disha
Nurturing Ambitions

FULLY SOLVED

NEET GUIDE for PHYSICS, CHEMISTRY & BIOLOGY

★ Brief Theory of Physics, Chemistry & Biology

★ 50-60 MCQs per chapter

★ Useful for NEET, AIIMS, JIPMER & Other Competitive Medical Entrance Exams

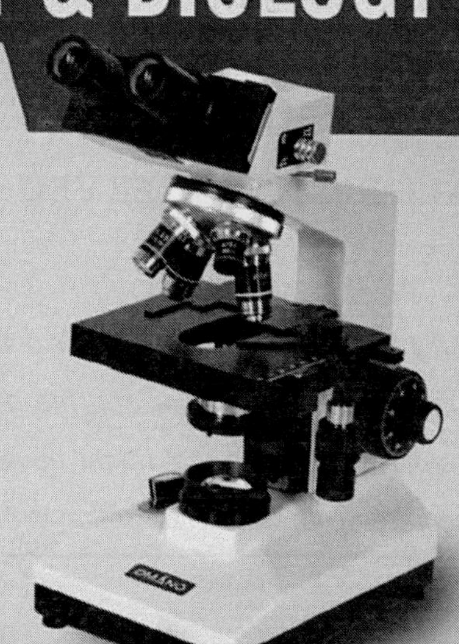

- **Head Office :** B-32, Shivalik Main Road, Malviya Nagar, New Delhi-110017
- **Sales Office :** B-48, Shivalik Main Road, Malviya Nagar, New Delhi-110017
  Tel. : 26692293 / 26692294

Typeset by Disha DTP Team

Printed at: Repro Knowledgecast Limited, Thane

DISHA PUBLICATION

For further information about the products from DISHA,

log on to **www.dishapublication.com** or **www.aiets.co.in** or email to **info@aiets.co.in**

INDEX

Physics

Chemistry

Biology

PHYSICS

1-286

PHYSICS

Physical World, Units & Measurements

CONCEPTS

WHAT IS PHYSICS?

Science refers to a system of acquiring knowledge. This system uses observation and experimentation to describe and explain natural phenomena. The term science also refers to the organized body of knowledge people have gained using that system.

Physics is the study of the interactions between physical systems. The physicist attempt to describe the interaction with the most fundamental and general law or principle possible.

The **purpose of science** is to produce useful models of reality. Most scientific investigations use some form of the scientific method.

The **scientific method** is a logical and rational order of steps by which scientists come to conclusions about the world around them. The Scientific method helps to organize thoughts and procedures so that scientists can be confident in the answers they find. Scientists use observations, hypotheses, and deductions to make these conclusions.

Science as defined above is sometimes called pure science to differentiate it from applied science, which is the application of research to human needs. Fields of science are commonly classified along two major lines:

Natural sciences, the study of the natural world, and *Social sciences*, the systematic study of human behavior and society. Natural science includes physics, chemistry, biology, etc.

SCOPE AND EXCITEMENT OF PHYSICS

The scope of physics is truly vast. It covers a tremendous range of magnitude of physical quantities like length, mass, time, energy, etc. Basically, there are two domains of interest : macroscopic and microscopic. The macroscopic domain includes phenomena at the laboratory, terrestrial and astronomical scales. The microscopic domain includes atomic, molecular and nuclear phenomena.

Classical physics deals mainly with macroscopic phenomena consisting of the study of heat, light, electricity, magnetism, optics, acoustics, and mechanics. Since the turn of the 20th century the study of quantum mechanics and relativity has become more important. Today, physics is often divided into branches such as nuclear physics, particle physics, quantum physics, theoretical physics, and solid-state physics. The study of the planets, stars, and their interactions is known as astrophysics, the physics of the Earth is called geophysics, and the study of the physical laws relating to living organisms is called biophysics.

Physics is exciting in many ways. Application and exploitation of physical laws to make useful devices is the most interesting and exciting part and requires great ingenuity and persistence of effort.

PHYSICS, TECHNOLOGY AND SOCIETY

We hear terms like science, engineering , technology all in same context though they are not exactly same.

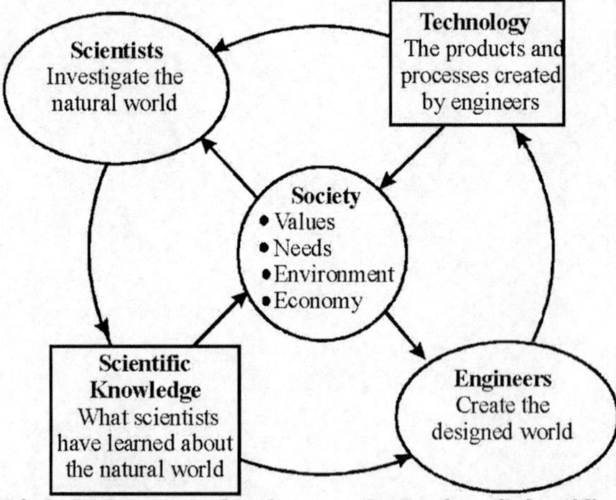

Link between technology and physics (Scientific Principle)

Technology	**Scientific principle(s)**
Steam engine	Laws of thermodynamics

Technology	Scientific principle(s)
Nuclear reactor	Controlled nuclear fission
Radio and Television	Generation, propagation and detection of em waves
Computers	Digital logic
Lasers	Light amplification by stimulated emission of radiation
Production of ultra high magnetic fields	Superconductivity
Rocket propulsion	Newton's laws of motion
Electric generator	EMI
Hydroelectric power	gravitational potential energy into electrical energy
Aeroplane	Bernoulli's principle in fluid
Particle accelerators	Motion of charged particles in electromagnetic fields
Sonar	Reflection of ultrasonic waves
Optical fibres	TIR of Light
Non-reflecting coatings	Thin film optical interference
Electron microscope	Wave nature of electrons
Photocell	Photoelectric effect
Fusion test reactor (Tokamak)	Magnetic confinement of plasma
Giant Metrewave Radio Telescope (GMRT)	Detection of cosmic radio waves
Bose-Einstein condensate	Trapping and cooling of atoms by laser beams and magnetic fields.

FUNDAMENTAL FORCES IN NATURE

Gravity is by far the weakest of the forces, but because the 2 nuclear forces have such a short range and the Universe is generally electricity neutral, gravity in fact rules the Universe.

Fundamental force	Range	Relative strength	Exchange particles	Major role
Strong nuclear force (between quarks)	Short, nuclear size ($\sim10^{-15}$m)	1	Gluons	Holding nuclei together
Electromagnetic force	Infinite	10^{-2}	Photons	Chemistry and biology
Weak nuclear force (that changes quark types)	Very short, Sub-nuclear size ($\sim10^{-1}$m)	10^{-13}	Weak bosons	Nuclear reactions
Gravitational force (between particles with mass)	Infinite	10^{-39}	Gravitons	Large-scale structure

Physicists are trying to derive a unified theory that would describe all the forces in nature as a single fundamental law. So far, they have succeeded in producing a unified description of the weak and electromagnetic forces, but a deeper understanding of the strong and gravitational forces has not yet been achieved. Theories that postulate the unification of the strong, weak, and electromagnetic forces are called Grand Unified Theories (often known by the acronym GUTs). Theories that add gravity to the mix and try to unify all four fundamental forces into a single force are called Superunified Theories. The theory that describes the unified electromagnetic and weak interactions is called the Standard Electroweak Theory, or sometimes just the Standard Model.

NATURE OF PHYSICAL LAWS

A physical law, scientific law, or a law of nature is a scientific generalization based on empirical observations of physical behavior. Empirical laws are typically conclusions based on repeated scientific experiments and simple observations, over many years, and which have become accepted universally within the scientific community. The production of a summary description of nature in the form of such laws is a fundamental aim of science.

Physical laws are distinguished from scientific theories by their simplicity. Scientific theories are generally more complex than laws; they have many component parts, and are more likely to be changed as the body of available experimental data and analysis develops. This is because a physical law is a summary observation of strictly empirical matters, whereas a theory is a model that accounts for the observation, explains it, relates it to other observations, and makes testable predictions based upon it. Simply stated, while a law notes that something happens, a theory explains why and how something happens.

Physical quantities (anything that can be measured), can be classified as fundamental or base quantities (such as length, mass, time, electric current, thermodynamic temperature, amount of substance, and luminous intensity) and derived quantities (e.g., velocity, acceleration, density etc.) which are derived from the base quantities, can be expressed as a combination of the base quantities.

UNITS

The chosen standard of measurement of a physical quantity, which has the same nature as that of the quantity is called the unit of that quantity.

Fundamental and Derived Units

Fundamental units are those, which are independent of each other and cannot be further resolved into any other units.

All units other than fundamental are derived units (which are dependent on fundamental units).

A complete set of units, both fundamental and derived, is called a **system of units.**

If the unit of a physical quantity is u and its numerical value in this unit is n then,

$$n \propto \frac{1}{u} \quad \text{or} \quad nu = \text{constant}$$

$$\text{or} \quad n_1 u_1 = n_2 u_2 = \dots\dots\dots = \text{constant}$$

when n_1, n_2 are magnitudes of the same physical quantity in units u_1 and u_2 respectively.

The International System of Units

This system (SI) is based on seven base units is at present internationally accepted and is widely used throughout the world.

The SI units are used in all physical measurements, for both the base quantities and the derived quantities obtained from them. Certain derived units are expressed by means of SI units with special names (such as joule, newton, watt, etc).

The SI units have well defined and internationally accepted unit symbols (such as m for metre, kg for kilogram, s for second, A for ampere, N for newton etc.).

Definitions of the 7 SI Base Units and 2 Supplementary Units

(i) **metre (17th CGPM, 1983) (m):** The metre (unit of length) is the length of the path travelled by light in vacuum during a time interval of 1/ 299,792,458 of a second.

(ii) **kilogram (3rd CGPM, 1901) (kg) :** The kilogram is the unit of mass; it is equal to the mass of the international prototype of the kilogram. The International Prototype of the Kilogram is kept at the International Bureau for Weights and Measures (BIPM), at Severe, near Paris, France, and is made of a platinum (90%) and iridium (10%) alloy. It has a density of approximately 21,500 kg/ m^3 and is shaped as a cylinder, with height and diameter equal to 39 mm.

(iii) **second (13th CGPM, 1967) (s) :** The second is the duration of 9,192,631,770 periods of the radiation corresponding to the transition between the two hyperfine levels of the ground state of the cesium-133 atom.

It is also 1/86,400 of a mean solar day. (There are 60 seconds in a minute, 60 minutes in an hour, and 24 hours in a day = $60 \times 60 \times 24 = 86{,}400$s)

(iv) **ampere (9th CGPM, 1948) (A) :** The ampere is that constant current which, if maintained in two straight parallel conductors of infinite length, of negligible circular cross section, and placed 1 metre apart in vacuum, would produce between these conductors a force equal to 2×10^{-7} newton per metre of length.

(v) **kelvin (13th CGPM, 1967) (K) :** The kelvin, unit of thermodynamic temperature, is the fraction 1/ 273.16 of the thermodynamic temperature of the triple point of water.

(vi) **mole (14th CGPM, 1971) (mol):**

(a) The mole is the amount of substance of a system which contains as many elementary entities as there are atoms in 0.012 kilogram of carbon 12.

(b) When the mole is used, the elementary entities must be specified and may be atoms, molecules, ions, electrons, other particles, or specified groups of such particles.

(vii) **Candela (16th CGPM, 1979) (cd) :** It is the luminous intensity in a perpendicular direction, of a surface of 1/600,000 square metre of a black body at the temperature of freezing platinum under a pressure of 1.013×10^5 N/m^2.

Two supplementary units

(viii) **Radian (rad) :** The radian is the plane angle between two radii of a circle that cut off on the circumference of an arc equal in length to the radius.

(ix) **Steradian (sr) :** The steradian is the solid angle that, having its vertex at the centre of a sphere, cuts off an area of the surface of the sphere equal to that of a square with sides of length equal to the radius of the sphere.

The solid angle is very similar to the plane angle with the only difference being that it is three-dimensional rather than just two-dimensional.

Imagine a sphere with a radius of 1 metre. A cone that covers an area of $1 m^2$ on the surface of the sphere encloses a solid angle of 1 steradian. A full sphere has a solid angle of 4π steradian.

PREFIXES FOR SI UNITS

In Physics we have to deal from very small (micro) to very large (macro) magnitudes. To express such large and small magnitudes simultaneously we use following prefixes:

Prefixes for powers of ten:

Multiple of 10	Prefix	Symbol
10^{24}	yotta	Y
10^{21}	zetta	Z
10^{18}	exa	E
10^{15}	peta	P
10^{12}	tera	T
10^{9}	giga	G
10^{6}	mega	M
10^{3}	kilo	k
10^{2}	hecto	h
10	deca	da
Sub -multiple	**Prefix**	**Symbol**
10^{-1}	deci	d
10^{-2}	centi	c
10^{-3}	milli	m
10^{-6}	micro	m
10^{-9}	nano	n
10^{-12}	pico	p
10^{-15}	femto	f
10^{-18}	atto	a
10^{-21}	zepto	z
10^{-24}	yocto	y

When a prefix is placed before the symbol of unit, the combined prefix and symbol should be considered as one new symbol which can be raised to a positive or negative power without any bracket, e.g., km^3 means $(10^3 m)^3$ but never $10^3 m^3$.

DIMENSIONS OF PHYSICAL QUANTITIES

All physical quantities can be expressed in terms of the fundamental quantities, e.g. consider the physical quantity force.

$$\text{Force} = \text{mass} \times \text{acceleration} = \text{mass} \times \frac{\text{velocity}}{\text{time}}$$

$$= \text{mass} \times \frac{\text{length / time}}{\text{time}} = \text{mass} \times \text{length} \times \text{time}^{-2}$$

$\therefore$ Unit of force = unit of mass × unit of length × (unit of time)$^{-2}$

Thus we can express the unit of force as product of different powers of the fundamental units of mass, length and time. These powers are called the **dimensions**. The dimensions of force are $[MLT^{-2}]$

For example, the dimensional equations of momentum and torque are : momentum = $[MLT^{-1}]$ and torque = $[ML^2T^{-2}]$

Uses of Dimensional Equations

(a) Conversion of one system of unit into another

(b) Checking the accuracy of various formulae

(c) Establish relation among different physical quantities.

A quantity can have unit without dimensions. e.g., plane angle and solid angle. But a quantity cannot have dimensions without unit.

Some pair which have same dimensions : (a) Torque and work (b) Angular momentum and Planck's constant (c) Latent heat and gravitational potential (d) Momentum and impulse.

ABSOLUTE ERROR, RELATIVE ERROR, AND PERCENTAGE ERROR

The magnitude of the difference between the individual measurement and the true value of the uantity is called **absolute error Δa** i.e., $\Delta a = a_1 - a_{mean}$

The ratio of the mean absolute error Δa_{mean} to the mean value a_{mean} of the quantity measured is called **relative error**.

Relative error = $\Delta a_{mean}/a_{mean}$

Percentage error : $\delta_a = (\Delta a_{mean}/a_{mean}) \times 100\%$

Combination of Errors (Maximum Permissible Error)

(a) Error of a sum or a difference : When two quantities are added or subtracted, the absolute error in the final result is the sum of the absolute errors in the individual quantities i.e. if $q = x \pm y$ then the maximum value of the error Δq is given by

$$\Delta q = \Delta x + \Delta y$$

(b) **Error of a product or a quotient :** When two or more quantities are multiplied or divided, the fractional error in the result is the sum of the fractional errors in the multipliers i.e. if $q = xy/z$, then the maximum permissible error in the measurement of q is given by

$$\frac{\Delta q}{q} = \frac{\Delta x}{x} + \frac{\Delta y}{y} + \frac{\Delta z}{z}$$

(c) **Error due to the power of a measured quantity :** The fractional error in a physical quantity raised to the power is the power times the fractional error in the individual quantity i.e. if $q = x^a y^b z^{-c}$ then the maximum permissible error in the measurement of q is given by

$$\frac{\Delta q}{q} = a\frac{\Delta x}{x} + b\frac{\Delta y}{y} + c\frac{\Delta z}{z}$$

(d) The quotient rule is not applicable if the numerator and denominator are dependent on each other.

e.g. if $R = \frac{XY}{X+Y}$. We cannot apply quotient rule to find the error in R. Instead we write the equation as follows

$\frac{1}{R} = \frac{1}{X} + \frac{1}{Y}$. Differentiating both the sides, we get

$$-\frac{dR}{R^2} = -\frac{dX}{X^2} - \frac{dY}{Y^2}$$

Thus $\frac{r}{R^2} = \frac{x}{X^2} + \frac{y}{Y^2}$ or $\frac{r}{R} = \frac{R}{X} \times \frac{x}{X} + \frac{R}{Y} \times \frac{y}{Y}$

PARALLAX METHOD

This method is used to measure large distance

$$D = \frac{b \text{ (distance between observatories)}}{\theta \text{ (parallax angle)}}$$

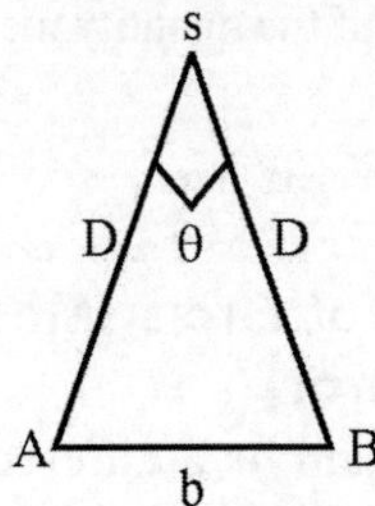

ACCURACY AND PRECISION

Accuracy refers to the closeness of observed values to its true value of the quantity while **precision** refers to closeness between the different observed values of the same quantity. High precision does not mean high accuracy.

SIGNIFICANT FIGURES

Significant figures indicate the precision of measurement which depends on the least count of the measuring instruments.

The reliable digits plus the first uncertain digit in the measurement are known as significant figures.

For **example,** if the length measured is 264.5 cm. here digits 2, 6 and 4 are certain digits while 5 is uncertain. Hence number of significant figure is four.

Rules to Determine the Numbers of Significant Figures

- All non-zero digits are significant. 235.75 has five significant figures.
- All zeroes between two non-zero digits are significant. 2016.008 has seven significant figures)
- All zeroes occurring between the decimal point and the non-zero digits are not significant. provided there is only a zero to left of the decimal point. 0.00652 has three significant figures.
- All zeroes written to the right of a non-zero digit in a number written without a decimal point are not significant. This rule does not work if the zero is a result of measurement. 54000 two significant figures. 54000m five significant figures.
- All zeroes occurring to the right of a non-zero digit in a number written with a decimal point are significant. 32.2000 has six significant figures.
- When a number is written in the exponential form, the exponential term does not contribute towards the significant figures. 2.465 x 10^5 has four significant figures.

Rules of Arithmetic Operations with Significant Figures

In arithmetical operations involving significant figures, the answer is reported in such a way that it reflects the reliability of the least precise operation.

In Addition and Subtraction: For addition and subtraction, look at the decimal portion (i.e., to the right of the decimal point) of the numbers only follow the following procedure :

(a) Count the number of significant figures in the decimal portion of each number in the problem. (The digits to the left of the decimal place are not used to determine the number of decimal places in the final answer)

(b) Add or subtract in the normal fashion.

(c) Round off the answer to the least number of places in the decimal portion of any number in the problem.

In Multiplication and Division: In multiplication or division the number of significant figures in the final answer is the same as the minimum number of the significant figures in the physical quantities being operated.

In Exponential notation: Sometimes this would required expressing the answer in exponential notation.

ROUNDING OFF

Rules of Rounding off Uncertain Digits

(a) The preceding digit is raised by 1 if the uncertain digit to be dropped is more than 5 and is left unchanged if the latter is less than 5.

Example : x = 5.68<u>6</u> is rounded off to 5.69 (as 6 > 5)

x = 3.46<u>2</u> is rounded off to 3.46 (as 2 < 5)

(b) If the uncertain digit to be dropped is 5, the preceeding digit raised by 1 if it is odd and is left unchanged if it is even digit.

Example : 7.735 is rounded off to three significant figures becomes 7.74 as preceeding digit is odd.

7.745 is rounded off to 7.74 as preceeding digit is even.

Exercise

1. The sum of the numbers 436.32, 227.2 and 0.301 in appropriate significant figures is
   (a) 6663.821 (b) 664
   (c) 663.8 (d) 663.8
2. Which of the following pairs of physical quantities does not have same dimensional formula?
   (a) Work and torque
   (b) Angular momentum and Planck's constant
   (c) Tension and surface tension
   (d) Impulse and linear momentum
3. On the basis of dimensions, decide which of the following relations for the displacement of a particle undergoing simple harmonic motion is not correct.
   (a) $y = a \sin 2\pi t/T - \cos 2\pi t/T$
   (b) $y = a \sin vt$
   (c) $y = \frac{a}{T}\sin\left(\frac{kt}{a}\right)$
   (d) $y = a\sqrt{2}\left(\sin\frac{2\pi t}{T} - \cos\frac{2\pi t}{T}\right)$
4. If P, Q, R are physical quantities, having different dimensions, which of the following combinations can never be a meaningful quantity?
   (a) $(P - Q) / R$ (b) $PQ - R$
   (c) PQ / R (d) $(PR - Q^2) / R$
5. Which of the following quantities has a unit but dimensionless?
   (a) Strain
   (b) Reynolds number
   (c) Angular displacement
   (d) Poisson's ratio
6. The speed of sound in a gas is given by $V = \sqrt{\frac{\gamma P}{d}}$ where P is pressure and d is density. The dimensional formula for γ is
   (a) $M^{-1} L^1 T^2$ (b) $M^0 L^0 T^0$
   (c) $M^0 L^{-1} T^0$ (d) $M^0 L^0 T^1$
7. Which of the following pair of quantities does not have the same dimensions?
   (a) Couple, linear momentum
   (b) Energy, work
   (c) Surface tension, spring factor
   (d) Torque, kinetic energy of rotation.
8. The dimensions of $(\mu_0 \in_0)^{\frac{-1}{2}}$ are
   (a) $[L^{1/2} T^{-1/2}]$ (b) $[L^{-1} T]$
   (c) $[L T^{-1}]$ (d) $[L^{-1/2} T^{1/2}]$
9. Which of the following sets can't be taken as fundamental quantities in any system of units :
   (a) Length, mass, velocity
   (b) Length, time, velocity
   (c) Mass, time, acceleration
   (d) Energy, force, time

10. The dimensional formula for relative density is
(a) $[ML^{-3}]$ (b) $[M^0 L^{-3}]$
(c) $[M^0 L^0 T^{-1}]$ (d) $[M^0 L^0 T^0]$

11. The percentage error in the measurement of length of a cube is 2%, what will be percentage error in the measurement of its volume
(a) 2% (b) 1%
(c) 4% (d) 6%

12. The solid angle sustended by the total surface area of a sphere, at the centre is
(a) 4π (b) 2π
(c) π (d) 3π

13. Mass of a body is 210 gm and its density is 7.981 g/cm^3 what will be its volume, with regard to significant digits
(a) 26.312 cm^3 (b) 26 cm^3
(c) 27 cm^3 (d) 26.3 cm^3

14. Random error can be eliminated by :
(a) careful observation
(b) eliminating the cause
(c) measuring the quantity with more than one instrument
(d) taking large number of observations and then their mean

15. Which of the following is the most accurate
(a) 200.0 m (b) 20×10^1 m
(c) 2×10^2 m (d) data is inadequate

16. If $f = x^2$, then the relative error in f is
(a) $\frac{2\Delta x}{x}$ (b) $\frac{(\Delta x)^2}{x}$
(c) $\frac{\Delta x}{x}$ (d) $(\Delta x)^2$

17. The least count of a stop watch is 0.2 second. The time of 20 oscillations of a pendulum is measured to be 25 second. The percentage error in the measurement of time will be
(a) 8% (b) 1.8%
(c) 0.8% (d) 0.1%

18. Two quantities A and B have different dimensions which mathematical operation given below is physically meaningful?
(a) A/B (b) $A+B$
(c) $A-B$ (d) $A=B$

19. The dimensional formula for angular momentum is
(a) $[M^0L^2T^{-2}]$ (b) $[ML^2T^{-1}]$
(c) $[MLT^{-1}]$ (d) $[ML^2T^{-2}]$

20. If C and R denote capacitance and resistance, the dimensional formula of CR is
(a) $[M^0L^0T^1]$
(b) $[M^0L^0T^0]$
(c) $[M^0L^0T^{-1}]$
(d) not expressible in terms of $M.L.T.$

21. The dimensional formula of torque is
(a) $[ML^2T^{-2}]$ (b) $[MLT^{-2}]$
(c) $[ML^{-1}T^{-2}]$ (d) $[ML^{-2}T^{-2}]$

22. Dimensional formula of self inductance is
(a) $[MLT^{-2}A^{-2}]$ (b) $[ML^2T^{-1}A^{-2}]$
(c) $[ML^2T^{-2}A^{-2}]$ (d) $[ML^2T^{-2}A^{-1}]$

23. Of the following quantities, which one has dimension different from the remaining three?
(a) Energy per unit volume
(b) Force per unit area
(c) Product of voltage and charge per unit volume
(d) Angular momentum

24. If $x = at + bt^2$, where x is the distance travelled by the body in kilometers while t is the time in seconds, then the units of b is
(a) km/s (b) kms
(c) km/s^2 (d) kms^2

25. According to Newton, the viscous force acting between liquid layers of area A and velocity gradient ΔV/ΔZ is given by $F = -\eta A \frac{\Delta V}{\Delta Z}$ where η is constant called coefficient of viscosity. The dimensional formula of η is
(a) $ML^{-2}T^{-2}$ (b) $M^0L^0T^0$
(c) ML^2T^{-2} (d) $ML^{-1}T^{-1}$

26. The frequency of vibration f of a mass m suspended from a spring of spring constant k is given by a relation of the type $f = c\, m^x k^y$, where c is a dimensionless constant. The values of x and y are
(a) $x = \frac{1}{2}, y = \frac{1}{2}$ (b) $x = -\frac{1}{2}, y = -\frac{1}{2}$
(c) $x = \frac{1}{2}, y = -\frac{1}{2}$ (d) $x = -\frac{1}{2}, y = \frac{1}{2}$

27. The dimensional formula of pressure is
(a) $[MLT^{-2}]$ (b) $[ML^{-1}T^2]$
(c) $[ML^{-1}T^{-2}]$ (d) $[MLT^2]$

28. The dimensional formula for permeability μ is given by

(a) $MLT^{-2}A^{-2}$
(b) M^0L^1T
(c) $M^0L^2T^{-1}A^2$
(d) None of the above

29. A certain body weighs 22.42 gm and has a measured volume of 4.7 cc. The possible error in the measurement of mass and volume are 0.01 gm and 0.1 cc. Then maximum error in the density will be
(a) 22% (b) 2%
(c) 0.2% (d) 0.02%

30. P represents radiation pressure, c represents speed of light and S represents radiation energy striking unit area per sec. The non zero integers x, y, z such that $P^x S^y c^z$ is dimensionless are
(a) $x=1, y=1, z=1$ (b) $x=-1, y=1, z=1$
(c) $x=1, y=-1, z=1$ (d) $x=1, y=1, z=-1$

31. The time dependence of a physical quantity p is given by $p = p_0 \exp(-\alpha t^2)$, where α is a constant and t is the time. The constant α
(a) is dimensionless
(b) has dimensions T^{-2}
(c) has dimensions T^2
(d) has dimensions of p

32. Turpentine oil is flowing through a tube of length ℓ and radius r. The pressure difference between the two ends of the tube is p. The viscosity of oil is given by

$$\eta = \frac{p\left(r^2 - x^2\right)}{4vl}$$

where v is the velocity of oil at a distance x from the axis of the tube. The dimensions of η are
(a) $[M^0L^0T^0]$ (b) $[MLT^{-1}]$
(c) $[ML^2T^{-2}]$ (d) $[ML^{-1}T^{-1}]$

33. In a vernier calliper N divisions of vernier scale coincides with (N – 1) divisions of main scale (in which length of one division is 1 mm). The least count of the instrument should be
(a) N (b) N – 1
(c) 1/10 N (d) 1/N – 1

34. In a particular system, the unit of length, mass and time are chosen to be 10 cm, 10 g and 0.1 s respectively. The unit of force in this system will be equivalent to
(a) 0.1 N (b) 1 N
(c) 10 N (d) 100 N

35. Which of the following is a dimensional constant?
(a) Refractive index
(b) Poissons ratio
(c) Relative density
(d) Gravitational constant

36. The percentage errors in the measurement of mass and speed are 2% and 3% respectively. The error in kinetic energy obtained by measuring mass and speed will be
(a) 12% (b) 10%
(c) 8% (d) 2%

37. An equation is given as : $\left(P + \frac{a}{V^2}\right) = b\frac{\theta}{V}$
where P = Pressure, V = Volume & θ = Absolute temperature. If a and b are constants, then dimensions of a will be
(a) $[ML^5T^{-2}]$ (b) $[M^{-1}L^5T^2]$
(c) $[ML^{-5}T^{-1}]$ (d) $[ML^5T^1]$

38. The density of a cube is measured by measuring its mass and length of its sides. If the maximum error in the measurement of mass and length are 4% and 3% respectively, the maximum error in the measurement of density will be
(a) 7% (b) 9%
(c) 12% (d) 13%

39. Which of the following will have the dimensions of time
(a) LC (b) $\frac{R}{L}$
(c) $\frac{L}{R}$ (d) $\frac{C}{L}$

40. The force F on a sphere of radius a moving in a medium with velocity v is given by $F = 6\pi\eta av$. The dimensions of η are
(a) $[ML^{-3}]$ (b) $[ML^{-2}]$
(c) $[ML^{-1}]$ (d) $[ML^{-1}T^{-1}]$

41. The dimensional formula for magnetic flux is
(a) $[ML^2T^{-2}A^{-1}]$ (b) $[ML^3T^{-2}A^{-2}]$
(c) $[M^0L^{-2}T^2A^{-2}]$ (d) $[ML^2T^{-1}A^2]$

42. Which one of the following groups have quantities that do not have the same dimensions?
(a) Pressure, stress (b) Velocity, speed
(c) Force, impulse (d) Work, energy

43. The dimensions of Planck's constant are same as
(a) energy (b) power
(c) momentum
(d) angular momentum

44. $5.4 \times 0.125 =$ ________ with due regards to significant figures
(a) 0.68 (b) 0.67
(c) 0.667 (d) None

45. How many significant figures are there in 0.030100×10^6
(a) 1 (b) 3
(c) 5 (d) 7

46. How many significant figures are there in 301.00?
(a) 1 (b) 3
(c) 5 (d) None

47. How many significant figures are there in 30100 m?
(a) 1 (b) 3
(c) 5 (d) None

48. How many significant figures are these in 0.30100?
(a) 1 (b) 3
(c) 5 (d) None

49. Which is most accurate
(a) 3 m (b) 3.0 m
(c) 3.00 m (d) 3.000 m

50. The dimensional formula for strain is same as that for
(a) Thrust
(b) Angle
(c) Modulus of elasticity
(d) Stress

51. The dimensional formula of $\frac{L}{R}$ is same as that of (where L is inductance and R is resistance)
(a) $(\text{Frequency})^2$ (b) Time period
(c) Frequency (d) $(\text{Time period})^2$

52. The division of energy and time is X. The dimensional formula of X is same as that of
(a) Momentum (b) Power
(c) Torque (d) Electric field

53. The Solar constant is defined as the energy incident per unit area per second. The dimensional formula for solar constant is
(a) $M^0L^0T^0$ (b) MLT^{-2}
(c) ML^2T^{-2} (d) MT^{-3}

54. The dimensional formula of latent heat is identical to that of
(a) Internal Energy
(b) Angular momentum
(c) Gravitational potential
(d) Electric potential

55. $\sqrt{\mu_0 \varepsilon_0}$ has the dimension
(a) LT^{-1} (b) $L^{-1}T^1$
(c) $M^2L^3T^{-1}$ (d) $M^{-1}L^3T^{-2}$

56. The dimension of $\sqrt{\mu rg}$ where μ = coefficient of friction, r = radius of circular path, g = acceleration due to gravity is same as that of
(a) $\frac{F}{Bq}$ (b) μF
(c) $\frac{F}{m}$ (d) qvB

57. Which of the following is a dimensional constant?
(a) Refractive index
(b) Dielectric constant
(c) Relative density
(d) Gravitational constant

58. The dimensional formula of modulus of rigidity is
(a) $[ML^2T^{-2}]$ (b) $[ML^{-1}T^{-3}]$
(c) $[ML^{-2}T^{-2}]$ (d) $[ML^{-1}T^{-2}]$

59. Which of the following has the same dimensions?
(a) Impulse and momentum
(b) Specific heat and latent heat
(c) Moment of Inertia and Force
(d) Thrust and surface tension

60. Measurement of a physical quantity results in a number (n) and unit (u). The relationship between these two quantities are
(a) $n \propto \frac{1}{u}$ (b) $n \propto u$
(c) $n \propto \frac{1}{u^2}$ (d) $n \propto u^2$

CHAPTER 2

Motion in a Straight Line

CONCEPTS

Mechanics is the study of motion of material objects.

Rectilinear motion is the motion of objects along a straight line.

REST AND MOTION

State of rest : When position of object does not change with time object is said to be in state of rest.

State of motion : When position of object changes with times object is said to be in state of motion. Rest and motion are relative terms.

DISTANCE AND DISPLACEMENT

Path length or distance is the total length of the path traversed by an object. It is a scalar quantity.

Displacement is the change in position when the object is moving in a straight line : $\Delta x = x_2 - x_1$. Path length is either greater or equal to the magnitude of the displacement between the same points. The displacement depends only on the end points; the path length (as the name implies) depends on the actual path. It has both magnitude as well as direction.

An object is said to be in **uniform motion in a straight** line if its displacement is equal in equal intervals of time. Otherwise, the motion is said to be non-uniform.

SPEED, VELOCITY AND ACCELERATION

Average speed is the ratio of total path length traversed and the corresponding time interval. The average speed of an object is lither greater than or equal to the magnitude of the average velocity over a given time interval. The two are equal only if the path length is equal to the magnitude of displacement.

$$\text{Average speed} = \frac{\text{Total distance travelled}}{\text{Total time taken}}$$

Average velocity is the displacement divided by the time interval in which the displacement occurs. On an x-t graph, the average velocity over a time interval is the slope of the line connecting the initial and final positions corresponding to the time interval.

$$\text{Average velocity} = \frac{\text{Displacement}}{\text{Total time taken}}$$

Instantaneous velocity or simply velocity is the limit of the average velocity as the time interval Δt becomes infinitesimally small. The velocity at a particular instant is equal to the slope of the tangent drawn on position-time graph at that instant.

$$\text{Instantaneous velocity, } \vec{V}_{inst} = \underset{\Delta t \to 0}{Lt} \frac{\Delta \vec{x}}{\Delta t} = \frac{\partial \vec{x}}{dt}$$

Uniform velocity : A body is said to move with uniform velocity, if it covers equal displacements in equal intervals of time, howsoever small these intervals may be.

Non-uniform velocity : The object is said to move with non-uniform velocity if it covers in equal displacements in equal intervals of time howsoever small these time intervals may be.

Acceleration is the rate of change of velocity of an object with time.

$$\text{Acceleration} = \frac{\text{Change in velocity}}{\text{Time interval}}$$

Average acceleration is the change in velocity divided by the time interval during which the change occurs.

$$\text{Average acceleration, } \vec{a}_{av} = \frac{\Delta \vec{v}}{\Delta t}$$

Instantaneous acceleration is defined as the limit of the average acceleration as the time interval Δt goes to zero.

$$\text{Instantaneous acceleration, } a_{inst} = \underset{\Delta t \to 0}{Lt} \frac{\Delta \vec{v}}{\Delta t} = \frac{\partial \vec{v}}{dt}$$

The acceleration of an object at a particular time is the slope of the velocity-time graph at that instant of time. For uniform motion, acceleration is zero and the x-t graph is a straight line inclined to the time axis and the v-t graph is a straight line parallel to the time axis. For motion with uniform acceleration, x-t graph is a parabola while the v-t graph is a straight line inclined to the time axis. The sign of acceleration does not tell us whether the particle's speed is increasing or decreasing. The sign of acceleration depends on the choice of the positive direction of the axis. For example, if the vertically upward direction is chosen to be the positive direction of the axis, the acceleration due to gravity is negative. If a particle is falling under gravity, this acceleration, though negative, results in increase in speed. For a particle thrown upward, the same negative acceleration (of gravity) results in decrease in speed. The area under the velocity-time curve between times t_1 and t_2 is equal to the displacement of the object during that interval of time.

KINEMATIC EQUATIONS OF MOTION

For objects in uniformly accelerated rectilinear motion, the five quantities displacement x, time taken t, initial velocity v_0, final velocity v and acceleration a are related by a set of simple equations called **kinematic equations of motion**. The definitions of instantaneous velocity and acceleration are exact and are always correct while the kinematic equations are true only for motion in which the magnitude and the direction of acceleration are constant during the course of motion. If acceleration (constant or variable) in backward motion and deceleration (constant or variable) in forward motion are identical then time of backward and that of forward motion is same and velocity at the same position in forward as well as in backward motion are same in magnitude but opposite in direction.

KINEMATIC EQUATIONS FOR UNIFORMLY ACCELERATED MOTION

Velocity time relation: $v = u + a_t\,(t)$

Velocity position relation: $v^2 = u^2 + 2a_t\,(s)$

Position time relation: $s = ut + \frac{1}{2}a_t\,(t)^2$

applicable only **when $|\vec{a}_t| = a_t$ is constant.**

a_t = magnitude of tangential acceleration, s = distance

If a_t is variable

(i) $a_t = f(t)$ $\qquad$ $v = f(t)$

$$\frac{d|\vec{v}|}{dt} = \frac{dv}{dt} = f(t) \qquad \frac{ds}{dt} = f(t)$$

$$\int dv = \int f(t)\,dt \qquad \int ds = \int f(t)\,dt$$

(ii) $\int_t a = f(s)$ $\qquad$ $v = f(s)$

$$v\frac{dv}{ds} = f(s) \qquad \frac{ds}{dt} = f(s)$$

$$\int vdv = \int f(s)\,ds \qquad \int \frac{ds}{f(s)} = \int dt$$

(iii) $a_t = f(v)$ $\qquad$ $a_t = f(v)$

$$\frac{dv}{dt} = f(v) \qquad v\frac{dv}{ds} = f(v)$$

$$\int \frac{dv}{f(v)} = \int dt \qquad \int v\frac{dv}{f(v)} = \int ds$$

Displacement of Particle in nth Second ($\vec{s}_n$)

$$\vec{s}_n = \vec{u} + \frac{1}{2}\vec{a}\,(2n-1)$$

Motion of Body Under Gravity (Free Fall)

If a body is dropped from some height

$$u = 0;\ a = +g,\ v = gt,\ h = \frac{1}{2}gt^2$$

$$v^2 = 2gh,\ h_n = \frac{g}{2}(2n-1)$$

Reaction time is the time taken by a person to observe, think and act.

REASONING STRATEGY IN APPLYING THE EQUATIONS OF KINEMATICS

(a) Make a drawing to represent the situation being studied. A drawing helps us to "see" what's happening.

(b) Decide which directions are to be called positive (+) and negative (–) relative to a conveniently chosen coordinate origin. Do not change your decision during the course of a calculation.

(c) In an organized way, write down the values (with appropriate plus and minus signs) that are given for any of the five kinematic variables (x, a, v, v_0, and t). Be on alert for "implied data" such as the phrase "starts from rest", which means that the value of the initial velocity is $v_0 = 0$ m/s. Identify the variables that you are being asked to determine.

(d) Before attempting to solve a problem, verify that the given information contains values for at least three of the five kinematic variables. Once the three known variables are identified along with the desired unknown variable, the appropriate relation can be selected. Remember that the motion of two objects may be interrelated, so they may share a piece of information. In such cases, data for only two variables need be specified for each object.

(e) When the motion of an object is divided into segments, remember that the final velocity of one segment is the initial velocity for the next segment.

(f) Keep in mind that there may be two possible answers to a kinematics problem. Try to visualize the different physical situations to which the answers correspond.

GRAPH

- **Reference point or origin point** is the point of intersection of three mutually perpendicular axes x, y and z.
- **Frame of reference** is the co-ordinate system along with a clock.
- **Point object :** An object whose dimension is negligible as compared to the distance travelled by it.

(a) **For a particle at rest**

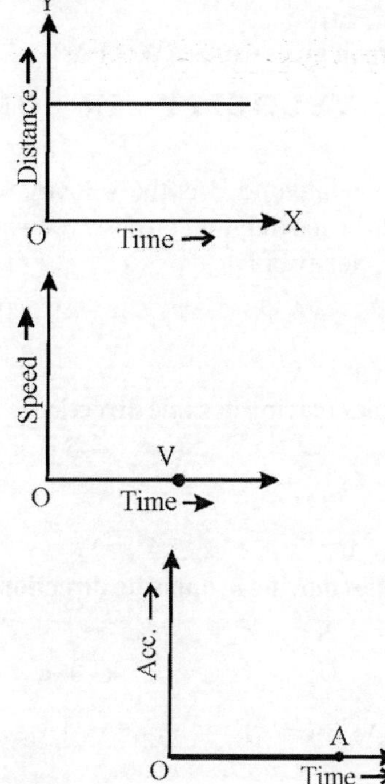

(b) **For a particle travelling with uniform speed**

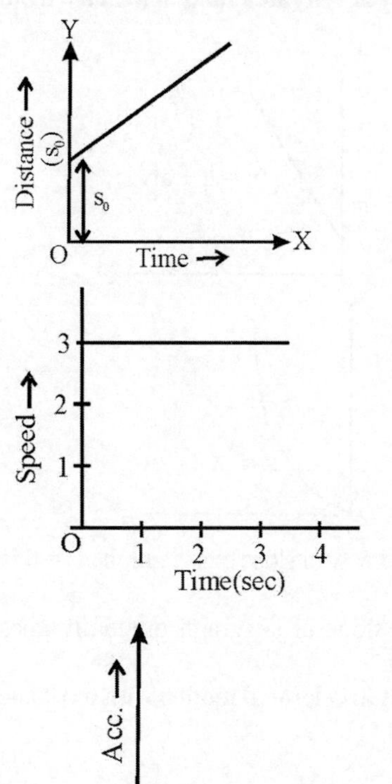

(c) **For one-dimensional motion with constant acceleration.**

Uniformly accelerated motion with u = 0 and s = 0 at t = 0

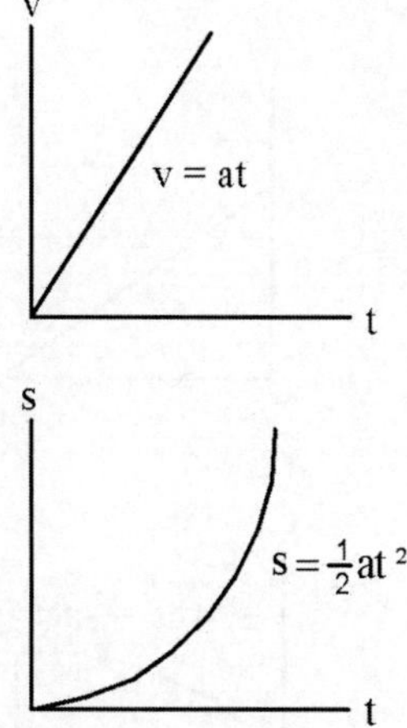

(i) $u = 0$, i.e., $v = 0$ at $t = 0$

(ii) $u = 0$, i.e., slope of s-t graph at $t = 0$, should be zero

(iii) a or slope of v-t graph is constant

Uniformly accelerated motion with $u \neq 0$ but $s = 0$ at $t = 0$

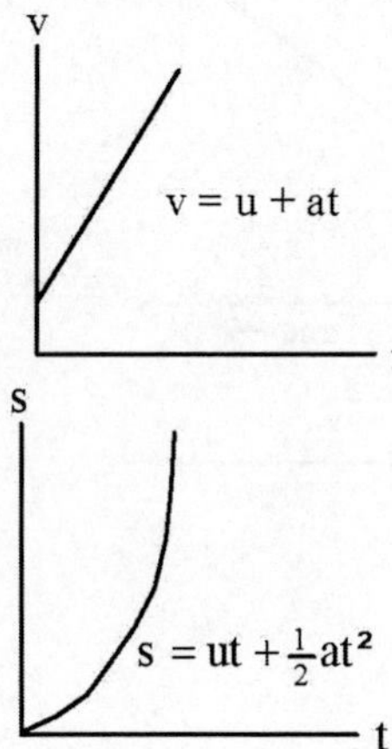

(i) $u \neq 0$, i.e., v or slope of s - t graph at $t = 0$ is not zero.

(ii) v or slope of s-t graph gradually goes on increasing

Uniformly accelerated motion with $u \neq 0$ and $s = s_0$ at $t = 0$

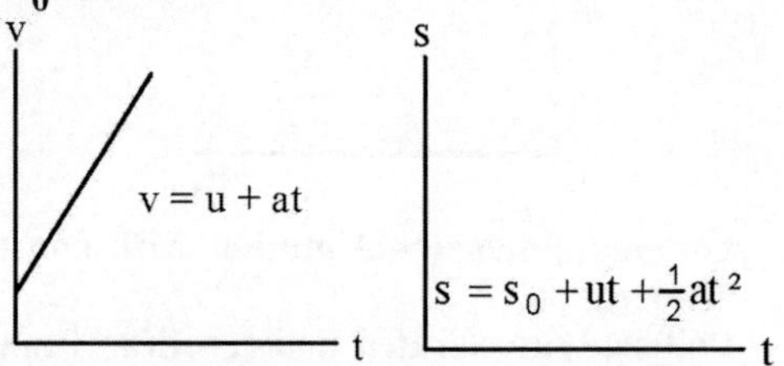

Uniformly retarded motion till velocity becomes zero

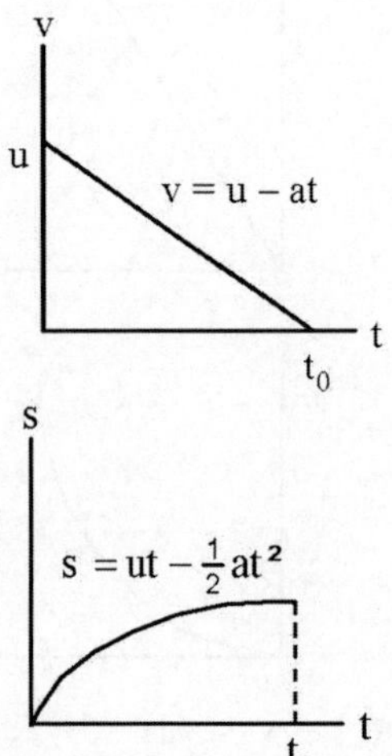

(i) Slope of s-t graph at $t = 0$ gives u

(ii) Slope of s-t graph at $t = t_0$ becomes zero

(iii) In this case u can't be zero

Uniformly retarded then accelerated in opposite direction

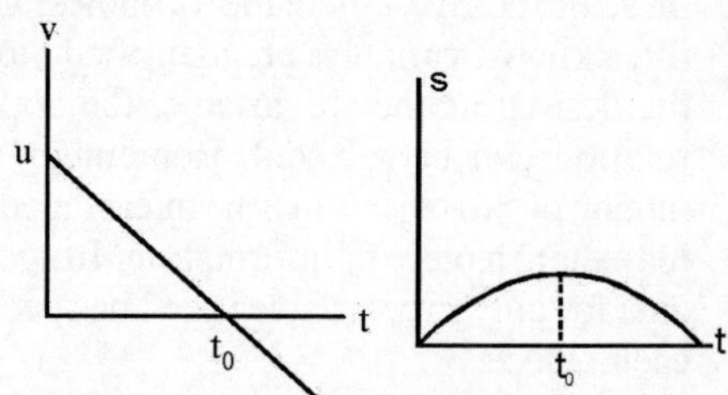

(i) At time $t = t_0$, $v = 0$ or slope of s-t graph is zero

(ii) In s-t graph, slope or velocity first decreases then increases with opposite sign.

Slope of displacement - time graph gives velocity $\left(v = \frac{ds}{dt}\right)$

Slope of velocity - time graph gives acceleration $\left(a = \frac{dv}{dt}\right)$

Slope of momentum - time graph gives force $\left(F = \frac{dp}{dt}\right)$

Area under v-t graph gives displacement ($ds = vdt$)

Area under a-t graph gives change in velocity ($dv = adt$)

Area under F-t graph gives impulse (change in momentum) ($\Delta p = F\Delta t$)

Area under F-s graph gives work ($\Delta W = F\Delta s$)

RELATIVE VELOCITY IN ONE DIMENSION

The velocity of A relative to B is the velocity with which A appears to be moving w.r.t an observer who is moving with the velocity of B

$$\vec{v}_{AB} = \vec{v}_A - \vec{v}_B$$

Similarly, $\vec{v}_{BA} = \vec{v}_B - \vec{v}_A$

Case 1: Bodies moving in same direction

$\vec{v}_A$ → $\vec{v}_B$ → (← – , → +)

$$\vec{v}_{AB} = \vec{v}_A - \vec{v}_B \Rightarrow v_{AB} = v_A - v_B$$

Case 2: Bodies moving in opposite direction

$\vec{v}_A$ → (← – , → +) ← $\vec{v}_B$

$$\vec{v}_{AB} = \vec{v}_A - (-\vec{v}_B) \Rightarrow v_{AB} = v_A + v_B$$

Exercise

1. Which one out of the following statement is false?
   (a) A body can have zero velocity and still be accelerated
   (b) A body can have a constant velocity and still have a varying speed
   (c) A body can have a constant speed and still have a varying velocity
   (d) The direction of the velocity of a body can change when its acceleration is constant
2. The displacement of a particle starting from rest (at $t = 0$) is given by $s = 6t^2 - t^3$. The time in seconds at which the particle will attain zero velocity again, is :
   (a) 2 (b) 4
   (c) 6 (d) 8
3. The study of motion, without consideration of its cause is studied in :
   (a) statistics (b) kinematics
   (c) mechanics (d) modern physics
4. The ratio of the numerical values of the average velocity and average speed of a body is always:
   (a) unity (b) unity or less
   (c) unity or more (d) less than unity
5. A particle has moved from one position to another position.
   (a) Its distance is zero
   (b) Its displacement is zero
   (c) Neither distance nor displacement is zero
   (d) Average velocity is zero
6. A vehicle travels half the distance L with speed V_1 and the other half with speed V_2, then its average speed is :
   (a) $\frac{V_1+V_2}{2}$ (b) $\frac{2V_1+V_2}{V_1+V_2}$
   (c) $\frac{2V_1 V_2}{V_1+V_2}$ (d) $\frac{2(V_1+V_2)}{V_1 V_2}$
7. The displacement of a body is zero. The distance covered :
   (a) is zero
   (b) is not zero
   (c) may or may not be zero
   (d) depends upon the acceleration
8. Which of the following changes when a particle is moving with uniform velocity?
   (a) speed (b) velocity
   (c) acceleration (d) position vector
9. A lift is coming from 8th floor and is just about to reach 4th floor. Taking ground floor as origin and positive direction upwards for all quantities, which one of the following is correct?
   (a) $X<0, v<0, a>0$ (b) $X>0, v<0, a<0$
   (c) $X>0, v<0, a>0$ (d) $X>0, v>0, a<0$
10. The given figure represents the displacement (x) time (t) graph for a particle in one dimensional motion.

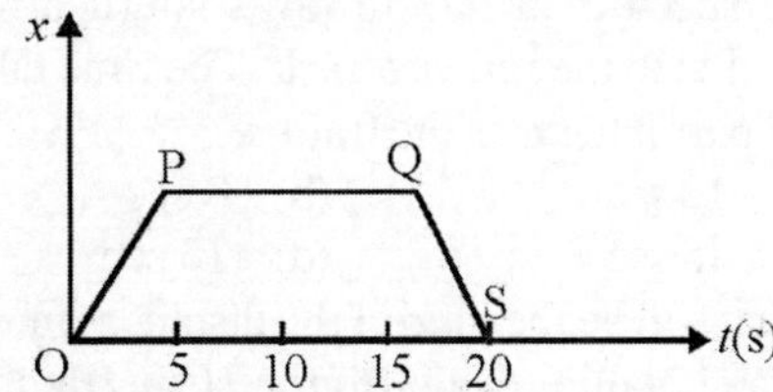

Which of the figure given below represents the variation in velocity of the particle with time?

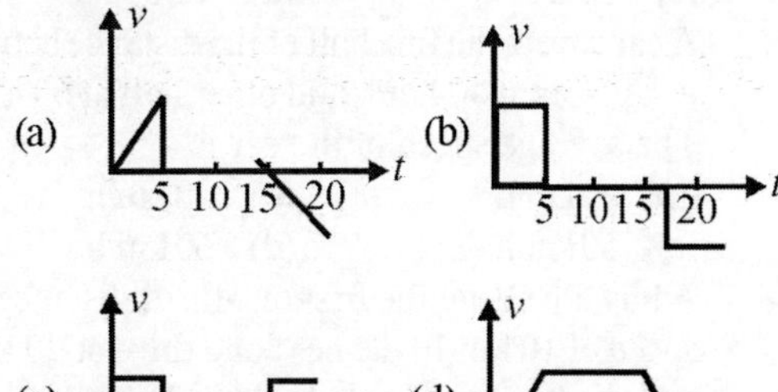

11. The slope of the velocity time graph for retarded motion is
    (a) positive
    (b) negative
    (c) zero
    (d) can be +ve, –ve or zero
12. The area of the acceleration-displacement curve of a body gives
    (a) impulse
    (b) change in momentum per unit mass
    (c) change in KE per unit mass
    (d) total change in energy

13. At a metro station, a girl walks up a stationary escalator in time t_1. If she ramains stationary on the escalator, then the escalator take her up in time t_2. The time taken by her to walk up on the moving escalator will be
(a) $(t_1 + t_2)/2$ (b) $t_1 t_2 / (t_2 - t_1)$
(c) $t_1 t_2 / (t_2 + t_1)$ (d) $t_1 - t_2$

14. A car is moving along a straight road with a uniform acceleration. It passes through two points P and Q separated by a distance with velocity 30 km/h and 40 km/h respectively. The velocity of the car midway between P and Q is
(a) 33.3 km/h (b) $20\sqrt{2}$ km/h
(c) $25\sqrt{2}$ km/h (d) 35 km/h

15. A train of 150 metre long is going towards north direction at a speed of 10 m/s. A parrot flies at the speed of 5 m/s towards south direction parallel to the railway track. The time taken by the parrot to cross the train is
(a) 12 sec (b) 8 sec
(c) 15 sec (d) 10 sec

16. What will be the ratio of the distances moved by a freely falling body from rest in 4th and 5th seconds of journey?
(a) 4 : 5 (b) 7 : 9
(c) 16 : 25 (d) 1 : 1

17. A car covers the first half of the distance between two places at 40 km/h and other half at 60 km/h. The average speed of the car is
(a) 40 km/h (b) 48 km/h
(c) 50 km/h (d) 60 km/h

18. A bus travelling the first one third distance at a speed of 10 km/h, the next one third at 20 km/h and the last one-third at 60 km/h. The average speed of the bus is
(a) 9 km/h (b) 16 km/h
(c) 18 km/h (d) 48 km/h

19. A car moves a distance of 200 m. It covers the first half of the distance at speed 40 km/h and the second half of distance at speed v. The average speed is 48 km/h. Find the value of v
(a) 56 km/h (b) 60 km/h
(c) 50 km/h (d) 48 km/h

20. A body dropped from top of a tower fall through 40 m during the last two seconds of its fall. The height of tower is ($g = 10$ m/s^2)
(a) 60 m (b) 45 m
(c) 80 m (d) 50 m

21. Which of the following curve does not represent motion in one dimension?

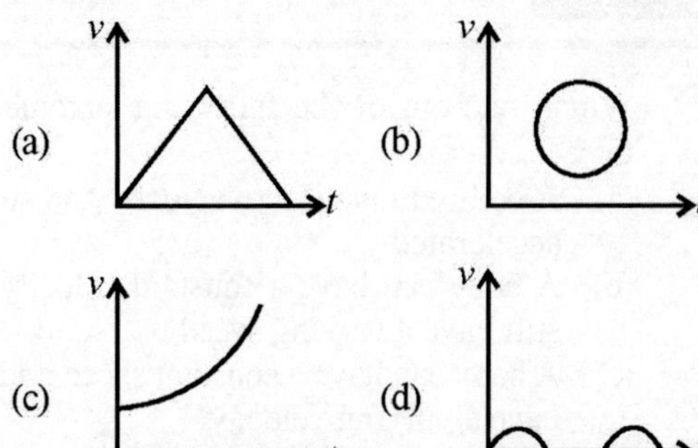

22. A body starts from rest, what is the ratio of the distance travelled by the body during the 4th and 3rd seconds?
(a) $\frac{7}{5}$ (b) $\frac{5}{7}$
(c) $\frac{7}{3}$ (d) $\frac{3}{7}$

23. A car accelerates from rest at a constant rate α for some time, after which it decelerates at a constant rate β and comes to rest. If the total time elapsed is t, then the maximum velocity acquired by the car is
(a) $\left(\frac{\alpha^2 + \beta^2}{\alpha\beta}\right)t$ (b) $\left(\frac{\alpha^2 - \beta^2}{\alpha\beta}\right)t$
(c) $\frac{(\alpha + \beta)t}{\alpha\beta}$ (d) $\frac{\alpha\beta t}{\alpha + \beta}$

24. The displacement time graph of a moving particle is shown below

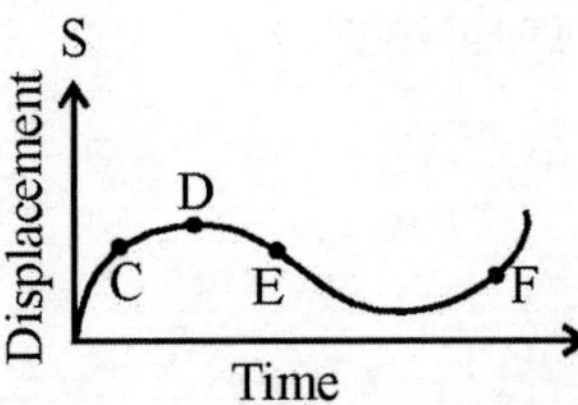

The instantaneous velocity of the particle is negative at the point
(a) D (b) F
(c) C (d) E

25. A particle moves along a straight line such that its displacement at any time t is given by $s = (t^3 - 6t^2 + 3t + 4)$ metres

The velocity when the acceleration is zero is
(a) 3 ms^{-1} (b) – 12 ms^{-1}
(c) 42 ms^{-2} (d) – 9 ms^{-1}

26. A body is thrown vertically upward from the ground. It reaches a maximum height of 20 m in 5 sec. After what time, it will reach the ground from its maximum height position?
(a) 2.5 sec (b) 5 sec
(c) 10 sec (d) 25 sec

27. A stone released with zero velocity from the top of a tower, reaches the ground in 4 sec. The height of the tower is $(g = 10 m/s^2)$
(a) 20 m (b) 40 m
(c) 80 m (d) 160 m

28. Three different objects of masses m_1, m_2 and m_3 are allowed to fall from rest and from the same point O along three different frictionless paths. The speeds of the three objects on reaching the ground will be in the ratio of
(a) $m_1 : m_2 : m_3$ (b) $m_1 : 2m_2 : 3m_3$
(c) 1:1:1 (d) $\frac{1}{m_1}:\frac{1}{m_2}:\frac{1}{m_3}$

29. The water drops fall at regular intervals from a tap 5 m above the ground. The third drop is leaving the tap at an instant when the first drop touches the ground. How far above the ground is the second drop at that instant? (Take $g = 10$ m/s^2)
(a) 1.25 m (b) 2.50 m
(c) 3.75 m (d) 5.00 m

30. If a ball is thrown vertically upwards with a velocity of 40 m/s, then velocity of the ball after two seconds will be ($g = 10$ m/s^2)
(a) 15 m/s (b) 20 m/s
(c) 25 m/s (d) 28 m/s

31. The displacement of a particle varies with time (t) as: $s = at^2 - bt^3$. The acceleration of the particle at any given time (t) will be equal to
(a) $\frac{a}{b}$ (b) $\frac{a}{3b}$
(c) $\frac{3b}{a}$ (d) $\frac{2a}{3b}$

32. If a car at rest accelerates uniformly to a speed of 144 km/h in 20 s, it covers a distance of
(a) 2880 m (b) 1440 m
(c) 400 m (d) 20 m

33. A car moving with a speed of 40 km/h can be stopped by applying brakes at least after 2 m. If the same car is moving with a speed of 80 km/h, what is the minimum stopping distance?
(a) 8m (b) 6m
(c) 4m (d) 2m

34. The displacement of a particle is represented by the following equation : $s = 3t^3 + 7t^2 + 5t + 8$ where s is in metre and t in second. The acceleration of the particle at $t = 1$s is
(a) 14 m/s^2 (b) 18 m/s^2
(c) 32 m/s^2 (d) zero

35. A stone is thrown vertically upwards with kinetic energy K. The kinetic energy at the highest point is
(a) $\frac{K}{2}$ (b) $\frac{\sqrt{K}}{2}$
(c) K (d) zero

36. If a ball is thrown vertically upwards with speed u, the distance covered during the last t seconds of its ascent is
(a) $(u + gt)t$ (b) ut
(c) $\frac{1}{2}gt^2$ (d) $ut - \frac{1}{2}gt^2$

37. Choose the correct statements from the following
(a) The magnitude of instantaneous velocity of a particle is equal to its instantaneous speed.
(b) The magnitude of the average velocity in an interval is equal to its average speed in that interval.
(c) It is possible to have a situation in which the speed of the particle is never zero but the average speed in an interval is zero.
(d) It is possible to have a situation in which the speed of particle is zero but the average speed is not zero.

38. Which of the following can be zero, when a particle is in motion for some time?
(a) distance (b) displacement
(c) speed (d) none of them

39. Which of the following decreases in motion along a straight line with constant retardation while the body is moving away from the origin
(a) speed (b) acceleration
(c) displacement (d) none of the above

40. If distance covered by a particle is zero, what can you say about its displacement?
    (a) It may or may not be zero
    (b) It cannot be zero
    (c) It is negative
    (d) It must be zero
41. If displacement of a particle is zero, distance covered by it
    (a) may be zero or may not be zero
    (b) must be zero
    (c) must not be zero
    (d) all are true
42. The numerical ratio of displacement to distance is
    (a) always less than one
    (b) always equal to one
    (c) always more than one
    (d) equal to or less than one
43. The numerical ratio of average velocity to average speed is
    (a) always less than one
    (b) always equal to one
    (c) always more than one
    (d) equal to or less than one.
44. A body moving with uniform velocity is in equilibrium.
    (a) True
    (b) False
    (c) May be true or false
    (d) can't say
45. The location of a particle has changed. What can we say about the displacement and the distance covered by the particle.
    (a) Neither can be zero
    (b) One may be zero
    (c) Both may be zero
    (d) One is +ve, other is –ve
46. Which of the following is not an example of linear motion
    (a) an aeroplane moving on a straight track with increasing speed
    (b) a body in uniform circular motion
    (c) wheel rotating at uniform speed on road
    (d) a body rolling down an inclined plane
47. A body cannot have a
    (a) zero speed and non zero acceleration
    (b) non zero speed and zero acceleration
    (c) constant velocity and a varying speed
    (d) constant speed and a varying velocity
48. A passenger in a moving train tosses a coin. If the coin falls behind him, the train must be moving with
    (a) an acceleration (b) a deceleration
    (c) a uniform speed (d) any of the above
49. A body is imparted motion from rest to move in a straight line. If it is then obstructed, by an opposite force, then
    (a) the body may necessarily change direction
    (b) the body is sure to slow down
    (c) the body will necessarily continue to move in the same direction at the same speed
    (d) None of the above
50. At the top of the trajectory of a projectile, the acceleration is
    (a) the maximum (b) the minimum
    (c) zero (d) g
51. *Assertion*: A body can have an acceleration even if its velocity is zero at a given instant of time.
    Reason: A body is momentarily at rest when it reverses its direction of motion.
    (a) If both Assertion and Reason are true and the Reason is correct explanation of the Assertion
    (b) If both Assertion and Reason are true, but Reason is not correct explanation of the Assertion
    (c) The Assertion is true, but the Reason is false
    (d) The Assertion is false and Reason is true
52. Fig shows the displacement (x) – time (t) graph of a particle moving on the X-axis.

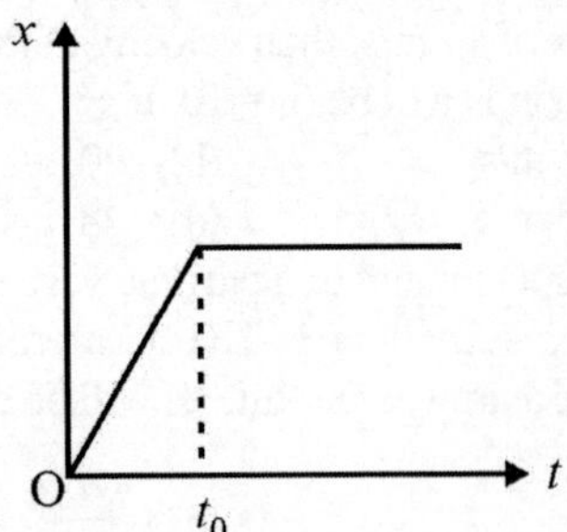

    (a) The particle is at rest
    (b) The particle is continuously going along X-direction
    (c) The velocity of particle increases upto time t_0 and then becomes constant
    (d) The particle moves at a constant velocity up to a time t_0, and then stops

53. The slope of velocity-time graph for motion with uniform velocity is equal to
(a) final velocity (b) initial velocity
(c) zero (d) none of the above

54. Which of the following speed time graphs is not possible

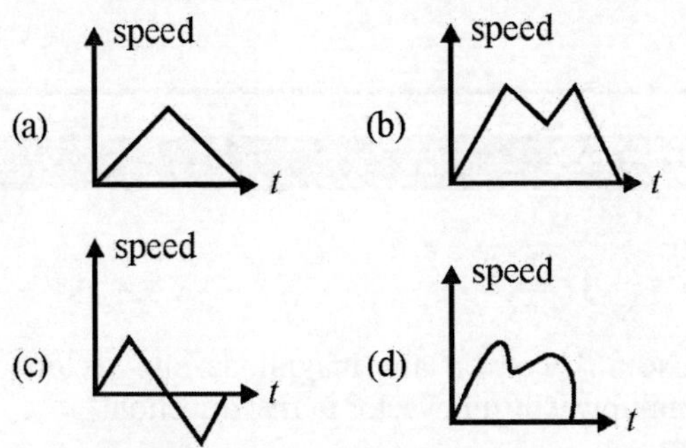

55. Which of the following displacement time graphs is not possible?

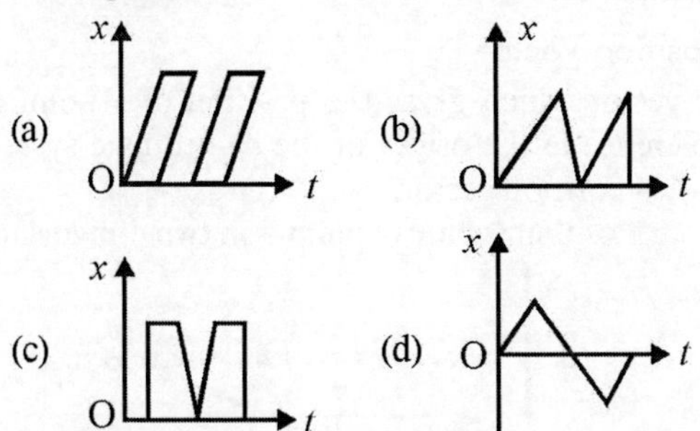

56. Which of the following velocity time graph is not possible?

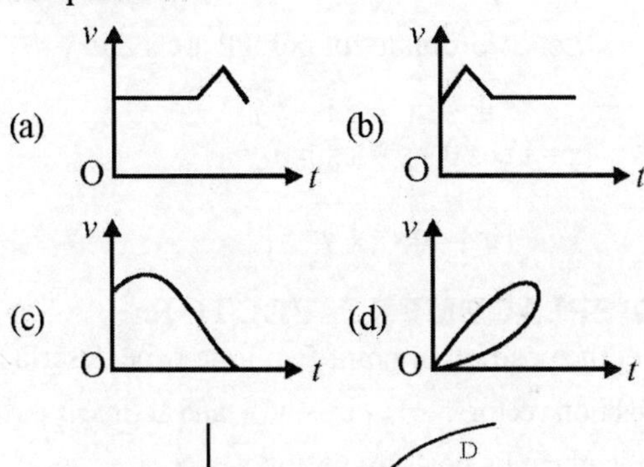

33.

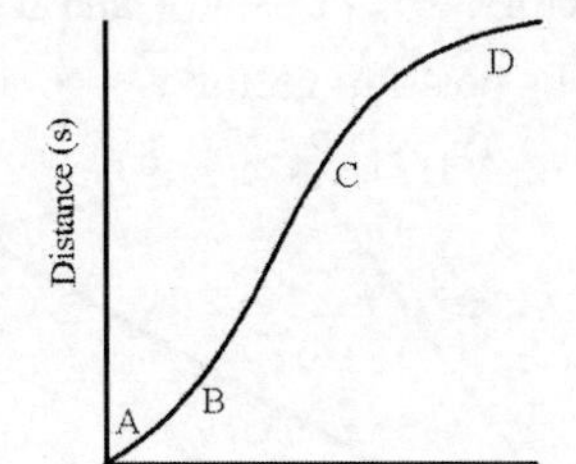

A particle shows distance - time curve as given in this figure. The maximum instantaneous velocity of the particle is around the point:
(a) B (b) C
(c) D (d) A

58. Which of the following time-displacement graph is not possible in nature

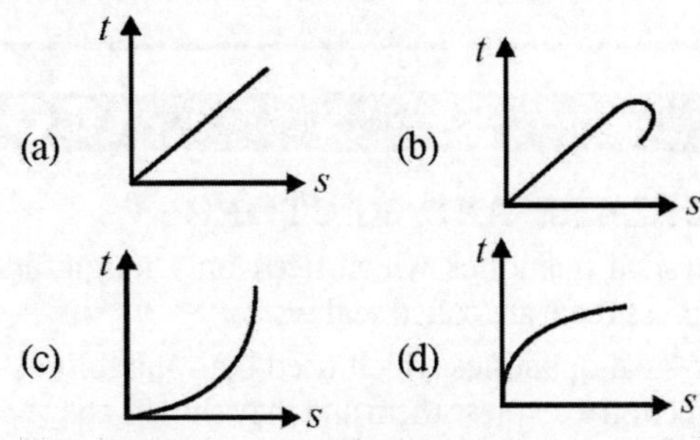

59. Fig shows the time-displacement curve of the particles P and Q. Which of the following statement is correct?

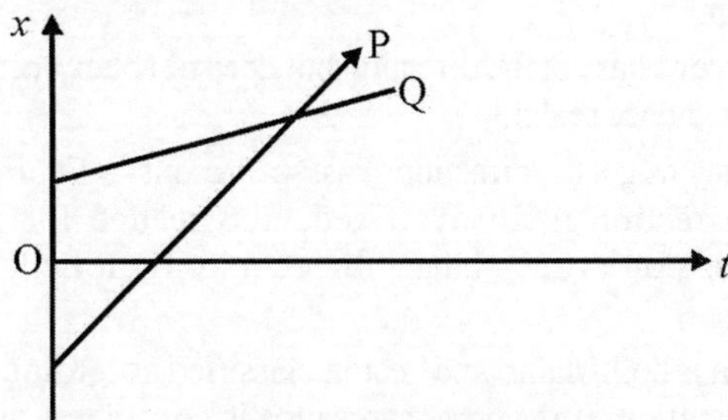

(a) Both P and Q move with uniform equal speed
(b) P is accelerated Q is retarded
(c) Both P and Q move with uniform speeds but the speed of P is more than the speed of Q
(d) Both P and Q move with uniform speeds but the speed of Q is more than the speed of P

60. A body is moving with velocity 30 m/s towards east. After 10 seconds its velocity becomes 40 m/s towards north. The average acceleration of the body is
(a) 1 m/s^2 (b) 7 m/s^2
(c) 7 m/s^2 (d) 5 m/s^2

Motion in a Plane

CONCEPTS

SCALARS AND VECTORS

Physical quantities which need only magnitude to express them are called **scalars**.

Physical quantities which need both magnitude and direction to express them and they should obey vector rules are called **vectors**.

Pressure has got the direction (always perpendicular to surface) but not needed for its expression hence scalar.

Current has got the direction but does not obey vector rules, hence scalar.

Time has got the direction (Past $\rightarrow$ Present $\rightarrow$ Future), as direction is always fixed, not needed for its expression (Time change direction only in fictious stories)

Area is both scalar and vector, classified according to application. (In general cases area is considered as a scalar but in electric field and magnetic field, it is considered as a vector).

Physical quantities which are neither considered as scalar nor vector are called **tensors.** Tensors have different values in different directions at the same point e.g., moment of inertia, stress etc.

Types of Vector

Unit vector : A vector divided by its magnitude is a unit vector. Unit vector for $\vec{A}$ is $\hat{A}$ (read as A cap or A hat)

Since, $\hat{A} = \frac{\vec{A}}{A} \Rightarrow \vec{A} = A\hat{A}$. Thus we can say that unit vector gives us the direction.

Null vector (0) : A vector having zero magnitude and arbitrary direction is called a zero vector or null vector.

Orthogonal unit vectors : $\hat{i}, \hat{j}$ and $\hat{k}$ are called orthogonal unit vectors. These vectors must form a Right Handed Triad (It is a coordinate system such that when we curl the fingers of right hand from x to y then we must get the direction of z along thumb). Then

$$\hat{i} = \frac{\vec{x}}{x}, \hat{j} = \frac{\vec{y}}{y}, \hat{k} = \frac{\vec{z}}{z} \qquad \therefore \vec{x} = x\hat{i},\ \vec{y} = y\hat{j},\ \vec{z} = z\hat{k}$$

Use : To convert any magnitude into vector form, multiply it by unit vector in that direction.

POSITION AND DISPLACEMENT VECTOR

Position Vector

A vector which gives the position of a point with reference to the origin of the co-ordinate system is called position vector.

Position vector of point P in two dimensions.

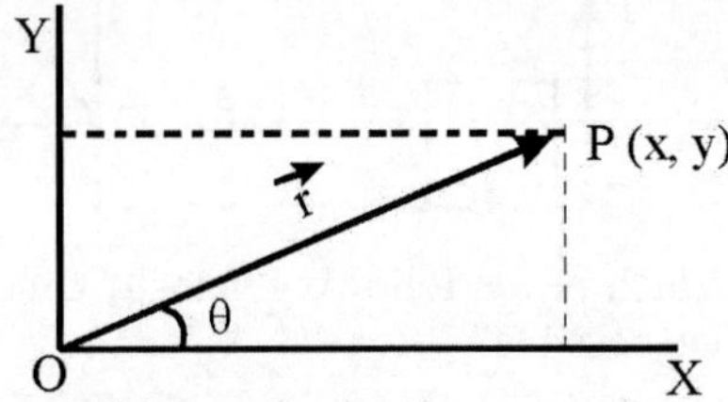

- Let co-ordinates of point P are x and y

$$\vec{OP} = \vec{r} = x\hat{i} + y\hat{j}$$

$$x = r\cos\theta,\ y = r\sin\theta$$

$$|\vec{r}| = \sqrt{x^2 + y^2}$$

DISPLACEMENT VECTOR

Let the position of point P at time t_1 be described by position vector $\vec{r_1} = x_1\hat{i} + y_1\hat{j}$ and at time t_2 position Q is given by position vector $\vec{r_2} = x_2\hat{i} + y_2\hat{j}$

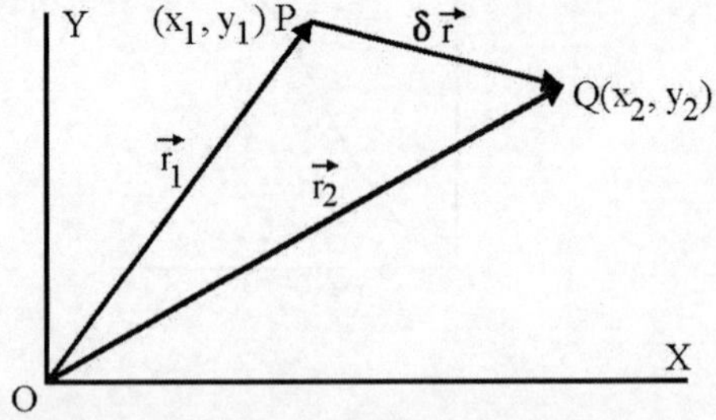

From $\Delta OPQ \Rightarrow \overrightarrow{OP} + \overrightarrow{PQ} = \overrightarrow{OQ}$

or $\quad \overrightarrow{PQ} = \overrightarrow{OQ} - \overrightarrow{OP}$

$\therefore$ Displacement $\overrightarrow{PQ} = \delta \vec{r} = \vec{r}_2 - \vec{r}_1$ in time interval

$\delta t = (t_2 - t_1)$

And similarly component of displacement are

$$s_x = u_x t + \frac{1}{2} a_x t^2 \quad \text{and} \quad s_y = u_y t + \frac{1}{2} a_y t^2$$

Hence, $s_x \hat{i} + s_y \hat{j} = (u_x \hat{i} + u_y \hat{j}) + \frac{1}{2}(a_x \hat{i} + a_y \hat{j}) t^2$

or, $\quad \vec{s} = \vec{u} t + \frac{1}{2} \vec{a} t^2$

Equality of vectors : Two vectors A and B are said to be equal if and only if they have the same magnitude and same direction.

LAWS OF VECTOR ADDITION

Triangle Law of Vector Addition

It states that if two vectors acting on a particle at the same time are represented in magnitude and direction by the two sides of a triangle taken in one order, their resultant vector is represented in magnitude and direction by the third side of the triangle taken in opposite order.

Let the two vectors $\vec{A}$ and $\vec{B}$, inclined at an angle θ be represented in magnitude and direction by two sides $\overrightarrow{OP}$ and $\overrightarrow{PQ}$ of triangle OPQ, taken in order. Then resultant R is represented by the third side $\overrightarrow{OQ}$ of the triangle taken in opposite order. ($\overrightarrow{OQ} \rightarrow$ directed from tail of first vector to head of second vector.)

Magnitude of $\vec{R}$: Draw QN perpendicular to OP produced. In Δ QNP, PN = PQ cos θ = B cos θ, QN = PQ sin θ = B sin θ

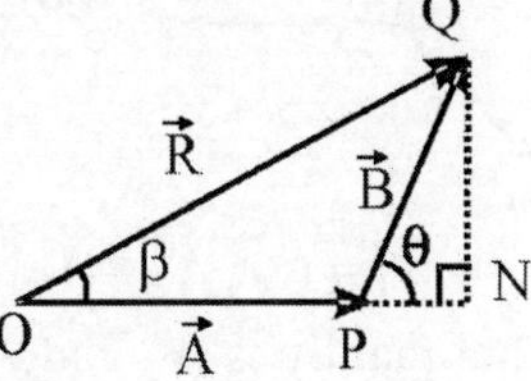

In right angled triangle ONQ, we have

$OQ^2 = ON^2 + NQ^2$

or $\quad OQ^2 = (OP + PN)^2 + NQ^2$

or $\quad R^2 = (A + B\cos\theta)^2 + (B\sin\theta)^2$

or $\quad R = \sqrt{A^2 + B^2 + 2AB\cos\theta}$

Direction of $\vec{R}$: Let the resultant $\vec{R}$ makes an angle β with the direction of $\vec{A}$. Then from right angled triangle QNO,

$$\tan\beta = \frac{QN}{ON} = \frac{QN}{OP + PN} = \frac{B\sin\theta}{A + B\cos\theta}$$

(i) Vector does not obey ordinary algebra.

(ii) In above formulation remember that θ is the angle between the vectors you have to add & angle between vectors means the smaller angle when tails of both vectors are joined.

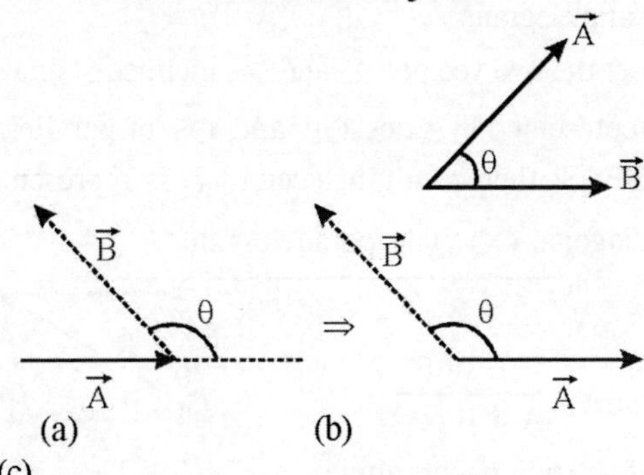

(a) (b)

(c)

In above diagrams (b) & (c) you can see θ by shifting $\vec{B}$ parallel to itself & joining its tail with $\vec{A}$.

Value of $\vec{R}$ ($\vec{R} = \vec{A} + \vec{B}$) depends on angle between vectors.

$|\vec{R}|$ is maximum if $\cos\theta = 1$ *i.e.*, $\theta = 0°$ (parallel vectors)

$\vec{A} \longrightarrow \vec{B} \longrightarrow$

$$R_{max} = \sqrt{A^2 + B^2 + 2AB} = A + B$$

$|\vec{R}|$ is minimum if $\cos\theta = -1$ *i.e.*, $\theta = 180°$ (antiparallel vectors)

$\longleftarrow \vec{B} \quad \vec{A} \longrightarrow$

$$R_{min} = \sqrt{A^2 + B^2 - 2AB} = A - B$$

Thus $R \in [A - B, A + B]$ i.e. $|\vec{R}|$ can have value between (A – B) & (A + B) including both. Like, the vector sum of the forces of 10N and 6N can be between [4, 16].

(a) To a vector only a vector of same type can be added that represents the same physical quantity and the resultant is a vector of the same type.

(b) Vector addition is commutative, i.e.,

$\vec{A} + \vec{B} = \vec{B} + \vec{A}$

(c) Vector addition is associative, i.e.,

$\vec{A} + (\vec{B} + \vec{C}) = (\vec{A} + \vec{B}) + \vec{C}$

(d) If the vectors A and B are orthogonal, *i.e.*, $\theta = 90°$ then $R = \sqrt{A^2 + B^2}$

Parallelogram Law of Vector Addition

It states that if two vectors are represented in magnitude and direction by the two adjacent sides of a parallelogram then their resultant is represented in magnitude and direction by the diagonal of the parallelogram.

Let the two vectors $\vec{A}$ and $\vec{B}$, inclined at angle θ be represented by sides $\overrightarrow{OP}$ and $\overrightarrow{OS}$ of parallelogram OPQS, then resultant vector $\vec{R}$ is represented by diagonal $\overrightarrow{OQ}$ of the parallelogram.

$$R = \sqrt{A^2 + B^2 + 2AB\cos\theta};$$

$$\tan\beta = \frac{B\sin\theta}{A + B\cos\theta}$$

If $\theta < 90°$ (Acute angle),

$\vec{R} = \vec{A} + \vec{B}$, $\vec{R}$ is called main (major) diagonal of parallelogram.

If $\theta > 90°$ (Obtuse angle), $\vec{R} = \vec{A} + \vec{B}$, $\vec{R}$ is called minor diagonal.

Polygon Law of Vector Addition

If a number of non-zero vectors are represented by the (n – 1) sides of an n-sided polygon in order then the resultant is given by the closing side or the nth side of the polygon taken in opposite order. So,

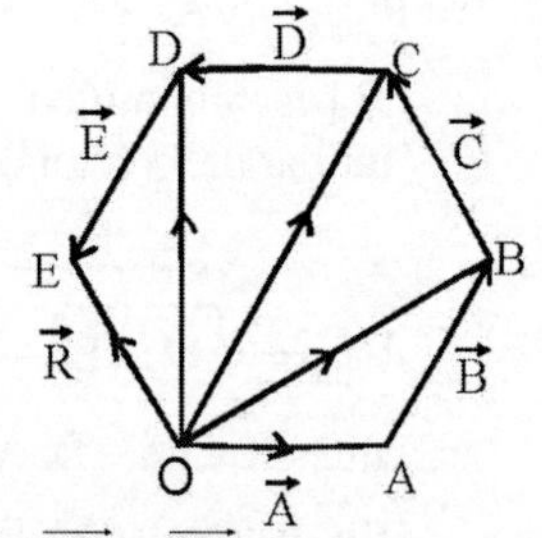

$$\vec{R} = \vec{A} + \vec{B} + \vec{C} + \vec{D} + \vec{E}$$

$$\overrightarrow{OA} + \overrightarrow{AB} + \overrightarrow{BC} + \overrightarrow{CD} + \overrightarrow{DE} = \overrightarrow{OE}$$

LAWS OF VECTOR ALGEBRA

- $\vec{A} + \vec{B} = \vec{B} + \vec{A}$ (Commutative law of addition)
- $\vec{A} + (\vec{B} + \vec{C}) = (\vec{A} + \vec{B}) + \vec{C}$ (Associative law of addition)
- $m\vec{A} = \vec{A}m$
- $m(n\vec{A}) = (mn)\vec{A}$
- $(m+n)\vec{A} = m\vec{A} + n\vec{A}$
- $m(\vec{A} + \vec{B}) = m\vec{A} + m\vec{B}$

SUBTRACTION OF VECTORS

Subtraction of a vector $\vec{B}$ from a vector $\vec{A}$ is defined as the addition of vectors $-\vec{B}$ (negative of vector $\vec{B}$) to vector $\vec{A}$.

Thus $\vec{A} + (-\vec{B}) = \vec{A} - \vec{B}$

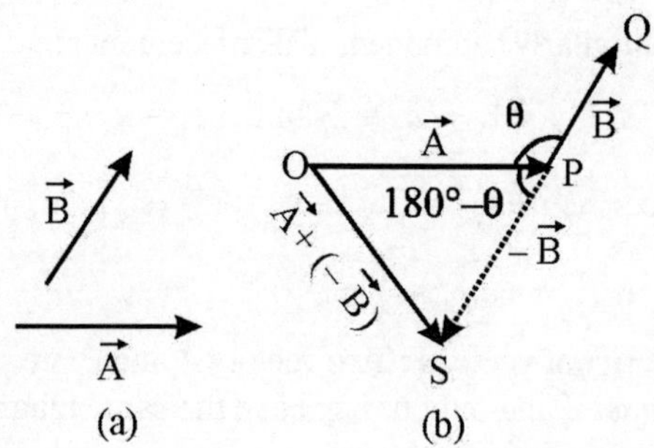

If θ is the angle between $\vec{A}$ and $\vec{B}$, then angle between $\vec{A}$ and $-\vec{B}$ is $(180° - \theta)$.

(a) Addition of a vector to its own negative vector is a null vector $\vec{A} + (-\vec{A}) = \vec{0}$ i.e. a vector with zero magnitude and an arbitrary direction.

(b) Subtraction is not commutative, i.e., $\vec{A} - \vec{B} \neq \vec{B} - \vec{A}$

RESOLUTION OF VECTORS

Consider a vector $\vec{A}$ that lies in xy plane, $\vec{A} = \vec{A}_1 + \vec{A}_2$ as shown.

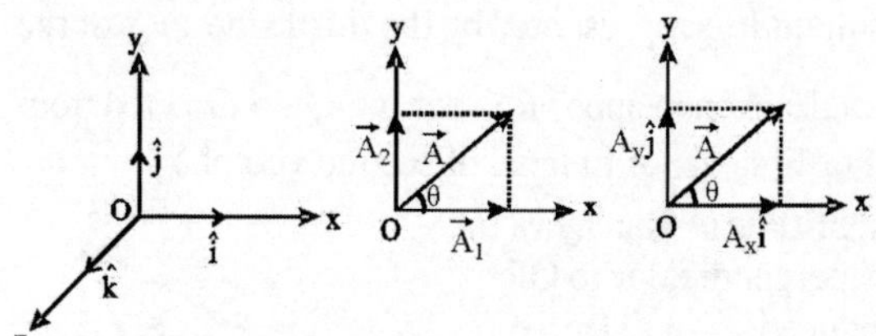

$$\vec{A}_1 = A_x\hat{i},\ \vec{A}_2 = A_y\hat{j} \Rightarrow \vec{A} = A_x\hat{i} + A_y\hat{j}$$

The quantities A_x and A_y are called x- and y-components of the vector $\vec{A}$. A_x is itself not a vector but $A_x\hat{i}$ is a vector and so is $A_y\hat{j}$.

$$A_x = A\cos\theta \quad \text{and} \quad A_y = A\sin\theta$$

VELOCITY AND ACCELERATION IN VECTOR FORM

Average velocity ($\overline{v}$): It is the ratio of the displacement and the corresponding time interval

$$\bar{v} = \frac{\Delta r}{\Delta t} = \frac{\Delta x\hat{i} + \Delta y\hat{j}}{\Delta t} = \hat{i}\frac{\Delta x}{\Delta t} + \hat{j}\frac{\Delta y}{\Delta t}$$

or $\quad \bar{v} = \bar{v}_x\hat{i} + \bar{v}_y\hat{j}$

Instantaneous velocity : If is the limiting value of the average velocity as the time interval approaches zero

$$v = \lim_{\Delta t \to 0} \frac{\Delta r}{\Delta t} + \frac{dr}{dt}$$

Average acceleration : If is the change in velocity of an object divided by the time interval.

$$\bar{a} = \frac{\Delta v}{\Delta t} = \frac{\Delta v_x}{\Delta t}\hat{i} + \frac{\Delta v_y}{\Delta t}\hat{j}$$

In terms of x and y

$$a_x = \frac{d^2x}{dt^2}\ ;\ a_y = \frac{d^2y}{dt^2}$$

Instantaneous accelesation: It is the limiting value of the average acceleration as the time interval approaches zero.

$$a = \lim_{\Delta t \to 0} \frac{\Delta v}{\Delta t}\ ; \qquad a = \bar{a}_x\hat{i} + \bar{a}_y\hat{j}$$

SCALAR PRODUCT AND VECTOR PRODUCT

Scalar or Dot Product

The scalar product or dot product of any two vectors $\vec{A}$ and $\vec{B}$, denoted as $\vec{A}.\vec{B}$ (read $\vec{A}$ dot $\vec{B}$) is defined as the product of their magnitude with cosine of angle between them. Thus, $\vec{A}.\vec{B} = AB\cos\theta$ (here θ is the angle between the vectors)

Properties of Scalar Product

- It is always a scalar.
- It is commutative, *i.e.*, $\vec{A}.\vec{B} = \vec{B}.\vec{A}$
- It is distributive, *i.e.* $\vec{A}.(\vec{B}+\vec{C}) = \vec{A}.\vec{B} + \vec{A}.\vec{C}$

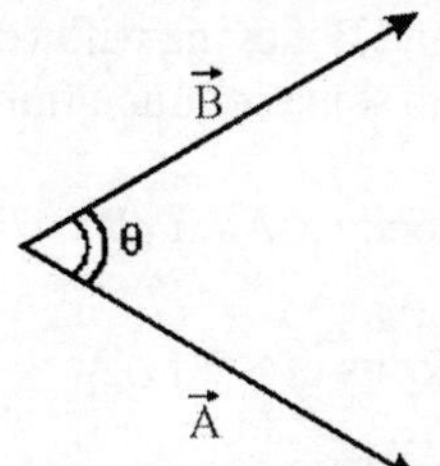

- As by definition $\vec{A}.\vec{B} = AB\cos\theta$
  The angle between the vectors

$$\theta = \cos^{-1}\left[\frac{\vec{A}.\vec{B}}{AB}\right]$$

- $\vec{A}.\vec{B} = A\,(B\cos\theta) = B\,(A\cos\theta)$
  Geometrically, $B\cos\theta$ is the projection of $\vec{B}$ onto $\vec{A}$ and $A\cos\theta$ is the projection of $\vec{A}$ onto $\vec{B}$ as shown. So $\vec{A}.\vec{B}$ is the product of the magnitude of $\vec{A}$ and the component of $\vec{B}$ along $\vec{A}$ and vice versa.

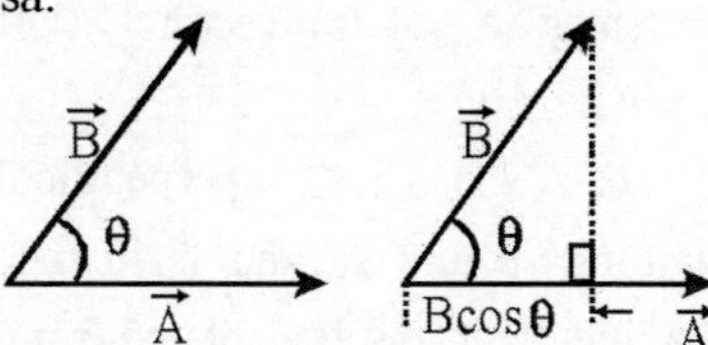

Component of $\vec{B}$ along

$$\vec{A} = B\cos\theta = \frac{\vec{A}.\vec{B}}{A} = \hat{A}.\vec{B}$$

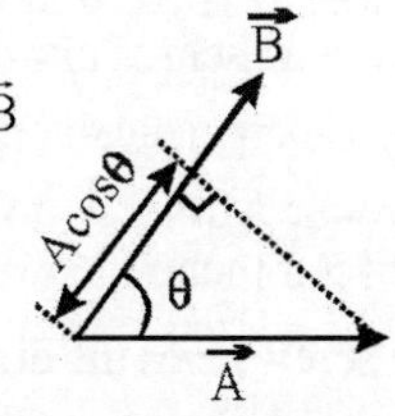

Component of $\vec{A}$ along

$$\vec{B} = A\cos\theta = \frac{\vec{A}.\vec{B}}{B} = \vec{A}.\hat{B}$$

- Scalar product of two vectors will be maximum when $\cos\theta$ is maximum *i.e.*, $\cos\theta = 1$, *i.e.* $\theta = 0°$,
  i.e., vectors are parallel $\Rightarrow (\vec{A}.\vec{B})_{max} = AB$
- If the scalar product of two non-zero vectors vanishes then the vectors are orthogonal.
- The scalar product of a vector by itself is termed as self dot product and is given by
  $(\vec{A})^2 = \vec{A}.\vec{A} = AA\cos\theta = A^2 \Rightarrow A = \sqrt{\vec{A}.\vec{A}}$
- In case of unit vector $\hat{n}$, $\hat{n}.\hat{n} = 1 \times 1 \times \cos 0° = 1$
  $\Rightarrow \quad \hat{n}.\hat{n} = \hat{i}.\hat{i} = \hat{j}.\hat{j} = \hat{k}.\hat{k} = 1$
- In case of orthogonal unit vectors, $\hat{i}$, $\hat{j}$ and $\hat{k}$;
  $\hat{i}.\hat{j} = \hat{j}.\hat{k} = \hat{k}.\hat{i} = 0\frac{n!}{r!(n-r)!}$
- $\vec{A}.\vec{B} = (\hat{i}A_x + \hat{j}A_y + \hat{k}A_z).(\hat{i}B_x + \hat{j}B_y + \hat{k}B_z)$
  $= [A_xB_x + A_yB_y + A_zB_z]$

Uses of Scalar Product

(i) To find angle between two vectors.

(ii) To find condition for two vectors to be perpendicular to each other $(\vec{A}\cdot\vec{B} = 0)$

(iii) To find physical quantities based on scalar product.

Vector or Cross Product

The vector product or cross product of two vectors is defined as a vector having a magnitude equal to the product of the magnitudes of two vectors with the sine of angle between them, and direction perpendicular to the plane containing the two vectors in accordance with right hand screw rule.

Thus, if $\vec{A}$ and $\vec{B}$ are two vectors, then their vector product written as $\vec{A}\times\vec{B}$ is a vector $\vec{C}$ defined by

$\vec{C}=\vec{A}\times\vec{B}=AB\sin\theta\hat{n}$

The direction of $\vec{A}\times\vec{B}$, i.e. $\vec{C}$ is perpendicular to the plane containing vectors $\vec{A}$ and $\vec{B}$ in the sense of advance of a right handed screw rotated from $\vec{A}$ (first vector) to $\vec{B}$ (second vector) through the small angle between them. Thus, if a right handed screw whose axis is perpendicular to the plane formed by $\vec{A}$ and $\vec{B}$ is rotated from $\vec{A}$ to $\vec{B}$ through the smaller angle between them, then the direction of advancement of the screw gives the direction of $\vec{A}\times\vec{B}$ i.e. $\vec{C}$.

$$\vec{C}\perp\vec{A},\ \vec{C}\perp\vec{B} \quad \text{Hence } \vec{C}.\vec{A}=0,\ \vec{C}.\vec{B}=0$$

Properties of Vector Product

- $\vec{A}\times\vec{B}\neq\vec{B}\times\vec{A}$ (Not Commutative)
  $\vec{A}\times\vec{B}=-\vec{B}\times\vec{A}$
- $\vec{A}\times(\vec{B}+\vec{C})=(\vec{A}\times\vec{B})+(\vec{A}\times\vec{C})$ (Follows Distributive law)
- $\vec{A}\times\vec{A}=0$
- The cross product of two vectors represents the area of the parallelogram formed by them.
- $\hat{i}\times\hat{i}=\hat{j}\times\hat{j}=\hat{k}\times\hat{k}=0$

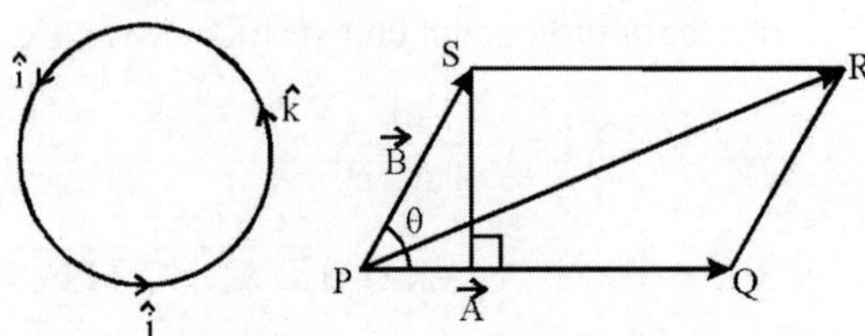

- $\hat{i}\times\hat{j}=\hat{k};\ \hat{j}\times\hat{k}=\hat{i};\ \hat{k}\times\hat{i}=\hat{j}$
- $\vec{A}\times\vec{B}=(A_x\hat{i}+A_y\hat{j}+A_z\hat{k})\times(B_x\hat{i}+B_y\hat{j}+B_z\hat{k})$

$$=\begin{vmatrix}\hat{i} & \hat{j} & \hat{k}\\ A_x & A_y & A_z\\ B_x & B_y & B_z\end{vmatrix}$$

$$=(A_yB_z-A_zB_y)\hat{i}+(A_zB_x-A_xB_z)\hat{j}+(A_xB_y-A_yB_x)\hat{k}$$

- $|\vec{A}\times\vec{B}|=A(B\sin\theta)$
  $=$ area of parallelogram PQRS
  $=2$ (area of ΔPQR)
- A unit vector which is perpendicular to $\vec{A}$ as well as $\vec{B}$ is $\frac{\vec{A}\times\vec{B}}{|\vec{A}\times\vec{B}|}$

Uses of Vector Product

(i) To find condition for two vectors to be parallel to each other $\vec{A}\times\vec{B}=\vec{0}$

(ii) To find physical quantities based on cross product.

(iii) To find area of parallelogram/triangle.

LAMI'S THEOREM

It states that if three forces acting at a point are in equilibrium, then each force is proportional to the sine of the angle between the other two forces.

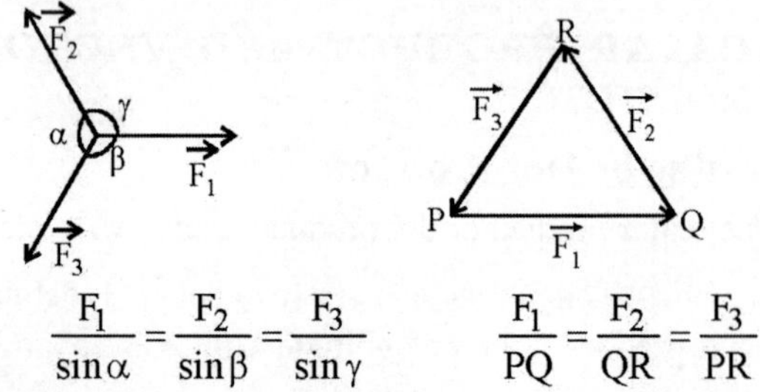

$$\frac{F_1}{\sin\alpha}=\frac{F_2}{\sin\beta}=\frac{F_3}{\sin\gamma} \qquad \frac{F_1}{PQ}=\frac{F_2}{QR}=\frac{F_3}{PR}$$

MOTION IN A PLANE

Motion in a plane can be treated as two separate simultaneous one-dimensional motions with constant acceleration along two perpendicular directions.

Relative Velocity in Two - Dimensional Motion

If two objects A and B moving with velocities V_A and V_B with respect to some common frame of reference, then

(i) relative velocity of A w.r.t B

$\vec{v}_{AB}=\vec{v}_A-\vec{v}_B$

(ii) relative velocity of B w.r.t. A

$\vec{v}_{BA}=\vec{v}_B-\vec{v}_A$

Therefore, $\vec{v}_{AB}=-\vec{v}_{BA}$ and $|\vec{v}_{AB}|=|\vec{v}_{BA}|$

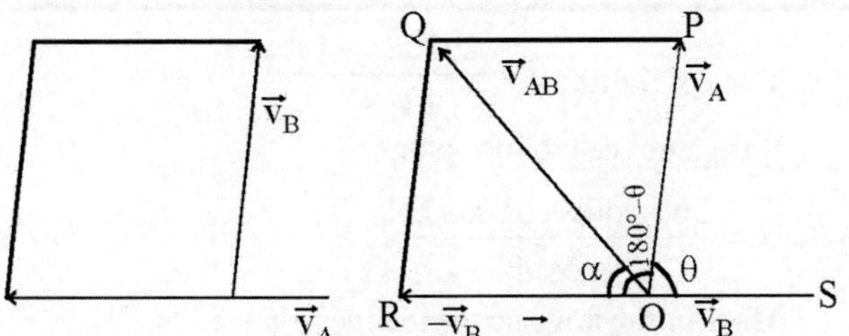

Let $\vec{v}_A$ and $\vec{v}_B$ act at a point at an angle θ. The resultant of $\vec{v}_A$ (along OP) and $-\vec{v}_B$ (along OR) will be represented by the diagonal OQ. Therefore the relative velocity $\vec{v}_{AB}$ is represented by OQ in magnitude and direction,

$$v_{AB} = \sqrt{v_A^2 + v_B^2 + 2v_A.v_B.\cos(180° - \theta)}$$

$$= \sqrt{v_A^2 + v_B^2 - 2v_A.v_B.\cos\theta}$$

and, $\tan\alpha = \dfrac{v_A \sin(180° - \alpha)}{v_B + v_A \cos(180° - \alpha)}$

$$= \frac{v_A \sin\alpha}{v_B - v_A \cos\alpha}$$

Swimmer Crossing a River :

(a) **For crossing the river through minimum distance**

(i) Starting from point P (in fig.) on one of the bank AB of the river a person has to move in a direction PR making angle θ from PQ in order to reach the point Q just opposite P.

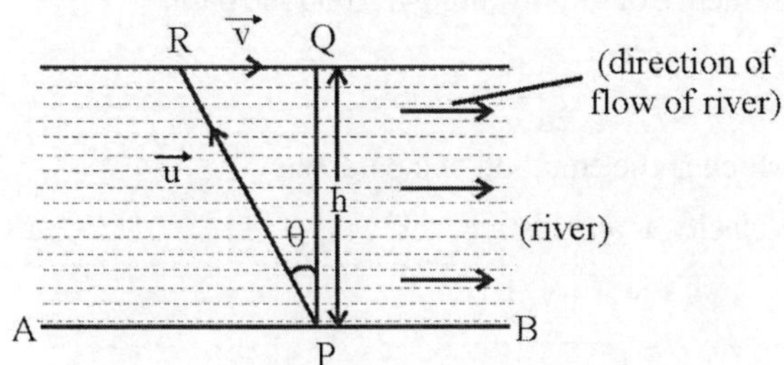

(ii) From the figure $\sin\theta = \dfrac{v}{u}$, $\cos\theta = \dfrac{h}{ut}$,

$$t = \frac{h}{u\cos\theta} = \frac{h}{\sqrt{u^2 - v^2}}$$

where $\vec{v}$ is velocity of the flow of river and $\vec{u}$ is the velocity of person.

(b) **For crossing the river in minimum time**

If the person moves in a direction perpendicular to the direction of river flow :

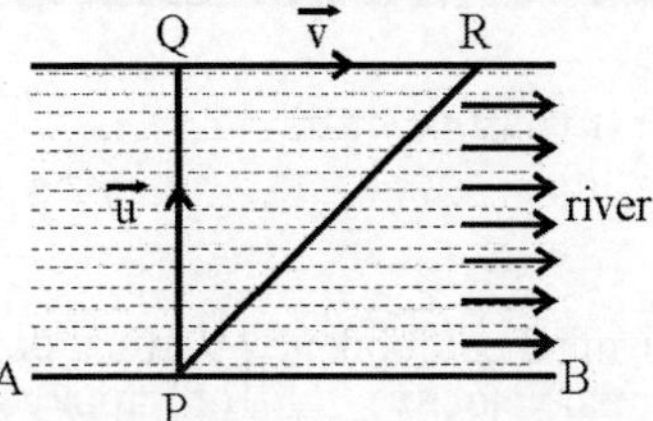

(i) In this case θ = 0° and cos θ will be maximum and consequently t will be minimum, i.e. $t_{min} = \dfrac{h}{u}$

(ii) The resultant velocity of the person $= \sqrt{u^2 + v^2}$ in the direction of $\overrightarrow{PR}$

(iii) The distance covered in the direction of river flow $= \dfrac{hv}{u} = QR$

(iv) The actual distance covered by the person $= t_{min}\sqrt{u^2 + v^2} = PR$

PROJECTILE MOTION

An object in flight after being projected is called a **projectile**. If an object is projected with initial velocity v_0 making an angle θ_0 with x-axis and assume its initial position to coincide with the origin of the coordinate system, then to find the position, velocity, height, trajectory equation, time of flight and range divide two dimensional motion in x and y directions and use motion in 1-D equation in both the direction and solve them simultaneously.

The shape of the trajectory of the motion of an object is not determined by the acceleration alone but also depends on the initial conditions of motion (initial position and initial velocity). For example, the trajectory of an object moving under the same acceleration due to gravity can be a straight line or a parabola depending on the initial conditions.

Consider a particle which is projected from a point O on level ground with a velocity u at an angle α to the horizontal, reaching ground level again at a point A.

The time of flight is $\dfrac{2u\sin\alpha}{g}$

The maximum height is

$$h = \frac{u^2 \sin^2\alpha}{2g}$$

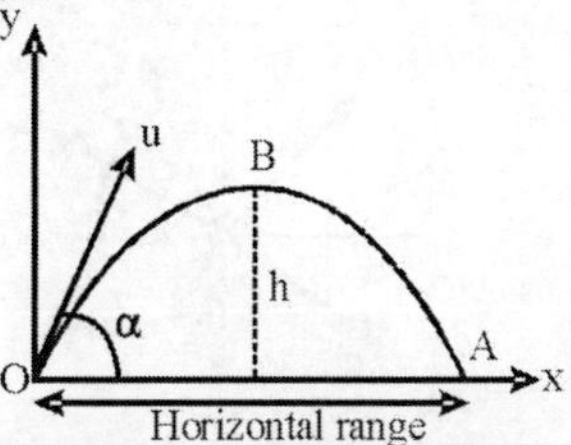

Range R is given by :

$$R = \frac{2}{g}(u\cos\theta)(u\sin\theta)$$

$$= \frac{2}{g}\begin{pmatrix}\text{Horizontal component} \\ \text{of initial velocity}\end{pmatrix}\begin{pmatrix}\text{Vertical component} \\ \text{of initial velocity}\end{pmatrix}$$

$$R = \frac{u^2 \sin 2\alpha}{g}$$

Trajectory equation : $y = x\tan\theta - \frac{gx^2}{2u^2\cos^2\theta}$ represents a parabola

Another form : $y = x\tan\theta - \frac{x^2\tan\theta}{R} = x\tan\theta\left(1 - \frac{x}{R}\right)$

Maximum range : For given speed, range varies with angle of projection and it is maximum at $\theta = 45^\circ$.

$$R = \frac{u^2 \sin 2\theta}{g}$$

$u \rightarrow$ constant, $R \rightarrow$ maximum, $\sin 2\theta \rightarrow$ maximum

$$(\sin 2\theta)_{max} = 1 \ ; \ 2\theta = 90^\circ \Rightarrow \theta = 45^\circ$$

The equation to the parabola traced by a body projected horizontally from the top of a tower of height y, with a velocity u is $y = gx^2/2u^2$, where x is the horizontal distance covered by it from the foot of the tower.

Projectile Motion up an inclined plane

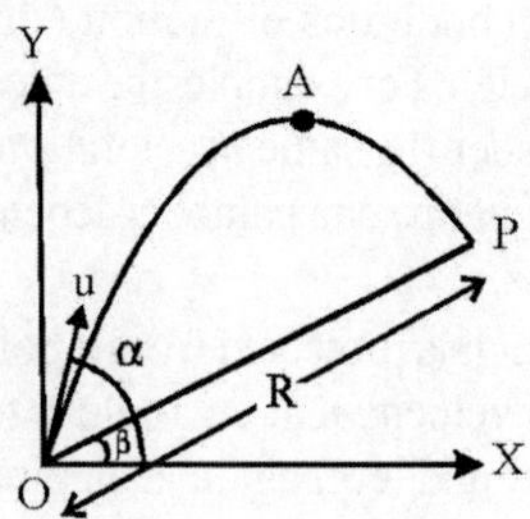

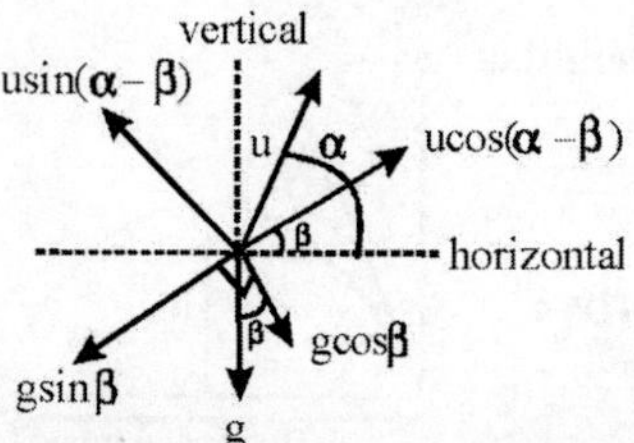

Time of flight : $T = \frac{2u\sin(\alpha - \beta)}{g\cos\beta}$

Range up the inclined plane :

$$R = \frac{2u^2\sin(\alpha - \beta).\cos\alpha}{g\cos^2\beta}$$

Maximum range up the inclined plane :

$$R_{max} = \frac{u^2}{g(1 + \sin\beta)}$$

Projectile Motion Down an Inclined Plane

Time of flight : $(T) = \frac{2u\sin(\alpha + \beta)}{g\cos\beta}$

Range down an inclined plane

$$R = \frac{2u^2\sin(\alpha + \beta)\cos\alpha}{g\cos^2\beta}$$

Direction for maximum range down an inclined plane

$$\alpha = \frac{90^\circ - \beta}{2} \Rightarrow 45^\circ - \frac{\beta}{2}$$

Maximum range down an inclined plane

$$R_{max} = \frac{u^2}{g(1 - \sin\beta)}$$

HORIZONTAL PROJECTILE

In this, the body is given an initial velocity directed along the horizontal and then it moves under the influence of gravity along a parabolic path.

$$y = \left(\frac{g}{2u^2}\right)x^2$$

which is the equation of a parabola

Velocity at any instant

$$\vec{v} = v_x\hat{i} + v_y\hat{j}$$

$$v = \sqrt{u^2 + g^2t^2}$$

If β is the angle made by $\vec{v}$ with the horizontal, then

$$\tan\beta = \frac{v_y}{v_x} = \frac{gt}{u}$$

If h is the distance of the ground from the point of projection, T is the time taken to strike the ground and R is the horizontal range of the projectile then

$$T = \sqrt{\frac{2h}{g}} \text{ and } R = u\sqrt{\frac{2h}{g}}$$

Case 1: If the projectile is projected from top of tower of height 'h', in horizontal direction, then the height of tower, range and time of flight are related as :

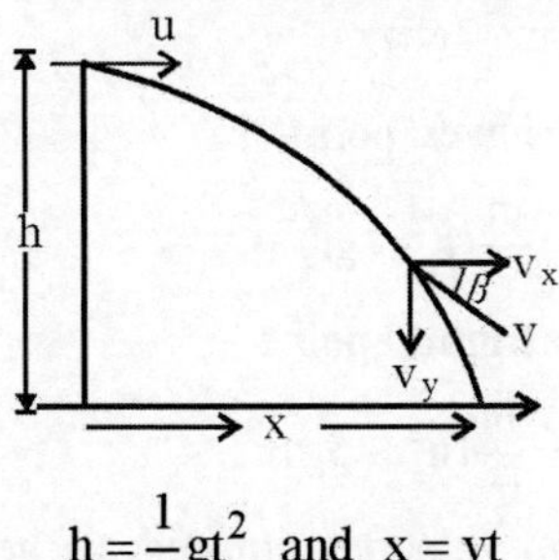

$$h = \frac{1}{2}gt^2 \text{ and } x = vt$$

Case 2: If a particle is projected at an angle (θ) in upward direction from the top of a tower of height h with velocity u, then

$u_y = u \sin\theta$

$a_y = -g$

$u_x = u \cos\theta$

$a_x = 0$

$h = u \sin\theta . t - \frac{1}{2}gt^2$

and $x = u \cos\theta . t$

+ − u sin θ, u, A, θ, u cos θ, h, x, B

Case 3: If a body is projected at an angle (θ) from the top of a tower in downward direction then

$u_y = -u \sin\theta,\ u_x = u \cos\theta,\ a_x = 0$

$a_y = g - h = -u \sin\theta . t - \frac{1}{2}gt^2$ and $x = u \cos\theta . t$

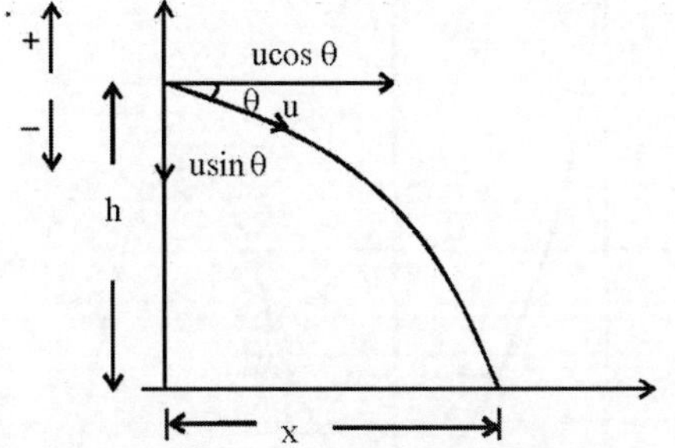

UNIFORM CIRCULAR MOTION

When an object follows a circular path at constant speed, the motion of the object is called **uniform circular motion**. The magnitude of its acceleration i.e., centripetal acceleration $a_c = v^2/R$. The direction of a_c is always towards the centre of the circle. The resultant acceleration of an object in circular motion is towards the centre only if the speed is constant. The angular speed ω, is the rate of change of angular distance. It is related to velocity v by $v = \omega R$. The acceleration is $a_c = \omega^2 R$.

There can be no circular motion without centripetal force

$$F_c = \frac{mv^2}{R}$$

In a uniform circular motion, velocity and acceleration are constants only in magnitude. Their directions change.

In a uniform circular motion, the kinetic energy of the body is a constant.

In a uniform circular motion, the work done by centripetal force in moving from any point in the circle to any other point will be zero. The centripetal force **F** is along the radius and the displacement **s** (which is along velocity) is along the tangent. This makes

$$W = \mathbf{F.s} = Fs\cos 90° = 0.$$

$a_r = \omega^2 r = \frac{v^2}{r} = \omega v$ (Always applicable)

$a_r = 4\pi^2 n^2 r = \frac{4\pi^2}{T^2} r$ (Applicable in uniform circular motion)

n = frequency of rotation, T = time period of rotation.

$$\vec{a}_r = \vec{\omega} \times \vec{v}$$

NON-UNIFORM CIRCULAR MOTION

An object moving in a circle with variable speed is said to be in non-uniform circular motion. If the angular velocity varies with time, the object has two accelerations possessed by it, centripetal acceleration (a_c) and tangential accelaration (a_T) and both are perpendicular to each other.

Net acceleration

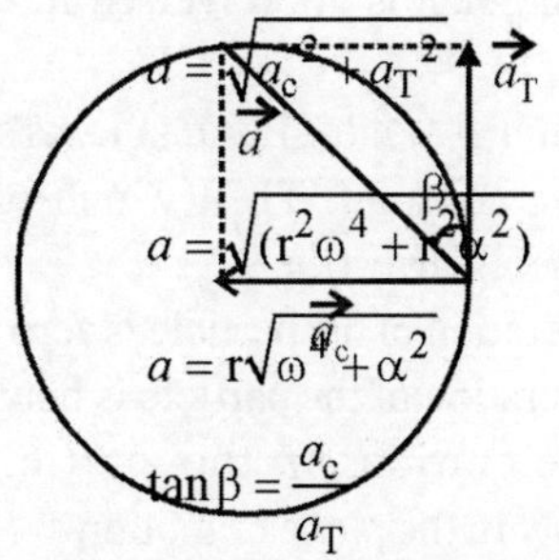

MOTION IN A VERTICAL CIRCLE

When an object moves in a vertical circle, acceleration due to gravity acts at every point and changes the speed at every point.

Consider an object of mass m attached to a string of length r to be whirled in a vertical circle.

Velocity at point P at vertical height *h*.

Loss in K.E = Gain in P.E.

$$\frac{1}{2}mu^2 - \frac{1}{2}mv^2 = mgh - 0$$

$$\therefore \quad v = \sqrt{u^2 - 2gh}$$

Tension at point P

$$T - mg\cos\theta = \frac{mv^2}{r}$$

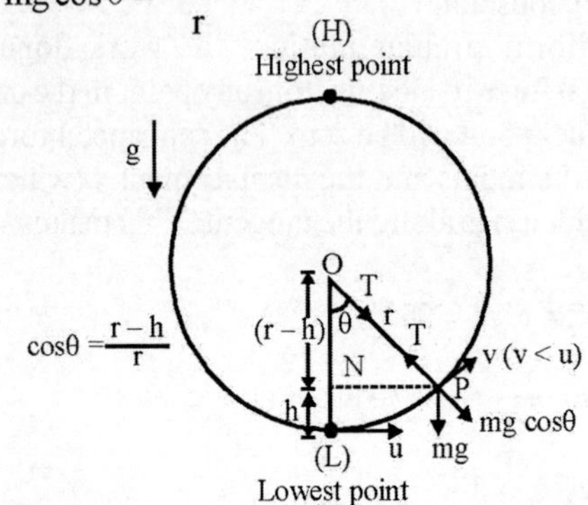

$$T = mg\cos\theta + \frac{mv^2}{r}$$

$$T = \frac{m}{r}(u^2 + gr - 3gh)$$

Tension at the lowest point

$$T_L = \frac{m}{r}(u^2 + gr)$$

Tension at the highest point

$$T_H = \frac{m}{r}(u^2 - 5gr)$$

The minimum velocity required at the bottom to complete the circle is given by

$$v_{min} \geq \sqrt{5gr}$$

The velocity at the highest point is

Exercise

1. In a two dimensional motion, instantaneous speed v_0 is a positive constant. Then which of the following are necessarily true?
   (a) The average velocity is not zero at any time
   (b) Average acceleration must always vanish
   (c) Displacements in equal time intervals are equal
   (d) Equal plan lengths are traversed in equal intervals

2. In a two dimensional motion, instantaneous speed v_0 is a positive constant. Then which of the following are necessarily true?
   (a) The acceleration of the particle is zero
   (b) The acceleration of the particle is bounded
   (c) The acceleration of the particle is necessarily in the plane of motion
   (d) The particle must be undergoing a uniform circular motion

3. A particle slides down a frictionless parabolic ($y = x^2$) track ($A - B - C$) starting from rest at point A figure. Point B is at the vertex of parabola and point C is at a height less than that of point A. After C, the particle moves freely in air as a projectile. If the particle reaches highest point at P, then

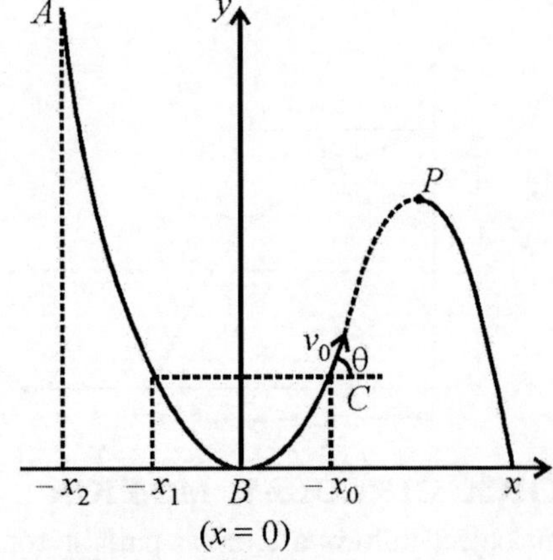

   (a) KE at P = KE at B
   (b) height at P = height at A
   (c) total energy at P = total energy at A
   (d) time of travel from A to B = time of travel from B to P.

4. It is found that $|A + B| = |A|$. This necessarily implies,

(a) $B = 0$
(b) A, B are antiparallel
(c) A, B are perpendicular
(d) $A.B \leq 0$

5. 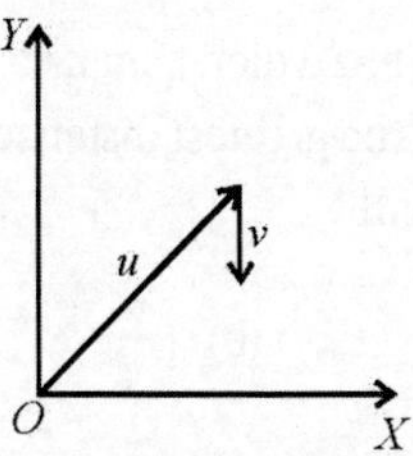

Figure shows the orientation of two vectors u and v in the XY plane.

If $u = a\hat{i} + b\hat{j}$ and $v = p\hat{i} + q\hat{j}$

which of the following is correct?
(a) a and p are positive while b and q are negative
(b) a, p and b are positive while q is negative
(c) a, q and b are positive while p is negative
(d) a, b, p and q are all positive

6. Which one of the following statements is true?
(a) A scalar quantity is the one that is conserved in a process
(b) A scalar quantity is the one that can never take negative values
(c) A scalar quantity is the one that does not vary from one point to another in space
(d) A scalar quantity has the same value for observers with different orientations of the axes

7. The angle between $A = \hat{i} + \hat{j}$ and $B = \hat{i} - \hat{j}$
(a) 45° (b) 90°
(c) −45° (d) 180°

8. Which of the following is true regarding projectile motion?
(a) horizontal velocity of projectile is constant
(b) vertical velocity of projectile is constant
(c) acceleration is not constant
(d) momentum is constant

9. The angular speed of a fly-wheel making 120 revolutions/minute is
(a) π rad/sec (b) 4π rad/sec
(c) 2π rad/sec (d) $4\pi^2$ rad/sec

10. Figure along side shows a body of mass m moving with a uniform speed v along a circle of radius r.

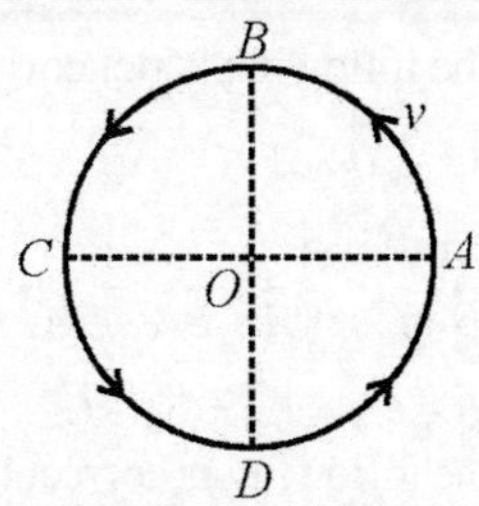

The change in velocity in going from A to B is
(a) $v\sqrt{2}$ (b) $v/\sqrt{2}$
(c) v (d) zero

11. If a unit vector is represented by $0.5\hat{i} + 0.8\hat{j} + c\hat{k}$, then the value of c is
(a) 1 (b) $\sqrt{0.8}$
(c) $\sqrt{0.11}$ (d) $\sqrt{0.01}$

12. Two vectors are perpendicular if
(a) $\vec{A}.\vec{B} = 1$ (b) $\vec{A} \times \vec{B} = 0$
(c) $\vec{A}.\vec{B} = 0$ (d) $\vec{A}.\vec{B} = AB$

13. A particle A is dropped from a height and another particle B is projected in horizontal direction with speed of 5 m/sec from the same height, then correct statement is
(a) particle A will reach at ground first with respect to particle B
(b) particle B will reach at ground first with respect to particle A
(c) both particles will reach at ground simultaneously
(d) both particles will reach at ground with same speed

14. The resultant of $\vec{A} \times 0$ will be equal to
(a) zero (b) A
(c) zero vector (d) unit vector

15. If $|\vec{a} + \vec{b}| = |\vec{a} - \vec{b}|$ then angle between $\vec{a}$ & $\vec{b}$ is
(a) 45° (b) 30°
(c) 90° (d) 180°

16. Which of the following is correct :
(a) $\vec{A}.\vec{B} \neq \vec{B}.\vec{A}$
(b) $\vec{A}.(\vec{B} + \vec{C}) = A.B + \vec{A}.\vec{C}$
(c) $\vec{A} \times \vec{B} = \vec{B} \times A$
(d) $\vec{A}.(\vec{B} + \vec{C}) \neq \vec{A}.\vec{B} + \vec{A}.\vec{C}$

17. Which of the following is not correct :

(a) $\vec{A}\times\vec{B}=-\vec{B}\times A$

(b) $\vec{A}\times\vec{B}\neq\vec{B}\times A$

(c) $\vec{A}\times(\vec{B}+\vec{C})=\vec{A}\times\vec{B}+\vec{A}\times\vec{C}$

(d) $\vec{A}\times(\vec{B}+\vec{C})=(\vec{A}\times\vec{B})+\vec{C}$

18. Which of the following is correct :

(a) $\vec{A}\times(\vec{B}\times\vec{C})=(\vec{A}\times\vec{B})\times\vec{C}$

(b) $\vec{A}(\vec{B}.\vec{C})=(\vec{A}.\vec{B})C$

(c) $\vec{A}\times(\vec{B}\times\vec{C})\neq(\vec{A}\times\vec{B})\times\vec{C}$

(d) $\vec{A}\times\vec{B}\neq\vec{B}\times\vec{A}$

19. In question (54) if the observer is on the road then the path will be

(a) parabola

(b) straight line vertically downwards

(c) hyperbola

(d) none

20. The magnitude of the vector product of two vectors is $\sqrt{3}$ times then scalar product. The angle between vectors is

(a) $\frac{\pi}{4}$ (b) $\frac{\pi}{6}$

(c) $\frac{\pi}{3}$ (d) $\frac{\pi}{2}$

21. Two vectors $\vec{A}$ and $\vec{B}$ are such that $\vec{A}+\vec{B}=\vec{C}$ and $A^2+B^2=C^2$. If θ is the angle between positive direction of $\vec{A}$ and $\vec{B}$, then the correct statement is :

(a) $\theta=\pi$ (b) $\theta=\frac{\pi}{2}$

(c) $\theta=0$ (d) $\theta=\frac{2\pi}{3}$

22. If $\vec{a}$ and $\vec{b}$ are two unit vectors such that $\vec{a}+\vec{b}$ and $\vec{a}-2\vec{b}$ are perpendicular then the angle between $\vec{a}$ and $\vec{b}$ is

(a) $\frac{\pi}{4}$ (b) $\frac{\pi}{3}$

(c) π (d) $\frac{\pi}{6}$

23. If $\vec{a}$, $\vec{b}$ and $\vec{c}$ are three unit vectors such that $\vec{a}+\vec{b}+\vec{c}=0$ than the value of $\vec{a}.\vec{b}+\vec{b}.\vec{c}+\vec{c}.\vec{a}$ is

(a) $\frac{3}{2}$ (b) -1

(c) 0 (d) $-\frac{3}{2}$

24. The greatest height to which a man can through a ball is h. What is the greatest distance to which he can throw the ball?

(a) $2h$ (b) $\frac{h}{4}$

(c) $\frac{h}{2}$ (d) none

25. The magnitudes of vectors $\vec{A}$, $\vec{B}$ and $\vec{C}$ are 3, 4 and 5 units respectively. If $\vec{A}+\vec{B}=\vec{C}$, then the angle between $\vec{A}$ and $\vec{A}$ is

(a) $\pi/2$ (b) $\cos^{-1} 0.6$

(c) $\tan^{-1}7/5$ (d) $\pi/4$

26. The angle between $\vec{A}$ and $\vec{A}$ is θ. The value of the triple product $\vec{A}.(\vec{B}\times\vec{A})$ is

(a) A^2B (b) zero

(c) $A^2B\sin\theta$ (d) $A^2B\cos\theta$

27. A bus is moving on a straight road towards north with a uniform speed of 50 km/hour turns through 90°. If the speed remains unchanged after turning, the increase in the velocity of bus in the turning process is

(a) 70.7 km/hour along south-west direction

(b) zero

(c) 50 km/hour along west

(d) 70.7 km/hour along north-west direction.

28. The maximum range of a gun of horizontal terrain is 16 km. If $g = 10\ ms^{-2}$, then muzzle velocity of a shell must be

(a) $160\ ms^{-1}$ (b) $200\sqrt{2}\ ms^{-1}$

(c) $400\ ms^{-1}$ (d) $800\ ms^{-1}$

29. Two bodies of same mass are projected with the same velocity at an angle 30° and 60° respectively. The ratio of their horizontal ranges will be

(a) 1 : 1 (b) 1 : 2

(c) 1 : 3 (d) $2:\sqrt{2}$

30. An electric fan has blades of length 30 cm measured from the axis of rotation. If the fan is rotating at 120 rpm, the acceleration of a point on the tip of the blade is
(a) $1600\ ms^{-2}$ (b) $47.4\ ms^{-2}$
(c) $23.7\ ms^{-2}$ (d) $50.55\ ms^{-2}$

31. When milk is churned, cream gets separated due to
(a) centripetal force
(b) centrifugal force
(c) frictional force
(d) gravitational force

32. Six vectors, $\vec{a}$ through $\vec{f}$ have the magnitudes and directions indicated in the figure. Which of the following statements is true?

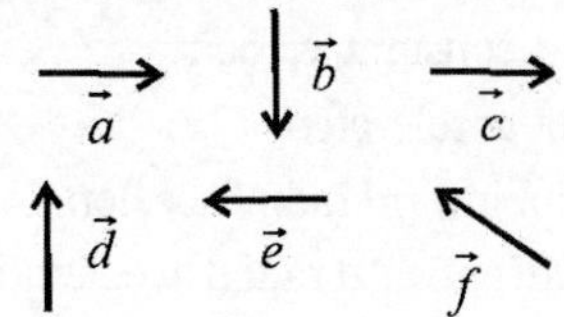

(a) $\vec{b}+\vec{c}=\vec{f}$ (b) $\vec{d}+\vec{c}=\vec{f}$
(c) $\vec{d}+\vec{e}=\vec{f}$ (d) $\vec{b}+\vec{e}=\vec{f}$

33. If a body A of mass M is thrown with velocity v at an angle of 30° to the horizontal and another body B of the same mass is thrown with the same speed at an angle of 60° to the horizontal, the ratio of horizontal range of A to B will be
(a) 1 : 3 (b) 1 : 1
(c) $1:\sqrt{3}$ (d) $\sqrt{3}:1$

34. A boat is sent across a river with a velocity of 8 km h^{-1}. If the resultant velocity of boat is 10 km h^{-1}, then the velocity of the river is
(a) 12.8 km h^{-1} (b) 6 km h^{-1}
(c) 8 km h^{-1} (d) 10 km h^{-1}

35. A body constrained to move in y-direction, is subjected to a force given by $\vec{F}=(-2\hat{i}+15\hat{j}+6\hat{k})N$. What is the workdone by this force in moving the body through a distance of 10m along y-axis?
(a) 190 J (b) 160 J
(c) 150 J (d) 20 J

36. When a body moves with a constant speed along a circle
(a) its velocity remains constant
(b) no force acts on it
(c) no work is done on it
(d) no acceleration is produced in it

37. The position vector of a particle is $\vec{r}=(a\cos\omega t)\hat{i}+(a\sin\omega t)\hat{j}$. The velocity of the particle is
(a) directed towards the origin
(b) directed away from the origin
(c) parallel to the position vector
(d) perpendicular to the position vector

38. Which of the following is not a vector quantity?
(a) speed (b) velocity
(c) torque (d) displacement

39. A bullet is fired from a gun with a speed of 1000 m/s in order to hit a target 100 m away. At what height above the target should the gun be aimed? (The resistance of air is negligible and $g = 10\ m/s^2$)
(a) 5 cm (b) 10 cm
(c) 15 cm (d) 20 cm

40. A body is whirled in a horizontal circle of radius 20 cm. It has an angular velocity of 10 rad/s. What is its linear velocity at any point on circular path
(a) $\sqrt{2}\ m/s$ (b) 2 m/s
(c) 10 m/s (d) 20 m/s

41. A ball whose kinetic energy is E, is thrown at an angle of 45° with the horizontal. Its kinetic energy at the highest point of its flight will be
(a) E (b) $\frac{E}{2}$
(c) $\frac{E}{\sqrt{2}}$ (d) 0

42. Find the torque of a force $\vec{F}=-3\hat{i}+\hat{j}+5\hat{k}$ acting at the point $\vec{r}=7\hat{i}+3\hat{j}+\hat{k}$.
(a) $-21\hat{i}+3\hat{j}+5\hat{k}$ (b) $-14\hat{i}+3\hat{j}+16\hat{k}$
(c) $4\hat{i}+4\hat{j}+6\hat{k}$ (d) $14\hat{i}-38\hat{j}+16\hat{k}$

43. Two particles A and B are connected by a rigid rod AB. The rod slides along perpendicular rails as shown here. The velocity of A to the left is 10 m/s. What is the velocity of B when angle $\alpha = 60°$?

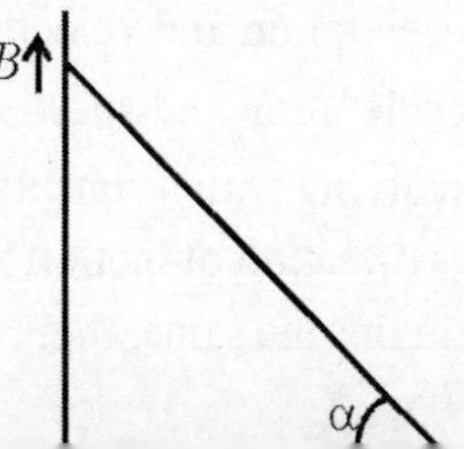

(a) 5.8 m/s (b) 9.8 m/s
(c) 10 m/s (d) 17.3 m/s

44. A ball of mass 0.25 kg attached to the end of a string of length 1.96 m is moving in a horizontal circle. The string will break if the tension is more than 25 N. What is the maximum speed with which the ball can be moved?
(a) 14 m/s (b) 3 m/s
(c) 5 m/s (d) 3.92 m/s

45. A 500 kg car takes a round turn of radius 50 m with a velocity of 36 km/h. The centripetal force is
(a) 250 N (b) 750 N
(c) 1000 N (d) 1200 N

46. What is the linear velocity if angular velocity vector $\vec{\omega} = 3\hat{i} - 4\hat{j} + \hat{k}$ and position vector $\vec{r} = 5\hat{i} - 6\hat{j} + 6\hat{k}$
(a) $6\hat{i} + 2\hat{j} - 3\hat{k}$ (b) $-18\hat{i} - 13\hat{j} + 2\hat{k}$
(c) $18\hat{i} + 13\hat{j} - 2\hat{k}$ (d) $6\hat{i} - 2\hat{j} + 8\hat{k}$

47. The angle between two vectors of magnitude 12 and 18 units when their resultant is 24 units, is
(a) 63° 51′ (b) 75° 52′
(c) 82° 31′ (d) 89° 16′

48. A body of mass 0.4 kg is whirled in a vertical circle making 2 rev/sec. If the radius of the circle is 1.2 m, then tension in the string when the body is at the top of the circle, is
(a) 41.56 N (b) 89.86 N
(c) 109.86 N (d) 115.86 N

49. In uniform circular motion
(a) both velocity and acceleration are constant
(b) acceleration and speed are constant but velocity changes
(c) both acceleration and velocity change
(d) both acceleration and speed are constant

50. A body is revolving with a constant speed along a circle. If its direction of motion is reversed but the speed remains the same, then,
(a) the centripetal force will not suffer any change in magnitude
(b) the centripetal force will have its direction reversed
(c) the centripetal force will suffer any change in direction
(d) the centripetal force would be doubled

51. When a body moves with a constant speed along a circle
(a) no work is done on it
(b) no acceleration is produced in the body
(c) no force acts on the body
(d) its velocity remains constant

52. A body is travelling in a circle at a constant speed. It
(a) has a constant velocity
(b) is not accelerated
(c) has an inward radial acceleration
(d) has an outward radial acceleration

53. When a particle is in uniform circular motion it does not have
(a) radial velocity and radial acceleration
(b) radial velocity and tangential acceleration
(c) tangential velocity and radial acceleration
(d) tangential velocity and transverse acceleration

54. A body is moving with a constant speed v in a circle of radius r. Its angular acceleration is
(a) vr (b) v/r
(c) zero (d) vr^2

55. A body of mass m is moving in a circle of radius r with a constant speed v. The work done by the centripetal force in moving the body over half the circumference of the circle is
(a) mv^2r (b) zero
(c) mv^2/r (d) $r^2/m\, v^2$

56. What happens to the centripetal acceleration of a revolving body if you double the orbital speed v and halve the angular velocity ω ?
(a) the centripetal acceleration remains unchanged
(b) the centripetal acceleration is halved
(c) the centripetal acceleration is doubled
(d) the centripetal acceleration is quadrupled.

57. A stone of mass m is tied to a string of length ℓ and rotated in a circle with a constant speed v, if the string is released the stone flies
(a) radially outward
(b) radially inward
(c) tangentially outward
(d) with an acceleration mv^2/ℓ

58. If a particle moves in a circle describing equal angles in equal interval of times, its velocity vector
(a) remains constant
(b) changes in magnitude
(c) changes in direction
(d) changes both in magnitude and direction

59. A particle of mass M is moving in a horizontal circle of radius R with uniform speed V. When it moves from one point to a diametrically opposite point, its
(a) kinetic energy changes by $MV^2/4$
(b) momentum does not change
(c) momentum changes by $2MV$
(d) kinetic energy changes by MV^2

60. A particle moves in a circular orbit under the action of a central attractive force inversely proportional to the distance 'r'. The speed of the particle is
(a) proportional to r^2
(b) independent of r
(c) proportional to r
(d) proportional to $1/r$

Laws of Motion

CONCEPTS

NEWTON'S LAWS OF MOTION

Newton's First Law of Motion

It gives qualitative definition of force. "Eeverybody continues to be in its state of rest or of uniform motion in a straight line, unless compelled by some external force to act otherwise". This law is also known as **law of inertia**.

- Inertia of a body is defined as, its ability to oppose the changes in its state of rest or motion.
- Inertia of a body, is determined by the matter content of the body.
- Inertia of a body is independent of its geometrical and dynamic conditions.

Newton's Second Law of Motion

It gives quantitative definition of force. "The rate of change of momentum of a body is proportional to the applied force and takes place in the direction in which the force acts".

i.e., Force $F \propto \frac{dp}{dt}$ or F = ma

- The second law is consistent with the first law (F = 0 implies a = 0)
- It is applicable to a particle, and also to a body or a system of particles, provided F is the total external force on the system and a is the acceleration of the system as a whole.
- F at a point at a certain instant determines a at the same point at that instant.
- Newton's second law is general. The first and third laws can be deduced from it.
- Not all forces produce acceleration. Only unbalanced forces do.

Momentum

It is the product of mass and velocity of the body.

Momentum P = mv

It is a vector quantity.

Newton's Third Law of Motion

It defines nature of force. "To every action, there is always an equal and opposite reaction that acts on different body".

- We cannot have an isolated force.
- Forces exist in pair which are equal and opposite.
- The action and reaction forces are always of the same nature.
- The action and the reaction always act on different bodies.

IMPULSE

It is the product of force and time which equals change in momentum.

Impulse, $I = F.\Delta t$

The notion of impulse is useful when a large force acts for a short time to produce a measurable change in momentum. Since the time of action of the force is very short, one can assume that there is no appreciable change in the position of the body during the action of the impulsive force.

You should be able to apply Newton's laws of motion to various kind of problems like in equilibrium condition, pulley system, wedge system, constrained motion and in accelerated frame of reference using pseudo force.

LAW OF CONSERVATION OF MOMENTUM

The total momentum of an isolated system of particles is conserved. The law follows from the second and third law of motion.

You should be able to apply momentum conservation and Newton's laws of motion to variable mass system.

Gun Firing a Bullet

If a gun of mass M fires a bullet of mass m with velocity v. Then from law of conservation of momentum, as initially bullet & gun are at rest position i.e., initial momentum is zero, so final momentum (gun + bullet) must also be zero.

Since on firing, the bullet moves with velocity $\vec{v}_b$ in forward direction, then from Newton's third law, the gun move in backward direction $\vec{v}_g$. So

Initial momentum = Final momentum

$$0 = \underset{\text{momentum of bullet}}{m\vec{v}_b} + \underset{\text{momentum of gun}}{M\vec{V}_g} \quad \therefore \vec{V} = \frac{-m\vec{v}}{M}$$

(–ve sign shows that the vel. of gun will have the opposite direction to that of bullet)

Rocket Propulsion (A Case of System of Variable Mass)

It is based on principle of conservation of linear momentum. In rocket, the fuel burns and produces gases at high temperature. These gases are ejected out of the rocket from nozzle at the backside of rocket and the ejecting gas exerts a forward force on the rocket which accelerates it. Let the gas ejects at a rate $r = -\frac{dM}{dt}$ and at constant velocity u w.r.t. rocket then from the conservation of linear momentum $\frac{dv}{dt} = \frac{ru}{M} = \frac{ru}{M_0 - rt}$ where $M = M_0$ - rt and M_0 is mass of rocket with fuel and solving this equation, we get $v = u \log_e\left(\frac{M_0}{M_0 - rt}\right)$

where v = velocity of rocket w.r.t. ground

EQUILIBRIUM OF A PARTICLE

A body is said to be in equilibrium, when no net force acts on the body.

i.e., $\Sigma\vec{F} = 0$

Then $\Sigma F_x = 0,\ \Sigma F_y = 0,\ \Sigma F_z = 0$

In this body is not in acceleration.

States of Equilibrium

(a) **Stable equilibrium :** If a body is slightly displaced from equlibrium position, it has the tendency to regain its original position, it is said to be in stable equilibrium.

In this case, P.E. is minimum. $\left(\frac{d^2u}{dr^2} = +ve\right)$

So, the centre of gravity is lowest.

(b) **Unstable equilibrium :** If a body, after being displaced from the equilibrium position, moves in the direction of displacement, it is said to be in unstable equilibrium.

In this case P.E. is maximum. $\left(\frac{d^2u}{dr^2} = -ve\right)$

So, the centre of gravity is highest.

(c) **Neutral equilibrium :** If a body, after being slightly displaced from the equilibrium position has no tendency to come back or to move in the direction of displacement the equilibrium is known to be neutral.

In this case P.E. is constant

$\left(\frac{d^2u}{dr^2} = \text{constant}\right)$

The centre of gravity remains at constant height.

SOME COMMON FORCES IN MECHANICS

Remember different kind of forces like 'friction', 'normal reaction' 'tension', 'air resistance', 'viscous drag', 'thrust', 'buoyancy' 'weight' 'centripetal force'.

If an object is connected by spring and spring is stretched or compressed by a distance x, then **spring** or **restoring** force on the object, F = – kx

A force which opposes relative motion between the surfaces in contact is called **frictional force** F = μN

This will be discussed in detail in later section.

Friction

Frictional force/ Friction is a force which opposes (impending or actual) relative linear motion between two surfaces in contact. It is the component of the contact force along the common tangent to the surface in contact. Static friction opposes impending relative motion; kinetic friction opposes actual relative motion. They are independent of the area of surface in contact and satisfy the following approximate laws :

$f_s \le (f_s)_{max} = \mu_s R$

$f_k = \mu_k R$

μ_s (co-efficient of static friction) and μ_k (co-efficient of kinetic friction) are constants characteristic of the pair of surfaces in contact. It is found experimentally that μ_k is less than μ_s and the rolling friction μ_R is much smaller than μ_R and μ_k.

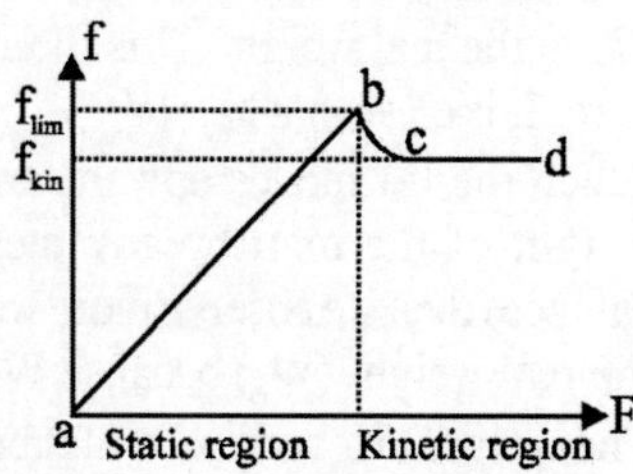

Angle of friction : At a point of rough contact, where slipping is about to occur, the two forces acting on each object are the normal reaction N and frictional force μN.

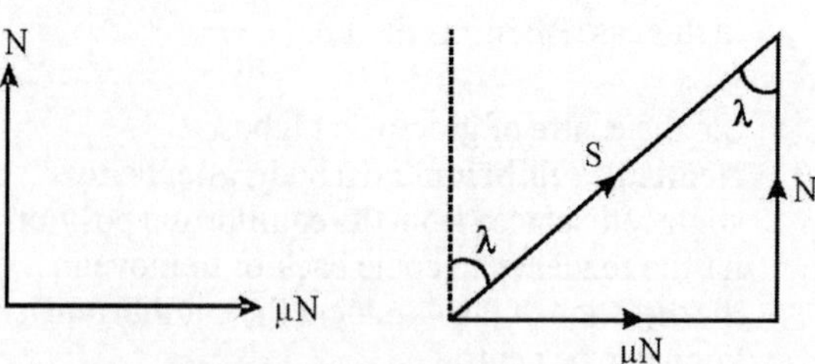

The resultant of these two forces is S, and S makes an angle λ with the normal, where

$$\tan\lambda = \frac{\mu N}{N} = \mu$$

The angle λ is called the angle of friction.

Retardation produced by friction on a car in horizontal motion = $\mu_s g$

APPARENT WEIGHT OF A BODY IN A LIFT

Lift is Accelerated Upwards

In this case the man also moves in upward direction with an acceleration $\vec{a}$.

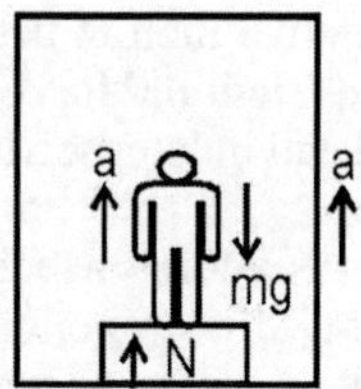

Then from Newton' Second Law

$N - mg = ma$ or $N = m(g+a)$

or $W_{app} = m(g+a) = W_o(1+a/g)$ (as W = mg)

where W_{app} is apparent weight of the man in the lift, W_o is the real weight, N is the reaction of lift on the man. It is clear that $N = W_{app}$.

When the lift moves upward and if we measure the weight of the man by any means (such as spring balance) then we observe more weight (i.e., W_{app}) than the real weight (W_o) because $W_{app} > W_o$.

The weight of a body W = mg, but is measured by the resistance, reaction or tension. If the resistance, reaction or tension becomes zero, the body fells weightlessness. When a body falls freely, only force of gravity acts on it. There is no resistance, reaction or tension. Hence, the body feels weightlessness.

Lift is Accelerated Downward

In this case from Newton's second law

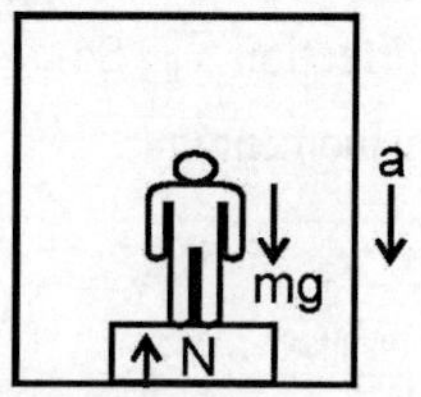

$mg - N = ma$

or $N = m(g-a) = W_o(1-a/g)$

or $W'_{app} = W_o(1-a/g)$ $\{\because W_o = mg\}$

In this case, if we measure the weight of man by spring balance, we observe deficiency because $W_{app} < W_o$.

Lift is at Rest or Moving with Constant Velocity

From Newton's second Law $N - mg = 0$ or $N = mg$

In this case spring balance gives the true weight of the man.

Pseudo Force

If a body of mass m is placed in a non-inertial frame having aceleration $\vec{a}$, then it experiences a Pseudo force acting in a direction opposite to the direction of $\vec{a}$.

$$\vec{F}_{pseudo} = m\vec{a}$$

The pseudo force is always directed in a direction opposite to the direction of the acceleration of the frame.

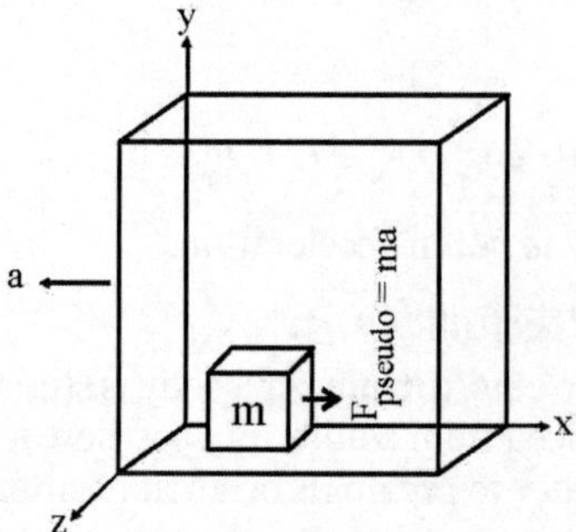

Pseudo force

- Is identified by an accelerating observer.
- Is an imaginary force.
- Does not obey Newton's third law.

- Can do work.
- Can accelerate a body.
- Acceleration of a body when it slides down a rough inclined plane = $g\sin\alpha - \mu_s g\cos\alpha$

A force is pseudo if (1) we cannot know where it is coming from, (2) it cannot be classified into one of the four forces in nature and (3) the force disappears when we change the frame to reference (observer)

FREE BODY DIAGRAM (FBD)

- Free body diagram shows the various forces acting on a body directly.
- The contact force between the two surfaces may be at an angle with the surface.
- The tension in a string always pulls a body.
- While drawing a FBD, tension in each branch of the string must form an Action-Reaction pair.
- Free body diagram is always drawn with respect to an observer.
- A body is said to be in equilibrium, if it is not accelerating.

Exercise

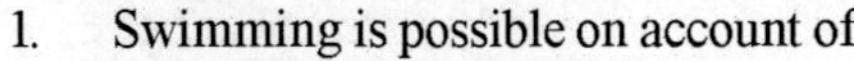

1. Swimming is possible on account of
   (a) first law of motion
   (b) second law of motion
   (c) third law of motion
   (d) Newton's law of gravitation
2. Inertia is that property of a body by virtue of which the body is
   (a) unable to change by itself the state of rest
   (b) unable to change by itself the state of uniform motion
   (c) unable to change by itself the direction of motion
   (d) unable to change by itself the state of rest and of uniform linear motion
3. An object will continue moving uniformly when
   (a) The resultant force on it is increasing continuously
   (b) the resultant force is at right angles to its rotation
   (c) The resultant force on it is zero
   (d) The resultant force on it begins to decrease
4. A man getting down a running bus falls forward because:
   (a) Due to inertia of rest, road is left behind and man reaches forward
   (b) Due to inertia of motion upper part of body continues to be in motion in forward direction while feet come to rest as soon as they touch the road
   (c) He leans forward as a matter of habit
   (d) Of the combined effect of all the three factors stated in (a), (b) and (c)
5. A man is at rest in the middle of a pond of perfectly smooth ice. He can get himself to the shore by making use of Newton's :
   (a) First law (b) Second law
   (c) Third law (d) All the laws
6. A cannon after firing recoils due to :
   (a) Conservation of energy
   (b) Backward thrust of gases produced
   (c) Newton's third law of motion
   (d) Newton's first law of motion
7. Newton's second law give the measure of :
   (a) Acceleration
   (b) Force
   (c) Momentum
   (d) Angular momentum
8. We can derive Newton's :
   (a) Second and third laws from the first law
   (b) First and second laws from the third law
   (c) Third and first laws from the second law
   (d) All the three laws are independent of each other
9. A jet plane moves up in air because :
   (a) The gravity does not act on bodies moving with high speeds
   (b) The thrust of the jet compensates for the force of gravity
   (c) The flow of air around the wings causes an upward force, which compensates for the force of gravity
   (d) The weight of air whose volume is equal to the volume of the plane is more than the weight of the plane

10. You are on a frictionless horizontal plane. How can you get off if no horizontal force is exerted by pushing against the surface :
    (a) By jumping
    (b) By spitting or sneezing
    (c) By rolling your body on the surface
    (d) By running on the plane
11. Pulling is easier than pushing because :
    (a) When we pull a roller, the vertical component of the pulling force acts in the direction of weight
    (b) The vertical component of the pulling force acts in the opposite direction of weight
    (c) Force of friction is in opposite direction
    (d) It is possible in the case of roller only
12. When a body is stationary :
    (a) There is no force acting on it
    (b) The force acting on it is not in contact with it
    (c) The combination of forces acting on it balances each other
    (d) The body is in vacuum
13. When a particle in motion is always acted upon by a force in a direction perpendicular to direction of motion, the path followed by particle is
    (a) Straight line (b) Parabolic
    (c) Circular (d) Hyperbolic
14. Which of the following statements about friction is true?
    (a) Friction can be reduced to zero
    (b) Frictional force cannot accelerate a body
    (c) Frictional force is proportional to the area of contact between the two surfaces
    (d) Kinetic friction is always greater than rolling friction
15. Which of the following is a self adjusting force?
    (a) Static friction (b) Limiting friction
    (c) Dynamic friction (d) Sliding friction
16. The force required to just move a body up the inclined plane is double the force required to just prevent the body from sliding down the plane. The coefficient of friction is μ. The inclination θ of the plane is
    (a) $\tan^{-1}\mu$ (b) $\tan^{-1}(\mu/2)$
    (c) $\tan^{-1}2\mu$ (d) $\tan^{-1}3\mu$
17. If μ_s, μ_k and μ_r are coefficients of static friction, sliding friction and rolling friction, then
    (a) $\mu_s < \mu_k < \mu_f$ (b) $\mu_k < \mu_r < \mu_s$
    (c) $\mu_r < \mu_k < \mu_s$ (d) $\mu_r = \mu_k = \mu_s$
18. A 30 kg block rests on a rough horizontal surface. A force of 200 N is applied on the body. The block acquires a speed of 4 m/sec, starting from rest, in 2 sec. What is the value of coefficient of friction?
    (a) $10/\sqrt{3}$ (b) $\sqrt{3}/10$
    (c) 0.47 (d) 0.185
19. A block is at rest on an inclined plane making an angle α with the horizontal. As the angle α of the inclination is increased, the block just starts slipping when the angle of inclination becomes θ. Then the coefficient of static friction between the block and the surface of the inclined plane is
    (a) $\sin\theta$ (b) $\cos\theta$
    (c) $\tan\theta$ (d) independent of θ
20. Which of the following statements is correct, when a person walks on a rough surface :
    (a) The frictional force exerted by the surface keeps him moving
    (b) The force which the man exerts on the floor keeps him moving
    (c) The reaction of the force which the man exerts on floor keeps him moving
    (d) None of the above
21. It is difficult to move a cycle with brakes on because
    (a) Rolling friction opposes motion on road
    (b) Sliding friction opposes motion on road
    (c) Rolling friction is more than sliding friction
    (d) Sliding friction is more than rolling friction
22. Which is a suitable method to decrease friction?
    (a) Polishing (b) Lubrication
    (c) Ball bearing (d) All of these
23. Centrifugal force is
    (a) a real force
    (b) balances centripetal force
    (c) always acts away from centre
    (d) all of these
24. A block of mass 2 kg is resting on a rough inclined plane of inclination 30°. The coefficient of friction is 0.8, the contact force between incline and the block is ($g = 10\ ms^{-2}$)
    (a) 20 N (b) 10 N
    (c) $10\sqrt{3}$ N (d) none of these

25. A rectangular block is placed on a rough horizontal surface in two different ways as shown, then

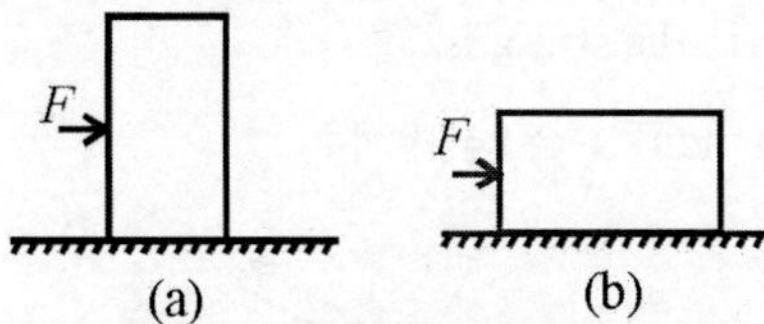

(a) friction will be more in case (a)
(b) friction will be more in case (b)
(c) friction will be equal in both the cases
(d) depends on the relations among its dimensions

26. A block of mass m is placed on a smooth horizontal surface as shown. The weight (mg) of the block and normal reaction (N) exerted by the surface on the block

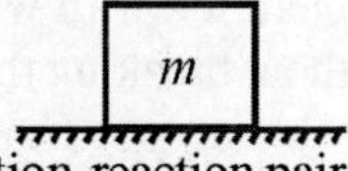

(a) form action-reaction pair
(b) balance each other
(c) act in same direction
(d) both (a) and (b)

27. On increasing the smoothness of surfaces in contact,
(a) frictional force b/w them must decrease
(b) frictional force b/w them must increase
(c) frictional force b/w them may increase or decrease
(d) frictional force does not change as it does not depend on area of contact

28. Centripetal force :
(a) can change speed of the body
(b) is always perpendicular to direction of motion
(c) is constant for uniform circular motion
(d) all of these

29. When a horse pulls a cart, the horse moves down to
(a) the force exerted by the horse on the cart
(b) the force exerted by the cart on the horse
(c) the force exerted by the horse on the earth
(d) the force exerted by the earth on the horse

30. Forces of action and reaction do not balance each other because
(a) they act in same direction
(b) their magnitudes are not equal
(c) they act on different bodies
(d) act perpendicular to each other

31. The force of action and reaction
(a) must be of same nature
(b) must be of different nature
(c) may be of different nature
(d) may not have equal magnitude

32. Which of the following forces is self-adjusting?
(a) Static friction (b) Kinetic friction
(c) Limiting friction (d) Rolling friction

33. Frictional force
(a) always acts opposite to direction of motion
(b) always acts in the direction of motion
(c) may act in the direction of motion
(d) always acts perpendicular to the direction of motion

34. A body is moving with uniform velocity, then
(a) no force must be acting on the body
(b) exactly two forces must be acting on the body
(c) body is not acted upon by a single force
(d) the number of forces acting on the body must be even

35. The direction of impulse is
(a) same as that of the net force
(b) opposite to that of the net force
(c) same as that of the final velocity
(d) same as that of the initial velocity

36. A monkey is climbing up a rope, then the tension in the rope.
(a) must be equal to the force applied by the monkey on the rope
(b) must be less than the force applied by the monkey on the rope
(c) must be greater than the force applied by the monkey on the rope
(d) may be equal to, less than or greater the force applied by the monkey on the rope

37. A coin is placed on a rotating disc and is stationary w.r.t. the disc, then the direction of friction is
(a) along the direction of motion of the coin w.r.t. ground
(b) opposite to the direction of motion of the coin w.r.t. ground
(c) towards the centre of the disc
(d) away from the centre of the disc

38. Starting from rest, a body slides down a 45° inclined plane in twice the time it takes to slide down the same distance in the absence of friction. The coefficient of friction between the body and the inclined plane is

(a) 0.80 (b) 0.75
(c) 0.25 (d) 0.33

39. Two bodies of masses m and 4m are moving with equal kinetic energies. The ratio of their linear momenta will be
(a) 1 : 4 (b) 4 : 1
(c) 1 : 2 (d) 2 : 1

40. A body of mass 5 kg explodes at rest into three fragments with masses in the ratio 1 : 1 : 3. The fragments with equal masses fly in mutually perpendicular directions with speeds of 21 m/s. The velocity of heaviest fragment in m/s will be
(a) $7\sqrt{2}$ (b) $5\sqrt{2}$
(c) $3\sqrt{2}$ (d) $\sqrt{2}$

41. A 4 kg mass and 1 kg are moving with equal kinetic energies. The ratio of the magnitudes of their linear momenta is
(a) 1 : 2 (b) 1 : 1
(c) 2 : 1 (d) 4 : 1

42. A 600 kg rocket is set for a vertical firing. If the exhaust speed is 1000 ms^{-1}, the mass of the gas ejected per second to supply the thrust needed to overcome the weight of rocket is
(a) 117.6 kg s^{-1} (b) 58.6 kg s^{-1}
(c) 6 kg s^{-1} (d) 76.4 kg s^{-1}

43. A particle of mass m is moving with a uniform velocity v_1. It is given an impulse such that its velocity becomes v_2. The impulse is equal to
(a) $m[|v_2| - |v_1|]$ (b) $\frac{1}{2}m[v_2^2 - v_1^2]$
(c) $m[v_1 + v_2]$ (d) $m[v_2 - v_1]$

44. A heavy uniform chain lies on horizontal table top. If the coefficient of friction between the chain and the table surface is 0.25, then the maximum fraction of the length of the chain that can hang over one edge of the table is
(a) 20% (b) 25%
(c) 35% (d) 15%

45. Physical independence of force is a consequence of
(a) third law of motion
(b) second law of motion
(c) first law of motion
(d) all of these laws

46. Consider a car moving along a straight horizontal road with a speed of 72 km/h. If the coefficient of static friction between the tyres and the road is 0.5, the shortest distance in which the car can be stopped is (taking g = 10 m/s^2)
(a) 30 m (b) 40 m
(c) 72 m (d) 20 m

47. For the arrangement shown in the Figure the tension in the string is
[Given : $\tan^{-1}(0.8) = 39°$]

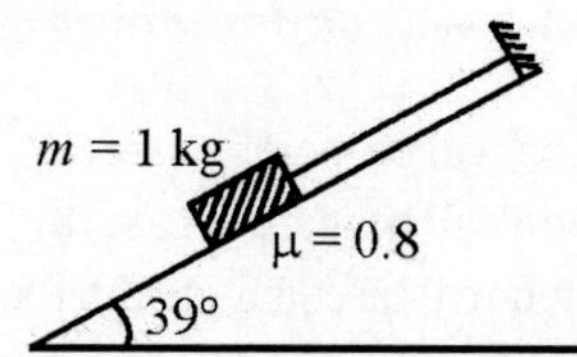

(a) 6 N (b) 6.4 N
(c) 0.4 N (d) zero.

48. A monkey is decending from the branch of a tree with constant acceleration. If the breaking strength is 75% of the weight of the monkey, the minimum acceleration with which monkey can slide down without breaking the branch is
(a) g (b) $\frac{3g}{4}$
(c) $\frac{g}{4}$ (d) $\frac{g}{2}$

49. A block has been placed on an inclined plane with the slope angle θ, block slides down the plane at constant speed. The coefficient of kinetic friction is equal to
(a) sin θ (b) cos θ
(c) g (d) tan θ

50. A shell is fired from a cannon, it explodes in mid air, its total
(a) momentum increases
(b) momentum decreases
(c) K.E. increases
(d) K.E. decreases

51. A satellite in a force free space sweeps stationary interplanetary dust at a rate $(dM/dt) = \alpha v$. The acceleration of satellite is
(a) $\frac{-2\alpha v^2}{M}$ (b) $\frac{-\alpha v^2}{M}$
(c) $\frac{-\alpha v^2}{2M}$ (d) $-\alpha v^2$

52. If the force on a rocket moving with a velocity of 300 m/sec is 345 N, then the rate of combustion of the fuel, is
(a) 0.55 kg/sec (b) 0.75 kg/sec
(c) 1.15 kg/sec (d) 2.25 kg/sec

53. What will be the maximum speed of a car on a road turn of radius 30 m if the coefficient of friction between the tyres and the road is 0.4
(Take $g = 9.8$ m/s^2)
(a) 10.84 m/s (b) 9.84 m/s
(c) 8.84 m/s (d) 6.84 m/s

54. A 10 N force is applied on a body produces an acceleration of 1 m/s^2. The mass of the body is
(a) 5 kg (b) 10 kg
(c) 15 kg (d) 20 kg

55. A ball of mass 150 g, moving with an acceleration 20 m/s^2, is hit by a force, which acts on it for 0.1 sec. The impulsive force is
(a) 0.5 N (b) 0.1 N
(c) 0.3 N (d) 1.2 N

56. A mass of 1 kg is suspended by a thread. It is
(i) lifted up with an acceleration 4.9 m/s^2,
(ii) lowered with an acceleration 4.9 m/s^2.
The ratio of the tensions is
(a) 3 : 1 (b) 1 : 2
(c) 1 : 3 (d) 2 : 1

57. A bullet is fired from a gun. The force on the bullet is given by $F = 600 - 2 \times 10^5\, t$ where, F is in newton and t in second. The force on the bullet becomes zero as soon as it leaves the barrel. What is the average impulse imparted to the bullet?
(a) 1.8 N-s (b) zero
(c) 9 N-s (d) 0.9 N-s

58. A 5000 kg rocket is set for vertical firing. The exhaust speed is 800 ms^{-1}. To give an initial upward acceleration of 20 ms^{-2}, the amount of gas ejected per second to supply the needed thrust will be ($g = 10$ ms^{-2})
(a) 127.5 kg s^{-1} (b) 187.5 kg s^{-1}
(c) 185.5 kg s^{-1} (d) 137.5 kg s^{-1}

59. Two blocks $m_1 = 5$ gm and $m_2 = 10$ gm are hung vertically over a light frictionless pulley as shown here. What is the acceleration of the masses when they are left free?

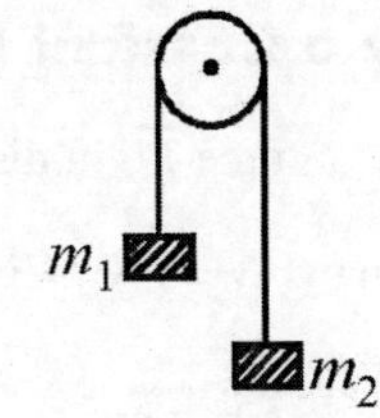

(a) $g/3$ (b) $g/2$
(c) g (d) $g/5$
(where g is acceleration due to gravity)

60. The effective acceleration of a body when thrown upwards with acceleration a is
(a) $(g + a)$ (b) $(a - g)$
(c) $\sqrt{(g^2 + a^2)}$ (d) $\{g - a^2/g\}$

Work, Energy & Power

CONCEPTS

WORK

Work Done by a Constant Force

Work done (W) by a force $\vec{F}$ in displacing a body through $\vec{x}$ is given by $W = \vec{F}.\vec{x} = Fx\cos\theta$

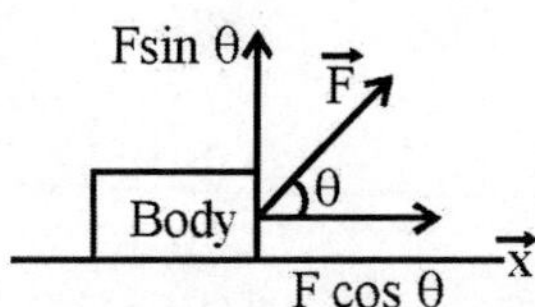

where θ is the angle between the applied force $\vec{F}$ and displacement $\vec{x}$. The S.I. unit of work is joule and its C.G.S. unit is erg. 1 joule = 10^7 erg. The dimensions of work are $[ML^2T^{-2}]$. It is a scalar quantity.

Work Done by a Variable Force

The work done by a force F_x from an initial point A to final point B is $W_{A\to B} = \int_{x_A}^{x_B} F_x dx$

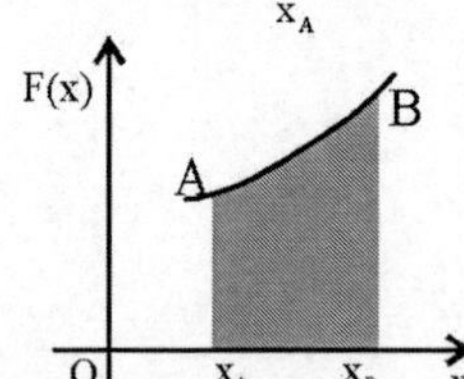

(a) The work done by variable force in displacing a particle from x_1 to x_2

$$W = \int_{x_1}^{x_2} F dx = \text{Area under force displacement graph}$$

(b) The work done in stretching or compressing a spring (against restoring force) by a distance x is

$W = \frac{1}{2}kx^2$ where K is spring constant or force constant. This work done is stored as elastic potential energy in spring.

For a particular displacement, work done by a force is independent of type of motion *i.e.* whether it moves with constant velocity, constant acceleration or retardation etc.

For a particular displacement work is independent of time. Work will be same for same displacement whether the time taken is small or large.

ENERGY

It is the capacity of doing work. The unit and dimensions of energy are same as that of work.

Potential Energy

The energy possessed by a body by virtue of its position or configuration is called potential energy. Potential energy is defined only for conservative forces. It does not exist for non conservative forces.

$$E_p = mgh$$

Potential energy depends on frame of reference but change in potential energy is independent of frame of reference .

Potential energy should be considered to be a property of the entire system, rather than assigning it to any specific particle.

For a conservative force in one dimension, we may define a potential energy function U (x) such that

$$F(x) = -\frac{dU(x)}{dx}$$

$$\text{or} \quad U_i - U_f = \int_{x_i}^{x_f} F(x)\,dx$$

The gravitational potential energy of a particle of mass m at a height x about the earth's surface is : $U(x) = mgx$,

The potential energy of a body subjected to a conservative force is always undetermined upto a constant. For example, the point where the potential energy is zero is a matter of choice. For the gravitational potential energy mgh, the zero of the potential energy is chosen to be the ground.
For the spring potential energy $kx^2/2$, the zero of the potential energy is the equilibrium position of the oscillating mass.
Every force encountered in mechanics does not have an associated potential energy. For example, work done by friction over a closed path is not zero and no potential energy can be associated with friction.

Kinetic Energy

The energy possessed by a body by virtue of its motion is called kinetic energy. The kinetic energy E_k is defined as

$$E_k = \frac{1}{2}mv^2 \quad ...(i)$$

Where m is mass of body, which is moving with velocity v.

Relation of Kinetic Energy with Linear Momentum

We know that linear momentum p of a body which is moving with a velocity v is given by

$$p = mv \quad ...(ii)$$

From eq(i) & (ii), we have

$$E_k = \frac{p^2}{2m} \quad ...(iii)$$

If total work is due to real forces only, then change in kinetic energy is defined with the help of velocity relative to inertial frame of reference.
If total work is due to real as well as pseudo forces, then change in kinetic energy is defined with the help of velocity relative to non-inertial reference frame.
Energy produced when rest mass is completely converted into energy. $E = m_0c^2$.
A force is conservative if (a) work done by it on an object is path independent and depends only on the end points $\{x_i, x_j\}$, or (b) the work done by the force is zero for an arbitrary closed path taken by the object such that it returns to its initial position.

$\Sigma W_{conservative} = -\Sigma\,\Delta U$; $\Sigma W_{conservative}$ ⇒ Total work due to all kinds of conservative forces.
$\Sigma\,\Delta U$ ⇒ Total change in all kinds of potential energy.

$$W_{nonconservative} + W_{pseudo} + W_{other} = \Delta K - W_{cons.} = \Delta K + \Delta U$$

$$W_{nonconservative} + W_{pseudo} + W_{other} = \Delta E$$

[ΔE = change in mechanical energy)

$$W_{conservative} + W_{nonconservative} + W_{pseudo} + W_{other} = \Delta K$$

If $W_{nonconservative} + W_{pseudo} + W_{other} = 0$ then

$W_{conservative} = \Delta K - \Delta U = \Delta K$; $\Delta U + \Delta K = 0$; U + K = constant;
E = constant (Conservation of mechanical energy)

THE WORK-ENERGY THEOREM

It states that the change in kinetic energy of a body is the work done by the net force on the body.

$$K_f - K_i = W_{net}$$

The WE theorem is not independent of Newton's second law. The WE theorem may be viewed as a scalar form of the Second Law.
The WE theorem holds in all inertial frames. It can also be extended to non-inertial frames provided we include the pseudo forces in the calculation of the net force acting on the body under consideration.

THE PRINCIPLE OF CONSERVATION OF MECHANICAL ENERGY

It states that the total mechanical energy of a body remains constant if the only forces that act on the body are conservative. The principle of conservation of mechanical energy may be viewed as a consequence of the WE theorem for conservative forces.

POWER

Power of the body is defined as the time rate of doing work by the body. The average power P_{av} over the time interval Δt is defined by

$$P_{av} = \frac{\Delta W}{\Delta t} \quad ...(i)$$

The instantaneous power P is defined by

$$P = \lim_{\delta t \to 0} \frac{\Delta W}{\Delta t} = \frac{dW}{dt} \quad ...(ii)$$

$$P = \frac{dW}{dt} = \frac{d}{dt}(\vec{F}.\vec{S}) = \vec{F}.\frac{d\vec{S}}{dt} = \vec{F}.\vec{v}$$

Power is a scalar quantity.
The unit of power is joule per second or watt

The horse-power (hp) is another unit of power

1hp = 746 W.

Electrical energy consumption is measured in kilowatt hour (kwh)

1 kwh = 3.6×10^6 J

The dimensions of power are $[ML^2T^{-3}]$

COLLISIONS

Elastic collision : The collision in which both the momentum and kinetic energy of the system remains conserved is called elastic collision.

Forces involved in the interaction of elastic collision are conservative in nature.

Inelastic collision : The collision in which only the momentum of the system is conserved but kinetic energy is not conserved is called inelastic collision.

Perfectly inelastic collision is one in which the two bodies stick together after the collision.

Velocity after colllision, $v = \dfrac{m_1u_1 + m_2u_2}{m_1 + m_2}$

Forces involved in the interaction of inelastic collision are non-conservative in nature.

Coefficient of Restitution : $e = |v_1 - v_2| / |u_1 - u_2|$

(i) $0 < e < 1$

(ii) For perfectly elastic collision, e = 1

(iii) For a perfectly inelastic collision, e = 0

Velocities After Collision

Two bodies of mass m_1 and m_2 moving unidirectionally collide (head on collision) such that their relative velocity is along the line of action. Let the velocities before the collision be u_1 and u_2 and that just after the collision be v_1 and v_2 respectively. Assume that the direction of motion of the bodies remains the same just before and after the collision.

$$v_1 = \frac{(m_1 - em_2)}{(m_1 + m_2)}u_1 + \frac{m_2(1+e)}{(m_1 + m_2)}u_2 \text{ and}$$

$$v_2 = \frac{m_1(1+e)}{(m_1 + m_2)}u_1 + \frac{(m_2 - em_1)}{(m_1 + m_2)}u_2$$

where e is coefficient of restitution.

Loss in Kinetic Energy in Collision

$$\Delta KE = \frac{1}{2} \cdot \frac{m_1m_2}{m_1 + m_2}(1 - e^2)(u_1 - u_2)^2$$

Oblique Collision

(i) This is the case of collision in two dimensions. After the collision, the particles move at different angle.

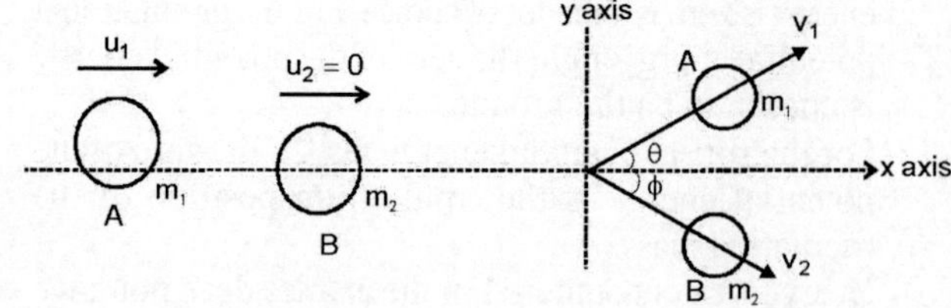

(ii) We will apply the principle of conservation of momentum in the mutually perpendicular direction.

Along x-axis, $m_1u_1 = m_1v_1\cos\theta + m_2v_2\cos\phi$

Along y-axis, $0 = m_1v_1\sin\theta - m_2v_2\sin\phi$

1. Work done by a force is zero if, F = 0 or S = 0 or $F \perp S$
2. Work done by a force is positive if angle between F and S is < 90°.
3. Work done by a force is negative if angle between F and S is > 90°.
4. When kinetic energy of a body is made n times, its momentum increases to $\sqrt{n}$ times.
5. When momentum of a body is made n times, its kinetic energy increases to n^2 times.
6. If a light body and a heavy body have equal momentum, then light body has greater kinetic energy.
7. If a light body and a heavy body have equal kinetic energy, then heavy body has greater momentum.
8. If a light body and a heavy body have equal kinetic energy, then heavy body has greater momentum.
9. When water flows through a uniform tube with a constant velocity v, the power to maintain the flow is proportional to cube of velocity.
10. $W_{spring} = -\left(\frac{1}{2}kx_2^2 - \frac{1}{2}kx_1^2\right)$
11. A sphere of mass m moving with a constant velocity u hits another stationary sphere of the same mass. If e is the coefficient of restitution, then ratio of velocities of the two spheres (First sphere/Second sphere) after collision will be : $\left(\frac{1-e}{1+e}\right)$.

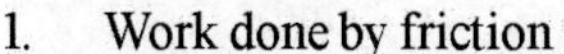

Exercise

1. Work done by friction :
   (a) is always negative (b) may be positive
   (c) is always zero (d) is always positive
2. The magnitude of work done by a force :
   (a) depends on frame of reference
   (b) does not depend on frame of reference
   (c) cannot be calculated in non-inertial frames.
   (d) both (a) and (b)
3. Work done by all non-conservative forces is equal to :
   (a) change in K.E. of the body
   (b) change in P.E. of the body
   (c) change in total mechanical energy of the body
   (d) none of these
4. Work done by a conservative force is positive if
   (a) P.E. of the body increases
   (b) P.E. of the body decreases
   (c) K.E. of the body increases
   (d) K.E. of the body decreases
5. A vehicle is moving with a uniform velocity on a smooth horizontal road, then power delivered by its engine must be :
   (a) uniform (b) increasing
   (c) decreasing (d) zero
6. Which of the following forces is/are non-conservative?
   (a) Frictional force (b) Spring force
   (c) Elastic force (d) All of these
7. Work done by frictional force in a round trip.
   (a) must be positive (b) must be negative
   (c) may be zero (d) must be none zero
8. A car and a truck moving with equal kinetic energies are stopped by applying equal retarding forces, then :
   (a) car will cover more distance before coming to rest
   (b) truck will cover more distance before coming to rest
   (c) both will cover equal distance before coming to rest
   (d) data insufficient
9. A ball of mass m and a ball B of mass $2m$ are projected with equal kinetic energies. Then at the highest point of their respective trajectories.
   (a) P.E. of A will be more than that of B
   (b) P.E. of B will be more than that of B
   (c) P.E. of A will be equal to that of B
   (d) can't be predicted
10. In case of elastic collision, at the time of impact.
   (a) total K.E. of colliding bodies is conserved
   (b) total K.E. of colliding bodies increases
   (c) total K.E. of colliding bodies decreases
   (d) total momentum of colliding bodies decreases
11. A ball A of mass m_A collide with a stationary ball B of mass m_B. The ball A can come to rest after collision *if*.
   (a) $m_A = m_B$ (b) $m_A < m_B$
   (c) $m_A > m_B$ (d) either (a) or (b)
12. A ball is projected from ground at a certain angle. After striking the ground, horizontal component of its velocity.
   (a) must change
   (b) must remain same
   (c) will change if surface is smooth
   (d) will change if surface is rough
13. In case of perfectly inelastic collision in two dimensions
   (a) Total K.E. remains conserved
   (b) Both the bodies stick together after the collision
   (c) Total momentum remains conserved
   (d) All of these
14. The engine of a vehicle delivers constant power. If the vehicle is moving up the inclined plane then, its velocity
   (a) must remain constant
   (b) must increase
   (c) must decrease
   (d) may increase, decrease or remain same
15. A ball projected from ground at a certain angle collides a smooth inclined plane at the highest point of its trajectory. If the collision is perfectly inelastic then after the collision.
   (a) ball will comes to rest
   (b) ball will move along the incline
   (c) ball will retrace its path
   (d) ball will bounce back but will strike the ground at a point other than point of projection

16. The co-efficient of restitution e for a perfectly elastic collision is
(a) 1 (b) 0
(c) ∞ (d) -1

17. A bullet of mass 10g leaves a rifle at an initial velocity of 1000 m/s and strikes the earth at the same level with a velocity of 500 m/s. The work done in joules overcoming the resistance of air will be
(a) 375 (b) 3750
(c) 5000 (d) 500

18. How much water, a pump of 2 kW can raise in one minute to a height of 10 m, take $g = 10\ m/s^2$?
(a) 1000 litres (b) 1200 litres
(c) 100 litres (d) 2000 litres

19. Two identical balls A and B moving with velocities + 0.5 m/s and –0.3 m/s respectively, collide head on elastically. The velocities of the balls A and B after collision, will be, respectively
(a) +0.5 m/s and +0.3 m/s
(b) – 0.3 m/s and +0.5 m/s
(c) +0.3 m/s and 0.5 m/s
(d) –0.5 m/s and +0.3 m/s

20. A position dependent force, $F = (7 - 2x + 3x^2)\ N$ acts on a small body of mass 2 kg and displaces it from $x = 0$ to $x = 5\ m$. Work done in joule is
(a) 35 (b) 70
(c) 135 (d) 270

21. Two masses of 1g and 9g are moving with equal kinetic energies. The ratio of the magnitudes of their respective linear momenta is
(a) 1 : 9 (b) 9 : 1
(c) 1 : 3 (d) 3 : 1

22. Consider a car moving along a straight horizontal road with a speed of 72 km/h. If the coefficient of static friction between road and tyres is 0.5, the shortest distance in which the car can be stopped is
(a) 30 m (b) 40 m
(c) 72 m (d) 20 m

23. If the momentum of a body is increased by 50%, then the percentage increase in its kinetic energy is
(a) 50% (b) 100%
(c) 125% (d) 200%

24. The kinetic energy acquired by a mass (m) in travelling distance (s) starting from rest under the action of a constant force is directly proportional to
(a) $1/\sqrt{m}$ (b) $1/m$
(c) $\sqrt{m}$ (d) m^0

25. A body of mass m moving with velocity 3 km/h collides with a body of mass 2 m at rest. Now the coalesced mass starts to move with a velocity
(a) 1 km/h (b) 2 km/h
(c) 3 km/h (d) 4 km/h

26. Which one of the following physical quantities is represented by the shaded area in the given graph?

Force
Distance

(a) Torque (b) Impulse
(c) Power (d) Work done

27. A molecule of mass m of an ideal gas collides with the wall of a vessel with a velocity v and returns back with the same velocity. The change in linear momentum of molecule is
(a) 2 mv (b) 4 mv
(c) 8 mv (d) 10 mv

28. From a stationary tank of mass 125000 pound a small shell of mass 25 pound is fired with a muzzle velocity of 1000 ft/sec. The tank recoils with a velocity of
(a) 0.1 ft/sec (b) 0.2 ft/sec
(c) 0.4 ft/sec (d) 0.8 ft/sec

29. Two bodies of masses 2 m and m have their K.E. in the ratio 8 : 1, then their ratio of momenta is
(a) 1 : 1 (b) 2 : 1
(c) 4 : 1 (d) 8 : 1

30. An object of 1 kg mass has a momentum of 10 kg m/sec then the kinetic energy of the object will be
(a) 100 J (b) 50 J
(c) 1000 J (d) 200 J

31. The kinetic energy of a body of mass 3 kg and momentum 2 Ns is
(a) 1 J (b) $\frac{2}{3}$ J
(c) $\frac{3}{2}$ J (d) 4J

32. If the momentum of a body increases by 0.01%, its kinetic energy will increase by
(a) 0.01% (b) 0.02%
(c) 0.04% (d) 0.08%

33. An electric motor exerts a force of 40 N on a cable and pulls it by a distance of 30 m in one minute. The power supplied by the motor (in watts) is
(a) 20 (b) 200
(c) 2 (d) 10

34. The bob A of a simple pendulum is released when the string makes an angle of 45° with the vertical. It hits another bob B of the same material and same mass kept at rest on the table. If the collision is elastic

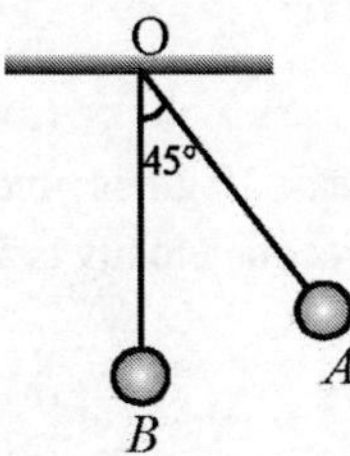

(a) Both A and B rise to the same height
(b) Both A and B come to rest at B
(c) Both A and B move with the same velocity of A
(d) A comes to rest and B moves with the velocity of A

35. An inelastic ball is dropped from a height of 100 m. Due to earth, 20% of its energy is lost. To what height the ball will rise?
(a) 80 m (b) 40 m
(c) 60 m (d) 20 m

36. A tennis ball is released from height h above ground level. If the ball makes inelastic collision with the gound, to what height will it rise after third collision
(a) he^6 (b) e^2h
(c) e^3h (d) None of these

37. A body of mass m moving with a constant velocity v hits another body of the same mass moving with the same velocity v but in the opposite direction and sticks to it. Velocity of the compound body after collision is
(a) v (b) $2v$
(c) zero (d) $v/2$

38. A force $\vec{F} = (5\hat{i} + 3\hat{j})\,N$ is applied over a particle which displace it from its origin to the point $\vec{r} = (2\hat{i} - 1\hat{j})$ m. Then the work done on the particle is
(a) 13 J (b) 10 J
(c) 7 J (d) −7 J

39. A stone is released from a tower, its total energy during its fall
(a) increases
(b) decreases
(c) remains constant
(d) first increases then remains constant

40. An engine pumps water through a hose pipe. Water passes through the pipe and leaves it with a velocity of 2 m/s. The mass per unit length of water in the pipe is 100 kg/m. What is the power of the engine?
(a) 400 W (b) 200 W
(c) 100 W (d) 800 W

41. A mass of 0.5 kg moving with a speed of 1.5 m/s on a horizontal smooth surface, collides with a nearly weightless spring of force constant $k = 50$ N/m. The maximum compression of the spring would be

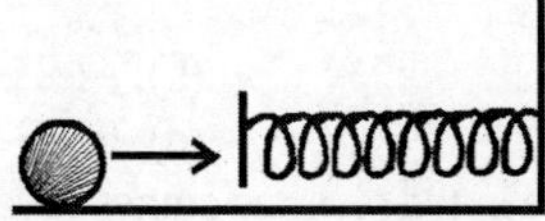

(a) 0.5 m (b) 0.15 m
(c) 0.12 m (d) 1.5 m

42. The block of mass M moving on the frictionless horizontal surface collides with the spring of spring constant k and compresses it by length L. The maximum momentum of the block after collision is

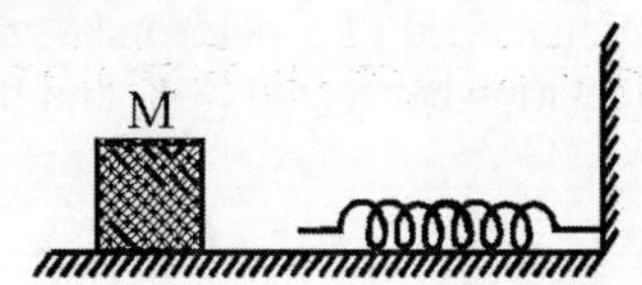

(a) $\frac{kL^2}{2M}$ (b) $\sqrt{Mk}\,L$
(c) $\frac{ML^2}{k}$ (d) zero

43. When a U^{238} nucleus, originally at rest, decays by emitting an α-particle, say with speed of v m/sec, the recoil speed of the residual nucleus is (in m/sec.)
(a) $-4v/234$ (b) $-4v/238$
(c) $4v/238$ (d) $-v/4$

44. A body is moved along a straight line by a machine delivering a constant power. The distance moved by the body in time 't' is proportional to
(a) $t^{3/4}$ (b) $t^{3/2}$
(c) $t^{1/4}$ (d) $t^{1/2}$

45. A particle moves in a straight line with retardation proportional to its displacement. Its loss of kinetic energy for any displacement x is proportional to
(a) x (b) e^x
(c) x^2 (d) $\log_e x$

46. A particle is acted upon by a force of constant magnitude which is always perpendicular to the velocity of the particle, the motion of the particles takes place in a plane. It follows that
(a) its kinetic energy is constant
(b) its acceleration is constant
(c) its velocity is constant
(d) it moves in a straight line

47. A particle of mass 100 g is thrown vertically upwards with a speed of 5 m/s. The work done by the force of gravity during the time the particle goes up is
(a) –0.5 J (b) –1.25 J
(c) 1.25 J (d) 0.5 J

48. An athlete in the olympic games covers a distance of 100 m in 10 s. His kinetic energy can be estimated to be in the range
(a) 200 J - 500 J (b) 2×10^5 J - 3×10^5 J
(c) 20,000 J - 50,000 J (d) 2,000 J - 5, 000 J

49. If W_1, W_2 and W_3 represent the work done in moving a particle from A to B along three different paths 1, 2 and 3 respectively (as shown) in the gravitational field of a point mass m, find the correct relation between W_1, W_2 and W_3

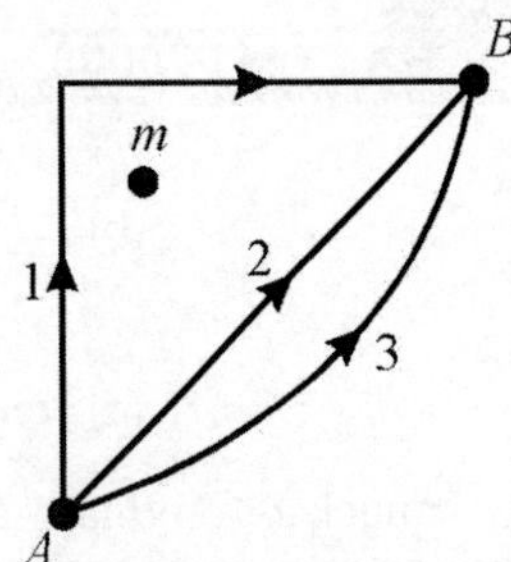

(a) $W_1 > W_2 > W_3$ (b) $W_1 = W_2 = W_3$
(c) $W_1 < W_2 < W_3$ (d) $W_2 > W_1 > W_3$

50. A rocket has a mass of 100 kg 90% of this is fuel. It ejects fuel vapours at the rate of 1 kg/sec with a velocity of 500 m/sec relative to the rocket. It is supposed that the rocket is outside the gravitational field. The initial upthrust on the rocket when it just starts moving upwards is
(a) zero (b) 500 N
(c) 1000 N (d) 2000 N

51. A wagon weighing 1000 kg is moving with a velocity 50 km/h on smooth horizontal rails. A mass of 250 kg is dropped into it. The velocity with which it moves now is
(a) 2.5 km/hour (b) 20 km/hour
(c) 40 km/hour (d) 50 km/hour

52. A rope of length 5 m is kept on frictionless surface and a force of 5 N is applied to one of its end. Find tension in the rope at 1m from this end
(a) 1 N (b) 3 N
(c) 4 N (d) 5 N

53. A bullet of mass 10 gm is fired from a gun of mass 1 kg. If the recoil velocity is 5 m/s, the velocity of the muzzle is
(a) 0.05 m/s (b) 5 m/s
(c) 50 m/s (d) 500 m/s

54. You lift a heavy book from the floor of the room and keep it in the book-shelf having a height 2 m. In this process you take 5 seconds. The work done by you will depend upon
(a) mass of the book and time taken
(b) weight of the book and height of the book-shelf
(c) height of the book-shelf and time taken
(d) mass of the book, height of the book-shelf and time taken

55. A body of mass 10 kg is dropped to the ground from a height of 10 metres. The work done by the gravitational force is (g = 9.8 m/sec^2)
(a) – 490 Joules (b) + 490 Joules
(c) – 980 Joules (d) + 980 Joules

56. A force of 5 N making an angle θ with the horizontal, acting on an object displaces it by 0.4 m along the horizontal direction. If the object gains kinetic energy of 1J, the horizontal component of the force is
(a) 1.5 N (b) 2.5 N
(c) 3.5 N (d) 4.5 N

57. A particle moves from position
$\vec{r}_1 = 3\hat{i} + 2\hat{j} + 9\hat{k}$ to position $\vec{r}_2 = 14\hat{i} + 13\hat{j} + 9\hat{k}$
under the action of force $4\hat{i} + \hat{j} + 3\hat{k}$ N. The work done will be
(a) 100 J (b) 50 J
(c) 200 J (d) 75 J

58. In an explosion, a body breaks up into two pieces of unequal masses. In this
    (a) both parts will have numerically equal momentum
    (b) lighter part will have more momentum
    (c) heavier part will have more momentum
    (d) both parts will have equal kinetic energy

59. A man starts walking from a point on the surface of earth (assumed smooth) and reaches diagonally opposite point. What is the work done by him?
    (a) Zero (b) Positive
    (c) Negative (d) Nothing can be said

60. The spring extends by x on loading, then energy stored by the spring is: (if T is the tension in spring and k is spring constant)
    (a) $\frac{T^2}{2k}$ (b) $\frac{T^2}{2k^2}$
    (c) $\frac{2k}{T^2}$ (d) $\frac{2T^2}{k}$

System of Particles and Rotational Motion

CONCEPTS

CENTRE OF MASS

The centre of mass of a system of particles is defined as the point whose position vector is $\vec{R}_{cm} = \frac{\sum m_i r_i}{M}$.

For uniform body position vector of centre of mass

$$\vec{R}_{cm} = \frac{\int dm\vec{r}}{\int dm}$$

The centre of gravity of an extended body is that point where the total gravitational torque on the body is zero. The **centre of gravity** of a body coincides with its centre of mass only if the gravitational field does not vary from one part of the body to the other.

Remember

(a) $M\vec{V}_{cm} = m_1\vec{v}_1 + m_2\vec{v}_2 + ... + m_n\vec{v}_n$

(b) $M\vec{a}_{cm} = \vec{F}_{ext}$

(c) If $\vec{F}_{ext} = 0$ then $\vec{v}_{cm}$ = constant

If no external force is acting on the system, net momentum of the system must remain constant. (momentum conservation)

A **rigid body** is one for which the distances between different particles of the body do not change, even though there are forces on them. A rigid body fixed at one point or along a line can have only rotational motion. A rigid body not fixed in some way can have either pure translation or a combination of translation and rotation.

In rotation about a fixed axis, every particle of the rigid body moves in a circle which lies in a plane perpendicular to the axis and has its centre on the axis. Every point in the rotating rigid body has the same angular velocity at any instant of time.

ANGULAR VELOCITY

The **angular velocity** is defined as the angle traversed by the radius vector per unit time.

Therefore average angular velocity $\bar{\omega} = \frac{\Delta\theta}{\Delta t}$

The **average angular acceleration** $\alpha_{avg.}$ of a rotating body is defined as

$$\alpha_{avg.} = \frac{\omega_2 - \omega_1}{t_2 - t_1} = \frac{\Delta\omega}{\Delta t}$$

Relationship Between Angular and Linear Velocities

$$\vec{v} = \vec{\omega} \times \vec{r} \Rightarrow v = r\omega \sin\theta$$

where θ is the angle between ω and r.

The linear velocity of a particle of a rigid body rotating about a fixed axis is given by $\vec{v} = \vec{\omega} \times \vec{r}$, where r is the position vector of the particle with respect to an origin along the fixed axis. The relation applies even to more general rotation of a rigid body with one point fixed. In that case, $\vec{r}$ is the position vector of the particle with respect to the fixed point taken as the origin.

TORQUE AND ANGULAR MOMENTUM

The moment of force is called **torque**. It is also defined as the product of force and the perpendicular distance of the force from the axis of rotation.

It is also called **couple**. A couple produces rotation without translation.

It is given as $\vec{\tau} = \vec{r} \times \vec{F} \Rightarrow \tau = rF\sin\theta$

where θ is the angle between $\vec{r}$ and $\vec{F}$.

The torque or moment of force on a system of n particles about the origin is $\tau = \sum_{i}^{n} \vec{r}_i \times \vec{F}_i$

The force F_i acting on the ith particle includes the external as well as internal forces. Assuming Newton's third law and that forces between any two particles act along the line joining the particles we can show $\tau_{int} = 0$ and $\frac{dL}{dt} = \tau_{ext}$

Angular momentum :

Angular momentum of a particle about an arbitary point is the moment of linear momentum taken about that point.

It is given as $\vec{L} = \vec{r} \times \vec{p}$

$L = rp\sin\theta$ where θ is the angle between $\vec{r}$ and $\vec{F}$.

The angular momentum of a system of n particles about the origin is $L = \sum_{i=1}^{n} r_i \times p_i$

Angular impulse :

$$J = \int_{t_1}^{t_2} \tau dt = L_2 - L_1 \text{ (Change in angular momentum)}$$

ANGULAR ACCELERATION

The angular acceleration of a rigid body rotating about a fixed axis is given by $I\alpha = \tau$. If the external torque τ is zero, the component of angular momentum about the fixed axis $I\alpha$ of such a rotating body is constant. The angular momentum L and the angular velocity ω are not necessarily parallel vectors. However, for the simpler situations when rotation is about a fixed axis which is an axis of symmetry of the rigid body, the relation $L = I\omega$ holds good, where I is the moment of the inertia of the body about the rotational axis.

The Kinematic equations : For constant angular acceleration

$$\vec{\omega}_{(t)} = \vec{\omega}_0 + \vec{\alpha}t \qquad \vec{\theta}_{(t)} = \vec{\theta}_0 + \vec{\omega}_0 t + \frac{1}{2}\vec{\alpha}t^2$$

$$\omega_{(t)}^2 = \omega_0^2 + 2\alpha\theta \; ; \; \theta = \left(\frac{\omega + \omega_0}{2}\right)t$$

$$\theta_{nth} = \omega_0 + \frac{\alpha}{2}(2n - 1)$$

Here ω_0 = initial angular velocity, $\omega_{(t)}$ = angular velocity after time t, θ_0 = initial angular position

$\theta_{(t)}$ = angular position after time t,

θ_{nth} = angular displacement in n^{th} second.

These equations are valid if α is a constant. If angular acceleration is variable then use calculus approach.

PRINCIPLE OF MOMENTS

According to **Principle of moments** for a lever, load arm × load = effortarm × effort

or $\quad d_1F_1 = d_2F_2$

This gives Mechanical advantage of lever is M.A. =

$$\frac{F_1}{F_2} = \frac{d_2}{d_1}$$

Thus if M.A. > 1 means a small effort can lift a larger load.

MOMENT OF INERTIA

The moment of inertia of a rigid body about an axis is defined by the formula $I = \sum m_i r_i^2$ where r_i is the perpendicular distance of the ith point of the body from the axis.

The kinetic energy of rotation is $K = \frac{1}{2}I\omega^2$

Moment of inertia is the analogue of mass in rotational motion.

Moment of Inertia of Diatomic Molecule

$$I = m_1r_1^2 + m_2r_2^2 = \left(\frac{m_1m_2}{m_1 + m_2}\right)r^2 = \mu r^2$$

where m_1 and m_2 are masses of any diatomic molecule and distance between them is r.

r_1 & r_2 are the distances of atoms from the centre of mass & μ is called the reduced mass.

Radius of Gyration

It is the distance from the axis of a mass point whose mass is equal to the mass of the whole body and whose moment of inertia is equal to the moment of inertia of the body about the axis.

Radius of Gyration K = $K = \sqrt{\frac{r_1^2 + r_2^2 + ... + r_n^2}{n}}$

Where, r_1, r_2 ......r_n are the distance of constituent particles from the axis of rotation.

Also, $K = \sqrt{\frac{I}{M}}$

THEOREMS OF PERPENDICULAR AND PARALLEL AXES

Theorem of Perpendicular Axes (Applicable only for two dimensional bodies or plane laminas)

The moment of inertia of a plane lamina about an axis perpendicular to its plane is equal to the sum of the moments of inertia of the lamina about any two mutually perpendicular axes in its own plane intersecting each other at the point through which the perpendicular axis passes. $\quad I_z = I_x + I_y$

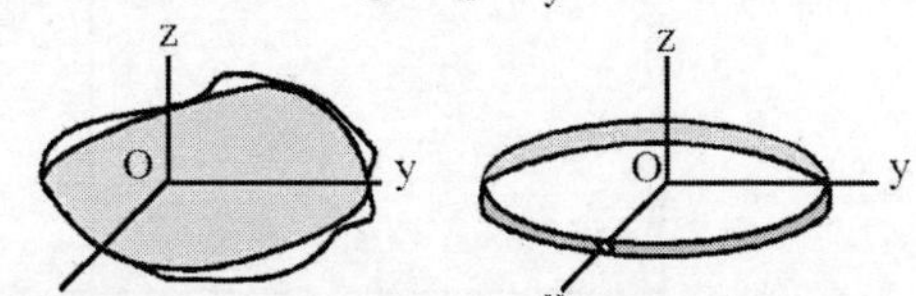

Theorem of Parallel Axes (Applicable for all Types of Bodies)

Moment of inertia of a body about any axis is equal to the moment of inertia about a parallel axis passing through the centre of mass plus product of mass of the body and the square of distance between these two parallel axes

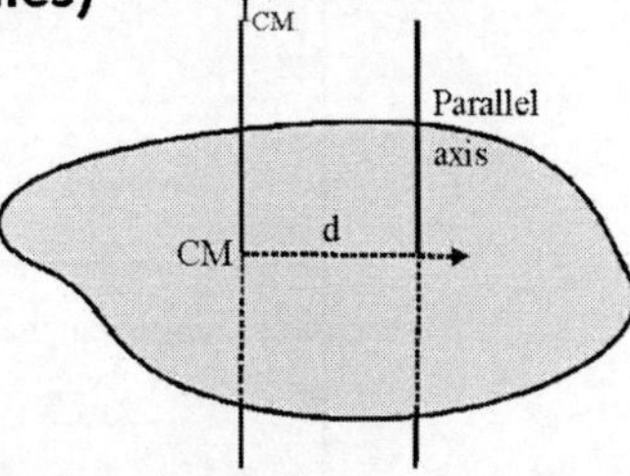

$$I = I_{CM} + Md^2$$

I_{CM} = moment of inertia about the axis passing through centre of mass.

Moment of inertia and radius of gyration of different objects

Shape of body	Rotational axis	Figure	Moment of inertia	Radius of gyration
(1) Ring M = mass R = radius	(a) Perpendicular to plane passing through centre of mass	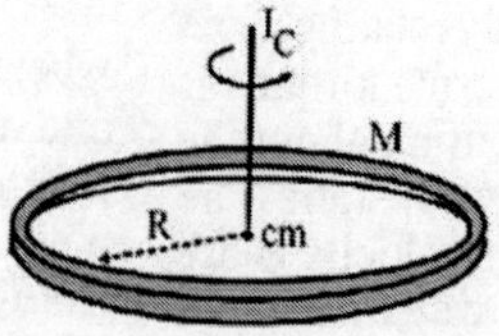	MR^2	R
	(b) Diameter in the plane	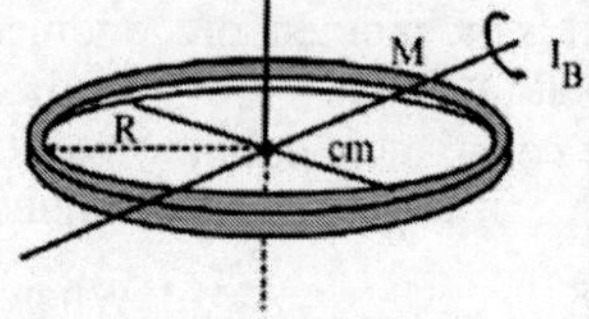	$\frac{1}{2}MR^2$	$\frac{R}{\sqrt{2}}$
	(c) Tangent perpendicular to plane	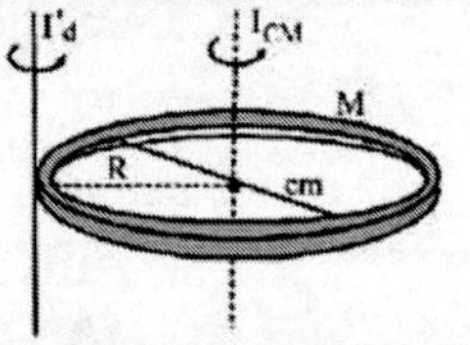	$2MR^2$	$\sqrt{2}R$
	(d) Tangent in the plane	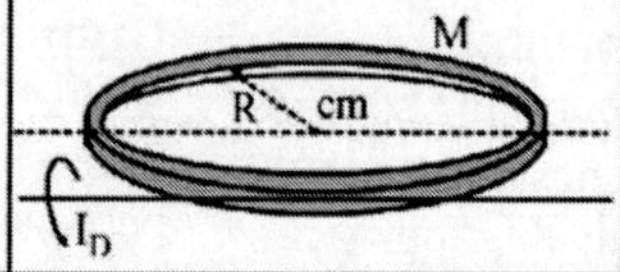	$\frac{3}{2}MR^2$	$\sqrt{\frac{3}{2}}R$
(2) Disc	(a) Perpendicular to plane passing through centre of mass	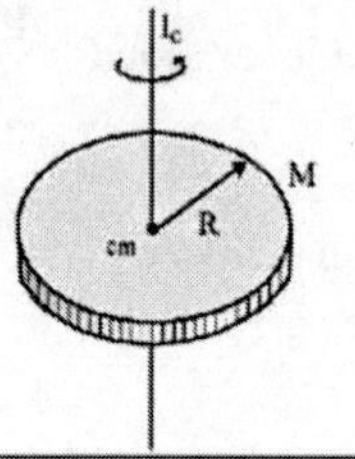	$\frac{1}{2}MR^2$	$\frac{R}{\sqrt{2}}$
	(b) Diameter in the plane	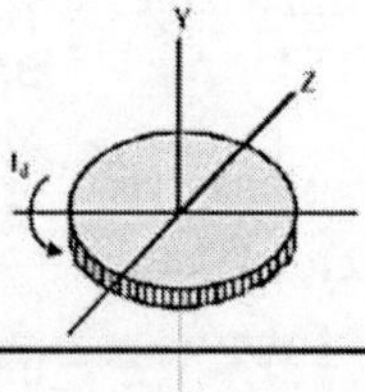	$\frac{MR^2}{4}$	$\frac{R}{2}$
	(c) Tangent in the plane	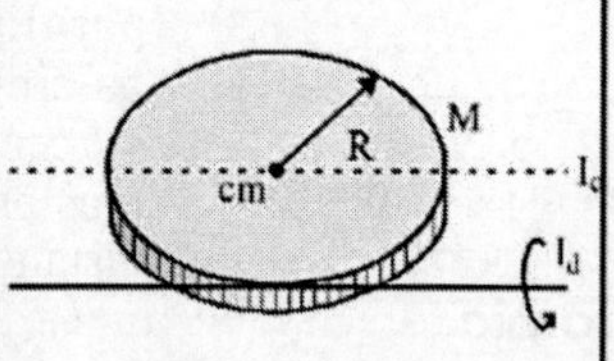	$\frac{5}{4}MR^2$	$\frac{\sqrt{5}}{2}R$

	(d) Tangent perpendicular to plane	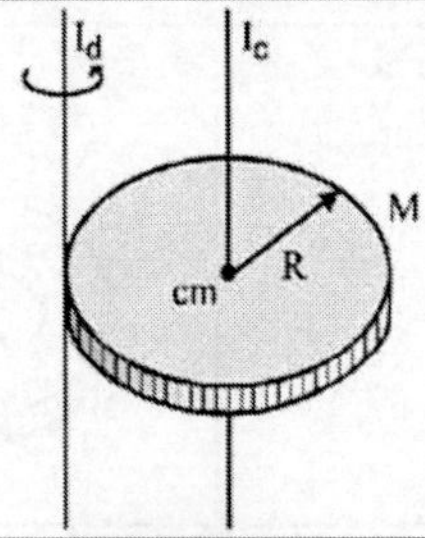	$\frac{3}{2}MR^2$	$\sqrt{\frac{3}{2}}\,R$
(3) Thin walled cylinder	(a) Geometrical axis	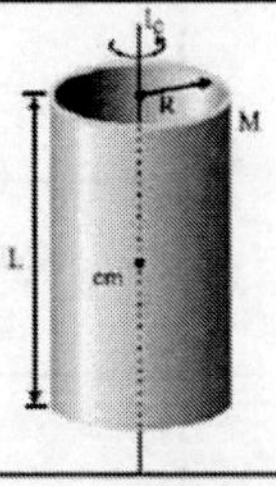	MR^2	R
	(b) Perpendicular to length passing through centre of mass	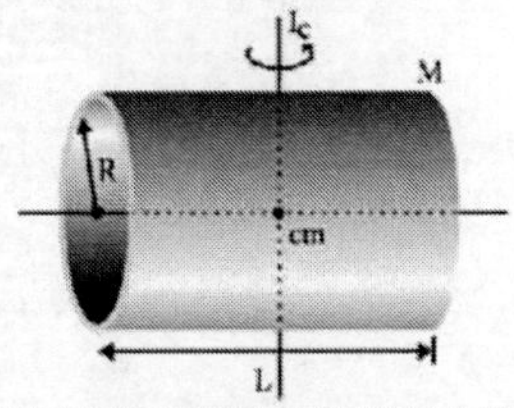	$M\left(\frac{R^2}{2}+\frac{L^2}{12}\right)$	$\sqrt{\frac{R^2}{2}+\frac{L^2}{12}}$
	(c) Perpendicular to length passing through one end	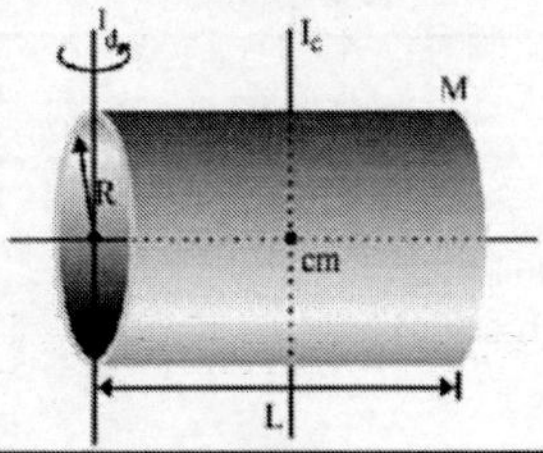	$M\left(\frac{R^2}{2}+\frac{L^2}{3}\right)$	$\sqrt{\frac{R^2}{2}+\frac{L^2}{3}}$
(4) Solid cylinder	(a) Geometrical axis	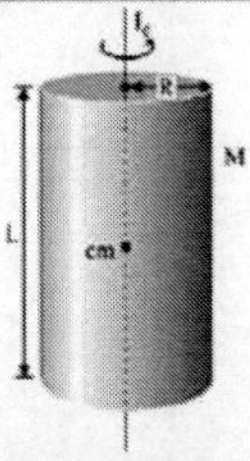	$\frac{MR^2}{2}$	$\frac{R}{\sqrt{2}}$
	(b) Perpendicular to length passing through centre of mass	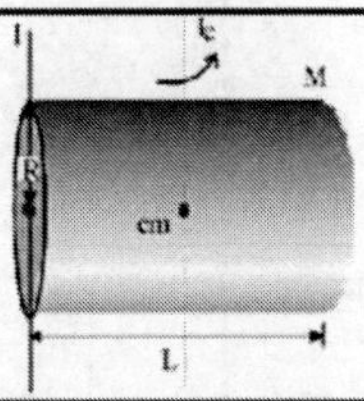	$M\left(\frac{R^2}{4}+\frac{L^2}{12}\right)$	$\sqrt{\frac{R^2}{4}+\frac{L^2}{12}}$
	(c) Perpendicular to length passing through one end	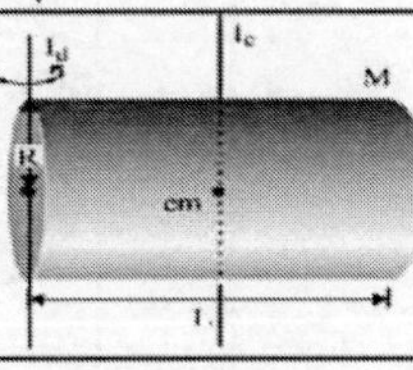	$M\left[\frac{R^2}{4}+\frac{L^2}{3}\right]$	$\sqrt{\frac{R^2}{4}+\frac{L^2}{3}}$

(5) Annular disc	(a) Perpendicular to plane passing through centre of mass	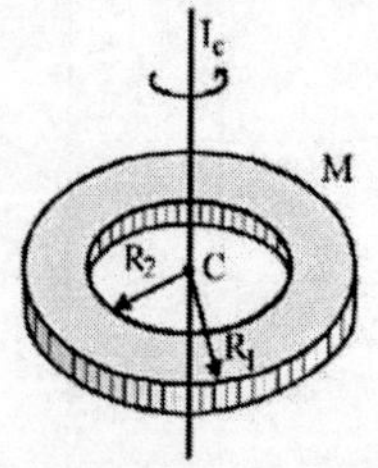	$\frac{M}{2}[R_1^2+R_2^2]$	$\sqrt{\frac{R_1^2+R_2^2}{2}}$
	(b) Diameter in the plane	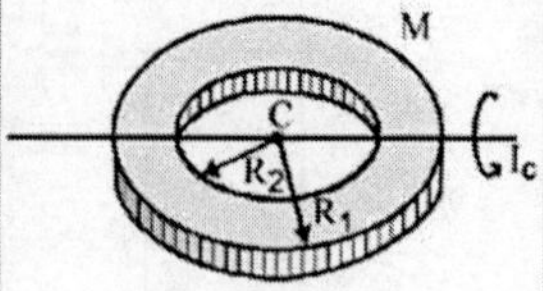	$\frac{M[R_1^2+R_2^2]}{4}$	$\sqrt{\frac{R_1^2+R_2^2}{4}}$
(6) Hollow cylinder	(a) Geometrical axis	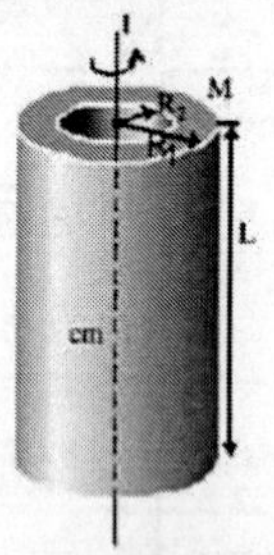	$M\left[\frac{R_1^2+R_2^2}{2}\right]$	$\sqrt{\frac{R_1^2+R_2^2}{2}}$
	(b) Perpendicular to length passing through centre of mass	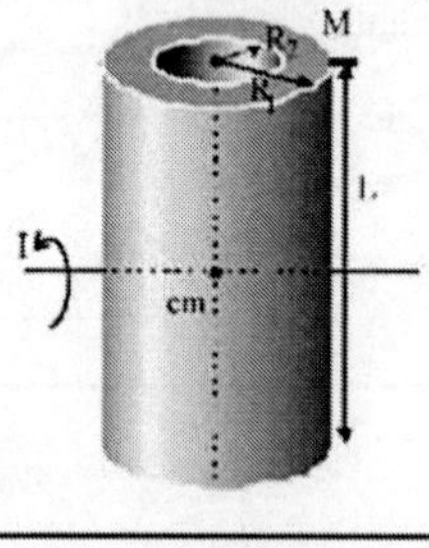	$M\left[\frac{L^2}{12}+\frac{(R_1^2+R_2^2)}{4}\right]$	$\sqrt{\frac{L^2}{12}+\frac{R_1^2+R_2^2}{4}}$
(7) Solid sphere	(a) Along the diameter	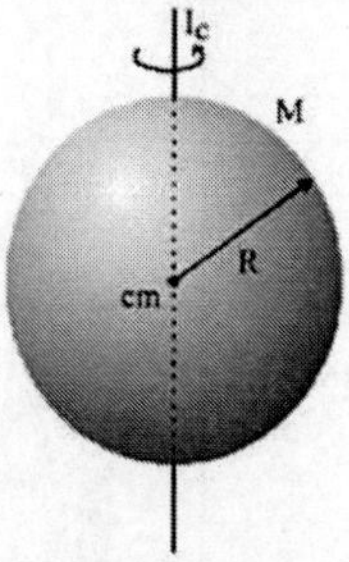	$\frac{2}{5}MR^2$	$\sqrt{\frac{2}{5}}R$
	(b) Along the tangent	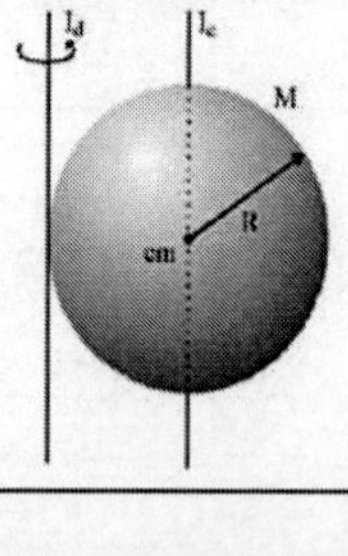	$\frac{7}{5}MR^2$	$\sqrt{\frac{7}{5}}R$

(8) Thin spherical shell	(a) Along the diameter	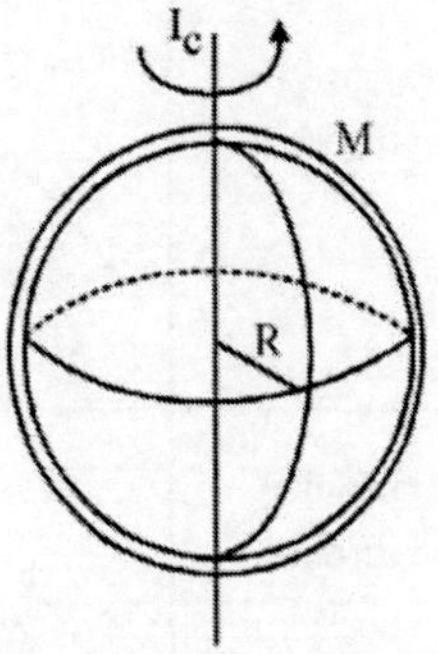	$\frac{2}{3}MR^2$	$\sqrt{\frac{2}{3}}R$
	(b) Along the tangent	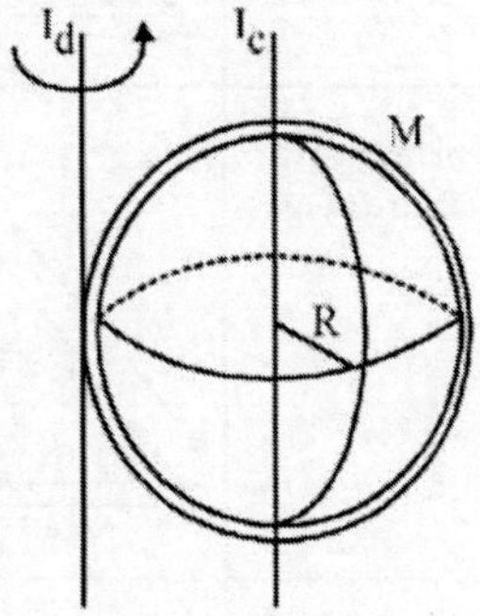	$\frac{5}{3}MR^2$	$\sqrt{\frac{5}{3}}R$
(9) Hollow sphere	Along the diameter	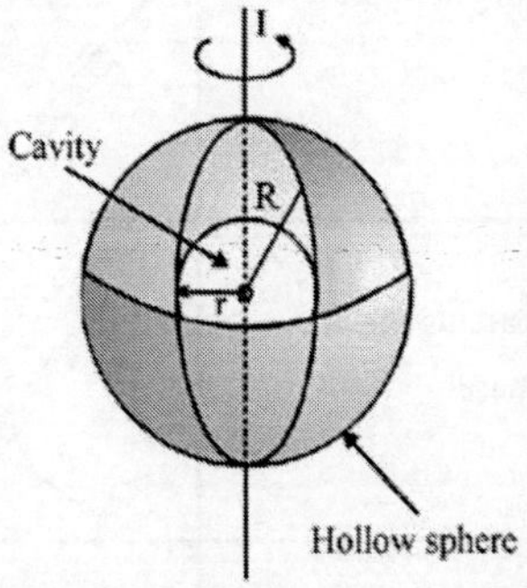	$\frac{2}{5}M\left[\frac{R^5-r^5}{R^3-r^3}\right]$	$\sqrt{\frac{2}{5}\frac{(R^5-r^5)}{(R^3-r^3)}}$
(10) Thin rod	(a) Perpendicular to length passing through centre of mass	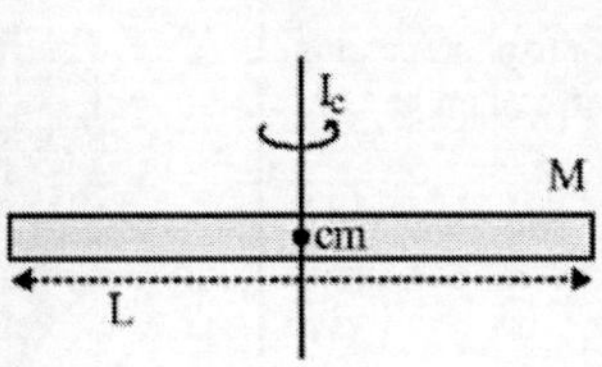	$\frac{ML^2}{12}$	$\frac{L}{2\sqrt{3}}$
	(b) Perpendicular to length passing through one end	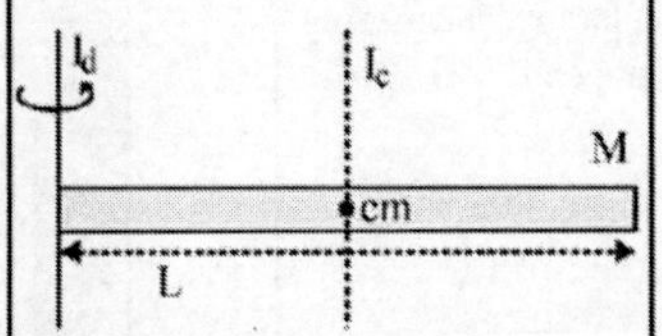	$\frac{ML^2}{3}$	$\frac{L}{\sqrt{3}}$

(11) Rectangular plate	(a) Perpendicular to length in the plane passing through centre of mass	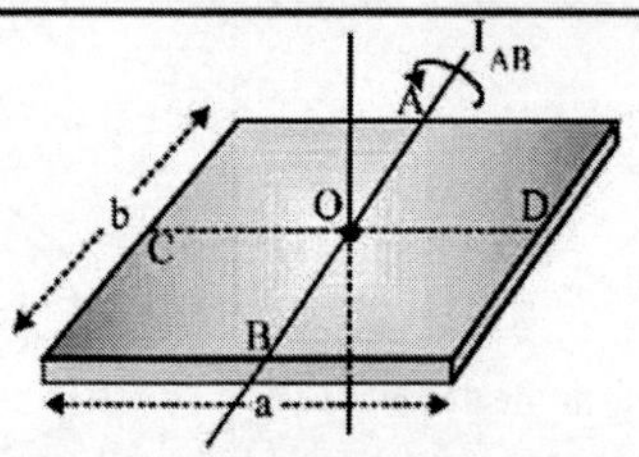	$\frac{Ma^2}{12}$	$\frac{a}{2\sqrt{3}}$
	(b) Perpendicular to breadth in the plane passing through centre of mass	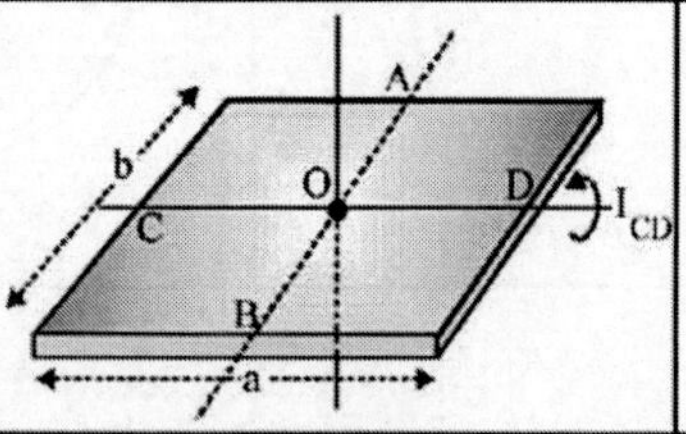	$\frac{Mb^2}{12}$	$\frac{b}{2\sqrt{3}}$
	(c) Perpendicular to plane passing through centre of mass	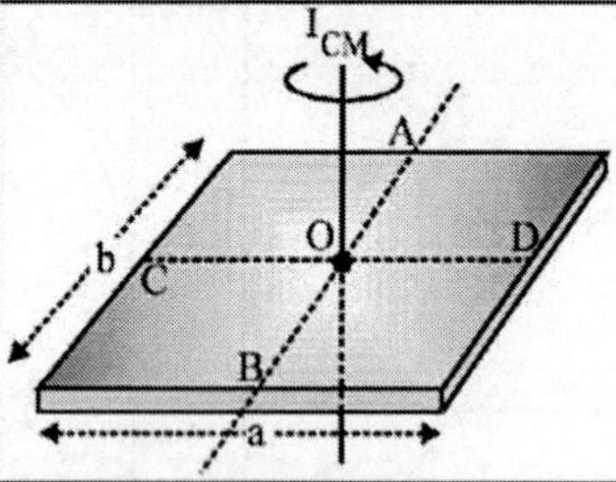	$\frac{M(a^2+b^2)}{12}$	$\frac{\sqrt{a^2+b^2}}{2\sqrt{3}}$
(12) Square Plate	(a) Perpendicular to plane passing through centre of mass	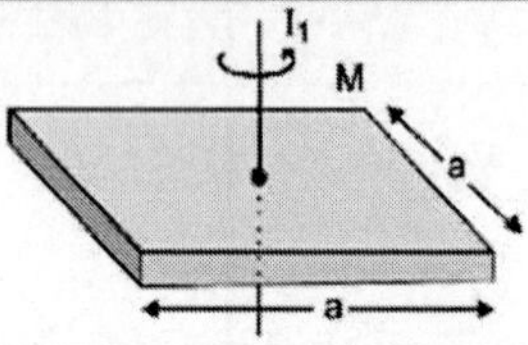	$I_1 = \frac{Ma^2}{6}$	$\frac{a}{\sqrt{6}}$
	(b) Diagonal passing through centre of mass	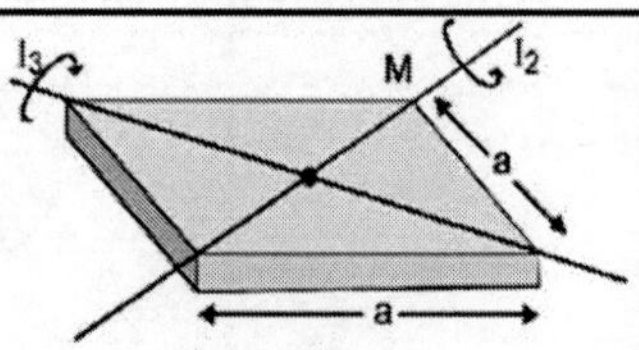	$I_2 = I_3 \frac{Ma^2}{12}$	$\frac{a}{2\sqrt{3}}$
(13) Cube	(a) Perpendicular to plane passing through centre of mass	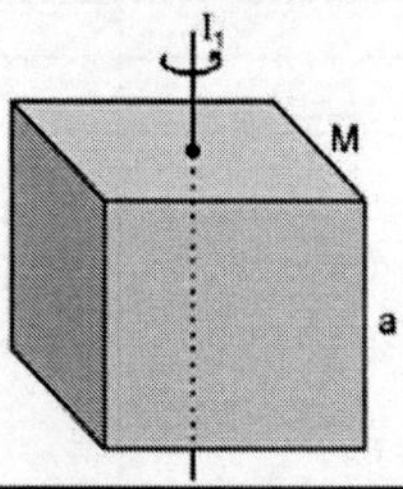	$I_1 = \frac{Ma^2}{6}$	$\frac{a}{\sqrt{6}}$
	(b) Perpendicular to plane passing through one end	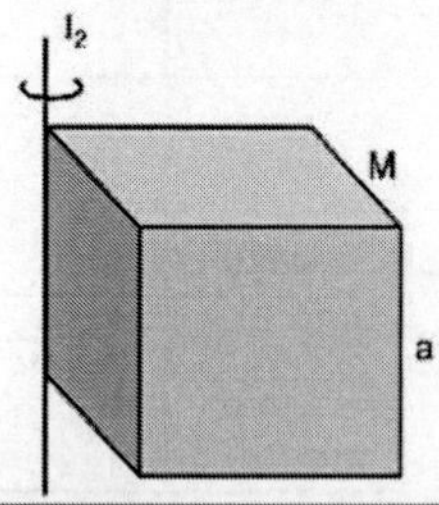	$I_2 = \frac{2Ma^2}{3}$	$\sqrt{\frac{2}{3}}\,a$

Exercise

1. Centre of mass of the earth and the moon system lies
   (a) closer to the earth
   (b) closer to the moon
   (c) at the mid-point of line joining the earth and the moon
   (d) can't be predicted

2. Four particles of masses m_1, m_2, m_3 and m_4 are placed at the vertices A, B, C and D as respectively of a square shown. The COM of the system will lie at diagonal AC if :

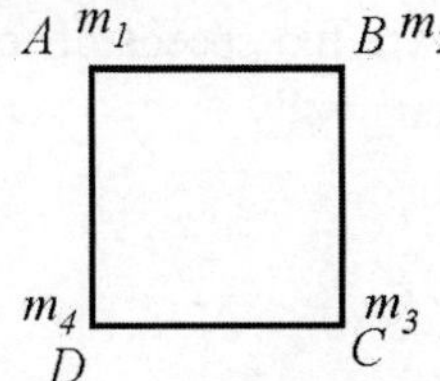

   (a) $m_1 = m_3$ (b) $m_2 = m_4$
   (c) $m_1 = m_2$ (d) $m_3 = m_4$

3. If COM of a system of particles lies at the origin of the coordinate system then
   (a) x-coordinate of all the particles may be positive
   (b) x-coordinate of all the particles may be negative
   (c) x-coordinate of all the particles may be zero
   (d) y-coordinate of all the particles may be positive

4. A nucleus at rest emits an α particle. As a result the nucleus recoils backwards, then K.E. of recoil of the nucleus will be :
   (a) greater than that of the α-particle
   (b) less than that of the α-particle
   (c) equal to that of the α-particle
   (d) depends on mass number of the nucleus

5. Two spheres A and B of masses m and $2m$ and radii $2R$ and R respectively are placed in contact as shown. The COM of the system lies

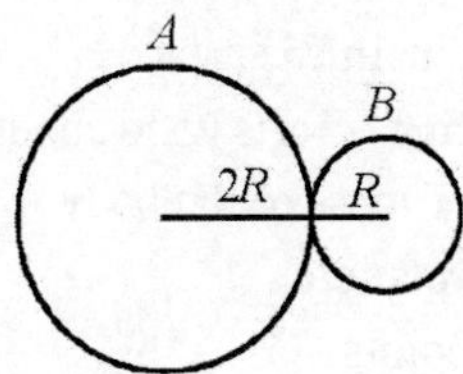

   (a) inside A
   (b) inside B
   (c) at the point of contact
   (d) none of these

6. In the system shown in the figure $m_1 > m_2$. If another block is attached below the block of mass m_1, then the acceleration of COM will :

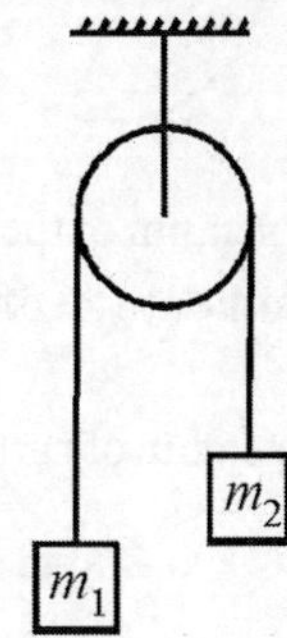

   (a) decrease
   (b) increase
   (c) remains some
   (d) may increase or decrease

7. Moment of inertia does not depend upon :
   (a) distribution of mass
   (b) axis of rotation
   (c) point of application of force
   (d) none of these

8. Angular momentum of a body is constant if the torque acting on the body is :
   (a) constant (b) positive
   (c) negative (d) zero

9. A disc is given a linear velocity on a rough horizontal surface then its angular momentum is :
   (a) conserved about COM only
   (b) conserved about the point of contact only
   (c) conserved about all the points
   (d) not conserved about any point

10. A body cannot roll without slipping on a
    (a) rough horizontal surface
    (b) smooth horizontal surface
    (c) rough inclined surface
    (d) smooth inclined surface
11. The equation $\tau = I\alpha$ is valid about
    (a) centre of mass of the body
    (b) axis of rotation
    (c) all the points
    (d) both (a) and (b)
12. A particle moving horizontally hits a stationary ring then about point of collision :

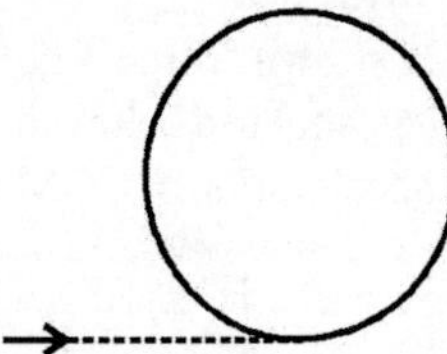

    (a) angular momentum of the ring is conserved
    (b) angular momentum of the particle is conserved
    (c) angular momentum of particle-ring system is conserved
    (d) all of these
13. A uniform rod is hanging with the help of the unintelligible strings as shown. Then:
    (a) $T_1 = T_2$
    (b) $T_1 > T_2$
    (c) $T_1 < T_2$
    (d) $T_1 \le T_2$

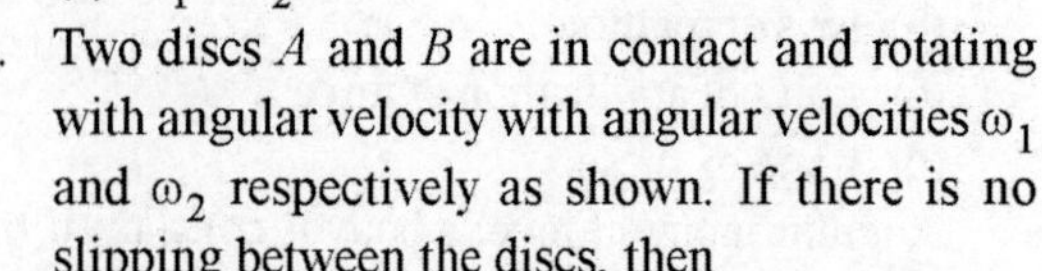

14. Two discs A and B are in contact and rotating with angular velocity with angular velocities ω_1 and ω_2 respectively as shown. If there is no slipping between the discs, then
    (a) $\omega_1 = \omega_2$
    (b) $\omega_1 > \omega_2$
    (c) $\omega_1 < \omega_2$
    (d) date insufficient

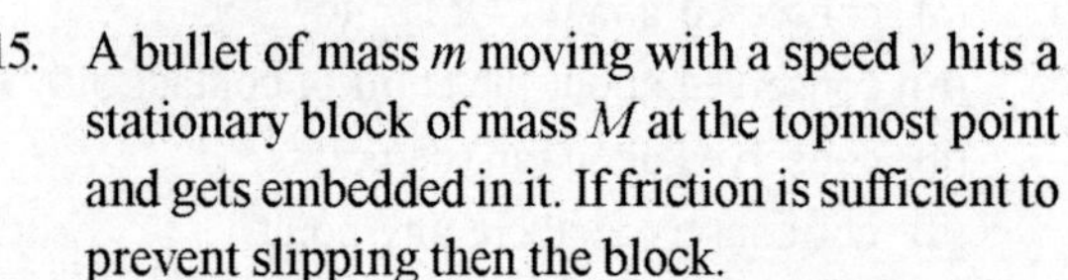

15. A bullet of mass m moving with a speed v hits a stationary block of mass M at the topmost point and gets embedded in it. If friction is sufficient to prevent slipping then the block.

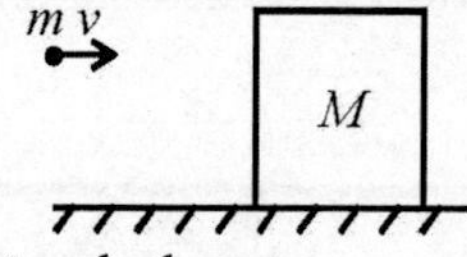

    (a) must topple down
    (b) must not topple down
    (c) may topple down
    (d) will remain at rest
16. A solid cylinder is rolling without shipping down an inclined plane. Then its angular momentum is :
    (a) conserved about COM of the cylinder
    (b) conserved about point of contact
    (c) conserved about all the points
    (d) not conserved about any point
17. In a bicycle, the radius of rear wheel is twice the radius of front wheel. It r_f and r_r are the radii and v_f and v_r are the speeds of topmost points of wheels then
    (a) $v_r = 2v_f$
    (b) $v_f = 2v_r$
    (c) $v_f = v_r$
    (d) $v_f = 4v_r$
18. If a person standing on a rotating disc stretches out his hands, the angular speed will :
    (a) increase (b) decrease
    (c) remains same (c) none of these
19. In an orbital motion, the direction of angular momentum vector is :
    (a) along the radius vector
    (b) parallel to the linear momentum
    (c) in the orbital plane
    (d) perpendicular to the orbital plane
20. A body is projected from ground with some angle to the horizontal. The angular momentum about the initial position will :
    (a) decrease
    (b) increase
    (c) remains same
    (d) first increase then decrease
21. A ring of mass m and radius r rotates about an axis passing through its centre and perpendicular to its plane with angular velocity ω. Its kinetic energy is

(a) $\frac{1}{2}mr^2\omega^2$ (b) $mr\omega^2$

(c) $mr^2\omega^2$ (d) $\frac{1}{2}mr\omega^2$

22. A solid homogeneous sphere of mass M and radius R is moving on a rough horizontal surface, partly rolling and partly sliding. During this kind of motion of the sphere
    (a) total kinetic energy is conserved
    (b) the angular momentum of the sphere about the point of contact with the plane is conserved
    (c) only the rotational kinetic energy about the centre of mass is conserved
    (d) angular momentum about the centre of mass is conserved

23. A solid cylinder of mass m and radius R rolls down an inclined plane of height h without slipping. The speed of its centre of mass when it reaches the bottom is

    (a) $\sqrt{(2gh)}$ (b) $\sqrt{4gh/3}$

    (c) $\sqrt{3gh/4}$ (d) $\sqrt{4g/h}$

24. A hollow sphere is held suspended. Sand is now poured into it in stages. The centre of gravity of the sphere with the sand

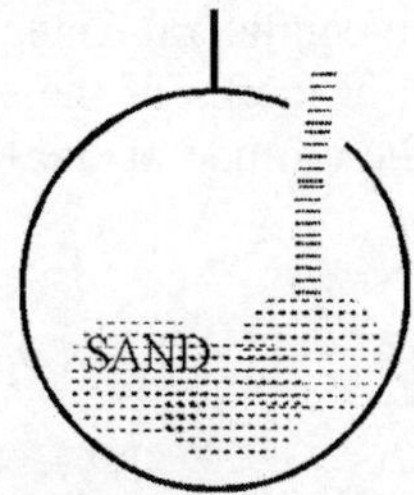

    (a) rises continuously
    (b) remains unchanged in the process
    (c) First rises and then falls to the original position
    (d) First falls and then rises to the original position

25. Moment of inertia of a uniform circular disc about a diameter is I. Its moment of inertia about an axis $\perp$ to its plane and passing through a point on its rim will be

    (a) $5I$ (b) $3I$
    (c) $6I$ (d) $4I$

26. The moment of inertia of a body about a given axis is 1.2 kg m^2. Initially, the body is at rest. In order to produce a rotational kinetic energy of 1500 joule, an angular acceleration of 25 radian/sec^2 must be applied about that axis for a duration of
    (a) 4 seconds (b) 2 seconds
    (c) 8 seconds (d) 10 seconds

27. A fly wheel rotating about a fixed axis has a kinetic energy of 360 joule when its angular speed is 30 radian/sec. The moment of inertia of the wheel about the axis of rotation is
    (a) 0.6 kg/m^2 (b) 0.15 kg m^2
    (c) 0.8 kg m^2 (d) 0.75 kg m^2

28. A particle of mass $m = 5$ is moving with a uniform speed $v = 3\sqrt{2}$ in the XOY plane along the line $y = x + 4$. The magnitude of the angular momentum of the particle about the origin is

    (a) 60 units (b) $40\sqrt{2}$ units
    (c) zero (d) 7.5 units

29. If a sphere is rolling, the ratio of the translational energy to total kinetic energy is given by
    (a) 7: 10 (b) 2 : 5
    (c) 10 : 7 (d) 5 : 7

30. In carbon monoxide molecule, the carbon and the oxygen atoms are separated by a distance 1.12×10^{-10} m. The distance of the centre of mass, from the carbon atom is
    (a) 0.64×10^{-10} m (b) 0.56×10^{-10} m
    (c) 0.51×10^{-10} m (d) 0.48×10^{-10} m

31. A thin circular ring of mass M and radius r is rotating about its axis with a constant angular velocity ω. Two objects each of mass m are attached gently to the opposite ends of a diameter of the ring. The ring will now rotate with an angular velocity
    (a) $\omega (M-2m)/(M+2m)$
    (b) $\omega M/(M+2m)$
    (c) $\omega (M+2m)/M$
    (d) $\omega M/(M+m)$

32. O is the centre of an equilateral triangle ABC. F_1, F_2 and F_3 are three forces acting along the sides AB, BC and AC as shown here. What should be the magnitude of F_3, so that the total torque about O is zero?

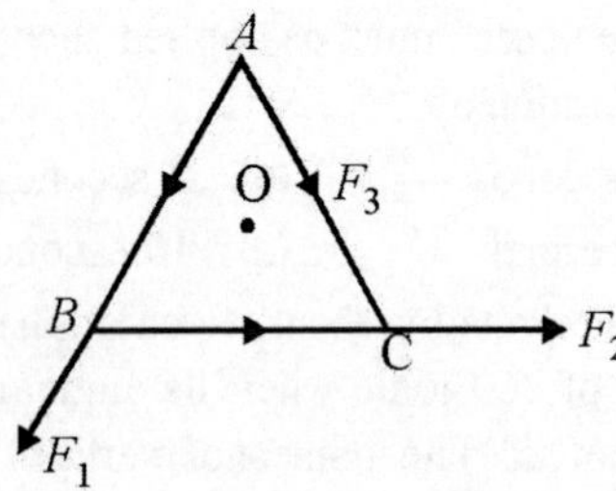

(a) $(F_1 + F_2)/2$ (b) $2(F_1 + F_2)$
(c) $(F_1 + F_2)$ (d) $[F_1 - F_2]$

33. A weightless ladder 20 ft long rests against a frictionless wall at an angle of 60° from the horizontal. A 150 pound man is 4 ft from the top of the ladder. A horizontal force is needed to keep it from slipping. Choose the correct magnitude of the force from the following
(a) 175 lb (b) 100 lb
(c) 120 lb (d) 69.2 lb

34. Two racing cars of masses m_1 and m_2 are moving in circles of radii r_1 and r_2 respectively. Their speeds are such that each makes a complete circle in the same time t. The ratio of the angular speeds of the first to the second car is
(a) 1 : 1 (b) $m_1 : m_2$
(c) $r_1 : r_2$ (d) $m_1 m_2 : r_1 r_2$

35. The moment of inertia of a disc of mass M and radius R about an axis, which is tangential to the circumference of the disc and parallel to its diameter, is
(a) $\frac{3}{2}MR^2$ (b) $\frac{2}{3}MR^2$
(c) $\frac{5}{4}MR^2$ (d) $\frac{4}{5}MR^2$

36. There is a flat uniform triangular plate ABC such that $AB = 4$ cm, $BC = 3$ cm and angle $ABC = 90°$. The moment of inertia of the plate about AB, BC and CA as axis is respectively I_1, I_2 and I_3. Which one of the following is true?

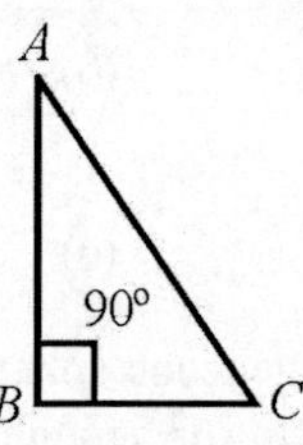

(a) $I_3 > I_2$ (b) $I_2 > I_1$
(c) $I_3 > I_1$ (d) $I_1 > I_2$

37. A solid cylinder and a hollow cylinder both of the same mass and same external diameter are released from the same height at the same time on an inclined plane. Both roll down without slipping. Which one will reach the bottom first?
(a) Both together
(b) Solid cylinder
(c) One with higher density
(d) Hollow cylinder

38. A constant torque of 1000 N-m turns a wheel of moment of inertia 200 kg-m^2 about an axis through its centre. Its angular velocity after 3 seconds is
(a) 1 rad/s (b) 5 rad/s
(c) 10 rad/s (d) 15 rad/s

39. A wheel of radius 1m rolls forward half a revolution on a horizontal ground. The magnitude of the displacement of the point of the wheel initially in contact with the ground is
(a) π (b) 2π
(c) $\sqrt{2}\pi$ (d) $\sqrt{\pi^2 + 4}$

40. Three point masses m_1, m_2, m_3 are located at the vertices of an equilateral triangle of length 'a'. The moment of inertia of the system about an axis along the altitude of the triangle passing through m_1 is
(a) $(m_2 + m_3)\frac{a^2}{4}$ (b) $(m_1 + m_2 + m_3)a^2$
(c) $(m_1 + m_2)\frac{a^2}{4}$ (d) $(m_2 + m_3)a^2$

41. The moment of inertia of uniform rectangular plate about an axis passing through its mid-point and parallel to its length l is (b = breadth of rectangular plate)
(a) $\frac{Mb^2}{4}$ (b) $\frac{Mb^3}{6}$
(c) $\frac{Mb^3}{12}$ (d) $\frac{Mb^2}{12}$

42. Point masses 1, 2, 3 and 4 kg are lying at the point (0, 0, 0), (2, 0, 0), (0, 3, 0) and (–2, –2, 0) respectively. The moment of inertia of this system about x-axis will be
(a) $43\ kg-m^2$ (b) $34\ kg-m^2$
(c) $27\ kg-m^2$ (d) $72\ kg-m^2$

43. A wheel has moment of inertia $5 \times 10^{-3}\ kg\ m^2$ and is making 20 rev/s. The torque needed to stop it in 10 s is .... $\times 10^{-2}$ N-m.
(a) 2π (b) 2.5π
(c) 4π (d) 4.5π

44. A torque of 30 N-m is applied on a 5 kg wheel whose moment of inertia is 2 kg-m^2 for 10 sec. The angle covered by the wheel in 10 sec will be
(a) 750 rad (b) 1500 rad
(c) 3000 rad (d) 6000 rad

45. A ballet dancer, dancing on a smooth floor is spinning about a vertical axis with her arms folded with an angular velocity of 20 rad/s. When she stretches her arms fully, the spinning speed decrease in 10 rad/s. If I is the initial moment of inertia of the dancer, the new moment of inertia is
(a) $2I$ (b) $3I$
(c) $I/2$ (d) $I/3$

46. The rotational kinetic energy of a body is E and its moment of inertia is I. The angular momentum is
(a) EI (b) $2\sqrt{EI}$
(c) $\sqrt{2EI}$ (d) E/I

47. The instantaneous angular position of a point on a rotating wheel is given by the equation $\theta(t) = 2t^3 - 6t^2$. The torque on the wheel becomes zero at
(a) $t = 1s$ (b) $t = 0.5\,s$
(c) $t = 0.25\,s$ (d) $t = 2s$

48. The moment of inertia of a thin uniform rod of mass M and length L about an axis passing through its mid point and perpendicular to its length is I_0. Its moment of inertia about an axis passing through one of its ends and perpendicular to its length is
(a) $I_0 + ML^2/2$ (b) $I_0 + ML^2/4$
(c) $I_0 + 2ML^2$ (d) $I_0 + ML^2$

49. A small mass attached to a string rotates on frictionless table top as shown. If the tension in the string is increased by pulling the string causing the radius of the circular motion to decrease by a factor of 2, the kinetic energy of the mass will

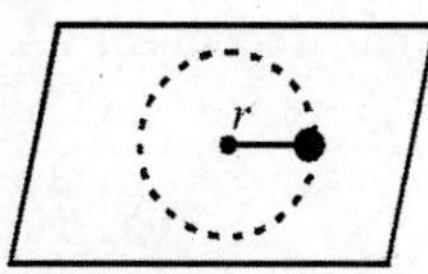

(a) remain constant
(b) increase by a factor of 2
(c) increase by a factor of 4
(d) decrease by a factor of 2

50. Two particles which are initially at rest, move towards each other under the action of their internal attraction. If their speeds are v and $2v$ at any instant, then the speed of centre of mass of the system will be:
(a) $2v$ (b) zero
(c) 1.5 (d) v

51. A circular disk of moment of inertia I_t is rotating in a horizontal plane, its symmetry axis, with a constant angular speed ω_i. Another disk of moment of inertia I_b is dropped coaxially onto the rotating disk. Initially the second disk has zero angular speed. Eventually both the disks rotate with a constant angular speed ω_f. The energy lost by the initially rotating disk to friction is:
(a) $\frac{1}{2}\frac{I_b^2}{(I_t + I_b)}\omega_i^2$ (b) $\frac{I_t^2}{(I_t + I_b)}\omega_i^2$
(c) $\frac{I_b - I_t}{(I_t + I_b)}\omega_i^2$ (d) $\frac{1}{2}\frac{I_b I_t}{(I_t + I_b)}\omega_i^2$

52. The speed of a homogenous solid sphere after rolling down an inclined plane of vertical height h from rest without sliding is
(a) $\sqrt{\frac{10}{7}gh}$ (b) $\sqrt{gh}$
(c) $\sqrt{\frac{6}{5}gh}$ (d) $\sqrt{\frac{4}{3}gh}$

53. A solid sphere, disc and solid cylinder all of the same mass and made of the same material are allowed to roll down (from rest) on the inclined plane, then
(a) Solid sphere reaches the bottom first
(b) Solid sphere reaches the bottom last
(c) Disc will reach the bottom first
(d) All reach the bottom at the same time

54. If a flywheel makes 120 revolutions/minute, then its angular speed will be
(a) 8π rad/sec (b) 6π rad/sec
(c) 4π rad/sec (d) 2π rad/s

55. A sphere rolls down an inclined plane without slipping. What fraction of its total energy is rotational?

(a) $\frac{2}{7}$ (b) $\frac{3}{7}$

(c) $\frac{4}{7}$ (d) $\frac{5}{7}$

56. What is the torque of force $\vec{F} = -3\hat{i} + \hat{j} + 5\hat{k}$ acting at point whose position vector is $\vec{r} = 7\hat{i} + 3\hat{j} + \hat{k}$?

(a) $14\hat{i} - 38\hat{j} + 16\hat{k}$ (b) $4\hat{i} + 4\hat{j} + 6\hat{k}$

(c) $-14\hat{i} + 38\hat{j} - 16\hat{k}$ (d) $-21\hat{i} + 3\hat{j} + 5\hat{k}$

57. Two bodies of masses 2 kg and 4 kg are moving with velocities 2 m/s and 10 m/s respectively along same direction. Then the velocity of their centre of mass will be

(a) 8.1 m/s (b) 7.3 m/s

(c) 6.4 m/s (d) 5.3 m/s

58. Masses 8, 2, 4, 2 kg are placed at the corners *A*, *B*, *C*, *D* respectively of a square *ABCD* of diagonal 80 cm. The distance of centre of mass from *A* will be

(a) 20 cm (b) 30 cm

(c) 40 cm (d) 60 cm

59. Two particles of masses m_1 and m_2 initially at rest start moving towards each other under their mutual force of attraction. The speed of the centre of mass at any time *t*, when they are at a distance *r* apart, is

(a) zero (b) $\left(G\frac{m_1 m_2}{r^2}.\frac{1}{m_1}\right)t$

(c) $\left(G\frac{m_1 m_2}{r^2}.\frac{1}{m_2}\right)t$ (d) $\left(G\frac{m_1 m_2}{r^2}.\frac{1}{m_1 + m_2}\right)t$

60. The centre of mass of triangle shown in figure has coordinates

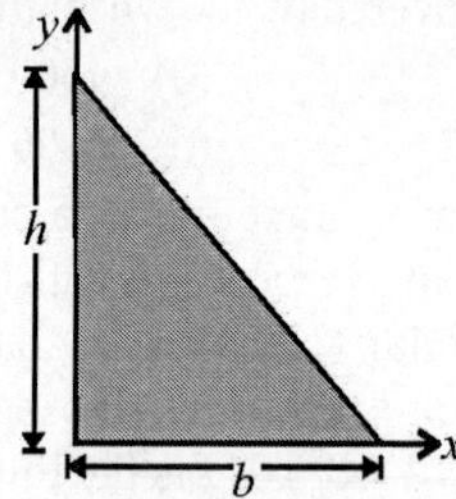

(a) $x = \frac{h}{2}, y = \frac{b}{2}$ (b) $x = \frac{b}{2}, y = \frac{h}{2}$

(c) $x = \frac{b}{3}, y = \frac{h}{3}$ (d) $x = \frac{h}{3}, y = \frac{b}{3}$

Gravitation

CONCEPTS

Gravitation is the force of attraction between any two bodies. If out of two bodies one is comparatively very heavy, for example celestial bodies- moon, earth, or any other planets then it is called **Gravity**.

KEPLER'S LAWS

Kepler's three laws of planetary motion :

(i) **Law of orbits :** All planets move in elliptical orbits with the Sun at one of the focal points

(ii) **Law of Areas :** The radius vector drawn from the sun to a planet sweeps out equal areas in equal time intervals. This follows from the fact that the force of gravitation on the planet is central and hence angular momentum is conserved.

(iii) **Law of periods :** The square of the orbital period of a planet is proportional to the cube of the semi-major axis of the elliptical orbit of the planet. i.e.,

$$T^2 \propto a^3 \text{ or } \left(\frac{T_1}{T_2}\right)^2 = \left(\frac{R_1}{R_2}\right)^3$$

The period T and radius R of the circular orbit of a planet about the Sun are related as

$$T^2 = \left(\frac{4\pi^2}{GM_S}\right) R^3$$

where M_S is the mass of the Sun. Most planets have nearly circular orbits about the Sun. For elliptical orbits, the above equation is valid if R is replaced by the semi-major axis, a.

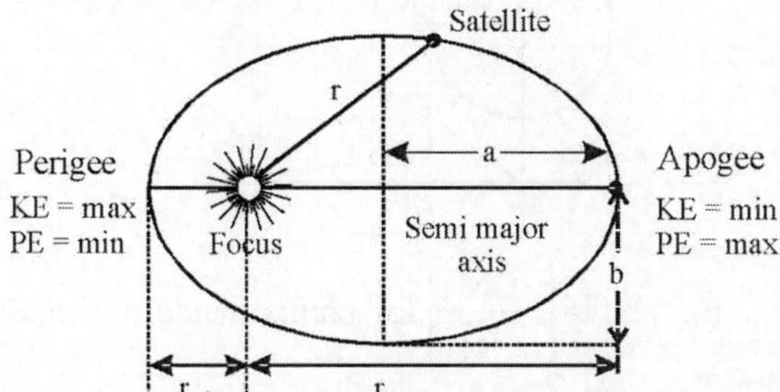

Conservation of angular momentum leads to Kepler's second law. However, it is not special to the inverse square law of gravitation. It holds for any central force. In Kepler's third law, $T^2 = K_s R^3$. The constant K_s is the same for all planets in circular orbits. This applies to satellites orbiting the Earth.

NEWTON'S UNIVERSAL LAW OF GRAVITATION

According to this law - Every two objects in the universe attract each other and the force of attraction between the two bodies is directly proportional to the product of their masses and inversely proportional to the square of distance between the two masses i.e., if two masses m_1 and m_2 are separated from each other by a distance r then

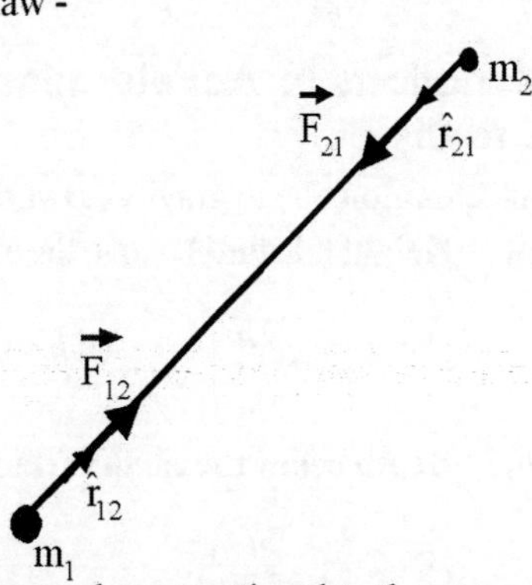

$$F \propto \frac{m_1 m_2}{r^2} \quad \text{or} \quad F = G\frac{m_1 m_2}{r^2}$$

where G = **universal gravitational constant**

The numerical value of G determined by scientists Henry cavendish in 1798,

$G = 6.67 \times 10^{-11}$ N-m^2 kg^{-2} **in SI system**

$= 6.67 \times 10^{-8}$ dyne cm^2 gm^{-2} **in CGS system**

Its **dimensions :** $[M^{-1}L^3 T^{-2}]$

In vector form, gravitational force $\vec{F} = \frac{Gm_1m_2}{r^3}\vec{r}$

Newton's Shell Theorem : A uniform shell of matter exerts no net gravitational force on a particle located inside it.

This statement does not mean that the gravitational forces on the particle from the various elements of the shell magically disappear. Rather, it means that the sum of the force vectors on the particle from all the elements is zero.

ACCELERATION DUE TO GRAVITY OF THE EARTH

The force of attraction exerted by earth on a body of mass m is the force of gravity. This force accelerates the body and the acceleration produced in the body under the influence of gravitational force is called acceleration due to gravity, denoted by 'g'.

$$g = \frac{GM_e}{r^2} \quad ...(i)$$

$$g = \frac{GM_e}{R_e^2} = \frac{G(4/3\pi R_e^3 \rho)}{R_e^2} = \frac{4}{3}\pi\rho GR \quad ...(ii)$$

Variations in Acceleration Due to Gravity

Acceleration due to gravity (g) varies with :

(a) **Height (altitude) above the earth's surface -**

$$g' = g\left(1 - \frac{2h}{R_e}\right) \qquad (\text{if } h << R_e)$$

(b) **Depth below the earth surface -**

$$g' = g\left(1 - \frac{d}{R_e}\right)$$

(c) **Shape of the earth** - The earth is elliptical in shape. It is flatter at the poles and bulged out at the equator. Now, we know that $g \propto 1/R^2$,

$R_P < R_E$

$\therefore \quad g_E < g_P$

Therefore the value of g at the equator is minimum and the value of g at the poles is maximum.

(d) **Axial rotation (latitude) of the earth** - If the observed value of g at the latitude λ is represented by g_λ ,then

$$g_\lambda = g_e - R_e\omega^2 \cos^2 \lambda$$

GRAVITATIONAL AND INERTIAL MASS

When mass is defined as the property of inertia, it is termed as **inertial mass**.

Gravitational mass : When mass is defined as the property of gravity, it is called gravitational mass.

Properties of gravitational mass : It is found that the ratio of two gravitational masses is same as the ratio of their inertial masses.

GRAVITATIONAL FIELD

The region of space around a body in which its gravitational influence is experienced by other bodies is the gravitational field of that body.

$$E_g = \frac{F}{m} = \left(\frac{GMm}{r^2}\right)/m = \frac{GM}{r^2}$$

(a) ***Gravitational Field Intensity (E_g) for spherical shell :***

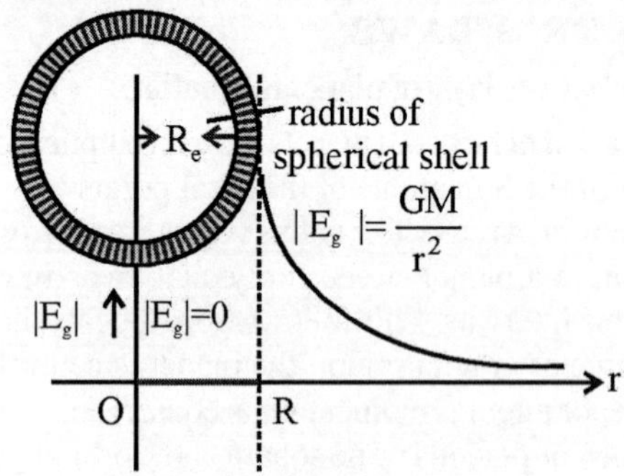

r = Radius of spherical shell

(i) $|E_g|=0$ at points inside the spherical shell

(ii) $|E_g| = \left|\frac{-GM}{R^2}\right|$ at the surface of the shell

(iii) $|E_g| = \left|\frac{-GM}{r^2}\right|$ for outside the spherical shell (r>R).

It is clear that as r increases, E_g reduces to zero.

(b) ***Gravitational Field Intensity for uniform solid sphere :***

E_g

Uniform solid sphere

r

$|E_g|$

R

r

(i) $|E_g| = \left|-\frac{GM.r}{R^3}\right|$ for points inside the solid sphere (r < R)

(ii) At the surface of solid sphere (r = R)

$$|E_g| = \left|-\frac{GM}{R^2}\right|$$

(iii) Outside the solid sphere (r > R)

$$|E_g| = \left|-\frac{GM}{r^2}\right|$$

GRAVITATIONAL POTENTIAL

Gravitational potential at any point is the work done by gravitational force in carrying a body of unit mass from infinity to that point in gravitational influence of source mass M. So

$$V_A = \frac{W}{m_o} = \frac{-GM}{r}$$

where r is the distance of point from source mass M.

Gravitational Potential Due to a Uniform Solid Sphere

Potential at a point, distant r from the centre of the sphere

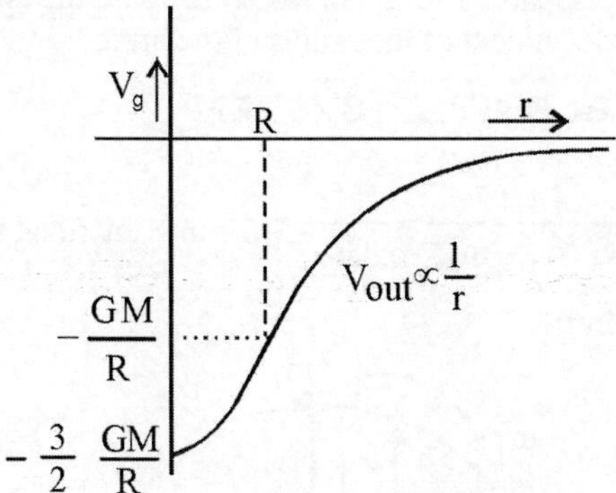

At internal point (r < R), $V = \frac{-GM(3R^2 - r^2)}{2R^3}$

At the surface (r = R), $V = -\frac{GM}{R}$

At the centre (r = 0), $V = -\frac{3GM}{2R}$

Gravitational Potential due to a Uniform Thin Spherical Shell

For an external point, spherical shell behaves as if whole of its mass is concentrated at the centre, i.e.,

$$V(r) = -\frac{GM}{r} \text{ for } [r \geq R]$$

For an internal point, (r < R) the potential everywhere is same and equal to its value at the surface, i.e.,

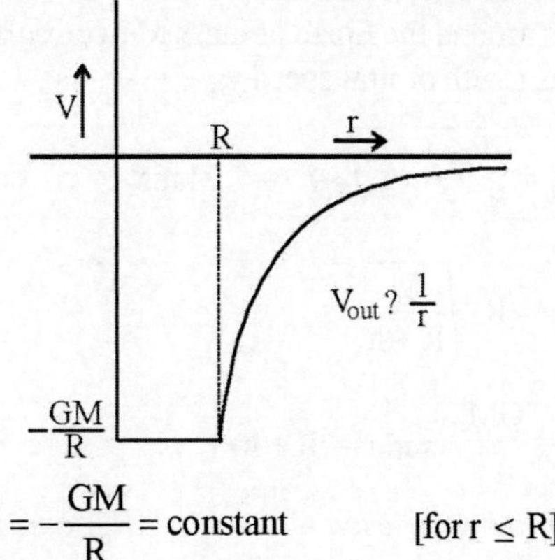

$$V = -\frac{GM}{R} = \text{constant} \qquad [\text{for } r \leq R]$$

GRAVITATIONAL POTENTIAL ENERGY

The work done against the field in bringing a body from infinity to a point in the gravitational field is called its gravitational potential energy. It is a scalar quantity. Its **SI unit** is joule.

The gravitational potential energy of a particle of mass 'm' at a point distant 'r' from the centre of earth is

$$U_g = -\frac{GM_e m}{r} = -\frac{GM_e m}{(R_e + h)} \quad (\text{if } r > R_e)$$

$$U_g = -\frac{GM_e m\,(3R_e^2 - r^2)}{2R_e^3} \quad (\text{if } r < R_e)$$

$$U_g = -\frac{GM_e m}{R_e} = -mgR_e \quad (\text{if } r = R_e)$$

Increase in potential energy on going to a height h above the surface

$$\Delta U_g = \frac{mgh}{(1 + h/R_e)} = mgh, \quad (\text{if } h << R_e)$$

Force between two particles, if their potential energy is U,

$$F = -\frac{dU}{dr} = -\frac{d}{dr}\left(-\frac{GMm}{r}\right) = -\frac{GMm}{r^2}$$

ESCAPE VELOCITY AND ORBITAL VELOCITY

The escape speed is the minimum speed with which a body must be projected up so that it overcomes the gravitational pull of the Earth.

$v_e = \sqrt{\frac{2GM_e}{R_e}} = \sqrt{2gR_e}$ and has a value of 11.2 km s^{-1}.

Orbital Velocity : It is the velocity required to put the satellite into its orbit around the earth. Let a satellite revolves around the Earth of mass M in circular orbit of radius r with orbital speed v_0

$$v_0 = \sqrt{\frac{GM}{r}} = \sqrt{gR} = 7.9\ \text{kms}^{-1}$$

or, $$v_0 = R\sqrt{\frac{g}{R+h}}$$

$\because$ $g = \frac{GM}{R^2}$ and $r = (R+h)$

Relation between escape velocity (V_e) and orbital velocity (V_o)

$$Ve = \sqrt{2}V_o$$

Time period of satellite : The time taken by satellite to go once around the planet.

$$T = 2\pi\sqrt{\frac{r^3}{GM}}$$

Energy of an orbiting satellite : If a body of mass m moves in a circular orbit of radius *a* mass *M*, where M>>m, the total energy of the system is

$E = -\frac{GMm}{2a}$ with the choice of the arbitrary constant in the potential energy . The total energy is negative for any bound system, that is, one in which the orbit is closed, such as an elliptical orbit. The kinetic (K) and potential(V) energies are

$$K = \frac{GMm}{2a}\ ;\ V = -\frac{GMm}{a}$$

GEOSTATIONARY AND POLAR SATELLITE

A **geostationary** (geosynchronous communication) or parking **satellite** moves in a circular orbit in the equatorial plane at an approximate distance of 4.22×10^4 km from the Earth's centre. The time period of a geostationary satellite is 24 hours. INSAT group of satellites are geostationary satellites widely used for telecommunications.

Polar satellites are low altitude (h $\approx$ 500 to 800 km) satellites. Its time period is around 100 minutes.

WEIGHTLESSNESS

The phenomenon of "weightlessness" occurs when there is no force of support on your body. When your body is effectively in "free fall", accelerating downward at the acceleration of gravity, then you are not being supported. The sensation of apparent weight comes from the support that you feel from the floor, from the seat, etc. Different sensations of apparent weight can occur on a roller-coaster or in an aircraft because they can accelerate either upward or downward.

The "weightlessness" you may feel in an aircraft occurs any time the aircraft is accelerating downward with acceleration 1g.

The satellite is moving in a circular orbit, it has a radial acceleration

$$a = \frac{v_0^2}{r} = \frac{GM}{r^2} \quad \left[\text{as } v_0 = \sqrt{\left(\frac{GM}{r}\right)}\right]$$

i.e., it is falling towards earth's centre with acceleration a, so apparent weight of the body in it

$$W_{ap} = m(g' - a)$$

where g' is the acceleration due to gravity of earth at the position (height) of satellite, i.e. $g' = (GM/r^2)$, so that

$$W_{ap} = m\left[\frac{GM}{r^2} - \frac{GM}{r^2}\right] = 0$$

i.e., the apparent weight of a body in a satellite is zero and independent of the radius of the orbit.

DOUBLE STAR SYSTEM

$T^2 = \frac{4\pi^2 r^3}{G(m_1+m_2)} = \frac{4\pi^2 r^3}{GM}$ [$M = m_1 + m_2$ (total mass of system)]

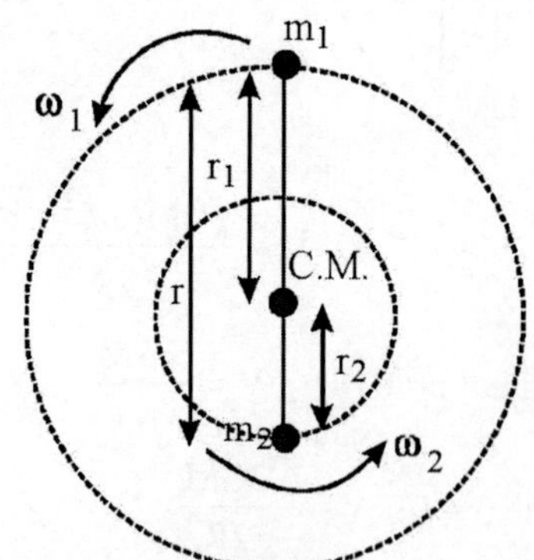

Angular momentum of system about centre of mass,

$$A.M. = \left(\frac{m_1 m_2}{m_1+m_2}\right) r^2\omega = \mu r^2\omega = I\omega$$

Kinetic energy of system,

$$K.E. = \frac{1}{2}\frac{m_1 m_2}{m_1+m_2} r^2\omega^2 = \frac{1}{2}\mu r^2\omega^2 = \frac{1}{2}I\omega^2$$

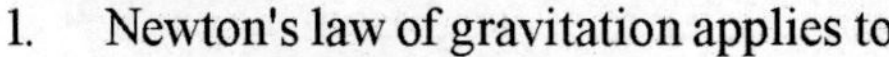

Exercise

1. Newton's law of gravitation applies to
   (a) small bodies only
   (b) planets only
   (c) both small and big bodies
   (d) only valid for solar system
2. The weight of a body at the centre of the earth is
   (a) zero
   (b) infinite
   (c) same as on the surface on earth
   (d) none of the above
3. The tides in sea are primarily due to
   (a) the atmospheric effect of the earth
   (b) the gravitational effect of venus on the earth
   (c) the gravitational effect of the sun on the earth
   (d) the gravitational effect of the moon on the earth
4. The magnitude of gravitational potential energy of earth-moon system is U which is zero at infinite separation. If K is the K.E. of the moon with respect to earth, then
   (a) $|U| = K$ (b) $|U| < K$
   (c) $|U| > K$ (d) Either B or C
5. Let V and E represent the gravitational potential and field at a distance r from the centre of a uniform solid sphere. Consider the two statements:
   (A) The plot of V against r is discontinuous.
   (B) The plot of E against r is discontinuous.
   (a) Both A and B are correct
   (b) A is correct but B is wrong
   (c) B is correct but A is wrong
   (d) Both A and B are wrong
6. Figure shows the elliptical path of a planet around the sun. The two shaded parts have equal area. If t_1 and t_2 be the time taken by the planet to go from a to b and from c to d respectively,

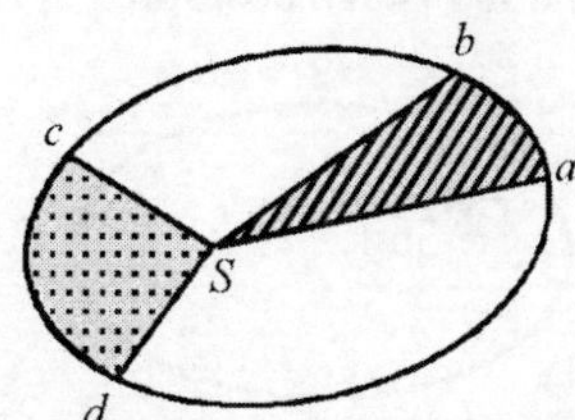

   (a) $t_1 < t_2$
   (b) $t_1 = t_2$
   (c) $t_1 > t_2$
   (d) insufficient information to deduce the relation between t_1 and t_2.
7. Geo-stationary satellite is one which
   (a) remains stationary at a fixed height from the earth's surface
   (b) revolves like other satellites but in the opposite direction of earth's rotation
   (c) revolves round the earth at a suitable height with same angular velocity and in the same direction as earth does about its own axis
   (d) none of the above
8. Weightlessness experienced while orbiting the earth in space ships is the result of
   (a) inertia (b) acceleration
   (c) zero gravity (d) centre of gravity
9. There is no atmosphere on the moon because
   (a) it is closer to the earth
   (b) it revolves round the earth
   (c) it gets light from the sun
   (d) the escape velocity of gas molecules is lesser than their root mean square velocity
10. To an astronaut in a space ship the sky appears black due to
    (a) absence of atmosphere in his neighbourhood
    (b) light from the sky is absorbed by the medium surrounding him
    (c) the fact that at height, sky radiations are only in the infra-red and the ultraviolet region
    (d) none of the above
11. An artificial satellite orbiting the earth does not fall down because the earth's attraction
    (a) is balanced by the attraction of the moon
    (b) vanishes at such distances
    (c) is balanced by the viscous drag produced by the atmosphere
    (d) produces the necessary acceleration of its motion in a curved path
12. The period of a satellite in a circular orbit near a planet is independent of
    (a) the mass of the planet
    (b) the radius of the planet
    (c) the mass of the satellite
    (d) all the above

13. Intensity of the gravitational field inside the solid sphere is
(a) variable (proportional to distance from centre)
(b) constant
(c) variable (does not depend on distance from the centre)
(d) zero

14. A reference frame attached to the earth
(a) is an inertial frame by definition
(b) can be an inertial frame because the earth is revolving round the sun
(c) considered a is an inertial frame because Newton's laws are applicable in this frame
(d) can not be inertial frame because the earth is rotating about its own axis.

15. Two satellites of earth, S_1 and S_2 are moving in the same orbit. The mass of S_1 is four times the mass of S_2. Which one of the following statements is true?
(a) The potential energies of earth satellites in the two cases are equal
(b) S_1 and S_2 are moving with the same speed
(c) The kinetic energies of the two satellites are equal
(d) The time period of S_1 is four times that of S_2

16. The kinetic energy needed to project a body of mass m from the earth surface (radius R) to infinity is
(a) $mgR/2$ (b) $2mgR$
(c) mgR (d) $mgR/4$

17. If suddenly the gravitational force of attraction between earth and a satellite revolving around it becomes zero, then the satellite will
(a) continue to move in its orbit with same velocity
(b) move tangentially to the original orbit in the same velocity
(c) become stationary in its orbit
(d) move towards the earth

18. Two satellites of masses $3M$ and M orbit the earth in circular orbits of radii r and $3r$ respectively. The ratio of their speeds is
(a) $1:1$ (b) $\sqrt{3}:1$
(c) $3:1$ (c) $9:1$

19. Which of the following is an evidence to show that there must be a force acting on the earth and directed towards the sun?
(a) Deviation of the falling bodies towards east
(b) Revolution of the earth round the sun
(c) Phenomenon of day & night
(d) Apparent motion of the sun round the earth

20. The orbital speed of Jupiter is :
(a) greater than the orbital speed of the earth
(b) less than the orbital speed of the earth
(c) equal to the orbital speed of the earth
(d) proportional to distance from the earth

21. Kepler's second law is based on :
(a) Newton's second law of motion
(b) Newton's law of gravitation
(c) Special theory of relativity
(d) Principle of conservation of angular momentum

22. A satellite is orbiting around the earth near its surface. If its kinetic energy is doubled, then
(a) it will remain in the same orbit
(b) it will fall on the earth
(c) it will revolve with greater speed
(d) it will escape out of the gravitational field of the earth

23. There is no atmosphere on the moon because
(a) it is closer to the earth and also it has the inactive inert gases in it
(b) it is too for from the sun and has very low pressure in its outer surface
(c) escape velocity of gas molecules is greater than their root mean square velocity
(d) escape velocity of gas molecules is less than their root mean square velocity

24. Escape velocity of a particle depends upon its mass as :
(a) $v_e \alpha m^2$ (b) $v_e \alpha m$
(c) $v_e \alpha m^0$ (d) $v_e \alpha m^{-1}$

25. A ball is dropped from a spacecraft revolving around the earth at a height of 100 km, then the ball
(a) fall down to the earth gradually
(b) go very far in the space
(c) continue to move with the same speed along the original orbit of the spacecraft
(d) move with the same speed, tangentially to the spacecraft

26. The maximum kinetic energy of a planet moving around the sun is at a position :

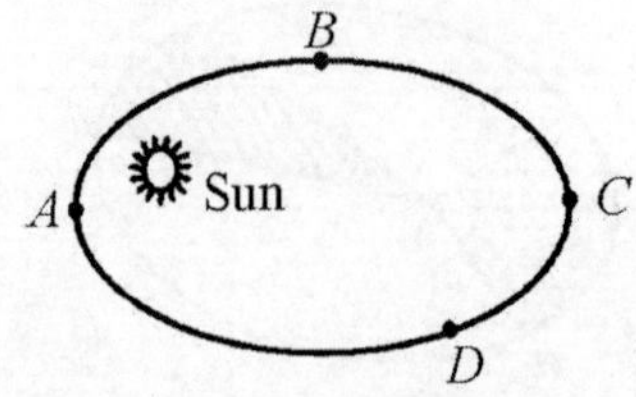

(a) A (b) B
(c) C (d) D

27. If radius of the earth is reduced, then :
(a) duration of the day reduce
(b) earth rotates slower
(c) time period of the rotation of earth decreases
(d) duration of the day increases

28. The value of escape velocity of a certain planet is 2 km/s. Then the value of orbital speed for a satellite orbiting closer to its surface is :
(a) 2 km/s (b) 1 km/s
(c) $\sqrt{2}$ km/s (d) $2\sqrt{2}$ km/s

29. When a body is taken from poles to equator on the earth, its weight
(a) increases
(b) decreases
(c) remains same
(d) increases at south pole and decreases at north pole

30. Which of the following quantities does not depend upon the orbital radius of the satellite?
(a) $\frac{T}{R}$ (b) $\frac{T^2}{R}$
(c) $\frac{T^2}{R^2}$ (d) $\frac{T^2}{R^3}$

31. With what minimum velocity a body must be projected from the surface of the earth making an angle 45° with the surface so that it escapes along from the gravitational field of the earth?
(a) 11.2 kms^{-1} (b) $11.2\sqrt{2}\ \text{kms}^{-1}$
(c) $5.6\sqrt{2}\ \text{kms}^{-1}$ (d) 22.4 kms^{-1}

32. A man waves his arms while walking. This is to :
(a) keep constant velocity
(b) ease the tension
(c) increase the velocity
(d) balance the effect of earth's gravity

33. A missile is launched with a velocity less than escape velocity. The sum of its kinetic and potential energies is :
(a) Zero
(b) Negative
(c) Positive
(d) May be positive, negative or zero

34. The distance of two planets from the sun are 10^{13} and 10^{12} metres respectively. The ratio of time periods of these two planets is
(a) $\frac{1}{\sqrt{10}}$ (b) 100
(c) $10\sqrt{10}$ (d) $\sqrt{10}$

35. The largest and the shortest distance of the earth from the sun are r_1 and r_2. Its distance from the sun when it is at perpendicular to the major-axis of the orbit drawn from the sun is
(a) $\frac{r_1+r_2}{4}$ (b) $\frac{r_1+r_2}{r_1-r_2}$
(c) $\frac{2r_1r_2}{r_1+r_2}$ (d) $\frac{r_1+r_2}{3}$

36. If the gravitational force between two objects were proportional to $1/R$ (and not as $1/R^2$) where R is separation between them, then a particle in circular orbit under such a force would have its orbital speed v proportional to
(a) $1/R^2$ (b) R^0
(c) R^1 (d) $1/R$

37. For a satellite escape velocity is 11 km/s. If the satellite is launched at an angle of 60° with the vertical, then escape velocity will be
(a) 11 km/s (b) $11\sqrt{3}$ km/s
(c) $\frac{11}{\sqrt{3}}$ km/s (d) 33 km/s

38. A planet is moving in an elliptical orbit around the sun. If T, V, E and L stand respectively for its kinetic energy, gravitational potential energy, total energy and magnitude of angular momentum about the centre of force, which of the following is correct?
(a) T is conserved
(b) V is always positive
(c) E is always negative
(d) L is conserved but direction of vector L changes continuously

39. A satellite of mass m is orbiting around the earth in a circular orbit with a velocity v. What will be its total energy?
(a) $(3/4)\,mv^2$ (b) $(1/2)\,mv^2$
(c) mv^2 (d) $-(1/2)mv^2$

40. A seconds pendulum is mounted in a rocket. Its period of oscillation decreases when the rocket
(a) comes down with uniform acceleration
(b) moves round the earth in a geostationary orbit
(c) moves up with a uniform velocity
(d) moves up with uniform acceleration

41. The mean radius of earth is R, its angular speed on its own axis is ω and the acceleration due to gravity at earth's surface is g. What will be the radius of the orbit of a geostationary satellite?
(a) $(R^2 g/\omega^2)^{1/3}$ (b) $(Rg/\omega^2)^{1/3}$
(c) $(R^2\omega^2/g)^{1/3}$ (d) $(R^2 g/\omega)^{1/3}$

42. A satellite A of mass m is at a distance of r from the surface of the earth. Another satellite B of mass 2m is at a distance of $2r$ from the earth's centre. Their time periods are in the ratio of
(a) 1 : 2 (b) 1 : 16
(c) 1 : 32 (d) $1 : 2\sqrt{2}$

43. The escape velocity from earth is 11.2 km/s. If a body is to be projected in a direction making an angle 45° to the vertical, then the escape velocity is
(a) 11.2×2 km/s (b) 11.2 km/s
(c) $11.2/\sqrt{2}$ km/s (d) $11.2\sqrt{2}$ km/s

44. The distance of Neptune and Saturn from the sun is nearly 10^{13} and 10^{12} meter respectively. Assuming that they move in circular orbits, their periodic times will be in the ratio
(a) 10 (b) 100
(c) $10\sqrt{10}$ (d) 1000

45. A satellite in force free space sweeps stationary interplanetary dust at a rate $dM/dt = \alpha v$ where M is the mass and v is the velocity of the satellite and α is a constant. What is the deceleration of the satellite?
(a) $-\alpha v^2$ (b) $-\alpha v^2/2M$
(c) $-\alpha v^2/M$ (d) $-2\alpha v^2/M$

46. The escape velocity from the surface of the earth is v_e. The escape velocity from the surface of a planet whose mass and radius are three times those of the earth, will be
(a) v_e (b) $3v_e$
(c) $9v_e$ (d) $1/3v_e$

47. What will be the formula of the mass in terms of g, R and G (R = radius of earth)
(a) $g^2\dfrac{R}{G}$ (b) $G\dfrac{R^2}{g}$
(c) $G\dfrac{R}{g}$ (d) $g\dfrac{R^2}{G}$

48. A ball is dropped from a satellite revolving around the earth at a height of 120 km. The ball will
(a) continue to move with same speed along a straight line tangentially to the satellite at that time
(b) continue to move with the same speed along the original orbit of satellite
(c) fall down to earth gradually
(d) go far away in space

49. The period of revolution of planet A around the Sun is 8 times that of B. The distance of A from the Sun is how many times greater than that of B from the Sun?
(a) 2 (b) 3
(c) 4 (d) 5

50. The escape velocity of a body on the surface of the earth is 11.2 km/s. If the earth's mass increases to twice its present value and the radius of the earth becomes half, the escape velocity would become
(a) 44.8 km/s
(b) 22.4 km/s
(c) 11.2 km/s (remains unchanged)
(d) 5.6 km/s

51. The escape velocity of a sphere of mass m is given by (G = Universal gravitational constant; M = Mass of the earth and R_e = Radius of the earth)
(a) $\sqrt{\dfrac{GM}{R_e}}$ (b) $\sqrt{\dfrac{2GM}{R_e}}$
(c) $\sqrt{\dfrac{2GMm}{R_e}}$ (d) $\sqrt{\dfrac{2GM+R_e}{R_e}}$

52. The escape velocity on the surface of earth is 11.2 km/s. What would be the escape velocity on the surface of another planet of the same mass but 1/4 times the radius of the earth?
(a) 22.4 km/s (b) 44.8 km/s
(c) 5.6 km/s (d) 11.2 km/s

53. A body weighs 72 N on the surface of the earth. What is the gravitational force on it due to earth at a height equal to half the radius of the earth from the surface?
(a) 32 N (b) 28 N
(c) 16 N (d) 72 N

54. Assuming earth to be a sphere of uniform density, what is the value of 'g' in a mine 100 km below the earth's surface? (Given, $R = 6400$ km)
(a) 9.65 m/s^2 (b) 7.65 m/s^2
(c) 5.06 m/s^2 (d) 3.10 m/s^2

55. The potential energy of a satellite, having mass m and rotating at a height of 6.4×10^6 m from the earth surface, is
(a) $-mgR_e$ (b) $-0.67\, mgR_e$
(c) $-0.5\, mgR_e$ (d) $-0.33\, mgR_e$

56. Energy required to move a body of mass m from an orbit of radius $2R$ to $3R$ is (M is the mass of the earth)
(a) $GMm/12R^2$ (b) $GMm/3R^2$
(c) $GMm/8R$ (d) $GMm/6R$

57. The escape velocity of a body depends upon mass as
(a) m^0 (b) m^1
(c) m^2 (d) m^3

58. The time period of a satellite of earth is 5 hours. If the separation between the earth and the satellite is increased to 4 times the previous value, the new time period will become
(a) 10 hours (b) 80 hours
(c) 40 hours (d) 20 hours

59. A particle of mass 10 g is kept on the surface of a uniform sphere of mass 100 kg and radius 10 cm. Find the work to be done against the gravitational force between them to take the particle far away from the sphere (you may take $G = 6.67 \times 10^{-11}\ \text{Nm}^2/\text{kg}^2$)
(a) 3.33×10^{-10} J (b) 13.34×10^{-10} J
(c) 6.67×10^{-10} J (d) 6.67×10^{-9} J

60. A planet in a distant solar system is 10 times more massive than the earth and its radius is 10 times smaller. Given that the escape velocity from the earth is 11 km s^{-1}, the escape velocity from the surface of the planet would be
(a) 1.1 km s^{-1} (b) 11 km s^{-1}
(c) 110 km s^{-1} (d) 0.11 km s^{-1}

Mechanical Properties of Solids

CONCEPTS

ELASTICITY AND PLASTICITY

Elasticity is the property of a body, by virtue of which it tends to regain its original shape and size when the applied force is removed. And if the body is permanently deformed such substances are called plastic and this property is called **Plasticity**.

STRESS

When a deforming force is applied to a body, an internal restoring force comes into play.
The restoring force per unit area is called **stress**.

$$\text{Stress} = \frac{\text{Restoring force}}{\text{Area}}$$

Its S.I. unit is Nm^{-2}.
Stress is a tensor as its value changes when direction changes.

Types of Stress

Longitudinal stress or tensile stress, volumetric stress and tangential stress.

STRAIN

It is defined as the ratio of the change in shape or size to the original shape or size of the body.

$$\text{Strain} = \frac{\text{Change in dimension}}{\text{Original dimension}}$$

Strain has no units or dimensions.

Types of Strain

Longitudinal strain : It is defined as the ratio of the change in length to the original length.

$$\text{Longitudinal strain} = \frac{\Delta \ell}{\ell}$$

Volume strain : It is the ratio of the change in volume to the original volume.

$$\text{Volume strain} = \frac{\Delta V}{V}$$

Shearing strain : It is the angular deformation produced in a body.

$$\text{Shearing strain } (\theta) = \frac{\Delta x}{h}$$

HOOKE'S LAW

It is the fundamental law of elasticity given by Robert Hooke in 1679. It states that "the stress is proportional to strain provided the strain is small".

i.e., Stress $\propto$ strain $\Rightarrow \frac{\text{Stress}}{\text{Strain}} = \text{constant} = E$

This proportionality constant is called **modulus of elasticity** (name given by Thomas Young) or coefficient of elasticity (E). There are three types of deformations and therefore three types of modulus of elasticity.

ELASTIC MODULI

(i) The **Young's modulus**

$$Y = \frac{\text{Tensile stress}}{\text{Tensile strain}} = \frac{F_\ell / A}{\Delta\ell / \ell}$$

where Δl is change in the length of bar, when we apply force F_ℓ.

(ii) The **shear modulus** or **modulus of rigidity**

$$\eta = \frac{\text{Shearing stress}}{\text{Shearing strain}} = \frac{F_t / A}{\Delta x / h}$$

(a) As we see, there is no change in volume under this deformation, but shape change.

(b) $\frac{\Delta x}{h} = \tan\theta \approx \theta$, where θ is shear angle.

(iii) The **bulk modulus** B is defined as

$$B = \frac{\text{Volume stress}}{\text{Volume stain}} = \frac{F_n / A}{\Delta V / V} = \frac{-\Delta P}{\Delta V / V}$$

Negative sign comes to make B positive,

because with the increase of pressure, the volume of body decreases or vice versa. The reciprocal of the bulk modulus is called compressibility of material i.e.,

Compressibility = 1/Bulk modulus.

POISSON'S RATIO

Lateral strain/longitudinal strain =Poisson's ratio (σ), the theoretical value of σ lies between –1 to 0.5 and practical value of σ lies between 0 to 0.5.

Relation between Y, K, η and σ

(i) $Y = 3K(1-2\sigma)$ (ii) $Y = 2\eta(1+\sigma)$

(iii) $\sigma = \dfrac{3K-2\eta}{2\eta+6K}$ (iv) $\dfrac{9}{Y} = \dfrac{1}{K} + \dfrac{3}{\eta}$

STRESS-STRAIN CURVE

The stress-strain graph of a ductile metal is shown in figure. Initially, the stress strain graph is linear and it obeys the Hooke's law upto the point called the proportional limit. After the proportional limit the graph is non-linear but it still remains elastic upto the yield point where the slope of the curve is zero. At the yield point the material starts deforming under constant stress and it behaves like a viscous liquid.

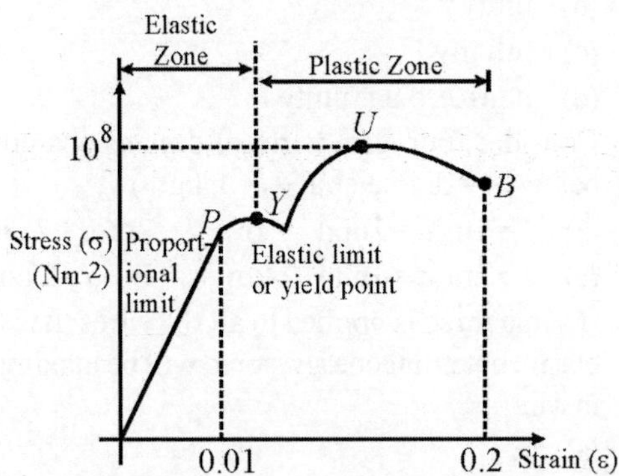

There are some materials which break as soon as the stress is increased beyond the elastic limit. They are called **brittle**, e.g. glass, ceramics etc.

Materials which have large plastic range of extension are called **ductile**. Using this property, materials can be drawn into thin wires, e.g. copper, aluminium etc.

Materials which can be hammered into thin sheets are called **malleable** e.g. gold, silver, lead, etc.

ELASTOMERS

Rubber has a large elastic region. It can be stretched several times its original length. On the removal of stress it returns to its original state. But the stress-strain graph is not a straight line. This means, it does not obey Hooke's law.

A pair of forces when applied parallel to the upper and lower faces, the solid deforms so that the upper face moves sideways with respect to the lower. The horizontal displacement ΔL of the upper face is perpendicular to the vertical height L. This type of deformation is called shear and the corresponding stress is the shearing stress. This type of stress is possible only in solids.

In this kind of deformation the Hooke's law takes the form

$$F/A = G \times \Delta L/L$$

where ΔL is the displacement of one end of object in the direction of the applied force F, and G is the shear modulus.

When an object undergoes hydraulic compression due to a stress exerted by a surrounding fluid, the Hooke's law takes the form $p = B(\Delta V/V)$, where p is the pressure (hydraulic stress) on the object due to the fluid, $\Delta V/V$ (the volume strain) is the absolute fractional change in the object's volume due to that pressure and B is the bulk modulus of the object.

In the case of a wire, suspended from ceiling and stretched under the action of a weight (F) suspended from its other end, the force exerted by the ceiling on it is equal and opposite to the weight. However, the tension at any cross-section A of the wire is just F and not 2F. Hence, tensile stress which is equal to the tension per unit area is equal to F/A.

The Young's modulus and shear modulus are relevant only for solids since only solids have lengths and shapes.

Metals have larger values of Young's modulus than alloys and elastomers. A material with large value of Young's modulus requires a large force to produce small changes in its length.

Bulk modulus is relevant for solids, liquid and gases. It refers to the change in volume when every part of the body is under the uniform stress so that the shape of the body remains unchanged.

THERMAL STRESS

Consider a bar which is heated and then prevented from contracting as it cools. A considerable force will be exerted at the ends of the bar, if the bar were free to contract, then decrease in length,

$$\Delta\ell = \ell\alpha\theta \quad \text{... (i)}$$

where ℓ is the original length, α is the coefficient of linear expansion and θ is the fall of temperature.

Again, $Y = \frac{F/A}{\Delta\ell/\ell} = \frac{F\ell}{A\Delta\ell}$ or $\Delta\ell = \frac{F\ell}{AY}$ ...(ii)

where, F is the force developed in the bar and A is the cross-sectional area of the bar.

From (i) and (ii), $\frac{F\ell}{AY} = \ell\alpha\theta$ or $F = YA\alpha\theta$

Thermal stress $= \frac{F}{A} = Y\alpha\theta$.

WORK DONE IN A STRETCHED WIRE

Elastic potential energy stored in the wire.

$U = \frac{1}{2}F \times \ell = \frac{1}{2}$ (stress) × (strain) × volume of the wire

∴ Elastic potential energy per unit volume of the wire (energy density)

$U = \frac{1}{2}$ (stress) × (strain)

$= \frac{1}{2}$ (Young's modulus × strain) × strain

(∵ Young's modulus = stress/strain)

∴ $U = \frac{1}{2}$ (Young's modulus) × $(\text{strain})^2$

Area under stress-strain diagram gives energy density. Material which stretches to a lesser extent for a given load is considered to be more elastic and its vice-versa.

Exercise

1. Modulus of rigidity of liquids is :
   (a) zero
   (b) unity
   (c) infinity
   (d) non-zero but finite
2. Gases possess
   (a) Young's modulus (b) Bulk modulus
   (c) Shear modulus (d) All
3. A steel wire is stretched to double its length, then its Young's modulus :
   (a) becomes half
   (b) becomes double
   (c) remains same
   (d) becomes one-fourth
4. If the tension in a wire is doubled, the elastic potential energy stored in the wire will :
   (a) become twice (b) become half
   (c) remains same (d) become four times
5. Elastomers are the materials which :
   (a) are not elastic at all
   (b) have very small elastic range
   (c) do not obey Hooke's law
   (d) none of these
6. Shear modulus of a perfectly rigid body is :
   (a) zero
   (b) unity
   (c) infinity
   (d) non-zero but unity
7. Consider four steel wires of dimensions given below (d = diameter and l = length) :
   (a) l = 1m, d = 1mm (b) l = 2m, d = 2 mm
   (c) l = 2m, d = 1mm (d) l = 1m, d = 2 mm
   If same force is applied to all the wires then the elastic potential energy stored will be maximum in wire :
   (a) A (b) B
   (c) C (d) D
8. Which of the following is responsible for change in shape?
   (a) Tensile stress
   (b) Compressive stress
   (c) Tangential stress
   (d) All of these
9. Poisson's ratio is the ratio of :
   (a) longitudinal strain to lateral strain
   (b) lateral strain to longitudinal strain
   (c) longitudinal strain to volumetric strain
   (d) volumetric strain to longitudinal strain

10. Steel is :
(a) perfectly elastic
(b) perfectly plastic
(c) partially plastic and partially elastic
(d) depends on applied force
11. Which of the following affects the elasticity of a substance?
(a) Hammering and annealing
(b) Change in temperature
(c) Impurity
(d) All of these
12. Which of the following materials is most elastic?
(a) Rubber (b) Lead
(c) Wood (d) Steel
13. If large deformation takes place between the elastic limit and fracture point then the material is
(a) ductile (b) elastomer
(c) brittle (d) none
14. SI unit of shear stress is
(a) $\frac{N}{m^2}$ (b) $\frac{N}{cm^2}$
(c) $\frac{Dy}{M^2}$ (d) $\frac{Dy}{cm^2}$
15. Hooke's law states that
(a) Stress is directly proportional to the strain
(b) Stress is inversely proportional to the strain
(c) Stress is proportional to Young's modulus
(d) Stress and strain are independent of each other
16. Hooke's law essentially defines
(a) Stress (b) Strain
(c) Yield point (d) Elastic limit
17. Young's modulus is defined as
(a) The ratio of linear strain to the normal stress
(b) The ratio of normal stress to strain
(c) Product of linear strain and normal stress
(d) Square of the ratio of normal stress to linear strain
18. The longitudinal extension of any elastic material is very small. In order to have an appreciable change, the material must be in the form of
(a) Thin block of any cross section
(b) Thick block of any cross section
(c) Long thin wire
(d) Short thin wire
19. M.K.S. unit of Young's modulus is
(a) N/m (b) N/m^2
(c) $N \times m$ (d) $N \times m^2$
20. From a steel wire of density ρ is suspended a brass block of density ρ_b. The extension of steel wire comes to e. If the brass block is now fully immersed in a liquid of density ρ_l, the extension becomes e'. The ratio $\frac{e}{e'}$ will be
(a) $\frac{\rho_b}{\rho_b - \rho_l}$ (b) $\frac{\rho_b - \rho_l}{\rho_b}$
(c) $\frac{\rho_b - \rho}{\rho_l - \rho}$ (d) $\frac{\rho_l}{\rho_b - \rho_l}$
21. The effect of temperature on the value of Young's modulus of elasticity for various substances in general is
(a) It increases with increase in temperature
(b) Remains constant
(c) Decrease with rise in temperature
(d) Sometimes increases and sometimes decreases with temperature
22. The breaking stress for a wire of unit cross section is called its
(a) Yield point (b) Tensile strength
(c) Elastic fatigue (d) Young's modulus
23. An iron bar of length ℓ cm and cross section A cm^2 is pulled by a force of F dynes from ends so as to produce an elongation $\Delta\ell$ cm. Which of the following statement is correct?
(a) Elongation is inversely proportional to length
(b) Elongation is directly proportional to cross section A
(c) Elongation is inversely proportional to A
(d) Elongation is directly proportional to Young's modulus
24. When a body undergoes a linear tensile strain it experiences a lateral contraction also. The ratio of lateral contraction to longitudinal strain is known as
(a) Young's modulus (b) Bulk modulus
(c) Poisson's ratio (d) Hooke's law

25. Energy per unit volume in a stretched wire is
(a) half of load × strain
(b) load × strain
(c) stress × strain
(d) half of stress × strain
26. A stretched rubber has
(a) Increased kinetic energy
(b) Increased potential energy
(c) Decreased kinetic energy
(d) Decreased potential energy
27. The Young's modulus of a perfectly rigid body is
(a) Unity
(b) Zero
(c) Infinity
(d) Some finite non-zero constant
28. The bulk modulus of a perfectly rigid body, is equal to
(a) Infinity
(b) Zero
(c) Some finite value
(d) Non-zero constant
29. The shear modulus of elasticity of a liquid is
(a) Infinity
(b) Unity
(c) Some finite, non-zero constant
(d) Zero
30. Identical springs of steel and copper ($Y_s > Y_{cu}$) are equally stretched
(a) Less work is done on steel spring
(b) Less work is done on copper spring
(c) Equal work is done on both the springs
(d) Data not complete
31. A spiral spring is stretched by a weight attached to it. The strain is
(a) Shear (b) Elastic
(c) Tensile (d) Bulk
32. A thick rope of density ρ and length L is hung from a rigid support. The Young's modulus of the material of rope is Y. The increase in length of the rope due to its own weight is
(a) $(1/4)\rho g L^2/Y$ (b) $(1/2)\rho g L^2/Y$
(c) $\rho g L^2/Y$ (d) $\rho g L/Y$
33. A metallic rod breaks when strain produced is 0.2%. The Young's modulus of the material of the rod is 7×10^9 N/m². What should be its area of cross-section to support a load of 10^4 N ?
(a) 7.1×10^{-8} m² (b) 7.1×10^{-6} m²
(c) 7.1×10^{-4} m² (d) 7.1×10^{-2} m²
34. A spherical ball contracts in volume by 0.02% when subjected to a pressure of 100 atmosphere. Assuming one atmosphere = 10^5 N m^{-2}, the bulk modulus of the material of the ball is
(a) 0.02×10^5 N/m² (b) 0.02×10^7 N/m²
(c) 50×10^7 N/m² (d) 50×10^9 N/m²
35. A wire is stretched by a force F. If s is the strain developed and Y is Young's modulus of material of wire then work done per unit volume is
(a) $\frac{Ys^2}{2}$ (b) $\frac{s^2}{2Y}$
(c) $\frac{1}{2}Fs$ (d) $\frac{Y}{2s^2}$
36. 111. Stress Vs strain for the elastic tissue of the aorta, the large tube (vessel) carrying blood from the heart, will be : [stress is proportional to square of the strain for the elastic tissue of the aorta]

(a) 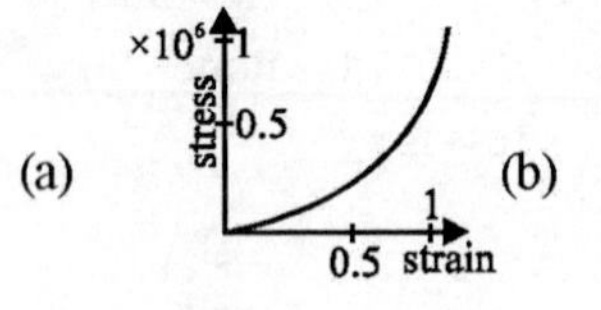

(b) 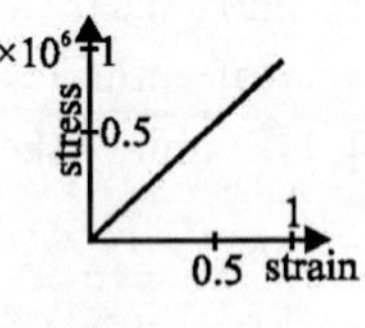

(c) 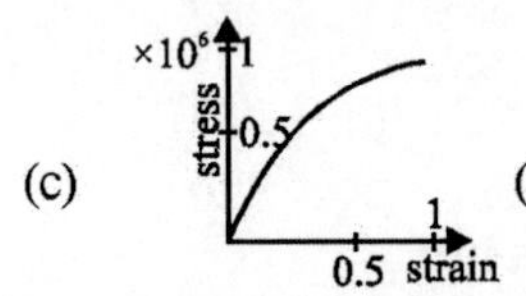

(d) 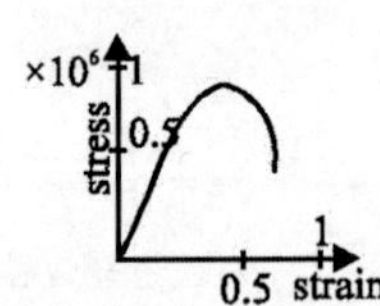

37. The potential energy U between two atoms in a diatomic molecules as a function of the distance x between atoms has been shown in the figure. The atoms are

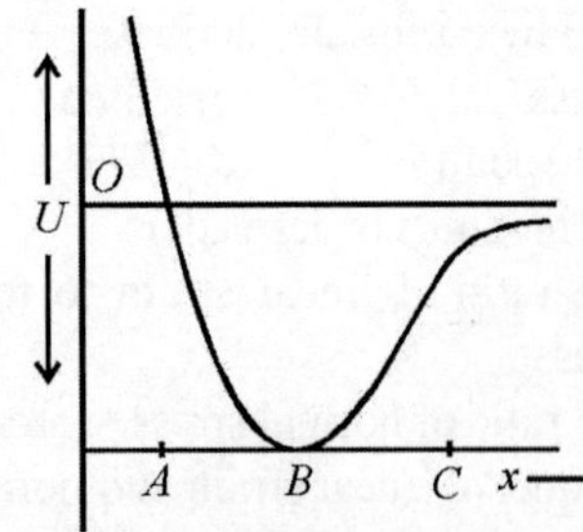

(a) attracted when x lies between A and B and are repelled when x lies between B and C
(b) attracted when x lies between B and C and are repelled when x lies between A and B
(c) are attracted when they reach B from C
(d) are repelled when they reach B from A

38. The amount of workdone in increasing the length of a wire through 1 cm will be

(a) $\frac{YA}{2L}$ (b) $\frac{YL}{2A}$

(c) $\frac{YL^2}{2A}$ (d) None of these

39. If in a wire of Young's modulus Y, longitudinal strain X is produced, then the value of potential energy stored in its unit volume will be

(a) YX^2 (b) $2YX^2$

(c) $Y^2X/2$ (d) $YX^2/2$

40. What is relation between Y, K, η for some isotropic solid material ?

(a) $\eta = \frac{3KY}{3K+Y}$ (b) $\eta = \frac{3KY}{9K-Y}$

(c) $\eta = \frac{3KY}{K-Y}$ (d) $Y = \frac{3K\eta}{3K-\eta}$

41. The value of tan $(90 - \theta)$ in the graph gives

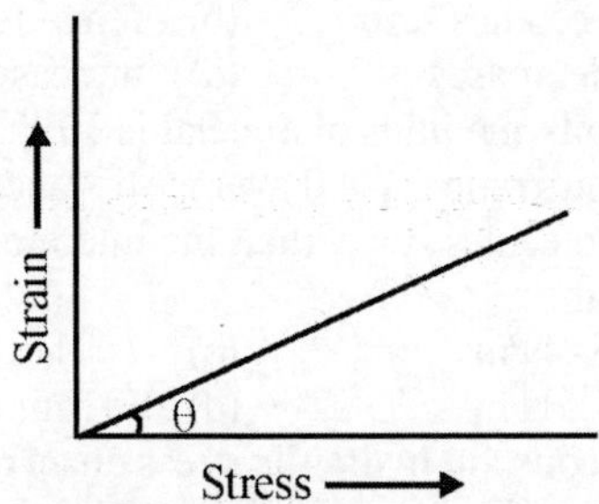

(a) Young's modulus of elasticity
(b) compressibility
(c) shear strain
(d) tensile strength

42. Two wires of same material and length but cross-sections in the ratio 1 : 2 are used to suspend the same loads. The extensions in them will be in the ratio

(a) 1:2 (b) 2:1

(c) 4:1 (d) 1:4

43. A steel ring of radius r and cross sectional area A is fitted onto a wooden disc of radius R $(R > r)$. If the Young's modulus of steel is Y, then the force with which the steel ring is expanded is

(a) $AY(R/r)$ (b) $AY(R-r)/r$

(c) $(Y/A)[(R-r)/r]$ (d) Yr/AR

44. A cube at temperature 0°C is compressed equally from all sides by an external pressure P. By what amount should its temperature be raised to bring it back to the size it had before the external pressure was applied. The bulk modulus of the material of the cube is B and the coefficient of linear expansion is α.

(a) $P/B\alpha$ (b) $P/3B\alpha$

(c) $3\pi\alpha/B$ (d) $3B/P$

45. A body of mass 10 kg is attached to a wire of radius 3 cm. It's breaking stress is 4.8×10^7 Nm^{-2}, the area of cross-section of the wire is 10^{-6} m^2. What is the maximum angular velocity with which it can be rotated in the horizontal circle?

(a) 1 rad sec^{-1} (b) 2 rad sec^{-1}

(c) 4 rad sec^{-1} (d) 8 rad sec^{-1}

46. The Young's modulus of brass and steel are respectively 10^{10} N/m^2 and 2×10^{10} N/m^2. A brass wire and a steel wire of the same length are extended by 1 mm under the same force, the radii of brass and steel wires are R_B and R_S respectively. Then

(a) $R_S = \sqrt{2}\,R_B$ (b) $R_S = R_B/\sqrt{2}$

(c) $R_S = 4R_B$ (d) $R_S = R_B/4$

47. A steel wire is suspended vertically from a rigid support. When loaded with a weight in air, it extends by ℓ_a and when the weight is immersed completely in water, the extension is reduced to ℓ_w. Then the relative density of material of the weight is

(a) ℓ_a/ℓ_w (b) $\frac{\ell_a}{\ell_a - \ell_w}$

(c) $\ell_w/(\ell_a - \ell_w)$ (d) ℓ_w/ℓ_a

48. Two rods of different materials having coefficients of linear expansion α_1, α_2 and Young's moduli Y_1 and Y_2 respectively are fixed between two rigid massive walls. The rods are heated such that they undergo the same increase in temperature. There is no bending of rods. If $\alpha_1 : \alpha_2 = 2 : 3$, the thermal stresses developed in the two rods are equal provided $Y_1 : Y_2$ is equal to

(a) 2:3 (b) 1:1

(c) 3:2 (d) 4:9

49. The diagram shown below represents the applied forces per unit area with the corresponding change X (per unit length) produced in a thin wire of uniform cross section in the curve shown. The region in which the wire behaves like a liquid is

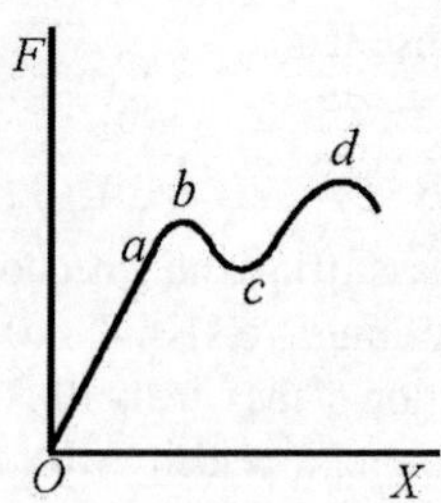

(a) ab (b) bc
(c) cd (d) Oa

50. A steel wire of length 20 cm and uniform cross-section 1 mm^2 is tied rigidly at both the ends. The temperature of the wire is altered from 40°C to 20°C. Coefficient of linear expansion for steel $\alpha = 1.1 \times 10^{-5}$/°C and Y for steel is 2.0×10^{11} N/m^2. The change in tension of the wire is
(a) 2.2×10^6 newton (b) 16 newton
(c) 8 newton (d) 44 newton

51. For the same cross-sectional area and for a given load, the ratio of depressions for the beam of a square cross-section and circular cross-section is
(a) $3:\pi$ (b) $\pi:3$
(c) $1:\pi$ (d) $\pi:1$

52. Two rods A and B of the same material and length have their radii r_1 and r_2 respectively. When they are rigidly fixed at one end and twisted by the same couple applied at the other end, the ratio $\left(\frac{\text{the angle of twist at the end of } A}{\text{the angle of twist at the end of } B}\right)$ is
(a) r_1^2/r_2^2 (b) r_1^3/r_2^3
(c) r_2^4/r_1^4 (d) r_1^4/r_2^4

53. A rod of length 20 cm is made of metal. It expands by 0.075 cm when its temperature is raised from 0°C to 100°C. Another rod of a different metal B having the same length expands by 0.045 cm for the same change in temperature. A third rod of the same length is composed of two parts, one of metal A and the other of metal B. This rod expands by 0.060 cm for the same change in temperature. The portion made of metal A has the length
(a) 20 cm (b) 10 cm
(c) 15 cm (d) 18 cm

54. Longitudinal strain is possible in :
(a) Liquid (b) Gases
(c) Solid (d) All of these

55. A beam of metal supported at the two edges is loaded at the centre. The depression at the centre is proportional to
(a) Y^2 (b) Y
(c) $1/Y$ (d) $1/Y^2$

56. A student performs an experiment to determine the Young's modulus of a wire, exactly 2 m long, by Searle's method. In a particular reading, the student measures the extension in the length of the wire to be 0.8 mm with an uncertainty of ± 0.05 mm at a load of exactly 1.0 kg. The student also measures the diameter of the wire to be 0.4 mm with an uncertainty of ± 0.01 mm. Take g = 9.8 m/s^2 (exact). The Young's modulus obtained from the reading is
(a) $(2.0 \pm 0.3) \times 10^{11}$ N/m^2
(b) $(2.0 \pm 0.2) \times 10^{11}$ N/m^2
(c) $(2.0 \pm 0.1) \times 10^{11}$ N/m^2
(d) $(2.0 \pm 0.05) \times 10^{11}$ N/m^2

57. According to Hook's law of elasticity, if stress is increased, then the ratio of stress to strain :
(a) becomes zero (b) remains constant
(c) decreases (d) increases

58. The bulk modulus of a metal is 10^{10} N/m^2 and Poisson's ratio 0.20. If average distance between the molecules is 3Å then the interatomic force constant :
(a) 5.4 N/m (b) 7.5 N/m
(c) 7.5 N/m (d) 30 N/m

59. For a constant hydraulic stress on an object, the fractional change in the object's volume ($\Delta V/V$) and its bulk modulus (B) are related as:
(a) $\frac{\Delta V}{V} \propto B$ (b) $\frac{\Delta V}{V} \propto \frac{1}{B}$
(c) $\frac{\Delta V}{V} \propto B^2$ (d) $\frac{\Delta V}{V} \propto B^{-2}$

60. Shear modulus is zero for
(a) solids (b) liquids
(c) gases (d) liquids and gases

CHAPTER 9

Mechanical Properties of Fluids

CONCEPTS

PRESSURE

The basic property of a fluid is that it can flow. The fluid does not have any resistance to change of its shape. Thus, the shape of a fluid is governed by the shape of its container. A liquid is incompressible and has a free surface of its own. A gas is compressible and it expands to occupy all the space available to it. One should not think of pressure of a fluid as being exerted only on a solid like the walls of a container or a piece of solid matter immersed in the fluid. Pressure exists at all points in a fluid. An element of a fluid is in equilibrium because the pressure exerted on the various faces are equal.

If F is the normal force exerted by a fluid on an area A then the average pressure P_{av} is defined as the ratio of the force to area. $P_{av} = \frac{F}{A}$. The unit of the pressure is the pascal (Pa). It is the same as N m^{-2}. Other common units of pressure are

1 atm = 1.01×10^5 Pa, 1 bar = 10^5 Pa, 1 torr = 133 Pa = 0.133 kPa,

1 mm of Hg = 1 torr = 133 Pa.

Pressure is a scalar quantity.

Atmospheric pressure : The atmosphere exerts pressure on the earth's surface. The atmospheric pressure at sea level is given by

$$P_0 = 1.01 \times 10^5 \text{ Pa}.$$

Barometer is used to measure atmospheric pressure.

Hydrostatic pressure : The hydrostatic pressure at a depth h below the surface of a fluid is given by

$$P = h\rho g$$

where ρ is the density of the fluid and g, the acceleration due to gravity.

Gauge pressure : The pressure at any point in a fluid is equal to the sum of the atmospheric pressure acting on its surface and the hydrostatic pressure due to the weight of the fluid above that point which is at a depth h below the surface of the fluid.

The gauge pressure at a point is given by : $P - P_a$

$$P = P_a + h\rho g$$

Manometer is used to measure gauge pressure

Archimedes' Principle

A body immersed in a fluid partially or wholly experiences an upward buoyant force equivalent to the weight of the fluid displaced by it.

Pascal's Law

It states that pressure in a fluid partially or wholly at rest is same at all points which are at the same height. A change in pressure applied to an enclosed fluid is transmitted undiminished to every point of the fluid and the walls of the containing vessel. Devices like hydraulic lift and hydraulic brakes are based on the pascal's law.

MOTION OF FLUIDS

The motion of fluids are of following four types :

(i) **Streamline motion :** In a fluid in motion, if fluid particles preceeding or succeeding a fluid particle follow the same path, then the path is called streamline; and then the motion of the fluid is called streamline motion. This type of motion takes place in non-viscous fluids having very small speed.

Principle of continuity : When incompressible, non-viscous liquid flows in non-uniform tube then in streamline flow product of area and velocity at any section remains same. The mass of liquid flowing in equals the mass flowing out.

i.e., $m_1 = m_2$

or, $v_1 A_1 \rho_1 \Delta t = v_2 A_2 \rho_2 \Delta t$ ..... (1)

As we have considered the fluid incompressible thus, $v_1 A_1 = v_2 A_2$

or $Av = constant$ ....(2)

Equations (1) and (2) are said to be as **equation of continuity**.

(ii) **Steady state motion :** In a liquid in motion, when liquid particles, crossing a point, cross it with same velocity, then the motion of the liquid is called steady state motion. This type of motion takes place in non-viscous liquids having very small speed.

(iii) **Laminar motion :** Viscous liquids flow in bounded region or in a pipe, in layers and when viscous liquid is in motion, different layers have different velocities. The layers in contact with the fixed surface has least velocity and the velocity of other parallel layers increases uniformly and continuously with the distance from the fixed surface to the free surface of the liquid. This is called laminar motion of the liquid.

(iv) **Turbulent motion :** When the velocity of a liquid is irregular, haphazard and large, i.e. Beyond a limiting value called critical velocity the flow of liquid loses steadiness then the motion of the liquid is called turbulent motion.

Critical velocity $V_c = \dfrac{K\eta}{\rho r}$

Here η is called coefficient of viscosity.

BERNOULLI'S PRINCIPLE

It states that as we move along a streamline, the sum of the pressure (P), the kinetic energy per unit volume ($\rho v^2/2$) and the potential energy per unit volume (ρgy) remains a constant.

i.e., $\quad P + \rho v^2/2 + \rho gy = \text{constant}$

The equation is basically the conservation of energy applied to non viscous fluid motion in steady state. There is no fluid which have zero viscosity, so the above statement is true only approximately. The viscosity is like friction and converts the kinetic energy to heat energy. Bernoulli's principle does not hold in presence of viscous drag on the fluid. The work done by this dissipative viscous force must be taken into account.

VELOCITY OF EFFLUX

It is the velocity of fluid out flow from a open tank. This equation is known as **Toricelli's law**.

$$v = \sqrt{2gh}$$

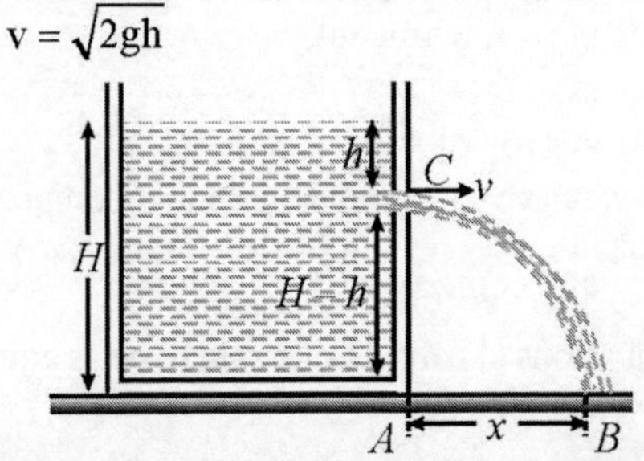

The speed of the liquid coming out of the orifice is independent of the nature and quantity of liquid in the container or the area of the orifice.

Greater is the distance of the hole from the free surface of liquid greater will be the velocity of efflux (i.e., $v \propto \sqrt{h}$). This is why liquid gush-out with maximum velocity from the orifice which is at maximum vertical distance from the free surface of the liquid.

As the vertical velocity of liquid at the orifice is zero and it is at a height (H – h) from the base, the time taken by the liquid to reach the base-level.

$$t = \sqrt{\frac{2(H-h)}{g}}$$

$$x = vt = \sqrt{2gh} \times \sqrt{[2(H-h)/g} = 2\sqrt{h(H-h)}$$

At $h = \dfrac{H}{2}$, $\quad x_{max} = 2\sqrt{\dfrac{H}{2}\left[H - \dfrac{H}{2}\right]} = H$ i.e.,

range x will be maximum (= H) when h = H/2.

Venturimeter

The device is used for the measurement of rate of flow of fluid through a tube.

The working of venturimeter is based on Bernoulli's principle.

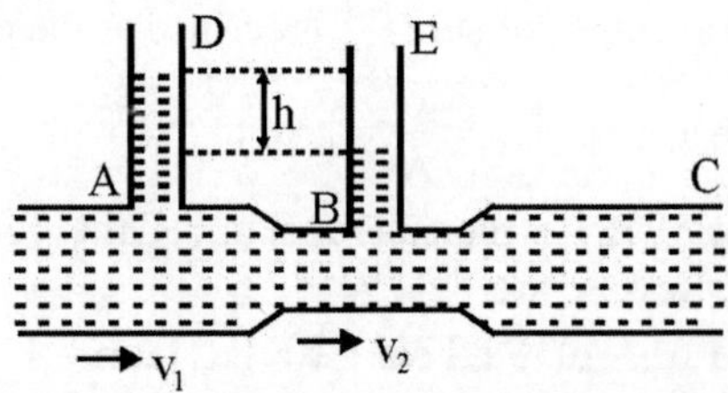

Rapidly moving fluids sustain less pressure than slowly moving fluids. Volume flowing per unit time is

$$v = a_1 a_2 \sqrt{\frac{2gh}{a_1^2 - a_2^2}}$$

(a_1 and a_2 are the area of cross-sections at points A and B)

VISCOSITY

The property of a fluid due to which it opposes the relative motion between its different layers is called viscosity.

Viscous force : The force between the layers of a liquid opposing the relative motion is called viscous force

$F = -\eta A \frac{dv}{dx}$; where η is a constant depending upon the nature of the liquid and is called the **coefficient of viscosity** and velocity gradient $= \frac{dv}{dx}$

The **coefficient of viscosity** has the dimension $[ML^{-1}T^{-1}]$ and its unit is newton second per square metre (Nsm^{-2}) or kilogram per metre per second ($kgm^{-1}s^{-1}$). In CGS the unit of viscosity is poise.

Stoke's Law

It states that the viscous drag force F on a spherical body of radius *a* moving with velocity *v* through a fluid of viscosity η is, $F = -6\pi\eta av$.

Terminal Velocity

When a spherical body is allowed to fall through viscous medium, its velocity increases till the viscous drag plus upthrust is equal to the weight of the body. After the body moves with constant velocity, called terminal velocity.

The terminal velocity is given

$$v = \frac{2R^2}{9\eta}(d_{body} - \sigma_{med})g$$

SURFACE TENSION

It is a force per unit lengths (or surface energy per unit area) acting in the plane of interface between the liquid and the bounding surface. It is the extra energy that the molecules at the interface have as compared to the interior. Surface tension arises due to excess potential energy of the molecules on the surface in comparison to their potential energy in the interior. Such a surface energy is present at the interface separating two substances at least one of which is a fluid. It is not the property of a single fluid alone.

or Surface tension, $T = \frac{F}{l}$

Angle of Contact

The angle between the tangent to the liquid surface and the tangent to the solid surface at the point of contact (inside the liquid) is known as angle of contact.

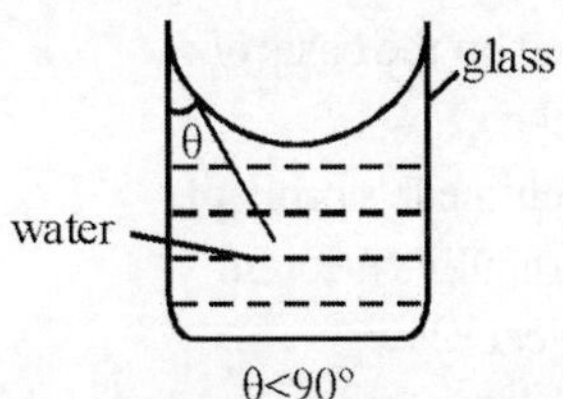

$\theta < 90°$

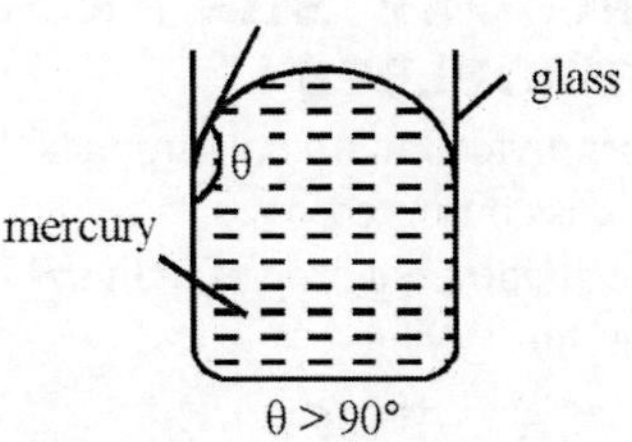

$\theta > 90°$

Some values of angle of contact of solid and liquid -

Pure water & Glass	0°
Normal water & Glass	8°
Silver & Glass	90°
Mercury & Glass	135°
Alcohol & Glass	138°

The angle of contact of solid and the liquid that wet solids (water) is acute angle and its meniscus is concave and the angle of contact of solid and liquid that do not wet solids (mercury) is obtuse angle and the meniscus is convex.

SURFACE ENERGY

In order to increase the surface area, the work has to be done over the surface of the liquid. This work done is stored in the liquid surface as its potential energy. Hence the surface energy of a liquid can be defined as the excess potential energy per unit area of the liquid surface.

$W = T\Delta A$, where ΔA = increase in surface area,

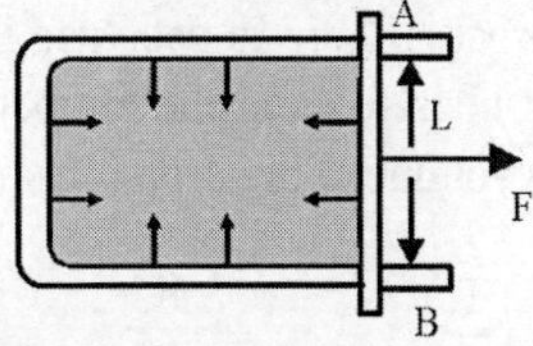

T = Surface tension of the liquid

Some Problems related to Surface Energy

(1) For a liquid drop (Only one free surface) :

For increasing radius 0 to R

Initial surface area $= 0$

Final surface area $= 4\pi R^2$

So, surface energy $= T\Delta A = 4\pi T(R^2 - 0)$

$= 4\pi R^2 T$

(2) For a soap bubble (For two free surfaces) :

For increasing radius R_1 to R_2 where $R_2 > R_1$

So, surface energy

$= T\Delta A = T \times 2(4\pi R_2^2 - 4\pi R_1^2) = 8\pi T(R_2^2 - R_1^2)$

DIVIDING ONE BIG DROP INTO SMALL DROPLETS

If a big drop is divided into small droplets then volume always remains conserved.

So, when we divide a big drop of radius *R* into *n* small droplets of radius *r*, then

$$\frac{4}{3}\pi R^3 = n\,\frac{4}{3}\pi r^3$$

or $$R^3 = nr^3 \Rightarrow n = \frac{R^3}{r^3} \Rightarrow n = \left(\frac{R}{r}\right)^3$$

Initial area of drop, $4\pi R^2$

Final area of drop, $n(4\pi r^2)$

Change in area $\Delta A = n4\pi r^2 - 4\pi R^2 = 4\pi(nr^2 - R^2)$

So, work done against surface tension

$$W = 4\pi T\,(nr^2 - R^2) \text{ or } W = 4\pi R^3 T\left(\frac{1}{r} - \frac{1}{R}\right)$$

In this process there is increase in surface area of the liquid. So energy is absorbed and decrease in temperature takes place. Due to this reason we feel cool near fountain.

Decrease in temperature $\Delta\theta = \frac{3T}{Jsd}\left(\frac{1}{r} - \frac{1}{R}\right)$

s = specific heat of liquid, d = density of liquid

SMALL DROPLETS COALESCE TO FORM A BIG DROP

In this process there is decrease in surface area so energy is liberated and the temperature of big drop increases volume of big drop = n small drops.

$$n = \left(\frac{R}{r}\right)^3 \Rightarrow r = \frac{R}{n^{1/3}}$$

Increase in temperature of big drop $\Delta\theta = \frac{3T}{Jsd}\left(\frac{1}{r} - \frac{1}{R}\right)$;

s = specific heat of liquid, d = density of liquid

Liberated energy $= \frac{4\pi T}{J}\,(nr^2 - R^2)$ cal.

$$W = 4\pi T\,(nr^2 - R^2) \text{ joules}$$

Excess pressure in liquid drop : P = 2T/R

Excess pressure in the bubble formed in a liquid : P = 2T/R

Excess pressure in soap bubble : There are two free surfaces in a soap bubble so P = 4T/R

CAPILLARITY

It is the rise or fall of liquid in a capillary (a narrow bore tube) relative to the surrounding liquid when dipped in liquid.

$$T = \frac{r\left[h + \left(\frac{r}{3}\right)\right]dg}{2\cos\theta}$$

If the tube is very narrow, r/3 can be neglected in comparison with h. Hence $h = \frac{2T\cos\theta}{rdg}$ or $T = \frac{hrdg}{2\cos\theta}$

Jurin's Law :

The height h of a liquid rised in a capillary tube is inversely proportional to the radius of the capillary tube.

i.e., $h \propto 1/r$

Exercise

1. The constant velocity attained by a body while falling through a viscous medium is termed as :
   (a) critical velocity (b) terminal velocity
   (c) threshold velocity (d) none of these
2. S.I. unit of Reynold's number is
   (a) $[ML^{-1}T^{-1}]$ (b) $[ML^{-3}T^{-1}]$
   (c) $[M^0L^0T^0]$ (d) $[M^0LT^{-3}]$
3. When a train crosses a platform with high velocity, passenger on the platform tend to fall towards the train. This phenomenon can be explained on the basis of :
   (a) Stoke's law
   (b) Archimede's principle
   (c) Bernoulli's theorem
   (d) Pascal's law

4. The difference is between viscosity and solid friction is/are :
   (a) viscosity depends on area while solid friction does not.
   (b) viscosity depends on nature of material but solid friction does not.
   (c) both (a) and (b)
   (d) neither (a) nor (b)
5. Water is not used in thermometer because
   (a) it sticks to glass
   (b) its shows anamalous expansion
   (c) both (a) and (b)
   (d) neither (a) nor (b)
6. Angle of contact depends on
   (a) temperature
   (b) presence of impurities
   (c) nature of liquid and the container
   (d) all of these
7. If r is the radius of influence of molecules of a certain liquid, then thickness of its surface film is
   (a) r (b) $\frac{r}{2}$
   (c) $2r$ (d) none of these
8. Bernoulli's theorems explains the phenomenon of
   (a) blowing of roob's due to wind
   (b) spinning of ball
   (c) dynamic lift of aeroplane
   (d) all of these
9. Toricelli's theorem is used to find :
   (a) the velocity of efflux through an orifice
   (b) the velocity of flow of liquid through a pipe
   (c) terminal velocity
   (d) critical velocity
10. A pin of density greater than that of water can float on the surface of water. It is due to :
    (a) viscosity (b) buoyancy
    (c) surface tension (d) none of these
11. With increase in temperature, the viscosity of gases :
    (a) increases
    (b) decrease
    (c) remains same
    (d) depends on the nature of the gas.
12. Gases do not posses
    (a) density (b) surface tension
    (c) volume (d) viscosity
13. Hydraulic lift is based on the principle of
    (a) Pascal's law (b) Bernoulli's theorem
    (c) Toricelli's theorem (d) Stoke's law
14. Paint-gun is based on :
    (a) Bernoulli's theorem
    (b) Archimede's principle
    (c) Boyle's law
    (d) Pascal's law
15. Water is flowing through a horizontal pipe having a restriction, then
    (a) Pressure will be greater at the restriction
    (b) Pressure will be greater in the wider portion
    (c) Pressure will be same through out the length of the pipe
    (d) none of the above
16. To get the maximum flight, the ball must be thrown as :

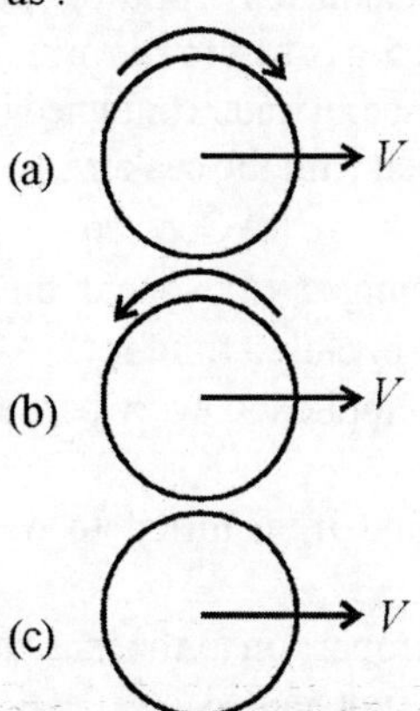

    (d) any of (a), (b) and (c)
17. When a liquid rises in a capillary tube, the gain in P.E. of the liquid is :
    (a) less than the work done by force of surface tension
    (b) greater than the work done by force of surface tension
    (c) equal to the work done by force of surface tension
    (d) may be less than or equal to the force of surface tension
18. Fevicol is added to point to be painted on the walls, because
    (a) it increases adhesive force between paint & wall
    (b) it decreases adhesive force between paint & wall molecules
    (c) it decreases cohesive force between paint molecules
    (d) none of the above

19. A sample of metal weighs 210 g in air, 140 g in water and 120 g in an unknown liquid. Then
   (a) The density of the metal is 3 g/cm^3
   (b) The density of the metal is 7 g/cm^3
   (c) Density of the metal is 4 times the density of unknown liquid
   (d) The metal still float in water
20. In order that a floating object be in a stable equilibrium, its centre of buoyancy should be
   (a) Vertically above its centre of gravity
   (b) Below its centre of gravity
   (c) Horizontally in a line with its centre of gravity
   (d) May be anywhere
21. A beaker containing water is balanced on the pan of a common balance. A solid of specific gravity 1 and mass 5 g is tied to the arm of the balance and immersed in water contained in the beaker. The scale pan with the beaker
   (a) Goes down (b) Goes up
   (c) Remains unchanged (d) None of these
22. Pressure applied to enclosed fluid is
   (a) Increased and applied to every part of the fluid
   (b) Diminished and transmitted to wall of container
   (c) Increased in proportion to the mass of the fluid and then transmitted
   (d) Transmitted unchanged to every portion of the fluid and wall of containing vessel
23. A common hydrometer reads specific gravity of liquids. Compared to the 1.6 mark of the stem the mark 1.5 will be
   (a) Upwards
   (b) Downwards
   (c) In the same place
   (d) May be upward or downward depending upon the hydrometer
24. An ice block floats in a liquid whose density is less than water. A part of block is outside the liquid. When whole of ice has melted, the liquid level will
   (a) Rise
   (b) Go down
   (c) Remain same
   (d) First rise then go down
25. A body partly immersed, floats in a liquid contained in a beaker. The beaker is kept inside a lift falling freely under gravity. The upthrust on the body due to liquid is :
   (a) Zero
   (b) Equal to weight of body in air
   (c) Equal to weight of liquid displaced
   (d) Equal to weight of immersed part of body
26. A small lead shot is embedded in a big lump of ice floating in a jar of water. The level of water in the jar is noted. When all the ice melts down, the level of water in the jar would:
   (a) Be raised
   (b) Go down
   (c) Remain unchanged
   (d) None of the above
27. To what height should a cylindrical vessel be filled with a homogeneous liquid to make the sides of the vessel equal to the force exerted by the liquid on the bottom of the vessel.
   (a) Equal to the radius
   (b) Less than radius
   (c) More than radius
   (d) Four times of the radius
28. The spring balance A reads 2 kg with a block of mass m suspended from it. A balance B reads 5 kg when a beaker with liquid is put on the pan of the balance. The two balances are now so arranged that the hanging mass is inside the liquid in a beaker as shown in fig.

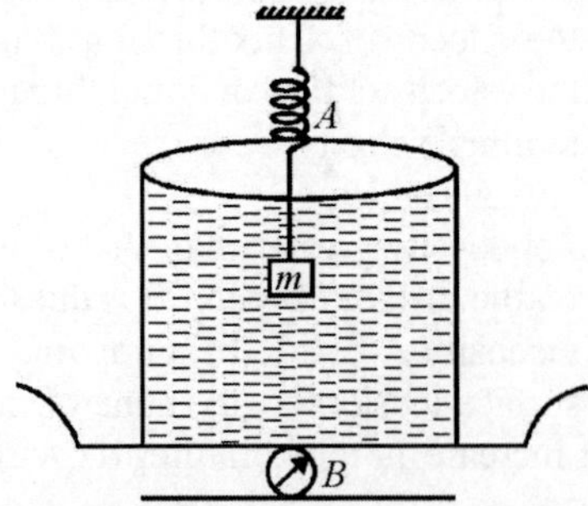

   (a) The balance A will read more than 2 kg
   (b) The balance B will read less than 5 kg
   (c) The balance A will read less than 2 kg and B will read more than 5 kg
   (d) The balance A will read more than 2 kg and B will read less than 5 kg

29. Surface tension of a liquid is due to
    (a) Gravitational force between molecules
    (b) Electrical force between molecules
    (c) Adhesive force between molecules
    (d) Cohesive force between molecules
30. Surface tension may be defined as
    (a) The work done per unit area in increasing the surface area of a liquid under isothermal conditions
    (b) The work done per unit area in increasing the surface area of a liquid under adiabatic conditions
    (c) The work done per unit area in increasing the surface area of a liquid under adiabatic conditions
    (d) Free surface energy per unit volume
31. Kerosene oil rises up in a wick of a lantern because of
    (a) Diffusion of the oil through the wick
    (b) Surface tension
    (c) Buoyant force of air
    (d) The gravitational pull of the wick
32. A drop of oil is placed on the surface of water. Which of the following statements is correct
    (a) It will remain on it as a sphere
    (b) It will spread as a thin layer
    (c) It will partly be a spherical droplet and partly a thin film
    (d) It will float as a distorted drop on the water surface
33. With the increase in temperature, the angle of contact
    (a) Decreases
    (b) Increases
    (c) Remains constant
    (d) Sometimes increases and sometimes decreases
34. At critical temperature, the surface tension of a liquid is
    (a) Zero
    (b) Infinity
    (c) The same as that at any other temperature
    (d) Cannot be determined
35. If two soap bubbles of different radii are in communication with each other
    (a) Air flows from the larger bubble into the smaller one until the two bubbles are of equal size
    (b) The size of the bubbles remains the same
    (c) Air flows from the smaller bubble into the larger one and the larger one grows at the expense of the smaller one
    (d) The air flows from the larger into the smaller bubble until the radius of the smaller one becomes equal to that of the larger one and of the larger one equal to that of the smaller one
36. Two water droplets merge with each other to form a larger droplet. In this process
    (a) Energy is liberated
    (b) Energy is absorbed
    (c) Energy is neither liberated nor absorbed
    (d) Some mass is converted into energy
37. Which graph represents the variation of surface tension with temperature over small temperature ranges for water?

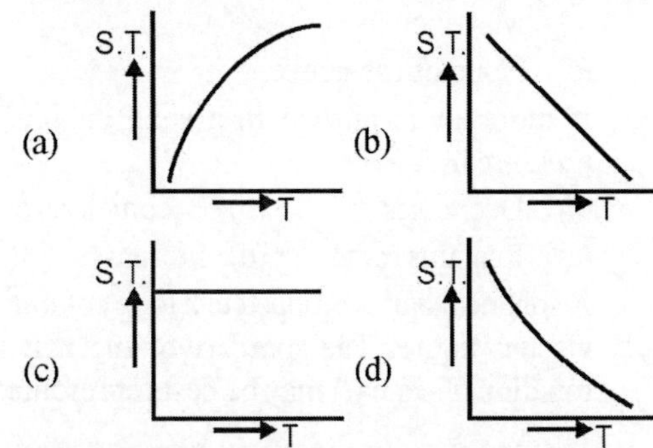

38. Due to capillary action, a liquid will rise in a tube if angle of contact is
    (a) Acute (b) Obtuse
    (c) 90° (d) Zero
39. Water rises in a capillary tube to a height h. It will rise to a height more than h
    (a) On the surface of sun
    (b) In a lift moving down with an acceleration
    (c) At the poles
    (d) In a lift moving up with an acceleration
40. Which molecule of a liquid has higher potential energy
    (a) One at the centre of gravity of the liquid
    (b) One at maximum distance from centre of gravity of liquid
    (c) One in the surface film
    (d) One at the bottom of the vessel
41. A body floats in a liquid containing in a beaker. The whole system as shown in Fig. is falling under gravity. The upthrust on the body due to liquid is

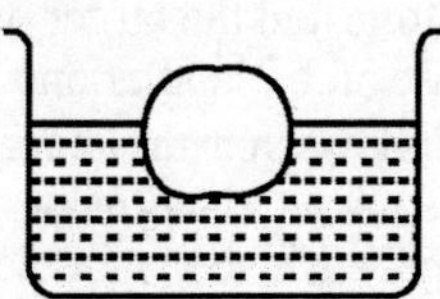

(a) zero
(b) equal to weight of body in air
(c) equal to weight of liquid displaced
(d) equal to weight of immersed part of the body

42. When a small quantity of impurity which is partially soluble, is added to water the surface tension of water
(a) Increases
(b) Decreases
(c) May increase or decrease
(d) None of the above

43. If more air is pushed in a soap bubble, the pressure in it
(a) Decreases (b) Becomes zero
(c) Remains same (d) Increases

44. A spherical ball is dropped in a long column of a viscous liquid. The speed (v) of the ball as a function of time (t) may be best represented by

(a) (b)

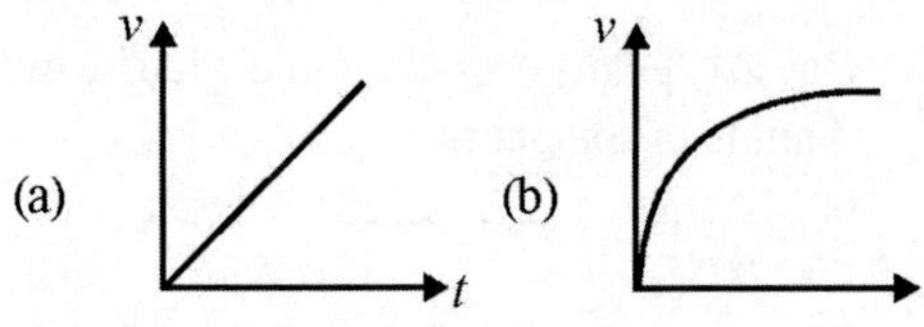

(c) (d)

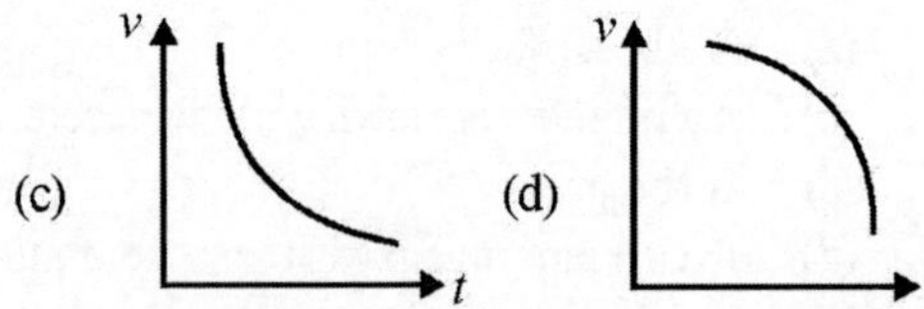

45. The adiabatic bulk modulus of a perfect gas at pressure P is given by
(a) P (b) $2P$
(c) $P/2$ (d) γP

46. At what speed, the velocity head of water is equal to pressure head of 40 cm of Hg ?
(a) 10.3 m/s (b) 2.8 m/s
(c) 5.6 m/s (d) 8.4 m/s

47. A hole is in the bottom of the tank having water. If total pressure at the bottom is 3 atm ($1 \text{ atm} = 10^5 \text{ Nm}^{-2}$), then velocity of water flowing from hole is :
(a) $\sqrt{400} \text{ ms}^{-1}$ (b) $\sqrt{600} \text{ ms}^{-1}$
(c) $\sqrt{60} \text{ ms}^{-1}$ (d) none of these

48. At which of the following temperature, the value of surface tension of water is minimum ?
(a) 4°C (b) 25°C
(c) 50°C (d) 75°C

49. The terminal velocity of a small object of radius r falling through a viscous liquid is proportional to
(a) $\frac{1}{r^2}$ (b) $\frac{1}{r}$
(c) r (d) r^2

50. Water from inside the earth rises through the trunk of a big tree to leaf high up. The main reason for this is
(a) capillary action
(b) high viscosity of water
(c) gravitational force
(d) evaporation of water

51. When a certain weight is suspended to a long uniform wire its length increases by one cm. If the same weight is suspended to another wire of the same material and length but having a diameter half of the first one, the increase in its length will be
(a) 0.5 cm (b) 2 cm
(c) 4 cm (d) 8 cm

52. On putting a capillary tube in a pot filled with water, the level of water rises up to a height of 4 cm in the tube. If a tube of half the diameter is used instead the water will rise to a height of nearly
(a) 2 cm (b) 5 cm
(c) 8 cm (d) 11 cm

53. If a small sphere is let fall vertically in a large quantity of a still liquid of density smaller than that of the material of the sphere, then
(a) at the first its velocity increases, but after sometime it approaches a constant value
(b) it falls with constant velocity all along from the very beginning
(c) at first it falls with constant velocity which after sometime goes on decreasing
(d) nothing can be said about its motion

54. A ball is falling freely in the air. After some time it achieves a terminal velocity. This is
(a) Boltzmann's law (b) Stokes' law
(c) Newton's law (d) Millikan's law

55. Terminal velocity of water drop depends upon the
(a) radius of drop
(b) charge on drop
(c) temperature of drop
(d) velocity of light in that medium

56. In the figure below is shown the flow of the liquid through a horizontal pipe. Three tubes *A*, *B* and *C* are connected to the pipe. The radii of the tubes *A*, *B* and *C* at the junction are respectively 2 cm, 1 cm and 2 cms. It can be said that

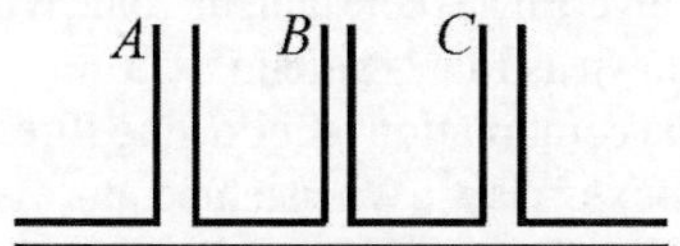

(a) the height of the liquid in the tube *A* is maximum
(b) height of the liquid in the tube *A* and tube *B* is the same
(c) height of the liquid in all the three tubes is the same
(d) height of the liquid in tube *A* and *C* is the same

57. The quantity on which the rise of liquid in a capillary tube does not depend on
(a) density of the liquid
(b) atmospheric pressure
(c) radius of the capillary tube
(d) angle of contact

58. A liquid rises in a capillary tube higher than water does. The difference may be due to the fact that
(a) diameter of the capillary may be different in the two cases
(b) temperature of liquid is higher than that of water
(c) surface tension of water is less than that of the liquid
(d) both (a) and (c)

59. A steel sphere dropping through a viscous medium attain some terminal velocity. This illustrates
(a) Bernoulli's principle
(b) Stoke's law
(c) Archimedes's principle
(d) Newton's third law

60. Figure shows a capillary rise H. If the ir is blown through the horizontal tube in the direction as shown then rise in capillary tube will be

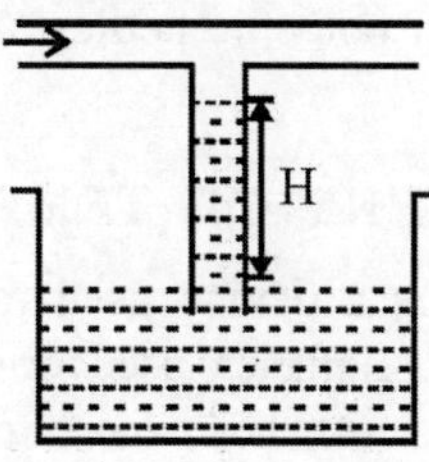

(a) =H (b) >H
(c) <H (d) zero

Thermal Properties of Matter

CONCEPTS

TEMPERATURE

The temperature indicates the degree of hotness or coldness of a body. It is a macroscopic concept. The flow of heat is always from higher temperature to lower temperature.

Thermal Equilibrium

Two bodies are said to be in thermal equilibrium with each other. When no heat flows from one body to the other, it is when both the bodies are at the same temperature.

MEASUREMENT OF TEMPERATURE

Thermometer is a device used to measure the temperature. Thermometers used for measuring very high temperatures are called **pyrometers**.

Different Types of Temperature Scales

Name of the Scale	Symbol for each degree	Lower Fixed Point (LFP)	Upper Fixed Point (UFP)	No. of division on the scales
Celsius	°C	0°C	100°C	100
Fahrenheit	°F	32°F	212°F	180
Kelvin	K	273.16 K	373.16 K	100
Reaumer	°R	0°R	80°R	80
Rankine	°Ra	460Ra	672Ra	212

Relationship between different scales of the temperature:

$$\frac{C-0}{100}=\frac{F-32}{212-32}=\frac{K-273.16}{373.16-273.16}$$

$$=\frac{R-0}{80-0}=\frac{Ra-460}{672-460}$$

T°K = (t°C + 273.16)

IDEAL-GAS EQUATION & ABSOLUTE TEMPERATURE

The equation, PV = nRT

where, n = number of moles in the sample of gas

R = Universal gas constant; its value is 8.31 J mol^{-1} K^{-1} is known as **Ideal-gas equation**

It is the combination of following three laws

(i) **Boyle's law :** When temperature is held constant, the pressure is inversely proportional to volume.

i.e., $P \propto \frac{1}{V}$ (at constant temperature)

(ii) **Charle's law :** When the pressure is held constant, the volume of the gas is directly porportional to the absolute temperature.

i.e., $V \propto T$ (at constant pressure)

(iii) **Avogadro's law :** When the pressure and temperature are kept constant, the volume is directly proportional to the number of moles of the ideal gas in the container.

i.e., $V \propto n$ (at constant pressure and temperature)

Absolute Temperature

The lowest temperature of –273.16 °C at which a gas is supposed to have zero volume and zero pressure and at which entire molecular motion stops is called absolute zero temperature. A new scale of temperature starting with –273.16°C by Lord Kelvin as Zero. This is called Kelvin scale or absolute scale of temperature.

T(K) = t°C + 273.16

THERMAL EXPANSION

Linear Expansion of Solid (Expansion of solid in one dimension)

Coefficient of linear expansion is defined as fractional increase in length per °C rise in temperature. If ℓ is the length of the rod at T K and as the temperature is changed to T + ΔT its length becomes $\ell + \Delta\ell$, so coefficient of linear expansion is given by

$$\alpha = \frac{\left(\frac{\Delta\ell}{\ell}\right)}{\Delta T} = \frac{d\ell}{\ell.dT}$$

It is positive for metals except carbon. The value of α is negative for plastic because in plastic when the temperature increases, length decreases. The numerical value of α is same in both the units i.e. in Per Kelvin or per °C

If α is coefficient of linear expansion at t_1°C.

ℓ_1 = length of the rod at t_1°C

ℓ_2 = length of the rod at t_2°C

$\ell_2 = \ell_1[1 + \alpha(t_2 - t_1)]$

t may be in any unit °C or K because in the formula there is a difference of temperature which remains same for °C or K.

Applications of Thermal Expansion in Solids

(i) Bimetallic strip : If two strips of different metals are welded together to form a bimetallic strip, when heated uniformly it bends in the form of an arc, the metal with greater coefficient of linear expansion lies on convex side.

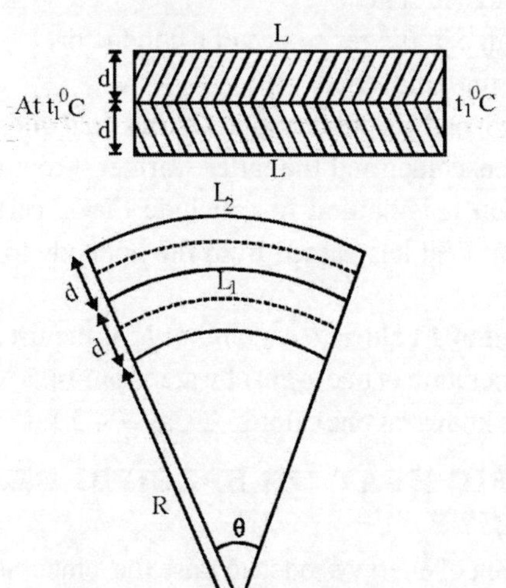

The radius of arc thus formed by bimetal is,

$$R = \frac{d}{(\alpha_2 - \alpha_1)(t_2 - t_1)} \text{ or } R = \frac{d}{(\alpha_2 - \alpha_1)\Delta t}$$

where Δt = temp. difference between the two ends.

d = thickness of each strip

α_1 and α_2 are coefficients of linear expansion of two different metals.

(ii) Thermal stress in a rigidly fixed rod :

$$\textit{Thermal strain} = \frac{\ell_2 - \ell_1}{\ell_2} = \alpha\,(t_2 - t_1)$$

$$\text{In elasticity, Young's modulus} = \frac{\text{stress}}{\text{strain}}$$

∴ Thermal stress = $Y\alpha\,(t_2 - t_1)$

Tensile strain $\simeq \alpha\Delta t$

(iii) Change of time period of pendulum clock :

$$\text{Change in time period}, \Delta T = T_2 - T_1 = \frac{1}{2}\alpha.\Delta\theta.T_1$$

The above expression represents the time lost per oscillation. Thus a pendulum clock loses time in summer and gains time in winter.

If a pendulum clock is giving correct time when time period is T then if T increases, clock becomes slow and vice-versa

Superficial Expansion of Solid

(Expansion of solid in two dimensions)

On increasing the temperature of solid, its area increases. This change in area is referred as superficial expansion of solids. If the area of solid at temperature t_1°C is A_1 and on heating the rod, the area becomes A_2 at t_2°C then $A_2 = A_1[1 + \beta(t_2 - t_1)]$

where β is coefficient of superficial expansion at t_1°C.

Coefficient of superficial expansion is defined as fractional increase in area per °C rise in temperature.

$$B = \frac{\Delta A / A}{\Delta T} = \frac{dA}{A.dT}$$

Volume Expansion of Solid

(Expansion of solid in three dimensions)

On increasing the temperature of rod, its volume changes.

If V_1 is the volume of solid at t_1°C and on increasing the temperature to t_2°C the volume becomes V_2 then,

$$V_2 = V_1\,[1 + \gamma(t_2 - t_1)]$$

where γ is coefficient of volume expansion at t_1°C.

Coefficient of volume expansion is defined as the fractional increase in volume per °C rise in temperature.

$$\gamma = \frac{\Delta V / V}{\Delta T} = \frac{dV}{V.dT}$$

Relation between coefficient of linear expansion (α), coefficient of superficial expansion (β) and coefficient of cubical expansion (γ)

$$\beta = 2\alpha \;;\; \gamma = 3\alpha;\; \beta = \frac{2}{3}\gamma$$

CHANGE IN DENSITY OF SOLID WITH TEMPERATURE

Suppose m is the mass of a solid which at a given temperature occupies a volume V so that density at 0°C is $d_0 = \frac{m}{V}$

Now if the temperature is increased by t°C, mass will remain unchanged but due to thermal expansion volume increases so that, $V' = V(1+\gamma t)$

Now density,

$$d_t = \frac{m}{V'} = \frac{m}{V(1+\gamma t)} = \frac{d_0}{(1+\gamma t)}$$

Here γ is coefficient of cubical expansion at 0°C.

THERMAL EXPANSION IN LIQUIDS

There is only volume expansion in liquids. Liquids have two coefficients of volume expansion

(i) **Coefficient of apparent expansion:** When the expansion of vessel containing the liquid is not taken into account

$$\gamma_a = \frac{(\Delta V)_a}{V_0 \times \Delta t} = \frac{\text{Apparent expansion in volume}}{\text{Initial volume } \times \Delta\theta}$$

(ii) **Coefficient of real expansion :** It is due to the actual increase in volume of liquid due to heating.

$$\gamma_r = \frac{(\Delta V)_r}{V_0 \times \Delta t} = \frac{\text{Real increase in volume}}{\text{Initial volume } \times \Delta\theta}$$

$$\gamma_{Real} = \gamma_{Apparent} + \gamma_{vessel}$$

Thermal Expansion of Water

Water shows anomalous expansion in the range from 0°C to 4°C. Water contracts on heating and expands on cooling i.e., γ is negative. At 4°C, density of water is maximum while its specific volume is minimum.

THERMAL EXPANSION OF GASES

Gases have only volume expansion as they have no definite shape. In case of gas, expansion of container is negligible in comparision to the gas hence gases have only real expansion.

$$\gamma_{gas} = \frac{\Delta V}{V_0} \times \frac{1}{\Delta t} \Rightarrow \text{ Final volume } V' = V_0(1 + \gamma\,\Delta t)$$

Pressure coefficient of gas

$$\gamma_p = \frac{P_t - P_0}{P_0 \Delta t}$$

For an ideal gas, coefficient of volume expansion is equal to coefficient of pressure expansion.

i.e., $\gamma_{gas} = \gamma_p$

HEAT

Calorimetry is the branch of thermodynamics that deals with measurement of heat.

Heat is a form of energy. It is measured in joule. It is also measured in the unit calorie

Heat is felt by its effects. Some of the effects of heat are :

(a) Change in the degree of hotness

(b) Expansion in length, surface area and volume

(c) Change in state

(d) Change in the resistance of a conductor

(e) Thermo e.m.f. effect

When a hot body is put in contact with a cold one, the former gets colder and the latter warmer. From this observation it is natural to conclude that a certain quantity of heat has passed from the hot body to the cold one.

Definition of 1 calorie : Amount of heat required to raise temperature of one gram of water from 14.5 °C to 15.5 °C is known as one calorie. 1 Cal $\simeq$ 4.2 J

SPECIFIC HEAT OR SPECIFIC HEAT CAPACITY

The amount of energy needed to raise the temperature of unit mass of that substance by 1°C (or 1K).

It is denoted by s or c.

If the temperature of a substance of mass m changes from T to T + dT when, it exchanges an amount of heat dQ with its surroundings then its specific heat is

$$c = \frac{1}{m.}\frac{dQ}{dT} \Rightarrow dQ = mcdT$$

The SI unit of specific heat is joule/kg K

Gram Specific Heat "c"

The amount of heat required to raise the temperature of the 1 gram substance by 1°C is known as "gram specific heat".

Heat required to raise the temperature of m gram substance by $\Delta\theta$°C, $Q = m\,c\,\Delta\theta$.

Unit : $\frac{\text{cal}}{\text{gm°C}}$ or $\frac{\text{cal}}{\text{gm kelvin}}$

Molar Specific Heat "C"

"Molar heat capacity C is the heat required to raise the temperature of 1 mole of a gas by 1°C (or 1 K)." Heat required to raise the temperature of μ gram-mole substance by $\Delta\theta$°C,

$$Q = \mu M_w c\,\Delta\theta$$

If $\mu = 1$ and $\Delta\theta = 1$°C

then $Q = C$, so $C = M_w c$

For a gas the value of C depends on the process through which its temperature is raised.

For example, in an isothermal process $\Delta T = 0$ or $C_{iso} = \infty$.

In an adiabatic process $\Delta Q = 0$. Hence, $C_{adi} = 0$.

Thus, molar heat capacity of a gas varies from 0 to ∞ depending on the process.

Molar Heat Capacity

The amount of energy needed to raise the temperature of one mole of a substance by 1°C (or 1K).

The molar heat capacity is the product of molecular weight and specific heat i.e.,

Molar heat capacity

C = (Molecular weight M) × (Specific heat c)

$$C = \frac{1}{\mu}.\frac{dQ}{dT}$$

where μ is the number of moles of the substance. If M is the molecular mass of the substance, then $\mu = \frac{m}{M}$

where m is the mass of the substance and, $C = \frac{M}{m}.\frac{dQ}{dT}$

Unit : SI - J/mol K.

THERMAL CAPACITY

The quantity of heat required to raise the temperature of the whole of that substance through 1°C. The thermal capacity of mass *m* of the whole of substance of specific heat *s* is equal to *ms* kilocalorie i.e.,

Thermal capacity = mass × specific heat

Water equivalent of a body : The thermal capacity of a body *ms* also represents its water equivalent, since the specific heat of water is unity.

The water equivalent of a body is the amount of water that absorbs or gives out the same amount of heat as is done by the body when heated or cooled through 1°C.

Mass of water having the same thermal capacity as the body is called the water equivalent of the body

W = mass of body × specific heat

Latent Heat or Hidden Heat

When state of a body changes, change of state takes place at constant temperature [m.pt. or b.pt.] and heat released or absorbed is given by,

$$Q = mL$$

where L is latent heat. Heat is absorbed, if solid converts into liquid (at m.pt.) or liquid converts into vapours (at b.pt.) and is released if liquid converts into solid or vapours converts into liquid.

Latent heat of fusion : The quantity of heat (in kilocalories) required to change its 1 kg mass from solid to liquid state at its melting point

For ice latent heat of fusion = 80 kilocal/kg.

Latent heat of vaporization : The quantity of heat required to change its 1 kg mass from liquid to vapour state at its boiling point. For water latent heat of vaporisation = 536 kilocal/kg.

LAW OF MIXTURES

When two bodies (one being solid and other liquid or both being liquid) at different temperatures are mixed, heat will be transferred from body at higher temperature to a body at lower temperature till both acquire same temperature. The body at higher temperature released heat while body at lower temperature absorbs it, so that :

Heat lost = Heat gained

This is also known as **laws of calorimetry.**

CHANGE OF STATE

When the exchange of heat takes place between the substance and its surrounding, then the state of substance changes

Thermal capacity : It is the amount of heat required to raise the temperature of substance through 1°C or 1K. Thermal capacity = mass × specific heat. Its unit is cal/°C, kcal/°C, J/K.

The latent heat of a substance may be defined as the quantity of heat required to change the unit mass of the substance completely from its one state to another at constant temperature.

Different processes of phase change

(i) **Sublimation :** It is the process of conversion of solids to gaseous state on heating. On cooling the vapours get converted back into solids.

(ii) **Condensaiton :** It is the process of conversion of gases/vapours to liquid state on cooling. On heating, these liquids are converted into vapours/gases.

(iii) **Boiling :** It is the process of conversion of liquid to gaseous state.

(iv) **Melting :** It is the process in which solid gets converted into liquid.

(v) Melting and boiling occur at definite temperature called melting point and boiling point respectively. The liquid boils at a temperature, at which its **vapour pressure is equal to the atmospheric pressure.**

Evaporation : Conversion of liquid into gaseous state at all temperatures is called evaporation. It is a phenomenon that occurs at the surface of the liquid.

HEAT TRANSFER

Conduction

Conduction is that mode of transmission of heat in which heat is transferred from a region of higher temperature to a region of lower temperature by the aid of particles of the body without their actual migration.

In conduction atoms or molecules near the hot end absorb energy and hence their amplitude of vibration about their mean position increases due to which they collide with other neighbouring atoms or molecules and thus transfer a part of energy, to them and so on.

In solids, heat transfer is mainly by conduction.

In metals the conduction of heat takes place by means of free electrons present in the metal.

$$\text{Temperature gradient} = \frac{\text{change in temperature}}{\text{distance}}$$

Unit: °C/metre

Thermal Conductivity

If a steady temperature difference $(\theta_1 - \theta_2)$ is to be maintained between the ends of a rod, heat must be supplied at a steady rate at one end and the same must be taken out at the other end. Suppose an amount of heat Q flows through the rod in time t then rate of heat flow or heat current $H = \frac{Q}{t}$

It has been found experimentally that in steady state.

$$H = \frac{KA(\theta_1 - \theta_2)}{\ell}$$

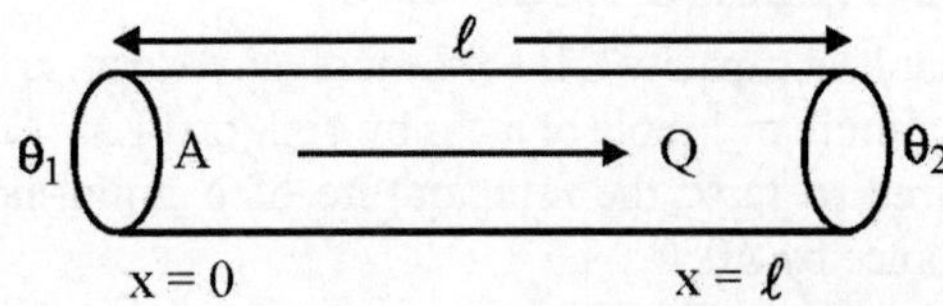

where K is a constant of proportionality called the thermal conductivity of the substance. It is a measure of how quickly heat energy can conduct (flow) through the substance.

Unit of K : CGS = cal/cm-sec-°C

MKS = kcal/m-sec-K SI = watt/m-K

Dimensions : $[M L T^{-3} \theta^{-1}]$

For a perfect conductor of heat K is infinite and for a perfect insulator K is zero.

Thermal Resistance : The thermal resistance of a body is a measure of its opposition to the flow of heat through it.

It is defined as the ratio of temperature difference to the heat current (= rate of flow of heat)

∴

$$R_{Th} = \frac{\theta_1 - \theta_2}{H} = \frac{\theta_1 - \theta_2}{(Q/t)} = \frac{\theta_1 - \theta_2}{KA(\theta_1 - \theta_2)/\ell} = \frac{\ell}{KA}$$

$$\text{or} \quad R_{Th} = \frac{\Delta\theta}{H} = \frac{\ell}{KA}$$

Unit : °C sec/cal or K sec/k-calorie.

Dimensions : $[M^{-1} L^{-2} T^3 \theta]$

Series Combination of Conductors

$$\frac{\ell_{eq.}}{K_{eq.} A_{eq.}} = \frac{\ell_1}{K_1 A_1} + \frac{\ell_2}{K_2 A_2}$$

If $A_1 = A_2 = A$

$$\therefore \frac{\ell_1 + \ell_2}{K_{eq.}} = \frac{\ell_1}{K_1} + \frac{\ell_2}{K_2} \quad \text{or} \quad K_{eq.} = \frac{\ell_1 + \ell_2}{\frac{\ell_1}{K_1} + \frac{\ell_2}{K_2}}$$

If $\ell_1 = \ell_2 = \ell$; $K_{eq.} = \dfrac{2K_1 K_2}{K_1 + K_2}$

If we have more than two slabs in the compound slab the total thermal resistance of the slabs connected in series.

$$R_{eq.} = R_1 + R_2 + R_3 + ... + R_n$$

Parallel Combination of Conductors

$$K_p = \frac{K_1 A_1 + K_2 A_2}{A_1 + A_2}$$

If $A_1 = A_2$, then $K_p = \dfrac{K_1 + K_2}{2}$

Growth of Ice on Ponds

When atmospheric temperature falls below 0°C the water in the lake will start freezing. Let at any time t, the thickness of ice in the lake be y and atmospheric temperature is $-\theta$°C. The temperature of water in contact with the lower surface of ice will be 0°C.

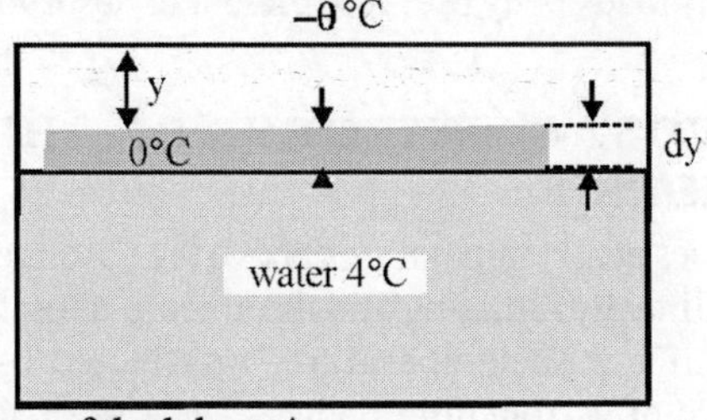

The area of the lake = A

Heat escaping through ice in time dt is

$$dQ_1 = KA\frac{[0-(-\theta)]}{y}dt = \frac{KA\theta}{y}dt$$

Now due to escaping of this heat if dy thickness of water in contact with lower surface of ice freezes,

$$dQ_2 = mL = \rho(dy\,A)L \quad [\text{as } m = \rho V = \rho A\,dy]$$

But as $dQ_1 = dQ_2$, the rate of growth of ice will be

$$\frac{dy}{dt} = \frac{K\theta}{\rho L} \times \frac{1}{y}$$

and so time taken by ice to grow a thickness y,

$$t = \frac{\rho L}{K\theta}\int_0^y y\,dy = \frac{1}{2}\frac{\rho L}{K\theta}y^2$$

Time taken to double and triple the thickness will be in the ratio $t_1 : t_2 : t_3 :: 1^2 : 2^2 : 3^2$

Convection

Convection requires a medium and is the process in which heat is transferred from one place to other by actual movement of heated substance (usually fluid). The rate of heat convection from an object is proportional to the temperature difference ($\Delta\theta$) between the object and convective fluid and the area of contact A, i.e., $\left[\dfrac{dQ}{dt}\right]_{convection} = hA\Delta\theta$ where, h represents a constant of proportionality called convection coefficient and depends on the properties of fluid such as density, viscosity, specific heat and thermal conductivity, etc.

Radiation

When a body is heated and placed in vacuum, it loses heat even when there is no medium surrounding it. The heat can not go out from the body by the process of conduction or convection since both of these process require the presence of a material medium between source and surrounding objects. The process by which heat is lost in this case is called radiation. This does not require the presence of any material medium. It is by radiation that the heat from the sun reaches the earth.

Radiation has the following properties :

(a) Radiant energy travels in straight lines and when some object is placed in the path, it's shadow is formed at the detector.

(b) It is reflected and refracted or can be made to interfere. The reflection or refraction are exactly as in case of light.

(c) It can travel through vacuum.

(d) Intensity of radiation follows the law of inverse square.

(e) Thermal radiation can be polarised in the same way as light by transmission through a nicol.

Let Q be the quantity of radiant heat incident onthe surface of the body and Q_a, Q_r and Q_t be the quantities of heat absobed, reflectd and transmitted by the body respectively.

Absorption Coefficient

$$a = \frac{Q_a}{Q} = \frac{\text{Quantity of radiant energy absorbed by the body}}{\text{Quantity of radiant energy incident on the surface of the body}}$$

Reflection Coefficient

$$r = \frac{Q_r}{Q} = \frac{\text{Quantity of radiant energy reflected}}{\text{Quantity of radiant energy incident on the surface of the body}}$$

Transmission Coefficient

$$t = \frac{Q_t}{Q}$$

$$= \frac{\text{Quantity of radiant energy transmitted by the body}}{\text{Quantity of radiant energy incident on the surface of the body}}$$

Then $Q_a + Q_r + Q_t = Q$

$$\therefore \quad \frac{Q_a}{Q} + \frac{Q_r}{Q} + \frac{Q_t}{Q} = 1 \quad \therefore a + r + t = 1$$

Emissive Power (e) : For a given surface it is defined as the radiant energy emitted per sec per unit area of the surface.

Spectral Emissive Power (e_λ) : If we consider emissive power of a surface for a particular wavelength instead of all wavelengths, it is called spectral emissive power and is represented by e_λ

Total emissive power $e = \int_0^\infty e_\lambda d\lambda$.

Emissive power of a surface depends on its nature and temperature.

It is maximum for a perfectly black body and minimum for a smooth shining body.

Unit : W/m^2

Absorptive Power (a) : Absorptive power of a surface is defined as the ratio of the radiant energy absorbed by it in a given time to the total radiant energy incident on it in the same time. If Q calories fall on the surface and Q′ calories out of it are absorbed, the absorptive power of the surface is given by $a = \frac{Q'}{Q}$

For a perfectly black body, absorptive power is maximum and unity. It has no units and dimensions.

Spectral absorptive power "a_λ" : Spectral absorptive power (a_λ) is defined as the ratio of the radiant energy of a given wavelength absorbed by a given surface in a given time to the total radiant energy of that wavelength incident in the same time on the same surface within a unit wavelength range.

Total absorptive power, $a = \int_0^\infty a_\lambda d\lambda$.

ABSOLUTE EMISSIVITY (e)

The emissivity e of a surface is defined as the quantity of heat radiated per second by one unit area of its surface when its temperature is 1°C above the surroundings.

Relative Emissivity (e) : It is defined as the ratio of radiant energy emitted per unit time per unit surface area of a body at a given temperature to the radiant energy emitted per unit time per unit surface area of a perfectly black body at the same temperature

i.e. $\quad e = \frac{E}{E_b}$

A perfectly black body : A black body is defined as one that will completely absorb all the radiation of whatever wavelength which falls on it. For perfectly black body, $a_\lambda = 1$.

Properties of a perfectly black body

(a) A perfectly black body absorbs all the radiant heat incident upon it. (i.e. a = 1)

(b) A perfectly black body does not reflect or transmit the radiant heat incident upon it. (i.e. t = 0, r = 0)

(c) The coefficient of emission of a perfectly black body is 1. It is very good emitter of heat.

(d) A perfectly black body does not exist in nature.

For practical purposes lamp black is treated as a perfectly black body as its coefficient of absorption is 0.98.

PREVOST'S THEORY OF HEAT EXCHANGE

All bodies at all temperatures above the absolute zero are continually radiating and the amount of radiation increasing with temperature, unaffected by the presence of surrounding bodies. The rise or fall of temperature which is observed in a body is due to its exchange of radiant energy with the surrounding bodies. This is known as Prevost's theory of heat exchange.

KIRCHHOFF'S LAW OF HEAT RADIATION

At any given temperature and for radiations of the same wavelength the ratio of the emissive power (e_λ) to the absorptive power (a_λ) is the same for all substances and is equal to the emissive power of a perfectly black body (E_λ).

i.e., $\quad \frac{e_\lambda}{a_\lambda} = \text{const.} = E_\lambda$

STEFAN BOLTZMANN LAW OF RADIATION

A perfectly black body is the best possible emitter of radiation at a given temperature. Stefan's Law gives the total amount of radiation emitted by a perfectly black body.

According to Stefan's law, the quantity of energy radiated in one second per unit area of a perfectly black body is directly proportional to the fourth power of its absolute temperature.

$$E_b \propto T^4 \quad \text{or} \quad E_b = \sigma T^4$$

where σ is a universal constant known as Stefan-Boltzmann constant and T is its temperature on absolute scale. The measured value of σ is

$\sigma = 5.67 \times 10^{-8}\ W/m^2\ K^4$ or $\sigma = 5.67 \times 10^{-5}$ erg/ $cm^2 sK^4$.

Eq^n. $E_b = \sigma T^4$ is called the Stefan-Boltzmann law. Stefan has suggested this law from experimental data available on radiation and Boltzmann derived it from thermodynamical considerations. The law is also quoted as Stefan's law and the constant σ as Stefan constant.

$$\therefore \frac{Q_b}{A.t} = \sigma T^4$$

Q_b = total energy radiated by the surface of a black body in time t.

or $Q_b = \sigma A t T^4$

The emissive power E of a non-black body is given by :

$$E = e E_b \quad \text{or} \quad E = e \sigma T^4$$

$\therefore$ Total energy radiated by the surface of a body in time t

$$Q = e \sigma A t T^4$$

NEWTON'S LAW OF COOLING

The rate of cooling of a body (rate of loss of heat) is directly proportional to the excess of temperature of the body over the surroundings, provided that this excess is small and loss of heat by radiation only.

If θ and θ_0 are the temperatures of the body and its surroundings respectively,

Rate of loss of heat $R = \left(-\frac{dQ}{dt}\right)$

The negative sign indicates that the amount of heat is decreasing with time.

$$-\frac{dQ}{dt} \propto (\theta - \theta_0) \quad \text{or} \quad -\frac{dQ}{dt} = k(\theta - \theta_0)$$

where k is the constant of proportionality.

Cooling Curves

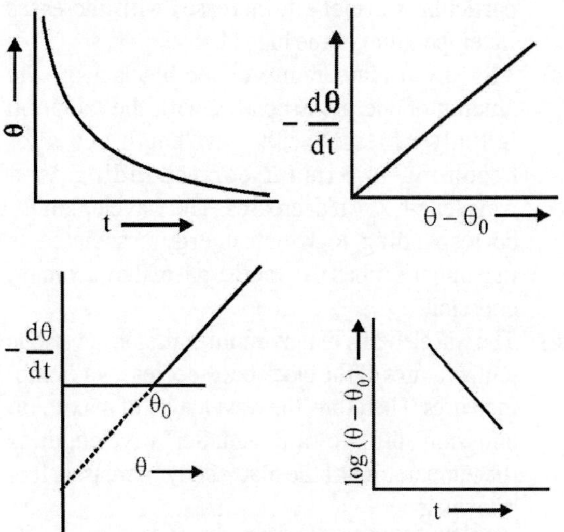

DISTRIBUTION OF ENERGY IN A BLACK BODY RADIATION

Figure shows how the energy of a black body radiation varies with temperature and wavelength.

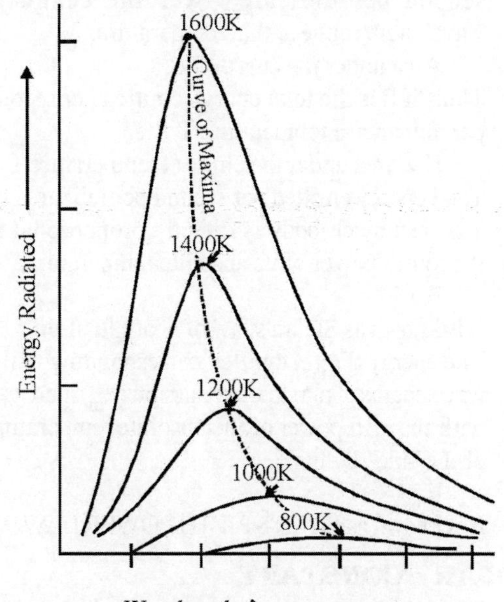

Each curve represents the variation of monochromatic emittance E_λ with wavelength λ for a certain temperature.

The following conclusions can be drawn from these curves :

(a) At each temperature, the black body emits continuous heat radiation spectrum.

(b) The energy associated with the radiation of a particular wavelength increases with increases in temperature of the black body.

(c) At a given temperature of the black body, the amount of energy associated with the radiation initially increases with wavelength and after becoming maximum corresponding to a wavelength λ_m, it decreases. The wavelength λ_m corresponding to which energy associated is maximum is called the wavelength of the maximum emission.

(d) The wavelengths of maximum emission at various temperatures of the black body decreases as temp. increases. Therefore, the wavelength of maximum emission shifts towards smaller wavelength as the temperature of the black body increases. It is known as
WIEN'S DISPLACEMENT LAW. $\lambda_m \times T = b$; b is known as Wien's constant
Numerical value of $b = 2.93 \times 10^{-3}$ mK

(e) The area under each curve represents the total energy emitted by the perfect black body per second per unit area over the complete wavelength range at that temperature.
Area under the curve $\propto T^4$
Thus, if E is the total energy emitted per second per unit area at temperature T, then
$E \propto$ area under the curve at temperature T.
Total energy emitted per second per unit area by a perfect black body is directly proportional to the fourth power of its absolute temperature.
$E \propto T^4$
It is known as Stefan's law of heat radiation.

(f) The energy (E_{max}) emitted corresponding to the wavelengths of maximum emission (λ_m) increases with the fifth power of the absolute temperature of the black body i.e.
$E_{max} \propto T^5$
It is known as WIEN'S FIFTH POWER LAW.

SOLAR CONSTANT

The sun emits radiant energy continuously in space of which an insignificant part reaches the earth. The solar radiant energy received per unit area per unit time by a black surface held at right angles to the sun's rays and placed at the mean distance of the earth (in the absence of atmosphere) is called Solar Constant. The solar constant S is taken to be 1400 watt/m^2.

1. At the following temperatures different temperature scales have the same reading
(i) Fahrenheit and Reaumur at –25.6°F = –25.6°R
(ii) Reaumur and Celsius at 0°R = 0°C
(iii) Celsius and Fahrenheit at –40°C = –40°F
(iv) Fahrenheit and Kelvin at 574.25°F = 574.25K
The Celsius scale and the Kelvin cannot have the same reading at any temperature.

2. In a compensated pendulum, the change in the length of the two rods should be same. This gives $L_1\alpha_1 = L_2\alpha_2$ where L_1, L_2 are their lengths and α_1 and α_2 are their linear expansivity respectively.

3. During change in temperature, the mass does not change but the density decreases due to increase in volume therefore

$$\rho = \frac{\rho_0}{1+\gamma\Delta T}$$

4. If equal mass of ice at 0°C and steam at 100°C are mixed, the resulting temperature will be 100°C.

5. If equal mass of ice at 0°C is mixed with water at or below 80°C, the resulting temperature will be 0°C.

6. If x g of ice at 0°C is mixed with x g of water above 80°C but below 100°C, the resulting temperature will always be less than 10°C. If θ is the temperature of water, (between 80°C and 100°C) then final temperature of the mixture will be

$$\frac{\theta - 80}{2x} \text{ in °C.}$$

7. If x g of ice at 0°C melts when y g of dry steam at 100°C is added to it, then x = 8y.

8. An isosceles triangle is formed with a rod of length ℓ_1 and coefficient of linear expansion α_1 for the base and two thin rods each of length ℓ_2 and coefficient of linear expansion α_2 for the two pieces, if the distance between the apex and the mid point of the base remains unchanged as the temperatures varied if $\frac{\ell_1}{\ell_2} = 2\sqrt{\frac{\alpha_2}{\alpha_1}}$.

Exercise

1. The coefficient of thermal expansion of water at 4°C is :
   (a) zero (b) maximum
   (c) minimum (d) infinity
2. The coefficient of thermal conductivity depends upon:
   (a) temperature difference between the two surfaces
   (b) area of the plate
   (c) material of the plate
   (d) all of these
3. Woolen clothes are used in winter season because woolen clothes :
   (a) are good sources for producing heat
   (b) absorb heat from surrounding
   (c) are bad conductors of heat
   (d) none of these
4. Infrared radiation is detected by
   (a) spectrometer (b) pyrometer
   (c) nanometer (d) photometer
5. Which of the following qualities are useful for a cooking utensil?
   (a) High specific heat and low thermal conductivity
   (b) High specific heat and high thermal conductivity
   (c) Low specific heat and low thermal conductivity
   (d) Low specific heat and high thermal conductivity
6. The wavelength of radiation emitted by a body depends upon:
   (a) the nature of its surface
   (b) the area of its surface
   (c) the temperature of its surface
   (d) all of the above
7. Consider the following statements :
   (A) Some bodies may contract on heating
   (B) Water shows anamolous expansion
   (a) (A) is true and (B) is false
   (b) (A) is false and (B) is true
   (c) Both (A) and (B) are true
   (d) Both (A) and (B) are false
8. The rate of flow of heat through a rod depends on:
   (a) thermal conductivity of the rod
   (b) length of the rod
   (c) temperature difference across the rod
   (d) all of the above
9. Which of the following represent convection?
   (a) Land breeze
   (b) Exhaust fan
   (c) Heating water by keeping the vessel on a flame
   (d) All of these
10. A vessel completely filled with a liquid is heated. If α & γ represent coefficient of linear expansion of material of vessel and coefficient of cubical expansion of liquid respectively, then the liquid will not overflow if:
    (a) $\gamma = 3\alpha$ (b) $\gamma > 3\alpha$
    (c) $\gamma < 3\alpha$ (d) $\gamma \leq 3\alpha$
11. The volume coefficient of a gas :
    (a) is constant for a given gas
    (b) always depends on temperature of the gas
    (c) may depend on temperature of the gas
    (d) does not depend on the type of process
12. Some ice is added to a glass of water, then
    (a) the equilibrium temperature must be positive
    (b) the equilibrium temperature may be negative
    (c) the equilibrium temperature may be zero
    (d) both (b) and (c)
13. Triple point is the temperature at which :
    (a) matter may simultaneously exist in liquid and gaseous state
    (b) matter may simultaneously exist in liquid and solid state
    (c) matter may simultaneously exist in solid and gaseous state
    (d) matter may simultaneously exist in all the three forms
14. The fastest mode of transfer of heat is :
    (a) conduction (b) convection
    (c) radiation (d) none of these

15. A perfectly black body when heated to extremely high temperature, it appears:
(a) black (b) red
(c) white (d) blue
16. At which temperature, the centrigrade and Fahrenheit scales are equal?
(a) 40° (b) – 40°
(c) 37° (d) – 80°
17. In order that the heat flows from one part of a solid to another part, what is required?
(a) Uniform density
(b) Temperature gradient
(c) Density gradient
(d) Uniform temperature
18. At a common temperature, a block of wood and a block of metal feel equally cold or hot. The temperatures of block and wood are
(a) equal to the temperature of the body
(b) less than the temperature of the body
(c) greater than temperature of the body
(d) either (b) or (c)
19. The bulb of one thermometer is spherical, while that of other is cylindrical. If both of them have equal amount of mercury, which one will respond quickly to the temperature?
(a) spherical (b) cylindrical
(c) elliptical (d) both (a) and (c)
20. Calorimeters are made of which of the following?
(a) Glass (b) Metal
(c) Wood (d) Either (a) or (c)
21. The sprinkling of water slightly reduces the temperature of a closed room because
(a) temperature of water is less than that of the room
(b) specific heat of water is high
(c) water has large latent heat of vaporisation
(d) water is a bad conductor of heat
22. Which of the following is not close to a black body?
(a) Black board paint (b) Green leaves
(c) Black holes (d) Red roses
23. According to Newton's law of cooling, the rate of cooling of a body is proportional to $(\Delta\theta)^n$, where $\Delta\theta$ is the difference of the temperature of the body and the surroundings, and n is equal to
(a) two (b) three
(c) four (d) one
24. N molecules each of mass m of a gas A and 2N molecules each of mass 2m of gas B are contained in the same vessel which is maintained at temperature T. The mean square velocity of molecules of B type is v^2 and the mean square rectangular component of the velocity of A type is denoted by ω^2. Then ω^2/v^2
(a) 2 (b) 1
(c) 1/3 (d) 2/3
25. 10 gm of ice cubes at 0°C are released in a tumbler (water equivalent 55 g) at 40°C. Assuming that negligible heat is taken from the surroundings, the temperature of water in the tumbler becomes nearly (L = 80 cal/g)
(a) 31°C (b) 22°C
(c) 19°C (d) 15°C
26. A centigrade and a Fahrenheit thermometer are dipped in boiling water. The water temperature is lowered until the Fahrenheit thermometer registers 140°. What is the fall in temperature as registered by the Centigrade thermometer?
(a) 80° (b) 60°
(c) 40° (d) 30°
27. Thermal capacity of 40 g of aluminium (s = 0.2 cal/g K) is
(a) 168 joule /°C (b) 672 joule/°C
(c) 840 joule/°C (d) 33.6 joule/°C
28. Mercury thermometer can be used to measure temperature upto
(a) 260°C (b) 100°C
(c) 357°C (d) 500°C
29. If the temperature of the sun is doubled, the rate of energy received on earth will be increased by a factor of
(a) 2 (b) 4
(c) 8 (d) 16
30. A body cools from 50.0°C to 48°C in 5s. How long will it take to cool from 40.0°C to 39°C? Assume the temperature of surroundings to be 30.0°C and Newton's law of cooling to be valid.
(a) 2.5 s (b) 10 s
(c) 20 s (d) 5 s
31. A beaker full of hot water is kept in a room. If it cools from 80°C to 75°C in t_1 minutes, from 75° C to 70°C in t_2 minutes and from 70°C to 65°C in t_3 minutes, then
(a) $t_1 = t_2 = t_3$ (b) $t_1 < t_2 = t_3$
(c) $t_1 < t_2 < t_3$ (d) $t_1 > t_2 > t_3$
32. A black body is at temperature of 500 K. It emits energy at rate which is proportional to
(a) $(500)^4$ (b) $(500)^3$
(c) $(500)^2$ (d) 500

33. The radiant energy from the sun, incident normally at the surface of earth is 20 k cal/m^2 min. What would have been the radiant energy, incident normally on the earth, if the sun had a temperature, twice of the present one?
(a) 160 k cal/m^2 min (b) 40 k cal/m^2 min
(c) 320 k cal/m^2 min (d) 80 k cal/m^2 min

34. If 1 g of steam is mixed with 1 g of ice, then the resultant temperature of the mixture is
(a) 270°C (b) 230°C
(c) 100°C (d) 50°C

35. The coefficient of linear expansions of brass and steel are α_1 and α_2 respectively. When we take a brass rod of length ℓ_1 and a steel rod of length ℓ_2 at 0°C, then the difference in their lengths $(\ell_2 - \ell_1)$ will remain the same at all temperatures if
(a) $\alpha_1 \ell_1 = \alpha_2 \ell_2$ (b) $\alpha_1 \ell_2 = \alpha_2 \ell_1$
(c) $\alpha_1^2 \ell_2 = \alpha_2^2 \ell_1$ (d) $\alpha_1 \ell_2^2 = \alpha_2 \ell_1^2$

36. The presence of gravitational field is required for the heat transfer by
(a) conduction (b) stirring of liquids
(c) natural convection (d) radiation

37. A black body has maximum wavelength λ_m at temperature 2000 K. Its corresponding wavelength at temperature 3000 K will be
(a) $\frac{3}{2}\lambda_m$ (b) $\frac{2}{3}\lambda_m$
(c) $\frac{4}{9}\lambda_m$ (d) $\frac{9}{4}\lambda_m$

38. A cylindrical rod having temperature T_1 and T_2 at its end. The rate of flow of heat is Q_1 cal/sec. If all the linear dimensions are doubled keeping temperature constant, then the rate of flow of heat Q_2 will be
(a) $4Q_1$ (b) $2Q_1$
(c) $Q_1/4$ (d) $Q_1/2$

39. Wien's law is concerned with
(a) relation between emissivity and absorptivity of a radiating surface
(b) total radiation, emitted by a hot surface
(c) an expression for spectral distribution of energy of a radiation from any source
(d) a relation between the temperature of a black body and the wavelength at which there is maximum radiant energy per unit wavelength

40. Radiation from which of the following sources, approximates black body radiation best?
(a) A tungsten lamp
(b) Sodium flame
(c) Hot lamp black
(d) A hole in a cavity, maintained at constant temperature

41. Two rods of thermal conductivities K_1 and K_2, cross-sections A_1 and A_2 and specific heats S_1 and S_2 are of equal lengths. The temperatures of two ends of each rod are T_1 and T_2. The rate of flow of heat at the steady state will be equal if
(a) $\frac{K_1}{A_1S_1} = \frac{K_2}{A_2S_2}$ (b) $K_1A_1 = K_2A_2$
(c) $K_1S_1 = K_2S_2$ (d) $A_1S_1 = A_2S_2$

42. Consider a compound slab consisting of two different materials having equal thicknesses and thermal conductivities K and $2K$, respectively. The equivalent thermal conductivity of the slab is
(a) $\frac{4}{3}K$ (b) $\frac{2}{3}K$
(c) $\sqrt{3}\,K$ (d) $3\,K$

43. If λ_m denotes the wavelength at which the radiative emission from a black body at a temperature T K is maximum, then
(a) $\lambda_m \propto T^{-1}$
(b) $\lambda_m \propto T^4$
(c) λ_m is independent of T
(d) $\lambda_m \propto T$

44. Which of the following circular rods (given radius r and length l), each made of the same material and whose ends are maintained at the same temperature will conduct most heat?
(a) $r = r_0; l = l_0$ (b) $r = 2r_0; l = l_0$
(c) $r = r_0; l = 2l_0$ (d) $r = 2r_0; l = 2l_0$

45. The temperature of inversion of a thermocouple is 620°C and the neutral temperature is 300°C. What is the temperature of cold junction?
(a) 320°C (b) 20°C
(c) –20°C (d) 40°C

46. A black body at 1227°C emits radiations with maximum intensity at a wavelength of 5000Å. If the temperature of the body is increased by 1000°C, the maximum intensity will be observed at

(a) 5000Å (b) 6000Å
(c) 3000Å (d) 4000Å

47. A black body is at 727° C. It emits energy at a rate which is proportional to

(a) $(1000)^4$ (b) $(1000)^2$
(c) $(727)^4$ (d) $(727)^2$

48. If the cold junction of a thermo-couple is kept at 0°C and the hot junction is kept at T°C then the relation between neutral temperature (T_n) and temperature of inversion (T_i) is

(a) $T_n = 2T_i$ (b) $T_n = T_i - T$
(c) $T_n = T_i + T$ (d) $T_n = T_i/2$

49. The filament of an evacuated light bulb has a length 10 cm, diameter 0.2 mm and emissivity 0.2, calculate the power it radiates at 2000 K. ($\sigma = 5.67 \times 10^{-8}$ W/m^{2}K^4)

(a) 21.5 W (b) 15.5 W
(c) 8.9 W (d) 11.4 W

50. The energy radiated from a black body at a temperature of 727°C is E. By what factor the radiated energy shall increase if the temperature is raised to 2227°C?

(a) 39 (b) 49
(c) 29 (d) 19

51. What temperature is the same on celsius scale as well as on Fahrenheit scale?

(a) −212°C (b) −40°C
(c) −32°C (d) 32°C

52. The temperature of a piece of metal is increased from 27°C to 84°C. The rate of radiated energy emitted from the metal will increase approximately:

(a) 16 times (b) 8 times
(c) 4 times (d) 2 times

53. The length of a metallic rod is 5 m at 0°C and becomes 5.01 m, on heating upto 100°C. The linear expansion of the metal will be :

(a) 2.33×10^{-5}/°C (b) 6.0×10^{-5}/°C
(c) 4.0×10^{-5}/°C (d) 2.0×10^{-5}/°C

54. The temperature of two bodies A and B are respecitvely 727°C and 327°C the ratio of the rates of heat radiated ($H_A : H_B$) by them, is :

(a) 625 : 81 (b) 25 : 9
(c) 5 : 3 (d) 727 : 372

55. A body initially at 80°C cools to 64°C in 5 minutes and to 52°C in 10 minutes. The temperature of the body after 15 minutes will be :

(a) 42.7°C (b) 35°C
(c) 47°C (d) 40°C

56. In a room where the temperature is 30°C, a body cools from 61°C to 59°C in 4 minutes. The time (in minutes) taken by the body to cool from 51°C to 49°C will be :

(a) 8 (b) 5
(c) 6 (d) 4

57. Solar radiation emitted by sun resembles that emitted by a black body at a temperature of 6000 K. Maximum intensity is emitted at a wavelength of about 4800 Å. If the sun were to cool down from 6000 K to 3000K, then the peak intensity would occur at a wavelength

(a) 4800 Å (b) 9600 Å
(c) 7200 Å (d) 6400 Å

58. There rods of the same dimensions have thermal conductivities 3K, 2K and K. They are arranged as shown in Fig. with their ends at 100°C, 50°C and 20°C. The temperature of their junction is

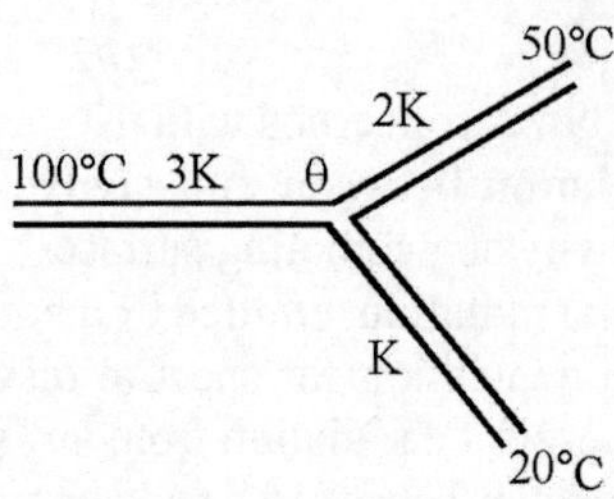

(a) 60° (b) 70°
(c) 50° (d) 35°

59. The earth radiates in the infrared region of the spectrum. The spectrum is correctly given by
(a) Rayleigh Jeans law
(b) Planck's law of radiation
(c) Stefan's law of radiation
(d) Wien's law

60. Two similar thermocouples made up of similar metals A and B are connected through a key K and sensitive galvanometer G as shown in the figure. One of the thermocouple is dipped in a hot bath maintained at temperature t_2 and other one in a cold bath at a temperature t_1. When the key K is pressed a deflection is seen in the galvanometer because :

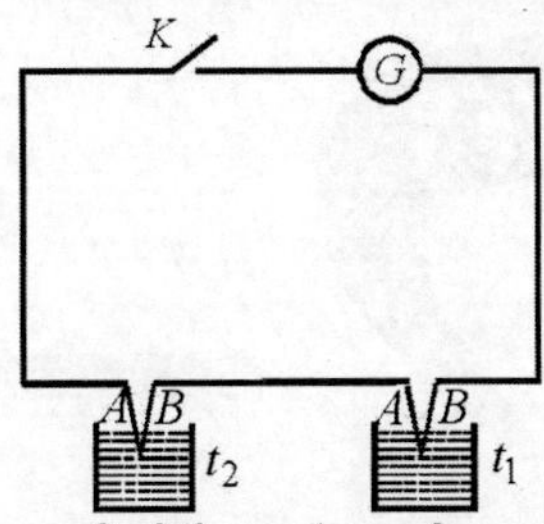

(a) an emf of the order a few of microvolt is generated which is proportional to $(t_2 - t_1)$
(b) an emf is generated the value of which will depend upon the temperature of the hot bath only
(c) an emf of about 1 volt is generated which will be proportional to $(t_2 - t_1)$
(d) an emf of a few microvolt is generated which will be proportional to t_2 only

CHAPTER 11

Thermodynamics

CONCEPTS

Thermodynamics originated as the study of the relationship between mechanical and thermal energy-the energy associated with the disordered motions of the atoms and molecules within a substance.

THERMODYNAMIC STATE

(a) The state of a system can be described completely by (i) composition (ii) temperature (iii) volume and (iv) pressure. These parameters are called the **thermodynamic parameters** or **variables of the system**.

(b) If a system is homogeneous and has definite mass and composition, then the state of the system can be described by the remaining three variables namely temperature, pressure and volume.

(c) These three variables are interrelated by equation $PV = \mu RT$ (Ideal gas equation)

(d) Thus, the thermodynamic state of the system is its condition as identified by **two independent thermodynamic** co-ordinates.

THERMAL EQUILIBRIUM

It refers to the situation when macroscopic variables describing the thermodynamic state of a system do not depend on time. Equilibrium of a system in mechanics means the net external force and torque on the system are zero. In a state of thermodynamic equilibrium, the microscopic constituents of a system are not in equilibrium (in the sense of mechanics). Equilibrium states of a thermodynamic system are described by state variables. Examples of state variables are pressure (P), volume (V), temperature (T), and mass (m). Heat and work are not state variables. An equation of state (like the ideal gas equation $PV = nRT$) is a relation connecting different state variables.

Mechanical equivalent of heat : According to Joule, work may be converted into heat and vice-versa. The ratio of work done to heat produced is always constant

$$\text{i.e., } \frac{W}{H} = J \text{ (constant) or } W = JH$$

where J is called mechanical equivalent of heat.

Its value is 4.2 joule/cal or 4.2×10^3 joule per kilo-cal. It is not a physical quantity but simply a conversion factor. It converts unit of work into that of heat and vice-versa.

In eq^n. W = JH, W must be in joule, irrespective of nature of energy or work and H must be in calorie or kilo-calorie.

Adiabatic wall : An insulating wall (can be movable) that does not allow flow of energy (heat) from one side to another.

Diathermic wall : A conducting wall that allows energy flow (heat) from one side to another.

Cyclic process : Cyclic process is that thermodynamic process in which the system returns to its initial state after undergoing a series of changes.

Non-cyclic process : Non-cyclic process is that process in which the system does not return to its initial state.

Quasi-static or equilibrium process : Quasi-static is a thermodynamic process which proceeds extremely slowly such that at every instant of time, the temperature and pressure are the same in all parts of the system.

HEAT, INTERNAL ENERGY AND WORK

Heat

It is energy in transit between two objects or system due to temperature difference between them. It exists in the form of translational, rotational and vibrational motion of molecules of a substance. It depends on process.

For constant pressure process $Q = nC_P \Delta T$

And for constant volume process $Q = nC_v \Delta T$

For any other process $Q = nC \Delta T$, where C is called molar specific heat for that process.

Internal Energy

Internal energy of a system is the sum of kinetic energies and potential energies of the molecular constituents of the system. It does not include the over-all kinetic energy of the system. If C_v is molar heat capacity ($C_v = Mc_v$, M being molecular weight), then for μ-moles of ideal gas

$$dU = \mu C_v dT = \frac{m}{M} C_v dT$$

It is obvious that internal energy in the absence of inter-molecular forces is simply the function of temperature and state only, it is independent of path followed.

$dU = U_f - U_i$

where U_i and U_f are internal energies in initial and final states.

Heat and work are two modes of energy transfer to the system. Heat is the energy transfer arising due to temperature difference between the system and the surroundings.

Work

Work is energy transfer brought about by other means, such as moving the piston of a cylinder containing the gas, by raising or lowering some weight connected to it etc.

ZEROTH LAW OF THERMODYNAMICS

It states that 'two systems are in thermal equilibrium with a third system separately are in thermal equilibrium with each other'. The zeroth law leads to the concept of temperature.

Temperature of a body is related to its average internal energy, not to the kinetic energy of motion of its centre of mass. A bullet fired from a gun is not at a higher temperature because of its high speed.

FIRST LAW OF THERMODYNAMICS

It is the general law of conservation of energy applied to any system in which energy transfer from or to the surroundings (through heat and work) is taken into account. It states that

$\Delta Q = \Delta U + \Delta W$,

where ΔQ is the heat supplied to the system, ΔW is the work done by the system and ΔU is the change in internal energy of the system.

If $Q > 0$, heat is added to the system, if $Q < 0$, heat is removed to the system, if $W > 0$, Work is done by the system, if $W < 0$, Work is done on the system.

Limitations of First Law of Thermodynamics :

(a) It does not explain the direction of heat flow.

(b) It does not explain how much amount of heat given will be converted into work.

Specific Heat Capacity

It is the amount of heat required to raise the temperature of unit mass of substance through 1 degree. It may be expressed in cal/g°C, kcal/kg°C, specific heat capacity, J/kgK.

$$c = \frac{Q}{m \times \Delta t}$$

For gases : During the **isothermal** process

$\Delta t = 0 \Rightarrow c = \infty$ (infinite) during the **adiabatic** process

$Q = 0 \Rightarrow c = 0$ (zero)

Specific heat of gas : The specific heat of a gas varies from zero to infinity. It may have any positive or negative value.

Following two specific heats are more significant.

(a) **Molar specific heat at constant volume** (C_V). It is the amount of heat required to raise the temperature of 1 mole of a gas through 1K or 1°C at constant volume.

$$C_V = \frac{1}{n}\left(\frac{dQ}{dT}\right)$$

(b) **Molar specific heat at constant pressure** (C_P) It is the amount of heat required to raise the temperature of 1 mole of a gas through 1K or 1°C at constant pressure.

$$C_P = \frac{1}{n}\left(\frac{dQ}{dT}\right)$$

Mayer's relation : $C_P - C_V = R$

where R is a gas constant. ($R = 8.314\ \text{Jmol}^{-1}\text{K}^{-1}$)

THERMODYNAMIC PROCESS

Thermodynamic process is said to take place if some change occurs in the state of a thermodynamic system, i.e. the thermodynamic variables of the system pressure, volume, temperature and entropy change with time.

Sign Conventions for the study of thermodynamic processes

The following sign conventions are adopted in the study of thermodynamical process :

(i) Heat gained by a system is taken as positive while that lost by a system is taken as negative.

(ii) The work done by a system is taken as positive while that done on the system is taken as negative.

(iii) Increase in the internal energy of a system is taken as positive. Decrease in the internal energy of system is taken as negative.

Some Thermodynamic Processes

(i) **Isothermal process :** If a thermodynamic system is perfectly conducting to surroundings and undergoes a physical change in such a way that temperature remains constant throughout, then process is said to be isothermal process.

For isothermal process, the equation of state is PV = nRT = constant, where *n* is no. of moles.

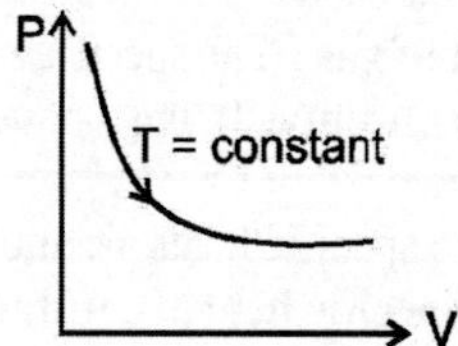

For ideal gas, since internal energy depends only on temperature

$$dU = 0 \Rightarrow dQ = dW = \int_{V_1}^{V_2} PdV = nRT\int_{V_1}^{V_2}\frac{dV}{V}$$

$$\text{or } dQ = nRT\log_e\frac{V_2}{V_1} = 2.303nRT\log_{10}\frac{V_2}{V_1}$$

(ii) **Adiabatic process :** If system is completely isolated from the surroundings so that no heat flows in or out of it, then any change that the system undergoes is called an adiabatic process.

For ideal gas, dQ = 0

$dU = \mu C_V dT$ (for any process)

$$dW = \int_{V_1}^{V_2} PdV = \int_{V_1}^{V_2}\frac{K}{V^\gamma}dV$$

(where $PV^\gamma = K$ = constant)

$$= \frac{K}{1-\gamma}\left(\frac{1}{V_2^{\gamma-1}} - \frac{1}{V_1^{\gamma-1}}\right) = \frac{(P_2V_2 - P_1V_1)}{1-\gamma}$$

where PV^γ = constant is applicable only in adiabatic process. Adiabatic process is called isoentropic process (in these process entropy is constant).

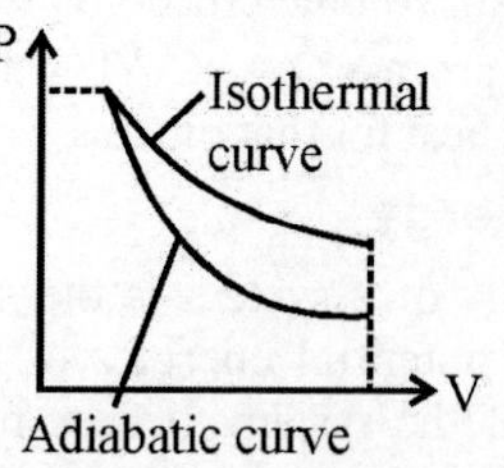

The slope of adiabatic curve is more than isthermal curve.

(iii) **Isobaric process :** A process taking place at constant pressure is called an isobaric process. In this process

$dQ = n C_p dT$, $dU = n C_V dT$, $dW = P(V_2 - V_1)$

(iv) **Isochoric process :** A process taking place at constant volume is called isochoric process. In this process, $dQ = dU = n C_V dT$, $dW = 0$

(v) **Cyclic process :** In this process the inital state and final state after traversing a cycle (shown in fig.) are same.

In cyclic process, $dU = 0 = U_f - U_i$,

dW = area of cycle = area (abcd)

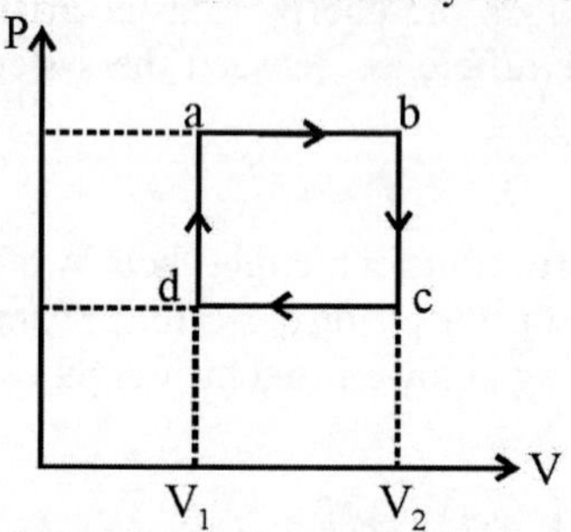

Equations of state for adiabatic processes

(i) PV^γ = constant

(ii) $TV^{\gamma-1}$ = constant

(iii) $\frac{P^{\gamma-1}}{T^\gamma}$ = Constant where $\gamma = C_p/C_v$

Indicator Diagram

The indicator diagram helps us to calculate the amount of work done by the gas or on the gas during expansion or compression.

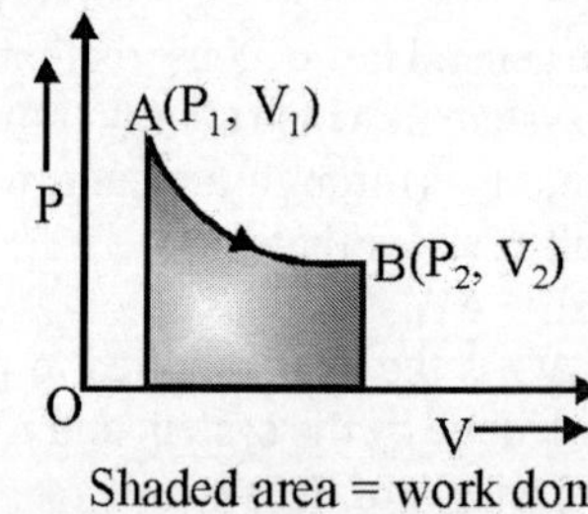

Shaded area = work done

Work done by thermodynamic system : Work is energy that is transmitted from one system to another by a force moving its points of application. The expression of work done on a gas or by a gas is :

$$W = \int dW = \int_{V_1}^{V_2} P dV$$

where V_1 is volume of gas in initial state and V_2 in final state.

(i) It is also macroscopic and path dependent function.

(ii) By sign convention it is +ive if system does work (i.e., expands against surrounding) and it is –ive, if work is done on system (i.e., contracts).

(iii) In cyclic process the work done is equal to area under the cycle and is negative if cycle is anticlockwise and +ive if cycle is clockwise (shown in fig.(a) and (b)).

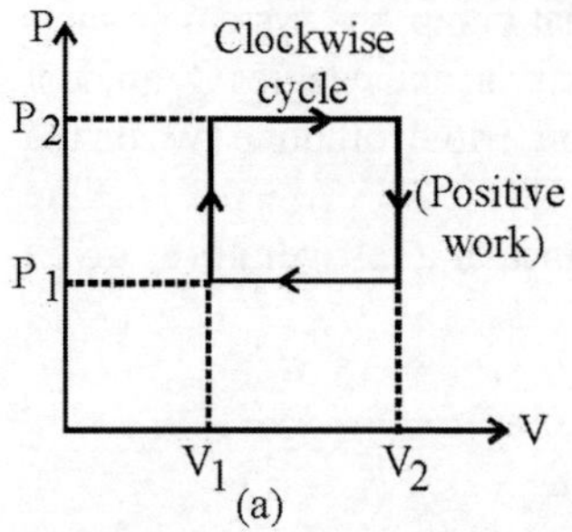

(a)

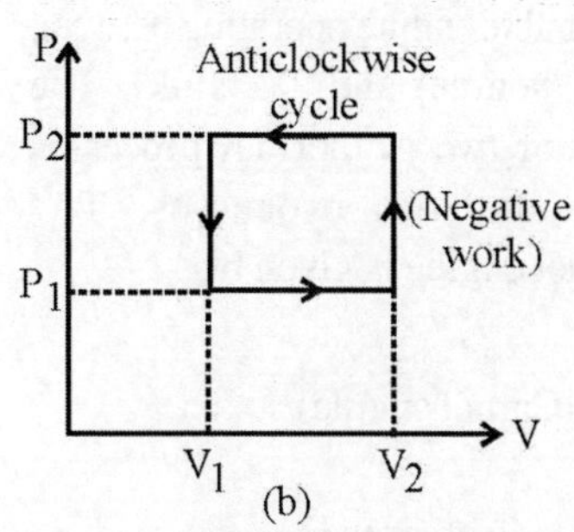

(b)

Following different cases are possible.

When volume is constant : V = constant , $W_{AB} = 0$

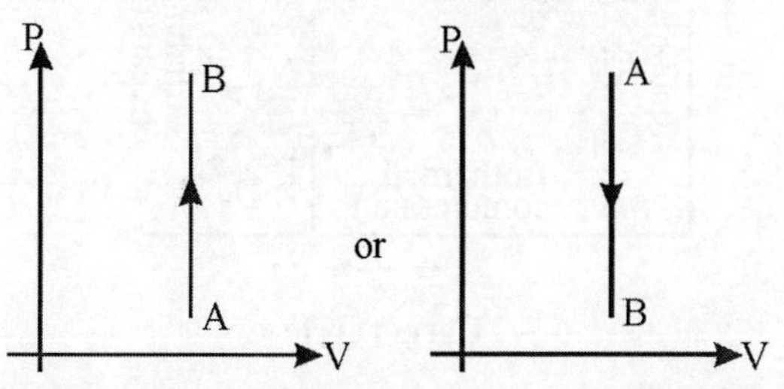

When volume is increasing (expansion) : V is increasing, $W_{AB} > 0$, W_{AB} = + Shaded area

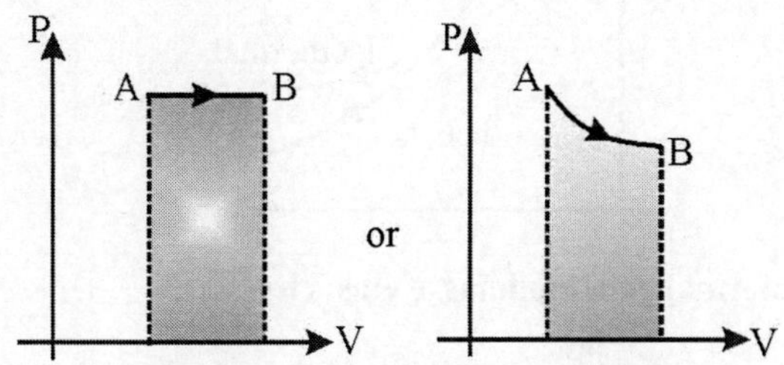

When volume is decreasing (compression)

V is decreasing, $W_{AB} < 0$, W_{AB} = – Shaded area

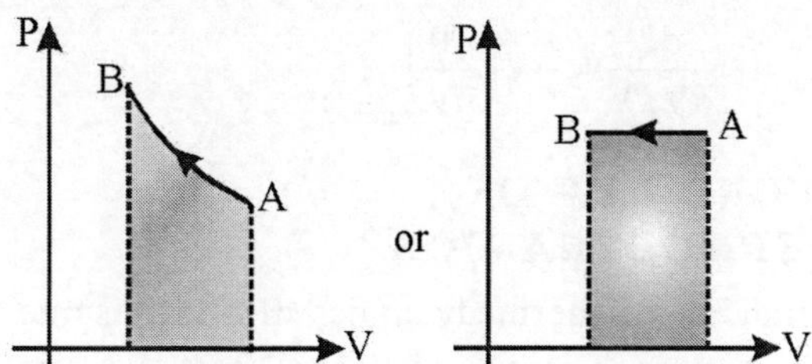

Slope of the PV curve $\left(\frac{dP}{dV}\right)$**:**

Isobaric process : $\frac{dP}{dV} = 0$

Isochoric or Isometric Process : $\frac{dP}{dV} = \infty$

Isothermal Process : $\frac{dP}{dV} = -\frac{P}{V}$

Adiabatic Process : $\frac{dP}{dV} = -\frac{\gamma P}{V}$

Adiabatic expansion of mono, dia and polyatomic gases

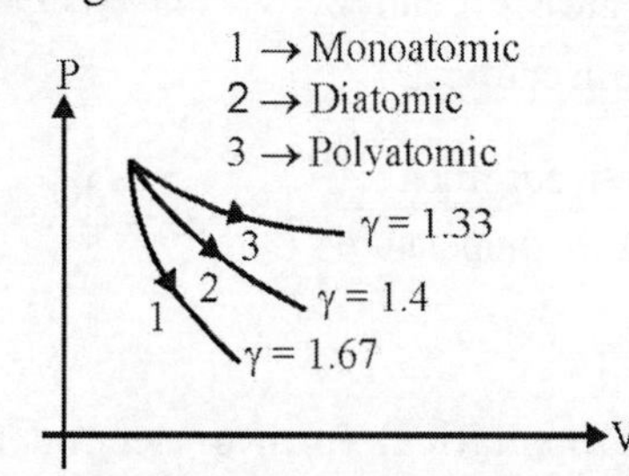

Adiabatic and isothermal expansion of an ideal gas

$$\frac{\text{slope of adiabatic changes}}{\text{slope of isothermal changes}} = \gamma$$

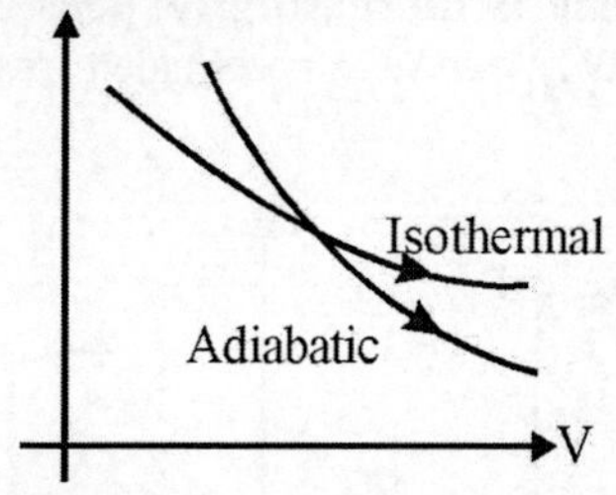

Isothermal bulk modulus of elasticity

$$E_{isotherm} = -\frac{dP}{dV/V} = -V\left(\frac{\partial P}{\partial V}\right)_{isotherm}$$

Adiabatic bulk modulus of elasticity

$$E_{adia} = -\frac{dP}{dV/V} = -V\left(\frac{\partial P}{\partial V}\right)_{adia}$$

SECOND LAW OF THERMODYNAMICS

Second law of thermodynamics disallows some processes consistent with the first law of thermodynamics.

Kelvin-Planck statement : No process is possible whose sole result is the absorption of heat from a reservoir and complete conversion of the heat into work.

Clausius statement : No process is possible whose sole result is the transfer of heat from a colder object to a hotter object. But simply, the second law implies that no heat engine can have efficiency η equal to 1 or no refrigerator can have co-efficient of performance β equal to infinity.

Entropy

It is a measure of molecular disorder of a system. Greater is disorder, greater is entropy.

The change in entropy,

$$dS = \frac{\text{heat absorbed}}{\text{absolute temperature}} \quad \text{i.e.,} \quad dS = \frac{dQ}{T}$$

or, $dQ = TdS$.

This is mathematical form of second law of thermodynamics.

REVERSIBLE AND IRREVERSIBLE PROCESS

A reversible process is one in which the changes in heat and work of direct process from initial to a final state are exactly retraced in opposite sense in the reverse process and the system and surroundings are left in their initial states. The reversibility is an ideal concept and can not be realised in practice.

The process which is not reversible is the irreversible process.

Heat Engine

Heat engine is a device in which a system undergoes a cyclic process resulting in conversion of heat into work. If Q_1 is the heat absorbed from the source, Q_2 is the heat released to the sink, and the work output in one cycle is W, the efficiency η of the engine is

$$\eta = \frac{W}{Q_1} = 1 - \frac{Q_2}{Q_1}.$$

Refrigerator and Heat Pump

In a refrigerator or a heat pump, the system extracts heat Q_2 from the cold reservoir and releases Q_1 amount of heat to the hot reservoir, with workdone (W) on the system. The co-efficient of performance of a refrigerator is given by

$$\beta = \frac{Q_2}{W} = \frac{Q_2}{Q_1 - Q_2}$$

Carnot Engine

Carnot engine is a reversible engine operating between two temperatures T_1 (source) and T_2 (sink). The Carnot cycle consists of two isothermal processes connected by two adiabatic processes. The efficiency(n) of a Carnot engine is given by

$$\eta = 1 - \frac{T_2}{T_1} \text{ (Carnot engine)}$$

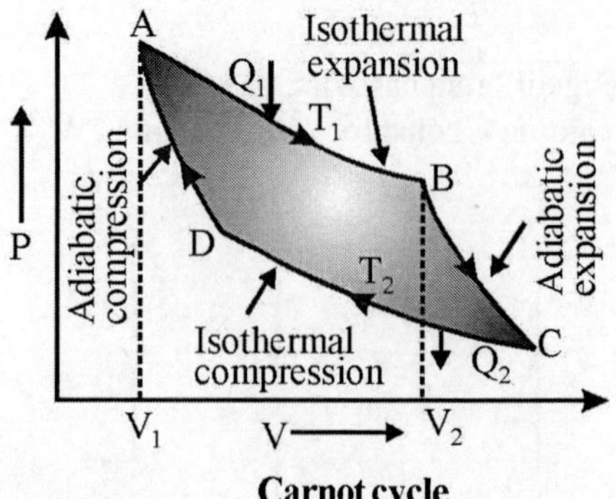

Carnot cycle

No engine operating between two temperatures can have efficiency greater than that of the Carnot engine.

1. Specific heat is zero during an adiabatic change since no heat is supplied. Specific heat will be infinite during an isothermal change since there is no temperature rise.
2. The ratio of increase in internal energy to total energy supplied to a gas is equal to $1/\gamma$. (γ is ratio of specific heats)
3. The ratio of external work to the total energy supplied to a gas $= \dfrac{\gamma - 1}{\gamma}$.
4. If earth had no atmosphere, it would have been terribly cool.
5. A gas undergoes a process in which the pressure and volume are related by VP^n = constant. The bulk modulus of the gas is P/n.
6. An ideal gas has a molar heat capacity C_V at constant volume. The molar heat capacity of this gas as a function of volume, if the gas undergoes the process :

   $T = T_0 e^{\alpha V}$ is $C_V + \dfrac{R}{\alpha V}$
7. Molar specific heat for a given process

   $$C = \frac{f}{2}R + \frac{RPdV}{PdV + VdP} = C_V + \frac{RPdV}{PdV + VdP}$$
8. For polytropic process PV^n = constant.

   $$P = \frac{K}{V^n} \Rightarrow \frac{dP}{dV} = n\left|\frac{K}{V^{n+1}}\right|, \quad C = \frac{R}{\gamma - 1} + \frac{R}{1-n}$$

   So C is constant for polytropic process.
9. When a gas expands its volume increases, then final pressure is less for adiabatic expansion. But, when a gas compresses its volume decreases, then the final pressure is more in case of adiabatic compression.

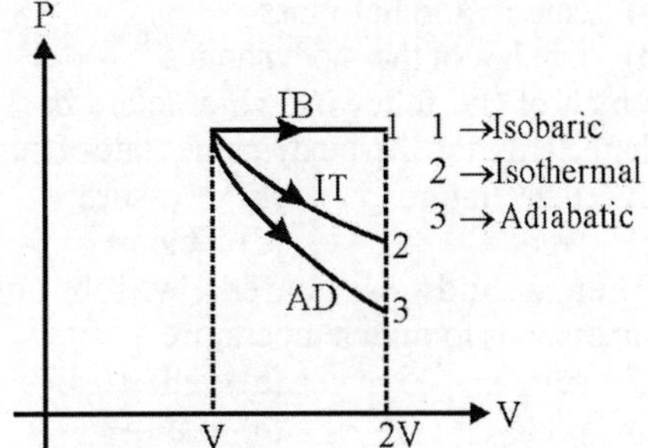

$W_{IB} > W_{IT} > W_{AD}$
$P_f \rightarrow P_{IB} > P_{IT} > P_{AD}$
$\Delta P \rightarrow \Delta P_{IB} > \Delta P_{IT} > \Delta P_{AD}$

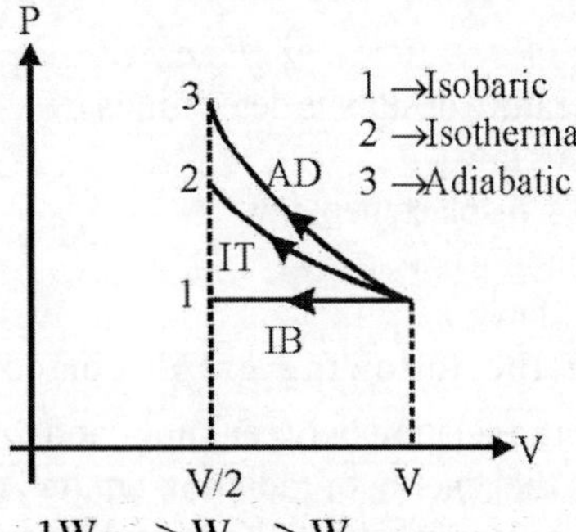

$1W_{AD} > W_{IT} > W_{IB}$
$P_f \rightarrow P_{AD} > P_{IT} > P_{IB}$
$\Delta P \rightarrow \Delta P_{AD} > \Delta P_{IT} > \Delta P_{IB}$

Exercise

1. Infrared radiation is detected by
   (a) spectrometer (b) pyrometer
   (c) nanometer (d) photometer
2. Which of the following is more close to a black body?
   (a) black board paint (b) green leaves
   (c) black holes (d) red roses
3. Even Carnot engine cannot give 100% efficiency because we cannot
   (a) prevent radiation
   (b) find ideal sources
   (c) reach absolute zero temperature
   (d) eliminate friction
4. Two spheres of the same material have radii 1 m and 4 m and temperatures 4000 K and 2000 K respectively. The ratio of the energy radiated per second by the first sphere to that by the second is
   (a) 1 : 1 (b) 16 : 1
   (c) 4 : 1 (d) 1 : 9

5. "Heat cannot by itself flow from a body at lower temperature to a body at higher temperature" is a statement or consequence of
(a) second law of thermodynamics
(b) conservation of momentum
(c) conservation of mass
(d) first law of thermodynamics

6. Which of the following parameters does not characterize the thermodynamic state of matter?
(a) Temperature (b) Pressure
(c) Work (d) Volume

7. When a body is heated, which colour corresponds to high temperature
(a) red (b) yellow
(c) white (d) orange

8. A radiation of energy E falls normally on a perfectly reflecting surface. The momentum transferred to the surface is
(a) Ec (b) $2E/c$
(c) E/c (d) E/c^2

9. The temperature of stars is determined by
(a) Stefan's law
(b) Wien's displacement law
(c) Kirchhoff's law
(d) Ohm's law

10. Which of the following graph correctly represents the relation between $\ln E$ and $\ln T$ where E is the amount of radiation emitted per unit time from unit area of body and T is the absolute temperature

(a) 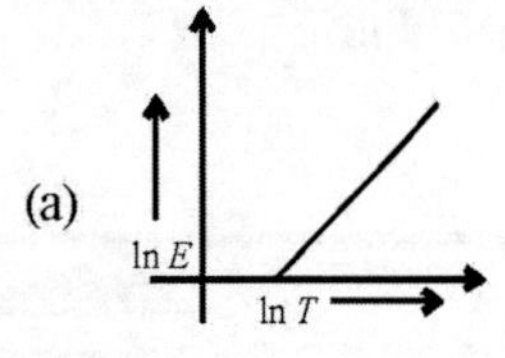

(b) 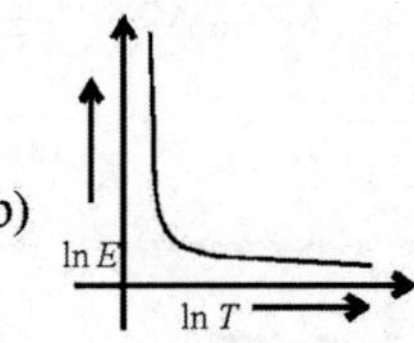

(c) 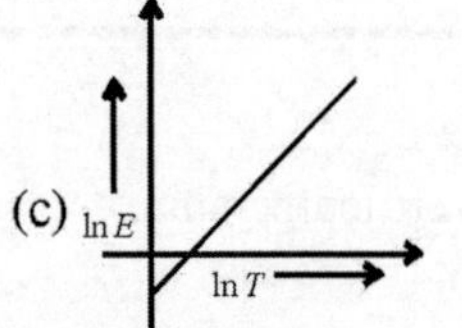

(d) 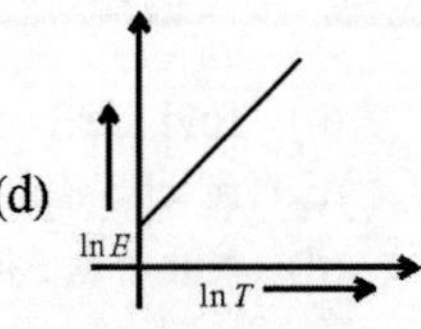

11. In the equation PV^γ = constant, the value of γ is unity. Then the process is
(a) isothermal (b) adiabatic
(c) isobaric (d) irreversible

12. Which of the following is the $\nu_m - T$ graph for a perfectly black body

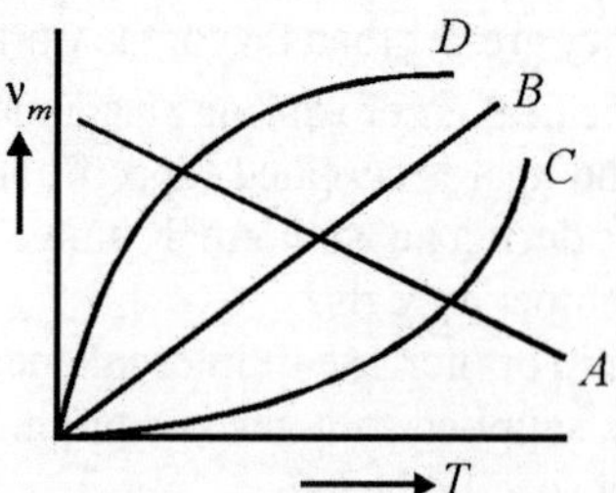

(a) A (b) B
(c) C (d) D

13. If a liquid is heated in weightlessness, the heat is transmitted through
(a) conduction
(b) convection
(c) radiation
(d) none of these, because the liquid cannot be heated in weightlessness

14. A system X is neither in thermal equilibrium with Y nor with Z. The systems Y and Z
(a) must be in thermal equilibrium
(b) cannot be in thermal equilibrium
(c) may be in thermal equilibrium
(d) none of these

15. Which of the curves in figure represents the relation between Celsius and Fahrenheit temperatures?

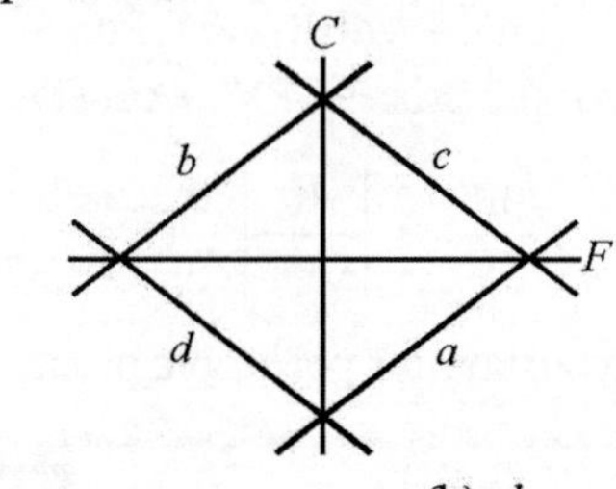

(a) a (b) b
(c) c (d) d

16. Which of the following pairs may give equal numerical values of the temperature of a body?
(a) Fahrenheit and Celsius
(b) Celsius and Kelvin
(c) Kelvin and Platinum
(d) None of these

17. For a constant volume gas thermometer, one should fill the gas at
(a) low temperature and low pressure
(b) low temperature and high pressure
(c) high temperature and low pressure
(d) high temperature and high pressure

18. Consider the following statements :
(A) The coefficient of linear expansion has dimension K^{-1}.
(B) The coefficient of volume expansion has dimension K^{-1}.
(a) A and B are both correct
(b) A is correct but B is wrong
(c) B is a correct but A is wrong
(d) A and B are both wrong

19. A metal sheet with a circular hole is heated. The hole
(a) gets larger
(b) gets smaller
(c) remains of the same size
(d) gets deformed

20. Two identical rectangular strips, one of copper and the other of steel, are rivetted together to form a bimetallic strip ($\alpha_{copper} > \alpha_{steel}$). On heating, this strip will
(a) remain straight
(b) bend with copper on convex side
(c) bend with steel on convex side
(d) get twisted

21. The temperature of water at the surface of a deep lake is 2°C. The temperature expected at the bottom is
(a) 0°C (b) 2°C
(c) 4°C (d) –6°C

22. In which of the following pairs of temperature scales, the size of a degree is identical?
(a) Mercury scale and ideal gas scale
(b) Celsius scale and mercury scale
(c) Celsius scale and ideal gas scale
(d) Real gas scale and absolute scale

23. As the temperature is increased, the time period of a pendulum
(a) increases proportionately with temperature
(b) increases
(c) decreases
(d) remains constant

24. The gas thermometers are more sensitive than liquid thermometers because
(a) gases expand more than liquids
(b) gases are easily obtained
(c) gases are much lighter
(d) gases do not easily change their states

25. Thermoelectric thermometer is based on
(a) Photoelectric effect (b) Seebeck effect
(c) Compton effect (d) Joule effect

26. If C_P and C_V denote the specific heats of nitrogen per unit mass at constant pressure and constant volume respectively, then
(a) $C_P - C_V = 28R$ (b) $C_P - C_V = R/28$
(c) $C_P - C_V = R/14$ (d) $C_P - C_V = R$

27. A solid ball of metal has a spherical cavity inside it. The ball is heated. The volume of cavity will
(a) decrease
(b) increase
(c) remain unchanged
(d) have its shape changed

28. On heating a liquid of coefficient of cubical expansion α in a container having coefficient of linear expansion $\alpha/3$, the level of liquid in the container will
(a) rise
(b) fall
(c) will remain almost stationary
(d) it is difficult to say

29. When water is heated from 0°C to 10°C, its volume
(a) increases
(b) decreases
(c) does not change
(d) first decreases and then increases

30. A liquid with coefficient of volume expansion γ is filled in a container of a material having the coefficient of linear expansion α. If the liquid overflows on heating
(a) $\gamma = 3\alpha$ (b) $\gamma > 3\alpha$
(c) $\gamma < 3\alpha$ (d) $\gamma = 3\alpha^3$

31. By increasing the temperature of liquid, its
(a) volume and density decreases
(b) volume and density increases
(c) volume increases and density decreases
(d) volume decreases and density increases

32. If for a gas, $\frac{R}{C_v} = 0.67$, the gas is made up of molecules which are
(a) diatomic
(b) mixture of diatomic and polyatomic molecules
(c) monoatomic
(d) polyatomic

33. A thermodynamic system is taken from state A to B along ACB and is brought back to A along BDA as shown in the PV diagram. The net work done during the complete cycle is given by the area

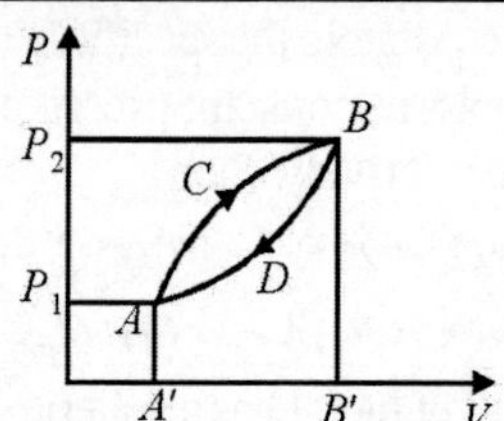

(a) $P_1ACBP_2P_1$ (b) $ACBB'A'A$
(c) $ACBDA$ (d) $ADBB'A'A$

34. An ideal gas A and a real gas B have their volumes increased from V to $2V$ under isothermal conditions. The increase in internal energy
(a) will be same in both A and B
(b) will be zero in both the gases
(c) of B will be more than that of A
(d) of A will be more than that of B

35. A gas at 27°C temperature and 30 atmospheric pressure is allowed to expand to the atmospheric pressure. If the volume becomes 10 times its initial volume, then the final temperature becomes
(a) 100°C (b) 173°C
(c) 273°C (d) –173°C

36. A Carnot engine whose efficiency is 50% has an exhaust temperature of 500 K. If the efficiency is to be 60% with the same intake temperature, the exhaust temperature must be (in K)
(a) 800 (b) 200
(c) 400 (d) 600

37. An ideal gas heat engine operates in a Carnot cycle between 227°C and 127°C. It absorbs 6 kcal at the higher temperature. The amount of heat (in kcal) converted into work is equal to
(a) 1.2 (b) 4.8
(c) 3.5 (d) 1.6

38. One mole of an ideal gas at an initial temperature of T K does 6R joules of work adiabatically. If the ratio of specific heats of this gas at constant pressure and at constant volume is 5/3, the final temperature of gas will be
(a) $(T-4)K$ (b) $(T+2.4)K$
(c) $(T-2.4)K$ (d) $(T+4)K$

39. Which of the following processes is reversible?
(a) Transfer of heat by conduction
(b) Transfer of heat by radiation
(c) Isothermal compression
(d) Electrical heating of a nichrome wire

40. An ideal gas heat engine operates in Carnot cycle between 227°C and 127°C. It absorbs 6×10^4 cals of heat at higher temperature. Amount of heat converted to work is
(a) 4.8×10^4 cals (b) 6×10^4 cals
(c) 2.4×10^4 cals (d) 1.2×10^4 cals

41. A Carnot engine whose sink is at 300 K has an efficiency of 40%. By how much should the temperature of source be increased so as to increase, its efficiency by 50% of original efficiency?
(a) 325 K (b) 250 K
(c) 380 K (d) 275 K

42. The molar specific heat at constant pressure of an ideal gas is (7/2) R. The ratio of specific heat at constant pressure to that at constant volume is
(a) 8/7 (b) 5/7
(c) 9/7 (d) 7/5

43. An engine has an efficiency of 1/6. When the temperature of sink is reduced by 62°C, its efficiency is doubled. Temperature of the source is
(a) 37°C (b) 62°C
(c) 99°C (d) 124°C

44. If Q, E and W denote respectively the heat added, change in internal energy and the work done in a closed cyclic process, then:
(a) $W=0$ (b) $Q=W=0$
(c) $E=0$ (d) $Q=0$

45. In thermodynamic processes which of the following statements is not true?
(a) In an isochoric process pressure remains constant
(b) In an isothermal process the temperature remains constant
(c) In an adiabatic process PV^γ = constant
(d) In an adiabatic process the system is insulated from the surroundings

46. The internal energy change in a system that has absorbed 2 kcals of heat and done 500 J of work is:
(a) 6400 J (b) 5400 J
(c) 7900 J (d) 8900 J

47. If ΔU and ΔW represent the increase in internal energy and work done by the system respectively in a thermodynamical process, which of the following is true?
(a) $\Delta U = -\Delta W$, in an adiabatic process
(b) $\Delta U = \Delta W$, in an isothermal process
(c) $\Delta U = \Delta W$, in an adiabatic process
(d) $\Delta U = -\Delta W$, in an isothermal process

48. During an isothermal expansion, a confined ideal gas does –150 J of work against its surroundings. This implies that
(a) 150 J heat has been removed from the gas
(b) 300 J of heat has been added to the gas
(c) no heat is transferred because the process is isothermal
(d) 150 J of heat has been added to the gas

49. When 1 kg of ice at 0°C melts to water at 0°C, the resulting change in its entropy, taking latent heat of ice to be 80 cal/°C, is
(a) 273 cal/K (b) 8 × 104 cal/K
(c) 80 cal/K (d) 293 cal/K

50. A mass of diatomic gas ($\gamma = 1.4$) at a pressure of 2 atmospheres is compressed adiabatically so that its temperature rises from 27°C to 927°C. The pressure of the gas in final state is
(a) 28 atm (b) 68.7 atm
(c) 256 atm (d) 8 atm

51. Which of the following is incorrect regarding first law of thermodynamics?
(a) It is a restatement of principle of conservation of energy
(b) It is applicable to cyclic processes
(c) It introduces the concept of entropy
(d) It introduces the concept of internal energy

52. For an isochoric thermodynamic process.
(a) $\Delta Q = \Delta W$ (b) $\Delta Q = \Delta U$
(c) $\Delta U = \Delta W$ (d) $\Delta U = 0$

53. In an adiabatic expansion, the product of pressure and volume :
(a) decreases
(b) increases
(c) remains constant
(d) first increases then decreases

54. Choose the incorrect statement related to an isobaric process.
(a) $\frac{V}{T}$ = constant
(b) $W = P\Delta V$
(c) heat given to a system is used up in raising the temperature only
(d) $\Delta Q > W$

55. At a given temperature, the pressure applied by the ideal gas is :
(a) directly proportional to the density of the gas
(b) directly proportional to the square of the density of the gas
(c) inversly proportional to the square of the density of the gas
(d) inversly proportional to the density of the gas

56. $(C_p - C_v)$ will be :
(a) maximum for oxygen
(b) maximum for helium
(c) maximum for carbon-dioxide
(d) same for all gases

57. The internal energy of an ideal gas does not depend upon:
(a) temperature of the gas
(b) pressure of the gas
(c) atomicity of the gas
(d) number of moles of the gas

58. Mark the incorrect statement related to a cyclic process.
(a) the initial and final conditions always coincide
(b) $Q = W$
(c) $W > 0$
(d) none of these

59. Which of the following is a very fast process?
(a) Isothermal (b) Adiabatic
(c) Isobaric (d) Isochoric

60. During isothermal expansion, the slope of P-V graph :
(a) decreases
(b) increases
(c) remains same
(d) may increase or decrease

CHAPTER 12 Kinetic Theory

CONCEPTS

Kinetic theory explains the behaviour of gases based on the idea that the gas consists of rapidly moving atoms or molecules.

BEHAVIOUR OF GASES

Ideal Gas : In an ideal gas, we assumed that molecules are point masses and there is no mutual attraction between them.

The ideal gas obey following laws :

(i) **Boyle's Law :** According to Boyle's law for a given mass of ideal gas, the pressure of a ideal gas is inversely proportional to the volume at constant temperature

i.e., $P \propto \frac{1}{V} \Rightarrow PV = \text{constant}$ or $P_1 V_1 = P_2 V_2$

(ii) **Charles' Law :** According to Charle's law for a given mass, the volume of a ideal gas is proportional to temperature at a constant pressure

i.e., $V \propto T \Rightarrow \frac{V}{T} = \text{constant}$ or, $\frac{V_1}{T_1} = \frac{V_2}{T_2}$

(iii) **Gay Lussac's Law :** According to Gay-lussac's Law for a given mass of ideal gas, the pressure is proportional to temperature at constant volume

i.e., $P \propto T \Rightarrow \frac{P}{T} = \text{constant}$ or, $\frac{P_1}{T} = \frac{P_2}{T_2}$

(iv) **Avogadro's Law :** According to Avogadro's law, the number of molecules of all gases are same at same temperature, pressure and volume

i.e., $N_{A_1} = N_{A_2}$ for same P, V & T.

The value of Avagadro number is 6.02×10^{23} molecules.

(v) **Graham's Law :**

(i) At constant temperature and pressure, the rms speed of diffusion of two gases is inversely proportional to the square root of the relative density.

i.e., $v_{rms} \propto \frac{1}{\sqrt{d}} \Rightarrow \frac{(v_{rms})_1}{(v_{rms})_2} = \sqrt{\frac{d_2}{d_1}}$

(ii) According to Graham's law, the rate of diffusion of a gas is inversely proportional to the square root of its density, provided pressure & temperature are constant.

Rate of diffusion $r \propto \frac{1}{\sqrt{\rho}}$

(vi) **Dalton's Law :** According to Dalton's law, the pressure exerted by a gaseous mixture is equal to sum of partial pressure of each component gases present in the mixture,

i.e., $P = P_1 + P_2 + P_3 + \ldots\ldots$

A relation connecting macroscopic properties P, V and T of a gas describing the state of the system is called equation of state. The equation of state for an ideal gas of n mole is

$$\frac{PV}{T} = nR \text{ or } PV = nRT$$

where R is universal gas constant with R = 8.31 J/mol K and $R = N_A k$, where k is Boltzmann's constant (N_A is Avagadro number).

n is the number of moles of a gas and

$n = \frac{m}{M} = \frac{N}{N_A}$, N_A is Avagadro number

m is the mass of a gas, N is the number of molecule and M is the molecular weight of a gas.

Equation of Real Gas

The real gas follows Vander Wall's law. According to this : $\left(p + \frac{n^2 a}{V^2}\right)(V - nb) = nRT$; a and b are Vander Wall's constant

KINETIC THEORY OF AN IDEAL GAS

The basic assumptions of kinetic theory are :

(i) A gas consists of particles called molecules which move randomly in all directions.

(ii) The volume of molecule is very small in comparison to the volume occupied by gas i.e., the size of molecule is infinitesimely small.

(iii) The collision between two molecules or between a molecule and wall are perfectely elastic and collision time is (duration of collision) very small.

(iv) The molecule exert no force on each other or on the walls of containers except during collision.

(v) The total number of molecules are large and they obey Newtonian Mechanics.

$$PV = \frac{1}{3} mn\,\overline{C^2} \quad ....(i)$$

Where $\overline{C^2}$ is called mean square velocity and $\sqrt{\overline{C^2}}$ is called root mean square velocity i.e,

$$\overline{C^2} = \frac{C_1^2 + C_2^2 + ...C_n^2}{n} \quad ....(ii)$$

and $v_{r.m.s.} = \sqrt{\overline{C^2}}$ ....(iii)

Where m = mass of one molecule, n = number of molecules

$$P = \frac{M}{3V}\overline{C^2} = \frac{1}{3}\rho v_{r.m.s}^2 \quad ....(iv)$$

$$\text{or } P = \frac{2}{3}\left[\frac{1}{2}\rho v_{r.m.s}^2\right]\left(\text{where } M = m \times n \ \& \ \rho = \frac{M}{V}\right)$$

$$\Rightarrow \quad P = \frac{2}{3}E \quad ....(v)$$

Where E is translational kinetic energy per unit volume of the gas. It is clear that pressure of ideal gas is equal to 2/3 of translation kinetic energy per unit volume.

Kinetic Interpretation of Temperature : From eq. (iv), we get

$$PV = \frac{1}{3}M\overline{C^2}$$

For 1 mole gas at temperature T :

$$PV = RT \text{ so } \frac{1}{3}m\overline{C^2} = RT$$

$$\frac{1}{3}m\overline{C^2} = \frac{R}{N}T = kT \quad \left(k = \frac{R}{N} \text{ is Boltzmann constant}\right)$$

or $\frac{1}{2}m\overline{C^2} = \frac{3}{2}kT$ ....(vi)

or $\frac{1}{2}mv_{r.m.s}^2 = \frac{3}{2}kT$

or $v_{r.m.s}^2 = \sqrt{\frac{3kT}{m}}$ ....(vii)

or $v_{r.m.s} = \sqrt{\frac{3nkT}{nm}} = \sqrt{\frac{3RT}{M}}$

It is clear from eq.(vi) that at a given temperature, the average translational kinetic energy of any gas molecules are equal i.e., it depends only on temperature.

From eq.(vii) It is clear that

$v_{r.m.s} \propto \sqrt{T}$ and $v_{r.m.s} \propto \frac{1}{\sqrt{M}}$ where M is moleculer weight of the gas.

RMS SPEED, MOST PROBABLE SPEED AND AVERAGE SPEED

RMS Speed V_{rms}

The square root of mean square speed is called root-mean-square speed or rms speed (v_{rms}). It is given by

$v_{rms} = \sqrt{\frac{3P}{\rho}}$, P = pressure exerted by the gas

ρ = density of the gas

Also, $v_{rms} = \frac{\sqrt{3K_BT}}{m}$, K_B = Boltzmann's constant

m = mass of each molecule of gas

T = temperature in the Kelvin scale

And $v_{rms} = \sqrt{\frac{3RT}{M_o}}$, M_o = molecular weight of gas

The temperature of a gas is a measure of the average kinetic energy of a molecule, independent of the nature of the gas or molecule.

Most Probable Speed V_{mp}

It is the speed which maximum number of molecules in a gas have at constant temperature and is given by

$$v_{mp} = \sqrt{\frac{2RT}{M_o}} = \left(\sqrt{\frac{2}{3}}\right) v_{rms} = 0.816\, v_{rms}$$

Average Speed V_{av}

It is the arithmetic mean of the speeds of molecules in a gas at a given temperature,

$$\text{i.e., } v_{av} = \frac{v_1 + v_2 + v_3 + ... + v_N}{N}$$

And according to kinetic theory of gases,

$$v_{av} = \sqrt{\frac{8RT}{\pi M_o}} = \left(\sqrt{\frac{8}{3\pi}}\right) v_{rms} = 0.92\ v_{rms}$$

elearly, $v_{rms} > v_{av} > v_{mp}$

THE LAW OF EQUIPARTITION OF ENERGY

It is stated as the energy for each degree of freedom in thermal equilibrium is $\frac{1}{2}k_BT$.

The total translational kinetic energy $E = \frac{3}{2}k_BNT$

$$= \frac{3}{2}nRT \cdot$$

This leads to a relation $PV = \frac{2}{3}E$.

But average translational kinetic energy, $K_{av} = \frac{3}{2}K_BT$

The term degree of freedom refers to the number of possible independent ways in which a system can have energy.

Degree of freedom (f) in different gas molecules

Molecules	Translational	Rotational
Monatomic	3	0
Diatomic	3	2
Polyatomic	3	2(linear molecule)
	3	(non-linear molecule)

SPECIFIC HEAT CAPACITY

It is the amount of heat required to raise the temperature of unit mass of substance through 1 degree. It may be expressed in cal/g°C, kcal/kg°C, BTU/1b°f, J/kgK.

$$\text{specific heat, } c = \frac{Q}{m \times \Delta t}$$

For gases : During the **isothermal** process

$\Delta t = 0 \Rightarrow c = \infty$ (infinite) during the **adiabatic** process

$Q = 0 \Rightarrow c = 0$ (zero)

Specific heat of gas : The specific heat of a gas varies from zero to infinity. It may have any positive or negative value.

Following of two specific heats are more significant.

Molar specific heat at constant volume (C_V). It is the amount of heat required to raise the temperature of 1 mole of a gas through 1K or 1°C at constant volume

$$\text{i.e., } C_V = \frac{1}{n}\left(\frac{dQ}{dT}\right)$$

Molar specific heat at constant pressure (C_P) It is the amount of heat required to raise the temperature of 1 mole of a gas through 1K or 1°C at constant pressure.

$$\text{i.e., } C_P = \frac{1}{n}\left(\frac{dQ}{dT}\right)$$

Mayer's relation : $C_P - C_V = R$

where R is a gas constant. ($R = 8.314\ Jmol^{-1}K^{-1}$)

Specific heat of gases :

As we know $U = \mu\frac{f}{2}RT$

For 1 gm mole gas $U = \frac{f}{2}RT \quad \therefore \quad \frac{dU}{dT} = \frac{f}{2}R$

$dU = \mu\, C_v\, dT.$ If $\mu = 1$ mole then $C_v = \frac{dU}{dT}$

So $C_v = \frac{1}{2}fR$ and hence $C_p = \frac{1}{2}Rf + R$

$$\text{Now } \gamma = \frac{C_p}{C_v} = \frac{\frac{1}{2}fR + R}{\frac{1}{2}fR} \text{ so } \gamma = 1 + \frac{2}{f}$$

$$\text{and } f = \frac{2}{\gamma - 1};\quad C_v = \frac{R}{\gamma - 1};\quad C_p = \frac{\gamma R}{\gamma - 1}$$

If P is the pressure of a gas, V its volume, T its temperature and ρ its density, then

PV = constant at constant T (Boyle's law),

$\frac{P}{\rho}$ = constant at contant T is another form of Boyle's law.

$\frac{P}{T}$ = constant at constant volume.

$\frac{V}{T}$ = constant at constant pressure (Charles law)

Vander Wall's Equation :

For μ gram mole, $\left(P+\mu^2\frac{a}{V^2}\right)(V-\mu b)=\mu RT$

Boyle temperature : The specific temperature at which the real gases obey Boyle's law, is defined as Boyle temperature.

$T_B = a/Rb$

Critical temperature (T_c) : The limiting temperature at which a gas can be liquified only by increasing the pressure and above which it cannot be liquified, how large the pressure may be increased, is defined as critical temperature.

$T_c = \frac{8a}{27Rb}$; $T_c = \frac{8}{27}T_B$

For CO_2, $T_c = 304.3$ K For O_2, $T_c = -118°C$

For N_2, $T_c = -147.1°C$ For steam, $T_c = 365°C$

Critical Pressure (P_c) : The minimum pressure necessary to liquify a gas at critical temperature is defined as critical pressure (P_c).

i.e., $P_c = \frac{a}{27b^2}$

For CO_2, $P_c = 73.87$ bar and for O_2, 49.7 atmosphere.

Critical Volume (V_c) : The volume of 1 mol of gas at critical pressure and critical temperature is defined as critical volume (V_c).

$V_c = 3b$; For CO_2, $V_c = 95 \times 10^{-6}$ m^3.

Relation between T_c, V_c and P_c : $\frac{P_cV_c}{T_c} = \frac{3}{8}R$

Values of a and b in terms of T_c, V_c and P_c.

$a = \frac{27R^2\,T_c^2}{64\,P_c}$ and $b = \frac{RT_c}{8P_c}$

MIXTURE OF GASES

Equivalent molar mass : When μ_1 moles of a gas with molar mass M_1 are mixed with μ_2 moles of a gas with molar mass M_2, the equivalent molar mass of the mixture is given by

$$M = \frac{\mu_1M_1+\mu_2M_2}{\mu_1+\mu_2}$$

C_V of the mixture : $C_V = \frac{\mu_1C_{V_1}+\mu_2C_{V_2}}{\mu_1+\mu_2}$

C_P of the mixture : $C_P = \frac{\mu_1C_{P_1}+\mu_2C_{P_2}}{\mu_1+\mu_2}$

γ of the mixture : $\frac{(\mu_1+\mu_2)}{\gamma_{mix}-1} = \frac{\mu_1}{\gamma_1-1}+\frac{\mu_2}{\gamma_2-1}$

MEAN FREE PATH

The distance covered by the molecules between two successive collisions is called the **free path**.

Mean free path : The average distance covered by the molecules between two successive collisions is called the **mean free path.**

i.e. $\lambda = \frac{1}{\sqrt{2}\,.\,\pi nd^2} = \frac{K_BT}{\sqrt{2}\pi d^2P}$

where, n = number of molecules per unit volume
d = diameter of each molecule
K_B = Boltzmann's constant

Mean free path depends on the diameter of molecule (d) and the number of molecules per unit volume n.

At N.T.P., mean free path, λ for air molecules is 0.01 μm.

Exercise

1. According to kinetic theory of matter, a molecule is the smallest particle of a substance and it possesses
   (a) all the properties of the substance
   (b) some of the properties of the substance
   (c) none of the properties of the substance
   (d) both (a) and (b) are true
2. According to the kinetic theory of gases, the pressure exerted by a gas on the walls is measured as
   (a) rate of change of momentum imparted to the walls per second per unit area

(b) momentum imparted to the walls per unit area
(c) change of momentum imparted to the walls per unit area.
(d) change in momentum per unit volume

3. Pressure exerted by a perfect gas is equal to
(a) mean kinetic energy per unit volume
(b) half of mean kinetic energy per unti volume
(c) two third of mean kinetic energy per unit volume
(d) one third of mean kinetic energy per unit volume

4. The internal energy of a gram-molecule of an ideal gas depends on
(a) pressure alone
(b) volume alone
(c) temperature alone
(d) both on pressure as well as temperature

5. The average velocity of the molecules in a gas in equilibrium is
(a) proportional to $\sqrt{T}$ (b) proportional to T
(c) proportional to T^2 (d) equal to zero

6. The absolute temperature of a gas determines
(a) the average momentum of the molecule
(b) the velocity of sound in the gas
(c) the number of molecules in the gas
(d) the mean square velocity of the molecules

7. At a given temperature the force between molecules of a gas as a function of intermolecular distance is
(a) always constant
(b) always decreases
(c) always increases
(d) first decreases and then increases

8. At 0K which of the following properties of a gas will be zero?
(a) Kinetic energy (b) Potential energy
(c) Vibrational energy (d) Density

9. A gas in a container A is in thermal equilibrium with antoher gas of the same mass in container B. If we denote the corresponding pressures and volumes by the suffixes A and B, then which of the following statement is most likely to be true.
(a) $P_A = P_B,\ V_A \neq V_B$
(b) $P_A \neq P_B,\ V_A = V_B$
(c) $P_A/V_A = P_B/V_B$
(d) $P_A V_A = P_B V_B$

10. In a cubical vessel are enclosed n molecules of a gas each having a mass m and an average speed v. If ℓ is the length of each edge of the cube, the pressure exerted by the gas will be
(a) $\frac{nmv^2}{\ell^3}$ (b) $\frac{nm^2v}{2\ell^3}$
(c) $\frac{mnv^2}{3\ell^3}$ (d) $\frac{nmv}{2\ell}$

11 The average kinetic energy of a molecule of a perfect gas is
(a) $(2/3)kT$ (b) $1.5kT$
(c) $2.5kT$ (d) none

12. The speed of sound in a gas is v.. The r.m.s. velocity of the gas molecules is C. The ratio of v to C is
(a) $3/\gamma$ (b) $\gamma/3$
(c) $\sqrt{(3/\gamma)}$ (d) $\sqrt{(\gamma/3)}$

13. In two vessels of same volume atomic hydrogen and helium at pressure 1 atm. and 2 atm. are filled. If temperature of both the samples is same, then average speed of hydrogen atom $\bar{v}_H$ will be related to helium $\bar{v}_{He}$ as
(a) $\bar{v}_H = \sqrt{2}\bar{v}_{He}$ (b) $\bar{v}_H = \bar{v}_{He}$
(c) $\bar{v}_H = 2\bar{v}He$ (d) $v_H = v_{He}/2$

14. The number of degrees of freedom for each atom of a monatomic gas is
(a) 3 (b) 5
(c) 6 (d) 1

15. At the same temperature, the ratio of energy per gm. mole of O_2 and H_2 is
(a) 1 (b) 1.4
(c) 2 (d) 3

16. The root mean square value of the speed of the molecules in a fixed mass of an ideal gas is increased by increasing
(a) the temperature while keeping the volume and pressure constant
(b) the pressure while keeping the volume constant
(c) the temperature while keeping the pressure constant
(d) the pressure while keeping the temperature constant

17. If a gas has 'n' degrees of freedom, the ratio of the specific heats γ of the gas is

(a) $\frac{1+n}{2}$ (b) $1+\frac{n}{2}$

(c) $1+\frac{1}{n}$ (d) $1+\frac{2}{n}$

18. The mean kinetic energy of a perfect gas monatomic gas molecule at the temperature T°K is

(a) $\frac{1}{2}kT$ (b) kT

(c) $\frac{3}{2}kT$ (d) $2kT$

19. The mean K.E. per gram mole of a gas is

(a) $\frac{3}{2}RT$ (b) $\frac{1}{2}RT$

(c) $\frac{2}{3}RT$ (d) $\frac{2}{3RT}$

20. The pressure P of an ideal gas and its mean K.E.E per unit volume are related as

(a) $P=\frac{E}{2}$ (b) $P=E$

(c) $P=\frac{3E}{2}$ (d) $P=\frac{2E}{3}$

21. At a given temperature which of the following gases possesses maximum r.m.s. velocity?
(a) hydrogen
(b) oxygen
(c) nitrogen
(d) carbon dioxide

22. The temperature of gas is produced by
(a) the potential energy of its molecules
(b) the kinetic energy of its molecules
(c) the attractive force between its molecules
(d) the repulsive force between its molecules

23. The root mean square speed of a group of n gas molecules, having speed $v_1, v_2, v_3, ..., v_n$ is

(a) $\frac{1}{n}\sqrt{(v_1+v_2+v_3+...+v_n)^2}$

(b) $\frac{1}{n}\sqrt{(v_1^2+v_2^2+v_3^2+...+v_n^2)}$

(c) $\sqrt{\frac{1}{n}(v_1^2+v_2^2+v_3^2+...+v_n^2)}$

(d) $\sqrt{\left[\frac{(v_1+v_2+v_3+...v_n)^2}{n}\right]}$

24. The kinetic theory of gases :
(a) explains the behaviour of an ideal gas
(b) describes the motion of a single atom or molecule
(c) relates the temperature of the gas with K.E. of atoms of the gas
(d) all of the above

25. The ratio of principal molar heat capacities of a gas is maximum for :
(a) a diatomic gas
(b) a monoatomic gas
(c) a polyatomic gas having linear molecules
(d) a polyatomic gas having non-linear molecules

26. The correct statement of the law of equipartition of energy is
(a) the total energy of a gas is equally divided among all the molecules
(b) the gas possess equal energies in all the three directions x, y and z-axis
(c) the total energy of a gas is equally divided between kinetic and potential energies
(d) the total kinetic energy of a gas molecules is equally divided among translational and rotational kinetic energies

27. The internal energy of an ideal gas is :
(a) The sum of total kinetic and potential energies
(b) The total translational kinetic energy
(c) The total kinetic energy of randomly moving molecules
(d) The total kinetic energy of gas molecules

28. The total degree of freedom of a CO_2 gas molecule is
(a) 3 (b) 6
(c) 5 (d) 4

29. The correct relations is/are

(a) $C_v=\frac{1}{2}fR$ (b) $C_v=\frac{R}{\gamma-1}$

(c) $C_v=\frac{C_p}{\gamma}$ (d) all of these

30. In kinetic theory of gases, it is assumed that molecules
    (a) have same mass but can have different volume
    (b) have same volume but mass can be different
    (c) have different mass as well as volume
    (d) have same mass but negligible volume
31. Gases exert pressure on the walls of the container because the gas molecules
    (a) possess momentum
    (b) collide with each other
    (c) have finite volume
    (d) obey gas laws
32. The average velocity of molecules of gas is given by:
    (a) $\sqrt{\frac{3RT}{M}}$ (b) $\sqrt{\frac{2RT}{M}}$
    (c) $\sqrt{\frac{8RT}{\pi M}}$ (d) zero
33. Hydrogen gas and oxygen gas have volume $1 cm^3$ each at NTP
    (i) Number of molecules is same in both the gases
    (ii) The rms velocity of molecules of both the gases is same
    (iii) The internal energy of each gas is same
    (iv) The average velocity of molecules of each gas is same
    The correct options are :
    (a) (i) and (ii) (b) (i), (iii) and (iv)
    (c) (ii) and (iii) (d) (i), (ii) and (iii)
34. Let v denote the rms speed of the molecules in an ideal diatomic gas at absolute temperature T. The mass of a molecule is 'm' Neglecting vibrational energy terms, the false statement is :
    (a) A molecule can have a speed greater than $\sqrt{2}v$
    (b) v is proportional to $\sqrt{T}$
    (c) The average rotational K.E. of a molecule is $\frac{1}{4}mv^2$
    (d) The average K.E. of a molecule is $\frac{5}{6}mv^2$
35. Two containers A and B are partly filled with water and closed. The volume of A is twice that of B and it contains half the amount of water in B. If both are at the same temperature, the water vapour in the containers will have pressure in the ratio of
    (a) 1 : 2 (b) 1 : 1
    (c) 2 : 1 (d) 4 : 1
36. At constant volume, temperature is increased then
    (a) collision on walls will be less
    (b) number of collisions per unit time will increase
    (c) collisions will be in straight lines
    (d) collisions will not change
37. A polyatomic gas with n degrees of freedom has a mean energy per molecule given by
    (a) $\frac{nkT}{N}$ (b) $\frac{nkT}{2N}$
    (c) $\frac{nkT}{2}$ (d) $\frac{3kT}{2}$
38. For a certain gas the ratio of specific heats is given to be $\gamma = 1.5$. For this gas
    (a) $C_V = 3R/J$ (b) $C_P = 3R/J$
    (c) $C_P = 5R/J$ (d) $C_V = 5R/J$
39. According to kinetic theory of gases, at absolute zero temperature
    (a) water freezes
    (b) liquid helium freezes
    (c) molecular motion stops
    (d) liquid hydrogen freezes
40. One mole of an ideal monoatomic gas requires 207 J heat to raise the temperature by 10 K when heated at constant pressure. If the same gas is heated at constant volume to raise the temperature by the same 10 K, the heat required is
    [Given the gas constant $R = 8.3$ J/ mol. K]
    (a) 198.7 J (b) 29 J
    (c) 215.3 J (d) 124 J
41. Relation between pressure (P) and energy (E) of a gas is
    (a) $P = \frac{2}{3}E$ (b) $P = \frac{1}{3}E$
    (c) $P = \frac{1}{2}E$ (d) $P = 3E$

42. Three containers of the same volume contain three different gases. The masses of the molecules are m_1, m_2 and m_3 and the number of molecules in their respective containers are N_1, N_2 and N_3. The gas pressure in the containers are P_1, P_2 and P_3 respectively. All the gases are now mixed and put in one of these containers. The pressure P of the mixture will be

(a) $P<(P_1+P_2+P_3)$ (b) $P=\frac{P_1+P_2+P_3}{3}$

(c) $P=P_1+P_2+P_3$ (d) $P>(P_1+P_2+P_3)$

43. The number of translational degrees of freedom for a diatomic gas is

(a) 2 (b) 3

(c) 5 (d) 6

44. The pressure of a gas is raised from 27°C to 927°C. The root mean square speed is

(a) $\sqrt{(927/27)}$ times the earlier value

(b) remain the same

(c) gets halved

(d) get doubled

45. The equation of state, corresponding to 8g of O_2 is

(a) $PV=8RT$ (b) $PV=RT/4$

(c) $PV=RT$ (d) $PV=RT/2$

46. At 0 K, which of the following properties of a gas will be zero?

(a) Kinetic energy (b) Potential energy

(c) Density (d) Mass

47. The degree of freedom of a molecule of a triatomic gas is

(a) 2 (b) 4

(c) 6 (d) 8

48. The equation of state for 5 g of oxygen at a pressure P and temperature T, when occupying a volume V, will be

(a) $PV=(5/16)RT$ (b) $PV=(5/32)RT$

(c) $PV=5RT$ (d) $PV=(5/2)RT$

where R is the gas constant.

49. At 10° C the value of the density of a fixed mass of an ideal gas divided by its pressure is x. At 110°C this ratio is:

(a) x (b) $\frac{383}{283}x$

(c) $\frac{10}{110}x$ (d) $\frac{283}{383}x$

50. If pressure of a gas contained in a closed vessel is increased by 0.4% when heated by 1°C, the initial temperature must be

(a) 250 K (b) 250°C

(c) 2500 K (d) 25°C

51. The r.m.s. velocity of oxygen molecule at 16°C is 474 m/sec. The r.m.s. velocity in m/s of hydrogen molecule at 127°C is

(a) 1603 (b) 1896

(c) 2230.59 (d) 2730

52. The gases are at absolute temperature 300°K and 350°K respectively. The ratio of average kinetic energy of their molecules is

(a) 7 : 6 (b) 6 : 7

(c) 36 : 49 (d) 49 : 36

53. N molecules, each of mass m, of gas A and 2 N molecules, each of mass 2 m, of gas B are contained in the same vessel which is maintained at a temperature T. The mean square of the velocity of molecules of B type is denoted by v^2 and the mean square of the X component of the velocity of A type is dentoed by ω^2, ω^2/v^2 is

(a) 2 (b) 1

(c) 1/3 (d) 2/3

54. A fixed mass of gas at constant pressure occupies a volume V. The gas undergoes a rise in temperature so that the root mean square velocity of its molecules is doubled. The new volume will be

(a) $V/2$ (b) $V/\sqrt{2}$

(c) $2V$ (d) $4V$

55. If one mole of a monatomic gas ($\gamma = 5/3$) is mixed with one mole of a diatomic gas ($\gamma = 7/3$), the value of γ for the mixture is
(a) 1.40 (b) 1.50
(c) 1.53 (d) 3.07

56. By what percentage should the pressure of a given mass of a gas be increased so as to decrease its volume by 10% at a constant temperature?
(a) 8.1% (b) 9.1%
(c) 10.1% (d) 11.1%

57. The density of a gas is 6×10^{-2} kg/m^3 and the root mean square velocity of the gas molecules is 500 m/s. The pressure exerted by the gas on the walls of the vessel is
(a) 5×10^3 N/m^2 (b) 1.2×10^{-4} N/m^2
(c) 0.83×10^{-4} N/m^2 (d) 30 N/m^2

58. At what temperature, pressure remaining constant, will the r.m.s. velocity of a gas be half of its value at 0°C?
(a) 0°C (b) −273°C
(c) 32°C (d) −204°C

59. A gas has pressure P and volume V. It is now compressed adiabatically to 1/32 times the original volume. Given that $(32)^{1.4} = 128$, the final pressure is ($\gamma = 1.4$)
(a) $P/128$ (b) $P/32$
(c) $32P$ (d) $128P$

60. At 27°C a gas is compressed suddenly such that its pressure becomes (1/8) of original pressure. Final temperature will be ($\gamma = 5/3$)
(a) 450 K (b) 300 K
(c) −142°C (d) 327°C

CHAPTER 13

Oscillations

CONCEPTS

PERIODIC AND OSCILLATORY MOTIONS

Any motion which repeats itself after regular interval of time is called **periodic motion** or **harmonic motion**. The constant interval of time after which the motion is repeated is called **time period**.

Examples :

(i) Motion of planets around the sun.

(ii) Motion of the pendulum of wall clock.

The motion of body is said to be **oscillatory** or **vibratory** motion if it moves back and forth (to and fro) about a fixed point after regular interval of time. The fixed point about which the body oscillates is called mean position or equilibrium position.

Examples:

(i) Vibration of the wire of 'Sitar'.

(ii) Oscillation of the mass suspended from spring.

Note : Every oscillatory motion is periodic but every periodic motion is not oscillatory.

SIMPLE HARMONIC MOTION (S.H.M.)

Simple harmonic motion is the simplest form of vibratory or oscillatory motion.

Types of S.H.M.

Linear S.H.M. : When a particle moves to and fro about a fixed point (called equilibrium position) along a straight line then its motion is called linear simple harmonic motion.

Example : Motion of a mass connected to a spring.

Angular S.H.M. : When a system oscillates angularly with respect to a fixed axis then its motion is called angular simple harmonic motion.

Example : Motion of a bob of a simple pendulum.

Necessary Condition to Execute S.H.M.

In linear S.H.M. : The restoring force (or acceleration) acting on the particle should always be proportional to the displacement of the particle and directed towards the equilibrium position.

i.e $F \propto x$ or $a \propto -x$

Negative sign shows that direction of force and acceleration is towards equilibrium position and x is displacement of particle from equilibrium position.

In angular S.H.M. : The restoring torque (or angular acceleration) acting on the particle should always be proportional to the angular displacement of the particle and directed towards the equilibrium position

i.e., $\tau \propto -\theta$ or $\alpha \propto -\theta$

EQUATION OF SIMPLE HARMONIC MOTION IN LINEAR AND ANGULAR S.H.M

In Linear S.H.M.

Restoring force acting on the particle, F = ma = – kx where k is a spring constant or force constant and x is displacement.

$$\text{Accelesation, } a = -\frac{kx}{m} \Rightarrow \frac{dv}{dt} = -\frac{kx}{m}$$

$$\Rightarrow \quad v\frac{dv}{dt} = -\frac{kx}{m}\frac{dx}{dt} \quad \left[\because v = \frac{dx}{dt}\right]$$

$$\int v\,dv = -\int \frac{kx}{m}\,dx \Rightarrow \frac{v^2}{2} = -\frac{kx^2}{2m} + C$$

At $x = 0, v = v_0$

$$C = \frac{v_0^2}{2} \Rightarrow \frac{v^2}{2} = -\frac{\omega^2 x^2}{2} + \frac{v_0^2}{2} \quad \left[\because \omega^2 = \frac{k}{m}\right]$$

$\Rightarrow \quad v = \sqrt{v_0^2 - \omega^2 x^2}\,, \quad \frac{dx}{dt} = \sqrt{v_0^2 - \omega^2 x^2}$

$\Rightarrow \quad \frac{dx}{\sqrt{v_0^2 - \omega^2 x^2}} = \int dt$

$\Rightarrow \quad \frac{1}{\omega}\sin^{-1}\left(\frac{\omega x}{v_0}\right) = t + c_1 \quad [\because v_0 = \omega A]$

$\Rightarrow \quad x = A\sin(\omega t + \omega c_1)$

$\because$ At $t = 0$, $x = 0$

If velocity is in + (ive) x direction then, $x = A\sin\omega t$

If velocity is in – (ive) x direction then, $x = -A\sin\omega t$.

In Angular S.H.M

Restoring torque acting on the particle $\tau = -C\theta$ where C is a constant which can be defined as torque per unit angular displacement.

Mathematically, $I\alpha = -C\theta$,

where I is the moment of inertia of the system about the axis of rotation and α is the angular acceleration.

$\Rightarrow \quad I\frac{d^2\theta}{dt^2} + C\theta = 0 \Rightarrow \frac{d^2\theta}{dt^2} + \left(\frac{C}{I}\right)\theta = 0$

Since, $\frac{d^2\theta}{dt^2} + \omega^2\theta = 0 \Rightarrow \omega = \sqrt{\left(\frac{C}{I}\right)}$

where C = restoring torque constant and
I = moment of inertia

Comparison of Linear S.H.M. and Angular S.H.M.

S.No.	Linear S.H.M.	Angular S.H.M.
(i)	$F \propto -y$ $F = -ky$ where k is restoring force constant	$\tau \propto -\theta$ $\tau = -C\theta$ where C is restoring torque constant.
(ii)	$a = -\frac{k}{m}y$	$\alpha = -\frac{C}{I}\theta$
(iii)	Equation of motion $\frac{d^2y}{dt^2} + \frac{k}{m}y = 0$	Equation of motion $\frac{d^2\theta}{dt^2} + \frac{C}{I}\theta = 0$
(iv)	$a = -\omega^2 y$	$\alpha = -\omega^2\theta$
(v)	$\omega^2 = \frac{k}{m} \Rightarrow \omega = \sqrt{\frac{k}{m}} = \frac{2\pi}{T} = 2\pi n$	$\omega = \sqrt{\frac{C}{I}} = 2\pi n = \frac{2\pi}{T}$

SOME BASIC TERMS

Mean position : It is the point at which the restoring force on the particle is zero and potential energy is minimum.

Restoring force : The force acting on the particle which tends to bring the particle towards its mean position, is known as restoring force. It always acts in a direction opposite to that of displacement. Displacement is measured from the mean position. It is given by formula, $F = -kx$

Amplitude (A) : The maximum (positive or negative) value of displacement of particle from mean position is defined as amplitude.

Time period (T) : The minimum time after which the particle keeps on repeating its motion is known as time period.

The smallest time taken to complete one oscillation or vibration is also defined as time period.

It is given by, $T = \frac{2\pi}{\omega} = \frac{1}{n}$ where ω is angular frequency and n is frequency.

Oscillation or vibration : When a particle goes on one side from mean position and returns back and then it goes to other side and again returns back to mean position, then this process is known as one oscillation.

Frequency (n or f) : The number of oscillations per second is defined as frequency.

It is given by $n = \frac{1}{T}$ or, $n = \frac{\omega}{2\pi}$

SI unit : hertz (Hz), 1 hertz = 1 cycle per second

Dimension: $[M^0L^0T^{-1}]$

Phase: The phase of a vibrating particle at any instant is the state of the vibrating particle regarding its displacement and direction of vibration at that particular instant.

In the equation $x = A\sin(\omega t + \phi)$, $(\omega t + \phi)$ is the phase of the particle.

- The phase angle at time t = 0 is known as **initial phase** or **epoch**.
- The difference of total phase angles of two particles executing S.H.M. with respect to the mean position is known as **phase difference** $\Delta\phi$.
- Two vibrating particles are said to be **in same phase** if the phase difference between them is an even multiple of π, i.e., $\Delta\phi = 2n\pi$, where n = 0, 1, 2, 3,....
- Two vibrating particle are said to be **in opposite phase** if the phase difference between them is an odd multiple of π, i.e., $\Delta\phi = (2n+1)\pi$, where n = 0, 1, 2, 3,....
- **Angular frequency (ω) :** The rate of change of phase angle of a particle with respect to time is defined as its angular frequency.

  Angular frequency $\omega = \sqrt{\dfrac{k}{m}}$

  where k = spring constant and m = mass.

  SI unit: radian/second,

  Dimension : $[M^0L^0T^{-1}]$

PROJECTION OF UNIFORM CIRCULAR MOTION ON A DIAMETER

Let at t = 0 the particle is at the point A and after time t it is at B. The foot of the perpendicular from the point on the diameter oscillates about O which is S.H.M. in nature. The displacement of the projection from centre is given by

$$y = R\sin(\omega t + \phi_0) \quad \text{...(i)}$$

The maximum displacement of the projection from the centre is called amplitude. If amplitude is denoted by A then R = A

$$\Rightarrow \quad y = A\sin(\omega t + \phi_0) \quad \text{...(ii)}$$

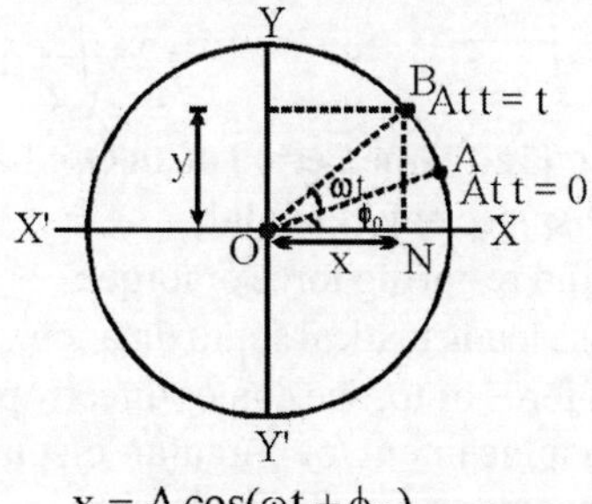

and $$x = A\cos(\omega t + \phi_0) \quad \text{...(iii)}$$

PARAMETERS RELATED TO SIMPLE HARMONIC MOTION

Displacement (x) in S.H.M.

$$x = A\cos(\omega t + \phi_0)$$

Velocity (v) in S.H.M. :

$$v = \frac{dx}{dt} = A\omega\cos(\omega t + \phi) = \omega\sqrt{(A^2 - x^2)},$$

$v_{max} = A\omega$ (in mean position or x = 0),

$v_{min} = 0$ (in extreme position or $x = \pm A$)

Acceleration (a) in S.H.M.

$$a = \frac{dv}{dt} = \frac{d^2x}{dt^2} = -\omega^2 A\sin(\omega t + \phi) = -\omega^2 x$$

$a_{max} = \pm\omega^2 A$ (in extreme position or $x = \pm A$)

$a_{min} = 0$ (in mean position or x = 0)

Force acting on the particle in S.H.M.

$$F(t) = -Kx(t)$$

$$K = m\omega^2 \text{ and } \omega = \sqrt{\frac{K}{m}}$$

Kinetic energy in S.H.M. (K.E.)

$$K.E. = \frac{1}{2}mv^2 = \frac{1}{2}m\omega^2A^2\cos^2(\omega t + \phi)$$

$$= \frac{1}{2}kA^2\cos^2(\omega t + \phi)$$

or $$K.E. = \frac{1}{2}m\omega^2(A^2 - x^2) = \frac{1}{2}k(A^2 - x^2)$$

$$(K.E.)_{max} = \frac{1}{2}kA^2 = \frac{1}{2}m\omega^2A^2 = 2\pi^2 mA^2n^2,$$

when x = 0

$(K.E.)_{min} = 0$, when $x = \pm A$.

Average K.E. with respect to displacement

$\langle K.E.\rangle_x = 1/3kA^2$

Average K.E. with respect to time $\langle K.E.\rangle_t = 1/4kA^2$

Potential energy in S.H.M (P.E.)

$$P.E. = U(x) = \frac{1}{2}kx^2 = \frac{1}{2}m\omega^2x^2 = \frac{1}{2}kA^2\sin^2(\omega t + \phi)$$

$$(P.E.)_{max} = \frac{1}{2}kA^2 = \frac{1}{2}m\omega^2A^2 = 2\pi^2mn^2A^2, \text{ when } x = \pm A$$

$(P.E.)_{min} = 0$, when x = 0

Average P.E. with respect to displacement $\langle P.E.\rangle_x = 1/6kA^2$

Average P.E. with respect to time $\langle P.E.\rangle_t = 1/4kA^2$

Total energy in S.H.M. (T.E.):

Total energy $E = KE + PE = 1/2kA^2 = 1/2\, m\omega^2A^2$

$= 2\pi^2mn^2A^2 = \text{constant}.$

DIFFERENT CURVES OF SIMPLE HARMONIC MOTION WHEN T = 0, INITIAL PHASE ϕ = 0.

(i) Displacement-Time

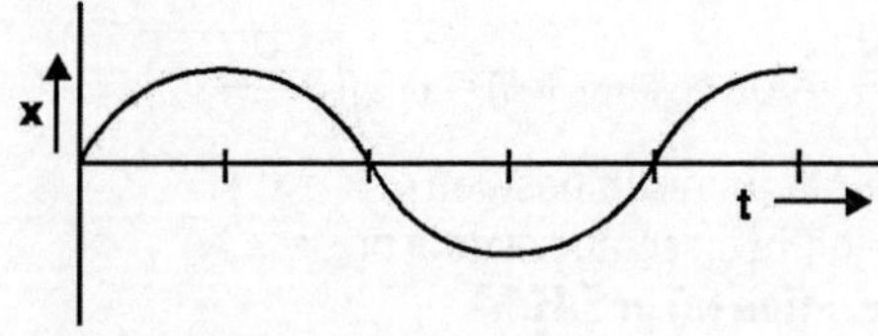

(ii) Velocity -Time

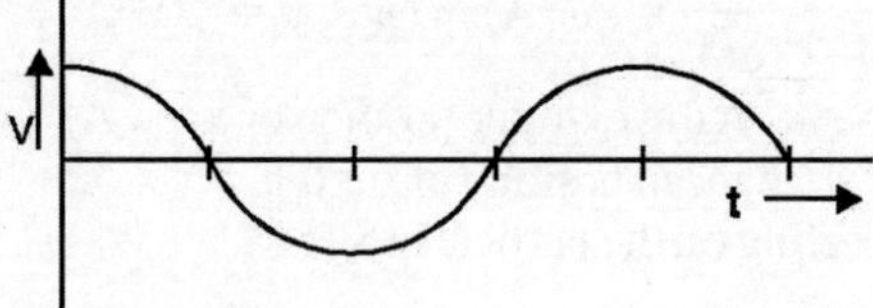

(iii) Acceleration -Time

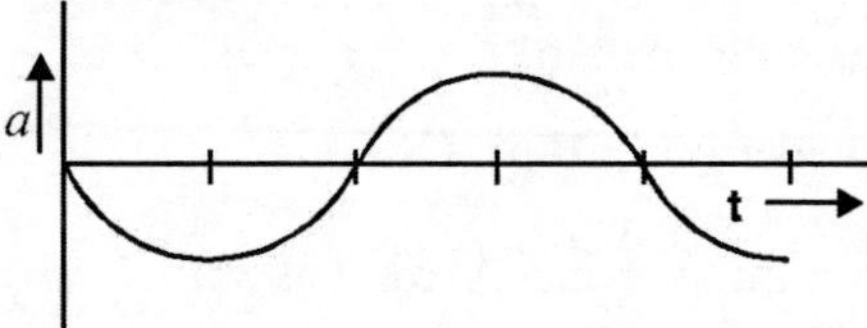

(iv) Force -Time

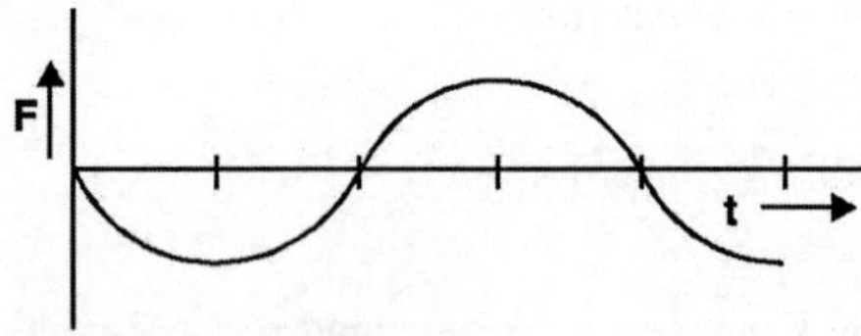

(v) Potential energy -Time

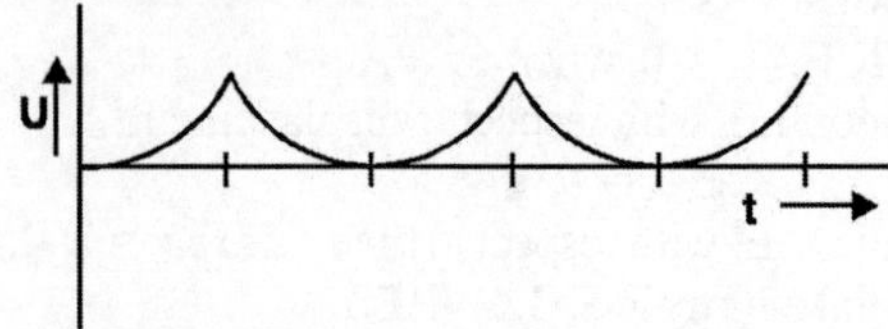

(vi) Kinetic energy -Time

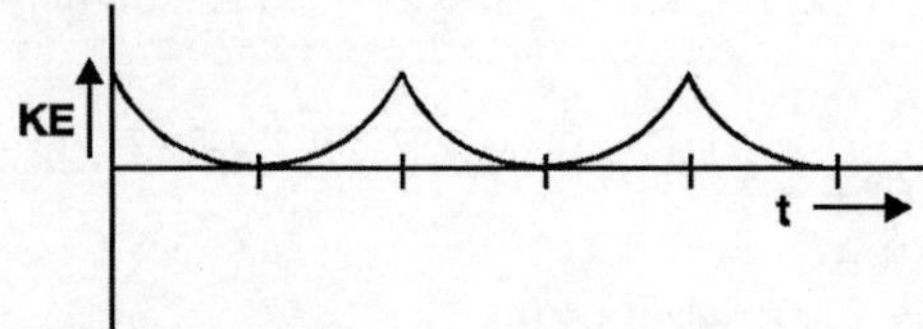

(vii) Velocity - Displacement

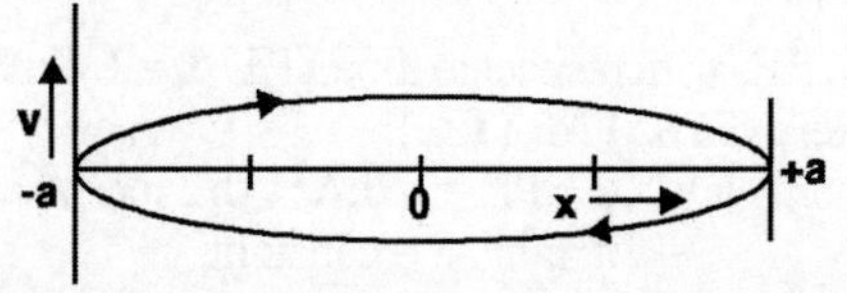

(viii) Acceleration - Displacement

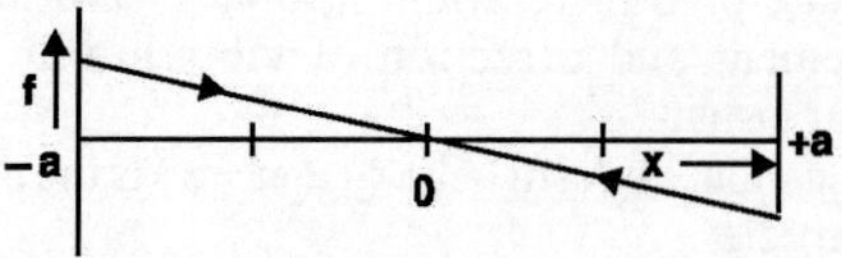

(ix) Force - Displacement

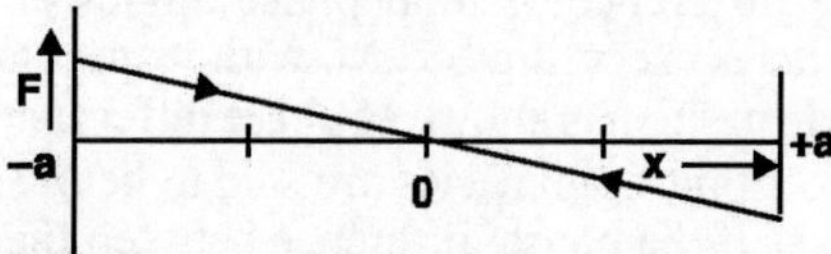

(x) Potential energy - Displacement

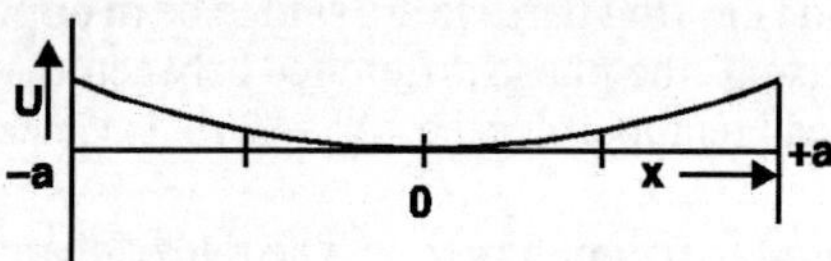

(xi) Kinetic energy - Displacement

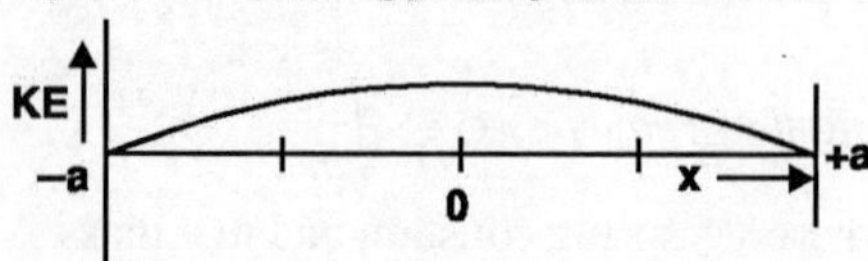

(xii) Total energy - Displacement

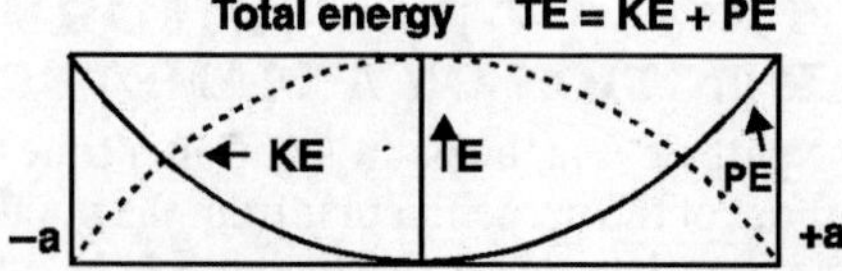

The simple pendulum : An ideal simple pendulum consists of a point-mass suspended by a flexible, inelastic and weightless string from a rigid support of infinite mass. In practice we obtain it by suspending a small metal bob by a long and fine cotton thread from a rigid support. Time period of simple pendulum,

$$T = 2\pi\sqrt{\frac{I}{mgL}} \quad \text{or} \quad T = 2\pi\sqrt{\frac{\ell}{g}} \quad [\because I = mL^2]$$

Steps to Find Time Period of an Oscillation

(i) Displace object slightly

(ii) Find restoring force or torque

(iii) Use mathematical approximation

(iv) If force or torque comes directly proportional to displacement or angular displacement then consider it as SHM and write

$T = 2\pi\sqrt{\frac{m}{k}}$ or $T = 2\pi\sqrt{\frac{I}{c}}$, where k or c are proportionality constant for linear or angular SHM in force equation.

(v) If oscillation is not SHM then see kinematic condition for the problem and solve accordingly.

(vi) For SHM time period can also be calculated from energy method.
Write total energy at any position, differentiate it w.r.t. time and use $v = \frac{dx}{dt}, a = \frac{dv}{dt} = v\frac{dv}{dx} = \frac{d^2x}{dt^2}$ to reduce equation in the form of $\frac{d^2x}{dt^2} + \omega^2 x = 0$ and compare it to get ω.

FORCE LAWS FOR SIMPLE HARMONIC MOTION

A particle of mass m oscillating under the influence of a Hooke's law, restoring force is given by F = – k x exhibits simple harmonic motion with $\omega = \sqrt{\frac{k}{m}}$ (angular frequency),

and $T = 2\pi\sqrt{\frac{m}{k}}$ (time period)

Such a system is also called a linear oscillator.

The time-period of simple pendulum is,

$$T = 2\pi\sqrt{\left(\frac{\ell}{g}\right)}$$

If a simple pendulum is made to oscillate in a non-inertial frame of reference, the torque due to pseudo force must be taken into account.

If ℓ is large of the order of R_e (radius of earth)

$$T = 2\pi\sqrt{\frac{R_e}{g\left(1+\frac{R_e}{\ell}\right)}}$$

when $\ell \to \infty$ then $T = 2\pi\sqrt{\frac{R_e}{g}}$ = 84.4 minutes

The time period of a simple pendulum placed inside a satellite is infinite i.e., it does not oscillate, because g = 0 there.

If a tunnel is dug across the earth along its diameter and a body is allowed to fall in it, then it will oscillate simple harmonically.

Its time period, $T = 2\pi\sqrt{\frac{R_e}{g}}$ = 84.4 minutes.

The time period of any simple pendulum cannot be equal to one day.

Graphs of Simple Pendulum

(i) T versus $\sqrt{\ell}$ **(ii) T^2 versus ℓ**

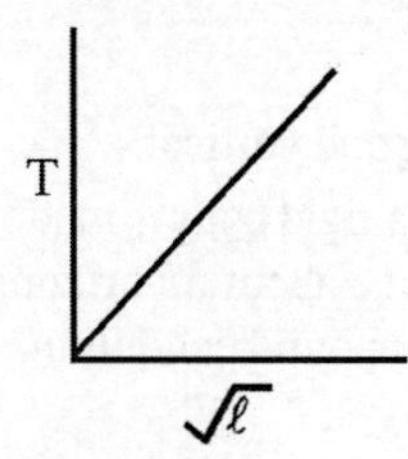

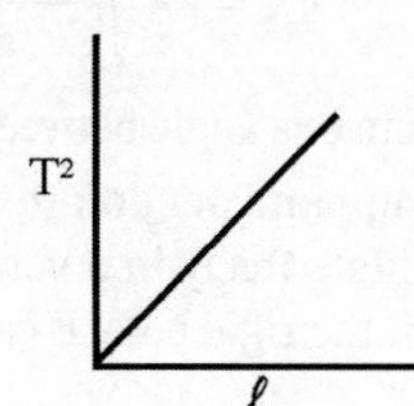

(iii) T versus ℓ

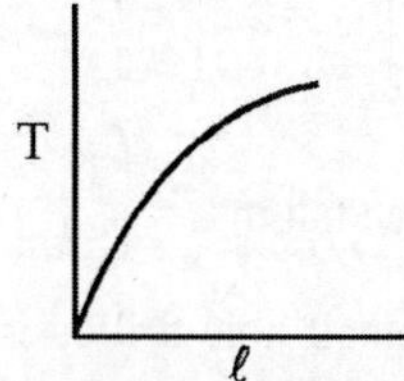

The time period in a reference frame moving with an acceleration *a* in a horizontal plane

$$T = 2\pi\left[\frac{\ell}{\sqrt{g^2+a^2}}\right]^{1/2}$$

OSCILLATIONS OF A PENDULUM IN A LIFT

(i) Lift moving up with an acceleration a, $T = 2\pi\sqrt{\frac{\ell}{g+a}}$

(ii) Lift moving down with an acceleration a, $T = 2\pi\sqrt{\frac{\ell}{g-a}}$

(iii) Lift falling freely (a = g), ∴ T = ∞.

(iv) Lift moving with a uniform speed (a = 0), $T = 2\pi\sqrt{\frac{\ell}{g}}$

Oscillations of a pendulum having bob of density ρ in a liquid of density d

$$T = 2\pi\sqrt{\frac{\ell}{g\left(1-\frac{d}{\rho}\right)}}$$

Second's pendulum

A simple pendulum whose period is 2 seconds, is called second's pendulum. Its length is nearly 100 cm on the earth.

Time period of Conical Pendulum

$$T = 2\pi\sqrt{\left(\frac{\ell\cos\theta}{g}\right)}$$

where θ is angle between string and vertical.

Compound pendulum : When a rigid body is made to oscillate freely in a vertical plane about a horizontal axis, the rigid body is called compound pendulum.

$$\text{Periodic time } T = 2\pi\left[\frac{\frac{K^2}{\ell}+\ell}{g}\right]^{1/2} = 2\pi\left[\frac{L}{g}\right]^{1/2}$$

$$\text{Length of equivalent simple pendulum } L = \left(\frac{K^2}{\ell}+\ell\right)$$

If a hollow bob with a very small hole drilled in its bottom, is filled with water and it is made to oscillate then first of all its period starts increasing but when more than half liquid is leaked out its period starts decreasing. This is due to the displacement of centre of mass.

Motion of the particle is bound between two points. This bound region is called potential well. When a potential well is parabolic, the motion of a particle in it is simple harmonic and its period $T = 2\pi\sqrt{m/k}$

A uniform cubic plank of side L is floating in a liquid of density n times its density (n > 1). If it is immersed slightly and released, it will execute simple harmonic motion of period $T = 2\pi\sqrt{\frac{L}{ng}}$.

If a spring of force constant k is cut into n equal pieces then the effective force constant becomes $1/\sqrt{n}$ times and time period becomes $\sqrt{n}$ times.

SERIES AND PARALLEL COMBINATION OF SPRINGS

Series Combination of Springs

When two spring are joined in series, the equivalent stiffness of the combination may be obtained as

$$\frac{1}{k} = \frac{1}{k_1} + \frac{1}{k_2}$$

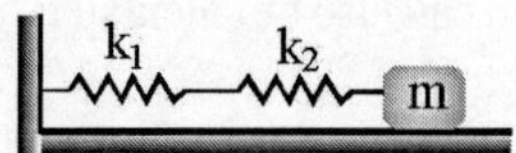

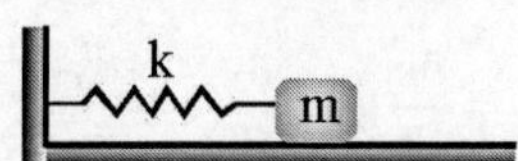

Parallel Combination of Springs

When two springs are joined in parallel, the equivalent stiffness of the combination is given by

$$k = k_1 + k_2$$

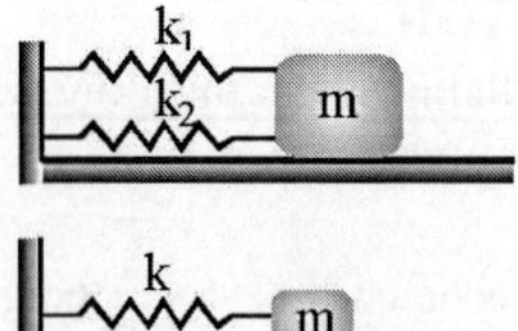

Two Particles System

$$\text{Time period } T = 2\pi\sqrt{\frac{\mu}{k}}$$

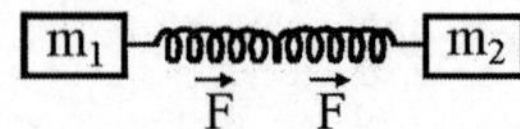

$$\text{where } \mu = \frac{m_1 m_2}{m_1 + m_2} = \text{reduced mass}$$

Two particles system with only m_1 oscillating

$$T = 2\pi\sqrt{\frac{\text{mass of oscillating particle}}{k}}$$

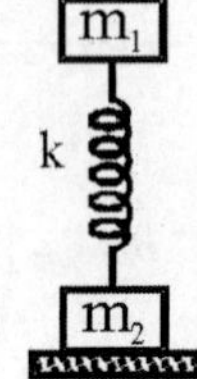

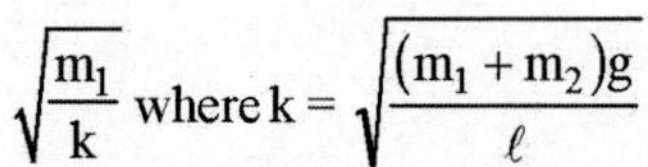

$$\sqrt{\frac{m_1}{k}} \text{ where } k = \sqrt{\frac{(m_1+m_2)g}{\ell}}$$

Vertical oscillations of a body suspended from a wire of Young's modulus Y

$$T = 2\pi\sqrt{\frac{mL}{YA}}$$

Oscillations of a frictionless piston fitted in a cylinder filled with ideal gas

$$T = 2\pi\sqrt{\frac{VM}{\gamma PA^2}}$$

where $\gamma = \frac{C_P}{C_V}$

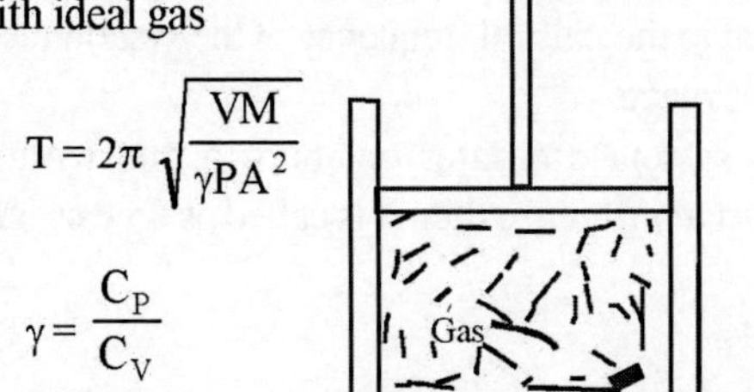

Oscillation of a small ball of radius r placed in a concave glass of radius R

$$T = 2\pi\sqrt{\frac{(R-r)}{g}} \;;\; T = 2\pi\sqrt{\frac{R}{g}} \quad (\text{if } r << R)$$

Oscillations of a liquid in a U-tube

If the liquid in one limb of a U tube is first pressed and then released, the motion of liquid is found to be simple harmonic.

If a = area of cross-section of limbs ; y = displacement; d = density of liquid

Restoring force F = – 2y a dg

Therefore, k = 2adg

and m = 2h ad

$$T = 2\pi\sqrt{\frac{m}{k}} = 2\pi\sqrt{\frac{h}{g}}$$, h is vertical height of liquid column.

Oscillations of a floating body

Cylindrical body

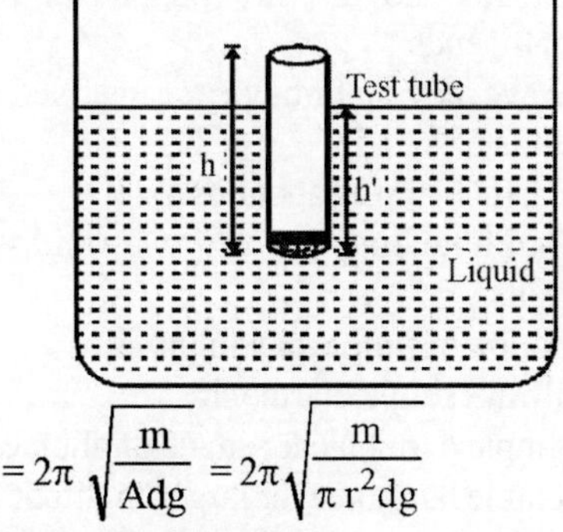

$$T = 2\pi\sqrt{\frac{m}{Adg}} = 2\pi\sqrt{\frac{m}{\pi r^2 dg}}$$

Rectangular block in liquid

Rectangular block floating in a liquid $T = 2\pi\sqrt{\frac{h'd'}{gd}}$

where d = density of liquid, d′ = density of block, h = height of block and h′ = height of block immersed in liquid

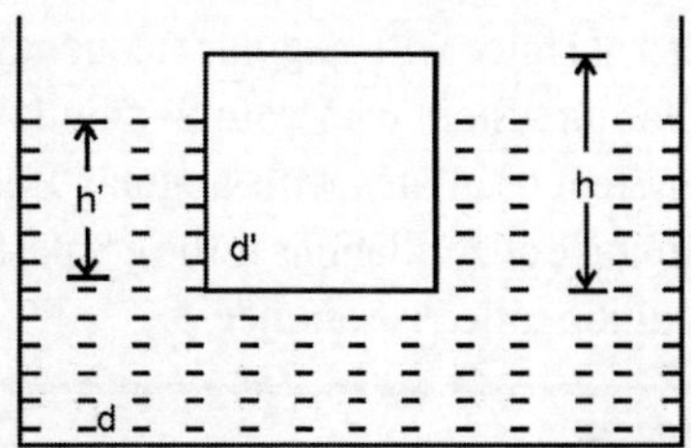

Period of a freely suspended magnet in a magnetic field B

When a freely suspended magnet is released after displacing it through an angle θ from the earth's magnetic meridian, then it starts oscillating.

$T = 2\pi\sqrt{\frac{I}{MB}}$, M is magnetic moment and I is moment of inertia.

A combination of two simple harmonic motions with arbitrary amplitudes and phases is not necessarily periodic. It is periodic only if frequency of one motion is an integral multiple of the other's frequency. However, a periodic motion can always be expressed as a sum of infinite number of harmonic motions with appropriate amplitudes.

For motion of a particle to be simple harmonic, its displacement x must be expressible in either of the following forms :

$x = A\cos\omega t + B\sin\omega t$;

$x = A\cos(\omega t + \alpha)$, $x = B\sin(\omega t + \beta)$

The three forms are completely equivalent (any one can be expressed in terms of any other two forms).

Damped Simple Harmonic Motion

The **damped simple harmonic motion** is not strictly simple harmonic, it is approximately so only for time intervals much less than 2m/b where b is the damping constant.

Damping force $f_d = -bv$

Oscillation performed under the influence of frictional force is called as damped oscillation.

FORCED OSCILLATIONS AND RESONANCE

In **forced oscillations**, the steady state motion of the particle (after the force oscillations die out) is simple harmonic motion whose frequency is the frequency of the driving frequency ω_d, not the natural frequency ω of the particle.

If an external force with angular frequency ω_d acts on an oscillating system with natural angular frequency ω, the system oscillates with angular frequency ω_d. The amplitude of oscillations is the greatest when $\omega_d = \omega$ a condition called **resonance**.

When the frequency of the external periodic force is equal to the natural frequency of the system resonance takes place.

If a system oscillates on its own and without any external influence then it is called as **free oscillation**.

Exercise

1. A child swinging on swing in sitting position stands up. The the time period of the swing will :
   (a) increase
   (b) decrease
   (c) remains same
   (d) increase if the child is tall and decrease if the child is short

2. The graph of time period (T) of simple pendulum verus its length (l) is

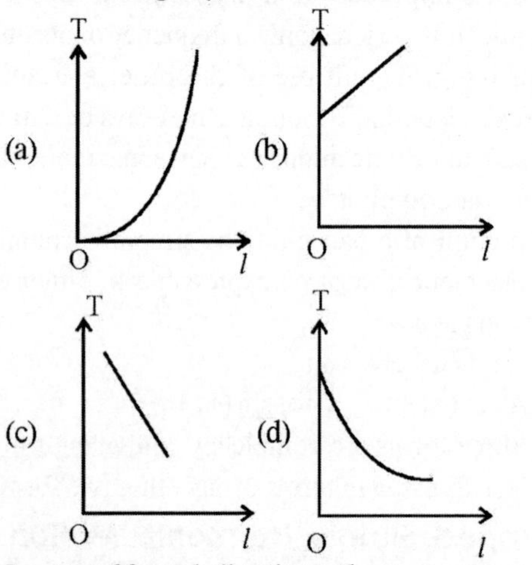

3. In case of forced vibrations, the resonance wave become very sharp when the
   (a) applied periodic force is small
   (b) quality factor is small
   (c) damping force is small
   (d) restoring force is small

4. What is the time period of a pendulum hanged in a satellite?
   (T is time period of the pendulum on the earth)
   (a) Zero (b) T
   (c) $\frac{T}{\sqrt{6}}$ (d) Infinity

5. Which of the following functions represents a simple harmonic oscillation?
   (a) $y = \sin \omega t - \cos \omega t$
   (b) $y = \sin \omega t + \sin 2\omega t$
   (c) $y = \sin \omega t + \cos 2\omega t$
   (d) $y = \sin^2 \omega t + \sin \omega t$

6. For a particle executing simple harmonic motion, which of the following statements is not correct?
   (a) Total energy of the particle always remains same
   (b) Restoring force is always directed towards a fixed point
   (c) Restoring force is maximum at the extreme position
   (d) Acceleration of the particle is maximum at the equilibrium position

7. If the metal bob of a simple pendulum is replaced by wooden bob of same size, then its time period will
   (a) increase
   (b) decrease
   (c) remains same
   (d) may increase or decrease

8. Which of the following is a simple harmonic motion?
   (a) Particle moving through a string fixed at both ends
   (b) Wave moving through a string fixed at both ends
   (c) Earth spinning about its axis
   (d) Ball bouncing between two rigid vertical walls

9. The motion of simple pendulum is :
   (a) always simple harmonic
   (b) simple harmonic for small amplitude
   (c) simple harmonic for large amplitude
   (d) simple harmonic for small length

10. The time period of second's pendulum is :
    (a) 1s (b) 2s
    (c) 4s (d) None of these

11. The phase difference between displacement and acceleration of simple harmonic oscillator is :
(a) $\frac{\pi}{2}$ (b) π
(c) $\frac{\pi}{4}$ (d) 2π

12. A simple pendulumn oscillates with a frequency f. The frequency with which its K.E. varies is :
(a) f (b) $\frac{f}{2}$
(c) $2f$ (d) $4f$

13. At resonance, the frequency of periodic force :
(a) must be equal to the frequency of oscillations
(b) must be less than the frequency of oscillations
(c) must be greater than the frequency of oscillations
(d) may be equal to the frequency of oscillations

14. A rod is hinged vertically at one end and is forced to oscillate in a vertical plane with hinged end at the top, the motion of the rod:
(a) is simple harmonic
(b) is oscillatory but not simple harmonic
(c) is pericolic but not oscillatory
(d) may be simple harmonic

15. The superposition of two mutually perpendicular simple harmonic motions having same amplitude and a phase difference of $\frac{\pi}{2}$ results into :
(a) a circle
(b) an ellipse
(c) simple harmonic motion
(d) none of these

16. The necessary and sufficient condition for simple harmonic motion is :
(a) constant acceleration
(b) constant time period
(c) constant amplitude
(d) restoring force directly proportional to displacement

17. The graph plotted between the velocity and displacement from mean position of a particle executing SHM is :
(a) circle (b) ellipse
(c) parabola (d) straight line

18. A particle moving along the X-axis, executes simple harmonic motion then the force acting on it is given by
(a) $-A\,kx$ (b) $A\cos(kx)$
(c) $A\exp(-kx)$ (d) Akx
where, A and k are positive constants.

19. A particle, with restoring force proportional to displacement and resistive force proportional to velocity is subjected to a force $F\sin\omega_0$. If the amplitude of the particle is maximum for $\omega = \omega_1$ and the energy of the particle is maximum for $\omega = \omega_2$, then
(a) $\omega_1 = \omega_0$ and $\omega_2 \neq \omega_0$
(b) $\omega_1 = \omega_0$ and $\omega_2 = \omega_0$
(c) $\omega_1 \neq \omega_0$ and $\omega_2 = \omega_0$
(d) $\omega_1 \neq \omega_0$ and $\omega_2 \neq \omega_0$

20. A mass m is suspended from a two coupled springs, connected in series. The force constant for springs are k_1 and k_2. The time period of the suspended mass will be
(a) $T = 2\pi\sqrt{\frac{m}{k_1 - k_2}}$
(b) $T = 2\pi\sqrt{\frac{mk_1k_2}{k_1 + k_2}}$
(c) $T = 2\pi\sqrt{\frac{m}{k_1 + k_2}}$
(d) $T = 2\pi\sqrt{\frac{m(k_1 + k_2)}{k_1k_2}}$

21. Select the correct statements
(a) A simple harmonic motion is necessarily periodic.
(b) A simple harmonic motion may be oscillatory
(c) An oscillatory motion is necessarily periodic
(d) A periodic motion is necessarily oscillatory

22. A particle moves in a circular path with a uniform speed. Its motion is
(a) Periodic
(b) Oscillatory
(c) Simple harmonic
(d) Angular simple harmonic

23. A particle moves in a circular path with a continuously increasing speed. Its motion is
(a) Periodic (b) Oscillatory
(c) Simple harmonic (d) None of them

24. Which of the following quantities are always negative in a simple harmonic motion?

(a) $\vec{F}.\vec{r}$. (b) $\vec{v}.\vec{r}$.

(c) $\vec{a}.\vec{r}$. (d) (a) & (c)

25. Which of the following quantities are always positive in a simple harmonic motion?

(a) $\vec{F}.\vec{a}$. (b) $\vec{v}.\vec{r}$.

(c) $\vec{a}.\vec{r}$. (d) $\vec{F}.\vec{r}$.

26. Which of the following quantities are always zero in a simple harmonic motion?

(a) $\vec{F}\times\vec{a}$ (b) $\vec{v}\times\vec{r}$

(c) $\vec{a}\times\vec{r}$ (d) All of these

27. Suppose a tunnel is dug along a diameter of the earth. A particle is dropped from a point, a distance h directly above the tunnel, the motion of the particle is

(a) Simple harmonic (b) Parabolic

(c) Oscillatory (d) Linear

28. Which of the following will change the time period as they are taken to moon?

(a) A simple pendulum

(b) A physical balance

(c) A torsional pendulum

(d) A spring-mass system

29. The displacement of a particle in simple harmonic motion in one time period is

(a) A (b) $2A$

(c) $4A$ (d) zero

30. The distance moved by a particle in simple harmonic motion in one time period is

(a) A (b) $2A$

(c) $4A$ (d) zero

31. The average acceleration in one time period in a simple harmonic motion is

(a) $A\omega^2$ (b) $A\omega^2/2$

(c) $A\omega^2/\sqrt{2}$ (d) zero

32. The motion of a particle is given by $x = A \sin\omega t + B \cos \omega t$. The motion of the particle is

(a) Not simple harmonic

(b) Simple harmonic with amplitude $(A - B)/2$

(c) Simple harmonic with amplitude $(A + B)/2$

(d) Simple harmonic with amplitude $\sqrt{A^2 + B^2}$

33. The displacement of a particle is given by $\vec{r} = A(\vec{i}\cos\omega t + \vec{j}\sin\omega t)$. The motion of the particle is

(a) Simple harmonic

(b) On a straight line

(c) On a circle

(d) With constant acceleration.

34. A particle moves on the X-axis according to the equation $x = A + B \sin \omega t$. The motion is simple harmonic with amplitude

(a) A (b) B

(c) $A+B$ (d) $\sqrt{A^2+B^2}$

35. The average energy in one time period in simple harmonic motion is

(a) $\frac{1}{2}m\omega^2a^2$ (b) $m\omega^2a^2$

(c) $1.5\, m\omega^2a^2$ (d) None

36. A spring-mass system oscillates with a frequency ν. If it is taken in an elevator slowly accelerating upward, the frequency will

(a) increase (b) decrease

(c) remain same (d) become zero

37. A simple pendulum clock that keeps correct time on the earth is taken to the moon. It will run

(a) at correct rate (b) 6 times faster

(c) $\sqrt{6}$ times faster (d) $\sqrt{6}$ times slower

38. A wall clock uses a vertical spring-mass system to measure the time. Each time the mass reaches an extreme position, the clock advances by a second. The clock gives correct time at the equator. If the clock is taken to the poles it will

(a) run slow (b) run fast

(c) stop working (d) give correct time

39. In a simple harmonic oscillator, at the mean position

(a) Kinetic energy is minimum, potential energy is maximum

(b) Both kinetic and potential energies are maximum

(c) Kinetic energy is maximum, potential energy is minimum

(d) Both kinetic and potential energies are minimum

40. The composition of two simple harmonic motions of equal periods at right angle to each other and with a phase difference of π results in the displacement of the particle along

(a) circle (b) figures of eight

(c) straight line (d) ellipse

41. The angular velocity and the amplitude of a simple pendulum is ω and a respectively. At a displacement x from the mean position if its kinetic energy is T and potential energy is V, then the ratio of T to V is

(a) $\frac{(a^2 - x^2\omega^2)}{x^2\omega^2}$ (b) $\frac{x^2\omega^2}{(a^2 - x^2\omega^2)}$

(c) $\frac{(a^2 - x^2)}{x^2}$ (d) $\frac{x^2}{(a^2 - x^2)}$

42. A body is executing S.H.M. When the displacements from the mean position are 4 cm and 5 cm, the corresponding velocities of the body are 10 cm per sec and 8 cm per sec. Then the time period of the body is

(a) 2π sec (b) π/2 sec
(c) π sec (d) (3π/2) sec

43. A simple pendulum is suspended from the roof of a trolley which moves in a horizontal direction with an acceleration a, then the time period is given by $T = 2\pi\sqrt{(l/g)}$, where g is equal to

(a) g (b) $g - a$
(c) $g + a$ (d) $\sqrt{(g^2 + a^2)}$

44. If a simple harmonic oscillator has got a displacement of 0.02 m and acceleration equal to 2.0 m/s^2 at any time, the angular frequency of the oscillator is equal to

(a) 10 rad/s (b) 0.1 rad/s
(c) 100 rad/s (d) 1 rad/s

45. A simple harmonic oscillator has an amplitude A and time period T. The time required by it to travel from $x = A$ to $x = A/2$ is

(a) $T/6$ (b) $T/4$
(c) $T/3$ (d) $T/2$

46. A wave has SHM (Simple Harmonic Motion) whose period is 4 seconds while another wave which also possess SHM has its period 3 seconds. If both are combined, then the resultant wave will have the period equal to

(a) 4 seconds (b) 5 seconds
(c) 12 seconds (d) 3 seconds

47. A body executes SHM with an amplitude A. At what displacement from the mean position is the potential energy of the body is one fourth of its total energy?

(a) $A/4$
(b) $A/2$
(c) $3A/4$
(d) Some other fraction of A.

48. Which one of the following is a simple harmonic motion?

(a) Ball bouncing between two rigid vertical walls
(b) Particle moving in a circle with uniform speed
(c) Wave moving through a string fixed at both ends
(d) Earth spinning about its own axis

49. A particle is subjected to two mutually perpendicular simple harmonic motions such that its x and y coordinates are given by $x = 2\sin\omega t$; $y = 2\sin\left(\omega t + \frac{\pi}{4}\right)$ The path of the particle will be

(a) a straight line (b) a circle
(c) an ellipse (d) a parabola

50. In a simple harmonic motion, when the displacement is one-half the amplitude, what fraction of the total energy is kinetic?

(a) 0 (b) $\frac{1}{4}$

(c) $\frac{1}{2}$ (d) $\frac{3}{4}$

51. A linear harmonic oscillator of force constant 2×10^6 N/m and amplitude 0.01 m has a total mechanical energy of 160 J. Its

(a) maximum potential energy is 160 J
(b) maximum potential energy is not 100 J
(c) minimum potential energy is zero
(d) minimum potential energy is 100 J

52. A particle starts simple harmonic motion from the mean position. Its amplitude is A and time period is T. What is its displacement when its speed is half of its maximum speed

(a) $\frac{\sqrt{2}}{3}A$ (b) $\frac{\sqrt{3}}{2}A$

(c) $\frac{2}{\sqrt{3}}A$ (d) $\frac{A}{\sqrt{2}}$

53. If the length of a simple pendulum is increased by 2%, then the time period

(a) increases by 2% (b) decreases by 2%
(c) increases by 1% (d) decreases by 1%

54. Two simple harmonic motions with the same frequency act on a particle at right angles i.e., along x and y axis. If the two amplitudes are equal and the phase difference is $\pi/2$, the resultant motion will be

(a) a circle
(b) an ellipse with the major axis along y-axis
(c) an ellipse with the major axis along x-axis
(d) a straight line inclined at 45° to the x-axis

55. A hollow sphere is filled with water. It is hung by a long thread. As the water flows out of a hole at the bottom, the period of oscillation will

(a) first increase and then decrease
(b) first decrease and then increase
(c) go on increasing
(d) go on decreasing

56. Two simple pendulums of length 5 m and 20 m respectively are given small linear displacement in one direction at the same time. They will again be in the phase when the pendulum of shorter length has completed .... oscillations

(a) 5 (b) 1
(c) 2 (d) 3

57. A mass m is vertically suspended from a spring of negligible mass; the system oscillates with a frequency n. What will be the frequency of the system, if a mass 4 m is suspended from the same spring?

(a) $\frac{n}{4}$ (b) $4n$
(c) $\frac{n}{2}$ (d) $2n$

58. The potential energy of a particle with displacement x is V(x). If the motion is simple harmonic, then (K is the spring constant)

(a) $V = \frac{Kx^2}{2}$ (b) $V = Kx^2$
(c) $V = K/x$ (d) $V = Kx$

59. The period of a simple pendulum of length ℓ at a place where the acceleration due to gravity is g is T. What is the period of a simple pendulum of the same length at a place where the acceleration due to gravity is 1.02 g?

(a) T (b) $1.02\,T$
(c) $0.99\,T$ (d) $1.01\,T$

60. A particle executes simple harmonic motion along a straight line with an amplitude A. The potential energy is maximum when the displacement is

(a) $\pm A$ (b) zero
(c) $\pm A/2$ (d) $\pm A/\sqrt{2}$

Waves

CONCEPTS

A wave is not a motion of matter as a whole in a medium. A wind is different from the sound wave in air. The former involves motion of air from one place to the other. The later involves compressions and rarefactions of layers of air. In a wave, energy and not the matter is transferred from one point to the other. Energy transfer takes place because of the coupling through elastic forces between neighbouring oscillating parts of the medium.

TRANSVERSE AND LONGITUDINAL WAVES

Transverse waves are waves in which the particles of the medium oscillate perpendicular to the direction of wave propagation. **Longitudinal waves** are waves in which the particles of the medium oscillate along the direction of wave propagation. Transverse waves can propagate only in medium with shear modulus of elasticity, Longitudinal waves need bulk modulus of elasticity and are therefore, possible in all media, solids, liquids and gases. Transverse waves can be produced in solids and liquids not in gases.

DISPLACEMENT RELATION IN A PROGRESSIVE OR TRAVELLING WAVE

The displacement in a sinusoidal wave propagating in the positive x direction is given by $y(x, t) = a \sin(kx - \omega t + \phi)$ where a is the amplitude of the wave, k is the angular wave number, ω is the angular frequency, $(kx - \omega t + \phi)$ is the phase, and ϕ is the phase constant or phase angle.

The travelling wave moving with constant speed v towards positive x direction must satisfy the following wave function.

i.e., $$\frac{d^2y}{dt^2} = v^2 \frac{d^2y}{dx^2}$$

Velocity of particle $v = \dfrac{dy}{dt} = a\omega \cos(\omega t - kx)$

Slope of wave $= \dfrac{dy}{dx} = -ak \cos(\omega t - kx)$

Acceleration (a_p) in a sinusoidal wave

$$a_p = \frac{d^2y}{dt^2} = \omega^2 A \sin(kx - \omega t) = -\omega^2 y$$

Equation of a Plane Progressive, Simple Harmonic Wave

Displacement, $y = a \sin 2\pi\left(\dfrac{t}{T} - \dfrac{x}{\lambda}\right)$

or $$y = a \sin \frac{2\pi}{\lambda}(vt - x)$$

THE SPEED OF A TRAVELLING WAVE

The speed of a progressive or travelling wave

$$v = \frac{\omega}{K} = \frac{\lambda}{T} = \lambda v$$

Wavelength (λ) of a progressive wave is the distance between two consecutive points of the same phase at a given time.

It is twice the distance between two consecutive nodes or antinodes in a stationary wave.

A stationary wave is formed when two identical waves travelling in the opposite directions along a stratight line interfere. It is possible only in bounded medium.

The period of oscillation of a wave, $T = \dfrac{2\pi}{\omega}$

The frequency of a wave ν or $f = \dfrac{1}{T} = \dfrac{\omega}{2\pi}$

Sound waves are longitudinal mechanical waves that can travel through solids, liquids, or gases. The speed v of sound wave in a fluid having bulk modulus B and density ρ is $v = \sqrt{B/\rho}$.

The speed of longitudinal waves in a metallic bar or solid is $v = \sqrt{Y/\rho}$.

For gases, since B = γ P, the speed of sound is $v = \sqrt{\gamma P/\rho}$

This is the modification of Newton's formula

$$v = \sqrt{\frac{p}{\rho}}$$ and is called Laplace correction.

The Speed of a Transverse Wave on a Stretched String

The speed of a transverse wave on a stretched string is set by the properties of the string. The speed on a string with tension T and linear mass density μ (i.e., mass per unit length) is $v = \sqrt{T/\mu}$

THE PRINCIPLE OF SUPERPOSITION OF WAVES

It is the addition of individual waveforms to determine the net wave form.

$$\gamma(x,t) = \gamma_1(x,t) + \gamma_2(x,t) + ... + \gamma_n(x,t)$$

REFLECTION OF WAVES

An **echo** is an example of the reflection of sound waves from the rigid boundary. It is repetition of sound heard after reflection from a distant object.

The incident and reflected waves of sound obey the usual laws of reflection.

i.e., (i) Angle of incident wave is equal to the angle of reflected wave.

(ii) The incident wave, the reflected wave and the normal all lie in the same plane.

Reverberation

Multiple reflections which are responsible for a series of waves falling on listener's ears, giving the impression of a persistence or prolongation of the sound are called reverberations.

The time gap between the initial direct note and the reflected note upto the minimum audibility level is called **reverberation time**.

INTENSITY AND ENERGY DENSITY OF A WAVE

Intensity of wave : In a medium, propagation energy perpendicular to per unit area per second is called intensity of wave.

$$I = 2\pi^2 n^2 a^2 \rho v$$

where ρ = density of the medium,

v = wave velocity,

a = amplitude, n = wave frequency

Unit : W/m^2

If v & ρ is constant, then I ∝ $(\text{frequency})^2$ and I ∝ $(\text{amplitude})^2$

Energy density : Energy per unit volume is called energy density.

$$E = \text{Energy density} = \frac{\text{wave intensity}}{\text{wave velocity}} = \frac{2\pi^2 n^2 a^2 \rho v}{v}$$

$$= 2\pi^2 n^2 a^2 \rho$$

CHARACTERISTICS OF SOUND

Loudness: Human ear is sensitive for extremely large range of intensity. So a logarithmic rather than an arithmetic scale is convenient. Accordingly, intensity level β of a sound wave is defined by the equation

$$\beta = 10 \log\left(\frac{I}{I_0}\right) \text{ decibel}$$

where $I_0 = 10^{-12} W/m^2$ is the reference or threshold intensity level to which any intensity I is compared.

Pitch : The pitch is a sound characteristic that depends on the frequency of the fundamental tone. Higher pitch means higher frequency.

Quality : The quality (or timbre) of a sound depends on the number and intensities of overtones. It is that characteristic of musical sounds by which we can distinguish sounds of the same pitch and loudness when produced by different instruments.

SONIC AND SHOCK WAVE

Infrasonic : The sound frequency less than 20 Hz is called infrasonic.

Ultrasonic : The sound frequency more than 20,000 Hz is called ultrasonic.

Supersonic : It is used with speed. The supersonic velocity is the one that is greater than the local velocity of sound.

Shock wave : It is the wave motion accompanying objects travelling at supersonic speed.

The ratio of the velocity of an object or of a shock wave front, to the local velocity of sound is called **mach number**.

INTERFERENCE

Superposition of two waves of nearly same amplitude having a constant phase difference (or no phase difference) travelling in the same direction and with the same wavelength show the phenomenon of interference.

The intensity of resultant wave in interference is

$$I = I_1 + I_2 + 2\sqrt{I_1}\sqrt{I_2}\cos\phi$$

Constructive and Destructive Interference

Constructive interference :

When waves meet in same phase

Phase difference $= \Delta\phi = 0, 2\pi, 4\pi$ ...., $2n\pi$

Path difference $= \Delta x = 0, \lambda, 2\lambda$ ...., $n\lambda$

[n = 0, 1, 2, ...]

$$\cos\phi = +1,\ I_{max} = \left(\sqrt{I_1} + \sqrt{I_2}\right)^2$$

$A_{max} = A_1 + A_2$

If $A_1 = A_2 = A$; $I_1 = I_2 = I$

$I_{max} = 4I$; $A_{max} = 2A$

Destructive Interference :

Phase difference $= \Delta\phi = \pi, 3\pi$ ...., $(2n+1)\pi$

Path difference $= \Delta x = \frac{\lambda}{2}, \frac{3\lambda}{2}$ ...., $(2n+1)\frac{\lambda}{2}$ [n = 0, 1, 2, ...]

$$A^2 = A_1^2 + A_2^2 + 2A_1A_2\cos\phi$$

$$I = I_1 + I_2 + 2\sqrt{I_1}\sqrt{I_2}\cos\phi\ ;\ \cos\phi = -1$$

$$A_{min} = A_1 - A_2\ ;\ I_{min} = \left(\sqrt{I_1} - \sqrt{I_2}\right)^2$$

If $A_1 = A_2 = A$; $I_1 = I_2 = I$; $A_{min} = 0$; $I_{min} = 0$

OPEN AND CLOSED ORGAN PIPE

Open Organ Pipe

If both ends of a pipe are open and a system of air is directed against an edge, standing longitudinal waves can be set up in the tube. The open ends are displacement antinodes and pressure nodes.

$$\text{Frequency} = \frac{v}{\lambda} = \frac{nv}{2L} \text{ where } n = 1, 2, ...$$

The oscillation mode with lowest frequency is called the **fundamental mode** or the **first harmonic**. The second harmonic is the oscillation mode with n = 2 and so on.

Maximum and minimum intensities in any interference wave form

$$\frac{I_{max}}{I_{min}} = \left(\frac{\sqrt{I_1} + \sqrt{I_2}}{\sqrt{I_1} - \sqrt{I_2}}\right)^2 = \left(\frac{a_1 + a_2}{a_1 - a_2}\right)^2$$

Average intensity of interference wave form is

$$\langle I \rangle \text{ or } I_{av} = \frac{I_{max} + I_{min}}{2}$$

Putting the value of I_{max} & I_{min}, $I_{av} = I_1 + I_2$

If $a = a_1 = a_2$ and $I_1 = I_2 = I$ then $I_{max} = 4I$, $I_{min} = 0$ and $I_{av} = 2I$

The interference of two identical waves moving in opposite directions produces standing waves. For a string with fixed ends, the standing wave is given by

$y(x, t) = [2a \sin kx] \cos \omega t$

Standing waves are characterised by fixed locations of zero displacement called **nodes** and fixed locations of maximum displacements called **antinodes**. The separation between two consecutive nodes or antinodes is $\lambda/2$.

VIBRATION IN A STRETCHED STRING

If a string fixed at two ends, vibrates in its fundamental mode, then

$$f = \frac{1}{2l}\sqrt{\frac{T}{\mu}}$$

μ = mass per unit length

= volume of unit length × density

$$= \pi r^2 \times 1 \times \rho = \pi \times \frac{D^2}{4} \times \rho$$

where D = diameter of the wire,

ρ = density.

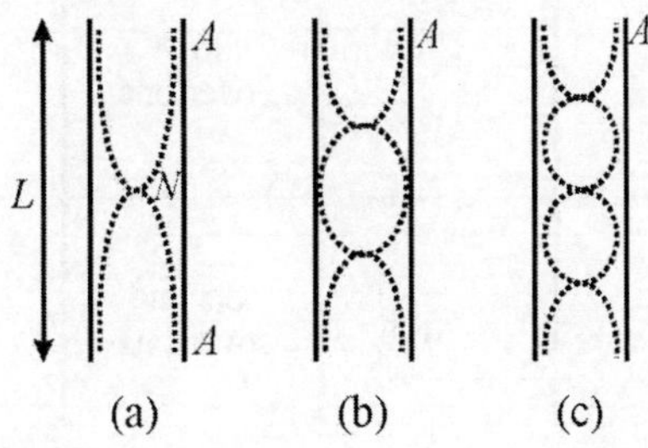

(a) (b) (c)

For vibrating string/ open organ pipe

Mode of vibration	Harmonic	Tone	Frequency	Wavelength		Shape for string
First or Fundamental	First	Fundamental tone	$\nu = \frac{v}{2l}$	$2l$		N A N
Second	Second	First overtone	2ν	$\frac{2l}{2}$		N A N A N
Third	Third	Second overtone	3ν	$\frac{2l}{3}$		N A N A N A N

Closed Organ Pipe

If one end of a pipe is closed, the reflected wave is 180° out of phase with the incoming wave. Thus the displacement of the small volume elements at the closed end must always be zero. Hence the closed end must be a displacement node and pressure antinode.

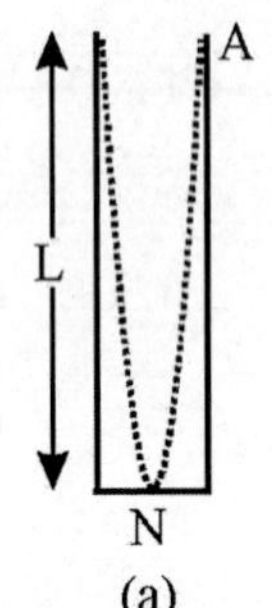

(a)

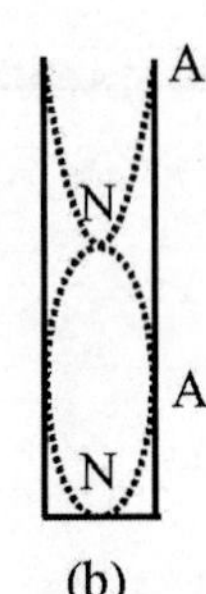

(b)

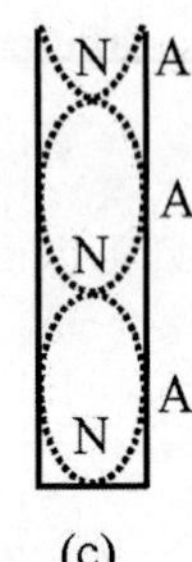

(c)

$$\text{Frequency} = \frac{(2n+1)v}{4L} \quad \text{where } n = 0, 1, 2 \ldots\ldots$$

For closed organ pipe

Mode of vibration	Harmonic	Tone	Frequency	Wavelength	Shape for string
First or Fundamental	First	Fundamental tone	$\nu = \frac{v}{4l}$	$4l$	
Second	Third	First overtone	3ν	$\frac{4l}{3}$	
Third	fifth	Second overtone	5ν	$\frac{4l}{5}$	

Due to finite momentum of air molecules in organ pipes reflection takes place not exactly at open end but somewhat it so in an organ pipe antinode is not formed exactly at free–end but above it at a distance $e = 0.6r$ (called end correction or Rayleigh correction) with r being the radius of pipe. So for closed organ pipe $L \rightarrow L + 0.6r$

while for open organ pipe L → L + 2 × 0.6r (as both ends are open) so that

$$f_{closed} = \frac{v}{4(L+0.6r)}$$

while $f_{open} = \frac{v}{2(L+1.2r)}$

This is why for a given v and L narrower the pipe higher will the frequency or pitch and shriller will be the sound.

A travelling wave, at a rigid boundary or a closed end, is reflected with a phase reversal but the reflection at an open boundary takes place without any phase change.

For an incident wave y (x, t) = a sin (kx – ωt) the reflected wave at a rigid boundary is

$$y_r(x, t) = -a\sin(kx + \omega t)$$

For reflection at an open boundary $y_r(x,t) = a\sin(kx + \omega t)$

BEATS

Beats arise when two waves having slightly different frequencies, f_1 and f_2 and comparable amplitudes, are superposed. The beat frequency is $f_{beat} = f_1 - f_2$.

Determination of frequency - If we know the frequency n_1 of a tuning fork, then we can determine the exact frequency of another fork of nearly equal frequency by the phenomenon of beats. For this, both the tuning forks are sounded together and the beats are heard. Suppose, x beats are heard in 1 second, then the frequency of the second fork will be either $(n_1 + x)$ or $(n_1 - x)$. Now one prong of this fork is loaded with a small wax so that its frequency is slightly lowered. Again, the two forks are sounded together and beats are heard. If the number of beats per second decreases then it means that the new (lowered) frequency of the second tuning forks is more nearer to the frequency of the first tuning fork. This would happen if the frequency of the second tuning fork is higher than the frequency of the first fork. Hence the frequency of the second fork is $(n_1 + x)$. On the other hand, if on loading with wax, the number of beats per second increases, then the frequency of the second fork is $(n_1 - x)$.

DOPPLER'S EFFECT

The Doppler effect is a change in the observed frequency of a wave when the source 's' and the observer 'O' moves relative to the medium. For sound the observed frequency n′ is given in terms of the source frequency n by $n' = n\left(\frac{v + v_0}{v + v_s}\right)$

Here v is the speed of sound through the medium, v_0 is the velocity of observer relative to the medium, and v_s is the source velocity relative to the medium. In using this formula, velocities in the direction OS (observer to source) should be treated as positive and those opposite to it should be taken to be negative.

When moving source crosses a stationary observer

$$n' = \frac{nv}{(v - v_s)} \quad ; \quad n'' = \frac{nv}{(n + v_s)}$$

$$\Delta n = n' - n'' = \frac{2nv_s v}{v\left[\frac{1 - v_s^2}{v^2}\right]} \quad ; \quad \Delta n = \frac{2nv_s}{v}$$

$$\left[\because \left(\frac{v_s}{v}\right)^2 << 1\right]$$

Exercise

1. Frequencies of sound produced from an organ pipe open at both ends are :
(a) only fundamental note
(b) only even harmonics
(c) only odd harmonics
(d) even and odd harmonics
2. The fundamental frequency of an organ pipe is 512 Hz. If its length is increased, then frequency will :
(a) decrease (b) increase
(c) remains same (d) can't be predicted
3. In a wave motion $y = a\sin(\omega t + kx)$ y can represent :
(a) electric field (b) displacement
(c) pressure (d) all of the above
4. A sounding horn is rotating rapidly in a horizontal circle, the apparent frequency of the horn observed at the centre of the circle will be :

(a) same as the actual frequency
(b) less than the actual frequency
(c) more than the actual frequency
(d) sometimes more than and sometimes less than the actual frequency

5. A star emitting light of green colour is approaching the earth with a uniform acceleration, its colour as seen from earth will :
(a) appear green
(b) turn gradually yellow
(c) turn gradually blue
(d) turn gradually red

6. The property of a medium necessary for wave propagation is :
(a) inertia (b) elasticity
(c) low resistance (d) all

7. The property of liquid which enables transverse wave to propagate on the surface of the liquid is
(a) bulk modulus (b) viscosity
(c) surface tension (d) all of these

8. Doppler's effect is not applicable for:
(a) Audio waves
(b) Electromagnetic waves
(c) Shock waves
(d) Sone of these

9. A wave $y = a \sin(\omega t - kx)$ on a string meets with another wave producing a node at $x = 0$. Then the equation of the unknown wave is :
(a) $y = a \sin(\omega t + kx)$
(b) $y = -a \sin(\omega t + kx)$
(c) $y = a \sin(\omega t - kx)$
(d) $y = -a \sin(\omega t - kx)$

10. Two sound waves of equal intensity I produce beats. The maximum intensity of sound produced in beats will be :
(a) I (b) $2I$
(c) $3I$ (d) $4I$

11. A source of frequency f is stationary and an observer starts moving towards it at $t = 0$ with constant small acceleration. Then the variation of observed frequency f registered by the observer with time is best represented by :

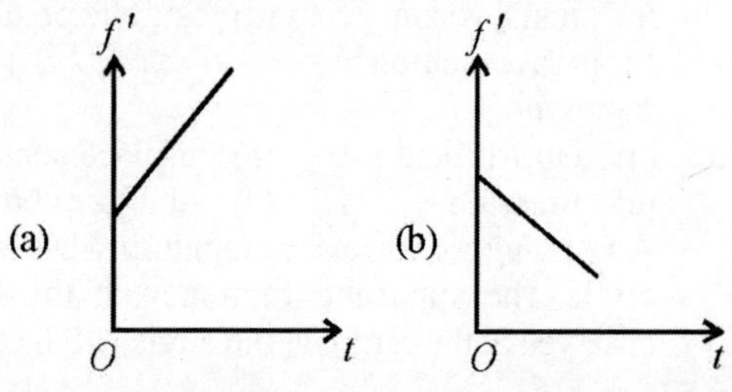

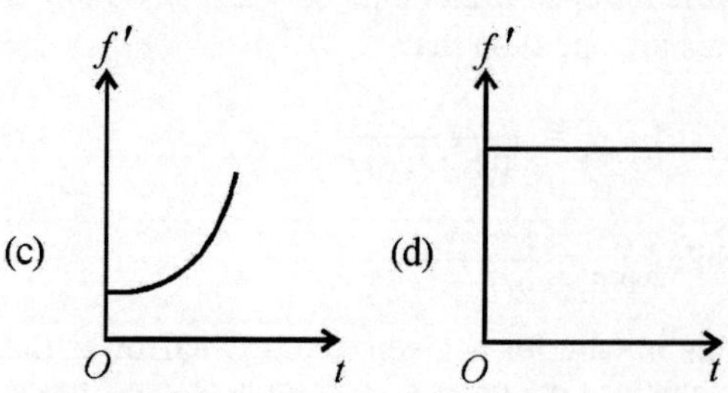

12. The velocity of sound in any gas depends upon
(a) wavelength of sound only
(b) density and elasticity of gas
(c) intensity of sound waves only
(d) amplitude and frequency of sound

13. If the amplitude of sound is doubled and the frequency is reduced to one fourth, the intensity of sound at the same point will be
(a) increasing by a factor of 2
(b) decreasing by a factor of 2
(c) decreasing by a factor of 4
(d) unchanged

14. A closed organ pipe (closed at one end) is excited to support the third overtone. It is found that air in the pipe has
(a) three nodes and three antinodes
(b) three nodes and four antinodes
(c) four nodes and three antinodes
(d) four nodes and four antinodes

15. With the propagation of a longitudinal wave through a material medium, the quantities transmitted in the propagation direction are
(a) Energy, momentum and mass
(b) Energy
(c) Energy and mass
(d) Energy and linear momentum

16. Which one of the following statements is true?
(a) The sound waves in air are longitudinal while the light waves are transverse
(b) Both light and sound waves in air are longitudinal
(c) Both light and sound waves can travel in vacuum
(d) Both light and sound waves in air are transverse

17. When a sound wave goes from one medium to another, the quantity that remains unchanged is
(a) Frequency (b) Amplitude
(c) Wavelength (d) Speed

18. Water waves are
(a) Longitudinal (b) Transverse
(c) Both longitudinal and transverse
(d) Neither longitudinal nor transverse

19. The ratio of the speed of a body to the speed of sound is called
(a) Mach number (b) Doppler ratio
(c) Sonic index (d) Refractive index

20. Of the following, the equation of plane progressive wave is
(a) $y = r \sin \omega t$
(b) $y = r \sin (\omega t - kx)$
(c) $y = \frac{a}{\sqrt{r}} \sin (\omega t - kx)$
(d) $y = \frac{a}{r} \sin (\omega t - kr)$

21. In ordinary talk, the amplitude of vibration is approximately
(a) 10^{-12} m (b) 10^{-11} m
(c) 10^{-8} m (d) 10^{-7} m

22. A balloon is filled with hydrogen. For sound waves, this balloon behaves like
(a) A converging lens (b) A diverging lens
(c) A concave mirror (d) Nothing can be said

23. Energy is not propagated by
(a) Stationary waves
(b) Electromagnetic waves
(c) Longitudinal progressive waves
(d) Transverse progressive waves

24. When a tuning fork vibrates, the vibrations of its two prongs have a phase difference of
(a) π (b) $\pi/2$
(c) Zero (d) 2π

25. On account of damping, the frequency of a vibrating body
(a) Remains unaffected
(b) Increases
(c) Decreases
(d) Changes erratically

26. A cylindrical tube open at both ends has a fundamental frequency n in air. The tube is dipped vertically in water so that half of it is immersed in water. The fundamental frequency of air column is
(a) $n/2$ (b) n
(c) $2n$ (d) $4n$

27. A closed pipe has a certain frequency. Now its length is halved. Considering the end correction, its frequency will now become
(a) Double (b) More than double
(c) Less than double (d) Four times

28. The fundamental frequency of a closed end organ pipe is n. Its length is doubled and radius is halved. Its frequency will become nearly
(a) $n/2$ (b) $n/3$
(c) n (d) $2n$

29. A source of sound is moving with a uniform speed along a circle. The frequency of sound as heard by listener stationed at the centre of the path
(a) Increases
(b) Decreases
(c) Remains the same
(d) May increase and decrease alternately

30. Light from a source travels out with a velocity c. If the source moves away from the observer with a velocity v, then relative velocity of light w.r.t. observer is
(a) c (b) $c + v$
(c) $c - v$ (d) $\sqrt{c^2 + v^2}$

31. The relation between velocity of sound in a gas (v) and r.m.s. velocity of molecules of gas (v_{rms}) is
(a) $v = v_{rms}(\gamma/3)^{1/2}$ (b) $v_{rms} = v(2/3)^{1/2}$
(c) $v = v_{rms}$ (d) $v = v_{rms}(3/\gamma)^{1/2}$

32. The wavelength of ultrasonics in air is of the order of
(a) 10^2 m (b) 10^1 m
(c) 10^{-2} m (d) 10^0 m

33. The wavelength of infrasonics in air is of the order of
(a) 10^0 m (b) 10^1 m
(c) 10^{-1} m (d) 10^{-2} m

34. When a wave is reflected from a denser medium, the change in phase is
(a) 0 (b) π
(c) 2π (d) 3π

35. On reflection of a wave from a rarer medium, change in phase is
(a) Zero (b) $\pi/2$
(c) π (d) $3\pi/4$

36. The speed of sound in a perfectly rigid rod is
(a) infinite (b) zero
(c) 332 m/s (d) 3×10^8 m/s

37. From a point source, if amplitude of waves at a distance r is A, its amplitude at a distance $2r$ will be
(a) A (b) 2 A
(c) A/2 (d) A/4

38. The differential equation of a wave is
(a) $d^2y/dt^2 = v^2 d^2y/dx^2$
(b) $d^2y/dx^2 = v^2 d^2y/dt^2$
(c) $d^2y/dx^2 = \frac{1}{v} d^2y/dt^2$
(d) $d^2y/dx^2 = -v\, d^2y/dt^2$

39. Equation of a progressive wave is given by
$$y = 4\sin\left[\pi\left(\frac{t}{5} - \frac{x}{9}\right) + \frac{\pi}{6}\right]$$
Then which of the following is correct?
(a) $v = 5$ cm (b) $\lambda = 18$ cm
(c) $a = 0.04$ cm (d) $f = 50$ Hz

40. A 5.5 metre long string has a mass of 0.035 kg. If the tension in the string is 77N, the speed of a wave on the string is
(a) 110 ms^{-1} (b) 165 ms^{-1}
(c) 77 ms^{-1} (d) 102 ms^{-1}

41. Velocity of sound waves in air is 330 m/s. For a particular sound wave in air, a path difference of 40 cm is equivalent to phase difference of 1.6 π. The frequency of this wave is
(a) 165 Hz (b) 150 Hz
(c) 660 Hz (d) 330 Hz

42. The transverse wave represented by the equation $y = 4\sin\left(\frac{\pi}{6}\right)\sin(3x - 15t)$ has
(a) amplitude = 4
(b) wavelength = $4\frac{\pi}{3}$
(c) speed of propagation = 5
(d) period $= \frac{\pi}{15}$

43. Two trains move towards each other with the same speed. The speed of sound is 340 m/s. If the height of the tone of the whistle of one of them heard on the other changes 9/8 times, then the speed of each train should be
(a) 20 m/s (b) 2 m/s
(c) 200 m/s (d) 2000 m/s

44. The frequency of sinusoidal wave $\xi = 0.40 \cos [2000\,t + 0.80]$ would be
(a) 1000 π Hz (b) 2000 Hz
(c) 20 Hz (d) $\frac{1000}{\pi}$ Hz

45. A transverse wave propagating along x-axis is represented by $y(x, t) = 8.0 \sin(0.5\pi x - 4\pi t - \frac{\pi}{4})$ where x is in metres and t is in seconds. The speed of the wave is
(a) 0.5 π m/s (b) $\frac{\pi}{4}$ m/s
(c) 8 m/s (d) 4π m/s

46. The time of reverberation of a room A is one second. What will be the time (in seconds) of reverberation of a room, having all the dimensions double of those of room A?
(a) 4 (b) $\frac{1}{2}$
(c) 1 (d) 2

47. Two sound waves with wavelengths 5.0 m and 5.5 m respectively, each propagate in a gas with velocity 330 m/s. We expect the following number of beats per second
(a) 0 (b) 1
(c) 6 (d) 12

48. Two points are located at a distance of 10 m and 15 m from the source of oscillation. The period of oscillation is 0.05 sec and the velocity of the wave is 300 m/sec. What is the phase difference between the oscillations of two points?
(a) $\frac{\pi}{3}$ (b) $\frac{2\pi}{3}$
(c) π (d) $\frac{\pi}{6}$

49. The wave described by $y = 0.25 \sin(10\pi x - 2\pi t)$, where x and y are in meters and t in seconds, is a wave travelling along the:
(a) –ve x direction with frequency 1 Hz.
(b) +ve x direction with frequency π Hz and wavelength $\lambda = 0.2$ m.
(c) +ve x direction with frequency 1 Hz and wavelength $\lambda = 0.2$ m
(d) –ve x direction with amplitude 0.25 m and wavelength $\lambda = 0.2$ m

50. Each of the two strings of length 51.6 cm and 49.1 cm are tensioned separately by 20 N force. Mass per unit length of both the strings is same and equal to 1 g/m. When both the strings vibrate simultaneously the number of beats is
(a) 7 (b) 8 (c) 3 (d) 5

51. The driver of a car travelling with speed 30 m/sec towards a hill sounds a horn of frequency 600 Hz. If the velocity of sound in air is 330 m/s, the frequency of reflected sound as heard by driver is
(a) 555.5 Hz (b) 720 Hz
(c) 500 Hz (d) 550 Hz

52. A wave in a string has an amplitude of 2 cm. The wave travels in the + ve direction of x axis with a speed of 128 m/sec and it is noted that 5 complete waves fit in 4 m length of the string. The equation describing the wave is
(a) $y = (0.02)\text{ m} \sin(15.7x - 2010t)$
(b) $y = (0.02)\text{ m} \sin(15.7x + 2010t)$
s(c) $y = (0.02)\text{ m} \sin(7.85x - 1005t)$
(d) $y = (0.02)\text{ m} \sin(7.85x + 1005t)$

53. A transverse wave is represented by $y = A \sin(\omega t - kx)$. For what value of the wavelength is the wave velocity equal to the maximum particle velocity?
(a) $\frac{\pi A}{2}$ (b) πA
(c) $2\pi A$ (d) A

54. A tuning fork of freqqency 512 Hz makes 4 beats per second with the vibrating string of a piano. The beat frequency decreases to 2 beats per sec when the tension in the piano string is slightly increased. The frequency of the piano string before increasing the tension was
(a) 510 Hz (b) 514 Hz
(c) 516 Hz (d) 508 Hz

55. Two tones of frequencies n_1 and n_2 are sounded together. The beats can be heard distinctly when
(a) $10 < (n_1 - n_2) < 20$ (b) $5 < (n_1 - n_2) > 20$
(c) $5 < (n_1 - n_2) < 20$ (d) $0 < (n_1 - n_2) < 10$

56. The equation of a spherical progressive wave is
(a) $y = a \sin \omega t$
(b) $y = a \sin(\omega t - kr)$
(c) $y = \frac{a}{\sqrt{2}} \sin(\omega t - kr)$
(d) $y = \frac{a}{r} \sin(\omega t - kr)$

57. The equation of a cylindrical progressive wave is
(a) $y = a \sin \omega t$
(b) $y = a \sin(\omega t - kr)$
(c) $y = \frac{a}{\sqrt{r}} \sin(\omega t - kr)$
(d) $y = \frac{a}{r} \sin(\omega t - kr)$

58. A body is executing S.H.M. When its displacement from the mean position are 4 cm and 5 cm, it has velocities 10 cm s^{-1} and 8 cm s^{-1} respectively. Its periodic time is
(a) $\pi/2$ s (b) π s
(c) $3\pi/2$ s (d) 2π s

59. A body of mass 5 gram is executing S.H.M. about a fixed point O. With an amplitude of 10 cm, its maximum velocity is 100 cm/s. Its velocity will be 50 cm s^{-1} at a distance (in cm)
(a) 5 (b) $5\sqrt{2}$
(c) $5\sqrt{3}$ (d) $10\sqrt{2}$

60. A simple pendulum performs S.H.M. about $x = 0$ with an amplitude a, and time period T. The speed of the pendulum at $x = a/2$ will be
(a) $\pi \frac{a\sqrt{3}}{T}$ (b) $\frac{\pi a\sqrt{3}}{2T}$
(c) $\frac{\pi a}{T}$ (d) $\frac{3\pi^2 a}{T}$

Electric Charges and Fields

ELECTRIC CHARGES AND THEIR CONSERVATION

Electric charge is something that the material objects have and makes it possible for them to exert electrical force and to respond to electrical force.

Electric and magnetic forces determine the properties of atoms, molecules and bulk matter. Coulomb force and gravitational force follow the same inverse-square law. But gravitational force has only one sign (always attractive), while Coulomb force can be of both signs (attractive and repulsive), allowing possibility of cancellation of electric forces. This is how gravity, despite being a much weaker force, can be a dominating and more pervasive force in nature.

Charge is not only a scalar (or invariant) under rotation; it is also invariant for frames of reference in relative motion. This is not always true for every scalar. For example, kinetic energy is a scalar under rotation, but is not invariant for frames of reference in relative motion.

Conservation of total charge of an isolated system is a property independent of the scalar nature of charge. It is not possible to create or destroy net charge carried by any isolated system. Conservation refers to invariance in time in a given frame of reference. A quantity may be scalar but not conserved (like kinetic energy in an inelastic collision). On the other hand, one can have conserved vector quantity (e.g., angular momentum of an isolated system). **Quantisation of electric charge** is a basic (unexplained) law of nature; interestingly, there is no analogous law on quantisation of mass.

Electric charge is quantized and charge on a body can be expressed as, $q = \pm ne$ where 'n' is an integer and 'e' charge of one electron

$e = 1.6 \times 10^{-19}C$

COULOMB'S LAW

The mutual electrostatic force between two point charges q_1 and q_2 is proportional to the product q_1q_2 and inversely proportional to the square of the distance r_{21} separating them.

Mathematically, $\vec{F}_{21}$ = force on q_2 due to

$$q_1 = \frac{k\,(q_1q_2)}{r_{21}^2}\hat{r}_{21}$$

where $\hat{r}_{21}$ is a unit vector in the direction from q_1 to q_2 and $k = 1/4\pi\varepsilon_0$ is the constant of proportionality.

In SI units, the unit of charge is coulomb. The experimental value of the constant ε_0 (absolute electrical permittivity of the free space) is

$\varepsilon_0 = 8.854 \times 10^{-12} C^2N^{-1}m^{-2}$.

The approximate value of k is $= 9 \times 10^9$ Nm^{2}C^{-2}.

SUPERPOSITION PRINCIPLE : FORCES BETWEEN MULTIPLE CHARGES

The electric force $\vec{F}_1$ on q_1 due to a number of charges placed in air or vacuum is given by

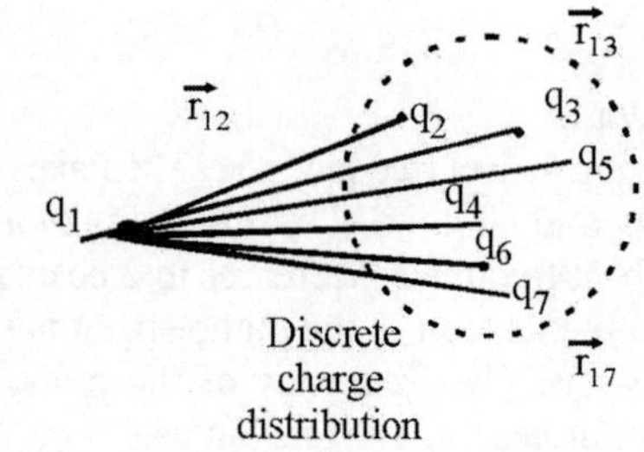

Discrete charge distribution

$$\vec{F}_1 = \frac{1}{4\pi\varepsilon_0}\sum_{i=1}^{n}\frac{q_1q_i}{r_{1i}^2}\hat{r}_{1i}$$

$$= \frac{1}{4\pi\varepsilon_0}\left[\frac{q_1q_2}{r_{12}^2}\hat{r}_{12} + \frac{q_1q_3}{r_{13}^2}\hat{r}_{13} + ...\right]$$

ELECTRIC FIELD INTENSITY DUE TO CONTINUOUS CHARGE DISTRIBUTION

The electric field E at a point due to a charge configuration is the force on a small positive test charge q placed at the point divided by the magnitude of the charge.

Electric field due to a point charge q has a magnitude $|q|/4\pi\varepsilon_0 r^2$, it is radially outwards from q, if q is positive, and radially inwards if q is negative. Like Coulomb force, electric field also satisfies superposition principle.

The electric field due to a discrete charge configuration is not defined at the locations of the discrete charges. For continuous volume charge distribution, it is defined at any point in the distribution. For a surface charge distribution, electric field is discontinuous across the surface.

Name/Type	Formula	Particular	Graph
Point charge	$\frac{kq}{\|\vec{r}\|^2}.\hat{r} = \frac{kq}{r^3}\vec{r}$	q is source charge. $\vec{r}$ is vector drawn from source charge to the test point. Electric field is non-uniform, radially outwards due to +ve charges & inwards due to –ve charges.	E, r
Infinitely long line charge	$\frac{\lambda}{2\pi\varepsilon_0 r}\hat{r} = \frac{2k\lambda\hat{r}}{r}$	λ is a linear charge density (assumed uniform) r is perpendicular distance of point from line charge. $\hat{r}$ is radial unit vector drawn from the charge to test point.	E, r
Finite line of charge (λ, Ey, P, Ex, α, β, r)	$E_x = \frac{k\lambda}{r}[\sin\beta + \sin\alpha]$ $E_y = \frac{k\lambda}{r}[\cos\alpha - \cos\beta]$ If $\alpha = \beta$ $E_y = 0, E_x = \frac{\lambda}{2\pi\varepsilon_0 r}$	λ is the linear charge density	
Infinite non-conducting thin sheet	$\frac{\sigma}{2\varepsilon_0}\hat{n}$	σ is surface charge density (assumed uniform) $\hat{n}$ is unit normal vector. Electric field intensity is independent of distance.	E, $\frac{\sigma}{2\varepsilon_o}$, r
Uniformly charged ring	$E = \frac{kQx}{(R^2+x^2)^{3/2}}$ $E_{centre} = 0$	Q is total charge of the ring. x = distance of point on the axis from centre of the ring. Electric field is always along the axis. E is maximum at $x = \pm R/\sqrt{2}$	E, E_{max}, –x, $-\frac{R}{\sqrt{2}}$, O, $\frac{R}{\sqrt{2}}$, X
Infinitely large charged conducting sheet	$\frac{\sigma}{\varepsilon_0}\hat{n}$	σ is the surface charge density $\hat{n}$ is unit normal vector perpendicular to the surface. Electric field intensity is independent of distance.	E, $\frac{\sigma}{\varepsilon_0}$, r

Uniformly charged hollow/ solid conducting sphere	(i) for $r \geq R$ $\vec{E} = \frac{kQ}{\|\vec{r}\|^2}\hat{r}$ (ii) for $r < R$ $\vec{E} = 0$	R is radius of the sphere $\vec{r}$ is a vector drawn from centre of sphere to the point. Sphere acts like a point charge, placed at centre for points outside the sphere. $\vec{E}$ is always along radial direction. Q is total charge ($= \sigma 4\pi R^2$) (σ = surface charge density)	E, $\frac{kQ}{R^2}$, R, r
Uniformly charged solid non-conducting sphere (insulating material)	(i) for $r \geq R$ $\vec{E} = \frac{kQ}{\|\vec{r}\|^2}\hat{r}$ (ii) for $r < R$ $\vec{E} = \frac{kQ\vec{r}}{R^3} = \frac{\rho\vec{r}}{3\varepsilon_0}$	$\vec{r}$ is a vector drawn from centre of sphere to the point. Sphere acts like a point charge, placed at centre for points outside the sphere. $\vec{E}$ is always along radial direction. Q is total charge $(\rho.\frac{4}{3}\pi R^3)$ (ρ = volume charge density) Inside the sphere $E \propto r$ Outside the sphere $E \propto 1/r^2$	E, $\frac{kQ}{R^2}$, R, r
Uniformly charged cylinder (non-conducting) with a linear charge density λ (R= radius of cylinder)	(i) for $r<R$, $E_{in} = \frac{\lambda r}{2\pi\varepsilon_0 R^2}$ (ii) for $r>R$, $E = \frac{\lambda}{2\pi \epsilon_0 r}$ (iii) for $r=R$, $E = \frac{\lambda}{2\pi \epsilon_0 R}$	r is the distance from the axis of cylinder	E, $E \propto r$, $E \propto \frac{1}{r}$, r
Uniformly charged (conducting) cylindrical shell with surface charge density σ is	(i) for $r<R$, $E_{in} = 0$, (ii) for $r>R$, $E = \frac{\lambda}{2\pi \epsilon_0 r}$ (iii) for $r=R$, $E = \frac{\lambda}{2\pi \epsilon_0 R}$	*R* is radius of cylinder *r* is the disance from the axis of cylinder	E, $\frac{kQ}{R^2}$, R, r

Electric Field Intensities Due to Various Charge Distributions

ELECTRIC FIELD LINES

An electric field line is a curve drawn in such a way that the tangent at each point on the curve gives the direction of electric field at that point. The relative closeness of field lines indicates the relative strength of electric field at different points; they crowd near each other in regions of strong electric field and are far apart where the electric field is weak. In regions of constant electric field, the field lines are uniformly spaced parallel straight lines.

Properties of Field Lines

Some of the important properties of field lines are:

(a) Field lines are continuous curves without any break.

(b) Two field lines cannot cross each other.

(c) Electrostatic field lines start at positive charges and end at negative charges —they cannot form closed loops.

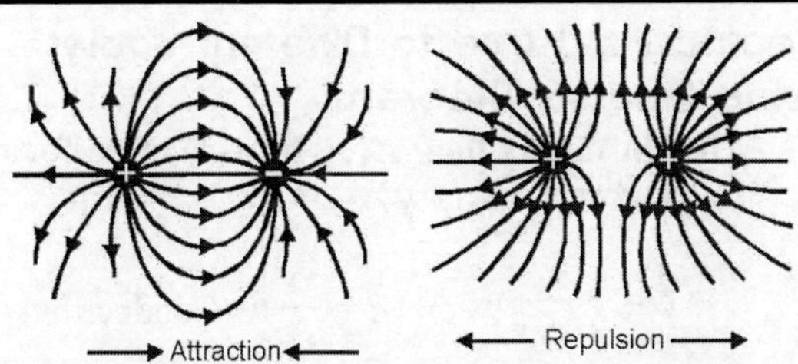

MOTION OF A CHARGED PARTICLE IN A UNIFORM ELECTRIC FIELD

If force of gravity does not exist or is balanced by some other force say reaction or neglected then

$$\vec{a} = \frac{\vec{F}}{m} = \frac{q\vec{E}}{m} = \text{constant} \qquad [\text{as } \vec{F} = q\vec{E}\,]$$

Here equations of motion are valid.

(a) **If the particle is initially at rest** then from $v = u + at$,

we get, $\quad v = at = \frac{qE}{m}t$

And from eqn. $s = ut + \frac{1}{2}at^2$ we get,

$$s = \frac{1}{2}at^2 = \frac{1}{2}\frac{qE}{m}t^2$$

The motion is accelerated translatory with $a \propto t^{\circ}$; $v \propto t$ and $s \propto t^2$

Here $W = \Delta K.E = \frac{1}{2}mv^2 = \frac{1}{2}m\left[\frac{qE}{m}t\right]^2$

Also $W = qEd = qV$

(b) **If the particle is projected perpendicular to the field** with an initial velocity v_0

From eqn. $v = u + at$ and $s = ut + \frac{1}{2}at^2$, $u_x = v_0$ and $a_x = 0$,

$u_x = v_0$ = constant and $x = v_0 t$

For motion along y-axis as $u_y = 0$ and $a_y = \frac{qE}{m}$,

$$v_y = \left[\frac{qE}{m}\right]t \text{ and } y = \frac{1}{2}\left[\frac{qE}{m}\right]t^2$$

So eliminating t between equation for x and y, we have $y = \frac{qE}{2m}\left[\frac{x}{v_0}\right]^2 = \frac{qE}{2mv_0^2}x^2$

If particle is projected perpendicular to field, the path is a parabola.

ELECTRIC DIPOLE

An electric dipole is a pair of equal and opposite charges q and – q separated by some distance 2a. Its dipole moment vector $\vec{p}$ has magnitude 2qa and is in the direction of the dipole axis from – q to q.

Electric Field Due to a Dipole

(a) **Field of an electric dipole in its equatorial plane** (i.e., the plane perpendicular to its axis and passing through its centre) at a distance r from the centre:

$$E = \frac{-\vec{p}}{4\pi\varepsilon_0}\frac{1}{(a^2+r^2)^{3/2}} \cong \frac{-\vec{p}}{4\pi\varepsilon_0 r^3}, \text{ for } r >> a$$

(b) **Dipole electric field on the axis at a distance r from the centre :** $\vec{E} = \frac{2\vec{p}r}{4\pi\varepsilon_0(r^2-a^2)^2} \cong \frac{2\vec{p}}{4\pi\varepsilon_0 r^3}$

for $r >> a$

The $1/r^3$ dependence of dipole electric fields should be noted in contrast to the $1/r^2$ dependence of electric field due to a point charge.

Torque on a Dipole

In a uniform electric field $\vec{E}$, a dipole experiences a **torque** $\vec{\tau}$ given by $\vec{\tau} = \vec{p} \times \vec{E}$ but experiences no net force.

Work done in rotating an electric dipole from θ_1 to θ_2 (uniform field)

$$W = \int_{\theta_1}^{\theta_2} pE \sin\theta \, d\theta = pE(\cos\theta_1 - \cos\theta_2)$$

Electrostatic potential energy : $U = -\vec{p}.\vec{E}$

Potential energy of a system of two charges Q_1 and Q_2 is,

$$U = \frac{1}{4\pi\epsilon_0}\frac{Q_1 Q_2}{r^2}$$

Q_1 ← Q_2 (separation r)

[Please note that in this formula we have to write charges with sign]

Potential energy of a system of three charges Q_1, Q_2 and Q_3 is

$$U = \frac{1}{4\pi\epsilon_0}\left[\frac{Q_1 Q_2}{r_1} + \frac{Q_2 Q_3}{r_2} + \frac{Q_3 Q_1}{r_3}\right]$$

Angular SHM :

When a dipole is suspended in uniform field, it will align itself parallel to the field. Now if it is given a small angular displacement θ about its angular position, the restoring couple will be

$\tau = -pE\sin\theta$. if θ is small $\Rightarrow \sin\theta \approx \theta$.

$\Rightarrow \tau = -pE\theta \Rightarrow \tau \propto -\theta$

(Angular SHM).

For balanced condition : $\tau_{deflecting} = \tau_{restoring}$

$I\alpha = -pE\theta$

$$\Rightarrow \alpha = -\left(\frac{pE}{I}\right)\theta = -\omega^2\theta$$

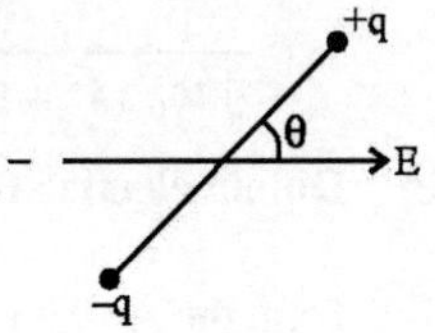

$$\Rightarrow \omega = \sqrt{\frac{pE}{I}}$$

$$\Rightarrow T = \frac{2\pi}{\omega} = 2\pi\sqrt{\frac{I}{pE}} \quad [I \to \text{moment of inertia}]$$

ELECTRIC FLUX

The flux Δϕ of electric field $\vec{E}$ through a small area element $\Delta\vec{S}$ is given by $\Delta\phi = \vec{E}.\Delta\vec{S}$

The vector area element ΔS is $\Delta\vec{S} = \Delta S\,\hat{n}$

where ΔS is the magnitude of the area element and $\hat{n}$ is a unit vector normal to the area element, which can be considered planar for sufficiently small ΔS. For an area element of a closed surface, $\hat{n}$ is taken to be the direction of outward normal, by convention.

GAUSS'S THEOREM

The flux of electric field through any closed surface S is $1/\varepsilon_0$ times the total charge enclosed by S. The law is especially useful in determining electric field $\vec{E}$, when the source distribution has simple symmetry:

$$\oint \vec{E}.d\vec{s} = \frac{q_{net}}{\varepsilon_0}$$

Electric Field Due to Different Bodies Using Gauss's Theorem

(a) Thin infinitely long straight wire of uniform linear charge density λ :

$$\vec{E} = \frac{\lambda}{2\pi\varepsilon_0 r}\hat{n} \quad \text{or} \quad E = \frac{\sigma}{\varepsilon_0}\hat{n} \text{ (Conducting)}$$

where r is the perpendicular distance of the point from the wire and $\hat{n}$ is the radial unit vector in the plane normal to the wire passing through the point.

(b) Infinite thin plane sheet of uniform surface charge density σ :

$$\vec{E} = \frac{\sigma}{2\varepsilon_0}\hat{n} \quad \text{(Non-conducting sheet)}$$

$$E = \frac{\sigma}{\epsilon_0}\hat{n} \quad \text{(Non-conducting sheet)}$$

(c) Thin spherical shell of uniform surface charge density σ :

$$\vec{E} = \frac{\sigma}{4\pi\varepsilon_0 r^2}\hat{r} \quad (r \ge R) \text{ (Outside of the shell);}$$

$$\vec{E} = 0 \quad (r < R) \text{ (Inside of the shell)}$$

ELECTROSTATIC PRESSURE

$$\frac{dF}{dA} = \frac{\sigma^2}{2\varepsilon_0}$$

A soap bubble or rubber balloon expands on providing charge to it (charge of any kind + or –).

Energy associated per unit volume of electric field of intensity E is defined as energy density.

$$u = \frac{dW}{dV} = \frac{\varepsilon_0 E^2}{2} = \frac{\sigma^2}{2\varepsilon_0} \text{ J/m}^3$$

Soap Bubble : In state of equilibrium $\frac{4T}{r} = \frac{\sigma^2}{2\varepsilon_0}$

(where T = surface tension).

Exercise

1. When a body is charged by induction, then the body
   (a) becomes neutral
   (b) does not lose any charge
   (c) loses whole of the charge on it
   (d) loses part of the charge on it
2. If a body is positively charged, then it has
   (a) excess of electrons
   (b) excess of protons
   (c) deficiency of electrons
   (d) deficiency of neutrons
3. On charging by conduction, mass of a body may
   (a) increase
   (b) decreases
   (c) increase or decrease
   (d) none
4. Coulomb's law is true for
   (a) atomic distances (= 10^{-11} m)
   (b) nuclear distances (= 10^{-15} m)
   (c) charged as well as uncharged particles
   (d) all the distances
5. Two charges are placed a certain distance apart. A metallic sheet is placed between them. What will happen to the force between the charges?
   (a) Increase
   (b) Decrease
   (c) Remains unchanged
   (d) Either (a) or (b)
6. Which of the following is best insulator?
   (a) Carbon (b) Paper
   (c) Graphite (d) Ebonite
7. The distinction between conductors, insulators and semiconductor is largely concerned with
   (a) their ability to conduct current
   (b) the type of crystal lattice
   (c) binding energy of their electrons
   (d) relative widths of their energy gaps
8. A cylindrical conductor is placed near another positively charged conductor. The net charge acquired by the cylindrical conductor will be
   (a) positive only
   (b) negative only
   (c) zero
   (d) either positive or negative
9. Two spheres A and B of exactly same mass are given equal positive and negative charges respectively. Their masses after charging
   (a) remains unaffected
   (b) mass of $A >$ mass of B
   (c) mass of $A <$ mass of B
   (d) nothing can be said
10. A positive point charge Q is brought near an isolated metal cube.
    (a) The cube becomes negatively charged
    (b) The cube becomes positively charged
    (c) The interior becomes positively charged and the surface becomes negatively charged
    (d) The interior remains charge free and the surface gets nonuniform charge distribution
11. A large nonconducting sheet M is given a uniform charge density. Two uncharged small metal rods A and B are placed near the sheet as shown in figure.

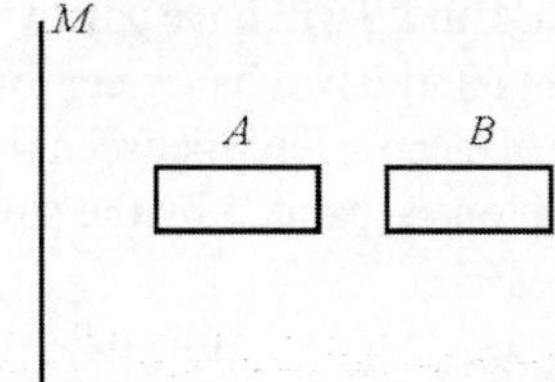

    (a) M attracts A (b) M attracts B
    (c) A attracts B (d) All of these
12. What happens when some charge is placed on a soap bubble?
    (a) Its radius decreases
    (b) Its radius increases
    (c) Ihe bubble collapses
    (d) None of the above
13. Quantisation of charge implies
    (a) charge cannot be destroyed
    (b) charge exists on particles
    (c) there is a minimum permissible charge on a particle
    (d) charge, which is a fraction of a Coulomb is not possible
14. Two charges are placed a certain distance apart in air. If a glass slab is introduced between them, the force between them will
    (a) increase (b) decrease
    (c) remains the same (d) be zero

15. A charge q_1 exerts some force on a second charge q_2. If third charge q_3 is brought near, the force of q_1 exerted on q_2
(a) decreases
(b) increases
(c) remains unchanged
(d) increases if q_3 is of same sign as q_1 and decreases if q_3 is of opposite sign.

16. A positively charged rod is brought near an uncharged conductor. If the rod is then suddenly withdrawn, the charge left on the conductor will be
(a) positive (b) negative
(c) zero (d) not sure

17. Two identical metal spheres *A* and *B* are supported on insulating stands and placed in contact. What kind of charges will *A* and *B* develop when a negatively charged ebonite rod is brought near *A*?
(a) *A* will have a positive charge and *B* will have a negative charge
(b) A will have a negative charge and *B* will have a positive charge
(c) Both *A* and *B* will have positive charges
(d) Both *A* and *B* will have negative charges

18. The ratio of electric force between two electrons to two protons separated by the same distance in air is
(a) 10^0 (b) 10^6
(c) 10^4 (d) none

19. Out of gravitational, electrostatic, vander waal and nuclear forces, which are able to provide attractive force between two neutrons
(a) electrostatic and gravitational
(b) electrostatic and nuclear
(c) van der Waal and nuclear
(d) nuclear and gravitational

20. The dielectric constant of a metal is
(a) ∞ (b) 0
(c) 1 (d) none

21. A metallic particle having no net charge is placed near a finite metal plate carrying a positive charge. The electric force on the particle will be
(a) towards the plate
(b) away from the plate
(c) parallel to the plate
(d) zero

22. If a body is charged by rubbing it, its weight
(a) remains precisely constant
(b) increases slightly
(c) decreases slightly
(d) may increase slightly or may decrease slightly

23. Positive electric flux indicates that electric lines of force are directed
(a) outwards
(b) inwards
(c) outwards or inwards
(d) none of these

24. Negative electric flux indicates that electric lines of force are directed
(a) outwards
(b) inwards
(c) outwards or inwards
(d) none of these

25. If the flux of the electric field through a closed surface is zero.
(a) the electric field must be zero every where on the surface
(b) the electric field may be zero everywhere on the surface
(c) the charge inside the surface must not be zero
(d) the charge in the vicinity of the surface must be zero

26. Electric flux at a point in an electric field is
(a) positive (b) negative
(c) zero (d) none of these

27. Electric flux over a surface in an electric field may be
(a) positive
(b) negative
(c) zero
(d) positive, negative, zero

28. Mark the correct options
(a) Gauss's law is valid only for assymmetrical charge distributions.
(b) Gauss's law is valid only for charge placed in vacuum.
(c) The electric field can not be calculated by Gauss's law is the field due to the charges inside the Gaussian surface.
(d) The flux of the electric field through a closed surface due to all the charges is equal to the flux due to the charges enclosed by the surface.

29. A closed surface S is constructed around a conducting wire connected to a battery and a switch. As the switch is closed, the free electrons in the wire start moving along the wire. In any time interval, the number of electrons entering the closed surface S is equal to the number of electrons leaving it. On closing the switch, the flux of the electric field through the closed surface

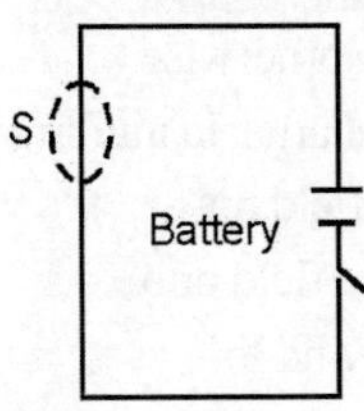

(a) is increased
(b) is decreased
(c) remains unchanged
(d) None of these

30. Figure below shows a closed surface which intersects a conducting sphere. If a positive charge is placed at the point P, the flux of the electric field through the closed surface

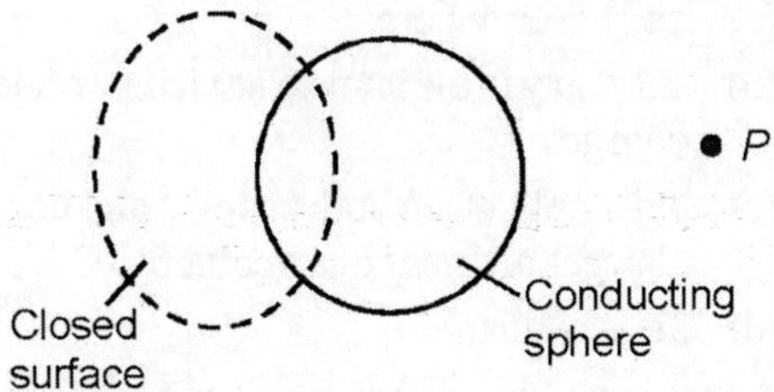

(a) will remain zero
(b) will become positive
(c) will become negative
(d) will become undefined

31. Figure shows a charge q placed at the centre of a hemisphere. A second charge Q is placed at one of the positions A, B, C and D. In which position(s) of this second charge, the flux of the electric field through the hemisphere remains unchanged?

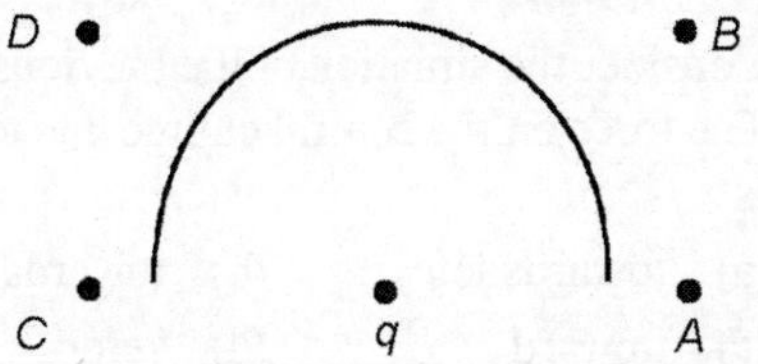

(a) A (b) B
(c) D (d) None of these

32. Debye is the unit of
(a) electric flux
(b) electric dipole moment
(c) electric potential
(d) electric field intensity

33. An electric dipole will experience a net force when it is placed in
(a) a uniform electric field
(b) a non-uniform electric field
(c) both the above cases
(d) none of these

34. An electric dipole is kept in a uniform electric field. It experiences
(a) a force and a torque
(b) a force, but no torque
(c) a torque but no force
(d) neither a force nor a torque

35. An electric dipole is kept in a non-uniform electric field. It experiences
(a) a force and a torque
(b) a force but not a torque
(c) a torque but no force
(d) neither a force nor a torque

36. Figure shows the electric lines of force emerging from a charged body. If the electric field at A and B are E_A and E_B respectively and if the displacement between A and B is r, then

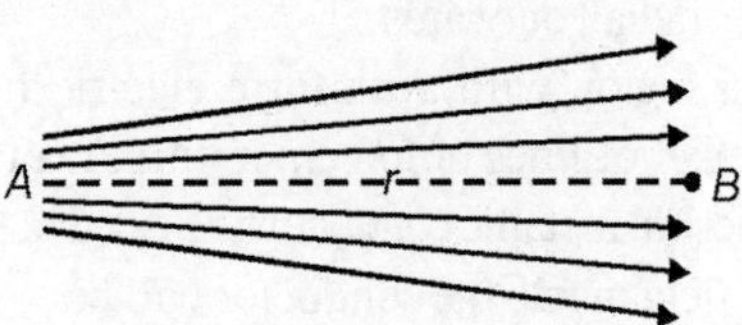

(a) $E_A > E_B$ (b) $E_A < E_B$
(c) $E_A = E_B/r$ (d) $E_A = E_B/r^2$

37. An electric field can deflect
(a) neutrons (b) X-rays
(c) γ-rays (d) α-particles

38. If an electron has an initial velocity in a direction different from that of an electric field, the path of the electron is
(a) a straight line (b) a circle
(c) an ellipse (d) a parabola

39. If a charge is moving along the direction of coulomb's force of an electric field,
(a) work is done by the electric field
(b) energy is used from some outside source
(c) strength of field decreases
(d) energy of the system is decreased

40. A hollow sphere of charge does not have electric field at
(a) outer point (b) interior point
(c) beyond 2 m (d) beyond 100 m

41. If one penetrates a uniformly charged solid sphere, the electric field E
(a) increases
(b) decreases
(c) is zero at all points
(d) remains same as at the surface

42. If one penetrates a uniformly charged spherical cloud, electric field strength
(a) decreases directly as the distance from the centre
(b) increases directly as the distance from the centre
(c) remains constant
(d) none of the above

43. Electric lines of force about a negative point charge are
(a) circular anticlockwise
(b) circular clockwise
(c) radial, inwards
(d) radial, outwards

44. In a region with a uniform electric field, the number of lines of force per unit area is E. If a spherical metallic conductor is placed in the area, the field inside the conductor will be
(a) zero (b) E
(c) more than E (d) less than E

45. Figure shows some of the electric field lines corresponding to an electric field. The figure suggests that

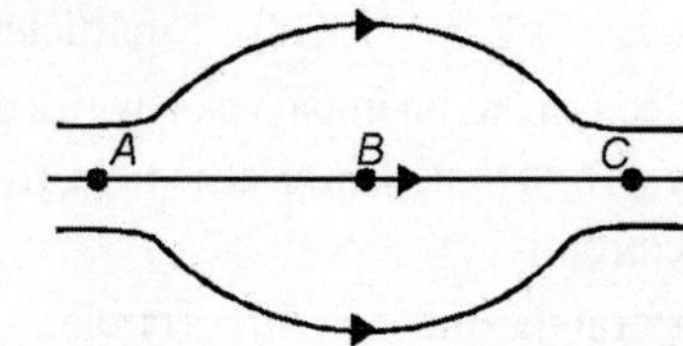

(a) $E_A > E_B > E_C$ (b) $E_A = E_B = E_C$
(c) $E_A = E_C > E_B$ (d) $E_A = E_C < E_B$

46. A point charge q is rotated along a circle in the electric field generated by another point charge Q. The work done by the electric field on the rotating charge in one complete revolution is
(a) zero
(b) positive
(c) negative
(d) zero if the charge Q is at the centre and nonzero otherwise

47. The electric charge, in uniform motion, produces
(a) electric field only
(b) magnetic field only
(c) both 'a' and 'b'
(d) neither 'a' and 'b'

48. An electron and a proton are situated in a uniform electric field. The ratio of their acceleration will be
(a) zero
(b) unity
(c) ratio of masses of proton and electron
(d) ratio of masses of electron and proton

49. Electric lines of force
(a) exist everywhere
(b) exist only in the immediate vicinity of electric charges
(c) exist only when both positive and negative charges are near one another
(d) are imaginary

50. A thin, metallic spherical shell contains a charge Q on it. A point charge q is placed at the centre of the shell and another charge q_1 is placed outside it as shown in figure. All the three charges are positive. The force on the charge at the centre is

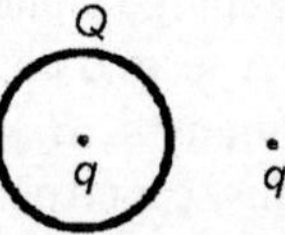

(a) towards left (b) towards right
(c) upward (d) zero

51. Consider the situation of the previous problem. The force on the central charge due to the shell is
(a) towards left (b) towards right
(c) upward (d) zero

52. An electric charge, $+Q$ is placed on the surface of a solid, conduction sphere of radius a. The distance measured from the centre of the sphere is denoted as r. Then
    (a) the charge gets distributed uniformly through the volume of the sphere
    (b) The electrostatic potential has the same value for $r < a$
    (c) an equal and opposite charge gets induce in the bottom half of the sphere
    (d) the electric field is given by $1/(4\pi \in_0 r^2)$ for $r < a$

53. A dielectric sphere is placed in a uniform electric field directed along the positive y-axis. Which one of the following represents the correct equipotential surface?

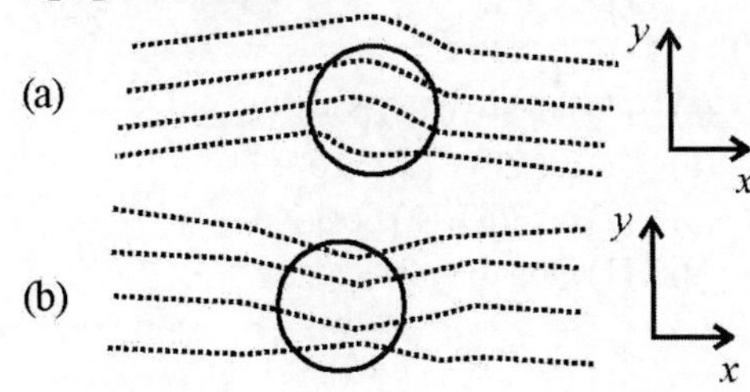

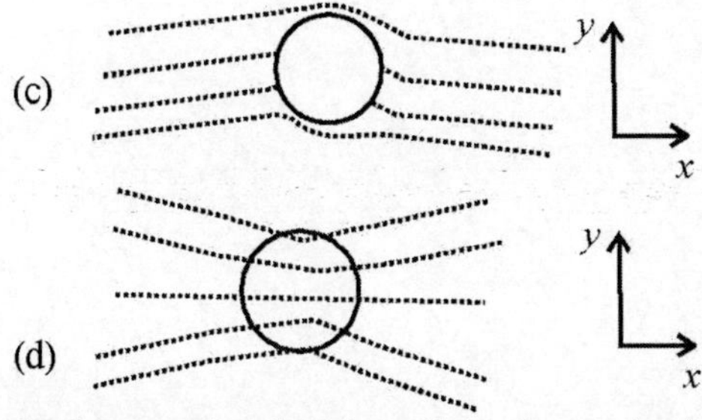

54. A charge q is placed at O in the cavity in a spherical uncharged conductor. Point S is outside the conductor.
If the charge is displaced from O towards S, still remaining within the cavity,

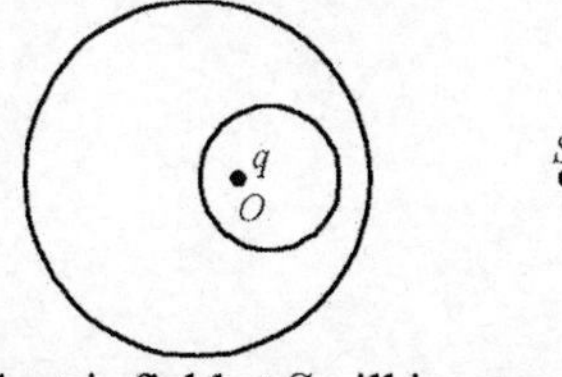

    (a) electric field at S will increase
    (b) electric field at S will decrease
    (c) electric field at S will first increase and then decrease
    (d) electric field at S will not change

55. Five styrofoam balls are suspended from insulating threads. Several experiments are performed on the balls and the following observations are made –
    (i) Ball A repels C and attracts B
    (ii) Ball D attracts B and has no effect on E
    (iii) A negatively charged rod attracts both A and E.
An electrically neutral styrofoam ball gets attracted if placed nearby a charged body due to induced charge. What are the charges, if any, on each ball?

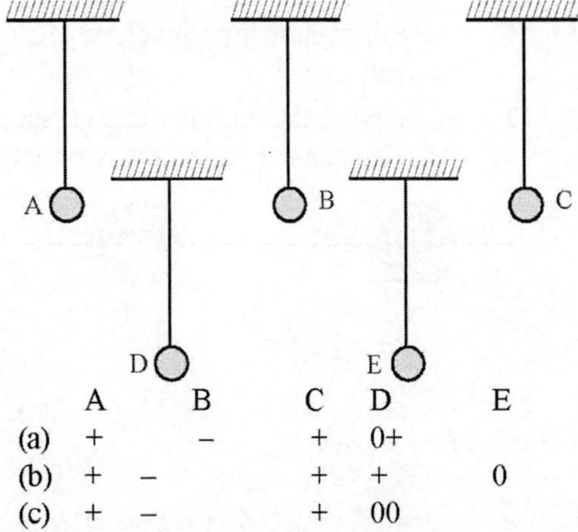

	A	B	C	D	E
(a)	+	–	+	0+	
(b)	+	–	+	+	0
(c)	+	–	+	00	
(d)	–	+	–	00	

56. Four charges are rigidly fixed along the Y-axis as shown. A positive charge approaches the system along the X-axis with initial speed just enough to cross the origin. Then its total energy at the origin is

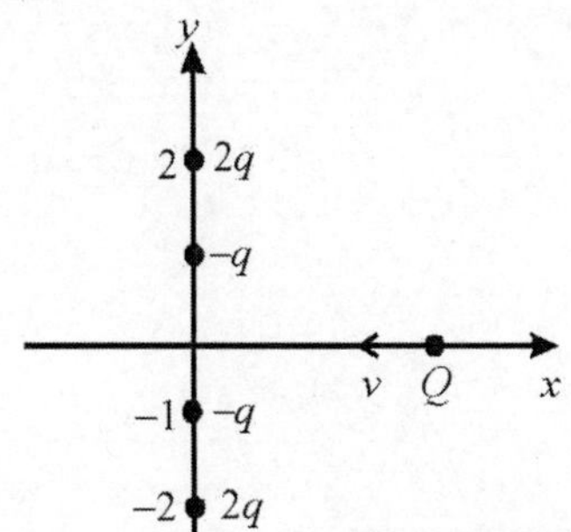

    (a) zero (b) positive
    (c) negative (d) data insufficient

57. A ball of charge $+50$ e lies at the centre of a hollow spherical metal shell that has a net charge of -150 e. The charge on the inner surface and outer surface of the spherical shell is :
    (a) $+50\,e, -200\,e$ (b) $-50\,e, -100\,e$
    (c) $-50\,e, -150\,e$ (d) $-50\,e, +100\,e$

58. The total positive charge in a glass of water containing 180 gm of water is approximately :
    (a) 1.0×10^7 coulomb
    (b) 6.03×10^7 coulomb
    (c) 1.80×10^7 coulomb
    (d) 1.20×10^7 coulomb

59. The figure shows two charged particles on x-axis. The particles are free to move. However, at one point, a third charged particle can be placed such that all three particles would be in equilibrium.
    1. Is the third particle positively or negatively charged?
    2. Is the point the third particle is placed to the left, right or between the two charged?

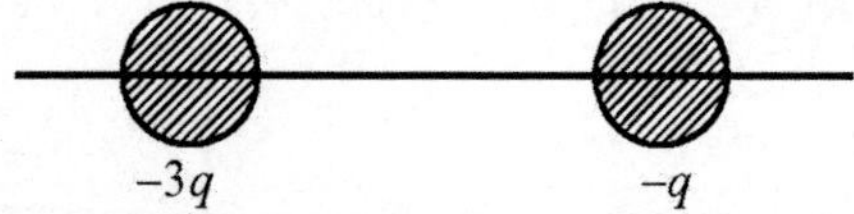

    (a) negative, to their right
    (b) positive, between them
    (c) negative, to their left
    (d) positive, to their right

60. The figure shows two situations in which a Gaussian cube sits in an electric field. The arrows and values indicate the directions and magnitudes (in N–m^2/C) of the electric fields. What is the net charge (in the two situations) inside the cube?

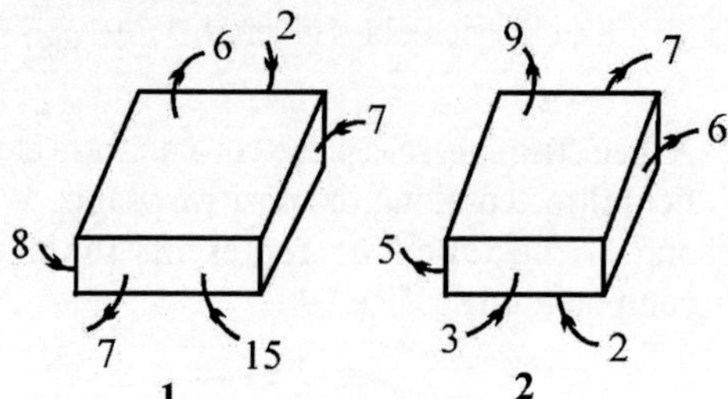

    (a) (1) negative (2) positive
    (b) (1) negative (2) zero
    (c) (1) positive (2) positive
    (d) (1) negative (2) negative

CHAPTER 16

Electrostatic Potential and Capacitance

CONCEPTS

ELECTRIC POTENTIAL

Electric potential at a point is the work done per unit charge (by an external agency) in bringing a charge from infinity to that point. Potential at a point is arbitrary to within an additive constant, since it is the potential difference between two points which is physically significant.

Electric Potential Due to a Point Charge

If potential at infinity is chosen to be zero; potential at a point with position vector $\vec{r}$ due to a point charge Q placed at the origin is given by $V(\vec{r}) = \frac{1}{4\pi\varepsilon_0}\frac{Q}{r}$.

Electric potential due to system of charges

For a charge configuration $q_1, q_2, ..., q_n$ with position vectors $r_1, r_2, ... r_n$, the potential at a point P is given by the superposition principle

$$V = \frac{1}{4\pi\varepsilon_0}\left(\frac{q_1}{r_{1P}} + \frac{q_2}{r_{2P}} + ... + \frac{q_n}{r_{nP}}\right)$$, where r_{1P} is the distance between q_1 and P, and so on.

In rectangular components, the electric field $\vec{E} = E_x\hat{i} + E_y\hat{j} + E_z\hat{k}$ and an infinitesimal displacement is $d\vec{r} = dx\,\hat{i} + dy\,\hat{j} + dz\,\hat{k}$

$$\int dV = -\int \vec{E}.d\vec{r} = -[E_x\ dx + E_y\ dy + E_z\ dz]$$

$$\vec{E} = -\frac{\partial V}{\partial x}\hat{i} - \frac{\partial V}{\partial y}\hat{j} - \frac{\partial V}{\partial z}\hat{k}$$

The electrostatic potential at a point with position vector $\vec{r}$ due to a point dipole of dipole moment $\vec{p}$ placed at the origin is $V(r) = \frac{1}{4\pi\varepsilon_0}\frac{\vec{p}.\hat{r}}{r^2}$.

Electric potential due to a charged ring

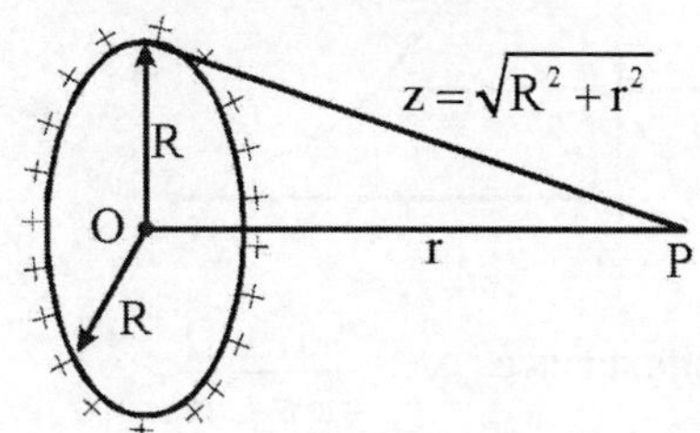

$$\text{At point P} = \frac{1}{4\pi\varepsilon_0}\frac{Q}{\sqrt{(R^2 + r^2)}}$$

$$\text{At centre } V = \frac{Q}{4\pi \in_0 R}$$

At a distance $r = \pm\frac{R}{\sqrt{2}}$, V is max, $V_{max} = \frac{Q}{2\sqrt{6}\pi \in_0}$

Electric potential due to a charged disc at a point on the axis

$$V = \frac{\sigma}{2\varepsilon_0}\left[(R^2 + r^2)^{1/2} - r\right]$$

Electric potential due to a Shell

(a) **At outside point :** $V = \frac{1}{4\pi\varepsilon_0}\frac{Q}{r}$

(b) **At an internal point :** $V = \frac{1}{4\pi\varepsilon_0}\frac{Q}{R}$ = V at surface

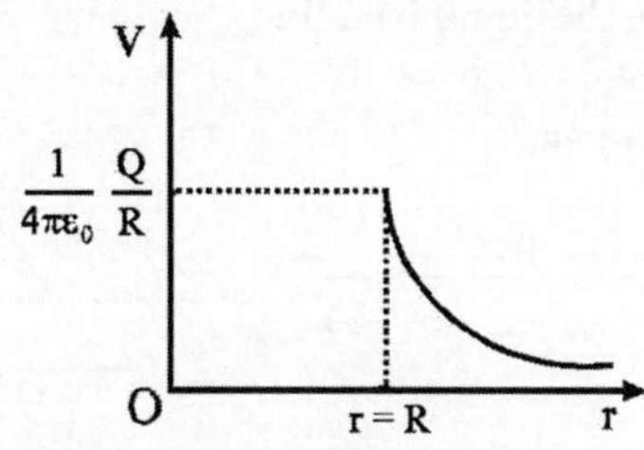

Electric potential due to a non-conducting charged sphere

(a) **At an external point :** $V = \frac{1}{4\pi\varepsilon_0}\frac{Q}{r}$

(b) **Potential at an internal point :**

$$V = \frac{1}{4\pi\varepsilon_0}\frac{Q}{2R^3}\left[3R^2 - r^2\right]$$

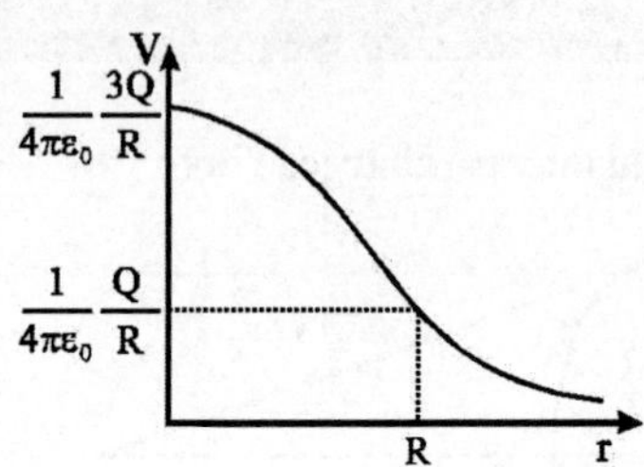

(c) **At the surface :** $V = \frac{1}{4\pi\in_0}\cdot\frac{Q}{R}$

Electric potential due to a finite length of charge with linear charge density λ

$$V = k\lambda \log_e\left(\frac{x+L}{x}\right)$$

Electric potential due to a dipole

(a) **Along the axial line**

$$V_{axial} = k\frac{p}{x^2 - l^2}$$

when $x >> l$ $V_{axial} = \frac{kp}{x^2}$

(b) **Along the equatorial line**

$V_{eq} = \text{zero}$

(c) **At any point**

$$V = k\frac{p\cos\theta}{(x^2 - l^2\cos^2\theta)}$$

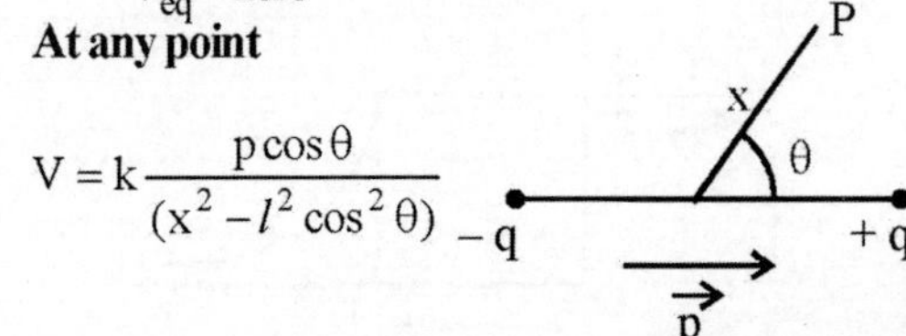

EQUIPOTENTIAL SURFACE

An equipotential surface is a surface over which potential has a constant value. For a point charge, concentric spheres centered at a location of the charge are equipotential surfaces. The electric field $\vec{E}$ at a point is perpendicular to the equipotential surface through the point. $\vec{E}$ is in the direction of the steepest decrease of potential.

Dielectric and Electric Polarisation

If the medium between the plates of a capacitor is filled with an insulating substance or **dielectric**, the electric field due to the charged plates induces a net dipole moment in the dielectric. This effect, called **polarisation**, gives rise to a field in the opposite direction. The net electric field inside the dielectric and hence the potential difference between the plates is thus reduced. Consequently, the capacitance C increases from its value C_0 when there is no medium (vacuum),
$C = KC_0$, where K is the **dielectric constant** of the insulating substance.

CAPACITANCE OF CAPACITORS

Capacitance is defined by C = Q/V, where Q and – Q are the charges on the two conductors and V is the potential difference between them. C is determined purely geometrically, by the shapes, sizes and relative positions of the two conductors.
The unit of capacitance is farad:, $1\text{ F} = 1\text{ C V}^{-1}$. For a parallel plate capacitor (with vacuum between the plates), $C = \varepsilon_0\frac{A}{d}$, where A is the area of each plate and d the separation between them.

Combination of Capacitors

For capacitors **in the series combination,** the total capacitance C

$$\frac{1}{C} = \frac{1}{C_1} + \frac{1}{C_2} + \frac{1}{C_3} + ...$$

For capacitors **in the parallel combination,** the total capacitance C :
$C = C_1 + C_2 + C_3 + ...$
where C_1, C_2, C_3... are individual capacitances.

Capacitance of Different Types of Capacitors

Capacitance of a praralled plate capacitors

(i) Capacitance of a parallel plate capacitor with a conducting slab

$$C = \frac{C_0}{\left(1 - \frac{t}{d}\right)}$$

Where, t = thickness of conducting slab
d = separation between the plates
And C_0 is the capacitance of parallel plate capacitor with air or vacuum inbetween.

(ii) Capacitance of a parallel plate capacitor with a dielectric slab.

$$C = \frac{\epsilon_0 A}{(d - t)\left(1 - \frac{1}{k}\right)}$$

Where, t = thickness of dielectric slab
k = dielectric constant of the slab

Capacitance of a spherical capacitor :

(i) If outer sphere is earthed

$$C = \frac{q}{v} = \frac{4\pi\varepsilon_0 .ab}{(b - a)}.$$

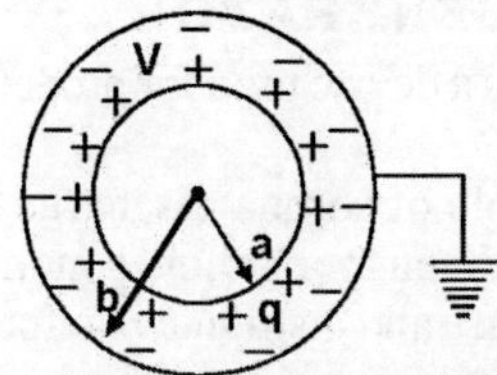

(ii) **If the inner sphere is earthed** and the outer sphere is charged, the capacity of the condenser becomes

$$C' = 4\pi\varepsilon_0 b + \frac{4\pi\varepsilon_0 .ab}{(b - a)} = \frac{4\pi\varepsilon_0 b^2}{(b - a)} = \frac{b}{a}C \Rightarrow C' > C.$$

Capacitance of cylindrical capacitor :

$$C = \frac{2\pi\varepsilon_0\varepsilon_r}{2.303\log_{10}(b/a)} \text{ farad}$$

The energy U stored in a capacitor of capacitance C, with charge Q and voltage V is

$$U = \frac{1}{2}QV = \frac{1}{2}CV^2 = \frac{1}{2}\frac{Q^2}{C}$$

VARIATION IN PARALLEL PLATE CAPACITOR UNDER TWO CONDITIONS

(a) Battery disconnected (after charging) $\rightarrow$ Q = const.

(b) Battery remains connected $\Rightarrow V \rightarrow$ const.

$$Q = CV, C = \frac{\varepsilon_0\varepsilon_r A}{d}, \varepsilon_0 = \frac{Q}{A\varepsilon_0\varepsilon_r} = \frac{V}{d} = \frac{\sigma}{\varepsilon_0\varepsilon_r}$$

(a) **Battery removed (Q $\rightarrow$ const.)**
(i) Initially Q_0 V_0 E_0
C_0 u_0
(ii) Insert a $\rightarrow$ $\downarrow$K $\downarrow$K
$\uparrow$K $\downarrow$K
dielectric slab.
(iii) Separation d $\uparrow$ $\rightarrow$ $\uparrow$
$\rightarrow\left(\frac{\sigma}{\varepsilon_0}\right)$ $\downarrow$ $\uparrow$
(iv) Eff. area A' $\uparrow$ $\rightarrow$ $\downarrow$ $\downarrow$
$\uparrow$ $\downarrow$

(b) **Battery remains connected (V $\rightarrow$ const.)**
(i) Initially Q_0 V_0 E_0
C_0 u_0
(ii) Insert a $\uparrow$K $\rightarrow$ $\rightarrow$
$\uparrow$K $\uparrow$K dielectric slab.
(iii) Separation d$\uparrow$ $\downarrow$ $\rightarrow$ $\downarrow$
$\downarrow$ $\downarrow$
(iv) Eff. area A' $\uparrow$ $\uparrow$ $\rightarrow$ $\rightarrow$
$\uparrow$ $\uparrow$

CHARGING AND DISCHARGING OF A CAPACITOR

Charging of a Capacitor

When a voltage is placed across the terminals of a capacitor the potential cannot rise to its final value instantaneously. As the charge builds up it tends to repel the addition of further charge.

$$q = Q_0(1 - e^{-t/RC}) \text{ and } V = V_0(1 - e^{-t/RC})$$

The current in circuit during charging is given by

$$I = I_0 e^{-t/RC}, \text{ where } I_0 = V_0/R.$$

the growth of q and V and decay of I is exponential, and depends on

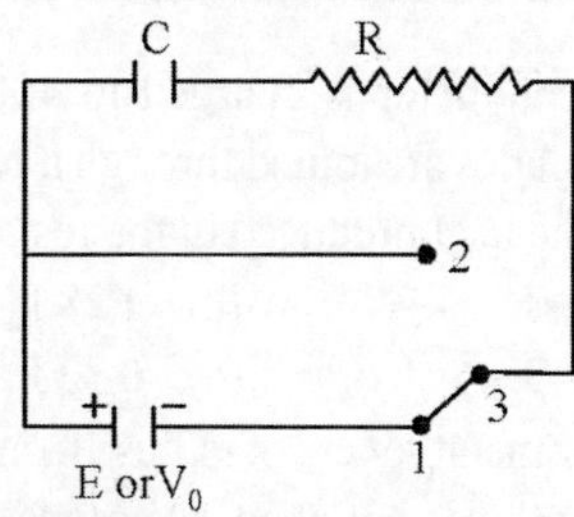

(a) The capacitance of the capacitor, and
(b) The resistance of the circuit through which it is being charged

Time constant of R – C circuits :
time const. $\tau = RC$.

If $t = RC, \Rightarrow q = Q_0(1 - 1/e) = 0.632\,Q_0$
$V = 0.632\,V_0$, and $I = 0.368\,I_0$

It is that time in which the charge on the plates grows to 63.2% of the final value.
It is the time in which the potential difference across the capacitor rises to 63.2% of its final value.
It is the time in which current falls to 36.8% of the initial value.

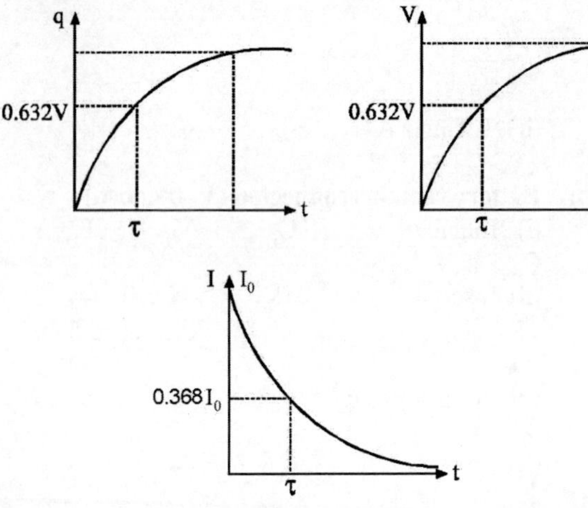

Current in R – C circuit while charging decreases exponentially (to 0.368 I in τ time, to 0.135 I_0 in time 2τ. and to 0.05 I_0 in time 3τ). Thus after same time the current in the RC circuit becomes unmeassurable (zero). Thus a capacitor blocks and direct current (d.c.) flow.

Discharging of a Capacitors

Let at time t = 0, the switch S in circuit of given fig. (fully charged capacitor) is changed to position 2 then the capacitor discharges through resistance R. The charge and voltage across the capacitor and the current in the circuit falls exponentially

$\Rightarrow q = Q_0 e^{-t/RC}$, $V = V_0 e^{-t/RC}$ and $I = -I_0 e^{-t/RC}$ at, time constant $\tau = RC \Rightarrow q = 0.368\, Q_0$, $V = 0.368\, V_0$

Half life (T_n):

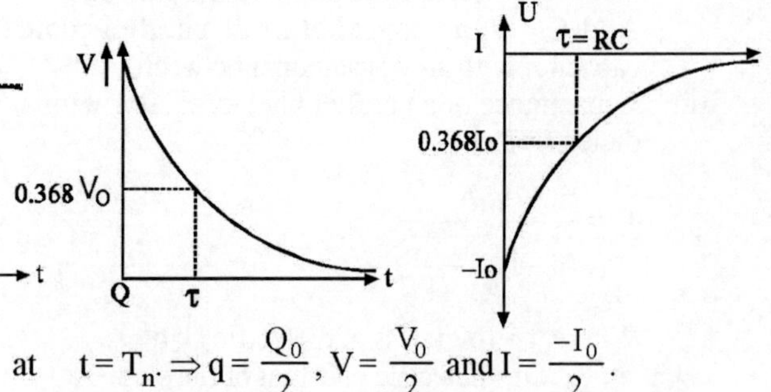

at $t = T_n \Rightarrow q = \frac{Q_0}{2}$, $V = \frac{V_0}{2}$ and $I = \frac{-I_0}{2}$.

Heat produced across resistance of connecting wire is equal to energy stored in capacitor

$$\text{Heat} = u_{cap.} = \frac{1}{2} CV^2$$

VAN DE GRAFF GENERATOR

A Van de graff generator is a device used for producing very high voltage.
It is based on the principle of corona discharge and the property that charge given to a hollow conductor is transferred to outer surface and distributed uniformly over the sphere.

Use of Van De Graff Generator

It is used to accelerate charged particles like- electrons, protons ions etc.

Exercise

1. A 4μF conductor is charged to 400 volts and then its plates are joined through a resistance of 1 kΩ. The heat produced in the resistance is
   (a) 0.16 J (b) 1.28 J
   (c) 0.64 J (d) 0.32 J
2. A hollow metal sphere of radius 10 cm is charged such that the potential on its surface is 80 V. The potential at the centre of the sphere is
   (a) zero (b) 80 V
   (c) 800 V (d) 8 V
3. The four capacitors, each of 25μ *F* are connected as shown in fig. The dc voltmeter reads 200 *V*. The charge on each plate of capacitor is

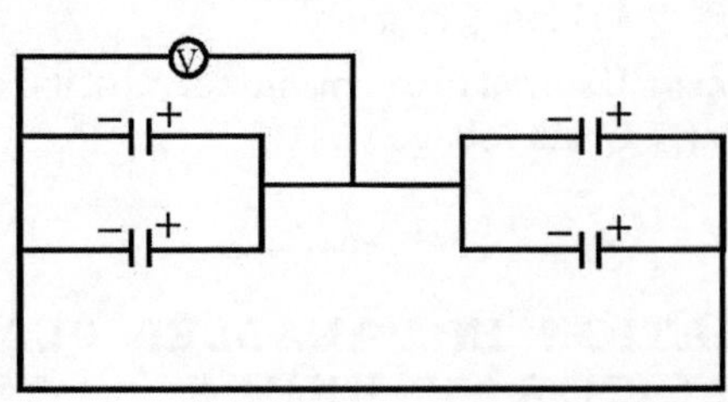

   (a) $\pm 2 \times 10^{-3}$C (b) $\pm 5 \times 10^{-3}$C
   (c) $\pm 2 \times 10^{-2}$C (d) $\pm 5 \times 10^{-2}$C
4. If the potential of a capacitor having capacity 6 μ*F* is increased from 10 *V* to 20 *V*, then increase in its energy will be

(a) 4×10^{-4} J (b) 4×10^{-4} J
(c) 9×10^{-4} J (d) 12×10^{-6} J

5. A parallel plate condenser with oil between the plates (dielectric constant of oil $K = 2$) has a capacitance C. If the oil is removed, then capacitance of the capacitor becomes
(a) $\sqrt{2}C$ (b) $2C$
(c) $\frac{C}{\sqrt{2}}$ (d) $\frac{C}{2}$

6. What is the effective capacitance between points X and Y?

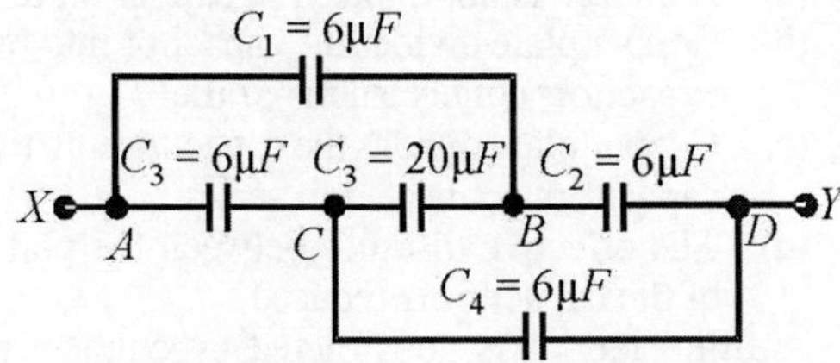

(a) 24 μF (b) 18 μF
(c) 12 μF (d) 6 μF

7. A capacitor is charged to store an energy U. The charging battery is disconnected. An identical capacitor is now connected to the first capacitor in parallel. The energy in each of the capacitor is
(a) $U/2$ (b) $3U/2$
(c) U (d) $U/4$

8. In a parallel plate capacitor, the distance between the plates is d and potential difference across the plates is V. Energy stored per unit volume between the plates of capacitor is
(a) $\frac{Q^2}{2V^2}$ (b) $\frac{1}{2}\varepsilon_0\frac{V^2}{d^2}$
(c) $\frac{1}{2}\frac{V^2}{\varepsilon_0 d^2}$ (d) $\frac{1}{2}\varepsilon_0\frac{V^2}{d}$

9. The capacity of a parallel plate condenser is 10 μF when the distance between its plates is 8 cm. If the distance between the plates is reduced to 4 cm then the capacity of this parallel plate condenser will be
(a) 5 μF (b) 10 μF
(c) 20 μF (d) 40 μF

10. Energy stored in a capacitor is
(a) $\frac{1}{2}QV$ (b) QV
(c) $\frac{1}{QV}$ (d) $\frac{2}{QV}$

11. A capacitor C_1 is charged to a potential difference V. The charging battery is then removed and the capacitor is connected to an uncharged capacitor C_2. The potential difference across the combination is
(a) $\frac{VC_1}{(C_1+C_2)}$ (b) $V\left(1+\frac{C_2}{C_1}\right)$
(c) $V\left(1+\frac{C_1}{C_2}\right)$ (d) $\frac{VC_2}{(C_1+C_2)}$

12. A solid spherical conductor is given a charge. The electrostatic potential of the conductor is
(a) constant throughout the conductor
(b) largest at the centre
(c) largest on the surface
(d) largest somewhere between the centre and the surface

13. Each corner of a cube of side l has a negative charge, $-q$. The electrostatic potential energy of a charge q at the centre of the cube is
(a) $-\frac{4q^2}{\sqrt{2}\pi\varepsilon_0 l}$ (b) $\frac{\sqrt{3}q^2}{4\pi\varepsilon_0 l}$
(c) $\frac{4q^2}{\sqrt{2}\pi\varepsilon_0 l}$ (d) $-\frac{4q^2}{\sqrt{3}\pi\varepsilon_0 l}$

14. Three capacitors each of capacity $4\mu F$ are to be connected in such a way that the effective capacitance is 6 μF. This can be done by
(a) connecting two in parallel and one in series
(b) connecting all of them in series
(c) connecting them in parallel
(d) connecting two in series and one in parallel

15. An electric dipole has the magnitude of its charge as q and its dipole moment is p. It is placed in uniform electric field E. If its dipole moment is along the direction of the field, the force on it and its potential energy are respectively
(a) $q.E$ and max. (b) 2 q.E and min.
(c) $q.E$ and p.E (d) zero and min.

16. A bullet of mass 2 g is having a charge of 2μC. Through what potential difference must it be accelerated, starting from rest, to acquire a speed of 10 m/s?
(a) 50 V (b) 5kV
(c) 50kV (d) 5V

17. An electron initially at rest is accelerated through a potential difference of one volt. The energy acquired by electron is :
(a) 10^{-19} J (b) 1.6×10^{-19} erg
(c) 1.6×10^{-19} J (d) 1 J

18. Let V be the electric potential at a given point. Then the electric field E_x along x-direction at that point is given by :
(a) $\int_0^{\infty} V dx$ (b) $\frac{dV}{dx}$
(c) $-\frac{dV}{dx}$ (d) $-V\frac{dV}{dx}$

19. Potential at any point inside a charged hollow sphere :
(a) increases with distance
(b) is a constant
(c) decreases with distance from centre
(d) is zero

20. Four charges $q_1 = 2 \times 10^{-8}$ C, $q_2 = -2 \times 10^{-8}$ C, $q_3 = -3 \times 10^{-8}$ C, and $q_4 = 6 \times 10^{-8}$ C are placed at four corners of a square of side $\sqrt{2}$ m. What is the potential at the centre of the square?
(a) 270 V (b) 300 V
(c) Zero (d) 100 V

21. A parallel plate air capacitor has a capacitance C. When it is half filled with a dielectric of dielectric constant 5, the percentage increase in the capacitance will be
(a) 400% (b) 66.6%
(c) 33.3% (d) 200%

22. The potential at a point P due to an electric dipole is 1.8×10^5V. If P is at a distance of 50 cm apart from the centre O of the dipole and if CP makes an angle 60° with the positive side of the axial line of the dipole, what is the moment of the dipole?
(a) 10 C-m (b) 10^{-3} C-m
(c) 10^{-4} C-m (d) 10^{-5} C-m

23. A copper plate is introduced between the plates A and B of a parallel plate capacitor as shown in Fig. The copper plate is not in contact with A or B. Which one of following statements is correct?

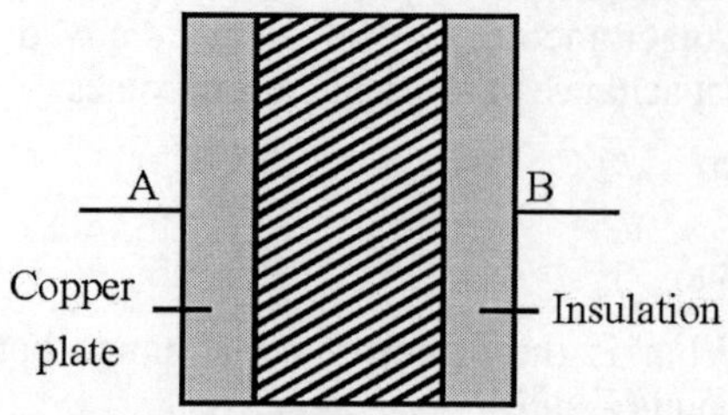

(a) A charge is not induced in copper plate
(b) Copper plate divides the capacitor into two capacitors connected in parallel
(c) Copper plate divides the capacitor into two capacitors connected in series
(d) The effective distance between the plates of the capacitor is reduced

24. A finite ladder is constructed by connecting several sections of 2 μF, 4 μF capacitor combinations as shown in the figure. It is terminated by a capacitor of capacitance C. What value should be chosen for C, such that the equivalent capacitance of the ladder between the point A and B becomes independent of the number of sections in between?

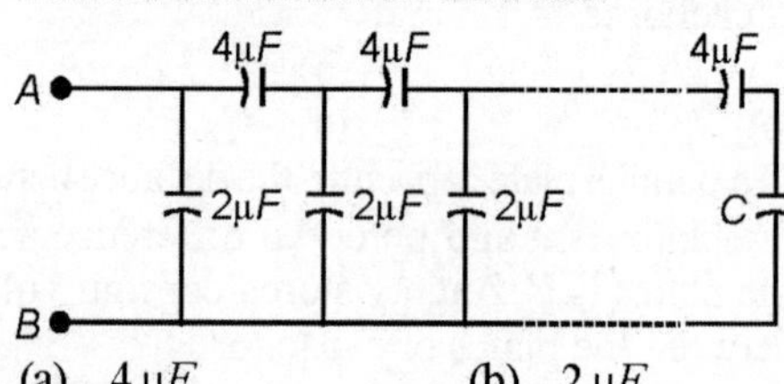

(a) 4 μF (b) 2 μF
(c) 18 μF (d) 6 μF

25. As per the diagram, a point charge $+q$ is placed at the origin O. Work done in taking another point charge $-Q$ from the point A [coordinates (0, a)] to another point B [coordinates (a, 0)] along the straight path AB is :

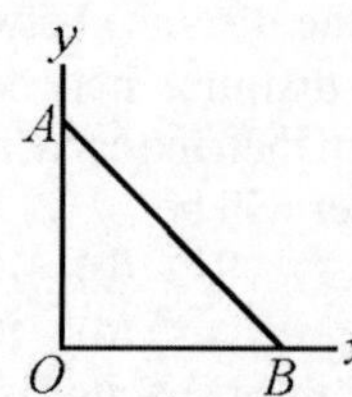

(a) zero (b) $\left(\frac{-qQ}{4\pi\varepsilon_0}\frac{1}{a^2}\right)\sqrt{2}a$
(c) $\left(\frac{qQ}{4\pi\varepsilon_0}\frac{1}{a^2}\right)\cdot\frac{a}{\sqrt{2}}$ (d) $\left(\frac{qQ}{4\pi\varepsilon_0}\frac{1}{a^2}\right)\cdot\sqrt{2}a$

26. Two charges q_1 and q_2 are placed 30 cm apart, as shown in the figure. A third charge q_3 is moved along the arc of a circle of radius 40 cm from C to D. The change in the potential energy of the system is $\frac{q_3}{4\pi \epsilon_0}k$, where k is

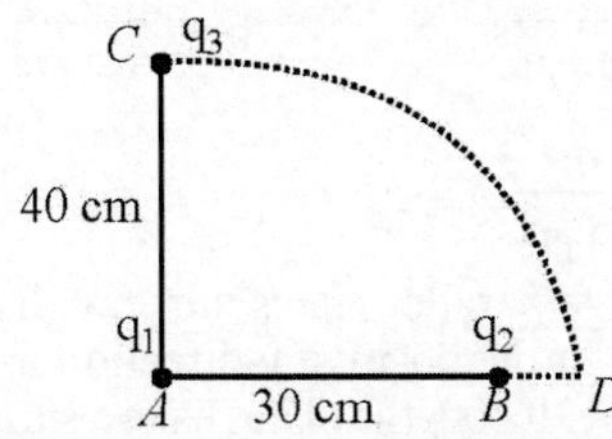

(a) $8q_1$ (b) $6q_1$
(c) $8q_2$ (d) $6q_2$

27. A network of four capacitors of capacity equal to $C_1 = C$, $C_2 = 2C$, $C_3 = 3C$ and $C_4 = 4C$ are conducted to a battery as shown in the figure. The ratio of the charges on C_2 and C_4 is:

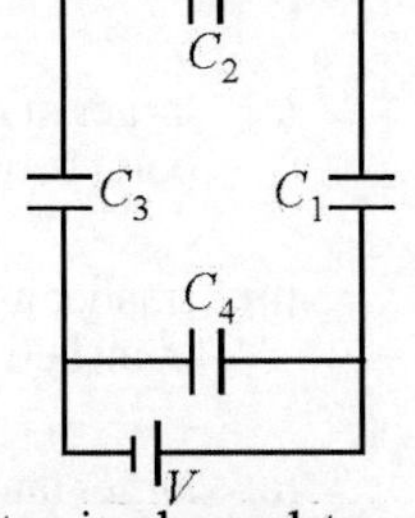

(a) $\frac{4}{7}$
(b) $\frac{3}{22}$
(c) $\frac{7}{4}$
(d) $\frac{22}{3}$

28. A parallel plate air capacitor is charged to a potential difference of V volts. After disconnecting the charging battery the distance between the plates of the capacitor is increased using an insulating handle. As a result the potential difference between the plates
(a) does not change (b) becomes zero
(c) increases (d) decreases

29. Charges $+q$ and $-q$ are placed at points A and B respectively which are a distance $2L$ apart, C is the midpoint between A and B. The work done in moving a charge $+Q$ along the semicircle CRD is

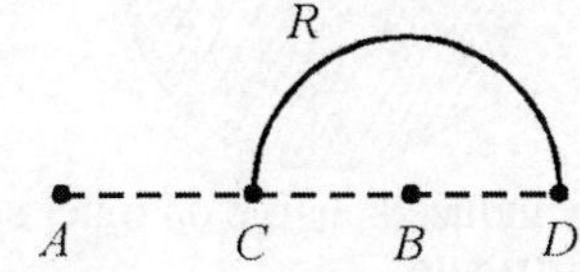

(a) $\frac{qQ}{2\pi\varepsilon_0 L}$ (b) $\frac{qQ}{6\pi\varepsilon_0 L}$
(c) $-\frac{qQ}{6\pi\varepsilon_0 L}$ (d) $\frac{qQ}{4\pi\varepsilon_0 L}$

30. Two condensers, one of capacity C and other of capacity $C/2$ are connected to a V-volt battery, as shown.

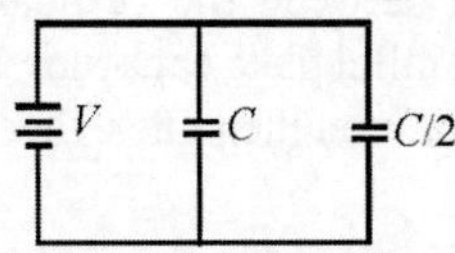

The work done in charging fully both the condensers is
(a) $\frac{1}{4}CV^2$ (b) $\frac{3}{4}CV^2$
(c) $\frac{1}{2}CV^2$ (d) $2CV^2$

31. The capacitance between points P and Q in circuit shown in fig is

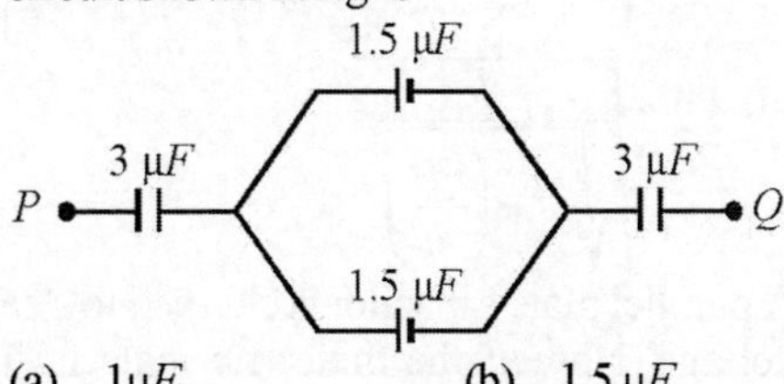

(a) 1μF (b) 1.5 μF
(c) 9 μF (d) 6.75 μF

32. The maximum electric field that can be held in air without producing ionisation of air is 10^7 V/m. The maximum potential therefore, to which a conducting sphere of radius 0.10 m can be charged in air is
(a) 10^9 V (b) 10^8 V
(c) 10^7 V (d) 10^6 V

33. Three capacitors C_1, C_2 and C_3 are connected as shown in the figure to a battery of V volt. If the capacitor C_3 breaks down electrically the change in total charge on the combination of capacitors is

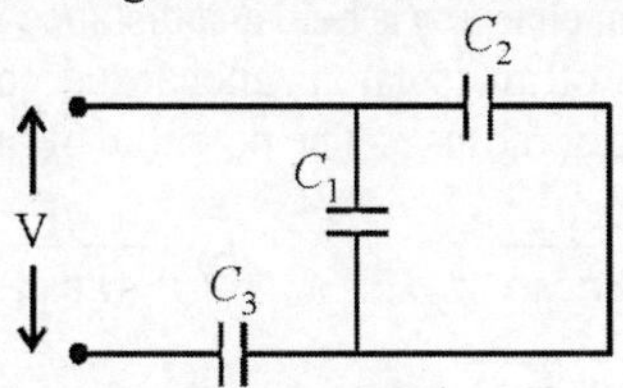

(a) $(C_1 + C_2) V[1 - C_3/(C_1 + C_2 + C_3)]$
(b) $(C_1 + C_2) V[1-(C_1 + C_2)/(C_1 + C_2 + C_3)]$
(c) $(C_1 + C_2) V[1+ C_3/(C_1 + C_2 + C_3)]$
(d) $(C_1 + C_2) V[1 - C_2/(C_1 + C_2 + C_3)]$

34. Two charges of magnitude $+q$ and $-3q$ are placed 100 cm apart. The distance from $+q$ between the charges where the electrostatic potential is zero is :
(a) 25 cm (b) 50 cm
(c) 75 cm (d) 80 cm

35. A simple pendulum has a length l, mass of bob m. The bob is given a charge q. The pendulum is suspended between the vertical plates of the charged parallel plate capacitor. If E is the field strength between the plates, then time period T equals

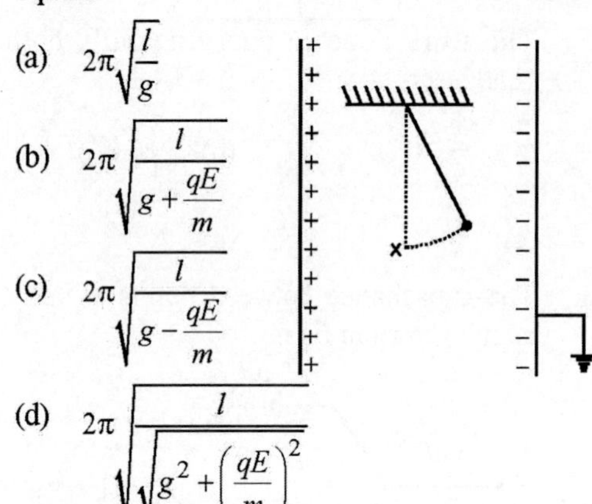

(a) $2\pi\sqrt{\frac{l}{g}}$

(b) $2\pi\sqrt{\frac{l}{g+\frac{qE}{m}}}$

(c) $2\pi\sqrt{\frac{l}{g-\frac{qE}{m}}}$

(d) $2\pi\sqrt{\frac{l}{\sqrt{g^2+\left(\frac{qE}{m}\right)^2}}}$

36. A parallel plate capacitor is charged to a certain voltage. Now, if the dielectric material (with dielectric constant k) is removed then the
(a) capacitance increases by a factor of k
(b) electric field reduces by a factor k
(c) voltage across the capacitor decreases by a factor k
(d) none of these

37. Two identical conducting balls having positive charges q_1 and q_2 are separated by a distance r. If they are made to touch each other and then separated to the same distance, the force between them will be
(a) less than before (b) same as before
(c) more than before (d) zero

38. A point charge q is held at a distance $2a$ from the centre of an isolated, uncharged conductance sphere of radius a. The potential of the sphere is

(a) $\frac{q}{8\pi\varepsilon_0 a}$ (b) $\frac{q}{4\pi\varepsilon_0 a}$

(c) $\frac{q}{2\pi\varepsilon_0 a}$ (d) zero

39. A sphere of radius R has uniform volume charge density. The electric potential at a points ($r<R$) is
(a) due to the charge inside a sphere of radius r only
(b) due to the entire charge of the sphere
(c) due to the charge in the spherical shell of inner and outer radii r and R, only
(d) independent of r

40. A point charge q is held at a distance $\frac{a}{2}$ from the centre of a thin, conducting, uncharged, spherical shell of radius a. The potential of the shell is

(a) $\frac{q}{2\pi\varepsilon_0 a}$ (b) $\frac{q}{4\pi\varepsilon_0 a}$

(c) $\frac{q}{8\pi\varepsilon_0 a}$ (d) zero

41. Two large parallel plates move with a constant speed v in the positive y-direction as shown in the figure. If both the plates have a surface charge density $\sigma > 0$, the magnetic field at the point P just above the top plate will have

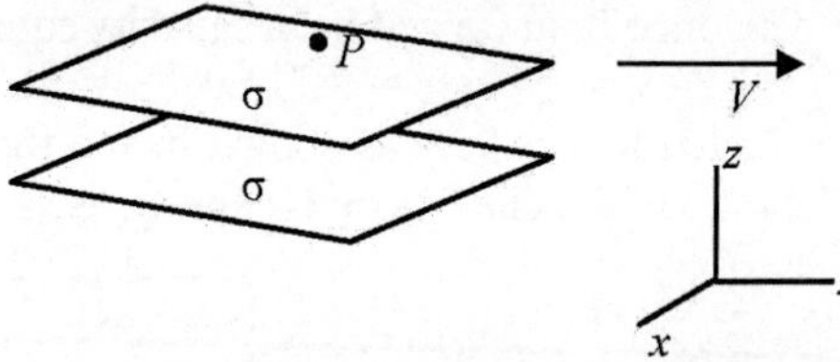

(a) larger magnitude than the field at the mid-point between the plates and point towards $-\hat{x}$
(b) smaller magnitude than the field at the mid-point between the plates and point towards $+\hat{x}$
(c) larger magnitude than the field at the mid-point between the plates and point towards $+\hat{x}$
(d) smaller magnitude than the field at the mid-point between the plates and point towards $-\hat{x}$

42. The figure shows a charge q placed inside a cavity in an uncharged conductor. Now if an external electric field is switched on

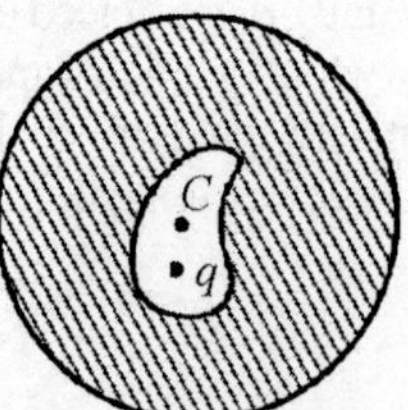

(a) only induced charge on outer surface will redistribute
(b) only induced charge on inner surface will redistribute
(c) both induced charge on outer and inner surface will redistribute
(d) force on charge q placed inside the cavity will change

43. In the figure two conducting concentric spherical shells are shown.

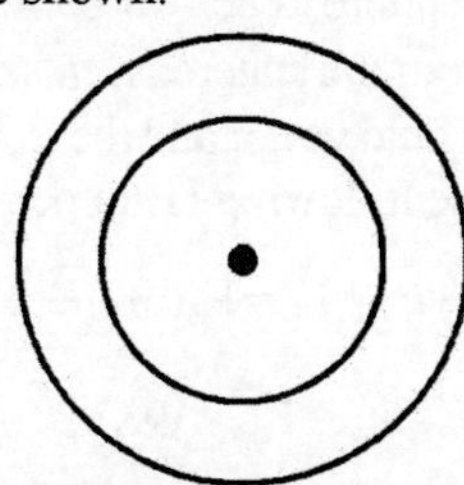

If the electric potential at the centre is 20V and the electric potential of the outer shell is 5V, then the potential of the inner shell is
(a) 5V (b) 15V
(c) 20V (d) 25V

44. Two point charge of 8 μC and 12 μC are kept in air at a distance of 10 cm from each other. The work required to change the distance between them to 6 cm will be
(a) 5.8 J (b) 4.8 J
(c) 3.8 J (d) 2.8 J

45. A condenser of capacity C is charged to a potential difference of V_1. The plates of the condenser are then connected to an ideal inductor of inductance L. The current through the inductor when the potential difference across the condenser reduces to V_2 is
(a) $\left(\frac{C(V_1^2-V_2^2)}{L}\right)^{1/2}$ (b) $\left(\frac{C(V_1-V_2)^2}{L}\right)^{1/2}$
(c) $\frac{C(V_1^2-V_2^2)}{L}$ (d) $\frac{C(V_1-V_2)}{L}$

46. Two conducting spheres of radii 5.0 cm and 10.0 cm are given a charge of 15.0 μC each. After connective the spheres by a copper wire, the charge on the smaller sphere is equal to :
(a) 20.0 μC (b) 5.0 μC
(c) 10 μC (d) none of the above

47. Four points a, b, c and d are set at equal distance from the centre of a dipole as shown the figure. The electrostatic potential V_a, V_b, V_c, and V_d would satisfy the following relation:
(a) $V_a > V_b > V_c > V_d$
(b) $V_a > V_b = V_d > V_c$
(c) $V_a > V_c = V_b = V_d$
(d) $V_b = V_d > V_a > V_c$

48. N small drops of a liquid each having a charge $+q$ are combined to form a larger drop. The new electrostatic potential V_0 is related to the potential V on each small drop by the equation :
(a) $V_0 = VN^3$ (b) $V_0 = VN^{2/3}$
(c) $V_0 = VN^{3/2}$ (d) $V_0 = VN$

49. Consider the network shown here, B is a 4V battery of zero internal resistance; S_1 and S_2 are switches and C_1 and C_2 are ideal capacitors of values shown. Suppose S_1 is closed keeping S_2 open till the capacitor C_1 is fully charged. Then S_1 is opened and S_2 is closed so that C_2 gets charged. The energy stored in the capacitors is equal to :

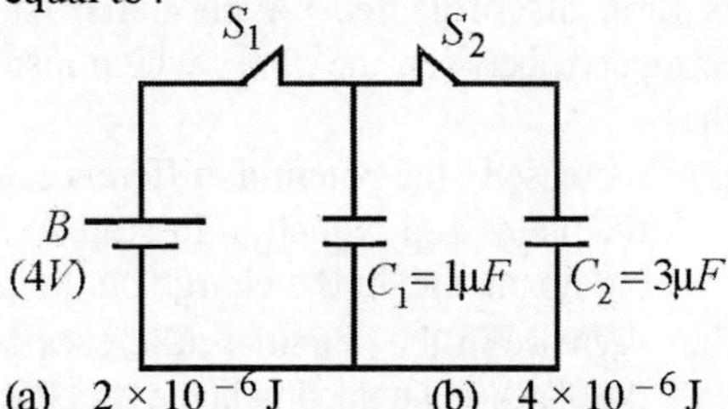

(a) 2×10^{-6} J (b) 4×10^{-6} J
(c) 8×10^{-6} J (d) 32×10^{-6} J

50. The figure given below shows four arrangements of charged particles, all at the same distance from the origin. Rank the situation according to the net electric potential (V_1, V_2, V_3, V_4) at the origin, most positive first :

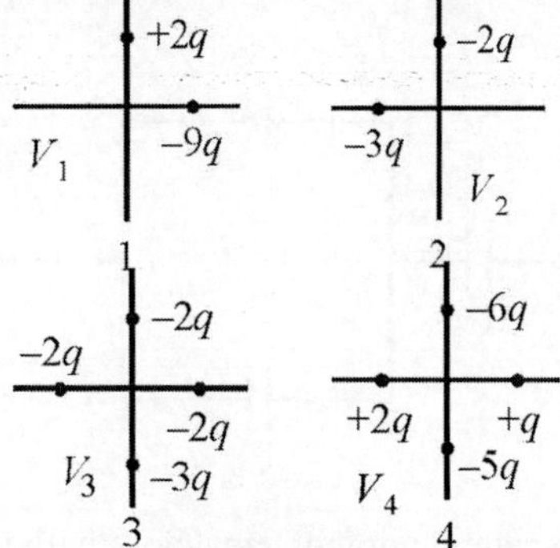

(a) $V_1 > V_2 > V_3 > V_4$ (b) $V_2 > V_1 > V_3 > V_4$
(c) $V_2 > V_1 > V_4 > V_3$ (d) $V_4 > V_1 > V_3 > V_2$

51. The electric potentials (in volts) and four points A, B, C and D are depicted in the figure on which of the following parts with the work done on a charge of -2 C, be maximum
(+2V) (–3V)
D • C •
(+1V) (–1V)
B • A •
(a) BC (b) BAD
(c) AD (d) ABC

52. A parallel plate capacitor has plate area A and separation d. It is charged to a potential difference V_0. The charging battery is disconnected and the plates are pulled apart to three times the initial separation. The work required to separate the plates is

(a) $\frac{3\varepsilon_0 A V_0^2}{d}$ (b) $\frac{\varepsilon_0 A V_0^2}{2d}$

(c) $\frac{\varepsilon_0 A V_0^2}{3d}$ (d) $\frac{\varepsilon_0 A V_0^2}{d}$

53. A capacitor is charged by using a battery which is then disconnected. A dielectric slab is introduced between the plates which results in the :
(a) increase in the potential difference across the plates and reduction in stored energy but no change in the charge on the plates
(b) decrease in the potential difference across the plates and reduction in the stored energy but no change in the charge on the plates
(c) reduction of charge on the plates and increase of potential difference across the plates
(d) increase in stored energy but no change in potential difference across the plates

54.

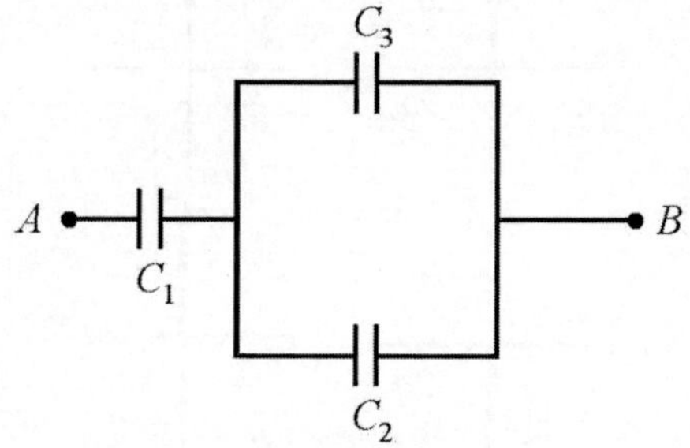

The combination of capacitors with $C_1 = 3\mu F$, $C_2 = 4\mu F$ and $C_3 = 2\mu F$ is charged by connecting AB to a battery. Consider the following statements :

I. Energy stored in C_1 = Energy stored in C_2 + Energy stored in C_3
II. Charge on C_1 = Charge on C_2 + Charge on C_3
III. Potential drop across C_1 = Potential drop across C_2 = Potential drop across C_3

Which of these is / are correct ?
(a) I and II (b) II only
(c) I and III (d) III only

55. Consider two points 1 and 3 in a region outside a charged sphere. The two points are not very far away from the sphere . If $\vec{E}$ and V represent the electric field vector and the electric potential, which of the following is not possible :

(a) $|\vec{E}_1| = |\vec{E}_2|, V_1 = V_2$ (b) $\vec{E}_1 \neq \vec{E}_2, V_1 \neq V_2$

(c) $\vec{E}_1 \neq \vec{E}_2, V_1 = V_2$ (d) $|\vec{E}_1| = |\vec{E}_2|, V_1 \neq V_2$

56. A parallel plate capacitor of plate area A and plates separation distance d is charged by applying a potential V_0 between the plates. The dielectric constant of the medium between the plates is K. What is the uniform electric field E between the plates of the capacitor?

(a) $E = \epsilon_0 \frac{CV}{kA}$ (b) $E = \frac{V_0}{kd}$

(c) $E = \frac{V_0}{ka}$ (d) $E = \frac{kv_0 d}{\epsilon_0 A}$

57. When a metal plate is introduced between the two plates of a charged capacitor and insulated from them, then
(a) The metal plate divides the capacitor into two capacitors connected in parallel to each other
(b) The metal plate divides the capacitors into two capacitors connected in series with each other
(c) The metal plate is equivalent to a dielectric of zero dielectric constant
(d) The metal plate is not equivalent to a dielectric of infinite dielectric constant

58. A capacitor works in
(a) AC. circuits (b) DC. circuits
(c) both 'a' and 'b' (d) neither 'a' nor 'b'

59. In a charged capacitor, the energy is stored in
(a) the negative charges
(b) the positive charges
(c) the field between the plates
(d) both 'a' and 'b'

60. A parallel plate capacitor is charged and then isolated. What is the effect of increasing the plate separation on charge, potential, capacitance, respectively?
(a) constant, decreases, decreases
(b) increases, decreases, decreases
(c) constant, decreases, increases
(d) constant, increases, decreases

Current Electricity

CONCEPTS

ELECTRIC CURRENT

It is the rate of flow of charge, conventionally, the direction of flow of positive charge is taken as the direction of electric current. It is a scalar quantity and its unit is ampere (A).

Electric current, $I = \frac{dq}{dt}$

The net charge in a current carrying conductor is zero.

Current through a given area of a conductor is the net charge passing per unit time through the area. Current is a scalar although we represent current with an arrow. Currents do not obey the law of vector addition. The current I through an area of cross-section is given by the scalar product of two vectors.

$$I = \vec{j} \cdot \Delta\vec{A}, \text{ where } \vec{j} \text{ and } \Delta\vec{A} \text{ are vectors.}$$

j gives the amount of charge flowing per second per unit area normal to the flow, $\vec{j} = nqv_d$

where n is the number density (number per unit volume) of charge carriers each of charge q, and v_d is the drift velocity of the charge carriers. For electrons q = – e. If $\vec{j}$ is normal to a cross-sectional area A and is constant over the area, **the magnitude of the current** I through the area is $\mathbf{nev_dA}$.

DRIFT VELOCITY

It is the average uniform velocity acquired by free electrons inside a metal by the application of an electric field which is responsible for current through it. Drift velocity is very small. It is of the order of 10^{-4} m/s as compared to thermal speed ($\approx 10^5$m/s) of electrons at room temperature.

Using, $\vec{v} = \vec{u} + \vec{a}t$ we have $\vec{v}_d = \frac{-e\vec{E}}{m}\tau$

($v_d \sim 10^{-4}$ m/sec) where τ is **relaxation time**.
The direction of drift velocity for electrons in a metal is opposite to that of applied field $\vec{E}$.

Mobility

mobility of a charge carrier is defined as drift velocity acquired per unit electric field.

$$\text{Mobility } \mu = \frac{v_d}{E} = \frac{e\tau}{m} = \frac{e}{m}\frac{m\sigma}{ne^2} = \frac{\sigma}{ne}$$

Its unit is $m^2 V^{-1} s^{-1}$ and dimensions are $[M^{-1} T^2 A^1]$. The mobility depends on applied potential difference, length of conductor, number density of charge carriers, current in conductor and area of cross-section of conductor.

Current due to translatory motion of charge : If n particles per unit volume each having charge q and moving with velocity v, the current through cross-section A is I = nqvA.

Current due to rotatory motion of charge : If a point charge q is moving in a circle of radius r with speed v (frequency ν, angular speed ω and time period T) then

corresponding current $I = q\nu = \frac{q}{T} = \frac{qv}{2\pi r} = \frac{q\omega}{2\pi}$

OHM'S LAW

The electric current I flowing through a conductor is proportional to the voltage V across its ends, i.e., $V \propto I$ or V = RI, where R is called the resistance of the substance. The **resistance** is the obstruction offered to the flow of electric current. The unit of resistance is ohm (Ω) : $1\Omega = 1\ VA^{-1}$.

Homogeneous conductors like silver or semiconductors like pure germanium or germanium containing impurities obey Ohm's law within some range of electric field values. They are called **ohmic** and if the field becomes too strong, there are departures from Ohm's law in all cases. The conductors that do not obey ohm's law called **non-ohmic**.

Equation $\vec{E} = \rho\vec{j}$ is microscopic form of Ohm's law, i.e., a conducting material obeys Ohm's law when the resistivity of the material does not depend on the magnitude and direction of applied electric field.

V-i Characteristics

It is the graph between current (i) and potential difference (V) V-i characteristics of ohmic substance is linear whereas of non-ohmic substance it is non-linear.

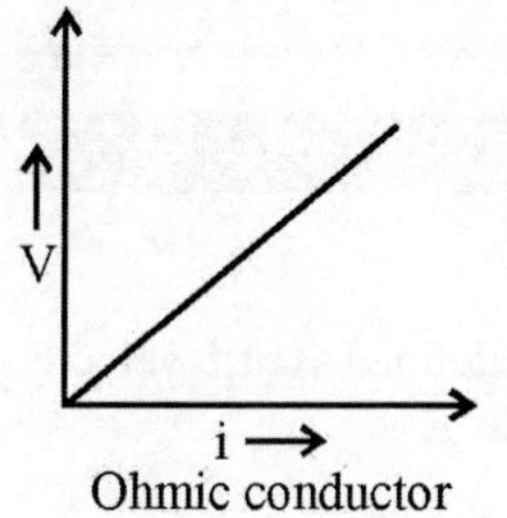

Ohmic conductor

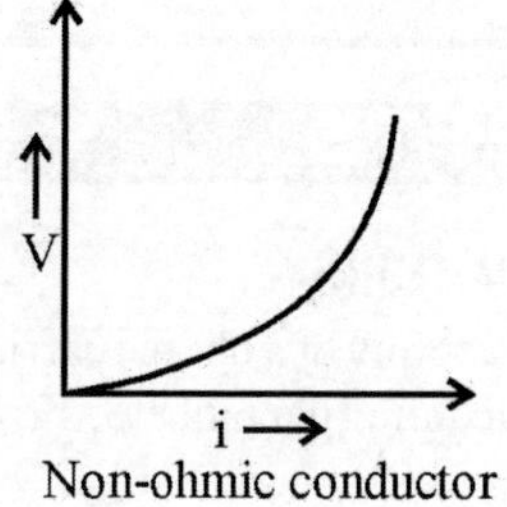

Non-ohmic conductor

Resistivity

The resistance R of a conductor depends on its length ℓ and constant cross-sectional area A through the relation,

$R = \frac{\rho\ell}{A}$, where ρ, called **resistivity** is a property of the material and depends on temperature and pressure. Electrical resistivity of substances varies over a very wide range. Metals have low resistivity, in the range of $10^{-8}\ \Omega$ m to $10^{-6}\ \Omega$ m. Insulators like glass and rubber have 10^{22} to 10^{24} times greater resistivity. Semiconductors like Si and Ge lie roughly in the middle range of resistivity on a logarithmic scale.

Temperature Dependence of Resistance

The value of resistivity or specific resistance changes with temperature. This change is due to change in resistance of the conductor with temperature. Dependence of resistivity on temperature is given by following equation :

$$\rho_t = \rho_0 (1 + \alpha t).$$

Thus, $R_t = R_0 (1 + \alpha t)$ or $\alpha = \frac{R_t - R_0}{R_0 \times t}$

Constant α is called the **temperature coefficient of resistance** of the material.

Conductivity and Conductance

The reciprocal of resistivity is called conductivity (σ) i.e. $\sigma = \frac{1}{\rho}$. Its unit is mho/m. The reciprocal of resistance is called conductance (C) i.e. $C = \frac{1}{R}$. Its unit is Ω^{-1} or "Siemen".

Stretching of Wire

If a conducting wire stretches, it's length increases, area of cross-section decreases, so resistance increases but volume remains constant.

$$\therefore\ A_1\ell_1 = A_2\ell_2 \quad \Rightarrow \frac{\ell_1}{\ell_2} = \frac{A_2}{A_1}$$

Ratio of resistances before and after stretching

$$\frac{R_1}{R_2} = \frac{\ell_1}{\ell_2} \times \frac{A_2}{A_1} = \left(\frac{\ell_1}{\ell_2}\right)^2 = \left(\frac{A_2}{A_1}\right)^2 = \left(\frac{r_2}{r_1}\right)^4 = \left(\frac{d_2}{d_1}\right)^4$$

Combination of Resistances

(a) Total resistance R of n resistors connected **in series** is given by $R = R_1 + R_2 + ... + R_n$

(b) Total resistance R of n resistors connected **in parallel** is given by $\frac{1}{R} = \frac{1}{R_1} + \frac{1}{R_2} + ... + \frac{1}{R_n}$.

COLOUR CODES FOR CARBON RESISTORS

Colour	Digit	Multiplier	Tolerance
Black	0	$10^0 = 1$	
Brown	1	$10^1 = 10$	
Red	2	10^2	
Orange	3	10^3	
Yellow	4	10^4	
Green	5	10^5	
Blue	6	10^6	
Violet	7	10^7	
Grey	8	10^8	
White	9	10^9	
Gold	–	0.4	± 5%
Silver	–	0.01	± 10%
No colour	–	–	± 20%

To learn the above table of colour codes of resistors let us learn this interesting sentence : BB ROY of Great Britain has a Very Good Wife.

ELECTRIC CELL

It is the device used to convert chemical energy into electrical energy.

Emf of cell (E) : It is the potential difference across the terminals of a cell when it is not in used.

Internal resistance of cell : It is the resistance offered by the electrolyte.

Terminal Potential Difference

The potential difference between the two electrodes of a cell in a closed circuit i.e. when current is being drawn from the cell is called terminal potential difference.

(a) When cell is discharging

When cell is discharging, current inside the cell is from cathode to anode.

Current $I = \frac{E}{r+R}$

or $E = IR + Ir = V + Ir$

or $\mathbf{V = E - Ir}$

When current is drawn from the cell, potential difference is less than emf of cell. Greater is the current drawn from the cell smaller is the terminal voltage. When a large current is drawn from a cell its terminal voltage is reduced.

(b) When cell is charging

When cell is charging, current inside the cell is from anode to cathode.

Current $I = \frac{V-E}{r}$

or $\mathbf{V = E + Ir}$

During charging, terminal potential difference is greater than emf of cell.

(c) When cell is in open circuit

In open circuit $R = \infty$ $\therefore$ $I = \frac{E}{R+r} = 0$

So $\mathbf{V = E}$

In open circuit terminal potential difference is equal to emf and is the maximum potential difference which a cell can provide.

(d) When cell is short circuited

In short circuit, $R = 0$ so

$I = \frac{E}{R+r} = \frac{E}{r}$ and $\mathbf{V = IR = 0}$

In short circuit, current from cell is maximum and terminal potential difference is zero.

(e) Power transferred to load by cell

$$P = I^2 R = \frac{E^2 R}{(r+R)^2}$$

so $P = P_{max}$

if $\frac{dP}{dR} = 0$

$\Rightarrow P = P_{max}$ if $r = R$

Power transferred by cell to load is maximum when

$r = R$ and $P_{max} = \frac{E^2}{4r} = \frac{E^2}{4R}$

Combination of Cells

(A) Series grouping : In series grouping anode of one cell is connected to cathode of other cell and so on. If identical cells are connected in series

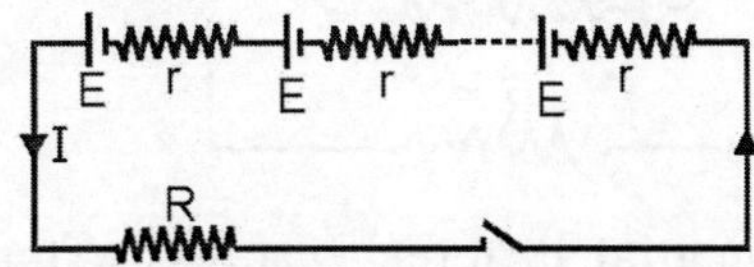

(a) Equivalent emf of the combination $E_{eq} = nE$

(b) Equivalent internal resistance $r_{eq} = nr$

(c) Main current = Current from each cell =

$$i = \frac{nE}{R + nr}$$

(d) If out of n cells in series m cells be connected in an opposite manner, then the current will be given by

$$i = \frac{\left[\sum_{i=1}^{(n-m)} E_i - \sum_{j=1}^{m} E_j\right]}{\sum_{i=1}^{n} r_i + R}$$

(e) Potential difference across external resistance $V = iR$

(f) Potential difference across each cell $V' = V/n$

(g) Power dissipated in the external circuit =

$$\left(\frac{nE}{R+nr}\right)^2 .R$$

(h) condition for maximum power $R = nr$ and

$$P_{max} = n\left(\frac{E^2}{4r}\right)$$

$\left(\frac{dP}{dR} = 0, \text{ Maximum power transfer theorem}\right)$

(B) Parallel grouping : In parallel grouping all anodes are connected at one point and all cathodes are connected together at other point. If n identical cells are connected in parallel as shown. Then,

(a) Equivalent emf $E_{eq} = E$

(b) Equivalent internal resistance $R_{eq} = r/n$

(c) Main current $i = \dfrac{E}{R + r/n}$

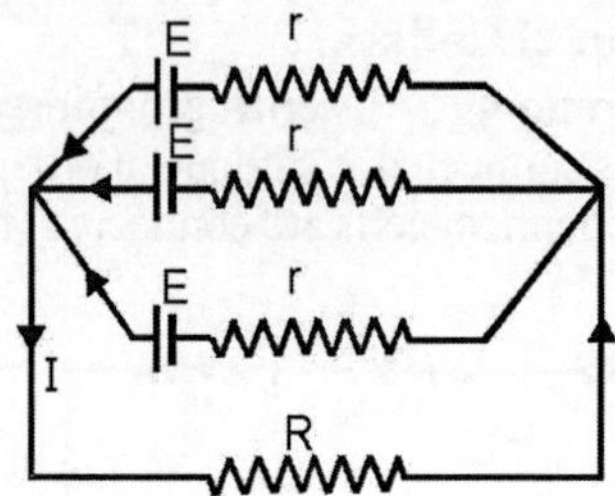

(d) Potential difference across external resistance = p.d. across each cell = V = iR

(e) Current from each cell $i' = \dfrac{i}{n}$

(f) Power dissipated in the circuit

$$P = \left(\frac{E}{R + r/n}\right)^2 .R$$

(g) Condition for maximum power is Conservation of energy R= r/n and

$$P_{max} = n\left(\frac{E^2}{4r}\right)$$

(h) Dissimilar cells in parallel : Two dissimilar cells in parallel are equivalent to a single cell of internal resistance

$$r_{eq} = \frac{r_1 r_2}{r_1 + r_2} \text{ and}$$

$$\text{emf } E_{eq} = \frac{E_1 r_2 + E_2 r_1}{r_1 + r_2} = \frac{r_1 r_2}{r_1 + r_2}\left[\frac{E_1}{r_1} + \frac{E_2}{r_2}\right]$$

(C) Mixed Grouping : If n identical cells are connected in m rows and such m rows are connected in parallel as shown. Then,

(a) Equivalent emf of the combination $E_{eq} = nE$

(b) Equivalent internal resistance of the combination

$r_{eq} = nr/m$

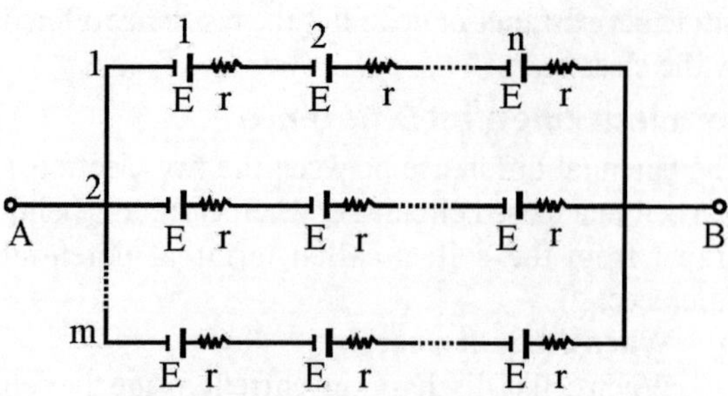

(c) Main current flowing through the load

$$i = \frac{nE}{R + (nr/m)} = \frac{mnE}{mR + nr}$$

(d) Potential difference across load V = iR

(e) Potential difference across each cell V' = V/n

(f) Current from each cell i' = i/m

(g) Condition for maximum power

$$R = \frac{nr}{m} \text{ and } P_{max} = (mn)\frac{E^2}{4r}$$

(h) In mixed grouping' as both current in the circuit and power transferred to the load are maximum under same condition it is preferred over series or parallel grouping of cells.

(i) Total number of cells = mn

(j) If N (= mn) and r/R = (m/n) are given then m = no. of rows and n = no. of cells contained in each row can be calculated for maximum current through R. It may be mentioned here, that the values of m and n should necessarily be positive integers. If, upon solving for m and n, the values come out to be fractions, then get the two set of integral values of m and n, one immediately lesser and the other greater than the obtained fractions. Next, check the value of i for these two set of m and n values, and then decide the values of m and n for i to be maximum.

KIRCHHOFF'S RULES

(a) **Junction rule**

At any junction of circuit elements, the sum of currents entering the junction must be equal to the sum of currents leaving it.

(b) **Loop rule**

The algebraic sum of changes in potential be around any closed loop must be zero.

Kirchhoff's junction rule is based on conservation of charge and the loop rule represents. Bending or reorienting the wire does not change the validity of Kirchhoff's junction rule.

SOME MEASURING INSTRUMENTS

Galvanometer : It is an instrument used to detect small current passing through it by showing deflection.

Moving coil type galvanometer

(a) Principle : When a current carrying coil is placed in magnetic field, it experiences a couple and the coil rotates.

(b) The torque of this couple is given by : $\tau = NiAB \sin\alpha$
where N = no. of turns in the coil, i = current through coil,
B = intensity of magnetic field
α = angle between magnetic field and plane of the coil.

(c) For radial magnetic field $\alpha = 90°$ $\therefore \tau = NiAB$

(d) In the state of equilibrium of coil
deflecting torque = restoring torque $\Rightarrow NiAB = C\theta$
where C is restoring torque per unit deflection, (C depends on elastic properties of spring)

or $i = \frac{C}{NAB}\theta$ or $i \propto \theta$

The current is proportional to deflection.

The constant $\frac{C}{NAB}$ is called galvanometer constant.

Ammeter : It is an instrument used to measure currents. It is put in series with the branch in which current is to be measured. An ideal ammeter has zero resistance.

A galvanometer with resistance G and current rating i_g can be converted into an ammeter of rating I by connecting a suitable resistance S in parallel to it. The resistance connected in parallel to the ammeter is called a **shunt. Shunt required for conversion of Galvanometer into ammeter.**

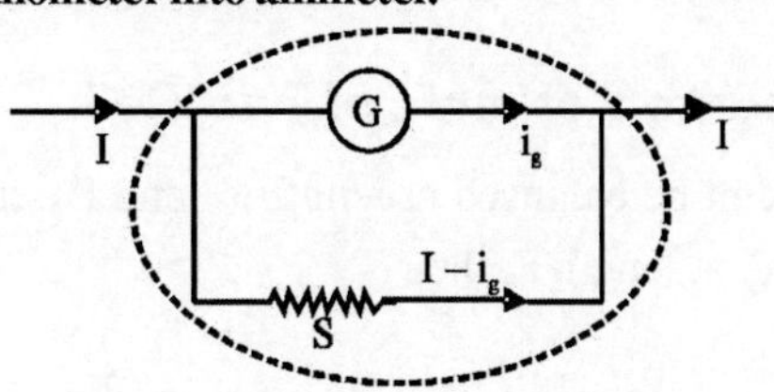

Thus $S(I - i_g) = i_g G \Rightarrow S = \frac{i_g G}{I - i_g}$

Voltmeter : It is an instrument to find the potential difference across two points in a circuit.

It is essential that the resistance R_v of a voltmeter be very large compared to the resistance of any circuit element with which the voltmeter is connected. Otherwise, the voltmeter itself becomes an important circuit element and alters the potential difference that is measured.

$$R_v >> R$$

For an ideal voltmeter $R_v = \infty$.

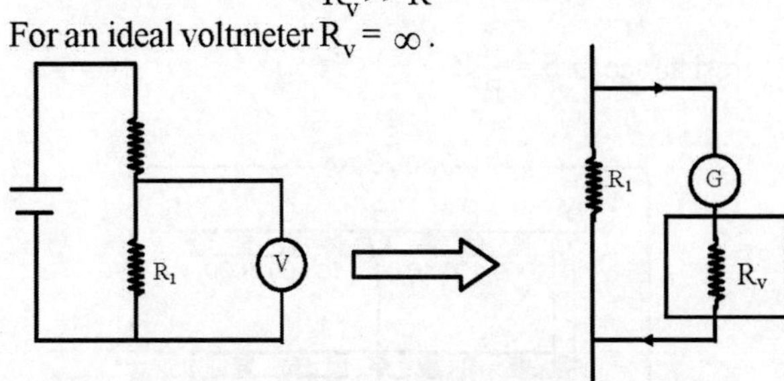

A galvanometer with current rating i_g can be converted into a voltmeter by connecting a large resistance R_V in series. **Large resistance in series required for conversion of galvanometer into voltmeter.**

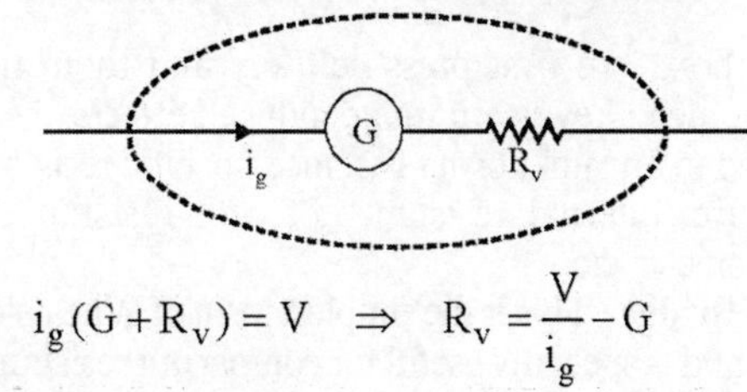

$$i_g(G + R_v) = V \Rightarrow R_v = \frac{V}{i_g} - G$$

WHEATSTONE BRIDGE AND ITS APPLICATIONS

Wheatstone Bridge

The Wheatstone bridge is an arrangement of four resistances R_1, R_2, R_3 and R_4.

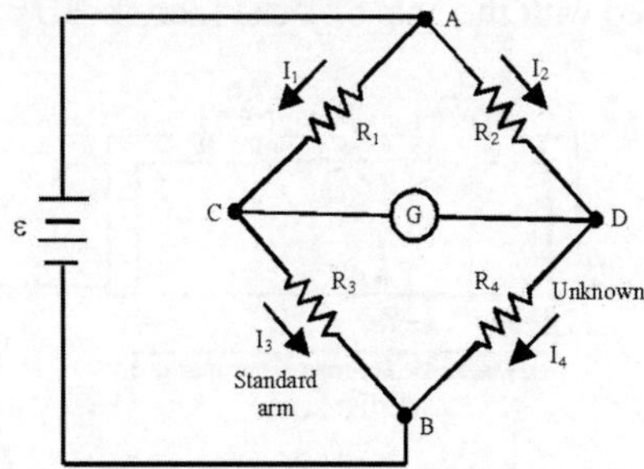

When the bridge is in balanced condition.

The null-point condition is given by $\frac{R_1}{R_2} = \frac{R_3}{R_4}$ using which the value of one resistance can be determined, knowing the other three resistances. Post office box and metre bridge are work on the principle of balanced Wheatstone bridge.

Post Office Box : It is so named because it has a shape of box and was designed to find resistance of electric cables and telegraph wires. It was used in post offices to determine resistance of transmission lines. It is based on the principle of Wheatstone bridge. Unknown resistance is $S = \frac{Q}{P}R$.

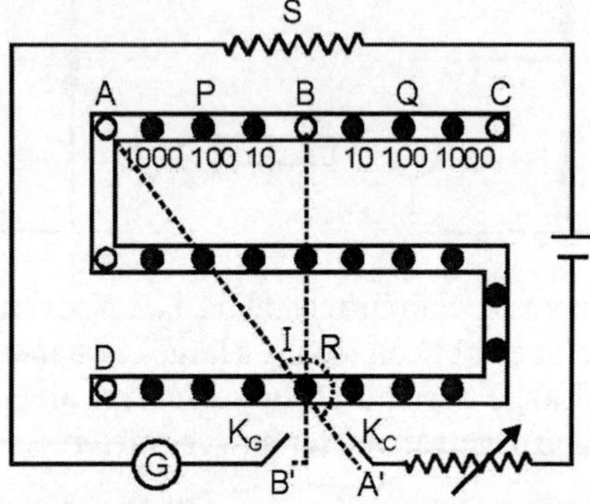

In PO box, we first press cell key and then press galvanometer key to eliminate induced effects.

It is used to find unknown resistance, specific resistance of a wire, internal resistance of cell, resistance of galvanometer, etc.

Metre Bridge : This is the simplest form of Wheatstone bridge and is specially useful for comparing resistances more accurately. The construction of the metre bridge is shown in the Figure. It consists of one metre resistance wire clamped between two metallic strips bent at right angles and it has two points for connection. There are two gaps; in one of them a known resistance and in second an unknown resistance whose value is to be determined is connected. The galvanometer is connected with the help of jockey across BD

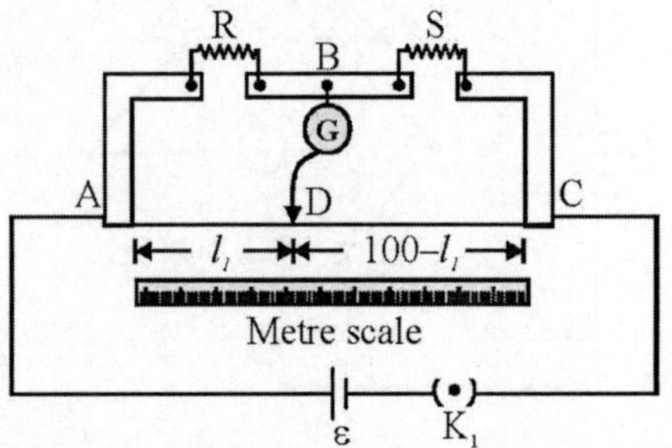

and the cell is connected across AC. After making connections, the jockey is moved along the wire and the null point is obtained, wire used is of uniform material and cross-section.

The resistance can be found with the help of the following relation:

$$\frac{R}{S} = \frac{\sigma \ell_1}{\sigma(100-\ell_1)} \text{ or } \frac{R}{S} = \frac{\ell_1}{100-\ell_1} \text{ or } R = S\frac{\ell_1}{100-\ell_1}$$

where σ is the resistance per unit length of the wire and ℓ_1 is the length of the wire from one end where null point is obtained. The bridge is most sensitive when null point is somewhere near the middle point of the wire. This is due to end resistances.

POTENTIOMETER

The potentiometer is a device to compare potential differences. Since the method involves a condition of no current flow, the device can be used to measure potential difference; internal resistance of a cell and compare emf's of two sources.

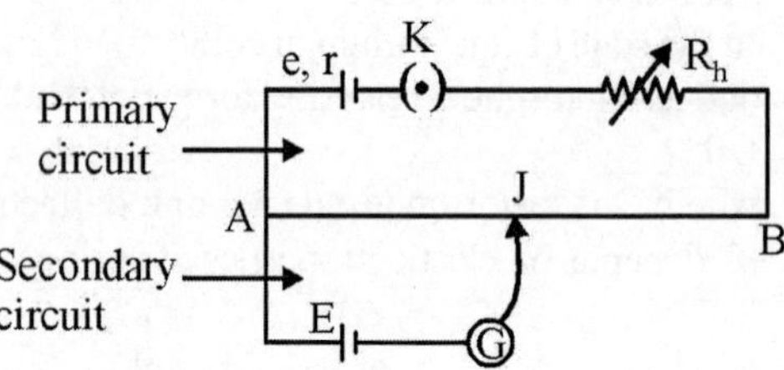

J = Jockey

K = Key

R = Resistance of potentiometer wire,

ρ = Specific resistance of potentiometer wire.

R_h = Variable resistance which controls the current through the wire AB

Potential gradient : The fall of potential per unit length of potentiometer wire is called potential gradient.

$$x = \frac{IR}{L} = \frac{e}{r + R_h + R}\left(\frac{R}{L}\right)$$

Comparison of emfs of Two Cells

Let E_1 emf be balanced at length ℓ_1 and E_2 emf be balanced at length ℓ_2 then

$$E_1 = x\ell_1 \text{ and } E_2 = x\ell_2 \quad \text{so } \frac{E_1}{E_2} = \frac{\ell_1}{\ell_2}$$

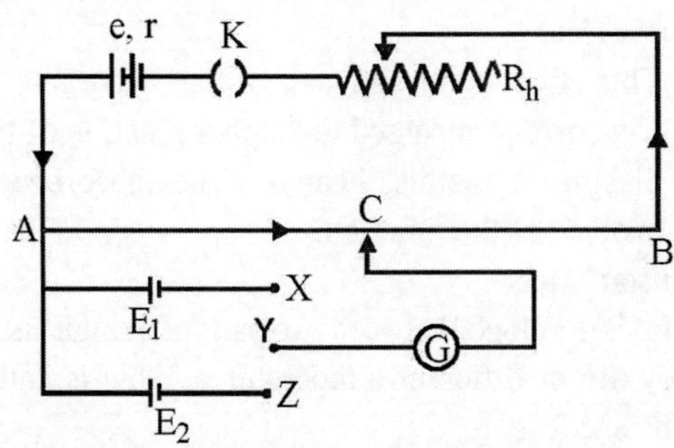

If two cells joined in series support each other (E_1 E_2) then the balancing length is ℓ_1 so $E_1 + E_2 = x\ell_1$

If two cells joined in series oppose each other (E_1 E_2) then the balancing length is ℓ_2 so $E_1 - E_2 = x\ell_2$.

$$\frac{E_1+E_2}{E_1-E_2}=\frac{\ell_1}{\ell_2} \quad \text{or} \quad \frac{E_1}{E_2}=\frac{\ell_1+\ell_2}{\ell_1-\ell_2}$$

Determination of Internal Resistance of a Cell

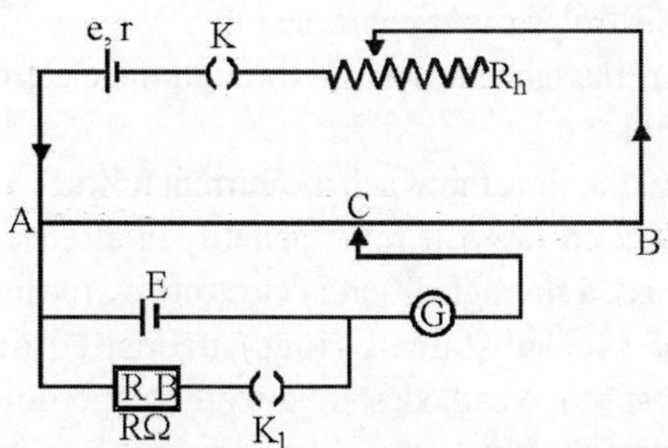

Keeping K_1 open the balancing length ℓ_1 gives emf of cell so $E = x\ell_1$.

Keeping K_1 closed the balancing length ℓ_2 for some resistance R gives potential difference so $V = x\ell_2$.

Internal resistance $r = \dfrac{E-V}{V}$

$$R = \frac{x\ell_1 - x\ell_2}{x\ell_2} \quad \text{or} \quad R = \left(\frac{\ell_1-\ell_2}{\ell_2}\right)R$$

HEATING EFFECTS OF ELECTRIC CURRENT

Joule's Laws of Heating Effects of Current

(i) The heat developed in a conductor is given by $H = i^2Rt$ joule $= 0.24\, i^2Rt$ cal.

(ii) The amount of heat developed in a conductor, in a given time, is directly proportional to the square of the current, i.e. $H \propto i^2$, when R and t are constants.

(iii) The amount of heat developed in a conductor by a given current in a given time is directly proportional to the resistance of the conductor, i.e. $H \propto R$, when i and t are constants.

(iv) The amount of heat developed in a given conductor due to a given current is proportional to the time of flow of the current.

i.e. $H \propto t$, when i and R are constants.

Rated power (of an electric device) : The voltage of operation needed is known as the rated voltage and the corresponding power consumed is called as the rated power.

Suppose, R is the electrical resistance of an electric device, with V_r and P_r is rated voltage and wattage respectively.

Then, $$P_r = \frac{V_r^2}{R} \quad \text{........ (1)}$$

Now, if the device is subjected to a voltage of V, then, the power consumption P will be

$$P = \frac{V^2}{R} \quad \text{.......... (2)}$$

The resistance R being the property of the electric device, is independent of the applied voltage.

Dividing eq. (2) by eq. (1), $P = \left(\dfrac{V}{V_r}\right)^2 P_r$

Thus, if the voltage be n times the rated one, then the actual power consumption will be n^2 times the rated one.

Electric fuse : Commercially, it is a device employed to save the different electrical appliances used in a house such as fan, bulb, T.V., tape recorder etc. in circumstances of abrupt increase of currents entering through the main supply.

An electric fuse consists of a simple wire made of an alloy of tin and lead, having low melting point and high resistivity. Properties of fuse wire is independent of its length.

Thermocouple

If two wires of different metals are joined at their ends so as to form two junctions, then the resulting arrangement is called a thermocouple.

It is used to measure very high temperature upto 2000°C as well as very low temperature – 200°C.

Seebeck Effect : When the two junctions of a thermocouple are maintained at different temperatures, then a current starts flowing through the wires. This is called seebeck effect and the e.m.f. developed in the circuit is called thermo emf.

If a graph be plotted for the thermoelectric emf e against the temperature T of the hot junction, then the curve obtained is a parabola as shown in figure.

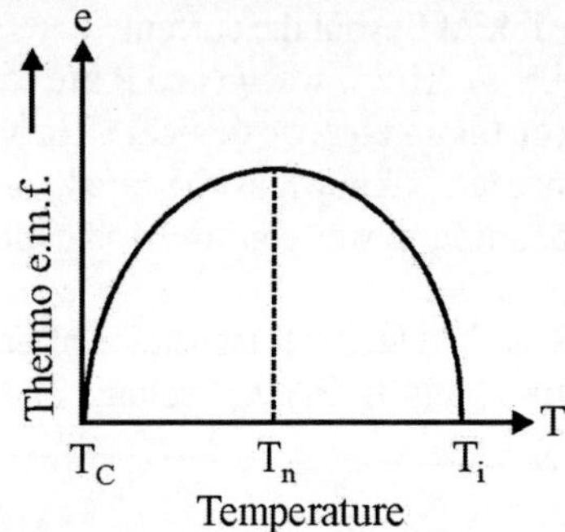

$T_n - T_c = T_i - T_n$

($T_n \rightarrow$ Neutral temperature at which e is maximum)

($T_i \rightarrow$ Inversion temperature or temperature of hot junction)

$e = aT + \frac{1}{2}bT^2$; $\frac{de}{dT} = a + bT$

The quantity $\frac{de}{dT}$ is called the **thermoelectric power** or the **Seebeck's coefficient** at $T = T_n, \frac{de}{dT} = 0$

$\Rightarrow a + bT_n = 0$

$T_n = \frac{-a}{b}, \; T_i = \frac{-2a}{b}$

Peltier Effect

(a) This effect is the converse of Seebeck effect.

(b) If a current is passed through a junction of two dissimilar metals, heat is either absorbed or evolved at the junction.

Thomson Effect

An emf is developed between two parts of a single metal if they are at different temperatures. This is called Thomson effect.

FARADAY'S LAWS OF ELECTROLYSIS

(i) **First law :** The total mass of ions liberated at an electrode, during electrolysis, is proportional to the quantity of electricity which passes through the electrolyte.

i.e. $m \propto Q$ But $Q = it$ $\therefore$ $m \propto it$ or $m = Zit$.

where Z is the constant of proportionality known as the **electro-chemical equivalent** of the substance. It is numerically equal to the mass in grams of the element deposited when a unit current flows in unit time.

Hence the first law may also be stated as follows. The mass of ions liberated at an electrode during electrolysis is proportional to

(a) the current flowing through the electrolyte, and

(b) the time for which the current flows.

(ii) **Second law :** If same quantity of electricity is passed through different electrolytes, the masses of the substances (ions) deposited at the respective cathodes are directly proportional to their chemical equivalents (equivalent weights).

i.e. $m \propto E$ (chemical equivalent) or

$\frac{m_1}{E_1} = \frac{m_2}{E_2} = \text{Constant}$

Exercise

1. In an electric circuit containing a battery, the charge (assumed positive) inside the battery
   (a) always goes from the positive terminal to the negative terminal
   (b) may go from the positive terminal to the negative terminal
   (c) always goes from the negative terminal to the positive terminal
   (d) does not move
2. A resistor of resistance R is connected to an ideal battery. If the value of R is decreased, the power dissipated in the resistor will
   (a) increase (b) decrease
   (c) remain unchanged (d) none of these
3. Two non-ideal batteries are connected in series. Consider the following statements:
   (A) The equivalent emf is larger than either of the two emfs.
   (B) The equivalent internal resistance is smaller than either of the two internal resistances.
   (a) Each of A and B is correct
   (b) A is correct but B is wrong
   (c) B is correct but A is wrong
   (d) Each of A and B is wrong
4. Two non-ideal batteries are connected in parallel. Consider the following statements
   (A) The equivalent emf is smaller than either of the two emfs.
   (B) The equivalent internal resistance is smaller than either of the two internal resistances.
   (a) Both A and B are correct
   (b) A is correct but B is wrong
   (c) B is correct but A is wrong
   (d) Both A and B are wrong
5. In a closed circuit the e.m.f. and internal resistance of the generator are E and r respectively. If the external resistance in the circuit is R, then the Ohm's law has the form
   (a) $I=\frac{E}{(R+r)}$ (b) $I=\frac{E}{R}$
   (c) $I=\frac{E}{r}$ (d) $I=\frac{E}{Rr}$
6. To draw a maximum current from a combination of cells, how should the cells be grouped?
   (a) Parallel
   (b) Series
   (c) Mixed grouping
   (d) Depends upon the relative values of internal and external resistances
7. Under what condition will the strength of current in a wire of resistance R be the same for connection is series and in parallel of n identical cells each of the internal resistance r, when
   (a) $R=nr$ (b) $R=r/n$
   (c) $R=r$ (d) $R\to\infty, r\to 0$
8. A cell of internal resistance r is connected to an external resistance R. The current will be maximum in R, if
   (a) $R=r$ (b) $R<r$
   (c) $R>r$ (d) $R=r/2$
9. Consider the following two statements :
   (A) Kirchoff's junction law follows from conservation of charge.
   (B) Kirchoff's loop law follows from conservative nature of electric field.
   (a) Both A and B are correct
   (b) A is correct but B is wrong
   (c) B is correct but A is wrong
   (d) Both A and B are wrong
10. In the network shown, each resistance is equal to R. The equivalent resistance between adjacent corners A and D is

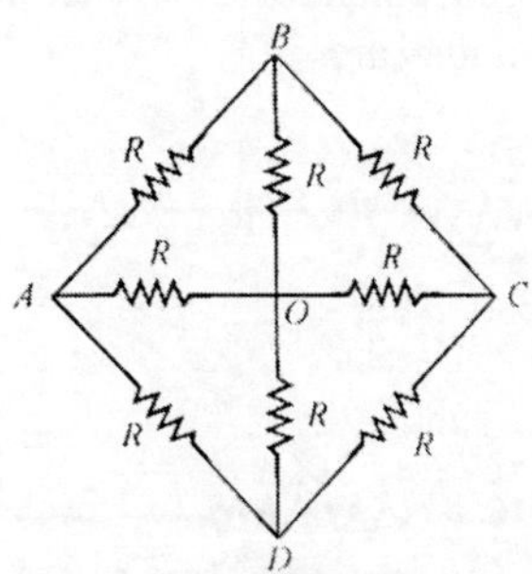

(a) R (b) $\frac{2}{3}R$

(c) $\frac{3}{7}R$ (d) $\frac{8}{15}R$

11. In a Wheatstone bridge all the four arms have equal resistance R. If the resistance of galvanometer arm is also R, the equivalent resistance of combination as seen by battery is
(a) $2R$ (b) $R/4$
(c) $R/$ (d) R

12. Wheatstone bridge is used to measure
(a) potential (b) e.m.f.
(c) current (d) resistance

13. Why is the Wheatstone bridge better than the other methods of measuring resistances?
(a) It does not involve Ohm's law
(b) It is based on Kirchoff's law
(c) It has four resistor arms
(d) It is a null method

14. If in the experiment of Wheatstone's bridge, the positions of cells and galvanometer are interchanged, then balance point will
(a) change
(b) remain unchanged
(c) depend on the internal resistance of cell and resistance of cell and resistance of galvanometer
(d) none of these

15. The masses of the three wires of copper are in the ratio of 1 : 3 : 5 and their lengths are in the ratio of 5 : 3 : 1. The ratio of their electrical resistance is
(a) 1:3:5 (b) 5:3:1
(c) 1:25:125 (d) 125:15:1

16. Two batteries of emf 4 V and 8 V with internal resistance 1 Ω and 2 Ω are connected in a circuit with a resistance of 9 Ω as shown in figure. The current and potential difference between the points P and Q are

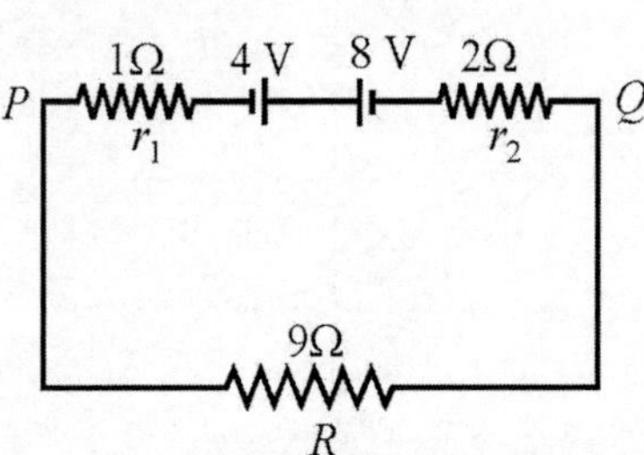

(a) $\frac{1}{3}A$ and 3 V (b) $\frac{1}{6}A$ and 4 V

(c) $\frac{1}{9}A$ and 9 V (d) $\frac{1}{12}A$ and 12V

17. n equal resistors are first connected in series and then connected in parallel. What is the ratio of the maximum to the minimum resistance?
(a) n (b) $1/n^2$
(c) n^2 (d) $1/n$

18. Forty electric bulbs are connected in series across a 220 V supply. After one bulb is fused the remaining 39 are connected again in series across the same supply. The illumination will be
(a) more with 40 bulbs than with 39
(b) more with 39 bulbs than with 40
(c) equal in both the cases
(d) in the ratio $40^2 : 39^2$

19. A current of 2 A, passing through a conductor produces 80 J of heat in 10 seconds. The resistance of the conductor in ohm is
(a) 0.5 (b) 2
(c) 4 (d) 20

20. Two identical batteries each of e.m.f . 2V and internal resistance 1Ω are available to produce heat in an external resistance by passing a current through it. The maximum power that can be developed across R using these batteries is
(a) 3.2 W (b) 2.0 W

(c) 1.28 W (d) $\frac{8}{9}W$

21. You are given several identical resistances each of value R = 10Ω and each capable of carrying a maximum current of one ampere. It is required to make a suitable combination of these resistances of 5Ω which can carry a current of 4 ampere. The minimum number of resistances of the type R that will be required for this job is
(a) 4 (b) 10
(c) 8 (d) 20

22. In the network shown in the Fig, each resistance is 1Ω. The effective resistance between A and B is

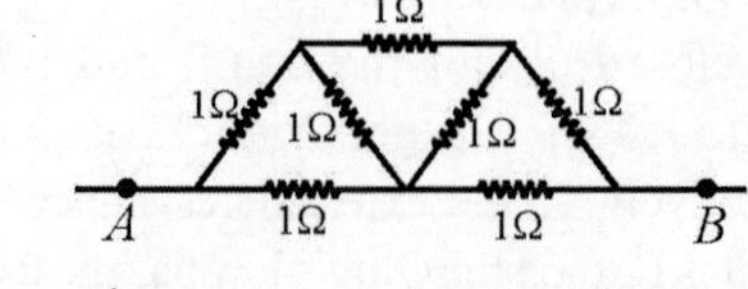

(a) $\frac{4}{3}\Omega$ (b) $\frac{3}{2}\Omega$

(c) 7 Ω (d) $\frac{8}{7}\Omega$

23. Faraday's laws are consequence of conservation of
    (a) energy
    (b) energy and magnetic field
    (c) charge
    (d) magnetic field
24. The velocity of charge carriers of current (about 1 amp) in a metal under normal conditions is of the order of
    (a) a fraction of mm/sec
    (b) velocity of light
    (c) several thousand metres/second
    (d) a few hundred metres per second
25. Direct current is passed through a copper sulphate solution using platinum electrodes. The elements liberated at the electrodes are
    (a) copper at anode and sulphur at cathode
    (b) sulphur at anode and copper at cathode
    (c) oxygen at anode and copper at cathode
    (d) copper at anode and oxygen at cathode
26. Three resistances each of 4 Ω are connected to form a triangle. The resistance between any two terminals is
    (a) 12 Ω (b) 2 Ω
    (c) 6 Ω (d) 8/3Ω
27. Which of the four resistances generate the greatest amount of heat when a current flows from A to B?

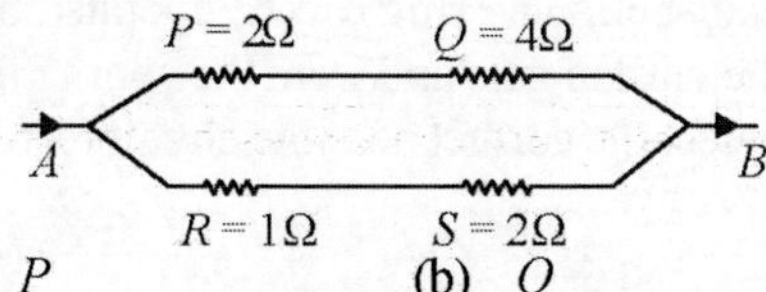

    (a) P (b) Q
    (c) R (d) S
28. The emf developed by a thermocouple is measured with the help of a potentiometer and not by a moving coil millivoltmeter because
    (a) the potentiometer is more accurate than the voltmeter
    (b) the potentiometer is more sensitive than voltmeter
    (c) the potentiometer makes measurement without drawing any current from the thermocouple
    (d) measurement using a potentiometer is simpler than with a voltmeter
29. You have been provided with four 100 ohm resistors each with a tolerance of 2%. The number of ways in which these can be combined to have different equivalent resistances is
    (a) seven different combinations and seven different equivalents
    (b) eight different combinations and seven different equivalents resistances
    (c) nine different combinations and eight different resistances
    (d) ten different combinations and nine different resistances
30. The equivalent resistance between points A and B is

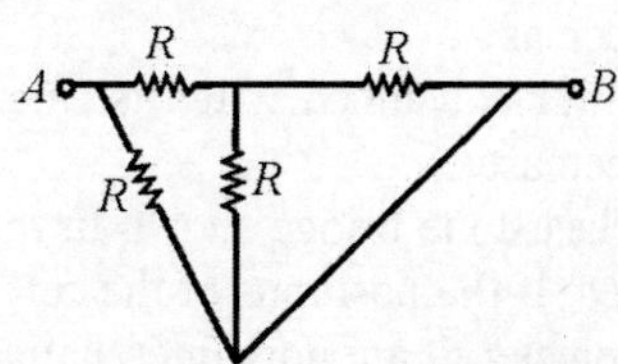

    (a) $2R$ (b) $(3/4)R$
    (c) $(4/3)R$ (d) $(3/5)R$
31. Current passing through 3Ω resistance is

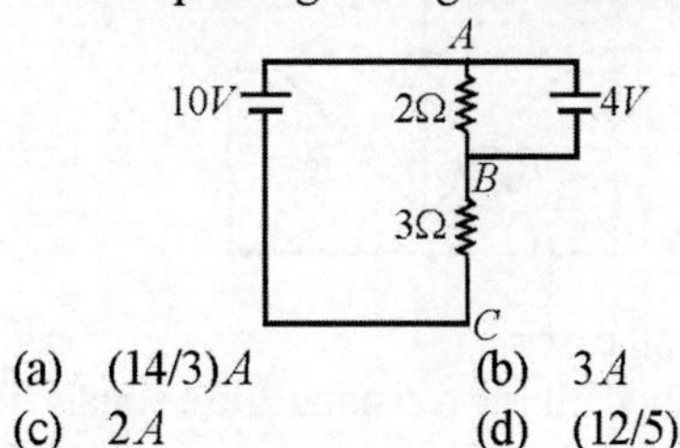

    (a) $(14/3)A$ (b) $3A$
    (c) $2A$ (d) $(12/5)A$
32. Circuit shown in figure is

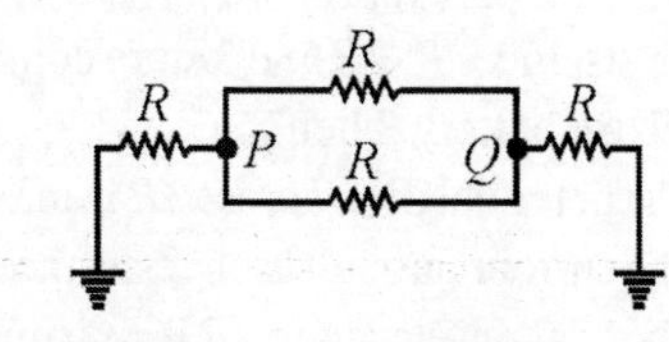

    (a) $R/2$ (b) $2R/5$
    (c) $3R/5$ (d) $R/3$
33. A 100 W bulb B_1 and two 60W bulbs B_2 and B_3, are connected to a 250V source, as shown in the figure. Now W_1, W_2 and W_3 are the output powers of the bulbs B_1, B_2 and B_3 respectively. Then –

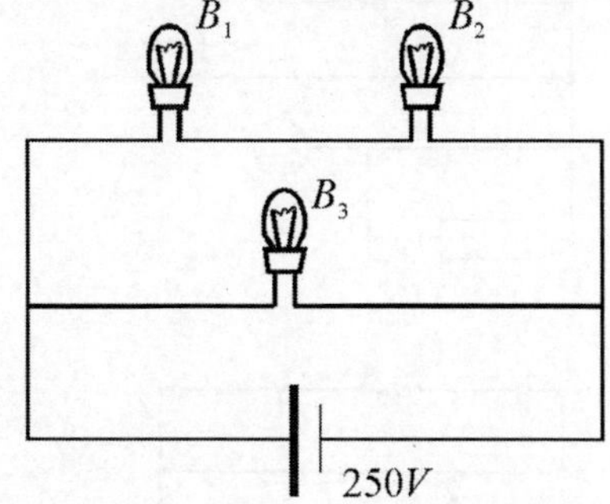

    (a) $W_1 > W_2 = W_3$ (b) $W_1 > W_2 > W_3$
    (c) $W_1 < W_2 = W_3$ (d) $W_1 < W_2 < W_3$

34. A battery of internal resistance 2Ω is connected to a variable resistor whose value can vary from 4Ω to 10Ω. The resistance is initially set at 4Ω. If the resistance is now increased then
(a) power consumed by it will decrease
(b) power consumed by it will increase
(c) power consumed by it may increase or may decrease
(d) power consumed will first increase then decrease

35. The Wheatstone bridge shown in the figure is balanced. If the positions of the cell C and the galvanometer G are now interchanged, G will show zero deflection –

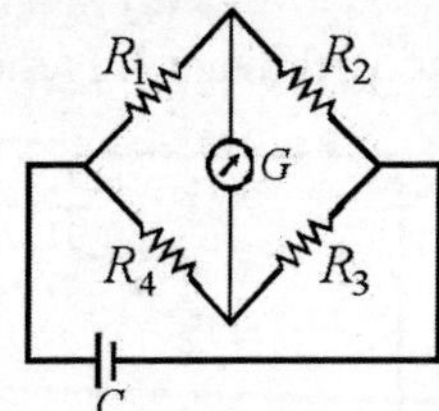

(a) in all cases
(b) only if all the resistances are equal
(c) only if $R_1 = R_3$ and $R_2 = R_4$
(d) only if $R_1/R_3/R_2/R_4$

36. Three resistances R, $2R$ and $3R$ are connected in parallel to a battery. Then
(a) The potential drop across $3R$ is maximum
(b) The current through each resistance is same
(c) The heat developed in $3R$ is maximum
(d) The heat developed in R is maximum

37. The correct circuit for the determination of internal resistance of a primary cell by using potentiometer is:

(a)
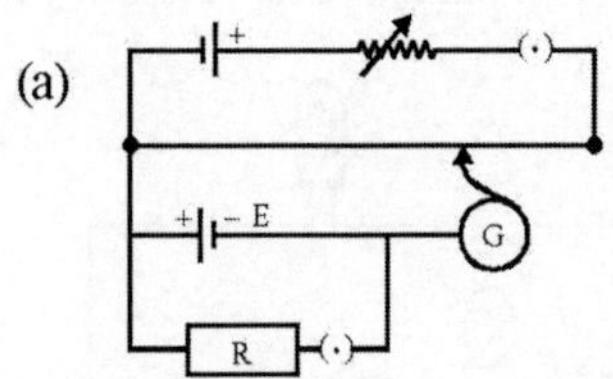

(b)
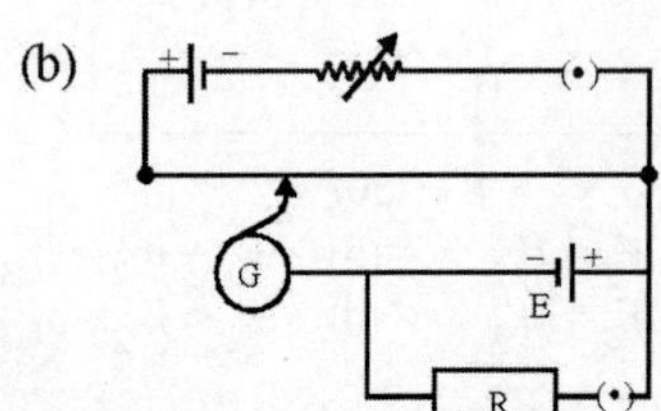

(c)
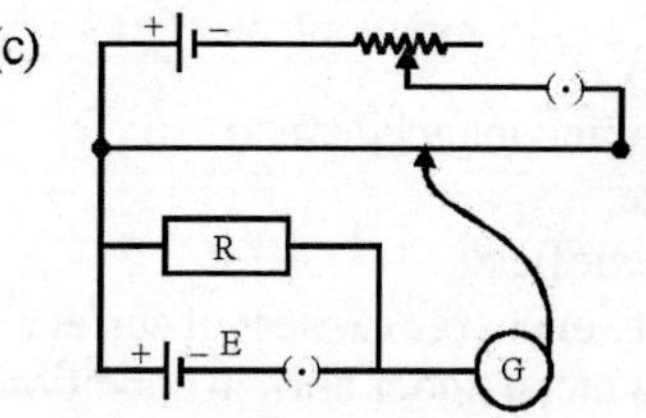

(d)
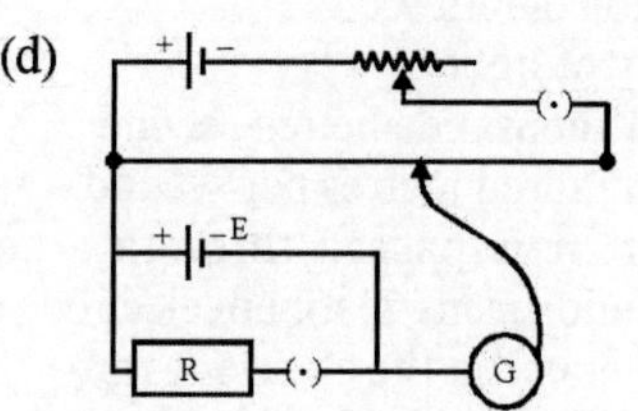

38. A piece of copper and another of germanium are cooled from room temperature to 80 K. The resistance of
(a) Each of them increases
(b) Each of them decreases
(c) Copper increases and germanium decreases
(d) Copper decreases and germanium increases

39. In the house of a person who is weak of hearing, a light bulb is also lit when somebody rings the door bell. The ring can be operated both from the garden gate and from the door of the house. Select the correct possible circuit required.

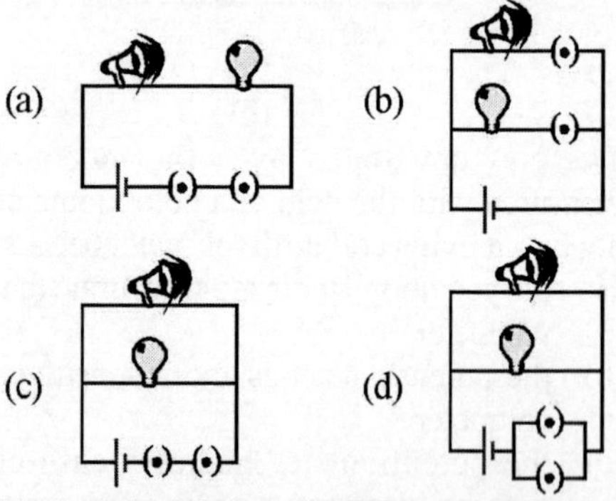

40. If two bulbs of power 25 W and 100 W respectively each rated at 220 V are connected in series with the supply of 440 V. Which bulb will fuse?
(a) 25 W bulb
(b) 40 watts
(c) none of these
(d) both (a) & (b)

41. Which of the following graphs represent the variation of thermo emf (E) of a thermocouple with temperature θ of hot junction (the cold junction being kept at 0°C)

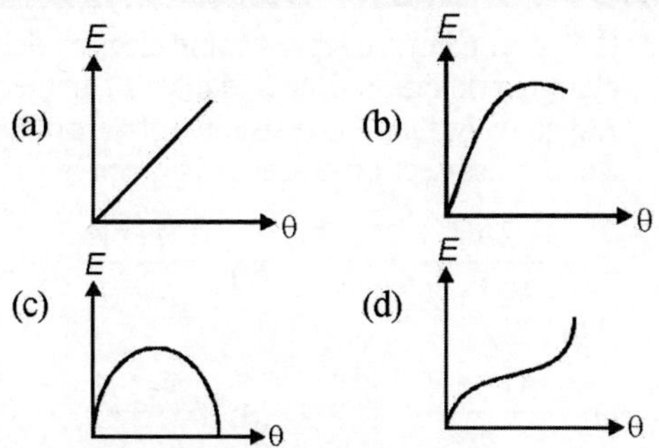

42. Consider the following devices :
I Copper voltameter
II Water voltameter
III Semiconductor diode
Which of these obey Ohm's law ?
(a) I only (b) I and II
(c) III only (d) I, II and III

43. A copper wire of length l and radius r is to be used as fuse. If the fuse wire attains, then the heat lost H per unit surface area per second is :
(a) $H \propto \frac{a}{r^3}$ (b) $H \propto r^2$
(c) $H \propto l$ (d) $H \propto r^3$

44. Five resistance, each of 5 Ω, are connected as shown in figure. Find the equivalent resistance between points (1) A and B, (2) A and C.

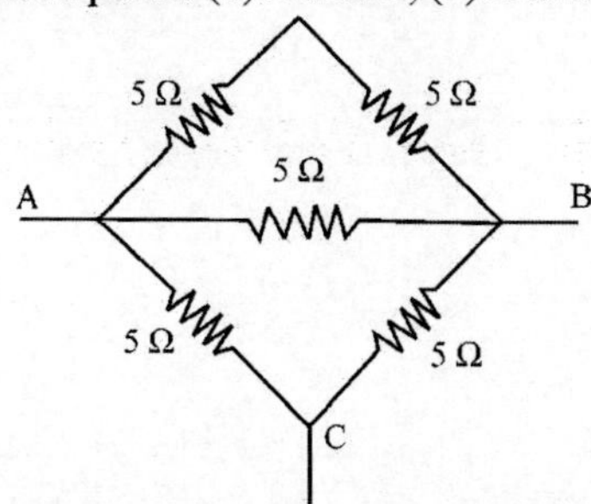

(a) (1) 7.5Ω (2) 2.25Ω (b) (1) 5Ω (2) 2.50Ω
(c) (1) 2.5Ω (2) 3.13Ω (d) (1) 3Ω (2) 2.50Ω

45. In which of the following the carriers of electric current are electrons only?
(a) a super conductor
(b) a voltaic cell
(c) a semiconductor
(d) a hydrogen discharge tube

46. It takes t minutes for a kettle of water to start boiling when placed on an electric heater. If the voltage drops to half its value, the time taken for this water to boil will be:
(a) $\sqrt{2}t$ minutes (b) $2t$ minutes
(c) $4t$ minutes (d) $8t$ minutes

47. Two electric bulbs of the same power, but with different marked voltage are connected in series across a power line. Their brightness will be :
(a) directly proportional to their marked voltages
(b) inversely proportional to their marked voltages
(c) directly proportional to their squares of their marked voltages
(d) inversely proportional to the square of their marked voltages

48. The resistance of a resistor with the following colour code

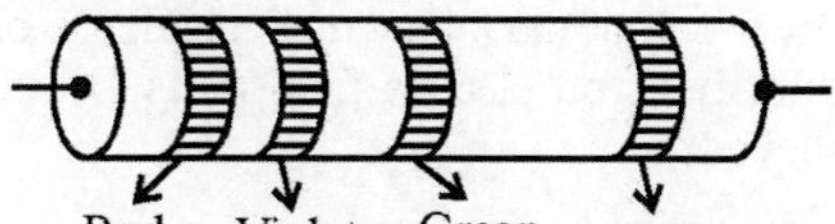

is equal to :
(a) $26 \times 10^4\,\Omega \pm 5\%$ (b) $25 \times 10^4\,\Omega \pm 10\%$
(c) $35 \times 10^5\,\Omega \pm 5\%$ (d) $27 \times 10^5\,\Omega \pm 5\%$

49. Two cells of the same emf E have different internal resistances r_1 and r_2. They are connected in series with an external resistance R and the potential difference across the first cell is found to be zero. Therefore, the external resistance R must be
(a) $r_1 - r_2$ (b) $r + r_2$
(c) $2r_1 - r_2$ (d) $r_1 - 2r_2$

50. The current which is assumed to be flowing in a circuit from positive terminal to negative, is called
(a) direct current
(b) pulsating current
(c) conventional current
(d) alternating current

51. Relation between drift velocity (v_d) of electron and thermal velocity (v_t) of electron is expressed as
(a) $v_d = v_t$ (b) $v_d > v_t$
(c) $v_d < v_t$ (d) $v_d = v_t = 0$

52. The speed at which the current travels, in conductor, is nearly
(a) 3×10^4 m/s (b) 3×10^5 m/s
(c) 4×10^6 m/s (d) 3×10^8 m/s

53. In the absence of an electric field, the mean velocity of free electrons in a conductor at absolute temperature (T) is
(a) zero (b) independent of T
(c) proportional to T (d) proportional to T^2

54. When a potential difference V is applied across a conductor at temperature T, the drift velocity of electrons is proportional to

(a) $\sqrt{V}$ (b) V

(c) $\sqrt{T}$ (d) T

55. An electronic charge e is revolving in a circular orbit of radius r around a nucleus with speed v. The equivalent current is

(a) $ev/2\pi r$ (b) $2ev/\pi r$

(c) $ev/4\pi r$ (d) none

56. The electric intensity E, current density J and specific resistance k are related to each other through the relation

(a) $E = J/k$ (b) $E = Jk$

(c) $E = k/J$ (d) $k = JE$

57. For which of the following dependence of drift velocity v_d on electric field E, is Ohm's law obeyed?

(a) $v_d \propto E^2$ (b) $v_d = E^{1/2}$

(c) v_d = constant (d) $v_d = E$

58. If N, e, τ and m are representing electron density, charge, relaxation time and mass of an electron respectively, then the resistance of wire of length ℓ and cross-sectional area A is given by

(a) $\dfrac{2m\ell}{Ne^2\tau A}$ (b) $\dfrac{2m\tau A}{Ne^2\ell}$

(c) $\dfrac{Ne^2\tau A}{2m\ell}$ (d) $\dfrac{Ne^2 A}{2m\tau\ell}$

59. The current density (number of free electrons per m^3) in metallic conductor is of the order of

(a) 10^{22} (b) 10^{24}

(c) 10^{26} (d) 10^{28}

60. A straigh t conductor of uniform cross-section carries a current I. If s is the specific charge of an electron, the momentum of all the free electrons per unit length of the conductor, due to their drift velocity only is

(a) Is (b) $\sqrt{I/s}$

(c) I/s (d) $(I/s)^2$

Moving Charges and Magnetism

CONCEPTS

BIOT-SAVART'S LAW

The Biot-Savart law asserts that the magnetic field $d\vec{B}$ due to an element $d\vec{\ell}$ carrying a steady current I at any point P at a distance r from the current element is

$$d\vec{B} = \frac{\mu_0}{4\pi} I \frac{d\vec{\ell} \times \vec{r}}{r^3}$$

To obtain the total field at P, we must integrate this vector expression over the entire length of the conductor. This law is analogus to coulomb's law in electrostatics.

Field due to a straight conductor

$\vec{B} = \frac{\mu_0 I}{4\pi a}(\sin\theta_1 + \sin\theta_2)$, where θ_1 and θ_2 are the angles between the line joining the point to the ends of conductor and perpendicular through the point to the conductor.

The magnitude of the magnetic field at a distance R from a long, straight wire carrying a current I is given by : $B = \frac{\mu_0 I}{2R}$

The field lines are circles concentric with the wire.

The magnitude of the magnetic field due to a circular coil of radius R carrying a current I at an axial distance x from the centre is

$$B = \frac{\mu_0 I R^2}{2(x^2 + R^2)^{3/2}}.$$

At the center this reduces to $B = \frac{\mu_0 I}{2R}$

MAGNETIC FIELD

Magnetic field is the space around a magnet/current carrying conductor where magnetic influence can be experienced.

Lorentz force.

$$\vec{F} = q(\vec{v} \times \vec{B})$$

Its direction can be determined by Fleming's left hand rule. Magnetic field is also called magnetic induction or magnetic flux density.

The **S.I. unit** of magnetic field $\vec{B}$ is tesla (T) or Wbm^{-2}

Magnetic field due to a circular arc at the centre (Subtending an angle θ at the centre) :

Consider a current element that subtends an angle θ as shown in figure.

Magnetic field due to this element is

$$B = \frac{\mu_0 I\theta}{4\pi R}$$

I θ R

Special Case :

Magnetic field at the centre of a circular loop :

Here the loop makes an angle $\theta = 2\pi$ at the centre

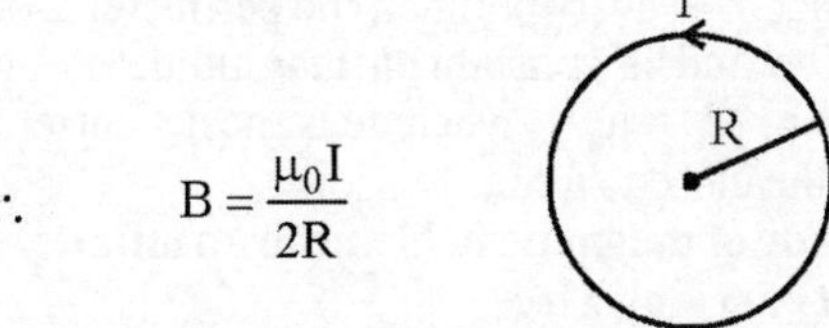

$$\therefore \quad B = \frac{\mu_0 I}{2R}$$

DIRECTION OF MAGNETIC FIELD

Right Hand Palm Rule

If we hold the thumb of right hand mutually perpendicular to the grip of the fingers such that the curvature of the finger represents the direction of current in the wire loop, then the thumb of the right hand will point in the direction of magnetic field near the centre of the current loop.

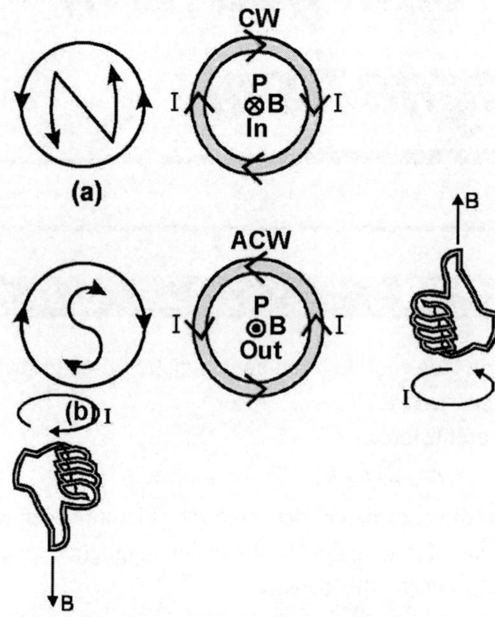

Here cross ⊗ represents the direction of magnetic field perpendicular and into the plane of the paper.
And dot ⊙ represents the direction of magnetic field perpendicular and out of the plane of the paper.
In: Magnetic field is away from the observer or perpendicular inwards.
Out : Magnetic field is towards the observers or perpendicular outwards.

AMPERE'S CIRCUITAL LAW

Let an open surface S be bounded by a loop C. Then the Ampere's law states that $\oint_C \vec{B}.d\vec{\ell} = \mu_0 I$, where I refers to the current passing through S. The sign of I is determined from the right-hand rule. If $\vec{B}$ is directed along the tangent to every point on the perimeter L of a closed curve and is constant in magnitude along perimeter then, $BL = \mu_0 I_e$, where Ie is the net current enclosed by the closed circuit.

The magnitude of magnetic field due to an infinitely long current carrying wire.

$$B = \frac{\mu_0 i}{4\pi r}$$

The magnitude of the field B inside a long solenoid carrying a current I.
$B = \mu_0 nI$, where n is the number of turns per unit length.
For a **toroid** one obtains,

$B = \frac{\mu_0 NI}{2\pi r}$, where N is the total number of turns and r is the average radius.

The magnetic field inside a solenoid of finite length at a point P is given by

$$B = \frac{\mu_0 ni}{2}[\sin\theta_1 + \sin\theta_2]$$

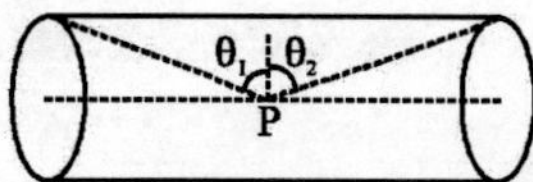

which is directed along the axis.
Where n is the number of turns per unit length.
Helmholtz coils : The two coaxial coils of equal radii placed at distance equal to the radius of either of the coils and in which same current in same direction is flowing are known as Helmholtz coils.
Resultant magnetic field in between two coils

$$B = \frac{8\mu_0 NI}{5\sqrt{5}\,a} = 1.432\,B, \text{ where } B = \frac{\mu_0 NI}{2R}$$

Magnetic field created by a long current carrying conducting cylinder

From Ampere's law

For r ≥ R :

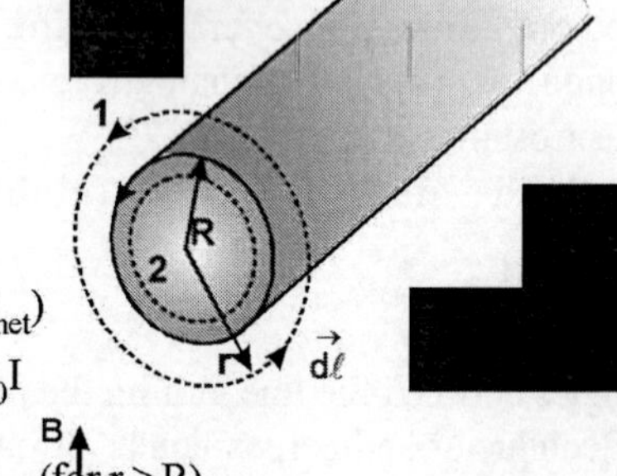

$$\oint \vec{B}.\vec{d\ell} = \mu_0 (I_{net})$$

or $B(2\pi r) = \mu_0 I$

and $B = \frac{\mu_0}{2\pi}\frac{I}{r}$ (for $r \geq R$)

For r < R :

$$B = \left(\frac{\mu_0 I}{2\pi R^2}\right) r$$

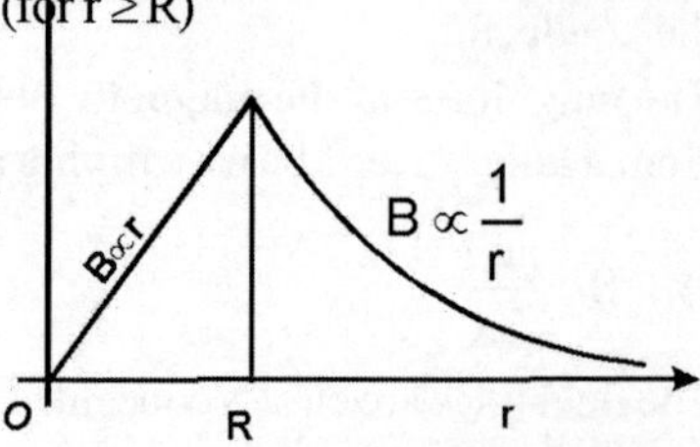

Magnetic field due to a current carrying hollow cylinder

(a) **For $r < r_1$**

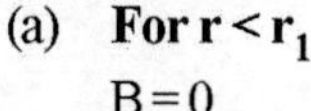

$B = 0$

(b) **For**

$r_1 < r < r_2$

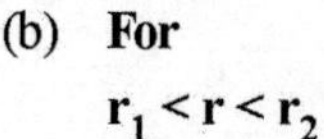

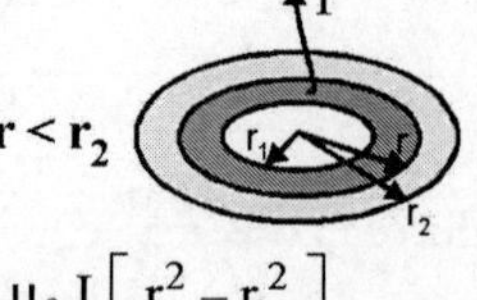

$$B = \frac{\mu_0 I}{2\pi r}\left[\frac{r^2 - r_1^2}{r_2^2 - r_1^2}\right].$$

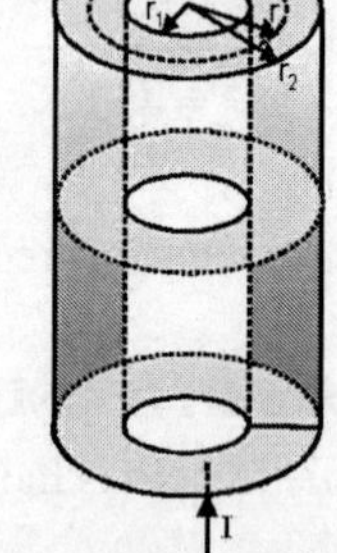

CYCLOTRON

The cyclotron, invented by Ernest O. Lawrence (1901-1958) in 1930, was the first machine developed to accelerate positively charged particles (like, α - particles, deutrons, etc.) to high velocities by causing them to pass repeatedly through the same accelerating region. Its operation depends on the remarkable fact that the period or the time required for a charged particle to complete one circular orbit in a uniform magnetic field $\vec{B}$ is independent of the speed of the particle v.

Cyclotron Frequency and Time Period

Under the action of the given magnetic field (B fixed), the given positive ion (e/m fixed) will cover the semi-circular path in a fixed time only, if t in equation is equal to T/2

where T is the time period of the electric field.

$$\therefore \quad \frac{T}{2} = \frac{\pi m}{Bq} \quad \text{or} \quad T = \frac{2\pi m}{Bq}$$

The cyclotron angular frequency, $\omega = \frac{Bq}{m}$.

Also, cyclotron frequency, $\nu = \frac{Bq}{2\pi m}$

Maximum energy attained by the charged particle,

$$E_{max} = \left(\frac{q^2B^2}{2m}\right) r_O^2 \text{ where, } r_O = \text{maximum radius}$$

FORCE ACTING ON A CHARGED PARTICLE MOVING IN A UNIFORM MAGNETIC FIELD

The force acting on a particle having a charge q and moving with velocity $\vec{v}$ in a uniform magnetic field $\vec{B}$ is given by

$\vec{F} = q(\vec{v} \times \vec{B}) \Rightarrow F = qvB \sin\theta$, where θ is the angle between $\vec{v}$ and $\vec{B}$

The Direction of $\vec{F}$ **:** We can use the rule of cross product. The direction of $\vec{F}$ is perpendicular to the plane containing $\vec{v}$ and $\vec{B}$ and can be found by right hand thumb rule.

Motion of a Charged Particle q in a Uniform Magnetic Field

Case 1 : (Straight line motion) : If a charged particle q is projected into a uniform magnetic field B with a velocity which is parallel to the field lines, the force experienced by the charge is zero and hence it travels in a straight line with uniform velocity.

$$\vec{F} = q(\vec{v} \times \vec{B}) = \vec{0} \text{ for } \vec{v} \parallel \vec{B}$$

Case II : (Uniform circular motion) : If a charged particle q is projected into a uniform magnetic field B with an initial velocity perpendicular to the lines of force, it gets trapped in a circular path. The force exerted by the field provides the necessary centripetal force.

Mathematically, we have

$$qvB = \frac{mv^2}{r} \Rightarrow r = \frac{mv}{qB}$$

The time period of revolution is

$$T = \frac{2\pi r}{v} \Rightarrow T = \frac{2\pi m}{qB}$$

Note that the plane of the circular path is perpendicular to the lines of force.

Case III : (Motion in a helical path).

If the charged particle's initial projection velocity makes an angle θ ($\theta \neq 0, 90°$) with the magnetic field B, it moves in a helical path. The axis of the helix is parallel to the lines of force.

Resolving the velocity of particle parallel and perpendicular to the field direction.

$$v_{\parallel} = v\cos\theta \text{ and } v_{\perp} = v\sin\theta$$

The v sinθ component is responsible for rotation of particle and hence,

(a) The radius of the helix is $r = \frac{mv_{\perp}}{qB}$

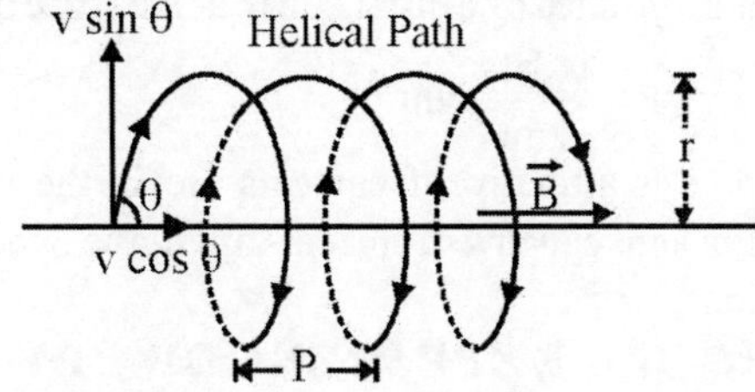

(b) The time period of revolution is $T = \frac{2\pi m}{qB}$

(c) The pitch of the helix (the displacement parallel to axis during one circular revolution).

$$P = v_{\|} T \Rightarrow P = \left(\frac{2\pi m}{qB}\right) v_{\|}$$

LORENTZ FORCE

The total force on a charge q moving with velocity v in the presence of magnetic and electric fields $\vec{B}$ and $\vec{E}$, respectively is called the **Lorentz force.** It is given by the expression: $\vec{F} = q(\vec{v} \times \vec{B} + \vec{E})$. The magnetic force $q(\vec{v} \times \vec{B})$ is normal to $\vec{v}$ and work done by it is zero. This velocity dependent force has occupied the attention of some of the greatest scientific thinkers. If one switches to a frame with instantaneous velocity $\vec{v}$, the magnetic part of the force vanishes. The motion of the charged particle is then explained by arguing that there exists an appropriate electric field in the new frame. The resolution of this paradox implies that electricity and magnetism are linked phenomena (electromagnetism) and that the Lorentz force expression does not imply a universal preferred frame of reference in nature.

Unit of $\vec{B}$: MKS weber/metre^2, SI tesla, CGS maxwell/ cm^2 or gauss.

One tesla = one (weber/m^2) = 10^4 (maxwell/cm^2) = 10^4 gauss.

A straight conductor of length ℓ and carrying a steady current I experiences a force $\vec{F}$ in a uniform external magnetic field $\vec{B}$, $\vec{F} = I\vec{\ell} \times \vec{B}$, the direction of ℓ is given by the direction of the current.

FORCE BETWEEN TWO PARALLEL CURRENT CARRYING CONDUCTORS

Force between two long parallel wires carrying currents I_1, I_2 and separated by a distance a in a free space or air

$$F = \frac{\mu_0 I_1 I_2}{2\pi a} \text{Nm}^{-1}.$$

The force is attractive if currents are in the same direction and repulsive if currents are in the opposite directions.

TORQUE EXPERIENCED BY A CURRENT CARRYING LOOP IN A MAGNETIC FIELD

A planar loop carrying a current I, having N closely wound turns, and an area A possesses a magnetic moment $\vec{M}$ where,

$\vec{M} = NI\vec{A}$ and the direction of M is given by the right-hand thumb rule : curl the palm of your right hand along the loop with the fingers pointing in the direction of the current. The thumb sticking out gives the direction of $\vec{M}$ (and $\vec{A}$).

When this loop is placed in a uniform magnetic field $\vec{B}$, the force $\vec{F}$ on it is: F = 0. And the torque on it is,

$\vec{\tau} = \vec{M} \times \vec{B}$

Work done = MB (1– cosθ)

$$\text{P.E.} = -\vec{M} \cdot \vec{B}$$

DIRECTION OF FORCE

Fleming's Left Hand Rule

Stretch the fore-finger, central finger and thumb of left hand mutually perpendicular.

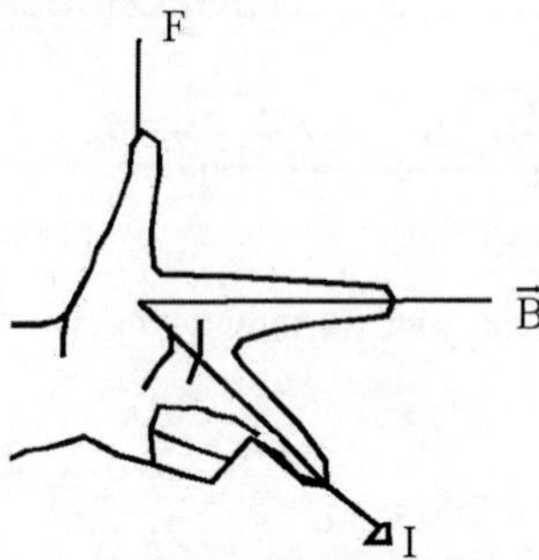

Then if the fore-finger points in the direction of field $(\vec{B})$, the central finger in the direction of current I, the thumb will point in the direction of force.

HALL EFFECT

In 1879 Edwin Hall discovered that when a current-carrying conductor is placed in a magnetic field, a voltage is generated in a direction perpendicular to both the current and the magnetic field. This observation, known as the **Hall Effect**, arises from the deflection of charge carriers to one side of the conductor as a result of the magnetic force experienced by the charge carriers.

MOVING COIL GALVANOMETER

For a moving coil instrument, we have $I = \left(\frac{C}{NAB}\right)\theta$

where C = spring constant, N = number of turns in the coil

A = area of the coil, B = flux density of the field

The **current sensitivity** is given by, $\frac{\theta}{I} = \frac{NAB}{C}$

The **voltage sensitivity** is given by, $\frac{\theta}{V} = \frac{\theta}{IR}$

Exercise

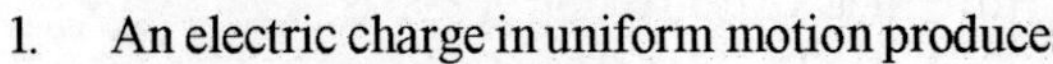

1. An electric charge in uniform motion produces
   (a) an electric field only
   (b) a magnetic field only
   (c) both electric and magnetic fields
   (d) no such field at all
2. The magnetic field at a point due to a current carrying conductor is directly proportional to
   (a) resistance of the conductor
   (b) thickness of the conductor
   (c) current flowing through the conductor
   (d) distance from the conductor
3. A current is passed through a straight wire. The magnetic field established around it has its lines of force
   (a) circular and endless
   (b) oval in shape and endless
   (c) straight
   (d) all are true
4. The strength of the magnetic field around a straight conductor
   (a) is same everywhere around the conductor
   (b) obeys inverse square law
   (c) is directly proportional to the square of the distance from the conductor
   (d) none of the above
5. Which of the following graphs, shows the variation of magnetic induction B with distance r from a long wire carrying a current

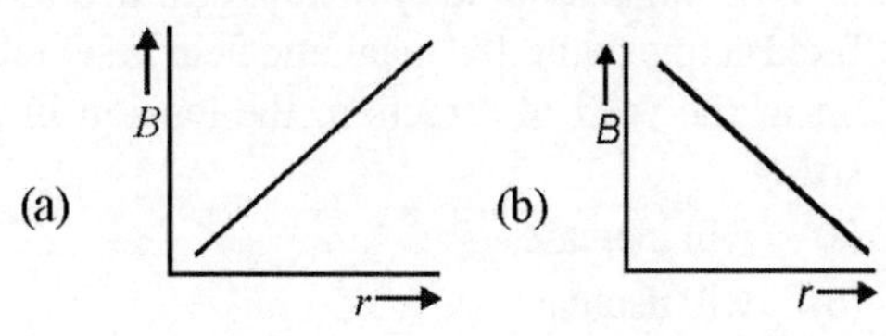

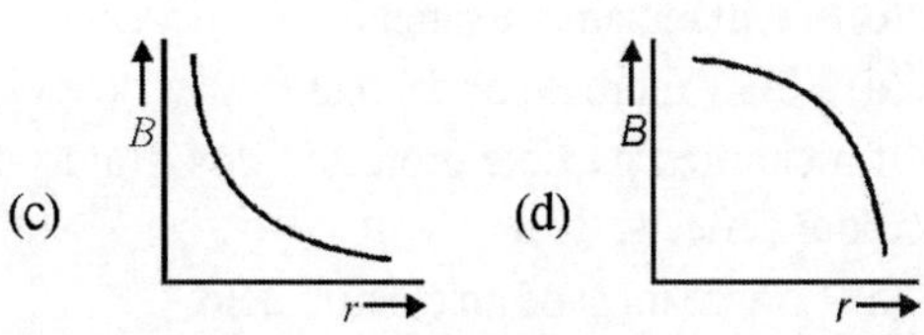

6. A long straight conductor carrying a current lies along the axis of a ring. The conductor will exert a force on the ring if the ring
   (a) carries a current
   (b) has uniformly distributed charge
   (c) has non-uniformly distributed charge
   (d) none of the above
7. A length of wire carries a steady current. It is bent first to form a circular plane coil of one turn. The same length is now bent more sharply to give a double loop of smaller radius. The magnetic field at the centre caused by the same current is
   (a) a quarter of its first value
   (b) unaltered
   (c) four times of its first value
   (d) a half of its first value
8. If a copper rod carries a direct current, the magnetic field associated with the current will be
   (a) only inside the rod
   (b) only outside the rod
   (c) both inside and outside the rod
   (d) neither inside nor outside the rod
9. If a long hollow copper pipe carries a direct current, the magnetic field associated with the current will be
   (a) only inside the pipe
   (b) only outside the pipe
   (c) neither inside nor outside the pipe
   (d) both inside and outside the pipe
10. A current i ampere flows along an infinitely long straight thin walled tube, then the magnetic induction at any point inside the tube is
    (a) ∞ (b) zero
    (c) $\frac{\mu_0}{4\pi}.\frac{2i}{r}$ tesla (d) $\frac{2i}{r}$ tesla
11. A steady current I flows down a hollow cylindrical tube of radius a and is uniformly distributed around the tube. Let r be the distance from the axis of symmetry of the tube to a given point. What is the magnitude of the magnetic field, B, at a point inside the tube? (cgs units)
    (a) 0 (b) $2\,I/r\,c$
    (c) $2\,I\,r/a^2 c$ (d) $4\,\pi\,(r-a)\,I/\,r^2$

12. A current i ampere flows in a circular arc of wire which subtends an angle $(3\ \pi/2)$ radians at its centre, whose radius is R. The magnetic induction B at the centre is
(a) $\mu_0\ i/R$ (b) $\mu_0\ i/2R$
(c) $2\ \mu_0\ i/R$ (d) $3\ \mu_0\ i/8R$
13. A current is flowing north along a power line. The direction of the magnetic field above it, neglecting the earth's field is towards
(a) North (b) East
(c) South (d) West
14. Field inside a solenoid is
(a) directly proportional to its length
(b) directly proportional to current
(c) inversely proportional to number of turns
(d) inversely proportional to current
15. Energy in a current carrying coil is stored in the form of
(a) electric field (b) magnetic field
(c) dielectric strength (d) heat
16. A charge q is moving with a velocity v parallel to a magnetic field B. Force on the charge due to magnetic field is
(a) $q\,v\,B$ (b) $q\,B/v$
(c) zero (d) $B\,v/q$
17. A proton moving with a constant velocity passes through a region of space without any change in its velocity. If E and B represent the electric and magnetic fields respectively, this region of space may not have
(a) $E=0, B=0$ (b) $E=0, B\neq 0$
(c) $E\neq 0, B=0$ (d) $E\neq 0, B\neq 0$
18. A proton (mass m and charge $+e$) and an alpha particle (mass $4\,m$ and charge $+2e$) are projected with the same kinetic energy at right angles to a uniform magnetic field. Which one of the following statements will be true?
(a) the alpha-particle will be bent in a circular path with a smaller radius than that of the proton
(b) the radius of the path of the alpha-particle will be greater than that of the proton
(c) the alpha-particle and the proton will be bent in a circular path with the same radius
(d) the alpha-particle and the proton will go through the field in a straight line
19. An electron having a charge e moves with a velocity v in X-direction. An electric field acts on it in Y-direction? The force on the electron acts in
(a) positive direction of Y-axis
(b) negative direction of Y-axis
(c) positive direction of Z-axis
(d) negative direction of Z-axis
20. A charged particle enters into a magnetic field with a velocity vector making an angle of 30° with respect to the direction of magnetic field. The path of the particle is
(a) circular (b) helical
(c) elliptical (d) straight line
21. If electron velocity is $2\,i+3\,j$ and it is subjected to magnetic field of $4\,k$, then its
(a) speed will change (b) path will change
(c) both a and b (c) none of the above
22. A positively charged particle moving due east enters a region of uniform magnetic field directed vertically upwards. This particle will
(a) get deflected in vertically upward direction
(b) move in circular path with an increased speed
(c) move in a circular path with decreased speed
(d) move in a circular path with uniform speed
23. The work done by a magnetic field, on a moving charge is
(a) zero because $\vec{F}$ acts parallel to $\vec{v}$
(b) positive because $\vec{F}$ acts perpendicular to $\vec{v}$
(c) zero because $\vec{F}$ acts perpendicular to $\vec{v}$
(d) negative because $\vec{F}$ acts parallel to $\vec{v}$
24. A charged particle is whirled in a horizontal circle on a frictionless table by attaching it to a string fixed at one point. If a magnetic field is switched on in the vertical direction, the tension in the string
(a) will increase
(b) will decrease
(c) will remain the same
(d) may increase or decrease
25. If a charged particle projected in a gravity-free room deflects, then
(a) there must be an electric field
(b) there must be a magnetic field
(c) both fields cannot be zero
(d) none of these

26. A charged particle goes undeflected in a region containing electric and magnetic fields. It is possible that
(a) $\vec{E} \parallel \vec{B}, \vec{v} \parallel \vec{E}$
(b) $\vec{E}$ is not parallel to $\vec{B}$
(c) $\vec{v} \parallel \vec{B}$ but $\vec{E}$ is not parallel to $\vec{B}$
(d) $\vec{E} \parallel \vec{B}$ but $\vec{v}$ is not parallel to $\vec{E}$

27. If a charged particle goes unaccelerated in a region containing electric and magnetic fields,
(a) $\vec{E}$ must be perpendicular to $\vec{B}$
(b) $\vec{v}$ must not be perpendicular to $\vec{B}$
(c) $\vec{v}$ must be perpendicular to $\vec{E}$
(d) $\vec{E}$ must be perpendicular to $v\,\vec{B}$

28. Two free parallel wires carrying currents in the opposite directions
(a) attract each other
(b) repel each other
(c) do not affect each other
(d) get rotated to be perpendicular to each other

29. Two parallel wires carrying currents in the same direction attract each other because of
(a) potential difference between them
(b) mutual inductance between them
(c) electric forces between them
(d) magnetic forces between them

30. A wire is placed parallel to the lines of force in a magnetic field and a current flows in the wire. Then
(a) the wire will experience a force in the direction of the magnetic field
(b) the wire will not experience any force at all
(c) the wire will experience a force in a direction opposite to the field
(d) it experiences a force in a direction perpendicular to lines of force

31. A conducting circular loop of radius r carries a constant current i. It is placed in a uniform magnetic field B such that B is perpendicular to the plane of the loop. The magnetic force acting on the loop is
(a) irB (b) $2\pi riB$
(c) zero (d) πriB

32. A straight steel wire of length l has a magnetic moment M. When it is bent in the form of a semi-circle its magnetic moment will be
(a) M (b) M/π
(c) $2M/\pi$ (d) $M\pi$

33. A current carrying coil is subjected to a uniform magnetic field. The coil will orient so that its plane becomes
(a) inclined at 45° to the magneic field
(b) inclined at any arbitrary angle to the magnetic field
(c) parallel to the magnetic field
(d) perpendicular to magnetic field

34. Two long wires are hanging freely. They are joined first in parallel and then in series and then are connected with a battery. In both cases which type of force acts between the two wires
(a) attraction force when in parallel and repulsion force when in series
(b) repulsion force when in parallel and attraction force when in series
(c) repulsion force in both cases
(d) attraction force in both cases

35. The magnetic dipole moment of current loop is independent of
(a) number of turns
(b) area of loop
(c) current in the loop
(d) magnetic field in which it is lying

36. A coil carrying a heavy current and having large number of turns is mounted in a *N-S* vertical plane. A current flows in the clockwise direction. A small magnetic needle at its centre will have its north pole in
(a) east-north direction
(b) west-north direction
(c) east-south direction
(d) west-south direction

37. If E and B denote electric and magnetic fields respectivley, which of the following is dimensionless?
(a) $\sqrt{\mu_0 \varepsilon_0}\,\frac{E}{B}$ (b) $\mu_0 \varepsilon_0 \frac{E}{B}$
(c) $\mu_0 \varepsilon_0 \left(\frac{B}{E}\right)^2$ (d) $\frac{E}{\varepsilon_0} \times \frac{\mu_0}{B}$

38. $ABCD$ is a square loop made of a uniform conducting wire. A current enters the loop at A and leaves at D. The magnetic field is

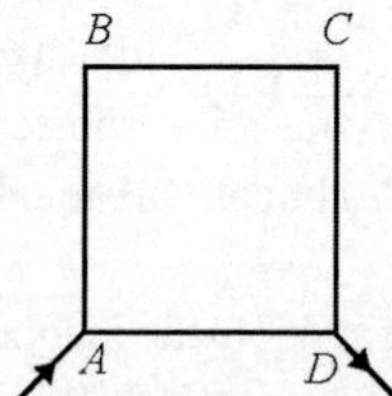

(a) zero only at the centre of the loop
(b) maximum at the centre of the loop
(c) zero at all points outside the loop
(d) zero at all points inside the loop

39. An electron having energy 10 eV is circulating in path having magnetic field 1.0×10^{-4} Tesla, the speed of the electron will be
(a) 1.9×10^6 m/s (b) 3.8×10^6 m/s
(c) 1.9×10^{12} m/s (d) 3.8×10^{12} m/s

40. The magnetic induction at a point P which is at a distance of 4 cm from a long current carrying wire is 10^{-3} T. The field of induction at a distance 12 cm from the current will be
(a) $3.33 \times 10^{-4}\,T$ (b) $1.11 \times 10^{-4}\,T$
(c) $3 \times 10^{-3}\,T$ (d) $9 \times 10^{-3}\,T$

41. A uniform magnetic field acts at right angles to the direction of motion of electron. As a result, the electron moves in a circular path of radius 2 cm. If the speed of electron is doubled, then the radius of the circular path will be
(a) 2.0 cm (b) 0.5 cm
(c) 4.0 cm (d) 1.0 cm

42. The magnetic field at a distance r from a long wire carrying current i is 0.4 tesla. The magnetic field at a distance $2r$ is
(a) 0.2 tesla (b) 0.8 tesla
(c) 0.1 tesla (d) 1.6 tesla

43. A straight wire of length 0.5 metre and carrying a current of 1.2 ampere is placed in uniform magnetic field of induction 2 tesla. The magnetic field is perpendicular to the length of the wire. The force on the wire is
(a) 2.4 N (b) 1.2 N
(c) 3.0 N (d) 2.0 N

44. A coil carrying electric current is placed in uniform magnetic field
(a) torque is formed
(b) e.m.f. is induced
(c) both (a) and (b) are correct
(d) none of the above

45. A charge moving with velocity v in X-direction is subjected to a field of magnetic induction in negative X-direction. As a result, the charge will
(a) remain unaffected
(b) start moving in a circular path Y–Z plane
(c) retard along X-axis
(d) move along a helical path around X-axis

46. A current of I ampere flows in a wire forming a circular arc of radius r metres subtending an angle θ at the centre as shown. The magnetic field at the centre O in tesla is

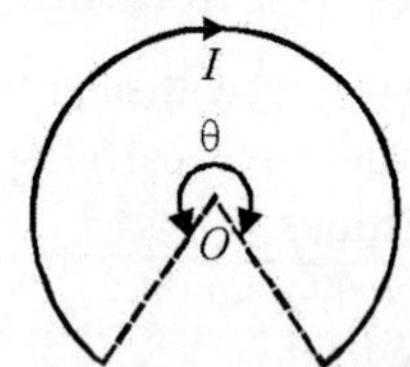

(a) $\frac{\mu_0 I\theta}{4\pi r}$ (b) $\frac{\mu_0 I\theta}{2\pi r}$
(c) $\frac{\mu_0 I\theta}{2r}$ (d) $\frac{\mu_0 I\theta}{4r}$

47. The wire loop $PQRS$ formed by joining two semi circular wires of radii R_1 and R_2 carries a current I as shown. The magnitude of magnetic induction at the centre O is

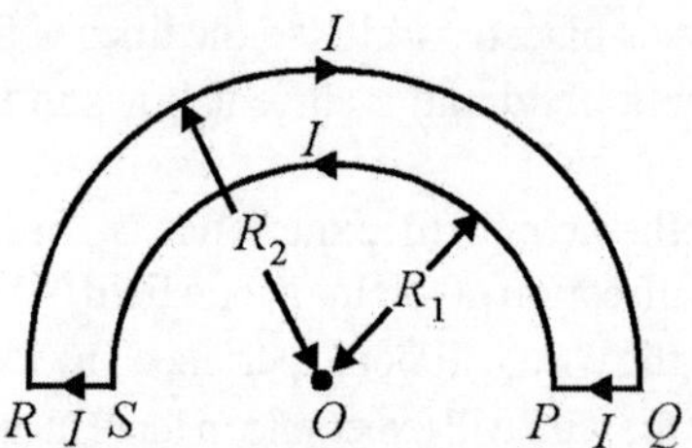

(a) $(\mu_0/4)I(R_1/R_2 - 1/R_1)$
(b) $(\mu_0/4)I(1/R_2 - 1/R_2)$
(c) $\mu_0 I(1/R_2 - 1/R_1)$
(d) $\mu_0 I(1/R_1)$.

48. At what distance from a long straight wire carrying a current of 12 A will the magnetic field be equal to 3×10^{-6} Wb/m^2?
(a) 8×10^{-2} m (b) 12×10^{-2} m
(c) 18×10^{-2} m (d) 24×10^{-2} m

49. A beam of electron passes undeflected through mutually perpendicular electric and magnetic fields. If the electric field is switched off, and the same magnetic field is maintained, the electrons move
(a) in a circular orbit
(b) along a parabolic path
(c) along a straight line
(d) in an elliptical orbit

50. A straight wire of diameter 0.5 mm carrying a current of 1 A is replaced by another wire of 1 mm diameter carrying same current. The strength of magnetic field far away is
(a) twice the earlier value
(b) same as the earlier value
(c) one-half of the earlier value
(d) one-quarter of the earlier value

51. An electron enters a region where magnetic field (B) and electric field (E) are mutually perpendicular, then
(a) it will always move in the direction of B
(b) it will always move in the direction of E
(c) it always possesses circular motion
(d) it can go undeflected also

52. When a charged particle moving with velocity $\vec{v}$ is subjected to a magnetic field of induction $\vec{B}$, the force on it is non-zero. This implies that
(a) angle between $\vec{v}$ and $\vec{B}$ can have any value other than 90°
(b) angle between $\vec{v}$ and $\vec{B}$ can have any value other than zero and 180°
(c) angle between $\vec{v}$ and $\vec{B}$ is either zero or 180°
(d) angle between $\vec{v}$ and $\vec{B}$ is necessarily 90°

53. A charged paritcle (charge q) is moving in a circle of radius R with uniform speed v. The associated magnetic moment μ is given by
(a) qvR^2 (b) $qvR^2/2$
(c) qvR (d) $qvR/2$

54. A very long straight wire carries a current I. At the instant when a charge $+Q$ at point P has velocity $\vec{v}$, as shown, the force on the charge is

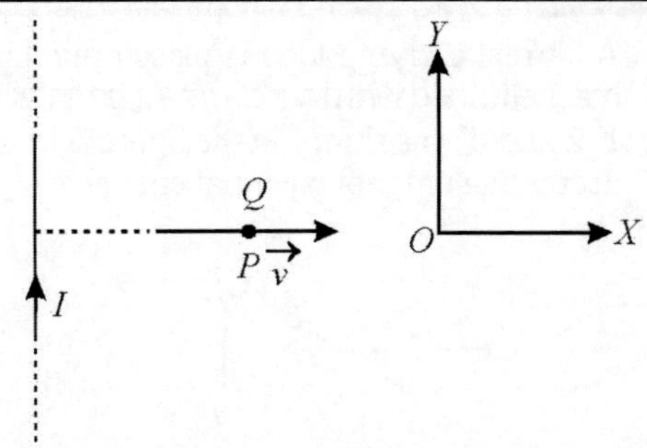

(a) along OY (b) opposite to OY
(c) along OX (d) opposite to OX

55. Current i is flowing in a coil of area A and number of turns N, then magnetic moment of the coil, M is
(a) NiA (b) $\frac{Ni}{A}$
(c) $\frac{Ni}{\sqrt{A}}$ (d) N^2Ai

56. A coil in the shape of an equilateral triangle of side l is suspended between the pole pieces of a permanent magnet such that $\vec{B}$ is in the plane of the coil. If due to a current i in the triangle a torque τ acts on it, the side l of the triangle is
(a) $\frac{2}{\sqrt{3}}\left(\frac{\tau}{B.i}\right)^{\frac{1}{2}}$ (b) $2\left(\frac{\tau}{\sqrt{3}B.i}\right)^{\frac{1}{2}}$
(c) $\frac{2}{\sqrt{3}}\left(\frac{\tau}{B.i}\right)$ (d) $\frac{1}{\sqrt{3}}\frac{\tau}{B.i}$

57. Two flat circular coils have a common center, but their planes are at right angles to each other. The inner coil has 150 turns and radius of π cm. The outer coil has 400 turns and a radius of 2π cm. The magnitude of the resultant magnetic induction at the common centers of the coils when a current of 200 mA is sent through each of them is
(a) 10^{-3} Wb/m^2 (b) 2×10^{-3} Wb/m^2
(c) 5×10^{-3} Wb/m^2 (d) 7×10^{-3} Wb/m^2

58. Two long parallel wires carry currents I_1 and I_2 ($I_1 > I_2$). When the currents are in the same direction, the magnetic field at a point midway between the wires is 10 μT. If the direction of I_2 is reversed, the field becomes 40 μT. The ratio I_1/I_2 is
(a) 5/2 (b) 5/3
(c) 4/3 (d) 3/2

59. A current carrying loop is placed in a uniform magnetic field with four different orientations X, Y, Z and W was shown in the figure. The correct decreasing order of potential energy is

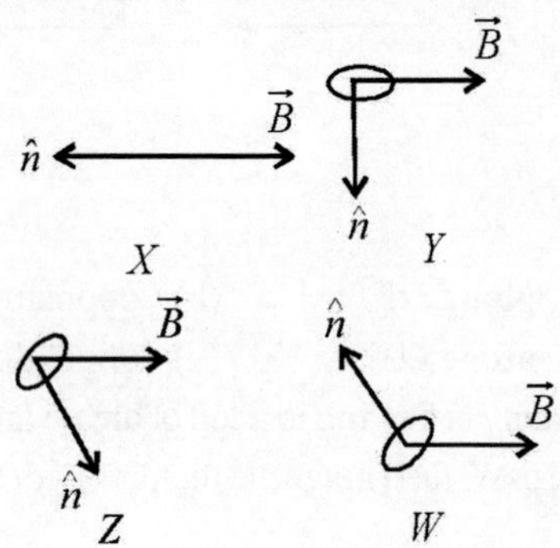

(a) $X > Y > Z > W$ (b) $Z > W > X > Y$
(c) $X > W > Y > Z$ (d) $X > Y > W > Z$

60. A cylindrical conductor having radius of cross section R carries a steady current I. If the distance from the axis of the conductor is r, then the magnetic field B varies as
(a) $1/r$
(b) r for $r < R$ and as $1/r$ for $r \geq R$
(c) r
(d) $1/r^2$

CHAPTER 19 Magnetism and Matter

CONCEPTS

BAR MAGNET

A bar magnet consists of two equal and opposite magnetic poles separated by a small distance. Poles are not exactly at the ends. The shortest distance between two poles is called effective length (L_e) and is less than its geometric length (L_g). For bar magnet $L_e = 2l$ and $L_e = (5/6)\, L_g$. For semicircular magnet $L_g = \pi R$ and $L_e = 2R$.

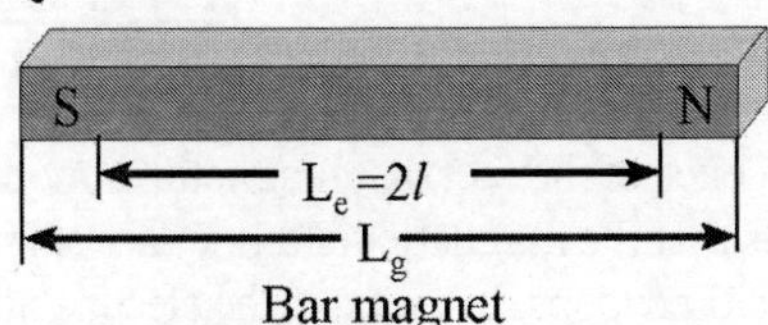

Bar magnet

Properties of Magnet

(i) Attractive property
(ii) Directive property
(iii) Poles of a magnet always exist in pairs
(iv) Repulsive Property

MAGNETIC DIPOLE

Magnetic dipole consists of two unlike poles of equal strength and separated by a small distance.
The distance between the two poles of a bar magnet is called the **magnetic length** of the magnet.

It is represented as $\overrightarrow{2\ell}$ as it is a vector, directed from S-pole to N-Pole.

Pole Strength (m)

The strength of a magnetic pole to attract magnetic materials towards itself is known as pole strength.

Magnetic Dipole Moment

The product of strength of either pole and the magnetic length of the magnet is called magnetic dipole moment.

$$\vec{M} = m(2\vec{\ell})$$

It is a vector quantity whose direction is from south pole to north pole of magnet. The unit of magnetic dipole moment is ampere metre^2 (Am^2) and joule/tesla (J/T). The dimensions are ($M^0L^2T^0A^1$.)

As magnetic moment is a vector, in case of two magnets having magnetic moments M_1 and M_2 with angle θ between them, the resulting magnetic moment

$$M = \left[M_1^2 + M_2^2 + 2M_1M_2\cos\theta\right]^{1/2} \text{ with}$$

$$\tan\phi = \left[\frac{M_2\sin\theta}{M_1 + M_2\cos\theta}\right]$$

Current Loop as a Magnetic Dipole

A current carrying loop behaves as a magnetic dipole. The two faces of the loop behaves as two poles of a magnet with equal strength.

Magnetic dipole moment of the current Loop

$M \propto I$ (Strength of current)
$M \propto A$ (area enclosed by the loop)
$\therefore\quad M \propto IA$ or, $M = KIA = IA\ [\because k = 1]$
Its SI unit is ampere metre^2 (Am^2)

Atom as a Magnetic Dipole

Magnetic dipole moment

$$M = IA = -\frac{ev}{2\pi r}\,.\,\pi r^2 = -\frac{evr}{2}$$

In vector form $\vec{M} = -\frac{e}{2m_e}\vec{L}$

The direction of magnetic dipole moment vector is opposite to angular momentum vector.

According to Bohr's theory $L = \frac{nh}{2\pi}$ n = 0, 1, 2.........

$$\text{So}\quad M = \left(\frac{e}{2m_e}\right)\frac{nh}{2\pi} = n\left(\frac{eh}{4\pi m_e}\right) = n\mu_B$$

where

$$\mu_B = \frac{eh}{4\pi m_e} = \frac{(1.6 \times 10^{-19} C)\ (6.62 \times 10^{-34} Js)}{4 \times 3.14 \times (9.1 \times 10^{-31} kg)}$$

$$= 9.27 \times 10^{-24}\ Am^2$$

It is called Bohr Magneton.
This is natural unit of magnetic moment.

MAGNETIC FIELD INTENSITY DUE TO A MAGNETIC DIPOLE

(i) **Along the axis of a magnetic dipole (or end on position):**

$$B = \frac{\mu_0}{4\pi} \frac{2M}{r^3}$$

This expression is also for the axial field of a **finite solenoid carrying current**.

(ii) **Perpendicular to the axis of a magnetic dipole : (or at the equatorial position)**

$$B = \frac{\mu_0 M}{4\pi r^3}$$

MAGNETIC LINES OF FORCE AND ITS PROPERTIES

(a) A line of force is an imaginary curve the tangent to which at a point gives the direction of magnetic field at that point.

(b) The magnetic field line is the imaginary path along which an isolated north pole will tend to move if it is free to do so.

(c) The magnetic lines of force are closed curves. They appear to converge or diverge at poles outside the magnet they run from north to south pole and inside from south to north.

(d) The number of lines originating or terminating on a pole is proportional to its pole strength.
Magnetic flux = number of magnetic lines of force $= \mu_0 \times m$
Where μ_0 is number of lines associated with unit pole.

(d) Magnetic lines of force do not intersect each other because if they do so there will be two directions of magnetic field which is not possible.

(f) The magnetic lines of force may enter or come out of surface at any angle.

(g) The number of lines of force per unit area at a point gives magnitude of field at that point. The crowded lines show a strong field while distant lines represent a weak field.

(h) The magnetic lines of force have a tendency to contract longitudinally like a stretched elastic string producing attraction between opposite poles.

(i) The magnetic lines of force have a tendency to repel each other laterally resulting in repulsion between similar poles.

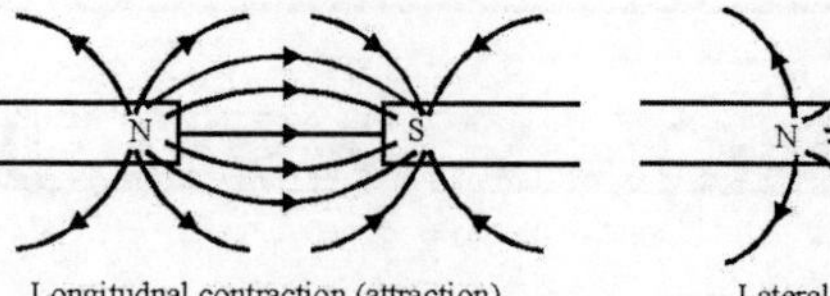

Longitudnal contraction (attraction)

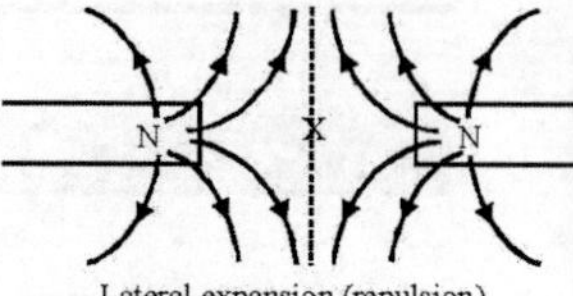

Lateral expansion (repulsion)

(j) The region of space with no magnetic field has no lines of force. At neutral point where resultant field is zero there cannot be any line of force.

(k) Magnetic lines of force exist inside every magnetised material.

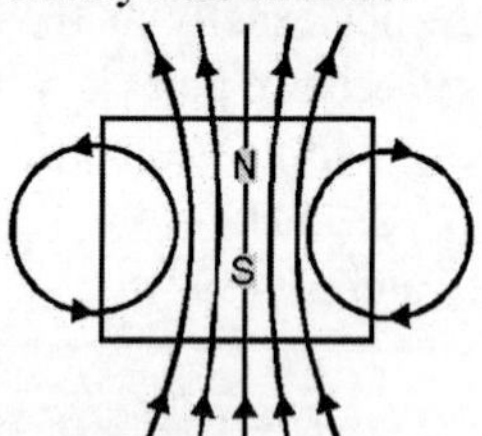

A consequence of the fact that magnetic monopoles do not exist is that the magnetic field lines are continuous and form closed loops. In contrast, the electrostatic lines of force begin on a positive charge and terminate on the negative charge (or fade out at infinity).

COULOMB'S LAW IN MAGNETISM

$$F = \frac{\mu_0}{4\pi} \frac{m_1 m_2}{r^2}$$

where $F = \frac{\mu_0}{4\pi} = 10^{-7}\ Wb\ A^{-1}\ m^{-1} = 10^{-7}$ henry/m

where μ_0 is permeability of free space and m_1 and m_2 are pole strength.

Bar magnet in a magnetic field :

Torque : $\vec{\tau} = \vec{M} \times \vec{B}$ **Potential energy :**

$$U = -\vec{M} . \vec{B}$$

Time period $= 2\pi\sqrt{\frac{I}{MB}}$

GAUSS'S LAW IN MAGNETISM

The surface integral of magnetic field $\vec{B}$ over a closed surface S is always zero.

Mathematically $\oint_S \vec{B} . \vec{ds} = 0$

(i) Isolated magnetic poles do not exist. It is a direct consequence of gauss law in magnetism.

(ii) The total magnetic flux linked with a closed surface is always zero.

(iii) If a number of magnetic field lines are leaving a closed surface, an equal number of field lines must also be entering the surface.

When a bar magnet of dipole moment M is placed in a uniform magnetic field $\vec{B}$,

(a) the force on it is zero,

(b) the torque on it is $\vec{M} \times \vec{B}$,

(c) its potential energy is $-\vec{M}.\vec{B}$, where we choose the zero of energy at the orientation when $\vec{M}$ is perpendicular to $\vec{B}$.

COMPARISON BETWEEN ELECTRIC DIPOLE AND MAGNETIC DIPOLE

Electric dipole	Magnetic dipole
(i) $\vec{p} = q \times 2\vec{d}$	(i) $\vec{M} = m2\vec{d}$, where m is pole strength.
(ii) directed from negative charge toward positive charge.	(ii) directed from south pole toward north pole. For current carrying loop of N turns $\vec{M} = Ni\vec{A}$
(iii) Torque $\vec{\tau} = \vec{p} \times \vec{E}$ (−q to +q, separation 2d)	(iii) Torque $\vec{\tau} = \vec{M} \times \vec{B}$ (S to N, separation 2d)
(iv) Potential energy $U = -\vec{p}.\vec{E}$	(iv) Potential energy $U = -\vec{M}.\vec{B}$
(v) Time period for small oscillation in uniform EF, $T = 2\pi\sqrt{\frac{I}{PE}}$	(v) Time period for small oscillation of small bar magnet in uniform MF, $T = 2\pi\sqrt{\frac{I}{MB}}$, where I = moment of inertia
(vi) Electric field at the axis $E = \frac{1}{4\pi\varepsilon_0}\frac{2p}{r^3}$	(vi) Magnetic field induction at the axis of current loop, $B = \frac{\mu_0}{4\pi}\frac{2M}{r^3}$.
(vii) Electric field at equatorial position $E = \frac{1}{4\pi\varepsilon_0}\frac{p}{r^3}$	(vii) Magnetic field induction at the equatorial position $B = \frac{\mu_0 M}{4\pi r^3}$

EARTH'S MAGNETIC FIELD AND MAGNETIC ELEMENTS

The **earth's magnetic field** resembles that of a (hypothetical) magnetic dipole located at the centre of the earth. The pole near the geographic north pole of the earth is called the north magnetic pole. Similarly, the pole near the geographic south pole is called the south magnetic pole. This dipole is aligned making a small angle with the rotation axis of the earth. The magnitude of the field on the earth's surface = 4×10^{-5} T. The earth's magnetic field is not due to a huge bar magnet inside it. The earth's core is hot and molten. Perhaps convective currents in this core are responsible for the earth's magnetic field. As to what 'dynamo' effect sustains this current, and why the earth's field reverses polarity every million years or so, we do not know.

Three quantities are needed to specify the magnetic field of the earth on its surface – the horizontal component, the magnetic declination, and the magnetic dip. These are known as the **elements of the earth's magnetic field**.

(a) **Angle of declination (ϕ) :** The angle between the magnetic meridian and geographic meridian at a place is called angle of declination.

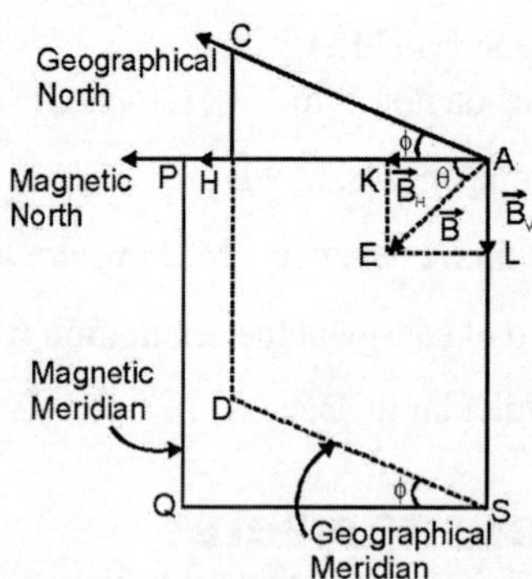

(b) **Angle of dip or inclination (θ) :** The angle through which the N pole dips down with reference to horizontal is called the angle of dip. At magnetic north and south poles, angle of dip is 90°. At magnetic equator, the angle of dip is zero.
In other words, the angle which the direction of resultant field of earth makes with the horizontal line of magnetic meridian is called angle of dip.

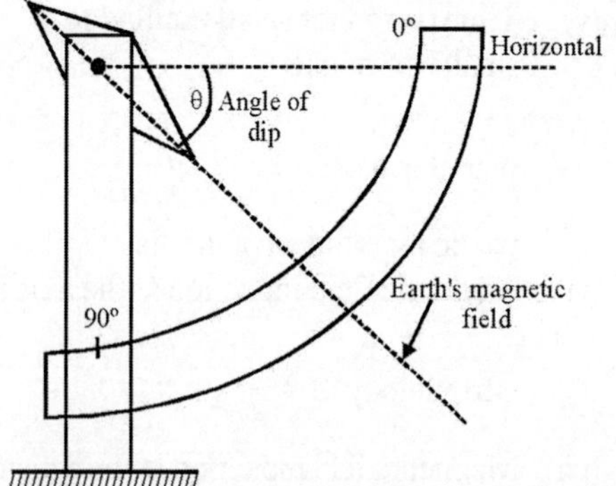

(c) **Horizontal component of earth's magnetic field** : The total intensity of the earth's magnetic field makes an angle θ with horizontal. It has
(i) component in horizontal plane called horizontal component B_H.
(ii) component in vertical plane called vertical component B_V.

$$B_V = B\sin\theta;\ B_H = B\cos\theta$$

$$\text{So } \frac{B_V}{B_H} = \tan\theta \text{ and } B = \sqrt{B_H^2 + B_V^2}$$

NEUTRAL POINTS

Neutral points are the points where the net field intensity due to the field of the bar magnet and field of earth is zero. When magnet is placed with its north pole towards geographic north, neutral points lie on equatorial line of the magnet. At each neutral point,

$$B_1 = \frac{\mu_0}{4\pi}\frac{M}{(d^2 + \ell^2)^{3/2}} = H$$

Where H = horizontal component of earth's magnetic field.
When the bar magnet is placed with its north pole towards geographic south, the neutral points lie on the axial line of the magnet. At each neutral point,

$$B_2 = \frac{\mu_0}{4\pi}\frac{M}{(d^2 - \ell^2)^2} = H$$

TANGENT LAW

If a small bar magnet is suspended in two mutually perpendicular uniform magnetic fields B and B_H such that it comes to rest making angle θ with direction of field B_H then $B = B_H \tan\theta$.

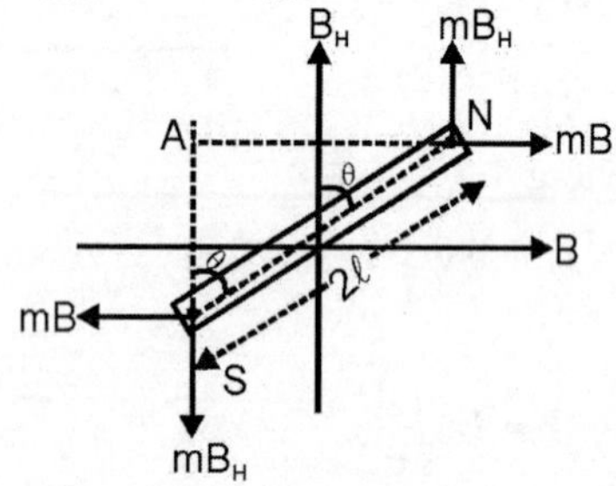

TANGENT GALVANOMETER

It is an instrument used to measure very small currents. It is a galvanometer with a fixed coil and moving magnet. It is based on tangent law.
According to **Tangent law**

$$B = B_H \tan\theta \text{ or } \frac{\mu_0 nI}{2a} = B_H \tan\theta$$

$$\text{or } I = \frac{2a}{n\mu_0}B_H\tan\theta = \frac{B_H}{G_1}\tan\theta = K\tan\theta$$

where $G_1 = \frac{n\mu_0}{2a}$ is galvanometer constant

and $K = \frac{B_H}{G_1}$ is reduction factor of galvanometer.

VIBRATION MAGNETOMETER

It is an instrument used to compare magnetic moments of two bar magnets, comparison of horizontal components of earth's magnetic field at two places and for measuring the horizontal component of earth's magnetic field.

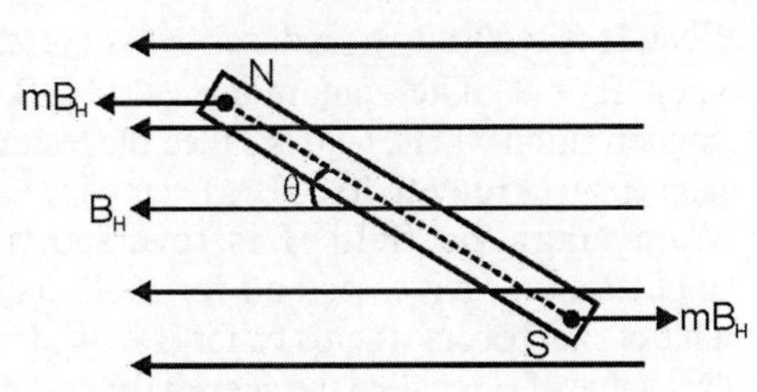

Time period,

$$T = 2\pi\sqrt{\frac{\text{Angular displacement}}{\text{Angular acceleration}}} = 2\pi\sqrt{\frac{I}{MB_H}}$$

(a) Comparison of Magnetic Moments of Magnets of same size $\frac{T_1^2}{T_2^2} = \frac{M_2}{M_1} = \frac{m_2}{m_1}$

(b) Comparison of Magnetic Moments of Magnets of different sizes : $\frac{M_1}{M_2} = \frac{T_1^2 + T_2^2}{T_2^2 - T_1^2}$

(c) Comparison of horizontal component of earth's magnetic field at different places : $\frac{B_H}{B'_H} = \frac{T'^2}{T^2}$

Consider a material placed in an external magnetic field $\vec{B}_0$. The magnetic intensity is defined as, $\vec{H} = \frac{\vec{B}_0}{\mu_0}$.

The magnetisation $\vec{M}$ of the material is its dipole moment per unit volume. The magnetic field B in the material is, $\vec{B} = \mu_0(\vec{H} + \vec{M})$

INTENSITY OF MAGNETISATION, MAGNETIC PERMEABILITY AND MAGNETIC SUSCEPTI-BILITY

Intensity of magnetisation :

It is defined as the magnetic dipole moment developed per unit volume when a magnetic material is subjected to magnetising field.

Intensity of magnetisation,

$$I = \frac{\text{Magnetic dipole moment}(M)}{\text{Volume}(V)}$$

It is a vector quantity whose direction is along the direction of magnetising field.

Its **SI unit** is ampere/m (A/m) and **dimensions** are [$M^0 L^{-1} T^0 A$]

The intensity of magnetisation shows the extent to which the substance is magnetised.

Intensity of magnetisation depends on nature of material, temperature etc.

For a linear material $\vec{M} = \chi\vec{H}$. So that $\vec{B} = \mu\vec{H}$ and χ is called the magnetic susceptibility of the material. The three quantities, χ , the relative magnetic permeability μ_r, and the magnetic permeability μ are related as follows:

$$\mu = \mu_0\mu_r \ ; \ \mu_r = 1 + \chi .$$

CLASSIFICATION OF MAGNETIC MATERIALS

Magnetic materials are broadly classified as: **diamagnetic**, **paramagnetic**, and **ferromagnetic**. For diamagnetic materials χ is negative and small and for paramagnetic materials it is positive and small. Ferromagnetic materials have large χ and are characterised by non-linear relation between $\vec{B}$ and $\vec{H}$. They show the property of hysteresis. A miniscule difference in the value of χ, the magnetic susceptibility, yields radically different behaviour: For diamagnetic materials $\chi = -10^{-5}$ whereas $\chi = +10^{-5}$ for paramagnetic materials. Diamagnetism is universal. It is present in all materials. But it is weak and hard to detect if the substance is para- or ferromagnetic.

The paramagnetic substances obey **Curie's law** according to which magnetic susceptibility is inversely proportional to its absolute temperature.

$$\chi_m = \frac{I}{H} = \frac{C}{T}$$

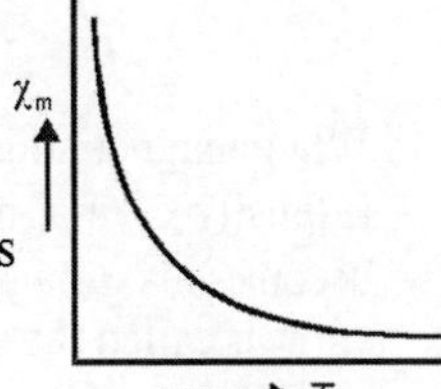

The ferromagnetic substances obey Curie-Weiss law

$$\chi_m = \frac{C}{T - T_C} \quad (\text{for } T > T_C)$$

T_C is Curie temperature which depends on nature of material

(T_C of iron = 1043K = 770°C).

Above Curie temperature a ferromagnetic material behaves as paramagnetic.

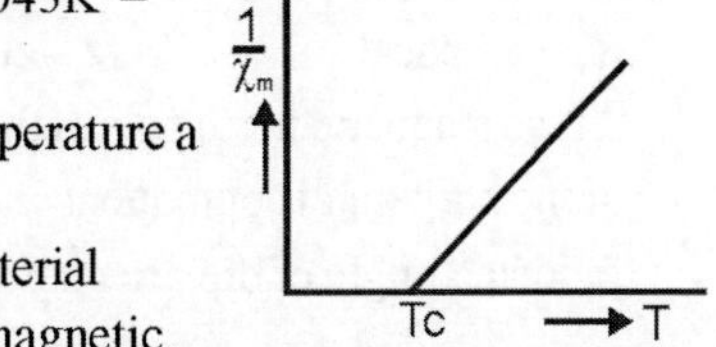

PERMANENT MAGNETS AND ELECTROMAGNETS

Permanent Magnet

It is the material which retain at room temperature its ferromagnetic properties for a long time.

The materials selected for permanent magnet should have high retentivity, high coercivity and high permeability.

Electromagnet

It is a temporary magnet. The core of electromagnets is made of ferromagnetic materials which have high permeability and low retentivity. Soft iron is suitable to make as electromagnet.

The strength of an electromagnet increases by increasing number of turns and current.

HYSTERESIS

When a bar of ferromagnetic material is magnetised by a varying magnetic field and the intensity of magnetisation I_m induced is measured for different values of magnetising field H, the graph of I versus H is as shown in fig. The graph shows :

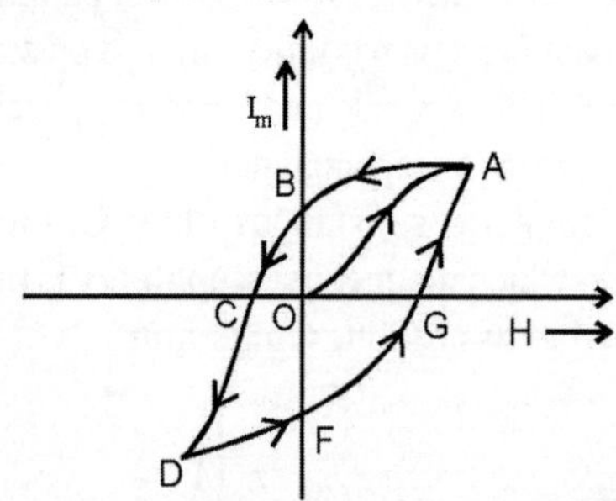

(i) When magnetising field is increased from O the intensity of magnetisation I_m increases and becomes maximum (i.e. point A). This maximum value is called the **saturation value**.

(ii) When H is reduced, I_m reduces but is not zero when H = 0. The remainder value OB of magnetisation when H = 0 is called the residual magnetism or **retentivity**. OB is retentivity.

(iii) When magnetic field H is reversed, the magnetisaiton decreases and for a particular value of H, it becomes zero i.e., for H = OC, I = 0. This value of H is called the **coercivity**.

(iv) When field H is further increased in reverse direction, the intensity of magnetisation attains saturation value in reverse direction (i.e., point D).

(v) When H is decreased to zero and changed direction in steps, we get the part DFGA.

COMPARISON OF PROPERTIES OF SOFT IRON AND STEEL

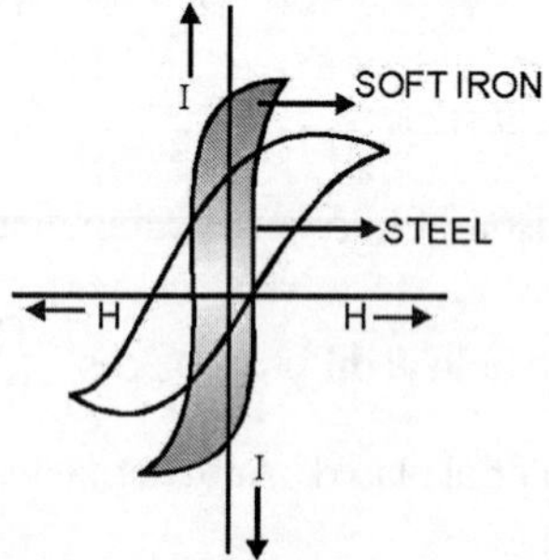

(i) The area of hysteresis loop for soft iron is much smaller than for steel so energy loss per unit volume per cycle of soft iron is smaller than steel.

(ii) The retentivity of soft iron is greater than that of steel.

(iii) The coercivity of steel is much larger than that of soft iron.

(iv) The magnetisation and demagnetisation is easier in soft iron than steel.

Exercise

1. The lines of force due to earth's horizontal magnetic field are
   (a) parallel and straight (b) concentric circles
   (c) elliptical (d) curved lines
2. At the magnetic north pole of the earth, the value of the horizontal component of earth's magnetic field and angle of dip are respectively
   (a) zero, maximum
   (b) maximum, minimum
   (c) maximum, maximum
   (d) minimum, minimum
3. The magnetic compass is not useful for navigation near the magnetic poles. Since
   (a) $R = 0$ (b) $V = 0$
   (c) $H = 0$ (d) $\theta = 0°$

4. The earth's magnetic field always has a vertical component except at the
   (a) magnetic equator
   (b) magnetic poles
   (c) geographic north pole
   (d) latitude 45°
5. A magnetic needle suspended horizontally by an unspun silk fibre oscillates in the horizontal plane, because of the restoring force originating mainly from
   (a) the torsion of the silk fibre
   (b) the horizontal component of earth's magnetic field
   (c) the force of gravity
   (d) all the above factors
6. At magnetic poles, the angle of dip is
   (a) 45° (b) 30°
   (c) zero (d) 90°
7. Magnetic meridian is
   (a) a point
   (b) a line along north-south
   (c) a horizontal plane
   (d) a vertical plane
8. A compass needle which is allowed to move in a horizontal plane is taken to a geomagnetic pole. It
   (a) will stay in north-south direction only
   (b) will stay in east-west direction only
   (c) will become rigid showing no movement
   (d) will stay in any position
9. A dip circle is taken to geomagnetic equator. The needle is allowed to move in a vertical plane perpendicular to the magnetic meridian. The needle will stay
   (a) in horizontal direction only
   (b) in vertical direction only
   (c) in any direction except vertical and horizontal
   (d) in any direction it is released
10. Which of the following is not a magnetic substance?
    (a) brass (b) iron
    (c) cobalt (d) nickel
11. Metals getting magnetised by orientation of atomic magnetic moments in external magnetic field are called
    (a) diamagnetics (b) paramagnetics
    (c) ferromagnetics (d) antimagnetics
12. The magnetic susceptibility for diamagnetic materials is
    (a) small and negative (b) small and positive
    (c) large and positive (d) large and negative
13. The ratio of intensity of magnetisation and magnetising field is called
    (a) permeability
    (b) magnetic intensity
    (c) diamagnetics
    (d) magnetic susceptibility
14. The meniscus of a liquid contained in one of the limbs of a narrow U-tube is held in an electromagnet with the meniscus in line with the field. The liquid is seen to rise. This indicates that the liquid is
    (a) ferromagnetic (b) paramagnetic
    (c) diamagnetic (d) non-magnetic
15. Which of the following instruments is used to measure magnetic field
    (a) thermometer (b) pyrometer
    (c) hygrometer (d) flux meter
16. If a diamagnetic solution is poured into a U-tube and one arm of this U-tube is placed between the poles of a strong magnet, with the meniscus in line with the field, then the level of solution will
    (a) rise (b) fall
    (c) oscillate slowly (d) remain as such
17. A magnetic needle suspended by a silk thread is vibrating in the earth's magnetic field. If the temperature of the needle is increased by 700°C, then
    (a) time period decreases
    (b) time period increases
    (c) time period remains unchanged
    (d) the needle stops vibrating
18. Susceptibility is positive and large for a
    (a) paramagnetic substance
    (b) ferromagnetic substance
    (c) diamagnetic substance
    (d) non-magnetic substance
19. A ferromagnetic material is heated above its curie temperature. Which one is a correct statement
    (a) ferromagnetic domains are perfectly arranged
    (b) ferromagnetic domains become random
    (c) ferromagnetic domains are not influenced
    (d) ferromagnetic material changes into diamagnetic material

20. The direction of magnetic lines of force of a bar magnet is
    (a) from south to north pole
    (b) from north to south pole
    (c) across the bar magnet
    (d) from south to north pole inside the magnet and from north to south pole outside the magnet
21. Which magnetic materials have negative susceptibility?
    (a) diamagnetic materials
    (b) paramagnetic materials
    (c) ferromagnetic materials
    (d) all the above
22. Of the following Fig., the lines of magnetic induction due to a magnet SN, are given by

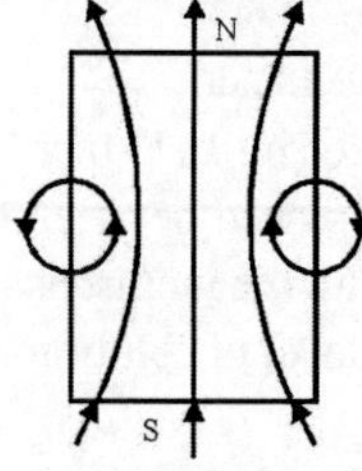

(1)

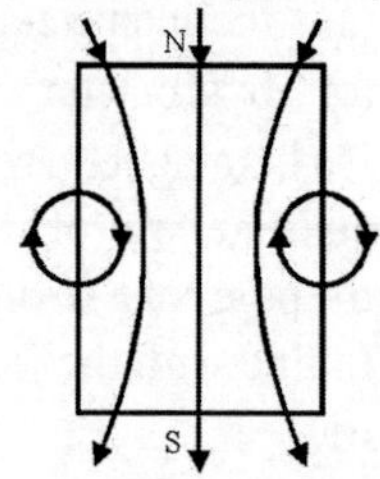

(2)

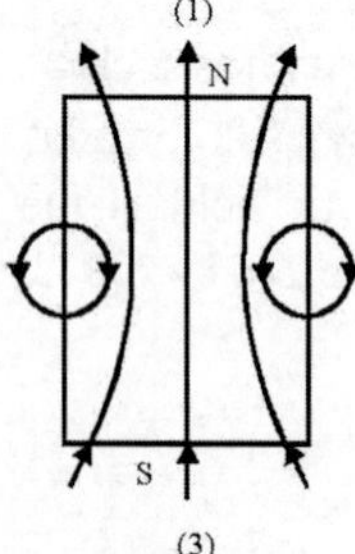

(3)

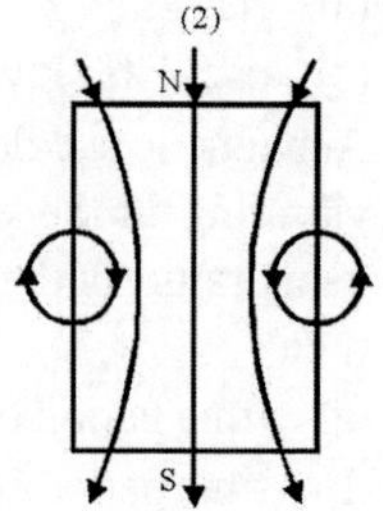

(4)

    (a) 1 (b) 2
    (c) 3 (d) 4
23. If a diamagnetic substance is brought near north or south pole of a bar magnet, it is
    (a) attracted by poles
    (b) repelled by poles
    (c) replaced by north pole and attracted by south pole
    (d) attracted by north pole and repelled by south pole
24. The magnetic lines of force inside a bar magnet
    (a) are from N-pole to S-pole of magnet
    (b) do not exist
    (c) depend upon the area of cross section of bar magnet
    (d) are from S-pole to N-pole of magnet
25. A bar magnet of magnetic moment M, is placed in magnetic field of induction B. The torque exerted on it is
    (a) $\vec{M}.\vec{B}$ (b) $-\vec{M}.\vec{B}$
    (c) $\vec{M}\times\vec{B}$ (d) $-\vec{B}\times\vec{M}$
26. The dimensional formula for magnetic flux is :
    (a) $ML^2T^{-2}A^{-1}$ (b) $ML^3T^{-2}A^{-2}$
    (c) $M^0L^{-2}T^{+2}A^{-2}$ (d) $ML^2T^{-1}A^2$
27. Current I is flowing in a coil of area A & number of turns is N, then magnetic moment of the coil is M equal to
    (a) NIA (b) NI/A
    (c) $NI/\sqrt{A}$ (d) N^2AI
28. A bar magnet is cut into two equal halves by a plane parallel to the magnetic axis. Of the following physical quantities the one which remains unchanged is
    (a) pole strength
    (b) magnetic moment
    (c) intensity of magnetisation
    (d) moment of inertia
29. A magnetic needle suspended freely orientates itself
    (a) in a definite direction
    (b) in no direction
    (c) upward
    (d) downward
30. Which of these is correct in regard to a magnet?
    (a) Geometric length = 0.8 times the magnetic length
    (b) Magnetic length = 0.8 times the geometric length
    (c) Magnetic length = Geometric length
    (d) Geometric length = 10/9 of magnetic length
31. Magnetic dipole moment is a vector quantity directed from
    (a) south pole to north pole
    (b) north pole to south pole
    (c) east to west
    (d) west to east
32. The ratio of magnetic fields due to a smaller bar magnet in the end on position to broadside on position is
    (a) 1/4 (b) 1/2
    (c) 1 (d) 2

33. A circular loop carrying a current is replaced by an equivalent magnetic dipole. A point on the axis of the loop is in
(a) end-on position
(b) broadside-on position
(c) both
(d) none

34. When a current in a circular loop is equivalently replaced by a magnetic dipole
(a) the pole strength *m* of each pole is fixed
(b) the distance *d* between the poles is fixed
(c) the product *md* is fixed
(d) none of the above

35. The B – H curve (i) and (ii) shown in fig. associated with

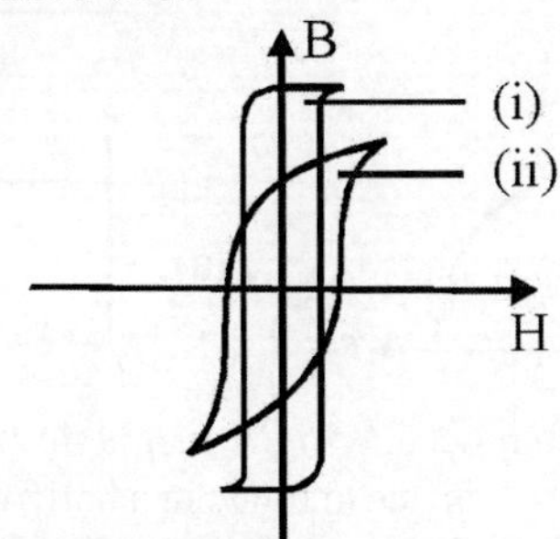

(a) (i) diamagnetic and (ii) paramagnetic substance
(b) (i) paramagnetic and (ii) ferromagnetic substance
(c) (i) soft iron and (ii) steel respectively
(d) (i) steel and (ii) soft iron respectively

36. A material is placed in a magnetic field and it is thrown out of it. Then the material is
(a) paramagnetic (b) diamagnetic
(c) ferromagnetic (d) non-magnetic

37. Among which the magnetic susceptibility does not depend on the temperature?
(a) diamagnetism (b) paramagnetism
(c) ferromagnetism (d) ferrite

38. If μ_0 is absolute permeability of vacuum and μ_r is relative magnetic permeability of another medium, then permeability μ of the medium is
(a) $\mu_0 \mu_r$ (b) μ_0/μ_r
(c) μ_r/μ_0 (d) $1/\mu_0 \mu_r$

39. The relation between *B*, *H* and *I* in S.I. units is
(a) $B = \mu_0(H + I)$ (b) $B = H + 4\pi I$
(c) $H = \mu_0(B + I)$ (d) none of these

40. A temporary magnet is made of
(a) cast iron (b) steel
(c) soft iron (d) stainless steel

41. Hysteresis is the phenomenon of lagging of
(a) *I* behind *B* (b) *B* behind *I*
(c) *I* and *B* behind *H* (d) *H* behind *I*

42. The narrowest hysteresis loop is for
(a) cobalt steel (b) alnico
(c) perm alloy (d) stainless steel

43. The hysteresis curve is studied generally for
(a) ferromagnetic materials
(b) paramagnetic materials
(c) diamagnetic materials
(d) all of these

44. In a cassette player, materials used for coating magnetic tapes are
(a) cobalt (b) $CoFe_2O_4$
(c) $NiFe_2O_4$ (d) Nickel

45. Of dia, para and ferromagnetism, the universal property of all substances is
(a) diamagnetism (b) paramagnetism
(c) ferromagnetism (d) all the above

46. The hysteresis cycle for the material of a transformer core is
(a) short and wide (b) tall and narrow
(c) tall and wide (d) short and narrow

47. The hysteresis cycle for the material of permanent magnet is
(a) short and wide (b) tall and narrow
(c) tall and wide (d) short and narrow

48. A diamagnetic material in a magnetic field moves
(a) perpendicular to field
(b) from stronger to weaker parts of field
(c) from weaker to stronger parts of the field
(d) none of the above directions

49. Curie temperature is the temperature above which
(a) a ferromagnetic material becomes paramagenetic
(b) a paramagnetic material becomes diamagnetic
(c) a ferromagnetic material becomes diamagnetic
(d) a paramagnetic material becomes ferromagnetic

50. A bar magnet is oscillating in the earth's magnetic field with a period *T*. What happens to its period of motion, if its mass is quadrupled
(a) motion remains simple harmonic with new period = T/2
(b) motion remains simple harmonic with new period = 2 T
(c) motion remains simple harmonic with new period = 4T
(d) motion remains simple harmonic and the period stays nearly constant

51. The work done in turning a magnet of magnetic moment M by an angle of 90° from the meridian, is n times the corresponding work done to turn it through an angle of 60°. The value of n is given by
(a) 2 (b) 1
(c) 0.5 (d) 0.25

52. For protecting a sensitive equipment from the external electric arc, it should be
(a) wrapped with insulation around it when a current is passing through it
(b) placed inside an iron can
(c) surrounded with fine copper sheet
(d) placed inside an aluminium can

53. The net magnetic moment of two identical magnets each of magnetic moment M_0, inclined at 60° with each other is

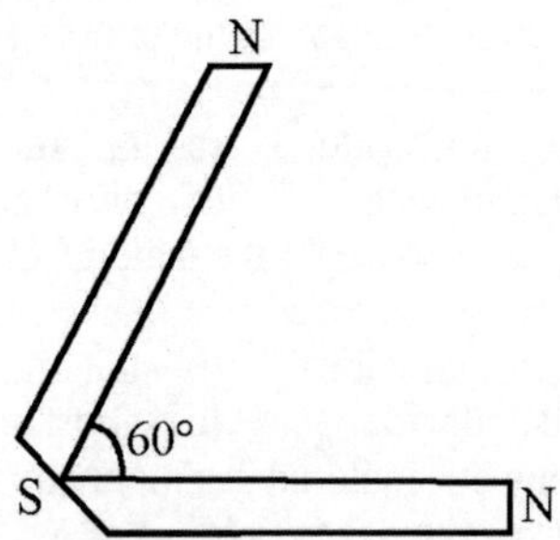

(a) M_0 (b) $\sqrt{2}\,M_0$
(c) $\sqrt{3}\,M_0$ (d) $2M_0$

54. If the magnetic dipole moment of an atom of diamagnetic material, paramagnetic material and ferromagnetic material are denoted by μ_d, μ_p and μ_f respectively, then
(a) $\mu_d = 0$ and $\mu_p \neq 0$ (b) $\mu_d \neq 0$ and $\mu_p = 0$
(c) $\mu_p = 0$ and $\mu_f \neq 0$ (d) $\mu_d \neq 0$ and $\mu_f \neq 0$

55. Above Curie temperature
(a) a paramagnetic substance becomes diamagnetic
(b) a diamagnetic substance becomes paramagnetic
(c) a paramagnetic substance becomes ferromagnetic
(d) a ferromagnetic substance becomes paramagnetic

56. Nickel shows ferromagnetic property at room temperature. If the temperature is increased beyond Curie temperature, then it will show
(a) anti ferromagnetism
(b) no magnetic property
(c) diamagnetism
(d) paramagnetism

57. A curve between magnetic moment and temperature of magnet is

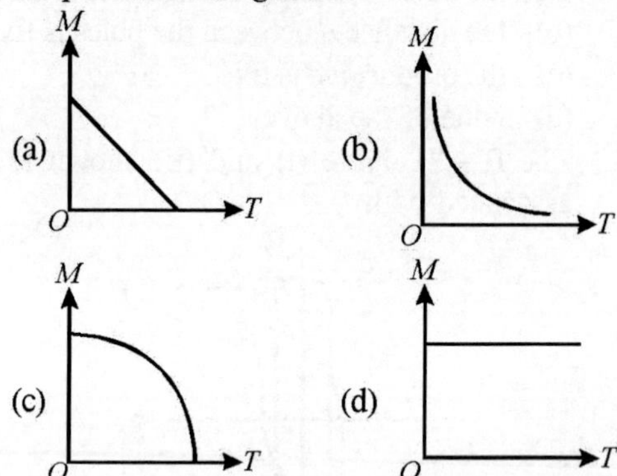

58. A bar magnet having a magnetic moment of $2 \times 10^4\ JT^{-1}$ is free to rotate in a horizontal plane. A horizontal magnetic field $B = 6 \times 10^{-4}\ T$ exists in the space. The work done in taking the magnet slowly from a direction parallel to the field to a direction 60° from the field is
(a) 12 J (b) 6 J
(c) 2 J (d) 0.6 J

59. A steel wire of length ℓ has a magnetic moment M. It is bent in L-shape (Figure).

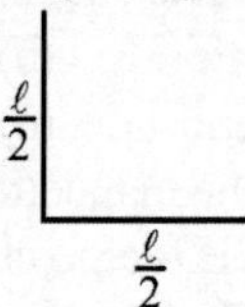

The new magnetic moment is
(a) M (b) $\frac{M}{\sqrt{2}}$
(c) $\frac{M}{2}$ (d) $2M$

60. A vibration magnetometer placed in magnetic meridian has a small bar magnet. The magnet executes oscillations with a time period of 2 sec in earth's horizontal magnetic field of 24 microtesla. When a horizontal field of 18 microtesla is produced opposite to the earth's field by placing a current carrying wire, the new time period of magnet will be

(a) 1 s (b) 2 s
(c) 3 s (d) 4 s

61. Electromagnets are made of soft iron because soft iron has
(a) low retentivity and high coercive force
(b) high retentivity and high coercive force
(c) low retentivity and low coercive force
(d) high retentivity and low coercive force

62. Two identical bar magnets are fixed with their centres at a distance d apart. A stationary charge Q is placed at P in between the gap of the two magnets at a distance D from the centre O as shown in the Figure

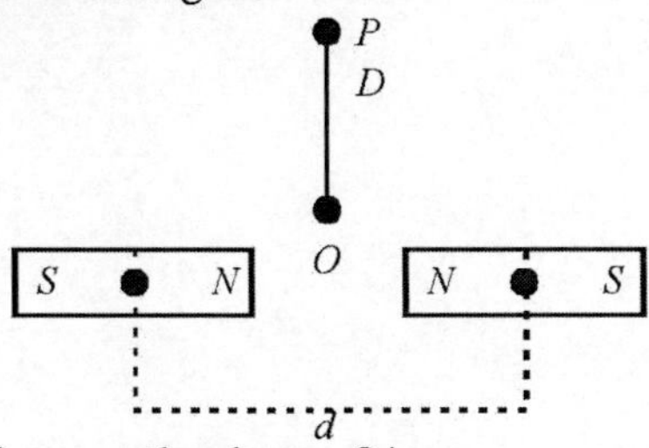

The force on the charge Q is
(a) directed perpendicular to the plane of paper
(b) zero
(c) directed along OP
(d) directed along PO

63. The magnetic moment of a diamagnetic atom is
(a) equal to zero
(b) much greater than one
(c) 1
(d) between zero and one

64. There are four light weigh rod samples A, B, C, D separately suspended by threads. A bar magnet is slowly brought near each sample and the following observations are noted
(i) A is feebly repelled
(ii) B is feebly attracted
(iii) C is strongly attracted
(iv) D remains unaffected
Which one of the following is true?
(a) B is of a paramagnetic material
(b) C is of a diamagnetic material
(c) D is of a ferromagnetic material
(d) A is of a nonmagnetic material

65. A short bar magnet of magnetic moment $0.4 \mathrm{J\,T^{-1}}$ is placed in a uniform magnetic field of 0.16 T. The magnet is in stable equilibrium when the potential energy is
(a) -0.64 J (b) zero
(c) -0.082 J (d) 0.064 J

66. A point near equator have
(a) $H >> V$ (b) $V = H \neq 0$
(c) $V >> H$ (d) $V = H$

67. The universal property of all substances is
(a) diamagnetism (b) ferromagnetism
(c) paramagnetism (d) all of these

68. Two magnets held together in earth's magnetic field, when same polarity together causes 12 vib /min and when opposite poles 4 vib /min. The ratio of magnetic moments is
(a) $9:1$ (b) $1:3$
(c) $1:9$ (d) $10:8$

69. A bar magnet passes through a wire coil as shown in the figure. The acceleration of magnet is

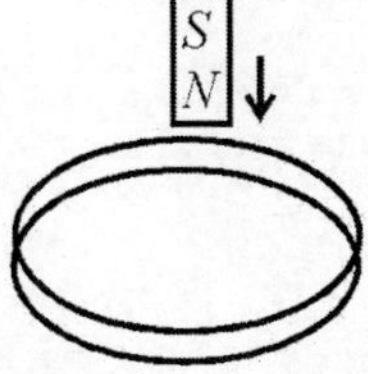

(a) less than g (b) equal to g
(c) more than g (d) none of the above

70. To shield an instrument from external magnetic field, it is placed inside a cabin made from
(a) wood
(b) ebonite
(c) iron
(d) diamagnetic substance

71. A hydrogen atom is paramagnetic. A hydrogen molecule is
(a) diamagnetic (b) paramagnetic
(c) ferromagnetic (d) none of these

72. Lines of force of bar magnet are
(a) parallel and straight
(b) concentric circle
(c) curved lines
(d) curved closed loops.

73. If r be the distance of a point on the axis of a bar magnet from its centre, the magnetic field at this point is proportional to
(a) $\frac{1}{r}$ (b) $\frac{1}{r^2}$
(c) $\frac{1}{r^3}$ (d) $\frac{1}{r^4}$

74. At a certain place, the angle of dip is 30° and the horizontal component of earth's magnetic field is 0.50 oersted. The earth's total magnetic field (in oersted) is

(a) $\sqrt{3}$ (b) 1

(c) $1/\sqrt{3}$ (d) $\frac{1}{2}$

75. Torques τ_1 and τ_2 are required for a magnetic needle to remain perpendicular to the magnetic field B_1 and B_2 at two different places. The ratio B_1 / B_2 is

(a) $\frac{\tau_2}{\tau_1}$ (b) $\frac{\tau_1}{\tau_2}$

(c) $\frac{\tau_1 + \tau_2}{\tau_1 - \tau_2}$ (d) $\frac{\tau_1 - \tau_2}{\tau_1 + \tau_2}$

CHAPTER 20

Electromagnetic Induction

CONCEPTS

The phenomenon of production of e.m.f. in a conductor due to change in magnetic flux linked with it is called electromagnetic induction.

MAGNETIC FLUX

The magnetic flux through a surface of area A placed in a uniform magnetic field $\vec{B}$ is defined as,

$\phi_B = \vec{B}.\vec{A} = BA\cos\theta$, where θ is the angle between $\vec{B}$ and $\vec{A}$.

The S.I. unit of magnetic flux is weber (wb).

FARADAY'S LAWS OF ELECTROMAGNETIC INDUCTION

Faraday gave two laws of electromagnetic induction.

First law : Whenever there is change in the magnetic flux associated with a circuit, an e.m.f. is induced in the circuit. This is also known as **Neumann's law**

Second law : The magnitude of the induced e.m.f. (e) is directly proportional to the time rate of change of the magnetic flux through the circuit.

i.e., $e \propto \dfrac{\Delta\phi}{\Delta t}$ or, $e = k\dfrac{\Delta\phi}{\Delta t}$

In the **S.I. system**, emf 'e' is measured in volt and $\dfrac{d\phi}{dt}$ in Wb/sec.

If the coil has N turns, then the emf will be induced in each turn and the emf's of all the turns will be added up. If the turns of coil are very close to each other, the magnetic flux passing through each turn will be same.

So, the induced emf in the whole coil $e = N\dfrac{\Delta\phi}{\Delta t}$

$N\phi$ = number of 'flux linkages' in the coil.

LENZ'S LAW

It states that the polarity of the induced emf is such that it tends to produce a current which opposes the change in magnetic flux that produces it. The negative sign in the expression for Faraday's law indicates this fact.

When a metal rod of length ℓ is placed normal to a uniform magnetic field B and moved with a velocity v perpendicular to the field, the induced emf (called motional emf) across its ends is $\varepsilon = B\ell v$

When the conductor is not straight, moving with uniform velocity $\vec{v}$ in a uniform magnetic field $\vec{B}$. If length MN = ℓ, then the induced emf between the ends is $e = \vec{\ell}.(\vec{v}\times\vec{B})$

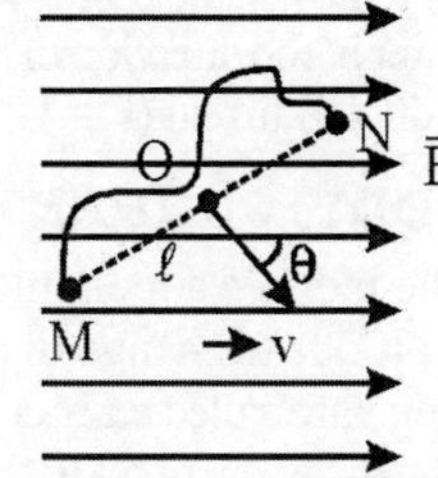

Motional EMF : In a conducting rod moving perpendicular through a uniform magnetic field as shown.

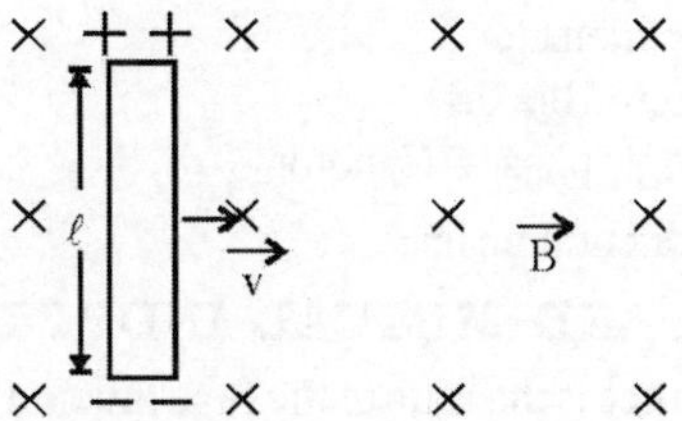

The induced emf produced across the rod

$$e = B\ell v = \int_0^{\ell}(\vec{v}\times\vec{B}).d\ell$$

This is also called **motional emf** and it develops when a metal rod cuts magnetic lines of force.

Special Case : If the rod moves in the magnetic field making an angle θ with it, then induced emf $e = B_n v\ell = Bv\ell\sin\theta$.

Motional EMF in a rod rotating about its one end with constant angular velocity ω in uniform magnetic field

$$e = \frac{1}{2}B\omega\ell^2$$

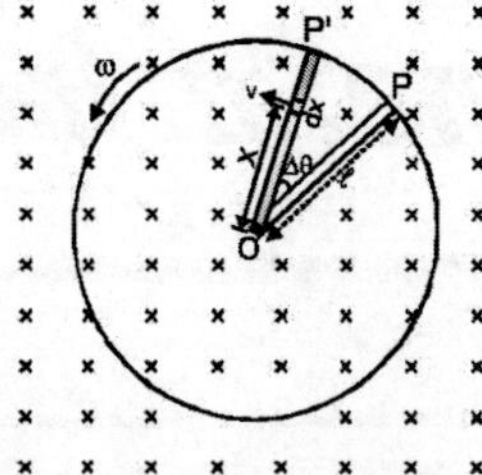

FLEMING'S RIGHT HAND RULE

"If we stretch the right-hand thumb and two nearby fingers perpendicular to one another, and the first finger points in the direction of magnetic field and the thumb in the direction of motion of the conductor, then the middle finger will point in the direction of the induced current (Fig.)."

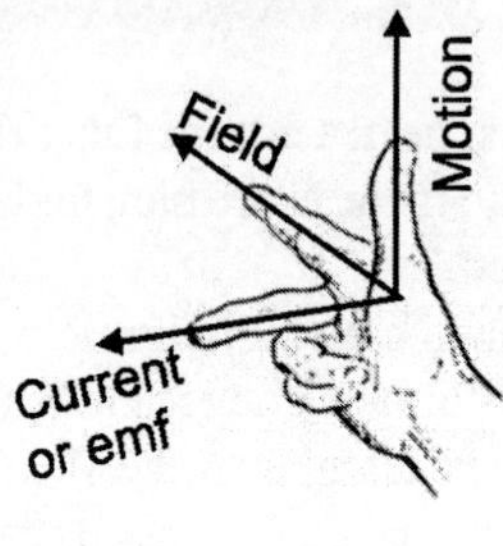

EDDY CURRENTS

The induced circulating current produced in a metal itself due to change in magnetic flux linked with the metal are called eddy current.

The direction of eddy currents is given by Lenz's law.

Applications of Eddy Currents

Eddy currents are useful in :

(1) Dead beat galvanometer.
(2) Energy meter.
(3) Speedometer.
(4) Electric brakes.
(5) Single phase AC motor.
(6) Induction furnace.

SELF AND MUTUAL INDUCTANCE

Inductance is the ratio of the flux-linkage to current. It is equal to $N\phi/L$.

When a current in a coil changes, it induces a back emf in the same coil.

The **self-induced emf** is given by, $\varepsilon = -L\frac{dI}{dt}$

L is the **self-inductance** of the coil. It is a measure of the inertia of the coil against the change of current through it.

The self-inductance of a long solenoid, the core of which consists of a magnetic material of permeability μ_r, is given by $L = \mu_r\,\mu_0\,n^2 A\ell$,

where A is the area of cross-section of the solenoid, ℓ its length and n the number of turns per unit length.

Self inductance of a plane coil

$$L = \frac{\mu_0\,\mu_r\,\pi N^2 R}{2}$$

Self inductance of a toroid, $L = \frac{\mu_0 N^2 R}{2}$

Self inductance of a square coil,

$$L = \frac{2\sqrt{2}\mu_0 N^2 a}{\pi} \quad \text{(a is side of square)}$$

Self inductance of a coaxial cylinders,

$$L = \frac{\mu_0}{2\pi R}\log_e\left(\frac{R_2}{R_1}\right)\left[R = \frac{R_1 + R_2}{2}\right]$$

Coefficient of self inductance of two coils in series

If M is the coefficient of mutual inductance between the two coils when they have flux linkage in same sense, then

$$L = L_1 + L_2 + 2M$$

For flux linkage in opposite sense $L = L_1 + L_2 - 2M$

When the coils are separated far then $L = L_1 + L_2$

Coefficient of self inductance of two coils in parallel

$$L = \frac{L_1 L_2}{L_1 + L_2}$$

When the coils are close together, $L = \frac{L_1 L_2 - M^2}{L_1 + L_2 \pm 2M}$

The coefficient of coupling between two coils having self inductances L_1 and L_2 and coefficient of mutual inductance M is $K = \frac{\pm M}{\sqrt{L_1 L_2}}$ $L ; 0 \le K \le 1$.

Generally, the value of K is less than 1.

If K is 1, then the coupling of two coils is tight while if $K < 1$, then coupling is loose.

Energy stored in a self inductance coil

$$U = \frac{L i_0^2}{2}$$

A changing current in a coil (coil 2) can induce an emf in a nearby coil (coil 1). This relation is given by,

$\varepsilon_1 = -M_{12}\frac{dI_2}{dt}$, The quantity M_{12} is called **mutual inductance** of coil 1 with respect to coil 2. One can similarly define M_{21}. There exists a general equality, $M_{12} = M_{21}$.

Mutual inductance between two plane coils

$$M = \frac{\mu_r \mu_0 N_1 N_2 \pi R_1^2}{2R_2}$$

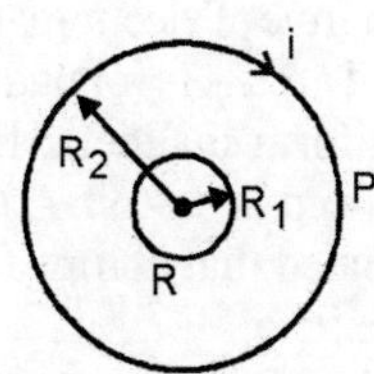

For a solenoid having a primary coil of N_1 turns and a secondary coil of N_2 turns, the coefficient of mutual inductance is given by

$$M = \frac{\mu_r \mu_0 N_1 N_2 A}{\ell}$$

where ℓ is the length of solenoid and A is the area of one turn of the secondary.

For two concentric coplanar square coils, the coefficient of mutual inductance is given by

$$M = \frac{\mu_0 \mu_r 2\sqrt{2} N_1 N_2 L_1^2}{\pi L_2}$$

L_1 and L_2 are the sides of the smaller and bigger square respectively.

GROWTH AND DECAY OF CURRENT IN LR CIRCUIT

Growth of Current in LR Circuit :

$$I = I_0 (1 - e^{-\frac{R}{L}t})$$

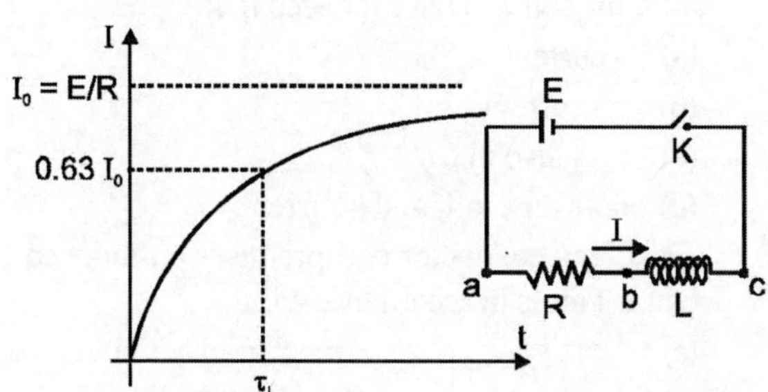

$\frac{L}{R}$ has dimensions of time. It is called inductive time constant (τ) of LR-circuit.

At $t = \frac{L}{R}$; $\quad I = I_0(1 - e^{-\frac{R}{L}\cdot\frac{L}{R}}) = I_0(1 - e^{-1})$

$$= I_0\left(\frac{e-1}{e}\right) = I_0\left(\frac{2.71-1}{2.71}\right) = 0.632\, I_0$$

The inductive time constant of an LR-circuit is the time in which the current grows from zero to 0.632 (or 63.2%) of its maximum value. When $t \to \infty$,

$$I = I_0(1 - e^{-\frac{R}{L}.\infty}) = I_0(1 - e^{-\infty}) = I_0(1-0) = I_0$$

Decay of Current in LR Circuit

$$I = I_0\, e^{-\frac{R}{L}t}$$

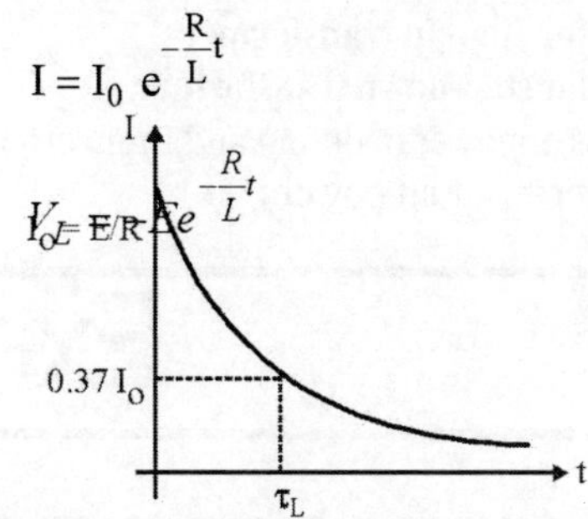

LC OSCILLATIONS

If a charged capacitor C is short-circuited through an inductor L, the charge and current in the circuit start oscillating simple harmonically. If the resistance of the circuit is zero, no energy is dissipated as heat.

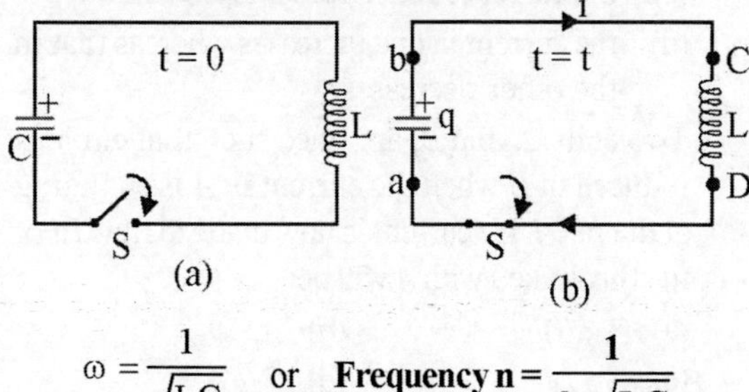

$$\omega = \frac{1}{\sqrt{LC}} \quad \text{or} \quad \textbf{Frequency } n = \frac{1}{2\pi\sqrt{LC}}$$

In L-C oscillations, q, i and $\frac{di}{dt}$ all oscillate harmonically with same angular frequency ω. But the phase difference between q and i or between i and $\frac{di}{dt}$ is $\pi/2$.

Their amplitudes are q_0, $q_0\omega$ and $\omega^2 q_0$ respectively.

So, $q = q_0 \cos\omega t$, then $i = \frac{dq}{dt} = -q_0\omega \sin\omega t$

$$\frac{di}{dt} = -q_0\omega^2 \cos\omega t$$

Similarly, potential energy across capacitor (U_C) and across inductor (U_L) also oscillate with double the frequency 2ω.

TRANSFORMER AND AC GENERATOR

It works on the phenomenon of mutual induction. It is a device for transforming a low alternating voltage of high current into a high alternating voltage of lower current and vice versa, without increasing power or changing frequency.

Transformation ratio of the transformer (K)

$$= \frac{\text{Number of turns in secondary } (N_s)}{\text{Number of turns in primary } (N_p)}$$

$K > 1$, **for step-up transformer**

$K < 1$, **for step-down transformer.**

If there is no loss of power in output and input circuits then, input power = output power

$$E_p \times I_p = E_s \times I_s \text{ or } \frac{I_p}{I_s} = \frac{E_s}{E_p} = \frac{N_s}{N_p}$$

In an **ac generator**, mechanical energy is converted to electrical energy by virtue of electromagnetic induction. If coil of N turns and area A is rotated at ν revolutions per second in a uniform magnetic field B, then the motional emf produced is $\varepsilon = NBA\,(2\pi)\sin(2\pi\nu t)$, where we have assumed that at time t = 0 s, the coil is perpendicular to the field.

Exercise

1. Two identical coaxial circular loops carry a current *i* each circulating in the same direction. If the loops approach each other, you will observe that
   (a) the current in each increases
   (b) the current in each decreases
   (c) the current in each remains the same
   (d) the current in one increases whereas that in the other decreases
2. Two coils, *A* and *B*, are lined such that emf ε is induced in *B* when the current in *A* is changing at the rate *I*. If *i* current is now made to flow in *B*, the flux linked with *A* will be
   (a) $(\varepsilon/I)i$ (b) $\varepsilon\, iI$
   (c) $(\varepsilon I)i$ (d) iI/ε
3. An induced e.m.f. is produced when a magnet is plunged into a coil. The strength of the induced e.m.f. is independent of
   (a) the strength of the magnet
   (b) number of turns of coil
   (c) the resistivity of the wire of the coil
   (d) speed with which the magnet is moved
4. Whenever the magnetic flux linked with a coil changes, an induced e.m.f. is produced in the circuit. The e.m.f. lasts
   (a) for a short time
   (b) for a long time
   (c) for ever
   (d) so long as the change in flux takes place
5. A magnet is moved towards a coil (i) quickly (ii) slowly, then the induced e.m.f. is
   (a) larger in case (i)
   (b) smaller in case (i)
   (c) equal to both the cases
   (d) larger or smaller depending upon the radius of the coil
6. Lenz's law is a consequence of the law of conservation of
   (a) charge (b) mass
   (c) energy (d) momentum
7. The laws of electromagnetic induction have been used in the construction of a
   (a) galvanometer (b) voltmeter
   (c) electric motor (d) generator
8. A cylindrical bar magnet is kept along the axis of a circular coil. On rotating the magnet about its axis, the coil will have induced in it
   (a) a current
   (b) no current
   (c) only an e.m.f.
   (d) both an e.m.f. and a current
9. A moving conductor coil produces an induced e.m.f. This is in accordance with
   (a) Lenz's law (b) Faraday's law
   (c) Coulomb's law (d) Ampere's law
10. A coil of insulated wire is connected to a battery. If it is taken to galvanometer, its pointer is deflected, because
    (a) induced current is produced
    (b) the coil acts like a magnet
    (c) the number of turns in the coil of the galvanometer are changed
    (d) none of the above

11. A conducting rod AB moves parallel to X-axis in a uniform magnetic field, pointing in the positive X-direction. The end A of the rod gets

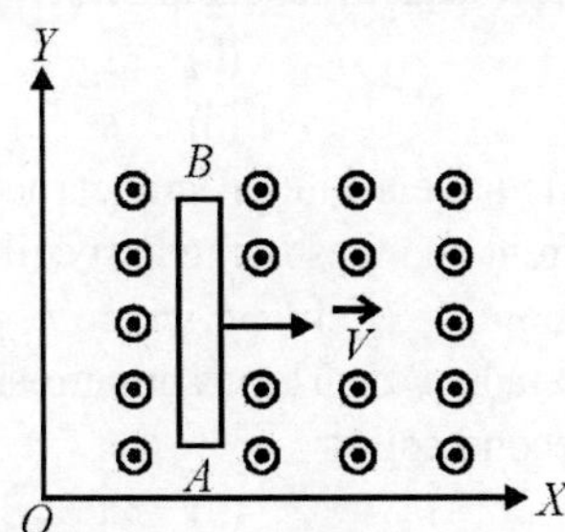

(a) positively charged
(b) negatively charged
(c) neutral
(d) first positively charged and then negatively charged

12. A metal ring is held horizontally and bar magnet is dropped through the ring with its length along the axis of the ring. The acceleration of the falling magnet is
(a) equal to g
(b) less than g
(c) more than g
(d) depends on the diameter of ring and length of magnet

13. Whenever, current is changed in a coil, an induced e.m.f. is produced in the same coil. This property of the coil is due to
(a) mutual induction (b) self induction
(c) eddy currents (d) hysteresis

14. When current i passes through an inductor of self inductance L, energy stored in it is $1/2. Li^2$. This is stored in the
(a) current (b) voltage
(c) magnetic field (d) electric field

15. A series combination of L and R is connected to a battery of E.M.F., E having negligible internal resistance. The final value of current depends upon
(a) L and R (b) E and L
(c) E and R (d) L, R and E

16. Two pure inductors, each of self inductance L are connected in parallel but are well separated from each other, then the total inductance is
(a) L (b) $2L$
(c) $L/2$ (d) $L/4$

17. Two coils of self inductances L_1 and L_2 are placed so close together that effective flux in one coil is completely linked with the other. If M is the mutual inductance between them, then
(a) $M = L_1 L_2$ (b) $M = L_1/L_2$
(c) $M = (L_1 L_2)^2$ (d) $M = \sqrt{(L_1 L_2)}$

18. A coil of resistance R and inductance L is connected to a battery of e.m.f. E volt. The final current in the coil is
(a) E/R (b) E/L
(c) $\sqrt{[E/(R^2 + L^2)]}$ (d) $\sqrt{[E\,L/(R^2 + L^2)]}$

19. An inductor may store energy in
(a) its electric field
(b) its coils
(c) its magnetic field
(d) both in electric and magnetic fields

20. Which of the following units denotes the dimension $\frac{ML^2}{Q^2}$, where Q denotes the electric charge?
(a) Wb/m^2 (b) Henry (H)
(c) H/m^2 (d) Weber (Wb)

21. If N is the number of turns is a coil, the value of self inductance varies as
(a) N^0 (b) N
(c) N^2 (d) N^{-2}

22. The SI unit of inductance, the henry, can be written as
(a) weber/ampere
(b) volt second/ampere
(c) joule/ampere^2
(d) all of these

23. Two coils are placed close to each other. The mutual inductance of the pair of coils depends upon
(a) relative position and orientation of the two coils.
(b) the materials of the wires of the coils
(c) the currents in the two coils
(d) the rates at which currents are changing in the two coils

24. A conductor of length l is moving perpendicular to a uniform magnetic field of induction $\vec{B}$ and perpendicular to its length with a uniform velocity $\vec{v}$. As a result, an emf $\varepsilon = Blv$ is induced between its ends. Under this condition free electrons in the conductor.

(a) drift from higher potential end to lower potential end
(b) drift from lower potential end to higher potential end
(c) do not drift along its length, since there is no electric field inside the conductor
(d) do not drift along its length, since the force on each electron due to the electric field is balanced by the force produced on it by the magnetic field.

25. A short-circuited coil is placed in a time-varying magnetic field. Electrical power is dissipated due to the current induced in the coil. If the number of turns were to be quadrupled and the wire radius halved, the electrical power dissipated would be
(a) halved (b) the same
(c) doubled (d) quadrupled

26. If rotational velocity of a dynamo armature is doubled, then induced emf will become
(a) unchanged (b) four times
(c) half (d) two times

27. The coil is in a uniform magnetic field, perpendicular to the plane of the coil, as shown.

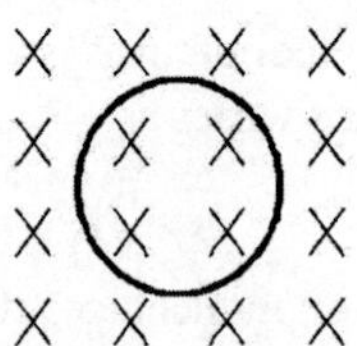

Consider now the following :
I. Rotation of the coil about an axis perpendicular to the coil and passing through its centre
II. Displacement of coil at constant speed towards right
III. Rotation of coil about an axis passing through one of its diameters
In which of these an emf will be induced in the coil :
(a) III only (b) I and III
(c) I and II (d) I, II and III

28. Two coils of self inductances 2 mH and 8 mH are placed so close together that the effective flux in one coil is completely linked with the other. The mutual inductance between these coils is
(a) 6 mH (b) 4 mH
(c) 16 mH (d) 10 mH

29. In an inductor of self-inductance L = 2 mH, current changes with time according to relation $i = t^2e^{-t}$. At what time emf is zero?
(a) 4s (b) 3s
(c) 2s (d) 1s

30. Two coils have a mutual inductance 0.005 H. The current changes in the first coil according to equation $I = I_0 \sin \omega t$, where $I_0 = 10A$ and $\omega = 100\pi$ radian/sec. The maximum value of e.m.f. in the second coil is
(a) 2π (b) 5π
(c) π (d) 4π

31. A varying current in a coil changes from 10 A to zero in 0.5 sec. If the average e.m.f. induced in the coil is 220 V, the self-inductance of the coil is
(a) 5 H (b) 6 H
(c) 11 H (d) 12 H

32. The total charge induced in a conducting loop when it is moved in a magnetic field depend on
(a) the rate of change of magnetic flux
(b) initial magnetic flux only
(c) the total change in magnetic flux
(d) final magnetic flux only

33. The magnetic flux through a circuit of resistance R changes by an amount $\Delta\phi$ in a time Δt. Then the total quantity of electric charge Q that passes any point in the circuit during the time Δt is represented by
(a) $Q = R.\frac{\Delta\phi}{\Delta t}$ (b) $Q = \frac{1}{R}.\frac{\Delta\phi}{\Delta t}$
(c) $Q = R.\frac{\Delta\phi}{\Delta t}$ (d) $Q = \frac{\Delta\phi}{\Delta t}$

34. A coil having an area A_0 is placed in a magnetic field which changes from B_0 to 4 B_0 in time interval t. The e.m.f. induced in the coil will be
(a) $3A_0 B_0/t$ (b) $4A_0 B_0/t$
(c) $3B_0/A_0t$ (d) $4A_0/B_0t$

35. A conducting square loop of side L and resistance R moves in its plane with a uniform velocity v perpendicular to one of its side. A magnetic induction B constant in

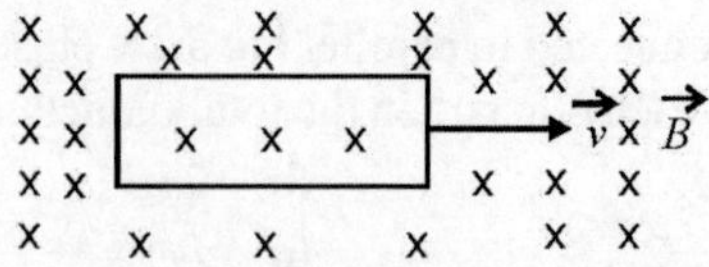

time and space, pointing perpendicular and into the plane of the loop exists everywhere. The current induced in the loop in

(a) $\frac{Blv}{R}$ clockwise

(b) $\frac{Blv}{R}$ anticlockwise

(c) $\frac{2Blv}{R}$ anticlockwise

(d) zero

36. If a current increases from zero to one ampere in 0.1 second in a coil of 5 mH, then the magnitude of the induced e.m.f. will be

(a) 0.005 volt (b) 0.5 volt
(c) 0.05 volt (d) 5 volt

37. A generator has an e.m.f. of 440 Volt and internal resistance of 4000 hm. Its terminals are connected to a load of 4000 ohm. The voltage across the load is

(a) 220 volt (b) 440 volt
(c) 200 volt (d) 400 volt

38. An e.m.f. of 2 volt is produced in a coil when the current changes at a steady rate from 3 to 2 amperes in 1 milli-second. The value of self inductance is

(a) zero (b) 2 mH
(b) 2 H (d) 200 mH

39. A coil has 200 turns and area of 70 cm^2. The magnetic field perpendicular to the plane of the coil is 0.3 Wb/m^2 and take 0.1 sec to rotate through 180°. The value of the induced e.m.f. will be

(a) 8.4 V (b) 84 V
(c) 42 V (d) 4.2 V

40. A thin circular ring of area A is held perpendicular to a uniform magnetic field of induction B. A small cut is made in the ring and a galvanometer is connected across the ends such that the total resistance of the circuit is R. When the ring is suddenly squeezed to zero area, the charge flowing through the galvanometer is

(a) $\frac{BR}{A}$ (b) $\frac{AB}{R}$

(c) ABR (d) $\frac{B^2A}{R^2}$

41. A small bar magnet is being slowly inserted with constant velocity inside a solenoid as shown in figure. Which graph best represents the relationship between emf induced with time

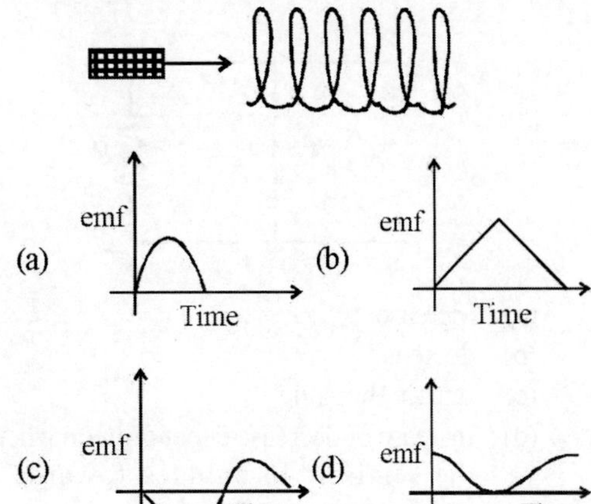

42. An infinitely long cylinder is kept parallel to an uniform magnetic field B directed along positive z axis. The direction of induced current as seen from the z axis will be

(a) zero
(b) anticlockwise of the +ve z axis
(c) clockwise of the +ve z axis
(d) along the magnetic field

43. The armature of a dc motor has 20W resistance. It draws a current of 1.5 A when run by a 220 V dc supply. The value of the back emf induced in it is

(a) 150 V (b) 170 V
(c) 180 V (d) 190 V

44. A small coil of radius r is placed at the centre of a large coil of radius R, where $R >> r$. The two coils are coplanar. The mutual induction between the coils is proportional to

(a) r/R (b) r^2/R
(c) r^2/R^2 (d) r/R^2

45. A coil having n turns and resistance $R\ \Omega$ is connected with a galvanometer of resistance 4R Ω. This combination is moved in time t seconds from a magnetic field W_1 weber to W_2 weber. The induced current in the circuit is

(a) $-\frac{(W_1 - W_2)}{Rnt}$ (b) $-\frac{n(W_2 - W_1)}{5\,Rt}$

(c) $-\frac{(W_2 - W_1)}{5\,Rnt}$ (d) $-\frac{n(W_2 - W_1)}{Rt}$

46. Consider the situation shown. The wire AB is sliding on fixed rails with a constant velocity. If the wire AB is replaced by semi-circular wire, the magnitude of induced e.m.f. will

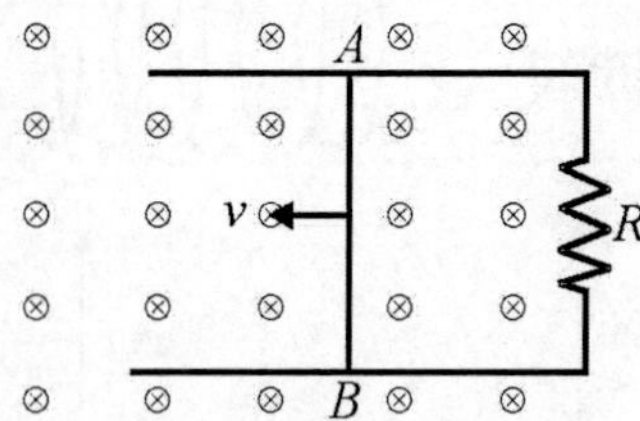

(a) increase
(b) decrease
(c) remain the same
(d) increase or decrease depending on whether the semi-circle buldges towards the resistance or away from it.

47. Fig represents an area A = 0.5 m^2 situated in a uniform magnetic field

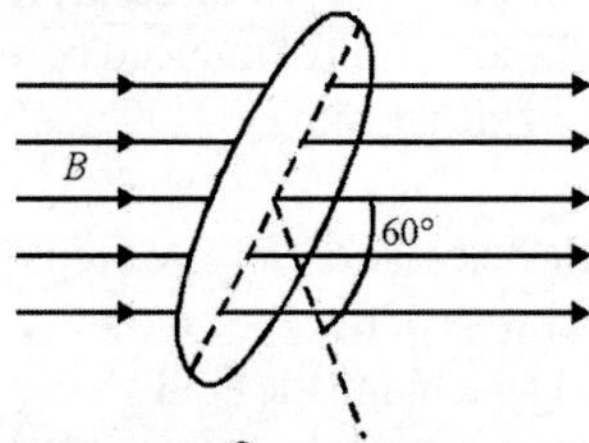

B = 2.0 weber/m^2 and making an angle of 60° with respect to magnetic field. The value of the magnetic flux through the area would be equal to

(a) 2.0 weber (b) $\sqrt{3}$ weber
(c) $\sqrt{3}/2$ weber (d) 0.5 weber

48. In a coil of area 10 cm^2 and 10 turns with magnetic field directed perpendicular to the plane and is changing at the rate of 10^8 Gauss/second. The resistance of the coil is 20Ω. The current in the coil will be
(a) 0.5 A (b) 5 A
(c) 50 A (d) 5×10^8 A

49. The flux linked with a coil at any instant 't' is given by $\phi = 10t^2 - 50t + 250$. The induced emf at t = 3s is
(a) −190 V (b) −10 V
(c) 10 V (d) 190 V

50. A horizontal telegraph wire 0.5 km long running east and west in a part of a circuit whose resistance is 2.5 Ω. The wire falls to g = 10.0 m/s^2 and $B = 2 \times 10^{-5}$ weber/m^2, then the current induced in the circuit is
(a) 0.7 amp (b) 0.04 amp
(c) 0.02 amp (d) 0.01 amp

51. A conducting square loop of side L and resistance R moves in its plane with a uniform velocity v perpendicular to one of its sides. A magnetic induction B constant in time and space, pointing perpendicular and into the plane at the loop exists everywhere with half the loop outside the field, as shown in figure. The induced emf is

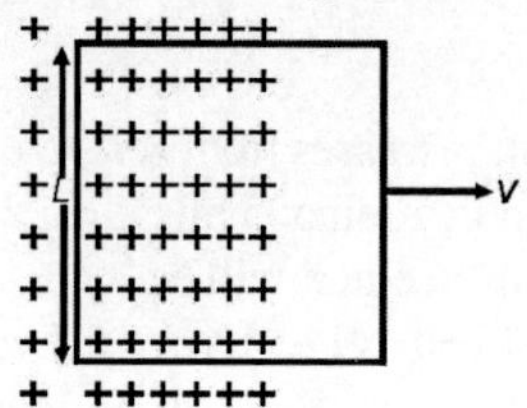

(a) zero (b) RvB
(c) vBL/R (d) vBL

52. If in a transformer the the number of turns of primary coil and secondary coil are 5 and 4 respectively and if 240 V is applied on primary coil then the ratio of current in primary and secondary coil is
(a) 4 : 5 (b) 5 : 4
(c) 5 : 10 (d) 8 : 12

53. The two rails of a railway track, insulated from each other and the ground, are connected to millivoltmeter. What is the reading of the millivoltmeter when a train passes at a speed of 180 km/hr along the track, given that the vertical component of earth's magnetic field is 0.2×10^{-4} wb/m^2 and rails are separated by 1 metre
(a) 10^{-2} volt (b) 10mV
(c) 1 volt (d) 1mV

54. When the current in a coil changes from 2 amp. to 4 amp. in 0.05 sec., an e.m.f. of 8 volt is induced in the coil. The coefficient of self inductance of the coil is
(a) 0.1 henry (b) 0.2 henry
(c) 0.4 henry (d) 0.8 henry

55. In an AC generator, a coil with N turns, all of the same area A and total resistance R, rotates with frequency ω in a magnetic field B. The maximum value of emf generated in the coil is
(a) $N.A.B.R.\omega$ (b) $N.A.B.$
(c) $N.A.B.R.$ (d) $N.A.B.\omega$

56. A copper disc of radius 0.1 m rotated about its centre with 10 revolutions per second in a uniform magnetic field of 0.1 tesla with its plane perpendicular to the field. The e.m.f. induced across the radius of disc is

(a) $\frac{\pi}{10}$ volts (b) $\frac{2\pi}{10}$ volts

(c) $\pi \times 10^{-2}$ volt (d) $2\pi \times 10^{-2}$ volt

57. A long solenoid having 200 turns per cm carries a current of 1.5 amp. At the centre of it is placed a coil of 100 turns of cross-sectional area 3.14×10^{-4} m^2 having its axis parallel to the field produced by the solenoid. When the direction of current in the solenoid is reversed within 0.05 sec, the induced em.f. in the coil is

(a) 0.48 V (b) 0.048 V

(c) 0.0048 V (d) 48 V

58. Eddy currents are produced when

(a) a metal is kept in varying magnetic field

(b) a metal is kept in steady magnetic field

(c) a circular coil is placed in a magnetic field

(d) through a circular coil, current is passed

59. In a uniform magnetic field of induction B, a wire in the form of a semicircle of radius r rotates about the diameter of the circle with an angular frequency ω. The axis of rotation is perpendicular to the field. If the total resistance of the circuit is R, the mean power generated per period of rotation is

(a) $\frac{(B\pi r\omega)^2}{2R}$ (b) $\frac{(B\pi r^2\omega)^2}{2R}$

(c) $\frac{B\pi r^2\omega}{2R}$ (d) $\frac{(B\pi r\omega^2)^2}{8R}$

60. The current in self inductance $L = 40$ mH is to be increased uniformly from 1 amp to 11 amp in 4 milliseconds. The e.m.f. induced in the inductor during the process is

(a) 100 volt (b) 0.4 volt

(c) 4.0 volt (d) 440 volt

CHAPTER 21

Alternating Current

CONCEPTS

ALTERNATING CURRENT

An alternating quantity such as current or voltage is that quantity whose magnitude varies with time between zero and a maximum value and reverses its direction periodically.

The instantaneous value of alternating current is expressed by, $I = I_0 \sin \omega t$

where I_0 = peak value or the amplitude of the alternating current and ωt = phase of alternating current.

The alternating current continuously varies in magnitude and periodically reverses its direction.

The current flowing in only one direction in a circuit is called direct current (D.C.). Batteries, thermocouples and solar cells are some of the sources of direct current.

In case of D.C. frequency $f = 0$

Magnitude of e.m.f. and current does not varies with time in D.C. circuits.

PEAK, AVERAGE AND RMS VALUE OF ALTERNATING CURRENT AND VOLTAGE

An alternating voltage $V = V_m \sin\omega t$ applied to a resistor R drives a current $i = i_m \sin\omega t$ in the resistor, $i_m = \frac{V_m}{R}$.

The current is in phase with the applied voltage.

For an alternating current $i = i_m \sin \omega t$ passing through a resistor R, the average power loss P (averaged over a cycle) due to joule heating is $(1/2)i^2_m R$. To express it in the same form as the dc power ($P = I^2R$), a special value of current is used. It is called **root mean square (rms) current** and is denoted by $I_{rms} = \frac{i_0}{\sqrt{2}} = 0.707\, i_0$

where I_0 = peak value of current

The mean or average value of alternating e.m.f. over positive half cycle (0 to T/2) is $E_m = \frac{2E_0}{\pi} = 0.637\, E_0$ where E_0 = peak value of voltage

The **average power loss** over a complete cycle is given by $P = VI\cos\phi$. The term $\cos\phi$ is called the **power factor**.

When a value is given for ac voltage or current, it is ordinarily the rms value.

The usual domestic power supply "220 - volt ac" has an rms voltage of 220 V. The voltage amplitude or peak value is $E_m = \sqrt{2}\, E_{rms} = \sqrt{2}(220V) = 311V$

REACTANCE AND IMPEDANCE

An A.C. voltage $V = V_m \sin\omega t$ applied to a pure inductor L, drives a current in the inductor $i = i_m \sin(\omega t - \pi/2)$, where

$i_m = V_m / X_L$. $X_L = \omega L$ is called **inductive reactance**.

The current in the inductor lags the voltage by $\pi/2$. The average power supplied to an inductor over one complete cycle is zero.

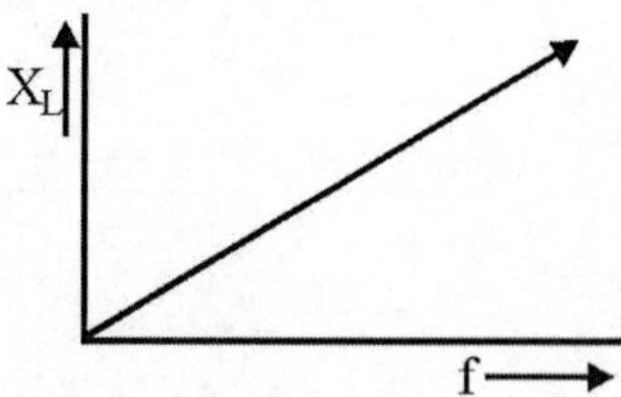

Inductor offers no opposition to the flow of d.c. or whereas a resistive path to a.c.

An ac voltage $V = V_m \sin\omega t$ applied to a capacitor drives a current in the capacitor: $i = i_m \sin(\omega t + \pi/2)$. Here,

$$i_m = \frac{V_m}{X_C},\ X_C = \frac{1}{\omega C}$$

is called **capacitive reactance X_C**.

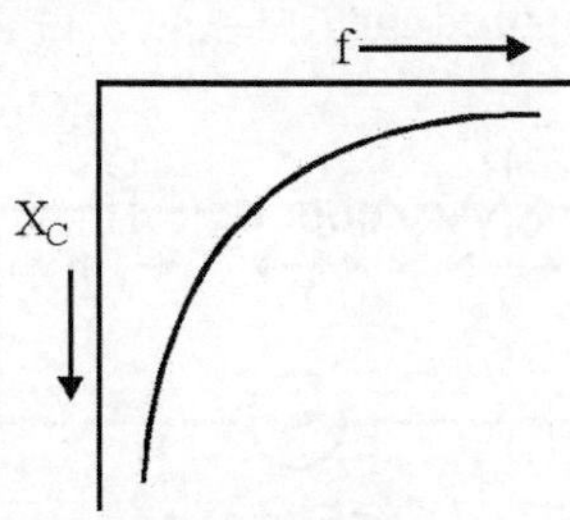

The current through the capacitor is $\pi/2$ ahead of the applied voltage. As in the case of inductor, the average power supplied to a capacitor over one complete cycle is zero.

In a high frequency A.C. circuit, the capacitor acts like a conductor.

In an ac circuit, while adding voltages across different elements, one should take care of their phases properly. For example, if V_R and V_C are voltages across R and C, respectively in an RC circuit, then the total voltage across

RC combination is $V_{RC} = \sqrt{V_R^2 + V_C^2}$ and not $V_R + V_C$ since V_C is $\pi/2$ out of phase of V_R.

For a series RLC circuit driven by voltage $V = V_m \sin\omega t$, the current is given by

$$i = i_m \sin(\omega t + \phi), \text{ where } i_m = \frac{V_m}{\sqrt{R^2 + (X_C - X_L)^2}}$$

and $\phi = \tan^{-1}\dfrac{X_C - X_L}{R}$

L C R

$Z = \sqrt{R^2 + (X_C - X_L)^2}$

is called the **impedance** of the circuit.

A.C. source

$E = E_0 \sin\omega t$

The reciprocal of impedance is called **admittance**

LR AND CR SERIES CIRCUIT

Resistance and Inductance in Series (L-R Circuit)

V_R is in phase with the current while V_L leads the current by $\dfrac{\pi}{2}$ so, V_R and V_L are mutually perpendicular (The sum of V_R and V_L cannot be put equal to applied e.m.f. E).

Phasor diagram for L-R series circuit :

The resultant of V_R and V_L = the magnitude of vector,

$$E = \sqrt{V_R^2 + V_L^2}$$

Thus, $E^2 = V_R^2 + V_L^2 = I^2(R^2 + X_L^2)$

or $I = \dfrac{E}{\sqrt{R^2 + X_L^2}}$

In L-R circuit **the applied emf E leads the current I or conversely the current I lags behind the e.m.f. E′ by a phase angle ϕ.**

$$\tan\phi = \frac{V_L}{V_R} = \frac{I X_L}{I R} = \frac{X_L}{R} = \frac{\omega L}{R}$$

$$\tan\phi = \frac{\omega L}{R} \quad \text{or} \quad \phi = \tan^{-1}\left(\frac{\omega L}{R}\right)$$

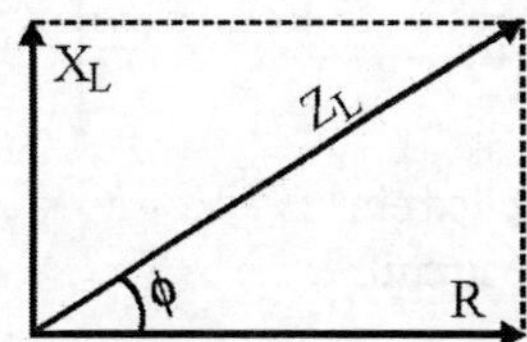

Inductive impedance Z_L :

$$Z_L = \sqrt{R^2 + \omega^2 L^2} \text{ or } Z_L = \sqrt{R^2 + (2\pi f L)^2}$$

The reciprocal of impedance is called **admittance**

$$\text{Admittance } Y_L = \frac{1}{Z_L} = \frac{1}{\sqrt{R^2 + \omega^2 L^2}}$$

Resistance and Capacitor in Series (RC-Series Circuit)

V_R is in phase with I, while V_C lags behind I by 90°.

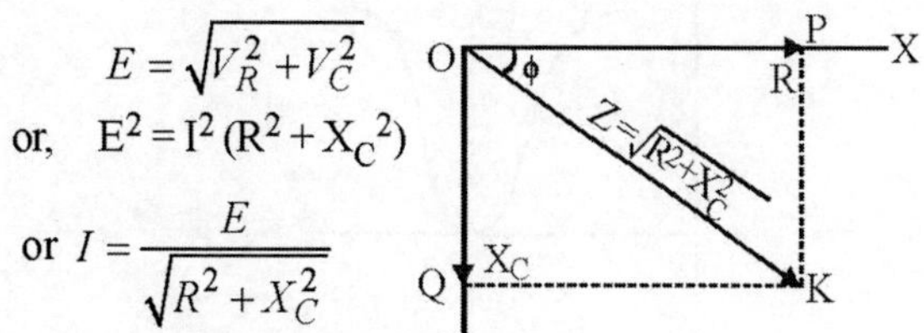

$$E = \sqrt{V_R^2 + V_C^2}$$

or, $E^2 = I^2(R^2 + X_C^2)$

or $I = \dfrac{E}{\sqrt{R^2 + X_C^2}}$

Capacitive impedance

$$Z_C = \sqrt{R^2 + X_C^2} = \sqrt{R^2 + \left(\frac{1}{\omega C}\right)^2}$$

The phasor diagram also shows that **in C-R circuit the applied e.m.f. lags behind the current I or the current I leads the emf E by a phase angle ϕ given by**

$$\tan\phi = \frac{V_C}{V_R} = \frac{X_C}{R} = \frac{1/\omega C}{R} = \frac{1}{\omega CR}$$

or, $\tan\phi = \frac{X_C}{R} = \frac{1}{\omega CR}$ or, $\phi = \tan^{-1}\left(\frac{1}{\omega CR}\right)$

LC OSCILLATIONS

Circuit Containing Inducatance and Capacitance in Series (LC Circuit)

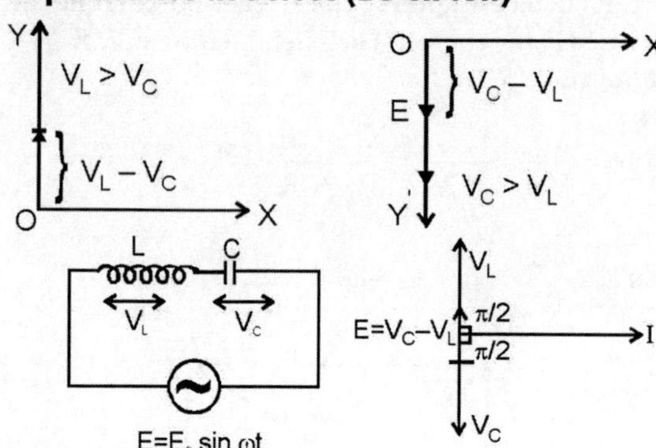

∴ Resultant applied emf, $E = V_C - V_L = X_C\mathbf{I} - X_L\mathbf{I}$

∴ Reactance of circuit,

$$X = E/\mathbf{I} = X_C - X_L = \left(\frac{1}{\omega C} - \omega L\right)$$

The current leads applied emf by $\phi = \pi/2$.

In case $X_C = X_L$, $Z = 0$, then $\frac{1}{\omega C} = \omega L$ or $\omega = \frac{1}{\sqrt{(LC)}}$

∴ **Frequency** $\mathbf{f = \omega/2\pi = \frac{1}{2\pi\sqrt{(LC)}}}$

At certain frequency the impedance of the circuit is minimum and the current is maximum.

This frequency is called the **resonant frequency**.

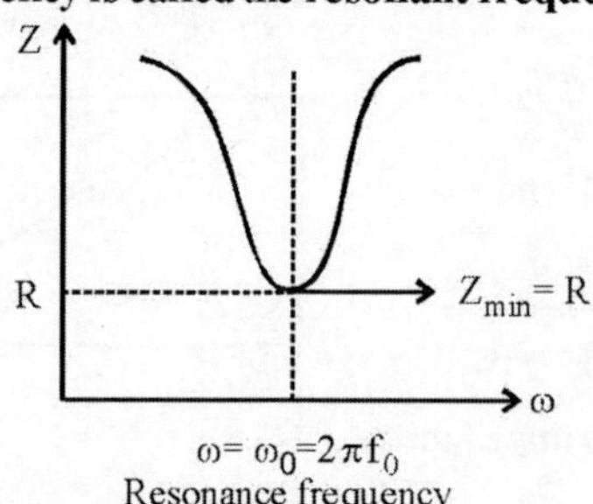

$\omega = \omega_0 = 2\pi f_0$
Resonance frequency

LCR SERIES CIRCUIT

Circuit Containing Resistance, Inductance and Capacitance in Series (Series LCR circuit)

V_R is in phase with current I but V_L leads I by 90° while V_C lags behind I by 90°.

∴ Resultant applied emf, $E = \sqrt{[V_R^2 + (V_C - V_L)^2]}$

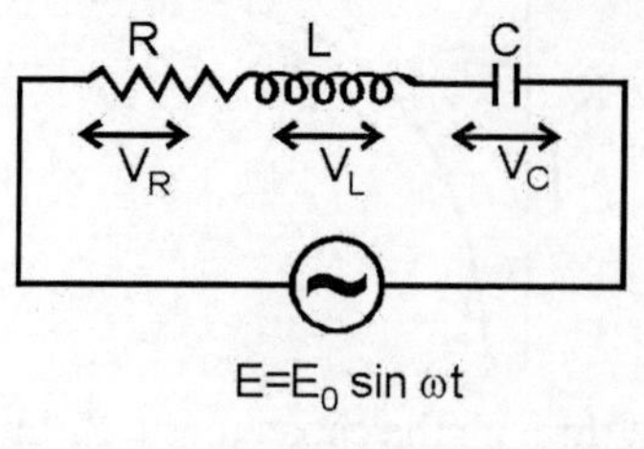

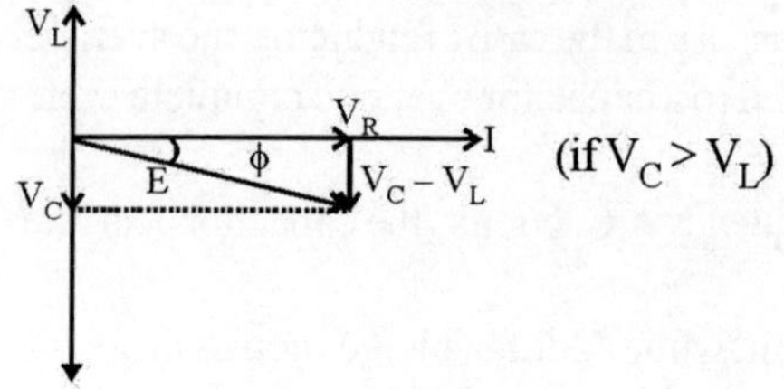

i.e., $E = \sqrt{\{(R\mathbf{I})^2 + (\mathbf{I}X_C - \mathbf{I}X_L)^2\}}$

∴ **Impedance**, $Z = \frac{E}{\mathbf{I}} = \sqrt{\{R^2 + (X_C - X_L)^2\}}$

The phase leads of current over applied emf is given by

$$\tan\phi = \frac{V_C - V_L}{V_R} = \frac{\mathbf{I}X_C - \mathbf{I}X_L}{R\mathbf{I}} = \frac{X_C - X_L}{R}$$

i.e. $\phi = \tan^{-1}\left(\frac{X_C - X_L}{R}\right)$

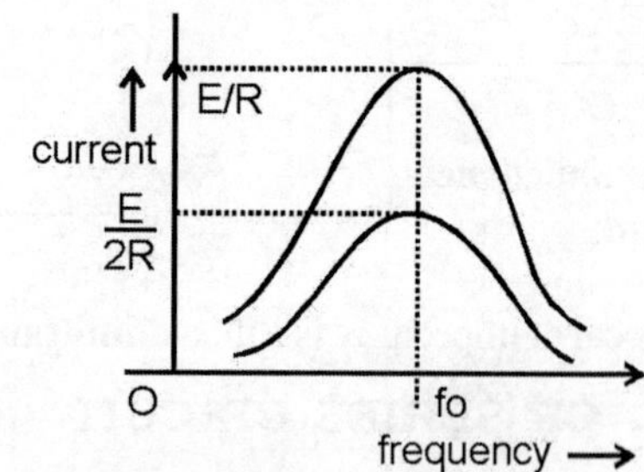

It is concluded that

(a) If $X_C > X_L$, the value of ϕ is positive, i.e., current leads the applied emf.

(b) If $X_C < X_L$, the value of ϕ is negative, i.e., current lags behind the applied emf.

(c) If $X_C = X_L$, the value of ϕ is zero, i.e., current and emf are in same phase. This is called the case of **resonance** and resonant frequency for condition $X_C = X_L$, is given by

$\frac{1}{\omega C} = \omega L$ i.e., $\omega = \frac{1}{\sqrt{LC}}$

$\therefore\ f_0 = \omega / 2\pi = \frac{1}{2\pi\sqrt{(LC)}}$

Thus the resonant frequency depends on the product of L and C and is independent of R.

At resonance, impedance is minimum, $Z_{min} = R$ and current is maximum $\mathbf{I}_{max} = \frac{E}{Z_{min}} = \frac{E}{R}$

(a) When $\omega L > \frac{1}{\omega C}$, then $\tan\phi$ is +ve i.e., ϕ is positive. In this case, the alternating emf leads the current I by a phase angle ϕ. The a.c. circuit is then inductance dominated circuit.

(b) When $\omega L < \frac{1}{\omega C}$, then $\tan\phi$ is –ve i.e., ϕ is negative. In this case, the alternating e.m.f. lags behind the current I by a phase angle ϕ. The a.c. circuit is then capacitance dominated circuit.

(c) When $\omega L = \frac{1}{\omega C}$, then $\tan\phi = 0$ i.e., $\phi = 0$. In the case, the alternating e.m.f. and current I are in phase. The a.c. circuit is then non-inductive.

RESONANCE AND QUALITY FACTOR (Q)

An interesting characteristic of a series LCR circuit is the phenomenon of **resonance**. The circuit exhibits resonance, i.e., the amplitude of the current is maximum at the **resonant frequency,**

$$\omega_0 = \frac{1}{\sqrt{LC}}\ (X_L = X_C). \Rightarrow \nu_0 = \frac{1}{2\pi\sqrt{LC}}$$

The **quality factor Q** defined by $Q = \frac{I \times L}{R} = \frac{I \times C}{R} = \frac{\omega_0 L}{R} = \frac{1}{\omega_0 CR} = \frac{1}{R}\sqrt{\frac{L}{C}}$ is an indicator of the sharpness of the **resonance,** the higher value of Q indicating sharper peak in the current.

Half power points

On the resonance curve there are two points where power in the circuit is half of the power at resonance.

The $\omega_1 = \omega_0 - \frac{R}{2L}$ and $\omega_2 = \omega_0 + \frac{R}{2L}$ (if $\frac{R}{2L} << \omega_0$).

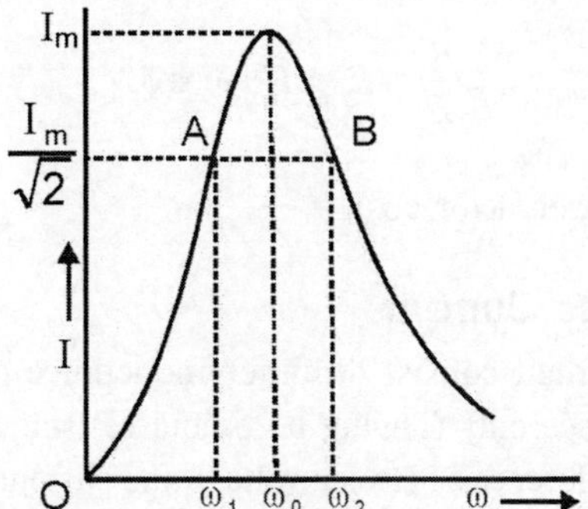

At lower half power frequency ω_1, the circuit is capacitive $X_C > X_L$.

At upper half power frequency ω_2, the circuit is inductive $X_L > X_C$.

Bandwidth :

$$Bw = w_2 - w_1 = \left(\omega_0 + \frac{R}{2L}\right) - \left(\omega_0 - \frac{R}{2L}\right) = \frac{R}{L}$$

The bandwidth is large for large R and small for small R.

The bandwidth Bw does not depend on C in the circuit (note that resonance frequency ω_r does not depend on R while bandwidth depends on R).

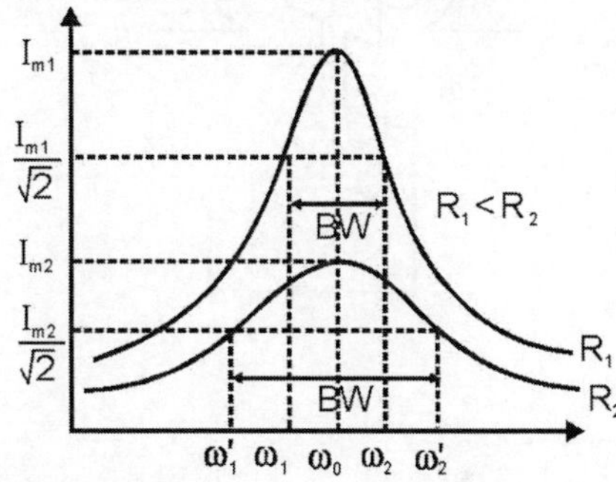

POWER IN A.C. CIRCUIT : THE POWER FACTOR

Average power $P_{av} = E_{rms}\ I_{rms} . \cos\phi$

It is also known as true power.

The quantity $E_{rms}\ I_{rms}$ is known as the **apparent power** or **virtual power.**

Power Factor

In the equation $P_{av} = E_{rms}. I_{rms}. \cos\phi$,. $\cos\phi$ is known as Power factor.

Average power = apparent power × power factor

It is customary to express true power in KW and apparent power in KVA.

$$P_{av} = E_{rms}\, I_{rms}.\frac{R}{Z} = E_{rms} I_{rms} \cos\phi$$

Here, power factor, $\cos\phi = \dfrac{R}{Z}$

Wattless Current

If the circuit consist of either inductance only or capacitance only (having no ohmic resistance), the phase difference ϕ between voltage and current is 90°.

The average power consumed,

$P = E_{rms}\, I_{rms} \times \cos 90° = 0$ $(\cos 90° = 0)$

Thus, if the resistance of an a.c. circuit is zero, although current flows in the circuit the average power consumed in the circuit remains zero. **The current in such a circuit is called wattless current.**

A.C. GENERATOR / DYNAMO

It is a device which converts mechanical energy into electric energy (A.C.) Its working is based on electromagnetic induction.

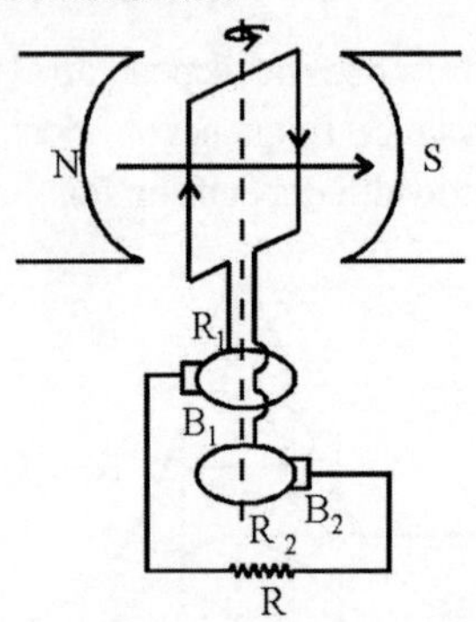

When a coil mounted on a rotor is rotated in a uniform magnetic field, a sinusoidal current is obtained according to electromagnetic induction.

$$E = nBA\omega \sin\omega t$$

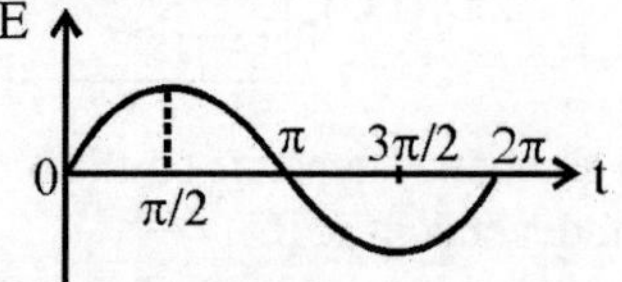

Choke Coil

A choke coil is simply an inductor with a large self-inductance and negligible resistance (zero in ideal case). It is used in A.C. circuits for reducing current without consuming power.

The average power consumed by the choke coil

$$P_{av} = E_v I_v \cos\frac{\pi}{2} = 0.$$

However, a practical inductance possesses a small resistance i.e., a practical inductance may be treated as a series combination of inductance L and a small resistance r.

Average power consumed in a practical inductance

$$P_{av} = E_v I_v \times \frac{r}{\sqrt{r^2 + \omega^2 L^2}}$$

For practical inductance power factor ($\cos\phi$)

$$= \frac{r}{\sqrt{r^2 + \omega^2 L^2}}$$

Exercise

1. For long distance transmission, the ac is stepped up because of high voltage, the transmission is
(a) faster (b) economical
(c) undamped (d) less dangerous

2. The transformer voltage induced in the secondary coil of a transformer is mainly due to
(a) a varying electric field
(b) a varying magnetic field
(c) the vibrations of the primary coil
(d) the iron core of the transformer

3. Eddy currents in the core of transformer can't be developed
(a) by increasing the number of turns in secondary coil
(b) by taking laminated transformer
(c) by making step down transformer
(d) by using a weak a.c. at high potential

4. The frequency of A.C. mains in India is
(a) 30 c/s (b) 50 c/s
(c) 60 c/s (d) 120 c/s

5. Hot wire ammeters are used for measuring
(a) A.C. only (b) D.C. only
(c) both A.C. and D.C. (d) none of these

6. Alternating current cannot be measured by D.C. ammeter because
(a) A.C. cannot pass through D.C. ammeter
(b) average value of current for complete cycle is zero
(c) A.C. is virtual
(d) A.C. changes its direction

7. Alternating current is converted to direct current by
(a) rectifier (b) dynamo
(c) transformer (d) motor

8. A transformer is employed to
(a) convert A.C. into D.C.
(b) convert D.C. into A.C.
(c) obtain a suitable A.C. voltage
(d) obtain a suitable D.C. voltage

9. To convert mechanical energy into electrical energy, one can use
(a) DC dynamo (b) AC dynamo
(c) motor (d) (a) & (b)

10. The AC voltage across a resistance can be meausred using
(a) a potentiometer
(b) a hot-wire voltmeter
(c) a moving-coil galvanometer
(d) a moving-magnet galvanometer

11. Conduction current flows
(a) only through resistance
(b) through wires and resistance
(c) only through capacitor
(d) through wires resistance and capacitor

12. Displacement current is
(a) continuous when electric field is changing in the circuit
(b) continuous when magnetic field is changing in the circuit
(c) continuous in both types of fields
(d) continuous through wires and resistance only

13. The nature of electric field in the region between the two plates of a capacitor, while charging is
(a) always constant
(b) varying
(c) depends on values of C & R
(d) depends on e.m.f. of battery

14. If $\vec{E}$ and $\vec{B}$ are the electric and magnetic field vectors of e.m. waves then the direction of propagation of e.m. wave is along the direction of
(a) $\vec{E}$ (b) $\vec{B}$
(c) $\vec{E}\times\vec{B}$ (d) none of these

15. Maxwell's modified form of Ampere's circuital law is
(a) $\oint_s \vec{B}.\vec{ds} = 0$
(b) $\oint_s \vec{B}.\vec{dl} = \mu_0 I$
(c) $\oint_s \vec{B}.\vec{dl} = \mu_0 I + \frac{1}{\varepsilon_o}\frac{dq}{dt}$
(d) $\oint_s \vec{B}.\vec{dl} = \mu_0 I + \mu_0 \varepsilon_o \frac{d\phi_E}{dt}$

16. According to Maxwell's equation the velocity of light in any medium is expressed as

(a) $\frac{1}{\sqrt{\mu_0 \varepsilon_o}}$ (b) $\frac{1}{\sqrt{\mu\varepsilon}}$

(c) $\sqrt{\mu/\varepsilon}$ (d) $\sqrt{\frac{\mu_0}{\varepsilon}}$

17. The displacement current is

(a) $\varepsilon_o\, d\phi_E/dt$ (b) $\frac{\varepsilon_o}{R} d\phi_E/dt$

(c) $\varepsilon_o E/R$ (d) $\varepsilon_o q\, C/R$

18. The electromagnetic waves do not transport
(a) energy (b) charge
(c) momentum (d) information

19. The amplitudes of electric and magnetic fields related to each other are

(a) $E_0 = B_0$ (b) $E_0 = c\,B_0$

(c) $E_0 = \frac{B_0}{c}$ (d) $E_0 = \frac{c}{B_0}$

20. In an electromagnetic wave, the direction of the magnetic induction $\vec{B}$ is
(a) parallel to the electric field $\vec{E}$
(b) perpendicular to the electric field $\vec{E}$
(c) antiparallel to the Poynting vector $\vec{S}$
(d) random

21. Which of the following electromagnetic waves have the longest wavelength?
(a) heat waves
(b) visible light
(c) radio frequency waves
(d) microwaves

22. Which of the following is of shortest wavelength?
(a) X-rays (b) γ-rays
(c) microwaves (d) radiowaves

23. Radiowaves with frequencies higher than television signals are
(a) ultrasonic waves (b) sound waves
(c) light waves (d) microwaves

24. Maxwell's equation describe the fundamental laws of
(a) electricity only (b) magnetism only
(c) mechanics only (d) both (a) and (b)

25. In a circuit inductance L and capacitance C are connected as shown in figure. A_1 and A_2

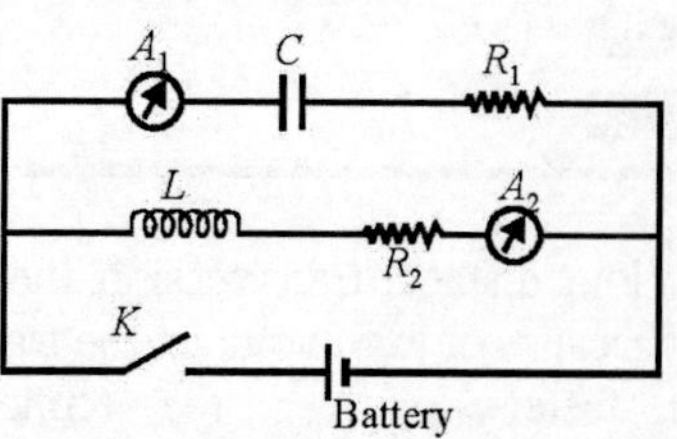

are ammeters. When key K is pressed to complete the circuit, then just after closing key (K), the readings of A_1 and A_2 will be
(a) zero in both A_1 and A_2
(b) maximum in both A_1 and A_2
(c) zero in A_1 and maximum in A_2
(d) maximum in A_1 and zero in A_2.

26. In an A.C. circuit, the current flowing in inductance is $I = 5\sin(100\,t - \pi/2)$ amperes and the potential difference is $V = 200\sin(100\,t)$ volts. The power consumption is equal to
(a) 1000 watts (b) 40 watts
(c) 20 watts (d) 0 watt

27. In an a.c. circuit V and I are given by
$V = 100\sin(100\,t)$ volts
$I = 100\sin(100\,t + \pi/3)$ mA
the power dissipated in the circuit is
(a) 10^4 watt (b) 10 watt
(c) 2.5 watt (d) 5.0 watt

28. The primary winding of a transformer has 100 turns and its secondary winding has 200 turns. The primary is connected to an A.C. supply of 120 V and the current flowing in it is 10 A. The voltage and the current in the secondary are
(a) 240 V, 5 A (b) 240 V, 10 A
(c) 60 V, 20 A (d) 120 V, 20 A

29. A step down transformer is connected to 2400 volts line and 80 amperes of current is found to flow in output load. The ratio of the turns in primary and secondary coil is 20 : 1. If transformer efficiency is 100%, then the current flowing in the primary coil will be
(a) 1600 amp (b) 20 amp
(c) 4 amp (d) 1.5 amp

30. In the circuit shown in fig, the resonant frequency is

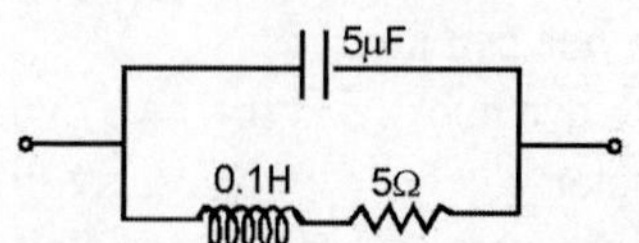

(a) 75 kc/s (b) 750 kc/s
(c) 7.5 kc/s (d) 75 mc/s

31. The power factor of an AC circuit having resistance (R) and inductance (L) connected in series and an angular velocity ω is
(a) $R/\omega L$ (b) $R/(R^2\omega^2L^2)$
(c) $\omega L/R$ (d) $R/(R^2\omega^2L^2)^{1/2}$

32. In an experiment, 200V A.C. is applied at the ends of an LCR circuit. The circuit consists of an inductive reactance (X_L) = 50Ω, capacitive reactance (X_C) = 50Ω and ohmic resistance (R) = 10Ω. The impedance of the circuit is
(a) 10Ω (b) 20Ω
(c) 30Ω (d) 40Ω

33. In an A.C. circuit with phase voltage V and current I, the power dissipated is
(a) VI (b) V^2I
(c) VI^2 (d) V^2I^2

34. In a transformers, number of turns in primary coil are 140 and that in secondary coil are 280. If current in primary coil is 4A, then that in secondary coil is
(a) 4 A (b) 2 A
(c) 6 A (d) 10 A

35. The core of any transformer is laminated so as to
(a) reduce the energy loss due to eddy currents
(b) make it light weight
(c) make it robust and sturdy
(d) increase secondary voltage

36. In an oscillating LC circuit the max. charge on the capacitor is Q. The charge on capacitor when the energy is stored equally between electric and magnetic field is
(a) $Q/2$ (b) $Q/\sqrt{3}$
(c) $Q/\sqrt{2}$ (d) Q

37. A capacitor acts as an infinite resistance for
(a) DC (b) AC
(c) DC as well as AC (d) neither AC nor DC

38. An inductor, a resistor and a capacitor are joined in series with an AC source. As the frequency of the source is slightly increased from a very low value, the reactance
(a) of the inductor increases
(b) of the resistor increases
(c) of the capacitor increases
(d) of the circuit increases

39. The reactance of a circuit is zero. It is possible that the circuit contains
(a) an inductor and a capacitor
(b) an inductor but no capacitor
(c) a capacitor but no inductor
(d) neither an resistor nor a capacitor

40. In an AC series circuit, the instantaneous current is zero when the instantaneous voltage is maximum. Connected to the source may be a
(a) pure inductor & pure resistor
(b) pure capacitor & pure resistor
(c) pure resistor
(d) combination of an inductor and a capacitor

41. An inductor-coil having some resistance is connected to an AC source. Which of the following quantities have zero average value over a cycle?
(a) current
(b) resistor
(c) joule heat
(d) magnetic energy stored in the inductor

42. An A.C. source is connected to a resistive circuit. What is true of the following :
(a) current leads ahead of voltage in phase
(b) current lags behind voltage in phase
(c) current and voltage are in same phase
(d) any of the above may be true depending upon the value of resistance

43. The capacitive reactance in an A.C. circuit is
(a) effective resistance due to capacity
(b) effective wattage
(c) effective voltage
(d) none of the above

44. The power factor in a circuit connected to an A.C. The value of power factor is :
(a) unity when the circuit contains an ideal inductance only
(b) unity when the circuit contains an ideal resistance only
(c) zero when the circuit contains an ideal resistance only
(d) unity when the circuit contains an ideal capacitance only

45. Current in a circuit is wattless if
(a) inductance in the circuit is zero
(b) resistance in the circuit is zero
(c) current is alternating
(d) resistance and inductance both are zero

46. Power factor is one for
(a) pure inductor
(b) pure capacitor
(c) pure resistor
(d) either an inductor or a capacitor

47. Of the following about capacitive reactance which is correct
(a) the reactance of the capacitor is directly proportional to its ability to store charge
(b) capacitive reactance is inversely proportional to the frequency of the current
(c) capacitive reactance is mesured in farad
(d) the reactance of a capacitor in an A.C. circuit is similar to the resistance of a capacitor in a D.C. circuit

48. With increase in frequency of an A.C. supply, the inductive reactance
(a) decreases
(b) increases directly with frequency
(c) increases as square of frequency
(d) decreases inversely with frequency

49. With increase in frequency of an A.C. supply, the impedance of an L-C-R series circuit
(a) remains constant
(b) increases
(c) decreases
(d) decreases at first, becomes minimum and then increases

50. An alternating voltage of frequency ω is induced in electric circuit consisting of an inductance L and capacitance C, connected in parallel. Then across the inductance coil the
(i) current is maximum when $\omega^2 = 1/(LC)$
(ii) current is minimum when $\omega^2 = 1/(LC)$
(iii) voltage is minimum when $\omega^2 = 1/(LC)$
(iv) voltage is maximum when $\omega^2 = 1/(LC)$
State if
(a) if (i) and (iii) are both correct
(b) if (i) and (iv) are both correct
(c) if (ii) and (iii) are both correct
(d) if (ii) and (iv) are both correct

51. The current in resistance R at resonance is

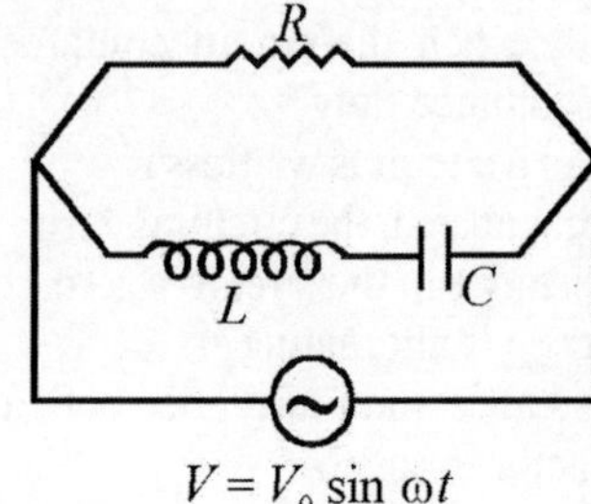

(a) zero
(b) minimum but finite
(c) maximum but finite
(d) infinite

52. In an L.C.R. series a.c. circuit, the current
(a) is always in phase with the voltage
(b) always lags the generator voltage
(c) always leads the generator voltage
(d) none of the above statements is true

53. The root mean square value of alternating current is equal to
(a) twice the peak value
(b) half the peak value
(c) $1/\sqrt{2}$ times the peak value
(d) equal to peak value

54. Same current is flowing in two alternating circuits. The first circuit contains only inductance and the other contains only a capacitor. If the frequency of the e.m.f. of a.c. is increased, the effect on the value of the current will be
(a) increases in the first circuit and decreases in the other
(b) increases in both the circuits
(c) decreases in both the circuits
(d) decreases in first circuit and increases in the other

55. The coil of a dynamo is rotating in a magnetic field. The developed induced e.m.f. changes and the number of magnetic lines of force also changes. Which of the following condition is correct
(a) lines of force minimum but induced e.m.f. is zero
(b) lines of force maximum but induced e.m.f. is zero
(c) lines of force maximum but induced e.m.f. is not zero
(d) lines of force maximum but induced e.m.f. is also maximum

56. The average power dissipated in a pure inductance is
(a) $\frac{1}{2}LI^2$ (b) LI^2
(c) $LI^2/4$ (d) zero

57. If a current I given by $I = I_0 \sin(\omega t - \pi/2)$ flows in inductance in an A.C. circuit across which an A.C. potential $E = E_0 \sin \omega t$ has been applied, then power consumption P in the circuit will be

(a) $P = E_0 I_0 / \sqrt{2}$ (b) $P = E I / \sqrt{2}$

(c) $P = E_0 I_0 / 2$ (d) zero

58. In series L-C-R circuit, the voltages across R, L and C are V_R, V_L and V_C respectively. Then the voltage of applied a.c. source must be

(a) $V_R + V_L + V_C$

(b) $\sqrt{[(V_R)^2 + (V_L - V_C)^2]}$

(c) $V_R + V_C - V_L$

(d) $[(V_R + V_L)^2 + (V_C)^2]^{1/2}$

59. The heat produced in a given resistance in a given time by the sinusoidal current $I_0 \sin \omega t$ will be the same as that of a steady current of magnitude nearly

(a) $0.71\, I_0$ (b) $1.412\, I_0$

(c) I_0 (d) $\sqrt{I_0}$

60. The sinusoidal A.C. current flows through a resistor of resistance R. If the peak current is I_p, then power dissipated is

(a) $I_p^2 R \cos\theta$ (b) $\frac{1}{2} I_p^2 R$

(c) $\frac{4}{\pi} I_p^2 R$ (d) $\frac{1}{\pi^2} I_p^2 R$

CHAPTER 22

Electromagnetic Waves

CONCEPTS

ELECTROMAGNETIC WAVES

An accelerated charge produces a sinusoidal time varying magnetic field, which in turn produces a sinusoidal time varying electric field. The two fields so produced are mutually perpendicular and are sources to each other. *The mutually perpendicular time varying electric and magnetic fields constitute electromagnetic waves* which can propagate through empty space.

Characteristics of Electromagnetic Waves

(i) The electric and magnetic fields satisfy the following wave equations, which can be obtained from Maxwell's third and fourth equations.

$$\frac{\partial^2 E}{\partial x^2} = \mu_0 \varepsilon_0 \frac{\partial^2 E}{\partial t^2} \text{ and } \frac{\partial^2 B}{\partial x^2} = \mu_0 \varepsilon_0 \frac{\partial^2 B}{\partial t^2}$$

where E and B are electric and magnetic field respectively.

(ii) Electromagnetic waves travel through vacuum with the speed of light.

(iii) The electric and magnetic fields of an electromagnetic wave are perpendicular to each other and also perpendicular to the direction of wave propagation. Hence, these are **transverse waves**.

(iv) The instantaneous magnitudes of $\vec{E}$ and $\vec{B}$ in an electromagnetic wave are related by the expression $\frac{\vec{E}}{\vec{B}} = c$

(v) Electromagnetic waves carry energy. The rate of flow of energy crossing a unit area is described by the Poynting vector $\vec{S}$, where $\vec{S} = \frac{1}{\mu_0}\vec{E} \times \vec{B}$

(vi) Electromagnetic waves carry momentum and hence can exert pressure (P) on surfaces, which is called **radiation pressure**. For an electromagnetic wave with poynting vector $\vec{S}$, incident on a perfectly absorbing surface then, $P = \frac{S}{c}$ and if incident on a perfectly reflecting surface then, $P = \frac{2S}{c}$

(vii) The electric and magnetic field of a sinusoidal plane electromagnetic wave propagating in the positive x-direction can also be written as

$$E = E_m \sin(kx - \omega t);\ B = B_m \sin(kx - \omega t)$$

where ω **is the angular frequency of the wave** and **k is wave number** which are given by

$$\omega = 2\pi f \text{ and } k = \frac{2\pi}{\lambda}$$

(viii) The intensity of a sinusoidal plane electromagnetic wave is defined as the average value of Poynting vector taken over one cycle.

$$S_{av} = \frac{E_m B_m}{2\mu_0} = \frac{E^2_m}{2\mu_0 c} = \frac{c}{2\mu_0} B^2_m$$

(ix) Electromagnetic waves obey the principle of superposition.

(x) The electric field vector of an electromagnetic field is responsible for all optical effects . For this reason electric field vector is also called a light vector.

CONDUCTION CURRENT AND DISPLACEMENT CURRENT

Conduction Current

It is a current in the electric circuit which arises due to the flow of electrons in the connecting wires of the circuit, in a definite closed path.

Displacement Current

It is that current which comes into play in the region, whenever the electric field and hence, the electric flux is changing with it. Displacement current was predicted by Maxwell.

1. The displacement current is given by the relation.

$$i_D = \varepsilon_0 \frac{d\phi_E}{dt}$$

where ε_0 = absolute permittivity of space, $d\phi_E / dt$ = rate of change of electric flux.

2. In case of a steady electric flux linked with a region, the displacement current is zero.

DIFFERENT TYPES OF ELECTROMAGNETIC WAVES (ELECTROMAGNETIC SPECTRUM)

The array obtained on arranging all the electromagnetic waves in an order on the basis of their wavelength is called electromagnetic spectrum.

Wave type	Wavelength	Production	Detection
Radio	>0.1 m	Rapid acceleration and decelerations of electrons in aerials	Receiver's aerials
Microwave	0.1m to 1 mm	Klystron valve or magnetron valve	Point contact diodes
Infra-red	1mm to 700 nm	Vibration of atoms and molecules photographic film	Thermopiles, Bolometer, Infrared
Visible	700 nm to 400 nm	Electrons in atoms emit light when they move from one energy level to a lower energy level	The eye Photocells Photographic film
Ultraviolet	400 nm to 1nm	Inner shell electrons in atoms moving from one energy level to a lower level	Photocells Photographic film
X-rays	1nm to 10^{-3} nm	X-ray tubes or inner shell electrons	Photographic film Geiger tubes Ionisation chamber
Gamma rays	<10^{-3} nm	Radioactive decay of the nucleus	Photographic film Geiger tubes Ionisation chamber

Uses of Electromagnetic Spectrum

Radiation	Uses
γ-rays	To get information on nuclear structure, medical treatment etc.
X-rays	Medical diagnosis and treatment, study of crystal structure, industrial radiograph etc.
UV-rays	Preserve food, sterilizing the surgical instruments, detecting the invisible writings, finger prints, etc.
Visible light	To see objects.
Infrared rays	To treat muscular strain, for taking photography during the fog haze etc.
Microwave	In microwave oven (microwave), in radar.
Radio wave	In telecommunication.

Exercise

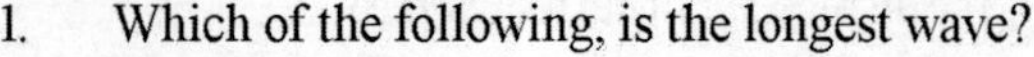

1. Which of the following, is the longest wave?
(a) X-rays (b) γ-rays
(c) microwaves (d) radiowaves

2. Pick out the longest wavelength from the following types of radiation.
(a) blue light (b) gamma rays
(c) X-rays (d) red light

3. The frequency of electromagnetic wave, which is best suited to observe a particle of radius 3×10^{-4} cm is of the order of
(a) 10^{15} (b) 10^{14}
(c) 10^{13} (d) 10^{12}

4. The structure of solids is investigated by using
(a) cosmic rays (b) X-rays
(c) γ-rays
(d) infra-red radiations

5. A signal emitted by an antenna from a certain point can be received at another point of the surface in the form of
(a) sky wave (b) ground wave
(c) sea wave (d) both a and b

6. The oscillating electric and magnetic field vectors of electromagnetic wave are oriented along
(a) the same direction and in phase
(b) the same direction but have a phase difference of 90°
(c) mutually perpendicular directions and are in phase
(d) mutually perpendicular directions but has a phase difference of 90°

7. Which one of the following electromagnetic radiations has the smallest wavelength?
(a) ultraviolet waves (b) X-rays
(c) γ-rays (d) microwaves

8. If ε_0 and μ_0 are the electric permittivity and magnetic permeability in vacuum, ε and μ are corresponding quantities in medium, then refractive index of the medium is
(a) $\sqrt{\frac{\varepsilon}{\varepsilon_0}}$ (b) $\sqrt{\frac{\varepsilon_0 \mu}{\varepsilon \mu_0}}$
(c) $\sqrt{\frac{\varepsilon_0 \mu_0}{\varepsilon \mu}}$ (d) $\sqrt{\frac{\varepsilon \mu}{\varepsilon_0 \mu_0}}$

9. In a circuit, inductance L and capacitance C are connected as shown in the figure. A_1 and A_2 are ammeters.

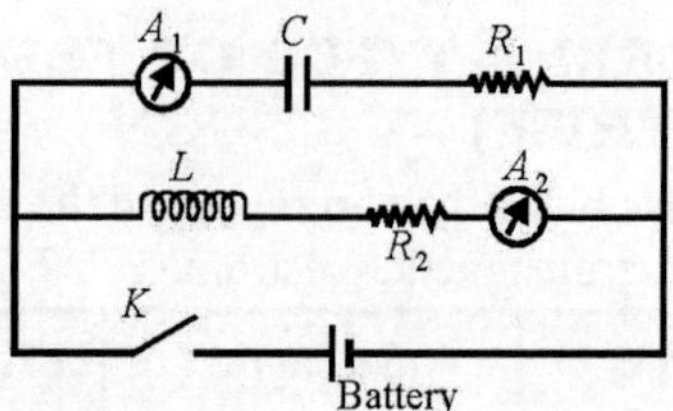

When key K is pressed to complete the circuit, then just after closing key (K), the readings of A_1 and A_2 will be
(a) zero in both A_1 and A_2
(b) maximum in both A_1 and A_2
(c) zero in A_1 and maximum in A_2
(d) maximum in A_1 and zero in A_2

10. The electromagnetic radiations are caused by
(a) a stationary charge
(b) uniformly moving charges
(c) accelerated charges
(d) all of above

11. The frequencies of X-rays, γ-rays and ultraviolet rays are respectively a, b, and c. Then
(a) $a < b, b < c$ (b) $a < b, b > c$
(c) $a > b, b > c$ (d) $a > b, b < c$

12. Which of the following is positively charged?
(a) α-particle (b) β-particle
(c) γ-rays (d) X-rays

13. If $\vec{E}$ and $\vec{B}$ represent electric and magnetic field vectors of the electromagnetic waves, then the direction of propagation of the waves will be along
(a) $\vec{B} \times \vec{E}$ (b) $\vec{E}$
(c) $\vec{B}$ (d) $\vec{E} \times \vec{B}$

14. Frequencies of sideband is
(a) 2.50005×10^6 Hz, 2.49995×10^6 Hz
(b) 2.505×10^6 Hz, 2.495×10^6 Hz
(c) 2.505×10^6 kHz, 2.495×10^6 kHz
(d) 2.505 MHz, 2.495 kHz

15. Green-house effect is the heating up of earth's atmosphere due to
(a) green plants (b) infrared rays
(c) X-rays (d) ultraviolet rays

16. Which one of the following rays is not electromagnetic wave ?
(a) heat rays (b) X-rays
(c) γ-rays (d) β-rays

17. We consider the radiation emitted by the human body. Which of the following statements is true?
(a) the radiation emitted lies in the ultraviolet region and hence is not visible
(b) the radiation emitted is in the infrared region
(c) the radiation is emitted only during the day
(d) the radiation is emitted during the summers and absorbed during the winters

18. If λ_v, λ_x and λ_m represent the wavelengths of visible light, X–rays and microwaves respectively, then
(a) $\lambda_m > \lambda_x > \lambda_v$ (b) $\lambda_m > \lambda_v > \lambda_x$
(c) $\lambda_v > \lambda_x > \lambda_m$ (d) $\lambda_v > \lambda_m > \lambda_x$

19. The electric and magnetic field of an electromagnetic wave are
(a) in opposite phase and perpendicular to each other
(b) in opposite phase and parallel to each other
(c) in phase and perpendicular to each other
(d) in phase and parallel to each other

20. The velocity of electromagnetic radiation in a medium of permittivity ε_0 and permeability μ_0 is given by
(a) $\sqrt{\frac{\varepsilon_0}{\mu_0}}$ (b) $\sqrt{\mu_0 \varepsilon_0}$
(c) $\frac{1}{\sqrt{\mu_0 \varepsilon_0}}$ (d) $\sqrt{\frac{\mu_0}{\varepsilon_0}}$

21. The electric field part of an electromagnetic wave in a medium is represented by $E_x=0$;

$$E_y = 2.5\frac{N}{C} \cos\left[\left(2\pi\times 10^6 \frac{\text{rad}}{m}\right) t - \left(\pi\times 10^{-2} \frac{\text{rad}}{s}\right) x\right];$$

$E_z = 0$. The wave is :
(a) moving along x direction with frequency 10^6 Hz and wave length 100 m
(b) moving along x direction with frequency 10^6 Hz and wave length 200 m
(c) moving along $-x$ direction with frequency 10^6 Hz and wave length 200 m
(d) moving along y direction with frequency $2\pi \times 10^6$ Hz and wave length 200 m

22. Which of the following statement is false for the properties of electromagnetic waves?
(a) Both electric and magnetic field vectors attain the maxima and minima at the same place and same time.
(b) The energy in electromagnetic wave is divided equally between electric and magnetic vectors
(c) Both electric and magnetic field vectors are parallel to each other and perpendicular to the direction of propagation of wave
(d) These waves do not require any material medium for propagation

23. The electric and the magnetic field associated with an E.M. wave, propagating along the +z-axis, can be represented by
(a) $\left[\vec{E} = E_0\hat{i}, \vec{B} = B_0\hat{j}\right]$
(b) $\left[\vec{E} = E_0\hat{k}, \vec{B} = B_0\hat{i}\right]$
(c) $\left[\vec{E} = E_0\hat{j}, \vec{B} = B_0\hat{i}\right]$
(d) $\left[\vec{E} = E_0\hat{j}, \vec{B} = B_0\hat{k}\right]$

24. The decreasing order of wavelength of infrared, microwave, ultraviolet and gamma rays is
(a) microwave, infrared, ultraviolet, gamma rays
(b) gamma rays, ultraviolet, infrared, microwaves
(c) microwaves, gamma rays, infrared, ultraviolet
(d) infrared, microwave, ultraviolet, gamma rays

25. The electromagnetic radiation used in food processing sterilizing agent is:
(a) microwaves (b) UV rays
(c) gamma rays (d) radio waves

26. An electromagnetic wave is made up of joint electric (E) and magnetic (b) fields. Its direction of propagation is:
(a) parallel to E
(b) perpendicular to E but parallel to B
(c) parallel to B
(d) perpendicular to both E and B

27. Electromagnetic radiation will be emitted in the case of a
    (a) neutron moving in a straight line with a constant speed
    (b) proton moving in a straight line with a constant speed
    (c) is constant inside and decays as I r outside the solenoid
    (d) is constant inside and decays as $\exp\left(-\frac{1}{r}\right)$ outside

28. Intensity of electromagnetic wave will be :
    (a) $I = c\mu_0 B_0^2 / 2$ (b) $I = c\varepsilon_0 B_0^2 / 2$
    (c) $I = B_0^2 / c\mu_0$ (d) $I = E_0^2 / 2c\varepsilon_0$

29. The electric (E) and magnetic (B) field amplitudes associated with an electromagnetic radiation from a point source behave at a distance r from the source as
    (a) E = constant, B = constant
    (b) $E \propto \frac{1}{r}$, $B \propto \frac{1}{r}$
    (c) $E \propto \frac{1}{r^2}$, $B \propto \frac{1}{r^2}$
    (d) $E \propto \frac{1}{r^3}$, $B \propto \frac{1}{r^3}$

30. In 1825, the existence of electromagnetic waves was predicted by
    (a) Maxwell (b) Hertz
    (c) Bose (d) Marconi

31. The electric field of a plane electromagnetic wave in free space travelling along x-direction is given by : $E_z = 60 \sin(kx + \omega t)$ V/m.
    The magnetic field component will be given as :
    (a) $B_z = \frac{60}{c} \sin(kz + \omega t) T$
    (b) $B_z = \frac{60}{c} \sin(kx + \omega t) T$
    (c) $B_y = \frac{60}{c} \sin(ky + \omega t) T$
    (d) $B_y = \frac{60}{c} \sin(kx + \omega t) T$

    Here c represents the speed of light in free space

32. If a source is transmitting electromagnetic wave of frequency 8.2×10^6 Hz, then wavelength of the electromagnetic waves transmitted from the source will be
    (a) 36.6 m (b) 40.5 m
    (c) 42.3 m (d) 50.9 m

33. In an electromagnetic wave, the electric and magnetic fields are 100 V m^{-1} and 0.265 A m^{-1}. The maximum energy flow is
    (a) 26.5 W/m^2 (b) 36.5 W/m^2
    (c) 46.7 W/m^2 (d) 765 W/m^2

34. Ionosphere as a whole is
    (a) positively charges (b) –vely charges
    (c) electrically neutral (d) can't be said

35. A TV tower has a height of 100 m. How much population is covered by the TV broadcast if the average population density around the tower is 100 km^{-2} (radius of the earth = 6.37×10^6 m)
    (a) 4 lakh (b) 4 billion
    (c) 40,000 (d) 40 lakh

36. In a plane electromagnetic wave propagating in space has an electric field of amplitude 9×10^3 V/m, then the amplitude of the magnetic field is
    (a) 2.7×10^{12} T (b) 9.0×10^{-3} T
    (c) 3.0×10^{-4} T (d) 3.0×10^{-5} T

37. Electromagnetic waves are transverse in nature is evident by
    (a) polarization (b) interference
    (c) reflection (d) diffraction

38. The electron density in all the layers of ionosphere
    (a) is the same
    (b) decreases with altitude
    (c) increases with altitude
    (d) sometimes decreases sometimes increases

39. Which of the following radiations has the least wavelength ?
    (a) γ-rays (b) β-rays
    (c) α-rays (d) X-rays

40. An electromagnetic wave of frequency $\nu = 3.0$ MHz passes from vacuum into a dielectric medium with permittivity $\varepsilon = 4.0$. Then
    (a) wave length is halved and frequency remains unchanged
    (b) wave length is doubled and frequency becomes half
    (c) wave length is doubled and the frequency remains unchanged
    (d) wave length and frequency both remain unchanged.

41. If $\lambda = 10 Å$ then it corresponds to
(a) infrared (b) microwaves
(c) ultraviolet (d) X-rays

42. 10 cm is a wavelength corresponding to the spectrum of
(a) infrared rays (b) ultraviolet rays
(c) microwaves (d) γ-rays

43. In an electromagnetic wave, the amplitudes of magnetic field H_0 and electric field E_0 in free space are related as :
(a) $H_0 = E_0$ (b) $H = \frac{E_0}{c}$
(c) $H_0 = E_0\sqrt{\mu_0\varepsilon_0}$ (d) $H_0 = E_0\sqrt{\frac{\varepsilon_0}{\mu_0}}$

44. In an electromagnetic wave
(a) power is transmitted along the magnetic field
(b) power is transmitted along the electric field
(c) power is equally transferred along the electric and magnetic fields
(d) power is transmitted in a direction perpendicular to both the fields

45. The frequency modulated waves are
(a) reflected by atmosphere
(b) absorbed by atmosphere
(c) bend by atmosphere
(d) radiowaves

46. The energy of electromagnetic wave in vacuum is given by the relation
(a) $\frac{E^2}{2\varepsilon_0} + \frac{B^2}{2\mu_0}$ (b) $\frac{1}{2}\varepsilon_0 E^2 + \frac{1}{2}\mu_0 B^2$
(c) $\frac{E^2 + B^2}{c}$ (d) $\frac{1}{2}\varepsilon_0 E^2 + \frac{B^2}{2\mu_0}$

47. An emf of 15 volt is applied in a circuit containing $5H$ inductance and 10Ω resistance. The ratio of currents at $t = \infty$ and $t = 1$ sec is
(a) $\frac{e^2}{e^2 - 1}$ (b) $\frac{e^{\frac{1}{2}}}{e^{\frac{1}{2}} - 1}$
(c) $1 - e$ (d) e^{-1}

48. Which of the following waves have the maximum wavelength ?
(a) Infrared rays (b) UV rays
(c) Radiowaves (d) X-rays

49. Which scientist experimentally proved the existence of electromagnetic waves?
(a) Sir J.C. Bose (b) Maxwell
(c) Marconi (d) Hertz

50. Which of the following is not transducer?
(a) Loudspeaker (b) Amplifier
(c) Microphone (d) All of these

51. In an LCR series circuit, the voltage across each of the components, L, C and R is 50 V. The voltage across the LC combination will be :
(a) 50 V (b) 0 V
(c) $50\sqrt{2}$ V (d) 100 V

52. Use of eddy currents is done in the following except
(a) moving coil galvanometer
(b) electric brakes
(c) induction motor
(d) dynamo

53. In a LCR series circuit, the potential difference between the terminals of the inductance is 60 V, between the terminals of the capacitor is 30 V and that across the resistance is 40V. Then, supply voltage will be equal to
(a) 50 V (b) 70 V
(c) 130 V (d) 10 V

54. A choke is preferred to a resistance for limiting current in AC circuit because
(a) choke is cheap
(b) there is no wastage of power
(c) choke is compact in size
(d) choke is a good absorber of heat

55. A solenoid 1.5 m long and 4.0 cm in diameter possesses 10 turns/cm. A current of 5.0 ampere is flowing through it. Calculate the magnetic induction (i) inside and (ii) at one end, on the axis of solenoid:
(a) $2\pi \times 10^{-3}$ tesla, $\pi \times 10^{-3}$ tesla
(b) $0.5\pi \times 10^{-3}$ tesla, $\frac{\pi}{2} \times 10^{-3}$ tesla
(c) $\frac{3}{2}\pi \times 10^{-3}$ tesla, $\frac{\pi}{3} \times 10^{-3}$ tesla
(d) none of these

56. Which one of the graphs shows variation of magnetic induction $\vec{B}$ with distance r from a long wire carrying current?

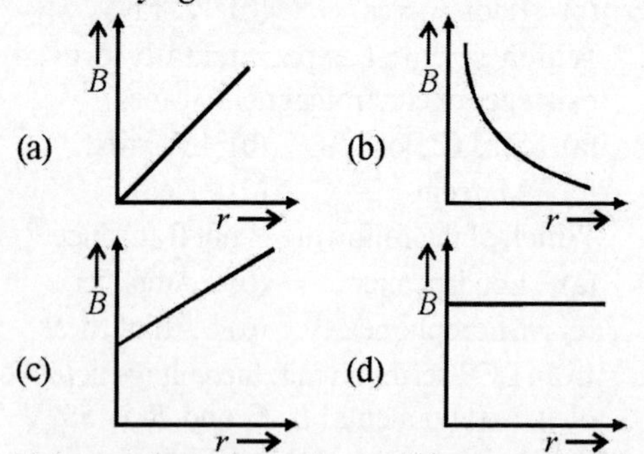

57. If the number of turns in a coil becomes doubled then its self inductances will be:
(a) unchanged (b) four times
(c) halved (d) doubled

58. The numbers of turns of primary and secondary coils of transformer are 5 and 10 respectively and the mutual inductance of the transformer is 25 H. Now the numbers of turns in the primary and secondary of the transformer are made 10 and 5 respectively. The mutual inductance of the transformer will be:
(a) 50 henry (b) 25 henry
(c) 12.5 henry (d) 6.25 henry

59. An aircraft with a wing span of 20 m flies at a horizontal speed of 360 km/h in a region where the vertical component of the earth field is 5×10^{-5} tesla. The potential difference between the tips of the wing will be:
(a) 0.30 eV (b) 0.20 V
(c) 0.10 V (d) 0.15 V

60. The magnitude of flux linked with the circuit of resistance 2 kΩ varies with t second according to equation:

$$\phi = 5t^2 - 4t + 12$$

The induced current in the circuit at $t = 0.2$ second is:
(a) 2 A (b) 1 A
(c) 2 mA (d) 1 mA

CHAPTER 23 Ray Optics and Optical Instruments

CONCEPTS

REFLECTION OF LIGHT

It is the turning back back of light in the same medium.

Laws of Reflection

There are three laws of reflection :

(i) The angle of incidence is equal to the angle of reflection.
i.e., $\angle i = \angle r$

(ii) The incident ray, the normal and the reflected ray lie in the same plane.

(iii) There is a phase change of π radians when light wave is reflected by permeable denser medium surface but no phase change occurs if it is reflected by rarer medium surface.

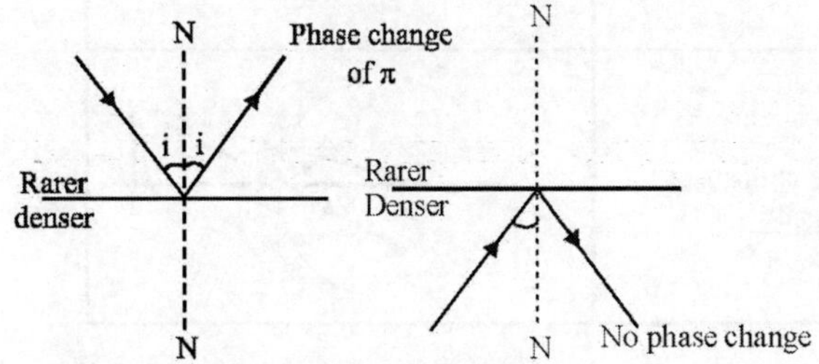

Reflection by Plane Mirror

(a) The image formed by the plane mirror is always erect, of the same size and at the same distance as the object is.

(b) To see the full image in a plane mirror, its length is just half the height of the man and it has to be kept in specific position.

(c) When the plane mirror or any reflecting surface is turned through an angle θ and if incident ray remains stationary, then the reflected ray will turn through 2θ.

(d) Image of an object formed by a plane mirror is perverted & if object is real then virtual image formed.

(e) When the two plane mirrors are parallel to each other, the number of images is infinity.

(f) When two plane mirrors are held at angle θ with their reflecting surfaces facing each other and an object is placed between them, images are formed by successive reflections.

First of all we will calculate, $n = \dfrac{360°}{\theta}$ then

(1) If n is fractional number, number of images will be whole part of this number.
Ex. : If n = 7.2 number of images = 7

(2) If n is whole number
(i) Even whole number, Number of images = n – 1
(ii) Odd whole number
(a) For symmetrical object, number of images = n – 1
(b) For asymmetrical object, number of images = n

Reflection at Curved Surfaces

Image formed by the concave mirror :

Position of object	Position of image	Nature	Figure
At infinity	At the focus	Real, inverted and diminished	

Between infinity and centre of curvature	Between focus & centre of curvature	Real, inverted and small in size	
At centre of curvature	At centre of curvature	Real, inverted and of the same size	
Between focus and centre of curvature	Beyond centre of curvature	Real, inverted and enlarged	
At focus	At infinity	Real, inverted and very large	
Between focus and Pole	Behind the mirror	Erect, virtual and enlarged	

Image formed by convex mirror :

A convex mirror forms only virtual images for all positions of the real object. The image is always virtual, erect, smaller than the object and is located between the pole and the focus.

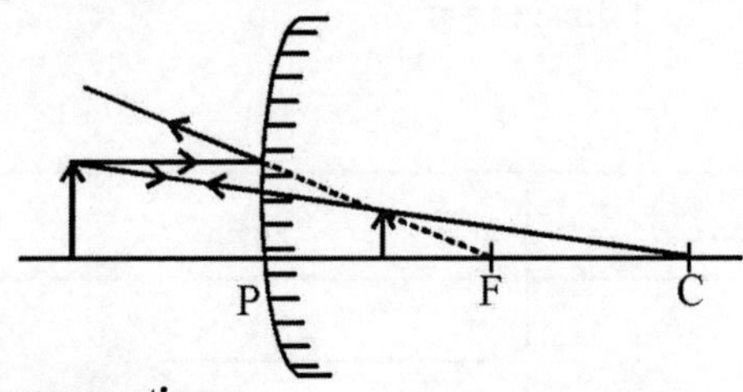

Sign convention :

(1) All distances are measured from the pole.

(2) The distance measured along the direction of propagation of light are taken as positive and the direction opposite to the propagation of light is taken as negative.

(3) The distance (heights) measured above the principal axis i.e. along positive Y axis, are taken as positive while distances below the principal axis i.e. along negative Y axis are taken as negative.

Mirror Formulae

1. If an object is placed at a distance u from the pole of a mirror and its image is formed at a distance v (from the pole) then

$$\frac{1}{v}+\frac{1}{u}=\frac{1}{f}$$

This is known as mirror formula.

In this formula to calculate unknown, known quantities are substituted with proper sign.

2. If a thin object linear size O situated vertically on the axis of a mirror at a distance u from the pole and its image of size I is formed at a distance v (from the pole), magnification (transverse) is defined as

$$m = \left[\frac{I}{O}\right] = -\left[\frac{v}{u}\right]$$

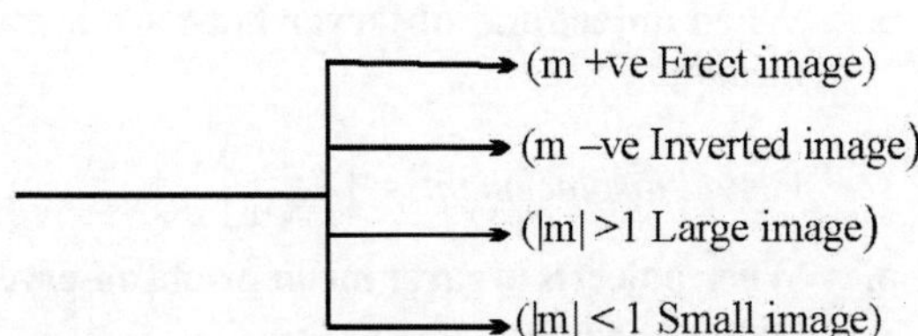

Here –ve magnification implies that image is inverted with respect to object while +ve magnification means that image is erect with respect to object.

3. **Other formulae of magnification**

$$m = \frac{f}{f-u}, \quad m = \frac{f-v}{f}$$

4. **The focal-length of a spherical mirror of radius R is given by** $f = (R/2)$

In sign convention, f (or R) is negative for concave or converging mirror and positive for convex or diverging mirror.

5. **Newton's formula** $f^2 = x_1 x_2$

f = focal length of mirror] x_1 = position of object with respect to focus] x_2 = position of image with respect to focus.

USES OF SPHERICAL MIRROR

Concave Mirror : Makeup and shaving mirrors are concave mirrors. When you place your face between the mirror and its focal point, you see an enlarged virtual image of yourself. Concave mirrors are also used in a new method for displaying the speed of a car.

Convex Mirror : It is used as rear view mirror. A convex mirror form a virtual image of the object, no matter where in front of the mirror the object is placed. Because of the shape of convex mirrors, they give a wider view than do other types of mirrors.

REFRACTION OF LIGHT

The bending of the light ray from its path in passing from one medium to the other medium is called refraction of light.

If the refracted ray bends towards the normal relative to the incident ray, then the second medium is said to be denser than the first medium. But if the refracted ray bends away from the normal, then the second medium is said to be rarer than the first medium.

Relative Refractive Index

When light passes from one medium to the other, the refractive index of medium 2 relative to 1 is written as ${}_1\mu_2$ and is defined as

$${}_1\mu_2 = \frac{\mu_2}{\mu_1} = \frac{(c/v_2)}{(c/v_1)} = \frac{v_1}{v_2}$$

Regarding refractive index it is worthnoting that

(a) It is a scalar and has no units and dimensions.

(b) If ϵ_0 and μ_0 are electric permittivity and magnetic permeability respectively of free space while ε and μ those of a given medium, then according to electromagnetic theory

$$c = \frac{1}{\sqrt{\varepsilon_0\mu_0}} \text{ and } v = \frac{1}{\sqrt{\varepsilon\mu}},$$

$$n = \frac{c}{v} = \sqrt{\frac{\varepsilon\mu}{\varepsilon_0\mu_0}} = \sqrt{\varepsilon_r\mu_r}$$

(c) For a given medium (other than free-space), the speed of light of different wavelengths is different, i.e., $v \propto \lambda$ and $\mu = (c/v)$, $\mu \propto (1/\lambda)$, i.e. greater the wavelength of light lesser will be the refractive index e.g. $\lambda_R > \lambda_B$, so in water or glass $\mu_R < \mu_B$, i.e., for a given medium (other than free space) refractive index depends on wavelength of light.

Laws of Refraction (At any Refracting Surface)

(i) The incident ray (AB), the normal to the refracting surface at the point of incidence (B) and the refracted ray (BC) all lie in the same plane called the plane of incidence or plane of refraction.

(ii) For any two given media and for light of a given wavelength

$$\frac{\sin i}{\sin r} = \text{constant}.$$

This is known as **Snell's law.**

Also, $\frac{\sin i}{\sin r} = {}_1\mu_2 = \frac{\mu_2}{\mu_1} = \frac{v_1}{v_2} = \frac{\lambda_1}{\lambda_2}$

where ${}_1\mu_2$ = refractive index of the second medium with respect to the first medium.

μ_1 = R.I. of the 1st medium

with respect to air or absolute R.I. $= \frac{c}{v_1}$

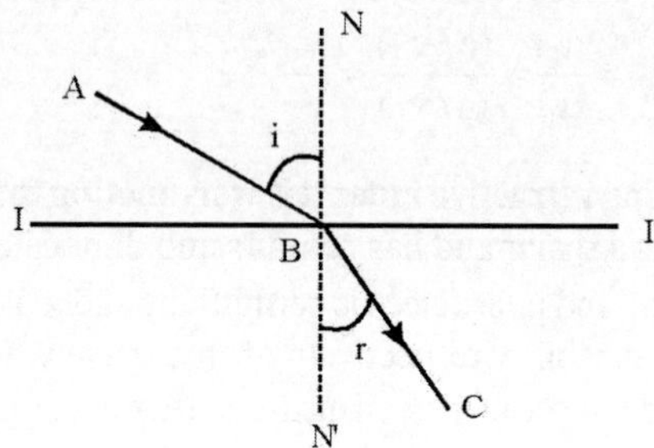

μ_2 = R.I. of the 2nd medium with respect

to air or absolute R.I. $= \frac{c}{v_2}$

v_1, v_2 are speed of light in first and second medium respectively. λ_1, λ_2 are wavelength of the light in 1st and 2nd medium respectively.

c = speed of light in air (or vacuum) = 3×10^8 m/s

(i) Higher the value of R.I. denser (optically) is the medium.

(ii) Frequency (colour), phase of light does not change during refraction.

Refraction Through Glass Slab

When one light ray incidents on slab then displacement produced in emergent ray due to refraction

$$x = \frac{t\sin(i-r)}{\cos r} = t\sec r\sin(i-r)$$

where, μ = refractive index of slab,

t = thickness of slab

1. **When object in denser medium and observer in rarer medium:**

Thickness of denser medium is t, in which a object is at a point.

Refractive index, $\mu = \frac{\textit{Real depth}}{\textit{Virtual depth}}$

Virtual depth $= \frac{t}{\mu}$

Virtual displacement $= t\left(1-\frac{1}{\mu}\right)$

2. **Refraction through successive slab of different thickness and refractive index:**

Virtual depth $= \frac{t_1}{\mu_1} + \frac{t_2}{\mu_2} + \frac{t_3}{\mu_3} + \ldots\ldots\ldots\ldots$

Virtual displacement

$$= t_1\left(1-\frac{1}{\mu_1}\right) + t_2\left(1-\frac{1}{\mu_2}\right) + t_3\left(1-\frac{1}{\mu_3}\right) + \ldots\ldots\ldots\ldots$$

3. **When object and observer both are in rarer medium:**

Virtual displacement $= \left(t - \frac{1}{\mu}\right)$

4. **When object is in rarer medium and observer in denser medium.**

$\frac{\textit{Real height}}{\textit{Virtual height}} = \frac{1}{\mu}$

Virtual displacement = (μ – 1) real height.

TOTAL INTERNAL REFLECTION (TIR)

In case of refraction of light, from snell's law, we have

$\mu_1 \sin i = \mu_2 \sin r$

$\sin i = \frac{\sin r}{\mu}$

The value of angle of incidence for which angle of refraction r = 90° is called **critical angle** and is denoted by θ_c.

$\sin\theta_C = \frac{\sin 90°}{\mu}$

i.e., $\sin\theta_C = \frac{1}{\mu}$ or $\sin i = \sin r \times \sin\theta_c$

i.e., $\sin r = \frac{\sin i}{\sin\theta_C}$

Conditions for Total Internal Reflection

(i) For total internal reflection to take place light must pass from denser to rarer medium thus for light from air to water (or glass) and from water to glass, total internal reflection can-never takes place.

(ii) Total internal reflection will take place only if angle of incidence is greater than critical angle i.e., $\angle i > \theta_c$

Some examples of total internal reflection :

(1) Shining of air bubble

(2) Sparking of diamond

(3) Optical - fibre

(4) Action of 'Porro' prism

(5) Mirage and looming

REFRACTION AT SPHERICAL SURFACE

Let the object is placed in a medium of refractive index μ_1. The spherical curved surface (convex or concave) separate it from another medium of refractive index μ_2. Then if v, u and R are respectively, the object distance, the image distance and the radius of curvature of the refracting surface, then the formula connecting u, v and R is

$$-\frac{\mu_1}{u}+\frac{\mu_2}{v}=\frac{\mu_2-\mu_1}{R}$$

This formula is applicable for convex as well as concave spherical curved surfaces.

LENS

A lens is a piece of transparent material with two refracting surface such that least one is curved and refractive index of use material is different from that of the surroundings.

Sign - Convention

1. Whenever and where possible, rays of light are taken to travel from left to right.
2. Transverse distance measured from optical centre and are taken to be positive while those below it negative.
3. Longitudinal distances are measured from optical centre and are taken to be positive if in the direction of light propagation and negative if opposite to it.

Image Formation by a Lens

For convergent or Convex lens

Position of object	Nature and Position of image	Figure
At infinity	Real, inverted, diminished (m<<–1) at F	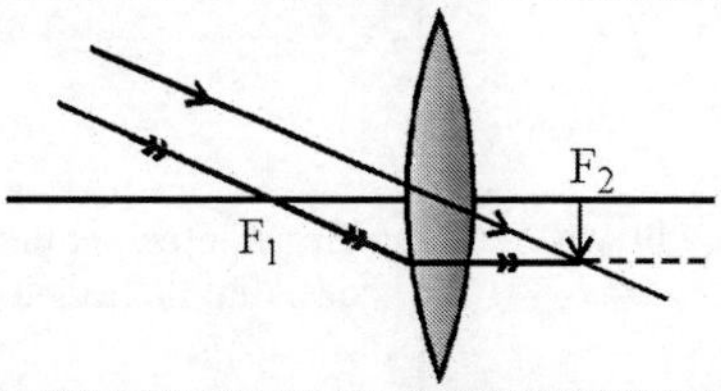 F_1, F_2
Between infinity and 2F	Real, inverted, diminished (m<–1) between F and 2F	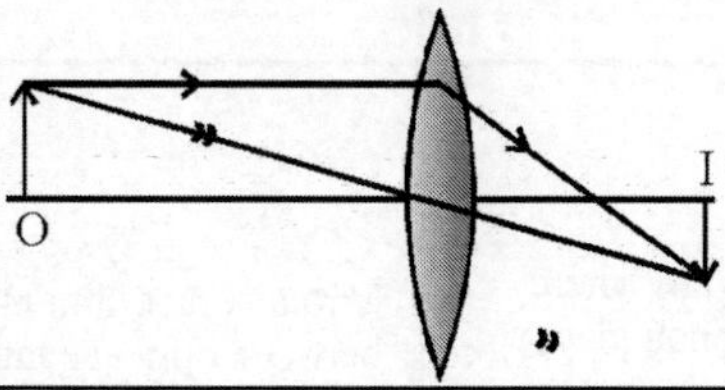 O, I
At 2F	Real, inverted diminished m = – at 2F	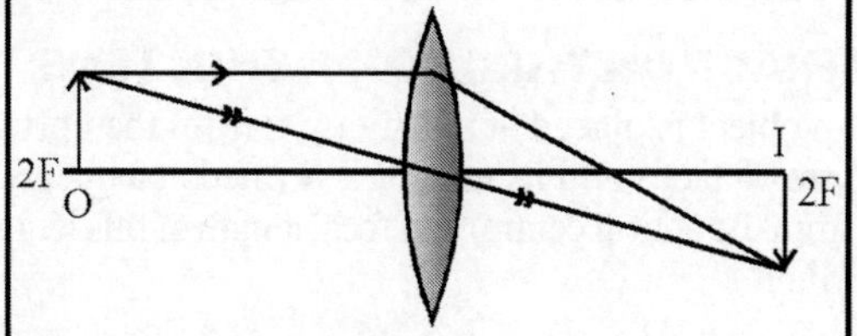 2F, O, I, 2F
Between 2F and F	Real, inverted, enlarged (m>–1) between 2F and ∞	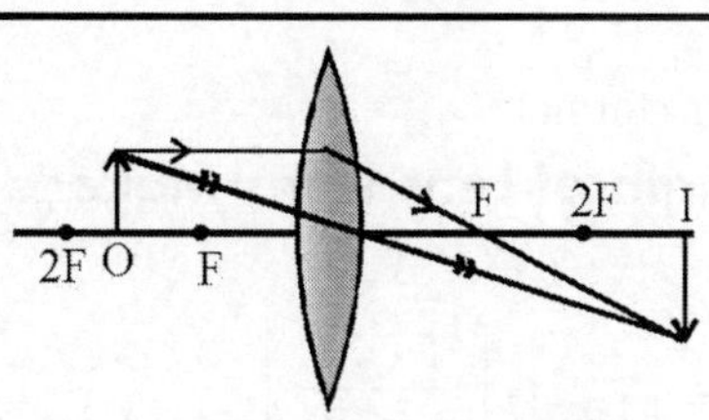 2F, O, F, F, 2F, I

At F	Real, inverted enlarged (m>>–1) At infinity	
Between focus and pole	Virtual, erect enlarged (m > +1) Between ∞ and object on same side	

For divergent or Concave lens

Position of object	Nature and position of image	Figure
Object is at infinity	Virtual, erect, point size, diminished, at F	
Object is anywhere on the principal axis	Virtual, erect, diminished, between optical centre and F,	

REFRACTION THROUGH A THIN LENS

If an object is placed at a distance u from the optical centre of a lens and its images is formed at a distance v (from the optical centre) and focal length of this length is f then

$$\frac{1}{v}-\frac{1}{u}=\frac{1}{f}$$

This is the lens formula.

Focal Length of Lens (Lens Maker's Formula)

$$\frac{1}{f}=({}_{m}\mu_{\ell}-1)\left[\frac{1}{R_1}-\frac{1}{R_2}\right]$$

where ${}_{m}\mu_{\ell}$ refractive index of lens with respect to medium.

R_1 = radius of curvature of first surface of lens,
R_2 = radius of curvature of second surface of lens

The power of a lens

$$P=\frac{1}{f(\text{in m})}=\frac{100}{f(\text{in cm})}$$

The unit of power is **diopter**.

If a thin object linear size O situated vertically on the axis of a lens at a distance u from the optical centre and its image of size I is formed at a distance v (from the optical centre)then **magnification(m)** (transverse) is defined as

$$m=\left[\frac{I}{O}\right]=\left[\frac{v}{u}\right]$$

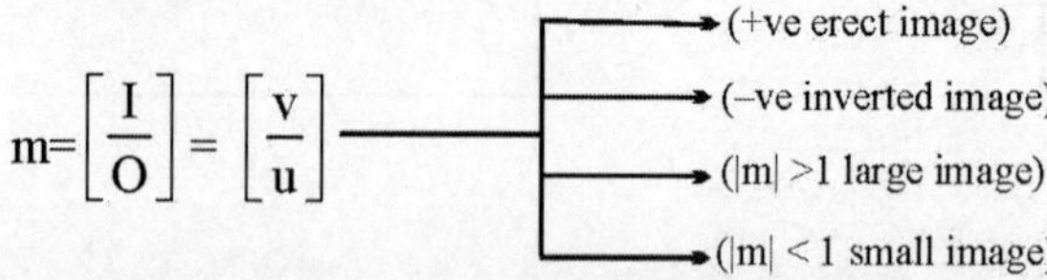

Here –ve magnification implies that image is inverted with respect to object while +ve magnification means that image is erect with respect to object.

Other formulae of magnification

$$m = \frac{f}{f+u}, \quad m = \frac{f-v}{f}$$

COMBINATION OF LENSES

When two lenses of focal lengths f_1 and f_2 are placed in contact, the combination behave as a single lens of equivalent focal length F given by –

$$\frac{I}{F} = \frac{1}{f_1} + \frac{1}{f_2}$$

and equivalent power, $P = P_1 + P_2$

If the two thin lens are separated by a distance d apart F is given by

$$\frac{I}{F} = \frac{1}{f_1} + \frac{1}{f_2} - \frac{d}{f_1 f_2}$$

and power $P = P_1 + P_2 - P_1 P_2 d.$

NOTE :

1. If a lens of focal length f is divided into two equal parts as in figure (A) each part has a focal length f' then

   $$\frac{1}{f} = \frac{1}{f'} + \frac{1}{f'} \; ; \; f' = 2f$$

   i.e. each part have focal length 2f now if these parts are put in contact as in (B) or (C) the resultant focal length of the combination will be

   $$\frac{1}{F} = \frac{1}{2f} + \frac{1}{2f} \quad \text{i.e. } F = f\,(=\text{initial value})$$

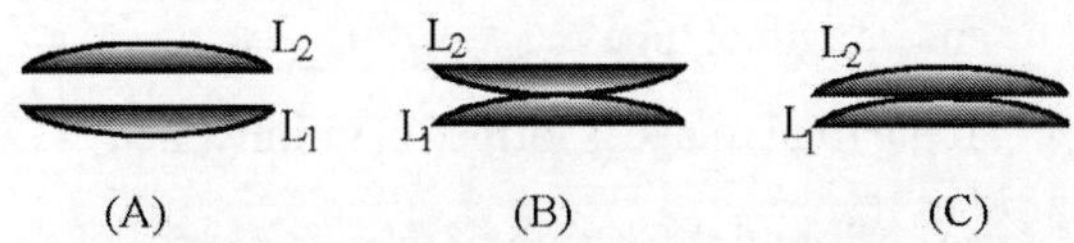

(A) (B) (C)

2. If a lens of focal length f is cut in two equal parts as shown,(A) in each will have focal length f. Now if these parts are put in contact as shown in (B) the resultant length will be

   $$\frac{1}{F} = \frac{1}{f} + \frac{1}{f} \quad \text{i.e. } F = (f/2)$$

L1 L2 L1 L2 L1 L2

(A) (B) (C)

However if the two parts are put in contact as shown in (C) first will behave as convergent lens of focal length f while the other divergent of same focal length (being thinner near the axis) so in this situation

$$\frac{1}{F} = \frac{1}{+f} + \frac{1}{-f} \quad \text{i.e. } -F = \infty \text{ and power } P = 0$$

REFRACTION BY A PRISM

1. **Formula for the refractive index of the prism**

   $$n = \frac{\sin i}{\sin r} = \frac{\sin\left(\frac{A+\delta_m}{2}\right)}{\sin\left(\frac{A}{2}\right)}$$

   Where A is prism angle and δ_m is the angle of minimum deviation.

   (a) When prism is thin, then value of A will be small ($\leq 10°$)

   $$\delta_m = (n-1)A$$

   (b) Condition for maximum deviation i_1 or $i_2 = 90°$.

2. **Angular dispersion for prism :** White light splits into its seven constituent colours (spectrum), on passing through a prism. This is known as **dispersion**. The angle between the emergent rays of any two colours is called angular dispersion between those colours.

   If deviation angle for violet and red are δ_v & δ_R respectively, then angular dispersion.

   $\theta = \delta_v - \delta_R \quad (\delta_v > \delta_R)$

   $\delta_R = (n_R - 1)A,$

   $\delta_v = (n_v - 1)A$

   $\theta = (n_v - 1)A - (n_R - 1)A = (n_v - n_R)A.$

3. **Dispersive power of prism**

   $$\omega = \frac{\delta_V - \delta_R}{\delta_Y} = \frac{(n_V - n_R)A}{(n_Y - 1)A}$$

   Where S_y is deviation angle for yellow colour.

   $$\omega = \frac{n_V - n_R}{n_Y - 1}$$

4. **Deviationless dispersion :** Yellow (mean ray) emergent through the combination of prism of different material will be parallel to the incident ray, but other emerging rays of different colours are not parallel to each other. Then,

   $$\frac{\text{Prism angle (2)}}{\text{Prism angle (1)}} = \frac{A_2}{A_1} = \frac{(n_1 - 1)}{(n_2 - 1)}$$

5. **Dispersionless deviation :** Emerging rays of different colour through the combination of prism of different material will be parallel to each other but not parallel to incident ray. Rays of different

colours are parallel, therefore angle between them is zero. It means angular dispersion is zero. Although rays of different colours are parallel but these are not coincident.

$$\frac{\text{Prism angle (2)}}{\text{Prism angle (1)}} = \frac{A_2}{A_1} = \frac{n_{1v} - n_{1r}}{n_{2v} - n_{2r}}$$

SCATTERING OF LIGHT

According to Rayleigh, the intensity of scattered light $I \propto \frac{1}{\lambda^4}$. Some phenomena like – reddish appearance of the sun at sunrise and sunset, blue colour of sky etc. based on scattering of light.

OPTICAL INSTRUMENTS

Human Eye

The normal range of vision for a healthy human eye is 25 cm (least distance of distinct vision D to ∞ (far point).

Defects of Vision

Nearsightedness: If the eyeball is too long or the lens too spherical, the distant objects cannot be seen clearly but nearby objects can be seen more easily. Eye glasses with concave lenses correct this problem by diverging the light rays before they enter the eye. Nearsightedness is called **myopia**.

If deflected far point is at a distance d from eye then focal length of used lens is f = – d = – (deflected far point)

A person can see upto distance $\to$ x, wants to see distance $\to y (y > x)$ so $f = \frac{xy}{x - y}$ or power of the lens $P = \frac{x - y}{xy}$.

Farsightedness : If the eyeball is too short or the lens too flat or inflexible, the nearby objects cannot be seen clearly but distant objects can be seen clearly. Eyeglasses with convex lenses can correct the problem. Farsightedness is called **hypermetropia**. If a person cannot see objects before distance d but wants to see the object placed at distance D from eye so $f = \frac{dD}{d - D}$

and power of the lens $P = \frac{d - D}{dD}$

Astigmatism: Astigmatism is the most common refractive problem responsible for blurry vision. Cylindrical lens is used to correct astigmatism.

Presbyopia ("after 40" vision) : After age 40, and most noticeably after age 45, the human eye is affected by presbyopia, which results in greater difficulty maintaining a clear focus at a near distance with an eye which sees clearly at a far away distance. This is due to a lessening of flexibility of the crystalline lens, as well as to a weakening of the ciliary muscles which control lens focusing, both attributable to the aging process.

Simple Microscope

1. Use of single lens of small focal length
2. Magnifying power

   Image formed at D: $M = \left(1 + \frac{D}{f}\right)$

   Image formed at infinite : $M = \frac{D}{f}$
3. Object is placed between optical centre and focus

Compound Microscope

1. Use of two lenses $f_0 < f_e$
2. Magnifying power

   Image formed at D

   $$M = -\frac{v_0}{u_0}\left(1 + \frac{D}{f_e}\right)$$

   Image formed at infinite, $M = -\frac{v_0 D}{u_0 f_e}$
3. Object is placed just outside the principl focus

Astronomical Telescope

1. If the final image is formed at a distance D,

   $$M = \frac{-f_0}{f_e}\left(1 + \frac{f_e}{D}\right)$$

   and length of tube is $L = f_0 + u_e = f_0 + \frac{f_e D}{f_e + D}$
2. If the final image is formed at infinity then

   $M = \frac{f_0}{f_e}$ and the length of tube is $\ell = f_0 + f_e$

Terrestrial Telescope

1. If image is formed at D

   $$M = \frac{f_0}{f_e}\left(1 + \frac{f_e}{D}\right)$$

   Length of tube = $f_0 + 4f + u_e$
2. If final image is formed at infinity, $M = \frac{f_0}{f_e}$;

   Length of tube = $f_o + f_e + 4f$, where f is the focal length of erecting lens, which is used in this Telescope.

RESOLVING POWER

(i) The resolving power of microscope is $R = \frac{2\mu \sin\theta}{\lambda}$

where μ = refractive index of medium between object and objective of microscope and θ = angle subtended by a radius of the objective on one of the objects. (When both objects are not self illuminous).

(ii) The resolving power of a telescope is $R = \frac{a}{1.22\lambda}$ where

a = diameter of objective of telescope.

Exercise

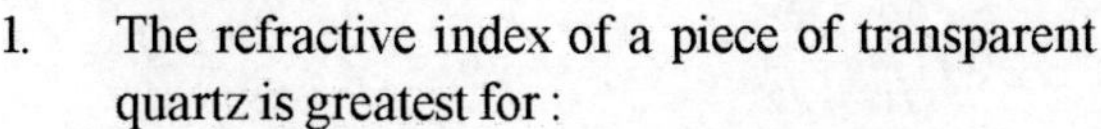

1. The refractive index of a piece of transparent quartz is greatest for :
   (a) violet light (b) green light
   (c) yellow light (d) red light
2. Two mirrors are placed at right angles to each other. A man is standing between them combing his hair. How many image will he see ?
   (a) 2 (b) 3
   (c) 1 (d) zero
3. In a room containing smoke particles, the intensity due to a source of light will :
   (a) obey the inverse square law
   (b) be constant at all distances
   (c) increase with distance from the source than the inverse fourth power law
   (d) fall faster with distance from the source than the inverse fourth power law
4. The image formed by a concave mirror
   (a) is always real
   (b) is always virtual
   (c) is certainly real if the object is virtual
   (d) is certainly virtual
5. A lens made of glass whose index of refraction is 1.60 has a focal length of + 20 cm in air. Its focal length in water, whose refractive index is 1.33, will be
   (a) three times longer than in air
   (b) two times longer than in air
   (c) same as in air
   (d) None of the above
6. Which of the following is not the case with the image formed by a concave lens?
   (a) It may be erect or inverted
   (b) It may be magnified or diminished
   (c) It may be real or virtual
   (d) Real image may be between the pole and focus or beyond focus
7. Two thin lenses of focal lengths f_1 and f_2 are made of materials of dispersive powers ω_1 and ω_2 respectively. The lenses are kept in contact. The combination will show no chromatic aberration if
   (a) $\omega_1 = \omega_2$ (b) $(f_1/f_2) = -(\omega_1/\omega_2)$
   (c) $f_1\omega_1 = f_2\omega_2$ (d) $f_1\omega_1 + f_2\omega_2 = 0$
8. The focal length of a normal eye-lens is about
   (a) 1 mm (b) 2 cm
   (c) 25 cm (d) 1 m
9. Which of the following is used in optical fibres?
   (a) Total internal reflection
   (b) Scattering
   (c) Diffraction
   (d) Refraction
10. A parallel beam of light is incident on a converging lens parallel to its principal axis. As one moves away from the lens on the other side on its principal axis, the intensity of light
    (a) remains constant
    (b) continuously increases
    (c) continuously decreases
    (d) first increases then decreases
11. The rays of different colours fail to converge at a point after going through a converging lens. This defect is called
    (a) spherical aberration
    (b) distortion
    (c) coma
    (d) chromatic aberration
12. If the light moving in a straight line bends by a small but fixed angle, it may be a case of
    (a) reflection (b) refraction
    (c) diffraction (d) (a) & (b)

13. What causes chromatic aberration?
(a) Marginal rays
(b) Central rays
(c) Difference in radii of curvature of its surfaces
(d) Variation of focal length of lens with colour

14. Consider the following two statements :
A. Line spectra contain information about atoms
B. Band spectra contain information about molecules
(a) Both A and B are wrong
(b) A is correct but B is wrong
(c) B is correct but A is wrong
(d) Both A and B are correct

15. The focal length of a converging lens are f_V and f_R for violet and red light respectively. Then
(a) $f_V > f_R$
(b) $f_V = f_R$
(c) $f_V < f_R$
(d) any of the three is possible depending on the value of the average refractive index m

16. A narrow beam of white light goes through a slab having parallel faces
(a) the light never splits in different colours
(b) the emergent beam is white
(c) the light inside the slab is split into different colours
(d) the light inside the slab is white

17. By properly combining two prisms made of different materials, it is not possible to
(a) have dispersion without average deviation
(b) have deviation without dispersion
(c) have both dispersion and average deviation
(d) have either dispersion or average deviation

18. A double convex thin lens made of refractive index 1.6 has radii of curvature 15 cm each. The focal length of this lens when immersed in a fluid of refractive index 1.63, is :
(a) 25 cm (b) 125 cm
(c) 250 cm (d) – 407.5 cm

19. When a beam of light ($\lambda = 6000\mathring{A}$) travelling in air enters a glass medium ($\mu = 1.5$), the wavelength of light in glass, would be :
(a) $2000\mathring{A}$ (b) $3000\mathring{A}$
(c) $4000\mathring{A}$ (d) $5000\mathring{A}$

20. An object is placed at a distance of 20 cm from a convex lens of focal length 10 cm. The image is formed on the other side of the lens at a distance of :
(a) 10 cm (b) 40 cm
(c) 25 cm (d) 20 cm

21. The condition for an achromatic doublet, is (here f_1 and f_2 are the focal length of individual lenses):
(a) $f_1 - f_2 = \frac{\omega_1\omega_2}{2} f_2$ (b) $\omega_1 f_1 + \omega_2 f_2 = 0$
(c) $\frac{\omega_1}{f_1} + \frac{\omega_2}{f_2} = 0$ (d) $\frac{f_1}{f_2} + \frac{\omega_1}{\omega_1} = 0$

22. Focal length of a convex lens will be maximum for
(a) blue light (b) yellow light
(c) green light (d) red light

23. An achromatic combination of lenses is formed by joining
(a) 2 convex lenses
(b) 2 concave lenses
(c) 1 convex and 1 concave lens
(d) 1 convex and 1 plane mirror

24. The hypermetropia is a
(a) short-sight defect
(b) long-sight defect
(c) bad vision due to old age
(d) none of these

25. If two mirrors are kept inclined at 60° to each other and a body is placed at the middle, then total number of images formed is
(a) six (b) five
(c) four (d) three

26. A lens is placed between a source of light and a wall. It forms images of area A_1 and A_2 on the wall for its two different positions. The area of the source of light is
(a) $\sqrt{A_1 A_2}$ (b) $\frac{A_1 + A_2}{2}$
(c) $\frac{A_1 - A_2}{2}$ (d) $\frac{1}{A_1} + \frac{1}{A_2}$

27. If f_V and f_R are the focal lengths of a convex lens for violet and red light respectively and F_V and F_R are the focal lengths of concave lens for violet and red light respectively, then we have

(a) $f_V < f_R$ and $F_V > F_R$
(b) $f_V < f_R$ and $F_V < F_R$
(c) $f_V > f_R$ and $F_V > F_R$
(d) $f_V > f_R$ and $F_V < F_R$

28. A ray of light from a denser medium strike a rarer medium at an angle of incidence i (see Fig). The reflected and refracted rays make an angle of 90° with each other. The angles of reflection and refraction are r and r' The critical angle is

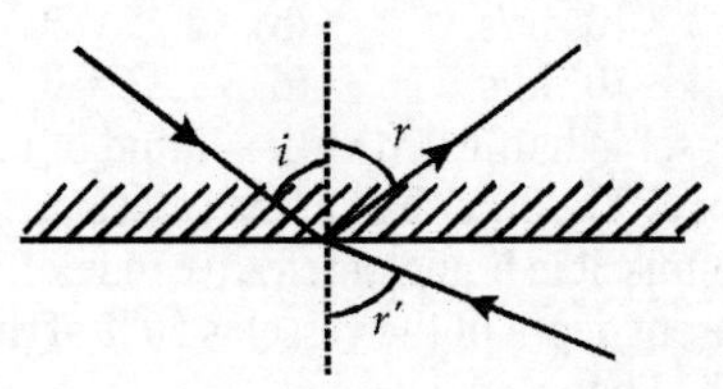

(a) $\sin^{-1}(\tan r)$
(b) $\sin^{-1}(\tan i)$
(c) $\sin^{-1}(\tan r')$
(d) $\tan^{-1}(\sin i)$

29. One face of a rectangular glass plate 6 cm thick is silvered. An object held 8 cm in front of the first face forms an image 12 cm behind the silvered face. The refractive index of the glass is
(a) 0.4
(b) 0.8
(c) 1.2
(d) 1.6

30. The focal length of converging lens is measured for violet, green and red colours. It is respectively f_v, f_g, f_r. We will get
(a) $f_v = f_g$
(b) $f_g > f_r$
(c) $f_v < f_r$
(d) $f_v > f_r$

31. If a glass prism is dipped in water, its dispersive power
(a) increases
(b) decreases
(c) does not change
(d) may increase or decrease depending on whether the angle of the prism is less than or greater than 60°

32. When a drop of oil is spread on a water surface, it displays beautiful colours in daylight because of
(a) dispersion of light
(b) reflection of light
(c) polarization of light
(d) interference of light

33. A bird in air looks at a fish vertically below it and inside water h_1 is the height of the bird above the surface of water and h_2, the depth of the fish below the surface of water. If refractive index of water with respect to air be μ, then the distance of the fish as observed by the bird is
(a) $h_1 + h_2$
(b) $h_1 + \frac{h_2}{\mu}$
(c) $\mu h_1 + h_2$
(d) $\mu h_1 + \mu h_2$

34. An electromagnetic radiation of frequency n, wavelength λ, travelling with velocity v in air, enters a glass slab of refractive index μ. The frequency, wavelength and velocity of light in the glass slab will be respectively
(a) $\frac{n}{\mu}, \frac{\lambda}{\mu}$ and $\frac{v}{\mu}$
(b) $n, \frac{\lambda}{\mu}$ and $\frac{v}{\mu}$
(c) $n, 2\lambda$ and $\frac{v}{\mu}$
(d) $\frac{2n}{\mu}, \frac{\lambda}{\mu}$ and v

35. Yellow light is refracted through a prism producing minimum deviation. If i_1 and i_2 denote the angle of incidence and emergence for this prism, then
(a) $i_1 = i_2$
(b) $i_1 > i_2$
(c) $i_1 < i_2$
(d) $i_1 + i_2 = 90°$

36. A luminous object is placed at a distance of 30 cm from the convex lens of focal length 20 cm. On the other side of the lens, at what distance from the lens a convex mirror of radius of curvature 10 cm be placed in order to have an upright image of the object coincident with it?
(a) 12 cm
(b) 30 cm
(c) 50 cm
(d) 60 cm

37. Light enters at an angle of incidence in a transparent rod of refractive index n. For what value of the refractive index of the material of the rod the light once entered into it will not leave it through its lateral face whatsoever be the value of angle of incidence?
(a) $n > \sqrt{2}$
(b) $n = 1$
(c) $n = 1.1$
(d) $n = 1.3$

38. A plano-convex lens is made of material of refractive index 1.6. The radius of curvature of the curved surface is 60 cm. The focal length of the lens is
(a) 50 cm
(b) 100 cm
(c) 200 cm
(d) 400 cm

39. Wavelength of light of frequency 100 Hz
(a) 2×16^6 m
(b) 3×10^6 m
(c) 4×10^6 m
(d) 5×10^6 m

40. The refractive index of the material of the prism is $\sqrt{3}$; then the angle of minimum deviation of the prism is
(a) 30°
(b) 45°
(c) 60°
(d) 75°

41. The radius of curvature of a thin plano-convex lens is 10 cm (of curved surface) and the refractive index is 1.5. If the plane surface is silvered, then it behaves like a concave mirror of focal length
(a) 10 cm (b) 15 cm
(c) 20 cm (d) 5 cm

42. The angular dispersion produced by a prism
(a) increases if the average refractive index increases
(b) increases if the average refractive index decreases
(c) remains constant whether the average refractive index increases or decreases
(d) has no relation with average refractive index

43. An astronomical telescope has a length of 44 cm and tenfold magnification. The focal length of the objective lens is
(a) 4 cm (b) 40 cm
(c) 44 cm (d) 440 cm

44. An electromagnetic radiation of frequency n, wavelength λ, travelling with velocity v in air enters in a glass slab of refractive index (μ). The frequency, wavelength and velocity of light in the glass slab will be respectively
(a) $n, \frac{\lambda}{\mu}$ and $\frac{v}{\mu}$ (b) $n, 2\lambda$ and $\frac{v}{\mu}$
(c) $\frac{n}{\mu}, \frac{\lambda}{\mu}$ and $\frac{v}{\mu}$ (d) $\frac{2\pi}{\mu}, \frac{\lambda}{\mu}$ and v

45. A man's near point is 0.5 m and far point is 3 m. Power of spectacle lenses required for (i) reading purposes, (ii) seeing distant objects, respectively, are
(a) –2 D and +3 D (b) +2 D and –3 D
(c) +2 D and –0.33 D (d) –2 D and +0.33 D

46. An equilateral prism is placed on a horizontal surface. A ray PQ is incident onto it. For minimum deviation

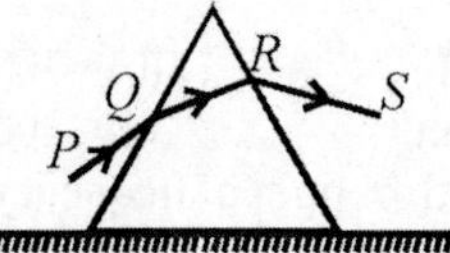

(a) PQ is horizontal
(b) QR is horizontal
(c) RS is horizontal
(d) Any one will be horizontal

47. Light travels through a glass plate of thickness t and having refractive index μ. If c be the velocity of light in vacuum, the time taken by the light to travel this thickness of glass is
(a) $\frac{t}{\mu c}$ (b) $t\mu c$
(c) $\frac{\mu t}{c}$ (d) $\frac{tc}{\mu}$

48. The refractive index of water is 1.33. What will be speed of light in water?
(a) 3×10^8 m/s (b) 2.25×10^8 m/s
(c) 4×10^8 m/s (d) 1.33×10^8 m/s

49. A vessel is half filled with a liquid of refractive index μ. The other half of the vessel is filled with an immiscible liquid of refrative index 1.5 μ. The apparent depth of the vessel is 50% of the actual depth. Then μ is
(a) 1.4 (b) 1.5
(c) 1.6 (d) 1.67

50. Two thin lenses of focal lengths f_1 and f_2 are in contact and coaxial. Its power is same as power of a single lens given by
(a) $\frac{f_1+f_2}{f_1 f_2}$ (b) $\sqrt{\left(\frac{f_1}{f_2}\right)}$
(c) $\sqrt{\left(\frac{f_2}{f_1}\right)}$ (d) $\frac{f_1+f_2}{2}$

51. Focal length of a convex lens of refractive index 1.5 is 2 cm. Focal length of the lens when immersed in a liquid of refractive index of 1.25 will be
(a) 10 cm (b) 2.5 cm
(c) 5 cm (d) 7.5 cm

52. Ray optics is valid, when characteristic dimensions are
(a) of the same order as the wavelength of light
(b) much smaller than the wavelength of light
(c) of the order of one millimetre
(d) much larger than the wavelength of light

53. A ray is incident at an angle of incidence i on one surface of a prism of small angle A and emerges normally from the opposite surface. If the refractive index of the material of prism is μ, the angle of incidence i is nearly equal to
(a) $\frac{A}{\mu}$ (b) $\frac{A}{2\mu}$
(c) μA (d) $\frac{\mu A}{2}$

54. A beam of monochromatic light is refracted from vacuum into a medium of refractive index 1.5, the wavelength of refracted light will be
(a) dependent on intensity of refracted light
(b) same
(c) smaller
(d) larger

55. Green light of wavelength 5460 $\mathring{A}$ is incident on an air-glass interface. If the refractive index of glass is 1.5, the wave length of light in glass would be ($c = 3 \times 10^8\ ms^{-1}$)
(a) 3640$\mathring{A}$ (b) 5460$\mathring{A}$
(c) 4861$\mathring{A}$ (d) none of the above

56. There is a prism with refractive index equal to $\sqrt{2}$ and the refracting angle equal to 30°. One of the refracting surfaces of the prism is polished. A beam of monochromatic light will retrace its path if its angle of incidence over the refracting surface of the prism is
(a) 0° (b) 30°
(c) 45° (d) 60°

57. Time taken by sunlight to pass through a window of thickness 4 mm whose refractive index is $\frac{3}{2}$ is
(a) 2×10^{-4} sec (b) 2×10^{8} sec
(c) 2×10^{-11} sec (d) 2×10^{11} sec

58. A point source of light is placed 4 m below the surface of water of refractive index $\frac{5}{3}$. The minimum diameter of a disc, which should be placed over the source, on the surface of water to cut off all light coming out of water is
(a) ∞ (b) 6m
(c) 4m (d) 3m

59. Angle of deviation (δ) by a prism (refractive index $= \mu$ and supposing the angle of prism A to be small) can be given by
(a) $\delta = (\mu - 1)A$ (b) $\delta = (\mu + 1)A$
(c) $\delta = \dfrac{\sin\frac{A+\delta}{2}}{\sin\frac{A}{2}}$ (d) $\delta = \dfrac{\mu - 1}{\mu + 1}A$

60. A convex lens of focal length 80 cm and a concave lens of focal length 50 cm are combined together. What will be their resulting power?
(a) +6.5 D (b) −6.5 D
(c) +7.5 D (d) −0.75 D

CHAPTER 24

Wave Optics

CONCEPTS

WAVEFRONT

The locus of all particles vibrating in the same phase is called wavefront.

(i) Each point of wavefront is considered as source of secondary wavelets.

(ii) The ray of light is considered in the direction of outward normal to the wavefront.

Waves from a point source spread out in all directions, while light was seen to travel along narrow rays. It required the insight and experiment of Huygens, Young and Fresnel to understand how a wave theory could explain all aspects of the behaviour of light.

HUYGEN'S PRINCIPLE

(1) Huygen proposed the wave theory of light. Light travels in the form of wavefront in the medium.

(2) Energy is emitted in the form of waves from light source.

(3) It need medium for propagation of light, that is why a universal hypothetical medium called ether is proposed.

(4) Each point on the wavefront acts as a centre of new disturbance and emits its own set of spherical waves called **secondary wavelets**.

(5) The envelope or the locus of these wavelets in the forward direction gives the position of new wavefront at any subsequent time.

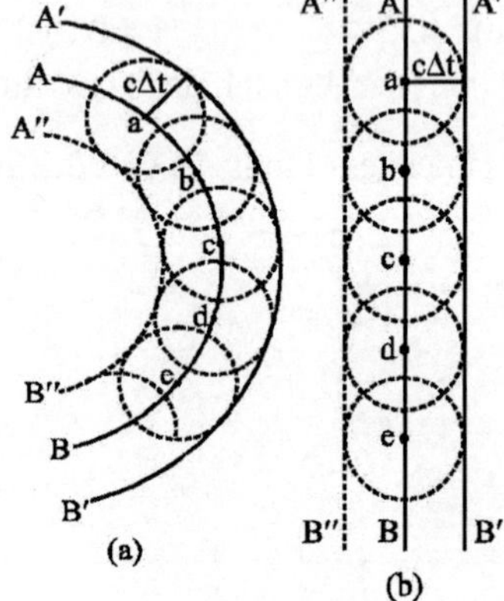

(6) Direction of wave propagation and wavefront are mutually perpendicular to each other.

(7) At large distances from the source, all wavefronts appear to be plane.

(8) The theory explained successfully the reflection, refraction, interference and diffraction.

(9) Huygen's wave theory cannot distinguish between longitudinal and transverse waves. Hence, the theory could not explain polarisation. However, later on it could interpret the phenomenon on the basis of light being assumed to be transverse.

(10) This principle is valid for all types of wave, e.g. light waves, sound waves etc.

Drawbacks of Huygen's Wave Theory

(a) This theory cannot explain photo-electric effect, compton, and Raman effect.

(b) Hypothetical medium in vacuum is not true imagination.

(c) The theory predicted the presence of back wave, which proved to be failure.

REFLECTION AND REFRACTION OF WAVEFRONT

Reflection :

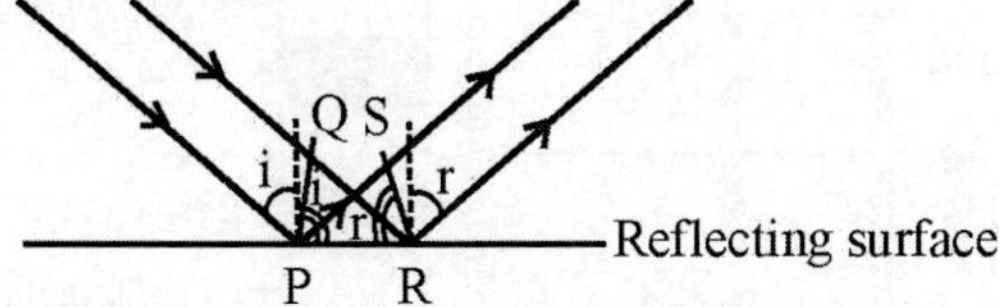

PS = QR and $\angle i = \angle r$ Law of reflection

Refraction :

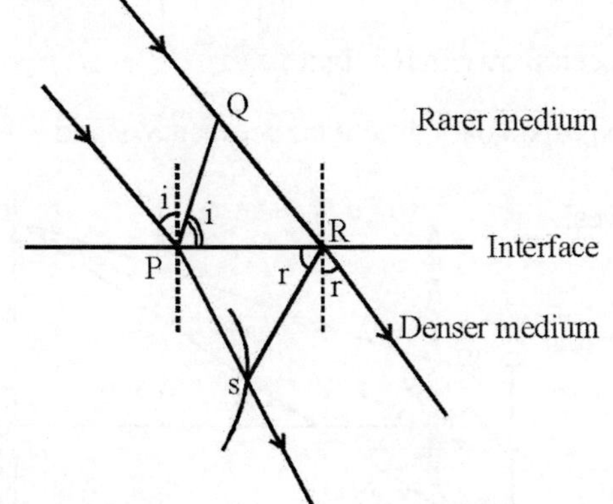

$\frac{QR}{PS} = \frac{V_1}{V_2} = \frac{\text{Sin } i}{\text{Sin } r} = \frac{\mu_2}{\mu_1}$ or $\mu = \frac{\sin i}{\sin r}$ Law of refraction

INTERFERENCE AND SUPERPOSITION OF WAVES

When we consider the intensity of light due to these sources at the given point, there is an interference term in addition to the sum of the individual intensities. But this term is important only if it has a non-zero average, which occurs only if the sources have the same frequency and a stable phase difference. Two waves arriving at same point produce brightness **(constructive interference)** if their path difference is equal to $n\lambda$, where n is an integer = 0, 1, 2, 3, ... etc. They produce darkness **(destructive interference)** when path difference is equal to $(2n+1)\frac{\lambda}{2}$ where n = 0, 1, 2, 3, ... etc.

Interference of Two Waves

When two waves of same frequency travel in a medium simultaneously in the same direction then, due to their **superposition**, the resultant intensity at any point of the medium is different from the sum of intensities of the two waves. At some points the intensity of the resultant wave is very large while at some other points it is very small or zero .

Mathematical interpretation of interference of two waves :

Let us consider two simple harmonic progressive waves of the same frequency travelling in the same direction, Let a_1 and a_2 be the amplitudes of the waves at that point be y_1 and y_2 then.

$y_1 = a_1 \sin \omega t$ .... (i)

and $y_2 = a \sin(\omega t + \phi)$ ....(ii)

$\omega/2\pi$ is the frequency of each wave. By the principle of superposition, the resultant displacement at the point is given by $y = y_1 + y_2$

Let $a_1 + a_2 \cos\phi = R\cos\theta$ ....(iii)

and $a_2 \sin\phi = R\sin\theta$ ....(iv)

where R and θ are new constants.

then $y = R\sin(\omega t + \theta)$

To determine R, we square eq.(iii) and (iv) and then add:

$R^2\cos^2\theta + R^2\sin^2\theta = (a_1 + a_2\cos\phi)^2 + (a_2\sin\phi)^2$

or $R^2 = a_1^2 + a_2^2 + 2a_1a_2\cos\phi$

The intensity is directly proportional to the square of the amplitude. Hence the resultant intensity I is given by

$I \propto (a_1^2 + a_2^2 + 2a_1a_2\cos\phi)$ and phase angle ϕ

$$= \tan^{-1}\frac{a_2\sin\phi}{a_1 + a_2\cos\phi}$$

Thus the resultant intensity at any point depends upon the phase difference ϕ between the two waves at that point.

Interference Due to Transmitted Light

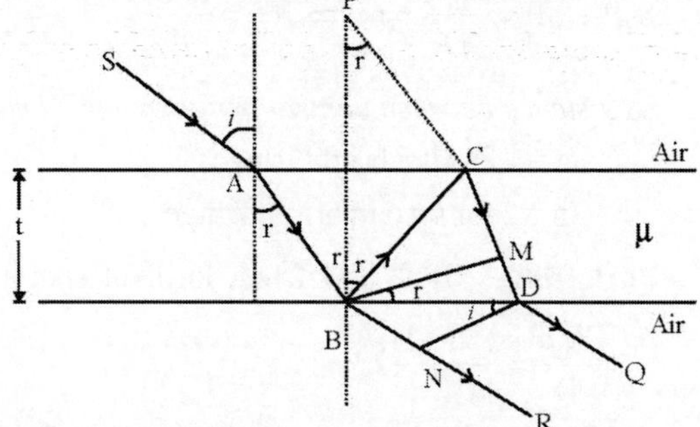

The optical path difference between the reflected ray (DQ) and the transmitted ray (NR) is given by

$x = \mu(BC + CD) - BN$

On simplification, we get $x = 2\mu t\cos r$

If $2\mu t\cos r = n\lambda$, where n = 0, 1, 2, 3, ...then constructive interference takes place and the film appears bright.

If $2\mu t\cos r = (2n+1)\frac{\lambda}{2}$, n = 0, 1, 2, ... then destructive interference takes place and the film appears dark.

YOUNG'S DOUBLE SLIT EXPERIMENT (YDSE)

- This experiment shows interference of light.
- Slits S_1 and S_2 behaves like two coherent sources.
- On screen, bright and dark portion are alternatively found.
- Bright portion is called bright fringe and dark portion is called dark fringe.
- Central fringe is always bright.
- Energy conserves in interference of light.
- It is explained on the basis of Huygen's principle.
- This experiment verifies the wave nature of light.

The arrangement of Young's double slit experiment has been shown in the figure.

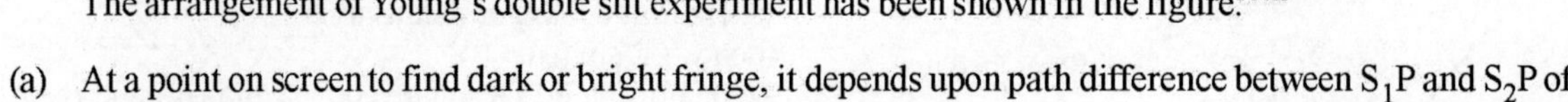

(a) At a point on screen to find dark or bright fringe, it depends upon path difference between S_1P and S_2P of light waves.

(b) There are two types of path difference between light waves:
(i) Geometrical path difference.
(ii) Optical path difference.

(c) In above experiment optical difference S_1P and S_2P has geometrical path difference so,
total path difference = geometrical difference

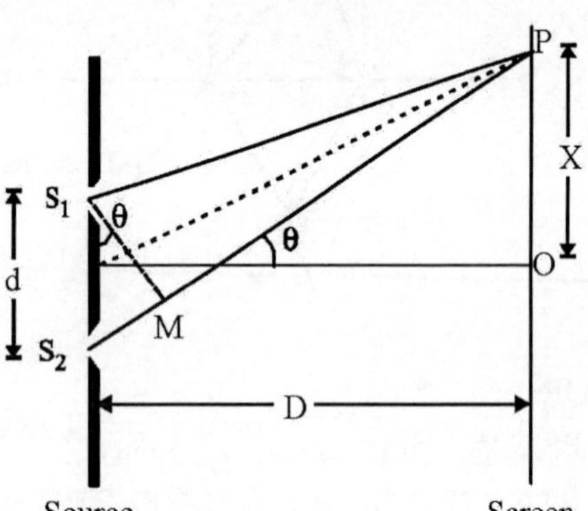

$$S_1P - S_2P = \frac{xd}{D} \qquad (\sin\theta \approx \tan\theta \approx Q)$$

Bright fringes: If n^{th} bright fringe forms at point P, then for bright fringe:

$$S_2P - S_1P = \frac{X_n d}{D} = n\lambda \Rightarrow X_n = \frac{n\lambda D}{d}$$

So *distance between central fringe and n^{th} bright fringe*: $X_n = \frac{n\lambda D}{d}$

n = 1 for first bright fringe.

⇒ n = 2 for second bright fringe.

Dark fringes : If n^{th} dark fringe forms at a point P, then for dark fringe :

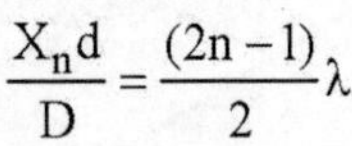

$$\frac{X_n d}{D} = \frac{(2n-1)}{2}\lambda$$

⇒ *Distance between central fringe and n^{th} dark fringe.*

$$X_n = \frac{(2n-1)\lambda D}{2d}$$

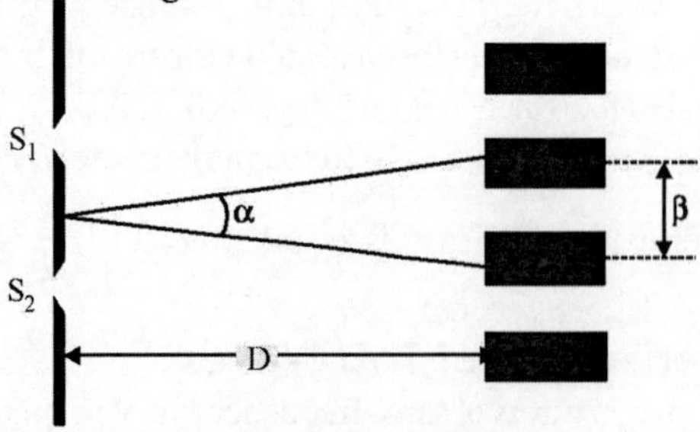

n = 1 for first dark fringe.
n = 2 for second dark fringe.

Distance between dark and bright fringe which are in coming order is called **fringe width**.

In interference, fringe width of dark and bright fringes are same $\beta = \frac{\lambda D}{d}$

Angular width of fringe $\alpha D = \beta$ or, $\alpha = \beta/D = \frac{\lambda}{d}$

Comments on Young's interference experiment :

(i) Energy is conserved in interference. This indicates that energy is redistributed from destructive interference region to the constructive interference region.

(ii) If the entire arrangement of young's double slit experiment is immersed in water then fringe width decreases.

$$\frac{\beta_{water}}{\beta_{air}} = \frac{\lambda_{water}}{\lambda_{air}} = \frac{1}{{}_a\mu_w} = \frac{1}{(4/3)}$$

(iii) If white light is used in place of monochromatic light in young's double slit experiment then,

(a) Central fringe is white.

(b) Coloured fringe is around the central white fringe.

(c) Inner edge of the dark fringe is red, while the outer edge is violet or blue.

(d) Inner edge of bright fringe is violet or blue and the outer edge is red.

(iv) If a filter allowing only λ_{red} is placed in front of slit s_1 and filter allowing only λ_{blue} is placed in fornt of slit s_2 then there is no interference pattern.

(v) If a thin glass plate or mica sheet is placed in front of one of the slit, then central fringe shifts towards that slit, refractive index of glass is μ and the thickness of sheet t, then the *optical path* = μt so, *extra path difference is* $(\mu - 1)t$.

If the central fringe now appears at the location of previously formed nth bright fringe then, $(\mu - 1)t = n\lambda$

If the central fringe appears at the position of previously formed nth dark fringe then, $(\mu - 1)t = (2n-1)\frac{\lambda}{2}$

(vi) If the width of the slit S increased then degree of spatial coherence decreases. As a result the interference pattern gradually disappears and occurs if distance between S_1 and S_2 is increased.

(vii) The fringe visibility :

$$V = \frac{I_{max} - I_{min}}{I_{max} + I_{min}}$$

V_{max}, if $I_1 = I_2 = I_0$ or $I_{min} = 0$

If widths of slits S_1 and S_2 are unequal the brightness of the bright fringe and the darkness of the dark fringe decreases.

If $I_1 >> I_2$ then $I_{max} = I_{min}$

(viii) When waves from two coherent sources S_1 and S_2 are interfere in space, the shape of the fringe is hyperbolic with foci at S_1 and S_2.

Fresnel Biprism

Biprism is actually a prism, whose refracting angle α is extremely 1/2° or 1°. When a biprism is used in front of a slit illuminated by a monochromatic source of light, then due to refraction by the biprism, two virtual sources S_1 and S_2 are formed. There occurs division of wavefront. The two virtual images act as two coherent sources.

The waves refracted from the upper and lower portions of the biprism superimpose to interfere and produce interference pattern.

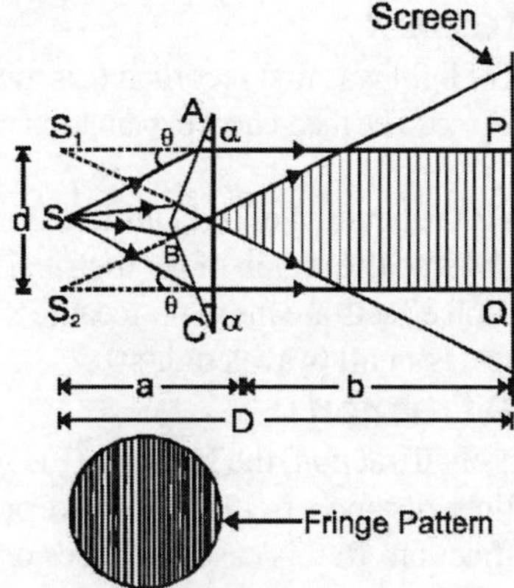

If d is the distance between the two virtual sources, then it can be proved that $d = 2a(\mu - 1)\alpha$ where a is the distance between the biprism and the slit, μ is the refractive index of the material of the prism and α is the prism angle (the refracting angle).

Use α in radians $\left[\alpha = \frac{1}{2} \times \frac{\pi^c}{180}\right]$.

Fringe Width β : It is given by $\beta = \frac{\lambda D}{d}$

where λ is the wavelength used, D is the distance between slit S (i.e. position of the two virtual sources) and the screen, and d is the separation between the virtual sources.

When Fresnel's biprism experimental arrangement is immersed in water then d = decreases ($d_{water} < d_{air}$)

$\lambda_w < \lambda_{air}$ (wavelength decreases) ;

$\beta_{water} > \beta_{air}$ (fringe width increases) ;

$$\beta_{water} = \beta_{air}\frac{(\mu_g - 1)}{(\mu_g - \mu_w)}$$

A thin film of thickness t and refractive index μ appears dark by reflection when viewed at an angle of refraction r if $2\mu t \cos r = n\lambda$ (n = 1, 2, 3, etc.)

And bright if $2\mu t \cos r = (2n+1)\frac{\lambda}{2}$; n = 0, 1, 2.

The minimum thickness (n = 1) of a film which appears dark by reflection at normal incidence (r = 0°) is $2\mu t = \lambda$.

The minimum thickness of a film, which appears bright under normal incidence of monochromatic light of wavelength λ is $2\mu t = \frac{\lambda}{2}$

COHERENT SOURCES

The two sources of light, whose frequencies are same and the phase difference between the waves emitted by which remains constant with respect to time are defined as coherent sources.

DIFFRACTION

Diffraction of light was first recorded (discovered) by Grimaldi in 1665. The theoretical explanation was given by Fresnel.

Diffraction is the bending of light around an obstacle and spreading into the region of geometrical shadow. The diffraction effects are more noticeable if the size of the obstacle is small (wavelengh-size).

Fresnel Diffraction

In this class of diffraction, the source or the screen or both are at finite distances from the aperture or obstacle causing diffraction. In this case no lenses or mirrors are used.

Fraunhoffer Diffraction

In this class of diffraction, the source and the screen on which the pattern is observed are at infinite distances from the aperture or the obstacle causing diffraction. Fraunhoffer diffraction pattern can be easily observed by using lenses. The incident light beam is made parallel with a lens and the diffracted beam is focussed on the screen with another lens.

Fraunhoffe Diffraction Due to Single Slit

In single slit diffraction, imagine aperature to be divided into two equal halves. Secondary sources in these two halves give first minima at $b\sin\theta = \lambda$.

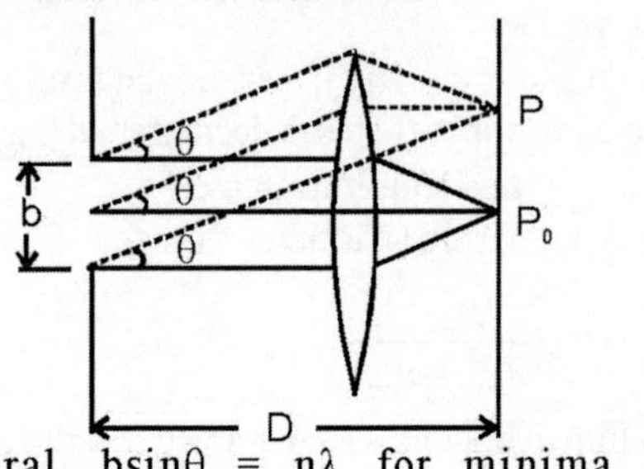

In general, $b\sin\theta = n\lambda$ for minima and,

$b\sin\theta = (2n+1)\frac{\lambda}{2}$ for maxima.

(i) The points of the maximum intensity lie nearly midway between the successive minima. The amplitude E_0' of the electric field at a general point P is

$E_0' = E_0 \frac{\sin\beta}{\beta}$ where $\beta = \frac{\pi b \sin\theta}{\lambda}$ and

E_0 = amplitude at the point P_0 i.e. at θ = 0

The intensity at a general point P is given as

$$I = I_0 \frac{\sin^2\beta}{\beta^2}$$

(ii) The graph for the variation of intensity as a function of sinθ is as follows :

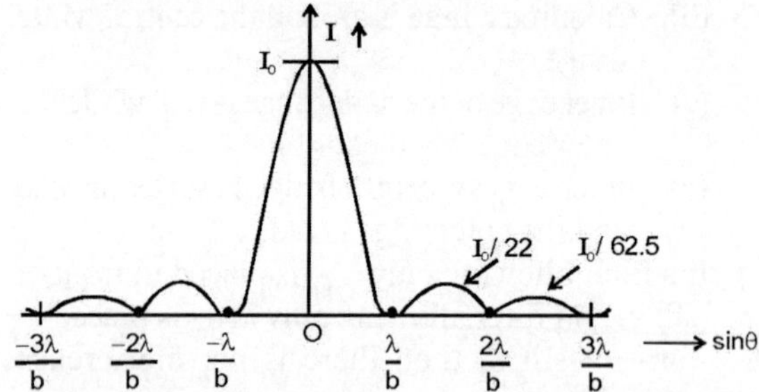

(iii) The **width of the central maxima** is $\left(\frac{2\lambda D}{b}\right)$ and **angular width of central maxima** is $\left(\frac{2\lambda}{b}\right)$.

Fraunhoffer Diffraction by a Circular Aperture

(i) The 1st dark ring is formed by the light diffracted from the circular aperture at an angle q with the axis where

$\sin\theta \approx \frac{1.22\lambda}{b}$ where T = wavelength of light used, b = diameter of circular aperture

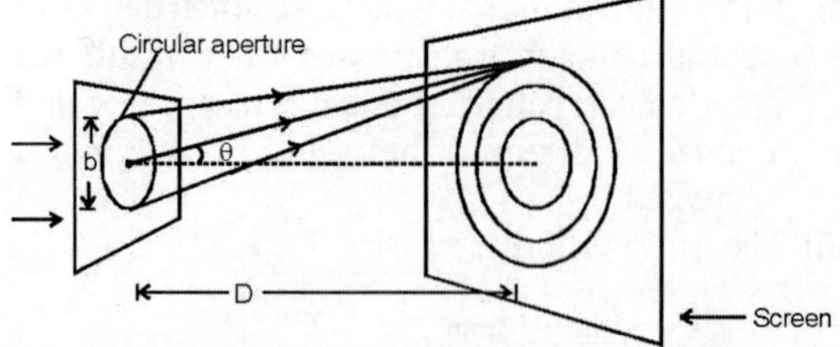

(ii) If the screen is at a distance D (D >> b) from the circular aperture, the radius of the 1st dark ring is,

$$R \approx \frac{1.22\lambda D}{b}$$

(iii) If the light transmitted by the hole is converged by a converging lens at the screen placed at the focal plane of the lens, the radius of the 1st dark ring is $R = \frac{1.22\lambda f}{b}$

This radius is also called the **radius of diffraction disc.**

For plane transmission diffraction grating $(a + b) \sin q_n = n\lambda$ – for maxima, where a = width of transparent portion, b – width of opaque portion.

A diffraction grating is made by a number of parallel slits ruled on a glass plate using a fine diamond point. The nth maximum appears at an angle θ_n given by equation $\sin \theta_n = Nn\lambda$, where N is the number of slits in unit length of the grating. If a is the width of the slit and b distance between the slits, then (a + b) is called grating element. $N = \frac{1}{a+b}$

RESOLVING POWER OF TELESCOPES AND MICROSCOPES

A large number of images are formed as a consequence of light diffraction from a source. If two sources are separated such that their central maxima do not overlap, their images can be distinguished and are said to be resolved R.P. of an optical instrument is its ability to distinguish two neighbouring points.

Linear R.P. = $d/\lambda D$

Angular R.P. = d/λ

D = Observed distance

d = Distance between two points

Telescope : Limit of resolution $= \theta = \sin^{-1} \frac{1.22\lambda}{a}$;

For small angles $\theta = \frac{1.22\lambda}{a}$

Resolving power $= \frac{1}{\text{limit of resolution}}$

Microscope : Limit of resolution (the smallest distance between two objects) $= x_{min} = \frac{1.22\lambda}{2\mu \sin\theta}$

Prism : R.P. = $t(d\mu/d\lambda) = \lambda/d\lambda$.

Diffraction Grating :

R.P. = $\lambda/d\lambda = N \times n$

(N is total number of lines and n is the order of spectrum)

Eye : The limit of resolution of human eye is 1' of arc (One minute of arc)

POLARISATION

An ordinary source such as bulb consists of a large number of waves emitted by atoms or molecules in all directions symmetrically. Such light is called **unpolarized light** (fig - a)

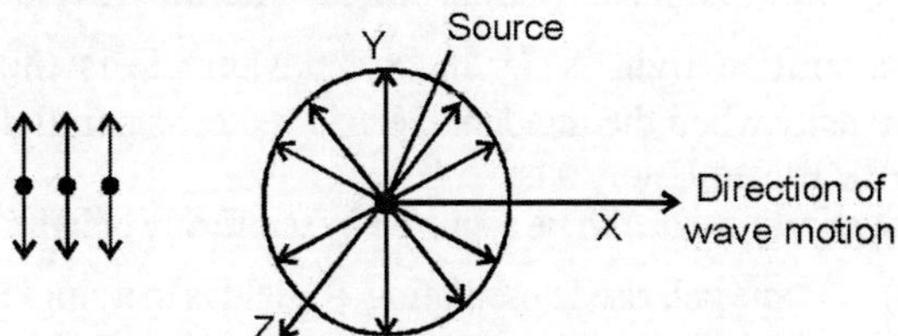

Fig (a) Unpolarised light

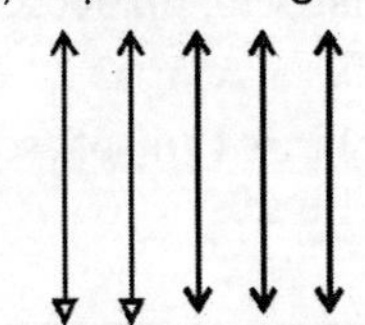

Fig (b) Polarised light

If we confine the direction of wave vibration of electric vector in one direction perpendicular to direction of wave propagation with the help of **polaroids or Nicol prism** then such type of light is called **plane polarised light** or linearly polarised. The phenomenon by which, we restrict the vibrations of wave in a particular direction Fig.(b) ⊥ to direction of wave propagation is called **polarisation.**

The plane of vibration is that which contains the vibrations of electric vector $\vec{E}$ and plane of polarisation is perpendicular to the plane of vibration Tourmaline and calcite polarizes an e.m. wave passing through it.

Polarisation by Reflection (Brewster's Law)

During reflection of a wave, we obtain a particular angle called angle of polarisation, for which the reflected light is completely plane polarised.

$\mu = \tan(i_p)$

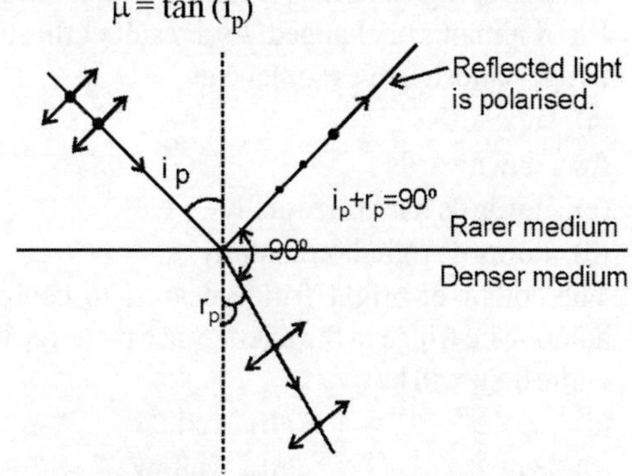

i_p : angle of incidence, such that the reflected and refracted waves are perpendicular to each other.

Law of Malus

If the electric vector is at angle θ with the transmission axis, light is partially transmitted. The intensity of transmitted light is $I = I_0 \cos^2\theta$ where I_0 is the intensity when the incident electric vector is parallel to the transmission axis.

Polarisation can also be achieved by scattering of light.

(a) Plane polarised : oscillating $\vec{E}$ field is in a single plane.

(b) Circularly polarised : tip of oscillating $\vec{E}$ field describes a circle.

(c) Elliptically polarised : tip of oscillating $\vec{E}$ field describes an ellipse.

DOPPLER'S EFFECT FOR LIGHT WAVES

(a) When the source moves towards the stationary observer or the observer moves towards the source, the apparent frequency.

$$\nu' = \nu\left(1 + \frac{v}{c}\right) \text{ (Blue shift)}$$

(b) When the source moves away from the stationary observer or vice-versa, $\nu' = \nu\left(1 - \frac{v}{c}\right)$ (Red shift)

where ν' = Apparent frequency, ν = Active frequency

v = Velocity of source, c = Velocity of light

But in both cases, the relative velocity v is small.

Exercise

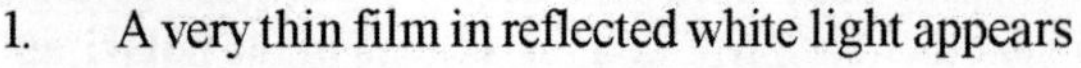

1. A very thin film in reflected white light appears
 (a) coloured (b) white
 (c) black (d) red
2. The fringe width for red colours as compared to that for violet colour is approximately
 (a) 3 times (b) double
 (c) 4 times (d) 8 times
3. Intensity of light depends on
 (a) amplitude (b) frequency
 (c) wavelength (d) velocity
4. In Young's double slit experiment, the source S and two slits A and B are lying in a horizontal plane. The slit A is above slit B. the fringes are obtained on a vertical screen K. The optical path from S to B is increased by putting a transparent material of higher refractive index. The path from S to A remains unchanged. As a result of this the fringe pattern moves somewhat
 (a) upwards
 (b) downwards
 (c) towards left horizontally
 (d) towards right horizontally
5. The colour of bright fringe nearest to central achromatic fringe in the interference pattern with white light will be
 (a) violet (b) red
 (c) green (d) yellow
6. When a thin film of thickness t is placed in the path of light wave emerging out of S_1 then increase in the length of optical path will be
 (a) $(\mu - 1)t$ (b) $(\mu + 1)t$
 (c) μt (d) $\frac{\mu}{t}$
7. The device which produces highly coherent sources is
 (a) Fresnel biprism (b) Young's double slit
 (c) Laser (d) Lloyd's mirror
8. Which out of following, cannot produce two coherent sources?
 (a) Lloyd's mirror (b) Fresnel biprism
 (c) Young's double slit (d) Prism
9. Coherence is a measure of
 (a) capability of producing interference by wave
 (b) waves being diffracted
 (c) waves being reflected
 (d) waves being refracted
10. In Young's double slit experiment, one slit is covered with red filter and another slit is covered by green filter, then interference pattern will be
 (a) red (b) green
 (c) yellow (d) invisible

11. Light waves can be polarised because they
(a) have high frequencies
(b) have short wavelength
(c) are transverse
(d) can be reflected

12. When unpolarised light is incident on a plane glass plate at Brewster's angle, then which of the following statements is correct?
(a) Reflected and refracted rays are completely polarised with their planes of polarization parallel to each other
(b) Reflected and refracted rays are completely polarised with their planes of polarization perpendicular to each other
(c) Reflected light is plane polarised but transmitted light is partially polarised
(d) Reflected light is partially polarised but refracted light is plane polarised

13. Light transmitted by Nicol prism is
(a) unpolarised
(b) plane polarised
(c) circularly polarised
(d) elliptically polarised

14. If the light is polarised by reflection, then the angle between reflected & refracted light is
(a) 180° (b) 90°
(c) 45° (d) 36°

15. From Brewster's law of polarisation, it follows that the anlge of polarisaiton depends upon
(a) the wavelength of light
(b) plane of polarisation's orientation
(c) plane of vibration's orientation
(d) none of the above

16. Optically active substances are those substances which
(a) produces polarised light
(b) produces double refraction
(c) rotate the plane of polarisation of polarised light
(d) converts a plane polarised light into circularly polarised light.

17. Polaroid glass is used in sun glasses because
(a) it reduces the light intensity to half on account of polarisation
(b) it is fashionable
(c) it has good colour
(d) it is cheaper

18. In the propagation of light waves, the angle between the plane of vibration & plane of polarisaiton is
(a) 0° (b) 90°
(c) 45° (d) 80°

19. In the propagation of electromagnetic waves, the angle between the direction of propagation and plane of polarisation is
(a) 0° (b) 45°
(c) 90° (d) 180°

20. If for a calcite crystal, μ_0 and μ_e are the refractive indices of the crystal for O-ray and E-ray respectively, then along the optic axis of the crystal
(a) $\mu_0 = \mu_e$ (b) $\mu_0 > \mu_e$
(c) $\mu_0 < \mu_e$ (d) None of the above

21. A star is emitting yellow light. If it is accelerated towards earth then to an observer on earth, it will appear
(a) shinning yellow
(b) gradually changing to violet
(c) gradually changing to red
(d) unchanged

22. When a source of light is receding away from an observer, then the spectral lines will get displaced towards
(a) violet region (b) red region
(c) ultraviolet region (d) blue region

23. An observer measures speed of light to be c, when he is stationary w.r.t. the source. If observer moves with velocity v towards the source, then the velocity of light observed will be
(a) c (b) $c+v$
(c) $c-v$ (d) $\sqrt{1-v^2/c^2}$

24. An observer on earth observes that wavelength of spectral line in spectrum of a milky way shifts towards red end of spectrum. According to the observer, the milky-way is
(a) approaching the earth
(b) receding away from earth
(c) stationary
(d) rotating about its own axis

25. The limit of resolution of eye is approximately
(a) 1° angle (b) 1' angle
(c) 1 mm (d) 1 cm

26. What should be refractive index of a transparent medium to be invisible in vacuum?
(a) 1 (b) <1
(c) >1 (d) None of the above

27. The ability of an optical instrument to show the images of two adjacent point objects as separate is called
(a) dispersive power (b) magnifying power
(c) resolving power (d) none of the above

28. Angular width (B) of central maxima of a diffraction pattern of a single slit does not depend upon
(a) distance between slit and source
(b) wavelength of the light used
(c) width of slit
(d) frequency of light used

29. Two waves are propagating with same amplitude and nearly same frequency, they result in
(a) beats (b) stationary wave
(c) resonance (d) wave packet

30. In diffraction at a single slit the angular width of central maxima does not depend upon
(a) distance between the slits
(b) wavelength of light
(c) distance between slit and the screen
(d) none of these

31. Which of the following does not support the wave nature of light :
(a) interference
(b) diffraction
(c) polarisation
(d) Photoelectric effect

32. When a compact disc is illuminated by small source of white light coloured bands are observed. This is due to
(a) dispersion (b) diffraction
(c) interference (d) reflection

33. The colours seen in the reflected white light from a thin oil film are due to
(a) Diffraction (b) Interference
(c) Polarisation (d) Dispersion

34. The reddish appearance of rising and setting sun is due to
(a) reflection of light
(b) diffraction of light
(c) scattering of light
(d) interference of light

35. A beam of light of wave length 600 nm from a distance source falls on a single slit 1 mm wide and a resulting diffraction pattern is observed on a screen 2 m away. The distance between the first dark fringes on either side of central bright fringe is
(a) 1.2 cm (b) 1.2 mm
(c) 2.4 cm (d) 2.4 mm

36. Monochromatic light of wavelength 400 nm and 560 nm are incident simultaneously and normally on double slits apparatus whose slits separation is 0.1 mm and screen distance is 1m. Distance between areas of total darkness will be
(a) 4mm (b) 5.6 mm
(c) 14mm (d) 28mm

37. A source emits sound of frequency 600 Hz inside water. The frequency heard in air will be equal to (velocity of sound in water = 1500 m/s, velocity of sound in air = 300 m/s)
(a) 3000 Hz (b) 120 Hz
(c) 600 Hz (d) 6000 Hz

38. In Young's double slit experiment intensity at a point is (1/4) of the maximum intensity. Angular position of this point is
(a) $\sin^{-1}(\lambda/d)$ (b) $\sin^{-1}(\lambda/2d)$
(c) $\sin^{-1}(\lambda/3d)$ (d) $\sin^{-1}(\lambda/4d)$

39. A nicol prism is based on the action of
(a) refraction (b) double refraction
(c) dichroism (d) both (b) and (c)

40. Which of the following cannot be polarised ?
(a) Radio waves (b) β rays
(c) Infrared rays (d) γ rays

41. The resolution of the human eye is 1. The resolving power of the human eye is nearly
(a) 360 (b) 3600
(c) 36000 (d) 360000

42. Ray optics is valid when characteristic dimensions are
(a) of the same order as the wavelength of light
(b) much smaller than the wavelength of light
(c) much larger than the wavelength of light
(d) of the order of 1 mm

43. Fluorescent tubes give more light than a filament bulb of same power because
(a) the tube contains gas at low temperature
(b) ultraviolet light is converted into visible light by fluorescence
(c) light is diffused through the walls of the tube
(d) it produces more heat than bulb

44. The phenomenon of interference is shown by
(a) longitudinal mechanical waves only
(b) transverse mechanical waves only
(c) non-mechanical transverse waves only
(d) all the above types of waves

45. The deflection of light in a gravitational field was predicted first by
(a) Einstein (b) Newton
(c) Max Planck (d) Maxwell

46. The transverse nature of light is shown by
(a) interference of light
(b) refraction of light
(c) polarization of light
(d) dispersion of light

47. Wavelength of light of frequency 100 Hz
(a) 2×10^6 m (b) 3×10^6 m
(c) 4×10^6 m (d) 5×10^6 m

48. A paper, with two marks having separation d, is held normal to the line of sight of an observer at a distance of 50 m. The diameter of the eye-lens of the observer is 2 mm. Which of the following is the least value of *d*, so that the marks can be seen as separate ? The mean wavelength of visible light may be taken as 5000 Å
(a) 1.25 m (b) 12.5 cm
(c) 1.25 cm (d) 2.5 mm

49. To demonstrate the phenomenon of interference, we require two sources which emit radiation
(a) of nearly the same frequency
(b) of the same frequency
(c) of different wavelengths
(d) of the same frequency and having a definite phase relationship

50. The angle of incidence at which reflected light is totally polarized for reflection from air to glass (refractive index *n*), is
(a) $\tan^{-1}(1/n)$ (b) $\sin^{-1}(1/n)$
(c) $\sin^{-1}(n)$ (d) $\tan^{-1}(n)$

51. The maximum number of possible interference maxima for slit-separation equal to twice the wavelength in Young's double-slit experiment is
(a) three (b) five
(c) infinite (d) zero

52. An electromagnetic wave of frequency $\nu = 3.0$ MHz passes from vacuum into a dielectric medium with permittivity $\in = 4.0$. Then
(a) wavelength is halved and frequency remains unchanged
(b) wavelength is doubled and frequency becomes half
(c) wavelength is doubled and the frequency remains unchanged
(d) wavelength and frequency both remain unchanged.

53. A Young's double slit experiment uses a monochromatic source. The shape of the interference fringes formed on a screen is
(a) circle (b) hyperbola
(c) parabola (d) straight line

54. When an unpolarized light of intensity I_0 is incident on a polarizing sheet, the intensity of the light which does not get transmitted is
(a) $\frac{1}{4}I_0$ (b) $\frac{1}{2}I_0$
(c) I_0 (d) zero

55. In a Young's double slit experiment the intensity at a point where the path difference is $\frac{\lambda}{6}$ (λ being the wavelength of light used) is *I*. If I_0 denotes the maximum intensity, $\frac{I}{I_0}$ is equal to
(a) $\frac{3}{4}$ (b) $\frac{1}{\sqrt{2}}$
(c) $\frac{\sqrt{3}}{2}$ (d) $\frac{1}{2}$

56. In an experiment, electrons are made to pass through a narrow slit of width '*d*' comparable to their de-Broglie wavelength. They are detected on a screen at a distance '*D*' from the slit (see figure).

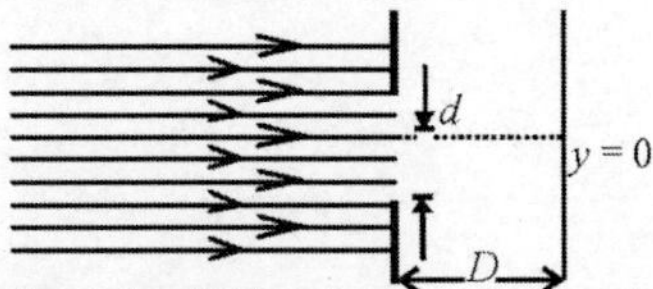

Which of the following graphs can be expected to represent the number of electrons '*N*' detected as a function of the detector position '*y*'(*y* = 0 corresponds to the middle of the slit)

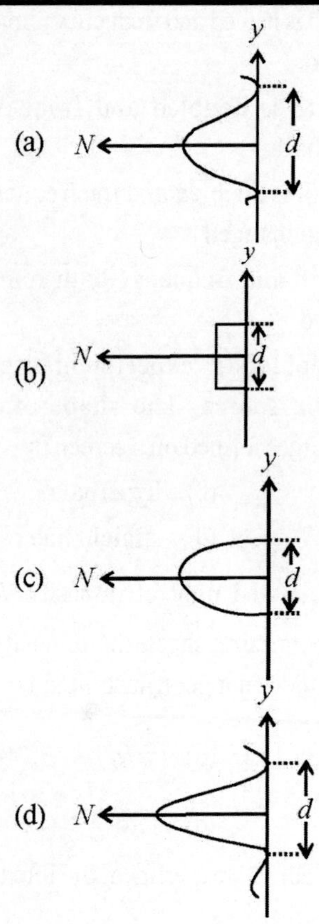

57. Which one of the following phenomena is not explained by Huygen's construction of wavefront?
(a) Refraction (b) Reflection
(c) Diffraction (d) Origin of spectra

58. Which of the following phenomenon is not common to sound and light waves ?
(a) Interference (b) Diffraction
(c) Coherence (d) Polarisation

59. Interference is possible in
(a) light waves only
(b) sound waves only
(c) both light and sound waves
(d) neither light nor sound waves

60. The Young's double slit experiment is performed with blue and with green light of wavelengths 4360Å and 5460Å respectively. If x is the distance of 4th maxima from the central one, then
(a) x (blue) = x (green)
(b) x (blue) > x (green)
(c) x (blue) < x (green)
(d) $\frac{x(\text{blue})}{x(\text{green})} = \frac{5460}{4360}$

CHAPTER 25

Dual Nature of Radiation and Matter

CONCEPTS

The dualism of matter is inherent in the de-Broglie relation which contains a wave concept (λ) and a particle concept (P). Radiation has dual nature i.e. it behaves both as particle and as wave.
But the wave nature and particle nature both cannot be possible simultaneously.

PHOTONS

Photon is a packet of energy emitted from a source of radiation. Photons are carrier particle of electromagnetic interaction. Photons travel in straight lines with speed of light.
The **energy of photons** is given as

$$E = h\nu = \frac{hc}{\lambda} = mc^2$$

where ν is frequency, λ is wavelength, h is Planck's constant.
The effective or motional mass of photon is given as

$$m = \frac{E}{c^2} = \frac{h\nu}{c^2} = \frac{h}{\lambda c}$$

The **momentum of a photon** is given as

$$p = mc = \frac{E}{c} = \frac{h\nu}{c} = \frac{h}{\lambda}$$

Photons are electrically neutral. They are not deflected by electric and magnetic fields.
If E is the energy of source in joule then **number of photons emitted**

$$n = \frac{\text{total energy radiated}}{\text{energy of each photon}} = \frac{E}{h\nu} = \frac{E\lambda}{hc}$$

Intensity of photons is defined as amount of energy carried per unit area per unit time, or power carried per unit area

$$\text{Intensity } (I_p) = \frac{\text{energy}}{\text{area} \times \text{time}} = \frac{\text{power}}{\text{area}},$$

$$I_p = nh\nu = \frac{N}{4\pi r^2}P$$

where, n = number of photons per unit area per unit time,

N = number of photons, P = power of source.

(a) For a point source $I_p = nh\nu = \frac{N}{4\pi r^2}P$

(b) For a line source $I_p = nh\nu = \frac{N}{2\pi r\ell}P$

PHOTOELECTRIC EFFECT

It is the phenomenon of emission of electrons by metals when illuminated by light of suitable frequency. Certain metals respond to ultraviolet light while others are sensitive even to the visible light. Photoelectric effect involves conversion of light energy into electrical energy. It follows the law of conservation of energy. The photoelectric emission is an instantaneous process and possesses certain special features. Below a certain frequency (threshold frequency) ν_0, characteristic of the metal, no photoelectric emission takes place, no matter how large the intensity may be. Observations on photoelectric effect imply that in the event of matter-light interaction, absorption of energy takes place in discrete units of $h\nu$. This is not quite the same as saying that light consists of particles, each of energy $h\nu$.

Photoelectric Current Depends on :

(i) the intensity of incident light,

(ii) the potential difference applied between the two electrodes,

(iii) the nature of the emitter material.

The Stopping Potential (V_0) Depends on :

(i) the frequency of incident light, and

(ii) the nature of the emitter material. For a given frequency of incident light, it is independent of its intensity. The stopping potential is directly related to the maximum kinetic energy of electrons emitted:

$e V_0 = (1/2) m v^2_{max} = K_{max}$.

EINSTEIN'S PHOTOELECTRIC EQUATION-PARTICLE NATURE OF LIGHT

Each metal has its own characteristic work function. According to Einstein, if the photon energy (hν) is at least equal to the work function, then absorption of a photon can eject an electron. If the photon energy is greater than the work function, some or all of the extra energy can appear as the ejected electron's kinetic energy. The maximum kinetic energy of an electron is the difference between the photon energy and the work function (ϕ).

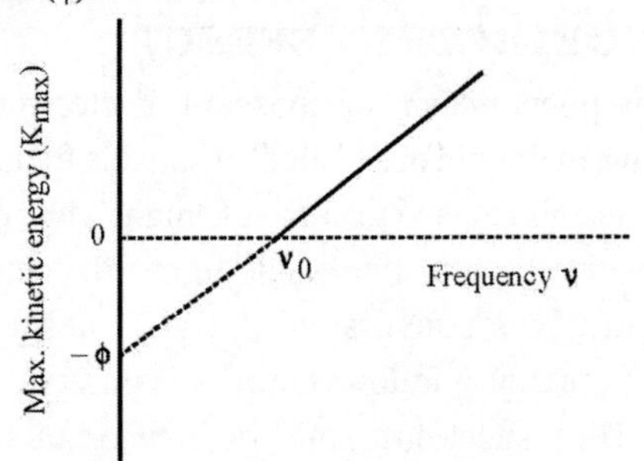

Einstein's Photoelectric equation

$$K_{max} = h\nu - \phi \qquad ...(1)$$

$$\text{Also, } k_{max} = h(\nu - \nu_0) = hc\left(\frac{1}{\lambda} - \frac{1}{\lambda_0}\right) = ev_s ...(2)$$

According to eq. (1), a graph of K_{max} versus ν is a straight line with a slope of h and a "y" -intercept of $-\phi$ this equation fits the experimental relationship (Fig.). The intercept on the frequency axis is the threshold frequency ν_0. Setting

$K_{max} = h\nu_0 - \phi = 0$ predicts a threshold frequency

$$\nu_0 = \frac{\phi}{h}$$

The four puzzling results of photoelectric effect experiments are easily explained using the photon concept:

(i) Light of greater intensity (but constant frequency) delivers more photons per unit time to the metal surface, so the number of electrons ejected per second increases as the intensity of the light increases. However, the energy of each photon remains the same. The maximum kinetic energy of the emitted electrons does not depend on the number of photons striking the metal per second because each emitted electron is the result of the absorption of one photon.

(ii) Higher-frequency light has larger energy photons. As the frequency of the light increases, the photon have more excess energy that can potentially become the electron's kinetic energy. Thus, K_{max} increases with increasing frequency.

(iii) Below the threshold frequency, a photon does not have enough energy to free an electron from the metal, so no electrons are emitted.

(iv) At low intensities, the number of photons per second is small, but the energy is still delivered in discrete packets. Just after the light is turned on, some photons hit the surface; some of them are absorbed and eject electrons from the metal. There is no time delay because the electrons cannot gradually accumulate energy; they either absorb a photon or they don't.

Laws of Photoelectric Effect:

(i) The number of photoelectrons emitted from a metal surface in a unit time (the rate of emission of photoelectrons) is proportional to the intensity of incident light.

(ii) The maximum kinetic energy of the emitted photoelectrons increases with increasing frequency or decreases with increasing wavelength of the incident light. The kinetic energy of the electrons does not depend upon the intensity of the incident light.

(iii) The emission of photoelectrons occurs upto a definite minimum frequency (maximum wavelength) of the incident light. This minimum frequency is called threshold frequency and the maximum wavelength. Threshold wavelength depends upon the nature of the substance and it is different for different metals.

(iv) Within the limits (~ 10^{-9} second) there is no time lag between the incidence of light at the metal surface and the emission of electrons from the metal surface, whatever is the intensity of incident light.

(v) The photoelectric effect takes place only when photons strike bound electrons because for free electrons energy and momentum conservations do not hold together.

Cesium is the best photo sensitive material.

Efficiency of photoemission,

$$\eta = \frac{\begin{array}{c}\textit{Number of photoelectrons emitted}\\ \textit{per unit area per unit time}\end{array}}{\begin{array}{c}\textit{Number of photons incident}\\ \textit{per unit area per unit time}\end{array}} = \frac{n_e}{n_p}$$

$$\eta = \frac{\textit{Intensity of emitted electrons}}{\textit{Intensity of incident radiation}} = \frac{I_e}{I_p}$$

Therefore, $\eta = \frac{n_e}{n_p} = \frac{I_e}{I_p}$

MATTER WAVES OR dE-BROGLIE WAVES

They are defined as the waves associated with every moving matter particles. Louis Victor de Broglie attributed a wave-like character to matter (material particles). The waves associated with the moving material particles are called matter waves or de Broglie waves.

The de Broglie wavelength (λ) associated with a moving particle is related to its momentum p as $\boldsymbol{\lambda = h/p}$. This is de Broglie relation. The de Broglie wavelength is independent of the charge and nature of the material particle. It is significantly measurable (of the order of the atomic-planes spacing in crystals) only in case of sub-atomic particles like electrons, protons, etc. (due to smallness of their masses and hence, momenta). However, it is indeed very small, quite beyond measurement, in case of macroscopic objects, commonly encountered in everyday life.

Electron diffraction experiments by Davisson and Germer, and by G. P. Thomson, as well as many later experiments, have verified and confirmed the wave-nature of electrons. The de Broglie hypothesis of matter waves supports the Bohr's concept of stationary orbits.

Wavelength of Matter Waves Associated with Accelerated Charged Particles :

If V is the accelerating voltage applied then :

(a) For the charged particle :

Energy $E = qV$,

Velocity $v = \sqrt{\frac{2qV}{m}} = \sqrt{\frac{2E}{m}}$

Momentum $p = \sqrt{2mE} = \sqrt{2mqV}$;

Wavelength $\lambda = \frac{h}{\sqrt{2mqV}} = \frac{h}{\sqrt{2mE}}$

(b) For electron $\lambda_e = \frac{12.27}{\sqrt{V}}$ Å

(c) For proton $\lambda_p = \frac{0.286}{\sqrt{V}}$ Å

(d) For alpha particle $\lambda_\alpha = \frac{0.101}{\sqrt{V}}$ Å

(e) For deuteron $\lambda_d = \frac{0.202}{\sqrt{V}}$ Å

(f) For neutral particles (neutron, atom or molecule)

(i) If E is the energy of the particle, then, $\frac{h}{\lambda} = \frac{h}{\sqrt{2mE}}$

(ii) If T is the temperature, then, $\lambda = \frac{h}{\sqrt{3mkT}}$

X-RAYS

X-rays are electromagnetic radiations of very short wavelength (0.1Å to 100 Å) and high energy which are emitted when fast moving electrons or cathode rays strike a target of high atomic mass.

Properties of X-Rays :

(i) These are highly penetrating rays and can pass through several materials which are opaque to ordinary light.

(ii) They ionize the gas through which they pass. While passing through a gas, they knock out electrons from several of the neutral atoms, leaving these atoms with +ve charge.

(iii) They cause fluorescence in several materials. A plate coated with barium platinocyanide, ZnS (zinc sulphide) etc becomes luminous when exposed to X-rays.

(iv) They affect photographic plates especially designed for the purpose.

(v) They are not deflected by electric and magnetic fields, showing that they are not charged particles.

(vi) They show all the properties of the waves except refraction. They show diffraction patterns when passed through a crystal which behaves like a grating.

Experimental observation and studies of spectra of X-rays reveal that X-rays are of two types and so are their respective spectras : Characteristic X-rays and Continuous X-rays.

(a) **Characteristic X-rays:**
The spectra of this group consists of several radiations with specific sharp wavelengths and frequency similar to the spectrum (line) of atoms like hydrogen. The wavelengths of this group show characteristic discrete radiations emitted by the atoms of the target material. The characteristic X-rays spectra helps us to identify the element of target material.
The frequency of emitted radiation (i.e. of photon) is given by $\nu = RZ_e^2\left(\frac{1}{n_1^2}-\frac{1}{n_2^2}\right)$; where R is constant and Z_e is effective atomic number.
Generally Z_e is taken to be equal to $Z - \sigma$, where Z is proton number or atomic number of the element and σ is called the screening constant.
Due to the presence of the other electrons, the charge of the nucleus as seen by the electron will be different in different shells.

(b) **Continuous X-rays :**
In addition to characteristic X-rays, tubes emit a continuous spectrum also.
The characteristic line spectra is superimposed on a continuous X-rays spectra of varying intensities. The wavelength of the continuous X-rays spectra are independent of material. One important feature of continuous X-rays is that they end abruptly at a certain lower wavelength for a given voltage.

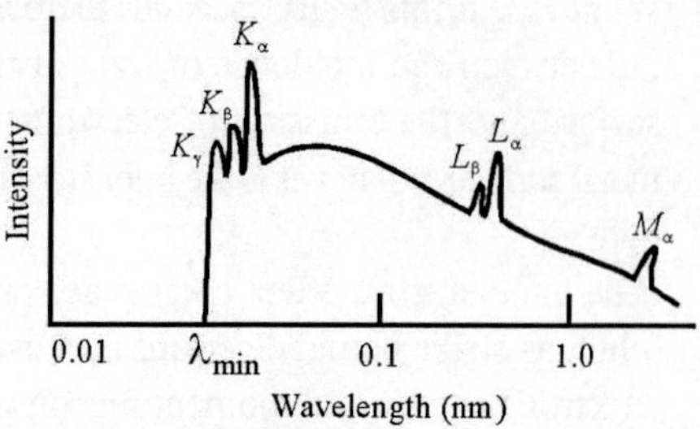

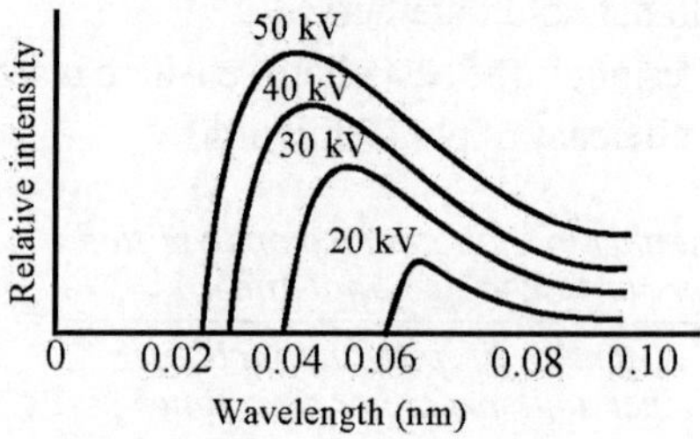

Bragg's condition for X-ray diffraction :
$n\lambda = 2d \sin \phi$.
Diffraction maximum of electrons accelerated with 54 volt is obtained at $\theta = 50^\circ$ for the Nickel crystal.

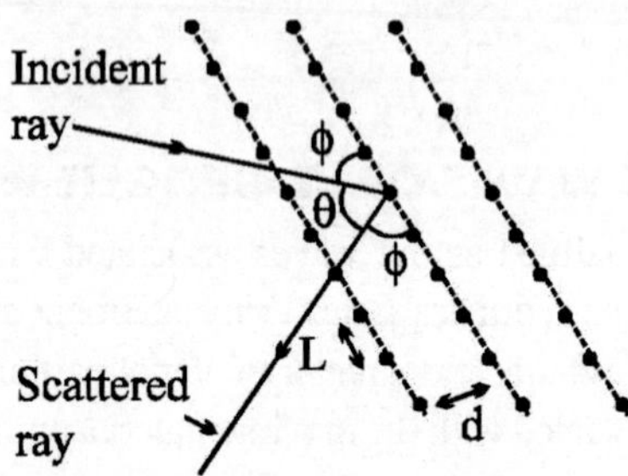

MOSELEY'S LAW :

If we plot $\sqrt{\nu}$ of K_α X-rays as a function of Z, the atomic number of material, we obtain a straight line. The relation of straight line is expressed as

$\sqrt{\nu} = a\,(Z - b)$, where a and b are constants.

This relation is called as Moseley's law. It helps to determine the atomic number Z of an atom. Here b is the screening constant.

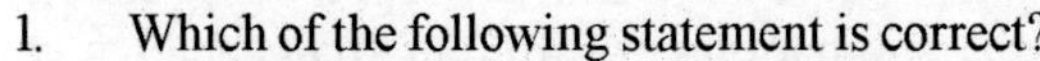

Exercise

1. Which of the following statement is correct?
   (a) Photocurrent increases with intensity of light
   (b) Photocurrent is proportional to the applied voltage
   (c) Current in photocell increases with increasing frequency
   (d) Stopping potential increases with increase of incident light
2. Cosmic rays are
   (a) high energy radiations
   (b) low energy radiations
   (c) ultra high energy radiations
   (d) very low energy radiations
3. The X-rays cannot be diffracted by means of an ordinary grating because of
   (a) high speed (b) short wavelength
   (c) large wavelength (d) none of these
4. As the intensity of incident light increases
   (a) photoelectric current increases
   (b) K. E. of emitted photoelectrons increases
   (c) photoelectric current decreases
   (d) K.E. of emitted photoelectrons decreases
5. Einstein work on the photoelectric effect provided support for the equation
   (a) $E = h\nu$ (b) $E = mc^2$
   (c) $E = -\frac{Rhc}{n^2}$ (d) $K.E. = \frac{1}{2}mv^2$
6. Which of the following moving particles (moving with same velocity) has largest wavelength of matter waves?
   (a) Electron (b) α-particle
   (c) Proton (d) Neutron
7. Which one of the following statements is not true for cathode rays?
   (a) Cathode rays produce heat when incident on metals
   (b) Cathode rays travel in a straight line
   (c) Cathode rays do not deflect in electric field
   (d) Cathode rays produce fluorescence when they fall on certain materials
8. When ultraviolet radiation is incident on a surface, no photoelectrons are emitted. If a second beam causes photoelectrons to be ejected, it may consists of
   (a) infrared waves (b) X-rays
   (c) visible light rays (d) radiowaves
9. A photoelectric cell is illuminated by a point source of light 1 m away. When the source is shifted to 2 m then
   (a) number of electrons emitted is a quarter of the initial number
   (b) each emitted electron carries one quarter of the initial energy
   (c) number of electrons emitted is half the initial number
   (d) each emitted electron carries half the initial energy
10. According to Einstein's photoelectric equation, the graph between the kinetic energy of photoelectrons ejected and the frequency of incident radiation is

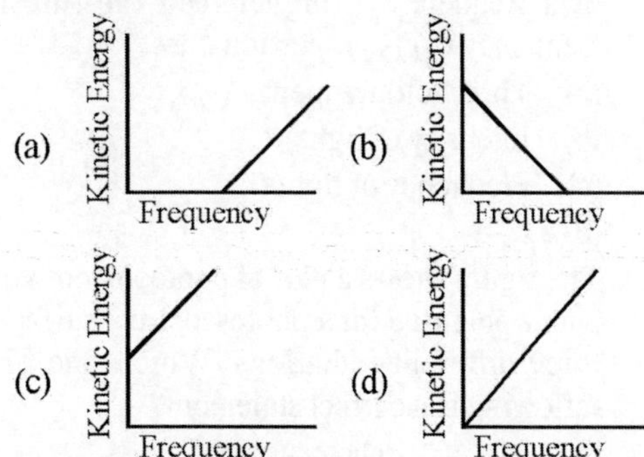

11. The work functions for metals A, B and C are respectively 1.92 eV, 2.0 eV and 5 eV. According to Einstein's equation, the metals which will emit photoelectrons for a radiation of wavelength 4100 Å is/are
   (a) none (b) A only
   (c) A and B only (d) all three metals
12. In a discharge tube ionization of enclosed gas is produced due to collisions between
   (a) negative electrons and neutral atoms/molecules
   (b) photons and neutral atoms/molecules
   (c) neutral gas atoms/molecules
   (d) positive ions and neutral atoms/molecules

13. A photo-cell employs photoelectric effect to convert
    (a) change in the intensity of illumination into a change in photoelectric current
    (b) change in the intensity of illumination into a change in the work function of the photocathode
    (c) change in the frequency of light into a change in the electric current
    (d) change in the frequency of light into a change in electric voltage
14. In the phenomenon of electric discharge through gases at low pressure, the coloured glow in the tube appears as a result of:
    (a) excitation of electrons in the atoms
    (b) collision between the atoms of the gas
    (c) collisions between the charged particles emitted from the cathode and the atoms of the gas
    (d) collision between different electrons of the atoms of the gas
15. The number of photoelectrons emitted for light of a frequency v (higher than the threshold frequency v_0) is proportional to:
    (a) Threshold frequency (v_0)
    (b) Intensity of light
    (c) Frequency of light (v)
    (d) $v - v_0$
16. The figure shows a plot of photocurrent versus anode potential for a photosensitive surface for three different radiations. Which one of the following is a correct statement?

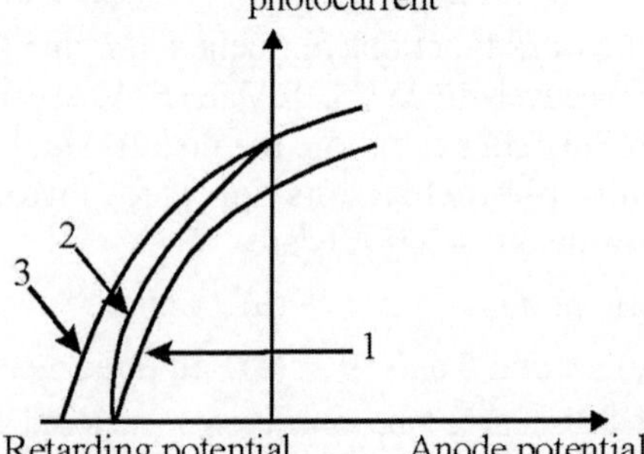

    (a) Curves (1) and (2) represent incident radiations of same frequency but of different intensities.
    (b) Curves (2) and (3) represent incident radiations of different frequencies and different intensities.
    (c) Curves (2) and (3) represent incident radiations of same frequency having same intensity.
    (d) Curves (1) and (2) represent incident radiations of different frequencies and different intensities.
17. In the Davisson and Germer experiment, the velocity of electrons emitted from the electron gun can be increased by
    (a) increasing the potential difference between the anode and filament
    (b) increasing the filament current
    (c) decreasing the filament current
    (d) decreasing the potential difference between the anode and filament
18. Electrons used in an electron microscope are accelerated by a voltage of 25 kV. If the voltage is increased to 100 kV then the de–Broglie wavelength associated with the electrons would
    (a) increase by 2 times
    (b) decrease by 2 times
    (c) decrease by 4 times
    (d) increase by 4 times
19. In photoelectric emission process from a metal of work function 1.8 eV, the kinetic energy of most energetic electrons is 0.5 eV. The corresponding stopping potential is
    (a) 1.8 V (b) 1.2 V
    (c) 0.5 V (d) 2.3 V
20. Photoelectric emission occurs only when the incident light has more than a certain minimum
    (a) power (b) wavelength
    (c) intensity (d) frequency
21. The phenomenon which can be explained only by quantum nature of light is
    (a) photoelectric effect (b) reflection
    (c) interference (d) polarization
22. In Millikan's oil drop experiment, an oil drop is observed to move vertically upward. The upward motion of the drop is due to
    (a) gravity (b) viscosity
    (c) buoyancy (d) electric field
23. Number of ejected photoelectron increases with increase
    (a) in intensity of light
    (b) in wavelength of light
    (c) in frequency of light
    (d) never

24. In photoelectric effect the work function of a metal is 3.5 eV. The emitted electrons can be stopped by applying a potential of –1.2 V. Then
(a) the energy of the incident photon is 4.7 eV
(b) the energy of the incident photon is 2.3 eV
(c) if higher frequency photon be used, the photoelectric current will rise
(d) when the energy of photon is 3.5 eV, the photoelectric current will be maximum

25. Momentum of a photon of wavelength λ is
(a) $\frac{h}{\lambda}$ (b) zero
(c) $\frac{h\lambda}{c^2}$ (d) $\frac{h\lambda}{c}$

26. Gases begin to conduct electricity at low pressure because
(a) at low pressures gases turn of plasma
(b) colliding electrons can acquire higher kinetic energy due to increased mean free path leading to ionisation of atoms
(c) atoms break up into electrons and protons
(d) the electrons in atoms can move freely at low pressures

27. According to Einstein's photoelectric equation, the plot of the kinetic energy of the emitted photo electrons from a metal Vs the frequency, of the incident radiation gives a straight line whose slope
(a) depends both on the intensity of the radiation and the metal used
(b) depends on the intensity of the radiation
(c) depends on the nature of the metal used
(d) is the same for all metals and independent of the intensity of the radiation

28. In gamma ray emission from a nucleus
(a) only the proton number changes
(b) both the neutron number and the proton number change
(c) there is no change in the proton number and the neutron number
(d) only the neutron number changes

29. In which of the following, emission of electrons does not take place
(a) thermionic emission
(b) X-rays emission
(c) photoelectric emission
(d) secondary emission

30. In Davison-Germer experiment, an electron beam is incident on a crystal. The reflected beam consists of
(a) photons (b) protons
(c) X-rays (d) electrons

31. Suppose an electron is attracted towards the origin by a force $\frac{k}{r}$ where 'k' is a constant and 'r' is the distance of the electron from the origin. By applying Bohr model to this system, the radius of the nth orbital of the electron is found to be 'r_n' and the kinetic energy of the electron to be 'T_n'. Then which of the following is true?
(a) $T_n \propto \frac{1}{n^2}, r_n \propto n^2$
(b) T_n independent of n, $r_n \propto n$
(c) $T_n \propto \frac{1}{n}, r_n \propto n$
(d) $T_n \propto \frac{1}{n}, r_n \propto n^2$

32. A radioactive nucleus of mass number A, initially at rest, emits an α-particle with a speed v. The recoil speed of daughter nucleus is
(a) $\frac{2v}{A-4}$ (b) $\frac{2v}{A+4}$
(c) $\frac{4v}{A-4}$ (d) $\frac{4v}{A+4}$

33. What is the kinetic energy gained by an electron due to acceleration through a potential difference of 1 V?
(a) 1 eV (b) 1 joule
(c) 5 Nm (d) 10 Nm

34. Light of wavelength 3500Å is incident on two metals A and B whose work functions are 3.2 eV and 1.9 eV respectively. Which metal will emit photoelectrons?
(a) A (b) B
(c) Both A and B (d) Neither A nor B

35. A steel ball of mass m is moving with a kinetic energy K. The de-Broglie wavelength associated with the ball is
(a) $\frac{h}{2mK}$ (b) $\sqrt{\frac{h}{2mK}}$
(c) $\frac{h}{\sqrt{2mK}}$ (d) meaningless

36. The half-life of a radioactive material is 77 days. The decay constant λ is numerically equal to (in days^{-1})
(a) 27×10^{-3} (b) 9×10^{-3}
(c) 3×10^{-3} (d) 1×10^{-3}

37. The de-Broglie wavelength of an electron moving in the nth Bohr orbit of radius r is given by
(a) $\frac{2\pi r}{n}$ (b) $n\pi r$
(c) $\frac{nr}{2\pi}$ (d) $\frac{nr}{\pi}$

38. Which of the following series in the spectrum of hydrogen atom lies in the visible region of the electromagnetic spectrum?
(a) Paschen series (b) Balmer series
(c) Lyman series (d) Brackett series

39. The anode voltage of a photocell is kept fixed. The wavelength λ of the light falling on the cathode is gradually changed. The plate current I of the photocell varies as follows

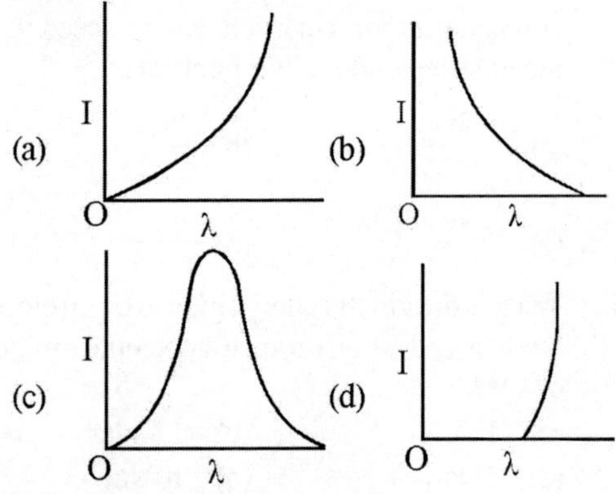

40. The maximum kinetic energy (E_{max}) of photoelectrons emitted in a photoelectric cell varies with frequency (ν) as shown in the graph. The slope of the graph is equal to

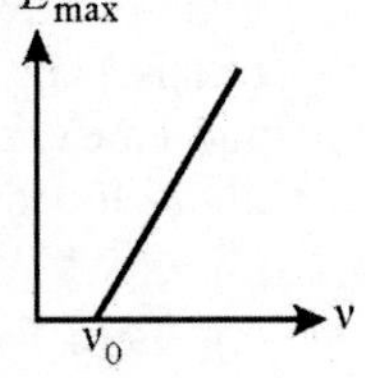

(a) charge of the electron
(b) $\frac{e}{m}$ of the electron
(c) work function of the emitter
(d) Planck's constant

41. The quantity $\frac{e^2}{2h\varepsilon_0 c}$ has a value
(a) $\frac{1}{137}$ ms^{-1} (b) $\frac{2}{137}$ ms^{-1}
(c) $\frac{1}{137}$ (d) $\frac{2}{137}$

42. The maximum velocity of an electron emitted by light of wavelength λ incident on the surface of a metal of work-function ϕ is
(a) $\sqrt{\frac{2(hc+\lambda\phi)}{m\lambda}}$ (c) $\frac{2(hc+\lambda\phi)}{m\lambda}$
(c) $\sqrt{\frac{2(hc-\lambda\phi)}{m\lambda}}$ (d) $\sqrt{\frac{2(h\lambda-\phi)}{m}}$

43. The threshold frequency for photoelectric effect on sodium corresponds to a wavelength of 5000 Å. Its work function is
(a) 4×10^{-19} J (b) 1 J
(c) 2×10^{-19} J (d) 3×10^{-19} J

44. The de-Broglie wave corresponding to a particle of mass m and velocity v has a wavelength associated with it
(a) $\frac{h}{mv}$ (b) $h\,m\,v$
(c) $\frac{mh}{v}$ (d) $\frac{m}{hv}$

45. Energy levels A, B, C of a certain atom correspond to increasing values of energy i.e., $E_A < E_B < E_C$. If $\lambda_1, \lambda_2, \lambda_3$ are the wavelengths of radiation corresponding to the transitions C to B, B to A and C to A respectively, which of the following relation is correct?
(a) $\lambda_3 = \lambda_1 + \lambda_2$ (b) $\lambda_3 = \frac{\lambda_1\lambda_2}{\lambda_1+\lambda_2}$
(c) $\lambda_1 + \lambda_2 + \lambda_3 = 0$ (d) $\lambda_3^2 = \lambda_1^2 + \lambda_2^2$

46. A radio transmitter operates at a frequency 880 kHz and a power of 10 kW. The number of photons emitted per second is
(a) 1.72×10^{31} (b) 1.327×10^{25}
(c) 1.327×10^{37} (d) 1.327×10^{45}

47. The momentum of a photon of an electromagnetic radiation is 3.3×10^{-29} kgms^{-1}. What is the frequency of the associated waves ?
($h = 6.6 \times 10^{-34}$ Js; $c = 3 \times 10^8$ ms^{-1})
(a) 1.5×10^{13} Hz (b) 7.5×10^{12} Hz
(c) 6.0×10^3 Hz (d) 3.0×10^3 Hz

48. Consider an electron in the nth orbit of a hydrogen atom in the Bohr model. The circumference of the orbit can be expressed in terms of de-Broglie wavelength λ of that electron as

(a) $(0.529)n\lambda$ (b) $\sqrt{n\lambda}$

(c) $(13.6)\lambda$ (d) $n\lambda$

49. The wavelength of a 1 keV photon is 1.24×10^{-9} m. What is the frequency of 1 MeV photon?

(a) 1.24×10^{15} (b) 2.4×10^{20}

(c) 1.24×10^{18} (d) $2 \times 4 \times 10^{23}$

50. Photoelectric work function of a metal is 1eV. Light of wavelength $\lambda = 3000$ Å falls on it. The photo electrons come out with velocity

(a) 10 metres/sec (b) 10^2 metres/sec

(c) 10^4 metres/sec (d) 10^6 metres/sec

51. The cathode of a photoelectric cell is changed such that the work function changes from W_1 to W_2 ($W_2 > W_1$). If the current before and after changes are I_1 and I_2, all other conditions remaining unchanged, then (assuming $h\nu > W_2$)

(a) $I_1 = I_2$ (b) $I_1 < I_2$

(c) $I_1 > I_2$ (d) $I_1 < I_2 < 2I_1$

52. A proton has kinetic energy $E = 100$ keV which is equal to that of a photon. The wavelength of photon is λ_2 and that of proton is λ_1. The ratio of λ_2/λ_1 is proportional to

(a) E^2 (b) $E^{1/2}$

(c) E^{-1} (d) $E^{-1/2}$

53. When light of wavelength 300 nm (nanometer) falls on a photoelectric emitter, photoelectrons are liberated. For another emitter, however, light of 600 nm wavelength is sufficient for creating photoemission. What is the ratio of the work functions of the two emitters?

(a) 1 : 2 (b) 2 : 1

(c) 4 : 1 (d) 1 : 4

54. Doubly ionised helium atoms and hydrogen ions are accelerated from rest through the same potential drop. The ratio of the final velocities of the helium and the hydrogen ion is

(a) 1/2 (b) 2

(c) $1/\sqrt{2}$ (d) $\sqrt{2}$

55. Kinetic energy of an electron, which is accelerated in a potential difference of 100 V is

(a) 1.6×10^{-17} J (b) 1.6×10^{-19} J

(c) 1.6×10^{-21} J (d) 1.6×10^{-25} J

56. The wavelength associated with an electron, accelerated through a potential difference of 100 V, is of the order of

(a) 1000 Å (b) 100 Å

(c) 10.5 Å (d) 1.2 Å

57. An electron of mass m and charge e is accelerated from rest through a potential difference of V volt in vacuum. Its final speed will be

(a) $\frac{eV}{2m}$ (b) $\frac{eV}{m}$

(c) $\sqrt{\frac{2eV}{m}}$ (d) $\sqrt{\frac{eV}{2m}}$

58. In a photo-emissive cell, with exciting wavelength λ, the fastest electron has speed v. If the exciting wavelength is changed to $\frac{3\lambda}{4}$, the speed of the fastest emitted electron will be

(a) $(3/4)^{1/2}.v$

(b) $(4/3)^{1/2}.v$

(c) less than $(4/3)^{1/2}.v$

(d) greater than $(4/3)^{1/2}.v$

59. The 21 cm radiowave emitted by hydrogen in interstellar space is due to the interaction called the hyperfine interaction in atomic hydrogen. The energy of the emitted wave is nearly

(a) 10^{-17} J (b) 1 J

(c) 7×10^{-8} J (d) 10^{-24} J

60. Light of wavelength 5000 Å falls on a sensitive plate with photoelectric work function of 1.9 eV. The kinetic energy of the photoelectrons emitted will be

(a) 0.58 eV (b) 2.48 eV

(c) 1.24 eV (d) 1.16 eV

CHAPTER 26

Atoms

CONCEPTS

RUTHERFORD MODEL OF ATOM

When a beam of α-particles was passed through a thin gold foil, it was found that

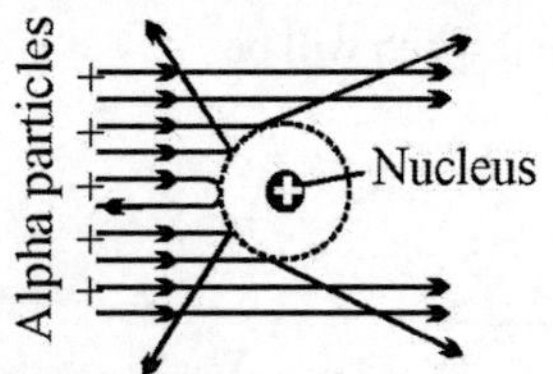

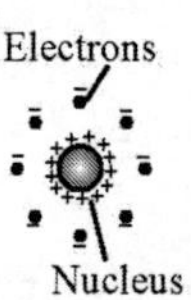

(i) most of the α-particles went straight through the gold foil and produced flashes on the screen as if there were nothing inside gold foil. Thus the atom is hollow.

(ii) few particles collided with the atoms of the foil which have scattered or deflected through considerable large angles. Few particles even turned back towards source itself.

(iii) the entire positive charge and almost whole mass of the atom is concentrated in small centre called a nucleus.

(iv) the electrons could not deflected the path of a α-particles i.e. electrons are very light.

(v) electrons revolve round the nucleus in circular orbits.

According to Rutherford scattering formula, the number of α-particles scattered at angle θ by a target is :

$$N(\theta) \propto \operatorname{cosec}^4 (\theta/2)$$

Impact parameter $b = \dfrac{2Ze^2 \cot(\theta/2)}{4\pi\varepsilon_0 m v_0^2}$

Distance of closest approach

$$r_0 = \frac{2Zeq}{4\pi\varepsilon_0 m v_0^2} = \frac{2(Ze)(2e)}{4\pi\varepsilon_0 m v_0^2}$$

BOHR'S MODEL

(i) Attractive coulomb force between electron and nucleus provides necessary centripetal force

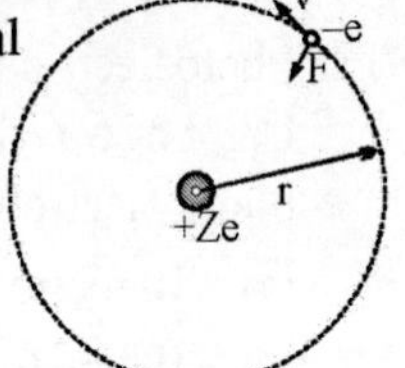

i.e, $\dfrac{mv^2}{r} = k\dfrac{Ze^2}{r^2}$

(ii) Rule of stable orbits : Electron orbits around nucleus in only those orbit where angular momentum is an integral multiple of $\dfrac{h}{2\pi}$

i.e, $mvr = n\dfrac{h}{2\pi}\left[\because \hbar = \dfrac{h}{2\pi}\right]$

(iii) Electromagnetic radiations are emitted if an electron jumps from stationary orbit of higher energy E_2 to another stationary orbit of lower energy E_1. The frequency ν of the emitted radiation is related by the equation.

$$E_2 - E_1 = h\nu$$

Drawbacks of Bohr model

(a) This model could not explain the fine structure of spectral lines, Zeeman effect and Stark effect.

(b) This model is valid only for single electron systems.

(c) This model is based on circular orbits of electrons whereas in reality the orbits are elliptical.

VARIOUS PARAMETER OF ELECTRON ACCORDING TO BOHR'S THEORY

(i) The radius of nth orbit :

(a) $r_n = \dfrac{n^2h^2}{4\pi^2 kZe^2 m}$ (b) $r_n \propto \dfrac{n^2}{mZ}$

(c) For hydrogen] $Z = 1, r_n = 0.529\, n^2$ Å

$r_1 : r_2 : r_3 = 1 : 4 : 9$

$r_H : r_{He^+} : r_{Li} = 1 : \frac{1}{2} : \frac{1}{3} = 6 : 3 : 1$

n - Order of orbit or principal quantum number
Z - Atomic number of element
m - Mass of particle like electron, meuon, etc. rotating about nucleus

(ii) The velocity of electron in n^{th} orbit :

(a) $V_n = \frac{2\pi kZe^2}{nh}$ (b) $V_n \propto \frac{Z}{n}$

(c) $V_n = \frac{c}{137} \frac{Z}{n}$

(d) For hydrogen]

$$v_n = \frac{2.188 \times 10^6}{n} = \frac{c}{137\,n} \text{ m/s}$$

$$v_1 : v_2 : v_3 = 1 : \frac{1}{2} : \frac{1}{3} = 6 : 3 : 2$$

(iii) Potential energy (U_n) in n-th orbit :

$$U_n = \frac{-kZe^2}{r_n} = \frac{-27.2}{n^2} Z^2 \text{ eV}$$

For Hydrogen atom $U_n = \frac{-ke^2}{r_n}$

(iv) Kinetic energy (E_{kn}) in n-th orbit :

$$E_{kn} = \frac{kZe^2}{2r_n} = \frac{13.6Z^2}{n^2} \text{ eV}$$

For Hydrogen atom, $E_{kn} = \frac{ke^2}{2r_n}$

(v) Total energy in n-th orbit :

= Kinetic energy + Potential energy

$E_n = U_n + E_{kn}$

(a) $E_n = \frac{-kZe^2}{2r_n}$

(b) $E_n = \frac{-2\pi^2 k^2 me^4 Z^2}{n^2 h^2}$

(c) $E_n = \frac{-RchZ^2}{n^2}$,

R = Rydberg constant $= \frac{2\pi^2 k^2 me^4}{ch^3}$

$= 1.1 \times 10^7 \text{ m}^{-1}$

(d) $E_n = -\frac{13.6Z^2}{n^2}$ eV

(e) For H atom $E_1 = -13.6$ eV, $E_2 = -3.40$ eV, $E_3 = -1.51$ eV

SERIES OF HYDROGEN SPECTRUM ENERGY LEVELS

The wavelength of radiation emitted due to transition of an electron between stages of quantum number n_1 and quantum number n_2 is given by

$$\bar{v} = \frac{1}{\lambda} = R\left(\frac{1}{n_1^2} - \frac{1}{n_2^2}\right)$$

where $R = 1.097 \times 10^7 \text{ m}^{-1}$ is called Rydberg constant.

If $n_1 = 1$, $n_2 = 2, 3, 4$, etc. Lyman series } u-v region

If $n_1 = 2$, $n_2 = 3, 4, 5$ etc. Balmer series } visible region

If $n_1 = 3$, $n_2 = 4, 5, 6$ etc. Pachen series ;
If $n_1 = 4$, $n_2 = 5, 6, 7$, etc. Bracket series } Infra red region

If $n_1 = 5$, $n_2 = 6, 7, 8$, etc. Pfund series.

When there is the transition from high energy state to low energy state it gives **emission spectrum.** When there is the transition from low energy state to high energy state it gives **absorption spectrum.**

In Bohr model, contrary to ordinary classical expectation, the frequency of revolution of an electron in its orbit is not connected to the frequency of spectral line. The later is the difference between two orbital energies divided by h. For transitions between large quantum numbers (n to n – 1, n very large), however, the two coincide as expected.

Quantum Numbers for the Hydrogen Atom :

Name	Symbol	Allowed Values
Principal quantum number	n	1, 2, 3, ...
Orbital quantum number	ℓ	0, 1, 2, ..., (n – 1)
Magnetic quantum number	m_ℓ	$-\ell$, ...–2, –1, 0, +1, +2, ...$+\ell$
Spin quantum no.	m_s	$-\frac{1}{2}, +\frac{1}{2}$

Paulli's exclusion principle : The state of an electron in an atom can be completely specified by the four quantum numbers n, ℓ, m_ℓ, m_s. Pauli's exclusion principle states that "no two electrons in an isolated atom can have the same values for all the four quantum numbers."

Exercise

1. The following statements are all true. Which one did Rutherford consider to be supported by the results of experiments in which α-particles were scattered by gold foil?
   (a) The nucleus of an atom is held together by forces which are much stronger than electrical or gravitational forces.
   (b) The force of repulsion between an atomic nucleus and an α-particle varies with distance according to inverse square law.
   (c) α-particles are nuclei of Helium atoms.
   (d) Atoms can exist with a series of discrete energy levels.
2. The proof of quantization of energy states in an atom is obtained by the experiment performed by:
   (a) Thomson (b) Millikan
   (c) Rutherford (d) Franck & Hertz
3. Which of the following types of radiation is not emitted by the electronic structure of atoms:
   (a) Ultraviolet light (b) X-rays
   (c) Visible light (d) γ-rays
4. The angular speed of the electron in the nth orbit of Bohr hydrogen atom is:
   (a) directly proportional to n
   (b) inversely proportional to $\sqrt{n}$
   (c) inversely proportional to n^2
   (d) inversely proportional to n^3
5. A spectral line is emitted when an electron:
   (a) rotates in the circular orbit
   (b) rotates in the elliptical orbit
   (c) jumps from lower orbit to higher orbit
   (d) jumps from higher orbit to lower orbit
6. Lines of Blamer series are emitted by the hydrogen atom when the electron jumps:
   (a) from higher orbits to first orbit
   (b) from higher orbit to second orbit
   (c) from second orbit to any other orbit
   (d) from third orbit to higher orbits
7. According to Bohr's model of hydrogen atom:
   (a) the linear velocity of the electron is quantised.
   (b) the angular velocity of the electron is quantised.
   (c) the linear momentum of the electron is quantised.
   (d) the angular momentum of the electron is quantised.
8. Consider the spectral line resulting from the transition $n = 2 \rightarrow n = 1$, in the atoms and ions given below, the shortest wavelength is produced by:
   (a) hydrogen atom
   (b) deuterium atom
   (c) singly ionised helium
   (d) doubly ionised lithium
9. Of the various series of the hydrogen spectrum, the one which lies wholly in the ultraviolet region is:
   (a) Lyman series (b) Balmer series
   (c) Paschen series (d) Bracket series
10. As the quantum number increases, the difference of energy between consecutive energy levels:
    (a) remain the same
    (b) increases
    (c) decreases
    (d) sometimes increases and sometimes decreases
11. Of the following transitions in hydrogen atom, the one which gives an absorption line of highest frequency is:
    (a) $n = 1$ to $n = 2$ (b) $n = 3$ to $n = 8$
    (c) $n = 2$ to $n = 1$ (d) $n = 8$ to $n = 3$
12. If the radius of the first orbit of hydrogen atom is 5.29×10^{-11} meter, the radius of the second orbit will be:
    (a) 21.16×10^{-11} metre
    (b) 15.87×10^{-11} metre
    (c) 10.58×10^{-11} metre
    (d) 2.64×10^{-11} metre
13. The minimum energy required to excite a hydrogen atom from its ground state is:
    (a) 3.4 eV (b) 13.6 eV
    (c) –13.6 eV (d) 10.2 eV
14. Line spectrum is obtained whenever the incandescent vapours at low pressure of the excited substance are in their
    (a) atomic state (b) molecular state
    (c) nuclear state (d) none of these

15. According to the Bohr model of hydrogen atom, radiation is emitted when the electron:
(a) revolves in an orbit
(b) jumps from an orbit to the nucleus
(c) jumps from a larger orbit to a smaller orbit
(d) jumps from a smaller orbit to a larger orbit

16. According to Bohr's theory of hydrogen atom, the angular momentum of an electron in any orbit of hydrogen atom is:
(a) directly proportional to the radius of the orbit
(b) inversely proportional to the radius of the orbit
(c) directly proportional to the square of the radius of the orbit
(d) directly proportional to the square root of the radius of the orbit

17. According to Bohr's model of the hydrogen atom, the radius of a stationary orbit characterised by the principal quantum number *n* is proportional to:
(a) n^{-1} (b) n
(c) n^{-2} (d) n^2

18. Who gave the principle that any given quantum orbit in an atom cannot be occupied by more than two electrons?
(a) Planck (b) Thomson
(c) Rutherford (d) Pauli

19. Fraunhofer lines are observed in the spectrum of:
(a) a hydrogen discharge tube
(b) a carbon arc
(c) the sun
(d) sodium vapour lamp

20. In which region of electromagnetic spectrum does the Lyman series of hydrogen atom lie:
(a) ultraviolet (b) infrared
(c) visible (d) X-ray

21. According to the Bohr theory of H-atom, the speed of the electron, its energy and the radius of its orbit varies with the principal quantum number *n*, respectively, as :
(a) $\frac{1}{n}, n^2, \frac{1}{n^2}$ (b) $n, \frac{1}{n^2}, n^2$
(c) $n, \frac{1}{n^2}, \frac{1}{n^2}$ (d) $\frac{1}{n}, \frac{1}{n^2}, n^2$

22. What is the wavelength of the most energetic photon emitted in the Balmer series of the Hydrogen atom?
(a) 654 nm (b) 580 nm
(c) 435 nm (d) 365 nm

23. In a hydrogen atom following the Bohr's postulates the product of linear momentum and angular momentum is proportional to $(n)^x$ where '*n*' is the orbit number. Then '*x*' is
(a) 0 (b) 2
(c) -2 (d) 1

24. One of the lines in the emission spectrum of Li^{2+} has the same wavelength as that of the 2nd line of Balmer series in hydrogen spectrum. The electronic transition corresponding to this line is –
(a) $n = 4 \rightarrow n = 2$ (b) $n = 8 \rightarrow n = 2$
(c) $n = 8 \rightarrow n = 4$ (d) $n = 12 \rightarrow n = 6$

25. Which one of the following is used as a moderator in nuclear reaction?
(a) Uranium (b) Heavy water
(c) Cadmium (d) Plutonium

26. The reaction responsible for the production of light energy from the sun will be:
(a) fission (b) fusion
(c) nuclear (d) none of these

27. Particle nature and wave nature of electromagnetic waves and electrons can be represented by :
(a) photoelectricity and electron microscopy
(b) light is refracted and diffracted
(c) X-rays is diffracted, reflected by thick metal sheet
(d) electrons have small mass, deflected by the metal sheet

28. Half life of a substance is 20 minutes, then the time between 33 % decay and 67 % decay will be
(a) 20 minutes (b) 40 minutes
(c) 50 minutes (d) 10 minutes

29. Who indirectly determined the mass of the electron by measuring the charge of the electron?
(a) Millikan (b) Rutherford
(c) Einstein (d) Thomson

30. When an electron jumps from the fourth orbit to the second orbit, one gets the
(a) second line of Lyman series
(b) second line of Paschen series
(c) second line of Balmer series
(d) first line of Pfund series

31. In which of the following systems will the radius of the first orbit ($n = 1$) be minimum?
(a) Hydrogen atom
(b) Doubly ionized lithium
(c) Singly ionized helium
(d) Deuterium atom

32. The Bohr model of atoms
(a) predicts the same emission spectra for all types of atoms
(b) assumes that the angular momentum of electrons is quantised
(c) uses Einstein's photoelectric equation
(d) predicts continuous emission spectra for atoms

33. Energy E of a hydrogen atom with principal quantum number n is given by $E = -13.6/n^2$ eV. The energy of photon ejected when the electron jumps from $n = 3$ state to $n = 2$ state of hydrogen is approximately
(a) 1.9 eV (b) 1.5 eV
(c) 0.85 eV (d) 3.4 eV

34. Ionization potential of hydrogen atom is 13.6 eV. Hydrogen atoms in the ground state are excited by monochromatic radiation of photon energy 12.1 eV. According to Bohr's theory, the spectral lines emitted by hydrogen will be
(a) three (b) Four
(c) One (d) Two

35. J.J. Thomson's cathode-ray tube experiment demonstrated that
(a) the e/m ratio of the cathode-ray particles changes when a different gas is placed in the discharge tube
(b) cathode rays are streams of negatively charged ions
(c) all the mass of an atom is essentially in the nucleus
(d) the e/m of electrons is much greater than the e/m of protons

36. To explain his theory, Bohr used
(a) conservation of linear momentum
(b) conservation of angular momentum
(c) conservation of quantum frequency
(d) conservation of energy

37. Which source is associated with a line emission spectrum?
(a) Electric fire (b) Neon street sign
(c) Red traffic light (d) sun

38. The spectrum obtained from a sodium vapour lamp is an example of
(a) band spectrum
(b) continuous spectrum
(c) emission spectrum
(d) absorption spectrum

39. When a hydrogen atom is raised from the ground state to an excited state,
(a) P.E. decreases and K.E. increases
(b) P.E. increases and K.E. decreases
(c) both K.E. and P.E. decrease
(d) absorption spectrum

40. When hydrogen atom is in its first excited level, its radius is
(a) four times its ground state radius
(b) twice
(c) same
(d) half

41. What is the radius of iodine atom (At. no. 53, mass no. 126)
(a) 2.5×10^{-11} m (b) 2.5×10^{-9} m
(c) 7×10^{-9} m (d) 7×10^{-6} m

42. The ionisation energy of hydrogen atom is 13.6 eV, the ionisation energy of helium atom would be
(a) 13.6 eV (b) 27.2 eV
(c) 6.8 eV (d) 54.4 eV

43. The ground state energy of H-atom 13.6 eV. The energy needed to ionize H-atom from its second excited state.
(a) 1.51 eV (b) 3.4 eV
(c) 13.6 eV (d) 12.1 eV

44. In terms of Bohr radius a_0, the radius of the second Bohr orbit of a hydrogen atom is given by
(a) $4\,a_0$ (b) $8\,a_0$
(c) $\sqrt{2}\,a_0$ (d) $2\,a_0$

45. The ionization energy of hydrogen atom is 13.6 eV. Following Bohr's theory, the energy corresponding to a transition between 3rd and 4th orbit is
(a) 3.40 eV (b) 1.51 eV
(c) 0.85 eV (d) 0.66 eV

46. In Rutherford scattering experiment, what will be the correct angle for α-scattering for an impact parameter, $b = 0$?
(a) 90° (b) 270°
(c) 0° (d) 180°

47. The radius of hydrogen atom in its ground state is 5.3×10^{-11} m. After collision with an electron it is found to have a radius of 21.2×10^{-11} m. What is the principal quantum number n of the final state of the atom
(a) $n = 4$ (b) $n = 2$
(c) $n = 16$ (d) $n = 3$

48. In the Bohr model of a hydrogen atom, the centripetal force is furnished by the coulomb attraction between the proton and the electron. If a_0 is the radius of the ground state orbit, m is the mass, e is the charge on the electron and ε_0 is the vacuum permittivity, the speed of the electron is
(a) 0 (b) $\frac{e}{\sqrt{\varepsilon_0 a_0 m}}$
(c) $\frac{e}{\sqrt{4\pi\varepsilon_0 a_0 m}}$ (d) $\frac{\sqrt{4\pi\varepsilon_0 a_0 m}}{e}$

49. Cathode rays moving with same velocity v describe an approximate circular path of radius r metre in an electric field of strength x volt metre^{-1}. If the speed of the cathode rays is doubled to $2v$, the value of electric field needed so that the rays describe the same approximate circular path (volt metre^{-1}) is
(a) $2x$ (b) $3x$
(c) $4x$ (d) $6x$

50. An electron changes its position from orbit $n = 2$ to the orbit $n = 4$ of an atom. The wavelength of the emitted radiations is (R = Rydberg's constant)
(a) $\frac{16}{R}$ (b) $\frac{16}{3R}$
(c) $\frac{16}{5R}$ (d) $\frac{16}{7R}$

51. The energy of hydrogen atom in nth orbit is E_n, then the energy in nth orbit of single ionised helium atom will be
(a) $4E_n$ (b) $E_n/4$
(c) $2E_n$ (d) $E_n/2$

52. The half life of radium is about 1600 years. Of 100 g of radium existing now, 25 g will remain unchanged after
(a) 3200 years (b) 4800 years
(c) 6400 years (d) 2400 years

53. The total energy of an electron in the first excited state of hydrogen atom is about –3.4 eV. Its kinetic energy in this state is
(a) 3.4 eV (b) 6.8 eV
(c) –3.4 eV (d) –6.8 eV

54. The total energy of electron in the ground state of hydrogen atom is – 13.6 eV. The kinetic energy of an electron in the first excited state is
(a) 6.8 eV (b) 13.6 eV
(c) 1.7 eV (d) 3.4 eV

55. The ground state energy of hydrogen atom is 13.6eV. When its electron is in the first excited state, its excitation energy is
(a) 3.4 eV (b) 6.8 eV
(c) 10.2 eV (d) 0

56. In a Rutherford scattering experiment when a projectile of charge Z_1 and mass M_1 approaches a target nucleus of charge Z_2 and mass M_2, the distance of closest approach is r_0. The energy of the projectile is
(a) directly proportional to $Z_1 Z_2$
(b) inversely proportional to Z_1
(c) directly proportional to mass M_1
(d) directly proportional to $M_1 \times M_2$

57. The ionization energy of the electron in the hydrogen atom in its ground state is 13.6 eV. The atoms are excited to higher energy levels to emit radiations of 6 wavelengths. Maximum wavelength of emitted radiation corresponds to the transition between
(a) $n = 3$ to $n = 1$ states
(b) $n = 2$ to $n = 1$ states
(c) $n = 4$ to $n = 3$ states
(d) $n = 3$ to $n = 2$ states

58. The energy of a hydrogen atom in the ground state is – 13.6 eV. The energy of a He^+ ion in the first excited state will be
(a) –13.6 eV (b) –27.2 eV
(c) –54.4 eV (d) –6.8 eV

59. An alpha nucleus of energy $\frac{1}{2}mv^2$ bombards a heavy nuclear target of charge Ze. Then the distance of closest approach for the alpha nucleus will be proportional to
(a) $\frac{1}{Ze}$ (b) v^2
(c) $\frac{1}{m}$ (d) $\frac{1}{v^4}$

60. The wavelength of the first line of Lyman series for hydrogen atom is equal to that of the second line of Balmer series for a hydrogen like ion. The atomic number Z of hydrogen like ion is
(a) 3 (b) 4
(c) 1 (d) 2

CHAPTER 27

Nuclei

CONCEPTS

COMPOSITION AND SIZE OF NUCLEUS

An atom has a nucleus. The nucleus is positively charged. The radius of the nucleus is smaller than the radius of an atom by a factor of 10^4. More than 99.9% mass of the atom is concentrated in the nucleus. The density of nuclear matter is independent of the size of the nucleus. The mass density of the atom does not follow this rule.

A nucleus can be considered to be spherical in shape and assigned a radius. Electron scattering experiments allow determination of the nuclear radius; it is found that radii of nuclei fit the formula $\mathbf{R = R_0 A^{1/3}}$, where R_0 = a constant = 1.2 fm and A = mass number. This implies that the **nuclear density** is independent of A. It is of the order of 10^{17} kg/m^3.

Neutrons and protons are bound in a nucleus by the short-range strong nuclear force. The nuclear force does not distinguish between neutron (n) and proton (p). A = n + p

TYPES OF NUCLEI

Isotopes

These are nuclei of same element having same atomic number Z but different mass number A

e.g. ${}_8O^{16}, {}_8O^{17}, {}_8O^{18}$; ${}_1H^1, {}_1H^2, {}_1H^3$; ${}_{92}U^{234}, {}_{92}U^{235}, {}_{92}U^{238}$

Isobars

These are nuclei of different elements having same A but different N and Z.

e.g ${}_6C^{14}$ and ${}_7N^{14}$; ${}_{18}Ar^{40}$ and ${}_{20}Ca^{40}$

Isotones

These are nuclei of different elements having same neutron N but different A.

e.g. ${}_6C_7^{13}$ & ${}_7N_7^{14}$; ${}_1H_2^3$ & ${}_2He_2^4$; ${}_2Be_5^9$ & ${}_5B_5^{10}$

MASS DEFECT AND BINDING ENERGY

The nuclear mass M is always less than the total mass, Σ m, of its constituents. The difference in mass of a nucleus and its constituents is called the **mass defect,**

$\Delta M = (Z m_p + (A - Z)m_n) - M$

Using **Einstein's mass energy relation**, $E = mc^2$ we express this mass difference in terms of energy as $\Delta E_b = \Delta M c^2$

The energy ΔE_b represents the **binding energy** of the nucleus.

Binding energy per nucleon $\overline{B} = \dfrac{\Delta E_b}{A}$

In the mass number range A = 30 to 170, the binding energy per nucleon is nearly constant, about 8 MeV/nucleon. The nature of the binding energy (per nucleon) curve shows that exothermic nuclear reactions are possible, when two light nuclei fuse or when a heavy nucleus undergoes fission into nuclei with intermediate mass. Although the binding energy (per nucleon) curve is smooth and slowly varying, it shows peaks at nuclides like ^{4}He, ^{16}O etc. This is considered as evidence of atom-like shell structure in nuclei.

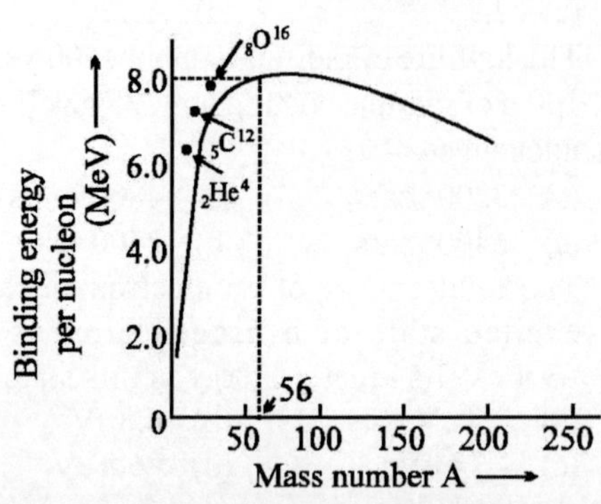

The **Q-value** of a nuclear process is Q = final kinetic energy – initial kinetic energy.
Due to conservation of mass-energy, this is also,
Q = (sum of initial masses – sum of final masses)c^2

RADIOACTIVITY

It is the phenomenon in which nuclei of a given species transform by giving out α or β or γ rays; α-rays are helium nuclei; β-rays are electrons and γ-rays are electromagnetic radiation of wavelengths shorter than X-rays.

Activity : The number of decays per unit time or decay rate is called activity.

Activity $A = -\frac{dN}{dt} = N_0 \lambda e^{-\lambda t} = A_0 e^{-\lambda t} = N\lambda$

where $N_0 \lambda = A_0$ is initial activity

$A = A_0 e^{-\lambda t}$ is the activity law which shows activity decreases exponentially with time. The activity of one gram of radioactive substance is called **specific activity**.

Units of activity :

Curie : The specific activity of 1 gm of Radium – 226 is called **one curie**.

1 curie = 3.7×10^{10} disintegrations per second

Rutherford : 1 Rutherford = 10^6 disintegrations per second

Becquerel : 1 Becquerel = 1 disintegration per second

Law of radioactive decay : $N(t) = N(0) e^{-\lambda t}$; where λ is the decay constant or disintegration constant.

PROPERTIES OF α, β AND γ RAYS

Property	α – particle	β – particle	γ – ray
1.Nature	These are doubly ionized helium atom $_2He^4$ charge $q = +2e = 3.2 \times 10^{-19}C$ mass $m = 2p + 2n = 4amu$ $= 4 \times 1.6 \times 10^{-27}$ kg	These are beam of fast moving electrons (β^-) and positrons (β^+) charge $\beta^- = -e = -1.6 \times 10^{-19}C$ $\beta^+ = +e = 1.6 \times 10^{-19}C$ $m(\beta^-) = m(\beta^+) = 9.1 \times 10^{-31}$ kg	These are electromagnetic radiations of high frequency and travel in form of photons charge q = 0 (chargeless) rest mass = 0 effective mass $= \frac{h\nu}{c^2} = \frac{h}{\lambda c}$
2.Velocity	Speed ranges between 1.4×10^7 to 2.20×10^7 m/s $v_\alpha \sim 0.05$ c	Speed ranges from 1% to 90% of velocity of light $v_\beta \sim 0.9c$	Speed equals velocity of light $v_\gamma = c$
3.Ionizing power	These have maximum ionizing power (1000)	Their ionizing power is less than α particles and more than γ rays (100)	Their ionizing power is least (1)
4.Penetrating power	The penetration power is smallest. Can only penetrate through 0.01 mm thick Al sheet (1)	Penetration power is about 100 times that of α rays, can penetrate through 1 mm thick Al sheet (100)	Penetration power is very large. Can penetrale about 30 cm thick Al sheet (10000)
5.Range	Range is very small (few cms in air)	Range is more than α rays. (few metres in air)	Range is very large (many hundreds of metre in air)
6.Nature of spectrum	Line spectrum	Continuous spectrum	Line spectrum
7.Interaction with matter	Produces heat	Produces heat	Produces photoelectric effect Compton effect and pair production
8.Effect of electric and magnetic field	Suffers small deflection	Suffers large deflection	Suffers null deflection
9.Effect on photo graphic plate and ZnS	Affects photographic plate and produces fluorescense	Affects photographic plate and produces fluorescence	Affects photographic plate and produces fluorescence.

HALF LIFE AND MEAN LIFE OF RADIO-ACTIVE SUBSTANCE

The half-life $T_{1/2}$ of a radionuclide is the time in which N has been reduced to one-half of its initial value. The mean life τ is the time at which N has been reduced to e^{-1} of its initial value $\quad T_{1/2} = \frac{\ln 2}{\lambda} = \tau \ln 2$

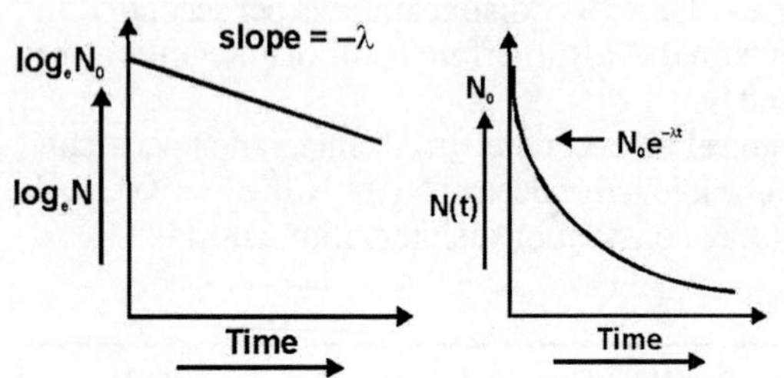

Mean life (τ) of an atom in a radioactive substance is called average life of radioactive substance.

$$\text{Mean life } \tau = \frac{\text{The sum of lives of all active atoms}}{\text{Total number of active atoms}}$$

or $\quad \tau = \frac{\int_0^{N_0} t_\lambda \, dN_{dt}}{N_0} = \frac{N_0\lambda \int_0^{N_0} t\, e^{-\lambda t} dt}{N_0} = \frac{1}{\lambda}$

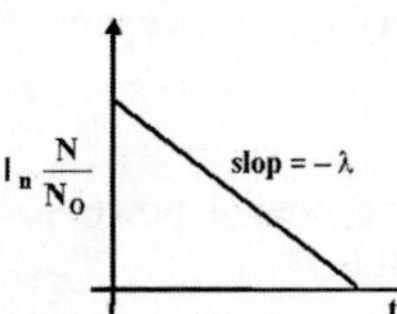

Mean life is equal to magnitude of reciprocal of slope of $\log \frac{N}{N_0}$ v/s t curve.

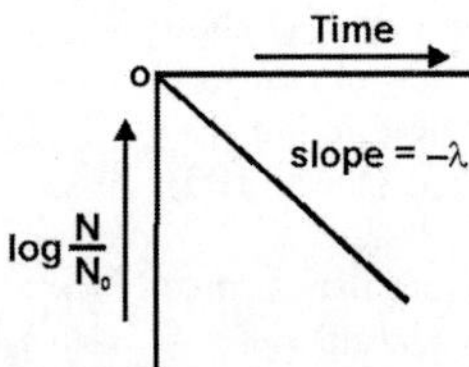

Carbon dating : Carbon dating is the process of determination of time interval which has passed by making use of radioactive decay of a sample containing radioactive substance. It helps in calculating age of geological specimens like rocks, biological specimens like bones of animals or trunk of trees and age of earth. The isotope of carbon, ${}_6C^{14}$ is radioactive. It is formed in atmosphere by bombardment of nitrogen atoms with cosmic rays

${}_7N^{14} + {}_0n^1 \rightarrow {}_6C^{14} + {}_1H^1$

The ${}_6C^{14}$ combines with oxygen to form carbondioxide which is absorbed by plants so concentration of ${}_6C^{14}$ is constant with time. The living plants and animals have a fixed ratio of ${}_6C^{14}$ to ordinary carbon ${}_6C^{12}$. When a plant or animal dies the content of ${}_6C^{14}$ decreases while that of ${}_6C^{12}$ remains constant. The ratio of two indicates the time that has passed since death of plant or animal. The time interval is calculated from the laws of radioactive disintegration

$$t = \frac{1}{\lambda}\log_e \frac{N_0}{N} = \frac{2.303}{\lambda}\log_{10}\frac{N_0}{N} \quad \left(\frac{N_0}{N} = \frac{A_0}{A}\right)$$

where N_o is number of ${}_6C^{14}$ nuclei at time of death, λ is decay constant of ${}_6C^{14}$ and N is number of ${}_6C^{14}$ nuclei currently present in sample.

NUCLEAR REACTIONS

Nuclear Fission

The process of splitting of a heavy nucleus into two nuclei of comparable size and release of large energy is called fission.

$${}_{92}U^{235} + {}_0n^1 \rightarrow {}_{56}Ba^{141} + {}_{36}Kr^{92} + 3\,{}_0n^1 + 200\,\text{MeV}$$

U^{235} nucleus captures a thermal neutron. This forms a compound nucleus U^{236} in excited state.

Nuclear fission was discovered by Otto Hahn and Strassman.

Nuclear fission can be explained on the basis of liquid drop model. The fission process can take place at normal pressure and temperature.

Nuclear Fusion :

The process in which two or more lighter nuclei combine to form a heavy nucleus is known as nuclear fusion. $\quad 4\,{}_1H^1 \longrightarrow {}_2He^4 + 2\,{}_{+1}e^0 + 2\nu + Q$

The binding energy per nucleon of product is greater than the reactants. The energy released per nucleon is large ~ 6.75 MeV. Fusion is possible at high pressure (~ 10^6 atom) and high temperature (~ 10^8 °C).

The proton-proton cycle happens at lower temperature as compared to carbon-nitrogen cycle.

Nuclear fusion is possible at a place which has reactants in large quantity. Hydrogen bomb works on principle of nuclear fusion. The explosion of a hydrogen bomb needs an explosion of atom bomb to generate required temperature. No harmful radiations are produced in fusion.

Moderator: An important aspect of the fission process is the speed of the neutrons. Slow neutrons hit uranium-235 nuclei more efficiently than do fast ones.

Because fission reactors are highly exothermic, the neutrons produced usually move with high velocities. For greater collision efficiency, neutrons must be slowed down. For this purposes a substance is used that can reduce the kinetic energy of neutrons. Such a substance is called as a moderator. A good moderator should be a nontoxic and inexpensive substance. And it should be resist conversion into radioactive substance by neutron bombardment. Graphite (C) or heavy water (D_2O) are commonly used as moderators.

Control rods: In principle, the main difference between an atomic bomb and nuclear reactor is that the chain reaction that takes place in a nuclear reactor is kept under controlled conditions at all the times. The factor limiting the rate of the reaction is the number of neutrons present. This can be controlled by lowering cadmium or boron rods between the fuel elements.

Exercise

1. The idea of the quantum nature of light has emerged in an attempt to explain
   (a) the thermal radiations of a black body
   (b) the interference of light
   (c) radioactivity
   (d) thermionic emission
2. A photon is
   (a) a quantum of light
   (b) a quantum of matter
   (c) a positively charged paticle
   (d) an instrument for measuring light intensity
3. Of the following properties, the photon does not possess
   (a) rest mass (b) momentum
   (c) energy (d) frequency
4. A photon will have less energy, if its
   (a) amplitude is higher
   (b) frequency is higher
   (c) wavelength is longer
   (d) wavelength is shorter
5. The energy spectrum of β-particles [number N(E) as a function of β-energy E] emitted from a radioactive source is

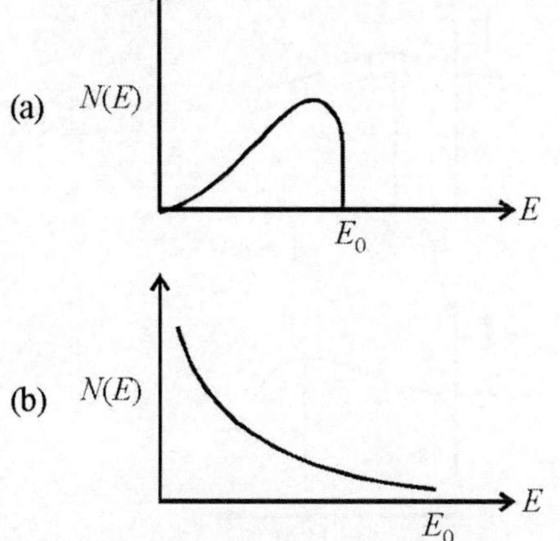

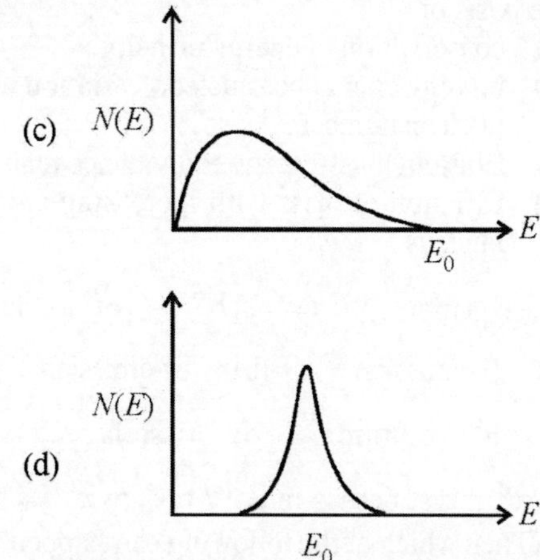

6. A nuclear transformation is denoted by $X(n, \alpha)$ ^7_3Li. Which of the following is the nucleus of element X?
   (a) $^{10}_5\text{Be}$ (b) $^{12}\text{C}_6$
   (c) $^{11}_4\text{Be}$ (d) ^9_5B
7. The speed of cathode rays is ($c = 3 \times 10^8$ m/s)
   (a) less than c (b) greater than c
   (c) equal to c (d) either 'b' or 'c'
8. A radioactive substance contains 10000 nuclei and its half-life period is 20 days. The number of nuclei present at the end of 10 days is
   (a) 7070 (b) 9000
   (c) 8000 (d) 7500
9. In fission of uranium, a chain reaction is possible because
   (a) huge amount of energy is released
   (b) more than 1 neutron is given out in each fission.
   (c) fragments in fission are radio-active
   (d) two fragments are formed in each fission.

10. Nuclear force is of
(a) short range and charge dependent
(b) long range and charge independent
(c) short range and charge independent
(b) long range like electrostatic type

11. Radioactive substance emits
(a) α -rays (b) γ -rays
(c) β -rays (d) all of these

12. Which of the following is deflected by electric field?
(a) X -rays (b) neutrons
(c) γ -rays (d) α -particles

13. Nuclear fission and fusion can be explained on the basis of
(a) conservation of energy principle
(b) binding energy per nucleon variation with nucleon number
(c) Einstein mass-energy equivalence relation
(d) variation of mass with increasing atomic nucleus

14. The equation ${}_{z}X^{A} \rightarrow {}_{z+1}Y^{A} + {}_{-1}e^{0} + \overline{\nu}$ is
(a) β -emission (b) α -emission
(c) e^{-} – capture (d) fission

15. A π^0 at rest decays into 2 γ rays as $\pi^0 \rightarrow \gamma + \gamma$. Then which of the following can happen?
(a) The two γ's move in same direction
(b) The two γ's move in opposite direction
(c) Both repel each other
(d) Both attract each other

16. Two protons exert a nuclear force on each other, the distance between them is
(a) 10^{-14} m (b) 10^{-10} m
(c) 10^{-12} m (d) 10^{-8} m

17. An element has binding energy 8 eV/nucleon. If it has total binding energy 128 eV, then the number of nucleons are
(a) 8 (b) 14
(c) 16 (d) 32

18. The specific charge for cathode rays is
(a) constant
(b) variable
(c) dependent upon the material of the cathode
(d) dependent upon the nature of gas in the discharge tube

19. Cathode rays are made to pass between the poles of a magnet perpendicular to axis, the effect of the magnetic field is
(a) to increase the velocity of rays
(b) to deflect them towards the north pole
(c) to deflect them towards the south pole
(d) to deflect them upwards above the plane of paper

20. The canal rays are
(a) a stream of electrons
(b) a stream of positively charged particles
(c) light rays
(d) a stream of uncharged particles

21. A strong argument for the particle nature of cathode rays is that they
(a) produce fluoroscence
(b) travel through vacuum
(c) get deflected by electric and magnetic fields
(d) cast shadow

22. Air becomes conducting when the pressure ranges between
(a) 76 cm and 10 cm
(b) 10 cm and 1 cm
(c) 1 cm and 10^{-3} cm
(d) 10^{-4} cm and 10^{-7} cm

23. In J.J. Thomson's method, electric field E, magnetic field B and velocity v of the electrons were in mutually perpendicular direction. This velocity selector allows particles of velocity v to pass undeflected when
(a) $v = BE$ (b) $v = E/B$
(c) $v = B/E$ (d) $v = B^2/E$

24. The magnetic Resonance Imaging (MRI) is based on the phenomenon of
(a) nuclear magnetic resonance
(b) electron spin resonance
(c) electron paramagnetic resonance
(d) diamagnetism of human tissues

25. The dependence of binding energy per nucleon, B_N, on the mass number A, is represented by

(a) 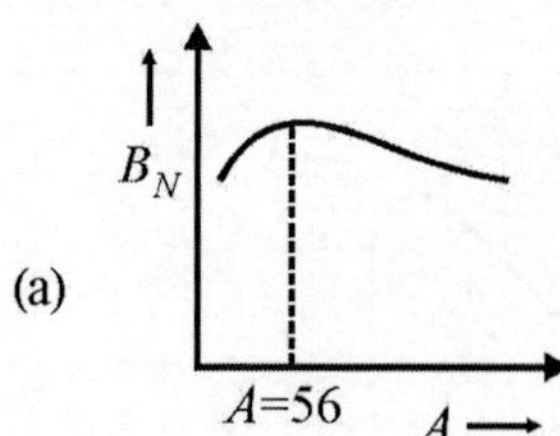

(b) 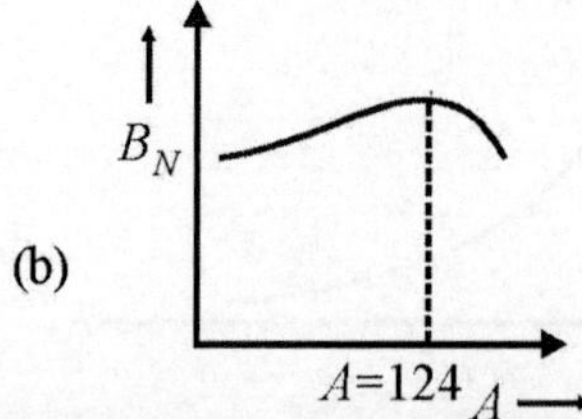

(c) 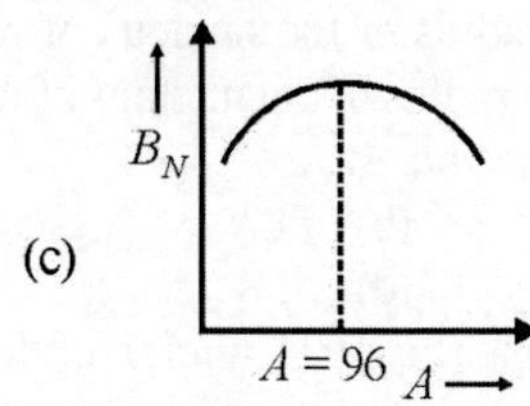

(d) 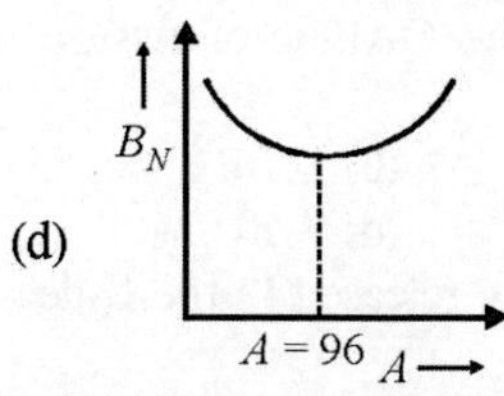

26. If an electron and a photon propagate in the form of waves having the same wavelength, it implies that they have the same
(a) energy
(b) momentum
(c) velocity
(d) angular momentum

27. Characteristic X-rays are produced due to
(a) transfer of momentum in collision of electrons with target atoms
(b) transition of electrons from higher to lower electronic orbits of an atom
(c) heating of the target
(d) transfer of energy in collision of electrons with atoms in the target

28. Nuclear fusion is possible
(a) only between light nuclei
(b) only between heavy nuclei
(c) between both light and heavy nuclei
(d) only between nuclei which are stable against β-decay

29. Radioactive nuclei that are injected into a potient collected at certain sites within its body, undergoing radioactive decay and emitting electromagnetic radiation. These radiations can then be recorded by a detector. This procedure provides an important diagnostic tools called
(a) Gamma camera
(b) CAT can
(c) Radiotracer technique
(d) Gamma ray spectroscopy

30. In a material medium, when a positron meets an electron both the particles annihilate leading to the emission of two gamma ray photons. This process forms the basis of an important diagnostic procedure called
(a) MRI (b) PET
(c) CAT (d) SPECT

31. The nuclei ${}_6C^{13}$ and ${}_7N^{14}$ can be described as
(a) isotones
(b) isobars
(c) isotopes of carbon
(d) isotopes of nitrogen

32. Which of the following statements is true for nuclear forces?
(a) they obey the inverse square law of distance
(b) they obey the inverse third power law of distance
(c) they are short range forces
(d) they are equal in strength to electromagnetic forces

33. Curie is a unit of
(a) energy of gamma-rays
(b) half-life
(c) radioactivity
(d) intensity of gamma-rays

34. Which one is possible?
(a) ${}_7N^{19} + {}_0n^1 \rightarrow {}_7N^{16} + {}_1H^1$
(b) ${}_{16}S^{32} + {}_1H^1 \rightarrow {}_{17}Cl^{35} + {}_2He^4$
(c) ${}_8O^{16} + {}_0n^1 \rightarrow {}_7N^{14} + 3\,{}_1H^1 + 2\,{}_{-1}\beta^0$
(d) ${}_1H^1 + {}_1H^1 \rightarrow {}_2He^4$

35. Carbon dating is best suited for determining the age of fossils if their age in years is of the order of
(a) 10^3 (b) 10^4
(c) 10^5 (d) 10^6

36. In nucleus of mass number A, originally at rest, emits an α-particle with speed v. The daughter nucleus recoils with a speed :
(a) $\frac{2v}{A+4}$ (b) $\frac{4v}{A+4}$
(c) $\frac{4v}{A-4}$ (d) $\frac{2v}{A-4}$

37. The operation of a nuclear reactor is said to be critical, if the multiplication factor (K) has a value
(a) 1 (b) 1.5
(c) 2.1 (d) 2.5

38. The ratio of the specific charge (e) of a proton to that of an α particle is
(a) 1 : 2 (b) 1 : 1/4
(c) 1 : 4 (d) 1 : 1/2

39. The radius of first orbit of hydrogen atom is 0.53Å. The radius of its fourth orbit will be
(a) 2.12Å (b) 8.48 Å
(c) 0.13Å (d) 0.03 Å

40. The energy obtained when 1 mg mass is completely annihilated is
(a) 9×10^{13} MeV (b) 9×10^{10} J
(c) 9×10^{13} J (d) 9×10^{16} J

41. Mosley law regarding characteristic X-rays states that (v = frequency, Z = atomic number and a, b are constants)
(a) $v^2 = aZ$ (b) $v = a(Z-b)$
(c) $\sqrt{v} = a(Z-b)$ (d) $v^2 = a(Z-b)$

42. If one gm of hydrogen is converted into 0.993 g of helium, in a thermonuclear reaction, the energy released in the reaction is
(a) 63×10^7 J (b) 63×10^{14} J
(c) 63×10^{10} J (d) 36×10^{20} J

43. Plutonium decays with a half-life of 24000 years. If plutonium is stored for 72,000 years, then the fraction of plutonium that remains is
(a) 1/2 (b) 1/4
(c) 1/3 (d) 1/8

44. An element A decays into an element C by a two step process
$A \rightarrow B + 2\text{He}^4$ and $B \rightarrow C + 2e^-$. Then,
(a) A and C are isotopes
(b) A and C are isobars
(c) B and C are isotopes
(d) A and B are isobars

45. In a nuclear fusion reaction involving deuterium and tritium, the released energy is typically of the order of:
(a) 9.10 MeV (b) 278 keV
(c) 1.78 GeV (d) 17.6 MeV

46. Nuclear forces are
(a) spin dependent and have no non-central part
(b) spin dependent and have a non-central part
(c) spin independent and have no non-central part
(d) spin independent and have a non-central part

47. Which one of the following nuclear reactions is possible?
(a) ${}^{14}_{7}N \longrightarrow {}^{13}_{6}C + \beta^+ + \nu_c$
(b) ${}^{13}_{7}N \longrightarrow {}^{13}_{6}C + \beta^+ + \nu_c$
(c) ${}^{13}_{7}N \longrightarrow {}^{13}_{6}C + \beta^+$
(d) ${}^{13}_{7}N \longrightarrow {}^{13}_{7} + \beta^+ + \nu_c$

48. The ratio of the radius of the nucleus of mass number 216 to the radius of the nucleus of mass number 64 is approximately:
(a) 1.0 (b) 1.2
(c) 1.5 (d) 1.8

49. At what rate must the ^{235}U nuclei undergo fission reaction to generate energy at the rate of 3.0 W. Assume the Q-value of fission to be 200 MeV:
(a) 10^{23} (b) 10^{19}
(c) 10^{11} (d) 10^{6}

50. What is the energy released in the β-decay of
$^{32}P \longrightarrow {}^{32}S$?
(Given : Atomic masses : 31.97391 u for ^{32}P and 31.97207 u for ^{32}S
(a) −1.2 MeV (b) +1.7 MeV
(c) +2.1 MeV (d) −0.9 MeV

51. In a nuclear reaction between a deuteron and ${}_6C^{12}$, ${}_7N^{13}$ is produced. The other liberated particle is a :
(a) proton (b) electron
(c) neutron (d) positron

52. How many alpha α – and β– decays does U^{238} experiences before turning finally into stable Pb^{206} isotope ?
(a) 12, 6 (b) 10, 4
(c) 8, 6 (d) 8, 8

53. The activity of a certain radioactive sample is 128 mCi. If the decay constant of the material be 6.93 per hour then its activity after 30 minutes will reduced.
(a) 4 mCi (b) 12.8 mCi
(c) 16 mCi (d) 32 mCi

54. Actinium 231, $^{231}AC_{89}$, emit in succession two β particles, four alphas, one β and one alpha plus several γ rays. What is the resultant isotope:
(a) $^{221}Au_{79}$ (b) $^{211}Au_{79}$
(c) $^{221}Pb_{82}$ (d) $^{211}Pb_{82}$

55. The order of nuclear density is :
(a) 10^{13} kg/m³ (b) 10^{15} kg/m³
(c) 10^{17} kg/m³ (d) 10^{19} kg/m³

56. In which sequence the radioactive radiation are emitted in the following nuclear reaction
${}_ZX^A \longrightarrow {}_{Z+1}Y^A \longrightarrow$
${}_{Z-1}K^{A-4} \longrightarrow {}_{Z-1}K^{A-4}$
(a) γ, α, β (b) α, β, γ
(c) β, γ, α (d) β, α, γ

57. The radioactive decay of uranium into thorium is expressed by the equation

$_{92}U^{238} \longrightarrow Th + X$, what X is?

(a) A deuteron (b) An alpha particle
(c) A proton (d) An electron

58. The activity of a radioactive substance is R_1 at time t_1 and R_2 at a later time instant t_2. Its decay constant is λ. Then,

(a) $R_1 t_1 = R_2 t_2$
(b) $R_2 \exp(\lambda t_2) = R_1 \exp(\lambda t_1)$
(c) $R_2 = R_1 \exp[\lambda(t_2 - t_1)]$
(d) $R_1 - R_2 =$ a constant $(t_2 - t_1)$

59. An element X decays, first by positron emission and then two α-particles are emitted in successive radioactive decay. If the product nuclei has a mass number 229 and atomic number 89, the mass number and atomic number of element X are

(a) 237, 93 (b) 237, 94
(c) 221, 84 (d) 237, 92

60. A heavy nucleus having mass number 200 gets disintegrated into two small fragments of mass number 80 and 120. If binding energy per nucleon for parent atom is 6.5 MeV and for daughter nuclei is 7 MeV and 8 MeV respectively, then the energy released in the decay will be –

(a) 200 MeV (b) –220 MeV
(c) 220 MeV (d) 180 MeV

Semiconductor Electronics : Material, Devices and Simple Circuits

CONCEPTS

ENERGY BANDS IN SOLIDS

There are two distinct band of energies **valence band** and **conduction band** in which the electrons in a material lie. Valence band energies are low as compared to conduction band energies. All energy levels in the valence band are filled while energy levels in the conduction band may be fully empty or partially filled.

The electrons in the conduction band are free to move in a solid and are responsible for the conductivity. The extent of conductivity depends upon the energy gap (E_g) between the top of valence band (E_V) and the bottom of the conduction band E_C. The electrons from valence band can be excited by heat, light or electrical energy to the conduction band and thus, produce a change in the current flowing in a semiconductor.

Comparison Between Conductors, Semiconductors and Insulators

Property	Conductors	Semiconductors	Insulators
Resistivity range	$10^{-6} - 10^{-8}$ Ωm	$10^{-4} - 0.5$ Ωm	$10^{7} - 10^{16}$ Ωm
Conductivity range	$10^{-6} - 10^{-8}$ mho/m	$10^{4} - 0.5$ mho/m	$10^{-7} - 10^{-16}$ mho/m
Temp. coefficient of resistance (α)	Positive	Negative	Negative
Flow of current	Due to free electrons	Due to electrons and holes	No current flow
Energy band diagram	Electron energy →; Conduction band (CB); No gap; Overlapping region; Valence band (VB)	Electron energy →; Conduction band (CB); Forbidden gap; $\Delta E_g \cong 1$eV; Valence band (VB)	Electron energy →; Conduction band (CB); Forbidden gap; $\Delta E_g \geq 6$eV; Valence band (VB)
Forbidden energy gap	$\cong$ 0eV	$\cong 0.1 - 3$eV	$\geq$ 6eV
Examples	Pt, Al, Cu, Ag etc.	Ge, Si, C, Ga, As GaF_2 etc.	Wood, plastic, diamond, mica

SEMICONDUCTOR ELECTRONICS

Semiconductors are the basic materials used in the present solid state electronic devices like diode, transistor, ICs, etc.

Semiconductors are elemental (Si, Ge) as well as compound (GaAs, CdS etc.).

Pure semiconductors are called **'intrinsic semiconductors'.** The presence of charge carriers (electrons and holes) is an 'intrinsic' property of the material and these are obtained as a result of thermal excitation. The number of electrons (n_e) is equal to the number of holes (n_h) in intrinsic semiconductors. Holes are essentially electron vacancies with an effective positive charge.

The number of charge carriers can be changed by 'doping' of a suitable impurity in pure semiconductors. Such semiconductors are known as **extrinsic semiconductors.**

Types of Extrinsic Semiconductors

There are two types of extrinsic semiconductors (n-type and p-type).

(i) **In n-type Semiconductors,** $n_e >> n_h$

n-type semiconduct or is obtained by doping Si or Ge with pentavalent atoms (donors) like As, Sb, P, etc.,

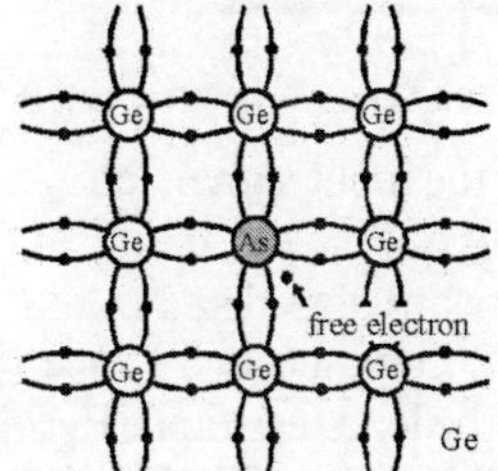

n-type semiconductor

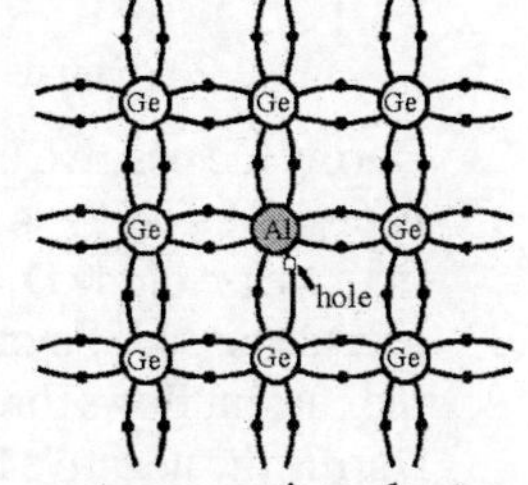

p-type semiconductor

(ii) **In p-type semiconductors,** $n_h >> n_e$

p-type semiconductor can be obtained by doping Si or Ge with trivalent atom (acceptors) like B, Al, In etc.

$n_e n_h = n_i^2$ in all cases. Further, the material possesses an overall charge neutrality.

P-N JUNCTION OR P-N JUNTION DIODE

p-n junction is the 'key' to all semiconductor devices. When such a junction is made, a 'depletion layer' is formed consisting of immobile ion-cores devoid of their electrons or holes. This is responsible for a junction potential barrier.

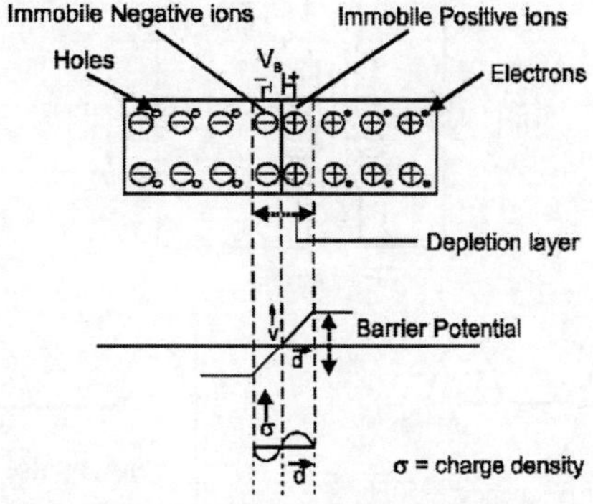

The p-n junction can be assumed as a capacitor having the depletion layer acting as a capacitor.

FORWARD AND REVERSE BIAS

By changing the external applied voltage, junction barriers can be changed. In forward bias (n-side is connected to negative terminal of the battery and p-side is connected to the positive), the barrier is decreased while the barrier increases in reverse bias. Hence, forward bias current is more (mA) while it is very small (μA) in a p-n junction diode.

Difference Between Forward Bias and Reverse Bias

	Forward bias		Reverse bias
1.	Potential barrier reduces.	1.	Potential barrier increases.
2.	Width of depletion layer decreases.	2.	Width of depletion layer increases.
3.	P-N Jn. provide very small resistance.	3.	P-N Jn. provide high resistance.
4.	Forward current flow in circuit.	4.	Very small current flow in circuit.
5.	Order of forward current in milli amp.	5.	Order of current in micro amp.
6.	Mainly majority current flows.	6.	Mainly minority current flows.
7.	Forward characteristic curve	7.	Reverse characteristic curve

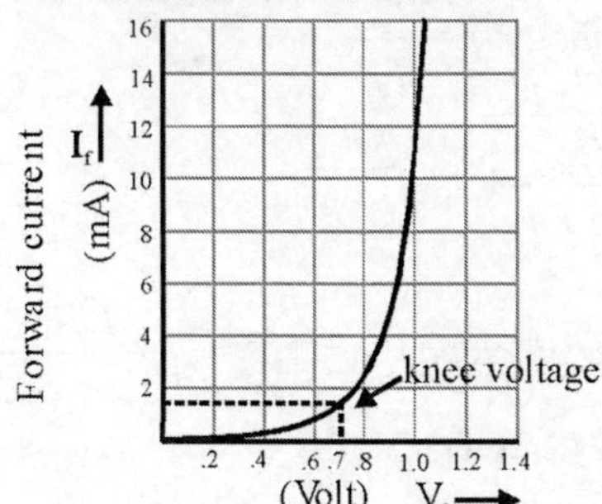

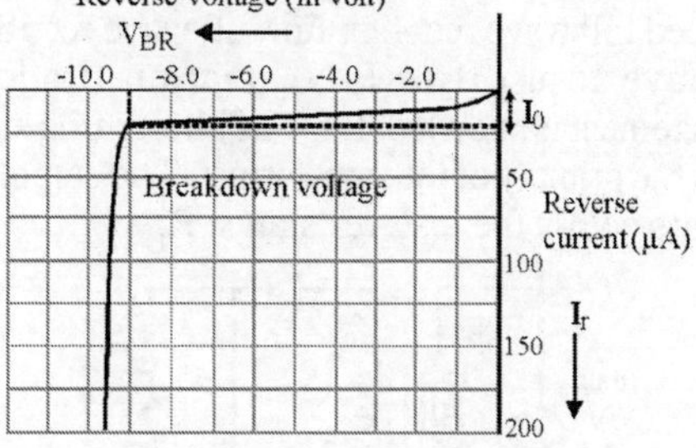

8.	Forward resistance $R_f = \dfrac{\Delta V_f}{\Delta I_f} \cong 100\ \Omega$	8.	Reverse resistance $R_r = \dfrac{\Delta V_r}{\Delta I_r} \cong 10^6\ \Omega$
9.	Knee or cut voltage Ge $\rightarrow$ 0.3 V, Si $\rightarrow$ 0.7 V	9.	Breakdown voltage Ge $\rightarrow$ 25 V, Si $\rightarrow$ 35 V

P–N JUNCTION DIODE AS A RECTIFIER

Rectifire is a device which is used for converting alternating current (a.c.)*into direct current* (d.c).

P-N Jn. Diode as a Half Wave Rectifier

For half wave rectification we have to use only one P-N junction. The arrangement is shown in fig.

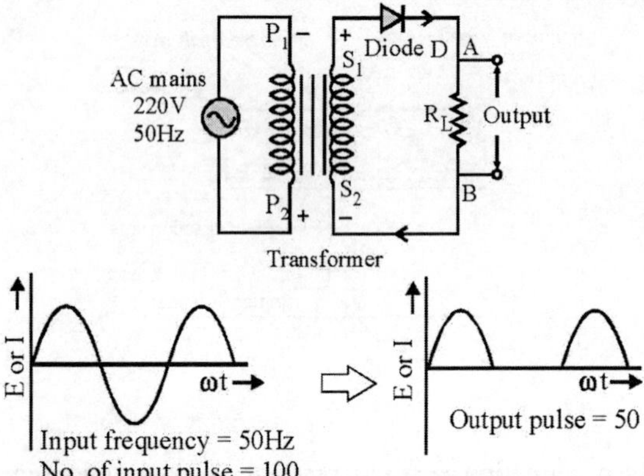

During the first half (positive) of the input signal, let S_1 is at positive and S_2 is at negative potential. So, the PN junction diode D is forward biased. The current flows through the load resistance R_L and output voltage is obtained.

During the second half (negative) of the input signal, S_1 and S_2 would be negative and positive respectively. The PN junction diode will be reversed biased. In this case, practically no current would flow through the load resistance. So, there will be no output voltage.

Thus, corresponding to an alternating input signal, we get a unidirectional pulsating output.

Peak voltage $V_s = V_{in}$

In half wave rectifier peak voltage = maximum voltage across secondary coil of transformer.

P-N Junction Diode as a Full Wave Rectifier

When the diode rectifies the whole of the AC wave, it is called full wave rectifier. For full wave rectification we have to use two P-N junction diode. The arrangement is shown in fig. The alternating signal is fed to the primary of the transformer. The output signal appears across the load resistance R_L.

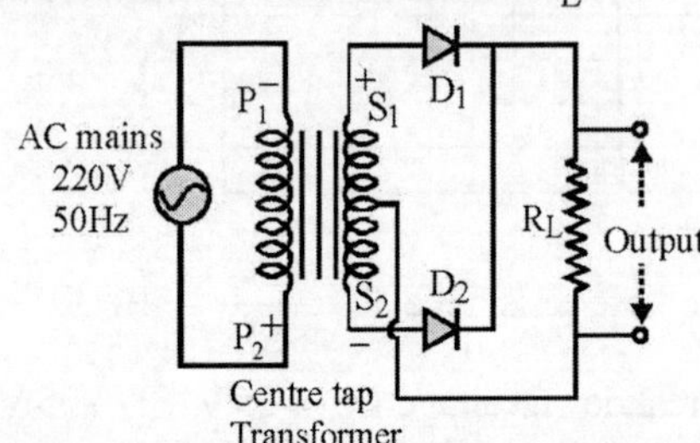

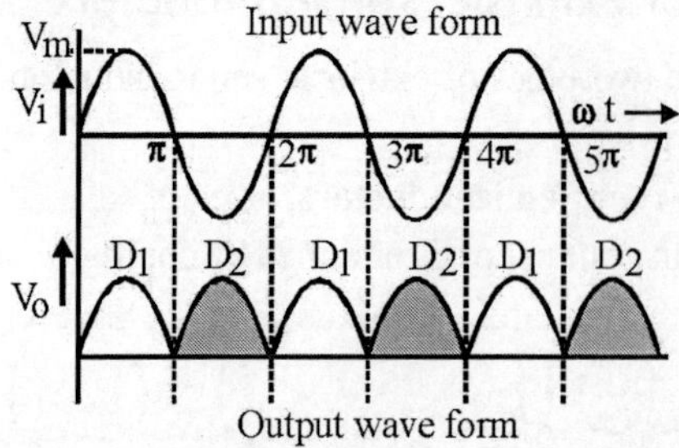

During the positive half of the input signal, let S_1 is positive and S_2 is negative.

In this case diode D_1 is forward biased and D_2 is reverse biased. So only D_1 conducts and hence the current flows through the load resistance R_L.

During the negative half of the input signal, S_1 is negative and S_2 is positive. So D_1 is reverse-biased and D_2 is forward biased. So only D_2 conducts and hence the current flows through the load resistance R_L.

It is clear that whether the input signal is positive or negative, the current always flows through the load resistance in the same direction and full wave rectification is obtained.

For full wave rectifier, form factor $F = \frac{\pi}{2\sqrt{2}}$

For half wave rectifier, form factor $F = \frac{\pi}{2}$

Ripple and ripple factor : In the output of rectifier some A. C. components are present. They are called ripple and their measurement is given by a factor so it is called ripple factor. For good rectifier ripple factor must be very low.

Total output current

$I_{rms} = \sqrt{I_{ac}^2 + I_{dc}^2}$, where I_{ac} = rms value of AC component and I_{dc} = rms value of DC component.

$$\text{Ripple factor } r = \frac{I_{ac}}{I_{dc}}$$

$$\Rightarrow r = \sqrt{\frac{I_{rms}^2}{I_{dc}^2} - 1} = \sqrt{F^2 - 1},$$

$$\text{Rectifier efficiency } \eta = \frac{P_{dc}}{P_{ac}} = \frac{I_{dc}^2 R_L}{I_{rms}^2 (R_F + R_L)}$$

SOME SPECIAL DIODES

There are some special purpose diodes.

Zener diode is one such special purpose diode. In reverse bias, after a certain voltage, the current suddenly increases (breakdown voltage) in a Zener diode. This property has been used to obtain **voltage**

regulation.

Zener diode

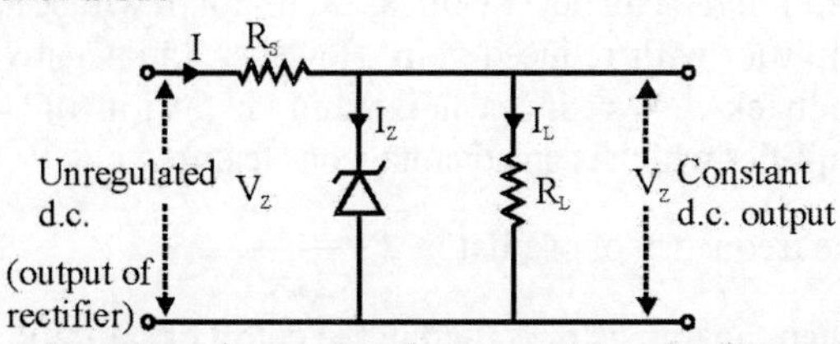

p-n junctions have also been used to obtain many photonic or optoelectronic devices where one of the participating entity is 'photon':

(a) **Photodiodes** in which photon excitation results in a change of reverse saturation current which helps us to measure light intensity.

(b) **Solar cells** which convert solar energy into electricity;

(c) **Light Emitting Diode (LED) and Diode Laser** in which electron excitation by a bias voltage results in the generation of light.

JUNCTION TRANSISTOR

Transistor is an n-p-n or p-n-p junction device.

The central block (thin and lightly doped) is called 'Base' while the other electrodes are 'Emitter' and 'Collector'. The emitter-base junction is forward biased while collector-base junction is reverse biased.

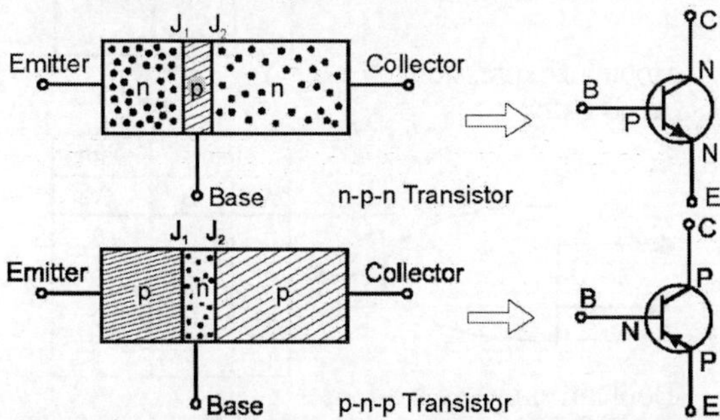

In transistors, the base region is both narrow and lightly doped, otherwise the electrons or holes coming from the input side (say, emitter in CE-configuration) will not be able to reach the collector

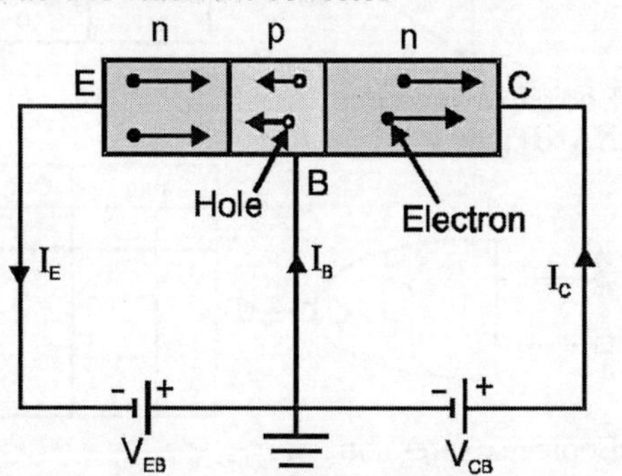

Working of transistor : Fig. shows a common base configuration of p-n-p transistor. The forward biasing of emitter junction lowers the emitter base potential barrier height, whereas the reverse biasing of collector junction increases the collector-base potential barrier height. Hence holes (majority carriers in p-type) flows through emitter to base and constitutes an emitter current I_E.

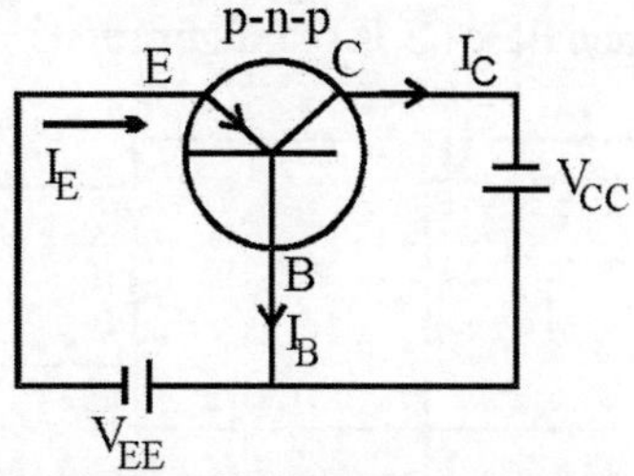

Since emitter is heavily doped in comparison to base, so approximately (only 5% holes recombine with electrons in base region and constitute base current I_B) 95% holes reach to collector and constitute collector current I_C.

From Kirchhoff's current Law,

$$I_E = I_C + I_B \quad ...(1)$$

Eq. (1) holds true regardless of circuit configuration or transistor type (p-n-p or n-p-n) that is used.

The **current gain** of transistor is defined as ratio of collector current I_C to base current I_B

i.e., $$\beta = \frac{I_c}{I_B} \quad ...(2)$$

The value of β lies between 10 and 100.

Since $I_E \approx I_C$ and exactly

$$I_E = \alpha I_C \quad ...(3)$$

Whereas α is defined as the ratio of collector current I_C to emitter current I_E. The value of α is always less than unity. In terms of β, α is

$$\alpha = \frac{\beta}{1+\beta} \Rightarrow \beta = \frac{\alpha}{1-\alpha} \quad ...(4)$$

The transistors can be connected in such a manner that either C or E or B is common to both the input and output. This gives the three configurations in which a transistor is used:

Common Emitter (CE), Common Collector (CC) and Common Base (CB). The plot between I_C and V_{CE} for fixed I_B is called **output characteristics** while the plot between I_B and V_{BE} with fixed V_{CE} is called **input characteristics**.

(i) **Common Collector (C.C) configuration :** Here the collector terminal is common to both input as well as output terminals as shown in fig. The base terminal is input and emitter is output terminal.

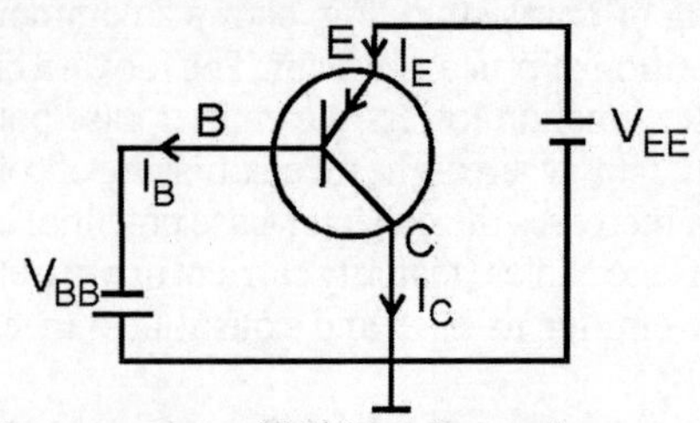

(ii) **Common Base (C.B.) configuration**

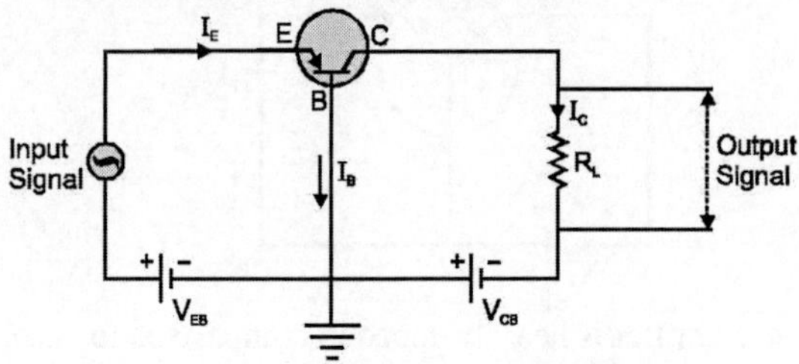

(iii) **Common Emitter (C.E.) configuration**

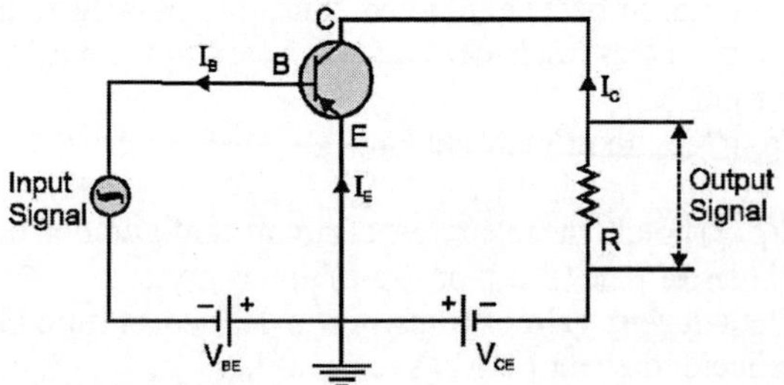

The important transistor parameters for CE-configuration are:

Input resistance, $r_i = \left(\dfrac{\Delta V_{BE}}{\Delta I_B}\right)_{V_{CE}}$

Output resistance, $r_0 = \left(\dfrac{\Delta V_{CE}}{\Delta I_C}\right)_{I_B}$

Current amplification factor, $\beta = \left(\dfrac{\Delta I_C}{\Delta I_B}\right)_{V_{CE}}$

TRANSISTOR AS AN AMPLIFIER, AN OSCILLATOR AND A SWITCH

Transistor can be used as an **amplifier** and **oscillator**. In fact, an oscillator can also be considered as a self-sustained amplifier in which a part of output is feed-back to the input in the same phase (positive feed back). The voltage gain of a transistor amplifier in common emitter configuration is:

$A_v = \left(\dfrac{V_0}{V_i}\right) = \beta\dfrac{R_C}{R_B}$, where R_C and R_B are respectively the resistances in collector and base sides of the circuit.

In an **oscillator**, the feedback is in the same phase (positive feedback). If the feedback voltage is in opposite phase (negative feedback), the gain is less than 1 and it can never work as oscillator. It will be an amplifier with reduced gain. However, the negative feedback also reduces noise and distortion in an amplifier which is an advantageous feature.

The frequency of oscillator $f = \dfrac{1}{2\pi\sqrt{LC}}$

When the transistor is used in the cutoff or saturation state, it acts as a **switch**.

LOGIC GATES

There are some special circuits which handle the digital data consisting of 0 and 1 levels. This forms the subject of **Digital Electronics.**

The important digital circuits performing special logic operations are called **logic gates**. These are: OR, AND, NOT, NAND, NOR and XOR gates. OR, AND and NOT are basic gates and NAND, NOR, XOR combination of basic gates.

(a) **OR gate :**
Symbol **Truth table**

Input		Output
A	B	Y
0	0	0
0	1	1
1	0	1
1	1	1

Boolean expression : $A + B = Y$

(b) **AND gate :**

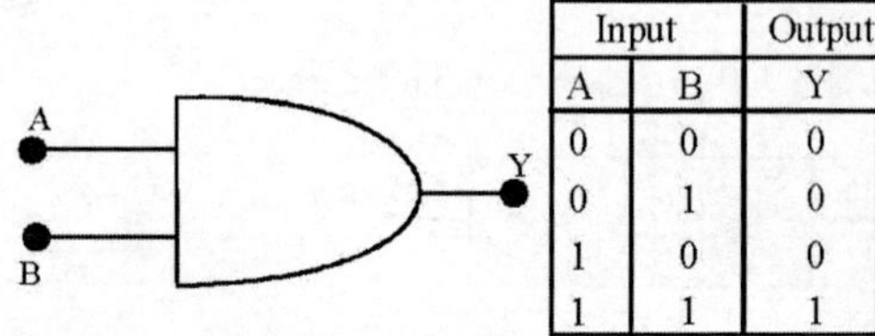

Input		Output
A	B	Y
0	0	0
0	1	0
1	0	0
1	1	1

Boolean expression : $A \cdot B = Y$

(c) **NOT gate :**

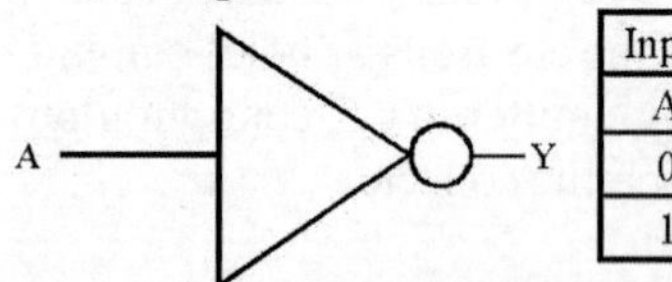

Input	Output
A	Y
0	1
1	0

Boolean expression : $\overline{A} = Y$

(d) **NAND gate :**

Input		Output
A	B	Y
0	0	1
0	1	1
1	0	1
1	1	0

Boolean expression : $\overline{A \cdot B} = Y$

(e) **NOR gate :**

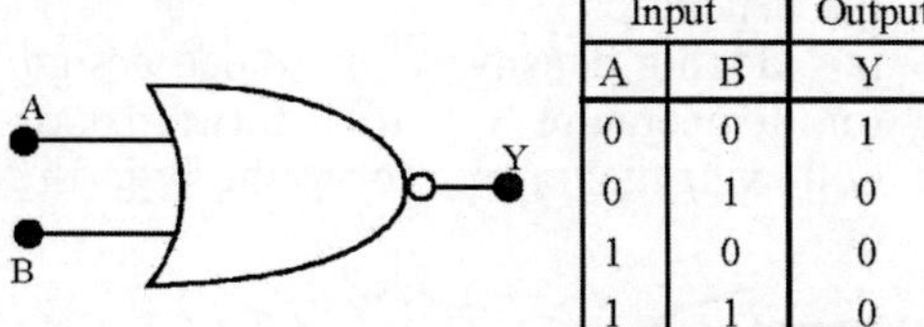

Input		Output
A	B	Y
0	0	1
0	1	0
1	0	0
1	1	0

Boolean expression : $\overline{A+B}=Y$

(f) **The XOR gate :** The output is high, when all inputs are low.

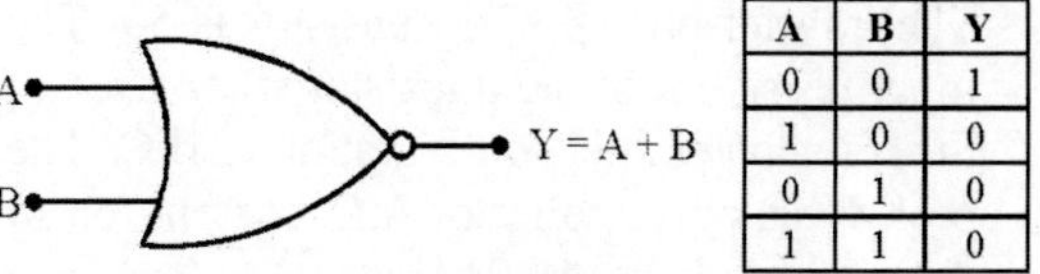

A	B	Y
0	0	1
1	0	0
0	1	0
1	1	0

Boolean expression : $\overline{A+B}=Y$

In modern day circuit, many logical gates or circuits are integrated in one single 'Chip'. These are known as **Integrated circuits (IC)**.

Exercise

1. Which of the following is NOT an amorphous substance ?
   (a) Copper (b) Rubber
   (c) Glass (d) Polymers
2. The number of atoms per unit cell in a sc, bcc and fcc system are
   (a) 1, 2 and 4 respectively
   (b) 8, 6 and 10 respectively
   (c) 1, 4 and 2 respectively
   (d) 2, 4 and 1 respectively
3. The electrical conductivity of pure germanium can be increased by
   (a) increasing the temperature
   (b) doping acceptor impurities
   (c) doping donor impurities
   (d) all of these
4. A semiconductor is doped with a donor impurity
   (a) the hole concentration increases
   (b) the hole concentration is not affected
   (c) the electron concentration increases
   (d) the electron concentration decreases
5. The resistivity of a semiconductor at room temperature is in between
   (a) 10^{-2} to 10^{-5} Ω cm (b) 10^{-3} to 10^{6} Ω cm
   (c) 10^{6} to 10^{8} Ω cm (d) 10^{10} to 10^{12} Ω cm
6. Number of electrons in the valence shell of a semiconductor is
   (a) 1 (b) 2
   (c) 3 (d) 4
7. What is the resistivity of a pure semiconductor at absolute zero?
   (a) Zero
   (b) Infinity
   (c) Same as that of conductors at room temperature
   (d) Same as that of insulators at room temperature
8. Temperature coefficient of resistance of semiconductor is
   (a) zero (b) constant
   (c) positive (d) negative
9. In a *p*-type semiconductor, the acceptor valence band is
   (a) close to the valence band of the host crystal
   (b) close to conduction band of the host crystal
   (c) below the conduction band of the host crystal
   (d) above the conduction band of the host crystal
10. In an n-type semiconductor, donor valence band is
    (a) above the conduction band of the host crystal
    (b) close to the valence band of the host crystal
    (c) close to the conduction band of the host crystal
    (d) below the valence band of the host crystal
11. In p-type semiconductor major current carriers are:
    (a) negative ions (b) holes
    (c) electrons (d) all of these
12. In a diode, when there is a saturation current, the plate resistance will be
    (a) data insufficient
    (b) zero
    (c) some finite quantity
    (d) infinite quantity
13. When the two semiconductors *p*- and *n*-type are brought into contact they form a *p-n* junction, which acts like a/an :
    (a) rectifier (b) amplifier
    (c) conductor (d) oscillator

14. The transfer ratio β of a transistor is 50. The input resistance of the transistor when used in the common emitter configuration is 1kΩ. The peak value of the collector A.C. current for an A.C. input voltage of 0.01 V, is
(a) 500 μA (b) 0.25 μA
(c) 0.01 μA (d) 100 μA
15. Boolean algebra is essentially based on:
(a) Numbers (b) Symbol
(c) Logic (d) Truth
16. A triode valve has an amplification factor of 20 and its plate is given a potential of 300 V. The grid voltage to reduce the plate current to zero, is
(a) 25 V (b) 15 V
(c) 12 V (d) 10 V
17. The relation between number of free electrons (n) in a semiconductor and temperature (T) is given by
(a) $n \propto T$ (b) $n \propto T^2$
(c) $n \propto \sqrt{T}$ (d) $n \propto T^{3/2}$
18. The mobility of free electrons is greater than that of free holes because
(a) they are light
(b) they carry negative charge
(c) they mutually colllide less
(d) they require low energy to continue their motion
19. According to Bravis the number of possible space lattice is
(a) 18 (b) 16
(c) 14 (d) 10
20. Identify the system of crystal structure, if $a = b \neq c$ and $\alpha = \beta = 90°$ and $\gamma = 120°$
(a) monoclinic (b) triclinic
(c) hexagonal (d) rhombohedral
21. The space lattice of diamond is
(a) fcc
(b) bcc with two atoms per unit cell
(c) bcc with 4 atoms per unit cell
(d) simple cubic
22. An n-p-n transistor conducts when
(a) both collector and emitter are negative with respect to the base
(b) both collector and emitter are positive with respect to the base
(c) collector is positive and emitter is negative with respect to the base
(d) collector is positive and emitter is at same potential as the base
23. Barrier potential of a *p-n* junction diode does not depend on
(a) doping density (b) diode design
(c) temperature (d) forward bias
24. Following diagram performs the logic function of

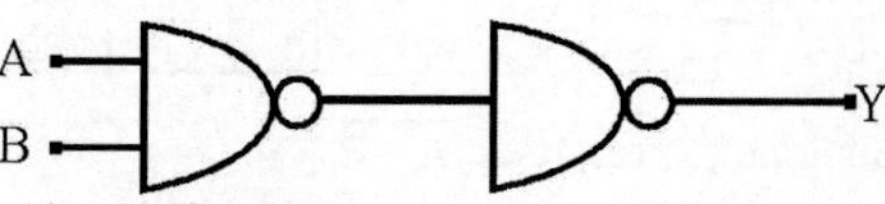

(a) XOR gate (b) AND gate
(c) NAND gate (d) OR gate
25. The time variations of signals are given as in *A*, *B* and *C*. Point out the true statement from the following :

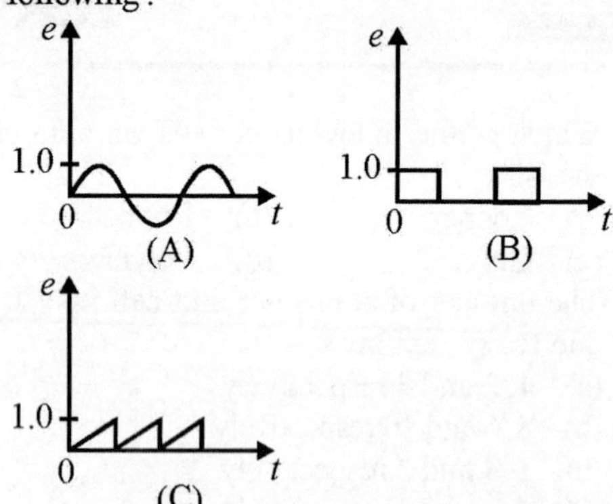

(a) *A*, *B* and *C* are analogue signals
(b) *A* and *B* are analogue, but *C* is digital signal
(c) *A* and *C* are digital, but B is analogue signal
(d) *A* and *C* are analogue, but *B* is digital signal
26. Transmission of light in optical fibre is due to
(a) scattering
(b) diffraction
(c) polarisation
(d) multiple total internal reflections
27. A d.c. battery of *V* volt is connected to a series combination of a resistor *R* and an ideal diode *D* as shown in the figure below. The potential difference across *R* will be

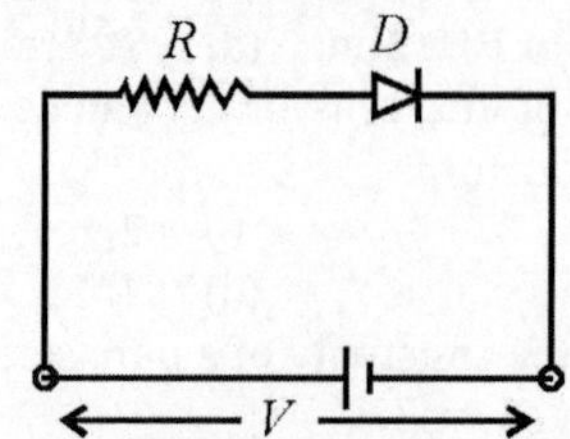

(a) 2*V* when diode is forward biased
(b) Zero when diode is forward biased
(c) *V* when diode is reverse biased
(d) *V* when diode is forward biased

28. The intrinsic semiconductor becomes an insulator at
(a) 0ºC (b) 0 K
(c) 300 K (d) –100ºC

29. For a common emitter circuit if $\frac{I_C}{I_E} = 0.98$ then current gain for common emitter circuit will be
(a) 49 (b) 98
(c) 4.9 (d) 25.5

30. An oscillator is nothing but an amplifer with
(a) positive feedback (b) negative feedback
(c) large gain (d) no feedback

31. When a *p-n* junction diode is reverse biased the flow of current across the junction is mainly due to
(a) diffusion of charges
(b) drift of charges
(c) depends on the nature of material
(d) both drift and diffusion of charges

32. In Fig., the input is across the terminals *A* and *C* and the output is across *B* and *D*. Then the output is

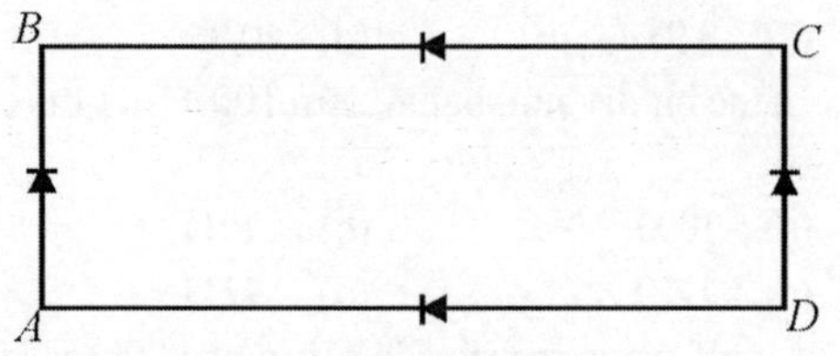

(a) zero
(b) same as the input
(c) half wave rectified
(d) full wave rectified

33. Which of the following gates corresponds to the truth table given below?

A	*B*	*Y*
1	1	0
1	0	1
0	1	1
0	0	1

(a) NAND (b) OR
(c) AND (d) XOR

34. In the diagram, the input is across the terminals *A* and *C* and the output is across *B* and *D*. Then the output is

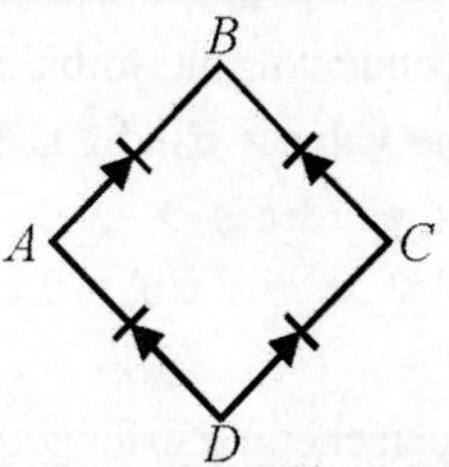

(a) zero (b) same as the input
(c) full wave rectifier (d) half wave rectifier

35. Which of the following, when added as an impurity, into the silicon, produces *n*-type semiconductor?
(a) Phosphorus (b) Aluminium
(c) Magnesium (d) Both b and c

36. The current gain for a transistor working as common-base amplifier is 0.96. If the emitter current is 7.2 mA, then the base current is
(a) 0.29 mA (b) 0.35 mA
(c) 0.39 mA (d) 0.43 mA

37. When an *n-p-n* transistor is used as an amplifier then
(a) the electrons flow from emitter to collector
(b) the holes flow from emitter to collector
(c) the electrons flow from collector to emitter
(d) the electrons flow from battery to emitter

38. When arsenic is added as an impurity to silicon, the resulting material is
(a) *n*-type semiconductor
(b) *p*-type semiconductor
(c) *n*-type conductor
(d) insulator

39. To obtain a *p*-type germanium semiconductor, it must be doped with
(a) arsenic (b) antimony
(c) indium (d) phosphorus

40. The following truth table belongs to which of the following four gates?

A	B	Y
1	1	0
1	0	0
0	1	0
0	0	1

(a) NOR (b) XOR
(c) NAND (d) OR

41. The forward biased diode is
(a) –2V +2V
(b) –4V –3V
(c) 3V 5V
(d) 0V –2V

42. In a semiconductor, the forbidden energy gap between the valence band and the conduction band is of the order is
(a) 1 MeV (b) 0.1 MeV
(c) 1 eV (d) 5 eV

43. The forbidden energy gap for germanium crystal at 0 K is
(a) 0.071 eV (b) 0.71 eV
(c) 2.57 eV (d) 6.57 eV

44. In an insulator, the forbidden energy gap between the valence band and conduction band is of the order of
(a) 1 MeV (b) 0.1 MeV
(c) 1 eV (d) 5 eV

45. Which of the following is a crystalline solid?
(a) Graphite (b) Calcite
(c) Gallium (d) Oxygen

46. Crystalline solids do not have
(a) a definite latent heat of fusion
(b) irregular atomic arrangement
(c) incompressibility
(d) free movement of atoms

47. In the case of a common emitter transistor amplifier, the ratio of the collector current to the emitter current I_c/I_e is 0.96. The current gain of the amplifier is
(a) 6 (b) 48
(c) 24 (d) 12

48. If in a p-n junction diode, a square input signal of 10 V is applied as shown

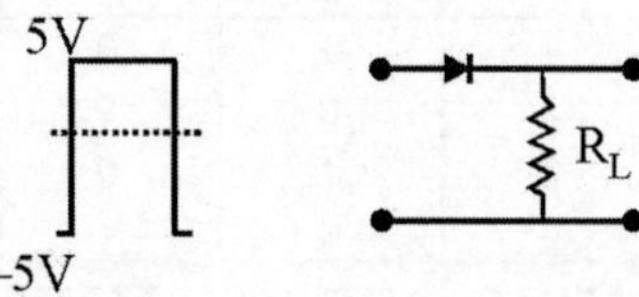

Then the output signal across R_L will be

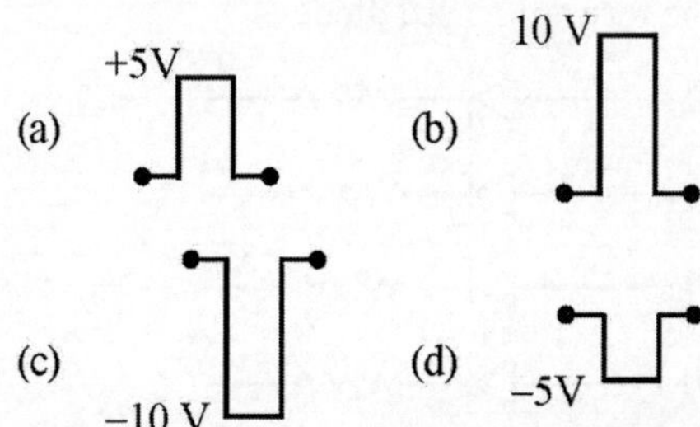

49. In a full wave rectifier circuit operating from 50 Hz mains frequency, the fundamental frequency in the ripple would be
(a) 25 Hz (b) 50 Hz
(c) 70.7 Hz (d) 100 Hz

50. If the forward voltage in a semiconductor diode is changed from 0.5 V to 0.7 V, then the forward current changes by 1.0 mA. The forward resistance of diode junction will be
(a) 100 Ω (b) 120 Ω
(c) 200 Ω (d) 240 Ω

51. Current gain of a transistor in common base mode is 0.95. Its value in common emitter mode is
(a) 0.95 (b) 1.5
(c) 19 (d) $(19)^{-1}$

52. A transistor has a base current of 1 mA and emitter current 90 mA. The collector current will be
(a) 90 mA (b) 1 mA
(c) 89 mA (d) 91 mA

53. For a common base amplifier, the values of resistance gain and voltage gain are 3000 and 2800 respectively. The current gain will be
(a) 1.1 (b) 0.98
(c) 0.93 (d) 0.83

54. In the binary number system 100 + 1011 is equal to
(a) 1000 (b) 1011
(c) 1110 (d) 1111

55. A half-wave rectifier is being used to rectify an alternating voltage of frequency 50 Hz. The number of pulses of rectified current obtained in one second is
(a) 50 (b) 25
(c) 100 (d) 2000

56. Which of the following gates will have an output of 1?

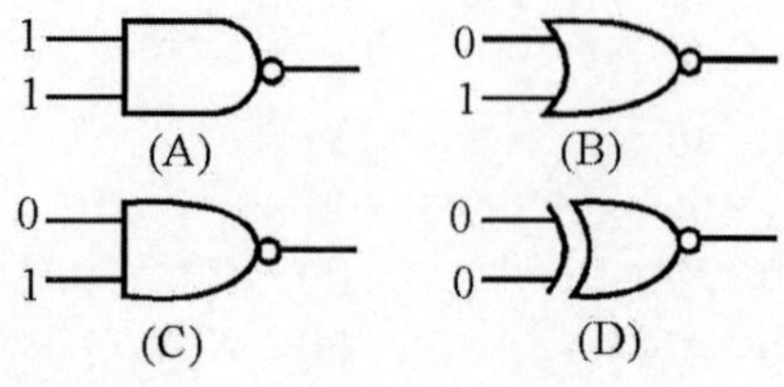

(a) D (b) A
(c) B (d) C

57. In a P-N junction
    (a) the potential of P & N sides becomes higher alternately
    (b) the P side is at higher electrical potential than N side
    (c) the N side is at higher electric potential than P side
    (d) both P & N sides are at same potential.

58. The frequency response curve of RC coupled amplifier is shown in figure. The band with of the amplifier will be

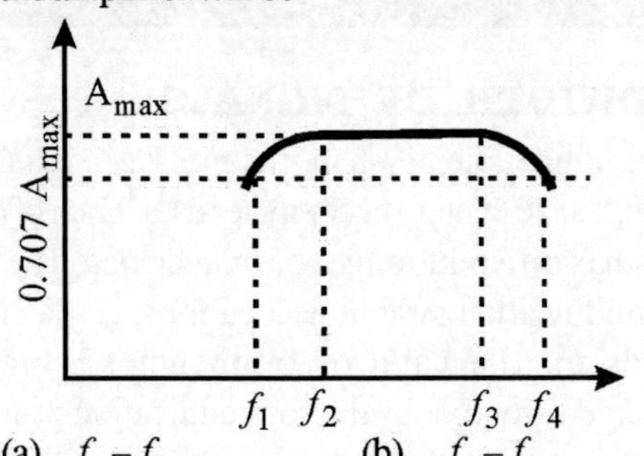

(a) $f_3 - f_2$
(b) $f_4 - f_1$
(c) $\frac{f_4 - f_2}{2}$
(d) $f_3 - f_1$

59. The energy band gap is maximum in
    (a) metals
    (b) superconductors
    (c) insulators
    (d) semiconductors.

60. Of the diodes shown in the following diagrams, which one is reverse biased ?

(a)

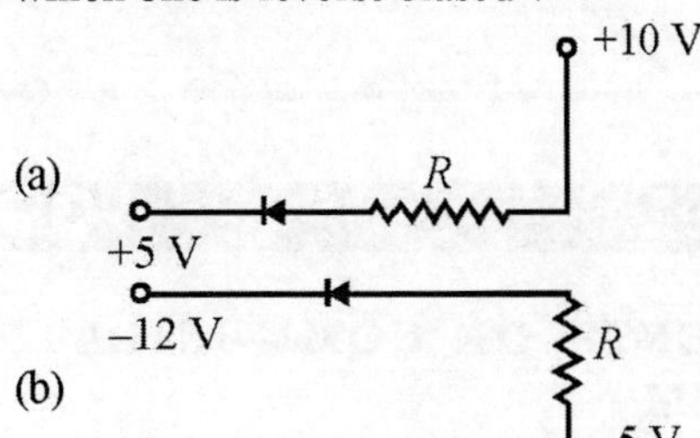

(b)

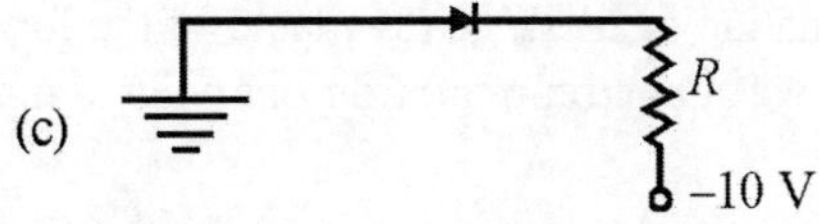

(c)

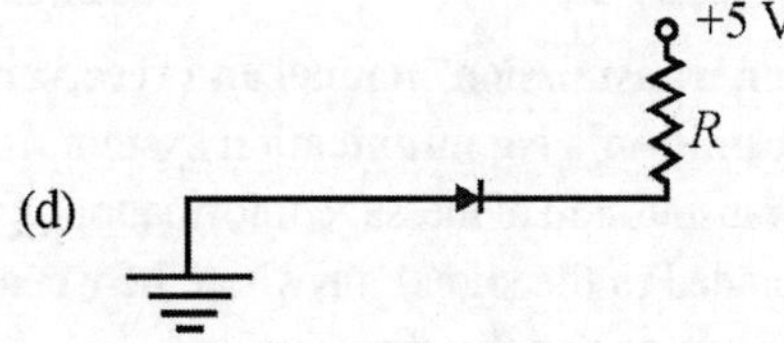

(d)

Communication Systems

CONCEPTS

ELEMENTS OF COMMUNICATION SYSTEM

Electronic communication refers to the faithful transfer of information or message (available in the form of electrical voltage and current) from one point to another point.

Transmitter, transmission channel and receiver are three basic units of a communication system. In the process of transmission of message/information signal, noise gets added to the signal anywhere between the information source and the receiving end.

Transmitter :

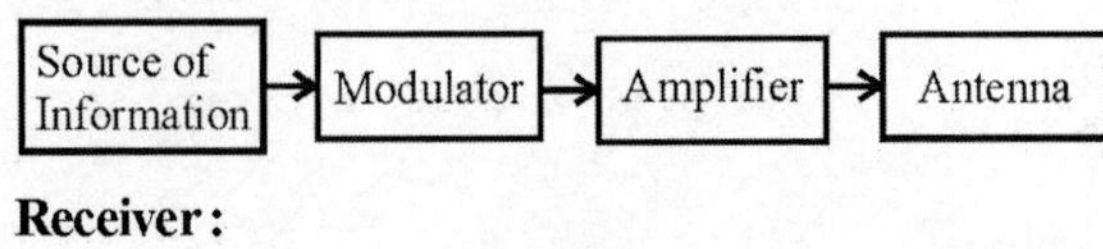

Receiver :

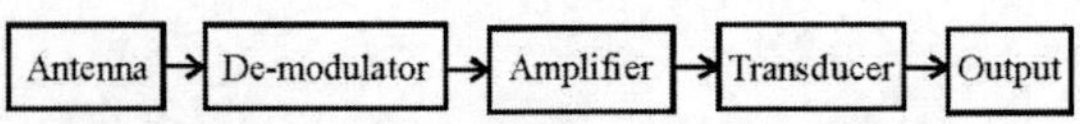

Two important forms of communication system are : **Analog and Digital.** The information to be transmitted is generally in continuous waveform for the former while for the latter it has only discrete or quantised levels.

There are two basic modes of communication:

(i) **Point-to-point communication mode :** In this mode, communication takes place over a link between a single transmitter and a receiver as in telephon

(ii) **Broadcast mode of communication :** In this mode, there are a large number of receivers corresponding to a single transmitter. Radio and television are most common examples of broadcast mode of communication.

BANDWIDTH OF SIGNALS

In a communication system, the message signal can be voice, music, picture or computer data. Each of these signals has different ranges of frequencies. The type of communication system needed for a given signal depends on the band of frequencies which is considered essential for the communication process. *The difference of maximum and minimum frequency in the range of each signal is called bandwidth of that signal.*

Bandwidth can be of message signal as well as of transmission medium.

(i) **Bandwidth for analog signals**

Bandwidth for some analog signals are listed below :

Signal required	Frequency range	Bandwidth
Speech	300-3100 Hz	3100 – 300 = 2800 Hz
Music	High frequencies produced by musical instrumentAudible range =20 Hz - 20 kHz	≈ 20 kHz
TV	Contains both voice and picture (76–88 MHz)	6 MHz

(ii) **Bandwidth for digital signal :** Basically digital signals are rectangular waves and these can be splitted into a superposition of sinusoidal waves of frequencies ν_0, $2\nu_0$, $3\nu_0$, $4\nu_0$, $n\nu_0$, where n is an integer extending to infinity. This implies that the infinite bandwidth is required to reproduce the rectangular waves. However, for practical purposes, higher harmonics are neglected for limiting the bandwidth.

MODULATION

Low frequencies cannot be transmitted to long distances. Therefore, they are superimposed on a high frequency carrier signal by a process known as **modulation.**

In modulation, some characteristic of the carrier signal like amplitude, frequency or phase varies in accordance with the modulating or message signal. Correspondingly, they are called Amplitude Modulated (AM), Frequency Modulated (FM) and Phase Modulated (PM) waves. In the process of modulation, new frequencies called sidebands are generated on either side (higher and lower than the carrier frequency) of the carrier by an amount equal to the highest modulating frequency.

Amplitude Modulation

It is the process of producing a modulated wave having the same frequency as that of the carrier wave but its amplitude is modified in accordance with that of the modulating wave.

Frequency Modulation

(i) In this type of modulation the frequency of the carrier voltage is varied in accordance with the instantaneous value of the modulating voltage.

(ii) Let modulating signal $e_m = E_m \cos w_m t$
Carrier signal $e_c = E_c \sin(w_c t + q)$
$\therefore$ Modulated signal

$$(e)_{FM} = E_c \sin\left[\omega_c t + \frac{K_f E_m}{\omega_m} \sin \omega_m t\right]$$

or $(e)_{FM} = E_c \sin(w_c t + m_f \sin w_m t)$.

(iii) m_f is known as the modulation index/depth of modulation

(iv) The ratio of frequency deviation to modulation frequency is known as δ.

$$\delta = m_f = \frac{k_f E_m}{\omega_m}.$$

(v) The total variation in frequency from the lowest to the highest is called **carrier swing (CS).**
$CS = 2 \times$ frequency deviation $\Rightarrow CS = 2 \times \delta$

Pulse Modulation

Pulse modulation could be classified as: Pulse Amplitude Modulation (PAM), Pulse Duration Modulation (PDM) or Pulse Width Modulation (PWM) and Pulse Position Modulation (PPM).

PROPAGATION OF EM WAVES

For transmission over long distances, signals are radiated into space using devices called antennas. The radiated signals propagate as electromagnetic waves and the mode of propagation is influenced by the presence of the earth and its atmosphere. Near the surface of the earth, electromagnetic waves propagate as surface waves. Surface wave propagation is useful up to a few MHz frequencies.

SKY and Space Wave Propagation

Long distance communication between two points on the earth is achieved through reflection of electromagnetic waves by ionosphere. Such waves are called **sky waves**. Sky wave propagation takes place up to frequency of about 30 MHz. Above this frequency, electromagnetic waves essentially propagate as **space waves**. **Space waves** are used for line-of-sight communication and satellite communication.

If an **antenna** radiates electromagnetic waves from a height h_T, then the **range** d_T is given by $\sqrt{2Rh_T}$ where R is the radius of the earth.

PRODUCTION AND DETECTION OF AMPLITUDE-MODULATED WAVE

Amplitude modulated signal contains frequencies $(\omega_c - \omega_m)$, ω_c and $(\omega_c + \omega_m)$.

Amplitude modulated waves can be produced by application of the message signal and the carrier wave to a non-linear device, followed by a band pass filter.

AM detection is the process of recovering the modulating signal from an AM waveform, is carried out using a rectifier and an envelope detector.

Exercise

1. The science of communication involving long distances is known as
   (a) mass-communication
   (b) distant-communication
   (c) tele-communication
   (d) trans-communication
2. The part of the total message that is conveyed during communication is known as
   (a) text (b) electric signal
   (c) bytes (d) information
3. A transducer used at the transmitting end, serves the purpose of converting
   (a) electrical signal to sound form
   (b) sound signal to electrical form
   (c) electrical signal to magnetic form
   (d) sound signal to magnetic form
4. During the process of transmission & reception the signal gets deteriorated due to
   (a) noise introduced in the system
   (b) distortion in the system
   (c) both (a) & (b)
   (d) neither (a) nor (b)
5. Buffer amplifier is used at the transmitting end to
   (a) feed carrier frequency to Master oscillator
   (b) to amplify carrier frequency
   (c) to mix modulating signal with carrier frequency
   (d) to isolate master oscillator from other stages of transmitter
6. Loudspeaker is a transducer used it
   (a) receiving end (b) transmitting end
   (c) both (a) & (b) (d) neither (a) nor (b)
7. The purpose of a detector at the receiving end is
   (a) to amplify signal
   (b) to reduce its frequency level
   (c) to modulate signal
   (d) to demodulate signal
8. For transmission of TV- signal, sound-part is
   (a) amplitude modulated
   (b) frequency modulated
   (c) phase modulated
   (d) pulse modulated
9. Reception of information involves
   (a) decoding of signal
   (b) storage of signal
   (c) interpretation of signal
   (d) all the above
10. Collation is a process of
    (a) sorting at transmitting end
    (b) decoding at receiving end
    (c) encoding at transmitting end
    (d) storing at receiving end
11. Encoding of signal is required for
    (a) modulation at transmitting end
    (b) modulation at receiving end
    (c) demodulation at receiving end
    (d) demodulation at transmitting end
12. Picture signal of TV-signal is
    (a) amplitude modulated
    (b) frequency modulated
    (c) phase modulated
    (d) pulse modulated
13. The term channel is after used to indicate
    (a) the amplitude range allocated to a given source
    (b) the frequency range allocated to a given source
    (c) the voltage-range allocated to a given source
    (d) all the above
14. For transmission of speeches, talks, music, dramas etc-------- is used
    (a) Radio broadcast transmitter
    (b) Radio telegraph transmitter
    (c) Radio transmitter
    (d) Navigation transmitter
15. Long wave broadcast transmitters are used in
    (a) temperate countries
    (b) tropical countries
    (c) everywhere countries
    (d) nowhere
16. A broadcast radio transmitter radiates 12 kW when percentage of modulation is 50%, then the unmodulated carrier power is
    (a) 5.67 kW (b) 7.15 kW
    (c) 9.6 kW (d) 12 kW

17. Consider the following amplitude modulated (AM) signal, where $f_m < B$
$x_{AM}(t) = 10(1 + 0.5 \sin 2\pi f_m t) \cos 2\pi f_c t$
The average side-band power for the AM signal given above is
(a) 25 (b) 12.5
(c) 6.25 (d) 3.125

18. Buffer amplifier is used between one master oscillator and Harmonic generator in AM-transmitter because
(a) it reduces S/N ratio
(b) it amplifies signal
(c) it increases signal frequency
(d) it avoids frequency deviation in tank-circuit

19. Magnetic-coil are used in camera tube to
(a) increase electron-beam intensity
(b) decrease electron-beam intensity
(c) give horizontal and vertical motion to the beam
(d) none of these

20. An *AM*- signal is given as
$x_{AM}(t) = 100\,[p(t) + 0.5g(t)] \cos \omega_c t$
in interval $0 \le t \le 1$. One set of possible values of the modulating signal and modulation index would be
(a) $t, 0.5$ (b) $t, 1.0$
(c) $t, 1.5$ (d) $t^2, 2.0$

21. The maximum value of induced emf in loop antenna occurs
(a) when plane of coil is along direction of transmitting signal
(b) when plane of coil is in perpendicular direction of transmitting signal
(c) when plane of coil is perpendicular to 45°
(d) none of these

22. Folded dipole is preferred over a simple dipole because it provides
(a) high input impedance for best matching
(b) increased bandwidth
(c) both (a) and (c)
(d) none of these

23. Main function of the RF amplifiers in a superheterodyne receiver is to
(a) amplify signal
(b) reject unwanted signal
(c) discriminate against image frequency signal and IF-signal
(d) all the above

24. The maximum range for the tropospheric transmission of radio wave of wavelength 3m using the transmitting antenna and receiving antenna of heights 100 m and 60 m respectively is
(a) 8m (b) 800m
(c) 8km (d) 80km

25. In a radar system, peak transmitted power is increased a factor of 81 and the antenna diameter is increased by a factor of 3, then the maximum-range will increase by a factor of
(a) 81 (b) 27
(c) 9 (d) 3

26. Optical fibre are used for long distance communication because
(a) it amplifies signals to be transmitted
(b) it transfer signals faster than electrical cables
(c) it pre-emphasise weak signals
(d) it provide little attenuation as compared to electrical cable for light propagation

27. Optical fibres transmits light along its axis, by the process of
(a) total internal reflection
(b) refraction
(c) interference
(d) diffraction

28. In an AM wave, the information is contained within
(a) r.f. carrier wave
(b) only lower and upper side frequencies
(c) both r.f. carrier and side frequencies
(d) none of these

29. The frequency deviation in a FM transmission is 18.75 KHz. If it broadcasts in 88-108 MHz band, then the percent modulation is
(a) 10% (b) 25%
(c) 50% (d) 75%

30. In FM, when frequency deviation is doubled, then
(a) modulation is halved
(b) carrier swing is halved
(c) modulation is doubled
(d) modulation index is decreased

31. Television signals are
(a) frequency modulated
(b) amplitude modulated
(c) both frequency and amplitude modulated
(d) phase modulated

32. The height of a television tower is 100 m. If radius of earth is 6.4×10^6 m and average-population density surrounding the tower is 1000 per km^2, then the population covered by the television transmission is
(a) 2.06×10^6 (b) 4.02×10^6
(c) 5.18×10^9 (d) 6.04×10^9
33. Pre-emphasis in FM system is done to
(a) compress modulating signal
(b) expand modulating signal
(c) amplify lower frequency component of the modulating signal
(d) amplify higher frequency component of the modulating signal
34. A 10 kW carrier is sinusoidally modulated by two carriers corresponding to a modulation index of 30% and 40% respectively then total power radiated by the modulator is
(a) 10.25 kW (b) 11.25 kW
(c) 12.75 kW (d) 17 kW
35. In a superheterodyne receiver, the IF is 455 kHz. If it is tuned 1200 kHz, the image frequency will be
(a) 1562 kHz (b) 1675 kHz
(c) 2110 kHz (d) 2025 kHz
36. Space wave communication is limited
(a) to the line of sight distance
(b) by earth's curvature
(c) none of these
(d) both (a) and (b)
37. The service area of space wave communication increases
(a) increasing the height of transmitting antenna
(b) decreasing the height of receiving antenna
(c) increasing the height of both transmitting and receiving antenna
(d) decreasing the distance between transmitting and receiving antenna
38. In a transistor the base is
(a) an insulator
(b) a conductor of low resistance
(c) a conductor of high resistance
(d) an extrinsic semiconductor
39. Which range of frequencies is suitable for sky wave propagation?
(a) 1 kHz to 500 kHz (b) 1 MHz to 2 MHz
(c) 2 MHz to 20 MHz (d) Above 30 MHz
40. In an optical fibre, the diameter of the core is nearly
(a) 10^{-2} m (b) 10^{-4} m
(c) 10^{-8} m (d) 1 Å
41. The waves which are electromagnetic in nature are
(a) sound waves and light waves
(b) water waves and radio waves
(c) light waves and X-rays
(d) sound waves and water waves
42. In communication with help of antenna if height is double then the range covered which was initially r would become
(a) $\sqrt{2}r$ (b) $3r$
(c) $4r$ (d) $5r$
43. CO_2 laser uses
(a) microwaves (b) infrared
(c) ultraviolet (d) visible light
44. An LED (Light Emitting Diode) is constructed from a p–n junction based on a certain Ga-As-P semi–conducting material whose energy gap is 1.9 eV. What is the wavelength of the emitted light?
(a) 650 nm (b) 65Å
(c) 800 nm (d) 8000 Å
45. A sinusoidal carrier voltage of frequency 10 MHz and amplitude 200 volts is amplitude modulated by a sinusoidal voltage of frequency 10 kHz producing 40% modulation. Calculate the frequency of upper and lower sidebands.
(a) 10010kHz, 9990kHz
(b) 1010 kHz, 990 kHz
(c) 10100 Hz, 9990 Hz
(d) 1010 MHz, 990 MHz
46. A transmitter radiates 10 kW of power with the carrier unmodulated and 11.8 kW with the carrier sinusoidally modulated. The modulation factor is
(a) 56% (b) 60%
(c) 72% (d) 84%
47. An FM signal has a resting frequency of 105 MHz and highest frequency of 105.03 MHz when modulated by a signal of frequency 5 kHz. The carrier swing is
(a) 25 kHz (b) 54 kHz
(c) 60 kHz (d) 75 kHz
48. Depth of modulation in terms of E_{max} and E_{min} is
(a) $m_a = E_{max} + E_{min}/E_{min}$
(b) $ma = E_{max} - E_{min}/E_{max}$
(c) $m_a = E_{max} - E_{min}/E_{max} + E_{min}$
(d) $ma = E_{max} + E_{min}/E_{max} - E_{min}$

49. An antenna has a radiation resistance of 68 Ω, a load resistance of 10 Ω and power gain of 16. The directive gain of the antenna is
(a) 15 (b) 16.02
(c) 17.08 (d) 18.35

50. What will be the image frequency of an FM radio receiver that is tuned to 98.6 MHz broadcast station
(a) 111.8 MHz (b) 108 MHz
(c) 121.6 MHz (d) 132 MHz

51. Which of the following frequencies will be suitable for beyond-the horizon communication using sky waves?
(a) 10 KHz (b) 10 MHz
(c) 1 GHz (d) 1000 GHz

52. Which mode of communication is most suitable for carrier wave of frequencies around 100 MHz:
(a) Satellite (b) Ground wave
(c) Line of sight (d) Ionospheric

53. Communication on ground is through electromagnetic waves of wavelength :
(a) larger than 600 m
(b) between 200 and 600 m
(c) between 1 and 5 m
(d) between 10^{-3} and 0.1

54. The maximum range for the tropospheric transmission of radiowave of wavelength 3m using the transmitting antenna and receiving antenna of heights 100 m and 60 m respectively is
(a) 8 m (b) 800 m
(c) 8 km (d) 80 km

55. Ground waves are polarised
(a) parallel to the earth's surface
(b) normal to the earth's surface
(c) at an angle 45° from earth's surface
(d) in any direction

56. Critical frequency that gets reflected back from ionosphere is
(a) same for all layers of the ionosphere
(b) different for different layers of the ionosphere
(c) not dependent on layers of the ionosphere
(d) none of these

57. Intensity of electric field obtained at receiver antenna for a space wave propagation is
(a) directly proportional to the perpendicular-distance from transmitter to antenna
(b) inversely proportional to the perpendicular-distance from transmitter to antenna
(c) directly proportional to the square perpendicular-distance from transmitter to antenna
(d) inversely proportional to the square perpendicular-distance from transmitter to antenna

58. In sky-wave propagation, skip-distance depends on :
(a) frequency of e.m. waves transmitted
(b) critical frequency of the layer
(c) height of layer above earth's surface
(d) all the above

59. For a single side band transmission a balanced modulator is used to
(a) increase power of carrier wave
(b) increase amplitude of carrier wave
(c) suppress audio signal
(d) suppress carrier component

60. The tank circuit used in a radio transmitter should have
(a) high effective Q
(b) low effective Q
(c) loosely coupled load
(d) Both (a) and (c)

1. PHYSICAL WORLD, UNITS & MEASUREMENTS

ANSWER KEY

1	(b)	**11**	(d)	**21**	(a)	**31**	(b)	**41**	(a)	**51**	(b)
2	(c)	**12**	(a)	**22**	(c)	**32**	(d)	**42**	(c)	**52**	(b)
3	(b)	**13**	(b)	**23**	(d)	**33**	(c)	**43**	(d)	**53**	(d)
4	(a)	**14**	(d)	**24**	(c)	**34**	(a)	**44**	(a)	**54**	(c)
5	(c)	**15**	(a)	**25**	(d)	**35**	(d)	**45**	(c)	**55**	(b)
6	(b)	**16**	(a)	**26**	(d)	**36**	(c)	**46**	(b)	**56**	(a)
7	(a)	**17**	(c)	**27**	(c)	**37**	(a)	**47**	(c)	**57**	(d)
8	(c)	**18**	(a)	**28**	(a)	**38**	(d)	**48**	(c)	**58**	(d)
9	(b)	**19**	(b)	**29**	(b)	**39**	(c)	**49**	(d)	**59**	(a)
10	(d)	**20**	(a)	**30**	(c)	**40**	(d)	**50**	(b)	**60**	(a)

Hints & Solutions

5. (c) Angular displacement has unit (degree or radian) but it is dimensionless.
Note : vice-versa is not possible.

6. (b) γ is ratio of specific heats.

8. (c) $c = \dfrac{1}{\sqrt{\mu_0 \in_0}}$, so, dimensions are $[LT^{-1}]$.

9. (b) Because by using length, time and velocity we can't find mass.

13. (b) Least no. of significant digits are 2, so round off the final answer to two significant digits.

17. (c) $\dfrac{\Delta T}{T} \times 100 = \dfrac{0.2}{25} \times 100.$

18. (a) Use principle of Homogeneity.

19. (b) [Angular momentum]
$= $ [Momentum of inertia] $\times$ [Angular velocity]
$= ML^2 \times T^{-1}$
$= ML^2T^{-1}$

20. (a) $CR = [M^{-1} L^{-2} T^4 A^2][M^1 L^2 T^{-3} A^{-2}]$

$$= [T] = [M^0 L^0 T^1]$$

21. (a) $\tau = [\text{Force} \times \text{distance}]$
$= [MLT^{-2}]\,[L] = ML^2T^{-2}$

22. (c) $L = \varepsilon\left(\dfrac{dt}{dI}\right) = \dfrac{W}{q}\left(\dfrac{T}{I}\right)$

or, $[L] = \dfrac{[ML^2T^{-2}][T]}{[AT][A]} = ML^2T^{-2}A^{-2}$

23. (d) For angular momentum, the dimensional formula is ML^2T^{-1}. For other three, it is ML^2T^{-2}.

24. (c) $[x] = [bt^2]$. Hence $[b] = [x/t^2] = \text{km/s}^2$.

25. (d) Substitute the dimensional formula of F, A, ΔV and ΔZ on both sides and find that for η.

26. (d) $f = c\, m^x k^y$;
Spring constant k = force/length.
$[M^0L^0T^{-1}] = [M^x (MT^{-2})^y] = [M^{x+y} T^{-2y}]$

$\Rightarrow x + y = 0,\ -2y = -1$ or $y = \dfrac{1}{2}$

Therefore, $x = -\dfrac{1}{2}$

27. (c) [Pressure] = [Force] / [Area]

$$= \frac{MLT^{-2}}{L^2} = ML^{-1}T^{-2}$$

28. (a) $\mu = \dfrac{B}{nI}$; $[B] = MT^{-2}A^{-1}$,
$[n] = L^{-1}$, $[I] = A$

$$\left[\frac{B}{nI}\right] = [MLT^{-2}A^{-2}]$$

29. (b) $D = M/V$

$$\therefore \frac{\Delta D}{D} = \frac{\Delta M}{M} + \frac{\Delta V}{V} = \left(\frac{0.01}{22.42} + \frac{0.1}{4.7}\right) \times 100$$
$= 2\,\%$

30. (c) Try out the given alternatives.
When $x = 1, y = -1, z = 1$

$$P^x S^y c^z = P^1 S^{-1} c^1 = \frac{Pc}{S}$$

$$= \frac{[ML^{-1}T^{-2}]\,[LT^{-1}]}{[ML^2T^{-2}/L^2T]} = [M^0L^0T^0]$$

31. (b) In $p = p_0 \exp(-\alpha t^2)$, αt^2 dimensionless

$$\therefore \alpha = \frac{1}{t^2} = \frac{1}{T^2} = [T^{-2}]$$

32. (d) $\eta = \dfrac{p(r^2 - x^2)}{4vl} = \dfrac{[ML^{-1}T^{-2}][L^2]}{[LT^{-1}][L]}$

$$= [ML^{-1}T^{-1}]$$

33. (c) Least count = 1MSD – 1 VSD

$$= 1\text{MSD} - \left(\frac{N-1}{N}\right)\text{MSD}$$

$$(\because N\ \text{VSD} = (N-1)\text{MSD} \therefore 1\ \text{VSD} = \frac{N-1}{N}\text{MSD})$$

$$= \frac{1}{N}\text{MSD}$$

$$= \frac{1}{N} \times \frac{1}{10}\text{cm} = \frac{1}{10N}$$

34. (a) $[F] = MLT^{-2} = (10g)\,(10\text{ cm})\,(0.1s)^{-2}$
$= (10^{-2}\text{ kg})\,(10^{-1}\text{m})\,(10^{-1}\text{s})^{-2} = 10^{-1}\text{N}.$

35. (d) A quantity which has dimensions and a constant value is called dimensional constant. Therefore, gravitational constant (G) is a dimensional constant.

36. (c) Percentage error in mass $\left(\dfrac{\Delta m}{m} \times 100\right) = 2$

and percentage error in speed $\left(\dfrac{\Delta v}{v} \times 100\right) = 3$.

$$\Rightarrow \frac{\Delta v}{v} = \frac{3}{100} \text{ and } \frac{\Delta m}{m} = \frac{2}{100}$$

Kinetic energy $= \dfrac{1}{2}mv^2 \propto mv^2$.

$\therefore$ Error in measurement of kinetic energy

$$= \frac{\Delta m}{m} + 2\left(\frac{\Delta v}{v}\right)$$

By Binomial Function,

$$\text{Reqd. error} = \left(\frac{\Delta m}{m}\right) + \left(2 \times \frac{\Delta v}{v}\right)$$

$$= \left(\frac{2}{100}\right) + \left(2 \times \frac{3}{100}\right) = \frac{8}{100} = 8\%$$

$\therefore$ %age error = 8%.

37. (a) $\left(P + \dfrac{a}{V^2}\right) = b\dfrac{Q}{V}$

According to the principle of homogeinity quantity with same dimension can be added or subtracted.

Hence, Dimension of P = Dimension of $\dfrac{a}{V^2}$

$\Rightarrow$Dimension of $\dfrac{\text{Force}}{\text{Area}}$ = Dimension of $\dfrac{a}{V^2}$

$$\Rightarrow \left[\frac{MLT^{-2}}{L^2}\right] = \frac{a}{[L^3]^2} \Rightarrow a = [M\,L^5\,T^{-2}]$$

38. (d) Density $= \dfrac{\text{Mass}}{\text{Volume}}$

$$\rho = \frac{M}{L^3} \qquad \therefore \frac{\Delta\rho}{\rho} = \frac{\Delta M}{M} + 3\frac{\Delta L}{L}$$

% error in density = % error in Mass
+ 3 (% error in length]
= 4 + 3(3) = 13%

39. (c) $e = -L\dfrac{di}{dt}$ ....(1)

$e = iR$ ....(2)

From (1) & (2), $iR = -L\dfrac{di}{dt}$

$\therefore$ Dimension of L.H.S. = Dimension of R.H.S.

$$[A]\,R = L\,[AT^{-1}] \Rightarrow \frac{L}{R} = [T]$$

40. (d) $F = 6\pi\eta\, av$

$$\eta = \frac{F}{6\pi\,av} = \frac{[MLT^{-2}]}{[L][LT^{-1}]} = [ML^{-1}T^{-1}]$$

41. (a) Dimension of magnetic flux
= Dimension of voltage × Dimension of time
$= [ML^2T^{-3}A^{-1}]\,[T] = [ML^2T^{-2}A^{-1}]$

$$\because \text{Voltage} = \frac{\text{work}}{\text{charge}}$$

42. (c) Force has dimension $[MLT^{-2}]$ while impulse has dimension $[MLT^{-1}]$, both have different dimensions.

43. (d) We know that $E = h\nu$

$$h = \frac{E}{\nu} = \frac{[ML^2T^{-2}]}{[T^{-1}]} = [ML^2T^{-1}]$$

Angular momentum = $I\omega$
$= [ML^2][T^{-1}] = [ML^2T^{-1}]$

44. (a) $5.4 \times 0.125 = 0.675 \approx 0.68$

49. (d) It has maximum number of significant figures.

50. (b) Both are dimensionless

51. (b) $\frac{L}{R}$ has same dimension as that of time

52. (b) Power $= \frac{\text{Energy}}{\text{time}}$

53. (d) Energy incident per unit area per second

$$= \frac{\text{Energy}}{\text{area} \times \text{second}} = \frac{ML^2T^{-2}}{L^2T} = MT^{-3}$$

54. (c) $Q = mL \Rightarrow L = \frac{Q}{m}$ = Energy per unit mass

Gravitational potential

$$V_g = \frac{W}{m} = \text{work per unit mass}$$

55. (b) $c = \frac{1}{\sqrt{\mu_0 \epsilon_0}}$ where c = speed of light

56. (a) $v = \sqrt{\mu r g}$ and $v = \frac{F}{Bq}$

58. (d) Modulus of rigidity, $\eta = \frac{F/a}{\theta}$

$$[\eta] = \frac{MLT^{-2}}{L^2} = ML^{-1}T^{-2}$$

(Since θ is a dimension less quantity)

59. (a) Impulse = change into momentum

60. (a) $n \times u$ = magnitude of physical quantity = constant

2. MOTION IN A STRAIGHT LINE

ANSWER KEY											
1	(b)	11	(b)	21	(b)	31	(b)	41	(a)	51	(a)
2	(b)	12	(c)	22	(a)	32	(c)	42	(d)	52	(d)
3	(b)	13	(c)	23	(d)	33	(a)	43	(d)	53	(c)
4	(b)	14	(c)	24	(d)	34	(c)	44	(a)	54	(c)
5	(c)	15	(d)	25	(d)	35	(d)	45	(a)	55	(a)
6	(c)	16	(b)	26	(b)	36	(c)	46	(b)	56	(d)
7	(c)	17	(b)	27	(c)	37	(a)	47	(c)	57	(b)
8	(d)	18	(c)	28	(c)	38	(b)	48	(a)	58	(b)
9	(a)	19	(b)	29	(c)	39	(a)	49	(b)	59	(c)
10	(b)	20	(b)	30	(b)	40	(d)	50	(d)	60	(d)

Hints & Solutions

2. (b) $s = 6t^2 - t^3$

$$v = \frac{ds}{dt} = 12t - 3t^2$$

for $v = 0$ $\quad 12t = 3t^2$

$$t = \frac{12}{3} = 4\,\text{s}$$

6. (c) $$\text{Average speed} = \frac{\text{Total distance travelled}}{\text{Total time taken}}$$

If L is the total distance travelled.

Time taken to cover half the distance $= t_1 = \dfrac{\frac{L}{2}}{V_1}$

time taken to cover test half the distance

$$= t_2 = \frac{\frac{L}{2}}{V_2}$$

$\therefore$ Average speed $$= \frac{\frac{L}{2} + \frac{L}{2}}{\frac{L}{2V_1} + \frac{L}{2V_2}}$$

$$= \frac{2\frac{L}{2}}{\frac{L}{2}\left(\frac{V_1 + V_2}{V_1V_2}\right)} = \frac{2V_1V_2}{V_1 + V_2}$$

10. (b) is the correct v-t graph. This is because from t = 0 to 5 sec, displacement increases linearly with time, velocity is uniform. From t = 5 sec to 15 sec, displacement is constant, velocity is zero. From t = 15 sec to 20 sec, displacement decreases linearly with time. Therefore, velocity is negative, but constant.

11. (b) The slope of velocity-time graph denotes acceleration. For a retarded motion acceleration is negative.

12. (c) Area of acceleration-displacement curve

$$A = \int_{x_1}^{x_2} a\,dx = \int_{x_1}^{x_2} \frac{dv}{dt}\,dx \quad \left[a = \frac{dv}{dt}, v = \frac{dx}{dt}\right]$$

$$A = \int_u^v v\,dv = \frac{1}{2}[v^2 - u^2]$$

$$A = \frac{1}{2}\frac{m[v^2 - u^2]}{m}$$

= Change in K.E. per unit mass.

14. (c) Let $PQ = x$, then

$$a = \frac{40^2 - 30^2}{2x} = \frac{350}{x} \quad [\because v^2 = u^2 + 2as]$$

Also, velocity at mid point is given by

$$v^2 - 30^2 = 2 \times \frac{350}{x} \times \frac{x}{2}$$

This gives $v = 25\sqrt{2}$ km/h

15. (d) Relative velocity of parrot w.r.t the train $= 10 - (-5) = 15\ \text{ms}^{-1}$.
Time taken by parrot to cross the train
$$= \frac{150}{15} = 10\ \text{s}$$

16. (b) $$\frac{x(4)}{x(5)} = \frac{\frac{g}{2}(2\times4-1)}{\frac{g}{2}(2\times5-1)} = \frac{7}{9}$$

$[\because S_{n^{th}} = u + \frac{a}{2}(2n-1)$ and $u = 0, a = g]$

17. (b) Total distance = s;

Total time taken $= \frac{s/2}{40} + \frac{s/2}{60} = \frac{5s}{240} = \frac{s}{48}$

$$\therefore \text{Average speed} = \frac{\text{total distance}}{\text{total time}}$$

$$= \frac{s}{s/48} = 48\ \text{km/h}$$

$$\left[\textbf{Alt: } v_{av} = \frac{2v_1v_2}{v_1+v_2} = \frac{2\times40\times60}{40+60} = 48\ \text{km/h}\right]$$

18. (c) Average speed $= \frac{s}{\frac{s/3}{10} + \frac{s/3}{20} + \frac{s/3}{60}}$

$$= \frac{s}{s/18} = 18\ \text{km/h}$$

19. (b) $$48 = \frac{200\times10^{-3}}{\left(\frac{100\times10^{-3}}{40}\right) + \left(\frac{100\times10^{-3}}{v}\right)}$$

or $\frac{1}{40} + \frac{1}{v} = \frac{2}{48} = \frac{1}{24}$

or $\frac{1}{v} = \frac{1}{24} - \frac{1}{40} = \frac{2}{120} = \frac{1}{60}$

or $v = 60$ km/h

$$\left[\textbf{Alt: } v_{av} = \frac{2v_1v_2}{v_1+v_2} \Rightarrow 48 = \frac{2\times40\times v}{40+v}\right.$$

$\left.\Rightarrow v = 60\ \text{km/h}\right]$

20. (b) Let the body fall through the height of tower in t seconds. From,

$D_n = u + \frac{a}{2}(2n-1)$ we have, total distance travelled in last 2 seconds of fall is

$$D = D_t + D_{(t-1)}$$

$$= \left[0 + \frac{g}{2}(2t-1)\right] + \left[0 + \frac{g}{2}\{2(t-1)-1\}\right]$$

$$= \frac{g}{2}(2t-1) + \frac{g}{2}(2t-3) = \frac{g}{2}(4t-4)$$

$$= \frac{10}{2}\times 4(t-1)$$

or, $40 = 20\,(t-1)$ or $t = 2 + 1 = 3$s
Distance travelled in t seconds is

$$s = ut + \frac{1}{2}at^2 = 0 + \frac{1}{2}\times10\times3^2 = 45\text{m}$$

21. (b) In one dimensional motion, the body can have at a time one velocity but not two values of velocities.

22. (a) $$\frac{D_4}{D_3} = \frac{0 + \frac{a}{2}(2\times4-1)}{0 + \frac{a}{2}(2\times3-1)} = \frac{7}{5}$$

23. (d)

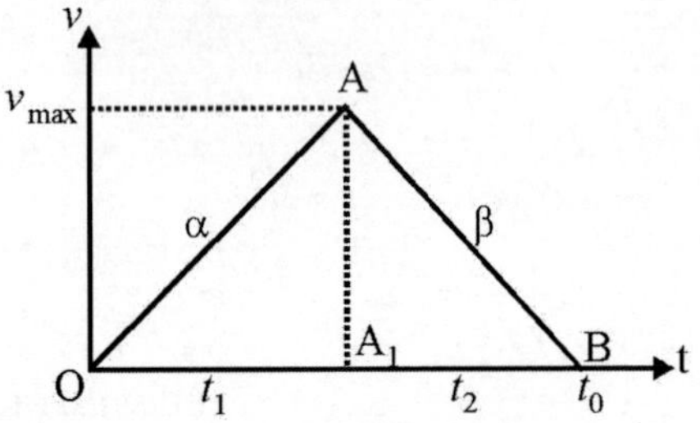

In Fig.
$AA_1 = v_{max.} = \alpha t_1 = \beta t_2$

But $t = t_1 + t_2 = \frac{v_{max}}{\alpha} + \frac{v_{max}}{\beta}$

$$= v_{max}\left(\frac{1}{\alpha} + \frac{1}{\beta}\right) = v_{max}\left(\frac{\alpha+\beta}{\alpha\beta}\right)$$

or, $v_{max} = t\left(\frac{\alpha\beta}{\alpha+\beta}\right)$

24. (d) At E, the slope of the curve is negative.

25. (d) Velocity, $v = \frac{ds}{dt} = 3t^2 - 12t + 3$

Acceleration, $a = \frac{dv}{dt} = 6t - 12$; For $a = 0$, we have, $0 = 6\,t - 12$ or $t = 2$s. Hence, at $t = 2$ s the velocity will be

$$v = 3\times2^2 - 12\times2 + 3 = -9\ \text{ms}^{-1}$$

26. (b) h_{max} = 20 m and t = 5 sec. Time taken by the body to reach the ground from some height is the same as taken to reach that height. Therefore, time to reach the ground from its maximum height is 5 sec.

27. (c) Initial velocity (u) = 0; Time (t) = 4 sec and gravitational acceleration (g) = 10 m/s^2.
Height of tower

$$h = ut + \frac{1}{2}gt^2 = (0 \times 4) + \frac{1}{2} \times 10 \times (4)^2$$

= 80 m.

28. (c) The speed of an object, falling freely due to gravity, depends only on its height and not on its mass. Since the paths are frictionless and all the objects fall through the same height, therefore, their speeds on reaching the ground will be in the ratio of 1 : 1 : 1

29. (c) Height of tap = 5m and (g) = 10 m/sec^2.

For the first drop, $5 = ut + \frac{1}{2}gt^2$

$$= (0 \times t) + \frac{1}{2} \times 10t^2 = 5t^2 \text{ or } t^2 = 1$$

or t = 1 sec.
It means that the third drop leaves after one second of the first drop. Or, each drop leaves after every 0.5 sec.
Distance covered by the second drop in 0.5 sec

$$= ut + \frac{1}{2}gt^2 = (0 \times 0.5) + \frac{1}{2} \times 10 \times (0.5)^2$$

= 1.25 m.
Therefore, distance of the second drop above the ground = 5 – 1.25 = 3.75 m.

30. (b) Initial velocity (u) = 40 m/s
Acceleration a = –g m/s^2 = –10 m/s^2
Time = 2 seconds
By Ist equation of motion,
$v = u + at$
v = 40 – 10 (2) = 20 m/s

31. (b) $s = at^2 - bt^3$

$$v = \frac{ds}{dt} = 2at - 3bt^2$$

$$a = \frac{dv}{dt} = 2a - 6bt$$

$$2a - 6bt = 0 \Rightarrow t = \frac{a}{3b}$$

32. (c) Initial velocity of car (u) = 0
Final velocity of car (v) = 144 km/hr = 40 m/s
Time taken = 20 s
We know that, $v = u + at$
$40 = a \times 20 \Rightarrow a = 2$ m/s^2
Also, $v^2 - u^2$ = 2as

$$\Rightarrow s = \frac{v^2 - u^2}{2a}$$

$$\Rightarrow s = \frac{(40)^2 - (0)^2}{2 \times 2} = \frac{1600}{4} = 400 \text{ m.}$$

33. (a) $v^2 - u^2 = 2as$

$$\Rightarrow a = \frac{v^2 - u^2}{2s}$$

$$= -\frac{u_1^2}{2s}, \quad \text{where } u_1 = 40 \text{ km/h}$$

For same retarding force $s \propto u^2$

$$\therefore \frac{s_2}{s_1} = \frac{u_2^2}{u_1^2}$$

$$\Rightarrow \frac{s_2}{s_1} = \left(\frac{80}{40}\right)^2 = 4$$

$\therefore s_2 = 4s_1$
$= 8$ m

If F is retarding force and s the stopping distance, then $\frac{1}{2}mv^2 = Fs$

For same retarding force, $s \alpha v^2$

$$\therefore \frac{s_2}{s_1} = \left(\frac{v_2}{v_1}\right)^2 = \left(\frac{80 \text{ km/h}}{40 \text{ km/h}}\right)^2 = 4$$

$$\therefore \quad s_2 = 4s_1 = 4 \times 2 = 8\text{m}$$

34. (c) Displacement
$s = 3t^3 + 7t^2 + 5t + 8$;

$$\text{Velocity} = \frac{ds}{dt} = 9t^2 + 14t + 5$$

$$\text{Acceleration} = \frac{d^2s}{dt^2} = 18t + 14$$

Acceleration at (t = 1s)
= 18×1+14 = 18 + 14 = 32 m/s^2

35. (d) At highest point, velocity is zero.
Hence, Kinetic energy is zero.

36. (c) Let body takes T sec to reach maximum height.
Then $v = u - gT$
v = 0, at highest point.

$$T = \frac{u}{g} \quad .....(1)$$

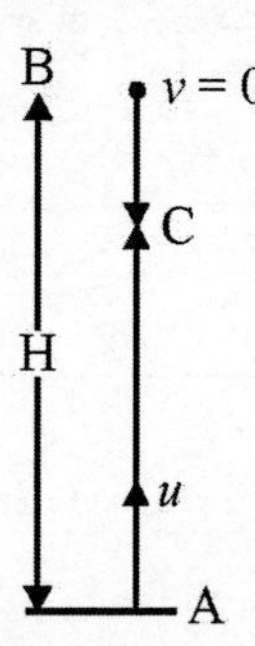

Velocity attained by body in $(T - t)$ sec

$v = u - g(T - t)$

$= u - gT + gt = u - g\frac{u}{g} + gt$

or $v = gt$ ....(2)

$\therefore$ Distance travelled in last t sec of its ascent

$$s = (gt)t - \frac{1}{2}gt^2 = \frac{1}{2}gt^2$$

37. (a) Instantaneous speed is the distance being covered by the particle per unit time at the given instant. It is equal to the magnitude of the instantaneous velocity at the given instant.
38. (b) Displacement may be zero because final position of the particle may coincide with its initial position.
39. (a) When a body moves along a straight line with constant retardation, its speed goes on decreasing.
40. (d) Distance covered by a particle is zero only when it is at rest. Therefore, its displacement must be zero.
41. (a) When displacement is zero, particle may be at rest, therefore, distance travelled = 0.

    Again, when displacement is zero, final position coincides with the initial position, distance travelled is not zero.
42. (d) As displacement $\leq$ distance, therefore, the ratio is equal to or less than one.
43. (d) From the previous question, it follows that average velocity is equal to or less than one.
44. (a) No net force acts on a particle moving with uniform velocity. Therefore, it is in equilibrium.
45. (a) When location of a particle has changed, it must have covered some distance and undergone some displacement.
46. (b) A body in uniform circular motion is moving in a plane and is two dimensional motion.
47. (c) Constant velocity means constant speed in a particular direction.
48. (a) As the coin falls behind him, force due to air must be backwards. Therefore, the train must be accelerating forward.
49. (b) The opposing force must cause retardation, due to which the body is sure to slow down.
50. (d) The acceleration due to gravity is acting throughout the motion of projectile.
51. (a) The reason is correct explanation of assertion.
52. (d) The displacement-time graph is a straight line inclined to time axis upto time t_0, indicates a uniform velocity. After time t_0, the displacement-time graph is a straight line parallel to time axis indicates the particle is at rest.
53. (c) The velocity-time graph for a uniform motion is a st. line parallel to time axis. Its slope is zero.
54. (c) This is because speed can never be negative.
55. (a) is not possible, because at a particular time t, displacement cannot have two values.
56. (d) is not possible, because at a particular time t, displacement cannot have two values.
57. (b) The slope of the graph $\frac{ds}{dt}$ is maximum at C and hence the instantaneous velocity is maximum at C.
58. (b) At a particular time, two values of displacement are not possible.
59. (c) As $x - t$ graph is a straight line in either case, velocity of both is uniform. As the slope of $x - t$ graph for P is greater, therefore, velocity of P is greater than that of Q.
60. (d) Average acceleration

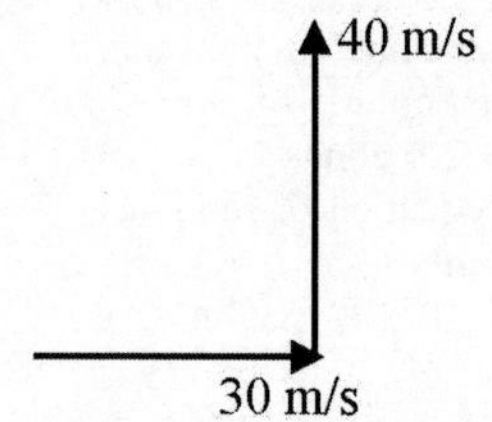

$$< a > = \frac{\text{Change in velocity}}{\text{Total Time}}$$

$$< a > = \frac{|40\hat{j} - 30\hat{i}|}{10 - 0}$$

$$< a > = \sqrt{4^2 + (-3)^2}$$

$$< a > = 5 \text{ m/sec}^2$$

3. MOTION IN A PLANE

ANSWER KEY											
1	(d)	11	(c)	21	(b)	31	(b)	41	(b)	51	(a)
2	(c)	12	(c)	22	(c)	32	(c)	42	(d)	52	(c)
3	(c)	13	(c)	23	(d)	33	(b)	43	(d)	53	(c)
4	(b)	14	(c)	24	(a)	34	(b)	44	(a)	54	(c)
5	(b)	15	(c)	25	(a)	35	(c)	45	(c)	55	(b)
6	(d)	16	(b)	26	(b)	36	(c)	46	(b)	56	(a)
7	(b)	17	(d)	27	(a)	37	(d)	47	(b)	57	(c)
8	(a)	18	(c)	28	(c)	38	(a)	48	(a)	58	(c)
9	(b)	19	(a)	29	(a)	39	(a)	49	(c)	59	(c)
10	(a)	20	(c)	30	(c)	40	(b)	50	(a)	60	(b)

Hints & Solutions

8. (a) Horizontal velocity of a projectile is not affected by gravity.

9. (b) $\omega = \dfrac{2\pi n}{t} = \dfrac{2\pi \times 120}{60\text{ s}}$

10. (a) Velocity at A, $\vec{v} = v\,\hat{j}$

Velocity at B, $\vec{v}_2 = -v\,\hat{i}$

Change in velocity,

$$\Delta\vec{v} = \vec{v}_2 - \vec{v}_1 = -v\,\hat{i} - v\,\hat{j}$$

$$|\Delta\vec{v}| = \sqrt{(-v)^2 + (-v)^2} = v\sqrt{2}.$$

11. (c) $\sqrt{0.5^2 + 0.8^2 + c^2} = 1$

or $c = \sqrt{1 - 0.89} = \sqrt{0.11}$

12. (c) For two perpendicular vectors,

$\vec{A}.\vec{B} = AB\cos 90° = 0.$

13. (c) Initially, both particles have zero vertical velocity. So, both particles take same time to fall through same height.

14. (c) The resultant of $\vec{A} \times 0$ is a vector of zero magnitude. The product of a vector with a scalar gives a vector.

15. (c) $|\vec{a} + \vec{b}|^2 = |\vec{a} - \vec{b}|^2$

$\Rightarrow |a|^2 + |b|^2 + 2\vec{a}.\vec{b} = |a|^2 + |b|^2 - 2\vec{a}.\vec{b}$

16. (b) Scalar product is distributive over addition.

19. (a) As the dropped ball has horizontal velocity wrt to the stationary observer.

23. (d) Use $(\vec{a} + \vec{b} + \vec{c}).(\vec{a} + \vec{b} + \vec{c}) = 0$

24. (a) $h = \dfrac{u^2}{2g} \Rightarrow u = \sqrt{2gh}$

now $R_{max} = \dfrac{u^2}{g} = \dfrac{2gh}{g} = 2h$

25. (a) $(\vec{A} + \vec{B})^2 = (\vec{C})^2$

$\Rightarrow A^2 + B^2 + 2\vec{A}.\vec{B} = C^2$

$\Rightarrow 3^2 + 4^2 + 2\vec{A}.\vec{B} = 5^2$

$\Rightarrow 2\vec{A}.\vec{B} = 0$

or $\Rightarrow \vec{A}.\vec{B} = 0$

$\therefore \vec{A} \perp \vec{B}$

Here $A^2 + B^2 = C^2$. Hence, $\vec{A} \perp \vec{B}$

26. (b) Note that $(\vec{B} \times \vec{A}) \perp \vec{A}$. Hence their dot product is zero.

27. (a) $\vec{v_1} = 50\,\text{km h}^{-1}$ due North;

$\vec{v_2} = 50\,\text{km h}^{-1}$ due West. Angle between $\vec{v_1}$ and $\vec{v_2} = 90°$

$-\overrightarrow{v_1} = 50 \text{ km h}^{-1}$ due South

$\therefore$ Change in velocity

$= |\overrightarrow{v_2} - \overrightarrow{v_1}| = |\overrightarrow{v_2} + (-\overrightarrow{v_1})|$

$= \sqrt{v_2^2 + v_1^2} = \sqrt{50^2 + 50^2} = 70.7 \text{ km/h}$

The direction of this change in velocity is in South-West.

28. (c) $R_{max} = \frac{v^2}{g} = 16000 \ [16\,\text{km} = 16000\,\text{m}]$

or $v = (16000g)^{1/2} = (16000 \times 10)^{1/2}$

$= 400 \text{ ms}^{-1}$

29. (a) Horizontal range is same when angle of projection is θ or $(90° - \theta)$.

30. (c) Centripetal acc. $= \omega^2 r = 4\pi^2 v^2 r$

$= 4 \times (3.14)^2 \times \frac{120}{60} \times \frac{30}{100} = 23.7 \text{ ms}^{-2}$

$[\because \omega = 2\pi v]$

31. (b) Cream gets separated from a churned milk due to centrifugal force.

32. (c) Using the law of vector addition, $(\vec{d} + \vec{e})$ is as shown in the fig.

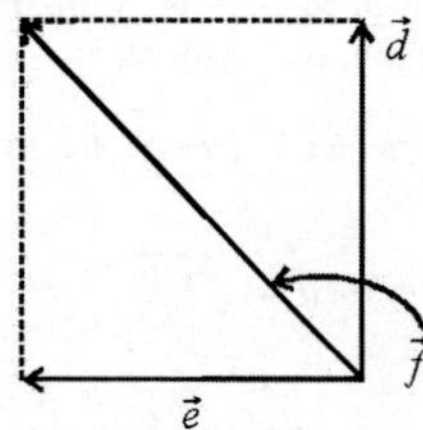

$\therefore \quad \vec{d} + \vec{e} = \vec{f}$

33. (b) Horizontal range is same when angle of projection with the horizonatal is θ and $(90° - \theta)$.

34. (b) $v_r = \sqrt{v_R^2 - v_B^2} = \sqrt{10^2 - 8^2} = 6 \text{ km h}^{-1}$

35. (c) Since displacement is along the y-direction, hence displacement $\vec{s} = 10\hat{j}$.

Work done $= \overrightarrow{F}.\overrightarrow{s}$

$= (-2\hat{i} + 15\hat{j} + 6\hat{k}).10\hat{j} = 150 \text{ J}$

36. (c) On circular motion, the force acts along the radius and displacement at a location is perpendicular to the radius i.e., $\theta = 90°$

As work done $= \overrightarrow{F}.\overrightarrow{S} = FS\cos 90° = 0$

37. (d) $\vec{r} = (a\cos\omega t)\hat{i} + (a\sin\omega t)\hat{j}$

$\vec{v} = \frac{d(\vec{r})}{dt} = \frac{d}{dt}\{(a\cos\omega t)\hat{i} + (a\sin\omega t)\hat{j}\}$

$= (-a\omega \sin\omega t)\hat{i} + (a\omega\cos\omega t)\hat{j}$

$= \omega[(-a\sin\omega t)\hat{i} + (a\cos\omega t)\hat{j}]$

Slope of position vector $= \frac{a\sin\omega t}{a\cos\omega t} = \tan\omega t$ &

slope of velocity vector

$= \frac{-a\cos\omega t}{a\sin\omega t} = \frac{-1}{\tan\omega t}$

$\therefore$ velocity is perpendicular to the displacement.

38. (a) A vector quantity has both magnitude and direction. In the given options, speed has only magnitude, therefore, it is non-vector or scalar quantity.

39. (a) Speed of the bullet $(v) = 1000$ m/s and horizontal distance of the target $(s) = 100$ m.

Time taken to cover the horizontal distance (t)

$= \frac{100}{1000} = 0.1 \text{ sec}$.

During this time, the bullet will fall down vertically due to gravitational acceleration.

$\therefore$ height $(h) = ut + \frac{1}{2}gt^2$

$= (0 \times 0.1) + \frac{1}{2} \times 10(0.1)^2 = 0.05 \text{ m} = 5 \text{ cm}$

40. (b) Radius of circular path $= 20$ cm $= \frac{2}{10}$ m

Angular speed of body $= 10$ rad/s

Linear velocity $=$ radius $\times$ Angular speed

$= \frac{2}{10} \times 10 = 2 \text{ m/s}$

41. (b) Kinetic energy is E

$\frac{1}{2}mv^2 = E \Rightarrow v = \sqrt{\frac{2E}{m}}$

Velocity at highest point

$= v\cos 45° = \sqrt{\frac{2E}{m}}.\frac{1}{\sqrt{2}} = \sqrt{\frac{E}{m}}$

[$\because$ at the highest point there is only horizontal velocity]

K.E. at highest point $= \frac{1}{2} \times m\frac{E}{m} = \frac{E}{2}$

42. (d) $\overrightarrow{F} = -3\hat{i} + \hat{j} + 5\hat{k}$; $\overrightarrow{r} = 7\hat{i} + 3\hat{j} + \hat{k}$

Torque $(\vec{\tau}) = \overrightarrow{r} \times \overrightarrow{F}$

$= (7\hat{i} + 3\hat{j} + \hat{k}) \times (-3\hat{i} + \hat{j} + 5\hat{k})$

$= 7\hat{k} + 35(-\hat{j}) - 9(-\hat{k}) + 15\hat{i} - 3\hat{j} + (-\hat{i})$

$= 14\hat{i} - 38\hat{j} + 16\hat{k}$

43. (d) Let after 1 sec angle become 60°. When the end A moves by 10 m left, the end B moves upward by

$BB' = 10 \times \sqrt{3} = 10 \times 1.73 = 17.3$ m/s

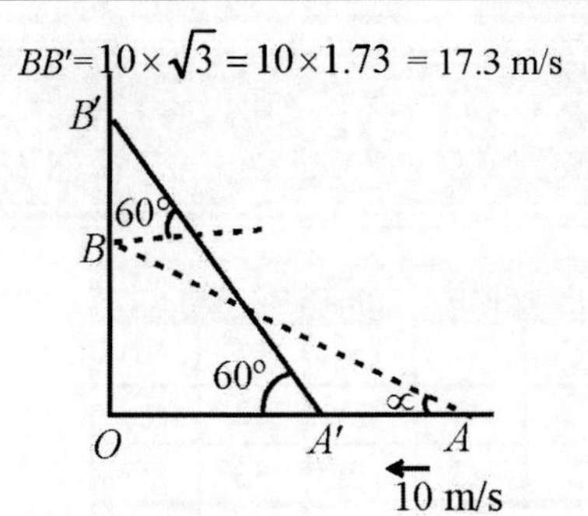

44. (a) $T = \dfrac{mv^2}{R}$

$$v = \sqrt{\frac{TR}{m}} = \sqrt{\frac{25 \times 1.96}{0.25}} = \frac{5 \times 14}{5} = 14 \text{ m/s}$$

45. (c) Centripetal force $= \dfrac{mv^2}{r} = \dfrac{500 \times (10)^2}{50}$
$= 1000$ N
[$\because$ 36 km/hr = 10 m/s]

46. (b) As we know that

$$\vec{v} = \vec{\omega} \times \vec{r} = (3\hat{i} - 4\hat{j} + \hat{k}) \times (5\hat{i} - 6\hat{j} + 6\hat{k})$$
$$= -18\hat{i} - 13\hat{j} + 2\hat{k}$$

47. (b) We know that, $R^2 = A^2 + B^2 + 2AB\cos\theta$

$$(24)^2 = (12)^2 + (18)^2 + 2(12)(18)\cos\theta$$

$$\cos\theta = \frac{108}{432} \Rightarrow \theta = 75°52'$$

48. (a) Given : Mass (m) = 0.4 kg
Its frequency (n) = 2 rev/sec
Radius (r) = 1.2 m. We know that linear velocity of the body (v) = $\omega r = (2\pi n)r$
$= 2 \times 3.14 \times 1.2 \times 2 = 15.08$ m/s.
Therefore, tension in the string when the body is at the top of the circle (T)

$$= \frac{mv^2}{r} - mg = \frac{0.4 \times (15.08)^2}{2} - (0.4 \times 9.8)$$
$$= 45.78 - 3.92 = 41.56\text{N}$$

49. (c) In circular motion with constant speed, acceleration is always inward, its magnitude is constant but direction changes, hence acceleration changes, so does velocity. K.E. is constant.

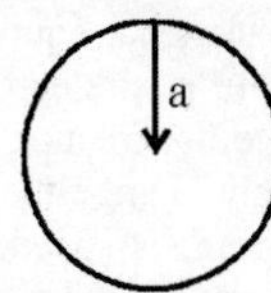

51. (a) Body moves with constant speed, it means that it performs uniform circular motion. In this motion the direction of motion is always perpendicular to centripetal force. Hence the work done by centripetal force is always zero

($dW = \vec{F}_C . \vec{dr} = F_C\, dr \cos\theta = 0 \Rightarrow \theta = 90°$)

52. (c) Body moves with constant speed it means that tangential acceleration $a_T = 0$ & only centripetal acceleration a_C exists whose direction is always towards the centre or inward (along the radius of the circle).

53. (c) It has only tangential velocity v_T & radial acceleration or centripetal acceleration a_C.

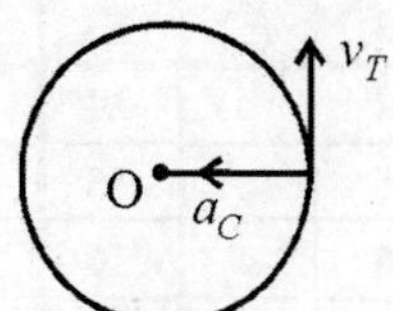

54. (c) Since the circular motion is uniform, therfore there is no change of angular velocity. Thus angular acceleration is zero.

56. (a) Since centripetal acceleration

$$a_C = \frac{v^2}{r} = r\omega^2 = v\omega \text{ because } v = r\omega$$

So there is no change if we double v & halve ω.

57. (c) In uniform circular motion, the body move with v_T (tangential velocity) & a_C. If $a_C = 0$ then it implies that the body is no longer bound to rotate in circle & so no change in the direction of velocity. Hence it move tangentially to the circle outward with velocity v_T.

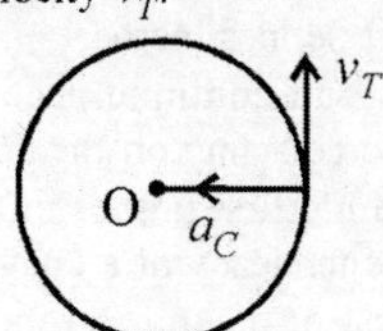

58. (c) Since velocity is defined as $v = \dfrac{\Delta s}{\Delta t} = r\dfrac{\Delta\theta}{\Delta t}$
Where $s = P_1P_2 = r.\theta$, is an arc to circle, which is swept by the particle in Δt time, r is radius of circle which is constant in uniform circular motion & θ is angular displacement in Δt time.
Hence if particle in a circle describe equal angles in equal intervals of time, its speed (magnitude of velocity vector) remains same but the direction changes due to centripetal acceleration a_C.

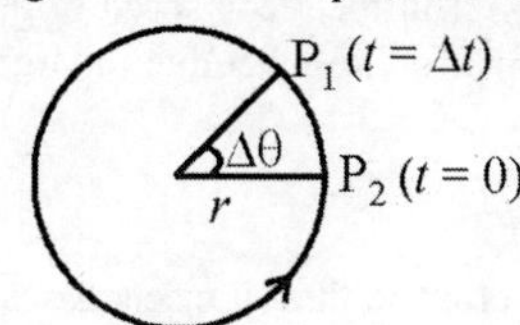

60. (b) Centripetal force, $F = \dfrac{mv^2}{r}$, so, $F \propto \dfrac{1}{r}$ so v is independent of r.

4. LAWS OF MOTION

ANSWER KEY											
1	(c)	**11**	(b)	**21**	(d)	**31**	(a)	**41**	(c)	**51**	(b)
2	(d)	**12**	(c)	**22**	(d)	**32**	(a)	**42**	(c)	**52**	(c)
3	(c)	**13**	(c)	**23**	(c)	**33**	(c)	**43**	(d)	**53**	(a)
4	(b)	**14**	(d)	**24**	(a)	**34**	(c)	**44**	(a)	**54**	(b)
5	(c)	**15**	(a)	**25**	(c)	**35**	(a)	**45**	(c)	**55**	(c)
6	(c)	**16**	(d)	**26**	(b)	**36**	(a)	**46**	(b)	**56**	(a)
7	(b)	**17**	(c)	**27**	(c)	**37**	(c)	**47**	(d)	**57**	(d)
8	(c)	**18**	(c)	**28**	(b)	**38**	(b)	**48**	(c)	**58**	(b)
9	(b)	**19**	(c)	**29**	(d)	**39**	(c)	**49**	(d)	**59**	(a)
10	(b)	**20**	(c)	**30**	(c)	**40**	(a)	**50**	(c)	**60**	(b)

Hints & Solutions

1. (c) When the swimmer push some water in backward direction, then he get some momentum in forward direction from water & starts to swim. This is according to Newton third Law. (action-reaction force)

2. (d) Newton' first law of motion is also called law of inertia as it defines inertia.

3. (c) The body will continue accelerating until the resultant force acting on the body becomes zero.

4. (b) See Newton's first law of motion according to which "the tendency of a body to continue in its state of rest or of uniform motion in a straight line, is called law of inertia".

5. (c) He can come at shore by making use of Newton's third law. In this case man push the ice backward & ice reacts back to the man in forward direction due to friction between ice & man. If friction is very small between him & the ice, then he come out from this pond only by taking very small steps.

6. (c) The gun applied a force F_{12} on the bullet in forward direction & according to Newton's third law bullet applies a reaction force on gun F_{21} in backward direction. But the recoil speed of gun is very low in comparison to bullet due to large mass.

7. (b) $F = \frac{dp}{dt}$

9. (b) When jet plane flies, it ejects gases in back ward direction at very high velocity. From Newton's third law, these gases provides the momentum to jet plane in forward direction plus compensates the force of gravity.

10. (b) By spitting or sneezing we get a momentum in opposite direction which will help us in getting off the plane. In all other cases we will slip on ice as there is no friction.

11. (b)

R
pushing → Block
f
W

Here $f = \mu R = \mu W$

Pushing

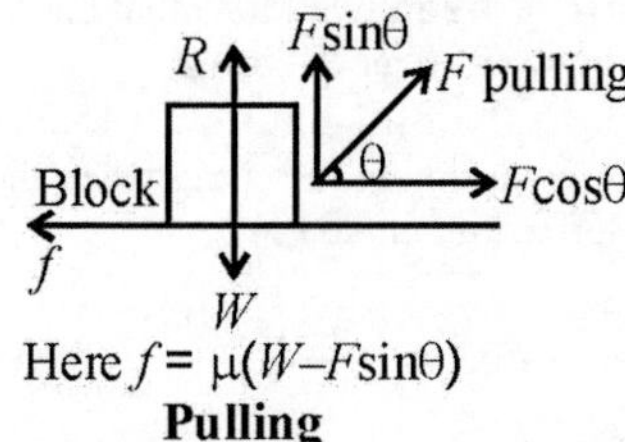

Here $f = \mu(W - F\sin\theta)$

Pulling

Since we required less force in pulling in comparison of pushing it. Hence pulling is easier then pushing.

12. (c) From Newton's second law if $\Sigma F_i = 0$ then the body is in translational equilibrium.

13. (c) When a particle is subjected to a force acting perpendicular to its direction of motion, the force applied will only change the direction of motion without changing its speed. Hence the particle will describe circular path.

14. (d) $\mu_{static} > \mu_{kinetic} > \mu_{rolling}$

15. (a) Static friction is a self adjusting force in magnitude and direction.

16. (d) In case (a) In case (b)

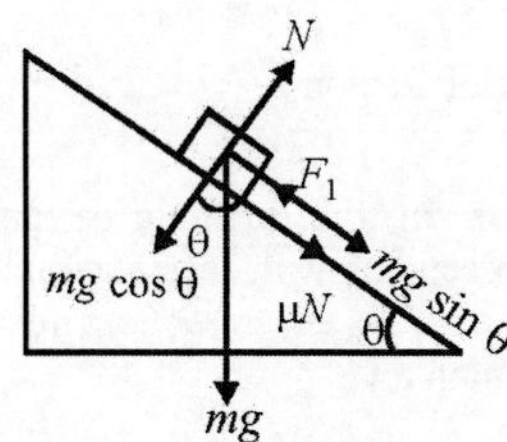

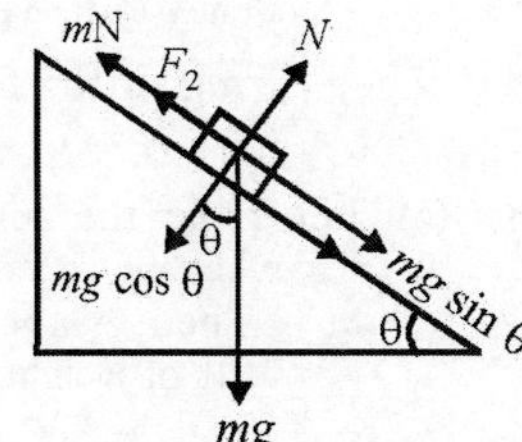

$mg \sin\theta = F_1 - \mu N$
$N = mg\cos\theta$
$\therefore\ mg\sin\theta + \mu\, mg\cos\theta = F_1$
In second case (b)
$\mu N + F_2 = mg\sin\theta$
$\mu\, mg\cos\theta - F_2 = mg\sin\theta$
or $F_2 = mg\sin\theta - \mu mg\cos\theta$
but $F_1 = 2F_2$
therefore $mg\sin\theta + \mu\, mg\cos\theta$
$= 2(mg\sin\theta - \mu\, mg\cos\theta)$
$mg\sin\theta = 3\,\mu\, mg\cos\theta$
or $\tan\theta = 3\mu$ or $\theta = \tan^{-1}(3\mu)$

17. (c) $\mu_s > \mu_k > \mu_r$

18. (c)

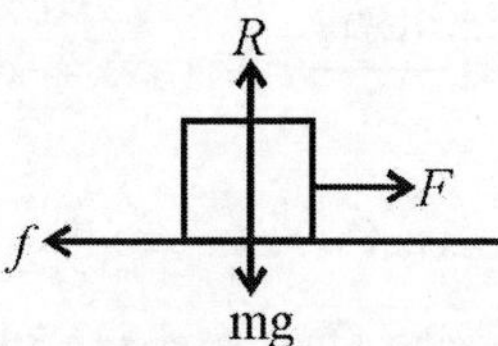

So equation of motion is $F - f = ma$
where $f = \mu R = \mu Mg = \mu \times 30 \times 9.8 = 294\mu$
Since initial velocity of the body is zero & its velocity increases from zero to 4 m/sec in 2 seconds, so acceleration of the body is
$v = u + at$
$\Rightarrow\quad 4 = 0 + a \times 2$ or $a = 2\text{ m/sec}^2$
So $F - f = 30 \times 2$
$\Rightarrow\quad 200 - \mu \times 294 = 60$

$$\Rightarrow\quad \mu = \frac{140}{294} = 0.476$$

19. (c) θ = Angle of repose $\therefore\ \mu = \tan\theta$.
20. (c) When the men push the rough surface on walking, then surface (from Newton' third Law) applies reaction force in forward direction. It occurs because there is friction between men & surface. If surface is frictionless (such as ice), then it is very difficult to move on it.
21. (d) When brakes are on, the wheels of the cycle will slide on the road instead of rolling there. It means the sliding friction will come into play instead of rolling friction. The value of sliding friction is more than that of rolling friction.
22. (d) Friction can be decreased by all the given methods.
23. (c) Centrifugal force is a pseudo force which appears to act on an object when observed from a rotating frame of reference. It always acts away from the centre of circular path.
24. (a) As the block is at rest, the net contact force on the block i.e. resultant of friction and normal reaction must balance the weight of the block. Hence $R = mg = 20\text{N}$

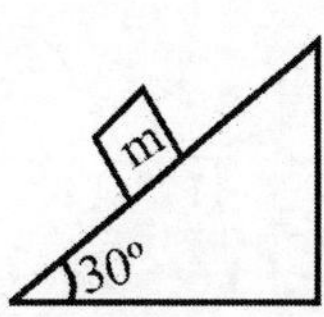

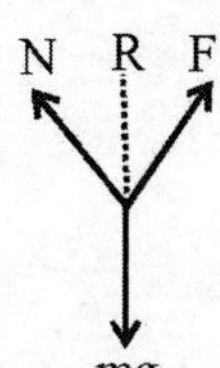

25. (c) Friction does not depend on the area of contact.
26. (b) Mg and N cannot form action-reaction pair as they are acting on same body.
27. (c) On increasing smoothness, friction first decreases and then increases.
28. (b) Centripetal force always acts perpendicular to the direction of motion, hence it can change direction of motion, of the body but not its speed. Again as direction of centripetal force is different at different positions, it is never constant.
29. (d) The horse exerts a backward force on the earth as a result, earth exerts a forward force on the horse and cause horse to move forward.
30. (c) Action and reaction always act on different bodies.
31. (a) Action and reaction forces are always of same nature.
32. (a) Static friction is self adjusting in nature.
33. (c) Friction always apposes relative motion between surfaces in contact and hence can act in any direction to oppose the relative motion.
34. (c) As the net force acting on the body is zero, either no force should act on the body or at least two forces should be acting on it.
35. (a) Direction of impulse is same as that of the net force.
36. (a) By Newton's third law of motion, the tension in the rope must be equal to the force applied by the monkey on the rope.
37. (c) Friction provides centripetal force to the coin, hence acts towards the centre of the disc.
38. (b) In presence of friction $a = (g\sin\theta - \mu g\cos\theta)$
$\therefore$ Time taken to slide down the plane

$$t_1 = \sqrt{\frac{2s}{a}} = \sqrt{\frac{2s}{g(\sin\theta - \mu\cos\theta)}}$$

In absence of friction $t_2 = \sqrt{\dfrac{2s}{g\sin\theta}}$

$t_1 = 2t_2$

$\therefore\ t_1^2 = 4t_2^2$

or $\dfrac{2s}{g(\sin\theta - \mu\cos\theta)} = \dfrac{2s \times 4}{g\sin\theta}$

$\sin\theta = 4\sin\theta - 4\mu\cos\theta$

$\mu = \dfrac{3}{4}\tan\theta = \dfrac{3}{4} = 0.75$

39. (c) $\dfrac{K_1}{K_2} = \dfrac{p_1^2}{m_1} \times \dfrac{m_2}{p_2^2}$ $[\because p = mv \Rightarrow K = \dfrac{p^2}{2m}]$

Hence, $\dfrac{p_1}{p_2} = \sqrt{\dfrac{M_1}{M_2}} = \sqrt{\dfrac{1}{4}} = \dfrac{1}{2}$

40. (a) Masses of the pieces are 1, 1, 3 kg. Hence

$(1 \times 21)^2 + (1 \times 21)^2 = (3 \times V)^2$

That is, $V = 7\sqrt{2}$.

41. (c) $E = \dfrac{1}{2}mv^2$. Hence, $mv = (2\ mE)^{1/2}$. For same KE, momentum $\propto \sqrt{m}$. Hence, the ratio is 2 : 1.

42. (c) Thrust $= \dfrac{udM}{dt} = mg \Rightarrow \dfrac{dM}{dt} = \dfrac{mg}{u}$

$= \dfrac{600 \times 10}{1000} = 6\ \text{kg s}^{-1}$

43. (d) Impulse = final momentum – initial momentum = $m(v_2 - v_1)$

44. (a) The force of friction on the chain lying on the table should be equal to the weight of the hanging chain. Let

ρ = mass per unit length of the chain

μ = coefficient of friction

l = length of the total chain

x = length of hanging chain

Now, $\mu(l - x)\rho g = x\rho g$ or $\mu(l - x) = x$

or $\mu l = (\mu + 1)x$ or $x = \mu l/(\mu + 1)$

$\therefore\ x = \dfrac{0.25l}{(0.25+1)} = \dfrac{0.25l}{1.25} = 0.2l$

$\dfrac{x}{l} = 0.2 = 20\%$

45. (c) Newton's first law of motion is related to physical independence of force.

46. (b) Here $u = 72$ km/h = 20 m/s; $v = 0$;

$a = -\mu g = -0.5 \times 10 = -5\ \text{m/s}^2$

As $v^2 = u^2 + 2as$,

$\therefore\ s = \dfrac{(v^2 - u^2)}{2a} = \dfrac{(0) - (20)^2}{2 \times (-5)} = 40\,\text{m}$

47. (d) Here $\tan\theta = 0.8$

where θ is angle of repose

$\theta = \tan^{-1}(0.8) = 39°$

The given angle of inclination is equal to the angle of repose. So the 1 kg block has no tendency to move.

$\therefore\ mg\sin\theta$ = force of friction

$\Rightarrow T = 0$

48. (c) Let T be the tension in the branch of a tree when monkey is descending with acceleration a. Then $mg - T = ma$; and T = 75% of weight of monkey

$\therefore\quad ma = mg - \left(\dfrac{75}{100}\right)mg = \left(\dfrac{1}{4}\right)mg$

or $\quad a = \dfrac{g}{4}$.

49. (d) When the block slides down the plane with a constant speed, then the inclination of the plane is equal to angle of repose (θ).

Coeff. of friction = tan of the angle of repose = $\tan\theta$.

50. (c) When shell explodes in mid air its chemical energy is partly converted into mechanical energy, hence K.E. increases.

51. (b) Thrust on the satellite,

$F = \dfrac{-vdM}{dt} = -v(\alpha v) = -\alpha v^2$

Acceleration $= \dfrac{F}{M} = \dfrac{-\alpha v^2}{M}$

52. (c) Velocity of the rocket (u) = 300 m/s and force (F) = 345N. Rate of combustion of fuel

$\left(\dfrac{dm}{dt}\right) = \dfrac{F}{u} = 1.15\text{kg/sec}$

53. (a) $r = 30\ m$ and $\mu = 0.4$.

$v_{max} = \sqrt{\mu rg} = \sqrt{0.4 \times 30 \times 9.8} = 10.84\ \text{m/s}$

54. (b) By Newton's IInd law of motion, $F = ma$

$\Rightarrow 10 = m(1) \Rightarrow m = 10$ kg.

55. (c) Mass = 150 gm = $\dfrac{150}{1000}$kg

Force = Mass × acceleration

$= \dfrac{150}{1000} \times 20\text{N} = 3\,\text{N}$

Impulsive force = $F.\Delta t = 3 \times 0.1 = 0.3\ N$

56. (a) In case (i) we have

$T_1 - (1 \times g) = 1 \times 4.9$

$\Rightarrow T_1 = 9.8 + 4.9 = 14.7$ N

In case (ii), $1 \times g - T_2 = 1 \times 4.9$

$\Rightarrow T_2 = 9.8 - 4.9 = 4.9$ N

$\therefore\ \dfrac{T_1}{T_2} = \dfrac{14.7}{4.9} = \dfrac{3}{1}$

57. (d) Given $F = 600-(2\times10^5 t)$
The force is zero at time t, given by

$$0 = 600-2\times10^5 t$$

$$\Rightarrow \quad t=\frac{600}{2\times10^5}=3\times10^{-3} \text{ seconds}$$

$$\therefore \text{ Impulse} = \int_0^t F dt = \int_0^{3\times10^{-3}} (600-2\times10^5 t)\, dt$$

$$= \left[600t-\frac{2\times10^5 t^2}{2}\right]_0^{3\times10^{-3}}$$

$$= 600\times3\times10^{-3}-10^5(3\times10^{-3})^2$$

$$= 1.8-0.9 = 0.9\text{Ns}$$

58. (b) Given : Mass of rocket (m) = 5000 kg
Exhaust speed (v) = 800 m/s
Acceleration of rocket (a) = 20 m/s^2
Gravitational acceleration (g) = 10 m/s^2

We know that upward force
$F = m\,(g + a) = 5000\,(10 + 20)$
$= 5000 \times 30 = 150000$ N.
We also know that amount of gas ejected

$$\left(\frac{dm}{dt}\right)=\frac{F}{v}=\frac{150000}{800}=187.5 \text{ kg/s}$$

59. (a)

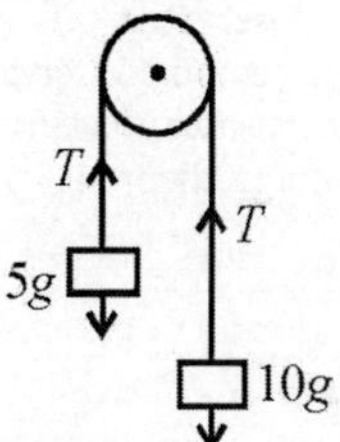

Let T be the tension in the string.
$\therefore\ 10g - T = 10a$ ....(i)
$T - 5g = 5a$ ....(ii)
Adding (i) and (ii),

$$5g = 15a \Rightarrow a = \frac{g}{3} \text{ m/s}^2$$

60. (b) Effective acceleration of a body = $[a - g]$.

5. WORK, ENERGY & POWER

ANSWER KEY											
1	(b)	**11**	(d)	**21**	(d)	**31**	(b)	**41**	(b)	**51**	(c)
2	(a)	**12**	(d)	**22**	(b)	**32**	(b)	**42**	(b)	**52**	(c)
3	(c)	**13**	(c)	**23**	(c)	**33**	(a)	**43**	(a)	**53**	(d)
4	(b)	**14**	(a)	**24**	(d)	**34**	(d)	**44**	(b)	**54**	(b)
5	(d)	**15**	(b)	**25**	(a)	**35**	(a)	**45**	(c)	**55**	(d)
6	(a)	**16**	(a)	**26**	(d)	**36**	(a)	**46**	(a)	**56**	(b)
7	(c)	**17**	(b)	**27**	(a)	**37**	(c)	**47**	(b)	**57**	(a)
8	(c)	**18**	(b)	**28**	(b)	**38**	(c)	**48**	(d)	**58**	(a)
9	(c)	**19**	(b)	**29**	(c)	**39**	(c)	**49**	(b)	**59**	(a)
10	(c)	**20**	(c)	**30**	(b)	**40**	(d)	**50**	(b)	**60**	(a)

Hints & Solutions

1. (b) Work done by friction may be positive, negative or zero.
2. (a) The magnitude of work done by a force depends on frame of reference.
3. (c) By work-energy theorem
Work done by all the forces,
$W = \Delta k$
or $W_C + W_{NC} = \Delta k$
but $W_C = -\Delta U \quad (U = P.E)$
$\therefore\ -\Delta U + W_{NC} = \Delta k$
or $W_{NC} = \Delta k + \Delta U = \Delta E$
$\therefore$ Work done by all non-conservative forces is equal to the change in total mechanical energy of the body.

4. (b) Work done by a conservative force is negative of change in P.E. of the body. Hence if work done by a conservative force is positive, its P.E. must decrease.

5. (d) $P = Fv$
as v = constant, $f = 0$
$\therefore P = 0$

6. (a) Friction force is non-conservative while all elastic forces are conservative.

7. (c) Work done by a non-conservative force in a round trip may be positive, negative or zero.

8. (c) By work-energy theorem
$W = \Delta k$
or $-Fs = -k$
or $S = \dfrac{k}{F}$
as both kinetic energy and retarding force are equal for the car and the truck, they must cover equal distance before coming to rest.

9. (c) K.E. at the highest point of trajectory,
$k' = k\cos^2\theta$
will be equal for both the balls hence their P.E. will also be equal.

10. (c) During collision, the P.E. of the colliding bodies increases due to deformation in the bodies, hence their K.E. decreases during the collision.

11. (d) Velocity of A after the collision is given by
$$V_1 = \frac{(m_1 - em_2)u_1}{m_1 + m_2} + \frac{(1+e)m_2u_2}{m_1 + m_2}$$
Here e is coefficient of restitution.
Here, $m_1 = m_p = m_2 = m_3$ $u = n$ $u_2 = 0$
$$\therefore \frac{(m_A - em_B)u}{m_A + m_B} = 0$$
or $m_A = em_B$ but $e \le 1$
$\therefore m_p \le m_B$

12. (d) If the surface is rough, friction will act along the horizontal and hence horizontal component of its velocity will change.

13. (c) In inelastic collision, K.E. does not remain conserved. Also bodies may or may not stick to each other if the collision is obrtque but the total momentum remains conserved.

14. (a) $P = Fv$
For the inclined plane,
$F = mg(\sin\theta + m\cos\theta)$ is constant, hence P = constant.

15. (b) Common normal component of velocity of the ball will become zero, hence the ball will move along the incline.

16. (a) $e = |v_1 - v_2| / |u_1 - u_2|$ which is 1 for a perfectly elastic collision.

17. (b) $W = \Delta E = \dfrac{1}{2}m(v_1^2 - v_2^2)$
$$= \frac{1}{2} \times 0.01[(1000)^2 - (500)^2] = 3750\text{ J}.$$

18. (b) $P = \dfrac{W}{t}$. Here, $P = 2$ kW $= 2000$ W.
$W = Mgh = M \times 10 \times 10 = 100$ M
and $t = 60$ s.
This gives, $M = 1200$ kg
Its volume = 1200 litre as 1 litre of water contains 1 kg of its mass.

19. (b) When the identical balls collide head-on, their velocities are exchanged.

20. (c) $W = \displaystyle\int_0^s F dx = \int_0^s (7 - 2x + 3x^2)dx$
$$= [7x - x^2 + x^3]_0^5 = 135\text{ J}$$

21. (c) $\dfrac{1}{2}(1)v_1^2 = \dfrac{1}{2}(9)v_2^2 \Rightarrow \dfrac{v_1^2}{v_2^2} = 9$ or $\dfrac{v_1}{v_2} = 3$
Ratio of their linear momenta
$$= \frac{m_1v_1}{m_2v_2} = \frac{1}{9} \times (3) = \frac{1}{3}$$

22. (b) Force due to friction = kinetic energy
$\mu mg s = \dfrac{1}{2}mv^2$ [Here, $v = 72$ km/h
$= \dfrac{72000}{60 \times 60} = 20\text{m/s}$]
or, $s = \dfrac{v^2}{2\mu g} = \dfrac{20 \times 20}{2 \times 0.5 \times 10} = 40\text{m}$

23. (c) Initial momentum $(p_1) = p$; Final momentum (p_2) $= 1.5\,p$ and initial kinetic energy $(K_1) = K$.
$$\text{Kinetic energy } (K) = \frac{p^2}{2m} \propto p^2$$
or, $\dfrac{K_1}{K_2} = \left(\dfrac{p_1}{p_2}\right)^2 = \left(\dfrac{p}{1.5p}\right)^2 = \dfrac{1}{2.25}$
or, $K_2 = 2.25\,K$.
Therefore, increase in kinetic energy is $2.25\,K - K = 1.25\,K$ or 125%.

24. (d) K.E. $= \dfrac{1}{2}mv^2$
Further, $v^2 = u^2 + 2as = 0 + 2ad = 2ad$
$= 2(F/m)d$
Hence, K.E. $= \dfrac{1}{2}m \times 2(F/m)d = Fd$
or, K.E. acquired = Work done
$= F \times d$ = constant.
i.e., it is independent of mass m.

25. (a) Applying law of conservation of momentum,
$m_1u_1 + m_2u_2 = (m_1 + m_2)v$
or, $m_1u_1 = (m_1 + m_2)v$ $(\because u_2 = 0)$
$\Rightarrow m\frac{(3\times1000)}{3600} = 3m(v)$
$\Rightarrow v = \frac{1000}{3600}$ m/s = 1 km/hr

26. (d) Work done $= \int F dx$

27. (a) mv → ×
Initial momentum $= mv$
Final momentum $= m(-v)$
change in momentum
$= mv - m(-v) = 2mv$

28. (b) According to conservation of momentum
Momentum of tank = Momentum of shell
$125000 \times v_{tank} = 25 \times 1000$
$\Rightarrow v_{tank} = 0.2$ ft/sec.

29. (c) $p = \sqrt{2mE} \therefore \frac{p_1}{p_2} = \sqrt{\frac{m_1}{m_2}\frac{E_1}{E_2}} = \sqrt{\frac{2}{1}\times\frac{8}{1}} = \frac{4}{1}$

30. (b) $E = \frac{p^2}{2m} = \frac{(10)^2}{2\times1} = 50\text{J}$

31. (b) $E = \frac{P^2}{2m} = \frac{4}{2\times3} = \frac{2}{3}\text{J}$

32. (b) $E \propto p^2$ (if m = constant)
Percentage increase in E
= 2 (percentage increase in P)
= 2 × 0.01% = 0.02%

33. (a) $p = Fv = F\times\frac{s}{t} = 40\times\frac{30}{60} = 20\text{ W}$

34. (d) Due to the same mass of A and B as well as due to elastic collision, velocities of spheres get interchanged after the collision.

35. (a) $mgh = \frac{80}{100}\times mg\times100 \Rightarrow h = 80\text{m}$

36. (a) $h_n = he^{2n}$, after third collision $h_3 = he^6$
[as $n = 3$]

37. (c) Initial momentum of the system = mv-mv = 0
As body sticks together
∴ final momentum = $2mV$
By conservation of momentum $2mV = 0$
∴ $V = 0$

38. (c) Work done = $\vec{F}\cdot\vec{r}$
= $(5i + 3j).(2i - j) = 7$ joules.

39. (c) From law of conservation of energy the total energy remains constant throughout the fall. However, as it falls potential energy changes into kinetic, due to which speed increases.

40. (d) Amount of water flowing per second from the pipe
$= \frac{m}{time} = \frac{m}{\ell}.\frac{\ell}{t} = \left(\frac{m}{\ell}\right)v$
Power = K.E. of water flowing per second
$= \frac{1}{2}\left(\frac{m}{\ell}\right)v.v^2 = \frac{1}{2}\left(\frac{m}{\ell}\right)v^3$
$= \frac{1}{2}\times100\times8 = 400\text{ W}$

41. (b) $\frac{1}{2}mv^2 = \frac{1}{2}kx^2 \Rightarrow mv^2 = kx^2$
or $0.5\times(1.5)^2 = 50\times x^2$
∴ $x = 0.15$ m

42. (b)

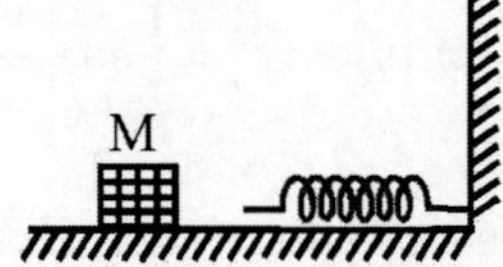

$\frac{1}{2}Mv^2 = \frac{1}{2}kL^2 : v = \sqrt{\frac{k}{M}}.L$

Momentum $= M\times v = M\times\sqrt{\frac{k}{M}}.L$
$= \sqrt{\text{kM}}.\text{L}$

43. (a) From $m_1v_1 + m_2v_2 = 0$,
$v_2 = \frac{-m_1v_1}{m_2} = -\frac{4v}{234}$

44. (b) We know that $F\times v$ = Power
$\therefore F\times v = c$ where c = constant
$\therefore m\frac{dv}{dt}\times v = c$ $\left(\therefore F = ma = \frac{mdv}{dt}\right)$
$\therefore m\int_0^v vdv = c\int_0^t dt$ $\therefore \frac{1}{2}mv^2 = ct$
$\therefore v = \sqrt{\frac{2c}{m}}\times t^{1/2}$

$\therefore \quad \frac{dx}{dt} = \sqrt{\frac{2c}{m}} \times t^{1/2} \quad$ where $v = \frac{dx}{dt}$

$\therefore \int_0^x dx = \sqrt{\frac{2c}{m}} \times \int_0^t t^{1/2} dt$

$x = \sqrt{\frac{2c}{m}} \times \frac{2t^{3/2}}{3} \Rightarrow x \propto t^{3/2}$

45. (c) Given : retardation $\propto$ displacement

i.e., $a = -x$ But $a = v\frac{dv}{dx}$

$\therefore \frac{vdv}{dx} = -x \Rightarrow \int_{v_1}^{v_2} v\, dv = -\int_0^x xdx$

$\left(v_2^2 - v_1^2\right) = -\frac{x^2}{2}$

$\Rightarrow \frac{1}{2}m\left(v_2^2 - v_1^2\right) = \frac{1}{2}m\left(\frac{-x^2}{2}\right)$

$\therefore$ Loss in kinetic energy, $\therefore \Delta K \propto x^2$

46. (a) Work done by such force is always zero since force is acting in a direction perpendicular to velocity.
$\therefore$ from work-energy theorem $= \Delta K = 0$
K remains constant.

47. (b) $\text{K.E.} = \frac{1}{2}mv^2 = \frac{1}{2} \times 0.1 \times 25 = 1.25$ J

$W = -mgh = -\left(\frac{1}{2}mv^2\right) = -1.25$ J

$\left[\because mgh = \frac{1}{2}mv^2 \text{ by energy conservation}\right]$

48. (d) The average speed of the athelete

$v = \frac{100}{10} = 10$ m/s $\quad \therefore \text{K.E.} = \frac{1}{2}mv^2$

If mass is 40 kg then,

$\text{K.E.} = \frac{1}{2} \times 40 \times (10)^2 = 2000$ J

If mass is 100 kg then,

$\text{K.E.} = \frac{1}{2} \times 100 \times (10)^2 = 5000$ J

49. (b) In a conservative field work done does not depend on the path. The gravitational field is a conservative field.
$\therefore \quad W_1 = W_2 = W_3$

50. (b) $F = u\left(\frac{dm}{dt}\right) = 500 \times 1 = 500\text{N}$

51. (c) According to principle of conservation of linear momentum
$1000 \times 50 = 1250 \times v \Rightarrow v = 40$ km/hr

52. (c) $T = \frac{F(L-x)}{L} = \frac{5(5-1)}{5} = 4\text{N}$

53. (d) $m_G v_G = m_B v_B$

$\Rightarrow v_B = \frac{m_G v_G}{m_B} = \frac{1 \times 5}{10 \times 10^{-3}} = 500$ m/s

54. (b) Work done = force × displacement
= Weight of the book × Height of the book shelf

55. (d) As the body moves in the direction of force therefore work done by gravitational force will be positive.
$W = Fs = mgh = 10 \times 9.8 \times 10 = 980$ J

56. (b) Work done on the body = K.E. gained by the body

$Fs\cos\theta = 1 \Rightarrow F\cos\theta = \frac{1}{s} = \frac{1}{0.4} = 2.5\text{N}$

57. (a) $W = \vec{F}.(\vec{r}_2 - \vec{r}_1)$

$= \left(4\hat{i} + \hat{j} + 3\hat{k}\right)\left(11\hat{i} + 11\hat{j} + 15\hat{k}\right)$

$W = 44 + 11 + 45 = 100$ Joule

58 (a) Both part will have numerically equal momentum and lighter part will have more velocity.

59. (a) As surface is smooth so work done against friction is zero. Also the displacement and force of gravity are perpendicular so work done against gravity is zero.

60. (a) $U = \frac{F^2}{2k} = \frac{T^2}{2k}$

6. SYSTEM OF PARTICLES & ROTATIONAL MOTION

ANSWER KEY											
1	(a)	11	(d)	21	(a)	31	(b)	41	(d)	51	(d)
2	(b)	12	(d)	22	(b)	32	(c)	42	(a)	52	(a)
3	(c)	13	(b)	23	(b)	33	(d)	43	(a)	53	(a)
4	(d)	14	(c)	24	(d)	34	(a)	44	(a)	54	(c)
5	(c)	15	(c)	25	(c)	35	(c)	45	(a)	55	(a)
6	(b)	16	(d)	26	(b)	36	(b)	46	(c)	56	(c)
7	(c)	17	(c)	27	(c)	37	(b)	47	(a)	57	(b)
8	(d)	18	(b)	28	(a)	38	(d)	48	(b)	58	(b)
9	(b)	19	(d)	29	(d)	39	(d)	49	(c)	59	(a)
10	(d)	20	(b)	30	(a)	40	(a)	50	(b)	60	(c)

Hints & Solutions

1. (a) Centre of mass of a two particle system always lies closer to the heavier particle.
2. (b) COM will lie on diagonal AC, i.e. COM will be equidistant from B and D if $m_2 = m_4$.
3. (c) If COM lies the origin x-coordinate of all the particles may be zero.
4. (d) By conservation of momentum, $p_\alpha = p_n$

$$\sqrt{2mk\alpha} = \sqrt{2m_n k_n} \quad \text{or} \quad \frac{k_n}{k_\alpha} = \frac{m_2}{m_n}$$

5. (c) Distance of COM from centre of sphere A,

$$r_1 = \frac{m_B r}{m_A + m_B} = \frac{2m \times 3R}{m + 2m} = 2R$$

6. (b) As $m_1 > m_2$, with increase in mass of m_1 acceleration of each block, hence that of COM increases.
7. (c) Moment of inertial depends on distribution of mass and the axis of rotation.
8. (d) $\tau = \frac{dL}{dt}$ $\therefore$ If L = constant, $\tau = 0$
9. (b) Situation is shown in the figure.

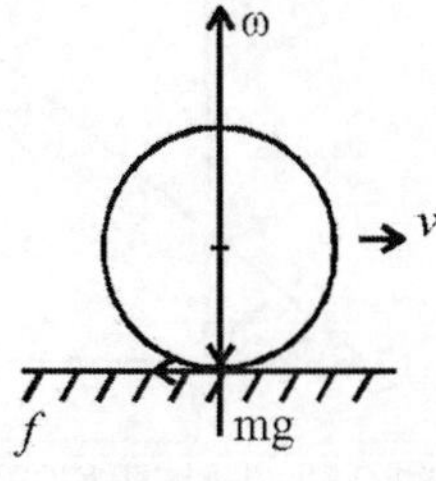

Here line of action of all the forces passes through the point of contact hence, net force about point of contact is zero. So, the angular momentum about point of contact is conserved.

10. (d) On smooth inclined plane, the speed of body will change due to gravity but there is not torque to change angular velocity of the body. Hence a body cannot roll without slipping on a smooth inclined surface.
11. (d) The equation $\tau = I\alpha$ is valid about COM or axis of rotation but not about any other point.
12. (d) No torque is acting on either object about point of contact, hence angular momentum of the ring as well as the particle and hence that of the system is conserved about point of contact.
13. (b) $T_1 r_1 = T_2 r_2$ but $r_1 < r_2$ $\therefore$ $T_1 > T_2$

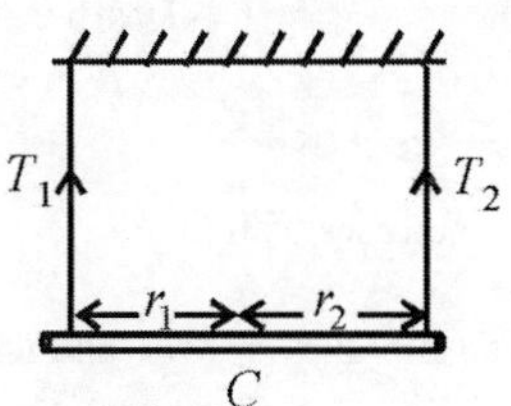

14. (c) For no slipping.

$$\omega_1 R_1 = \omega_2 R_2 \quad \text{or} \quad \frac{\omega_1}{\omega_2} = \frac{R_2}{R_1}$$

but $R_1 > R_2$ $\therefore$ $\omega_1 < \omega_2$

15. (c) If the K.E. of the system just after the collision is sufficient to rise the block as shown, the block will topple, otherwise it will not topple down.

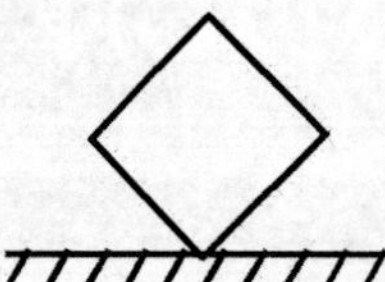

16. (d) The forces are not concurrent, hence a finite torque will act about any of the point and hence angular momentum is not conserved about any point.

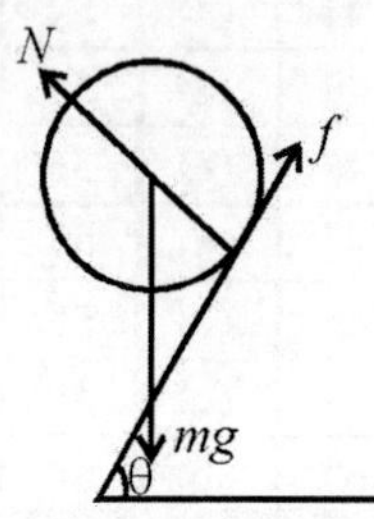

17. (c) Speeds of centre of wheels act hence that of topmost points must be equal.

18. (b) When the person stretches his hands, his moment of inertia increase, hence by conservation of angular momentum, its angular speed decreases.

19. (d) Direction of angular momentum is perpendicular to the orbital plane.

20. (b) At any instant, angular momentum about point of projection,

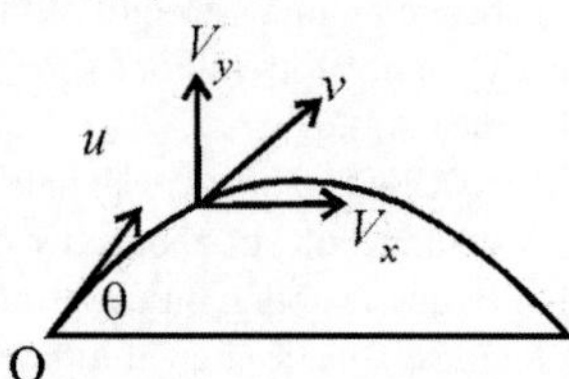

$$L = v_x y - V_y x = (v\cos\theta)\left(u\sin gt - \frac{1}{2}gt^2\right)$$
$$-(u\sin\theta - gt)(n\cos\theta t)$$

$$= \frac{1}{2}gu^2\cos\theta t^2$$

$\therefore$ L increases with time.

21. (a) Kinetic energy $= \frac{1}{2}I\omega^2$ and for ring

$I = mr^2$

Hence, K.E. $= \frac{1}{2}mr^2\omega^2$

22. (b) Angular momentum about the point of contact with the surface includes the angular momentum about the centre. Due to friction linear momentum is conserved.

23. (b) $\text{K.E.} = \frac{1}{2}I\omega^2 + \frac{1}{2}mv^2$

$$\text{K.E.} = \frac{1}{2}\left(\frac{1}{2}mr^2\right)\omega^2 + \frac{1}{2}mv^2$$

$$= \frac{1}{4}mv^2 + \frac{1}{2}mv^2 = \frac{3}{4}mv^2$$

Now, gain in K.E. = Loss in P.E.

$$\frac{3}{4}mv^2 = mgh \Rightarrow v = \sqrt{\left(\frac{4}{3}\right)gh}$$

24. (d) Initially centre of gravity is at the centre. When sand is poured it will fall and again after a limit, centre of gravity will rise.

25. (c) M.I. of uniform circular disc about its diameter $= I$

According to theorem of perpendicular axes,

M.I. of disc about its axis $= \frac{1}{2}mr^2 = 2I$

Applying theorem of || axes,

$$\left(\because I = \frac{1}{4}mr^2\right)$$

M.I. of disc about the given axis
$= 2I + mr^2 = 2I + 4I = 6I$

26. (b) $I = 1.2$ kg m^2, $E_r = 1500$ J,
$\alpha = 25$ rad/sec^2, $\omega_1 = 0$, $t = ?$

As $E_r = \frac{1}{2}I\omega^2$, $\omega = \sqrt{\frac{2E_r}{I}}$

$$= \sqrt{\frac{2\times1500}{1.2}} = 50 \text{ rad/sec}$$

From $\omega_2 = \omega_1 + \alpha t$
$50 = 0 + 25t$, $t = 2$ seconds

27. (c) $E_r = \frac{1}{2}I\omega^2$

$$I = \frac{2E_r}{\omega^2} = \frac{2\times360}{30\times30} = 0.8 \text{ kg m}^2$$

28. (a)

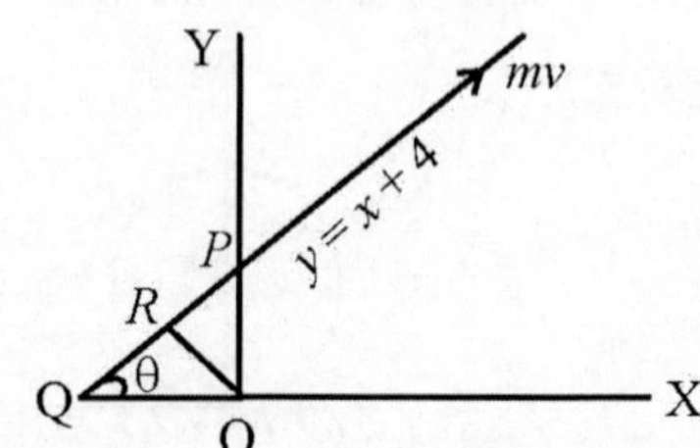

$y = x + 4$ line has been shown in the figure when $x = 0, y = 4$. So, $OP = 4$.

The slope of the line can be obtained by comparing with the equation of the straight line

$y = mx + c$

$m = \tan\theta = 1$

$\Rightarrow \quad \theta = 45°$

$\angle OQP = \angle OPQ = 45°$

If we draw a line perpendicular to this line, length of the perpendicular = OR

In ΔOPR, $\dfrac{OR}{OP} = \sin 45°$

$\Rightarrow OR = OP \sin 45°$

$= 4 \times \dfrac{1}{\sqrt{2}} = \dfrac{4}{\sqrt{2}} = 2\sqrt{2}$

Angular momentum of particle going along this line

$= r \times mv = 2\sqrt{2} \times 5 \times 3\sqrt{2} = 60$ units

29. (d) $E = E_t + E_r = \dfrac{1}{2}mv^2 + \dfrac{1}{2}I\omega^2$

$= \dfrac{1}{2}mv^2 + \dfrac{1}{2} \times \left(\dfrac{2}{5}mr^2\right)\omega^2$

$= \dfrac{1}{2}mv^2 + \dfrac{1}{5}mv^2 = \dfrac{7}{10}mv^2$

$\therefore \dfrac{E_t}{E} = \dfrac{\frac{1}{2}mv^2}{\frac{7}{10}mv^2} = \dfrac{5}{7}$

30. (a) (12 amu) Ⓒ ← 1.12×10^{-10} m → Ⓞ (16 amu)

c.m.

← d → ← $1.12\times10^{-10} - d$ →

From definition of centre of mass.

$d = \dfrac{16 \times 1.12 \times 10^{-10} + 12 \times 0}{16 + 12}$

$= \dfrac{16 \times 1.12 \times 10^{-10}}{28} = 0.64 \times 10^{-10}$ m.

31. (b) By the principle of conservation of angular momentum, $I_1\omega_1 = I_2\omega_2$

$MR^2\omega = (M + 2m)R^2\omega'$

$\omega' = \dfrac{M\omega}{(M + 2m)}$

32. (c) Perpendicular distance of each line of action of force from the centre is same, say x. Now taking moments about O,

$F_1 \times x + F_2 \times x - F_3 \times x = 0$

$\therefore F_3 = F_1 + F_2$

33. (d) AB is the ladder, let F be the horizontal force and W is the weigth of man. Let N_1 and N_2 be normal reactions of ground and wall, respectively. Then for vertical equilibrium

$W = N_1$ ..... (1)

For horizontal equilibrium, $N_2 = F$ ..... (2)

Taking moments about A,

$N_2(AB \sin 60°) - W(AC \cos 60°) = 0$ ..... (3)

Using (2) and AB = 20 ft, BC = 4 ft, we get

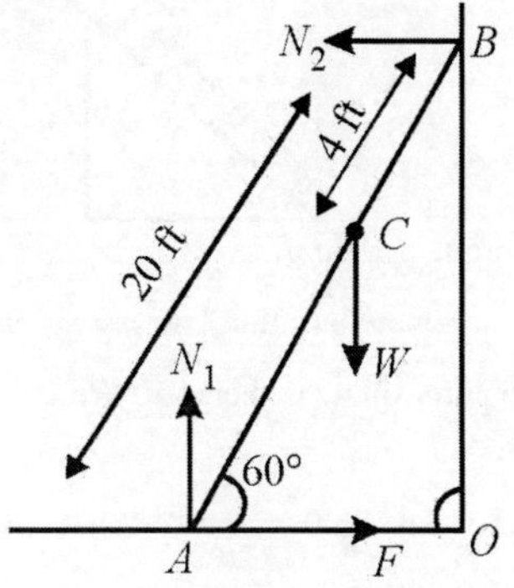

$F\left(20 \times \dfrac{\sqrt{3}}{2}\right) - W\left(16 \times \dfrac{1}{2}\right) = 0$

$\Rightarrow F = \dfrac{8W \times 2}{20\sqrt{3}} = \dfrac{4W}{5\sqrt{3}} = \dfrac{150 \times 4}{5\sqrt{3}}$ pound

$= 40\sqrt{3} = 40 \times 1.73 = 69.2$ pound

34. (a) As time taken by both car to complete one revolution is same.

As $\omega = \dfrac{2\pi}{T} \Rightarrow \omega \propto \dfrac{1}{T}$, as T is same in both cases. Hence 'ω' will also be same.

35. (c) Moment of inertia of disc about its diameter is

$I_d = \dfrac{1}{4}MR^2$

MI of disc about a tangent passing through rim and in the plane of disc is

$I = I_d + MR^2 = \dfrac{1}{4}MR^2 + MR^2 = \dfrac{5}{4}MR^2$

36. (b)

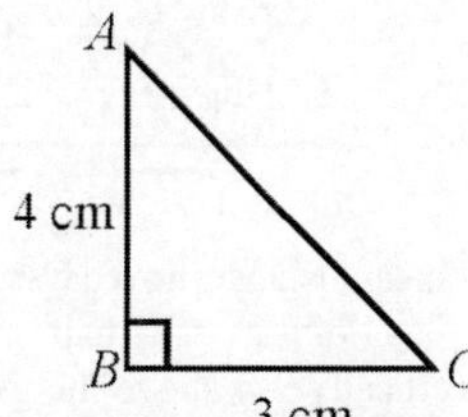

Moment of Inertia depend upon mass and distribution of masses as $I = \Sigma mr^2$.
Further, as the distance of masses is more, more is the moment of Inertia.
If we choose BC as axis. Distance is maximum. Hence, Moment of Inertia is maximum.

$\therefore \quad I_2 > I_1, I_2 > I_3$

37. (b)

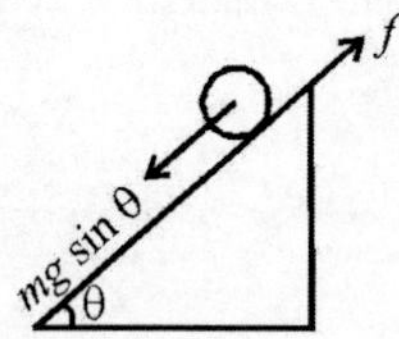

Torque, $I\alpha = f.R$.

Using Newton's IInd law, $mg \sin\theta - f = ma$

$\because$ pure rolling is there, $a = R\alpha$

$$\Rightarrow \quad mg \sin\theta - \frac{I\alpha}{R} = ma$$

$$\Rightarrow \quad mg \sin\theta - \frac{Ia}{R^2} = ma \qquad \left(\because \alpha = \frac{a}{R}\right)$$

or, acceleration, $a = \dfrac{mg\sin\theta}{(I/R^2 + m)}$

Using, $s = ut + \dfrac{1}{2}at^2$

or, $s = \dfrac{1}{2}at^2 \Rightarrow t \propto \dfrac{1}{\sqrt{a}}$

t minimum means a should be more. This is possible when I is minimum which is the case for solid cylinder. Therefore, solid cylinder will reach the bottom first.

38. (d) $\tau = 1000$ N - m, $I = 200$ kg-m^2

$\therefore I\alpha = 1000$

$\Rightarrow \alpha = \dfrac{1000}{200} = 5$ rad/sec^2

$\omega = \omega_0 + \alpha t = 0 + 3 \times 5 = 15$ rad/s

39. (d)

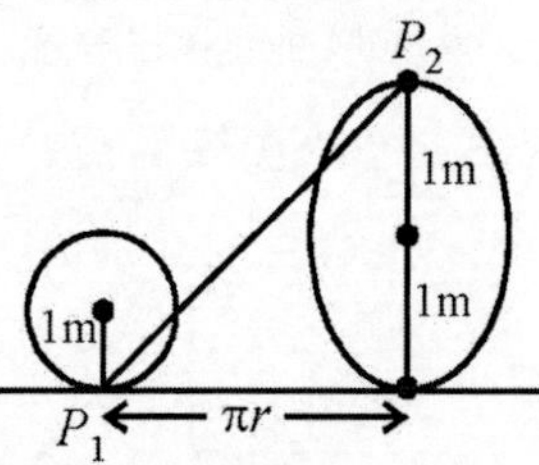

Linear distance moved by wheel in half revolution $= \pi r$. Point P_1 after half revolution reaches at P_2 vertically 2m above the ground.

$\therefore$ Displacement P_1P_2

$$= \sqrt{\pi^2 r^2 + 2^2} = \sqrt{\pi^2 + 4}$$

$$[\because r = 1m]$$

40. (a) M.I. of system about the axis which passing through m_1

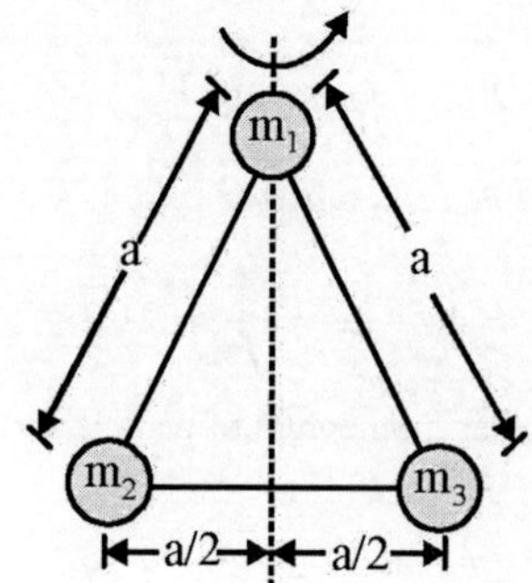

$$I_{system} = m_1(0)^2 + m_2\left(\frac{a}{2}\right)^2 + m_3\left(\frac{a}{2}\right)^2$$

$$I_{system} = (m_2 + m_3)\frac{a^2}{4}$$

41. (d)

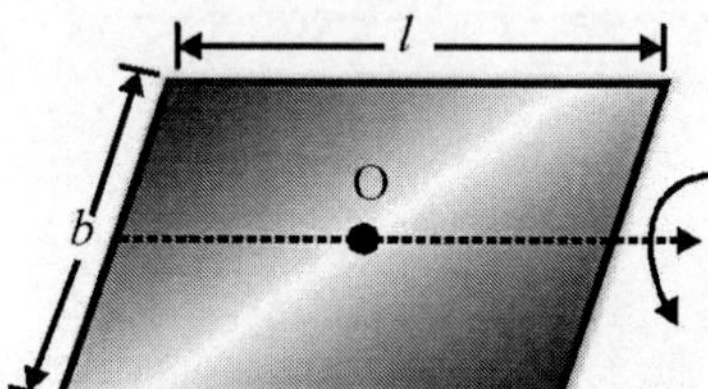

M.I. of plate about O and parallel to length

$$= \frac{Mb^2}{12}$$

42. (a)

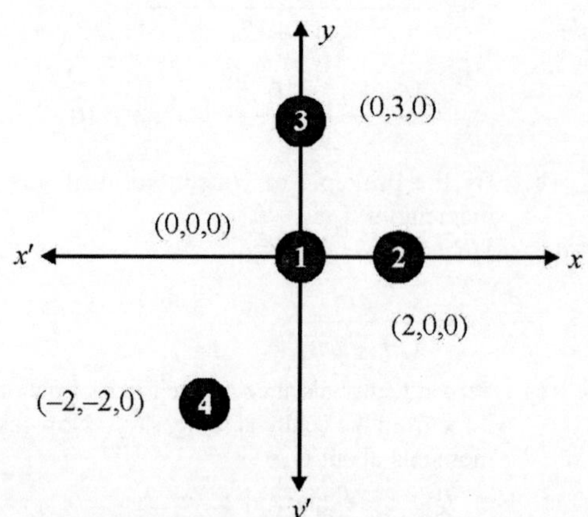

$I_1 = I_2 = 0$, because these particles are placed on x-axis

The M.I. of system about x-axis,

$= I_1 + I_2 + I_3 + I_4$

$= 0 + 0 + 3 \times (3)^2 + 4 \times (-2)^2$

$= 27 + 16 = 43$ kg – m^2

43. (a) $\alpha = \dfrac{2\pi(n_2 - n_1)}{t}$

$= \dfrac{2\pi(0-20)}{10} = -4\pi \text{ rad/sec}^2$

$\tau = I\alpha = 5 \times 10^{-3} \times 4\pi$

$= 20\pi \times 10^{-3} = 2\pi \times 10^{-2}$ N-m

44. (a) $\alpha = \dfrac{\tau}{I} = \dfrac{20}{2} = 15 \text{ rad/s}^2$

$\therefore \theta = \omega_0 t + \dfrac{1}{2}\alpha t^2$

$= 0 + \dfrac{1}{2} \times (15) \times (10)^2 = 750 \text{ rad}$

45. (a) Angular momentum of system remains constant

$I \propto \dfrac{1}{\omega} \Rightarrow \dfrac{I_2}{I_1} = \dfrac{\omega_1}{\omega_2} = \dfrac{20}{10} \Rightarrow I_2 = 2I_1 = 2I$

46. (c) $E = \dfrac{L^2}{2I} \Rightarrow L = \sqrt{2EI}$

47. (a) When angular acceleration (α) is zero then torque on the wheel becomes zero.

$\theta(t) = 2t^3 - 6t^2$

$\Rightarrow \dfrac{d\theta}{dt} = 6t^2 - 12t$

$\Rightarrow \alpha = \dfrac{d^2\theta}{dt^2} = 12t - 12 = 0$

$\therefore \quad t = 1$ sec.

48. (b) By theorem of parallel axes,

$I = I_{cm} + Md^2$

$I = I_0 + M(L/2)^2 = I_0 + ML^2/4$

49. (c) K.E. $= \dfrac{L^2}{2I}$

The angular momentum L remains conserved about the centre.

That is, L = constant.

$I = mr^2$

$\therefore$ K.E. $= \dfrac{L^2}{2mr^2}$

In 2nd case,

K.E. $= \dfrac{L^2}{2(mr'^2)}$

But $r' = \dfrac{r}{2}$

$\therefore K.E' = \dfrac{L^2}{2m.\dfrac{r^2}{4}} = \dfrac{4L^2}{2mr^2}$

$\Rightarrow$ K.E.$'$ = 4 K.E.

$\therefore$ K.E. is increased by a factor of 4.

50. (b) If no external force actson a system of particles, the centre of mass remains at rest. So, speed of centre of mass is zero.

51. (d) By conservation of angular momentum, $I_t\omega_i = (I_t + I_b)\,\omega_f$

where ω_f is the final angular velocity of disks

$\therefore \quad \omega_f = \left(\dfrac{I_t}{I_t + I_b}\right)\omega_i$

Loss in K.E., ΔK = Initial K.E. – Final K.E.

$= \dfrac{1}{2}I_t\omega_i^2 - \dfrac{1}{2}(I_t + I_b)\omega_f^2$

$\Rightarrow \Delta K = \dfrac{1}{2}I_t\omega_i^2 - \dfrac{1}{2}(I_t + I_b)\dfrac{I_t^2}{(I_t + I_b)^2}\omega_i^2$

$= \dfrac{1}{2}\omega_i^2 \dfrac{I_t}{I_t + I_b}(I_t + I_b - I_t)$

$= \dfrac{1}{2}\omega_i^2 \dfrac{I_t I_b}{I_t + I_b}$

52. (a) P.E. = total K.E.

$mgh = \dfrac{7}{10}mv^2, v = \sqrt{\dfrac{10gh}{7}}$

53. (a) For solid sphere, $\dfrac{K^2}{R^2} = \dfrac{2}{5}$

For disc and solid cylinder, $\dfrac{K^2}{R^2} = \dfrac{1}{2}$

As $\dfrac{K^2}{R^2}$ for solid sphere is smallest, it takes minimum time to reach the bottom of the incline.

54. (c) Angular speed, $\omega = \dfrac{120 \times 2\pi}{60} = 4\pi$ rad/sec

55. (a) Rotational energy

$$= \frac{1}{2}I(\omega)^2 = \frac{1}{2}(mK^2)\omega^2$$

Linear kinetic energy $= \frac{1}{2}m\omega^2R^2$

$\therefore$ Required fraction

$$= \frac{\frac{1}{2}(mK^2)\omega^2}{\frac{1}{2}(mK^2)\omega^2 + \frac{1}{2}m\omega^2R^2} = \frac{K^2}{K^2+R^2}$$

$$= \frac{\frac{2}{5}R^2}{\frac{2}{5}R^2 + R^2} = \frac{2}{7}$$

($\because$ For a solid sphere, $K^2 = \frac{2}{5}R^2$)

56. (c) $\vec{F} = -3\hat{i} + \hat{j} + 5\hat{k}$

$\vec{r} = 7\hat{i} + 3\hat{j} + \hat{k}$

Torque, $\vec{\tau} = \vec{r} \times \vec{F}$

$$= \left(-3\hat{i} + \hat{j} + 5\hat{k}\right) \times \left(7\hat{i} + 3\hat{j} + \hat{k}\right)$$

$$= \begin{vmatrix} \hat{i} & \hat{j} & \hat{k} \\ -3 & 1 & 5 \\ 7 & 3 & 1 \end{vmatrix}$$

$$= \hat{i}(1-15) - \hat{j}(-3-35) + \hat{k}(-9-7)$$

$$= -14\hat{i} + 38\hat{j} - 16\hat{k}$$

57. (b) $\vec{v}_{cm} = \frac{m_1\vec{v}_1 + m_2\vec{v}_2}{m_1 + m_2}$

$$= \frac{2\times 2 + 4\times 10}{2+4} = 7.3 \text{ m/s}$$

58. (b) According to figure let A is the origin and co-ordinates of centre of mass be (x, y), then,

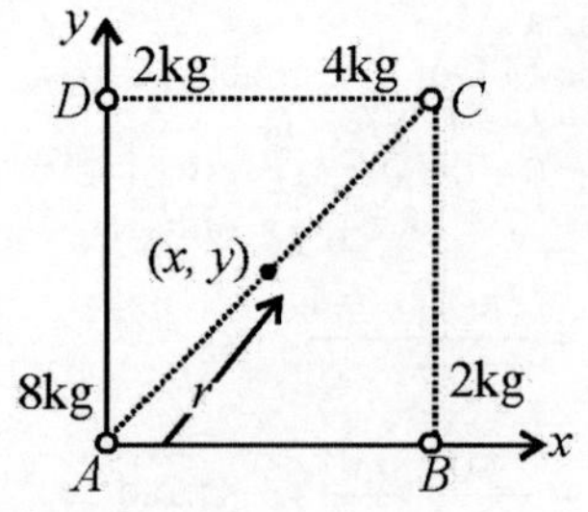

$$x = \frac{m_1x_1 + m_2x_2 + m_3x_3 + m_4x_4}{m_1 + m_2 + m_3 + m_4}$$

$$= \frac{0 + 2\times\frac{80}{\sqrt{2}} + 4\times\frac{80}{\sqrt{2}} + 0}{16} = \frac{30}{\sqrt{2}}$$

Similarly, $y = \frac{30}{\sqrt{2}}$ so, $r = \sqrt{x^2 + y^2} = 30$ cm

59. (a) As initially both the particles were at rest therefore velocity of centre of mass was zero and there is no external force on the system so speed of centre of mass remains constant i.e. it should be equal to zero.

60. (c) We can assume that three particles of equal mass m are placed at the corners of triangle

$\vec{r}_1 = 0\hat{i} + 0\hat{j}, \vec{r}_2 = b\hat{i} + 0\hat{j}$

and $\vec{r}_3 = 0\hat{i} + h\hat{j}$

$$\therefore \overrightarrow{r_{cm}} = \frac{m_1\vec{r}_1 + m_2\vec{r}_2 + m_3\vec{r}_3}{m_1 + m_2 + m_3} = \frac{b}{3}\hat{i} + \frac{h}{3}\hat{j}$$

i.e. coordinates of centre of mass is $\left(\frac{b}{3}, \frac{h}{3}\right)$

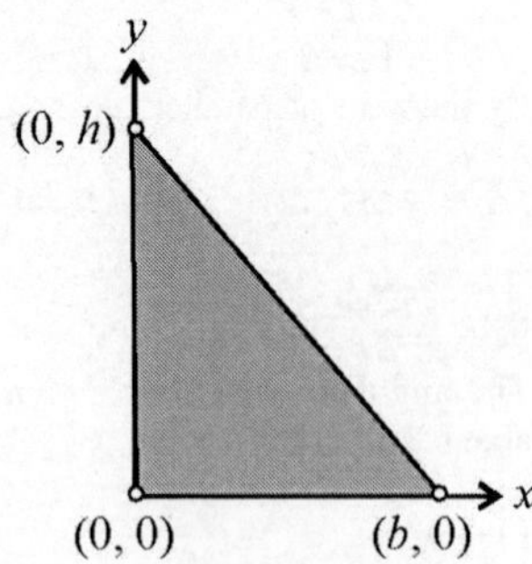

7. GRAVITATION

ANSWER KEY											
1	(c)	11	(d)	21	(d)	31	(a)	41	(a)	51	(b)
2	(a)	12	(c)	22	(d)	32	(d)	42	(d)	52	(a)
3	(d)	13	(a)	23	(d)	33	(b)	43	(b)	53	(a)
4	(c)	14	(c)	24	(c)	34	(c)	44	(c)	54	(a)
5	(d)	15	(b)	25	(c)	35	(c)	45	(c)	55	(c)
6	(b)	16	(c)	26	(a)	36	(b)	46	(a)	56	(d)
7	(c)	17	(b)	27	(c)	37	(a)	47	(d)	57	(a)
8	(c)	18	(b)	28	(c)	38	(c)	48	(b)	58	(c)
9	(d)	19	(b)	29	(b)	39	(d)	49	(a)	59	(c)
10	(a)	20	(b)	30	(d)	40	(d)	50	(b)	60	(c)

Hints & Solutions

1. (c) It is applicable to both small & big bodies.
2. (a) The weight (= mg) of the body at the centre of the earth is zero, because the value of g at centre is zero.
3. (d) Tides occurs due to gravitational pull of moon on the earth & they have regular character. On new moon & full moon high tides occur.
4. (c) The orbital velocity of moon is

$$v_{0m} = \sqrt{\frac{GM_e}{r}} \quad ....(i)$$

So kinetic energy of moon is $K = ½\, M_m\, v_{0m}^2$

$$\frac{GM_e M_m}{2r} \quad ....(ii)$$

$$U = -\frac{GM_e M_m}{r} \quad ....(iii)$$

where r is distance between the centres of earth & moon.

It is clear from (ii) & (iii) that $U>K$ (in magnitudes term)

5. (d)

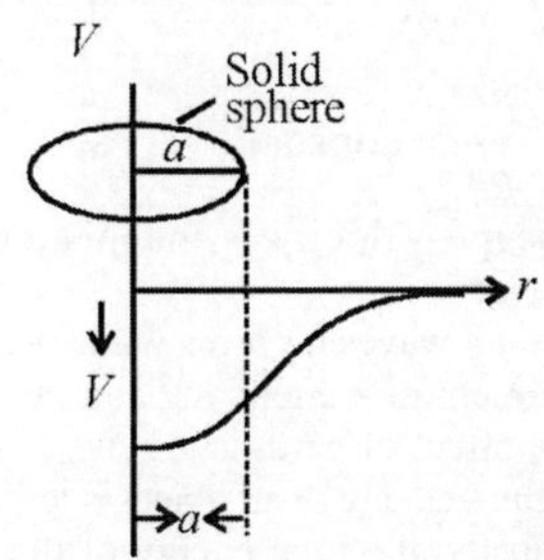

(i) For inside the (r < a) solid sphere,

$$V = \frac{3GM}{2a}\left(1 - \frac{r^2}{3a^2}\right)$$

(ii) At surface ($r = a$) of solid sphere,

$$V = \frac{-GM}{a}$$

(iii) Outside the solid sphere (r > a),

$$V = \frac{-GM}{r}$$

(B)

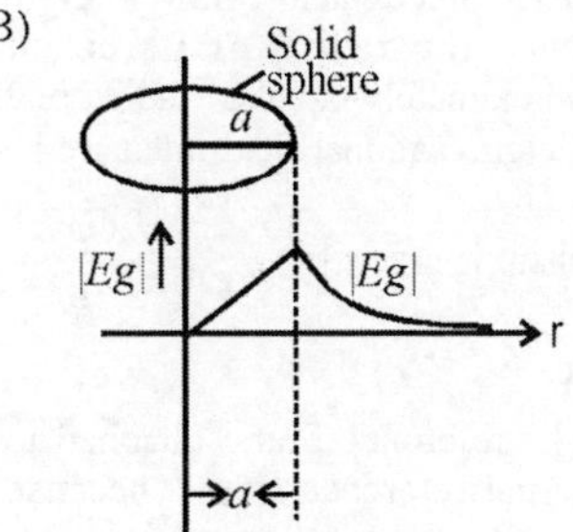

(i) $Eg = \frac{-GM}{R^3} r$ for inside the sphere ($r < a$)

(ii) At surface ($r = a$) of solid sphere,

$$Eg = \frac{-GM}{R^2}$$, where R is radius of solid sphere (in generalised way)

(iii) Outside the solid sphere $Eg = \frac{-GM}{r^2}$ ($r > a$)

6. (b) Since areas are equal, when a planet revolves around the sun, areal velocity is constant ie,

$$\frac{d\vec{A}}{dt} = \text{constant} \quad \text{or} \quad t_1 = t_2$$

7. (c) Geo-stationary satellites are also called synchronous satellite. They always remain about the same path on equater, i.e., it has a period of exactly one day (86400 sec)

So orbit radius $\left(T = 2\pi\sqrt{\frac{r^3}{GM}}\right)$ comes out to be 42400 km, which is nearly equal to the circumterence of eath. So height of Geo-stationary satellite from the earth surface is 42,400 – 6400 = 36,000 km.

8. (c) Weightlessness means that there is no reaction on a body from the floor. Since both the artificial satellite & the astronaut have same centripetal acceleration (as in a lift; which is falling freely, we does not feel any weight, because both lift & we fall with same acceleration). So the astronaut does not feel any weight inside the space craft.

9. (d) There is no atmosphere at moon, because escape velocity is less than the root mean square velocity of the molecules at moon. Hence all molecules escape.

12. (c) The orbital velocity of satellite moving in circular orbit near a planet is $v_0 = \sqrt{\frac{GM}{R}}$

so period $T = \frac{2\pi R}{v_0} = 2\pi R\sqrt{\frac{R}{GM}} = 2\pi\sqrt{\frac{R^3}{GM}}$

Where R & M are the radius & mass of that planet. Since satellite moves very near to planet, hence the radius of its circular orbit is approximately equal to radius of the planet.

13. (a) The gravitational field intensity inside the solid sphere is given by $E_g = -\frac{GMr}{R^3}$.

i.e, $E_g \propto r$

14. (c) The reference frame attached to earth is not inertial reference frame, because the earth is rotating about its own axis.

15. (b) Since orbital velocity of satellite is

$v = \sqrt{\frac{GM}{r}}$, it does not depend upon the mass of the satellite.

Therefore, both satellites will move with same speed.

16. (c) $K.E. = \frac{1}{2}mv_e^2$ where v_e = escape velocity $= \sqrt{2gR}$

$$\therefore K.E. = \frac{1}{2}m \times 2gR = mgR$$

17. (b) Due to inertia of motion it will move tangentially to the original orbit in the same velocity.

18. (b) Orbital speed of a satellite is given by

$$v_0 = \sqrt{\frac{GM_e}{r}}$$

$\Rightarrow$ v_0 is independent of mass of satellite

$$\therefore \quad \frac{v_{0_1}}{v_{0_2}} = \sqrt{\frac{r_2}{r_1}} = \sqrt{\frac{3r}{r}} = \sqrt{3}:1$$

19. (b) Gravitational force of the sun provides necessary centripetal force to the earth.

20. (b) $V \alpha \frac{1}{\sqrt{r}}$

where r is distance from the sun. As distance of Jupiter from the sun is more than that of the earth, its orbital speed is less than that of the earth.

21. (d) Kepler's second law is based on the principle of conservation of angular momentum.

22. (d) If K.E. of the satellite is doubled, its speed will become sufficient to escape out of the gravitational field of the earth.

23. (d) Escape velocity from the surface of the moon is much less than the rms velocity of gas molecules, hence there is no atmosphere on the moon.

24. (c) Escape velocity is independent of mass.

25. (c) The ball will have velocity equal to that of the satellite hence will revolve around the earth in the same orbit.

26. (a) Speed of the planet is maximum when it is nearest to the sun.

27. (c) If the radius of the earth is reduced, its moment of inertia will decrease, hence by conservation of angular momentum, angular speed of the earth will increase causing a decrease in the time period of its rotation.

28. (c) When the satellite revolves close to the surface of the planet,

$$V_o = \frac{V_e}{\sqrt{2}} = \frac{2}{\sqrt{2}} = \sqrt{2}\text{ km/s.}$$

29. (b) The value of acceleration due to gravity is more at poles than at equator.

30. (d) $T^2 \propto R^3$

$$\therefore \frac{T^2}{R^3} = \text{constant}$$

31. (a) Escape velocity is independent of angle of projection.

32. (d) A man waves his arms while walking so that the moment of weights remains zero i.e. to balance the effect of earth is gravity.

33. (b) If the velocity of an object is less than the escape velocity, the total energy of the object must be negative.

34. (c) $T^2 \propto R^3$ (Kepler's law)

$$\frac{T_1^2}{T_2^2} = \left(\frac{10^{13}}{10^{12}}\right)^3 \Rightarrow \frac{T_1}{T_2} = 10\sqrt{10}$$

35. (c) Applying the properties of ellipse, we have

$$\frac{2}{R} = \frac{1}{r_1} + \frac{1}{r_2} = \frac{r_1 + r_2}{r_1 r_2}$$

Instant position of satellite

Sun

R

major axis

r_1 r_2

$$R = \frac{2r_1 r_2}{r_1 + r_2}$$

36. (b) $F = \frac{k}{R} = \frac{Mv^2}{R}$. Hence $v \propto R^0$

37. (a) Since, escape velocity $\left(v_e = \sqrt{2gR_e}\right)$ is independent of angle of projection, so it will not change.

38. (c) In a circular or elliptical orbital motion, torque is always acting parallel to displacement or velocity. So, angular momentum is conserved. In attractive field, potential energy is negative. Kinetic energy changes as velocity increase when distance is less. So, option (c) is correct.

39. (d) Total energy = – K.E. = $\frac{\text{P.E.}}{2}$

K.E. = $-\frac{1}{2}mv^2$

40. (d) $T = 2\pi\sqrt{l/g}$. When the rocket accelerates upwards g increases to $(g + a)$.

41. (a) $T = \frac{2\pi r}{v_0} = \frac{2\pi r}{(gR^2/r)^{1/2}} = \frac{2\pi r^{3/2}}{\sqrt{gR^2}} = \frac{2\pi}{\omega}$

Hence, $r^{3/2} = \frac{\sqrt{gR^2}}{\omega}$ or $r^3 = \frac{gR^2}{\omega^2}$

or, $r = (gR^2/\omega^2)^{1/3}$

42. (d) Time period does not depend on the mass. Also, $T^2 \propto r^3$.

43. (b) Escape velocity does not depend on the angle of projection.

44. (c) $T^2 \propto R^3$ (According to Kepler's law)

$T_1^2 \propto (10^{13})^3$ and $T_2^2 \propto (10^{12})^3$

$$\therefore \frac{T_1^2}{T_2^2} = (10)^3 \text{ or } \frac{T_1}{T_2} = 10\sqrt{10}$$

45. (c) $F = \left(\frac{dM}{dt}\right)v = \alpha v^2 \qquad \left(\because \frac{dM}{dt} = \alpha v\right)$

$\therefore$ Retardation $= \frac{-F}{M} = -\frac{\alpha v^2}{M}$

46. (a) Escape velocity on surface of earth

$$(v_e) = \sqrt{\frac{2GM_e}{R_e}} \propto \sqrt{\frac{M_e}{R_e}}.$$

$$\therefore \frac{v_e}{v_P} = \sqrt{\frac{M_e}{R_e}} \times \sqrt{\frac{R_P}{M_P}}$$

$$= \sqrt{\frac{M_e}{R_e}} \times \sqrt{\frac{3R_e}{3M_e}} = \frac{1}{1} = 1$$

or, $v_P = v_e$.

47. (d) We know that $mg = \frac{GMm}{R^2}$

$$\therefore g = \frac{GM}{R^2} \Rightarrow M = \frac{gR^2}{G}$$

48. (b) The orbital speed of satellite is independent of mass of satellite, so the ball will behave as a satellite and will continue to move with the same speed in the original orbit.

49. (a) Let T_A and T_B be time period of A and B about sun.

$T_A = 8T_B$

$\frac{T_A}{T_B} = 8$ ...(1)

According to Kepler's Law $T^2 \propto r^3$

$$\frac{T_A}{T_B} = \frac{(r_A)^3}{(r_B)^3} \Rightarrow \left(\frac{r_A}{r_B}\right)^3 = 8 \Rightarrow \frac{r_A}{r_B} = 2$$

50. (b) Escape velcocity

$$v_e = \sqrt{\frac{2GM_e}{R_e}},\ v'_e = \sqrt{\frac{2GM'_e}{R'_e}}$$

$$\therefore \frac{v'_e}{v_e} = \sqrt{\frac{M'_e}{M_e} \times \frac{R_e}{R'_e}}$$

Given $M'_e = 2M_e$ and $R'_e = \frac{R_e}{2}$

$$\therefore \frac{v'_e}{v_e} = \sqrt{\frac{2M_e}{M_e} \times \frac{R_e}{R_e/2}} = \sqrt{4} = 2$$

$v'_e = 2v_e = 2 \times 11.2 = 22.4$ km/s

51. (b) Escape velocity is the minimum velocity with which a body is projected to escape from earth's gravitational field

$$\frac{1}{2}mv_e^2 = \frac{GMm}{R_e} \Rightarrow v_e = \sqrt{\frac{2GM}{R_e}}$$

52. (a) $v_{\text{earth}} = \sqrt{\frac{2GM_e}{R_e}}$

$$v_{\text{planet}} = \sqrt{\frac{2GM_p}{R_p}} = \sqrt{\frac{2GM_e}{R_e/4}} = \sqrt{\frac{8GM_e}{R_e}}$$

$$\frac{v_{\text{planet}}}{v_{\text{earth}}} = \sqrt{\frac{8GM_e}{R_e}} \times \sqrt{\frac{R_e}{2GM_e}} = 2$$

$$\therefore v_{\text{planet}} = 2 \times v_{\text{earth}} = 2 \times 11.2 = 22.4 \text{ km/s}$$

53. (a) $mg = 72$ N (body weight on the surface)

$$g = \frac{GM}{R^2}$$

At a height $H = \frac{R}{2}$,

$$g' = \frac{GM}{\left(R + \frac{R}{2}\right)^2} = \frac{4\ GM}{9\ R^2}$$

Body weight at height $H = \frac{R}{2}$,

$$mg' = m \times \frac{4\ GM}{9\ R^2}$$

$$= m \times \frac{4}{9} \times g = \frac{4}{9}mg$$

$$= \frac{4}{9} \times 72 = 32 \text{ N}$$

54. (a) We know that effective gravity g' at depth below earth surface is given by

$$g' = g\left(1 - \frac{d}{R}\right)$$

Here, $d = 100$ km, $R = 6400$ km,

$$\therefore g' = 9.8\left(1 - \frac{100}{6400}\right) = 9.65 \text{ m/s}^2$$

55. (c) Mass of the satellite = m and height of satellite from earth $(h) = 6.4 \times 10^6$ m.

We know that gravitational potential energy of the satellite at height

$$h = -\frac{GM_e m}{R_e + h} = -\frac{gR_e^2 m}{2R_e}$$

$$= -\frac{gR_e m}{2} = -0.5\, mgR_e$$

(where, $GM_e = gR_e^2$ and $h = R_e$)

56. (d) Energy required = (Potential energy of the Earth -mass system when mass is at distance $3R$) – (Potential energy of the Earth -mass system when mass is at distance $2R$)

$$= \frac{-GMm}{3R} - \left(\frac{-GMm}{2R}\right) = \frac{-GMm}{3R} + \frac{GMm}{2R}$$

$$= \frac{-2GMm + 3GMm}{6R} = \frac{GMm}{6R}$$

57. (a) Escape velocity, $v_e = \sqrt{2gR} = \sqrt{\frac{2GM}{R}}$

$$\Rightarrow v_e \propto m^0$$

Where M, R are the mass and radius of the planet respectively. In this expression the mass of the body (m) is not present showing that the escape velocity is independent of the mass.

58. (c) According to Kepler's law of planetary motion, $T^2 \propto R^3$

$$\therefore T_2 = T_1\left(\frac{R_2}{R_1}\right)^{3/2} = 5 \times \left[\frac{4R}{R}\right]^{3/2}$$

$$= 5 \times 2^3 = 40 \text{ hours}$$

59. (c) Workdone,

$$W = \Delta U = U_f - U_i = 0 - \left[\frac{-GMm}{R}\right]$$

$$W = \frac{6.67 \times 10^{-11} \times 100}{0.1} \times \frac{10}{1000}$$

$$= 6.67 \times 10^{-10} \text{ J}$$

60. (c) $$\frac{(v_e)_p}{(v_e)_e} = \frac{\sqrt{\frac{2GM_p}{R_p}}}{\sqrt{\frac{2GM_e}{R_e}}} = \sqrt{\frac{M_p}{M_e} \times \frac{R_e}{R_p}}$$

$$= \sqrt{\frac{10M_e}{M_e} \times \frac{R_e}{R_e/10}} = 10$$

$$\therefore (v_e)_p = 10 \times (v_e)_e = 10 \times 11 = 110 \text{ km/s}$$

8. MECHANICAL PROPERTIES OF SOLIDS

ANSWER KEY

1	(a)	11	(d)	21	(c)	31	(a)	41	(a)	51	(a)
2	(b)	12	(d)	22	(b)	32	(b)	42	(b)	52	(c)
3	(c)	13	(a)	23	(c)	33	(c)	43	(b)	53	(b)
4	(d)	14	(a)	24	(c)	34	(d)	44	(b)	54	(c)
5	(c)	15	(a)	25	(d)	35	(a)	45	(c)	55	(c)
6	(c)	16	(d)	26	(b)	36	(a)	46	(b)	56	(b)
7	(c)	17	(b)	27	(c)	37	(b)	47	(b)	57	(b)
8	(d)	18	(c)	28	(a)	38	(a)	48	(c)	58	(a)
9	(b)	19	(b)	29	(d)	39	(d)	49	(b)	59	(b)
10	(d)	20	(a)	30	(b)	40	(b)	50	(d)	60	(d)

Hints & Solutions

1. (a) Liquids do not have finite shape hence their modulus of rigidity is zero.
2. (b) Gases & liquids possess only Bulk modulus.
3. (c) Young's modulus of a given material is constant and does not depend on dimensions of the wire.
4. (d) $U = \frac{1}{2} \times T \times \Delta l$

   but $\Delta l \propto T$

   $\therefore \quad U \propto T^2$

   $\therefore$ If tension is doubled, elastic P.E. stored in the wire will becomes four times.
5. (c) Elastomers do not obey Hooke's law.
6. (c) Shear modulus is given by

   $$\eta = \frac{F}{A\theta}$$

   for a perfectly rigid body,

   $\theta = 0$

   $\therefore \eta = \infty$
7. (c) We have, $U = \frac{F^2}{2k}$

   where $k = \frac{Yl}{A} = \frac{Yl}{\frac{1}{4}\pi d^2}$

   $\therefore \; U \propto \frac{d^2}{l}$
8. (d) Shape of a body changes when a any type stress is applied on it.
9. (b) Poisson's ratio is the ratio of lateral strain to longitudinal strain.
10. (d) Steel may behave as perfectly elastic, perfectly plastic or partially elastic depending on the stress developed in the steel wire object.
11. (d) All these factor affect elasticity of a substance.
12. (d) Among the given materials steel is most elastic.
15. (a) According to Hooke's Law–for metals within the limit of proportionality, stress applied to the body is proportional to resultant strain i.e., stress α strain ⇒ stress/strain = E=const. Where the proportionality constant E, is called modulus of elasticity or simply elasticity.
16. (d) Hooke's Law applies only in elastic region. In the elastic region stress is proportional to strain. Beyond the elastic region, plastic region starts, in which stress is not proportional to strain. So Hooke's Law defines elastic limit of the material.

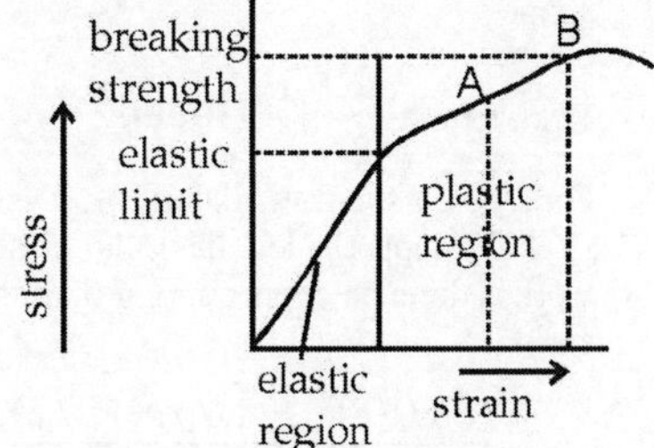

17. (b) Young's modulus Y is defined as

    $$Y = \frac{\text{Normal stress}}{\text{strain}} = \frac{F/A}{\Delta L/L}$$

    So the unit of Y is, N/m^2
18. (c) $Y = FL/A\,l$ or $l = FL/A\,Y$
20. (a) Weights without and with liquid proportional to ρ_b and $\rho_b - \rho_l$.
21. (c) We know that interatomic binding energy decreases with temperature. So, more strain is produced at higher temperature. As a result, Y decreases.

22. (b) Tensile strength = breaking stress × area of cross section

23. (c) According to Hook' Law

Stress α strain i.e., $\frac{F}{A} \propto \frac{\Delta l}{l}$

⇒ For same F & l, $\Delta l \propto \frac{1}{A}$

24. (c) Poisson's ratio σ is defined as :

$$\sigma = \frac{\text{lateral strain}}{\text{longitudinal strain}} = -\left(\frac{\Delta r}{r} / \frac{l}{\Delta l}\right)$$

The theoritical value of σ lies between -1 to 0.5 & practical value of σ lies between - zero to 0.5.

25. (d) $U = \begin{pmatrix}\text{energy per} \\ \text{unit volume}\end{pmatrix} = \frac{1}{2} \times \text{stress} \times \text{strain}$

$= ½ (F/A)×(\Delta l/l) = ½ × Y×(\Delta l/l)^2$

26. (b) When rubber is stretched, some work is done due to elasticity on the rubber. This work is stored in the form of elastic potential energy. As a result, the potential energy of the rubber increases.

27. (c) For a perfectly rigid body strain produced is zero for the given force applied, so

Y = stress/strain = ∞

28. (a) For a perfectly rigid body $\Delta V = 0$ so

$$B = -\frac{\Delta P}{\Delta V / V} = \infty$$

29. (d) Shear modulus of elasticity of a liquid is zero because a liquid does not resist change in its shape.

30. (b) Workdone = $\frac{1}{2} \times Y \times (\text{strain})^2$

For a given strain, workdone ∝ Y

As Y for copper is less than that of steel, so less work is done on copper spring than that on steel spring.

32. (b) $\Delta \ell = \frac{F(L/2)}{AY} = \frac{(AL\rho g)(L/2)}{AY}$

$= \left(\frac{1}{2}\right) \rho g L^2 / Y$

33. (c) Maximum possible strain = 0.2/100

$\therefore \quad A = \frac{F}{Y \times \text{strain}}$

$= \frac{10^4 \times 100}{(7 \times 10^9) \times 0.2} = 7.1 \times 10^{-4} \text{ m}^2$

34. (d) $B = -\frac{\Delta P}{\Delta V / V}$

$= \frac{100 \times 10^5}{(0.02/100)} = 50 \times 10^9 \text{ N/m}^2$

35. (a) Workdone per unit volume

$= \frac{1}{2} \times \text{stress} \times \text{strain}$

$= \frac{1}{2}(\text{Young's modulus} \times \text{strain}) \text{ strain}$

$= \frac{1}{2}(Y \times s)s = \frac{Y s^2}{2}$

37. (b) The atoms when brought from infinity are attracted due to interatomic electrostatic force of attraction. At point B, the potential energy is minimum and force of attraction is maximum. But if we bring atoms closer than $x = B$, force of repulsion between two nucleii starts and P.E. increases.

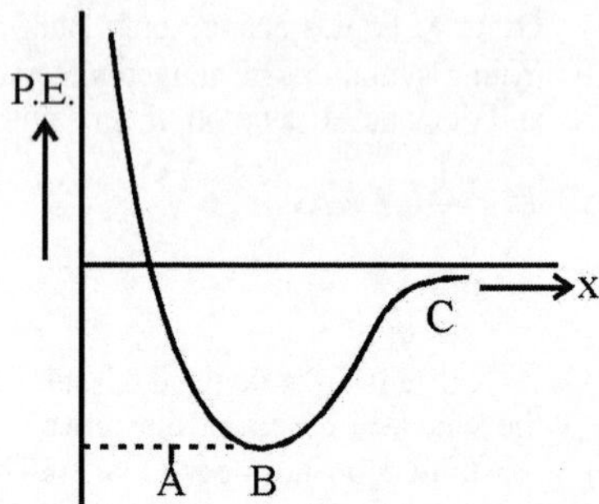

38. (a) $W = \frac{1}{2} \times \text{stress} \times \text{strain} \times \text{volume},$

$= \frac{1}{2} \times (Y \times \text{strain}) \times \text{strain} \times \text{volume}$

$= \frac{1}{2} Y (\ell / L)^2 AL$

$= \frac{1}{2}(YA/L)\ell^2 = \frac{1}{2}(YA/L)(1)^2 = \frac{\text{YA}}{\text{2L}}$

39. (d) P.E. $= \frac{1}{2} \times \text{stress} \times \text{strain} \times \text{volume}$

or (P.E./volume) $= \frac{1}{2} \times (Y \times \text{strain})\,(\text{strain})$

$= \frac{1}{2} Y (\text{strain})^2 = \frac{1}{2} Y X^2$

40. (b) The relation between Y, k, η for some isotropic solid material is given by

$$Y = \frac{9K\eta}{3K+\eta} = \frac{9K}{\frac{3K}{\eta}+1} \Rightarrow \frac{3K}{\eta}+1 = \frac{9K}{Y}$$

$$\Rightarrow \frac{3K}{\eta} = \frac{9K}{Y} - 1 = \frac{9K-Y}{Y} \Rightarrow \eta = \frac{3KY}{9K-Y}.$$

41. (a) $\tan(90-\theta) = \frac{\text{stress}}{\text{strain}}$

42. (b) Let W newton be the load suspended. Then

$$Y = \frac{(W/A_1)}{(\ell_1/L)} = \frac{WL}{A_1\ell_1} \quad \text{...(1)}$$

and $$Y = \frac{(W/A_2)}{(\ell_2/L)} = \frac{WL}{A_2\ell_2} \quad \text{....(2)}$$

Dividing equation (1) by equation (2), we get

$$1 = \left(\frac{\ell_2}{\ell_1}\right)\left(\frac{A_2}{A_1}\right) = \left(\frac{\ell_2}{\ell_1}\right)\left(\frac{2}{1}\right)$$

$$\therefore \frac{\ell_1}{\ell_2} = \frac{2}{1} \text{ or } \ell_1 : \ell_2 = 2:1$$

43. (b) Let T be the tension in the ring, then

$$Y = \frac{T.2\pi r}{A.2\pi(R-r)} = \frac{Tr}{A(R-r)}$$

$$\therefore T = \frac{YA(R-r)}{r}$$

44. (b) Bulk modulus $B = \frac{-P}{(\Delta V/V)} = \frac{-PV}{\Delta V}$ ....(1)

and $\Delta V = \gamma V \Delta T = 3\alpha.V.T$

or $\frac{-V}{\Delta V} = \frac{1}{3\alpha.T.}$ ...(2)

From eqs. (1) and (2), $B = P/(3\alpha.T)$

or $T = \frac{P}{3\alpha B}$

45. (c) Given that $F/A = 4.8 \times 10^7$ Nm^{-2}

$\therefore$ $F = 4.8 \times 10^7 \times A$ or

$$\frac{mv^2}{r} = 4.8 \times 10^7 \times 10^{-6} = 48$$

or $\frac{mr^2\omega^2}{r} = 48$ or $\omega^2 = \frac{48}{mr}$

$$\omega = \sqrt{\left(\frac{48}{10 \times 0.3}\right)} = \sqrt{16} = 4 \text{ rad/sec}$$

46. (b) We know that $Y = FL/\pi r^2 \ell$ or $r^2 = FL/(Y\pi\ell)$

$\therefore$ $R_B^2 = FL/(Y_B \pi \ell)$

and $R_S^2 = FL/(Y_S \pi \ell)$

or $\frac{R_B^2}{R_S^2} = \frac{Y_S}{Y_B} = \frac{2 \times 10^{10}}{10^{10}} = 2$

or $R_B^2 = 2R_S^2$ or $R_B = \sqrt{2}\, R_S$

$\therefore R_S = R_B/\sqrt{2}$

47. (b) Let V be the volume of the load and ρ its relative density

So, $Y = \frac{FL}{A\ell_a} = \frac{V\rho g L}{A\ell_a}$ .....(1)

When the load is immersed in the liquid, then

$$Y = \frac{F'L}{A\ell_w} = \frac{(V\rho g - V \times 1 \times g)L}{A\ell_w} \quad \text{....(2)}$$

($\because$ Now net weight = weight – upthrust)

From eqs. (1) and (2), we get

$$\frac{\rho}{\ell_a} = \frac{(\rho-1)}{\ell_w} \text{ or } \rho = \frac{\ell_a}{(\ell_a - \ell_w)}$$

48. (c) Thermal stress = $Y\alpha(\Delta T)$

(Thermal stress)$_1$ = $Y_1\alpha_1\Delta T$,

(Thermal stress)$_2$ = $Y_2\alpha_2\Delta T$,

$\therefore$ $Y_1\alpha_1\Delta T = Y_2\alpha_2\Delta T$

$$\frac{Y_1}{Y_2} = \frac{\alpha_2}{\alpha_1} = \frac{3}{2}$$

49. (b) The wire starts behaving like a liquid at point b. It behaves like a viscous liquid in the region bc of the graph.

50. (d) $F = YA\alpha t = (2.0 \times 10^{11})(10^{-6})(1.1 \times 10^{-5})(20) = 44$ newton

51. (a) $\delta = \frac{W\ell^3}{3YI}$, where W = load, ℓ = length of beam

and I is geometrical moment of inertia for rectangular beam,

$I = \frac{bd^3}{12}$ where b = breadth and d = depth

For square beam $b = d$

$$\therefore I_1 = \frac{b^4}{12}$$

For a beam of circular cross-section, $I_2 = \left(\frac{\pi r^4}{4}\right)$

$$\therefore \delta_1 = \frac{W\ell^3 \times 12}{3Yb^4} = \frac{4W\ell^3}{Yb^4}$$

(for sq. cross section)

and $\delta_2 = \frac{W\ell^3}{3Y(\pi r^4/4)} = \frac{4W\ell^3}{3Y(\pi r^4)}$

(for circular cross-section)

Now $\frac{\delta_1}{\delta_2} = \frac{3\pi r^4}{b^4} = \frac{3\pi r^4}{(\pi r^2)^2} = \frac{3}{\pi}$

($\because b^2 = \pi r^2$ i.e., they have same cross-sectional area)

52. (c) Couple per unit angle of twist, $C = \frac{\pi \eta r^4}{2\ell}$

$\therefore$ Couple $\tau = C\theta = \frac{\pi\eta r^4\theta}{2\ell}$

Here η, ℓ, C & τ are same. So, $r^4\theta$ = constant

$\therefore \frac{\theta_1}{\theta_2} = \left(\frac{r_2^4}{r_1^4}\right)$

53. (b) **Rod A**

$L_A = 20\text{cm}$
$\Delta L_A = 0.075$ cm
$(\Delta\theta)_A = (100-0)°C = 100°C$

Rod B

$L_B = L_A = 20$ cm
$\Delta L_B = 0.045$ cm
$(\Delta\theta)_B = (\Delta\theta)_A = 100°C$

Composite Rod

$L_3 = L'_A + L'_B = 20$ cm
$\Delta L_3 = 0.060$ cm
$(\Delta\theta)_3 = (\Delta\theta)_A = 100°C$
$L_A = ?$

Clearly

$$\alpha_A = \frac{\Delta L_A}{L_A(\Delta\theta)_A} = \frac{0.075}{20\times 100}$$

$$\alpha_B = \frac{\Delta L_B}{L_B(\Delta\theta)_B} = \frac{0.045}{20\times 100}$$

If $\Delta L'_A$ and $\Delta L'_B$ be the increase in length of the part made by A and B respectively, then

$$\Delta L_B = \Delta L'_B + \Delta L'_B$$

$$= L'_A\alpha_A(\Delta\theta)_A + L'_B\alpha_B(\Delta\theta)_B$$

$$= L'_A(\Delta\theta)_A(\alpha_A + \alpha_B)$$

$$\therefore \quad L'_A = \frac{\Delta L_3}{(\Delta\theta)_A(\alpha_A+\alpha_B)}$$

$$= \frac{0.060}{100\times\left(\frac{0.075}{20\times 100} + \frac{0.045}{20\times 100}\right)}$$

$$= \frac{0.060\times 20}{0.075+0.045} = \frac{0.06\times 20}{0.120} = 10.$$

$\therefore$ Length of the rod made by A is 10 cm.

54. (c) Longitudinal strain is possible only in solids because only solids can have length which can be stretched by applying force.

55. (c)

For a beam, the depression at the centre is given by, $\delta = \left(\frac{fL}{4Ybd^3}\right)$

[f, L, b, d are constants for a particular beam]

56. (b) We know that $Y = \frac{mg}{\pi\frac{D^2}{4}} \times \frac{L}{\ell}$

$$\therefore \frac{\Delta Y}{Y} = \frac{2\Delta D}{D} + \frac{\Delta\ell}{\ell}$$

[$\therefore$ value of m,g and L are exact]

$$= 2\frac{0.01}{0.4} + \frac{0.05}{0.8} = 2\times 0.025 + 0.0625$$

$= 0.05 + 0.0625 = 0.1125$
$\Delta Y = 2\times 10^{11} \times 0.1125$
$= 0.225 \times 10^{11}$.

[$\therefore$ the value of $Y = 2\times 10^{11}$ in all the four options]

57. (b) The Ratio of stress to strain is always constant. If stress is increased, strain will also increase so that their ratio remains constant.

58. (a) Young's modulus,

$Y = 3K(1-2\sigma) = 3\times 10^{10}(1-2\times 0.2)$
$= 1.8\times 10^{10}$ N/m^2

$\therefore$ Interatomic force constant is
$K = Yr = 1.8\times 10^{10}\times 3\times 10^{-10} = 5.4$ N/m

59. (b) $B = \frac{\text{Stress}}{\text{Volume strain}} = \frac{\text{Stress}}{\Delta V/V}$

$$\frac{\Delta V}{V} = \frac{\text{Stress}}{B}$$

As stress is constant. So, $\frac{\Delta V}{V} \propto \frac{1}{B}$

60. (d) Shear modulus is applicable to solids where deforming force causes change in shape of body. For fluids it is not possible since they have no fixed shape.

9. MECHANICAL PROPERTIES OF FLUIDS

ANSWER KEY											
1	(b)	11	(a)	21	(c)	31	(b)	41	(a)	51	(c)
2	(c)	12	(b)	22	(d)	32	(b)	42	(b)	52	(c)
3	(c)	13	(a)	23	(a)	33	(a)	43	(a)	53	(a)
4	(a)	14	(a)	24	(b)	34	(a)	44	(b)	54	(b)
5	(c)	15	(b)	25	(a)	35	(c)	45	(d)	55	(a)
6	(d)	16	(b)	26	(b)	36	(a)	46	(a)	56	(d)
7	(a)	17	(a)	27	(a)	37	(b)	47	(a)	57	(b)
8	(d)	18	(a)	28	(c)	38	(a)	48	(d)	58	(c)
9	(a)	19	(a)	29	(d)	39	(b)	49	(d)	59	(b)
10	(c)	20	(a)	30	(a)	40	(c)	50	(a)	60	(b)

Hints & Solutions

1. (b) When a body falls through a viscous medium, it attains a constant velocity called terminal velocity.
2. (c) Reynold's number is a dimensionless quantity.
3. (c) When the train crosses the platform, velocity of air molecules near the train become greater hence the passenger standing on the platform experience a pull towards the train as fast moving air molecules exert less pressure.
4. (a) Both friction and viscosity depend on nature of material but friction does not depend on area of contact.
5. (c) Water shows anamalous expansion and it sticks to glass as well.
6. (d) Angle of contact depends on temperature, nature of liquid and the container and on the presence of impuraties.
7. (a) Thickness of the surface film is equal to the radius of influence.
8. (d) All the given phenomena can be explained by Bernoulli's theorem.
9. (a) Toricelli's theorem is used to find velocity of efflux of an ideal liquid through on orifice.
10. (c) A pin of density greater than that of water can float on the surface of water due to surface tension.
11. (a) With increase in temperature, the viscosity of gases increases while those of liquids decreases.
12. (b) Surface tension is possessed by liquids only.
13. (a) Hydraulic lift is based on the principle of Pascal's law.
14. (a) Paint gun is based on Bernoulli's theorem.
15. (b) By equation of continuity, the velocity of liquid will be more at narrower portion and less at wider portion, hence pressure will be more at the wider portion, as according to Bernoulli's theorem, pressure is more at a point where velocity is less.
16. (b) Apply Bernoulli's theorem.
17. (a) Some of the work done by the force of surface tension is used in change in P.E. of the liquid and remaining is loss in the form of heat.
18. (a) Fevicol increases adhesive force between paint and wall molecules.
19. (a) Specific gravity $= \dfrac{\text{weight of body in air}}{\text{weight of water displaced}}$

$$= \frac{210}{210-140} = \frac{210}{70} = 3$$

Density of body = sp. gravity × density of water $= 3 \times 1\ g/cm^3 = 3\ g/cm^3$

21. (c) Effective weight of solid of specific gravity 1 when immersed in water will be zero.
22. (d) Pressure applied to enclosed fluid is transmitted equally in all direction according to Pascal law.
23. (a) In hydrometer lower density mark is upwards and higher density mark is downwards.
24. (b) Ice is lighter than water. When ice melts, the volume occupied by water is less than that of ice. Due to which the level of water goes down.

25. (a) When the lift is falling down freely, the apparent weight of body in a lift becomes zero. Due to which the upthrust on the body due to liquid is zero.

26. (b) When ice having lead balls melts, the volume of water formed and lead balls left will be smaller than that of ice. Due to which the level of water will go down.

27. (a) Let h be the desired height of liquid in cylinder for which the force on the bottom and sides of the vessel is equal.

Force on bottom = $\rho g h \times \pi R^2$

Force on the walls of vessel

$= \rho g (h/2) \times 2 \pi Rh$

$\therefore \quad \rho g h \pi R^2 = \rho g h \pi R h \quad$ or $R = h$

28. (c) The effective weight of the block in liquid will become less than 2 kg due to buoyancy of liquid. As a result of which A will read less than 2 kg.

As the body immersed in liquid has some effective weight acting downwards so the reading of B will be more than 5 kg.

29. (d) Surface tension of a liquid is due to force of attraction between like molecules of a liquid i.e. cohesive force between the molecules.

30. (a) Surface tension = work done per unit area in increasing the surface area of a liquid under isothermal conditions.

31. (b) Kerosene oil rises up in wick of a lantern because of capillary action. If the surface tension of oil is zero, then it will not rise, so oil rises up in a wick of a lantern due to surface tension.

32. (b) The surface tension of oil is less than that of water, so the oil spreads as a thin layer.

33. (a) With the increase in temperature, the surface tension of liquid decreases and angle of contact also decreases.

34. (a) The surface tension of liquid at critical temperature is zero.

35. (c) Excess of pressure in a soap bubble,

$P = 4T/r$ i.e. $P \propto \frac{1}{r}$ therefore pressure in a smaller bubble is more than that of a bigger bubble. When two bubbles of different radii are in communication, then the air flows from higher pressure to lower pressure i.e. from smaller bubble into larger one.

36. (a) When two drops merge together to form one drop, the surface area of drop will decrease, due to which the energy of bigger drop is less than the sum of the energy of two smaller drops. Due to it, the energy is released.

37. (b) Over a small temperature ranges, S.T. of water decreases linearly with rise of temperature.

38. (a) $h = \frac{2\,T\cos\theta}{r\rho g}$. The liquid will rise i.e. h is positive if $\cos\theta$ is +ve. It is so if $\theta < 90°$ or θ is acute.

39. (b) When lift is accelerated downwards, the observed weight of body in a lift decreases. Hence to counter balance the upward pull due to S.T. on the liquid meniscus, the height through which the liquid rises must increase.

40. (c) A molecule in the surface film has maximum potential energy because the molecule when brought in surface film, the work is done against both gravity and surface tension.

42. (b) When a sparingly soluble salt (like detergent) added to water, the surface tension of water decreases.

43. (a) When air is pushed in soap bubble, its radius increases. As excess pressure $P \propto \frac{1}{r}$, so pressure in it decreases.

44. (b) When the ball is dropped in a viscous liquid, the speed of the ball, first increases due to gravity pull but the viscous force acting on the ball also increases with the increase in speed of ball. When viscous force balances gravity pull on ball, then the ball moves with constant velocity.

45. (d) Adiabatic bulk modulus $K = \gamma P$

Where γ is the ratio of specific heat at constant pressure to the specific heat at constant volume.

46. (a) $\frac{1}{2}\rho v^2 = 13.6\, gh$

[Density of mercury is 13.6 times water]

$\frac{1}{2} \times v^2 = 13.6 \times 10 \times \frac{40}{100}$ 1144

$v^2 = 2 \times 4 \times 13.6$ or $v = 10.3$ m/sec

47. (a) Pressure difference created = 3 – 1 = 2 atm $= 2 \times 10^5\ \text{Nm}^{-2}$

Pressure energy created in unit mass

$= \frac{2\times10^5}{\rho} = \frac{2\times10^5}{10^3} = 2\times10^2$

Now K.E. of unit mass $= \frac{1}{2}v^2$

$\frac{1}{2}v^2 = 2\times10^2 \Rightarrow v^2 = 4\times10^2$

$v = \sqrt{400}$ m/sec

48. (d) Surface tension is the tensile force that occurs on the free surface of a liquid due to the cohesion effect between the particles of liquid. The cohesion is due to the molecular attraction between the molecules of a liquid. With the rise in temperature the liquid molecules become more mobile and the cohesion force decreases. Hence the surface tension of water at 75°C will be the minimum.

49. (d) We know that the Terminal velocity

$$v=\frac{2r^2(\rho-\rho_0)g}{9\eta} \quad \therefore v\propto r^2$$

(Where η = viscosity of medium
ρ = density of body
ρ_0 = density of liquid)

50. (a) A lot of water evaporates through the stomata present along the lower surface of leaf. The loss of water in vapour form from the leaf into the atmosphere is called transpiration. Due to this phenomenon leaf creates a suction which pulls water from xylem cells by capillary action.

51. (c) $Y=\frac{FL}{(\pi r^2)\Delta L}$.

As material and length are same $\Rightarrow$ Y, L are same. Also load F is same.

$$\Rightarrow \pi r_1^2\Delta L_1=\pi r_2^2\Delta L_2$$

$$r_1^2\times 1\text{ cm}=\frac{1}{4}r_1^2\Delta L_2$$

$$\Delta L_2=4\text{ cm.}$$

52. (c) The height to which water rises in a capillary is given by, $h=\frac{2T}{R\rho g}$.

When radius is halved, the height to which water level rises is doubled.

53. (a) Whenever a sphere of radius 'r' falls through a fluid of viscosity 'η', a frictional force given by Stokes law, $F=6\pi\eta rv$ acts. At the start of motion $v=0$, so frictional force, $F=0$. But as velocity increases friction force also increase till terminal velocity v_T is reached where upward force (friction) balances downward force of weight, and body falls with constant velocity.

54. (b) According to Stoke's law the backward retarding force of friction, $f=6\pi\eta rv$ increases with velocity, till it becomes equal and opposite to downward force. As no resultant force acts on the body, it falls with constant velocity.

55. (a) The expression for $v_T=\frac{2}{9}\frac{r^2(\rho-\sigma)}{\eta}g$

Thus, terminal velocity depends on radius of drop.

56. (d) The rise of water in capillary tube is inversely proportional to radius, $h=\frac{2T}{r\rho g}$. It is given that radius of A and C are same. So, height of the liquid in tube A and C is the same.

57. (b) Rise of liquid does not depend on atmospheric pressure.

58. (c) The height to which liquid rises in a capillary is given by, $h=\frac{2T}{r\rho g}$. Thus, if height to which liquid rises is more, the liquid have greater surface tension than water. Option (a) is not suitable because diameter of capillary tube containing liquid should be smaller than the capillary in which water rises. This is not mentioned clearly in the option (a).

59. (b) According to Stoke's law, the retarding upward frictional force increases till it becomes equal to downward force. Net force being zero, body stops accelerating & starts falling with constant velocity.

60. (b) Due to increase in velocity, pressure will be low above the surface of water.

10. THERMAL PROPERTIES OF MATTER

ANSWER KEY

1	(a)	11	(c)	21	(c)	31	(c)	41	(b)	51	(b)
2	(c)	12	(c)	22	(a)	32	(a)	42	(a)	52	(d)
3	(c)	13	(d)	23	(d)	33	(c)	43	(a)	53	(d)
4	(b)	14	(c)	24	(d)	34	(c)	44	(b)	54	(a)
5	(d)	15	(c)	25	(b)	35	(a)	45	(c)	55	(a)
6	(d)	16	(b)	26	(c)	36	(c)	46	(c)	56	(c)
7	(c)	17	(b)	27	(d)	37	(b)	47	(a)	57	(b)
8	(d)	18	(a)	28	(c)	38	(b)	48	(d)	58	(b)
9	(d)	19	(b)	29	(d)	39	(d)	49	(d)	59	(d)
10	(d)	20	(b)	30	(b)	40	(d)	50	(a)	60	(b)

Hints & Solutions

1. (a) At 4°C, density of water is maximum, hence $\frac{dv}{v} = 0$ or $\alpha = \frac{1}{T}\frac{dv}{v} = 0$
2. (c) Coefficient of thermal conductivity depends only on the material of the plate.
3. (c) Wool is a bad conductor hence it does not allow heat from our body to lose to the surroundings.
5. (d) A cooking utensil must heat up quickly and must transmit heat rapidly. Hence, it must have low specific heat and high thermal conductivity.
6. (d) $P = e\sigma A(T^4 - T_0^{\ 4})$
7. (c) Some bodies like wood, rubber etc. have negative value of coefficient of thermal expansion, hence they contract on heating, hence (A) is true. (B) is also true.
8. (d) Rate of flow of heat through a rod,

$$H = \frac{kA}{2}\Delta T$$

9. (d) All the given phenomena are caused due to convection.
10. (d) Liquid does not over flow on heating, when apparent value of coefficient of volume expansion,

$\gamma_\alpha = \gamma - 3\alpha \leq 0$
or $\gamma \leq 3\alpha$

11. (c) Volume coefficient of a gas varies from process to process and may depend upon temperature of the gas.
12. (c) When ice is added to water, the equilibrium temperature may be zero or positive depending on amount of ice and water and their initial temperatures.
13. (d) Triple point is the temperature where three states of matter can exist simultaneously.
14. (c) Radiation is the fastest mode of transfer of heat.
15. (c) A body at higher temperature emits radiation of all wavelengths absorbed by it a lower temperature.
16. (b) $\frac{C}{5} = \frac{F-32}{9}$ or $\frac{C}{5} = \frac{C-32}{9}$

$4C = -160$ or $C = F = -40°$.

17. (b) As heat flows from higher temperature to lower temperature, so a temperature gradient is required.
18. (a) The temperatures of the block and wood are equal to the temperature of the body as both feel equally hot or cold.
19. (b) As the surface area of cylindrical bulb is larger than a spherical bulb, heat will be transmitted quickly through a cylindrical bulb and it will respond quickly to the temperature.
20. (b) In order that the heat exchange is quick, the calorimeters are made of metals.
21. (c) Water has a large latent heat of vaporisation. When it is sprinked over a large area, its evaporation occurs which, in turn, causes cooling.
22. (a) Black body also emits radiation whereas nothing escapes a black hole.
23. (d) $\frac{dH}{dt} \propto (\theta_2 - \theta_1) = (\Delta\theta)^n \Rightarrow n = 1$
24. (d) Mean kinetic energy of the two types of molecules should be equal. The mean square velocity of A type molecules

$= \omega^2 + \omega^2 + \omega^2 = 3\omega^2$

Therefore, $\frac{1}{2}m(3\omega^2) = \frac{1}{2}(2m)v^2$

This gives $\omega^2 / v^2 = 2/3$

25. (b) Let the final temperature be T

Heat gained by ice $= mL + m \times s \times (T - 0)$
$= 10 \times 80 + 10 \times 1 \times T$

Heat lost by water $= 55 \times 1 \times (40 - T)$

By using law of calorimetery,

$800 + 10\,T = 55 \times (40 - T)$

$\Rightarrow T = 21.54°C = 22°C$

26. (c) Using $\frac{F-32}{180} = \frac{C}{100}$

$$\Rightarrow \frac{140-32}{180} = \frac{C}{100} \Rightarrow C = 60$$

Temperature of boiling water = 100°C

We get, fall in temperature = 100 – 60 = 40°C

27. (d) Thermal capacity = ms = 40 × 0.2 = 8 cal/°C = 4.2 × 8 J = 33.6 joules/°C
28. (c) Mercury thermometer is based on the principle of change of volume with rise of temperature and can measure temperatures ranging from – 30°C to 357°C.
29. (d) Amount of energy radiated $\propto T^4$
30. (b) Rate of cooling $\propto$ temperature difference between system and surrounding.

As the temperature difference is halved, so the rate of cooling will also be halved

31. (c) Let θ_0 be the temperature of the surro-unding. Then

$$\frac{80-75}{t_1} = k\left(\frac{80+75}{2} - \theta_0\right)$$

or, $$\frac{5}{t_1} = k(77.5 - \theta_0)$$

or, $$t_1 = \frac{5}{k(77.5 - \theta_0)} \quad \text{... (1)}$$

Similarly, $t_2 = \dfrac{5}{k(72.5-\theta_0)}$ ...(2)

and $t_3 = \dfrac{5}{k(67.5-\theta_0)}$ ... (3)

From (1), (2) & (3), it is obvious that $t_1 < t_2 < t_3$

32. (a) According to Stefan's Law $E = \sigma e A T^4$

$E \propto T^4$; so, $E \propto (500)^4$

33. (c) According to Stefan's law

$E \propto T^4$

$$\frac{E_1}{E_2} = \frac{T_1^4}{T_2^4} \text{ or } \frac{20}{E_2} = \frac{T^4}{2^4 T^4}$$

$\Rightarrow E_2 = 320$ kcal/m^2 min.

34. (c) Heat required by ice at 0°C to reach a temperature of 100°C $= mL + mc\Delta\theta$
$= 1 \times 80 + 1 \times 1 \times (100 - 0) = 180$ cal
Heat available with 1 g steam to condense into 1 g of water at 100°C = 536 cal.
Obviously the whole steam will not be condensed and ice will attain a temperature of 100°C; so the temperature of mixture = 100°C.

35. (a) Let ΔT be increase in the temperature of brass wire.Then length of brass wire,
$(l') = l_1(1 + \alpha_1 \Delta T)$
Similarly, length of steel wire when temperature is increased by ΔT
$(l'') = l_2(1 + \alpha_2 \Delta T)$
According to question, $l' - l'' = l_1 - l_2$
$l_1 + l_1\alpha_1\Delta T - l_2 + l_2\alpha_2\Delta T = l_1 - l_2$
$l_1\alpha_1 = l_2\alpha_2$, as $\Delta T \neq 0$

36. (c) In convection, the temperature gradient exists in the vertical direction and not in the horizontal direction. So, up and down movement of particles takes place which depends on the weight and gravity.

37. (b) According to Wien's displacement law,
$\lambda_m T = 2.88 \times 10^{-3}$
When $T = 2000$ K,
$\lambda_m (2000) = 2.88 \times 10^{-3}$ ....(1)
When $T = 3000$ K,
$\lambda'_m (3000) = 2.88 \times 10^{-3}$ ....(2)
Dividing (1) by (2),

$$\frac{2}{3}\frac{\lambda_m}{\lambda'_m} = 1 \Rightarrow \frac{\lambda_m}{\lambda'_m} = \frac{3}{2} \Rightarrow \lambda'_m = \frac{2}{3}\lambda_m$$

38. (b) $Q = \dfrac{KA(\theta_1 - \theta_2)t}{l}$

Rate of heat flow

$$H = \frac{Q}{t} = \frac{KA(\theta_1 - \theta_2)}{l} \quad i.e., \ H \propto \frac{A}{l}$$

Dimensions of area $A = [L^2]$, dimensions of distance $l = [L]$

$$\therefore H \propto L \Rightarrow \frac{H_2}{H_1} = \frac{L_2}{L_1} = 2 \Rightarrow H_2 = 2H_1$$

39. (d) According to Wien's displacement law, product of wavelength belonging to maximum intensity and temperature is constant i.e., $\lambda_m T$ = constant.

41. (b) Rate of heat flow for one rod

$$= \frac{K_1 A_1 (T_1 - T_2)}{d} (d \to \text{Length})$$

Rate of heat flow for other rod

$$= \frac{K_2 A_2 (T_1 - T_2)}{d}$$

In steady state, $\dfrac{K_1 A_1 (T_1 - T_2)}{d}$

$$= \frac{K_2 A_2 (T_1 - T_2)}{d} \Rightarrow K_1 A_1 = K_2 A_2$$

42. (a) In series, equivalent thermal conductivity

$$K_{eq} = \frac{2K_1 K_2}{K_1 + K_2}$$

or, $K_{eq} = \dfrac{2 \times K \times 2K}{K + 2K} = \dfrac{4}{3}K$

43. (a) From Wien's displacement law
$\lambda_m T$ = constant
$\Rightarrow \lambda_m \propto T^{-1}$

44. (b) We know that $Q = \dfrac{T_H - T_L}{R}$

Also, Thermal resistance $R = \dfrac{l}{KA} = \dfrac{l}{K\pi r^2}$

Heat flow will be maximum when thermal resistance is minimum.
From given option
(i) $r = 2r_0, l = 2l_0$

$$\therefore R = \frac{2l_0}{K\pi(2r_0)^2} = \frac{l_0}{2K\pi r_0^2}$$

(ii) $r = 2r_0, l = l_0$

$$\therefore R = \frac{l_0}{K\pi(2r_0)^2} = \frac{l_0}{4K\pi r_0^2}$$

(iii) $r = r_0, l = 2l_0$

$$\therefore R = \frac{2l_0}{K\pi r_0^2} = \frac{2l_0}{K\pi r_0^2}$$

(iv) $r = r_0, l = l_0$

$$\therefore R = \frac{l_0}{K\pi r_0^2} = \frac{l_0}{K\pi r_0^2}$$

It is clear that for option (2) resistance is minimum, hence, heat flow will be maximum.
[Alt : (i) Rate of heat flow is directly proportional to area (ii) inversely proportional to length.
$\therefore$ Heat flow will be maximum when r is maximum and ℓ is minimum.]

45. (c) $\theta_n = \frac{\theta_c + \theta_i}{2}$

$\therefore \theta_c = 2\theta_n - \theta_i = 2(300) - 620 = -20°C$

46. (c) Applying Wien's displacement law,

$\lambda_m T$ = constant

5000 Å × (1227 + 273) = (2227 + 273) × λ_m

$$\lambda_m = \frac{5000 \times 1500}{1000} = 3000\text{Å}$$

47. (a) According to Stefan's law,

$E \propto T^4$

$\propto (t + 273)^4 K$ [727°C = (727 + 273)K]

$\propto (727 + 273)^4 K$

$\propto (1000)^4 K$

48. (d) Since $T_n = \frac{T_i + T_c}{2}$ =Neutral temperature

$$T_n = \frac{T_i + 0}{2} = \frac{T_i}{2}$$

[T_c = 0°C = temperature of cold junction]

49. (d) l = 10 cm = 0.1 m, d = 0.2 mm,

r = 0.1 mm = 1 × 10^{-4} m,

e = 0.2, T = 2000 K, $\sigma = 5.67 \times 10^{-8}$ W/m² K⁴

According to stefan's law of radiation, rate of emission of heat for an ordinary body,

$E = \sigma A e T^4 = \sigma(2\pi r l) e T^4$

$= 5.67 \times 10^{-8} \times 2 \times 3.14 \times 1 \times 10^{-4} \times 0.1 \times 0.2 \times (2000)^4$

= 11.4 W

∴ Power radiated by the filament = 11.4 W

[$A = 2\pi r l$]

50. (a) $\frac{E_2}{E_1} = \left[\frac{T_2}{T_1}\right]^4 = \left[\frac{2227 + 273}{727 + 273}\right]^4$

$$= \left[\frac{2500}{1000}\right]^4 = 39$$

51. (b) $\frac{C}{5} = \frac{F - 32}{9}$ Here C = F

$\frac{C}{5} = \frac{C - 32}{9} \Rightarrow 9C = 5C - 160$

4C = – 160 ⇒ C = – 40°C.

Thus at –40°C and –40° F the temperture is same.

52. (d) $E = \sigma T^4$ $\therefore \frac{E}{E'} = \frac{T^4}{T'^4}$

$\Rightarrow \frac{E}{E'} = \left(\frac{273+27}{273+84}\right)^4 \Rightarrow \frac{E'}{E} = \left(\frac{357}{300}\right)^4 = 2$

$\therefore E' = 2E$

Rate of radiated energy will be 2 times.

53. (d) $\ell = 5$m $t_1 = 0°C$

$\ell_2 = 5.01$m $t_2 = 100°C$

$$\alpha = \frac{\ell_2 - \ell_1}{\ell_1(t_2 - t_1)} = \frac{5.01 - 5}{5 \times 100} = 2 \times 10^{-5}/°C.$$

54. (a) $\frac{E_1}{E_2} = \frac{T_1^4}{T_2^4} = \left(\frac{T_1}{T_2}\right)^4$

or, $\frac{E_1}{E_2} = \left(\frac{273 + 727}{273 + 327}\right)^4 = \left(\frac{1000}{600}\right)^4 = \frac{625}{81}$

55. (a) From Newton's law of cooling :–

$\frac{\theta_1 - \theta_2}{t} = k\left(\frac{\theta_1 + \theta_2}{2} - \theta_0\right)$ where θ_1 is higher temperature θ_2, is lower temperature.

$$\frac{80 - 64}{5} = k\,(72 - \theta_0) \quad \text{.... (i)}$$

Where θ_0 is temperature of surroundings

$$\frac{64 - 52}{10} = k\,(58 - \theta_0) \quad \text{.... (ii)}$$

Dividing (i) and (ii) we get θ_0

$$\frac{52 - \theta}{15} = k\left(\frac{52 + \theta}{2} - \theta_0\right) \quad \text{.... (iii)}$$

Thus θ is found.

56. (c) From Newton's law of cooling

$$\frac{dQ}{dt} = -KA\frac{dT}{dx}$$

Area of cross-section A and thickness dx is the same.

Also $dQ = mCd\theta$

Thus in first case $\frac{m \times C \times (61° - 59°)}{4}$

$$= \frac{-KA}{dx}\left[\left(\frac{61° + 51°}{2}\right) - 30°\right]$$

...(i)

In second case, $\frac{m \times C \times (51° - 49°)}{t}$

$$= \frac{-KA}{dx}\left[\left(\frac{51° + 49°}{2}\right) - 30°\right] \quad \text{... (ii)}$$

Dividing equation (i) by equation (ii)

$$\frac{t}{4} = \frac{30}{20}$$

or t = 6 minutes.

57. (b) From Wien's displacement law,
$\lambda_{max} T$ = constant
$\therefore \quad (\lambda_{max})_1 T_1 = (\lambda_{max})_2 T_2$.
$\therefore \quad (\lambda_{max})_2 = \frac{4800 \times 6000}{3000} \text{Å} = 9600\text{Å}$

58. (b) $\frac{dQ}{dt} = KA\frac{\Delta T}{L}$

For the first rod, $\left(\frac{dQ}{dt}\right)_1 = \frac{3KA}{L}(100 - \theta)$

Similarly, $\left(\frac{dQ}{dt}\right)_2 = 2K\frac{A}{L}(\theta - 50)$

$\left(\frac{dQ}{dt}\right)_3 = K\frac{A}{L}(\theta - 20)$

Now, $\left(\frac{dQ}{dt}\right)_1 = \left(\frac{dQ}{dt}\right)_2 + \left(\frac{dQ}{dt}\right)_3$

$\Rightarrow$ 3 (100 – θ) = 2 (θ – 50) + (θ – 20)
$\Rightarrow$ θ = 70°

60. (b) Seebeck discovered that when the junctions of two dissimilar metals are kept at different temperature, an emf results which will depend on the temperature of hot bath only.

11. THERMODYNAMICS

ANSWER KEY

1	(b)	11	(a)	21	(c)	31	(c)	41	(b)	51	(c)
2	(a)	12	(b)	22	(c)	32	(c)	42	(d)	52	(b)
3	(c)	13	(a)	23	(b)	33	(c)	43	(c)	53	(a)
4	(a)	14	(c)	24	(a)	34	(b)	44	(c)	54	(c)
5	(a)	15	(a)	25	(b)	35	(d)	45	(a)	55	(a)
6	(c)	16	(a)	26	(b)	36	(c)	46	(c)	56	(d)
7	(c)	17	(c)	27	(b)	37	(a)	47	(a)	57	(b)
8	(b)	18	(a)	28	(c)	38	(a)	48	(a)	58	(c)
9	(b)	19	(a)	29	(d)	39	(c)	49	(d)	59	(b)
10	(c)	20	(b)	30	(b)	40	(d)	50	(c)	60	(a)

Hints & Solutions

2. (a) Black body also emits radiation whereas nothing escapes a black hole.
3. (c) Absolute zero temperature is practically not reachable.
4. (a) Energy radiated ∞ R^2T^4.
5. (a) External amount of work must be done in order to flow heat from lower temperature to higher temperature. This is according to second law of thermodynamics.
7. (c) When a body is heated then relation between colours and temperature is according to Prevost's theory of radiation which states that everybody emitting radiant energy in all directions at a rate depending only on the nature of its surface and its temperature e.g., when a body is placed in an enclosure (furnace) it would acquire the temperature of furnace and seem white means radiate white light. So it becomes first dark and then white.

8. (b) Momentum of photon $= \frac{E}{c}$

Change in momentum $= \frac{2E}{c}$
= momentum transferred to the surface
(the photon will reflect with same magnitude of momentum in opposite direction)

9. (b) Temperature of stars can be determined by Winn's displacement law
λ_m = constant

10. (c) We know

$E = eT^4 \Rightarrow \ln E = \ln e + 4\ln T$

$\therefore$ $\ln e$ is negative because $e < 1$

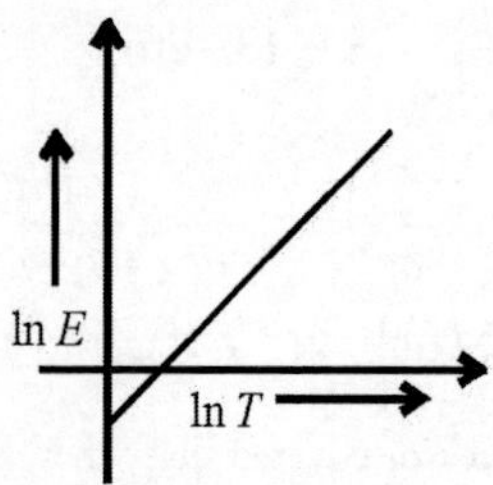

11. (a) PV = constant represents isothermal process.

12. (b) $\lambda_m T$ = constant

or $T \propto \frac{1}{\lambda_m}$ or $T \propto \nu_m$.

13. (a) In condition of weightlessness, convection is not possible.

14. (c) Since system X is neither in thermal equilibrium with Y nor with Z, it means

$$(X)_{\text{temperature}} \neq (Y)_{\text{temperature}}$$

& $(X)_{\text{temperature}} \neq (Z)_{\text{temperature}}$

but it may happen that

$$(Y)_{\text{temperature}} = (Z)_{\text{temperature}}$$

(see the Zeroth' law of thermodynamics)

16. (a) The normal temperature of a person is 98.6°F or 37°C.

18. (a) $L=L_0(1+\alpha\Delta T)$

$V=V_0(1+\gamma\Delta T)$ where $\gamma = 3\alpha$

19. (a) The area of circular hole increases when we heat the metal sheet & expansion of metal sheet will be independent of shape & size of the hole.

20. (b)

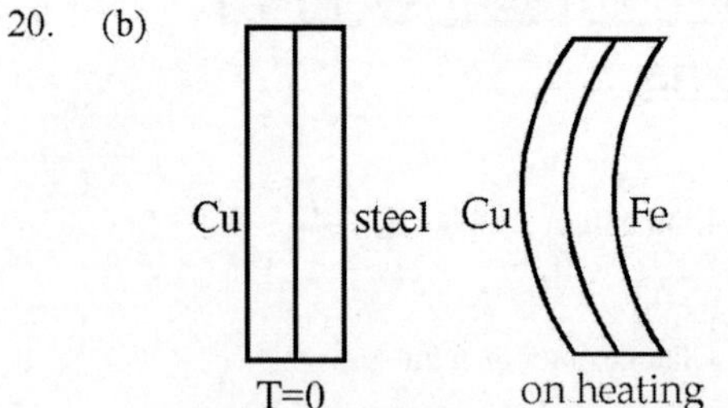

23. (b) As temperature increases, length of pendulum increases & hence time period also increases.

$$T = 2\pi\sqrt{l/g}$$

25. (b) Thermoelectric thermometer is based on seebeck effect

26. (b) According to Mayer's relationship

$$C_p - C_v = R$$

$$\therefore \quad \frac{C_p}{M} - \frac{C_v}{M} = \frac{R}{M}$$

Here M = 28.

28. (c) In case of expansion of volume V of liquid

$$V_L' = V(1+\gamma_\ell \Delta\theta) = V(1+\alpha\Delta\theta)$$

Where α is cubical expansion of liquid

While V is volume of container

$$V_S' = V(1+\gamma_s \Delta\theta)$$

$$= V(1+(\gamma/3)\Delta\theta) = V(1+\alpha\Delta\theta)$$

because of volume expansion

$\gamma_S = 3 \times$ (linear coefficient of expansion)

$$3 \times \frac{\alpha}{3} = \alpha$$

So change in volume of liquid relative to container is

$$V_L' - V_S' = 0$$

So level of liquid in container will remain almost stationary.

30. (b) Coefficient of cubical expansion of container = 3α.

When $\gamma > 3\alpha$, increase in the volume of the liquid is more than that of vessel.

31. (c) Since density $\rho = m/V$

due to thermal expansion volume increase & so density decreases i.e., $V'=V(1+\gamma\Delta\theta)$ so

$$\rho' = \frac{m}{V'} = \frac{\rho}{(1+\gamma\Delta\theta)}$$

where γ is coefficient of volume expansion.

32. (c) Since $\frac{R}{Cv} = 0.67 \Rightarrow Cv = \frac{3}{2}R$, hence gas is monoatomic.

33. (c) Work done = Area under curve $ACBDA$

34. (b) Under isothermal conditions, there is no change in internal energy.

35. (d) Given : Initial temperature of gas

(T_1) = 27°C = 300 K

Initial pressure (P_1) = 30 atm

Initial volume $(V_1) = V$

Final pressure (P_2) = 1 atm

Final volume $(V_2) = 10\,V$.

We know from the general gas equation that

$$\frac{P_1V_1}{T_1} = \frac{P_2V_2}{T_2}$$

or, $\frac{30 \times V}{300} = \frac{1 \times 10V}{T_2}$

or, T_2 = 100 K = –173°C.

36. (c) $\eta = 1 - \frac{T_2}{T_1}$ or $\frac{50}{100} = 1 - \frac{500}{T_1}$

$\Rightarrow T_1 = 1000\ K$

Also, $\frac{60}{100} = 1 - \frac{T_2}{1000} \Rightarrow T_2 = 400\ K$

37. (a) Efficiency $= \frac{T_1 - T_2}{T_1}$

$T_1 = 227 + 273 = 500$ K
$T_2 = 127 + 273 = 400$ K

$\eta = \frac{500 - 400}{500} = \frac{1}{5}$

Hence, output work

$= (\eta) \times \text{Heat input} = \frac{1}{5} \times 6 = 1.2$ kcal

38. (a) $T_1 = T$, $W = 6R$ joules, $\gamma = \frac{5}{3}$

$W = \frac{P_1V_1 - P_2V_2}{\gamma - 1} = \frac{nRT_1 - nRT_2}{\gamma - 1}$

$= \frac{nR(T_1 - T_2)}{\gamma - 1}$

$n = 1, T_1 = T \Rightarrow \frac{R(T - T_2)}{5/3 - 1} = 6R$

$\Rightarrow \quad T_2 = (T - 4)\ K$

39. (c) For a process to be reversible, it must be quasi-static. For quasi-static process, all changes take place infinitely slowly. Isothermal process occur very slowly so it is quasi-static and hence it is reversible.

40. (d) We know that efficiency of Carnot engine

$= 1 - \frac{T_2}{T_1} = 1 - \frac{400}{500} = \frac{1}{5}$

[$\because$ $T_1 = (273 + 227)$K = 500 K
and $T_2 = (273 + 127)$K = 400 K]

Efficiency of Heat engine $= \frac{\text{Work output}}{\text{Heat input}}$

or, $\frac{1}{5} = \frac{\text{work output}}{6 \times 10^4}$

$\Rightarrow$ work output $= 1.2 \times 10^4$ cals

41. (b) We know that efficiency of Carnot Engine

$= \frac{T_1 - T_2}{T_1}$

where, T_1 is temp. of source & T_2 is temp. of sink

$\therefore\ 0.40 = \frac{T_1 - 300}{T_1} \Rightarrow T_1 - 300 = 0.40T_1$

$0.6T_1 = 300 \Rightarrow T_1 = \frac{300}{.6} = \frac{3000}{6} = 500$K

Now efficiency to be increased by 50%

$\therefore \quad 0.60 = \frac{T_1 - 300}{T_1} \Rightarrow T_1 - 300 = 0.6T_1$

$0.4T_1 = 300 \Rightarrow T_1 = \frac{300}{.4} = \frac{300 \times 10}{4} = 750$

Increase in temp = 750 – 500 = 250 K

42. (d) $C_P = \frac{7}{2}R;\ C_V = C_P - R = \frac{7}{2}R - R = \frac{5}{2}R$

$\frac{C_P}{C_V} = \frac{7/2\,R}{5/2\,R} = \frac{7}{5}$

43. (c) Since efficiency of engine is $\eta = 1 - \frac{T_2}{T_1}$

According to problem,

$\frac{1}{6} = 1 - \frac{T_2}{T_1}$ ... (1)

When the temperature of the sink is reduced by 62°C, its efficiency is doubled

$2\left(\frac{1}{6}\right) = 1 - \frac{T_2 - 62}{T_1}$ ... (2)

Solving (1) and (2) $T_2 = 372$ K
$T_1 = 99°\text{C}$ = Temperature of source.

44. (c) In a cyclic process, the initial state coincides with the final state. Hence, the change in internal energy is zero, as it depends only on the initial and final states. But Q & W are non-zero during a cycle process.

45. (a) In an isochoric process volume remains constant whereas pressure remains constant in isobaric process.

46. (c) According to first law of thermodynamics
$Q = \Delta U + W$
$\Delta U = Q - W$
$= 2 \times 4.2 \times 1000 - 500 = 8400 - 500$
$= 7900$ J

47. (a) By first law of thermodynamics,

$$\Delta Q = \Delta U + \Delta W$$

In adiabatic process, $\Delta Q = 0$

$\therefore \quad \Delta U = -\Delta W$

In isothermal process, $\Delta U = 0$

$\therefore \quad \Delta Q = \Delta W$

48. (a) or (d)

If a process is expansion then work done is positive so answer will be (a).

But in question work done by gas is given –150J so that according to it answer will be (d).

49. (d) Change in entropy is given by

$$dS = \frac{dQ}{T} \text{ or } \Delta S = \frac{\Delta Q}{T} = \frac{mL_f}{273}$$

$$\Delta S = \frac{1000 \times 80}{273} = 293 \text{cal/K}.$$

50. (c) $T_1 = 273 + 27 = 300\text{K}$

$T_2 = 273 + 927 = 1200\text{K}$

For adiabatic process,

$P^{1-\gamma} T^{\gamma} = \text{constant}$

$\Rightarrow P_1^{1-\gamma} T_1^{\gamma} = P_2^{1-\gamma} T_2^{\gamma}$

$$\Rightarrow \left(\frac{P_2}{P_1}\right)^{1-\gamma} = \left(\frac{T_1}{T_2}\right)^{\gamma}$$

$$\Rightarrow \left(\frac{P_1}{T_2}\right)^{1-\gamma} = \left(\frac{T_2}{T_1}\right)^{\gamma}$$

$$\left(\frac{P_1}{P_2}\right)^{1-1.4} = \left(\frac{1200}{300}\right)^{1.4}$$

$$\left(\frac{P_1}{P_2}\right)^{-0.4} = (4)^{1.4}$$

$$\left(\frac{P_2}{P_1}\right)^{0.4} = 4^{1.4}$$

$$P_2 = P_1 4^{\left(\frac{1.4}{0.4}\right)} = P_1 4^{\left(\frac{7}{2}\right)}$$

$$= P_1 (2^7) = 2 \times 128 = 256 \text{ atm}$$

51. (c) The concept of entropy is introduced by the second law of thermodynamics.

52. (b) For an isochoric process, $W = 0$

$\therefore \quad \Delta Q = \Delta U.$

53. (a) In adiabatic expansion,

$\Delta Q = OL$

$\therefore \quad \Delta U = -W$

as W is positive, ΔU must be negative.

but $\Delta U = nCu\Delta T$

$\therefore$ In adiabatic expansion, temperature decrease and as

$PV = nRT$

the product of pressure and volume also decreases.

54. (c) In isobaric process, volume of system changes, hence heat supplied is used to increase temperature and also in work done. Hence (c) is incorrect.

55. (a) According to Boyle's law, at constant temperature,

$$P \propto \frac{1}{V} \text{ or } V \propto \rho \qquad \left[\because P = \frac{m}{v}\right]$$

56. (d) $C_p - C_v = R$ is always a constant.

57. (b) Internal energy of an ideal gas

$U = nWT$

depends of temperature number of moles and atomicity (as C_v depends on atomicity) of the gas, but it does not depend on pressure of the gas.

58. (c) For a cyclic process, work done can be positive as well as negative.

59. (b) Adiabatic processes are very fast processes.

60. (a) During isothermal expansion, the slope of P = V graph,

$$\frac{dP}{dP} = -\frac{p}{V}, \text{ always decreases.}$$

12. KINETIC THEORY

ANSWER KEY

1	(a)	11	(b)	21	(a)	31	(a)	41	(a)	51	(c)
2	(a)	12	(d)	22	(b)	32	(d)	42	(c)	52	(b)
3	(c)	13	(c)	23	(c)	33	(b)	43	(b)	53	(d)
4	(c)	14	(a)	24	(d)	34	(c)	44	(d)	54	(d)
5	(a)	15	(a)	25	(b)	35	(b)	45	(b)	55	(c)
6	(d)	16	(c)	26	(b)	36	(b)	46	(a)	56	(d)
7	(d)	17	(d)	27	(d)	37	(c)	47	(c)	57	(a)
8	(a)	18	(c)	28	(c)	38	(b)	48	(b)	58	(d)
9	(d)	19	(a)	29	(d)	39	(c)	49	(d)	59	(d)
10	(c)	20	(d)	30	(d)	40	(d)	50	(a)	60	(c)

Hints & Solutions

3. (c) $P = \frac{2/3(\frac{1}{2}mN\, v_{rms}^2)}{V}$ where m is mass of molecule of gas, N is number of molecules in gas & v_{rms} is root mean square velocity of the molecule.

5. (a) $v_{avg.} = \sqrt{\frac{8kT}{\pi m}}$

8. (a) The kinetic energy is directly proportional to temperature.

9. (d) Apply Boyle's Law

13. (c) The average speed is given by

$$\bar{v} = \sqrt{\frac{8kT}{\pi m}} = 1.6\sqrt{\frac{kT}{m}}$$

here $(M)_{\substack{atomic\\hydrogen}} = 1.0$ gm,

$(m)_{\substack{atomic\\helium}} = 4.0$ gm,

so, $\bar{v}(H)_{atomic} = 2\bar{v}(He)_{atomic}$

17. (d) Let us consider 1 mole of an ideal gas as a Kelvin temperature T. It has N_A molecules (Avogadro's number). The internal energy of an ideal gas is entirely kinetic. The average K.E. per molecule of an ideal gas $= \frac{1}{2}\, n\, k\, T$, where n is degrees of freedom. Therefore the internal energy of one mole of an ideal gas would be

$$U = N\left(\frac{1}{2}nkT\right) = \frac{1}{2}nRT \quad (\because k = R/N_A)$$

Now, $C_v = \frac{dU}{dT} = \frac{n}{2}R$

and $C_P = \frac{n}{2}R + R = \left(\frac{n}{2}+1\right)R$

$$\frac{C_P}{C_v} = \frac{\left(\frac{n}{2}+1\right)R}{\frac{n}{2}R} = \left(1+\frac{2}{n}\right) = \gamma$$

20. (d) $P = \frac{1}{3}\rho\bar{v}^2 = \frac{2}{3}\times\left(\frac{1}{2}\rho\bar{v}^2\right) = \frac{2}{3}E$

24. (d) All the given phenomena are explained by kinetic theory of gases.

25. (b) $\gamma = 1 + \frac{2}{F}$

where F is number of degree of freedom. As F is least for a monoatomic gas molecules, γ is maximum for monoatomic gas molecule.

26. (b) According to the law of equipartition of energies, the gas possesses equal energies in all the three direction x, y and z-axis.

27. (d) There is no interaction among the molecules of a gas, hence potential energy of an ideal gas is zero.

28. (c) CO_2 gases a linear molecule, hence for CO_2, $F = 5$.

30. (d) In kinetic theory of gas, if is assumed that molecules/ atoms have negligible volume.

31. (a) The pressure is exerted on walls due to collision of molecules with the walls of the container, i.e. due to change in momentum.

32. (d) In a gas molecules move randomly in all possible directions, hence their average velocity is zero.

33. (b) As both the gases have equal volume, they must have equal number of moles, hence (i) is correct. R.M.S. velocity depends on molar mass, hence is different for both the gases (ii) is incorrect. Both the gases are diatomic, hence, they must have some internal energy. (iii) is correct. Average velocity of molecules is zero for both the gases. (iv) is correct.

34. (c) Total K.E. $= \frac{1}{2} fkT = \frac{1}{2}(5)k\left(\frac{mv^2}{3k}\right) = \frac{5}{6}mv^2$

and rotational K.E. $= \frac{1}{2} \times 2 \times k\,k = \frac{1}{3}mv^2$.

35. (b) Vapour pressure does not depend on the amount of substance. It depends on the temperature alone.

36. (b) As the temperature increases, the average velocity increases. So, the collisions are faster.

37. (c) According to law of equipartition of energy, the energy per degree of freedom is $\frac{1}{2}kT$.

For a polyatomic gas with n degrees of freedom, the mean energy per molecule $= \frac{1}{2}nkT$

38. (b) $\gamma = \frac{C_P}{C_V} = \frac{15}{10} = \frac{3}{2} \Rightarrow C_V = \frac{2}{3}C_P$

$$C_P - C_V = \frac{R}{J} \Rightarrow C_P - \frac{2}{3}C_P = \frac{R}{J}$$

$$\Rightarrow \frac{C_P}{3} = \frac{R}{J} \Rightarrow C_P = \frac{3R}{J}$$

40. (d) $C_p = \frac{5}{2}R$ and $C_v = \frac{3}{2}R$

We know that $Q_v = nC_v\Delta T$ and $Q_p = nC_p\Delta T$

$$\Rightarrow \frac{Q_v}{Q_p} = \frac{3}{5}.$$

Given $Q_p = 207$ J $\Rightarrow Q_v \cong 124$ J

41. (a) $P = \frac{1}{3}Nmc^2 = \frac{2}{3} \times \left(\frac{1}{2}Nm\right)c^2 = \frac{2}{3}K.E.$

42. (c) According to Dalton's law of partial pressures, we have $P = P_1 + P_2 + P_3$

43. (b) Number of translational degrees of freedom are same for all types of gases.

44. (d) $c_{rms} \propto \sqrt{T}$

As temperature increases from 300 K to 1200 K that is four times, so, c_{rms} will be doubled.

45. (b) 8g of oxygen is equivalent to (1/4) mole.

46. (a) At 0 K, molecular motion stops. Hence, kinetic energy of molecules becomes zero.

47. (c) No. of degree of freedom $= 3K - N$ where K is no. of atom and N is the number of relations between atoms. For triatomic gas, $K = 3$, $N = 3$

No. of degree of freedom $= 3(3) - 3 = 6$

48. (b) $n = \frac{5}{32}$ $\therefore PV = \frac{5}{32}RT$ $[\because PV = nRT]$

49. (d) Let the mass of the gas be m.

At a fixed temperature and pressure, volume is fixed.

Density of the gas, $\rho = \frac{m}{V}$

Now $\frac{\rho}{P} = \frac{m}{PV} = \frac{m}{nRT}$

$\Rightarrow \frac{m}{nRT} = x$ (By question)

$\Rightarrow xT =$ constant $\Rightarrow x_1T_1 = x_2T_2$

$$\Rightarrow x_2 \Rightarrow \frac{x_1T_1}{T_2} = \frac{283}{383}x \quad \left[\begin{array}{l}\therefore \\ T_1 = 283K \\ T_2 = 383K\end{array}\right]$$

50. (a) $\frac{P_1V_1}{T_1} =$ constant for one mole

and $\frac{P_1}{T_1} =$ constant for constant volume

so $\frac{P_1}{T_1} = \frac{1.004P_1}{(T_1+1)} \Rightarrow T_1 = 250K$

51. (c) $v_{oxg.} = \sqrt{\frac{3R \times 289}{32}}$ $\left(v_{rms} = \sqrt{\frac{3RT}{M}}\right)$

$v_H = \sqrt{\frac{3R \times 400}{2}}$ so $v_H = 2230.59$ m/sec

53. (d) Total K.E. of A type molecules $=\frac{3}{2}m\omega^2$

Total K.E. of A type of molecule is

$$K.E_A=\frac{1}{2}m\left[(v_{r.ms})_x^2+(v_{r.ms})_y^2+(v_{r.ms})_z^2\right]$$

but $(v_{r.ms})_x=\omega$

so $(v_{r.ms})_y=(v_{r.ms})_z=\omega$

Total K.E. of B type molecules

$$=\frac{1}{2}\times 2mv^2=mv^2$$

Now, $\frac{3}{2}\times m\omega^2=mv^2$ or $(\omega^2/v^2)=\frac{2}{3}$

54. (d) Since v_{rms} is doubled by increasing the temp. so by $v_{rms}=\sqrt{\frac{3KT}{m}}$, the temp. increase by four times.

Now for constant pressure $\frac{V_1}{T_1}=\frac{V_2}{T_2}$

$V_1=V, T_1=T°K, T_2=4T°K, V_2=?$

$V_2=4V$

55. (c) Let γ_1 and γ_2 denote the ratio of two specific heats of gas 1 and gas 2 respectively when they are mixed in the ratio of n_1 and n_2 mole. The resultant value of γ is the weighted mean of γ_1 and γ_2 i.e.,

$$\gamma=\frac{n_1\gamma_1+n_2\gamma_2}{n_1+n_2}=\frac{1(5/3)+1(7/5)}{1+1}$$

$$=\frac{25+21}{15\times 2}=\frac{46}{30}=1.53$$

56. (d) $\frac{PV}{T}=\frac{P'(90/100)V}{T}$

or $\frac{P'}{P}=\frac{100}{90}=1+\frac{10}{90}$

or $\frac{P'-P}{P}=\frac{10}{90}=11.1\%$

57. (a) $P=\frac{1}{3}\rho\bar{\upsilon}^2=\frac{1}{3}\times(6\times10^{-2})\times(500)^2$

$=5\times10^3\ N/m^2$

58. (d) $v_{rms}\propto\sqrt{T}$

At 0°C, $v_{rms}\propto\sqrt{273}$

At temp. T, $\frac{v_{rms}}{2}\propto\sqrt{T}$

$\therefore\ \frac{1}{2}=\frac{\sqrt{T}}{\sqrt{273}}$ or $T=\frac{273}{4}=68.2K\simeq 69K$

$T=69-273=-204°C$

59. (d) $PV^\gamma=P_1\left(\frac{V}{32}\right)^\gamma$

or $P_1=(32)^\gamma,\ P=(32)^{1.4},\ P=128P$

60. (c) $T_1^\gamma P_1^{1-\gamma}=T_2^\gamma P_2^{1-\gamma}$

13. OSCILLATIONS

ANSWER KEY											
1	(b)	11	(b)	21	(a)	31	(d)	41	(c)	51	(a)
2	(a)	12	(c)	22	(a)	32	(d)	42	(c)	52	(b)
3	(c)	13	(a)	23	(d)	33	(c)	43	(d)	53	(c)
4	(d)	14	(d)	24	(d)	34	(b)	44	(a)	54	(a)
5	(a)	15	(b)	25	(a)	35	(a)	45	(a)	55	(a)
6	(d)	16	(d)	26	(d)	36	(c)	46	(c)	56	(c)
7	(c)	17	(b)	27	(c, d)	37	(d)	47	(b)	57	(c)
8	(b)	18	(a)	28	(a)	38	(d)	48	(c)	58	(a)
9	(b)	19	(c)	29	(d)	39	(c)	49	(c)	59	(c)
10	(b)	20	(d)	30	(c)	40	(c)	50	(d)	60	(a)

Hints & Solutions

1. (b) As the child stands up, centre of mass of the system shifts up decreasing the effective length of the pendulum, hence time period of the swing will decrease.

2. (a) For a simple pendulum

$$T = 2\pi\sqrt{\frac{l}{g}}$$

Hence graph between T and l is a parabola

3. (c) The resonance becomes sharp when damping force is small

4. (d) In a satellite, effective value of acceleration due to gravity.

$g' = 0$

$$\therefore\ T' = 2\pi\sqrt{\frac{l}{g'}} = \infty$$

5. (a) Superposition of two S.H.M's gives a S.H.M if both have equal frequency.

6. (d) At equilibrium position, net force and hence acceleration of particle executing simple harmonic motion is zero.

7. (c) Time period of a simple pendulum is independent of mass of the bob.

8. (b) The particles of wave moving through a string fixed at both ends execute simple harmonic motion.

9. (b) The motion of simple pendulum is simple harmonic only for small amplitude.

10. (b) A second's pendulum crosses mean position at the interval of is, hence its time period is $2s$.

11. (b) For a simple harmonic motion, if

$x = A \sin \omega t$

then $a = -\omega^2 x = \omega^2 A \sin \omega t = \omega^2 A \sin (\omega t + \pi)$

hence phase difference between x and a is π.

12. (c) Let the S.H.M is described by the equation

$x = A \sin \omega t$

$$v = \frac{dx}{dt} = \omega A \cos\omega t$$

$\therefore$ kinetic energy

$$k = \frac{1}{2}mv^2 = \frac{1}{2}m\omega^2 A^2 \cos^2 \omega t$$

$$= \frac{1}{4}m\omega^2 A^2 (1 + \cos 2\omega t)$$

$\therefore$ K.E. of simple harmonic oscillator, oscillates with frequency $2f$.

13. (a) At resonance, the frequency of periodic force must be equal to the frequency of oscillations.

14. (d) The oscillations of the rod are simple harmonic for small amplitude only.

15. (b) Let the two SHM's are described

$x = A \sin \omega t$ ...(1)

and $y = A \sin (\omega t + \frac{\pi}{2}) = A \cos \omega t$ ...(2)

Squaring and adding,

$x^2 + y^2 = A^2$, which is equation of circle.

16. (d) The necessary and sufficient condition for simple harmonic motion is :

$F = -kx$

i.e. a restoring force proportional to displacement.

17. (b) For a simple harmonic motion,

$x = A \sin \omega t$ ...(1)

$$\therefore\ v = \frac{dx}{dt} = \omega A \cos \omega t \quad ...(2)$$

From (1) and (2)

$\frac{x^2}{A^2} + \frac{v^2}{\omega^2 A^2} = 1$, which is equation of an ellipse.

18. (a) For simple harmonic motion, $F = -kx$.

Here, $k = Ak$.

19. (c) At maximum energy of the particle, velocity resonance takes place, which occurs when frequency of external periodic force is equal to natural frequency of undamped vibrations, i.e. $\omega_2 = \omega_0$.

Further, amplitude resonance takes place at a frequency of external force which is less than the frequency of undamped natural vibrations, i.e. $\omega_1 \neq \omega_0$.

20. (d) The effective spring constant of two springs in series; $K = \frac{k_1 k_2}{k_1 + k_2}$.

Time period,

$$T = 2\pi\sqrt{\frac{m}{K}} = 2\pi\sqrt{\frac{m(k_1 + k_2)}{k_1 k_2}}$$

21. (a) Every simple harmonic motion (SHM) is necessarily periodic, but a periodic motion may or may not be simple harmonic motion.

22. (a) The motion of the particle is periodic, (not oscillatory), beacuse it returns to its starting point after a fixed time (called time period).

23. (d) Its motion is complex, but not periodic, oscillatory or S.H.M.

24. (d) In case of simple pendulum, the restoring force always points towards the equilibrium position i.e., always opposite to displacement. Hence acceleraion is in the negative direction of displacement & it is a characteristic of S.H.M.

$F = -kx$ by Hooke's Law

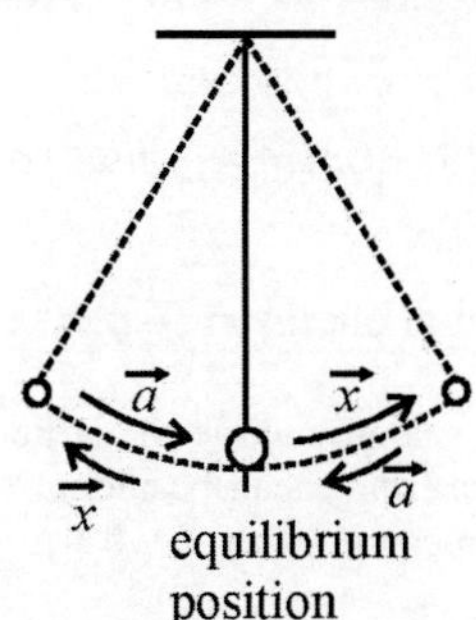

or $\frac{md^2x}{dt^2} = -kx$ by Newton's second Law

or $a = \frac{d^2x}{dt^2} = -\left(\frac{k}{m}\right)x$ ...(i)

where x is displacement and equation (i) is followed by S.H.M. So, $\vec{F}.\vec{r}$ and $\vec{a}.\vec{r}\left(\frac{\vec{F}.\vec{r}}{m}\right)$ are always negative in S.H.M.

25. (a) $\vec{F}.\vec{a}$ is always positive, because both have same direction.

26. (d) $\vec{F}\times\vec{a}, \vec{a}\times\vec{r}$ and $\vec{v}\times\vec{r}$ are always zero, because each quantity is parallel or anti parallel to other.

27. (c,d) When a particle is dropped from a height h above the centre of tunnel.

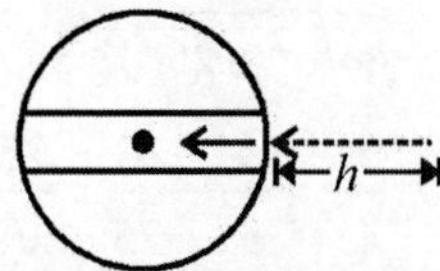

(i) It will oscillate, through the earth to a height h on both sides.
(ii) The motion of particle is periodic.
(iii) The motion of particle will not be S.H.M.

28. (a) (i) For simple pendulum $T = 2\pi\sqrt{L/g}$
(ii) For physical pendulum

$$T = 2\pi\sqrt{I/mgL}$$

So in both above case, time peroid is changed if they are taken to the moon.

(iii) For torsional pendulum $T = 2\pi\sqrt{I/C}$
(iv) For spring-mass system

$$T = 2\pi\sqrt{m/k}$$

In above case T does not depend on 'g', so it will not change if we take them to moon.

29. (d) As seen from figure after one time period the bob return to its equilibrium position, so diplacement of the particle is zero, but distance covered by the particle in one time period is $4A$ (where A is amplitude of bob, when it does S.H.M.)

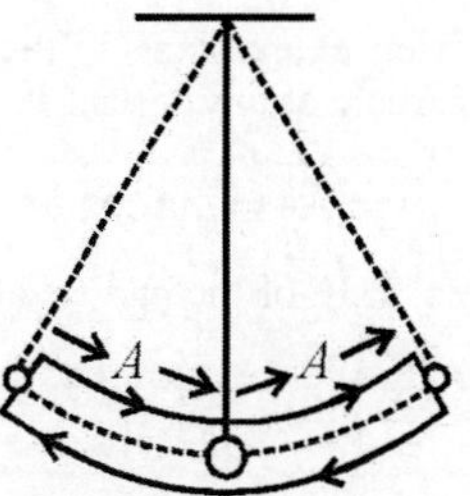

30. (c) The distance moved by a particle in S.H.M. in one time period = $4A$

32. (d) The motion of particle is S.H.M. with
$x = A \sin \omega t + B \cos \omega t$
$= a \sin (\omega t + \theta)$

where $a = \sqrt{A^2 + B^2}$, $A = a \cos\theta$,
$B = a \sin \theta$, $\theta = \tan^{-1} B/A$.

33. (c) The motion of particle is circular or elliptic, when two S.H.M. which are perpendicular to each other superimpose on the particle. The particle moves on a
(i) Ellipse if amplitudes of two S.H.M. are different
(ii) Circle, if amplitudes of two S.H.M. are same.

34. (b) The motion is S.H.M. (as seen from fig.) with an amplitude B.

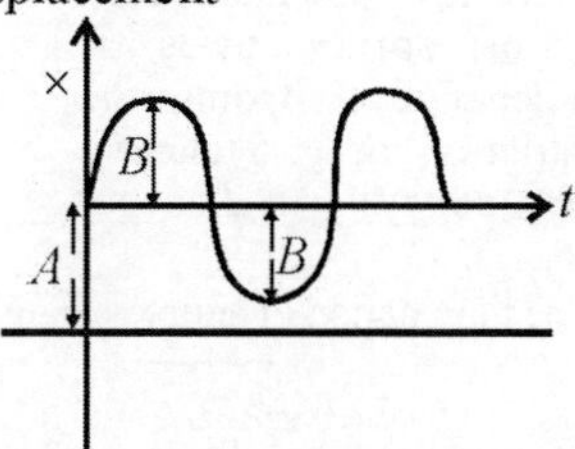

35. (a) Let at any instant the displacement of particle in S.H.M. is

$y = a \sin \omega t$ ...(1)

where a is amplitude & ω is the angular frequency of the particle
The P.E. at any instant is

$$\text{P.E} = \frac{1}{2}ky^2 = \frac{1}{2}m\omega^2 y^2 \quad ...(2)$$

where k is constant & m is the mass of particle. The average potential energy of the particle over a whole period

$$T = \frac{1}{T}\int_0^T \frac{1}{2} m\omega^2 y^2 dt$$

$$= \frac{1m\omega^2 a^2}{4} \quad ...(3)$$

From expression (1) the velocity of the particle at any instant is

$$v = \frac{dy}{dt} = a\omega \cos\omega t$$

So, K.E. of the particle at any instant is

$$= \frac{1}{T}\int_0^T \frac{1}{2} mv^2 dt$$

$$= \frac{1}{T}\int_0^T \frac{1}{2} ma^2\omega^2 \cos^2 \omega t\, dt = \frac{1}{4} ma^2\omega^2$$

So total average energy of the particle in S.H.M. over a whole time period is

= average P.E. over a time period T + average K.E. over a time period T

$$= \frac{1}{4} ma^2\omega^2 + \frac{1}{4} ma^2\omega^2 = \frac{1}{2} ma^2\omega^2$$

36. (c) For spring mass m system, the time period of the oscillation of mas m is defined as

$$T = 2\pi\sqrt{m/k} = 2\pi\sqrt{y/g}$$

where m = mass of particle

k = spring constant

y = extension in spring

L = natural length of spring

If g is changed, then y also changes so that y/g is constant, so the time period T of spring mass is independent from the variation in g. Hence ν (frequency) will also not change.

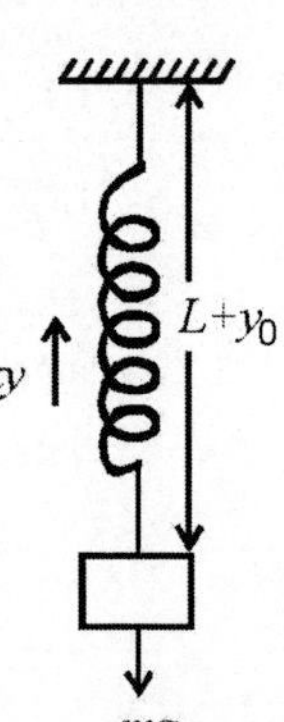

37. (d) The time period of simple pendulum on earth is

$$T_E = 2\pi\sqrt{L/g_E}$$

where L is the length of simple pendulum. Now, if take it to the moon, then

$$T_m = 2\pi\sqrt{L/g_m} \text{ but } g_m = \frac{g_E}{6}$$

So $T_m = 2\pi\sqrt{\frac{L \times 6}{g_E}} = \sqrt{6}T_E$

It means that it is $\sqrt{6}$ times slower on moon.

38. (d) If it gives correct time at equator, it will give correct time at poles also because the time period of spring-mass system is independent of g.

39. (c) In S.H.M., kinetic energy of particle at any point P is

$$\text{Kinetic energy} = \frac{1}{2} m\omega^2 (a^2 - x^2)$$

$$\text{Potential energy} = \left(\frac{1}{2} m\omega^2 x^2\right)$$

where a is amplitude of particle and x is the distance from mean position.

So, at mean position, $x = 0$

$$\text{K.E.} = \frac{1}{2} m\omega^2 a^2 \quad \text{(maximum)}$$

P.E. = 0

40. (c) $x = a \sin \omega t$

and $y = b \sin (\omega t + \pi) = -b \sin \omega t$

or, $\frac{x}{a} = -\frac{y}{b}$ or $y = -\frac{b}{a}x$

It is an equation of a straight line.

41. (c) P.E., $V = \frac{1}{2} m\omega^2 x^2$

and K.E., $T = \frac{1}{2} m\omega^2 (a^2 - x^2)$

$$\therefore \quad \frac{T}{V} = \frac{a^2 - x^2}{x^2}$$

42. (c) For S.H.M., Velocity,

$v = \omega\sqrt{a^2 - x^2}$ at displacement x.

$$\Rightarrow \quad 10 = \omega\sqrt{a^2 - 16} \quad .....(1)$$

$$\text{and} \quad 8 = \omega\sqrt{a^2 - 25} \quad .....(2)$$

Dividing, $\frac{5^2}{4^2} = \frac{a^2 - 16}{a^2 - 25} = \frac{25}{16}$

or, $16a^2 - 256 = 25a^2 - 625 \Rightarrow 9a^2 = 369$

$$\therefore \quad a^2 = \frac{369}{9}$$

Putting this value in equation (2) mentioned above,

$$10 = \omega\sqrt{\frac{369}{9} - 16} \Rightarrow 10 = \omega\sqrt{\frac{225}{9}}$$

or, $\omega = \frac{10 \times 3}{15} = 2 \text{ radian/sec}.$

Time period $= \frac{2\pi}{\omega} = \frac{2\pi}{2} = \pi$ sec.

43. (d) The effective value of acceleration due to gravity is $\sqrt{(a^2 + g^2)}$

44. (a) $\omega^2 = \frac{\text{acceleration}}{\text{displacement}} = \frac{2.0}{0.02}$

$\omega^2 = 100$ or $\omega = 10$ rad/s

45. (a) For S.H.M., $x = A\sin\left(\frac{2\pi}{T}t\right)$

When $x = A$, $A = A\sin\left(\frac{2\pi}{T}.t\right)$

$\therefore \sin\left(\frac{2\pi}{T}.t\right) = 1$

$\Rightarrow \sin\left(\frac{2\pi}{T}.t\right) = \sin\left(\frac{\pi}{2}\right) \Rightarrow t = (T/4)$

When $x = \frac{A}{2}, \frac{A}{2} = A\sin\left(\frac{2\pi}{T}.t\right)$

or, $\sin\frac{\pi}{6} = \sin\left(\frac{2\pi}{T}t\right)$ or $t = (T/12)$

Now, time taken to travel from $x = A$ to $x = A/2$ is $(T/4 - T/12) = T/6$

46. (c) Beats are produced. Frequency of beats will be $\frac{1}{3} - \frac{1}{4} = \frac{1}{12}$. Hence time period = 12 s.

[**Alt** : Frequency of combined S.H.M

$= \frac{T_1T_2}{T_2 - T_1} = \frac{3\times4}{4-3} = 12\text{s}$]

47. (b) $P.E. = \frac{1}{2}m\omega^2x^2$.

Total energy $E = \frac{1}{2}m\omega^2A^2$

P.E. $= \frac{1}{4}E \Rightarrow \frac{1}{2}m\omega^2x^2 = \frac{1}{8}m\omega^2A^2$

$\therefore \quad x = \frac{1}{2}A.$

[**Alt** : The displacement at which the P.E. becomes n times the total energy is given by $x = \sqrt{n}A$, where A is the amplitude.]

48. (c) A wave moving through a string fixed at both ends, is a transverse wave formed as a result of simple harmonic motion of particles of the string.

49. (c) As phase difference $= \frac{\pi}{4}$, the resultant path of particle is an ellipse.

50. (d) Total energy of particle executing S.H.M. of amplitude (A).

$E = \frac{1}{2}m\omega^2A^2$

K.E. of the particle

$= \frac{1}{2}m\omega^2\left(A^2 - \frac{A^2}{4}\right) \quad \left(\text{when } x = \frac{A}{2}\right)$

$= \frac{1}{2}m\omega^2 \times \frac{3}{4}A^2 = \frac{1}{2}\times\frac{3}{4}m\omega^2A^2$

Clearly, $\frac{\text{K.E.}}{\text{Total Energy}} = \frac{3}{4}$

[**Alt** : If the displacement be n times the amplitude, then $\frac{K.E.}{T.E.} = 1 - n$]

51. (a) $k = 2 \times 10^6$ N/m

Amplitude (A) = 0.01 m

Total mechanical energy = 160 J

As we know that total mechanical energy is equal to total maximum potential energy.

Maximum K.E.

$= \frac{1}{2}ka^2 = \frac{1}{2}\times(2\times10^6)\times(0.01)^2 = 100\text{J}$

52. (b) $v_{max} = A\omega$ when $v = \frac{v_{max}}{2} = \frac{A\omega}{2}$

$v = \omega\sqrt{A^2 - y^2}$

$\Rightarrow \quad \frac{A\omega}{2} = \omega\sqrt{A^2 - y^2} \Rightarrow y = \pm\frac{\sqrt{3}}{2}A$

[**Alt** : The displacement at which the speed is n times the maximum speed is given by $y = a\sqrt{1 - n^2}$]

53. (c) We know that $T = 2\pi\sqrt{\frac{l}{g}}$

$\frac{\Delta T}{T}\times100 = \frac{1}{2}\frac{\Delta l}{l}\times100$

If length is increased by 2%, time period increases by 1%.

54. (a) Equation of two simple harmonic motions

$y = A\sin(\omega t + \phi)$ ....(1)

$x = A\sin\left(\omega t + \phi + \frac{\pi}{2}\right)$

$\Rightarrow \quad x = A\cos(\omega t + \phi)$ ....(2)

On squaring and adding equations (1) and (2),

$x^2 + y^2 = A^2$

This is an equation of a circle. Hence, resulting motion will be a circular motion.

55. (a) Time period of simple pendulum

$T = 2\pi\sqrt{\left(\frac{l}{g}\right)} \propto \sqrt{l}$ where l is effective length.

[i.e distance between centre of suspension and centre of gravity of bob]

Initially, centre of gravity is at the centre of sphere. When water leaks the centre of gravity goes down until it is half filled; then it begins to go up and finally it again goes at the centre. That is effective length first increases and then decreases. As $T \propto \sqrt{l}$, so time period first increases and then decreases.

56. (c) Let the pendulums be in phase after t sec of start. Within this time, if the bigger pendulum executes n oscillations, the smaller one will have executed $(n + 1)$ oscillations.

Now, the time of n oscillation = $2\pi\sqrt{\frac{20}{g}} \times n$

and the time of $(n + 1)$ oscillation

$= 2\pi\sqrt{\frac{5}{g}} \times (n+1)$

To be in phase

$$2\pi\sqrt{\frac{20}{g}} \times n = 2\pi\sqrt{\frac{5}{g}} \times (n+1)$$

or, $2n = n + 1$

or, $n = 1$

Hence, the no. of oscillations executed by shorter pendulum $= n + 1 = 1 + 1 = 2$

57. (c) $n = \frac{1}{2\pi}\sqrt{\frac{k}{m}}$

$$n' = \frac{1}{2\pi}\sqrt{\frac{k}{4m}} = \frac{1}{2} \times \frac{1}{2\pi}\sqrt{\frac{k}{m}}$$

On putting the value of n we get $n' = \frac{n}{2}$

58. (a) The potential energy of a spring executing SHM is given by $V = \frac{Kx^2}{2}$

59. (c) Time period of a simple pendulum is given by,

$T = 2\pi\sqrt{\frac{\ell}{g}}$ when acceleration is 1.02 g,

$$T_1 = 2\pi\sqrt{\frac{\ell}{1.02g}}$$

$$\Rightarrow T_1 = \frac{1}{\sqrt{1.02}}\left[2\pi\sqrt{\frac{\ell}{g}}\right] = \frac{T}{\sqrt{1.02}} = 0.99T$$

60. (a) Potential energy of a particle executing SHM is maximum when K.E. is zero. So,

$$\frac{1}{2}m\omega^2(A^2 - y^2) = 0 \text{ or } y = \pm A$$

14. WAVES

ANSWER KEY

1	(d)	11	(a)	21	(c)	31	(a)	41	(c)	51	(b)
2	(a)	12	(b)	22	(b)	32	(c)	42	(c)	52	(c)
3	(d)	13	(c)	23	(a)	33	(b)	43	(a)	53	(c)
4	(a)	14	(d)	24	(c)	34	(b)	44	(d)	54	(d)
5	(c)	15	(b)	25	(c)	35	(a)	45	(c)	55	(d)
6	(d)	16	(a)	26	(b)	36	(a)	46	(d)	56	(d)
7	(c)	17	(a)	27	(c)	37	(c)	47	(c)	57	(c)
8	(c)	18	(c)	28	(a)	38	(a)	48	(b)	58	(b)
9	(b)	19	(a)	29	(c)	39	(b)	49	(c)	59	(c)
10	(d)	20	(b)	30	(a)	40	(a)	50	(a)	60	(a)

Hints & Solutions

1. (d) An open organ pipe produces both even and odd harmonics.
2. (a) Fundamental frequency of an organ pipe, $\nu \propto \frac{1}{L}$
3. (d) A wave function can be represented by electric field, magnetic field, displacement, pressure, density, etc.
4. (a) The direction of motion is always perpendicular to the line joining the source and the listener, hence apparent frequency is equal to the actual frequency.
5. (c) As the star is moving towards the earth, the apparent frequency of light must be greater than its actual frequency.
6. (d) For propagation of a wave, a medium must possess elasticity, inertia and low resistance.
7. (c) Free surface of liquids tends to regain its shape due to its surface tension and causes a transverse wave to propagate on the surface.
8. (c) Shock waves do not show Doppler's effect.
9. (b) For producing standing waves, the superposition waves must have same amplitude & speed, a phase difference of π and must be travelling in opposite direction.
10. (d) $I_{non} = \left(\sqrt{I_1}+\sqrt{I_2}\right)^2 = \left(2\sqrt{I}\right)^2 = 4I$
11. (a) $v' = \frac{v+v_0}{v}\nu$

    where v = velocity of sound

    and v_0 = velocity of observer

    Let a_0 be the acceleration of observer, then, $v_0 = a_0 t$

    $\therefore\ v' = \frac{v+a_0 t}{v}\nu$
12. (b) Velocity of sound in any gas depends upon density and elasticity of gas.
13. (c) Intensity is proportional to $(\text{amplitude})^2$ and also intensity α $(\text{frequency})^2$. Therefore, intensity becomes $\frac{2^2}{4^2} = \frac{1}{4}$th
14. (d) Third overtone has a frequency $7n$, which means $L = \frac{7\lambda}{4}$ = three full loops + one half loop, which would make four nodes and four antinodes.
15. (b) With the propagation of a longitudinal wave, energy alone is propagated.
16. (a) Sound waves in air are longitudinal and the light waves are transverse.
17. (a) Frequency of wave is a function of the source of waves. Therefore, it remains unchanged.
18. (c) Water waves are both longitudinal and transverse.
19. (a) $\frac{\text{Speed of body}}{\text{Speed of sound}} = \text{Mach number}$
20. (b) The position of such a wave changes in two dimensional plane with time.
21. (c) In ordinary talk, the amplitude of vibration is about 10^{-8} m
22. (b) As density of hydrogen is less than density of air, velocity of sound in hydrogen is greater than vel. of sound in air. Therefore hydrogen acts as a rarer medium and the balloon behaves as a diverging lens.
23. (a) Energy is not propagated in stationary waves
24. (c) There is no phase difference between the vibrations of two prongs of a fork.
25. (c) Damping is caused by opposing force, which decreases the frequency.
26. (b) When tube is open, $n = \frac{v}{2\ell}$, where n is fundamental frequency of open organ pipe in air When half of tube is dipped vertically in water, it behaves as a closed pipe of length $\frac{\ell}{2}$, so fundamental frequency n in this case is

    $\therefore\ n' = \frac{v}{4(\ell/2)} = \frac{v}{2\ell} = n$
27. (c) $n_1 = \frac{v}{4(\ell+x)}$, ($n_1$ is fundamental frequency of closed organ pipe)

    $n_2 = \frac{v}{4\left(\frac{\ell}{2}+x\right)} = \frac{2v}{4(\ell+2x)}$, where x is end correction.

    Clearly, n_2 is less than double of n_1.
28. (a) Frequency does not depend upon radius. As length is doubled, fundamental frequency becomes half.
29. (c) The apparent frequency remains the same because distance between the source and observer does not change.
30. (a) The speed of light in free space is always a constant. It is independent of the relative motion of the source and the observer.
31. (a) Velocity of sound in a gas is

    $v = \sqrt{\gamma P/\rho}$ and from $P = \frac{1}{3}\rho v_{rms}^2$

    $v_{rms} = \sqrt{3P/\rho}\ \therefore\ \frac{v}{v_{rms}} = \sqrt{\frac{\gamma}{3}}$

32. (c) For ultrasonics, freq. $n > 20000$ c/s

$$\lambda = \frac{v}{n} < \frac{330}{20000} \approx 10^{-2}\text{ m}$$

33. (b) For infrasonics, freq. $n < 20$ c/s

$$\lambda = \frac{v}{n} > \frac{330}{20} \approx 15\text{m} \approx 10^{1}\text{ m}$$

34. (b) On reflection at a denser medium, change of phase = π (radian)

35. (a) On reflection at a rarer medium, no change of phase occures

36. (a) For a perfectly rigid rod, $\eta = \infty$

$$\therefore \quad v = \sqrt{\eta/\rho} \rightarrow \text{infinite}$$

37. (c) Intensity = energy/sec/area = power/area
From a point source, energy spreads over the surface of a sphere of radius r.

$$\therefore \quad \text{Intensity} = \frac{P}{A} = \frac{P}{4\pi r^2} \propto \frac{1}{r^2}$$

But Intensity = (Amplitude)^2

$$\therefore \quad (\text{Amplitude})^2 \propto \frac{1}{r^2}$$

At a distance 2 r, amplitude becomes $A/2$

38. (a) The correct form of differential eqn. of a wave is

$$\frac{d^2y}{dt^2} = v^2\frac{d^2y}{dx^2}$$

39. (b) The standard equation of a progressive wave is

$$y = a\sin\left[2\pi\left(\frac{t}{T} - \frac{x}{\lambda}\right) + \phi\right]$$

The given equation can be written as

$$y = 4\sin\left[2\pi\left(\frac{t}{10} - \frac{x}{18}\right) + \frac{\pi}{6}\right]$$

∴ $a = 4$ cm, $T = 10$ s, $\lambda = 18$ cm and $\phi = \pi/6$
Hence, (b) is correct.

40. (a) $m = \dfrac{0.035}{5.5}\text{ kg/m}, T = 77N$

$$v = \sqrt{\frac{T}{m}} = \sqrt{\frac{77 \times 5.5}{0.035}} = 110\text{ m/s}$$

41. (c) From $\Delta x = \dfrac{\lambda}{2\pi}\Delta\phi,$

$$\lambda = 2\pi\frac{\Delta x}{\Delta\phi} = \frac{2\pi(0.4)}{1.6\pi} = 0.5\text{ m}$$

$$f = \frac{v}{\lambda} = \frac{330}{0.5} = 660\text{ Hz}$$

42. (c) Compare the given equation with standard form

$$y = r\sin\left[\frac{2\pi x}{\lambda} - \frac{2\pi t}{T}\right]$$

$$\frac{2\pi}{\lambda} = 3, \lambda = \frac{2\pi}{3} \text{ and } \frac{2\pi}{T} = 15$$

$$T = \frac{2\pi}{15}$$

Speed of propagation,

$$v = \frac{\lambda}{T} = \frac{2\pi/3}{2\pi/15} = 5$$

43. (a) Here, $\nu' = \dfrac{9}{8}\nu$

Source and observer are moving in opposite direction, therefore, apparent frequency

$$\nu' = \nu \times \frac{(v+u)}{(v-u)}$$

$$\frac{9}{8}\nu = \nu \times \frac{340+u}{340-u}$$

$$\Rightarrow \quad 9 \times 340 - 9u = 8 \times 340 + 8u$$

$$\Rightarrow 17u = 340 \times 1 \Rightarrow u = \frac{340}{17} = 20\text{ m/sec.}$$

44. (d) Comparing with the equation

$$\xi = A\cos(2\pi\nu t + \phi)$$

We get $2\pi\nu = 2000$ $\nu = \dfrac{2000}{2\pi} = \dfrac{1000}{\pi}\text{Hz}$

45. (c) Speed of a wave represented by the equation

$y(x, t) = A\sin(kx - \omega t + \phi)$ is $v = \dfrac{\omega}{k}$

By comparison, $\omega = 4\pi$, $k = 0.5\pi$

$$v = \frac{\omega}{k} = \frac{4\pi}{0.5\pi} = 8\text{ m/sec}$$

46. (d) Reverberation time is defined as the time during which the intensity of sound in the auditorium becomes one millionth of the initial intensity. Sabine has shown that standard reverberation time for an auditorium is given by the formula

$$T_R = K\frac{V}{\alpha S}$$

Here, V is volume of the auditorium, S is the surface area. So, $T_R = \dfrac{K.V}{\alpha\, S} = 1$ (given)

$$T_R = \frac{K \cdot l^3}{6\alpha l^2}$$

(Assuming auditorium to be cubic in shape)

$$= \frac{K}{6\alpha} l$$

So, $T_R \alpha\, l$

If dimension is doubled, reverberation time t will be doubled. So,

New $T_R = 2$ sec.

47. (c) Frequencies of sound waves are $\frac{330}{5}$ & $\frac{330}{5.5}$

i.e., 66 Hz and 60 Hz

Frequencies of beat = 66 – 60 = 6 per second

48. (b) Here, $T = 0.05$ sec, $v = 300\ ms^{-1}$.

Now $\lambda = \frac{v}{v} = vT = (300 \times 0.05)$m

or, $\lambda = 15$ m

Phase of the point at 10 m from the source

$$= \frac{2\pi}{\lambda} \times x = \frac{2\pi}{15} \times 10 = \frac{4\pi}{3} \text{rad}$$

Phase of the point at 15 m from the source

$$\frac{2\pi}{\lambda} \times x = \frac{2\pi}{15} \times 15 = 2\pi \text{rad}$$

∴ The phase difference between the points

$$= 2\pi - \frac{4\pi}{3} = \frac{2\pi}{3} \text{rad}$$

49. (c) $y = 0.25 \sin(10\pi x - 2\pi t)$

Comparing this equation with the standard wave equation

$$y = a \sin(kx - \omega t)$$

We get, $k = 10\pi$

$$\Rightarrow \frac{2\pi}{\lambda} = 10\pi \Rightarrow \lambda = 0.2 \text{ m}$$

And $\omega = 2\pi$ or, $2\pi v = 2\pi \Rightarrow v = 1$Hz.

The sign inside the bracket is negative, hence the wave travels in + ve x-direction.

50. (a) The frequency of vibration of a string is given by, $f = \frac{1}{2l}\sqrt{\frac{T}{m}}$ where m is mass per unit length.

$$f_1 = \frac{1}{2l_1}\sqrt{\frac{T}{m}},\ f_2 = \frac{1}{2l_2}\sqrt{\frac{T}{m}},$$

$$f_2 - f_1 = \frac{1}{2}\sqrt{\frac{T}{m}}\frac{(l_1 - l_2)}{l_1 l_2}$$

$$\sqrt{\frac{T}{m}} = \sqrt{\frac{20}{10^{-3}}} = \sqrt{2} \times 10^2 = 1.414 \times 100$$

$$= 141.4$$

$$\frac{l_1 - l_2}{l_1 l_2} = \frac{(51.6 - 49.1) \times 10^2}{51.6 \times 49.1}$$

$$= \frac{2.5 \times 10^2}{50 \times 50} = \frac{1}{10}$$

$$\therefore f_2 - f_1 = \frac{1}{2} \times 141.4 \times \frac{1}{10} = 7 \text{ beats}$$

51. (b) f (S) → 30 ms^{-1} → $|f'$ O; f'' (O) → 30 ms^{-1} ← $|f'$ S

f' is the apparent frequency received by an observer at the hill. f'' is the frequency of the reflected sound as heard by driver.

$$f' = \frac{v}{v - 30} f,$$

$$f'' = \frac{v + 30}{v} f' = \frac{v + 30}{v - 30} f = \frac{360}{300} \times 600$$

$$= 720 \text{ Hz}$$

52. (c) $A = 2$ cm, $\frac{\omega}{k} = 128\ ms^{-1}$, $5\lambda = 4$, $\lambda = \frac{4}{5}$m

$y = A \sin(kx - \omega t)$,

$$k = \frac{2\pi}{\lambda} = \frac{2\pi \times 5}{4} = \frac{31.4}{4} = 7.85$$

$y = 0.02$ m $\sin(7.857 - 1005\ t)$

$\omega = 128 \times 7.85 = 1005$

53. (c) $y = A \sin(\omega t - kx)$

Particle velocity,

$$v_p = \frac{dy}{dt} = A\,\omega \cos(\omega t - kx)$$

$$\therefore v_{p\,max} = A\,\omega$$

$$\text{wave velocity} = \frac{\omega}{k}$$

$$\therefore A\omega = \frac{\omega}{k}$$

i. e., $A = \frac{1}{k}$ But $k = \frac{2\pi}{\lambda}$

$$\therefore \lambda = 2\pi A$$

54. (d) The frequency of the piano string $= 512 \pm 4 = 516$ or 508. When the tension is increased, beat frequency decreases to 2, it means that frequency of the string is 508 as frequency of string increases with tension.

55. (d) As number of beats/sec = diff. in freq-uencies has to be less than 10, therefore $0 < (n_1 - n_2) < 10$

56. (d) In the spherical source, the amplitude A of wave is inversely proportional to the distance r i.e.,

$$A \propto \frac{1}{r}$$

where r is distance of source from the point of consideration.

57. (c) The surface area of cylindrical surface is $2\pi r\ell$, where r is radius of base, ℓ is length of cylinder. We know that with increasing distance from source, the total energy or power transmitted remains same, but intensity decreases. For any source of power P, intensity I at distance r from it will be $I = P/S$, where S is surface area.

(i) For spherical wave front $\Rightarrow S = 4\pi r^2$(surface)

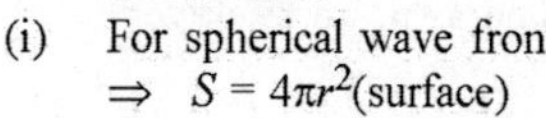

So $I = \left(\frac{P}{4\pi}\right)\left(\frac{1}{r^2}\right) \Rightarrow I \propto \frac{1}{r^2}$

Since $I \propto (\text{Amplitude } A)^2 \Rightarrow A \propto \frac{1}{r}$

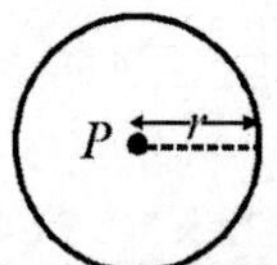

(ii) For cylindrical wave front $S = 2\pi r\ell$

So $I = \left(\frac{P}{2\pi\ell}\right)\left(\frac{1}{r}\right) \Rightarrow I \propto \frac{1}{r}$ So $A \propto \frac{1}{\sqrt{r}}$

58. (b) Using $v^2 = \omega^2 (a^2 - y^2)$ we have

$10^2 = \omega^2 (a^2 - 4^2)$

and $8^2 = \omega^2 (a^2 - 5^2)$;

so $10^2 - 8^2 = \omega^2 (5^2 - 4^2) = (3\omega)^2$

or $6 = 3\omega$ or $\omega = 2$

or $T = 2\pi/\omega = 2\pi/2 = \pi$ s.

59. (c) $v_{max} = 100 = a\omega$;

$\omega = 100/a = 100/10 = 10$ rad/s

$v^2 = \omega^2 (a^2 - y^2)$ or $50^2 = 10^2 (10^2 - y^2)$ or $25 = 100 - y^2$

or $y = \sqrt{75} = 5\sqrt{3}$ cm.

60. (a) $v = \omega\sqrt{a^2 - y^2}$

At $x = 0$, $v = \omega\sqrt{a^2 - o^2} = \omega a$.

At $x = \frac{a}{2}, v' = \omega\sqrt{a^2 - \left(\frac{a}{2}\right)^2} = \omega\sqrt{\frac{3a^2}{4}}$

$\therefore \frac{v'}{v} = \frac{\sqrt{3}}{2}$

or $v' = \frac{\sqrt{3}}{2}\omega a = \frac{\sqrt{3}\pi a}{2}\omega a = \frac{\sqrt{3}\pi a}{T}$

$\left(\because \omega = \frac{2\pi}{T}\right)$

15. ELECTRIC CHARGES AND FIELDS

ANSWER KEY

1	(b)	11	(d)	21	(d)	31	(a)	41	(b)	51	(b)
2	(c)	12	(b)	22	(d)	32	(b)	42	(a)	52	(b)
3	(c)	13	(d)	23	(a)	33	(b)	43	(c)	53	(b)
4	(d)	14	(b)	24	(b)	34	(c)	44	(a)	54	(d)
5	(b)	15	(c)	25	(b)	35	(a)	45	(c)	55	(c)
6	(d)	16	(c)	26	(c)	36	(a)	46	(a)	56	(b)
7	(d)	17	(a)	27	(d)	37	(d)	47	(c)	57	(b)
8	(c)	18	(a)	28	(c)	38	(d)	48	(c)	58	(a)
9	(c)	19	(d)	29	(c)	39	(a)	49	(d)	59	(b)
10	(d)	20	(a)	30	(b)	40	(b)	50	(d)	60	(a)

Hints & Solutions

1. (b) Charging by induction involves transfer of charges from one part to the other of the body. No loss of charge is involved.
2. (c) Positive charge is due to deficiency of electrons.
3. (c) On charging by conduction, body may gain mass, if it acquires negative charge. It may lose mass, if it acquires positive charge.
4. (d) Coulomb's law is true for all distances small and large. Hence it is called a long range force.
5. (b) $F_m = \frac{F_0}{K}$. For metals $K = \infty$, therefore, $F_m \to 0$, in the thickness of sheet. Hence force between the charges will decrease.
6. (d) Ebonite is the best insulator.
7. (d) According to band theory, distinction between conductors, insulators and semiconductors is based on relative width of energy gaps between valence band and conduction band.
8. (c) Net charge acquired by induction is zero, as there is only transfer of electrons from one part of body to the other.
9. (c) A loses electrons and B gains electrons. Therefore, mass of A < mass of B.
11. (d) There may be two case

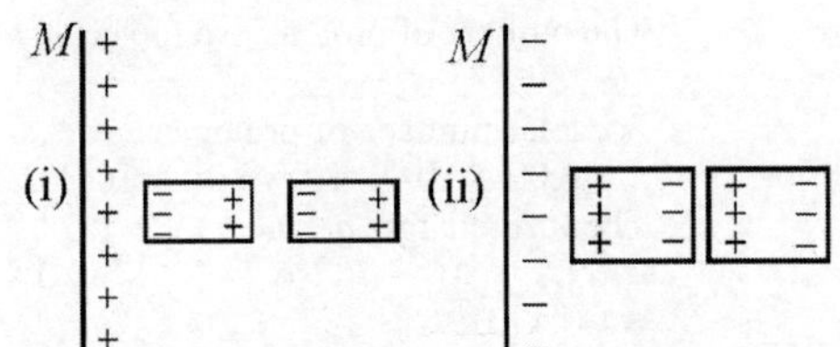

12. (b) The radius of soap bubble increases because of outward force acting on the bubble due to charging.
13. (d) $q = \pm ne$ shows that minimum value of $q = \pm 1e$.
where e = 1.6×10^{-19} Coulomb = Charge of one electron
14. (b) $F_g = \frac{1 q_1 q_2}{4\pi\epsilon_0 r^2} = \frac{F_\upsilon}{K}$

As K (for glass) > 1. $\therefore F_g < F_\upsilon$
15. (c) The force of q_1 or q_2 is not affected by q_3. However, resultant force on q_2 will be affected.
16. (c) Positively charged rod induces negative charge on the side of the sphere closer to the rod and an equal positive charge on the side away from the rod. When the rod is withdrawn, the negative charges move to neutralise the equal positive charge. The charge left is zero.
17. (a) In contact, A and B behave as one body. Therefore, by induction, positive charge developes on A and negative charge developes on farther side of B.
18. (a) $\frac{F_e}{F_p} = 10^0 = 1$, because force depends only on charges, distance and nature of medium.
19. (d) Attractive force between two neutrons may be due to gravitational forces or due to nuclear forces.
20. (a) Metal is a good conductor. So, its $K = \infty$.
22. (d) The weight can be increased slightly, if it acquire negative charge & weight can be decreased slightly, if it acquires positive charge.
23. (a) Positive electric flux is due to + charge. So lines of force are directed outwards.
24. (b) Negative electric flux is due to negative charge. Therefore, lines of force are directed inwards.
26. (c) As area of a point is zero,

$\therefore \phi = E\,(ds)\cos\theta = E\cos\theta \times 0 = \text{Zero}.$
27. (d) Though electric flux is scalar, yet its value may be positive, negative or zero.
30. (b) The positive & negative charges developed on different parts of conducting sphere due to induction. Hence the flux through the closed surface is positive.

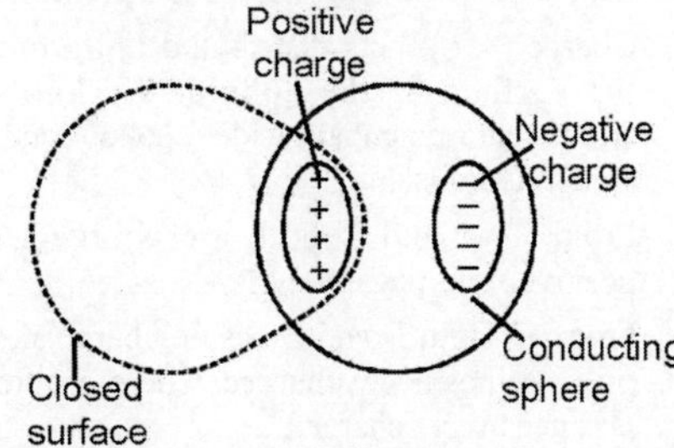

32. (b) Debye is the unit of electric dipole moment.
33. (b) Net force (as well as torque) are experienced by the dipole in a nonuniform electric field.
34. (c) In a uniform electric field, net force = 0, but torque $\neq 0$.
35. (a) In a nonuniform electric field, net force $\neq 0$ and net torque $\neq 0$.
36. (a) Figure indicates the presence of some positive charge to the left of A.

$\therefore E_A > E_B \quad (\because r_A < r_B)$
37. (d) An electric field can deflect only a charged particle.

38. (d) The path is a parabola, because initial velocity can be resolved into two rectangular components, one along $\vec{E}$ and other $\perp$ to $\vec{E}$. The former decreases at a constant rate and latter is unaffected. The resultant path is therefore a parabola.

39. (a) Work is done by the electric field because $\theta = 0°$, $W = F s \cos\theta \rightarrow$ positive.

40. (b) $E = 0$, at any point inside the sphere.

41. (b) E decreases as we move from the surface of the sphere to its centre.

42. (a) A spherical cloud behaves as a solid sphere. Therefore, E decreases directly with the decreasing distance from the centre.

43. (c) For a single negative point charge, electric lines of force are radial and inwards.

44. (a) Electric field inside a conductor is zero.

46. (a) Because it is moving in a equipotential surface, hence the work done in complete rotation is zero.

47. (c) An electric charge in uniform motion produces both an electric and a magnetic field.

48. (c) $\dfrac{a_e}{a_m} = \dfrac{F_e/m_e}{F_p/m_p} = \dfrac{m_p(eE)}{m_e(eE)} = \dfrac{m_p}{m_e}$.

49. (d) Electric lines of force are imaginary.

50. (d) The charge q, which is kept at the centre of metallic spherical shell transfered to the outer surface of shell and inside the shell the electric field is zero and hence force is also zero.

52. (b) Charge $(+Q)$ gets distributed uniformly over the surface of the sphere. We know that electrostatic potential inside a solid, conducting sphere is constant.

55. (c) From (i) A and C both are charged, either positively or negatively.
From (ii) Both D and E has no charge and from (iii), A is positively charged. Therefore from (i), B is negatively charged.

56. (b) There exists a point P on the x-axis (other than the origin), where net electric field is zero. Once the charge Q reaches point P, attractive forces of the two –ve charge will dominate and automatically cause the charge Q to cross the origin.
Now if Q is projected with just enough velocity to reach P, its K.E. at P is zero, but while being attracted towards origin it acquires K.E. and hence its net energy at the origin is positive. (P.E. at origin = zero)

57. (b)

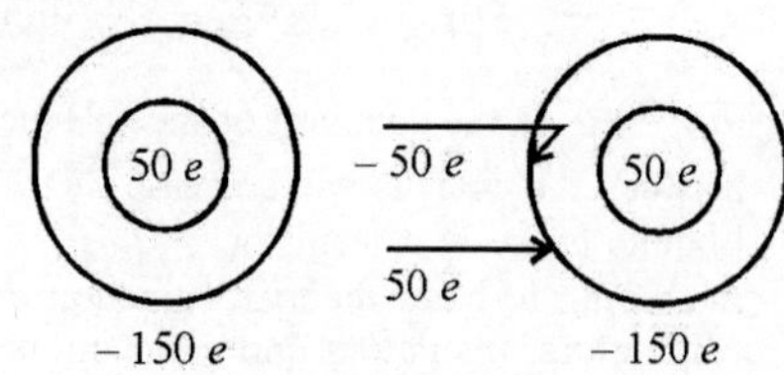

The charge on inner and outer surfaces $= -50\,e, -100\,e$

58. (a) The number of molecules in 18 g of water $= 6.02 \times 10^{23}$

$\therefore$ Number of molecules in 180 g of water

$$= 6.02 \times 10^{23} \times \frac{180}{18} = 6.02 \times 10^{24}$$

The number of protons in a molecule of water = 10
So, total number of protons,
$n = 6.02 \times 10^{24} \times 10 = 6.02 \times 10^{25}$
The total charge, $q = ne$
$= 6.02 \times 10^{25} \times 1.6 \times 10^{-19} = 0.96 \times 10^{7}$
$\approx 1.0 \times 10^{7}$ C.

59. (b) For the equilibrium the net force on the third charged particle must be zero. This can be possible only when positively charged particle is placed between the given charges.

60. (a) The field lines in 1 are :
(6 + 7) out – (2 + 7+ 15 + 8) in = 19 (in)
It represents negative charge.
The field lines in 2 are :
(9 + 5 + 6) out – (7+ 3 + 2) in = 8 (out)
It represents positive charge.

16. ELECTROSTATIC POTENTIAL AND CAPACITANCE

ANSWER KEY

1	(d)	11	(a)	21	(b)	31	(a)	41	(c)	51	(a)
2	(b)	12	(a)	22	(d)	32	(d)	42	(a)	52	(c)
3	(b)	13	(d)	23	(c)	33	(a)	43	(c)	53	(b)
4	(c)	14	(d)	24	(a)	34	(a)	44	(a)	54	(b)
5	(d)	15	(d)	25	(a)	35	(d)	45	(a)	55	(d)
6	(d)	16	(c)	26	(c)	36	(d)	46	(c)	56	(b)
7	(a)	17	(c)	27	(b)	37	(c)	47	(b)	57	(b)
8	(b)	18	(c)	28	(c)	38	(b)	48	(b)	58	(a)
9	(c)	19	(b)	29	(c)	39	(a)	49	(c)	59	(c)
10	(a)	20	(a)	30	(b)	40	(b)	50	(c)	60	(d)

Hints & Solutions

1. (d) The energy stored in the capacitor $=\frac{1}{2}CV^2$

$$=\frac{1}{2}\times 4\times 10^{-6}\times(400)^2=0.325$$

This energy will be converted into heat in the resistor.

2. (b) Potential at the centre of the sphere = potential on the surface = 80 V.

3. (b) Charge on each plate of each capacitor

$$Q=\pm CV=\pm 25\times 10^{-6}\times 200$$

$$=\pm 5\times 10^{-3}\,C$$

4. (c) Capacitance of capacitor (C) = 6 μF = 6×10^{-6} F, Initial potential (V_1) = 10 V and final potential (V_2) = 20 V.
The increase in energy (ΔU)

$$=\frac{1}{2}C(V_2^2-V_1^2)$$

$$=\frac{1}{2}\times(6\times 10^{-6})\times[(20)^2-(10)^2]$$

$$=(3\times 10^{-6})\times 300=9\times 10^{-4}J.$$

5. (d) When oil is placed between space of plates

$$C=\frac{2A\varepsilon_0}{d} \quad ...(1)$$

$$\left[\because C=\frac{KA\varepsilon_0}{d},\ \text{if } K=2\right]$$

When oil is removed $C'=\frac{A\varepsilon_0}{d}$ ...(2)

On comparing both equations, we get $C'=C/2$

6. (d)

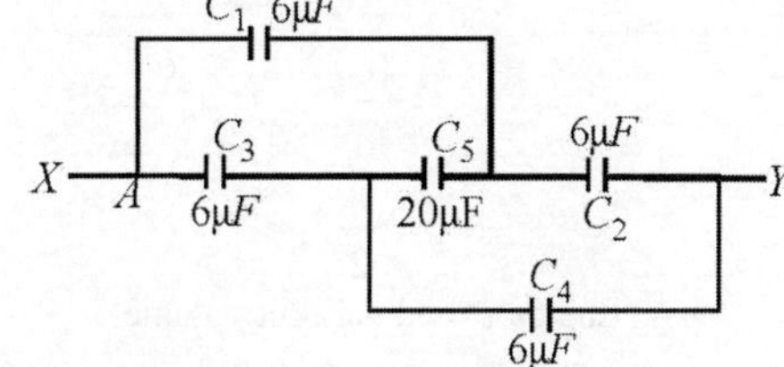

Equivalent circuit

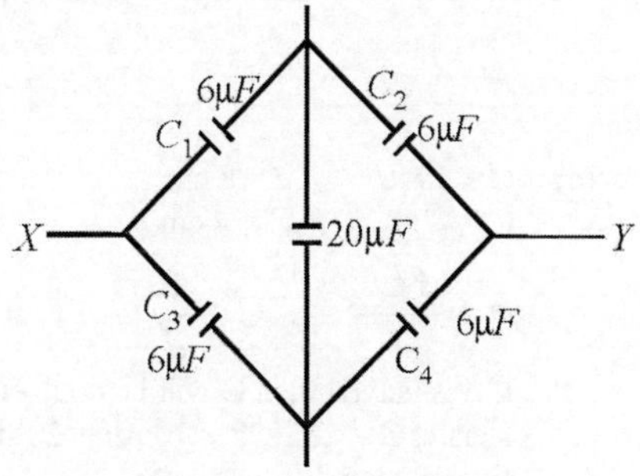

Here, $\frac{C_1}{C_3}=\frac{C_2}{C_4}$

Hence, no charge will flow through 20μF

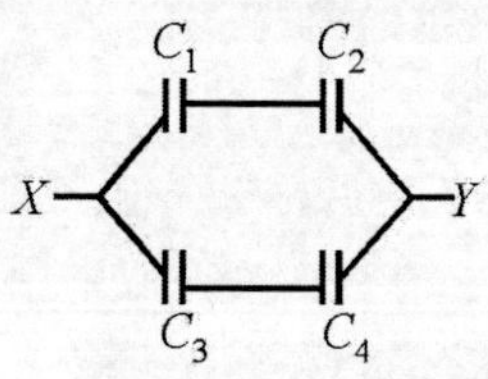

$\Rightarrow$ 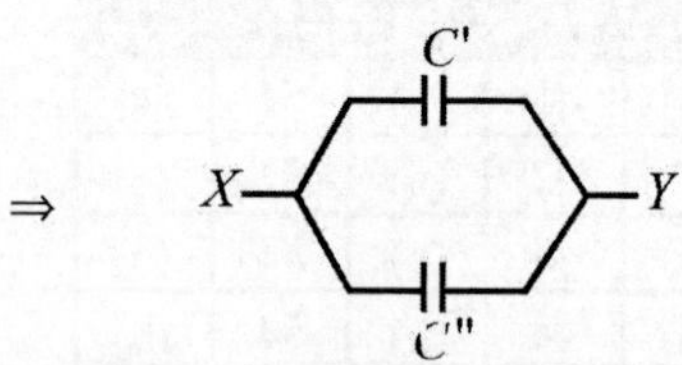

C_1 and C_2 are in series, also C_3 and C_4 are in series.
Hence, $C' = 3\ \mu F, C'' = 3\ \mu F$
C' and C'' are in parallel.
Hence net capacitance $= C' + C'' = 3 + 3$
$= 6\ \mu F$

7. (a) In Ist case when capacitor C connected with battery charged with the energy,
$U_1 = U$ (stored energy on capacitor).
In IInd case, after disconnection of battery similar capacitor is connected in parallel with Ist capacitor then $C_{eq} = C' = 2C$.

Now, $\dfrac{U_1}{U_2} = \dfrac{\frac{1}{2}\frac{q^2}{C}}{\frac{1}{2}\frac{q^2}{C'}} = \dfrac{C'}{C}$

$= \dfrac{2C}{C} \quad (\because C' = 2C)$

$U_2 = \dfrac{U}{2}$

8. (b) Energy stored per unit volume

$$= \frac{1}{2}\varepsilon_0 E^2 = \frac{1}{2}\varepsilon_0\left(\frac{V}{d}\right)^2 = \frac{1}{2}\varepsilon_0\frac{V^2}{d^2}$$

$$\left(\because E = \frac{V}{d}\right)$$

9. (c) $C = 10\ \mu F \quad d = 8$ cm
$C' = ? \quad d' = 4$ cm

$C = \dfrac{A\epsilon_0}{d} \Rightarrow C \alpha \dfrac{1}{d}$

If d is halved then C will be doubled.
Hence, $C' = 2C = 2 \times 10\ \mu F = 20\ \mu F$

10. (a) Energy stored in capacitor

$$= \frac{1}{2}CV^2 = \frac{1}{2}\frac{Q}{V}V^2 \quad (Q = CV)$$

$$= \frac{1}{2}QV$$

11. (a) Charge $Q = C_1 V$
Total capacity of combination (parallel)
$C = C_1 + C_2$

P.D. $= \dfrac{Q}{C} = \dfrac{C_1 V}{C_1 + C_2}$

12. (a) Electric potential is constant
(equal to $\dfrac{kq}{R}$, where $k = \dfrac{1}{4\pi\epsilon_0}$)
within or on the surface of conductor.

13. (d) Length of body diagonal $= \sqrt{3}\,l$
$\therefore$ Distance of centre of cube from each corner

$r = \dfrac{\sqrt{3}}{2}l$

P.E. at centre
$= 8 \times$ Potential Energy due to A

$= 8 \times \dfrac{Kq \times (-q)}{r}$

$= 8 \times \dfrac{1}{4\pi\varepsilon_0\sqrt{3}\,l} \times 2 \times q \times (-q) = \dfrac{-4q^2}{\sqrt{3}\pi\varepsilon_0 l}$

14. (d) For series, $C' = \dfrac{C_1 \times C_2}{C_1 + C_2} = \dfrac{4 \times 4}{4+4} = 2\mu F$

For parallel, $C_{eq} = C' + C_3 = 2 + 4 = 6\mu F$

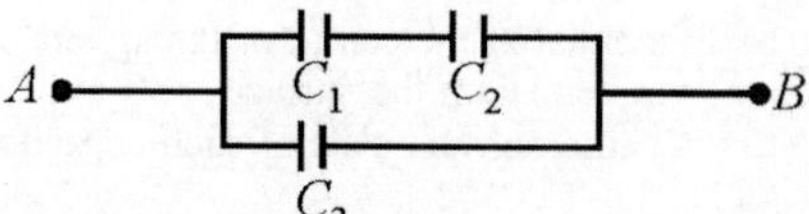

15. (d) When the dipole is in the direction of field then net force is $qE + (-qE) = 0$

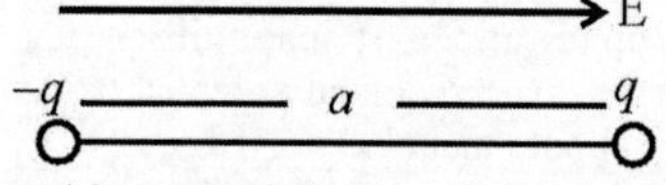

and its potential energy is minimum
$= -$ P.E. $= -qaE$

16. (c) $qV = \dfrac{1}{2}mv^2$

$\Rightarrow 2 \times 10^{-6} \times V = \dfrac{1}{2} \times \dfrac{2}{1000} \times 10 \times 10$

$\therefore V = 50\ kV$

17. (c) Energy $= eV = 1.6 \times 10^{-19} \times 1$

18. (c) The component of electric field in any direction is negative of the rate of change of electric potential with distance in that direction.

$\therefore E_x = -\dfrac{dV}{dx}$

19. (b) As, $E = -\dfrac{dV}{dr}$ or, $0 = -\dfrac{dV}{dr}$
because electric field inside a charged hollow sphere is zero.
or, dv = constant

20. (a) $AC = BD = \sqrt{(\sqrt{2})^2 + (\sqrt{2})^2} = 2$ m

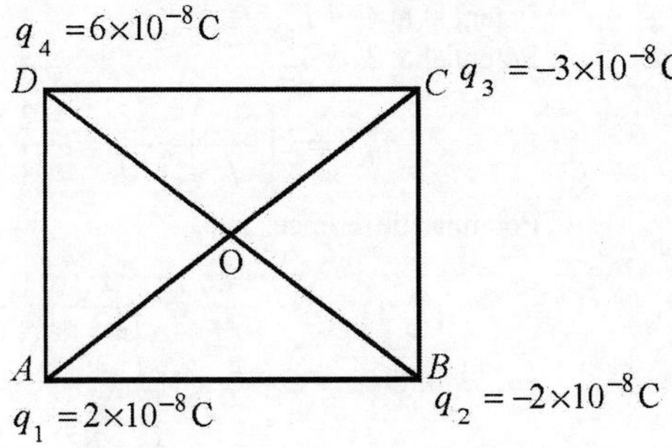

$\therefore$ $DO = OB = AO = OC = \dfrac{2}{2} = 1\text{m}$

$\therefore$ Potential at the centre O, $V = k\dfrac{q}{r}$

$$V = k\left[\frac{2\times10^{-8}}{1} + \frac{-2\times10^{-8}}{1} + \frac{-3\times10^{-8}}{1} + \frac{6\times10^{-8}}{1}\right]$$

$$V = k\times3\times10^{-8} = 9\times10^{9}\times3\times10^{-8} \text{ volt}$$
$= 27 \times 10 = 270\ V$

21. (b) Initial capacitance $= \dfrac{\epsilon_0 A}{d}$
When it is half filled by a dielectric of dielectric constant K, then

$$C_1 = \frac{K\varepsilon_0 A}{d/2} = 2K\frac{\varepsilon_0 A}{d}$$

and $C_2 = \dfrac{\varepsilon_0 A}{d/2} = \dfrac{2\varepsilon_0 A}{d}$

$$\therefore \frac{1}{C'} = \frac{1}{C_1} + \frac{1}{C_2} = \frac{d}{2\varepsilon_0 A}\left(\frac{1}{K} + 1\right)$$

$$= \frac{d}{2\varepsilon_0 A}\left(\frac{1}{5} + 1\right) = \frac{6}{10}\frac{d}{\varepsilon_0 A}$$

$$C' = \frac{5\varepsilon_0 A}{3d}$$

Hence, % increase in capacitance

$$= \left(\frac{\frac{5}{3}\frac{\varepsilon_0 A}{d} - \frac{\varepsilon_0 A}{d}}{\frac{\varepsilon_0 A}{d}}\right) \times 100$$

$$= \left(\frac{5}{3} - 1\right)\times100 = \frac{2}{3}\times100 = 66.6\%$$

22. (d) $V = \dfrac{1}{4\pi\varepsilon_0}\dfrac{p\cos\theta}{r^2}$
Here, $V = 1.8 \times 10^5$V, $\theta = 60°$,
$r = 50 \times 10^{-2} = 0.5$ m

$$\therefore 1.8\times10^5 = 9\times10^9 \times \frac{p\cos60°}{(0.5)^2}$$

or $p = \dfrac{1.8\times10^5\times0.25\times2}{9\times10^9} = 10^{-5}$C-m

24. (a) From left hand side, the effective capacitance between the terminals is C and the ladder is infinite.
The effective network is shown in figure. Now, the capacitance between A and B is

$$C = \frac{4C}{C+4} + 2 = \frac{6C+8}{C+4}$$

or $C^2 + 4C - 6C - 8 = 0$
Solving $C = 4\ \mu F$

25. (a) We know that potential energy of two charge system is given by

$$U = \frac{1}{4\pi\epsilon_0}\frac{q_1 q_2}{r}$$

According to question,

$$U_A = \frac{1}{4\pi\epsilon_0}\frac{(+q)(-Q)}{a} = -\frac{1}{4\pi\varepsilon_0}\frac{Qq}{a}$$

and $U_B = \dfrac{1}{4\pi\epsilon_0}\dfrac{(+q)(-Q)}{a}$

$$= -\frac{1}{4\pi\varepsilon_0}\frac{Qq}{a}$$

$\Delta U = U_B - U_A = 0$
We know that for conservative force,
$W = -\Delta U = 0$

26. (c) We know that potential energy of discrete system of charges is given by

$$U=\frac{1}{4\pi\in_0}\left(\frac{q_1q_2}{r_{12}}+\frac{q_2q_3}{r_{23}}+\frac{q_3q_1}{r_{31}}\right)$$

According to question,

$$U_{\text{initial}}=\frac{1}{4\pi\in_0}\left(\frac{q_1q_2}{0.3}+\frac{q_2q_3}{0.5}+\frac{q_3q_1}{0.4}\right)$$

$$U_{\text{final}}=\frac{1}{4\pi\in_0}\left(\frac{q_1q_2}{0.3}+\frac{q_2q_3}{0.1}+\frac{q_3q_1}{0.4}\right)$$

$$U_{\text{final}}-U_{\text{initial}}=\frac{1}{4\pi\in_0}\left(\frac{q_1q_2}{0.1}-\frac{q_2q_3}{0.5}\right)$$

$$=\frac{1}{4\pi\in_0}[10q_2q_3-2q_2q_3]=\frac{q_3}{4\pi\in_0}(8q_2)$$

27. (b)

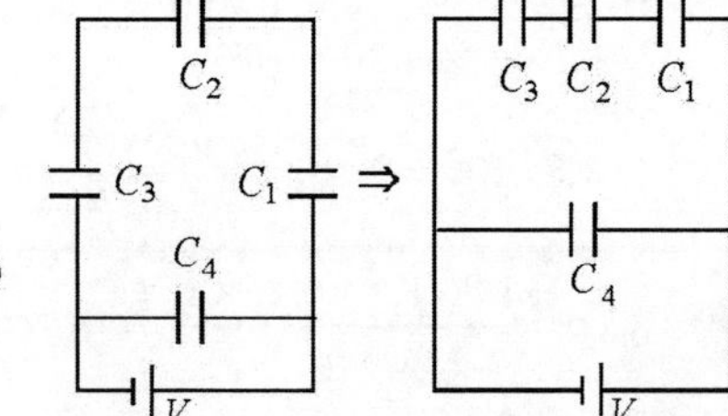

Equivalent capacitance for three capacitors (C_1, C_2 & C_3) in series is given by

$$\frac{1}{C_{eq.}}=\frac{1}{C_1}+\frac{1}{C_2}+\frac{1}{C_3}=\frac{C_2C_3+C_3C_1+C_1C_2}{C_1C_2C_3}$$

$$\Rightarrow \quad C_{eq.}=\frac{C_1C_2C_3}{C_1C_2+C_2C_3+C_3C_1}$$

$$\Rightarrow \quad C_{eq.}=\frac{C(2C)(3C)}{C(2C)+(2C)(3C)+(3C)C}$$

$$=\frac{6}{11}C$$

$\Rightarrow$ Charge on capacitors (C_1, C_2 & C_3) in series

$$=C_{eq}V=\frac{6C}{11}V$$

Charge on capacitor $C_4=C_4V=4C\,V$

$$\frac{\text{Charge on } C_2}{\text{Charge on } C_4}=\frac{\frac{6C}{11}V}{4CV}=\frac{6}{11}\times\frac{1}{4}=\frac{3}{22}$$

28. (c) If we increase the distance between the plates its capacity decreases resulting in higher potential as we know $Q = CV$. Since Q is constant (battery has been disconnected), on decreasing C, V will increase.

29. (c)

Potential at $C = V_C = 0$

Potential at $D = V_D$

$$=k\left(\frac{-q}{L}\right)+\frac{kq}{3L}=-\frac{2}{3}\frac{kq}{L}$$

Potential difference

$$V_D-V_C=\frac{-2}{3}\frac{kq}{L}=\frac{1}{4\pi\in_0}\left(-\frac{2}{3}.\frac{q}{L}\right)$$

$\Rightarrow$ Work done $= Q\,(V_D - V_C)$

$$=-\frac{2}{3}\times\frac{1}{4\pi\varepsilon_0}\frac{qQ}{L}=\frac{-qQ}{6\pi\varepsilon_0 L}$$

30. (b) Work done = Change in energy

$$=\frac{1}{2}\left(C+\frac{C}{2}\right)V^2=\frac{1}{2}\left(\frac{3C}{2}\right)V^2=\frac{3CV^2}{4}$$

31. (a)

$$\frac{1}{C_{\text{effective}}}=\frac{1}{3}+\frac{1}{3}+\frac{1}{3}=1$$

$\Rightarrow C_{\text{effective}} = 1\ \mu F$

32. (d) $E = 10^7$ V/m

$r = 0.10$ m

$$E=\frac{V}{r}$$

$$V=Er=10^7\times\frac{10}{100}$$

$V = 10^6\ V$

33. (a) Equivalent capacitance of circuit,

$$\frac{1}{C_{eq}}=\frac{1}{C_3}+\frac{1}{C_1+C_2}$$

(Since C_1 and C_2 are in parallel and which is in series with C_3).

i.e., $$\frac{1}{C_{eq}}=\frac{C_1+C_2+C_3}{C_3(C_1+C_2)}$$

$$\therefore \quad C_{eq}=\frac{C_3(C_1+C_2)}{C_1+C_2+C_3}$$

Since V is the voltage of battery,
charge, $q = C_{eq}V$

$$= \frac{C_3(C_1+C_2)V}{C_1+C_2+C_3}$$

If the capacitor C_3 breaks down, then effective capacitance,

$$C'_{eq} = C_1 + C_2$$

$\therefore$ New charge $q' = C'_{eq}V = (C_1+C_2)V$

Change in total charge $= q' - q$

$$= (C_1+C_2)V - \frac{C_3(C_1+C_2)V}{C_1+C_2+C_3}$$

$$= (C_1+C_2)V\left[1 - \frac{C_3}{C_1+C_2+C_3}\right]$$

34. (a) q ——— – 3q (1m)

Suppose x is the required distance from q, then

$$\frac{1}{4\pi\varepsilon_0}\left[\frac{9}{x} + \frac{(-3q)}{1-x}\right] = 0$$

After solving for x, we get $x = \frac{1}{4}$ m

36. (d) All (a), (b), (c) are not true. When dielectric is removed C decreases by a factor k. Hence (b) is not true.
$q = CV$ = constant. $\therefore$ V increases by a factor k. (c) is not true.
$E = V/d$. $\therefore$ E increases by a factor k. Hence (b) is not true.

37. (c) Before touching, force $\propto q_1 q_2$.
After touching and separating, force $\propto \left(\frac{q_1+q_2}{2}\right)^2$
After touching and then separating to the same distance, the force is greater than before.

38. (b) $V = \frac{q}{4\pi\varepsilon_0 a}$

39. (a) Due to the change inside a sphere of radius r only.

41. (c) At mid-points the field due to plates are oppositely directed.

42. (a) The distribution of charge on the outer surface, depends only on the charges outside and it distributes itself such that the net, electric field inside the outer surface due to the charge on outer surface and all the outer charges is zero. Similarly, the distribution of charge on the inner surface, depends only on the charges inside the inner surface, and it distributes itself such that the net, electric field outside the inner surface due to the charge on inner surface and all the inner charges is zero. Also the force on charge inside the cavity is due to the charge on the inner surface.

43. (c) The electric field inside the inner shell is zero. So, the potential on inner shell and all the points inside it will be constant.

44. (a) Work = Increase in potential energy

$$= Kq_1q_2\left(\frac{1}{r_2} - \frac{1}{r_1}\right)$$

$$= 9 \times 10^9 \times 8 \times 10^{-6} \times 12 \times 10^{-6}\left(\frac{1}{6\times10^{-2}} - \frac{1}{10\times10^{-2}}\right)$$

$$= 9 \times 8 \times 12\left(\frac{1}{6} - \frac{1}{10}\right) \times 10^{-1} = 5.8 \text{ J}$$

45. (a) $q = CV_1 \cos\omega t$

$$\Rightarrow \quad i = \frac{dq}{dt} = -\omega C v_1 \sin\omega t$$

Also, $\omega^2 = \frac{1}{LC}$ and $V = V_1 \cos\omega t$

At $t = t_1$, $V = V_2$ and $i = -\omega CV_1 \sin\omega t_1$

$\therefore$ $\cos\omega t_1 = \frac{V_2}{V_1}$ (–ve sign gives direction)

Hence, $i = V_1\sqrt{\frac{C}{L}}\left(1 - \frac{V_2^2}{V_1^2}\right)^{1/2}$

$$= \left(\frac{C(V_1^2 - V_2^2)}{L}\right)^{1/2}$$

46. (c) If q_1 and q_2 are the final charges, then
$q_1 + q_2 = 15 + 15 = 30$ ...(1)

and $\frac{1}{4\pi\varepsilon_0}\frac{q_1}{r_1} = \frac{1}{4\pi\varepsilon_0}\frac{q_2}{r_2}$

or $\frac{q_1}{q_2} = \frac{r_1}{r_2} = \frac{5}{10} = \frac{1}{2}$ ...(2)

Solving (i) & (ii), $q_1 = 10\mu C$

47. (b) The potential of A, V_a = + ve
$V_b = 0$
$V_d = 0$
V_c = – ve
Thus $V_a > V_b = V_d > V_c$

48. (b) By conservation of charge, the charge on bigger drop $Q = Nq$
If R is the radius of the bigger drop, then

$$\frac{4}{3}\pi R^3 = N \times \frac{4}{3}\pi r^3 \Rightarrow R = N^{1/3} r$$

For small drop $V = \frac{1}{4\pi\varepsilon_0}\frac{q}{r}$

The potential of bigger drop,

$$V_0 = \frac{1}{4\pi\varepsilon_0}\frac{Q}{R}$$

$$= \frac{1}{4\pi\varepsilon_0}\frac{Nq}{N^{1/3} r} = N^{2/3}\frac{1}{4\pi\varepsilon_0}\frac{q}{r}$$

$= N^{2/3} V$

49. (c) $U = \frac{1}{2}C_1 V^2 = \frac{1}{2} 1 \times 10^{-6} \times 4^2$
$= 8 \times 10^{-6}$ J

50. (c) V_1 V_2 V_3 V_4
–7 –5 –9 –8

51. (a) Work done = potential difference × charge
For BC,
$W = (V_C - V_B) \times Q = (-3 - 1) \times -2 = 8$J
(Work done depends only on initial and final position)
For BAD,
$W = (V_D - V_B) \times Q = (2 - 1) \times -2 = -2$J
For AD,
$W = (V_D - V_A) \times Q = (2 + 1) \times -2 = -6$J
For ABC,
$W = (V_C - V_A) \times Q = (-3) - (-1) \times -2 = +4$ J
Thus work is maximum for path BC

52. (c) $\frac{\varepsilon_0 A}{d}$, charge on plate is $Q = CV = \frac{\varepsilon_0 A V_0}{d}$
Energy stored is $W = Q^2/2C$
As d is increased 3 times, C decreases 3 times. Battery is disconnected, so Q remains same. The difference in the energy is the work done.

53. (b) If a dielectric slab of dielectirc constant K is filled in between the plates of a capacitor after charging the capacitor (i.e., after removing the connection of battery with the plates of capacitor) the potential difference between the plates reduces to $\frac{1}{K}$ times and the potential energy of capacitor reduces to $\frac{1}{K}$ times but there is no change in the charge on the plates.

54. (b) By Kirchoff first rule,
$q_1 = q_2 + q_3$

55. (d) At two different points, outside the sphere if $|E_1| = |E_2|$, then their potential must be equal.

56. (b) $E = \frac{V}{d} = \frac{V_0/k}{d} = \frac{V_0}{kd}$

57. (b) The capacitor is divided into two capacitors joined in series.

58. (a) A capacitor works in a.c. circuits only as it blocks d.c.

59. (c) Electrostatic energy of a condenser lies in the field in between the plates of the condenser.

60. (d) As the capacitor is isolated after charging, charge on it remains constant. Plate separation increases d, decreases $C = \epsilon_0 A/d$ and hence increases potential $V = Q/C$.

17. CURRENT ELECTRICITY

ANSWERKEY

1	(c)	11	(d)	21	(c)	31	(c)	41	(c)	51	(c)
2	(a)	12	(d)	22	(d)	32	(b)	42	(b)	52	(d)
3	(b)	13	(d)	23	(a)	33	(d)	43	(a)	53	(a)
4	(c)	14	(b)	24	(a)	34	(a)	44	(c)	54	(b)
5	(a)	15	(d)	25	(c)	35	(a)	45	(a)	55	(a)
6	(d)	16	(a)	26	(d)	36	(d)	46	(c)	56	(b)
7	(c)	17	(c)	27	(d)	37	(d)	47	(c)	57	(d)
8	(a)	18	(b)	28	(c)	38	(d)	48	(d)	58	(a)
9	(b)	19	(b)	29	(d)	39	(d)	49	(a)	59	(d)
10	(d)	20	(b)	30	(d)	40	(a)	50	(c)	60	(c)

Hints & Solutions

1. (c) Inside the battery or cell, the positive charge move from negative terminal (lower potential) to positive terminal (higher potential) of the battery. While in external circuit the positive charge move from positive terminal to negative terminal of the battery.

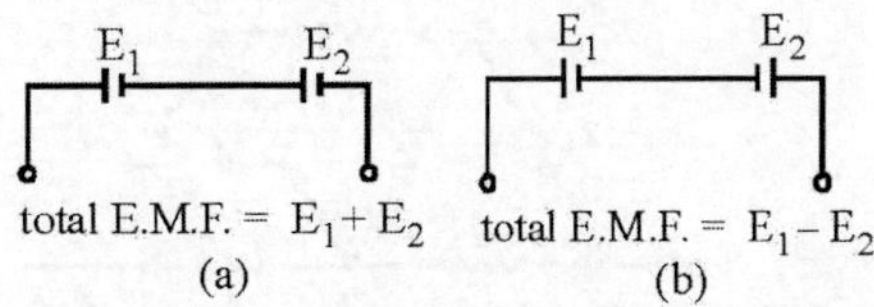

2. (a) $P = I^2R$, If we decrease R, then I increases hence P increases.

3. (b) If we add two non ideal batteries in series then

E_1 E_2	E_1 E_2
total E.M.F. = $E_1 + E_2$	total E.M.F. = $E_1 - E_2$
(a)	(b)

so in case (a) the equivalent e.m.f may be larger than either of two e.m.f, but in case (b) the equivalent e.m.f may be smaller than either of two e.m.f.

In series total resistance is always greater than individual resistance, whether batteries are connected in any way [i.e., either according to case (a) or case (b)].

7. (c) In series combination of cells, Current

$$I = \frac{nE}{nr + R'}$$

In parallel combination of cells,

$$I' = \frac{E}{(r/n) + R}$$

If $I = I'$ then

$$\frac{nE}{nr + R} = \frac{E}{(r/n) + R} = \frac{nE}{r + nR}$$

It will be so if $r = R$

8. (a) $I = \dfrac{E}{R + r}$

$$= \frac{E}{[(\sqrt{R})^2 + (\sqrt{r})^2 - 2\sqrt{2r}] + 2\sqrt{Rr}}$$

$$= \frac{E}{(\sqrt{R} - \sqrt{r})^2 + 2\sqrt{Rr}}$$

I will be maximum if $(\sqrt{R} - \sqrt{r}) = 0$ or $R = r$

10. (d) The equivalent circuit is as shown in figure. The resistance of arm AOD (= R + R) is in parallel to the resistance R of arm AD. Their effective resistance

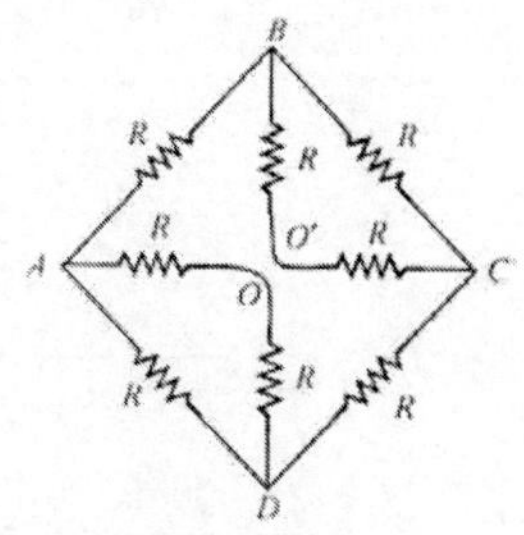

$$R_1 = \frac{2R \times R}{2R \times R} = \frac{2}{3}R$$

The resistance of arms AB, BC and CD is

$$R_2 = R + \frac{2}{3}R + R = \frac{8}{3}R$$

The resistance R_1 and R_2 are in parallel. The effective resistance between A and D is

$$R_3 = \frac{R_1 \times R_2}{R_1 + R_2} = \frac{\frac{2}{3}R \times \frac{8}{3}R}{\frac{2}{3}R + \frac{8}{3}R} = \frac{8}{15}R.$$

11. (d)

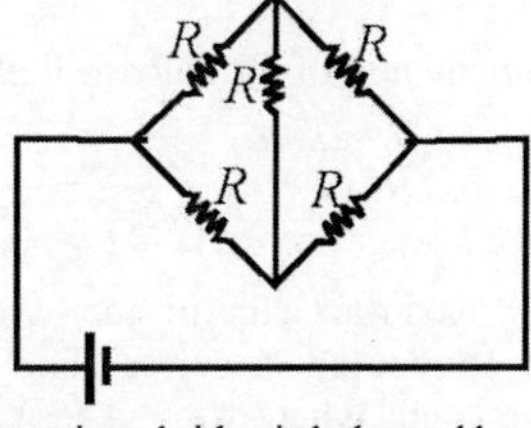

Hint ⇒ since bridge is balanced hence equivalent resistance is $\frac{1}{R'}$ = (½R + ½R)

⇒ $R' = R$

14. (b) When Wheatstone bridge is balanced, then $\frac{P}{Q} = \frac{R}{S}$ or $\frac{P}{R} = \frac{Q}{S}$. If the galvanometer is replaced with cell in balanced Wheatstone bridge, then the condition for balanced bridge will be $\frac{P}{R} = \frac{Q}{S}$, which is there: Hence balance point will remain unchanged, where galvanometer shows no current.

15. (d) $R=\frac{\rho l}{\pi r^2}$. But $m=\pi r^2\, ld \therefore \pi r^2=\frac{m}{ld}$

$$\therefore R=\frac{\rho l^2 d}{m},\ R_1=\frac{\rho l_1^2 d}{m_1},\ R_2=\frac{\rho l_2^2 d}{m_2}$$

$$R_3=\frac{\rho l_3^2 d}{m_3}$$

$$R_1:R_2:R_3=\frac{l_1^2}{m_1}:\frac{l_2^2}{m_2}:\frac{l_3^2}{m_3}$$

$$R_1:R_2:R_3=\frac{25}{1}:\frac{9}{3}:\frac{1}{5}=125:15:1$$

16. (a) $I=\frac{8-4}{1+2+9}=\frac{4}{12}=\frac{1}{3}A;$

$$V_P-V_Q=4-\frac{1}{3}\times 3=3\,\text{volt}$$

17. (c) In series, $R_s=nR$

In parallel, $\frac{1}{R_p}=\frac{1}{R}+\frac{1}{R}+...n$ terms

$\therefore R_s/R_p=n^2/1=n^2$

18. (b) Since, the voltage is same for the two combinations, therefore $H\propto\frac{1}{R}$. Hence, the combination of 39 bulbs will glow more.

19. (b) $H=I^2Rt$ or $R=\frac{H}{(I^2t)}=\frac{80}{(2^2\times 10)}=2\Omega$

20. (b) For maximum current, the two batteries should be connected in series. The current will be maximum when external resistance is equal to the total internal resistance of cells i.e. 2Ω. Hence power developed across the resistance R will be

$$I^2R=\left(\frac{2E}{R+2r}\right)^2 R=\left(\frac{2\times 2}{2+2}\right)\times 2=2W$$

21. (c) To carry a current of 4 ampere, we need four paths, each carrying a current of one ampere. Let r be the resistance of each path. These are connected in parallel. Hence, their equivalent resistance will be $r/4$. According to the given problem $\frac{r}{4}=5$ or $r=20\Omega$.

For this propose two resistances should be connected. There are four such combinations. Hence, the total number of resistance $=4\times 2=8$.

22. (d) At A current is distributed and at B currents are collected. Between A and B, the distribution is symmetrical. It has been shown in the figure. It appears that current in AO and OB remains same. At O, current i_4 returns back without any change. If we detach O from AB there will not be any change in distribution.

Now, CO & OD will be in series hence its total resistance = 2Ω

It is in parallel with CD, so, equivalent resistance

$$=\frac{2\times 1}{2+1}=\frac{2}{3}\Omega$$

This equivalent resistance is in series with AC & DB, so, total resistance

$$=\frac{2}{3}+1+1=\frac{8}{3}\Omega$$

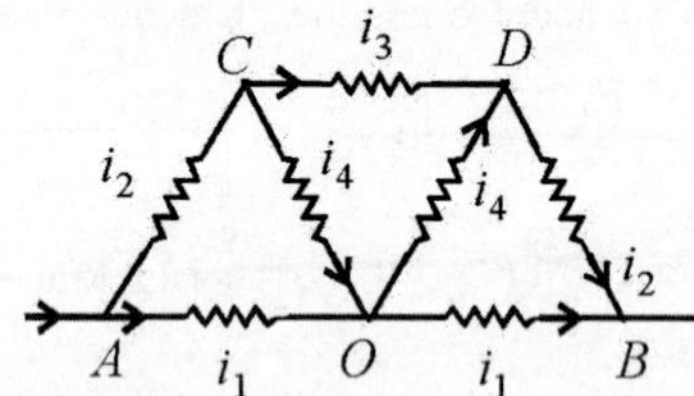

Now $\frac{8}{3}\Omega$ is parallel to AB, that is, 2Ω, so total resistance

$$=\frac{8/3\times 2}{8/3+2}=\frac{16/3}{14/3}=\frac{16}{14}=\frac{8}{7}\Omega$$

[Alt :

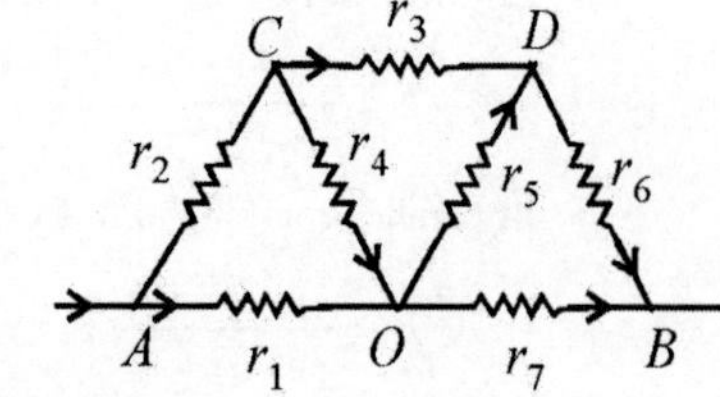

Between C & D, the equivalent resistance is given by

$$1/r=\frac{1}{r_3}+\frac{1}{(r_4+r_5)}=1+\frac{1}{2}=\frac{3}{2}$$

Equivalent resistance along

$$ACDB=1+\frac{2}{3}+1=\frac{8}{3}$$

$\therefore$ Effective resistance between A and B is

$$\frac{1}{R}=\frac{3}{8}+\frac{1}{2}=\frac{7}{8}\text{ or } R=\frac{8}{7}\Omega]$$

23. (a) Faraday's laws are based on the conversion of electrical energy into mechanical energy; which is in accordance with the law of conservation of energy.

25. (c) In the electrolysis of $CuSO_4$, oxygen is liberated at anode and copper is deposited at cathode.

26. (d) The two resistances are connected in series and the resultant is connected in parallel with the third resistance.

$\therefore R' = 4\ \Omega + 4\ \Omega = 8\Omega$ and $\frac{1}{R''} = \frac{1}{8} + \frac{1}{4} = \frac{3}{8}$

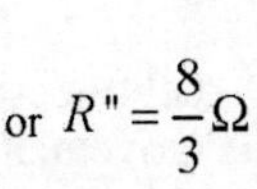

or $R'' = \frac{8}{3}\Omega$

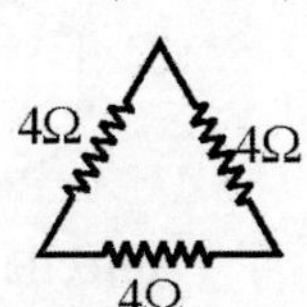

27. (d) If current in P and $Q = I_1$ then current in R and S will be $2I$.
Current^2 × resistance is greatest for S.

28. (c) Null methods are always better.

29. (d) Arrange the resistors in all possible combinations and get the number of combinations as ten with nine different values. The combinations can be all 4 in series, all 4 in parallel, 3 in series with 1 in parallel, 3 in parallel with 1 in series etc.

30. (d) The circuit can be redrawn as and finally

$R_{AB} = \frac{3}{5}R$

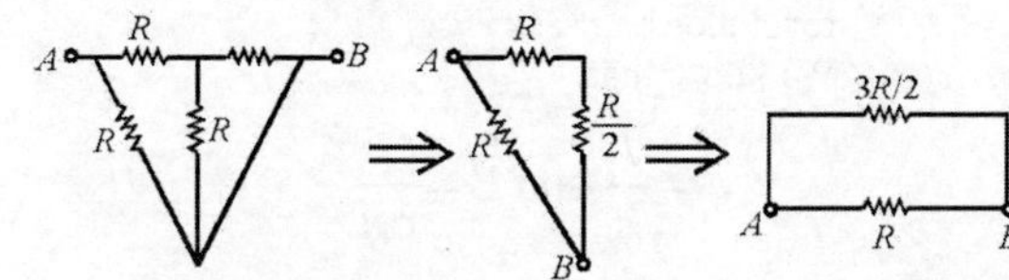

31. (c) $V_A - V_B = 4V$ ...(i)
$V_A - V_C = 10\ V$ ...(ii)

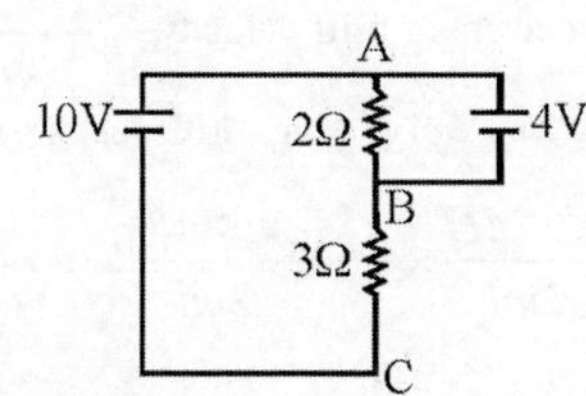

$\therefore V_B - V_C = 6\ V$

32. (b) The circuit can be drawn as :

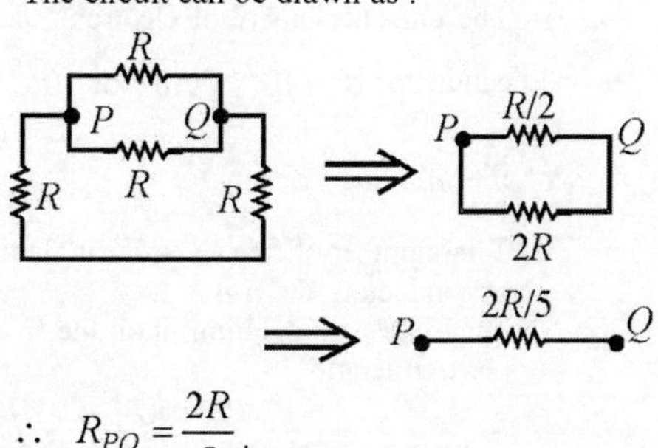

$\therefore\ R_{PQ} = \frac{2R}{5}.$

33. (d) Voltage across bulb B_2 will be less than that across B_3. $\Rightarrow W_2 < W_3$.

34. (a) Power maximum when $r = R$.
So, power consumed by it will decrease for $R > r$.

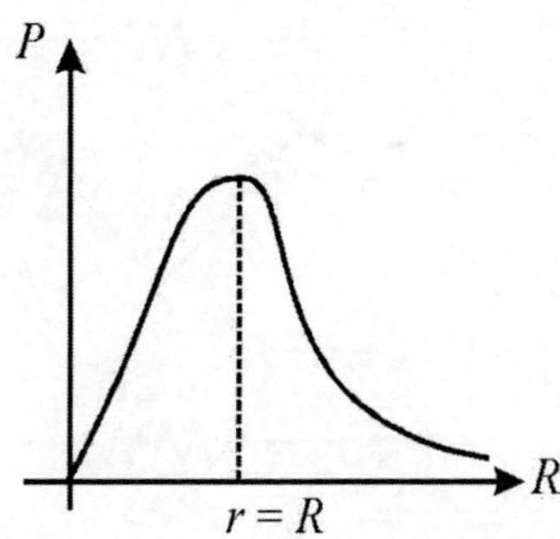

35. (a) In a Wheatstone bridge, the deflection in the galvanometer does not change if the battery and galvanometer are interchanged.

36. (d) In parallel combination, potential drop across each resistance is same.

Heat developed $= \frac{V^2}{R}$

$$H_1 = \frac{V^2}{R},\ H_2 = \frac{V^2}{2R},\ H_3 = \frac{V^2}{3R}$$

$\therefore$ Heat developed in resistance R is maximum.

38. (d) Copper is a metal whereas Germanium is semi-conductor.

43. (a) $H = \frac{i^2R}{2\pi rl} = \frac{i^2 \frac{\rho l}{\pi r^2}}{2\pi rl}$

or $H \propto \frac{1}{r^3}$

44. (c) RAB = 2.5 Ω
$R_{AC} \Rightarrow$

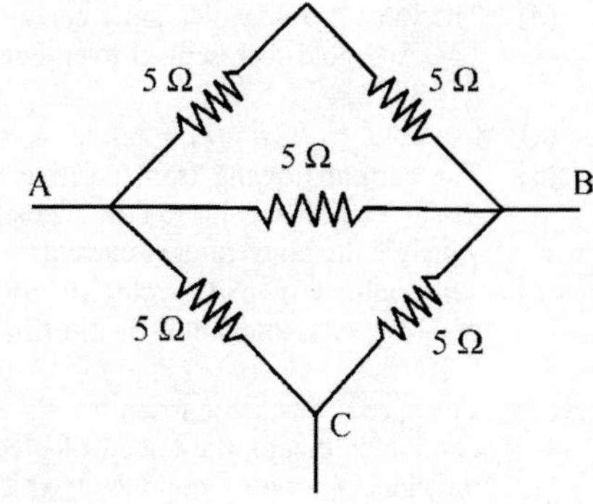

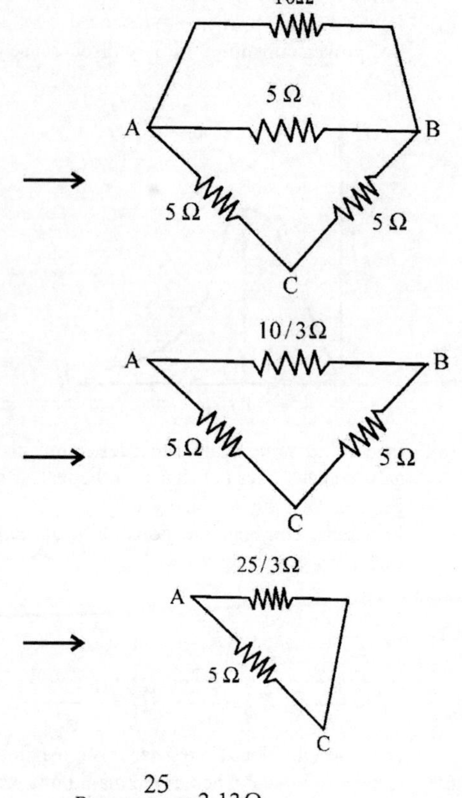

$R_{AC} = \frac{25}{8} = 3.13\Omega$

45. (a) In conductors, the charge carriers are free electrons.

46. (c) $H = \frac{V^2}{R}t = \frac{(V/2)^2}{\rho}t'$

$\therefore \quad t' = 4t$

47. (c) $P = \frac{V^2}{R} = \frac{V'^2}{R'}$

48. (d) The value of red, violet and green are : 2, 7 and 10^5. And gold represents a tolerance 5%. Thus $R = 27 \times 10^5 \pm 5\%\ \Omega$.

49. (a) Current $I = 2E/(R + r_1 + r_2)$.

50. (c) The current flowing from positive terminal to negative terminal is due to flow of positive charge, which is the conventional current.

51. (c) The value of drift velocity of electrons $= 10^{-5}$ m/s and that of thermal velocity $= 10^5$ m/s.

52. (d) The speed at which current travels through the conductor means the speed of electric effect travelling through a conductor which is at the speed of light.

53. (a) In the absence of electric field, the electron moves in zig-zag direction as shown in figure. So net displacement is zero, hence mean velocity of free electrons in a conductor at absolute temperature is zero

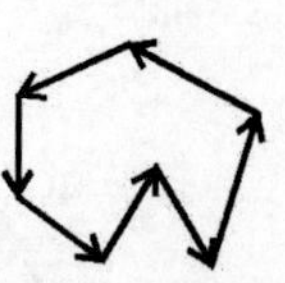

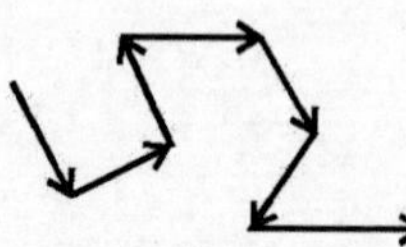

direction of electron (In the absence of $\vec{E}$)

net direction or movement (In presence of electric field)

[v_d = drift velocity $= \frac{I}{neA}$, where n is number of free electrons per unit volume, I is current that flows through a cross section A & e is electronic charge (1.6×10^{-19} coulomb)].

54. (b) Drift velocity $v_d = \frac{eE}{m}\tau = \frac{e}{m}\tau\left(\frac{V}{l}\right)$;

so $v_d \propto V$

55. (a) Electric current, I = charge /time period of revolution = $e/(2\pi r / v)$.

56. (b) We know that

$$J = \frac{I}{A} \text{ and } E = \frac{I}{\sigma A};$$

$$\therefore \ E = \frac{J}{\sigma} = \rho J = kJ.$$

58. (a) Since average drift velocity $= \frac{1}{2}\frac{eE}{m} \times (\tau)$

Now $I = NeA \times$ (avg. drift velocity)

$$= \frac{Ne^2AE}{2ml} \times \tau = \frac{Ne^2AV}{2ml} \times \tau$$

$R = \frac{V}{I} = \frac{2ml}{Ne^2\tau A}$, where N is electron density.

59. (d) The current density of electrons in a metallic conductor is $\approx 10^{22}/\text{cm}^3$ or $10^{28}/\text{m}^3$.

60. (c) $I = nAev_d$ or $v_d = \frac{I}{nAe}$

Total number of free electrons in the unit length of conductor, $N = nA \times 1$.

Total linear momentum of all the free electrons per unit length

$$= (Nm)v_d = nAm \times \frac{I}{nAe} = \frac{I}{(e/m)} = \frac{I}{s}$$

18. MOVING CHARGES AND MAGNETISM

ANSWERKEY											
1	(c)	**11**	(a)	**21**	(b)	**31**	(c)	**41**	(c)	**51**	(d)
2	(c)	**12**	(d)	**22**	(d)	**32**	(c)	**42**	(a)	**52**	(b)
3	(a)	**13**	(b)	**23**	(c)	**33**	(d)	**43**	(b)	**53**	(d)
4	(d)	**14**	(b)	**24**	(d)	**34**	(a)	**44**	(a)	**54**	(a)
5	(c)	**15**	(b)	**25**	(c)	**35**	(d)	**45**	(a)	**55**	(a)
6	(d)	**16**	(c)	**26**	(a)	**36**	(b)	**46**	(a)	**56**	(b)
7	(c)	**17**	(c)	**27**	(a)	**37**	(a)	**47**	(b)	**57**	(a)
8	(c)	**18**	(c)	**28**	(b)	**38**	(a)	**48**	(a)	**58**	(b)
9	(b)	**19**	(b)	**29**	(d)	**39**	(a)	**49**	(a)	**59**	(c)
10	(b)	**20**	(b)	**30**	(b)	**40**	(a)	**50**	(b)	**60**	(b)

Hints & Solutions

3. (a)

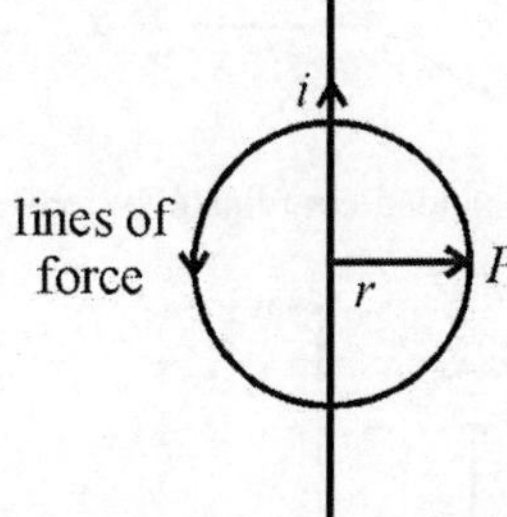

4. (d) $B = \frac{\mu_o i}{2\pi r}$, where r is distance of any point P from the straight conductor, at which we want to calculate the magnetic field.

5. (c) $B = \frac{\mu_0 i}{2\pi r}$ or $B \propto 1/r$.

7. (c) The magnetic field at the centre of a circular coil is

$$B_{\text{centre}} = \frac{\mu_o \pi i \times n}{2\pi r}$$

where $2\pi r = l$ = length of wire & n is number of turns

It is given that $2\pi r_1 = l = 2(2\pi r_2) \Rightarrow r_2 = r_1/2$

So B_1 at centre $= \frac{\mu_o \pi i \times 1}{2\pi r_1}$

B_2 at centre $= \frac{\mu_o i \times \pi \times 2}{\pi r_1}$

so $\frac{B_2}{B_1} = \frac{2 \times 2}{1} \Rightarrow B_2 = 4B_1$

8. (c) In a metallic rod, charge carriers flow through whole of the cross-section, the magnetic field exists both inside as well as outside.

9. (b) According to Ampere's law $\oint \vec{B} \cdot \vec{dl} = \mu_0 i$

To any closed path inside pipe, there will be no current enclosed. So B inside pipe is zero. Only, there would be field outside the pipe.

10. (b) Magnetic induction inside a thin walled tube is zero. (According to Ampere's Law)

11. (a) Ampere's Law states that the line integral of magnetic field induction $\vec{B}$ around a closed path in vacuum is equal to μ_0 times the total current I threading the closed path.

12. (d) For a loop, magnetic induction at centre

$$B = \frac{\mu_0}{4\pi} \times \frac{2\pi i}{r}$$

When loop subtends angle θ at centre, then

$$B = \frac{\mu_0}{4\pi} \times \frac{\theta i}{r} \text{ (Here angle is } \theta \text{ instead of } 2\pi)$$

In the given problem, $\theta = 3\pi/2$ and $r = R$

$$\therefore B = \frac{\mu_0}{4\pi} \times \frac{3\pi}{2} \times \frac{i}{R} = \frac{3\mu_0 i}{8R}$$

13. (b) Apply Right hand thumb rule.
According to Right-Hand Rule, the direction of $\vec{F}$ on positively charged particle will be perpendicular to the plane of paper upward.

16. (c) $F_n = q(\vec{v} \times \vec{B})$

$= qvB \sin\theta = 0$ (because $\theta = 0°$)

17. (c) (i) When no field is present $E = 0, B = 0$, the proton experiences no force. Thus it moves with a constant velocity.

(ii) When $E = 0$ and $B \neq 0$, then there will be a probability that proton may move parallel to magnetic field. In this situation, there will be no force acting on proton.

(iii) When both fields are present $E \neq 0, B \neq 0$, then let E, B and v may be mutually perpendicular to each other. In this case, the electric and magnetic forces acting on the proton may be equal and opposite. Thus, there will be no resultant force on the proton.

18. (c) Given that $K_p = K_\alpha$

We know that $r = \dfrac{mv}{qB} = \dfrac{(2mK)^{1/2}}{q\text{B}}$

$\therefore \quad r_p = \dfrac{(2m_p K_p)^{1/2}}{q_p B}$

and $\quad r_\alpha = \dfrac{(2m_\alpha K_\alpha)^{1/2}}{q_\alpha B}$

Now $\dfrac{r_p}{r_\alpha} = \sqrt{\left(\dfrac{m_p}{m_\alpha}\right)} \times \dfrac{q_\alpha}{q_p}$

$= \sqrt{\left(\dfrac{m_p}{4m_p}\right)} \times \dfrac{2e}{e} = 1$

19. (b) $\vec{F}_e = -q\vec{E}$, where – ve sign comes due to electron, which has negative charge. It is clear from expression that force on electron is opposite to electric field i.e., in negative direction.

20. (b) Here velocity vector have two components

(i) $v \cos\theta$, parallel to magnetic field

(ii) $v \sin\theta$, perpendicular to magnetic field. Due to component $v \cos\theta$, the particle will have a linear motion but due to $v \sin\theta$, the particle will have simultaneously a circular motion. The resultant of the two is a helical path.

22. (d) When a positively charged particle enters in a region of uniform magnetic field directed vertically upwards, it experiences a centripetal force which moves the particle in circular path with a uniform speed (in clockwise direction).

23. (c) Force on moving charge while moving in magnetic field is; $\vec{F} = q(\vec{v} \times \vec{B})$ where $\vec{F}$ is perpendicular to $\vec{v}$.

Work done/sec $= \vec{F}.\vec{v} = Fv \cos 90° = 0$.

24. (d)

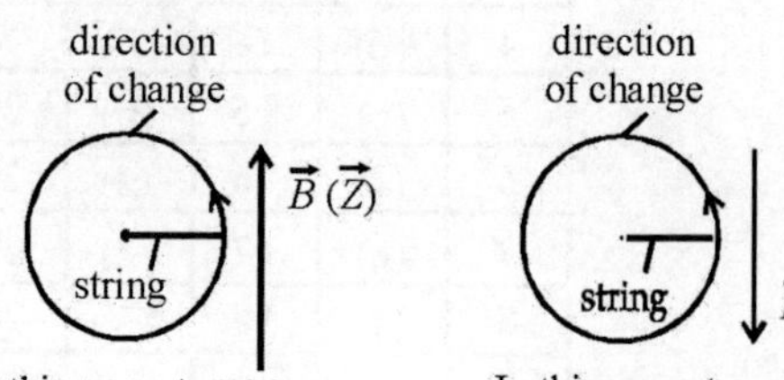

In this case at every point tension decreases

In this case at every point tension increases

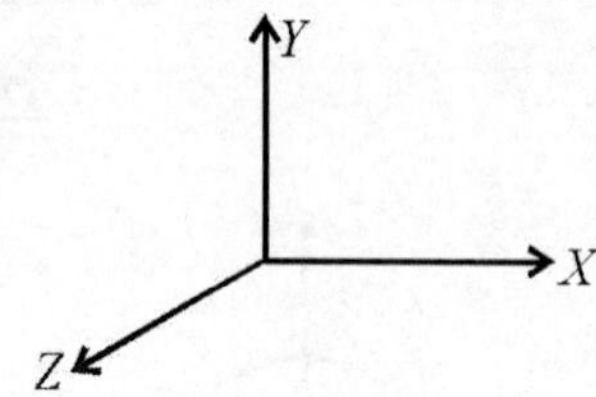

(right handed coordinate system)

30. (b)

lines of force

wire

$\vec{B}$

i

$F_m = I(\vec{l} \times \vec{B}) = 0$ because the directions of current I & magnetic field $\vec{B}$ are parallel.

31. (c) The force acting on various small current carrying elements of the circumference of the loop will be distributed randomly in all possible directions. The vector addition of such randomly distributed forces will be zero.

33. (d) $\vec{\tau} = (\vec{M} \times \vec{B})$, where $|\vec{M}| = i.A$

$= MB \sin\theta$

where θ is angle between Magnetic moment & $\vec{B}$.

For $\theta = 0$ $\vec{\tau} = 0$ & coil is in stable equilibrium. Hence plane of coil must be perpendicular to magnetic field.

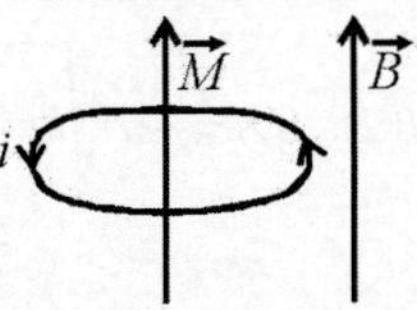

35. (d) Magnetic dipole moment of current loop $M = niA$, it is independent of B.

36. (b) When the plane of coil is in N-S vertical plane and current is passed in clockwise direction, the magnetic field at the centre of coil due to current is directed East to West. Earth's magnetic field is south to North. Since the magnetic needle will be in the direction of resultant magnetic field, which will be in the west-north direction.

37. (a) $\frac{E}{B} = c$ (velocity light if we consider free space)

& $\frac{1}{\sqrt{\mu_o \varepsilon_o}} = c$ so $\sqrt{\mu_o \varepsilon_o}\,\frac{E}{B}$ is dimensionless.

38. (a) The resistance of AD is one-third of the resistance along the parallel path $AB + BC + CD$. Hence if current i enters at A, $\frac{3i}{4}$ will flow along AD and $\frac{i}{4}$, along AB. The magnetic field at the centre due to AD is equal and opposite to the combined effects of AB, BC and CD.

39. (a) An electron which has potential energy is converted in kinetic energy.
According to question,

$$10\text{ eV} = \frac{1}{2}mv^2 = \text{K.E. of } e^-$$

$$v = \sqrt{\frac{2\times10\times1.6\times10^{-19}}{9\times10^{-31}}} = 1.9\times10^6\text{ m/s}$$

[1 electron volt is the amount of kinetic energy an electron acquires when it is accelerated in an electric field produced by a potential difference of 1 volt.
1eV = charge of an electron × 1 volt

$= 1.6\times10^{-19}$ coulomb × 1 volt

$= 1.6\times10^{-19}$ joule (M of $e^- = 9\times10^{-31}$ kg)]

40. (a) $B \propto \frac{1}{r} \Rightarrow r' = 3r$

$$\therefore B' = \frac{1}{3}B = \frac{1}{3}\times10^{-3} = 3.33\times10^{-4}T$$

41. (c) $r = \frac{mv}{qB}$ or $r \propto v$

As v is doubled, the radius also becomes double.
Hence radius = 2 × 2 = 4 cm

42. (a) $B = \frac{\mu_0 i}{2\pi r}$ or $B \propto \frac{1}{r}$

When r is doubled, the magnetic field becomes half, i.e., now the magnetic field will be 0.2 T.

43. (b) $F = Bil = 2\times1.2\times0.5 = 1.2\ N$

44. (a) A current carrying coil has magnetic dipole moment. Hence a torque $\vec{p}_m \times \vec{B}$ acts on it in magnetic field.

45. (a) The force acting on a charged particle in magnetic field is given by
$F = q\,(v \times B)$ or $F = qvB\sin\theta$
when angle between v and B is 180°,
$F = 0$

46. (a) $B = \frac{\mu_0 I}{2r}\times\frac{\theta}{2\pi} = \frac{\mu_0 I\theta}{4\pi r}$

48. (a) Current (I) = 12 A and magnetic field (B) = 3×10^{-5} Wb/m^2. Consider magnetic field $\vec{B}$ at distance r.

Magnetic field, $B = \frac{\mu_0 I}{2\pi r}$

$$\Rightarrow r = \frac{\mu_0 I}{2\pi B} = \frac{(4\pi\times10^{-7})\times12}{2\times\pi\times(3\times10^{-5})} = 8\times10^{-2}m$$

49. (a) If the electric field is switched off, and the same magnetic field is maintained, the electrons move in a circular orbit and electron will travel a magnetic field $\perp$ to its velocity.

50. (b) The current is same in both the wire, hence magnetic field induced will be same.

51. (d) When the deflection produced by electric field is equal to the deflection produced by magnetic field, then the electron can go undeflected.

$B = \frac{\mu_0 i}{2\pi r}$ and so it is independent of thickness.

52. (b) Force on a particle moving with velocity v in a magnetic field B is $F = q\,(\vec{v}\times\vec{B})$

If angle between $\vec{v}$ & $\vec{B}$ is either zero or 180°, then value of F will be zero as cross product of $\vec{v}$ and $\vec{B}$ will be zero.

53. (d) Magnetic moment, $m = IA$

$$= \frac{qv}{2\pi R}(\pi R^2) = \frac{qvR}{2}$$

$$\left[\because I = \frac{q}{T} \text{ and } T = \frac{2\pi R}{v}\right]$$

54. (a) The direction of $\vec{B}$ is along $(-\hat{k})$

$\therefore$ The magnetic force

$$\vec{F} = Q(\vec{v} \times \vec{B}) = Q(v\hat{i}) \times B(-\hat{k}) = QvB\hat{j}$$

$\Rightarrow \vec{F}$ is along OY.

55. (a) Magnetic moment linked with one turn $= iA$

Magnetic moment linked with N turn $= iNA$ amp-m^2. Here, A = Area of current loop.

56. (b) $\tau = mB \sin\theta$

$\tau = iAB \sin 90°$

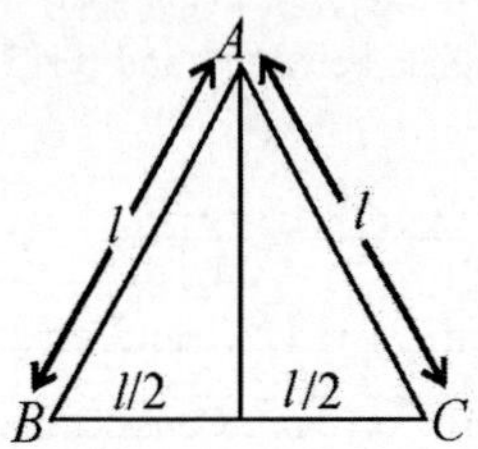

$\therefore A = \dfrac{\tau}{iB}$; Also, $A = 1/2\,(BC)\,(AD)$

But $\dfrac{1}{2}(BC)(AD) = \dfrac{1}{2}(l)\sqrt{l^2 - \left(\dfrac{l}{2}\right)^2} = \dfrac{\sqrt{3}}{4}l^2$

$$\Rightarrow \frac{\sqrt{3}}{4}(l)^2 = \frac{\tau}{Bi} \therefore l = 2\left(\frac{\tau}{\sqrt{3}\,B.i}\right)^{\frac{1}{2}}$$

57. (a) $B = \dfrac{\mu_0 In}{2\pi a}$ Add vectorially.

58. (b) Use the expression for magnetic field due to a current carrying straight conductor and

$$\frac{B_1 - B_2}{B_1 + B_2} = \frac{I_1 - I_2}{I_1 + I_2}$$

59. (c) Potential energy is given by $U = -\mu B$

60. (b) Using Ampere's law and noting that the steady current is uniformly distributed over the cross section of the conductor, the magnetic field is proportional to the distance r from the centre for $r < R$ and outside the conductor proportional to $1/r$.

19. MAGNETISM AND MATTER

ANSWERKEY

1	(a)	11	(b)	21	(a)	31	(a)	41	(c)	51	(a)	61	(d)	71	(a)
2	(a)	12	(a)	22	(a)	32	(d)	42	(c)	52	(b)	62	(b)	72	(d)
3	(c)	13	(d)	23	(b)	33	(a)	43	(a)	53	(c)	63	(a)	73	(c)
4	(a)	14	(b)	24	(d)	34	(c)	44	(c)	54	(a)	64	(a)	74	(c)
5	(b)	15	(d)	25	(c)	35	(c)	45	(a)	55	(d)	65	(a)	75	(b)
6	(d)	16	(b)	26	(a)	36	(b)	46	(b)	56	(d)	66	(a)		
7	(d)	17	(d)	27	(a)	37	(a)	47	(c)	57	(c)	67	(a)		
8	(d)	18	(b)	28	(c)	38	(a)	48	(b)	58	(b)	68	(d)		
9	(d)	19	(b)	29	(a)	39	(a)	49	(a)	59	(b)	69	(a)		
10	(a)	20	(d)	30	(b)	40	(a)	50	(b)	60	(d)	70	(c)		

Hints & Solutions

1. (a) Due to horizontal component of earth's field, the magnetic lines of force are parallel and straight.

2. (a) At magnetic north pole of earth, $H = 0$ and $\delta = 90°$, maximum.

3. (c) Near the magnetic poles, $H = 0$, therefore, magnetic compass will not work.

4. (a) At magnetic equator, $R = H$; $V = 0$.

5. (b) The restoring torque is due to horizontal component of earth's field.

6. (d) At poles, $\delta = 90°$.

10. (a) Brass is non magnetic material.

11. (b) In paramagnetic metals, magnetism is acquired by orientation of atomic dipoles.

12. (a) For diamagnetic material, χ_m is low and negative.

13. (d) $I_m/H = \chi_m$.

14. (b) Paramagnetic liquid tends to flow from region of weaker magnetic fields to stronger magnetic fields.
15. (d) A flux meter is used for measuring a magnetic field.
16. (b) A diamagnetic liquid moves from stronger parts of magnetic field to weaker parts.
17. (d) On increasing the temperature by 700°C, the magnetic needle is demagnetised. Therefore, the needle stops vibrating.
18. (b) For ferromagnetic substances, χ_m is large and positive.
19. (b) Beyond curie temperature, ferromagnetic material turns into paramagnetic material, as if ferromagnetic domains become random.
20. (d) The direction of magnetic lines of force of a bar magnet is from north to south pole outside the magnet and from south pole to north pole inside the magnet.
22. (a) As lines of magnetic induction B are continuous curves, they run continuously through the bar and outside, as shown in Fig. (1).
23. (b) Diagmagnetic substances are repelled by the poles of a bar magnet.
28. (c) For each half $M = m \times 2l$ becomes half and volume $V = a \times 2l$ also becomes half therefore, $I = M/V$, remains constant.
29. (a) A magnetic needle suspended freely orientates itself in a definite direction.
30. (b) Magnetic length of a magnet is roughly 0.8 times the geometric length.
31. (a) Magnetic dipole moment is directed from south pole to north pole of magnetic dipole.
32. (d) End on position corresponds to axial line and Broadside on position corresponds to equatorial line. Therefore, $B_1 / B_2 = 2$.
35. (c) The loop (i) is for soft iron and the loop (ii) is for steel in Fig.
36. (b) Diagmagnetic materials are repelled by the magnetic field.
37. (a) In diamagnetism, susceptibility does not depend on the temperature.
38. (a) $\mu = \mu_r \mu_0$, as $\mu_r = \mu/\mu_0$.
39. (a) In S.I. units, we have $B = \mu_0 (H + I)$
50. (b) The time period of a bar magnet in a magnetic field is given by.

$$T = 2\pi\sqrt{\frac{I}{MB}};$$

Here, I = moment of inertia $\propto m$,

M = moment of magnet,
B = magnetic field.

$T \propto \sqrt{I} \propto \sqrt{m}$; so, T becomes twice as mass becomes four times

51. (a) Magnetic moment = M; Initial angle through which magnet is turned (θ_1) = 90° and final angle through which magnet is turned (θ_2) = 60°. Work done in turning the magnet through 90° (W_1)
$= MB(\cos 0° - \cos 90°) = MB(1 - 0) = MB$.
Similarly, $W_2 = MB(\cos 0° - \cos 60°)$

$$= MB\left(1 - \frac{1}{2}\right) = \frac{MB}{2}$$

$\therefore W_1 = 2W_2$ or $n = 2$.
52. (b) The iron can produces a magnetic screening for the equipment as lines of magnetic force cannot enter iron enclosure.
53. (c) $$M_{net} = \sqrt{M_0^2 + M_0^2 + 2M_0^2 \cos 60°}$$
$$= \sqrt{3M_0^2} = \sqrt{3}M_0$$
54. (a) The magnetic dipole moment of diamagnetic material is zero as each of its pair of electrons have opposite spins, i.e., $\mu_d = 0$.
Paramagnetic substances have dipole moment > 0, i.e. $\mu_p \neq 0$, because of excess of electrons in its molecules spinning in the same direction.
Ferromagnetic substances are very strong magnets and they also have permanent magnetic moment, i.e. $\mu_f \neq 0$.
55. (d) Above Curie Temperature, a ferromagnetic substance becomes paramagnetic.
56. (d) Beyond Curie temperature, ferromagnetic substances behaves like a paramagnetic substance.
58. (b) Work done
$= MB(\cos\theta_1 - \cos\theta_2)$
$= MB(\cos 0° - \cos 60°)$

$$= MB\left(1 - \frac{1}{2}\right) = \frac{2\times10^4 \times 6\times10^{-4}}{2} = 6 \text{ J}$$

59. (d) DMagnetic moment, $M = m\,l$

$\frac{M}{l} = m$, where m is the polestrength.

Therefore distance between poles

$$= \sqrt{(l/2)^2 + (l/2)^2} = \frac{l}{\sqrt{2}}$$

So, $M' = \frac{ml}{\sqrt{2}} = \frac{M}{\sqrt{2}}$

60. (d) Time period of a vibration magnetometer,

$$T \propto \frac{1}{\sqrt{B}}$$

$$\Rightarrow \frac{T_1}{T_2} = \sqrt{\frac{B_2}{B_1}}$$

$\Rightarrow T_2 = T_1\sqrt{\frac{B_1}{B_2}}$

$= 2\sqrt{\frac{24\times10^{-6}}{6\times10^{-6}}} = 4\,\text{s}$

61. (d) Soft iron has high retentivity and low coercive force.

62. (b) Force on a charged particle is given by $F = qvB$. Here $v = 0$ and also resultant B is zero.
$\therefore$ Force = 0.

63. (a) The magnetic moment of a diamagnetic atom is equal to zero.

64. (a) A $\rightarrow$ diamagnetic
B $\rightarrow$ paramagnetic
C $\rightarrow$ Ferromagnetic
D $\rightarrow$ Non-magnetic

65. (a) For stable equilibrium
$U = -MB$
$= -(0.4)(0.16)$
$= -0.064$ J

66. (a) As $\tan\delta = V/H$
At equator, $\delta = 0$,
so, $H >> V$.

67. (a) Universal property of all substances is Diamagnetism.

68. (d) Here we use sum and difference method.

$$\frac{M_2}{M_1} = \frac{T_2^2 - T_1^2}{T_2^2 + T_1^2}$$

Where T_1 and T_2 are time periods

$T \propto \frac{1}{n}$ $(\because n$ is frequency.)

$$\Rightarrow \frac{M_1}{M_2} = \frac{n_1^2 + n_2^2}{n_1^2 - n_2^2}$$

$$= \frac{(12)^2 + (4)^2}{(12)^2 - (4)^2} = \frac{144+16}{144-16} = \frac{10}{8}$$

$M_1 : M_2 = 10 : 8$

69. (a) Induced current is produced in the wire coil, due to change of flux linked with it, when a bar magnet passes through the coil in the downward direction.
According to Lenz's law the direction of the induced current is such as to oppose the motion of the magnet. So the acceleration of the magnet will be less than g.

70. (c) The cabin is made up of iron because maximum magnetic lines of forces pass through it.

71. (a) A hydrogen molecule consisting of two hydrogen atoms is diamagnetic in nature.

72. (d) Lines of force of a bar magnet are curved closed loops, as it starts from north pole and ends at south pole.

73. (c) Magnetic field at a point at a distance r on the axis of the bar magnet $B = \frac{\mu_0}{4\pi}.\frac{2M}{r^3}$

$\therefore B \propto \frac{1}{r^3}$

74. (c) $R = \frac{H}{\cos\delta} = \frac{0.50}{\cos 30°} = \frac{0.50\times2}{\sqrt{3}} = 1/\sqrt{3}$

75. (b) As $\tau = MB\sin\theta$; $\therefore \tau_1 = MB_1 \sin 90° = MB_1$, $\tau_2 = MB_2 \sin 90° = MB_2$

$\therefore \frac{MB_1}{MB_2} = \frac{\tau_1}{\tau_2}$ or $\frac{B_1}{B_2} = \frac{\tau_1}{t_2}$.

20. ELECTROMAGNETIC INDUCTION

ANSWERKEY

1	(b)	11	(a)	21	(c)	31	(c)	41	(c)	51	(d).
2	(a)	12	(b)	22	(d)	32	(c)	42	(a)	52	(a)
3	(c)	13	(b)	23	(a)	33	(c)	43	(d)	53	(d)
4	(d)	14	(c)	24	(a)	34	(a)	44	(b)	54	(b)
5	(a)	15	(c)	25	(b)	35	(d)	45	(b)	55	(d)
6	(c)	16	(c)	26	(d)	36	(c)	46	(c)	56	(c)
7	(d)	17	(d)	27	(a)	37	(d)	47	(d)	57	(b)
8	(b)	18	(a)	28	(b)	38	(b)	48	(b)	58	(a)
9	(b)	19	(c)	29	(c)	39	(a)	49	(b)	59	(b)
10	(a)	20	(b)	30	(b)	40	(b)	50	(c)	60	(a)

Hints & Solutions

2. (a) Let M = mutual inductance between A and B.

$$\varepsilon_B = M\frac{di_A}{dt} \text{ or } \varepsilon_B = M\dot{I} \text{ or } M = \varepsilon_B/I$$

where $I = di_A/dt$

$$\phi_A = Mi_B = (\varepsilon_B/I)i_B$$

$$\therefore \phi_A = (\varepsilon/I)i \;\; (i = i_B)$$

3. (c) $\varepsilon = -\dfrac{Nd\phi}{dt} \Rightarrow \varepsilon \infty N \text{ and } \varepsilon \infty \dfrac{d\phi}{dt}$

If so the speed of magnet is fast then correspondingly rate of change of flux is fast & ε is maximum. It does not depend on the resistance of coil.

5. (a) Since $\varepsilon = -\dfrac{Nd\phi}{dt}$ if $\dfrac{d\phi}{dt}$ is fast, so ε is large.

9. (b) According to Faraday's law, the changing magnetic flux causes induced e.m.f.

11. (a) According to right hand palm rule, the Lorentz force on free electrons in the conductor will be directed towards end B. Hence, the end A gets positively charged.

12. (b) Induced e.m.f. in the ring opposes the motion of the magnet.

16. (c) When the coils are connected in parallel, let the currents in the two coils be i_1 and i_2 respectively. Total induced current

$$i = i_1 + i_2 \text{ or } \frac{di}{dt} = \frac{di_1}{dt} + \frac{di_2}{dt} \quad ...(1)$$

Now $e = -L_1\left(\dfrac{di_1}{dt}\right) = -L_2\left(\dfrac{di_2}{dt}\right)$

($\because$ In parallel, induced e.m.f. across each coil will be same)

Hence $\dfrac{di_1}{dt} = -\dfrac{e}{L_1}$ and $\dfrac{di_2}{dt} = -\dfrac{e}{L_2}$ ... (2)

Let L' be the equivalent inductance.

Then $e = -L'\dfrac{di}{dt}$ or $\dfrac{di}{dt} = -\dfrac{e}{L'}$ ... (3)

From eqs. (1), (2) and (3), we get

$$-\frac{e}{L'} = -\frac{e}{L_1} - \frac{e}{L_2} \text{ or } \frac{1}{L'} = \frac{1}{L_1} + \frac{1}{L_2}$$

$$\therefore L' = \frac{L_1 L_2}{L_1 + L_2}$$

Here $L_1 = L_2 = L$

$$\therefore L' = \frac{L \times L}{L + L} = \frac{L}{2}$$

20. (b) Mutual inductance = $\dfrac{\phi}{I} = \dfrac{BA}{I}$

$$[\text{Henry}] = \frac{[MT^{-1}Q^{-1}L^2]}{[QT^{-1}]} = ML^2Q^{-2}$$

21. (c) $L = (\mu_0 N^2 \pi r^2/2) \;\therefore\; L \propto N^2$

24. (a) $F = qvB; B = \mu_0 I/2\pi a$

25. (b) As number of turns are quadrupled, the induced emf will increase four times. Also, resistance of coil increases sixteen times. Hence power $\left(\dfrac{V^2}{R}\right)$ will not change.

26. (d) Induced emf, $E = NAB\omega \sin \omega t$

$\Rightarrow$ max emf ($\sin \omega t = 1$)

$E_{max} = NAB\omega$

If $\omega' = 2\omega$ then

$E' = 2E_{max}$

27. (a) When coil is rotated about the diametre, the flux will change. In other cases no flux will change.

28. (b) Mutual Inductor of two coils

$$M = \sqrt{M_1 M_2} = \sqrt{2\text{mh} \times 8\text{mh}} = 4\text{mH}$$

29. (c) $L = 2$ mH, $i = t^2 e^{-t}$

$$E = -L\frac{di}{dt} = -L[-t^2e^{-t} + 2te^{-t}]$$

when $E = 0$,

$-e^{-t}t^2 + 2te^{-t} = 0$ or, $2t\,e^{-t} = e^{-t}t^2$

$\Rightarrow t = 2$ sec.

30. (b) $e = -M\dfrac{di}{dt} = -0.005 \times \dfrac{d(i_0 \sin \omega t)}{dt}$

$= -0.005 \times i_0 \times (\omega \cos \omega t)$

$e_{max.} = 0.005 \times i_0 \times \omega$ (when $\cos \omega t = -1$)

$= 0.005 \times 10 \times 100\pi = 5\pi$ V

31. (c) Initial current (I_1) = 10 A; Final current (I_2) = 0; Time (t) = 0.5 sec and induced e.m.f. (ε) = 220 V.

Induced e.m.f. (ε)

$$= -L\frac{dI}{dt} = -L\frac{(I_2 - I_1)}{t} = -L\frac{(0 - 10)}{0.5} = 20L$$

or, $L = \dfrac{220}{20} = 11\text{H}$

[where L = Self inductance of coil]

32. (c) $q = \int i dt = \frac{1}{R}\int e dt = \frac{1}{R}\int\left(\frac{-d\phi}{dt}\right)dt$

$= \frac{1}{R}\int d\phi$ (taking only magnitude of e)

Hence, total charge induced in the conducting loop depend upon the total change in magnetic flux.

33. (c) $\frac{\Delta\phi}{\Delta t} = \varepsilon = iR \Rightarrow \Delta\phi = (i\Delta t)R = QR$

$\Rightarrow Q = \frac{\Delta\phi}{R}$

34. (a) Induced e.m.f. $\varepsilon = \frac{d\phi}{dt} = \frac{dBA}{dt} = A_0\frac{dB}{dt}$

$= A_0\left(\frac{4B_0 - B_0}{t}\right) = 3A_0 B_0/t$

35. (d) Since the magnetic field is uniform the flux f through the square loop at any time t is constant, because

$f = B \times A = B \times L^2 =$ constant

$\therefore\ \varepsilon = -\frac{d\phi}{dt} =$ zero

36. (c) $\varepsilon = (5\times10^{-3})(1/0.1) = 0.05$V.

37. (d) Total resistance of the circuit = 4000 + 400 = 4400 W

Current flowing $i = \frac{V}{R} = \frac{440}{4400} = 0.1$ amp.

Voltage across load = $R\,i$ = 4000 × 0.1 = 400 volt.

38. (b) $\varepsilon = L\frac{di}{dt}$ or $2 = L\left[\frac{(3-2)}{10^{-3}}\right]$

$\therefore\ L = \frac{2\times10^{-3}}{1} = 2$mH.

39. (a) Change in flux = 2 B A N

$\therefore$ Induced e.m.f.

$= \frac{2\times0.3\times200\times70\times10^{-4}}{0.1}$

40. (b) The individual emf produced in the coil

$e = \frac{-d\phi}{dt}$

$\therefore$ The current induced will be

$i = \frac{|e|}{R} \Rightarrow i = \frac{1}{R}\frac{d\phi}{dt}$

But $i = \frac{dq}{dt}$

$\Rightarrow \frac{dq}{dt} = \frac{1}{R}\frac{d\varphi}{dt} \Rightarrow \int dq = \frac{1}{R}\int d\varphi \Rightarrow q = \frac{BA}{R}$

41. (c) Initially ϕ_B increases as magnet approaches the solenoid

$\therefore \varepsilon = -ve$ and increasing in magnitude. When magnet is moving inside the solenoid, increases in ϕ_B slows down and finally ϕ_B starts decreasing.

$\therefore$ emf is positive and increasing.

Only graph (c) shows these charateristic.

42. (a) For a current to induce in the cylindrical conducting rod.

(i) The cylindrical rod should cut magnetic lines of force which will happen only when the cylindrical conducting rod is moving. Since conducting rod is at rest, no current will be induced.

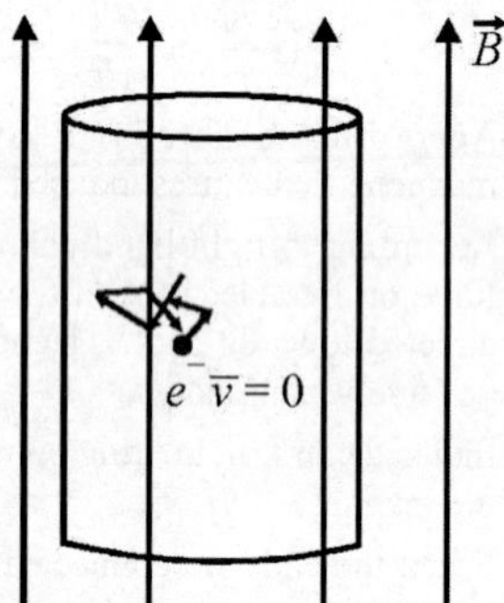

(ii) The magnitude / direction of the magnetic field changes. A changing magnetic field will create an electric field which can apply force on the free electrons of the conducting rod and a current will get induced.

But since the magnetic field is constant, no current will be induced.

44. (b) Let I current flow in the larger coil.

Magnetic field at the centre = $B = \frac{\mu_0 I}{2R}$

Flux linked with the smaller coil

$= \phi = \pi r^2 B = (\pi r^2)\frac{\mu_0 I}{2R}$

But, $\mu = MI$ or $M = \frac{\phi}{I} = \left(\frac{\mu_0\pi}{2}\right)\frac{r^2}{R}$.

45. (b) $\frac{d\phi}{dt} = \frac{(W_2 - W_1)}{t}$ $R_{tot} = (R + 4R)\Omega = 5R\ \Omega$

$i = \frac{n d\phi}{R_{tot} dt} = \frac{-n(W_2 - W_1)}{5Rt}$

($\because$ W_2 & W_1 are magnetic flux)

46. (c) E.M.F. will remain same because change in area per unit time will be same in both cases.

47. (d) $\phi = BA\cos\theta = 2.0 \times 0.5 \times \cos 60^\circ$

$= \frac{2.0 \times 0.5}{2} = 0.5$ Weber.

48. (b) $\varepsilon = \frac{d\phi}{dt} = nA\frac{dB}{dt}$

$\therefore \varepsilon = 10 \times (10 \times 10^{-4})(10^4)$

(10^8 Gauss/sec=10^4 T/s) = 100 V.

$I = (\varepsilon/R) = (100/20) = 5$amp.

49. (b) $\phi = 10t^2 - 50t + 250$

$e = -\frac{d\phi}{dt} = -(20t - 50)$

$e_{t=3} = -10\,V$.

50. (c) $i = \frac{\varepsilon}{R} = \frac{1}{R}\frac{d\phi}{dt}$

Here $d\phi = B \times A = (2 \times 10^{-5}) \times (0.5 \times 10^{+3} \times 5)$

dt = time taken by the wire to fall at ground

$= (2h/g)^{1/2} = (10/10)^{1/2} = 1$ sec.

$\therefore i = \frac{1}{2.5}\left[\frac{(2 \times 10^{-5}) \times (0.5 \times 10^3 \times 5)}{1}\right]$

= 0.02 amp.

53. (d) $\varepsilon = Blv$

$= (0.2 \times 10^{-4})(1)(180 \times 5/18)$

$= 10^{-3}\,V = 1\,mV$

54. (b) $\varepsilon = M\frac{di}{dt}$ or $8 = M\left[\frac{(4-2)}{0.05}\right]$

$\therefore M = \frac{8 \times 0.05}{2} = 0.2$ henry

55. (d) $e = -\frac{d\phi}{dt} = -\frac{d(N\vec{B}.\vec{A})}{dt}$

$= -N\frac{d}{dt}(BA\cos\omega t) = NBA\omega\sin\omega t$

$\Rightarrow e_{max} = NBA\omega$

56. (c) e.m.f. induced $= \frac{1}{2}BR^2\omega = \frac{1}{2}BR^2(2\pi n)$

$= \frac{1}{2} \times (0.1) \times (0.1)^2 \times 2\pi \times 10$

$= (0.1)^2\ \pi$ volts

57. (b) $B = \mu_0 ni$

$= (4\pi \times 10^{-7})\ (200 \times 10^{-2}) \times 1.5$

$= 3.8 \times 10^{-2}$ Wb/m^2

Magnetic flux through each turn of the coil

$\phi = BA = (3.8 \times 10^{-2})(3.14 \times 10^{-4})$

$= 1.2 \times 10^{-5}$ weber

When the current in the solenoid is reversed, the change in magnetic flux

$= 2 \times (1.2 \times 10^{-5}) = 2.4 \times 10^{-5}$ weber

Induced e.m.f.

$= N\frac{d\phi}{dt} = 100 \times \frac{2.4 \times 10^{-5}}{0.05} = 0.048\,V.$

58. (a) Eddy currents are produced when a metal is kept in a varying magnetic field.

59. (b) $\phi = \vec{B}.\vec{A}$ $\qquad \phi = BA\cos\omega t$

$\varepsilon = -\frac{d\phi}{dt} = \omega BA\sin\omega t$; $i = \frac{\omega BA}{R}\sin\omega t$

$P_{inst} = i^2R = \left(\frac{\omega BA}{R}\right)^2 \times R\sin^2\omega t$

$P_{avg} = \frac{\int_0^T P_{inst} \times dt}{\int_0^T dt} = \frac{(\omega BA)^2}{R}\frac{\int_0^T \sin^2\omega t\,dt}{\int_0^T dt}$

$= \frac{(\omega BA)^2}{R}\left(\frac{T}{2.T}\right)$

$\therefore \quad P_{avg} = \frac{(\omega B\pi r^2)^2}{2R}$

60. (a) $e = L\frac{di}{dt}$

Given that $L = 40 \times 10^{-3}$ H,

$di = 11\,A - 1\,A = 10\,A$

and $dt = 4 \times 10^{-3}$ s

$\therefore e = 40 \times 10^{-3} \times \left(\frac{10}{4 \times 10^{-3}}\right) = 100V$

21. ALTERNATING CURRENT

ANSWERKEY											
1	(c)	11	(d)	21	(b)	31	(b)	41	(a)	51	(c)
2	(b)	12	(a)	22	(b)	32	(a)	42	(c)	52	(d)
3	(b)	13	(b)	23	(d)	33	(a)	43	(a)	53	(c)
4	(b)	14	(c)	24	(d)	34	(b)	44	(b)	54	(d)
5	(c)	15	(d)	25	(d)	35	(a)	45	(b)	55	(b)
6	(b)	16	(b)	26	(d)	36	(c)	46	(c)	56	(d)
7	(a)	17	(a)	27	(c)	37	(a)	47	(b)	57	(d)
8	(c)	18	(b)	28	(a)	38	(a)	48	(b)	58	(b)
9	(d)	19	(b)	29	(c)	39	(a)	49	(d)	59	(a)
10	(b)	20	(b)	30	(a)	40	(d)	50	(d)	60	(b)

Hints & Solutions

1. (c) The transmission is undamped, because energy loss becomes negligible.

4. (b) In India the frequency of A.C. current is 50 Hz.

7. (a)

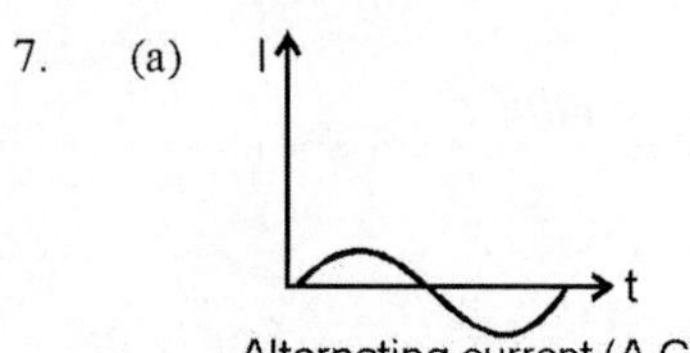

Alternating current (A.C)

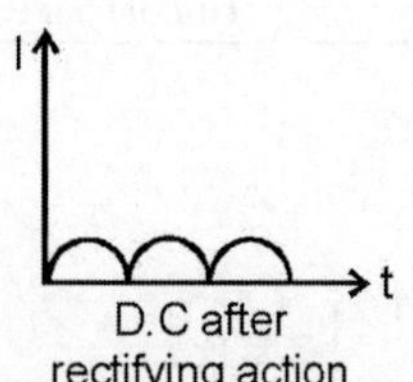

D.C after rectifying action

12. (a) Displacement current is set up in a region where the electric field is changing with time.

13. (b) During charging a capacitor, there will be different amount of charges on the plates of capacitor at different instants. Due to which the electric field in the region between the plates of capacitor is varying.

14. (c) The direction of propagation of electromagnetic wave is perpendicular to the variation of electric field $\vec{E}$ as well as to the magnetic field $\vec{B}$

16. (b) Velocity of light in a medium,

$$c = \frac{1}{\sqrt{\mu_0 \varepsilon_o \mu_r \varepsilon_r}} = \frac{1}{\sqrt{\mu \varepsilon}}$$

17. (a) $I_D = \varepsilon_o \, d\phi_E / dt$.

23. (d) The frequency of television signal is smaller than that of microwave and visible waves.

24. (d) Maxwell's equations describe the fundamental laws of electricity and magnetism.

25. (d) Initially there is no D.C. current in inductive circuit and maximum D.C.current is in capacitive current. Hence, the current is zero in A_2 and maximum in A_1.

26. (d) Power, $P = I_{r.m.s} \times V_{r.m.s} \times \cos\phi$
In the given problem, the phase difference between voltage and current is $p/2$. Hence
$$P = I_{r.m.s} \times V_{r.m.s} \times \cos(\pi/2) = 0.$$

27. (c) $P = V_{r.m.s} \times I_{r.m.s} \times \cos\phi = \frac{1}{2} V_0 I_0 \cos\phi$
$$= \frac{1}{2} \times 100 \times (100 \times 10^{-3}) \cos \pi/3 = 2.5\,\text{W}$$

28. (a) $\frac{E_s}{E_p} = \frac{n_s}{n_p}$ or $E_s = E_p \times \left(\frac{n_s}{n_p}\right)$
$$\therefore\ E_s = 120 \times \left(\frac{200}{100}\right) = 240V$$
$$\frac{I_p}{I_s} = \frac{n_s}{n_p} \text{ or } I_s = I_p \left(\frac{n_p}{n_s}\right)$$
$$\therefore\ I_s = 10\left(\frac{100}{200}\right) = 5 \text{ amp}$$

29. (c) $\frac{I_s}{I_p} = \frac{n_p}{n_s}$; $\frac{80}{I_p} = \frac{20}{1}$ or $I_p = 4$ amp.

32. (a) Impedance of circuit is

$$Z = \sqrt{R^2 + (X_L - X_C)^2}$$

36. (c) In the case of maximum charge on capacitor, the whole energy is, stored in capacitor in the form of electric field which is

$$U = \frac{1}{2}\frac{Q^2}{C}$$

When energy in distributed equally between electric and magnetic field , then energy stored in electric field i.e. in capacitor is

$$U_1 = \frac{U}{2} = \left(\frac{1}{2}\frac{Q^2}{C}\right)\frac{1}{2}$$

At that time if charge on capacitor is Q_1, then

$$U_1 = \frac{1}{2}\frac{Q_1^2}{C} = \frac{U}{2} = \frac{1}{4}\frac{Q^2}{C} \Rightarrow Q_1 = Q/\sqrt{2}$$

37. (a) X_C (reactance of capacitor) $= \frac{1}{\omega C}$ for D.C.,

$\omega = 0 \quad \Rightarrow X_C = \infty$

38. (a) The reactance of inductor, $X_L = \omega L$

The reactance of capacitor, $X_C = \frac{1}{\omega C}$

where $\omega = 2\pi n$ & n is the frequency of A.C source.

39. (a) The reactance of circuit is zero in two cases
(i) At resonance, $X_L = X_C$
So total reactance, $X = (X_L - X_C) = 0$ because they are in opposite phase.

42. (c) When resistance is connected to A.C. source, then current & voltage are in same phase.

43. (a) Capacitive reactance in an A.C. circuit is

$X_C = \frac{1}{\omega C}$ ohm , where C is the capacitance of capacitor & $\omega = 2\pi n$ (n is the frequency of A.C source).

44. (b) $\cos\phi = \frac{R}{Z}$, where Z is the impedance &

$Z = \sqrt{R^2 + (X_L - X_C)^2}$, if there is only resistance then $Z = R \Rightarrow \cos\phi = 1$

45. (b) If $R = 0 \Rightarrow \cos\phi = 0 \Rightarrow \phi = 90°$ so $P = 0$, in this case power loss is zero & current flowing in the circuit is called wattless current.

47. (b) $X_C = \frac{1}{\omega C} \Rightarrow X_C \propto \frac{1}{\omega}$ for given C.

48. (b) $X_L = \omega L \Rightarrow X_L \propto \omega$

49. (d)

Z

$Z_{min} = R$

$\omega_0 \rightarrow \omega$

51. (c) At resonance $X_L = X_C \Rightarrow Z = R$ & current is maximum but finite, which is $I_{max} = \frac{E}{R}$, where E is applied voltage.

52. (d)

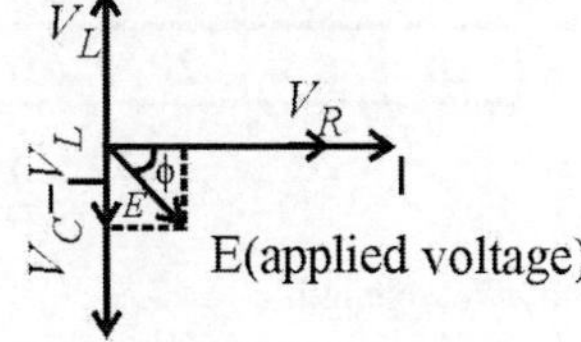

$$\tan\phi = \frac{V_C - V_L}{V_R} \text{ (if } V_C > V_L)$$

$$= \frac{V_L - V_C}{R} \text{ (if } V_L > V_C)$$

where ϕ is angle between current & applied voltage.

54. (d) In first circuit $I_1 = \frac{E}{\omega L}$

In second circuit $I_2 = \left(\frac{E}{1/\omega C}\right)$

56. (d) In case of pure inductance $\cos\phi = 0$, so no power dissipates.

57. (d) $P = V_{r.m.s.} \times I_{r.m.s.} \times \cos\phi$

$$= \frac{1}{2}E_0 \times I_0 \cos\pi/2 = 0$$

60. (b) The value of r.m.s current is

$$I_{rms} = \frac{I_P}{\sqrt{2}}$$

so power dissipated is $P = I_{rms}^2 R = \frac{1}{2} I_P^2 R$

22. ELECTROMAGNETIC WAVES

ANSWERKEY											
1	(d)	11	(b)	21	(b)	31	(d)	41	(d)	51	(b)
2	(d)	12	(a)	22	(c)	32	(a)	42	(c)	52	(d)
3	(b)	13	(d)	23	(a)	33	(a)	43	(c)	53	(a)
4	(b)	14	(a)	24	(a)	34	(c)	44	(d)	54	(b)
5	(d)	15	(b)	25	(b)	35	(d)	45	(c)	55	(a)
6	(c)	16	(d)	26	(a)	36	(d)	46	(d)	56	(b)
7	(c)	17	(b)	27	(c)	37	(a)	47	(a)	57	(b)
8	(d)	18	(b)	28	(b)	38	(c)	48	(c)	58	(b)
9	(d)	19	(c)	29	(c)	39	(a)	49	(d)	59	(c)
10	(c)	20	(c)	30	(a)	40	(a)	50	(b)	60	(d)

Hints & Solutions

1. (d) RaysWavelength

	[Range in m]
X-rays	1×10^{-11} to 3×10^{-8}
γ-rays	6×10^{-14} to 1×10^{-11}
Microwaves	10^{-3} to 0.3
Radiowaves	10 to 10^4

2. (d) Wavelength of red light is longest.

3. (b) Size of particle $= \lambda = \dfrac{c}{\nu}$

$$3 \times 10^{-6} = \frac{3 \times 10^8}{\nu}$$

$\nu = 10^{14}$ Hz

However, when frequency is higher than this, wavelength is still smaller. Resolution becomes better.

4. (b) X-rays are used for the investigation of structure of solids.

6. (c) The direction of oscillations of E and B fields are perpendicular to each other as well as to the direction of propagation. So, electromagnetic waves are transverse in nature.
The electric and magnetic fields oscillate in same phase.

7. (c) RaysWavelength

	[Range in m]
X-rays	1×10^{-11} to 3×10^{-8}
γ-rays	6×10^{-14} to 1×10^{-11}
Microwaves	10^{-3} to 0.3
Radiowaves	10 to 10^4

Wavelength of U.V. Rays ranges from 6×10^{-8} to 4×10^{-7}.

8. (d) We know that velocity of electromagnetic wave in vacuum $(v_0) = \dfrac{1}{\sqrt{\mu_0 \varepsilon_0}}$ and velocity of electromagnetic wave in medium is $(v) = \dfrac{1}{\sqrt{\mu\varepsilon}}$.
Therefore refractive index of the medium

$$(\mu) = \frac{\text{Vel. of E.M. wave in vacuum}(v_0)}{\text{Vel. of E.M. wave in medium}(v)}$$

$$= \frac{1/\sqrt{\mu_0\varepsilon_0}}{1/\sqrt{\mu\varepsilon}} = \sqrt{\frac{\mu\varepsilon}{\mu_0\varepsilon_0}}$$

9. (d) Initially there is no D.C. current in inductive circuit and maximum D.C. current is in capacitive current. Hence, the current is zero in A_2 and maximum in A_1.

10. (c) A stationary charge produces electric field only; an uniformly moving charge produces localised electromagnetic field; an accelerated charge produces electro-magnetic radiations.

11. (b) γ rays has lowest wavelength and highest frequency among them while ultraviolet ray has highest wavelength and lowest frequency.
Order of frequency : $b > a > c$

12. (a) α rays contain Helium nuclei which contains 2 unit of positive charge.

13. (d) Direction of propagation of electromagnetic waves is perpendicular to Electric field and Magnetic field. Hence, direction is given by Poynting vector

$$\vec{S} = \vec{E} \times \vec{H} = \frac{\vec{E} \times \vec{B}}{\mu_0}.$$

14. (a) $\omega_c + \omega_m$
$= 2.5 \times 10^6 + 0.0005 \times 10^6 Hz$
$\omega_c - \omega_m$
$= 2.5 \times 10^6 - 0.0005 \times 10^6$ Hz
$= 2.4995 \times 10^6$ Hz

15. (b) Infrared rays is the cause of Green house effect. The glass transmits visible light and short infrared rays which are absorbed by plants. Then it emits long infrared rays, which are reflected back by glass.

16. (d) β ray is not electromagnetic ray

17. (b) Depends on the magnitude of frequency

18. (b) We know $E = \frac{hc}{\lambda} \Rightarrow E \propto \frac{1}{\lambda}$

$\Rightarrow E_m < E_v < E_x$
$\therefore \quad \lambda_m > \lambda_v > \lambda_x$

19. (c) Variation in magnetic field causes electric field and vice-versa.

In electromagnetic waves, $\vec{E} \perp \vec{B}$. Both $\vec{E}$ and $\vec{B}$ are in the same phase.
In electromagnetic waves

$E = E_0 \sin(\omega t - kx)$

$B = B_0 \sin(\omega t - kx)$

The electromagnetic waves travel in the direction of $(\vec{E} \times \vec{B})$.

20. (c) The velocity of electromagnetic radiation in a medium of permittivity ϵ_0 and permeability μ_0 is equal to $\frac{1}{\sqrt{\mu_0 \epsilon_0}}$.

21. (b) Comparing with the equation of wave.
$E_y = E_0 \cos(\omega t - kx)$
$\omega = 2\pi f = 2\pi \times 10^6 \quad \therefore \; f = 10^6$ Hz
$\frac{2\pi}{\lambda} = k = \pi \times 10^{-2}\, m^{-1}, \lambda = 200$ m

22. (c) Electromagnetic waves are the combination of mutually perpendicular electric and magnetic fields. So, option (c) is false.

23. (a) E.M. wave always propagates in a direction perpendicular to both electric and magnetic fields. So, electric and magnetic fields should be along + X– and + Y–directions respectively. Therefore, option (a) is the correct option.

24. (a) The decreasing order of the wavelengths is as given below :
microwave, infrared, ultraviolet, gamma rays.

25. (b) Ultraviolet radiation is used in food processing as sterilizing agent.

27. (c) Accelerating charge produces electro-magnetic radiation which is called Bremstrahlung radiation.

30. (a) The existence of electromagnetic wave was first predicted by Maxwell.

31. (d) Given $E_z = 60\sin(kx + \omega t)$ V/m. The magnetic field component must be perpendicular to E_z and x.

Also $B_0 = \frac{E_0}{c} = \frac{60}{c}$. Thus

$B_y = \frac{60}{c}\sin(kx + \omega t)$.

32. (a) Here, $\lambda = \frac{c}{\nu} = \frac{3 \times 10^8}{8.2 \times 10^6} = 36.6\,\text{m}$.

33. (a) Here, amplitude of electric field, $E_0 = 100$ V/m; amplitude of magnetic field, $H_0 = 0.265$ A/m. We know that the maximum rate of energy flow, $S = E_0 \times H_0 = 100 \times 0.265 = 26.5\ W/m^2$.

34. (c) Ionosphere contain free electron & + ve ions. In equillibrium, the no. of free electron is equal to the number of positive ions. So as a whole it is electrically neutral.

35. (d) $d = \sqrt{2hR}$.
Population covered
$= \pi d^2 \times$ population density

36. (d) $B_0 = \frac{E_0}{c} = \frac{9 \times 10^3}{3 \times 10^8} = 3 \times 10^{-5} T$.

38. (c) Electron density of each layer of ionosphere is different from the other. i.e. they are stratified. As we move upward density increases.

39. (a) $\xrightarrow[\text{RMIVUXGC}]{\lambda \text{ decreasing}}$

R → Radio waves M → Micro waves
I → Infra red rays V → Visible rays
U → Ultraviolet rays X → X rays
G → γ rays C → Cosmic rays
⇒ γ rays has least wavelength

40. (a) Velocity of light,

$v = \frac{1}{\sqrt{\mu_0 \varepsilon_0 \varepsilon_r}} = \frac{1}{\sqrt{\mu_0 \varepsilon_0 . 4}} = \frac{c}{2}$

v remains constant hence

$\lambda = \frac{v}{\nu} = \frac{\lambda_0}{2}$ since $v = \frac{c}{2}$

∴ wavelength is halved and frequency remains unchanged

41. (d) $\lambda = 10Å = 10 \times 10^{-10} m = 10^{-9} m$
X- ray wavelength is of the order of 1Å

42. (c) Microwave region wavelength $= 10^{-3}$ m to 1m

43. (c) Relation between E_0 and H_0

$$\frac{E_0}{H_0} = \frac{1}{C} = \sqrt{\mu_0 \varepsilon_0}$$

i.e. $H_0 = E_0\sqrt{\varepsilon_0 \mu_0}$

46. (d) $\frac{1}{2}\varepsilon_0 E_0^2$ is electric energy density.

$\frac{B^2}{2\mu_0}$ is magnetic energy density.

So, total energy $= \frac{1}{2}\varepsilon_0 E_0^2 + \frac{B_0^2}{2\mu_0}$

47. (a) For growth of current in LR circuit

$$i = i_0\left(1 - e^{-\frac{R}{L}t}\right) = i_0\left(1 - \frac{1}{e^{Rt/L}}\right)$$

for $t = \infty, i = i_0$

So current after one second

$$i = i_0\left(1 - \frac{1}{e^{\frac{10}{5}\times 1}}\right)$$

$$= i_0\left(1 - \frac{1}{e^2}\right) = i_0\left(\frac{e^2 - 1}{e^2}\right)$$

$$\Rightarrow \frac{i_0}{i} = \frac{e^2}{(e^2 - 1)}$$

ratio of current at $t = \infty$ and t = 1 is

$$= \frac{e^2}{(e^2 - 1)}.$$

48. (c) Radio waves are low frequency electro-magnetic waves. So, their wave length will be comparatively large.

49. (d) Hertz experiment is known for existence of electromagnetic waves.

50. (b) A transducer is a device which produce electrical signal from other form of energy such as pressure. This is the phenomena which has been used in Loudspeaker, Amplifier, Microphone etc. All are called transducer.

51. (b) Voltage across LC combination

$$= \left(\omega L - \frac{1}{\omega C}\right) I_{rms}$$

$$= \omega L I_{rms} - \frac{1}{\omega C} I_{rms} = (50 - 50)V = 0V$$

In other words, as their potential are in opposite phase, they will cancel out each other.

52. (d) Eddy current is the current produced in the body of a conductor due to electromagnetic induction. Moving coil galvanometer, electric brakes and induction motor based on electromagnetic induction. Hence they use eddy current. Dynamo doesn't use it.

53. (a)

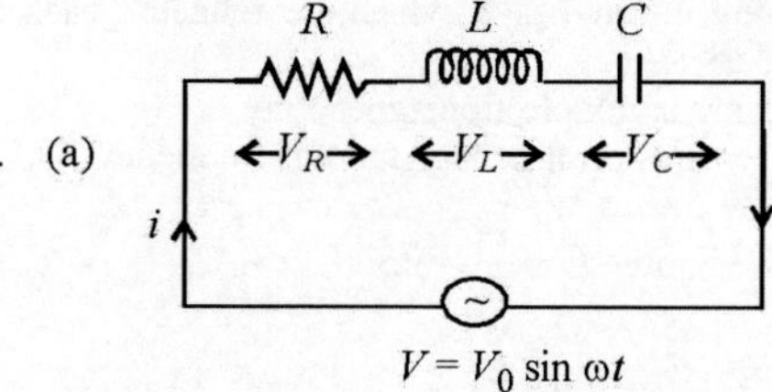

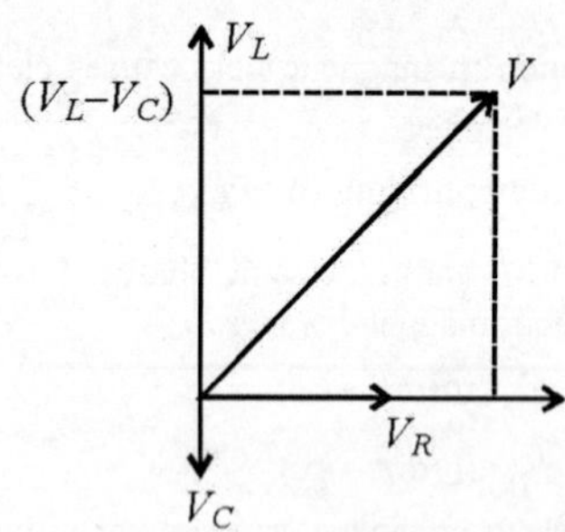

Voltage $V = \sqrt{V_R^2 + (V_L - V_C)^2}$

$$= \sqrt{40^2 + (60 - 30)^2}$$

$$= \sqrt{40^2 + 30^2} = 50 \text{ volt.}$$

54. (b) Choke coil is a device having high inductance and negligible resistance. It is used to control current in ac circuits and the power loss in a circuit containing choke coil is least.

55. (a) $l = 1.5$ m;

$r = \frac{4}{2} = 2$ cm

$n = 10$ turns/cm $= 1000$ turn/m

$I = 5$A

(i) $B_{inside} = \mu_0 nI = 4\pi \times 10^{-7} \times 1000 \times 5 = 2\pi \times 10^{-3}\, T$

(ii) $B_{end} = \frac{\mu_0 nI}{2} = \pi \times 10^{-3} T$

56. (b) The magnetic field at a point, distant r from the straight wire conductor carrying current I is given by,

$$B = \frac{\mu_0}{4\pi}.\frac{21}{r}$$

$$B \propto \frac{1}{r}$$

i.e., graph is rectangular hyperbolic.

57. (b) Self inductance of a small circular coil of radius r,

$$L = \frac{\mu_0 N^2 A}{2r} = \frac{\mu_0 N^2 (\pi r^2)}{2r}$$

i.e. $L \propto N^2$

If no. of turns are doubled then self inductance will be four times.

58. (b) $n_1 = 5, n_2 = 10$

$M = 25\ H$

$M \propto n_1 n_2$

$25 \propto 5 \times 10$ ... (i)

$n_1 = 10; n_2 = 5$

$M \propto 10 \times 5$ ... (ii)

From (i) and (ii), we get

$M = 25$ H.

59. (c) Potential difference e = Bvl

$$e = 5\times10^{-5}\times360\times\frac{5}{18}\times20$$

$e = 400 \times 25 \times 10^{-5}$

$e = 10^4 \times 10^{-5}$

$e = 0.10$ V

60. (d) $R = 2k\ \Omega$

$\phi = 5t^2 - 4t + 12$

$t = 0.2$ sec.

Induced emf $e = \frac{-d\phi}{dt}$

$e = -10t + 4 = -10 \times 0.2 + 4 = +2$ V

$$I = \frac{e}{R} = \frac{+2}{2\times10^3}$$

$I = 10^{-3}$ A

$I = 1$ mA

23. RAY OPTICS AND OPTICAL INSTRUMENTS

ANSWERKEY

1	(a)	**11**	(d)	**21**	(c)	**31**	(b)	**41**	(a)	**51**	(c)
2	(b)	**12**	(d)	**22**	(d)	**32**	(d)	**42**	(a)	**52**	(d)
3	(d)	**13**	(d)	**23**	(c)	**33**	(b)	**43**	(b)	**53**	(c)
4	(c)	**14**	(d)	**24**	(b)	**34**	(b)	**44**	(a)	**54**	(c)
5	(a)	**15**	(c)	**25**	(b)	**35**	(a)	**45**	(c)	**55**	(a)
6	(d)	**16**	(b)	**26**	(a)	**36**	(c)	**46**	(b)	**56**	(c)
7	(b)	**17**	(d)	**27**	(a)	**37**	(a)	**47**	(c)	**57**	(c)
8	(b)	**18**	(d)	**28**	(a)	**38**	(b)	**48**	(b)	**58**	(b)
9	(a)	**19**	(c)	**29**	(c)	**39**	(b)	**49**	(d)	**59**	(a)
10	(a)	**20**	(d)	**30**	(c)	**40**	(c)	**50**	(a)	**60**	(d)

Hints & Solutions

1. (a) In terms of wavelength,

$$\mu \propto \frac{1}{\lambda}.$$

So shorter the wavelength, greater is the refractive index. λ is minimum for violet so μ is maximum.

2. (b) When two plane mirrors are inclined at an angle θ, the number of image obtained is

$$n = \left(\frac{360°}{\theta} - 1\right) = \frac{360°}{90°} - 1 = 3$$

3. (d) The inverse square law and inverse fourth power law do not hold in smoked filled room. So, the intensity of source falls faster.

4. (c) The image formed by a concave mirror is always real when the object is placed between infinity and focus. It forms a virtual image when the object is placed between focus and pole. This means that when the object is virtual, the image is certainly real.

5. (a) ${}_an_l = 1.6, {}_wn_l = 1.33$
$f = 20$ cm
We have,

$$\frac{1}{f} = ({}_an_l - 1)\left(\frac{1}{R_1} - \frac{1}{R_2}\right)$$

$$\frac{1}{20} = (1.6 - 1)\left(\frac{1}{R_1} - \frac{1}{R_2}\right) \quad \text{... (1)}$$

Also, $\frac{1}{f'} = ({}_wn_l - 1)\left(\frac{1}{R_1} - \frac{1}{R_2}\right)$

$$= \left(\frac{{}_an_l}{{}_wn_l} - 1\right)\left(\frac{1}{R_1} - \frac{1}{R_2}\right)$$

$$\frac{1}{f'} = \left(\frac{1.6}{1.33} - 1\right)\left(\frac{1}{R_1} - \frac{1}{R_2}\right) \quad \text{.... (2)}$$

Dividing equation (1) by (2)

$$\Rightarrow \frac{f'}{20} = \frac{0.6}{(1.2 - 1)}$$

$$f' = \frac{0.6 \times 20}{0.2} = 60 \text{ cm.}$$

Hence it's focal length is three times longer than in air.

6. (d) The image formed by a concave lens is always virtual, erect and diminished. When object is placed any where on the principal axis, image is formed between optical centre and focus.

7. (b) $\frac{\omega_1}{f_1} + \frac{\omega_2}{f_2} = 0$ or $\frac{f_1}{f_2} = -\frac{\omega_1}{\omega_2}$

9. (a) The basic principle of communication in fibre optics is based on the phenomenon of total internal reflection.

11. (d) Since refractive index for different wave length of light is different. Hence the different colours of light forms images at different position. This phenomenon is called chromatic aberration.

12. (d) It occurs in both reflection & refraction & the angle of bending rays is constant.

13. (d) The cause of chromatic aberration is that lens focusses different colours at different points

14. (d) **(i) Line spectra :** The line spectra consists of distinct bright lines & is produced by an excited source in atomic state. These lines are the characterstic of element & are used for its identification.
(ii) Band Spectra : The band spectra consists of distinct bright bands & is produced by an excited source in molecular state.

15. (c) Since $\lambda_R > \lambda_V \quad \mu_R < \mu_V$

$$\left(\because \mu \propto \frac{1}{\lambda}\right) \Rightarrow f_V < f_R \left(\because \frac{1}{f} \propto (\mu - 1)\right)$$

16. (b) If faces of prism on which light is incident & from which it emerge are parallel, then angle of prism will be zero & so deviation will also be zero i.e., the prism will act as a transparent prism.

17. (d) We can combine two prisms in such a way
(i) deviation is zero but dispersion not
(ii) dispersion is zero but deviation is not.
But in any situation both deviation & dispersion cannot be zero simultaneously.

18. (d) Using the formula for focal length,

$$\frac{1}{f} = \left(\frac{\mu_g}{\mu_l} - 1\right)\left(\frac{1}{R_1} - \frac{1}{R_2}\right)$$

$$\frac{1}{f} = \left(\frac{1.6}{1.63} - 1\right)\left(\frac{1}{15} + \frac{1}{15}\right)$$

As refractive index of surrounding medium is greater, the nature of lens is changed and we get a negative focal length.

$$\frac{1}{f} = -\frac{0.03}{1.63} \times \frac{2}{15}$$

$$\Rightarrow f = -\frac{815}{2} = -407.5 \text{ cm}$$

19. (c) The relation for refractive index and wavelength,

$$\mu = \frac{\lambda\,(\text{air})}{\lambda\,(\text{medium})}$$

$$\lambda_{\text{glass}} = \frac{\lambda_{\text{air}}}{\mu} = \lambda_{\text{glass}} = \frac{6000}{3/2} = 4000\text{Å}$$

20. (d) $u = -20$ cm, $f = 10$ cm

$$\frac{1}{f} = \frac{-1}{u} + \frac{1}{v} \Rightarrow \frac{1}{v} = \frac{1}{10} - \frac{1}{20}$$

$\Rightarrow v = 20$ cm.

21. (c) The condition for an achromatic doublet is

$$\frac{\omega_1}{f_1} + \frac{\omega_2}{f_2} = 0$$

This gives the combination of two lenses which will be free from chromatic aberration.

22. (d) For red light, focal length of lens is maximum because $f \propto \lambda$ and λ is maximum for red light.

23. (c) If two or more lenses are combined together in such a way that this combination produces images of different colours at the same point and of the same size, then this property is called 'achromatism'. Concave and convex type of lenses are used for this combination.

24. (b) A person suffering from hypermetropia can see objects beyond a particular point called the near point. If the object lies at a point nearer than this point, then image is not formed at the retina. This is also known as long-sight defect.

25. (b) Angle between two mirrors (θ) = 60°. Number of images formed by the inclined mirror $(n) = \frac{360°}{\theta} - 1 = \frac{360°}{60°} - 1 = 6 - 1 = 5$

26. (a) Size of images = A_1 and A_2. From the displacement method, the area of the source of light = $\sqrt{A_1A_2}$.

27. (a) $\frac{1}{f} = (\mu - 1)\left(\frac{1}{R_1} - \frac{1}{R_2}\right)$

According to Cauchy relation

$\mu = A + \frac{B}{\lambda^2} + \frac{C}{\lambda^4} \cdots$ Hence $f \propto \lambda$.

Hence, red light having maximum wavelength has maximum focal length.

$\therefore\ f_v < f_r$ and also $F_v > F_r$ as focal length is negative for a concave lens.

28. (a) ${}_1^{}\mu_2 = \frac{\sin 90°}{\sin C} = \frac{1}{\sin C}$ [For critical angle]

$\therefore \quad C = \sin^{-1}\left(\frac{1}{{}_1\mu_2}\right)$ ... (i)

Applying Snell's law at P we get

${}_1\mu_2 = \frac{\sin r'}{\sin i} = \frac{\sin(90 - r)}{\sin r}$

$[\because i = r; r' + r = 90°]$

${}_1\mu_2 = \frac{\cos r}{\sin r}$ ... (ii)

From (i) and (ii)

$C = \sin^{-1}(\tan r)$

29. (c) Thickness of glass plate (t) = 6 cm;
Distance of the object (u) = 8 cm
And distance of the image (v) = 12 cm
Let x = Apparent position of the silvered surface in cm
Since the image is formed due to reflection at the silvered face and by the property of mirror image
Distance of object from the mirror = Distance of image from the mirror
or, $x + 8 = 12 + 6 - x \Rightarrow x = 5$ cm
Therefore, refractive index of glass

$$= \frac{\text{Real depth}}{\text{Apparent depth}} = \frac{6}{5} = 1.2$$

30. (c) According to Cauchy relation

$\mu = A + \frac{B}{\lambda^2} + \frac{C}{\lambda^4} \cdots$ Hence $f \propto \lambda$.

$\therefore$ Red light has wavelength greater than violet light. Therefore focal length of lens for red is greater than for violet. ($f_r > f_v$)

31. (b) Dispersive power of a prism

$$\omega = \frac{\mu_V - \mu_R}{\mu_y - 1} = \frac{d\mu}{\mu - 1},$$

where $\mu = \mu_y = \frac{\mu_V + \mu_R}{2}$

32. (d) The colours are seen due to interference of light. The colours seen in reflected light are complementary with the colours seen in transmitted light.

33. (b) Apparent position of fish as seen by bird $= [h_1 + (h_2/\mu)]$

34. (b) We know that frequency of electromagnetic radiation remains the same when it changes the medium. Further

$$\mu = \frac{\text{wavelength of light in vacuum}}{\text{wavelength of light in medium}} = \frac{\lambda_v}{\lambda_m}$$

$$\lambda_m = \frac{\lambda_v}{\mu} = \frac{\lambda}{\mu}$$

Similarly, $\mu = \frac{\text{velocity of light in vacuum}}{\text{velocity of light in medium}}$

$$\lambda_m = \frac{v}{\mu}$$

35. (a) In the position of minimum deviation, $i_1 = i_2$

36. (c)

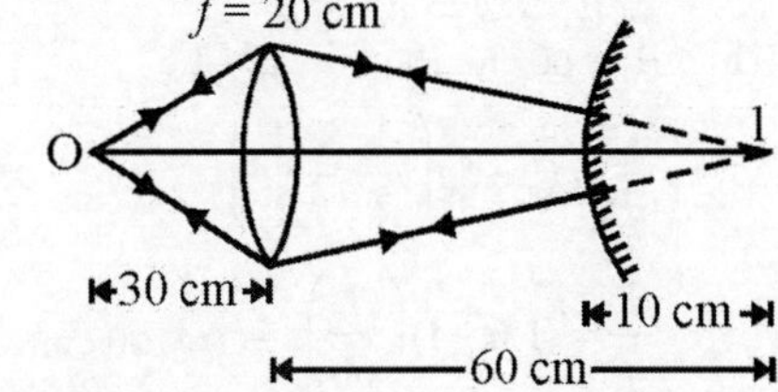

For the lens,

$\frac{1}{v} - \frac{1}{u} = \frac{1}{f};\ \frac{1}{v} - \frac{1}{-30} = \frac{1}{20} \Rightarrow v = 60$ cm

Coincidence is possible when the image is formed at the centre of curvature of the mirror. Only then the rays refracting through the lens will fall normally on the convex mirror and retrace their path to form the image at O. So, the distance between lens and mirror = 60 – 10 = 50 cm.

37. (a) Let a ray of light enter at A and the refracted beam is AB. This is incident at an angle θ. For no refraction at the lateral face, $\theta > C$

or, $\sin\theta > \sin C$ But $\theta + r = 90°$

$\Rightarrow \theta = (90° - r)$

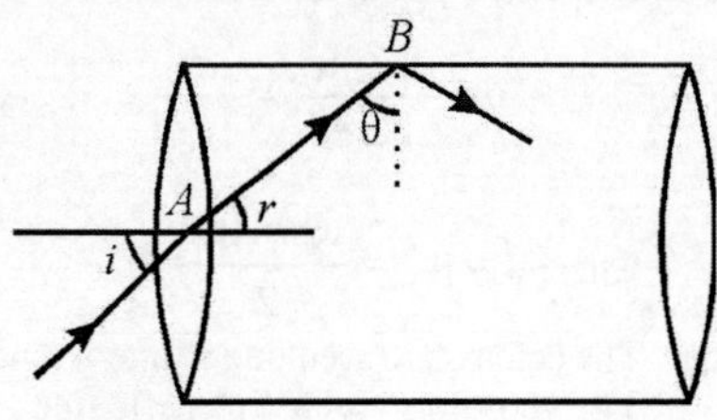

$\therefore\ \sin(90° - r) > \sin C$

or $\cos r > \sin C$ ...(1)

From Snell's law, $n = \dfrac{\sin i}{\sin r} \Rightarrow \sin r = \dfrac{\sin i}{n}$

$$\therefore \cos r = \sqrt{1-\sin^2 r} = \sqrt{\left(1-\frac{\sin^2 i}{n^2}\right)}$$

$\therefore$ equation (1) gives

$$\sqrt{1-\frac{\sin^2 i}{n^2}} > \sin C \Rightarrow 1-\frac{\sin^2 i}{n^2} > \sin^2 C$$

Also, $\sin C = \dfrac{1}{n}$

$$\therefore 1-\frac{\sin^2 i}{n^2} > \frac{1}{n^2} \text{ or } 1 > \frac{\sin^2 i}{n^2}+\frac{1}{n^2}$$

or $\dfrac{1}{n^2}(\sin^2 i + 1) < 1$ or $n^2 > (\sin^2 i + 1)$

Maximum value of $\sin i = 1$

$$\therefore n^2 > 2 \Rightarrow n > \sqrt{2}$$

38. (b) $R_1 = 60$ cm, $R_2 = \infty$, $\mu = 1.6$

$$\frac{1}{f} = (\mu - 1)\left(\frac{1}{R_1} - \frac{1}{R_2}\right)$$

$$\frac{1}{f} = (1.6 - 1)\left(\frac{1}{60}\right) \Rightarrow f = 100 \text{ cm.}$$

39. (b) Frequency $(n) = 100$ Hz

$v = n\lambda$

$$\lambda = \frac{3 \times 10^8}{100}$$

[where, velocity of light $(v) = 3 \times 10^8$ m/s]

$$\lambda = 3 \times 10^6 \text{ m}$$

40. (c) Angle of minimum deviation

$$\mu = \frac{\sin\left(\frac{A+\delta_m}{2}\right)}{\sin\left(\frac{A}{2}\right)} \Rightarrow \sqrt{3} = \frac{\sin\left(\frac{60°+\delta_m}{2}\right)}{\sin\left(\frac{60°}{2}\right)}$$

$$\Rightarrow \sin\left(30° + \frac{\delta_m}{2}\right) = \frac{\sqrt{3}}{2} \Rightarrow 30° + \frac{\delta_m}{2} = 60°$$

$\Rightarrow \delta_m = 60°$.

41. (a) The silvered plano convex lens behaves as a concave mirror; whose focal length is given by

$$\frac{1}{F} = \frac{2}{f_1} + \frac{1}{f_m}$$

If plane surface is silvered

$$f_m = \frac{R_2}{2} = \frac{\infty}{2} = \infty$$

$$\therefore \frac{1}{f_1} = (\mu - 1)\left(\frac{1}{R_1} - \frac{1}{R_2}\right)$$

$$= (\mu - 1)\left(\frac{1}{R} - \frac{1}{\infty}\right) = \frac{\mu - 1}{R}$$

$$\therefore \frac{1}{F} = \frac{2(\mu - 1)}{R} + \frac{1}{\infty} = \frac{2(\mu - 1)}{R}$$

$$F = \frac{R}{2(\mu - 1)}$$

Here, $R = 10$ cm, $\mu = 1.5$

$$\therefore F = \frac{10}{2(1.5 - 1)} = 10 \text{ cm}$$

42. (a) The angular dispersion θ i.e., the angle between the extreme rays of light,

$\theta = (\delta_V - \delta_R)$ where $\delta_V = (\mu_V - 1)A$, $\delta_R = (\mu_V - 1)A$ & A is angle of prism.

So if refractive index increases, then δ increases & hence θ increases.

43. (b) Given : Length of astronomical telescope $(f_0 + f_e) = 44$ cm and magnification $\left(\dfrac{f_0}{f_e}\right) = 10$.

From the given magnification, we find that $f_0 = 10f_e$. Therefore, $10f_e + f_e = 44$ or $11f_e = 44$ or $f_e = 4$. And focal length of the objective $(f_0) = 44 - f_e = 44 - 4 = 40$ cm.

44. (a) When electromagnetic wave enters in other medium, frequency remains unchanged while wavelength and velocity become $\dfrac{1}{\mu}$ times.

So, For e.m. wave entering from air to glass slab (μ), frequency remains n, wavelength, $\lambda' = \dfrac{\lambda}{\mu}$

and velocity, $v' = \dfrac{v}{\mu}$

45. (c) For reading purposes :

$u = -25$ cm, $v = -50$ cm, $f = ?$

$$\frac{1}{f} = \frac{1}{v} - \frac{1}{u} = -\frac{1}{50} + \frac{1}{25} = \frac{1}{50};$$

$$P = \frac{100}{f} = +2D$$

For distant vision, f = distance of far point = -3 m

$$P = \frac{1}{f'} = -\frac{1}{3}D = -0.33\ D$$

46. (b) For minimum deviation, incident angle is equal to emerging angle.
$\therefore$ QR is horizontal.

47. (c) $\mu = \dfrac{\text{velocity of light in vacuum}}{\text{velocity of light in glass plate}}$

or $\mu = \dfrac{c}{c'}$ or $c' = \dfrac{c}{\mu}$

Time taken = distance/velocity

$$= t/(c/\mu) = \frac{\mu t}{c}$$

48. (b) $\dfrac{v_2}{v_1} = \dfrac{\mu_1}{\mu_2} = \dfrac{1}{1.33}$

or $v_2 = \dfrac{v_1}{1.33} = 2.25 \times 10^8 \text{m/s}$

49. (d) Let d be the depth of two liquids.
Then apparant depth

$$\frac{(d/2)}{\mu} + \frac{(d/2)}{1.5\mu} = \frac{d}{2} \text{ or } \frac{1}{\mu} + \frac{2}{3\mu} = 1$$

Solving we get $\mu = 1.671$

50. (a) $\dfrac{1}{F} = \dfrac{1}{f_1} + \dfrac{1}{f_2} = \dfrac{f_2 + f_1}{f_1 f_2}$; $P = \dfrac{1}{F} = \dfrac{f_1 + f_2}{f_1 f_2}$

51. (c) $\dfrac{f_a}{f_l} = \dfrac{\left(\dfrac{\mu_g}{\mu_l} - 1\right)}{(\mu_g - 1)} = \dfrac{\left(\dfrac{1.5}{1.25} - 1\right)}{1.5 - 1} = \dfrac{\frac{1}{5}}{\frac{1}{2}} = \dfrac{2}{5}$

$$f_l = \frac{5}{2} f_a = \frac{5}{2} \times 2 = 5 \text{ cm}$$

52. (d) Characteristic dimensions must be much larger than the wavelength of light.

53. (c) As refracted ray emerges normally from oppostite surface, $r_2 = 0$
As $A = r_1 + r_2$ $\therefore$ $r_1 = A$

Now, $\mu = \dfrac{\sin i_1}{\sin r_1} = \dfrac{i_1}{r_1} = \dfrac{i}{A}; i = \mu A$

54. (c) From $\mu = \dfrac{c}{v} = \dfrac{n\lambda_v}{n\lambda_m}$, $\lambda_m = \dfrac{\lambda_v}{\mu}$

Here, c = velocity of light in medium and v = velocity of light in vacuum;
m = refractive index of the medium.

Hence, wavelength in medium $(\lambda_m) < \lambda_a$
($\because \mu > 1$, given)

So, the required wavelength decreases.

Alternatively,

$c = v\lambda$. On refraction, the frequency, v do not change. When light is refracted from vacuum to a medium, the velocity, c decreases. Therefore, λ also decreases.

55. (a) $\lambda_g = \dfrac{\lambda_a}{\mu} = \dfrac{5460}{1.5} = 3640\text{Å}$

56. (c)

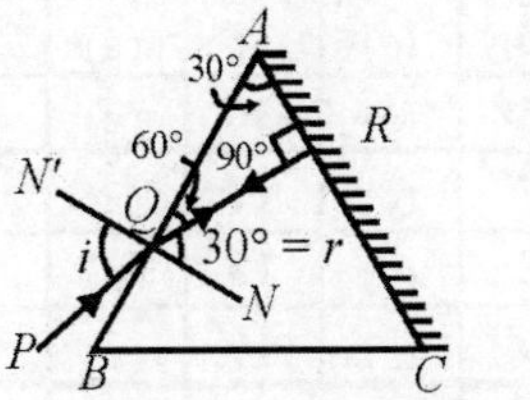

A
30°
R
60°
90°
N'
O
i
30° = r
N
P
B
C

It is clear from the figure that the ray will retrace the path when the refracted ray QR is incident normally on the polished surface AC. Thus, angle of refraction $r = 30°$

We know that $\mu = \dfrac{\sin i}{\sin r}$

$\therefore$ $\sin i = \mu \times \sin r = \sqrt{2} \times \sin 30°$

$$= \sqrt{2} \times \frac{1}{2} = \frac{1}{\sqrt{2}} \text{ or } i = \sin^{-1}\frac{1}{\sqrt{2}} = 45°$$

57. (c) $v_g = \dfrac{c}{\mu} = \dfrac{3 \times 10^8}{\frac{3}{2}} = 2 \times 10^8$ m/s

$$t = \frac{x}{v_g} = \frac{4 \times 10^{-3}}{2 \times 10^8} = 2 \times 10^{-11} \text{ s}$$

58. (b) $\sin C = \dfrac{1}{\mu} = \dfrac{1}{\frac{5}{3}} = \dfrac{3}{5} \Rightarrow \tan C = \dfrac{3}{4}$

Now, $\tan C = \dfrac{r}{h}; r = h\tan C = 4 \times \dfrac{3}{4} = 3\text{m}$

Diameter of disc = $2r$ = 6 m

59. (a) When the angle of prism is small,
$\delta = (\mu - 1)A$

60. (d) We know that $\dfrac{1}{f} = \sum_{i=1}^{n} \dfrac{1}{f_i}$

$\dfrac{1}{f} = \dfrac{1}{f_1} + \dfrac{1}{f_2}$; $f_1 = 80$ cm, $f_2 = -50$ cm

$$\frac{1}{f} = \frac{1}{\frac{80}{100}} - \frac{1}{\frac{50}{100}}$$

$\Rightarrow P = \dfrac{1}{f} = 1.25 - 2 = -0.75\text{D}$

24. WAVE OPTICS

ANSWERKEY											
1	(c)	**11**	(c)	**21**	(c)	**31**	(d)	**41**	(b)	**51**	(b)
2	(b)	**12**	(c)	**22**	(b)	**32**	(b)	**42**	(c)	**52**	(a)
3	(a)	**13**	(b)	**23**	(a)	**33**	(b)	**43**	(b)	**53**	(d)
4	(b)	**14**	(b)	**24**	(b)	**34**	(c)	**44**	(d)	**54**	(b)
5	(a)	**15**	(a)	**25**	(b)	**35**	(d)	**45**	(b)	**55**	(a)
6	(a)	**16**	(c)	**26**	(a)	**36**	(d)	**46**	(c)	**56**	(d)
7	(c)	**17**	(a)	**27**	(c)	**37**	(c)	**47**	(b)	**57**	(d)
8	(d)	**18**	(b)	**28**	(a)	**38**	(c)	**48**	(b)	**58**	(d)
9	(a)	**19**	(a)	**29**	(b)	**39**	(d)	**49**	(d)	**59**	(c)
10	(d)	**20**	(a)	**30**	(c)	**40**	(b)	**50**	(d)	**60**	(c)

Hints & Solutions

1. (c) When the film is thin, $t \to 0$, path diff. $= \lambda/2$. Therefore, in reflected light, the film appears black.

2. (b) As $\lambda_r \approx 8000 A^\circ$ and $\lambda_v \approx 4000 \mathring{A}$ and

$\beta = \dfrac{\lambda D}{d}$ i.e.

$\beta \propto \lambda$, therefore β_{red} is roughly double of β_{violet}.

3. (a) According to wave theory, intensity of light is directly proportional to square of amplitude.

4. (b) As optical path *SB* of lower slit is increased, therefore, fringe pattern shifts somewhat downwards.

5. (a) As $\beta = \dfrac{\lambda D}{d} \therefore \beta \propto \lambda$.

As λ for violet is least, therefore, fringe nearest to central achromatic fringe will be violet.

6. (a) Increase in length of optical path $= \mu t - t = (\mu - 1)t$.

7. (c) Highly coherent sources are produced using laser.

8. (d) A prism cannot produce coherent sources.

9. (a) Coherence is a measure of capability of producing interference by waves.

10. (d) Interference pattern will be invisible, because red and green are complimentary colours.

11. (c) Light waves can be polarized because they are transverse in nature.

12. (c) At Brewster's angle, only the reflected light is plane polarised, but transmitted light is partially polarised.

13. (b) Light transmitted by nicol prism is plane polarised.

14. (b) On polarisation by reflection, the reflected and refracted waves are at 90° to each other.

16. (c) Such substances rotate the plane of polarised light.

17. (a) Polaroid glass polarises light reducing the light intensity to half its original value.

18. (b) Angle between plane of vibration and plane of polarisation is 90°.

19. (a) Plane of vibration is $\perp^r$ to direction of propagation and also $\perp^r$ to plane of polarisation. Therefore, angle between plane of polarisation and direction of propagation is 0°.

20. (a) Along the optic axis, $\mu_0 = \mu_e$.

22. (b) When source is receding away, apparent wavelength increases. Displacement is towards red region.

23. (a) Velocity of light (in vacuum) is an absolute constant and does not depend upon the motion of the observer.

24. (b) Shifting towards red end means wavelength is increasing. Therefore, milky way is receding away from earth.

25. (b) Limit of resolution of eye is roughly 1 minute (angle).

26. (a) To be invisible in vacuum, μ of medium must be equal to μ of vacuum, which is 1.

27. (c) This ability refers to resolving power of the instrument.

28. (a) For single slit diffraction pattern $e\sin\theta = \lambda$
Angular width, e = slit width

$\therefore \quad \theta = \sin^{-1}\left(\frac{\lambda}{e}\right)$

it is independent of D, distance between screen and slit.

29. (b) When two waves of same amplitude and nearly same frequency in opposite direction are propagated. They produce stationary waves. For beats, there must be a difference in frequency.

30. (c) $e\sin\theta = n\lambda$
e = slit width
it is independent of the distance between slit and screen.

31. (d) Photoelectric effect does not support the wave nature of light.

32. (b) The line rulings, each of 0.5 μm width, on a compact disc function as a diffraction grating. When a small source of light illuminates a disc, diffracted light forms coloured 'lanes' that are the composite of the diffraction patterns from the ruling.

34. (c) Scattering of light

35. (d) The distance between the first dark fringe on either side of the central bright fringe
= width of central maximum

$= \frac{2D\lambda}{a}$

$= \frac{2 \times 2 \times 600 \times 10^{-9}}{10^{-3}}$

$= 2.4 \times 10^{-3}$ m = 2.4 mm

36. (d) At the area of total darkness minima will occur for both the wavelength.

$\therefore \quad \frac{(2n+1)}{2}\lambda_1 = \frac{(2m+1)}{2}\lambda_2$

$\Rightarrow \quad (2n+1)\lambda_1 = 2(m+1)\lambda_2$

or $\quad \frac{(2n+1)}{(2m+1)} = \frac{560}{400} = \frac{7}{5}$

or $\quad 10\,n = 14\,m + 2$
by inspection for $m = 2, n = 3$
for $m = 7, n = 10$
the distance between them will be the distance between such points

i.e., $\quad \Delta s = \frac{D\lambda_1}{d}\left\{\frac{(2n_2+1)-(2n_1-1)}{2}\right\}$

put $n_2 = 10, n_1 = 3$
on solving we get, $\Delta s = 28$ mm

37. (c) Frequency does not change with change of medium.

38. (c) Let P be the point on the central maxima whose intensity is one fourth of the maximum intensity.
For interference we know that

$$I = I_1 + I_2 + 2\sqrt{I_1 I_2}\cos\phi$$

Where I is the intensity at P and I_1, I_2 are the intensity of light originating from A and B respectively and ϕ is the phase difference at P.

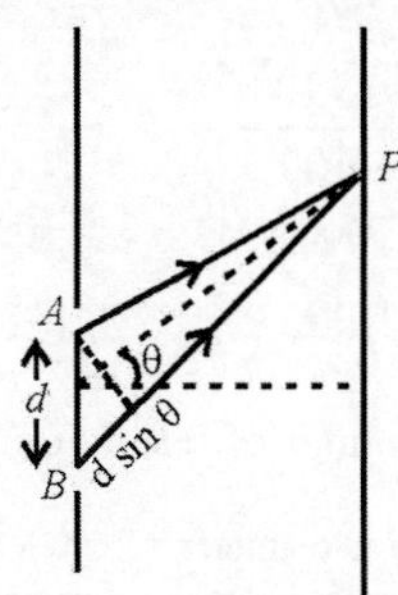

In $\quad YDSE, I_1 = I_2 = I$ and $I_{max} = 4\,I$
We are concentrating at a point where the intensity is one fourth of the max intensity.

$\therefore \quad I = I + I + 2\,I\cos\phi$

$\Rightarrow \quad -\frac{1}{2}\cos\phi \Rightarrow \phi = \frac{2\pi}{3}$

[P1 note that we take the least value of the angle as the point is in central maxima]
For a phase difference of 2π, the path difference is λ

For a phase difference of $\frac{2\pi}{3}$, the path difference is $\frac{\lambda}{2\pi} \times \frac{2\pi}{3} = \frac{\lambda}{3}$

But the path difference (in terms of P and Q) is d sin θ as shown in fig.

$\therefore \quad d\sin\theta = \frac{\lambda}{3}$

$\Rightarrow \quad \sin\theta = \frac{\lambda}{3d}$

$\Rightarrow \quad \theta = \sin^{-1}\left(\frac{\lambda}{3d}\right)$

39. (d) When a ray of light enters nicol prism, it splits into two plane polarised light in mutually perpendicular direction. One of this light undergoes total reflection and absorption whereas other comes out as a plane polarised light.

40. (b) Longitudinal waves cannot be polarised.

41. (b) Resolving power = $\frac{1}{\theta}$

43. (b) The fluorescent material present in the tube converts u.v. rays into visible light.

45. (b) Newton first predicted deflection of light by gravitational field.

47. (b) Frequency (n) = 100 Hz

$v = n\lambda$

$$\lambda = \frac{3\times10^8}{100}$$

[where, velocity (v) = 3 × 10^8 m/s of light]

$\lambda = 3\times10^6 \text{m}$.

48. (b) Angular limit of resolution of eye, $\theta = \frac{\lambda}{d}$,

where, d is diameter of eye lens.

Also, if y is the minimum separation between two objects at distance D from eye then

$$\theta = \frac{y}{D}$$

$$\Rightarrow \frac{y}{D} = \frac{\lambda}{d} \Rightarrow y = \frac{\lambda D}{d} \quad ...(1)$$

Here, wavelength $\lambda = 5000\text{Å} = 5\times10^{-7}\text{m}$

D = 50 m

Diameter of eye lens = 2 mm = 2×10^{-3} m

From eq. (1) minimum separation is

$$y = \frac{5\times10^{-7}\times50}{2\times10^{-3}} = 12.5\times10^{-3}\ m = 12.5\,cm$$

49. (d) For the phenomenon of interference we require two sources of light of same frequency and having a definite phase relationship (a phase relationship that does not change with time)

50. (d) The angle of incidence for total polarization is given by $\tan\theta = n \Rightarrow \theta = \tan^{-1} n$

where n is the refractive index of the glass.

51. (b) For constructive interference $d\sin\theta = n\lambda$

given $d = 2\lambda \Rightarrow \sin\theta = \frac{n}{2}$

n = 0, 1, – 1, 2, – 2 hence five maxima are possible

52. (a) Frequency remains constant during refraction

$$v_{med} = \frac{1}{\sqrt{\mu_0 \in_0 \times 4}} = \frac{c}{2}$$

$$\frac{\lambda_{med}}{\lambda_{air}} = \frac{v_{med}}{v_{air}} = \frac{c/2}{c} = \frac{1}{2}$$

∴ wavelength is halved and frequency remains unchanged.

53. (d) The shape of interference fringes formed on a screen in case of a monochromatic source is a straight line.

54. (b) $I = I_0 \cos^2\theta$

Intensity of polarized light = $\frac{I_0}{2}$

⇒ Intensity of untransmitted light

$$= I_0 - \frac{I_0}{2} = \frac{I_0}{2}$$

55. (a) For path difference of λ, the phase difference is 2π

For path difference of $\frac{\lambda}{6}$, the phase difference is

$$\frac{2\pi\times\lambda/6}{\lambda} = \frac{\pi}{3}$$

∴ Intensity $I = I_1 + I_2 + 2\sqrt{I_1}\sqrt{I_2}\cos\frac{\pi}{3}$

∴ $I = I_1 + I_2 + \sqrt{I_1}\sqrt{I_2}$

when $I_1 = I_2 = I'$ (say) then $I = 3I'$

$I_{max} = \left(\sqrt{I_1} + \sqrt{I_2}\right)^2$

$= \left(\sqrt{I'} + \sqrt{I'}\right)^2 = \left(2\sqrt{I'}\right)^2 = 4I'$

∴ $\frac{I}{I_{max}} = \frac{3}{4}$

56. (d) The electron beam will be diffracted and the maxima is obtained at y = 0. Also the distance between the first minima on both side will be greater than d.

57. (d) Huygen's construction of wavefront does not apply to origin of spectra which is explained by quantum theory.

58. (d) Sound waves cannot be polarised as they are longitudinal. Light waves can be polarised as they are transverse.

59. (c) Interference is a wave phenomenon shown by both the light waves and sound waves.

60. (c) Distance of nth maxima, $x = n\lambda\frac{D}{d} \propto \lambda$

As $\lambda_b < \lambda_g$

∴ $x_{blue} < x_{green}$

25. DUAL NATURE OF RADIATION AND MATTER

ANSWERKEY

1	(a)	11	(c)	21	(a)	31	(b)	41	(c)	51	(a)
2	(a)	12	(a)	22	(d)	32	(c)	42	(c)	52	(d)
3	(b)	13	(a)	23	(a)	33	(a)	43	(a)	53	(b)
4	(a)	14	(a)	24	(a)	34	(c)	44	(a)	54	(d)
5	(a)	15	(b)	25	(a)	35	(c)	45	(b)	55	(a)
6	(a)	16	(a)	26	(b)	36	(b)	46	(a)	56	(d)
7	(c)	17	(a)	27	(d)	37	(a)	47	(a)	57	(c)
8	(b)	18	(b)	28	(c)	38	(b)	48	(d)	58	(d)
9	(a)	19	(c)	29	(b)	39	(b)	49	(b)	59	(d)
10	(a)	20	(d)	30	(d)	40	(d)	50	(d)	60	(a)

Hints & Solutions

1. (a) According to photoelectric effect, speed of electron (kinetic energy) emitted depends upon frequency of incident light while number of photoelectrons emitted depends upon intensity of incident light. Hence, as the intensity of light increases, the photocurrent increases. In a photo-cell, the photocurrent has no relation with the applied voltage.
Stopping potential is the (negative) potential at which the current is just reduced to zero. It is independent of intensity of light but depends on the frequency of light similar to K.E.

2. (a) Cosmic rays have low wavelength and high frequency hence these rays emit high energy radiation.

3. (b) We know that the X-rays are of short wavelength as compared to grating constant of optical grating. As a result of this, it makes difficult to observe X-rays diffraction with ordinary grating.

4. (a) K.E. of electrons emitted depends upon the frequency of incident rays rather than the intensity. While number of photoelectrons emitted depends upon intensity of radiation.

5. (a) Einstein work on photoelectric effect supports the equation $E = h\nu$. It is based on quantum theory of light.

6. (a) de-Broglie wavelength $\lambda = \dfrac{h}{mv}$

For same velocity, $\lambda \propto \dfrac{1}{m}$

Out of given particles, the mass of electron is minimum, so the associated de-Broglie wavelength is maximum for electron.

7. (c) Cathode rays are negatively charged rays. Hence, when electric field is applied to them, then they deflect in direction opposite to electric field.

8. (b) Energy of photon of X-rays is more than energy of photon of ultraviolet rays. Because frequency of X- rays is more than ultraviolet rays.

9. (a) Power $\propto$ No. of electrons emitted (N)

$$P \propto \frac{1}{r^2} \Rightarrow N \propto \frac{1}{r^2}$$

10. (a) K.E. $= h\nu - \phi$
$\therefore$ graph of K.E. versus ν is a straight line with +ve slope h and K.E. > 0 for ν such that $h\nu > \phi$.

11. (c) $E = \dfrac{hc}{\lambda}$

$$\Rightarrow E = \frac{12375}{\lambda(in\ \AA)}\text{eV}$$

$$\Rightarrow E = \frac{12375}{4100}\text{eV} \qquad \Rightarrow E \approx 3\text{ eV}$$

So, only metals having work function less than 3eV can emit photoelectrons for the incident radiation of wavelength 4100Å.

12. (a) When electrons emitted from cathode collide with gas molecules or atoms, they knock out outer electrons and produce positively charged ions. They become part of positive rays.

13. (a) A photo-cell employs photoelectric effect to convert light energy into photoelectric current.

14. (a) The coloured glow in the tube appears as a result of excitations of electrons in the atoms.

15. (b) The number of photoelectrons emitted is proportional to the intensity of incident light. Saturation current $\propto$ intensity.
16. (a) Retarding potential depends on the frequency of incident radiation but is independent of intensity.
17. (a) In the Davisson and Germer experiment, the velocity of electrons emitted from the electron gun can be increased by increasing the potential difference between the anode and filament.
18. (b) $\lambda \propto \frac{1}{\sqrt{V}}$

$$\Rightarrow \frac{\lambda_1}{\lambda_2} = \sqrt{\frac{V_2}{V_1}} = \sqrt{\frac{100keV}{25keV}} = 2$$

$$\Rightarrow \lambda_2 = \frac{\lambda_1}{2}$$

19. (c) The stopping potential is equal to maximum kinetic energy.
20. (d) For occurence of photoelectric effect, the incident light should have frequency more than a certain minimum which is called the threshold frequency (v_0).

We have, $\frac{1}{2}mv^2 = hv - hv_0$

For photoelectric effect emission $\nu > \nu_0$ where ν is the frequency of the incident light.

21. (a) Photoelectric effect can be explained only by quantum nature of light.
22. (d) The oil drop moves upward due to applied electric field.
23. (a) Photoelectric current is directly proportional to the intensity of incident light.
24. (a) $hv = W_0 + E_k = 3.5 + 1.2 = 4.7$ eV
25. (a) According to de-Brogie wave equation,

$$\lambda = \frac{h}{mv} = \frac{h}{p}$$

$$\therefore p = \frac{h}{\lambda}$$

26. (b) The ionisation requires high energy electrons.
27. (d) From Equation $K.E. = h\nu - \phi$

slope of graph of K.E. & ν is h (Planck's constant) which is same for all metals

28. (c) There is no change in the proton number and the neutron number as the γ-emission takes place as a result of excitation or de-excitation of nuclei. γ-rays have no charge or mass.
29. (b) **Thermoionic emission :** When a metal is heated to a high temperature, the free electrons gain kinetic energy and escape from the surface of the metal.

Secondary emission : When an electron strikes the surface of a metallic plate, it emits other electrons from the surface.

Photoelectric emission : Emission of electrons from the metal surface on irradiation with radiation of suitable frequency.

X-rays emission : High speed electrons (emitted from a filament) striking a target (a copper block) to emit X-rays.

30. (d) The reflected beam consists of electrons.
31. (b) When

$$F = \frac{k}{r} = \text{centripetal force, then} \frac{k}{r} = \frac{mv^2}{r}$$

$\Rightarrow mv^2 = $ constant

$\Rightarrow$ kinetic energy is constant

$\Rightarrow T$ is independent of n.

32. (c) We assume that mass number of nucleus when it was at rest = A

$\because$ mass number of α-particle = 4

$\therefore$ mass number of remaining nucleus = $A - 4$

As there is no external force, so momentum of the system will remain conserved.

$$\Rightarrow 0 = (A-4)v' + 4v \quad \Rightarrow v' = -\frac{4v}{(A-4)}$$

negative sign represents that direction is opposite to the direction of motion of α-particle.

33. (a) Gain in K.E. = $qV = (1.6 \times 10^{-19} \times 1)$ J = 1 eV.
34. (c) Work function, $W_0 = \frac{hc}{\lambda_0}$

In electron volt,

$$W_0(\text{in } eV) = \frac{hc}{e\lambda_0} = \frac{12375}{\lambda_0(\mathring{A})}$$

$$\therefore \quad \lambda_0(\mathring{A}) = \frac{12375}{W_0(eV)}$$

$\Rightarrow$ For metal A, $\lambda_0 = \frac{12375}{3.2} = 3867\text{Å}$

For metal B, $\lambda_0 = \frac{12375}{1.9} = 6513\text{Å}$

$\because$ Incident wavelength (λ) < threshold wavelength λ_0 for both the metals, hence photoelectrons will emit for both.

35. (c) de-Broglie's relation, $\lambda = \frac{h}{p}$

momentum $p = \sqrt{2mE}$

$$\Rightarrow \lambda = \frac{h}{\sqrt{2mE}} = \frac{h}{\sqrt{2mK}} \quad (\because E = K)$$

36. (b) The decay constant,

$$\lambda = \frac{0.693}{T}; \text{ where } T = 1/2 \text{ life}$$

$$\lambda = \frac{0.693}{77} = 9 \times 10^{-3}/\text{day}$$

37. (a) Using the relation,

$2\pi r_n = n\lambda$

where λ is the de-Broglie wavelength of the electron moving in nth orbit.

$$\therefore\ \lambda = \frac{2\pi r}{n}$$

38. (b) Transition from higher states to $n = 2$ lead to emission of radiation with wavelengths 656.3 nm and 365.0 nm. These wavelengths fall in the visible region and constitute the Balmer series.

39. (b) As λ decreases, v increases and hence the speed of photoelectron increases. The chances of photo electron to meet the anode increases and hence photo electric current increases.

41. (c)

$$\frac{e^2}{2h\varepsilon_0 c} = \frac{(1.6\times10^{-19})^2}{2\times6.26\times10^{-34}\times8.857\times10^{-12}\times3\times10^8}$$

$$= \frac{1}{137} \text{ it is unit less quantity.}$$

42. (c) $\frac{1}{2}mv^2 = \frac{hc}{\lambda} - \phi \Rightarrow v = \sqrt{\frac{2(hc-\lambda\phi)}{\lambda m}}$

43. (a) $W_0 = \frac{hc}{\lambda_0}$

$$= \frac{6.63\times10^{-34}\times3\times10^8}{5000\times10^{-10}} = 4\times10^{-19}\text{ J}$$

44. (a) De-Broglie wavelength, $\lambda = \frac{h}{p} = \frac{h}{mv}$

45. (b) $(E_2 - E_1) = h\nu = \frac{hc}{\lambda}$

$$\therefore\ \frac{hc}{\lambda_1} = (E_C - E_B),\ \frac{hc}{\lambda_2} = (E_B - E_A)$$

and $\frac{hc}{\lambda_3} = (E_C - E_A)$

Now, $(E_C - E_A) = (E_C - E_B) + (E_B - E_A)$

or, $\frac{hc}{\lambda_3} = \frac{hc}{\lambda_1} + \frac{hc}{\lambda_2}$ or $\frac{1}{\lambda_3} = \frac{1}{\lambda_1} + \frac{1}{\lambda_2}$

$$\therefore\ \frac{1}{\lambda_3} = \frac{\lambda_1+\lambda_2}{\lambda_1\lambda_2} \text{ or } \lambda_3 = \frac{\lambda_1\lambda_2}{\lambda_1+\lambda_2}$$

46. (a) No. of photons emitted per sec,

$$n = \frac{\text{Power}}{\text{Energy of photon}}$$

$$= \frac{P}{h\nu} = \frac{10000}{6.6\times10^{-34}\times880\times10^3}$$

$$= 1.72\times10^{31}$$

47. (a) As $\lambda = \frac{h}{p}$ and $\lambda = \frac{c}{\nu}$; so

$$\nu = \frac{c}{\lambda} = \frac{cP}{h} = 3\times10^8\times\frac{3.3\times10^{-29}}{6.6\times10^{-34}}$$

$$= 1.5\times10^{13}\text{ Hz}$$

48. (d) The circumference of an orbit in an atom in terms of wavelength of wave associated with electron is given by the relation, circumference = $n\lambda$ where, n = 1, 2, 3, ...

49. (b) Here, $\frac{hc}{\lambda} = 10^3$ eV and $h\nu = 10^6$ eV

$$\text{Hence, } \nu = \frac{10^3 c}{\lambda} = \frac{10^3\times3\times10^8}{1.24\times10^{-9}}$$

$$= 2.4\times10^{20}\text{ Hz}$$

50. (d) $h\nu = W + \frac{1}{2}mv^2$ or $\frac{hc}{\lambda} = W + \frac{1}{2}mv^2$

Here λ = 3000 Å = 3000×10^{-10} m
and W = 1 eV = 1.6×10^{-19} joule

$$\therefore\ \frac{(6.6\times10^{-34})(3\times10^8)}{3000\times10^{-10}}$$

$$= (1.6\times10^{-19}) + \frac{1}{2}\times(9.1\times10^{-31})v^2$$

Solving we get, $v \cong 10^6$ m/s

51. (a) The work function has no effect on photoelectric current so long as $h\nu > W_0$. The photoelectric current is proportional to the intensity of incident light. Since there is no change in the intensity of light, hence $I_1 = I_2$.

52. (d) For photon $E = h\nu$

$$E = \frac{hc}{\lambda} \Rightarrow \lambda_2 = \frac{hc}{E} \quad ...(i)$$

for proton $E = \frac{1}{2}m_p v_p^2$

$$E = \frac{1}{2}\frac{m_p^2 v_p^2}{m} \Rightarrow p = \sqrt{2mE}$$

From De-Broglie Eqn.

$$p = \frac{h}{\lambda_1} \Rightarrow \lambda_1 = \frac{h}{p} = \frac{h}{\sqrt{2mE}} \quad ...(ii)$$

$$\frac{\lambda_2}{\lambda_1} = \frac{hc}{E \times \frac{h}{\sqrt{2mE}}} \propto E^{-1/2}$$

53. (b) $W_0 = \frac{hc}{\lambda_0}$ or $W_0 \propto \frac{1}{\lambda_0}$;

$$\Rightarrow \frac{W_1}{W_2} = \frac{\lambda_2}{\lambda_1} = \frac{600}{300} = 2$$

54. (d) $\frac{1}{2}mv^2 = \text{eV}$, or, $v = (2\text{eV/m})^{1/2}$

the mass of helium ion is four times that of hydrogen ion. Further, the charge on helium ion is twice that of hydrogen ion.

$\therefore v_{He}/v_H = \sqrt{2}$

55. (a) Potential difference (V) = 100 V. The kinetic energy of an electron = 1 eV = 1 × (1.6 × 10^{-19}) = 1.6 × 10^{-19} J. Therefore kinetic energy in 100 volts = (1.6 × 10^{-19}) × 100 = 1.6 × 10^{-17}J.

[Alt : K.E. = qV
= 1.6 × 10^{-19} × 100 J = 1.6 × 10^{-17} J]

56. (d) Potential difference = 100 V

K.E. acquired by electron = e (100)

$$\frac{1}{2}mv^2 = e(100) \Rightarrow v = \sqrt{\frac{2e(100)}{m}}$$

According to de-Broglie's concept

$$\lambda = \frac{h}{mv}$$

$$\Rightarrow \lambda = \frac{h}{m\sqrt{\frac{2e(100)}{m}}}$$

$$= \frac{h}{\sqrt{2me(100)}} = 1.2 \times 10^{-10} = 1.2\text{Å}$$

57. (c) Kinetic energy of electron accelerated through a potential V= eV

$$\Rightarrow \frac{1}{2}mv^2 = \text{eV}$$

$$\Rightarrow v^2 = \frac{2\text{eV}}{m} \Rightarrow v = \sqrt{\frac{2\text{eV}}{m}}$$

58. (d) $\frac{1}{2}mv^2 = \frac{hc}{\lambda} - W_0$ or $\frac{hc}{\lambda} = \frac{1}{2}mv^2 + W_0$ and

$$\frac{1}{2}mv_1^2 = \frac{hc}{(3\lambda/4)} - W_0$$

$$= \frac{4}{3}\left(\frac{1}{2}mv^2 + W_0\right) - W_0$$

So, v_1 is greater than $v\left(\frac{4}{3}\right)^{1/2}$

59. (d) $E = h\nu = \frac{hc}{\lambda} = \frac{6.6 \times 10^{-34} \times 3 \times 10^8}{0.21}$

$$= \frac{6.6}{7} \times 10^{-24} = 0.94 \times 10^{-24} \text{ J}$$

60. (a) K.E. = $h\nu - h\nu_0$

$$= \frac{hc}{\lambda} - h\nu_0 = \frac{12375}{5000} - 1.9$$

= 2.48 – 1.9 = 0.58 eV

26. ATOMS

ANSWERKEY											
1	(b)	11	(a)	21	(d)	31	(b)	41	(a)	51	(a)
2	(d)	12	(a)	22	(d)	32	(b)	42	(d)	52	(a)
3	(d)	13	(d)	23	(a)	33	(a)	43	(a)	53	(a)
4	(d)	14	(a)	24	(d)	34	(a)	44	(a)	54	(d)
5	(d)	15	(c)	25	(b)	35	(b)	45	(d)	55	(c)
6	(b)	16	(d)	26	(b)	36	(b)	46	(d)	56	(a)
7	(d)	17	(d)	27	(a)	37	(b)	47	(b)	57	(c)
8	(d)	18	(d)	28	(a)	38	(c)	48	(c)	58	(a)
9	(a)	19	(c)	29	(d)	39	(b)	49	(a)	59	(c)
10	(c)	20	(a)	30	(c)	40	(a)	50	(b)	60	(d)

Hints & Solutions

1. (b) Some of α-particles was found to be scattered at very large angles inspite of having very high kinetic energy. This shows that these are the α-particles which will be passing very close to Nucleus. Rutherford confirmed that the repulsive force on α-particle due to nucleus varies with distance according to inverse square law.

3. (d) Ultraviolet light is emitted when electron makes transition from any higher energy level to $n = 1$ level. Similarly, visible light is emitted when electron jumps from any higher level to $n = 2$ level. X-rays are also emitted when one of the electrons from some inner shell of atom is knocked out and the vacancy so created is filled up by jumping of electron from higher energy shell. On the other hand, γ-rays are emitted from the nucleus of an atom.

4. (d) $\omega = \frac{v}{r}$ As $v \propto \frac{1}{n}$ and $r \propto n^2$

hence $\omega \propto \frac{1}{n^3}$

7. (d) According to Bohr's second postulate:

$mvr = \frac{nh}{2\pi}, n = 1, 2, ...$

8. (d) $\frac{1}{\lambda_z} = Z_2 R\left(\frac{1}{1^2} - \frac{1}{2^2}\right)$; λ_z is minimum.

when Z is maximum. Z is maximum (= 3) for doubly ionised lithium.

9. (a) The range of wavelengths of the lines of Lyman series varies from 912 A° to 1216 A°.

11. (a) As $\Delta E = hv$, hence v is highest when ΔE is maximum. ΔE is maximum for $n = 1$ to $n = 2$ transition. For absorption, it should be a transition from lower energy level to higher energy level.

14. (a) Line spectrum is related with atomic state of matter because every atom is found to emit a radiations of certain fixed discrete wavelengths only.

16. (d) $L = mvr = \frac{nh}{2\pi}$, As $r \propto n^2$, hence $L \propto \sqrt{r}$.

19. (c) When white light from the photosphere (central portion of the sun) passes through vapours of various elements present in the outer chromosphere, then these elements absorb those wavelengths which they themselves emit on being incandescent. Hence dark lines (absence of light) appear at those places in the continuous solar spectrum

20. (a) The wavelengths of different lines in Lyman series varies from 912 A° to 1216 A°.

21. (d) Speed of electro, $v = \frac{C}{137}\frac{z}{n}$,

Energy $= \frac{-13.6}{n^2}$ and radius $= 0.53\, n^2$Å

22. (d) $\frac{1}{\lambda}=R\left(\frac{1}{2^2}-\frac{1}{\infty^2}\right)$

$\therefore \lambda=\frac{4}{R}=4\times 912=365 nm$.

23. (a) Linear momentum $\Rightarrow mv \propto 1/n$
angular momentum $\Rightarrow mv \propto n$
$\therefore$ product of linear momentum and angular momentum $\propto n^0$.

24. (d) For 2nd line of Balmer series in hydrogen spectrum

$$\frac{1}{\lambda}=R(1)\left(\frac{1}{2^2}-\frac{1}{4^2}\right)=\frac{3}{16}R$$

For Li^{2+} which is satisfied by only (d).

25. (b) Heavy water (D_2O) is used as a moderator in nuclear reaction.

26. (b) Fusion is the reaction responsible for the production of light energy from the Sun. In it two hydrogen molecules fuse to form helium.

$$_1H^1 + _1H^1 \longrightarrow {_2He^4} + \text{Energy}$$

27. (a) In photoelectricity, photon is acting as particles which knocks out electron from the surface of the metal. In electron microscopy electron behaves as waves so we can take image of objects with the help of ray of electrons.

28. (a) When the body is already 33% decayed to be decayed to a further 33%, it will require a period equal to its half life. So achieve level of decay from 33% to 67% it takes time equal to half life or 20 minutes.

30. (c) When the electron drops from any orbit to second orbit, then wavelength of line obtained belongs to Balmer series.

31. (b) $r \propto \frac{1}{Z}$; $Z(=3)$ is maximum for Li^{2+}.

32. (b) In Bohr's model, angular momentum is quantised i.e $l=n\left(\frac{h}{2\pi}\right)$

33. (a) $\Delta E=E_3-E_2$

$$=\frac{-13.6}{3^2}+\frac{13.6}{2^2}$$

$$=13.6\left\{\frac{1}{4}-\frac{1}{9}\right\}\text{eV}=1.9\text{ eV}$$

34. (a) Energy of ground state 13.6 eV
Energy of first excited state

$$=-\frac{13.6}{4}=-3.4\text{ eV}$$

Energy of second excited state

$$=-\frac{13.6}{9}=-1.5\text{ eV}$$

Difference between ground state and 2nd excited state = 13.6 – 1.5 = 12.1 eV
So, electron can be excited upto 3rd orbit
No. of possible transition
$1\rightarrow 2, 1\rightarrow 3, 2\rightarrow 3$
So, three lines are possible.

35. (b) Cathode rays are streams of negatively charged ions

36. (b) Bohr used conservation of angular momentum.

37. (b) Neon street sign is a source of line emission spectrum.

38. (c) A spectrum is observed, when light coming directly from a source is examined with a spectroscope. Therefore spectrum obtained from a sodium vapour lamp is emission spectrum.

39. (b) $P.E.=-\frac{KZe^2}{r}$ and $K.E.=\frac{KZe^2}{2r}$, where, r is the radius of orbit which increases as we move from ground to an excited state. Therefore, when a hydrogen atom is raised from the ground state, it increases the value of r. As a result of this, P.E. increases (decreases in negative) and K.E. decreases.

40. (a) $R=\frac{R_0 n^2}{Z}$

Radius in ground state $=\frac{R_0}{Z}$

Radius in first excited state $=\frac{R_0\times 4}{Z}$

$(\because n=2)$

Hence, radius of first excited state is four times the radius in ground state.

41. (a) 53 electrons in iodine atom are distributed as 2, 8, 18, 18, 7
$\therefore n=5$

$$r_n=(0.53\times 10^{-10})\frac{n^2}{Z}$$

$$=\frac{0.53\times 10^{-10}\times 5^2}{53}=2.5\times 10^{-11}\text{m}$$

42. (d) $E\propto Z^2$ and Z for helium = 2
$\therefore (E)_{He}=4\times 13.6=54.4$ eV

43. (a) Second excited state corresponds to $n=3$

$$\therefore E=\frac{13.6}{3^2}\text{eV}=1.51\text{ eV}$$

44. (a) As $r \propto n^2$, therefore, radius of 2nd Bohr's orbit = 4 a_0.

45. (d) $E = E_4 - E_3$

$$= -\frac{13.6}{4^2} - \left(-\frac{13.6}{3^2}\right) = -0.85 + 1.51$$

= 0.66 eV

46. (d) Impact parameter for Rutherford scattering experiment,

$$b = \frac{Ze^2 \cot\left(\frac{\theta}{2}\right)}{4\pi \epsilon_0 k_i} = 0 \Rightarrow \cot\left(\frac{\theta}{2}\right) = 0$$

$$\Rightarrow \frac{\theta}{2} = 90° \text{ or } \theta = 180°$$

47. (b) $r \propto n^2$

$$\therefore \frac{\text{radius of final state}}{\text{radius of initial state}} = n^2$$

$$\frac{21.2 \times 10^{-11}}{5.3 \times 10^{-11}} = n^2$$

$\therefore n^2 = 4$ or $n = 2$

48. (c) Centripetal force = Coulombian force

$$\frac{mv^2}{a_0} = \frac{1}{4\pi\varepsilon_0} \cdot \frac{e \times e}{a_0^2}$$

$$\Rightarrow v^2 = \frac{e^2}{4\pi\varepsilon_0 . a_0 . m} \Rightarrow v = \frac{e}{\sqrt{4\pi\varepsilon_0 . a_0 . m}}$$

49. (a) According to relation,

$$Bqv = \frac{Mv^2}{r} \Rightarrow r = \frac{Mv}{Bq}$$

$$\Rightarrow r = \frac{m2v}{B'q} \Rightarrow B' = 2B = 2x$$

50. (b) $\bar{v} = R\left(\frac{1}{n_1^2} - \frac{1}{n_2^2}\right)$, where $n_1 = 2, n_2 = 4$

$$\bar{v} = R\left(\frac{1}{4} - \frac{1}{16}\right)$$

$$\frac{1}{\lambda} = R\left(\frac{12}{4 \times 16}\right) \quad \Rightarrow \quad \lambda = \frac{16}{3R}$$

51. (a) We have $E_n = \frac{-2\pi^2 mK^2 Z^2 e^4}{n^2 h^2}$. For helium $Z = 2$. Hence requisite answer is $4E_n$

52. (a) 100 g will become 25 g in two half lives, so, it is 3200 years.

53. (a) $\text{K.E.} = \frac{Z^2}{n^2}(13.6 \text{ eV})$

$$\text{Mechanical energy} = \frac{-Z^2}{n^2}(13.6 \text{ eV})$$

$\therefore$ K.E. in 2nd orbital for hydrogen = – Mechanical energy

$$= -\frac{(1)^2}{(2)^2}(13.6) = +3.4\text{eV}$$

54. (d) Energy in the first excited state

$$= \frac{-13.6}{n^2} = \frac{-13.6}{2^2} = -3.4\text{eV}$$

But K.E. = –(Total energy) = +3.4 eV.

55. (c) When the electron is in first excited state ($n = 2$), the excitation energy is given by

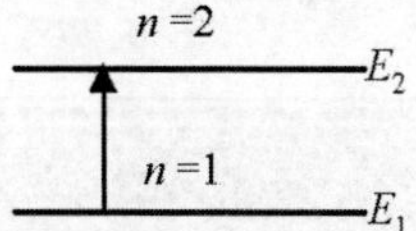

$$\Delta E = E_2 - E_1$$

We have, $E_n = -\frac{13.6}{n^2} eV$

$$\therefore E_2 = -\frac{13.6}{2^2} \text{eV} = -3.4\text{eV}$$

Given $E_1 = -13.6$ eV

$\therefore \Delta E = (-3.4) - (-13.6) = 10.2$ eV.

56. (a) The kinetic energy of the projectile is given by

$$\frac{1}{2}mv^2 = \frac{Ze(2e)}{4\pi\varepsilon_0 r_0}$$

$$= \frac{Z_1 Z_2}{4\pi\varepsilon_0 r_0}$$

Thus energy of the projectile is directly proportional to Z_1, Z_2

57. (c) $\frac{n(n-1)}{2} = 6$

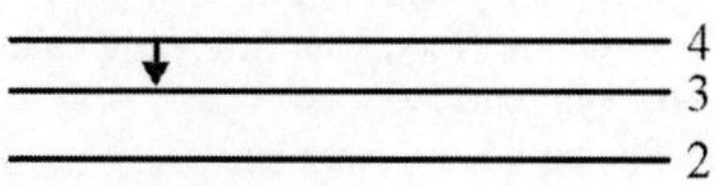

$n^2 - n - 12 = 0$

$(n - 4)(n + 3) = 0$ or $n = 4$

58. (a) Energy of a H-like atom in it's n^{th} state is given by

$$E_n = -Z^2 \times \frac{13.6}{n^2} \text{ eV}$$

For, first excited state of He^+, $n = 2, Z = 2$

$$\therefore E_{He^+} = -\frac{4}{2^2} \times 13.6 = -13.6 \text{ eV}$$

59. (c) Kinetic energy of alpha nucleus is equal to electrostatic potential energy of the system of the alpha particle and the heavy nucleus. That is,

$$\frac{1}{2}mv^2 = \frac{1}{4\pi\varepsilon_0}\frac{q_\alpha Ze}{r_0}$$

where r_0 is the distance of closest approach

$$r_0 = \frac{2}{4\pi\varepsilon_0}\frac{q_\alpha Ze}{mv^2}$$

$$\Rightarrow r_0 \propto Ze \propto q_\alpha \propto \frac{1}{m} \propto \frac{1}{v^2}$$

Hence, correct option is (c).

60. (d) For first line of Lyman series of hydrogen

$$\frac{hc}{\lambda_1} = Rhc\left(\frac{1}{1^2} - \frac{1}{2^2}\right)$$

For second line of Balmer series of hydrogen like ion

$$\frac{hc}{\lambda_2} = Z^2 Rhc\left(\frac{1}{2^2} - \frac{1}{4^2}\right)$$

By question, $\lambda_1 = \lambda_2$

$$\Rightarrow \left(\frac{1}{1} - \frac{1}{2}\right) = Z^2\left(\frac{1}{4} - \frac{1}{16}\right) \text{ or } Z = 2$$

27. NUCLEI

ANSWERKEY											
1	(a)	**11**	(d)	**21**	(c)	**31**	(a)	**41**	(c)	**51**	(c)
2	(a)	**12**	(d)	**22**	(c)	**32**	(c)	**42**	(c)	**52**	(c)
3	(a)	**13**	(c)	**23**	(b)	**33**	(c)	**43**	(d)	**53**	(a)
4	(c)	**14**	(a)	**24**	(a)	**34**	(c)	**44**	(a)	**54**	(d)
5	(c)	**15**	(b)	**25**	(a)	**35**	(b)	**45**	(d)	**55**	(c)
6	(a)	**16**	(a)	**26**	(b)	**36**	(c)	**46**	(b)	**56**	(d)
7	(d)	**17**	(c)	**27**	(b)	**37**	(a)	**47**	(b)	**57**	(b)
8	(a)	**18**	(a)	**28**	(a)	**38**	(d)	**48**	(c)	**58**	(b)
9	(b)	**19**	(d)	**29**	(c)	**39**	(b)	**49**	(c)	**59**	(b)
10	(c)	**20**	(b)	**30**	(b)	**40**	(b)	**50**	(b)	**60**	(c)

Hints & Solutions

1. (a) Quantum theory of light came to explain the thermal radiations of a black body
2. (a) A photon is a packet of energy called quantum of light energy
3. (a) Photon has no rest mass
4. (c) Energy of a photon $E = \frac{hc}{\lambda}$; E is less if λ is longer
5. (c) The range of energy of β-particles is from zero to some maximum value.
6. (a) $_ZX^A + _0n^1 \longrightarrow _3Li^7 + _2He^4$

On comparison,

$A = 7 + 4 - 1 = 10, z = 3 + 2 - 0 = 5$

It is boron $_5B^{10}$

8. (a) $\frac{N}{N_0} = \left(\frac{1}{2}\right)^{\frac{t}{T}}$ or $\frac{N}{10000} = \left(\frac{1}{2}\right)^{\frac{10}{20}}$

or $N = \frac{10000 \times 1}{\sqrt{2}} = \frac{10000}{1.414} = 7070.$

9. (b) In fission of uranium, a chain reaction is possible because in each fission of uranium, 2 or 3 neutrons are produced. Hence option (b) is correct.

10. (c) The nuclear force is of short range i.e. it acts over a very short distance. The force operates between the interacting particles when they are separated by a distance not more than 10^{-15} m. The nuclear force is also charge independent. This means that forces between a proton and proton (pp force) or between a proton and a neutron (pn force) or between a neutron and neutron (nn force) are all attractive in nature and are basically the same.

11. (d) All Radioactive substances emit α-rays, γ-rays and β-rays.

12. (d) α-particles have charge, therefore it is deflected by electric field. Other particles like X-rays, neutrons, γ-rays are neutral towards electric field. Option (d) is correct.

13. (c) Nuclear fission and fusion are based on Einstein equation
$E = mc^2$ or Einstein mass energy equivalence relation.

14. (a) Option (a) is correct because from emission of β-particle, atomic number increases.

15. (b) By conservation of linear momentum, the two γ's move in opposite direction.

16. (a) Distance between protons $\approx$ size of nucleus $\approx$ 10^{-14} m

17. (c) Total binding energy
= Binding energy per nucleon × Number of molecules
$\Rightarrow 128 = 8 \times N \Rightarrow N = 16$

18. (a) Specific charge (e/m) for cathode rays is constant as e and m for cathode rays are fixed.

19. (d) The force on cathode rays (i.e. charge e) moving with velocity $\vec{v}$ due to a perpen-dicular magnetic field $\vec{B}$ is $-e(\vec{v} \times \vec{B})$, which acts perpendicular to plane of $\vec{v}$ and $\vec{B}$.

20. (b) Canal rays are the stream of positively charged particles.

21. (c) As cathode rays are deflected by electric and mangetic fields, it shows that cathode rays carry charged particles (i.e. electrons).

22. (c) When pressure in a tube is reduced in the range 1 cm and 10^{-3} cm ; the mean free path of moving electron in the discharge tube increases. As a result of which the electron gets higher KE while moving towards anode and then cause ionisation of the atoms with which it will collide on its ways causing excitation phenomenon.

23. (b) $eE = evB$ or $v = E/B$

24. (a) The nuclei of many atoms are magnetic dipoles. This fact helps us in studying nuclear structure. Magnetic resonance imaging is an important device which is based upon determining the location and density of particular atoms with the help of magnetic resonance.

25. (a) Binding energy per nucleon is maximum for atomic number (A) = 56, so figure (a) is correct alternative.

26. (b) A photon and electron will have same momentum if their wavelength are equal.

For electron, momentum = $\frac{h}{\lambda}$.

For photon, momentum

$$= \frac{mc^2}{c} = \frac{E}{c} = \frac{h\nu}{c} = \frac{h}{\lambda}$$

27. (b) When we bombard electrons on a target consisting of heavier atoms, the electron of inner orbitals of target atoms get knocked out. So to fill their position, other electrons from the higher energy orbitals make transition to lower energy orbitals. Its results is emission of X-rays.

28. (a) Nuclear fusion is possible only between light nuclei as they become more stable by acquiring greater atomic no.

29. (c) In radiotracer technique we trace the existence of an atom by detecting the radiation emitted by it if atom is a radioactive one. The case as stated in the question confirms to it so it is an example of radiotracer technique.

30. (b) Positron emission tomography (PET) is a nuclear medicine medical imaging technique which produces a three dimensional image are map of functional process in the body. The positron annihilates with an electron producing a pair of annihilation photons (gamma rays) moving in opposite direction

31. (a) As ${}_6C^{13}$ and ${}_7N^{14}$ have same no. of neutrons (13 – 6 =7 for C and 14 – 7 =7 for N), so they are isotones.

32. (c) Nuclear forces are short range attractive forces which balance the repulsive forces between the protons inside the nucleus.

33. (c) Curie is a unit of radioactivity.

34. (c) By law of lepton numbers, mass number and atomic number is to be conserved on both sides.

35. (b) Carbon dating is best suited for determining the age of fossils if their age in years is of the order of 10,000 years. This is because fossil older than this are destroyed due to time factor. The tracks that are built by the rays are destroyed. For fossil earlier then this, the tracks are so small that it cannot be measured with accuracy.

36. (c) Applying law of conservation of momentum

$$MV = mv$$

Let mass of each proton = m

Mass of daughter nucleus = (Am – $4m$)

Mass of α-particle = $4m$

So, $(Am - 4m)\,V = 4m \times v$

$$V = \frac{4mv}{m(A-4)} = \frac{4v}{A-4}$$

37. (a) The operation of a nuclear reactor is said to be critical if the multiplication factor is

$$k = \frac{\text{rate of production of neutrons}}{\text{rate of loss of neutrons}}$$

If $k > 1$ explosion occurs
If $k < 1$ the chain reaction comes to half.

38. (d) $\dfrac{\text{Specific charge of proton}}{\text{Specific charge of } \alpha \text{ particle}}$

$$= \frac{e/m}{2e/4m} = 1/(1/2) = 2$$

39. (b) Radius of Bohr's nth orbit for hydrogen atom

$r_n = 0.53\ n^2 Å$

For Ist orbit, $n = 1$

So, $r_n = 0.53 \times (1)^2 = 0.53$ Å

For IInd orbit, $n = 2$, $r_n = 0.53 \times 4 = 2.12$ Å

For fourth orbit, $n = 4$, $r_n = 0.53 \times (4)^2 = 8.48$ Å

40. (b) According to Einstein's mass=energy equivalence relation
$E = mc^2$

$m = 1\ mg = 1 \times 10^{-6}$ kg, $c = 3 \times 10^8$ m/s

$E = mc^2 = 1 \times 10^{-6} \times (3 \times 10^8)^2$
$= 9 \times 10^{10}$ joule

41. (c) According to Mosley law regarding characteristic X-rays

$\sqrt{\nu} = a(Z - b)$

Where ν = frequency, Z = atomic number; a and b are constants.

42. (c) Here $\Delta m = (1 - 0.993)\,\text{gm} = 0.007$ gm
$= 7 \times 10^{-6}$ kg.

Using the Einstein's formula

$E = \Delta mc^2$, $c = 3 \times 10^8\,\text{m/s}$

$\Rightarrow E = 7 \times 10^{-6} \times (3 \times 10^8)^2$
$= 63 \times 10^{10}$ joule.

43. (d) $n = \dfrac{\text{Total time}}{\text{half life}} = \dfrac{T}{t_{1/2}} = \dfrac{72000}{24000} = 3$

$$N_{\text{remained}} = N_0\left(\frac{1}{2}\right)^n = N_0\left(\frac{1}{2}\right)^3 = \frac{1}{8}N_0$$

Remaining fraction $= \dfrac{N_{\text{remained}}}{N_0} = \dfrac{1}{8}$

44. (a) ${}^{X}_{Y}A \rightarrow {}^{Y}_{X}A \rightarrow {}^{Y-4}_{X-2}B + {}_2He^4$,

${}^{Y-4}_{X-2}B \rightarrow {}^{Y-2}_{X}C + 2e^-\ {}^{Y}_{X}A$ and ${}_XC^{Y-2}$

Hence, A and C are isotopes.

45. (d) In a nuclear fusion reaction involving Deuterium and Tritium, the released energy is typically of the order of 17.6 MeV.

48. (c) We know, $r = r_0 A^{1/3}$

$\therefore \dfrac{r_1}{r_2} = \left(\dfrac{A_1}{A_2}\right)^{1/3}$

$$= \left[\frac{216}{64}\right]^{1/3} = \frac{6}{4} = 1.5$$

49. (c) The rate of fission = P/E

$$= \frac{3}{200 \times 1.6 \times 10^{-13}} = 10^{11}$$

50. (b) The mass defect,
$\Delta m = 31.97391 - 31.97207 = 0.00184$u
The energy released $E = \Delta mc^2$
$= 1.7$ MeV.

51. (c) ${}_6C^{12} + {}_1H^1 = {}_7N^{13} + {}_0n^1$

52. (c) Let n-alpha and m-beta particles be emitted, then
$238 = 206 + n \times 4 \quad \Rightarrow n = 8$
and $92 - 8 \times 2 + m \times 1 = 82 \Rightarrow m = 6$.

53. (a) $\dfrac{A}{A_0} = e^{-\lambda t}$

$A = A_0 e^{-\lambda t} = 128\ \text{e}^{-6.95 \times 1/2}$
$128\ \text{e}^{-3.475} = 4$m Cu

OR

$$T = \frac{0.693}{6.93} = \frac{1}{10}$$

$$n = \frac{t}{T} = \frac{1/2}{1/10} = 5,$$

$$A = 128\left(\frac{1}{2}\right)^5 = 4$$

54. (d) Five alpha and three beta particles results

$${}^{231}AC_{89} \longrightarrow \underset{(89-5\times2+3\times1)=82}{\overset{231-5\times4=211}{\text{Pb}}}$$

55. (c) The order of nuclear density is 10^{17} kg/m^3.

56. (d) ${}^{A}_{Z}X \longrightarrow {}_{Z+1}Y^A$ (since $Z \rightarrow Z + 1$)

$\therefore$ β emission

${}^{A}_{Z+1}Y \longrightarrow {}_{Z-1}Y^{A-4}$

(decrease in $A = 4$, decrease in $Z = 2$)

$\therefore$ α particle emission

$${}_{Z-1}K^{A-4} \longrightarrow {}_{Z-1}K^{A-4}$$

(decrease in $A = 0$, decrease in $Z = 0$)

$\therefore$ γ emission.

57. (b) Thorium has atomic mass 234 and atomic number 90 hence

$${}^{235}_{92}\text{U} \longrightarrow {}^{234}_{90}\text{Th} + \text{X}$$

For X, A = 4, Z = 2

which means Z = alpha particle.

58. (b) If N_0 is the initial number of nuclei present, then the activities at the time instants t_1 and t_2 can be written as $R_1 = N_0 \exp(-\lambda t_1)$ and $R_2 = N_0 \exp(-\lambda t_2)$ respectively. Equate for N_0 to get the result.

59. (b) $${}_{Z}X^{A} \longrightarrow {}_{Z-1}Y^{A} + e^{+} + \nu$$
$$\downarrow$$
$$2\alpha + {}_{Z-5}X'^{A-B}$$

Given $A - 8 = 224$

& $Z - 5 = 89 \quad \Rightarrow \quad A = 237, Z = 94$

60. (c) Energy released = $(80 \times 7 + 120 \times 8 - 200 \times 6.5)$ = 220 MeV.

28. SEMICONDUCTOR ELECTRONICS : MATERIAL, DEVICES AND SIMPLE CIRCUITS

ANSWERKEY											
1	(a)	11	(b)	21	(a)	31	(b)	41	(d)	51	(c)
2	(a)	12	(d)	22	(c)	32	(d)	42	(c)	52	(c)
3	(d)	13	(a)	23	(b)	33	(a)	43	(b)	53	(c)
4	(c)	14	(a)	24	(b)	34	(c)	44	(d)	54	(d)
5	(b)	15	(c)	25	(d)	35	(a)	45	(b)	55	(b)
6	(d)	16	(b)	26	(d)	36	(a)	46	(d)	56	(d)
7	(b)	17	(d)	27	(b)	37	(a)	47	(c)	57	(b)
8	(d)	18	(a)	28	(b)	38	(a)	48	(a)	58	(b)
9	(a)	19	(c)	29	(a)	39	(c)	49	(d)	59	(c)
10	(c)	20	(c)	30	(a)	40	(a)	50	(c)	60	(d)

Hints & Solutions

2. (a) Number of atoms per unit cell,

$$n = \frac{N_c}{8} + \frac{N_f}{2} + \frac{N_i}{1}$$

For scc system, $N_c = 8$, $N_f = 0$ and $N_i = 2$, so $n = 2$

For f.c.c. system, $N_c = 8$, $N_f = 0$ and $N_i = 4$, so $n = 4$

5. (b) Resistivity of a semiconductor at room temp. is in between 10^{-5} Ωm to 10^{4} Ωm i.e. 10^{-3} to 10^{6} Ωcm

6. (d) The valency of semiconductor (Ge or Si) is four, hence it has 4 valence electrons in the outermost orbit of the Ge or Si-atom

7. (b) The electrical conductivity of a semiconductor at 0 K is zero. Hence resistivity (= 1/electrical conductivity) is infinity.

8. (d) The temperature coefficient of resistance of a semiconductor is negative. It means that resistance decrease with increase of temperature.

9. (a) The acceptor valence band is close to the valence band of host crystal

10. (c) The donor valence band lies little below the conduction band of the host crystal

11. (b) In p-type semiconductor, holes predominates, so, holes are major current carriers.

12. (d) Dynamic plate resistance = $\frac{\Delta V}{\Delta I}$

In case of saturation current, $\Delta I = 0$

Plate resistance = ∞

13. (a) In case of diode (formed by p-n junction) it acts as rectifier. It means it conducts only in case of forward biasing & not in reverse biasing.

14. (a) $\beta = \frac{I_c}{I_b} \Rightarrow I_b = \frac{0.01}{1 \times 10^3} = 10^{-5}$

or $50 = \frac{I_c}{10^{-5}} \Rightarrow I_c = 50 \times 10^{-5}$

$= 500 \times 10^{-6}$

$I_c = 500\,\mu A$

15. (c) Boolean algebra is based on logic.

16. (b) Amplification factor = 20 = $\frac{\text{Plate potential}}{\text{Grid potential}}$

$20 = \frac{300}{V} \Rightarrow V = \frac{300}{20} = 15V$

17. (d) For semiconductor, $n = AT^{3/2} e^{-\frac{E_g}{2KT}}$;

so $n \propto T^{3/2}$

19. (c) Number of possible space lattice according to Bravis is 14

20. (c) For hexagonal crystal structure, $a = b \neq c$ and $\alpha = \beta = 90^o$ but $\gamma = 120^o$.

21. (a) The diamond is fcc structure

22. (c) When the collector is positive and emitter is negative w.r.t. base, it causes the forward biasing for each junction, which causes conduction of current.

23. (b) Barrier potential does not depend on diode design while barrier potential depends upon temperature, doping density, and forward biasing.

24. (b)

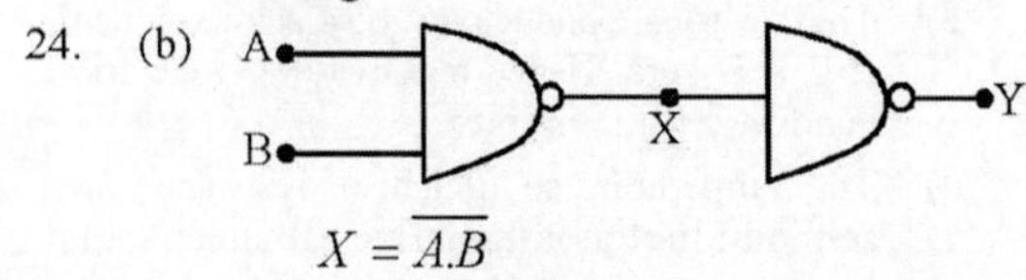

$X = \overline{A.B}$

$\therefore Y = \overline{X} = \overline{\overline{A.B}}$

$Y = A.B$ by Demorgan therorem

$\therefore$ This diagram performs the function of AND gate.

25. (d) A and C are analogue but B is digital signal.

26. (d) Light is transmitted through optical fibre due to multiple total internal reflections.

27. (b) In forward biasing, the diode conducts. For ideal junction diode, the forward resistance is zero; therefore, entire applied voltage occurs across external resistance R i.e., there occurs no potential drop, so potential across R is V in forward biased.

28. (b) At 0K, motion of free electrons stop. Hence conductivity becomes zero. Therefore, at 0K intrinsic semiconductor becomes an insulator.

29. (a) We have, $\beta = \frac{\alpha}{1-\alpha} = \frac{0.98}{1-0.98} = \frac{0.98}{0.02} = 49$

30. (a) A positive feedback from output to input in an amplifier provides oscillations of constant amplitude.

31. (b) When p-n junction is reverse biased, the flow of current is due to drifting of minority charge carriers across the junction.

32. (d) It is the circuit of full wave rectifier.

33. (a) This truth table is of the identity, $Y = \overline{A.B}$, hence it is NAND gate.

Here, the output is high even if all inputs are low or one input is low.

34. (c) The given circuit is a circuit of full wave rectifier.

35. (a) Phosphorus (P) is pentavalent and silicon is tetravalent. Therefore, when silicon is doped with pentavalent impurity, it forms a n-type semiconductor.

36. (a) Current gain (α) = 0.96

$I_e = 7.2$ mA

$\frac{I_c}{I_e} = \alpha = 0.96$

$I_c = 0.96 \times 7.2\,\text{mA} = 6.91$ mA

$I_e = I_c + I_b$

$\Rightarrow I_b = I_e - I_c = 7.2 - 6.91 = 0.29$ mA

37. (a) In an n-p-n transistor, the charge carriers, are free electrons in the transistor as well as in external circuit; these electrons flow from emitter to collector.

38. (a) Arsenic contains 5 electrons in its outermost shell. When Arsenic is mixed with silicon there is one electron extra in silicon crystal. Hence, such type of semiconductor is n-type semiconductor.

39. (c) p-type germanium semiconductor is formed when it is doped with a trivalent impurity atom.

40. (a) The given truth table is of (OR gate + NOT gate) $\equiv$ NOR gate

41. (d) A (0V) —▷|—ᴡᴡ— B (–2V)

As A is at higher potential than B, hence current will flow from A to B i.e. from p junction to n junction and diode is forward biased.

43. (b) The forbidden energy gap for germanium crystal is 0.71 eV.

45. (b) Calcite is a crystalline solid.

46. (d) In a crystalline solid, there is no free movement of atoms.

47. (c) $\frac{I_c}{I_e} = 0.96$

$\Rightarrow \quad I_c = 0.96 I_e$

But $I_e = I_c + I_b = 0.96 I_e + I_b$

$\Rightarrow \; I_b = 0.04 I_e$

$\therefore$ Current gain, $\beta = \frac{I_c}{I_b} = \frac{0.96 I_e}{0.04 I_e} = 24$

48. (a) The current will flow through R_L when the diode is forward biased.

49. (d) In full wave rectifier, we get the output for the positive and negative cycle of input a.c. Hence the frequency of the ripple of the output is twice than that of input a.c. i.e. 100 Hz

50. (c) Forward resistance

$$= \frac{\Delta V}{\Delta I} = \frac{0.7 - 0.5}{1.0 \times 10^{-3}} = 200\,\Omega$$

51. (c) $\beta = \frac{\alpha}{1-\alpha} = \frac{0.95}{1-0.95} = \frac{0.95}{0.05} = 19$

52. (c) $I_C = I_E - I_B = 90 - 1 = 89$ m A

53. (c) Current gain, $\alpha = \frac{A_V}{A_R} = \frac{2800}{3000} = 0.93$

54. (d) $(100)_2 + (1011)_2 = (0 \times 2^0 + 0 \times 2^1 + 1 \times 2^2) + (1 \times 2^0 + 1 \times 2^1 + 0 \times 2^2 + 1 \times 2^3)$

$= (0 + 0 + 4) + (1 + 2 + 0 + 8) = (15)_{10}$

2	15		
2	7	– 1	
2	3	– 1	↑ readup
2	1	– 1	the remainders.
	0	– 1	or $(15)_{10} = (1111)_2$

ALTERNATIVELY :

$0100 + 1011 = (1111)_2$

$(1111)_2 = 1 \times 2^3 + 1 \times 2^2 + 1 \times 2^1 + 1 \times 2^0 = (15)_{10}$

55. (b) In halfwave rectifier only half of the wave is rectified

56. (d) (A) is a NAND gate so output is $\overline{1 \times 1} = \bar{1} = 0$

(B) is a NOR gate so output is $\overline{0+1} = \bar{1} = 0$

(C) is a NAND gate so output is $\overline{0 \times 1} = \bar{0} = 1$

(D) is a XOR gate so output is $0 \oplus 0 = 0$

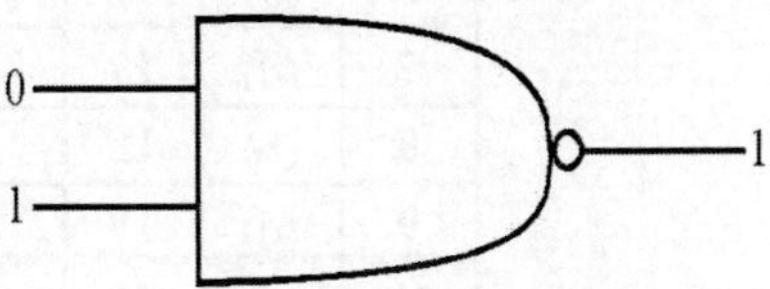

Following is NAND Gate $Y = \overline{AB}$

57. (b) [Hint : For easy flow of current the P side must be connected to +ive terminal of battery i.e., it is connected to higher potential in comparison to N. This connection is called forward biased. In this case the input resistance is very low. In reverse-biased, the P-side is connected to–ive terminal & N side to (+ive) terminal to battery. In this case input resistance is very high.

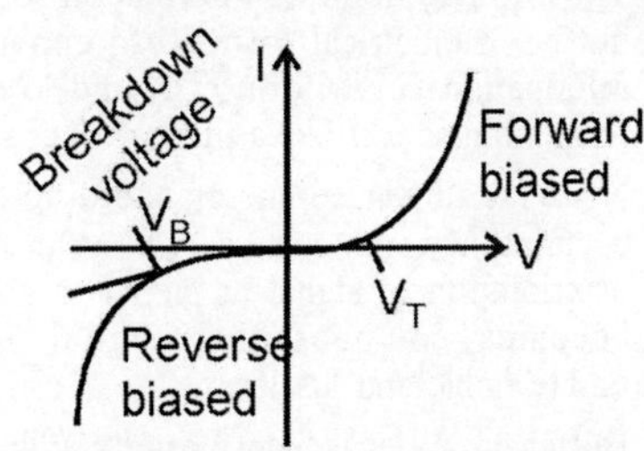

59. (c) Maximum in insulators and overlaping in metals

60. (d) Positive terminal is at lower potential (0V) and negative terminal is at higher potential 5V.

29. COMMUNICATION SYSTEMS

ANSWERKEY											
1	(c)	**11**	(a)	**21**	(a)	**31**	(c)	**41**	(c)	**51**	(b)
2	(d)	**12**	(a)	**22**	(c)	**32**	(b)	**42**	(a)	**52**	(c)
3	(b)	**13**	(b)	**23**	(d)	**33**	(d)	**43**	(b)	**53**	(d)
4	(c)	**14**	(a)	**24**	(c)	**34**	(b)	**44**	(a)	**54**	(c)
5	(d)	**15**	(a)	**25**	(c)	**35**	(c)	**45**	(a)	**55**	(b)
6	(a)	**16**	(c)	**26**	(d)	**36**	(d)	**46**	(b)	**56**	(b)
7	(d)	**17**	(c)	**27**	(a)	**37**	(c)	**47**	(c)	**57**	(d)
8	(b)	**18**	(d)	**28**	(b)	**38**	(d)	**48**	(c)	**58**	(d)
9	(d)	**19**	(c)	**29**	(b)	**39**	(c)	**49**	(d)	**59**	(d)
10	(d)	**20**	(a)	**30**	(c)	**40**	(b)	**50**	(c)	**60**	(d)

Hints & Solutions

1. (c) Tele means long distance.

2. (d) Out of the total message usually a part is conveyed. This part of message in form of bits is called information.

3. (b) The message from the information source may not be in electrical form so to convert these information i.e. in form of sound to electrical form a transducer like a microphone is used.

4. (c) Noise the unwanted energy and distortion both occuring at various stage of a system leads to deterioration of signal as signal to noise ratio becomes so poor that signal becomes unintelligible and useless.

5. (d) Buffer amplifier isolate master oscillator from the influence of modulation done at a later stage.

6. (a) Loudspeaker is a transducer which converts electrical signal transmitted by the transmitter to sound form at the receiving end.

7. (d) The RF-signal coming from the transmitter needs to be demodulated by a detector inorder to remove the carrier frequency and receive back the low-frequency original signal.

8. (b) Due to several advantage of FM over AM, to get better quality signal the sound part of TV-signal is frequency modulated.

9. (d) The received signals is either AM or FM so it needs to be demodulated i.e. decoded to get back original signal. It also needs to be stored and interpreted at receiving end.

10. (d) Collation is a process of storing information done before transmitting the required signal.

11. (a) Encoding modulation of signal i.e. to be transmitted is done with carrier frequency at transmitting end to avoid interference with other signals that are also transmitted.

12. (a) Picture signal in amplitude modulated to avoid complication in development of transmitter & receiver structure.

13. (b) Channel indicate frequency range at which different R.F. signals all transmitted.

14. (a) Radio broadcast transmitter are used for above purpose. AM transmitter operating on long wave, medium & short wave are used.

15. (a) Long wave broadcast transmitters are used in temperate countries because there atmospheric disturbance on long wave are not severe.

16. (c) $P_c = \dfrac{P_t}{1+\dfrac{m_a^2}{2}} = \dfrac{12}{1+\dfrac{(0.5)^2}{2}} = \dfrac{12}{1.25} = 9.6\text{ kW}$

17. (c) Average side-band power $P_{av} = \dfrac{m_a}{4} P_c^{\,2}$

Here $m_a = 0.5$

$P_c = 10$

$\therefore \quad P_{av} = \dfrac{0.5 \times 10 \times 10}{4} = 6.25$

19. (c) The electron-beam coming out of the electron gun needs to scan each element on the photoconductive layer from top to bottom so for scanning entire target electron beam is deflected by these magnetic deflecting coils.

20. (a) Comparing $(x_{AM})\,t = 100\,[1 + 0.5\,t]\cos\omega_c t$ for $0 < t < 1$
with standard AM signal $x_{AM} = E_c\,[1 + m_a \cos \omega_m t]\cos\omega_c t$
We have modulating signal t and $m_a = 0.5$.

21. (a) Induced emf $e = \dfrac{2\pi NA\cos\theta}{\lambda} E_m$ volts

$\Rightarrow \quad e \propto \theta$

maximum value of e occurs when

$$\theta = 0, \pi, 2\pi.........$$

$\Rightarrow$ when plane of the coil is along the direction of transmitting signal.

24. (c) $\lambda = 3\text{m},\ \text{H} = 100\ \text{m},\ \text{h} = 60\ \text{m}$
Path difference between direct wave and wave obtained after reflection from earth surface is

$$\frac{2\pi}{\lambda}\left(\frac{2\text{Hh}}{\text{d}}\right) + \pi = 2\pi$$

or $\dfrac{4\text{Hh}}{\lambda\text{d}} = 1$ or $\text{d} = \dfrac{4\text{Hh}}{\lambda}$

$$\Rightarrow \text{d} = \frac{4 \times 100 \times 60}{3} = 8 \times 10^3\,\text{m} \quad \text{or} \quad 8\text{km}$$

25. (c) $R_{max} \propto d \propto P^{\frac{1}{4}}$

$$\Rightarrow \frac{R_2\ (\text{new range})}{R_1\ (\text{original range})}$$

$$= \frac{D_2\ (\text{new diameter})}{D_1\ (\text{original diameter})}$$

$$= \frac{P_2^{\frac{1}{4}}\ (\text{new transmitted power})}{P_1^{\frac{1}{4}}\ (\text{original transmitted power})}$$

$$= \frac{3D_1}{D_1} \times \frac{(81\,P_1)^{\frac{1}{4}}}{\frac{1}{4}} = 3 \times 3 = 9$$

27. (a) The side-wall of optical fibre provides total internal reflection of beam of light incident on it.

28. (b) The modulated voltage comprises of

(i) Carrier wave of frequency ω_c

(ii) Lower side frequency $(\omega_c - \omega_m)$ wave

(iii) Upper side frequency $(\omega_c + \omega_m)$ wave

Thus in an AM wave information is contained in lower $(\omega_c - \omega_m)$ and upper $(\omega_c + \omega_m)$ side frequencies.

29. (b) For given transmission band 88-108 MHz
$(\Delta f)_{max} = 75$ kHz

given $(\Delta f)_{actual} = 18.75$ kHz

$\therefore$ % modulation

$$m = \frac{(\Delta f)_{actual}}{(\Delta f)_{max}} \times 100 = \frac{18.75}{75} = 25\%$$

30. (c) $m = \dfrac{(\Delta f)_{actual}}{(\Delta f)_{max}} \times 100 \Rightarrow m \propto (\Delta f)_{actual}$

i.e. if frequency deviation is doubled then modulation is doubled.

31. (c) TV signal comprises of video and audio signals. Video signal is AM and sound signal is FM.

32. (b) $h = 100m,\ R = 6.4 \times 10^6\,m$

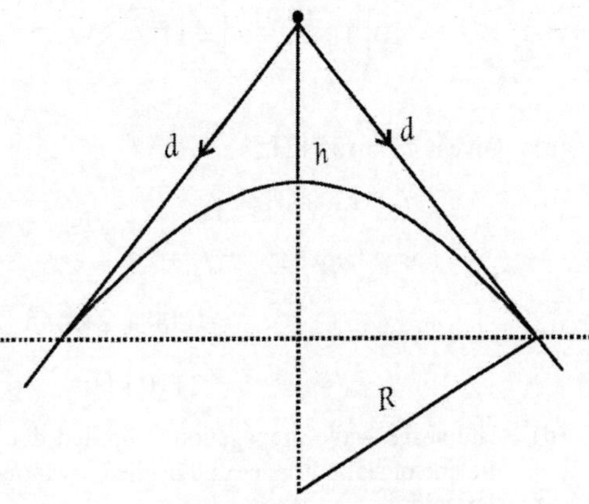

Average population density $= 1000\ \text{km}^{-2}$

$$= 10^{-3}\ \text{m}^{-2}$$

Maximum range of TV transmission

$$d = \sqrt{2hR} = (2 \times 100 \times 6.4 \times 10^6)^{1/2}$$

∴ Maximum area covered by TV transmission

$$= \pi d^2 = \pi \times 2hR$$

$$= 2 \times 3.14 \times 100 \times 6.4 \times 10^6$$

$$= 4.02 \times 10^9 \text{ m}^2$$

∴ the population covered by TV transmission

$$= (4.02 \times 10^9) \times 10^{-3} = 4.02 \times 10^6$$

33. (d) Pre-emphasis of higher frequency component is required in FM-system because high frequency terms of modulating signal have small amplitude and therefore small power relative to those of low frequency term.
In the reproduced program at the o/p, these high frequency terms have poor S/N ratio and at time noise may completely mask the signal at these high frequencies, so it is necessary to provide pre-emphasis of high frequencies.

34. (b) $P_c = \frac{E_c^2}{2} = 10 \text{ kW}$

$$m_a = \sqrt{m_1^2 + m_2^2} = \sqrt{0.30^2 + 0.40^2} = 0.50$$

$$\therefore P_t = P_c\left(1 + \frac{m_a^2}{2}\right)$$

$$= 10\left(1 + \frac{(0.5)^2}{2}\right) = 11.25 \text{ kW}$$

35. (c) Given $f_i = 455 \text{ kHz}$

$f_s = 1200$ kHz

∴ Image frequency $= f_s + 2f_i$

$$= 1200 + 2 \times 455$$

$$= 2110 \text{ kHz}$$

36. (d) The space wave propagation is limited due to the line of sight distance and by the curvature of the earth.

37. (c) Maximum range of space wave propagation

$$d = \frac{4}{3} \times 1.23\left[\sqrt{H_t} + \sqrt{H_r}\right]$$

$\Rightarrow$ $d \propto H_t$ and $d \propto H_r$

∴ d increases if H_t and H_r i.e. height of transmitting and receiving antenna increases.

38. (d) A junction transistor is formed by sandwitching a p-type material between two layers of n-type materials or an n-type materials between two layers of p-type materials. The material between the two same layers is termed as the base of the transistor. Therefore, the base of a transistor is an extrinsic semiconductor (a semiconductor doped with impurities).

39. (c) Sky wave propagation is suited for frequency range 2 MHz to 20 MHz.

40. (b) The diameter of the core in an optical fibre is nearly 10^{-4} m.

41. (c) Light waves and X-rays are electromagnetic waves.

42. (a) Range of antenna $= r\sqrt{2hr}$, h = height of antenna,
R = radius of earth
if h is doubled i.e., $h' = 2h$, then new range

$$r' = \sqrt{2h'r}$$

$$\Rightarrow \quad r' = \sqrt{2.2hr} = \sqrt{2}\sqrt{2hr} = \sqrt{2}\,\text{r}$$

43. (b) CO_2 laser uses infrared waves.

44. (a) The wavelength of emitted light

$$\lambda = \frac{hc}{E_g}$$

where E_g = energy gap of semiconductor
= 1.9eV
$= 1.9 \times 1.6 \times 10^{-19}$V

$$\lambda = \frac{6.6 \times 10^{-34} \times 3 \times 10^8}{1.9 \times 1.6 \times 10^{-19}} \text{m}$$

$= 6.5 \times 10^{-7}$ m
$= 650 \times 10^{-9}$ m
= 650 nm

45. (a) Modulating signal frequency → 10kHz
Carrier signal frequency→ 10 MHz
∴ Side band frequency are
USB = 10 MHz + 10 kHz = 10010 kHz
LSB = 10MHz – 10 kHz = 9990 kHz

46. (b) $P_c = P_t\left(1 + \frac{m^2}{2}\right)$

$$\Rightarrow \quad 11.8 = 10\left(1 + \frac{m^2}{2}\right)$$

$\Rightarrow$ $m = 0.6$ $\Rightarrow$ % modulation = 60%.

47. (c) Carrier Swing $= 2\times\Delta f$

$$= 2\times 105.03 - 105 = 2\times 0.03 \text{ MHz}$$

$$= 0.06 \text{ MHz} = 60 \text{ kHz}$$

49. (d) $\eta = \dfrac{R}{R+R_L} = \dfrac{68}{68+10} = 0.872$

$\Rightarrow$ Power gain $= \eta G$

$\Rightarrow$ directive gain

$$G = \frac{\text{Power gain}}{\eta} = \frac{16}{0.872} = 18.35$$

50. (c) FM (I.F) = 11.5 MHz

F(image) $= f_s + 2f_{IF}$

$$= 98.6 + 2\times 11.5 = 121.6 \text{ MHz}$$

51. (b) Medium wavelength is suitable for sky waves.

52. (c) Line of sight mode of communication for carrier waves around 100 MHz.

53. (d) The required wavelength should be from 10^{-3}m to 0.1m.

54. (c) $\lambda = 3$ m, $H = 100$ m, $h = 60$ m

Path difference between direct wave and wave obtained after reflection from earth surface is

$$\frac{2\pi}{\lambda}\left(\frac{2Hh}{d}\right) + \pi = 2\pi$$

or $\dfrac{4Hh}{\lambda d} = 1$ or $d = \dfrac{4Hh}{\lambda}$

$$\Rightarrow d = \frac{4\times 100\times 60}{3} = 8\times 10^3\, m \text{ or } 8 \text{ km}$$

55. (b) Ground waves have the property of being polarized normal to the earth's surface.

56. (b) Critical frequency $f_c = 9\sqrt{N_m}$

where N_m represents election density of layers

$\because f_c \propto \sqrt{N_m}$

$\Rightarrow f_c$ is different for different layers.

57. (d) $\text{Electric Field} = -\dfrac{4\pi HhE}{\lambda d^2}$

$\Rightarrow E.F. \propto \dfrac{1}{d^2}$

58. (d) Skip distance is the minimum distance on earth's surface from the transmitter where e.m. wave of a definite frequency can reach after reflection from the ionosphere

It is given by $D_{\text{skip}} = 2h\sqrt{\dfrac{f^2}{f_c^2} - 1}$

$\Rightarrow D_{\text{skip}}$ is dependent on h, f and fc.

59. (d) Since maximum part of the power of modulated wave is contained with the carrier wave which does not transmit any desired information, hence to avoid wastage of power to suppress carrier balanced modulator is used.

60. (d) Harmonics and fundamental frequency voltages cross-modulate with each other producing fundamental frequency currents whose relation with the normal fundamental current vary with amount of harmonics in the oscillator circuit. The phase angle of the resultant fundamental frequency current thus varies resulting in frequency variation. Such frequency variation can be minimized by using a tank circuit having high effective Q having loosely coupled load.

CHEMISTRY

1-306

Some Basic Concepts of Chemistry

CONCEPTS

MATTER :

Anything that occupies space is called matter.

States of Matter :

The Solid State :
The solid state is characterised by, having a definite shape, distinct boundaries, rigidity and incompressibility.

The Liquid State :
The liquid state is characterised as having fluidity, low compressibility and no definite shape. Liquids take the shape of container in which they are stored.

The Gaseous State :
The gaseous state is characterised as having fluidity, high compressibility, no definite boundaries, no definite volume and no definite shape.

Classification of Matter

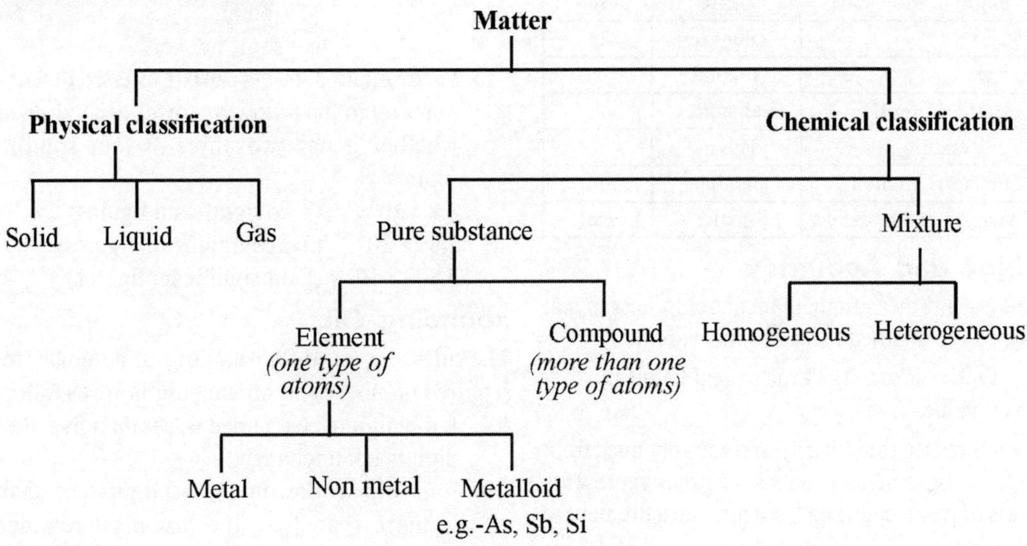

MIXTURE :

A material containing two or more substances in any proportion is called a mixture. There are two types of mixtures :

Homogeneous Mixture

It is a mixture that has the same composition throughout. e.g. a solution of sugar in water.

Heterogeneous Mixture

In such a mixture the particles of each component of the mixture remain separate and can be observed as individual grains under a microscope. e.g. mixture of grains and sand.

PURE SUBSTANCE

A pure substance is one that has a uniform composition and always have the same texture, colour, taste at a given temperature and pressure.

Elements

Pure substances which are made up of only one kind of atoms are known as elements. *e.g.* Pure gold is an element it contains only gold atoms.

Compounds

A chemical compound is a pure substance formed from chemical combination of two or more elements.

MEASUREMENTS IN CHEMISTRY

Measurement of fundamental physical quantities like mass (amount), length, temperature and time have important place in description and analysis of matter for practical purposes in chemistry. To represent the physical quality of matter, two aspects are important (i) quantitative value (digit) and (ii) unit.

S.I. Units :

In Standard International system, there are 7 fundamental units from which other derivatives or units are obtained.

No.	Physical quantity	Name of unit	Symbol
1	Length	metre	m
2	Mass	kilogram	kg
3	Time	second	s
4	Electric current	ampere	A
5	Temperature	kelvin	K
6	Luminous intensity	candela	cd
7	Amount of substance	mole	mol

Precision and Accuracy

Accuracy of any measurement means closeness to real value or result obtained from that measurement.

E = T – O, E = Error, T = True or real value and O = Observed value.

Precision is related to similarity in results obtained from repeated measurements whereas accuracy represents closeness of result obtained from any measurement to its real value.

SIGNIFICANT FIGURES

Significant figures are total number of digits in any number including the last digits whose value is uncertain are called as significant number.

Rules for Determining Significant Figures:

(i) All non-zero digits are significant for example, 768 cm has three significant figures, 0.768 has also three significant figures.

(ii) Zeros to the left of the first non-zero digit in the number are not significant for example, 0.008 gram has only one significant figure; 0.036 has two significant figures.

(iii) Zeros between non-zero digits are significant. For example, 3.05 has three significant figures.

(iv) Zeros to the right of the decimal point are significant. For example, 0.600, 0.060, 0.6000 have three, two and four significant figures respectively.

(v) If a number ends in zeros that are not to right of a decimal, the zeros may or may not be significant. For example, 3800 cm may have two, three or four significant figures.

(vi) Scientific notation : The ambiguity in rule (v) can be removed by expressing the number in scientific notation. In scientific notation the number is written in the standard exponential form as $N \times 10^n$.

Where N = a number with a single non-zero digit to the left of the decimal point.

n = some integer

For example, 3800 cm can be expressed in scientific notation in the following forms depending upon whether it has two, three or four significant figures.

3.8×10^3 (Two significant figures)
3.80×10^3 (Three significant figures)
3.800×10^3 (Four significant figures)

Rounding Off

The rules employed for rounding off a number to the required number of significant digits are as follows –

(i) If the digit to be retained is less than five, the last digit is left unchanged.

(ii) If the digit following the last digit to be retained is more than five, the last digit retained is increased by one.

(iii) If the digit following the last digit to be retained is equal to five, the last digit is left unchanged if it is even and is increased by one if it is odd.

LAWS OF CHEMICAL COMBINATION :

Law of Conservation of Mass (by Lavoisier 1789)

During any physical or chemical change, the total mass of reactants is equal to total mass of products.

Law of Definite Proportions on Constant Compositions (by Proust 1779)

Elements combine in fixed ratio of their weights to form compounds or a compound always has same elements combine together in same fixed ratio by weight.

Law of Multiple Proportions (by Dalton 1803)

When two elements combine to form two or more compounds, then the weight of one element which combine with a fixed weight of other bear a simple ratio to one another.

Examples : Oxides of nitrogen, CO and CO_2, SO_2 and SO_3.

Law of Reciprocal Proportions or Law of Equivalent Proportions (by Richter 1792)

When two elements combine with fixed weight of third element, then it is either the same or simple multiple of ratio of weights of two elements which directly combine with each other.

Examples : CH_4, CO_2 and H_2O.

Law of Combining Volumes (by Gay-Lussac)

When gases react together, they always do so in volumes which bear a simple ratio to one another and to volume of products, when all measurements are made under same conditions of temperature and pressure. **This law is applicable only for gases.**

Avogadro's Law :

This law states that "equal volumes of all gases under similar conditions of temperature and pressure contain equal number of molecules".

Example : $2H_2(g) + O_2(g) \longrightarrow 2H_2O(g)$

2 unit vol. 1 unit vol. 2 unit vol.

Ratio of number of molecules = 2 : 1 : 2

It provides a relationship between vapour density and molecular weight of substances.

By definition : Vapour density (VD)

$$= \frac{\text{Mass of same volume of gas at STP}}{\text{Mass of same volume of } H_2 \text{ at STP}}$$

or molecular weight of gas = 2 × vapour density.

DALTON'S ATOMIC THEORY :

(i) All matter is made up of atoms.

(ii) Atoms are very tiny, indivisible and indestructible ultimate particles.

(iii) Atoms of a given element are identical in mass and chemical properties.

(iv) Atoms of different elements have different masses and chemical properties.

(v) Atoms combine in small whole numbers to form compounds.

(vi) Atoms can neither be created nor destroyed.

ATOMIC AND MOLECULAR MASS

Atomic Mass

The modern reference standard for atomic weight is carbon isotope of mass number 12.

It is the number of times an atom of an element is heavier than 1/12 th of an atom of C–12.

Atomic weight of an element

$$= \frac{\text{Weight of 1 atom of element}}{1/12 \times \text{weight of 1 atom of C-12}}$$

One atomic mass unit is defined as a mass exactly equal to one twelfth the mass of one carbon-12 atom.

And 1 amu = 1.66056×10^{-24} g

The ***'average atomic mass'*** is obtained by taking into account the natural abundance of different isotopes of that element.

Average Atomic mass

$$= \frac{\text{R.A(1)} \times \text{mass number} + \text{R.A.(2)} \times \text{mass number}}{\text{R.A.(1)} + \text{R.A(2)}}$$

Molecular Weight/Mass

Molecular mass is the sum of atomic masses of the elements present in a molecule. It is obtained by multiplying the atomic mass of each element by the number of its atoms and adding them together.

The ***'formula mass'*** of a substance (ionic substances) is the sum of the atomic masses of all the atoms in a formula unit of a compound. e.g., Formula mass of NaCl = 23 + 35.5 = 58.5 u).

Equivalent Weight

The part of a substance that combines with or displaces directly or indirectly **1.0008 parts by mass of hydrogen or 8 parts by mass of oxygen or 35.5 parts by mass of chlorine.**

(i) Eq. wt. of element = $\frac{\text{Atomic weight.}}{\text{Valency}}$

(ii) Eq. wt. of acid or base = $\frac{\text{Molecular weight}}{\text{Basicity / Acidity}}$

$\left(\begin{array}{l}\text{Basicity = no. of replaceable } H^+ \\ \text{Acidity = no. of replaceable } OH^-\end{array}\right)$

(iii) Eq. wt. of salt

$$= \frac{\text{Molecular weight}}{\text{Total positive or total negative charges}}$$

(iv) Eq. wt. of oxidising agent or reducing agent

$$= \frac{\text{Molecular wt. of substance}}{\text{No. of } e^- \text{s gained or No. of } e^- \text{s lost by one molecule}}$$

METHODS FOR DETERMINING ATOMIC WEIGHTS:

(i) Atomic Weight = Equivalent weight × Valency

(ii) **Dulong and Petits method**

- **Used to determine atomic weight of solids except Be, B, C and Si.**

Approximate atomic weight

$$= \frac{6.4}{\text{Specific heat (in cal/gm)}}$$

To know the exact atomic weight, we use valency because valency is always whole number.

$$\text{Valency} = \frac{\text{Approximate atomic weight}}{\text{Equivalent weight}}$$

∴ Exact atomic weight = Exact valency × Eq. wt.

(iii) **Specific heat method**

- **Used to determine atomic weight of elementary gases only**

$$\text{At. wt. of gas} = \frac{\text{Molecular weight}}{\text{Atomicity}}$$

(Atomicity = No. of atoms present in 1 molecule of element)

(iv) **Volatile Chloride method**

- **Used to determine atomic weights of elements which form volatile chloride.**

Valency of metal whose chloride is volatile

$$= \frac{2 \times \text{V.D. of metal chloride}}{\text{Eq. wt. of metal} + 35.5}$$

Atomic weight = Eq. wt. × Valency.

METHODS FOR DETERMINING MOLECULAR WEIGHT:

(i) Molecular wt. = 2 × vapour density

(ii) Molecular wt. = 22.4 L at STP

(iii) $\frac{r_1}{r_2} = \sqrt{\frac{M_2}{M_1}}$ $\left[\begin{array}{l}r_1 \text{ and } r_2 \text{ are rates of diffusion of gases} \\ M_1 \text{ and } M_2 \text{ are molecular wts.}\end{array}\right]$

(iv) **Cryoscopic method :** Molecular Wt.

$$= \frac{K_f \times \text{wt. of solute} \times 1000}{\text{wt. of solvent} \times \text{depression in freezing point}}$$

(v) **Ebulloscopic method :** Molecular Wt.

$$= \frac{K_b \times \text{wt. of solute} \times 1000}{\text{wt. of solvent} \times \text{elevation in boiling point}}$$

(vi) **Victor Meyer Method**

- **Used only for volatile organic compound only**

Molecular Mass of volatile organic molecules

$$= \frac{\text{Mass of organic compound} \times 22400}{\text{Volume of organic compound}}$$

METHODS FOR DETERMINING EQUIVALENT WEIGHT:

(i) **Hydrogen displacement method**

Eq. wt. of metal

$$= \frac{\text{Mass of metal}}{\text{Mass of Hydrogen displaced}} \times 1.008$$

(ii) **Oxide formation method**

Eq. wt. of metal

$$= \frac{\text{Mass of metal}}{\text{Mass of Oxygen combined}} \times 8.0$$

(iii) **Chloride formation method**

Eq. wt. of metal

$$= \frac{\text{Mass of metal}}{\text{Mass of Chlorine combined}} \times 35.5$$

(iv) Eq. wt. = $\frac{\text{Strength}}{\text{Normality}}$

(v) **Metal displacement method**

$$\frac{\text{Mass of metal added}}{\text{Mass of metal displaced}} = \frac{\text{Eq. wt. of metal added}}{\text{Eq. wt. of metal displaced}}$$

(vi) **Silver salt method**

- **Used for determination of molecular wt. of organic acids only.**

$$\text{Eq. wt. of acid} = \frac{108 \times \text{wt. of silver salt}}{\text{wt. of silver metal}} \times 107$$

Molecular wt. of acid = Eq. wt. of acid × basicity

(vii) **Electrolytic method**

Gram eq. wt. = 96500 × electrochemical equivalent

$$\frac{W_1}{W_2} = \frac{E_1}{E_2} \left[\begin{array}{l} \text{where } w_1 \text{ and } w_2 \text{ are weights of metal} \\ \text{deposited and } E_1 \text{ and } E_2 \text{ are equivalent} \\ \text{weights} \end{array} \right]$$

(viii) **Double decomposition method**

For a chemical reaction

$$AB + CD \longrightarrow AD\downarrow + CB$$

$$\frac{\text{Wt. of comp. AB}}{\text{Wt. of comp. AD}} = \frac{\text{Eq. wt. of A + Eq. wt. of B}}{\text{Eq. wt. of salt in ppt}}$$

and

$$\frac{\text{Wt. of salt taken}}{\text{Mass of ppt obtained}} = \frac{\text{Eq. wt. of salt}}{\text{Eq. wt. of salt in ppt}}$$

MOLE CONCEPT

The number of moles can be calculated by a number of formulae depending upon the data given :

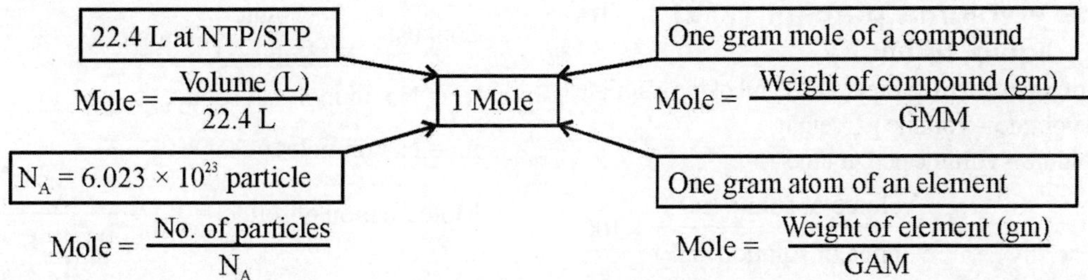

Relation of Mole with Gas Volume :

* Volume of one mole of a gas at NTP = 22.4 Litre.

* Since 1 mole gas contain 6.023×10^{23} molecules So, 6.023×10^{23} (N_A) molecule have volume at NTP = 22.4 lit.

* Moles at NTP = $\frac{\text{Volume (Litre)}}{22.4}$

* Mass of one mole gas = (Molecular mass of gas) gm = 1 GMM

* Molecular mass of gas = 2 × vapour density of gas.

* Vapour density of gas

$$= \frac{\text{Weight of a certain vol. of gas at NTP}}{\text{Weight of the same vol. of } H_2 \text{ gas at NTP}}$$

* Weight of 1 ml of H_2 gas at NTP

$$= \frac{\text{Weight of one mole of } H_2 \text{ gas}}{\text{Volume of one mole of } H_2 \text{ gas at NTP.}}$$

$$= \frac{2\,\text{gm}}{22400\,\text{ml}} = 0.00009\,\text{gm}$$

EMPIRICAL AND MOLECULAR FORMULA

Empirical Formula :

The simplest whole number ratio of atoms of elements present in one molecule of substance.

Molecular Formula :

The actual number of atoms of different elements present in its one molecule.

Molecular formula = n × Empirical formula

$n = \frac{\text{Molecular formula mass}}{\text{Empirical formula mass}}$ (n is always a whole number)

STOICHIOMETRY :

Calculation based on chemical equations

Chemical Equations :

A shorthand representation of a chemical reaction in terms of symbols and formulae of the substances involved in the reaction is called a chemical equation.

Limiting Reagent :

The reactant which is present in least quantity and is completely consumed during the reaction. It decides the amount of products formed during the reaction. While solving problems based on chemical equations, always find if any of the reagent is limiting reagent or not before marking correct answer.

EXPRESSION OF STRENGTH/ CONCENTRATION OF SOLUTION :

The amount of solute which dissolved in unit volume of solution is called concentration of solution.

$$\text{Concentration} = \frac{\text{amount of solute}}{\text{volume of solution}}$$

Weight – weight percent (w/W) : Weight of solute present in 100 gm of the solution.

Weight – Weight percentage

$$= \frac{\text{weight of solute (gm)}}{\text{weight of solution (gm)}} \times 100$$

Volume – volume percent (v/V) : (In liquid – liquid solution)

Volume of solute in ml. present in 100 ml of the solution is called volume – volume percentage.

Volume – volume percentage

$$= \frac{\text{Volume of solute (ml.)}}{\text{volume of solution (ml.)}} \times 100$$

Weight – volume percentage (w/V) :–

Weight of solute in gm. present in 100 ml of the solution is called weight — volume percentage.

Weight – volume percentage

$$= \frac{\text{weight of solute (gm)}}{\text{volume of solution (ml)}} \times 100$$

Normality :

$$\text{Normality} = \frac{\text{Number of gram equivalents of solute}}{\text{Volume of solution (lit.)}}$$

Molarity :

$$\text{Molarity} = \frac{\text{Number of gram moles of solute}}{\text{Volume of solution (lit.)}}$$

Molarity of dilution :

Before dilution After dilution

$$M_1V_1 = M_2V_2$$

Molarity of mixing :

$M_1V_1 + M_2V_2 + M_3V_3 = M_R(V_1 + V_2 + V_3)$

M_R = resultant molarity

Specific gravity of solution :

$\frac{\text{weight of solution}}{\text{volume of solution}}$ i.e., weight of 1 mL solution.

Molality :

$$\text{Molality} = \frac{\text{Gram moles of solute}}{\text{Weight of solvent (kg)}}$$

Formality :

$$\text{Formality} = \frac{\text{Weight of solute (gm)}}{\text{Formula weight of solute}} \times \frac{1}{\text{Volume of solution (lit.)}}$$

Mole Fraction :

Suppose $\left.\begin{matrix}A - \text{Solute} \\ B - \text{Solvent}\end{matrix}\right\}$ Solution ;

n_A = No. of moles of solute,

n_B = No. of moles of solvent

Mole fraction of solute $= X_A = \frac{n_A}{n_A + n_B}$

Mole fraction of solvent $= X_B = \frac{n_B}{n_A + n_B}$

; $X_A + X_B = 1$

For gaseous mixture :

A binary system of two gases A & B

P_A = Partial pressure of A

P_B = Partial pressure of B

$P = P_A + P_B$ = Total pressure of gaseous mixture

Mole fraction of gas A : $X_A = \frac{P_A}{P_A + P_B} = \frac{P_A}{P}$

Mole fraction of gas B : $X_B = \frac{P_B}{P_A + P_B} = \frac{P_B}{P}$

Mole percentage :

Mole percentage = Mole fraction × 100

Mole percent of A = $X_A \times 100$

Mole percent of B = $X_B \times 100$

ppm. (Parts per million) : The parts of the component per million parts (10^6) of the solution.

$$\text{ppm} = \frac{w}{w + W} \times 10^6 ;$$

where, w = weight of solute,
W = weight of solvent

Exercise

1. Given the numbers : 161 cm, 0.161 cm, 0.0161 cm. The number of significant figures for the three numbers are
(a) 3, 4 and 5 respectively
(b) 3, 3 and 4 respectively
(c) 3, 3 and 3 respectively
(d) 3, 4 and 4 respectively
2. If the density of a solution is 3.12 g mL^{-1}, the mass of 1.5 mL solution in significant figures is______.
(a) 4.7 g (b) 4680×10^{-3} g
(c) 4.680 g (d) 46.80 g
3. In which of the following number all zeros are significant?
(a) 0.0005 (b) 0.0500
(c) 50.000 (d) 0.0050
4. Two students performed the same experiment separately and each one of them recorded two readings of mass which are given below. Correct reading of mass is 3.0 g. On the basis of given data, mark the correct option out of the following statements.

Students	Readings (i)	Readings (ii)
A	3.01	2.99
B	3.05	2.95

(a) Results of both the students are neither accurate nor precise.
(b) Results of student A are both precise and accurate.
(c) Results of student B are neither precise nor accurate.
(d) Results of student B are both precise and accurate.
5. In the final answer of the expression
$$\frac{(29.2-20.2)(1.79\times10^{5})}{1.37}$$
the number of significant figures is :
(a) 1 (b) 2
(c) 3 (d) 4
6. A measured temperature on Fahrenheit scale is 200 °F. What will this reading be on Celsius scale ?
(a) 40° C (b) 94° C
(c) 93.3 °C (d) 30° C
7. Which of the following is an example of a heterogeneous substance?
(a) Bottled water (b) Table salt
(c) Pieces of copper (d) Candle
8. Which of the following statements about a compound is incorrect ?
(a) A molecule of a compound has atoms of different elements.
(b) A compound cannot be separated into its constituent elements by physical methods of separation.
(c) A compound retains the physical properties of its constituent elements.
(d) The ratio of atoms of different elements in a compound is fixed.
9. The prefix 10^{18} is
(a) giga (b) kilo
(c) exa (d) nano
10. Which of the following statements is correct about the reaction given below ?
$$4Fe(s) + 3O_2(g) \longrightarrow 2Fe_2O_3(g)$$
(a) Total mass of iron and oxygen in reactants = total mass of iron and oxygen in product therefore, it follows law of conservation of mass.
(b) Total mass of reactants = total mass of product; therefore, law of multiple proportions is followed.
(c) Amount of Fe_2O_3 can be increased by reducing the amount of any one of the reactants (iron or oxygen).
(d) Amount of Fe_2O_3 produced will decrease if the amount of any one of the reactants (iron or oxygen) is taken in excess.
11. Two samples of lead oxide were separately reduced to metallic lead by heating in a current of hydrogen. The weight of lead from one oxide was half the weight of lead obtained from the other oxide. The data illustrates –
(a) law of reciprocal proportions
(b) law of constant proportions
(c) law of multiple proportions
(d) law of equivalent proportions

12. In compound A, 1.00g of nitrogen unites with 0.57g of oxygen. In compound B, 2.00g of nitrogen combines with 2.24g of oxygen. In compound C, 3.00g of nitrogen combines with 5.11g of oxygen. These results obey the following law
(a) law of constant proportion
(b) law of multiple proportion
(c) law of reciprocal proportion
(d) Dalton's law of partial pressure

13. One mole of a gas occupies a volume of 22.4 L. This is derived from
(a) Berzelius' hypothesis
(b) Gay-Lussac's law
(c) Avogadro's law
(d) Dalton's law

14. The percentage of copper and oxygen in samples of CuO obtained by different methods were found to be the same. The illustrate the law of
(a) constant proportions
(b) conservation of mass
(c) multiple proportions
(d) reciprocal proportions

15. The law of definite proportions was given by –
(a) John Dalton (b) Humphry Davy
(c) Proust (d) Michael Faraday

16. Which one of the following pairs of compounds illustrate the law of multiple proportions ?
(a) H_2O, Na_2O (b) MgO, Na_2O
(c) Na_2O, BaO (d) $SnCl_2, SnCl_4$

17. Molecular mass is defined as the
(a) mass of one atom compared with the mass of one molecule
(b) mass of one atom compared with the mass of one atom of hydrogen
(c) mass of one molecule of any substance compared with the mass of one atom of C-12
(d) None of the above

18. 1 amu is equal to
(a) $\frac{1}{14}$ of O-16 (b) $\frac{1}{12}$ of C-12
(c) 1 g of H_2 (d) 1.66×10^{-23} kg

19. The modern atomic weight scale is based on
(a) O^{16} (b) C^{12}
(c) H^1 (d) C^{13}

20. Two containers P and Q of equal volume (1 litre each) contain 6 g of O_2 and SO_2 respectively at 300 K and 1 atmosphere, then
(a) Number of molecules in P is less than that in Q
(b) Number of molecules in P and Q is same
(c) Number of molecules in Q is less than that in P
(d) Either (a) or (b)

21. In a chemical reaction
$$K_2Cr_2O_7 + xH_2SO_4 + ySO_2 \longrightarrow K_2SO_4 + zCr_2(SO_4)_3 + H_2O$$
the values of x, y, z are
(a) 4, 1, 4 (b) 1, 3, 1
(c) 3, 2, 3 (d) 2, 1, 2

22. 1 c.c. N_2O at NTP contains :
(a) $\frac{1.8}{224} \times 10^{22}$ atoms
(b) $\frac{6.02}{22400} \times 10^{23}$ molecules
(c) $\frac{1.32}{224} \times 10^{23}$ electrons
(d) all the above

23. The weight of one molecule of a compound $C_{60}H_{122}$ is
(a) 1.2×10^{-20} gram (b) 1.4×10^{-21} gram
(c) 5.025×10^{23} gram (d) 6.023×10^{23} gram

24. With increase of temperature, which of these changes?
(a) molality
(b) weight fraction of solute
(c) fraction of solute present in water
(d) mole fraction.

25. A molal solution is one that contains 1 mole of a solute in
(a) one litre of the solvent
(b) 1000 g of the solvent
(c) one lit re of the solution
(d) 22.4 litres of the solution

26. Equivalent weight of crystalline oxalic acid is
(a) 90 (b) 53
(c) 63 (d) 45

27. The total number of atoms represented by the compound $CuSO_4.5H_2O$ is
(a) 27 (b) 21
(c) 5 (d) 8

28. The molecular formula P_2O_5 means that
   (a) a molecule contains 2 atoms of P and 5 atoms of O
   (b) the ratio of the mass of P to the mass of O in the molecule is 2:5
   (c) there are twice as many P atoms in the molecule as there are O atoms
   (d) the ratio of the mass of P to the mass of O in the molecule is 5 : 2
29. In complete combustion of C_4H_{10}, total molecules of O_2 required are
   (a) $\frac{13}{2}O_2$ (b) $8O_2$
   (c) $\frac{15}{2}O_2$ (d) $\frac{11}{2}O_2$
30. The empirical formula of a compound is CH_2. One mole of this compound has a mass of 42 grams. Its molecular formula is :
   (a) C_3H_6 (b) C_3H_8
   (c) CH_2 (d) C_2H_2
31. 50 ml 10 N H_2SO_4, 25 ml 12 N HCl and 40 ml 5 N HNO_3 were mixed together and the volume of the mixture was made 1000 ml by adding water. The normality of the resultant solution will be
   (a) 2 N (b) 1 N
   (c) 3 N (d) 4 N
32. 25.4 g of I_2 and 14.2 g of Cl_2 are made to react completely to yield a mixture of ICl and ICl_3. Calculate moles of ICl and ICl_3 formed
   (a) 0.1, 0.1 (b) 0.2, 0.2
   (c) 0.1, 0.2 (d) 0.2, 0.1
33. The Statue of Liberty is made of 2.0×10^5 lbs of copper sheets bolted to a framework. (1lb = 454 g) How many atoms of copper are on the statue?
   (Atomic weight: Cu = 63.5).
   (a) 2.1×10^{27} (b) 8.6×10^{29}
   (c) 4.3×10^{26} (d) 8.6×10^{26}
34. The empirical formula and molecular mass of a compound are CH_2O and 180 g respectively. What will be the molecular formula of the compound ?
   (a) $C_9H_{18}O_9$ (b) CH_2O
   (c) $C_6H_{12}O_6$ (d) $C_2H_4O_2$
35. What is the mass percent of carbon in carbon dioxide ?
   (a) 0.034% (b) 27.27%
   (c) 3.4% (d) 28.7%
36. At S.T.P. the density of CCl_4 vapours in g/L will be nearest to :
   (a) 6.87 (b) 3.42
   (c) 10.26 (d) 4.57
37. One litre hard water contains 12.00 mg of Mg^{2+}. Mili-equivalents of washing soda required to remove its hardness is :
   (a) 1 (b) 12. 16
   (c) 1×10^{-3} (d) 12.16×10^{-3}
38. What is the weight of oxygen required for the complete combustion of 2.8 kg of ethylene ?
   (a) 2.8 kg (b) 6.4 kg
   (c) 9.6 kg (d) 96 kg
39. The number of gram molecules of oxygen in 6.02×10^{24} CO molecules is
   (a) 10 gm molecules (b) 5 gm molecules
   (c) 1 gm molecules (d) 0.5 gm molelcules
40. The number of oxygen atoms in 4.4 g of CO_2 is
   (a) 1.2×10^{23} (b) 6×10^{22}
   (c) 6×10^{23} (d) 12×10^{23}
41. The molecular weight of O_2 and SO_2 are 32 and 64 respectively. At 15°C and 150 mm Hg pressure, one litre of O_2 contains 'N' molecules. The number of molecules in two litres of SO_2 under the same conditions of temperature and pressure will be :
   (a) N/2 (b) N
   (c) 2N (d) 4N
42. A 5 molar solution of H_2SO_4 is diluted from 1 litre to a volume of 10 litres, the normality of the solution will be :
   (a) 1N (b) 0.1N
   (c) 5N (d) 0.5N
43. If N_A is Avogadro's number then number of valence electrons in 4.2g of nitride ions (N^{3-}) is
   (a) $2.4 N_A$ (b) $4.2 N_A$
   (c) $1.6 N_A$ (d) $3.2 N_A$
44. The percentage weight of Zn in white vitriol $[ZnSO_4.7H_2O]$ is approximately equal to (Zn = 65, S = 32, O = 16 and H = 1)
   (a) 33.65 % (b) 32.56 %
   (c) 23.65 % (d) 22.65 %

45. Liquid benzene (C_6H_6) burns in oxygen according to the equation

$$2C_6H_6(l) + 15O_2(g) \longrightarrow 12CO_2(g) + 6H_2O(g)$$

How many litres of O_2 at STP are needed to complete the combustion of 39 g of liquid benzene?(Mol. wt. of O_2 = 32, C_6H_6 = 78)
(a) 74 L (b) 11.2 L
(c) 22.4 L (d) 84 L

46. An organic compound containing C, H and N gave the following analysis :
C = 40% ; H = 13.33% ; N = 46.67%
Its empirical formula would be
(a) $C_2H_7N_2$ (b) CH_5N
(c) CH_4N (d) C_2H_7N

47. Haemoglobin contains 0.334% of iron by weight. The molecular weight of haemoglobin is approximately 67200. The number of iron atoms (at. wt. of Fe is 56) present in one molecule of haemoglobin are
(a) 1 (b) 6
(c) 4 (d) 2

48. An organic compound containing C, H and O gave on analysis C – 40% and H – 6.66%. Its empirical formula would be
(a) C_3H_6O (b) CHO
(c) CH_2O (d) CH_4O

49. Assuming fully decomposed, the volume of CO_2 released at STP on heating 9.85 g of $BaCO_3$ (Atomic mass, Ba = 137) will be
(a) 2.24 L (b) 4.96 L
(c) 1.12 L (d) 0.84 L

50. Percentage of Se in peroxidase anhydrase enzyme is 0.5% by weight (at. wt. = 78.4) then minimum molecular weight of peroxidase anhydrase enzyme is
(a) 1.568×10^3 (b) 15.68
(c) 2.136×10^4 (d) 1.568×10^4

51. Which has maximum number of molecules?
(a) 7 gm N_2 (b) 2 gm H_2
(c) 16 gm NO_2 (d) 16 gm O_2

52. In Haber process 30 litres of dihydrogen and 30 litres of dinitrogen were taken for reaction which yielded only 50% of the expected product. What will be the composition of gaseous mixture under the aforesaid condition in the end?
(a) 20 litres ammonia, 25 litres nitrogen, 15 litres hydrogen
(b) 20 litres ammonia, 20 litres nitrogen, 20 litres hydrogen
(c) 10 litres ammonia, 25 litres nitrogen, 15 litres hydrogen
(d) 20 litres ammonia, 10 litres nitrogen, 30 litres hydrogen

53. The maximum number of molecules is present in
(a) 15 L of H_2 gas at STP
(b) 5 L of N_2 gas at STP
(c) 0.5 g of H_2 gas
(d) 10 g of O_2 gas

54. The number of moles of $KMnO_4$ reduced by one mole of KI in alkaline medium is:
(a) one (b) two
(c) five (d) one fifth

55. Concentrated aqueous sulphuric acid is 98% H_2SO_4 by mass and has a density of 1.80 g mL^{-1}. Volume of acid required to make one litre of 0.1M H_2SO_4 solution is
(a) 16.65 mL (b) 22.20 mL
(c) 5.55 mL (d) 11.10 mL

56. The number of moles of $KMnO_4$ that will be needed to react with one mole of sulphite ion in acidic solution is
(a) 4/5 (b) 2/5
(c) 1 (d) 3/5

57. Volume occupied by one molecule of water (density = 1 g cm^{-3}) is :
(a) 9.0×10^{-23} cm^3 (b) 6.023×10^{-23} cm^3
(c) 3.0×10^{-23} cm^3 (d) 5.5×10^{-23} cm^3

58. Number of moles of MnO_4^- required to oxidize one mole of ferrous oxalate completely in acidic medium will be :
(a) 0.6 moles (b) 0.4 moles
(c) 7.5 moles (d) 0.2 moles

59. What volume of oxygen gas (O_2) measured at 0°C and 1 atm, is needed to burn completely 1L of propane gas (C_3H_8) measured under the same conditions ?
(a) 7L (b) 6L
(c) 5L (d) 10L

60. An organic compound contains carbon, hydrogen and oxygen. Its elemental analysis gave C, 38.71% and H, 9.67%. The empirical formula of the compound would be :
(a) CH_3O (b) CH_2O
(c) CHO (d) CH_4O

CHAPTER 2

Structure of Atom

CONCEPTS

- The ratio of charge to mass (i.e., e/m) and the value of the charge of the particles in cathode rays does not depends upon the nature of the gas whereas in anode rays. If depends upon the nature of the gas taken in the discharge tube.

FUNDAMENTAL PARTICLES OF ATOM :

S.No.	Properties	Electron	Proton	Neutron
(i)	Mass	9.1×10^{-31} kg	1.673×10^{-27} kg	1.675×10^{-27} kg = 1/1837 times of H atom
(ii)	Mass in amu	5.485×10^{-4}	1.007	1.008
(iii)	Actual charge (coulomb)	1.602×10^{-19}	1.602×10^{-19}	0
(iv)	Actual charge (e.s.u.)	4.8×10^{-10}	$+4.8 \times 10^{-10}$	0
(v)	Charge	-1.6×10^{-19} C or	$+1.6 \times 10^{-19}$ C or	neutral
(vi)	e/m	1.76×10^{8} Coulomb/g	9.48×10^{4} C/g	zero
(vii)	Discovered in	Cathode ray exp.	Anode ray exp.	bombardment of α particles on Be atom.
(viii)	Discovered by	J.J. Thomson	E.Goldstein	James Chadwick

NUCLEUS (DISCOVERED BY RUTHERFORD BY BOMBARDMENT OF α-PARTICLE WITH THIN GOLD SHEET):

(i) It has neutrons and protons.

Neutrons + Protons = Nucleons.

(ii) **Radius of nucleus (r_n)**

$r_n = r_0 \times A^{1/3}$

where r_0 = constant = 1.4×10^{-13} cm

A = mass no.

Size of nucleus is measured in Fermi (1 Fermi = 10^{-13} cm = 10^{-15}m)

The **nuclear radii** lie in **range of 1.5 – 6.5 Fermi**.

(iii) **Volume of nucleus 10^{-39} cm^3** (about 10^{-15} times smaller than that of atom).

(iv) **Density of nucleus is of 10^{14} g cm^{-3}** = 10^8 tonnes cm^{-3} = 10^{11}kg/cc.

Atomic number (Z)

= Number of protons = Number of electrons

Mass number (A)

= Number of proton + Number of neutrons

DIFFERENT TYPES OF ATOMIC SPECIES :

(i) **Isotopes** - Same atomic number and different atomic masses.

(ii) **Isobars** - Atoms of different elements with same atomic masses and different atomic numbers.

(iii) **Isotones** - Atoms of different elements having different atomic masses, different atomic no. but same no. of neutrons. e.g.- ${}^{16}_{8}O$ and ${}^{15}_{7}N$.

(iv) **Isodiaphers** - They have different atomic numbers and mass numbers but the same difference between the number of neutrons and protons are electrons. For example, for both $_{90}Th^{234}$ and $_{92}U^{238}$ the difference between neutron number (N) and proton number (Z) is N – Z = 54.

(v) **Isosters** - They have **same number of atoms and electrons**. e.g.- N_2 and CO (both have two atoms and 14 electrons each).

(vi) **Isoelectronic** - These species contains same of electrons. For example O^{2-} and F^{-}.

ELECTROMAGNETIC RADIATIONS :

These are energy transporters in form of waves. They travel in space with velocity of light.

(i) $\nu = \frac{c}{\lambda}$ where ν = frequency, c = velocity, λ = wavelength.

(ii) **Wave number ($\bar{\nu}$)** gives number of wavelength in 1 cm. $\bar{\nu} = \frac{1}{\lambda}$

(iii) **Electromagnetic spectrum** is arrangement of different electromagnetic radiations in order of increasing wavelengths or decreasing frequencies.

Cosmic rays < γ-rays < X-rays < UV rays < Visible < IR rays < Micro waves < Radio waves

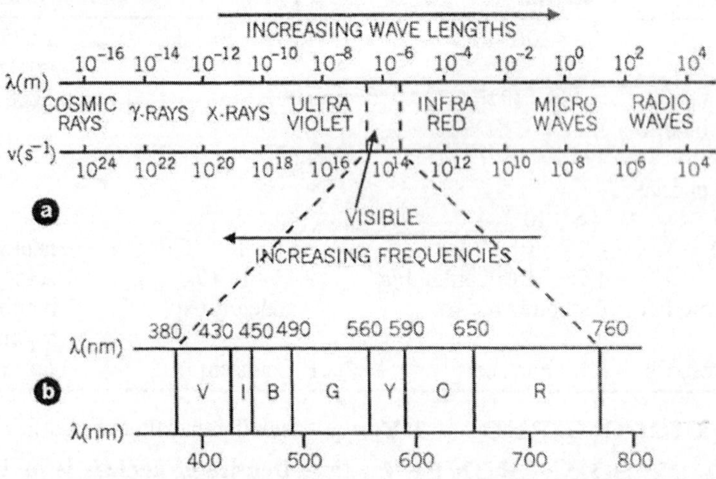

QUANTUM THEORY (PROPOSED BY MAX PLANCK)

(i) **Energy is absorbed or evolved in small packets called quanta. One quanta of light is called photon**. It is never in fractions.

(ii) E ∝ ν or E = hν. Where E = energy of radiation, h = Planck's constant = 6.625×10^{-34} Js.

$$\frac{E_1}{E_2} = \frac{\lambda_1}{\lambda_1}\left(\because E = \frac{hc}{\lambda}\right).$$

PHOTOELECTRIC EFFECT

(i) When radiations with certain minimum frequency (called **threshold frequency**) fall on metal surface, electrons are ejected from metal surface. It is called **photoelectric effect**. Ejected electrons are called **photoelectrons**.

(ii) For photoelectric effect to be observed, frequency of incident radiations must be more than threshold frequency.

(iii) **Intensity of incident radiation ∝ No. of photoelectrons ejected**.

(iv) **Total energy of ejected electron = K.E. + W**

or, $h\nu = \frac{1}{2}mv^2 + h\nu_0$.

or, K.E. of ejected photon = $\frac{1}{2}mv^2 = h\nu - h\nu_0$

where ν_0 = Threshold frequency

ν = Frequency of incident radiation.

W = work function.

ATOMIC SPECTRUM

Photograph of process of dispersion of light and pattern of bands obtained when light passes through a substance.

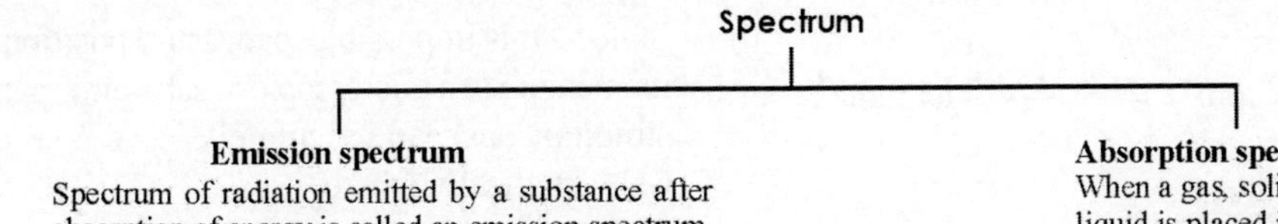

Spectrum

Emission spectrum
Spectrum of radiation emitted by a substance after absorption of energy is called an emission spectrum.

Continuous spectrum
A spectrum having no lines or bonds, especially a spectrum of radiation distributed over an uninterrupted range of wavelengths e.g.- spectra of bulb

Line or atomic spectrum
A **line spectrum** is that spectrum in which only certain wave lengths or bright coloured lines appear on a dark background. Atomic spectra of most elements is line spectrum.

Absorption spectrum
When a gas, solid or transparent liquid is placed in the path of bright light emitted from any source and then analysed by the spectroscope, dark lines are observed in the spectrum. This type of spectrum is called **absorption spectrum.**

LINE SPECTRUM OF HYDROGEN

It has 6 series **Lyman, Balmer, Paschen, Brackett, Pfund and Humphry**.

(i) Rydberg's formula :

$$\overline{v} = R\left(\frac{1}{n_1^2} - \frac{1}{n_2^2}\right)$$

where $\overline{v}$ = wave no.,
R = Rydberg's constant = 109677 cm^{-1}, n_1 and n_2 are electronic levels involved in transition.

For hydrogen like atom

$$\overline{v} = R \times Z^2\left[\frac{1}{n_1^2} - \frac{1}{n_2^2}\right]$$

where Z = atomic number of atom.

(ii)

S.No.	Series	Region	n_1	n_2
1.	Lyman	Ultraviolet	1	$2, 3, 4 \ldots\ldots \infty$.
2.	Balmer	Visible	2	$3, 4, 5 \ldots\ldots \infty$.
3.	Paschen	Infrared	3	$4, 5, 6 \ldots\ldots \infty$.
4.	Brackett	Infrared	4	$5, 6, 7 \ldots\ldots \infty$.
5.	Pfund	Infrared	5	$6, 7, 8 \ldots\ldots \infty$.
6.	Humphry	Infrared	6	$7, 8, 9 \ldots\ldots \infty$.

(iii) **No. of spectral lines** when electron jumps from n^{th} level to ground level = $\mathbf{\frac{n(n-1)}{2}}$

(iv) The intensities of spectral lines decreases with increase in value of n.

(v) **The limiting line** for any series of line in hydrogen spectrum is **when $n_2 = \infty$** in Rydbergs equation.

BOHR'S ATOMIC MODEL

It is based on **Planck's quantum theory**. The postulates of this model are as follows.

(i) Electron moves on certain definite paths only, around nucleus without losing or gaining energy. These paths are called orbits and represented as K, L, M, N etc.

(ii) **Angular momentum (mvr) of an electron in orbit**

$$= n\frac{h}{2\pi}$$

where n = whole number.

Angular momentum of electron in any orbital

$$= \sqrt{\ell(\ell+1)}\frac{h}{2\pi}$$

$\because$ It depends only on value of *l* and not on n.
$\therefore$ Angular momentum of electrons in 1s, 2s, 3s, 4s will be same.

(iii) Energy is absorbed or emitted only when electron jumps from one orbit to other.

$$\Delta E = E_2 - E_1 = h\nu.$$

(iv) **Radius of nth orbit of H like atom**

$$r_n = \frac{5.29 \times 10^{-11} n^2}{Z} \text{meter} = \frac{0.529 \times n^2}{Z} A^o$$

where n = no. of orbit, Z = atomic number.

(v) **Velocity of electron in n^{th} orbit (v_n)**

$$v_n = \frac{2.188 \times 10^6 Z}{n} ms^{-1}$$

(vi) No. of revolutions of electron per second in n^{th} orbit

$$= \frac{\text{velocity}}{\text{circumferance}} = \frac{v_n}{2\pi r_n}.$$

(vii) **Energy of electron in n^{th} orbit of H like atom (E_n)**

$$E_n = \frac{-13.6\,Z^2}{n^2}\,\text{eV atm}^{-1} = \frac{-1312\,Z^2}{n^2}\,\text{kJ mol}^{-1}$$

$$= \frac{-2.172\times10^{-18}\,Z^2}{n^2}\,\text{J atm}^{-1}$$

Electrical energy is negative because when electron is at infinite distance from nucleus, interaction between nucleus and electrons is zero. When electron comes closer to nucleus, attraction increases, energy is released and it becomes negative as it was already zero.

LIMITATION'S OF BOHR'S ATOMIC MODEL:

(i) **Failed to explain line spectra of multielectron atoms.**

(ii) **Could not explain Zeeman effect** (splitting of spectral lines **in magnetic field**) and **Stark effect** (splitting of spectral lines in **electric field**)

(iii) It is **against Heisenberg's uncertainty principle**.

(iv) Could not explain the presence of multiple spectral lines

(v) Unable to explain de-Broglie's concept of dual nature.

(vi) No conclusion was given for complete quantization.

DUAL NATURE OF LIGHT :

(i) **Photoelectric effect and black body radiations support particle nature of light.**

(ii) **Interference and diffraction support wave nature of light.**

∴ light has dual nature (particle and wave).

DUAL NATURE OF MATTER (BY DE-BROGLIE).

(i) All matter has dual nature-particle and wave both.

(ii) Waves associated with matter are called **matter waves**.

(iii) **de-Broglie equation :** $\lambda = \frac{h}{mv} = \frac{h}{p} = \frac{h}{\sqrt{2mE}}$

where, p = momentum.

(iv) **It is not significant for large objects.**

HEISENBERG'S UNCERTAINTY PRINCIPLE:

(i) States, "It is impossible to measure position and momentum of a microscopic or subatomic particle simultaneously and accurately."
Mathematically, where,
Δx = uncertainty in position,
Δp = uncertainty in momentum.

(ii) It is not applicable to large bodies.

QUANTUM OR WAVE MECHANICAL MODEL OF ATOM (BY SCHRODINGER)

(i) **It is based on de-Broglie's principle, Heisenberg's uncertainty principle and dual nature of matter.**

(ii) Motion of electron in three dimentional space is represented by Schrodinger wave equation.

$$\frac{\partial^2\psi}{\partial x^2}+\frac{\partial^2\psi}{\partial y^2}+\frac{\partial^2\psi}{\partial z^2}+\frac{8\pi^2 m}{h^2}(E-V)\psi = 0$$

where ψ = amplitude of electron wave at point with x, y, z co-ordinates.
E = Total energy
V = Potential energy
ψ = Wave function
ψ^2 = Probability of finding electron.

Quantum efficiency

$$= \frac{\text{No. of moles of substance reacted}}{\text{No. of quanta absorbed}}.$$

PROBABILITY DISTRIBUTION :

(i) **ψ is wave function** and has no physical significance.

(ii) **ψ^2 is probability density of electron and has physical significance.**

(iii) **Atomic orbital** is three dimentional space around nucleus where the probability of finding electron is high.

(iv) **Radial probability distribution curve** gives probability of finding electron in a given volume.

(v) **Node** is a spherical surface within an orbital on which the probability of finding the electron is zero. **No. of nodes in an orbital = $n - l - 1$.**

(vi) **Nodal plane** is the plane passing through the nucleus where the probability of finding an electron is zero. **No. of nodal planes in an orbital = l.**

DIFFERENT ORBITALS OF ATOM

S.No.	Property	s-orbital	p-orbital	d-orbital	f-orbital
(i)	shape	spherical	3 dumb-bells	5 dumb-bells	7 dumb-bells
(ii)	structure	symmetrical	2-lobes in each p_x, p_y, p_z axis	2 lobes each d_{xy}, d_{yz}, d_{xz}, $d_{x^2-y^2}, d_{z^2}$	2-lobes each $f_{z^2}, f_{xz^2}, f_{xyz}$, $f_{x(x^2-3y^2)}, f_{z(x^2-y^2)}$, $f_{y(3x^2-y^2)}$

QUANTUM NUMBERS

- Used to define position of electron in an atom. There are 4 quantum numbers.

S.No.	Property	Principal quantum number	Azimuthal quantum number	Magnetic quantum number	Spin quantum number
(i)	Symbol	n	l	m	s
(ii)	Tells about	**size of orbit** and energy of orbit	**shape of orbital**	**orientation of orbital**	**direction of spin of electron**
(iii)	Possible	from 1 to ∞ any values	**from 0 to (n – 1)** whole no. (1 is nearest to nucleus)	**from $-l$ to** $l = 0$ is s orbital	**+ 1/2 and – 1/2** **zero to + l** = 1 is p orbital = 2 is d orbital = 3 is f orbital
(iv)	Discoverer	Bohr	Sommerfield	Lande	Goudsmit

- No. of subshells in n^{th} shell = n
  No. **of orbitals in subshell** = $(2l + 1)$
  Maximum no. of electrons in subshell = $2(2l + 1)$
  Maximum **no. of electrons in n^{th} shell** = $2n^2$.

RULES AND PRINCIPLES TO WRITE ELECTRONIC CONFIGURATIONS

(a) **Aufbau principle :** It states that electrons are filled in increasing order of energy in an atom.

(b) **n + l rule**
   (i) Electrons are filled in order of increasing n + l value
   (ii) If two orbitals have same value of n + l then orbital having lower value of n is filled first.

(c) **Hund's rule :** It states that pairing of electrons in orbitals belonging to same subshell (degenerate orbitals) does not take place till each orbital of that subshell is singly occupied.

(d) **Pauli's exclusion principle :** No two electrons of same atom can have all 4 quantum number values identical.

(e) **Exactly half filled and fully filled orbitals are more stable than nearly half filled and nearly fully filled orbitals** due to greater symmetry and greater exchange energy.

Exercise

1. Which of the following statement is not correct about the characteristics of cathode rays?

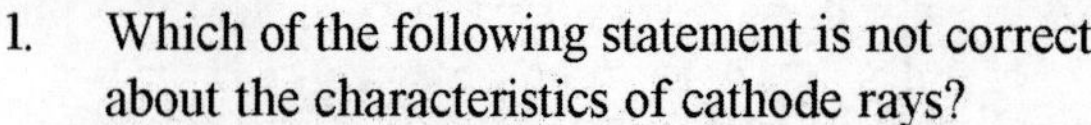

   (a) They start from the cathode and move towards the anode.
   (b) They travel in straight line in the absence of an external electrical or magnetic field.
   (c) Characteristics of cathode rays do not depend upon the material of electrodes in cathode ray tube.
   (d) Characteristics of cathode rays depend upon the nature of gas present in the cathode ray tube.
2. Which of the following is never true for cathode rays ?
   (a) They possess kinetic energy
   (b) They are electromagnetic waves
   (c) They produce heat
   (d) They produce mechanical pressure.
3. Which of the following statements about the electron is incorrect?
   (a) It is negatively charged particle
   (b) The mass of electron is equal to the mass of neutron.
   (c) It is a basic constituent of all atoms.
   (d) It is a constituent of cathode rays.
4. Which of the following pairs have identical values of e/m?
   (a) A proton and a neutron
   (b) A proton and deuterium
   (c) Deuterium and an α-particle
   (d) An electron and γ-rays
5. While performing cathode ray experiments, it was observed that there was no passage of electric current under normal conditions. Which of the following can account for this observation ?
   (a) Dust particles are present in air
   (b) Carbon dioxide is present in air
   (c) Air is a poor conductor of electricity under normal conditions
   (d) None of the above
6. When beryllium is bombarded with alpha particles (Chadwick's experiment) extremely penetrating radiations, which cannot be deflected by electrical or magnetic field are given out. These are :
   (a) A beam of protons
   (b) Alpha rays
   (c) A beam of neutrons
   (d) A beam of neutrons and protons
7. Which of the following properties of atom could be explained correctly by Thomson Model of atom?
   (a) Overall neutrality of atom.
   (b) Spectra of hydrogen atom.
   (c) Position of electrons, protons and neutrons in atom.
   (d) Stability of atom.
8. Rutherford's α-particle dispersion experiment concludes
   (a) all positive ions are deposited at small part
   (b) all negative ions are deposited at small part
   (c) proton moves around the electron
   (d) neutrons are charged particles.
9. The diagram given below represent an atom of

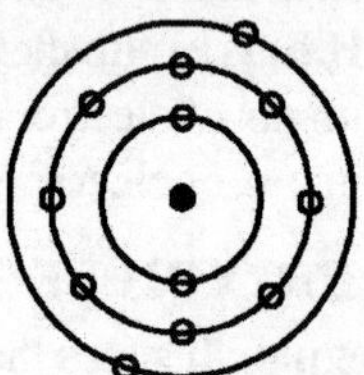

   (a) Chlorine (b) Magnesium
   (c) Calcium (d) Helium
10. Number of protons, neutrons and electrons in the element ${}_{89}X^{231}$ is
   (a) 89, 89, 242 (b) 89, 142, 89
   (c) 89, 71, 89 (d) 89, 231, 89
11. Chlorine exists in two isotopic forms, C1-37 and C1-35 but its atomic mass is 35.5. This indicates the ratio of C1-37 and C1-35 is approximately
   (a) 1 : 2 (b) 1 : 1
   (c) 1 : 3 (d) 3 : 1
12. Boron has two stable isotopes, ^{10}B (19%) and ^{11}B (81%). Average atomic weight for boron in the periodic table is
   (a) 10.8 (b) 10.2
   (c) 11.2 (d) 10.0
13. An element, X has the following isotopic composition :

$^{200}X : 90\%$ $^{199}X : 8.0\%$ $^{202}X : 2.0\%$

The weighted average atomic mass of the naturally occuring element X is closest to

(a) 201 amu (b) 202 amu

(c) 199 amu (d) 200 amu

14. Which one of the following constitutes a group of the isoelectronic species?

(a) C_2^{2-}, O_2^-, CO, NO

(b) $CN^-, N_2, O_2^{2-}, C_2^{2-}$

(c) $NO^+, C_2^{2-}, CN^-, N_2$

(d) N_2, O_2^-, NO^+, CO

15. If travelling at same speeds, which of the following matter waves have the shortest wavelength?

(a) Electron

(b) Alpha particle (He^{2+})

(c) Neutron

(d) Proton

16. Which of the following statements do not form a part of Bohr's model of hydrogen atom ?

(a) Energy of the electrons in the orbits are quantized

(b) The electron in the orbit nearest the nucleus has the lowest energy

(c) Electrons revolve in different orbits around the nucleus

(d) The position and velocity of the electrons in the orbit cannot be determined simultaneously.

17. Number of unpaired electrons in N^{2+} is

(a) 2 (b) 0

(c) 1 (d) 3

18 An ion has 18 electrons in the outermost shell, it is

(a) Cu^+ (b) Th^{4+}

(c) Cs^+ (d) K^+

19. In a given atom no two electrons can have the same values for all the four quantum numbers. This is called

(a) Hund's Rule

(b) Aufbau principle

(c) Uncertainty principle

(d) Pauli's Exclusion principle.

20. The energy of an electron in the nth Bohr orbit of hydrogen atom is

(a) $-\frac{13.6}{n^4}eV$ (b) $-\frac{13.6}{n^3}eV$

(c) $-\frac{13.6}{n^2}eV$ (d) $-\frac{13.6}{n}eV$

21. The electron was shown experimentally to have wave properties by

(a) de Broglie

(b) Davisson and Germer

(c) N. Bohr

(d) Schrodinger.

22. The orbitals are called degenerate when

(a) they have the same wave functions

(b) they have the same wave functions but different energies

(c) they have different wave functions but same energy

(d) they have the same energy

23. The ion that is isoelectronic with CO is

(a) CN^- (b) O_2^+

(c) O_2^- (d) N_2^+

24. The orientation of an atomic orbital is governed by

(a) Spin quantum number

(b) Magnetic quantum number

(c) Principal quantum number

(d) Azimuthal quantum number

25. Maximum number of electrons in a subshell of an atom is determined by the following:

(a) $2l+1$ (b) $4l-2$

(c) $2n^2$ (d) $4l+2$

26. Which of the following electronic configurations is wrong?

(a) Be(3) = 2, 1 (b) O(8) = 2, 6

(c) S(16) = 2, 6, 8 (d) P(15) = 2, 8, 5

27. The orbital diagram in which both the Pauli's exclusion principle and Hund's rule are violated, is

(a)  (b)  

(c)  (d)

28. Which one of the following set of quantum numbers is not possible for 4p electron?

(a) $n=4, l=1, m=-1, m_s=+\frac{1}{2}$

(b) $n=4, l=1, m=0, m_s=+\frac{1}{2}$

(c) $n=4, l=1, m=2, m_s=+\frac{1}{2}$

(d) $n=4, l=1, m=-1, m_s=-\frac{1}{2}$

29. In the photo-electron emission, the energy of the emitted electron is
(a) greater than the incident photon
(b) same as than of the incident photon
(c) smaller than the incident photon
(d) proportional to the intensity of incident photon.

30. The spectrum of He is expected to be similar to that
(a) H (b) Li^+
(c) Na (d) He^+

31. If E_1, E_2, and E_3 represent respectively the kinetic energies of an electron and an alpha particle and a proton each having same de-broglie wavelength then
(a) $E_1 > E_3 > E_2$ (b) $E_2 > E_3 > E_1$
(c) $E_1 > E_2 > E_3$ (d) $E_1 = E_2 = E_3$

32. The first emission line in the atomic spectrum of hydrogen in the Balmer series appears at

(a) $\frac{9R}{400}$ cm^{-1} (b) $\frac{7R}{144}$ cm^{-1}

(c) $\frac{3R}{4}$ cm^{-1} (d) $\frac{5R}{36}$ cm^{-1}

33. The third line of the Balmer series, in the emission spectrum of the hydrogen atom, is due to the transition from the
(a) fourth Bohr orbit to the first Bohr orbit
(b) fifth Bohr orbit to the second Bohr orbit
(c) sixth Bohr orbit to the third Bohr orbit
(d) seventh Bohr orbit to the third Bohr orbit

34. In a Bohr model of an atom, when an electron jumps from $n = 3$ to $n = 1$, how much energy will be emitted?
(a) 2.15×10^{-11} ergs (b) 2.389×10^{-12} ergs
(c) 0.239×10^{-10} ergs (d) 0.1936×10^{-10} ergs

35. The radius of hydrogen atom in the ground state is 0.53Å. The radius of Li^{2+} ion (atomic number = 3) in a similar state is
(a) 0.17 Å (b) 1.06 Å
(c) 0.53 Å (d) 0.265 Å

36. The uncertainty in the momentum of an electron is 1.0×10^{-5} kg ms^{-1}. The uncertainty in its position will be ($h = 6.62 \times 10^{-34}$ kg m^{2}s^{-1})
(a) 1.05×10^{-26} m (b) 1.05×10^{-28} m
(c) 5.27×10^{-30} m (d) 5.25×10^{-28} m

37. The electrons, identified by quantum numbers n and l (i) $n=4, l=1$ (ii) $n=4, l=0$ (iii) $n=3, l=2$ (iv) $n=3, l=1$ can be placed in order of increasing energy, from the lowest to highest, as
(a) (iv) < (ii) < (iii) < (i) (b) (ii) < (iv) < (i) < (iii)
(c) (i) < (iii) < (ii) < (iv) (d) (iii) < (i) < (iv) < (ii)

38. A 5f orbital has
(a) one spherical node
(b) two spherical nodes
(c) three spherical nodes
(d) four spherical nodes

39. The correct order of increasing energy of atomic orbitals is
(a) $5p < 4f < 6s < 5d$ (b) $5p < 6s < 4f < 5d$
(c) $5p < 5d < 4f < 6s$ (d) None of these

40. If the nitrogen atom had electronic configuration $1s^7$ it would have energy lower than that of the normal ground state configuration $1s^2\ 2s^2\ 2p^3$ because the electrons would be closer to the nucleus. Yet $1s^7$ is not observed. It violates
(a) Heisenberg's uncertainty principle
(b) Hund's rule
(c) Pauli exclusion principle
(d) Bohr postulate of stationary orbits

41. The number of spherical nodes in 3p orbitals are
(a) one (b) three
(c) none (d) two

42. If r is the radius of the first orbit, the radius of nth orbit of H-atom is given by
(a) rn^2 (b) rn
(c) r/n (d) $r^2 n^2$

43. The maximum number of electrons in a subshell is given by the expression
(a) $4l-2$ (b) $4l+2$
(c) $2l+2$ (d) $2n^2$

44. The total number of electrons that can be accommodated in all the orbitals having principal quantum number 2 and azimuthal quantum number 1 is
(a) 2 (b) 4
(c) 6 (d) 8

45. For azimuthal quantum number ℓ = 3, the maximum number of electrons will be
(a) 2 (b) 6
(c) 0 (d) 14.

46. The order of filling of electrons in the orbitals of an atom will be
(a) 3d, 4s, 4p, 4d, 5s (b) 4s, 3d, 4p, 5s, 4d
(c) 5s, 4p, 3d, 4d, 5s (d) 3d, 4p, 4s, 4d, 5s

47. Which of the following species has four lone pairs of electrons ?
(a) I (b) O^-
(c) Cl^- (d) He

48. If ionization potential for hydrogen atom is 13.6 eV, then ionization potential for He^+ will be
(a) 54.4 eV (b) 6.8 eV
(c) 13.6 eV (d) 24.5 eV

49. For which one of the following sets of four quantum numbers, an electron will have the heighst energy?

	n	l	m	s
(a)	3	2	1	1/2
(b)	4	2	–1	1/2
(c)	4	1	0	–1/2
(d)	5	0	0	–1/2

50. If electron has spin quantum number + 1/2 and a magnetic quantum number – 1, it cannot be present in
(a) d-orbital (b) f-orbital
(c) p-orbital (d) s-orbital.

51. Uncertainty in position of an electron (mass = 9.1×10^{-28} g) moving with a velocity of 3×10^4 cm/s accurate upto 0.001% will be (use $h/4\pi$) in uncertainty expression where h = 6.626×10^{-27} erg-second)
(a) 1.93 cm (b) 3.84 cm
(c) 5.76 cm (d) 7.68 cm

52. The momentum of a particle having a de Broglie wavelength of 10^{-17} metres is (Given $h = 6.625 \times 10^{-34}$ Js)
(a) 3.3125×10^{-7} kg ms^{-1}
(b) 26.5×10^{-7} kg ms^{-1}
(c) 6.625×10^{-17} kg ms^{-1}
(d) 13.25×10^{-17} kg ms^{-1}

53. The Bohr orbit radius for the hydrogen atom (n = 1) is approximately 0.530 Å. The radius for the first excited state (n = 2) orbit is (in Å)
(a) 0.13 (b) 1.06
(c) 4.77 (d) 2.12

54. The position of both, an electron and a helium atom is known within 1.0 nm. Further the momentum of the electron is known within 5.0×10^{-26} kg ms^{-1}. The minimum uncertainty in the measurement of the momentum of the helium atom is
(a) 50 kg ms^{-1} (b) 80 kg ms^{-1}
(c) 8.0×10^{-26} kg ms^{-1} (d) 5.0×10^{-26} kg ms^{-1}

55. According to Bohr's theory the energy required for an electron in the Li^{2+} ion to be emitted from n = 2 state is (given that the ground state ionization energy of hydrogen atom is 13.6 eV)
(a) 61.2 eV (b) 13.6 eV
(c) 30.6 eV (d) 10.2 eV

56. If the energy of a photon is given as : = 3.03×10^{-19} J then, the wavelength (λ) of the photon is :
(a) 6.56 nm (b) 65.6 nm
(c) 656 nm (d) 0.656 nm

57. Set of isoelectronic species is
(a) N_2, CO_2, CN^-, O (b) N_2, H_2S, CO
(c) N_2, CO, CN^-, O_2^{+2} (d) Ca, Mg, Cl

58. The following quantum numbers are possible for how many orbital (s) n = 3, l = 2, m = +2 ?
(a) 1 (b) 3
(c) 2 (d) 4

59. In hydrogen atom, energy of first excited state is –3.4 eV. Find out KE of the same orbit of Hydrogen atom
(a) +3.4 eV (b) +6.8 eV
(c) – 13.6 eV (d) + 13.6 eV

60. The value of Planck's constant is 6.63×10^{-34} Js. The velocity of light is 3.0×10^8 m s^{-1}. Which value is closest to the wavelength in nanometers of a quantum of light with frequency of 8×10^{15} s^{-1}?
(a) 3×10^7 (b) 2×10^{-25}
(c) 5×10^{-18} (d) 4×10^1

Classification of Elements and Periodicity in Properties

CONCEPTS

DEVELOPMENT OF PERIODIC TABLE

Dobereiner's Law of Triads

When elements having similar properties are placed in increasing order of atomic weight in group of three, then atomic weight of middle element is arithmetic mean of other two elements. e.g.- **Ca, Sr, Ba** and **Cl, Br, I**. But all elements can't be classified like this.

Newlands Law of Octaves

When elements are arranged in increasing order of atomic weights, every eighth element has properties similar to those of first element just like every eighth note of an octave of music.

Lother Mayer Volume Curve

When a **graph** is plotted **between atomic volumes and atomic weights** of elements, the **element with similar properties occupy similar position on the curve**.

Telluric Helix

It is the idea of periodic arrangements of elements (was given by **Chancourtois**).

Mendeleev's Periodic Table :

(i) **Based on atomic weights.**

(ii) **Mendeleev's periodic law :** The physical and chemical properties of elements are **periodic function** of their **atomic weights**.

(iii) Only **63 elements** were known at that time.

(iv) It has **9 vertical columns** called **group I – VIII and zero**. Except VIII group all groups are sub divided into 2 subgroups A and B.

(v) It has 7 **horizontal columns** called **periods 1 – 7**.

(vi) Order of atomic weights was changed at places to justify properties of element.

(vii) It had vacant places for elements which were discovered later. e.g.- **eka aluminium (now known as gallium) and eka silicon (now known as germanium).**

Moseley gave **modern periodic law** which states that **chemical and physical properties of elements** are **periodic function of their atomic numbers.**

MODERN OR LONG FORM OF PERIODIC TABLE (BOHR'S TABLE)

Old no.	IA	IIA											IIIA	IVA	VA	VIA	VIIA	VIIIA
New no.	1	2											13	14	15	16	17	18
			IIIB	IVB	VB	VIB	VIIB	VIII	VIII	VIII	IB	IIB						
			3	4	5	6	7	8	9	10	11	12						
	s-block (ns^{1-2})		d-block [$(n-1)d^{1-10}$, (ns^{1-2})]										p-block (ns^2, np^{1-6})					

f-block $(n-2)f^{1-14}, (n-1)d^{0-1}, ns^2$

(i) It has 7 – **horizontal rows** – called **periods**.

(ii) It has **18 vertical rows** – called **groups**.

(iii) **Representative elements :** s and p block.

(iv) **Transition elements :** d-block.

(v) **Inner transition elements :** f-block.

(vi) Alkali metals - group 1
Alkaline earth metals - group 2
Pnicogens - group 3
Chalcogens - group 16
Halogens - group 17
Noble gases or **aerogens - group 18**

(vii) **Typical element** - members **third period**.

(viii) **Bridge element** - members of second period.

(ix) **Lanthanides** - Elements of at. no. 58-71

(x) **Actanides** - Elements of at. no. 90-103

(xi) **Transuranics** - All elements after Uranium (Atomic Number 92).

NOMENCLATURE OF ELEMENTS WITH Z > 100

(I)

DIGIT	0	1	2	3	4	5	6	7	8	9
Root	nil	un.	bi.	tri.	quad.	pent.	hex.	sept.	oct.	enn.
Abbre-viation	n	u	b	t	q	p	h	s	o	e

e.g.- element with atomic no. 102 - Unnilbium (Unb)

(ii) **Common names** of some elements
101 - Mendelevium, 104 - Rutherfordium, 107- Bohrium

(iii) **Name of elements on countries :**
Ru - Ruthenium (Country - Ruthenia)
Ge - Germanium (Country - Germany)
Po - Polonium (Country - Poland)
Am - Americium (Country - America)

PERIODIC TRENDS IN PHYSICAL PROPERTIES:

Atomic Size

Distance between nucleus of the atom and outermost electron

(i) Absolute value of atomic size cannot be determined because exact position of electron cannot located.

(ii) **Covalent radius** (r_{cov})

(a) For a homoatomic molecule, $r_{(cov)}$ = 1/2 (distance between two covalently bonded atoms)

(b) For hetrodiatomic molecule in which electronegativity remains approximately same. $d_{A-B} = \gamma_A + \gamma_B$

(c) For a heteronuclear diatomic molecule AB, while difference between the electronegativity values of atom A and B is relatively larger,

$$d_{A-B} = \gamma_A + \gamma_B - 0.09(X_A - X_B)$$

where X_A and X_B are electronegativity values of high electronegative element A and less electro negative element B respectively.

(iii) **van der Waal's radius** ($r_{v.\ Waal}$)

$r_{(v.\ Waal)}$ = 1/2 (distance between nuclei of two non bonded isolated atoms or adjacent atoms belonging to neighbouring molecules of an element)

(iv) **Metallic radius** is half of distance between centres of nuclei of two adjacent atoms in metallic crystal.

(v) vander Waal's radius > Metallic radius > Covalent radius

(vi) Atomic radius decreases along period and increases along a group.

(vii) Ionic radius :
It may be defined as the effective distance from the centre of the nucleus of the ion up to which it exerts its influence on its electron cloud. Similar to the covalent radii ionic radii of cation and anion increases on moving down the group.

➢ Size of anion > Size of atom > Size of cation

(a) **Cation is always smaller in size than corresponding neutral atom.**
$M > M^+ > M^{2+} > M^{3+}$

(b) **Anion is always larger in size than corresponding neutral atom.**
$A^{3-} > A^{2-} > A^- > A$

(c) **Isoelectronic species** have same no. of electrons but different number of protons. For isoelectronic species.

(a) Cation with highest positive charge is smallest in size.

(b) Anion with highest negative charge is largest in size.

Ionisation Energy/Ionisation Enthalpy (I.E.)

(i) It is amount of energy required to take out most loosely bounded electron from isolated gaseous atom to change it into gaseous cation. **Its value is positive** (because energy is supplied).

$$A(g) + I.E. \longrightarrow A^{+}(g) + e^{-}$$

(ii) **$I.E_3 > I.E_2 > I.E_1$**. (∵ of increased force of attraction after removal of electron)

(iii) **Units – eV/atom or kcal.mol^{-1} or $kJmol^{-1}$**

(iv) **Factors effecting ionisation enthalpy**

(a) I.E ∝ Effective nuclear charge

(b) I.E ∝ ^{1}/ atomic radius

(c) Penetrating power of electron (s > p > d > f)

(d) I.E. ∝ ^{1}/Shielding effect.

(e) Stable configurations (half and fulfilled orbitals)

(v) For second period elements correct increasing order of ionisation energies is as follows :

$IE_1 = Li < B < Be < N < C < O < N < F < Ne$

$IE_2 = Be < C < B < N < F < O < Ne < Li$

(vi) For third period elements correct increasing order of ionisation energies is as follows

$IE_1 = Na < Al < Mg < Si < S < P < Cl < Ar$

$IE_2 = Mg < Si < Al < P < S < Cl < Ar < Na$

Electron Affinity (EA)

(i) It is amount of energy released when gaseous atom accepts an electron to form gaseous anion.

$$X(g) + e^{-} \longrightarrow X^{-}(g) + E.A.$$

(ii) EA_1 is **negative** but EA_2 and successive electron affinities are **positive** because it becomes difficult to add electron to anion.

$$O(g) + e^{-} \longrightarrow O^{-}(g), \ EA_1 = -141\ kJmol^{-1}$$

$$O(g) + e^{-} \longrightarrow O^{2-}(g), \ EA_2 = +780\ kJmol^{-1}.$$

(iii) **Factors effecting electron affinity :**

(a) E.A. ∝ ^{1}/Atomic size.

(b) E.A. ∝ ^{1}/Nuclear charge.

(c) Stable electronic configuration. (half filled and filled orbitals)

(d) E.A. ∝ ^{1}/Shielding effect.

Electronegativity (EN)

(i) It is **tendency of an atom to attract shared pair of electrons in a molecule**.

(ii) The value of electronegativity depends on

(a) Charge on the nucleus

(b) Size of atom

(c) Shielding effect.

(iii) **Measurement of electronegativity :**

(a) **Mullikan scale**

Electronegativity $= \frac{IE + EA}{2}$ where I.E. and EA are expressed in eV.

(b) **Pauling scale :** If two atoms, A and B, having different electronegativity values, get bonded to form a molecule, AB, then the bond between A and B in A – B will have both covalent and ionic properties.

Δ_{A-B} = Observed bond energy – Energy of 100% covalent or $\Delta_{A-B} = D - E_{A-B}$

where, D = Observed bond energy

E_{A-B} = Bond energy of pure covalent bond of A – B.

The value of E_{A-A} and E_{B-B} is $E_{A-B} = 1/2\ [E_{A-A} + E_{B-B}]$

$= 0.208\sqrt{\Delta_{A-B}} = X_A - X_B$

where $X_A > X_B$ or $0.043 \times \Delta_{A-B} = (X_A - X_B)^2$

PERIODIC TRENDS IN CHEMICAL PROPERTIES:

It is defined as the number of atoms of hydrogen (or twice the number of oxygen) required by one atom of the element to form a stable compound.

The valency increases from 1 to 8 along a period.On moving down a group valency remain same.

- Among representative elements, the valence is either equal to the number of electrons in the outermost orbitals or eight minus the number of outermost electrons.
- **Digonal relationship:** It is observed that first element of a group on the left has similar properties (physical and chemical) to second element to its right diagonally placed. This is referred as diagonal relationship.

First member of every group differs from rest of the members due to

(i) very small size.

(ii) high electronegativity.

(iii) absence of vacant d-orbitals.

- For the same alkali metal, the melting and boiling points decrease regularly in the order Fluoride > chloride > bromide > iodide
- For the same metal lattice energy decreases with the decrease in electronegativity of the halogen.

For example,

Metal halide	NaF	NaCl	NaBr	NaI
Lattice energy (kJ/mole)	910	769	732	682
Melting point (K)	1298	1081	1028	934

- Amongst ionic halides, melting point decrease as lattice energy decreases as we move down the group i.e.,

NaCl	>	KCl	>	RbCl	>	CsCl
1081K		1045		990		918 K

- The solubilities of alkali metal fluorides in water increase regularly from lithium to caesium

Metal fluoride	LiF	NaF	KF	RbF	CsF
Solubility in water at 298 K (gm/litre)	2.7	42	1020	1310	3700

- LiCl has much higher solubility in water than NaCl. This is due to small size of Li^+ ion and much higher hydration energy. However from NaCl to CsCl, solubility in water increase regularly due to decrease in their lattice energy.
- The carbonates and bicarbonates of alkali metals are quite soluble in water and their solubility increase as we move down the group from Li to Cs.
- Amongst alkali carbonates, except Li_2CO_3, all are quite stable to heat.
- Stabilities of the carbonates of alkaline earth metals incrases from $BeCO_3$ to $BaCO_3$ primarily due of alkaline earth metal carbonates.
- Like carbonates, the solubilities of sulphates of alkaline earth metals decrease down the group from Be to Ba.

$BeSO_4$	$MgSO_4$	$CaSO_4$	$SrSO_4$	$BaSO_4$
(Fairly soluble)	soluble	sparingly	Insoluble	Insoluble

- Hydroxide of alkali metals (MOH) are appreciably soluble in water. Their solubility further increase as we move down the group due to decrease in lattice energy. Thus,

LiOH NaOH KOH RbOH CsOH

$\xrightarrow{\hspace{6cm}}$

Solubility in water increase

- Basic strength of alkaline earth metal hydroxides increases as we move down the group from $Be(OH)_2$ to $Ba(OH)_2$.

$Be(OH)_2$	$Mg(OH)_2$	$Ca(OH)_2$	$Sr(OH)_2$	$Ba(OH)_2$
Amphoteric	Weak base	moderate base	Moderate base	Strong base

$\xrightarrow{\hspace{6cm}}$

Base strength increase

Exercise

1. Cl, Br, I, if this is Dobereiner's triad and the atomic masses of Cl and I are 35.5 and 127 respectively the atomic mass of Br is –
(a) 162.5 (b) 91.5
(c) 81.25 (d) 45.625
2. Newlands could classify elements only upto –
(a) copper (b) chlorine
(c) calcium (d) chromium
3. If the two members of a Dobereiner triad are phosphorus and antimony, the third member of this triad is –
(a) arsenic (b) sulphur
(c) iodine (d) calcium
4. The most significant contribution towards the development of periodic table was made by
(a) Mendeleev (b) Avogadro
(c) Dalton (d) Cavendish
5. The statement that is not correct for the periodic classification of elements is
(a) The properties of elements are the periodic functions of their atomic number.
(b) Non-metallic elements are lesser in number than metallic elements.
(c) The first ionization energies of elements along a period do not vary in a regular manner with increase in atomic number.
(d) For transition elements the d-sub shells are filled with electrons monotonically with increase in atomic number.
6. The statement that is not correct for periodic classification of elements is :
(a) The properties of elements are periodic function of their atomic numbers.
(b) Non-metallic elements are less in number than metallic elements.
(c) For transition elements, the $3d$-orbitals are filled with electrons after $3p$-orbitals and before $4s$-orbitals.
(d) The first ionisation enthalpies of elements generally increase with increase in atomic number as we go along a period.
7. According to the Periodic Law of elements, the variation in properties of elements is related to their
(a) nuclear masses
(b) atomic numbers
(c) nuclear neutron-proton number ratios
(d) atomic masses
8. Mendeleev classified elements in –
(a) increasing order of atomic groups
(b) eight periods and eight groups
(c) seven periods and nine groups
(d) eight periods and seven groups
9. Noble gases were included in Mendeleev's periodic table in the –
(a) 1st group (b) 7th group
(c) 8th group (d) none of these
10. In the modern periodic table one of the following does not have appropriate position –
(a) transition elements
(b) inert gases
(c) inner transition elements
(d) halogens
11. The screening effect of 'd' electrons is
(a) Much less than s- electrons
(b) Much more than s- electrons
(c) Equal to s- electrons
(d) Equal to p- electrons
12. Consider the following changes
$$A \rightarrow A^+ + e^- : E_1 \text{ and } A^+ \rightarrow A^{2+} + e^- : E_2$$
The energy required to pull out the two electrons are E_1 and E_2 respectively. The correct relationship between two energies would be
(a) $E_1 < E_2$ (b) $E_1 = E_2$
(c) $E_1 > E_2$ (d) $E_1 \geq E_2$
13. Pauling's electronegativity values for elements are useful in predicting
(a) Polarity of the molecules
(b) Position in the E.M.F. series
(c) Coordination numbers
(d) Dipole moments.
14. In the periodic table, with the increase in atomic number, the metallic character of an element
(a) Decreases in a period and increases in a group
(b) Increases in a period and decreases in a group
(c) Increases both in a period and the group
(d) Decreases in a period and the group.

15. If the atomic number of an element is 33, it will be placed in the periodic table in the
    (a) First group (b) Third group
    (c) Fifth group (d) Seventh group.
16. In the periodic table from left to right in a period, the atomic volume
    (a) Decreases
    (b) Increases
    (c) Remains same
    (d) First decrease then increases
17. An atom has electronic configuration $1s^2\ 2s^2\ 2p^6\ 3s^2\ 3p^6\ 3d^3\ 4s^2$, you will place it in which group?
    (a) Fifth (b) Fifteenth
    (c) Second (d) Third
18. Ionic radii are
    (a) inversely proportional to effective nuclear charge
    (b) inversely proportional to square of effective nuclear charge
    (c) directly proportional to effective nuclear charge
    (d) directly proportional to square of effective nuclear charge
19. Which of the following electronic configuration an atom has the lowest ionisation enthalpy?
    (a) $1s^2\ 2s^2\ sp^3$ (b) $1s^2\ 2s^2\ 2p^5\ 3s^1$
    (c) $1s^2\ 2s^2\ 2p^6$ (d) $1s^2\ 2s^2\ 2p^5$
20. The atoms of elements belonging to the same group of periodic table have the same –
    (a) number of protons
    (b) number of electrons
    (c) number of neutrons
    (d) number of electrons in the outermost shell
21. The elements with atomic numbers 2, 10, 18, 36, 54 and 86 are all –
    (a) halogens (b) noble gases
    (c) noble metals (d) light metals
22. The number of elements in each of the inner transition series is –
    (a) 2 (b) 8
    (c) 10 (d) 14
23. The noble gases are unreactive because
    (a) they react with sodium
    (b) they have a full outer shell of electrons
    (c) they have a half outer shell of neutrons
    (d) they are too thin
24. Which of the following element is not in the liquid state?
    (a) Hg (b) Li
    (c) Ga (d) Br
25. The electron affinity for the inert gases is –
    (a) zero (b) high
    (c) begative (d) positive
26. Which is metalloid –
    (a) Pb (b) Sb
    (c) Si (d) Zn
27. To which block is related an element having electronic configuration $1s^2\ 2s^22p^6\ 3s^23p^63d^{10}\ 4s^1$ in the periodic table–
    (a) s–block (b) p–block
    (c) d–block (d) f–block
28. Ionisation energy decreases down the group due to
    (a) increase in charge
    (b) increase in atomic size
    (c) decrease in size
    (d) decreasing in shielding effect
29. Which form coloured salts?
    (a) Non-metals
    (b) Metals
    (c) p-block elements
    (d) Transitional elements
30. The element with the outer electronic configuration $3d^64s^2$ is a :
    (a) Metalloid (b) Non-metal
    (c) Transition metal (d) Noble gas
31. Consider the isoelectronic species, Na^+, Mg^{2+}, F^- and O^{2-}. The correct order of increasing length of their radii is ______.
    (a) $F^- < O^{2-} < Mg^{2+} < Na^+$
    (b) $Mg^{2+} < Na^+ < F^- < O^{2-}$
    (c) $O^{2-} < F^- < Na^+ < Mg^{2+}$
    (d) $O^{2-} < F^- < Mg^{2+} < Na^+$
32. Consider the following statements
    I. The radius of an anion is larger than that of the parent atom.
    II. The ionization energy generally increases with increasing atomic number in a period.
    III. The electronegativity of elements increases on moving down across a group.

    Which of the above statements is/are correct?
    (a) I alone (b) II alone
    (c) I and II (d) II and III
33. Which of the following is not an actinoid ?
    (a) Curium ($Z = 96$) (b) Californium ($Z = 98$)
    (c) Uranium ($Z = 92$) (d) Terbium ($Z = 65$)

34. The formation of the oxide ion, $O^{2-}(g)$, from oxygen atom requires first an exothermic and then an endothermic step as shown below :

$$O(g) + e^- \longrightarrow O^-(g); \Delta H^\ominus = -141 \text{ kJ mol}^{-1}$$

$$O^-(g) + e^- \longrightarrow O^{2-}(g); \Delta H^\ominus = +780 \text{ kJ mol}^{-1}$$

Thus process of formation of O^{2-} in gas phase is unfavourable even though O^{2-} is isoelectronic with neon. It is due to the fact that,
(a) oxygen is more electronegative.
(b) addition of electron in oxygen results in larger size of the ion.
(c) electron repulsion outweighs the stability gained by achieving noble gas configuration.
(d) O^- ion has comparatively smaller size than oxygen atom.

35. Electronic configurations of four elements A, B, C and D are given below :
(i) $1s^2\,2s^2\,2p^6$ (ii) $1s^2\,2s^2\,2p^4$
(iii) $1s^2\,2s^2\,2p^6\,3s^1$ (iv) $1s^2\,2s^2\,2p^5$
Which of the following is the correct order of increasing tendency to gain electron ?
(a) A<C<B<D (b) A<B<C<D
(c) D<B<C<A (d) D<A<B<C

36. The period number in the long form of the periodic table is equal to
(a) magnetic quantum number of any element of the period.
(b) atomic number of any element of the period.
(c) maximum Principal quantum number of any element of the period.
(d) maximum Azimuthal quantum number of any element of the period.

37. Which of the following option is incorrect with respect to ionic radii ?
(a) $Ti^{4+} < Mn^{2+}$ (b) $^{35}Cl^- < {}^{37}Cl^-$
(c) $K^+ > Cl^{-1}$ (d) $P^{3+} > P^{5+}$

38. The incorrect statement among the following is
(a) the first ionization potential of Al is less than the first ionization potential of Mg
(b) The second ionization potential of Mg is greater than the second ionization potential of Na
(c) The first ionization potential of Na is less than the first ionization potential of Mg
(d) The third ionization potential of Mg is greater than the third ionization potential of Al.

39. Which of the following order is wrong?
(a) $NH_3 < PH_3 < AsH_3$ — Acidic
(b) Li < Be < B < C — IE_1
(c) $Al_2O_3 < MgO < Na_2O < K_2O$ — Basic
(d) $Li^+ < Na^+ < K^+ < Cs^+$ — Ionic radius

40. The valence shell of element A contains 3 electrons while the valence shell of element B contains 6 electrons . If A combines with B, the probable formula of the compound formed will be
(a) AB_2 (b) A_2B
(c) A_2B_3 (d) A_3B_2

41. An element having electronic configuration $1s^2\,2s^2\,2p^6\,3s^2\,3p^6\,4s^1$ forms
(a) Acidic oxide (b) Basic oxide
(c) Amphoteric oxide (d) Neutral oxide

42. Among Al_2O_3, SiO_2, P_2O_3 and SO_2 the correct order of acidic strength is
(a) $Al_2O_3 < SiO_2 < SO_2 < P_2O_3$
(b) $SiO_2 < SO_2 < Al_2O_3 < P_2O_3$
(c) $SO_2 < P_2O_3 < SiO_2 < Al_2O_3$
(d) $Al_2O_3 < SiO_2 < P_2O_3 < SO_2$

43. Following statements regarding the periodic trends of chemical reactivity of the alkali metals and the halogens are given. Which of these statements gives the correct picture?
(a) Chemical reactivity increases with increase in atomic number down the group in both the alkali metals and halogens
(b) In alkali metals the reactivity increases but in the halogens it decreases with increase in atomic number down the group
(c) The reactivity decreases in the alkali metals but increases in the halogens with increase in atomic number down the group
(d) In both the alkali metals and the halogens the chemical reactivity decreases with increase in atomic number down the group

44. The electronic configuration of four elements are given below. Which elements does not belong to the same family as others ?
(a) $[Xe]4f^{14}5d^{10}1s^2$ (b) $[Kr]4d^{10}5s^2$
(c) $[Ne]3s^23p^5$ (d) $[Ar]\,3d^{10}\,4s^2$

45. Elements of which of the following groups will form anions most readily ?
(a) Oxygen family (b) Nitrogen family
(c) Halogens (d) Alkali metals

46. Which of the following sets has strongest tendency to form anions ?
(a) Ga, In, Tl (b) Na, Mg, Al
(c) N, O, F (d) V, Cr, Mn

47. Na^+, Mg^{++}, Al^{3+} and Si^{4+} are isoelectronic. The order of their ionic size is
(a) $Na^+ > Mg^{++} < Al^{3+} < Si^{4+}$
(b) $Na^+ < Mg^{++} > Al^{3+} > Si^{4+}$
(c) $Na^+ > Mg^{++} > Al^{3+} > Si^{4+}$
(d) $Na^+ < Mg^{++} > Al^{3+} < Si^{4+}$

48. One of the characteristic properties of non-metals is that they
(a) Are reducing agents
(b) Form basic oxides
(c) Form cations by electron gain
(d) Are electronegative

49. The electronic configuration of an element is $1s^2 2s^2 2p^6 3s^2 3p^3$. What is the atomic number of the element, which is just below the above element in the periodic table?
(a) 33 (b) 34
(c) 36 (d) 49

50. The element, with atomic number 118, will be
(a) alkali (b) noble gas
(c) lanthanide (d) transition element

51. Which one of the following ions will be the smallest in size?
(a) Na^+ (b) Mg^{2+}
(c) F^- (d) O_2^-

52. Which of the following does not represent the correct order of the properties indicated?
(a) $Ni^{2+} > Cr^{2+} > Fe^{2+} > Mn^{2+}$ (size)
(b) Sc > Ti > Cr > Mn (size)
(c) $Mn^{2+} > Ni^{2+} < Co^{2+} < Fe^{2+}$ (unpaired electron)
(d) $Fe^{2+} > Co^{2+} > Ni^{2+} > Cu^{2+}$ (unpaired electron)

53. The first ionization potentials (eV) of Be and B respectively are
(a) 8.29, 9.32 (b) 9.32, 9.32
(c) 8.29, 8.29 (d) 9.32, 8.29

54. Of the given electronic configurations for the elements, which electronic configuration indicates that there will be abnormally high difference in the second and third ionization energy for the element?
(a) $1s^2 2s^2 2p^6 3s^2$ (b) $1s^2 2s^2 2p^6 3s^1$
(c) $1s^2 2s^2 2p^6 3s^2 3p^1$ (d) $1s^2 2s^2 2p^6 3s^2 3p^2$

55. Correct order of first IP among following elements Be, B, C, N, O is
(a) B < Be < C < O < N (b) B < Be < C < N < O
(c) Be < B < C < N < O (d) Be < B < C < O < N

56. Among K, Ca, Fe and Zn the element which can form more than one binary compound with chlorine is
(a) Fe (b) Zn
(c) K (d) Ca

57. Which one of the following oxides is expected to exhibit paramagnetic behaviour?
(a) CO_2 (b) SiO_2
(c) SO_2 (d) ClO_2

58. Which one of the following arrangements represents the correct order of electron gain enthalpy (with negative sign) of the given atomic species?
(a) S < O < Cl < F (b) Cl < F < S < O
(c) F < Cl < O < S (d) O < S < F < Cl

59. Identify the correct order of the size of the following:
(a) $Ca^{2+} < K^+ < Ar < Cl^- < S^{2-}$
(b) $Ar < Ca^{2+} < K^+ < Cl^- < S^{2-}$
(c) $Ca^{2+} < Ar < K^+ < Cl^- < S^{2-}$
(d) $Ca^{2+} < K^+ < Ar < S^{2-} < Cl^-$

60. Which one of the following ionic species has the greatest proton affinity to form stable compound?
(a) NH_2^- (b) F^-
(c) I^- (d) HS^-

Chemical Bonding and Molecular Structure

CONCEPTS

CHEMICAL BOND :

Force which holds the atoms of molecule together. Atoms form **bond** to **decrease** the **energy and increase the stability**.

VALENCY :

Valency is termed as the number of chemicals bonds formed by an atom in a molecule.

LEWIS STRUCTURE :

The chemical symbols in which the valence electrons in molecules are represented as dots, are called Lewis structures. e.g. $_6C$ has electronic configuration [He] $2s^2$ $2p^2$. Its electron dot symbol shows 4 valence electrons as shown. $\cdot\dot{\underset{\cdot}{C}}\cdot$

CLASSIFICATION OF CHEMICAL BONDS :

On the basis of electronic valency theory and structure, chemical bonds can be classified as follows.

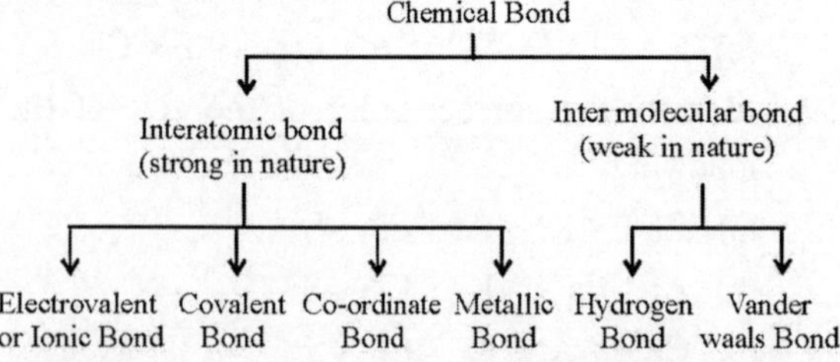

IONIC BONDS (EXPLAINED BY KOSSEL) :

(i) They are formed by transfer of electron from one atom to other.

(ii) **Favourable conditions for formation of ionic bond.**

(a) low *ionisation energy* of atom forming cation.

(b) high *electron affinity* of atom forming anion.

(c) high *lattice energy* of crystal formed.

(d) Factors influencing magnitude of lattice energy are as follows.

1. **Size of ions :** Smaller the size of the ions, lesser is the internuclear distance consequently the inter ionic attraction will be high and lattice energy will also be large.

2. **Charge on ions :** Larger the magnitudes of charge on the ions, greater will be the attractive forces between the ions. Consequently the lattice energy will be high.

Energy Steps Involved in Ionic Bond Formation:

(i) **Formation of cation :**

(a) $M_{(s)} + \text{Sublimation energy} \rightarrow M_{(g)}$
$\Delta H = + SE$

(b) $M_{(g)} + IP \rightarrow M^+_{(g)} + e^-$
$\Delta H = + IP$

(ii) **Formation of anion :**

(a) $\frac{1}{2}X_{2(g)} + \frac{1}{2}\text{Dissociation energy} \rightarrow X_{(g)}$
$\Delta H = +D/2$

(b) $X_{(g)} + e^- \rightarrow X^-_{(g)} + \text{Electron affinity}$
$\Delta H = -EA$

(iii) **Formation of ionic bond :**

$M^+ + X^- \rightarrow M^+X^- + \text{Lattice energy}$
$\Delta H = -\text{Lattice energy}$

Born – Haber equation :

Total

$$\Delta H = \underbrace{S.E. + I.P. + D/2}_{\text{Energy absorbed}} - \underbrace{E.A. - L.E.}_{\text{Energy released}}$$

Ionic bond is formed if : Total $\Delta H = -ve$ (energy released)

More the –ve value, more will be the strength of ionic bond

Characteristics of Ionic Compounds

(i) **Ionic bonds are non directional**

(ii) Good conductors of electricity in molten and aqueous state.

(iii) They have **high melting and boiling points**.

(iv) Highly soluble in water (polar solvent)

COVALENT BOND (EXPLAINED BY LEWIS)

(i) Formed by sharing of electrons. Covalent bond is further classified as polar and non-polar depending upon the fact whether the electron pair is shared unequally between the atoms or shared equally.

(ii) **Characteristics of covalent compounds :**

(a) The covalent bonds are directional and rigid.

(b) They have low boiling and melting points.

(c) Covalent compounds do not conduct electricity in fused state.

(d) Covalent substances are generally insoluble in polar solvents like water but soluble in non-polar solvents like benzene, ether, chloroform etc.

LEWIS STRUCTURE AND COVALENT BOND :

Electron dot structures, also known **Lewis structures of covalent molecules,** are written in accordance with octet rule. Accordingly all atoms in a formula will have a total of eight electrons by sharing in the valence shell except the H-atom (hydrogen will have only two electrons).

$$:\underset{\cdot\cdot}{\overset{\cdot\cdot}{Cl}}. + {}_{\times}\underset{\times\times}{\overset{\times\times}{Cl}}{}^{\times}_{\times} \longrightarrow :\underset{\cdot\cdot}{\overset{\cdot\cdot}{Cl}}{}^{\cdot}_{\times}\,\underset{\times\times}{\overset{\times\times}{Cl}}{}^{\times}_{\times} \text{ or } Cl-Cl$$

Chlorine

Formal Charge

Formal charge (F.C.) on an atom in a Lewis structure = total number of valence electrons in the free atom – total number of non bonding (lone pair) electrons – (1/2) [total number of bonding (shared) electrons]

OCTET RULE :

The *octet rule* is defined as the tendency of atoms of different elements to acquire 8 electrons in their outermost shell.

Exception to Octet Rule:

Violation of octet rule is observed in the formation of stable compounds like PCl_5, SF_6, IF_7, etc. In each of these molecules the central atom P,S and I have 10, 12 and 14 electrons respectively as shown below.

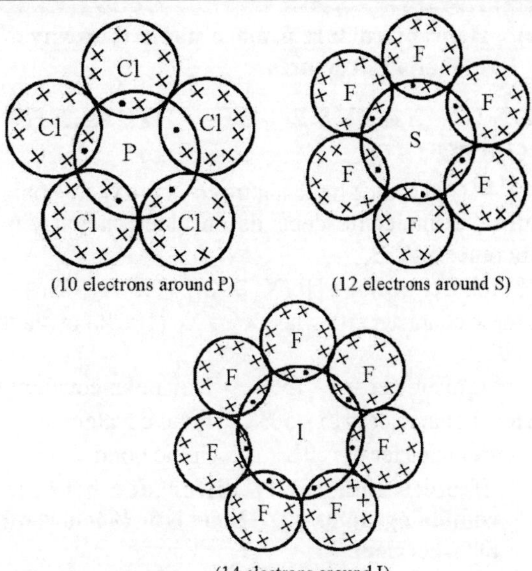

(10 electrons around P) (12 electrons around S)

(14 electrons around I)

CO-ORDINATE OR DATIVE BOND (PROPOSED BY SIDWICK)

(i) It is one sided sharing of electrons in which one atom donates (donor) electrons and other (acceptor) simply shares it.

(ii) Directional bond and exhibits stereo-isomerism.

(iii) Do not conduct electricity.

(iv) Their m.p. and b.p. are higher than purely covalent compounds and lower than ionic compounds.

(v) These are sparingly soluble in polar solvents like water but readily soluble in non-polar solvents.

METALLIC BOND

The simultaneous force of attraction between mobile electrons and kernels which hold the metal atoms together. Due to presence of **these bonds metals are lustrous, malleable, ductile and conductors**.

RESONANCE :

(i) When all the properties of molecule cannot be explained by one structure, then compound is said to be having more than one structure. All these structures are called **resonating structures or canonical structure or contributing structures**. The real structure of compound is intermediate of all resonating structures and is called resonance hybrid.

(ii) **Resonance energy** = Energy of resonance hybrid – Energy of most stable contributing structure.

(iii) **Hybrid structure is more stable than any of resonating structures.**

IONIC NATURE OF COVALENT COMPOUND.

In A—B molecule : Ionic nature $\propto X_A - X_B$ i.e. ionic nature of molecule depends on electronegativity difference of A–B.

if % Ionic character = $[16(X_A - X_B) + 3.5(X_A - X_B)^2$

% Ionic character = 0 when $X_A = X_B$ (100% covalent bond)

% Ionic character = 0 – 15%	Non polar covalent
% Ionic character = 15 – 50%	Polar covalent
% Ionic character = > 50%	Ionic bond

(i) If **electronegativity difference** between **combining atoms = 1.7. Bond is 50% ionic and 50% covalent.**

(ii) If electronegativity **difference > 1.7 bond is ionic.**

(iii) If electronegativity **difference < 1.7 bond is covalent.**

DIPOLE MOMENT (μ)

Dipole moment is defined as the product of charge on any one of the atom and the distance between the centre of two charges.

(i) $\mu = q \times d$

where, q = charge on any one covalently bonded atom, d = distance between covalently bonded atoms (bond length)

(ii) **Unit of dipole moment is Debye.**

$1\,D = 10^{-8}$ esu.

(iii) Dipole moment is a vector quantity (where its head is towards more electronegative atom).

$\overset{\longmapsto}{\text{HCl}}$

(iv) For polyatomic molecule dipole moment is resultant of vectors representing various bond moments.

(v) **Dipole moment helps in**

(a) **Predicting weather molecule is polar or non polar.**

(b) Predicting geometry of molecule ($\because$ molecule may have polar bonds but its dipole moment still may be zero).

(c) **Calculation of % of ionic character in a covalent bond**

% of ionic character in covalent bond

$$= \frac{\text{Actual dipole moment of bond}}{\text{Dipole moment of pure ionic bond}} \times 100$$

(d) **Used to distinguish between cis and trans isomers**. Cis isomers have higher dipole moment (due to less symmetry) as compared to trans isomer.

FAJANS RULE :

(i) All ionic compounds have some covalent character because of tendency of cation to polarize the anion.

(ii) Magnitude of covalent character depends on extent of polarisation caused by cation.

(iii) Polarizing power of cation $\propto \frac{1}{\text{size}} \propto$ charge on ion.

$\therefore$ smaller the size of cation more is its polarising power.

(iv) **Polarisability of anions $\propto$ size.**

$\therefore$ larger the size of anion more is its polarisability.

(v) Ions of d-block elements have more polarising power than that of corresponding s block element.

(vi) Polarisation capacity of pseudo inert gas config > Polarisation capacity of inert gas config.

VSEPR THEORY

The basic concept of this theory as suggested by Sidwick and Powell (1940) has provided a highly useful idea for predicting geometries of molecules . According to this model, the arrangement of bonds around the central atom depends upon the repulsions operating between pairs of electrons (bonded or non-bonded) arround the central atom; The main postulates of VSEPR theory are

(i) For polyatomic molecules containing 3 or more atoms, one of the atoms is called the central atom to which other atoms are linked.

(ii) The geometry of molecule depends upon the total number of valence shell electron pairs (bonded or non-bonded) present around the central atom and their repulsion due to relative sizes and shapes.

(iii) If the central atom is linked to similar atoms the repulsion between them are also similar. It gives the symmetrical shape to the molecule.

(iv) If the central atom is lined to atoms of different sizes or is surrounded by lone pairs as well as bond pairs of electrons, then the molecule has a distorted geometry. The relative order of repulsion between electron pairs is as follows.

Lone pair – Lone pair > Lone pair – Bond pair > Bond pair – Bond pair

Molecule type	No. of bonding pairs	No. of lone pairs	Total electron pairs	Arrangement of electron pairs	Shape (geometry)	Example
AB_2L	2	1	3	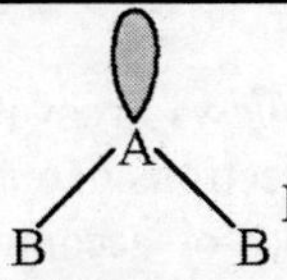 Trigonal planar	Bent or Vshaped	SO_2, O_3, $SnCl_2$
AB_3L	3	1	4	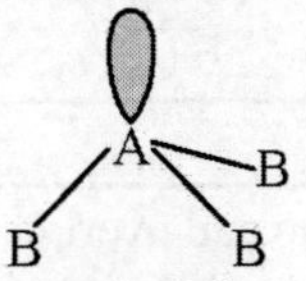 Tetrahedral	Pyramidal	NH_3
AB_2L_2	2	2	4	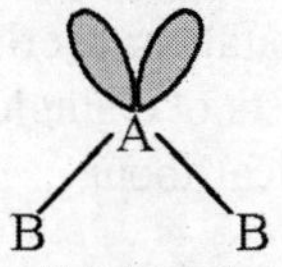 Tetrahedral	V shaped	H_2O
AB_4L	4	1	5	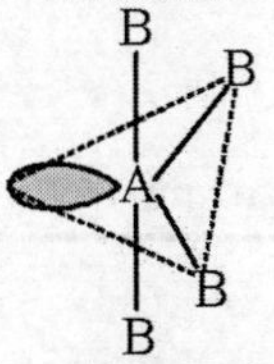 Trigonal bipyramidal	See saw	SF_4
AB_3L_2	3	2	5	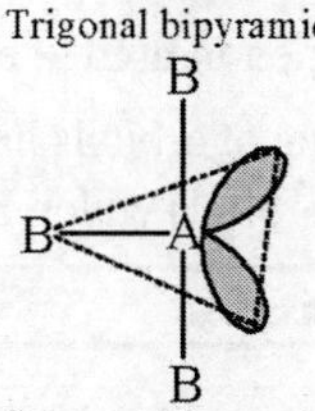 Trigonal bipyramidal	T shaped	ClF_3
AB_2L_3	2	3	5	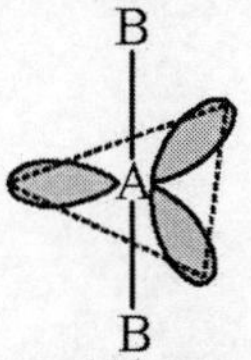 Trigonal bipyramidal	Linear	XeF_2
AB_5L	5	1	6	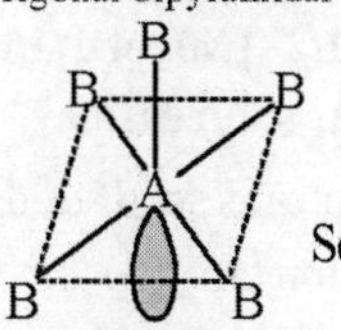 Octahedral	Square pyramidal	BrF_5
AB_4L_2	4	2	6	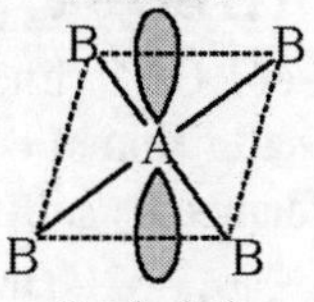 Octahedral	Square planar	XeF_4

There are other theories also to predict shapes or geometry of various molecules which you will study in further higher classes.

VALENCE BOND THEORY (VBT) :

It was presented by Heitler & London to explain how a covalent bond is formed and it was extended by Pauling & Others.

According to *Valence Bond Theory*, a covalent bond is formed by the partial overlapping of two half filled atomic orbitals containing electrons in opposite spins.

Two types of bonds are formed on account of overlapping.

(a) Sigma (σ) bond (b) Pi (π) bond

Difference in σ and π bonds

Sigma (σ) bond	Pi (π) bond
(1) It results from the end to end overlapping of two s orbitals or two p–orbitals or overlapping of two p orbitals.	(1) It result from the sidewise (lateral) one s and one p–orbital.
(2) Its bonded orbital consists of a single electron cloud symmetrical about internuclear axis.	(2) Its bonded orbital consists of two electron clouds one above and the other below the plane of participating atoms
(3) Stronger	(3) Less strong
(4) Bond energy 80 Kcals	(4) Bond energy 65 Kcals
(5) Less reactive	(5) More reactive
(6) Can be independently existent	(6) Always exist along with a σ bond

HYBRIDISATION :

(i) It is hypothetical phenomenon used to **explain shapes of molecules**.

(ii) It is process of intermixing of orbitals having slight difference in energies, so as to redistribute their energies to form hybrid orbitals which have same shape, size and energy.

S.No.	Type of hybridisation	Shape of molecule/ion
1.	sp	linear
2.	sp^2	trigonal planar
3.	sp^3	tetrahedral
4.	sp^3d	trigonal bipyramidal
5.	sp^3d^2	octahedral
6.	sp^3d^3	pentagonal bipyramidal

(iii) **Hybridisation (H) = 1/2 [No. of e^- in valence shell of central atom + No. of monovalent atoms + charge on anion – charge on cation]**

H = 7 means sp^3d^3, 6 means sp^3d^2 or d^2sp^3, 5 means sp^3d or dsp^3, 4 means sp^3 or dsp^2, 3 means sp^2, 2 means sp hybridisation etc.

MOLECULAR ORBITAL THEORY (PROPOSED BY MULLIKAN) :

(i) **According to principle of LCAO** (linear combination of atomic orbitals) **BMO** (bonding molecular orbitals) **are formed by addition of wave functions of atomic orbitals and ABMO** (anti bonding molecular orbitals) **are formed by subtraction of atomic orbitals**.

$$\psi_A + \psi_B \longrightarrow \text{BMO}$$
$$\psi_A - \psi_B \longrightarrow \text{ABMO}$$

(ii)

	Properties	Bonding M.O.	Antibonding M.O.
1.	Formed by	Constructive interferance $\psi_b = \psi_A + \psi_B$	Destructive interferance $\psi_a = \psi_A - \psi_B$
2.	Energy	less than individual atomic orbital	more than individual atomic orbital
3.	Maximum no. of e^-	2	2

(iii) Applicable to elements with atomic no. > 7	Applicable to element with atomic no. < 7
(e.g.- O_2, F_2, CO)	(e.g.-B_2, C_2, N_2, Li_2, Be_2)
$\sigma 1s^2, \sigma^* 1s^2, \sigma 2s^2, \sigma^* 2s^2, \sigma 2p_z^2,$ $\pi 2p_x^2 \equiv \pi 2p_y^2, \pi^* 2p_x^2 \equiv \pi^* 2p_y^2, \sigma^* 2p_z^2$	$\sigma 1s^2, \sigma^* 1s^2, \sigma 2s^2, \sigma^* 2s^2, \pi 2p_x^2 \equiv \pi 2p_y^2$ $\sigma 2p_z^2, \pi^* 2p_x^2 \equiv \pi^* 2p_y^2, \sigma^* 2p_z^2$

The pictorial representation of energy diagram for simple homonuclear diatomic molecules specially like O_2, F_2 or Ne_2 is given in the following diagram.

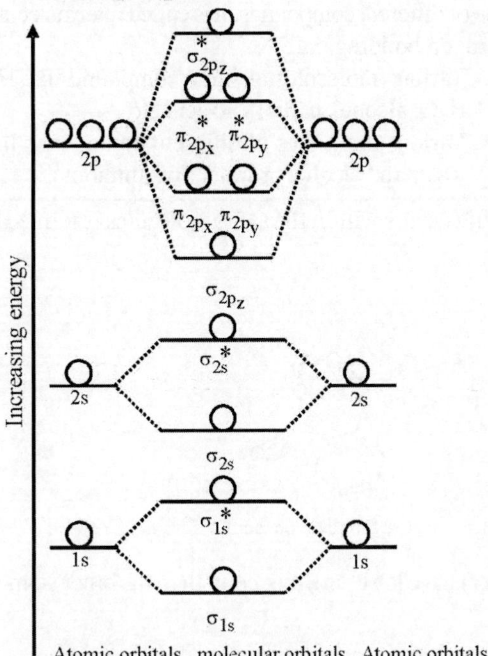

SOME IMPORTANT CHARACTERISTICS OF BONDS:

Bond Length

Average distance between centres of nuclei of two bonded atoms.

(i) **Bond length $\propto$ size of atom**

(ii) **Bond length decreases with multiplicity of the bond**. So, $C \equiv C < C = C < C - C$.

(iii) **Greater the s character smaller the bond length**.
$sp^3C - H > sp^2C - H > spC - H$

Bond Energy or Bond Dissociation Energy

The amount of energy required to break one mole of the bond and separate the bonded atoms as gaseous atoms. Its units are **kJmol^{-1}**.

(i) Bond energy $\propto \dfrac{1}{\textbf{atomic size}}$

$\therefore$ smaller the size of bonded atom more will be bond energy.

(ii) **Bond energy $\propto$ difference in electronegativities of bonded atoms.**

$\therefore$ bond energy of halogen acids are HF > HCl > HBr > HI

(iii) Bond energy $\propto$ % of s character in hybrid orbital

	sp^3 <	sp^2 <	sp
s character	25%	33.3%	50%

(iv) Bond energy $\propto \dfrac{1}{\text{lone pair of } e^-}$

Bond Angle (θ) :

Average angle between bonded atoms.

(i) **Bond angle $\propto$ % of s character of hybrid orbital**

	sp^3 <	sp^2 <	sp
	25% s	33.3% s	50% s
Bond angle	109.28	120°	180°

(ii) **Bond angle $\propto$ electronegativity of central atom**

Molecule	H_2Te	H_2O	H_2S	H_2Se
Bond angle	89.5°	104.5°	92°	91°

Bond Order:

It is defined as the number of covalent bonds by which the two atoms are joined in the molecule. It is calculated as given below.

$$\text{Bond order} = \frac{1}{2}\left[\begin{array}{ccc}\text{Number of electrons} & & \text{Number of electrons} \\ \text{in bonding M.O.} & - & \text{in antibonding M.O.} \\ \text{i.e.} N_b & & \text{i.e.} N_a\end{array}\right]$$

$$\therefore \text{Bond order} = \frac{1}{2}[N_b - N_a]$$

Bond order helps in

(i) **Predicting formation of molecule/ion :** If bond order = 0 or negative, it means no bond is there and molecule does not exist.

(ii) **Bond order $\propto$ stability**

Higher bond order means high stability.

(iii) **Bond order** $\propto \frac{1}{\text{bond length}}$

(iv) Bond order $\propto$ bond energy

HYDROGEN BONDING :

When a hydrogen atom linked to a highly electronegative atom (like F, O or N) comes under the influence of another strongly electronegative atom, then a weak bond is developed between them. This type of bond is called hydrogen bond and is represented by dotted lines as shown below:

$$\overset{\delta_+}{—H}—\overset{\delta_-}{F} \;......\; \overset{\delta_+}{H}—\overset{\delta_-}{F} \;.......\; \overset{\delta_+}{H}—\overset{\delta_-}{F}$$

Hydrogen bond

Types of Hydrogen Bond:

Hydrogen bonding are of two types

(a) Intermolecular hydrogen bonding

(b) Intramolecular hydrogen bonding

Intermolecular hydrogen bonding: When hydrogen bonding takes place between various molecules of the same or different compounds, it is called intermolecular hydrogen bonding e.g.

(i) Various molecules of same compound like HF, H_2O, alcohol, carboxylic acid etc.

(ii) Various molecules of different compounds like water and alcohol or water and ammonia.

Intramolecular hydrogen bonding : When hydrogen bonding occurs within the same molecules is termed as intramolecular hydrogen bonding.

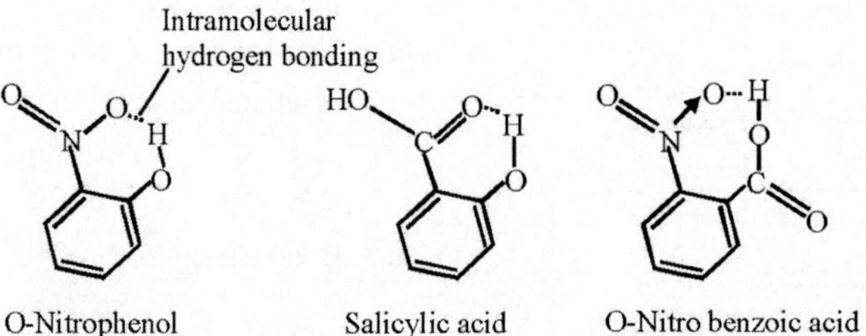

O-Nitrophenol Salicylic acid O-Nitro benzoic acid

Compound having intramolecular hydrogen bond (i.e. o-isomer) have lower melting point than the other isomers (meta or para isomers).

VAN DER WAAL'S FORCES :

Weak forces of attraction between covalent molecules. They are of three types.

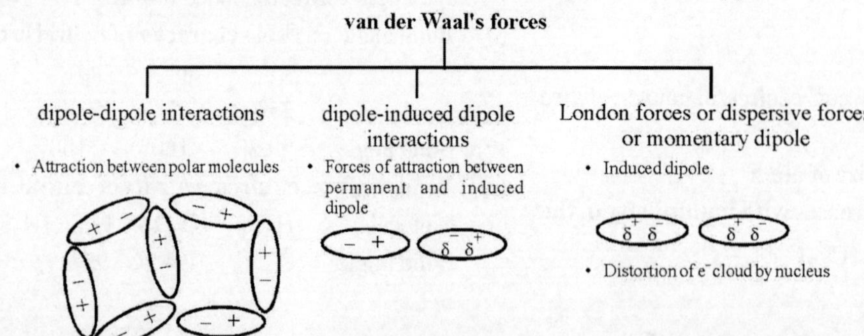

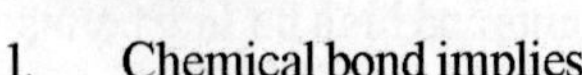

Exercise

1. Chemical bond implies
   (a) repulsion
   (b) attraction
   (c) attraction and repulsion balanced at a particular distance
   (d) attraction and repulsion
2. Hydrogen chloride molecule contains
   (a) Polar covalent bond
   (b) Double bond
   (c) Co-ordinate bond
   (d) Electrovalent bond
3. Sodium chloride is an ionic compound whereas hydrogen chloride is mainly covalent because
   (a) Sodium is less reactive
   (b) Hydrogen is non-metal
   (c) Hydrogen chloride is a gas
   (d) Electronegativity difference in the case of hydrogen and chlorine is less than 2.1.
4. In PO_4^{3-}, the formal charge on each oxygen atom and the P—O bond order respectively are
   (a) −0.75, 0.6 (b) −0.75, 1.0
   (c) −0.75, 1.25 (d) −3, 1.25
5. Which one of the following has the shortest carbon carbon bond length ?
   (a) Benzene (b) Ethene
   (c) Ethyne (d) Ethane
6. Which of the following pairs will form the most stable ionic bond ?
   (a) Na and Cl (b) Mg and F
   (c) Li and F (d) Na and F
7. In which of the following the bond angle is maximum?
   (a) NH_3 (b) SCl_2
   (c) NH_4^+ (d) PCl_3
8. The relationship between the dissociation energy of N_2 and N_2^+ is :
   (a) Dissociation energy of N_2^+ > dissociation energy of N_2
   (b) Dissociation energy of N_2 = dissociation energy of N_2^+
   (c) Dissociation energy of N_2 > dissociation energy of N_2^+
   (d) Dissociation energy of N_2 can either be lower or higher than the dissociation energy of N_2^+
9. Which does not show resonance ?
   (a) Benzene (b) Aniline
   (c) Ethylamine (d) Toluene
10. Which of the following bonds will be most polar?
    (a) N − Cl (b) O − F
    (c) N − F (d) N − N
11. The geometry of ClO_3^- ion according to Valence Shell Electron Pair Repulsion (VSEPR) theory will be
    (a) Planar triangular (b) Pyramidal
    (c) Tetrahedral (d) Square planar
12. Which of the following overlap is correct ?
    (a)
    (b) 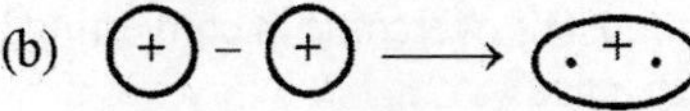
    (c) 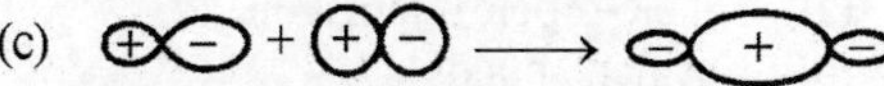
    (d) None of the above
13. The angle between the overlapping of one s-orbital and one p-orbital is
    (a) 180° (b) 120°
    (c) 109°28' (d) 120°, 60°
14. Linear combination of two hybridized orbitals belonging to two atoms and each having one electron leads to a
    (a) Sigma bond
    (b) Double bond
    (c) Co-ordinate covalent bond
    (d) Pi bond.
15. Which statement is NOT correct ?
    (a) A sigma bond is weaker than a π-bond.
    (b) A sigma bond is stronger than a π-bond.
    (c) A double bond is stronger than a single bond.
    (d) A double bond is shorter than a single bond.
16. Which one shows maximum hydrogen bonding?
    (a) H_2O (b) H_2Se
    (c) H_2S (d) HF.
17. The weakest among the following types of bonds is
    (a) ionic (b) covalent
    (c) metallic (d) H–bond.

18. The ground state electronic configuration of valence shell electrons in nitrogen molecule (N_2) is written as KK $\sigma 2s^2, \sigma^* 2s^2, \pi 2p_x^2, \pi 2p_y^2 \sigma 2p_z^2$
Bond order in nitrogen molecule is
(a) 0 (b) 1
(c) 2 (d) 3
19. H_2O is dipolar, whereas BeF_2 is not. It is because
(a) the electronegativity of F is greater than that of O
(b) H_2O involves hydrogen bonding whereas BeF_2 is a discrete molecule
(c) H_2O is linear and BeF_2 is angular
(d) H_2O is angular and BeF_2 is linear
20. The number of unpaired electrons in a paramagnetic diatomic molecule of an element with atomic number 16 is
(a) 3 (b) 4
(c) 1 (d) 2
21. What is the dominant intermolecular force or bond that must be overcome in converting liquid CH_3OH to a gas?
(a) Dipole-dipole interaction
(b) Covalent bonds
(c) London dispersion force
(d) Hydrogen bonding
22. Experiment shows that H_2O has a dipole moment while CO_2 has not. Point out the structures which best illustrates these facts
(a) O = C = O, O (bonded to H and H, angular: H / O \ H)
(b) O = C = O, H – O – H
(c) C (double bonded to O and O, angular: O // C \\ O), H – O – H
(d) O = C = O (vertical: O ‖ C = O); H – O – H (vertical: H | O – H)
23. In the N_2 molecule, number of electron pairs shared by nitrogen atoms is :
(a) 1 (b) 2
(c) 3 (d) 5
24. The bond between two identical non-metal atoms has a pair of electrons
(a) Unequally shared between the two
(b) Transferred fully from one atom to another
(c) With identical spins
(d) Equally shared between them
25. As compared to covalent compounds, electrovalent compounds generally have
(a) Low melting points and low boiling points
(b) Low melting points and high boiling points
(c) High melting points and low boiling points
(d) High melting points and high boiling points
26. Which of the following statements is true for an ion having sp^3 hybridisation?
(a) all bonds are ionic
(b) H-bonds are situated at the corners of a square
(c) all bonds are co-ordinate covalent
(d) H-atoms are situated at the corners of tetrahedron
27. Which of the following is insoluble in water?
(a) H_2S (b) $HgCl_2$
(c) $Ca(NO_3)_2$ (d) CaF_2
28. Electron deficient molecule is :
(a) CCl_4 (b) PCl_5
(c) BF_3 (d) SF_6
29. Which one of the following does not follow octate rule?
(a) PF_3 (b) BF_3
(c) CO_2 (d) CCl_4
30. The boiling point of *p*-nitrophenol is higher than that of *o*-nitrophenol because
(a) NO_2 group at *p*-position behave in a different way from that at *o*-position.
(b) intramolecular hydrogen bonding exists in *p*-nitrophenol
(c) there is intermolecular hydrogen bonding in *p*-nitrophenol
(d) *p*-nitrophenol has a higher molecular weight than *o*-nitrophenol.
31. Sequence of bond length of the following is
A. ethane B. ethene
C. ethyne
(a) A > B > C (b) B > A > C
(c) C > B > A (d) C > A > B
32. Which one of the following molecules will have unequal M – F bond lengths ?
(a) NF_3 (b) BF_3
(c) PF_5 (d) SF_4

33. Match List I (Molecules) with List II (Bond order) and select the correct answer using the codes

List I		List II	
I.	Li_2	A.	3
II	N_2	B.	1.5
III	Be_2	C.	1.0
IV	O_2	D.	0
		E.	2

Codes

(a) I - B, II - C, III - A, IV - E
(b) I - C, II - A, III - D, IV - E
(c) I - D, II - A, III - E, IV - C
(d) I - C, II - B, III - E, IV - A

34. The molecular shapes of SF_4, CF_4 and XeF_4 are
(a) different with 1, 0 and 2 lone pairs of electrons on the central atom, respectively
(b) different with 0, 1 and 2 lone pairs of electrons on the central atom, respectively
(c) the same with 1, 1 and 1 lone pair of electrons on the central atoms, respectively
(d) the same with 2, 0 and 1 lone pairs of electrons on the central atom, respectively

35. Which of the following statements is/are true?
1. PH_5 and $BiCl_5$ do not exist
2. $p\pi - d\pi$ bond is present in SO_2
3. I_3^+ has bent geometry
4. SeF_4 and CH_4 have same shape

(a) 1, 2, 3 (b) 1, 3
(c) 1, 3, 4 (d) 1, 2, 4

36. Which of the following molecule does not have a linear arrangement of atoms ?
(a) H_2S (b) C_2H_2
(c) BeH_2 (d) CO_2

37. In which one of the following molecules the central atom said to adopt sp^2 hybridization?
(a) BeF_2 (b) BF_3
(c) C_2H_2 (d) NH_3

38. Which one of the following formulae does not correctly represent the bonding capacities of the two atoms involved ?

(a) $\left[\begin{array}{c} H \\ | \\ H-P-H \\ | \\ H \end{array}\right]^+$

(b) F–O–F (bent: F \ O / F)

(c) $O \leftarrow N$ (=O) (–O–H)

(d) $H-C=C$ (=O) (–O–H)

39. Which of the following statements is not correct ?
(a) Double bond is shorter than a single bond
(b) Sigma bond is weaker than a π (pi) bond
(c) Double bond is stronger than a single bond
(d) Covalent bond is stronger than hydrogen bond.

40. Which one of the following is the correct order of interactions ?
(a) Covalent < hydrogen bonding < vander Waals < dipole-dipole
(b) vander Waals < hydrogen bonding < dipole < covalent
(c) vander Waals < dipole-dipole < hydrogen bonding < covalent
(d) Dipole-dipole < vander Waals < hydrogen bonding < covalent.

41. Mark the incorrect statement in the following
(a) The bond order in the species O_2, O_2^+ and O_2^- decreases as $O_2^+ > O_2 > O_2^-$
(b) The bond energy in a diatomic molecule always increases when an electron is lost
(c) Electrons in antibonding M.O. contribute to repulsion between two atoms.
(d) With increase in bond order, bond length decreases and bond strength increases.

42. The distance between the two adjacent carbon atoms is largest in
(a) benzene (b) ethene
(c) butane (d) ethyne

43. The BCl_3 is a planar molecule whereas NCl_3 is pyramidal because
(a) B-Cl bond is more polar than N-Cl bond
(b) N-Cl bond is more covalent than B-Cl bond
(c) nitrogen atom is smaller than boron atom
(d) BCl_3 has no lone pair but NCl_3 has a lone pair of electrons

44. The correct order of N–O bond lengths in NO, NO_2^-, NO_3^- and N_2O_4 is
(a) $N_2O_4 > NO_2^- > NO_3^- > NO$
(b) $NO > NO_3^- > N_2O_4 > NO_2^-$
(c) $NO_3^- > NO_2^- > N_2O_4 > NO$
(d) $NO > N_2O_4 > NO_2^- > NO_3^-$

45. N_2 and O_2 are converted into monocations, N_2^+ and O_2^+ respectively. Which of the following statements is wrong ?
(a) In N_2, the N—N bond weakens
(b) In O_2, the O—O bond order increases
(c) In O_2, paramagnetism decreases
(d) N_2^+ becomes diamagnetic

46. The number of anti-bonding electron pairs in O_2^{-2} molecular ion on the basis of molecular orbital theory is, (Atomic number of O is 8)
(a) 5 (b) 2
(c) 3 (d) 4

47. Among the following which one is not paramagnetic? [Atomic numbers : Be = 4, Ne = 10, As = 33, Cl = 17]
(a) Cl^- (b) Be^+
(c) Ne^{+2} (d) As^+

48. The dipole moments of diatomic molecules AB and CD are 10.41D and 10.27 D, respectively while their bond distances are 2.82 and 2.67 Å, respectively. This indicates that
(a) bonding is 100% ionic in both the molecules
(b) AB has more ionic bond character than CD
(c) AB has lesser ionic bond character than CD
(d) bonding is nearly covalent in both the molecules

49. Which of the following molecules is planar?
(a) SF_4 (b) XeF_4
(c) NF_3 (d) SiF_4

50. Which one of the following molecules will form a linear polymeric structure due to hydrogen bonding?
(a) NH_3 (b) H_2O
(c) HCl (d) HF

51. Among the following ions the pπ–dπ overlap could be present in
(a) NO_2^- (b) NO_3^-
(c) PO_4^{3-} (d) CO_3^{2-}

52. Which of the following two are isostructural?
(a) NH_3, BF_3 (b) PCl_5, ICl_5
(c) XeF_2, IF_2^- (d) CO_3^{-2}, SO_3^{-2}

53. In X—H --- Y, X and Y both are electronegative elements
(a) Electron density on X will increase and on H will decrease
(b) In both electron density will decrease
(c) In both electron density will increase
(d) Electron density will decrease on X and will increase on H

54. Which of the following statements is not correct for sigma and pi-bonds formed between two carbon atoms?
(a) Sigma-bond determines the direction between carbon atoms but a pi-bond has no primary effect in this regard
(b) Sigma-bond is stronger than a pi-bond
(c) Bond energies of sigma- and pi-bonds are of the order of 264 kJ/mol and 347 kJ/mol, respectively
(d) Free rotation of atoms about a sigma-bond is allowed but not in case of a pi-bond

55. In a regular octahedral molecule, MX_6 the number of X - M - X bonds at 180° is
(a) three (b) two
(c) six (d) four

56. In BrF_3 molecule, the lone pairs occupy equatorial positions to minimize
(a) lone pair - bond pair repulsion only
(b) bond pair - bond pair repulsion only
(c) lone pair - lone pair repulsion and lone pair - bond pair repulsion
(d) lone pair - lone pair repulsion only

57. The correct sequence of increasing covalent character is represented by
(a) $LiCl < NaCl < BeCl_2$
(b) $BeCl_2 < LiCl < NaCl$
(c) $NaCl < LiCl < BeCl_2$
(d) $BeCl_2 < NaCl < LiCl$

58. Which of the following molecules has trigonal planar geometry?
(a) BF_3 (b) NH_3
(c) PCl_3 (d) IF_3

59. Which of the following is not a correct statement?
(a) The canonical structures have no real existence
(b) Every AB_5 molecule does in fact have square pyramidal structure
(c) Multiple bonds are always shorter than corresponding single bonds
(d) The electron-deficient molecules can act as Lewis acids

60. The electronegativity difference between N and F is greater than that between N and H yet the dipole moment of NH_3 (1.5 D) is larger than that of NF_3 (0.2D). This is because
(a) in NH_3 the atomic dipole and bond dipole are in the same direction whereas in NF_3 these are in opposite directions
(b) in NH_3 as well as NF_3 the atomic dipole and bond dipole are in opposite directions
(c) in NH_3 the atomic dipole and bond dipole are in the opposite directions whereas in NF_3 these are in the same direction
(d) in NH_3 as well as in NF_3 the atomic dipole and bond dipole are in the same direction

States of Matter

CONCEPTS

INTERMOLECULAR FORCES :

Intermolecular forces are the forces of attraction and repulsion between interacting particles. Attractive intermolecular forces are known as van der Waals forces. Different types of van der walls forces are as follows:

(i) **Dispersion forces or London forces :** The forces of attraction between the induced momentary dipoles (due to instantaneous distortion of their electron clouds) are called London forces.

(ii) **Dipole-dipole forces :** These forces act between molecules possessing permanent dipole. They occur among polar molecules. For example:

$$\ldots\ldots \overset{\delta+}{H}-\overset{\delta-}{Cl}\ldots\ldots \overset{\delta+}{H}-\overset{\delta-}{Cl}\ldots\ldots \overset{\delta+}{H}-\overset{\delta-}{Cl}\ldots\ldots$$

(iii) **Dipole induced dipole forces :** These forces operate between molecules having permanent dipole and molecules lacking permanent dipole. Permanent dipole of the polar molecule induces dipole on the electrically neutral molecule by deforming its electron cloud. Thus an induced dipole is developed in the other molecule.

(iv) **Hydrogen bond :** It is a special case of dipole-dipole interaction.

GASEOUS STATE

(i) Gases or their mixture are homogeneous in composition

(ii) Gases have very low density

(iii) Gases exert pressure

(iv) Gases possess high diffusibility

(v) Gases have infinite expansibility and high compressibility

(vi) Gases can be liquified by cooling under high pressure

(vii) All gases undergo similar change with the change of temperature and pressure. This behaviour of gases can be described by certain laws known as **gas laws**.

GAS LAWS

Include Boyle's law, Charle's law, Amontons law, Dalton's law of partial pressure and Graham's law of diffusion.

Boyle's Law

$V \propto 1/P$ (constant T and n)

or $PV = \text{constant}$ or $P_1V_1 = P_1V_2$

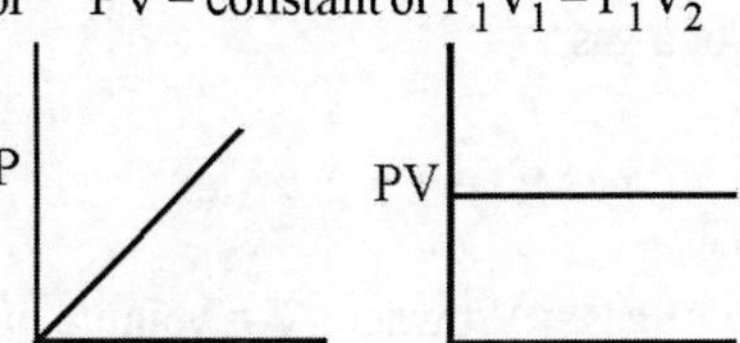

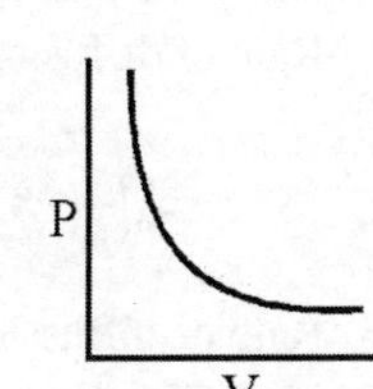

Charle's Law

$V \propto T$ (at constant P and n)

or $V/T = \text{constant}$

or $\dfrac{V_1}{T_1} = \dfrac{V_2}{T_2}$

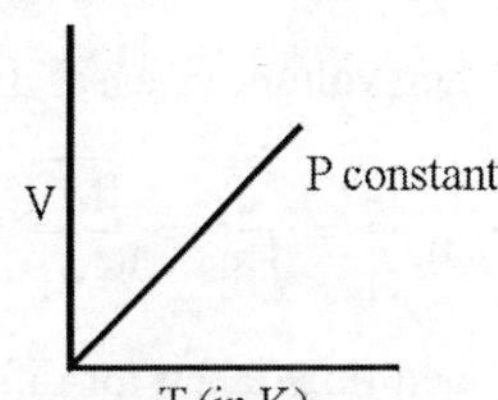

The law does not hold good for liquids.

Amontons Law or Gay Lussac's Law (Pressure Temperature Law)

$P \propto T$ (at constant V and n)

$P/T = \text{Constant}$

$$\frac{P_1}{T_1} = \frac{P_2}{T_2}$$

Dalton's Law of Partial Pressure

When a mixture of two or more non reacting gases is kept in a closed container then the total pressure exerted by gaseous mixture is equal to sum of all partial pressures, that each gas would exert when present alone in the same vessel at the same temperature.

$$P = p_1 + p_2 + p_3 + ...$$

The pressure exerted by gas in a mixture when it occupies whole volume is called its partial pressure.

partial pressure (p) = P × x

(Total pressure × mole fraction)

Relationship between partial pressure and number of moles :

(a) $P_1 = \left(\dfrac{n_1}{n_1 + n_2}\right) P = x_1 P$

where x_1 = mole fraction of gas

(b) Partial pressure of a gas in the mixture

$$= \frac{\text{Volume of the gas}}{\text{Total Volume}} \times P$$

Graham's Law of Diffussion :

Under similar conditions of temperature and pressure. Rate of diffusion of a gas

$$\propto \sqrt{\frac{1}{M}} \propto \sqrt{\frac{1}{d}} \quad (\because \text{Mol. wt.} = 2 \times \text{v.d.})$$

Rate of diffusion of gas = V/t where V = volume of gas diffused and t = time taken

$$\therefore \frac{r_1}{r_2} = \sqrt{\frac{d_2}{d_1}} = \sqrt{\frac{M_2}{M_1}} \text{ or } \frac{V_1/t_1}{V_2/t_2} = \sqrt{\frac{d_2}{d_1}} = \sqrt{\frac{M_2}{M_1}}$$

When volume of gases diffused are equal ($V_1 = V_2$),

then $\dfrac{t_2}{t_1} = \sqrt{\dfrac{d_2}{d_1}} = \sqrt{\dfrac{M_2}{M_1}}$

When time taken for diffusion of gases is same ($t_1 = t_2$), then $\dfrac{V_1}{V_2} = \sqrt{\dfrac{d_2}{d_1}} = \sqrt{\dfrac{M_2}{M_1}}$

Diffusion : It is intermixing of gases without help of external forces.

Rate of diffusion or effusion $\propto$ u

$$\frac{u_1}{u_2} = \sqrt{\frac{M_{W_2}}{M_{W_1}}}$$

Effusion : It is escape of gas through small hole.

Avogadro's Law :

In 1812, Amadeo Avogadro stated that samples of different gases which contain the same number of molecules (any complexity, size, shape) occupy the same volume at the same temperature and pressure. It follows from Avogadro's hypothesis that $V \propto n$ (T and P are constant).

$$V \propto n \text{ (T, P constant)} \Rightarrow \frac{V_1}{n_1} = \frac{V_2}{n_2}$$

IDEAL GAS EQUATION

$$\frac{P_1V_1}{T_1} = \frac{P_2V_2}{T_2}$$

PV = nRT (for n moles of gas)

$PV = N_AkT$ ($N_Ak = R$)

or $n = \dfrac{PV}{RT}$ ($\because$ n = mass/molecular mass)

or $P = \dfrac{m}{V}\dfrac{RT}{M}$

or $P = \rho\dfrac{RT}{M}$ (ρ is density and $= \dfrac{\text{mass}}{\text{volume}}$)

$\therefore \rho = \dfrac{PM}{RT}$

KINETIC MOLECULAR THEORY OF GASES :

This theory is based on seven assumptions which are as follows —

(i) Every gas is made of large number of extremely small particles which are called molecules. All the molecules of gas are identical in mass and size.

(ii) The molecules are very small and are at very large distance from each other, so the **volume occupied by molecules is very small as compared to total volume occupied by gas.**

(iii) Since the distance between molecules is very large, **the attractive forces existing between particles are negligible.**

(iv) The molecules of gas are in the constant state of motion. During motion they collide with each other as well as with walls of container.

(v) The pressure exerted by gas is due to the bombardment of moving molecules against the wall of container.

(vi) The molecules of gas are perfectly elastic and spherical. Therefore during collision, energy is not lost and it is transferred from one molecule to other.

(vii) Average kinetic energy of molecule $\propto$ absolute temperature of gas. This is called **Maxwell's generalisation.** K.E. $\propto$ T

KINETIC GAS EQUATION

Mathematical expression derived from an ideal gas on the basis of assumption of kinetic gas equation.

$$PV = \frac{1}{3}mNu^2 = \frac{1}{3}Mu^2 \quad \begin{pmatrix} \text{where m = mass of 1 molecule} \\ \text{N = total no. of molecules} \\ \text{m} \times \text{N = M (molecular mass)} \end{pmatrix}$$

or $$P = \frac{1}{3}\frac{M}{V}u^2 = \frac{1}{3}du^2$$

$$u = \sqrt{\frac{3P}{d}}.$$

KINETIC ENERGY AND MOLECULAR SPEED:

(i) $$K.E. = \frac{3}{2}PV = \frac{3}{2}RT/mol$$

$$= \frac{3}{2}nRT \text{ for n moles.}$$

(ii) $$\text{Average} = \frac{3}{2}kT \quad \begin{pmatrix} \text{where } k = \dfrac{R}{N_0} \text{ and is called} \\ \textbf{Boltzmann constant} \end{pmatrix}$$

∴ **Average K.E. of gas molecule is independent of mass of gas molecule.**

∴ At absolute zero, K.E. of gas molecule is zero.

The continuous motion of gas molecules results in the change in their speed and the redistribution of energy. If their individual speeds are u_1, u_2, u_3 ..... u_n (where 1, 2, ...n is the number of molecules), then the average speed of molecules is given as:

$$u_{av} = \frac{u_1 + u_2 + ... + u_n}{n}$$

The distribution of speeds can be shown with the help of a plot called the Maxwell-Boltsmann distribution of speeds, as shown below :

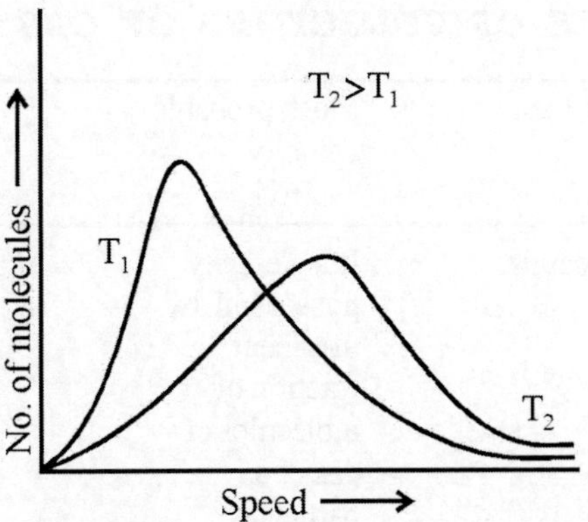

The figure is a schematic plot of number of molecules Vs. Molecular speed at two different temperatures T_1 and T_2 ($T_2 > T_1$).

AVERAGE, MEAN, MEAN SQUARE AND r.m.s. VELOCITY :

Average Velocity :

Since gas molecules moves randomly in all possible directions so average velocity of gas molecules is zero.

Mean Speed :

Mean speed of average speed is the arithmetic mean of the speeds of molecules in a gas at a given temperature.

$$\bar{v} = \frac{v_1 + v_2 + ... + v_N}{N}$$

Mean Square Velocity :

Mean square velocity is defined as the average of the square of the velocities of all molecules.

Let v_1, v_2, ..., v_N be the velocities of N molecules of a gas, then mean square velocity is,

$$\overline{v^2} = \frac{v_1^2 + v_2^2 + ... + v_N^2}{N}$$

Root Mean Square Velocity (R.M.S. Velocity):

The square root of the average of the squares of the velocities of gas molecules is called the R.M.S. velocity.

$$v_{rms} = \sqrt{\frac{v_1^2 + v_2^2 + ... + v_N^2}{N}} = \sqrt{\overline{v^2}}$$

TYPES OF VELOCITIES OF GAS MOLECULES

S. No.	Velocity	Most probable velocity (α)	Average velocity (v)	Root mean square velocity (u)
1.	Feature	It is velocity possessed by maximum fraction of molecules of gas at a particular temperature	It is arithematic different velocities possessed by gas molecules at particular temperature	It is square root of mean of the squares of velocities of the molecule.
2.	Expression	$\alpha = \sqrt{2RT/M}$ $= \sqrt{2PV/M}$	$v = \sqrt{8RT/\pi M}$ $= \sqrt{8PV/\pi M}$	$u = \sqrt{3RT/M}$ $= \sqrt{3PV/M}$

RELATIONSHIP BETWEEN α, v AND u

$$\alpha : v : u = \sqrt{\frac{2RT}{M}} : \sqrt{\frac{8RT}{\pi M}} : \sqrt{\frac{3RT}{M}}$$

$$v = 0.921\,u$$

$$\alpha = 0.816\,u$$

$$\therefore \alpha : v : u = 1 : 1.28 : 1.224$$

$$\therefore \alpha < v < u$$

Root mean square speed :

$$v_{rms} = \sqrt{\frac{3PV}{M}} = \sqrt{\frac{3P}{\rho}} = \sqrt{\frac{3RT}{M_w}} = \sqrt{\frac{3kT}{m}}$$

$$(\because R = N_A k \text{ and } M_w = m \,.\, N_A)$$

(i) For two different temperatures of a gas

$$\frac{v_{rms_2}}{v_{rms_1}} = \sqrt{\frac{T_2}{T_1}}$$

(ii) For two different gases $\frac{v_{rms_2}}{v_{rms_1}} = \sqrt{\frac{T_2}{T_1} \times \frac{M_{w_1}}{M_{w_2}}}$

(iii) For given temp. lighter the gas, larger the root mean square velocity of the molecules.

$$v_{rms} \propto \sqrt{\frac{1}{m}}$$

REAL GAS VS IDEAL GAS :

The real gases do not behave as ideal gas because the two assumptions of kinetic molecular theory are wrong.

(i) The forces of attraction between molecules of gases are negligible.

(ii) The actual volume occupied by gas molecules is negligible as compared to total volume of gas.

VAN DER WAAL'S EQUATION :

In order to explain behaviour of real gas, van der Waal introduced volume correction and pressure correction in ideal gas equation.

$$\left(P + \frac{a}{V^2}\right)(V - b) = RT \quad \text{(For 1 mole of gas)}$$

$$\left(P + \frac{an^2}{V^2}\right)(V - nb) = nRT \quad \text{(For n moles of gas)}$$

a and b are called van der Waal's constants.

a- gives idea about **magnitude of attractive forces** existing between gas molecules. Greater the value of 'a' more easily gas can be liquified.

b- gives idea about effective size of gas molecules. Greater the value of **'b'** larger will be size of atom of gas and smaller is compressible volume. **b** is also called as **co-volume or excluded volume.**

COMPRESSIBILITY FACTOR (Z)

(i) It gives the extent of deviation of real gas from ideal behaviour.

$Z = \frac{PV}{nRT}$ when Z = 1, gas is real

Z > 1 shows **positive deviation** and gas is less compressible than ideal behaviour.

Z < 1 shows **negative deviation** and gas is more compressible than ideal behaviour.

(ii) At **low pressure** all gases show ideal behaviour

(iii) **Boyle's temperature** is the temperature at which real gas behaves as ideal gas. eg- for N_2 it is 50°C. $T_b = \frac{a}{Rb}$

VAN DER WAAL'S EQUATION AT DIFFERENT TEMPERATURES AND PRESSURES

(i) **At high temperature :** Both V and P will be large, so both 'a' and 'b' can be neglected thus, the equation changes as follows

$PV = RT$

(ii) **At moderate pressure :** Since V is large. Thus b can be neglected and equation becomes as follows

$$\left(P+\frac{a}{V^2}\right)V = RT$$

$$PV = RT - \frac{a}{V}; \quad \Rightarrow Z = \frac{PV}{RT} = 1 - \frac{a}{VRT}$$

(iii) **At very lower pressure :** V becomes so large that **'b'** can be comparitively neglected. At very high value of V correction term $\frac{a}{V^2}$ is also very small that it can be neglected. Thus, equation becomes as

$$PV = RT; Z = \frac{PV}{RT} = 1$$

(iv) **At high pressure :** Since P is very large, pressure correction is neglected but volume correction cannot be neglected, so the equation changes as follows

$$P(V-b) = RT$$

or $PV = RT + Pb \Rightarrow Z = \frac{PV}{RT} = 1 + \frac{Pb}{RT}$

CRITICAL CONSTANT :

Critical Pressure (P_c)

The maximum pressure required to liquify the gas at critical temperature.

$$P_c = \frac{a}{27b^2}$$

Critical Temperature (T_c)

The temperature above which a gas cannot be liquified even by applying any amount of pressure.

$$T_c = \frac{8a}{27\,Rb}$$

Critical Volume (V_c)

The volume occupied by one mole of gas at critical temperature and pressure. $V_c = 3\,b$

Relationship between P_c, T_c and V_c : $P_cV_c = 3/8\,RT_c$

Joule-Thomson Effect :

When a real gas is allowed to expand adiabatically through a porous plug or a fine hole into region of low pressure, it is accompanied by cooling, but **H_2 and He warm up**. Ideal gases do not show cooling or heating effect because they do not have inter molecular forces of attraction. **Inversion temperature** is the temperature at which gas shows neither cooling nor heating.

Triple point is the temperature at which a substance exists in all three states.

LIQUID STATE :

Properties of The Liquid :

(i) **Shape**: Liquids have no definite shape of thier own. It acquires the shape of the container in which it is put.

(ii) **Volume:** Liquids have a definite volume.

(iii) **Density:** The densities of liquids are nearly 1000 times more than the densities of the gases under identical conditions.

(iv) **Compressibility:** Liquids are very less compressible compared to gases.

(v) **Diffusion :** Liquids diffuse like gases but the process of diffusion (also known as intermixing) is much slower.

(vi) **Vapour pressure**

It is pressure exerted by vapours in atmosphere in equilibrium with liquid at particular temperature.

It increases with temperature.

(vii) **Evaporation**

Spontaneous phenomenon occuring at all temperatures.

During the process liquid changes into vapour and leaves surface.

It results in cooling.

Rate of evaporation $\propto$ temperature

(viii) **Boiling point :** Temperature at which vapour pressure becomes equal to atmospheric pressure.

(ix) **Surface tension**

The surface tension is defined as the force per cm acting perpendicular to the tangential line on the surface of the liquid which tend to compress the surface area.

Surface tension $\propto$ 1/temperature

Surface tension is measured by **stalagmometer**

Liquid drops are spherical due to surface tension.

The units of surface tension are force per unit length i.e. dynes cm^{-1} or Nm^{-1} (in SI units)

Stronger the forces of intermolecular attraction (cohesive forces) greater is the surface tension.

(x) **Viscosity:** Viscosity may be defined as the force of friction between two layers of a liquid moving past one another with different velocities.

The viscosities of liquids are compared in terms of coefficient of viscosity which is defined as the force per unit area needed to maintain a unit difference in velocities between two consecutive parallel layers which are one cm apart.

The units of viscosity are poise (*P*) where $1p = 1g\ cm^{-1}s^{-1}$. In SI unit of viscosity is $1p = 0.1$ N sec m^{-2}.

The liquid having stronger forces of attraction has a higher viscosity.

Exercise

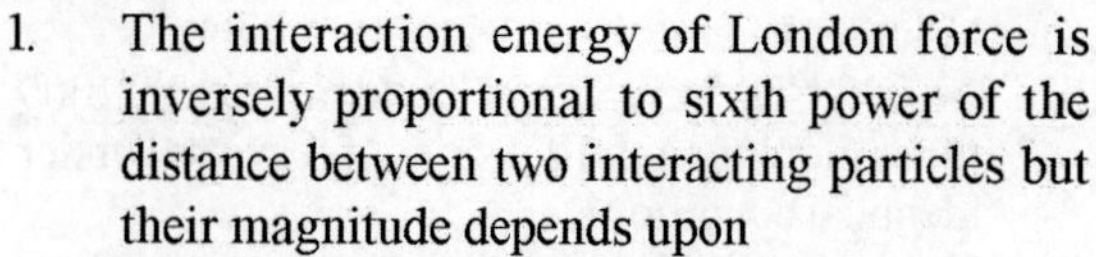

1. The interaction energy of London force is inversely proportional to sixth power of the distance between two interacting particles but their magnitude depends upon
   (a) charge of interacting particles
   (b) mass of interacting particles
   (c) polarisability of interacting particles
   (d) strength of permanent dipoles in the particles.
2. Dipole-dipole forces act between the molecules possessing permanent dipole. Ends of dipoles possess 'partial charges'. The partial charge is
   (a) more than unit electronic charge
   (b) equal to unit electronic charge
   (c) less than unit electronic charge
   (d) double the unit electronic charge
3. Boyle's law states that the
   (a) pressure of a gas is directly proportional to the temperature at constant volume
   (b) pressure of a gas is inversely proportional the volume at constant temperature
   (c) volume is directly proportional to the temperature at constant pressure
   (d) None of the above
4. Which of the following expression at constant pressure represents Charle's law?
   (a) $V \propto \frac{1}{T}$ (b) $V \propto \frac{1}{T^2}$
   (c) $V \propto T$ (d) $V \propto d$
5. Equal volumes of all gases under similar conditions of temperature and pressure contain equal numbers of molecules. This statement was made by
   (a) Gay-Lussac (b) Avogadro
   (c) Berzilius (d) John Dalton
6. Which is not true in case of an ideal gas ?
   (a) It cannot be converted into a liquid
   (b) There is no interaction between the molecules
   (c) All molecules of the gas move with same speed
   (d) At a given temperature, PV is proportional to the amount of the gas
7. Select one correct statement. In the gas equation, PV = nRT
   (a) n is the number of molecules of a gas
   (b) V denotes volume of one mole of the gas
   (c) n moles of the gas have a volume V
   (d) P is the pressure of the gas when only one mole of gas is present.
8. Correct gas equation is :
   (a) $\frac{V_1T_2}{P_1} = \frac{V_2T_1}{P_2}$ (b) $\frac{P_1V_1}{P_2V_2} = \frac{T_1}{T_2}$
   (c) $\frac{P_1T_2}{V_1} = \frac{P_2V_2}{T_2}$ (d) $\frac{V_1V_2}{T_1T_2} = P_1P_2$
9. If P, V, M, T and R are pressure, Volume, molar mass, temperature and gas constant respectively, then for an ideal gas, the density is given by
   (a) $\frac{RT}{PM}$ (b) $\frac{P}{RT}$
   (c) $\frac{M}{V}$ (d) $\frac{PM}{RT}$
10. The total pressure exerted by a number of non-reacting gases is equal to the sum of the partial pressures of the gases under the same conditions is known as
    (a) Boyle's law (b) Charle's law
    (c) Avogadro's law (d) Dalton's law
11. Kinetic theory of gases proves
    (a) only Boyle's law
    (b) only Charles' law
    (c) only Avogadro's law
    (d) All of these
12. According to the kinetic theory of gases, in an ideal gas, between two successive collisions a gas molecule travels
    (a) in a wavy path
    (b) in a straight line path
    (c) with an accelerated velocity
    (d) in a circular path

13. The compressibility factor for an ideal gas is
(a) 1.5 (b) 1.0
(c) 2.0 (d) ∞
14. Gases possess characteristic critical temperature which depends upon the magnitude of intermolecular forces between the particles. Following are the critical temperature of some gases.

Gases	H_2	He	O_2	N_2
Critical temperature in Kelvin	33.2	5.3	154.3	126

From the above data what would be the order of liquefaction of these gases ?
Start writing the order from the gas liquefying first
(a) H_2, He, O_2, N_2 (b) He, O_2, H_2, N_2
(c) N_2, O_2, He, H_2 (d) O_2, N_2, H_2, He
15. Generally, liquid drops assume spherical shape because:
(a) a sphere has maximum surface area
(b) a sphere has minimum surface area
(c) sphere is symmetrical in shape
(d) None of these
16. In van der Waal's equation of state for a non-ideal gas, the term that accounts for intermolecular forces is :
(a) $(V-b)$ (b) $(RT)^{-1}$
(c) $\left(P+\frac{a}{V^2}\right)$ (d) RT
17. A gas is said to behave like an ideal gas when the relation PV/T = constant. When do you expect a real gas to behave like an ideal gas ?
(a) When the temperature is low
(b) When both the temperature and pressure are low
(c) When both the temperature and pressure are high
(d) When the temperature is high and pressure is low
18. The correct value of the gas constant 'R' is close to :
(a) 0.082 litre-atmosphere K
(b) 0.082 litre-atmosphere K^{-1} mol^{-1}
(c) 0.082 litre – atmosphere^{-1} K mol^{-1}
(d) 0.082 litre^{-1} atmosphere^{-1} K mol
19. An ideal gas can't be liquefied because
(a) its critical temperature is always above 0°C
(b) its molecules are relatively smaller in size
(c) it solidifies before becoming a liquid
(d) forces operated between its molecules are negligible
20. A liquid can exist only :
(a) between triple point and critical temperature
(b) at any temperature above the melting point
(c) between melting point and critical temperature
(d) between boiling and melting temperature.
21. The kinetic energy of molecules in gaseous state is :
(a) More than those in the liquid state
(b) Less than those in the liquid state
(c) Equal to those in the liquid state
(d) Less than those in solid state
22. The total pressure of a mixture of two gases is :
(a) The sum of the partial pressures
(b) The difference between the partial pressures
(c) The product of the partial pressures
(d) The ratio of the partial pressures.
23. Value of universal gas constant (R) depends upon
(a) Number of moles of gas
(b) Volume of gas
(c) Temperature of gas
(d) None of these
24. Graph between P and V at constant temperature is
(a) straight
(b) curved increasing
(c) straight line with slope
(d) none of these
25. At higher altitudes
(a) boiling point of a liquid increases
(b) boiling point of a liquid decreases
(c) no change in boiling point
(d) melting point of solid increases
26. Which of the following describes the liquid phase ?
(a) It has a definite shape and a definite volume
(b) It has a definite shape but not a definite volume
(c) It has a definite volume but not a definite shape
(d) It has neither a definite shape nor a definite volume
27. The change of a liquid into vapour is called
(a) vapourization (b) solidification
(c) sublimation (d) None of these

28. Select the correct order of evaporation for water, alcohol, petrol and kerosene oil
   (a) water > alcohol > kerosene oil > petrol
   (b) alcohol > petrol > water > kerosene oil
   (c) petrol > alcohol > water > kerosene oil
   (d) petrol > alcohol > kerosene oil > water
29. In van der Waals equation of state of the gas law, the constant 'b' is a measure of
   (a) volume occupied by the molecules
   (b) intermolecular attraction
   (c) intermolecular repulsions
   (d) intermolecular collisions per unit volume
30. Select the one that when used would be considered as best condition for liquification of a gas.
   (a) Increasing the temperature.
   (b) Decreasing the pressure.
   (c) Increasing the pressure and decreasing the temperature.
   (d) Decreasing the pressure and increasing the temperature.
31. A plot of volume (V) versus temperature (T) for a gas at constant pressure is a straight line passing through the origin. The plots at different values of pressure are shown in figure. Which of the following order pressure is correct for this gas ?

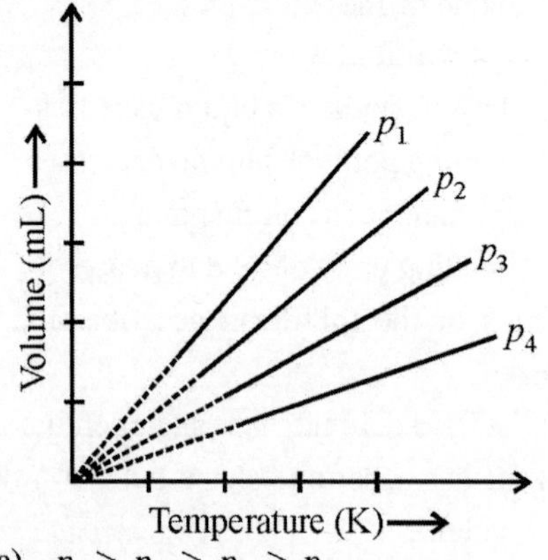

   (a) $p_1 > p_2 > p_3 > p_4$
   (b) $p_1 = p_2 = p_3 = p_4$
   (c) $p_1 < p_2 < p_3 < p_4$
   (d) $p_1 < p_2 = p_3 < p_4$
32. Pressure remaining the same, the volume of a given mass of an ideal gas increases for every degree centigrade rise in temperature by definite fraction of its volume at
   (a) 0°C
   (b) its critical temperature
   (c) absolute zero
   (d) its Boyle temperature
33. Root mean square velocity of a gas molecule is proportional to
   (a) $m^{1/2}$ (b) m^0
   (c) $m^{-1/2}$ (d) m
34. In a closed flask of 5 litres, 1.0 g of H_2 is heated from 300 to 600 K. Which statement is not correct?
   (a) Pressure of the gas increases
   (b) The rate of collision increases
   (c) The number of moles of gas increases
   (d) The energy of gaseous molecules increases
35. At constant temperature, for a given mass of an ideal gas
   (a) The ratio of pressure and volume always remains constant.
   (b) Volume always remains constant.
   (c) Pressure always remains constant.
   (d) The product of pressure and volume always remains constant.
36. The root mean square speeds at STP for the gases H_2, N_2, O_2 and HBr are in the order :
   (a) $H_2 < N_2 < O_2 < HBr$
   (b) $HBr < O_2 < N_2 < H_2$
   (c) $H_2 < N_2 = O_2 < HBr$
   (d) $HBr < O_2 < H_2 < N_2$.
37. A closed flask contains water in all its three states solid, liquid and vapour at 0°C. In this situation, the average kinetic energy of water molecules will be
   (a) the greatest in all the three states
   (b) the greatest in vapour state
   (c) the greatest in the liquid state
   (d) the greatest in the solid state
38. At STP, 0.50 mol H_2 gas and 1.0 mol He gas
   (a) have equal average kinetic energies
   (b) have equal molecular speeds
   (c) occupy equal volumes
   (d) have equal effusion rates
39. Under what conditions will a pure sample of an ideal gas not only exhibit a pressure of 1 atm but also a concentration of 1 mole litre^{-1} ?
   (R = 0.082 litre atm mol^{-1}deg^{-1})
   (a) At STP
   (b) When V = 22.4 litres
   (c) When T = 12 K
   (d) Impossible under any conditions

40. When is deviation more in the behaviour of a gas from the ideal gas equation PV = nRT ?
(a) At high temperature and low pressure
(b) At low temperature and high pressure
(c) At high temperature and high pressure
(d) At low temperature and low high pressure

41. The temperature of the gas is raised from 27°C to 927°C, the root mean square velocity is
(a) $\sqrt{927/27}$ time the earlier value
(b) same as before
(c) halved
(d) doubled

42. 500 ml of nitrogen at 27°C is cooled to –5°C at the same pressure. The new volume becomes
(a) 326.32 ml (b) 446.66 ml
(c) 546.32 ml (d) 771.56 ml

43. 600 c.c. of a gas at a pressure of 750 mm is compressed to 500 c.c. Taking the temperature to remain constant, the increase in pressure, is
(a) 150 mm (b) 250 mm
(c) 350 mm (d) 450 mm

44. Cyclopropane and oxygen at partial pressures 170 torr and 570 torr respectively are mixed in a gas cylinder. What is the ratio of the number of moles of cyclopropane to the number of moles of oxygen (nC_3H_6/nO_2)?
(a) $\frac{170 \times 42}{570 \times 32} = 0.39$
(b) $\frac{170}{42} \bigg/ \left(\frac{170}{42} + \frac{570}{32}\right) \approx 0.19$
(c) $\frac{170}{740} = 0.23$
(d) $\frac{170}{570} = 0.30$

45. At which one of the following temperature - pressure conditions the deviation of a gas from ideal behaviour is expected to be minimum?
(a) 350 K and 3 atm. (b) 550 K and 1 atm.
(c) 250 K and 4 atm. (d) 450 K and 2 atm.

46. If 500 ml of gas A at 400 torr and 666.6 ml of B at 600 torr are placed in a 3 litre flask, the pressure of the system will be
(a) 200 torr (b) 100 torr
(c) 550 torr (d) 366 torr

47. A gaseous mixture contains H_2 and O_2 in the molar ratio 8 : 1. The ratio of $H_2 : O_2$ by weight in this mixture would be
(a) 4 : 1 (b) 1 : 8
(c) 8 : 1 (d) 1 : 2

48. The surface tension of which of the following liquid is maximum?
(a) C_2H_5OH (b) CH_3OH
(c) H_2O (d) C_6H_6

49. If a gas expands at constant temperature, it indicates that :
(a) kinetic energy of molecules decreases
(b) pressure of the gas increases
(c) kinetic energy of molecules remains the same
(d) number of the molecules of gas increases

50. The pressure exerted by 6.0g of methane gas in a 0.03 m^3 vessel at 129°C is (Atomic masses : C = 12.01, H = 1.01 and R = 8.314 JK^{-1} mol^{-1})
(a) 31684 Pa (b) 215216 Pa
(c) 13409 Pa (d) 41648 Pa

51. Three moles of an ideal gas expanded spontaneously into vacuum. The work done will be :
(a) Zero (b) Infinite
(c) 3 Joules (d) 9 Joules

52. A gaseous mixture was prepared by taking equal mole of CO and N_2. If the total pressure of the mixture was found 1 atmosphere, the partial pressure of the nitrogen (N_2) in the mixture is :
(a) 0.5 atm (b) 0.8 atm
(c) 0.9 atm (d) 1 atm

53. A bubble of air is underwater at temperature 15°C and the pressure 1.5 bar. If the bubble rises to the surface where the temperature is 25°C and the pressure is 1.0 bar, what will happen to the volume of the bubble ?
(a) Volume will become greater by a factor of 1.6.
(b) Volume will become greater by a factor of 1.1.
(c) Volume will become smaller by a factor of 0.70.
(d) Volume will become greater by a factor of 2.5.

54. At a certain temperature the volume of gas sample is 120 ml at one atmosphere pressure. Keeping the temperature same the pressure is raised to five atmospheres. The volume of the gas would change to :
(a) 24 ml (b) 125 ml
(c) 300 ml (d) 600 ml

55. When helium is allowed to expand into vacuum, heating effect is observed. Its reason is that :
(a) Helium is an ideal gas
(b) Helium is an inert gas
(c) The inversion temperature of helium is very low
(d) The boiling point of helium is the lowest amongst the elements

56. Pure hydrogen sulphide is stored in a tank of 100 litre capacity at 20°C and 2 atm. pressure. The weight of gas in grams will be :
(a) 34 g (b) 340 g
(c) 282.4 g (d) 28.24 g

57. For a gas, maximum deviation from ideal gas behaviour is observed at :
(a) 0°C and 1.0 atm (b) 100°C and 2.0 atm
(c) –100°C and 2.0 atm (d) –100°C and 5.0 atm

58. The approximate density of methane at 27°C and 6 atm. pressure (R = 0·0821 lit atm mol^{-1} K^{-1}) :
(a) 2 gm/lit (b) 4 gm/lit
(c) 16 gm/lit (d) 32 gm/lit

59. At STP, the density of a gas (molecular weight 45) is
(a) 2 g/litre (b) 22.4 g/litre
(c) 11.2 g/litre (d) 26.6 g/litre

60. A certain sample of gas has a volume of 0.2 litres measured at 1 atm. pressure and 0°C. At the same pressure but at 273°C, its volume will be
(a) 0.4 litre (b) 1.4 litre
(c) 0.8 litre (d) 1.8 litre

Thermodynamics

CONCEPTS

- **Thermodynamics** deals with quantitative relationship between different forms of energy.
- **Chemical energetics** is study of energy changes during chemical reaction.

THERMODYNAMIC TERMS

System and Surroundings

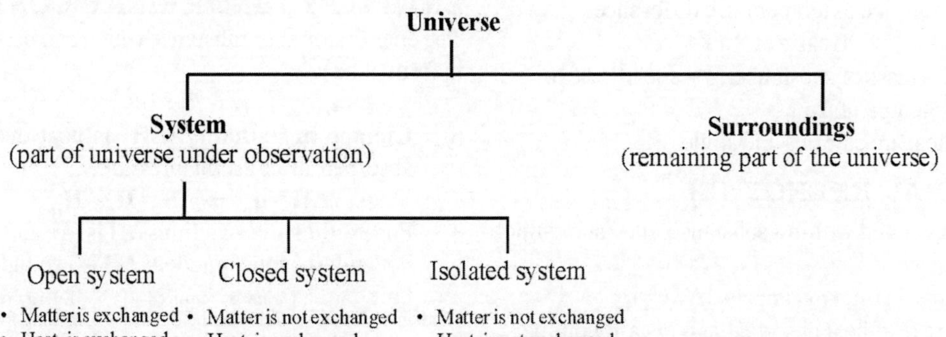

State Functions

The thermodynamic parameters which depend upon the initial and final states of the system only and are independent of how the change is accomplished are called state functions. Examples of the state functions are internal energy (E), enthalpy (H), entropy (S), free energy (G), pressure (P), temperature (T) and volume (V), etc. Heat (q) and work (W) are not the state functions being path dependent.

A system in which the state functions (variables) have constant values throughout is said to be in a state of thermodynamic equilibrium.

The state functions are of two types

- **Intensive variables:** The state functions of the system which are independent of the quantity of the substance present in the system are called intensive variables. For example, pressure, temperature, concentration, density, viscosity, surface tension, refractive index and specific heat, etc.
- **Extensive Variables:** The state variables which depend upon the quantity of the substance present in the system are termed as **extensive variables.** Volume, mass, internal energy, enthalpy, entropy and work are examples of extensive variables.

THERMODYNAMIC PROCESSES

(i) **Isothermal process** (T = constant, $\Delta T = 0$)

(ii) **Isobaric process** (P = constant, $\Delta P = 0$)

(iii) **Isochoric process** (V = constant, $\Delta V = 0$)

(iv) **Adiabatic process** ($dq = 0$, no heat is exchanged)

(v) **Reversible process** takes place at very slow speed and can be reversed at any stage.

(vi) **Irreversible process** can not be reversed.

(vii) **Cyclic process :** When a system returns to original position after completing a series of changes, then it is known that a cycle is completed.

WORK

Transference of energy may occur as work, when the system and surroundings have different pressure. If pressure inside the system is higher, work is said to be done by the system. If pressure on the surrounding is higher, the work is said to be done on the system and the work done = P Δ V.

Work done by a gas for expansion from volume V_1 to volume V_2 against a constant external pressure P is w = $P(V_2 - V_1) = P\Delta V$

MKS unit of work is newton and CGS unit is erg.

1 KJ = 10^3J, 1J = 10^7ergs = 0.2390 cal

1 cal = 4.184 J and 1 K cal = 4.184 KJ

HEAT (Q) :

Amount of energy that flows between system and surroundings due to temperature difference.

Heat = m × s × t

where m = mass of substance, s = specific heat, t = temperature.

Units of heat are calories and Joule.

INTERNAL ENERGY (E)

The energy stored within a substance. Its value cannot be determined.

(i) Change in **internal energy (ΔE) = q_v**
where q_v = heat absorbed at constant volume.

(ii) **ΔE can be calculated by using Bomb calorimeter.** The formula used is as follows

$$\Delta E = C \times \Delta t \times \frac{M}{m}$$

where
C = heat capacity of calorimeter, Δt = rise in temperature, M = molecular mass of substance, m = mass of substance taken.

(iii) Internal energy depends on T, P, V and quantity of matter.

(iv) Internal energy of element in most stable form is taken as zero.

FIRST LAW OF THERMODYNAMICS

States that energy can neither be created nor destroyed but can be transformed from one form to another

(i) **Mathematical form of first law of thermodynamics**

$$\Delta E = q + w$$

where
ΔE = change in internal energy, q = heat change, w = work done.

(ii) **Sign conventions**

Work	Heat
work done by system –ve	Absorbed +ve
work done on system +ve	Evolved –ve

(iii) **Expression for work of expansion**

$$W = -P\Delta V$$

$$W = -2.303\, nRT \log\frac{V_2}{V_1} = -2.303\, nRT \log\frac{P_1}{P_2}$$

(–sign indicates that during expansion work is done by system)

(iv) For isothermal reversible change

$$q = -w = 2.303\, nRT \log\frac{V_2}{V_1} = 2.303\, nRT \log\frac{P_1}{P_2}$$

(v) For adiabatic change, q = 0

$$\Delta E = W_{ad}$$

ENTHALPY (H) OR HEAT CONTENT

The energy stored in substance which can be available in form of heat.

$$H = E + PV$$

(i) **Change in enthalpy (ΔH)** is heat evolved or absorbed at constant pressure.

$$\Delta H = q_p \Rightarrow \Delta H = H_p - H_R$$

For **exothermic reactions ΔH is –ve** and $H_R > H_P$

For **endothermic reactions ΔH is +ve** and $H_P > H_R$

$$\text{Enthalpy change of a reaction} = \begin{bmatrix}\text{Energy required}\\ \text{to break the bonds}\\ \text{of reactants}\end{bmatrix} - \begin{bmatrix}\text{Energy released}\\ \text{during formation}\\ \text{of bonds in products}\end{bmatrix}$$

i.e. $\Delta H_{reaction}$ = [Bond energy of reactants] – [Bond energy of products]

$$\Delta H_{Reaction} = \Sigma\Delta H_f^\circ(\text{Products}) - \Sigma\Delta H_f^\circ(\text{Reactants}).$$

(ii) $\mathbf{\Delta H = \Delta E + P\Delta V = \Delta E + \Delta n_{(g)} RT}$,
where $\Delta n_{(g)} = (n_p - n_r)_{gaseous}$

(iii) **ΔH is calculated by calorimeter**

$$\Delta H = Z \times \Delta T \times M/m$$

where Z = Heat capacity of calorimeter.

MOLAR HEAT CAPACITY

Heat required to raise temperature of system of one mole pure substance by 1°C and for 1 g of substance it is called **specific heat**.

(i) $C = \frac{q}{\Delta T}$

Heat capacities are of two types

Heat capacity at constant volume (C_V)

$$C_v = \left[\frac{\partial E}{\partial T}\right]_v$$

Heat capacity at constant pressure (C_P)

$$C_p = \left[\frac{\partial H}{\partial T}\right]_p$$

(ii) $C_p - C_v = R$ (R = 2 cal. = 8.314)

(iii) $\gamma = C_p/C_v$

γ **= 1.66 for monoatomic gas**

γ **= 1.40 for diatomic gas**

γ **= 1.30 for triatomic gas**

HEAT OF REACTION

It is heat change taking place when number of moles of reactants and products are same as indicated in balance chemical equation.

Standard state of substance means its most stable form at 298 K and 1 atm pressure.

Different Forms of Heat of Reaction

(i) **Heat of formation**
- It is the heat change when one mole of substance is formed from its elements.
- **Standard heat of formation** of compound is heat of formation under standard conditions.
- Standard heat of formation of all elements is taken as zero.

(ii) **Heat of combustion**
- It is total heat liberated when 1 mole of substance undergoes complete combustion.
- **Combustion is always exothermic process** $\Delta H_{comb.} < 0$

(iii) **Heat of hydration :** It is total heat change when 1 mole of substance undergoes complete hydration (combine with required number of H_2O molecules) e.g.

$$CuSO_4(s) + 5H_2O_{(\ell)} \longrightarrow CuSO_4.5H_2O; \quad \Delta H = -78.2 \text{ kJ}$$

(iv) **Heat of dissolution :** It is total heat change when 1 mole of substance is completely dissolved in excess of solvent.

(v) **Heat of fusion :** It is total heat change when 1 mole of solid substance completely changes into liquid at its melting point.

$$H_2O_{(s)} \longrightarrow H_2O_{(\ell)}; \Delta H_{fus} = 6.0 \text{ kJ}$$

(vi) **Heat of vapourisation :** It is total heat change when 1 mole of liquid substance completely changes into vapours at its boiling point.

$$H_2O_{(\ell)} \longrightarrow H_2O_{(g)}; \quad \Delta H_{vap} = 40.6 \text{ kJ}$$

(vii) **Heat of sublimation :** It is heat required to convert one mole of solid directly into vapours below its boiling point.

$$I_{2(s)} \longrightarrow I_{2(g)}; \Delta H_{sub} = 62.4 \text{ kJ.}$$

Heat of sublimation

= Heat of fusion + Heat of vapourisation.

(viii) **Heat of neutralisation :** It is heat change taking place when **1 gm - equivalent of an acid is completely neutralised** by **base or vice-versa.**
- Heat of neutralization of strong acid by strong base or vice versa is always constant ($\because$ it is in fact heat of formation of water).

  $\Delta H_{neutralisation}$ (strong acid with strong base) **= –57.1 kJ = –13.7 k cals.**
- If either acid or base is weak then, the value of heat of neutralization is less than –57.1 kJ. ($\because$ some heat is required to dissociate the weak acid or weak base)
- The **smaller the value of heat of neutralization weaker is the acid or base.**

(ix) **Enthalpy of transition :** It is the enthalpy change when 1 mol of an element changes from one allotropic form to another

e.g.

$$C_{(diamond)} \longrightarrow C_{(graphite)} \quad \Delta H = +3.3 \text{ kcal/mol}$$

(x) **Enthalpy of ionisation :** It is defined as the amount of heat absorbed when 1 mole of an electrolyte completely dissociates into ions.

e.g.

$$CH_3COOH \longrightarrow CH_3COO^- + H^+ \quad \Delta H = 3 \text{ kcal}$$

HESS'S LAW OF CONSTANT HEAT SUMMATION :

Hess's law states that " the total amount of heat absorbed or evolved in a chemical reaction is same whether the reaction takes place in one step or in a number of steps."

BOND ENERGY AND BOND DISSOCIATION ENERGY

Bond energy	Bond dissociation energy
1. It is **average energy** required **to break** one **mole of bonds** of particular type	It is energy required to **break one mole of bonds of diatomic molecule** in gaseous state. in gaseous molecules.
2. It is an average of various bond dissociation energies of similar bonds and is expressed as kJ mol^{-1}.	It is a definite quantity and is expressed in kJ mol^{-1}.

SPONTANEOUS PROCESSES:

A process which can take place by itself under some given set of conditions, or by initiation if necessary, is called a spontaneous process.

A process that does not occur of its own i.e. a process which does not have any natural tendency to occur is termed as a non-spontaneous process.

ENTROPY (S)

(i) It is measure of randomness or disorderness of system.

(ii) $\Delta S = S_2 - S_1 = \frac{q_{(rev)}}{T}$

(iii) Units of entropy are $JK^{-1}\ mol^{-1}$

(iv) $\Delta S_{total} = \Delta S_{system} + \Delta S_{surroundings}$

for irreversible process $\Delta S_{uni} > 0$

for reversible process $\Delta S_{uni} = 0$

Entropy Changes in Processes :

(i) **During isothermal reversible expansion**

$$\Delta E = 0 \qquad \Delta S = nRT\ ln\frac{P_1}{P_2}$$

(ii) **During adiabatic reversible expansion**

$\Delta q = 0$ and $\therefore \Delta S = 0$

such process is called **isoentropic process.**

(iii) **During isobaric process**

$$\Delta S = C_p ln\frac{T_1}{T_2}$$

(iv) **During isochoric process**

$$\Delta S = C_p ln\frac{T_2}{T_1}$$

(v) **Entropy change during transition**

$$\Delta S = \frac{q_{(rev)}}{T} = \frac{\Delta H}{T}$$

$$S_{fus} = \frac{\Delta H_{fus}}{T_{fus}} \qquad \Delta S_{vap} = \frac{\Delta H_{vap}}{T_{vap}}$$

SECOND LAW OF THERMODYNAMICS :

Entropy of universe is always increasing or all spontaneous processes are accompanied by a net increase of entropy.

$$\Delta S_{total} = \Delta S_{system} + \Delta S_{surrounding} > 0.$$

- **During stretching of rubber band, entropy decreases** because when rubber band is stretched macromolecules are uncoiled and the arrangement of uncoiled macromolecule is more specific and hence entropy decreases.
- **Entropy of egg increases when egg is boiled** because on hard boiling denaturation of proteins occurs in which proteins change from helical form to random coiled form as a result entropy increases.

GIBBS FREE ENERGY OR GIBBS FUNCTION

Energy which can be converted into useful work done by system.

(i) $\Delta G = G_2 - G_1$

$\Delta G = \Delta H - T\Delta S$ **Gibbs Helmholtz equation**

(ii) When ΔG is positive – reaction is non spontaneous.

When ΔG is negative – reaction is spontaneous.

When ΔG is zero – reaction is in equilibrium.

(iii)

ΔH	ΔS	$\Delta G = \Delta H - T\Delta S$	Remarks
–	+	–	Reaction is spontaneous at all temperatures.
+	–	+	Reaction is non-spontaneous at all temperatures.
–	–	– (at low T)	Reaction is spontaneous at low temperatures.
–	–	+ (at high T)	Reaction is non-spontaneous at high temperatures.
+	+	+ (at low T)	Reaction is non-spontaneous at low temperatures.
+	+	– (at high T)	Reaction is spontaneous at high temperatures.

(iv) **Standard energy changes during chemical reaction (ΔG°)**

$$\Delta G^\circ_{reaction} = \Sigma G^\circ_{f(products)} - \Sigma G^\circ_{f(reactants)}$$

where G°_f are heats of formation.

(v) **Relationship between G° and K_c (equilibrium constant)**

$$\Delta G^\circ = -2.303\ RT \log K_c$$

(vi) $\Delta G^\circ = -nFE^\circ$

where F = Faraday, E° = standard emf of cell

n = no. of electrons in cell reaction.

(vii) Standard free energy of formation of all elementary substances is taken as zero.

THIRD LAW OF THE THERMODYNAMICS

States that entropy of all crystalline substances is zero at absolute zero temperature.

ZEROTH LAW OF THERMODYNAMICS

$$A \xrightarrow{\text{thermal equilibrium}} B \text{ and } A \xrightarrow{\text{thermal equilibrium}} C$$

$\therefore$ **A is in thermal equilibrium with C.**

$$\therefore A \xrightarrow{\text{thermal equilibrium}} B \xrightarrow{\text{thermal equilibrium}} C$$

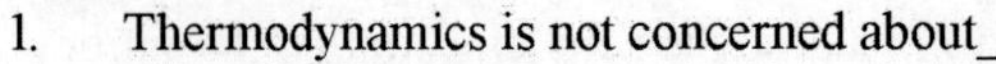

Exercise

1. Thermodynamics is not concerned about____.
   (a) energy changes involved in a chemical reaction.
   (b) the extent to which a chemical reaction proceeds.
   (c) the rate at which a reaction proceeds
   (d) the feasibility of a chemical reaction.
2. Which of the following statements is correct?
   (a) The presence of reacting species in a covered beaker is an example of open system.
   (b) There is an exchange of energy as well as matter between the system and the surroundings in a closed system.
   (c) The presence of reactants in a closed vessel made up of copper is an example of a closed system.
   (d) The presence of reactants in a thermos flask or any other closed insulated vessel is an example of a closed system.
3. The state of a gas can be described by quoting the relationship between _____.
   (a) pressure, volume, temperature
   (b) temperature, amount, pressure
   (c) amount, volume, temperature
   (d) pressure, volume, temperature, amount
4. Which one of the following statements is false?
   (a) Work is a state function.
   (b) Temperature is a state function.
   (c) Change in the state is completely defined when the initial and final states are specified
   (d) None of the above
5. In an adiabatic process, no transfer of heat takes place between system and surroundings. Choose the correct option for free expansion of an ideal gas under adiabatic condition from the following.
   (a) $q=0, \Delta T \neq 0, w=0$
   (b) $q \neq 0, \Delta T=0, w=0$
   (c) $q=0, \Delta T=0, w=0$
   (d) $q=0, \Delta T<0, w \neq 0$
6. During isothermal expansion of an ideal gas, its
   (a) internal energy increases
   (b) enthalpy decreases
   (c) enthalpy remains unaffected
   (d) enthalpy reduces to zero.
7. Adiabatic expansions of an ideal gas is accompanied by
   (a) decrease in ΔE
   (b) increase in temperature
   (c) decrease in ΔS
   (d) no change in any one of the above properties
8. 2 mole of an ideal gas at 27°C temperature is expanded reversibly from 2 lit to 20 lit. Find the entropy change (R = 2 cal/mol K)
   (a) 92.1 (b) 0
   (c) 4 (d) 9.2
9. If a reaction involves only solids and liquids which of the following is true ?
   (a) $\Delta H < \Delta E$ (b) $\Delta H = \Delta E$
   (c) $\Delta H > \Delta E$ (d) $\Delta H = \Delta E + RT\Delta n$
10. Which is an extensive property of the system ?
    (a) Volume (b) Viscosity
    (c) Temperature (d) Refractive index
11. The molar heat capacity of water at constant pressure is 75 $JK^{-1}\ mol^{-1}$. When 1kJ of heat is supplied to 100 g of water, which is free to expand, the increase in temperature of water is
    (a) 6.6 K (b) 1.2 K
    (c) 2.4 K (d) 4.8 K
12. The factor of ΔG values is important in metallurgy. The ΔG values for the following reactions at 800°C are given as :

    $$S_2(s) + 2O_2(g) \longrightarrow 2SO_2(g)\ ;\ \Delta G = -544\ kJ$$

    $$2Zn(s) + S_2(s) \longrightarrow 2ZnS(s)\ ;\ \Delta G = -293\ kJ$$

    $$2Zn(s) + O_2(g) \longrightarrow 2ZnO(s)\ ;\ \Delta G = -480\ kJ$$

    Then ΔG for the reaction :

    $$2ZnS(s) + 3O_2(g) \longrightarrow 2ZnO(s) + 2SO_2(g)$$

    will be :
    (a) –357 kJ (b) –731 kJ
    (c) –773 kJ (d) –229 kJ
13. Unit of entropy is
    (a) $JK^{-1}\ mol^{-1}$ (b) $J\ mol^{-1}$
    (c) $J^{-1}\ K^{-1}\ mol^{-1}$ (d) $JK\ mol^{-1}$

14. Considering entropy (S) as a thermodynamic parameter, the criterion for the spontaneity of any process is
(a) $\Delta S_{system} + \Delta S_{surroundings} > 0$
(b) $\Delta S_{system} - \Delta S_{surroundings} > 0$
(c) $\Delta S_{system} > 0$ only
(d) $\Delta S_{surroundings} > 0$ only

15. Identify the correct statement for change of Gibbs energy for a system (ΔG_{system}) at constant temperature and pressure
(a) If $\Delta G_{system} = 0$, the system has attained equilibrium
(b) If $\Delta G_{system} = 0$, the system is still moving in a particular direction
(c) If $\Delta G_{system} < 0$, the process is not spontaneous
(d) If $\Delta G_{system} > 0$, the process is not spontaneous

16. During complete combustion of one mole of butane, 2658 kJ of heat is released. The thermochemical reaction for above change is
(a) $2C_4H_{10}(g) + 13O_2(g) \rightarrow 8CO_2(g) + 10H_2O(l)$
$\Delta_c H = -2658.0\ kJ\ mol^{-1}$
(b) $C_4H_{10}(g) + \frac{13}{2}O_2(g) \rightarrow 4CO_2(g) + 5H_2O(g)$
$\Delta_c H = -1329.0\ kJ\ mol^{-1}$
(c) $C_4H_{10}(g) + \frac{13}{2}O_2(g) \rightarrow 4CO_2(g) + 5H_2O(l)$
$\Delta_c H = -2658.0\ kJ\ mol^{-1}$
(d) $C_4H_{10}(g) + \frac{13}{2}O_2(g) \rightarrow 4CO_2(g) + 5H_2O(l)$
$\Delta_c H = +2658.0\ kJ\ mol^{-1}$

17. Identify the correct statement regarding a spontaneous process:
(a) Lowering of energy in the process is the only criterion for spontaneity.
(b) For a spontaneous process in an isolated system, the change in entropy is positive.
(c) Endothermic processes are never spontaneous.
(d) Exothermic processes are always spontaneous.

18. Identify the correct statement regarding entropy.
(a) At absolute zero temperature, entropy of a perfectly crystalline substance is taken to be zero.
(b) At absolute zero temperature, the entropy of a perfectly crystalline substance is positive.
(c) Absolute entropy of a substance cannot be determined.
(d) At 0°C, the entropy of a perfectly crystalline substance is taken to be zero

19. Which of the following is not correct?
(a) ΔG is zero at equilibrium.
(b) ΔG is positive for a spontaneous reaction.
(c) ΔG is negative for a spontaneous reaction.
(d) ΔG is positive for a non-spontaneous reaction.

20. A chemical reaction will be spontaneous if it is accompanied by a decrease of
(a) entropy of the system.
(b) enthalpy of the system.
(c) internal energy of the system.
(d) free energy of the system.

21. According to the third law of thermodynamics which one of the following quantities for a perfectly crystalline solid is zero at absolute zero?
(a) Free energy (b) Entropy
(c) Enthalpy (d) Internal energy

22. For the reaction

$$C_3H_8(g) + 5O_2(g) \rightarrow 3CO_2(g) + 4H_2O_{(l)}$$

at constant temperature, $\Delta H - \Delta E$ is
(a) $-RT$ (b) $+RT$
(c) $-3\,RT$ (d) $+3\,RT$

23. Which of the following pairs of a chemical reaction is certain to result in a spontaneous reaction?
(a) Exothermic and increasing disorder
(b) Exothermic and decreasing disorder
(c) Endothermic and increasing disorder
(d) Endothermic and decreasing disorder

24. Assume each reaction is carried out in an open container. For which reaction will $\Delta H = \Delta E$?
(a) $C(s) + 2H_2O(g) \rightarrow 2H_2(g) + CO_2(g)$
(b) $PCl_5(g) \rightarrow PCl_3(g) + Cl_2(g)$
(c) $2CO(g) + O_2(g) \rightarrow 2CO_2(g)$
(d) $H_2(g) + Br_2(g) \rightarrow 2HBr(g)$

25. Which of the following are *not* state functions ?
(I) $q+w$ (II) q
(III) w (IV) $H-TS$
(a) (I) and (IV) (b) (II), (III) and (IV)
(c) (I), (II) and (III) (d) (II) and (III)

26. The heat of combustion of a substance is :
(a) Always positive
(b) Always negative
(c) Numerically equal to the heat of formation
(d) Unpredictable

27. In which of the following entropy decreases?
(a) Crystallization of sucrose solution
(b) Rusting of iron
(c) Melting of ice
(d) Vaporization of camphor

28. Calorie is equivalent to :
(a) 0.4184 Joule (b) 4.184 Joule
(c) 41.84 Joule (d) 418.4 Joule

29. A spontaneous reaction is impossible if
(a) both ΔH and ΔS are negative
(b) both ΔH and ΔS are positive
(c) ΔH is negative and ΔS is positive
(d) ΔH is positive and ΔS is negative

30. In an endothermic reaction, the value of change in enthalpy (ΔH) is
(a) positive (b) zero
(c) negative (d) either (b) or (c)

31. Enthalpy of an exothermic reaction is always
(a) positive (b) zero
(c) negative (d) may be (a) or (b)

32. Which of the following is combustion reaction?
(a) $C+O_2 \rightarrow CO_2$
(b) $CH_4+O_2 \rightarrow CO_2+H_2O$
(c) $Mg+O_2 \rightarrow MgO$
(d) All of these

33. A heat engine absorbs heat Q_1 at temperature T_1 and heat Q_2 at temperature T_2. Work done by the engine is J (Q_1+Q_2). This data
(a) violates 1st law of thermodynamics
(b) violates 1st law of thermodynamics if Q_1 is –ve
(c) violates 1st law of thermodynamics of Q_2 is –ve
(d) does not violate 1st law of thermodynamics

34. An ideal gas is allowed to expand both reversibly and irreversibly in an isolated system. If T_i is the initial temperature and T_f is the final temperature, which of the following statements is correct?
(a) $(T_f)_{rev} = (T_f)_{irrev}$
(b) $T_f = T_i$ for both reversible and irreversible processes
(c) $(T_f)_{irrev} > (T_f)_{rev}$
(d) $T_f > T_i$ for reversible process but $T_f = T_i$ for irreversible process

35. For the reaction $CO(g)+\frac{1}{2}O_2(g) \rightarrow CO_2(g)$
Which one of the statement is correct at constant T and P ?
(a) $\Delta H = \Delta E$
(b) $\Delta H < \Delta E$
(c) $\Delta H > \Delta E$
(d) ΔH is independent of physical state of the reactants

36. On the basis of thermochemical equations (a), (b) and (c), find out which of the algebric relationships given in options (i) to (iv) is correct.
(1) $C(\text{graphite})+O_2(g) \rightarrow CO_2(g);$ $\Delta_r H = x\ \text{kJ mol}^{-1}$
(2) $C(\text{graphite})+\frac{1}{2}O_2(g) \rightarrow CO(g);$ $\Delta_r H = y\ \text{kJ mol}^{-1}$
(3) $CO(g)+\frac{1}{2}O_2(g) \rightarrow CO_2(g);$ $\Delta_r H = z\ \text{kJ mol}^{-1}$
(a) $z = x+y$ (b) $x = y-z$
(c) $x = y+z$ (d) $y = 2z-x$

37. Oxidising power of chlorine in aqueous solution can be determined by the parameters indicated below:

$$\frac{1}{2}Cl_2(g) \xrightarrow{\frac{1}{2}\Delta_{diss}H^\ominus} Cl(g) \xrightarrow{\Delta_{eg}H^\ominus} Cl^-(g) \xrightarrow{\Delta_{Hyd}H^\ominus} Cl^-(aq)$$

using the data,
$\Delta_{diss} H^\ominus_{Cl_2} = 240\ \text{kJ mol}^{-1}$,
$\Delta_{eg} H^\ominus_{Cl} = -349\ \text{kJ mol}^{-1}$,
$\Delta_{hyd} H^\ominus_{Cl^-} = -381\ \text{kJ mol}^{-1}$, will be
(a) $+152\ \text{kJ mol}^{-1}$ (b) $-850\ \text{kJ mol}^{-1}$
(c) $-610\ \text{kJ mol}^{-1}$ (d) $+120\ \text{kJ mol}^{-1}$

38. The enthalpies of elements in their standard states are taken as zero. The enthalpy of formation of a compound
(a) is always negative
(b) is always positive
(c) may be positive or negative
(d) is never negative

39. If ΔH is the change in enthalpy and ΔE, the change in internal energy accompanying a gaseous reaction, then
(a) ΔH is always greater than ΔE
(b) $\Delta H < \Delta E$ only if the number of moles of the products is greater than the number of moles of the reactants
(c) ΔH is always less than ΔE
(d) $\Delta H < \Delta E$ only if the number of moles of products is less than the number of moles of the reactants

40. Consider the following reaction occurring in an automobile
$$2C_8H_{18}(g) + 25O_2(g) \longrightarrow 16CO_2(g) + 18H_2O(g)$$
the sign of ΔH, ΔS and ΔG would be
(a) +, –, + (b) –, +, –
(c) –, +, + (d) +, +, –

41. If enthalpies of formation of $C_2H_4(g)$, $CO_2(g)$ and $H_2O_{(l)}$ at 25°C and 1atm pressure are 52, – 394 and – 286 kJ/mol respectively, the change in ethalpy is equal to
(a) – 141.2 kJ/mol (b) – 1412 kJ/mol
(c) + 14.2 kJ/mol (d) + 1412 kJ/mol

42. The correct relationship between free energy and equilibrium constant K of a reaction is
(a) $\Delta G° = -RT\,ln K$ (b) $\Delta G = RT\,ln K$
(c) $\Delta G° = RT\,ln K$ (d) $\Delta G = -RT\,ln K$

43. Given the following entropy values (in $J K^{-1} mol^{-1}$) at 298 K and 1 atm : $H_2(g)$: 130.6, $Cl_2(g)$: 223.0, $HCl(g)$: 186.7. The entropy change (in $J K^{-1} mol^{-1}$) for the reaction
$H_2(g) + Cl_2(g) \longrightarrow 2HCl(g)$ is
(a) +540.3 (b) +727.0
(c) –166.9 (d) +19.8

44. One mole of an ideal gas at 300 K is expanded isothermally from an initial volume of 1 litre to 10 litres. The ΔE for this process is (R = 2 cal. $mol^{-1} K^{-1}$)
(a) 163.7 cal (b) zero
(c) 1381.1 cal (d) 9 lit. atm

45. For a cyclic process, which of the following is not true?
(a) $\Delta H = 0$ (b) $\Delta E = 0$
(c) $\Delta G = 0$ (d) Total W = 0

46. For a reaction in which all reactants and products are liquids, which one of the following equations is most applicable ?
(a) $\Delta H < \Delta E$ (b) $\Delta H = \Delta S$
(c) $\Delta H = \Delta E$ (d) $\Delta H = \Delta G$

47. The entropy change in the fusion of one mole of a solid melting at 27°C (Latent heat of fusion, 2930 J mol^{-1}) is :
(a) 9.77 $J K^{-1} mol^{-1}$ (b) 10.73 $J K^{-1} mol^{-1}$
(c) 2930 $J K^{-1} mol^{-1}$ (d) 108.5 $J K^{-1} mol^{-1}$

48. The values of heat of formation of SO_2 and SO_3 are –298.2 kJ and –98.2 kJ. The heat of formation of the reaction
$SO_2 + (1/2)O_2 \rightarrow SO_3$ will be
(a) –200 kJ (b) –356.2 kJ
(c) +200 kJ (d) –396.2 kJ

49. Enthalpy of $CH_4 + \frac{1}{2}O_2 \rightarrow CH_3OH$ is negative. If enthalpy of combustion of CH_4 and CH_3OH are x and y respectively, then which relation is correct
(a) $x > y$ (b) $x < y$
(c) $x = y$ (d) $x \geq y$

50. When 1 mol of a gas is heated at constant volume, temperature is raised from 298 to 308 K. If heat supplied to the gas is 500 J, then which statement is correct ?
(a) $q = w = 500$ J, $\Delta U = 0$
(b) $q = \Delta U = 500$ J, $w = 0$
(c) $q = -w = 500$ J, $\Delta U = 0$
(d) $\Delta U = 0$, $q = w = -500$ J

51. What is the enthalpy change for,
$2H_2O_2(l) \rightarrow 2H_2O(l) + O_2(g)$ if heat of formation of H_2O_2 (l) and H_2O (l) are –188 and –286 kJ/mol respectively?
(a) –196 kJ/mol (b) +948 kJ/mol
(c) + 196 kJ/mol (d) –948 kJ/mol

52. Heat of combustion $\Delta H°$ for C (s), H_2 (g) and CH_4 (g) are –94, –68 and –213 kcal/mol, then $\Delta H°$ for $C(s) + 2H_2(g) \rightarrow CH_4(g)$ is
(a) –17 kcal (b) – 111 kcal
(c) –170 kcal (d) –85 kcal

53. What is the entropy change (in $JK^{-1}\ mol^{-1}$) when one mole of ice is converted into water at 0° C? (The enthalpy change for the conversion of ice to liquid water is 6.0 kJ mol^{-1} at 0°C)
(a) 21.98 (b) 20.13
(c) 2.013 (d) 2.198

54. For which one of the following equations is ΔH°_{react} equal to ΔH_f° for the product?
(a) $2CO(g) + O_2(g) \rightarrow 2CO_2(g)$
(b) $N_2(g) + O_3(g) \rightarrow N_2O_3(g)$
(c) $CH_4(g) + 2Cl(g) \rightarrow CH_2Cl_{2(l)} + 2HCl(g)$
(d) $Xe(g) + 2F_2(g) \rightarrow XeF_4(g)$

55. If the bond energies of H–H, Br–Br, and H-Br are 433, 192 and 364 kJ mol^{-1} respectively, the ΔH° for the reaction
$H_2(g) + Br_2(g) \rightarrow 2HBr(g)$ is
(a) –261 kJ (b) +103 kJ
(c) +261kJ (d) – 103 kJ

56. Standard enthalpy and standard entropy changes for the oxidation of ammonia at 298 K are – 382.64 kJ mol^{-1} and –145.6 $JK^{-1}\ mol^{-1}$, respectively. Standard Gibb's energy change for the same reaction at 298 K is
(a) –22.1 kJ mol^{-1} (b) –339.3 kJ mol^{-1}
(c) –439.3 kJ mol^{-1} (d) –523.2 kJ mol^{-1}

57. The work done during the expansion of a gas from a volume of 4 dm^3 to 6 dm^3 against a constant external pressure of 3 atm is (1 L atm = 101.32 J)
(a) – 6 J (b) – 608 J
(c) + 304 J (d) – 304 J

58. A reaction occurs spontaneously if
(a) TΔS < ΔH and both ΔH and ΔS are + ve
(b) TΔS > ΔH and ΔH is + ve and ΔS is – ve
(c) TΔS > ΔH and both ΔH and ΔS are + ve
(d) TΔS = ΔH and both ΔH and ΔS are + ve

59. The enthalpy and entropy change for the reaction
$Br_2(l) + Cl_2(g) \rightarrow 2BrCl(g)$
are 30kJ mol^{-1} and 105 $JK^{-1}\ mol^{-1}$ respectively. The temperature at which the reaction will be in equilibrium is
(a) 273 K (b) 450 K
(c) 300 K (d) 285.7K

60. For the gas phase reaction,
$PCl_5(g) \rightleftharpoons PCl_3(g) + Cl_2(g)$
which of the following conditions are correct ?
(a) ΔH = 0 and ΔS < 0
(b) ΔH > 0 and ΔS > 0
(c) ΔH < 0 and ΔS < 0
(d) ΔH > 0 and ΔS < 0

CHAPTER 7

Equilibrium

CONCEPTS

TYPES OF REACTIONS

- **Irreversible reactions :** Such reactions which go only in the forward direction are called irreversible reactions, For example,

$$Pb(NO_3)_2(aq) + K_2CrO_4(aq) \longrightarrow PbCrO_4(s) + 2KNO_3(aq)$$

The reaction in which the products do not react to give back the reactants are called irreversible reactions.

- **Reversible reactions :** The reaction which can take place in both the directions under similar conditions of temperature and pressure are called reversible reaction. Such reactions are carried out in a closed vessel e.g.
  (i) $H_2(g) + I_2(g) \rightleftharpoons 2HI(g)$
  (ii) $N_2(g) + 3H_2(g) \rightleftharpoons 2NH_3(g)$

EQUILIBRIUM :

The state at which the observable properties of system do not change with time is called state of equilibrium.

(i) **Physical equilibrium :** The state of equilibrium attained between two physical phases of any chemical compound is termed as physical equilibrium. e.g.

(a) Equilibrium between solid and its liquid.
Solid $\rightleftharpoons$ liquid
For this equilibrium
Rate of melting = Rate of freezing.

(b) Equilibrium between liquid and its vapours.
Liquid $\rightleftharpoons$ Gas
For this equilibrium
Rate of evaporation = Rate of condensation.

(c) Equilibrium between solids which sublime to vapour phase
$I_2(\text{solid}) \rightleftharpoons I_2(\text{vapour})$
Rate of sublimation of solid = Rate of sublimation of vapours

- **Henry's law :** According to Henry's law the mass of gas dissolved per unit mass of the solvent at a specific temperature is proportional to the pressure of the gas in equilibrium with the solution. $m \propto p$ or $m = kp$
where m = mass of the gas dissolved in unit mass of solvent.

(ii) **Chemical equilibrium :** When a reversible reaction is carried out in a closed vessel a stage is reached when the speed of forward reaction equals the speed of the backward reaction and chemical equilibrium is said to be established.

- **General characteristics of physical and chemical equilibria :**
  (i) Equilibria can be attained only in a closed vessel.
  (ii) An equilibrium is **dynamic** in nature (achieved when forward and backward both reactions takes place at equal speed).
  (iii) It can be attained from either side of the reaction.
  (iv) Equilibrium can be established only if none of the products is allowed to escape.
  (v) The measurable properties at equilibrium are constant (concentration of reactants and products become constant).
  (vi) **Catalyst does not alter state of equilibrium.** It changes time required to attain equilibrium.

LAW OF MASS ACTION AND CHEMICAL EQUILIBRIUM

Law of Mass Action :

The rate of a chemical reaction is proportional to the product of the active masses of the reactants, each raised to the power given by its coefficient as represented in the balanced chemical equation.

$$aA + bB \longrightarrow$$
$$r = K[A]^a [B]^b$$

Law of Chemical Equilibrium

For reaction $aA + bB \rightleftharpoons xX + yY$

$$K = \frac{[X]^x[Y]^y}{[A]^a[B]^b}$$

where K = equilibrium constant,

$$\frac{[X]^x[Y]^y}{[A]^a[B]^b} = \text{concentration quotient (Q)}$$

EQUILIBRIUM CONSTANT AND ITS CHARACTERIST59ICS

Equilibrium Constant

Equilibrium constant in terms of concentration (K_c)	Equilibrium constant in terms of pressure (K_p)
$K_c = \frac{[X]^x[Y]^y}{[A]^a[B]^b}$	$K_p = \frac{[P_X]^x \times [P_Y]^y}{[P_A]^a \times [P_B]^b}$
Units K_c = (mol L^{-1}) $^{(x+y)-(a+b)}$	Units of K_p = (atm) $^{(x+y)-(a+b)}$

Characteristics of Equilibrium Constant

(i) Its value depends on temperature. It has fixed value for given temperature.

(ii) The value of K is independent of
- (a) Catalyst
- (b) Initial concentration of reactants
- (c) Direction from which equilibrium is attained
- (d) Presence of inert material.

(iii) When reaction is reversed value of K changes to K′

$$K' = \frac{1}{K}$$

(iv) When reaction is divided by 2

$$K' = \sqrt{K}$$

(v) When reaction is multiplied by 2

$$K' = (K)^2$$

(vi) When reaction is written in two steps each having equilibrium constant as K_1 and K_2 then

$$K = K_1 \times K_2$$

(vii) The magnitude of equilibrium constant is helpful in predicting the relative amount of the reactants and the products of a reversible reaction.

If $K_c > 10^3$, products predominate over reactants, i.e., if K_c is very large, the reaction proceeds nearly to completion.

If $K_c < 10^{-3}$, reactants predominate over products, i.e., if K_c is very small, the reaction proceeds rarely.

Relation between K_p and K_c

$$K_p = K_c RT^{(\Delta n)}$$

where $\Delta n = (n_p - n_r)_{gaseous}$

When $\Delta n = 0$ $K_p = K_c$

When $\Delta n > 0$ $K_p > K_c$

When $\Delta n < 0$ $K_p < K_c$

CONCENTRATION QUOTIENT OR REACTION QUOTIENT (Q)

$$aA + bB \rightleftharpoons xX + yY$$

$$Q = \frac{[X]^x[Y]^y}{[A]^a[B]^b}$$

(i) when Q = K, the reaction is in equilibrium.

(ii) when Q > K, the net reaction proceed in backward direction.

(iii) when Q < K, the net reaction proceed in forward direction.

- A relationship between the equilibrium constant K, reaction quotient and Gibbs energy.

  $\Delta G = \Delta G° + RT \ln Q$

  At equilibrium $\Delta G = 0$ and Q = K then $\Delta G° = -RT \ln K_c$

  $\therefore$ $\Delta G° = -RT \ln K_p$

LE CHATELIER'S PRINCIPLE

If a system in equilibrium is subjected to change in factors that determine the equilibrium conditions of a system, the equilibrium shifts in a direction as to nullify or undo the effect of change imposed.

FACTORS EFFECTING EQUILIBRIUM

- **Effect of Change of Concentration :**
  - (i) Increase in concentration of any of the reactants, shifts the equilibrium in the forward direction.
  - (ii) Increase in the concentration of any of the products, shifts the equilibrium in the backward direction.
- **Effect of Change of Temperature :**
  - (i) Endothermic reactions shift the equilibrium in the forward direction with increase of

temperature.

(ii) Exothermic reactions shift the equilibrium in the forward direction with decrease of temperature.

- **Effect of Change of Pressure :**
  (i) Increase of pressure shifts the equilibrium in the direction where total number of moles decreases or $\Delta n < 0$
  (ii) Decrease of pressure alters the equilibrium in the direction where total number of moles increases or $\Delta n > 0$
  (iii) Change of pressure has no effect on equilibrium, if the number of moles of gaseous products is equal to moles of gaseous reactants i.e. $\Delta n = 0$

- **Effect of Additon of Catalyst :**
  Addition of catalyst has no effect on equilibrium because the catalyst alters the rate of forward and backward reaction equally.

- **Effect of Addition of Inert Gas (Argon):**
  At constant volume :
  No effect, what so ever is the value of Δn
  At constant pressure :
  Addition of inert gas favours forward reaction when $\Delta n > 0$
  Addition of inert gas favours backward reaction when $\Delta n < 0$
  Addition of inert gas has no effect when $\Delta n = 0$

- **Effect of Temperature, Pressure, Concentration and Addition of Inert Gas on Equilibrium :**

	Effect of increase in pressure	Effect of increase in concentration of reactant	Effect of increase in temperature	Effect of addition of inert gas
$2SO_2 + O_2 \rightleftharpoons 2SO_3 + x$	Forward	Forward	Backward	Backward
$2NH_3 \rightleftharpoons N_2 + 3H_2 - 93.0kJ$	Backward	Forward	Forward	Forward
$N_2O_4 \rightleftharpoons 2NO_2 - 62\ kJ$	Backward	Forward	Forward	Forward
$N_2 + O_2 \rightleftharpoons 2NO + 180\ kJ$	No effect	Forward	Backward	No.effect
$PCl_5 \rightleftharpoons PCl_3 + Cl_2 - x\ kJ$	Backward	Forward	Forward	Forward
$H_2 + I_2 \rightleftharpoons 2HI - 12.4\ kJ$	No effect	Forward	Forward	No. effect
ice + heat $\rightleftharpoons$ water	Forward	Forward	Forward	–
water + heat $\rightleftharpoons$ water(vapours)	Backward	Forward	Forward	–

Electrolyte
(Conducts electricity in molten and aqueous state)

- Strong electrolyte dissociates completely e.g.- HCl, NaOH
- Weak electrolyte dissociate to small extent e.g.- NH_4Cl, CH_3COOH

CONCEPTS OF ACIDS AND BASES

Property	Arrhenius concept	Bronsted-Lowry concept	Lewis concept
1. **Acids**	give H^+ in aq. solution	H^+ donors	e^- acceptors
2. **Bases**	give OH^- in aq. solution	H^+ acceptors	e^- donors
3. **Fails to explain**	• Behaviour of acids and bases in non aqueous solution • Acidic character of salts like $AlCl_3$, BF_3 and basic character of NH_3, PH_3 etc.	• Acidic and basic properties of compounds which do not have H^+ ions	

Conjugate Acid-Base Pair :

(i) It is pair of acid and base which **differ by H^+ ion.**

conjugate acid $-H^+ \rightleftharpoons$ **conjugate base.**

Base $+ H^+ \rightleftharpoons$ Conjugate acid

(ii) **Weak acid has strong conjugate base** and vice-versa

- **Examples of Lewis acids**
  (i) **Molecules in which cation has incomplete octet** e.g. $AlCl_3$, BF_3, $FeCl_3$.
  (ii) **Cations** – Na^+, Fe^{3+}, Ag^+
  (iii) **Molecules having multiple bonds** between different atoms e.g. SO_2, CO_2
  (iv) **Molecules having empty d-orbitals.** e.g. $SnCl_4$
- **Examples of Lewis bases**
  (i) **Neutral molecules having lone pair of electrons on central atom** e.g. $\ddot{N}H_3, R\ddot{N}H_2$
  (ii) **Anions** - CN^-, Cl^-, OH^-

IONIC PRODUCT OF WATER (K_W)

Product of molar concentrations of H^+ and OH^-.

$$K_w = [H^+][OH^-]$$

$$K_w = 10^{-14} \text{ at } 25°C.$$

- Value of K_w increases with temperature ($\because$ dissociation of H_2O increases with increase in temperature)
- Value of **K_w is not effected by addition of salt or acid or base if temperature is constant.**

pH SCALE (GIVEN BY SORENSEN)

Used to express the acidic and basic strength of a solution.

0 ←— acidic —→ 7 ←— basic —→ 14

(i) $pH = -\log [H^+]$

(ii) $pOH = -\log [OH^-]$

(iii) $pH + pOH = 14$ at 298 K.

(iv) **If pH = 0**, then solution is **acidic**.

(v) pH < 7 for acidic solution
pH > 7 for basic solution
pH = 7 for neutral solution

(vi) $\mathbf{pH} \propto \dfrac{1}{\textbf{acidity}}$

(vii) **pH of water decreases** with **increase in temperature** because ionic product of water increases.

ACID-BASE INDICATORS (OSTWALD THEORY)

(i) Acid-base indicators are complex organic molecules that change colour in solution when pH changes.

(ii) They are either weak organic acids (e.g.- phenolphthalein) or weak organic-bases (e.g.- methyl orange)

$$\underset{\substack{\text{Phenolpthalein}\\\text{(colourless)}}}{HPh} \rightleftharpoons H^+ + \underset{\text{(Pink)}}{Ph^-}$$

$$\underset{\substack{\text{Methyl orange}\\\text{(yellow)}}}{MeOH} \rightleftharpoons \underset{\text{(red)}}{Me^+} + OH^-$$

IONISATION OF WEAK ACIDS

$$HX(aq) + H_2O(l) \rightleftharpoons H_3O^+(aq) + X^-(aq)$$

Initial	C	0	0
At equilibrium	C – Cα	Cα	Cα

$$K_a = \frac{C^2\alpha}{C(1-\alpha)} = \frac{C\alpha 2}{(1-2)}$$

$$pK_a = -\log (K_a)$$

Ionisation of weak bases

$$MOH(aq) \rightleftharpoons M^+(aq) + OH^-(aq)$$

$$K_b = \frac{C\alpha^2}{(1-\alpha)}$$

$$pK_b = -\log (K_b)$$

RELATIVE STRENGTH OF ACIDS AND BASES

$$\frac{\text{Strength of acid } HA_1}{\text{Strength of acid } HA_2} = \frac{\alpha_1}{\alpha_2} = \sqrt{\frac{K_{a1}}{K_{a_2}}}$$

$$PK_a + PK_b = PK_w = 14$$

$$\frac{\text{Strength of base } (BOH)_1}{\text{Strength of base } (BOH)_2} = \sqrt{\frac{K_{b_1}}{K_{b_2}}}$$

- **Greater the value of K_a stronger the acid.**
- For **polyprotic acids** $\mathbf{K_{a_1} > K_{a_2} > K_{a_3}}$
  (K_{a_1}, K_{a_2} and K_{a_3} are first, second and third dissociation constants).
- **Common ion effect** means that dissociation of weak electrolyte is further supressed in presence of strong electrolyte having common ion e.g.-

dissociation of CH_3COOH is supressed in presence of CH_3COONa.

- **Hydrolysis constant (K_h)**

$$\underset{(salt)}{BA} + H_2O \rightleftharpoons \underset{(acid)}{HA} + \underset{(base)}{BOH}$$

$$K_h = \frac{[HA][BOH]}{[BA]}$$

- Degree of hydrolysis (h)

$$= \frac{\text{No. of moles of salt hydrolysed}}{\text{Total no. of moles of salt taken}}$$

SALT HYDROLYSIS

Salt hydrolysis is defined as " a process in which a salt reacts with water to produce either acidic or basic or neutral solution."

(i) **Salts of strong acid and strong base** e.g., NaCl, KNO_3, Na_2SO_4, $KClO_4$, etc. Such salts on hydrolysis produce equally strong acids and bases. Thus, their solutions are almost neutral with pH = 7.

(ii) **Salts of weak acid and strong base** e.g., Na_2CO_3, KCN, CH_3COONa, K_2CO_3, etc. Hydrolysis of these salts produce weak acid and strong base and thus their aqueous solutions are basic (pH > 7).

$$K_h = \frac{K_w}{K_a};$$

$$\text{Degree of hydrolysis, } h = \frac{\sqrt{K_h}}{C} = \sqrt{\frac{K_w}{K_a C}}$$

$$pH = 7 + \frac{1}{2}pK_a + \frac{1}{2}\log C \text{ or } -\frac{1}{2}[\log K_w + \log K_a - \log C]$$

Here K_h is hydrolysis constant of salt K_w and K_a are ionisation constants of water and weak acid respectively. C is the concentration of salt in mole/litre.

(iii) **Salt of strong acid and weak base.** e.g., NH_4Cl, $CuSO_4$, NH_4NO_3, $AlCl_3$, etc. These produce strong acid and weak base on hydrolysis resulting in acidic nature of solution (pH < 7).

$$K_h = \frac{K_w}{K_b};\ h = \sqrt{\frac{K_w}{K_b C}}$$

$$pH = 7 - \frac{1}{2}pK_b - \frac{1}{2}\log C \text{ or } -\frac{1}{2}[\log K_w - \log K_b + \log C]$$

(iv) **Salts of weak acids and weak base** e.g., $(NH_4)_2CO_3$, $CH_3COO^- NH_4^+$, $AlPO_4$ etc. Their hydrolysis give weak acids and weak base.

$$K_h = \frac{K_w}{K_a K_b};$$

$$pH = 7 + \frac{1}{2}pK_a - \frac{1}{2}pK_b \text{ or } -\frac{1}{2}\log K_a + \log K_w - \log K_b$$

BUFFER SOLUTION

The solution which resists the change in hydrogen ion concentration on the addition of small amount of acid or base.

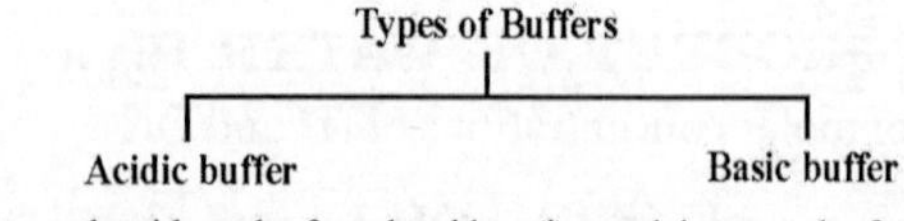

(i) **For acidic buffers**

$$pH = pK_a + \left[\frac{salt}{acid}\right] \text{ or } pH = pK_a + \log\left[\frac{\text{conjugate base}}{\text{acid}}\right]$$

(ii) **For basic buffers**

$$pOH = pK_b + \left[\frac{salt}{base}\right] \text{ or } pH = pK_b + \log\left[\frac{\text{conjugate acid}}{\text{base}}\right]$$

- **Buffer Capacity or Buffer index**

$$= \frac{\text{No. of moles of acid or base added per litre}}{\text{Change in pH}}$$

- **Buffer**

Blood and sea water are examples of natural buffer.

SOLUBILITY PRODUCT (K_{SP})

Defined as product of molar concentration of ions of an electrolyte in a concentrated solution, when each concentration term raised to the power equal to the number of ions produced on dissociation of one molecule of electrolyte.

$$A_xB_y \rightleftharpoons xA^+ + yB^-$$

$$K_{sp} = [A^{y+}]^x \times [B^{x-}]^y = (xs)^x (ys)^y$$

$$= x^x . y^y S^{(x+y)}$$

$$S^{(x+y)} = \frac{K_{sp}}{x^x . y^y}$$

$$S = \left(\frac{K_{sp}}{x^x . y^y}\right)^{\frac{1}{x+y}}$$

(i) **Solubility product** of a substance has **constant** value **for given temperature**.

(ii) Solubility product is **used in qualitative salt analysis.**

(iii) Ionic product > K_{sp} precipitation occurs

K_{sp} > ionic product no precipitation occurs

(iv) **Solubility product is used for purification of common salt :** When HCl is passed in solution of impure NaCl, pure NaCl precipitates when ionic product of NaCl exceeds its solubility product.

(v) Precipitation of soap by adding NaCl is called **salting out of soap**.

Exercise

1. Which of the following is not a general characteristic of equilibria involving physical processes ?
   (a) Equilibrium is possible only in a closed system at a given temperature.
   (b) All measurable properties of the system remain constant.
   (c) All the physical processes stop at equilibrium.
   (d) The opposing processes occur at the same rate and there is dynamic but stable condition.
2. According to Le-chatelier's principle, adding heat to a solid $\rightleftharpoons$ liquid equilibrium will cause the
   (a) temperature to increase
   (b) temperature to decrease
   (c) amount of liquid to decrease
   (d) amount of solid to decrease
3. When pressure is applied to the equilibrium system

   Ice $\rightleftharpoons$ Water

   Which of the following phenomenon will happen?
   (a) More ice will be formed
   (b) Water will evaporate
   (c) More water will be formed
   (d) Equilibrium will not be formed
4. K_1 and K_2 are equilibrium constant for reactions (1) and (2)

   $N_2(g) + O_2(g) \rightleftharpoons 2NO(g)$ ....(1)

   $NO(g) \rightleftharpoons \frac{1}{2}N_2(g) + \frac{1}{2}O_2(g)$ ......(2)

   Then,

   (a) $K_1 = \left(\frac{1}{K_2}\right)^2$ (b) $K_1 = K_2^2$

   (c) $K_1 = \frac{1}{K_2}$ (d) $K_1 = (K_2)^0$
5. The value of equilibrium constant of the reaction

   $HI(g) \rightleftharpoons \frac{1}{2}H_2(g) + \frac{1}{2}I_2$ is 8.0

   The equilibrium constant of the reaction

   $H_2(g) + I_2(g) \rightleftharpoons 2HI(g)$ will be:

   (a) $\frac{1}{16}$ (b) $\frac{1}{64}$

   (c) 16 (d) $\frac{1}{8}$
6. In which of the following equilibrium K_c and K_p are not equal?

   (a) $2NO(g) \rightleftharpoons N_2(g) + O_2(g)$

   (b) $SO_2(g) + NO_2(g) \rightleftharpoons SO_3(g) + NO(g)$

   (c) $H_2(g) + I_2(g) \rightleftharpoons 2HI(g)$

   (d) $2C(s) + O_2(g) \rightleftharpoons 2CO_2(g)$
7. The reaction quotient (Q) for the reaction

   $N_2(g) + 2H_2(g) \rightleftharpoons 2NH_3(g)$

   is given by $Q = \frac{[NH_3]^2}{[N_2][H_2]^3}$. The reaction will proceed from right to left if

(a) $Q=0$ (b) $Q=K_c$
(c) $Q<K_c$ (d) $Q>K_c$

where K_c is the equilibrium constant

8. Which one of the following information can be obtained on the basis of Le Chatelier principle?
(a) Dissociation constant of a weak acid
(b) Entropy change in a reaction
(c) Equilibrium constant of a chemical reaction
(d) Shift in equilibrium position on changing value of a constraint

9. Which equilibrium can be described as an acid-base reaction using the Lewis acid-base definition but not using the Bronsted-Lowry definition?
(a) $2NH_3 + H_2SO_4 \rightleftharpoons 2NH_4^+ + SO_4^{2-}$
(b) $NH_3 + CH_3COOH \rightleftharpoons NH_4^+ + CH_3COO^-$
(c) $H_2O + CH_3COOH \rightleftharpoons H_3O^+ + CH_3COO^-$
(d) $[Cu(H_2O)_4]^{2-} + 4NH_3 \rightleftharpoons [Cu(NH_3)_4]^{2+} + 4H_2O$

10. Conjugate acid of NH_2^- is :
(a) NH_4^+ (b) NH_3
(c) NH_2 (d) NH

11. The pH value of a 10 M solution of HCl is
(a) less than 0 (b) equal to 0
(c) equal to 1 (d) equal to 2

12. K_{a_1}, K_{a_2} and K_{a_3} are the respective ionisation constants for the following reactions.

$H_2S \rightleftharpoons H^+ + HS^-$

$HS^- \rightleftharpoons H^+ + S^{2-}$

$H_2S \rightleftharpoons 2H^+ + S^{2-}$

The correct relationship between K_{a_1}, K_{a_2} and K_{a_3} is

(a) $K_{a_3} = K_{a_1} \times K_{a_2}$
(b) $K_{a_3} = K_{a_1} + K_{a_2}$
(c) $K_{a_3} = K_{a_1} - K_{a_2}$
(d) $K_{a_3} = K_{a_1} / K_{a_2}$

13. When HCl gas is passed through a saturated solution of NaCl, the solubility of NaCl
(a) increases
(b) decreases
(c) does not change
(d) NaCl is decomposed

14. $FeCl_3$ solution is
(a) Acidic (b) Basic
(c) Neutral (d) Amphoteric

15. Which of the following pairs constitutes a buffer?
(a) NaOH and NaCl
(b) HNO_3 and NH_4NO_3
(c) HCl and KCl
(d) HNO_2 and $NaNO_2$

16. On increasing the pressure, the gas phase reaction proceed forward to re-establish equilibrium, as predicted by applying the Le Chatelier's principle. Consider the reaction.

$$N_2(g) + 3H_2(g) \rightleftharpoons 2NH_3(g)$$

Which of the following is correct, if the total pressure at which the equilibrium is established, is increased without changing the temperature?
(a) K will remain same
(b) K will decrease
(c) K will increase
(d) K will increase initially and decrease when pressure is very high

17. Which of the following can act both as Bronsted acid and Bronsted base?
(a) Na_2CO_3 (b) OH^-
(c) HCO_3^- (d) NH_3

18. The ionisation constant of an acid, K_a, is the measure of strength of an acid. The K_a values of acetic acid, hypochlorous acid and formic acid are 1.74×10^{-5}, 3.0×10^{-8} and 1.8×10^{-4} respectively. Which of the following orders of pH of 0.1 mol dm^{-3} solutions of these acids is correct?
(a) acetic acid > hypochlorous acid > formic acid
(b) hypochlorous acid > acetic acid > formic acid
(c) formic acid > hypochlorous acid > acetic acid
(d) formic acid > acetic acid > hypochlorous acid

19. Aqueous solution of acetic acid contains
(a) CH_3COO^- and H^+
(b) CH_3COO^-, H_3O^+ and CH_3COOH
(c) CH_3COO^-, H_3O^+ and H^+
(d) CH_3COOH, CH_3COO^- and H^+

20. The rate constant for forward and backward reaction of hydrolysis of ester are 1.1×10^{-2} and 1.5×10^{-3} per minute respectively. Equilibrium constant for the reaction
$CH_3COOC_2H_5 + H^+ \rightleftharpoons CH_3COOH + C_2H_5OH$ is
(a) 4.33 (b) 5.33
(c) 6.33 (d) 7.33

21. For dibasic acid correct order is
(a) $K_{a_1} < K_{a_2}$ (b) $K_{a_1} > K_{a_2}$
(c) $K_{a_1} = K_{a_2}$ (d) not certain

22. Which one of the following compounds is not a protonic acid?
(a) $SO_2(OH)_2$ (b) $B(OH)_3$
(c) $PO(OH)_3$ (d) $SO(OH)_2$

23. H_2S gas when passed through a solution of cations containing HCl precipitates the cations of second group of qualitative analysis but not those belonging to the fourth group. It is because
(a) presence of HCl decreases the sulphide ion concentration.
(b) solubility product of group II sulphides is more than that of group IV sulphides.
(c) presence of HCl increases the sulphide ion concentration.
(d) sulphides of group IV cations are unstable in HCl.

24. If the concentration of OH^- ions in the reaction $Fe(OH)_3(s) \rightleftharpoons Fe^{3+}(aq) + 3OH^-(aq)$ is decreased by $\frac{1}{4}$ times, then equilibrium concentration of Fe^{3+} will increase by :
(a) 8 times (b) 16 times
(c) 64 times (d) 4 times

25. Which of the following is least likely to behave as Lewis base ?
(a) H_2O (b) NH_3
(c) BF_3 (d) OH^-

26. In a reversible chemical reaction having two reactants in equilibrium, if the concentration of the reactants are doubled then the equilibrium constant will
(a) Also be doubled
(b) Be halved
(c) Become one-fourth
(d) Remain the same

27. For the manufacture of ammonia by the reaction
$N_2 + 3H_2 \rightleftharpoons 2NH_3 + 2\,kcal$
the favourable conditions are
(a) Low temperature, low pressure and catalyst
(b) Low temperature, high pressure and catalyst
(c) High temperature, low pressure and catalyst
(d) High temperature, high pressure and catalyst

28. Neutralization of an acid with a base, invariably results in the production of
(a) H_3O^+ (b) H_2O
(c) H^+ and OH^- (d) OH^-

29. Which of the following reaction will be favoured at low pressure ?
(a) $H_2 + I_2 \rightleftharpoons 2HI$
(b) $N_2 + 3H_2 \rightleftharpoons 2NH_3$
(c) $PCl_5 \rightleftharpoons PCl_3 + Cl_2$
(d) $N_2 + O_2 \rightleftharpoons 2NO$

30. About buffer solution which is incorrect ?
(a) It contains a weak acid and its conjugate base
(b) It contains a weak base and its conjugate acid
(c) It shows change in pH on adding small amount of acid or base
(d) All of the above

31. A salt 'X' is dissolved in water of pH = 7. The salt is made resulting solution becomes alkaline in nature. The salt is made
(a) A strong acid and strong base
(b) A strong acid and weak base
(c) A weak acid and weak base
(d) A weak acid and strong base

32. The equilibrium which remains unaffected by pressure change is
(a) $N_2(g) + O_2(g) \rightleftharpoons 2NO(g)$
(b) $2SO_2(g) + O_2(g) \rightleftharpoons 2SO_3(g)$
(c) $2O_3(g) \rightleftharpoons 3O_2(g)$
(d) $2NO_2(g) \rightleftharpoons N_2O_4(g)$

33. In $A + B \rightleftharpoons C$. The unit of equilibrium constant is :
(a) Litre $mole^{-1}$ (b) Mole litre
(c) Mole $litre^{-1}$ (d) No unit

34. If s and S are respectively solubility and solubility product of a sparingly soluble binary electrolyte then :
(a) $s = S$ (b) $s = S^2$
(c) $s = S^{1/2}$ (d) $s = \frac{1}{2}S$

35. Suitable conditions for melting of ice :
(a) high temperature and high pressure
(b) high temperature and low pressure
(c) low temperature and low pressure
(d) low temperature and high pressure

36. A reaction is $A + B \rightleftharpoons C + D$. Initially we start with equal concentrations of A and B. At equilibrium we find that the moles of C is two times of A. What is the equilibrium constant of the reaction?
(a) $\frac{1}{4}$ (b) $\frac{1}{2}$
(c) 4 (d) 2

37. For the reversible reaction,
$N_2(g) + 3H_2(g) \rightleftharpoons 2NH_3(g)$ at 500°C, the value of K_p is 1.44×10^{-5} when partial pressure is measured in atmospheres. The corresponding value of K_C, with concentration in mole litre^{-1}, is
(a) $\frac{1.44 \times 10^{-5}}{(0.082 \times 500)^{-2}}$ (b) $\frac{1.44 \times 10^{-5}}{(8.314 \times 773)^{-2}}$
(c) $\frac{1.44 \times 10^{-5}}{(0.082 \times 773)^{2}}$ (d) $\frac{1.44 \times 10^{-5}}{(0.082 \times 773)^{-2}}$

38. For the reaction $C(s) + CO_2(g) \rightleftharpoons 2CO(g)$, the partial pressures of CO_2 and CO are 2.0 and 4.0 atm respectively at equilibrium. The K_p for the reaction is.
(a) 0.5 (b) 4.0
(c) 8.0 (d) 32.0

39. $\Delta G°$ for the reaction $X + Y \rightleftharpoons Z$ is –ve 4.606 kcal. The equilibrium constant for the reaction at 227°C is
(a) 100 (b) 10
(c) 2 (d) 0.01

40. In which of the following reactions, the equilibrium remains unaffected on addition of small amount of argon at constant volume?
(a) $H_2(g) + I_2(g) \rightleftharpoons 2HI(g)$
(b) $PCl_5(g) \rightleftharpoons PCl_3(g) + Cl_2(g)$
(c) $N_2(g) + 3H_2(g) \rightleftharpoons 2NH_3(g)$
(d) The equilibrium will remain unaffected in all the three cases.

41. When hydrochloric acid is added to cobalt nitrate solution at room temperature, the following reaction takes place out the reaction mixture becomes blue. On cooling the mixture it becomes pink. On the basis of this information mark the correct answer.

$$\underset{\text{(pink)}}{\left[Co(H_2O)_6\right]^{3+}}(aq) + 4Cl^-(aq) \rightleftharpoons \underset{\text{(blue)}}{\left[CoCl_4\right]^{2-}}(aq) + 6H_2O(l)$$

(a) $\Delta H > 0$ for the reaction
(b) $\Delta H < 0$ for the reaction
(c) $\Delta H = 0$ for the reaction
(d) The sign of ΔH cannot be predicted on the basis of this information.

42. The pK_a of a weak acid, HA, is 4.80. The pK_b of a weak base, BOH, is 4.78. The pH of an aqueous solution of the corresponding salt, BA, will be
(a) 9.58 (b) 4.79
(c) 9.22 (d) 7.01

43. What will be the value of pH of 0.01 mol dm^{-3} CH_3COOH $\left(K_a = 1.74 \times 10^{-5}\right)$?
(a) 3.4 (b) 3.6
(c) 3.9 (d) 3.0

44. The pH of a solution which is 0.1 M in HA and 0.5 M in NaA. K_a for HA is 1.8×10^{-6}
(a) 5.44 (b) 6.44
(c) 6.0 (d) 4.73

45. Which of the following will produce a buffer solution when mixed in equal volumes?
(a) 0.1 mol dm^{-3} NH_4OH and 0.1 mol dm^{-3} HCl
(b) 0.05 mol dm^{-3} NH_4OH and 0.1 mol dm^{-3} HCl
(c) 0.1 mol dm^{-3} NH_4OH and 0.05 mol dm^{-3} HCl
(d) 0.1 mol dm^{-3} CH_4COONa and 0.1 mol dm^{-3} NaOH

46. A solution which is 10^{-3} M each in Mn^{2+}, Fe^{2+}, Zn^{2+} and Hg^{2+} is treated with 10^{-6} M sulphide ion. If K_{sp} of MnS, FeS, ZnS and HgS are 10^{-15}, 10^{-23}, 10^{-20} and 10^{-54} respectively which one will precipitate first
(a) FeS (b) MgS
(c) HgS (d) ZnS
47. The compound whose aqueous solution has the highest pH is
(a) NaCl (b) $NaHCO_3$
(c) Na_2CO_3 (d) NH_4Cl
48. 0.1 M solution of which one of these substances will be basic ?
(a) Sodium borate
(b) Ammonium chloride
(c) Calcium nitrate
(d) Sodium sulphate.
49. In which of the following solvents, AgBr will have the highest solubility ?
(a) 10^{-3}M NaBr (b) 10^{-3}M NH_4OH
(c) Pure water (d) 10^{-3}M HBr
50. Which of the following is most soluble ?
(a) Bi_2S_3 ($K_{sp} = 1 \times 10^{-17}$)
(b) MnS ($K_{sp} = 7 \times 10^{-16}$)
(c) CuS ($K_{sp} = 8 \times 10^{-37}$)
(d) Ag_2S ($K_{sp} = 6 \times 10^{-51}$).
51. The pH value of blood does not appreciably change by a small addition of an acid or a base, because the blood
(a) is a body fluid
(b) can be easily coagulated
(c) contains iron as a part of the molecule
(d) contains serum protein which acts as buffer
52. If α is the fraction of HI dissociated at equilibrium in the reaction, $2\,HI(g) \rightleftharpoons H_2(g) + I_2(g)$, starting with 2 moles of HI, the total number of moles of reactants and products at equilibrium are
(a) $2+2\alpha$ (b) 2
(c) $1+\alpha$ (d) $2-\alpha$
53. The solubility products of CuS, Ag_2S and HgS are 10^{-31}, 10^{-44}, 10^{-54} respectively. The solubilities of these sulphides are in the order
(a) $Ag_2S > HgS > CuS$
(b) $Ag_2S > CuS > HgS$
(c) $HgS > Ag_2S > CuS$
(d) $CuS > Ag_2S > HgS$
54. If K_1 and K_2 are the respective equilibrium constants for the two reactions
$XeF_6(g) + H_2O(g) \rightleftharpoons XeOF_4(g) + 2HF(g)$
$XeO_4(g) + XeF_6(g) \rightleftharpoons XeOF_4(g) + XeO_3F_2(g)$
the equilibrium constant of the reaction
$XeO_4(g) + 2HF(g) \rightleftharpoons XeO_3F_2(g) + H_2O(g)$
will be
(a) $K_1/(K_2)^2$ (b) $K_1 . K_2$
(c) K_1/K_2 (d) K_2/K_1
55. What is the H^+ ion concentration of a solution prepared by dissolving 4 g of NaOH (Atomic weight of Na = 23 amu) in 1000 ml?
(a) 10^{-10} M (b) 10^{-4} M
(c) 10^{-1} M (d) 10^{-13} M
56. Solubility of a M_2S salt is 3.5×10^{-6}, then its solubility product will be
(a) 1.7×10^{-16} (b) 1.7×10^{-6}
(c) 1.7×10^{-18} (d) 1.7×10^{-12}
57. In HS^-, I^-, RNH_2 and NH_3, order of proton accepting tendency will be
(a) $I^- > NH_3 > RNH_2 > HS^-$
(b) $HS^- > RNH_2 > NH_3 > I^-$
(c) $RNH_2 > NH_3 > HS^- > I^-$
(d) $NH_3 > RNH_2 > HS^- > I^-$
58. Solubility of MX_2-type eletrolytes is 0.5×10^{-4} mole/lit, then find out K_{sp} of electrolytes
(a) 5×10^{-12} (b) 25×10^{-10}
(c) 1×10^{-13} (d) 5×10^{-13}
59. Solution of 0.1 N NH_4OH and 0.1 N NH_4Cl has pH 9.25. Then find out pK_b of NH_4OH
(a) 9.25 (b) 4.75
(c) 3.75 (d) 8.25
60. For the reaction

$$CH_4(g) + 2O_2(g) \rightleftharpoons CO_2(g) + 2H_2O(l),$$
$$\Delta_r H = -170.8 \text{ kJ mol}^{-1}$$

Which of the following statements is not true ?
(a) The equilibrium constant for the reaction is given by $K_p = \frac{[CO_2]}{[CH_4][O_2]}$
(b) Addition of CH_4 (g) or O_2 (g) at equilibrium will cause a shift to the right
(c) The reaction is exothermic
(d) At equilibrium, the concentrations of CO_2 (g) and H_2O(l) are not equal

Redox Reactions

CONCEPTS

- **Redox reactions or electron transformation reactions :** Oxidation and reduction occur simultaneously.

OXIDATION AND REDUCTION

Parameter	Oxidation	Reduction
(i) Hydrogen	loss	gain
(ii) Oxygen	gain	loss
(iii) Electron	loss	gain
(iv) Oxidation no.	increase	decrease

Oxidising agent or Oxidant	Reducing agent or reductant
(i) Undergoes reduction	Undergoes oxidation
(ii) Gains electrons	Donates electrons
(iii) Oxidation no. of its atoms increases	Oxidation no. of its atoms decreases.

STOCK NOTATION

Representation of oxidation state of metal in roman numerals inside parenthesis.

e.g.- $Fe_2(SO_4)_3$ - iron (III) sulphate.

OXIDATION NUMBER AND RULES FOR ITS CALCULATION

Oxidation No.

The charge which an atom appears to have when present in combined state.

Rules for Calculation of Oxidation Number

(a) Type of atom	Oxidation number
(i) Atom in elementary state	zero
(ii) Monoatomic ion charge on it	
(iii) Hydrogen (except active metal hydride)	+1
(iv) Hydrogen in hydrides of active metals	–1
(v) Oxygen (except peroxides)	–2
(vi) Oxygen in peroxides	–1
(vii) Halogens in metal halides	–1
(viii) Alkali metals	+1
(ix) Alkaline earth metals	+2
(x) Metals in carbonyls and amalgams	zero

(b) Sum of oxidation states of all atoms in a neutral compound is equal to zero and in charged ions it is equal to charge on ion (complex).

(c) **Evaluation of oxidation no.**

(i) Oxidation state of **C in CH_2Cl_2 is zero.**

(ii) Oxidation state of **Fe in Fe_3O_4 is 8/3.**

(iii) Oxidation state of **Cr in CrO_5.**

Peroxide linkage ← [structure of CrO_5: Cr bonded to one O by a double bond and to two O–O peroxide groups] → Peroxide linkage

$$x + (1 \times -2) + (4 \times -1) = 0$$

($\because$ 4 O have peroxide linkage)

$\therefore x = +6.$

(iv) Oxidation state of S in H_2SO_5 (or permonosulphuric acid)

$$\text{H}-\text{O}-\underset{\underset{\text{O}}{\|}}{\overset{\overset{\text{O}}{\|}}{\text{S}}}-\text{O}-\text{O}-\text{H}$$

$$\underset{\text{(for H)}}{2 \times (+1)} + \underset{\text{(for S)}}{x} + \underset{\text{(for O)}}{3 \times (-2)} + \underset{\text{(for peroxide linkage)}}{2 \times (-1)} = 0$$

$\therefore x = +6.$

(v) Oxidation state of S in $H_2S_2O_8$

$$\begin{array}{c} \quad O \qquad\quad O \\ \quad \| \qquad\quad \| \\ H-O-S-O-O-S-O-H \\ \quad \| \qquad\quad \| \\ \quad O \qquad\quad O \end{array}$$

$$2 \times (+1) + 2 \times x + 6 \times (-2) + 2 \times (-1) = 0$$

(for peroxide linkage)

$\therefore\ x = +6.$

DISPROPORTIONATION:

There are some substances which function as both oxidising as well as reducing agents e.g., H_2O_2, SO_2, H_2SO_3, HNO_2, O_3, etc. This can be explained by the following example,

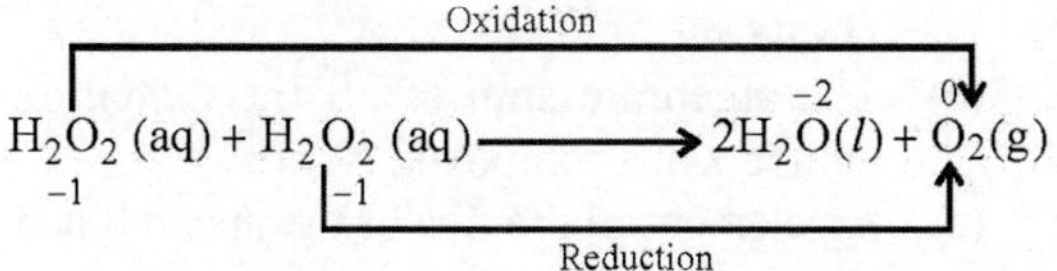

BALANCING OF CHEMICAL EQUATIONS :

For balancing a chemical equation the two important methods are

(1) Oxidation Number Method :

The certain rules are as follows

(a) Assign oxidation number to the atoms showing a change in oxidation state.

(b) Balance the total number of atoms undergoing change in oxidation state.

(c) Balance the number of electrons gained and lost.

(d) Balance [O] on both sides by adding H_2O.

(e) Balance H atoms by adding H^+ ions

(f) If the reaction proceeds in basic solution add sufficient number of $\overline{OH}$ ions on both sides.

(2) Ion Electron Method :

The rules are as follows:

(a) Split up the reaction into two half reactions showing oxidation and reduction separately.

(b) Balance number of atoms undergoing the change of oxidation state.

(c) Balance O on both sides by adding H_2O.

(d) Balance H atoms by adding H^+ ions.

(e) Balance charge by adding required number of electrons.

(f) Make the number of electrons equal in two half reactions by multiplying with suitable coefficient.

(g) Add the two half reactions.

ELECTROCHEMICAL SERIES :

The arrangement of metal ions in order of increasing value of standard reduction potentials is known as electrochemical series.

Elements	Electrode reactions	E^0_{red} (volts)
Li	$Li^+(aq) + e^- \longrightarrow Li(s)$	−3.05
K	$K^+(aq) + e^- \longrightarrow K(s)$	−2.93
Ba	$Ba^{2+}(aq) + 2e^- \longrightarrow Ba(s)$	−2.90
Ca	$Ca^{2+}(aq) + 2e^- \longrightarrow Ca(s)$	−2.87
Na	$Na^+(aq) + e^- \longrightarrow Na(s)$	−2.71
Mg	$Mg^{2+}(aq) + 2e^- \longrightarrow Mg(s)$	−2.37
Al	$Al^{3+}(aq) + 3e^- \longrightarrow Al(s)$	−1.66
Zn	$Zn^{2+}(aq) + 2e^- \longrightarrow Zn(s)$	−0.76
Cr	$Cr^{3+}(aq) + 3e^- \longrightarrow Cr(s)$	−0.74
Fe	$Fe^{2+}(aq) + 2e^- \longrightarrow Fe(s)$	−0.44
H_2	$H_2O(l) + 2e^- \longrightarrow \frac{1}{2}H_2(g) + OH^-(aq)$	−0.41
Cd	$Cd^{2+}(aq) + 2e^- \longrightarrow Cd(s)$	−0.40
Pb	$PbSO_4(s) + 2e^- \longrightarrow Pb(s) + SO_4^{2-}(aq)$	−0.31
Co	$Co^{2+}(aq) + 2e^- \longrightarrow Co(s)$	−0.28
Ni	$Ni^{2+}(aq) + 2e^- \longrightarrow Ni(s)$	−0.25
Sn	$Sn^{2+}(aq) + 2e^- \longrightarrow Sn(s)$	−0.14
Pb	$Pb^{2+}(aq) + 2e^- \longrightarrow Pb(s)$	−0.13
	$2H^+ + 2e^- \rightarrow H_2(g)$	0.00
Cu	$Cu^{2+}(aq) + 2e^- \longrightarrow Cu(s)$	+0.34
I_2	$I_2(s) + 2e^- \longrightarrow 2I^-(s)$	+0.54
Fe	$Fe^{3+}(aq) + e^- \longrightarrow Fe^{2+}(aq)$	+0.77

Hg	$Hg_2^{2+}(aq) + 2e^- \longrightarrow 2Hg(\ell)$	+0.79
Ag	$Ag^+ (aq) + e^- \longrightarrow Ag(s)$	+0.80
Hg	$Hg^{2+}(aq) + 2e^- \longrightarrow Hg(\ell)$	+0.85
Br_2	$Br_2(aq) + 2e^- \longrightarrow 2Br^-(aq)$	+1.08
O_2	$1/2\,O_2(g) + 2H_3O^+(aq) + 2e^- \longrightarrow 3H_2O$	+1.08
Cr	$Cr_2O_7^- + 14H^+ + 6e^- \longrightarrow 2Cr^{3+} + 7H_2O$	+1.33
Cl_2	$Cl_2(g) + 2e^- \longrightarrow 2Cl^-(aq)$	+1.36
Au	$Au^{3+}(aq) + 3e^- \longrightarrow Au(s)$	+1.42
Mn	$MnO_4^-(aq) + 8H_3O^+(aq) + 5e^- \longrightarrow Mn^{2+}(aq) + 12H_2O(\ell)$	+1.51
F_2	$F_2(g) + 2e^- \longrightarrow 2F^-(g)$	+2.87

Exercise

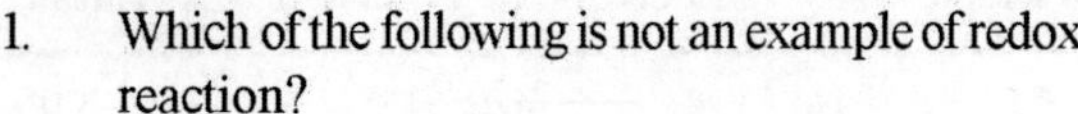

1. Which of the following is not an example of redox reaction?
   (a) $CuO + H_2 \rightarrow Cu + H_2O$
   (b) $Fe_2O_3 + 3CO \rightarrow 2Fe + 3CO_2$
   (c) $2K + F_2 \rightarrow 2KF$
   (d) $BaCl_2 + H_2SO_4 \rightarrow BaSO_4 + 2HCl$
2. Of the following reactions, only one is a redox reaction. Identify it
   (a) $Ca(OH)_2 + 2HCl \rightarrow CaCl_2 + 2H_2O$
   (b) $BaCl_2 + MgSO_4 \rightarrow BaSO_4 + MgCl_2$
   (c) $2S_2O_7^{2-} + 2H_2O \rightarrow 4SO_4^{2-} + 4H^+$
   (d) $Cu_2S + 2FeO \rightarrow 2Cu + 2Fe + SO_2$
3. The loss of electron is termed as
   (a) oxidation (b) reduction
   (c) combustion (d) neutralization
4. The oxide, which cannot act as a reducing agent, is
   (a) NO_2 (b) SO_2 (c) CO_2 (d) ClO_2
5. $Zn^{2+}(aq.) + 2e^- \longrightarrow Zn(s)$. This is
   (a) Oxidation
   (b) Reduction
   (c) Redox reaction
   (d) None of the above
6. $Co(s) + Cu^{2+}(aq) \longrightarrow Co^{2+}(aq) + Cu(s)$
   The above reaction is :
   (a) Oxidation reaction
   (b) Reduction reaction
   (c) Redox reaction
   (d) None of these
7. The oxidation number of an element in a compound is evaluated on the basis of certian rules. Which of the following rules is not correct in this respect?
   (a) The oxidation number of hydrogen is always + 1.
   (b) The algebraic sum of all the oxidation numbers in a compound is zero.
   (c) An element in the free or the uncombined state bears oxidation number zero.
   (d) In all its compounds, the oxidation number of fluorine is – 1.
8. In which of the following compounds, the oxidation number of iodine is fractional ?
   (a) IF_7 (b) I_3^-
   (c) IF_5 (d) IF_3
9. Oxidation numbers of P in PO_4^{3-}, of S in SO_4^{2-} and that of Cr in $Cr_2O_7^{2-}$ are respectively
   (a) +3, +6 and +5 (b) +5, +3 and +6
   (c) −3, +6 and +6 (d) +5, +6 and +6
10. The O.N. of nitrogen in NO_3^- is
    (a) −1 (b) +2
    (c) +3 (d) +5
11. What products are expected from the disproportionation reaction of hypochlorous acid?
    (a) HCl and Cl_2O (b) HCl and $HClO_3$
    (c) $HClO_3$ and Cl_2O (d) $HClO_2$ and $HClO_4$
12. $E^\ominus$ Values of some redox couples are given below. On the basis of these values choose the correct option.
    $E^\ominus$ values : $Br_2/Br^- = +1.90$; $Ag^+/Ag(s) = +0.80$
    $Cu^{2+}/Cu(s) = +0.34$; $I_2(s)/I^- = 0.54$
    (a) Cu will reduce Br^- (b) Cu will reduce Ag
    (c) Cu will reduce I^- (d) Cu will reduce Br_2

13. For the redox reaction

$MnO_4^- + C_2O_4^{2-} + H^+ \rightarrow Mn^{2+} + CO_2 + H_2O$

the correct coefficients of the reactants for the balanced reaction are:

	MnO_4^-	$C_2O_4^{2-}$	H^+
(a)	2	5	16
(b)	16	5	2
(c)	5	16	2
(d)	2	16	5

14. Arrange the following in the order of their decreasing electrode potentials : Mg, K, Ba and Ca
(a) K, Ca, Ba, Mg (b) Ba, Ca, K, Mg
(c) Ca, Mg, K, Ba (d) Mg, Ca, Ba, K

15. In the following reaction

$4P + 3KOH + 3H_2O \longrightarrow 3KH_2PO_2 + PH_3$

(a) only phosphorus is oxidised and reduced.
(b) only phosphorus is reduced.
(c) phosphorus is both oxidised and reduced.
(d) phosphorus is neither oxidised nor reduced.

16. The oxidation number of chromium in potassium dichromate is
(a) $+6$ (b) -5
(c) -2 (d) $+2$

17. The oxidation number of Mn in $KMnO_4$ is
(a) $+7$ (b) -7
(c) $+1$ (d) -1

18. Oxidation number of iodine varies from
(a) -1 to $+1$ (b) -1 to $+7$
(c) $+3$ to $+5$ (d) -1 to $+5$

19. Iodine shows the highest oxidation state in the compound :
(a) KI (b) KI_3
(c) IF_5 (d) KIO_4

20. The oxidation number of carbon in $H_2C_2O_4$:
(a) $+4$ (b) $+3$
(c) $+2$ (d) -2

21. Oxidation number of oxygen atom in O_2 molecule is
(a) 0 (b) -2
(c) $+2$ (d) $-\frac{1}{2}$

22. O.N. of P in KH_2PO_2 is
(a) $+1$ (b) $+3$
(c) $+5$ (d) -4

23. Which of the following reactions depict the oxidising behaviour of H_2SO_4:
(a) $2PCl_5 + H_2SO_4 \longrightarrow 2POCl_3 + 2HCl + SO_2Cl_2$
(b) $2NaOH + H_2SO_4 \longrightarrow Na_2SO_4 + H_2O$
(c) $NaCl + H_2SO_4 \longrightarrow NaHSO_4 + HCl$
(d) $2HI + H_2SO_4 \longrightarrow I_2 + SO_2 + 2H_2O$

24. Oxidation number of S in Na_2SO_4 is
(a) -2 (b) $+2$
(c) -6 (d) $+6$

25. Which of the following involves transfer of five electrons ?
(a) $MnO_4^- \rightarrow Mn^{2+}$
(b) $CrO_4^{2-} \rightarrow Cr^{3+}$
(c) $MnO_4^{2-} \rightarrow MnO_2$
(d) $Cr_2O_7^{2-} \rightarrow 2Cr^{3+}$

26. Which of the following arrangements represent increasing oxidation number of the central atom?
(a) $CrO_2^-, ClO_3^-, CrO_4^{2-}, MnO_4^-$
(b) $ClO_3^-, CrO_4^{2-}, MnO_4^-, CrO_2^-$
(c) $CrO_2^-, ClO_3^-, MnO_4^-, CrO_4^{2-}$
(d) $CrO_4^{2-}, MnO_4^-, CrO_2^-, ClO_3^-$

27. When $KMnO_4$ acts as an oxidising agent and ultimately forms MnO_4^{-2}, MnO_2, Mn_2O_3 and Mn^{+2}, then the number of electrons transferred in each case respectively is
(a) 4, 3, 1, 5 (b) 1, 5, 3, 7
(c) 1, 3, 4, 5 (d) 3, 5, 7, 1

28. Nitric oxide acts as a reducing agent in the reaction
(a) $4NH_3 + 5O_2 \rightarrow 4NO + 6H_2O$
(b) $2NO + 3I_2 + 4H_2O \rightarrow 2NO_3^- + 6I^- + 8H^+$
(c) $2NO + H_2SO_3 \rightarrow N_2O + H_2SO_4$
(d) $2NO + H_2S \rightarrow N_2O + S + H_2O$

29. In the reaction

$3Br_2 + 6CO_3^{2-} + 3H_2O \rightarrow 5Br^- + BrO_3^- + 6HCO_3^-$

(a) Bromine is oxidised and carbonate is reduced.
(b) Bromine is reduced and water is oxidised
(c) Bromine is neither reduced nor oxidised
(d) Bromine is both reduced and oxidised

30. The oxidation number of sulphur in $Na_2S_4O_6$ is
(a) 1.5 (b) 2.5
(c) 3 (d) 2

31. Thiosulphate reacts differently with iodine and bromine in the reactions given below:

$2S_2O_3^{2-} + I_2 \rightarrow S_4O_6^{2-} + 2I^-$

$S_2O_3^{2-} + Br_2 + 5H_2O \rightarrow 2SO_4^{2-} + 2Br^- + 10H^+$

Which of the following statements justifies the above dual behaviour of thiosulphate?
(a) Bromine is a stronger oxidant than iodine.
(b) Bromine is a weaker oxidant than iodine.
(c) Thiosulphate undergoes oxidation by bromine and reduction by iodine in these reactions.
(d) Bromine undergoes oxidation and iodine undergoes reduction in these reactions.

32. The more positive the value of $E^\ominus$, the greater is the tendency of the species to get reduced. Using the standard electrode potential of redox couples given below find out which of the following is the strongest oxidising agent.

$E^\ominus$ Values : $Fe^{3+}/Fe^{2+} = +0.77; I_2(s)/I^- = +0.54;$
$Cu^{2+}/Cu = +0.34; Ag^+/Ag = +0.80V$
(a) Fe^{3+} (b) I_2
(c) Cu^{2+} (d) Ag^+

33. Using the standard electrode potential, find out the pair between which redox reaction is not feasible.

$E^\ominus$ values : $Fe^{3+}/Fe^{2+} = +0.77; I_2/I^- = +0.54;$
$Cu^{2+}/Cu = +0.34; Ag^+/Ag = +0.80V$
(a) Fe^{3+} and I^- (b) Ag^+ and Cu
(c) Fe^{3+} and Cu (d) Ag and Fe^{3+}

34. Which of the following elements does not show disproportionation tendency?
(a) Cl (b) Br
(c) F (d) I

35. $(x)MnO_4^- + (y)H_2O_2 \rightarrow$

$2Mn^{+2} + 5H_2O + 13/2O_2 + (z)e^-$

In this reaction, value of (x), (y) and (z) are
(a) 2, 5, 6 (b) 5, 2, 9
(c) 3, 5, 5 (d) 2, 6, 6

36. $KMnO_4$ oxidises oxalic acid in acidic medium. The number of CO_2 molecules produced as per the balanced equation is
(a) 10 (b) 8
(c) 6 (d) 3

37. In the reaction

$Cr_2O_7^{2-} + 14H^+ + 6I^- \longrightarrow 2Cr^{3+} + 7H_2O + 3I_2$

Which element is reduced
(a) I (b) O
(c) H (d) Cr

38. Which substance serves as reducing agent in the following reaction ?

$14H^+ + Cr_2O_7^2 + 3Ni \longrightarrow 2Cr^{3+} + 7H_2O + 3Ni^{2+}$

(a) H_2O (b) Ni
(c) H^+ (d) $Cr_2O_7^2$

39. Phosphorus has the oxidation state of + 3 in
(a) Phosphorous acid
(b) Orthophosphoric acid
(c) Hypophosphorous acid
(d) Metaphosphoric acid.

40. In which of the following reactions, there is no change in valency ?
(a) $4KClO_3 \longrightarrow 3KClO_4 + KCl$
(b) $SO_2 + 2H_2S \longrightarrow 2H_2O + 3S$
(c) $BaO_2 + H_2SO_4 \longrightarrow BaSO_4 + H_2O_2$
(d) $3BaO + O_2 \longrightarrow 2BaO_2$.

41. Which of the following involves a redox reaction?
(a) Reaction of H_2SO_4 with NaOH
(b) Production of ozone from oxygen in the atmosphere by lightning
(c) Production of nitrogen oxides from nitrogen and oxygen in the atmosphere by lightning
(d) Evaporation of water

42. The oxidation number of phosphorus in pyrophosphoric acid is
(a) +3 (b) +1
(c) +4 (d) +5

43. The following redox reaction is balanced by which set of coefficients ?

$$aZn + bNO_3^- + cH^+ \rightarrow dNH_4^+ + eH_2O + fZn^{2+}$$

	a	b	c	d	e	f
(a)	1	1	10	1	3	1
(b)	2	2	10	2	3	2
(c)	4	2	10	1	3	4
(d)	4	1	10	1	3	4

44. A compound contains atoms of three elements A, B and C. If the oxidation number of A is +2, B is +5, and that of C is –2, the possible formula of the compound is :
(a) $A_2(BC_3)_2$ (b) $A_3(BC_4)_2$
(c) $A_3(B_4C)_2$ (d) ABC_2

45. Zn gives H_2 gas with H_2SO_4 and HCl but not with HNO_3 because
(a) Zn acts as an oxidising agent when it reacts with HNO_3
(b) HNO_3 is weaker acid than H_2SO_4 and HCl
(c) In electrochemical series, Zn is above hydrogen
(d) $NO_3^{\ominus}$ is reduced in preference to hydronium ion

46. The oxidation states of sulphur in the anions SO_3^{2-}, $S_2O_4^{2-}$ and $S_2O_6^{2-}$ follow the order
(a) $S_2O_6^{2-} < S_2O_4^{2-} < SO_3^{2-}$
(b) $S_2O_4^{2-} < SO_3^{2-} < S_2O_6^{2-}$
(c) $SO_3^{2-} < S_2O_4^{2-} < S_2O_6^{2-}$
(d) $S_2O_4^{2-} < S_2O_6^{2-} < SO_3^{2-}$

47. Oxygen has an oxidation state of +2 in the compound :
(a) H_2O_2 (b) CO_2
(c) H_2O (d) F_2O

48. In the compounds $KMnO_4$ and $K_2Cr_2O_7$ the highest oxidation state is of the element :
(a) Potassium (b) Manganese
(c) Chromium (d) Oxygen

49. $SnCl_2$ gives a precipitate with a solution of $HgCl_2$. In this process $HgCl_2$ is
(a) Reduced
(b) Oxidised
(c) Converted into a complex compound containing both Sn and Hg
(d) Converted into a chloro complex of Hg.

50. If three electrons are lost by a metal ion M^{+3}, its final oxidation number would be
(a) zero (b) +6
(c) +2 (d) +4

51. The standard electrode potentials of four elements A, B, C and D are –3.05, –1.66, –0.40 and +0.80. The highest chemical reactivity will be exhibited by :
(a) A (b) B
(c) C (d) D

52. Which one of the following metals can deposit copper from copper sulphate solution?
(a) Mercury (b) Iron
(c) Gold (d) Platinum

53. What is the valency of Manganese in K_2MnO_4?
(a) 2 (b) 4
(c) 6 (d) 7

54. The charge on cobalt in $[Co(CN)_6]^{3-}$ is
(a) – 6 (b) – 3
(c) +3 (d) +6

55. The violent reaction between sodium and water is an example of :
(a) Reduction
(b) Oxidation
(c) Redox reaction
(d) Neutralization reaction

56. The compound that can work both as an oxidizing and reducing agent is :
(a) $KMnO_4$ (b) H_2O_2
(c) BaO_2 (d) $K_2Cr_2O_7$

57. Br^- is converted into Br_2 by using :
(a) Cl_2 (b) conc. HCl
(c) HBr (d) H_2S

58. Which of these substances is a good reducing agent?
(a) HI (b) KBr
(c) $FeCl_3$ (d) KClO

59. Oxidation state of oxygen in hydrogen peroxide is
(a) – 1 (b) + 1
(c) 0 (d) – 2

60. In ferrous ammonium sulphate oxidation number of Fe is
(a) +3 (b) +2
(c) +1 (d) – 2

CHAPTER 9

Hydrogen

CONCEPTS

- **Hydrogen was discovered by Henry Cavandish.**

ISOTOPES OF HYDROGEN :

Hydrogen has three isotopes

Name	$_1H^1$ Protium	$_1H^2$ or D Deuterium	$_1H^3$ or T Tritium
Abundance	99.985%	0.016%	Traces (10^{-15}%)

Deuterium is known as heavy hydrogen. Tritium is formed in upper atmosphere and is β-radio active.

$$_2He^3 + {}_{-1}e^0 \rightarrow {}_1^3H\ (t_{1/2} = 12.34\ \text{year})$$

$$_7N^{14} \rightarrow {}_6C^{12} + {}_1H^3\ \text{(formation)}$$

PREPARATION AND PROPERTIES OF HYDROGEN

Methods of Preparation :

Laboratory method of preparation of hydrogen (H_2)

$$Zn + \text{dil. } H_2SO_4 \longrightarrow ZnSO_4 + H_2$$

Commercial processes (for manufacture of hydrogen)

(a) Electrolysis of acidified water using Pt electrodes give hydrogen,

$$2H_2O(l) \xrightarrow[\text{traces of acids/bases}]{\text{Electolysis}} 2H_2(g) + O_2(g)$$

(b) Reaction of steam on hydrocarbons or coke.

$$C + H_2O \xrightarrow{\Delta} \underbrace{CO + H_2}_{\text{water gas}}$$

$$\text{Hydrocarbon} + H_2O \xrightarrow[\text{catalyst}]{1270\ K} CO + H_2$$

this mixture of CO and H_2 is used for the synthesis of methanol thus also know as syn gas. Water gas is now mixed with steam and passed over a catalyst Fe_2O_3 with suitable promoter kept at 773 K, where CO is oxidised to CO_2

$$\underbrace{CO + H_2}_{\text{water gas}} + H_2O \xrightarrow[777K]{Fe_2O_3 + Cr_2O_3} CO_2 + 2H_2$$

Properties of Hydrogen :

1. **Physical properties :** Hydrogen is tasteless, odourless, colourless gas. It is non poisonous but presence of AsH_3 makes it poisonous. Its critical temperature is –234.5°C which makes its liquefaction difficult. It is slightly soluble in water.

2. **Chemical properties :** Its chemical properties are

   (a) **Reaction with non-metals :**

   $$H_2 + X_2 \rightarrow 2HX\ (X = F, Cl, Br, I)$$

   $$2H_2 + O_2 \rightarrow 2H_2O;\ 3H_2 + N_2 \rightarrow 2NH_3$$

   (b) **Reaction with metals :**

   $$2Na + H_2 \rightarrow NaH$$

   $$Ca + H_2 \rightarrow CaH_2\ \text{(Hydrolith)}$$

   Metals like Fe, Ni and Pd form interstitial or metallic hydrides.

   (c) **Reducing property :**

   $$CuO + H_2 \rightarrow Cu + H_2O$$

   $$Ag_2O + H_2 \rightarrow 2Ag + H_2O$$

   $$Fe_3O_4 + 4H_2 \rightarrow 3Fe + 4H_2O$$

   (d) **Hydrogenation :**

   $$C_nH_{2n} + H_2 \xrightarrow{Ni} C_nH_{2n+2}$$

USES OF HYDROGEN

(i) It is used in the manufacture of ammonia, methyl alcohol (from CO), gasoline (synthetic petrol), HCl, etc.

(ii) It is used in the manufacture of vegetable ghee (hydrogenation of oils of fats).

(iii) Oxy-hydrogen flame (oxy-hygrogen torch) and atomic hydrogen flame (atomic hydrogen torch) are usd for cutting and welding steel and other hard metals.

(iv) It is used for providing reducing atmosphere is furnaces.

(v) Due to its low density, hydrogen in used in filling baloons and gas bags for air ships. However, it is now replaced by helium.

(vi) Several fuel gases (e.g., coal gas, water gas, oil gas) contain hydrogen as an important constitutent.

FORMS OF HYDROGEN :

(i) **Atomic hydrogen :**

$H_2 \xrightarrow{\text{Electric arc}} 2H \quad \Delta H = 105.4 \text{ k cal mol}^{-1}$

It is very reactive and its half life period is 0.33 second.

(ii) **Occluded hydrogen :** Hydrogen adsorbed by certain metals e.g. Pt, Pd, Fe, Ni etc is known as occluded **hydrogen**.

(iii) **Nascent hydrogen :** Freshly prepared hydrogen is known as ***nascent hydrogen*** and is more reactive than ordinary hydrogen.

HYDRIDES :

Hydrogen combines with a number of metal and nonmetal to form hydrides.

(a) Ionic Hydrides :

These are formed by metals which have low electronegative values and are electropositive with respect to hydrogen, i.e., alkali and alkaline earth metals (except Be and Mg).

(i) $2Na + H_2 \longrightarrow 2NaH$,

(ii) $Ca + H_2 \longrightarrow CaH_2$

(b) Interstitial Hydrides

Also called metallic or non-stoichiometric hydrides. These are formed by Be, Mg (both are s-block elements) and transitions (d-block) elements. These are interstitial compounds as the hydrogen atoms occupy interstitial places in the metal lattices.

(c) Covalent Hydrides :

(i) Electron precise e.g., BH_3.

(ii) Electron precise e.g., CH_4.

(iii) Electron rich e.g., NH_3.

WATER (H_2O) :

Water is a colourless liquid and possesses a high dielectric constant. It has strong inter molecular H-bonding due to which it has high enthalpy of vaporization, high enthalpy of fusion and high polarity (good solvent). Water is a good solvent for ionic compounds but a poor solvent for covalent compounds.

Ice is less dense than water because its 3D structure has large open spaces. In a H_2O, molecule each O atom is bonded to 4 H atoms, with 2 H atoms it's covalently bonded & with other 2 it has H-bond.

Soft and Hard Water :

(i) Soft water produces lather with soap and hard water does not.

(ii) **Types of hardness of water**

Temporary hardness

(i) Caused by **bicarbonates of Ca and Mg**.

(ii) Removed by

(a) **Boiling**

$Ca(HCO_3)_2 \xrightarrow{\Delta} CaCO_3 \downarrow + CO_2 + H_2O$

(b) **Clark's method** - by adding $Ca(OH)_2$

- $Ca(HCO_3)_2 + Ca(OH)_2 \longrightarrow 2CaCO_3 \downarrow + 2H_2O$
- $Mg(HCO_3)_2 + Ca(OH)_2 \longrightarrow 2CaCO_3 \downarrow + 2H_2O + MgCO_3 \downarrow$

Permanent hardness

(i) Caused by **soluble chlorides and sulphates of Ca and Mg**.

(ii) Removed by

(a) **Soda lime method** (adding Na_2CO_3)

Bicarbonates and chlorides of Ca & Mg + $Na_2CO_3 \rightarrow$ Carbonates of Ca & Mg. $\downarrow + Na_2SO_4$

(b) **Ion exchange method**

Ca^{++} and Mg^{++} ions of hard water are exchanged with those present in exchanger

Inorganic ion exchanger

- **Zeolite** [(Na_2Z – where Z is $Al_2Si_2O_8xH_2O$)] is used to exchange Ca^{2+} and Mg^{2+} with Na^+.

Organic exchanger

- **Resins** are used

Cation exchange resin

- Hydrocarbon has acidic group. $-SO_3H$

Anion exchange resin

- Hydrocarbon has basic group. $-NH_2$

HEAVY WATER (D_2O) :

It is discovered by Urey. ordinary water contains a fraction (about 1 part in 6000 parts) of heavy water. On a large scale, it is produced by repeated electrolysis of ordinary water containing alkali.

It is made up of two deuterium atoms and one oxygen atom. Its molecular weight is 20. Physical properties of D_2O is different from that H_2O due to greater nuclear mass of deuterium.

HYDROGEN PEROXIDE :

It was discovered by **Thennard.**

Preparation

(i) **Lab method** - By action of H_2SO_4 on hydrated barium peroxide ($BaO_2.8H_2O$)

$$BaO_2.8H_2O + H_2SO_4 \longrightarrow BaSO_4 + H_2O_2 + 8H_2O$$

(ii) **Merck's method**

$$BaO_2 + H_2O + CO_2 \rightarrow BaCO_3 + H_2O_2$$

(iii) **By electrolysis of 50% ice cold H_2SO_4**

$$2H_2SO_4 \xrightarrow{\text{electricity}} 2H^+ + 2HSO_4^-$$

At cathode $\quad 2H^+ + 2e^- \longrightarrow H_2 \uparrow$

At anode $\quad 2HSO_4^- \longrightarrow 2HSO_4 + 2e^-$

$$2HSO_4 \longrightarrow \underset{\text{perdisulphuric acid}}{H_2S_2O_8} + 2e^-$$

$$H_2S_2O_8 + 2H_2O \longrightarrow 2H_2SO_4 + H_2O_2$$

(iv) **By auto oxidation of 2-ethyl anthraquinol**

$$\underset{\text{2-ethylanthraquinol}}{\text{2-ethylanthraquinol}} \underset{H_2/Pd}{\overset{O_2}{\rightleftharpoons}} \underset{\text{2-ethylanthraquinone}}{\text{2-ethylanthraquinone}} + H_2O_2$$

Properties

1. **Physical properties:**

(i) It is thick syrupy pale blue liquid.

(ii) The dipole moment is slightly more than water.

(iii) It is more viscous than water.

(iv) It is completely miscible with water, alcohol and ether.

(v) It is diamagnetic in nature.

(vi) Pure H_2O_2 is a weak dibasic acid. It is stored in coloured, plastic bottles away from sunlight otherwise it disintegrates.

(vii) **Structure of H_2O_2 : It has open book structure**

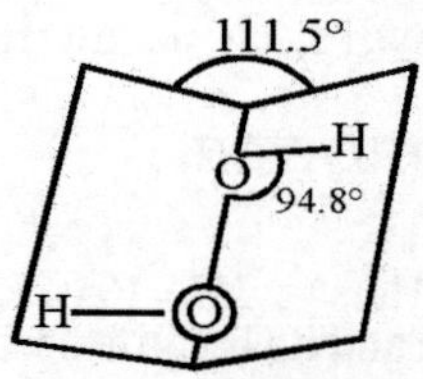

(viii) **Strength of H_2O_2 solution**

(a) **When expressed as percentage** means % of H_2O_2 in solution w/v.

(b) **When expressed in volume** strength means the volume of O_2 (in L) obtained by decomposition of 1 L sample of H_2O_2.

(c) Volume strength of H_2O_2 = Normality × 5.6

(d) Normality of H_2O_2

$$= \frac{\text{percentage strength}}{\text{Eq. wt. of } H_2O_2} \times 10$$

(Equivalent weight of H_2O_2 = 17)

2. **Chemical properties**

In acidic medium :

The basic reaction is H_2O_2 is used as a bleaching agent due to its ability to oxidise the coloured matter,

$$H_2O_2 \longrightarrow H_2O + [O], \; H_2O_2 \longrightarrow H_2O + O$$

e.g.

i. $2FeSO_4 + H_2SO_4 + H_2O_2 \longrightarrow Fe_2(SO_4)_3 + 2H_2O$

ii. $H_2SO_3 + H_2O_2 \longrightarrow H_2SO_4 + H_2O$

iii. $PbS + 4H_2O_2 \longrightarrow PbSO_4 + 4H_2O$

In basic medium :

The basic reaction is

$$H_2O_2 + 2OH^- + 2e^- \longrightarrow 2H_2O + 2O^-$$

e.g.

i. $Mn^{2+} + H_2O_2 + 2OH \longrightarrow Mn^{4+} + 2H_2O$

ii. $2HCHO + H_2O_2 \xrightarrow{\text{pyrogallol}} 2HCOOH + H_2$

3. **Reducing action :**

It acts as a reducing agent towards oxidising agents in acidic as well as alkaline medium. Reducing action in acidic solution in somewhat slow that in alkaline solution.

In acidic solution :

$$H_2O_2 \longrightarrow O_2 + 2H^+ + 2e^-$$

e.g.

i. $2KMnO_4 + 5H_2O_2 + 3H_2SO_4 \longrightarrow K_2SO_4 + 2MnSO_4 + 8H_2O + 5O_2$

ii. $HOCl + H_2O_2 \longrightarrow Cl^- + O_2 + H_3O^+$

Alkaline medium:

$$H_2O_2 + 2OH^- \longrightarrow 2H_2O + O_2 + 2e^-$$

i. $Cl_2 + H_2O_2 + 2OH^- \longrightarrow 2Cl^- + 2H_2O + O_2$

ii. $2Fe^{3+} + H_2O_2 + 2OH^- \longrightarrow 2Fe^{2+} + 2H_2O + O_2$

- **Bleaching action:** H_2O_2 is used as bleaching agent due to its ability to oxidise the coloured matter

$$H_2O_2 \rightarrow H_2O + [O]$$

Colouring matter + [O] → Oxidised coloured matter.

Uses of H_2O_2.

It is used

(i) For bleaching delicate articles like wool, hair, leather ivory, etc.

(ii) In production of epoxide, propylene oxide, etc.

(iii) As an antiseptic for washing wounds, teeth, ears under the name of perhydrol.

(iv) As a propellent for torpendose.

(v) As an oxidant for rocket.

H_2O_2 is used to remove excess chlorine from textiles so this property is called *antichlor*.

- **Tests of H_2O_2**

(i) Changes **black PbS to white $PbSO_4$**.

(ii) $\underset{\text{(orange)}}{K_2Cr_2O_7} + H_2O_2 \longrightarrow \underset{\text{(blue)}}{CrO_5}$

- **Hydrogen economy** means use of liquid hydrogen as alternate source of fuel.

Exercise

1. Hydrogen resembles halogens in many respects for which several factors are responsible. Of the following factors which one is most important in this respect?
   (a) Its tendency to lose an electron to form a cation.
   (b) Its tendency to gain a single electron in its valence shell to attain stable electronic configuration.
   (c) Its low negative electron gain enthalpy value.
   (d) Its small size.
2. Radioactive elements emit α, β and γ-rays and are characterised by their half-lives. The radioactive isotope of hydrogen is
   (a) Protium (b) Deuterium
   (c) Tritium (d) Hydronium
3. Which of the following reactions is an example of use of water gas in the synthesis of other compounds?
   (a) $CH_4(g) + H_2O(g) \xrightarrow[\text{Ni}]{1270K} CO(g) + H_2(g)$
   (b) $CO(g) + H_2O(g) \xrightarrow[\text{Catalyst}]{673K} CO_2(g) + H_2(g)$
   (c) $C_nH_{2n+2} + nH_2O\,(g) \xrightarrow[\text{Ni}]{1270K} nCO + (2n+1)H_2$
   (d) $CO(g) + 2H_2(g) \xrightarrow[\text{Catalyst}]{\text{Cobalt}} CH_3OH(l)$
4. Hydrogen is not obtained when zinc reacts with
   (a) Cold water (b) dil. HCl
   (c) dil. H_2SO_4 (d) Hot NaOH (20%)
5. In aqueous solution, H_2 will not reduce
   (a) Fe^{3+} (b) Cu^{2+}
   (c) Ag^+ (d) Zn^{2+}
6. Metal hydrides are ionic, covalent or molecular in nature. Among LiH, NaH, KH, RbH, CsH, the correct order of increasing ionic character is
   (a) LiH > NaH > CsH > KH > RbH
   (b) LiH < NaH < KH < RbH < CsH
   (c) RbH > CsH > NaH > KH > LiH
   (d) NaH > CsH > RbH > LiH > KH
7. $LiAlH_4$ is used as :
   (a) An oxidizing agent
   (b) A reducing agent
   (c) A mordant
   (d) A water softener
8. At its melting point ice is lighter than water because
   (a) H_2O molecules are more closely packed in solid state
   (b) Ice crystals have hollow hexagonal arrangement of H_2O molecules.
   (c) On melting of ice the H_2O molecule shrinks in size
   (d) Ice froms mostly heavy water on first melting.
9. The H–O–H angle in water molecule is about
   (a) 90° (b) 180°
   (c) 102° (d) 105°

10. The boiling point of water is exceptionally high because
    (a) There is covalent bond between H and O
    (b) Water molecule is linear
    (c) Water molecules associate due to hydrogen bonding
    (d) Water molecule is not linear
11. When a substance A reacts with water it produces a combustible gas B and a solution of substance C in water. When another substance D reacts with this solution of C, it also produces the same gas B on warming but D can produce gas B on reaction with dilute sulphuric acid at room temperature. A imparts a deep golden yellow colour to a smokeless flame of Bunsen burner. A, B, C and D respectively are
    (a) Na , H_2, NaOH, Zn
    (b) K, H_2, KOH, Al
    (c) Ca, H_2, $Ca(OH)_2$, Sn
    (d) CaC_2, C_2H_2, $Ca(OH)_2$, Fe
12. Permanent hardness of water can be removed by adding Calgon $(NaPO_3)_n$. This is an example of
    (a) Adsorption (b) Exchange of ion
    (c) Precipitation (d) None
13. The volume strength of 1.5 N H_2O_2 solution is
    (a) 4.8 L (b) 5.2 L
    (c) 8.4 L (d) 8.8 L
14. The reaction
    $H_2S + H_2O_2 \longrightarrow S + 2H_2O$ manifests
    (a) Acidic nature of H_2O_2
    (b) Alkaline nature of H_2O_2
    (c) Oxidising action of H_2O_2
    (d) Reducing action of H_2O_2.
15. Heavy water is
    (a) H_2O (b) D_2O
    (c) H_2O_2 (d) none of these
16. Which of the following statements is correct ?
    (a) Hydrogen has same IP as alkali metals
    (b) Hydrogen has same electronegativity as halogens
    (c) It has oxidation number of –1 and +1
    (d) It will not be liberated at anode
17. Which of the following is the true structure of H_2O_2 ?
    (a) H – O – O – H (b) 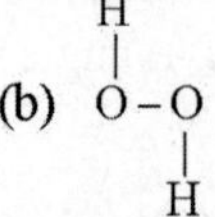
    (c) $\mathrm{H_2}$>O = O (d) $\mathrm{H_2}$>O ← O.
18. Calgon used as a water softener is
    (a) $Na_2[Na_4(PO_3)_6]$
    (b) $Na_4[Na_2(PO_3)_6]$
    (c) $Na_4[Na_4(PO_4)_5]$
    (d) $Na_4[Na_2(PO_4)_6]$
19. The ionization of hydrogen atom would give rise to
    (a) hydride ion (b) hydronium ion
    (c) proton (d) hydroxyl ion
20. Which of the following groups of ions makes the water hard ?
    (a) Sodium and bicarbonate
    (b) Magnesium and chloride
    (c) Potassium and sulphate
    (d) Ammonium and chloride
21. H_2 from HCl can be prepared from :
    (a) Mg (b) Cu
    (c) P (d) Hg
22. Which of the following statements is incorrect ?
    (a) H_2O_2 can act as an oxidising agent
    (b) H_2O_2 can act as a reducing agent
    (c) H_2O_2 has acidic properties
    (d) H_2O_2 has basic properties
23. Which one the following removes temporary hardness of water ?
    (a) Slaked lime (b) Plaster of Paris
    (c) Epsom (d) Hydrolith
24. Heavy water is used in nuclear reactors as
    (a) Source of α -particles
    (b) Slowing down the speed of high energy neutrons
    (c) Transporting heat of the reactor
    (d) Heating purpose
25. The structure of H_2O_2 is
    (a) Planar (b) Non planar
    (c) Spherical (d) Linear
26. Hydrogen molecules differs from chlorine molecule in the following respect
    (a) Hydrogen molecule is non-polar but chlorine molecule is polar
    (b) Hydrogen molecule is polar while chlorine molecule is non-polar
    (c) Hydrogen molecule can form intermolecular hydrogen bonds but chlorine molecule does not
    (d) Hydrogen molecule cannot participate in coordination bond formation but chlorine molecule can

27. The property of hydrogen which distinguishes it from alkali metals is
    (a) Its electropositive character
    (b) Its affinity for non metal
    (c) Its reducing character
    (d) Its non-metallic character
28. Hydrogen accepts an electron to form inert gas configuration. In this it resembles
    (a) Halogen (b) Alkali metals
    (c) Chalcogens (d) Alkaline earth metals
29. Water possesses a high dielectric constant, therefore :
    (a) it always contains ions
    (b) it is a universal solvent
    (c) can dissolve covalent compounds
    (d) can conduct electricity
30. Chemical A is used for water softening to remove temporary hardness. A reacts with sodium carbonate to generate caustic soda. When CO_2 is bubbled through a solution of A, it turns cloudy. What is the chemical formula of A.
    (a) $CaCO_3$ (b) CaO
    (c) $Ca(OH)_2$ (d) $Ca(HCO_3)_2$
31. Which one of the following processes will produce hard water ?
    (a) Saturation of water with $MgCO_3$
    (b) Saturation of water with $CaSO_4$
    (c) Addition of Na_2SO_4 to water
    (d) Saturation of water with $CaCO_3$
32. Which of the following reactions increases production of dihydrogen from synthesis gas?
    (a) $CH_4(g)+H_2O(g)\xrightarrow[Ni]{1270K}CO(g)+3H_2(g)$
    (b) $C(s)+H_2O(g)\xrightarrow{1270K}CO(g)+H_2(g)$
    (c) $CO(g)+H_2O(g)\xrightarrow[Catalyst]{673K}CO_2(g)+H_2(g)$
    (d) $C_2H_6+2H_2O\xrightarrow[Ni]{1270K}2CO+5H_2$
33. Consider the following statements :
    1. Atomic hydrogen is obtained by passing hydrogen through an electric arc.
    2. Hydrogen gas will not reduce heated aluminium oxide.
    3. Finely divided palladium absorbs large volume of hydrogen gas.
    4. Pure nascent hydrogen is best obtained by reacting Na with C_2H_5OH.

    Which of the above statements is/are correct ?
    (a) 1 alone (b) 2 alone
    (c) 1, 2 and 3 (d) 2, 3 and 4
34. Elements of which of the following group(s) of periodic table do not form hydrides.
    (a) Groups 7, 8, 9 (b) Group 13
    (c) Groups 15, 16, 17 (d) Group 14
35. When two ice cubes are pressed over each other, they unite to form one cube. Which of the following forces is responsible to hold them together ?
    (a) Hydrogen bond formation
    (b) Van der Waals forces
    (c) Covalent attraction
    (d) Ionic interaction
36. Water contracts on heating
    (a) to 100°C (b) from 0°C to 4°C
    (c) to 273 K (d) from 10°C to 20°C
37. The m.p. of most of the solid substances increase with an increase of pressure . However ice melts at a temperature lower than its usual melting point when pressure is increased . This is because
    (a) ice is less denser than H_2O
    (b) pressure generates heat
    (c) the chemical bonds break under pressure
    (d) ice is not a true solid
38. Match list I with list II and select the correct answer using the codes given below the lists :

List I	List II
1. Heavy water	A. Bicarbonates of Mg and Ca in water
2. Temporary hard water	B. No foreign ions in water
3. Soft water	C. D_2O
4. Permanent hard water	D. Sulphates & chlorides of Mg & Ca in water

    Codes
    (a) 1–C, 2–D, 3–B, 4–A
    (b) 1–B, 2–A, 3–C, 4–D
    (c) 1–B, 2–D, 3–C, 4–A
    (d) 1–C, 2–A, 3–B, 4–D
39. Consider the reactions
    (A) $H_2O_2+2HI\rightarrow I_2+2H_2O$
    (B) $HOCl+H_2O_2\rightarrow H_3O^++Cl^-+O_2$
    Which of the following statements is correct about H_2O_2 with reference to these reactions? Hydrogen peroxide is ______ .
    (a) an oxidising agent in both (A) and (B)
    (b) an oxidising agent in (A) and reducing agent in (B)
    (c) a reducing agent in (A) and oxidising agent in (B)
    (d) a reducing agent in both (A) and (B)

40. The oxide that gives H_2O_2 on treatment with dilute H_2SO_4 is–
(a) PbO_2 (b) BaO_2
(c) MnO_2 (d) TiO_2

41. Which of the following equations depict the oxidising nature of H_2O_2?
(a) $2MnO_4^- + 6H^+ + 5H_2O_2 \rightarrow 2Mn^{2+} + 8H_2O + 5O_2$
(b) $2Fe^{3+} + 2H^+ + H_2O_2 \rightarrow 2Fe^{2+} + 2H_2O + O_2$
(c) $2I^- + 2H^+ + H_2O_2 \rightarrow I_2 + 2H_2O$
(d) $KIO_4 + H_2O_2 \rightarrow KIO_3 + H_2O + O_2$

42. Which of the following equation depicts reducing nature of H_2O_2?
(a) $2[Fe(CN)_6]^{4-} + 2H^+ + H_2O_2 \rightarrow 2[Fe(CN)_6]^{3-} + 2H_2O$
(b) $I_2 + H_2O_2 + 2OH^- \rightarrow 2I^- + 2H_2O + O_2$
(c) $Mn^{2+} + H_2O_2 \rightarrow Mn^{4+} + 2OH^-$
(d) $PbS + 4H_2O_2 \rightarrow PbSO_4 + 4H_2O$

43. Decomposition of H_2O_2 is retarded by :
(a) H_3PO_4 (b) Alcohol
(c) Acetanilide (d) All

44. Why does H^+ ion always get associated with other atoms or molecules?
(a) Ionisation enthalpy of hydrogen resembles that of alkali metals.
(b) Its reactivity is similar to halogens.
(c) It resembles both alkali metals and halogens.
(d) Loss of an electron from hydrogen atom results in a nucleus of very small size as compared to other atoms or ions. Due to small size it cannot exist free.

45. The reaction of H_2O_2 with hydrogen sulphide is an example of ....reaction
(a) Addition (b) Oxidation
(c) Reduction (d) Redox

46. The O – O – H bond angle in H_2O_2 is
(a) 106° (b) 109°28'
(c) 120° (d) 97°

47. The hydride ion, H^-, is a stronger base than the hydroxide ion, OH^-. Which one of the following reactions will occur if sodium hydride (NaH) is dissolved in water?
(a) $H^-(aq) + H_2O(l) \rightarrow H_3O^-(aq)$
(b) $H^-(aq) + H_2O(l) \rightarrow OH^-(aq) + H_2(g)$
(c) $H^-(aq) + H_2O(l) \rightarrow OH^-(aq) + 2H^+(aq) + 2e$
(d) $H^-(aq) + H_2O(l) \rightarrow$ No reaction

48. Which one of the following pairs of substances on reaction will not evolve H_2 gas?
(a) Iron and H_2SO_4 (aqueous)
(b) Iron and steam
(c) Copper and HCl (aqueous)
(d) Sodium and ethyl alcohol

49. When H_2O_2 is oxidised the product is
(a) OH^- (b) O_2
(c) O^{2-} (d) HO_2^-

50. Some statements about heavy water are given below:
(A) Heavy water is used as a moderator in nuclear reactors.
(B) Heavy water is more associated than ordinary water.
(C) Heavy water is more effective solvent than ordinary water.
Which of the above statements are correct?
(a) (A) and (C) (b) (A) and (B)
(c) (A), (B) and (C) (d) (B) and (C)

51. Commercial "10 volume" H_2O_2 is a solution with a strength of approximately :
(a) 30% (b) 3%
(c) 1% (d) 10%

52. Which of the following statements is not true for hydrogen peroxide?
(a) Pure H_2O_2 is fairly stable
(b) It sometimes acts as a reducing agent
(c) It acts as an oxidising agent
(d) Aqueous solution of H_2O_2 is weakly basic

53. Which of the following metals reacts with H_2O at room temp?
(a) Ag (b) Fe
(c) Al (d) Na

54. The compound that can work both as an oxidising agent and a reducing agent is
(a) $KMnO_4$ (b) H_2S
(c) BaO_2 (d) H_2O_2.

55. Hard water when passed through ion exchange resin containing R'COOH groups, becomes free from :
(a) Cl^- (b) SO_4^{2-}
(c) H_3O^+ (d) Ca^{2+}

56. H_2O_2 when added to a solution containing $KMnO_4$ and H_2SO_4 :
(a) Acts as an oxidizing agent
(b) Acts as a reducing agent
(c) Acts both as an oxidizing as well as a reducing agent
(d) Producing no reaction

57. $H_2O_2 \longrightarrow 2H^+ + O_2 + 2e^-; E° = -0.68V.$
The above equation represents which of the following behaviour of H_2O_2 ?
(a) Reducing (b) Oxidising
(c) Acidic (d) Catalytic.

58. Moist hydrogen peroxide can not be dried over conc. H_2SO_4 because
(a) It can catch fire
(b) It reduced by H_2SO_4
(c) It is oxidised by H_2SO_4
(d) It decomposes H_2O_2.

59. Which does not replace hydrogen from HCl ?
(a) Cu (b) Mg
(c) Na (d) Al

60. The number of moles of H_2 in 0.224 litre of hydrogen gas at STP (273 K, 1 atm) is :
(a) 0.1 (b) 0.01
(c) 0.001 (d) 1

CHAPTER 10

The s-Block Elements

CONCEPTS

GROUP-1 (ALKALI METALS)

- **Members** Li, Na, K, Rb, Cs, Fr.
- **General electronic configuration** – ns^1.

Properties of Alkali Metals

(A) Physical properties

1. **Atomic size :** largest in period and increases down the group.
2. **Density : In general low** and increases down the group. **Density of K is lesser than Na** due to abnormal increase in size while moving from Na to K.
3. **Melting and boiling points are low in general** and decreases down the group (because of large size, strength of metallic bond is low).
4. **Ionisation enthalpies :** Lowest in period and decrease down the group. The second ionisation enthalpies are very high because after loss of 1 electron they acquire noble gas configuration.
5. **Electropositive character/metallic nature :** Strongly electropositive elements and it increases down group as ionisation energy decreases.
6. **Oxidation state** +1
7. **Flame colouration : All alkali metals and their salts impart characteristic colour to Bunsen burner flame.** As their ionisation energies are low, the electrons absorb energy from flame and are excited to higher state. This excited state is unstable so the electrons fall to ground state emitting light in visible region.

Element	Li	Na	K	Rb	Cs
Colour of flame	Crimson red	Golden yellow	Pale violet	violet	Sky blue

8. **Photoelectric effect :** Due to low ionisation enthalpy all alkali metals show photoelectric effect. The ease increases down group (due to decrease in IP). **K and Cs are used in photoelectric cells.**
9. **Electrical conductivity :** Alkali metals are **good conductors** of electricity. Electrical conductance increases down group as distance between valence electrons and nucleus increases.
10. **Hydration of ions :**
    (i) Hydration is exothermic process.
    (ii) **Smaller the size of cation greater is hydration.**
    ∴ Li^+ is most extensively hydrated. (due to smallest size among alkali metals)
    ∴ **Hydration energy decreases from Li^+ to Cs^+.**
11. Mobility of ion $\propto \dfrac{1}{\text{size of hydrated ion}}$

    ∴ **Li^+** (being largest in size due to maximum hydration) **has least mobility**.
12. **Reducing character decreases down group : Li is strongest reducing agent**. This property depends on ionisation potential and hydration energy. **Hydration energy of Li is sufficiently high to compensate for its high ionisation energy**. Li is strongest reducing agent.

(B) Chemical properties

1. **Action of air :** They tarnish on exposure to air due to formation of oxides and hydroxides on surface. Therefore **alkali metals are kept in kerosene oil**.
2. **Reaction with oxygen**

$$Li + O_2 \longrightarrow Li_2O \text{ (Oxide only)}$$
$$2Na + O_2 \longrightarrow \underset{\text{(peroxide)}}{Na_2O_2}$$
$$\text{K or Rb or Cs} + O_2 \longrightarrow \text{Superoxide}$$

Li forms monoxide, Na forms peroxide and other metals form superoxides. Li being a small cation has a strong positive field around it and can combine with oxide ion (O^{2-}) which has a strong negative field around it. Na^+ being larger has weaker positive field and can stabilise peroxide ion (O_2^{2-}). The even larger cations like K^+, Rb^+, Cs^+ have still weaker positive field and can stabilise only very weak negative field anions like superoxide (O_2^-).

All oxides, peroxides and superoxides are basic in nature.

The solubility and basic strength of oxides increase in the order

$Li_2O < Na_2O_2 < KO_2 < RbO_2 < CsO_2$

3. **Reaction with water:**

$$2M + 2H_2O \longrightarrow 2MOH + H_2$$

Basicity of hydroxides

LiOH < NaOH < KOH < RbOH < CsOH (On moving down the group size of alkali metal atom increases and M – OH bond get weakens.

The reactivity with water increases down the group from Li to Cs as Li < Na < K < Rb < Cs due to increase in electropositive character in the same order.

4. **Reaction with halogens**
   - Alkali metals form ionic halides (M^+X^-)
   - For particular halogen order of reactivity is Li < Na < K < Rb < Cs
   - For particular alkali metal order of reactivity $F_2 > Cl_2 > Br_2 > I_2$
   - Melting point for same alkali metal Fluorides > Chlorides > Bromides > Iodides.

5. **Reaction with hydrogen :** Form **ionic hydride** (M^+H^-)
   - Order of reactivity is Li > Na > K > Rb > Cs
   - Stability of hydrides is LiH > NaH > KH> RbH > CsH

6. **Solubility in liquid ammonia:**

(i) All alkali metals dissolve in NH_3 to produce ammoniated electrons.

$$M + (y+x)NH_3 \longrightarrow \underset{\text{Ammoniated cation}}{[M(NH_3)_x]^+} + \underset{\text{Ammoniated electron}}{[e(NH_3)_y]^-}$$

(ii) Solution dark blue in colour.

(iii) Conducting due to presence of ammoniated electron.

(iv) Powerful reducing agent, and used inorganic chemistry under the name Birch reaction.

(v) Paramagnetic.

7. **Reaction with sulphur and phosphorous**

$$16M + S_8 \xrightarrow{\Delta} \underset{\text{Metal Sulphide}}{8M_2S}; \quad 12M + P_4 \xrightarrow{\Delta} \underset{\text{Metal Phosphide}}{4M_3P}$$

8. **Nature of carbonates and bicarbonates:**

(i) Stability of carbonates of alkali metals increases down the group.

(ii) **$LiHCO_3$ does not exist in solid state.**

(iii) The solubility of carbonates and bicarbonates increases on moving down the group.

Anomalous Behaviour of Li

(a) It has the smallest size in the group

(b) It has very high ionization energy and highest electronegativity in the group.

(c) It has no vacant *d*-orbitals in the valence shell.

As a result, it differs from the other member of the alkali metals family in following respects:

(i) **Forms only monoxide** while **other group members form per and superoxides** also.

(ii) **Li reacts with N_2** to form Li_3N, **other alkali metals do not react with N_2**.

$$6Li + N_2 \xrightarrow{\Delta} \underset{\text{Lithium nitride}}{2Li_3N}$$

(iii) $Li_2CO_3 \xrightarrow{\Delta} Li_2O + CO_2$, other carbonates do not give CO_2.

(iv) **LiOH** is **weak base** which **decomposes on heating** while hydroxides of other group members are stable.

(v) Salts of lithium are insoluble in water but that of other alkali metals are soluble in water.

(vi) **$LiNO_3$ on heating evolves NO_2 and O_2** while **other nitrates give only O_2**.

$$4LiNO_3 \xrightarrow{\Delta} 2Li_2O + 4NO_2 + O_2$$

$$2NaNO_3 \xrightarrow{\Delta} 2NaNO_2 + O_2$$

Li_2SO_4 is insoluble in water, rest are soluble.

Diagonal Relationship Between Li and Mg

(i) Both are hard and ductile metals.

(ii) LiOH and $Mg(OH)_2$ both are weak bases.

(iii) **Nitrates of both on heating evolve NO_2 and O_2.**

(iv) Both **react with N_2 to form nitride (Li_3N and Mg_3N_2)**

(v) Both have almost similar atomic radius, ionic radius and electronegativity.

(vi) Both are predominantly covalent.

Some Important Compounds of Sodium

Sodium Chloride, NaCl : To obtain pure sodium chloride, the crude salt is dissolved in minimum amount of the water and filtered to remove insoluble impurities. Hydrogen chloride gas is passed through the filtrate where crystals of pure NaCl separate out due to common ion effect, leaving behind calcium chloride and magnesium chloride in solution.

$$NaCl \rightleftharpoons Na^+ + Cl^-$$

$$HCl \rightleftharpoons H^+ + Cl^-$$

- **Properties**

(i) It is a while crystalline solid.

(ii) Pure sodium chloride is non-hygroscopic but the usual impurities of $CaCl_2 . 2H_2O$ and $MgCl_2 . 2H_2O$ make it hygroscopic.

- **Uses**

(i) It is used as a table salt, an essential constituent of our food.

(ii) It is used in the manufacture of Na_2CO_3, NaOH, HCl, Cl_2 etc.

(iii) It is used in soap industry for salting out of soap

Sodium Hydroxide

- **Manufacture**

(i) **Causticing process** (or **Gossage process**)

$$Na_2CO_3 + Ca(OH)_2 \longrightarrow CaCO_3 \downarrow + 2NaOH$$

(ii) **Electrolytic process :** Electrolysis of aq. NaCl,

$$NaCl + H_2O \xrightarrow{\text{electricity}} NaOH + H_2 + Cl_2$$

Electrolytic process

Nelson Cell

- Asbestos diaphram is used to separate cathode and anode.

Castner - Kellner cell

Anode - graphite Cathode - Hg Electrolyte - NaCl	Anode - Hg Cathode - Fe rods Electrolyte - dil NaOH	Anode - graphite Cathode - Hg Electrolyte - NaCl
Outer compartment	Central compartment	Outer compartment

Reactions in outer compartments

At Anode $Cl^- \rightarrow Cl + e^-$,

$2Cl \rightarrow Cl_2 \uparrow$

At Cathode $Na^+ + e^- \rightarrow Na$,

$Na + Hg \rightarrow \underset{\text{(amalgam)}}{Na/Hg}$

Reaction in central compartment

At Anode $OH^- \rightarrow OH + e^-$,

$Na/Hg + OH \rightarrow NaOH + Hg$

At Cathode $H^+ + e^- \rightarrow H$,

$2H \rightarrow H_2$

- **Properties**

(i) It is a hygroscopic, deliquescent white solid, absorbs CO_2 and moisture from the atmosphere.

(ii) **Reaction with salts**

(a) $FeCl_3 + 3NaOH \rightarrow Fe(OH)_3 \downarrow + 3NaCl$

(b) $2AgNO_3 + 2NaOH \rightarrow 2NaNO_3 + 2AgOH$

$2AgOH \rightarrow \underset{\text{Brown}}{Ag_2O} \downarrow + H_2O$

(c) $ZnSO_4 + 2NaOH \rightarrow Zn(OH)_2 \downarrow + Na_2SO_4$

$Zn(OH)_2 + 2NaOH \rightarrow \underset{\text{Sodium zincate}}{Na_2ZnO_2} + 2H_2O$

(d) $NH_4Cl + NaOH \xrightarrow{\Delta} NaCl + NH_3 \uparrow + H_2O$

(iii) **Reaction with halogens**

$X_2 + 2NaOH(\text{cold}) \rightarrow NaX + \underset{\text{Sodium hypohalite}}{NaXO} + H_2O$

$3X_2 + 6NaOH(\text{hot}) \rightarrow 5NaX + \underset{\text{Sodium halate}}{NaXO_3} + 3H_2O$

(where X= Cl, Br or I)

(iv) **Reaction with metals**

$$Zn + 2NaOH \rightarrow Na_2ZnO_2 + H_2$$

(v) **Reaction with sand**

$$2NaOH + SiO_2 \rightarrow \underset{\text{sodium silicate (glass)}}{Na_2SiO_3} + H_2O$$

Washing Soda

$Na_2CO_3.10H_2O$

- **Preparation**

(i) **Manufactured by Solvay process**

- Raw material used is $NaCl + CO_2 + NH_3$

$$NH_3 + H_2O + CO_2 \longrightarrow NH_4HCO_3$$

$$NH_4HCO_3 + NaCl \longrightarrow NaHCO_3 \downarrow + NH_4Cl$$

$NaHCO_3$ is separated and heated

$$NaHCO_3 \xrightarrow{250°C} Na_2CO_3 + H_2O + CO_2$$

- NH_4Cl is treated with milk of lime to give valuable by products – NH_3 and $CaCl_2$.

$$2NH_4Cl + Ca(OH)_2 \longrightarrow CaCl_2 + 2H_2O + 2NH_3$$

(ii) **Electrolytic Process :-** In the Nelson cell used for the manufacture of NaOH, CO_2 under Pressure is blown with steam

$$2NaOH + CO_2 \rightarrow Na_2CO_3 + H_2O$$

- **Properties**

(i) Sodium Carbonate crystallizes from water as decahydrate which effloresces on exposure to dry air forming monohyrate which on heating change to anhydrous salt (soda-ash).

$$Na_2CO_3.10H_2O \xrightarrow{\text{dry air}} Na_2CO_3.H_2O + 9H_2O$$

$$Na_2CO_3.H_2O \xrightarrow{\Delta} \underset{\text{soda ash}}{Na_2CO_3} + H_2O$$

(ii) On hydrolysis it forms an alkaline solution

$$Na_2CO_3 + 2H_2O \rightarrow \underset{\text{(weak acid)}}{H_2CO_3} + \underset{\text{(strong base)}}{2NaOH}$$

(iii) Aqueous sodium carbonate solution react with CO_2 gas and forms sodium bicarbonate.

$$Na_2CO_3 + H_2O + CO_2 \rightarrow 2NaHCO_3$$

- **Uses**

Sodium carbonate is used

(i) in laundries and in softening of hard water as washing soda,

(ii) in the manufacture of glass, caustic soda, soap powders, etc,

(iii) as laboratory reagent (Na_2CO_3 + K_2CO_3 mixture is known as fusion mixture), and

(iv) in textile and petroleum refining.

Sodium Bicarbonate or Baking Soda, $NaHCO_3$

- **Preparation**

Commercially, it is obtained by Solvay process, described in Na_2CO_3. However, in the laboratory it is prepared by saturating a cold solution of Na_2CO_3 with CO_2.

$$Na_2CO_3 + H_2O + CO_2 \longrightarrow 2NaHCO_3$$

- **Properties**

(i) Sodium bicarbonate is a white crystalline substance, sparingly soluble in water. The aqueous solution is alkaline in nature.

$$NaHCO_3 + H_2O \rightleftharpoons H_2CO_3 + NaOH$$

It is called baking soda because it decomposes on heating to generate bubbles of CO_2 gas.

- **Uses**

Sodium bicarbonate is used

(i) in the preparation of baking powder,

(ii) in the preparation of effervescent drinks and fruit salts,

(iii) in medicines to remove acidity of the stomach (i.e. as antacid), and

(iv) in fire extinguishers.

GROUP-2 : ALKALINE EARTH METALS

- They are Be, Mg, Ca, Sr, Ba, Ra.
- **General electronic configuration** - ns^2.
- They are called alkaline earth metals because their oxides are alkaline in nature and are found in earth crust.

Properties of Alkaline Earth Metals

(A) **Physical properties**

1. **Atomic size :** In general large and increases down the group.
2. **Ionisation potential :** In general low and decreases down the group.
3. **Density :** Greater than alkali metals.
4. **Melting and boiling points :** Higher than alkali metals because of more number of electrons in valence shell.
5. **Electropositive character :** Less than alkali metals (due to their smaller size and higher I.E.)
6. **Oxidation state** +2.
7. **Flame colouration :**

Element	Be	Mg	Ca	Sr	Ba	Ra
Colour of	–	–	Brick red	Crimson red	Apple green	Crimson

Be and Mg do not impart colour to flame due to high ionisation potential.

8. **Heat and electrical conductivity :** Due to the presence of two loosely held valence electrons, the alkaline earth metals are good conductor of heat and electricity.

9. **Reducting properties**. Due to their large size and low ionisation energy, they are strong reducing agents.

(B) **Chemical properties**

1. **Nature of compounds**
Be and Mg compounds - covalent
Other group element compounds - ionic.

2. **Reaction with oxygen**

$$\underset{}{2M + O_2} \longrightarrow \underset{\text{metal oxide}}{2MO} \quad (M = \text{all group 2 members})$$

$$M + O_2 \longrightarrow \underset{\text{metal peroxide}}{MO_2} \quad (M = \text{Ba, Sr})$$

3. **Reaction with water**

$$M + 2H_2O \xrightarrow{\Delta} 2M(OH)_2 + H_2$$

- Reaction becomes vigorous on moving down the group.
Order of reactivity Ba > Sr > Ca > Mg > Be.

4. **Reaction with H_2**

$$M + H_2 \xrightarrow{\Delta} MH_2 \quad (M = \text{group 2 member except Be})$$

- BeH_2 is covalent, rest of group hydrides are ionic.
- **BeH_2 and MgH_2 are polymeric**.
- BeH_2 can be prepared by other methods.
- CaH_2 is called hydrolith.

5. **Reaction with halogens**

$$M + X_2 \longrightarrow MX_2$$

- BeX_2 are covalent and rest of halides are ionic.
- **The halides are soluble in water and their solubility decreases in the order**
$MgX_2 > CaX_2 > SrX_2 > BaX_2$
- $BeCl_2$ has a polymeric structure in solid state but exists as dimer in the vapour state and as a monomer at 1200K.

6. **Reaction with N_2**

$$3M + N_2 \longrightarrow M_3N_2$$

$$M_3N_2 + 6H_2O \longrightarrow 3M(OH)_2 + 2NH_3$$

7. **Reaction with acids** - All members of group 2 **except Be displace H_2 from acids**.

$$M + H_2SO_4 \longrightarrow MSO_4 + H_2$$

(M = group 2 element except Be)

8. **Reaction with carbon**
(i) $Be + C \longrightarrow Be_2C$
- **Be_2C has** C^{4-} **anion and is called** methanide.
- **It gives** CH_4 **on reaction with H_2O.**
$Be_2C + 2H_2O \longrightarrow 2BeO + CH_4$

(ii) Other elements of group except Be form **acetylide**

$$M + 2C \longrightarrow \underset{\text{acetylide}}{MC_2}$$

- MC_2 has C_2^{2-} **ion.**
- **It gives** acetylene **on reaction with H_2O.**

$$CaC_2 + 2H_2O \longrightarrow Ca(OH)_2 + \underset{\text{acetylene}}{HC \equiv CH}$$

- **General characteristics of compounds of the alkaline Earth metals.**

(i) **Basic strength of oxides and hydroxides increases** down the group. **BeO and $Be(OH)_2$ are amphoteric and rest are basic.**

(ii) **Solubility of sulphates and carbonates**
$BeSO_4 > MgSO_4 > CaSO_4 > SrSO_4 > BaSO_4$
$BeCO_3 > MgCO_3 > CaCO_3 > SrCO_3 > BaCO_3$
($\because$ lattice energy remains constant and hydration energy decreases)

(iii) **Solubility of hydroxides**
$Ba(OH)_2 > Sr(OH)_2 > Ca(OH)_2 > Mg(OH)_2 > Be(OH)_2$
($\because$ lattice energy decreases more rapidly than hydration energy)

(iv) Stability of carbonates and bicarbonates increases down the group.
$BaCO_3 > SrCO_3 > MgCO_3 > BeCO_3$

Anomalous Behaviour of Berilium (Be)

(i) **BeO and $Be(OH)_2$ are amphoteric** but **oxides and hydroxides of other alkaline earth metals are basic.**

(ii) **Be does not liberate H_2** from acids but other members of group do so.

(iii) **Be does not react with H_2O** even at high temperature but other group members react with water and liberate H_2.

(iv) **Be forms covalent compounds** but other group members form ionic compounds.

(v) $\mathbf{BeC_2 + H_2O \longrightarrow CH_4}$,
carbide of other member + $\mathbf{H_2O \longrightarrow C_2H_2}$.

(vi) Coordination no. of Be = 4
Coordination no. of other members of group = 6.

Diagonal Relationship - Be and Al

(i) Both do not impart colour to bunsen burner flame.
(ii) Carbides of both on reaction with H_2O give CH_4.
(iii) Both are rendered passive by conc. HNO_3.
(iv) Oxides and hydroxides of both are amphoteric.
(v) Both have strong tendency to form covalent compounds.
(vi) Chlorides of both act as Lewis acid.

Some Important Compounds of Calcium

Calcium Oxide, Quick lime, Burnt lime, Lime CaO

- **Preparation**
  By the thermal decomposition of calcium carbonate.

$$CaCO_3 \xrightarrow{1070-1270K} CaO + CO_2;\ \Delta H = +1080\,kg$$

(ii) When exposed to air, it absorbs moisture and carbon dioxide forming slaked lime ($CaCO_3$).
(iii) It reacts readily with water releasing large amount of heat . This process is called slaking of lime.

$$CaO + H_2O \longrightarrow Ca(OH)_2 + heat.$$

- **Uses**
  Calcium oxide is used
  (i) in the manufacture of bleaching powder, slaked lime and lime colours,
  (ii) in the manufacture of calcium carbide, cement, glass, mortar (slaked like + sand + water), etc.,
  (iii) in the conversion of Na_2CO_3 into caustic soda,

Gypsum or Albastar ($CaSO_4.2H_2O$)

$$\underset{\text{Gypsum}}{CaSO_4.2H_2O} \xrightarrow{390\,K} \underset{\text{Plaster of Paris}}{CaSO_4.\tfrac{1}{2}H_2O} + 1\tfrac{1}{2}H_2O$$

- **Uses**
  Plaster of Paris is used
  (i) is making moulds for casting.
  (ii) It is also used for preparing black board chalks.
  (iii) Plaster of Paris is used in surgical bandages for plastering broken or fractured bones of the body.
  (iv) It is also used for making statues, models and other decorative materials.

Milk of Lime or Calcium Hydroxide (Slaked Lime)

Suspension of $Ca(OH)_2$ in water.

- **Preparation**
  By dissolving quick lime in water.

$$CaO + H_2O \longrightarrow Ca(OH)_2;\quad \Delta H = -63kJ$$

- **Properties**
  (i) Its suspension in water is known as milk of lime.
  (ii) It gives $CaCO_3$(milky) and then $Ca(HCO_3)_2$ with CO_2.

$$\underset{\text{Lime water}}{Ca(OH)_2 + CO_2} \longrightarrow \underset{\text{(Milkiness)}}{CaCO_3} + H_2O$$

$$CaCO_3 + H_2O + \underset{\text{(Excess)}}{CO_2} \longrightarrow \underset{\text{(Soluble)}}{Ca(HCO_3)_2}$$

  (iii) It reacts with Cl_2 to give bleaching powder $CaOCl_2$.

$$Ca(OH)_2 + Cl_2 \longrightarrow CaOCl_2 + H_2O$$

- **Uses**
  Calcium hydroxide is used
  (i) in the manufacture of bleaching powder and caustic soda, (ii) for white washing buildings and for softening of water,

Bleaching Powder, ($CaOCl_2$) (Calcium chloro hypochlorite, Chloride of Lime):

- **Preparation**
  By passing a current of chlorine over dry slaked lime.

$$Ca(OH)_2 + Cl_2 \longrightarrow CaOCl_2 + H_2O$$

- **Properties**
  (i) It is a mixture (mixed salt) of calcium hypochlorite ($Ca.(OCl)_2.4H_2O$) and basic calcium chloride ($CaCl_2.Ca(OH)_2.H_2O$).
  (ii) Its aqueous solution gives Ca^{2+}, Cl^- and OCl^- ions.
  (iii) With dil. H_2SO_4, it gives nascent oxygen which cause its oxidising and bleaching power.

$$2CaOCl_2 + H_2SO_4 \longrightarrow CaCl_2 + CaSO_4 + 2HClO$$
$$HClO \longrightarrow HCl + O$$

  (iv) With excess of dil. H_2SO_4 (or CO_2), it forms Cl_2 known as **available chlorine**.

$$CaOCl_2 + H_2SO_4 \longrightarrow CaSO_4 + H_2O + Cl_2$$
$$CaOCl_2 + CO_2 \longrightarrow CaCO_3 + Cl_2$$

Cement

(i) **Cement is chemically :** Calcium silicate + Calcium aluminate + 1% gypsum.

(ii) Raw material

- **Limestone** (Provides CaO)
- **Clay** (Provides SiO_2, Fe_2O_3, Al_2O_3)

(iii) **Approximate composition of cement:**

Calcium oxide (CaO)	62%
Silica (SiO_2)	22%
Alumina (Al_2O_3)	7.5%
Magnesia (MgO)	2.5%
Ferric oxide (Fe_2O_3)	2.5%

(iv) If cement has
(a) **excess of** lime-cracks appear
(b) **excess of** silica-slow setting
(c) **less of lime**-weak cement
(d) **excess of** Al_2O_3 - fast setting

(v) Cement with Fe is **white cement.**

(vi) **Setting of cement is exothermic process.**

(vii) **Concrete** = cement + sand + gravel.

(viii) **Reinforced cement** = cement + sand + iron rods.

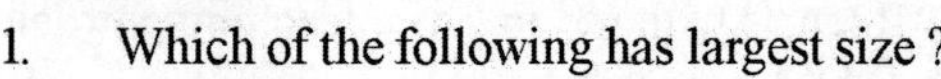

Exercise

1. Which of the following has largest size ?
(a) Na (b) Na^+
(c) Na^- (d) Can't be predicted

2. The sequence of ionic mobility in aqueous solution is :
(a) $K^+ > Na^+ > Rb^+ > Cs^+$
(b) $Cs^+ > Rb^+ > K^+ > Na^+$
(c) $Rb^+ > K^+ > Cs^+ > Na^+$
(d) $Na^+ > K^+ > Rb^+ > Cs^+$

3. Which of the following is not correct ?
(a) $2Li_2O \xrightarrow[673k]{heat} Li_2O_2 + 2Li$
(b) $2K_2O \xrightarrow[673k]{heat} K_2O_2 + 2K$
(c) $2Na_2O \xrightarrow[673k]{heat} Na_2O_2 + 2Na$
(d) $2Rb_2O \xrightarrow[673k]{heat} Rb_2O_2 + 2Rb$

4. Which of the following bicarbonates does not exist as solid?
(a) $KHCO_3$ (b) $NaHCO_3$
(c) $CsHCO_3$ (d) $LiHCO_3$

5. Which of the following does not illustrate the anomalous properties of lithium?
(a) The melting point and boiling point of Li are comparatively high
(b) Li is much softer than the other group I metals
(c) Li forms a nitride Li_3N unlike group I metals
(d) The ion of Li and its compounds are more heavily hydrated than those of the rest of the group

6. Washing soda has formula
(a) $Na_2CO_3.7H_2O$ (b) $Na_2CO_3.10H_2O$
(c) $Na_2CO_3.3H_2O$ (d) Na_2CO_3

7. Which of the following metal ions plays an important role in muscle contraction ?
(a) K^+ (b) Na^+
(c) Mg^{2+} (d) Ca^{2+}

8. The first ionization energy of magnesium is lower than the first ionization energy of
(a) Lithium (b) Sodium
(c) Calcium (d) Beryllium

9. Dehydration of hydrates of halides of calcium, barium and strontium i.e., $CaCl_2.6H_2O$, $BaCl_2.2H_2O$, $SrCl_2 . 2H_2O$, can be achieved by heating. These become wet on keeping in air. Which of the following statements is correct about these halides?
(a) act as dehydrating agent
(b) can absorb moisture from air
(c) tendency to form hydrate decreases from calcium to barium
(d) All of the above

10. Magnesium keeps on burning in
(a) N_2 (b) CO_2
(c) NO_2 (d) N_2 as well as CO_2

11. Which of the following has correct increasing basic strength ?
(a) MgO < BeO < CaO < BaO
(b) BeO < MgO < CaO < BaO
(c) BaO < CaO < MgO < BeO
(d) CaO < BaO < BeO < MgO

12. Thermal stability of alkaline earth metal carbonates decreases in order
(a) $BaCO_3 > SrCO_3 > CaCO_3 > MgCO_3$
(b) $BaCO_3 > SrCO_3 > MgCO_3 > CaCO_3$
(c) $CaCO_3 > SrCO_3 > MgCO_3 > BaCO_3$
(d) None of these

13. Which of the following statements is true about $Ca(OH)_2$?
(a) It is used in the preparation of bleaching powder.
(b) It is a light blue solid.
(c) It does not possess disinfectant property.
(d) It is used in the manufacture of cement.

14. Gypsum on heating to 300°C gives
(a) $CaSO_4 . 2H_2O$ (b) $CaSO_4$
(c) $CaSO_4 . \frac{1}{2}H_2O$ (d) SO_3 and CaO

15. Raw materials used in the manufacture of cement are :
(a) Limestone, gypsum and clay
(b) Limestone, clay and sand
(c) Quick lime and gypsum
(d) Lime, gypsum and alumina

16. The alkali metals have low melting point. Which of the following alkali metal is expected to melt if the room temperature rises to 30°C?
(a) Na (b) K
(c) Rb (d) Cs

17. Lithium is strongest reducing agent among alkali metals due to which of the following factor?
(a) Ionization energy
(b) Electron affinity
(c) Hydration energy
(d) Lattice energy

18. Which statement is false for alkali metals?
(a) Lithium is the strongest reducing agent
(b) Sodium is amphoteric in nature
(c) Li^+ is exceptionally small
(d) All alkali metals give blue solution in liquid ammonia

19. The formula of soda ash is
(a) $Na_2CO_3 . 10H_2O$ (b) $Na_2CO_3 . 2H_2O$
(c) $Na_2CO_3 . H_2O$ (d) Na_2CO_3

20. Dead burnt plaster is
(a) $CaSO_4$ (b) $CaSO_4 . \frac{1}{2}H_2O$
(c) $CaSO_4 . H_2O$ (d) $CaSO_4 . 2H_2O$

21. Which one of the following properties of alkali metals increases in magnitude as the atomic number rises ?
(a) Ionic radius
(b) Melting point
(c) Electronegativity
(d) First ionization energy

22. Compared with the alkaline earth metals, the alkali metals exhibit
(a) Smaller ionic radii
(b) Highest boiling points
(c) Greater hardness
(d) Lower ionization energies

23. Among the following oxides, the one which is most basic is
(a) ZnO (b) MgO
(c) Al_2O_3 (d) N_2O_5

24. In the case of the alkali metals
(a) The cation is less stable than the atom
(b) The cation is smaller than the atom
(c) The cation and the atom have about the same size
(d) The cation is larger than the atom

25. Sodium carbonate is manufactured by :
(a) Castner's process
(b) Solvay's process
(c) Electrolysis of brine in the presence of CO_2
(d) Ostwald's process

26. Na metal is stored in :
(a) C_6H_6 (b) Kerosene
(c) Alcohol (d) Toluene

27. Which of the following salts becomes plaster of Paris on being appropriately hydrated ?
(a) $CaCO_3$ (b) $MgSO_4$
(c) $CaSO_4$ (d) $ZnCO_3$

28. Plaster of paris hardens by
(a) giving off CO_2
(b) changing into $CaCO_3$
(c) combining with water
(d) giving out water

29. Baking powder has one of the following constituents
(a) Na_2CO_3 (b) Na_2CO_4
(c) $NaHCO_3$ (d) K_2CO_3

30. The gas evolved on heating Na_2CO_3 is
(a) CO_2 (b) Water vapour
(c) CO (d) No gas

31. Which of the following properties is not true for an alkali metal ?
(a) Low atomic volume
(b) Low ionization energy
(c) Low density
(d) Low electronegativity

32. Among the alkaline earth elements, covalent character is maximum in the compounds of :
(a) Mg (b) Zinc
(c) Strontium (d) Barium

33. Which of the following metal is found in green colouring pigment chlorophyll of plants?
(a) Fe (b) Mg
(c) Na (d) Al

34. Of the metals Be, Mg, Ca and Sr of group II A. In the perodic table the least ionic chloride would be formed by
(a) Be (b) Mg
(c) Ca (d) Sr

35. Plaster of paris is used
(a) in surgery and dentary
(b) as a whitewash
(c) as a constituent of toothpastes
(d) for the preparation of blackboard chalks

36. The solubility of metal halides depends on their nature, lattice enthalpy and hydration enthalpy of the individual ions. Amongst fluorides of alkali metals, the lowest solubility of LiF in water is due to
(a) Ionic nature of lithium fluoride
(b) High lattice enthalpy
(c) High hydration enthalpy for lithium ion.
(d) Low ionisation enthalpy of lithium atom

37. Which of the following alkali metal ions has lowest ionic mobility in aqueous solution ?
(a) Rb^+ (b) Cs^+
(c) Li^+ (d) Na^+

38. On heating sodium metal in a current of dry ammonia gas the compound formed is
(a) Sodium nitrate (b) Sodium hydride
(c) Sodium amide (d) Sodium azide

39. In the synthesis of sodium carbonate, the recovery of ammonia is done by treating NH_4Cl with $Ca(OH)_2$. The by-product obtained in this process is
(a) $CaCl_2$ (b) NaCl
(c) NaOH (d) $NaHCO_3$

40. Sodium cannot be extracted by the electrolysis of brine solution because :
(a) Electrolysis cannot take place with brine solution.
(b) Sodium reacts with water to produce NaOH + H_2
(c) Sodium being more electropositive than hydrogen, H_2 is liberated at cathode not sodium.
(d) None.

41. Some of the Group 2 metal halides are covalent and soluble in organic solvents. Among the following metal halides, the one which is soluble in ethanol is
(a) $BeCl_2$ (b) $MgCl_2$
(c) $CaCl_2$ (d) $SrCl_2$

42. Chemical A is used for water softening to remove temporary hardness. A reacts with Na_2CO_3 to generate caustic soda. When CO_2 is bubbled through A, it turns cloudy. What is the chemical formula of A
(a) $CaCO_3$ (b) CaO
(c) $Ca(OH)_2$ (d) $Ca(HCO_3)_2$

43. Amphoteric hydroxides react with both alkalies and acids. Which of the following Group 2 metal hydroxides is soluble in sodium hydroxide?
(a) $Be(OH)_2$ (b) $Mg(OH)_2$
(c) $Ca(OH)_2$ (d) $Ba(OH)_2$

44. Which of the following atoms will have the smallest size ?
(a) Mg (b) Na
(c) Be (d) Li

45. Which one of the following has minimum value of cation/anion ratio.
(a) NaCl (b) KCl
(c) $MgCl_2$ (d) CaF_2

46. All the following substances react with water. The pair that gives the same gaseous product is
(a) K and KO_2 (b) Na and Na_2O_2
(c) Ca and CaH_2 (d) Ba and BaO_2

47. Which of the following statement is false ?
(a) Strontium decomposes water readily than beryllium
(b) Barium carbonate melts at a higher temperature than calcium carbonate
(c) Barium hydroxide is more soluble in water than magnesium hydroxide
(d) Beryllium hydroxide is more basic than barium hydroxide.

48. Identify the correct statement
    (a) gypsum is obtained by heating plaster of Paris
    (b) plaster of Paris can be obtained by hydration of gypsum
    (c) plaster of paris is obtained by partial oxidation of gypsum
    (d) gypsum contains a lower percentage of calcium than plaster of Paris
49. Which one is the correct statement with reference to solubility of $MgSO_4$ in water?
    (a) SO_4^{2-} ion mainly contributes towards hydration energy
    (b) Sizes of Mg^{2+} and SO_4^{2-} are similar
    (c) Hydration energy of $MgSO_4$ is higher in comparison to its lattice energy
    (d) Ionic potential (charge/radius ratio) of Mg^{2+} is very low
50. In crystals of which one of the following ionic compounds would you expect maximum distance between centres of cations and anions?
    (a) LiF (b) CsF
    (c) CsI (d) LiI
51. A solid compound 'X' on heating gives CO_2 gas and a residue. The residue mixed with water forms 'Y'. On passing an excess of CO_2 through 'Y' in water, a clear solution 'Z', is obtained. On boiling 'Z', a compound 'X' is reformed. The compound 'X' is
    (a) $Ca(HCO_3)_2$ (b) $CaCO_3$
    (c) Na_2CO_3 (d) K_2CO_3
52. In which of the following the hydration energy is higher than the lattice energy?
    (a) $MgSO_4$ (b) $RaSO_4$
    (c) $SrSO_4$ (d) $BaSO_4$
53. The alkali metals form salt-like hydrides by the direct synthesis at elevated temperature. The thermal stability of these hydrides decreases in which of the following orders ?
    (a) CsH > RbH > KH > NaH > LiH
    (b) KH > NaH > LiH > CsH > RbH
    (c) NaH > LiH > KH > RbH > CsH
    (d) LiH > NaH > KH > RbH > CsH
54. Which of the following oxides is not expected to react with sodium hydroxide?
    (a) CaO (b) SiO_2
    (c) BeO (d) B_2O_3
55. Property of the alkaline earth metals that increases with their atomic number :
    (a) Solubility of their hydroxides in water
    (b) Solubility of their sulphates in water
    (c) Ionization energy
    (d) Electronegativity
56. Apart from lime, the major constituent of Portland cement is :
    (a) Alumina (b) Silica
    (c) Magnesia (d) Iron oxide
57. Which is manufactured by electrolysis of fused NaCl ?
    (a) NaOH (b) Na
    (c) NaClO (d) $NaClO_3$.
58. Out of the following elements which one do you expect to be most reactive chemically ?
    (a) Be (b) Ca
    (c) Ba (d) Sr.
59. The formula of calcium cyanamide is
    (a) $Ca(CN)_2$ (b) CaC_2N
    (c) CaNCN (d) $CaCHNH_2$
60. The process associated with sodium carbonate manufacture is known as :
    (a) Chamber (b) Haber
    (c) LeBlanc (d) Castner

The p-Block Elements (Group 13 & 14)

CONCEPTS

GROUP 13 (BORON FAMILY)

- Members of group - B , Al , Ga, In, Tl.
- General electronic configuration - ns^2, np^1

Properties of Group 13 Elements

A. Physical Properties

(a) **Atomic radii** (or atomic size)

(i) Increases down the group, with an exception at gallium.

(ii) **Radii (or size) of Ga (135pm) is less than Al(143pm) due to poor screening effect of d-electrons.**

(b) **Ionisation energy**

First ionisation energies of group members follow the following order (due to intervening d and f electrons)

B > Al < Ga > In < Tl

(c) **Oxidation state**

(i) **Inert pair effect** is reluctance of s-electrons of valence shell to take part in bonding

(ii) +1 and +3 are two oxidation states shown by group members.

(iii) Stability order of +1 oxidation state is B < Al < Ga < In < Tl.

(iv) Stability of +3 oxidation state decreases down the group.

$\therefore$ **$T\ell^+$ is more stable than $T\ell^{+3}$**

(v) B and Al show an oxidation state of +3 only while Ga, In, and Tl show oxidation state of both +1 and +3.

(d) **Boiling and melting points** show no regular trend.

(i) B has exceptionally high m.p due to its three dimentional network structure.

(ii) Ga has low melting point as it consists of only Ga_2 molecules.

(e) **Density :** It increases regularly on moving down the group from B to Tl.

(f) **Reducing Character :** It decreases down the group from Al to Tl because of the increase in electrode potential values for M^{3+}/M. Therefore, it follows the order.

$$Al > Ga > In > Tl$$

B. Chemical Properties

(a) **Complex formation :** Due to their smaller size and greater charge, these elements have greater tendency to form complexes than the s-block elements.

(b) **Reaction with air (Formation of oxides)**

(i) All members of group react with air or oxygen to form oxide (M_2O_3).

(ii) Basic character of oxide increases down group.

B_2O_3	Al_2O_3, Ga_2O_3	In_2O_3, Tl_2O_3
Acidic	Amphoteric	Basic

Besides Tl_2O_3, Tl also forms Tl_2O due to inert pair effect.

(c) **Reaction with alkalies** - All members of this group dissolve in alkalies with the evolution of H_2.

$$2B + 6NaOH \longrightarrow 2Na_3BO_3 + 3H_2$$

$$2Al + 2NaOH + 2H_2O \longrightarrow 2NaAlO_2 + 3H_2$$

(d) **Reaction with N_2 :** All members of this group on reaction with nitrogen form corresponding nitrides.

$$2B + N_2 \xrightarrow{\Delta} \underset{\text{(boron nitride)}}{2BN}$$

(e) **Action of halogen :** $2M + 3X_2 \longrightarrow 2MX_3$, All the group 13 elements form trihalide except Tl. Tl form TlX.

(f) **Action of acids :** $2M + 6H^+ \longrightarrow 2M^{3+} + 3H_2$. Boron is not affected by non-oxidizing acids like HCl and dilute H_2SO_4 while other elements dissolve to form trivalent salts. Conc. HNO_3 and conc. H_2SO_4 reacts with all the group 13 elements, but HNO_3 makes Al passive.

$$2B + 3H_2SO_4 \longrightarrow 2H_3BO_3 + 3SO_2$$
$$B + 3HNO_3 \longrightarrow H_3BO_3 + 3NO_2$$
$$2Al + 6H_2SO_4 \longrightarrow Al_2(SO_4)_3 + 3SO_2 + 6H_2O$$

General characteristics of the compounds of group 13 elements

- **Hydrides**
  (i) Hydrides are not formed by direct reaction with H_2.
  (ii) **B** form two series of hydrides called as **Boranes**. General formula of boranes are B_nH_{n+4} and B_nH_{n+6}.
  (iii) **Al forms polymeric hydrides** of formula $(AlH_3)_x$.
  (iv) Rest of the group members form single hydrides (MH_3) which are unstable.
- **Halides**
  (i) All elements form trihalide (MX_3) except Tl (due to inert pair effect).
  (ii) Halides of boron are electron deficient compounds (Lewis acids). **The acidic strength of boron halides is $BF_3 < BCl_3 < BBr_3 < BI_3$.**
  The order is reverse of as expected by electronegativity data. This order is possible due to **pπ - pπ back bonding** between B and F
  (iii) All trihalides of boron have sp^2 hybridisation and **trigonal planar** shape.
  (iv) BCl_3 does not form dimer because of its small size can't hold large bridged chlorine atom. **$AlCl_3$ exists in form of dimer**.
  (v) Order of acidic strength when central atoms is different is $BX_3 > AlX_3 > GaX_3 > InX_3$.

Anomalous Behaviour of B

(i) B shows allotropy but other group members don't show allotropy.
(ii) B shows maximum covalency of 4 eg $[BH_4^-]$ while other group members show maximum co-valency of 6 eg. $[AlF_6]^3$.
(iii) Boron forms series of hydrides (boranes).
(iv) B does not react with steam.
(v) It is a non-metal while other members of the group are metallic in character.
(vi) The halides of boron exist as monomers while $AlCl_3$ exists as a dimer.
(vii) The oxides and hydroxides of boron are weak acidic while those of aluminium are amphoteric and those of other elements are basic.

Diagonal Relationship Between B and Si

(i) Both are non metals and semiconductors.
(ii) Both form covalent , volatile and hygroscopic halides.
(iii) Their oxides are acidic in nature and dissolve in alkalis.
(iv) Both do not form ions *i.e.* B^{3+} and Si^{4+}.

Some Important Compounds of Boron

Boric acid (H_3BO_3) or Orthoboric acid

- **Preparation**
  From borax :-($Na_2B_4O_7 . 10H_2O$):

$$Na_2B_4O_7 + H_2SO_4 + 5H_2O \longrightarrow Na_2SO_4 + \underset{\text{(boric acid)}}{4H_3BO_3}$$

or

$$Na_2B_4O_7 + 2HCl + 5H_2O \longrightarrow 2NaCl + 4B(OH)_3$$

- **Properties**
  Action of heat:-

$$\underset{\text{boric acid}}{H_3BO_3} \xrightarrow{373\,K} \underset{\text{metaboric acid}}{HBO_2} \xrightarrow{433\,K} \underset{\text{tetraboric acid}}{H_2B_4O_7} \xrightarrow[-H_2O]{\text{red hot}} \underset{\text{boron trioxide}}{B_2O_3}$$

Boric acid behaves as a weak monobasic acid.

- **Uses**
  (i) It is used as an antiseptic, as eye-wash (eye-lotion) and a food preservative.
  (ii) It is used in the preparation of glass, glazes and enamels.

Borax ($Na_2B_4O_7 . 10H_2O$)

- **Preparation**
  From colemanite

$$Ca_2B_6O_{11} + 2Na_2CO_3 \longrightarrow 2CaCO_3 \downarrow + Na_2B_4O_7 + 2NaBO_2$$

- **Properties**
  (i) Its solution is basic in nature due to hydrolysis.

$$Na_2B_4O_7 + 7H_2O \rightleftharpoons \underset{\text{strong base}}{2NaOH} + \underset{\text{weak acid}}{4H_3BO_3}$$

  (ii) **Action of heat:**

$$Na_2B_4O_7.10H_2O \xrightarrow[-10H_2O]{} \underset{\text{anhydrous}}{Na_2B_4O_7} \xrightarrow{740°C} \uparrow 2NaBO_2 + \underset{\text{transparent bead}}{B_2O_3} \uparrow$$

- **Borax bead test :** Borax on strong heating forms B_2O_3 which forms coloured glassy bead with coloured compounds of certain metals. It is called borax bead test.
- **Uses**
  (i) Borax is used in the manufacture of enamels, glazes, pyrex glass which in turn is used in the production of crystal glass and optical glasses, soap and drying oils.
  (ii) It is also used as a cleansing and stiffening agent in laundry work and in candle industry.
  (iii) It is used for glazing paper, playing cards and also as an antiseptic.

Diborane (B_2H_6)

- **Preparation**

(i) $2NaBH_4 + I_2 \xrightarrow{\text{Polyether}} B_2H_6 + 2NaI + H_2$

(ii) $2BF_3 + 6NaH \xrightarrow{450K} B_2H_6 + 6NaF$

- **Properties**

(i) $B_2H_6 + 3O_2 \longrightarrow B_2O_3 + 3H_2O;$
$\Delta H_c = -1976\,kJ/mol.$

As it has high heat of combustion per unit mass, therefore used as a high energy fuel.

(ii) $B_2H_6 + 6H_2O \longrightarrow \underset{\text{Boric acid}}{2H_3BO_3} + 6H_2$

(iii) $2LiH + B_2H_6 \longrightarrow \underset{\text{Lithium Borohydride}}{2Li(BH_4)}$

(iv) Reaction with ammonia

$$2NH_3 + B_2H_6 \longrightarrow [BH_2(NH_3)_2]^+BH_4^-$$

$$\underset{\text{Diammoniate of diborane}}{3[BH_2(NH_3)_2]^+BH_4^-} \xrightarrow{450K} \underset{\text{Borazine}}{2B_3N_3H_6} + 12H_2$$

$B_3N_3H_6$ resembles benzene in structure and is called inorganic benzene or borazole or borazine. Boron nitride $(BN)_x$ has structure similar to graphite and is called Inorganic graphite.

- **Structure of Diborane**

The four terminal hydrogen atoms and the two Boron atoms lie in one plane. Above and below this plane, there are two bridging H-atoms. The four terminal B–H bonds are regular two centre – two electron bonds while the two bridge (B–H–B) bonds are different and can be described in terms of three centre – two electron bonds as shown in below.

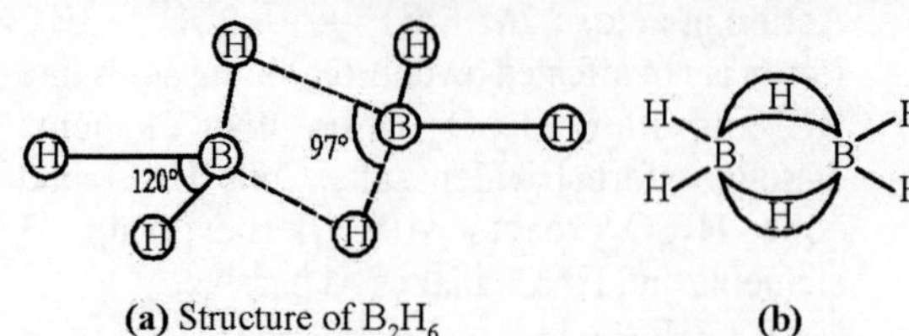

(a) Structure of B_2H_6 (b)

Aluminium (Al) and Its Compounds

(a) Ores of Aluminium

(i) Bauxite – $Al_2O_3.2H_2O$
(ii) Corundum – Al_2O_3
(iii) Cryolite – Na_3AlF_6
(iv) Feldspar – $K_2O.Al_2O_3.6SiO_2$.

(b) Extraction of Al from Bauxite

(i) **Concentration**

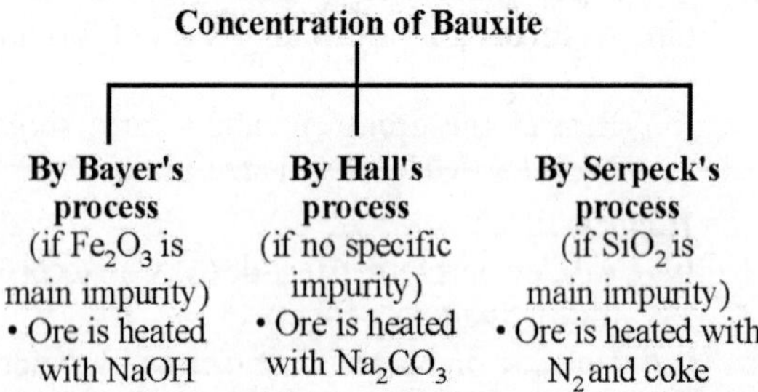

(ii) **Electrolysis of pure Al_2O_3 (Hall and Heroult's process)** Cathode made of Fe, Anode made of graphite are used in this method. **Cryolite is used as electrolyte to decrease temperature of reaction and increase the conductivity**.

(iii) **Refining of Alumina (Hoop's electrolysis) :** The refining cell has three fused layers of different densities as follows

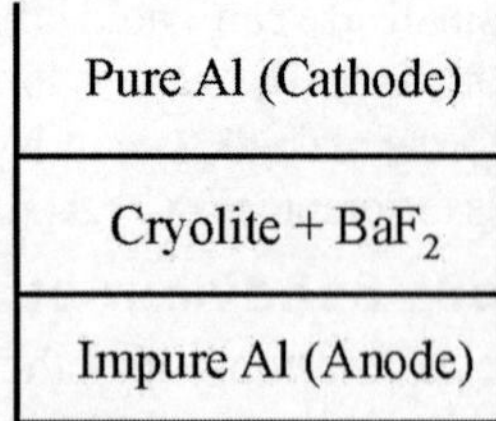

Pure Al is collected at cathode.

(c) Anhydrous $AlCl_3$

It fumes in moist air due to hydrolysis and the resulting solution is acidic in nature

$$AlCl_3 + 3H_2O \longrightarrow Al(OH)_3 + 3HCl$$

It exists as dimer Al_2Cl_6 in which each Al atom is

terahedrally surrounded by four Cl atoms as below.

$$\begin{array}{c}Cl\diagdown\quad\;Cl\quad\;\diagup Cl\\ \;Al\quad\;Al\\ Cl\diagup\quad\;Cl\quad\;\diagdown Cl\end{array}$$

(d) Alums : These are double salts.

(i) **General formula of alum is**
- $M_2SO_4 . M_2'(SO_4)_3 . 24H_2O$
where M = monovalent metal cation.
M′ = trivalent metal cation.

(ii) Alums are used in tanning of leather, antiseptic to stop bleeding and purification of water.

GROUP 14 (CARBON FAMILY)

- Members of group - C, Si, Ge, Sn, Pb.
- General electronic configuration - ns^2, np^2.

- **Properties of group 14 elements**

1. **Atomic radii** - Increases down the group (C to Pb).
2. **Ionisation enthalpy C > Si > Ge > Sn < Pb** (Due to intervening effect of d and f electrons)
3. **Electronegativity** decreases down the group but not in a regular manner. Carbon is most electronegative (2.4 on Pauling scale) element of group.
4. **Catenation** is property of self linkage. It is maximum for carbon due to high stability of C–C bond. It decreases down group. **Sn and Pb do not show catenation.**
5. **Metallic character :** Carbon is a non–metal, carbon is a typical non-metal forming covalent bonds in the vast majority of its compounds using all its four valence electrons. Silicon is also a non-metal but in some physical properties it behaves as a semimetal, germanium is a metalloid, tin and lead are metals.
6. **Oxidation state :** All the elements have, in common, the oxidation state of +4. Since the inert pair effect increases from Ge to Pb, the stability of M^{4+} ions decreases whereas that of M^{2+} ion increases. Thus the stability order is Ge^{4+} . > $Sn^{4+} > Pb^{4+}$ and $Ge^{2+} < Sn^{2+} < Pb^{2+}$ and $< Sn^{2+} < Sn^{4+}$, $Pb^{2+} > Pb^{4+}$
7. **Melting and boiling point :** On moving down the group, the melting and boiling point decreases.

GENERAL CHARACTERISTICS OF THE COMPOUOND OF GROUP 14

- **Hydrides**

(i) All group members form covalent hydrides.
(ii) Hydrides of carbon are called **hydrocarbons** (alkane, alkene, alkyne and arene).
(iii) Hydrides of Si are called **Silane**, general formula Si_nH_{2n+2}.
(iv) Hydrides of Ge are called **germanes**, only 3 germanes are known.
(v) Hydrides of Sn are called **stannane**, two stannane are known SnH_4 and Sn_2H_6.
(vi) Lead forms only one hydride **plumbane** (PbH_4).

- **Oxides**

(i) All group members form two oxides - monoxide and dioxide.
(ii) **Monoxides (MO)** - CO is neutral and rest of monoxides are basic.
(iii) **Preparation of CO**

$C + H_2O \longrightarrow CO + H_2$ (water gas)
$C + O_2 + 4N_2 \longrightarrow 2CO + 4N_2$ (producer gas)

(iv) **Uses**

(a) It is used in making carbonyls among which nickel tetracarbonyl is used in the extraction of nickel (Mond's Process).

(b) It is used in the manufacture of methanol, synthetic petrol and phosgene.

CO + haemoglobin $\longrightarrow$ Carboxyhaemoglobin (stops supply of O_2)

Antidote for CO poisoning is carbogen (95 % O_2 and 5% CO_2)

(v) **Characteristics of Oxides**

(a) CO_2 is gas at room temperature and forms $p\pi$ - $p\pi$ bonding.

(b) SiO_2 and other dioxides exists in form of network solid and does not form $p\pi$ - $p\pi$ bond . Si is in sp^3 hybrid state in SiO_2. It exists in three **crystalline forms which are quartz, cristobalite and tridymite.** Amorphous form of silica are silica gel and fumed silica.

(c) **Basicity of dioxides increases down the group.**

$\underbrace{CO_2 \quad SiO_2}_{\text{acidic}} \quad \underbrace{GeO_2 \quad SnO_2}_{\text{amphoteric}} \quad \underbrace{PbO_2}_{\text{basic}}$

(d) All the dioxide of group members dissolve in alkalies.

- **Dry ice :** Solid carbon dioxide (obtained on cooling at – 40°C) is called as dry ice, or drikold. It can be obtained if liquid CO_2 is allowed to escape through a nozzle from a high pressure container.
- **Action of water :** C, S, Ge and Pb are unaffected by H_2O.

$$Sn + 2H_2O \longrightarrow SnO_2 + 2H_2 \uparrow$$

Halides

(i) General formula of halides of group is MX_4. Halides are covalent in nature and have tetrahedral structure.

(ii) **Stability of tetrahalides**

$CX_4 > SiX_4 > GeX_4 > SnX_4 > PbX_4$

$MF_4 > MCl_4 > MBr_4 > MI_4$

(iii) **$PbBr_4$ and PbI_4 does not exist** because Pb^{4+} is strong oxidant and does not survive in presence of strong reductants like Br^- and I^- and is immediately reduced to Pb^{2+}.

(iv) **CCl_4 (or CX_4) cannot be hydrolysed** (due to absence of vacant d-orbitals) but other halides of group are easily hydrolysed.

Anomalous Behaviour of Carbon

Carbon behaves differently from rest of the members of the group due to its smaller size, higher electronegtivity, higher I.E. and unavailability of d-orbitals.

(i) due to non-anvailability of d-orbitals, it cannot extend its covalence beyond four.

(ii) carbon has a unique ability to form pπ – pπ multiple bonds with itself and with other atoms of small size and high electronegativity. Heavier elements do not form pπ – pπ bonds due to large size of their orbitals.

(iii) C-shows the tendency of catenation i.e. carbon atoms link with one another through covalent bonds to form rings and chains. This is because the C–C bonds are very strong. Down the group, size increase, electronegativity decreases and thereby tendency to show catenation decreases.

Allotropy

Existence of same substance in two or more forms in same physical state having different properties. All the elements except Pb show allotropy.

(i) **Allotrops of Carbon**

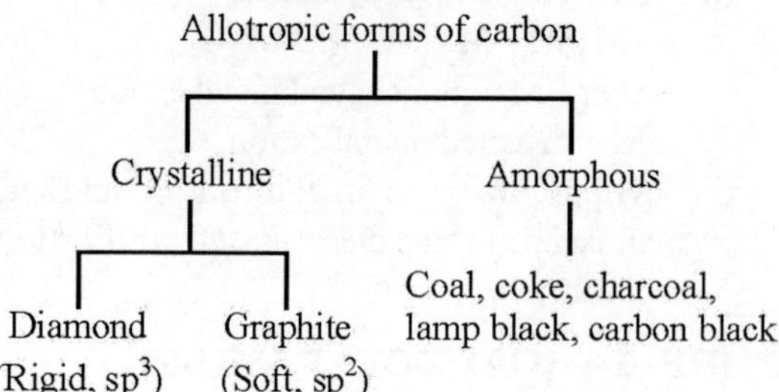

(ii) Allotrops of Si - crystalline and amorphous

(iii) Allotrops of Ge - two crystalline forms

(iv) **Allotrops of Sn - Grey tin, white tin, rhombic tin.**

Carbon does not form any complex due to absence of d-orbital, whereas other elements can form.

Compounds of Carbon

(a) **Fullrenes** are members of spheroidal carbon cage molecules with hexagonal and pentagonal patterns.

(b) **Carbides** are compounds of carbon with more electro positive elements than itself.

(c) **Calcium cyanide ($CaCN_2$) + C (graphite)** are called **nitrolium** and is used as nitrogen fertilizer.

(d) **Gaseous fuels**

(i) **Water gas or synthesis gas** - $CO + H_2$

$$\underset{\text{Steam}}{C + H_2O} \xrightarrow{\Delta} \underbrace{CO + H_2}_{\text{water gas}}$$

(ii) **Producer gas** - $CO + N_2$

(iii) **Oil gas** - $CH_4 + H_2 + CO + CO_2$

(iv) **Semi water gas** - $CO + H_2 + N_2$ (with CH_4 1 – 5%, CO_2 4 – 5%)

(v) **LPG** - Butane + isobutane

(vi) **Natural gas** - mainly CH_4 ($C_2H_6 + H_2 + CO + CO_2$)

(vii) **Biogas or gobar gas** - CH_4 = 55%, H_2 = 7.4 %, CO_2 = 35%, N_2 = 2.6 %, H_2O in traces.

(viii) **Coal Gas** - CH_4 = 32%, H_2 = 40%, CO = 7%, C_2H_2 = 2%, C_2H_4 = 3%, N_2 = 4%, CO_2 = 1%.

Compounds of Silicon

(a) Silicates

- All silicates have basic repeating unit $(SiO_4)^{4-}$ and has tetrahedron shape.

Types of silicates :

S. No.	Name of type	Basic unit	Examples
1	Orthosilicate	SiO_4^{2-}	Zircon ($ZrSiO_4$)
2	Pyrosilicates	$Si_2O_7^{6-}$	Thortveitite ($Sc_2Si_2O_7$)
3.	Cyclic silicates	$Si_3O_9^{6-}$	Beryl ($Be_3Al_2Si_6O_{18}$)
4.	Chain silicates	$(SiO_3)^{2-}$ or $(Si_4O_{11})_n^{6-}$	Asbestos
5.	Sheet silicates	$(Si_2O_5)_n^{2-}$	Mica, Clay, Kaolin , Talc
6.	Three dimentional or Frame work silicate	$(SiO_3)_n$	Zeolite , Quartz.

(b) **Silicones**

- Organosilicon compounds have repeating R_2SiO units and **Si– O – Si linkage**.
- They are water repellent, heat resistant and inert.
- They are used as electrical insulators and lubricants.

(c) **Silica (SiO_2) :** It exists in three crystalline forms Quartz, Trydimite and Crystobalite.

(d) **Zeolites** -They have honey comb like structure and have the general formula $\frac{Mx}{n}(AlO_2)_x.(SiO_2)_y.zH_2O$

$M = Na^+, K^+$ or Ca^{++}

n = Charge on metal

They act as ion exchanger and molecular sieves. They can be artificially perpared by heating **China clay, Silica** and **Na_2CO_3.**

Glass

An amorphous transparent or translucent super cooled liquid which is mixture of silicates and borates of K, Ca and Pb.

Type of glass -	Batch	Properties	Uses
(i) Soft or soda glass	$Na_2O.CaO.SiO_2$	Soft	Window glass and bottles
(ii) Hard or potash glass	Sand + lime stone + K_2CO_3	Hard	Hard glass apparatus
(iii) Flint glass	Sand +Red lead + K_2CO_3	High refractive index	Optical instrument, prisms
(iv) Pyrex glass	$Na_2O.B_2O_3.Al_2O_3.SiO_2$	Low coefficient	Glass ware of expansion
(v) Jena glass	Zn and Barium Borosilicate	Low coefficient of expansion	Glass wares
(vi) Crookes glass	$K_2O.PbO.CeO_2.SiO_2$	Cuts off U.V. rays	Eye lenses
(vii) Quartz glass or silica glass	Pure SiO_2	Low coefficient of expansion	Scientific wares
(viii) Safety glass or shatter proof glass	Between two glass layers a transparent plastic layer of vinyl acetate resin is placed		Automobile wind shields, goggles etc.
(ix) Reinforced glass	Has network of wires	Does not shatter	
(x) Ground glass	Soft glass with ground surface		

Tin

(a) (i) **Ore - Cassiterite or tin stone SnO_2.**

(ii) Tinpyrites, Sn, Cu_2S, FeS.

(b) **Tinning** is process of covering surface of utensils made of copper or brass by tin because tin is not attacked by organic acids present in food stuff.

Soft silvery white metal, ductile and malleable. It has maximum number of isotopes and three allotropic forms.

$$\text{Grey} \underset{}{\overset{18^\circ}{\rightleftharpoons}} \text{White} \underset{}{\overset{161^\circ}{\rightleftharpoons}} \text{Rhombic}$$

Tin cry : It produces a peculiar cracking sound on

bending which is known as tin cry.

Tin plague : It is the conversion of white tin to grey tin at low temperature which crumbles into powder.

Tin plating : It involves the depositing of thin protective layer of tin over sheets of iron electrolytically.

(c) **$SnCl_2 . 5H_2O$ is called butter of tin** and used as mordant.

(d) $Sn + 4HNO_{3(conc)} \longrightarrow \underset{\text{metastanic acid}}{H_2SnO_3} + 4NO_2 + H_2O$

(e) $SnCl_2$ is ionic and reducing in nature
$SnCl_4$ is covalent and oxidising in nature

Lead

(a) **Ore**

(i) **Galena** - PbS
(ii) **Cerussite** - $PbCO_3$
(iii) **Anglesite** - $PbSO_4$

(b) **PbO_2**

- **Preparation**

$$Pb_3O_4 + 4HNO_3 \xrightarrow{\Delta} 2Pb(NO_3)_2 + 2H_2O + PbO_2$$

- **Properties :** It is powerful oxidising agent and is amphoteric in nature.

(c) **PbO (Litharge)**

- **Preparation** :

$$2Pb(NO_3)_2 \xrightarrow{\Delta} 2PbO + 4NO_2 + O_2$$

$$6PbO + O_2 \xrightarrow{470°C} \underset{\text{Read lead}}{2Pb_3O_4} \xrightarrow{775K} 6PbO + O_2$$

- **Properties :** It is used in glazing pottery.

(d) **TEL :** $Pb(C_2H_5)_4$ is used as antiknock agent.

(e) **Lithophone** is ($BaSO_4$ + ZnS) is substitute of white lead.

Exercise

1. Which one of the following has the lowest m.p.?
(a) B (b) Al
(c) Ga (d) Tl

2. Which one of the following is strongest Lewis acid ?
(a) BF_3 (b) BCl_3
(c) BBr_3 (d) BI_3

3. Thallium shows different oxidation states because
(a) It is transition element
(b) Of inert pair effect
(c) Of its amphoteric character
(d) Of its higher reactivity

4. Orthoboric acid –
(a) donate proton to form $H_2BO_3^-$
(b) accept proton of form $H_4BO_3^+$
(c) donate OH^- to form $H_2BO_2^+$
(d) accept OH^- to form $[B(OH)_4]^-$

5. Inorganic benzene is :
(a) $B_3H_3N_3$ (b) BH_3NH_3
(c) $B_3H_6N_3$ (d) $H_3B_3N_6$

6. Chemically borax is
(a) Sodium metaborate
(b) Sodium orthoborate
(c) Sodium tetraborate
(d) Sodium tetraborate decahydrate

7. Which metal is protected by a layer of its own oxide?
(a) Al (b) Ag
(c) Au (d) Fe

8. Which does not exist ?
(a) $[SiCl_6]^{2-}$ (b) $[GeF_6]^{2-}$
(c) $[CCl_6]^{2-}$ (d) $[SnCl_6]^{2-}$

9. Least thermally stable is –
(a) CCl_4 (b) $SiCl_4$
(c) $GeCl_4$ (d) $GeBr_4$

10. Catenation i.e., linking of similar atoms depends on size and electronic configuration of atoms. The tendency of catenation in Group 14 elements follows the order :
(a) C > Si > Ge > Sn (b) C >> Si > Ge ≈ Sn
(c) Si > C > Sn > Ge (d) Ge > Sn > Si > C

11. Graphite conducts electricity because of :
(a) Weak van der Waal's forces between layers
(b) Covalent bonding between carbon atoms of layers
(c) Delocalized electrons in each layer
(d) sp^2 hybridisation of carbon atoms in a layer

12. Gas A is bubbled through slaked lime when a white precipitate is formed. On prolonged bubbling the precipitate is dissolved. On heating the resultant solution, the white precipitate

appears with evolution of gas B. The gases A and B respectively are

(a) CO and CO (b) CO_2 and CO

(c) CO and CO_2 (d) CO_2 and CO_2

13. Quartz is the crystalline variety of
(a) silicon (b) sodium silicate
(c) silica (d) silicon carbide

14. The straight chain polymer (silicones) is formed by:
(a) hydrolysis of CH_3SiCl_3 followed by condensation polymerisation
(b) hydrolysis of $(CH_3)_4Si$ by addition polymerisation
(c) hydrolysis of $(CH_3)_2SiCl_2$ followed by condensation polymerisation
(d) hydrolysis of $(CH_3)_3SiCl$ followed by condensation polymerisation

15. Name the type of the structure of silicate in which one oxygen atom of $[SiO_4]^{4-}$ is shared ?
(a) Linear chain silicate
(b) Sheet silicate
(c) Pyrosilicate
(d) Three dimensional

16. Aluminium chloride is a/an
(a) Bronsted - Lowery acid
(b) Arhenius acid
(c) Lewis acid
(d) Lewis base

17. Which is not correct?
(a) Al acts as a reducing agent
(b) Al does not react with steam even at higher temperature
(c) Al forms a number of alloys with other metals
(d) Al is ionic in all its compounds

18. Dry ice is
(a) solid NH_3 (b) solid SO_2
(c) solid CO_2 (d) solid N_2

19. Which of the following types of forces bind together the carbon atoms in diamond ?
(a) Ionic (b) Covalent
(c) Dipolar (d) van der Waals

20. Water gas is produced by
(a) Passing steam through a red hot coke bed
(b) Saturating hydrogen with moisture
(c) Mixing oxygen and hydrogen in the ratio of 1 : 2
(d) Heating a mixture of CO_2 and CH_4 in petroleum refineries.

21. Synthesis gas is a mixture of
(a) Steam and CO (b) CO and N_2
(c) H_2 and CO (d) H_2 and CH_4

22. Which of the following is a semiconductor ?
(a) C (b) Pb
(c) Ge (d) Sn

23. Ordinary glass is :
(a) Sodium silicate
(b) Calcium silicate
(c) Sodium and calcium silicate
(d) Mixed salt of Na and Ca

24. Which of the following is most abundant in the earth crust ?
(a) Boron (b) Aluminium
(c) Gallium (d) Thallium

25. Nitrolim is a mixture of
(a) CaC_2 and Ga (b) $CaCN_2$ and C
(c) $Ca(CN)_2$ and Ga (d) $CaCN_2$ and N_2

26. In graphite C atom is in .... state
(a) sp^3 (b) sp
(c) sp^2 (d) None of these

27. Thermite is a mixture of :
(a) $Cr_2O_3 + Al_2O_3$ (b) $Fe_2O_3 + Al$
(c) $Fe_2O_3 + Al_2O_3$ (d) $Al_2O_3 + 2Cr$

28. Producer gas is the mixture of :
(a) $CO + N_2$ (b) $CO + H_2$
(c) CO + water vapour (d) $N_2 + CH_4$

29. Coal gas is a mixture of :
(a) H_2O and CO (b) H_2, CO and CH_4
(c) H_2 and CO (d) CH_4 and CO

30. Amphoteric oxide among the following is –
(a) B_2O_3 (b) Ga_2O_3
(c) In_2O_3 (d) Tl_2O_3

31. The exhibition of highest co-ordination number depends on the availability of vacant orbitals in the central atom. Which of the following elements is not likely to act as central atom in MF_6^{3-} ?
(a) B (b) Al (c) Ga (d) In

32. Identify the statement that is not correct as far as structure of diborane is concerned
(a) There are two bridging hydrogen atoms and four terminal hydrogen atoms in diborane
(b) Each boron atom forms four bonds in diborane
(c) The hydrogen atoms are not in the same plane in diborane
(d) All, B – H bonds in diborane are similar

33. Aluminium vessels should not be washed with materials containing washing soda because
(a) washing soda is expensive
(b) washing soda is easily decomposed
(c) washing soda reacts with aluminium to form soluble aluminate
(d) washing soda reacts with aluminium to form insoluble aluminium oxide

34. Which is false in case of boric acid H_3BO_3?
(a) It acts as a tribasic acid.
(b) It has a planar structure.
(c) It acts as a monobasic acid.
(d) It is soluble in hot water.

35. A compound X, of boron reacts with NH_3 on heating to give another compound Y which is called inorganic benzene. The compound X can be prepared by treating BF_3 with Lithium aluminium hydride. The compounds X and Y are represented by the formulas.
(a) $B_2H_6, B_3N_3H_6$ (b) $B_2O_3, B_3N_3H_6$
(c) $BF_3, B_3N_3H_6$ (d) $B_3N_3H_6, B_2H_6$

36. $AlCl_3$ on hydrolysis gives
(a) $Al_2O_3.H_2O$ (b) $Al(OH)_3$
(c) Al_2O_3 (d) $AlCl_3.6H_2O$

37. Ionisation enthalpy (Δ_iH_1kJ mol^{-1}) for the elements of Group 13 follows the order.
(a) B > Al > Ga > In > Tl
(b) B < Al < Ga < In < Tl
(c) B < Al > Ga < In > Tl
(d) B > Al < Ga > In < Tl

38. Which of the following is most stable?
(a) Sn^{2+} (b) Ge^{2+}
(c) Si^{2+} (d) Pb^{2+}

39. Cement, the important building material is a mixture of oxides of several elements. Besides calcium, iron and sulphur, oxides of elements of which of the group (s) are present in the mixture?
(a) Group 2
(b) Groups 2, 13 and 14
(c) Groups 2 and 13
(d) Groups 2 and 14

40. The most commonly used reducing agent is
(a) $AlCl_3$ (b) $PbCl_2$
(c) $SnCl_4$ (d) $SnCl_2$

41. Lead pipes are not suitable for drinking water because
(a) lead forms basic lead carbonate
(b) lead reacts with water containing air to form $Pb(OH)_2$
(c) a layer of lead dioxide is deposited over lipes
(d) lead reacts with air to form litharge

42. Graphite is a soft solid lubricant extremely difficult to melt. The reason for this anomalous behaviour is that graphite
(a) is an allotropic form of diamond
(b) has molecules of variable molecular masses like polymers
(c) has carbon atoms arranged in large plates of rings of strongly bound carbon atoms with weak inter plate bonds
(d) is a non-crystalline substance

43. Which statement is false?
(a) Water gas is a mixture of hydrogen and carbon dioxide.
(b) Producer gas is a mixture of CO and nitrogen.
(c) Water gas is a mixture of carbon monoxide and hydrogen.
(d) Natural gas consists of methane, ethane and gaseous hydrocarbons.

44. Silicon has a strong tendency to form polymers like silicones. The chain length of silicone polymer can be controlled by adding
(a) $MeSiCl_3$ (b) Me_2SiCl_2
(c) Me_3SiCl (d) Me_4Si

45. Glass is a
(a) Liquid
(b) Solid
(c) Supercooled liquid
(d) Transparent organic polymer

46. Which of the following elements is extracted commercially by the electrolysis of an aqueous solution of its compound ?
(a) Cl (b) Br
(c) Al (d) Na

47. Which of the following statements about H_3BO_3 is not correct ?
(a) It is a strong tribasic acid.
(b) It is prepared by acidifying an aqueous solution of borax.
(c) It has a layer structure in which planar BO_3 units are joined by hydrogen bonds.
(d) It does not act as proton donor but acts as a Lewis acid by accepting a lone pair of electrons.

48. In graphite, electrons are
(a) Localised on every third C-atom
(b) Present in anti-bonding orbital
(c) Localised on each C-atom
(d) Spread out between the structure

49. Glass reacts with HF to produce
(a) SiF_4 (b) H_2SiF_6
(c) H_2SiO_3 (d) Na_3AlF_6

50. Which one of the following statements about the zeolites is false ?
    (a) They are used as cation exchangers
    (b) They have open structure which enables them to take up small molecules
    (c) Zeolites are aluminosilicates having three dimensional network
    (d) Some of the SiO_4^{4-} units are replaced by AlO_4^{5-} and AlO_6^{9-} ions in zeolites

51. The element which is exclusively applied as semi-conductor :
    (a) Au (b) Ge
    (c) Pt (d) Si

52. The oxide of lead used in lead accumulators is :
    (a) PbO (b) Pb_2O_3
    (c) Pb_3O_4 (d) PbO_2

53. Silicon is an important constituent of
    (a) Chlorophyll (b) Haemoglobin
    (c) Rocks (d) Amalgams

54. When a mixture of air and steam is passed over red hot coke, the outgoing gas is
    (a) Producer gas
    (b) Water gas
    (c) Coal gas
    (d) None of these gases

55. White lead is
    (a) Basic lead aeetate
    (b) Acidic lead carbonate
    (c) Basic lead carbonate
    (d) Basic lead hydroxide

56. Alum is not used :
    (a) As a mordant in dyeing
    (b) As an insecticide
    (c) In the purification of water
    (d) In tanning of leather.

57. When vapours of $SiCl_4$ are passed over hot Mg, the product formed is :
    (a) $MgSiCl_6$ (b) $SiCl_2 + MgCl_2$
    (c) $Si + MgCl_2$ (d) $MgSi + Cl_2$

58. Lead Pencil contains
    (a) PbS (b) FeS
    (c) Pb_3O_4 (d) Graphite

59. Anhydrous $AlCl_3$ is prepared from
    (a) conc. HCl and aluminium metal
    (b) aluminium and Cl_2
    (c) dry HCl gas + heated aluminium metal
    (d) none of above

60. Inorganic graphite is :
    (a) BN (b) BF_3
    (c) B_2H_6 (d) none of these

Organic Chemistry - Some Basic Principles & Techniques

CONCEPTS

CARBON UNDERGOES THREE TYPES OF HYBRIDIZATION

(i) sp^3 in saturated compounds such as alkanes and cycloalkanes.

(ii) sp^2 in all unsaturated compounds containing a double bond such as alkenes, arenes, etc.

(iii) sp in all unsaturated compounds containing a triple bond i. e., $-C \equiv C-$

Electronegativity of carbon atom varies with the state of hybridization. For example, electronegativity of carbon increases as the s-character of the hybrid orbital increases.

For example electronegativity increases in the order

$$\mathbf{C\,(sp^3) < C\,(sp^2) < C(sp)}$$
$$\mathbf{25\% \quad 33\% \quad 50\%}$$

The bond length decreases as the size of the orbitals forming the bond decreases.

CLASSIFICATION OF ORGANIC COMPOUNDS

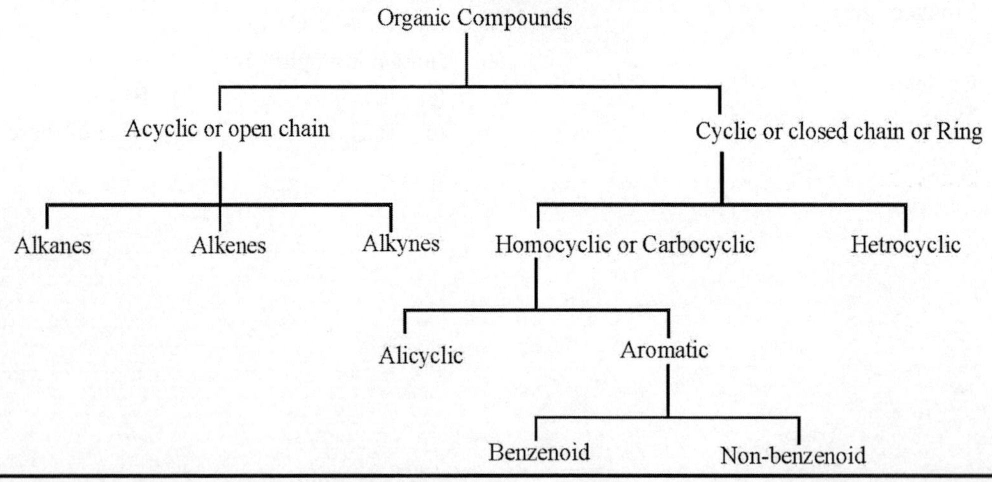

SOME COMMON ORGANIC COMPOUNDS

- **Acyclic or Open Chain Hydrocarbons**

$$H_3C-C \equiv C-CH_2-CH_3$$
Pentyne - 2

$$H_3C-\underset{|}{\overset{CH_3}{HC}}-CH_3$$
2 Methyl propane

- **Cyclic or Closed Ring Hydrocarbons**

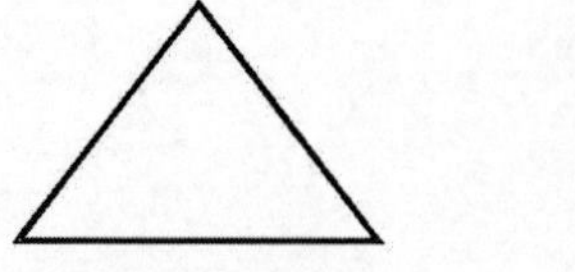
Cyclopropane

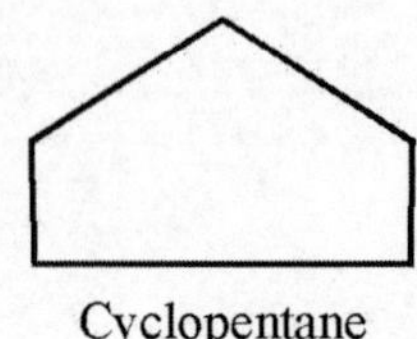
Cyclopentane

- **Aromatic Hydrocarbons**

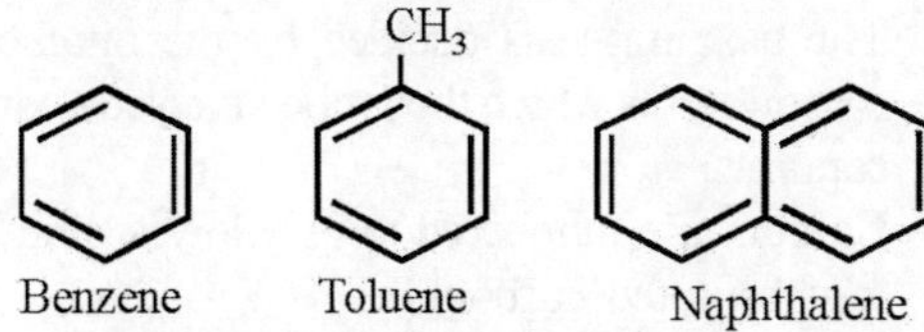

Benzene Toluene Naphthalene

Benzenoids contain at least one benzene ring in their molecules, but if a non-benzene ring is present it is termed as non-benzenoid compound e.g.

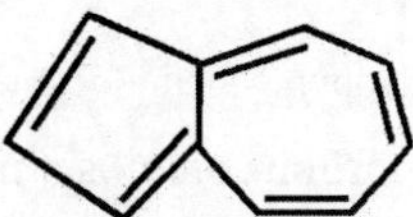

non-benzenoid compound

- **Heterocyclic Compounds**

The ring system is made up of two or more than two types of atoms.

Aromatic : eg.

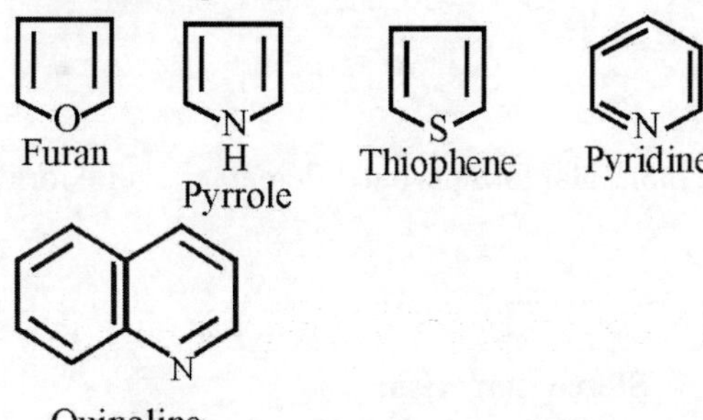

Furan Pyrrole Thiophene Pyridine

Quinoline

IUPAC RULES FOR NOMENCULATURE OF ORGANIC COMPOUNDS

(i) **Longest chain rule :** Select longest possible carbon atom chain which may or may not be in one plane. It is called **parent chain.**

(ii) **Numbering of carbon atoms on longest carbon chain** is done in such a way that the carbon atom carrying the substituent group gets the lowest possible number. The number which indicates the position of substituent on parent chain is called **position number or locant.** It is decided by following rule

Lowest set of locants rule : When two or more substituents are present, the carbon atom of parent chain are numbered in such a way that the sum of locants is lowest. This is called lowest set of locants rule.

$$\begin{array}{c} \overset{2}{}\;\overset{1}{}\;\overset{7'}{} \\ {}^{6'}CH_2\,CH_3 \\ |^{5'}\quad{}^{4'} \\ CH_3-\underset{3}{CH}-\underset{4}{CH}-CH_2\,CH_3 \\ |^{3'}\;{}^{2'}\;{}^{1'} \\ \underset{5}{CH_2}\underset{6}{CH_2}\underset{7}{CH_3} \end{array}$$

Sum of locants $= 3 + 4$ correct

$= 4' + 5'$ incorrect

(iii) **Preference list of functional groups**

carboxylic acid > sulphonic acid > anhydrides > esters > acid chlorides > acid amides > nitriles > aldehydes > ketones > alcohols >amines > ethers > alkenes >alkynes

Functional group	Prefix	Suffix
Sulphonic acid $-SO_3H$	sulpho	
Carboxylic acid $-COOH$	carboxy	– oic acid
Anhydride $(RCOO)_2O$		– alkanoic anhydride
Ester $-COOR$	carboxy	alkyl – alkanoate
Acid halide $-COX$	haloformyl	– oyl halide
Acid amide $-CONH_2$	carbamoyl	– amide
Cyanide $-C\equiv N$	cyano	– nitrile
Aldehyde $\begin{matrix}R\\H\end{matrix}\!\!>C=O$	formyl	– al
Ketone $\begin{matrix}R\\R'\end{matrix}\!\!>C=O$	oxo	– one
Alcohol $-OH$	hydroxy	– ol
Amine $-NH_2$	amino	alkanamines

(iv) When using **prefixes such as formyl, cyano, carboxy, carbamoyl** etc, the **carbon of substituent is not counted in principal chain**.

(v) **Some important names :**

Phenyl – (Benzene – H) **Benzyl –** (Toluene – H of side chain) **Tolyl –** (Toluene – H of ring)

(vi) When **two like carbon containing functional groups are present at two ends of chain, then carbon of both functional groups is counted in chain**.

$$\overset{3}{HOOC}-\overset{2}{CH_2}-\overset{1}{COOH}$$

propane-1,3-dioic acid

(vii) When **two different carbon atom containing functional groups are present** at two ends of carbon chain, only the **carbon atom of principal functional group is counted towards parent carbon.**

$$\overset{4}{NC}-\overset{3}{CH_2}-\overset{2}{CH_2}-\overset{1}{CH_2}-COOH$$

4–cyanobutanoic acid

(viii) If a **hydrocarbon has both double and triple bond, it is named as alkyne.**

$$\overset{5}{CH_3}-\overset{4}{C}\equiv\overset{3}{C}-\overset{2}{CH}=\overset{1}{CH_2}$$

pent – 1–en–3– yne

While **numbering double bond is always prefered over triple bond.**

(ix) If **two atoms or groups of same preference occupy identical positions from either end of the parent chain, the lower number must be given to atom/group whose prefix comes first in alphabetic order.**

$$\overset{5}{Cl-CH_2}-\overset{4}{CH_2}-\overset{3}{CH_2}-\overset{2}{CH_2}-\overset{1}{CH_2}-Br$$

1–bromo–5–chloropentane

NOMENCLATURE OF CYCLOALKANES

(i) The base name is decided by the number of carbon atoms which the cyclic or acyclic portion contains.

(ii) Carbons are numbered to give lowest numbers to substituted carbons.

(iii) When there are more acyclic than cyclic carbons the cyclic part becomes cycloalkyl substituent.

4-Cyclopropyl-3-methyloctane

(iv) When acyclic portion contains a multiple bond or a functional group, the cyclic portion is treated as substituent.

$$\overset{4}{CH}=\overset{3}{CH}-\overset{2}{\underset{}{C}}(=O)-\overset{1}{CH_3}$$

4-Cyclohexylbut-3-en-2-one

ISOMERISM

It is shown by different compounds which have same molecular formula but different structural formula.

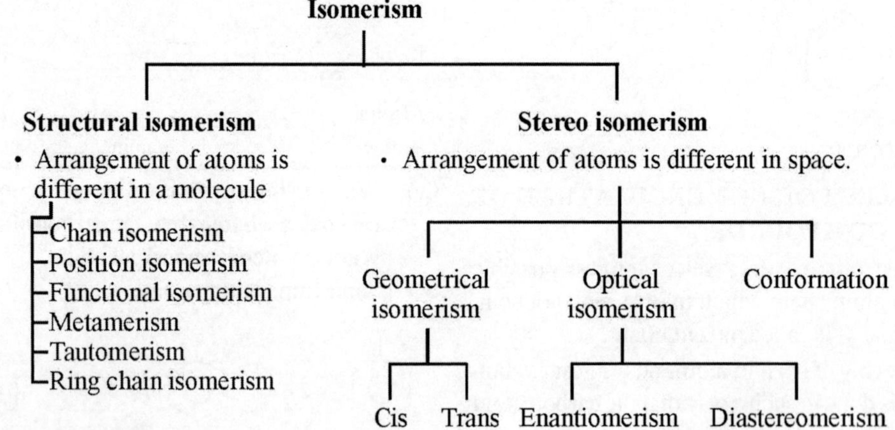

Structural Isomerism

1. **Chain isomerism :** Different skeleton of carbon chains is present in different isomers. e.g.- C_5H_{12} has three isomers normal, iso and neo pentane.

$CH_3CH_2.CH_2.CH_2–CH_3$

n-Pentane

$(CH_3)_2>CH–CH_2–CH_3$

Iso-Pentane

$CH_3-C(CH_3)_2-CH_3$

Neo-Pentane

2. **Position isomerism :** Position of functional group or multiple bond is different in isomers. e.g.-

(i) **o, p and m isomers of same compound.**

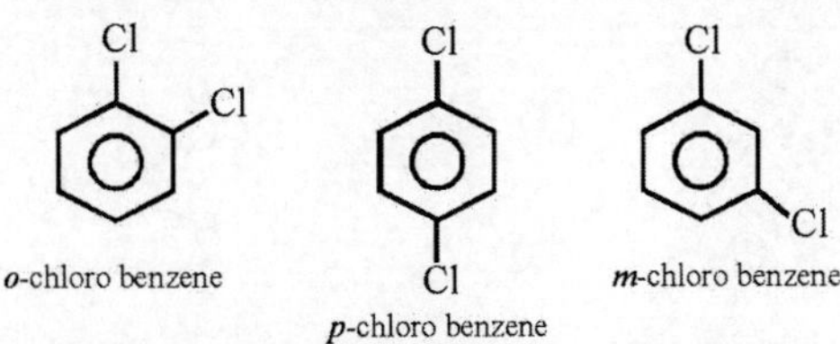

(ii) **butanol-1 and butanol-2**.

3. **Functional isomerism :** Different functional groups are possible by same molecular formula in different isomers.

e.g.-

C_2H_5OH ethyl alcohol	CH_3OCH_3 dimethyl ether
CH_3COOH Acetic acid	$HCOOCH_3$ Methyl formate
CH_3CH_2CHO Propionaldehyde	CH_3COCH_3 Acetone

4. **Metamerism :** Different alkyl groups are attached to same functional groups in different isomers.
   - **Ethers, esters, ketones, 2°-amines shows metamerism** because they have two alkyl groups.

   e.g.-

   $C_2H_5-O-C_2H_5$ (diethyl ether) and $CH_3-O-C_3H_7$ (methyl propyl ether)

5. **Tautomerism or demotropism or alleotropism or kryptomerism :**
   - It is **dynamic isomerism**.
   - Two isomers can never be separated.
   - It is **caused due to presence of labile H in compound**.
   -

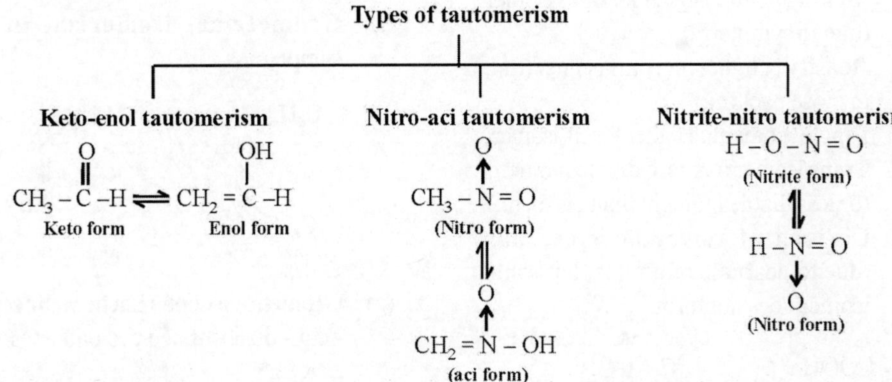

6. **Keto enol tautomerism**
   (i) In **simple forms, keto form is more stable** due to greater strength of C – O π bond as compared to C – C π bond.
   (ii) In some cases, **enolic form is stablised by intramolecular hydrogen bonding (Chelation)**.

$$CH_3-\overset{O}{\overset{\|}{C}}-CH_2-\overset{O}{\overset{\|}{C}}-OC_2H_5 \rightleftharpoons CH_3-\overset{O\cdots H-O}{\overset{\|\qquad\quad |}{C-CH=C}}-OC_2H_5$$

Acetoacetic ester (keto form) ⇌ Enolic form (intramolecular H-bonding)

   (iii) When benzene ring is present in enolic form, double bond is in conjugation with the π cloud of benzene ring, the **stablisation of enolic form further increases by conjugation**.

$$C_6H_5-\overset{O}{\overset{\|}{C}}-CH_2-\overset{O}{\overset{\|}{C}}-C_6H_5 \rightleftharpoons C_6H_5-\overset{O\cdots H-O}{\overset{\|\qquad\quad |}{C-CH=C}}-C_6H_5$$

keto form (11%) Benzyleacetophenone Enol form (89%)

   (iv) **Polar protic solvents** (e.g.- H_2O, CH_3OH) form hydrogen bonding with keto form and **decrease** the **enol content. Aprotic solvents increase the enol content.**

7. **Ring-chain isomerism**
   - Cyclic and open chain both structures are possible by the same formula.

   e.g.- **Cycloalkane and alkenes**.

   △ (cyclopropane) and $CH_3CH=CH_2$ (propene)

Stereo Isomerism

1. **Geometrical isomerism (cis-trans isomerism) or configurational isomerism :**
   (a) It is **caused due to hindered rotation across C = C bond** (or C = N or N = N bond).
   (b) **Conditions necessary for compound to exhibit geometrical isomerism.**

(i) The compound must have **at least one C=C**.

(ii) The **two groups attached to same carbon atom must be different.**

$$\begin{matrix} a & & a \\ & C=C & \\ b & & b \end{matrix} \qquad \begin{matrix} a & & b \\ & C=C & \\ b & & a \end{matrix}$$

Cis [similar groups on same side of double bond] **Trans** [similar groups on opposite sides of double bond]

$$\begin{matrix} HOOC & & COOH \\ & C=C & \\ H & & H \end{matrix} \qquad \begin{matrix} HOOC & & H \\ & C=C & \\ H & & COOH \end{matrix}$$

Maleic acid (Cis) **Fumaric acid (Trans)**

(c) **Difference in properties of cis and trans isomer**

(i) **Boiling point :** Higher for cis isomer.

(ii) **Melting point :** Higher for trans isomer (due to symmetry)

(iii) **Density :** Higher for trans isomer (due to close packing)

(iv) **Dipole moment :** Higher for cis isomer. Trans isomer has zero dipole moment (or less dipole moment than cis form.)

(v) **Cis-isomer forms cyclic compounds** (due to nearness of groups) but trans isomer does not form.

$$\begin{matrix} H-C-COOH \\ \| \\ H-C-COOH \end{matrix} \xrightarrow{\Delta} \begin{matrix} H & \\ & C-CO \\ & \| \quad\;\; O + H_2O \\ & C-CO \\ H & \end{matrix}$$

Maleic acid (Cis) **Maleic anhydride**

$$\begin{matrix} H-C-COOH \\ \| \\ HOOC-C-H \end{matrix} \xrightarrow{\Delta} \text{No reaction}$$

Fumaric acid (trans)

(d) **E and Z system of nomenclature of geometrical isomers**

(i) The groups on each carbon atom are assigned priority no. 1 and 2 **as per sequence rules**. (Sequnce rules states that group or atom with higher atomic mass has higher priority, double and triple bonds are treated as if they have duplicate or triplicate single bonds)

(ii) **If two groups of higher priority are on same side of double bond**, it is assigned **Z configuration and if on opposite side** it is assigned **E configuration.**

$$\begin{matrix} \overset{②}{H_3C} & & \overset{②}{CH_2CH_3} \\ & C=C & \\ \underset{①}{I} & & \underset{①}{Br} \end{matrix} \qquad \begin{matrix} \overset{①}{HCO} & & \overset{②}{CH_3} \\ & C=C & \\ \underset{②}{H} & & \underset{①}{COOH} \end{matrix}$$

Z **E**

(e) **Geometrical isomerism in compound having C = N**

Syn - if H (or alkyl group) are on same side of C = N.

anti - if H (or alkyl group) are on opposite side C = N.

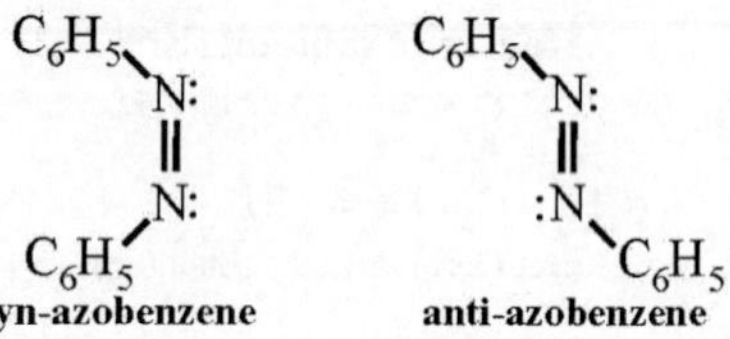

(f) **Geometrical isomerism in N = N compounds.**

$$\begin{matrix} C_6H_5 & \\ & N: \\ & \| \\ & N: \\ C_6H_5 & \end{matrix} \qquad \begin{matrix} C_6H_5 & & \\ & N: & \\ & \| & \\ & :N & \\ & & C_6H_5 \end{matrix}$$

syn-azobenzene **anti-azobenzene**

(g) **Geometrical isomerism in cyclic compounds** - e.g.- disubstituted cycloalkanes.

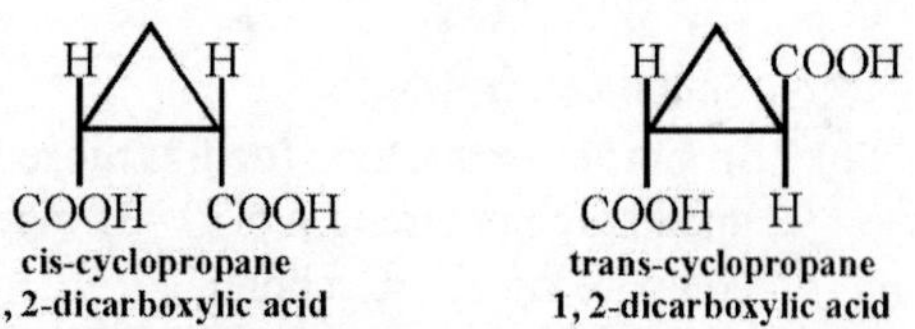

2. Optical isomerism

(a) **Plane polarised** light has vibrations only in one plane.

(b) **Optically active compounds rotate plane polarised light.**

(c) **Angle of rotation** (α) is the angle through which plane polarised light is rotated when passed through solution of optically active substance. It is measured by **polarimeter**.

(d) **Specific rotation** is angle of rotation produced by solution of length 10 cm (or 1 dm) and having unit concentration (1 g/ cm^3) for given wavelength (λ) of light at given temperature (t).

$$[\alpha]_\lambda^t = \frac{100 \times \text{observed angle of rotation } (\theta)}{\text{length in dm} \times \text{conc. in g/cm}^3 \text{ solution}}$$

(e) Optical isomers have identical physical and chemical properties but have different behaviour towards plane polarised light.

(f) **Conditions for compound to show optical isomerism**

 (i) The compound **must have** at least one **asymmetric or chiral carbon atom** (all four groups attached to chiral carbon atom are different).

 (ii) The compound **must not have plane of symmetry**.

(g) **Enantiomers** are pair of optical isomers which are **related as non-superimposable mirror images of each other.** They **have identical properties** but differ in their behaviour towards plane polarised light.

(h) **Diastereomers** are pair of optical isomers which **cannot be related as non-superimposable mirror images of each other**. They have **different physical properties.**

(i) **Meso compounds** do not show optical activity inspite of presence of chiral carbon atom due to presence of molecular symmetry. It is called **internal compensation.**

```
      COOH         Mirror         COOH
       |             |             |
   H - C - OH        |        HO - C - H
       |             |             |
  OH - C - H         |         H - C - OH
       |             |             |
      COOH           |            COOH
      (I)            |            (II)
  d-tartaric acid    |      l-tartaric acid

        COOH
         |
     H - C - OH       Plane of symmetry
  ---+--------------------↓----
     H - C - OH
         |
        COOH
        (III)
   meso tartaric acid
```

I and II are **enantiomers**.
I and III are **diasteromers**.
III - **meso compound**.

(j) **Substituted allenes and substituted biphenyls** do not have asymmetric carbon atom but **they show optical isomerism due to chirality (dissymmetry) in molecule**.

(k) **Asymmetric synthesis** is synthesis of optically active compound from an optically inactive molecule.

(l) **Walden inversion** is conversion of d-form of an optically active compound into its ℓ-form or vice-versa.

(m) **Resolution** is separation of d and l forms from racemic mixture.

(n) **Racemisation** is process of formation of recamic mixture by adding 50% d and 50% ℓ-form of same compound.

- **R and S configuration of optical isomers** (developed by **Cahn, Ingold-Prelog**) :

 (i) Assign priority to groups attached to chiral carbon atom (as 1, 2, 3, 4) in accordance with sequence rules.

 (ii) Arrange groups/atoms in such a way that group having lowest priority is directed away from us.

 (iii) If in going from group of highest priority to lowest priority, the eye moves in **clockwise direction**, it is **assigned configuration R** (right) and if it **moves in anti clockwise direction**, it is **assigned** S (sinister or left).

TYPES OF BOND CLEVAGES

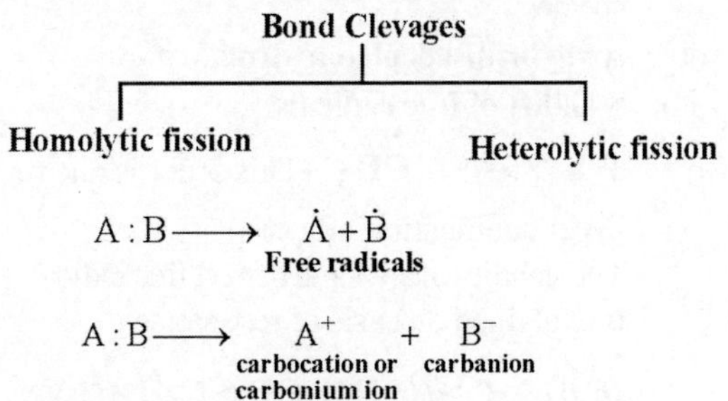

REACTION INTERMEDIATES

(a) Carbocation or Carbonium Ion

(i) It has **6e⁻** (electron deficient species).

(ii) **sp^2 hybridisation, planar shape**.

(iii) **Stability of carbocations**

- **3° > 2° > 1°** (This order is due to + I and hyperconjugation effect)

- Allyl and benzyl carbocation are stabilised by resonance.
benzyl > isopropyl > allyl
- Greater the no. of phenyl groups attached to carbocation more is its stability.

$$(C_6H_5)_3C^{\oplus} > (C_6H_5)_2CH^{\oplus} > (C_6H_5)CH_2^{\oplus}$$

- Stability of carbocation $\propto$ 1/s-character

$$CH \equiv \overset{\oplus}{C} < CH_2 = \overset{\oplus}{C}H < CH_3 - \overset{\oplus}{C}H_2$$

(b) Carbanion

(i) It has 8e⁻ and is electron rich species.
(ii) **sp^3 hybridised and pyramidal in shape.**
(iii) **Stability of carbanions**

- CH_3 **> 1° > 2° > 3°** (This order is due to inductive effect)
- Allyl and benzyl carbanions are stabilised by resonance.
- Electron donating group destabilise carbanion and electron withdrawing group stabilize carbanion.
- **Stability of carbanion increases as s-character of carbon holding negative charge increases**

$$\underset{sp(50\%\,s)}{H-C \equiv \ddot{C}^-} > \underset{sp^2(33.3\%\,s)}{CH_2 = \ddot{C}H^-} > \underset{sp^3(25\%\,s)}{CH_3 - \ddot{C}H_2^-}$$

(c) Free Radical

(i) It has 7e⁻, **paramagnetic**, electron deficient species.
(ii) **sp^2 hybridised, planar structure.**
(iii) **Stability of free radicals**

- **3° > 2° > 1° > $\dot{C}H_3$** (This order is due to hyper conjugation effect)
- The stability of allyl and benzyl free radicals is explained on basis of resonance.

$$(C_6H_5)_3\dot{C} > (C_6H_5)_2\dot{C}H > C_6H_5\dot{C}H_2$$
$$> 3° > 2° > 1° > \dot{C}H_3 > \text{vinyl}$$

(d) Carbene

(i) They are neutral divalent carbon species bearing no charge and surrounded by sextet of electrons.
(ii) They behave as **electrophile or Lewis acids.**
(iii) Carbenes are of two types

- **Singlet carbene** ⇅ e.g.- $:CH_2$, sp^2 hybridised and V shaped.
- **Triplet carbene** ⇅ sp hybridised and linear. **Triplet carbene is more stable than singlet carbene.**

(e) Nitrenes. These are neutral monovalent nitrogen species and are analogus to carbene.

TYPES OF REAGENTS

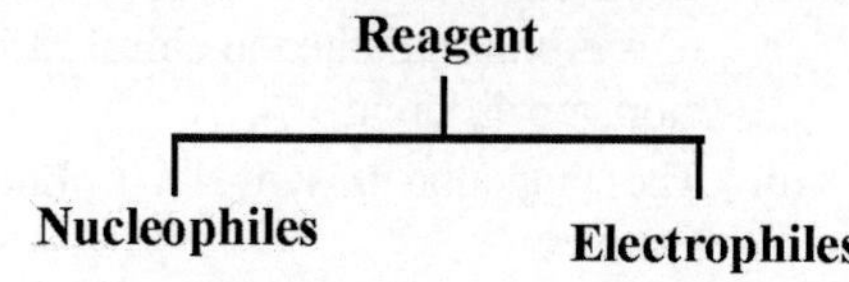

Nucleophiles	Electrophiles
(i) Negatively charged or e⁻ **rich species**	(i) Positively charged or e⁻ **deficient species.**
(ii) *e.g.*- OH^-, $\ddot{N}H_3$,	(ii) e.g.- $AlCl_3$, NO_2^+, $NaBH_4$
(iii) **Generally Lewis bases**	(iii) **Generally Lewis acids**

Nucleophilicity is defined by the rate of attack on an electrophilic carbon atom.

1. Species with negative charge are stronger nucleophiles than analogous species without a negative charge.
$OH^- > H_2O$, $SH^- > H_2S$, $NH_2^- > NH_3$
2. Nucleophilicity decreases from left to right across the periodic table eg.
$OH^- > F^-$, $NH_3 > H_2O$
3. Nucleophilicity increases down the periodic table.
$I^- > Br^- > Cl^- > F^-$, $HSe^- > HS^- > HO^-$

TYPES OF DISPLACEMENTS IN COVALENT BONDS

(a) Inductive Effect or Transmission Effect

(i) **It is permanent displacement of σ-electrons** towards more electronegative group/atom.
(ii) It is of two types

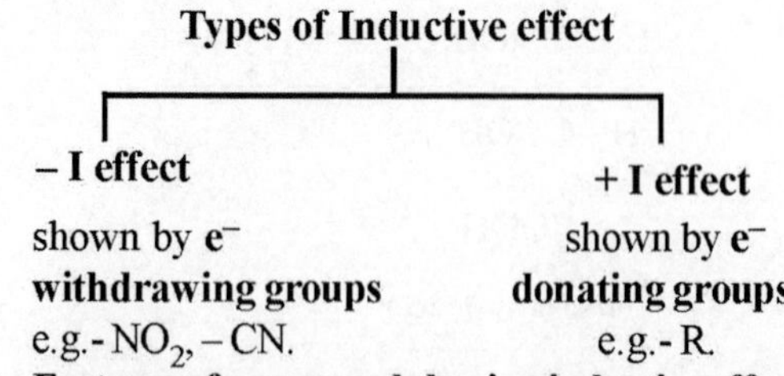

– I effect	+ I effect
shown by e⁻ **withdrawing groups**	shown by e⁻ **donating groups**
e.g.- NO_2, – CN.	e.g.- R.

(iii) **Features of compounds having inductive effect**
- Responsible for greater reactivity of alkyl halides.
- Responsible for increase in acidity of carboxylic acid.

- It effects basicity of amines.

The inductive effect in di-and tri-halogens substituted acids are progressively more stronger than the corresponding monohalogen substituted acid.

Cl_3CCOOH > $Cl_2CHCOOH$
Trichloroacetic acid Dichloroacetic acid
> $ClCH_2COOH$ > CH_3COOH
Chloroacetic acid Acetic acid

$\overset{+}{N}R_3 > \overset{+}{N}H_3 > NO_2 >$

$CN > COOH > F > Cl > Br > I$
$> OAr > OR > OH > C \equiv CR > Ar > H$

⟵ Increasing – I effect (Electrons attracting)

$O^- > COO^- > CR_3 > CHR_2$
$> CH_2R > CH_3 > D > H$

⟵ Increasing + I effect (electrons repelling)

(b) Electromeric Effect

(i) **It is temporary effect of π electron displacement in a multiple bond in presence of reagent.**

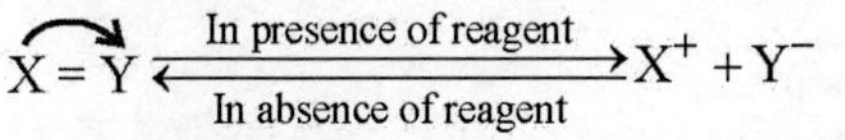

$$X=Y \underset{\text{In absence of reagent}}{\overset{\text{In presence of reagent}}{\rightleftharpoons}} X^+ + Y^-$$

(ii) **+E - effect** transfer of π electrons is towards attacking species.

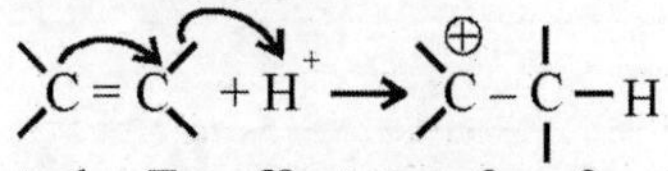

$$>C=C< + H^+ \rightarrow >\overset{\oplus}{C}-\overset{|}{\underset{|}{C}}-H$$

and – **E - effect** transfer of π electrons is away from attacking species

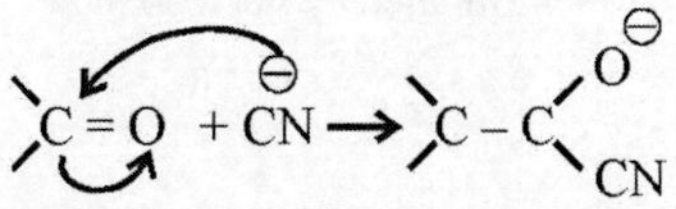

$$>C=O + \overset{\ominus}{C}N \rightarrow >C-C\begin{matrix}O^{\ominus}\\CN\end{matrix}$$

(c) Resonance or Mesomeric effect (R or M Effect)

(i) The generation of centres of low and high electron density by **transfer of π electrons from one part of conjugated π system** (having alternate single and double bond) to other due to resonance is called **resonance effect**.

(ii) It is permanent effect.

(iii) **+ M (or + R) effect** is shown by groups which donate e^- towards conjugate system.

e.g. – OH, – OR, – NH_2, – X.

– M (or – R) effect is shown by groups which withdraw e^- from conjugate system.

e.g. $>C=O$, – CN, – NO_2, – COOR

(iv) **Conditions for resonance**

- The relative position of nuclei should remain same.
- All resonating structures must have same number of unpaired electrons.

(v) **Resonance explains**

- Stability of reaction intermediates.
- Low reactivity of vinyl and aryl halides.
- Acidic nature of acids.
- Order of basicity of amines.

(d) Hyperconjugation or Baker - Nathan Effect

Baker and Nathan suggested that alkyl groups with at least with one hydrogen atom on the α–carbon atom, attached to an unsaturated carbon atom, are able to release electrons by a mechanism similar to that of the electromeric effect, e.g.

$$-\overset{H}{\overset{|}{C}}-C=C \longrightarrow -\overset{\overset{+}{H}}{C}=C-\overset{-}{C}$$

the delocalisation involves σ and π bond orbitals (or orbitals in case of free radicals); thus it also known as **σ — π conjugation**. The type of electron release due to the presence of the system H – C = C is known as hyperconjugation.

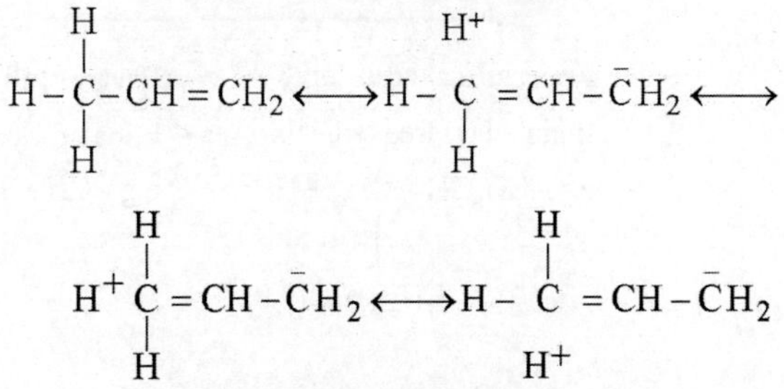

$$H-\underset{H}{\overset{H}{C}}-CH=CH_2 \longleftrightarrow H-\underset{H}{\overset{H^+}{C}}=CH-\overset{-}{C}H_2 \longleftrightarrow$$

$$H^+\underset{H}{\overset{H}{C}}=CH-\overset{-}{C}H_2 \longleftrightarrow H-\underset{H^+}{\overset{H}{C}}=CH-\overset{-}{C}H_2$$

- **Effects of hyperconjugation**

(i) **Stability of alkenes:** Hyperconjugation explains the stability of certain olefins over other alkenes. Hyperconjugation is also known as no–bond resonance. Greater the number of alkyl groups attached to the doubly–bonded carbons, the more stable the alkene is.

(ii) **Stability of alkyl free–radicals :** The concept of hyperconjugation can also be extended to explain the following relative stabilities of alkyl radicals.

tert – Alkyl > sec – Alkyl > prim – Alkyl > Methyl radical

e.g., $(CH_3)_3C^+ > (CH_3)_2\overset{+}{C}H > CH_3\overset{+}{C}H_2 > \overset{+}{C}H_3$

Stability order of carbocations

Triphenyl > benzyl > allyl > 3° > 2° > 1° > methyl

Stability order of carboanions

$(C_6H_5)_3\bar{C} > (C_6H_5)_2\bar{C}H > C_6H_5\bar{C}H_2 > CH_2 = CH - \bar{C}H_2 >$

$\bar{C}H_3 > CH_3 \;\; CH_2 > (CH_3)_2\bar{C}H > (CH_3)_3\bar{C}$

TYPES OF ORGANIC REACTIONS

(a) Substitution reactions
(b) Addition reactions
(c) Elimination reactions
(d) Rearrangement reactions
(e) Condensation reactions
(f) Isomerisation reactions

Substitution Reactions

- During such reaction atom/group replaces other atom / group.
- Characteristic of stable molecules.

Substitution reactions

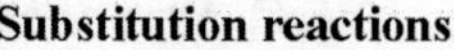

Free radical substitution
- Take place in presence of light
- *e.g.* - Alkanes.
- $CH_4 + Cl_2 \xrightarrow{U.V.} CH_3Cl + HCl$

Necleophilic substitution
- *e.g.* - Alkyl Halides

$RX + OH^- \longrightarrow ROH + X^-$

- S_N1 — *e.g.* 3°RX
- S_N2 — *e.g.* 1°RX

Electrophilic substitution
- *e.g.*-Benzene and its derivatives

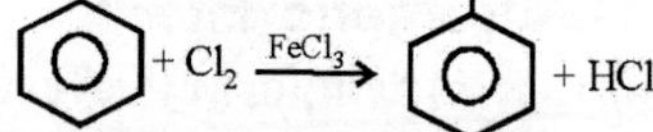

Addition Reactions

- Characteristic of **multiple bonded compounds.**

Addition reactions

Free radical addition
- Initiated by free radicals

e.g.- $CH_3CH = CH_2 + HBr \xrightarrow{\text{peroxide}} CH_3CH_2CH_2Br$

Electrophilic addition
- Initiated by electrophile

e.g.- $CH_3CH = CH_2 + HBr \xrightarrow{\text{No peroxide}} CH_3CHBrCH_3$

Nucleophilic addition
- Initiated by nucleophile

e.g.- $>C = O + HC \equiv N \longrightarrow >C(OH)(CN)$

Elimination Reactions

- Loss of molecules/atoms occur during reaction.
- **Multiple bonds are formed during this reaction.**
-

Types of elimination reactions

α-elimination
- Two atoms/groups are lost from same carbon atom.

e.g.- **Reimer-Tiemann reaction.**

β-elimination
- Two atoms/groups are lost from adjacent carbon atoms.

e.g.- $CH_3CH_2OH \xrightarrow{H_2SO_4} H_2C = CH_2 + H_2O$

γ-elimination
- Two atoms/groups are lost from α- and γ-positions (i.e.- 3 bonds away)

Condensation Reaction

Two or more molecules of same or different type combine to form product with or without loss of small molecules (like H_2O, NH_3). *Example* - **Aldol condensation.**

Rearrangement Reaction

It involves migration of an atom/group from one position to other within same molecule. Example - **Hofmann bromamide reaction.**

Isomerisation Reaction :

Interconversion of isomers take place.

PURIFICATION OF ORGANIC COMPOUNDS

1. Methods of Purification of Solids.

(i) Crystallisation

- **Principle :** Certain organic substances dissolve partly in solvent and their solubility in this solvent increases with increase in temperature.
- A **suitable solvent** is selected to dissolve solute. The **suitable solvent is one in which substance is appreciably soluble at its boiling point, which is not inflammable and which does not react with compound to be crystalised.** e.g.-
- **Sugar having impurity of NaCl is crystalised by using ethyl alcohol (C_2H_5OH)** because sugar is soluble in it and **NaCl is insoluble.**
- Mixture of **benzoic acid (soluble) and napthalene (insoluble) are separated by hot H_2O.**

(ii) Fractional crystallisation

- Used to **separate mixtures of solids.**
- **Principle :** The **crystals of less soluble component separates out first.** Following mixtures are separated by this method using water as solvent.
  - (a) **Mixture of Na_2SO_4 (less soluble) and $Na_2Cr_2O_4$ (more soluble)**
  - (b) **$KClO_3$ (less soluble) and KCl (more soluble).**

(iii) Sublimation

- Sublimating substances which directly get converted from solid to vapour state can be purified by this method.
- **Examples - Camphor, napthalene, benzoic acid, NH_4Cl, I_2, anthracene, pthalic anhydride.**

Methods of Purification of Liquids

(i) Distillation (simple distillation)

- Purifies liquids which do not decompose at their boiling points and contain only non volatile impurities.
- It is **process of simultaneous vapourisation and condensation.**

(ii) Fractional distillation :

- It is used when two liquids to be separated differ in boiling point by more than 10-15 K.
- **Examples -**
  - (a) **Separation of mixture of acetone (b.p. 325 K) and CH_3OH (b.p. 338 K)**
  - (b) **Separation of crude petroleum into gasoline, kerosene oil, diesel oil.**

(iii) Vaccum distillation (distillation under reduced pressure) :

- **Used for liquids which decompose on or before their boiling points.**
- **Examples -**
  - (a) **glycerol**
  - (b) **concentration of sugar cane juice** in sugar industry.

(iv) Steam distillation

- Substances which are volatile at the temperature of steam and immiscible in water are purified by this method.
- Examples - **Aniline, nitrobenzene, sandalwood oil, terpentine oil.**

(v) Differential extraction

- This method is used to separate an organic compound present in aq. solution by shaking it with suitable organic solvent which is immiscible with water and in which organic compound is more soluble as compared to water.
- **Greater the number of extractions more will be the amount of organic substance extracted.**
- **Examples - Benzoic acid is separated from water using benzene.**

CHROMATOGRAPHY (DISCOVERED BY BOTANIST TSWETT)

(i) The method is based on principle of selective distribution of components of mixture between fixed (stationary phase) and moving phase. **Stationary phase can be solid or liquid and mobile phase can be liquid or gas.**

(ii)	Types of chromatography	Mobile phase	Stationary phase	Uses
1.	Column chromatography	liquid	solid	Large scale separations
2.	Thin layer chromatography	liquid	solid	Qualitative analysis of organic compound
3.	High performance liquid chromatography (HPLC)	liquid	solid	Qualitative and quantitative analysis of organic compounds
4.	Gas liquid chromatography (GLC)	gas	liquid	Qualitative and quantitative analysis of organic compound
5.	Paper chromatography	liquid	liquid	Analysis of polar organic compounds

QUALITATIVE ANALYSIS OF ORGANIC COMPOUNDS

1. Detection of Carbon and Hydrogen

$$\text{Org. comp.} + \text{dry CuO} \xrightarrow[\text{Strong}]{\Delta} CO_2 + H_2O$$

CO_2 → turns lime water milky → shows C present

H_2O → turns anhydrous $CuSO_4$ blue → shows H present

2. Detection of Nitrogen :

Take Lassaigne's extract add few drops of freshly prepared solution of ferrous sulphate. The contents are warmed a little and then acidified with dil. H_2SO_4. Appearance of prussian blue colour indicates the presence of nitrogen.

$Na + C + N \longrightarrow NaCN$ (formed in Lassaigne's extract)

$$2NaCN + FeSO_4 \longrightarrow Na_2SO_4 + Fe(CN)_2$$

$$Fe(CN)_2 + 4NaCN \longrightarrow Na_4[Fe(CN)_6]$$

$$3Na_4[Fe(CN)_6] + 4Fe^{3+} \longrightarrow \underset{\text{Prussian blue}}{Fe_4[Fe(CN)_6]_3} + 12Na^+$$

But if sulphur is also present along with nitrogen one gets blood red colouration due to formation of ferric thiocyanate.

$$Na + C + S + N \longrightarrow NaSCN$$

$$Fe^{3+} + 2NaSCN \longrightarrow \underset{\text{Blood red}}{Fe(SCN)_3} + 3Na^+$$

3. Detection of Sulphur

(i) Lassaigne's extract + sod. nitroprusside $\longrightarrow$ violet colouration.

$$\underset{\text{(from Lassaigne's extract)}}{Na_2S} + \underset{\textbf{sod. nitroprusside}}{Na_2[Fe(NO)(CN)_5]} \longrightarrow \underset{\textbf{sod. thionitroprusside (violet colouration)}}{Na_4[Fe(CN)_5NOS]}$$

(ii)

$$\underset{\text{(from Lassaigene's extract)}}{Na_2S} + \underset{\text{lead acetate}}{(CH_3COO)_2Pb} \longrightarrow \underset{\text{black ppt of lead sulphide}}{PbS} + 2CH_3COONa$$

4. Detection of Halogens

(i) **Beilstein test**

$$\text{Org. comp.} \xrightarrow{\text{Heat over copper wire}}$$

green flame means any halogen is presentin organic compound.

(ii) **Lassaigne's test**

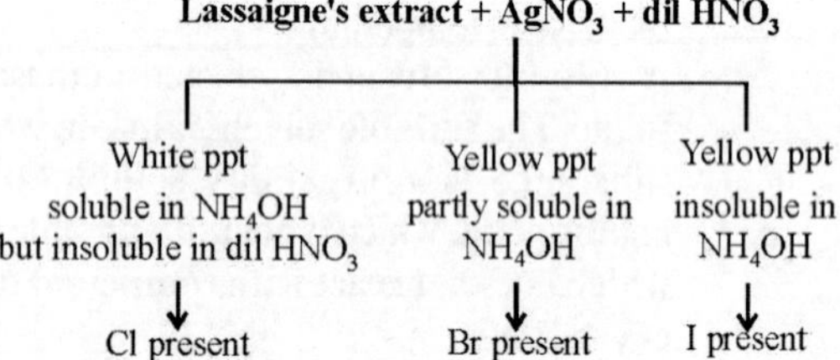

5. Detection of Phosphorous

Phosphorous is detected by fusing the organic compound with Na_2O_2 (oxidizing agent) and then extracting with water. The aqueous solution is boiled with HNO_3 and then few drops of ammonium molybdate solution is added. The appearance of yellow ppt or coloration indicates the presence of phosphorous.

$$5Na_2O_2 + 2P \longrightarrow 2Na_3PO_4 + 2Na_2O$$

$$Na_3PO_4 + 3HNO_3 \longrightarrow H_3PO_4 + 3NaNO_3$$

$$H_3PO_4 + 12(NH_4)_2MoO_4 + 21\,HNO_3 \longrightarrow$$

$$(NH_4)_3PO_4.12MoO_3 + 21NH_4NO_3 + 12H_2O$$

Ammonium phospomolybdate (yellow)

(C) QUANTITATIVE ANALYSIS OF ORGANIC COMPOUNDS

(i) Percentage of carbon (Liebig Method)

$$= \frac{12}{44} \times \frac{\text{Mass of } CO_2 \text{ formed}}{\text{Mass of substance taken}} \times 100$$

(ii) Percentage of hydrogen (Liebig Method)

$$= \frac{2}{18} \times \frac{\text{Mass of water formed}}{\text{Mass of substance taken}} \times 100$$

(iii) Percentage of nitrogen (Duma method)

$$= \frac{28}{22400} \times \frac{\text{Volume of } N_2 \text{ at STP}}{\text{Mass of substance taken}} \times 100$$

(Kjeldahl's method)

$$= \frac{1.4 \times \text{Normality of acid} \times \text{Volume of acid used}}{\text{Mass of substance taken}}$$

(iv) Percentage of chlorine (Carlus Method)

$$= \frac{35.5 \times \text{Mass of AgCl formed}}{143.5 \times \text{Mass of substane taken}} \times 100$$

(v) Percentage of bromine (Carlus Method)

$$= \frac{80 \times \text{Mass of AgBr formed}}{188 \times \text{Mass of substance taken}} \times 100$$

(vi) Percentage of iodine (Carlus Method)

$$= \frac{127 \times \text{Mass of AgI formed}}{235 \times \text{Mass of substance taken}} \times 100$$

(vii) Percentage of sulphur (Carlus Method)

$$= \frac{32 \times \text{Mass of } BaSO_4 \text{ formed}}{233 \times \text{Mass of substance taken}} \times 100$$

(viii) Percentage of phosphorous (Carlus Method)

$$= \frac{31 \times \text{Mass of ammonium phosphomolybdate}}{1877 \times \text{Mass of substance taken}} \times 100$$

(ix) Percentage of oxygen

$$= \frac{32 \times \text{Mass of carbon dioxide}}{88 \times \text{Mass of organic compound}} \times 100$$

Exercise

1. Electronegativity of carbon atoms depends upon their state of hybridisation. In which of the following compounds, the carbon marked with asterisk is most electronegative?
   (a) $CH_3 - CH_2 - {}^*CH_2 - CH_3$
   (b) $CH_3 - {}^*CH = CH - CH_3$
   (c) $CH_3 - CH_2 - C \equiv {}^*CH$
   (d) $CH_3 - CH_2 - CH = {}^*CH_2$
2. Correct IUPAC name of

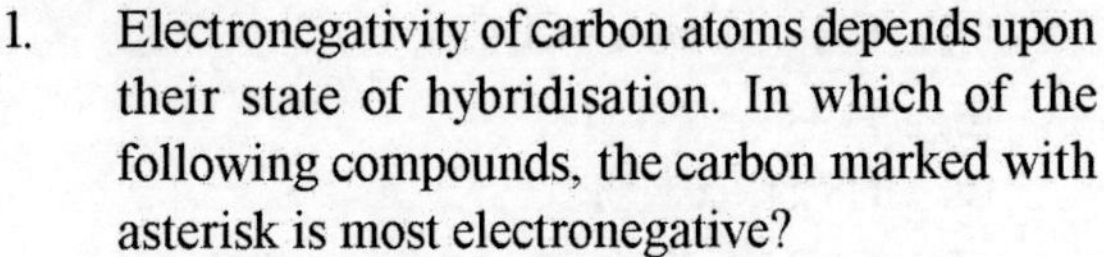

   (a) 2–ethyl-3-methylpentane
   (b) 3,4-dimethylhexane
   (c) 2-sec-butylbutane
   (d) 2, 3-dimethylbutane
3. The IUPAC name of

$$CH_3 - CH(OH) - CH = C(CH_3) - CHO$$

   (a) 4-Hydroxy-1-methylpentanal
   (b) 4-Hydroxy-2-methylpent-2-en-1-al
   (c) 2-Hydroxy-4-methylpent-3-en-5-al
   (d) 2-Hydroxy-3-methylpent-2-en-5-al
4. The IUPAC name for

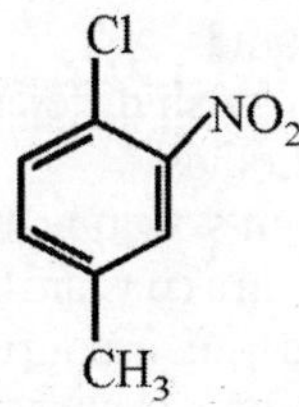

   (a) 1-Chloro-2-nitro-4-methyl benzene
   (b) 1-Chloro-4-methyl-2-nitrobenzene
   (c) 2-Chloro-1-nitro-5-methyl benzene
   (d) *m*-Nitro-*p*-chlorotoluene
5. Which of the following is the most stable carbocation (carbonium ion) ?
   (a) $CH_3CH_2^+$ (b) $(CH_3)_2\overset{+}{C}H$
   (c) $(CH_3)_3\overset{+}{C}$ (d) $C_6H_5\overset{+}{C}H_2$
6. The order of decreasing stability of the carbanions
   $(CH_3)_3C^-$ (I); $(CH_3)_2CH^-$ (II); $CH_3CH_2^-$ (III); $C_6H_5CH_2^-$ (IV) is
   (a) I > II > III > IV (b) IV > III > II > I
   (c) IV > I > II > III (d) I > II > IV > III

7. The most stable free radical among the following is
(a) $C_6H_5CH_2\dot{C}H_2$ (b) $C_6H_5\dot{C}HCH_3$
(c) $CH_3\dot{C}H_2$ (d) $CH_3\dot{C}HCH_3$

8. The kind of delocalization involving sigma bond orbitals is called
(a) Inductive effect
(b) Hyperconjugation effect
(c) Electromeric effect
(d) Mesomeric effect

9. Which of the following does not have a resonating structure ?
(a) Benzene (b) Benzaldehyde
(c) Acetaldehyde (d) Benzylamine

10. A bottle containing two immiscible liquids is given to you. They may be separated by using a
(a) Fractionating column
(b) Separating funnel
(c) Vacuum distillation
(d) Steam distillation

11. Fractional distillation is used when :
(a) There is a large difference in the boiling point of liquids
(b) There is a small difference in the boiling points of liquids
(c) Boiling points of liquids are same
(d) Liquids form a constant boiling mixture

12. During hearing of a court case, the judge suspected that some changes in the documents had been carried out. He asked the forensic department to check the ink used at two different places. According to you which technique can give the best results?
(a) Column chromatography
(b) Solvent extraction
(c) Distillation
(d) Thin layer chromatography

13. Lassaigne's test is used in qualitative analysis to detect
(a) Nitrogen (b) Sulphur
(c) Chlorine (d) All of these

14. In Lassaigne's test for N, S and halogens, organic compound is
(a) Fused with sodium
(b) Dissolved with sodomite
(c) Extracted with sodomite
(d) Fused with calcium

15. Kjeldahl's method is used in the estimation of
(a) Nitrogen (b) Halogens
(c) Sulphur (d) Oxygen

16. Which of the following compounds will show metamerism?
(a) $CH_3–CO–C_2H_5$ (b) $C_2H_5–S–C_2H_5$
(c) $CH_3–O–CH_3$ (d) $CH_3–O–C_2H_5$

17. The principle involved in paper chromatography is
(a) Adsorption (b) Partition
(c) Solubility (d) Volatility

18. Which are isomers ?
(a) ethyl alcohol and dimethyl ether
(b) acetone and acetaldehyde
(c) propionic acid and propanone
(d) methyl alcohol and dimethyl ether

19. $C_6H_5C \equiv N$ and $C_6H_5N \overset{\rightarrow}{=} C$ are which type of isomers?
(a) Position (b) Functional
(c) Tautomerism (d) Linkage

20. Which of the following is *not* a nucleophile?
(a) CN^- (b) OH^-
(c) NH_3 (d) BF_3

21. Which one of the following can exhibit cis-trans isomerism ?
(a) $CH_3 - CHCl - COOH$
(b) $H - C \equiv C - Cl$
(c) $ClCH = CHCl$
(d) $ClCH_2 - CH_2Cl$

22. In sodium fusion test of organic compounds, the nitrogen of the organic compound is converted into
(a) Sodamide (b) Sodium cyanide
(c) Sodium nitrite (d) Sodium nitrate

23. Isomers of a substance must have the same
(a) Structural formula
(b) Physical properties
(c) Chemical properties
(d) Molecular formula

24. The most reactive compound for electrophilic nitration is
(a) Benzene (b) Nitrobenzene
(c) Benzoic acid (d) Toluene

25. IUPAC name of the following is
$CH_2 = CH - CH_2 - CH_2 - C \equiv CH$
(a) 1, 5-hexenyne (b) 1-hexene- 5-yne
(c) 1-hexyne- 5-ene (d) 1, 5-hexynene

26. Which one of the following pairs represents stereoisomerism?
(a) Structural isomerism and Geometrical isomerism
(b) Optical isomerism and Geometrical isomerism
(c) Chain isomerism and Rotational isomerism
(d) Linkage isomerism and Geometrical isomerism

27. Which of the following species is not electrophilic in nature?
(a) $\overset{\oplus}{NO_2}$ (b) $\overset{\oplus}{Cl}$
(c) BH_3 (d) $H_3\overset{\oplus}{O}$

28. In Duma's method, nitrogen present in organic compound is estimated as
(a) N_2 (b) NH_3
(c) NO_2 (d) None

29. Sodium extract of an organic compound gives blood red colour with $FeCl_3$. It contains
(a) N (b) S
(c) N and S (d) S and Cl

30. The homolytic fission of a covalent bond liberates
(a) Carbonium ions (b) Carbanions
(c) Free radicals (d) Carbenes

31. Which does not show resonance ?
(a) Benzene (b) Aniline
(c) Ethyl amine (d) Toluene

32. Which of the following is the correct IUPAC name?
(a) 3-Ethyl-4, 4-dimethylheptane
(b) 4,4-Dimethyl-3-ethylheptane
(c) 5-Ethyl-4, 4-dimethylheptane
(d) 4, 4-Bis(methyl)-3-ethylheptane

33. The IUPAC name for

$$CH_3-\overset{\overset{\large O}{\|}}{C}-CH_2-CH_2-\overset{\overset{\large O}{\|}}{C}-OH$$

is ___.
(a) 1-hydroxypentane-1, 4-dione
(b) 1,4-dioxopentanol
(c) 1-carboxybutan-3-one
(d) 4-oxopentanoic acid

34. In which of the following, functional group isomerism is not possible?
(a) Alcohols (b) Aldehydes
(c) Alkyl halides (d) Cyanides

35. The number of possible open chain (acyclic) isomeric compounds for molecular formula C_5H_{10} would be
(a) 8 (b) 7
(c) 6 (d) 5

36. Which of the following compounds exhibits stereoisomerism ?
(a) 2-methylbutene-1
(b) 3-methylbutyne-1
(c) 3-methylbutanoic acid
(d) 2-methylbutanoic acid

37. Select the most stable carbocation from amongst the following

(a) 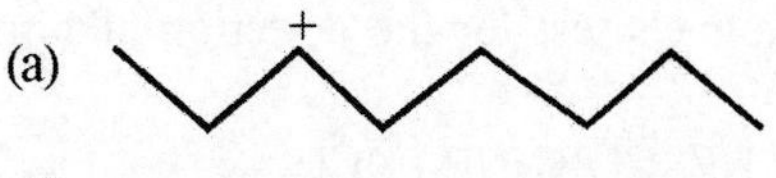

(b) 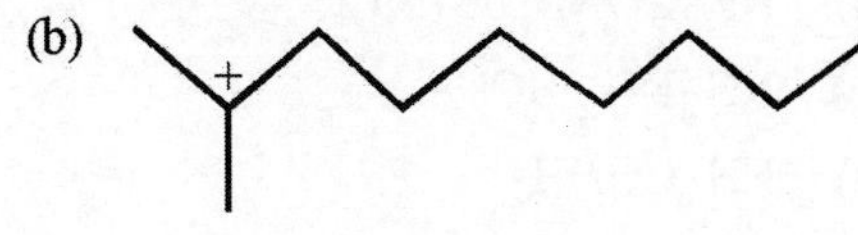

(c) 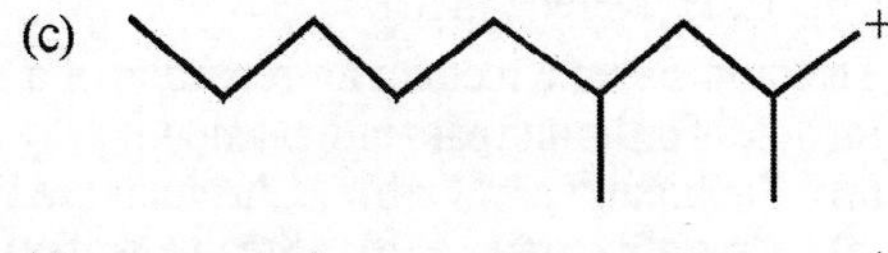

(d)

38. What is the correct order of decreasing stability of the following cations.

I. $CH_3-\overset{\oplus}{C}H-CH_3$

II. $CH_3-\overset{\oplus}{C}H-OCH_3$

III. $CH_3-\overset{\oplus}{C}H-CH_2-OCH_3$

(a) II > I > III (b) II > III > I
(c) III > I > II (d) I > II > III

39. The fragrance of flowers is due to the presence of some steam volatile organic compounds called essential oils. These are generally insoluble in water at room temperature but are miscible with water vapour in vapour phase. A suitable method for the extraction of these oils from the flowers is:
(a) Distillation
(b) Crystallisation
(c) Distillation under reduced pressure
(d) Steam distillation

40. How many chain isomers could be obtained from the alkane C_6H_{14} ?
(a) Four (b) Five
(c) Six (d) Seven

41. Which of the following IUPAC names is correct for the compound?

$$H_3C-\underset{\underset{CH_3}{|}}{CH}-\underset{\underset{CH_2CH_3}{|}}{CH}-CH_2-CH_3$$

(a) 2-Methyl-3-ethylpentane
(b) 3-Ethyl-2-methylpentane
(c) 2-Ethyl-3-methylpentane
(d) 3-Methyl-2-ethylpentane

42. Lassaigne's test for the detection of nitrogen fails in
(a) $NH_2CONHNH_2.HCl$
(b) $NH_2NH_2.HCl$
(c) NH_2CONH_2
(d) $C_6H_5NHNH_2.HCl$

43. The most suitable method for separtion of a 1 : 1 mixture of ortho and para nitrophenols is
(a) Sublimation (b) Chromatography
(c) Crystallization (d) Steam distillation

44. The number of possible isomers of the compound with molecular formula C_7H_8O is
(a) 3 (b) 5
(c) 7 (d) 9

45. The IUPAC name of

$$H_3C-\underset{\underset{CH_3}{|}}{CH}-\underset{\underset{CH_3}{|}}{CH}-CH_2-CH_2-\underset{\underset{CH_3}{|}}{CH}-CH_3$$

is
(a) 1, 3-isopropyl-3-methylpropane
(b) 2, 3, 6-trimethylheptane
(c) 2, 5, 6-trimethylheptane
(d) 2, 6, 3-trimethylheptane

46. Decreasing order of reactivity towards nucleophilic addition to carbonyl group among cyclopentanone, 3-pentanone and n-pentanal is
(a) 3-pentanone, cyclopentanone, n-pentanal
(b) n-pentanal, 3-pentanone, cyclopentanone
(c) n-pentanal, cyclopentanone, 3-pentanone
(d) cyclopentanone, 3-pentanone, n-pentanal

47. Among the following compounds (I - III), the ease of their reaction with electrophiles is,

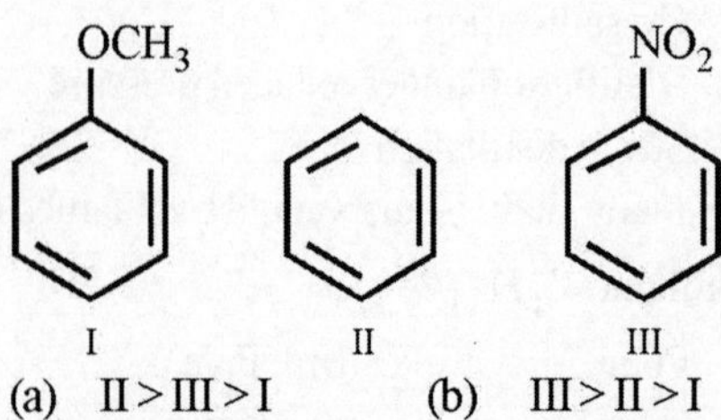

(a) II > III > I (b) III > II > I
(c) II > I > III (d) I > II > III

48. Which one of the following compounds will be most easily attacked by an electrophile?

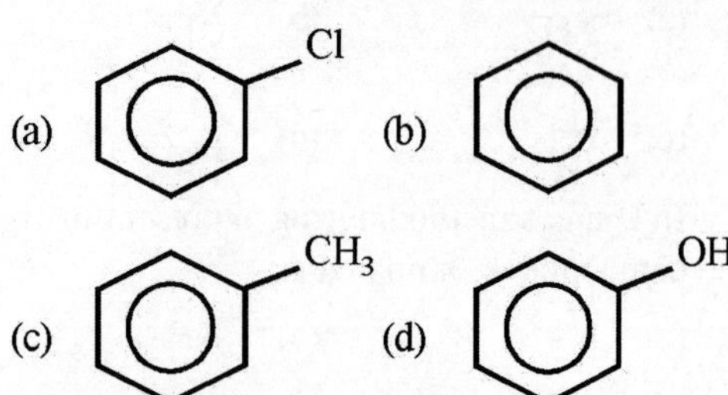

49. Which one of the following orders is correct regarding the —I effect of the substituents ?

(a) $—NR_2 < —OR > —F$
(b) $—NR_2 > —OR > —F$
(c) $—NR_2 < —OR < —F$
(d) $—NR_2 > —OR < —F$

50. Consider the following phenols :

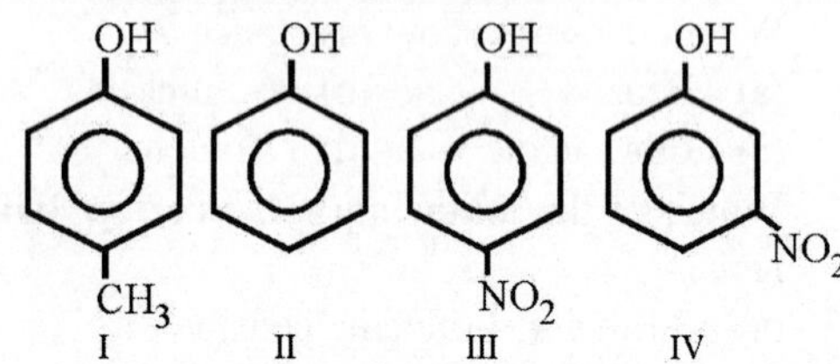

The decreasing order of acidity of the above phenols is
(a) III > IV > II > I (b) II > I > IV > III
(c) I > IV > II > III (d) III > IV > I > II

51. Correct order of stability is :
(a) cis -2- butene > 1-butene > trans -2-butene
(b) trans-2-butene > cis-2-butene > 1-butene
(c) 1-butene > cis-2-butene > trans-2- butene
(d) cis-2-butene > trans-2-butene > 1-butene

52. Which of the following compounds reacts slower in electrophilic substitution?
(a) $C_6H_5NO_2$ (b) C_6H_5OH
(c) $C_6H_5CH_3$ (d) $C_6H_5NH_2$

53. In steam distillation of toluene, the pressure for toluene in vapour is
(a) Equal to pressure of barometer
(b) Less than pressure of barometer
(c) Equal to vapour pressure of toluene in simple distillation
(d) More than vapour pressure of toluene in simple distillation

54. The incorrect IUPAC name is

(a) $CH_3-\underset{\underset{O}{||}}{C}-\underset{\underset{CH_3}{|}}{CH}-CH_3$

2-Methyl-3-butanone

(b) $CH_3-\underset{\underset{CH_3}{|}}{CH}-\underset{\underset{CH_2CH_3}{|}}{CH}-CH_3$

2,3-Dimethylpentane

(c) $CH_3\underset{\underset{Cl}{|}}{CH}-\underset{\underset{Br}{|}}{CH}-CH_3$

2-Bromo-3-chlorobutane

(d) $CH_3-C\equiv CCH(CH_3)_2$

4-Methyl-2-pentyne

55. Name of the compound given below is

(skeletal structure: H_3C–CH(CH_3)–CH(CH_2CH_3)–$CH_2CH_2CH_2CH_3$)

(a) 5-ethyl-6-methyloctane
(b) 4-ethyl-3-methyloctane
(c) 3-methyl-4-ethyloctane
(d) 2, 3-diethylheptane

56. Which one of the following is a free-radical substitution reaction?

(a) $CH_3CHO + HCN \longrightarrow CH_3CH(OH)CN$

(b) $C_6H_5CH_3 + Cl_2 \xrightarrow{\text{Boiling}} C_6H_5CH_2Cl$

(c) $C_6H_6 + CH_3Cl \xrightarrow{\text{Anh. AlCl}_3} C_6H_5CH_3$

(d) $C_6H_5CH_2Cl + AgNO_2 \longrightarrow C_6H_5CH_2NO_2$

57. The molecular formula of diphenyl methane, $C_6H_5-CH_2-C_6H_5$, is $C_{13}H_{12}$.

How many structural isomers are possible when one of the hydrogens is replaced by a chlorine atom?

(a) 6 (b) 4
(c) 8 (d) 7

58. Which one of the following compounds is most acidic?

(a) $Cl-CH_2-CH_2-OH$ (b) C_6H_5OH (phenol)

(c) o-nitrophenol (OH, NO_2) (d) o-cresol (OH, CH_3)

59. The best method for the separation of naphthalene and benzoic acid from their mixture is:

(a) distillation (b) sublimation
(c) chromatography (d) crystallisation

60. The IUPAC name of $CH_3CH_2CH(CH_3)CH(CH_3)COCl$ (skeletal structure) is

(a) 1- chloro -1-oxo - 2, 3- dimethyl pentane
(b) 2-ethyl -3- methylbutanoyl chloride
(c) 2, 3-dimethylpentanoyl chloride
(d) 3, 4-dimethyl pentanoyl chloride

Hydrocarbons

CONCEPTS

SOURCES OF HYDROCARBONS

- **Petroleum** is main **source of aliphatic hydrocarbons.**
- **Coal is source of aromatic hydrocarbons.**

PETROLEUM (MAIN OR WHITE GOLD)

- Viscous liquid found inside earth crust made of dead remains of plants and animals is called petroleum.
- **Refining of petroleum** is a process of separating it into different useful fractions.

Products of Petroleum Fractionation

	Fractions	Composition	Boiling Range	Uses
(a)	**Uncondensed gases**	C_1–C_4	0-30°C	
	(a) Cymogene			Used in the manufacture of Ice
	(b) Rhigolene			Used as local anaesthetic
(b)	**Crude Naphtha** Refractionation	C_5–C_{10}	30-150°C	
	(a) Petroleum ether	C_5–C_6	30-70°C	Solvent for oils, fats and rubber, dry cleaning
	(b) Gasoline or Petrol	C_6–C_8	70-90°C	Fuel and dry cleaning
	(c) Ligroin or Light Petroleum	C_6–C_8	90-120°C	In dry cleaning as solvent
	(d) Benzene or Benzolene	C_8–C_{10}	120-150°C	Solvent (oil and paints industry), dry cleaning
(c)	**Kerosene Oil**	C_{11}–C_{18}	150-300°C	As fuel, for illumination, making oil gas
(d)	**Diesel or Heavy Oil**	C_{18}–C_{42}	Above 300°C	Fuel in heavy automobiles
(e)	**Residue** Refractionation under reduced pressure	C_{30}–C_{45}	Above 400°C	
	(a) Paraffin Wax			For Candles and boot polishes
	(d) Lubricating Oils			For lubrications
	(c) Vaseline			In ointments and toilet goods
	(d) Pitch			In paints and varnishes, as fuel
	(e) Petroleum Coke			As fuel.

- **Natural gas** has about 90% methane and rest is ethane, propane and butanes.
- **LPG** is – propane + butane.
- **CNG** is compressed natural gas.

OCTANE AND CETANE NUMBER

Octane No.

Used to determine quality of gasoline.

(i) It is the **% of iso octane in mixture of iso octane and n-heptane** which produces same knocking effect as the fuel under examination.

$CH_3CH_2CH_2CH_2CH_2CH_2CH_3$ (n-heptane, octane no = 0)

$CH_3-C(CH_3)_2-CH_2-CH(CH_3)-CH_3$ (isooctane, octane no = 100)

(ii) **Higher octane no. means better fuel.**

(iii) Straight chain hydrocarbons have low octane number.

(iv) Octane number increases with branching.

(v) Unsaturated hydrocarbons and cycloalkanes have higher octane number than corresponding n-alkanes.

(vi) **Octane number of aromatic compounds is high**.

(vii) Octane number of fuel is increased by adding TEL (tetra ethyl lead).

(viii)
- Octane number of ordinary gasoline = 74 (used in car/scooter).
- Octane number of premium gasoline = 84 and above.
- Octane number of aviation gasoline = 100 and above.

CETANE No.

Used to determine quality of diesel. It is the % by volume of cetane in a mixture of cetane (cetane no = 100) and α-methyl napthalene (cetane no = 0) which has same spark as fuel under consideration.

CRACKING AND REFORMING

Used to increase the quality of petrol, diesel and kerosene.

It is the conversion of less valuable higher fraction to the more valuable lower fraction by the application of heat.

- Cracking can be vapour phase, thermal cracking or catalytic cracking.
- **Reforming or aromatisation or platforming** is a process of converting straight chain hydrocarbon into aromatic compound.

$$n-\text{hexane} \xrightarrow[Pt]{873K} \text{benzene}$$

- In reforming antiknock properties increases by special type of cracking which includes alkylation, isomerisation, aromatisation, cyclisation, dehydrogenation, fractionation etc.

SYNTHETIC PETROL

Manufactured by two methods :

(i) **Bergius process**

Powdered Coal + Heavy Oil $\xrightarrow[400-500°C,\ 200-700\ Atm]{H_2/Sn,\ Pb\ or\ Mo\ Compounds}$

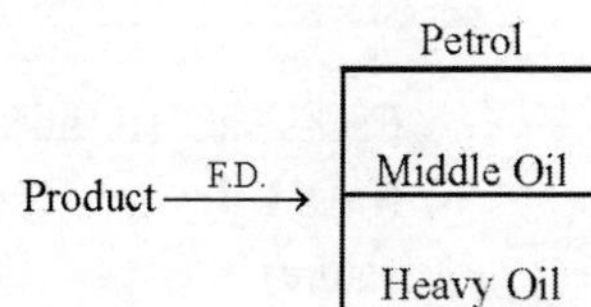

Heavy oil is reused.

(ii) **Fischer - Tropsch process.**

$$\text{Steam} + \text{Coke (Red hot)} \longrightarrow \underset{\text{water gas}}{CO + H_2}$$

$$n(CO+H_2)+nH_2 \xrightarrow[200-250\ °C]{Catalyst} C_nH_{2n} + nH_2O \text{ (Olefin)}$$

$$nCO+(2n+1)H_2 \xrightarrow{910\ atm.} C_nH_{2n+2} + nH_2O \text{ (Paraffin)}$$

Catalyst Cobalt (100 parts), thoria (5 parts), magnesia (8 parts), Kieselguhr (200 parts).

- **Petrochemicals** are organic compounds obtained directly or indirectly from natural gas or petroleum.

COAL :

(a)

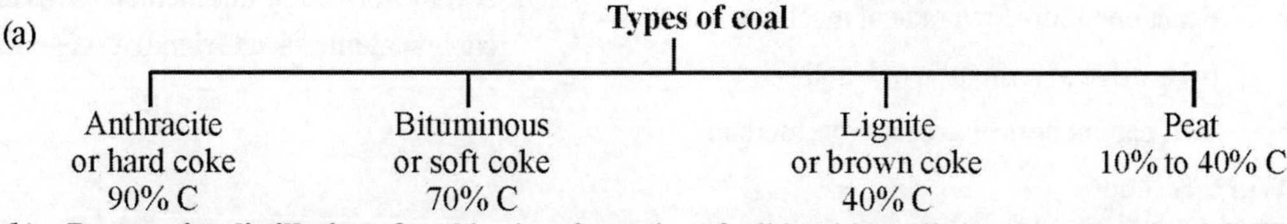

(b) **Destructive distillation of coal** leads to formation of solid residue called **coal tar**. Fractional distillation of coal-tar gives different aromatic compounds.

S.No.	Name of fraction	Temperature range (K)	Major aromatic compounds
1.	Light oil	upto 443	Benzene, toluene, xylenes
2.	Middle oil or carbolic acid	443 – 503	Phenol, pyridine, napthalene
3.	Heavy oil or creosote oil	503 – 543	Methyl and lighter alkyl phenols
4.	Green oil or Anthracene oil	543 – 633	Anthracene, phenanthrene
5.	Pitch (residue)	Non volatile	92 – 94% C

ALKANES (OR PARAFFINS)

- General formula C_nH_{2n+2}, carbon is sp^3 hybridised.
- **Main source** - Crude petroleum and natural gas.

Preparation of Alkanes :

(i) Decarboxylation of fatty acids

$$\underset{\text{sod. salt of carboxylic acid}}{RCOONa} + NaOH \xrightarrow{CaO} \underset{\text{alkane}}{R-H} + Na_2CO_3$$

- CaO – NaOH is called **soda lime**. **CaO keeps NaOH dry** ($\because$ NaOH is hygroscopic in nature).
- Reaction is used in descending of homologous series ($\because$ alkane has 1 carbon less than carboxylic acid)

(ii) Kolbe's electrolysis

$$\underset{\text{Pot. salt of carboxylic acid}}{2RCOOK} + 2H_2O \xrightarrow{\text{electricity}} \underset{\text{alkane}}{R-R} + 2CO_2 + 2KOH + H_2$$

At anode	At cathode
$2RCOO^- \longrightarrow 2RCOO + 2e^-$	$2H_2O \longrightarrow 2H^+ + 2OH^-$
$2RCOO \longrightarrow R-R + 2CO_2$	$2H^+ + 2e^- \longrightarrow H_2$
$\therefore$ **At anode CO_2 is liberated and alkane is produced.**	$\therefore$ **At cathode H_2 is liberated.**

- Reaction follows free radical mechanism.
- pH > 7 due to formation of NaOH.
- **CH_4 cannot be prepared by this method.**

(iii) Wurtz reaction

$$\underset{\text{alkyl halide}}{R-X} + 2Na + X-R \xrightarrow{\text{dry ether}} \underset{\text{alkane}}{R-R} + 2NaX$$

Only symmetrical alkane are prepared by this method.

- Alkane has double the carbon atoms as in alkyl halide.
- **CH_4 cannot be prepared by this method.**
- **3° alkyl halides do not undergo Wurtz reaction because** they prefer to undergo elemination reaction to give alkene.

  Unsymmetrical alkane cannot be prepared as it requires two different alkyl halides. This reaction will result into formation by three different alkanes

$$R-X + 2Na + R'-X \xrightarrow{\text{dry ether}} R-R' + 2NaX$$

$$R-X + 2Na + R-X \xrightarrow{\text{dry ether}} R-R + 2NaX$$

$$R'-X + 2Na + R'-X \xrightarrow{\text{dry ether}} R'-R' + 2NaX$$

(iv) Frankland reaction

$$\underset{\text{alkyl halide}}{2RX} + Zn \longrightarrow \underset{\text{alkane}}{R-R} + ZnX_2$$

(v) Hydrolysis of Grignard reagent

$$RMgX + \underset{\text{water}}{HOH} \longrightarrow \underset{\text{alkane}}{RH} + Mg(OH)X$$

$$RMgX + \underset{\text{alcohol}}{R'-OH} \longrightarrow \underset{\text{alkane}}{RH} + Mg(OR')X$$

$$RMgX + \underset{\text{1° amine}}{R'-NH_2} \longrightarrow \underset{\text{alkane}}{RH} + Mg(R'NH)X$$

- **Alkane** formed by this method has **as many carbon atoms as in Grignard reagent.**

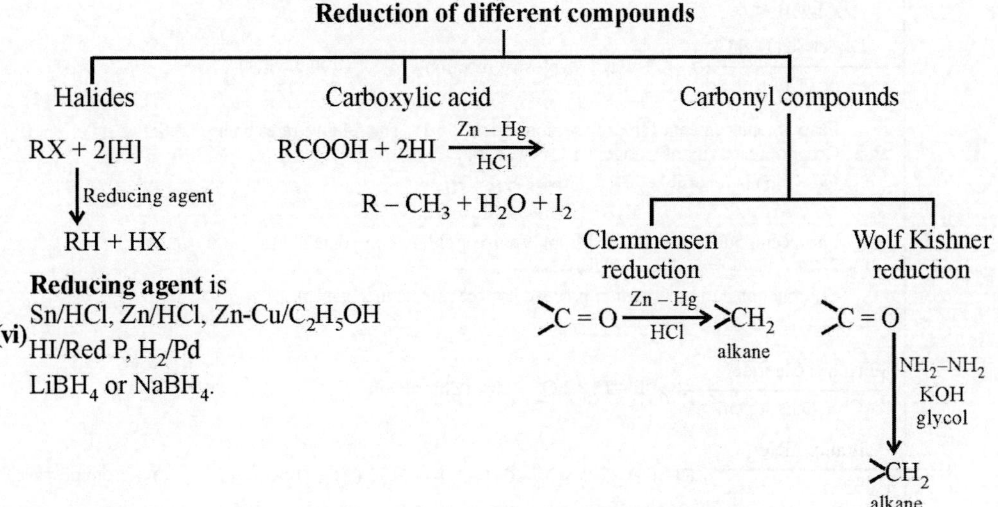

Reducing agent is
(vi) Sn/HCl, Zn/HCl, Zn-Cu/C_2H_5OH
HI/Red P, H_2/Pd
$LiBH_4$ or $NaBH_4$.

(vii) Sebatier Sandern's reaction

$$\text{alkene/alkyne} + H_2 \xrightarrow{Ni/\Delta} \text{alkane}$$

$$\underset{\text{alkene}}{R-CH=CH_2} + H_2 \xrightarrow{Ni/\Delta} \underset{\text{alkane}}{R-CH_2-CH_3}$$

$$\underset{\text{alkyne}}{R-C\equiv CH} + H_2 \xrightarrow{Ni/\Delta} \underset{\text{alkane}}{R-CH_2-CH_3}$$

(viii) Hydrolysis of methanides (Al_4C_3 and Be_2C)

$$\underset{\text{Aluminium carbide}}{Al_4C_3} + 12H_2O \longrightarrow 4Al(OH)_3 + \underset{\text{methane}}{3CH_4}$$

$$\underset{\text{Berylium carbide}}{Be_2C} + 4H_2O \longrightarrow 2Be(OH)_2 + \underset{\text{methane}}{CH_4}$$

(ix) Corey-House reaction is for synthesis of unsymmetrical alkanes.

$$\underset{\text{dialkyl copper lithium}}{R_2CuLi} + R'X \xrightarrow{\text{ether}} \underset{\text{alkane (unsymmetrical)}}{R-R'} + RCu + LiX$$

(x) By Berthlot's reaction

$$C + 2H_2 \xrightarrow[\text{arc}]{\text{Electric}} CH_4$$

Physical Properties of Alkanes

(i) Boiling point increases with increase in molecular mass.

(ii) Boiling point decreases as branching increases. This is because with increase in number of branched chains, molecule attains spherical shape and surface area of a sphere is minimum.

(iii) Melting point increases with increase in molecular mass, but increase is not regular.

- n-alkane having even number of C atom have higher melting point than next lower n-alkane with odd no. of carbon atoms.

Chemical Properties of Alkanes

RH (Alkane)

- Halogenation (Cl_2), UV light / or Δ → CH_3Cl . CH_2Cl_2 . $CHCl_3$. CCl_4 (from methane)
- Direct Sun Light, $2Cl_2$ → C + 4HCl (explosive reaction) (from methane)
  - Ease of replacement of hydrogen atom 3° > 2° > 1°. The relative rates being **5:3.8:1**
  - Order of reactivity of halogens $F_2 > Cl_2 > Br_2 > I_2$
  - Iodination is reversible $CH_4 + I_2 \rightleftharpoons CH_3I + HI$

    $5HI + HIO_3 \longrightarrow 3I_2 + 3H_2O$
  - Fluro compounds are obtained from bromo or chloro compounds

    $2RBr + HgF_2 \longrightarrow 2RF + HgBr_2$
  - Chlorination and bromination proceed by free radical mechanism.
- Sulphuryl Chloride, SO_2Cl_2+ light or peroxide → $R-Cl + SO_2 + HCl$ (Chlorination)
- Nitration HNO_3, 450°C vapour phase → $CH_3CH_2CH_2NO_2 + CH_3\underset{}{\overset{NO_2}{\overset{|}{C}}}H-CH_3 + CH_3CH_2NO_2 + CH_3NO_2$ (from propane)

  $2CH_3COOH + CO_2 + H_2O$ (from neopentane)
- Sulphonation, H_2SO_4 400°C → RSO_3H (Alkyl sulphonic acid) $+ H_2O$
- Oxidation O_2, Cu tube 100 atm., 200°C → CH_3OH (From methane)
- MoO / O_2 → $HCHO + H_2O$ (From methane)
- $(CH_3COO)_2 Mn/O_2$ → $HCOOH + H_2O$ (From methane)
- Alk. $KMnO_4$ [O] → R_3COH (From Isobutane)

  Alkanes containing 2° and 1° H atoms are generally not oxidised by $K_2Cr_2O_7$ and $KMnO_4$
- Combustion, O_2 Excess Δ → $CO_2 + H_2O$
- Isomerisation, Anhy. $AlCl_3$ + Conc. HCl → $CH_3-\overset{CH_3}{\overset{|}{C}}H-CH_3$ (from n – butane)
- Aromatisation Cr_2O_3, 450–500°C → (benzene) + $4H_2$ (from n-Hexane)
- Dehydrogenation, Cr_2O_3 / Al_2O_3 Δ → $CH_3-CH=CH_2 + H_2$ (from propane)
- Alkylation, Isobutane + Isobutylene → $CH_3-\underset{CH_3}{\underset{|}{\overset{CH_3}{\overset{|}{C}}}}-CH_2-\overset{CH_3}{\overset{|}{C}}H-CH_3$

  2,2,4–trimethyl pentane (common name isoctane)
- Cracking Δ, n-butane gives → $CH_3-CH=CH_2 + CH_4 + CH_2=CH_2 + CH_3-CH_3$
- $(CH_3)_3CH + CH_3COCH_3$, Conc. H_2SO_4 → $CH_3-\underset{CH_3}{\underset{|}{\overset{CH_3}{\overset{|}{C}}}}-\underset{OH}{\underset{|}{\overset{CH_3}{\overset{|}{C}}}}-CH_3$

CONFORMATIONAL ANALYSIS

Different arrangements of atoms in a molecule convertible into one another by rotation about single bonds are called **conformers** or **conformations**. Their study is known as **conformational analysis**. Ethane can exist in an infinite number of conformations. They are

Eclipsed $\theta = 60°$ Staggered $\theta < 60 > 0$ Skew

θ = It is dihedral angle between C–H bonds on the front of a Newman projection and those on the back.
Order of stability : Staggered > Skew > Eclipsed

- **Conformations of Propane C_3H_8**

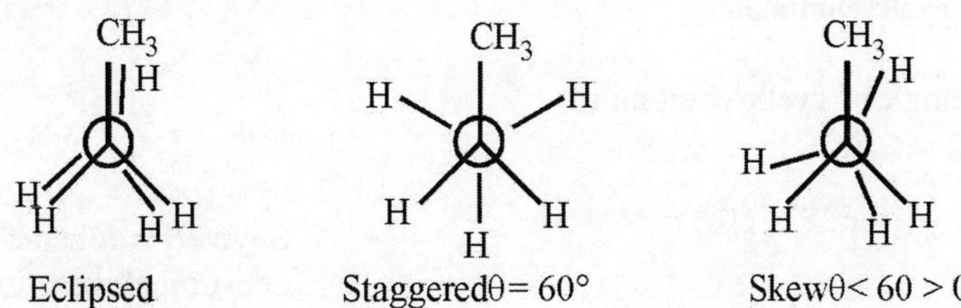

Eclipsed Staggered$\theta = 60°$ Skew$\theta < 60 > 0$

Order of stability : Staggered > Skew > Eclipsed

- **Conformations of n-Butane C_4H_{10}**

$\theta = 0°$
Fully Eclipsed

$\theta < 60 > 0$
Skew

$\theta = 0°$
Fully Eclipsed

$\theta < 60 > 0$
Skew

$\theta = 60°$ $\theta = 180°$ Anti
Gauche

$\theta = 120°$ Eclipsed

Order of stability : Anti > Gauche > Skew > Eclipsed > Fully Eclipsed

- **Conformations of Cyclohexane :** It exists in two non-planar, strainless forms, the boat and the chair form.

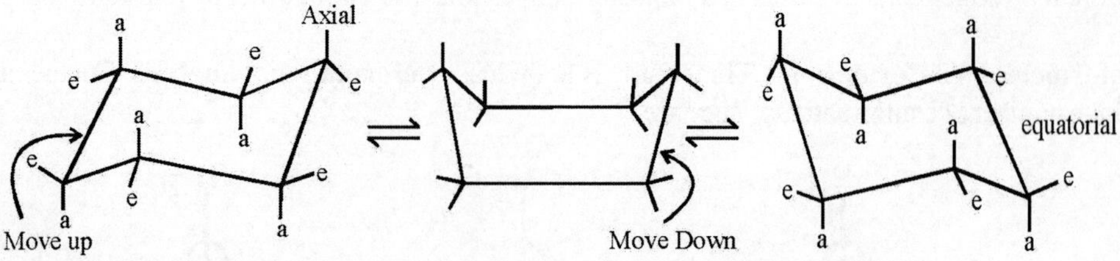

Chair form (Most stable) *Boat form* (Least stable) *Chair form*

Baeyer's Strain Theory :

When ring compounds are formed, the bonds deviate from normal positions which produces a condition of strain in the molecule. The strain is directly proportional to angle of deviation and can be calculated as follows. The more the angle of deviation, the less is the stability.

Angle of deviation of cyclic compound

$$= \frac{1}{2}(109°28' - \text{angle of cyclic compound})$$

For Cyclo propane $= \frac{1}{2}(109°28' - 60°) = 24°44'$

Cyclo pentane $= \frac{1}{2}(109°28' - 108°) = 0.44'$

Cyclo hexane $= \frac{1}{2}(109°28' - 120°)$

$= -5°16'$ (deviation outside)

- **Limitations :** Baeyer's strain theory fails to explain the stability of large ring alicyclic compounds.

ALKENES (OLEFINS)

- **General formula, C_nH_{2n}, carbon has sp^2 hybridisation.**

Methods of Preparation

(i) **By dehydration of alcohols**

- $\underset{\text{ethanol}}{CH_3CH_2OH} \xrightarrow[\text{or } Al_2O_3,\ 623\ K]{\text{Conc. } H_2SO_4,\ 443K} \underset{\text{ethene}}{CH_2 = CH_2} + H_2O$

- **Ease of dehydration** among alcohols 3° > 2° > 1°

- $\underset{2° \text{ alcohol}}{H_3C - \overset{OH}{\overset{|}{C}H} - CH_2 - CH_3} + \text{conc. } H_2SO_4$

$\longrightarrow \underset{\text{But-2-ene (80\%) major product}}{H_3C - CH = CH - CH_3}$

$\longrightarrow \underset{\text{But-1-ene (20\%) minor product}}{H_3C - CH_2 - CH = CH_2}$

Product is unexpected due to rearrangement of 1° carbonium ion to more stable 2° carbonium ion.

(ii) Dehydrohalogenation of alkyl halides (β-elimination)

$CH_3 - \underset{X}{\underset{|}{C}H} - CH_2 - CH_3 \xrightarrow{\text{alc KOH}}$

$\underset{\text{(major product)}}{\underset{\text{but-2-ene}}{CH_3CH = CH - CH_3}} + \underset{\text{(minor product)}}{CH_3CH_2CH = CH_2}$

- **Saytzeff rule** states during dehydration or dehydrohalogenation reaction hydrogen is preferably taken from carbon atom having lesser number of hydrogen atoms.
- **Order of reactivity** towards dehydrohalogenation

  3° > 2° > 1° (alkyl halides)

  RI > RBr > RCl
- More substituted alkenes are more stable.

$R_2C = CR_2 > R_2C = CHR > R_2C = CH_2 >$

$RCH = CHR > RCH = CH_2 > CH_2 = CH_2$

(iii) **Dehalogenation of vicinal and gem dihalides**

- gem.- (2 halogen atoms are present on the same carbon atom)
- vic.- (2 halogen atoms are present on neighbouring carbon atoms)

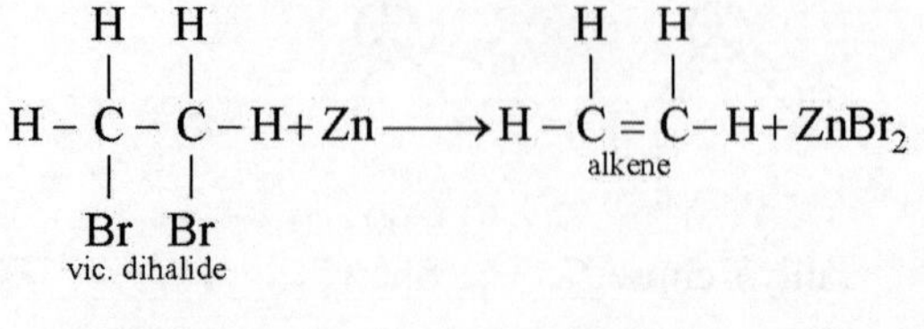

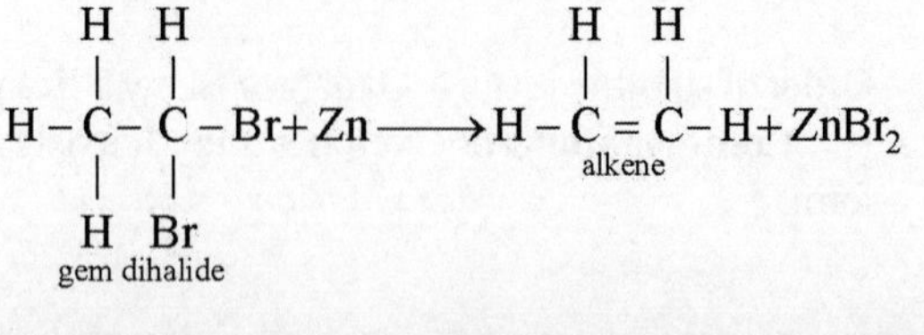

$\underset{\text{vic. dihalide}}{H - \overset{H}{\overset{|}{\underset{Br}{\underset{|}{C}}}} - \overset{H}{\overset{|}{\underset{Br}{\underset{|}{C}}}} - H} + Zn \longrightarrow \underset{\text{alkene}}{H - \overset{H}{\overset{|}{C}} = \overset{H}{\overset{|}{C}} - H} + ZnBr_2$

$\underset{\text{gem dihalide}}{H - \overset{H}{\overset{|}{\underset{H}{\underset{|}{C}}}} - \overset{H}{\overset{|}{\underset{Br}{\underset{|}{C}}}} - Br} + Zn \longrightarrow \underset{\text{alkene}}{H - \overset{H}{\overset{|}{C}} = \overset{H}{\overset{|}{C}} - H} + ZnBr_2$

(iv) Kolbe's electrolysis

$$\underset{\text{sod. succinate}}{\begin{array}{c}CH_2COONa\\|\\CH_2COONa\end{array}} \xrightarrow{\text{electricity}} \underset{\text{ethene}}{CH_2 = CH_2} + 2CO_2 + H_2 + 2NaOH$$

(v) Partial reduction of alkynes

$$\underset{\substack{\text{trans alkene}\\\text{(always)}}}{\overset{H}{R}{>}C = C{<}\overset{R}{H}} \xleftarrow[\text{Birch reduction}]{Na/liq\,NH_3} \underset{\text{alkyne}}{R-C\equiv C-R} \xrightarrow[H_2/Pd/CaCO_3/S]{\text{Lindlar's catalyst}} \underset{\substack{\text{cis alkene}\\\text{(always)}}}{\overset{R}{H}{>}C = C{<}\overset{R}{H}}$$

Properties of Alkenes :

Physical properties : C_2–C_4 gases; C_4-C_{15} liquids, C_{16} onwards solids. Less volatile than alkanes and possess anaesthetic properties.

Chemical properties :

$R–CH=CH_2$

- $\xrightarrow{H_2/Ni\ 200-300°C}$ CH_3-CH_3 **(Sabatier - Senderen reaction) (Syn addition)**
- $\xrightarrow{Br_2/CCl_4 \text{ or } CHCl_3}$ $CH_2Br.CH_2.Br$ (Colour of bromine discharged) (**Anti addition**)
- $\xrightarrow{\text{HOX hypohalous acid}}$ $CH_2OH.CH_2Cl$ ethylene chlorohydrin
- $\xrightarrow{\text{Cold conc. } H_2SO_4}$ $\underset{\textbf{Ethylhydrogen sulphate}}{CH_3.CH_2HSO_4} \xrightarrow{HOH} \underset{\textbf{Ethyl alcohol}}{CH_3CH_2OH}$
- $\xrightarrow{\text{NOCl Nitrosyl Chloride}}$ $CH_2Cl\,.\,CH_2NO$ Ethylene nitroso chloride
- $\xrightarrow[\text{Ozone}]{O_3}$ ozonide (ring: CH_2–O–CH_2, O—O) $\xrightarrow{H_2O}$ $2HCHO + H_2O_2 \longrightarrow 2HCOOH$; ozonide $\xrightarrow{H_2/Zn}$ $2HCHO + H_2O$

Addition of Zn powder removes H_2O_2 and aldehyde is obtained

$$Zn + H_2O_2 \longrightarrow ZnO + H_2O$$

- $\xrightarrow{\text{Addition of } HNO_3}$ $CH_2(OH)CH_2NO_2$ β-Nitro ethyl alcohol
- $\xrightarrow[\text{e.g. } CH_3COCl]{\text{Addition of acyl halide}}$ $CH_2(Cl)CH_2COCH_3$ Methyl β-chloro ethyl Ketone
- $\xrightarrow[200-400°C]{\text{Addition of } O_2 \text{ Ag catalyst}}$ $\begin{array}{c}CH_2\\|\\CH_2\end{array}{>}O$

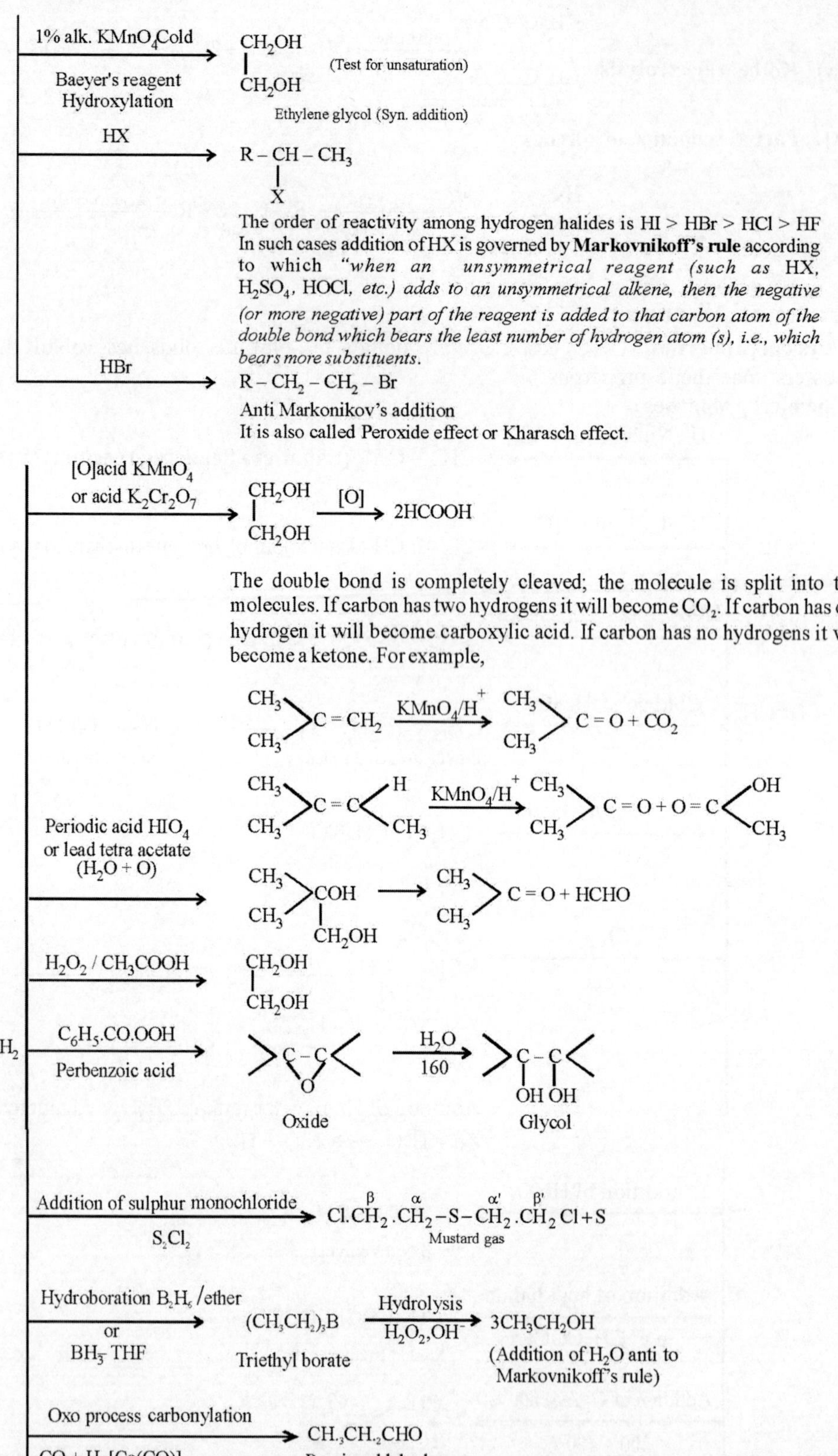
1% alk. KMnO4 Cold
Baeyer's reagent
Hydroxylation
CH2OH | CH2OH
(Test for unsaturation)
Ethylene glycol (Syn. addition)
HX
R – CH – CH3 | X
The order of reactivity among hydrogen halides is HI > HBr > HCl > HF
In such cases addition of HX is governed by Markovnikoff's rule according to which "when an unsymmetrical reagent (such as HX, H2SO4, HOCl, etc.) adds to an unsymmetrical alkene, then the negative (or more negative) part of the reagent is added to that carbon atom of the double bond which bears the least number of hydrogen atom (s), i.e., which bears more substituents.
HBr
R – CH2 – CH2 – Br
Anti Markonikov's addition
It is also called Peroxide effect or Kharasch effect.
[O]acid KMnO4
or acid K2Cr2O7
CH2OH | CH2OH
[O]
2HCOOH
The double bond is completely cleaved; the molecule is split into two molecules. If carbon has two hydrogens it will become CO2. If carbon has one hydrogen it will become carboxylic acid. If carbon has no hydrogens it will become a ketone. For example,
CH3 CH3 C = CH2
KMnO4/H+
CH3 CH3 C = O + CO2
CH3 CH3 C = C H CH3
KMnO4/H+
CH3 CH3 C = O + O = C OH CH3
Periodic acid HIO4
or lead tetra acetate
(H2O + O)
CH3 CH3 COH | CH2OH
CH3 CH3 C = O + HCHO
H2O2 / CH3COOH
CH2OH | CH2OH
R – CH = CH2
C6H5.CO.OOH
Perbenzoic acid
H2O
160
OH OH
Oxide
Glycol
Addition of sulphur monochloride
S2Cl2
β α α' β'
Cl.CH2.CH2–S–CH2.CH2 Cl+S
Mustard gas
Hydroboration B2H6 /ether
or
BH3-THF
(CH3CH2)3B
Triethyl borate
Hydrolysis
H2O2,OH–
3CH3CH2OH
(Addition of H2O anti to
Markovnikoff's rule)
Oxo process carbonylation
CO + H2 [Co(CO)]4
CH3CH2CHO
Propionaldehyde

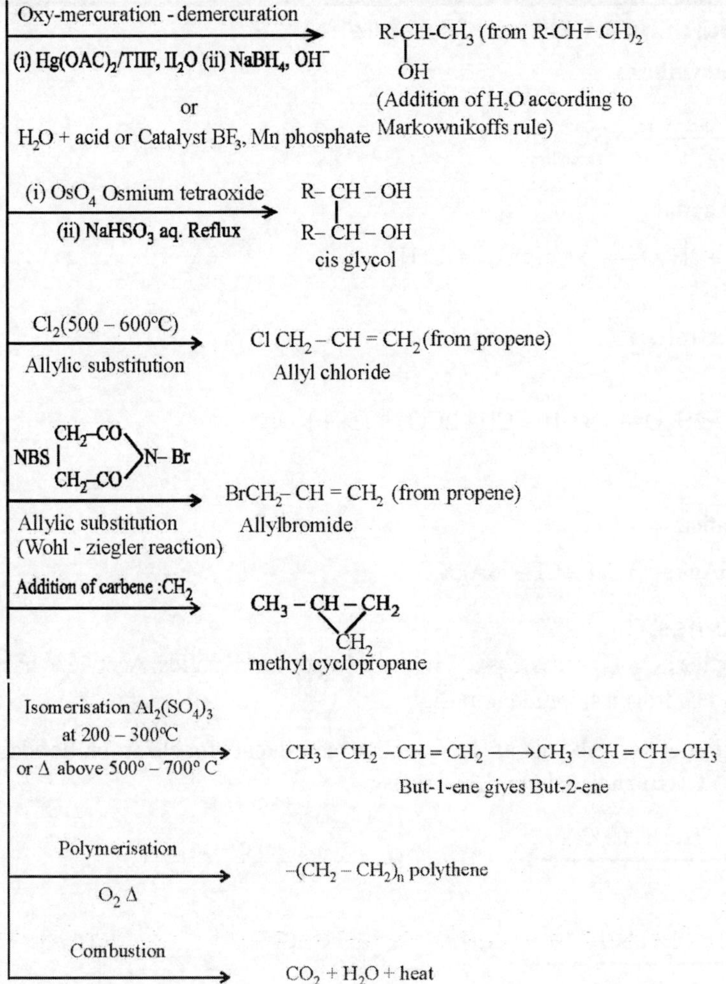

ALKYNES

- **General formula** C_nH_{2n-2}.

Methods of Preparation of Alkynes

(i) Dehydrohalogenation of vic. and gem dihalides

$$\underset{\text{vic dibromide}}{Br-CH_2-CH_2-Br} + 2KOH(alc) \xrightarrow{\Delta} \underset{\text{ethyne}}{HC \equiv CH} + 2KBr + 2H_2O$$

$$\underset{\text{gem dibromide}}{CH_3CHBr_2} + 2KOH(alc) \xrightarrow{\Delta} \underset{\text{ethyne}}{HC \equiv CH} + 2KBr + 2H_2O$$

(ii) Dehalogenation of 1, 1, 2, 2 tetrahaloalkanes

$$R-CBr_2-CHBr_2 + 2Zn \xrightarrow{CH_3OH,\ \Delta} RC \equiv CH + 2ZnBr_2$$

(iii) Alkylation of acetylides

$$\underset{\text{sod. acetilide}}{H-C \equiv CNa} + RX \longrightarrow \underset{\text{alkyne}}{HC \equiv C-R} + NaX$$

Methods of Preparation of Ethyne (Acetylene)

(i) Berthelot's synthesis

$$2C + H_2 \xrightarrow[1475\text{ K}]{\text{Electric arc}} \underset{\text{acetylene}}{HC \equiv CH}$$

(ii) Wohler's reaction

$$\underset{\text{calcium carbide}}{CaC_2} + 2H_2O \longrightarrow Ca(OH)_2 + \underset{\text{acetylene}}{C_2H_2}$$

(iii) Kolbe's Electrolysis

$$\underset{\text{pot. fumarate}}{\begin{matrix} CHCOOK \\ \| \\ CHCOOK \end{matrix}} + 2H_2O \longrightarrow \underset{\text{acetylene}}{CH \equiv CH} + 2CO_2 + H_2 + KOH$$

(iv) Dehalogenation

$$2CHX_3 + 6Ag \longrightarrow CH \equiv CH + 6AgX$$

Properties of Alkynes

Physical properties : C_2–C_4 gases C_5–C_{12} liquids. C_{13} onwards solids. Acetylene has garlic odour due to phosphene and hydrogen sulphide impurity.

- **Chemical properties:** Alkynes are less reactive then alkenes for electrophilic addition reactions.

(A) Electrophilic Addition Reactions :

$R - C \equiv CH$

$\xrightarrow{H_2/\text{Ni } 300°C}$ $CH_2 = CH_2 \xrightarrow{H_2/\text{Ni } 200°C} CH_3 - CH_3$

$\xrightarrow[\text{Kiesulguhr}]{Cl_2 / SbCl_5}$ $\underset{\text{Acetylene dichloride}}{\begin{matrix} CHCl \\ \| \\ CHCl \end{matrix}} \xrightarrow{Cl_2} \underset{\text{Acetylene tetrachloride (Westron)}}{\begin{matrix} CHCl_2 \\ | \\ CHCl_2 \end{matrix}} \xrightarrow[\text{Lime}]{BaCl_2 \text{ or}} \underset{\text{Trichloro ethylene (Westrosol)}}{\begin{matrix} CCl_2 \\ \| \\ CH.Cl \end{matrix}} + HCl$

Reactivity of halogens $Cl > Br > I$

$\xrightarrow{\text{Addition of HX}}$ $\begin{matrix} CH_2 \\ \| \\ CHX \end{matrix} \xrightarrow{HX} \underset{\text{Ethylidene halide}}{\begin{matrix} CH_3 \\ | \\ CHX_2 \end{matrix}}$

$\xrightarrow{\text{Addition of 2HOCl}}$ $\underset{\text{(Unstable)}}{\begin{matrix} CHCl_2 \\ | \\ CH(OH)_2 \end{matrix}} \xrightarrow{-H_2O} \underset{\text{Dichloro acetaldehyde}}{Cl_2CH.CHO}$

$\xrightarrow{\text{Addition of conc. } H_2SO_4}$ $\underset{\text{Ethylidene hydrogen Sulphate}}{\begin{matrix} CH_3 \\ | \\ CH(HSO_4)_2 \end{matrix}} \xrightarrow{H_2O} \underset{\text{(Acetaldehyde)}}{\begin{matrix} CH_3 \\ | \\ CHO \end{matrix}}$

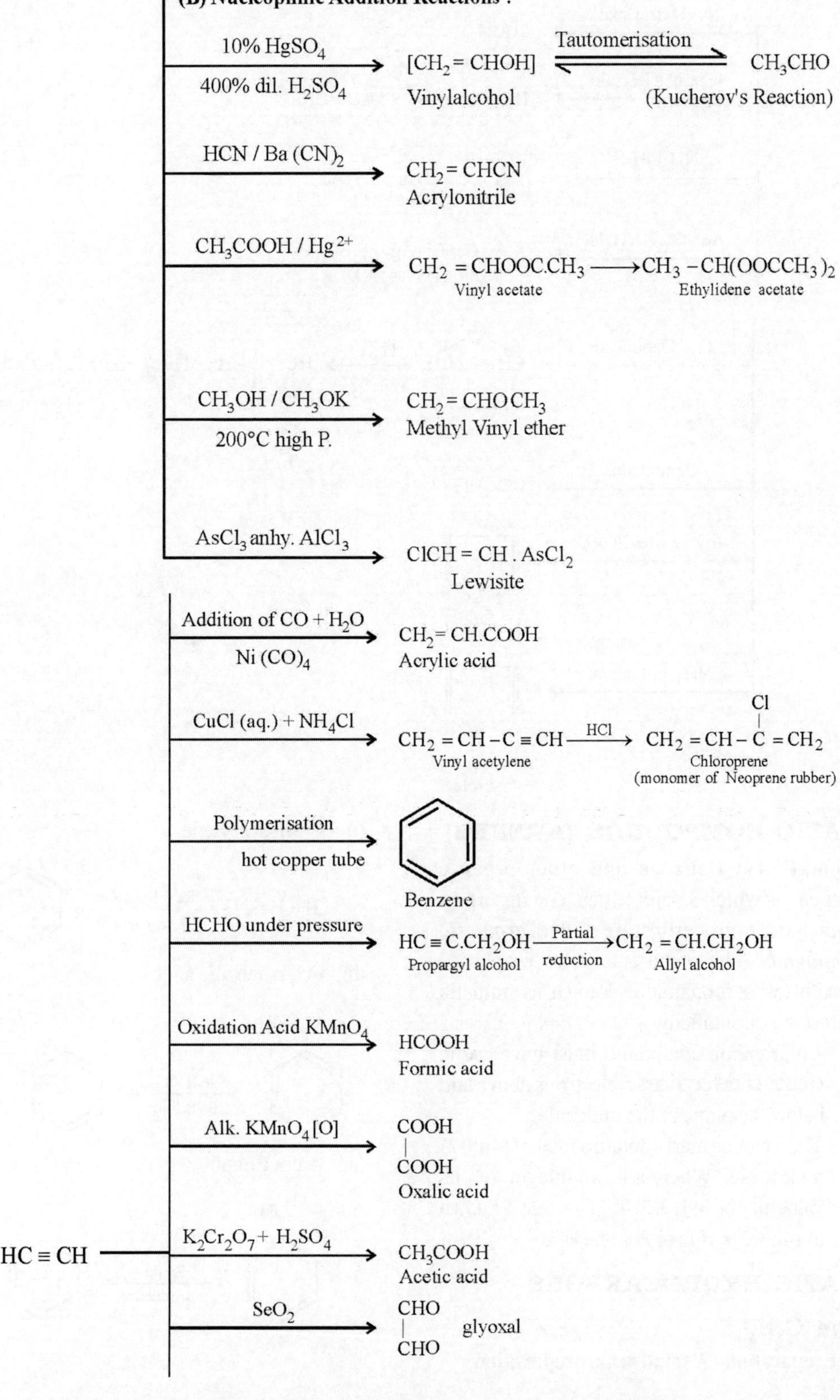
(B) Nucleophilic Addition Reactions :
10% $HgSO_4$
400% dil. H_2SO_4
$[CH_2=CHOH]$
Vinylalcohol
Tautomerisation
CH_3CHO
(Kucherov's Reaction)
HCN / Ba $(CN)_2$
$CH_2=CHCN$
Acrylonitrile
CH_3COOH / Hg^{2+}
$CH_2=CHOOC.CH_3 \longrightarrow CH_3-CH(OOCCH_3)_2$
Vinyl acetate
Ethylidene acetate
CH_3OH / CH_3OK
200°C high P.
$CH_2=CHOCH_3$
Methyl Vinyl ether
$AsCl_3$ anhy. $AlCl_3$
$ClCH=CH.AsCl_2$
Lewisite
Addition of $CO+H_2O$
Ni $(CO)_4$
$CH_2=CH.COOH$
Acrylic acid
CuCl (aq.) + NH_4Cl
$CH_2=CH-C\equiv CH \xrightarrow{HCl} CH_2=CH-C(Cl)=CH_2$
Vinyl acetylene
Chloroprene
(monomer of Neoprene rubber)
Polymerisation
hot copper tube
Benzene
HCHO under pressure
$HC\equiv C.CH_2OH \xrightarrow[\text{reduction}]{\text{Partial}} CH_2=CH.CH_2OH$
Propargyl alcohol
Allyl alcohol
Oxidation Acid $KMnO_4$
HCOOH
Formic acid
Alk. $KMnO_4$[O]
COOH
COOH
Oxalic acid
$HC\equiv CH$
$K_2Cr_2O_7+H_2SO_4$
CH_3COOH
Acetic acid
SeO_2
CHO
CHO
glyoxal

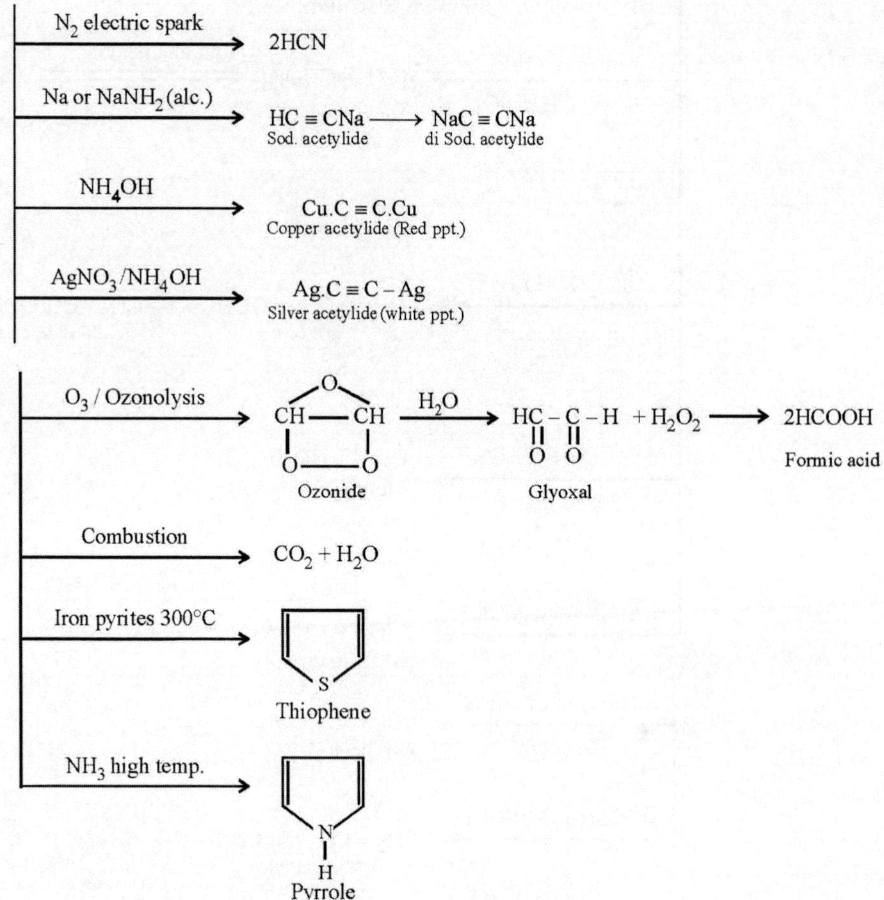

AROMATIC COMPOUNDS (ARENES)

Aromaticity: Benzene and other organic compounds which resemble benzene in certain characteristic properties are called ***aromatic compounds***. These characteristics properties constitute what is commonly known as aromatic character or aromaticity.

(a) An aromatic compound must have cyclic clouds of delocalised π electrons above and below the plane of the molecule.

(b) The π clouds must contain a total of $(4n + 2)$ π electrons. Where π is an integer i.e., its value may be 0, 1, 2, 3, 4... This rule is known as $(4n + 2)\pi$ rule or ***Huckel rule***.

AROMATIC HYDROCARBONS

Benzene C_6H_6

(A) Preparation - A small scale preparation

(i) **From acetylene**

$$3HC \equiv CH \xrightarrow[500°C]{\text{Red hot tube}} C_6H_6$$

(ii) **From benzoic acid**

$$C_6H_5COOH \xrightarrow[\text{Soda lime}]{NaOH + CaO} C_6H_6 + Na_2CO_3$$

(iii) **From Phenol**

$$C_6H_5OH \xrightarrow[\text{distillation}]{\text{Zn dust}} C_6H_6 + ZnO$$

(iv) **From Chlorobenzene**

$$C_6H_5Cl + 2H \xrightarrow{Mg/dil.\ HCl} C_6H_6 + HCl$$

(v) **From benzene sulphonic acid**

$$C_6H_5SO_3H + 2H_2O \text{ (Steam)} \xrightarrow{200°C} C_6H_6 + H_2SO_4$$

(vi) **From benzene diazonium chloride**

$$C_6H_5N_2Cl \xrightarrow[Cu^+]{H_3PO_2} C_6H_6 + N_2 + HCl$$

(B) Large Scale preparation
From Petroleum : n-hexane fraction of petroleum

$$\underset{\text{(n-Hexane)}}{CH_3(CH_2)_4CH_3} \xrightarrow[550°C]{Cr_2O_3/Al_2O_3} C_6H_6 + 4H_2$$

Properties : Colourless Liquid : bpt 80.1°C. Insoluble in H_2O.

(A) Electrophilic Substitution Reactions

Benzene →

- **Conc. HNO_3 + H_2SO_4**, **nitration, electrophile $\overset{+}{NO_2}$** → Nitrobenzene ($C_6H_5NO_2$) + H_2O
- Sulphonation Conc. H_2SO_4, Electrophile SO_3 or $^+HSO_3$ → Benzene Sulphonic acid ($C_6H_5SO_3H$)
- Halogenation $Cl_2/FeCl_3$, Electrophile $\overset{+}{Cl}$ → Chlorobenzene (C_6H_5Cl)
- Friedel Craft's Alkylation CH_3Cl/Anhy $AlCl_3$, Electrophile $\overset{+}{C}H_3$ (Carbonium ion) → Toluene ($C_6H_5CH_3$)
- Friedel Craft's Acylation CH_3COCl/Anhy $AlCl_3$, Electrophile $CH_3-\overset{O}{\overset{\|}{C}}-^+$ acylium ion → Acetophenone ($C_6H_5COCH_3$)

(B) Addition reaction

- $3H_2/Ni$ → Cyclohexane
- $3Cl_2$/UV light → $C_6H_6Cl_6$ — BHC - Benzene Hexa Chloride, Gammexane, Lindane used as insecticide.

- (I) O_3 (II) H_2O/Zn, Ozonolysis → $3\ \begin{matrix} HC=O \\ | \\ HC=O \end{matrix}$ glycoxal
- Oxidation V_2O_5, 450°C → Maleic anhydride + $4H_2O + 4CO_2$
- 700 – 800°C, Red hot iron tube → Biphenyl

Exercise

1. The correct IUPAC name of following alkane is

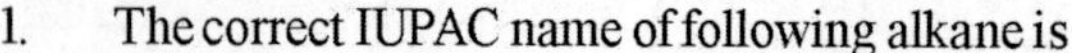

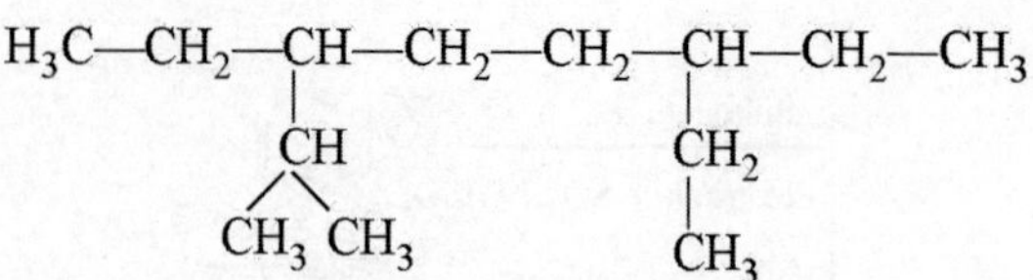

(a) 3,6 – Diethyl – 2 – methyloctane
(b) 5 – Isopropyl – 3 – ethyloctane
(c) 3 – Ethyl – 5 – isopropyloctane
(d) 3 – Isopropyl – 6 – ethyloctane

2. Which one of the following cannot be prepared by Wurtz reaction ?
(a) CH_4 (b) C_2H_6
(c) C_3H_8 (d) C_4H_{10}

3. Arrange the following in decreasing order of their boiling points.
(A) n–butane
(B) 2-methylbutane
(C) n-pentane
(D) 2, 2–dimethylpropane
(a) $A > B > C > D$ (b) $B > C > D > A$
(c) $D > C > B > A$ (d) $C > B > D > A$

4. Arrange the halogens F_2, Cl_2, Br_2, I_2, in order of their increasing reactivity with alkanes.
(a) $I_2 < Br_2 < Cl_2 < F_2$
(b) $Br_2 < Cl_2 < F_2 < I_2$
(c) $F_2 < Cl_2 < Br_2 < I_2$
(d) $F_2 < I_2 < Cl_2 < F_2$

5. The most stable conformation of n-butane is
(a) skew boat (b) gauche
(c) staggered-anti (d) eclipsed

6. Dehydration of ethanol gives
(a) Acetic acid (b) Ethane
(c) Ethylene (d) Acetylene

7. Ethylene from ethyl bromide is obtained by treating it with
(a) hydrogen
(b) alcoholic caustic potash
(c) aqueous caustic potash
(d) aqueous caustic soda

8. Alcoholic solution of KOH is used for
(a) Dehydrogenation
(b) Dehalogenation
(c) Dehydration
(d) Dehydrohalogenation

9. An alkene having molecular formula C_7H_{14} was subjected to ozonolysis in the presence of zinc dust. An equimolar amount of the following two compounds was obtained

$$\begin{matrix} CH_3 \\ CH_3 \end{matrix}\!\!>C=O \text{ and } \begin{matrix} CH_3 \\ CH_3CH_2 \end{matrix}\!\!>C=O$$

The IUPAC name of the alkene is
(a) 3, 4-dimethyl-3-pentene
(b) 3, 4-dimethyl-2-pentene
(c) 2, 3-dimethyl-3-pentene
(d) 2, 3-dimethyl-2-pentene

10. Polythene is a resin obtained by polymerisation of
(a) Butadiene (b) Ethylene
(c) Methane (d) Ethyne

11. $R-CH_2-CCl_2-R \xrightarrow{\text{Reagent}} R-C \equiv C-R$
The reagent is
(a) Na (b) HCl in H_2O
(c) KOH in C_2H_5OH (d) Zn in alcohol.

12. Acetylene when passed through 20% H_2SO_4 at 80°C gives acetaldehyde. The catalyst required for this conversion is
(a) Anhydrous $AlCl_3$ (b) $HgSO_4$
(c) Pd (d) Pt

13. Huckel's rule states that a monocyclic conjugated compound will be aromatic if it contains
(a) $(4n + 2\pi)$ electrons
(b) $(4\pi + 2n)$ electrons
(c) 4π electrons
(d) $(4n + 2)\pi$ electrons

14. $C_6H_6 + CH_3Cl \xrightarrow[AlCl_3]{\text{anhydrous}} C_6H_5CH_3 + HCl$
is an example of
(a) Friedel - Craft's reaction
(b) Kolbe's synthesis
(c) Wurtz reaction
(d) Grignard reaction

15. Which of the following reactions takes place when a mixture of concentrated HNO_3 and H_2SO_4 reacts on benzene at 350 K ?
(a) Sulphonation (b) Nitration
(c) Chlorination (d) Diazotisation

16. Which of the following will be most easily attacked by an electrophile?

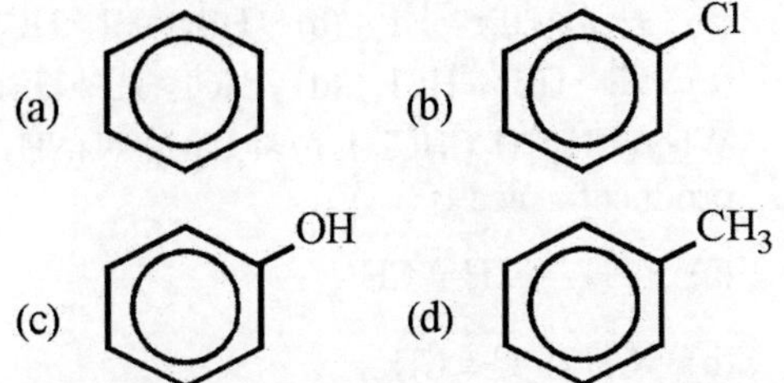

17. Select the true statement about benzene amongst the following
(a) Because of unsaturation benzene easily undergoes addition
(b) There are two types of C – C bonds in benzene molecule
(c) There is cyclic delocalisation of pi-electrons in benzene
(d) Monosubstitution of benzene gives three isomeric products.

18. A compound is treated with $NaNH_2$ to give sodium salt. Identify the compound
(a) C_2H_2 (b) C_6H_6
(c) C_2H_6 (d) C_2H_4.

19. When CH_3Cl and $AlCl_3$ are used in Friedel-Crafts reaction, the electrophile is
(a) Cl^+ (b) $AlCl_4^-$
(c) CH_3^+ (d) $AlCl_2^+$

20. The alkene $R - CH = CH_2$ reacts readily with B_2H_6 and formed the product B which on oxidation with alkaline hydrogen peroxides produces
(a) $R - CH_2 - CHO$
(b) $R - CH_2 - CH_2 - OH$
(c) $R - \underset{\substack{| \\ CH_3}}{C} = O$ (d) $R - \underset{\substack{| \\ OH}}{CH} - \underset{\substack{| \\ OH}}{CH_2}$

21. The reaction of ethyl magnesium bromide with water would give
(a) Ethane (b) Ethyl alcohol
(c) Ethyl bromide (d) Ethyl ether

22. For the formation of toluene by Friedal Craft reaction, reactants used in presence of anhydrous $AlCl_3$ are
(a) C_2H_2 and CCl_4 (b) CH_4 and $CaCN_2$
(c) C_6H_6 and CH_3Cl (d) C_2H_5 OH and Zn

23. Reaction of HBr with propene in the presence of peroxide gives
(a) isopropyl bromide (b) 3-bromo propane
(c) allyl bromide (d) n-propyl bromide

24. The IUPAC name of the compound $CH_3CH = CHC \equiv CH$ is
(a) Pent-l-yn-3-ene (b) Pent-4-yn-2-ene
(c) Pent-3-en-1-yne (d) Pent-2-en-4-yne

25. Which one of these is not compatible with arenes?
(a) Greater stability
(b) Delocalisation of π-electrons
(c) Electrophilic additions
(d) Resonance

26. The ring structure of benzene was proposed by
(a) Faraday (b) Davy
(c) Kekule (d) Wohler

27. Sodium ethoxide is a specific reagent for
(a) dehydration
(b) dehydrogenation
(c) dehydrohalogenation
(d) dehalogenation

28. Ethyl hydrogen sulphate is obtained by reaction of H_2SO_4 on :
(a) Ethylene (b) Ethane
(c) Ethyl chloride (d) Ethanol

29. The negative part of an addendum adds on to the carbon atom joined to the least number of hydrogen atoms. This statement is called
(a) Thiele's theory
(b) Peroxide effect
(c) Markownikoff's rule
(d) Baeyer's strain theory

30. The hydrogen atoms are most acidic in nature in:
(a) Ethane (b) Ethene
(c) Ethyne (d) Benzene

31. The acetylene molecule contains :
(a) 5 sigma bonds
(b) 4 sigma and 1 pi bonds
(c) 3 sigma and 2 pi bonds
(d) 2 sigma and 3 pi bonds

32. The disappearance of the characteristic purple colour of $KMnO_4$ in its reaction with an alkene is the test for unsaturation. It is known as
(a) Markownikoff test
(b) Baeyer test
(c) Wurtz test
(d) Grignard test

33. The reaction,

$CH_3 - Br + 2Na + Br - CH_3 \rightarrow$ the product, is called
(a) Wurtz reaction
(b) Perkin's reaction
(c) Aldol condensation
(d) Levit reaction

34. Which of the following is the most stable alkene?
(a) $CH_2 = CH_2$ (b) $RCH = CH_2$
(c) $R_2C = CR_2$ (d) $RCH = CHR$

35. Ethyl alcohol is heated with conc. H_2SO_4. The product formed is :
(a) $H_3C-\underset{\underset{O}{\|}}{C}-OC_2H_5$ (b) C_2H_6
(c) C_2H_4 (d) C_2H_2

36. Pure methane can be produced by
(a) Wurtz reaction
(b) Kolbe's electrolytic method
(c) Soda-lime decarboxylation
(d) Reduction with H_2

37. Which of the following reactions of methane is incomplete combustion ?
(a) $2CH_4 + O_2 \xrightarrow{Cu/523\,K/100\,atm} 2CH_3OH$
(b) $CH_4 + O_2 \xrightarrow{Mo_2O_3} HCHO + H_2O$
(c) $CH_4 + O_2 \rightarrow C(s) + 2H_2O(l)$
(d) $CH_4 + 2O_2 \rightarrow CO_2(g) + 2H_2O(l)$

38. Arrange the following alkyl halides in decreasing order of the rate of β–elimination reaction with alcoholic KOH.
(A) $CH_3-\underset{\underset{CH_3}{|}}{\overset{\overset{H}{|}}{C}}-CH_2Br$
(B) CH_3-CH_2-Br
(C) $CH_3-CH_2-CH_2-Br$
(a) A>B>C (b) C>B>A
(c) B>C>A (d) A>C>B

39. The addition of HBr to 1-butene gives a mixture of products A, B and C
(A) $H_5C_2-\underset{\underset{H}{|}}{\overset{\overset{Br}{|}}{C}}-CH_3$ (B) $H-\underset{\underset{Br}{|}}{\overset{\overset{C_2H_5}{|}}{C}}-CH_3$
(C) $CH_3-CH_2-CH_2-CH_2-Br$
The mixture consists of
(a) A and B as major and C as minor products
(b) B as major, A and C as minor products
(c) B as minor, A and C as major products
(d) A and B as minor and C as major products

40. Arrange the following hydrogen halides in order of their decreasing reactivity with propane.
(a) HCl > HBr > HI (b) HBr > HI > HCl
(c) HI > HBr > HCl (d) HCl > HI > HBr

41. When $CH_3CH_2CHCl_2$ is treated with $NaNH_2$, the product formed is
(a) $CH_3-CH=CH_2$
(b) $CH_3-C\equiv CH$
(c) $CH_3CH_2CH\begin{smallmatrix}NH_2\\NH_2\end{smallmatrix}$
(d) $CH_3CH_2CH\begin{smallmatrix}Cl\\NH_2\end{smallmatrix}$

42. Choose the correct alkyne and reagents for the preparation of

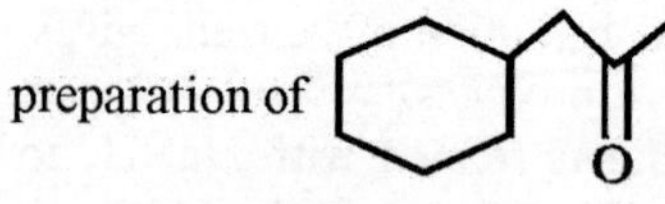

(a) , $HgSO_4$, H_2SO_4, H_2O
(b) , $HgSO_4$, H_2SO_4, H_2O
(c) , BH_3, H_2O_2, NaOH
(d) , BH_3, H_2O_2, NaOH

43. 3-Hexyne reacts with Na/liq. NH_3 to produce
(a) cis-3-Hexene (b) trans-3-Hexene
(c) 3-Hexylamine (d) 2-Hexylamine

44. Propyne and propene can be distinguished by
(a) conc. H_2SO_4
(b) Br_2 in CCl_4
(c) dil. $KMnO_4$
(d) $AgNO_3$ in ammonia

45. Which of the following is the most reactive towards ring nitration ?
(a) Benzene (b) Mesitylene
(c) Toluene (d) m-Xylene

46. The order of activity of the various o- and p-director is
(a) $-O^- > -OH > -OCOCH_3 > -COCH_3$
(b) $-OH > -O^- > -OCOCH_3 > -COCH_3$
(c) $-OH > -O^- > -COCH_3 > -OCOCH_3$
(d) $-O^- > -COCH_3 > -OCOCH_3 > -OH$

47. Among the following compounds, the decreasing order of reactivity towards electrophilic substitution is

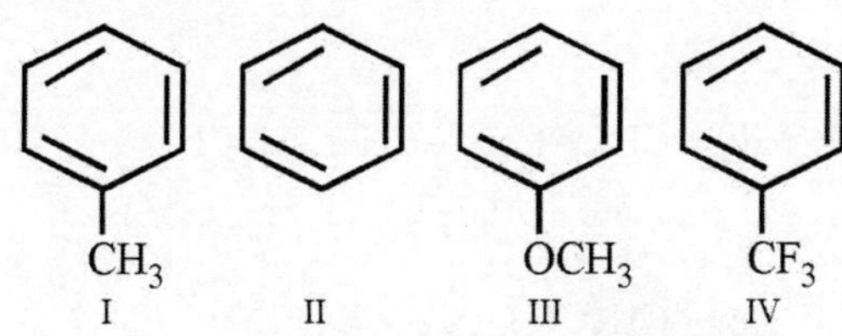

(a) III > I > II > IV (b) IV > I > II > III
(c) I > II > III > IV (d) II > I > III > IV

48. Acetylenic hydrogens are acidic because
(a) Sigma electron density of C – H bond in acetylene is nearer to carbon, which has 50% s-character
(b) Acetylene has only open hydrogen in each carbon
(c) Acetylene contains least number of hydrogens among the possible hydrocarbons having two carbons
(d) Acetylene belongs to the class of alkynes with molecular formula, C_nH_{2n-2}.

49. Which is the most suitable reagent among the following to distinguish compound (3) from rest of the compounds ?
1. $CH_3-C\equiv C-CH_3$
2. $CH_3-CH_2-CH_2-CH_3$
3. $CH_3-CH_2C\equiv CH$
4. $CH_3-CH=CH_2$.
(a) Bromine in carbon tetrachloride
(b) Bromine in acetic acid
(c) Alk $KMnO_4$
(d) Ammonical silver nitrate.

50. When hydrochloric acid gas is treated with propene in presence of benzoyl peroxide, it gives
(a) 2-Chloropropane (b) Allyl chloride
(c) No reaction (d) n-Propyl chloride.

51. Which of the following compounds has the lowest boiling point ?
(a) $CH_3CH_2CH_2CH_2CH_3$
(b) $CH_3CH=CHCH_2CH_3$
(c) $CH_3CH=CH-CH=CH_2$
(d) $CH_3CH_2CH_2CH_3$

52. In the free radical chlorination of methane, the chain initiating step involves the formation of
(a) Chlorine free radical
(b) Hydrogen chloride
(c) Methyl radical
(d) Chloromethyl radical.

53. In the presence of platinum catalyst, hydrocarbon A adds hydrogen to form n-hexane. When hydrogen bromide is added to A instead of hydrogen, only a single bromo compound is formed. Which of the following is A?
(a) $CH_3-CH_2-CH=CH-CH_2-CH_3$
(b) $CH_3-CH_2-CH_2-CH=CH-CH_3$
(c) $CH_3-CH=CH-CH_2-CH_2-CH_3$
(d) $CH_2=CH-CH_2-CH_2-CH_2-CH_3$

54. Which one of the following reactions is expected to readily give a hydrocarbon product in good yields ?
(a) $RCOOK \xrightarrow[\text{oxidation}]{\text{Electrolytic}}$
(b) $RCOO^-Ag^+ \xrightarrow{Br_2}$
(c) $CH_3CH_3 \xrightarrow[h\upsilon]{Cl_2}$
(d) $(CH_3)_3CCl \xrightarrow{C_2H_5OH}$

55. A hydrocarbon A on chlorination gives B which on heating with alcoholic potassium hydroxide changes into another hydrocarbon C. The latter decolourises Baeyer's reagent and on ozonolysis forms formaldehyde only. A is
(a) Ethane (b) Butane
(c) Methane (d) Ethene

56. Which of the following reagents convert propene to 1-propanol?
(a) H_2O, H_2SO_4
(b) aqueous KOH
(c) $MgSO_4$, $NaBH_4/H_2O$
(d) B_2H_6, H_2O_2, OH^-

57. Which alkene on ozonolysis gives CH_3CH_2CHO and $CH_3\overset{\underset{\|}{O}}{C}CH_3$
(a) $CH_3CH_2CH=C(CH_3)_2$
(b) $CH_3CH_2CH=CHCH_2CH_3$
(c) $CH_3CH_2CH=CHCH_3$
(d) $CH_3-\underset{\underset{CH_3}{|}}{C}=CHCH_3$

58. Using anhydrous $AlCl_3$ as catalyst, which one of the following reactions produces ethylbenzene (PhEt)?

(a) $H_3C-CH_2OH+C_6H_6$

(b) $CH_3-CH=CH_2+C_6H_6$

(c) $H_2C=CH_2+C_6H_6$

(d) $H_3C-CH_3+C_6H_6$

59. Which one of the following alkenes will react faster with H_2 under catalytic hydrogenation conditions?
(R = Alkyl Substituent)

(a) 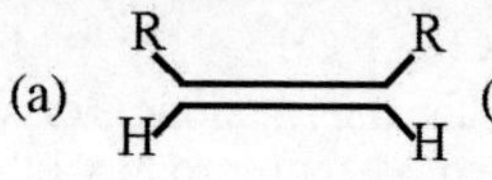
(b) 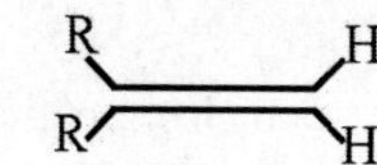

(c) 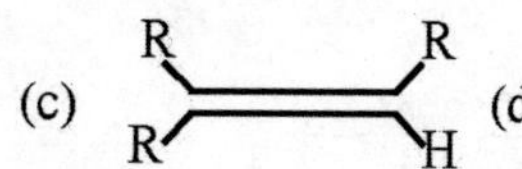
(d) 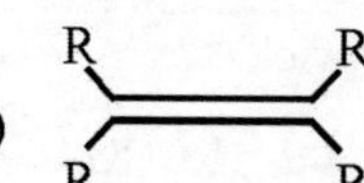

60. Predict the product C obtained in the following reaction of butyne-1.

$$CH_3CH_2-C\equiv CH+HCl\longrightarrow B\xrightarrow{HI}C$$

(a) $CH_3-CH_2-CH_2-\overset{I}{\underset{Cl}{C}}-H$

(b) $CH_3-CH_2-\overset{I}{CH}-CH_2Cl$

(c) $CH_3CH_2-\overset{I}{\underset{Cl}{C}}-CH_3$

(d) $CH_3-\underset{Cl}{CH}-CH_2CH_2I$

Environmental Chemistry

CONCEPTS

POLLUTANT

A pollutant is substance or energy introduced into the environment that has undesired effects, or adversly affects the usefulness of a resource. The pollutants may be inorganic, biological or radiological in nature.

Pollutants

- Primary pollutants (H_2S, SO_2, CO, NO, HF, NH_3 Smoke, dust, Fumes)
- Secondary pollutants (SO_3, NO_2, CH_4, aldehydes, Ketones, nitrates, Sulphates)
- Bio-degradable Pollutants (domestic waste)
- Non-Biodegradable Pollutants (Mercuric salt, lead compounds, pesticides.etc.)

ENVIRONMENT

The conditions existing around animal or human life. It is further classified as

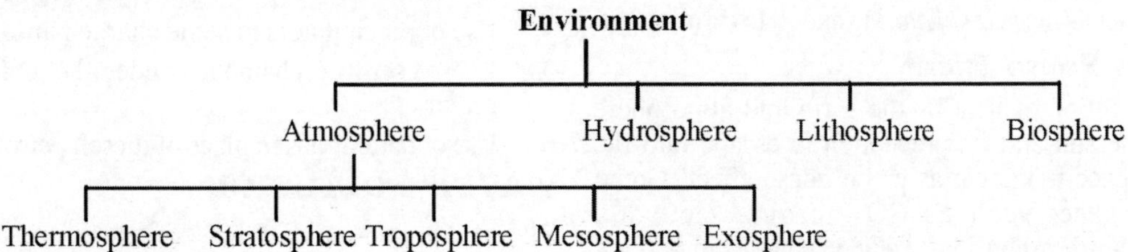

AIR POLLUTION

Major Air Pollutants

The major air pollutants are following :

(a) **Carbon monoxide (CO) :** It is produced by incomplete combustion of gasoline in motor vehicles, wood, coal, inceneration and forest fires.

(b) **Sulphur dioxide (SO_2) :** It is produced by petrol combustion, coal combustion, petrol refining and smelting operations.

(c) **Oxides of nitrogen (NO_2 and NO) :** It is produced by combustion of coal, gasoline, natural gas, petroleum refining, chemical plants manufacturing explosives and fertilizers, tobacco smoke.

(d) **Smoke, dust :** Sources foundries, cement works, iron and steel works, gas works, power generating stations.

(e) **Ammonia :** Ammonia works

(f) **Chlorine and hydrogen chloride :** Chlorine works

(g) **Chlorinated hydrocarbons :** Dry cleaning works

(h) **Mercaptans :** Oil refineries, coke ovens

(i) **Zn and Cd :** Zinc industries

(j) **Freon :** Refrigeration works.

- **Photochemical pollutants :** The nitrogen dioxide by absorbing sunlight in blue and U. V. region decomposes into nitric oxide and atomic oxygen followed by a series of other reactions producing O_3, formaldehyde, acrolein and peroxyacylnitrates.

Smog

It is a mixture of smoke and fog in suspended droplet form. It is of two types :

(a) **London smog or classical smog :** It is coal smoke plus fog. The fog part is mainly SO_2 and SO_3. It has sulphuric acid aerosol. It causes bronchial irritation and acid rain. It is reducing in nature.

(b) **Photochemical smog or Los Angeles smog :** The oxidised hydrocarbons and ozone in presence of humidity cause photochemical smog. It is oxidising in nature and causes irritation to eyes, lungs, nose, asthamatic attack and damage plants. Hydrocarbons + O_2, NO_2, NO, O, O_3 $\rightarrow$ Peroxides, formaldehyde, peroxyacetylnitrate (PAN), acrolein etc.

Acid Rain

The oxides of C, N and S present in the atmosphere, dissolve in water and produce acids and which lowers the pH of water below 5.5.

$$H_2O + CO_2 \longrightarrow H_2CO_3 \rightleftharpoons 2H^+ + CO_3^{2-}$$

$$2H_2O + 2SO_2 + O_2 \rightarrow 2H_2SO_4 \rightleftharpoons 2H^+ + SO_4^{2-}$$

$$2H_2O + 4NO_2 + O_2 \rightarrow 4HNO_3 \rightleftharpoons H^+ + NO_3^-$$

The acids are toxic to vegetation, react with marble and damage buildings.

$$CaCO_3 + H_2SO_4 \longrightarrow CaSO_4 + H_2O + CO_2$$

Acids corrode water pipes and produce salts with heavy metals ions viz Cu, Pb, Hg and Al toxic in nature.

Green House Effect

The retention of heat by the earth and atmosphere from the sun and its prevention to escape into the outer space is known as green house effect. Green house gases such as CO_2, ozone, methane, chlorofluorocarbon and water vapour form a thick cover around the earth which prevents the IR rays emitted by the earth to escape. It gradually leads to increase in temperature of atmosphere.

Consequences of green house effect

(i) Global warming would result in rise in sea level due to increased rate of melting of glaciers and floods.

(ii) Increase in infectious diseases like Malaria, Dengue etc.

Ozone Layer and its Depletion :

The ozone layer, existing between 20 to 35 km above the earth's surface, shield the earth from the harmful U. V. radiations from the sun. The U. V. radiations cause skin cancer, cataract of eye, and harmful to vegetation. Depletion of ozone is caused by oxides of nitrogen

$$N_2O + h\upsilon \longrightarrow NO + N$$

reactive nitric oxide

$$NO + O_3 \longrightarrow NO_2 + O_2,$$

$$O_3 + h\upsilon \longrightarrow O_2 + O,\ NO_2 + O \longrightarrow NO + O_2$$

$$2O_3 + h\upsilon \longrightarrow 3O_2 \text{ (Net reaction)}$$

The presence of chlorofluorocarbons also increase the decomposition of O_3.

$$CF_2Cl_2 + h\upsilon \longrightarrow CF_2Cl + Cl$$

$$CFCl_3 + h\upsilon \longrightarrow CFCl_2 + Cl,$$

$$Cl + O_3 \longrightarrow ClO + O,$$

$$ClO + O \longrightarrow Cl + O_2$$

$$O_3 + O \longrightarrow 2O_2 \text{ (Net reaction)}$$

Control of Air Pollution

It can be controlled by

1. Dissolving HCl, HF, SiF_4 in water and SO_2, Cl_2, H_2S in alkaline solution.
2. Adsorbing gas and liquids molecules over activated charcoal and silica gel.
3. Chemical reactions

$$2CuO + O_2 + 2SO_2 \longrightarrow 2CuSO_4$$

$$2CaCO_3 + O_2 + 2SO_2 \longrightarrow 2CaSO_4 + 2CO_2$$

$$4NO + 4NH_3 + O_2 \longrightarrow 2N_2 + 6H_2O$$

4. Use of precipitators to settle charge particles
5. Use of settling chambers under the action of gravity
6. Use of natural gas in place of diesel, petrol etc.

WATER POLLUTION

The contamination of water by foreign substances which would constitute a health hazard and make it harmful for all purposes (domestic, industrial or agriculture etc.) is known as water pollution. The polluted water may have affensive odour, bad taste, unpleasant colour, murky oily etc.

Sources of Water Pollution

(i) **Domestic sewage :** Discharges from kitchens, baths, lavotories etc.

(ii) **Industrial waters :** Wastes from manufacturing processes which includes acids, alkalies, pesticides, insecticides, metals like copper, Zinc, lead, mercury, fungicides etc.

(iii) **Oil :** From oil spills or washings of automobiles.

(iv) **Atomic explosion** and processing of radioactive materials

(v) **Suspended particles** (organic or inorganic) viruses, bacterials algae protozoa etc.

(vi) **Wastes from fertilizer** plants such as phosphates, nitrates ammonia etc.

(vii) **Clay :** Ores, minerals, fine particles of soil.

Effects of Impurities in Water

Dissolved substances

(i) **Hardness :** Corrossive effect on boils, alkalinity, laxative effect.

(ii) **Fluorides :** Motting of teeth enamel, nervous and skeleton disorders, above 1 mg/litre causes fluorosis.

(iii) **Sulphates :** Sulphates of Na, K, Mg cause **diarrihoea**.

(iv) **Sodium chloride :** It imparts bad taste to water.

(v) **Iron and manganese :** Stain fabrics, bad taste, modify colours.

(vi) **Lead :** It damages kidney, liver, brain and central nervous system.

(vii) **Cadmium and Mercury :** Cause kidney damage.

(viii) **Zn :** Causes vomitting, dizziness and diarrhoea.

(ix) **Arsenic** can cause cramps, paralysis and death.

(x) **Phosphates from fertilizers :** They promote algae growth and reduce D.O. concentration of water. This process is known as *eutrophication*.

(xi) **Anionic detergents** (eg. alkyl benzene sulphonates, ABS) : They produce stable foam, stabilise colloidal impurities and inhibit oxidation of organic compounds like phenol. ABS is not bio-degradable.

(xii) **Hydrogen sulphide :** Acidic, rotten-egg odour and corrosive to metals.

(xiii) **Polychlorinated biphenyls :** They are resistant to oxidation and cause skin disorders and are carcinogenic.

(xiv) **Acid polluted water (pH < 3) :** H_2SO_4 produced by oxidation of Iron pyrites (FeS_2) harmful to life.

Suspended Impurites

(i) **Parasitic worms :** They cause infections

(ii) **Bacterias :** Cause dysentery, typhoid, cholera

(iii) **Viruses :** Cause entroviral infections

(iv) **Algae :** Cause foul odour, taste, turbidity

International Standards for Drinking Water

S.No.	Characteristics (mg/l)	Acceptable limit	Rejection limit
1.	pH value	7 - 8.5	6.5 - 9.5
2.	Total dissolved solids	500	1500
3.	Total hardness (as $CaCO_3$)	200	600
4.	Fluorides	1.0	1.5
5.	Chlorides	200	1000
6.	Sulphates	200	400
7.	Nitrates	45	45
8.	Magnesium	30	150
9.	Calcium	75	200
10.	Zinc	5.0	15.0
11.	Anionic detergents	0.2	1.0
12.	Iron	0.1	1.0
13.	Manganese / Copper	0.05	0.5
14.	Phenolic compounds	0.001	0.002
15.	Toxic Materials eg. As, Cd, Cr, CN^-, Pb, Se	0.05 - 0.001	0.05 - 0.001

Control of Water Pollution

The water pollution can be reduced by following techniques

(i) **Recycling of waste water** : by aeration and use of trickling filter.

(ii) **Use of chemicals :** Effective filtration and chlorination.

(iii) **Special techniques** : Such as adsorption, ion-exchangers, reverse osomosis, electrodialysis, etc.

(iv) **Waste-water reclamation :** Sewage water can be directly used for irrigation and fish farms. Since it contains N, P and K.

Aerobic and Anaerobic Oxidation

The oxidation of organic compounds present in sewage in presence of good amount of dissolved or free oxygen (approx. 8.5 ml/l) by aerobic bacterials is called *aerobic oxidation*. When dissolved or free oxygen is below a certain value the sewage is called *stale* anaerobic bacterias bring out purification producing H_2S, NH_3, CH_4, $(NH_4)_2S$ etc. This type of oxidation is called *anaerobic* oxidation. The optimum value of D.O. for good quality of water is 4-6 ppm (4-6 mg/l). The lower the concentration of D.O., the more polluted is the water.

Biological Oxygen Demand (BOD)

It is defined as the amount of free oxygen required for biological oxidation of the organic matter by aerobic conditions at 20°C for a period of five days. Its unit is mg/l or ppm. An average sewage has BOD of 100 to 150 mg/l.

Chemical Oxygen Demand (COD)

It is a measure of all types of oxidisable impurities (biologically oxidisable and biologically inert organic matter such as cellulose) present in the sewage. COD values are higher than BOD values.

- **COD determination :** A known volume of sample is refluxed with known volume of standard $K_2Cr_2O_7$ + dil. H_2SO_4 in presence of Ag_2SO_4 (catalyst) for 1½ hours. The unreacted $K_2Cr_2O_7$ is then titrated against $FeSO_4(NH_4)_2SO_4.6H_2O$ solution. The oxygen equivalent of $K_2Cr_2O_7$ consumed is taken as a measure of COD.

  1 ml. of 1 N $K_2Cr_2O_7 \equiv 0.008$ g oxygen

SOIL OR LAND POLLUTION

The addition of substances in an indefenite proportion changing the productivity of the soil is known as soil or land pollution.

Sources of Soil Pollution

(i) **Agricultural pollutants** Chemicals like pesticides, fertilizers, fumigants, insecticides, herbicides, fungicides.

(ii) **Domestic refuge** and industrial wastes.

(iii) **Radioactive wastes** from research centres, and hospitals.

(iv) **Soil conditioners** containing toxic metals like Hg, Pb, As, Cd etc.

(v) **Farm wastes** from poultries, dairies and piggery forms.

(vi) **Improper disposal** of human and animal excreta.

(vii) **Pollutants** present in air from chemical works.

Pesticides

The chemical substances used to kill or stop the growth of unwanted organisms are called pesticides. They are further classified as

(a) **Insecticides :** They are used to kill insects. The most common insecticides are

(i) 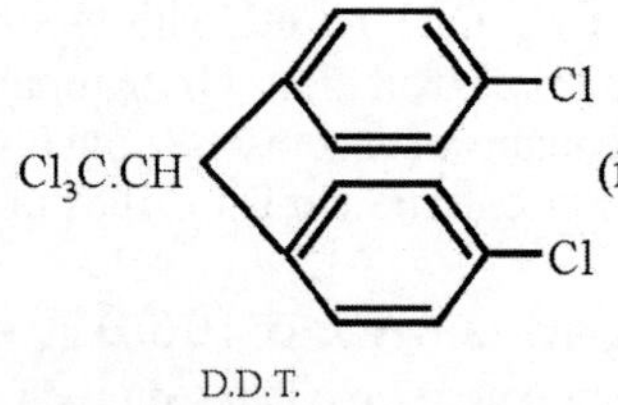 $Cl_3C.CH$ (with two p-chlorophenyl rings, -Cl)

D.D.T.

(ii) 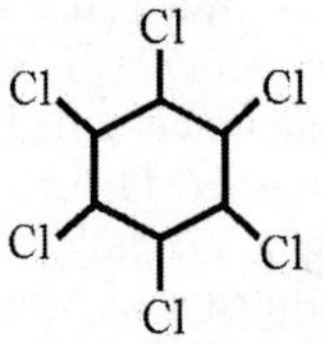 (cyclohexane ring with six Cl)

BHC, 666, gamexane

(iii) 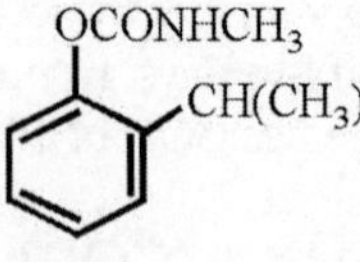 (benzene ring with $OCONHCH_3$ and $CH(CH_3)_2$)

Baygon

(iv) 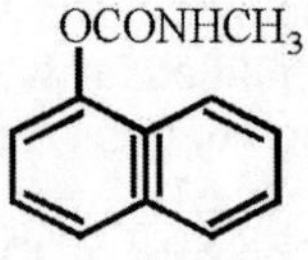 (naphthalene ring with $OCONHCH_3$)

Sevin, Carbaryl

(v) 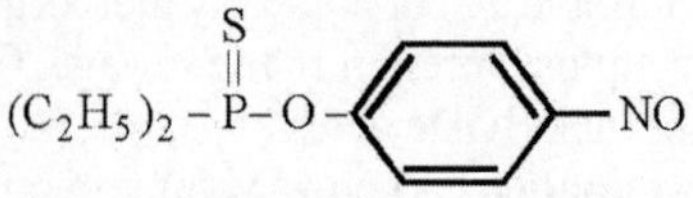 $(C_2H_5)_2-\overset{\overset{S}{||}}{P}-O-C_6H_4-NO$

Parathion

(vi) 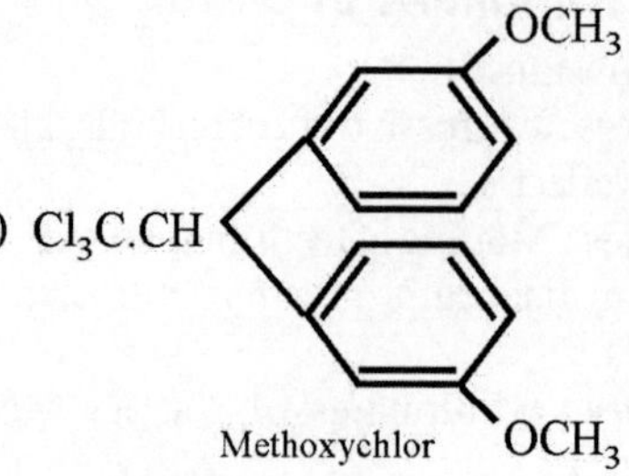 $Cl_3C.CH$ (with two p-methoxyphenyl rings, OCH_3)

Methoxychlor

(vii) Aldrin (Cl, Cl, CCl_2, CH_2, Cl, Cl)

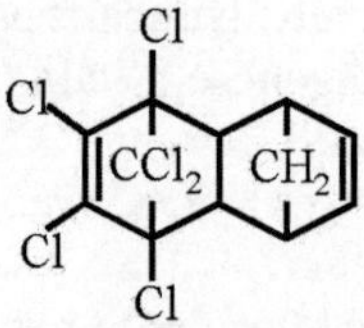

(b) **Herbicides :** They are used to kill weeds e.g.

(i) 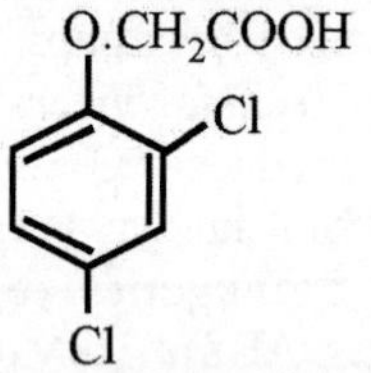 (benzene ring with $O.CH_2COOH$, Cl, Cl)

2,4-dichlorophenory acetic acid (2,4D)

(ii) Triazines

(iii) $NaClO_3$
Sodium chlorate

(iv) Na_3AsO_3
Sodium arsenite

The (iii) and (iv) are not used now a days.

(c) **Fungicides :** They are used to stop or kill fungus e.g.

(i) $CuSO_4$
copper sulphate

(ii) 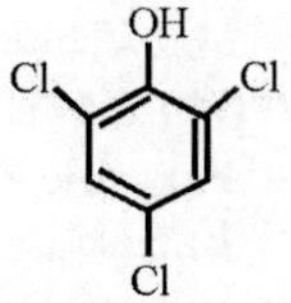 (benzene ring with OH, Cl, Cl, Cl)
2,4,6-trichlorophenol

(d) **Rodenticides :** They are used to kill rodents e.g.

(i) Zn_3P_2
Zincphosph ide

(ii) $TlSO_4$
Thalium sulphate

(iii) $ClCH_2COONa$
Sodium monochloro acetate

Exercise

1. The region which is greatly affected by air pollution is
(a) Thermosphere (b) Stratosphere
(c) Troposphere (d) Mesosphere
2. The uppermost region of the atmosphere is called
(a) Ionosphere (b) Stratosphere
(c) Troposphere (d) Exosphere
3. SO_2 is one of the air pollutants. SO_2
(a) is a lung irritant
(b) dissolves in water to form acid rain
(c) both (a) and (b)
(d) none of the above
4. Dinitrogen and dioxygen are main constituents of air but these do not react with each other to form oxides of nitrogen because __________.
(a) the reaction is endothermic and requires very high temperature.
(b) the reaction can be initiated only in presence of a catalyst.
(c) oxides of nitrogen are unstable.
(d) N_2 and O_2 are unreactive
5. The green house effect is caused by
(a) CO_2 (b) NO_2 (c) NO (d) CO
6. Which is related to 'Green House Effect'?
(a) Farming of Green plants
(b) Farming of Vegetables in Houses
(c) Global Warming
(d) Biodegradable pollutant
7. The substance having the largest concentration in acid rain?
(a) H_2CO_3 (b) HNO_3
(c) HCl (d) H_2SO_4
8. The non-viable particulate among the following is
(a) Dust (b) Bacteria
(c) Moulds (d) Fungi
9. Photochemical smog occurs in warm, dry and sunny climate. One of the following is not amongst the components of photochemical smog, identify it.
(a) NO_2
(b) O_3
(c) SO_2
(d) Unsaturated hydrocarbon
10. The gas responsible for ozone depletion :
(a) NO and freons (b) SO_2
(c) CO_2 (d) CO
11. Water polluton is caused by
(a) Pesticides (b) SO_2
(c) O_2 (d) CO_2
12. Biochemical Oxygen Demand, (BOD) is a measure of organic material present in water. BOD value less than 5 ppm indicates a water sample to be __________.
(a) rich in dissolved oxygen
(b) poor in dissolved oxygen
(c) highly polluted
(d) not suitable for aquatic life
13. Phosphate fertilizers when added to water leads to
(a) Increased growth of decomposers
(b) Reduced algal growth
(c) Increased algal growth
(d) Nutrient enrichment (eutrophication)
14. Chief source of soil and water pollution is
(a) Mining
(b) Agro industry
(c) Thermal power plant
(d) All the above
15. Green chemistry means such reactions which :
(a) produce colour during reactions
(b) reduce the use and production of hazardous chemicals
(c) are related to the depletion of ozone layer
(d) study the reactions in plants
16. The pollutants which came directly in the air from sources are called primary pollutants. Primary pollutants are sometimes converted into secondary pollutants. Which of the following belongs to secondary air pollutants?
(a) CO
(b) Hydrocarbon
(c) Peroxyacetyl nitrate
(d) NO
17. Which of the following gases is not a green house gas?
(a) CO (b) O_3
(c) CH_4 (d) H_2O vapour
18. The type of pollution caused by spraying of DDT
(a) Air and soil (b) Air and water
(c) Air (d) Air, water and soil

19. Water is often treated with chlorine to
(a) remove hardness
(b) increase oxygen content
(c) kill germs
(d) remove suspended particles

20. The greenhouse effect is because of the
(a) presence of gases, which in general are strong infrared absorbers, in the atmosphere
(b) presence of CO_2 only in the atmosphere
(c) pressure of O_3 and CH_4 in the atmosphere
(d) N_2O and chlorofluorohydrocarbons in the atmosphere

21. The quantity of CO_2 in atmosphere is
(a) 3.34% (d) 6.5%
(c) 0.034% (d) 0.34%

22. The main element of smog is
(a) O_3 and PAN (b) O_3
(c) PAN (d) PPN and PBN

23. Which pollution is harmful for 'Tajmahal'?
(a) Hydrogen (b) O_2
(c) SO_2 (d) Chlorine

24. Which gas is responsible for ' Bhopal Gas Tragedy' in 1984?
(a) CO (b) Methyl isocynate
(c) SO_2 and NO_2 (d) Ethyl Isocynate

25. The quantity of DDT in food chain
(a) Decreases (b) Remains same
(c) Increases (d) Changes

26. The substance which is a primary pollutant?
(a) H_2SO_4 (b) CO
(c) PAN (d) Aldehydes

27. Depletion of ozone layer causes
(a) Breast cancer (b) Blood cancer
(c) Lung cancer (d) Skin cancer

28. The substance which is not regarded as a pollutant?
(a) NO_2 (b) CO_2
(c) O_3 (d) Hydrocarbons

29. Thermal pollution affects mainly –
(a) vegetation (b) aquatic creature
(c) rocks (d) air

30. Select the one that has an adverse effect on ozone layer.
(a) Carbon dioxide
(b) Chlorofluorocarbons
(c) Soil
(d) Dust particles

31. The effect of polluted water on soil is, that
(a) it decreases fertility
(b) it contaminates ground water
(c) it renders soil acidic or basic
(d) all of the above

32. A green house
(a) is an enclosure of glass in which plants are kept to protect them from cold air outside
(b) is that "body" which allows the short wavelength incoming solar radiation to enter in, but does not allow long wavelength outgoing infra-red radiations to escape out
(c) both the above
(d) none of the above

33. Which of the following causes lung cancer?
(a) Textiles (b) Asbestos
(c) Silica (d) Paper

34. Ozone hole refers to
(a) Increase in concentration of ozone
(b) Hole in ozone layer
(c) Reduction in thickness of ozone layer in troposphere
(d) Reduction in thickness of ozone layer in stratsophere

35. Sewage mostly constitutes
(a) Non-biodegradable pollutants
(b) Biodegradable pollutants
(c) Effluents
(d) Air pollutants

36. Which of the following is not a consequence of greenhouse effect?
(a) Climatic conditions will be changed resulting in
(b) Plants in warmer climates with adequate rainfall would grow faster
(c) The incidence of infectious diseases is likely to increase
(d) Malaria will be controlled as the mosquitoes will not survive.

37. Which of the following statements is not true about classical smog?
(a) Its main components are produced by the action of sunlight on emissions of automobiles and factories.
(b) Produced in cold and humid climate.
(c) It contains compounds of reducing nature.
(d) It contains smoke fog and sulphur dioxide

38. Which of the following statements about photochemical smog is wrong?
    (a) It has high concentration of oxidising agents
    (b) It has low concentration of oxidising agent
    (c) It can be controlled by controlling the release of NO_2, hydrocarbons ozone, etc.
    (d) Plantation of some plants like pinus helps in controlling photochemical smog.
39. Which of the following statements is wrong?
    (a) Ozone is not responsible for green house effect.
    (b) Ozone can oxidise sulphur dioxide present in the atmosphere to sulphur trioxide.
    (c) Ozone hole is thinning of ozone layer present in stratosphere.
    (d) Ozone is produced in upper stratosphere by the action of UV rays on oxygen.
40. Which of the following statements is correct?
    (a) Ozone hole is a hole formed in stratosphere from which ozone oozes out.
    (b) Ozone hole is a hole formed in the troposphere from which ozone oozes out.
    (c) Ozone hole is thinning of ozone layer of stratosphere at some places.
    (d) Ozone hole means vanishing of ozone layer around the earth completely.
41. Phosphate pollution is caused by
    (a) sewage and agricultural fertilizers
    (b) weathering of phosphate rocks only
    (c) agricultral fertilizers only
    (d) phosphate rocks and sewage
42. Which among the following statements is *false*?
    (a) Oil slick in sea water increases D.O. value.
    (b) The main reason for river water pollution is industrial and domestic sewage discharge.
    (c) Surface water contains a lot of organic matter mineral nutrients and radioactive materials.
    (d) Oil spill in sea water causes heavy damage to fishery.
43. Sewage containing organic waste should not be disposed in water bodies because it causes major water pollution. Fishes in such a polluted water die because of
    (a) large number of mosquitoes
    (b) increase in the amount of dissolved oxygen
    (c) decrease in the amount of dissolved oxygen in water
    (d) clogging of gills by mud.
44. Which of the following practices will not come under green chemistry?
    (a) If possible, making use of soap made of vegetable oils instead of using synthetic detergents.
    (b) Using H_2O_2 for bleaching purpose instead of using chlorine based bleaching agents.
    (c) Using bicycle for travelling small distances instead of using petrol/ diesel based vehicles.
    (d) Using plastic cans for neatly storing substances.
45. Identify the wrong statement in the following:
    (a) Chlorofluorocarbons are responsible for ozone layer depletion.
    (b) Greenhouse effect is responsible for global warming.
    (c) Acid rain is mostly because of oxides of nitrogen and sulphur.
    (d) Ozone layer does not permit infrared radiation from the sun to reach the earth.
46. Which of the following is/are the hazardous pollutant(s) present in automobile exhaust gases?
    (i) N_2 (ii) CO
    (iii) CH_4 (iv) Oxides of nitrogen
    (a) (ii) and (iii) (b) (i) and (ii)
    (c) (ii) and (iv) (d) (i) and (iii)
47. Which one of the following statement is not true ?
    (a) pH of drinking water should be between 5.5 – 9.5.
    (b) Concentration of DO below 6 ppm is good for the growth of fish.
    (c) Clean water would have a BOD value of less than 5 ppm.
    (d) Oxides of sulphur, nitrogen and carbon are the most widespread air pollutant.

48. BOD of pond is connected with
   (a) Microbes & organic matter
   (b) Organic matter
   (c) Microbes
   (d) None of these
49. When rain is accompanied by a thunderstorm, the collected rain water will have a pH value
   (a) slightly lower than that of rain water without thunderstorm
   (b) slightly higher than that when the thunderstorm is not there
   (c) uninfluenced by occurrence of thunderstorm
   (d) which depends upon the amount of dust in air
50. The reaction caused by U.V. radiations from sun produces
   (a) Ozone (b) Carbon monoxide
   (c) Sulphur dioxide (d) Fluorides
51. Lead is
   (a) Radiological pollutant
   (b) Sound pollutant
   (c) Soil pollutant
   (d) Air pollutant
52. CFC which is a main reason behind air pollution, is produced by?
   (a) Sewage pollutant (b) Aerosols
   (c) Industrial remains (d) Above all
53. Which of the following is the coldest region of atmosphere?
   (a) Thermosphere (b) Mesosphere
   (c) Troposphere (d) Stratosphere
54. Formation of London smog takes place in
   (a) Winter during day time
   (b) Summer during day time
   (c) Summer during morning time
   (d) Winter during morning time
55. Green house gases –
   (a) allow shorter wavelength to enter earth's atmosphere while doesn't allow longer wavelength to leave the earth's atmosphere.
   (b) allow longer wavelength to enter earth atmosphere while doesn't allow shorter wavelength to leave the surface
   (c) don't have wavelength specific character.
   (d) show wavelength specific behaviour near the earth while far from earth these have wavelength independent behaviour.
56. Select the process that does not add particulate materials to air.
   (a) Use of air conditioner
   (b) Burning of fosssil fuels
   (c) Paper industry
   (d) Incomplete combustion of coal
57. Which of the following is the major cause of global warming?
   (a) re-radiation of U.V. rays by CO_2 and H_2O
   (b) re-radiation of I.R. rays by CO_2 and H_2O
   (c) re-radiation of I.R. rays by O_2 and N_2
   (d) re-radiation of U.V. rays by O_2 and N_2
58. Select the one that is an oxygen demanding waste.
   (a) Grease (b) Oil
   (c) Pesticides (d) Domestic sevage
59. Which is the function of oxygen present in air ?
   (a) Respiration (b) Burning of fuels
   (c) Fermentation (d) All of these
60. The process of respiration
   (a) involves oxidation of food
   (b) involves oxidation of glucose to carbon dioxide and water
   (c) is accompanied by release a large amount of energy
   (d) all the above

The Solid State

CONCEPTS

SOLIDS

Solids are the form of matter which have high density, low compressibility, definite shape & volume, considerable mechanical strength, rigidity, high force of attraction, no translatory motion and vibrating about fixed position.

Types of Solids

- **True solids :** Solids which retain their shape against mild distorting forces.
- **Pseudo solids :** They can be easily distorted by bending or compressing forces. They are also called as super cooled liquid i.e., glass.

Types of solids

Crystalline solids

(a) They have a definite geometrical shape due to the definite and orderly arrangement of particles in the three dimensional space.
(b) They have sharp melting point.
(c) When cut with a sharp edged tool, they split into two pieces and the newly generated surfaces are plain and smooth.
(d) They have crystal symmetry.
(e) They possess anisotropic properties, *i.e.*, the magnitude of some of the physical properties of crystalline solids like refractive index, coefficient of thermal expansion and electrical and thermal conductivities is different in different directions within the crystal.

Amorphous solids

(a) They do not have an orderly arrangement of particles and therefore, do not possess a definite geometrical shape.
(b) They do not have sharp melting point.
(c) When they are cut, they split into two pieces and the newly generated surfaces are ir regular in shape.
(d) They do not have crystal symmetry.
(e) They have isotropic properties, *i.e.*, the magnitude of physical properties is same in all directions.

Types of Crystalline Solids

S. No.	Property	Ionic solid	Metallic solid	Covalent solid	Molecular solid
1.	Constituent particle	ions	cations, kernels	atoms	molecules
2.	Type of attractive force	electrostatic	metallic bonds	covalent bonds	van-der waals
3.	Examples	NaCl, LiF, KNO_3	All metals	SiO_2, S	dry CO_2, I_2, Ar

Type of Molecular Solids :

- **Non-polar molecular solids:** In these solids, atoms are held together by weak dispersion forces or London forces.
- **Polar molecular solids:** In these solids, molecules are held together by relatively stronger dipole-dipole interactions.
- **Hydrogen bonded molecular solids:** The molecules are held together by hydrogen bonds.

CRYSTAL LATTICES AND UNIT CELLS

Basic Terms :

- **Crystal :** A crystal is a homogeneous portion of a solid substance made by regular pattern of structural units bonded by plane surface making definite angles with each other.
- **Space lattice :** The arrangement of constituents like atom, ions and molecules in different sites in three dimensional space is called space lattice.
- **Unit cell :** The smallest repeating unit in space lattice which when repeats over and over again, results in a crystal of the given substance called unit cell.
- **Crystallographic axis :** The lines drawn parallel to the lines of intersection of any three faces of the unit cell which do not lie in the same plane OX, OY, OZ.
- **Interfacial angle** : The angle between the three crystallographic axis α, β, γ.
- **Primitives :** The three sides of a unit cell a, b, c.

Types of Unit Cubic Cells

There are three types of unit cells based on the cube. These are:

(*a*) **Primitive** or **simple cube** which has one constituent at each corner.

(*b*) **Body-centred cube** in which one constituent at the centre of the cube as well as one at each corner.

(*c*) **Face-centred cube** in which there is one constituent at the centre of each face as well as one at each corner.

S. No.	Property	Simple cubic lattice	Face centre cubic lattice	Body centre cubic lattice
1.	Position of particles	At corners of cube only $8 \times 1/8 = 1$	At corners of cube and centre of each face $8 \times 1/8 + 6 \times 1/2 = 4$	At corners of cube and of unit cell $8 \times 1/8 + 1 \times 1 = 2$
2.	Packing fraction	52%	74%	74%
3.	Radius of atoms/ particle (r)	$a/2$	$r = \sqrt{2}a/4$ where a = edge length	$r = \sqrt{3}a/4$

Types of Crystal Systems

System	Maximum symmetry elements	Axes and Angles	Examples
1. Cubic or Regular	9 planes 13 axis	$a = b = c$ $\alpha = \beta = \gamma = 90°$	NaCl, KCl, CaF_2, ZnS, Cu_2O, Diamond, Alums, Pb, Ag, Au, Hg
2. Tetragonal	5 planes 5 axis	$a = b \neq c$ $\alpha = \beta = \gamma = 90°$	SnO_2, ZnO_2, TiO_2, $NiSO_4$, $ZrSiO_4$, $PbWO_4$, White Sn
3. Hexagonal	7 planes 7 axis	$a = b \neq c$ $\alpha = \beta = 90°, \gamma = 120°$	ZnO, PbI_2, CdS, HgS, Graphite, Ice, Beryl, Mg, Zn, Cd
4. Trigonal or Rhombohedral	7 planes 7 axis	$a = b = c$ $\alpha = \beta = \gamma \neq 90°$	$NaNO_3$, $CaSO_4$, Calcite, ICl, Quartz, As, Sb, Bi
5. Orthorhombic (Rhombic)	3 planes 3 axis	$a \neq b \neq c$ $\alpha = \beta = \gamma = 90°$	KNO_3, $K_2SO_4 PbCO_3$, $BaSO_4$ Rhombic sulphur, $MgSO_4.7H_2O$
6. Monoclinic	1 plane 1 axis	$a \neq b \neq c$ $\alpha = \gamma = 90°$ $\beta \neq 90°$	$Na_2SO_4.10H_2O$, $Na_2B_4O_7.10H_2O$ $CaSO_4.2H_2O$, Monoclinic sulphur,
7. Triclinic	No plane No axis	$a \neq b \neq c$ $\alpha \neq \beta \neq \gamma \neq 90°$	$CuSO_4.5H_2O$, $K_2Cr_2O_7$, H_3BO_3

DETERMINATION OF STRUCTURE OF CRYSTAL (BRAGG's EQUATION)

$$2d\sin\theta = n\lambda$$

where d = distance between two layers of solids

θ = angle of diffraction, n = order of diffraction, λ = wavelength of x-ray.

- **Co-ordination number:** is the number of particles with which a particle is in contact with.

PACKING FRACTION (P.F.)

It is defined as ratio of the volume of the unit cell that is occupied by spheres of the unit cell to the total volume of the unit cell.

Simple cubic unit cell P.F. = 48%

Body centered unit cell P.F = 32%

Face centered unit cell P.F = 26%

Hence, the order of P.F is fcc < bcc < sc.

RADIUS RATIO

The ratio of cation to that of anion (r^+/r^-).

Radius Ratio Rule

r^+/r^- (radius ratio)	Co-ordination no.	Arrangement of particles
0.155 – 0.225	3	Triangular planar
0.225 – 0.414	4	Tetrahedral
0.414 – 0.732	6	Octahedral
0.732 – 1.0	8	Body-centred cubic

CLOSE PACKING IN SOLIDS

Close Packing in Two Dimension

(It is arrangement of particles in 2 dimensions to have maximum force of attraction among them and best way to occupy available space)

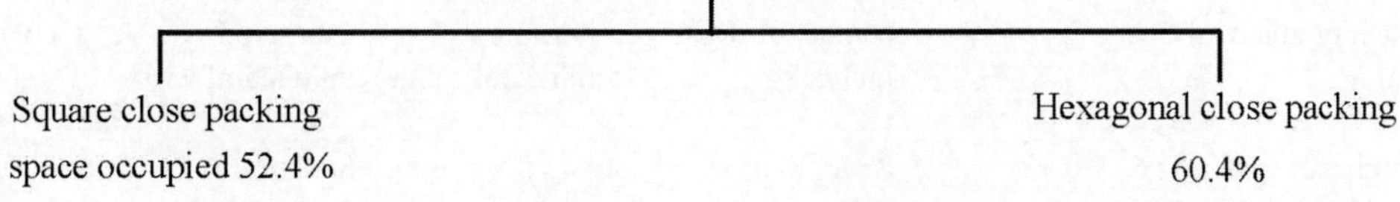

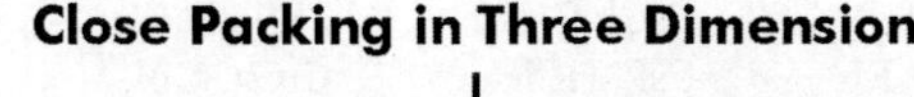

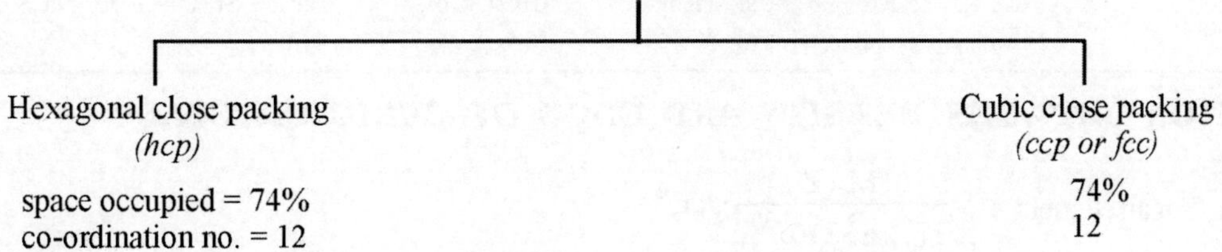

INTERSTITIAL VOIDS

In *hcp* as well *ccp* only 74% of the available space is occupied, rest space is called interstitial voids.

Classification of Voids

Voids
(Interstitial sites left in crystal lattice)

Tetrahedral voids

- It is a triangular void surrounded by 4 spheres (atoms)
- No. of tetrahedral voids = 2 × no. of atoms

Radius ratio $= \frac{r_{void}}{r_{sphere}} = 0.225$

Octahedral voids

- It is combination of two triangular voids and 6 spheres (atoms) surround it.
- No. of octahedral voids = No. of atoms in close packed arrangement

Radius ratio $= \frac{r_{void}}{r_{sphere}} = 0.414$

Structure of Simple Ionic Compounds

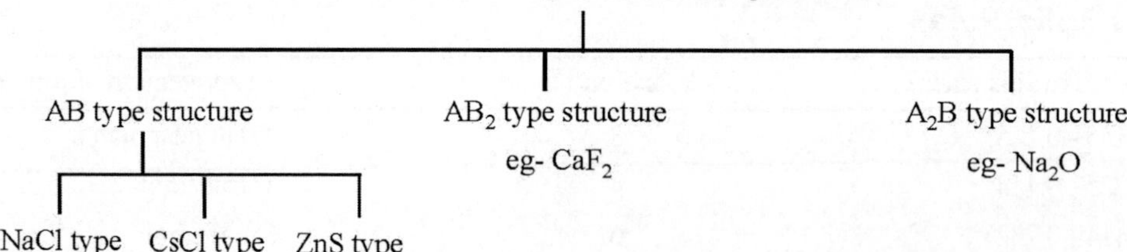

S. No.	Property	NaCl type (AB)	CsCl type (AB) structure	ZnS type (AB) structure (CaF_2 type)	Fluorite (AB_2) (Na_2O type)	Antifluorite (A_2B)
1.	Constituting ions	Na^+ and Cl^-	Cs^+ and Cl^-	Zn^{2+} and S^{2-}	Ca^{2+} and F^-	$2Na^+$ and O^{2-}
2.	Position of cation	Octahedral voids	Centre of unit cell	Zn^{++} hcp	Corner and cubic sites	Tetrahedral voids
3.	Position of anion	FCC	corners of BCC unit cell	Alternate tetrahedral voids	Centre of tetrahedral voids	A^-(CCP)
4.	Co-ordination no.	6 : 6	8 : 8	4 : 4	8 : 4	4 : 8
5.	No. of formula units/unit cell	4	1	4	4	4
6.	Examples	Halides of Li, Na, K, Rb, AgF, AgBr, NH_4Cl	CsCl, CsBr, CsI, TiCl, TiCN	ZnS, CuCl, CuBr, CuI, BeS, AgI	BeF_2, CaF_2, $BaCl_2$, SrF_2, PbF_2	Na_2O, Li_2O, K_2O, Na_2S

RELATION BETWEEN DENSITY AND EDGE OF CUBIC CRYSTALS

(i) For an element $\rho = \frac{M \times Z}{N_0 \times a^3 \times 10^{-30}}$ g/cm^3

where, ρ = density of cubic crystal, M = At. mass of element

Z = No. of atoms/unit cell 1 for simple, 2 for BCC and 4 for FCC.

(for ionic compounds its no. of formula units/cell e.g. 4 for NaCl and ZnS, 1 for CsCl)

a = edge length of cubic cell, N_0 = Avogadro's number

(ii) $\rho = \dfrac{Z \times M}{a^3 \times N_0}$ where M = molar mass in kg mol^{-1}

a = edge in metres, ρ = density in kg m^{-3}
If M = g mol^{-1}, a = edge in cm

$$\rho = \frac{Z \times M}{a^3 \times N_0} \text{ g/cm}^3$$

IMPERFECTIONS IN SOLIDS

•

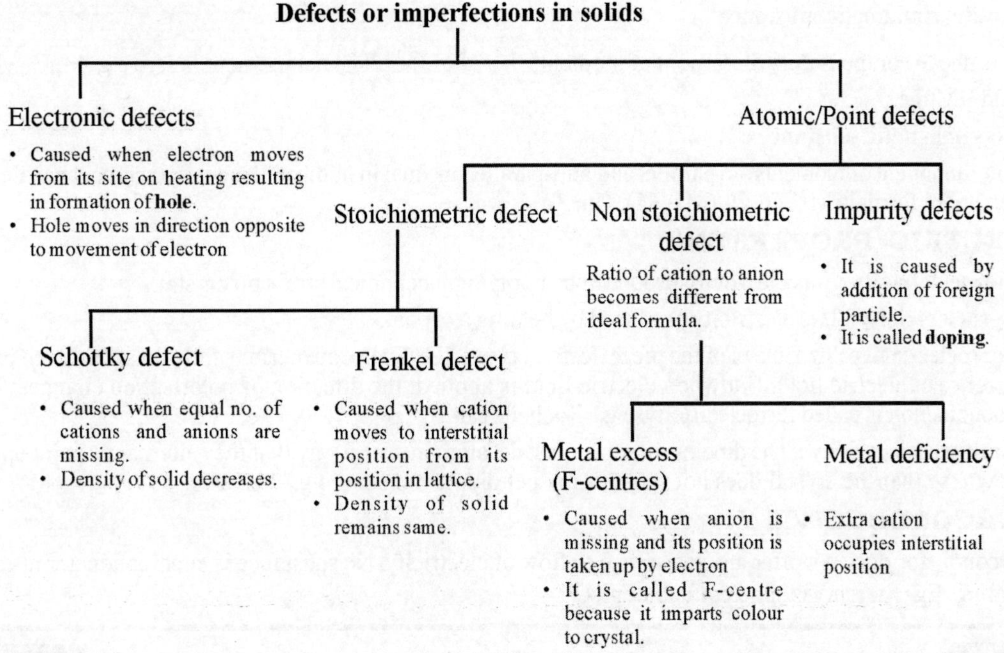

CLASSIFICATION OF SOLIDS (BASED ON THEIR ELECTRICAL CONDUCTIVITY)

(a) **Conductors:** Metals are good conductors of electricity having conductivities of the order of $10^7 \Omega^{-1}$ m^{-1}.

(b) **Insulators:** They have very low conductivities, ranging between 10^{-20} to $10^{-10} \Omega^{-1}$ m^{-1}.

(c) **Semi conductors:** Semiconductors having conductivities intermediate between those of conductors and insulators, varying between 10^{-6} to $10^4 \Omega^{-1}$ m^{-1}.

Types of semiconductors

- Pure substances that show conducting behaviour like that of silicon and germanium are called **intrinsic semiconductors.**
- **Extrinsic semiconductors** are obtained by adding an impurity to a pure substance, *i.e.*, to an intrinsic semiconductors.
- An ***n*-type semiconductor** is obtained by doping silicon with a group 15 element (P, As, Sb). The impurity provides an excess of electrons.
- An ***p*-type semiconductor** is obtained by doping silicon with a group 13 element (B, Al, Ca, In). The impurity provides an excess of positive charge which are termed as holes.

MAGNETIC PROPERTIES OF SOLIDS

Related to their behaviour in magnetic fields.

(1) **Paramagnetic substances**
- Attracted by magnetic field.
- Unpaired electrons present.

(2) **Diamagnetic substances**
- Weakly repelled in magnetic field.
- No unpaired electron (all paired electrons).

(3) **Ferromagnetic substances**
- Show permanent magnetism even in absence of magnetic field.
- Example - Fe, Co, Ni.

(4) **Antiferrimagnetic substance**

It is due to compensatory alignment of moments $\uparrow\downarrow\ \uparrow\downarrow\ \uparrow\downarrow$. The net moment is zero e.g. MnO, MnSe and $KMnFe_3$.

(5) **Ferrimagnetic substances**

The alignment of moments are parallel and anti-parallel unequal in number, there is net moment e.g. Fe_3O_4, ferrites of formula $M^{2+}Fe_2O_4$ (M = Mg, Cu, Zn)

DIELECTRIC PROPERTIES

(i) **Piezoelectricity :** It is electricity produced by applying mechanical stress on crystal.

(ii) **Pyroelectricity :** It is electricity produced by heating a crystal.

(iii) **Ferroelectricity :** In some of of the piezoelectric crystal, the dipoles are permanently polarized even in the absence of electric field. But when electric field is applied, the direction of polarisation changes. This phenomenon is called ferroelectricity. e.g.- Rochelle salt.

(iv) **Antiferroelectricity :** The dipoles, in some crystals, align in such a way that they alternately point up and down, so that the crystal does not possess any net dipole moment. e.g.- $PbZnO_3$.

SUPERCONDUCTIVITY

A superconductor does not offer any resistance to flow of electricity. No substance is superconductor at room temperature. Eg- Mercury is superconductor at 4 K.

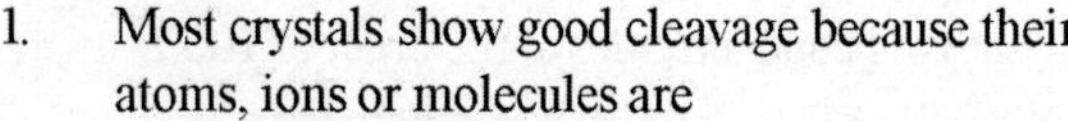

Exercise

1. Most crystals show good cleavage because their atoms, ions or molecules are
   (a) weakly bonded together
   (b) strongly bonded together
   (c) spherically symmetrical
   (d) arranged in planes.
2. Which of the following compounds is a good conductor of electricity in solution state?
   (a) Covalent
   (b) Molecular solid
   (c) Metallic solid
   (d) Ionic compounds
3. A compound formed by elements X and Y crystallizes in a cubic structure in which the X atoms are at the corners of a cube and the Y atoms are at the face centres. The formula of the compound is
   (a) XY_3 (b) X_3Y
   (c) XY (d) XY_2
4. Potassium crystallizes with a
   (a) Body-centred cubic lattice
   (b) Face-centred cubic lattice
   (c) Simple cubic lattice
   (d) Orthorhombic lattice
5. Solid CH_4 is
   (a) ionic solid (b) covalent solid
   (c) molecular solid (d) does not exist

6. Cells present in a cube-shaped ideal crystal of NaCl of mass 1.00 g are [Atomic masses : Na = 23, Cl = 35.5]
(a) 5.14×10^{21} unit cells
(b) 1.28×10^{21} unit cells
(c) 1.71×10^{21} unit cells
(d) 2.57×10^{21} unit cells

7. The fraction of total volume occupied by the atoms present in a simple cube is
(a) $\frac{\pi}{3\sqrt{2}}$ (b) $\frac{\pi}{4\sqrt{2}}$
(c) $\frac{\pi}{4}$ (d) $\frac{\pi}{6}$

8. Percentage of free space in a body centred cubic unit cell is :
(a) 30% (b) 32%
(c) 34% (d) 28%

9. Lithium metal crystallises in a body centred cubic crystal. If the length of the side of the unit cell of lithium is 351 pm, the atomic radius of the lithium will be:
(a) 151.8 pm (b) 75.5 pm
(c) 300.5 pm (d) 240.8 pm

10. What type of crystal defect is indicated in the diagram below ?

$Na^+Cl^-Na^+Cl^-Na^+Cl^-$
Cl^- □ Cl^- Na^+ □ Na^+
Na^+Cl^- □Cl^- Na^+ Cl^-
Cl^- Na^+ Cl^-Na^+ □ Na^+

(a) Interstitial defect
(b) Schottky defect
(c) Frenkel defect
(d) Frenkel & Schottky defects

11. When electrons are trapped into the crystal in anion vacancy, the defect is known as :
(a) Schottky defect
(b) Frenkel defect
(c) Stoichiometric defect
(d) F-centres

12. On doping Ge metal with a little of In or Ga, one gets
(a) p-type semi conductor
(b) n-type semi conductor
(c) insulator
(d) rectifier

13. With which one of the following elements silicon should be doped so as to give p-type of semiconductor ?
(a) Germanium (b) Arsenic
(c) Selenium (d) Boron

14. Schottky defect defines imperfection in the lattice structure of
(a) solid (b) gas
(c) liquid (d) plasma

15. Schottky defect in crystals is observed when
(a) an ion leaves its normal site and occupies an interstitial site
(b) unequal number of cations and anions are missing from the lattice
(c) density of the crystal is increased
(d) equal number of cations and anions are missing from the lattice

16. The appearance of colour in solid alkali metal halides is generally due to
(a) Schottky defect
(b) Frenkel defect
(c) Interstitial positions
(d) F-centres

17. The ability of a substances to assume two or more crystalline structures is called
(a) Isomerism (b) Polymorphism
(c) Isomorphism (d) Amorphism

18. In graphite electrons are :
(a) localised on each carbon atom
(b) spread out between the sheets
(c) localised on every third carbon atom
(d) present in antibonding orbital.

19. A solid with high electrical and thermal conductivity from the following is
(a) Si (b) Li
(c) NaCl (d) Ice

20. Among the following which one has the highest cation to anion size ratio?
(a) NaF (b) CsI
(c) CsF (d) LiF

21. How many Cl^- ions are there around Na^+ ion in NaCl crystal ?
(a) 3 (b) 4
(c) 6 (d) 8

22. Doping of silicon (Si) with boron (B) leads to :
(a) *n*-type semiconductor
(b) *p*-type semiconductor
(c) metal
(d) insulator

23. Each of the following solids show, the Frenkel defect except
(a) ZnS (b) AgBr
(c) AgI (d) KCl

24. Lattice energy of a solid increases if :
(a) size of ions are small
(b) charges of ions are small
(c) ions are neutral
(d) none of the above

25. The radius ratio in CsCl is 0.93. The expected lattice structure is
(a) tetrahedral
(b) square planar
(c) octahedral
(d) body-centred cubic

26. In Alternate tetrahedral voids are occupied in crystals
(a) NaCl (b) ZnS
(c) CaF_2 (d) Na_2O

27. The co-ordination number of a metal crystallising in a hexagonal close packed strcuture is
(a) 12 (b) 4
(c) 8 (d) 6

28. The limiting radius ratio for tetrahedral shape is:
(a) 0 to 0.155 (b) 0.225 to 0.414
(c) 0.155 to 0.225 (d) 0.414 to 0.732

29. Which of the following has maximum value of cation/anion ratio?
(a) KCl (b) NaCl
(c) CaF_2 (d) $MgCl_2$

30. In the fluorite structure, the coordination number of Ca^{2+} ion is :
(a) 4 (b) 6
(c) 8 (d) 3

31. In the solid state, MgO has the same structure as that of sodium chloride. The number of oxygens surrounding each magnesium in MgO is
(a) 6 (b) 1
(c) 2 (d) 4

32. The intermetallic compound LiAg crystallizes in a cubic lattice in which both lithium and silver atoms have coordination number of eight. To what crystal class does the unit cell belong ?
(a) Simple cubic
(b) Face-centred cubic
(c) Body-centred cubic
(d) None

33. CsBr crystallises in a body centered cubic lattice. The unit cell length is 436.6 pm. Given that the atomic mass of Cs = 133 and that of Br = 80 amu and Avogadro number being 6.02×10^{23} mol^{-1}, the density of CsBr is
(a) 0.425 g/cm^3 (b) 8.25 g/cm^3
(c) 4.25 g/cm^3 (d) 42.5 g/cm^3

34. Which of the following is the incorrect statement?
(a) NaCl has 6 : 6 coordination and CsCl has 8 : 8 coordination.
(b) In Na_2O each oxide ion is coordinated by $8Na^+$ ions and each Na^+ ion by 4 oxide ions
(c) NaCl structure transform to CsCl structure on heating
(d) In CaF_2 structure each F^- ion is coordinated by 4 Ca^{2+} and vice-versa.

35. Which of the following is a correct statement ?
(a) Bonding in metallic crystals is directional
(b) Diamond has two dimensional crystal lattice
(c) Co-ordination number of bcc lattice is 12
(d) A ccp structure has 8 tetrahedral and 4 Octahedral interstices.

36. To get n-type doped semiconductor, impurity to be added to silicon should have the following number of valence electrons
(a) 2 (b) 5
(c) 3 (d) 1

37. The r_+/r_- ratio of ZnS is 0.402. Pick out the false statements of the following
(a) ZnS is 4 : 4 coordination compound.
(b) ZnS does not crystallize in rock salt type lattice because r_+ / r_- is too small to avoid overlapping of S^{2-} ions.
(c) Zn^{2+} ion is too small to fit precisely into the octahedral voids of S^{2-} ions.
(d) Zn^{2+} ion is too large to fit into the octahedral voids of S^{2-} ions.

38. The unit cell dimensions of a cubic lattice (edges a, b, c and the angles between them, α, β and γ) are
(a) $a = b = c, \alpha = \beta = \gamma = 90^0$
(b) $a = b \neq c, \alpha = \beta = \gamma = 90^0$
(c) $a = b = c, \alpha = \gamma = 90^0, \beta \neq 90^0$
(d) $a \neq b \neq c, \alpha = \beta = 90^0, \gamma \neq 90^0$

39. The second order Bragg's diffraction of X-rays of wavelength 100 pm from a set of parallel lattice planes in a metal occurs at a grazing angle of 30°. The spacing between the successive scattering planes in the cystal is
(a) 50 pm (b) 100 pm
(c) 150 pm (d) 200 pm

40. A NaCl crystal is changed into CsCl type structure by:
(a) Increasing temperature and decreasing pressure.
(b) Increasing pressure and decreasing temperature.
(c) Increasing both temperature and pressure
(d) None of these.

41. Non stoichiometric defects are formed by
(a) s- block elements
(b) p-block elements
(c) either s-block elements or d-block elements
(d) only d-block elements.

42. M_2X has anti fluorite structure. In such structure
(a) X^{--} ions occupy all the 8 octahedral voids
(b) Each X^{--} is surrounded by 4 M^+ in tetrahedral arrangement
(c) Larger cations occupy the position of F^- ions and smaller anions that of C^{++} ions
(d) Smaller cations occupy the position of F^- ions and larger anions that of C^{++} ions

43. A metallic element exists as cubic lattice. Each edge of the unit cell is 2.88 Å. The density of the metal is 7.20 g cm^{-3}. How many unit cell will be present in 100 g of the metal?
(a) 6.85×10^2 (b) 5.82×10^{23}
(c) 4.37×10^5 (d) 2.12×10^6

44. For orthorhombic system axial ratios are $a \neq b \neq c$ and the axial angles are
(a) $\alpha = \beta = \gamma \neq 90°$
(b) $\alpha = \beta = \gamma = 90°$
(c) $\alpha = \beta = \gamma = 90°, \beta = 90°$
(d) $\alpha \neq \beta \neq \gamma = 90°$

45. The pure crystalline substance on being heated gradually first forms a turbid liquid at constant temperature and still at higher temperature turbidity completely disappears. The behaviour is a characteristic of substance forming.
(a) Allotropic crystals
(b) Liquid crystals
(c) Isomeric crystals
(d) Isomorphous crystals

46. The edge length of a centred unit cubic cell is 508 pm. If the radius of the cation is 100 pm, the radius of the anion is
(a) 288 pm (b) 398 pm
(c) 154 pm (d) 618 pm

47. For two ionic solids CaO and KI, identify the wrong statement amongst the following :
(a) The lattice energy of CaO is much large than that of KI
(b) KI is more soluble in water
(c) KI has higher melting point
(d) CaO has higher melting point

48. The second order Bragg diffraction of X-rays with = 1.00 Å from a set of parallel planes in a metal occurs at an angle 60°. The distance between the scattering planes in the crystal is
(a) 0.575 Å (b) 1.00 Å
(c) 2.00 Å (d) 1.15 Å

49. When molten zinc is converted into solid state, it acquires HCP structure. The number of nearest neighbours will be
(a) 6 (b) 12
(c) 8 (d) 4

50. The pyknometric density of sodium chloride crystal is 2.165×10^3 kg m^{-3} while its X-ray density is 2.178×10^3 kg m^{-3}. The fraction of unoccupied sites in sodium chloride crystal is
(a) 5.96×10^{-3} (b) 5.96
(c) 5.96×10^{-2} (d) 5.96×10^{-1}

51. If NaCl is doped with 10^{-4} mol % of $SrCl_2$, the concentration of cation vacancies will be ($N_A = 6.02 \times 10^{23}$ mol^{-1})
(a) 6.02×10^{16} mol^{-1} (b) 6.02×10^{17} mol^{-1}
(c) 6.02×10^{14} mol^{-1} (d) 6.02×10^{15} mol^{-1}

52. Which of the folloiwng anions is present in the chain structure of silicates?
(a) $(Si_2O_5^{2-})_n$ (b) $(SiO_3^{2-})_n$
(c) SiO_4^{4-} (d) $Si_2O_7^{6-}$

53. If 'a' stands for the edge length of the cubic systems : simple cubic, body centred cubic and face centred cubic, then the ratio of radii of the spheres in these systems will be respectively,

(a) $\frac{1}{2}a : \frac{\sqrt{3}}{4}a : \frac{1}{2\sqrt{2}}a$

(b) $\frac{1}{2}a : \sqrt{3}a : \frac{1}{\sqrt{2}}a$

(c) $\frac{1}{2}a : \frac{\sqrt{3}}{2}a : \frac{\sqrt{3}}{2}a$

(d) $1a : \sqrt{3}a : \sqrt{2}a$

54. Copper crystallises in a face-centred cubic lattice with a unit cell length of 361 pm. What is the radius of copper atom in pm?

(a) 157 (b) 181
(c) 108 (d) 128

55. AB; crystallizes in a body centred cubic lattice with edge length '*a*' equal to 387 pm. The distance between two oppositely charged ions in the lattice is :

(a) 335 pm (b) 250 pm
(c) 200 pm (d) 300 pm

56. A solid compound XY has NaCl structure. If the radius of the cation is 100 pm, the radius of the anion (Y^-) will be :

(a) 275.1 pm (b) 322.5 pm
(c) 241.5 pm (d) 165.7 pm

57. The number of atoms in a face centred cubic unit cell are

(a) 4 (b) 6
(c) 8 (d) 12

58. For FCC if AB is just like the rock salt like structure then, A^+ and B^- are located at –

(a) A^+ - Tetrahedral voids ; B^- - Corner
(b) A^+ - Corner and faces ; B^- - Octahedral voids
(c) A^+ - Octahedral voids ; B^- - Corner and faces
(d) A^+ - Corner and faces ; B^- - Octahedral and tetrahedral voids

59. The ionic radii of X^+ and Y^- ions are 146 and 216 pm respectively. The probable type of structure shown by it is

(a) CsCl type (b) Rock salt type
(c) Zinc blend type (d) CaF_2 type

60. A solid is made of two elements X and Z. The atoms Z are in CCP arrangement while the atoms X occupy all the tetrahedral sites. What is the formula of the compound?

(a) XZ (b) XZ_2
(c) X_2Z (d) X_2Z_3

Solutions

CONCEPTS

SOLUTIONS

- A solution is a homogeneous mixture of two or more chemically non-reacting substance whose composition can be varied within certain limit. Binary solution has 2 components : Solute and Solvent
- Solute is component present in lesser quantity and solvent is present in larger proportion.

Types of Solutions

(a) On the basis of phase of solute or solvent

S.No.	Solute	Solvent	Type of solution	Examples
1	Solid	Solid	Solid in solid	Alloys
2	Liquid	Solid	Liquid in solid	Mercury in Zinc amalgam
3	Gas	Solid	Gas in solid	Dissolved gases in mineral
4	Solid	Liquid	Solid in liquid	Salt in water
5	Liquid	Liquid	Liquid in Liquid	Alcohol in water
6	Gas	Liquid	Gas in Liquid	Aerated drinks
7	Solid	Gas	Solid in gas	Iodine vapour in air
8	Liquid	Gas	Liquid in gas	Humidity in air
9	Gas	Gas	Gas in gas	Air

(b) On the basis of concentration of solution :

(i) **Dilute solution :** A solution in which relatively a small amount of solute is dissolved in large amount of solvent is called a dilute solution.

(ii) **Concentrated solution :** A solution in which relatively a large amount of solute is dissolved in small amount of solvent is called a concentrated solution.

(iii) **Saturated solution :** A solution which remains in contact with excess of the solute is said to be saturated.

(iv) **Supersaturated solution :** A solution containing more amount of solute than that required for saturation of a given amount of solvent at a particular temperature, is called a supersaturated solution. It is unstable system.

SOLID SOLUTION

- Both components are solid

Solid solution

Substitutional solid solution	**Interstitial solid solution**
• One element takes position of other substance in crystal lattice.	• The other substance enters into smaller interstitial sites of host lattice
e.g., - Bronze	e.g., - Tungston carbide

SOLUBILITY

(i) It is amount of solute dissolved in 100g of a solvent.

(ii) A solid dissolves only when
Hydration energy > Lattice energy
Lattice energy is energy required to dissociate one mole of an ionic crystal into ions and **hydration energy** is amount of energy released when 1 mole of ions are hydrated.

$$\Delta H_{(solution)} = \Delta H_{(lattice)} - \Delta H_{(hydration)}$$

(iii) Molecular solids containing polar groups (e.g. OH) are soluble in polar solvents like water, methanol etc. Others are soluble in non polar solvents.

Factors Influencing the Solubility of a Gas

(i) Nature of the gas : Easily liquifiable gases are more soluble. e.g. CO_2 is more soluble than O_2 and H_2

(ii) Nature of solvent : Gases capable of forming ions in aqueous solution are more soluble in water than in other solvents. eg. HCl, NH_3 etc.

(iii) Temperature : Solubility decreases with rise of temperature at constant pressure.

(iv) Pressure of the gas (Henry's Law)
Henry's law : The mass of a gas dissolved per unit volume of solvent is proportional to the pressure of the gas at constant temperature.

$m \propto p$ or
$m = k.P$ where k is Henry's constant.

Henry's Law in terms of volume : The volume of a gas dissolved in a solvent at a given temperature is independent of the pressure.

(v) **Effect of temperature on the solubility of gases :** Dissolution of a gas in a liquid is generally an exothermic process i.e., heat is liberated during the dissolution. Thus, according to the Le-chatelier principle, it is expected that the solubility of a gas will decrease with increase of temperature. Experimentally, it is found to be so for most gases. The quantitative variation of solubility with temperature at constant pressure can be derived thermodynamically. Assuming the gas to be ideal the relation is :

$\frac{d \ln C}{dT} = \frac{\Delta H}{RT^2}$; where C is the molar concentration of gas in the liquid phase, and ΔH is the differential heat of solution of 1 mol of gas in a saturated solution at temperature T. If ΔH is assumed to be independent of temperature, then integration of above equation gives :

$$\ln \frac{C_2}{C_1} = -\frac{\Delta H}{R}\left(\frac{1}{T_2} - \frac{1}{T_1}\right)$$

where C_1 and C_2 are the concentration of the solution at T_1 and T_2 respectively.

METHODS OF EXPRESSING CONCENTRATION

(i) **Weight percentage** : It is wt. of solute per 100 g of solution (W/W)

(ii) **Volume percentage :** It is the volume of solute / 100 ml of solution (V/V)

(iii) **Weight-volume percentage :** It is wt of solute / 100 ml of solution (W/V)

(iv) **Parts per million (ppm)**

$$\text{ppm} = \frac{\text{mass of solute}}{\text{total mass of solution}} \times 10^6$$

(iii) **Normality (N) :** Number of gram equivalents of the solute dissolved per litre of the solution.
Normality(N)

$$= \frac{\text{wt. of solute} \times 1000}{\text{E. wt of solute} \times \text{volume in mL}}$$

(iv) **Molarity (M) :** Number of moles of the solute dissolved in one litre of solution.
Molarity (M)

$$= \frac{\text{wt. of solute} \times 1000}{\text{M. wt of solute} \times \text{volume in mL}}$$

(v) **Molality (m) :** Number of moles of solute dissolved in 1000 g of the solvent.
Molality (m)

$$= \frac{\text{wt. of solute} \times 1000 \text{ g solvent}}{\text{M. wt of solute} \times \text{wt. of solvent in grams}}$$

(vi) **Mole fraction :** The ratio of the number of moles of one component to the total number of moles of solution. For components A and B.

$$x_A = \frac{n_A}{n_A + n_B};\ x_B = \frac{n_B}{n_A + n_B};\ x_A + x_B = 1 \text{ always.}$$

(vii) Formality

$$= \frac{\text{weight of solute (gm)}}{\text{formula weight of solute}} \times \frac{1}{\text{volume of solution}}$$

(viii) Mole percentage = Mole fraction × 100

Some Important Relations

(i) $M = \frac{md}{1 + m\frac{M_2}{1000}}$ or $d = M\left(\frac{1}{m} + \frac{M_2}{100}\right)$

(ii) $m = \frac{1000x_2}{x_1 M_1}$

(iii) $M = \frac{1000\, dx_2}{x_1 M_1 + x_2 M_2}$

(iv) Mole fraction $(x_2) = \frac{m}{m + \frac{1000}{M_1}}$

M = molarity
m = molality
d = density of solution
x_1 = mole fraction of solvent
x_2 = mole fraction of solute
M_1 = molecular mass of solvent
M_2 = molecular mass of solute
Normality = n × Molarity

where $n = \frac{\text{Molecular Mass}}{\text{Equivalent mass}}$

- **Molality and mole fraction are preferred to express concentration** of solution because they do not involve volume terms and do not change with temperature.

VAPOUR PRESSURE

- It is pressure exerted by vapour in equilibrium with liquid at a given temperature.
- Vapour pressure

$$\propto \frac{1}{\text{forces of attraction between molecules}}$$

- Vapour pressure depends on following factors :
  (a) Nature of solvent
  (b) Temperature
  (c) %age purity of liquid or surface area of liquid
- **Variation of Vapour Pressure with Temperature:**
  It is given by Clapeyron-Clausius Equation

$$\log\frac{P_2}{P_1} = \frac{\Delta H_V}{2.303R}\left[\frac{1}{T_1} - \frac{1}{T_2}\right]$$

ΔH_v = Enthalpy of vapourisation of liquid;
R = gas constant;
P_2 = vapour pressure at T_2 and
P_1 = vapour pressure at T_1.

RAOULT's LAW

(i) The vapour pressure of the solvent in a solution is equal to the product of its mole fraction and its vapour pressure in the pure state at the same temperature.
$P_S = P_S° \times x$ $x \rightarrow$ mole fraction

(ii) Vapour pressure of pure solvent decreases when solute is added into it.

(iii) **Raoult's Law for volatile solutes**
For solution having two component A & B
$P_{total} = p_A + p_B$

$$= p_A^\circ x_A + p_B^\circ x_B = (p_B^\circ - p_A^\circ)x_B + p_A^\circ$$

(iv) **Raoults Law for non volatile solute**

$$\frac{p_A^\circ - p_A}{p_A^\circ} = x_B = \frac{n}{n+N}$$

where x_B = mole fraction of solute
p_A° = Vapour pressure of pure solvent
p_A = Vapour pressure of solution.
$n \rightarrow$ moles of B
$N \rightarrow$ moles of A

(v) **Limitations of Raoult's law :**
(a) As described earlier, Raoult's law is applicable only to very dilute solutions.
(b) Raoult's law is applicable to solutions containing non-volatile solute only.
(c) Raoult's law is not applicable to solutes which dissociates or associate in the particular solution

IDEAL AND NON-IDEAL SOLUTIONS

Ideal solutions	Non ideal solutions
1. Follows Raoult's law at all temperature and pressure	Does not follow Raoults law at all temperature and pressure
2. $\Delta V_{mix} = 0$	$\Delta V_{mix} \neq 0$
3. $\Delta H_{mix} = 0$	$\Delta H_{mix} \neq 0$
4. e.g. - hexane and heptane $C_2H_5Br + C_2H_5Cl$	e.g., water + C_2H_5OH, HNO_3 + water

Types of Non-ideal Solution

Non ideal solution showing positive deviation	Non ideal solution showing negative deviation
1. $p_A > p_A^\circ x_A,\ p_B > p_B^\circ x_B$ $P > p_A^\circ x_A + p_B^\circ x_B$	1. $p_A > p_A^\circ x_A,\ p_B > p_B^\circ x_B$ $P < p_A^\circ x_A + p_B^\circ x_B$
2. ΔH_{mix} and ΔV_{mix} are positive	2. ΔH_{mix} and ΔV_{mix} are negative
3. New interaction are weaker than those existing in pure components.	3. New interactions are stronger than those existing in pure components.
4. e.g., CH_3COCH_3 + C_2H_5OH, $H_2O + C_2H_5OH$	4. e.g., $H_2SO_4 + H_2O$, $CH_3COCH_3 + CHCl_3$

Graphical representation of ideal and non ideal solutions

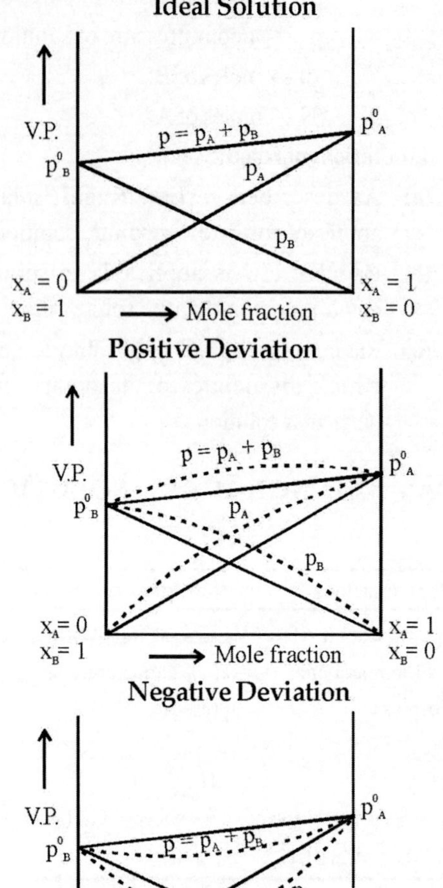

AZEOTROPIC MIXTURES

Mixtures of two liquids which boil at constant temperature and are distilled without changing composition. Non ideal solutions form azeotropic mixture.

Types of Azeotrops

S.No	Properties	Minimum boiling Azeotrops	Maximum boiling Azeotrops
1	Boiling point	less than either of two pure components	more than either of two pure components.
2	Composition	Non ideal solution showing positive deviation	Non ideal solution showing negative deviation.
3	e.g.	95.5% C_2H_5OH+4.5% H_2O boils at 351.15K	68% HNO_3 + 32% H_2O boils at 393.5 K

COLLIGATIVE PROPERTIES

Those properties of solution which **depend only on the number of solute particles** and **not on nature of solute particles**. These are

(i) Relative lowering of vapour pressure
(ii) Osmotic pressure
(iii) Elevation in boiling point
(iv) Depression in freezing point

Osmosis

(a) It is movement of solvent particles from region of their higher concentration to lower concentration across semi-permeable membrane.

(b) **Osmotic pressure** is the minimum pressure which must be applied on the solution to prevent the entry of solvent into the solution through semipermeable membrane. It is **measured** by **Berkeley and Hartly method**.

(c) **Types of solutions**

S.No.	Isotonic Solution	Hypotonic Solution	Hypertonic solution
1	The solutions have same osmotic pressure	The solution has lower osmotic pressure than other across semipermeable membrane	The solution has higher osmotic pressure than other across semipermeable membrane
2	e.g Human RBC and 0.9% NaCl	Human RBC and < 0.91% NaCl	Human RBC and > 0.91% NaCl
3	The object is not effected	The object swells and even bursts	The object shrinks

(d) (i) **Exosmosis** - Water comes out from cell into solution.

(ii) **Endosmosis** - Water flows inside cell from solution.

(e) **Van't Hoff law for dilute solution :** The osmotic pressure (P or π) of a solution is directly proportional to its concentration (C) when the temperature is kept constant. The concentration of the solution contain one gram mole in V litres is equal to 1/V.

thus $\pi \propto C$ (when temp. is const.)

or $\pi \propto \frac{1}{V}$; $\pi V = \text{constant}$

(f) **Calculation of osmotic pressure (π)**

$$\pi = \frac{n}{v}RT = CRT = \frac{wRT}{MV} \qquad \left(n = \frac{w}{M}\right)$$

$\left(C = \frac{n}{V}\right)$

For isotonic solution, $\pi_1 = \pi_2$

$$\therefore \quad \frac{n_1}{v_1} = \frac{n_2}{v_2} \quad \text{or} \quad \frac{w_1}{M_1V_1} = \frac{w_2}{M_2V_2}$$

- **Reverse Osmosis :** If a pressure higher than osmotic pressure is applied on the solution, the solvent will flow from the solution into the pure solvent through the semipermeable membrane. Since here the flow of solvent is in the reverse direction to that observed in the usual osmosis, the process is called reverse osmosis.

Relative Lowering of Vapour Pressure

When a non-volatile solute is dissolved in a pure solvents, the vapour pressure of the solvent is lowered i.e. the vapour pressure of a solution is always lower than that of pure solvents, because the escaping tendency of solvents molecules decreases (due to lesser surface area of solutions available for evaporation).

If at a certain temperature P^0 is the pressure of pure solvent, and P_S is the vapour pressure of solution then

Lowering of vapour pressure $= P^0 - P_S$;

Relative lowering of vapour pressure $= \frac{P^0 - P_S}{P^0}$

; From Raoult's law, $\frac{P^0 - P_S}{P^0} = \frac{n_A}{n_A + n_B}$

For a more dilute solution $n_A <<< n_B$; so

$$\frac{P^0 - P_S}{P^0} = \frac{n_A}{n_B}; \quad \frac{\Delta P}{P^0} = \frac{P^0 - P_S}{P^0} = \frac{n_A}{n_B};$$

$$\frac{\Delta P}{P^0} = \frac{n_A}{n_B}$$

or relative lowering of vapour pressure =

$$\frac{P^0 - P_S}{P^0} = \frac{n_A}{n_B}$$

Elevation in Boiling Point

(i) Boiling point is the temperature at which vapour pressure = atmospheric pressure

(ii) $\Delta T_b = T_b - T_0$

where ΔT_b = elevation in boiling point

T_b = Boiling point of solution

T_0 = Boiling point of pure solvent

For a dilute solution $\Delta T_b \propto m$

$\Rightarrow \Delta T_b = K_b \times m$ (m = molality)

(iii) $\Delta T_b = \frac{K_b \times w_B \times 1000}{M_B \times W_A}$

K_b = molal elevation constant or ebullioscopic constant

w_B = wt. of solute

w_A = wt. of solvent

M_B = molecular wt. of solute

m = molality

$$K_b = \frac{RT_o^2}{1000L_v} = \frac{M_A RT_o^2}{\Delta H_{vap} \times 1000}$$

where L_v = latent heat of vapourisation per gram of the solvent

Depression in Freezing Point

(i) **Freezing point** of a substance is the temperature at which the liquid and solid phases are in equilibrium and hence have same vapour pressure.

(ii) $\Delta T_f = T_f^\circ - T_f$

where ΔT_f = depression in freezing point

T_f° = freezing point of pure solvent

T_f = Freezing point of solution

(iii) $\Delta T_f \propto m$

$$\Delta T_f = K_f \times m = \frac{K_f \times w_B \times 100}{M_B \times w_A}$$

where K_f = molal depression constant or cryoscopic constant.

(iv) $K_f = \frac{RT_o^2}{1000L_f} = \frac{M_A RT_o^2}{\Delta H_{Fusion} \times 1000}$

where L_f = latent heat of fusion pen gram of the solvent

VAN'T HOFF FACTOR

$$i = \frac{\text{Observed value of colligative property}}{\text{Normal value of colligative property}}$$

$$\text{Molecular mass} \propto \frac{1}{\text{colligative property}}$$

$$\therefore i = \frac{\text{Normal molecular mass}}{\text{Observed molecular mass}}$$

i = solute remain uneffected

i > 1 for dissociation

i < 1 for association

Modified Formula for Substances which Associate or Dissociate

$\pi = iCRT$

$\Delta T_b = iK_b m$

$\Delta T_f = iK_f m$

Van't Hoff Factor and Degree of Dissociation

$$(A)_n \longrightarrow nA$$

Initial moles 1 0

At equilibrium $1-\alpha$ $n\alpha$

Total moles at equilibrium $= 1-\alpha+n\alpha$

$$\therefore \quad i = \frac{1-\alpha+n\alpha}{1} \quad \text{or} \quad \alpha = \frac{i-1}{n-1}$$

Van't Hoff Factor and Degree of Association

$$nA \longrightarrow (A)_n$$

Initial moles 1 0

At equilibrium $1-\alpha$ $\frac{\alpha}{n}$

Total moles at equilibrium $= 1-\alpha+\frac{\alpha}{n}$

$$i = \frac{1-\alpha+\alpha/n}{1}$$

$$\therefore \quad \alpha = \frac{i-1}{\frac{1}{n}-1}$$

DISTRIBUTION LAW (BY NERNST)

(i) A solute distributes itself between two immiscible solvents in such a way that the ratio of the concentrations of the solute in the two solvents is constant at constant temperature provided that the solute does not undergo any dissociation or association in any of the solvents.

Mathematically

$$\frac{\text{concentration of solute in solvent A}}{\text{concentration of solute in solvent B}} = \frac{C_1}{C_2} = K_D$$

where K_D = distribution coefficient or partition coefficient

(ii) **Distribution law is used in the process of solvent intraction** Effeciency of process of extraction depends upon -

- **Distribution ratio** and **number of extractions. Multiple extractions are more effective than single step extraction.**

Exercise

1. Which of the following concentration unit is independent of temperature ?
   (a) Normality (b) Molarity
   (c) Formality (d) Molality
2. Isotonic solutions have same
   (a) Molar concentration
   (b) Molality
   (c) Normality
   (d) None of these
3. By increasing the temperature, the vapour pressure of a substance
   (a) always increases
   (b) does not depend on temperature
   (c) always decreases
   (d) partially depends on temperature
4. The vapour pressure of two liquids 'P' and 'Q' are 80 and 60 torr, respectively. The total vapour pressure of solution obtained by mixing 3 mole of P and 2 mole of Q would be
   (a) 72 torr (b) 140 torr
   (c) 68 torr (d) 20 torr
5. When mercuric iodide is added to the aqueous solution of potassium iodide then
   (a) freezing point is raised.
   (b) freezing point is lowered.
   (c) freezing point does not change.
   (d) boiling point does not change.
6. An ideal solution is formed when its components
   (a) have no volume change on mixing
   (b) have no enthalpy change on mixing
   (c) have both the above characteristics
   (d) have high solubility.
7. Colligative properties of the solution depend on
   (a) Nature of solute
   (b) Nature of solvent
   (c) Number of particles present in the solution
   (d) Number of moles of solvent only
8. Which is not a colligative property ?
   (a) Freezing point
   (b) Lowering of vapour pressure
   (c) Depression of freezing point
   (d) Elevation of boiling point
9. A solution containing 10 g per dm^3 of urea (molecular mass = 60 g mol^{-1}) is isotonic with a

5% solution of a non-volatile solute. The molecular mass of this nonvolatile solute is

(a) $300\ g\ mol^{-1}$ (b) $350\ g\ mol^{-1}$
(c) $200\ g\ mol^{-1}$ (d) $250\ g\ mol^{-1}$

10. The van't Hoff factor i for a compound which undergoes dissociation in one solvent and association in other solvent is respectively :
(a) less than one and greater than one.
(b) less than one and less than one.
(c) greater than one and less than one.
(d) greater than one and greater than one.

11. A solution of acetone in ethanol
(a) shows a positive deviation from Raoult's law
(b) behaves like a non ideal solution
(c) obeys Raoult's law
(d) shows a negative deviation from Raoult's law

12. During osmosis, flow of water through a semipermeable membrane is
(a) from both sides of semipermeable membrane with equal flow rates
(b) from both sides of semipermeable membrane with unequal flow rates
(c) from solution having lower concentration only
(d) from solution having higher concentration only

13. Which one of the following salts will have the same value of van't Hoff factor (i) as that of $K_4[Fe(CN)_6]$.
(a) $Al_2(SO_4)_3$ (b) NaCl
(c) $Al(NO_3)_3$ (d) Na_2SO_4.

14. For a dilute solution, Raoult's law states that
(a) the lowering of vapour pressure is equal to the mole fraction of the solute
(b) the relative lowering of vapour pressure is equal to the mole fraction of the solute
(c) the relative lowering of vapour pressure is proportional to the amount of solute in solution
(d) the vapour pressure of the solution is equal to the mole fraction of the solvent

15. The osmotic pressures of equimolar solution of $BaCl_2$, NaCl and sucrose will be in the order
(a) Sucrose > NaCl > $BaCl_2$
(b) Sucrose > $BaCl_2$ > NaCl
(c) NaCl > $BaCl_2$ > Sucrose
(d) $BaCl_2$ > NaCl > Sucrose

16. Pressure cooker reduces cooking time because
(a) the heat is more easily distributed
(b) the higher pressure tenderizes the food
(c) the boiling point of the water inside is elevated
(d) a larger flame is used

17. All form ideal solution except
(a) C_6H_6 and $C_6H_5CH_3$
(b) C_2H_6 and C_2H_5I
(c) C_6H_5Cl and C_6H_5Br
(d) C_2H_5I and C_2H_5OH.

18. Which of the following aqueous solution has minimum freezing point ?
(a) 0.01 m NaCl (b) 0.005 m C_2H_5OH
(c) 0.005 m MgI_2 (d) 0.005 m $MgSO_4$.

19. Which one is a colligative property ?
(a) boiling point (b) vapour pressure
(c) osmotic pressure (d) freezing point

20. A solution contains non-volatile solute of molecular mass M_2. Which of the following can be used to calculate the molecular mass of solute in terms of osmotic pressure ?

(a) $M_2 = \left(\frac{m_2}{\pi}\right) VRT$

(b) $M_2 = \left(\frac{m_2}{V}\right) \frac{RT}{\pi}$

(c) $M_2 = \left(\frac{m_2}{V}\right) \pi RT$

(d) $M_2 = \left(\frac{m_2}{V}\right) \frac{\pi}{RT}$

21. Camphor is often used in molecular mass determination because
(a) it is readily available
(b) it has a very high cryoscopic constant
(c) it is volatile
(d) it is solvent for organic substances

22. The mole fraction of the solute in one molal aqueous solution is:
(a) 0.009 (b) 0.018
(c) 0.027 (d) 0.036

23. In mixture A and B components show -ve deviation as
(a) $\Delta V_{mix} > 0$
(b) $\Delta H_{mix} < 0$
(c) $A - B$ interaction is weaker than $A - A$ and $B - B$ interaction
(d) $A - B$ interaction is stronger than $A - A$ and $B - B$ interaction.

24. Which of the following liquid pairs shows a positive deviation from Raoult's law ?
(a) Water - nitric acid
(b) Benzene - methanol
(c) Water - hydrochloric acid
(d) Acetone - chloroform

25. Equimolar solutions in the same solvent have
(a) Different boiling and different freezing points
(b) Same boiling and same freezing points
(c) Same freezing point but different boiling points
(d) Same boiling point but different freezing points

26. The molecular weight of benzoic acid in benzene as determined by depression in freezing point method corresponds to :
(a) ionization of benzoic acid.
(b) dimerization of benzoic acid.
(c) trimerization of benzoic acid.
(d) solvation of benzoic acid.

27. As a result of osmosis, the volume of the solution
(a) gradually decreases
(b) gradually increases
(c) is not affected
(d) suddenly increases

28. If A contains 2% NaCl and is separated by a semipermeable membrane from B which contains 10% NaCl, then which event will occur ?
(a) NaCl will flow from A to B
(b) NaCl will flow from B to A
(c) Water will net flow from A to B
(d) Water will net flow from B to A

29. Which of the following pairs of solution are isotonic at the same temperature ?
(a) 0.1 M $Ca(NO_3)_2$ and 0.1 M Na_2SO_4
(b) 0.1 M NaCl and 0.1 M Na_2SO_4
(c) 0.1 M urea and 0.1 M $MgCl_2$
(d) 0.2 M urea and 0.1 M NaCl

30. Which one of the following is not correct for an ideal solution?
(a) It must obey Raoult's law
(b) $\Delta H = 0$
(c) $\Delta C = 0$
(d) $\Delta H = \Delta V \neq 0$

31. Osmotic pressure of a solution at a given temperature
(a) increases with concentration
(b) decreases with concentration
(c) remains same
(d) initially increases and then decreases

32. For an electrolyte, elevation of B.P. is directly proportional to :
(a) molarity (b) molality
(c) mole fraction (d) all of these

33. Formation of a solution from two components can be considered as
(i) Pure solvent $\rightarrow$ separated solvent molecules, ΔH_1
(ii) Pure solute $\rightarrow$ separated solute molecules, ΔH_2
(iii) Separted solvent & solute molecules $\rightarrow$ Solution, ΔH_3
Solution so formed will be ideal if
(a) $\Delta H_{soln} = \Delta H_3 - \Delta H_1 - \Delta H_2$
(b) $\Delta H_{soln} = \Delta H_1 + \Delta H_2 + \Delta H_3$
(c) $\Delta H_{soln} = \Delta H_1 + \Delta H_2 - \Delta H_3$
(d) $\Delta H_{soln} = \Delta H_1 - \Delta H_2 - \Delta H_3$

34. Which of the following colligative property can provide molar mass of proteins (or polymers or colloids) with greatest precision ?
(a) Osmotic pressure
(b) Elevation of boiling point
(c) Depression of freezing point
(d) Relative lowering of vapour pressure

35. The vapour pressure of a solvent decreased by 10mm of mercury when a non-volatile solute was added to the solvent. The mole fraction of the solute in the solution is 0.2. What should be the mole fraction of the solvent if the decrease in the vapour pressure is to be 20mm of mercury?
(a) 0.8 (b) 0.6
(c) 0.4 (d) 0.2

36. A solution of urea (mol. mass 56 g mol^{-1}) boils at 100.18°C at the atmospheric pressure. If K_f and K_b for water are 1.86 and 0.512 K kg mol^{-1} respectively, the above solution will freeze at
(a) 0.654°C (b) –0.654°C
(c) 6.54°C (d) –6.54°C

37. The normality of 10% (weight/volume) acetic acid is
(a) 1 N (b) 10 N
(c) 1.7 N (d) 0.83 N

38. Osmotic pressure of a sugar solution at 24°C is 2.5 atmosphere. The concentration of the solution in gm mole per litre is
(a) 10.25 (b) 1.025
(c) 102.5 (d) 0.1025

39. The freezing point of a solution prepared from 1.25 gm of a non electrolyte and 20 gm of water is 271.94 *K*. If K_f is 1.86 $Kmol^{-1}$. Then the molar mass of the solute will be nearly :
(a) 109 (b) 106
(c) 115 (d) 93

40. A solution is obtained by dissolving 12 gm of urea (mol. wt. 60) in a litre of water. Another solution is obtained by dissolving 68.4 gm of cane sugar (mol. wt. 342) in a litre of water at the same temperature. The lowering of vapour pressure in the first solution is
(a) same as that of 2nd solution
(b) nearly one-fifth of the 2nd solution
(c) double that of 2nd solution
(d) nearly five times that of 2nd solution

41. The 5.85 g of NaCl and one kg of water is added to prepare a solution. What is the strength of NaCl in this solution ?
(Molecular weight of NaCl = 58.5)
(a) 0.1 Normal (b) 0.1 Molal
(c) 0.1 Molar (d) 0.1 Formal

42. The relative lowering of the vapour pressure is equal to the ratio between the number of
(a) solute molecules to the solvent molecules
(b) solute molecules to the total molecules in the solution
(c) solvent molecules to the total molecules in the solution
(d) solvent molecules to the total number of ions of the solute.

43. Blood cells retain their normal shape in solution which are
(a) hypotonic to blood
(b) isotonic to blood
(c) hypertonic to blood
(d) equinormal to blood.

44. If 0.1 M solution of glucose and 0.1 M solution of urea are placed on two sides of the semipermeable membrane to equal heights, then it will be correct to say that
(a) there will be no net movement across the membrane
(b) glucose will flow towards urea solution
(c) urea will flow towards glucose solution
(d) water will flow from urea solution to glucose

45. At 25°C, the highest osmotic pressure is exhibited by 0.1 M solution of
(a) $CaCl_2$ (b) KCl
(c) Glucose (d) Urea

46. Vapour pressure of benzene at 30°C is 121.8 mm. When 15 g of a non-volatile solute is dissolved in 250 g of benzene its vapour pressure decreased to 120.2 mm. The molecular weight of the solute (Mo. wt. of solvent = 78)
(a) 356.2 (b) 456.8
(c) 530.1 (d) 656.7

47. The vapour pressure at a given temperature of an ideal solution containing 0.2 mol of a non-volatile solute and 0.8 mol of solvent is 60 mm of Hg. The vapour pressure of the pure solvent at the same temperature is
(a) 150 mm of Hg (b) 60 mm of Hg
(c) 75 mm of Hg (d) 120 mm of Hg

48. Which of the following 0.10 m aqueous solutions will have the lowest freezing point ?
(a) $Al_2(SO_4)_3$ (b) $C_6H_{12}O_6$
(c) KCl (d) $C_{12}H_{22}O_{11}$

49. Which of the following statements, regarding the mole fraction (x) of a component in solution, is incorrect?
(a) $0 \le x \le 1$
(b) $x \le 1$
(c) x is always non-negative
(d) $-2 \le x \le 2$

50. Molarity of liquid HCl will be, if density of solution is 1.17 gm/cc
(a) 36.5 (b) 32.05
(c) 18.25 (d) 42.10

51. 1 M, 2.5 litre NaOH solution is mixed with another 0.5 M, 3 litre NaOH solution. Then find out the molarity of resultant solution
(a) 0.80 M (b) 1.0 M
(c) 0.73 M (d) 0.50 M

52. 1.00 g of a non-electrolyte solute (molar mass 250 g mol^{-1}) was dissolved in 51.2 g of benzene. If the freezing point depression constant, K_f of benzene is 5.12 K kg mol^{-1}, the freezing point of benzene will be lowered by
(a) 0.3 K (b) 0.5 K
(c) 0.4 K (d) 0.2

53. 0.5 molal aqueous solution of a weak acid (HX) is 20% ionised. If K_f for water is 1.86 K kg mol^{-1}, the lowering in freezing point of the solution is
(a) 0.56 K (b) 1.12 K
(c) −0.56 K (d) −1.12 K

54. A 0.0020 m aqueous solution of an ionic compound $Co(NH_3)_5(NO_2)Cl$ freezes at – 0.00732 °C. Number of moles of ions which 1mol of ionic compound produces on being dissolved in water will be (K_f=– 1.86°C/m)
(a) 3 (b) 4
(c) 1 (d) 2
55. An aqueous solution is 1.00 molal in KI. Which change will cause the vapour pressure of the solution to increase?
(a) Addition of NaCl
(b) Addition of Na_2SO_4
(c) Addition of 1.00 molal KI
(d) Addition of water
56. A solution of sucrose (molar mass = 342 g mol^{-1}) has been prepared by dissolving 68.5 g of sucrose in 1000 g of water. The freezing point of the solution obtained will be (K_f for water = 1.86 K kg mol^{-1}).
(a) – 0.372°C (b) – 0.520°C
(d) + 0.372°C (d) – 0.570°C
57 25.3 g of sodium carbonate, Na_2CO_3 is dissolved in enough water to make 250 mL of solution. If sodium carbonate dissociates completely, molar concentration of sodium ions, Na^+ and carbonate ions, CO_3^{2-} are respectively (Molar mass of Na_2CO_3 = 106 g mol^{-1})
(a) 0.955 M and 1.910 M
(b) 1.910 M and 0.955 M
(c) 1.90 M and 1.910 M
(d) 0.477 M and 0.477 M
58. The freezing point depression constant for water is – 1.86°C m^{-1}. If 5.00 g Na_2SO_4 is dissolved in 45.0 g H_2O, the freezing point is changed by –3.82°C. Calculate the van't Hoff factor for Na_2SO_4.
(a) 2.05 (b) 2.63
(c) 3.11 (d) 0.381
59. Mole fraction of the solute in a 1.00 molal aqueous solution is :
(a) 0.1770 (b) 0.0177
(c) 0.0344 (d) 1.7700
60. A 0.1 molal aqueous solution of a weak acid is 30% ionized. If K_f for water is 1.86°C/m, the freezing point of the solution will be :
(a) –0.18°C (b) –0.54°C
(c) –0.36°C (d) –0.24°C

Electrochemistry

CONCEPTS

Electrochemistry is the branch of chemistry which deals with relationship between electrical and chemical energy

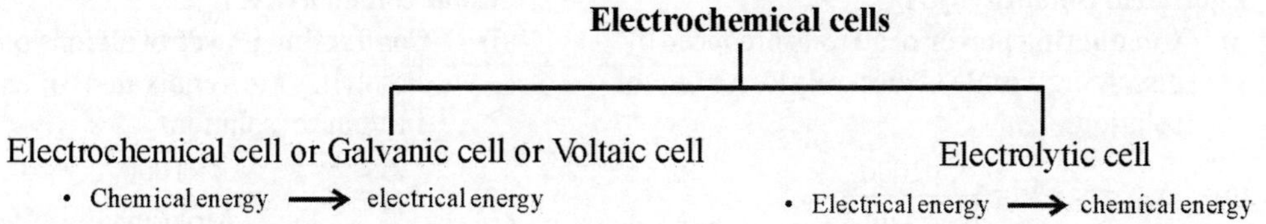

ELECTROLYTIC CELL :

The apparatus used to carry out electrolysis is known as electrolytic cell. The main features of electrolytic cell are :

Feature	*Cathode*	*Anode*
Sign	Negative since attached to negative end of external battery	Positive since attached to positive end of external battery
Direction of movement of electrons	into the cell	out of the cell
Direction of movement of ions	Cations	Anions
Half-Reaction	Reduction	Oxidation

ELECTROLYSIS

Process of decomposition of an electrolyte by passage of electricity through molten or aqueous state. It is carried out in **electrolytic cell.**

Faraday's law of electrolysis

(i) **Faraday's First law of electrolysis**

$W \propto Q \Rightarrow W = ZQ = Zit$

where Q = quantity of charge

Z = electrochemical equivalent

i = current strength in amp. t = time in sec.

(ii) **Faraday's Second law of electrolysis** states when same quantity of electricity is passed in solutions of different electrolytes, the weight of ions deposited on electrodes is proportional to their equivalent weights.

$$\frac{\text{Weight of substance} - 1}{\text{Weight of substance} - 2} = \frac{\text{Equivalent weight of} - 1}{\text{Equivalent weight of} - 2}$$

or $$\frac{w_1}{w_2} = \frac{Eq_1}{Eq_2} = \frac{Z_1 Q}{Z_2 Q} = \frac{Z_1}{Z_2}$$

- **1 Faraday is charge carried by 1 mole of electrons**
- **1 F = 96500 coulomb**
- **$Z \times 96500$ = Eq. wt. of substance.**

OHM'S LAW :

$I \propto E \Rightarrow I = \dfrac{E}{R}$ $\qquad$ $R = E/I$

where I = current, E = potential difference, R = resistance. Units of R is ohm.

Resistance (R) or resistivity (ρ)	Conductance (C) or conductivity (κ)	Specific resistance	Specific conductivity
• Opposition to flow of e^-	• ease of flow of e^- • C = 1/R mho	• It is resistance of conductor having length 1 cm and area of cross section 1 cm^2	• reciprocal of specific resistance. It is defined as conductance of cell having length 1 cm and area of cross section 1 cm^2.
• R = E/I		• $\rho = R\,a/l$ ohm cm	• $\kappa = 1/\rho = C \times l/a$
• ohm or (Ω)	• mho (Ω^{-1})	• ohm cm	• $ohm^{-1}\,cm^{-1}$

• Cell constant $= \dfrac{\ell}{a} = \dfrac{R}{\rho} = \dfrac{\text{observed resistance}}{\text{specific resistance}} = \dfrac{\text{specific conductance}}{\text{observed conductance}}$

EQUIVALENT AND MOLAR CONDUCTIVITY

Equivalent conductivity $\Lambda_{(eq)}$	Molar conductivity $\Lambda_{(m)}$
(i) **Conducting power** of all ions produced by dissolving **1 mole of electrolyte** in Vcm^3 of solution.	(i) **Conducting power** of all ions produced by dissolving **1 gm equivalent of electrolyte** in Vcm^3 of solution.
(ii) $\Lambda_{eq} = \kappa \times V = \dfrac{\kappa \times 1000}{\text{Normality}}$	(ii) $\Lambda_m = \kappa \times V = \dfrac{\kappa \times 1000}{\text{Molarity}}$
(iii) Unit is S $cm^2 eq^{-1}$	(iii) Unit is S $cm^2 mol^{-1}$

• Equivalent conductivity increases and specific conductivity decreases on dilution.

Relation between Λ_{eq} and Λ_m

$\Lambda_m = n_{Factor} \times \Lambda_{eq}$

where n_{Factor} = total change carried by either ion = M /E

Variation of Molar Conductivity with Concentration

(i) **For strong electrolytes** (given by **Huckel Onsagar equation**),

$$\Lambda_m = \Lambda_m^0 - b\sqrt{c}$$

where b = constant (depends on nature of solvent and temperature)

Λ_m^0 = molar conductance at infinite dilution.

The **conductance of strong electrolyte increases with dilution** due to decrease in interionic attractions with dilution.

(ii) **For weak electrolytes conductance increases with dilution** due to increase in dissociation on dilution. The magnitude of increase in conductance is higher than that of strong electrolytes.

(ii) **Concentration of the solution :** The concentrated solutions of strong electolytes have significant inter-ionic attractions, which reduce the speed of ions and lower the value of Λ_m. The dilution decreases such attractions and increases the value of Λ_m. The limiting value Λ_m(Λ_m^∞ the molar conductivity at zero concentration or at infinite dilution) can be obtained by extrapolating the graph.

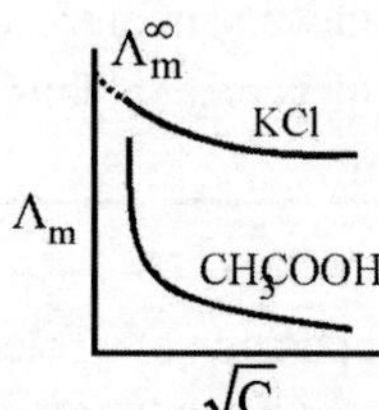

In case of weak electrolytes, the degree of ionisation increases which increases the value of Λ_m. The limiting value Λ_m^∞ cannot be obtained by extrapolating the graph. The limiting value, Λ_m^∞, for weak electrolytes is obtained by Kohlrausch law.

KOHLRAUSCH'S LAW :

At infinite dilution the molar conductivity of an electrolyte is the sum of the ionic conductivities of the cations and anions. e.g. for A_xB_y.

$$\Lambda_m^\infty(A_xB_y) = x\lambda^\infty(A^+) + y\lambda^\infty(B^-)$$

Uses of Kohlrausch's law

(i) **To find molar conductivity at infinite (∞) dilution for weak electrolytes.**

$$\Lambda^{\infty}_{CH_3COOH} = \Lambda^{\infty}_{CH_3COONa} + \Lambda^{\infty}_{HCl} - \Lambda^{\infty}_{NaCl}$$

(ii) **To find degree of dissociation (α)**

$$\alpha = \frac{\text{molar conductivity at conc. C}}{\text{molar conductivity at infinite dilution}}$$

(iii) **To calculate solubility of sparingly soluble salt**

$$\text{solubility (moles / L)} = \frac{\kappa \times 1000}{\Lambda^{\infty}_{m}}$$

ELECTROCHEMICAL CELL OR GALVANIC CELL OR VOLTAIC CELL

(i) Converts chemical energy into electrical energy.

(ii) **Daniel cell is electrochemical cell** having Zn and Cu electrodes.

Cell reactions in Daniel cell

$$Zn + CuSO_4 \longrightarrow ZnSO_4 + Cu$$

$$Zn - 2e^- \longrightarrow Zn^{2+} \text{ Oxidation half reaction}$$

$$Cu^{2+} + 2e^- \longrightarrow Cu \text{ Reduction half reaction}$$

$$Zn + Cu^{2+} \longrightarrow Zn^{2+} + Cu \text{ Cell reaction}$$

(iii) **Representation of cell**

⊖Anode | Anodic electrolyte (conc.) ||

Cathodic electrolyte (conc.) | Cathode ⊕

The vertical double line indicates salt bridge.

(iv) **Some facts about electrochemical cell**

(a) **Anode is negative electrode, cathode is positive electrode.**

(b) **Oxidation occurs at anode and reduction occurs at cathode.**

(c) Flow of electrons – anode ⟶ Cathode.

(d) Flow of current – cathode ⟶ Anode.

(e) **Salt bridge** helps in completing internal circuit of cell, maintains electrical neutrality between two cells and provides path for mobility of charges. Salt bridge is made of strong electrolytes.

ELECTRODE POTENTIAL

Tendency of an electrode to gain or loose electron when in contact with its own ions.

(i) **Oxidation potential - tendency to lose electrons**
Reduction potential - tendency to accept electrons
Standard electrode potential - is measurement at 298 K, 1 atm. P and 1 M conc.

(ii) Absolute value of electrode potential can't be determined because oxidation and reduction reaction can't occur alone. So we use **reference electrodes. Hydrogen electrode** or (NHE or SHE) **electrode having standard potential zero** is a reference electrode.

Representation of **hydrogen electrode**

$$Pt \mid H_2(1\ \text{atm.}) \mid H^+_{(1\,M)}$$

The electrode can undergo oxidation or reduction depending on nature of other electrode to which it is attached. **Calomel electrode** is another reference electrode.

(iii) The potential difference which causes a current to flow from electrode of higher negative potential called the **e.m.f.** It is measured by **potentiometer**.

(a) $\mathbf{E_{cell} = E_C - E_A}$ (when E_C and E_A both values are in terms of reduction potential)

(b) $E^{\circ}_{cell} = E^{\circ}_{C} - E^{\circ}_{A}$

(c) $E_{red} = -E_{oxid.}$ and $E_{oxid.} = -E_{red.}$

(iv) **Electrochemical series** is formed by arranging elements in the increasing order of their reduction potentials. It helps in comparing reactivity of metals and to know if H_2 will be displaced by metal from acid or not.

NERNST EQUATION

(i) It gives relationship between electrode potential and ionic concentration of electrolyte solution.

(ii) **Nernst equation for electrode (reduction)**

$$M^{n+} + ne^- \longrightarrow M_{(s)}$$

$$E_{M^{n+}/M_{(s)}} = E^{\circ}_{M^{n+}/M(s)} + \frac{2.303\,RT}{nF}\log[M^{n+}]$$

$$= E^{\circ}_{M^{n+}/M(s)} + \frac{0.0591}{n}\log[M^{n+}]\ \text{(at 298 K)}$$

where n = number of electrons in electrode reaction.

(iii) **Nernst equation and equilibrium constant (K_c):**

When E_{cell} drops to zero the concentration of ions will be equilibrium concentrations. We have for general cell reaction

$$aA + bB \rightleftharpoons cC + dD$$

$$E_{cell} = E^{o}_{cell} - \frac{2.303RT}{nF}\log\frac{[C]^{c}[D]^{d}}{[A]^{a}[B]^{b}}$$

$$E^{o}_{cell} = \frac{2.303RT}{nF}\log K_c$$ [Since at equilibrium $E_{cell} = 0$]

$$E^{o}_{cell} = \frac{0.0591}{n}\log K_c \text{ at 298K.}$$

- **Free energy change and EMF of cell**

$$\Delta G^{\circ} = -nFE^{o}_{cell} = -2.303RT\log K_C$$

COMMERCIAL CELLS

Battery refers to two or more electrochemical cells **connected in series to generate electricity.**

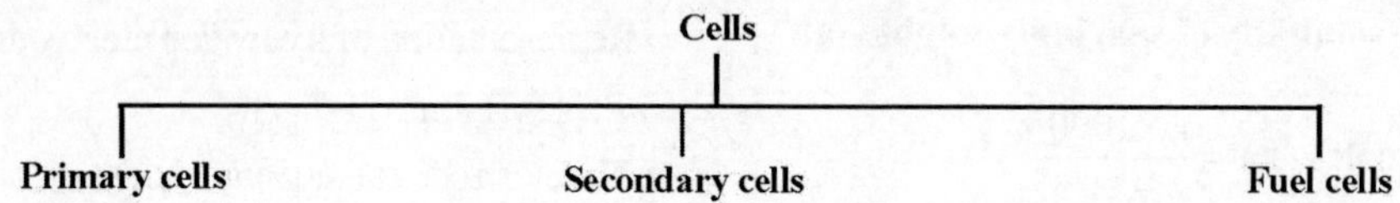

Primary cells
- not rechargable
- e.g.- Leclanche cell, Mercury cell

Secondary cells
- can be recharged,
- e.g.- Ni-Cd cell, lead storage battery

Fuel cells
- give continuous supply of energy untill supply of reactants is maintained.
- energy produced by combustion of fuel changes into electrical energy e.g.- $H_2 - O_2$ fuel cell.

Dry cell – Anode of Zn, Cathode of graphite,

Electrolyte touching anode - $NH_4Cl + ZnCl_2$

Electrolyte touching cathode - $MnO_2 + C$

Its life is not long because NH_4Cl corrodes Zn container even when the cell is not in use.

Lead - storage battery :

Anode – spongy lead

Cathode – grid of lead packed with PbO_2

Reaction occurring during discharge of cell

At anode $Pb + SO_4^{2-} \longrightarrow PbSO_4 + 2e^-$

At cathode $PbO_2 + SO_4^{2-} + 2e^- + 4H^+ \longrightarrow PbSO_4 + 2H_2O$

Cell

$Pb + PbO_2 + 2H_2SO_4 \longrightarrow 2PbSO_4 + 2H_2O$

Recharge reaction is reverse of cell reaction

Concentration cell :

Consists of **two solutions of the same electrolyte having different concentrations connected by salt bridge with electrodes of the same material dipped in them.**

e.g.- $Zn/Zn^{2+}_{(C_1)} \parallel Zn^{2+}_{(C_2)}/Zn$

$$E_{cell} = \frac{0.0591}{n}\log\frac{C_2}{C_1} \quad (\text{where } C_2 > C_1)$$

CORROSION

Wasteful conversion of metal into compounds by attack of atmospheric gases.

Corrosion of iron is called rusting. Rust is $Fe_2O_3.xH_2O$.

(i) **Process of rusting**

$$\underset{\text{layer on surface of iron}}{H_2O} + \underset{\text{from air}}{CO_2} \longrightarrow H_2CO_3$$

$$H_2CO_3 \rightleftharpoons 2H^+ + CO_3^{2-}$$

At anode $Fe \longrightarrow Fe^{2+} + 2e^-$

At cathode $H^+ + e^- \longrightarrow H$

$4H + O_2 \longrightarrow 2H_2O$

$$\underset{\text{(dissolved in water)}}{4Fe^{2+}} + O_2 + 4H_2O \longrightarrow \underset{\text{(at surface)}}{2Fe_2O_3} + 8H^+$$

$$Fe_2O_3 + xH_2O \longrightarrow \underset{\text{(Rust)}}{Fe_2.O_3xH_2O} + 4H^+$$

(ii) **Factors that promote rusting : Impurities** in iron, **moisture** (H_2O), **electrolyte**. (e.g.- NaCl in saline water)

(iii) **Prevention of rusting : Barrier protection** (e.g.- paint), **Sacrificial protection** by coating tin, **Galvanisation** by Zn, **electrical protection.**

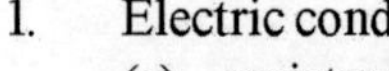

Exercise

1. Electric conductance is the direct measure of
(a) resistance (b) current
(c) concentration (d) None

2. In electrolytic cell, cathode acts as an/a
(a) oxidising agent (b) reducing agent
(c) either of the two (d) neither (a) nor (b)

3. The cell reaction $Cu + 2Ag^+ \rightarrow Cu^{+2} + Ag$ is best represented by
(a) $Cu(s) \mid Cu^{+2}(aq) \mid\mid Ag^+(aq) \mid Ag(s)$
(b) $Pt \mid Cu^{+2} \mid\mid Ag^+(aq) \mid Ag(s)$
(c) $Cu^{+2} \mid Cu \mid\mid Pt \mid Ag$
(d) None of the above representations

4. In the electrolytic cell, flow of electrons is from
(a) Cathode to anode in solution
(b) Cathode to anode through external supply
(c) Cathode to anode through internal supply
(d) Anode to cathode through internal supply

5. For the galvanic cell
$Zn \mid Zn^{2+}(0.1M) \mid\mid Cu^{2+}(1.0M) \mid Cu$ the cell potential increase if :
(a) $[Zn^{2+}]$ is increased
(b) $[Cu^{2+}]$ is increased
(c) $[Cu^{2+}]$ is decreased
(d) surface area of anode is increased

6. The e.m.f. of a Daniell cell at 298 K is E_1.

$$Zn \left| \begin{matrix} ZnSO_4 \\ (0.01\ M) \end{matrix} \right|\left| \begin{matrix} CuSO_4 \\ (1.0\ M) \end{matrix} \right| Cu$$

When the concentration of $ZnSO_4$ is 1.0 M and that of $CuSO_4$ is 0.01 M, the e.m.f. changed to E_2. What is the relationship between E_1 and E_2?
(a) $E_2 = 0 \neq E_1$ (b) $E_1 > E_2$
(c) $E_1 < E_2$ (d) $E_1 = E_2$

7. Equivalent conductances of NaCl, HCl and CH_3COONa at infinite dilution are 126.45, 426.16 and 91 $ohm^{-1}\ cm^2$ respectively. The equivalent conductance of CH_3COOH at infinite dilution would be
(a) 101.38 $ohm^{-1}\ cm^2$ (b) 253.62 $ohm^{-1}\ cm^2$
(c) 390.71 $ohm^{-1}\ cm^2$ (d) 678.90 $ohm^{-1}\ cm^2$

8. Which of the following is the use of electrolysis?
(a) Electrorefining (b) Electroplating
(c) Both (a) & (b) (d) None of these

9. In a hydrogen-oxygen fuel cell, combustion of hydrogen occurs to
(a) produce high purity water
(b) create potential difference between two electrodes
(c) generate heat
(d) remove adsorbed oxygen from elctrode surfaces

10. Several blocks of magnesium are fixed to the bottom of a ship to
(a) make the ship lighter
(b) prevent action of water and salt
(c) prevent puncturing by under-sea rocks
(d) keep away the sharks

11. The conductivity of strong electrolyte is
(a) Increases on dilution slightly
(b) Decreases on dilution
(c) Does not change with dilution
(d) Depends upon density of electrolytes itself

12. Which of the following is non-electrolyte?
(a) NaCl (b) $CaCl_2$
(c) $C_{12}H_{22}O_{11}$ (d) CH_3COOH

13. The unit of equivalent conductivity is
(a) $S\ cm^{-2}$
(b) ohm cm^2 (g - equivalent)
(c) ohm cm
(d) $ohm^{-1}\ cm^2\ (g\text{-equivalent})^{-1}$

14. The standard electrode potentials of four elements A, B, C and D are –3.05, –1.66, –0.40 and +0.80. The highest chemical reactivity will be exhibited by :
(a) A (b) B
(c) C (d) D

15. The standard reduction potentials at 25°C of Li^+/Li, Ba^{2+}/Ba, Na^+/Na and Mg^{2+}/Mg are – 3.03, – 2.73, – 2.71 and – 2.37 volt respectively. Which one of the following is the strongest oxidising agent?
(a) Na^+ (b) Li^+
(c) Ba^{2+} (d) Mg^{2+}

16. Which of the following is an insulator ?
(a) Graphite (b) Aluminium
(c) Diamond (d) Silicon.

17. The most durable metal plating on iron to protect against corrosion is
(a) nickel plating (b) copper plating
(c) tin plating (d) zinc plating.

18. The most convenient method to protect the bottom of ship made of iron is
(a) Coating it with red lead oxide
(b) White tin plating
(c) Connecting it with Mg block
(d) Connecting it with Pb block

19. Which reaction is not feasible?
(a) $2KI + Br_2 \rightarrow 2KBr + I_2$
(b) $2KBr + I_2 \rightarrow 2KI + Br_2$
(c) $2KBr + Cl_2 \rightarrow 2KCl + Br_2$
(d) $2H_2O + 2F_2 \rightarrow 4HF + O_2$

20. Kohlrausch's law states that at :
(a) finite dilution, each ion makes definite contribution to equivalent conductance of an electrolyte, whatever be the nature of the other ion of the electrolyte.
(b) infinite dilution each ion makes definite contribution to equivalent conductance of an electrolyte depending on the nature of the other ion of the electrolyte.
(c) infinite dilution, each ion makes definite contribution to conductance of an electrolyte whatever be the nature of the other ion of the electrolyte.
(d) infinite dilution, each ion makes definite contirubtion to equivalent conductance of an electrolyte, whatever be the nature of the other ion of the electrolyte.

21. An increase in equivalent conductance of a strong electrolyte with dilution is mainly due to:
(a) increase in ionic mobility of ions
(b) 100% ionisation of electrolyte at normal dilution
(c) increase in both i.e. number of ions and ionic mobility of ions
(d) increase in number of ions

22. Which of the following expressions correctly represents the equivalent conductance at infinite dilution of $Al_2(SO_4)_3$, Given that $\Lambda^{\circ}_{Al^{3+}}$ and $\Lambda^{\circ}_{SO_4^{2-}}$ are the equivalent conductances at infinite dilution of the respective ions?
(a) $\frac{1}{3}\Lambda^{\circ}_{Al^{3+}} + \frac{1}{2}\Lambda^{\circ}_{SO_4^{2-}}$
(b) $2\Lambda^{\circ}_{Al^{3+}} + 3\Lambda^{\circ}_{SO_4^{2-}}$
(c) $\Lambda^{\circ}_{Al^{3+}} + \Lambda^{\circ}_{SO_4^{2-}}$
(d) $\left(\Lambda^{\circ}_{Al^{3+}} + \Lambda^{\circ}_{SO_4^{2-}}\right) \times 6$

23. Standard electrode potential of three metals X, Y and Z are – 1.2 V, + 0.5 V and – 3.0 V, respectively. The reducing power of these metals will be :
(a) Y>Z>X (b) X>Y>Z
(c) Z>X>Y (d) X>Y>Z

24. Standard electrode potential for Sn^{4+} / Sn^{2+} couple is + 0.15 V and that for the Cr^{3+} / Cr couple is – 0.74 V. These two couples in their standard state are connected to make a cell. The cell potential will be :
(a) +1.19 V (b) +0.89 V
(c) +0.18 V (d) +1.83 V

25. Which organic compound is an electrolyte?
(a) CH_3OH (b) $CH_3—O—CH_3$
(c) $C_6H_{12}O_6$ (d) CH_3COONa

26. The best way to prevent rusting of iron is
(a) making it cathode
(b) putting in saline water
(c) both of these
(d) none of these

27. The standard hydrogen electrode potential is zero, because
(a) Hydrogen oxidized easily
(b) Electrode potential is considered as zero
(c) Hydrogen atom has only one electron
(d) Hydrogen is a very light element

28. H_2 cannot be displaced by :
(a) Li^+ (b) Sr^{2+}
(c) Al^{3+} (d) Ag^+

29. On passing C ampere of electricity through a electrolyte solution for t second. m gram metal deposits on cathode. The equivalent weight E of the metal is
(a) $E = \frac{C \times t}{m \times 96500}$ (b) $E = \frac{C \times m}{t \times 96500}$
(c) $E = \frac{96500 \times m}{C \times t}$ (d) $E = \frac{C \times t \times 96500}{m}$

30. The aqueous solution of which of the following decomposes on passing electric current
(a) Cane sugar (b) Urea
(c) Methanol (d) Potassium iodide

31. The standard electrode potential of the half cells are given below

$$Zn^{2+} + 2e^- \longrightarrow Zn \; ; E = -7.62 \text{ V}$$

$$Fe^{2+} + 2e^- \longrightarrow Fe; \; E = -7.81 \text{ V}$$

The emf of the cell, $Fe^{2+} + Zn \longrightarrow Zn^{2+} + Fe$ is
(a) 1.54 V (b) -1.54 V
(c) -0.19 V (d) $+0.19$ V

32. Aluminium displaces hydrogen from acids but copper does not. A galvanic cell prepared by combining Cu/Cu^{2+} and Al/Al^{3+} has an e.m.f of 2.0 V at 298°K. If the potential of copper electrode is $+0.34$ V, that of aluminium is
(a) $+1.66$ (b) -1.66
(c) $+2.34$ (d) -2.3 V

33. Two electrolytic cells, one containing acidified ferrous chloride and another acidified ferric chloride, are connected in series. The ratio of iron deposited at cathodes in the two cells will be :
(a) 3 : 1 (b) 2 : 1
(c) 1 : 1 (d) 3 : 2

34. An electrochemical cell is set up as :
Pt;H_2 (1atm)|HCl(0.1 M) || CH_3COOH (0.1 M)| H_2 (1atm); Pt. The e.m.f of this cell will not be zero, because
(a) the temperature is constant
(b) e.m.f depends on molarities of acids used
(c) acids used in two compartments are different
(d) pH of 0.1 M HCl and 0.1 M CH_3COOH is not same

35. On passing a current of 1.0 ampere for 16 min and 5 sec through one litre solution of $CuCl_2$, all copper of the solution was deposited at cathode. The strength of $CuCl_2$ solution was (Molar mass of Cu = 63.5; Faraday constant = 96,500 $Cmol^{-1}$)
(a) 0.01 N (b) 0.01 M
(c) 0.02 M (d) 0.2 N

36. E° for the cell, $Zn\left|Zn^{2+}(aq)\right|\left|Cu^{2+}(aq)\right|Cu$ is 1.10 V at 25°C. The equilibrium constant for the cell reaction:

$Zn + Cu^{2+}(aq) \rightleftharpoons Cu + Zn^{2+}(aq)$, is of the order of
(a) 10^{-18} (b) 10^{-37}
(c) 10^{18} (d) 10^{37}

37. Standard potentials (E°) for some half-reactions are given below :

(1) $Sn^{4+} + 2e \rightarrow Sn^{2+}$; E° = +0.15 V

(2) $2Hg^{2+} + 2e \rightarrow Hg_2^{2+}$; E° = +0.92 V

(3) $PbO_2 + 4H^+ + 2e \rightarrow Pb^{2+} + 2H_2O$; E° = +1.45 V

Based on the above, which one of the following statements is correct ?
(a) Sn^{4+} is a stronger oxidising agent than Pb^{4+}
(b) Sn^{2+} is a stronger reducing agent than Hg_2^{2+}
(c) Hg^{2+} is a stronger oxidising agent than Pb^{4+}
(d) Pb^{2+} is a stronger reducing agent than Sn^{2+}

38. For the cell reaction,
$Cu^{2+}(C_1, aq) + Zn(s) \rightarrow Zn^{2+}(C_2, aq) + Cu(s)$ of an electrochemical cell, the change in free energy, ΔG at a given temperature is a function of
(a) $ln(C_1)$ (b) $ln(C_2/C_1)$
(c) $ln(C_2)$ (d) $ln(C_1 + C_2)$

39. In the silver plating of copper, $K[Ag(CN)_2]$ is used instead of $AgNO_3$. The reason is
(a) A thin layer of Ag is formed on Cu
(b) More voltage is required
(c) Ag^+ ions are completely removed from solution
(d) Less availability of Ag^+ ions, as Cu cannot displace Ag from $[Ag(CN)_2]^-$ ion

40. The standard e.m.f. of a galvanic cell involving cell reaction with n = 2 is found to be 0.295 V at 25°C. The equilibrium constant of the reaction would be
(Given F = 96500 C mol^{-1}; R = 8.314 $JK^{-1}mol^{-1}$)
(a) 2.0×10^{11} (b) 4.0×10^{12}
(c) 1.0×10^2 (d) 1.0×10^{10}

41. A hypothetical electrochemical cell is shown below

$$\overset{\ominus}{A} \mid A^+(xM) \parallel B^+(yM) \mid \overset{\oplus}{B}$$

The emf measured is +0.20 V. The cell reaction is
(a) $A^+ + e^- \rightarrow A$; $B^+ + e^- \rightarrow B$
(b) The cell reaction cannot be predicted
(c) $A + B^+ \rightarrow A^+ + B$
(d) $A^+ + B \rightarrow A + B^+$

42. On electrolysis of dilute sulphuric acid using platinum electrodes, the product obtained at the anode will be
(a) hydrogen (b) oxygen
(c) hydrogen sulphide (d) sulphur dioxide.

43. On heating one end of a piece of a metal, the other end becomes hot because of
(a) resistance of the metal
(b) mobility of atoms in the metal
(c) energised electrons moving to the other end
(d) minor perturbation in the energy of atoms

44. If 0.5 amp current is passed through acidified silver nitrate solution for 100 minutes. The mass of silver deposited on cathode, is (eq.wt.of silver nitrate = 108)
(a) 2.3523 g (b) 3.3575 g
(c) 5.3578 g (d) 6.3575 g

45. If 0.01 M solution of an electrolyte has a resistance of 40 ohms in a cell having a cell constant of 0.4 cm^{-1}, then its molar conductance in $ohm^{-1}\ cm^2\ mol^{-1}$ is
(a) 10^2 (b) 10^4
(c) 10 (d) 10^3

46. Without losing its concentration $ZnCl_2$ solution cannot be kept in contact with
(a) Au (b) Al
(c) Pb (d) Ag

47. What is the E°_{cell} for the reaction

$$Cu^2\ (aq)\ \ Sn^2\ (aq) \rightleftharpoons Cu(s) + Sn^{4+}(aq)$$

at 25°C if the equilibrium constant for the reaction is 1×10^6?
(a) 0.5328 V (b) 0.3552 V
(c) 0.1773 V (d) 0.7104 V

48. The ionic conductance of Ba^{2+} and Cl^- are respectively 127 and 76 $ohm^{-1}\ cm^2$ at infinite dilution. The equivalent conductance (in $ohm^{-1}\ cm^2$) of $BaCl_2$ at infinite dilution will be :
(a) 139.5 (b) 203
(c) 279 (d) 101.5

49. $Cu^+(aq)$ is unstable in solution and undergoes simultaneous oxidation and reduction according to the reaction :

$$2Cu^+(aq) \rightleftharpoons Cu^{2+}(aq) + Cu(s)$$

choose correct E° for above reaction if $E^\circ_{Cu^{2+}/Cu}$ $= 0.34$ V and $E^\circ_{Cu^{2+}/Cu^+} = 0.15$ V
(a) –0.38 V (b) +0.49 V
(c) +0.38 V (d) –0.19 V

50. Specific conductance of a 0.1 N KCl solution at 23°C is 0.012 $ohm^{-1}\ cm^{-1}$. Resistance of cell containing the solution at same temperature was found to be 55 ohm. The cell constant is
(a) 0.918 cm^{-1} (b) 0.66 cm^{-1}
(c) 1.142 cm^{-1} (d) 1.12 cm^{-1}

51. Standard electrode potentials are : Fe^{+2}/Fe $E^\circ = -0.44$; Fe^{+3}/Fe^{+2} $E^\circ = +0.77$; If Fe^{+2}, Fe^{+3} and Fe blocks are kept together, then
(a) Fe^{+2} increases
(b) Fe^{+3} decreases
(c) $\frac{Fe^{+2}}{Fe^{+3}}$ remains unchanged
(d) Fe^{+2} decreases

52. In electrolysis of NaCl when Pt electrode is taken then H_2 is liberated at cathode while with Hg cathode it forms sodium amalgam. This is because
(a) Hg is more inert than Pt
(b) More voltage is required to reduce H^+ at Hg than at Pt
(c) Na is dissolved in Hg while it does not dissolve in Pt
(d) Conc. of H^+ ions is larger when Pt electrode is taken

53. On the basis of the information available from the reaction

$$\frac{4}{3}Al + O_2 \rightarrow \frac{2}{3}Al_2O_3, \Delta G = -827\ kJ\ mol^{-1}\ of\ O_2$$

the minimum e.m.f required to carry out an electrolysis of Al_2O_3 is (F = 96500 C mol^{-1})
(a) 8.56 V (b) 2.14 V
(c) 4.28 V (d) 6.42 V

54. 4.5 g of aluminium (at. mass 27 amu) is deposited at cathode from Al^{3+} solution by a certain quantity of electric charge. The volume of hydrogen produced at STP from H^+ ions in solution by the same quantity of electric charge will be
(a) 44.8 L (b) 22.4 L
(c) 11.2 L (d) 5.6 L

55. If $E^\circ_{Fe^{2+}/Fe} = -0.441V$ and $E^\circ_{Fe^{3+}/Fe^{2+}} = +771V$, the standard EMF of the reaction $Fe + 2Fe^{3+} \rightarrow 3Fe^{2+}$ will be
(a) 1.653 V (b) 1.212 V
(c) 0.111 V (d) 0.330 V

56. The equilibrium constant of the reaction:

$$Cu(s) + 2Ag^+(aq) \longrightarrow Cu^{2+}(aq) + 2Ag(s);$$

$E° = 0.46$ V at 298 K is

(a) 2.0×10^{10} (b) 4.0×10^{10}

(c) 4.0×10^{15} (d) 2.4×10^{10}

57. On the basis of the following E° values, the strongest oxidizing agent is :

$[Fe(CN)_6]^{4-} \rightarrow [Fe(CN)_6]^{3-} + e^{-1}; E° = -0.35$ V

$Fe^{2+} \rightarrow Fe^{3+} + e^{-1}; E° = -0.77$ V

(a) $[Fe(CN)_6]^{4-}$ (b) Fe^{2+}

(c) Fe^{3+} (d) $[Fe(CN)_6]^{3-}$

58. Given:

(i) $Cu^{2+} + 2e^- \rightarrow Cu, E° = 0.337$ V

(ii) $Cu^{2+} + e^- \rightarrow Cu^+, E° = 0.153$ V

Electrode potential, E° for the reaction,

$Cu^+ + e^- \rightarrow Cu$, will be :

(a) 0.90 V (b) 0.30 V

(c) 0.38 V (d) 0.52 V

59. Al_2O_3 is reduced by electrolysis at low potentials and high currents. If 4.0×10^4 amperes of current is passed through molten Al_2O_3 for 6 hours, what mass of aluminium is produced? (Assume 100% current efficiency. At. mass of Al = 27 g mol^{-1})

(a) 8.1×10^4 g (b) 2.4×10^5 g

(c) 1.3×10^4 g (d) 9.0×10^3 g

60. The equivalent conductance of $\frac{M}{32}$ solution of a weak monobasic acid is 8.0 mhos cm^2 and at infinite dilution is 400 mhos cm^2. The dissociation constant of this acid is:

(a) 1.25×10^{-6} (b) 6.25×10^{-4}

(c) 1.25×10^{-4} (d) 1.25×10^{-5}

CHAPTER 18

Chemical Kinetics

CONCEPTS

CHEMICAL KINETICS

Deals with rates of reactions, factors affecting it and mechanism of reaction.

RATE OF REACTION

It is rate of change of concentration of reactant or product with respect to time. This is in fact **average rate** of reaction.

$$A + 2B \longrightarrow 3C$$

$$\text{rate} = -\frac{\Delta[A]}{\Delta t} = -\frac{1}{2}\frac{\Delta[B]}{\Delta t} = \frac{1}{3}\frac{\Delta[C]}{\Delta t}$$

where –ve sign indicates that concentration of reactants is decreasing with time.

- Units of rate of reaction are mol L^{-1} s^{-1}.
- Rate of reaction decreases with time for reactants.

In case of gaseous reactants rate of reaction is given as atmosphere per second.

Instantaneous Rate of Reaction

Calculated for any particular instant of time

$$A + 2B \longrightarrow 3C$$

instantaneous rate

$$= \frac{dx}{dt} = -\frac{d[A]}{dt} = -\frac{1}{2}\frac{d[B]}{dt} = \frac{1}{3}\frac{d[C]}{dt}$$

where dt = infinestimal small time interval

dx = corresponding small change in concentration.

Relation between Average and Instantaneous Rate

Instantaneous rate = Average rate as Δt approaches to zero

$$\text{Instantaneous rate} = \left(\frac{\Delta x}{\Delta t}\right)_{\Delta t \to 0} = \frac{dx}{dt}$$

Factor Affecting the Rate of Reaction

(i) **Nature of the reactants :** Reactions involving lesser bond rearrangement proceed much faster than those which involve larger bond rearrangement.

(ii) **Temperature :** In most cases the rate of reaction in a homogeneous system is approximately doubled or even tripled by an increase in temeprature of 10°C. (See 13).

(iii) **Concentration of reactants :** At fixed temperature and in absence of catalyst, the rate increases with increased concentration of reactants.

(iv) **Surface area :** The more the surface area, the more is the rate of reaction.

(v) **Positive catalyst :** Increases the rate by lowering the energy of activation.

(vi) **Presence of light :** Photochemical reactions are influenced by radiations of specific wavelength.

LAW OF MASS ACTION

For hypothetical reaction $aA + bB \longrightarrow \text{products}$

rate $\propto [A]^a [B]^b$

rate = $k[A]^a[B]^b$, where k = rate constant or velocity constant or specific reaction rate.

Rate constant is rate of reaction when the concentration of reactants is unity.

Units of k

For n^{th} order reaction

Rate = $dx/dt = k(\text{conc.})^n$

$$\therefore k = \frac{dx}{dt} \times \frac{1}{(\text{conc.})^n} = \frac{\text{conc.}}{\text{time}} \times \frac{1}{(\text{conc.})^n}$$

$$= (\text{mol L}^{-1})^{1-n}\ \text{time}^{-1}$$

Order of reaction	Units of k
Zero (n = 0)	mol L^{-1} time^{-1}
First (n = 1)	time^{-1}
Second (n = 2)	L mol^{-1}time^{-1}

Characteristics of Rate Constant

(i) The value of k is different for different reactions.

(ii) At fixed temperature the value of k is constant.

(iii) It is independent of concentration but depends on temperature.

(iv) The larger the value of k the faster is the reaction and vice versa.

RATE LAW AND ORDER OF REACTION

For a reaction $aA + bB \longrightarrow$ product

If rate $= k[A]^{\alpha}[B]^{\beta}$ then the expression rate $= k[A]^{\alpha}[B]^{\beta}$ is called **rate law** and order with respect to A is α and with respect to B is β. Order of reaction $= \alpha + \beta$.

Order of Reaction

The sum of exponents to which the concentration terms are raised in rate equation of the reaction. It can be fraction, zero or any whole number.

Rate of determining step : The slowest step is called the rate determining step. The rate law equation is always written from slow step.

- **Fractional order reactions :**

  (I) Reaction between H_2 and D_2.

  Rate $= k\,P_{H_2}(P_{D_2})^{1/2}$. Order w.r.t. D_2 is 1/2.

  (II) Conversion of para hydrogen to ortho hydrogen at high temperature.

  Rate $\propto (P_{H_2})^{3/2}$

  $\therefore$ Rate $= k(P_{H_2})^{1.5}$. Rate $= k(P_{H_2})^{3/2}$.

  Order w.r.t. H_2 is 3/2.

- **Negative order reactions :**

  Conversion of ozone into oxygen.

  Rate $= K[O_3]^2[O_2]^{-1}$

  Order with respect to oxygen is -1.

- **Pseudo first order reactions** are those reactions which are not truly of the first order but are in fact second order reactions in which one of the reactant is present in large excess and its concentration does not change with time. e.g.- acid hydrolysis of esters, hydrolysis of cane sugar.

  $$\underset{\text{ester}}{RCOOR'} + \underset{\text{water}}{H_2O} \longrightarrow \underset{\text{acid}}{RCOOH} + \underset{\text{alcohol}}{R'OH}$$

- Reaction of higher order are rare because probability of collision of many molecules at one time is very small.

  $4KClO_3 \longrightarrow 3KClO_4 + KCl$ is 4th order reaction.

MOLECULARITY OF REACTION

The number of reactant species which must collide in a step leading to chemical reaction.

Order of reaction	Molecularity of reaction
1. Experimentally determined	Theoretically deduced
2. Can be in fractions	Cannot be in fractions
3. Can be zero	Can never be zero
4. Can change with time	Remains same throughout reaction.
5. Provides the mechanism of the reaction	does not give any clue about mechanism of the reaction
6. It is the sum of power of concentrations in the rate law	It is equal to the number of reactant molecules taking part in a reaction.

HALF LIFE PERIOD

Time required for half of the reaction to complete.

Half-life Period for the nth Order Reaction :

When the order of reaction is n, $t_{1/2}$ is given by

$$t_{1/2} = \frac{1}{k(n-1)}\left[\frac{2^{n-1}-1}{a^{n-1}}\right]$$

$$\therefore t_{1/2} \propto \frac{1}{a^{n-1}}$$

Unit of k $= (\text{conc})^{1-n}.\ \text{time}^{-1}$

Sl. No.	Reaction	Order	Rate law equation	Expression for rate constant	Half life period
(i)	$A \rightarrow Product$	0	Rate = k	$k = 1/t[(A^o)-(A)]$	
	$t_{1/2} = [A^o]/2k = a/2k$	(graph: r vs conc)			
(ii)	$A \rightarrow Product$	1	Rate = [A]	$k = \frac{2.303}{t}\log[A^o/A]$	$t_{1/2} = 0.693/k$
		(graph: r vs conc)			
(iii)	$2A \rightarrow product$	2	$Rate = k[A]^2$	$k = \frac{2.303}{t}\left[\frac{1}{A}-\frac{1}{A^o}\right]$	$t_{1/2} \propto 1/[A^o]$
		(graph: r vs $conc^2$)			

where A^o = Initial concentration
A = Concentration after time t
k = Rate constant
a = initial concentration

TEMPERATURE COEFFICIENT

It is the ratio of two rate constants differing by a temperature of 10°. Generally the temperature are 298K & 308K.

$$\text{Temperature Coefficient} = \frac{\text{Rate constant at 308K}}{\text{Rate constant at 298K}}$$

The value of temperature coefficient is generally 2 and 3. In rare cases the value is 4, 5 etc.

ARRHENIUS EQUATION

Mathematical expression to give quantitative relationship between rate constant and temperature. The Arrhenius equation is

$$k = A.e^{-E_a/RT} \Rightarrow \log k = \log A - E_a/2.303\,RT$$

where A = frequency factor, E_a = Activation energy, R = Gas constant, T = Temperature, k = Rate constant and $e^{-E_a/RT}$ is called **Boltzmann factor**.

$$\log\frac{k_2}{k_1} = \frac{E_a}{2.303\,R}\left(\frac{T_2 - T_1}{T_1 T_2}\right)$$

where k_1 and k_2 are rate constants, T_1 and T_2 are temperatures.

COLLISION THEORY

Chemical reaction occurs as a result of effective collisions between reacting molecules. For this two things are important.

(I) Proper orientation of reacting molecules.

(II) Possession of certain minimum amount of energy by reacting molecules

Threshold Energy

The minimum amount of energy possessed by the reacting molecules to have effective collisions, resulting in the formation of product, is called threshold energy.

Activation Energy (E_a)

The excess energy over and above the average potential energy possessed by reacting molecules to have effective collisions resulting in the formation of product is known as activation energy.

Activation Energy = threshold energy – average energy of reactants

For fast reaction the activation energy is low.

For slow reaction the activation energy is high.

Collision Frequency (z)

Total number of collisions which occur among the reacting molecules per second per unit volume is called collision frequency. Its value is given by

$$z = \sqrt{2}\,\pi\, v^{-}\, \sigma^2\, n^2$$

v^{-} = average velocity, σ = molecular diameter in cm., n = number of molecules per cc.

Rate of Reaction from Collision Theory

It is given by

Rate of reaction $= f \times z$

z = collision frequency, f = fraction of effective collisions

$$= \frac{\Delta n}{N} = e^{-E_a/RT}$$

$\therefore$ Rate (k) = $z\,e^{-Ea/RT}$

Evaluation of Energy of Activation

From Arrhenius equation

$$k = A\,e^{-E_a/RT}$$

$$\ln k = \ln A - \frac{E_a}{RT}$$

$$\log k = \log A - \frac{E_a}{2.303R}\cdot\frac{1}{T}$$

Note that as the value of Ea increases, the value of K decreases and rate of reaction decreases. Again as the value of T increases the value of k increases and rate of reaction increases.

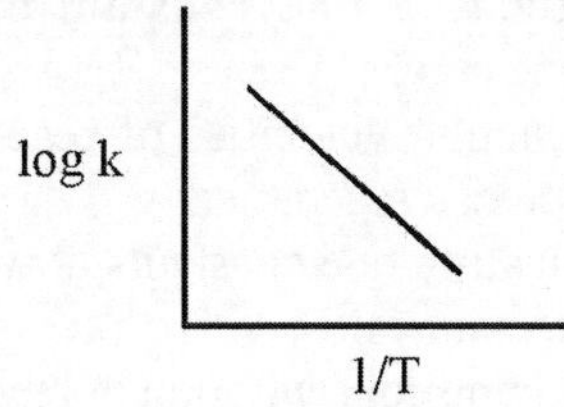

A graph between log k and $\frac{1}{T}$ is a straight line having slope equal to $-\frac{E_a}{2.303R}$.

$\therefore E_a = -2.303R \times \text{slope}$

ACTIVATED COMPLEX THEORY

(i) **Activation energy** = Threshold energy – average energy of reactants

(ii) According to this theory first activated complex is formed by reaction between reactant molecules.

The **energy of activated complex is more than that of the reactant molecules and it is unstable.** This complex dissociates to form more stable products.

PHOTOCHEMICAL REACTIONS

(i) The reaction takes place by the absorption of radiations of suitable wavelength.

(ii) Intensity of light affects rate of photochemical reaction.

(iii) **Photosensitiser** is a substance which starts a photochemical reaction without itself undergoing any change.

(iv) **Quantum yield or Quantum efficiency of photochemical reaction**

$$= \frac{\text{No. of molecules reacting in given time}}{\text{No. of quanta of light absorbed in the same time}}$$

(v) **Free energy change** of a photochemical reaction may be positive as a part of light energy absorbed by reactants changes into free energy of products.

(vi) **Vision** is a photochemical reaction.

Exercise

1. In the following reaction, how is the rate of appearance of the underlined product related to the rate of disappearance of the underlined reactant ?

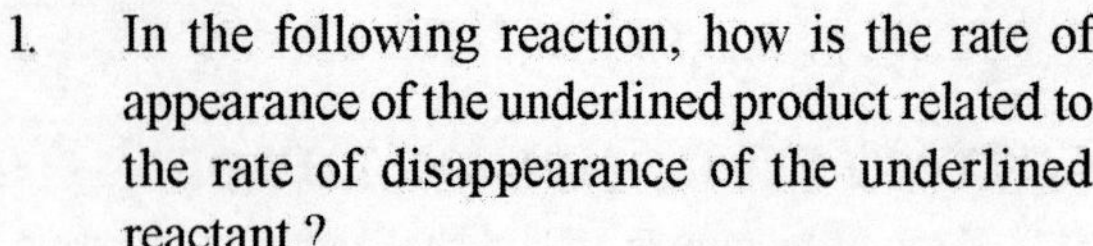

$BrO_3^-(aq) + \underline{5Br^-}(aq) + 6H^+(aq) \longrightarrow \underline{3Br_2(l)} + 3H_2O(l)$

(a) $\frac{d[Br_2]}{dt} = -\frac{5}{3}\frac{d[Br^-]}{dt}$

(b) $\frac{d[Br_2]}{dt} = -\frac{d[Br^-]}{dt}$

(c) $\frac{d[Br_2]}{dt} = \frac{3}{5}\frac{d[Br^-]}{dt}$

(d) $\frac{d[Br_2]}{dt} = -\frac{3}{5}\frac{d[Br^-]}{dt}$

2. Consider the reaction
$N_2(g) + 3H_2(g) \rightarrow 2\,NH_3(g)$

The equality relationship between $\frac{d[NH_3]}{dt}$ and $-\frac{d[H_2]}{dt}$ is

(a) $+\frac{d[NH_3]}{dt} = -\frac{2}{3}\frac{d[H_2]}{dt}$

(b) $+\frac{d[NH_3]}{dt} = -\frac{3}{2}\frac{d[H_2]}{dt}$

(c) $\frac{d[NH_3]}{dt} = -\frac{d[H_2]}{dt}$

(d) $\frac{d[NH_3]}{dt} = -\frac{1}{3}\frac{d[H_2]}{dt}$

3. The rate of reaction depends upon the
(a) volume (b) force
(c) pressure (d) conc. of reactants

4. The rate of a reaction does not depend upon
(a) Temperature (b) Concentration
(c) Catalyst (d) None of these

5. The plot of concentration of the reactant vs. time for a reaction is a straight line with a negative slope. The reaction follows a
(a) zero order rate equation
(b) first order rate equation
(c) second order rate equation
(d) third order rate equation

6. In a first-order reaction A $\rightarrow$ B, if k is rate constant and inital concentration of the reactant A is 0.5 M, then the half-life is

(a) $\frac{\log 2}{k}$ (b) $\frac{\log 2}{k\sqrt{0.5}}$

(c) $\frac{\ln 2}{k}$ (d) $\frac{0.693}{0.5k}$

7. The temperature dependence of rate constant (k) of a chemical reaction is written in terms of Arrhenius equation, $k = Ae^{-E_a^*/RT}$. Activation energy (E_a^*) of the reaction can be calculated by plotting

(a) $\log k$ vs $\frac{1}{\log T}$ (b) k vs T

(c) k vs $\frac{1}{\log T}$ (d) $\log k$ vs $\frac{1}{T}$

8. Activation energy of a chemical reaction can be determined by
(a) evaluating rate constant at standard temperature
(b) evaluating velocities of reaction at two different temperatures
(c) evaluating rate constants at two different temperatures
(d) changing concentration of reactants

9. According to the collision theory of reaction rates, the rate of reaction increases with temperature due to
(a) greater number of collision
(b) higher velocity of reacting molecules
(c) greater number of molecules having the activation energy
(d) decrease in the activation energy

10. The rate constant of a reaction depends on
   (a) temperature
   (b) initial concentration of the reactants
   (c) time of reaction
   (d) extent of reaction
11. Which one of the following statement for order of reaction is not correct?
   (a) Order can be determined experimentally
   (b) Order of reaction is equal to sum of the powers of concentration terms in differential rate law.
   (c) It is not affected with the stoichiometric coefficient of the reactants
   (d) Order cannot be fractional.
12. Which of the following statements is incorrect?
   (a) Activation energy for the forward reaction equals activation energy for the reverse reaction
   (b) For a reversible reaction, an increase in temperature increases the reaction rate for both the forward and the backward reaction
   (c) The larger the initial reactant concentration for a second order reaction, the shorter its half-life.
   (d) When Δt is infinitesimally small, the average rate equals the instantaneous rate
13. $3A \rightarrow 2B$, rate of reaction $\frac{d[B]}{dt}$ is equal to
   (a) $-\frac{3}{2}\frac{d[A]}{dt}$ (b) $-\frac{2}{3}\frac{d[A]}{dt}$
   (c) $-\frac{1}{3}\frac{d[A]}{dt}$ (d) $+2\frac{d[A]}{dt}$
14. If the rate of the reaction is equal to the rate constant, the order of the reaction is
   (a) 3 (b) 0
   (c) 1 (d) 2
15. For the reaction, $N_2 + 3H_2 \rightarrow 2NH_3$,
   $\frac{d[NH_3]}{dt} = 2 \times 10^{-4}$ mol L^{-1} s^{-1} , the value of
   $\frac{-d[H_2]}{dt}$ would be:
   (a) 4×10^{-4} mol L^{-1} s^{-1}
   (b) 6×10^{-4} mol L^{-1} s^{-1}
   (c) 1×10^{-4} mol L^{-1} s^{-1}
   (d) 3×10^{-4} mol L^{-1} s^{-1}
16. The unit of rate constant for a zero order reaction is
   (a) mol L^{-1} s^{-1} (b) L mol^{-1} s^{-1}
   (c) L^2 mol^{-2} s^{-1} (d) s^{-1}
17. The rate of a chemical reaction
   (a) Increases as the reaction proceeds
   (b) Decreases as the reaction proceeds
   (c) May increase or decrease during the reaction
   (d) Remains constant as the reaction proceeds
18. A catalyst is used
   (a) Only for increasing the velocity of the reaction
   (b) For altering the velocity of the reaction
   (c) Only for decreasing the velocity of the reaction
   (d) All (a), (b) and (c) are correct.
19. The enzyme which can catalyse the conversion of glucose to ethanol is
   (a) Zymase (b) Invertase
   (c) Maltase (d) Diastase
20. A zero order reaction is one whose rate is independent of
   (a) the concentration of the reactants
   (b) the temperature of reaction
   (c) the concentration of the product
   (d) the material of the vessel in which reaction is carried out
21. In the Haber process for the manufacture of ammonia the following catalyst is used
   (a) Platinized asbestos
   (b) Iron with molybdenum as promoter
   (c) Copper oxide
   (d) Alumina
22. Regarding criteria of catalysis which one of the following statements is not true ?
   (a) The catalyst is unchanged chemically at the end of the reaction
   (b) A small quantity of catalyst is often sufficient to bring about a considerable amount of reaction
   (c) In a reversible reaction the catalyst alters the equilibrium position
   (d) The catalyst accelerates the reaction.
23. $C_{12}H_{22}O_{11} + H_2O \xrightarrow{\text{dil } H_2SO_4} C_6H_{12}O_6(aq) + C_6H_{12}O_6(aq)$
   In this reaction, dil. H_2SO_4 is called.
   (a) homogeneous catalysis
   (b) heterogeneous catalysis
   (c) homogeneous catalyst
   (d) heterogeneous catalyst

24. A catalyst can affect reversible reaction by
(a) changing equilibrium constant
(b) slowing forward reaction
(c) attaining equilibria in both directions
(d) None of these

25. Order of reaction can be
(a) 0
(b) fraction
(c) whole number
(d) integer, fraction, zero

26. Units of rate constant of first and zero order reactions in terms of molarity M unit are respectively
(a) sec^{-1}, $Msec^{-1}$ (b) sec^{-1}, M
(c) $Msec^{-1}$, sec^{-1} (d) M, sec^{-1}.

27. In respect of the equation $k = Ae^{-E_a/RT}$ in chemical kinetics, which one of the following statements is correct ?
(a) A is adsorption factor
(b) E_a is energy of activation
(c) R is Rydberg's constant
(d) k is equilibrium constant

28. A reaction involving two different reactants can never be
(a) bimolecular reaction
(b) second order reaction
(c) first order reaction
(d) unimolecular reaction

29. Rate of a reaction can be expressed by Arrhenius equation as : $k = A\,e^{-E/RT}$. In this equation, E represents
(a) the total energy of the reacting molecules at a temperature, T
(b) the fraction of molecules with energy greater than the activation energy of the reaction
(c) the energy below which all the colliding molecules will react
(d) the energy below which colliding molecules will not react

30. The plot that represents the zero order reaction is :

(a) 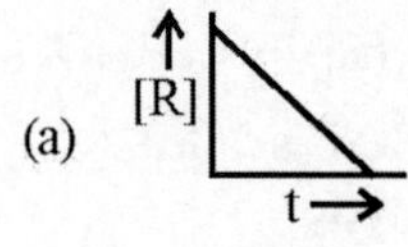

(b) 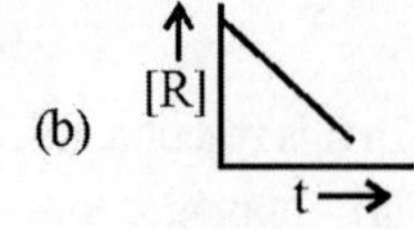

(c) 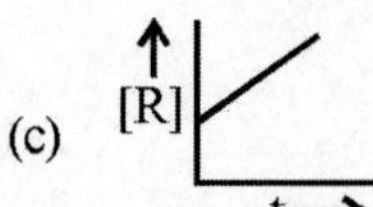

(d) 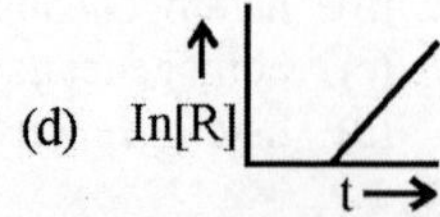

31. Select the rate law that corresponds to data shown for the following reaction

$A + B \longrightarrow$ products.

Exp.	[A]	[B]	Initial rate
1	0.012	0.035	0.1
2	0.024	0.070	0.8
3	0.024	0.035	0.1
4	0.012	0.070	0.8

(a) rate $= k[B]^3$ (b) rate $= k[B]^4$
(c) rate $= k[A][B]^3$ (d) rate $= k[A]^2[B]^2$

32. For an exothermic reaction, the energy of activation of the reactants is
(a) equal to the energy of activation of products
(b) less than the energy of activation of products
(c) greater than the energy of activation of products
(d) sometimes greater and sometimes less than that of the products

33. In a reversible reaction the energy of activation of the forward reaction is 50 kcal. The energy of activation for the reverse reaction will be
(a) < 50 kcal
(b) either greater than or less than 50 kcal
(c) 50 kcal
(d) > 50 kcal

34. $3A \rightarrow B + C$, it would be a zero order reaction when
(a) the rate of reaction is proportional to square of concentration of A
(b) the rate of reaction remains same at any concentration of A
(c) the rate remains unchanged at any concentration of B and C
(d) the rate of reaction doubles if concentration of B is increased to double

35. For the reaction $2A + B \rightarrow 3C + D$ which of the following does not express the reaction rate ?
(a) $-\frac{d[B]}{dt}$ (b) $\frac{d[D]}{dt}$
(c) $-\frac{1}{2}\frac{d[A]}{dt}$ (d) $-\frac{1}{3}\frac{d[C]}{dt}$

36. The reaction $A \rightarrow B$ follows first order kinetics. The time taken for 0.8 mole of A to produce 0.6 mole of B is 1 hour. What is the time taken for conversion of 0.9 mole of A to produce 0.675 mole of B?
(a) 2 hours (b) 1 hour
(c) 0.5 hour (d) 0.25 hour

37. The rate of reaction between two reactants A and B decreases by a factor of 4 if the concentration of reactant B is doubled. The order of this reaction with respect to reactant B is:
(a) 2 (b) –2
(c) 1 (d) –1

38. The reaction of hydrogen and iodine monochloride is given as:

$$H_2(g) + 2ICl(g) \longrightarrow 2HCl(g) + I_2(g)$$

The reaction is of first order with respect to $H_2(g)$ and ICI(g), following mechanisms were proposed.
Mechanism A:

$$H_2(g) + 2ICl(g) \longrightarrow 2HCl(g) + I_2(g)$$

Mechanism B:

$$H_2(g) + ICl(g) \longrightarrow HI(g); \text{slow}$$

$$HI(g) + ICl(g) \longrightarrow HCl(g) + I_2(g); \text{fast}$$

Which of the above mechanism(s) can be consistent with the given information about the reaction?
(a) A and B both (b) neither A nor B
(c) A only (d) B only

39. The rate of reaction between A and B increases by a factor of 100, when the concentration of A is increased 10 folds, the order of reaction with respect to A is
(a) 10 (b) 1
(c) 4 (d) 2

40. For the reaction $H_2(g) + Br_2(g) \rightleftharpoons 2HBr(g)$, the rate law is rate $= k[H_2][Br_2]^{1/2}$. Which of the following statement is true about this reaction
(a) The reaction is a second order one
(b) Molecularity of the reaction is 3/2
(c) The unit of k is s^{-1}
(d) Molecularity of the reaction is 2

41. Consider a reaction $aG + bH \rightarrow$ Products. When concentration of both the reactants G and H is doubled, the rate increases by eight times. However, when concentration of G is doubled keeping the concentration of H fixed, the rate is doubled. The overall order of the reaction is
(a) 0 (b) 1
(c) 2 (d) 3

42. For the reaction $A \longrightarrow B$, the rate law expression is : rate = k [A]. Which of the following statements is incorrect?
(a) The reaction follows first order kinetics
(b) The $t_{1/2}$ of reaction depends on initial concentration of reactants
(c) k is constant for the reaction at a constant temperature
(d) The rate law provides a simple way of predicting the conc. of reactants and products at any time after the start of the reaction.

43. The chemical reaction $2O_3 \longrightarrow 3O_2$ proceeds as follows :

$$O_3 \xrightarrow{\text{Fast}} O_2 + O\,; \quad O + O_3 \xrightarrow{\text{Slow}} 2O_2$$

the rate law expression should be
(a) $r = k[O_3]^2$ (b) $r = k[O_3]^2[O_2]^{-1}$
(c) $r = k^3[O_3][O_2]^2$ (d) $r = [O_3][O_2]^2$

44. A substance 'A' decomposes by a first order reaction starting initially with [A] = 2.00 m and after 200 min, [A] becomes 0.15 m. For this reaction $t_{1/2}$ is
(a) 53.72 min (b) 50.49 min
(c) 48.45 min (d) 46.45 min

45. Half life of a first order reaction is 4 s and the initial concentration of the reactants is 0.12 M. The concentration of the reactant left after 16 s is
(a) 0.0075 M (b) 0.06 M
(c) 0.03 M (d) 0.015 M

46. When a biochemical reaction is carried out in laboratory in the absence of enzyme then rate of reaction obtained is 10^{-6} times, then activation energy of reaction in the presence of enzyme is
(a) $\dfrac{6}{RT}$
(b) Different from E_a obtained in laboratory
(c) P is required
(d) Can't say anything

47. For the reaction $2N_2O_5 \rightarrow 4NO_2 + O_2$, rate and rate constant are 1.02×10^{-4} mol lit^{-1} sec^{-1} and 3.4×10^{-5} sec^{-1} respectively then concentration of N_2O_5 at that time will be
(a) 1.732M (b) 3M
(c) 3.4×10^5M (d) 1.02×10^{-4}M

48. The activation energy for a simple chemical reaction $A \rightarrow B$ is E_a in forward direction. The activation energy for reverse reaction
(a) is always double of E_a
(b) is negative of E_a
(c) is always less than E_a
(d) can be less than or more than E_a

49. The rate of a first order reaction is 1.5×10^{-2} mol L^{-1} min^{-1} at 0.5 M concentration of the reactant. The half life of the reaction is
(a) 0.383 min (b) 23.1 min
(c) 8.73 min (d) 7.53 min

50. For a first order reaction $A \longrightarrow B$ the reaction rate at reactant concentration of 0.01 M is found to be 2.0×10^{-5} mol L^{-1} S^{-1}. The half life period of the reaction is
(a) 30 s (b) 220 s
(c) 300 s (d) 347 s

51. If 60% of a first order reaction was completed in 60 minutes, 50% of the same reaction would be completed in aproximately
(a) 45 minutes (b) 60 minutes
(c) 40 minutes (d) 50 minutes
(log 4 = 0.60, log 5 = 0.69)

52. The rate constants k_1 and k_2 for two different reactions are $10^{16} . e^{-2000/T}$ and $10^{15} . e^{-1000/T}$, respectively. The temperature at which $k_1 = k_2$ is :
(a) 1000 K (b) $\frac{2000}{2.303}$K
(c) 2000 K (d) $\frac{1000}{2.303}$K

53. Half life period of a first-order reaction is 1386 seconds. The specific rate constant of the reaction is:
(a) 0.5×10^{-2} s^{-1} (b) 0.5×10^{-3} s^{-1}
(c) 5.0×10^{-2} s^{-1} (d) 5.0×10^{-3} s^{-1}

54. For the reaction $A + B \longrightarrow$ products, it is observed that:
(1) On doubling the initial concentration of A only, the rate of reaction is also doubled and
(2) On doubling the initial concentrations of both A and B, there is a change by a factor of 8 in the rate of the reaction.
The rate of this reaction is given by:
(a) rate = k [A] $[B]^2$ (b) rate = k $[A]^2$ $[B]^2$
(c) rate = k [A] [B] (d) rate = k $[A]^2$ [B]

55. For the reaction $[N_2O_5(g) \longrightarrow 2NO_2(g) + 1/2\, O_2(g)]$ the value of rate of disappearance of N_2O_5 is given as 6.25×10^{-3} mol $L^{-1}s^{-1}$. The rate of formation of NO_2 and O_2 is given respectively as :
(a) 6.25×10^{-3} mol $L^{-1}s^{-1}$ and 6.25×10^{-3} mol $L^{-1}s^{-1}$
(b) 1.25×10^{-2} mol $L^{-1}s^{-1}$ and 3.125×10^{-3} mol $L^{-1}s^{-1}$
(c) 6.25×10^{-3} mol $L^{-1}s^{-1}$ and 3.125×10^{-3} mol $L^{-1}s^{-1}$
(d) 1.25×10^{-2} mol $L^{-1}s^{-1}$ and 6.25×10^{-3} mol $L^{-1}s^{-1}$

56. For an endothermic reaction, energy of activation is E_a and enthalpy of reaction of ΔH (both of these in kJ/mol). Minimum value of E_a will be.
(a) less than ΔH (b) equal to ΔH
(c) more than ΔH (d) equal to zero

57. During the kinetic study of the reaction, $2A + B \rightarrow C + D$, following results were obtained:

Run	[A]/mol L^{-1}	[B]/mol L^{-1}	Initial rate of formation of D/mol L^{-1} min^{-1}
I	0.1	0.1	6.0×10^{-1}
II	0.3	0.2	7.2×10^{-1}
III	0.3	0.4	2.88×10^{-1}
IV	0.4	0.1	2.40×10^{-1}

Based on the above data which one of the following is correct?
(a) rate = $k\,[A]^2$ [B] (b) rate = k[A] [B]
(c) rate = $k[A]^2[B]^2$ (d) rate = k[A] $[B]^2$

58. The rate of the reaction

$2NO + Cl_2 \longrightarrow 2NOCl$ is given by the rate equation rate = k $[NO]^2$ $[Cl_2]$
The value of the rate constant can be increased by:
(a) increasing the concentration of NO.
(b) increasing the temperature.
(c) increasing the concentration of the Cl_2
(d) doing all of these

59. For the reaction $N_2(g) + O_2(g) \rightleftharpoons 2NO(g)$, the equilibrium constant is K_1. The equilibrium constant is K_2 for the reaction
$2NO(g) + O_2(g) \rightleftharpoons 2NO_2(g)$.
What is K for the reaction

$$NO_2(g) \rightleftharpoons \frac{1}{2}N_2(g) + O_2(g)\ ?$$

(a) $1/(2K_1K_2)$ (b) $1/(4K_1K_2)$
(c) $[1/K_1K_2]^{1/2}$ (d) $1/(K_1K_2)$

60. The rate of the reaction $2N_2O_5 \rightarrow 4NO_2 + O_2$ can be written in three ways :

$$\frac{-d[N_2O_5]}{dt} = k\,[N_2O_5]$$

$$\frac{d[NO_2]}{dt} = k'\,[N_2O_5]$$

$$\frac{d[O_2]}{dt} = k''\,[N_2O_5]$$

The relationship between k and k' and between k and k'' are :
(a) $k' = 2k\,;\, k' = k$ (b) $k' = 2k\,;\, k'' = k/2$
(c) $k' = 2k\,;\, k'' = 2k$ (d) $k' = k\,;\, k'' = k$

CHAPTER 19

Surface Chemistry

CONCEPTS

ADSORPTION :

- **Adsorption** is the phenomenon of higher in concentration of species on the surface of the solid, then in bulk.
- **Adsorbate** is molecule of solid which is adsorbed on surface and the substance on which it is adsorbed is called **adsorbent**.
- **Desorption** is removal of adsorbed substance from surface.
- **Occulation** is adsorption of gases on surface of solids.

Characteristics of Adsorption

(i) It is specific and selective in nature.

(ii) It is spontaneous process (ΔG is negative)

(iii) It is exothermic process.

Difference between Absorption and Adsorption

Absorption	Adsorption
1. Bulk phenomenon	1. Surface phenomenon
2. Rate of absorption is uniform	2. Rate varies
3. Concentration is same throughout.	3. Not same throughout the bulk

Sorption

When adsorption and absorption occur simultaneously and it is difficult to distinguish between the two processes, it is called sorption.

Types of adsorption

S.No. Property	Physical or van der Waal's adsorption	Chemical or Langmuir adsorption
(i) Forces of attraction	van der Waal	Chemical bonds
(ii) Heat of adsorption	20-40 kJ/mole	40-400 kJ/mole
(iii) Nature of process	Reversible	Irrversible
(iv) Effect of temperature	Decreases with increse in temperature	First increases and then decreases

Factors Effecting Extent of Adsorption :

(i) Nature of adsorbate and adsorbent.

(ii) **Temperature :** Adsorption is exothermic process.

Gas (adsorbate) + Solid (adsorbate) $\rightleftharpoons$ Gas adsorbed on solid + heat

$\therefore$ According to Le Chatelier's Principle, **increase in temperature decreases the rate of adsorption**.

(iii) **Pressure :** When T is constant, increase in pressure results in increase in rate of adsorption.

Adsorption Isotherm

Plot of the mass of gas adsorbed per gram of adsorbent (x/m) versus pressure at constant temperature.

- **Freundlich adsorption isotherm :** It is an empirical equation expressing the relationship between the extent of adsorption (x/m) and pressure (p).

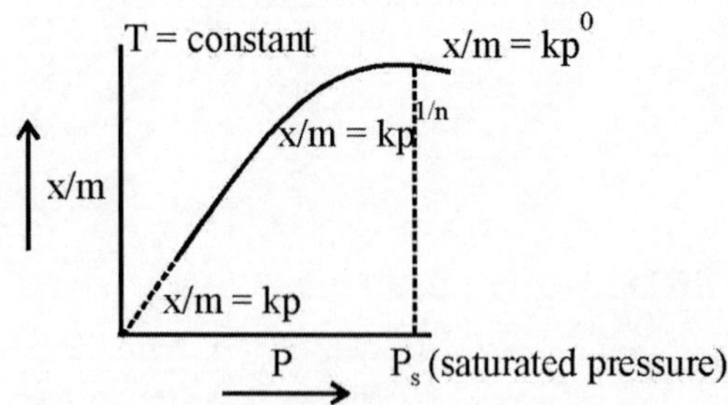

From graph, we can say

(i) at high pressure $x/m \propto p^0$

(ii) at low pressure $x/m \propto p$

(iii) at intermediate pressure $x/m \propto p^{1/n}$ $(n > 1)$

$$\log\frac{x}{m} = \log k + \frac{1}{n}\log p \text{ where } k = \text{constant.}$$

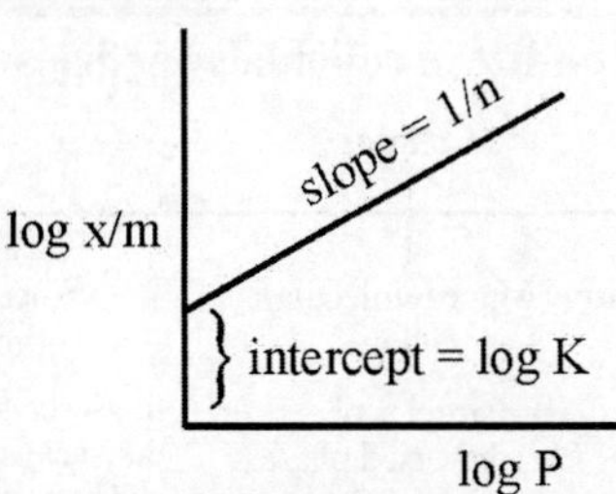

Above equation is called ***FREUNDLICH ISOTHERM***

Main Points of Langmuir Adsorption Isotherm

(a) The adsorption of a gas on the surface of a solid takes place till the whole of the surface is covered with a unimolecular layer of the gas molecules.

(b) There are two opposing processes occuring simultaneously. Adsorption of the gas molecules on the surface of the solid, and desorption of the adsorbed molecules from the surface of the solid.

(c) At equilibrium, the rate of adsorption is equal to the rate of desorption.
Based on above theory,
Langmuir derived the following relation,

$$\left(\frac{x}{m} = \frac{ap}{1+bp}\right) \qquad (i)$$

where a and b are Langmuir parameters.

- **Adsorption Isobar** is plot of x/m versus T at constant P.
- **Easily liquefiable gases** (like HCl, NH_3, H_2S and CO_2) are easily adsorbed as compared to H_2, N_2, O_2. **Activated charcoal** is most common adsorbent.

Applications of Adsorption

(i) In gas masks
(ii) For removing poisonous gases
(iii) In chromatography
(iv) For dyeing
(v) For water purification

CATALYSTS :

Catalyst is a substance which changes speed of reaction and process but itself remaining unchanged chemically. They are selective and specific.

- **Types of Catalysts**

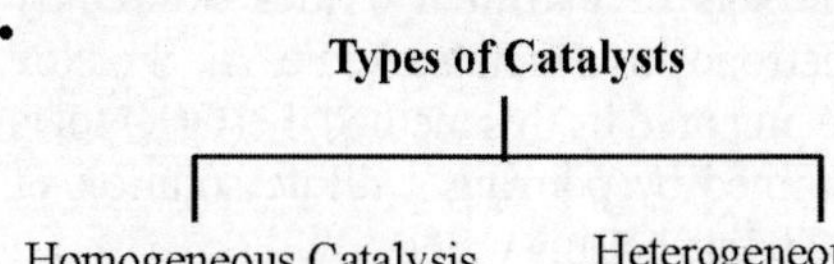

Note : The main function of catalyst is to lower the activation energy of reaction.

Catalytic Convertors

Fitted in automobiles which convert polluting exhaust gases into non-toxic compounds.

Autocatalysis

Phenomenon in which one of the products formed during reaction act as catalyst for reaction.

Positive Catalysis

When a substance increases the rate of reaction then it is called positive catalyst and process as positive catalysis.

Negative Catalysis

Phenomenon in which presence of foreign substance retards rate of chemical reaction.

(i) Addition of 1% C_2H_5OH to $CHCl_3$ retards rate of decompositon of $CHCl_3$

(ii) Addition of 1% H_3PO_3 to H_2O_2.retards rate of decompositon of H_2O_2

Induced Catalysis

In induced catalysis a chemical reaction acts as catalyst.

Catalytic Promotor

Substances which increase the activity of catalyst e.g..- Mo act as catalytic promotor for Fe in Haber's process.

Catalytic Poisons or Inhibitors

Substances which decreases the reactivity of catalyst. e.g..- S or quinoline act as poison for Pd in Rosenmund's reduction.

Enzymes

They are biological catalysts. Chemically they are **globular proteins**.

Main features of enzyme catalysis are:

(a) Their efficiency is very high even at low concentration.

(b) They are highly specific, *e.g.*, urease, an enzyme that catalyses the hydrolysis of urea.

$$NH_2CONH_2 + H_2O \xrightarrow{\text{Urease}} 2NH_3 + 2CO_2$$

None of the enzymes except urease, present in the cell catalyses the above reaction.

(c) They are highly sensitive to temperature changes.

(d) They are highly sensitive to acidity or alkalinity of the solution.

(e) Michaelis and Menten proposed the following mechanism for enzyme catalysis

$$\underset{\text{Enzyme}}{E} + \underset{\text{Substrate}}{S} \rightleftharpoons \underset{\text{Activated Complex}}{ES} \longrightarrow \underset{\text{Product}}{P} + \underset{\text{Enzyme}}{E}$$

COLLOIDS

It is heterogeneous system of two immiscible phases. - dispersion medium (It is continuous phase) and dispersed phase (It is discontinuous phase)

- The disperse phase constitutes the colloidal particles whereas the dispersion medium constitutes the medium in which particles are dispersed.

Types of Colloids :

S.No.	Dispersed phase	Dispersion medium	Name of colloid	Example
1	Solid	Solid	Solid sol.	Coloured glass, Precious stones
2	Solid	Liquid	Solution	Mud
3	Solid	Gas	Aerosol	Smoke, dust
4	Liquid	Solid	Gel	Cheese, Jellies, butter
5	Liquid	Liquid	Emulsion	Milk, creams
6	Liquid	Gas	Aerosol	Mist, fog, cloud
7	Gas	Solid	Solid foam	Pumic stone, bread rubber
8	Gas	Liquid	Foam	Whipped cream froth

Sols

Colloidal system in which the solid is desperse phase and the liquid is dispersion medium.

- **Alcosols** - have alcohol as dispersion medium.
- **Hydrosols** - have H_2O as dispersion medium

Classification of Colloids

(a) **Based on affinity between dispersion medium and dispersed phase**

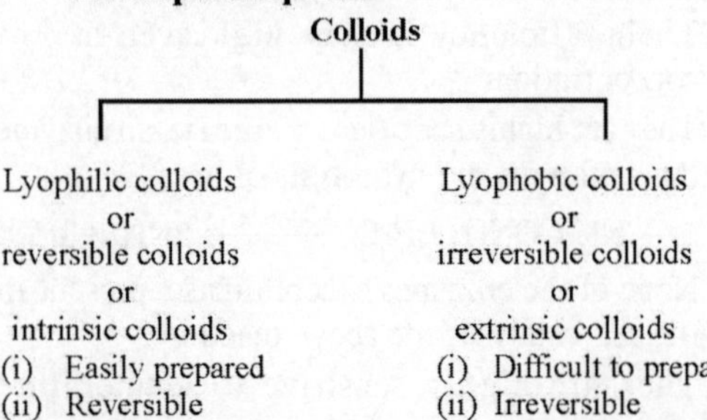

Lyophilic colloids	Lyophobic colloids
(i) Easily prepared	(i) Difficult to prepare
(ii) Reversible	(ii) Irreversible
(iii) Highly solvated	(iii) Not solvated
(iv) Quite stable and can't be precipitated or coagulated	(iv) Easily precipitated by adding small amount of electrolyte

(b) **Based on size of colloidals particles**

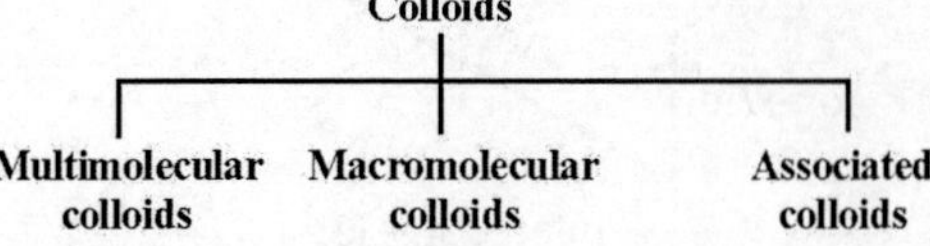

Multimolecular colloids	Macromolecular colloids	Associated colloids
(i) Formed by aggression of atoms / molecules	(i) Particles of dispersed phase are large and form colloid when dissolved in water	(i) In low concentration they act as electrolyte and at high concentration they associate to form aggregated particles called micells
(ii) Lyophilic	(ii) Lyophobic	(ii) Both lyophilic and lyophobic
(iii) Examples - sols of gold and sulphur	(iii) Examples - starch, nylon	(iii) Examples - detergents and soaps

Comparision of True Solution, Colloid and Suspension :

S.No.	Property	True solution	Colloidal solution	Suspension
1	Particle size	Less than 1nm (10Å)	between 1 – 1000 nm (10 Å – 10000Å)	more than 1000 nm > 10000Å
2	Settling	Do not settle	Do not settle	Settle on standing
3	Filterability	through filter paper and animal membrane	through filter paper only	Do not pass through either filter paper or animal membrane

Methods of Preparation of Colloids

(a) **Chemical methods :**

(i) **Oxidation,**

e.g., a sulphur sol is prepared by oxidation of H_2S.

$$SO_2 + 2H_2S \longrightarrow 2H_2O + 3S\ (sol)$$

(ii) **Reduction :**

The colloidal solutions of metals like gold silver, platinum, etc., are obtained by the reduction of their dilute salt solutions by formaldehyde or hydrazine.

$$2AuCl_3 + 3HCHO + 3H_2O \longrightarrow 2\ Au\ (sol) + 3HCOOH + 6HCl$$

$$PtCl_4 + 2HCHO + 2H_2O \longrightarrow Pt\ (sol) + 2HCOOH + 4HCl$$

(iii) **Hydrolysis:**

The sols of hydrated oxides of weakly electropositive metals like Fe, Al, Sn, etc., are prepared by this method. $Fe(OH_3)$ sol is obtained by pouring a dilute solution of $FeCl_3$ into boiling water.

$$FeCl_3 + 3H_2O \longrightarrow Fe(OH)_3 + 3HCl$$

(iv) **Double decomposition,**

e.g., $As_2O_3 + 3H_2S \longrightarrow As_2S_3\ (sol) + 3H_2O$

(b) **Physical methods:**

(i) **Exchange of solvent :**
A colloidal sol is obtained when a true solution is mixed with an excess of another solvent in which the solute does not dissolve but the solvent dissolves, *e.g.*, sulphur sol is obtained when solution of sulphur in alcohol is mixed with excess of water.

(ii) **Excessive cooling :**
A colloidal sol of ice in an organic solvent ($CHCl_3$ or ether) can be obtained by freezing a solution of water in the solvent.

(*c*) **Dispersion methods:**

(i) **Mechanical dispersion :**
In this method, the course suspension of the substance is brought into the colloidal state in the dispersion medium by grinding it in a colloid mill.

(ii) **Bredig's arc method :**
When an electric arc is struck between two electrodes of a metal (Au, Ag, Pt, Cu, etc.) in water containing traces of an alkali, the metal passes into the colloidal solution.

(iii) **Peptization method :**
The process of transferring a freshly precipitated substance into a colloidal state is called peptization. This is done by the addition of a small amount of a suitable electrolyte to a freshly precipitated substance. Thus, a sol of ferric hydroxide is obtained by treating fresh precipitate of $Fe(OH)_3$ with dilute $FeCl_3$ solution.

(iv) **Exchange of solvent**
When H_2O is added to solution of S or P in alcohol, they form colloidal solution.

Different Methods for Purification of Colloids

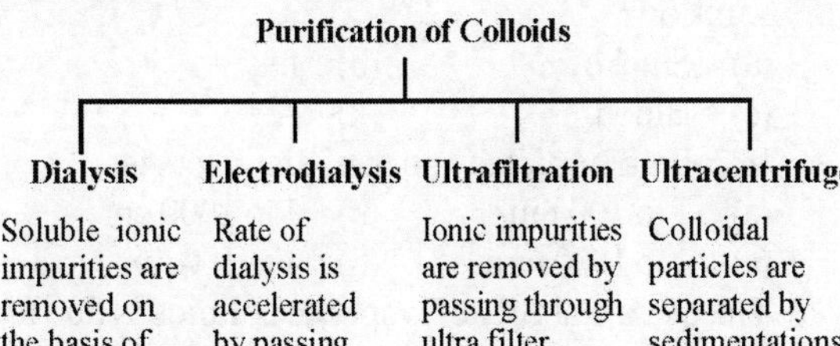

PROPERTIES OF COLLOIDS

Heterogeneous nature
Colloid has dispersion medium and dispersed phase.

Filterability
They can pass through filter paper.

Tyndall effect (Optical property)
Because of their large size they scatter light in all possible directions. True solution and suspension do not show Tyndall effect. Blue colour of sky and sea water, twinkling of stars and visibility of sharp rays of sunlight passing through a slit in dark room are examples of Tyndall effect.

Brownian movement - (Mechanical properties)
The particles of colloid are in state of continuous random movement called brownian movement and was first observed by Robert Brown. This random motion is due to collision between the colloidal particles.

Stability of colloids

(a) All particles of dispersed phase carry same type of charge (either +ve or –ve) while the dispersion medium has equal and opposite charge.

(b) Charge on colloids is due to perferential adsorption of ions. Positively charged sols are metal hydroxides [$Fe(OH)_3$, $Al(OH)_3$, $Ca(OH)_2$], haemoglobin. Negatively charged sols are metals like Au, Cu, At, Ag, starch, clay, As_2S_3, Sb_2S_3 and CdS.

Electrophoresis or cataphoresis (Electrical property)

- It is migration of colloidal particles towards oppositely charged electrode when electricity is applied.
- It shows that colloids are charged. On applying direct current colloidal particles move and coagulate.

Electro-Osmosis
Movement of dispersion medium under the influence of an electric field where as colloidal particles are not allowed to move.

Coagulation or Flocculation

- It is process of conversion of colloidal particle into precipitate by addition of electrolyte.
- **Coagulation value** (or precipitation value or flocculation value) is the minimum amount of the electrolyte in millimoles that must be added to one litre of colloidal sol to bring about its precipitation.

Flocculation value $\propto \frac{1}{\text{Coagulating power}}$

- **Hardy -Schulze rule** - The coagulating powers of different electrolytes are compared by this rule
  - (i) The ions having opposite charge to sol cause coagulation.
  - (ii) **Greater the valency more is coagulating power of electrolyte.**

    Decreasing order of coagulating power of negatively charged sols is

    $[Fe(CN)_6]^{4-} > PO_4^{3-} > SO_4^{2-} > Cl^-$

    Decreasing order for coagulating power of positively charge sols is

    $Al^{3+} > Ba^{++} > Na^+$

Protective Action of Lyophilic Colloids

- Some lyophilic colloids are used to make lyophobic colloids stable or to prevent their coagulation. Such lyophilic colloids are called **protective colloids.**
- **Gold number** is used to compare the protective powers of lyophilic colloids. Gold number is the no. of milligrams of the protective colloid required to just prevent the coagulation of 10 mL of gold sol when 1 ml of 10% of NaCl is added to it.

  Lower the gold no. more is the protective power of colloid

EMULSIONS

(i) They are colloidal solution of two immisible liquids

(ii) **Types of emulsions.**

S.No.	Property	Water in oil (W in O)	Oil in water (O in W)
1	Dispersed phase	water	oil
2	Dispersion medium	oil	water
3	examples	butter, cold cream	milk

GELS

- They are colloids in which liquids are dispersed phase and solids act as dispersion medium
- **Synerises** or weeping is process when gel shrinks by loosing some water.

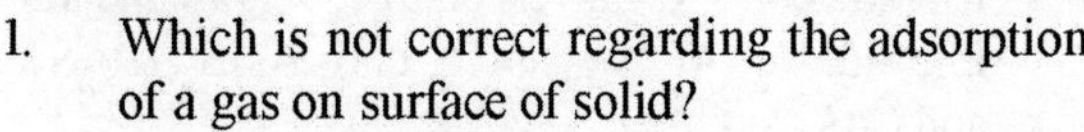

Exercise

1. Which is not correct regarding the adsorption of a gas on surface of solid?
   (a) On increasing temperature, adsorption increases contineuously
   (b) Enthalpy and entropy changes are –ve
   (c) Adsorption is more for some specific substance
   (d) This Phenomenon is reversible

2. In physical adsorption, gas molecules are bound on the solid surface by
   (a) Chemical forces
   (b) Electrostatic forces
   (c) Gravitational forces
   (d) Van der waal's forces

3. Which one of the following statements is incorrect in the case of heterogeneous catalysis ?
   (a) The catalyst lowers the energy of activation
   (b) The catalyst actually forms a compound with the reactant
   (c) The surface of the catalyst plays a very important role
   (d) There is no change in the energy of activation.

4. According to the adsorption theory of catalysis, the speed of the reaction increases because-
   (a) Adsorption lowers the activation energy of the reaction
   (b) The concentration of reactant molecules at the active centres of the catalyst becomes high due to strong adsorption
   (c) In the process of adsorption, the activation energy of the molecules becomes large
   (d) Adsorption produces heat which increases the speed of the reaction

5. Which one of the following in not a colloidal solution?
   (a) Smoke (b) Ink
   (c) Blood (d) Air

6. In colloid particles, range of diameters is
   (a) 1 to 100 nm (b) 1 to 1000 cm
   (c) 1 to 1000 mm (d) 1 to 100 km

7. The extra stability of lyophilic colloids is due to
   (a) Charge on their particles
   (b) A layer of medium of dispersion on their particles
   (c) The smaller size of their particles
   (d) The large size of their particles

8. Small liquid droplets dispersed in another liquid is called
   (a) gel (b) suspension
   (c) emulsion (d) true solution
9. Milk is a colloid in which a
   (a) Liquid is dispersed in a liquid
   (b) Solid is dispersed in a liquid
   (c) Gas is dispersed in a liquid
   (d) Sugar is dispersed in a liquid
10. Which of the following statements is *not* true about the oil-in-water type emulsion?
   (a) On addition of small amount of water, no separate layer of water appears
   (b) On addition of oil, separate layer of oil is formed
   (c) Addition of an electrolyte causes the conductivity of the emulsion to increase
   (d) Addition of small amount of oil soluble dye renders the entire emulsion coloured.
11. Which of the following statements regarding catalyst is not true ?
   (a) A catalyst remains unchanged in composition and quantity at the end of the reaction
   (b) A catalyst can initiate a reaction
   (c) A catalyst does not alter the equilibrium in a reversible reaction
   (d) Catalysts are sometimes very specific in respect of reaction.
12. When dispersed phase is liquid and dispersion medium is gas then the colloidal system is called
   (a) Smoke (b) Clouds
   (c) Jellies (d) Emulsions
13. Catalyst increases the rate of reaction by
   (a) decreasing threshold energy
   (b) decreasing activation energy
   (c) increasing activation energy
   (d) decreasing equilibrium constant
14. During dialysis
   (a) only solvent molecules can diffuse
   (b) solvent molecules, ions and colloidal particles can diffuse
   (c) all kinds of particles can diffuse through the semi-permeable membrane
   (d) solvent molecules and ions can diffuse
15. The ability of an ion to bring about coagulation of a given colloid depends upon
   (a) its size
   (b) the magnitude of its charge
   (c) the sign of the charge alone
   (d) both magnitude and sign of its charge
16. At the critical micelle concentration (CMC) the surfactant molecules
   (a) decompose
   (b) dissociate
   (c) associate
   (d) become completely soluble
17. Which is used for ending charge on colloidal solution?
   (a) Electrons
   (b) Electrolytes
   (c) Positively charged ions
   (d) Compounds
18. Which one is a colloid?
   (a) Sodium chloride (b) Urea
   (c) Cane sugar (d) Blood
19. An arsenious sulphide sol carries a negative charge. The maximum precipitating power for this sol is possessed by
   (a) K_2SO_4 (b) $CaCl_2$
   (c) Na_3PO_4 (d) $AlCl_3$
20. Smoke is an example of
   (a) Gas dispersed in liquid
   (b) Gas dispersed in solid
   (c) Solid dispersed in gas
   (d) Solid dispersed in solid
21. Which of the following is not a colloid ?
   (a) Chlorophyll (b) Smoke
   (c) Ruby glass (d) Milk
22. Suspensions are
   (a) Visible to naked eye
   (b) Not visible by any means
   (c) Invisible under electron microscope
   (d) Invisible through microscope
23. Brownian movement is discovered by
   (a) Robert Brown (b) Zsigmondy
   (c) Hardy-Schulzi (d) Grahm
24. Which of the following terms is not related with colloids ?
   (a) Brownian movement
   (b) Dialysis
   (c) Ultrafiltration
   (d) Wavelength
25. Tyndall effect in colloidal solutions is due to :
   (a) Scattering of light
   (b) Reflection of light
   (c) Absorption of light
   (d) Presence of electrically charged particles

26. Cloud or fog is an example of colloidal system of
(a) Liquid dispersed in gas
(b) Gas dispersed in gas
(c) Solid dispersed in gas
(d) Solid dispersed in liquid

27. Which of the following forms a colloidal solution in water ?
(a) NaCl (b) Glucose
(c) Starch (d) Barium nitrate

28. Colloidal solution can be purified by
(a) peptization (b) coagulation
(c) dialysis (d) Bredig Arc method

29. Tyndall effect is shown by
(a) Colloidal solution
(b) Osmotic solution
(c) Isotonic solution
(d) Hypertonic solution

30. A colloid always :
(a) Contains two phases
(b) Is a true solution
(c) Contains three phases
(d) Contains only water soluble particles

31. The electrolytic impurities of a sol can most easily be separated by
(a) dialysis (b) electrosmosis
(c) electrophoresis (d) electrodialysis

32. The simplest way to check whether a system is colloidal is by
(a) Tyndall effect
(b) Brownian movement
(c) Electrodialysis
(d) Measuring particle size

33. Substances whose solutions can readily diffuse through parchment membrane are
(a) Colloids (b) Crystalloids
(c) Electrolytes (d) Non electrolytes

34. Wood charcoal is used to decolourise sugar because it
(a) adsorbs coloured material
(b) absorbs decolourised material
(c) reduces coloured material
(d) None of these

35. A catalyst can affect reversible reaction by
(a) changing equilibrium constant
(b) slowing forward reaction
(c) attaining equilibria in both directions
(d) None of these

36. Hardy-Schulze rule explains the effect of electrolytes on the coagulation of colloidal solution. According to this rule, coagulation power of cations follow the order
(a) $Ba^{+2} > Na^{+} > Al^{+3}$ (b) $Al^{+3} > Na^{+} > Ba^{+2}$
(c) $Al^{+3} > Ba^{+2} > Na^{+}$ (d) $Ba^{+2} > Al^{+3} > Na^{+}$

37. Position of non-polar and polar part in micelle is
(a) Polar at outer surface and non-polar at inner surface
(b) Polar at inner surface and non-polar at outer surface
(c) Distributed all over the surface
(d) Present in the surface only

38. The Langmuir adsorption isotherm is deduced using the assumption
(a) the adsorption sites are equivalent in their ability to adsorb the particles
(b) the heat of adsorption varies with coverage
(c) the adsorbed molecules interact with each other
(d) the adsorption takes place in multilayers.

39. Gold numbers of protective colloids A, B, C and D are 0.50, 0.01, 0.10 amd 0.005, respectively. the correct order of their protective powers is
(a) $D < A < C < B$ (b) $C < B < D < A$
(c) $A < C < B < D$ (d) $B < D < A < C$

40. Which of the following statements is incorrect regarding physiosorption ?
(a) More easily liquefiable gases are adsorbed readily.
(b) Under high pressure it results into multi molecular layer on adsorbent surface.
(c) Enthalpy of adsorption ($\Delta H_{adsorption}$) is low and positive.
(d) It occurs because of van der Waal's forces.

41. The disperse phase in colloidal iron (III) hydroxide and colloidal gold is positively and negatively charged, respectively. Which of the following statements is NOT correct ?
(a) Coagulation in both sols can be brought about by electrophoresis.
(b) Mixing the sols has no effect.
(c) Sodium sulphate solution causes coagulation in both sols.
(d) Magnesium chloride solution coagulates, the gold sol more readily than the iron (III) hydroxide sol.

42. For adsorption of a gas on a solid, the plot of log x/m vs log P is linear with slope equal to (n being whole number)
(a) k (b) log k
(c) n (d) $\frac{1}{n}$

43. Which one of the following forms micelles in aqueous solution above certain concentration?
(a) Dodecyl trimethyl ammonium chloride
(b) Glucose
(c) Urea
(d) Pyridinium chloride

44. If x is amount of adsorbate and m is amount of adsorbent, which of the following relations is not related to adsorption process ?
(a) $x/m = f(p)$ at constant T.
(b) $x/m = f(\text{T})$ at constant p.
(c) $p = f(\text{T})$ at constant (x/m).
(d) $\frac{x}{m} = p \times T$

45. Which one of the following substances gives a positive charged sol?
(a) Gold (b) A metal sulphide
(c) Ferric hydroxide (d) An acidic dye

46. The colloidal solution of gelatin is known as
(a) Solvent loving sol (b) Reversible sol
(c) Hydrophilic sol (d) All the above

47. When a colloidal solution is observed under an ultramicroscope, we can see
(a) Light scattered by colloidal particles
(b) Size of the particle
(c) Shape of the particle
(d) Relative size

48. Tyndall phenomena is exhibited by
(a) NaCl solution (b) Starch solution
(c) Urea solution (d) $FeCl_3$ solution

49. In colloidal state, particle size ranges from
(a) 1 to 10 Å (b) 20 to 50 Å
(c) 10 to 1000 Å (d) 1 to 280 Å

50. Which of the following methods is used for the destruction of sol ?
(a) Condensation
(b) Dialysis
(c) Diffusion through animal membrane
(d) Addition of an electrolyte.

51. Colloidal solutions are not purified by
(a) Dialysis (b) Electrodialysis
(c) Ultrafiltration (d) Electrophoresis

52. Tyndall effect will be observed in
(a) solution (b) precipitate
(c) sol (d) vapour

53. Which one of the following colloidal solution is positive sol?
(a) Blood
(b) Clay soil
(c) Smoke
(d) Gelatin in strongly acidic solution

54. Butter is a colloid formed when
(a) water is dispersed in fat
(b) milk is dispersed in fat
(c) fat is dispersed in water
(d) milk is dispersed in water

55. Which of the following statements is true about a colloidal system?
(a) It carries a net electric charge
(b) It consists of one phase only
(c) It can be made out of two gases
(d) It is electrically neutral as a whole

56. One desires to prepare a positively charged sol of silver iodide. This can be achieved by
(a) adding small amount of $AgNO_3$ solution to KI solution in slight excess
(b) adding small amount of KI solution to $AgNO_3$ solution in slight excess
(c) mixing equal volumes of equimolar solutions of $AgNO_3$ and KI
(d) None of these

57. Which of the following ionic substances will be most effective in precipitating the sulphur sol?
(a) KCl (b) $BaCl_2$
(c) $Fe_2(SO_4)_3$ (d) Na_3PO_4

58. Identify the correct statement regarding enzymes
(a) Enzymes are specific biological catalysts that cannot be poisoned
(b) Enzymes are normally heterogeneous catalysts that are very specific in their action
(c) Enzymes are specific biological catalysts that can normally function at very high temperatures (T~1000K)
(d) Enzymes are specific biological catalysts that possess well-defined active sites

59. Blood may be purified by
(a) Dialysis (b) Electroosmosis
(c) Coagulation (d) Filtration

60. Surface tension of lyophillic sols is
(a) Lower than that of H_2O
(b) More than that of H_2O
(c) Equal to that of H_2O
(d) None of the above

General Principles and Processes of Isolation of Elements

CONCEPTS

OCCURRENCE OF ELEMENTS

Total **naturally occurring element-90 (about 70% are metal)**

Occurrence of elements

- **Native state or free state** (not attacked by moisture, oxygen and CO_2 of air) e.g.- Au, Ag, noble gases
- **Combined state** (attacked by moisture, oxygen and CO_2 of air) e.g.- Na, K etc

- **Minerals** are compounds of metals in which metal occurs in nature.
- **Ore** is mineral from which metal can be obtained conveniently and economically.
- **Ganguge or matrix** are **impurities** present in ore.
- **Weathering** is selective extraction of igneous rocks by H_2O, CO_2 and humic acid to remove Na^+, Ca^{2+}, Mg^{2+} and Fe^{2+} ions from it leaving behind insoluble residue.
- **Mining :** Process of taking out ores from earth's crust.

Metallurgy : Process of obtaining pure metal from its ores

It is done in following steps -

(a) Concentration or benefication or dressing of ore.

(b) Conversion of ore into oxide.

(c) Conversion of oxide into impure metal. (Reduction)

(d) Purification of metal.

CONCENTRATION OF ORE

Concentration of ore is achieved by

(i) physical methods, and

(ii) chemical methods

Physical Methods

(i) Hand-picking: It is used in the case when the impurities are quite distinct from the ore so that these may be differentiated by naked eye.

(ii) Levigation or gravity separation or hydraulic washing

- used to separate lighter ganguge particles from heavy ore particles.

(iii) Froth flotation process - Finely divided ore is mixed with oil (pine oil, eucalyptus oil or camphor oil) and agitated with water containing a detergent (foaming agent). When air is bubbled through into the mixture, the air bubbeles are stabilised by the detergent .These adsorb mineral particles wetted with oil and rise to the surface. The earthy matter wetted by water settles down at the bottom.

- **Collectors** - Which increase the non wettability of ore particles e.g. pine oil, xanthates and fatty acids.
- **Froth stabiliser** - Which stabilise the froth e.g. cresols and aniline.
- **Depressants** - Depressants prevent the formation of froth eg NaCN, when added to ore containing ZnS and PbS form a complex with ZnS as $Na_2[Zn(CN)_4]$ and prevent it from forming froth. PbS is then easily separated from ZnS.

(iv) Magnetic separation is used when either ore or impurity is magnetic in nature. e.g.- **Wolframite (ferromagnetic)** ore is separated **from cassiterite ore (SnO_2)** by this method.

Chemical Method

Leaching involves dissolving ore in such solvent in which only ore is soluble and impurities are not. e.g.- Ores of Al, Au and Ag.

$$\underset{\text{bauxite (ore of Al)}}{Al_2O_3.2H_2O} + 2NaOH \longrightarrow \underset{\text{soluble meta aluminate}}{2NaAlO_2} + 3H_2O$$

EXTRACTION OF CRUDE METAL

Conversion to Oxide

(i) **Calcination :** It is the heating of the ore in a suitable furnace in absence of air much below its melting point to cause decomposition and elimination of volatile products.

$PbCO_3 \rightarrow PbO + CO_2$

The process is generally applied to hydrated oxide or carbonate ores.

(ii) **Roasting :** It is the heating of the concentrated ore in a suitable furnace strongly in presence of air with or without certain substances below the melting point which causes chemical reaction to expel volatile impurities eg oxides of S, As and Sb.

$$2ZnS + 3O_2 \rightarrow 2ZnO + 2SO_2$$

$$2MoS_2 + 7O_2 \rightarrow 2MoO_3 + 4SO_2$$

- **Furnaces** are used to carry out these processes. e.g.- blast furnace.
- **Refractory material** used is that which bear high temperature without melting or becoming soft to line the furnaces.

  Acidic refractories - SiO_2.

  Basic refractories - CaO, MgO.

  Neutral refractories - graphite, bone ash.

Conversion of Oxide into Impure Metal (Reduction):

(i) **Smelting:** This involves the reduction of the ore to the molten metal at a high temperature. For the extraction of electropositive metals such as Pb, Fe, Sn, powerful reducing agent like C, H_2 CO, Al, Mg, etc., are used. Examples are

- **Carbon reduction process:** The oxides of less electropositive metals are reduced by strongly heating them with coke.

  $PbO + C \longrightarrow Pb + CO$

  $Fe_2O_3 + 3C \longrightarrow 2Fe + 3CO$

  $Fe_2O_3 + 3CO \longrightarrow 2Fe + 3CO_2$

- **Reduction with Aluminium (Gold - Schmidt Aluminothermic process):** This process is used in case of those metals which have very high melting points and are to be extrated from their oxides.

  $Cr_2O_3 + 2Al \longrightarrow 2Cr + Al_2O_3$

(ii) **Self reduction process:** Also called auto-reduction process. The sulphide ores of less electropositive metals like Pb, Cu, etc., are heated in air to convert part of the ore into oxide which then reacts with remaining sulphide ore to give the metal and SO_2. No external reducing agent is used.

$2Cu_2S + 3O_2 \longrightarrow 2Cu_2O + 2SO_2$

$Cu_2S + 2Cu_2O \longrightarrow 6\ Cu + SO_2$ (self reduction)

(iii) **Hydrometallurgy:** It is used for the reduction of metal oxide to free metal. It is based on the fact that more electropositive metal can displace less electropositive metal from its salt solution. The ore is treated with such chemical reagents which convert it into soluble compound. By the addition of more electropositive metal to the filtrate, the metal present in the ore can be precipitated, *e.g.*,in extraction of copper,

(iv) **Electrolytic reduction (Electro metallurgy)** Highly electropositive metals (Na, K, Ca, Mg) are obtained by this method by electrolysis of their oxides, hydroxides or chlorides in fused state.

REFINING OR PURIFICATION OF PURE METAL

(i) **Liquation**

- Used to purify **metals which are readily fusible** (have low melting point) such as **Pb, Sn, Bi, Hg** as compared to impurities.
- **Infusible matter during liquation is called dross.**

(ii) **Distillation**

- Used to refine **metals with low boiling point and are easily volatile**. e.g.- **Hg, Zn, Cd**.

(iii) **Polling**

- The molten metal is stirred with green poles of wood, which liberates gas like methane. The latter reduces any oxide present in the metal eg CuO in the blister copper is reduced to copper.

(iv) **Cupellation**

- Used to **remove impurity of Pb from Ag**. ($\because$ oxide of Pb is volatile)

(v) **Electro refining**

- In this method anode is made of impure metal and cathode of pure metal and electolyte is suitable salt.

- Cu, Ag, Au, Cr, Zn, Ni are purified by this method.
- Pure metal deposits at cathode.

(vi) Zone refining

- **Semiconductors (e.g.- Si, Ge, Ga) are purified by this method.**
- Method is **based on difference in solubilities in molten and solid state of metal.**

(vii) Vapour-phase refining

The crude metal is freed from impurities by first converting it into a suitable volatile compound by heating it with a specific reagent at a lower temperature and then decomposing the volatile compound at some higher temperature to give the pure metal.

- **Van Arkel method**

  Ti and Zr are to purified by this method.

$$Ti + 2I_2 \xrightarrow{500K} \underset{\text{volatile}}{TiI_4} \xrightarrow[\text{filament}]{1700K} Ti + 2I_2$$

- **Mond's process**

  Used for **refining of Ni** by using carbon monoxide.

$$\underset{\text{impure}}{Ni} + 4CO \xrightarrow{330-350K} \underset{\text{volatile}}{Ni(CO)_4}$$

$$\xrightarrow{450-470K} \underset{\text{pure}}{Ni} + 4CO$$

EXTRACTION OF IRON AND COPPER

Extraction of Fe from its Oxide

Oxide ores of iron are mixed with limestone and coke and fed into a **Blast furnace** from its top. Here, the oxide is reduced to the metal.

$$FeO\,(s) + C\,(s) \longrightarrow Fe\,(s/l) + CO\,(g)$$

In the Blast furnace, reduction of iron oxides takes place in different temperature ranges.

At 500 – 800 K (lower temperature range)

$$3Fe_2O_3 + CO \longrightarrow 2Fe_3O_4 + CO_2$$

$$Fe_3O_4 + 4CO \longrightarrow 3Fe + 4CO_2$$

$$Fe_2O_3 + CO \longrightarrow 2FeO + CO_2$$

At 900 – 1500 K (higher temperature range)

$$C + CO_2 \longrightarrow 2CO$$

$$FeO + CO \longrightarrow Fe + CO_2$$

Extraction of Copper from Cuprous Oxide (Cu_2O)

Most of the copper ores are sulphide ores. The sulphide ores are roasted/smelted to give oxides:

$$2Cu_2S + 3O_2 \longrightarrow 2Cu_2O + 2SO_2$$

The oxide can then be easily reduced to metallic copper using coke.

$$Cu_2O + C \longrightarrow 2Cu + CO$$

RELATIVE ABUNDANCE

Abundance of elements in the earth's crust (by weight)

$$O > Si > Al > Fe > Ca > Na > K > Mg > H > Ti > Cl > F$$

Abundance of elements in the earth's crust (in terms of number of atoms per 100 atoms)

$$O > Si > Al > H > Na > Ca > Fe > Mg >$$

$$K > Ti > C > Mn > S > F > Cl > Cr$$

THERMODYNAMIC PRINCIPLES OF METALLURY

Theory of metallurgical transformations can be interpreted by Gibb's free energy change at any specified temp.

$$\Delta G = \Delta H - T\Delta S \text{ or } \Delta G^o = -RT\,\ell nK$$

where, ΔH = enthalpy change and
ΔS = entropy change
K = equilibrium constt. at tempt.

The reducing agent is oxidised and metal oxide is reduced. The role of reducing agent is to provide ΔG^o negative.

During reduction the metal oxide decomposes.

$$M_xO(s) \longrightarrow xM \text{ (Solid or liquid)} + \frac{1}{2}O_2(g)$$

The reducing agent (C or CO) is oxidised

$$C(s) + \frac{1}{2}O_2(g) \longrightarrow CO(g)$$

$$CO(g) + \frac{1}{2}O_2(g) \longrightarrow CO_2(g)$$

$$C(s)O_2(g) \longrightarrow CO_2(g)$$

If net ΔG of two possible reactions (Reduction/Oxidation) is –ve, the overall reaction will occur.

H.J.T. Elligham diagram (plots of ΔG Vs T) provides a sound basis for considering the choice of reducing agent in the reduction of oxides.

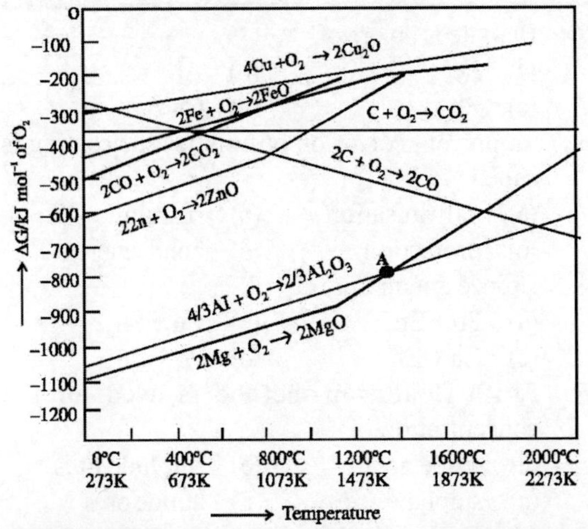

Fig. Gibbs energy (ΔG^o) vs T plots (schematic) for formation of some oxides (Ellingham diagram)

When alumina is reduced by magnesium, the two equations are

(i) $4\ Al + 3O_2 \longrightarrow 2\ Al_2O_3$ (ii)

$2Mg + O_2 \longrightarrow 2MgO$

ΔG becomes ZERO at the point of intersection of the Al_2O_3 and MgO curves (marked "A"). Above this point the magnesium can reduce alumina.

Note: Although thermodynamically feasible the magnesium metal, is not used for the reduction of aluminas. The temperature required would be so high and the process would be technologically difficult and uneconomic.

Exercise

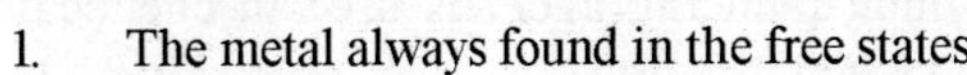

1. The metal always found in the free states is
   (a) Au (b) Ag
   (c) Cu (d) Na
2. Which of the following is not an ore of iron?
   (a) limonite (b) cassiterite
   (c) magnetite (d) None of these
3. The process to remove unwanted impurities from the ore is called –
   (a) purification (b) calcination
   (c) bassimerisation (d) concentration
4. Heating of concentrated ore in absence of air for conversion into oxide ore is known as –
   (a) roasting (b) calcination
   (c) reduction (d) none of these
5. The reducing agent in thermite process is
   (a) Al (b) Mg
   (c) BaO_2 (d) MnO_2
6. The commonest method for the extraction of metals from oxide ores involves :
   (a) Reduction with carbon
   (b) Reduction with aluminium
   (c) Reduction with hydrogen
   (d) Electrolytic method
7. Which of the following statements, about the advantage of roasting of sulphide ore before reduction is not true?
   (a) The ΔG_f^o of the sulphide is greater than those for CS_2 and H_2S.
   (b) The ΔG_f^o is negative for roasting of sulphide ore to oxide.
   (c) Roasting of the sulphide to the oxide is thermodynamically feasible.
   (d) Carbon and hydrogen are suitable reducing agents for metal sulphides.
8. In the extraction of aluminium from bauxite, cryolite is used to
   (a) decrease the boiling point of aluminium
   (b) purify the bauxite
   (c) make the mass lighter
   (d) decrease melting point of alumina and increase conductivity
9. During the process of electrolytic refining of copper, some metals present as impurity settle as 'anode mud'. These are
   (a) Fe and Ni (b) Ag and Au
   (c) Pb and Zn (d) Sn and Ag
10. The method of zone refining of metals is based on the principle of
    (a) Greater solubility of the impurity in the molten state than in the solid
    (b) Greater mobility of the pure metal than that of the impurity
    (c) Higher melting point of the impurity than that of the pure metal
    (d) Greater noble character of the solid metal than that of the impurity

11. Among the following statements, the incorrect one is
(a) calamine and siderite are carbonates
(b) argentite and cuprite are oxides
(c) zinc blende and pyrites are sulphides
(d) malachite and azurite are ores of copper
12. Calcination is the process of heating the ore
(a) In a blast furnace (b) In absence of air
(c) In presence of air (d) None of these
13. The chief ore of aluminium is :
(a) Cryolite (b) Feldspar
(c) Kaolin (d) Bauxite
14. A naturally occurring substance from which a metal can be profitably extracted is known as :
(a) mineral (b) gangue
(c) ore (d) flux
15. Which of the following is an ore of silver?
(a) Argentite (b) Stibnite
(c) Haematite (d) Bauxite
16. Which is not a basic flux ?
(a) $CaCO_3$ (b) Lime
(c) SiO_2 (d) CaO
17. Electrolytic reduction method is used in the extraction of
(a) Highly electronegative elements
(b) Highly electropositive elements
(c) Transition metals
(d) Noble metals
18. In alumino thermite process which of the following is reduced :
(a) Aluminium (b) Oxide
(c) Sulphide (d) Carbonate
19. Refractory metals are used in construction of furnaces because
(a) they can withstand high temperature
(b) they are chemically inert
(c) their melting point is high
(d) None of these
20. Which of the following pairs of metals is purified by van Arkel method ?
(a) Ga and In (b) Zr and Ti
(c) Ag and Au (d) Ni and Fe
21. Amalgams are :
(a) Highly coloured alloys
(b) Always in liquid state
(c) Alloys which contain mercury as one of their constituents
(d) Alloys which have a great resistance to abrasion
22. Bauxite is an ore of
(a) Fe (b) Al
(c) Pb (d) Cu
23. To prevent corrosion of iron by zinc coating is called
(a) galvanisation (b) rusting
(c) protection (d) bond energy
24. Bronze is a mixture of
(a) Pb + Sn (b) Cu + Sn
(c) Cu + Zn (d) Pb + Zn
25. Froth floatation method is used for the concentration of –
(a) oxide ores (b) sulphide ores
(c) sulphate ores (d) halide ores
26. Chemical method used in concentration of ore is known as–
(a) bleaching (b) leaching
(c) roasting (d) calcination
27. Which reducing agent is used in chemical reduction?
(a) C (b) CO
(c) Al (d) All of these
28. Which reagent is used in Bayer's process?
(a) Na_2CO_3 (b) Carbon
(c) NaOH (d) Silica
29. What is anode mud?
(a) Fan of anode
(b) Metal of anode
(c) Impurities collected at anode in electrolysis during purification of metals
(d) All of these
30. Matrix is defined as –
(a) the unwanted foreign material present in the ore
(b) the flux added to remove the unwanted impurities from ore
(c) the slag formed as a result of the reaction of flux with gangue
(d) the material used in the reduction of metal oxide to metal
31. Aluminium is used in thermite welding because–
(a) aluminium is a light metal
(b) aluminium has more affinity for oxygen
(c) aluminium is a strong oxidising agent
(d) aluminium is a reactive metal
32. The process of extraction of metal from its ores, is known as –
(a) concentration (b) calcination
(c) purification (d) metallurgy

33. The process to heat the ore in the presence of excess supply of air below its melting point is called –
(a) roasting (b) calcination
(c) smelting (d) liquation

34. The metal extracted by leaching with a cyanide is
(a) Mg (b) Ag
(c) Cu (d) Na

35. The role of calcination in metallurgical operations is
(a) to remove moisture
(b) to decompose carbonate
(c) to drive off organic matter
(d) to achieve all the above

36. Which one of the following ores is best concentrated by froth-flotation method ?
(a) Galena (b) Cassiterite
(c) Magnetite (d) Malachite

37. Which of the following factors is of *no significance* for roasting sulphide ores to the oxides and not subjecting the sulphide ores to carbon reduction directly?
(a) Metal sulphides are thermodynamically more stable than CS_2
(b) CO_2 is thermodynamically more stable than CS_2
(c) Metal sulphides are less stable than the corresponding oxides
(d) CO_2 is more volatile than CS_2

38. The chemical composition of 'slag' formed during the smelting process in the extraction of copper is
(a) $Cu_2O + FeS$
(b) $FeSiO_3$
(c) $CuFeS_2$
(d) $Cu_2S + FeO$

39. Sulphide ores of metals are usually concentrated by froth flotation process. Which one of the following sulphide ores offer an exception and is concentrated by chemical leaching?
(a) Galena (b) Copper pyrite
(c) Sphalerite (d) Argentite

40. In the froth floatation process for the purification of ores, the ore particles float because
(a) They are light
(b) Their surface is hydrodphobic i.e. not easily wetted by water
(c) They bear electrostatic charge
(d) They are insoluble

41. A compound is formed which is used in the metallurgy of Al and which is used to remove the impurities :
(a) Slag (b) Flux
(c) Acidic flux (d) Basic flux

42. Which one of the following metals is not obtained by electrolysis in their aq state?
(a) Cu (b) Mg
(c) Ni (d) Cr

43. In the extraction of Cu, the reaction which takes place in Bessemer converter is
(a) $2FeS + 3O_2 \longrightarrow 2FeO + 2SO_2$
(b) $2Cu_2O + Cu_2S \longrightarrow 6Cu + SO_2$
(c) $2Cu_2S + 3O_2 \longrightarrow 2Cu_2O + 2SO_2$
(d) $2CuFeS_2 + O_2 \longrightarrow Cu_2S + FeS + SO_2$

44. Method used for obtaining highly pure silicon used as a semiconductor material, is
(a) Oxidation (b) Electrochemical
(c) Crystallization (d) Zone refining

45. Which of the following elements is present as the impurity to the maximum extent in the pig iron ?
(a) Manganese (b) Carbon
(c) Silicon (d) Phosphorus

46. The following reactions take place in the blast furnace in the preparation of impure iron. Identify the reaction pertaining to the formation of the slag.
(a) $Fe_2O_3(s) + 3\,CO(g) \rightarrow 2\,Fe\,(l) + 3\,CO_2\,(g)$
(b) $CaCO_3\,(s) \rightarrow CaO\,(s) + CO_2\,(g)$
(c) $CaO\,(s) + SiO_2(s) \rightarrow CaSiO_3\,(s)$
(d) $2C(s) + O_2\,(g) \rightarrow 2\,CO(g)$

47. Duraluminium is an alloy of
(a) Al and Mg
(b) Al, Mg and Ni
(c) Al, Mg, Mn and Cu
(d) Al and Ni.

48. Thomas slag is
(a) $Ca_3(PO_4)_2$
(b) $CaSiO_3$
(c) Mixture of (a) and (b)
(d) $FeSiO_3$

49. During the extraction of Fe, slag obtained is
(a) FeO (b) $FeSiO_3$
(c) $MgSiO_3$ (d) $CaSiO_3$

50. Which of the following ore is chloride?
(a) Bauxite (b) Horn silver
(c) Magnesite (d) Zincite

51. Blister copper is
(a) Impure Cu
(b) Cu alloy
(c) Pure Cu
(d) Cu having 1% impurity

52. Which of the following methods is not involved in metallurgy of iron?
(a) Calcination
(b) Smelting
(c) Concentration of ore
(d) None of these

53. $Cu_2S + 2Cu_2O \longrightarrow 6Cu + SO_2$
In which process of metallurgy of copper, above equation is involved?
(a) Roasting (b) Self reduction
(c) Refining (d) Purification

54. $2CuFeS_2 + O_2 \longrightarrow Cu_2S + 2FeS + SO_2$
Which process of metallurgy of copper is represented by above equation?
(a) Concentration (b) Roasting
(c) Reduction (d) Purification

55. Cinnabar is an ore of –
(a) Hg (b) Cu
(c) Pb (d) Zn

56. Iron is obtained on large scale from Fe_2O_3 by
(a) Reduction with Na
(b) Reduction with Al
(c) Reduction with CO
(d) Passing H_2

57. The important oxide ore of iron is
(a) Siderite (b) Haematite
(c) Pyrites (d) Bauxite

58. Calcination is used in metallurgy for removal of?
(a) Water and sulphide
(b) Water and CO_2
(c) CO_2 and H_2S
(d) H_2O and H_2S

59. The process of zone refining is used in the purification of
(a) Si (b) Al
(c) Ag (d) Cu

60. Roasting is generally done in case of the following
(a) Oxide ores (b) Silicate ores
(c) Sulphide ores (d) Carbonate ores

CHAPTER 21

The p-Block Elements (Group 15, 16, 17 & 18)

CONCEPTS

GROUP 15 (NITROGEN FAMILY)

- Members of group - N, P, As, Sb, Bi. General electronic configuration - $ns^2 np^3$.
- The group members are known as **Pnicogens** meaning suffocation.

Properties of Group 15

Physical properties

(i) **Ionisation enthalpy** of group members are **exceptionally high** due to exactly half filled valence shell. Ionisation enthalpy decrease down to group due to increase in size.

(ii) **Electronegativity** decreases down the group.

(iii) **Non metallic character** increases down the group due to increases in size

N P (non-metals) As Sb (metalloid) Bi (metal)

(iv) **Catenation** decreases down the group- N > P > As > Sb > Bi.

(v) **Allotropy**

(a) Except Bi all elements show allotropy.

(b) N – two allotrops - α and β nitrogen.

(c) P – 3 allotrops - white (or yellow), red and black phosphorus.

(d) As – 3 allotrops - grey, yellow and black arsenic.

(e) Sb – 2 allotrops - metallic and yellow antimony.

(vi) **Elemental state**

(a) Nitrogen exist in form of N_2

(b) Nitrogen displays a great tendency to form $p\pi - p\pi$ multiple bonds with itself as well as with carbon and oxygen. The tendency to exhibit $p\pi - p\pi$ multiple bonding decreases as we move down the group.

(c) **P, As and Sb exist in form of P_4, As_4 and Sb_4** due to formation of 3 σ bonds.

Chemical properties

(i) **Oxidation state:** The elements of this group can exhibit various oxidation states ranging between –3 to + 5 Negative oxidation state will be exhibited when they combine with less electro negative element & positive oxidation state will be exhibited with more electronegative element. Positive oxidation state becomes more favourable as we move down the group due to increasing metallic character & electropositivity. Although due to inert pair effect the stability of +5 state will also decrease. The only stable compound of Bi (v) is BiF_5.

(ii) **Covalency :** This is the number of covalent bonds formed by an atom. Maximum covalency of Nitrogen is four because it does not have d-orbitals to expand its covalency. where as other elements can exhibit covalency of 5 & 6 as well by the use of their d-orbitals eg PCl_5, $[PCl_6]^-$

(iii) **Hydrides (MH_3)**

(a) **Bond angle** decreases due to decrease in bond pair - bond pair repulsion.

NH_3	PH_3	AsH_3	SbH_3	BiH_3
107°	94°	92°	91°	90°

(b) **Basic character** $NH_3 > PH_3 > AsH_3 > SbH_3 > BiH_3$ (due to decrease in availability of lone pair)

(c) **Boiling point** $PH_3 < AsH_3 < NH_3 < SbH_3 < BiH_3$ ($\because$ of intermolecular hydrogen bonding in nitrogen it shows exceptionally high boiling point)

(d) **Thermal stability** decreases down the group due to decrease in strength of M–H bond.

$NH_3 > PH_3 > AsH_3 > SbH_3 > BiH_3$

(e) **Reducing character** increases down the group at strength of M–H bond decreases.

$NH_3 < PH_3 < AsH_3 < SbH_3 < BiH_3$.

(iv) Halides

All the group members form trihalides (MX_3) and pentahalides (MX_5)

(a) All the elements of the group form trihalides. The ionic character of trihalides increases we move down the group. Except NCl_3, all the trihalides are hydrolysed by water. This is due to the absence of d-orbitals in nitrogen.

(b) PF_3 is not hydrolysed because fluorine being more electronegative than oxygen forms more stable bonds with phosphorus than P – O bonds.

(c) N cannot form NX_5 because of non-availability of d-orbitals. Bi cannot form a BiX_5 because of reluctance of 6s electrons of Bi to participate in bond formation.

(d) The hybridisation of M in MX_3 is sp^3 and shape is pyramidal. In MX_5 M is sp^3d hybridised and shape is trigonal bipyramid. The axial bonds in MX_5 are weaker and longer, So MX_5 are less stable and decompose on heating eg:

$$PCl_5 \xrightarrow{\Delta} PCl_3 + Cl_2$$

(v) Oxides

(a) Nitrogen forms a number of oxides. The rest of the members (P, As, Sb and Bi) of the group form two types of oxides: E_2O_3 and E_2O_5.

(b) The reluctance of P, As, Sb and Bi to enter into $p\pi - p\pi$ multiple bonding leads to cage structures of their oxides and they exist as dimers, E_4O_6 and E_5O_{10}.

(c) The basic nature of the oxides increases with increase in atomic number of the element. Thus, the oxides of nitrogen (except N_2O and NO), P (III) and As (III) are acidic, Sb (III) oxide is amphoteric and Bi (III) oxide is basic.

Anamolous Behaviours of Nitrogen

Nitrogen shows anomalous properties due to very small size, high electro negativity, and absence of d-orbitals in the valance shell some of its anomalous properties are:

(a) forms pπ-pπ bonds with itself, oxygen and carbon, others do not form pπ- pπ bonds due to their large size.

(b) Nitrogen exists as discrete N_2 molecule due to its ability of pπ-pπ bonding & so it is gas, others are solid

(c) It is highly uncreative due to very strong triple bond, $N \equiv N$, having bond energy of 941.4 kJM^{-1}

Nitrogen

(a) Found in 78% in free form and rest in form of nitrates and ammonium salts.

(b) **Lab method for preparation of N_2**

$$NH_4Cl + NaNO_2 \longrightarrow NH_4NO_2 + NaCl$$
$$NH_4NO_2 \xrightarrow{\Delta} N_2 + 2H_2O$$

(c) **Ammonia (NH_3),** Manufactured by **Haber's process.** $N_2 + 3H_2 \underset{\Delta}{\overset{Fe, Mo}{\rightleftharpoons}} 2NH_3$

(d) **Nessler's reagent** is **alk. solution of K_2HgI_4** It is used to detect presence of NH_4^+ due to formation of reddish brown ppt of Millon's base

$$\underset{\text{Nessler's reagent}}{K_2HgI_4} + NH_3 + H_2O \longrightarrow \underset{\substack{\text{Millon's base} \\ \text{(red-brown ppt)}}}{NH_2 - Hg - O - Hg - I}$$

Uses of Nitrogen

(i) Manufacture of NH_3 and other industrial chemicals containing nitrogen, *e.g.*, calcium cyanamide.

(ii) As inert diluent for reactive chemicals in iron and steel industry.

(iii) As a refrigerant to preserve biological materials, food items and in cryosurgery.

Oxides of Nitrogen

S.No.	Oxide	Oxidation state	Nature	Preparation	Properties
1.	N_2O (nitrous oxide) N – N – O	+1	neutral	$2NaNO_3 + (NH_4)_2SO_4$ ↓ $2N_2O + 4H_2O + Na_2SO_4$	• Diamagnetic • Called **laughing gas** • Used as anesthetic.

2.	NO (nitric oxide)	+2	neutral N – O	$2NaNO_2 + 2FeSO_4 + 2H_2SO_4$ ↓ $NO + Fe_2(SO_4)_3 + 2H_2O$	• Paramagnetic. • Dimer is diamagnetic
3.	NO_2 (nitrogen dioxide) O–N–O (N bonded to two O)	+3	acidic	$2Pb(NO_3)_2 \xrightarrow{\Delta} 2PbO + 4NO_2 + O_2$	• Paramagnetic. • $NO_2 \underset{Heat}{\overset{Cool}{\rightleftharpoons}} N_2O_4$ • Brown fumes • Pungent smell
4.	N_2O_3 $O=N-NO_2$	+4	acidic	$NO + NO_2 \xrightarrow{\Delta} N_2O_3$	• Anhydride of HNO_2 • Pale blue solid
5.	N_2O_5 $O_2N-O-NO_2$	+5	acidic	$2HNO_3 + P_2O_5 \rightarrow N_2O_5 + 2HPO_3$	• Anhydride of HNO_3

Oxoacids of Nitrogen - HNO_2, HNO_3 and $H_2N_2O_2$ (hyponitrous acid)

HNO_2 - Is oxidising as well as reducing agent.

HNO_3 - (Called Aqua fortis)

Nitric Acid

• **Preparation of Nitric Acid (HNO_3)**

• **Birkeland - Eyde process - Preparation of HNO_3 from air**

$$N_2 + O_2 \longrightarrow \underset{\text{Nitric oxide}}{2NO};$$

$$2NO + O_2 \longrightarrow \underset{\text{Nitrogen dioxide}}{2NO_2}$$

$$2NO_2 + H_2O \longrightarrow \underset{\text{Nitrous acid}}{HNO_2} + \underset{\text{Nitric acid}}{HNO_3};$$

$$3HNO_2 \longrightarrow HNO_3 + H_2O + NO$$

• **Ostwald process (oxidation of NH_3)**

$$4NH_3 + 5O_2 \xrightarrow[500k]{Pt/Rh\ gauge} 4NO + 6H_2O;$$

$$2NO + O_2 \rightleftharpoons 2NO_2$$

$$3NO_2 + H_2O \longrightarrow 2HNO_3 + NO$$

Properties : Yellow colour of HNO_3 is due to **dissolved NO_2** in it.

• Very powerful oxidising agent.

• Mg and Mn liberate H_2 on reaction with HNO_3.

• 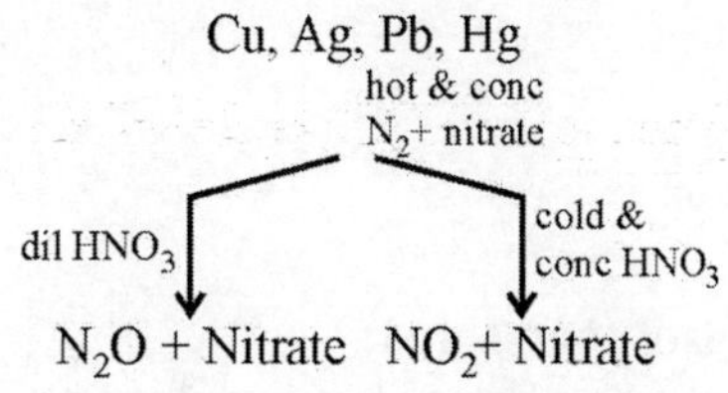

• 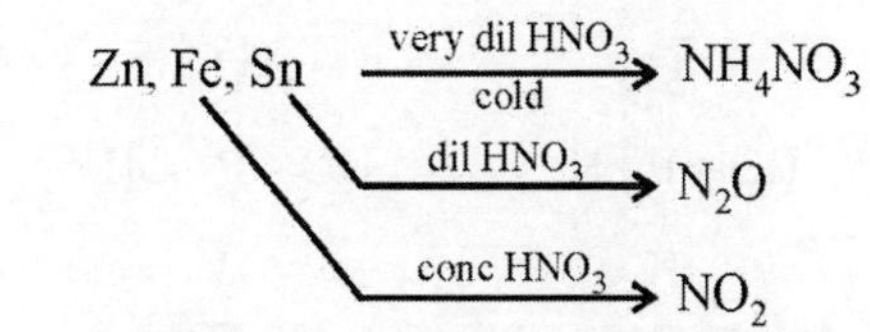

• Metals like, Fe, Cr, Ni, Al or Co become inactive or passive due to stable oxide layers. Noble Metals like Pt, Pd, Os, Ir and Au do not react with nitric acid. They react with **aqua regia** (1 vol. Conc. HNO_3 + 3 vol. Conc. HCl).

Phosphorus

(a) **Ores - Phosphorite** $Ca_3(PO_4)_2$, **Chlorapatite** - $3Ca_3(PO_4)_2.CaCl_2$.

(b) **Allotrops of Phosphorus**

	Properties	White phosphorus	Red phosphorus	Black Phosphorus
1.	Nature	Poisonous	Non-poisonous	Non-poisonous
2.	Solubility in CCl_4 and CS_2	soluble	insoluble	
3.	Smell	Garlic	Odourless	Colourless
4.	Catches fire	spontaneously (so it is kept in water)	does not catch fire easily	does not catch fire easily
5.	Stabililty	Less stable at ordinary	Most stable at temperature	Most stable ordinary trmperature
6.	Chemical reactivity	very reactive	less reactive	Least reactive
7.	Structure	P_4 unit	Polymeric	double layered

(c) **Phosphorus** forms two oxides - P_2O_3 and P_2O_5

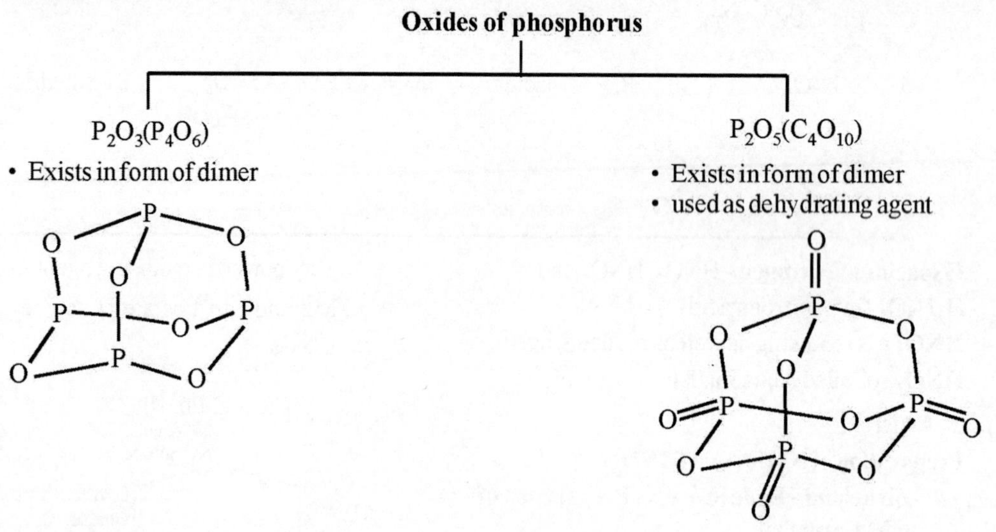

(d) **Oxoacids of phosphorus :**

(i) Each phosphorus atom is sp^3 hybridised.

(ii)

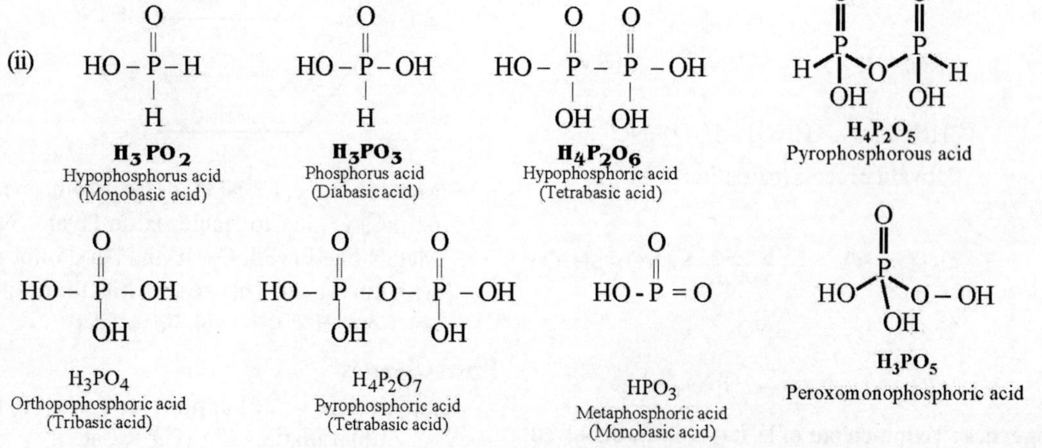

Phosphine (PH_3)

Preparation

(i) **Any phosphide** $+H_2O \rightarrow PH_3$

$$Ca_3P_2 + H_2O \rightarrow 2PH_3 + Ca(OH)_2$$

$$Na_3P + 3H_2O \rightarrow PH_3 + 3NaOH$$

$$2AlP + 3H_2SO_4 \rightarrow 2PH_3 + Al_2(SO_4)_3$$

(ii) ***Decomposition*** **of H_3PO_3**:

$$4H_3PO_3 \rightarrow 3H_3PO_4 + PH_3$$

(iii) ***Lab. method***:

$$4P + 3NaOH + 3H_2O \rightarrow 3NaH_2PO_2 + PH_3$$

(iv) ***Pure*** **PH_3** :

$$PH_4I + NaOH \rightarrow PH_3 + NaI + H_2O$$

Properties

Colourless, highly poisonous, with rotten fish odour gas, slightly soluble in water

Chemical properties :

(i) **Basic nature** : $PH_3 + HI \rightarrow PH_4I$

(ii) **Decomposition** : $4PH_3 \xrightarrow{311K} P_4 + 6H_2$

(iii) **Combustibility** :

$$4PH_3 + 8O_2 \rightarrow P_4O_{10} + 6H_2O$$

Pure PH_3 is not spontaneously inflammable. Ordinary PH_3 is spontaneously inflammable due to the presence of P_2H_4

(iv) **With metallic salts** :

$$3CuSO_4 + 2PH_3 \rightarrow Cu_3P_2 + 3H_2SO_4$$

(v) **With chlorine** : $PH_3 + 3Cl_2 \rightarrow PCl_3 + 3HCl$

Fertilisers

Required for growth of plants.

(i) **Nitrogenous fertilizers** : These provide nitrogen to the plants. They are
Ammonium sulphate (Sindri fertilizer) :

$$(NH_4)_2CO_3 + CaSO_4 \longrightarrow (NH_4)_2SO_4$$

Basic Calcium nitrate :
Calcium cyanamide :

$$CaC_2 + N_2 \xrightarrow{850°–100°C} \underset{\text{Nitrolim}}{[CaCN_2 + C]}$$

(ii) **Phosphatic fertilizers** : These provide phosphorous to plants
Super phosphate of lime :

$$Ca_3(PO_4)_2 + 2H_2SO_4 \longrightarrow \underline{Ca(H_2PO_4)_2 + 2CaSO_4}$$

Phosphatic slag or Thomas slag : $2Ca_3(PO_4)_2$

Triple super phosphate :

$$3Ca(H_2PO_4)_2 + 4H_3PO_4 + Ca_3(PO_4)_2$$

GROUP 16 (OXYGEN FAMILY)

- Members of group are O, S, Se, Te, Po. General electronic configuration $ns^2\ np^4$.
- O, S, Se and Te are called **Chalcogens** (Ore forming substances).

Properties of Group 16

Physical properties :

(i) **Ionisation potential** lower than group 15 members.

(ii) **Electron affinity** in general high except for oxygen and decreases down the group (EA of oxygen is low due to its small size).

(iii) **Allotropy :**

(a) Allotrops of oxygen - O_2 and O_3

(b) Allotrops of S - Rhombic, Monoclinic, Plastic.

(c) Allotrops of Se - Red and grey.

(d) Te - Non metallic and metallic

(e) Po - α and β both metallic

(iv) **Metallic and non metallic character**

$$\underbrace{O \quad S}_{\text{Non metals}} \quad \underbrace{Se \quad Te}_{\text{Metalloids}} \quad \underset{\text{(Metal)}}{Po}$$

(Radio active) $t_{1/2}$ 138.4 days

(v) **Abundance** $O > S > Se > Te > Po$
Oxygen is the most abundant element.

(vi) **Electronegativity** - It decreases down the group due to decrease in the effective nuclear charge down the group.

Chemical properties :

(i) **Oxidation states :**

(a) **–2 oxidation** state is shown by oxygen. Other elements do not show –2 oxidation state due to low ionization enthalpy and low electronegativity.

(b) **+2, +4 and +6**. oxidation states are shown by other members of that group. Stability of +4 oxidation state increases down the group due to inert pair effect.

O	S	Se	Te	Po
–1,–2	–2 to +6	–2 to +6	–2 to +6	–2 to +6

In OF_2 the oxidation state of oxygen is +2

(ii) **Catenation** - The tendency to form chains of identical atoms is known as catenation. It follows the order

S-S >	Se - Se >	O - O >	Te -Te
226	172	142	126 kJ/mole

The higher the bond strength, the higher is the catenation.

(iii) **Atomicity** - Oxygen is diatomic, sulphur and selenium octa atomic with puckered ring structure

S S S S
S S S S

(iv) **Molecular state :**

(a) **Oxygen exists in form of O_2** molecule due to pπ-pπ bond.

(b) **Other group members exist in form of octa atomic puckered 8 atom rings** as they are incapable of forming pπ-pπ bond.

(v) **Hydrides**

(a) H_2O is liquid (due to intermolecular hydrogen bonding) whereas H_2S, H_2Te, H_2Se, H_2Te are gases.

(b) The volatility increases markedly from water to hydrogen sulphide and then declines. This is evident in their boiling point. Increasing order of boiling points of hydrides is $H_2S < H_2Se < H2\,Te < H_2O$. Down the group boiling point increases because of increasing M.W which increases the vandrwaal interaction. H_2O has abnormally high b.p. due to hydrogen bonding.

(c) **Thermal stability** $H_2O > H_2S > H_2Se > H_2Te$.

(d) **Bond angles :** $H_2O > H_2S > H_2Se > H_2Te$.

(e) **Reducing character :** $H_2O < H_2S < H_2Se < H_2Te$.

(vi) **Halides**

(a) **Oxygen** forms OF_2, Cl_2O and Br_2O. All of them have tetrahedral shape.
Compounds of O and F are called **fluorides.** Compounds of O and Cl or Br or I are called **oxides** because F is more electronegative than O but other halogens are less negative than oxygen.

(b) **S forms** S_2F_2, S_2Cl_2, S_2Br_2, SF_2, SCl_2, SBr_2, SF_4, SCl_4 and SF_6.

(c) **Se and Te** forms SeF_6, TeF_6 and chlorides and bromides in +1, +2 and +4 oxidation states.

(d) As the size of halogen atom increases, maximum coordination number of elements of this group decreases. So no hexachloride is known, only three tetrabromides are known.

(e) **All elements excepts oxygen form hexafluorides :** They have sp^3d^2 hybridisation and octahedral shape.

(f) **All elements excepts oxygen form tetrafluoride :** They have sp^3d hybridisation.

(vii) **Oxides**

(a) The elements form oxides of +2 to +6 oxidation state.

(b) Acidic character - $\underbrace{SO_2 > SeO_2}_{\text{acidic}} > \underbrace{TeO_2 > PoO_2}_{\text{amphoteric}}$.

(c) Acidic character - $SO < SO_2 < SO_3$ (according to oxidation state).

(d) **SO_2** - gas at room temperature.
- Acts as both oxidising and reducing agent.
- Bleaching action of SO_2 is due to reduction
$$SO_2 + 2H_2O \longrightarrow H_2SO_4 + 2[H]$$
- **It is anhydride of H_2SO_3.**

(e) SeO_2 and TeO_2 are polymers having zig-zag structure.

(f) SeO_3 has tetrameric cyclic structure.

(g) **SO_3** - Acidic oxide.
- Oxidising agent.
- **It is anhydride of H_2SO_4.**

Oxoacids of Sulphur

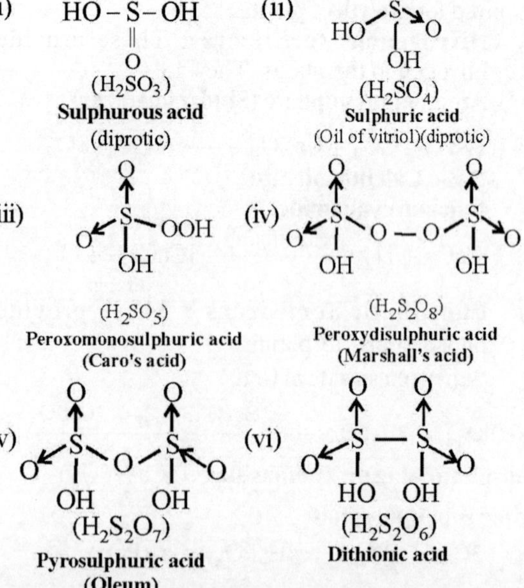

Dioxygen

Preparation:

(i) By heating oxygen containing salts, such as chlorates nitrates and permanganates.

$$2KClO_3 \xrightarrow[MnO_2]{\Delta} 2KCl + 3O_2$$

(ii) By thermal decomposition of the oxides of metals low in the electrochemical series and higher oxides of some metals.

$$2Ag_2O\,(s) \longrightarrow 4Ag\,(s) + O_2\,(g)$$

(iii) By catalytic decomposition of H_2O_2.

(iv) From water or air by electrolysis.

(v) By liquefaction of air and followed by its fractional distination when N_2 distils off at 77K and O_2 remains os liquid.

Properties:

(i) Colourless and odourless gas soluble in water.

(ii) It has three stable isotopes: ^{16}O, ^{17}O and ^{18}O.

(iii) It reacts with nearly all metals and non-metals except Au, Pt and some noble gases.

Uses:

(i) In respiration and combustion processes.

(ii) In oxyacetylene welding and manufacture of steel.

(iii) As oxygen cylinders in hospitals, high altitude flying and in mountaineering.

Ozone

Pale blue gas at room temperature.

(i) It is prepared by silent electric discharge in **Siemens's ozoniser**

$$3O_2 \rightleftharpoons 2O_3$$

(ii) It is strong oxidising agent. It oxidises -

(a) $PbS + 4O_3 \rightarrow PbSO_4 + 4O_2$

(b) $2HCl + O_3 \rightarrow Cl_2 + H_2O + O_2$

(c) $2KI + H_2O + O_3 \rightarrow 2\,KOH + I_2 + O_2$

Sulphur

(i) It is extracted by Frasch process.

(ii) **S is used in vulcanisation of rubber**.

(iii) H_2S is prepared in **Kipp's apparatus in lab**

$$FeS + H_2SO_4 \longrightarrow FeSO_4 + H_2S$$

(iv) **Allotropes of sulphur** – Yellow rhombic (α – sulphur) and monoclinic (β - sulphur).

- **Properties of yellow sulphur :**
  - Its mp is 385.8K
  - Its sp. gravity is 2.06
  - It is insoluble in water, somewhat soluble in benzene, alcohol, ether & readily soluble in CS_2.
- **Properties of monoclinic sulphur :**
  - Its mp is 393 k
  - Its sp gravity is 1.98
  - It is soluble in CS_2
  - It is stable only above 369K

Sulphuric Acid (H_2SO_4)

Preparation

(i) **Lead chamber process**

$$2SO_2 + O_2 \underset{\text{of Nitrogen}}{\overset{\text{oxides}}{\rightleftharpoons}} 2SO_3\ ;$$

$$SO_3 + H_2SO_4 \longrightarrow \underset{\text{Oleum}}{H_2S_2O_7}$$

$$H_2S_2O_7 + H_2O \longrightarrow \underset{\text{sulphuric acid}}{2H_2SO_4}$$

(ii) **Contact process** – V_2O_5 is used as catalyst is place of oxides of nitrogen

Properties

(i) Strong dibasic acid.

(ii) Dehydrating agent.

(iii) Oxidising agent.

Hypo (Sodium Thiosulphate) $Na_2S_2O_3.5H_2O$

(i) Used in iodometric titration and to remove I_2 stains from skin and clothes.

$$2Na_2S_2O_3 + \underset{\text{(brown)}}{I_2} \longrightarrow \underset{\substack{\text{Sod. tetrathionate}\\\text{(colourless)}}}{Na_2S_4O_6} + 2NaI$$

(ii) Used as **antichlor** for removing excess Cl_2 in bleaching industry.

$$Cl_2 + H_2O + Na_2S_2O_3 \longrightarrow 2HCl + Na_2SO_4 + S$$

(iii) Used in **photography** to remove undecomposed silver halide on photographic paper.

$$AgBr + 2Na_2S_2O_3 \longrightarrow \underset{\substack{\text{sod. argento thiosulphate}\\\text{(soluble complex)}}}{Na_3[Ag(S_2O_3)_2]} + NaBr$$

GROUP 17 (HALOGEN FAMILY)

- **Members of group :** F, Cl, Br, I, At. General electronic configuration $ns^2\,np^5$.
- They are called **halogens** due to their salt forming tendency.

Properties of Group 17

Physical properties :

(i) **Atomic size :** Smallest in period and increases down the group.

(ii) **Electronegativity** highest in period. F > Cl > Br > I.

(iii) **Electron affinities** are high, increases from F to Cl and then decreases down the group. The electron affinity of F is less than that of Cl due to its small size with high electron density. Electron affinity varies as Cl > F > Br > I

(iv) **Bond energy :** $Cl_2 > Br_2 > F_2 > I_2$. (because of greater inter electronic repulsion in F_2)

(v) **Bond length :** $I_2 > Br_2 > Cl_2 > F_2$.

(vi) **Atomicity :** diatomic covalent molecules.

(vii) **Colour :**

Halogen F_2 Cl_2 Br_2
I_2 colour light greenish
redish violet
yellow yellow brown

(viii) **Heat of hydration** decreases with increase in size of halogen. $F^- > Cl^- > Br^- > I^-$.

(ix) **Boiling and melting points** increase down group due to increase in van der Waal's forces of attraction. $F_2 < Cl_2 < Br_2 < I_2$.

Chemical properties :

(i) **Oxidation states :**
 (a) –1 exhibited by all halogens.
 (b) +3, +5, +7 are exhibited by halogens other than F.
 (c) +4 and +6 oxidation states are found in oxy acids.

(ii) **Oxidising power :** F_2 is strongest oxidising agent. $F_2 > Cl_2 > Br_2 > I_2$.

(iii) **Reactivity** decreases down group. F > Cl > Br > I.

(iv) **Hydrogen Halides**
 (a) **Boiling point :** HF > HI > HBr > HCl (due to H bonding in HF)
 (b) **Acidic strength :** HI > HBr > HCl > HF (due to decrease in bond dissociation energy)
 (c) **Reducing character :** HI > HBr > HCl > HF (due to weakest bond in HI)
 (d) **Thermal stability :** HF > HCl > HBr > HI.

(v) **Oxides**

(a) Covalent in nature due to lesser electronegativity difference between halogen and oxygen.

Oxidation state	F	Cl	Br	I
–1	OF_2, O_2F_2	–	–	–
+1	–	Cl_2O	Br_2O	–
+4	–	ClO_2	BrO_2	–
+5	–	–	–	I_2O_5
+6	–	ClO_3	BrO_3	–
+7	–	Cl_2O_7	–	–

(b) **Acidic strength of oxides ∝ oxidation state.**

Oxoacids of Halogens

(a) F forms only one oxo-acid HOF. It is known as fluoric (I) acid or hypofluorous acid. Other halogen form some of the oxo-acids of the series: HOX, HOXO, $HOXO_2$, $HOXO_3$.

(b) Hypohalous acids are all weak acids and exist in solution only. Acid strength decreases down the group. HOCl > HOBr > HOI

(c) All oxoacids are monobasic.

(d) Thermal stability and acidic character increases with increase in oxidation number of halogen.

$HClO < HClO_2 < HClO_3 < HClO_4$.

(e) Oxidising power $HClO > HClO_2 > HClO_3 > HClO_4$.

(f) Strength of conjugate base $ClO^- > ClO_2^- > ClO_3^- > ClO_4^-$.

(g) Stability of anions of oxoacids $ClO^- < ClO_2^- < ClO_3^- < ClO_4^-$.

(h) Acidity of oxoacids of different halogens having same oxidation no. $HClO_4 > HBrO_4 > HIO_4$.

Interhalogen Compounds

(i) The compound formed by reaction of halogens among themselves are called interhalogen compounds. They are formed due to large difference of size and electronegativity among halogens.

(ii) These compounds are covalent in nature due to less differrence in electronegativities.

(iii) They are diamagnetic.

(iv) They are strong oxidising agent.

(v) More electronegative halogen form negative part of interhalogen of molecule.

	Types of interhalogen comp.	AB type	AB_3 type	AB_5 type	AB_7 type
1.	Examples	ClF, BrF	ClF_3, ICl_3	ClF_5, BrF_5	IF_7
2.	Shape	Linear	T shaped	Square pyramidal	Pentagonal bipyramidal
3.	Hybridisation	sp	dsp^3	d^2sp^3	d^3sp^3

Pseudohalogens and Pseudohalides

Pseudohalogen molecule is made of two electronegative atoms or groups and they resemble halogens in properties.

Pseudohalogen	**Pseudohalide ion**
Cyanogen $(CN)_2$	Cyanide CN^-
Oxocyanogen $(OCN)_2$	Cyanate OCN^-
Thiocyanogen $(SCN)_2$	Thiocyanate SCN^-
Selenocynogen $(SeCN)_2$	SelenocyanateSeCN^-

Freons (CFC)

(a) They cause ozone layer depletion.

(b) **Freon-11** is CCl_3F, **Freon-12** is CCl_2F_2.

Fluorine (F_2) (Isolated by Moissan)

(a) **Ores** - Fluospar (CaF_2), Cryolite (Na_3AlF_6).

(b) **Isolation of F_2 is difficult** because it is very reactive and strong oxidising agent.

Dennis method :

$$KHF_{2(molten)} \longrightarrow KF + HF$$

$$KHF_2 \longrightarrow K^+ + 2F^+ + H^+$$

$$\text{At anode } 2F^- - 2e \longrightarrow F_2$$

$$\text{At cathode } 2H^+ + 2e^- \longrightarrow H_2$$

(c) **F_2 gives flames with moist air due to formation of liquid HF.**

(d) $$2F_2 + \underset{\text{(Hot and conc)}}{4KOH} \longrightarrow 4KF + O_2 + 2H_2O$$

Chlorine (Discovered by Scheele)

(a) **Manufacture by Deacon's process**

$$2HCl + O_2 \xrightarrow[400°C]{CuCl_2} 2Cl_2 + 2H_2O$$

(b) **In lab it is prepared by action of MnO_2 or $KMnO_4$ or $K_2Cr_2O_7$ with HCl.**

(c) **Reaction with NaOH**

$$Cl_2 + \text{dil. NaOH} \xrightarrow{\text{Cold}} NaCl + NaClO + H_2O$$

$$3Cl_2 + \text{conc. 6NaOH} \xrightarrow{\text{Hot}} 5NaCl + NaClO_3 + 3H_2O$$

(d) **Reaction with NH_3 :**

$$8NH_{3\,(excess)} + 3Cl_2 \longrightarrow 6NH_4Cl + N_2$$

$$3Cl_{2\,(excess)} + NH_3 \longrightarrow NCl_3 + 3HCl$$

(e) Chlorine is dried over conc. H_2SO_4. It can't be dried over CaO or KOH or NaOH because it reacts with them.

Bromine

It was discovered by Belard.

(a) Occurs in combined form in sea water.

(b) $$2KBr + 3H_2SO_4 + MnO_2 \longrightarrow 2KHSO_4 + MnSO_4 + 2H_2O + Br_2$$

Iodine

(a) **Occurance** - Deep sea weed **Laminaria** is main source of iodine. Its ash called as **kelp** has about 1% iodine in form of alkali metal iodides. **Caliche** (Crude chili salt peter, $NaNO_3$) has NaI.

(b) Solubility of I_2 in water increases on addition of KI due to formation I_3^- ion.

$$KI + I_2 \rightleftharpoons KI_3$$

$$I_2 + I^- \longrightarrow I_3^-$$

Bleaching Powder ($CaOCl_2$)

(a) It loses its chlorine by action of dil acids or CO_2

$$CaOCl_2 + CO_2 \longrightarrow CaCO_3 + Cl_2$$

$$CaOCl_2 + 2HCl \longrightarrow CaCl_2 + H_2O + Cl_2$$

Amount of chlorine liberated in this way is called **available chlorine**.

(b) It is **oxidising** and **bleaching** agent.

Hydrogen Chloride

(a) **Preparation:** By heating NaCl with conc. H_2SO_4.

(b) **Properties:**

(*i*) Colourless and pungent smelling gas, easily liquefied to a colourless liquid and freezes to a white crystalline solid.

(*ii*) Extremely soluble in water.

(c) **Uses:**

(*i*) Manufacture of Cl_2, NH_4Cl and glucose (from corn starch),

(*ii*) For extracting glue from bones and purifying bone black.

(*iii*) In medicine and as a laboratory reagent.

GROUP 18 (NOBLE ELEMENTS)

- Members of group - He, Ne, Ar, Kr, Xe. General electronic configuration ns^2np^6.
- They are called **aerogens** since they occur in air.

Properties of Group 18

(i) **Atomic radii** - largest in period and increases in group. **It is van der Waal's radii not covalent radii.**

(ii) **Ionisation enthalpies** highest in period due to stable configuration. It decreases down the group.

(iii) **Electron gain enthalpies** zero due to completely filled valence shell.

(iv) **Boiling point and melting point** increases down the group due to increase in magnitude of van der Waal's forces.

$$He < Ne < Ar < Kr < Xe < Rn$$

(v) **Polarizability** increases down group.

$$He < Ne < Ar < Kr < Xe$$

(vi) **Ease of liquation :**

They cannot be liquified easily at low temperature since atoms are held together by weak van der Waal's forces of attraction. Ease of liquification increases down group from He to Rn.

$$He < Ne < Ar < Kr < Xe < Rn.$$

($\because$ when attraction increases down the group, van der Waal's forces of attraction increases and hence liquification becomes easier).

(vii) **Solubility** - sparingly soluble in water.

$He < Ne < Ar < Kr < Xe.$

(viii) **Thermal conductivity :** $He > Ne > Ar > Kr > Xe.$

(ix) **Diffusion :** $He > Ne > Ar > Kr > Xe.$

Xenon

(a) **Xe forms compounds** with fluorine and oxygen at it has least ionisation potential in the group.

(b) Xe forms binary fluorides XeF_2, XeF_4 and XeF_6.

$Xe + F_2$ $\xrightarrow[\text{Ni vessel}]{\text{in sealed}}$
- $\xrightarrow{2:1,\ 673K,\ 1bar} XeF_2$
- $\xrightarrow{1:5,\ 873K,\ 7\ bar} XeF_4$
- $\xrightarrow{1:20,\ 573K,\ 60-70\ bar} XeF_6$

(c) XeF_2 on hydrolysis forms Xe, HF and O_2.

$$2XeF_2\,(s) + 2H_2O\,(l) \longrightarrow 2Xe\,(g) + 4HF\,(aq) + O_2\,(g)$$

(d) Hydrolysis of XeF_4 and XeF_6:

$$6XeF_4 + 12H_2O \longrightarrow 2XeO_3 + 4Xe + 3O_2 + 24HF$$

$$XeF_6 + 3H_2O \longrightarrow XeO_3 + 6HF$$

(e) $XeF_6 + H_2O \xrightarrow[\text{hydrolysis}]{\text{Partial}} XeOF_4 + 2HF$

$$XeF_6 + 2H_2O \xrightarrow[\text{hydrolysis}]{\text{Partial}} XeO_2F_2 + 4HF$$

	Molecule	Hybridisation	Structure
(1)	XeF_2	sp^3d	linear
(2)	XeF_4	sp^3d^2	square planar
(3)	XeF_6	sp^3d^3	distorted octahedral
(4)	$XeOF_2$	sp^3d	T-shpaed
(5)	$XeOF_4$	sp^3d^2	square pyramidal
(6)	XeO_2	sp^3	pyramidal

Clathrates

Molecules with large spaces in which small gas molecules are traped. They are non stoichiometric compounds. Only Ar, Kr and Xe form clathrates since their sizes match the size of cavities present in most of the organic and inorganic compounds. He and Ne do not form clathrates since they are too small to be entrapped within the cavities.

Uses of Noble Gases

(i) Mixture of 20% He + 80% O_2 is used by deep sea divers for artificial respiration.

(ii) Neon is used in discharge lamps and signs.

(iii) Radon is used in treatment of cancer (radiotherapy).

Exercise

1. Nitrogen is relatively inactive element because
   (a) its atom has a stable electronic configuration
   (b) it has low atomic radius
   (c) its electronegativity is fairly high
   (d) dissociation energy of its molecule is fairly high

2. Of the following which is paramagnetic and has three electron bond in its structure?
   (a) N_2O (b) NO
   (c) N_2O_3 (d) N_2O_5

3. Which of the following metal evolves hydrogen on reacting with cold dilute HNO_3?
   (a) Mg (b) Al
   (c) Fe (d) Cu

4. With respect to protonic acids, which of the following statements is correct ?
(a) PH_3 is more basic than NH_3
(b) PH_3 is less basic than NH_3
(c) PH_3 is equally basic as NH_3
(d) PH_3 is amphoteric while NH_3 is basic

5. PCl_5 is possible but NCl_5 does not exist :
(a) in N, d-sub-shell is absent
(b) ionization energy of N is very high
(c) it does not like Cl
(d) None of these

6. Which of the following oxy-acids has the maximum number of hydrogens directly attached to phosphorus?
(a) $H_4P_2O_7$ (b) H_3PO_2
(c) H_3PO_3 (d) H_3PO_4

7. The gases respectively absorbed by alkaline pyrogallol and oil of cinnamon are
(a) O_3, CH_4 (b) O_2, O_3
(c) SO_2, CH_4 (d) N_2O, O_3.

8. Bleaching action of SO_2 is due to
(a) reduction (b) oxidation
(c) hydrolysis (d) its acidic nature

9. The acid which has a peroxy linkage is
(a) Sulphurous acid
(b) Pyrosulphuric acid
(c) Dithionic acid
(d) Caro's acid

10. Among the following which is the strongest oxidising agent?
(a) Br_2 (b) I_2
(c) Cl_2 (d) F_2

11. The correct order of acid strength is:
(a) $HClO_4 < HClO_3 < HClO_2 < HClO$
(b) $HClO < HClO_2 < HClO_3 < HClO_4$
(c) $HClO_4 < HClO < HClO_2 < HClO_3$
(d) $HClO_2 < HClO_3 < HClO_4 < HClO$

12. Noble gases do not react with other elements because
(a) they are mono atomic
(b) they are found in abundance
(c) the size of their atoms is very small
(d) they are completely paired up and stable electron shells

13. Difference between S and S^{2-} as S^{2-} has
(a) larger radii and large size
(b) smaller radii and large size
(c) larger radii and small size
(d) smaller radii and small size

14. The correct order of acidic strength is
(a) $HF < HCl < HBr < HI$
(b) $HBr < HCl < HI < HF$
(c) $HCl < HBr < HF < HI$
(d) $HI < HBr < HCl < HF$

15. The structure of XeF_6 is
(a) distorted octahedral
(b) pyramidal
(c) tetrahedral
(d) None of the above.

16. Of the four elements, the one having maximum electron affinity is :
(a) Fluorine (b) Chlorine
(c) Bromine (d) Iodine

17. The structural formula of hypophosphorous acid is

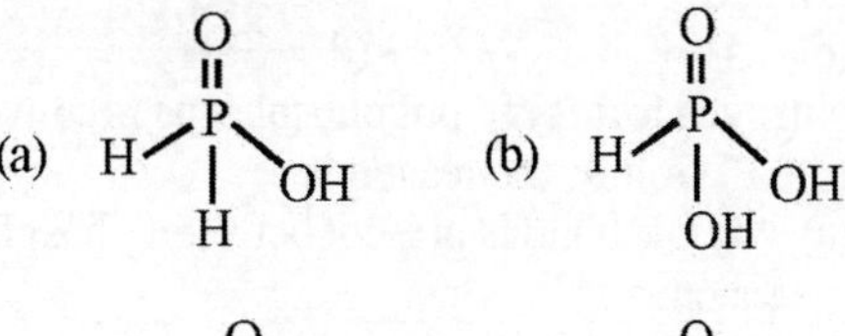

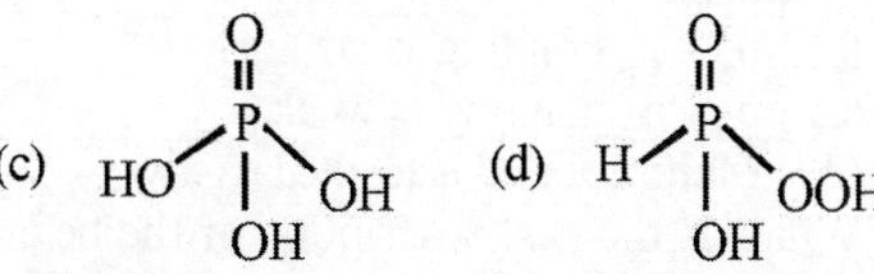

18. Oleum is
(a) Castor Oil (b) Oil of vitriol
(c) Fuming H_2SO_4 (d) None of them

19. Basicity of orthophosphoric acid is
(a) 2 (b) 3
(c) 4 (d) 5

20. Number of electrons shared in the formation of nitrogen molecule is
(a) 6 (b) 10
(c) 2 (d) 8

21. H_3PO_2 is the molecular formula of an acid of phosphorus. Its name and basicity respectively are
(a) Phosphorus acid and two
(b) Hypophosphorous acid and two
(c) Hypophosphorous acid and one
(d) Hypophosphoric acid and two

22. A solution of potassium bromide is treated with each of the following. Which one would liberate bromine ?
(a) Hydrogen iodide (b) Sulphur dioxide
(c) Chlorine (d) Iodine

23. The formula for calcium chlorite is
(a) $Ca(ClO_4)_2$ (b) $Ca(ClO_3)_2$
(c) $CaClO_2$ (d) $Ca(ClO_2)_2$
24. Brown ring test is used to detect
(a) Iodine (b) Nitrate
(c) Iron (d) Bromide
25. About 20 km above the earth, there is an ozone layer. Which one of the following statements about ozone and ozone layer is true?
(a) ozone has a triatomic linear molecule
(b) it is harmful as it stops useful radiation
(c) it is beneficial to us as it stops U.V radiation
(d) conversion of O_3 to O_2 is an endothermic reaction
26. Which of the following species has the highest dipole moment ?
(a) NH_3 (b) PH_3
(c) AsH_3 (d) SbH_3
27. Nitrogen forms N_2, but phosphorus is converted into P_4 from P, the reason is
(a) Triple bond is present between phosphorus atom
(b) $p_\pi - p_\pi$ bonding is strong
(c) $p_\pi - p_\pi$ bonding is weak
(d) Multiple bond is formed easily
28. Which is the best description of the behaviour of bromine in the reaction given below ?

$H_2O + Br_2 \rightarrow HOBr + HBr$

(a) Proton acceptor only
(b) Both oxidized and reduced
(c) Oxidized only
(d) Reduced only
29. How many bridging oxygen atoms are present in P_4O_{10}?
(a) 5 (b) 6
(c) 4 (d) 2
30. Which of the following oxide is amphoteric ?
(a) SnO_2 (b) CaO
(c) SiO_2 (d) CO_2
31. Which one of the following statements is true for HNO_2 ?
(a) It can not act as reducing agent.
(b) It can not act as oxidising agent.
(c) It can not act both as oxidant reductant.
(d) It is stable only in aqueous solution.
32. Which one of the following property is not correct for ozone ?
(a) It oxidises lead sulphide
(b) It oxidises potassium iodide
(c) It oxidises mercury
(d) It cannot act as bleaching agent
33. The sequence of acidic character is
(a) $SO_2 > CO_2 > CO > N_2O_5$
(b) $SO_2 > N_2O_5 > CO > CO_2$
(c) $N_2O_5 > SO_2 > CO > CO_2$
(d) $N_2O_5 > SO_2 > CO_2 > CO$
34. If you touch concentrated HNO_3 with your finger and immediately wash it with water, then the skin at the place where it came into contact with HNO_3 becomes yellow. This is because of :
(a) the formation of Xanthoprotein
(b) the absorption of HNO_3 by skin
(c) the blood absorbing HNO_3
(d) the absorption of yellow/brown NO_2 gas
35. The boiling points of the following hydrides follow the order of
(a) $SbH_3 > NH_3 > AsH_3 > PH_3$
(b) $NH_3 > AsH_3 > PH_3 > SbH_3$
(c) $SbH_3 > AsH_3 > PH_3 > NH_3$
(d) $NH_3 > PH_3 > AsH_3 > SbH_3$
36. Which of the following fluorides of xenon is impossible?
(a) XeF_2 (b) XeF_3
(c) XeF_4 (d) XeF_6
37. The order of reactivity of halogens is
(a) F > Cl > Br > I (b) F < Cl < Br < I
(c) F < Cl > Br < I (d) F < Cl < Br > I
38. Which of the following statements is correct ?
(a) Only iodine forms oxy acid.
(b) Only chlorine and bromine form oxy acid.
(c) All the halogens form oxy acid.
(d) All the halogens form oxy acids except fluorine.
39. Which of the following statements is not correct for nitrogen ?
(a) Its electronegativity is very high
(b) d-orbitals are available for bonding
(c) It is a typical non-metal
(d) Its molecular size is small
40. Which one of the following orders is not in accordance with the property stated against is ?
(a) HI > HBr > HCl > HF : Acidic property in water
(b) $F_2 > Cl_2 > Br_2 > I_2$: Electronegativity
(c) $F_2 > Cl_2 > Br_2 > I_2$: Bond dissociation energy
(d) $F_2 > Cl_2 > Br_2 > I_2$: Oxidising power

41. Which of the following statements is false ?
   (a) Radon is obtained from the decay of radium.
   (b) Helium is inert gas.
   (c) Xenon is the most reactive among the rare gases.
   (d) The most abundant rare gas found in the atmosphere is helium.
42. Hypo is used in photography to
   (a) Reduce AgBr grains to metallic silver
   (b) Convert metallic silver to silver salt
   (c) Remove undecomposed silver bromide as a soluble complex
   (d) Remove reduced silver
43. Which of the following is a nitric acid anhydride?
   (a) NO (b) NO_2
   (c) N_2O_5 (d) N_2O_3
44. Oxygen will directly react with each of the following elements except
   (a) P (b) Cl
   (c) Na (d) S.
45. Bleaching powder reacts with a few drops of dilute HCl to give
   (a) chlorine
   (b) hypochlorous acid
   (c) calcium oxide
   (d) oxygen
46. Which one of the following compounds does not exist ?
   (a) NCl_5 (b) AsF_5
   (c) $SbCl_5$ (d) PF_5
47. When orthophosphoric acid is heated to 600°C, the product formed is
   (a) PH_3 (b) P_2O_5
   (c) H_3PO_3 (d) HPO_3
48. It is possible to obtain oxygen from air by fractional distillation because
   (a) oxygen is in a different group of the periodic table from nitrogen
   (b) oxygen is more reactive than nitrogen
   (c) oxygen has higher b.p. than nitrogen
   (d) oxygen has a lower density than nitrogen.
49. The bleaching action of chlorine is due to
   (a) Reduction (b) Hydrogenation
   (c) Chlorination (d) Oxidation
50. Which would quickly absorb oxygen ?
   (a) Alkaline solution of pyrogallol
   (b) Conc. H_2SO_4
   (c) Lime water
   (d) Alkaline solution of $CuSO_4$
51. P_2O_5 is heated with water to give
   (a) Hypophosphorous acid
   (b) Phosphorous acid
   (c) Hypophosphoric acid
   (d) Orthophosphoric acid
52. PCl_3 reacts with water to form
   (a) PH_3 (b) H_3PO_3, HCl
   (c) $POCl_3$ (d) H_3PO_4
53. $PH_4I + NaOH$ forms
   (a) PH_3 (b) NH_3
   (c) P_4O_6 (d) P_4O_{10}
54. Pure nitrogen is prepared in the laboratory by heating a mixture of
   (a) $NH_4OH + NaCl$ (b) $NH_4NO_3 + NaCl$
   (c) $NH_4Cl + NaOH$ (d) $NH_4Cl + NaNO_2$.
55. Which one of the following substance is used in the laboratory for fast drying of neutral gases?
   (a) Phosphorus pentoxide
   (b) Active charcoal
   (c) Anhydrous calcium chloride
   (d) Na_3PO_4.
56. When chlorine is passed over dry slaked lime at room temperature, the main reaction product is
   (a) $Ca(ClO_2)_2$ (b) $CaCl_2$
   (c) $CaOCl_2$ (d) $Ca(OCl)_2$
57. Which of the following fertilizers has the highest nitrogen percentage ?
   (a) Ammonium sulphate
   (b) Calcium cyanamide
   (c) Urea
   (d) Ammonium nitrate
58. A certain compound (X) when treated with copper sulphate solution yields a brown precipitate. On adding hypo solution, the precipitate turns white. The compound is
   (a) K_2CO_3 (b) KI
   (c) KBr (d) K_3PO_4
59. HI can be prepared by all the following methods, except
   (a) $PI_3 + H_2O$ (b) $KI + H_2SO_4$
   (c) $H_2 + I_2 \xrightarrow{Pt}$ (d) $I_2 + H_2S$
60. Which among the following is paramagnetic ?
   (a) Cl_2O (b) ClO_2
   (c) Cl_2O_7 (d) Cl_2O_6

CHAPTER 22

The d and f-Block Elements

CONCEPTS

TRANSITION ELEMENTS

Elements where the last orbitals filled are the d orbitals known as transition elements. They have been placed in the middle of the periodic table between electropositive s-block and electronegative p-block elements.

- General electronic configuration $(n-1)d^{1-10}$, ns^{1-2}.
- There are three transition series in d-block.
  (i) First transition series – Sc (21) to Zn (30) [3d orbital is filled]. $3d^{1-10}\ 4s^{1-2}$
  (ii) Second transition series – Y (39) to Cd (48) [4d orbital is filled]. $4d^{1-10}\ 5s^{0-2}$
  (iii) Third transition series – La (57) and Hf (72) to Hg (80) [5d orbital is filled]. $5d^{1-10}\ 6s^{1-2}$
- **Zn, Cd and Hg** due to completely filled d-orbital are **not d-block elements by definition**.

PROPERTIES OF TRANSITION ELEMENTS

Atomic Size :

(i) In a series, atomic radii first decreases upto middle of series, then becomes constant and again increases at the end.
(ii) Down the group atomic radii increases.
(iii) Atomic radii of second and third transition series members are almost same due to **lanthanide contraction**.

- The electrons in the f-subshell are not very effective in shielding outer electrons from the nucleus, there is a decrease in atomic size in the series of elements in which the 4f-subshell is filled. This is known as **lanthanoid contraction.**

Metallic Character :

(i) All d-block elements are metals.
(ii) Greater the number of unpaired electrons in (n – 1)d, more is metallic character.

Melting and Boiling Points :

(i) In general high due to formation of metal bonds by (n – 1)d electrons also.
(ii) **B.P. of Zn, Cd and Hg are relatively low** due to completely filled (n – 1) orbital.

Ionisation Potential :

(i) IP greater than s-block elements but lesser than p-block elements.
(ii) IP decrease in series but trend is not regular.

Density :

(i) In general high due to strong metallic bonds.
(ii) **Density of Osmium is highest**.

Thermodynamic Stability of Compounds :

(i) Smaller the value of ionization enthalpy, more will be the stability of compound.
(ii) ∴ Ni (II) compounds are more stable than Pt (II) compounds.
but Pt (IV) compounds are more stable than Ni (IV) compounds.
∴ **$K_2[PtCl_6]$ having Pt (IV) is well known compound** but **$K_2[NiCl_6]$ is not known**.

Electrode Potential and Reducing Character :

(i) The **standard reduction potential of all transition elements** (except Cu and Hg) is lower (that is negative) **than that of hydrogen**.
∴ All the transition metals (except Hg and Cu) liberate H_2 from dilute acids
(ii) All the transition elements are good reducing agent (except Cu) but weaker reducing agent than s-block elements.

(iii) **Electrode potential** (E^0) ($E^0 M^{n+}/M$) is governed by three factors
  (a) Heat of sublimation
  (b) Heat of ionisation
  (c) Heat of hydration

For the 3d transition metals the $E^0 M^{2+}/M$ values are

V	Cr	Mn	Fe	Co	Ni	Cu
-1.18	-0.91	-1.18	-0.44	-0.28	-0.25	0.35

(Volts)

The irregular trend is due to variation in ionization energies and sublimation energies.

Oxidation States :

(i) All transition elements show variable oxidation states.
(ii) **Highest oxidation state of Mn is** +7 in MnO_4^-.
(iii) **Most common oxidation state of 3d-series is +2** (except Sc which shows +3 oxidation state).
(iv) Highest oxidation states of elements are found in fluorides and oxides. It is +8 for transition elements as electronegativity values of oxgen and fluorine are high and have small size.
(v) Transition metals exhibit zero oxidation state also (e.g.- in carbonyls).

Complex Formation :

Transition elements form complexes due to
(i) high nuclear charge
(ii) small size of cation
(iii) presence of empty d-orbitals to accept electrons from ligands.

Alloy Formation :

Due to their almost equal atomic size, transition elements substitute in each others lattice to form alloy.

Formation of Coloured Compounds (Both in Solid State as well as in Aqueous Solution)

Another very common characteristics of transition metals. This is due to absorption of some radiation from visible light to cause $d-d$ transition of electrons in transition metal atom. The d-orbitals do not have same energy and under the influence of ligands, the d-orbitals split into two sets of orbitals having different energies; transition of electrons can take place from one set of d-orbitals to another set within the same sub-shell. Such transitions are called $d-d$ transitions.

Magnetic Properties :

(i) Unpaired electron – **paramagnetic**, All electrons paired – **diamagnetic**.
(ii) Paramagnetic character $\propto$ no. of unpaired electrons.
paramagnetic character is measured in terms of magnetic movement.
(iii) Magnetic moment $(\mu) = \sqrt{n(n+2)}$ Bohr Magneton (B.M.) where n = no. of unpaired electrons.

Interstitial Compounds

Formed by traping of small atoms or molecules like H, C, B, N in interstitial sites of transition elements.
(i) Steel is interstitial compound having carbon trapped in it.
(ii) They have higher melting points and are harder than pure metals.

Non-stoichiometric Compounds

Have indefinite composition e.g.- $Fe_{0.98}O$.

Catalytic Properties :

Many of transition elements and their compounds are used as catalysts due to variable oxidation states and presence of free valencies on metal surface.

Oxides :

(i) of lower oxidation states +2 and +3 are basic (except Cr_2O_3-which is amphoteric).
(ii) of higher oxidation state ≥ 5 are acidic.
(iii) of intermediate oxidation state +4 are amphoteric.

Halides :

(i) Halides of transition metals in higher oxidation states fumes in air due to hydrolysis.

$$TiCl_4 \longrightarrow TiO_2 + 4HCl$$

(ii) Fluorides are ionic in nature and ionic nature decreases from fluorides to iodides.

Sulphides :

(i) Only the sulphides of lower oxidation state of metals are formed because sulphur is weak oxidising agent.
(ii) Sulphides of transition elements are usually coloured and are insoluble in water.
MnS - flesh coloured, CoS, NiS, CuS - black
ZnS - white, CdS - yellow.

SOME IMPORTANT TRANSITION ELEMENTS AND THEIR COMPOUNDS

Iron :

(a) **Ores** - Haematite - Fe_2O_3, Magnetite – Fe_3O_4.

(b) **Extraction of Fe from haematite :**

(i) Crushed ore is concentrated by **gravity separation.**

(ii) Ore is then **roasted** to remove volatile impurities and to make it porus.

(iii) **Smelting** - It is carried out in **blast furnace.** Roasted **ore is mixed with coke and limestone**. Temperature of furnace is from 200°C to 1600°C. Following reactions occur in furnace -

$$C + O_2 \longrightarrow CO_2 + 97kCal;$$

$$CO_2 + C \longrightarrow 2CO - 39kCal$$

$$\underset{\text{Lime stone}}{CaCO_3} \longrightarrow CaO + CO_2$$

$$\underset{\text{Impurity}}{SiO_2} + CaO \longrightarrow \underset{\text{slag}}{CaSiO_3}$$

$$FeO + CO \longrightarrow Fe + CO_2$$

$$Fe_2O_3 + 3CO \longrightarrow 2Fe + 3CO_2;$$

(∵ CO is reducing agent)

(iv) Spongy iron floats over slag.

(c) **Varieties of iron** - depends on % impurity of carbon in it.

Pig iron or cast iron	Steel	Wrought iron or malleable iron
About 3-4% impurity	0.1-1.5% impurity	0.2 to 0.5% of impurity

(d) **Steel is prepared by Bessemer process, Open Hearth process** and **Electric furnace process**.

(e) **Heat treatment of steel :** The hardness of steel depends on carbon content and heat treatment.

(i) **Annealing**

- Process of heating steel to redness followed by slow cooling.
- It makes steel soft.

(ii) **Quenching or hardening**

- Process of heating steel to redness and then cool suddenly by plunging in water or oil.
- It makes steel hard and brittle.

(iii) **Tempering**

- Process of heating the hardened or quenched steel to 504K-574K, and then cool slowly.
- It makes steel hard but brittleness disappears.

(iv) **Case hardening**

- Process of giving thin coating of hardened steel to wrought iron or steel by heating it in contact with charcoal followed by quenching in oil.

(v) **Nitriding**

- Process of heating steel at 700°C (973K) in an atmosphere of NH_3.
- It gives hard coating of iron nitride on surface of steel.

(f) **Mohr's salt** - $FeSO_4.(NH_4)_2SO_4.6H_2O$.

(i) Light green coloured double salt.

(ii) Used as primary standard in volumetric analysis because a standard solution of Fe^{++} ions can be obtained directly by weighing known amount of Mohr salt.

Copper :

(a) **Ores - Copper glance** Cu_2S, **Copper pyrite** $CuFeS_2$, **Basic copper carbonate** (Malachite) - $CuCO_3.Cu(OH)_2$.

(b) **Extraction of Cu from copper pyrite.**

(i) Concentration of pyrite is done by froth floatation process.

(ii) **Roasting**

Main reaction

$$2CuFeS_2 + O_2 \longrightarrow Cu_2S + 2FeS + SO_2$$

Side reactions

$$2Cu_2S + 3O_2 \longrightarrow 2Cu_2O + 2SO_2$$

$$2FeS + 3O_2 \longrightarrow 2FeO + 2SO_2$$

(iii) **Smelting** - Roasted ore is mixed with silica.

$$FeO + SiO_2 \longrightarrow \underset{\text{slag}}{FeSiO_3}$$

$$Cu_2O + FeS \longrightarrow FeO + Cu_2S$$

Matte is Cu_2S + FeS. Slag floats over matte.

(iv) **Bassemerisation** is done in **Bassemer converter** where Matte is heated in current of air.

$$2Cu_2S + 3O_2 \longrightarrow 2Cu_2O + 2SO_2$$

$$2Cu_2O + Cu_2S \longrightarrow \underset{\text{blister copper}}{6Cu} + SO_2$$

Blister copper has blister like appearance due to removal of SO_2 from it. **It is 99% pure Cu.**

(v) **Electrolytic refining :** Impure Cu - anode, Pure Cu - cathode. dil. $CuSO_4$ is electrolyte. Pure Cu deposits at cathode by electrolytic refining.

(c) **Oxidation states of copper :** Cu^+ and Cu^{2+}

(i) Cu (I) salts are colourless and Cu (II) salts are blue coloured.

(ii) **Cu^{2+} is more stable than Cu^+.**

$$2Cu^+ \longrightarrow Cu^{2+} + Cu$$

(disproportionation reaction)

(d) Tetrammine copper sulphate is known as **Schwitzer' Reagent** $[Cu(NH_3)_4]SO_4$.
It is used to dissolve cellulose in the manufacture of artificial silk.

(e) Mixture of $CuSO_4.5H_2O$ + Lime is known as Bordeaux mixture. It is used as fungicide, in the preparation of Fehling solution and electric batteries.

Silver :

It is best conductor of heat and electricity.

(a) **Ores - Argentite** Ag_2S, **Horn silver** - AgCl

(b) **Extraction by cyanide process (or Mac-Arthur-Forest cyanide process)**

(i) Powdered argentite + NaCN + O_2 $\longrightarrow$ soluble complex

(ii) Complete equation

$$2Ag_2S + 8NaCN + O_2 + 2H_2O \longrightarrow 4Na[Ag(CN)_2] + 4NaOH + 2S$$

(iii) Precipitation of silver

$$2Na[Ag(CN)_2] + Zn \longrightarrow Na_2[Zn(CN)_4] + 2Ag \downarrow \text{(ppt)}$$

(c) **Extraction of Ag from argentiferous lead** ($PbS + Ag_2S$)

(i) Argentiferous lead has 0.01% to 0.1% silver.

(ii) Method of extracting silver from argeniferous lead is called **Parke's process**.

(iii) It is based on following facts :

- Silver is more soluble in molten zinc.
- Molten Zn and Ag are immiscible.
- Zn-Ag alloy solidifies earlier then molten Pb.
- Ag is separated from Zn by distillation.

(iv) **Silver is purified by cupellation method.**

(d) **Fineness** is amount of Ag present in alloy of Ag. e.g.- **625 fine silver means alloy has 62.5% Ag.**

(e) **Silver halides**

- Only AgF is soluble in water. AgBr, AgCl, AgI are insoluble in water.
- AgBr, AgCl and AgI all are photosensitive.
- AgCl does not react with $NaNO_3$.

(f) **Photography**

(i) **Ripening** is process of coating photographic plate with AgBr.

(ii) During **exposure** AgBr is reduced to metallic Ag.

$$2AgBr \longrightarrow 2Ag + Br_2.$$

(iii) **Developing :**

$$2AgBr + \underset{\substack{\text{Quinol} \\ \text{(Developer)}}}{C_6H_4(OH)_2} \longrightarrow 2Ag + 2HBr + \underset{\text{Quinone}}{C_6H_4O_2}$$

Developer does not affect those parts of the film which are not affected by light.

(iv) **Fixing :** Hypo ($Na_2S_2O_3$) is used to remove unchanged AgBr.

$$AgBr + \underset{\text{Hypo}}{2Na_2S_2O_3} \longrightarrow \underset{\text{(soluble)}}{Na_3[Ag(S_2O_3)_2]} + NaBr$$

(v) After taking print of photograph it is dipped in solution of $K_2[PtCl_6]$ to give shining grey tinge.

(vi) Developing in photography involves decomposition of AgBr.

(vii) **Sterling silver** is 80% Ag and 20% copper and is used in making jewellery.

Gold (King of Metals) :

(a) **Ores :** (i) Claverite AuI_2 (ii) Sylvanite (Ag.Au) Te_2

(b) **Extraction :**

(i) Extraction of gold is same as of Ag by cyanide process.

(ii) **Plattners chlorine process :** From auriferous pyrites.

Auriferous pyrites (moistened) + Cl_2 $\longrightarrow$ $AuCl_3$

$$AuCl_3 + 3FeSO_4 \longrightarrow FeCl_3 + Fe_2(SO_4)_3 + Au$$

$$2AuCl_3 + 3H_2S \longrightarrow 6HCl + 3S + 2Au$$

(c) Removal of Ag and Cu from gold is called **parting**.

(d) Purity of gold is expressed in terms of **carats**. 24 carats is purest form of gold.

(e) **Purple of Cassius** is colloidal gold.

(f) Gold dissloves is aqua regia due to formation of $H[AuCl_4]$

Zinc :

(a) **Ores - Zincite** ZnO, **Calamine** $ZnCO_3$.

(b) **Extraction** - It is extracted by reduction process from ZnS (Zinc blende).

(i) **Concentration** - The ore is concentrated by froth flotation process.

(ii) **Roasting** - The concentrated Zinc blende is roasted in a current of air.

$2ZnS + 3O_2 \rightarrow 2ZnO + 2SO_2$

(iii) **Reduction** - The ZnO is reduced by mixing with carbon and heating in fire clay retort.

$ZnO + C \rightarrow Zn + CO$

(iv) **Purification** - Zinc so obtained contains the impurities of Fe, Pb, Cd, As or Sb. It is purified by

- Distillation or
- Electrolytic method

Anode impure (Zn) $\quad Zn - 2e^- \rightarrow Zn^{++}$

Cathode pure (Zn) $\quad Zn^{++} + 2e \rightarrow Zn$

Electrolyte - Acidic solution of $ZnSO_4$.

- Liquation - Molten Zn is allowed to flow down on sloping hearth when non fusible impurities are left behind.

(v) **Electrolytic method** - Pure $ZnSO_4$ is electrolysed when Zn is deposited on aluminium cathode. It is scrapped off and melted to obtain 99.95% pure metal.

(c) **Zinc dust** -It is prepared by atomising molten zinc with blast of air.

(d) **Granulated Zinc** - It is prepared by pouring molten Zinc into cold water.

(e) **Zn is volatile metal**.

(f) $Zn + 2NaOH \longrightarrow \underset{\text{sod. zincate}}{Na_2ZnO_2} + H_2$

(g) $Zn + H_2SO_4(\text{dil.}) \longrightarrow ZnSO_4 + H_2 \uparrow$

Mercury :

(a) **Ore - Cinnabar** - HgS, **Calomel** – Hg_2Cl_2

(b) **$HgCl_2$ Mercuric chloride (Corrosive sublimate)** (It has corrosive action on glass and it sublimates on heating) $Hg + Cl_2 \longrightarrow HgCl_2$

Properties of $HgCl_2$:

(i) $HgCl_2 + 2KI \longrightarrow \underset{\text{(Red ppt.)}}{HgI_2} + 2KCl$

(ii) $HgCl_2 + 2NH_4OH \longrightarrow \underset{\text{(White ppt.)}}{Hg(NH_2)Cl} + NH_4Cl + 2H_2O$

(iii) $HgCl_2$ **is poisonous and its antidote is white of an egg.**

(iv) $2HgCl_2 + SnCl_2 \longrightarrow \underset{\text{(White ppt.)}}{Hg_2Cl_2} + SnCl_4$

$Hg_2Cl_2 + SnCl_2 \longrightarrow \underset{\text{(Grey)}}{2Hg} + SnCl_4$

(c) **Hg_2Cl_2 (Calomel) :** $HgCl_2 + Hg \longrightarrow Hg_2Cl_2$

(i) $Hg_2Cl_2 + NH_4OH \longrightarrow \underbrace{Hg + Hg(NH_2)Cl}_{\text{Black mixture}} + 2H_2O + NH_4Cl$

(ii) Used as purgative in medicines.

(iii) Used in making **standard electrode** called **calomel electrode.**

(d) **Nessler's reagent** is alkaline solution of K_2HgI_4 and is used to detect presence of NH_4^+ ions.

$K_2Cr_2O_7$:

Chromate and dichromate can be inter converted to each other in aqueous solution by altering pH of solution.

(a) $\underset{\substack{\text{chromate ion} \\ \text{(yellow)}}}{Cr_2O_7^2} \quad H_2O \underset{H}{\overset{OH}{\rightleftharpoons}} \underset{\substack{\text{dichromate ion} \\ \text{(orange)}}}{2CrO_4^2} \quad 2H$

(b) It is powerful oxidising agent.

$Cr_2O_7^{2-} + 14H^+ + 6e^- \longrightarrow 2Cr^{3+} + 7H_2O$

$\text{Eq. wt. of } K_2Cr_2O_7^{2-} = \frac{\text{Molecular weight}}{6} = \frac{294}{6} = 49.$

(c) It oxidises

$I^- \longrightarrow I_2, Fe^{2+} \longrightarrow Fe^{3+}, S \longrightarrow SO_4^{2-}$

$KMnO_4$

(a) **Equivalent weight** of $KMnO_4$ in different medium.

(i) In acidic medium $= \frac{\text{Mol. wt.}}{5} = \frac{158}{5} = 31.68$

(ii) In basic medium $= \frac{\text{Mol. wt.}}{1} = \frac{158}{1} = 158$

(iii) In neutral medium $= \frac{\text{Mol. wt.}}{3} = \frac{158}{3} = 52.6$

(b) **Oxidation by $KMnO_4$ in acidic medium**

(i) $Fe^{2+} \longrightarrow Fe^{3+} + e^-$

(ii) $C_2O_4^{2-} \longrightarrow 2CO_2 + 2e^-$

(iii) $I^- \longrightarrow I_2$

(iv) $SO_3^{2-} \longrightarrow SO_4^{2-}$

(c) **Oxidation by $KMnO_4$ in alkaline medium**

(i) $I^- \longrightarrow IO_3^-$

(ii) $MnO_4^- \longrightarrow MnO_4^{--}$

(d) **Oxidation by $KMnO_4$ in neutral medium**

(i) $CH_2 = CH_2 \longrightarrow \underset{\text{ethylene glycol}}{CH_2OH - CH_2OH}$

(ii) Toulene ⟶ Benzoic acid

INNER TRANSITION ELEMENTS (f-BLOCK ELEMENTS)

The elements in which the filling of atomic orbitals by electrons take place in f subshells, two levels inside the outer subshell, are known as inner transition elements. Thus these elements form a series within the transition series. They are also known as f- block elements since the differentiating electron enters the f -subshell.

Classification of f-Block Elements

They have been classified into two series.

(i) **4 f-series (first inner transition series)** - The differentiating electron enters in 4 f orbitals. The elements belonging to this series are also known as ***Lanthanides or Lanthanones.***

(ii) **5 f -series (second inner transition series)** - The differentiating electron enters in 5 f orbitals. The elements belonging to this series are also known as ***Actinides or actinones.***

For the sake of symmetry of the periodic table they have been placed outside the periodic table.

LANTHANIDES

The fifteen elements from lanthanum (At. no. 57) to lutetium (At. no. 71) are known as ***lanthanides*** or ***rare earths*** (because they were obtained as earths (oxides) from relatively rare minerals).

Properties of Inner Transition Elements

(i) **Electronic configuration** - The general electronic configuration of these elements is $[Xe]4f^{0-14}5d^{0-1}6s^2$

The lanthanum electronic configuration $[Xe]4f^0 5d^1 6s^2$ and lutetium electronic configuration $[Xe]4f^{14}5d^1 6s^2$, have no partially filled 4 f orbital in their ground state, are considered as lanthanides due to their properties close to these elements.

(ii) **Oxidation state** - The common oxidation state of lanthanides is +3 but some elements also exhibit +2 and +4oxidation states in which they leave behind stable ions eg.

$Eu^{2+}[Xe]4f^7$, Yb^{2+} $[Xe]4f^{14}$

$Ce^{4+}[Xe]4f^0$, $Tb[Xe]4f^7$

An aqueous solution of Ce^{4+} is a good oxidising agent. The Eu^{2+} and Yb^{2+} can exist in aqueous solution and are good reducting agents. But there are exceptions also e.g.

$Sm^{2+}[Xe]4f^6$, $Tm^{2+}[Xe]4f^{13}$,

$Pr^{4+}[Xe]4f^1$, $Nb[Xe]4f^2$

(iii) **Magnetic properties** -All lanthanide ions with the exception of La^{3+}, Lu^{3+} and Ce^{4+} are paramagnetic in nature.

Lanthanide Contraction :

The decrease in size while moving from La to Lu is called lanthanide contraction. It is due to greater effect of the increased nuclear charge than that of screening effect.

Consequences of lanthanide contraction -

(i) Covalent character of cations increase.

(ii) The electronegativity of trivalent ions increase slightly.

(iii) There is decrease in basic strength of oxides and hydroxides.

(iv) There is small increase in standard electrode potential values.

(v) **Colour** - The species containing unpaired electrons are coloured and so is the case with lanthanide ions. The f-f transitions are possible due to absorption of light from the visible region.

(vi) **Melting and boiling point** - Lanthanides have high melting and boiling points but there is no regular trend.

(vii) **Density** - Lanthanides have densities varying from 6.77 to 9.74 g cm^{-3}. But there is no definite trend for these values.

(viii) **The element electronegatively** values of lanthanides are almost same as that of s-block elements. Lanthanides form ionic compounds.

(ix) **Ionisation energies** - The ionisation energy values of lanthanides are not very high due to their large size and are comparable with those of alkaline earth metals.

(x) **Complex compound** - Due to having large ionic size they have little tendency to form complexes.

(xi) **Reactivity** - Due to their low values of ionisation energies the lanthanides are very reactive.

(xii) **Alloys** - They form alloy especially with iron e.g. *MISCH METAL* rare earths 94 – 95% iron 5% and S, C, Ca and Al in traces.

- f-block elements have no tendency to form complexes.

ACTINIDES

The fifteen elements from Actinium (At. no. 89) to Lawrencium (At. no. 103) are known as actinides and constitute the 5f. series. From neptunium to onwards the elements are man made (artificially prepared) and also known as *trans uranium* elements.

Properties of Actinides

(i) **Electronic configuration :**

$[Rn]5f^{0-14}6d^{0-2}7s^2$

(ii) **Oxidation state** - The common oxidation state is +3 but other oxidation states are also exhibited by actinides the maximum being +7.

(iii) **Magnetic properties** - The magnetic moments of actinide ions are smaller than theoretical values. It is hard to interpret due to large spin orbit coupling.

(iv) **Actinide contraction** - It is similar to lanthanide contraction due to poor shielding of $5f$ electrons.

(v) **Melting and boiling points** - They have high values for melting and boiling points but there is no regular trend.

(vi) **Density** - The value of density vary from 7.0 g cm^{-3} to 19.84 g cm^{-3}. Again there is no regular trend.

(vii) **Reducing character** -They are strong reducing agents as they have high E^0 values approximately 2.0 volts.

(viii) **Reactivity** - Actinides are very reactive in nature and combine with oxygen and halogens like lanthanides.

(ix) **Coloured ions** - Actinide ions are coloured due to the presence of unpaired electrons and $f-f$ transitions.

(x) **Complex formation** - They have higher tendency to form complex compounds.

Exercise

1. The transition elements have a general electronic configuration
   (a) ns^2, np^6, nd^{1-10}
   (b) $(n-1)d^{1-10}, ns^{0-2}, np^{0-6}$
   (c) $(n-1)d^{1-10}, ns^{1-2}$
   (d) nd^{1-10}, ns^{-2}
2. The electronic configurations of four elements are given below. Which element does not belong to the same family as others ?
   (a) $[Xe]4f^{14}5d^{10}6s^2$ (b) $[Kr]4d^{10}5s^2$
   (c) $[Ne]3s^2 3p^5$ (d) $[Ar]3d^{10}4s^2$
3. Of the following transition metals, the maximum numbers of oxidation states are exhibited by:
   (a) Chromiun (Z = 24)
   (b) Manganese (Z = 25)
   (c) Iron (Z = 26)
   (d) Titanium (Z = 22)
4. Which one of the following characteristics of the transition metals is associated with their catalytic activity?
   (a) Variable oxidation states
   (b) High enthalpy of atomization
   (c) Parmagnetic behaviour
   (d) Colour of hydrated ions
5. Which forms coloured salts ?
   (a) Metals
   (b) Non-metals
   (c) p-Block elements
   (d) Transition elements
6. The oxidation state of Cr in $K_2Cr_2O_7$ is
   (a) +5 (b) +3
   (c) +6 (d) +7
7. When MnO_2 is fused with KOH, a coloured compound is formed, the product and its colour is:
   (a) K_2MnO_4, purple green
   (b) $KMnO_4$, purple
   (c) Mn_2O_3, brown
   (d) Mn_3O_4, black
8. The main reason for larger number of oxidation states exhibited by the actinoids than the corresponding lanthanoids, is

(a) more energy difference between 5f and 6d orbitals than between 4f and 5d orbitals.
(b) lesser energy difference between 5f and 6d orbitals than between 4f and 5d orbitals.
(c) larger atomic size of actinoids than the lanthanoids.
(d) greater reactive nature of the actinoids than the lanthanoids.

9. Identify the incorrect statement among the following:
(a) Lanthanoid contraction is the accumulation of successive shrinkages.
(b) As a result of lanthanoid contraction, the properties of 4d series of the transition elements have no similarities with the 5d series of elements.
(c) Shielding power of 4f electrons is quite weak.
(d) There is a decrease in the radii of the atoms or ions as one proceeds from La to Lu.

10. Photographic films and plates have an essential ingredient of
(a) Silver nitrate (b) Silver bromide
(c) Sodium chloride (d) Oleic acid

11. The complex ion $[Co(NH_3)_6]^{3+}$ is formed by sp^3d^2 hybridisation. Hence the ion should possess
(a) Octahedral geometry
(b) Tetrahedral geometry
(c) Square planar geometry
(d) Tetragonal geometry.

12. Which one of the following ionic species will impart colour to an aqueous solution?
(a) Ti^{4+} (b) Cu^+
(c) Zn^{2+} (d) Cr^{3+}

13. Four successive members of the first row transition elements are listed below with their atomic numbers. Which one of them is expected to have the highest third ionization enthalpy?
(a) Vanadium (Z = 23)
(b) Chromium (Z = 24)
(c) Manganese (Z = 25)
(d) Iron (Z = 26)

14. Which of the following metals corrodes readily in moist air ?
(a) Gold (b) Silver
(c) Nickel (d) Iron

15. The common oxidation states of Ti are
(a) +2, +3 (b) +3, +4
(c) –3, –4 (d) +2, +3, +4

16. Actinides
(a) Are all synthetic elements
(b) Include element 104
(c) Have any short lived isotopes
(d) Have variable valency

17. The electronic configuration of gadolinium (Atomic number 64) is
(a) $[Xe]\,4f^8\,5d^0\,6s^2$ (b) $[Xe]\,4f^3\,5d^5\,6s^2$
(c) $[Xe]\,4f^6\,5d^2\,6s^2$ (d) $[Xe]\,4f^7\,5d^1\,6s^2$

18. Which of the following forms colourless compound?
(a) Sc^{3+} (b) V^{3+}
(c) Ti^{3+} (d) Cr^{3+}

19. General electronic configuration of lanthanides is
(a) $(n-2)\,f^{1-14}\,(n-1)\,s^2p^6d^{0-1}\,ns^2$
(b) $(n-2)\,f^{10-14}\,(n-1)\,d^{0-1}\,ns^2$
(c) $(n-2)\,f^{0-14}\,(n-1)\,d^{10}\,ns^2$
(d) $(n-2)\,d^{0-1}\,(n-1)\,f^{1-14}\,ns^2$

20. Which of the following shows maximum number of oxidation states?
(a) Cr (b) Fe
(c) Mn (d) V

21. Lanthanoids are
(a) 14 elements in the sixth period (atomic no. = 90 to 103) that are filling 4f sublevel
(b) 14 elements in the seventh period (atomic no. = 90 to 103) that are filling 5f sublevel
(c) 14 elements in the sixth period (atomic no. = 58 to 71) that are filling 4f sublevel
(d) 14 elements in the seventh period (atomic no. = 58 to 71) that are filling 4f sublevel

22. Which of the following oxidation states are the most characteristic for lead and tin respectively?
(a) +2, +4 (b) +4, +4
(c) +2, +2 (d) +4, +2

23. Which one of the following ions has electronic configuration $[Ar]\,3d^6$?
(a) Ni^{3+} (b) Mn^{3+}
(c) Fe^{3+} (d) Co^{3+}
(At. Nos. Mn = 25, Fe = 26, Co = 27, Ni = 28)

24. Which of the following pairs has the same size?
(a) Fe^{2+}, Ni^{2+} (b) Zr^{4+}, Ti^{4+}
(c) Zr^{4+}, Hr^{4+} (d) Zn^{2+}, Hr^{4+}

25. Which of the following oxidation states is the most common among the lanthanoids?
(a) 3 (b) 4
(c) 2 (d) 5

26. Which element belongs to d-block ?
(a) Na (b) Ca
(c) Cu (d) Ar

27. Which of the following electronic configurations is that of a transition element ?
(a) $1s^2 2s^2 2p^6 3s^2 3p^6 3d^{10} 4s^2 4p^6$
(b) $1s^2 2s^2 2p^6 3s^2 3p^6 3d^{10} 4s^2 4p^1$
(c) $1s^2 2s^2 3p^6 3s^2 3p^6 3d^2 4s^2$
(d) $1s^2 2s^2 2p^6 3s^2 3p^6 4s^2$

28. Which of the following ions having following electronic structure would have maximum magnetic moment?
(a) $1s^2 2s^2 2p^6 3s^2 3p^6 3d^3$
(b) $1s^2 2s^2 2p^6 3s^2 3p^6 3d^5$
(c) $1s^2 2s^2 2p^6 3s^2 3p^6 3d^7$
(d) $1s^2 2s^2 2p^6 3s^2 3p^6 3d^9$

29. Which of the following is not correct about transition metals?
(a) Their melting and boiling points are high
(b) Their compounds are generally coloured
(c) They can form ionic or covalent compounds
(d) They do not exhibit varialbe valency

30. The transition metals have a less tendency to form ions due to
(a) High ionisation energy
(b) Low heat of hydration of ions
(c) High heat of sublimation
(d) All of the above

31. $(n-1)d^{10}ns^2$ is the general electronic configuration of
(a) Fe, Co, Ni (b) Cu, Ag, Au
(c) Zn, Cd, Hg (d) Se, Y, La

32. The last electron in d-block elements goes to
(a) (n-1) d (b) nd
(c) np (d) (n-1) s

33. Oxidation state of osmium (Os) in OsO_4 is
(a) +4 (b) +6
(c) +7 (d) +8

34. Which of the following is not property of the transition elements?
(a) Colour (b) Paramagnetism
(c) Fixed valency (d) None of these

35. Transition metals usually exhibit highest oxidation states in their
(a) Chlorides (b) Fluorides
(c) Bromides (d) Iodides

36. Cuprous ion is colourless while cupric ion is coloured because
(a) Both have half filled p-and d-orbitals
(b) Cuprous ion has incomplete d-orbitaland cupric ion has a complete d-orbital
(c) Both have unpaired electrons in the d-orbitals
(d) Cuprous ion has complete d-orbital and cupric ion has an imcomplete d-orbital.

37. A blue colouration is not obtained when
(a) Ammonium hydroxide dissolves in copper sulphate
(b) Copper sulphate solution reacts with $K_4[Fe(CN)_6]$
(c) Ferric chloride reacts with sod. ferrocyanide
(d) Anhydrous $CuSO_4$ is dissolved in water

38. The lanthanide contraction is responsible for the fact that
(a) Zr and Y have about the same radius
(b) Zr and Nb have similar oxidation state
(c) Zr and Hf have about the same radius
(d) Zr and Zn have the same oxidation states
(Atomic numbers : Zr = 40, Y = 39, Nb = 41, Hf = 72, Zn = 30)

39. Which of the following combines with Fe (II) ions to form a brown complex?
(a) NO (b) N_2O
(c) N_2O_3 (d) N_2O_5

40. The correct order of ionic radii of Y^{3+}, La^{3+}, Eu^{3+} and Lu^{3+} is
(a) $La^{3+} < Eu^{3+} < Lu^{3+} < Y^{3+}$
(b) $Y^{3+} < La^{3+} < Eu^{3+} < Lu^{3+}$
(c) $Y^{3+} < Lu^{3+} < Eu^{3+} < La^{3+}$
(d) $Lu^{3+} < Eu^{3+} < La^{3+} < Y^{3+}$
(Atomic nos. Y = 39, La = 57, Eu = 63, Lu = 71)

41. The basic character of the transition metal monoxides follows the order
(Atomic Nos., Ti = 22, V = 23, Cr = 24, Fe = 26)
(a) TiO > VO > CrO > FeO

(b) $VO > CrO > TiO > FeO$
(c) $CrO > VO > FeO > TiO$
(d) $TiO > FeO > VO > CrO$

42. Among the following series of transition metal ions, the one where all metal ions have $3d^2$ electronic configuration is (At. nos. Ti = 22; V = 23; Cr = 24; Mn = 25)
(a) $Ti^{3+}, V^{2+}, Cr^{3+}, Mn^{4+}$
(b) $Ti^{+}, V^{4+}, Cr^{6+}, Mn^{7+}$
(c) $Ti^{4+}, V^{3+}, Cr^{2+}, Mn^{3+}$
(d) $Ti^{2+}, V^{3+}, Cr^{4+}, Mn^{5+}$

43. In which of the following pairs are both the ions coloured in aqueous solutions ?
(a) Sc^{3+}, Ti^{3+} (b) Sc^{3+}, Co^{2+}
(c) Ni^{2+}, Cu^{+} (d) Ni^{2+}, Ti^{3+}
(At. no. : Sc = 21, Ti = 22, Ni = 28, Cu = 29, Co = 27)

44. Which of the following statement is incorrect ?
(a) Silver glance mainly contains silver sulphide.
(b) Gold is found in native state.
(c) Zinc blende mainly contains zinc chloride.
(d) Copper pyrites also contain Fe_2S_3.

45. In the extraction of silver from argentite ore. The ore is treated with dil. solution of NaCN in water in the presence of Y, whereby the following reaction takes place :
$Ag_2X + 4NaCN + 2Y \rightarrow 2Na[Ag(CN)_2] + Na_2XY_4$.
X and Y in this reaction are respectively :
(a) Sb and S (b) S and O_2
(c) O and O_2 (d) O and S

46. Which of the following statement is not true about Mohr's salt ?
(a) It decolourises $KMnO_4$
(b) It is a primary standard
(c) It is a double salt
(d) Oxidation state of iron is +3 in it

47. When excess of $SnCl_2$ is added to a solution of $HgCl_2$, a white precipitate turning to grey is obtained. This grey colour is due to the formation of
(a) Hg_2Cl_2 (b) $SnCl_4$
(c) Sn (d) Hg

48. Ammonium dichromate is used in some fireworks. The green coloured powder blown in the air is
(a) CrO_3 (b) Cr_2O_3
(c) Cr (d) $CrO(O_2)$

49. Prussian blue is formed when
(a) Ferrous sulphate reacts with $FeCl_3$
(b) Ferric sulphate reacts with $K_4[Fe(CN)_6]$
(c) Ferrous ammonium sulphate reacts with $FeCl_3$
(d) Ammonium sulphate reacts with $FeCl_3$

50. Nitriding is the process of surface hardening of steel by treating it in an atmosphere of
(a) NH_3 (b) O_3
(c) N_2 (d) H_2S

51. The composition of 'golden spangles' is
(a) $PbCrO_4$ (b) PbI_2
(c) As_2S_3 (d) $BaCrO_4$

52. When $(NH_4)_2Cr_2O_7$ is heated, the gas evolved is
(a) N_2 (b) NO_2
(c) O_2 (d) N_2O

53. Among the lanthanides the one obtained by synthetic method is
(a) Lu (b) Pm
(c) Pr (d) Gd

54. Stainless steel contains iron and
(a) Cr + Ni (b) Cr + Zn
(c) Zn + Pb (d) Fe + Cr + Ni

55. Cuprous compounds such as CuCl, CuCN and CuSCN are the only salts stable in water due to
(a) high hydration energy of Cu^{+} ions
(b) their inherent tendency not do disproportionate
(c) diamagnetic nature
(d) insolubility in water

56. $K_2Cr_2O_7$ on heating with aqueous NaOH gives
(a) CrO_4^{2-} (b) $Cr(OH)_3$
(c) $Cr_2O_7^{2-}$ (d) $Cr(OH)_2$

57. CrO_3 dissolves in aqueous NaOH to give
(a) $Cr_2O_7^{2-}$ (b) CrO_4^{2-}
(c) $Cr(OH)_3$ (d) $Cr(OH)_2$

58. Which one of the following elements shows maximum number of different oxidation states in its compounds?
(a) Eu (b) La
(c) Ge (d) Am

59. On heating chromite ($FeCr_2O_4$) with Na_2CO_3 in air, which of the following product is obtained?
(a) $Na_2Cr_2O_7$ (b) FeO
(c) Fe_3O_4 (d) Na_2CrO_4

60. The addition of excess of aqueous HNO_3 to a solution containing $[Cu(NH_3)_4]^{2+}$ produces
(a) Cu^{+} (b) $[Cu(H_2O)_4]^{2+}$
(c) $Cu(OH)_2$ (d) $Cu(NO_3)_2$

Coordination Compounds

CONCEPTS

ADDITION COMPOUNDS

Double salts	Co-ordination compounds or complex compounds
• Lose their identity in dissolved state e.g.- Mohr salt : $FeSO_4(NH_4)_2SO_4.6H_2O$	• Retain their identity even in dissolved state. e.g.- $K_4[Fe(CN)_6]$

CO-ORDINATION COMPOUNDS

Important terms in co-ordination compounds.

Complex Ion

It is formed by union of simple cation with one or more neutral molecules or simple ions.

Ligands

The ions or molecules which donate lone pair of electron to central atom are called **ligands**.

- **Types of ligands**

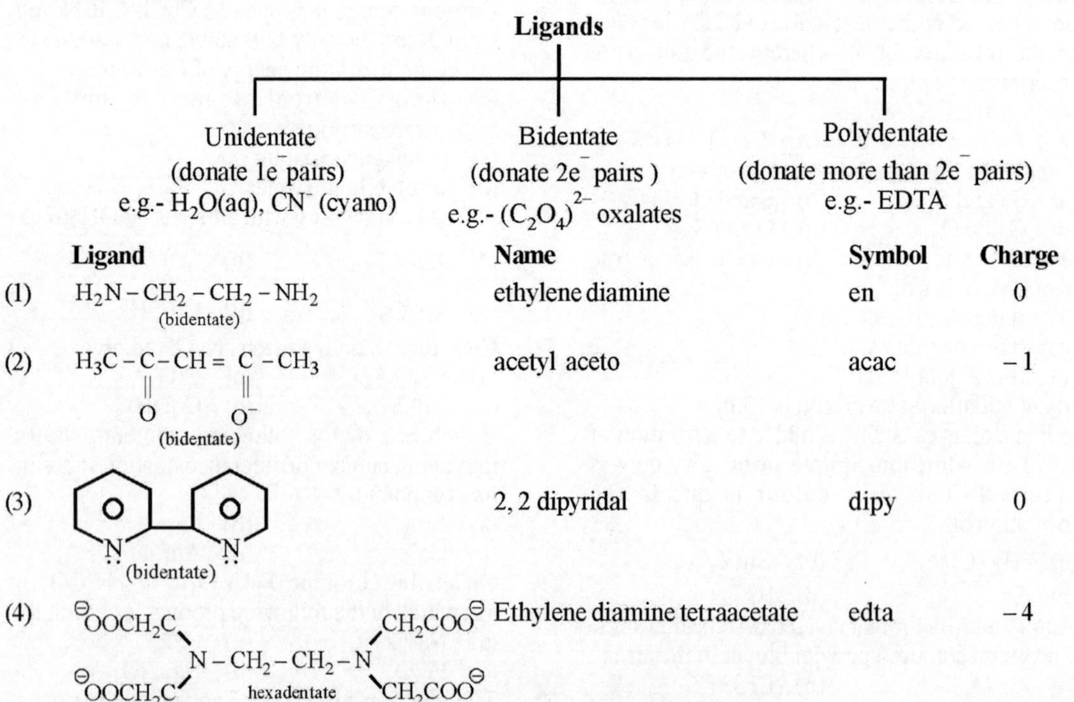

Ligands

Unidentate	Bidentate	Polydentate
(donate 1e$^-$ pairs)	(donate 2e$^-$ pairs)	(donate more than 2e$^-$ pairs)
e.g.- H_2O(aq), CN^- (cyano)	e.g.- $(C_2O_4)^{2-}$ oxalates	e.g.- EDTA

	Ligand	Name	Symbol	Charge
(1)	$H_2N-CH_2-CH_2-NH_2$ (bidentate)	ethylene diamine	en	0
(2)	$H_3C-\underset{\underset{O}{\parallel}}{C}-CH=\underset{\underset{O^-}{\parallel}}{C}-CH_3$ (bidentate)	acetyl aceto	acac	–1
(3)	(bidentate)	2, 2 dipyridal	dipy	0
(4)	$(^{\ominus}OOCH_2C)_2N-CH_2-CH_2-N(CH_2COO^{\ominus})_2$ hexadentate	Ethylene diamine tetraacetate	edta	–4

- **Bridging ligands -** Such ligands can bind to more than one metal atom.
- **Homoleptic ligand -** Metal bound to only one type of donor group.
- **Heterolepltic ligand -** Metal bound to more than one type donor group.

Co-ordination Number or Ligancy

Total number of coordinate bonds with which central atom is linked to the ligands in the complex.

Chelate and Chelating Agent

- When a multidentate ligand simultaneously forms more than two bonds with central atom, a ring like structure is formed which is called **chelate** and ligand is called **chelating agent**.
- **Stability of complex is increases by chelation** more the number of rings in complex, more will be stability of complex.

Co-ordination Sphere

The combination of central metal atom and ligands written in "square brackets" is called the coordination sphere.

Ionisation Sphere

The portion present outside the square bracket is called ionisation sphere.

K_4 [$Fe(CN)_6$]

Ionisation sphere (K_4) ↓ Central metal atom (Fe) ↓ Ligand (CN) — Coordination sphere ([$Fe(CN)_6$])

Oxidation Number

Charge carried by the central metal atom

Effective Atomic Number (EAN)

It can be obtained from the following simple expression

$$EAN = Z - O.N + 2\,(CN)$$

Where Z = Atomic number of central metal atom

O.N. = Oxidation number of central metal atom

CN = Coordination number of central metal atom.

It is equal to the number of monodentate ligands, twice the number of bidentate ligands and so on.

WERNERS THEORY

According to this theory metals possess two types of valencies -

(i) **Primary or ionisable valency :** These are non directional and are satisfied by negative ions only.

(ii) **Secondary or non ionisable valency :** Its number is fixed for a metal. They are **non ionisable and directional** in nature. It is satisfied by negative, neutral or positive species.

Formula for Writing Mononuclear Coordination Entities

(i) Ligands and polydentate ligands are listed alphabetically.

(ii) Formula of coordination entity is enclosed in square brackets.

(iii) There should be no space between ligands and metals.

(iv) When there is no counter ion charge is indicated outside the square bracket e.g. $[Co(CN)_6]^{3-}$.

Nomenclature of Complex Compounds

(i) The positive part of a coordinate compound is named first followed by name of negative part.

(ii) In co-ordination sphere ligands are named first followed by name of central atom. Name of ligands are arranged in alphabatic order.

(iii) Bi, tri, tetra are used for monodentate ligands and bis, tris, tetrakis are used for bi and poly dentate ligands to indicate the presence of similar kind of ligand when present more than one.

(iv) μ is used to show linkage in polynuclear complexes.

(v) The oxidation state of central metal atom is written in paranthesis in roman numerals.

(vi) The name of negative complex ends in-ate.

(vii) The name of complex compound should not start with capital letter.

(viii) Name of complex ion is one world.

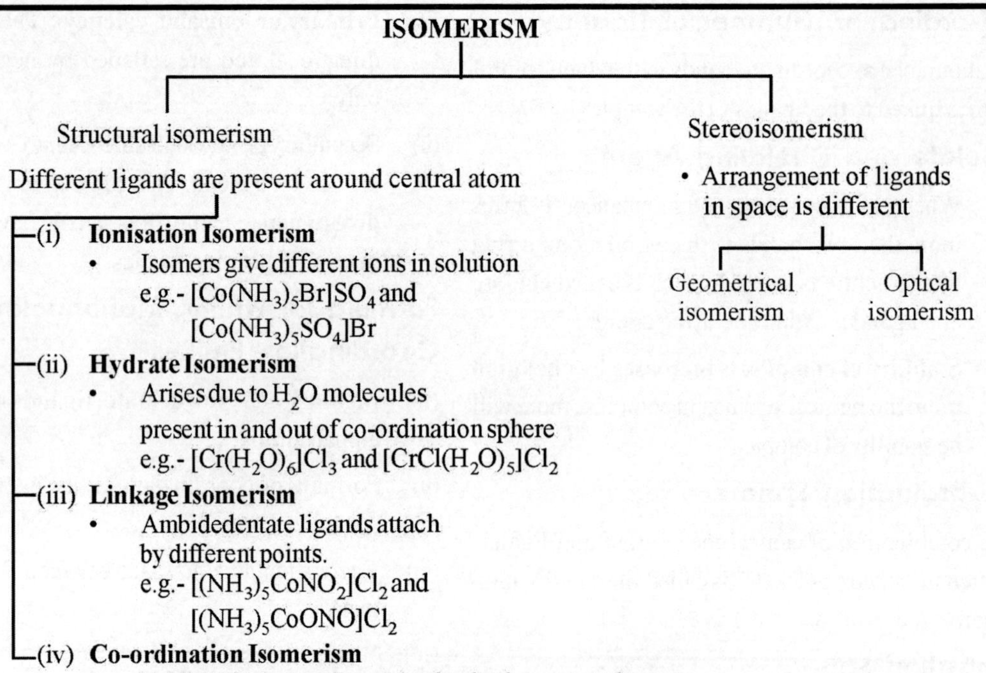

(v) **Ligand Isomerism** - Occurs when more than one isomer of the ligand is possible e.g., 1, 2 diamino propane and 1, 3-diaminopropane

(vi) **Coordination Position Isomerism** - In this coordinating groups occupy different positions and the isomerism occurs generally in bridged complex. e.g.

$$\left[(NH_3)_4Co\begin{matrix} NH_2 \\ O_2 \end{matrix}Co(Cl)_2(NH_3)_2\right]^{2+} \text{ and}$$

$$\left[(NH_3)_3(Cl)Co\begin{matrix} NH_2 \\ O_2 \end{matrix}Co(Cl)(NH_3)_3\right]^{2+}$$

STEREO ISOMERISM

- **Geometrical** — It is due to different spatial arrangements of ligands around central metal atom
  - Square planar: MA_2X_2, $MABX_2$, $MABXY$, $M(AA')_2$
  - Octahedral complex: MA_4X_2, MA_3X_3
- **Optical Isomers** — Shown by molecules which do not have plane of symmetry. These are non super imposable mirror images of each other
  - [Mabcdef], $[M(AA)_3]$, $[M(AA)_2B_2]$, $[M(AA)B_2C_2]$, $[MA_2B_2C_2]$

Geometrical Isomers

Square planar		Octahedral	
MA_2X_1	$[Co(NH_3)_2Cl_2][Pt(NH_3)_2Cl_2]$	MA_4X_2	$[Pt(NH_3)_4Cl_2]^{2+}$
$MABX_2$	$[Co(NH_3)_2ClBr]$	MA_3X_3	$[Co(NH_3)_3Cl_3]$
$MABXY$	$[Pt(NH_3)(NH_3OH)(NO_2)(PY)]NO_2$		
$M(AA')_2$	$[Pt(Gly)_2]$ (here Gly $H_2N.CH_2COO^-$)		

Optical Isomers :

Type	Examples
$[M(AA)_3]$	$[Co(en)_3]^{3+}$
$[M(AA)_2B_2]$	$[Co(en)_2Cl_2]^+$
$[M(AA)B_2C_2]$	$[Co(en)(NH_3)_2Cl_2]$
$[MA_2B_2C_2]$	$[Co(en)_2(NH_3)_2Cl_2]^+$
[Mabcdef]	$[Pt(Py)NH_3NO_2ClBrI]$

- **Optical isomerism** is given by octahedral complexes, square planar do not show optical isomerism.

VALENCE BOND THEORY

Features of This Theory are

1. Uses hybrid orbitals to hold the donated electron pairs for the formation of the coordinate bonds.
2. Can explain the structure and magnetic properties.

Limitations of Valence Bond Theory are Following

1. It does not explain the relative stability of complexes.
2. It can not explain colour and spectra
3. The relative stability of structural isomers.

CRYSTAL FIELD THEORY

It is based on assumption that metal ion and ligands act as point charges and interaction between them is purely electrostatic. Ligands cause splitting of d-orbital energies called crystal field splitting. Depending on the ability of ligand to cause crystal field splitting, ligands are arranged in increasing order of strength and series so produced is called **spectrochemical series.**

$I^- < Br^- < S^{2-} < Cl^- < F^- < Ox^{2-} < H_2O < py < NH_3 <$ en $< NO_2^- < CN^- < CO$

From I^- to NH_3 – weak ligands. (Cause weak splitting)
NH_3 to CO – strong ligands. (Cause large splitting)

Only strong ligands force pairing of unpaired e^-.

It accounts for both the colour and the magnetic properties of complexes. It is based on d- orbital energy level splitting

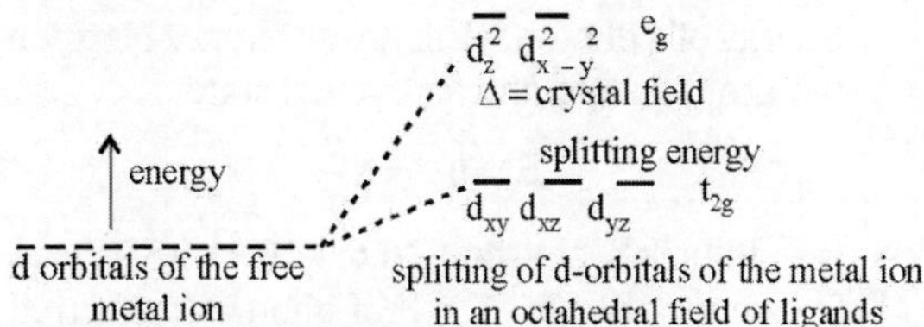

The Size of Δ Depends on -

(i) The nature of the ligand "Spectro chemical series" Δ decreases as shown below

$\bar{I} < \bar{Br} < \bar{Cl} < \bar{OH} < F^- < C_2O_4^{2-} < H_2O < \bar{CNS} <$
weak ligands
$< NH_3 <$ en $< \bar{NO_2} < \bar{CN} < CO$
strong ligands

(ii) The oxidation state of the metal Δ is greater for M^{3+} than for M^{2+}

(iii) The row of the metal in the periodic table. For a given ligand and oxidation state of the metal, Δ increases going down in a group eg.
Δ is greater in $[Ru(NH_3)_6]^{3+}$ than in $[Fe(NH_3)_6]^{3+}$

Effect of Strong Field Ligands and Weak Field Ligands

Strong field ligands forces the electrons of central metal for pairing and the complex formed is known as low spin complex. While weak field ligands do not forces the electron of central atom for pairing and the compelx formed is known as high spin complex. For example– In $[Ni(CN)_4]$, CN is a strong field ligand which forces electrons of Ni^{2+} for pairing

Ni^{2+}; $[Ar]\ 3d^8$ ↑↓ | ↑↓ | ↑↓ | ↑ | ↑

$[Ni(CN)_4]^{2-}$

3d 4s 4p

∴ [Ar] ↑↓ | ↑↓ | ↑↓ | ↑↓ | : | : | : | : |

electrons provided by $4CN^-$
Hybridisation: dsp^2 (square planar)

In $[NiCl_4]^{2-}$, Cl^- is a weak field ligand which do not forces electrons of Ni^{2+} for pairing

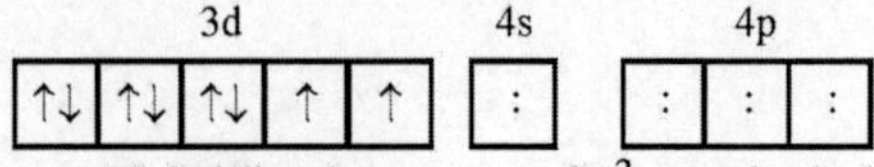

Hybridisation ∴ Sp^3, tetrahedral

Colours of Metal Complexes

It is due to electronic transitions between t_{2g} and e_g energy levels. The energy of an electron is increased by absorbing light energy and it moves to a higher energy level.

Energy of a photon = Energy difference between the ground state and an excited state

$$E = h\nu = h\frac{c}{\lambda}$$

h = Planck's constant (6.63×10^{-34} J.S.) ν = frequency of light. Δ is E of photon (measured with UV. or visible spectroscopy)

Magnetic Properties of Metal Complexes

- Paramagnetic - unpaired electrons
- Diamagnetic - no unpaired electrons
- Determined from crystal field splitting diagrams

Stability of Coordination Compounds in Solution

Consider the following equilibrium between undissociated complex ion and dissociated ion.

$$[MLn]^{b+} \rightleftharpoons M^{a+} + nL^{x-}$$

The equilibrium constant

$$K_c = \frac{[M^{a+}][L^{x-}]^n}{[(MLn)^{b+}]} = \frac{1}{K_s}$$

The smaller the value of K_c, the greater is the stability of complex ion and vice versa. The reciprocal of equilibrium constant is called stability constant (K_s).

Value of K_s depend on

(a) **Nature of central metal atom** - The more the polarizing power of the central metal ion the more is the stability of complex ion.

$$\text{The polarising power } \phi = \frac{\text{charge}}{\text{radius}}$$

Thus complex of Fe^{3+} is more stable then Fe^{2+}

(b) **Nature of ligand** - Since ligand is a Lewis base, the more the basic character of ligand the more is the stability of complex ion.

Drawbacks of Crystal Field Splitting

CFT is successful in explaining the formation, structures, colour and magnetic properties of coordination compounds.

(a) From the assumption that ligands are point charges, it follows that anionic ligands should exert greatest spliting effect. But anionic ligands actually are found at the low end of the spectro chemical series.

(b) It does not take into account for the covalent character of bonding between the ligand and the central atom.

ORGANOMETALLIC COMPOUNDS

These compounds have at least one metal-carbon bond

Organometallics or Organometallic Compounds

σ bonded organometallic comp.

- Formed by non-metals and metalloids except Zn, Cd and Hg.
- Bond between carbon and metalloid is normal covalent bond.
- e.g.- R–Mg–X, $(CH_3CH_2)_2$ Zn

π bonded organometallic comp.

- Formed by transition elements.
- Normal σ bond is formed through π cloud of e^- of organic molecule.
- e.g.- (i) **Zeisse's salt** $K[PtCl_3(\eta-C_2H_4)]$
  (ii) **Ferrocene** $Fe[\eta^5-(C_5H_5)_2]$

(Metal carbonyls)

- Metal-carbon bond possesses both σ and π bond character.
- Transition metal carbonyls form this type of bond
- e.g.- $Ni(CO)_4$

Applications of Organometallics and Co-ordination Compound

(i) **Wilkinson's Catalyst** $(Ph_3P)_3RhCl$ is used for hydrogenation of alkenes.

(ii) **Zeigler-Natta Catalyst** - $TiCl_4 + (C_2H_5)_3Al$ is used for polymerisation of olefines.

(iii) **TEL** - $[(C_2H_5)_2Pb]$ is used as anti knock compound.

(iv) **EDTA** - is used for estimation of hardness in water as Ca^{2+} and Mg^{2+}.

(v) **Cis platin** i.e. Cis$[PtCl_2(NH_3)_2]$ is **used in treatment of cancer.**

(vi) Chlorphyll and haemoglobin are co-ordination complexes having Mg^{2+} and Fe^{2+} ions respectively.

(vii) **Vitamin B_{12}** is co-ordination **complex of** Co^{2+}.

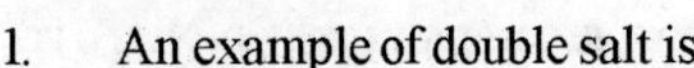

Exercise

1. An example of double salt is
   (a) Bleaching powder (b) $K_4[Fe(CN)_6]$
   (c) Hypo (d) Potash alum

2. $K_3[Al(C_2O_4)_3]$ is called
   (a) Potassium alumino oxalate
   (b) Potassium trioxalateoaluminate (III)
   (c) Potassium aluminium (III) oxalate
   (d) Potassium trioxalato aluminate (VI)

3. Correct formula of potassium ferrocyanide is :
   (a) $K_3[Fe(CN)_6]$
   (b) $K_4[Fe(CN)_6]$
   (c) $K_2[Fe(CN)_6]$
   (d) $K[Fe(CN)_6]H_2O$

4. The complexes $[Co(NH_3)_6]$ $[Cr(CN)_6]$ and $[Cr(NH_3)_6]$ $[Co(CN)_6]$ are the examples of which type of isomerism?
   (a) Linkage isomerism
   (b) Ionization isomerism
   (c) Coordination isomerism
   (d) Geometrical isomerism

5. Which of the following pairs represent linkage isomers?
   (a) $[Pd(P Ph_3)_2(NCS)_2]$ and $[Pd(P Ph_3)_2(SCN)_2]$
   (b) $[Co(NH_3)_5NO_3]SO_4$ and $[Co(NH_3)_5SO_4]NO_3$
   (c) $[PtCl_2(NH_3)_4]Br_2$ and $[PtBr_2(NH_3)_4]Cl_2$
   (d) $[Cu(NH_3)_4][PtCl_4]$ and $[Pt(NH_3)_4][CuCl_4]$

6. The coordination number of a central metal atom in a complex is determined by
   (a) the number of ligands around a metal ion bonded by sigma and pi-bonds both
   (b) the number of ligands around a metal ion bonded by pi-bonds
   (c) the number of ligands around a metal ion bonded by sigma bonds
   (d) the number of only anionic ligands bonded to the metal ion.

7. Which of the following is common donor atom in ligands?
   (a) Arsenic (b) Nitrogen
   (c) Oxygen (d) Both 'b' and 'c'

8. Which of the following may be considered to be an organometallic compound?
   (a) Nickel tetracarbonyl
   (b) Chlorophyll
   (c) $K_3[Fe(C_2O_4)_3]$
   (d) $[Co(en)_3]Cl_3$

9. Which of the following organometallic compound is σ and π bonded?
   (a) $[Fe(\eta^5-C_5H_5)_2]$
   (b) $Fe(CH_3)_3$
   (c) $K[PtCl_3(\eta^2-C_2H_4)]$
   (d) $[Co(CO)_5NH_3]^{2+}$

10. Coordination compounds have great importance in biological systems. In this context which of the following statements is **incorrect** ?
    (a) Cyanocobalamin is B_{12} and contains cobalt
    (b) Haemoglobin is the red pigment of blood and contains irons
    (c) Chlorophylls are green pigments in plants and contain calcium
    (d) Carboxypeptidase - A is an exzyme and contains zinc.

11. Among the following, the compound that is both paramagnetic and coloured is
(a) $K_2Cr_2O_7$ (b) $(NH_4)_2(TiCl_6)$
(c) $CoSO_4$ (d) $K_3[Cu(CN)_4]$
12. The complex ion which has no 'd' electron in the central metal atom is
(a) $[MnO_4]^-$ (b) $[Co(NH_3)_6]^{3+}$
(c) $[Fe(CN)_6]^{3-}$ (d) $[Cr(H_2O)_6]^{3+}$
13. Which kind of isomerism is exhibited by octahedral $Co(NH_3)_4Br_2Cl$?
(a) Geometrical and Ionization
(b) Geometrical and Optical
(c) Optical and Ionization
(d) Geometrical only
14. In which of the following compounds does iron exhibit zero oxidation state?
(a) $[Fe(H_2O)_6](NO_3)_3$ (b) $K_3[Fe(CN)_6]$
(c) $K_4[Fe(CN)_6]$ (d) $[Fe(CO)_5]$
15. CN^- is a strong field ligand. This is due to the fact that
(a) it carries negative charge
(b) it is a pseudohalide
(c) it can accept electrons from metal species
(d) it forms high spin complexes with metal species
16. The number of geometrical isomers for $[Pt(NH_3)_2Cl_2]$ is
(a) 2 (b) 1
(c) 3 (d) 4
17. Which of the following will exhibit maximum ionic conductivity?
(a) $K_4[Fe(CN)_6]$ (b) $[Co(NH_3)_6]Cl_3$
(c) $[Cu(CN_3)_4]Cl_2$ (d) $[Ni(CO)_4]$
18. Oxidation number of Ni in $[Ni(C_2O_4)_3]^{4-}$ is
(a) 3 (b) 4
(c) 2 (d) 6
19. The number of unpaired electrons in the complex ion $[CoF_6]^{3-}$ is (Atomic no.: Co = 27)
(a) zero (b) 2
(c) 3 (d) 4
20. Of the following complex ions, which is diamagnetic in nature ?
(a) $[NiCl_4]^{2-}$ (b) $[Ni(CN)_4]^{2-}$
(c) $[CuCl_4]^{2-}$ (d) $[CoF_6]^{3-}$
21. Hybridization of Ag in the linear complex $[Ag(NH_3)_2]^+$ is :
(a) dsp^2 (b) sp
(c) sp^2 (d) sp^3
22. Correct name of $K_4[Fe(CN)_6]$ is :
(a) Potassium ferricyanide
(b) Potassium ferrocyanide
(c) Potassium hexa cyanoferrate (II)
(d) Potassium hexa cyanoferrate (III)
23. Ammonia will not form complex ions with
(a) Ag^+ (b) Cd^{2+}
(c) Cu^{2+} (d) Pb^{2+}
24. $CH_3-Mg-Br$ is an organo metallic compound due to
(a) Mg – Br bond (b) C – Mg bond
(c) C – Br bond (d) C – H bond.
25. Ammonia forms the complex ion $[Cu(NH_3)_4]^{2+}$ with copper ions in alkaline solutions but not in acidic solutions. What is the reason for it ?
(a) In acidic solutions protons coordinate with ammonia molecules forming NH_4^+ ions and NH_3 molecules are not available
(b) In alkaline solutions insoluble $Cu(OH)_2$ is precipitated which is soluble in excess of any alkali
(c) Copper hydroxide is an amphoteric substance
(d) In acidic solutions hydration protects copper ions
26. $[EDTA]^{4-}$ is a :
(a) monodentate ligand
(b) bidentate ligand
(c) quadridentate ligand
(d) hexadentate ligand
27. $K_4[Fe(CN)_6]$ is a :
(a) double salt
(b) complex compound
(c) acid
(d) base
28. The IUPAC name of $K_2[PtCl_6]$ is
(a) hexachloroplatinate potassium
(b) potassium hexachloroplatinate (IV)
(c) potassium hexachloroplatinate
(d) potassium hexachloroplatinum (IV)
29. Which of the following is an organometallic compound?
(a) Lithium methoxide
(b) Lithium acetate
(c) Lithium dimethylamide
(d) Methyl lithium.

30. The species having tetrahedral shape is
(a) $[PdCl_4]^{2-}$ (b) $[Ni(CN)_4]^{2-}$
(c) $[Pd(CN)_4]^{2-}$ (d) $[NiCl_4]^{2-}$

31. The IUPAC name of $[Ni(NH_3)_4][NiCl_4]$ is
(a) Tetrachloronickel (II) - tetraamminenickel (II)
(b) Tetraamminenickel (II) - tetrachloronickel (II)
(c) Tetraamminenickel (II) - tetrachloronickelate (II)
(d) Tetrachloronickel (II) - tetrachloronickelate (0)

32. Both $[Ni(CO)_4]$ and $[Ni(CN)_4]^{2-}$ are diamagnetic. The hybridisations of nickel in these complexes, respectively, are
(a) sp^3, sp^3 (b) sp^3, dsp^2
(c) dsp^2, sp^3 (d) dsp^2, sp^2

33. Which of the following statements is correct ? (Atomic number of Ni = 28)
(a) $Ni(CO)_4$ is diamagnetic and $[NiCl_4]^{2-}$ and $[Ni(CN)_4]^{2-}$ are paramagnetic.
(b) $Ni(CO)_4$ and $[Ni(CN)_4]^{2-}$ are diamagnetic and $[NiCl_4]^{2-}$ is paramagnetic.
(c) $Ni(CO)_4$ and $[NiCl_4]^{2-}$ are diamagnetic and $[Ni(CN)_4]^{2-}$ is paramagnetic.
(d) $[NiCl_4]^{2-}$ and $[Ni(CN)_4]^{2-}$ are diamagnetic and $Ni(CO)_4$ is paramagnetic.

34. Among the following complexes, optical activity is possible in
(a) $[Co(NH_3)_6]^{3+}$
(b) $[Co(H_2O)_2(NH_3)_2Cl_2]^+$
(c) $[Cr(H_2O)_2Cl_2]^+$
(d) $[Co(CN)_5NC]$

35. The total number of possible isomers for the complex compound $[Cu^{II}(NH_3)_4][Pt^{II}Cl_4]$
(a) 3 (b) 6
(c) 5 (d) 4

36. IUPAC name of $[Pt(NH_3)_3(Br)(NO_2)Cl]Cl$ is
(a) Triamminechlorobromonitroplatinum (IV) chloride
(b) Triamminebromonitrochloroplatinum (IV) chloride
(c) Triamminebromochloronitroplatinum (IV) chloride
(d) Triamminenitrochlorobromoplatinum (IV) chloride

37. Which one of the following will show paramagnetism corresponding to 2 unpaired electrons? (Atomic numbers : Ni = 28, Fe = 26)
(a) $[FeF_6]^{3-}$ (b) $[NiCl_4]^{2-}$
(c) $[Fe(CN)_6]^{3-}$ (d) $[Ni(CN)_4]^{2-}$

38. Which one of the following complexes will have four different isomers ?
(a) $[Co(en)_2Cl_2]Cl$
(b) $[Co(en)(NH_3)Cl_2]Cl$
(c) $[Co(PPH_3)_2Cl_2]Cl$
(d) $[Co(en)_3]Cl_3$

39. The hypothetical complex chlorodiaquatriammine cobalt (III) chloride can be represented as
(a) $[CoCl(NH_3)_3(H_2O)_2]Cl_2$
(b) $[Co(NH_3)_3(H_2O)Cl_3]$
(c) $[Co(NH_2)_3(H_2O)_2Cl]$
(d) $[Co(NH_3)_3(H_2O)_3]Cl_3$

40. Which of the following does **not** have a metal-carbon bond?
(a) $Al(OC_2H_5)_3$ (b) C_2H_5MgBr
(c) $K[Pt(C_2H_4)Cl_3]$ (d) $Ni(CO)_4$

41. Which of the following is considered to be an anticancer species ?

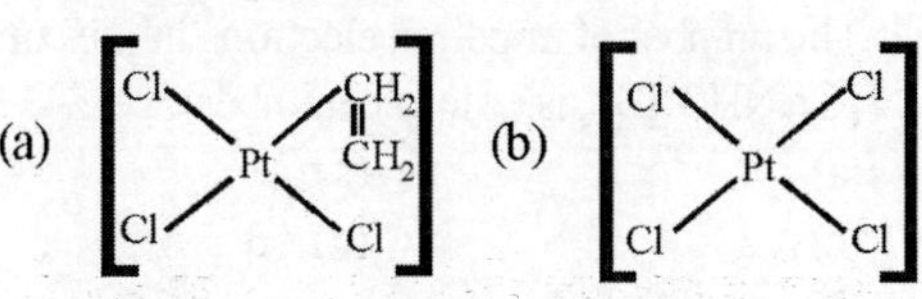

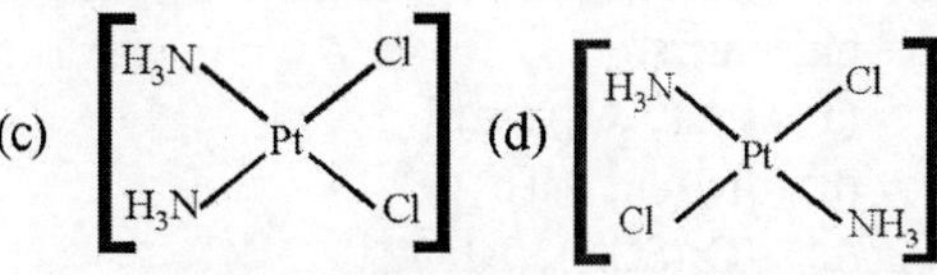

42. The d electron configurations of Cr^{2+}, Mn^{2+}, Fe^{2+} and Ni^{2+} are $3d^4$, $3d^5$, $3d^6$ and $3d^8$ respectively. Which one of the following aqua complexes will exhibit the minimum paramagnetic behaviour?
(a) $[Fe(H_2O)_6]^{2+}$ (b) $[Ni(H_2O)_6]^{2+}$
(c) $[Cr(H_2O)_6]^{2+}$ (d) $[Mn(H_2O)_6]^{2+}$
(At. No. Cr = 24, Mn = 25, Fe = 26, Ni = 28)

43. The complex ion $[Co(NH_3)_6]^{3+}$ is formed by sp^3d^2 hybridisation. Hence the ion should possess
(a) Octahedral geometry
(b) Tetrahedral geometry
(c) Square planar geometry
(d) Tetragonal geometry.

44. The formula for the complex, dichlorobis (urea) copper (II) is
(a) $[Cu\{O=C(NH_2)_2\}]Cl_2$
(b) $[Cu\{O=C(NH_2)_2\}Cl]Cl$
(c) $[CuCl_2\{O=C(NH_2)_2\}_2]$
(d) $[CuCl_2][\{O=C(NH_2)_2\}]_2$

45. The number of geometrical isomers of the complex $[Co(NO_2)_3(NH_3)_3]$ is
(a) 2 (b) 3
(c) 4 (d) zero

46. A co-ordination complex compound of cobalt has the molecular formula containing five ammonia molecules, one nitro group and two chlorine atoms for one cobalt atom. One mole of this compound produces three mole ions in an aqueous solution. On reacting this solution with excess of $AgNO_3$ solution, we get two moles of AgCl precipitate. The ionic formula for this complex would be
(a) $[Co(NH_3)_4(NO_2)Cl][(NH_3)Cl]$
(b) $[Co(NH_3)_5Cl][Cl(NO_2)]$
(c) $[Co(NH_3)_5(NO_2)]Cl_2$
(d) $[Co(NH_3)_5][(NO_2)_2Cl_2]$

47. The number of unpaired electrons in the complex $[Cr(NH_3)_6]Br_3$ is (Atomic number Cr = 24)
(a) 4 (b) 1
(c) 2 (d) 3

48. Which of the following will give maximum number of isomers?
(a) $[Ni(C_2O_4)(en)_2]^{2-}$
(b) $[Ni(en)(NH_3)_4]^{2+}$
(c) $[Cr(SCN)_2(NH_3)_4]^+$
(d) $[Co(NH_3)_4Cl_2]$

49. Which statement is incorrect?
(a) $Ni(CO)_4$ – Tetrahedral, paramagnetic
(b) $[Ni(CN)_4]^{2-}$ – Square planar, diamagnetic
(c) $Ni(CO)_4$ – Tetrahedral, diamagnetic
(d) $[NiCl_4]^{2-}$ – Tetrahedral, paramagnetic

50. Atomic number of Cr and Fe are respectively 25 and 26, which of the following is paramagnetic?
(a) $[Cr(CO)_6]$ (b) $[Fe(CO)_5]$
(c) $[Fe(CN)_6]^{-4}$ (d) $[Cr(NH_3)_6]^{+3}$

51. $CuSO_4$ when reacts with KCN forms CuCN, which is insoluble in water. It is soluble in excess of KCN due to formation of the following complex
(a) $K_2[Cu(CN)_4]$ (b) $K_3[Cu(CN)_4]$
(c) $CuCN_2$ (d) $Cu[KCu(CN)_4]$

52. Which one of the following octahedral complexes will not show geometric isomerism? (A and B are monodentate ligands)
(a) $[MA_5B]$ (b) $[MA_2B_4]$
(c) $[MA_3B_3]$ (d) $[MA_4B_2]$

53. According to IUPAC nomenclature sodium nitroprusside is named as
(a) Sodium pentacyanonitrosyl ferrate (III)
(b) Sodium nitroferrocyanide
(c) Sodium nitroferrocyanide
(d) Sodium pentacyanonitrosyl ferrate (II)

54. Among the following, which is **not** the π-bonded organometallic compound?
(a) $(CH_3)_4Sn$
(b) $K[PtCl_3(\eta^2-C_2H_4)]$
(c) $Fe(\eta^5-C_5H_5)_2$
(d) $Cr(\eta^6-C_6H_6)_2$

55. Which of the following coordination compounds would exhibit optical isomerism?
(a) pentamminenitrocobalt(III) iodide
(b) diamminedichloroplatinum(II)
(c) trans-dicyanobis (ethylenediamine) chromium (III) chloride
(d) tris-(ethylendiamine) cobalt (III) bromide

56. Considering H_2O as a weak field ligand, the number of unpaired electrons in $[Mn(H_2O)_6]^{2+}$ will be (At. no. of Mn = 25)
(a) three (b) five
(c) two (d) four

57. Which one of the following is an inner orbital complex as well as diamagnetic in behaviour? (Atomic number: Zn = 30, Cr = 24, Co = 27, Ni = 28)
(a) $[Zn(NH_3)_6]^{2+}$ (b) $[Cr(NH_3)_6]^{3+}$
(c) $[Co(NH_3)_6]^{3+}$ (d) $[Ni(NH_3)_6]^{2+}$

58. Which one of the following is expected to exhibit optical isomerism? (en = ethylenediamine)
(a) cis–$[Pt(NH3)_2Cl_2]$
(b) trans–$[Pt(NH_3)_2Cl_2]$
(c) cis–$[Co(en)_2Cl_2]$
(d) trans–$[Co(en)_2Cl_2]$

59. $[Co(NH_3)_4(NO_2)_2]Cl$ exhibits
   (a) linkage isomerism, ionization isomerism and geometrical isomerism
   (b) ionization isomerism, geometrical isomerism and optical isomerism
   (c) linkage isomerism, geometrical isomerism and optical isomerism
   (d) linkage isomerism, ionization isomerism and optical isomerism

60. $[Cr(H_2O)_6]Cl_3$ (at no. of Cr = 24) has a magnetic moment of 3.83 B. M. The correct distribution of 3d electrons in the Chromium of the complex is
   (a) $3d\,xy^1$, $(3dx^2-y^2)^1$, $3dyz^1$
   (b) $3d\,xy^1$, $3dyz^1$, $3dxz^1$
   (c) $3dxy^1$, $3dyz^1$, $3dz^2$
   (d) $(3dx^2-y^2)^1$, $3dz^2$, $3dxz^1$

Haloalkanes and Haloarenes

CONCEPTS

ALKYL HALIDES (RX) :

- General formula – $C_nH_{2n+1}X$.
- They can show chain, position and optical isomerism.
- Due to electronegativity difference between the carbon and the halogen, the shared pair of electron lies closer to the halogen atom. As a result, the halogen carries a small negative charge, while the carbon carries a small positive charge. Consequently, C – X bond is a polar covalent bond.

Methods of Preparation of Alkyl Halides :

- They can show chain, position and optical isomerism.

(a)

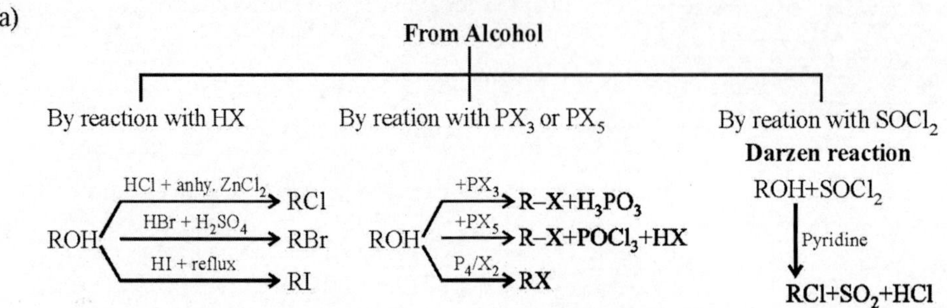

- $ZnCl_2$ act as dehydrating agent to promote C–O bond clevage.
- **Alkyl chrorides are best prepared** by $SOCl_2$ because SO_2 and HCl formed as by product are gases which can be easily removed.
- The **reactivity of alcohols** – $3° > 2° > 1°$.
  The **reactivity** of alkyl halides – HI > HBr > HCl.

(b)

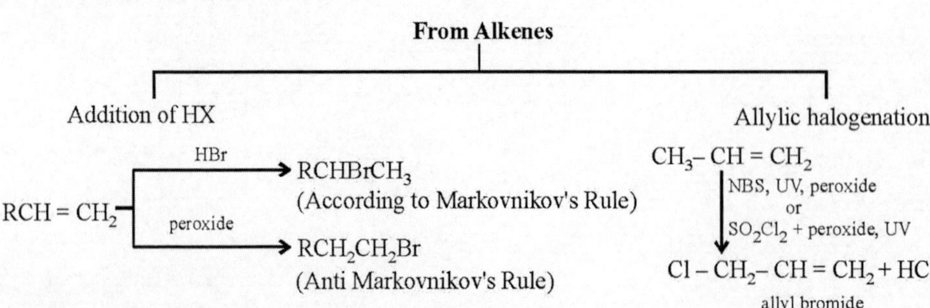

(c) **Halide exchange reaction (Finklestein reaction)**

$$RCl \text{ or } RBr + NaI \xrightarrow{\text{acetone}} R - I + NaCl / NaBr$$

Only iodoalkanes can be prepared by this method by displacing Cl or Br.

(d) **Borodine - Hunsdiecker reaction**

$$RCOOAg + Br_2 \xrightarrow{CCl_4 / reflux} R-Br + AgBr + CO_2$$

(i) Reaction **proceeds via free radical**.
(ii) Reaction is used in descending of homologus series.
(iii) Yield of alkyl bromides follow the order; 1° > 2° > 3°

Properties of Alkyl Halides :

Physical properties of alkyl halides :

(a) **Melting and boiling points :**
(i) RI > RBr > RCl > RF (when R is same)
(ii) $CH_3CH_2CH_2X > CH_3CH_2X > CH_3X$
(iii) For isomeric alkyl halides boiling point decreases with increase in branching
$CH_3CH_2CH_2CH_2Cl > CH_3CH_2CHClCH_3 > (CH_3)_2CHCH_2Cl > (CH_3)_3CCl$

(b) **Bond strength** decreases with increase in size of halogen
$CH_3F > CH_3Cl > CH_3Br > CH_3I$

(c) **Stability order** R–F > R–Cl > R–Br > R–I

(d) **Dipole moment** - $CH_3Cl > CH_3F > CH_3Br > CH_3I$

Chemical Properties :

They undergo **nucleophilic substitution and elimination reaction.**

(a) Nucleophilic substitution reaction proceeds via either S_N1 or by S_N2 type mechanism.

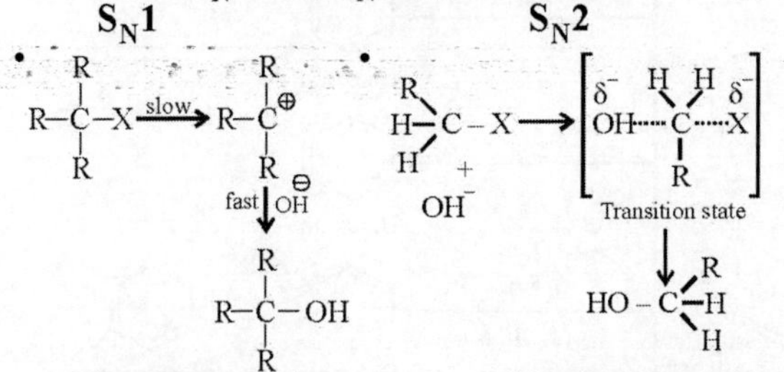

S_N1	S_N2
• Two step reaction	• Single step reaction
• r = k[RX]	• r = k[RX][Nu⁻]
• Polar solvents favour it.	• Non polar solvents favour it.
• Racemisation takes place.	• Inversion of configuration takes place.
• Low concentration of nucleophile favour it.	• High concentration of nucleophile favour it.

(b) **Elimination or Dehydrohalogenation reactions :**

• Reaction is governed Saytzeff rule (more highly substituted alkane is the major product of dehydro-halogenation).

$$CH_3-CH_2-\underset{Br}{\underset{|}{CH}}-CH_3 \xrightarrow[-KBr,\ -H_2O]{alc\ KOH,\ heat} \underset{(major)}{CH_3-CH=CH-CH_3} + \underset{(minor)}{CH_3-CH_2-CH=CH_2}$$

• **Ease of dehydrogenation** 3° > 2° > 1°.
• Types of eliminations – E_1 and E_2 reaction.

(c) **Reaction with KOH**

$$RX \xrightarrow{+KOH\ (aq)} ROH + KX$$
$$RX \xrightarrow{+KOH\ (alc)} alkene + H_2O + KX$$

(d) **Reaction with Ag_2O**

$$RX \xrightarrow{+Ag_2O\ (moist)} \underset{alcohol}{ROH} + AgX$$
$$RX \xrightarrow{+Ag_2O\ (dry)} \underset{ether}{R-O-R} + 2AgX$$

(e) **Williamson's synthesis**

$$RX + \underset{sod.\ alkoxide}{RONa} \longrightarrow \underset{ether}{R-O-R} + NaX$$

(f)

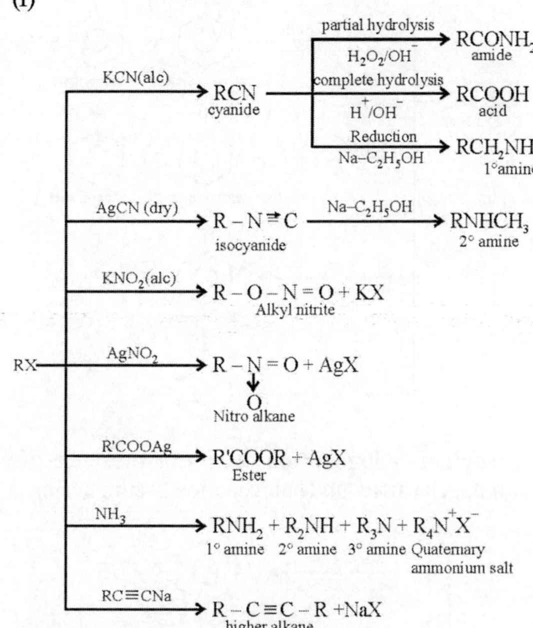

ARYL HALIDES (ArX)

Reactions of Haloarenes :

Aryl halides are extremely less reactive towards nucleophilic substitution reaction due to following reasons :

(a) **Resonance effect:** In haloarenes, the lone pair of electron on the halogen atom are delocalized on the benzene ring as:

As a result, C – Cl bond acquires some double bond character. Consequently, C – X bond in aryl halides is little stronger than in alkyl halides, and hence cannot be easily broken.

(b) In haloalkane, the halogen is attached to sp^3-hybridized carbon while in haloarenes the halogen is attached to sp^2-hybridized carbon. Since sp^2-hybridized orbital is smaller in sizes as compared to sp^3-carbital of carbon, therefore, the C – Cl bond in haloarenes should be shorter and hence stronger than in methyl chloride.

(c) In haloarenes, the phenyl cation formed as a result of self-ionisation will not be stabilised by resonance and therefore, S_N1 mechanism is ruled out.

Methods of Preparation :

(i) **From diazonium salt**

$C_6H_5\overset{+}{N}_2\overset{-}{Cl}$ (Diazonium salt)

- $\xrightarrow{\text{CuCl/HCl or CuBr/HBr}}$ (Sandmeyer reaction) chlorobenzene or Bromobenzene $+ N_2$
- $\xrightarrow{\text{Cu/HCl or Cu/HBr}}$ (Gattermann reaction) Chlorobenzene or Bromobenzene $+ N_2$
- $\xrightarrow{HBF_4}$ (Balz Schiemann reaction) Fluorobenzene $+ N_2 + BF_3$
- $\xrightarrow{\text{aq KI}/\Delta}$ Iodobenzene $+ N_2$

(ii) **Nuclear halogenation** occurs in presence of halogen carrier and halogenation occurs in ring.

Benzene

- $\xrightarrow{Cl_2/AlCl_3}$ Chlorobenzene + HCl
- $\xrightarrow{Br_2/AlBr_3}$ Bromobenzene + HBr
- $\xrightarrow{I_2/HIO_3}$ Iodobenzene + HI

(iii) **Side chain substitution** occurs in absence of Lewis acid and in presence of light. Side chain hydrogens are substituted by halogen atoms.

Toluene $\xrightarrow[hv]{Cl_2}$ Benzyl chloride ($C_6H_5CH_2Cl$) $\xrightarrow[hv]{Cl_2}$ Benzal chloride ($C_6H_5CHCl_2$) $\xrightarrow[hv]{Cl_2}$ Benzo chloride ($C_6H_5CCl_3$)

Properties of Aryl Halides :

Physical properties :

(i) Like alkyl halides, aryl halides are insoluble in water due to their incapability of forming H-bonds.

(ii) Aryl halides are less polar than alkyl halides.

(iii) Possibility of resonance in chlorobenzene makes the C–Cl bond shorter and hence, its dipole moment is less than that of cyclohexyl chloride.

(iv) The *p*-isomer has higher melting point than the *o*- and *m*-isomers.

Chemical properties :

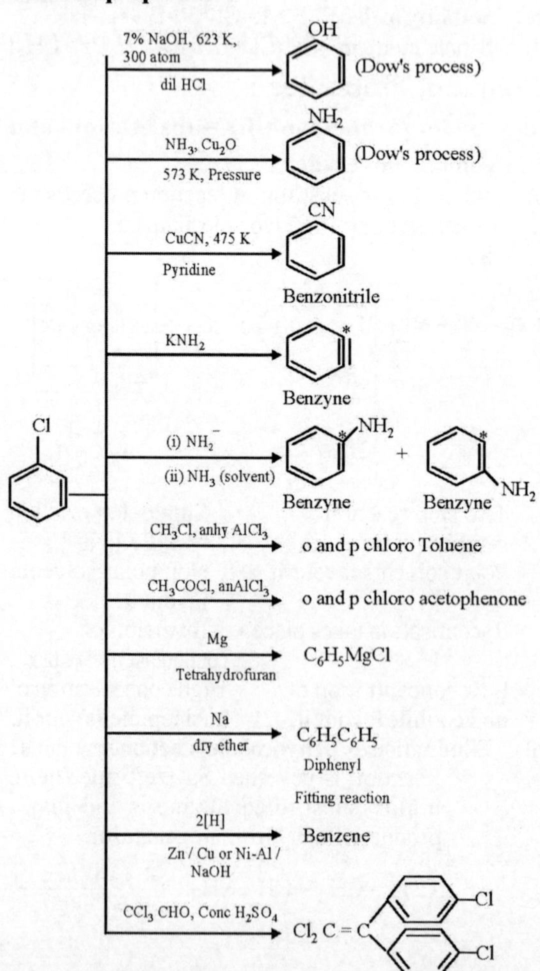

Dichloro Ethanes :

They are of two types -

- **Alkylidene or gem dihalide** - Both halogen atoms present on same carbon atom.
- **Alkylene or vic dihalide** - Both halogen atoms are present on adjacent carbon atoms.

Preparation

gem. dichloroethane

(i) $CH_3CHO + PCl_5 \longrightarrow CH_3CHCl_2 + POCl_3$ (acetaldehyde)

(ii) $HC \equiv CH + 2HCl \longrightarrow CH_3CHCl_2$ (acetylene)

vic dichloroethane

(i) $CH_2 = CH_2 + Cl_2 \longrightarrow CH_2Cl - CH_2Cl$ (ethene)

(ii) $\begin{matrix} CH_2OH \\ | \\ CH_2OH \end{matrix} + 2PCl_5 \xrightarrow[425]{HCl} \begin{matrix} CH_2Cl \\ | \\ CH_2Cl \end{matrix}$ (Glycol)

Properties

gem. dichloroethane

(i) $CH_3CHCl_2 + 2KOH(aq) \longrightarrow CH_3CH = O$ (acetaldehyde)

(ii) $CH_3CHCl_2 + Zn \longrightarrow \begin{matrix} CH_2 \\ \| \\ CH_2 \end{matrix}$ (ethene)

(iii) $CH_3CHCl_2 + KOH(alc.) \longrightarrow \begin{matrix} CH \\ ||| \\ CH \end{matrix}$ (acetylene)

vic dichloroethane

(i) $\begin{matrix} CH_2Cl \\ | \\ CH_2Cl \end{matrix} + aq.KOH \longrightarrow \begin{matrix} CH_2OH \\ | \\ CH_2OH \end{matrix}$ (glycol)

(ii) $\begin{matrix} CH_2Cl \\ | \\ CH_2Cl \end{matrix} + Zn \longrightarrow \begin{matrix} CH_2 \\ \| \\ CH_2 \end{matrix}$ (ethene)

(iii) $\begin{matrix} CH_2Cl \\ | \\ CH_2Cl \end{matrix} + KOH(alc.) \longrightarrow \begin{matrix} CH \\ ||| \\ CH \end{matrix}$ (acetylene)

Chloroform ($CHCl_3$)

Preparation :

(i) **From ethyl alcohol**

$CH_3CH_2OH \xrightarrow{CaOCl_2 + H_2O} CHCl_3 + (HCOO)_2Ca$ (chloroform, calcium formate)

(ii) **From acetone**

$CH_3COCH_3 \xrightarrow{CaOCl_2 + H_2O} CHCl_3 + (CH_3COO)_2Ca$ (acetone; chloroform, calcium acetate)

Properties :

(i) **Oxidation**

$CHCl_3 + O_2(air) \longrightarrow COCl_2 + HCl$

Phosgene (Poisonous gas)

∴ Chloroform is mixed with 1% C_2H_5OH (while storing) to change phosgene into non-poisonous diethyl carbonate.

$COCl_2 + 2C_2H_5OH \longrightarrow (C_2H_5)_2CO_3 + 2HCl$ (Phosgene; diethyl carbonate)

(ii) $CHCl_3 + HNO_3 \longrightarrow NO_2.CCl_3 + H_2O$

chloropicrin **(used as tear gas and insecticide)**

(iii) **Carbylamine reaction**

1° amine (aliphatic or aromatic) + KOH (alc) + $CHCl_3$ $\longrightarrow$ Bad smelling isocyanide

$C_6H_5NH_2 + CHCl_3 + 3KOH(alc)$ (Aniline)

$\longrightarrow C_6H_5N \overset{\rightarrow}{=} C + 3KCl + 3H_2O$

Phenyl isocyanide (bad smelling)

Reaction is used to test 1° amines.

Reimer Tiemann reaction :

Phenol (OH on benzene) $+ CHCl_3 + 3KOH \xrightarrow[(-3KCl,\ -2H_2O)]{heat}$ (O^-K^+, CHO on benzene) $\xrightarrow{H^+}$ (OH, CHO on benzene)

Salicylaldehyde

(iv) **Dehalogenation**

$CH\boxed{Cl_3 + 6Ag + Cl_3}CH \longrightarrow CH \equiv CH + 6AgCl$ (acetylene)

(v) **Condensation with acetone**

$CHCl_3 + CH_3COCH_3 \longrightarrow$ (acetone) $(CH_3)_2C(OH)CCl_3$

Chloretone (used as hypnotic and sleep inducing drug)

(vi) **Hydrolysis**

$$CHCl_3 + KOH_{(aq)} \longrightarrow \underset{\text{pot. formate}}{HCOOK} + H_2O$$

$$CCl_4 \xrightarrow{KOH\ (aq.)} C(OH)_4$$

$$\xrightarrow{-2H_2O} CO_2 \xrightarrow[(-H_2O)]{KOH} K_2CO_3$$

Due to non-availability of *d*-orbital in C, CCl_4 resists hydrolysis with boiling water.

IODOform (CHI_3)

(i) It is yellow solid at room temperature.

(ii) **Iodoform test** is used to detect presence of CH_3CO group or $CH_3CH(OH)$ group in organic compound.

Compound having

$$CH_3CO\ gr. + NaOH / KOH + I_2 \longrightarrow \text{Yellow ppt of } CHI_3$$

(iii) **It is used as antiseptic** because it liberates free I_2.

COMMERCIALLY IMPORTANT COMPOUNDS:

(i) **DDT** [2, 2-bis (p-chlorophenyl) 1, 1, 1-trichloroethane]

$$2\ Cl-C_6H_4-H\ (\text{Chlorobenzene}) + \underset{\text{chloral}}{CCl_3CHO} \xrightarrow{\text{conc. } H_2SO_4} \underset{\text{DDT}}{(Cl-C_6H_4)_2CHCCl_3} + H_2O$$

- Used **as insecticide**.
- It is non-biodegradable.

(ii) **Benzene hexachloride (BHC) or Gammaxane or 666 or Lindane:**

$$C_6H_6 + 3Cl_2 \xrightarrow{U.V} C_6H_6Cl_6$$

- Used **as pesticide** (to kill termites)

(iii) **Westron ($Cl_2HC.CHCl_2$)** used as solvent in paints and varnishes.

(iv) **Westrosol** ($Cl_2C = CHCl$) used as solvent for dry cleaning.

Exercise

1. Phosgene is a common name for
   (a) phosphoryl chloride
   (b) Thionyl chloride
   (c) Carbon dioxide and phosphine
   (d) Carbonyl chloride.

2. Which of the following reactions is an example of nucleophilic substitution reaction?
   (a) $2\,RX + 2\,Na \rightarrow R-R + 2\,NaX$
   (b) $RX + H_2 \rightarrow RH + HX$
   (c) $RX + Mg \rightarrow RMgX$
   (d) $RX + KOH \rightarrow ROH + KX$

3. Which of the following possesses highest melting point?
   (a) Chlorobenzene
   (b) m-dichlorobenzene
   (c) o-dichlorobenzene
   (d) p-dichlorobenzene

4. Which one is most reactive towards S_N1 reaction?
   (a) $C_6H_5CH(C_6H_5)Br$
   (b) $C_6H_5CH(CH_3)Br$
   (c) $C_6H_5C(CH_3)(C_6H_5)Br$
   (d) $C_6H_5CH_2Br$

5. In a S_N2 substitution reaction of the type

$$R-Br + Cl^- \xrightarrow{DMF} R-Cl + Br^-$$

which one of the following has the highest relative rate?

(a) $CH_3-CH_2-CH_2Br$

(b) $CH_3-CH(CH_3)-CH_2Br$

(c) $CH_3-C(CH_3)_2-CH_2Br$

(d) CH_3CH_2Br

6. $CH_3-CH_2-CH(Cl)-CH_3$ obtained by chlorination of n-butane, will be

(a) *l*-form (b) d-form

(c) Meso form (d) Racemic mixture

7. A Grignard reagent may be made by reacting magnesium with

(a) Methyl amine (b) Diethyl ether

(c) Ethyl iodide (d) Ethyl alcohol

8. Which chloro derivative of benzene among the following would undergo hydrolysis most readily with aqueous sodium hydroxide to furnish the corresponding hydroxy derivative?

(a) 2,4,6-trinitrochlorobenzene: O_2N–$C_6H_2(NO_2)_2$–Cl

(b) $O_2N-C_6H_4-Cl$

(c) $Me_2N-C_6H_4-Cl$

(d) C_6H_5Cl.

9. Benzene reacts with CH_3Cl in the presence of anhydrous $AlCl_3$ to form:

(a) chlorobenzene (b) benzylchloride

(c) xylene (d) toluene

10. Trichloroacetaldehyde, CCl_3CHO reacts with chlorobenzene in presence of sulphuric acid and produces:

(a) $Cl-C_6H_4-C(H)(C_6H_4Cl)-C_6H_4-Cl$

(b) $Cl-C_6H_4-C(OH)(Cl)-C_6H_4-Cl$

(c) $Cl-C_6H_4-CH(CCl_3)-C_6H_4-Cl$

(d) $Cl-C_6H_4-C(Cl)(CH_2Cl)-C_6H_4-Cl$

11. Chlorobenzene reacts with Mg in dry ether to give a compound (A) which further reacts with ethanol to yield

(a) Phenol (b) Benzene

(c) Ethylbenzene (d) Phenyl ether.

12. Chloroform on treatment with conc. HNO_3 gives

(a) Chloropicrin (b) Nitromethane

(c) Picric acid (d) Acetylene

13. Which of the following is gem-dihalide?

(a) CH_3CHBr_2

(b) $BrCH_2CH_2Br$

(c) $CH_3CHBrCH_2Br$

(d) $CH_3CHBrCH_2CH_2CH_2Br$

14. Benzene reacts with n-propyl chloride in the presence of anhydrous $AlCl_3$ to give

(a) 3 – Propyl – 1 – chlorobenzene

(b) n-Propylbenzene

(c) No reaction

(d) Isopropylbenzene.

15. The replacement of chlorine of chlorobenzene to give phenol requires drastic conditions, but the chlorine of 2,4-dinitrochlorobenzene is readily replaced since,

(a) nitro groups make the aromatic ring electron rich at ortho/para positions
(b) nitro groups withdraw electrons from the meta position of the aromatic ring
(c) nitro groups donate electrons at meta position
(d) nitro groups withdraw electrons from ortho/para positions of the aromatic ring

16. Which of the following is responsible for depletion of the ozone layer in the upper strata of the atmosphere?
(a) Polyhalogens (b) Ferrocene
(c) Fullerenes (d) Freons

17. Chloropicrin is obtained by the reaction of
(a) steam on carbon tetrachloride
(b) nitric acid on chlorobenzene
(c) chlorine on picric acid
(d) nitric acid on chloroform

18. Bromine gives substitution reaction products with benzene readily in presence of :
(a) Sunlight
(b) Platinum
(c) $FeBr_3$ at 35°C
(d) $FeBr_3$ at high temperature

19. Haloforms are trihalogen derivatives of
(a) Ethane (b) Methane
(c) Propane (d) Benzene

20. Chloropicrin is
(a) $C_2H_5C(NO)_5SH$ (b) CCl_3CHO
(c) CCl_3NO_2 (d) CCl_3NO_3

21. Chloroform cannot be prepared from which of the following?
(a) CH_3OH (b) C_2H_5OH
(c) CH_3CHO (d) $(CH_3)_2CO$

22. Which reagent cannot be used to prepare an alkyl halide from an alcohol ?
(a) $HCl + ZnCl_2$ (b) NaCl
(c) PCl_5 (d) $SOCl_2$

23. Which of the following compounds is used as a refrigerant ?
(a) Acetone (b) CCl_4
(c) CF_4 (d) CCl_2F_2

24. Which of the following compound will make precipitate most readily with $AgNO_3$?
(a) $CCl_3.CHO$ (b) $CHCl_3$
(c) $C_6H_5CH_2Cl$ (d) C_6H_5Cl

25. Chloroform is used as :
(a) Fire extinguisher (b) Industrial solvent
(c) Refrigerant (d) Insecticide

26. Which of the following is the example of Friedal Craft Reaction :
(a) $C_6H_6 + CH_3CH_2Cl \xrightarrow{AlCl_3} C_6H_5C_2H_5 + HCl$
(b) $C_6H_5OH + CH_3CH_2Cl \xrightarrow{AlCl_3} C_6H_5C_2H_5 + HOCl$
(c) $C_6H_5Cl + CH_3CH_2Cl \xrightarrow{[AlCl_3]} C_6H_5C_2H_5 + Cl_2$
(d) $C_6H_5CH_3 + CH_3CH_2Cl \xrightarrow{[AlCl_3]} C_6H_5C_2H_5 + CH_3Cl$

27. Which of the following does not form Ethyl chloride from ethyl alcohol ?
(a) Cl_2 (b) Cl_2/P
(c) conc HCl (d) PCl_3

28. Which does not react with I_2 & NaOH :
(a) Ethyl Alcohol (b) Acetaldehyde
(c) Acetone (d) Propanol

29. Oxidation of primary alcohols with chlorine yields :
(a) Acyl chloride (b) Alkyl chloride
(c) Aldehyde (d) Ketone

30. In which of the following conversions, phosphorus pentachloride is used as the reagent?
(a) $H_2C = CH_2 \longrightarrow CH_3CH_2Cl$
(b) $CH_3CH_2OH \longrightarrow CH_3CH_2Cl$
(c) $H_3C - O - CH_3 \longrightarrow CH_3Cl$
(d) $CH \equiv CH \longrightarrow CH_2 = CHCl$

31. $AgNO_3$ does not give precipitate with $CHCl_3$ because
(a) $CHCl_3$ does not ionise in water
(b) $CHCl_3$ does not react with $AgNO_3$
(c) $CHCl_3$ is chemically inert
(d) None of these

32. Iodoform test is not given by
(a) 2-pentanone (b) 3-pentanone
(c) Ethanol (d) None

33. When chlorine is passed through propene at 400°C, which of the following is formed ?
(a) PVC
(b) Allyl chloride
(c) Nikyl chloride
(d) 1, 2-Dichloroethane

34. An organic compound A (C_4H_9Cl) on reaction with Na/diethyl ether gives a hydrocarbon which on monochlorination gives only one chloro derivative, then A is
(a) tert-butyl chloride
(b) sec-butyl chloride
(c) isobutyl chloride
(d) n-butyl chloride

35. Reactivity order of halides for dehydrohalogenation is
(a) $R-F > R-Cl > R-Br > R-I$
(b) $R-I > R-Br > R-Cl > R-F$
(c) $R-I > R-Cl > R-Br > R-F$
(d) $R-F > R-I > R-Br > R-Cl$

36. $$CH_3CH_2Cl \xrightarrow{NaCN} X \xrightarrow{Ni/H_2} Y$$
$$Y \xrightarrow[\text{anhydride}]{\text{Acetic}} Z$$
Z in the above reaction sequence is
(a) $CH_3CH_2CH_2NHCOCH_3$
(b) $CH_3CH_2CH_2NH_2$
(c) $CH_3CH_2CH_2CONHCH_3$
(d) $CH_3CH_2CH_2CONHCOCH_3$

37. When $CH_3CH_2CHCl_2$ is treated with $NaNH_2$, the product formed is
(a) $CH_3-CH=CH_2$
(b) $CH_3-C\equiv CH$
(c) $CH_3CH_2CH\langle^{NH_2}_{NH_2}$
(d) $CH_3CH_2CH\langle^{Cl}_{NH_2}$

38. Identify X, Y, Z in the following series
$$C_2H_5I \xrightarrow{\text{alc. KOH}} X \xrightarrow{Br_2} Y \xrightarrow{KCN} Z$$
(a) $BrCH_2CH_2Br$
(b) $BrCH_2CH_2CN$
(c) $CN-CH_2-CH_2-CN$
(d) $Br-CH=CH-CN$

39. Under certain conditions an alkyl halide reacts with base to give an alkene and HCl [Elimination Reaction] for example
$R-CH_2-CH_2-Cl \rightarrow R-CH=CH_2+HCl$
The extent of these reactions depends on the structure of alkyl halides (e.g. primary, secondary or tertiary). The relative extent to which such reactions take place is in the order (of haloalkanes) :
(a) Primary < secondary < Tertiary
(b) Primary > Secondary > Tertiary
(c) Primary > Secondary < Tertiary
(d) Primary < Secondary > Tertiary

40. On sulphonation of C_6H_5Cl
(a) benzene sulphonic acid is formed
(b) metachlorobenzene sulphonic acid is formed
(c) orthochlorobenzene sulphonic acid is formed
(d) ortho and para chlorobenzene sulphonic acids are formed.

41. S_N2 mechanism is involved in the following substitution :
(a) $CH_3-CH_2-Cl+OH^-$
(b) $CH_3-C(Cl)(CH_3)-CH_3+OH^-$
(c) $CH_3-CH(Cl)-CH_3+OH^-$
(d) $CH_3-CH_2-C(Cl)(CH_3)-CH_3+OH^-$

42. An alkyl halide by formation of its Grignard reagent and heating with water yields propane. What is the original alkyl halide ?
(a) Methyl iodide (b) Ethyl iodide
(c) Ethyl bromide (d) Propyl bromide

43. Identify the set of reagent / reaction conditions 'X' and 'Y' in the following set of transformations
$$CH_3-CH_2-CH_2Br \xrightarrow{X} \text{Product} \xrightarrow{Y} CH_3-CH(Br)-CH_3$$

(a) X = dilute aqueous NaOH, 20°C; Y = HBr/acetic acid, 20°C
(b) X = concentrated alcoholic NaOH, 80°C; Y = HBr/acetic acid, 20°C
(c) X = dilute aqueous NaOH, 20°C ; Y = Br_2/$CHCl_3$, 0°C
(d) X = concentrated alcoholic NaOH, 80°C; Y = Br_2/$CHCl_3$, 0°C

44. The chief reaction product of reaction between *n*-butane and bromine at 130°C is :
(a) $CH_3CH_2CH_2CH_2Br$
(b) $CH_3CH_2CHBr(CH_3)$
(c) $CH_3-CH_2CHBr(CH_2Br)$
(d) $CH_3CH_2CBr_2(CH_3)$

45. The reaction conditions leading to the best yields of C_2H_5Cl are :
(a) $C_2H_6\text{ (excess)} + Cl_2 \xrightarrow{\text{uv light}}$
(b) $C_2H_6 + Cl_2 \xrightarrow[\text{room temperature}]{\text{dark}}$
(c) $C_2H_6 + Cl_2\text{ (excess)} \xrightarrow{\text{uv light}}$
(d) $C_2H_6 + Cl_2 \xrightarrow{\text{uv light}}$

46. The reaction of toluene with chlorine in presence of ferric chloride gives predominantly :
(a) benzoyl chloride
(b) *m*-chlorotoluene
(c) benzyl chloride
(d) *o*- and *p*-chlorotoluene

47. The number of structural and configurational isomers of a bromo compound, C_5H_9Br, formed by the addition of HBr to 2-pentyne respectively are
(a) 1 and 2 (b) 2 and 4
(c) 4 and 2 (d) 2 and 1

48. Industrial preparation of chloroform employs acetone and
(a) Phosgene
(b) Calcium hypochlorite
(c) Chlorine gas
(d) Sodium chloride.

49. Zerevitinov's determination of active hydrogen in a compound is based upon its reaction with
(a) Na (b) CH_3MgI
(c) Zn (d) Al.

50. 2-Bromopentane is heated with potassium ethoxide in ethanol. The major product obtained is
(a) 2-ethoxypentane (b) pentene-1
(c) trans-2-pentene (d) cis-pentene-2

51. The reaction of toluene with Cl_2 in presence of $FeCl_3$ gives 'X' and reaction in presence of light gives 'Y'. Thus, 'X' and 'Y' are :
(a) X = Benzal chloride, Y = *o* – Chlorotoluene
(b) X = *m* – Chlorotoluene, Y = *p* –Chlorotoluene
(c) X = o –and *p* – Chlorotoluene, Y = Trichloromethyl – benzene
(d) X = Benzyl chloride, Y = *m* – Chlorotoluene

52. Consider the reactions :
(i) $(CH_3)_2CH-CH_2Br \xrightarrow{C_2H_5OH} (CH_3)_2CH-CH_2OC_2H_5 + HBr$
(ii) $(CH_3)_2CH-CH_2Br \xrightarrow{C_2H_5O^-} (CH_3)_2CH-CH_2OC_2H_5 + Br^-$

The mechanisms of reactions (i) and (ii) are respectively :
(a) S_N1 and S_N2 (b) S_N1 and S_N1
(c) S_N2 and S_N2 (d) S_N2 and S_N1

53. A salt solution is treated with chloroform drops. Then it is shaken with chlorine water. Chloroform layer becomes violet. Solution contains.
(a) NO_2^- ion (b) NO_3^- ion
(c) Br^- ion (d) I^- ion

54. An alkyl halide reacts with metallic sodium in dry ether. The reaction is known as :
(a) Frankland's reaction
(b) Sandmeyer's reaction
(c) Wurtz reaction
(d) Kolbe's reaction

55. Vapour density of an organic compound is 23. It contains 52.17% of carbon and 13% of hydrogen. The compound gives iodoform test. The compound is :
(a) Ethanol (b) Dimethyl ether
(c) Acetone (d) Methanol

56. When chloroform reacts with acetone the product is
(a) Ethyliene chlorde (b) Mesitylene
(c) Chloretone (d) Chloral

57. Ethylene dichloride can be prepared by adding HCl to
(a) Ethane (b) Ethylene
(c) Acetylene (d) Ethylene glycol

58. Chloroform on warming with Ag powder gives
(a) C_6H_6 (b) C_2H_4
(c) C_2H_2 (d) None

59. When chloroform is exposed to light and air, it forms
(a) chlorine gas
(b) methyl chloride
(c) phosgene gas
(d) carbon tetrachloride

60. When 2-bromobutane reacts with alcoholic KOH, the reaction is called
(a) halogenation
(b) chlorination
(c) hydrogenation
(d) dehydro halogenation

Alcohols, Phenols and Ethers

CONCEPTS

ALCOHOLS

(R – OH)

- General formula $C_nH_{2n+1}OH$

Methods of Preparation of Alcohols

(a) From alkyl halides:

$$RX \xrightarrow{\text{aq KOH}} ROH + KX$$

$$RX \xrightarrow{\text{Moist AgOH}} ROH + AgX$$

(b) By reduction of carbonyl compounds, acids and esters:

(i) Aldehydes $\xrightarrow{LiAlH_4}$ 1° alcohol

$$\underset{\text{acetaldehyde}}{CH_3CHO} \xrightarrow{LiAlH_4} \underset{\text{ethanol}}{CH_3CH_2OH}$$

(ii) Ketone $\xrightarrow{LiAlH_4}$ 2° alcohol

$$\underset{\text{acetone}}{CH_3-\overset{\overset{\large O}{\|}}{C}-CH_3} \xrightarrow{LiAlH_4} \underset{\text{2-propanol}}{CH_3-\overset{OH}{\underset{H}{C}}-CH_3}$$

(iii) Acid $\xrightarrow{LiAlH_4 / H_2-Ni}$ 1° alcohol

(iv) Ester $\xrightarrow{LiAlH_4 / Na-alcohol}$ mixture of alcohols

$$RCOOR' \xrightarrow{LiAlH_4} RCH_2OH + R'OH$$

(c) By reaction of Grignard reagent and carbonyl compounds:

(i) $HCHO + RMgX \xrightarrow{H_2O/H^+}$ 1° alcohol

(ii) Any other aldehyde + RMgX $\xrightarrow{H_2O/H^+}$ 2° alcohol

(iii) Ketone + RMgX $\xrightarrow{H_2O/H^+}$ 3° alcohol

CH_3OH cannot be prepared by Grignard reagent.

(d) From esters (with Grignard reagent):

(i) Formic esters + 2RMgX $\xrightarrow{H_2O/H^+}$ 2° alcohol

$$\underset{\text{ethylformate}}{H-\overset{\overset{\large O}{\|}}{C}-OC_2H_5} + 2R'MgX \xrightarrow[\text{(ii) }-Mg(OH)X]{\text{(i) dry ether}} \underset{\text{2° alcohol}}{H-\overset{OH}{\underset{R'}{C}}-R'}$$

(ii) Esters other than formic ester + RMgX $\xrightarrow{H_2O/H^+}$ 3° alcohol

$$R-\overset{\overset{\large O}{\|}}{C}-OC_2H_5 + 2R'MgX \xrightarrow[\text{(ii) }Mg(OH)X]{\text{(i) dry ether}} \underset{\text{3° alcohol}}{R-\overset{OH}{\underset{R'}{C}}-R'}$$

(e) From alkenes:

(i) Hydration of alkene

$$RCH = CH_2 + H_2O \xrightarrow[\text{Anti Markownikoff rule}]{\text{conc } H_2SO_4} R-\underset{OH}{CH}-CH_3$$

- Rearrangement occurs during reaction because reaction must proceed via most stable carbonium ion.
- Trans product are formed during reaction.

1-methoxycyclohexene $\xrightarrow[THF]{BH_3}$ Trans-2-methylcyclohexanol

- The addition of water to alkene is syn.

(ii) **Hydroboration – Oxidation reaction**

$$\underset{\text{diborane}}{RCH = CH_2 + B_2H_6} \xrightarrow[\text{Anti Markownikoff addition}]{\text{dry ether}}$$

$$\underset{\text{Trialkylboron}}{2(RCH_2CH_2)_3B} \xrightarrow{NaOH/H_2O_2} RCH_2CH_2OH + 2H_3BO_3$$

- It is **anti-Markownikoff addition** of H_2O to alkene.
- The addition of water to alkene is syn.

(iii) **Oxymercuration – demercuration reaction**

$$RCH = CH_2 \xrightarrow[\text{Markownikoff addition}]{(CH_3COO)_2Hg,\ THF-H_2O}$$

$$R-\underset{OH}{CH}-\underset{HgOCOCH_3}{CH_2} \xrightarrow{NaBH_4/OH} R-\underset{OH}{CH}-CH_3$$

- **It is Markownikoff addition of water to alkene.**
- **The addition of water to alkene is anti.**

(iv) **Oxo process** (Carbomylation or hydroformylation reaction)

$$\underset{\text{ethene}}{CH_2 = CH_2} + CO + H_2O \xrightarrow[\Delta]{[Co(CO)_4]}$$

$$\underset{\text{propanal}}{CH_3CH_2CHO} \xrightarrow{H_2/Ni} \underset{\text{propanol-1}}{CH_3CH_2CH_2OH}$$

(f) **By fermentation of carbohydrates:** (Commercial method)

(i) **From molasses** - It is solution left after crystallisation of sugar from sugar cane juice.

$$\underset{\text{sucrose}}{C_{12}H_{22}O_{11}} + H_2O \xrightarrow{\text{Invertase}} \underset{\text{glu cose}}{C_6H_{12}O_6} + \underset{\text{fructose}}{C_6H_{12}O_6}$$

$$C_6H_{12}O_6 \xrightarrow{\text{Zymase}} \underset{\text{ethyl alcohol}}{2C_2H_5OH} + 2CO_2$$

(ii) **From starch** -

$$\underset{\text{Starch}}{2(C_6H_{10}O_5)_n} + nH_2O \xrightarrow{\text{Diastase}} \underset{\text{Maltose}}{nC_{12}H_{22}O_{11}}$$

$$\underset{\text{Maltose}}{C_{12}H_{22}O_{11}} \xrightarrow{H_2O/\text{Maltase}} \underset{\text{glu cos e}}{2C_6H_{12}O_6}$$

$$\underset{\text{glu cos e}}{C_6H_{12}O_6} \xrightarrow{\text{Zymase}} \underset{\text{ethyl alcohol}}{2C_2H_5OH} + 2CO_2$$

Properties of Alcohols

(i) Boiling point of alcohols are higher than those of hydrocarbons and isomeric ethers of comparable molecular mass due to intermolecular hydrogen bonding.

(ii) **Reactivity of alcohols**

- The reactions in which O – H bond cleaves $1° > 2° > 3°$
- The reactions in which C – O bond cleaves $3° > 2° > 1°$
- **Alcohols are weak acids.**

Order of acidic strength $1° > 2° > 3°$

(iii)

$$ROH \xrightarrow{Na} RONa + H_2\uparrow \text{ (shows acidic nature of alcohol)}$$

$$ROH \xrightarrow{R'COCl/Pyridine} \underset{\text{ester}}{RCOOR'} + HCl$$

$$ROH \xrightarrow{(RCO)_2O} \underset{\text{ester}}{RCOOR}$$

$$ROH \xrightarrow{R'MgX} \underset{\text{alkane}}{R-H} + Mg(OR')X$$

$$ROH \xrightarrow[\text{conc } H_2SO_4]{R-COOH} RCOOR' + H_2O$$

- The order of ease of formation of ester Primary alcohol > Secondary alcohol > Tertiary alcohol.

(iv) **Reaction with conc H_2SO_4**

$$C_2H_5OH \xrightarrow[H_2SO_4]{383\,K} \underset{\text{ethyl sulphate}}{(C_2H_5)_2SO_4}$$

$$2C_2H_5OH \xrightarrow[H_2SO_4]{413\,K} \underset{\text{diethyl ether}}{CH_3CH_2-O-CH_2CH_3} + H_2O$$

$$C_2H_5OH \xrightarrow[H_2SO_4]{443\,K} \underset{\text{ethene}}{H_2C = CH_2} + H_2O$$

(v) **Reaction with Cu/Δ** - (Dehydrogenation)

$$1° \text{ alcohol} \xrightarrow{Cu/\Delta} \text{Aldehyde}$$

$$2° \text{ alcohol} \xrightarrow{Cu/\Delta} \text{Ketone}$$

$$3° \text{ alcohol} \xrightarrow{Cu/\Delta} \text{Alkene} + H_2O \text{ (dehydration)}$$

(vi) **Reaction with phosphorous halides:** Phosphorous halides convert alcohols to alkyl halides

$$ROH + PCl_5 \rightarrow RCl + POCl_3 + HCl$$

$$3ROH + PBr_3 \xrightarrow{P/Br_2} 3RBr + H_3PO_3$$

$$3ROH + PI_3 \xrightarrow{P/I_2} 3RI + H_3PO_3$$

(vii) **Reaction with thionyl chloride:** Alkyl chlorides are obtained

$$ROH + SOCl_2 \longrightarrow RCl + SO_2\uparrow + HCl\uparrow$$

(viii) **Reaction with halogens:** The reaction with halogens is oxidation and halogenation.

$$CH_3CH_2OH + Cl_2 \longrightarrow CH_3CHO + 2HCl \text{ (oxidation)}$$

$$CH_3CHO + 3Cl_2 \longrightarrow \underset{\text{Chloral}}{CCl_3CHO} + 3HCl \text{ (halogenation)}$$

(ix) **Reaction with (Al_2O_3) :**

$$2CH_3CH_2OH \xrightarrow[513-523K]{Al_2O_3} CH_3CH_2-O-CH_2CH_3 + H_2O$$

Distinction between 1°, 2° and 3° Alcohols

(i) **Victor – Meyer Test**

Alcohol + P/I_2 + $AgNO_2$ + HNO_2 + NaOH

- Red colouration – 1° alcohol
- Blue colouration – 2° alcohol
- No colour – 3° alcohol

(ii) **Lucas test**

- Lucas reagent is **anhydrous** $ZnCl_2$ + conc HCl

Alcohol + Lucas reagent

- cloudiness appear immediately – **3° alcohol**
- cloudiness appear after 5-10 minutes – **2° alcohol**
- No cloudiness at room temperature – **1° alcohol**

PHENOLS

- Phenols are compounds having –OH group directly attached to aromatic ring system. C_6H_5OH is simplest phenol and is called **carbolic acid**.

Methods of Preparation of Phenols

(i) **From benzene diazonium chloride**

$$C_6H_5N_2Cl + H_2O \xrightarrow{\Delta} C_6H_5OH + N_2 + HCl$$

Diazonium salt → Phenol

(ii) **From Grignard reagent**

$$C_6H_5MgBr + 1/2\ O_2\ (air) \xrightarrow{H_2O/HCl} C_6H_5OH + MgBr(OH)$$

Phenol

(iii) **From sodium benzene sulphonic acid**

$$C_6H_5SO_3Na + NaOH \xrightarrow{\text{Fuse and HCl}} C_6H_5OH + Na_2SO_3$$

sod. benzene sulphonic acid → phenol

(iv) **From Cumene**

$$C_6H_6 + CH_3-CH=CH_2 \xrightarrow{H_3PO_4}$$

Benzene, Propene

$$C_6H_5-CH(CH_3)-CH_3 \xrightarrow[(ii)\ H_2O,\ H_2O_2]{(i)\ O_2} C_6H_5OH + CH_3COCH_3$$

Cumene → Phenol + Acetone

(v) **Dow's process**

$$C_6H_5Cl + NaOH \xrightarrow[200\ atms]{573-623\ K} C_6H_5ONa$$

Chlorobenzene

$$C_6H_5ONa \xrightarrow{HCl} C_6H_5OH + NaCl$$

Sod. phenoxide → Phenol

(vi) **Raschig's process:** Developed in Germany

$$C_6H_6 + HCl + ½\ O_2 \xrightarrow{Cu/Fe} C_6H_5Cl + H_2O$$

$$C_6H_5Cl + H_2O \xrightarrow[425°C]{Steam} C_6H_5OH + HCl$$

(vii) **Oxidation of benzene:**

$$2\ C_6H_6 + O_2 \xrightarrow[315°C]{V_2O_5} 2\ C_6H_5OH + HCl$$

Properties of Phenols

(i) **Physical properties:** Phenol is a colourless crystalline solid mpt. 42°C and bpt 183°C. It is deliquescent, becomes pinkish on exposure to air due to oxidation sparingly soluble in cold water.

(ii) **Chemical properties of phenols :**

(a) Acidity of phenols

- Phenols are weak acids. They turn blue litmus red. They are weaker acids than H_2CO_3.
- Phenols are stronger acids than alcohols (due to resonance) but weaker acids than carboxylic acids.
- Electron withdrawing groups ⟶ increase acidity of phenol
- Electron donating groups ⟶ decrease acidity of phenol
- Phenols form the intermolecular hydrogen binding and and their m.pt and b.pt are much higher than hydrocarbons of the compound of the comparable molecular weights

(b) **Alkylation of phenol**

Phenol (OH) $+ CH_3I \xrightarrow{\Delta}$ Anisole (ether) (OCH_3) $+ NaI$

Phenol $\xrightarrow{C_2H_5I}$ Phenetole (ether) (OC_2H_5)

(c) **Benzoylation**

Phenol (OH) $+ C_6H_5COCl$ (Benzoyl chloride) $\xrightarrow{NaOH}$ Phenyl benzoate ($COOC_6H_5$) $+ NaCl + H_2O$

(Schotten Beumann reaction)

(d) **Distillation with Zn**

Phenol (OH) $+ Zn \xrightarrow{distil}$ Benzene $+ ZnO$

(e) **Kolbe's reaction**

Phenol (OH) $+ NaOH \longrightarrow$ Sod. phenoxide (ONa) $\xrightarrow[4-7\ atm]{+CO_2\ 400\ K}$

Sod. salicylate (OCOONa) $\xrightarrow{HCl}$ Salicylic acid (OH, COOH)

Salicylic acid (OH, COOH) $\xrightarrow[+\ conc\ H_2SO_4]{(CH_3CO)_2O}$ Aspirin (2-acetoxy benzoic acid) ($OCOCH_3$, COOH) $+ CH_3COOH$

Salicylic acid $\xrightarrow{C_6H_5OH + POCl_3}$ Phenyl salicylate (salol) (OH, $COOC_6H_5$)

Salicylic acid $\xrightarrow{CH_3OH\ (reflux)}$ Methyl salicylate (oil of winter green) (OH, $COOCH_3$)

- **Aspirin** is used as **analgesic and antipyretic.**
- **Salol** is used as **antiseptic.**
- Oil of winter green is used in perfumery, as flavouring agent in food and drinks. Also used as analgesic in rheumatism and sciatica pain.

(f) **Reimer-Tiemann reaction**

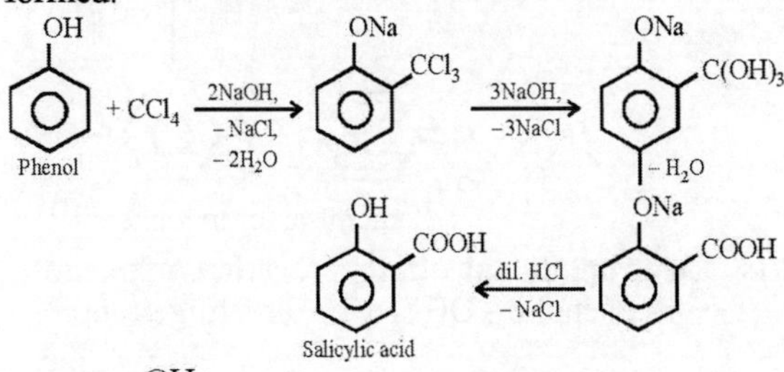

If CCl_4 is used instead of $CHCl_3$ salicylic acid is formed.

Phenol (OH) $+ CCl_4 \xrightarrow[-NaCl,\ -2H_2O]{2NaOH,}$ (ONa, CCl_3) $\xrightarrow[-3NaCl]{3NaOH,}$ (ONa, $C(OH)_3$) $\xrightarrow{-H_2O}$ (ONa, COOH) $\xrightarrow[-NaCl]{dil.\ HCl}$ Salicylic acid (OH, COOH)

(g) Phenol (OH) $\xrightarrow{O_2/CrO_3}$ O=⟨⟩=O Benzoquinone (pink colour)

Phenol $\xrightarrow[\text{Coupling reaction}]{C_6H_5N_2Cl}$ $C_6H_5-N=N-C_6H_4OH$

p – hydroxyazobenzene (Orange – red dye)

(h) **Elbs persulphate reaction :**

Phenol (OH) $\xrightarrow[\text{alkaline solution}]{K_2S_2O_8}$ Quinol (Major) (OH, OH) + Catechol (Minor) (OH, OH)

(i) **Phthalein reaction** :

Phenol (OH) + Phthalic anhydride (OH, CO–O–CO) $\xrightarrow{conc.\ H_2SO_4}$ Phenolphthalein

(j) **Libermann's nitroso reaction**

Phenol (OH) $\xrightarrow[\text{Conc } H_2SO_4]{NaNO_2}$ p-nitrosophenol (OH, NO) $\rightleftharpoons$ quinone oxime (O, NOH)

$O=C_6H_4=NOH + H-C_6H_4-OH \xrightarrow[H_2O]{H_2SO_4}$ $O=C_6H_4=N-C_6H_4-OH$

Indophenol (red)

$\downarrow H_2O \mid NaOH$

$O=C_6H_4=N-C_6H_4-ONa$

Sodium salt of Indophenol

(k) **Electrophilic substitution reactions of phenol**

- Phenolic – OH is o, p - directing group.

(1) **Bromination**

Phenol $+ Br_2 \xrightarrow{CS_2}$ p-bromo phenol (Major) + o-bromo phenol (Minor)

Phenol $\xrightarrow{+ Br_2(aq)}$ 2, 4, 6 – tri bromophenol **(white ppt)**

(2) **Sulphonation**

p-hydroxybenzenesulphonic acid (SO_3H) $\xleftarrow[373\ K]{H_2SO_4}$ Phenol $\xrightarrow[293\ K]{H_2SO_4}$ o-hydroxybenzenesulphonic acid (SO_3H)

(3) **Nitration**

p-nitrophenol (NO_2) + o-nitrophenol (NO_2) $\xleftarrow[H_2SO_4]{\text{dil. } HNO_3}$ Phenol

Phenol $\xrightarrow[H_2SO_4]{\text{conc } HNO_3}$ 2, 4, 6 – trinitrophenol **(picric acid)**

Distinction between Alcohol and Phenol

	Test	Phenol	Alcohol
(i)	Litmus test	Blue $\longrightarrow$ red	No change
(ii)	Neutral $FeCl_3$ test	Blue or violet colouration	No change
(iii)	$Br_2 + H_2O$	White ppt	No change
(iv)	$C_6H_5N_2Cl$	Orange red dye	No change
(v)	Liberman nitroso test	Red colouration changing to $(H_2SO_4 + NaOH)$	No change green and again red on dilution

GLYCOL : CH_2OH
|
CH_2OH

- **Viscous** liquid due to extensive hydrogen bonding due to presence of two – OH groups.

Preparation

$$CH_2 = CH_2 + H_2O + O \xrightarrow[\text{alkaline}]{1\%\ KMnO_4} \begin{array}{c} CH_2OH \\ | \\ CH_2OH \end{array}$$

Cis addition

$$H_2C=CH_2 \xrightarrow[\text{Peracid}]{RCOO.OH} \begin{matrix} CH_2 \\ | \\ CH_2 \end{matrix}\!\!>O \xrightarrow[H^+]{H_2O} \begin{matrix} CH_2OH \\ | \\ HO-CH_2 \end{matrix} \text{ Anti addition}$$

$$H_2C=CH_2 + HOCl \longrightarrow \underset{\text{Ethylene chlorhydrin}}{\begin{matrix} CH_2OH \\ | \\ CH_2Cl \end{matrix}} \xrightarrow{NaHCO_3} \begin{matrix} CH_2OH \\ | \\ CH_2OH \end{matrix} + NaCl + CO_2$$

Industrial method

$$\begin{matrix} CH_2 \\ \| \\ CH_2 \end{matrix} + \frac{1}{2}O_2 \xrightarrow[Ag]{\text{Catalyst}} \underset{\text{Ethylene oxide}}{\begin{matrix} CH_2 \\ | \\ CH_2 \end{matrix}\!\!>O} \xrightarrow[\text{dil. HCl}]{H_2O} \begin{matrix} CH_2OH \\ | \\ CH_2OH \end{matrix}$$

Properties

- **Physical properties :** It is colourless viscous liquid, b.p. = 197°C, m.p. = –11.5°C, miscible with water in all proportions due to intermolecular hydrogen bonding.
- **Chemical properties :**

(i) $$\underset{\text{Glycol}}{\begin{matrix} CH_2OH \\ | \\ CH_2OH \end{matrix}} + \underset{\text{phthalic acid}}{HOOC-C_6H_4-COOH} \xrightarrow{H_2SO_4} \text{Dacron (terylene)}$$

(ii) **1, 2 – splitting products of glycol by reaction (oxidation) with different compounds :**

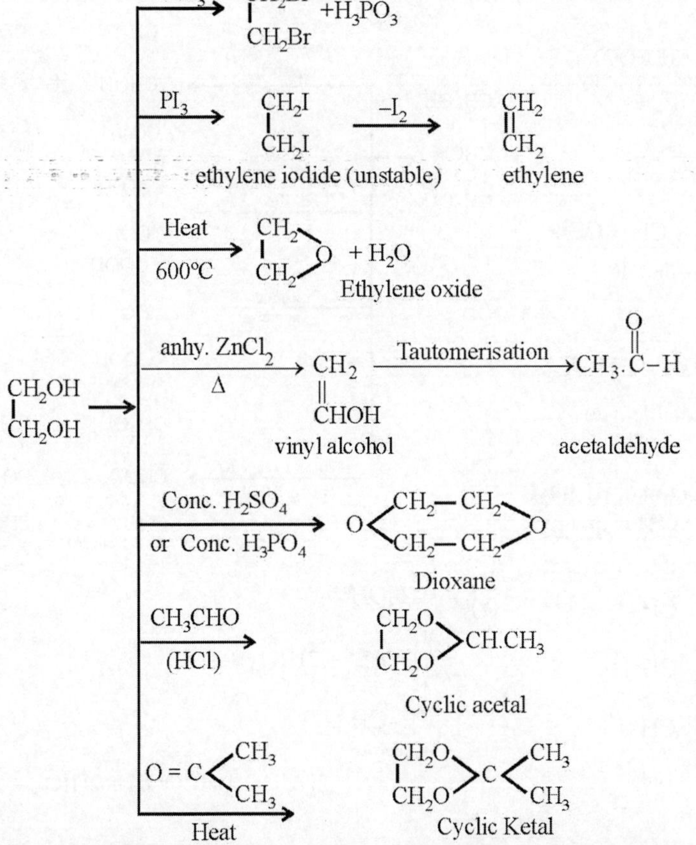

$\begin{matrix} CH_2OH \\ | \\ CH_2OH \end{matrix} \longrightarrow$

- $\xrightarrow{PBr_3} \begin{matrix} CH_2Br \\ | \\ CH_2Br \end{matrix} + H_3PO_3$
- $\xrightarrow{PI_3} \begin{matrix} CH_2I \\ | \\ CH_2I \end{matrix} \xrightarrow{-I_2} \begin{matrix} CH_2 \\ \| \\ CH_2 \end{matrix}$ — ethylene iodide (unstable); ethylene
- $\xrightarrow[600°C]{\text{Heat}} \begin{matrix} CH_2 \\ | \\ CH_2 \end{matrix}\!\!>O + H_2O$ — Ethylene oxide
- $\xrightarrow[\Delta]{\text{anhy. } ZnCl_2} \begin{matrix} CH_2 \\ \| \\ CHOH \end{matrix} \xrightarrow{\text{Tautomerisation}} CH_3.\overset{O}{\overset{\|}{C}}-H$ — vinyl alcohol; acetaldehyde
- $\xrightarrow[\text{or Conc. } H_3PO_4]{\text{Conc. } H_2SO_4} O\!<\!\begin{matrix} CH_2-CH_2 \\ CH_2-CH_2 \end{matrix}\!>\!O$ — Dioxane
- $\xrightarrow[(HCl)]{CH_3CHO} \begin{matrix} CH_2O \\ | \\ CH_2O \end{matrix}\!>CH.CH_3$ — Cyclic acetal
- $O=C\!<\!\begin{matrix} CH_3 \\ CH_3 \end{matrix} \xrightarrow{\text{Heat}} \begin{matrix} CH_2O \\ | \\ CH_2O \end{matrix}\!>C\!<\!\begin{matrix} CH_3 \\ CH_3 \end{matrix}$ — Cyclic Ketal

$$\underset{\text{glycol}}{\begin{array}{l}CH_2OH\\|\\CH_2OH\end{array}} \xrightarrow{HIO_4} \underset{\text{formaldehyde}}{2HCHO} + HIO_3$$

$$\xrightarrow{HNO_3} \underset{\text{glycollic acid}}{\begin{array}{l}COOH\\|\\CH_2OH\end{array}} + \underset{\text{oxalic acid}}{\begin{array}{l}COOH\\|\\COOH\end{array}}$$

$$\xrightarrow{KMnO_4/H_3O^+} \underset{\text{oxalic acid}}{\begin{array}{l}COOH\\|\\COOH\end{array}} + 2H_2O$$

$$\xrightarrow{(CH_3COO)_4Pb} \underset{\text{Formic acid}}{2HCHO} + \underset{\text{Acetic acid}}{2CH_3COOH} + \underset{\text{Lead acetate}}{(CH_3COO)_2Pb}$$

Glycol is used as an antifreeze for automobile radiators and as an coolant for aeroplane fuel.

GLYCEROL

$(CH_2OH-CHOH-CH_2OH)$

- Its common name is **glycerine**.

Preparation

(i) **By saponification of oils and fats :**

$$\underset{\text{oil or fat}}{\begin{array}{l}CH_2OCOR\\|\\CHOCOR\\|\\CH_2OCOR\end{array}} + 3NaOH \longrightarrow \underset{\text{glycerol}}{\begin{array}{l}CH_2OH\\|\\CHOH\\|\\CH_2OH\end{array}} + \underset{\substack{\text{sod. salt of fatty acid}\\\text{(soap)}}}{3RCOONa}$$

(ii) **From propene**

$$\underset{\text{Propene}}{CH_3-CH=CH_2} \xrightarrow[500^\circ C]{Cl_2} \underset{\text{Allyl Chloride}}{ClCH_2-CH=CH_2}$$

(iii) **By fermentation of sugars:**

$$\underset{\text{Glucose}}{C_6H_{12}O_6} \xrightarrow[Na_2SO_3]{\text{Yeast}} \underset{\text{Glycerol}}{C_3H_8O_3} + \underset{\text{acetaldehyde}}{CH_3CHO} + CO_2$$

Properties

(i) **Reaction with HI – with small amount of HI, allyl iodide** is formed but with **excess of HI isopropyl iodide** is formed.

$$\underset{\text{glycerol}}{\begin{array}{l}CH_2OH\\|\\CHOH\\|\\CH_2OH\end{array}} \xrightarrow{3HI} \begin{array}{l}CH_2I\\|\\CHI\\|\\CH_2I\end{array} \xrightarrow{-I_2} \underset{\text{allyl iodide}}{\begin{array}{l}CH_2\\||\\CH\\|\\CH_2I\end{array}}$$

$$\xrightarrow{+HI} \begin{array}{l}CH_3\\|\\CHI\\|\\CH_2I\end{array} \xrightarrow{-I_2} \underset{\text{propene}}{\begin{array}{l}CH_3\\|\\CH\\||\\CH_2\end{array}} \xrightarrow{HI} \underset{\text{Isopropyl Iodide}}{\begin{array}{l}CH_3\\|\\CHI\\|\\CH_3\end{array}}$$

(ii) **Reaction with oxalic acid**

$$\underset{\text{glycerol}}{\begin{array}{l}CH_2OH\\|\\CHOH\\|\\CH_2OH\end{array}} \xrightarrow[383\,K]{(COOH)_2} \underset{\text{formic acid}}{HCOOH} + \underset{\substack{\text{glycerol}\\\text{(It is regenerated)}}}{\begin{array}{l}CH_2OH\\|\\CHOH\\|\\CH_2OH\end{array}}$$

$$\xrightarrow[503\,K]{(COOH)_2} \underset{\text{allyl alcohol}}{CH_2=CH-CH_2OH} + 2H_2O + 2CO_2$$

(iii) **Reaction with H_2SO_4 and HNO_3**

$$\begin{array}{l}CH_2OH\\|\\CHOH\\|\\CH_2OH\end{array} \xrightarrow[+\text{ conc } HNO_3]{\text{conc } H_2SO_4} \begin{array}{l}CH_2ONO_2\\|\\CHONO_2\\|\\CH_2ONO_2\end{array}$$

Nitroglycerine
or glycerol trinitrate
or **Nobel's oil**

(iv)

$$\begin{array}{l}CH_2OH\\|\\CHOH\\|\\CH_2OH\end{array} \xrightarrow{\text{dil } HNO_3} \underset{\text{glyceric acid}}{\begin{array}{l}CH_2OH\\|\\CHOH\\|\\COOH\end{array}} + \underset{\text{tartonic acid}}{\begin{array}{l}COOH\\|\\CHOH\\|\\COOH\end{array}}$$

$$\xrightarrow{KMnO_4} \begin{array}{l}COOH\\|\\COOH\end{array} + CO_2 + H_2O$$

$$\xrightarrow{\text{Bismuth nitrate}} \underset{\text{Meso oxalic acid}}{\begin{array}{l}COOH\\|\\CO\\|\\COOH\end{array}}$$

$$\xrightarrow[\text{or } Br_2-NaOH]{FeSO_4+H_2O_2\ \text{(Fenton's reagent)}} \underset{\text{glycreralaldehyde}}{\begin{array}{l}CH_2OH\\|\\CHOH\\|\\CHO\end{array}} + \underset{\text{dihydroxy acetone}}{\begin{array}{l}CH_2OH\\|\\CO\\|\\CH_2OH\end{array}}$$

(v)

$$\underset{\text{glycerol}}{\begin{array}{l}CH_2OH\\|\\CHOH\\|\\CH_2OH\end{array}} + 2HIO_4 \longrightarrow$$

$$2HCHO + HCOOH + 2HIO_3 + H_2O$$

(vi)
$$\begin{array}{c} CH_2OH \\ | \\ CHOH \\ | \\ CH_2OH \end{array} \xrightarrow[\text{or only } ZnCl_2]{KHSO_4 \text{ or } H_2SO_4} \begin{array}{c} CH_2 \\ \| \\ CH \\ | \\ CH_2 \end{array}$$
glycerol — acrolein (Bad smelling)

The reaction is used as test for glycerol.

Uses of Glycerol

(i) Used in manufacture of dynamite, high quality soaps and cosmetics.
(ii) Sweetening agent in beverages and confectionary.
(iii) As medicines. In treatment of heat disease and asthma.

ETHERS (ROR OR ROR')

Methods of Preparation of Ethers

(i) **From alcohol**

$$\underset{\text{alcohol}}{2ROH} \xrightarrow[Al_2O_3,\ 633\ K]{\text{conc } H_2SO_4,\ 413K \text{ or}} \underset{\text{ether}}{R-O-R} + H_2O$$

(ii) $2RX + Ag_2O_{(dry)} \longrightarrow \underset{\text{ether}}{R-O-R} + 2AgX$

(iii) **Williamson's synthesis**

$$\underset{\text{alk. halide}}{RX} + \underset{\text{sod. alkoxide}}{RONa} \longrightarrow \underset{\text{simple ether}}{R-O-R} + NaX$$

$$\underset{\text{alk. halide}}{RX} + \underset{\text{sod. alkoxide}}{R'ONa} \longrightarrow \underset{\text{mixed ether}}{R-O-R'} + NaX$$

- The alkyl halide should be primary as 2° and 3° halides undergoes elimination to give alkenes.
- Mixed and simple both ethers are prepared by this method.
- To prepare **alkyl aryl ether, sodium phenoxide and alkyl halide should be used**.

No reaction takes place if aryl halide is reacted with sodium alkoxide due to lesser reactivity of aryl halides.

$$C_6H_5ONa + \underset{\text{methyl chloride}}{Cl-CH_3} \longrightarrow C_6H_5CH_3 + NaCl$$
sod. phenoxide — methyl phenyl ether

Properties of Ethers

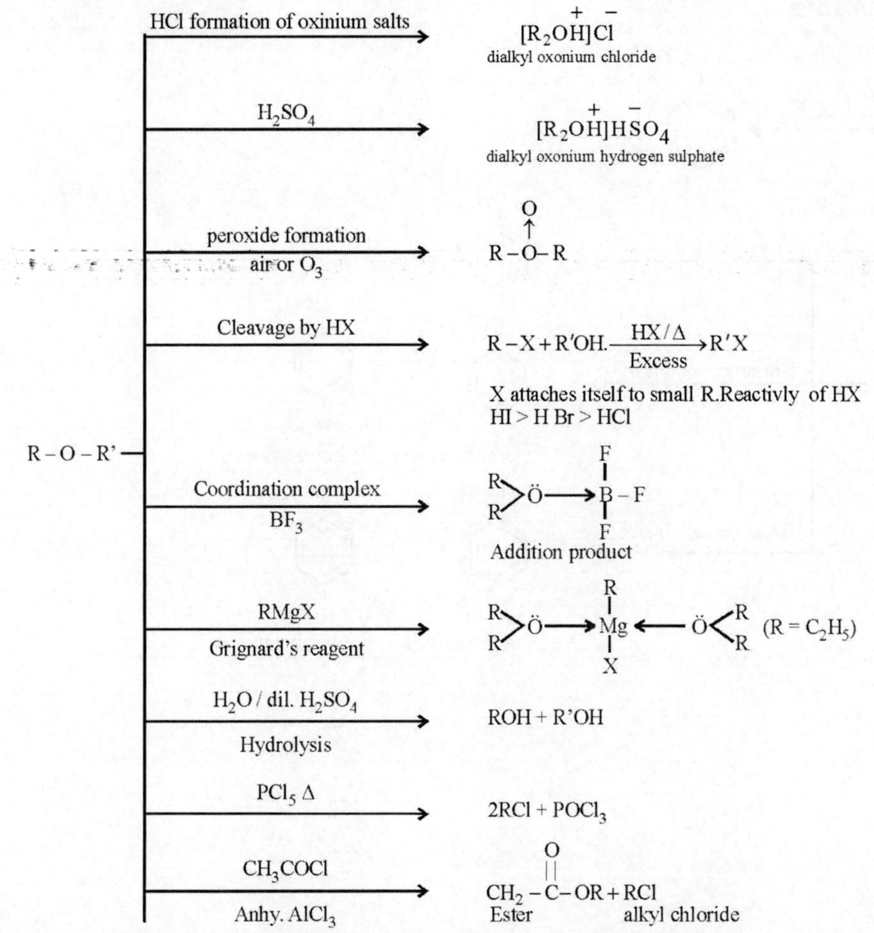

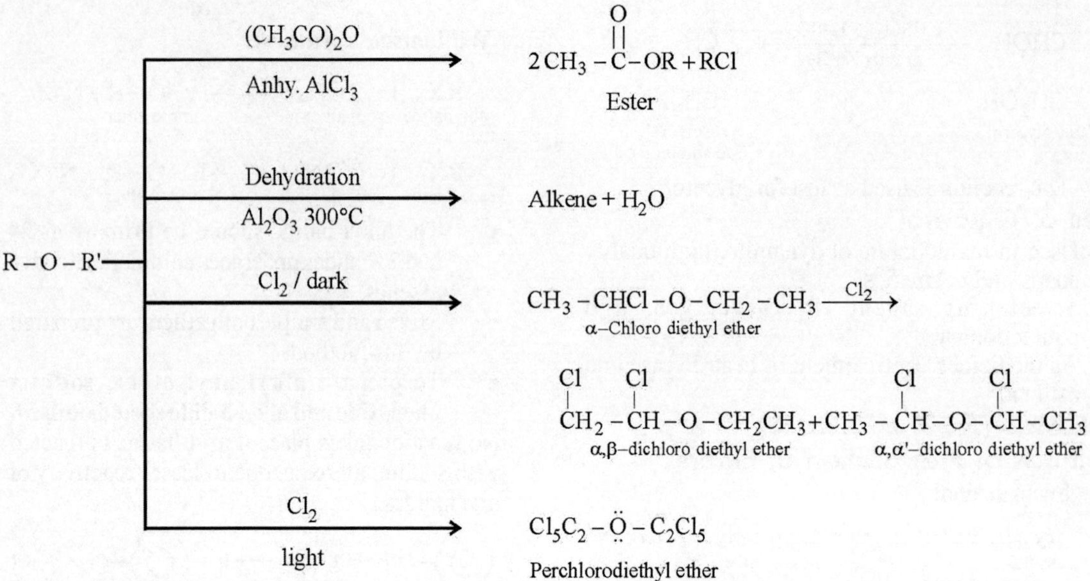

Aromatic Ether : Anisole

Preparation

$$C_6H_5OH \xrightarrow{CH_2N_2} C_6H_5OCH_3 + N_2$$

Properties

Anisole ($C_6H_5OCH_3$):

- $\xrightarrow{HI}$ $C_6H_5OH + CH_3I$
- $\xrightarrow{\text{Nitration Conc } HNO_3,\ H_2SO_4}$ o-nitro anisole + p-nitro anisole
- $\xrightarrow{\text{Bromination } Br_2/CS_2}$ o-bromo anisole + p-bromo anisole

Exercise

1. Propene, $CH_3CH = CH_2$ can be converted into 1-propanol by oxidation. Indicate which set of reagents amongst the following is ideal to effect the above conversion ?
(a) $KMnO_4$ (alkaline)
(b) Osmium tetraoxide (OsO_4/CH_2Cl_2)
(c) B_6H_6 and alk. H_2O_2
(d) O_3/Zn

2. Which of the following will not form Phenol or phenoxide :
(a) $C_6H_5N_2Cl$ (b) $C_6H_5SO_3Na$
(c) C_6H_5Cl (d) $C_6H_5CO_2H$

3. Ethanol and dimethyl ether form a pair of functional isomers. The boiling point of ethanol is higher than that of dimethyl ether, due to the presence of
(a) H-bonding in ethanol
(b) H-bonding in dimethyl ether
(c) CH_3 group in ethanol
(d) CH_3 group in dimethyl ether

4. The ionization constant of phenol is higher than that of ethanol because :
(a) phenoxide ion is bulkier than ethoxide
(b) phenoxide ion is stronger base than ethoxide
(c) phenoxide ion is stabilized through delocalization
(d) phenoxide ion is less stable than ethoxide

5. On reaction with hot conc. H_2SO_4, which one of the following compounds loses a molecule of water ?
(a) CH_3COCH_3 (b) CH_3COOH
(c) CH_3OCH_3 (d) CH_3CH_2OH

6. Lucas reagent is
(a) Conc. HCl and anhydrous $ZnCl_2$
(b) Conc. HNO_3 and hydrous $ZnCl_2$
(c) Conc. HCl and hydrous $ZnCl_2$
(d) Conc. HNO_3 and anhydrous $ZnCl_2$

7. When phenol is treated with excess bromine water, it gives:
(a) *m*-Bromophenol
(b) *o*- and *p*-Bromophenol
(c) 2, 4-Dibromophenol
(d) 2, 4, 6-Tribromophenol

8. When phenol is heated with $CHCl_3$ and alcoholic KOH when salicyladehyde is produced. This reaction is known as
(a) Rosenmund's reaction
(b) Reimer-Tiemann reaction
(c) Friedel-Crafts reaction
(d) Sommelet reaction

9. Which enzyme converts glucose and fructose both into ethanol ?
(a) Diastase (b) Invertase
(c) Zymase (d) Maltase

10. Williamson synthesis is used to prepare :
(a) Alcohols (b) Amines
(c) Aldehydes (d) Ethers

11. The major organic product in the reaction, $CH_3 — O — CH(CH_3)_2 + HI \rightarrow$ Product is
(a) $ICH_2OCH(CH_3)_2$
(b) $CH_3OC(CH_3)_2$ with I attached to the central C
(c) $CH_3I + (CH_3)_2CHOH$
(d) $CH_3OH + (CH_3)_2CHI$

12. Diethyl ether reacts, inspite of its usual inert nature, with :
(a) Dilute suphuric acid
(b) Dilute sodium hydroxide
(c) Boron trifluoride
(d) Metallic sodium

13. Benzyl alcohol is obtained from benzaldehyde by
(a) Fitting's reaction
(b) Cannizzaro's reaction
(c) Kolbe's reaction
(d) Wurtz's reaction

14. Formation of diethyl ether from ethanol is based on a :
(a) Dehydrogenation reaction
(b) Hydrogenation reaction
(c) Dehydration reaction
(d) Heterolytic fission reaction

15. In the reaction

$$\text{Phenol} \xrightarrow{\text{NaOH}} (A) \xrightarrow[140^\circ]{CO_2+HCl} (B), \text{ here B}$$

is
(a) benzaldehyde (b) chlorobenzene
(c) benzoic acid (d) salicylic acid

16. The compound which reacts fastest with Lucas reagent at room temperature is
(a) Butan-1-ol
(b) Butan-2-ol
(c) 2-Methyl propan-1-ol
(d) 2-Methylpropan-2-ol

17. Which one of the following compounds will be most readily attacked by an electrophile ?
(a) Chlorobenzene (b) Benzene
(c) Phenol (d) Toluene

18. Which one of the following on oxidation gives a ketone ?
(a) Primary alcohol
(b) Secondary alcohol
(c) Tertiary alcohol
(d) All of these

19. What is formed when a primary alcohol undergoes catalytic dehydrogenation ?
(a) Aldehyde (b) Ketone
(c) Alkene (d) Acid

20. Reaction of $CH_2 - CH_2$ (bridged by O, epoxide) with RMgX leads to formation of
(a) RCHOHR (b) $RCHOHCH_3$
(c) RCH_2CH_2OH (d) R_2CHCH_2OH (R, R > CHCH₂OH)

21. When phenol is treated with $CHCl_3$ and NaOH, the product formed is
(a) Benzaldehyde (b) Salicylaldehyde
(c) Salicylic acid (d) Benzoic acid

22. Consider the following reaction:

$$\text{Phenol} \xrightarrow{\text{Zn dust}} X \xrightarrow[\text{Anhydrous } AlCl_3]{CH_3Cl} Y \xrightarrow{\text{Alkaline } KMnO_4} Z$$

The product Z is:
(a) benzaldehyde (b) benzoic acid
(c) benzene (d) toluene

23. Among the following four compounds
(i) phenol
(ii) methylphenol
(iii) meta-nitrophenol
(iv) para-nitrophenol
the acidity order is :
(a) ii > i > iii > iv (b) iv > iii > i > ii
(c) iii > iv > i > ii (d) i > iv > iii > ii

24. Primary and secondary alcohols on action of reduced copper give
(a) Aldehydes and ketones respectively
(b) Ketones and aldehydes respectively
(c) Only aldehydes
(d) Only ketones

25. Williamson synthesis is used to prepare
(a) Acetone (b) Diethyl ether
(c) P.V.C. (d) Bakelite

26. An example of a compound with functional group –O– is :
(a) Acetic acid (b) Methyl alcohol
(c) Diethyl ether (d) Acetone

27. In the reduction

$$R-CHO + H_2 \longrightarrow RCH_2OH$$

the catalyst used is :
(a) Ni (b) Pd
(c) Pt (d) Any of these

28. When vapours of an alcohol are passed over hot reduced copper, alcohol is converted into alkene, the alcohol is
(a) Primary (b) Secondary
(c) Tertiary (d) None of above

29. Phenol can be converted into salicylic acid by :
(a) Etard reaction
(b) Kolbe reaction
(c) Reimer-Tiemann reaction
(d) Cannizzaro's reaction

30. Ethylene reacts with Baeyer's reagent to give
(a) ethane (b) ethyl alcohol
(c) ethylene glycol (d) none

31. Butane-2-ol is
(a) Primary alcohol
(b) Secondary alcohol
(c) Tertiary alcohol
(d) Aldehyde

32. Alcohols of low molecular weight are
(a) Soluble in water
(b) Soluble in water on heating
(c) Insoluble in water
(d) Insoluble in all solvents

33. When ethyl alcohol reacts with acetic acid, the products formed are
(a) Sodium ethoxide + hydrogen
(b) Ethyl acetate + water
(c) Ethyl acetate + soap
(d) Ethyl alcohol + water

34. The reaction given below is known as
$C_2H_5ONa + IC_2H_5 \longrightarrow C_2H_5OC_2H_5 + NaI$
(a) Kolbe's synthesis
(b) Wurtz' synthesis
(c) Williamson's synthesis
(d) Grignard's synthesis

35. On distilling phenol with Zn dust, one gets :
(a) Toluene
(b) Benzaldehyde + ZnO
(c) ZnO + benzene
(d) Benzoic acid

36. Which of the following compounds is oxidised to prepare methyl ethyl ketone?
(a) 2-Propanol (b) l-Butanol
(c) 2-Butanol (d) t-Butyl alcohol

37. An industrial method of preparation of methanol is :
(a) catalytic reduction of carbon monoxide in presence of $ZnO–Cr_2O_3$
(b) by reacting methane with steam at 900°C with a nickel catalyst
(c) by reducing formaldehyde with lithium aluminium hydride
(d) by reacting formaldehyde with aqueous sodium hydroxide solution

38. Diethyl ether on heating with conc. HI gives two moles of
(a) ethanol (b) iodoform
(c) ethyl iodide (d) methyl iodide

39. When phenol is reacted with $CHCl_3$ and NaOH followed by acidification, salicyladehyde is obtained. Which of the following species are involved in the above mentioned reaction as intermediate?

(a) O, H, $\overset{+}{C}Cl_2$ (b) OH, $CHCl_2$

(c) O, H, CHCl, OH (d) O^-, $CHCl_2$

40. If there be a compound of the formula $CH_3C(OH)_3$ which one of the following compounds would be obtained from it without reaction with any reagent?
(a) CH_3OH (b) C_2H_5OH
(c) CH_3COOH (d) HCHO

41. An alkene, obtained by the dehydration of an alcohol (A), on ozonolysis gives two molecules of acetaldehyde for every molecule of alkene. The alcohol (A) is
(a) $CH_3CH_2CH_2OH$
(b) CH_3CH_2OH
(c) $CH_3CH=CHCH_2OH$
(d) $CH_3CH_2CH(OH)CH_3$

42. Which statement is not correct about alcohol?
(a) Molecular weight of alcohol is higher than water.
(b) Alcohol of less no. of carbon atoms is less soluble in water than alcohol of more no. of carbon atoms
(c) Alcohol evaporates quickly.
(d) All of these

43. Most acidic is :

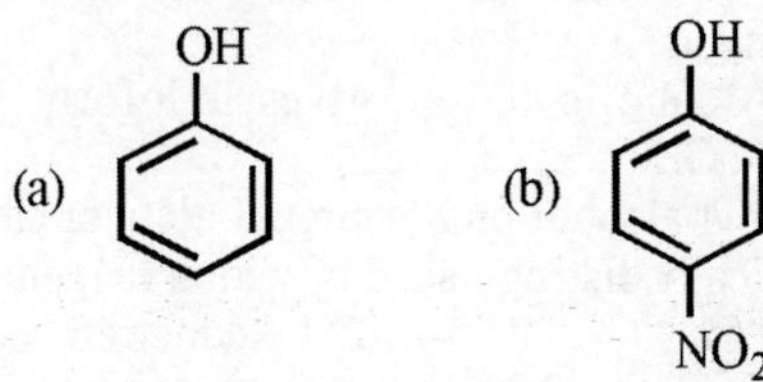

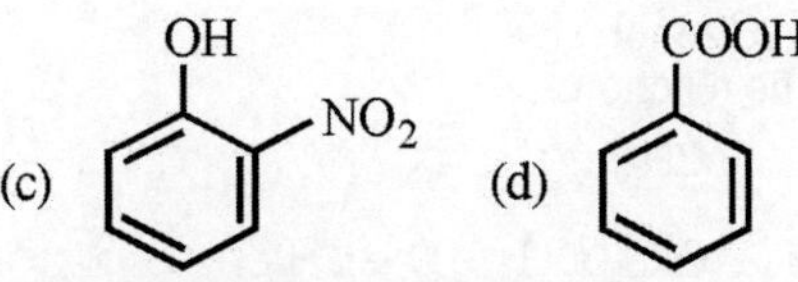

44. Which one is formed when sodium phenoxide is heated with ethyl iodide ?
(a) Phenetole
(b) Ethyl phenyl alcohol
(c) Phenol
(d) None of these

45. How many isomers of $C_5H_{11}OH$ will be primary alcohols ?
(a) 5 (b) 4
(c) 2 (d) 3

46. HBr reacts fastest with
(a) 2-Mehtylpropan-1-ol
(b) 2-Methylpropene-2-ol
(c) propan-2-ol
(d) propan-1-ol.

47. Increasing order of acid strength among p-methoxyphenol, p-methylphenol and p-nitrophenol is
(a) p-Nitrophenol, p-Methoxyphenol, p-Methylphenol
(b) p-Methylphenol, p-Methoxyphenol, p-Nitrophenol
(c) p-Nitrophenol, p-Methylphenol, p-Methoxyphenol.
(d) p-Methoxyphenol, p-Methylphenol, p-Nitrophenol

48. The stablest among the following is
(a) $CH_3CH(OH)_2$ (b) $ClCH_2CH(OH)_2$
(c) $(CH_3)_2C(OH)_2$ (d) $CCl_3CH(OH)_2$.

49. 1-Phenylethanol can be prepared by the reaction of benzaldehyde with
(a) ethyl iodide and magnesium
(b) methyl iodide and magnesium
(c) methyl bromide and aluminium bromide
(d) methyl bromide

50. Which of the following is correct ?
(a) On reduction of any aldehyde, secondary alcohol is formed
(b) Reaction of vegetable oil with H_2SO_4 gives glycerine
(c) Sucrose on reaction with NaCl gives invert sugar
(d) Alcoholic iodine gives iodoform with NaOH

51. *n*-Propyl alcohol and isopropyl alcohol can be chemically distinguished by which reagent?
(a) PCl_5 (b) Reduction
(c) Oxidation with potassium dichromate
(d) Ozonolysis

52. In the reaction:

$$CH_3-\underset{}{\overset{CH_3}{\overset{|}{C}H}}-CH_2-O-CH_2-CH_3+HI\xrightarrow{\text{Heated}}$$

Which of the following compounds will be formed?
(a) $CH_3-\underset{CH_3}{\underset{|}{C}H}-CH_3+CH_3CH_2OH$
(b) $CH_3-\underset{CH_3}{\underset{|}{C}H}-CH_2OH+CH_3CH_3$
(c) $CH_3-\overset{CH_3}{\overset{|}{C}H}-CH_2OH+CH_3-CH_2-I$
(d) $CH_3-\overset{CH_3}{\overset{|}{C}H}-CH_2-I+CH_3CH_2OH$

53. Consider the following reaction,

$$\text{ethanol}\xrightarrow{PBr_3}X\xrightarrow{\text{alc.KOH}}Y\xrightarrow[\text{(ii) }H_2O\text{, heat}]{\text{(i) }H_2SO_4\text{ room temperature}}Z;$$

the product Z is:
(a) $CH_3CH_2-O-CH_2-CH_3$
(b) $CH_3-CH_2-O-SO_3H$
(c) CH_3CH_2OH
(d) $CH_2=CH_2$

54. Which one of the following compounds has the most acidic nature?

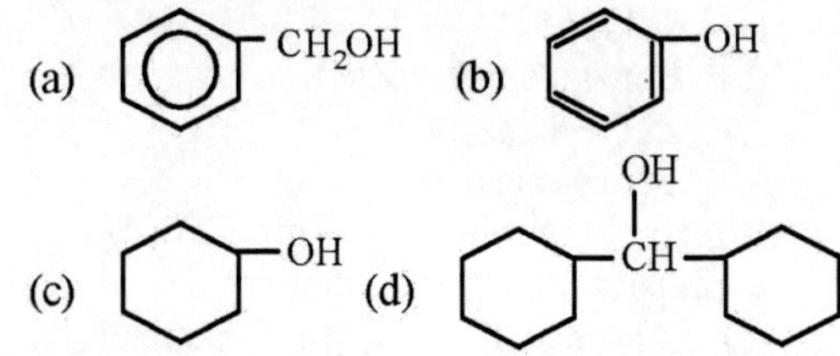

55. Ethylene glycol, on oxidation with periodic acid, gives
(a) Oxalic acid (b) Glycol
(c) Formaldehyde (d) Glycolic acid

56. Ether can be used
(a) as a general anaesthetic
(b) as a refrigerant
(c) in perfumery
(d) all of the above

57. Glycerol boils at 290°C with slight decomposition. Impure glycerine can be purified by
(a) Steam distillation
(b) Simple distillation
(c) Vacuum distillation
(d) Extraction with a solvent

58. When sodium ethoxide is heated with ethyl iodide the new compound formed is :
(a) Sodium ethoxy iodide
(b) Butyl ethyl ether
(c) Diethyl ether
(d) A mixed ether

59. Dehydration of glycerol gives
(a) Propane (b) Propene
(c) Acrolein (d) Benzene

60. Ethyl alcohol is industrially prepared from ethylene by
(a) Permanganate oxidation
(b) Catalytic reduction
(c) Absorbing in H_2SO_4 followed by hydrolysis
(d) All the three

Aldehydes, Ketones and Carboxylic acids

CONCEPTS

ALDEHYDES AND KETONES

- Aldehydes and ketones are functional isomers.

Methods of Preparation for Aldehydes and Ketones Both

(i) **Oxidation of alcohols** (oxidation by $K_2Cr_2O_7/H_2SO_4$ and $KMnO_4/KOH$)

$$1° \text{ alcohol} \longrightarrow \text{aldehyde}$$

$$2° \text{ alcohol} \longrightarrow \text{ketone}$$

(ii) **Dehydrogenation of alcohols**

$$1° \text{ alcohol} \xrightarrow{Cu/573\text{ K}} \text{aldehyde}$$

$$2° \text{ alcohol} \xrightarrow{Cu/573\text{ K}} \text{ketone}$$

(iii) **Dry distillation of calcium salt of fatty acids**

$$\underset{\text{calcium formate}}{2(HCOO)_2Ca} \xrightarrow{\text{distill}} \underset{\text{formaldehyde}}{HCHO} + CaCO_3$$

$$(RCOO)_2Ca + Ca(HCOO)_2 \xrightarrow{\Delta} \underset{\text{aldehyde}}{2RCHO} + 2CaCO_3$$

$$2(RCOO)_2Ca \xrightarrow{\Delta} \underset{\text{ketone}}{RCOR} + CaCO_3$$

(iv) **By passing vapours of fatty acids over MnO**

$$2HCOOH \xrightarrow[MnO]{573\text{ K}} \underset{\text{formaldehyde}}{HCHO} + CO_2 + H_2O$$

$$RCOOH + HCOOH \xrightarrow[MnO]{573\text{ K}} \underset{\text{aldehyde}}{RCHO} + CO_2 + H_2O$$

$$2RCOOH \xrightarrow[MnO]{573\text{ K}} \underset{\text{ketone}}{RCOR} + CO_2 + H_2O$$

(v) **By hydrolysis of gem-dihalides with aq KOH**

$$\underset{\text{1, 1-dichloroethane}}{CH_3CHCl_2} + 2KOH_{(aq)} \xrightarrow[-H_2O]{-2KCl} \underset{\text{acetaldehyde}}{CH_3CHO}$$

$$\underset{\text{2, 2-dichloroethane}}{(CH_3)_2CCl_2} + 2KOH_{(aq)} \xrightarrow[-H_2O]{-2KCl} \underset{\text{acetone}}{CH_3-\overset{O}{\overset{\|}{C}}-CH_3}$$

(vi) **Hydration of alkynes**

$$\underset{\text{alkyne}}{R-C\equiv CH} + H_2O \xrightarrow{Hg^{2+},\text{ dil }H_2SO_4} \underset{\text{ketone}}{R-\overset{O}{\overset{\|}{C}}-R}$$

(vii) **Reductive ozonolysis of alkenes**

$$\underset{\text{alkene}}{\overset{a}{\underset{b}{}}\!\!>C=C<\!\!\overset{x}{\underset{y}{}}} \xrightarrow{O_3} \underset{\text{ozonoid}}{\text{ozonoid (C–O–C ring with O–O)}} \xrightarrow[Zn]{H_2O/H^+} \underset{\text{carbonyl compounds}}{\overset{a}{\underset{b}{}}\!\!>C=O + \overset{x}{\underset{y}{}}\!\!>C=O} + ZnO$$

Methods of Preparation for Aldehydes Only

(i) $\underset{\text{acid chloride}}{R-\overset{O}{\overset{\|}{C}}-Cl} + H_2 \xrightarrow[\text{S or quinoline}]{Pd-BaSo_4} \underset{\text{aldehyde}}{R-\overset{O}{\overset{\|}{C}}-H} + HCl$

- Pd – $BaSO_4$ poisoned by S or quinoline is called Lindlar catalyst.

(ii) **Stephen's reduction**

$$\underset{\text{cyanide}}{R-C\equiv N} + 2[H] \xrightarrow[\text{ether}]{SnCl_2/HCl} \underset{\text{aldamine}}{RCH=NH} \xrightarrow{\text{Boiling } H_2O} \underset{\text{aldehyde}}{RCHO} + NH_3$$

(iii) **From Grignard reagent**

(a) $HCN \xrightarrow[\text{dry ether}]{RMgX} H-\underset{R}{\underset{|}{C}}=NMgX \xrightarrow[-MgX]{H_2O/H^+} \underset{\text{Aldehyde}}{H-\underset{R}{\underset{|}{C}}=O} + NH_3 + Mg(OH)Br$

(b) **By reaction of RMgX with ethyl formate**

$$\underset{\text{Ethyl formate}}{H-\overset{\overset{\displaystyle O}{\|}}{C}-OC_2H_5} \xrightarrow{RMgX} H-\underset{\underset{\displaystyle R}{|}}{\overset{\overset{\displaystyle OMgX}{|}}{C}}-OC_2H_5$$

$$\xrightarrow{-MgX(OC_2H_5)} \underset{\text{Aldehyde}}{H-\overset{\overset{\displaystyle O}{\|}}{C}-R}$$

This method is not good because aldehyde formed reacts further to form 2° alcohol.

(iv) **Wacker method**

$$\underset{\text{ethene}}{CH_2=CH_2} + PdCl_2 + H_2O \xrightarrow{CuCl_2} \underset{\text{acetaldehyde}}{CH_3CHO} + Pd + 2HCl$$

(v) **Etard reaction** [Oxidation of methyl benzene by chromyl chloride (CrO_2Cl_2)]

$$\underset{\text{toluene}}{C_6H_5CH_3} \xrightarrow[CCl_4]{2Cr_2O_2Cl_2} \underset{\text{Brown coloured complex}}{\left[C_6H_5CH(OCrCl_2OH)_2\right]} \xrightarrow{H_2O} \underset{\text{Benzaldehyde}}{C_6H_5CHO}$$

(vi) **Gattermann aldehyde synthesis**

$$C_6H_6 + HCN + HCl \xrightarrow{\text{anhyd } AlCl_3} C_6H_5CH{=}NH \xrightarrow{H_2O/H^+} C_6H_5CHO$$

(vii) **Gattermann Koch synthesis**

$$\underset{\text{Benzene}}{C_6H_6} + CO + HCl \xrightarrow{\text{anhyd } AlCl_3} \underset{\text{Benzaldehyde}}{C_6H_5CHO}$$

(viii) **Oxidation of methyl benzene by Cr_2O_3 and $(CH_3CO)_2O$**

$$\underset{\text{toluene}}{C_6H_5CH_3} \xrightarrow[(CH_3CO)_2O]{Cr_2O_3} \underset{\text{gem diacetate}}{C_6H_5-CH(OCOCH_3)_2} \xrightarrow{OH/H_2O} \underset{\text{Benzaldehyde}}{C_6H_5CHO} + 2CH_3COOH$$

- $(CH_3CO)_2O$ is used to trap benzaldehyde as benzylidene diacetate to avoid further oxidation to benzoic acid.

(ix) **Vilsmeyer reaction**

- Aromatic aldehydes are prepared by this method.

$$\underset{\text{Benzene}}{C_6H_6} \xrightarrow{HCOOH + 2^\circ \text{ amine} + POCl_3} \underset{\text{Benzaldehyde}}{C_6H_5CHO} + HCl$$

Methods for Preparation of Ketones Only

(i) **By Friedel Craft acylation**

$$\underset{\text{Benzene}}{C_6H_6} + \underset{\text{acid chloride}}{RCOCl} \xrightarrow{\text{Anhy. } AlCl_3} \underset{\text{Ketone}}{C_6H_5COR} + HCl$$

(ii) **From dialkyl cadmium**

$$2\underset{\text{acid chloride}}{R-\overset{\overset{\displaystyle O}{\|}}{C}-Cl} + \underset{\text{dialkyl cadmium}}{R'_2Cd} \xrightarrow{\text{dry ether}} 2\underset{\text{ketone}}{R-\overset{\overset{\displaystyle O}{\|}}{C}-R'} + CdCl_2$$

(iii) **From Grignard reagent**

$$R-C\equiv N + RMgX \xrightarrow{\text{dry ether}} R-\underset{\underset{\displaystyle R}{|}}{C}=NMgX$$

$$\xrightarrow{H^+/H_2O} R-\overset{\overset{\displaystyle O}{\|}}{C}-R + NH_3 + MgX(OH)$$

Properties of Aldehydes and Ketones

(a) Characteristic reactions – **Nucleophilic addition reactions.**

(b) **Ketones are less reactive as compared to aldehydes** due to + I effect of (two) alkyl groups and steric hinderance.

(c) **Reactions common for aldehydes and ketones**

(i)

$$>C=O \xrightarrow{HCN} \underset{\text{Cyanohydrin}}{>C(OH)(CN)}$$

$$>C=O \xrightarrow{NaHSO_3} \underset{\text{Bisulphite addition product (white ppt)}}{>C(OH)(SO_3Na)}$$

$$>C=O \xrightarrow{PCl_5/SOCl_2} \underset{\text{gem dichloride}}{>CCl_2}$$

(ii) **Addition of Grignard reagent** gives alcohols.

$$HCHO + RMgX \longrightarrow 1^\circ \text{ alcohol}$$

Any other aldehyde + RMgX ⟶ 2° alcohol

Ketone + RMgX ⟶ 3° alcohol

(iii) **Addition of alcohol**

- Aldehydes react with alcohol to form acetal

$$\underset{\text{aldehyde}}{\text{R(H)C=O}} + ROH \xrightarrow{\text{dry HCl}/H^+} \underset{\text{acetal}}{\text{R(H)C(OR)}_2}$$

- Ketones react with **dihydric alcohols only** to form cyclic ketals.

$$\underset{\text{acetone}}{(CH_3)_2C=O} + \underset{\text{glycol}}{\begin{matrix} HO-CH_2 \\ | \\ HO-CH_2 \end{matrix}} \xrightarrow[-H_2O]{\text{P-toluene sulphuric acid, Reflux}} \underset{\text{acetone ethylene glycol ketal}}{(CH_3)_2C\begin{matrix} O-CH_2 \\ | \\ O-CH_2 \end{matrix}} + H_2O$$

- Acetal and ketal are stable towards alkalies but are hydrolysed by dilute acids to give back original aldehyde and ketone.

(iv) **Addition of NH_3**

- $6HCHO + 4NH_3 \longrightarrow (CH_2)_6N_4 + 6H_2O$

  Hexamethylene tetramine (Trade name - **Urotropine**) (used as **Urinary antiseptic**)

- Any other aldehyde forms aldimine

$$\text{R(H)C=O} + NH_3 \longrightarrow \text{R(H)C(OH)(NH}_2) \xrightarrow{H_2O} \underset{\text{aldimine}}{\text{R(H)C=NH}}$$

- Ketones form complex addition product.

- $2CH_3COCH_3 + NH_3 \longrightarrow$

$$\underset{\text{diacetonamine}}{CH_3-\overset{NH_2}{\underset{CH_3}{C}}-CH_2-\overset{O}{\overset{||}{C}}-CH_3}$$

- $3C_6H_5CHO + 2NH_3 \longrightarrow \underset{\text{Hydrobenzamide}}{(C_6H_5CH=N)_2CHC_6H_5} + 3H_2O$

  ($C_6H_5CH{=}N$, $C_6H_5CH{<}{}^N_N$, $C_6H_5CH{=}N$)

(v) **Addition reaction with ammonia derivatives**

$$>C=\boxed{O + H_2}N-G \xrightarrow[-H_2O]{H^+} >C=NG$$ (Where NH_2– G is ammonia derivative)

S. No.	Ammonia derivative used Structure	Name	Product formed Structure	Name
1.	H_2N-OH	Hydroxylamine	$>C=N-OH$	Oxime
2.	H_2N-NH_2	Hydrazine	$>C=N-NH_2$	Hydrazone
3.	$H_2N-NHCONH_2$	Semicarbazide	$>C=NHCONH_2$	Semicarbazone
4.	$H_2N-NH-C_6H_5$	Phenyl hydrazine	$>C=N-NH-C_6H_5$	Phenyl hydrazone
5.	$H_2N-NH-C_6H_3(NO_2)_2$ (2,4-)	**2,4-dinitro phenyl hydrazine** (2, 4 DNP)	$>C=N-NH-C_6H_3(NO_2)_2$	2,4-dinitro phenyl hydrazone **(orange ppt)**

- All these derivatives are crystalline solids having sharp melting points and are used to identify carbonyl compounds.
- **The pH of reaction must be between 3 – 4**.

(d) **Oxidation**

(i) Aldehydes are easily oxidised to carboxylic acids

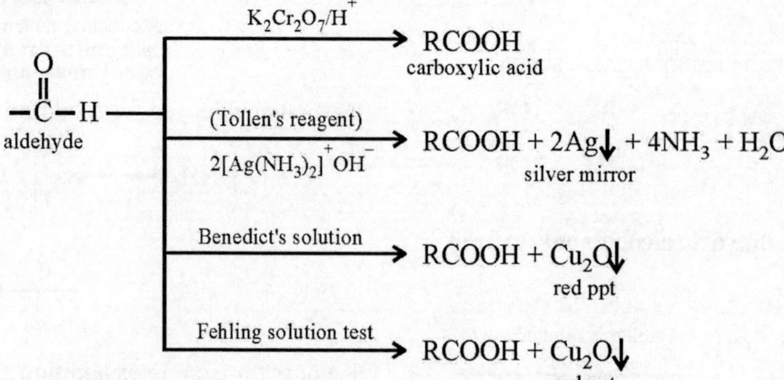

- **Fehling's solution is alkaline solution of a $CuSO_4$** containing **Rochelle salt** (sod. pot. tartarate).
- **Benendict solution is alkaline solution of copper acetate and sodium citrate**.
- Aldehydes give red ppt of Cu_2O with Fehling and Benedict solution.
- **Aromatic aldehydes** do not reduce Fehling solution and Benedict solution (∵ they are weak reagents).

(ii) **Ketones are not oxidised easily**. Under drastic conditions they are oxidised to give carboxylic acid having lesser number of carbon atoms. Strong oxidising agent (alk $KMnO_4$, conc HNO_3) are used to oxidise ketones.

Popoff's rule states that during oxidation of unsymmetrical ketones, the clevage of bonds take place in such a way that carbonyl group is retained by smaller alkyl group.

$$H_3C-\overset{O}{\overset{\|}{C}}-CH_2-CH_2-CH_3 \xrightarrow{\text{conc } HNO_3} CH_3-\overset{O}{\overset{\|}{C}}-OH + CH_3CH_2COOH$$

2-pentanone → ethanoic acid + propanoic acid

(e) **Reduction**

(i) **Reduction to alcohol**

$$>C=O \xrightarrow[\text{LiAlH}_4 \text{ or } NaBH_4]{H_2/Pt \text{ or}} >C<^{OH}_{H}$$

carbonyl comp → alcohol

(ii) **Reduction to alkane**

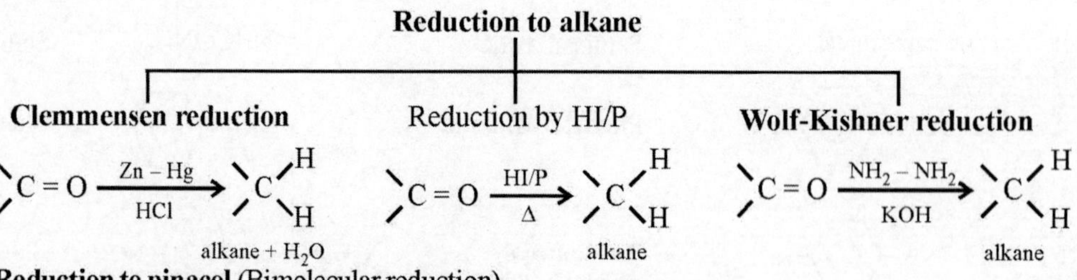

(iii) **Reduction to pinacol** (Bimolecular reduction)

- Aldehyde do not give this reaction

$$CH_3-\underset{\underset{O}{\|}}{\overset{CH_3}{\overset{|}{C}}} + \underset{\underset{O}{\|}}{\overset{CH_3}{\overset{|}{C}}}-CH_3 \xrightarrow{Mg-Hg/H_2O} CH_3-\underset{\underset{OH}{|}}{\overset{CH_3}{\overset{|}{C}}}-\underset{\underset{OH}{|}}{\overset{CH_3}{\overset{|}{C}}}-CH_3$$

acetone (two molecules) (Pinacolone) → 2, 3-dimethyl butane-2, 3-diol (Pinacol)

(iv) **Meerwein Ponndorf -Verley (MPV) reduction**

- Carbonyl group is reduced to alcohol without reducing – NO_2, $CH_2= CH_2$ or – $C \equiv C$ – groups.

$$\text{cyclohex-2-enone} + (CH_3)_2CHOH \text{ (isopropyl alcohol)} \xrightarrow[\text{Aluminium isopropoxide}]{[(CH_3)_2CHO]_3Al} \text{cyclohex-2-enol} + CH_3COCH_3$$

(f) **Aldol condensation**

(i) **Given by aldehydes and ketones which have α-H atom.**

(ii) The aldehydes and ketones having α-hydrogen atom undergo self condensation reaction in presence of aq. alkali [dil NaOH or Na_2CO_3 or Ba $(OH)_2$] to give β-hydroxy aldehyde or β-hydroxy ketone called as aldol or ketol. The aldol and ketol undergo dehydration to give α, β unsaturated aldehyde or ketone.

$$\underset{\text{acetaldehyde}}{2CH_3CHO} \xrightarrow{\text{dil NaOH}} \underset{\text{3-hydroxybutanal (aldol)}}{CH_3-CH(OH)-CH_2-CHO} \xrightarrow[-H_2O]{\Delta} \underset{\text{2-butenal (crotonaldehyde)}}{CH_3CH=CHO}$$

$$\underset{\text{acetone}}{2CH_3COCH_3} \xrightarrow{\text{dil NaOH}} \underset{\text{4-hydroxy 4-methyl 2-pentanone (diacetone alcohol)}}{CH_3-C(OH)(CH_3)-CH_2COCH_3} \xrightarrow[-H_2O]{\Delta} \underset{\text{Mesityl oxide}}{(CH_3)_2C=CHCOCH_3}$$

(iii) **Cross aldol condensation** takes place between two different aldehyde or ketone or one aldehyde and one ketone molecule.

$$\underset{\text{methanal}}{HCHO}+\underset{\text{ethanal}}{CH_3CHO} \xrightarrow{\text{dil NaOH}} \underset{\text{3-hydroxypropanal}}{H-CH(OH)-CH_2-C(=O)-H}$$

(g) **Cannizzaro reaction**

(i) Given by **aldehydes which do not have α-hydrogen atom.**

(ii) It is **disproportionation** reaction in which half of molecules are oxidised and other half are reduced.

$$\underset{\text{formaldehyde}}{2HCHO} + NaOH \longrightarrow \underset{\text{methanol}}{CH_3OH} + \underset{\text{sod. formate}}{HCOONa}$$

$$2\,\underset{\text{benzaldehyde}}{C_6H_5CHO} + NaOH \longrightarrow \underset{\text{benzyl alcohol}}{C_6H_5CH_2OH} + \underset{\text{sodium benzoate}}{C_6H_5COONa}$$

(iii) **In cross Cannizzaro reaction** two different aldehydes react. HCHO being smaller molecule is easily oxidised and other aldehyde is reduced.

$$\underset{\text{Formaldehyde}}{HCHO} + \underset{\text{benzaldehyde}}{C_6H_5CHO} \xrightarrow{\text{dil NaOH}} \underset{\text{sod. formate}}{HCOONa} + \underset{\text{benzyl alcohol}}{C_6H_5CH_2OH}$$

(h) **Polymerisation reactions**

(i) **Polymerisation of formaldehyde**

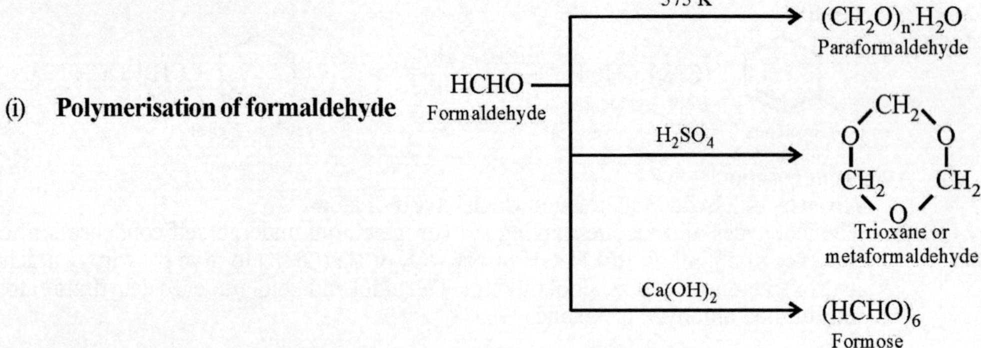

(ii) **Polymerisation of acetaldehyde**

CH_3CHO (Acetaldehyde) $\xrightarrow{\text{dil. } H_2SO_4 \text{(room temperature)}}$ Paraldehyde (cyclic trimer: six-membered ring of three O and three $CH(CH_3)$ units)

CH_3CHO (Acetaldehyde) $\xrightarrow{\text{conc. } H_2SO_4,\ 273 \text{ K}}$ Metaldehyde (eight-membered ring: $CH_3-CH-O-CH-CH_3$, each C bonded through O to $CH_3-CH-O-CH-CH_3$)

(iii) **Polymerisation of acetone**

Acetone $\xrightarrow[-H_2O,\ \Delta]{\text{HCl, Room Temp}}$ $(CH_3)_2C=CH-\overset{O}{\overset{||}{C}}-CH_3$ (Mesityl oxide)

Acetone $\xrightarrow[-2H_2O]{\text{HCl, } \Delta}$ $(CH_3)_2C=CH-\overset{O}{\overset{||}{C}}-CH=C(CH_3)_2$ (Phorone)

Acetone $\xrightarrow[-3H_2O]{\text{conc. } H_2SO_4}$ Mesitylene (1,3,5-trimethylbenzene: benzene ring with three CH_3 groups)

Electrophilic substitution reaction: It takes place at the ring in which the carbonyl group acts as a deactivating and meta-directing group

$C_6H_5-CHO \xrightarrow[273-283 \text{ K}]{HNO_3/H_2SO_4}$ m-$O_2N-C_6H_4-CHO$

(j) **Claisen-Schmidt or Claisen condensation**

Aromatic aldehyde + aliphatic aldehyde or ketone.

$\xrightarrow{\text{dil NaOH}}$ unsaturated carbonyl compound

$$\underset{\text{benzaldehyde}}{C_6H_5-CH=O}+\underset{\text{acetone}}{CH_3-\overset{O}{\overset{\|}{C}}-CH_3}$$

$$\xrightarrow{\text{dil NaOH}} \underset{\substack{\text{Benzalacetone} \\ \text{(4–Phenyl but-3-ene-2-one)}}}{C_6H_5-CH=CH-COCH_3}$$

(k) **Tischenko reaction**

Aldehyde + Aluminium ethoxide $\longrightarrow$ ester

$$\underset{\text{acetaldehyde}}{2CH_3CHO}\xrightarrow{Al(OC_2H_5)_3}\underset{\text{ethyl acetate}}{CH_3COOCH_2CH_3}$$

(l) **Benzoin condensation**

2 molecules of aromatic aldehydes $\xrightarrow{\text{alc KCN}}$ $\alpha-$hydroxyketone

$$\underset{\text{benzaldehyde}}{2\,C_6H_5CHO}\xrightarrow{\text{(alc) KCN}}\underset{\text{benzoin}}{C_6H_5-\overset{OH}{\overset{|}{CH}}-\overset{O}{\overset{\|}{C}}-C_6H_5}$$

$$\xrightarrow[\text{[O]}]{NaBrO_3} \underset{\text{Benzil}}{C_6H_5-\overset{O}{\overset{\|}{C}}-\overset{O}{\overset{\|}{C}}-C_6H_5}$$

(m) **Perkin condensation**

Aromatic aldehyde + aliphatic anhydride

$\xrightarrow[\text{(ii) Hydrolysis}]{\text{(i) salt of same acid}}$ $\alpha, \beta-$unsaturated acid

$$\underset{\text{benzaldehyde}}{C_6H_5-CHO}+\underset{\text{acetic anhydride}}{(CH_3CO)_2O}\xrightarrow[453\text{ K}]{CH_3COONa}$$

$$\underset{\text{condensation product}}{C_6H_5CH=CH-CO-O-COCH_3}$$

$$\xrightarrow{H_2O/H^+} \underset{\text{acetic acid}}{CH_3COOH}+\underset{\text{cinnamic acid}}{C_6H_5CH=CHCOOH}$$

(n) **Knoevenogel condensation**

aldehyde + active methylene compound.

$\xrightarrow[-H_2O,\,-CO_2]{\text{organic base}/\Delta}$ $\alpha, \beta-$unsaturated acid

Examples:

(i)

$$\underset{\text{benzaldehyde}}{(C_6H_5)(H)C=O}+\underset{\text{malonic acid}}{H_2C(COOH)_2}\xrightarrow[-H_2O,\,-CO_2]{\text{Pyridine, }\Delta}\underset{\text{cinnamic acid}}{C_6H_5-CH=CHCOOH}$$

(ii)

$$\underset{\text{acetaldehyde}}{(CH_3)(H)C=O}+\underset{\text{malonic acid}}{H_2C(COOH)_2}\xrightarrow[-H_2O,\,-CO_2]{\text{Pyridine, }\Delta}\underset{\text{Crotonic acid}}{CH_3-CH=CH-COOH}$$

- **Active methylene compounds** are - malonic acid, malonic ester.
- **Organic bases** are pyridine, piperidine.

(o) **Reformatsky reaction**

aldehyde or ketone + α-bromo ester

$\xrightarrow[\text{(ii) Hydrolysis}]{\text{(i) Zn-ether}}$ β-hydroxy esters

$$\underset{\text{acetaldehyde}}{CH_3CHO}+\underset{\text{ethyl bromo acetate}}{Br-CH_2COOC_2H_5}\xrightarrow[\text{(ii) Hydrolysis}]{\text{(i) Zn–ether}}$$

$$\underset{\substack{\text{ethyl }\beta\text{-hydroxybutyrate} \\ (\beta\text{-hydroxy ester})}}{CH_3(\beta)-\overset{OH}{\overset{|}{\underset{H}{\underset{|}{C}}}}-(\alpha)CH_2COOC_2H_5}$$

(p) **Benzil-Benzillic acid rearrangement**

Benzoin $\xrightarrow{\text{conc }HNO_3\text{(oxidation)}}$ Benzil $\xrightarrow[H^+/H_2O]{\text{alc KOH}}$ Benzilic acid

Example:

$$\underset{\text{Benzoin}}{C_6H_5-\overset{O}{\overset{\|}{C}}-\overset{OH}{\overset{|}{CH}}-C_6H_5}\xrightarrow[\text{oxidation}]{\text{conc }HNO_3}$$

$$\underset{\text{Benzil}}{C_6H_5-\overset{O}{\overset{\|}{C}}-\overset{O}{\overset{\|}{C}}-C_6H_5}\xrightarrow[H^+/H_2O]{\text{alc KOH}}\underset{\text{Benzilic acid}}{C_6H_5-\overset{OH}{\overset{|}{\underset{C_6H_5}{\underset{|}{C}}}}-COOH}$$

(q) **Schmidt reaction**

$$\text{Aldehyde} + \underset{\substack{\text{Hydrazoic} \\ \text{acid}}}{N_3H}\xrightarrow[330-340\text{ K}]{\text{conc }H_2SO_4}$$

cyanide + formyl derivative of 1° amine + N_2

Example:

$$RCHO + N_3H \xrightarrow[330-340K]{conc\ H_2SO_4} \underset{cyanide}{RCN} + RNHCHO + N_2$$

$$Ketone + N_3H \xrightarrow[330\text{-}340\ K]{conc\ H_2SO_4} Amide + N_2$$

Example:

$$\underset{ketone}{RCOR} + N_3H \xrightarrow[330-340\ K]{conc\ H_2SO_4} \underset{amide}{RCONHR} + N_2$$

(r) **Beckmann rearrangement**

$$Oxime \xrightarrow[\text{or Phosphoric acid (PPA)}]{conc.\ H_2SO_4\ or\ SOCl_2\ or\ PCl_5}$$

substituted amide

$$\underset{\text{phenyl acetophenone}}{(C_6H_5)(CH_3)C{=}N{-}OH} \xrightarrow[(ii)\ LiAlH_4]{(i)\ conc\ HNO_3} \underset{\text{N-Phenylacetamide or Acetanilide}}{CH_3-\overset{O}{\overset{||}{C}}-NHC_6H_5}$$

- During rearrangement trans or anti group w.r.t. OH group migrates.

(s) **Witting reaction**

$$\underset{\text{Aldehyde or ketone}}{>C=O} + \underset{\text{Phosphorus ylide}}{CH_2 = P(C_6H_5)_3} \xrightarrow[\text{or THF}]{\text{dry ether}}$$

$$\underset{\text{substituted alkene}}{>C = CH_2} + \underset{\text{Triphenyl phosphine oxide}}{(C_6H_5)_3P = O}$$

$$\underset{Benzaldehyde}{C_6H_5CH = O} + \underset{\text{methylenetriphenlyl phosphorane}}{CH_2 = P(C_6H_5)_3} \xrightarrow{\text{dry ether}}$$

$$\underset{styrene}{C_6H_5 - CH = CH_2} + \underset{\text{Triphenyl phosphine oxide}}{(C_6H_5)_3PO}$$

(t) **Pinacol-pinacolone rearrangement**

$$\text{Substituted glycol} \xrightarrow[-H_2O]{\text{Mineral acid}} \text{Ketone}$$

$$\underset{Pinacol}{CH_3-\overset{OH}{\underset{CH_3}{C}}-\overset{OH}{\underset{CH_3}{C}}-CH_3} \xrightarrow[-H_2O]{H_2SO_4}$$

$$\underset{Pinacolone}{CH_3-\overset{CH_3}{\underset{CH_3}{C}}-\overset{O}{\overset{||}{C}}-CH_3} + H_2O$$

(u) **Baeyer - Villiger reaction**

$$\text{Aliphatic ketones} \xrightarrow{peracid} \text{Ester}$$

$$\underset{ketone}{R-\overset{O}{\overset{||}{C}}-R} \xrightarrow{\text{Perbenzoic acid}} \text{Ester}$$

CARBOXYLIC ACID :

- General formula $C_nH_{2n+1}COOH$.

Methods of Preparation of Carboxylic Acids

(i) By Grignard reaction (Carbonation reaction)

$$RMgX + \underset{(or\ CO_2)}{O = C = O} \xrightarrow{\text{dry ether}}$$

$$R-\overset{O}{\overset{||}{C}}-OMgX \xrightarrow{H^+/H_2O} R-\overset{O}{\overset{||}{C}}-OH$$

Carboxylic acid formed by this method has one carbon atom more than Grignard reagent.

(ii) $$\underset{1^\circ\ alcohol}{RCH_2OH} \xrightarrow[KMnO_4\ (alk)]{K_2Cr_2O_7\ and\ H_2SO_4} \underset{acid}{RCOOH}$$

(iii) **Hydrolysis of nitriles**

$$\underset{nitrile}{R-C \equiv N} + 2H_2O \xrightarrow{H^+} R-COOH + NH_4Cl$$

(iv) **Hydrolysis of acid derivatives**

$$\underset{ester}{RCOOR'} + H_2O \xrightarrow{H^+/OH^-} \underset{\text{carboxylic acid}}{RCOOH} + \underset{alcohol}{R'OH}$$

$$\underset{\text{acid chloride}}{RCOCl} + H_2O \xrightarrow{H^+/OH^-} \underset{\text{carboxylic acid}}{RCOOH} + HCl$$

$$\underset{\text{acid amide}}{RCONH_2} + H_2O \xrightarrow{H^+/OH^-} \underset{\text{carboxylic acid}}{RCOOH} + NH_3$$

(v) **Hydrolysis of Anhydride**

$$(RCO)_2O + H_2O \xrightarrow{H^+} 2RCOOH$$

- Ease of hydrolysis

$$RCOCl > (RCO)_2O > RCOOR' > RCONH_2$$

- Alkaline hydrolysis of an ester is known as **saponification**

(vi) **Oxidation of alkyl benzenes**

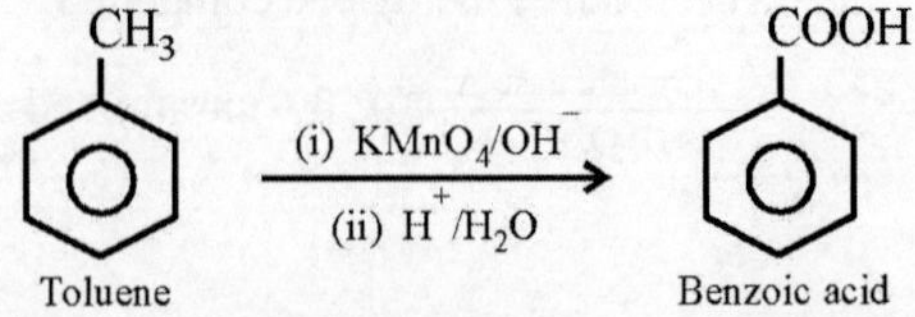

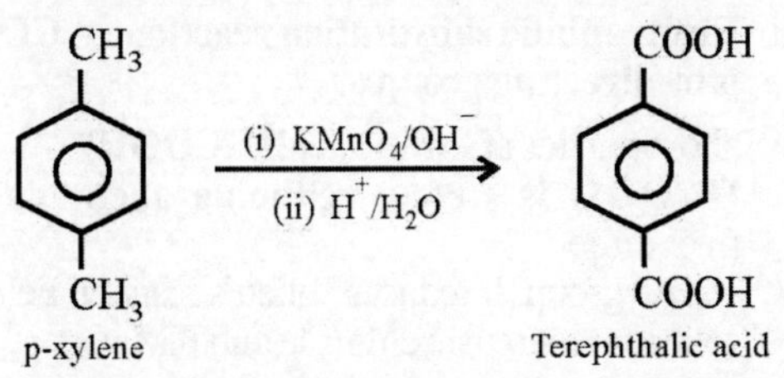

(vii) **Koch reaction :** It is acid catalysed hydrocarboxylation of olefins.

$$\underset{\text{ethene}}{CH_2 = CH_2} + CO + H_2O_{(steam)} \xrightarrow[\substack{573\,K \\ \text{under pressure}}]{H_3PO_4} \underset{\text{propionic acid}}{CH_3CH_2COOH}$$

(viii) **Hydrolysis of trihalogen derivative of alkanes**

$$R.CCl_3 + 3NaOH \longrightarrow RC(OH)_3 \xrightarrow{-H_2O} R-\overset{O}{\overset{||}{C}}-OH$$

(ix) **From Sodium alkyl and CO_2**

$$C_2H_5Na + CO_2 \longrightarrow C_2H_5COONa \xrightarrow{HCl} C_2H_5COOH + NaCl$$

(x) **From alkynes**

$$R-C \equiv C-R \xrightarrow[H_2O]{O_3/CCl_4} R-\overset{O}{\overset{||}{C}}-\overset{O}{\overset{||}{C}}-R + H_2O_2 \longrightarrow 2RCOOH$$

(xi) **By heating dicarboxylic acids eg oxalic acid or malonic acid**

$$(COOH)_2 \xrightarrow{\Delta} HCOOH + CO_2$$

(xii) **By acidic hydrolysis of Malonic or aceto acetic ester**

$$RCH(COOC_2H_5)_2 \xrightarrow{H^+} RCH(COOH)_2 \xrightarrow{\Delta} RCH_2.COOH$$

$$CH_3CO.CHR.COOC_2H_5 \xrightarrow{NaOH} CH_3COOH + RCH_2COOH + C_2H_5OH$$

Manufacture of Methanoic Acid

$$NaOH + CO \xrightarrow[\text{under 8 atm.}]{150°C} HCOONa \xrightarrow{H_2SO_4} HCOOH + NaHSO_4$$

Manufacture of Acetic Acid

(i) **By air oxidation of Acetaldehyde**

$$2CH_3CHO + O_2 \xrightarrow[50-70°C]{(CH_3COO)_2\,Mg} 2CH_3COOH$$

(ii) **By air oxidation of Butane**

$$CH_3CH_2\,CH_2.CH_3 + O_2 \xrightarrow[200°C]{(CH_3COO)_2\,Co} 2CH_3COOH$$

(iii) **Quick vinegar process**

$$2CH_3CH_2OH + O_2 \xrightarrow{\text{aceti. Bacterium}}$$

$$2CH_3CHO + 2H_2O \xrightarrow[\text{aceti. B}]{O_2} 2CH_3COOH$$

Properties of Carboxylic Acid

Physical properties :

(i) Lower acids are soluble in water due to H-bonding.

(ii) **Boiling point** in general high due to hydrogen bonding.

(iii) Exists in form of dimer due to hydrogen bonding

(iv) **Melting points** of aliphatic carboxylic acids show **alternation effect** i.e. the melting point of an acid with even number of C-atom is higher than the next lower and next higher homologue containing odd number of carbon atoms.

(v) **Acidity of carboxylic acids :**

(a) The carboxylate anion is stabilised by resonance.

$$R-C{\overset{\displaystyle O}{\underset{\displaystyle O^-}{\lessgtr}}} \longleftrightarrow R-C{\overset{\displaystyle O^-}{\underset{\displaystyle O}{\lessgtr}}}$$

The greater the stability of carboxylate anion, the stronger is the acid.

(b) Carboxylic acids are stronger acids than phenols and alcohols. This is because of resonance stabilized carboxylate ions.

(c)

Effect of substituents on acidity of aliphatic acids

Groups having + I effect	Groups having − I effect
• acidity decreases	• acidity increases.
$HCOOH > CH_3COOH > (CH_3)_2CHCOOH > (CH_3)_3COOH$	$Cl_3CCOOH > Cl_2CHCOOH > ClCH_2COOH > CH_3COOH$
• acidity decrease with increase in no. of alkyl groups.	• acidity increases with increase in no. ofhalogen atoms
	• $CH_3CH_2CHClCOOH > CH_3CHClCH_2COOH > ClCH_2CH_2COOH$ acidity decreases as distance of halogen from carboxyl group increases.

(vi) **Acidity of aromatic acids**

(a) **Aromatic acids are more acidic than aliphatic acids.**

(b) **Electron donating groups decrease acidity** and election withdrawing groups increase acidity.

(c) **Ortho effect :** Acidic strength increases by presence of any substituent at ortho position due to ortho effect, o-substituted benzoic acid are stronger acids than benzoic acid regardless of the nature of substituent.

Chemical properties :

RCOOH–

- $\xrightarrow{Na} RCOONa + 1/2\ H_2$
- $\xrightarrow{NaOH} RCOONa + H_2O$
- $\xrightarrow{NaHCO_3} RCOONa + H_2O + CO_2$ (Test for acids also)

(Salt formation)

- $\xrightarrow{PCl_5} RCOCl + POCl_3 + HCl$
- $\xrightarrow{PCl_3} RCOCl + H_3PO_3$
- $\xrightarrow{SOCl_2} RCOCl + SO_2 + HCl$

(Acid chloride)

- $\xrightarrow{NH_3} RCOO\,NH_4 \xrightarrow{\Delta} RCONH_2$
- $\xrightarrow{P_2O_5} RCN$
- $\xrightarrow{P_2O_5} (RCO)_2O + H_2O$ (Anhydride)
- $\xrightarrow{R'OH/H_2SO_4} RCOOR' + H_2O$ (Ester formation)
- $\xrightarrow{NaOH+CaO} RH + CO_2$ (Decarboxylation)
- $\xrightarrow{LiAlH_4} R.CH_2OH$ (Reduction)
- $\xrightarrow{HI/Red\,P} RCH_3$ (Reduction)
- $\xrightarrow{Cl_2/P \text{ or } Br_2/P} RCHCl.COOH$ or $RCHBrCOOH$ (HVZ reaction) α-halogenated acid (**Hell volhard zelinsky** reaction)
- $\xrightarrow{\text{Ca Salt distillation.}} RCOR$ (Ketone)
- $\xrightarrow{\text{K or Na Salt Electrolysed}} R-R$ (at anode) Kolbe's electrolytic method
- $\xrightarrow{Ag\ Salt/Br_2\ a\ Cl_2/CCl_4} RBr$ or RCl (**Hunsdiecker's** reaction)

- **Electrophilic substitution reaction : – COOH is m-directing group.**

Special properties of formic acid (HCOOH)

- **HCOOH is strong reducing agent** due to presence of –CHO group. It reduces Tollen's reagent, Fehling solution; mercuric chloride and $KMnO_4$.

$$\underset{\text{Formic acid}}{HCOOH} + \underset{\text{Tollen's reagent}}{Ag_2O} \xrightarrow{NH_4OH} CO_2 + H_2O + \underset{\text{silver mirror}}{2Ag\downarrow}$$

$$\underset{\text{Formic acid}}{HCOOH} + \underset{\text{Fehling solution}}{2CuO} \xrightarrow{NH_4OH} CO_2 + H_2O + \underset{\text{red ppt}}{Cu_2O\downarrow}$$

$$\underset{\text{Formic acid}}{HCOOH} + 2HgCl_2 \longrightarrow \underset{\text{(White ppt)}}{Hg_2Cl_2} + 2HCl + CO_2$$

Acid Chlorides (RCOCl)

Methods of preparation of acid chlorides

(i) $RCOOH + PCl_5 \longrightarrow RCOCl + POCl_3 + HCl$

$\quad \xrightarrow{+PCl_3} RCOCl + H_3PO_3$

(ii) $3RCOONa + PCl_3 \xrightarrow{distil} 3RCOCl + Na_3PO_3$

Properties of acid chloride

(i) Boiling points are lower than that of corresponding acids due to absence of hydrogen bonding.

(ii)

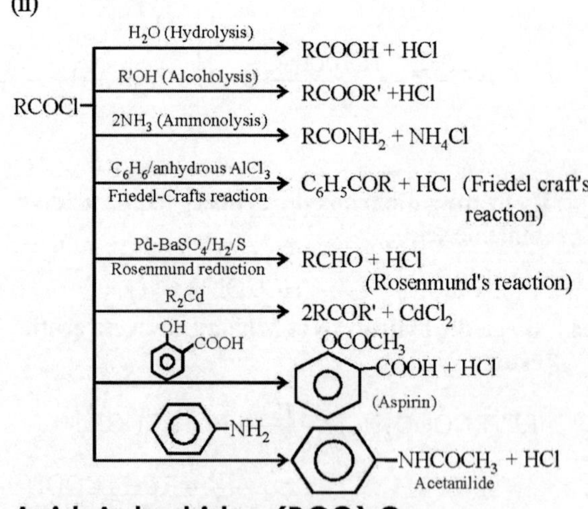

Acid Anhydrides $(RCO)_2O$

Methods of preparation of acid anhydrides

(i) $\underset{\text{Carboxylic acid}}{2RCOOH} \xrightarrow[\Delta]{P_2O_5} \underset{\text{Acid Anhydride}}{(RCO)_2O} + H_2O$

(ii) $\underset{\text{Acid chloride}}{RCOCl} + RCOOH \xrightarrow[\Delta]{P_2O_5} \underset{\text{Acid anhydride}}{(RCO)_2O} + HCl$

(iii) $RCOCl + RCOONa \xrightarrow{\Delta} (RCO)_2O + NaCl$

(iv) $\underset{\text{Ketene}}{CH_2 = C = O} + CH_3COOH \longrightarrow \underset{\text{acetic anhydride}}{(CH_3CO)_2O}$ (Industrial method)

Properties of acid anhydride

(i) The boiling points of anhydrides are higher than those of acids from which they are derived mainly due to larger size of molecule.

(ii)

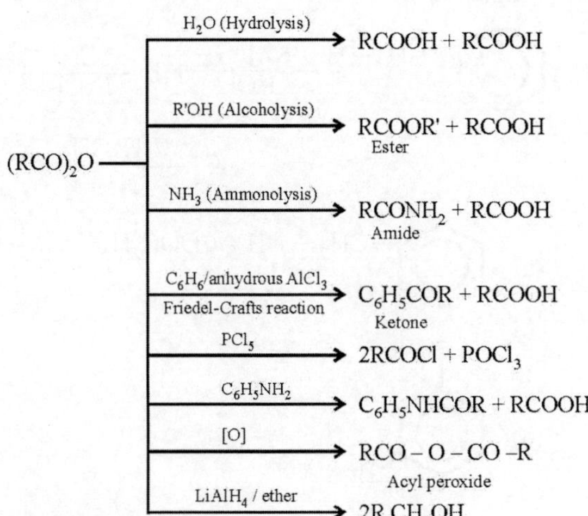

Esters (RCOOR) :

Methods of preparation of esters

(i) **Esterification**

$$RCOOH + ROH \xrightarrow{H^+} RCOOR + H_2O$$

(ii) $RCOOH + \underset{\text{diazomethane}}{CH_2N_2} \longrightarrow \underset{\text{methyl ester}}{RCOOCH_3} + N_2$

(iii) $RCOCl + R'OH \longrightarrow RCOOR' + HCl$

(iv) $R(CO)_2O + R'OH \longrightarrow RCOOR' + RCOOH$

Properties of esters

(i)

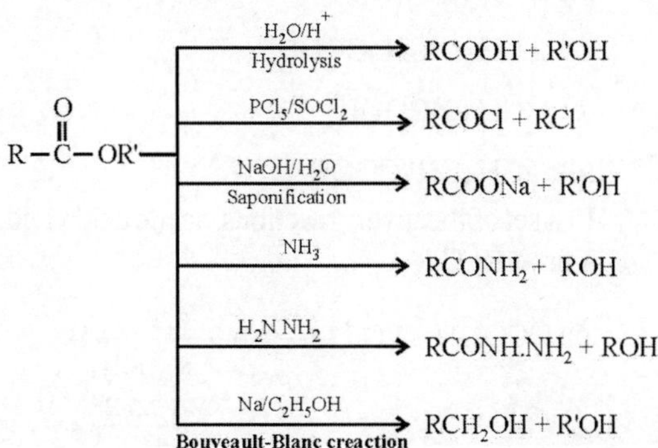

(ii) **Transesterification reaction is conversion of ester of one acid into ester of other acid.**

$$\underset{\text{ester}}{RCOOR'} + \underset{\text{alcohol}}{R''OH} \xrightarrow{\text{acid}} RCOOR'' + R'OH$$

Acid Amides ($RCONH_2$)

$$\underset{1^\circ\text{ amide}}{R-\overset{O}{\overset{\|}{C}}-NH_2} \qquad \underset{2^\circ\text{ amide}}{R-\overset{O}{\overset{\|}{C}}-NHR'} \qquad \underset{3^\circ\text{ amide}}{R-\overset{O}{\overset{\|}{C}}-NR'_2}$$

Methods of preparation of acid amides

(i) **By ammonolysis of acid derivatives**

$$RCOCl + NH_3 \longrightarrow \underset{\text{amide}}{RCONH_2} + HCl$$

$$RCOOR' + 2NH_3 \longrightarrow \underset{\text{amide}}{RCONH_2} + R'OH$$

(ii) **By thermal decomposition of ammonium salts**

$$RCOONH_4 \xrightarrow{\Delta} RCONH_2 + H_2O$$

(iii) **By partial hydrolysis of cyanides**

$$\underset{\text{cyanide}}{R-C \equiv N} + H_2O \xrightarrow[\text{or } H_2O_2-NaOH]{\text{conc HCl}} \underset{\text{amide}}{R-CONH_2}$$

Properties of acid amides

(i) They have high boiling and melting points (as compared to corresponding acids) due to strong intermolecular hydrogen bonding.

(ii) Hydrolysis

$$\underset{\text{amide}}{RCONH_2} \xrightarrow[\text{Acidic hydrolysis}]{H^+/H_2O} \underset{\text{Carboxylic acid}}{RCOOH} + NH_4^+$$

$$\underset{\text{amide}}{RCONH_2} \xrightarrow[\text{Basic hydrolysis}]{NaOH} \underset{\text{Carboxylic acid}}{RCOOH} + NH_3$$

(iii) **Amphoteric nature**

$$RCONH_2 + HgO \longrightarrow (RCONH)_2Hg + H_2O$$

$$RCONH_2 + \text{conc HCl} \longrightarrow RCONH_2.HCl \text{ (basic nature)}$$

(iv) **Dehydration**

$$\underset{\text{amide}}{RCONH_2} \xrightarrow{P_2O_5 \text{ or } SOCl_2} \underset{\text{cyanide}}{RC \equiv N} + H_2O$$

(v) **Reduction**

$$\underset{\text{amide}}{RCONH_2} \xrightarrow{LiAlH_4 / \text{dry ether}} \underset{1^\circ\text{ amine}}{RCH_2NH_2}$$

(vi) **Hoffmann Bromamide reaction**

$$\underset{\text{amide}}{RCONH_2} + Br_2 + 4KOH \longrightarrow \underset{1^\circ\text{ amine}}{RNH_2} + K_2CO_3 + 2KBr + H_2O$$

- **The 1° amine has one carbon less than parent amide.**
- The reaction is used in step down conversions.

Exercise

1. In the following reaction, product 'P' is

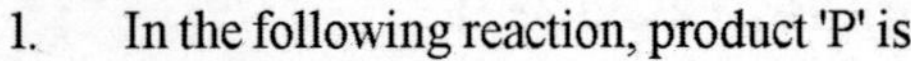

$R-\underset{\underset{O}{||}}{C}-Cl \xrightarrow[Pd-BaSO_4]{H_2} P$

(a) RCH_2OH (b) $RCOOH$
(c) $RCHO$ (d) RCH_3

2. In the Freidel Craft's acylation reaction, the effective electrophile is
(a) $RCOCl^{\oplus}$ (b) $AlCl_3$
(c) $RCOCl$ (d) $RCO^{\oplus}$

3. Acetaldehyde reacts with
(a) Electrophiles only
(b) Nucleophiles only
(c) Free radicals only
(d) Both electrophiles and nucleophiles

4. The product formed in Aldol condensation is
(a) a beta-hydroxy aldehyde or a beta-hydroxy ketone
(b) an alpha-hydroxy aldehyde or ketone
(c) an alpha, beta unsaturated ester
(d) a beta-hydroxy acid

5. When m-chlorobenzaldehyde is treated with 50% KOH solution, the product(s) obtained is (are)

(a) OH OH, –CH–CH– linking two benzene rings, each with OH (meta)

(b) COO^- (benzene ring with OH) + CH_2OH (benzene ring with OH)

(c) COO^- (benzene ring with Cl) + CH_2OH (benzene ring with Cl)

(d) OH OH, –CH–CH– linking two benzene rings, each with Cl (meta)

6. Formalin is an aqueous solution of
(a) Fluorescein (b) Formic acid
(c) Formaldehyde (d) Furfuraldehyde.

7. In a set of reactions, ethylbenzene yielded a product D.

$C_6H_5-CH_2CH_3 \xrightarrow[KOH]{KMnO_4} B \xrightarrow[FeCl_3]{Br_2} C \xrightarrow[H^+]{C_2H_5OH} D$

(a) $C_6H_5-CH_2-\underset{\underset{Br}{|}}{CH}-COOC_2H_5$

(b) Benzene ring with Br, Br and $CH_2COOC_2H_5$

(c) Benzene ring with COOH and OCH_2CH_3

(d) Benzene ring with $COOC_2H_5$ and Br

8. Among the following the strongest acid is
(a) CH_3COOH
(b) CH_2ClCH_2COOH
(c) $CH_2ClCOOH$
(d) CH_3CH_2COOH

9. In a set of the given reactions, acetic acid yielded a product C.

$CH_3COOH + PCl_5 \longrightarrow A \xrightarrow[Anh.\ AlCl_3]{C_6H_6} B \xrightarrow[Ether]{C_2H_5MgBr} C$

Product C would be -

(a) $CH_3-\underset{}{\overset{C_2H_5}{\overset{|}{C}}}(OH)C_6H_5$

(b) $CH_3CH(OH)C_2H_5$

(c) $CH_3COC_6H_5$

(d) $CH_3CH(OH)C_6H_5$

10. Propionic acid with $Br_2|P$ yields a dibromo product. Its structure would be:

(a) $H-\underset{Br}{\underset{|}{\overset{Br}{\overset{|}{C}}}}-CH_2COOH$

(b) CH_2Br-CH_2-COBr

(c) $CH_3-\underset{Br}{\underset{|}{\overset{Br}{\overset{|}{C}}}}-COOH$

(d) $CH_2Br-CHBr-COOH$

11. Benzoic acid gives benzene on being heated with X and phenol gives benzene on being heated with Y. Therefore X and Y are respectively

(a) Soda-lime and copper
(b) Zn dust and NaOH
(c) Zn dust and soda-lime
(d) Soda-lime and zinc dust

12. Among acetic acid, phenol and n-hexanol, which of the following compounds well react with $NaHCO_3$ solution to give sodium salt and carbon dioxide?

(a) Acetic acid
(b) n-Hexanol
(c) Acetic acid and phenol
(d) Phenol

13. $(CH_3)_3C—CHO$ does not undergo Aldol condensation due to

(a) three electron donating methyl groups
(b) cleavage taking place between —C— CHO bond
(c) absence of alpha hydrogen atom in the molecule
(d) bulky $(CH_3)_3$ C—group

14. Consider the following transformations :

$$CH_3COOH \xrightarrow{CaCO_3} \boxed{A} \xrightarrow{heat} \boxed{B} \xrightarrow[NaOH]{I_2} \boxed{C}$$

The molecular formula of $\boxed{C}$ is

(a) $CH_3-\underset{I}{\underset{|}{\overset{OH}{\overset{|}{C}}}}-CH_3$ (b) $ICH_2—COCH_3$

(c) CHI_3 (d) CH_3I

15. $C_6H_5MgBr \xrightarrow[(ii)\,H_3O^{\oplus}]{(i)\,CO_2} P$

In the above reaction product 'P' is

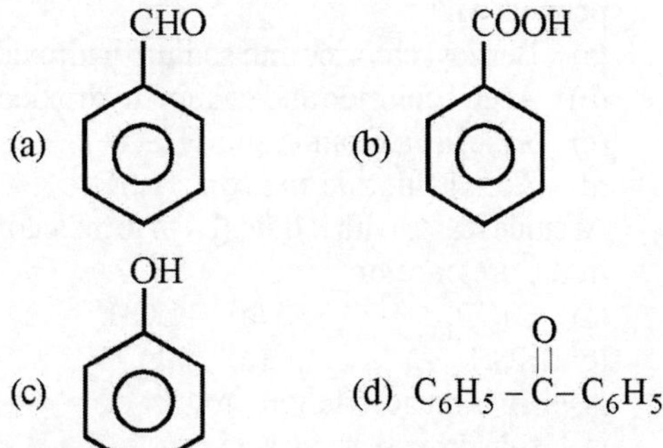

16. In a set of reactions acetic acid yielded a product D.

$$CH_3COOH \xrightarrow{SOCl_2} A \xrightarrow[Anhy.\,AlCl_3]{Benzene} B \xrightarrow{HCN} C \xrightarrow{H.OH} D$$

The structure of D would be:

(a) $C_6H_5-CH_2-\underset{OH}{\underset{|}{\overset{COOH}{\overset{|}{C}}}}-CH_3$

(b) $C_6H_5-\underset{OH}{\underset{|}{\overset{CN}{\overset{|}{C}}}}-CH_3$

(c) $C_6H_5-CH_2-\underset{CN}{\underset{|}{\overset{OH}{\overset{|}{C}}}}-CH_3$

(d) $C_6H_5-\underset{CH_3}{\underset{|}{\overset{OH}{\overset{|}{C}}}}-COOH$

17. Which one of the following on treatment with 50% aqueous sodium hydroxide yields the corresponding alcohol and acid?
(a) C_6H_5CHO
(b) $CH_3CH_2CH_2CHO$
(c) $CH_3-\overset{\overset{O}{\|}}{C}-CH_3$
(d) $C_6H_5CH_2CHO$

18. Which of the following is the correct decreasing order of acidic strength of ?
(i) Methanoic acid (ii) Ethanoic acid
(iii) Propanoic acid (iv) Butanoic acid
(a) (i) > (ii) > (iii) > (iv) (b) (ii) > (iii) > (ii) > (i)
(c) (i) > (iv) > (iii) > (ii) (d) (iv) > (i) > (iii) > (ii)

19. Schotten-Baumann reaction is a reaction of phenols with
(a) Benzoyl chloride and sodium hydroxide
(b) Acetyl chloride and sodium hydroxide
(c) Salicylic acid and conc. H_2SO_4
(d) Acetyl chloride and conc H_2SO_4

20. Acetone reacts with iodine (I_2) to form iodoform in the presence of
(a) $CaCO_3$ (b) NaOH
(c) KOH (d) $MgCO_3$

21. Aspirin is an acetylation product of
(a) o-hydroxybenzoic acid
(b) o-dihydroxybenzene
(c) m-hydroxybenzoic acid
(d) p-dihydroxybenzene

22. Iodoform test is not given by
(a) 2-Pentanone (b) Ethanol
(c) Ethanal (d) 3-Pentanone

23. Phenylmethyl ketone can be converted into ethylbezene in one step by which of the following reagents?
(a) $LiAlH_4$ (b) Zn-Hg/HCl
(c) $NaBH_4$ (d) CH_3MgI

24. Benzoic acid may be converted to ethyl benzoate by reaction with :
(a) sodium ethoxide
(b) ethyl chloride
(c) dry HCl—C_2H_5OH
(d) ethanol

25. The catalyst used in Rosenmund's reduction is
(a) $HgSO_4$ (b) $Pd/BaSO_4$
(c) anhydrous $AlCl_3$ (d) anhydrous $ZnCl_2$

26. The OH group of an alcohol or the –COOH group of a carboxylic acid can be replaced by –Cl using
(a) phosphorus pentachloride
(b) hypochlorous acid
(c) chlorine
(d) hydrochloric acid

27. Reduction of aldehydes and ketones into hydrocarbons using zinc amalgam and conc. HCl is called
(a) Cope reduction
(b) Dow reduction
(c) Wolf-Kishner reduction
(d) Clemmensen redcution

28. Which of the following reactions will not result in the formation of carbon-carbon bonds?
(a) Reimer-Tieman reaction
(b) Cannizaro reaction
(c) Wurtz reaction
(d) Friedel-Crafts acylation

29. Clemmensen reduction of a ketone is carried out in the presence of which of the following ?
(a) Glycol with KOH
(b) Zn-Hg with HCl
(c) $LiAlH_4$
(d) H_2 and Pt as catalyst

30. An organic compound 'A' on treatment with NH_3 gives 'B' which on heating gives 'C', 'C' when treated with Br_2 in the presence of KOH produces ethylamine. Compound 'A' is:
(a) CH_3COOH
(b) $CH_3CH_2CH_2COOH$
(c) $CH_3-\underset{\underset{CH_3}{|}}{CH}COOH$
(d) CH_3CH_2COOH

31. [Benzene ring with three CH_3 groups at 1,3,5 positions]

The above compound describes a condensation polymer which can be obtained in two ways : either treating 3 molecules of acetone (CH_3COCH_3) with conc. H_2SO_4 or passing propyne ($CH_3C\equiv CH$) through a red hot tube. The polymer is
(a) Phorone (b) Mesityl oxide
(c) Deacetonyl alcohol (d) Mesitylene

32. The reagent (s) which can be used to distinguish acetophenone from benzophenone is (are)
(a) 2,4-Dinitrophenylhydrazine
(b) Aqueous solution of $NaHSO_3$
(c) Benedict reagent
(d) I_2 and Na_2CO_3

33. Which of the following compound will undergo self aldol condensation in the presence of cold dilute alkali ?
(a) $CH_2 = CH - CHO$
(b) $CH \equiv C - CHO$
(c) C_6H_5CHO
(d) CH_3CH_2CHO

34. An ester is boiled with KOH. The product is cooled and acidified with concentrated HCl. A white crystalline acid separates. The ester is
(a) Methyl acetate (b) Ethyl acetate
(c) Ethyl formate (d) Ethyl benzoate

35. An ester (A) with molecular fomula, $C_9H_{10}O_2$ was treated with excess of CH_3MgBr and the complex so formed was treated with H_2SO_4 to give an olefin (B). Ozonolysis of (B) gave a ketone with molecular formula C_8H_8O which shows +ve iodoform test. The structure of (A) is
(a) $C_6H_5COOC_2H_5$
(b) $C_2H_5COOC_6H_5$
(c) $H_3COCH_2COC_6H_5$
(d) $p-H_3CO-C_6H_4-COCH_3$

36. Ketones
$[R-\underset{\underset{O}{\|}}{C}-R_1$, where $R = R_1 =$ alkyl groups$]$
can be obtained in one step by
(a) oxidation of primary alcohols
(b) hydrolysis of esters
(c) oxidation of tertiary alcohols
(d) reaction of acid halides with alcohols

37. The cyanohydrin of a compound on hydrolysis gives an optically active α-hydroxy acid. The compound is
(a) Diethyl ketone (b) Formaldehyde
(c) Acetaldehyde (d) Acetone

38. Which of the following is incorrect?
(a) $NaHSO_3$ is used in detection of carbonyl compound.
(b) $FeCl_3$ is used in detection of phenolic group.
(c) Tollen reagent is used in detection of unsaturation.
(d) Fehling solution is used in detection of glucose.

39. A and B in the following reactions are

$$R-\underset{\underset{O}{\|}}{C}-R' \xrightarrow[KCN]{HCN} A \xrightarrow{B} R-\underset{R'}{C}\begin{matrix} OH \\ CH_2NH_2 \end{matrix}$$

(a) $A = RR'C\begin{matrix} CN \\ OH \end{matrix}$, $B = LiAlH_4$
(b) $A = RR'C\begin{matrix} OH \\ COOH \end{matrix}$, $B = NH_3$
(c) $A = RR'C\begin{matrix} CN \\ OH \end{matrix}$, $B = H_3O^{\oplus}$
(d) $A = RR'CH_2CN$, $B = NaOH$

40. Which of the following represents the correct order of the acidity in the given compounds?
(a) $FCH_2COOH > CH_3COOH > BrCH_2COOH > ClCH_2COOH$
(b) $BrCH_2COOH > ClCH_2COOH > FCH_2COOH > CH_3COOH$
(c) $FCH_2COOH > ClCH_2COOH > BrCH_2COOH > CH_3COOH$
(d) $CH_3COOH > BrCH_2COOH > ClCH_2COOH > FCH_2COOH$

41. Which one of the following can be oxidised to the corresponding carbonyl compound?
(a) 2-hydroxy-propane
(b) Ortho-nitrophenol
(c) Phenol
(d) 2-methyl-2 hydroxy-propane

42. If formaldehyde and KOH are heated, then we get
(a) Methane (b) Methyl alcohol
(c) Ethyl formate (d) Acetylene

43. 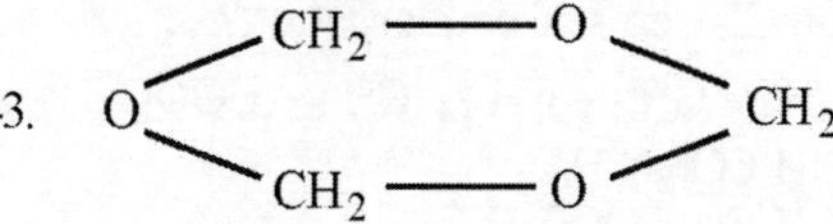

The above shown polymer is obtained when a carbonyl compound is allowed to stand. It is a white solid. The polymer is
(a) Trioxane (b) Formose
(c) Paraformaldehyde (d) Metaldehyde

44. In which of the following, the number of carbon atoms does not remain same when carboxylic acid is obtained by oxidation ?
(a) CH_3COCH_3 (b) CCl_3CH_2CHO
(c) $CH_3CH_2CH_2OH$ (d) CH_3CH_2CHO

45. Sodium formate on heating yields
(a) Oxalic acid and H_2
(b) Sodium oxalate and H_2
(c) CO_2 and NaOH
(d) Sodium oxalate

46. $(CH_3)_2C = CHCOCH_3$ can be oxidized to $(CH_3)_2C = CHCOOH$ by
(a) Chromic acid (b) NaOI
(c) Cu at 300°C (d) $KMnO_4$

47. Pinacolone is
(a) 2, 3-Dimethyl-2 3-butanediol
(b) 3, 3-Dimethyl-2 butanone
(c) I-Phenyl-2Propanone
(d) 1,1-Diphenyl-2-ethandiol

48. Benzaldehyde reacts with ethanoic KCN to give
(a) $C_6H_5CHOHCN$
(b) $C_6H_5CHOHCOC_6H_5$
(c) $C_6H_5CHOHCOOH$
(d) $C_6H_5CHOHCHOHC_6H_5$

49. Aldehydes and ketones will not form crystalline derivatives with
(a) Sodium bisulphite
(b) Phenylhydrazine
(c) Semicarbazide hydrochloride
(d) Dihydrogen sodium phosphate

50. Formic acid is obtained when
(a) Calcium acetate is heated with conc. H_2SO_4
(b) Calcium formate is heated with calcium acetate
(c) Glycerol is heated with oxalic acid at 373 K
(d) Acetaldehyde is oxidised with $K_2Cr_2O_7$ and H_2SO_4

51. The preparation of ethyl acetoacetate involves
(a) Witting reaction
(b) Cannizzaro's reaction
(c) Reformatsky reaction
(d) Claisen condensation

52. Which one of the following esters cannot undergo Claisen self-condensation?
(a) $CH_3-CH_2-CH_2-CH_2-COOC_2H_5$
(b) $C_6H_5COOC_2H_5$
(c) $C_6H_5CH_2COOC_2H_5$
(d) $C_6H_{11}CH_2COOC_2H_5$

53. Acetaldehyde reacts with semicarbazide and forms semicarbazone. Its structure is
(a) $CH_3CH=NNHCON=CHCH_3$
(b) $CH_3CH=NNHCONH_2$
(c) $CH_3CH=N-\underset{\displaystyle OH}{\underset{|}{N}}-CONH_2$
(d) $CH_3CH=N-CONHNH_2$

54. Reaction of phenylacetylene with dil. H_2SO_4 and $HgSO_4$ gives
(a) acetophenone
(b) 2-phenylethanol
(c) phenylacetaldehyde
(d) phenylacetic acid

55. Polarization of electrons in acrolein may be written as:
(a) $\overset{\delta-}{C}H_2=CH-CH=\overset{\delta+}{O}$
(b) $CH_2=\overset{\delta+}{C}H-\overset{\delta-}{C}H=O$
(c) $\overset{\delta+}{C}H_2=CH-CH=\overset{\delta-}{O}$
(d) $\overset{\delta-}{C}H_2=CH-\overset{\delta+}{C}H=O$

56. During reduction of aldehydes with hydrazine and potassium hydroxide, the first is the formation of :
(a) $R-CH=N-NH_2$
(b) $R-C\equiv N$
(c) $R-\underset{\displaystyle O}{\underset{||}{C}}-NH_2$
(d) $R-CH=NH$

57. In the reaction
$CH_3CHO+HCN \rightarrow CH_3CH(OH)CN$
$\xrightarrow{H.OH} CH_3CH(OH)COOH$
an asymmetric centre is generated. The acid obtained would be
(a) 20 % D + 80 % L-isomer
(b) D-isomer
(c) L-isomer
(d) 50% D + 50% L-isomer

58. A carbonyl compound reacts with hydrogen cyanide to form cyanohydrin which on hydrolysis forms a racemic mixture of α-hydroxy acid. The carbonyl compound is
(a) acetone (b) diethyl ketone
(c) formaldehyde (d) acetaldehyde

59. Nucleophilic addition reaction will be most favoured in
(a) $(CH_3)_2C=O$
(b) CH_3CH_2CHO
(c) CH_3CHO
(d) $CH_3-CH_2-CH_2\overset{\displaystyle O}{\overset{||}{C}}-CH_3$

60. Self condensation of two moles of ethyl acetate in presence of sodium ethoxide yields
(a) acetoacetic ester
(b) methyl acetoacetate
(c) ethyl propionate
(d) ethyl butyrate

Amines

CONCEPTS

AMINES (R – NH_2)

$R-NH_2$ — 1° amine (alkanamine)

$R-\underset{|}{\overset{R}{N}}-H$ — 2° amine (N-alkylamine)

$R-\underset{|}{\overset{R}{N}}-R$ — 3° amine (N, N-dialkylamine)

Methods of Preparation of Amines

(i) **Reduction of alkyl cyanides and isocyanides :**

$$\underset{\text{alkylcyanide}}{R-C\equiv N}\xrightarrow{LiAlH_4}\underset{1^\circ \text{ amine}}{RCH_2NH_2};$$

$$\underset{\text{isocyanide}}{R-N\equiv C}\xrightarrow{LiAlH_4}\underset{2^\circ \text{ amine}}{RNHCH_3}$$

Reduction of cyanides by Na $-C_2H_5OH$ is called **Mendius reaction or Mendius reduction.**

(ii) Reduction of Nitrocompounds

$$R-NO_2\xrightarrow{\text{Raney Ni / Pt}}R-NH_2$$

$$\underset{\text{Nitro benzene}}{C_6H_5NO_2}\xrightarrow{Sn/HCl}\underset{\text{Aniline}}{C_6H_5NH_2}$$

(iii) **Reduction of amides**

$$\underset{\text{amide}}{RCONH_2}\xrightarrow[LiAlH_4]{H_2/\text{Pt or}}\underset{1^\circ \text{ amine}}{RCH_2NH_2}$$

(iv) **Reduction of aldoximes and ketoximes**

$$\underset{\text{aldoxime}}{RCH=NOH}\xrightarrow[\text{Pt or } LiAlH_4]{\text{Na-alcohol}}\underset{1^\circ \text{ amine}}{RCH_2NH_2}+H_2O$$

$$\underset{\text{ketoxime}}{R-\overset{R}{\overset{|}{C}}=NOH}\xrightarrow[\text{Pt or } LiAlH_4]{\text{Na-alcohol}}\underset{1^\circ \text{ amine}}{R-\overset{R}{\overset{|}{C}}HNH_2}$$

(v) **Hoffmann's bromamide reaction**

$$\underset{\text{amide}}{RCONH_2}+Br_2+4KOH\longrightarrow\underset{1^\circ \text{ amine}}{RNH_2}+2KBr+K_2CO_3+2H_2O$$

- **1° amine prepared** by this method has **one carbon atom less** than parent amide.

(vi) **Hoffmann's ammonolysis**

$$\underset{\text{alkyl halide}}{RX}+2NH_3\xrightarrow{373\text{ K}}\underset{1^\circ \text{ amine}}{RNH_2}+NH_4X$$

$$\underset{1^\circ \text{ amine}}{RNH_2}\xrightarrow[373\text{ K}]{RX}\underset{2^\circ \text{ amine}}{R_2NH}\xrightarrow[373\text{ K}]{RX}\underset{3^\circ \text{ amine}}{R_3N}\xrightarrow[373\text{ K}]{RX}\underset{\text{quaternary ammonium salt}}{R_4N^+X^-}$$

- It gives mixture of 1°, 2°, 3° amines.
- When NH_3 is in excess $\longrightarrow$ 1° amine is main product.
  When RX is in excess $\longrightarrow$ Quarternary ammonium salt is main product.
- **Aryl amines are not prepared by this method.**

(vii) **Reductive amination of aldehydes and ketones**

$$\underset{\text{aldehyde}}{R-\overset{O}{\overset{||}{C}}-H}+NH_3\longrightarrow\underset{\text{aldimine}}{R-CH=NH}\xrightarrow{Ni/H_2}\underset{1^\circ \text{ amine}}{R-CH_2-NH_2}$$

$$\underset{\text{ketone}}{R-\overset{O}{\overset{||}{C}}-R}+NH_3\longrightarrow\underset{\text{ketimine}}{R-\overset{R'}{\overset{|}{C}}=NH}\xrightarrow{Ni/H_2}\underset{2^\circ \text{ amine}}{R-\overset{R'}{\overset{|}{C}}H-NH_2}$$

(viii) **Gabriel Phthalimide reaction**

$$\underset{\text{Phthalimide}}{C_6H_4(CO)_2NH}\xrightarrow[-H_2O]{\text{KOH (alc)}}\underset{\text{Pot. phthalimide}}{C_6H_4(CO)_2NK}$$

$$\xrightarrow[-KX]{RX}\underset{\text{N-alkyl phthalimide}}{C_6H_4(CO)_2N-R}\xrightarrow{H_2O/\overset{+}{H}}\underset{1^\circ \text{ amine}}{RNH_2}+\underset{\text{Phthalic acid}}{C_6H_4(COOH)_2}$$

- **Used for preparation of pure 1° aliphatic amines** but 1° **aryl amines can't be prepared** by this method because aryl halides do not undergo nucleophilic substitution reaction under mild conditions.

(ix) **From Chloramine and Grignard's reagent**

$$RMgX + ClNH_2 \longrightarrow RNH_2 + ClMgX$$

(x) **Reduction of azide**

$$\underset{}{R-X+NaN_3} \longrightarrow \underset{\text{Alkylazide}}{RN_3} \xrightarrow[H_2O]{NaBH_4} \underset{1°\text{-a min e}}{RNH_2}$$

(xi) **Ritter reaction for amines containing tertiary alkyl groups**

$$R_3COH \xrightarrow{H+} H_2O + R_3C^+ \xrightarrow{HCN}$$

$$R_3C\overset{+}{N} \equiv CH \xrightarrow{H_2O} R_3CNH.CHO$$

$$\xrightarrow{OH^-} R_3CNH_2 + HCO_2^-$$

(xii) **By Lossen rearrangement:** From Hydroxamic acid

$$RCONHOH \xrightarrow{HCl} RCONHOH_2^+ \xrightarrow{-H_2O}$$

$$R-\overset{O}{\overset{\|}{C}}-\overset{+}{\underset{..}{N}}-H \xrightarrow{-H^+} R-N=C=O$$

$$\xrightarrow{+H_2O} RNH_2 + CO_2$$

(xiii) **Schmidt reaction:**

$$\underset{\text{Acid}}{RCOOH} + \underset{\text{Hydrazoic acid}}{HN_3} \xrightarrow{H_2SO_4} RNH_2 + CO_2 + N_2$$

(xiv) **Curtius rearrangement**

$$RCOOH + SOCl_2 \longrightarrow \underset{\text{Acylchloride}}{RCOCl} \xrightarrow{NaN_3} \underset{\text{Acylazide}}{RCON_3}$$

$$\xrightarrow{-N_2} \underset{\text{Isocyanate}}{RNCO} \xrightarrow{2NaOH} RNH_2 + Na_2CO_3$$

(xv) **By Leuckart reaction:**

$$\underset{\text{Aldehyde or Ketone}}{>C=O} + \underset{\text{Amm. Formate}}{2HCOONH_4} \longrightarrow$$

$$>CHNH-\overset{O}{\overset{||}{C}}-H + 2H_2O + CO_2 + NH_3 \xrightarrow[H+]{HOH}$$

$$>\underset{1°.\text{amine}}{CH.NH_2} + H_2O + CO_2$$

$$\underset{\text{Aldehyde or Ketone}}{>C=O} + \underset{\text{Formamide}}{2HCONH_2} \longrightarrow$$

$$>CHNH-\overset{O}{\overset{||}{C}}H + CO_2 + NH_3$$

$$\xrightarrow{HOH/H+} >\underset{1°-\text{amine}}{CHNH_2} + H_2O + CO_2$$

Properties of Amines

(i) **Basic character**

- **Amines are basic** in nature due to presence of lone pair of electron on nitrogen.
- Aliphatic amines are stronger base than aromatic amines.
- Due to Resonance the lone pair of electrons is delocalised within benzene nucleus and not available for protonation and basic character is suppressed.
- Basicity of amines

  in gas state (or non aq. solution)

  $3° > 2° > 1° > NH_3$

  in aq. solution $2° > 1° > 3°$
- Basicity of amine depends on (i) **+I effect of alkyl groups** (ii) **Hydration energy** (iii) **Steric factor**.
- **Higher the value of K_b or lower the value of pK_b stronger is amine.**
- **Electron donating groups** (e.g.- CH_3, OCH_3) **- increase basicity.**

  Electron withdrawing groups (e.g.-NO_2, –CN) **- decrease basicity.**

(ii) **Reaction with aldehyde and ketones**

$$\underset{\text{Amine}}{RNH_2} + \underset{\text{Aldehyde}}{O=CHR'} \xrightarrow{acid} \underset{\text{Anil or Schiff's base}}{R-N=CHR'} + H_2O$$

$$RNH_2 + O=CR_2' \xrightarrow{acid} \underset{\text{Anil}}{R-N=CR_2'} + H_2O$$

- **Schiff's base on catalytic reduction gives 2°amine.**

(iii) **Hoffmann Mustard oil reaction**

$$R-NH_2 + S=C=S \xrightarrow{HgCl_2} R-N=C=S$$

(or $ArNH_2$) (Carbon disulphide) — Alkyl or aryl isothiocyanate (smells like mustard oil)

(iv)

$$RNH_2 + R'MgX \longrightarrow R'H + Mg(NHR)X$$

1° amine — alkane (This alkane have as many carbon atoms asin Grignard reagent)

(v)

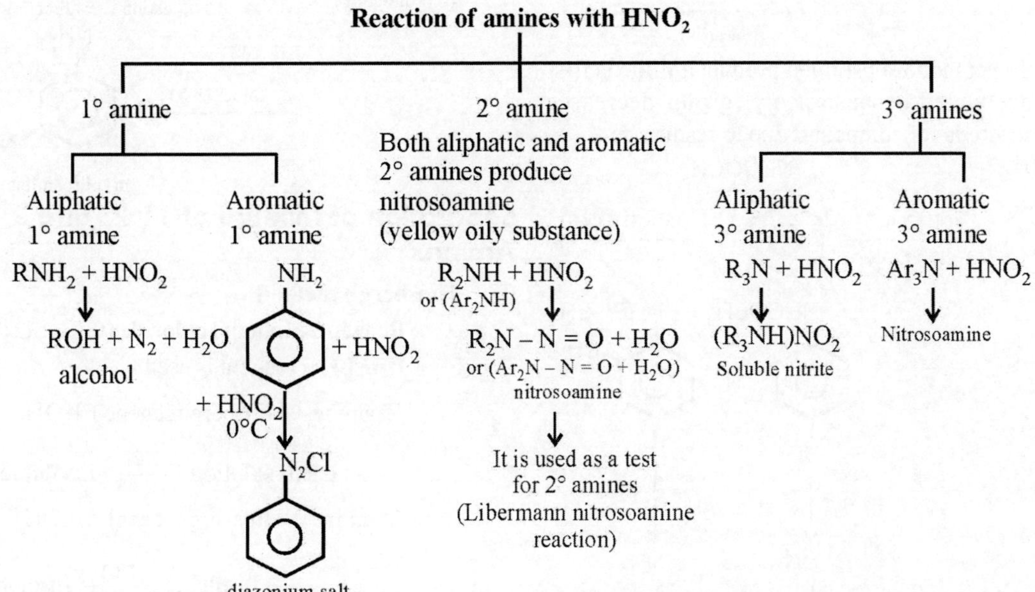

(vi) **Oxidation**

(a) **Oxidation of aromatic amines**

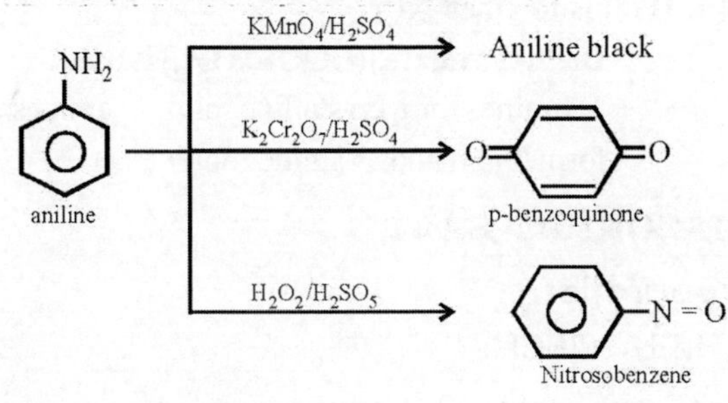

(b) **Oxidation of 1° amines**

RCH_2-NH_2 with $KMnO_4$ → $RCHO$ (aldehyde)

RCH_2-NH_2 with H_2SO_5 (caro's acid) → RCH_2NHOH (N-alkyl hydroxyl amine) $+ RCH=NOH$ (aldoxime) $+ RC(OH)=NOH$ (Hydroxamic acid)

(c) **Oxidation of 2° amines**

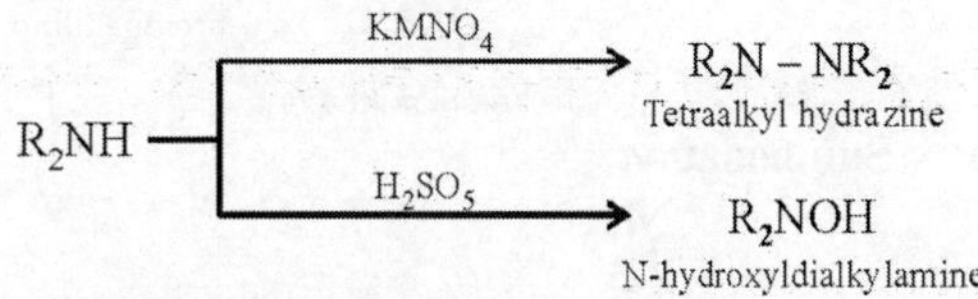

(d) **Oxidation of 3° amines**

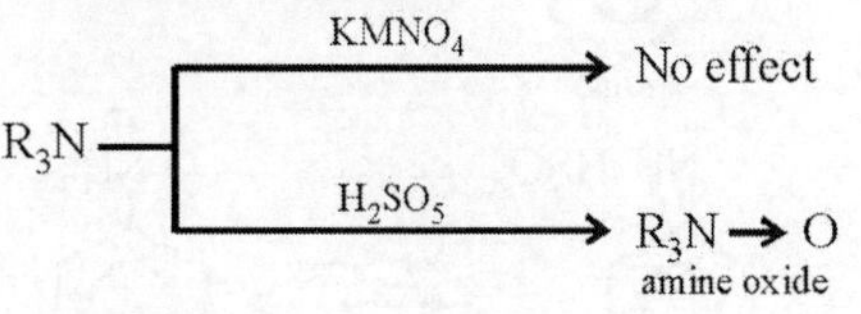

(vii)

$$R\ddot{N}H_2 \xrightarrow[\text{Benzoylation}]{C_6H_5COCl/NaOH} RNH.COC_6H_5 + NaCl + H_2O$$

(Schotten Baumann reaction)

(viii)

$$\ddot{R}NH_2 \xrightarrow{CHCl_3+KOH} R-N{\equiv}C + 3KCl + 3H_2O$$

Carbylamine reacition

(ix) **Electrophilic substitution reaction**

- $-NH_2$ group is *o*, *p* - directing group.

(a) **Halogenation**

aniline + $3Br_{2(aq)}$ → 2, 4, 6 - tribromoaniline (white ppt.) + 3HBr

To get mono substituted product aniline is first acetylated because acetyl group decreases reactivity of compound due to resonance.

Aniline $\xrightarrow[+\ CH_3COONa]{(CH_3CO)_2O}$ acetanilide $\xrightarrow{Br_2/CH_3COOH}$ p-bromoacetanilide + o-bromoacetanilide $\xrightarrow{H^+/H_2O}$ p- bromoaniline + o-bromoaniline

(b) **Sulphonation**

aniline + H_2SO_4(conc) ⟶ anilinium hydrogen sulphate ($\overset{+}{N}H_3\overset{-}{H}SO_4$) $\xrightarrow{455-473K}$ Sulphanilic acid ($\overset{\oplus}{N}H_3$, $SO_3^{(-)}$)

(Zwitter ion structure)

(c) **Nitration**

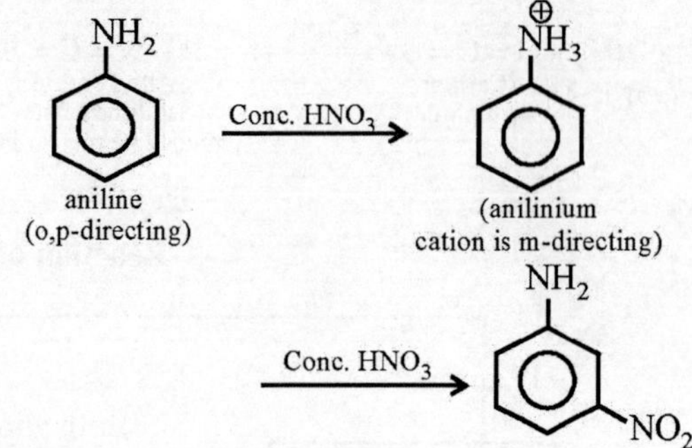

Conc. HNO_3 → m- nitroaniline

Separation of Mixture of 1°, 2° and 3° Amines

(i) **Hinsberg's method**

- **Benzenesulphonyl chloride ($C_6H_5SO_2Cl$) Hinsberg** reagent is used.

1° amine + Hinsberg reagent + KOH

$\longrightarrow$ Clear solution $\xrightarrow{HCl}$ Insoluble

2° amine + Hinsberg reagent + KOH

$\longrightarrow$ Insoluble $\xrightarrow{HCl}$ Insoluble

3° amine + Hinsberg reagent + KOH

$\xrightarrow{HCl}$ No reaction

(ii) **Hoffmann's method**

- **Diethyl oxalate [$(COOC_2H_5)_2$]** is used.
- 1° amines form crystalline solid, 2° amines form liquid and 3° amines don't react.

DIAZONIUM SALT

Preparation

$C_6H_5HN_2 + 2HCI$

$\xrightarrow{273-278k} C_6H_5N_2 + Cl- + NACI + 2H_20$

Chemical Reactions of Diazonium Salt

Replacement of - N_2Cl group is unimolecular nucleophilic in nature

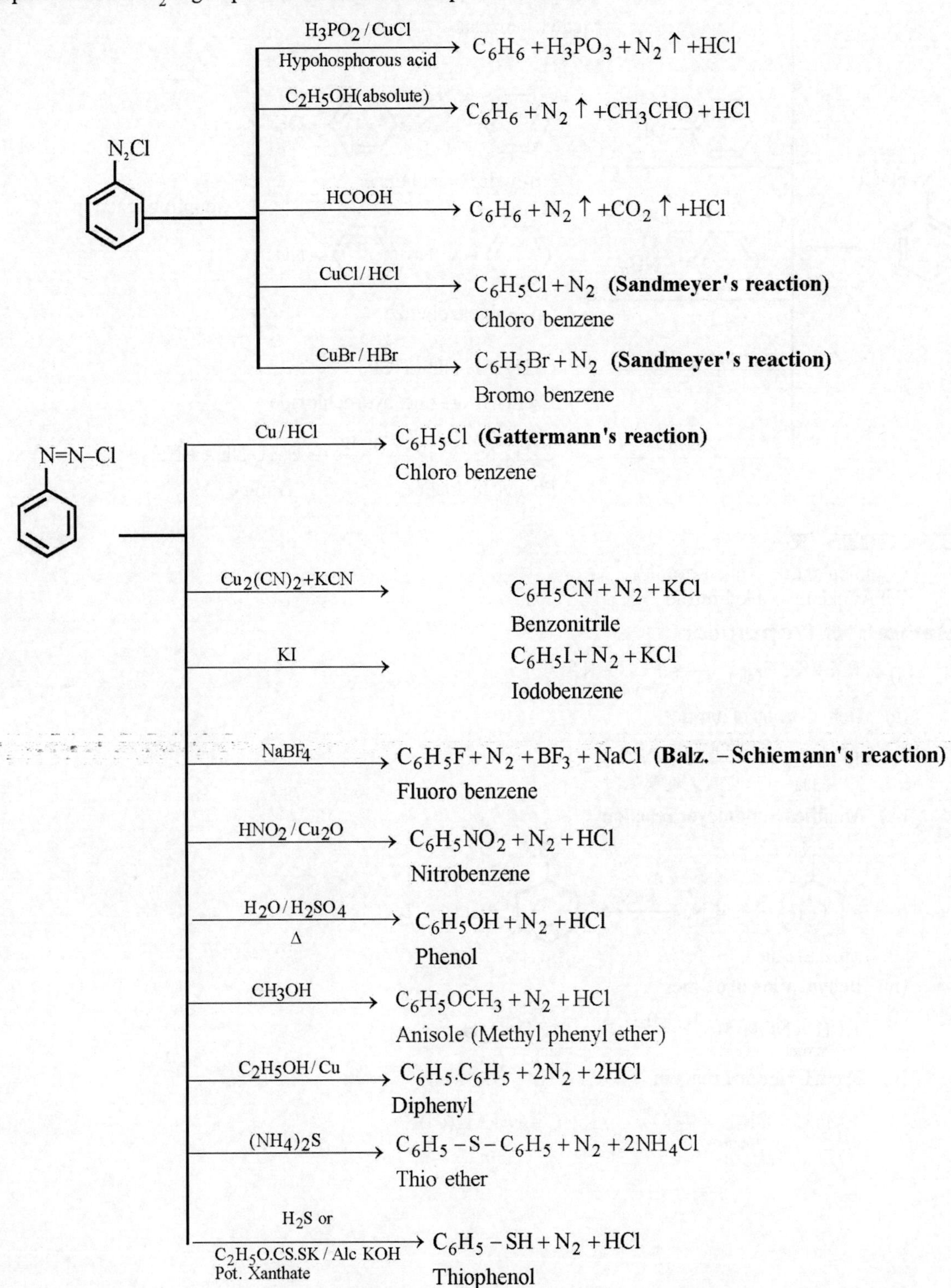

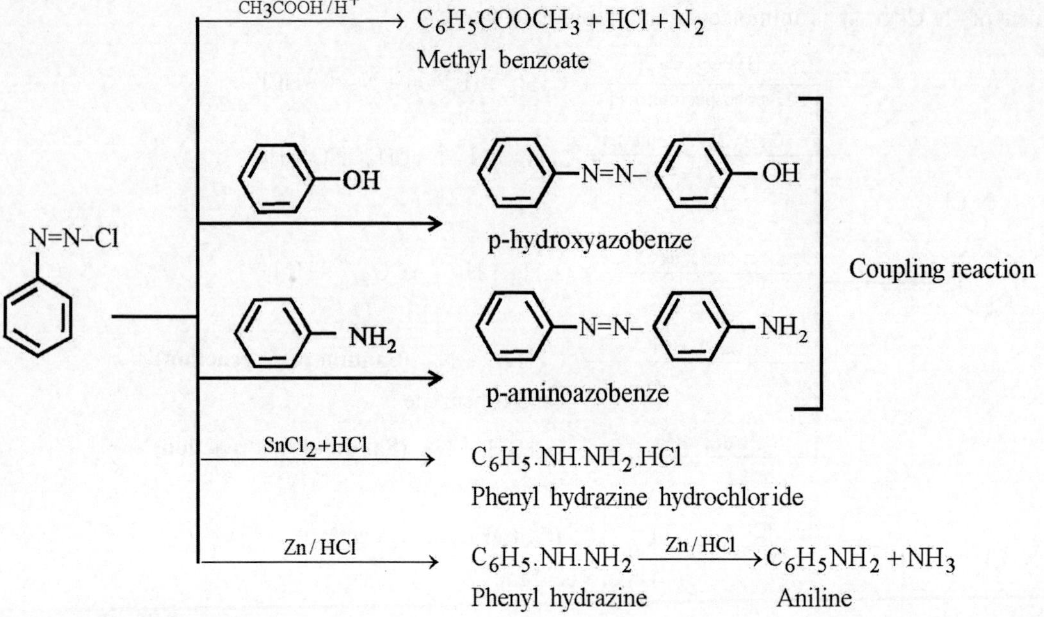

CYANIDES (R – C ≡ N)

- Common name → alkyl-cyanide
  IUPAC name → alkyl-nitrile

Methods of Preparation

(i) $\underset{}{RX + KCN_{(alc)}} \longrightarrow \underset{\text{cyanide}}{RCN} + KX$

(ii) **Dehydration of amides**

$$\underset{\text{amide}}{RCONH_2} \xrightarrow{P_2O_5} \underset{\text{cyanide}}{RCN} + H_2O$$

(iii) **Modified Sandmeyer reaction**

$$\underset{\text{Diazonium salt}}{C_6H_5N_2Cl} + CuCN \xrightarrow{KCN} \underset{\text{arylcyanide}}{C_6H_5CN} + CuCl + N_2$$

(iv) **Dehydration of oximes**

$$\underset{\text{oxime}}{RCH = NOH} \xrightarrow{(CH_3CO)_2O/\Delta} \underset{\text{cyanide}}{RCN} + H_2O$$

(v) **From Grignard reagent**

$$RMgX + \underset{\text{cyanogen chloride}}{ClCN} \xrightarrow{\text{dry ether}} R - C \equiv N + Mg(Cl)X$$

Chemical Reactions of Cynides

$$RCN \longrightarrow \begin{cases} \xrightarrow{H_2O/NaOH} RCONH_2 \xrightarrow{H_2O} RCOOH + NH_3 \\ \xrightarrow{H_2O/HCl} RCONH_2 \xrightarrow{H_2O} RCOOH + NH_3 \\ \xrightarrow[Na+C_2H_5OH]{Reduction} RCH_2NH_2 \text{ (reduction with } H_2/Pt \text{ or } LiAlH_4 \text{ is called Mandius reduction)} \\ \xrightarrow[R'MgX]{\text{with Grignard's Reagent}} R-\underset{R'}{\underset{|}{C}}=N-MgX \xrightarrow{HOH} R-\underset{R'}{\underset{|}{C}}=O+NH_3+HOMgX \text{ (Ketone)} \\ \xrightarrow[SnCl_2+HCl(gas)]{\text{Stephen's reaction}} R-CH=NH \text{ (Aldimine)} \xrightarrow{HOH} RCHO+NH_3 \text{ (Aldehyde)} \\ \xrightarrow{H_2S \text{ Addition}} RCS.NH_2 \text{ (Thioamide)} \\ \xrightarrow{NH_3 \text{ Addition}} R-C(=NH)NH_2 \text{ (Amidines)} \end{cases}$$

- **Cyanide (C N) is meta directing group.**

ISOCYANIDES (R - N$\rightleftarrows$C)

- They do not have IUPAC name. **Commonly called as carbylamines or Isonitriles.**

Methods of Preparation

(i) $$\underset{\text{alkyl halide}}{R-X} + \underset{\text{silver cyanide}}{AgCN} \longrightarrow RN\rightleftarrows C + AgX$$

(ii) Carbylamine reaction

$$\underset{\substack{1^\circ \text{ amine} \\ \text{(aliphatic or aromatic)}}}{RNH_2} + \underset{\text{chloroform}}{CHCl_3} + 3KOH \xrightarrow{\Delta} \underset{\substack{\text{isocyanide} \\ \text{(bad smelling compound)}}}{RNC} + 3KCl + 3H_2O$$

Properties

(i) **Hydrolysis**

$$\underset{\text{isocyanide}}{RNC} + 2H_2O \xrightarrow{H^+} \underset{1^\circ \text{ amine}}{RNH_2} + \underset{\text{formic acid}}{HCOOH}$$

(ii) **Reduction**

$$\underset{\text{isocyanide}}{RNC} \xrightarrow{Na-C_2H_5OH} \underset{2^\circ \text{ amine}}{R-NHCH_3}$$

(iii) They form addition products with Cl_2, S, HgO and O_3

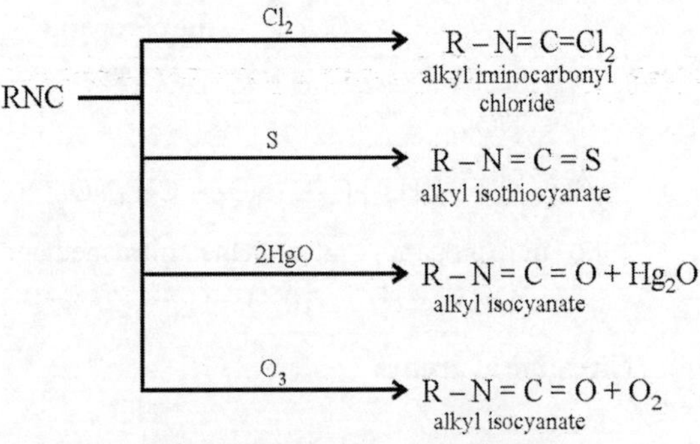

(iv) **Isocyanide is m-directing group.**

DIFFERENCE BETWEEN CYANIDES AND ISOCYANIDES

(i) **Reduction of KCN $\longrightarrow$ Cyanide**

reduction of AgCN $\longrightarrow$ isocyanide

(ii) **Alkyl cyanides have pleasant smell but isocyanides have highly unpleasant odour.**

(iii) Alkyl cyanides have lower dipole moment than isocyanides.

(iv) Boiling points of isocyanides are lower than that of cyanides.

(v) **Cyanides** $\xrightarrow{H^+/OH^-}$ $RCONH_2$ (amide)

Isocyanides $\xrightarrow{\text{acids}}$ 1° amine + HCOOH

(vi) **Cyanides** $\xrightarrow{\text{reduction}}$ 1° amine

Isocyanides $\xrightarrow{\text{reduction}}$ 2° amine

NITROALKANES AND ALKYL NITRITE

- **Nitroalkanes ($R-NO_2$) and alkyl nitrite ($R-O-N=O$) are functional isomers.**

Methods of Preparation of RNO_2 and RONO

(i) **From alkyl halide**

$$\underset{\substack{\text{Nitro alkane}\\\text{(Minor)}}}{RNO_2} + \underset{\substack{\text{Alkyl nitrite}\\\text{(Major)}}}{R-O-N=O} \xleftarrow{KNO_2} RX$$

$$\xrightarrow{2AgNO_2} \underset{\substack{\text{Nitro alkane}\\\text{(Major)}}}{RNO_2} + \underset{\substack{\text{Alkyl nitrite}\\\text{(Minor)}}}{R-O-N=O} + 2AgI$$

(ii) **Direct nitration of paraffins**

$$CH_3CH_2CH_3 \xrightarrow[200-600°C]{HNO_3} \underset{\substack{\text{1 – nitropropane}\\(25\%)}}{CH_3CH_2CH_2NO_2}$$

$$+ \underset{\substack{\text{2 – nitro propane}\\(40\%)}}{CH_3\overset{\overset{NO_2}{|}}{CH}-CH_3} + \underset{\substack{\text{nitro ethane}\\(10\%)}}{C_2H_5NO_2} + \underset{\substack{\text{nitro methane}\\(25\%)}}{CH_3NO_2}$$

(iii) **Nitration of arenes**

$$\underset{\text{benzene}}{C_6H_6} + HNO_3 \xrightarrow[333\text{ K}]{\text{conc }H_2SO_4} \underset{\text{nitroalkane}}{C_6H_5NO_2} + H_2O$$

- $HNO_3 + H_2SO_4$ is called **nitrating mixture.**
- H_2SO_4 acts on HNO_3 **to generate NO_2^+ ion which acts as electrophile.**
- Mixture of conc. nitric acid + acetic anydride gives better yield **(Pictet and Kholinsky's method)**
- $N_2O_4 + AlCl_3$ **(Schaar's Schmidt method)**

Chemical Properties of Nitro Compounds

(i) **Reaction with HNO_2**

1° nitro compound $+HNO_2 \longrightarrow$ Nitrolic acid $\xrightarrow{NaOH}$ Red colouration

2° nitro compound $+HNO_2 \longrightarrow$ Pseudonitrol $\xrightarrow{NaOH}$ Blue colouration

3° nitro compound $+HNO_2 \longrightarrow$ No reaction.

(ii) RNO_2 —

- $\xrightarrow[\text{Ni/Pt, or }LiAlH_4]{\text{Reduction Sn/HCl or}}$ $R-NH_2$ (1° amine)
- $\xrightarrow{Zn/NH_4Cl}$ $R-NHOH + H_2O$ (alkyl-hydroxylamine)
- $\xrightarrow{\text{Fe, Sn or Zn/HCl,}}$ RNH_2 (Primary amine)

$$RNO_2 \xrightarrow[\text{1° and 2° Nitro alkanes}]{\text{Tautomerisation by}}$$

$$\underset{\substack{\text{Nitro Compound}\\\text{or}\\\text{pseudo acid form}}}{R-\overset{\overset{H}{|}}{\underset{\underset{H}{|}}{C}}-N\begin{smallmatrix}\nearrow O\\ \searrow O\end{smallmatrix}} \rightleftharpoons \underset{\substack{\text{Nitrolic acid}\\\text{or}\\\text{acid form}}}{R-\underset{\underset{H}{|}}{C}=N\begin{smallmatrix}\nearrow \ddot{O}-H\\ \searrow \ddot{O}:\end{smallmatrix}}$$

$$R-\overset{\overset{H}{|}}{\underset{\underset{R}{|}}{C}}-N\begin{smallmatrix}\nearrow O\\ \searrow O\end{smallmatrix} \rightleftharpoons R-\underset{\underset{R}{|}}{C}-N\begin{smallmatrix}\nearrow O-H\\ \searrow O\end{smallmatrix}$$

(iii) **Chemical properties of Alkyl nitrites :**

R–O–N=O —

- $\xrightarrow[6H]{\text{Reduction}}$ $ROH + NH_3 + H_2O$
- $\xrightarrow[NaOH]{\text{Alk. Hydrolysis}}$ $ROH + NaNO_2$
- $\xrightarrow{\text{acid Hydrolysis}}$ $ROH + HNO_2$

(HNO_2 decomposes to HNO_3 and NO)

(iv) **Reduction of nitroarenes**

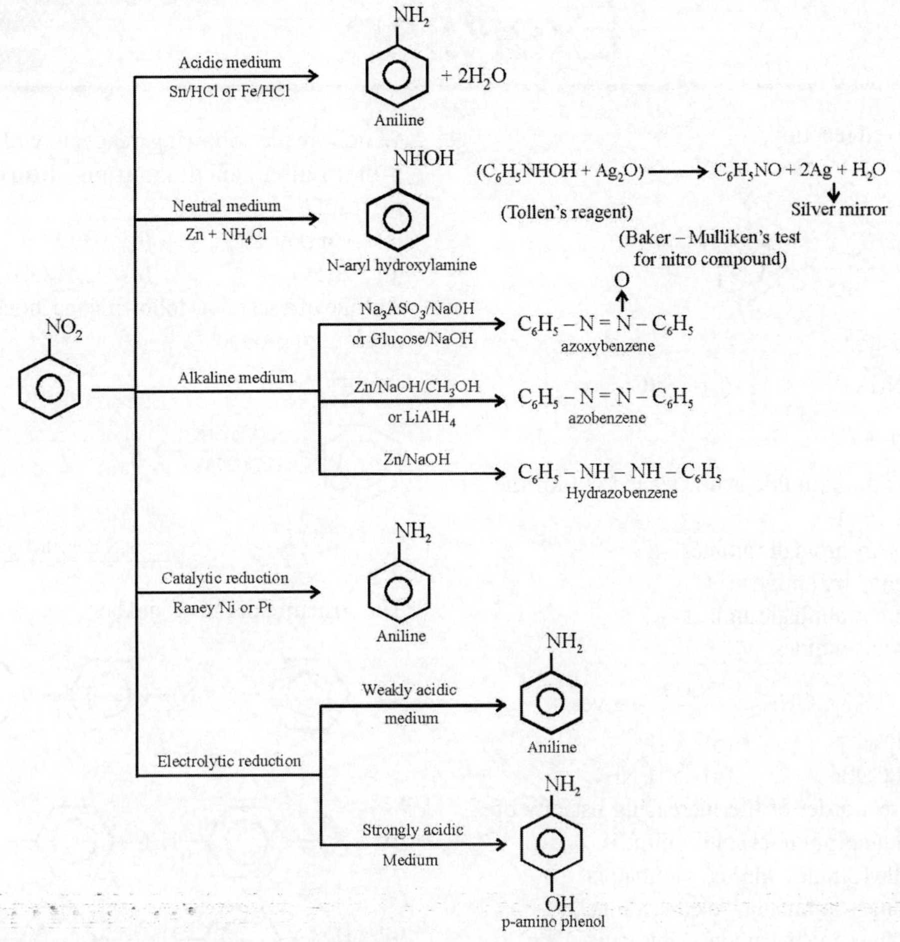

(v) $CH_3NO_2 + 3Cl_2 + 3NaOH \longrightarrow Cl_3CNO_2 + 3NaCl + 3H_2O$

Chloropicrin

(vi) **Nitro group is m-directing.**

Electrophilic Substitution Reactions

NO_2 (nitrobenzene)

Nitration, Conc HNO_3 + H_2SO_4 ⟶ m-dinitro benzene (NO_2, NO_2)

Sulphonation, Conc H_2SO_4 ⟶ m-nitro benzene sulphonic acid (NO_2, SO_3H)

Halogenation Cl_2 /Fe ⟶ **m-chloronitro benzene** (NO_2, Cl)

Exercise

1. In the given reaction,

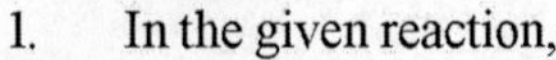

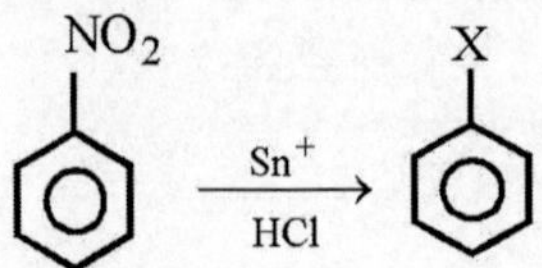

X- stands for

(a) $-NH_2$ (b) $SnCl_2$

(c) $-Cl$ (d) –

2. Gabriel's phthalimide synthesis is used for the preparation of
   (a) Primary aromatic amines
   (b) Secondary amines
   (c) Primary aliphatic amines
   (d) Tertiary amines

3. Reaction $CH_3CONH_2 \xrightarrow{NaOBr}$ gives
   (a) CH_3Br (b) CH_4
   (c) CH_3COBr (d) CH_3NH_2

4. The correct order of the increasing basicity of methyl amine, ammonia and aniline is
   (a) methyl amine < aniline < ammonia
   (b) aniline < ammonia < methyl amine
   (c) aniline < methyl amine < ammonia
   (d) ammonia < aniline < methyl amine

5. Which of the following is the most basic compound ?
   (a) Aniline (b) p-nitroaniline
   (c) Acetanilide (d) Benzylamine

6. An isocyanide is prepared by
   (a) Friedel-Crafts reaction
   (b) Perkin reaction
   (c) Carbylamine reaction
   (d) Wurtz reaction

7. The *reaction* of nitrous acid on ethyl amine gives
   (a) Ethane (b) Ethyl nitrite
   (c) Ethyl alcohol (d) Nitroethane

8. Which of the following reaction takes place when a mixture of conc. HNO_3 and H_2SO_4 reacts with benzene at 300 K ?
   (a) Sulphonation (b) Nitration
   (c) Hydrogenation (d) Dehydration

9. Which of the following reagents will convert p-methylbenzenediazonium chloride into p-cresol?
   (a) Cu powder (b) H_2O
   (c) H_3PO_2 (d) C_6H_5OH

10. Aniline in a set of the following reactions yielded a coloured product 'Y'

$C_6H_5NH_2 \xrightarrow[(273\text{-}278K)]{NaNO_2/HCl} X \xrightarrow{\text{N,N-dimethylaniline}} Y$

The structure of 'Y' would be ;

(a) $C_6H_5-N=N-C_6H_4-N(CH_3)_2$

(b) $CH_3-HN-C_6H_4-NH-C_6H_4-NH-CH_3$

(c) $H_3C-C_6H_4-N=N-C_6H_4-NH_2$

(d) $CH_3-HN-C_6H_4-N=N-C_6H_4-NH-CH_3$

11. Indicate which nitrogen compound amongst the following would undergo Hofmann reaction (i.e.., reaction with Br_2 and strong KOH) to furnish the primary amine ($R-NH_2$)
   (a) $RCONHCH_3$
   (b) $RCOONH_4$
   (c) $RCONH_2$
   (d) $R-CO-NHOH$.

12. Aniline is an activated system for electrophilic substitution. The compound formed on heating aniline with acetic anhydride is

(a) NH_2, $COCH_3$

(b) NH_2, $COCH_3$

(c) NH_2, $COCH_3$

(d) $NHCOCH_3$

13. Which of the following is more basic than aniline?
(a) Triphenylamine (b) p-Nitroaniline
(c) Benzylamine (d) Diphenylamine

14. In test for primary amines, the amine is treated with $CHCl_3$ and KOH and a bad smelling compound is formed. If the primary amine used is ethylamine, identify the bad smelling compound formed?
(a) CH_3CN (b) CH_3CNO
(c) CH_3CH_2NC (d) CH_3NCO

15. The correct order of increasing basic nature for the bases NH_3, CH_3NH_2 and $(CH_3)_2NH$ is
(a) $(CH_3)_2NH < NH_3 < CH_3NH_2$
(b) $NH_3 < CH_3NH_2 < (CH_3)_2NH$
(c) $CH_3NH_2 < (CH_3)_2NH < NH_3$
(d) $CH_3NH_2 < NH_3 < (CH_3)_2NH$

16. Which one of the following is the strongest base in aqueous solution ?
(a) Methylamine (b) Trimethylamine
(c) Aniline (d) Dimethylamine.

17. Consider the following sequence of reactions :

$$\text{Compound}[A] \xrightarrow{\text{Reduction}} [B] \xrightarrow{HNO_2} CH_3CH_2OH$$

The compound [A] is
(a) CH_3CH_2CN (b) CH_3NO_2
(c) CH_3NC (d) CH_3CN

18. The compound obtained by heating a mixture of a primary amine and chloroform with ethanolic potassium hydroxide (KOH) is
(a) an alkyl cyanide
(b) a nitro compound
(c) an alkyl isocyanide
(d) an amide

19. The final product C, obtained in this reaction

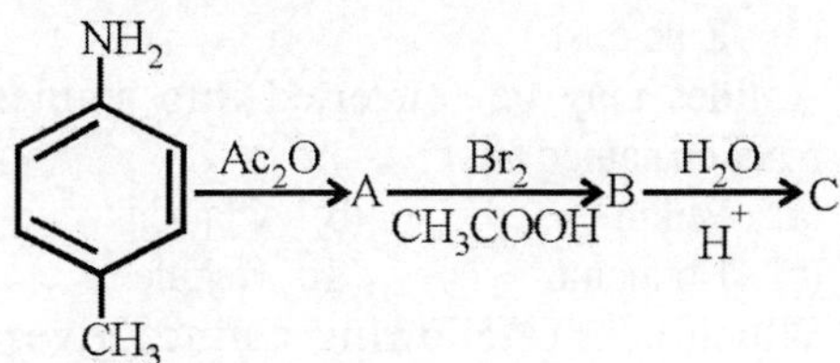

would be

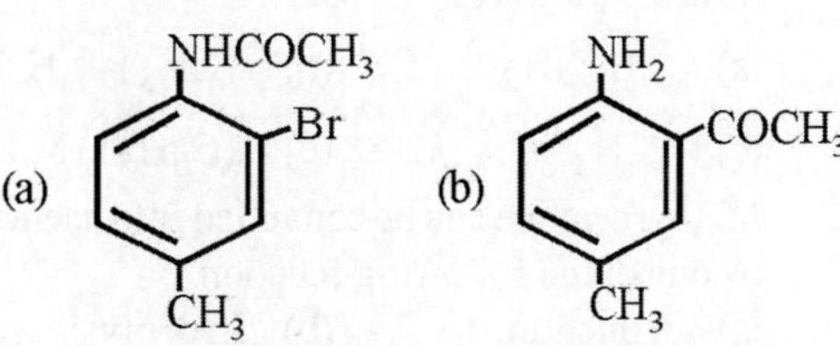

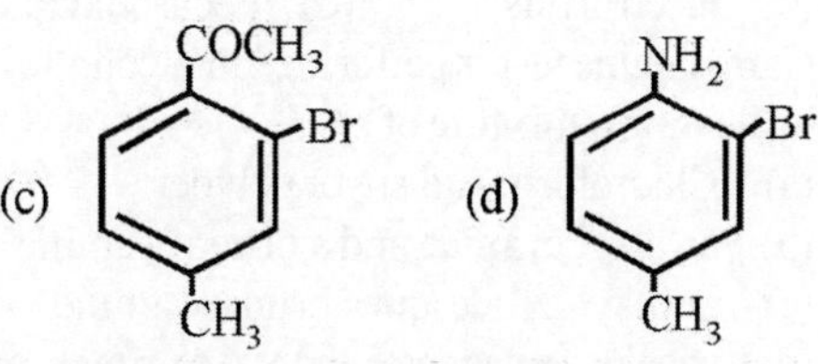

20. Nitrobenzene can be prepared from benzene by using a mixture of conc. HNO_3 and conc. H_2SO_4 in the mixture, nitric acid acts as a/an:
(a) acid (b) base
(c) catalyst (d) reducing agent

21. Acetamide is treated with the following reagents separately. Which one of these would yield methylamine?
(a) $NaOH - Br_2$ (b) Sodalime
(c) Hot conc. H_2SO_4 (d) PCl_5

22. Which of the following statements about primary amines is **'False'** ?
(a) Alkyl amines are stronger bases than aryl amines
(b) Alkyl amines react with nitrous acid to produce alcohols
(c) Aryl amines react with nitrous acid to produce phenols
(d) Alkyl amines are stronger bases than ammonia

23. Aniline on treatment with HCl and $NaNO_2$ gives
(a) Nitroaniline (b) Chloroaniline
(c) Diazo salt (d) Aminophenol

24. Primary amines can be distinguished from secondary and tertiary amines by reacting with
(a) Chloroform and alcoholic KOH
(b) Methyl iodide
(c) Chloroform alone
(d) Zinc dust

25. Amides may be converted into amines by reaction named after :
(a) Perkin (b) Claisen
(c) Hofmann (d) Kekule

26. Which of the following amines gives the characteristic offensive smell of isocyanide when heated with $CHCl_3$ & KOH ?
(a) CH_3NH_2 (b) $(CH_3)_2NH$
(c) $(CH_3)_3N$ (d) $(C_6H_5)_2NH$

27. Methyl cyanide can be converted into acetic acid by one of the following reaction :
(a) reduction (b) hydrolysis
(c) electrolysis (d) decarboxylation

28. Carbylamine test is performed in alcoholic KOH by heating a mixture of
(a) Chloroform and silver powder
(b) an alkyl cyanide and a primary amine
(c) an alkyl halide and a primary amine
(d) trihalogenated methane and a primary amine

29. $R \text{.......} \overset{\overset{O}{||}}{C} \text{.......} NH_2$ on reduction with Lithium aluminimum hydride gives :
(a) 4° ammonium salt (b) 1° amine
(c) 2° amine (d) 3° amine

30. Which of the following is most basic?
(a) Benzamide (b) Butamine
(c) Nitrobenzene (d) Benzene

31. Acetamide and ethylamine can be distinguished by reacting with
(a) Aqueous HCl and heat
(b) Aqueous NaOH and heat
(c) Acidified $KMnO_4$
(d) Bromine water

32. Aniline is reacted with bromine water and the resulting product is treated with an aqueous solution of sodium nitrite in presence of dilute hydrochloric acid. The compound so formed is converted into a tetrafluoroborate which is subsequently heated dry. The final product is
(a) 1,3, 5-tribromobenzene
(b) p-bromofluorobenzene
(c) p-bromoaniline
(d) 2,4, 6-tribromofluorobenzene

33. $[A] \xrightarrow{\text{reduction}} [B] \xrightarrow{CHCl_3 + KOH}$

$[C] \xrightarrow{\text{reduction}} \text{N} - \text{Methylaniline, A is}$

(a) Formaldehyde (b) Trichloromethane
(c) Nitrobenzene (d) Toluene

34. Aniline in a set of reactions yielded a product D.

$$C_6H_5NH_2 \xrightarrow[HCl]{NaNO_2} A \xrightarrow{CuCN} B$$

$$\xrightarrow[Ni]{H_2} C \xrightarrow{HNO_2} D$$

The structure of the product D would be:
(a) C_6H_5NHOH (b) $C_6H_5NHCH_2CH_3$
(c) $C_6H_5CH_2NH_2$ (d) $C_6H_5CH_2OH$

35. Which one of the following on reduction with lithium aluminium hydride yields a secondary amine?
(a) Methyl isocyanide (b) Acetamide
(c) Methyl cyanide (d) Nitroethane

36. On oxidation with sodium dichromate and sulphuric acid aniline gives aniline black. On controlled oxidation with the same reagent the product is
(a) Benzoic acid (b) Nitrobenzene
(c) p-Benzoquinone (d) Phenol

37. Mark the correct statement :
(a) Methyl amine is stronger base than NH_3.
(b) Methyl amine is more acidic than NH_3.
(c) Methyl amine forms salt with alkali.
(d) Methyl amine is less basic than ammonia.

38. Which of the following is not correct ?
(a) Ethyl amine and aniline both have $-NH_2$ group
(b) Ethyl amine and aniline dissolve in HCl
(c) Ethyl amine and aniline both react with $CHCl_3$ and KOH to form unpleasant smelling compound
(d) Ethyl amine and aniline both react with HNO_2 in cold to give hydroxy compounds

39. In the following reaction :

$$\text{Acetamide} \xrightarrow[\Delta]{P_2O_5} Y \xrightarrow[\Delta]{H_2O/HCl} Z \xrightarrow[CaO]{NaOH} CH_4, \text{ identify } Z:$$

(a) $CH_3CH_2NH_2$ (b) CH_3COOH
(c) CH_3CONH_2 (d) CH_3CHO

40. Which one of the following methods is neither meant for the synthesis nor for separation of amines?
(a) Curtius reaction (b) Wurtz reaction
(c) Hofmann method (d) Hinsberg method

41. Ethyl isocyanide on hydrolysis in acidic medium generates
(a) propanoic acid and ammonium salt
(b) ethanoic acid and ammonium salt
(c) methylamine salt and ethanoic acid
(d) ethylamine salt and methanoic acid

42. What is the decreasing order of basicity of primary, secondary and tertiary ethylamines and NH_3 ?
(a) $NH_3 > C_2H_5NH_2 > (C_2H_5)_2NH > (C_2H_5)_3N$
(b) $(C_2H_5)_3N > (C_2H_5)_2NH > C_2H_5NH_2 > NH_3$
(c) $(C_2H_5)_2NH > C_2H_5NH_2 > (C_2H_5)_3N > NH_3$
(d) $(C_2H_5)_2NH > (C_2H_5)_3N > C_2H_5NH_2 > NH_3$.

43. Which is formed, when acetonitrile is hydrolysed partially with cold concentrated HCl?
(a) Acetic acid (b) Acetamide
(c) Methyl cyanide (d) Acetic anhydrides

44. When aniline reacts with oil of bitter almonds (C_6H_5CHO) condensation takes place and benzal derivative is formed. This is known as
(a) Million's base (b) Schiff's reagent
(c) Schiff's base (d) Benedict's reagent

45. The consituent of the powerful explosive RDX is formed during the nitration of
(a) toluene (b) phenol
(c) glycerol (d) urotropine

46. Intermediates formed during reaction of $R\overset{O}{\overset{||}{C}}NH_2$ with Br_2 and KOH are
(a) RNHBr and RCONHBr
(b) RNHCOBr and RNCO
(c) RCONHBr and RNCO
(d) $RCONBr_2$

47. (3-methoxyphenyl with $C\equiv N$; OCH_3) $+ CH_3MgBr \xrightarrow{H_3O^{\oplus}} P$

Product 'P' in the above reaction is
(a) ring with $CH(OH)-CH_3$ and OCH_3 (b) ring with $\overset{O}{\overset{||}{C}}-CH_3$ and OCH_3
(c) ring with CHO and OCH_3 (d) ring with COOH and OCH_3

48. The major organic product formed from the following reaction :

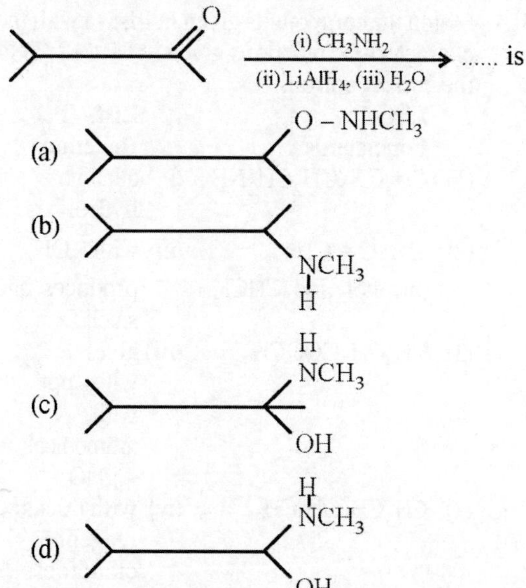

49. Electrolytic reduction of nitrobenzene in weakly acidic medium gives
(a) N–Phenylhydroxylamine
(b) Nitrosobenzene
(c) Aniline
(d) p–Hydroxyaniline

50. Predict the product:

$C_6H_5-NHCH_3 + NaNO_2 + HCl \longrightarrow$ Product

(a) $C_6H_5-N(CH_3)-NO_2$

(b) o-NO substituted $C_6H_4NHCH_3$ + p-NO substituted $C_6H_4NHCH_3$

(c) $C_6H_5-N(OH)-CH_3$

(d) $C_6H_5-N(CH_3)-N=O$

51. Match the compounds given in List - I with their characteristic reactions given in List - II. Select the correct option.

List - I Compounds	List - II Reactions
(1) $CH_3CH_2CH_2CH_2NH_2$	(i) alkaline hydrolysis
(2) $CH_3C \equiv CH$	(ii) with KOH (alcohol) and $CHCl_3$ produces bad smell
(3) $CH_3CH_2COOCH_3$	(iii) gives white ppt. with ammonical $AgNO_3$
(4) $CH_3CH(OH)CH_3$	(iv) with Lucas reagent cloudiness appears after 5 minutes

Options:

	(1)	(2)	(3)	(4)
(a)	(iv)	(ii)	(iii)	(i)
(b)	(ii)	(i)	(iv)	(iii)
(c)	(iii)	(ii)	(i)	(iv)
(d)	(ii)	(iii)	(i)	(iv)

52. In a set of reactions m-bromobenzoic acid gave a product D. Identify the product D.

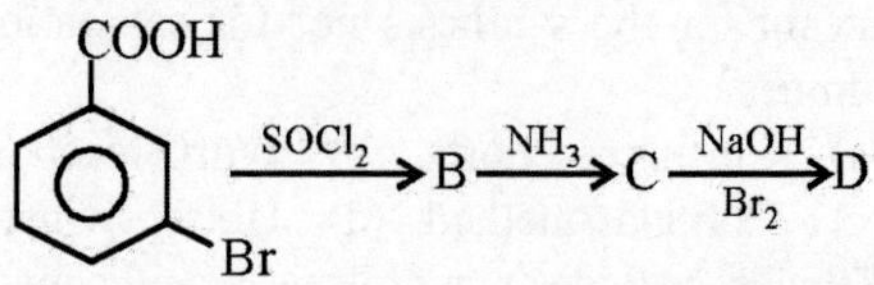

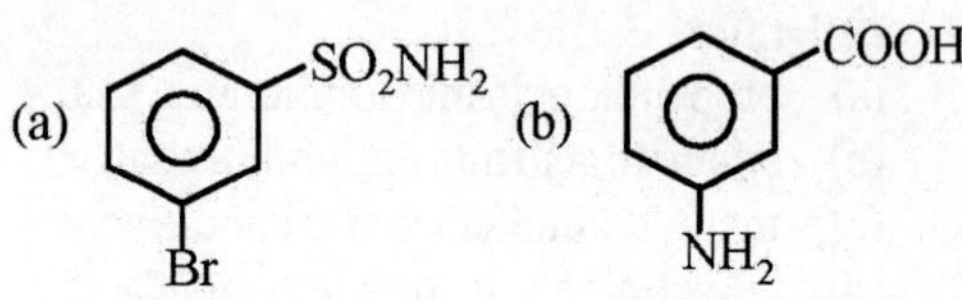

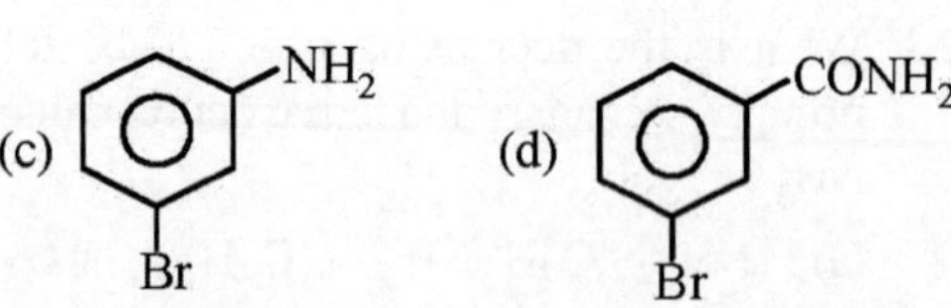

53. What is the product obtained in the following reaction ?

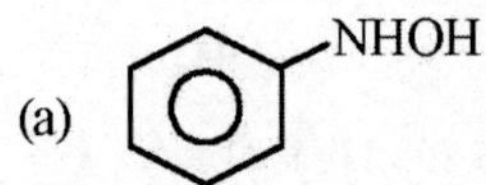

(a) C_6H_5-NHOH

(b)
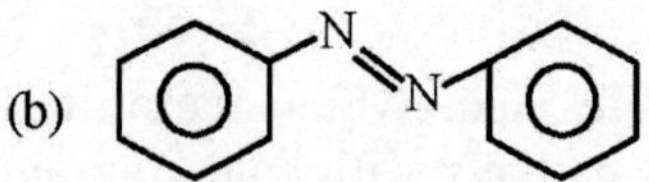

(c)
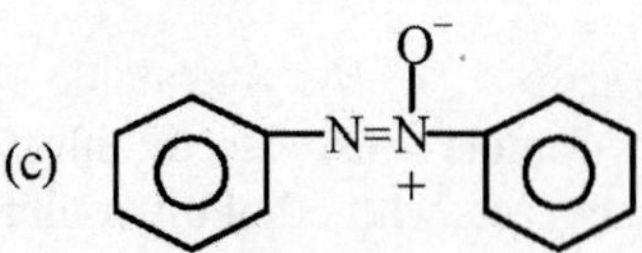

(d) $C_6H_5-NH_2$

54. Ethyl amine can be obtained by the
(a) Action of NH_3 on ethyl iodide.
(b) Action of NH_3 on ethyl alcohol.
(c) Both
(d) None

55. The end product of the reaction :

$$\text{Ethyl amine} \xrightarrow{HNO_2} A \xrightarrow{PCl_5} B \xrightarrow{HNH_2} C$$

is

(a) Ethyl amide (b) Ethyl amine
(c) Methyl amine (d) Acetamide

56. Acetamide on treatment with nitrous acid gives:
(a) Acetic acid (b) Acetone
(c) Methanol (d) Methyl amine

57. The bad smelling substance formed by the action of alcoholic caustic potash on chloroform and aniline is :
(a) Phenyl Isocyanide
(b) Phenyl Isocyanate
(c) Nitrobenzene
(d) Phenyl Cyanide

58. Acetanilide can be prepared from aniline with
(a) Ethanol (b) Acetaldehyde
(c) Acetic anhydride (d) None of the above

59. When ethylamine react with sodium metal, the gas evolved is
(a) H_2 (b) C_2H_5
(c) N_2 (d) NH_3

60. $C_6H_5C \equiv N$ and $C_6H_5N \rightleftarrows C$ exhibit which type of isomerism?
(a) Position
(b) Functional
(c) Dextroisomerism
(d) Position isomerism

CHAPTER 28

Biomolecules

CONCEPTS

Complex organic molecules which are essential for growth and maintenance of living organism. They include carbohydrates, amino acids and proteins, enzymes, nucleic acids, lipids, hormones, vitamins.

CARBOHYDRATES

(i) They are polyhydroxy aldehydes or ketones or compounds which give these on hydrolysis.

(ii) **Classification of carbohydrates**

(a) **Sugars** - Crystalline solids of sweet taste.

(b) **Non-sugar**-do not have sweet taste. e.g.- **starch**.

(c) **Monosaccharides** cannot by hydrolysed to simpler molecules. They are of two types – **aldoses** (having an aldehyde group) and **ketoses** (having keto group).

(d) **Disaccharides** give two molecules of same or different monosaccharides on hydrolysis e.g.- **sucrose (glucose + fructose), maltose (2 molecules of glucose), lactose (glucose + galactose).**

(e) **Oligosaccharides** give 2–9 molecules of monosaccharides on hydrolysis. i.e. di, tri, tetra saccharides and so on.

(f) **Polysaccharides** give large number of monosaccharide molecules on hydrolysis. e.g.- **Starch, glycogen, cellulose.**

(iii) **Reducing and non reducing sugars** : **Reducing sugars** are carbohydrates **having – CHO group** which are capable of reducing Tollen's reagent and Fehling's solution. Non reducing sugars do not reduce Tollen's reagent and Fehling solution e.g.-**glucose is reducing sugar** and **fructose is non reducing sugar**.

(iv) **Glycoside linkage** is the linkage which holds monosaccharides units in polysaccharides.

Glucose

- The molecular formula of glucose is $C_6H_{12}O_6$.

Preparation of Glucose

(a) **From sucrose (cane sugar):** By boiling sucrose with dil HCl or dil H_2SO_4 in alcoholic solution.

$$\underset{\text{Sucrose}}{C_{12}H_{22}O_{11}} + H_2O \xrightarrow{H^+} \underset{\text{Glucose}}{C_6H_{12}O_6} + \underset{\text{Fructose}}{C_6H_{12}O_6}$$

(b) **From starch :** By the hydrolysis of starch with dil H_2SO_4 at 393 K under pressure.

$$\underset{\text{Starch}}{(C_6H_{10}O_5)_n} + H_2O \xrightarrow[393K,\ 2-3bar]{H^+} n\,\underset{\text{Glucose}}{C_6H_{12}O_6}$$

Reactions of Glucose

Glucose is an aldohexose and is also known as dextrose. It gives following reactions:

(a)

$$\begin{array}{l} CHO \\ | \\ (CHOH)_4 \\ | \\ CH_2OH \end{array} \xrightarrow{HI,\ \Delta} \underset{\text{(n–Hexane)}}{CH_3-CH_2-CH_2-CH_2-CH_2-CH_3}$$

this reaction indicates that all six carbons of glucose are in a chain.

(b)

$$\begin{array}{l} CHO \\ | \\ (CHOH)_4 \\ | \\ CH_2OH \end{array} \xrightarrow{NH_2OH} \begin{array}{l} CH=N-OH \\ | \\ (CHOH)_4 \\ | \\ CH_2OH \\ \text{Oxime} \end{array}$$

this reaction indicates that glucose has a carbonyl group.

$$\begin{array}{l} CHO \\ | \\ (CHOH)_4 \\ | \\ CH_2OH \end{array} \xrightarrow{HCN} \begin{array}{l} CH\langle^{CN}_{OH} \\ | \\ (CHOH_4) \\ | \\ CH_2OH \\ \text{Cyanohydrin} \end{array}$$

this reaction indicates that glucose has a carbonyl group.

(c)
$$\begin{array}{l} CHO \\ | \\ (CHOH)_4 \\ | \\ CH_2OH \end{array} \xrightarrow{Br_2\ water} \begin{array}{l} COOH \\ | \\ (CHOH)_4 \\ | \\ CH_2OH \end{array}$$
Gluconic acid

this reaction indicates that carbonyl group is aldehydic in nature which is easily oxidized by bromine water.

(d)
$$\begin{array}{l} CHO \\ | \\ (CHOH)_4 \\ | \\ CH_2OH \end{array} \xrightarrow{Acetic\ anhydride} \begin{array}{l} CHO \\ | \\ (CH-O-\overset{O}{\overset{\|}{C}}-CH_3)_4 \\ | \\ CH_2-O-\overset{O}{\overset{\|}{C}}-CH_3 \end{array}$$
Glucose pentaacetate

this reaction indicates that there are five hydroxyl groups.

(e)
$$\begin{array}{l} CHO \\ | \\ (CHOH)_4 \\ | \\ CH_2OH \end{array} \xrightarrow{Oxidation} \begin{array}{l} COOH \\ | \\ (CHOH)_4 \\ | \\ COOH \end{array}$$
Saccharic acid

$$\xleftarrow{Oxidation} \begin{array}{l} COOH \\ | \\ (CHOH)_4 \\ | \\ CH_2OH \end{array}$$
Gluconic acid

D and L Configuration

A molecule is assigned D-configuration if – OH group attached to the carbon adjacent to the – CH_2OH group (last chiral carbon) is on right and it is assigned L-configuration if – OH group is on left.

The Fisher projection for D-and L-glucose are shown below:

D-glucose		L-glucose	
CHO		CHO	
H	OH	H	H
OH	H	H	OH
H	OH	OH	H
H	OH	OH	H
CH_2OH		CH_2OH	

Cyclic Structures

(i) All pentoses and hexoses exists in cyclic hemiacetal or hemiketal structures. When it is **six membered ring it is called pyranose form** and **when it is 5 membered ring it is called furanose form**.

(ii) The cyclic structure of glucose was proposed by R. D. Haworth. The six-membered cyclic structure of glucose is called as **pyranose structure (α or β)**.

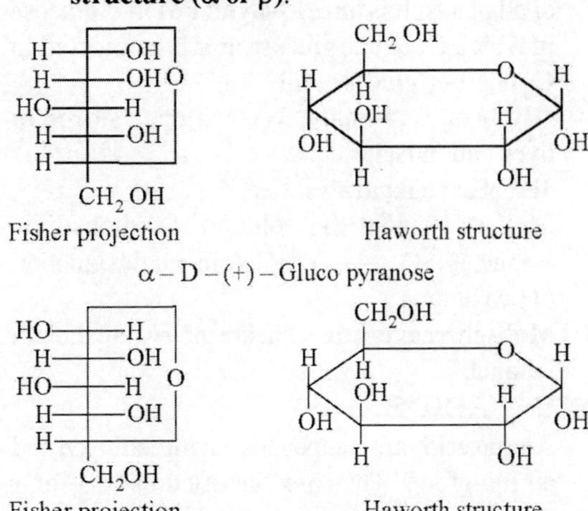

β – D – (+) – Gluco pyranose

- **Anomers differ in configuration around C–1 only**. e.g.- α-D glucose and β-D glucose.
- **Epimers differ in configuration around any one C atom other than** C–1. e.g.- D-glucose and D-manose.
- **Mutarotation :** When either α-D or β-D glucose is dissolved in water, it changes structure as follows.

$$\underset{36\%}{\alpha - D\ glucose} \rightleftharpoons \underset{1\%}{Open\ chain\ structure} \rightleftharpoons \underset{63\%}{\beta - D - glucose}$$

Specific rotation of α-D glucose is +112° and that of β-D glucose is +19°. The specific rotation of equilibrium mixture is +52.7°. **The change in specific rotation of an optically active compound in solution with time to an equilibrium value is called mutarotation**.

- **Inversion of sugar**

$$\underset{[\alpha]_D = +66.5°}{Sucrose} + H_2O \longrightarrow \underset{[\alpha]_D = +52.7°}{glucose} + \underset{[\alpha]_D = -92.4°}{fructose}$$

Sucrose is dextrorotatory, the sign of rotation changes to leavorotatory after hydrolysis. Such change in sign of optical rotation is called inversion of sugar.

- **Starch** is **polysaccharide** made of two components – **Amylose** (water soluble) (20%) and **amylopectin** (water insoluble) (80%). **Amylose is linear polymer of α-D glucose** and **amylopectin is branched chain polymer of α-D glucose**.

- **Glucose** is chief structural material of cell walls of all plants. It is **linear polymer of β-D glucose in which C_1 of one glucose unit is connected to C_4 of other glucose unit.**
- **Glycogen** is **animal polysaccharide stored in liver and muscles.**
- **Test of carbohydrates**
  Molisch reagent +aq. solution of carbohydrate + conc. $H_2SO_4 \longrightarrow$ violet ring at the junction of two liquids.z
- **Molisch reagent is solution of α-napthol in ethanol.**

AMINO ACIDS

- **Amino acids** are compounds having carboxyl and amino group. They are classified as α, β or γ depending upon position of – NH_2 group with respect to – COOH group.
- **α-amino acids** are building blocks of proteins. They have – COOH group present at α-carbon atom.

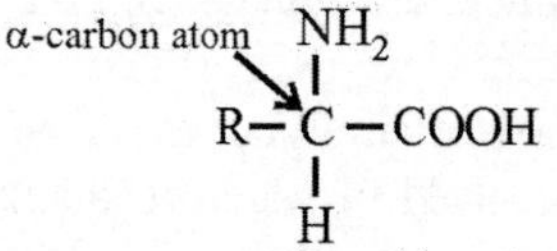

$$\text{R}-\overset{\overset{\text{NH}_2}{|}}{\underset{\underset{\text{H}}{|}}{\text{C}}}-\text{COOH}$$

α-carbon atom (arrow to C)

α-amino acid

- There are **26 amino acids** in nature. The amino acids differ from each other in nature of side chain groups R.

Structure of Amino Acids

$$\text{R}-\overset{\overset{\text{NH}_2}{|}}{\text{CH}}-\text{COOH} \rightleftharpoons \text{R}-\overset{\overset{\oplus\text{NH}_3}{|}}{\text{CH}}-\text{COO}^-$$

Zwitter ion or dipolar ion

The **pH of a solution of amino acids at which amino acids exists in form of dipolar ion is called iso-electric point**. Different amino acids have different iso electric points.

Classification of Amino Acids

(i) $\alpha, \beta, \gamma, \delta$ ***amino acids*** depending upon the position of - NH_2 w.r.t. – COOH group.

(ii) *Neutral* - Having one –NH_2 and one –COOH e.g. $NH_2.CH_2.COOH$ glycine

(iii) *Acidic* - Having one –NH_2 and two –COOH

e.g. $\text{HOOC.}\overset{\overset{\text{NH}_2}{|}}{\text{CH}}-\text{COOH}$

Aspartic acid.

(iv) *Basic* - Having two or more –NH_2 and one – COOH

e.g. $\text{H}_2\text{N(CH}_2)_4.\overset{\overset{\text{NH}_2}{|}}{\text{CH}}-\text{COOH}$

Lysine

Types of Amino Acids

(a) **Essential amino acids** are those which our body can't synthesize and we must take it from outside. Valine, leucine, isoleucine, phenylalanine, methionine, tryptophan, threonine, lysine, arginine and histidine are essential amino acids.

(b) **Non-essential amino acids** can be synthesised inside our body.

- **Peptides are** amides formed by condensation of NH_2 group of one α-amino acid with COOH group of other with elimination of water. The **CO – NH bond thus formed is called peptide bond or peptide linkage**. Each polypeptide chain has a free amino group at one end called, **N-Terminal end** and free carboxyl group at other end called **C-terminal end**.

PROTEINS

Chemically condensation polymers of α-amino acids. They are complex nitrogenous organic compounds found in animals and plants.

Structure of Proteins

(a) ***Primary structure*** - It simply reveals the sequence of amino acids.

(b) ***Secondary structure*** - α - helix structure maintained by hydrogen bonds or β - pleated sheet structure when size of R groups is small.

(c) ***Tertiary structure*** - The folding and super imposition of polypeptide chains forms a compact globular shape, termed as tertiary structure. It is stabilised by covalent, ionic, hydrogen and disulphide bridge.

(d) **Quaternary structure**-The precise arrangement constitutes the quaternary structure.

Types of Proteins

(a) **On the basis of hydrolysis products**

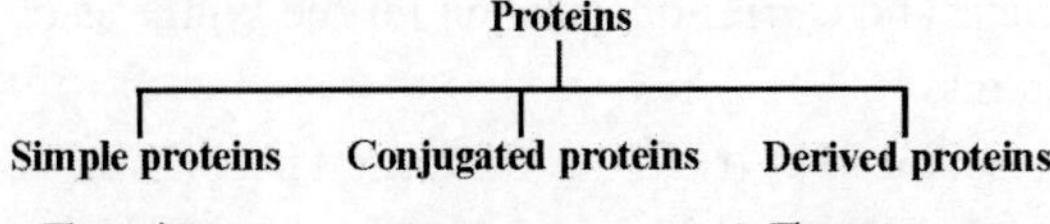

- **Simple proteins:** They give only α-amino acid on hydrolysis e.g.-**albumin, keratin.**
- **Conjugated proteins:** They give α-amino acid + non protein portion (called **prosthetic group**) e.g.- **Haemoglobin.**
- **Derived proteins:** They are degradation product obtained by partial hydrolysis of simple or conjugated proteins

Protein	**Prosthetic group**
Nucleoproteins	Nucleic acid
Phospho proteins	Phosphoric acid
Glyco proteins	Carbohydrates
Metallo proteins	Metals
Chromo proteins	Pigment
Lipo proteins	Lipids

(b) **On the basis of molecular structure**

Proteins

Fibrous proteins
- Linear thread like molecule which lie side by side to form fibres and held together by H-bonding.
- e.g.- keratin, myosin, fibrocin, collagen

Globular protein
- The polypeptide chain of protein coils around itself to give spherical structure.
- e.g.- **All enzymes,** many hormones (insulin) **antibodies,** haemoglobin.

Classification of Proteins According to Functions

(i) ***Structural proteins :*** Fibrous proteins

(ii) ***Enzymes :*** Serve as catalyst pepsin, trypsin etc.

(iii) ***Harmones :*** Insulin

(iv) ***Contractile Proteins :*** Found in muscles eg myosin, actin.

(v) ***Anti bodies :*** Gamma globulins present in blood.

(vi) ***Blood protein :*** Albumins, haemoglobin and fibrinogen.

Denaturation of Proteins

It refers to the loss of a biological activity of a protein brought about by changes in its secondary and tertiary structure due to heat, pH change, presence of salts, etc.

- Examples of denaturation of proteins are

(i) coagulation of albumin present in egg white when we boil egg.

(ii) Formation of cheeze from milk on addition of lemon juice.

Test for Proteins

(a) **Biuret test :**

alk. solution of protein + 1% $CuSO_4$ + NaOH $\longrightarrow$ violet colouration

(b) **Millon's test :**

aq. solution of protein + **Millions reagent $\longrightarrow$ Pink colour**

Million's reagent → solution of mercuric nitrate and nitrile in nitric acid containing traces of nitrous acid.

(c) **Ninhydrin test :**

Protein + water + ninhydrin $\xrightarrow{\text{boil}}$ blue colouration.

ENZYMES

Biological catalysts and chemically they are **globular proteins.**

(i) **Coenzymes** - Non proteinous components required for the activity of certain enzymes are known as coenzymes.

- **Coenzymes** include metal ions like Mn^{2+}, Mg^{2+}, K^+, Na^+, Zn^{++}, Co^{++} etc. hetero cyclic ring systems (pyrrole, purine, pyidine etc), a sugar residue, phosphoric acid residue or vitamins like thiamine, riboflavin, niacin etc. In such cases the protein part of enzymes is called **apoenzyme**

(ii) **Endoenzyme** - acts in the same cell in which it is synthesised.

(iii) **Exoenzyme** - acts outside the cell in which it is synthesised.

(iv) **Nomenclature** - They are usually named by adding the suffix - ase to the root name of the substrate eg urease, maltase.

(v) **Hydrolytic enzymes** - Catalyse hydrolysis and are mostly simple proteins.

(vi) **Oxidative enzymes** - Catalyse oxidation reduction reaction and are mostly conjugated proteins.

(vii) **Holoenzyme** (Protein-cofactor complex) (Active) $\rightleftharpoons$ **Apoenzyme** (inactive) + **cofactor**

Properties of Enzymes

- Highly specific, have very high effectivity, required in very small quantity.
- The enzyme action follows **Lock and Key mechanism :** Enzyme action is inhibited by certain organic molecules called as **inhibitors**.

Applications

(a) ***Treatment of diseases*** - The congenital disease phenyl ketonuria caused by phenylalanine - hydroxylase can be cured by diet of low phenylalanine content. Albinism is caused by deficiency of tyrosinase. Enzyme streptokinase is used for blood cloting to prevent heart disease.

(b) ***In industry*** - Tanning of leather, fermentation process.

NUCLEIC ACIDS

(i) **Nucleic acids** are polymers of complex repeating units called **nucleotides**. There are two types of nucleic acids: DNA (deoxyribonucleic acid) and RNA (ribonucleic acid).

- **Nucleotide** = pentose sugar + nitrogeneous base + phosphoric acid unit.
- Nucleoside = pentose sugar + nitrogeneous base.
- Pentose sugar is of 2 types **Ribose sugar (in RNA) and Deoxyribose sugar (in DNA)**

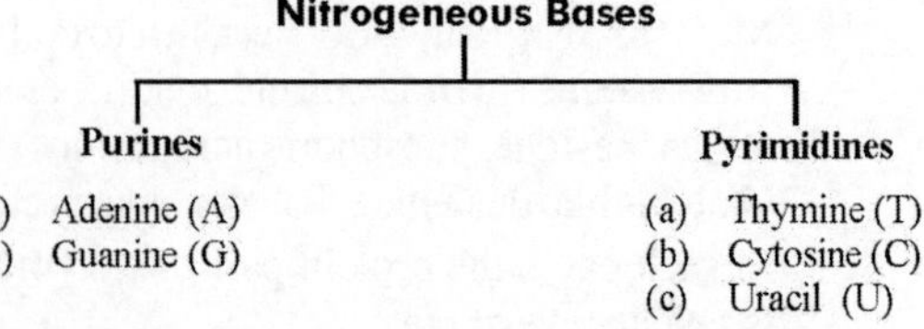

Types of Nucleic Acids – DNA and RNA

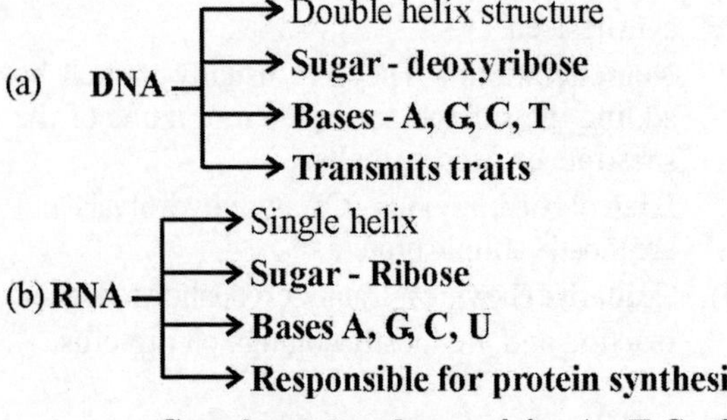

- **Complementary base pairing** A = T, C ≡ G

Types of RNA

Messenger RNA (m-RNA). It is produced in the nucleus and carries information for the synthesis of proteins.

Ribosomal RNA (r - RNA) - Its role yet not known.

Transfer RNA (Soluble or Adoptive RNA) (S.RNA, t - RNA) - It is found in cytoplasm. Its function is to collect amino acids from cytoplasm for protein synthesis.

Functions of Nucleic Acids

(i) To direct the synthesis of proteins.

(ii) To transfer the genetic information (hereditary characters).

- **Replication** is process of formation of new DNA molecules from DNA templet.
- **Protein synthesis :** Occurs in two steps.
  (a) **Transcription** is copying of DNA base sequence into m-RNA.
  (b) **Translation** – mRNA directs protein synthesis with help of t-RNA and r-RNA.
- **Gene** - The portion of DNA carrying information about a specific protein is called gene.
- **Genetic Code** - The relation between the amino acid and the nucleotide triplet is called genetic code.
- **Codons** - The nucleotide bases in RNA function in groups of three (triplet) in coding amino acids. These base triplets are called codons.
- **Mutation** - Error in genetic code is called mutation.

LIPIDS

(i) They are oily or waxy substances present in all living cells.

(ii) **Found in cell membrane and store energy for cell.**

(iii)

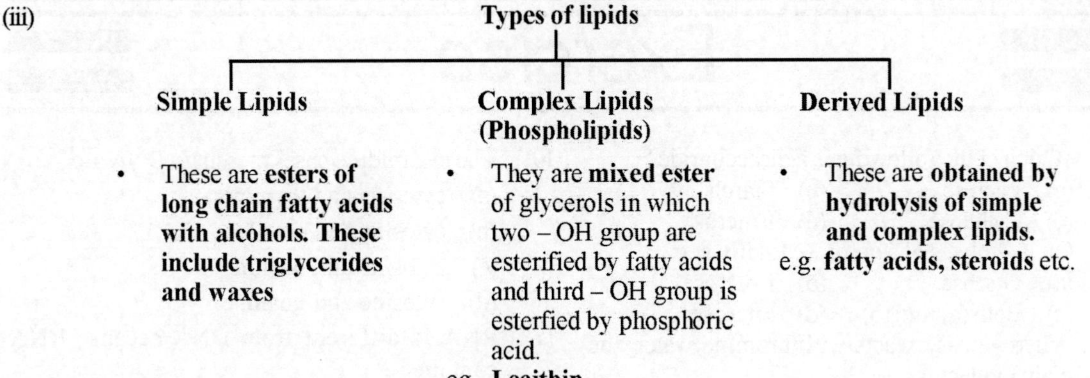

HORMONES

(i) These are biomolecules produced in **endocrine** (or ductless) **glands**. They reach all body parts via blood.

(ii) **Functions of hormones**

S.No.	Name	Organ of secretion	Function
		1. Steroidal hormone	
(a)	**Androgens**	Testes	control and development of male sex organs.
(b)	**Estrogens**	overies	control and development of female sex organs.
(c)	**Progesterone**	corpus luteum	controls development and maintenance of pregnancy.
(d)	**Corticoids**	adrenal cortex	Regulates fat, carbohydrate and protein metabolism.
		2. Peptide hormones	
(a)	**Oxytocin**	pituitary gland	controls contraction of uterus after the child birth.
(b)	**Vasopressin**	pituitary gland	controls reabsorption of water in the kidney.
(c)	**Insulin**	pancreas	controls metabolism of carbohydrates and maintains glucose level in blood.
		3. Amine hormone	
(a)	**Adrenaline**	Adrenal medulla	prepares animal for emergency situation.
(b)	**Thyroxin**	Thyroid gland	controls metabolism of proteins, lipids, carbohydrates.

VITAMINS

(i) They are biomolecules required in small amount for normal growth and health of normal human beings and animal organisms.

(ii) **Fat soluble vitamins – A, D, E, K.**
Water soluble vitamins – B, C.

(iii)

	Vitamins	Deficiency diseases
1.	**Vitamin A (Retinol)**	(i) Night blindness. (ii) **Xerophthalmia** (hardening of cornea)
2.	**Vitamin B_1 (Thiamin)**	Beriberi (disease of nervous system)
3.	**Vitamin B_2 (Riboflavin)**	Inflation of tongue, dermatitis
4.	**Vitamin B_6 (Pyridoxin)**	**Pellagra**
5.	**Vitamin B_{12} (Cyanocobalamin)**	**Pernicious anaemia**
6.	**Vitamin D (Ergocalciferol)**	**Rickets**
7.	**Vitamin E (Tocopherol)**	**Sterility**
8.	**Vitamin K** **(Phylloquinone** or **antihaemorrgagic vitamin)**	**Haemorrhage and lengthens time of blood clotting.**

Exercise

1. Which of the following is a disaccharide ?
(a) Lactose (b) Starch
(c) Cellulose (d) Fructose
2. On hydrolysis of starch, we finally get
(a) Glucose (b) Fructose
(c) Both (a) and (b) (d) Sucrose
3. When glucose reacts with bromine water, the main product is
(a) gluconic acid (b) glyceraldehyde
(c) saccharic acid (d) acetic acid
4. Complete hydrolysis of cellulose gives
(a) D-ribose (b) D-glucose
(c) L-glucose (d) D-fructose
5. The number of essential amino acids in man is
(a) 8 (b) 10
(c) 18 (d) 20
6. 
For $-\overset{O}{\overset{\|}{C}}-\ddot{N}H-$ (peptide bond)
Which statement is incorrect about peptide bond?
(a) C — N bond length in proteins is longer than usual bond length of the C — N bond
(b) Spectroscopic analysis shows planar structure of the $-\underset{O}{\underset{\|}{C}}-NH-$ group
(c) C — N bond length in proteins is smaller than usual bond length of the C—N bond
(d) None of the above
7. The helical structure of protein is stabilized by
(a) dipeptide bonds (b) hydrogen bonds
(c) ether bonds (d) peptide bonds
8. The function of enzymes in the living system is to
(a) transport oxygen
(b) provide energy
(c) provide immunity
(d) catalyse biochemical reactions
9. Which of the following statements about vitamin B_{12} is incorrect ?
(a) It has a cobalt atom
(b) It also occurs in plants
(c) It is also present in rain water
(d) It is needed for human body in very small amounts
10. The pyrimidine bases present in DNA are
(a) cytosine and thymine
(b) cytosine and uracil
(c) cytosine and adenine
(d) cytosine and guanine
11. RNA is different from DNA because RNA contains
(a) ribose sugar and thymine
(b) ribose sugar and uracil
(c) deoxyribose sugar and thymine
(d) deoxyribose sugar and uracil.
12. The reason for double helical structure of DNA is operation of
(a) dipole-dipole interaction
(b) hydrogen bonding
(c) electrostatic attractions
(d) van der Waals' forces
13. Which base is present in RNA but not in DNA ?
(a) Guanine (b) Cytosine
(c) Uracil (d) Thymine
14. The change in optical rotation (with time) of freshly prepared solutions of sugar is known as:
(a) Specific rotation (b) Inversion
(c) Rotatory motion (d) Mutarotation
15. Glucose gives silver mirror test with Tollen's reagent. It shows the presence of
(a) acidic group (b) alcoholic group
(c) ketonic group (d) aldehyde group
16. The linkage present in proteins and peptides is
(a) $-\overset{O}{\overset{\|}{C}}-NH-$ (b) $-\overset{O}{\overset{\|}{C}}-O$
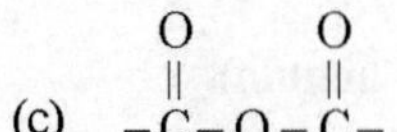
(c) $-\overset{O}{\overset{\|}{C}}-O-\overset{O}{\overset{\|}{C}}-$ (d) $-NH-$
17. Which carbohydrate is essential constituent of plant cell?
(a) Starch (b) Glucose
(c) Cellulose (d) Sucrose
18. Amino acids are the building blocks of
(a) fats (b) proteins
(c) vitamins (d) carbohydrates

19. The reactions of (a) oxygen and (b) carbon monoxide with haeme (the prosthetic group of haemoglobin) give
(a) only oxygen-haeme complex
(b) only carbon monoxide-haeme complex
(c) both oxygen-haeme and carbon monoxide-haeme complexes but oxygen-haeme complex is more stable
(d) both oxygen-haeme and carbon monoxide-haeme complexes but carbon monoxide-haeme complex is more stable

20. Which of the following is the sweetest sugar?
(a) Sucrose (b) Glucose
(c) Fructose (d) Maltose

21. Mg is present in which of the following :
(a) Starch (b) Chlorophyll
(c) Both (d) None

22. Cellulose is a polymer of
(a) Glucose (b) Fructose
(c) Ribose (d) Sucrose

23. Enzymes are made up of
(a) Edible proteins
(b) Proteins with specific structure
(c) Nitrogen containing carbohydrates
(d) Carbohydrates

24. Which one of the following structures represents the peptide chain?

```
      H              O
      |              ||
(a) – N – C – N – C – NH – C – NH –
          ||  |   |
          O   H

      H              H
      |      |   |   |   |   |   |
(b) –N–C–C–C–C–N–C–C–C–
        ||   |   |   |   |   |   |
        O

      H        H       H   O
      |   |    |   |   |   ||
(c) –N–C–C–N–C–C–N–C–C–
          |  ||    |  ||   |
             O        O

      H       O           H
      |   |   |   ||      |   |   |   |
(d) –N–C–C–C–N–C–C–N–C–C–C–
          |   |       |   |       ||  |   |
                      H           O
```

25. The cell membranes are mainly composed of
(a) fats (b) proteins
(c) phospholipids (d) carbohydrates

26. The human body does not produce
(a) Vitamins (b) Hormones
(c) Enzymes (d) DNA

27. Glucose contains
(a) one – CHO group
(b) one primary alcoholic group
(c) four secondary alcoholic groups
(d) all are correct

28. The commonest disaccharide has the molecular formula
(a) $C_{10}H_{18}O_9$ (b) $C_{10}H_{20}O_{10}$
(c) $C_{18}H_{22}O_{11}$ (d) $C_{12}H_{22}O_{11}$

29. Which one is a disaccharide ?
(a) Glucose (b) Fructose
(c) Xylose (d) Sucrose

30. Deficiency of vitamin A results in
(a) Lose in apetite (b) Skin diseases
(c) Sterility (d) Retarted growth

31. The pair of compounds in which both the compounds give positive test with Tollen's reagent is
(a) Glucose and Sucrose
(b) Fructose and Sucrose
(c) Acetophenone and Hexanal
(d) Glucose and Fructose

32. The two forms of D–glucopyranose obtained from the solution of D–glucose are called
(a) Isomers (b) Anomers
(c) Epimers (d) Enantiomers

33. Insulin production and its action in human body are responsible for the level of diabetes. This compound belongs to which of the following categories?
(a) An enzyme (b) A hormone
(c) A co-enzyme (d) An antibiotic

34. In both DNA and RNA, heterocylic base and phosphate ester linkages are at –
(a) C_5' and C_1' respectively of the sugar molecule
(b) C_1' and C_5' respectively of the sugar molecule
(c) C_2' and C_5' respectively of the sugar molecule
(d) C_5' and C_2' respectively of the sugar molecule

35. The secondary structure of a protein refers to
(a) fixed configuration of the polypeptide backbone
(b) α – helical backbone
(c) hydrophobic interactions
(d) sequence of α – amino acids.

36. α – D– (+)–glucose and β–D–(+)–glucose are
(a) conformers (b) epimers
(c) anomers (d) enatiomers

37. Proteins when heated with conc. HNO_3 give a yellow colour. This is
(a) Oxidizing test
(b) Xanthoproteic test
(c) Hoppe's test
(d) Acid base test

38. Which of the following reagent cannot distinguish between glucose and fructose?
(a) Fehling's solution
(b) Tollen's reagent
(c) Benedict's solution
(d) All of these

39. Enzymes take part in a reaction and
(a) decrease the rate of a chemical reaction
(b) increase the rate of a chemical reaction
(c) both (a) and (b)
(d) None of these

40. Chemically considering digestion is basically
(a) Anabolism
(b) Hydrogenation
(c) Hydrolysis
(d) Dehydrogenation.

41. The α-D glucose and β-D glucose differ from each other due to difference in carbon atom with respect to its
(a) conformation
(b) configuration
(c) number of OH groups
(d) size of hemiacetal ring

42. Which of the following protein destroy the antigen when it enters in body cell?
(a) Antibodies (b) Insulin
(c) Chromoprotein (d) Phosphoprotein

43. In reference to biological role, Ca^{2+} ions are important in
(a) triggering the contraction of muscles
(b) generating right electrical potential across cell membrane
(c) hydrolysis of ATP
(d) defence mechanism

44. The number of molecules of ATP produced in the lipid metabolism of a molecule of palmitic acid is
(a) 130 (b) 36
(c) 56 (d) 86

45. Glucose molecule reacts with 'X' number of molecules of phenylhydrazine to yield osazone. The value of 'X' is
(a) four (b) one
(c) two (d) three

46. In cells the net production of ATP molecules generated from one glucose molecule is
(a) 46 (b) 32
(c) 36 (d) 40

47. Which of the following is a steroid hormone?
(a) Cholesterol (b) Adrenaline
(c) Thyroxine (d) Progesterone

48. Which of the following gives positive Fehling solution test ?
(a) Protein (b) Sucrose
(c) Glucose (d) Fats

49. Which is correct statement?
(a) Starch is a polymer of α-glucose.
(b) In cyclic structure of fructose, there are four carbons and one oxygen atom.
(c) Amylose is a component of cellulose.
(d) Proteins are composed of only one type of amino acids.

50. Which is not a true statement?
(a) α-Carbon of α-amino acid is asymmetric.
(b) All proteins are found in L-form.
(c) Human body can synthesize all proteins they need.
(d) At pH = 7 both amino and carboxylic groups exist in ionised form.

51. Glycolysis is
(a) conversion of glucose to haem
(b) oxidation of glucose to glutamate
(c) conversion of pyruvate to citrate
(d) oxidation of glucose to pyruvate

52. Phospholipids are esters of glycerol with
(a) Three phosphate groups
(b) Three carboxylic acid residues
(c) Two carboxylic acid residues and one phosphate group
(d) One carboxylic acid residue and two phosphate groups

53. A sequence of how many nucleotides in messenger RNA makes a codon for an amino acid?
(a) Three (b) Four
(c) One (d) Two

54. The hormone that helps in the conversion of glucose to glycogen is
    (a) Cortisone (b) Bile acids
    (c) Adrenaline (d) Insulin
55. The enzyme which hydrolyses triglycerides to fatty acids and glycerol is called
    (a) Maltase (b) Lipase
    (c) Zymase (d) Pepsin
56. The correct statement in respect of protein haemoglobin is that it
    (a) functions as a catalyst for biological reactions
    (b) maintains blood sugar level
    (c) acts as an oxygen carrier in the blood
    (d) forms antibodies and offers resistance to dieases
57. Which one of the following is a peptide hormone ?
    (a) Testosterone (b) Thyroxin
    (c) Adrenaline (d) Glucagon
58. During the process of digestion, the proteins present in food materials are hydrolysed to amino acids. The two enzymes involved in the process

    $$\text{Proteins} \xrightarrow{\text{Enzyme(A)}} \text{Polypeptides}$$

    $$\xrightarrow{\text{Enzyme(B)}} \text{Amino acids}$$

    are respectively
    (a) Diastase and Lipase
    (b) Pepsin and Trypsin
    (c) Invertase and Zymase
    (d) Amylase and Maltase
59. Which one of the following is an amine hormone ?
    (a) Thyroxine (b) Oxypurin
    (c) Insulin (d) Progesterone
60. The segment of DNA which acts as the instrumental manual for the synthesis of the protein is:
    (a) ribose (b) gene
    (c) nucleoside (d) nucleotide

CHAPTER 29

Polymers

CONCEPTS

- Polymers have high molecular mass and formed by union of monomers.
- **Homopolymers :** Polymers made of same monomer.
- **Copolymers :** Polymers made of different types of monomers.

CLASSIFICATION OF POLYMERS

Classification of Polymers on the Basis of Source

(i) **Natural polymers :** Proteins, cellulose, starch, resins and rubber.

(ii) **Semi-synthetic polymers :** Cellulose derivatives as cellulose acetate (rayon) and cellulose nitrate, etc.

(iii) **Synthetic polymers :** Plastic (polythene), synthetic fibres (nylon 6, 6) and synthetic rubbers (Buna – S).

Classification Based on Mode of Polymerisation

(i) **Chain growth polymers or addition polymers :** They are formed by successive addition of monomers without loss of simple molecules like H_2O, NH_3 etc. e.g. polythene.

(ii) **Condensation polymers :** They are formed by the condensation reactions between two monomers, each monomers having two functional groups, with the elimination of small molecules such as water, alcohol and ammonia, *e.g.,* Nylon 66.

Classification Based on Molecular Forces

(i) **Elastomers :** The strands of polymer are held together by weak inter molecular forces (van der waal's). e.g.- **vulcanised rubber**.

(ii) **Fibres** are held together by hydrogen bonds. e.g.- **nylon, polyester, polyamide**.

(iii) **Thermoplastics** can be easily moulded on heating. They don't have cross-links, e.g.- **polyethene**.

(iv) **Thermosetting polymers** have cross links, cannot be remolded on heating. e.g.- **Bakelite**.

(v) **Plastisizer** are high boiling esters which are added to plastic to make it soft and rubber like.

MOLECULAR WEIGHT :

- It is expressed as number average molecular mass $(\overline{M}^n)$ or weight average molecular mass $(\overline{M}_W)$ of a polymer.

$$\overline{M}_n = \frac{\Sigma N_i M_i}{\Sigma N_i} \text{ and } \overline{M}w = \frac{\Sigma N_i M_i^2}{\Sigma N_i M_i}$$

where N^i is the number of molecules of molecular mass M^i.

POLY DISPERSITY INDEX (PDI)

- It is the ratio of the weight and number average molecular masses $(\overline{M}_W / \overline{M}_n)$.
- In some natural polymers, the PDI is unity, *i.e.,* $\overline{M}_W = \overline{M}_n$ because they are mono dispersed.
- In synthetic polymers, the PDI is always greater than unity, *i.e.,* $\overline{M}_W > \overline{M}_n$ because they are always poly dispersed.

NATURAL AND SYNTHETIC RUBBER

Natural Rubber

- It is cis 1, 4-polyisoprene. It is a linear 1,4-polymer of isoprene. It is manufactured from rubber latex which is a colloidal suspension of rubber in water.
- The process of heating a mixture of raw rubber and sulphur at 373 K to 415 K is known as **vulcanisation of rubber**. The process of vulcanisation is accelerated by additing additives such as ZnO.

Synthetic Rubber

- These are either homopolymers of 1, 3-butadiene derivatives or are copolymers in which one of the monomers is 1, 3-butadiene or its derivatives. Examples are Buna – S, Buna – N, neoprene and butyl rubber.

Examples of Synthetic Rubbers

(a) **Buna-S or styrene-Butadiene rubber (SBR)**
Addition copolymer of

$$\underset{\text{1, 3-butadiene}}{CH_2 = CH - CH = CH_2} \text{ and } \underset{\text{Styrene}}{C_6H_5CH = CH_2}$$

- Used in tyres, rubber soles, belts etc.

(b) **Buna-N or (Nitrile rubber)**
Addition copolymer of

$$\underset{\text{1, 3-butadiene}}{CH_2 = CH - CH = CH_2} \text{ and } \underset{\text{acrylonitrile}}{CH_2 = CHCN}$$

(c) **Neoprene**
Addition homopolymer of

$$\underset{\text{chloroprene}}{CH_2 = CH - \overset{\overset{\large Cl}{|}}{C} = CH_2}$$

- It is used for manufacturing conveyor belts, gaskets and hoses.

SOME IMPORTANT POLYMERS

Polyesters :

(a) **Terylene or dacron**

Copolymer of $\underset{\text{ethylene glycol}}{CH_2OH - CH_2OH}$ and

$$\underset{\text{terephthalic acid}}{COOH-C_6H_4-COOH}$$

- Used for making magnetic recording tapes.

(b) **Glyptal or alkydresin**

Copolymer of $\underset{\text{ethylene glycol}}{CH_2OH - CH_2OH}$ and

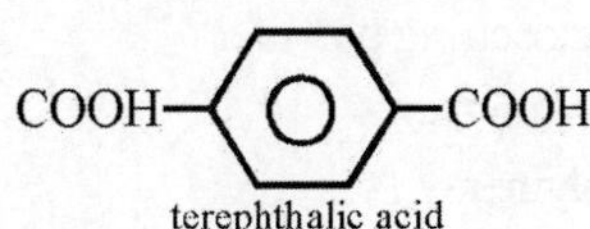

phthalic acid

- Used in paints and building materials.

Polyamides :

(a) **Nylon 66**

Copolymer of adipic acid + hexamethylenediamine

$COOH(CH_2)_4COOH + NH_2 - (CH_2)_6 - NH_2$

- It is used in making sheets, bristles for brushes and in textile industry.

(b) **Nylon 6**

Homopolymer of caprolactum

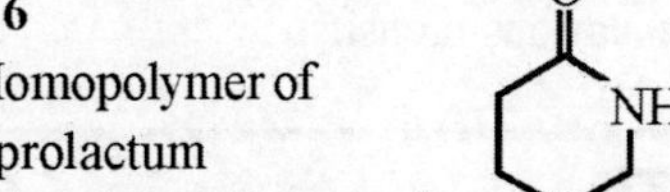

- It is used for the manufacture of tyre cords, fabrics and ropes.

(c) **Nylon 6, 10**

Copolymer of $\underset{\text{hexamethylenediamine}}{NH_2(CH_2)_6NH_2} + \underset{\text{sebacoylchlorids}}{ClOC(CH_2)_8COCl}$

Polyacrylates :

(a) **Poly methyl methacrylate (PMMA)**
Addition homopolymer of

$$\underset{\text{(Methyl methacrylate)}}{CH_2 = \overset{\overset{\large CH_3}{|}}{C} - COOCH_3}$$

- Used in lenses, aircraft windows, protective coatings.

(b) **Polyacrylonitrile (PAN) or Orlon**

Addition homopolymer of $\underset{\text{(acrylonitrile)}}{CH_2 = CHCN}$

- Used for making blankets, sweaters, synthetic carpets.

Polyolefins

(a) **Polystyrene (or styron)**

Addition homopolymer of $\underset{\textbf{(styrene)}}{C_6H_5CH = CH_2}$

- Used in plastic utensils, insulation.

(b) **Polythene or polyethylene**
Addition polymer of **ethylene** $CH_2 = CH_2$.
- It is used in plastic bags, bottes.

Polyhalolefins

(a) **Poly vinyl chloride (PVC)**
Chain growth homopolymer of **vinyl chloride** $(CH_2 = CH - Cl)$.
- Used in shower curtains, tubing.

(b) **Polytetrafluoroethylene or Teflon (PTFE)**
Chain growth homopolymer of **tetrafluoroethylene** $(F_2C = CF_2)$.
- Used in nonstick coating.

(c) **Polymonochlorotrichloroethylene** (PCTFE)
Homopolymer of $ClFC = CF_2$
(monochlorotrifluoroethylene)

Bakelite :

Copolymer of **phenol + formaldehyde**.

- It is used for making combs, photograph records, electrical switches and handles of various utensils.

Melamine :

Copolymer of **formaldehyde** and **melamine** (2, 4, 6-triamino-1, 3, 5-triazine)

- It is used in the manufacture of unbreakable crockery.

1. Which is not true about polymers?
   (a) Polymers do not carry any charge
   (b) Polymers have high viscosity
   (c) Polymers scatter light
   (d) Polymers have low molecular weight
2. Which of the following polymers do not involve cross linkages?
   (a) Melmac (b) Bakelite
   (c) Polythene (d) Vulcanised rubber
3. Which of the following is not an example of addition polymer ?
   (a) Polystyrene (b) Nylon
   (c) PVC (d) Polypropylene
4. In elastomer, intermolecular forces are
   (a) strong (b) weak
   (c) nil (d) none of these
5. Which is an example of thermosetting polymer?
   (a) Polythene (b) PVC
   (c) Neoprene (d) Bakelite
6. Nylon threads are made of
   (a) polyester polymer
   (b) polyamide polymer
   (c) polyethylene polymer
   (d) polyvinyl polymer
7. Of the following which one is classified as polyester polymer ?
   (a) Terylene (b) Bakelite
   (c) Melamine (d) Nylone-66
8. Bakelite is obtained from phenol by reacting with:
   (a) acetaldehyde (b) formaldehyde
   (c) baking soda (d) amino acid
9. Which one of the following monomers gives the polymer neoprene on polymerization ?
   (a) $CF_2 = CF_2$
   (b) $CH_2 = CHCl$
   (c) $CCl_2 = CCl_2$
   (d) $CH_2 = \underset{|}{\overset{Cl}{C}} - CH = CH_2$
10. Natural rubber is a polymer of
    (a) isoprene (b) acrylic acid
    (c) ethylene (d) 1, 4 butadiene
11. Nylon – 66 is made by using
    (a) succinic acid
    (b) benzaldehyde
    (c) adipic acid
    (d) benzyl chloride
12. PVC is polymer of
    (a) $CH_2 = CH_2$ (b) $CH_2 = CH{-}Cl$
    (c) $CH_2 = CH{-}CH_2Cl$ (d) $CH_3{-}CH = CH{-}Cl$
13. Which one of the following is used to make "non-stick" cook-wares ?
    (a) Polystyrene
    (b) Poly (ethylene terephthalate)
    (c) Polytetrafluoroethylene
    (d) Poly (vinyl chloride)
14. $\sim\sim[NH(CH_2)NHCO(CH_2)_4CO]_n\sim\sim$ is a
    (a) addition polymer
    (b) thermosetting polymer
    (c) homopolymer
    (d) copolymer
15. Which one of the following polymers is prepared by condensation polymerisation?
    (a) Teflon (b) Natural rubber
    (c) Styrene (d) Nylon-66
16. Which of the following is a polyamide?
    (a) Bakelite (b) Terylene
    (c) Nylon-66 (d) Teflon
17. Which of the following is fully fluorinated polymer ?
    (a) PVC (b) Thiokol
    (c) Teflon (d) Neoprene

18. $CF_2 = CF_2$ is a unit of
(a) Teflon (b) Buna – S
(c) Bakelite (d) Polyethene

19. Which of the following structures represents neoprene polymer?
(a) $-(CH_2-\underset{\underset{Cl}{|}}{C}=CH-CH_2)-_n$
(b) $-(CH_2-\overset{\overset{CN}{|}}{CH})-_n$
(c) $-(CH_2-\overset{\overset{Cl}{|}}{CH})-_n$
(d) $-(\underset{\underset{C_6H_5}{|}}{CH}-CH_2)-_n$

20. Polythene is a resin obtained by polymerisation of
(a) Butadiene (b) Ethylene
(c) Isoprene (d) Propylene

21. Teflon is a polymer of the monomer
(a) Monofluoroethene
(b) Difluoroethene
(c) Trifluoroethene
(d) Tetrafluoroethene

22. Buna-S is a polymer of
(a) Butadiene and styrene
(b) Butadiene
(c) Styrene
(d) Butadiene and chloroprene

23. Polymer obtained by condensation polymerization is
(a) Polythene
(b) Teflon
(c) Phenol-formaldehyde
(d) Nitrile rubber

24. A condensation polymer among the following is
(a) Dacron (b) PVC
(c) Polystyrene (d) Teflon

25. On the basis of mode of formation, polymers can be classified?
(a) as addition polymers only
(b) as condensation polymers only
(c) as copolymers
(d) Both as addition and condensation polymers

26. The process involving heating of rubber with sulphur is called
(a) Galvanisation (b) Vulcanization
(c) Bessemerisaion (d) Sulphonation

27. Terylene is a condensation polymer of ethylene glycol and
(a) Benzoic acid (b) Phthalic acid
(c) Salicyclic acid (d) Terephthalic acid

28. Teflon, styron and neoprene are all
(a) Copolymers
(b) Condensation polymers
(c) Homopolymers
(d) Monomers

29. An example of biopolymer is
(a) Teflon (b) Neoprene
(c) Nylon-66 (d) DNA

30. Which of the following is not correctly matched?
(a) Terylene $\left[-OCH_2-CH_2-\overset{\overset{O}{||}}{C}-C_6H_4-\overset{\overset{O}{||}}{C}-\right]_n$
(b) Neoprene $\left[-CH_2-\underset{\underset{Cl}{|}}{C}=CH-CH_2-\right]_n$
(c) Nylon-66 $\left[NH-(CH_2)_6-NH-\overset{\overset{O}{||}}{C}-(CH_2)_4-\overset{\overset{O}{||}}{C}-O\right]_n$
(d) PMMA $\left[-CH_2-\underset{\underset{COOCH_3}{|}}{\overset{\overset{CH_3}{|}}{C}}-\right]_n$

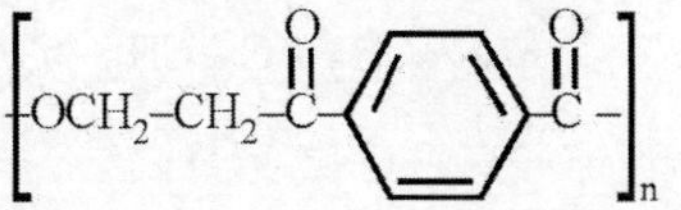

31. Acrolin is a hard, horny and a high melting material. Which of the following represents its structure ?

(a) 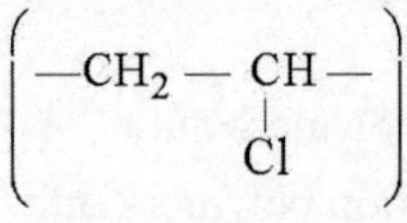 $\left(-CH_2-\underset{Cl}{\underset{|}{CH}}-\right)_n$

(b) 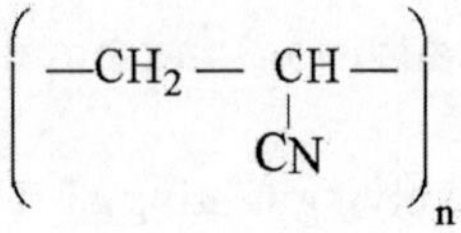 $\left(-CH_2-\underset{CN}{\underset{|}{CH}}-\right)_n$

(c) 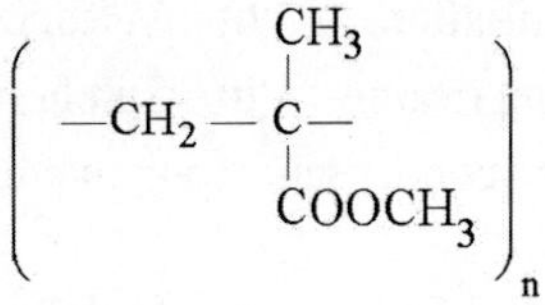 $\left(-CH_2-\overset{CH_3}{\overset{|}{\underset{COOCH_3}{\underset{|}{C}}}}-\right)_n$

(d) 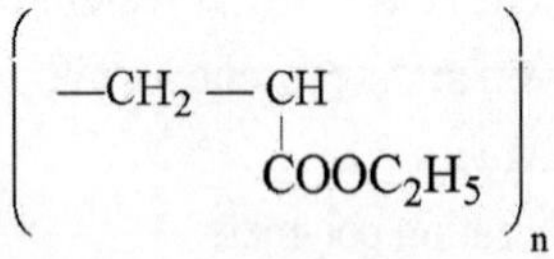 $\left(-CH_2-\underset{COOC_2H_5}{\underset{|}{CH}}\right)_n$

32. Which one of the following is a chain growth polymer?
(a) Starch (b) Nucleic acid
(c) Polystyrene (d) Protein

33. The monomer of the polymer;

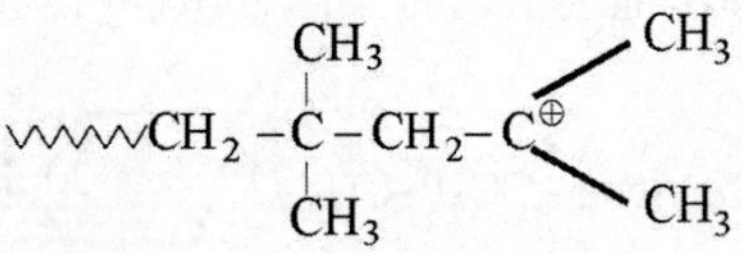

$\sim\sim CH_2-\overset{CH_3}{\overset{|}{\underset{CH_3}{\underset{|}{C}}}}-CH_2-\overset{\oplus}{C}(CH_3)_2$ is

(a) 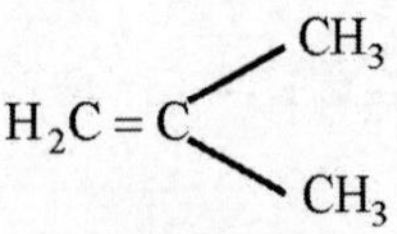 $H_2C=C(CH_3)_2$
(b) $CH_3CH{=}CHCH_3$
(c) $CH_3CH=CH_2$
(d) $(CH_3)_2C=C(CH_3)_2$

34. Polymer formation from monomers starts by
(a) condensation reaction between monomers
(b) coordinate reaction between monomers
(c) conversion of monomer to monomer ions by protons
(d) hydrolysis of monomers.

35. Buna-N synthetic rubber is a copolymer of :
(a) $H_2C=CH-CH=CH_2$ and $H_5C_6-CH=CH_2$
(b) $H_2C=CH-CN$ and $H_2C=CH-CHCH_2$
(c) $H_2C=CH-CN$ and $H_2C=CH-\underset{CH_3}{\underset{|}{C}}=CH_2$
(d) 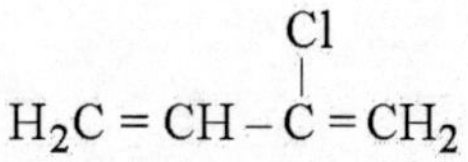 $H_2C=CH-\overset{Cl}{\overset{|}{C}}=CH_2$ and $H_2C=CH-CH=CH_2$

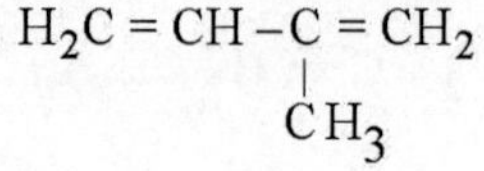

36. Monomer of $\left[-\overset{CH_3}{\overset{|}{\underset{CH_3}{\underset{|}{C}}}}-CH_2-\right]_n$ is
(a) 2-methylpropene (b) Styrene
(c) Propylene (d) Ethene

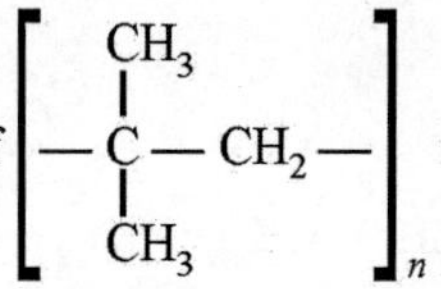

37. Vitamin B_{12} contains
(a) Ca(II) (b) Fe(II)
(c) Co(III) (d) Zn(II)

38. Which one of the following statement is *not true*?
(a) In vulcanization the formation of sulphur bridges between different chains make rubber harder and stronger.
(b) Natural rubber has the *trans* -configuration at every double bond.
(c) Buna-S is a copolymer of butadiene and styrene.
(d) Natural rubber is a 1, 4 - polymer of isoprene.

39. Polymer used in bullet proof glass is
(a) Lexan (b) PMMA
(c) Nomex (d) Kevlar

40. Orlon is a polymer of
(a) Styrene
(b) Tetrafluoroethylene
(c) Vinyl chloride
(d) Acrylonitrile

41. Caprolactam polymerises to give
(a) Terylene (b) Teflon
(c) Glyptal (d) Nylon-6

42. Interparticle forces present in nylon-66 are
(a) van der Waals'
(b) Hydrogen bonding
(c) Dipole-dipole interactions
(d) None of these

43. Ebonite is
(a) Natural rubber
(b) Synthetic rubber
(c) Highly vulcanized rubber
(d) Polypropene

44. Formation of polyethylene from calcium carbide takes place as follows

$CaC_2 + 2H_2O \longrightarrow Ca(OH)_2 + C_2H_2$

$C_2H_2 + H_2 \longrightarrow C_2H_4$

$nC_2H_4 \longrightarrow (-CH_2-CH_2-)_n$

The amount of polyethylene obtained from 64.1 kg of CaC_2 is
(a) 7 kg (b) 14 kg
(c) 21 kg (d) 28 kg

45. The turbidity of a polymer solution measures
(a) the light scattered by solution
(b) the light absorbed by a solution
(c) the light transmitted by a solution
(d) none of the above

46. The catalyst used for the polymerization of olefins is
(a) Ziegler-Natta catalyst
(b) Wilkinson's cataylst
(c) Pd-catalyst
(d) Zeise's salt complex

47. Which of the following is used in paints?
(a) Terylene (b) Nylon
(c) Glyptal (d) Chloroprene

48. Which one of the following is not an example of chain growth polymer?
(a) Neoprene (b) Buna-S
(c) PMMA (d) Glyptal

49. Which of the following is a polyamide molecule?
(a) Terylene (b) Rayon
(c) Nylon – 6 (d) Polystyrene

50. Polymer used in bullet proof glass is
(a) PMMA (b) Lexan
(c) Nomex (d) Kevlar

51. Dimethyl phthalate and ethylene glycol react to form
(a) Nylon-6 (b) Nylon-66
(c) Dacron (d) Neoprene

52. Glyptal polymer is obtained from glycol by reacting with
(a) Malonic acid (b) Phthalic acid
(c) Maleic acid (d) Acetic acid

53. Soft drinks and baby feeding bottles are generally made up of
(a) Polystyrene (b) Polyurethane
(c) Polyurea (d) Polyamide

54. Urethane is
(a) $H_2N-C\equiv N$
(b) $H_2N-\underset{\underset{O}{\|}}{C}-OH$
(c) $HO-C\equiv N$
(d) $H_2N-\underset{\underset{O}{\|}}{C}-OC_2H_5$

55. Match List-I (Monomer) with List II (Polymer) and select the correct answer using the codes given below the lists:

	List I		List II
I.	Hexamethylenediamine	A.	Bakelite
II.	Phenol	B.	Dacron
III.	Phthalic acid	C.	Glyptal
IV.	Terephthalic acid	D.	Melamine
		E.	Nylon

Codes:
(a) I-E, II-A, III-B, IV-C (b) I-E, II-A, III-C, IV-B
(c) I-D, II-C, III-A, IV-B (d) I-D, II-C, III-A, IV-B

56. Synthetic human hair wigs are made from a copolymer of vinyl chloride and acrylonitrile and is called
(a) PVC (b) Polyacrylonitrile
(c) Cellulose (d) Dynel

57. $n(CF_2 = CF_2) \xrightarrow[\Delta]{(NH_4)_2SO_4} X$

Here, X is :
(a) RMMA (b) PVC
(c) PAN (d) None of these

58. Polymer of chloroethylene is:
(a) teflon (b) nylon
(c) terylene (d) PVC

59. PVC is :
(a) thermoplastic polymer
(b) compound polymer
(c) thermosetting polymer
(d) simple polymer

60. Rayon is :
(a) synthetic plastic (b) natural rubber
(c) natural silk (d) artificial silk

Chemistry in Everyday life

CONCEPTS

CHEMOTHERAPY

Branch of chemistry dealing in treatment of diseases by suitable chemicals.

TYPES OF DRUGS

Analgesics

Drugs used for relieving pain.

(*a*) Aspirin (2-acetoxy benzoic acid) is the most common analgesic. It has antipyretic (temperature lowering) properties also.

(*b*) These days, aspirin is used for the prevention of heart attack as it prevents clotting of blood.

(*c*) In case of severe pain, narcotics like morphine, codeine and heroin can also be used.

Antipyretics

It bring down body temperature during high fever. e.g.-

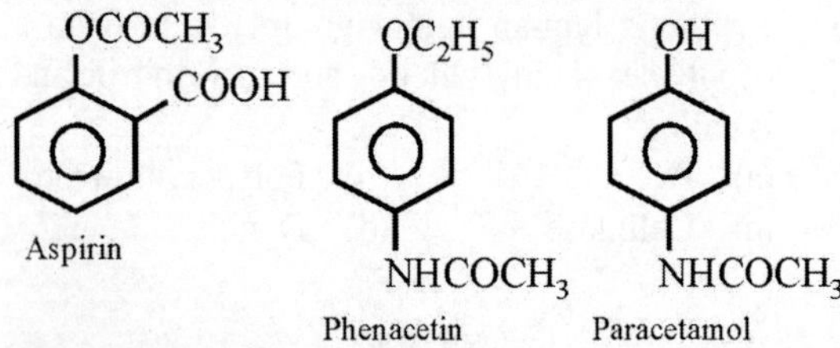

Antimalarials

These are drugs which cure malaria e.g.- **quinine, chloroquine**.

Antiseptics and Disinfectants

Antiseptics	Disinfectants
• prevents growth of microorganisms or may even kill them.	• kills and prevents growth of microorganisms.
•**Not harmful for humans**	• **Harmful for humans.**
Examples :	*Example :*
(i) **Dettol (chlorohexenol + terpeneol)**	**1 % of solution of phenol.**
(ii) **Bithional**	
(iii) **Salol**	

Tranquilizers

Used to treat mental diseases and stress. They act on central nervous system and reduce anxiety. e.g.- **Barbituric acid, luminal, seconal.**

Antibiotics

Chemicals which are produced by some specific micro-organism and are used to kill other micro-organism. e.g.- **Penicillin**, **Streptomycin** (for tuberculosis), **Chloroamphenicol** (for typhoid). **Broad spectrum antibiotics** can kill all range of micro-organisms. e.g.- Tetracyclin.

Sulphadrugs

These are derivative of sulphanilamide and have antibacterial powers.

Antihistamines

Drugs used to treat allergy such as skin rashes. They are called so because allergic reactions are caused due to the liberation of histamine in the body.

(*a*) These drugs are also used for treating conjuctivitis (inflammation of conjuctiva) and rhinitis (inflammation of nasal mucosa).

(*b*) The commonly used antihistamines drugs are diphenyl hydraamine, chloropheniramine, promethazine and cetrizine.

Antacids

Antacids are the drugs which neutralize excess acid in the gastric Juices and give relief from acid indigestion. They remove the excess acid and raise the pH to appropriate level in stomach. There are mainly weak bases.

Examples– $Mg(OH)_2$, $KHCO_3$

Antifertility Drug

The drugs which are used to control the pregnancy

are known as Antifertility drugs or oral Contraceptives. These are essentially a mixture of estrogen and progesterone derivatives.

Examples – Ormeloxifene, mix pristone

ARTIFICIAL SWEETENERS AND FOOD PRESERVATIVES

Artificial Sweeteners

The chemical compounds that are added to foods to make them sweet.

(*a*) Saccharin is the first artificial sweetener used as it is water soluble sodium or calcium salt.

(*b*) Nowadays aspartame is used as artificial sweetener, but it is unstable at cooking temperatures.

(*c*) Alitame, another sweetener is more stable than aspartame at cooking temperature.

Food Preservatives

The chemical substances that are added to the food to prevent their decaying and to retain their nutritive value for long periods.

- Sodium benzoate is the most commonly used preservative.

SOAPS AND DETERGENTS

Soaps

Sodium or potassium salts of long chain fatty acids, *e.g.*, stearic oleic and palmitic acids. Soaps containing sodium salts are formed by heating fat (*i.e.*, glyceryl ester of fatty acid) with aqueous sodium hydroxide solution. This process is called **saponification**.

- Soaps do not work in hard water. Hard water contains Ca^{+2} and Mg^{+2} ions. These ions form insoluble calcium and magnesium soaps respectively when sodium or potassium soaps are dissolved in hard water. These insoluble soaps separate as scum in water and are useless as cleansing agent.

Synthetic Detergents

Sodium salts of alkylbenzene sulphonic acids. They are better cleansing agents than soap. These are of three types:

(a) **Anion detergents** are those detergent which contain large part of anion.

   (i) For preparing anionic detergent, long chain alcohols are first treated with concentrated H_2S_4, giving alkyl hydrogen sulphates. These are neutralized with alkali to give anionic detergent.

   (ii) They are also effective in slightly acidic solutions. In acidic solution, they form alkyl hydrogen sulphate which is soluble in water whereas soaps form insoluble fatty acids.

(b) **Cationic detergents** are mostly acetates or chlorides of quaternary amines. These detergents have germicidal properties and are extensively used as germicides.

(c) **Non-ionic detergents** are esters of high molecular mass.

Exercise

1. The use of chemicals for treatment of diseases is called as
   (a) Homoeotherapy
   (b) Isothermotherapy
   (c) Angiotherapy
   (d) Chemotherapy

2. The function of enzymes in the living system is to
   (a) transport oxygen
   (b) provide energy
   (c) provide immunity
   (d) catalyse biochemical reactions

3. Which one of the following is employed as Antihistamine ?
   (a) Chloramphenicol
   (b) Diphenyl hydramine
   (c) Norothindrone
   (d) Omeprazole

4. Which one of the following is employed as a tranquilizer?
   (a) Naproxen
   (b) Tetracycline
   (c) Chlorpheninamine
   (d) Equanil

5. Which of the following drugs is an analgesic
(a) Sulphaguanidine (b) Paludrin
(c) Analgin (d) Iodex
6. An antibiotic with a broad spectrum
(a) Kills the antibodies
(b) Acts on a specific antigen
(c) Acts on different antigens
(d) Acts on both the antigens and antibodies
7. Alkaline hydrolysis of esters is known as :
(a) Esterification (b) Saponification
(c) dehydration (d) alkalination
8. Commercial detergent contains mainly
(a) RCOONa (b) RONa
(c) RSNa (d) RSO_3Na
9. Which of the following is used for inducing sleep ?
(a) Paracetamol
(b) Chloroquine
(c) Bithional
(d) Barbituric acid derivatives
10. Sodium alkyl benzene sulphonate is used as
(a) Soap (b) Fertilizers
(c) Pesticides (d) Detergents
11. Substances which bring body temperature down are known as
(a) Antipyretics (b) Analgesics
(c) Antobiotics (d) None
12. Pheromones are chemicals
(a) formed by fermentation process of fungi
(b) secreted by endocrine glands of man
(c) secreted outside the body of insects
(d) plant growth hormones
13. Which of these is a hypnotic ?
(a) Metaldehyde (b) Acetaldehyde
(c) Paraldehyde (d) None of these
14. Tranquillizers are substances used for the treatment of
(a) cancer (b) AIDS
(c) mental diseases (d) physical disorders
15. Which one of the following is employed as a tranquilizer drug?
(a) Promethazine (b) Valium
(c) Naproxen (d) Mifepristone
16. Penicillin was first discovered by
(a) A. Fleming (b) Tence and Salke
(c) S. A Waksna (d) Lewis Pasteur
17. Acetoxy benzoic acid is
(a) Antiseptic (b) Aspirin
(c) Antibiotic (d) Mordant dye
18. Aspirin is
(a) Antibiotic (b) Antipyretic
(c) Sedative (d) Psychedelic
19. Which of the following drugs is a tranquilizer and sedative
(a) Sulphadiazine (b) Papaverine
(c) Equanil (d) Mescaline
20. An antipyretic is
(a) Quinine (b) Paracetamol
(c) Luminal (d) Piperazine
21. The drug used as an antidepressant is
(a) Luminol (b) Tofranil
(c) Mescaline (d) Sulphadiazine
22. The correct structure of the drug paracetamol is
(a) benzene ring with OH (top) and $CONH_2$ (bottom) (b) benzene ring with OH (top) and $NHCOCH_3$ (bottom)
(c) benzene ring with Cl (top) and $CONH_2$ (bottom) (d) benzene ring with Cl (top) and $COCH_3$ (bottom)
23. Drug which helps to reduce anxiety and brings about calmness is
(a) tranquillizer (b) diuretic
(c) analgesic (d) antihistamine
24. Phenol is used as
(a) an antisptic (b) a disinfectant
(c) both (a) and (b) (d) a styptic
25. The drug given during hypertension is
(a) streptomycin (b) chloroxylenol
(c) equanil (d) aspirin
26. Commonly used antiseptic 'Dettol' is a mixture of
(a) o-chlorophenozylenol + terpeneol
(b) o-cresol + terpeneol
(c) phenol + terpeneol
(d) chlorozylenol + terpeneol
27. Which one of the following can possibly be used as analgesic without causing addiction and mood modification ?
(a) Diazepam
(b) Morphine
(c) N-Acetyl-para-aminophenol
(d) Tetrahydrocannabinol

28. Cobalt (60) isotope is used in the treatment of :
(a) Heart diseases (b) Skin diseases
(c) Diabetes (d) Cancer

29. Saccharine contains
(a) C and H (b) C, H, N and O
(c) C, H, N, O & S (d) C, O and H

30. Which of the following is not true ?
(a) Some disinfectants can be used as antispetics
(b) sulphadiazine is a synthetic antibacterial
(c) aspirin is analgesic as well as antipyretic
(d) polystyrene is used to make non-stick cookware

31. Structurally biodegrdable detergent should contain
(a) normal alkyl chain
(b) branched alkyl chain
(c) phenyl side chain
(d) cyclohexyl side chain

32. An ester used as medicine is
(a) ethyl acetate (b) methyl acetate
(c) methyl salicylate (d) ethyl benzoate

33. Streptomycin, well known antibiotic, is a derivative of
(a) peptides (b) carbohydrates
(c) purines (d) terpenes

34. A medicine which promotes the secretion of urine is called
(a) Uretic (b) Monouretic
(c) Diuretic (d) Triuretic

35. Veronal, a barbiturate drug is used as
(a) Anaesthetic (b) Sedative
(c) Antiseptic (d) None of these

36. Antiseptic chloroxylenol is
(a) 4-chloro-3, 5-dimethylphenol
(b) 3-chloro-4, 5-dimethylphenol
(c) 4-chloro-2, 5-dimethylphenol
(d) 5-chloro-3, 4-dimethylphenol

37. Which of the following is a hypnotic drug
(a) Luminal (b) Salol
(c) Catechol (d) Chemisol

38. Arsenic drugs are mainly used in the treatment of
(a) Jaundice (b) Typhoid
(c) Syphilis (d) Cholera

39. The following compound is used as

$$C_6H_4(COOH)(O-\overset{O}{\overset{\|}{C}}-CH_3)$$

(a) An anti-inflammatory compound
(b) Analgesic
(c) Hypnotic
(d) Antiseptic

40. Amoxillin is semi-synthetic modification of
(a) Penicillin (b) Streptomycin
(c) Tetracycline (d) Chloroampheniol

41. Which of the following is an antidiabatic drug
(a) Insulin (b) Penicillin
(c) Chloroquine (d) Aspirin

42. Detergents are prepared by the action of H_2SO_4 followed by neutralization by starting with
(a) chloesterol (b) lauryl alcohol
(b) cyclohexanol (d) p- nitrophenol

43. Streptomycin, well known antibiotic, is a derivative of
(a) peptides (b) carbohydrates
(c) purines (d) terpenes

44. Match List I with List II and selcet the correct answer using the codes gives below the lists :

List I	List II
I. Iodoform	A. Anaesthetic
II. Methyl salicylate	B. Antiseptic
III. Diethyl ether	C. Insecticide
IV. Hexachlorocyclohexane	D. Detergent
	E. Pain balm

Codes :
(a) I-B, II-E, III-C, IV-D
(b) I-D, II-B, III-A, IV-C
(c) I-B, II-E, III-A, IV-C
(d) I-C, II-A, III-D, IV-B

45. Washing soap can be prepared by saponification with alkali of which of the following oil
(a) Rose oil (b) Paraffin oil
(c) Groundnut oil (d) Kerosene oil

46. Palmitic acid and stearic acid are used as :
(a) medicine (b) soap
(c) antiseptic cream (d) pickle

47. Paracetamol is:
(a) an antipyretic (b) an antiseptic
(c) an antibiotic (d) a disinfectant

48. Which of the following compounds is used as broad spectrum antibiotics?
(a) Ampicillin (b) Penicilin G
(c) Penicillin K (d) Tetracyline
49. Bithional is an example of
(a) disinfectant (b) antiseptic
(c) antibiotic (d) analgesic
50. Chloramine-T is used as an :
(a) analgesic (b) antiseptic
(c) antipyretic (d) anti-inflamatory
51. Soaps can be classified as :
(a) esters (b) salts of fatty acids
(c) alcohols (d) phenols
52. The chloramphenicol is an example of :
(a) antiseptic (b) antibiotic
(c) analgesics (d) antipyretics
53. Penicillin is an :
(a) antibiotic (b) anaesthetic
(c) antiseptic (d) antipyretic
54. Which is used as medicine?
(a) PVC (b) Terylene
(c) Glyptal (d) Urotropine
55. Which of the following is a broad spectrum drug?
(a) Plasmoquine
(b) Chloroquine
(c) Chloramphenicol
(d) D.D.T.

1. SOME BASIC CONCEPTS OF CHEMISTRY

ANSWER KEY											
1	(c)	11	(c)	21	(b)	31	(b)	41	(c)	51	(b)
2	(a)	12	(b)	22	(d)	32	(a)	42	(a)	52	(c)
3	(c)	13	(c)	23	(b)	33	(b)	43	(a)	53	(a)
4	(b)	14	(a)	24	(c)	34	(c)	44	(d)	54	(a)
5	(c)	15	(c)	25	(b)	35	(b)	45	(d)	55	(c)
6	(c)	16	(d)	26	(c)	36	(a)	46	(c)	56	(b)
7	(d)	17	(c)	27	(b)	37	(a)	47	(c)	57	(c)
8	(c)	18	(b)	28	(a)	38	(c)	48	(c)	58	(b)
9	(c)	19	(b)	29	(a)	39	(b)	49	(c)	59	(c)
10	(a)	20	(c)	30	(a)	40	(a)	50	(d)	60	(a)

Hints & Solutions

1. (c) Each has three significant figures. When zero is used to locate the decimal point, it is not considered as significant figure.

3. (c) If zero is used to locate the decimal point it is considered as a significant figure. In 50.000 all zero are significant.

5. (c) On calculation we find

$$\frac{(29.2-20.2)(1.79\times10^5)}{1.37}=1.17\times10^6$$

As the least precise number contains 3 significant figures therefore answers should also contains 3 significant figures.

7. (d) Candle is a heterogeneous mixture of wax and threads. Copper is an element while bottled water and table salt are compounds.

9. (c) Exa = 10^{18}

11. (c) Law of multiple proportions.

12. (b) Law of multiple proportion. As the ratio of oxygen which combine with fix weights of 1 g of nitrogen bears a simple whole number ratio
0.57 : 1 : 12 : 1.7031 : 2 : 3

14. (a) Constant proportions according to which a pure chemical compound always contains same elements combined together in the same definite proportion of weight.

16. (d) $SnCl_2$ $SnCl_4$
119 : 2 × 35.5 119 : 4 × 35.5
Chlorine ratio in both compounds is
= 2 × 35.5 : 4 × 35.5 = 1 : 2

18. (b) 1 amu = $\frac{1}{12}$ of the mass of C-12.

19. (b) The modern atomic weight scale is based on C^{12} (see text).

20. (c) Mol. of $O_2=\frac{6}{32}$ & mol. of $SO_2=\frac{6}{64}$. The number of molecules of SO_2 will be less than that of O_2.

21. (b) $\underset{1}{K_2Cr_2O_7}+\underset{1}{H_2SO_4}+\underset{3}{3SO_2}\longrightarrow$

$$K_2SO_4+Cr_2(SO_4)_3+H_2O$$

22. (d) AT NTP 22400 cc of N_2O = 6.02×10^{23} molecules

$$\therefore 1\text{ cc } N_2O=\frac{6.02\times10^{23}}{22400}\text{ molecules}$$

$$=\frac{3\times6.02\times10^{23}}{22400}\text{ atoms}$$

$$=\frac{1.8}{224}\times10^{22}\text{ atoms}$$

No. of electrons in a molecule of N_2O
= 7 + 7 + 8 = 22
Hence no. of electrons

$$=\frac{6.02\times10^{23}}{22400}\times22\text{ electrons}=\frac{1.32\times10^{23}}{22400}$$

23. (b) Molecular weight of $C_{60}H_{122}$
= (12 × 60) + 122 = 842.
Therefore weight of one molecule

$$= \frac{\text{Molecular weight of } C_{60}H_{122}}{\text{Avagadro's number}}$$

$$= \frac{842}{6.023 \times 10^{23}} = 1.4 \times 10^{-21} g$$

24. (c) Volume increases with rise in temperature. Thus, some water molecule may be evaporated at high temp.

25. (b) Molal solution contains 1 mole of solute in 1000 g solvent.

26. (c) Eq. Wt of crystalline oxalic acid$(COOH)_2 . 2H_2O$

$$= \frac{\text{M.Wt.}}{2} = \frac{126}{2} = 63$$

27. (b) 1 atom of Cu + 1 atom of sulphur + 9 atoms of oxygen + 10 atoms of hydrogen. Total number of atoms in compound is 21.

29. (a) $C_4H_{10} + 13/2\ O_2 \longrightarrow 4CO_2 + 5H_2O$

30. (a) Empirical formula of compound = CH_2
Molecular mass of the compound = 42
$\therefore n = 42/14 = 3$
$\therefore$ Hence molecular formula = C_3H_6

31. (b) $N = \frac{N_1V_1 + N_2V_2 + N_3V_3}{\text{Total volume}}$

$$= \frac{50 \times 10 + 25 \times 12 + 40 \times 5}{1000} = 1N$$

32. (a)

	I_2	+ $2Cl_2$	$\longrightarrow$ ICl	+ ICl_3
Initially	$\frac{25.4}{254}$	$\frac{14.2}{71}$	0	0
After reaction	0.1	0.2	0	0
	0	0	0.1	0.1

34. (c) $n = \frac{180}{\text{Molecular mass of } CH_2O} = \frac{180}{30}$

$n = 6$

36. (a) 1 mol CCl_4 vapour = 12 + 4 × 35.5
= 154 g ≡ 22.4 L at STP

$\therefore$ Density = $\frac{154}{22.4} gL^{-1} = 6.875 gL^{-1}$

37. (a) $Mg^{++} + Na_2CO_3 \longrightarrow MgCO_3 + 2Na^+$
1 g eq. 1g eq.
1 g eq. of Mg^{2+} = 12 g of Mg^{2+} = 12000 mg
= 1000 milli eq. of Na_2CO_3
$\therefore$ 12 mg Mg^{++} = 1 milli eq. Na_2CO_3

38. (c) $C_2H_4 + 3\ O_2 \longrightarrow 2CO_2 + 2H_2O$
28 kg 96 kg
$\because$ 28 kg of C_2H_4 undergo complete combustion by = 96 kg of O_2
$\therefore$ 2.8 kg of C_2H_4 undergo complete combustion by = 9.6 g of O_2.

39. (b) 6.02×10^{23} molecules of CO = 1mole of CO 6.02×10^{24} CO molecules = 10 moles CO
= 10 g atoms of O = 5 g molecules of O_2

40. (a) 4.4 g $CO_2 = \frac{4.4}{44} = 0.1$ mol CO_2
(mol. wt. of CO_2 = 44)
= 6×10^{22} molecules
= $2 \times 6 \times 10^{22}$ atoms of O.

41. (c) According to Avogadro's law "equal volumes of all gases contain equal numbers of molecules under similar conditions of temperature and pressure". Thus if 1 L of one gas contains N molecules, 2 L of any gas under the same conditions will contain 2N molecules.

42. (a) 5 MH_2SO_4 = 10 N H_2SO_4,
($\because$ Basicity of H_2SO_4 = 2)
$N_1V_1 = N_2V_2$,
$10 \times 1 = N_2 \times 10$ or $N_2 = 1$ N

43. (a) No of moles of nitride ion

$$= \frac{4.2}{14} = 0.3 \text{ mol} = 0.3 \times N_A \text{ nitride ions.}$$

Valence electrons = 8 × 0.3 N_A = 2.4 N_A (5 + 3 due to charge). One N^{3-} ion contains 8 valence electrons.

44. (d) Molecular weight of $ZnSO_4.7H_2O$
= 65 + 32 + (4 × 16) + 7(2 × 1 + 16) = 287.
$\therefore$ percentage mass of zinc (Zn)

$$= \frac{65}{287} \times 100 = 22.65\%$$

45. (d) $\underset{2(78)}{2C_6H_6} + \underset{15(32)}{15O_2(g)} \rightarrow 12CO_2(g) + 6H_2O(g)$

$\because$ 156 gm of benzene required oxygen
= 15 × 22.4 litre
$\therefore$ 1 gm of benzene required oxygen

$$= \frac{15 \times 22.4}{156} \text{litre}$$

$\therefore$ 39 gm of Benzene required oxygen

$$= \frac{15 \times 22.4 \times 39}{156} = 84.0 \text{ litre}$$

46. (c) As the sum of the percentage of C, H & N is 100. Thus it does not contains O atom.

Table for empirical formula

Element	%	At. wt.	Rel. Number	Ratio
C	40.00	12	$\frac{40}{12}=3.66$	$\frac{3.66}{3.33}=1.09\sim 1$
H	13.33	1	$\frac{13.33}{1}=13.33$	$\frac{13.33}{3.33}=4$
N	46.67	14	$\frac{46.67}{14}=3.33$	$\frac{3.33}{3.33}=1$

Hence empirical formula = CH_4N

47. (c) Given : Percentage of the iron = 0.334%; Molecular weight of the haemoglobin = 67200 and atomic weight of the iron = 56. We know that the number of iron atoms

$$=\frac{\text{Molecular wt. of haemoglobin}\times\text{\% of iron}}{100\times\text{Atomic weight of iron}}$$

$$=\frac{67200\times 0.334}{100\times 56}=4$$

48. (c)

Element	%	At.wt	Relative number	Ratio
C	40	12	$\frac{40}{12}=3.33$	$\frac{3.33}{3.33}=1$
H	6.66	1	$\frac{6.66}{1}=6.66$	$\frac{6.66}{3.33}=2$
O	53.34	16	$\frac{53.34}{16}=3.33$	$\frac{3.33}{3.33}=1$

(% of O in organic compound
Table for empirical formula :
$=100-(40+6.66)=53.34\%$)

Empirical formula of organic compound = CH_2O.

49. (c) $BaCO_3 \rightarrow BaO + CO_2$
197 gm

$\because$ 197 gm of $BaCO_3$ released carbon dioxide = 22.4 litre at STP

$\therefore$ 1 gm of $BaCO_3$ released carbon dioxide

$=\frac{22.4}{197}$ litre

$\therefore$ 9.85 gm of $BaCO_3$ released carbon dioxide = $\frac{22.4}{197}\times 9.85=1.12$ litre

50. (d) Suppose the mol. wt. of enzyme = x
Given 100 g of enzyme wt of Se = 0.5 gm

$\therefore$ In x g of enzyme wt. of Se = $\frac{0.5}{100}\times x$

Hence $78.4=\frac{0.5\times x}{100}$

$\therefore x = 15680 = 1.568\times 10^4$

51. (b) 2g of H_2 means one mole of H_2, hence contains 6.023×10^{23} molecules. Others have less than one mole, so have less no. of molecules.

52. (c) $N_2 + 3H_2 \rightarrow 2NH_3$
1 vol. 3 vol. 2 vol.
10 litre 30 litre 20 litre

It is given that only 50% of the expected product is formed hence only 10 litre of NH_3 is formed
N_2 used = 5 litres, left = 30 – 5 = 25 litres
H_2 used = 15 litres, left = 30 – 15 = 15 litres

53. (a) No. of molecules in different cases

(a) $\because$ 22.4 litre at STP contains
$=6.023\times 10^{23}$ molecule of H_2

$\therefore$ 15 litre at STP contains $=\frac{15}{22.4}\times 6.023\times 10^{23}$

(b) $\because$ 22.4 litre at STP contains
$=6.023\times 10^{23}$ molecule of N_2

$\because$ 5 litre at STP contians $=\frac{5}{22.4}\times 6.023\times 10^{23}$

(c) $\because$ 2 gm of $H_2 = 6.023\times 10^{23}$ molecules of H_2

$\because$ 0.5 gm of $H_2 = \frac{0.5}{2}\times 6.023\times 10^{23}$

(d) Similarly 10 g of O_2 gas

$=\frac{10}{32}\times 6.023\times 10^{23}$ molecules

Thus (a) will have maximum number of molecules

54. (a) $\overset{+7}{KMnO_4}\xrightarrow{OH^-}K_2\overset{+6}{MnO_4}$

Change in oxidation number of Mn in basic medium is 1. Hence mole of KI is equal to mole of $KMnO_4$.

55. (c) Molarity of H_2SO_4 solution

$=\frac{98\times 1000}{98\times 100}\times 1.84=18.4$

Suppose V ml of this H_2SO_4 is used to prepare 1 lit. of 0.1M H_2SO_4

$\therefore V\times 18.02=1000\times 0.1$

or $V=\frac{1000\times 0.1}{18.02}=5.55$ ml.

56. (b) The balance chemical equation is :

$2MnO_4^- + 6H^+ + 5SO_3^{--} \longrightarrow$

$$2Mn^{++} + 5SO_4^{--} + 3H_2O$$

From the equation it is clear that

Moles of MnO_4^- require to oxidise 5 moles of SO_3^{--} are 2

Moles of MnO_4^- require to oxidise 1 mole of SO_3^{--} are 2/5.

57. (c) $\text{Density} = \frac{\text{Mass}}{\text{Volume}}$

$$1 \text{ gram cm}^{-3} = \frac{1 \text{ gram}}{\text{cm}^3}$$

$$\text{Volume} = \frac{\text{Mass}}{\text{Density}}$$

$$= \frac{1 \text{ gram}}{1 \text{ gram cm}^{-3}} = 1\text{cm}^3$$

$\therefore$ Volume occupied by 1 gram water = 1 cm^3

or Volume occupied by

$\frac{6.023 \times 10^{23}}{18}$ molecules of water = 1 cm^3

[$\therefore$ 1g water = $\frac{1}{18}$ moles of water]

Thus volume occupied by 1 molecules of water

$$= \frac{1 \times 18}{6.023 \times 10^{23}} \text{cm}^3 = 3.0 \times 10^{-23} \text{ cm}^3.$$

i.e. the correct answer is option (c).

58. (b) $5 \begin{matrix} COO^- \\ | \\ COO^- \end{matrix} + 2\,MnO_4^- + 16\,H^+ \rightarrow$

oxalate ion

$$2\,Mn^{2+} + 10CO_2 + 8H_2O$$

From above equation 2 moles MnO_4 required to oxide 5 moles of oxalate.

Thus number of moles of MnO_4 required to oxidise one mole of oxalate =2/5 =0.4

59. (c) Writing the equation of combustion of propane (C_3H_8), we get

$$\underset{\substack{1vol \\ 1L}}{C_3H_8} + \underset{\substack{5vol \\ 5L}}{5O_2} \rightarrow 3CO_2 + 4H_2O$$

From the above equation we find that we need 5 L of oxygen at NTP to completely burn 1 L of propane at N.T.P.

If we change the conditions for both the gases from N.T.P. to same conditions of temperature and pressure. The same results are obtained. i.e. 5 L is the correct answer.

60. (a)

Element	Percentage	Atomic weight	Atomic ratio	Simple ratio
C	38.71	12	$\frac{38.71}{12} = 3.23$	$\frac{3.23}{3.23} = 1$
H	9.67	1	$\frac{9.67}{1} = 9.67$	$\frac{9.67}{3.23} = 3$
O	100 − (38.71+9.67) = 51.62	16	$\frac{51.62}{16} = 3.23$	$\frac{3.23}{3.23} = 1$

Thus empirical formula is CH_3O.

2. STRUCTURE OF ATOM

ANSWER KEY

1	(d)	11	(c)	21	(b)	31	(a)	41	(a)	51	(a)
2	(b)	12	(a)	22	(d)	32	(d)	42	(a)	52	(c)
3	(b)	13	(d)	23	(a)	33	(b)	43	(b)	53	(d)
4	(c)	14	(c)	24	(b)	34	(d)	44	(c)	54	(d)
5	(c)	15	(b)	25	(d)	35	(a)	45	(d)	55	(c)
6	(c)	16	(d)	26	(c)	36	(c)	46	(b)	56	(c)
7	(a)	17	(c)	27	(d)	37	(a)	47	(b)	57	(c)
8	(a)	18	(a)	28	(c)	38	(a)	48	(a)	58	(a)
9	(b)	19	(d)	29	(c)	39	(b)	49	(b)	59	(a)
10	(b)	20	(c)	30	(b)	40	(c)	50	(d)	60	(d)

Hints & Solutions

2. (b) Cathode rays are never electromagnetic waves.
4. (c) Deuterium and an α-particle have identical values of *e*/*m*.
6. (c) As the neutron is a chargeless particle, hence, the beam of neutrons is not deflected by electrical or magnetic field.
8. (a) All positive ions are deposited at small part. (nucleus of atom).
10. (b) Number of p = number of e^- = 89 and neutrons 231 – 89 = 142.
12. (a) Average atomic mass

$$= \frac{R.A.(1) \times M.No + R.A.(2) \times M.No.}{R.A.(1) + R.A.(2)}$$

where R.A. = relative abundance
M.No = Mass number

$$= \frac{19 \times 10 + 81 \times 11}{100} = 10.81$$

13. (d) Average isotopic mass of

$$X = \frac{200 \times 90 + 199 \times 8 + 202 \times 2}{90 + 8 + 2}$$

$$= \frac{18000 + 1892 + 404}{100}$$

$$= \frac{19996}{100} = 199.96 \text{ amu}$$

14. (c) Species having same number of electrons are isoelectronic calculating the number of electrons in each species given here, we get.
CN^- (6 + 7 + 1 = 14); N_2 (7 + 7 = 14);
O_2^{2-} (8 + 8 + 2 = 18); C_2^{2-} (6 + 6 + 2 = 14);
O_2^- (8 + 8 + 1 = 17); NO^+ (7 + 8 – 1 = 14)
CO (6 + 8 = 14); NO (7 + 8 = 15)
From the above calculation we find that all the species listed in choice (c) have 14 electrons each so it is the correct answer.
16. (d) Uncertainty principle which was given by Hiesenberg and not Bohr's postulate.
17. (c) N(7) = $1s^2\ 2s^2\ 2p^3$

$$N^{2+(s)} = 1s^2, 2s^2 2p_x^1$$

Unpaired electrons = 1.
18. (a) Cu^+ = 29 – 1 = 28 e^-
thus the electronic confingration of Cu^+ is Cu^+

$$(28) = 1s^2\ 2s^2\ 2p^6\ \underbrace{3s^2\ 3p^6\ 3d^{10}}_{18e^-}$$

19. (d) This is as per the definition of Pauli's Exclusion principle.
20. (c) Energy of an electron in Bohr's orbit is given by the relationship. $E_n = -\frac{13.6}{n^2} eV.$
21. (b) Wave nature of electron was shown by Davisson and Germer. Davission and germer demonstrated the physical reality of the wave nature of electrons by showing that a beam of electrons could also diffracted by crystals just like light of x-rays.
22. (d) The orbitals which have same energy are called degenerate orbitals eg. p_x, p_y, p_z.
23. (a) We know that ions which have the same number of electrons are called isoelectronic. We also know that both CO and CN^- have 14 electrons, therefore these are isoelectronic.
24. (b) Magnetic quantum no. represents the orientation of atomic orbitals in an atom. For example p_x, p_y & p_z have orientation along X-axis, Y-axis & Z-axis
25. (d) The number of sub shell is $(2l + 1)$. The maximum number of electrons in the sub shell is $2(2l + 1)$ $= (4l + 2)$
28. (c) For 4*p* electron $n = 4$, $l = 1$, $m = -1, 0 + 1$ and $s = +\frac{1}{2}$ or $-\frac{1}{2}$
29. (c) K.E. of emitted electron
$= h\upsilon - h\upsilon_0$ (i.e. smaller than $h\upsilon$)
30. (b) Both He and Li^+ contain 2 electrons each therefore their spectrum will be similar.
31. (a) Since, K.E. $= \frac{1}{2}mv^2$ and $\lambda = \frac{h}{mv}$.

$$\therefore \text{ K.E.} = \frac{1}{2}m\ \frac{h^2}{m^2\lambda^2} = \frac{h^2}{2m\lambda^2}.$$

As λ is the same.

$$\therefore \text{ K.E.} \propto \frac{1}{m}$$

32. (d) For Balmer $n_1 = 2$ and $n_2 = 3$; v

$$= R\left(\frac{1}{2^2} - \frac{1}{3^2}\right) = \frac{5R}{36} cm^{-1}$$

34. (d) $E_1 = \frac{-13.6}{1}$; $E_3 - E_1 = -1.5 - (-13.6)$;
$= 12.1$ eV $= 12.1 \times 1.6 \times 10^{-12}$ ergs

$= 0.1936 \times 10^{-10}$ ergs

(1 ev = 1.6×10^{-12} ergs).

35. (a) $r_n = \frac{a_0 \times n^2}{Z} = \frac{0.53Å \times n^2}{Z}$;

For $Li^{++} = \frac{0.53Å \times (1)^2}{3} = 0.17Å$

36. (c) $\Delta x = \frac{h}{4\pi \times \Delta p} = \frac{6.63 \times 10^{-34}}{4 \times 3.14 \times 10^{-5}}$

$= 5.27 \times 10^{-30}$ m

37. (a) (n + l) rule the higher the value of (n + l), the higher is the energy. When (n + l) value is the same see value of n.

	I	II	III	IV
(n + l)	(4 + 1)	(4 + 0)	(3 + 2)	(3 + 1)
	5	4	5	4

∴ IV < II < III < I

38. (a) Angular nodes = l, spherical nodes (n – l – 1); Total (n – 1). Hence spherical nodes for 5f orbital. = (5 –3 –1) = 1

39. (b)

	5p	4f	6s	5d
(n + l)	5 + 1	4 + 3	6 + 0	5 + 2
	6	7	6	7

Hence the order is 5p < 6s < 4f < 5d

40. (c) Not more than two electrons can be present in same atomic orbital. This is Paulis exclusion principle.

41. (a) No. of radial nodes in 3p-orbital

$= (n - \ell - 1)$

[for p ortbital $\ell = 1$]

$= 3 - 1 - 1 = 1$

42. (a) Radius of nth orbit = $r_1 n^2$. (for H-atom)

43. (b) No. of orbitals in a sub-shell = $2\ell + 1$

⇒ No. of electrons = $2(2\ell + 1) = 4\ell + 2$

44. (c) n = 2, l = 1 means 2 p – orbital. Electrons that can be accommodated = 6 as p sub-shell has 3 orbital and each orbital contains 2 electrons.

45. (d) $\ell = 3$ means f-subshell. Maximum no. of electrons $= 4\ell + 2 = 4 \times 3 + 2 = 14$

46. (b) The sub-shell with lowest value of (n + ℓ) is filled up first. When two or more sub-shells have same (n + ℓ) value the subshell with lowest value of 'n' is filled up first therefore the correct order is

orbital	4s	3d	4p	5s	4d
n + ℓ	4 + 0	3 + 2	4 +1	5 + 0	4 + 2
value	= 4	5	5	5	6

47. (b) Outer electronic configuration of Cl

$= 3s^2 3p_x^2 3p_y^2 3p_z^1$

Outer electronic configuration of

$Cl^- = 3s^2 3p_x^2 3py^2 3pz^2$

hence Cl^- contain four lone pairs of electron

48. (a) The ionization energy of any hydrogen like species (having one electron only) is given by the equation

$$I.E = \frac{2\pi^2 Z^2 m e^4}{h^2} \text{ or } I.E \propto Z^2$$

Since the atomic number of H is 1 and that of He is 2, therefore, the I.E. of He^+ is four times (2^2) the I.E. of H i.e., $13.6 \times 4 = 54.4$ eV

49. (b) The sub-shell are 3d, 4d, 4p and 4s, 4d has highest energy as n + ℓ value is maximum for this.

50. (d) As m = (2ℓ + 1), hence m = – 1 means,

$-1 = 2\ell + 1.\ \ell = 1$

i.e least value of $\ell = 1$. So it cannot be present in s-orbital. Because for s orbital $\ell = 0$.

51. (a) Given mass of an electron(m)

$= 9.1 \times 10^{-28}$ g;

Velocity of electron $(v) = 3 \times 10^4$ cm/s;

Accuracy in velocity $= 0.001\% = \frac{0.001}{100}$;

Actual velocity of the electron

$(\Delta v) = 3 \times 10^4 \times \frac{0.001}{100} = 0.3$ cm/s.

Planck's constant (h) = 6.626×10^{-27} erg-sec.

∴ Uncertainty in the position of the electron

$$(\Delta x) = \frac{h}{4\pi m \Delta v}$$

$$= \frac{6.626 \times 10^{-27} \times 7}{4 \times 22 \times (9.1 \times 10^{-28}) \times 0.3}$$

=1.93cm

52. (c) Acc. to deBroglie

$$\lambda = \frac{h}{mv} \Rightarrow mv = \frac{h}{\lambda} = \frac{6.626 \times 10^{-34}}{10^{-17}}$$

⇒ p = 6.626×10^{-17} kg m/s

53. (d) **Given :** Radius of hydrogen atom = 0.530 Å, Number of excited state (n) = 2 and atomic number of hydrogen atom (Z) = 1. We know that the Bohr radius.

$$(r) = \frac{n^2}{Z} \times \text{Radius of atom} = \frac{(2)^2}{1} \times 0.530$$

$= 4 \times 0.530 = 2.12$ Å

54. (d) By Heisenberg uncertainty Principle

$$\Delta x \times \Delta p = \frac{h}{4\pi} \text{ (which is constant)}$$

As Δx for electron and helium atom is same thus momentum of electron and helium will also be same therefore the momentum of helium atom is equal to 5×10^{-26} kg. m.s^{-1}.

55. (c) Energy of electron in 2nd orbit of Li^{+2}

$$= -13.6\frac{z^2}{n^2} = \frac{-13.6 \times (3)^2}{(2)^2} = -30.6 \text{ eV}$$

Energy required = 0 – (–30.6) = 30.6 eV

56. (c) The energy of photon,

$$E = \frac{hc}{\lambda} = 3.03 \times 10^{-19}$$

$$\text{or } \lambda = \frac{6.626 \times 10^{-34} \times 3 \times 10^8}{3.03 \times 10^{-19}}$$

$$\text{or } \lambda = \frac{19.878}{3.03} \times 10^{-7} = 6.56 \times 10^{-7} \text{m}$$

$$= 656 \text{ nm}$$

57. (c) The molecule which contains same number of electrons are called isoelectronic. eg.

$$N_2 = CO = CN^- = O_2^{+2} = 14e$$

58. (a) Quantum number n = 3, l = 2, m = +2 represent an orbital with

$$s = \pm\frac{1}{2} \quad \left(3d_{xy} \text{ or } 3d_{x^2-y^2}\right)$$

which is possible only for one e^-.

59. (a) Suppose the nucleus of hydrogen atom have charge of one proton i.e. The electron revolves in a radius of r around it. Therefore the centripital force is supplied by electrostatic force of attraction i.e.

$$\frac{mv^2}{r} = \frac{ze^2}{r^2}$$

$$\text{or} \quad \frac{mv^2}{r} = \frac{ze^2}{r}$$

$$\text{or} \quad \frac{1}{2}mv^2 = \frac{1}{2}\frac{ze^2}{r} = \text{K.E}$$

now total energy (E_n) = K.E + P.E in first excited state

$$E = \frac{1}{2}mv^2 + \left[-\frac{ze^2}{r}\right]$$

$$= +\frac{1}{2}\frac{ze^2}{r} - \frac{ze^2}{r}$$

$$-3.4 \text{ eV} = -\frac{1}{2}\frac{ze^2}{r}$$

$$\therefore \text{K.E} = \frac{1}{2}\frac{ze^2}{r} = +3.4 \text{ eV}$$

60. (d) $E = h\nu = \frac{hc}{\lambda}$ or $\lambda = \frac{c}{\nu}$

$$\Rightarrow \lambda = \frac{3 \times 10^8}{8 \times 10^{15}} = 3.75 \times 10^{-8} \text{m}$$

In nanometer λ = 3.75 × 10

which is closest to 4 × 10^1

3. CLASSIFICATION OF ELEMENTS AND PERIODICITY IN PROPERTIES

ANSWER KEY											
1	(c)	11	(a)	21	(b)	31	(b)	41	(b)	51	(b)
2	(c)	12	(a)	22	(d)	32	(c)	42	(d)	52	(a)
3	(a)	13	(a)	23	(b)	33	(d)	43	(b)	53	(d)
4	(a)	14	(a)	24	(b)	34	(c)	44	(c)	54	(a)
5	(d)	15	(c)	25	(a)	35	(a)	45	(c)	55	(a)
6	(c)	16	(d)	26	(c)	36	(c)	46	(c)	56	(a)
7	(b)	17	(a)	27	(a)	37	(d)	47	(c)	57	(d)
8	(c)	18	(a)	28	(b)	38	(b)	48	(a)	58	(d)
9	(d)	19	(b)	29	(d)	39	(b)	49	(a)	59	(a)
10	(c)	20	(d)	30	(c)	40	(c)	50	(b)	60	(a)

Hints & Solutions

1. (c) According to Dobereneir's triad the atomic mass of Br will be average of the atomic masses of Cl & I

$$= \frac{35.5 + 127}{2} = 81.25$$

7. (b) The properties of elements change with a change in atomic number.
11. (a) The screening effect follows the order $s > p > d > f$.
12. (a) IE_1 is always less than IE_2.
13. (a) Pauling scale of electronegativity was helpful in predicting
(i) Nature of bond between two atoms
(ii) Stability of bond
by calculating the difference in electro-negativities polarity of bond can be calculated.
14. (a) Metallic character decreases in a period and increases in a group.
15. (c) Element with Z = 33
$(1s^2 2s^2 p^6 3s^2 p^6 d^{10} 4s^2 p^3)$ lies in fifth (or 15th) group.
16. (d) Atomic volume is the volume occupied by one gram of an element. Within a period from left to right, atomic volume first decreases and then increases due to increases of nuclear charge and increase in the number of electrons in the valence shell.
17. (a) The electronic configuration clearly suggest that it is a d-block element (having contiguration $(n-1)d^{1-10} ns^{0-2}$) which starts from III B and goes till II B. Hence with d^3 configuration it would be classified in the group.
18. (a) Ionic radii are inversely proportional to effective nuclear charge.
Ionic radii in the nth orbit is given as

$$r_n = \frac{n^2 a_0}{Z} \text{ or } r_n \propto \frac{1}{Z}$$

when n = principal quantum number
Z-effective nuclear change.
28. (b) When we move down the group, atomic size increases, distance between nucleus and valence shell electron increases and ionisation energy decreases.
29. (d) Most of the transition metal compounds (ionic as well as covalent) are coloured both in the solid state and in aqueous solution in contrast to the compounds of s and p-block elements due to the presence of incomplete d-subshell.
30. (c) Element with the outer electronic configuration $3d^6 4s^2$ is d-block element. It is a transition metal.
37. (d) P^{5+} has more effective nuclear charge and smaller size than P^{3+}.
38. (b) IE_2 of Mg is lower than that of Na because in case of Mg^+, 3s-electron has to be removed where as in case of Na^+, an electron is removed from the stable inert gas configuration which is difficult.
39. (b) The right sequence of $I.E_1$ of Li < B < Be < C.
40. (c) The element A is $ns^2 np^1$ and B is $ns^2 np^4$.
They can form compound of the type A_2B_3
41. (b) It is electronic configuration of alkali metal. Hence it will form basic oxide.
42. (d) $SO_2 > P_2O_3 > SiO_2 > Al_2O_3$
(SO_2, P_2O_3: Acidic; SiO_2: Weak acidic; Al_2O_3: Amphoteric)
43. (b) The alkali metals are highly reactive because their first ionisation potential is very low and hence they have great tendency to lose electrons to form unipositive ions. On moving down a group from Li to Cs ionisation enthalpy decreases hence the reactivity increases.
The halogens are most reactive elements due to their low bond dissociation energy, high electron affinity and high enthalpy of hydration of halide ion however their reactivity decreases with increase in atomic number. As the size increases, the attraction for an additional electron by the nucleus becomes less. Thus reactivity decreases.
44. (c) Elements (a), (b) and (d) belong to the same group since each one of them has two electrons in the s sub shell. In contrast, element (c) has seven electrons in the valence shell and hence does not lie in the same group in which elements (a), (b) and (d) lie.
45. (c) Electron affinity values are high in case of halogen because halogens have seven electrons $(ns^2 np^5)$ in the valence shell, they have a strong tendency to acquire the nearest inert gas configuration by gaining an electron from the metallic atom and form halide ions easily.
46. (c) N, O and F (p-block elements) are highly electronegative non metals and will have the strongest tendency to form anions by gaining electrons from metal atoms.
47. (c) Amongst isoelectronic ions, the size of the cation decreases as the magnitude of the charge increases.
48. (a) Non metals form oxides with oxygen and thus reduce oxides of metals behaving as reducing agents.
49. (a) Atomic number of the given element is 15 and it belongs to 5th group. Therefore atomic number of the element below the above element = 15 + 18 = 33.
50. (b) Electronic configuration of element with atomic number 118 will be $[Rn]5f^{14}6d^{10}7s^2 7p^6$. Since its elctronic configuration in the outer most orbit $(ns^2 np^6)$ resemble with that of inert or noble

gases, therefore it will be noble gas element.

51. (b) Greater is the positive charge on atom, large is effective nuclear charge. Hence smaller is the size

52. (a) In a period on moving from left to right ionic radii decreases.
(a) So order of cationic radii is
$Cr^{2+} > Mn^{2+} > Fe^{2+} > Ni^{2+}$ and
(b) $Sc > Ti > Cr > Mn$
(correct order of atomic radii)
(c) For unpaired electrons
$$Mn^{2+}\text{(Five)} > Ni^{2+}\text{(Two)} < Co^{2+}\text{(Three)} < Fe^{2+}\text{(Four)}$$
(d) For unpaired electrons
$$Fe^{2+}\text{(Four)} > Co^{2+}\text{(Three)} < Ni^{2+}\text{(Two)} < Cu^{2+}\text{(One)}$$

53. (d) First ionisation potential of Be is greater than boron due to following configuration
${}_4Be = 1s^2, 2s^2$ ${}_5B = 1s^2, 2s^2 2p^1$
Order of attraction of electrons towards nucleus 2s>2p, so more amount of energy is required to remove the electron with 2s-orbital in comparison to 2p orbital.

54. (a) $Mg = 1s^2 2s^2 2p^6 3s^2$
After removing of 2 electron, the magnesium acquired noble gas configuration hence removing of 3rd electron will require large amount of energy.

55. (a) Be – $1s^2 2s^2$; B – $1s^2 2s^2 2p^1$; C – $1s^2 2s^2 2p^2$; N – $1s^2 2s^2 2p^3$; O – $1s^2 2s^2 2p^4$. IP increases along the period. But IP of Be > B. Further IP of O < N because atoms with fully or partly filled orbitals are most stable and hence have high ionisation energy.

56. (a) Among the given options, only Fe shows variable oxidation states so it can form two chlorides, viz. $FeCl_2$ and $FeCl_3$.

57. (d) Due to odd number of electrons in ClO_2.

Therefore 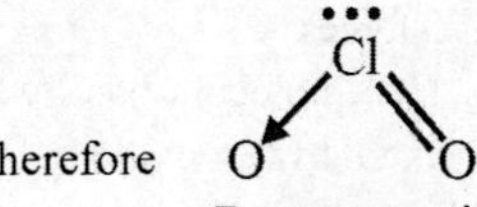

Paramagnetic

58. (d) The amount of energy released when an electron is added to an isolated gaseous atom to produce a monovalent anion is called electron gain enthalpy. Electron affinity value generally increase on moving from left to right in a period however there are exceptions of this rule in the case of those atoms which have stable configuration. These atoms resist the addition of extra electron, therefore the low value of electron affinity
$$\underset{-1.48}{O} < \underset{-2.0}{S} < \underset{-3.6}{F} < \underset{-3.8}{Cl}$$
On the other hand Cl because of its compariti very bigger size than F, allow the addition of an extra electron more easily.

59. (a) For isoelectronic species, size of anion increases as negative charge increases whereas size of cation decreases with increase in positive charge. Further ionic radii of anions is more than that of cations. Thus the correct order is
$$Ca^{++} < K^+ < Ar < Cl^- < S^{--}$$

60. (a) Proton affinity decreases in moving across the period from left to right due to increase in charge, within a group the proton affinities decreases from top to bottom.
Nitrogen family > Oxygen family > Halogens

4. CHEMICAL BONDING AND MOLECULAR STRUCTURE

ANSWER KEY

1	(c)	11	(b)	21	(d)	31	(a)	41	(b)	51	(c)
2	(a)	12	(a)	22	(a)	32	(c)	42	(c)	52	(c)
3	(b)	13	(a)	23	(c)	33	(b)	43	(d)	53	(a)
4	(c)	14	(a)	24	(d)	34	(a)	44	(c)	54	(c)
5	(c)	15	(a)	25	(d)	35	(a)	45	(b)	55	(a)
6	(b)	16	(d)	26	(d)	36	(a)	46	(d)	56	(c)
7	(c)	17	(d)	27	(d)	37	(b)	47	(a)	57	(c)
8	(c)	18	(d)	28	(c)	38	(d)	48	(c)	58	(a)
9	(c)	19	(d)	29	(b)	39	(b)	49	(b)	59	(b)
10	(c)	20	(d)	30	(c)	40	(b)	50	(d)	60	(a)

Hints & Solutions

1. (c) At bond distance the attractive forces overweigh the repulsive forces.
2. (a) A gaseous HCl molecule has hydrogen and chlorine linked by a covalent bond. Here electronegativity of chlorine is greater than that of hydrogen. Due to this the shared pair of electron is more attracted towards chlorine. Thus, chlorine end of molecule has higher electron density and becomes slightly negative and the hydrogen and slightly positive. Hence the covalent bond in HCl has a polar character as shown below

$$\overset{\delta^+}{H} \longrightarrow \overset{\delta^-}{Cl}$$

3. (b) Hydrogen is non metal and non metal atoms form covalent bond
4. (c)

$$\left[\begin{matrix} & O & \\ & | & \\ O- & P & -O \\ & \| & \\ & O & \end{matrix}\right]^{3-} \leftrightarrow \left[\begin{matrix} & O & \\ & | & \\ O- & P & =O \\ & | & \\ & O & \end{matrix}\right]^{3-} \leftrightarrow$$

$$\left[\begin{matrix} & O & \\ & \| & \\ O- & P & -O \\ & | & \\ & O & \end{matrix}\right]^{3-} \leftrightarrow \left[\begin{matrix} & O & \\ & | & \\ O= & P & -O \\ & | & \\ & O & \end{matrix}\right]^{3-}$$

Bond order

$$= \frac{\text{Number of bonds}}{\text{Number of Resonating structures}}$$

$$= \frac{5}{4} = 1.25$$

Three unit negative charge is being shared by four O atoms. Formal charge = –3/4

5. (c) The bond length decreases in the order $sp^3 > sp^2 > sp$.
Because of the triple bond, the carbon-carbon bond distance in ethyne is shortest.
6. (b) The stability of the ionic bond depends upon the lattice energy which is expected to be more between Mg and F due to +2 charge on Mg atom.
7. (c) We know that bond angles of NH_3 = 107°, NH_4^+ = 109.5°, PCl_3 = 100°. Therefore bond angle of NH_4^+ is maximum.
8. (c) Dissociation energy of any molecules depends upon bond order. Bond order in N_2 molecule is 3 while bond order in N_2^+ is 2.5. Further we know that more the Bond order, more is the stability and more is the BDE.
9. (c) Benzene and the compounds having benzene ring show resonance. Hence, ethylamine does not show resonance.
10. (c) Polarity of the bond depends upon the electronegativity difference of the two atoms forming the bond. Greater the electronegativity difference, more is the polarity of the bond.

N – Cl	O – F	N – F	N – N
3.0–3.0	3.5–4.0	3.0–4.0	3.0–3.0

As the electronegativity difference between N and F is maximum hence this bond is most polar.

11. (b) Hybridisation is sp^3 and shape pyramidal.
13. (a)

s orbital p orbital

The overlap between s- and p-orbitals occurs along internuclear axis and hence the angle is 180°.

14. (a) Linear combination of two hybridized orbitals leads to the formation of sigma bond.
15. (a) A σ – bond is stronger than a π -bond hence option (a) is not correct. Sigma (σ) bonds are formed by head on overlap of unhybridised s–s, p–p or s–p orbitals and hybridised orbitals (sp, sp^2, sp^3, sp^3d and sp^3d^2) hence σ bonds are strong bonds. Where as Pi (π)-bonds are formed by side ways overlap of unhybridised p- and d orbitals-hence π bonds are weak bonds.
16. (d) With the increase of electronegativity and decrease in size of the atom to which hydrogen is covalently linked, the strength of hydrogen bond increases. As F is most electronegative thus HF shows maximum strength of hydrogen bond.
17. (d) H-bond is the weakest.
18. (d) In this configuration, there are four completely filled bonding molecular orbitals and one completely filled antibonding molecular orbital.
So that $N_b = 8$ and $N_a = 2$.

$$\therefore \text{Bond order} = \frac{1}{2}(N_b - N_a) = \frac{1}{2}(8-2) = 3.$$

19. (d) In a linear symmetrical molecule like BeF_2, the bond angle between three atoms is 180°, hence the polarity due to one bond is cancelled by the other equal polarity due to other bond, while it is not so in angular molecules, like H_2O.
20. (d) Electronic configuration of the molecule according to molecular orbital theory, is
$\sigma 1s^2 \sigma^* 1s^2 \sigma 2s^2 \sigma^* 2s^2 \sigma 2p_z^2\ (\pi 2p_x^2 = \sigma 2p_y^2)$

$(\pi^*2p_x^2 = \pi 2p_y^2)$. $\sigma^*2p_z^2\sigma 3s^2\sigma^*3s^2\sigma 3p_z^2$
$(\pi 3p_x^2 = \pi 3p_y^2)$ $(\pi^*3p_x^1 = 3p_y^1)$
Last two electrons are unpaired. So no. of unpaired electron is 2.

21. (d) Due to intermolecular hydrogen bonding in methanol, it exist as assosiated molecule.

22. (a) $O \equiv C \equiv O$, $\mu = 0$ H–O–H

$\mu = 1.84D$

These structure are best illustrates the fact that CO_2 has zero dipole moment but H_2O has dipole moment

23. (c) In the N_2 molecule, 3 electron pairs are shared by nitrogen atoms as there are 5 electrons in its outermost shell.

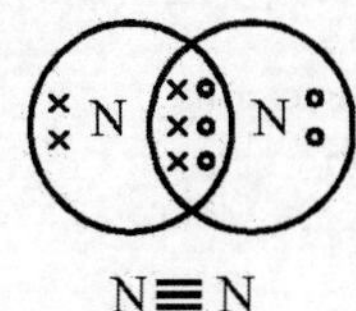

$N \equiv N$

24. (d) Electrons are equally shared in the bond formed between two identical non-metal atoms.

25. (d) Ionic compounds generally have high melting and boiling points because of the strong electrostatic force of attraction between oppositely charged ions. Consequently, a considerable amount of energy is required to overcome strong attractive inter-ionic forces and to break down the crystal lattice.

26. (d) Ions having sp^3 hybridisation contain hydrogen atoms at the corners of tetrahedron.

27. (d) CaF_2 is insoluble in water due to high lattice energy than others (which is given in options).

28. (c) Structure of BF_3 is

:F: :B: :F:

:F:

In this example octet is incomplete.

29. (b) BF_3 does not follow octate rule because central atom, boron lacks an electron pair. Thus, it also acts as Lewis acid.

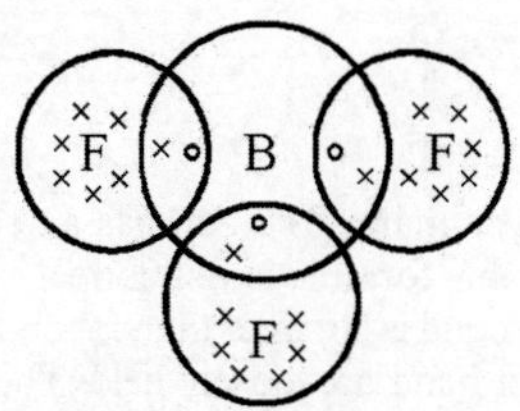

30. (c) The b.p. of p-nitrophenol is higher than that of o-nitrophenol because in p-nitrophenol there is intermolecular H-bonding but in o-nitrophenol it is intramolecular H-bonding.

31. (a) More the number of bonds between atoms, shorter will be the bond length.

$H_3C–CH_3$ (A) $H_2C=CH_2$ (B)

$H–C \equiv C–H$ (C)

$\therefore$ Sequence of bond order is A > B > C.

32. (c) In PF_5 the hybridisation is sp^3d P–F Axial

P – F bonds are longer than equatorial bonds.

33. (b) $B.O. = \frac{1}{2}[N_b - N_a]$ $Li_2 = \frac{4-2}{2} = 1; N_2 = 3;$ Be $= 0, O_2 = 2.$

34. (a) SF_4 – Configuration of excited S atom :

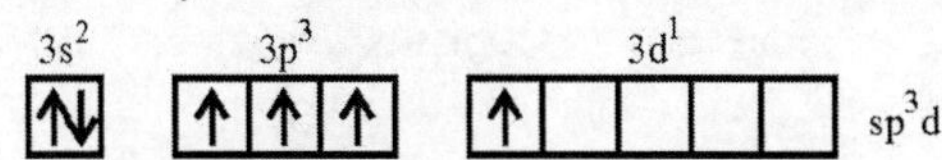

shape-square pyramidal, one lone pair

CF_4 – Configuration of excited C-atom :

$2s^1$ ↑ $2p^3$ ↑ ↑ ↑ sp^3

shape-tetrahedral; no lone pair

XeF_6 – have square planar geometry with 2 lone pairs.

36. (a) For linear arrangement of atoms the hybridisation should be sp(linear shape, 180° angle). Only H_2S has sp^3-hybridization and hence has angular shape while C_2H_2, BeH_2 and CO_2 all involve sp - hybridization and hence have linear arrangement of atoms.

37. (b) BF_3 involves sp^2-hybridization.

38. (d) $H-C=\overset{\overset{O}{\|}}{C^*}-O-H$

The star marked carbon has a valency of 5 and hence this formula is not correct.

39. (b) Sigma bond is stronger than π-bond. The electrons in the π bond are loosely held. The bond is easily broken and is more reactive than σ -bond. Energy released during sigma bond formation is always more than π bond because of greater extent of overlapping

40. (b) The strength of the interactions follows the order vander Waal's < hydrogen – bonding < dipole-dipole < covalent.

41. (b) The removal of an electron from a diatomic molecule may increase the bond order as in the conversion $O_2\ (2) \longrightarrow O_2^+(2.5)$ or decrease the bond order as in the conversion, $N_2(3.0) \longrightarrow N_2^+(2.5)$, As a result, the bond energy may increase or decrease. Thus, statement (b) is incorrect.

42. (c) The C–C bond distance decreases as the multiplicity of the bond increases. Thus, bond distance decreases in the order: butane (1.54 Å) > benzene (1.39 Å) > ethene (1.34) Å > ethyne (1.20 Å). Thus in butane, C – C bond distance is the largest.

43. (d) As there is no lone pair on boron in BCl_3 therefore no repulsion takes place. But there is a lone pair on nitrogen in NCl_3. Therefore repulsion takes place. Thus BCl_3 is planar molecule but NCl_3 is a pyramidal molecule.

44. (c) The N–O bond length decreases in the order

:O: (double bonded to N) :O:=N–O⁻ > :O:=N–O⁻ > :O:=N–N:=O (each N bearing O←) > ·N:=:O

45. (b) We know that in O_2 bond, the order is 2 and in O_2^- bond, the order is 1.5. Therefore the wrong statements is (b)

46. (d) Total no. of electrons in O_2^{2-} = 16 + 2 = 18

Distribution of electrons in molecular orbital

$\sigma 1s^2, \sigma^* 1s^2, \sigma 2s^2, \sigma^* 2s^2, \sigma 2p_x^2, \pi 2p_y^2$

$\pi 2p_z^2, \pi^* 2p_x^2\, \pi^* 2p_y^2$

Anti bonding electron = 8 (4 pairs)

47. (a) Paramagnetic character is based upon presence of unpaired electron

${}_{17}Cl^- = 1s^2, 2s^2\, 2p^6, 3s^2\, 3p_x^2\, 3p_y^2\, 3p_z^2$

${}_4Be = 1s^2, 2s^1$

${}_{10}Ne^{2+} = 1s^2, 2s^2\, 2p_x^2\, 2p_y^1\, 2p_z^1$

${}_{33}As^+ = 1s^2, 2s^2\, 2p^6, 3s^2\, 3p^6\, 3d^{10}, 4s^2\, 4p_x^1\, 4p_y^1\, 4p_z^0$

Hence only Cl^- do not have unpaired electrons.

48. (c) As dipole moment = electric charge × bond length

D. M. of AB molecules

$= 4.8 \times 10^{-10} \times 2.82 \times 10^{-8} = 13.53D$

D.M. of CD molecules

$= 4.8 \times 10^{-10} \times 2.67 \times 10^{-10} = 12.81D$

now% ionic character

$= \dfrac{\text{Actual dipole moment of the bond}}{\text{Dipole moment of pure ionic compound}}$

then % ionic character in

$AB = \dfrac{10.41}{13.53} \times 100 = 76.94\%$

% ionic character in

$CD = \dfrac{10.27}{12.81} \times 100 = 80.23\%$

49. (b) XeF_4 hybridisation is $= \dfrac{1}{2}(V + X - C + A)$

hence V = 8 (no. of valence e^-)

$X = 4$ (no of monovalent atom)

$\dfrac{1}{2}(8 + 4 + 0 - 0) = 6 \quad sp^3d^2$

C = 0 charge on cation

A = 0 (charge on anion). The shape is

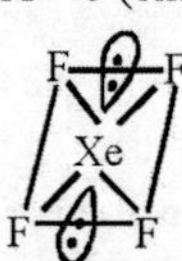

Square planar shape.

50. (d) HF form linear polymer structure due to hydrogen bonding.

51. (c) In P–O bond, π bond is formed by the sidewise overlapping of d-orbital of P and p-orbital of oxygen. Hence it is formed by $p\pi$ and $d\pi$ overlapping.

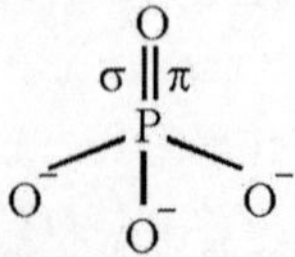

52. (c) In XeF_2 and IF_2^- . Both XeF_2 and IF_2^- are sp^3d hybridized and have planar shape.

53. (a) In X—H- - - Y, X and Y both are electronegative elements (i.e attracts the electron pair) then electron density on X will increase and on H will decrease.

54. (c) As sigma bond is stronger than the π (pi) bond, so it must be having higher bond energy than π (pi) bond.

55. (a)

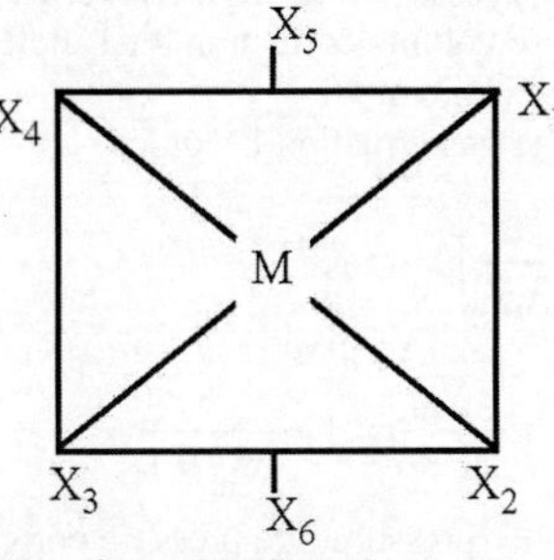

Thus here bond angles between

$$X_4 - M - X_2 = 180°$$

$$X_1 - M - X_3 = 180°$$

$$X_5 - M - X_6 = 180°$$

56. (c)

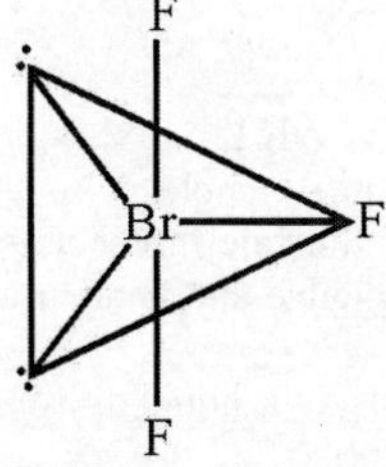

In BrF_3, both bond pairs as well as lone pairs of electrons are present. Due to the presence of lone pairs of electrons (lp) in the valence shell, the bond angle is contracted and the molecule takes the T-shape. This is due to greater repulsion between two lone pairs or between a lone pair and a bond pair than between the two bond pairs.

57. (c) As difference of electronegativity increases % ionic character increases and covalent character decreases i.e., negativity difference decreases covalent character increases.
Further greater the charge on the cation more will be its covalent character. Be has maximum (+2) charge.

58. (a) BF_3 is sp^2 hybridised. So, it is trigonal planner. NH_3, PCl_3 has sp^3 hybridisation hence has trigonal bipyramidal shape, IF_3, has sp^3d hydridization and has linear shape.

59. (b) Statement (a), (c), (d) are correct. Statement (b) is incorrect statement.
AB_5 may have two structures as follows :

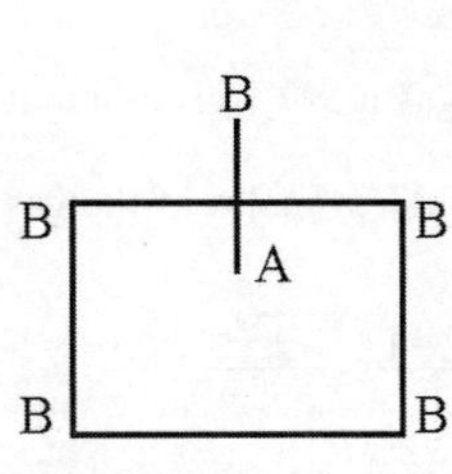

Square Pyramidal

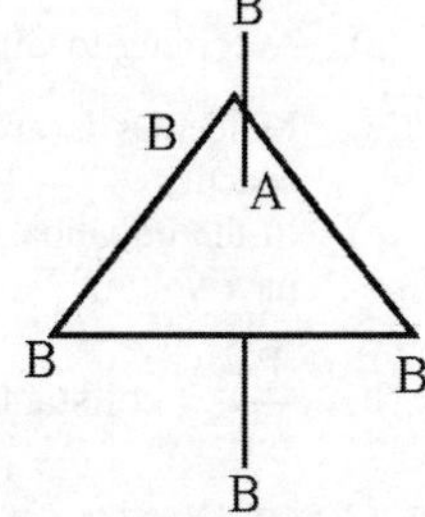

Trigonal Bipyramidal

60. (a) In NH_3 the atomic dipole and bond dipole are in the same direction whereas in NF_3 these are in opposite direction so in the former case they are added up whereas in the latter case net result is reduction of dipole moment. It has been shown in the following figure :

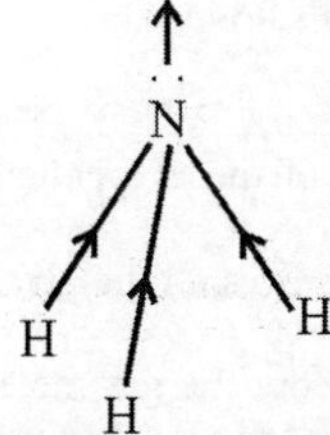

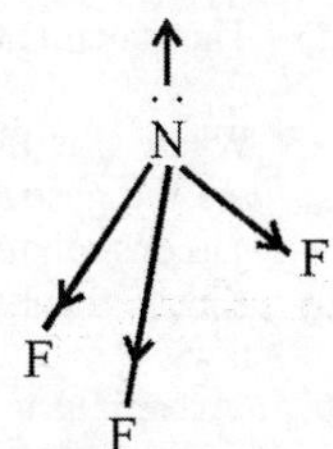

5. STATES OF MATTER

ANSWER KEY

1	(c)	11	(d)	21	(a)	31	(c)	41	(d)	51	(a)
2	(c)	12	(b)	22	(a)	32	(a)	42	(b)	52	(a)
3	(b)	13	(b)	23	(d)	33	(c)	43	(a)	53	(a)
4	(c)	14	(d)	24	(d)	34	(c)	44	(d)	54	(a)
5	(b)	15	(b)	25	(b)	35	(d)	45	(b)	55	(c)
6	(c)	16	(c)	26	(c)	36	(b)	46	(a)	56	(c)
7	(c)	17	(d)	27	(a)	37	(b)	47	(d)	57	(d)
8	(b)	18	(b)	28	(d)	38	(a)	48	(c)	58	(b)
9	(d)	19	(d)	29	(a)	39	(c)	49	(c)	59	(a)
10	(d)	20	(d)	30	(c)	40	(b)	50	(d)	60	(a)

Hints & Solutions

3. (b) Boyle's law

$P \propto \frac{1}{V}$

$P = \frac{K}{V}$

$PV = K$

4. (c) According to Charle's law $V \propto T$ or $\frac{V}{T} = k$

6. (c) Molecules in an ideal gas move with different speeds.

7. (c) In the equation PV = nRT, n moles of the gas have volume V.

8. (b) $\frac{PV}{T} = \text{constant}$ or $\frac{P_1V_1}{T_1} = \frac{P_2V_2}{T_2}$

$\Rightarrow \frac{P_1V_1}{P_2V_2} = \frac{T_1}{T_2}$

9. (d) $PV = nRT = \frac{m}{M}RT$

or $PM = \frac{m}{V}RT = dRT \Rightarrow d = \frac{PM}{RT}$

10. (d) The statement of Dalton's law is
$P = P_A^\circ + P_B^\circ + \ldots\ldots\ldots\ldots P_n^\circ$
Where P is the total pressure P_A°, P_B° .... are the partial pressure of the components of gaseous mixture.

11. (d) Kinetic theory of gases proves all the given gas laws.

12. (b) According to kinetic theory the gas molecules are in a state of constant rapid motion in all possible directions colloiding in a random manner with one another and with the walls of the container and between two successive collisions molecules travel in a straight line path but show haphazard motion due to collisions.

13. (b) The compressibility factor of a gas is defined as

$Z = \frac{pV_m}{RT}$

For an ideal gas, $pV_m = RT$. Hence $Z = 1$

15. (b) Liquid drops assume spherical shape because a sphere has minimum surface area.

16. (c) Van der Waal's equation for 1 mole :

$\left(P + \frac{a}{V^2}\right)(V - b) = RT$

Here, $\left(P + \frac{a}{V^2}\right)$ represent the intermolecular forces.
(V – b) is the corrected volume.

17. (d) *At low pressure and high temperature:* At low pressure volume correction for 1 mole of a gas in negligible, i.e b = 0
thus the gas equation becomes

$\left(P + \frac{a}{V^2}\right)V = RT$

or $Z = \frac{PV_m}{RT} = 1 - \frac{a}{V_mRT}$

At higher pressure, the pressure correction for 1 mole of gas in negligible i.e $\frac{a}{V^2} = 0$

or (P + 0) (V – b) = RT
or $P(V_m - b) = RT$
or $PV_m = RT + Pb$

or $Z = \frac{PV_m}{RT} = 1 + \frac{Pb}{RT}$

18. (b) R = 0.082 litre atm K^{-1} $mole^{-1}$.

19. (d) In the ideal gas, the intermolecular forces of attraction are negligible and hence it cannot be liquefied.

20. (d) A substance exists as a liquid above its m. pt. and below its b. pt.

21. (a) The kinetic energy of molecules in gaseous state is more than those in the liquid state as the molecules in gaseous state can move freely (with higher speed) as compared in liquid state.

22. (a) By Dalton's law of partial pressures, the total pressure of a mixture of two gases is the sum of the partial pressures.

23. (d) Value of gas constant depends only upon units of measurement.

24. (d) $P \propto \frac{1}{V}$ (at constant T)

$\therefore$ PV = constant.

25. (b) At higher altitudes atmospheric pressure decrease. Therefore at lower temperature vapour pressure becomes equal to atmospheric pressure. Thus boiling point of a liquid decreases.

27. (a) During vapourization energy is supplied to liquid in the form of heat. Liquid reaches its boiling point and get converted into gas (vapours).

29. (a) In van der waals equation 'b' is for volume correction

32. (a) **Charle's Law** : The volume of the given mass of a gas increases or decreases by $\frac{1}{273}$ of its volume at 0°C for each degree rise or fall of temperature at constant pressure.

$V_t = V_0\left(1 + \frac{t}{273}\right)$ at const. Pressure

33. (c) According to kinetic gas equation

$PV = \frac{1}{3}mNu^2$, u = root mean square velocity

$\Rightarrow u^2 = \frac{3PV}{mN}$ or $u \propto \frac{1}{\sqrt{m}}$ i.e $u \propto m^{-\frac{1}{2}}$

34. (c) Here volume is constant. Again the mass of H_2 is fixed so the number of moles of the gas do not change. As temperature increases the pressure also increases. The rate of collision among the gas molecules and their energy also increases.

35. (d) According to Boyle's law at constant temperature,

$V \propto \frac{1}{P}$ or PV = constant

36. (b) $PV = \frac{1}{3}mNu^2 = \frac{1}{3}Mu^2$ or $u = \sqrt{3PV/M}$.

At STP $u \propto \sqrt{\frac{1}{M}}$

i.e higher will be the molar mass lower will be the value of u_{rms}.

Molecular masses of H_2, N_2, O_2 and HBr are 2, 28, 32 and 81. Hence the correct order is HBr < O_2 < N_2 < H_2

37. (b) Velocity and hence the K.E is maximum in the gaseous state.

38. (a) Average kinetic energy depends only on temperature $\left(K.E. = \frac{3}{2}kT\right)$

39. (c) $PV = nRT$ or $P = \frac{n}{V}RT = CRT$.

Hence 1 = 1 × 0.082 × T is $T = \frac{1}{0.082} = 12K$

40. (b) At low temperature and high pressure.

41. (d) $u \alpha \sqrt{T}$ or $u_1/u_2 = \sqrt{T_1/T_2}$

$= \sqrt{\frac{27+273}{927+273}} = \sqrt{\frac{300}{1200}} = \frac{1}{2}$

$u_2 = 2u_1$

42. (b) Given initial volume (V_1) = 500 ml ; Initial temperature (T_1) = 27°C = 300 K and final temperature (T_2) = –5°C = 268 K.

From Charle's law :

$\frac{V_1}{T_1} = \frac{V_2}{T_2}$ or $\frac{500}{300} = \frac{V_2}{268}$

Where V_2 = New volume of gas

$V_2 = \frac{500}{300} \times 268 = 446.66$ml.

43. (a) Given initial volume (V_1) = 600 c.c.; Initial pressure (P_1) = 750 mm and final volume (V_2) = 500 c.c. according to Boyle's law,

$P_1V_1 = P_2V_2$

or 750 × 600 = P_2 × 500

or $P_2 = \frac{750 \times 600}{500} = 900\,mm$.

Therefore increase in pressure = (900 – 750) = 150 mm.

44. (d) By Ideal gas equation

$P_1V = n_1RT$

$n_1 \propto P_1$ and $n_2 \propto P_2$

$\frac{n_1}{n_2} = \frac{P_1}{P_2} \Rightarrow \frac{n_1}{n_2} = \frac{170}{570} = 0.30$

45. (b) At low pressure and high temperature real gas nearly behave like ideal gas. Hence deviation is minimum from ideal behaviour.

46. (a) Applying Boyle's law $P_1V_1 = P_2V_2$ for both gases

$\frac{500}{1000} \times 400 = P \times 3 \Rightarrow P = \frac{200}{3}$

$600 \times \frac{666.6}{1000} = P' \times 3 \Rightarrow P' = \frac{400}{3}$

$\Rightarrow P_T = P + P' = \frac{200}{3} + \frac{400}{3} = \frac{600}{3} = 200$ tor

47. (d) $\frac{\text{Moles of } H_2}{\text{Moles of } O_2} = \frac{8}{1}$ (Given)

$\left(\frac{\text{M.W. of } O_2}{\text{M.W. of } H_2}\right) \cdot \left(\frac{\text{weight of } H_2}{\text{weight of } O_2}\right) = \frac{8}{1}$

$\frac{\text{weight of } H_2}{\text{weight of } O_2} = \frac{8 \times 2}{32 \times 1} = \frac{1}{2}$

48. (c) Due to intermolecular H-bonding the surface tension of H_2O is more than other liquid. One H_2O molecule is joined with 4 another H_2O molecule through H–bond.

Hydrogen bonding is in order $H_2O > C_2H_5OH > CH_3OH$.

49. (c) At any constant temperature the K.E. of gaseous molecules remains same. (K.E. ∝ T). Thus option (c) is correct answer.

50. (d) $P = \frac{nRT}{V} = \frac{\frac{6}{16.02} \times 8.314 \times 402}{0.03} \simeq 41648\,Pa$

51. (a) Ideal gas during spontaneous expansion into vacuum does not do any external work.

52. (a) Given $n_{CO} = n_{N_2}$

$P_{CO} + P_{N_2} = 1$ atm

Partial pressure of a gas = mole fraction of gas ×

total pressure

$$\therefore\ P_{N_2} = \frac{n_{N_2}}{n_{CO}+n_{N_2}} \times 1$$

$$= \frac{n_{N_2}}{2n_{N_2}} \times 1 = \frac{1}{2} = 0.5 \text{ atm.}$$

53. (a) Given
$P_1 = 1.5$ bar $T_1 = 273 + 15 = 288$ K $V_1 = V$
$P_2 = 1.0$ bar $T_1 = 273 + 25 = 298$K $V_2 = ?$

$$\frac{P_1V_1}{T_1} = \frac{P_2V_2}{T_2}$$

$$\frac{1.5 \times V}{288} = \frac{1 \times V_2}{298}$$

$V_2 = 1.55$ V i.e., volume of bubble will be almost 1.6 time to initial volume of bubble.

54. (a) At a constant temperature,
$V_1 = 120$ ml & $V_2 = ?$
$P_1 = 1$ atm & $P_2 = 5$ atm
Applying Boyle's law
$P_1V_1 = P_2V_2$
$\Rightarrow 120 \times 1 = 5 \times V_2$
$\Rightarrow V_2 = 24$ ml

55. (c) As the inversion temperature of helium is very low, hence during the expansion into vacuum, heating effect is observed.

56. (c) V = 100 litre, T = 20°C + 273 = 293 K
P = 2 atm, $M(H_2S) = 2 + 32 = 34$
By ideal gas eqn.,

$$\Rightarrow PV = \frac{w}{M}RT \Rightarrow w = \frac{PVM}{RT}$$

$$\Rightarrow w = \frac{2 \times 100 \times 34}{0.0821 \times 293} = 282.4 \text{ g}$$

57. (d) For a gas, maximum deviation from ideal gas behaviour is observed at low temperature & high pressure.

58. (b) From ideal gas equation
PV = nRT
n = 1, P = 6 atm. T = 273 + 27 = 300 K
R = 0.0821 lit atm. $mol^{-1}K^{-1}$)

$$P\frac{M}{d} = RT \qquad \left(\because V = \frac{m}{d}\right)$$

$$d = \frac{P \times M_{CH_4}}{RT} = \frac{6 \times 16}{0.0821 \times 300} = 4 \text{ gm/lit.}$$

59. (a) M. wt. of a gas = 45 (given)
Volume of a gas at STP = 22.4 litre.
Density of a gas

$$= \frac{\text{M.wt. of a gas}}{\text{Volume of a gas at S.T.P.}}$$

$$= \frac{45}{22.4} = 2 \text{ gm/litre}$$

Hence, density of a gas at STP is 2 gm/litre.
Option (a) is correct.

60. (a) Initial volume of a gas $V_1 = 0.2$ liters (given)
Initial temperature of a gas
$T_1 = 0°\text{ C} + 273\ °\text{K}$ (given)
Final temperature of a gas at same pressure
$T_2 = 273° + 273°\text{ C} = 546$ K (given)
Final volume of a gas at same Pressure $V_2 = ?$
By Charle's law,

$$\frac{V_1}{V_2} = \frac{T_1}{T_2}$$

$$\Rightarrow V_2 = \frac{V_1 \times T_2}{T_1} = \frac{0.2 \times 546}{273} = 0.4 \text{ litre}$$

Hence, final volume of a gas at the same pressure is 0.4 litre.

6. THERMODYNAMICS

ANSWER KEY

1	(c)	11	(c)	21	(b)	31	(c)	41	(b)	51	(a)
2	(c)	12	(b)	22	(c)	32	(d)	42	(a)	52	(a)
3	(d)	13	(a)	23	(a)	33	(d)	43	(d)	53	(a)
4	(a)	14	(a)	24	(d)	34	(c)	44	(b)	54	(d)
5	(c)	15	(a)	25	(d)	35	(b)	45	(d)	55	(d)
6	(c)	16	(c)	26	(b)	36	(c)	46	(c)	56	(b)
7	(a)	17	(b)	27	(a)	37	(c)	47	(a)	57	(b)
8	(d)	18	(a)	28	(b)	38	(c)	48	(c)	58	(c)
9	(b)	19	(b)	29	(d)	39	(d)	49	(a)	59	(d)
10	(a)	20	(d)	30	(a)	40	(b)	50	(b)	60	(b)

Hints & Solutions

5. (c) Justification : free expansion w = 0
adiabatic process q = 0
$\Delta U = q + w = 0$, this means that internal energy remains constant. Therefore, $\Delta T = 0$.

6. (c) During isothermal expansion of ideal gas,
$\Delta T = 0$. Now H = E + PV
$\because \Delta H = \Delta E + \Delta(PV)$
$\therefore \Delta H = \Delta E + \Delta(nRT)$;
Thus if $\Delta T = 0.$, $\Delta H = \Delta E$
i.e., remain unaffected

7. (a) $\Delta E = \Delta Q - W$
For adiabatic expansion, $\Delta Q = 0$
$\Rightarrow \Delta E = -W$
The negative sign shows decrease in Internal energy, which is equal to the work done on the system by the surroundings.

8. (d) For isothermal reversible expansion

$$w = q = nRT \times 2.303 \log \frac{V_2}{V_1}$$

$$= 2RT \times 2.303 \log \frac{20}{2}$$

$$= 2 \times 2 \times T \times 2.303 \times 1 = 9.2\ T$$

Entropy change, $\Delta S = \frac{q}{T} = \frac{9.2T}{T} = 9.2$ cal.

9. (b) $\Delta H = \Delta E + P\Delta V$, for solid and liquid, ΔV = or $\Delta H = \Delta E + \Delta n\ RT$, for solids and liquids $\Delta n = 0$.

10. (a) Volume depends upon mass. Hence it is extensive property.

11. (c) Given $C_p = 75\ JK^{-1}\ mol^{-1}$

$n = \frac{100}{18}$ mole, Q = 1000 J ΔT = ?

$Q = nC_p\Delta T$ $\Rightarrow$

12. (b) For the reaction
$2ZnS \rightarrow 2Zn + S_2$; $\Delta G_1° = 293$ kJ ....(1)
$2Zn + O_2 \rightarrow 2ZnO$; $\Delta G_2° = -480$ kJ ....(2)
$S_2 + 2O_2 \rightarrow 2SO_2$; $\Delta G_3° = -544$ kJ ....(3)
$\Delta G°$ for the reaction
$2ZnS + 3O_2 \rightarrow 2ZnO + 2SO_2$
can be obtained by adding eqn. (1), (2) and (3)
$\Rightarrow \Delta G° = 293 - 480 - 544 = -731$ kJ

13. (a) $\Delta S = \frac{q}{T}$
q ⟶ required heat per mole
T ⟶ constant absolute temperature
Unit of entropy is $JK^{-1}\ mol^{-1}$

14. (a) For a spontaneous process, ΔS_{total} is always positive.

15. (a) If $\Delta G_{system} = 0$ the system has attained equilibrium is right choice.
In it alternative (d) is most confusing as when $\Delta G > 0$, the process may be spontaneous when it is coupled with a reaction which has $\Delta G < 0$ and total ΔG is negative, so right answer is (a).

17. (b) Spontaneity of reaction depends on tendency to acquire minimum energy state and maximum randomness. For a spontaneous process in an isolated system the change in entropy is positive.

18. (a) Third law of Thermodynamics.

20. (d) ΔG is negative for a spontaneous process.

21. (b) Entropy states the randomness or disorderness of the system. At absolute zero, the movement of molecules of the system or randomenss of the system is zero, hence entropy is also zero.

22. (c) $\Delta H = \Delta E + \Delta nRT$
$\Delta n = 3 - (1 + 5)$
$= 3 - 6 = -3$

$\Delta H - \Delta E = (-3RT)$

23. (a) Measure of disorder of a system is nothing but Entropy. For a spontaneous reaction, $\Delta G < 0$.
As per Gibbs Helmnoltz equation,
$\Delta G = \Delta H - T\Delta S$
Thus ΔG is –ve only
When ΔH = –ve (exothermic)
and ΔS = +ve (increasing disorder)

24. (d) We know that
$\Delta H = \Delta E + P\Delta V$
In the reactions, $H_2 + Br_2 \rightarrow 2HBr$ this is no change in volume or $\Delta V = 0$
So, $\Delta H = \Delta E$ for this reaction

25. (d) We know that q (heat and work (w) are not state functions but (q + w) is a state functions. H – TS (i.e. G) is also a state functions. Thus II and III are not state functions so the correct answer is option (d).

26. (b) Heat of combustion of a substance is always negative as it is the amount of heat evolved (i.e. decrease in enthalpy) when one mole of the substance is completely burnt in air or oxygen.

27. (a) Crystallization of sucrose solution. Entropy is a measure of randomness during the crystallisation of sucrose solution liquid state is changing into solid state hence entropy decreases.

28. (b) 1 calorie = 4.184 joule

29. (d) $\Delta G = \Delta H - T\Delta S$; ΔG is positive for a reaction to be non-spontaneous when ΔH is positive and ΔS is negative.

30. (a) In endothermic reaction , the enthalpy of reactants is less than the enthalpy of products, since heat is absorbed in endothermic reaction. Therefore, in endothermic reaction change in enthalpy is always positive.

$$\underset{H_R}{A + B} \rightleftharpoons \underset{H_P}{C + D} \quad \Delta H = +ve$$

$\Delta H = H_P - H_R = +ve$ $(\because H_P > H_R)$

31. (c) In exothermic reaction, the enthalpy of reactant is higher than enthalpy of product therefore, the enthalpy of an exothermic reaction is always negative. In exothermic reaction,

$$\underset{H_R}{A + B} \rightleftharpoons \underset{H_P}{C + D} \quad (\Delta H = -ve)$$

$\Delta H = (H_P - H_R) = -ve$ $(\because H_R > H_P)$

32. (d) Combustion is heating of substance in excess of oxygen so as to form carbon bioxide and water. Sometimes water is not formed during combustion.

33. (d) It violate 2nd law of thermodynamics, not 1st.

34. (c) In a reversible process the work done is greater than in irreversible process. Hence the heat absorbed in reversible process would be greater than in the latter case. So,

$T_f(\text{rev.}) < T_f(\text{irr.})$

35. (b) $\Delta n = -\frac{1}{2}; \Delta H = \Delta E - \frac{1}{2}RT; \Rightarrow \Delta E > \Delta H$

37. (c) The energy involved in the conversion of

$\frac{1}{2}Cl_2(g)$ to Cl^{-1} (aq) is given by

$$\Delta H = \frac{1}{2}\Delta_{diss}H^{(-)}_{Cl_2} + \Delta_{eg}H^{(-)}_{Cl} + \Delta_{hyd}H^{(-)}_{Cl}$$

Substituting various values from given data, we get

$$\Delta H = \left(\frac{1}{2}\times 240\right) + (-349) + (-381)\,kJmol^{-1}$$

$= (120 - 349 - 381)$ kJ mol^{-1}

$= -610$ kJ mol^{-1}

39. (d) As $\Delta H = \Delta E + \Delta n_g RT$

if $n_p < n_r$; $\Delta n_g = n_p - n_r = -$ ve.

Hence $\Delta H < \Delta E$.

40. (b) This is combustion reaction, which is always exothermic hence

$\Delta H = -$ve

As the no. of gaseous molecules are increasing hence entropy increases

now $\Delta G = \Delta H - T\Delta S$

For a spontaneous reaction

$\Delta G = -$ve

Which is possible in this case as $\Delta H = -$ve and $\Delta S = +$ve.

41. (b) Enthalpy of formation of C_2H_4, CO_2 and H_2O are 52, – 394 and – 286 kJ/ mol respectively. (Given)

The reaction is

$C_2H_4 + 3O_2 \rightarrow 2CO_2 + 2H_2O.$

change in enthalpy,

$(\Delta H) = \Delta H_{products} - \Delta H_{reactants}$

$= 2\times(-394) + 2\times(-286) - (52+0)$

$= -1412$ kJ/ mol.

42. (a) The relation between free energy change and equilibrium constant is given by Nernst equation

$$E = E^\circ - \frac{RT}{nF}\ln Q$$

At equilibrium, $E_C = 0$ and $Q = K_C$

$$\therefore E^\circ = \frac{RT}{nF}\ln K \quad .....(i)$$

Again $\Delta G^\circ = nFE^\circ$ .....(ii)

put in (i)

$$\frac{-\Delta G^\circ}{nF} = \frac{RT}{nF}\,ln K \;;\; \Delta G^\circ = -RT\,ln K$$

43. (d) Entropy change

$\Delta S = \Delta S_{product} - \Delta S_{reactant}$

$= 2\,(186.7) - (223 + 130.6)$

$= 373.4 - 353.6$

$= 19.8$ JK^{-1} mol^{-1}

44. (b) For an isothermal process $\Delta E = 0$

45. (d) For a cyclic process

$\Delta E = 0$, $\Delta H = 0$ & $\Delta G = 0$. As all depend upon final state and initial state, W doesn't depend on path followed.

46. (c) As all reactant and product are liquid $\Delta n_{(g)} = 0$

$\Delta H = \Delta E - \Delta nRT$

$\Delta H = \Delta E \quad (\because \Delta n = 0)$

47. (a) $$\Delta S = \frac{\text{Latent heat of fusion}}{\text{Melting point}} = \frac{\Delta H_y}{T}$$

$= \frac{2930}{300}$ J K^{-1} mol^{-1}

$= 9.77$ J K^{-1} mol^{-1}

48. (c) $SO_2 + \frac{1}{2}O_2 \longrightarrow SO_3$

$\Delta H = \Delta H^\circ_{f\,(SO_3)} - \Delta H^\circ_{f\,(SO_2)}$

$= -98.2 + 298.2 = 200$ kJ/Mole

49. (a) $CH_4\,(g) + \frac{1}{2}O_2\,(g) \rightarrow CH_3OH(l)\;\Delta H = ?$

$\therefore \Delta H = [(\Delta H$ of combustion of $CH_3OH)$

$-(\Delta H$ of combustion of $CH_4)]$

$= [(-y) - (-x)] = [-y + x] = x - y$

Given $\Delta H = -$ve

$\therefore x - y < 0$

hence $x < y$

50. (b) As volume is constant hence work done in this proces is zero hence heat supplied is equal to change in internal energy.

51. (a) $2H_2O_2(l) \longrightarrow 2H_2O(l) + O_2(g)\;\Delta H = ?$

$\Delta H = [2\times\Delta H_f$ of $H_2O(l) + (\Delta H_f$ of $O_2)]$

$-(2\times\Delta H_f$ of $H_2O_2(l))$

$= [(2\times -286) + (0) - (2\times -188)]$

$= [-572 + 376] = -196$ kJ/mole

52. (a) $C(s) + O_2(g) \rightarrow CO_2(g)$

$\Delta H = -94$ kCal/mole ...(i)

$H_2(g) + \frac{1}{2}O_2 \rightarrow H_2O(g)$

$\Delta H = -68$ kCal/mole ...(ii)

$CH_4 + 2O_2 \rightarrow CO_2 + 2H_2O,$

$\Delta H = -213$ k Cal/mole ...(iii)

$C(s) + 2H_2 \rightarrow CH_4(g),\ \Delta H = ?$ ....(iv)

Eqn. (iv) can be obtained by eq. (i) + eq. (ii) × 2 – eq.(iii)

$C(s) + O_2 \rightarrow CO_2(g)$

$2H_2 + O_2 \rightarrow 2H_2O(g)$

$CO_2 + 2H_2O \rightarrow CH_4(g) + 2O_2$

$C(s) + 2H_2(g) \rightarrow CH_4(g)$

So, $\Delta H_{CH_4} = -94 + 2(-68) - (-213)$

$= -94 - 136 + 213 = -17$ k Cal/mole

53. (a) $\Delta S = \dfrac{\Delta H}{T}$

$\Delta S(\text{per mole}) = \dfrac{\Delta H \text{ per mole}}{T} = \dfrac{6000}{273}$

$= 21.98\ JK^{-1}mol^{-1}$

54. (d) XeF_4 sublimes at room temp., while N_2O_3 does not exist at room temp. It decomposes even at – 30ºC.

55. (d) $H_2(g) + Br_2(g) \rightarrow 2HBr(g)$

$\Delta H° = (B\ E)_{reactant} - (B\ E)_{product}$

$= (433 + 192) - (2 \times 364)$

$= 625 - 728 = -103$ kJ

56. (b) $\Delta G = \Delta H - T\Delta S$

$= -382.64 - (298 \times 145.6 \times 10^{-3})$

$= -339.3\ kJ\ mol^{-1}$

57. (b) $W = -\ p\Delta V$

$= -3(6-4) = -6$ litre atmosphere

$= -6 \times 101.32 = -608J$

58. (c) For a spontaneous reaction $\Delta G(-ve)$, which is possible if ΔS = +ve, ΔH = +ve and $T\Delta S > \Delta H$ [As $\Delta G = \Delta H - T\Delta S$]

59. (d) We know that

$\Delta G = \Delta H - T\Delta S$

When the reaction is in equilibrium, $\Delta G = 0$

$0 = \Delta H - T\Delta S \Rightarrow T = \dfrac{\Delta H}{\Delta S}$

$T = \dfrac{2 \times 30 \times 1000}{2 \times 105} = 285.7\ K$

60. (b) For the reaction

$PCl_5(g) \rightleftharpoons PCl_3(g) + Cl_2(g)$

The reaction given is an example of decomposition reaction and we know that decomposition reactions are endothermic in nature, i.e, $\Delta H > 0$.

Further

$\Delta n = (1 + 1) - 1 = +1$

Hence more number of molecules are present in products which shows more randomness i.e. $\Delta S > 0$ (ΔS is positive)

7. EQUILIBRIUM

ANSWER KEY

1	(c)	11	(a)	21	(b)	31	(d)	41	(a)	51	(d)
2	(d)	12	(a)	22	(b)	32	(a)	42	(d)	52	(b)
3	(c)	13	(b)	23	(a)	33	(a)	43	(a)	53	(d)
4	(a)	14	(a)	24	(c)	34	(c)	44	(b)	54	(d)
5	(b)	15	(d)	25	(c)	35	(a)	45	(c)	55	(d)
6	(d)	16	(a)	26	(d)	36	(c)	46	(c)	56	(a)
7	(d)	17	(c)	27	(b)	37	(d)	47	(c)	57	(c)
8	(d)	18	(d)	28	(b)	38	(c)	48	(a)	58	(d)
9	(d)	19	(b)	29	(c)	39	(a)	49	(b)	59	(b)
10	(b)	20	(d)	30	(c)	40	(d)	50	(b)	60	(a)

Hints & Solutions

2. (d) Solid $\rightleftharpoons$ Liquid

It is an endothermic process. So when temperature is raised, more liquid is formed. Hence adding heat will shift the equilbrium in the forward direction.

3. (c) Ice melts with a reduction in volume. So Increase

in pressure shifts equilibrium to water side, result in meeting of ice Acc. to Lechatelier's principle.

4. (a) For reation (1)

$$K_1 = \frac{[NO]^2}{[N_2][O_2]}$$

and for reaction (2)

$$K_2 = \frac{[N_2]^{½}[O_2]^{½}}{[NO]} \text{ therefore } K_1 = \frac{1}{K_2^2}$$

5. (b) **Given:** Equilibrium constant (K_1) for the reaction

$$HI(g) \overset{K_1}{\rightleftharpoons} \frac{1}{2}H_2(g) + \frac{1}{2}I_2(g);\ K_1 = 8;$$ ....(i)

To find equilibrium constant for the following reaction

$H_2(g) + I_2(g) \rightleftharpoons 2HI(g);\ K_2 = ?$ .....(ii)

For this multiply (i) by 2, we get

$2HI(g) \rightleftharpoons H_2(g) + I_2(g);$

$K_1 = 8^2 = 64.$ .....(iii)

[**Note:** When the equation for an equilibrium is multiplied by a factor, the equilibrium constant must be raised to the power equal to the factor]

Now reverse equation (iii), we get

$$H_2(g) + I_2(g) \rightleftharpoons 2HI(g); K = \frac{1}{64}$$ .....(iv)

[**Note:** For a reversible reaction, the equilibrium constant of the backward reaction is inverse of the equilibrium constant for the forward reaction.]

Equation (iv) is the same as the required equation (ii), thus K_2 for equation (ii) is $\frac{1}{64}$ i.e. option (b) is correct.

6. (d) $2C(s) + O_2(g) \rightleftharpoons 2CO_2(g)$

$\Delta n = 2 - 1 = +1$

$\therefore$ K_c and K_p are not equal.

7. (d) For reaction to proceed from right to left

$$\underset{\begin{pmatrix}\text{backward}\\ \text{rate}\end{pmatrix}}{Q} > \underset{\begin{pmatrix}\text{forward}\\ \text{rate}\end{pmatrix}}{K_c}$$ i.e the reaction will be fast in backward direction i.e $r_b > r_f$.

8. (d) According to Le-chatelier's principle" whenever a constraint is applied to a system in equilibrium, the system tends to readjust so as to nullify the effect of the constraint.

9. (d) $[Cu(H_2O)_4]^{2+} + 4NH_3 \rightleftharpoons [Cu(NH_3)_4]^{2+} + 4H_2O$

involves lose and gain of electrons. H_2O is coordinated to Cu by donating electrons (LHS). It is then removed by withdrawing electrons.

10. (b) Because NH_3 after losing a proton (H^+) gives NH_2^-

$$NH_3 + H_2O \rightleftharpoons NH_2^- + H_3O^+$$

(Conjugate acid-base pair differ only by a proton)

11. (a) Molarity (M) = 10M. HCl is a strong acid and it is completely dissociated in aqueous solutions as : HCl (10) $\rightleftharpoons$ $H^+(10) + Cl^-$.

So, for every moles of HCl, there is one H^+.

Therefore $[H^+]$ = [HCl] or $[H^+]$ = 10.

$pH = -\log[H^+] = -\log[10] = -1.$

13. (b) The presence of common ion causes the solubility to decrease.

14. (a) $FeCl_3$ on hydrolysis produces weak base $Fe(OH)_3$ & a strong acid HCl thus $FeCl_3$ solution is acidic in nature

15. (d) HNO_2 is a weak acid and $NaNO_2$ is salt of that weak acid and strong base (NaOH).

16. (a) Justification : According to Le-Chatelier's principle, at constant temperature, the equilibrium composition will change but K will remain same.

17. (c) $H_2CO_3 \rightleftharpoons H^+ + HCO_3^-$

$HCO_3^- \longrightarrow H^+ + CO_3^-$

$HCO_3^- + H^+ \longrightarrow H_2CO_3$

HCO_3^- can donate and accept H^+.

19. (b) An aqueous solution of acetic acid dissociates as

$CH_3COOH + H_2O \rightleftharpoons CH_3COO^- + H_3O^+$

20. (d) Rate constant of forward reaction (K_f) = 1.1×10^{-2} and rate constant of backward reaction (K_b) = 1.5×10^{-3} per minute. Equilibrium constant (K_c)

$$= \frac{K_f}{K_b} = \frac{1.1 \times 10^{-2}}{1.5 \times 10^{-3}} = 7.33$$

21. (b) In polyprotic acids the loss of second proton occurs much less readily than the first. Usually the K_a values for successive loss of protons from these acids differ by at least a factor of 10^{-3} i.e.,

$$K_{a_1} > K_{a_2}$$

$$H_2X \rightleftharpoons H^+ + HX^- \left(K_{a_1}\right)$$

$$HX^- \rightleftharpoons H^+ + X^{2-} \left(K_{a_2}\right)$$

22. (b) $B(OH)_3$ does not provide H^+ ions in water instead it accepts OH^- ion and hence it is Lewis acid

$$B(OH)_3 + H_2O \rightleftharpoons [B(OH)_4]^- + H^+$$

23. (a) IVth group needs higher S^{2-} ion concentration. In presence of HCl, the dissociation of H_2S decreases hence produces less amount of sulphide ions due to common ion effect, thus HCl decreases the solubility of H_2S which is sufficient to precipitate IInd group radicals.

24. (c) For this reaction $K_{eq.}$ is given by

$$K = \frac{[Fe^{3+}][OH^-]^3}{[Fe(OH)_3]}$$

$= (Fe^{3+})(OH^-)^3$ $[\because [solid]=1]$.

If (OH^-) is decreased by $\frac{1}{4}$ times then for reaction equilibrium constant to remain constant, we have to increase the concentration of $[Fe^{3+}]$ by a factor of 4^3 i.e $4 \times 4 \times 4 = 64$.

25. (c) BF_3 behaves as Lewis acid.

26. (d) Equilibrium constant (K) is independent of concentrations of reactions and products.

27. (b) The most favourable conditions are :
(i) High pressure ($\Delta n < 0$)
(ii) Low temperature (Exothermic reaction)
(iii) Catalyst Fe is presence of Mo.

28. (b) Neutralization of an acid and base involves the combination of H^+ and OH^- to form water.

$$H^+ + OH^- \longrightarrow H_2O$$

29. (c) As in this no. of moles are increasing hence low pressure will favour the forward direction.

$\Delta n = (1 + 1) - 1 = 1$

30. (c) Weak acid + its conjugate base $\longrightarrow$ acid buffer
Weak base + its conjugate acid $\longrightarrow$ basic buffer
A buffer solution resist the change in pH on adding a drop of acid or a base.

31. (d) A salt of strong base with weak acid undergoes anionic hydrolysis to give basic solution.

32. (a) For $\Delta n = 0$, no effect of pressure.

33. (a) For $A + B \rightleftharpoons C$, $\Delta n = 1 - 2 = -1$

$$\text{Unit of } K_c = \left[\frac{mol}{litre}\right]^{\Delta n} = \left[\frac{mol}{litre}\right]^{-1}$$

$=$ Litre $mole^{-1}$

34. (c) Let binary electrolyte be AB

$$\underset{s}{AB} \rightleftharpoons \underset{s}{A^+} + \underset{s}{B^-}$$

Hence, solubility product of AB
$K_{sp} = [A^+][B^-]$
$S = [s.][s] \Rightarrow s = S^{1/2}$

35. (a) Melting of ice involve absorption of heat i.e Endothermic hence high temperature favour the process.
Further for a given mass volume of water is less than ice thus high pressure favour the process. High pressure and high temperature convert ice into liquid.

36. (c)

$$A + B \rightleftharpoons C + D$$

At eqb. x x 2x 2x

$$K_c = \frac{2x.2x}{x.x} = 4$$

37. (d) $K_c = \dfrac{K_p}{(RT)^{\Delta n}}$

$$= \frac{1.44 \times 10^{-5}}{(0.082 \times 773)^{-2}} \text{ (R in L. atm. } K^{-1} \text{ mole}^{-1}\text{)}.$$

38. (c) $K_p = \dfrac{P^2co}{Pco_2}; K_p = \dfrac{4 \times 4}{2} = 8; C(s) = 1;$

The concentration of solids and liquids are taken as unity.

39. (a) $\Delta G^0 = -2.303\ RT \log K$
$-4.606 \times 10^3 = -2.303 \times 2 \times 500 \log K_c$
$\therefore K_c = 100.$

42. (d) In aqueous solution BA(salt) hydrolyses to give

$$BA + H_2O \rightleftharpoons \underset{\text{Base}}{BOH} + \underset{\text{acid}}{HA}$$

Now pH is given by

$$pH = \frac{1}{2}pK_w + \frac{1}{2}pKa - \frac{1}{2}pK_b$$

substituting given values, we get

$$pH = \frac{1}{2}(14 + 4.80 - 4.78) = 7.01$$

44. (b) $[H^+] = \dfrac{K_a[ACID]}{[SALT]}; [H^+] = \dfrac{1.8 \times 10^{-6} \times 0.1}{0.5}$

$= 0.36 \times 10^{-6}; \log[H^+] = -\log[0.13 \times 10^{-6}]$
$\therefore pH = 6.44$

46. (c) Since k_{sp} of HgS is minimum among others, HgS will precipitate first.

47. (c) NaCl is a salt of strong acid and strong base hence its aqueous solution will be neutral ie pH = 7. $NaHCO_3$ is an acidic salt hence pH < 7 . Na_2CO_3

is a salt of weak acid and strong base. Hence its aqueous solution will be strongly basic ie. pH > 7. NH_4Cl is salt of weak base and strong acid, hence its aqueous solution will be strongly acidic i.e. pH < 7.

48. (a) Sodium borate is a salt of strong base (NaOH) and weak acid (H_3BO_3). Hence its aqueous solution will be basic

49. (b) AgBr has the highest solubility in 10^{-3}M NH_4OH. All other solvents will dissolve AgBr poorly. Moreover bromides of Ag^+, Hg_2^{2+} and Cu_2^{2+} are water insoluble

50. (b) For Bi_2S_3. $K_{sp-} = (2s)^2.(3s)^3$
$= 4s^2.\ 27s^3 = 108s^5$

or $$s = \sqrt[5]{\frac{K_{sp}}{108}} = \sqrt[5]{\frac{1\times10^{-17}}{108}}$$

For MnS. $K_{sp} = s^2$

or $$s = \sqrt{K_{sp}} = \sqrt{7\times10^{-16}}$$

for CuS $$s = \sqrt{K_{sp}} = \sqrt{8\times10^{-37}}$$

For Ag_2S $K_{sp} = 2s^2.\ s = 4s^3$

or $$s = \sqrt[3]{\frac{K_{sp}}{4}}$$

thus MnS has maximum solubility.

51. (d) The buffer system present in serum is $H_2CO_3 + NaHCO_3$ and as we known that a buffer solution resist the change in pH therefore pH value of blood does not change by a small addition of an acid or a base.

52. (b) According to equation

$$2HI \rightleftharpoons H_2 + I_2$$

At t = 0 (2 moles) 0 0
At equilibrium (2 – 2α) moles α mole α mole

Total moles at equilibrium = 2 – 2α + α + α= 2 mole

53. (d) $CuS \rightleftharpoons \underset{S}{Cu^{2+}} + \underset{S}{S^{2-}}$

$$K_{sp} = S^2 \text{ or } S = \sqrt{K_{sp}}$$

For Binary salts like CuS & HgS, solubilty,

$$S = \sqrt{K_{sp}}$$

$$\Rightarrow S_{CuS} = \sqrt{10^{-31}},\ S_{HgS} = \sqrt{10^{-54}}$$

For $Ag_2S \rightarrow \underset{2s}{2Ag^+} + \underset{s}{S^{2-}}$

$$K_{sp} = 4s^3 \text{ or } S_{Ag_2S} = \sqrt[3]{\frac{K_{sp}}{4}} = \sqrt[3]{\frac{10^{-44}}{4}}$$

∴ The order is CuS > Ag_2S > HgS

54. (d) For the reaction

$$XeF_6(g) + H_2O(g) \rightleftharpoons XeOF_4(g) + 2HF(g)$$

$$K_1 = \frac{[XeOF_4][HF]^2}{[XeF_6][H_2O]} \quad ....(1)$$

and for the reaction

$$XeO_4(g) + XeF_6(g) \rightleftharpoons XeOF_4(g) + XeO_3F_2(g)$$

$$K_2 = \frac{[XeOF_4][XeO_3F_2]}{[XeO_4][XeF_6]} \quad ....(2)$$

For reaction :

$$XeO_4(g) + 2HF(g) \rightarrow XeO_3F_{2(g)} + H_2O(g)$$

$$K = \frac{[XeO_3F_2][H_2O]}{[XeO_4][HF]^2}$$

∴ From eqs. no. (1) and (2)

$$K = K_2 / K_1$$

55. (d) No. of moles of NaOH = $\frac{4}{40} = 0.1$

[Molecular weight of NaOH = 40]

No. of moles of OH^- = 0.1

Concentration of OH^- = $\frac{0.1}{1 \text{ litre}} = 0.1 \text{Mole/L}$

As we know that, $[H^+][OH^-] = 10^{-14}$

$$\therefore [H^+] = 10^{-13} \quad (\because OH^- = 10^{-1})$$

56. (a) $M_2S \rightleftharpoons \underset{2x}{2M^+} + \underset{x}{S^{--}}$

Solubility product = $(2x)^2\ (x) = 4x^3$
$= 4\ (3.5 \times 10^{-6})^3 = 1.7 \times 10^{-16}$

57. (c) Strong base has higher tendency to accept the proton. Increasing order of base and hence the order of accepting tendency of proton is

$$I^- < HS^- < NH_3 < RNH_2$$

58. (d) Given s = 0.5 × 10^{-4} moles/lit

$[MX_2 \rightleftharpoons M^{2+} + 2X^-]$

∵ For MX_2, $K_{sp} = s \times (2s)^2 = 4s^3$
$K_{sp} = 4 \times (0.5 \times 10^{-4})^3 = 4 \times 0.125 \times 10^{-12}$
$= 0.5 \times 10^{-12} = 5 \times 10^{-13}$

59. (b) $$pOH = pK_b + \log\frac{[Salt]}{[Base]}$$

or $pK_b = pOH - \log\frac{[salt]}{[Base]}$

but pOH + pH = 14 or pOH = 14 – pH

$\therefore\ 14 - pH - \log\frac{[Salt]}{[Base]} = pK_b$

$14 - 9.25 - \log\frac{0.1}{0.1} = pK_b$

$14 - 9.25 - 0 = pK_b \qquad pK_b = 4.75$

60. (a) First option is incorrect as the value of K_P given is wrong. It should have been

$$K_P = \frac{P_{CO_2}}{P_{CH_4} \times [P_{O_2}]^2}$$

8. REDOX REACTIONS

ANSWER KEY

1	(d)	**11**	(b)	**21**	(a)	**31**	(a)	**41**	(c)	**51**	(a)
2	(d)	**12**	(d)	**22**	(a)	**32**	(d)	**42**	(d)	**52**	(b)
3	(a)	**13**	(a)	**23**	(d)	**33**	(d)	**43**	(d)	**53**	(c)
4	(c)	**14**	(d)	**24**	(d)	**34**	(c)	**44**	(b)	**54**	(c)
5	(b)	**15**	(c)	**25**	(a)	**35**	(a)	**45**	(d)	**55**	(c)
6	(c)	**16**	(a)	**26**	(a)	**36**	(a)	**46**	(b)	**56**	(b)
7	(a)	**17**	(a)	**27**	(c)	**37**	(d)	**47**	(d)	**57**	(a)
8	(b)	**18**	(b)	**28**	(b)	**38**	(b)	**48**	(b)	**58**	(d)
9	(d)	**19**	(d)	**29**	(d)	**39**	(a)	**49**	(a)	**59**	(a)
10	(d)	**20**	(b)	**30**	(b)	**40**	(c)	**50**	(b)	**60**	(b)

Hints & Solutions

2. (d) In redox reactions oxidation and reduction take place side by side.

$Cu_2S + 2FeO \rightarrow 2Cu + 2Fe + SO_2$.

O.N. of Cu changes from +1 to 0 (reduction) and O.N. of S changes from –2 to +4 (oxidation).

3. (a) Losing of electron is called oxidation,

4. (c) Carbon has the maximum oxidation state of + 4, therefore carbon dioxide (CO_2) cannot act as a reducing agent.

5. (b) $Zn^{2+} + 2e^- \rightarrow Zn(s)$

Here electrons are reducing from Zn^{2+} to Zn.

6. (c) $Co(s) + Cu^{2+}(aq) \longrightarrow Co^{2+}(aq) + Cu(s)$

This reaction is a redox reaction as Co undergoes oxidation whereas Cu^{+2} undergoes reduction.

8. (b) O.N. of iodine in I_3^- is –1/3

9. (d) $PO_4^{3-} = x + 4(-2) = -3;\ x - 8 = -3;\ x = +5$

$SO_4^{2-} = x + 4(-2) = -2;\ x - 8 = -2;\ x = +6$

$Cr_2O_7^{2-} = 2x + 7(-2) = -2;\ 2x - 14 = -2;$

$2x = 12;\ x = +6$

10. (d) NO_3^-

Let x be the oxidation numbers of nitrogen

$[x + 3 \times (-2)] = -1 \Rightarrow x = -1 + 6 = 5$

11. (b) During disproportionation same compound undergo simultaneous oxidation reduction.

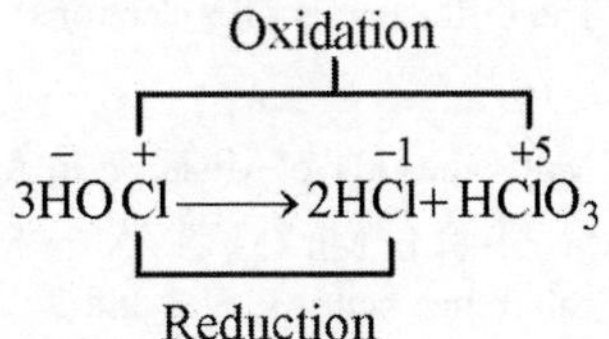

13. (a) The balanced equation is

$$2MnO_4^- + 5C_2O_4^{2-} + 16H^+ \rightarrow$$

$$2Mn^{2+} + 10CO_2 + 8H_2O$$

14. (d) Order of decreasing electrode potentials of Mg, K, Ba and Ca is

Mg > Ca > Ba > K

It can be explained by their standard reduction potentials.

$E^0_{K^+|K} = -2.925$

$E_{Ba^{2+}|Ba} = -2.90$

$E^0_{Ca^{2+}|Ca} = -2.87$

$E^0_{Mg^{2+}|Mg} = -2.37$

Highly negative value of E^0_{red} shows the least value of electrode potential.

15. (c) $4P + 3KOH + 3H_2O \rightarrow KH_2PO_2 + PH_3$

O.N of P = 0, In KH_2PO_2 it is + 1, In PH_3 it is –3. Hence P is oxidised and reduced.

16. (a) Let x = oxidation no. of Cr in $K_2Cr_2O_7$.
$\therefore (2 \times 1) + (2 \times x) + 7(-2) = 0$
or 2 + 2x – 14 = 0 or x = + 6.

17. (a) $1 + x + 4 \times (-2) = 0; x = +7$

18. (b) In KI it shows – 1 O.S & in IF_7 it shows +7 O.S hence O.S of I varies from – 1 to + 7.

19. (d) In KI : $+1 + x = 0 \Rightarrow x = -1$
In KI_3 : $+1 + 3x = 0 \Rightarrow x = -1/3$
In IF_5 : $x + 5(-1) = 0 \Rightarrow x = +5$
In KIO_4 : $+1 + x + 4(-2) = 0 \Rightarrow x = +7$

20. (b) Let O.N. of carbon is x in $H_2C_2O_4$
$\Rightarrow 2(+1) + 2x + 4(-2) = 0 \Rightarrow x = +3$

21. (a) Ox. No. of an element in uncombined state = 0.

22. (a) $1 + 2 \times 1 + x + 2(-2) = 0, x = +1$

23. (d) $\underset{\substack{\text{Oxidised} \\ \text{RA}}}{2H\overset{-1}{I}} + H_2\overset{+6}{S}O_4 \rightarrow \overset{0}{I_2} + \underset{\substack{\text{Reduced} \\ \text{OA}}}{\overset{+4}{S}O_2} + 2H_2O$

24. (d) $2 \times 1 + x + 4 \times (-2) = 0, \ x = +6$

25. (a) O.N. of Mn in MnO_4^- is +7 and in Mn^{2+} it is +2. The difference is of 5 electrons.

26. (a)

27. (c) In $KMnO_4$ the O.N. of Mn is +7, in MnO_4^{-2} +6, in MnO_2 + 4 in Mn_2O_3 +3 and in Mn^{2+} is +2. The difference being 1, 3, 4 and 5 respectively.

28. (b) O.N. of N changes from + 2 to + 5 hence NO is reducing.

29. (d) $3Br_2 + 6CO_3^{2-} + 3H_2O \rightarrow$

$$5Br^- + BrO_3^- + 6HCO_3^-$$

O.N. of Br_2 changes from 0 to –1 and +5 hence it is reduced as well as oxidised.

30. (b) $Na_2S_4O_6$ Let O.N. of S be x then
$2x(+1) + 4(x) + 6x(-2) = 0$
$\therefore$ x = 2.5. By chemical bonding method the two S atoms have O.N. + 5 and two S atoms have O.N. zero.

35. (a) $2MnO_4^- + 5H_2O_2 \rightarrow$

$$2Mn^{2+} + 5H_2O + 13/2O_2 + 6\bar{e}$$

$x = 2, \ y = 5$ and $z = 6$

36. (a) 10 moles of CO_2 are produced.

37. (d) $2I^- \rightarrow I_2$ is oxidation (loss of electrons) ; Cr (+6) changes to Cr (+3) by gain of electrons. Hence Cr is reduced.

38. (b) The element undergo oxidation itself and reduces others is known as reducing agent. In this reaction O. N. of Ni Changes from 0 to + 2 and hence Ni acts as a reducing agent.

39. (a) O.N. of P in H_3PO_3 (phosphorous acid)
$3 \times 1 + x + 3 \times (-2) = 0$ or $x = +3$
In orthophosphoric acid (H_3PO_4) O.N. of P is + 5, in hypophosphorous acid (H_3PO_2) it is + 1 while in metaphosphoric acid (HPO_3), it is + 5,

40. (c) $\overset{+2}{Ba}\overset{-1}{O_2} + \overset{+1}{H_2}\overset{+6}{S}\overset{-2}{O_4} \longrightarrow \overset{+2}{Ba}\overset{+6}{S}\overset{-2}{O_4} + \overset{+1}{H_2}\overset{-1}{O_2}$

In this reaction, none of the elements undergoes a change in oxidation number or valency.

41. (c) $2NaOH + H_2SO_4 \longrightarrow Na_2SO_4 + 2H_2O$ (neutralization)

(not redox)

$3\overset{0}{O_2} \xrightarrow{\text{Light}} 2\overset{0}{O_3}$ (not redox reaction)

$\underset{0}{N_2} + \underset{0}{O_2} \xrightarrow{\text{Light}} 2\underset{+2\,-2}{NO}$ (redox reaction)

here oxidation of N_2 & reduction of O_2 is taking place

$H_2O(l) \xrightarrow{\Delta} H_2O(g)$ (not redox reaction)

42. (d) Pyrophosphoric acid $H_4P_2O_7$
Let oxidation state of phosphorus is x
$(4 \times 1 + (-2) \times 7 + 2x) = 0$
$\therefore 2x = 10$ or $x = +5$

43. (d) $Zn \rightarrow Zn^{+2} + 2e^-$ ....(1)

$8e^- + 10H^+ + NO_3^- \rightarrow NH_4^+ + 3H_2O$ .....(2)

operate eq. (1) × 4 + eq. (2) × 1

$4Zn + 10H^+ + NO_3^- \rightarrow 4Zn^{+2} + NH_4^+ + 3H_2O$

44. (b) Oxidation number of a compound must be 0. Using the values for A, B and C in the four options we find that $A_3(BC_4)_2$ is the answer.
Check : (+2)3 + [(+5)+4(–2)]2 = 6 + (5–8)2 = 0

45. (d) Zinc gives H_2 gas with dil H_2SO_4/HCl but not with HNO_3 because in HNO_3, NO_3^- ion is reduced and give NH_4NO_3, N_2O, NO and NO_2 (based

upon the concentration of HNO_3)

$$[Zn + 2HNO_3 \longrightarrow Zn(NO_3)_2 + 2H] \times 4$$
(nearly 6%)

$$HNO_3 + 8H \longrightarrow NH_3 + 3H_2O$$

$$NH_3 + HNO_3 \longrightarrow NH_4NO_3$$

$$4Zn + 10HNO_3 \longrightarrow 4Zn(NO_3)_2 + NH_4NO_3 + 3H_2O$$

Zn is on the top position of hydrogen in electrochemical series. So Zn displaces H_2 from dilute H_2SO_4 and HCl with liberation of H_2.
$Zn + H_2SO_4 \rightarrow ZnSO_4 + H_2$

46. (b) $SO_3^{2-} \rightarrow$ S is in +4 oxidation state

$S_2O_4^{2-} \rightarrow$ S is in +3 oxidation state

$S_2O_6^{2-} \rightarrow$ S is in +5 oxidation state

47. (d) In $H_2O_2 : \Rightarrow 2 \times (+1) + 2 \times x = 0$
$\Rightarrow x = -1$
In $CO_2 : \Rightarrow 4 + 2x = 0 \Rightarrow x = -2$
In $H_2O : \Rightarrow 2 \times (+1) + x = 0 \Rightarrow x = -2$
In $F_2O : \Rightarrow 2 \times (-1) + x = 0 \Rightarrow x = +2$

48. (b) In $KMnO_4$: Let O.N. of Mn be x
$\Rightarrow +1 + x + 4(-2) = 0 \Rightarrow x = +7$
In $K_2Cr_2O_7$: Let O.N. of Cr be x
$\Rightarrow 2(1) + 2x + 7(-2) = 0 \Rightarrow x = +6$

49. (a)

oxidised

$$\overset{+2}{Sn}Cl_2 + 2\overset{+2}{Hg}Cl_2 \rightarrow \overset{+1}{Hg_2}Cl_2 + \overset{+4}{Sn}Cl_2$$

reduced

50. (b) On losing the three electrons the ion M^{3+} will further acquire a charge of + 3 units, hence the total + ve charge will be 6 units. Hence O.S. of M will be + 6.

51. (a) Standard electrode potential i.e. reduction potential of A is minimum (–3.05V) i.e. its oxidation potential is maximum which implies 'A' is most reactive chemically.

52. (b) Iron (Fe) can deposit copper from copper sulphate solution as it reduces $CuSO_4$ to Cu.

$$\underset{+2}{CuSO_4} + \underset{0}{Fe} \longrightarrow \underset{+2}{FeSO_4} + \underset{0}{Cu}$$
(Oxd. state)

53. (c) Let the valency of Mn be x in K_2MnO_4
$\Rightarrow 2 \times (+1) + x + 4 \times (-2) = 0 \Rightarrow x = +6$

54. (c) $x + 6 \times (-1) = -3, x = +3$

55. (c) The violent reaction between sodium and water is an example of redox reaction :

$$\underset{0}{2Na} + \underset{+1}{2H_2O} \longrightarrow \underset{+1}{2NaOH} + \underset{0}{H_2}$$
(Oxd. state)

In this reaction, sodium (Na) is oxidised to NaOH while H_2O is reduced to H_2.

56. (b) H_2O_2 can act both as an oxidizing and reducing agent, as oxygen has intermediate oxd. state in it.

57. (a) Remember : Any halogen placed above in periodic table will oxidize halide ion of halogen placed below.
$Cl_2 + 2\,Br^- \longrightarrow Br_2 + 2Cl^-$
Tendancy to undergo oxidation of halide ion increases down the group.

58. (d) All the given compounds ionise in solution as

$$HI \longrightarrow H^+ + I^-$$
$$KBr \longrightarrow K^+ + Br^-$$
$$FeCl_3 \longrightarrow Fe^{3+} + 3Cl^-$$
$$KClO \longrightarrow K^+ + ClO^-$$

A reducing agent is one which get oxidised easily and in this process, reducing agent provides the electrons for reduction of the other substanc. Out of these given ions I^-, Br^- and Cl^- are in –1 oxidation state whereas in ClO^-, Cl is in +1 oxidation state. ClO^- ion can get, oxidised in solution to ClO_2, $ClO_3^-ClO_4^-$ ions increasing the oxidation state to +4, +5 and +7 state which are more stable. Hence, it acts as a good reducing agent.

59. (a) $2 \times 1 + 2x = 0, x = -1$

60. (b) $FeSO_4.(NH_4)_2SO_4 \rightarrow FeSO_4 + (NH_4)_2SO_4;$

$$FeSO_4 \rightarrow Fe^{2+} + SO_4^{2-}$$

i.e. O.St of Fe = +2

9. HYDROGEN

ANSWER KEY

1	(b)	11	(a)	21	(a)	31	(b)	41	(c)	51	(b)
2	(c)	12	(b)	22	(d)	32	(c)	42	(b)	52	(d)
3	(d)	13	(c)	23	(a)	33	(c)	43	(d)	53	(d)
4	(a)	14	(c)	24	(b)	34	(a)	44	(d)	54	(d)
5	(d)	15	(b)	25	(b)	35	(a)	45	(d)	55	(d)
6	(b)	16	(c)	26	(d)	36	(b)	46	(d)	56	(b)
7	(b)	17	(b)	27	(d)	37	(a)	47	(b)	57	(a)
8	(b)	18	(a)	28	(a)	38	(d)	48	(c)	58	(c)
9	(d)	19	(c)	29	(b)	39	(b)	49	(b)	59	(a)
10	(c)	20	(b)	30	(c)	40	(b)	50	(b)	60	(b)

Hints & Solutions

4. (a) Zinc has no action on cold water.

5. (d) In aqueous solution, H_2 will not reduce Zn^{2+} ions because hydrogen is less electropositive than Zn.

7. (b) $LiH + AlCl_3 \longrightarrow (AlH_3)_n$

$$\xrightarrow[\text{LiH}]{\text{excess}} Li[AlH_4]$$

Lithuim aluminium hydride is a most useful organic reducing agent. It reduces functional groups but does not attack double bonds.

8. (b) In the structure of ice each molecule of H_2O is surrounded by three H_2O molecules in hexagonal honey comb manner. On the other in water, each molecule is surrounded by four neighbouring molecules randomly which results an open cage like structure. As a result there are a number of 'hole' or open spaces. In such a structure lesser number of molecules are packed per ml. When ice melts a large no. of hydrogen bonds are broken. The molecules therefore move into the holes or open spaces and come closer to each other than they were in solid state. This result sharp increase in the density. Therefore ice has lower density than water.

9. (d) The hybridisation in water is sp^3 and bond angle 105°.

10. (c) The high boiling point of water is due to H-bonding.

11. (a) $\underset{\text{'A'}}{2Na} + 2H_2O \rightarrow \underset{\text{'C'}}{2NaOH} + \underset{\text{'B'}}{H_2} \uparrow$

$$\underset{\text{'D'}}{Zn} + \underset{\text{'C'}}{2NaOH} \rightarrow Na_2ZnO_2 + \underset{\text{'B'}}{H_2} \uparrow$$

$$\underset{\text{'D'}}{Zn} + \text{dil. } H_2SO_4 \rightarrow ZnSO_4 + \underset{\text{'B'}}{H_2} \uparrow$$

Na produces golden yellow colour with smokeless flame of Bunsen burner.

13. (c) Volume strength = 5.6 × Normality
= 5.6 × 1.5 = 8.4 L

14. (c) H_2S is oxidised to S by H_2O_2.

15. (b) D_2O - heavy water, H_2O_2 - hydrogen peroxide, H_2O - water.

16. (c) In metal hydrides the O.S. of hydrogen –1 otherwise it is +1.

17. (b) H–O–O–H (skew) is the true structure of H_2O_2.

18. (a) The complex salt of metaphosphoric acid sodium hexametaphosphate $(NaPO_3)_6$, is known as calgon. It is represented as $Na_2[Na_4(PO_3)_6]$.

19. (c) $H(g) \longrightarrow H^+(g) + e^-$.

20. (b) Temporary hardness is due to presence of bicarbonates of calcium and magnesium and permanent hardness is due to the sulphates or chlorides of both of calcium and magnesium.

21. (a) Hydrogen can only be displaced from dilute acids if electrons are supplied to H^+ ions of the acid. Mg which is a s-block element can supply its electrons very quickly. Also Cu and Hg comes after H in the reactivity series. So, both Cu and Hg cannot reduce H^+ ion in HCl and phosphorus

which is a non-metal does not supply electrons from the reduction of H^+.

$$Mg(s) + 2HCl(aq) \longrightarrow MgCl_2(aq) + H_2\uparrow$$

22. (d) H_2O_2 does not have basic properties.
23. (a) This method is known as Clark's process. In this method temporary hardness is removed by adding lime water or milk of lime.

$$Ca(OH)_2 + Ca(HCO_3)_2 \longrightarrow \underset{\text{ppt.}}{2CaCO_3\downarrow} + 2H_2O$$

24. (b) Heavy water is used as a moderator in nuclear reactors to slow down the speed of fast moving (high energy) neutrons.
25. (b) Structure of H_2O_2 is nonplanar
26. (d) Chlorine has lone pair which it can donate to form coordinate bond while hydrogen cannot.
27. (d) Hydrogen is a non-metal while all other members of group 1 (alkali metals) are metals.
28. (a) $H + e^- \ (1s^1) \rightarrow H^- \ (1s^2 \text{ or } [He]^2)$
    $F + e^-([He]^2 2s^2 2p^2) \rightarrow F^- ([He]^2\ 2s^2\ 2p^6$ or $[Ne]^{10})$
29. (b) Due to high dielectric constant, water acts as a good solvent therefore it is also called a universal solvent.
30. (c) $Ca(OH)_2$ is used for the softening of temporary hard water.

$$Ca(OH)_2(aq) \longrightarrow \underset{\text{cloudiness}}{CaCO_3(s)} + H_2O(l)$$

31. (b) Permanent hardness of water is due to chlorides and sulphates of calcium and magnesium i.e $CaCl_2$, $CaSO_4$, $MgCl_2$ and $MgSO_4$.
32. (c) The production of dihydrogen can be increased by reacting CO of syngas mixtures with $H_2O(g)$ in the presence of iron chromate as catalyst.
33. (c) Pure hydrogen is evolved by reacting absolute alcohol and Na

$$C_2H_5OH + Na \rightarrow C_2H_5ONa + \frac{1}{2}H_2$$

other statements are correct.

35. (a) Two ice cubes stick to each other due to H-bonding.
36. (b) When water is heated from 0°C to 4°C, its density increases and volume decreases.

$$\left(d = \frac{m}{V}\right)$$

37. (a) ice occupy more volume than liquid water (Ice $\rightleftharpoons$ water). Increase of pressure favours forward reaction (Le-Chatelier's principle).
38. (d) Heavy water is D_2O (1 – C); Temporary hard water contains the bi-carbonates of Mg and Ca (2 – A); Soft water contains no foreign ions (3 – B); Permanent hard water contains the sulphates and chlorides of Mg and Ca (4 – D) therefore the answer is D.
43. (d) Acetanilide, alcohol and H_3PO_4 are negative catalyst and retard decomposition of H_2O_2.
45. (d)

$$H_2S + H_2O_2 \longrightarrow 2H_2O + S$$

(oxidation: $H_2S \rightarrow S$; reduction: $H_2O_2 \rightarrow H_2O$)

In this reaction H_2S is oxidised to sulphur and H_2O_2 is reduced to H_2O, hence this reaction show oxidation-reduction both i.e., redox reaction.

46. (d) O – O – H bond angle in H_2O_2 is 97°.
47. (b) $\underset{\text{base 1}}{H^-(aq)} + \underset{\text{acid 1}}{H_2O(l)} \longrightarrow \underset{\text{base 2}}{OH^-(aq)} + \underset{\text{acid2}}{H_2(g)}$

In this reaction H^- acts as bronsted base as it accepts one proton (H^+) from H_2O and for H_2.

48. (c) $Fe + dil.H_2SO_4 \rightarrow FeSO_4 + H_2\uparrow$

$$3Fe + \underset{\text{Steam}}{4H_2O} \rightarrow Fe_3O_4 + 4H_2\uparrow$$

Cu + dil. HCl → No reaction

Copper does not evolve H_2 from acid as it is below hydrogen in electro chemical series.

$$2Na + C_2H_5OH \rightarrow 2C_2H_5ONa + H_2\uparrow$$

49. (b) $H_2O_2 + [O] \xrightarrow{\text{Oxidation}} H_2O + O_2\uparrow$
51. (b) "10 vol" H_2O_2 solution means its 1 ml on decomposition gives 10 ml of O_2 gas at NTP.

∵ 1 ml of 10 vol H_2O_2 gives 10 ml O_2

∴ 100 ml of 10 vol H_2O_2 gives

= 10 × 100 ml O_2 at NTP

= 1000 ml O_2 at NTP

$$\underset{\substack{2\,\text{mole} \\ 2\times 34\,\text{gm}}}{2H_2O_2} \longrightarrow 2H_2 + \underset{\substack{1\text{mole} \\ 22400\,\text{ml at NTP}}}{O_2}$$

∴ 22400 ml of O_2 at NTP is obtained from 68 gm H_2O_2

∴ 1 ml of O_2 at NTP is obtained from

$$= \frac{68}{22400}\text{gm } H_2O_2$$

∴ 1000 ml of O_2 at NTP is obtained from

$$= \frac{68 \times 1000}{22400}\text{gm} = 3.035 \text{ gm} = 3.0 \text{ gm approx.}$$

Hence "10 vol" H_2O_2 solution is 3.0% approx.

52. (d) It is weakly acidic in nature and pure hydrogen peroxide turns blue litmus red. ($K_a = 1.57 \times 10^{-12}$ at 293 K). It ionises in two steps.

$H_2O_2 \rightleftharpoons H^+ + O_2$

$HO_2^- \rightleftharpoons H^+ + O_2^{2-}$

So, option (d) is not true.

53. (d) Sodium is most electro +ve element among those given.

$$2Na + 2H_2O \xrightarrow[\text{Room temp.}]{} 2NaOH + H_2$$

54. (d) $KNO_2 + H_2O_2 \longrightarrow KNO_3 + H_2O$

(oxidising nature)

$Cl_2 + H_2O_2 \longrightarrow 2HCl + O_2$

(reducing agent)

55. (d) An ion exchange resin containing R–COOH group exchange cations like Ca^{2+}, Mg^{2+}, Na^+, Fe^{2+} with H^+ when hard water is passed through it. This resin is called cation exchange resin.

56. (b) Acidified $KMnO_4$ (pink or purple) is reduced to colourless Mn^{++} ion :

$$2KMnO_4 + 3H_2SO_4 \longrightarrow K_2SO_4 + 2MnSO_4 + 3H_2O + 5[O]$$

$$[H_2O_2 + [O] \longrightarrow H_2O + O_2] \times 5$$

$$\underset{\text{Pink or purple}}{2\overset{+7}{KMnO_4}} + 3H_2SO_4 + 5\overset{-1}{H_2O_2} \longrightarrow$$

$$\underset{\text{Colourless}}{K_2SO_4 + 2MnSO_4 + 8H_2O + 5\overset{0}{O_2}}$$

57. (a) The equation represents reducing behaviour of H_2O_2.

58. (c) $H_2O_2 + \underset{\text{From } H_2SO_4}{|O|} \longrightarrow H_2O + O_2$

59. (a) Cu does not replace hydrogen from HCl because Cu is less electropositive than H as it comes below H^- in the electrochemical sereis.

60. (b) $\because$ 22.4 L of H_2 at STP = 1 mole of H_2

$\therefore$ 0.224 L of H_2 at STP $= \frac{1}{22.4} \times 0.224$

= 0.01 moles.

10. THE s-BLOCK ELEMENTS

ANSWER KEY

1	(c)	11	(b)	21	(a)	31	(a)	41	(a)	51	(b)
2	(b)	12	(a)	22	(d)	32	(a)	42	(c)	52	(a)
3	(a)	13	(a)	23	(b)	33	(b)	43	(a)	53	(d)
4	(d)	14	(b)	24	(b)	34	(a)	44	(c)	54	(a)
5	(b)	15	(a)	25	(b)	35	(a)	45	(c)	55	(a)
6	(b)	16	(d)	26	(b)	36	(b)	46	(c)	56	(b)
7	(d)	17	(c)	27	(c)	37	(c)	47	(d)	57	(b)
8	(d)	18	(b)	28	(c)	38	(c)	48	(d)	58	(c)
9	(d)	19	(d)	29	(c)	39	(a)	49	(c)	59	(c)
10	(d)	20	(a)	30	(d)	40	(c)	50	(c)	60	(c)

Hints & Solutions

1. (c) A cation is always much smaller than the corresponding atom, whereas an anion is always larger than the corresponding atom, hence the size decreases in the order

$Na^- > Na > Na^+$

2. (b) Smaller the ion more is its ionic mobility in aqueous solution. Ionic radii of the given alkali metals is in the order $Na^+ < K^+ < Rb^+ < Cs^+$ and thus expected ionic mobility will be in the order $Cs^+ < Rb^+ < K^+ < Na^+$. However due to high degree of solvation (or hydration) because of lower size or high charge density, the hydrated ion size follows the order $Cs^+ < Rb^+ < K^+ < Na^+$ and thus conductivity order is $Cs^+ > Rb^+ > K^+ > Na^+$.

3. (a) Lithium does not form peroxide.

4. (d) $LiHCO_3$ is unstable and exists only in solution

5. (b) Li is much softer than the other group I metals. Actually Li is harder then other alkali metals.
6. (b) Washing soda is $Na_2 CO_3 . 10 H_2O$.
7. (d) Ca^{2+} ions is an essential element for the contraction of muscles.

10. (d) $3Mg + N_2 \xrightarrow{Heat} Mg_3N_2$;

$$2Mg + CO_2 \xrightarrow{Heat} 2MgO + C.$$

11. (b) $BeO < MgO < CaO < BaO$.

The basic character of the oxides increases down the group. Among these BeO is amphoteric.

14. (b) Gypsum $(CaSO_4.2H_2O)$ on heating to 300°C gives $CaSO_4$ (anhydrite, dead burnt plaster).
15. (a) Lime stone provides CaO; clay gives silica (SiO_2) and alumina (Al_2O_3). They all react to give a mixture of silicates and aluminates of calcium. Gypsum is also added to prevent quick setting of cement.
17. (c) Li due to highest hydration energy among the alkali metals is the strongest reducing agent.
18. (b) Na is basic in nature.
21. (a) Within a group, ionic radius increases with increase in atomic number. The melting points decrease down the group due to weakning of metallic bond. The electronegativity and the 1st ionization energy also decreases down the group.
22. (d) Because of larger size and smaller nuclear charge, alkali metals have low ionization potential relative to alkaline earth metals.
23. (b) MgO. N_2O_5 is strongly acidic, ZnO and $Al_2 O_3$ are amphoteric, therefore, MgO is most basic.
24. (b) Gp I metals form cations M^+ by loss of electron from outermost shell. Electronic configuration of Gp I metals is ns^1. When the outer electron is removed to give a positive ion, the size decreases because the outermost shell is completely removed. After removal of an electron, the positive charge of the nucleus is greater on the remaining electrons so that each of the remaining electrons are attracted more strongly towards the nucleus. This further reduces the size.
25. (b) Sodium carbonate is produced by Solvay's process. The process involves the formation of a sparingly soluble sodium bicarbonate by the reaction of sodium chloride and ammonium bicarbonate in aqueous solution. Sodium bicarbonate on heating decomposes into sodium carbonate. The reactions taking place in the process are :

$$NH_3 + H_2O + CO_2 \longrightarrow \underset{\text{Ammonium bicarbonate}}{NH_4HCO_3}$$

$$NaCl + NH_4HCO_3 \longrightarrow \underset{\text{Sodium bicarbonate}}{NaHCO_3} + NH_4Cl$$

$$2NaHCO_3 \xrightarrow{\Delta} Na_2CO_3 + H_2O + CO_2$$

Castner's process is a method to synthesize NaOH.

HNO_3 is produced by ostwald's process.

26. (b) Na reacts violently and may catch fire on exposure to moisture (air + water). So it is always stored in kerosene. Na reacts with alcohol to produce H_2.
27. (c) $CaSO_4$ can exist as a hemihydrate $CaSO_4 . 1/2H_2O$ which is plaster of paris. This is prepared by partially dehydrating gypsum.

$$\underset{\text{gypsum}}{CaSO_4.2H_2O} \xrightarrow{150^\circ} CaSO_4.\frac{1}{2}H_2O + \frac{3}{2}H_2O$$

28. (c) Plaster of paris reacts with water to form gypsum which is hard. e.g.

$$\underset{\text{Plaster of Paris}}{CaSO_4.\frac{1}{2}H_2O} + H_2O \rightarrow \underset{\text{Gypsum}}{CaSO_4.2H_2O}$$

29. (c) Baking powder contains $NaHCO_3$, potassium hydrogen tartrate and starch.
30. (d) Na_2CO_3 does not decompose on heating even to redness.
31. (a) In any period, alkali metal (except noble gas) has the lowest density. Therefore, alkali metals have large atomic volumes.
32. (a) Zn is not alkaline earth element. Among alkaline earth elements (Mg, Sr, Ba), Mg has the maximum covalent character because of small size of cation (Mg^{2+}). Lattice energy decreases as the size of metal increases. Remember, higher the lattice energy, greater is the ionic character and greater is the solubility. Exceptional case is fluorides & hydroxides (due to their small ionic radii, they form covalent compouds with large size cations).
34. (a) Because of small atomic size and high I.E. Be forms covalent chloride.
35. (a) Plaster of Paris is used as a setting material for fractured bones and also in dentary.
37. (c) Li^+ ion due to excessive hydration has lowest ionic mobility in aqueous solution.
38. (c) $Na + NH_3 \longrightarrow \underset{\text{Sodium amide}}{NaNH_2} + 1/2\ H_2$
42. (c)

$$\underset{\text{temp. hardness}}{Ca(HCO_3)_2} + Ca(OH)_2 \longrightarrow 2CaCO_3 \downarrow + 2H_2O$$

$$\underset{A}{Ca(OH)_2} + Na_2CO_3 \longrightarrow \underset{\text{Caustic soda}}{2NaOH} + CaCO_3$$

$$\underset{A}{Ca(OH)_2} + CO_2 \longrightarrow \underset{\text{milkiness}}{CaCO_3 \downarrow} + H_2O$$

44. (c) Within a period, the atomic size decreases from left to right.Further atomic size increases down the group. Hence the correct order is i,e. Na > Mg > Li > Be.

45. (c) Atomic size of $K^+ > Ca^{2+} > Mg^{2+}$ and that of $Cl^- > F^-$. Therefore, Mg^{2+}/Cl^- ratio has the minimum value.

46. (c) Ca and CaH_2 both react with H_2O to form H_2 gas,

$$Ca + 2H_2O \longrightarrow Ca(OH)_2 + H_2$$

$$CaH_2 + 2H_2O \longrightarrow Ca(OH)_2 + 2H_2$$

whereas

K gives H_2 while KO_2 gives O_2 and H_2O_2

$$2K + 2H_2O \longrightarrow 2KOH + H_2$$

$$2KO_2 + 2H_2O \longrightarrow 2KOH + O_2 + H_2O_2$$

Similarly, Na gives H_2 while Na_2O_2 gives H_2O_2

$$2Na + 2H_2O \longrightarrow 2NaOH + H_2$$

$$Na_2O_2 + 2H_2O \longrightarrow 2NaOH + H_2O_2$$

Likewise Ba gives H_2 while BaO_2 gives H_2O_2

$$Ba + 2H_2O \longrightarrow Ba(OH)_2 + H_2$$

$$BaO_2 + 2H_2O \longrightarrow Ba(OH)_2 + H_2O_2$$

47. (d) $Be(OH)_2$ is amphoteric, but the hydroxides of other alkaline earth metals are basic. The basic strength increases gradually.

48. (d) Gypsum is $CaSO_4.2H_2O$ and plaster of Paris is $(CaSO_4)_2.H_2O$. Therefore gypsum contains a lower percentage of calcium than plaster of Paris.

49. (c) $MgSO_4$ is the only alkaline earth metal sulphate which is soluble in water and for solubility hydration energy should be greater than lattice energy i.e hydration energy > lattice energy

50. (c) As Cs^+ ion has larger size than Li^+ and I^- has larger size than F^-, therefore maximum distance between centres of cations and anions is in CsI.

51. (b)

$$\underset{\text{Solid}}{X} \xrightarrow{\text{heat}} CO_2 + \text{Residue}$$

Residue $\xrightarrow{H_2O}$ Y; Y $\xrightarrow{\text{excess } CO_2}$ $\underset{\text{Clear solution}}{Z}$; Z $\xrightarrow{\text{boil}}$ X

The given properties coincide with $CaCO_3$

$$\underset{X}{CaCO_3} \xrightarrow{\text{heat}} CO_2 + \underset{\text{Residue}}{CaO}$$

$$CaO \xrightarrow{H_2O} \underset{Y}{Ca(OH)_2} \xrightarrow{CO_2} CaCO_3 \xrightarrow[CO_2]{\text{excess}} Ca(HCO_3)_2 \xrightarrow{\text{heat}} CaCO_3$$

52. (a) The solubility of sulphates of alkaline earth metals decreases as we move down the group from Be to Ba due to the reason that ionic size increases down the group. The lattice energy remains constant because sulphate ion is so large, so that small change in cationic sizes do not make any difference. Thus the order:

$$BeSO_4 > MgSO_4 > CaSO_4 > SrSO_4 > BaSO_4$$

53. (d) The stability of alkali metal hydrides decreases from Li to Cs. It is due to the fact that M–H bonds becomes weaker with increase in size of alkali metals as we move down the group from Li to Cs. Thus the order of stability of hydrides is LiH > NaH > KH > RbH > CsH

54. (a) NaOH is a strong alkali. It combines with acidic and amphoteric oxides to form salts. Since CaO is a basic oxide hence does not reacts with NaOH.

55. (a) Lattice energy decreases more rapidly than hydration energy for alkaline earth metal hydroxides.

56. (b) Portland cement is made by heating correct mixture of limestone ($CaCO_3$) with sand (SiO_2) & clay (aluminosilicate) at a temperature of 1450 – 1600°C. The composition of portland cement is CaO (70%), SiO_2 (20%), Al_2O_3(5%), Fe_2O_3(3%), $CaSO_4.2H_2O$(2%).

57. (b) Na metal is manufactured by electrolysis of fused NaCl by two methods.

(i) Castner's process

(ii) Down's process

In both the above processes electrolysis of fused sodium chloride produces Na at cathode.

$$2NaOH \rightleftharpoons 2Na^+ + OH^-$$

at cathode $2Na^+ + 2e^- \longrightarrow 2Na$

at anode $4OH^- \longrightarrow 2H_2O + O_2 + 4e$

58. (c) Among alkaline earth metals. The electropositive character increases from Be to Ba. Reactivity and metallic character are directly linked with the tendency to lose electron or electrons i.e. with electropositive nature. Thus reactivity increase gradually from Be to Ba.

Be < Mg < Ca < Sr < Ba < Ra

thus Ba is most reactive.

59. (c) Calcium Cyanamide is CaNCN or CaN_2.

11. THE P-BLOCK ELEMENTS (Group 13 and 14)

ANSWER KEY

1	(c)	11	(c)	21	(c)	31	(a)	41	(b)	51	(b)
2	(d)	12	(d)	22	(c)	32	(d)	42	(c)	52	(d)
3	(b)	13	(c)	23	(c)	33	(c)	43	(a)	53	(c)
4	(d)	14	(c)	24	(b)	34	(a)	44	(c)	54	(d)
5	(c)	15	(c)	25	(b)	35	(a)	45	(c)	55	(c)
6	(d)	16	(c)	26	(c)	36	(b)	46	(c)	56	(b)
7	(a)	17	(d)	27	(b)	37	(d)	47	(a)	57	(c)
8	(c)	18	(c)	28	(a)	38	(d)	48	(d)	58	(d)
9	(d)	19	(b)	29	(b)	39	(b)	49	(b)	59	(c)
10	(b)	20	(a)	30	(b)	40	(d)	50	(c)	60	(a)

Hints & Solutions

1. (c) The m.p decreases from B to Ga , hence gallium (Ga) has least m.p. (303 K) among group of 13 element.

2. (d) The relative strength of Lewis acids of boron trihalide increases in the order
$BF_3 < BCl_3 < BBr_3 < BI_3$
This can be explained on the basis of the tendency of the halogen atom to back donte its electron to the boron atom and makes it less electron deficient.

3. (b) Thallium shows different (+1 and +3) oxidation states because of inert pair effect.

4. (d) H_3BO_3 acts as a Lewis acid and accepts an OH^- ions to form $[B(OH)_4]^-$.

5. (c)

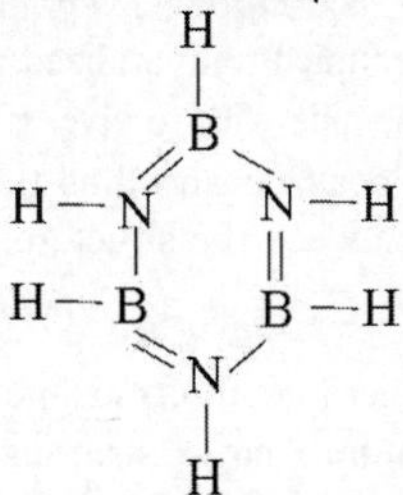

Inorganic benzene, $B_3N_3H_6$

It is isoelectronic with benzene.

6. (d) Chemically borax is sodium tetraborate decahydrate and it has the formula $Na_2B_4O_7.10H_2O$.

8. (c) Carbon cannot expand its coordination number beyond four due to the absence of d-orbitals, hence it cannot form $[CCl_6]^{2-}$ ion.

9. (d) The thermal stability of tetrahalides decreases in order
$CX_4 > SiX_4 > GeX_4 > SnX_4$ and in terms of same metal with different halides is in order of $MF_4 > MCl_4 > MBr_4 > MI_4$.

11. (c) Graphite is composed of flat two-dimensional sheets of carbon atoms. Each sheet is a hexagonal net of C atoms. The 3 e^- of C form σ bond & 4th electron is a $\pi\ e^-$ and is delocalized over the whole sheet & is thus mobile. Conduction of electricity is due to these delocalized electrons within each layer. Conduction does not occur from one sheet to another.

12. (d) $\underset{(A)}{CO_2} + Ca(OH)_2 \longrightarrow \underset{\text{white ppt}}{CaCO_3 \downarrow} + H_2O$

$CaCO_3 + CO_2 + H_2O \longrightarrow \underset{\substack{\text{Calcium bicarbonate} \\ \text{(Souble in water)}}}{Ca(HCO_3)_2}$

$Ca(HCO_3)_2 \xrightarrow{\text{Heat}} CaCO_3 + \underset{(B)}{CO_2} + H_2O$

A and B are CO_2 and CO_2.

13. (c) The formula of quartz is same as silica (SiO_2). It is hard and crystalline form, having polymeric cross linked structure.

```
  |         |         |
  Si – O – Si – O – Si
  |         |         |
  O         O         O
  |         |         |
  Si – O – Si – O – Si
  |         |         |
```

14. (c) Hydrolysis of substituted chlorosilanes yield corresponding silanols which undergo polymerisation.

$$\begin{matrix}CH_3\\CH_3\end{matrix}\!\!>Si<\!\!\begin{matrix}Cl\ H\,OH\\+\\Cl\ H\,OH\end{matrix}\xrightarrow{-2HCl}\begin{matrix}CH_3\\CH_3\end{matrix}\!\!>Si<\!\!\begin{matrix}OH\\OH\end{matrix}$$

Dialkyl silandiol

Polymerisation of dialkyl silandiol yields linear thermoplastic polymer.

$$HO-\underset{CH_3}{\overset{CH_3}{Si}}-OH+HO-\underset{CH_3}{\overset{CH_3}{Si}}-OH\longrightarrow$$

$$HO-\underset{CH_3}{\overset{CH_3}{Si}}-O-\underset{CH_3}{\overset{CH_3}{Si}}-OH$$

15. (c)

$$(O^-)_3Si-O-Si(O^-)_3$$

Pyrosilicate $[Si_2O_7]^{6-}$

17. (d) Al in its compounds forms covalent bonds.

19. (b) In diamond each carbon atom is sp^3 hybridized and thus forms covalent bonds with four other carbon atoms lying at the corners of a regular tetrahedron.

20. (a) Water gas is made by blowing steam through the layer of incandescent coal.

$$\underset{\text{Steam}}{H_2O} + \underset{\text{Red hot}}{C} \longrightarrow \underset{\text{water gas}}{H_2 + CO} - 28\text{ kcal}$$

21. (c) Mixture of H_2 and CO is known as synthesis gas or water gas.

22. (c) Ge is used as a semiconductor in transistors.

23. (c) Normal glass is calcium alkali silicate glass made by fusing the alkali metal carbonate, $CaCO_3$ and SiO_2.

24. (b) Aluminium does not occur in the free state in nature but is most abundant metal in the earth's crust.

25. (b) Nitrolim is a mixture of $CaCN_2$ and graphite (c) which is used as fertilizer.

Option (b) is correct.

26. (c) In graphite each carbon atom is in sp^2 hybridized state and is thus attached to three carbon atom by three σ bonds forming a hexagonal planar structure.

27. (b) Thermite is mixture of Fe_2O_3 + Al.

28. (a) Producer gas is a fuel gas and is mixture of CO and N_2.

29. (b) Coal gas is a mixture of $H_2 + CO + CH_4$

30. (b) Down the group basic character of oxides increases.

B_2O_3 - Acidc
Al_2O_3 - Amphoteric
Ga_2O_3 - Amphoteric
In_2O_3 - Basic
Tl_2O_3 - Basic

34. (a) H_3BO_3 is monobasic acid.

38. (d) Pb^{2+} due to inert pair effect.

41. (b) $2Pb + 2H_2O + O_2 \rightarrow 2Pb(OH)_2$

43. (a) Water gas is $CO + H_2$

45. (c) Glass is a super cooled liquid.

46. (c) Aluminium can be extracted by electrolysis pure alumina (Al_2O_3). Alumina ionises as

$$Al_2O_3 \rightleftharpoons \underset{\text{cathode}}{Al^{3+}} + \underset{\text{anode}}{AlO_3^{3-}}$$

$$Al^{3+} + 3e \longrightarrow Al \text{ (at cathode)}$$

$$4AlO_3^{3-} \longrightarrow 2Al_2O_3 + 3O_2 + 12e \text{ (at anode)}$$

Aluminium of 99.8% purity is obtained from this process.

47. (a) H_3BO_3 is a weak monobasic acid.

48. (d) In graphite, each carbon is sp^2 -hybridized and the single occupied unhybridized p-orbitals of C-atoms overlap side wise to give π -electron cloud which is delocalized and thus the electrons are spread out between the structure.

49. (b) $6HF + SiO_2 \rightarrow H_2SiF_6 + 2H_2O$

50. (c) Zeolite are microporous crystalline solid with well defined structure. They contain silicon, aluminium and oxygen in their framework and cations, water & other molecules within their pores. The general formula of zeolite is

$M_{x/y}[(AlO_2)_x(SiO_2)_y]$. mH_2O.

51. (b) Both Ge and Si are extensively used as semiconductors. Semiconductors are solids where there is only a small difference in energy, called band gap, between the filled valency band of electrons and a conduction band since the

band gap of Ge is less than Si, it is a better element to be used as semiconductor.

52. (d) PbO_2 is a strong oxidising agent and is produced in situ in lead storage batteries. The anode is oxidized to PbO_2 and cathode is reduced to spongy Pb.

53. (c) Rocks are mainly silicates.

54. (d) Producer gas ($CO + N_2$) is prepared by incomplete combustion of coal, coke or charcoal. Water gas ($CO + H_2$) is prepared by blowing steam through the layers of incandescent coal and coal gas (a mixture of H_2, CH_4, CO, C_2H_2, C_2H_4, CO_2, N_2 & O_2) is obtained by destructive distillation of coal at about 1000°C.

55. (c) White lead is also called basic lead carbonate,

$Pb(OH)_2.2PbCO_3$

56. (b) Alum is not used as an insecticide.

57. (c) $SiCl_4 + 2Mg \longrightarrow Si + 2MgCl_2$
(vapours) hot.

58. (d) Lead pencil has graphite.

59. (c) When dry hydrochloride gas is passed over heated aluminium metal, then anhydrous aluminium chloride is obtained.

$$\underset{(Dry)}{HCl(g)} + \underset{(Heated)}{Al} \longrightarrow \text{Anhydrous } AlCl_3$$

60. (a) Boron nitride is sometimes referred to as 'inorganic graphite' because its structure is like graphite with alternating boron and nitrogen atoms replacing the carbon atoms in the hexagonal structure.

12. ORGANIC CHEMISTRY - SOME BASIC PRINCIPLES AND TECHNIQUES

ANSWER KEY											
1	(c)	**11**	(b)	**21**	(c)	**31**	(c)	**41**	(b)	**51**	(b)
2	(b)	**12**	(d)	**22**	(b)	**32**	(b)	**42**	(b)	**52**	(a)
3	(b)	**13**	(d)	**23**	(d)	**33**	(d)	**43**	(d)	**53**	(b)
4	(b)	**14**	(a)	**24**	(d)	**34**	(c)	**44**	(b)	**54**	(a)
5	(c)	**15**	(a)	**25**	(b)	**35**	(c)	**45**	(b)	**55**	(b)
6	(b)	**16**	(b)	**26**	(b)	**36**	(d)	**46**	(c)	**56**	(b)
7	(b)	**17**	(b)	**27**	(d)	**37**	(b)	**47**	(d)	**57**	(b)
8	(b)	**18**	(a)	**28**	(a)	**38**	(a)	**48**	(d)	**58**	(c)
9	(c)	**19**	(b)	**29**	(c)	**39**	(d)	**49**	(c)	**59**	(b)
10	(b)	**20**	(d)	**30**	(c)	**40**	(b)	**50**	(a)	**60**	(c)

Hints & Solutions

3. (b) $\overset{5}{C}H_3 - \overset{4}{C}H(OH) - \overset{3}{C}H = \overset{2}{C}(CH_3) - \overset{1}{C}HO$

$${}^5CH_3 - {}^4\underset{\displaystyle OH}{\underset{|}{C}H} - {}^3CH = {}^2\underset{\displaystyle CH_3}{\underset{|}{C}} - {}^1CHO$$

4-Hydroxy-2-methylpent-2-en-1-al

5. (c) Higher the possibility of delocalisation of the positive charge, greater is stability of the species. Thus

$$C_6H_5\overset{+}{C}H_2 > (CH_3)_3C^+ > (CH_3)_2\overset{+}{C}H > CH_3\overset{+}{C}H_2$$

Benzyl carbocation is more stable than *tert*-butyl due to resonance in the former.

6. (b) $C_6H_5CH_2^- > CH_3CH_2^- > (CH_3)_2CH^- > (CH_3)_3C^-$
Dispersal of –ve Intensification of –ve charge due charge due to to + I effect of CH_3 gps. resonance and - I effect

7. (b) $C_6H_5\dot{C}HCH_3$ is a 2° benzylic free radical, hence stabilized most due to resonance.

9. (c) Does not contain benzene nucleus as in other options. Also delocalization of electrons is not possible.

10. (b) Being immiscible liquids, they form different layers hence can be separated with the help of separating funnel.

11. (b) If there is a small difference (10 or less) in the boiling points of liquids fractional distillation is used e.g. acetone b.p. 333 K and methanol b.p. 338 K.

13. (d) Nitrogen, sulphur and halogens are tested in an organic compound by Lassaigne's test. The organic compound is fused with sodium metal as to convert these elements into ionisable inorganic substances,

$$Na + C + N \longrightarrow NaCN$$

$$2Na + S \longrightarrow Na_2S$$

$$2Na + X_2 \longrightarrow 2NaX$$

The cyanide, sulphide or halide ions can be confirmed in aqueous solution by usual test.

14. (a) N, S and halogens present in organic compounds are converted from covalent form into the ionic form by fusing the compound with Na metal.

$Na + C + N \longrightarrow NaCN$ (if N is present)

$2Na + S + \longrightarrow Na_2S$ (if S is present)

$Na + X \longrightarrow NaX$ (if X is present)

$Na + C + N + S \xrightarrow{\text{fusion}} NaCNS$

(if both N & S are present)

15. (a) Kjeldal's method is suitable for estimating nitrogen in those compounds in which nitrogen is linked to carbon and hydrogen. The method is not used in case of nitro, azo and azoxy compound. This method is basically used for estimating nitrogen in food fertilizers and agricultural products.

20. (d) BF_3 has electron deficient B, hence it is an electrophile.

21. (c) Such isomers, which possess the same molecular and structural formula but differ in the arrangement of atoms around the double bonded carbon atoms are known as geometrical isomers.

```
H – C – Cl          H – C – Cl
    ||                  ||
H – C – Cl         Cl – C – H
cis-                trans-
```

22. (b) Sodium cyanide ($Na + C + N \rightarrow NaCN$). (Lassaigne's test)

23. (d) Organic compounds having same molecular formula but differ from each other in physical properties or chemical properties or both are known as isomers.

24. (d) Due to + I-effect of the CH_3 group, toluene has much higher electron density in the ring than benzene, nitrobenzene and benzoic acid as nitro and carboxylic group show- I-effect and hence toluene is most reactive towards nitration.

25. (b) The IUPAC name of

$$\overset{1}{C}H_2 = \overset{2}{C}H - \overset{3}{C}H_2 - \overset{4}{C}H_2 - \overset{5}{C} \equiv \overset{6}{C}H$$

is Hex-1-en-5-yne or 1-hexene-5-yne

The lowest number is given to the C = C double bond.

26. (b) Optical and geometrical isomerism pair up to exhibit stereoisomerism. This is because the isomers differ only in their orientation in space.

27. (d) $H_3\ddot{O}^+ \left(\begin{array}{c} H \\ H - \overset{+}{\underset{\cdot\cdot}{O}} - H \end{array} \right)$ has a lone pair of electrons on oxygen atom, thus not an electrophile. Also octet is complete.

28. (a) Duma's method is based upon the fact that nitrogenous compound when heated with CuO in an atmosphere of CO_2 yield free nitrogen.

$$C_xH_yN_z + (CuO) \longrightarrow xCO_2 + \frac{y}{2}H_2O + \frac{z}{2}N_2 + (Cu)$$

29. (c) Blood red colour is due to the formation of ferric thiocyanate, $Fe(CNS)_3$. Hence N and S both elements must be present in the compound.

30. (c) In homolytic fission each of the atoms acquires one of the bonding electrons producing free radicals (species having one unpaired electron).

$$A \overset{\bullet\bullet}{-} B \longrightarrow A\bullet + \bullet B$$

31. (c) In $C_2H_5 - NH_2$, the lone pair of electrons on nitrogen is not conjugated and hence resonance is not possible.

Whereas in benzene, toluene and aniline electrons localised, hence show resonance.

35. (c) C_5H_{10} has 1° degree of unsaturation since the isomers are acyclic, all of these are alkenes. For writing the isomers, first introduce the double bond at different possible positions, and then consider the possibility of branching in the alkyl group.

$CH_3CH_2CH_2CH = CH_2$

1-pentene (*i*)

$CH_3CH_2CH{=}CHCH_3$

2- pentene, (*cis,- trans*) (*ii*), (*iii*)

```
      CH3
      |
CH3 – CH – CH = CH2
```

3-methyl-1-butene,

```
         CH3
         |
CH3CH2 C = CH2
```

(*iv*) 2-methyl-1-butene, (*v*)

```
      CH3
      |
CH3 – C = CHCH3
```

2-methyl-2-butene, (*vi*)

37. (b) Structure (*b*) is a 3° carbocation, while (*a*) is 2° and (*c*) and (*d*) are 1° carbocations; thus (*b*) is the most stable.

40. (b) Five chain isomers are possible which are –

(i) $CH_3-CH_2-CH_2-CH_2-CH_2-CH_3$
n-hexane

(ii) $CH_3-CH(CH_3)-CH_2-CH_2-CH_3$
2 methyl pentane

(iii) $CH_3-C(CH_3)_2-CH_2-CH_3$
2, 2 dimethyl butane

(iv) $CH_3-CH(CH_3)-CH(CH_3)-CH_3$
2, 3 dimethyl butane

(v) $CH_3-CH_2-CH(CH_3)-CH_2-CH_3$
3 methyl pentane

41. (b) $^1CH_3-{}^2CH(CH_3)-{}^3CH(CH_2CH_3)-{}^4CH_2-{}^5CH_3$
3-Ethyl-2-methylpentane

42. (b) Hydrazine (NH_2NH_2) does not contain carbon and hence on fusion with Na metal, it cannot form NaCN; consequently hydrazine does not show Lassaigne's test for nitrogen.

43. (d) The boiling point of o-nitrophenol is less than para-nitrophenol due to presence of intramolecular hydrogen bonding. Since p-nitrophenol 'is less' volatile' in than o-nitrophenol due to presence of inter molecular hydrogen bonding hence they can be separated by steam distillation.

44. (b) The possible isomers of the compound with molecular formula C_7H_8O is 5. These are $C_6H_5OCH_3$, $C_6H_5CH_2OH$ and
(Anisol) benzyl alcohol

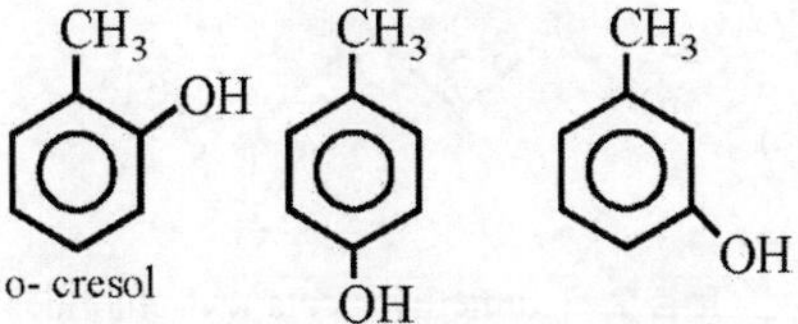

o- cresol, p-cresol, m-cresol

45. (b) When many substituents are present, the numbering is done from the end where the sum of locants is the lowest (lowest sum rule)

$$\overset{7}{C}H_3-\overset{6}{C}H(CH_3)-\overset{5}{C}H(CH_3)-\overset{4}{C}H_2-\overset{3}{C}H_2-\overset{2}{C}H(CH_3)-\overset{1}{C}H_3$$

2, 5, 6 trimethyl heptane (wrong)
2 + 5 + 6 = 13

$$\overset{1}{C}H_3-\overset{2}{C}H(CH_3)-\overset{3}{C}H(CH_3)-\overset{4}{C}H_2-\overset{5}{C}H_2-\overset{6}{C}H(CH_3)-\overset{7}{C}H_3$$

2 + 3 + 6 = 11
2, 3, 6 Trimethyl Heptane (correct)

46. (c) More is the electron - dificiency of the carbonyl carbon, greater will be the reactivity of the carbonyl compounds towards nucleophilic addition.

$CH_3CH_2CH_2CHO$ > (cyclopentanone) > $CH_3CH_2COCH_2CH_3$

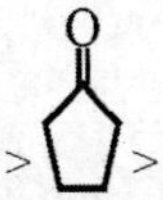

47. (d) $-OCH_3$ activates the benzene ring. $-NO_2$ deactivates the ring. Hence the reaction of the given compounds with electrophiles is in the order, I > II > III.

48. (d) —Cl atom shows o/p-directive influence but deactivate the benzene ring, while [—OH/—CH_3] groups show o/p influence but activate the benzene ring but in these —OH group is more activating than —CH_3.
Hence order of electrophilic substitution

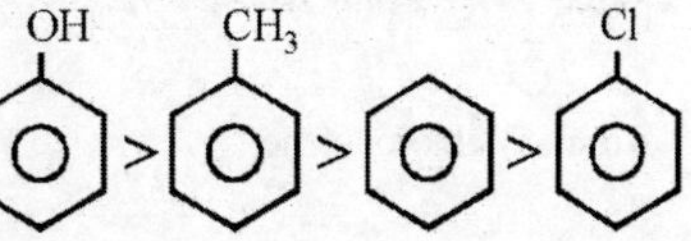

(phenol, OH) > (toluene, CH_3) > (benzene) > (chlorobenzene, Cl)

49. (c) The atom or group which has more power to attract electrons in comparision to hydrogen is said to have -I effect. Thus higher the electronegativity of atom stronger will be the -I effect. As electronegativity of N, O & F follow the order N < O < F hence based upon electronegative character order of-I effect is $-NR_2$ < – OR < – F.

50. (a) Electron withdrawing group ($-NO_2$) increases acidity while electron releasing group ($-CH_3$, –H) decreases acidity. Also effect will be more if functional group present at para position then ortho and meta position.

51. (b) Stability of an alkene depends upon the heat of hydrogenation of an alkene. The lower the heat of hydrogenation of an alkene higher will be stability.

Order of stability	Heat of hydrogenation (kJ/mol)
trans-2-butene	115.5
cis-2-butene	119.6 and
1-butene	126.8 respectively.

52. (a) CH_3 and NH_2 are electron repelling groups, whereas NO_2 and OH are electron attracting groups and they leave the benzene ring deactivated. Due to stronger electron attracting (–I effect) effect of NO_2 group $C_6H_5NO_2$ shows least reactivity towards electrophilic substitution.

53. (b) The principle of steam distillation is based on Dalton's law of partial pressures. Suppose p_1 and p_2 be the vapour pressures of water vapour and the toluene at the distillation temperature. Toluene boils when total pressure is equal to atmospheric pressure p i.e $p = p_1 + p_2$
or $p_2 = p - p_1$
as a result when toluene boil in the presence of steam. Its partial pressure p_2 is less than atmospheric pressure.

54. (a) $\overset{1}{CH_3}-\underset{\underset{O}{||}}{\overset{2}{C}}-\underset{\underset{CH_3}{|}}{\overset{3}{CH}}-\overset{4}{CH_3}$

3 Methyl-2- butanone

$\overset{1}{CH_3}-\underset{\underset{CH_3}{|}}{\overset{2}{CH}}-\underset{\underset{\overset{4}{CH_2}-\overset{5}{CH_3}}{|}}{\overset{3}{CH}}-CH_3$

2,3dimethyl pentane

$\overset{4}{CH_3}-\underset{\underset{Cl}{|}}{\overset{3}{CH}}-\underset{\underset{Br}{|}}{\overset{2}{CH}}-\overset{1}{CH_3}$

2 Bromo 3 chlorobutane

$\overset{1}{CH_3}-\overset{2}{C}\equiv\overset{3}{C}-\underset{\underset{CH_3}{|}}{\overset{\overset{^5CH_3}{|}}{\overset{4}{CH}}}$

4 - Methyl-2-pentyne

55. (b) CH_3 (1) – 2 – 3 (CH_3) – 4 (CH_2CH_3) – 5 – 6 – 7 – CH_3

4 ethyl - 3 methyl octane

56. (b) In the presence of UV rays or energy, by boiling chlorine, free radical is generated which attack the methyl carbon atom of the toluene.

$C_6H_5CH_3 \xrightarrow{\Delta} C_6H_5\dot{C}H_2 + H^{\bullet}$
(benzyl free radical)

$Cl_2 \xrightarrow{h\nu} 2Cl^{\bullet}$

$C_6H_5\dot{C}H_2 + Cl^{\bullet} \longrightarrow C_6H_5CH_2Cl$

57 (b) In diphenylmethane monochlorination at following positions will produce structured isomers

$C_6H_5-CH_2-C_6H_5$ (positions 1, 2, 3 on the ring and 4 on CH_2)

(1) p-$ClC_6H_4-CH_2-C_6H_5$

(2) o-$ClC_6H_4-CH_2-C_6H_5$

(3) m-$ClC_6H_4-CH_2-C_6H_5$

(4) $C_6H_5-CH(Cl)-C_6H_5$

58. (c) Phenols are more acidic than alcohol as they are resonance stabilised whereas alcohols are not .
Further nitro is an electron withdrawing which increases acidic character and facilitates release of proton, whereas $-CH_3$ is an electron donating group.Which decreases acidic character , thus removal of H^+ becomes very difficult.

59. (b) Among the given compounds naphthelene is volatile but benzoic acid is non-volatile (it forms a dimer). So, the best method for their separation is sublimation, which is applicable to compounds which can be converted directly into the vapour phase from its solid state on heating and back to the solid state on cooling. Hence it is the most appropriate method.

60. (c) $\overset{5}{CH_3}-\overset{4}{CH_2}-\underset{\underset{CH_3}{|}}{\overset{3}{CH}}-\underset{\underset{CH_3}{|}}{\overset{2}{CH}}-\overset{1}{C}(=O)-Cl$

It is 2, 3 dimethyl pentanoyl chloride.

13. HYDROCARBONS

ANSWER KEY

1	(a)	**11**	(c)	**21**	(a)	**31**	(c)	**41**	(b)	**51**	(d)
2	(a)	**12**	(b)	**22**	(c)	**32**	(b)	**42**	(b)	**52**	(a)
3	(d)	**13**	(d)	**23**	(d)	**33**	(a)	**43**	(b)	**53**	(a)
4	(a)	**14**	(a)	**24**	(c)	**34**	(c)	**44**	(d)	**54**	(a)
5	(c)	**15**	(b)	**25**	(c)	**35**	(c)	**45**	(b)	**55**	(a)
6	(c)	**16**	(c)	**26**	(c)	**36**	(c)	**46**	(a)	**56**	(d)
7	(c)	**17**	(c)	**27**	(c)	**37**	(c)	**47**	(a)	**57**	(a)
8	(d)	**18**	(a)	**28**	(a)	**38**	(d)	**48**	(a)	**58**	(c)
9	(d)	**19**	(c)	**29**	(c)	**39**	(a)	**49**	(d)	**59**	(a)
10	(b)	**20**	(b)	**30**	(c)	**40**	(c)	**50**	(a)	**60**	(c)

Hints & Solutions

2. (a) CH_4 has only one carbon atom, hence it can't be prepared by Wurtz reaction, which involves two molecules of alkyl halide.

5. (c) Order of stability : staggered anti > gauche > skew boat > eclipsed.
Newmann projection of n butane is given as

staggered (Anti) Eclipsed Skew or gauche

The maximum staggered conformation is most stable in which methyl groups are far apart as for as possible, due to minimum repulsion between methyl groups and is also called anti conformation.

6. (c) Dehydration of ethanol (alcohol) takes place in presence of acid.

$$CH_3CH_2OH \xrightarrow{H^+} CH_3CH_2\overset{\oplus}{O}H_2 \xrightarrow{-H_2O}$$

$$CH_2\overset{\oplus}{C}H_2 \xrightarrow{-H^+} CH_2 = CH_2$$

7. (c) Dehydrohalogenation takes place.

$$\underset{\text{Ethyl bromide}}{CH_3CH_2Br} \xrightarrow[\text{KOH}]{\text{Alcoholic}} \underset{\text{Ethylene}}{CH_2 = CH_2}$$

8. (d) Alcoholic KOH is used for dehydrohalogenation e.g.

$$CH_3 - CH_2 - CH_2 - CH_2Cl \xrightarrow[\text{KOH}]{\text{alc.}} CH_3 - CH_2 - CH = CH_2$$

9. (d) $$(CH_3)_2C = O + O = C(CH_3)(C_2H_5) \xrightarrow{\text{Ozonolysis}} \underset{\text{2,3-Dimethyl-2-pentene}}{(CH_3)_2C = C(CH_3)(CH_2CH_3)}$$

10. (b) Polythene is manufactured by heating ethylene to 473K under a pressure of 1500 atmosphere and in the presence of a trace of oxygen.

$$n(CH_2 = CH_2) \xrightarrow[\text{Trace of oxygen}]{\text{473K, 1500atm}} (-CH_2 - CH_2-)_n$$

The polythene manufactured in this way is called low density of polythene.

11. (c) On heating ethylene chloride (1, 1 dichloro ethane) with alcoholic potash followed by sodamide alkyne is obtained

$$R - CH_2 - CCl_2 - R \xrightarrow{\text{alc.KOH}} R - CH = CCl - R \xrightarrow{NaNH_2} R - C \equiv C - R$$

12. (b) Reaction of alkyne with dil H_2SO_4 follows Markownikoff's addition of H_2O which takes place in presence of $HgSO_4$. Pd and Pt are catalysts suitable for hydrogenation. Anhydrous $AlCl_3$ is used in Friedel Craft's reaction.

13. (d) Huckel's rule states that for aromaticity there must be (4n + 2) π electron present in compound where n is an integer

14. (a) This is an example of Friedel - Craft alkylation.

15. (b) $C_6H_6 \xrightarrow[H_2SO_4,\ 328K]{HNO_3} C_6H_5NO_2 \xrightarrow[H_2SO_4,\ 350K]{HNO_3}$ m-dinitrobenzene

Nitro group being deactivating group is m-directing, the second $-NO_2$ group goes at m-position and also nitration of nitrobenzene requires higher temperature.

16. (c) Phenol as well as toluene have electron-releasing groups, however –OH group, in phenol, is more electron-releasing (due to +R effect) than the $-CH_3$ group in toluene, so phenol is more easily attacked by an electrophile.

17. (c) Benzene do not show addition reaction like other unsaturated hydrocarbons. However it show substitution reactions. Due to resonance all the C – C bonds have the same nature, which is possible because of the cyclic delocalisation of π-electrons in benzene. Monosubstitution will give only a single product.

18. (a) Only C_2H_2 (acetylene) has acidic H-atoms and hence reacts with $NaNH_2$ to form sodium salt, i.e.,

$$HC \equiv CH + NaNH_2 \longrightarrow HC \equiv CNa + NH_3.$$

19. (c) $CH_3Cl + AlCl_3 \longrightarrow \underset{\text{Electrophile}}{CH_3^+} + AlCl_4^-$

20. (b) $6R-CH=CH_2 \xrightarrow[\text{Ether}, 0°C]{B_2H_6} 2(RCH_2CH_2)_3$

$$\xrightarrow[NaOH]{H_2O_2} 6RCH_2CH_2OH + 2H_3BO_3.$$

21. (a) $C_2H_5MgBr + H_2O \longrightarrow C_2H_6 + Mg(Br)(OH)$

22. (c) $C_6H_6 + CH_3Cl \xrightarrow{\text{anhydrous } AlCl_3} C_6H_5CH_3 + HCl$

23. (d) In presence of peroxide, HBr adds on alkenes in anti–markovnikov's way, thus

$$\underset{\text{Propene}}{H_3C-CH=CH_2} + HBr \xrightarrow{\text{Peroxide}} \underset{\text{n-propyl bromide}}{H_3C\ CH_2CH_2Br}$$

Kharasch observed that the addition of HBr to unsymmetrical alkene in the presence of organic peroxides follows an opposite course to that suggested by Markownikoff. This is termed anti-Markownikoff or peroxide effect.

24. (c) When both double and triple bonds are present, then triple bond is considered as the principal group.

25. (c) In arenes, π electrons are delocalised, hence arenes do not undergo addition reactions easily. Aromatic compounds (Arenes) are highly stable and show resorance. eg. Benzene is the simplest example.

26. (c) Kekule in 1865 suggested a ring structure of benzene in which the ring was composed of six carbon atoms, each of which carries one atom of hydrogen. To satisfy the fourth valency of the carbon atom, he suggested three alternate double bonds.

(Kekule structure: ring of six CH units — CH, HC, CH, HC, CH, CH)

27. (c) $CH_3CH_2Br \xrightarrow{C_2H_5O^-Na^+} CH_3\overset{\oplus}{C}H_2 + C_2H_5O^{\ominus}$

Ethoxide ion is a strong base which abstracts proton from carbocation to form alkene.

$$\underset{|\atop H}{CH_2}-\overset{\oplus}{C}H_2 + C_2H_5\overset{\ominus}{O} \longrightarrow CH_2 = CH_2 + C_2H_5OH$$

Thus, there is a loss of hydrogen and halogen both leading to dehydrohalogenation.

28. (a) $CH_2 = CH_2 + \underset{H_2SO_4}{H-O-SO_2OH} \longrightarrow$

$$CH_3\overset{\oplus}{C}H_2 + \overset{\ominus}{O}SO_2OH \longrightarrow \underset{\text{Ethyl hydrogen sulphate}}{CH_3CH_2OSO_2OH}$$

Addition of sulphuric acid takes place according to Markownikoff's rule. Alkanes do not absorb cold conc. H_2SO_4.

29. (c) According to Markownikoff's rule, "in case of addition of an unsymmetrical reagent (H– X), the positive part get attached to the C which is least substituted or which bears larger number of hydrogen atoms."

$$R(H)C=CH_2 + H-X \longrightarrow R-\underset{H}{\overset{H}{C}}-\overset{H}{C}-H$$

Markownikoff's rule is based on the stability of carbocations (3° > 2° > 1° > methyl).

30. (c) In ethyne both carbons are *sp* hybridised. – C – H bond becomes weak as s-character of C atom increases. The electrons of the C–H bond in ethyne are strongly held by carbon nuclei. This facilitates the removal of hydrogen as proton.

$\underset{sp\quad sp}{H-C\equiv C-H}$ (ethyne), $\underset{sp^2\quad sp^2}{CH_2=CH_2}$ (ethene),

benzene (all carbons sp^2), CH_3-CH_3 (ethane, sp^3)

31. (c) Acetylene molecule can be represented as,

$$H\overset{}{\underset{\sigma}{-}}C\overset{\pi}{\underset{\pi}{\equiv}}{}^{\sigma}C\underset{\sigma}{-}H$$

So, it contains 3 σ and 2 π bonds.

32. (b) Alkenes react with cold alkaline $KMnO_4$ solution to form glycols. During this reaction $KMnO_4$ (violet) gets converted to MnO_2 (brown ppt.) and therefore the colour of $KMnO_4$ is disappeared.

$$>C=C<+KMnO_4 \xrightarrow{OH^{\ominus}} -\underset{OH}{\overset{|}{C}}-\underset{OH}{\overset{|}{C}}-+MnO_2+KOH$$

1,2–diols

This test is known as Baeyer test.

33. (a) When alkyl halide is treated with sodium metal in presence of ether, alkane is obtained, this reaction is called as Wurtz reaction.

$$R-\mathbf{X}+\mathbf{2Na}+\mathbf{X}-R' \xrightarrow{\text{dry ether}} R-R'+2NaBr$$

$$CH_3-Br + 2Na+Br-CH_3 \xrightarrow{\text{dry ether}} CH_3-CH_3+2NaBr$$

34. (c) If alkyl group increases in alkene, then its stability also increases. So $R_2C=CR_2$ is more stable than others. Stability of alkene increases in the order

$$CH_2=CH_2 < RCH=CH_2 < RCH=CHR < R_2C=CR_2$$

35. (c) $C_2H_5OH \xrightarrow[KMnO_4]{conc.\ H_2SO_4} C_2H_4+H_2O$

Note : If ethyl alcohol is taken in excess and the reaction is carried out at a temperature of 433–443 K diethyl ether is formed.

$$\underset{(excess)}{2C_2H_5OH} \xrightarrow[433-443\ K]{conc.\ H_2SO_4} C_2H_5OC_2H_5+H_2O$$

36. (c) Other three methods can be used for the preparation of hydrocarbons having at least 2 carbon atoms.

42. (b) Cyclohexyl-propyne $\xrightarrow[H_2O]{HgSO_4\ H_2SO_4}$ [enol (OH)] $\xrightarrow{\text{Tautomerisation}}$ ketone (O)

43. (b) Reduction of alkynes with Na/liq. NH_3 gives trans-alkenes.

44. (d) Ammonical silver nitrate and ammonical cuprous chloride give precipitate only with terminal alkynes (propyne, $CH_3C\equiv CH$)

45. (b) Mesitylene (1,3,5-trimethylbenzene) has three electron-releasing ($-CH_3$) groups, while m-xylene, toluene and benzene have 2, 1 and 0 electron-releasing ($-CH_3$) groups

46. (a) $:\ddot{O}:^-$ (phenoxide) > $:\ddot{O}H$ (phenol)

Most electron-releasing due to –ve charge on O

Electron releasing due to –OH group

> $:\ddot{O}-\overset{O}{\overset{\|}{C}}-CH_3$ (phenyl acetate) > $\overset{O}{\overset{\|}{C}}-CH_3$ (acetophenone)

Less electron releasing due to presence of –CO grouping which shows resonance with the electrons on O

Electron-withdrawing group

47. (a) $:\ddot{O}CH_3$ (anisole) > CH_3 (toluene)

Strong electron-releasing group is present due to + R

Electron-releasing group is present

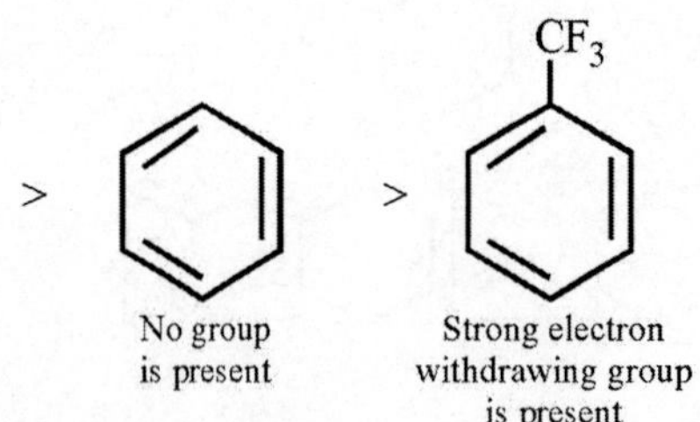

No group is present

Strong electron withdrawing group is present

48. (a) The acidity of acetylene or 1--alkynes can be explained on the basis of molecular orbital concept according to which formation of C—H bond in acetylene involves sp-hybridised carbon atom. Now since s electrons are closer to the nucleus than p electrons, the electrons present in a bond having more s character will be correspondingly more closer to the nucleus.
Thus owing to high s character of the C—H bond in alkynes (s = 50%), the electrons constituting this bond are more strongly held by the carbon nucleus (i.e., the acetylenic carbon atom or the sp orbital acts as more electronegative species than the sp^2 and sp^3 with the result the hydrogen present on such a carbon atom (≡C—H) can be esily removed as a proton.

49. (d) Br_2 in CCl_4 (a), Br_2 in CH_3 COOH (b) and alk. $KMnO_4$ (c) will react with all unsaturated compounds, i.e., 1, 3 and 4 while ammonical $AgNO_3$ (d) reacts only with terminal alkynes, i.e., 3 and hence compund 3 can be distinguished from 1, 2 and 4 by. ammonical $AgNO_3$ (d).

50. (a) Peroxide effect is observed only in case of HBr. Therefore, addition of HCl to propene even in the presence of benzyoyl peroxide occurs according to Markovnikov's rule :

$$CH_3-CH=CH_2 \xrightarrow[(C_6H_5CO)_2O_2]{HCl} CH_3-CHCl-CH_3$$

51. (d) Of all the options listed $CH_3CH_2CH_2CH_3$ has the least number of C-atoms and hence has the lowest b.p.

52. (a) $Cl_2 \xrightarrow[\text{Chain initiation}]{hv} 2Cl^{\bullet}$

53. (a) $CH_3-CH_2-CH=CH-CH_2-CH_3 \xrightarrow[\text{Platinum}]{H_2} CH_3-CH_2-CH_2-CH_2-\underset{\displaystyle CH_3}{\underset{|}{CH_2}}$

$\xrightarrow{HBr} CH_3-CH_2-\underset{\displaystyle Br}{\underset{|}{CH}}-CH_2-CH_2-CH_3$

54. (a) Electrolysis of a concentrated aqueous solution of either sodium or potassium salts of saturated mon-carboxylic acids yields higher alkane at anode.

$$2RCOOK \xrightarrow[\text{Oxidation}]{\text{Electrolytic}} \underset{\text{Anode}}{2RCOO^-} + \underset{\text{Cathode}}{2K^+}$$

At anode $2RCOO^- \rightarrow 2RCOO + 2e^- \longrightarrow R-R + 2CO_2$

At Cathode $2K^+ + 2e^- \rightarrow 2K$
$2K + H_2O \rightarrow 2KOH + H_2\uparrow$

(Kolbe's Method)

55. (a) Given

$$A \xrightarrow{Cl_2} B \xrightarrow{alc./KOH} \underset{\text{Hydrocarbon}}{C} \xrightarrow{O_3/H_2O} CH_2O$$

Since hydrocarbon C give only CH_2O, on ozonolysis, C should be $CH_2 = CH_2$ hence going backward A should be ethane. Thus the reactions are

$$\underset{(A)}{CH_3CH_3}\xrightarrow{Cl_2/h\nu}\underset{(B)}{CH_3CH_2Cl}\xrightarrow[KOH]{alc.}\ldots$$

$$\underset{(C)}{CH_2{=}CH_2}\xrightarrow[\Delta]{O_3/H_2O}\underset{(D)}{HCHO}$$

56. (d) We know that

$$\underset{\text{1, Propene}}{6\,CH_3-CH=CH_2}\xrightarrow[\text{ether, }0°C]{B_2H_6}$$

$$2(CH_3CH_2CH_2)_3\xrightarrow[OH^-]{H_2O_2}$$

$$6CH_3CH_2CH_2OH+2H_3BO_3$$
$$\text{Propanol}$$

57. (a)

$$CH_3-CH_2-\overset{H}{\overset{|}{C}}=\underset{CH_3}{\underset{|}{\overset{CH_3}{\overset{|}{C}}}}\xrightarrow{O_3}CH_3-CH_2-\underset{\underset{O-O}{|}}{\overset{H\;\;O}{\overset{\backslash\;/\;\backslash}{C}}}-C{\small\begin{matrix}CH_3\\CH_3\end{matrix}}$$

$$CH_3-\overset{O}{\overset{||}{C}}-CH_3+CH_3-CH_2-CHO\xleftarrow{(-H_2O)}$$

58. (c) $C_6H_6+CH_2=CH_2\xrightarrow{AlCl_3}C_6H_5CH_2CH_3$

59. (a) Heat of hydrogenation of alkene

$$\propto\frac{1}{\text{Stability of alkene}}$$

Hence the alkene which will react fastest with H_2 will be the least stable. Among the given options the compound having least number of alkyl groups (R) will be the least stable.

Further the relative rates of hydrogenation decrease with increase of steric hindrance

$$\begin{matrix}R\\R\end{matrix}{>}C=C{<}\begin{matrix}R\\R\end{matrix}\;>\;\begin{matrix}R\\R\end{matrix}{>}C=C{<}\begin{matrix}R\\H\end{matrix}$$

$$>\;\begin{matrix}R\\R\end{matrix}{>}C=C{<}\begin{matrix}H\\H\end{matrix}\;>\;\begin{matrix}R\\H\end{matrix}{>}C=C{<}\begin{matrix}R\\H\end{matrix}$$

Hence $\begin{matrix}R\\H\end{matrix}{>}C=C{<}\begin{matrix}R\\H\end{matrix}$ will react fast with H_2

60. (c) This reaction occurs according to Markownikoff's rule which states that when an unsymmetrical alkene undergo hydrohalogenation, the negative part goes to that C-atom which contain lesser no. of H-atom.

$$CH_3-CH_2-C\equiv CH+HCl$$

$$\longrightarrow CH_3-CH_2-\underset{Cl}{\underset{|}{C}}=CH_2$$

$$\xrightarrow{HI}CH_3-CH_2-\underset{Cl}{\underset{|}{\overset{I}{\overset{|}{C}}}}-CH_2$$

14. ENVIRONMENTAL CHEMISTRY

ANSWER KEY

1	(c)	**11**	(a)	**21**	(c)	**31**	(d)	**41**	(a)	**51**	(c)
2	(d)	**12**	(a)	**22**	(a)	**32**	(c)	**42**	(a)	**52**	(b)
3	(c)	**13**	(d)	**23**	(c)	**33**	(a)	**43**	(c)	**53**	(b)
4	(a)	**14**	(d)	**24**	(b)	**34**	(d)	**44**	(d)	**54**	(d)
5	(a)	**15**	(b)	**25**	(c)	**35**	(b)	**45**	(d)	**55**	(a)
6	(c)	**16**	(c)	**26**	(b)	**36**	(d)	**46**	(c)	**56**	(a)
7	(d)	**17**	(a)	**27**	(d)	**37**	(a)	**47**	(b)	**57**	(b)
8	(a)	**18**	(d)	**28**	(b)	**38**	(b)	**48**	(a)	**58**	(d)
9	(c)	**19**	(c)	**29**	(b)	**39**	(a)	**49**	(a)	**59**	(d)
10	(a)	**20**	(a)	**30**	(b)	**40**	(c)	**50**	(a)	**60**	(d)

Hints & Solutions

1. (c) Air pollution greatly affect the troposphere.
2. (d) The uppermost region of atmosphere is exosphere.
5. (a) CO_2 causes Green House Effect.
7. (d) Acid rain contains $H_2SO_4 > HNO_3 > HCl$.
8. (a) Dust
10. (a) NO and freons are responsible for ozone depletion.
11. (a) Pesticides cause water pollution.
13. (d) Addition of phosphate fertilizers to water leads to nutrient enrichment (eutrophication)
15. (b) Green chemistry may be defined as the programme of developing new chemical products and chemical processes or making improvements in the already existing compounds and processes so as to make less harmful to human health and environment. This means the same as to reduce the use and production of hazardous chemicals.
18. (d) DDT causes water, air and soil pollution.
19. (c) water is often treated with Cl_2 to kill germs.
20. (a) Green house gases such as CO_2, ozone, methane, the chlorofluoro carbon compounds and water vapour form a thick cover around the earth which prevents the IR rays emitted by the earth to escape. It gradually leads to increase in temperature of atmosphere.
26. (b) CO is primary pollutant.
27. (d) Depletion of ozone layer causes skin cancer.
28. (b) CO_2 is generally not regarded as pollutant.
29. (b) Thermal pollution is caused by power plants. Power plant requires a larger quantity of water for cooling. The water after cooling is left in the water body. The temperature of left water is generally very high and affects aquatic life.
30. (b) They create holes in ozone layer.
33. (a) Textiles cause lung cancer.
34. (d) Ozone hole is reduction in ozone layer in stratosphere.
35. (b) Domestic sewage constitute biodegradable pollutants.
36. (d) The mosquitoes will increase their population and spread malaria.
41. (a) Phosphate pollution is caused by sewage and agricultural fertilizers.
42. (a) Oil slick in sea water decreases D.O value.
45. (d) Ozone layer acts as a shield and does not allow ultraviolet radiation from sun to reach earth. It does not prevent infra-red radiation from sun to reach earth, thus option (d) is wrong statement and so it is the correct answer.
46. (c) CO and oxides of Nitrogen are poisnous gases present in automobile exhaust gases.
47. (b) The ideal value of D.O for growth of fishes is 8 mg/ℓ. 7mg ℓ is desirable range, below this value fishes get susceptible to desease. A value of 2 mg/ℓ or below is lethal for fishes.
48. (a) BOD of pond is connected with microbes and organic matter.
49. (a) Normal rain water has pH 5.6. Thunderstorm results in the formation of NO and HNO_3 which lowers the pH.
50. (a) $O_2 \longrightarrow O + O;\ O + O_2 \longrightarrow O_3$
53. (b) The coldest region is mesosphere (temp. –27°C to –92°C)
54. (d) London smog is formed in morning during winter.
55. (a) Radiation coming from sun or outerspace have high energy or short wavelength, which are allowed to enter by green house gases. However, radiation emitted by earth is in infrared region, having long wavelength, are reflected back by the envelope of green house gases.
58. (d) [Many organic substances break up into simpler substances by taking up dissolved oxygen in presence of some bacteria. As they continue, dissolved oxygen, runs short for aquatic life, which is then badly affected. Moreover many harmful products are formed]

15. THE SOLID STATE

ANSWER KEY

1	(d)	11	(d)	21	(c)	31	(a)	41	(d)	51	(b)
2	(d)	12	(a)	22	(d)	32	(c)	42	(d)	52	(b)
3	(a)	13	(d)	23	(d)	33	(d)	43	(b)	53	(a)
4	(a)	14	(a)	24	(a)	34	(c)	44	(b)	54	(d)
5	(c)	15	(d)	25	(d)	35	(d)	45	(b)	55	(a)
6	(d)	16	(d)	26	(b)	36	(b)	46	(c)	56	(c)
7	(d)	17	(b)	27	(a)	37	(d)	47	(c)	57	(a)
8	(b)	18	(b)	28	(b)	38	(a)	48	(d)	58	(c)
9	(a)	19	(b)	29	(c)	39	(d)	49	(b)	59	(b)
10	(c)	20	(c)	30	(c)	40	(b)	50	(a)	60	(c)

Hints & Solutions

1. (d) Crystals show good cleavage because their constituent particles are arranged in planes.
2. (d) Ionic compounds are dissociated in solution state and form ions. Ions are good carrier of charge. So it makes solution conducting.
3. (a) For a cubic structure,

No. of X atoms $= 8 \times \frac{1}{8} = 1$

No. of Y atoms $= 6 \times \frac{1}{2} = 3$

$\therefore$ Formula of the compound $= XY_3$

4. (a) Potassium crystallises in BCC lattice.
5. (c) Solid CH_4 is a molecular solid. In this, the constituent molecules are held together by vander Waal's forces.
6. (d) Since in NaCl type of structure 4 formula units form a cell. Number of formulas in cube shaped

crystals $= \frac{1.0}{58.5} \times 6.02 \times 10^{23} = 1.029 \times 10^{22}$

$\therefore$ units cells $= \frac{1.0 \times 6.02 \times 10^{23}}{58.5 \times 4} = 2.57 \times 10^{21}$

unit cells.

7. (d) Number of atoms per unit cell = 1

Atoms touch each other along edges.

Hence $r = \frac{a}{2}$

(r = radius of atom and a = edge length)

Therefore % fraction $= \frac{\frac{4}{3}\pi r^3}{(2r)^3} = \frac{\pi}{6} = 0.52$

8. (b) Percentage of occupied space in a *bcc* is 68%.

$\therefore$ Percentage of free space in a *bcc* is (100 – 68) = 32% i.e., option (b) is correct

9. (a) Since lithium metal crystallises in a body centred cubic crystal

$\therefore \; r = \frac{a\sqrt{3}}{4} = \frac{351 \times 1.732}{4} = 151.98$ pm

11. (d) When electrons are trapped in anion vacancies, these are called F-centres.

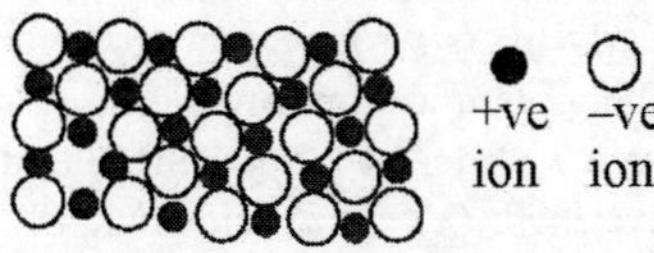

F- centre in crystal

12. (a) p-type of semiconductors are produced

(a) due to metal deficiency defects

(b)by adding impurity containing less electrons (i.e. atoms of group 13)

Ge belongs to Group 14 and In to Group 13. Hence on doping, p-type semicondutor is obtained.

13. (d) The semiconductors formed by the introduction of impurity atoms containing one elecron less than the parent atoms of insulators are termed as p-type semiconductors. Therefore silicon containing 14 electrons is to be doped with boron containing 13 electrons to give a p- type semi-conductor.
14. (a) Schottky defects are found in solid.
15. (d) If in an ionic crystal of the type A^+, B^-, equal number of cations and anions are missing from their lattice sites so that the electrical neutrality is maintained. The defect is called Schottky defect.
16. (d) The appearance of colour in solid alkali metal halide is due to presence of F-centres found as defect in the crystal structure.
17. (b) The phenomenon of occurrence of a substance in two or more crystalline structures is called polymorphism.
18. (b) In graphite, the electrons are spread out between the sheets.
19. (b) Out of the given substances, only Li has high electrical and thermal conductivity.
20. (c) $\frac{M^+}{X^-}$ is highest in CsF
21. (c) Each of the Na^+ and Cl^- ions has coordination number of 6.

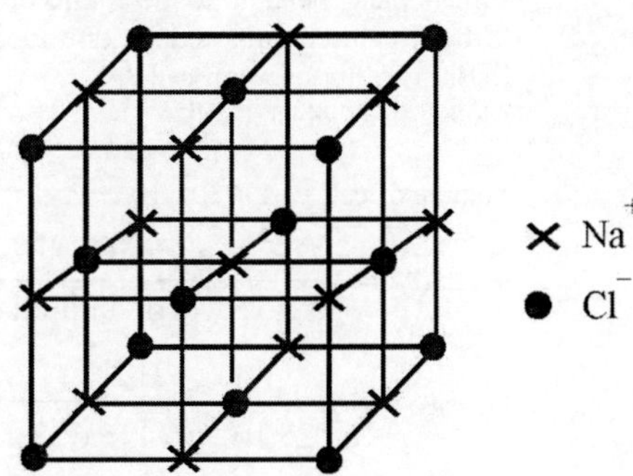

Cubical structure of sodium chloride crystal

23. (d) In KCl, co-ordination number of cation and anion is 6 and 6 respectively. KCl is highly ionic so

Schottky defect is common.
Note : Schottky defect is common in compounds having high coordination nummber while Frenkel defect is common in compounds with low coordination number.

24. (a) According to Born-Lande equation :

$$U = \frac{Z^+Z^-e^2AN}{r_{node}}\left(\frac{1}{n}-1\right)$$

Lattice energy is inversely proportional to inter-ionic distance (r_{node}) i.e., ions should be of small size to give high lattice energy.

25. (d) Since the radius ratio of CsCl is 0.93, it is expected to have a body centred cubic structure.

26. (b) In ZnS structure, sulphide ions occupy all FCC lattice points while Zn^{2+} ions are present in alternate tetrahedral voids.

27. (a) In hcp the coordination number is 12.

28. (b) For tetrahedral shape radius ratio is 0.225 – 0.414.

29. (c) Among the three options KCl, NaCl and $MgCl_2$, the size of anion is same. So larger the cation, larger will be the cation/anion ratio i.e., KCl will have larger cation/anion ratio among the three. So, we left with two options KCl and CaF_2. Among these two CaF_2 will have maximum value of cation/anion ratio because decrease in ionic radii of anion from Cl^- to F^- does not overcome the effect of decrease in ionic radii of cation from K^+ to Ca^{2+}.

30. (c) In fluorite structure each F^- ion is surrounded by four Ca^{++} ions whereas each Ca^{++} is surrounded by eight F^- ions, giving a body centred cubic arrangement. Thus the co-ordination number of $Ca^{++} = F^-$ are 8 and 4.

31. (a) Since MgO has a rock salt structure. In this structure each cation is surrounded by six anions and vice versa.

32. (c) A body-centred cubic system consists of all eight corners plus one atom at the centre of cube.

33. (d) In body centred cubic lattice one molecule of CsBr is within one unit cell.
Atomic mass of unit cell
= 133 + 80 = 213 a.m.u
Volume of cell = $(436.6 \times 10^{-12})^3$

$$\text{Density} = \frac{n \times \text{at.wt.}}{\text{Av.no.} \times \text{vol.of unit all cell}}$$

$$\text{Density} = \frac{213}{6.02 \times 10^{23} \times (436.6 \times 10^{-12})^3}$$

$$\left[\rho = \frac{ZM}{a^3 N_A}\right] = \frac{213 \times 10^7}{6.02 \times (436.6)^3} = 42.5 \text{ gm/cm}^3$$

34. (c) CsCl (8 : 8 coordination) transform to NaCl (6 : 6 coordination) on heating but reverse is not possible.

36. (b) To get n-type semi conductor the 14 group elements must be doped with 15 group elements (valence electrons 5).

37. (d) Zn^{2+} ion is too small to fit into the octahedral voids of S^{2-} ions.

38. (a) It is based on the definition of the cubic lattice.

39. (d) *d*-spacing

$$= \frac{n\lambda}{2\sin\theta} = \frac{2 \times 100}{2 \times \sin 30^\circ} = 200\text{pm}$$

40. (b) Low co-ordination compound can be changed into high co-ordination compound by increasing pressure and decreasing temperature.

$$\underset{(6,6)}{\text{NaCl}} \underset{\text{Temp.}}{\overset{\text{Pr.}}{\rightleftharpoons}} \underset{(8,8)}{\text{CsCl}}$$

41. (d) In nonstoichiometric defects ratio of cation and anion is not same that is represented by chemical ideal formula, this is due to variable oxidation nos of cation. Transition elements show variable oxidation nos.

43. (b) The volume of the unit cell
= $(2.88 \text{ Å})^3 = 23.9 \times 10^{-24}$ cm^3.
The volume of 100 g of the metal

$$= \frac{m}{\rho} = \frac{100}{7.20} = 13.9 \text{ cm}^3$$

Number of unit cells in this volume

$$= \frac{13.9 \text{ cm}^3}{23.9 \times 10^{-24} \text{cm}^3} = 5.82 \times 10^{23}$$

44. (b) For orthorhombic system $\alpha = \beta = \gamma = 90^\circ$

45. (b) Liquid crystals on heating first become turbid and then clear.

46. (c) For fcc

$$2(r_{cation} + r_{anion}) = 508$$

$$(r_{cation} + r_{anion}) = 254$$

$$r_{anion} + 100 = 254, \Rightarrow r_{anion} = 154$$

47. (c) CaO has higher lattice energy because of higher charge on Ca^{2+} and O^{2-}, which results in higher attraction. KI is more soluble in water because of low lattice energy and higher hydration energy. Clearly (c) is wrong because CaO has higher melting point as compared to KI.

48. (d) **Given :** Order of Bragg diffraction (n) = 2 ; Wavelength (λ) = 1 Å and angle (θ) = 60°. We know from the Bragg's equation $n\lambda = 2d \sin\theta$
or $2 \times 1 = 2d \sin 60^\circ$

$$\Rightarrow 2\times1 = 2.d.\frac{\sqrt{3}}{2} \Rightarrow d = \frac{2}{\sqrt{3}} = 1.15\text{Å}$$

(where d = Difference between the scattering planes)

49. (b) hcp is a closed packed arrangement in which the unit cell is hexagonal and coordination number is 12.

50. (a) Fraction of unoccupied sites in NaCl crystal

$$\frac{\text{x-ray density–pyknometric density}}{\text{x-ray density}}$$

$$= \frac{2.178\times10^3 - 2.165\times10^3}{2.178\times10^3} = \frac{0.13\times10^3}{2.178\times10^3}$$

$$= \frac{130}{2178} = 5.96\times10^{-3}$$

51. (b) Since each Sr^{++} ion provides one cation vacancy, hence

Concentration of cation vacancies = mole % of $SrCl_2$ added

$= 10^{-4}$ mole %

$$= \frac{10^{-4}}{100}\times6.023\times10^{23} = 6.023\times10^{17}$$

52. (b) $(SiO_3^{2-})_n$ Two oxygen atoms per SiO_4 tetrahedron are shared giving polymeric anion chains.

Example : Spodumene LiAl $(SiO_3)_2$, Diposide $CaMg(SiO_3)_2$.

53. (a) Following generalization can be easily derived for various types of lattice arrangements in cubic cells between the edge length (a) of the cell and r the radius of the sphere.

For simple cubic : a = 2r or $r = \frac{a}{2}$

For body centred cubic :

$$a = \frac{4}{\sqrt{3}}r \text{ or } r = \frac{\sqrt{3}}{4}a$$

For face centred cubic :

$$a = 2\sqrt{2}r \text{ or } r = \frac{1}{2\sqrt{2}}a$$

Thus the ratio of radii of spheres for these will be

simple : bcc : fcc

$$= \frac{a}{2} : \frac{\sqrt{3}}{4}a : \frac{1}{2\sqrt{2}}a$$

i.e. option (a) is correct answer.

54. (d) Since Cu metal crystallises in a face centred cubic lattice

$$\therefore r = \frac{a\sqrt{2}}{4}$$

$$r = \frac{a\times1.414}{4} = \frac{361\times1.414}{4} = 128\text{pm}$$

55. (a) For BCC lattice body diagonal = $a\sqrt{3}$.

The distance between the two oppositely charged ions $= \frac{a}{2}\sqrt{3}$

$$= \frac{387\times1.732}{2} = 335\text{pm}$$

56. (c) Radius ratio of NaCl like crystal

$$= \frac{r^+}{r^-} = 0.414$$

$$r^- = \frac{100}{0.414} = 241.5\text{pm}$$

57. (a) $\frac{1}{8}\times8(\text{at corners}) + \frac{1}{2}\times6$ (at the centres of six faces) = 4

58. (c) AB is just like NaCl. Thus twelve A^+ are at edges and 1 within body of fcc i.e. in octahedral voids and six B^- at faces and 8 at corner.

59. (b) The radius ratio $\frac{r_+}{r_-} = \frac{146}{216} = 0.675$

It lies between 0.414 – 0.732. Hence, it exhibits rock salt type structure.

60. (c) Let the number of atoms Z in the CCP arrangement = 100

Thus the number of tetrahedral sites = 200

Since all the tetrahedral sites are occupied by atoms X, the number of atoms X = 200

Hence ratio of X : Z = 2 : 1

Thus the formula is X_2Z

16. SOLUTIONS

ANSWER KEY											
1	(d)	**11**	(a)	**21**	(b)	**31**	(a)	**41**	(b)	**51**	(c)
2	(a)	**12**	(c)	**22**	(b)	**32**	(b)	**42**	(b)	**52**	(c)
3	(a)	**13**	(a)	**23**	(d)	**33**	(b)	**43**	(b)	**53**	(b)
4	(a)	**14**	(c)	**24**	(b)	**34**	(a)	**44**	(a)	**54**	(d)
5	(a)	**15**	(d)	**25**	(b)	**35**	(b)	**45**	(a)	**55**	(d)
6	(c)	**16**	(c)	**26**	(b)	**36**	(b)	**46**	(a)	**56**	(a)
7	(c)	**17**	(d)	**27**	(b)	**37**	(c)	**47**	(c)	**57**	(b)
8	(a)	**18**	(a)	**28**	(c)	**38**	(d)	**48**	(a)	**58**	(b)
9	(a)	**19**	(c)	**29**	(a)	**39**	(a)	**49**	(a)	**59**	(b)
10	(c)	**20**	(b)	**30**	(d)	**40**	(a)	**50**	(b)	**60**	(d)

Hints & Solutions

1. (d) Volume is temperature dependent, hence expression involving volume term (normality, molarity and formality) varies with temperature

$$\text{Normality} = \frac{\text{geq of solute}}{\text{vol. of solution in L}}$$

$$\text{Molarity} = \frac{\text{moles of solute}}{\text{vol. of solution in L}}$$

$$\text{Formality} = \frac{\text{formula mass}}{\text{vol. of solution in L}}$$

$$\text{Molality} = \frac{\text{moles of solute}}{\text{mass of solvent in kg}}$$

Since molality does not include the volume term, it is independent of temperature.

2. (a) Isotonic solutions have same molar concentration at given temperature provided the Van't Hoff factor (i) is same

3. (a) By increasing the temperature the vapour pressure of a substance always increases because more vapour will be formed at higher temperature.

4. (a) Given $V.P_P = 80$ torr

$V.P_Q = 60$ torr

$P_{total} = V{\cdot}P_P \times X_P + V{\cdot}P_q \times X_Q$

$$= \left[80 + \frac{3}{5} + 60 \times \frac{2}{5}\right] = 16 \times 3 + 12 \times 2$$

$P_{total} = 48 + 24 = 72$ torr

5. (a) Added HgI_2 forms a complex with KI in the solution as follows

$2KI + HgI_2 \rightarrow K_2[HgI_4]$

As a result, number of particles decreases and so ΔT_f increases.

6. (c) For ideal solution,

$$\Delta V_{mixing} = 0 \text{ and } \Delta H_{mixing} = 0.$$

7. (c) Colligative properties of dilute solution containing non volatile solute depends upon the number of particles of the solute present in the solution.

8. (a) Colligative property is strictly true only for dilute solutions which behaves as nearly ideal solution. These properties are : lowering of vapour pressure of the solvent, osmotic pressure of the solution, elevation in boiling point of the solvent, depression in freezing point of the solvent. Hence, option (a) is not true.

9. (a) Osmotic pressure of urea from the formula

$$\pi V = nRT \Rightarrow \pi = \frac{nRT}{V}$$

$$\Rightarrow \frac{10}{60} \times .0821 \times T \qquad [1\ dm^3 = 1 \text{ litre}]$$

5% solution means

100 ml ≡ 5gm

1000 ml ≡ 50 g/L

Osmotic pressure of solution having non-volatile solute $\pi_2 = \frac{50}{M} \times .0821 \times T$

For isotonic solution, $\pi_1 = \pi_2$

$$\frac{10}{60} \times .0821 \times T = \frac{50}{M} \times .0821 \times T$$

$M = 300$ gm mol^{-1}

10. (c) If compound dissociates in solvent $i > 1$ and on association $i < 1$.

11. (a) A solution of acetone in ethanol shows positive deviation from Raoult's law. It is because ethanol molecules are strongly hydrogen bonded. When acetone is added, these molecules break the hydrogen bonds and ethanol becomes more volatile. Therefore its vapour pressure is increased.

12. (c) During osmosis water flows through semipermeable membrane from lower concentration to higher concentration.

13. (a) $K_4[Fe(CN)_6]$ and $Al_2(SO_4)_3$ both dissociates to give 5 ions or i = 5

$$K_4[Fe(CN)_6] \rightleftharpoons 4K^+ + [Fe(CN)_6]^-$$

and

$$Al_2(SO_4)_3 \rightleftharpoons 2Al^{3+} + 3SO_4^{--}$$

14. (c) According to Raoult's law
Relative lowering of vapour pressure

$$= \frac{p^\circ - p}{p^\circ} = \frac{n_{solute}}{n_{solute} + N_{solvent}} = x_{solute}$$

15. (d) The number of mole particles of $BaCl_2$, NaCl and sucrose in the solution are 3, 2 and 1 respectively.

16. (c) The boiling point of water inside the cooker increases beyond 100°C due to accumulation of steam and increase in pressure. Thus making it possible to cook food faster.

17. (d) C_2H_5I and C_2H_5OH form non-ideal solution.

18. (a) $\Delta T_f = i \times K_f \times m$
Van't Hoff factor, i = 2 for NaCl,
hence $\Delta T_f = 0.02\ K_f$ which is maximum in the present case.

Hence ΔT_f is maximum or freezing point is minimum.

19. (c) Osmotic pressure is a colligative property.

20. (b) $\pi V = nRT = \frac{m}{M}RT$ or $M_2 = \left(\frac{m_2}{V}\right)\frac{RT}{\pi}$

21. (b) Solvent having high cryoscopic constant can be used in determination of molecular mass by cryoscopic method.

22. (b) One molal solution means one mole solute present in 1 kg (1000 g) solvent
i.e., mole of solute = 1

$$\text{Mole of solvent } (H_2O) = \frac{1000g}{18g} = \frac{1000}{18}$$

$$\text{Mole fraction of solute} = \frac{1}{\left(1 + \frac{1000}{18}\right)}$$

$$= \frac{18}{1018} = 0.018.$$

23. (d) In solution containing A and B component showing negative deviation A–A and B–B interactions are weaker than that of A–B interactions. For such solutions.
$\Delta H = -ve$ and $\Delta V = -ve$

24. (b) Positive deviations are shown by such solutions in which solvent-solvent and solute-solute interactions are stronger than the solvent interactions. In such solution, the interactions among molecules becomes weaker. Therefore their escaping tendency increases which results in the increase in their partial vapour pressures.
In a solutions of benzene and methanol there exists inter molecular H– bonding.

```
---O – H--- O—H --- O—H---
   |        |        |
  CH3      CH3      CH3
```

In this solution benzene molecules come between ethanol molecules which weaken intermolecular forces. This results in increase in vapour pressure.

25. (b) Equimolar solutions of normal solutes in the same solvent will have the same b. pts and same f. pts.

26. (b) Benzoic acid exists as dimer in benzene.

27. (b) Increases gradually.

28. (c) Solvent net flow from low concentration solution to high concentration solution.

29. (a) The solution which provide same number of ions are isotonic.
$Ca(NO_3)_2 \longrightarrow Ca^{2+} + 2NO_3^-$
Total ions produced = 3
$Na_2SO_4 \longrightarrow 2Na^+ + SO_4^{2-}$
Total ions produced = 3
∴ 0.1 M $Ca(NO_3)_2$ and 0.1 M Na_2SO_4 are isotonic.

30. (d) For an ideal solution, $\Delta H = 0$, $\Delta V = 0$
Hence, option (d) is incorrect.

31. (a) According to Boyle-van't Hoff law,
$\pi \propto c$ (at constant temp)

32. (b) $\Delta T_b = K_b \times i \times$ molality

where ΔT_b = elevation in boiling point
K_b = molal elevation constant
i = vant Hoff factor

∴ $\Delta T_b \propto$ molality.

33. (b) For an ideal solution, $\Delta H_{mixing} = 0$
$\Delta H = \Delta H_1 + \Delta H_2 + \Delta H_3$ (Accroding to Hess's law)
i.e., for ideal solutions there is no change in magnitude of the attractive forces in the two components present.

34. (a) Molecular masses of polymers are best determined by osmotic pressure method . Firstly because other colligative properties give so low values that they cannot be measured accurately and secondly, osmotic pressure measurements can be made at

room temperature and do not require heating which may change the nature of the polymer.

35. (b) According to Raoult's law

$$\frac{\Delta p}{p_0} = \frac{n}{n+N} \text{ (mole fraction of solute)}$$

$$\frac{10}{p_0} = 0.2 \therefore p_0 = 50 \text{ mm}$$

For other solution of same solvent

$$\frac{20}{p_0} = \frac{n}{n+N} \text{ (Mole fraction of solute)}$$

$$\Rightarrow \frac{20}{50} = \text{mole fraction of solute}$$

$\Rightarrow$ Mole fraction of solute = 0.4

As mole praction of solute + mole traction of solvent = 1

Hence, mole fraction of solvent = 1 – 0.4 = 0.6

36. (b) As $\Delta T_f = K_f\ m$

$\Delta T_b = K_b .\ m$

Hence, we have $m = \dfrac{\Delta T_f}{K_f} = \dfrac{\Delta T_b}{K_b}$

or $\Delta T_f = \Delta T_b \dfrac{K_f}{K_b}$

$$\Rightarrow [\Delta T_b = 100.18 - 100 = 0.18°C]$$

$$= 0.18 \times \frac{1.86}{0.512} = 0.654°C$$

As the Freezing Point of pure water is 0°C,

$\Delta T_f = 0 - T_f$

$0.654 = 0 - T_f$

$\therefore T_f = -0.654$

Thus the freezing point of solution will be – 0.654°C.

37. (c) Normality = $\dfrac{10 \times 1000}{60 \times 100} = 1.7$;

Mol. mass of CH_3COOH = 60.

38. (d) $C = \dfrac{\pi}{RT} = \dfrac{2.5}{0.082 \times 297} = 0.1025 \text{ mol } \ell^{-1}$

39. (a) Weight of solute or non electrolyte (w) = 1.25 gm

Weight of solvent or water (W) = 20 gm.

Freezing point of solution T = 271.94 K

Molal depression constant or cryoscopic constant

$K_f = 1.86 \text{ K mole}^{-1}$

Molar mass of solute (m) = $\dfrac{1000 \times K_f \times w}{(T_0 - T) \times W}$

$$= \frac{1000 \times 1.86 \times 1.25}{(273 - 271.94) \times 20} = \frac{50 \times 1.86 \times 1.25}{1.06}$$

= 109.67

40. (a) Given : In the Ist solution,

Weight of urea (w_u) = 12 g

$M_u = 60$; V_w = 1 litre, and

In second solution, w_c = 68.4 gm; M_c = 342

We know that according to Ist solution,

Number of the moles of urea (n_u)

$$= \frac{w_u}{M_u} = \frac{12}{60} = \frac{1}{5} = 0.2$$

According IInd solution,

Number of the moles of cane sugar n_c

$$= \frac{w_c}{M_c} = \frac{68.4}{342} = 0.2$$

We know that lowering in vapour pressure depends on the number of solute. Since both solutions have same number of moles, therefore, lowering in V. P. will be same.

41. (b) Molality = mol kg^{-1} = $\dfrac{5.85}{58.5} = 0.1$

42. (b) According to Raoult's law, the relative lowering in vapour pressure of a dilute solution is equal to the mole fraction of the solute present in the solution.

$$\frac{p^o - p}{p} = \text{mole fraction of solute} = \frac{n}{n+N}$$

43. (b) Blood cells neither swell nor shrink in isotonic solution. As isotonic solutions have equal concentration therefore there is no flow of solvent is found and hence solvent neither enters nor flow out of the blood cells.

44. (a) As both the solutions are isotonic hence there is no net movement of the solvent through the semipermeable membrane between two solutions.

45. (a) Conc. of particles in $CaCl_2$ sol. will be max. as i = 3 is max.

Note : Glucose and Urea do not dissociate into ions, as they are nonelectrolytes.

46. (a) Give vapour pressure of pure solute

$(P^0) = 121.8$ mm; Weight of solute (w) = 15 g

Weight of solvent (W) = 250 g; Vapour pressure of pure solvent (P) = 120.2 mm and Molecular weight of solvent (M) = 78

From Raoult's law

$$= \frac{P^o - P}{P^o} = \frac{w}{m} \times \frac{M}{W} = \frac{121.8 - 120.2}{121.8} = \frac{15}{m} \times \frac{78}{250}$$

$$\text{or } m = \frac{15 \times 78}{250} \times \frac{121.8}{1.6} = 356.2$$

47. (c) According to Raoult's law

$$\frac{P^o - P}{P^o} = X_B$$

$$\left[\begin{array}{l} X_B = \text{Mole fraction of solute} = \dfrac{.2}{.2+.8} = \dfrac{1}{5} \\ P = 60\text{mm of Hg} \end{array}\right]$$

$$\frac{P^o - P}{P^o} = \frac{1}{5} \text{ or } 4P^o = (P) \times 5$$

$$\Rightarrow P^o = \frac{60 \times 5}{4} = 75\text{mm of Hg}$$

48. (a) Depression in F.P. ∝ No. of particles.

$Al_2(SO_4)_3$ provides five ions on ionisation

$$Al_2(SO_4)_3 \longrightarrow 2Al^{3+} + 3SO_4^{2-}$$

while KCl provides two ions

$$KCl \rightleftharpoons K^+ + Cl^-$$

$C_6H_{12}O_6$ and $C_{12}H_{22}O_{11}$ are not ionised so they have single particle.

Hence, $Al_2(SO_4)_3$ have maximum value of depression in F.P or lowest F.P

49. (a) Mole fraction of any component A

$$x = \frac{\text{No. of moles of A}}{\text{Total No. of moles}}$$

As total no. of moles > no. of moles of A thus x can never be equal to one on zero.

50. (b) Density = 1.17 gm/cc (Given)

$$\text{As } d = \frac{\text{Mass}}{\text{Volume}}$$

volume = 1cc ∴ mass = d = 1.17g

$$\text{now molarity} = \frac{\text{No. of moles}}{\text{Volume in litre}} = \frac{11.7 \times 1000}{36.5 \times 1}$$

$$= \frac{1170}{36.5} = 32.05M$$

51. (c) From molarity equation

$M_1V_1 + M_2V_2 = MV$

$1 \times 2.5 + 0.5 \times 3 = M \times 5.5$

$$M = \frac{4}{5.5} = 0.73M$$

52. (c) $\Delta T = K_f m = 5.12 \times \dfrac{1}{250} \times \dfrac{1000}{51.2} = 0.4K$

53. (b) As $\Delta T_f = i\, K_f\, m$

For

	HX	$\rightleftharpoons$	H^+	+	X^-
t = 0	1		0		0
teq	(1 – 0.20)		0.20		0.20

Total no. of moles = 1 – 0.20 + 0.20 + 0.20
= 1 + 0.20 = 1.2

∴ $\Delta T_f = 1.2 \times 1.86 \times 0.5 = 1.1160 \approx 1.12$ K

54. (d) $\Delta T_f = 0 - (0.00732°) = 0.00732$

$\Delta T_f = i \times K_f \times m$

$$i = \frac{\Delta T_f}{K_f \times m} = \frac{0.00732}{1.86 \times 0.002} = 2$$

55. (d) When the aqueous solution of one molal KI is diluted with water, concentration decreases, therefore the vapour pressure of the resulting solution increases.

56. (a) $\Delta T_f = K_f \dfrac{1000 W_2}{M_2 W_1}$

$$= \frac{1.86 \times 1000 \times 68.5}{342 \times 1000} = 0.372$$

$T_f = -0.372°C$

57. (b) Concentration of

$$Na_2CO_3 = \frac{25.3}{106} \times \frac{1000}{250} = 0.955\text{ M}$$

$[Na^+] = 2 \times 0.955 = 1.91$ M

$\left[CO_3^{2-}\right] = 0.955$ M

58. (b) Given $K_f = -1.86°\ cm^{-1}$, mass of solute = 5.00 g, mass of solvent = 45.0 g

$\Delta T_f = i \times K_f\, m$

$$3.82 = i \times 1.86 \times \frac{5 \times 1000}{142 \times 45}$$

∴ i = 2.63 (Molecular mass of Na_2SO_4 = 142)

59. (b) 1 molal solution means 1 mole of solute dissolved in 1000 gm solvent.

∴ $n_{solute} = 1$ $w_{solvent} = 1000g$

$$\therefore n_{solvent} = \frac{1000}{18} = 55.56$$

$$X_{solute} = \frac{1}{1 + 55.56} = 0.0177$$

60. (d) Given α = 30% i.e., 0.3

$HA \longrightarrow H^+ + A^-$

$1 - \alpha \quad \alpha \quad \alpha$

$1 - 0.3 \quad 0.3 \quad 0.3$

i = 1 – 0.3 + 0.3 + 0.3

i = 1.3

$\Delta T_f = 1.3 \times 1.86 \times 0.1 = 0.2418$

$T_f = 0 - 0.2418 = -0.2418$ °C

17. ELECTROCHEMISTRY

ANSWER KEY											
1	(a)	**11**	(a)	**21**	(a)	**31**	(c)	**41**	(c)	**51**	(b)
2	(b)	**12**	(c)	**22**	(c)	**32**	(b)	**42**	(b)	**52**	(b)
3	(a)	**13**	(d)	**23**	(c)	**33**	(d)	**43**	(c)	**53**	(b)
4	(c)	**14**	(a)	**24**	(b)	**34**	(d)	**44**	(b)	**54**	(d)
5	(b)	**15**	(d)	**25**	(d)	**35**	(a)	**45**	(d)	**55**	(b)
6	(b)	**16**	(c)	**26**	(a)	**36**	(d)	**46**	(b)	**56**	(c)
7	(c)	**17**	(d)	**27**	(b)	**37**	(b)	**47**	(c)	**57**	(c)
8	(c)	**18**	(c)	**28**	(d)	**38**	(b)	**48**	(a)	**58**	(d)
9	(b)	**19**	(b)	**29**	(c)	**39**	(d)	**49**	(c)	**59**	(a)
10	(b)	**20**	(d)	**30**	(d)	**40**	(d)	**50**	(b)	**60**	(d)

Hints & Solutions

1. (a) We know that electric conductance is inversely proportional to resistance.

   Electric conductance $\propto \frac{1}{R}$

2. (b) In an electrolytic cell reduction occurs at cathode.

4. (c) In an electrolytic cell, electrons do not flow themselves. It is the migration of ions towards oppositely charged electrodes that indirectly constitutes the flow of electrons from cathode to anode through internal supply.

5. (b) For the given cell

   $$E_{cell} = E^{\circ}_{cell} - \frac{0.059V}{2}\log\frac{[Zn^{+}(aq)]}{[Cu^{+}(aq)]}$$

   The cell potential will decreases with increase in $[Zn^{2+}$ (aq)] and will increases with increase in $[Cu^{2+}$(aq)]. i.e. (c)

6. (b) Cell reaction is, $Zn + Cu^{2+} \rightarrow Zn^{2+} + Cu$

   $$E_{cell} = E^{o}{}_{cell} - \frac{RT}{nF}\ell n\frac{[Zn^{2+}]}{[Cu^{2+}]}$$

   Greater the factor $\left[\frac{(Zn^{2+})}{(Cu^{2+})}\right]$, less is the EMF

   Hence $E_1 > E_2$

7. (c) By Kohlraush's law, $\wedge^{0}_{eq}\,NaCl = 126.45$

   $\lambda^{0}_{Na^{+}} + \lambda^{0}_{Cl^{-}} = 126.45$ ....(1)

   $\lambda^{0}_{H^{+}} + \lambda^{0}_{Cl^{-}} = 426.16$ ...(2)

   $\lambda^{0}_{CH_3OO^{-}} + \lambda^{0}_{Na^{+}} = 91$ ....(3)

   on adding (2) and (3) then subtract (1) from it

   $\lambda^{0}_{CH_3COO^{-}} + \lambda^{0}_{Na^{+}} = 517.16 - 126.45$

   $\wedge^{0}_{(CH_3COOH)} = 390.71 ohm^{-1}\,cm^{2}$

8. (c) Electrorefining and electroplating are done by electrolysis.

9. (b) In $H_2 - O_2$ fuel cell, the combustion of H_2 occurs to create potential difference between the two electrodes

10. (b) Magnesium provides cathodic protection and prevent rusting or corrosion.

11. (a) We know that according to Debye Huckel rule, on dilution the ionisation of strong acid increases due to increasing volume. Since ionic mobility also increases. Therefore the conductivity of strong electrolyte increases on dilution slightly.

12. (c) The compounds which when dissolved in water, do not produce ions, are called as non-electrolytes. Therefore, $C_{12}H_{22}O_{11}$ is non-electrolyte because it does not give ions with water.

13. (d) The equivalent conductivity of a solution,

$$\Lambda_{eq} = \frac{1000}{C} \times \kappa$$

where,
κ = specific conductance = Unit $ohm^{-1}\ cm^{-1}$
C = normality of the solution unit gm eq/cm^3

Hence, the unit of Λ_{eq} is $Ohm^{-1} . cm^2$ (gm equivalent)$^{-1}$.

14. (a) Standard electrode potential i.e. reduction potential of A is minimum (–3.05V) i.e. its oxidation potential is maximum which implies 'A' is maximum reactive chemically.
15. (d) Higher the reduction potential, stronger is the oxidizing agent.
16. (c) Diamond is an insulator.
17. (d) This is because zinc has higher oxidation potential than Ni, Cu and Sn. The process of coating of iron surface with zinc is known as galvanization. Galvanized iron sheets maintain their lustrue due to the formation of protective layer of basic zinc carbonate.
18. (c) For bottom of ship to be protected it is connected with more reactive metal than iron like magnesium. This technique is called cathodic protection.
19. (b) $2KBr + I_2 \longrightarrow 2KI + Br_2$

    reaction is not possible because Br^- ion is not oxidised to Br_2 with I_2 due to higher electrode (oxidation) potential of I_2 than bromine.
20. (d) Kohlrausch's Law states that at infinite dilution, each ion migrates independently of its co-ion and contributes to the total equivalent conductance of an eletrolyte a definite share which depends only on its own nature.
21. (a) Dilution of strong electrolytes increases ionisation, hence ionic mobility of ions which in turn increases equivalent conductance of the solution.
22. (c) Equivalent conductance of an electrolyte at infinite dilution is given by the sum of equivalent conductances of the respective ions at infinite dilution.
23. (c) As the value of standard reduction potential decreases the reducing power increases i.e.,

    $$\underset{(-3.0)}{Z} > \underset{(-1.2)}{X} > \underset{(+0.5)}{Y}$$
24. (b) Given $E_{Sn^{+4}/Sn^{+2}} = +0.15$ V

    $E_{Cr^{+3}/Cr} = -0.74$ V

    $E^o_{cell} = E^o_{cathode} - E^o_{anode}$
    $= 0.15 - (-0.74)$
    $= +0.89$ V
25. (d) Sodium acetate ionises in aqueous solution hence (CH_3COONa) it is an electrolyte.
26. (a) Cathodic protection is best method to prevent iron from rusting. In this method iron is made cathode by application of external current. Saline water is highly conducting and hence accelerates the formation of rust.
27. (b) Electrode potential is considered as zero.
28. (d) Ag^+ comes after H^+ in the electrochemical series, hence it can't displace H_2.
30. (d) KI is an electrolyte.
31. (c) In the below reaction,

    $Zn^{+2} + 2e^- \longrightarrow Zn$; $E_1 = -7.62$ V ....(1)

    $Fe^{+2} + 2e^- \longrightarrow Fe$; $E_2 = -7.81$ V ....(2)

    On substracting the equation (1) and (2),

    $Fe^{++} + Zn \longrightarrow Fe + Zn^{++}$

    $E = E_2 - E_1 = -7.81 - (-7.62)$

    $\Rightarrow E = -0.19$ V
32. (b) The given statement explains that Al is placed above Cu in E.C.S. hence it can reduce Cu from its salt solution (reduction). Thus Al acts as a reducing agent now,
    Given : electrode potential of copper

    $(E^0_{cell}) = +0.34$ V
    e.m.f of cell = 2 V
    We know that,
    emf of a galvanic cell (E^0_{cell})
    = electrode potential of Cu (cathode) – electrode potential of Al (anode)

    $2.0 = 0.34 - E^0_{Al}$

    $E^0_{Al} = 0.34 - 2.00 = -1.66$ V
    Hence, electrode potential of Al is – 1.66 V.
33. (d) At cathodes : $Fe^+ + 2e^- \longrightarrow Fe$;

    $Fe^{3+} + 3e^- \longrightarrow Fe$

    $(E_{Fe})_1 = \frac{\text{At. wt.}}{2}$; $(E_{Fe})_2 = \frac{\text{At. wt.}}{3}$

    Hence, $\frac{(E_{Fe})_1}{(E_{Fe})_2} = \frac{3}{2}$

34. (d) For a concentration cell having different concentrations of ions.

$$E = \frac{0.0591}{n} \log \frac{c_1}{c_2}$$

If all the concentrations are identical then obviously the cell voltage is zero. But as the pH of 0.1 M HCl (strong acid) & pH of 0.1M CH_3 COOH is (weak acid) not same, therefore the cell voltage is not zero.

35. (a) By Faraday's Ist Law, $\frac{W}{E} = \frac{q}{96500}$

(where q = it = charge of ion)
We know that no of equivalent

$$= \frac{W}{E} = \frac{it}{96500} = \frac{1 \times 965}{96500} = \frac{1}{100}$$

(where i= 1 A, t = 16×60+5 = 965 sec.)
Since, we know that

$$\text{Normality} = \frac{\text{no. of equivalent}}{\text{Volume (in litre)}} = \frac{\frac{1}{100}}{1}$$

$$= 0.01 \text{ N}$$

36. (d) Given for the reaction

$$Zn(s) + Cu^{2+}(aq) \rightleftharpoons Cu(s) + Zn^{2+}(aq),$$

$E° = +1.10$ V.
At equilibrium

$$\therefore \quad E^{\circ}_{cell} = \frac{0.0591}{n} \log_{10} K_{eq}$$

here (n → number of exchange of electrons)

$$\text{or} \quad 1.10 = \frac{0.0591}{2} \log_{10} K_{eq}$$

$$\frac{2.20}{0.059} = \log_{10} k_{eq} = 37.22$$

or $K_{eq} = 1.66 \times 10^{37}$

37. (b) In electrochemical series, Sn is above hydrogen and Hg below hydrogen. Elements above hyrogen are reducing in nature while below hydrogen are oxidising in nature.

38. (b) $\Delta G = -nE^{\circ}F$

For concentration cell, $E = \frac{RT}{nF} ln \frac{C_2}{C_1}$

In it R, T, n and F are constant

So, E is based upon $ln \; C_2 / C_1$

Now, $\Delta G = -nEF = -nF \times \frac{RT}{nF} ln \, C_2 / C_1$

$= -RTlnC_2/C_1$

At constant temperature ΔG is based upon lnC_2/C_1.

39. (d) In the silver plating of copper, $K[Ag(CN)_2]$ is used instead of $AgNO_3$. Copper being more electropositive readily precipitate silver from their salt solution

$$Cu + 2AgNO_3 \longrightarrow Cu(NO_3)_2 + Ag$$

whereas in $K[Ag(CN)_2]$ solution a complex anion $[Ag(CN)_2]^-$ is formed and hence Ag^+ are less available in the solution and therefore copper cannot displace Ag from its complex ion.

40. (d) $E° = \frac{0.0591}{n} \log_{10} K$

Here, $n = 2, E° = 0.295$

$\therefore \log_{10} K = \frac{2 \times 0.295}{0.0591} = 10$ or $K = 10^{10}$

41. (c) The cell reacton is as follows :

$$A \rightarrow A^+ + e$$
$$B^+ + e \rightarrow B$$

Adding $A + B^+ \rightarrow A^+ + B$

(c) is correct answer.

42. (b) At anode :

$$2OH^- \rightleftharpoons H_2O + \frac{1}{2}O_2$$

43. (c) When one end of a metal is heated, the free electrons are energised and move to the other end. It heats up the other end of the metal.

44. (b) Given current (i) = 0.5 amp;
Time (t) = 100 minutes × 60 = 6000 sec
Equivalent weight of silver nitrate (E) = 108.
According to Faraday's first law of electrolysis

$$W = \frac{Eit}{96500} = \frac{108 \times 0.5 \times 6000}{96500} = 3.3575 \text{ g.}$$

45. (d) Given molarity = 0.01 M
Resistance = 40 ohm;

Cell constant $\frac{l}{A} = 0.4 \, cm^{-1}$.

Specific conductivity (κ)

$= \frac{\text{cell constant}}{\text{resistance}} = \frac{0.4}{40}$

$= 0.01\ \text{ohm}^{-1}\ \text{cm}^{-1}$

Molar conductance $(\wedge_m) = \frac{1000\kappa}{\text{Molarity}}$

$= \frac{1000 \times .01}{.01} = 10^3\ \text{ohm}^{-1}\ \text{cm}^2\ \text{mol}^{-1}$

46. (b) Without losing its concentration $ZnCl_2$ solution can not kept in contact with Al because Al is more reactive than Zn due to electrode (reduction) potential.

47. (c) $E^{\circ}_{cell} = \frac{2.303RT}{nF} \log K_{eq} = \frac{.0591}{n} \log K_{eq}$

$= \frac{.0591}{2} \log 10^6 = .0591 \times 3 = 0.1773$

48. (a) The equivalent conductance of $BaCl_2$ at infinite dilution

$\wedge_{\infty}$ of $BaCl_2 = \frac{1}{2} \lambda_{\infty}$ of $Ba^{++} + \lambda_{\infty}$ of Cl^-

$= \frac{127}{2} + 76 = 139.5\ \text{ohm}^{-1}\ \text{cm}^2$

49. (c) $2Cu^+ \longrightarrow Cu^{+2} + Cu$

$2e^- + Cu^{+2} \longrightarrow Cu;\ E^{\circ}_1 = 0.34V;$ ...(i)

$e^- + Cu^{+2} \longrightarrow Cu^+;$

$E^{\circ}_2 = 0.15V;$ ...(ii)

$Cu^+ + e^- \rightarrow Cu;\ E^{\circ}_3 = ?$ ... (iii)

Now, $\Delta G^{\circ}_1 = -nFE^{\circ}_1 = -2 \times 0.34F$

$\Delta G^{\circ}_2 = -1 \times 0.15F,\ \Delta G^{\circ}_3 = -1 \times E^{\circ}_3 F$

Again, $\Delta G^{\circ}_1 = \Delta G^{\circ}_2 + \Delta G^{\circ}_3$

$\Rightarrow -0.68F = -0.15F - E^{\circ}_3 F$

$\Rightarrow E^{\circ}_3 = 0.68 - 0.15 = 0.53V$

$E^{\circ}_{cell} = E^{\circ}_{cathode}(Cu^+/Cu) - E^{\circ}_{anode}(Cu^{+2}/Cu^+)$

$= 0.53 - 0.15 = 0.38$ V.

50. (b) Given specific conductance of the solution (κ) = 0.012 $\text{ohm}^{-1}\ \text{cm}^{-1}$ and resistance (R) = 55 ohm. We know that cell constant = Specific conductance × Observed resistance = 0.012 × 55= 0.66cm^{-1}.

51. (b) $Fe^{+2}/Fe \quad E^{\circ} = -0.44$

$Fe^{+3}/Fe^{+2} \quad E^{\circ} = 0.77$

The metals having higher negative electrode potential can displaced metals having lower values of negative electrode potential from their salt solutions.

52. (b) In electrolysis of NaCl when Pt electrode is taken then H_2 liberated at cathode while with Hg cathode it forms sodium amalgam because more voltage is required to reduce H^+ at Hg than Pt.

53. (b) $\Delta G = -nEF$

For 1 mol of Al, n = 3

$\therefore$ for $\frac{4}{3}$ mol of Al, $n = 3 \times \frac{4}{3} = 4$

According to question,

$827 \times 1000 = 4 \times E \times 96500$

$E = \frac{827 \times 1000}{4 \times 96500} = 2.14V$

54. (d) No. of gram equivalent of H^+ = No. of gm eq. of Al^{3+}

$= \frac{4.5}{9} = 0.5$

Eq. wt of $Al^{3+} = \frac{27}{3} = 9$

No. of gm eq. of H^+ = no. of mole of H^+

Hence Mass of $H^+ = 0.5 \times 1g = 0.5$ g

We know that, 2g H_2 at STP = 22.4 L

$\therefore$ 0.5 g H_2 at STP = $= \frac{22.4}{2} \times 0.5 = 5.6$ L

55. (b)

$Fe \longrightarrow Fe^{2+} + 2e$ [Anode] $E = -0.441V$

$[Fe^{3+} + e \longrightarrow Fe^{2+}] \times 2$ [Cathode] $E = +0.771V$

$Fe + 2Fe^{3+} \longrightarrow 3Fe^{2+}$

$E^{o}_{cell} = E^{o}_{cathode} - E^{o}_{Anode}$

$= .771 - (-.441) = .771 + .441 = 1.212$ V

56. (c) As $E^o_{cell} = \frac{0.0591}{n} \log K_c$

$\therefore 0.46 = \frac{0.0591}{2} \log K_c$

$\therefore \log K_c = \frac{2 \times 0.46}{0.0591} = 15.57$

or K_c = Antilog 15.57 = $3.7 \times 10^{15} \approx 4 \times 10^{15}$

57. (c) From the given data we find Fe^{3+} is strongest oxidizing agent. More the positive value of E^o_{OP}, more is the tendency to get oxidized.

58. (d) $Cu^{2+} + 2e^- \rightarrow Cu$; $\Delta G^o = -nE^o F$
$= -2 \times F \times 0.337$
$= -0.674$ F ....(i)

$Cu^+ \rightarrow Cu^{2+} + e^-$; $\Delta G^o = -nE^o F$
$= -1 \times F \times -0.153$
$= 0.153$ F ....(ii)

On adding eqns (i) & (ii)
$Cu^+ + e^- \rightarrow Cu$;
$\Delta G^o = -0.521 F = -nE^oF$,
Here n = 1 $\therefore E^o = +0.52$ V

59. (a) $\because Q = i \times t$
$\therefore Q = 4.0 \times 10^4 \times 6 \times 60 \times 60$ C
$= 8.64 \times 10^8$ C

Now, since 96500 C liberates 9 g of Al

8.64×10^8 C liberates $\frac{9}{96500} \times 8.64 \times 10^8$ g Al
$= 8.1 \times 10^4$ g of Al

60. (d) Degree of dissociation

$$\alpha = \frac{\Lambda}{\Lambda_\infty} = \frac{8.0}{400} = 2 \times 10^{-2}$$

$$K_a = \frac{C\alpha^2}{(1-\alpha)} \approx C\alpha^2 = \frac{1}{32} \times \left(2 \times 10^{-2}\right)^2$$

$= 1.25 \times 10^{-5}$

18. CHEMICAL KINETICS

ANSWER KEY

1	(d)	11	(d)	21	(b)	31	(a)	41	(d)	51	(a)
2	(a)	12	(a)	22	(c)	32	(b)	42	(b)	52	(d)
3	(d)	13	(b)	23	(c)	33	(b)	43	(b)	53	(b)
4	(d)	14	(b)	24	(c)	34	(b)	44	(a)	54	(a)
5	(b)	15	(d)	25	(d)	35	(d)	45	(a)	55	(b)
6	(c)	16	(a)	26	(a)	36	(b)	46	(b)	56	(c)
7	(d)	17	(b)	27	(b)	37	(b)	47	(b)	57	(d)
8	(c)	18	(b)	28	(d)	38	(d)	48	(d)	58	(b)
9	(a)	19	(a)	29	(d)	39	(d)	49	(b)	59	(c)
10	(a)	20	(a)	30	(c)	40	(d)	50	(d)	60	(b)

Hints & Solutions

1. (d) Rate of reaction

$$-\frac{1}{5}\frac{d[Br^-]}{dt} = +\frac{1}{3}\frac{d[Br_2]}{dt}$$

$$\frac{d[Br_2]}{dt} = -\frac{3}{5}\frac{d[Br^-]}{dt}$$

2. (a) If we write rate of reaction in terms of concentration of NH_3 and H_2,then

Rate of reaction $= \frac{1}{2}\frac{d[NH_3]}{dt} = -\frac{1}{3}\frac{d[H_2]}{dt}$

So, $\frac{d[NH_3]}{dt} = -\frac{2}{3}\frac{d[H_2]}{dt}$

3. (d) The rate of a reaction is the speed at which the reactants are converted into products. It depends upon the concentration of reactants. e.g for the reaction

$A + B \longrightarrow$ Product; $r \propto [A][B]$

4. (d) Rate of reaction depends upon all the three factor i.e., temperature, concentration and catalyst.

→ As the concentration of reactants increases, the

reaction rate increases.

→ Catalyst is a substance that increases the rate of reaction.

→ Rate of reaction increases with increase in temperature.

5. (b) For a first reaction $\frac{dx}{dt} = k(a-x)$

on intergration

$$\int \frac{dx}{(a-x)} = \int k\,dt$$

i.e $-\ell n\ (a-x) = kt + c$ or $kt = \ell n\ a - \ell n\ (a-x)$

or $kt = 2.303\ [\log a - \log (a-x)]$

Thus if we plot a graph between log a & t we get

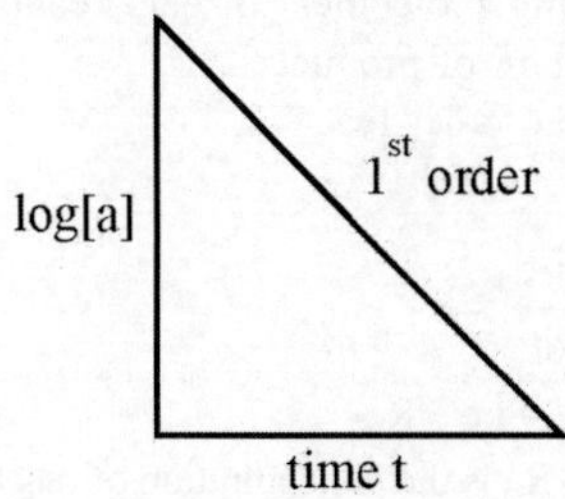

6. (c) For a first order reaction

$$k = \frac{2.303}{t}\log_{10}\frac{a}{a-x}$$

when $t = t_{1/2}$

$$k = \frac{2.303}{t_{1/2}}\log_{10}\frac{a}{a-a/2}$$

$$\text{or } t_{1/2} = \frac{2.303}{k}\log_{10}2 = \frac{\ln 2}{k}$$

7. (d) $k = Ae^{-Ea/RT}$

$$\text{or } \log k = \log A - \frac{E_a}{2.303\ RT}$$

Comparing the above equation with $y = mx + c$

$$y = \log k,\ x = \frac{1}{T}$$

Thus A plot of $\log_{10}k$ vs 1/T should be a straight line, with slope equal to $-$ Ea/2.303 RT and intercept equal to log A

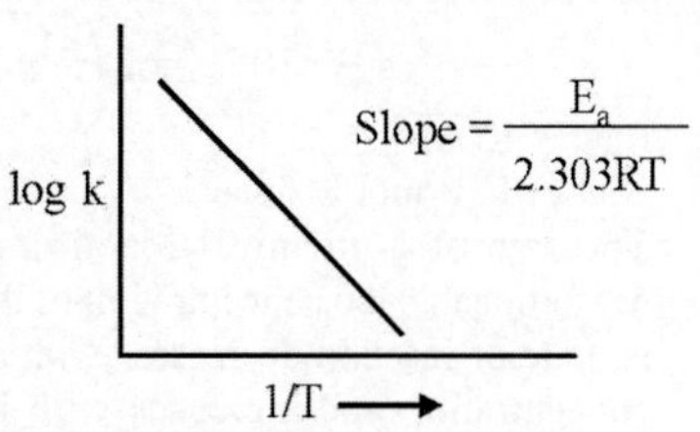

$$\therefore\ \text{Slope} = \frac{-Ea}{2.303R}$$

$$\text{or } E_a = -2.303R \times \text{Slope}$$

8. (c) We know that the activation energy of chemical reaction is given by formula

$$\log\frac{k_2}{k_1} = \frac{E_a}{2.303R}\left[\frac{T_2 - T_1}{T_1T_2}\right],$$ where k_1 is the rate constant at temperature T_1 and k_2 is the rate constant at temperature T_2 and E_a is the activation energy. Therefore activation energy of chemical reaction is determined by evaluating rate constant at two different temperatures.

10. (a) It is a constant of a particular reaction at a given temperature. It does not depend upon initial concentration of the reactants, time of reaction and extent of reaction.

11. (d) Order of a reaction can be fractional. Rest of all are true. Order of a reaction can be determined experimentally.

12. (a) $E_a(\text{F.R.}) \neq E_a(\text{B.R.})$

13. (b) $3A \longrightarrow 2B$

Rate of appearance of B is equal to rate of disappearance of A.

$$\frac{1}{2}\frac{d[B]}{dt} = -\frac{1}{3}\frac{d[A]}{dt} \Rightarrow \frac{d[B]}{dt} = -\frac{2}{3}\frac{d[A]}{dt}$$

14. (b) $\because\ r = k[A]^n$

if n = 0

$$r = k[A]^0$$

or r = k thus for zero order reactions rate is equal to the rate constant.

15. (d) Rate of disappearance of H_2 = rate of formation of NH_3.

$$-\frac{1}{3}\frac{d[H_2]}{dt} = \frac{1}{2}\frac{d[NH_3]}{dt}$$

$$\Rightarrow \frac{-d[H_2]}{dt} = \frac{3}{2}\frac{d[NH_3]}{dt} = \frac{3}{2}\times 2\times 10^{-4}$$

$$= 3\times10^{-4}\text{mol L}^{-1}\text{s}^{-1}$$

16. (a) Rate = $K(A)^0$
Unit of K = mol L^{-1} sec^{-1}

17. (b) The rate of a chemical reaction is directly proportional to the concentration of the reactants i.e. rate of reaction decreases with decrease of concentration and increases with increase in concentration. As the reaction progresses the reactant is converted into the product and thus the concentration of reactants decreases while that of the products increases.

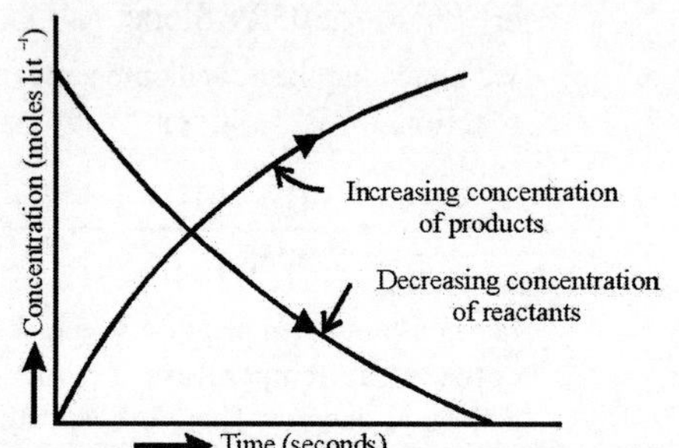

18. (b) A catalyst is used for increasing or decreasing the velocity of a reaction.

19. (a) Zymase catalyses the conversion of glucose to ethanol.

20. (a) For zero order reaction, $\frac{dx}{dt} = k\,[\text{reactant}]^\circ$
Thus the rate of zero order reaction is independent of concentration of reactants.

21. (b) In Haber's process, ammonia is manufactured from N_2 and H_2 using iron as catalyst with molybdenum as promoter at high temperature and pressure

$$N_2 + 3H_2 \xrightarrow[\text{Mo(catalytic promoter)}]{Fe_2O_3\text{(catalyst)}} 2NH_3$$

22. (c) Catalyst does not alter the position of equilibrium in a reversible reaction.

23. (c) The catalyst is in the same phase as the reactants and products.
Therefore, in this reaction, H_2SO_4 is called as homogeneous catalyst.

$$\underset{\text{(solid)}}{C_{12}H_{22}O_{11}} + H_2O \xrightarrow{\text{dil. } H_2SO_4}$$

$$C_6H_{12}O_6\text{(aq.)} + C_6H_{12}O_6\text{ (aq.)}$$

24. (c) A catalyst can affect reversible reaction by attaining equilibria in both directions.

25. (d) Order of reaction is equal to the number of molecules whose concentration is changing with time. It can be zero or in fractions or an integer.

26. (a) For a zero order reaction.
rate $=k[A]^\circ$ i.e. rate = k
hence unit of k = M.sec^{-1}
For a first order reaction.
rate = $k\,[A]$
k = M.sec^{-1}/M = sec^{-1}

27. (b) In equation $K = Ae^{-E_a/RT}$; A = Frequency factor
K = velocity constant, R = gas constant and E_a = energy of activation

28. (d) The reaction involving two different reactant can never be unimolecular.

29. (d) In Arrhenius equation $K = A\,e^{-E/RT}$, E is the energy of activation, which is required by the colliding molecules to react resulting in the formation of products.

30. (c) For zero order reaction,
rate, r = $k[R]^\circ$

$$\Rightarrow \frac{dR}{dt} = k$$

$$\Rightarrow R = kt + R_0$$

where R_0 is the concentration of reactant at time t = 0.
Thus [R] increases with time

31. (a) From data 1 and 3, it is clear that keeping (B) const, [A] is doubled, rate remains unaffected. Hence rate is independent of [A]. from 1 and 4, keeping [A] constant, [B] is doubled, rate become 8 times. Hence rate $\propto [B]^3$.

32. (b) $E_{a(\text{Forward})} + \Delta H = E_{a(\text{Backward})}$
For Exothermic reaction, ΔH = –ve and
$\therefore$ activation energy of reactant is less than the energy of activation of products.

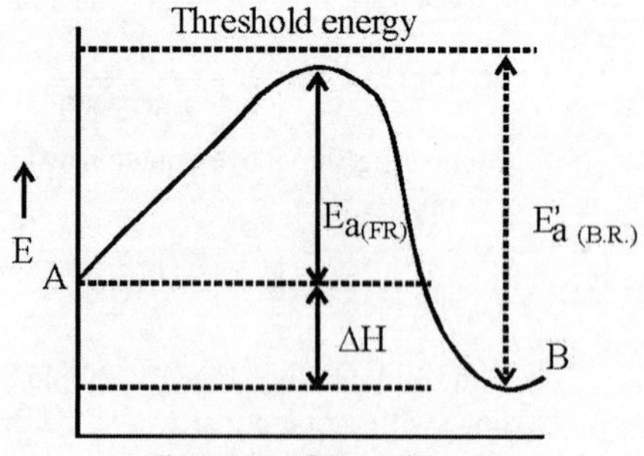

33. (b) $\Delta H = E_{a(f)} - E_{a(b)}$
Thus energy of activation for reverse reaction depend upon whether reaction is exothermic or endothermic.
If reaction is exothermic,

$\Delta H = -ve \quad E_{a(b)} > E_{a(f)}$

If reaction is endothermic

$\Delta H = +ve \quad E_{a(b)} < E_{a(f)}$

34. (b) For reaction $3A \longrightarrow B + C$
If it is zero order reaction r = k $[A]^0$, i.e the rate remains same at any concentration of 'A'. i.e independent upon concentration of A.

35. (d) In the given options $-\frac{d[C]}{3.dt}$ will not represent the reaction rate. It should not have –ve sign as it is product.

since $\frac{1}{3}\frac{dC}{dt}$ show the rate of formation of product C which will be positive.

36. (b) $A \rightarrow B$ For a first order reaction
given a = 0.8 mol, (a – x) = 0.8 – 0.6 = 0.2

$$k = \frac{2.303}{1}\log\frac{0.8}{0.2} \text{ or } k = 2.303 \log 4$$

Again, $a = 0.9, \; a - x = 0.9 - 0.675 = 0.225$

$$k = \frac{2.303}{t}\log\frac{0.9}{0.225}$$

$$2.303 \log 4 = \frac{2.303}{t}\log 4$$

37. (b) $Rate_1 = k[A]^x[B]^y$ ... (1)

$$\frac{Rate_1}{4} = k[A]^x[2B]^y \quad ... (2)$$

or $Rate_1 = 4k[A]^x[2B]^y$
From (1) and (2) we get

$$\frac{k[A]^x[B]^y}{4} = k[A]^x[2B]^y$$

$$\frac{[B]^y}{4} = [2B]^y$$

or $\frac{1}{4} = \left(\frac{2B}{B}\right)^y \Rightarrow \frac{1}{4} = 2^y$ or $2^{-2} = 2^y$

or y = –2.

38. (d) As the slowest step is the rate determining step thus the mechanism B will be more consistent with the given information also because it involve one molecule of H_2 and one molecule of ICl it can expressed as

$r = k[H_2][ICl]$

Which shows that the reaction is first order w.r.t. both H_2 & ICl.

39. (d) $(Rate)_1 = K[A]^l[B]^m$;

$$(Rate)_2 = K(10[Al])^l[B]^m$$

Hence, $\frac{(Rate)_2}{(Rate)_1} = 100 = 10^l$

or $10^2 = 10^l \Rightarrow l = 2$

40. (d) $H_2(g) + Br_2(g) \rightleftharpoons 2HBr(g)$
Rate law, $R = k[H_2][Br_2]^{1/2}$
Order of reaction = 1 + ½ = 3/2
Molecularity of reaction = 2

The unit of $k = \frac{R}{[H_2][Br_2]^{1/2}}$

$$= \frac{mole.\,lit^{-1}s^{-1}}{[mole.\,lit^{-1}][mole.\,lit^{-1}]^{1/2}}$$

$= mole^{-1/2}.\,lit^{1/2}.\,s^{-1}$

41. (d) Overall order = sum of orders w.r.t each reactant.
Let the order be *x* and *y* for G and H respectively

Exp.No.	[G]mole litre^{-1}	[H]mole litre^{-1}	rate(mole litre^{-}time^{-1})
1	*a*	*b*	*r*
2	2*a*	2*b*	8*r*
3	2*a*	*b*	2*r*

Applying $r = k[G]^x[H]^y$ we get,
$x = 1, y = 2$
∵ For (1) and (3), the rate is doubled when conc. of G is doubled keeping that of H constant
i.e., rate ∝ [G] ∴ $x = 1$
From (2) and (3), y = 2
∴ Overall order is 3.

42. (b) For 1st order rection $t_{1/2}$ independent of initial concentration.

43. (b) $O_3 \xrightarrow{Fast} O_2 + O;\; O + O_3 \xrightarrow{Slow} 2O_2$

$k = \frac{[O_2][O]}{[O_3]}$ (I) Rate = k' $[O_3][O]$ put [O] from (I)

$$r = \frac{k'[O_3]K[O_3]}{[O_2]} = k[O_3]^2[O_2]^{-1}$$

Note intermediates are never represented in rate law equation.

44. (a) Given initial concentration (a) = 2.00 m; Time taken (t) = 200 min and final concentration (a – x) = 0.15 m. For a first order reation rate constant,

$$k = \frac{2.303}{t}\log\frac{a}{a-x} = \frac{2.303}{200}\log\frac{2.00}{0.15}$$

$$= \frac{2.303}{200}\times(0.301+0.824) = 1.29\times10^{-2}\ \text{min}^{-1}.$$

Further

$$(t_{1/2}) = \frac{0.693}{k} = \frac{0.693}{1.29\times10^{-2}} = 53.49\ \text{min}.$$

45. (a) $t_{1/2} = 4$ s T = 16

$$n = \frac{T}{t_{1/2}} = \frac{16}{4} = 4 \qquad (\because T = n\times t_{½})$$

$$A = A^o\left(\frac{1}{2}\right)^n = 0.12\times\left(\frac{1}{2}\right)^4 = \frac{0.12}{16} = 0.0075\ \text{M}$$

where Å = initial concentration and A = concentration left after time t

46. (b) The presence of enzyme (catalyst) increases the speed of reaction by lowering the energy barrier, i.e. a new path is followed with lower activation energy.

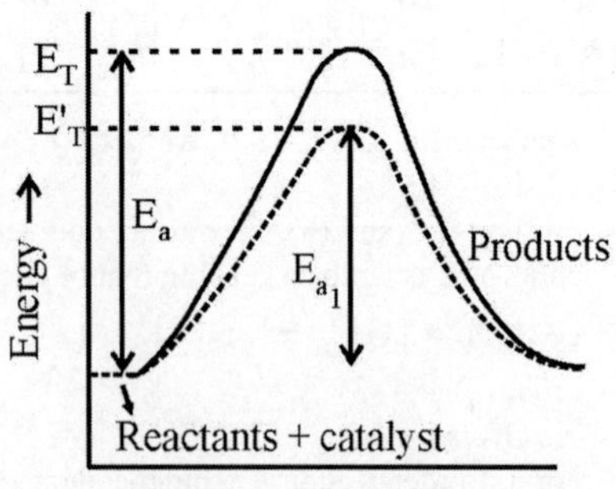

Here E_T is the threshold energy.
E_a and E_{a_1} is energy of activation of reaction in absence and presence of catalyst respectively.

47. (b) $2N_2O_5 \rightarrow 4NO_2 + O_2$

from the unit of rate constant it is clear that the reaction follow first order kinetics. Hence by rate law equation, $r = k[N_2O_5]$
where $r = 1.02\times10^{-4}$, $k = 3.4\times10^{-5}$

$1.02\times10^{-4} = 3.4\times10^{-5}[N_2O_5]$
$[N_2O_5] = 3M$

48. (d) The activation energy of reverse reaction will depend upon whether the forward reaction is exothermic or endothermic.
As $\Delta H = E_{a\ (forward\ reaction)} - E_{a(backward\ reaction)}$
For exothermic reaction
$\Delta H = -ve$
$\therefore -\Delta H = E_{a(f)} - E_{a(b)}$
or $E_{a(f)} = E_{a(b)} - \Delta H$
$\therefore\ E_{a(f)} < E_{a(b)}$
For endothermic reaction
$\Delta H = +ve$
$\therefore \Delta H = E_{a(f)} + E_{a(b)}$ or $E_{a(f)} = \Delta H + E_{a(b)}$

49. (b) For a first order reaction, A → products

$$r = k[A] \text{ or } k = \frac{r}{[A]} \Rightarrow k = \frac{1.5\times10^{-2}}{0.5}$$

$= 3\times10^{-2}$

$$\text{Further, } t_{1/2} = \frac{0.693}{k} = \frac{0.693}{3\times10^{-2}} = 23.1$$

50. (d) Given [A] = 0.01 M
Rate = 2.0×10^{-5} mol L^{-1} S^{-1}
For a first order reaction
Rate = k[A]

$$k = \frac{2.0\times10^{-5}}{[0.01]} = 2\times10^{-3}$$

$$t_{1/2} = \frac{0.693}{2\times10^{-3}} = 347\ \text{sec.}$$

51. (a) For a first order reaction

$$k = \frac{2.303}{t}\log\frac{a}{a-x}$$

when t = 60 and x = 60%

$$k = \frac{2.303}{60}\log\frac{100}{100-60} = \frac{2.303}{60}\log\frac{100}{40}$$

Now,

$$t_{½} = \frac{2.303}{0.0153}\log\frac{100}{100-50} = \frac{2.303}{0.0153}\times\log 2$$

$$= \frac{2.303}{0.0153}\times0.3010 = 45.31\ \text{min}.$$

52. (d) Given, $k_1 = 10^{16}.e^{-\frac{2000}{T}}$

and $k_2 = 10^{15}.e^{-\frac{1000}{T}}$

when k_1 and k_2 are equal at any temperature T, we have

$$10^{16}.e^{-\frac{2000}{T}} = 10^{15}.e^{-\frac{1000}{T}}$$

or $10 \times 10^{15}.e^{-\frac{2000}{T}} = 10^{15}.e^{-\frac{1000}{T}}$

or $10.e^{-\frac{2000}{T}} = e^{-\frac{1000}{T}}$

or $\ln 10 - \frac{2000}{T} = -\frac{1000}{T}$

or $\ln 10 = \frac{2000}{T} - \frac{1000}{T}$

or $2.303 \log 10 = \frac{1000}{T}$

or $2.303 \times 1 \times T = 1000$ $[\because \log 10 = 1]$

or $T = \frac{1000}{2.303} K$

53. (b) For a first order reaction

$t_{1/2} = \frac{0.693}{k}$; $k = \frac{0.693}{1386} = 0.5 \times 10^{-3} s^{-1}$

54. (a) When concentration A is doubled, rate is doubled. Hence order with respect to A is one.
When concentrations of both A and B are doubled, rate increases by 8 times hence order with respect to B is 2.
$\therefore$ rate = k $[A]^1$ $[B]^2$
Total order = 1 + 2 = 3

55. (b) $N_2O_5\ (g) \longrightarrow 2\ NO_2\ (g) + 1/2\ O_2(g)$

$$-\frac{d}{dt}[N_2O_5] = +\frac{1}{2}\frac{d}{dt}[NO_2] = 2\frac{d}{dt}[O_2]$$

$\frac{d}{dt}[NO_2] = 1.25 \times 10^{-2}$ mol $L^{-1}s^{-1}$ and

$\frac{d}{dt}[O_2] = 3.125 \times 10^{-3}$ mol $L^{-1}s^{-1}$

56. (c)

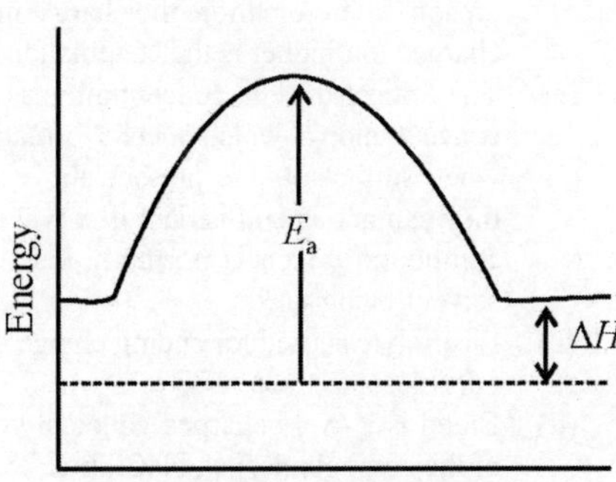

$E_a > \Delta H$

57. (d) In case of (II) and (III) Keeping concentration of [A] constant, when the concentration of [B] is doubled, the rate quadruples. Hence it is second order with respect to B. In case of I & IV Keeping the concentration of [B] constant. when the concentration of [A] is increased four times, rate also increases four times. Hence, the order with respect to A is one. hence

Rate = k [A]$[B]^2$

58. (b) $2\ NO(g) + Cl_2(g) \rightleftharpoons 2\ NOCl(g)$
Rate = k $[NO]^2$ [Cl]
The value of rate constant can be increased by increasing the temperature.

59. (c) For the reaction

$$N_2(g) + O_2(g) \rightleftharpoons 2NO(g) \quad ...(i)$$

$$K_1 = \frac{[NO_2]^2}{[N_2][O_2]}$$

For the reaction

$$2NO(g) + O_2(g) \rightleftharpoons 2NO_2(g) \quad ...(ii)$$

$$K_2 = \frac{[NO_2]^2}{[NO]^2[O_2]}$$

For the reaction

$$NO_2(g) \rightleftharpoons \frac{1}{2}N_2(g) + O_2(g) \quad ...(iii)$$

$$K = \frac{[N_2]^{\frac{1}{2}}[O_2]}{[NO_2]}$$

Hence, $K = \sqrt{\frac{1}{K_1} \times \frac{1}{K_2}}$

$$K = \frac{[N_2]^{\frac{1}{2}}[O_2]^{\frac{1}{2}}}{[NO]} \times \frac{[NO][O_2]^{\frac{1}{2}}}{[NO_2]}$$

$$= \frac{[N_2]^{\frac{1}{2}}[O_2]}{[NO_2]}$$

60. (b) Rate of disappearance of reactants = Rate of appearance of products

$$-\frac{1}{2}\frac{d(N_2O_5)}{dt} = \frac{1}{4}\frac{d(NO_2)}{dt} = \frac{d(O_2)}{dt}$$

$$\frac{1}{2}k(N_2O_5) = \frac{1}{4}k'(N_2O_5) = k''(N_2O_5)$$

$$\frac{k}{2} = \frac{k'}{4} = k''$$

$k' = 2k$, $k'' = \frac{k}{2}$

19. SURFACE CHEMISTRY

ANSWER KEY											
1	(a)	11	(b)	21	(a)	31	(d)	41	(b)	51	(d)
2	(d)	12	(b)	22	(a)	32	(a)	42	(d)	52	(c)
3	(c)	13	(b)	23	(a)	33	(b)	43	(a)	53	(d)
4	(a)	14	(d)	24	(d)	34	(a)	44	(d)	54	(a)
5	(d)	15	(d)	25	(a)	35	(c)	45	(c)	55	(d)
6	(a)	16	(c)	26	(a)	36	(c)	46	(d)	56	(b)
7	(b)	17	(b)	27	(c)	37	(a)	47	(a)	57	(c)
8	(c)	18	(d)	28	(c)	38	(a)	48	(b)	58	(d)
9	(a)	19	(d)	29	(a)	39	(c)	49	(c)	59	(a)
10	(d)	20	(c)	30	(a)	40	(c)	50	(d)	60	(a)

Hints & Solutions

1. (a) On increasing temperature adsorption of a gas on surface of solid decreases. Solid adsorb greater amount of substances at lower temperature.
2. (d) In physiosorption, gas molecule is held on the solid surface by weak van der Waal's forces.
3. (c) The theory of heterogeneous catalysis is based upon the phenomenon of adsorption. The activity of catalyst is due to the presence of free valencies on its sufrace due to which surface of catalyst has force of attraction.
4. (a) According to the adsorption theory of catalysis, the activity of catalysis is due to the presence of free valencies on its surface due to which surface of catalyst has chemical force of altraction. When a gas comes in contact with this surface molecules get attached attached through these valencies. Further the rate of reaction is always increases by decreases in activation energy. When any of the reactants is strongly adsorbed on the surface of catalyst, the rate becomes inversely proportional to the concentration of that reaction. The reaction is then said to be inhibited by such reactant. The lowering of activation energy always leads to the increase in speed of reaction.
5. (d) Air is a homogeneous mixture of gases, mainly nitrogen and oxygen.
6. (a) The colloidal particles have diameter 1-100 nm.
7. (b) The stability of a lyophilic sol is due to both the charge and solvation of particles, but largely to the latter factor.
8. (c) When the dispersed phase and dispersion medium both are liquid, the colloidal system is called as an emulsion like milk, vasnishing cream etc.
9. (a) Milk is liquid dispersed in liquid.
10. (d) Since, oil is the inner (dispersed) phase, the added dye will not be dissolved in the emulsion.
11. (b) A catalyst can not initiate a reaction.
12. (b) Cloud consists of fine droplets of water suspended in air.
13. (b) Catalyst decreases the activation energy of the reaction by forming an intermediate product. So no of molecules having activation energy increases hence rate of reaction increases.
14. (d) The use of membrane for separating colloidal particles is termed as dialysis. Hence it is clear that colloidal particle cannot pass through animal membrane. Hence only solvent molecules and ions (in case of electrodialysis) can diffuse.
15. (d) According to the Hardy Schulze rule the coagulating effect of an ion on dispersed phase of opposite charge increases with the valency of the ion. Therefore more the charge on oppositely charged ion higher is the coagulation value
16. (c) The critical micelle concentration is the lowest concentration at which micelle formation appears when surfactants are present above that CMC, they can act as emulsifiers that will solubilise a compound which is normally insoluble in the solvent being used.
17. (b) Electrolyte is used for ending charge on colloidal particles.
18. (d) Blood is a –vely charged colloidal system. Rest of the compounds, i.e., NaCl, urea & cane sugar form true solution in water.
19. (d) As_2S_3 sol. carries a negatively charged, so for coagulation, cations will be the effective ions, coagulation power of an effective ion increases with its charge.

20. (c) Smoke consists of carbon particles (solid) dispersed in air (gas).
21. (a) Chlorophyll. Smoke is an example of solid-gas colloid system Ruby glass is an example of solid-solid colloid system.
Milk is an liquid -liquid colloid system.
22. (a) Suspension particles are visible under a microscope and sometimes even to a naked eye
23. (a) Robert Brown, an English botanist observed that the particles in colloidal solution were in constant random motion, traversing no definite set paths, but travelling in zig-zag directions all over the field of view. This movement of colloidal particles is called Brownian movement.
24. (d) Dialysis and ultrafiltration are the methods of purification of colloidal solutions Brownian movement is a property of colloidal solution. Only wavelength is not related with colloids
25. (a) When a beam of light is passed through a colloidal solution, the path of the beam is illuminated and becomes visible as a blue bright streak when seen in a direction perpendicular to the incident beam. This phenomenon is called as Tyndall Effect. This phenomenon is due to scattering of light by the sol particles.
26. (a) Fog is a colloidal system consisting water droplets dispersed in air.
27. (c) Starch molecules have colloidal dimensions whereas NaCl, glucose and $Ba(NO_3)_2$ are crystalloids and soluble in water.
28. (c) Collidal solutions can be purified by dialysis. It may be defined as the process of separating a crystalloid from a colloid by diffusion or filtration through a fine membrane. The process of dialysis can be quickened by using hot water (hot dialysis) or by applying an electric field (electrodialysis).
29. (a) Tyndall effect is shown by colloidal solution. If a beam of light, such as that from a flashlight passes through a colloid, the light is reflected by the colloidal particles and the path of light can therefore be observed. The scattering of light by colloids is known as Tyndall effect.
30. (a) A collidal solution is biphasic and heterogeneous. It is composed of two phases :
1. Dispersed phase; 2. Dispersion medium.
31. (d) Electrolytic (Ionic) impurities can be most easily removed on application of electric field.
32. (a) Tyndall effect is the simplest way to check colloidal system since path becomes visible due to scattering of light.
33. (b) Crystalloids. can easily be obtained in the crystalline form and their solution can diffuse rapidly through a vegetable or animal membrane [i.e., parchment membrane]; e.g. sugar, urea, salts, acids, etc.
34. (a) Wood charcoal is used to decolourise sugar because it has a highly porous structure, providing a large surface area for adsorption. Therefore it adsorbs coloured material from sugar.
Option (a) is correct.
35. (c) A catalyst can affect reversible reaction by attaining equilibria in both directions.
Option (c) is correct.
36. (c) According to this law the coagulating effect of an ion on dispersed phase of opposite charge increases with the valency of the ion. The precipitating power of Al^{3+} , Ba^{++}, Na^{+} ions is in order $Al^{3+} > Ba^{2+} > Na^{+}$.
37. (a) o- Polarhead
n- Non-polar tail
(micelle)
38. (a) Langmuir adsorption isotherm is based on the assumption that every adsorption site is equivalent and the ability of a particle to bind there is independent of whether or not nearby sites are occupied.
39. (c) For a protective colloid lesser the value of gold number better is the protective power. Thus the correct order of protective power of A, B, C and D is

$\Rightarrow$	(A) <	(C) <	(B) <	(D)
Gold number	0.50	0.10	0.01	0.005

40. (c) Adsorption is an exothermic process, hence ΔH will always be negative.
41. (b) When oppositely charged sols are mixed their charges are neutralised. Both sols may be partially or completely precipitated.
42. (d) According to Freundlich adsorption isotherm. At intermediate pressure, extent of adsorption

$$\frac{x}{m} = kP^{1/n} \text{ or } \log\frac{x}{m} = \log k + \frac{1}{n}\log P;$$

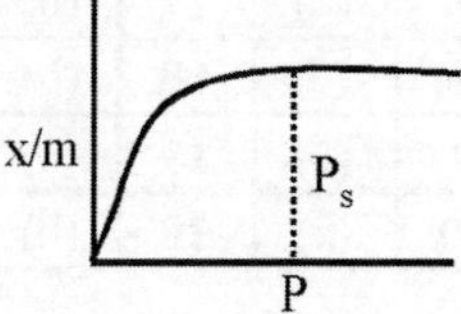

plot of $\log\frac{x}{m}$ vs log P is linear with slope

$= \frac{1}{n}$

43. (a) Micelle formation is shown by surfactants detergents (Dodecyl trimethyl ammonium chloride) in their aqueous solutions.
45. (c) $Fe(OH)_3$ particles absorb Fe^{3+} ions and get peptized to give a positively charged sol.
46. (d) Gelatin is a lyophilic solution (lyo-liquid philic -love). Lyophilic colloids are also known as

reversible colloids since on evaporating the dispersion medium (i.e. water), the residue can again be easily reconverted into colloidal state simply by addition of the liquid i.e. water.

47. (a) Scattering of light from the surface of colloidal particles–tyndall effect is observed. When a colloidal solution is observed under an ultramicroscope.

48. (b) Scattering of light (Tyndall effect) is exhibited by a colloidal solution. Among the given solutions starch solution is a colloid whereas other are example of true solution.

49. (c) Size of particles in colloidal state ranges from 10 to 1000 Å

Size of particles ------ 10Å ----- 1000Å ------
True solution | particles in colloidal state | Suspension

50. (d) Addition of electrolytes causes precipitation (coagulation) of a sol.

51. (d) Colloidal solutions are not purified by electrophoresis. Movement of colloidal particles under the influence of electric field is called electrophoresis. So, it can make easier. Electrophoresis is the property of colloids not the purification method.

52. (c) Sol is a colloidal solution so, it will show Tyndall effect.

54. (a) Butter is obtained from milk which itself is an emulsion of fat dispersed in water, reverse is true for butter.

56. (b) $KI + AgNO_3$ (slight excess) $\longrightarrow AgI + KNO_3$

$$AgNO_3 \longrightarrow Ag^+ + NO_3^-$$

$$AgI(s) + Ag^+ \longrightarrow [AgI]Ag^+$$

58. (d) Enzymes are very specific biological catalysts possessing well - defined active sites.

59. (a) Blood is purified by dialysis.

60. (a) Surface tension of lyophilic sols is lower than water (dispersion medium).

20. GENERAL PRINCIPLES AND PROCESSES OF ISOLATION OF ELEMENTS

ANSWER KEY

1	(a)	11	(b)	21	(c)	31	(b)	41	(d)	51	(d)
2	(b)	12	(b)	22	(b)	32	(d)	42	(b)	52	(a)
3	(d)	13	(d)	23	(a)	33	(a)	43	(b)	53	(b)
4	(b)	14	(c)	24	(b)	34	(b)	44	(d)	54	(b)
5	(a)	15	(a)	25	(b)	35	(d)	45	(b)	55	(a)
6	(a)	16	(c)	26	(b)	36	(c)	46	(c)	56	(c)
7	(d)	17	(b)	27	(d)	37	(c)	47	(c)	57	(b)
8	(d)	18	(b)	28	(c)	38	(b)	48	(c)	58	(b)
9	(b)	19	(a)	29	(c)	39	(d)	49	(d)	59	(a)
10	(a)	20	(b)	30	(a)	40	(b)	50	(b)	60	(c)

Hints & Solutions

1. (a) Gold being least reactive found native.

2. (b) Limonite and magnetite are ores of iron but cassiterite is not an ore of iron. Cassiterite (SnO_2) is an ore of tin (Sn).

4. (b) Calcination involves heating of the ore below its fusion temperature in absence of air.

5. (a) Aluminium works as reducing agent in thermite process because it displaces Fe from Fe_2O_3. Hence, option (a) is correct.

6. (a) Most metals occur in their combined state. They have to be reduced to become free metals. The choice of reducing agent is decided by the factors of energetics and economics. In this regard C is the most common reducing agent used. Carbon can be used in the form of coal, coke, charcoal and carbon monoxide. e.g :-

$$SnO_2 + 2C \xrightarrow{1473-1573K} Sn + 2CO$$

$$ZnO + CO \xrightarrow{1600K} Zn + CO_2$$

$$Fe_2O_3 + CO \xrightarrow{823K} 2FeO + CO_2$$

$$FeO + CO \xrightarrow{1123K} Fe + CO_2$$

7. (d) The sulphide ore is roasted to oxide before reduction because the ΔG_f^o of most of the sulphides are greater than those of CS_2 and H_2S, therefore neither C nor H can reduce metal sulphide to metal. Further, the standard free energies of formation of oxide are much less than those of SO_2. Hence oxidation of metal sulphides to metal oxide is thermodynamically favourable.
8. (d) The M.P of alumina is 2000° C, which is reduced to 900°C in the presence of cryolite (Na_3AlF_6) and mixture is also a good conductor of electricity. Thus, cryolite decreases the energy consumption during the process.
9. (b) During the process of electrolytic refining Ag and Au are obtained as anode mud.
10. (a) Zone refining is based on the difference in solubility of impurities in molten and solid state of the metal. This method is used for obtaining metals of very high purity.
11. (b) Argentite (Ag_2S) is a sulphide ore. It is not an acidic ore.
12. (b) **Calcination** is the heating of the ore in a suitable furnace in absence of air much below its melting point to cause decomposition and elimination of volatile products.

$$PbCO_3 \rightarrow PbO + CO_2$$

The process is generally applied to hydrated oxide or carbonate ores.
13. (d) Bauxite may be written as $Al_2O_3.2H_2O$ Cryolite is $Na_3[AlF_6]$. Feldspar are mock forming minerals. Kaolin is a Clay (China Clay).
14. (c) Minerals are naturally occurring chemical substances present in earth's crust which are obtained by mining. Minerals are source of *ores* from which metal can be commercialy recoverd. Gangue is undersirable impurity associated with ore. Flux is a substance mixed with metal to promote fusion.
15. (a) Silver ores are Ag_2S (argentite) and AgCl (horn silver). Haematite is ore of iron, bauxite of Aluminium and stibnite (Sb_2S_3) is an ore of antimony.
16. (c) SiO_2 is an acidic flux.
17. (b) Highly electropositive element Na is extracted by electrolytic reduction.
18. (b) Aluminium has the power of displacing elements less electropositive than itself from their oxides as it has great affinity for oxygen at high temperature. The reaction being so strongly exothermic that the metal set free is obtained in the molten condition and it is protected from oxidation by a layer of fluid slag consisting of Al_2O_3.

$$\text{Oxide} + Al \longrightarrow \underbrace{Al_2O_3 + \text{metal}}_{\text{Molten state}} + \text{heat energy}$$

19. (a) Refractory metals are used in the construction of furnaces because they can withstand high temperature e.g. silica, flint, lime, etc.
20. (b) Zr and Ti are purified by van Arkel method.

$$Zr(s) + 2I_2(g) \longrightarrow ZrI_4(g)$$

$$ZrI_4(g) \xrightarrow[\text{filament}]{\text{On the hot}} Zr(s) + 2I_2(g)$$

21. (c) Amalgams (alloys of mercury) are quite often used as catalysts in organic reactions e.g., zinc mercury amalgam is used in clemmensen reduction.
22. (b) Bauxite is an ore of aluminium (Al) consisting of hydrous aluminium oxide and aluminous laterite. It is a clay like amorphous material formed by the weathering of silicate rocks under tropical conditions.
Hence, option (b) is correct.
23. (a) The coating of zinc on iron is called as galvanization to prevent corrosion of iron.
24. (b) Bronze is a mixture of Cu + Sn
Option (b) is correct.
26. (b) Leaching is a chemical method of concentration of ores. Leaching involves the treatment of the ore with a suitable reagent as to make it soluble while impurities remain insoluble. The ore is recovered from the solution by suitable chemical method.
27. (d) Smelting involves the reduction of the ore to the molten metal at a high temperature. For the extraction of less electropositive metal powerful reducing agents such as C, H_2, CO water gas, Na, K, Mg, Al may be used.
34. (b) Silver ore forms a soluble complex with NaCN from which silver is precipitated using scrap zinc.

$$Ag_2S + 2NaCN \rightarrow Na[Ag(CN)_2] \xrightarrow{Zn}$$

$$Na_2[Zn(CN)_4] + Ag \downarrow$$

sodargento-cynanide
(soluble)

36. (c) Galena is PbS and thus purified by froth floatation method.

Froath floatation method is used to concentrate sulphide ores. This method is based on the preferential wetting properties with the froathing agent and water.

37. (c) The reduction of metal sulphides by carbon reduction process is not spontaneous because ΔG for such a process is positive. The reduction of metal oxide by carbon reduction process is spontaneous as ΔG for such a process is negative. From this we find that on thermodynamic considerations CO_2 is more stable than CS_2 and the metal sulphides are more stable than corresponding oxides.
In view of above the factor listed in choice (c) is incorrect and so is of no significance.
Hence the correct answer is (c)

38. (b) During the extraction of copper, iron is present in the ore as impurity (FeS).
The ore together with a little coke and silica is smelted; FeS present as impurity in the ore is oxidized to iron oxide, which then reacts with silica to form fusible ferrous silicate which is removed as slag.

$$2FeS + 3O_2 \longrightarrow 2FeO + 2SO_2\uparrow ;$$

$$FeO + SiO_2 \longrightarrow \underset{(Slag)}{FeSiO_3}$$

39. (d) Leaching is the selective dissolution of the desired mineral leaving behind the impurities in a suitable dissolving agent eg
Argentitie or Silver glance, Ag_2S is an ore of silver. Silver is extracted from argentite by the mac-Arthur and Forest process (leaching process).

$$Ag_2S + 4NaCN \rightarrow 2Na[Ag(CN)_2] + Na_2S$$

$$4Au + 8KCN + 2H_2O + O_2 \rightarrow 4K[Au(CN)_2] + 4KOH$$

40. (b) In froth floatation process, ore particles float because their surface is hydrophobic i.e., not easily wetted by oil.

41. (d) Basic flux is used to remove the impurities in the metallurgy of Al.

42. (b) Mg is highly electropositive.

43. (b) In the Bassemer converter the molten matter (Cu_2S and little FeS) is heated in blast of air. A part of cuprous sulphide is oxidised into cuprous oxide which combines with remaining Cu_2S to form copper metal.

$$2Cu_2S + 3O_2 \longrightarrow 2Cu_2O + 3SO_2$$

$$2Cu_2O + Cu_2S \longrightarrow 6Cu + SO_2$$

44. (d) Si obtained by reduction of $SiCl_4$ with H_2 is further purified by zone refining method to get Si of very high purity. Silicon is purified by zone-refining process because the impurities present in it are more soluble in the liquid phase than in the solid phase.

45. (b) Pig iron or cast iron contains 3 – 5% carbon and varying amounts of Mn, Si, P and S which makes the iron hard and brittle.

46. (c) In blast furnace at about 1270 K, calcium carbonate is almost completely decomposed to give CaO which acts as a flux and combines with SiO_2 present as impurity (gangue) in the ore to form calcium silicate (fusible slag)
CaO(s) (basic flux) + SiO_2 (s) (acidic flux) $\longrightarrow CaSiO_3$ (s) (slag)

47. (c) Duraluminium alloy contains Al (95%), Cu (3%), Mn (1%), Mg (1%).

48. (c) Calcium silicophosphate (a mixture of $Ca_3(PO_4)_2$ & Ca_2SiO_4) is called Thomas slag.

49. (d) The Iron ore contains impurities as silica gangue and the flux added to the furnace is lime stone. Hence, the slag is fromed as calcium silicate $(CaSiO_3)$.

$$CaCO_3 \longrightarrow CaO + CO_2$$

$$\underset{Flux}{CaO} + \underset{gangue}{SiO_2} \longrightarrow \underset{slag}{CaSiO_3}$$

50. (b) Silver chloride (AgCl) is known as horn silver in which chloride is present. Therefore, the ore of chloride is horn silver.

51. (d) Blister-Copper contains 1 – 2 % impurities. It is obtained after Bessemerisation of crude copper.

53. (b) This process is also called autoreduction process or air reduction process. The sulphide ores of less electropositive metals are heated in air to convert part of the ore into oxide or sulphate which then react with the remaining sulphide ore to give the metal and sulphur dioxide.

$$2Cu_2S + 3O_2 \longrightarrow 2Cu_2O + 2SO_2$$

$$Cu_2S + 2Cu_2O \longrightarrow 6Cu + SO_2$$

56. (c) $Fe_2O_3 + 3CO \rightarrow 2Fe + 3CO_2$ (in blast furnace)

57. (b) Important ore of iron is Haematite Fe_2O_3.

58. (b) Calcination is used for removal of volatile impurities and decompose carbonates.

59. (a) Si

60. (c) In this process sulphides ores are converted into oxide ores $2ZnS + 3O_2 \rightarrow 2ZnO + 2SO_2\uparrow$

21. THE P-BLOCK ELEMENTS (Group 15, 16, 17 & 18)

ANSWER KEY

1	(d)	11	(b)	21	(c)	31	(d)	41	(d)	51	(d)
2	(b)	12	(d)	22	(c)	32	(d)	42	(c)	52	(b)
3	(a)	13	(a)	23	(d)	33	(d)	43	(c)	53	(a)
4	(b)	14	(a)	24	(b)	34	(a)	44	(b)	54	(d)
5	(a)	15	(a)	25	(c)	35	(a)	45	(a)	55	(a)
6	(b)	16	(b)	26	(a)	36	(b)	46	(a)	56	(c)
7	(b)	17	(a)	27	(c)	37	(a)	47	(d)	57	(c)
8	(a)	18	(c)	28	(b)	38	(d)	48	(c)	58	(b)
9	(d)	19	(b)	29	(b)	39	(b)	49	(d)	59	(b)
10	(d)	20	(a)	30	(a)	40	(c)	50	(a)	60	(b)

Hints & Solutions

1. (d) N_2 molecule contains triple bond between N atoms having very high dissociation energy (946 kJ mol^{-1}) due to which it is relatively inactive.

2. (b) The structure of NO is $\overset{\cdots}{N} = O$. It is an odd electron molecule and is paramagnetic. It contains a three electron bond.

3. (a) Magnesium and manganese are the metals that produce hydrogen with dilute nitric acid
$Mg + 2HNO_3 \longrightarrow Mg(NO_3)_2 + H_2$

4. (b) The basic character decreases from NH_3 to BiH_3. The basic nature is due to the presence of lone pair of electrons on the central atom. NH_3 is the strongest electron pair donor due to its small size as the electron density of the electron pair is concentrated over a small region. As the size increases the electron density gets diffused over a large region and hence the ability to donate the electron pair (basic nature) decreases.

5. (a) NCl_5 in not possible because N does not contain d orbitals.
Only nitrogen has a tendency to form $P\pi — P\pi$ multiple bonds. Other forms $d\pi - p\pi$ multiple bonds easily.

6. (b)

(a) $H_4P_2O_7 \Rightarrow O{=}P(OH)_2{-}O{-}P(OH)_2{=}O$

Pyrophosphoric acid

(b) $H_3PO_2 \Rightarrow OH{-}P(=O)(H){-}H$

Hypophosphorous acid

(c) $H_3PO_3 \Rightarrow HO - P(=O)(H) - OH$

Phosphorous acid

(d) $H_3PO_4 \Rightarrow HO - P(=O)(OH) - OH$

orthophosphoric acid

7. (b) Alkaline pyrogallol absorbs O_2 and oil of cinnamon absorbs O_3.

8. (a) In the presence of moisture, SO_2 bleaches due to reduction

$$SO_2 + 2H_2O \longrightarrow H_2SO_4 + 2H$$

9. (d) Caro's acid is H_2SO_5 which contains one S – O – O – H peroxy linkage. It is also known as permonosulphuric acids.

$$H-O-O-S(=O)_2-OH$$

Caro's acid

10. (d) Standard reduction potential of halogens are positive and decreases from fluorine to iodine. Therefore halogens act as strong oxidising agent and their oxidising power decreases from fluorine to iodine.

11. (b) As oxidation number of central atom in oxy-acid increases strength increases. Hence the correct

order of acidic strength is

$$\overset{+1}{HClO} < \overset{+3}{HClO_2} < \overset{+5}{HClO_3} < \overset{+7}{HClO_4}$$

12. (d) On account of highly stable ns^2np^6 configuration in the valence shell. These elements have no tendency either to lose gain or share electrons with atoms of other elements i.e., their combining capacity or valency is zero. Further all the orbitals in the atoms of these elements are doubly occupied i.e electrons are not available for sharing.

13. (a) The size of anion is always larger than corresponding neutral atom. So S^{2-} has larger radii and larger size than S.

14. (a) HI is most acidic due to weak bonding between H and I. It breaks into H^+ and I^+. HF is weakest acid because of extensive hydrogen bonding H^+ does not ionise freely.

15. (a) sp^3d^3 hybridization will give pentagonal bipyramid geometry with one trans position occupied by a lone pair and shape of the molecule will be distorted octahedral.

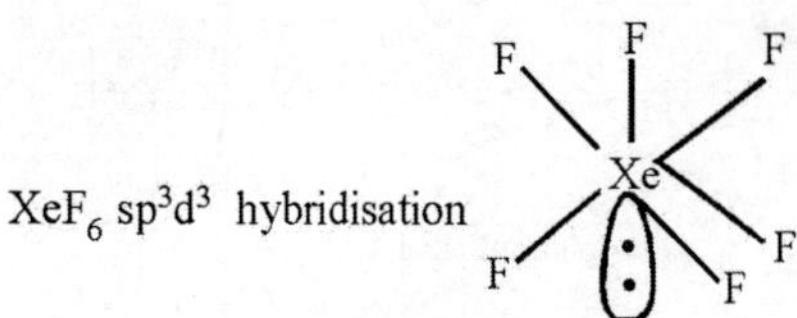

Distorted octahedral

16. (b) In halogens electron affinity decreases in group except chlorine
$Cl > F > Br > I$
The electron gain enthalpy of F is less negative than that of Cl because of its small size as a result of which inter-electronic repulsions present in its 2p subshell are comparatively large.

17. (a) We know that empirical formula of hypophosphorus acid is H_3PO_2. In this only one ionisable hydrogen atom is present i.e. it is monobasic. Therefore option (a) is correct structural formula of it.

18. (c) Oleum is $H_2S_2O_7$ $(H_2SO_4 + SO_3)$ which is obtained by dissolving SO_3 in H_2SO_4 and is called fuming sulphuric acid.

19. (b) It is a tribasic acid all the three hydrogen atoms are ionisable. It forms three series of salts.

$$H_3PO_4 \rightleftharpoons H^+ + H_2PO_4^-$$
$$\Updownarrow$$
$$3H^+ + PO_4^{3-} \rightleftharpoons 2H^+ + HPO_4^{2-}$$

20. (a) Nitrogen molecule is diatomic containing a triple bond between two N atoms, $\ddot{N} \equiv \ddot{N}$. Therefore, nitrogen molecule is formed by sharing six electrons.

21. (c) H_3PO_2 is named as hypophosphorous acid. As it contains only one P – OH group, its basicity is one.

22. (c) A stronger oxidising agent (Cl_2) displaces a weaker oxidising agent (Br_2) from its salt solution.

$$2KBr + Cl_2 \rightarrow 2KCl + Br_2$$

23. (d) Calcium chlorite is $Ca(ClO_2)_2$

24. (b) Brown ring test is done for confirmation of NO_3^- ions.

$$NaNO_3(aq) + H_2SO_4(aq) \longrightarrow NaHSO_4(aq) + HNO_3(aq)$$

$$6FeSO_4 + 2HNO_3 + 3H_2SO_4 \longrightarrow 3Fe_2(SO_4)_3 + 2NO + 4H_2O$$

$$FeSO_4 + NO \longrightarrow \underset{\text{Ferrous nitroso-sulphate (Brown ring)}}{[Fe(NO)]SO_4}$$

25. (c) Ozone layer is beneficial to us, because it stops harmful ultraviolet radiations to reach the earth.

26. (a) Order of dipole moment decreases as
$NH_3 > PH_3 > AsH_3 > SbH_3$
(Based upon electronegativity)

27. (c) Nitrogen form N_2 (i.e. $N \equiv N$) but phosphorus form P_4, because in P_2, $p_\pi - p_\pi$ bonding is present which is a weaker bonding.

28. (b) $$H_2O + \overset{0}{Br_2} \longrightarrow \overset{+1}{HOBr} + \overset{-1}{HBr}$$

Thus here oxidation number of Br increases from 0 to +1 and also decreases from 0 to –1. Thus it is oxidised as well as reduced.

29. (b)

bridging oxygen
bridging oxygen
bridging oxygen
bridging oxygen

30. (a) SnO_2 is an amphoteric oxide because it reacts with acids as well as bases to form corresponding salts.

$$SnO_2 + 2H_2SO_4(\text{conc}) \longrightarrow Sn(SO_4)_2 + 2H_2O$$

$$SnO_2 + 2NaOH \longrightarrow Na_2SnO_3 + H_2O$$

31. (d) HNO_2 is formed only in solution.

32. (d) Ozone can act as a bleaching agent and the process is called dry bleach.

33. (d) Out of N, S and C, nitrogen has the highest electronegativity and it decrease in the order N > S > C. Thus the oxide of nitrogen in its highest (+5) oxidation state will the most acidic.

Next to N_2O_5 in the decreasing order of acidity will be SO_2 then CO_2 and finally CO which is neutral.

34. (a) Nitric acid attacks proteins forming a yellow nitro compound called xanthoprotein. It therefore, stains skin and renders wool yellow. This property is utilised for the test of proteins.

35. (a) Boiling point of NH_3 is high due to intermolecular hydrogen bonding. In others it varies directly with molecular weight of molecule due to increased van der Wall's force. This force is least in PH_3 & highest in SbH_3.

36. (b) $Xe + F_2 \xrightarrow{673K} XeF_2$

2 : 1

$Xe + 2F_2 \xrightarrow[5-6 Alm]{673K} XeF_4$

1 : 5

$Xe + 3F_2 \xrightarrow[50-60 Atm]{573K} XeF_6$

1 : 20

Structures of Xenon fluorides

Xe F_2 : Hybridization sp^3d

Linear

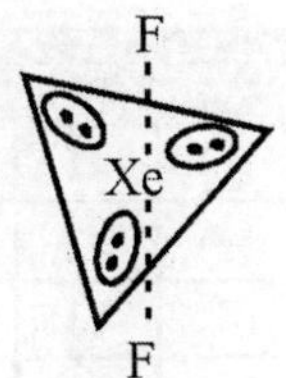

Xe F_4 : Hybridization sp^3d^2 Square planar

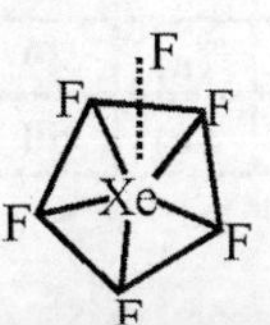

XeF_6 : Hybribisation sp^3d^3

Pentagonal pyramidal or distorted octahedral

37. (a) The reactivity of halogens are depend on electro negativity.

Since electronegativity of halogen decreases in groups. Therefore correct order of reactivity of halogen :

F > Cl > Br > I

38. (d) Fluorine does not form oxyacids because in oxyacids halogen has positive oxidation state. Fluorine is most electronegative element. It can not have positive oxidation state. So, it does not form oxyacids.

39. (b) In case of nitrogen, d-orbitals are not available.

40. (c) Bond dissociation energy of fluorine is less because of its small size and repulsion between electrons of two flourine atoms. So option (c) is wrong order. The correct order is $Cl_2 > Br_2 > F_2 > I_2$

41. (d) The most abundant rare gas found in the atmosphere is argon and not helium.

42. (c) Undecomposed AgBr forms a soluble complex with hypo

$$AgBr + 2Na_2S_2O_3 \rightarrow \underset{\text{soluble complex}}{Na_3[Ag(S_2O_3)_2]} + NaBr$$

It is washed with water and the image is fixed

43. (c) $N_2O_5 + H_2O \longrightarrow 2HNO_3$

44. (b) $S + O_2 \longrightarrow SO_2$ (burns with blue light)

$4Na + O_2 \longrightarrow 2NaO$

(burns with yellow light)

$P_4 + 3O_2 \longrightarrow P_4O_6$

$P_4 + 5O_2 \longrightarrow P_4O_{10}$

$Cl + O_2 \longrightarrow$ No reaction

Chlorine does not react directly with oxygen.

45. (a) $\underset{\text{Bleaching powder}}{CaOCl_2} + 2HCl \rightarrow CaCl_2 + H_2O + Cl_2$

46. (a) As Nitrogen does not have d-orbital in its valence shell its maximum covalency is 3 while in the case of other elements the maximum covalency is 5.

47. (d) $\underset{\text{ortho phosphoric acid}}{2H_3PO_4} \xrightarrow[-2H_2O]{600°C} \underset{\text{meta phosphoric acid}}{2HPO_3}$

48. (c) Air is liquified by making use of the joule-Thomson effect (cooling by expansion of the gas). Water vapour and CO_2 are removed by solidification. The remaining constituents of liquid air i.e., liquid oxygen and liquid nitrogen are separated by means of fractional distillation (b.p. of O_2 = –183°C : b. P. of N_2 = – 195.8°C).

49. (d) Bleaching action of chlorine is due to oxidation in presence of moisture. It is permanent

$H_2O + Cl_2 \rightarrow 2HCl + |O|$

Colouring matter + | O | → colourless matter

50. (a) Alkaline solution of pyrogallol absorbs oxygen quickly.

51. (d) P_2O_5 have great affinity for water. The final product is orthophosphoric acid.

$$P_4O_{10} \xrightarrow{2H_2O} \underset{\text{Metaphosphoric acid}}{4HPO_3} \xrightarrow{2H_2O} \underset{\text{Pyrophosphoric acid}}{2H_4P_2O_7} \xrightarrow{2H_2O} \underset{\text{orthophosphoric acid}}{4H_3PO_4}$$

52. (b) $PCl_3 + 3H_2O \rightarrow H_3PO_3 + 3HCl$

53. (a) $PH_4I + NaOH \rightarrow NaI + PH_3 + H_2O$

54. (d) Pure nitrogen in the lab can be obtained by heating ammonium nitrate. Ammonium nitrate is not a stable compound it dissociate to give nitrogen.

$$NH_4Cl + NaNO_2 \xrightarrow{\text{Heat}} NH_4NO_2 \xrightarrow{\text{Heat}} N_2 + 2H_2O.$$

55. (a) Phosphorus pentoxide has great affinity for water. It forms ortho phosphoric acid on absorbing water

i.e. $P_4O_{10} + 6H_2O \longrightarrow 4H_3PO_4$

It is thus used as a powerful dehydrating or drying agent.

56. (c) $Ca(OH)_2 + Cl_2 \rightarrow CaOCl_2 + H_2O$

57. (c) Urea (46.6%N). % of N in other compound are : $(NH_4)_2SO_4 = 21.2\%$;

$CaCN_2 = 35.0\%$ and

$NH_4NO_3 = 35.0\%$

58. (b) KI reacts with $CuSO_4$ solution to produce cuprous iodide (white precipitate) and I_2 (which gives brown colour) Iodine reacts with hypo ($Na_2S_2O_35H_2O$) solution. Decolourisaiton of solution shows the appearance of white precipitate.

$$2CuSO_4 + 4KI \rightarrow 2K_2SO_4 + \underset{\substack{\text{Cuprous iodide} \\ \text{(White ppt.)}}}{2CuI} + \underset{\substack{\text{(Brown colour} \\ \text{in solution)}}}{I_2}$$

$$2Na_2S_2O_3 + I_2 \longrightarrow \underset{\substack{\text{Sod. tetra} \\ \text{thionate} \\ \text{(colourless)}}}{Na_2S_4O_6} + 2NaI$$

59. (b) HI cannot be prepared by heating iodides with concentrated H_2SO_4 because HI is a strong reducing agent and reduce H_2SO_4 to SO_2 and get themselves oxidised to I_2.

$$2HI + H_2SO_4 \longrightarrow SO_2 + I_2 + 2H_2O$$

60. (b) ClO_2 contains 7 + 12 i.e. 19 electrons (valence) which is an odd number, i.e. there is (are) free electron(s). Hence it is paramagnetic in nature.

22. THE d AND f-BLOCK ELEMENTS

ANSWER KEY

1	(c)	11	(a)	21	(c)	31	(c)	41	(a)	51	(b)
2	(c)	12	(d)	22	(a)	32	(a)	42	(d)	52	(a)
3	(b)	13	(c)	23	(d)	33	(d)	43	(d)	53	(b)
4	(a)	14	(d)	24	(c)	34	(c)	44	(c)	54	(d)
5	(d)	15	(d)	25	(a)	35	(b)	45	(b)	55	(a)
6	(c)	16	(d)	26	(c)	36	(d)	46	(d)	56	(a)
7	(a)	17	(d)	27	(c)	37	(b)	47	(d)	57	(b)
8	(b)	18	(a)	28	(b)	38	(c)	48	(b)	58	(d)
9	(b)	19	(a)	29	(d)	39	(a)	49	(b)	59	(d)
10	(b)	20	(c)	30	(d)	40	(c)	50	(a)	60	(b)

Hints & Solutions

1. (c) General electronic configuration of transition elements is $(n-1)d^{1-10}ns^{1-2}$

2. (c) $[Ne]3s^2 3p^5$ is the electronic configuration of a p-block element whereas other configuration are those of d-block elements

3. (b) Manganese shows max. no. of oxidation states, +2, +3, + 4, + 5, + 6, + 7
Other metals shows the following oxidation states
Cr = + 2, +3, + 4, + 5, + 6
Fe = + 2, + 3
Ti = + 2, + 3, + 4

4. (a) Transition metals have the ability to utilize (n – 1)d -orbitals. Thus variable oxidation states enables the transition element to associate with the reactants in different forms.

5. (d) Transition elements form coloured salts due to presence of incompletely filled d-orbitals.

6. (c) $K_2^{+1}Cr_2^{x}O_7^{-2}$ = + 2 + 2x – 14 = 0 or x = + 6

7. (a) Stable oxidation state of Mn in alkaline medium is +6. So, MnO_2 is oxidised to K_2MnO_4 (purple green) by atmospheric oxygen in KOH medium.

$$2MnO_2 + 4KOH + O_2 \rightarrow \underset{\text{(Purple)}}{2K_2MnO_4} + 2H_2O$$

8. (b) The cause of showing different oxidation is due to the fact that there is only a small difference between the energies of electron in the *ns* orbitals and *(n – 1)d* orbitals with the result both *ns* as well *(n – 1)d* electrons can be used for compound formation. Lesser energy difference between 5f and 6d orbitals than between 4f and 5d orbitals result in larger no. of oxidation state.

9. (b) There is a steady decrease in the radii as the atomic number of the lanthanide elements increases. For every additional proton added in nucleus the corresponding electron goes to *4f* subshell.
The shape of f -orbitals is very much diffused and they have poor shielding effect. The effective nuclear charge increases which causes the contraction in the size of electron charge cloud. This contraction in size is **quite regular** and known as **Lanthanoid contraction.**
Since the change in the ionic radii in the lanthanide series is very small, thus their chemical properties are similar.

10. (b) AgBr is highly photosensitive and is used in photographic films and plates.

11. (a) According to VSEPR theory, a molecule with six bond pairs must be octahedral.

12. (d) We know that chromium (III) salts dissolve in water to give violet solution. The violet colour is due to the hydrated chromium (III) in $[Cr(H_2O)_6]^{3+}$.

13. (c) For third ionization enthalpy last configuration of

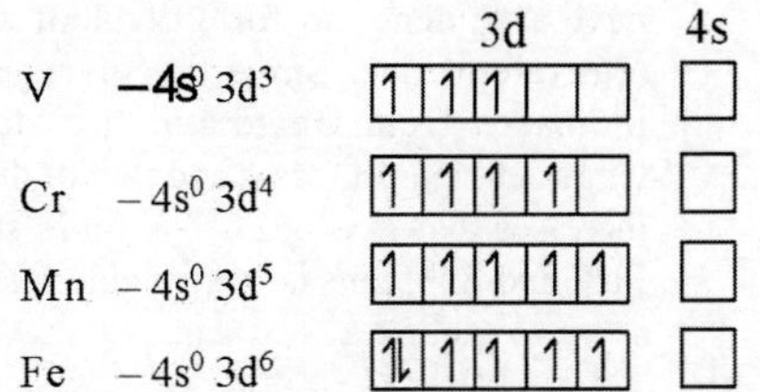

For third Ionization enthalpy Mn has stable configuration due to half filled d-orbital.

14. (d) In moist air, corrosion of iron takes place in the form of rust $(Fe_2O_3 + Fe(OH)_3)$. As iron is most reactive among all.

15. (d) The minimum oxidation state in transition metal is equal to the number of electrons in 4s shell and the maximum oxidation state is equal to the sum of the 4s and 3d electrons.
Ti $3d^2\ 4s^2$
hence minimum oxidation state is +2 and maximum oxidation state is +4. Thus the common oxidation states of Ti are +2, +3 & +4

16. (d) Actinides have variable valency due to very small difference in energies of 5f, 6d and 7s orbitals. Actinides are the elements from atomic number 89 to 103.

17. (d) We know that atomic number of gadolinium is 64. Therefore the electronic configuration of gadolinium is [Xe] $4f^7\ 5d^1\ 6s^2$. Because the half filled and fully filled orbitals are more stable.

18. (a) $Sc^{+3} \rightarrow [Ar]^{18}\ 3d^0;\ V^{+3} \rightarrow [Ar]^{18}\ 3d^2$

$Ti^{+3} \rightarrow [Ar]^{18}\ 3d^1;\ Cr^{+3} \rightarrow [Ar]^{18}\ 3d^3$

Scandium ion doesn't contain any unpaired electron in d orbitals hence it form colourless compound.

19. (a) The Lanthanides are transition metals from Atomic numbers 58 (Ce) to 71(Lu).
Hence the electron configuration becomes :
$(n-2)\,f^{1-14}\,(n-1)\,s^2p^6\,d^{0-1}\,ns^2$.

20. (c) Mn : [Ar] $3d^5\ \ 4s^2$
Shows +2, +3, +4, +5, +6 & +7 oxidation states

21. (c) Lanthanides are 4*f*-series elements starting from cerium (Z= 58) to lutetium (Z = 71). These are placed in the sixth period and in third group.

22. (a) All the group IV elements show tetravalency which can be explained on the basis of shifting of one e^- of ns to the vacant np orbital.These four orbitals give rise to four sp^3 hybrid orbitals. When ns^2 electrons of the outermost shell do not participate in bonding it is called inert pair and the effect is called inert pair effect. The last three elements of group 4 i.e. Ge, Sn and Pb, have a tendency to form divalent as well as tetravalent ions. Since the inert pair effect increases from Ge to Pb, the stability of tetravalent ion decreases and that of divalent ion increases that is why Pb^{++} is more stable than Pb^{4+} and Sn^{4+} ions is more stable than Sn^{2+}.

23. (d) Ni^{3+} : [Ar] $3d^7$

Mn^{3+} : [Ar] $3d^4$

Fe^{3+} : [Ar] $3d^5$

Co^{3+} : [Ar] $3d^6$

24. (c) Due to lanthanide contraction, the size of Zr and Hf (atom and ions) become nearly similar

26. (c) The electronic configuration of Cu(29) is $1s^2\ 2s^2\ 2p^6\ 3s^2\ 3p^6\ 4s^1\ 3d^{10}$ since the last electron goes to the d-orbital hence Cu belongs to d-block of the periodic table.

27. (c) The general electronic configuration of transition elements is $(n-1)d^{1-10}ns^{1-2}$, where n is outer most shell, hence $1s^2 2s^2 2p^6 3s^2 3p^6 3d^2 4s^2$ is the electronic configuration of transition element.

28. (b) The more the number of unpaired electrons, the more is magnetic moment.Therefore the answer is (b).

29. (d) Incorrect, transition metals exhibit variable valency.

30. (d) All statements are correct.

33. (d) O.S. of Os in OsO_4 is +8

34. (c) Transition metals exhibit variable valency.

35. (b) Since reduction potential of fluorine is highest transition metals exhibit highest Oxidation State with fluorine.

36. (d) In $Cu^+\ [Ar]3d^{10}$ there is no unpaired electron,

$Cu^{2+}\ [Ar]3d^9$ contains one unpaired electron hence coloured.

37. (b) $2CuSO_4 + K_4[Fe(CN)_6] \longrightarrow$

$$Cu_2[Fe(CN)_6] + 2K_2SO_4$$

Chocolate ppt.

38. (c) We know that regular decrease in the size of the atoms and ions is called lanthanide contraction. In vertical column of transition elements there is a very small change in size and some times size is found same from second member to third member.The similarity in size of the atoms of Zr and Hf is evident due to the object of lanthanide contraction. Therefore Zr and Hf both have same radius 160 pm.

39. (a) We know that when nitrogen oxide (NO) combines with Fe (II) ions, a brown complex is formed. This reaction is called brown ring test.

40. (c) In lanthanide series there is a regular decrease in the atomic as well as ionic radii of trivalent ions (M^{3+}) as the atomic number increases. Although the atomic radii do show some irregularities but ionic radii decreases from La(103 pm) to Lu (86pm).

41. (a) The basic character of the transition metal monoxide is TiO > VO > CrO > FeO because basic character of oxides decrease with increase in atomic number. Oxides of transitional metals in low oxidation state i.e., + 2 and + 3 are generally basic except Cr_2O_3.

42. (d) The electronic configuration of different species given in the question are :

(a) ${}_{22}Ti^{3+} : 1s^2 2s^2 p^6 3s^2 p^6 d^1$

(b) ${}_{22}Ti^{+} : 1s^2 2s^2 p^6 3s^2 .p^6 d^2 4s^1$

(c) ${}_{22}Ti^{4+} : 1s^2 2s^2 p^6 3s^2 p^6$

(d) ${}_{22}Ti^{2+} : 1s^2 2s^2 p^6 3s^2 p^6 d^2$

Thus options (a) and (c) are discarded; now let us observe the second point of difference.

${}_{23}V^{4+} : 1s^2 2s^2 p^6 3s^2 p^6 d^1$ Thus option (b) is discarded

$${}_{23}V^{3+} : 1s^2 2s^2 p^6 3s^2 p^6 d^2$$

$${}_{24}Cr^{4+} : 1s^2 2s^2 p^6 3s^2 p^6 d^2$$

$${}_{25}Mn^{5+} : 1s^2 2s^2 p^6 3s^2 p^6 d^2$$

43. (d) Sc^{3+} : $1s^2$, $2s^2p^6$, $3s^2p^6d^0$, $4s^0$; no unpaired electron.

Cu^+ : $1s^2$, $2s^2p^6$, $3s^2p^6d^{10}$, $4s^0$; no unpaired electron.

Ni^{2+}: $1s^2$, $2s^2p^6$, $3s^2p^6d^8$, $4s^0$; unpaired electron present.

Ti^{3+} : $1s^2$, $2s^2p^6$, $2s^2p^6d^1$, $4s^0$; unpaired electron present

Co^{2+} : $1s^2$, $2s^2p^6$, $3s^2p^6d^7$, $4s^0$; unpaired electron present

So from the given option the only correct combination is Ni^{2+} and Ti^{3+}

44. (c) Zinc blende mainly contains ZnS and not $ZnCl_2$.

45. (b) $Ag_2S + 4NaCN + 2O_2 \longrightarrow 2Na[Ag(CN)_2] + Na_2SO_4$

46. (d) Mohr's salt is $FeSO_4.(NH_4)_2SO_4.6H_2O$
The oxidation state of iron is +2 in it.

(a) $\overset{+7}{Mn}O_4^- + 5Fe^{2+} + 8H^+ \longrightarrow \underset{\text{Colourless}}{Mn^{2+}} + 5Fe^{3+} + 4H_2O$ (Coloured)

(b) It is a primary standard since it is available in pure state & can be stored without decomposition.

(c) It is a double salt since in solution it shows properties of Fe^{2+}, SO_4^{2-}, NH_4^+.

47. (d) $2HgCl_2 + SnCl_2 \longrightarrow \underset{\text{White ppt}}{Hg_2Cl_2} + SnCl_4$

$Hg_2Cl_2 + SnCl_2 \longrightarrow \underset{\text{grey}}{2Hg} + SnCl_4$

48. (b) $(NH_4)_2Cr_2O_7 \xrightarrow{\Delta} \underset{\text{(green)}}{Cr_2O_3} + N_2 + 4H_2O$

49. (b) Potassium ferrocyanide solution is added to Fe^{3+} ions in solution to give deep blue solution or precipitate.

$4FeCl_3 + 3K_4[Fe(CN)_6] \longrightarrow \underset{\text{Prussian blue}}{Fe_4[Fe(CN)_6]_3} + 12KCl$

50. (a) When steel is heated in presence of NH_3, iron nitride on the surface of steel is formed which imparts a hard coating. This process is called nitriding.

51. (b) PbI_2 is yellow and is called golden spangles.

52. (a) $(NH_4)_2Cr_2O_7 \xrightarrow{\text{Heat}} Cr_2O_3 + N_2 + 4H_2O$

53. (b) Pm is obtained by synthetic method.

54. (d) Stainless steel contains 73% Fe, 18% Cr and 8% Ni.

55. (a) The smaller is the size of cation, the more is heat of hydration. Since hydration is exothermic process, thus heat evolves and hence system having lower energy are stable.

56. (a) $K_2Cr_2O_7 + 2NaOH \longrightarrow K_2CrO_4 + Na_2CrO_4 + H_2O$

Hence CrO_4^{2-} ion is obtained.

57. (b) $CrO_3 + 2NaOH \rightarrow Na_2CrO_4 + H_2O$

58. (d) We know that lanthanides Eu, La, Gd shows +2, +3, +3 and +3 oxidation states respectively only. while actinides, Am shows +2, +3, +4, +5 and +6 oxidation state. Therefore American (Am) has maximum number of oxidation state.

59. (d) By heating chromite with Na_2CO_3, Na_2CrO_4 is obtained

$4FeCr_2O_4 + 8Na_2CO_3 + 7O_2 \longrightarrow 8Na_2CrO_4 + 2Fe_2O_3 + 8CO_2$

60. (b) $[Cu(NH_3)_4]^{2+}$ on addition of excess of aqueous HNO_3 gives $[Cu(H_2O)_4]^{2+}$

23. COORDINATION COMPOUNDS

ANSWER KEY											
1	(d)	11	(c)	21	(b)	31	(c)	41	(c)	51	(b)
2	(b)	12	(a)	22	(c)	32	(b)	42	(b)	52	(a)
3	(b)	13	(a)	23	(d)	33	(b)	43	(a)	53	(a)
4	(c)	14	(d)	24	(b)	34	(b)	44	(c)	54	(a)
5	(a)	15	(b)	25	(a)	35	(d)	45	(a)	55	(d)
6	(c)	16	(a)	26	(d)	36	(c)	46	(c)	56	(b)
7	(d)	17	(a)	27	(b)	37	(b)	47	(d)	57	(c)
8	(b)	18	(c)	28	(b)	38	(b)	48	(c)	58	(c)
9	(d)	19	(d)	29	(d)	39	(a)	49	(a)	59	(a)
10	(c)	20	(b)	30	(b)	40	(a)	50	(d)	60	(b)

Hints & Solutions

1. (d) Potash alum $K_2SO_4.Al_2(SO_4)_3.24H_2O$ is a double salt.
2. (b) IUPAC name is potassium trioxalatoaluminate (III).
3. (b) In potassium ferrocyanide, Fe is in +2 oxidation state. so, its correct formula is $K_4[Fe(CN)_6]$.
4. (c) Coordination isomerism occurs when cationic and anionic complexes of different metal ions are present in a salt. Interchange of ligand between the complexes give isomers e.g. $[Co(NH_3)_6][Cr(CN)_6]$ is an isomer of $[Co(CN)_6][Cr(NH_3)_6]$
5. (a) The SCN^- ion can coordinate through S or N atom giving rise to linkage isomerism

   $M \leftarrow SCN$ thiocyanato

   $M \leftarrow NCS$ isothiocyanato.
6. (c) The coordination number of central metal atom in a complex is equal to number of monovalent ligands, twice the number of bidentate ligands and so on, around the metal ion bonded by coordinate bonds.

   Hence coordination number = no. of σ bonds formed by metals with ligands
7. (d) In the formation of a coordinate bond, the ligands donate a pair of electrons to the metal atom. Futher nitrogen and oxygen has great tendency to donate the pair of electrons in most of the compounds. Therefore both nitrogen and oxygen are common donor atoms in ligands.
8. (b) Organometallic compounds are those compounds in which there is a bond which involve metal. In chlorophyll there is bond involving metal Mg
9. (d) $[Co(CO)_5NH_3]^{2+}$. In this complex. Co-atom attached with NH_3 through σ bonding with CO attached with dative π-bond.
10. (c) The chlorophyll molecule plays an important role in photosynthesis, contain porphyrin ring and the metal Mg not Ca.
11. (c) In $(NH_4)_2(TiCl_6)$, Ti^{4+} $(3d^0\ 4s^0)$ has no unpaired electrons.

    In $K_2Cr_2O_7$, Cr^{6+} $(3p^6\ d^0)$ has no unpaired electrons.

    In $CoSO_4$, Co^{2+} (d^7) has three unpaired electrons in d-orbitals, so it is both paramagnetic and coloured.

    In $K_3[Cu(CN)_4]$, Cu^+ $(3d^{10})$, no unpaired electron.
12. (a) In $[MnO_4]^-$, Mn is in +7 oxidation state.

    Electronic configuration of Mn (Z = 25) : [Ar] $3d^54s^2$

    Electronic configuration of Mn^{7+} :

    [Ar] $3d^04s^0$

    Central atom in other ions have definite number of d electrons

    No. of electrons

    $[Co(NH_3)_6]^{3+}$ in Co^{3+} = Six

    $[Fe(CN)6]^{3-}$ in Fe^{3+} = Five

    $[Cr(H_2O)_6]^{3+}$ in Cr^{3+} = three
13. (a) $Co(NH_3)_4Br_2Cl$ will show both geometrical and ionization isomerism.

    $[Co(NH_3)_4Br_2]Cl$ and $[Co(NH_3)_4BrCl]Br$ are ionization isomers and geometrical isomers are

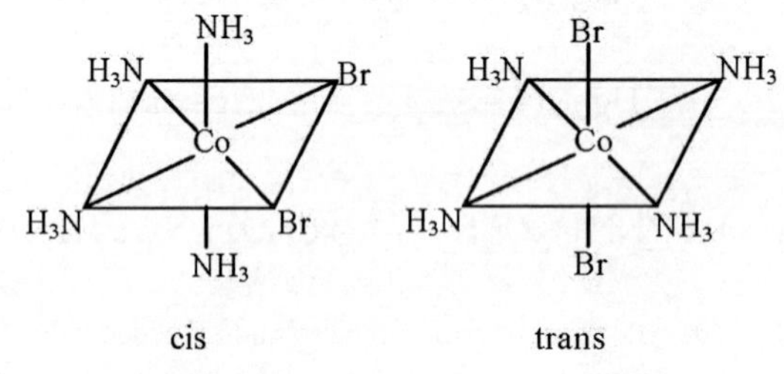

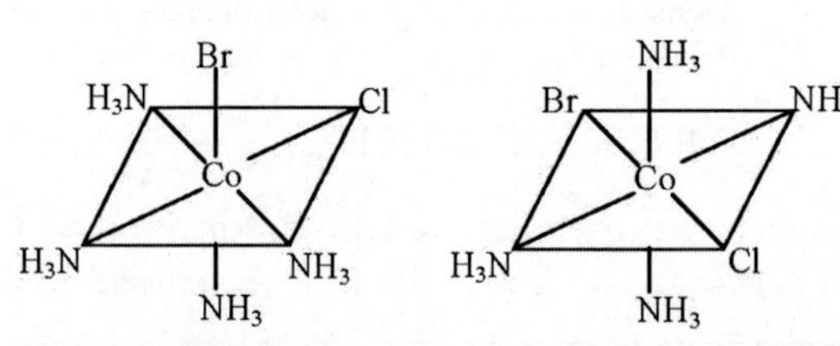

14. (d) In $[Fe(CO)_5]$ Iron exist in zero oxidation state
15. (b) CN^- is a strong field ligand as it is a psuedohalide ion. These ions are strong coordinating ligands and hence have the tendency to form σ-bond (from the pseudo halide to the metal) and π-bond. (from the metal to pseudo halide)
16. (a) Pt $(NH_3)_2\ Cl_2$ is a disubstituted complex and shows only *cis*-& *trans*-isomers

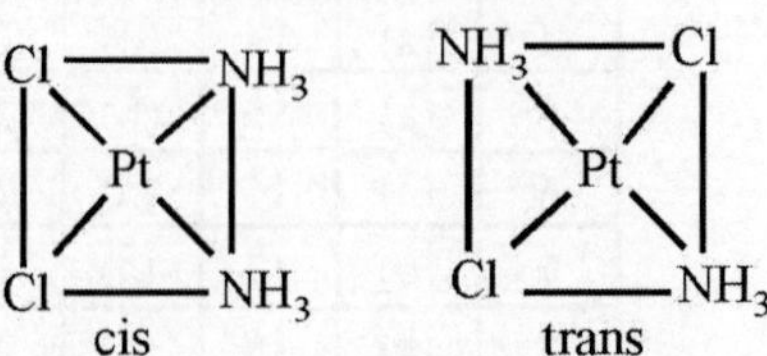

17. (a) The complex ions which give maximum ion in solution exhibit maximum ionic conductivity

$$K_4[Fe(CN)_6] \rightarrow [Fe(CN)_6]^{-4} + 4K^+ \rightarrow 5 \text{ ions}$$

$$[Co(NH_3)_6]Cl_3 \rightarrow [Co(NH_3)_6]^{3+} + 3Cl^-$$
$$\rightarrow 4 \text{ ions}$$

$$[Cu(NH_3)_4]Cl_2 \rightarrow [Cu(NH_3)_4]^{2+} + 2Cl^-$$
$$\rightarrow 3 \text{ ions}$$

$$[Ni(CO)_4] \rightarrow \text{No ions}$$

18. (c) O.N. of Ni in $[Ni(C_2O_4)_3]^{-4}$

$$= x + 3(-2) = -4 \Rightarrow x = -4 + 6 = 2$$

19. (d) Co here is in +3 oxidation state

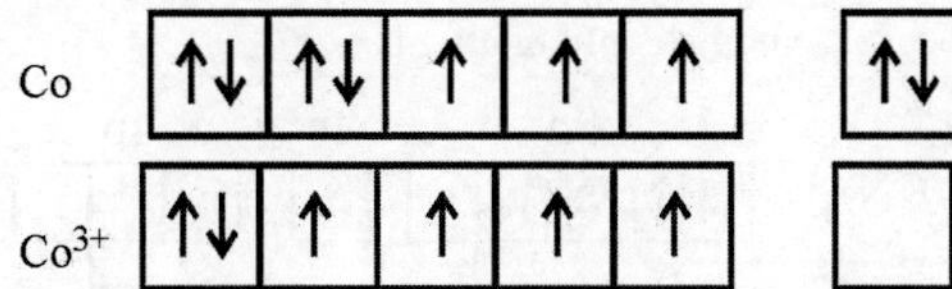

Unpaired electrons = 4 and sp^3d^2 hybridisation and octahedral shape.

20. (b) $Ni^{++} = 3d^8\, 4s^\circ$

1L	1L	1L	1	1

Since, the coordination number of Ni in this complex is 4, the configuration of Ni^{++} at first sight shows that the complex is paramagnetic with two unpaired electron. However, experiments show that the complex is diamagnetic. This is possible when the 3d electrons rearrange against the Hund's rule as shown below. This is in accordance with the fact that the ligand involved here is strong i.e., CN^- ion.

Ni^{++} (after rearrangement)

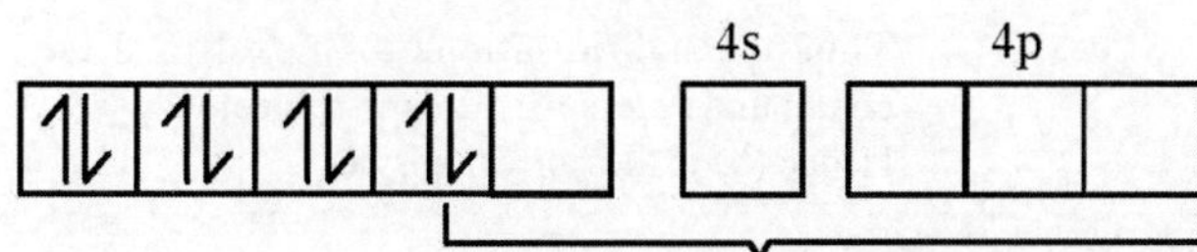

Hence, now dsp^2 hybridization involving one 3d, one 4s and two 4p orbitals, takes place leading to four dsp^2 hybrid orbitals, each of which accepts four electron pairs from CN^- ion forming $[Ni(CN)_4]^{2-}$ ion.

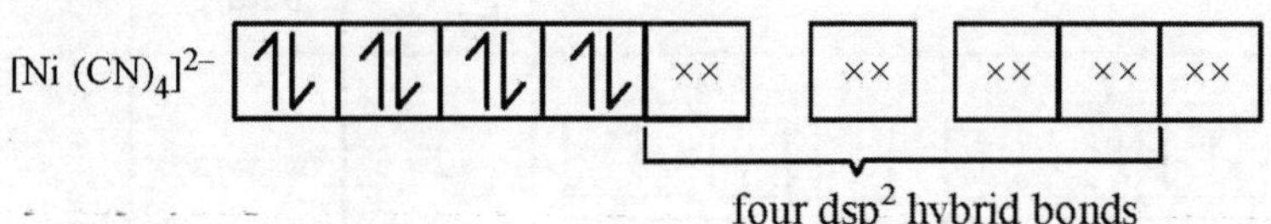

Thus, the complex is diamagnetic as it has no unpaired electron.

21. (b) In $[Ag(NH_3)_2]^+$ ground state configuration of Ag = $4d^{10}5s^1$ excited state configuration of Ag = $4d^{10}5s^0$.

sp hybridization in Ag^+ ion

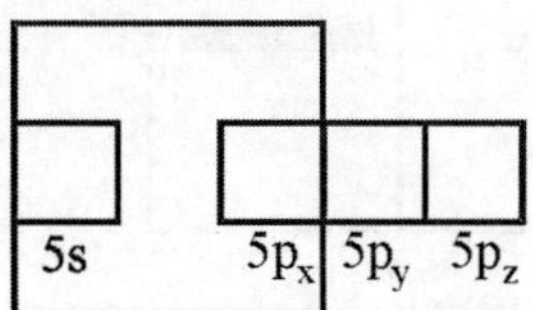

The two *sp* hybrid orbitals in Ag^+ overlap with the filled atomic orbitals of N in NH_3 molecules.

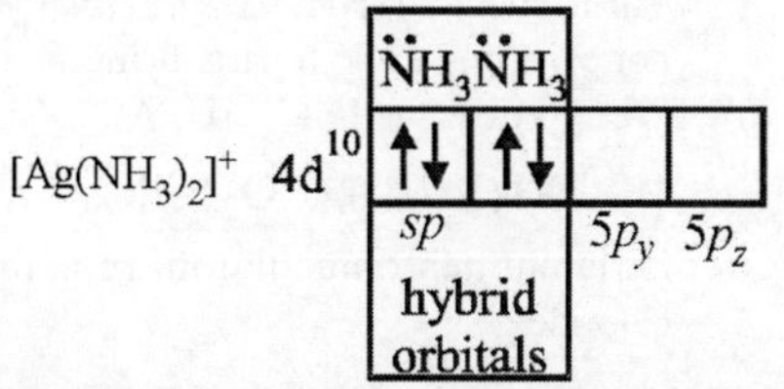

22. (c) Oxidation state of iron is +2 in

$$K_4[Fe(CN)_6]$$

so, its correct name is Potassium hexacyanoferrate (II)

23. (d) NH_3 does not form complex ion with Pb^{2+} as it is not a d-block element.

24. (b) Compounds that contain at least one carbon metal bond are known as organometallic compounds. In CH_3-Mg-Br (Grignard's reagent) a bond is present between carbon and Mg (Metal) hence it is an organometallic compound.

25. (a) $\ddot{N}H_3 + H^+ \text{ (acid medium)} \rightleftharpoons \overset{+}{N}H_4$

26. (d) $EDTA^{4-}$ is a hexadentate ligand, because it has six donor atoms and donate 6 pair of electrons to central metal atom in the complex.

27. (b) Complex compounds do not dissociate into constituent ions.

$K_4[Fe(CN)_6] \longrightarrow 4K^+ + [Fe(CN_6)]^{4-}$

$\therefore$ It is a complex because no CN^- is formed on dissociation.

28. (b) $K_2[Pt\ Cl_6]$
Potassium hexachloroplatinate (IV)
Oxidation state of Pt is + 4 in the complex and anion is present in form of complex.

29. (d) Organometallic compounds are those compounds in which metal atom is directly bonded with C-atom. $H_3C - Li$.

30. (b) The configuration of Ni^{2+} is $3d^8$. For the elements of the first transition series, Cl^- behaves as a weak field/high spin ligand. Hence Ni in $[NiCl_4]^{2-}$ is sp^3 hybridised leading to tetrahedral shape.

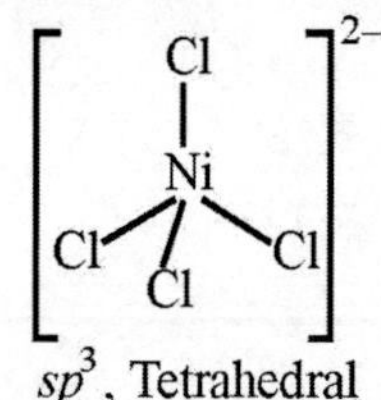

sp^3, Tetrahedral

31. (c) The correct IUPAC name of the given compound is tetramminenickel (II) - tetrachloronickelate (II) thus (c) is the correct answer.

32. (b) In carbonyls O.S. of metal is zero
In $[Ni(CO)_4]$, the oxidation state of nickel is zero. Its configuration in $Ni(CO)_4$ is

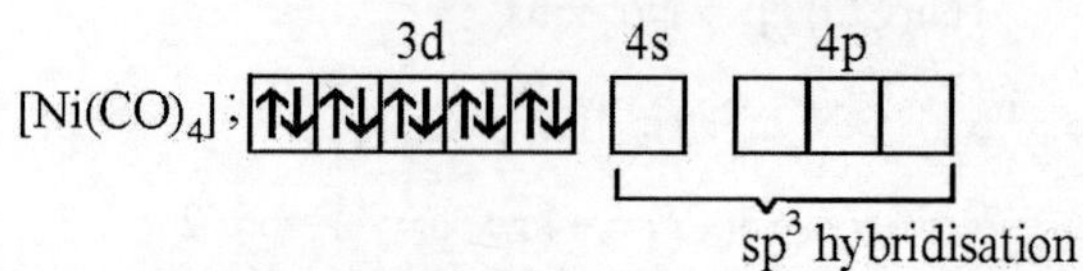

In $[Ni(CN)_4]^{2-}$ the oxidation state of Ni is 2+ and its configuration is

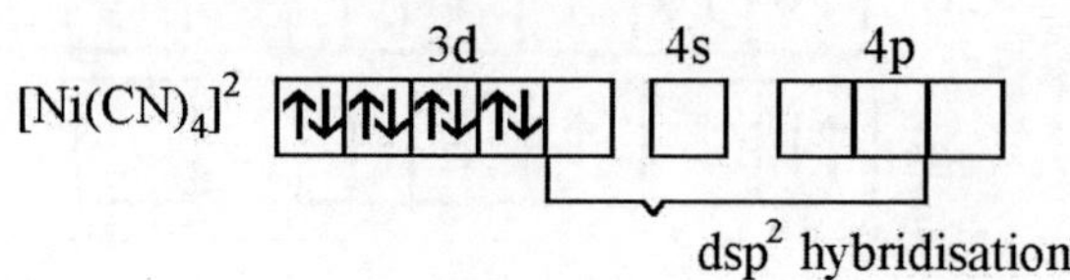

Thus the hybridisations of nickel in these compounds are sp^3 and dsp^2 respectively. Hence (b) is the correct answer.

33. (b)

Atom/Ion Complex	Configuration (3d, 4s, 4p)	No. of unpaired electrons	Magnetic nature
Ni^{2+} (d^8)	3d: ↑↓ ↑↓ ↑↓ ↑ ↑; 4s: □; 4p: □ □ □	2	Paramagnetic
$[NiCl_4]^{2-}$	3d: ↑↓ ↑↓ ↑↓ ↑ ↑; 4s: :; 4p: : : : (sp^3)	2	Paramagnetic
$[Ni(CN)_4]^{2-}$	3d: ↑↓ ↑↓ ↑↓ ↑↓ :; 4s: :; 4p: : : □ (Rearrangement, dsp^2)	0	Diamagnetic
Ni (d^8s^2)	3d: ↑↓ ↑↓ ↑↓ ↑ ↑; 4s: ↑↓; 4p: □ □ □	2	Paramagnetic
$[Ni(CO)_4]$	3d: ↑↓ ↑↓ ↑↓ ↑↓ ↑↓; 4s: :; 4p: : : : (Rearrangement, sp^3)	0	Diamagnetic

34. (b) It is optically active when two Cl atoms are in cis position.

35. (d) The total number of isomers for the complex compound

$[Cu^{II}(NH_3)_4][Pt^{II}Cl_4]$ is four.

These four isomers are

$[Cu(NH_3)_3Cl]\ [Pt(NH_3)Cl_3]$,

$[Cu(NH_3)Cl_3]\ [Pt(NH_3)_3Cl]$,

$[CuCl_4][Pt(NH_3)_4]$

and $[Cu(NH_3)_4][PtCl_4]$.

The isomer $[Cu\ (NH_3)_2\ Cl_2][Pt\ (NH_3)_2\ Cl_2]$ does not exist due to both parts being neutral.

36. (c) We know that IUPAC name of $[Pt(NH_3)_3\ (Br)\ (NO_2)Cl]Cl$ istriamminebromochloronitroplatinum (IV) chloride.

37. (b) As in $[NiCl_4]^{-2}$ Chloride ion being a weak ligand is not able to paired the electron in d orbital.

38. (b) Complex $Co(en)(NH_3)_2Cl_2]Cl$ will have four different isomers.

(i) **Geometrical isomers**

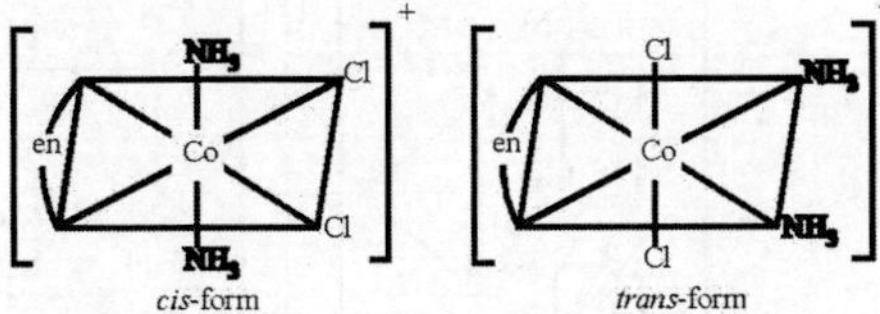

cis-form *trans*-form

(ii) **Optical isomers**

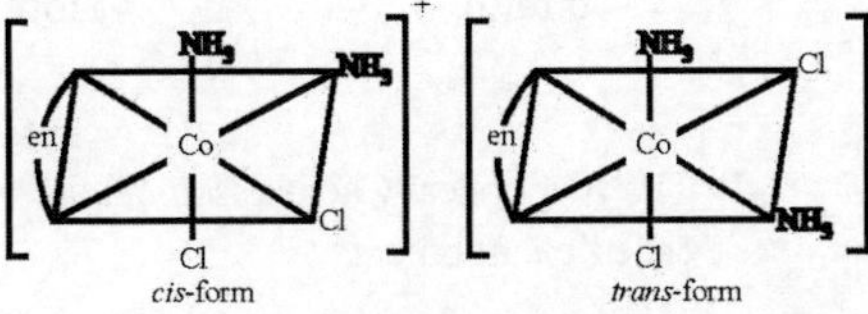

cis-form *trans*-form

39. (a) Chloro diaquatriammine cobalt (III) chloride is $[CoCl(NH_3)_3(H_2O)_2]Cl_2$

40. (a) Triethoxyaluminium has no Al – C linkage

$$Al \begin{cases} O-CH_2CH_3 \\ O-CH_2CH_3 \\ O-CH_2CH_3 \end{cases}$$

41. (c) Diaminodichloroplatinum (II) commonly known as *cis* - **platin** is found to have anticancer property.

42. (b) Lesser is the number of unpaired electrons smaller will be the paramagnetic behaviour. As Cr^{++}, Mn^{++}, Fe^{++} & Ni^{++} contains.

Cr^{++} $(3d^4)$ = | 1 | 1 | 1 | 1 | |

= 4 unpaired e^-.

Mn^{++} $(3d^5)$ = | 1 | 1 | 1 | 1 | 1 |

= 5 unpaired e^-.

Fe^{++} $(3d^6)$= | 1l | 1 | 1 | 1 | 1 |

= 4 unpaired e^-.

Ni^{++} $(3d^8)$ = | 1l | 1l | 1l | 1 | 1 |

= 2 unpaired e^-.

As Ni^{++} has minimum no. of unpaired e^- thus this is least paramagnetic.

43. (a) According to VSEPR theory, a molecule with six bond pairs must be octahedral.

44. (c) $[CuCl_2\{(O = C(NH_2)_2\}_2]$

45. (a) $[Co(NO_2)_3(NH_3)_3]$

Possible geometrical isomers

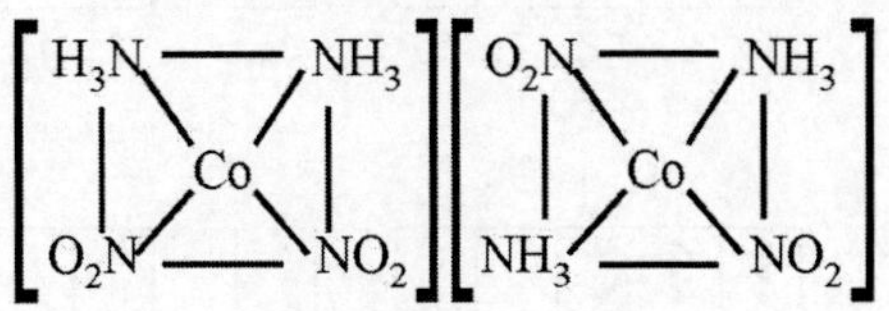

cis *trans*

Total geometrical isomers = 2

46. (c) As it forms two moles of silver chloride thus it has two moles of ionisable Cl.

$$[Co(NH_3)_5NO_2]Cl_2 \rightarrow [Co(NH_3)_5NO_2]^{++} + 2Cl^-$$

$$2Cl^- + 2AgNO_3 \rightarrow 2AgCl + 2NO_3^-$$

47. (d) In $[Cr(NH_3)_6]Br_3$, Cr is in +3 oxidation state etc.

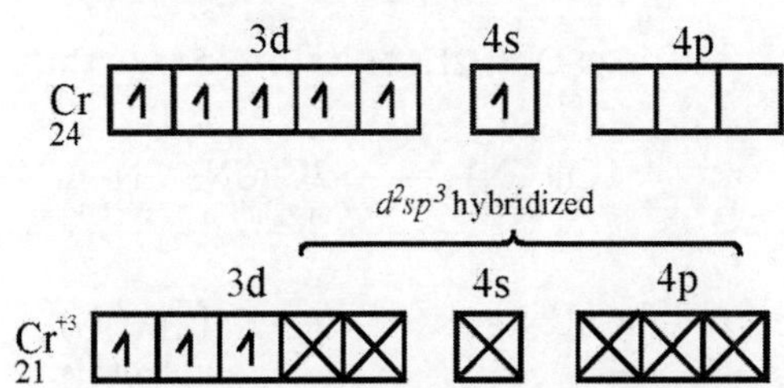

Its ion is octahedral in nature. Due to the presence of three unpaired electrons it is paramagnetic.

48. (c) $[Cr(SCN)_2(NH_3)_4]^+$ shows linkage, geometrical and optical isomerism. Hence produces maximum no. of isomers.

49. (a) (CO) carbonyl group being a strong ligand paired all electrons present in d orbital of Ni. Hence form dsp^2 hybrid orbitals and hence the shape of molecule is square planar.

In $Ni(CO)_4$ complex, Ni will have $3d^{10}$ configuration. It has tetrahedral geometry but diamagnetic as there are no unpaired electrons.

50. (d) Cr^{3+} has $4s^o$ $3d^3$ electronic configuration with 3 unpaired electrons, hence paramagnetic. In other cases pairing of d-electrons take place in presence of strong field ligands such as CO or CN^-.

In $Cr(CO)_6$ molecule 12 electrons are contributed by CO group and it contain no odd electron $[Cr(CO)_6]$

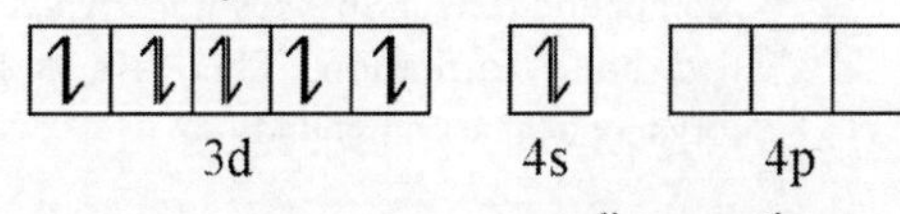

— diamagnetic

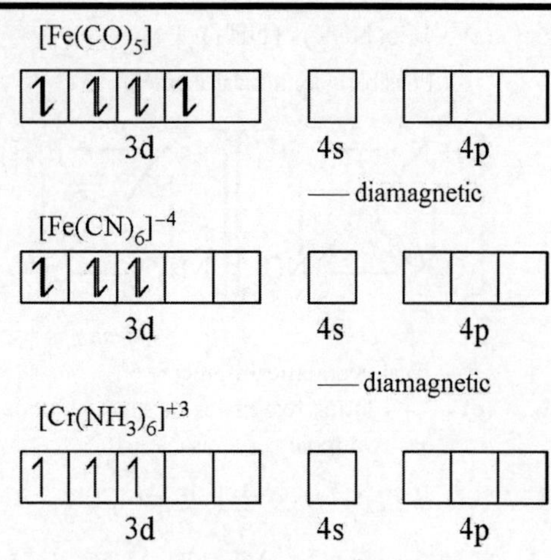

— Paramagnetic

51. (b) Copper sulphate react with KCN to give white ppt of $Cu(CN)_2$ and cyanogen gas. The insoluble copper cynaide dissolve in excess of KCN and give soluble potassium cuprocyanide

$$CuSO_4 + 2KCN \longrightarrow K_2SO_4 + Cu(CN)_2$$

$$2Cu(CN)_2 \longrightarrow \underset{\text{insoluble}}{2CuCN} + \underset{\text{cyanogen}}{CN-CN}$$

$$CuCN + 3KCN \longrightarrow \underset{\text{Soluble}}{K_3[Cu(CN)_4]}$$

52. (a) $MA_3\,B_3$ – 2 geometrical isomers
$MA_2\,B_4$ – 2 geometrical isomers
$MA_4\,B_2$ – 2 geometrical isomers
The complexes of general formula Ma_6 and Ma_5b octahedral geometry do not show geometrical isomerism.

53. (a) IUPAC name of sodium nitroprusside $Na_2[Fe(CN)_5NO]$ is sodium pentacyanonitrosoyl ferrate (III) because in it NO is neutral ligand. Hence
2×O.N. of Na + O.N. of Fe + 5 × O.N. of CN
1 × O.N. of NO = 0
$2 \times (+1) + \text{O.N. of Fe} + 5 \times (-1) + 1 \times 0 = 0$
O.N. of Fe = 5 – 2 = +3, Hence ferrate (III)

54. (a) The number of carbon atom found in π bonded organometallic compounds is indicated by the greek latter 'η' with a number. The prefixes η^2, η^5 and η^6 indicate that 2, 5 and 6 carbon atom are bound to the metal in the compound.$(CH_3)_4Sn$ does not involve any pi (π) bond formation. This is a σ bonded organometallic compound

55. (d) The optical isomers are pair of molecules which are non super imposable mirror images of each other

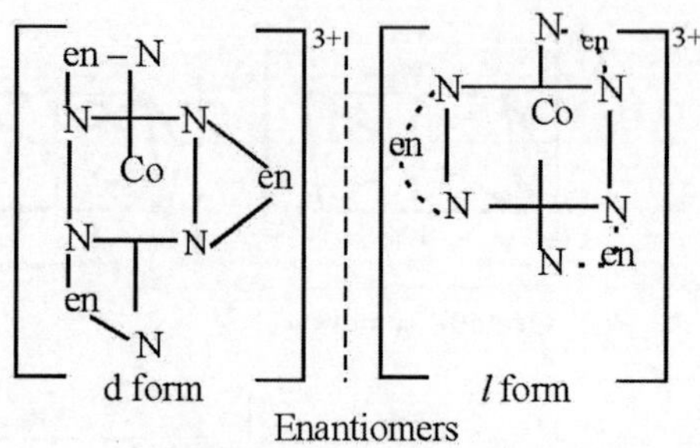

Enantiomers

The two optically active isomers are collectivity called enantiomers.

56. (b) Since H_2O is a weak ligand, it will not cause pairing of electrons in the metal ion Mn^{2+}. Thus electronic configuration of the metal (Mn^{2+}) in the complex will be

$$_{25}Mn^{2+} : 1s^2 2s^2 p^6 3s^2 3p^6 3d^5$$

i.e s unpaired electrons.

57. (c) $[Co(NH_3)_6]^{3+}$ $Co^{3+}(27 - 3 = 24)$

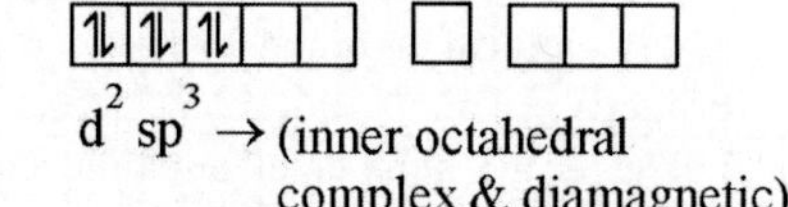

$d^2sp^3 \rightarrow$ (inner octahedral complex & diamagnetic)

$[Cr(NH_3)_6]^{3+}$ $Cr^{3+}(24 - 3 = 21)$

$d^2sp^3 \rightarrow$(inner octahedral complex & paramagnetic)

$[Ni(NH_3)_6]^{2+}$ Ni^{2+} $(28 - 2 = 26)$

$sp^3d^2 \rightarrow$(outer octahedral complex & paramagnetic)

$[Zn(NH_3)_6]^{2+}$ $Zn^{2+}(30 - 2 = 28)$

$sp^3d^2 \rightarrow$ (outer octahedral complex &diamagnetic)

58. (c)

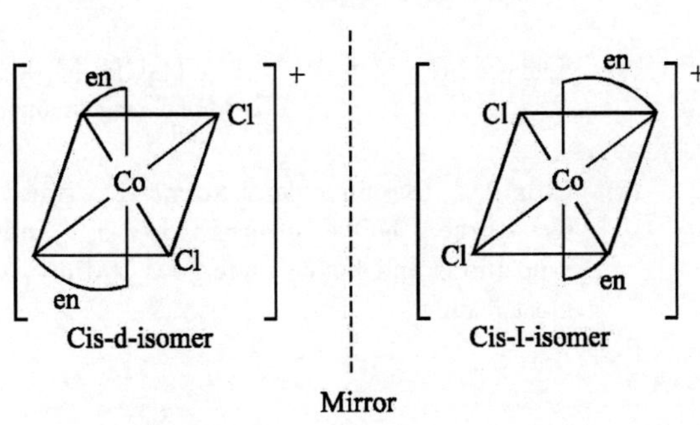

Transform of $[M(AA)_2 a_2]^{n\pm}$ does not shows optical isomerism.

59. (a) The given compound may have linkage isomerism due to presence of NO_2 group which may be in the form $-NO_2$ or –ONO.
It may have ionisation isomerism due to presence of two ionisable group $-NO_2$ & –Cl. It may have geometrical isomerism in the form of *cis-trans* form as follows :
$[Co(NH_3)_4Cl(NO_2)]NO_2$ & $[Co(NH_3)(NO_2)_2]Cl$ —— ionisation isomers.
$[Co(NH_3)_5(NO_2)_2]Cl$ & $[Co(NH_3)_5(ONO)_2Cl$ —— Linkage isomers

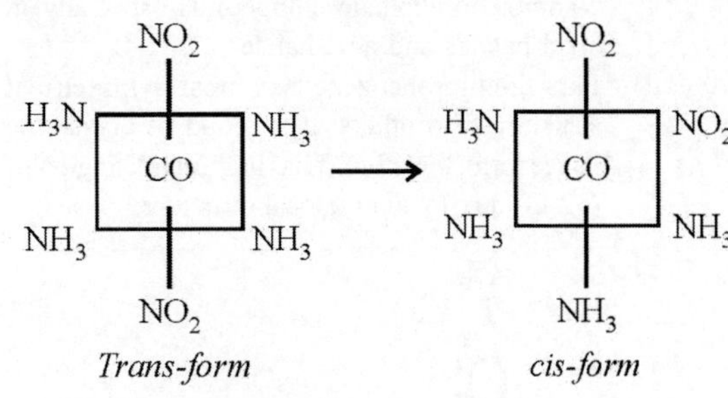

Geometrical isomers

60. (b) $\mu = \sqrt{n(n+2)}$

$3.83 = \sqrt{n(n+2)}$

on solving n = 3
as per magnetic moment, it has three unpaired electron.
Cr^{3+} will have configuration as
Cr $1s^2\ 2s^2\ 2p^6\ 3s^2\ 3p^6\ 3d^4\ 4s^2$
Cr^{3+} $1s^2\ 2s^2\ 2p^6\ 3s^2\ 3p^6\ 3d^3$

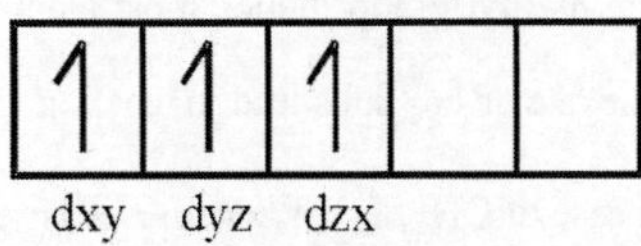

dxy dyz dzx

So 3dxy' 3dyz' 3dxz'

24. HALOALKANES AND HALOARENES

ANSWER KEY											
1	(d)	**11**	(b)	**21**	(a)	**31**	(a)	**41**	(a,c)	**51**	(c)
2	(d)	**12**	(a)	**22**	(b)	**32**	(b)	**42**	(d)	**52**	(a)
3	(d)	**13**	(a)	**23**	(d)	**33**	(b)	**43**	(b)	**53**	(d)
4	(c)	**14**	(d)	**24**	(c)	**34**	(a)	**44**	(b)	**54**	(c)
5	(d)	**15**	(d)	**25**	(b)	**35**	(b)	**45**	(a)	**55**	(a)
6	(d)	**16**	(d)	**26**	(a)	**36**	(a)	**46**	(d)	**56**	(c)
7	(c)	**17**	(d)	**27**	(a)	**37**	(b)	**47**	(b)	**57**	(d)
8	(a)	**18**	(c)	**28**	(d)	**38**	(b)	**48**	(b)	**58**	(c)
9	(d)	**19**	(b)	**29**	(c)	**39**	(a)	**49**	(b)	**59**	(c)
10	(c)	**20**	(c)	**30**	(b)	**40**	(d)	**50**	(c)	**60**	(d)

Hints & Solutions

1. (d) Carbonyl chloride $(COCl_2)$.
2. (d) In **nucleophilic substitution,** a nucleophile provides an electron pair to the substrate and the leaving group departs with an electron pair.

$$Nu: + R-X \longrightarrow R-Nu + X:^{\ominus}$$

These are usually written as S_N (S stands for substitution and N for nucleophilic) and are common in aliphatic compounds especially in alkyl halides and acyl halides.

3. (d) Para-di-chlorobenzene has most symmetrical structure than others. It is found as crystalline lattice form, therefore, it has highest melting point (52°C) due to symmetrical structure.

Para Chlorobenzene
Option (d) is correct.

4. (c) S_{N^1} reactions involve the formation of carbocations, hence higher the stability of carbocation, more will be reactivity of the parent alkyl halide. Thus tertiary carbocation formed from (c) is stabilized by two phenyl groups and one methyl group, hence most stable.

5. (d) The rate of S_N^2 substitution reaction is maximum in case of CH_3CH_2Br because S_N^2 mechanism is followed in case of primary and secondary halides i.e., S_N^2 reaction is favoured by small groups on the carbon atom attached to halogens so

$CH_3\,CH_2\,Br > CH_3\,CH_2\,CH_2\,Br >$

$CH_3-CH(CH_3)-Br > CH_3-C(CH_3)_2-Br$

6. (d) Chlorination of n-butane taken place via free radical formation i.e., $Cl_2 \xrightarrow{h\nu} Cl^{\bullet} + Cl^{\bullet}$

$CH_3-CH_2-CH_2-CH_3 \xrightarrow{Cl_2/h\nu}$

$H-C(CH_3)(C_2H_5)-Cl + Cl-C(CH_3)(C_2H_5)-H$
(d) (l)

Racemic mixture
50% d form + 50% *l* form

$Cl^{\bullet}$ may attack on either side and give a racemic mixture of 2 chloro butane which contain 50% d form and 50% *l*-form.

7. (c) $CH_3CH_2I + Mg \xrightarrow[\text{Ether}]{\text{Dry}} CH_3CH_2MgI$ (Ethyl magnesium iodide)

8. (a) Cl in 2, 4, 6-trinitrochlorobenzene is activated by three NO_2 groups at o, and p-positions and hence undergoes hydrolysis most readily.

9. (d)

$C_6H_6 + CH_3Cl \xrightarrow[AlCl_3]{\text{Anhyd.}} C_6H_5CH_3 + HCl_3$
Toluene

Friedel Craft's reaction.

10. (c)

$Cl_3C-CH{=}O + 2\,C_6H_5Cl \xrightarrow{\text{conc. } H_2SO_4} Cl_3C-CH(C_6H_4Cl)_2$

DDT

11. (b) $C_6H_5Cl \xrightarrow{Mg} C_6H_5MgBr$

$\xrightarrow{CH_3CH_2OH} C_6H_6 + CH_3CH_2OMgBr$

12. (a) $Cl_3C-\boxed{H + HO}-NO_2 \longrightarrow Cl_3C-NO_2$ Chloropicrin (used as an insecticide)

13. (a) When halogens are attached to same C atom then they are called gem-halides

$R-CH(X)-X$, $R-CH(X)-CH-X$
Vicinal halides

14. (d) $C_6H_6 + CH_3CH_2CH_2Cl \xrightarrow[AlCl_3]{\text{Anhyd.}}$

$C_6H_5-CH(CH_3)-CH_3$

Isopropyl benzene

15. (d) $-NO_2$ group is electron attractive group, so it is able to deactivate the benzene ring.

(Structure: benzene ring with Cl at top, NO_2 at ortho position and NO_2 at para position)

hence withdrawl of electrons from ortho & para position cause easy removal of –Cl atom due to development of +ve charge on o- and p positions.

16. (d) Chlorofluorocarbons, e.g. CF_2Cl_2, CHF_2Cl_2, HCF_2CHCl_2. These are non-inflammable colour less and stable upto 550°C. These are emitted as propellants in aerosol spray, cans refrigerators, fire fighting reagents etc. They are chemically inert and hence do not react with any substance with which they come in contact and therefore float through the atmosphere and as a result enter the stratosphere. There theyabsorb UV-rays and due to this they produce free atomic chlorine which results decomposition of ozone which cause depletion of ozone layer.

17. (d) Chloropicrin is nitrochloroform. It is obtained by the nitration of chloroform with HNO_3.

$$\underset{\text{Chloroform}}{HCCl_3} \xrightarrow{HNO_3} \underset{\text{Chloropicrin}}{O_2NCCl_3}$$

Chloropicrin is a liquids, poisonous and used as an insecticide and a war gas

18. (c) $C_6H_6 \xrightarrow[35°C]{Br_2/FeBr_3} C_6H_5Br + HBr + FeBr_3$

19. (b) Haloform compounds with the formula CHX_3, where X is a halogen atom.
Haloforms are trihalogen derivatives of methane.
Example : Chloroform $CHCl_3$.

20. (c) Chloropicrin is CCl_3NO_2 also known as nitrochloroform is a useful insecticide and lachrymatry substance.

21. (a) $\underset{\text{Ethanol}}{CH_3CH_2OH} + Cl_2 \xrightarrow{\text{oxidation}} \underset{\text{Ethanal}}{CH_3CHO} + 2HCl$

$$\underset{\text{Ethanal}}{CH_3CHO} + 3Cl_2 \longrightarrow \underset{\text{Chloral}}{CCl_3CHO} + 3HCl$$

$$2CCl_3CHO + Ca(OH)_2 \longrightarrow \underset{\text{Chloroform}}{2CHCl_3} + \underset{\text{Cal. formate}}{(HCOO)_2Ca}$$

$$\underset{\text{Acetone}}{CH_3COCH_3} + 3Cl_2 \longrightarrow \underset{\text{Trichloroacetone}}{CCl_3COCH_3} + 3HCl$$

$$2CCl_3CO.CH_3 + Ca(OH)_2 \longrightarrow 2CHCl_3 + (CH_3COO)_2Ca$$

22. (b) NaCl is an ionic compound cannot displace –OH by Cl. Rest all other reagents ($HCl + ZnCl_2$, $PCl_5, SOCl_2$) displaces –OH from alcohol and provide Cl^- as a nucleophile.

23. (d) Under ordinary conditions freon is a gas. Its boiling point is –29.8°C. It can easily be liquified. It is chemically intert. It is used in air-conditioning and in domestic refrigerators for cooling purposes (as refrigerant)

24. (c) Oxidation along with reaction with alcoholic silver nitrate can be used for differentiating between —Cl present on nucleus and on the side chain.

$$\underset{\text{Benzyl chloride}}{C_6H_5CH_2Cl} \xrightarrow{\text{alc. } AgNO_3} \underset{\textbf{(White)}}{AgCl\downarrow};$$

(—Cl on side chain, reactive)

$$\underset{\text{Chlorobenzene}}{C_6H_5Cl} \xrightarrow[AgNO_3]{\text{alc.}} \text{No AgCl}$$

(—Cl on nucleus inert)

25. (b) Chloroform ($CHCl_3$) is used as industrial solvent.

26. (a) Friedel-Craft's reaction is mainly applied on benzene.

$$C_6H_6 + CH_3CH_2Cl \xrightarrow{AlCl_3} C_6H_5C_2H_5 + HCl$$

27. (a) $C_2H_5OH \xrightarrow{PCl_3} CH_3CH_2Cl + H_3PO_3$

$$C_2H_5OH \xrightarrow{HCl} C_2H_5Cl + H_2O$$

$C_2H_5OH \xrightarrow{Cl_2/P}$ will give PCl_3 in situ and C_2H_5Cl is formed.

Molecular Cl_2 cannot displace —OH from the alcohol.

28. (d) CH_3CH_2OH – & CH_3CO – groups react with I_2 and NaOH to give Haloform (Iodoform Reaction) $CH_3CH_2 - CH_2 - OH$ (propanol) Lacks this grouping therefore does not react.

29. (c) Primary alcohol on oxidation produce an aldehyde.

$$\underset{\text{Pri. Chloride}}{ROH} \xrightarrow{X_2} RX \xrightarrow[-2HX]{\{O\}} \underset{\text{Aldehyde}}{RCHO}$$

30. (b) When ethyl alcohol is treated with PCl_5, then ethyl chloride is formed.

$$CH_3CH_2-OH + PCl_5 \xrightarrow{\Delta} CH_3CH_2-Cl + HCl + POCl_3$$

31. (a) Chloroform is an organic compound which does not ionise in water. Since it can not provide Cl^-, therefore, it is not precipitated with $AgNO_3$.

32. (b) Iodoform test is given that by compound which has α-CH_3 group. Since

$$CH_3-CH_2-\overset{\overset{O}{\|}}{C}-CH_2CH_3$$ does not contain α-CH_3 group. Therefore, it does not show Iodoform test.

33. (b) At high temp. i.e., 400°C substitution occurs in preference to addition.

$$CH_3CH=CH_2 \xrightarrow[-HCl]{Cl_2, 400°C} ClCH_2CH=CH_2$$

34. (a)

$$\underset{\text{t-Butyl chloride}}{CH_3-\underset{CH_3}{\overset{CH_3}{|}{C}}-Cl} + 2Na + Cl-\underset{CH_3}{\overset{CH_3}{C}}-CH_3 \xrightarrow{\text{Wurtz Rxn}} CH_3-\underset{CH_3}{\overset{CH_3}{C}}-\underset{CH_3}{\overset{CH_3}{C}}-CH_3 \xrightarrow[\text{Clorination}]{\text{Mono}} CH_3-\underset{CH_3}{\overset{CH_3}{C}}-\underset{CH_3}{\overset{CH_3}{C}}-CH_2Cl$$

35. (b) The order of atomic size of halogens decrease in the order I > Br > Cl > F
i.e on moving down a group atomic size increases. Further the bond length of C-X bond deccreases in the order
C – I > C – Br > C – Cl > C – F
and hence the bond dissociation energy decreases in the order
R – F > R – Cl > R – Br > R – I
hence B – I being a weakest bond break most easily.
hence R – I is most reactive.

36. (a) $CH_3CH_2Cl \xrightarrow{NaCN} \underset{(X)}{CH_3CH_2CN} + NaCl \xrightarrow{Ni/H_2}$

$$\underset{(Y)}{CH_3CH_2CH_2NH_2} \xrightarrow{(CH_3CO)_2O} CH_3CH_2CH_2NHCOCH_3(Z) + CH_3COOH$$

37. (b) $CH_3-CH_2-CHCl_2 \xrightarrow[\Delta]{NaNH_2}$

$$CH_3-CH=CHCl \xrightarrow[\Delta]{NaNH_2} \underset{\text{Final Product}}{CH_3-C\equiv CH}$$

38. (b) $C_2H_5I \xrightarrow{\text{alc. KOH}} \underset{(X)}{CH_2=CH_2} \xrightarrow{Br_2}$

$$\underset{(Y)}{\underset{Br}{CH_2}-\underset{Br}{CH_2}} \xrightarrow{KCN} \underset{(Z)}{\underset{Br}{CH_2}-CH_2CN}$$

39. (a) Primary halide < Secondary halide < Tertiary halide.

40. (d) On sulphonation of chlorobenzene, ortho and para chlorobenzene is formed because – Cl group is para and ortho directing.

$$C_6H_5Cl \xrightarrow[\text{Fuming } H_2SO_4]{H_2S_2O_7} o\text{-}ClC_6H_4SO_3H + p\text{-}ClC_6H_4SO_3H$$

41. (a,c) S_N2 mechanism is followed in the case of primary and secondary alkyl halides, but primary is more reactive than secondary i.e.

$$CH_3CH_2Cl > (H_3C)_2CH-Cl$$

In comparision to *p*-halides reactivity of *s*-hlides towards S_N2 reaction is negligible.

42. (d) $CH_3CH_2CH_2Br + Mg \longrightarrow CH_3CH_2CH_2MgBr$
(Propyl Bromide)

$CH_3CH_2CH_2MgBr + H_2O$

$\longrightarrow CH_3CH_2CH_3 + Mg(OH)Br$

43. (b) Dehydrobromination by strong base (alc. NaOH) followed by Markownikoff addition of HBr.

44. (b) The reaction proceeds via free radical mechanism. As 2° free radical is more stable than 1°, so $CH_3CH_2CH(Br)CH_3$ would be formed.

45. (a) Chlorination beyond monochlorination during the preparation of alkyl halides in presence of UV light can be suppressed by taking **alkane in excess**.

46. (d) The given reaction is an example of electrophilic substitution. Further, CH_3 group in toluene is *o*, *p*-directing

47. (b) Addition of HBr of 2-pentyne gives two structural isomers (I) and (II)

$CH_3-C\equiv C-CH_2CH_3 \xrightarrow{HBr}$

$\underset{(I)}{CH_3C(Br)=CHCH_2CH_3} + \underset{(II)}{CH_3CH=C(Br)CH_2CH_3}$

Each one of these will exist as a pair of geometrical isomers. Thus, there are two structural and four configurational isomers.

48. (b) By distilling ethanol or acetone with a paste of bleaching powder *(laboratory and commercial method)*.

$CaOCl_2 + H_2O \longrightarrow Ca(OH)_2 + Cl_2$

Cl_2, so obtained acts as a mild oxidising as well as chlorinating agent

(a) $CH_3COCH_3 + 3Cl_2 \longrightarrow$

$CCl_3.CO.CH_3 + 3HCl$ **(Chlorination)**

(b) $2CCl_3.CO.CH_3 + Ca(OH)_2 \longrightarrow$

$2CHCl_3 + (CH_3COO)_2Ca$

(Hydrolysis)

49. (b) Number of active hydrogen in a compound corresponds to the number of moles of CH_4 evolved per mole of the compound.

$-OH, -NH_2, -SH, -OH$ or $-C\equiv CH \xrightarrow{CH_3MgI}$

$CH_4\uparrow$ ($2CH_4$ from $-NH_2$)

50. (c) Potassium ethoxide is a strong base, and 2-bromopentane is a 2° bromide, so elimination raction predominates

$CH_3CH(Br)CH_2CH_2CH_3 \xrightarrow{OC_2H_5^-}$

$\underset{\text{Pentene - 2(major) } trans}{CH_3CH=CHCH_2CH_3} + \underset{\text{Pentene-1(minor) } cis}{CH_2=CHCH_2CH_2CH_3}$

Since *trans*- alkene is more stable than *cis*. Thus *trans*-pentene -2 is the main product.

51. (c) Toluene ($C_6H_5CH_3$) $\xrightarrow[FeCl_3]{Cl_2}$ o-chlorotoluene (CH_3, Cl) + p-chlorotoluene (CH_3, Cl)

Toluene ($C_6H_5CH_3$) $\xrightarrow{Cl_2, hv}$ $C_6H_5CH_2Cl$ $\xrightarrow{Cl_2, hv}$ $C_6H_5CHCl_2$ $\xrightarrow{Cl_2, hv}$ $C_6H_5CCl_3$

52. (a) A strong nucleophile favours the S_N2 reaction and a weak nucleophile favours the S_N1 reaction. First reaction is S_N1 reaction because C_2H_5OH used as solvent which is a weak nucleophile. Second reaction is S_N2 reaction because $C_2H_5O^-$ is strong nucleophile.

53. (d) Cl_2 water being aqueous and $CHCl_3$ being organic form two layers. Cl_2 water oxidizes $I^- \longrightarrow I_2$ which shows its colour (violet) in organic layer $(CHCl_3)$. This is layer test which is used for detection of I^- and Br^- in a solution. Br^- gives brown layer.

54. (c) Wurtz reaction : It involves the interaction of two molecules of an alkyl halide (preferably bromide or iodide) with metalic sodium in presence of dry ether to form symmetrical alkanes containing double the number of carbon atoms present in the alkyl halide. For example,

$$\underset{\text{Alkyl halide}}{R-X} + 2Na + X-R \xrightarrow{\text{Dry ether}} \underset{\text{Alkane}}{R-R} + 2NaX$$

55. (a) We know that, moleculer mass = 2 × v.d.
⇒ molecular mass of the given organic compound is 46.

	%composition	Atomic mass	% composition / atomic mass	Wh.one no.ratio
C	52.17	12	4.3475	2
H	13	1	13	6
O	34.83	16	2.176875	1

So, Formula of the compound = $(C_2H_6O)_n$

$n = \frac{46}{46} = 1 \Rightarrow$ molecular formula is C_2H_6O

The compound also gives iodoform test so, it would contain –OH group or $-\overset{\overset{O}{\|}}{C}-$

As our molecular formula contains two carbon atoms. So, the organic compound is ethanol.

56. (c) $CH_3-\underset{\underset{O}{\|}}{C}-CH_3 + H-CCl_3 \longrightarrow$

$H_3C-\overset{\overset{OH}{|}}{\underset{\underset{CCl_3}{|}}{C}}-CH_3$

Chloretone (used as an drug)

57. (d) Ethylene dichloride can be prepared by adding HCl to ethylene glycol ($CH_2OH.\ CH_2OH$).

58. (c) $CH\boxed{Cl_3 + 6Ag + Cl_3}CH \longrightarrow CH \equiv CH + 6AgCl$

59. (c) When chloroform is exposed to light it is oxidised to a poisonous gas known as Phosgene.

$$2CHCl_3 + 2O_2 \longrightarrow 2COCl_2 + Cl_2$$

60. (d) $CH_3-\overset{\overset{Br}{|}}{CH}-CH_2-CH_3 \xrightarrow{\text{alc KOH}} CH_2 = CH-CH_2-CH_3$

In this reaction both hydrogen and halogen atom has been removed so it is known as dehydro halogenation reaction.

25. ALCOHOLS, PHENOLS AND ETHERS

ANSWER KEY

1	(c)	**11**	(c)	**21**	(b)	**31**	(b)	**41**	(d)	**51**	(c)
2	(d)	**12**	(c)	**22**	(b)	**32**	(a)	**42**	(b)	**52**	(c)
3	(a)	**13**	(b)	**23**	(b)	**33**	(b)	**43**	(d)	**53**	(c)
4	(c)	**14**	(c)	**24**	(a)	**34**	(c)	**44**	(a)	**54**	(b)
5	(d)	**15**	(d)	**25**	(b)	**35**	(c)	**45**	(b)	**55**	(c)
6	(a)	**16**	(d)	**26**	(c)	**36**	(c)	**46**	(b)	**56**	(d)
7	(d)	**17**	(c)	**27**	(d)	**37**	(a)	**47**	(d)	**57**	(c)
8	(b)	**18**	(b)	**28**	(c)	**38**	(c)	**48**	(d)	**58**	(c)
9	(c)	**19**	(a)	**29**	(c)	**39**	(d)	**49**	(b)	**59**	(c)
10	(d)	**20**	(c)	**30**	(c)	**40**	(c)	**50**	(d)	**60**	(c)

Hints & Solutions

1. (c) $KMnO_4$ (alkaline) and OsO_4 / CH_2Cl_2 are used for hydroxylation of double bond while O_3 /Zn is used for ozonolysis. Therefore, the right option is (c), i.e.,

$$3CH_3CH = CH_2 \xrightarrow{BH_3 \text{ in THF}} (CH_3CH_2CH_2)_3B$$

$$\xrightarrow[\text{NaOH}]{3H_2O_2} 3CH_3CH_2CH_2OH + H_3BO_3$$
1-propanol

2. (d) Benzoic acid (C_6H_5COOH) will not form phenol or phenoxide.
3. (a) Due to H-bonding, the boiling point of ethanol is much higher than that of the isomeric diethyl ether.
4. (c) The acidic nature of phenol is due to the formation of stable phenoxide ion in solution

$$C_6H_5OH + H_2O \rightleftharpoons \underset{\text{Phenoxide ion}}{C_6H_5O^-} + H_3O^+$$

The phenoxide ion is stable due to resonance.

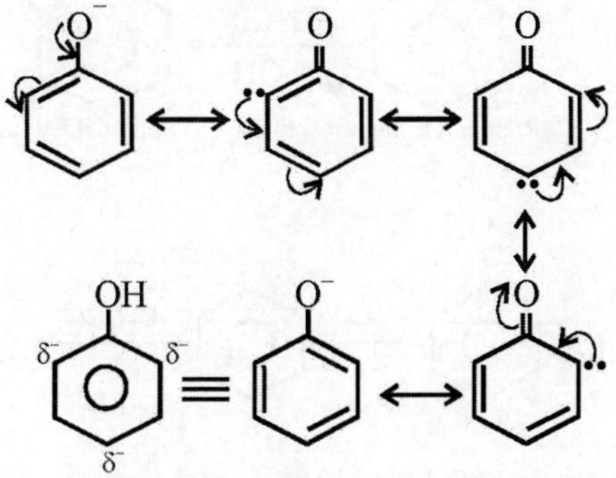

The negative charge is delocalized in the benzene ring which is a stabilizing factor in the phenoxide ion and increase acidity of phenol. wheras no resonance is possible in alkoxide ions (RO^-)derived from alcohol. The negative charge is localized on oxygen atom. Thus, alcohols are not acidic.

5. (d) $CH_3CH_2OH \xrightarrow[H_2SO_4]{\text{Conc.}} \underset{\text{alkene}}{CH_2 = CH_2} + H_2O$

Ketones do not react with hot conc. H_2SO_4 Acids on reaction with conc. H_2SO_4 decompose to give $CO_2 + H_2O$. Ethers with hot conc. H_2SO_4 yield alcohol and alkyl hydrogen sulphate. So only alcohols lose a H_2O molecule with hot conc. H_2SO_4 to yield alkene.

6. (a) Lucas reagent is conc. HCl + anhyd. $ZnCl_2$.
7. (d)

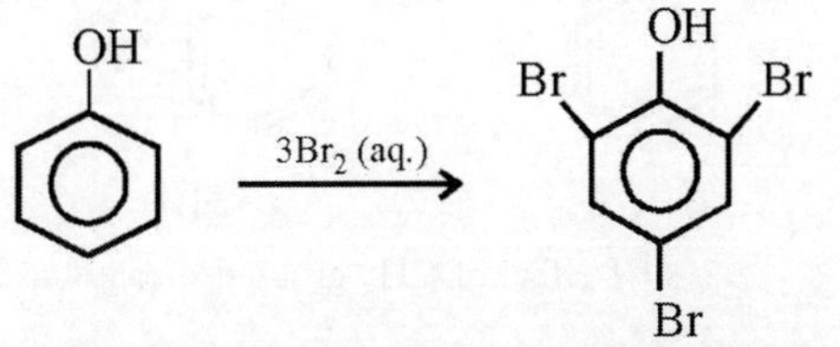

Phenol $\xrightarrow{3Br_2\ (aq.)}$ 2, 4,6-Tribromophenol

Note : The –OH group in phenol, being activating group, facilitates substitution in the *o*- and *p*-positions.

8. (b) Phenol $+ CHCl_3 \longrightarrow$ (o-$CHCl_2$ phenol) $\xrightarrow{2KOH}$ (o-$CH(OH)_2$ phenol) $\xrightarrow{-H_2O}$ (o-CHO phenol)

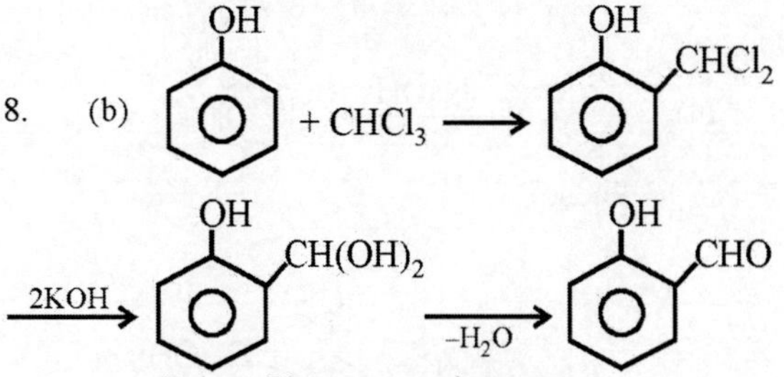

Reimer-Tiemann reaction.

9. (c) Glucose and fructose obtained by hydrolysis of sucrose, are converted into alcohol by enzyme zymase.

$$C_6H_{12}O_6 \xrightarrow{\text{zymase}} 2C_2H_5OH + 2CO_2$$

10. (d) The reaction of alkyl halides with sodium alkoxide or sodium phenoxide to form ethers is called Williamson synthesis. For example,

$$\underset{\text{Alkyl halide}}{R-X} + \underset{\text{Sod. alkoxide}}{R'-ONa} \longrightarrow \underset{\text{Ether}}{R-O-R'} + NaX$$

11. (c) In case of unsymmetrical ethers, the site of cleavage depends on the nature of alkyl group e.g.,

$$CH_3O-CH(CH_3)_2 + HI \xrightarrow{373K} \underset{\text{Methyl iodide}}{CH_3I} + \underset{\text{Isopropyl alcohol}}{(CH_3)_2CHOH}$$

The alkyl halide is always formed from the smaller alkyl group.

12. (c) Ethers readily form co-ordination complexes with lewis acids like BF_3, $AlCl_3$, RMgX, etc.

$$R-\ddot{O}\!:\!(R) + BF_3 \rightarrow [R-O(R) \rightarrow BF_3]$$

13. (b) By heating benzaldehyde with conc. NaOH or KOH (Cannizzaro reaction).

$$\underset{\text{Benzaldehyde}}{C_6H_5CHO} + NaOH \longrightarrow \underset{\text{Benzyl alcohol}}{C_6H_5CH_2OH} + \underset{\text{Sod. benzoate}}{C_6H_5COONa}$$

14. (c) Formation of diethylether from ethanol is based on dehydration.

$$C_2H_5\boxed{OH \;+\; H}-O-C_2H_5 \xrightarrow[-H_2O]{X,\ \Delta} C_2H_5-O-C_2H_5$$

X could be conc. H_2SO_4 at 413 K or Al_2O_3 at 523 K.

15. (d) Phenol (OH-benzene) $\xrightarrow{NaOH}$ sodium phenoxide (ONa-benzene) $\xrightarrow[140° C]{CO_2 + HCl}$ Salicylic acid (o-HO–C$_6$H$_4$–COOH)

16. (d) The rates of reaction with lucas reagent follows the order.
3° alcohol > 2° alcohol > 1° alcohol
since carbocations are formed as intermediate, more stable the carbocation, higher will be the reactivity of the parent compound (alcohol). 2-Methylpropan-2-ol generates a 3° carbocation, so it will react fastest; other three generates either 1° or 2° carbocations.

$$CH_3-\underset{CH_3}{\overset{CH_3}{C}}-OH \xrightarrow{H^+}$$

2-Methyl-2-propanol

$$CH_3-\underset{CH_3}{\overset{CH_3}{C^{\oplus}}} \xrightarrow{Br^-} CH_3-\underset{CH_3}{\overset{CH_3}{C}}-Br$$

17. (c) Due to strong electron-donating effect of the OH group, the electron density in phenol is much higher than that in toluene, benzene and chlorobenzene and hence phenol is readily attacked by the electrophile.

18. (b) Secondary alcohols on oxidation give ketones.
Note : – Primary alcohols from aldehydes.

$$R_2CHOH \xrightarrow{[O]} R_2C=O$$

Isopropyl alcohol → Ketone

19. (a) 1° Alcohols on catalytic dehydrogenation give aldehydes.

$$\underset{1°\ alcohol}{RCH_2OH} \xrightarrow[300°C]{Cu} \underset{Aldehyde}{RCHO} + H_2$$

20. (c) We know that

$$H_2C\underset{O}{-}CH_2 + RMgX \longrightarrow \underset{OMgX}{CH_2}-\underset{R}{CH_2} \xrightarrow[-Mg(OH)X]{H_2O} \underset{OH}{CH_2}-\underset{R}{CH_2}$$

21. (b)

Phenol $\xrightarrow[NaOH]{CHCl_3}$ Salicylaldehyde (o-HO–C$_6$H$_4$–CHO)

(Riemer Tiemann reaction) Salicylaldehyde

22. (b) Phenol $\xrightarrow{Zn}$ X (benzene) $\xrightarrow[anhy\ AlCl_3]{CH_3Cl}$ Y (toluene, C$_6$H$_5$CH$_3$) $\xrightarrow[KMnO_4]{alc.}$ Z (benzoic acid, C$_6$H$_5$COOH)

23. (b) p-nitrophenol (iv) > m-nitrophenol (iii) > phenol (i) > m-cresol (ii)

(iv) (– I and – M effects, both increase acidity)

(iii) (only – I effect)

(i) (ii)

(+ I effect of CH_3 group decreases acidity)

24. (a) Alcohols are oxidized by removal of H_2 in presence of a heated metal catalyst (Cu)

$$\underset{1^\circ \text{ alcohol}}{CH_3CH_2OH} \xrightarrow[300^\circ]{Cu} \underset{\text{Aldehyde}}{CH_3CHO} + H_2$$

$$\underset{2^\circ \text{ Alcohol}}{(H_3C)_2CH\text{—}OH} \xrightarrow[300^\circ]{Cu} \underset{\text{Ketone}}{CH_3\overset{\underset{\|}{O}}{C}CH_3} + H_2$$

25. (b) By heating alkyl halide with sod. or pot. alkoxides (Williamson synthesis) diethyl ether is formed. This is the most important industrial and laboratory method and may be used for preparing simple as well as mixed ethers. For example :

$$C_2H_5ONa + C_2H_5I \longrightarrow C_2H_5O\,.\,C_2H_5 + NaI$$

26. (c) Ethers contain the functional group – O –
27. (d) Any one of Ni, Pt or Pd can be used in the reduction of aldehydes.
28. (c) A tertiary alcohol having no oxidisable hydrogen linked to carbon atom bearing hydroxyl group, is stable to oxidation.

$$\underset{\text{Tert. butyl alcohol}}{2CH_3-C(CH_3)_2-OH} \xrightarrow[Cu,\ 300^\circ C]{[O]} \underset{\text{Isobutylene}}{CH_3-C(CH_3)=CH_2}$$

$$CH_3-C(CH_3)_2-OH \xrightarrow[Cu]{\text{hot reduced}} CH_3-C(=CH_2)-CH_3$$

29. (c) Conversion of phenol into salicyclic acid can be carried out by Riemer-Tiemann reaction.

$$C_6H_5OH + CCl_4 \xrightarrow[340K,\ -NaCl,\ -H_2O]{NaOH} C_6H_4(ONa)(CCl_3)$$

$$\underset{\text{Unstable}}{C_6H_4(ONa)C(OH)_3} \xrightarrow[-NaCl]{\text{dil. HCl}} C_6H_4(OH)COOH$$

30. (c) $CH_2 = CH_2 \xrightarrow[OH^-]{KMnO_4} \underset{\text{Glycol}}{CH_2(OH)-CH_2(OH)}$ ($KMnO_4/OH^-$: Baeyer's reagent)

This reaction is known as Baeyer's test for unsaturation.

31. (b) $CH_3CH_2CH(OH)CH_3$ – a secondary alcohol.
32. (a) The lower alcohols are readily soluble in water and the solubility decreases with the increase in molecular weight. The solubility of alcohols in water can be explained due to the formation of hydrogen bond between the highly polarised —OH groups present both in alcohol and water.
33. (b) $C_2H_5OH + CH_3COOH \longrightarrow CH_3COO.C_2H_5 + H_2O$
34. (c) Preparation of ethers by reacting sodium ethoxide with alkyl halide is called Williamson synthesis.
35. (c) When phenol reacts with Zinc dust, then benzene and zinc oxide are formed

$$\underset{\text{Phenol}}{C_6H_5OH} + Zn \longrightarrow \underset{\text{Benzene}}{C_6H_6} + ZnO$$

36. (c) Secondary alcohols oxidise to produce kenone.

$$\underset{\text{2-Butanol}}{CH_3CHOHCH_2CH_3} \xrightarrow{(O)} \underset{\text{Ethyl methyl ketone}}{CH_3COCH_2CH_3}$$

37. (a) $\underbrace{CO + H_2}_{\text{water gas}} + H_2 \xrightarrow[300^\circ C]{Cr_2O_3-ZnO} \underset{\text{Methanol}}{CH_3OH}$
38. (c) $C_2H_5OC_2H_5 + 2\ HI \xrightarrow{\Delta} 2\ C_2H_5I + H_2O$
39. (d) Riemer-Tiemann reaction involves electrophilic substitution on the highly reactive phenoxide ring.

$$HCCl_3 + OH^- \longrightarrow H_2O + {}^-{:}CCl_3 \quad :\bar{C}Cl_3 \longrightarrow Cl^- + :CCl_2$$

Note the C has only a sextet of electrons

$$\text{Phenoxide} \xrightarrow{:CCl_2} \text{(intermediate with } CCl_2\text{, H)} \longrightarrow$$

$$\text{o-}CHCl_2\text{ phenoxide} \xrightarrow{OH^-} \text{o-}CHO\text{ phenoxide}$$

A benzal chloride

40. (c) $CH_3C(OH)_3 \xrightarrow{-2H_2O} CH_3COOH$

41. (d) $CH_3.CH_2-\underset{OH}{\underset{|}{CH}}-CH_3 \xrightarrow[-H_2O]{}$

$$CH_3CH=CH-CH_3$$

$$\xrightarrow{O_3} \text{(ozonide) } CH_3CH\text{–O–}CH-CH_3 \longrightarrow 2CH_3CHO$$

42. (b) The solubility of alcohols depend on number of C-atoms of alcohols. The solubility of alcohols in water is decreased by increasing number of C-atoms of alcohol. As resulting molecular weight increases, the polar nature of O – H bond decreases and hence strength of hydrogen bond decreases.

43. (d) Phenol is acidic in nature and the acidic nature of phenol is increased by the presence of –I (electron attracting) groups at o – and p– position of the – OH groups. However, benzoic acid is more acidic than nitro-phenol.

44. (a) $C_6H_5ONa + C_2H_5I \xrightarrow{\Delta}$

$$\underset{\text{Phenetole}}{C_6H_5OC_2H_5} + NaI$$

45. (b) Four primary alcohols of $C_5H_{11}OH$ are possible. These are:

(i) $CH_3CH_2CH_2CH_2CH_2OH$

(ii) $CH_3CH_2-\underset{CH_3}{\underset{|}{CH}}-CH_2OH$

(iii) $CH_3-\underset{CH_3}{\underset{|}{CH}}-CH_2CH_2OH$

(iv) $CH_3-\overset{CH_3}{\overset{|}{\underset{CH_3}{\underset{|}{C}}}}-CH_2OH$

46. (b) Greater the stability of the intermediate carbocation, more reactive is the alcohol. Since 2-methylpropan-2-ol generates 3° carbocation, therefore, it reacts fastest with HBr.

47. (d) Electron-donating groups (– OCH_3, – CH_3 etc.) tend to decrease and electron withdrawing groups (– NO_2, – OCH_3 etc.) tend to increase the acidic character of phenols. Since – OCH_3 is a more powerful electron-donating group than – CH_3 group, therefore, p-methylphenol is slightly more acidic than p-methoxyphenol while p-nitrophenol is the strongest acid. Thus, option (d), i.e. p-methoxyphenol, p-methylphenol, p-nitrophenol is correct.

48. (d) Due to –I-effect of the three C–Cl-bonding between CI and C-atom of the OH group, CCl_3 CH $(OH)_2$ is most stable.

49. (b) $CH_3I + Mg \xrightarrow{\text{Dry ether}} CH_3MgI$

$$CH_3MgI + \underset{H}{\overset{C_6H_5}{>}}C{=}O \longrightarrow \underset{CH_3}{\overset{C_6H_5}{>}}CH-O-MgI$$

$$\xrightarrow[-Mg(OH)I]{H_2O} \underset{CH_3}{\overset{C_6H_5}{>}}CHOH$$

(1-Phenyl ethanol)

50. (d) $C_2H_5OH + 4I_2 + 6NaOH \longrightarrow$

$$\underset{\text{Iodoform}}{CHI_3}\downarrow + HCOONa + 5NaI + 5H_2O$$

51. (c) Primary alcohol on oxidation give aldehyde which on further oxidation give carboxylic acid whereas secondary alcohols give ketone.

$$\underset{\text{n-propyl alcohal}}{CH_3CH_2CH_2OH} \xrightarrow{[O]}$$

$$CH_3CH_2CHO \xrightarrow{[O]} CH_3CH_2COOH$$

$$\underset{\text{isopropyl alcohal}}{\underset{H_3C}{\overset{H_3C}{>}}CH-OH} \xrightarrow{[O]} \underset{\text{Ketone}}{\underset{H_3C}{\overset{H_3C}{>}}C=O}$$

52. (c) In the cleavage of mixed ethers with two different alkyl groups, the alcohol and alkyl iodide that

form depend on the nature of alkyl group. When primary or secondary alkyl groups are present, it is the lower alkyl group that forms alkyl iodide therefore

$$CH_3-\underset{\substack{|\\CH_3}}{CH}-CH_2-O-CH_2-CH_3+HI\xrightarrow{\Delta}$$

$$CH_3-\overset{\substack{CH_3\\|}}{CH}-CH_2OH+CH_3CH_2I$$

53. (c)

$$CH_3CH_2OH\xrightarrow{PBr_3}CH_3CH_2Br\xrightarrow{alc.KOH}CH_2=CH_2$$

$$\downarrow H_2SO_4$$

$$CH_3CH_2OH\xleftarrow[heat]{H_2O}CH_3-CH_2-HSO_4$$

54. (b) Phenol is most acidic because its conjugate base is stabilised due to resonance, while the rest three compounds are alcohols, hence, their corrosponding conjugate bases do not exhibit resonance

55. (c) When three –OH groups are attached with a single C atom the compound is unstable hence get dissociated and form carboxylic acid on leaving 2 –OH molecules of H_2O.
Oxidation with periodic acid (HIO_4 or H_5IO_6) gives formaldehyde, HCHO.

$$\underset{\text{Ethylene glycol}}{\begin{matrix}CH_2OH\\|\\CH_2OH\end{matrix}}\xrightarrow{HIO_4}\begin{matrix}CH_2O\\+\\CH_2O\end{matrix}$$

Remember that compounds containing two –OH groups on the two adjacent carbon atoms are easily oxidised by periodic acid forming aldehyde/ketone.

56. (d) Ether is used (i) as a general anaesthetic, (ii) as refrigerant since it produces cooling on evaporation, (iii) as solvent for oils, fats, resins etc. (iv) for providing inert medium in Wurtz reaction, (v) for preparing Grignard reagent, (vi) in perfumery.

57. (c) Glycerine decomposes at its boiling point so cannot be distilled by simple distillation. It is distilled at 180° (lower than its boiling point) at 12 mm pressure by vaccum distillation.

58. (c) $C_2H_5I+C_2H_5ONa\xrightarrow{\Delta}$

$$\underset{\text{Diethyether}}{C_2H_5OC_2H_5}+NaI$$

This reaction is known as Williamson synthesis.

59. (c)

$$\underset{\text{Glycerol}}{\begin{matrix}CH_2OH\\|\\CHOH\\|\\CH_2OH\end{matrix}}\xrightarrow[KHSO_4]{\Delta}\underset{\text{Acrolein}}{\begin{matrix}CH_2\\||\\CH\\|\\CHO\end{matrix}}+H_2O$$

This reaction is often used as a qualitative test for the presence of glycerol since acrolein is easily detected due to its peculiar odour.

60. (c) Ethylene is passed into concentrated sulphuric acid at 75–80°C under pressure when a mixture of ethyl hydrogen sulphate and diethyl sulphate is formed.

$$\underset{\text{Ethylene}}{H_2C=CH_2}+H_2SO_4\xrightarrow{100°C}C_2H_5HSO_4$$

$$\xrightarrow[H_2O]{\Delta}C_2H_5OH+H_2SO_4$$

26. ALDEHYDES, KETONES AND CARBOXYLIC ACIDS

ANSWER KEY

1	(c)	11	(d)	21	(a)	31	(d)	41	(a)	51	(d)
2	(d)	12	(a)	22	(d)	32	(d)	42	(b)	52	(b)
3	(b)	13	(c)	23	(b)	33	(d)	43	(a)	53	(b)
4	(a)	14	(c)	24	(c)	34	(d)	44	(a)	54	(a)
5	(c)	15	(b)	25	(b)	35	(a)	45	(b)	55	(c)
6	(c)	16	(d)	26	(a)	36	(c)	46	(b)	56	(a)
7	(d)	17	(a)	27	(d)	37	(c)	47	(b)	57	(d)
8	(c)	18	(a)	28	(b)	38	(c)	48	(b)	58	(d)
9	(a)	19	(a)	29	(b)	39	(a)	49	(d)	59	(c)
10	(c)	20	(b)	30	(d)	40	(c)	50	(c)	60	(a)

Hints & Solutions

1. (c) $R-\underset{\underset{O}{\|}}{C}-Cl \xrightarrow[Pd-BaSO_4]{H_2} R-\underset{\underset{O}{\|}}{C}-H + HCl$ 'P'

2. (d) $R\,COCl + AlCl_3 \longrightarrow R-\overset{\overset{O}{\|}}{C}^{\oplus} + AlCl_4^-$ (Electrophile)

3. (b) Acetaldehyde reacts only with nucleophiles. Since the mobile p electrons of carbon–oxygen double bond are strongly pulled towards oxygen, carbonyl carbon is electron-deficient and carbonyl oxygen is electron-rich. The electron deficient (acidic) carbonyl carbon is most susceptible to attack by electron rich nucleophilic reagents, that is, by base. Hence the typical reaction of aldehydes and ketones is nucleophilic addition.

4. (a) Aldehydes and ketones having at least one α-hydrogen atom in presence of dilute alkali give β-hydroxy aldehyde or β-hydroxy ketone

$$\underset{\text{Acetaldehyde}}{CH_3-\overset{\overset{O}{\|}}{\underset{\underset{H}{|}}{C}}} + HCH_2CHO \xrightarrow{\text{dil.NaOH}} \underset{\text{Aldol}}{CH_3-\overset{\overset{OH}{|}}{\underset{\underset{H}{|}}{C}}-CH_2-CHO}$$

$$\xrightarrow[-H_2O]{\Delta} \underset{\text{Crotonaldehyde}}{CH_3-CH=CH.CHO}$$

5. (c) It is a simple Cannizzro reaction.

$$2\ \text{(3-chlorobenzaldehyde, } C_6H_4(Cl)CHO) \xrightarrow{50\%\ KOH} C_6H_4(Cl)COO^- + C_6H_4(Cl)CH_2OH$$

6. (c) Formalin is an aqueous solution (40%) of formaldehyde.

7. (d)

$$C_6H_5CH_2-CH_3 \xrightarrow[\text{(ii) } KMnO_4, KOH \text{ (iii) } H^+]{\text{(i) [O]}} C_6H_5COOH$$

$$\text{(3-bromobenzoic acid) } C_6H_4(Br)COOH \xrightarrow[H^+]{C_2H_5OH} C_6H_4(Br)COOC_2H_5$$

8. (c) Chlorine is electron withdrawing group. Further Inductive effect is stronger at α position than β-position. i.e.,

$$\overset{\alpha}{C}H_2ClCOOH > \overset{\beta}{C}H_2Cl\overset{\alpha}{C}H_2COOH$$

9. (a)

$$CH_3COOH + PCl_5 \rightarrow \underset{[A]}{CH_3COCl} \xrightarrow[\text{Anhyd}AlCl_3]{C_6H_6} C_6H_5COCH_3 \text{ (Fridle Craft reaction)}$$

$$\xrightarrow{MgBrC_2H_5} C_2H_5-\overset{\overset{OMgBr}{|}}{\underset{\underset{C_6H_5}{|}}{C}}-CH_3 \xrightarrow[\text{ether hydrolysis}]{H^+} \underset{(C)}{C_2H_5-\overset{\overset{OH}{|}}{\underset{\underset{C_6H_5}{|}}{C}}-CH_3}$$

10. (c) This reaction is an example of Hell - Volhard Zelinsky reaction. In this reaction acids containing α – H on treatment with X_2 /P give di-halo substituted acid.

$$CH_3-CH_2COOH \xrightarrow{Br_2/P} CH_3-CBr_2-COOH$$

11. (d)

$$C_6H_5COOH \xrightarrow{\text{Soda-lime(X)}} C_6H_6 + Na_2CO_3 + H_2O$$

$$C_6H_5OH \xrightarrow{\text{Zn dust (Y)}} C_6H_6 + ZnO$$

12. (a) Among acetic acid, phenol and n-hexanol only CH_3COOH reacts with $NaHCO_3$ to evolve CO_2 gas.

$$CH_3COOH + NaHCO_3 \longrightarrow CH_3COONa + CO_2 + H_2O$$

13 (c) Aldol condensation is given by the compounds which contain α hydrogen atom. As the given compound does not contain α hydrogen atom. Hence it does not undergo aldol condensation.

14. (c) $CH_3COOH + CaCO_3 \rightarrow (CH_3COO)_2Ca$

$$\xrightarrow{\text{Heat}} \begin{matrix} CH_3 \\ CH_3 \end{matrix} > CO$$

$$\xrightarrow{I_2+NaOH} CHI_3 + NaI + CH_3COONa + 3H_2O$$

15. (b) Grignard reagent forms addition product with bubbled carbondioxide which on hydrolysis with HCl yields benzoic acid.

$$C_6H_5MgBr \xrightarrow{(i)\ CO_2} C_6H_5-\overset{O}{\overset{\|}{C}}-OMgBr \xrightarrow{H_3O^+} C_6H_5-\overset{O}{\overset{\|}{C}}-O-H + Mg(OH)Br$$

'P'

Benzoic acid

16. (d) $CH_3COOH \xrightarrow{SOCl_2} CH_3-COCl$

Friedel craft's reaction C_6H_6 / $AlCl_3$

$$C_6H_5-\overset{O}{\overset{\|}{C}}-CH_3 \xrightarrow{HCN} C_6H_5-C(OH)(CH_3)-CN$$

H_2O (hydrolysis of cyano group)

$$C_6H_5-C(OH)(CH_3)-COOH$$

17. (a) Aldehydes containing no α-hydrogen atom on warming with 50% NaOH or KOH undergo disproportionation i.e. self oxidation - reduction known as cannizzaro's reaction.

$$2C_6H_5CHO + NaOH \xrightarrow{50\%\ NaOH} C_6H_5COONa + C_6H_5CH_2OH$$

18. (a) An electron releasing substituent (+I effect) intensity the negative charge on the anion resulting in the decrease of stability and thus decreases the acidity of the acid. Hence acidic character decreases as the + I-effect of the alkyl group increases as

$$CH_3 < CH_3CH_2 < CH_3CH_2CH_2 < CH_3CH_2CH_2CH_3$$

Hence the order of acidic strength becomes:
(i) > (ii) > (iii) > (iv)

19. (a)

$$C_6H_5OH + C_6H_5COCl \xrightarrow{aq.\ NaOH} C_6H_5O.COC_6H_5$$

phenyl benzoate

The function of NaOH is
(i) To convert phenol to more stronger nucleophile PhO^-
(ii) To neutralize the acid formed

20. (b) $\underset{\text{Acetone}}{CH_3COCH_3} + 3I_2 + 4NaOH \longrightarrow$

$$\underset{\text{Iodoform}}{CHI_3} + 3NaI + CH_3COONa + 3H_2O$$

Thus acetone reacts with iodine to form iodoform in the presence of NaOH.

21. (a)

$$\underset{\substack{\text{O-hydroxy benzoic acid} \\ \text{(Salicylic acid)}}}{o\text{-}HOC_6H_4COOH} + \underset{\text{acetyl chloride}}{ClCOCH_3} \xrightarrow{\text{Pyridine}} \underset{\text{Aspirin}}{o\text{-}CH_3COOC_6H_4COOH}$$

22. (d) Iodoform test is exhibited by ethyl alcohol acetaldehyde, acetone methyl ketones and those alcohols which possess $CH_3CH(OH)$-group. As 3-pentanone does not contain CH_3CO-group as therefore it does not give iodoform test.

23. (b)

$$\underset{\text{Phenyl methyl ketone}}{C_6H_5-\overset{O}{\overset{\|}{C}}-CH_3} \xrightarrow{Zn\text{-}Hg/HCl} \underset{\text{ethyl benzene}}{C_6H_5-CH_2-CH_3}$$

This reaction is known as clemmensen's reduction.

24. (c) $$C_6H_5COOH + \underset{\text{Ethanol}}{C_2H_5OH} \xrightarrow[\text{dry}]{HCl} C_6H_5COOC_2H_5 + H_2O$$

This process is known as esterification.

25. (b) Catalyst used in Rosenmund reduction is Pd/$BaSO_4$. Rosenmund Reduction is used for reduction of acid chloride.

$$R-\overset{O}{\overset{\|}{C}}-Cl \xrightarrow{Pd/BaSO_4} R-\overset{O}{\overset{\|}{C}}-H$$

26. (a) Among the given options, only PCl_5 can convert an alcoholic group as well as a carboxyl group to chloride.

$$RCH_2OH \xrightarrow{PCl_5} RCH_2Cl$$

$$RCOOH \xrightarrow{PCl_5} RCOCl$$

27. (d) $$>C=O \xrightarrow{\text{Zn-Hg / conc. HCl}} >CH_2 + H_2O$$

Clemmensen reduction

e.g.

$$(CH_3)_2C=O \xrightarrow[\text{conc. HCl}]{\text{Zn-Hg /}} (CH_3)_2CH_2 + H_2O$$

28. (b) (a) $$C_6H_5OH \xrightarrow[NaOH]{CHCl_3} o\text{-}HOC_6H_4CHO$$

(b) $$H-\overset{O}{\overset{\|}{C}}-H \xrightarrow{NaOH} CH_3OH + HCOONa$$

(c) $$2CH_3Cl \xrightarrow[\text{dry ether}]{Na} CH_3-CH_3$$

(d) $$C_6H_6 \xrightarrow[\text{anh.}AlCl_3]{CH_3Cl} C_6H_5CH_3$$

Note that new C–C bond is formed is a, c and d.

29. (b) Clemmensen reduction is

$$>C=O \xrightarrow{\text{Zn-Hg/HCl}} >CH_2$$

30. (d) Since, C when heated with Br_2 in presence of KOH produces ethylamine, hence it must be propanamide and hence the organic compound (A) will be propanoic acid. The reactions follows.

$$\underset{(A)}{CH_3-CH_2-COOH} \xrightarrow{NH_3} \underset{(B)}{CH_3-CH_2-CO\overset{-}{O}\overset{+}{N}H_4} \xrightarrow{\Delta}$$

$$CH_3-CH_2-CONH_2 \xrightarrow[\text{Hoffmann bromamide reaction}]{KOH+Br_2} \underset{\text{(Ethylamine)}}{CH_3-CH_2-NH_2}$$

31. (d) Acetone forms mesitylene (1,3,5-trimethyl benzene) on distillation with conc. H_2SO_4.

32. (d) I_2 and Na_2CO_3 react with acetophenone ($C_6H_5COCH_3$) to give yellow ppt. of CHI_3 but benzophenone ($C_6H_5COC_6H_5$) does not and hence can be used to distinguish between them.

33. (d) Aldehydes which contain a α-hydrogen on a saturated carbon, i.e., CH_3CH_2CHO undergo aldol condensation.

$$\underset{\text{propanol}}{CH_3CH_2-\overset{H}{\overset{|}{C}}=O} + H-\underset{\alpha}{H\overset{CH_3}{\overset{|}{C}}}-CHO$$

$$\xrightarrow{OH^-} CH_3CH_2-\overset{CH_3}{\overset{\beta|}{\underset{OH}{\underset{|}{C}}}}-\overset{CH_3}{\overset{\alpha|}{C}}HCHO$$

3.hydroxy,2 methyl pentanal

34. (d) Methyl acetate and ethyl acetate on hydrolysis give CH_3COOH which is a liquid. Similarly ethyl formate on hydrolysis will give formic acid which is also a liquid. Only ethyl benzoate on hydrolysis will give benzoic acid which is a solid.

35. (a)

$$C_6H_5COOC_2H_5 \xrightarrow{CH_3MgBr} C_6H_5-\overset{OMgBr}{\overset{|}{\underset{CH_3}{\underset{|}{C}}}}-OC_2H_5$$

$$\xrightarrow{-Mg(OC_2H_5)Br} C_6H_5-\overset{O}{\overset{\|}{C}}-CH_3 \xrightarrow[CH_3MgBr]{\text{Excess}}$$

$$C_6H_5-\overset{OMgBr}{\overset{|}{\underset{CH_3}{\underset{|}{C}}}}-CH_3 \xrightarrow{H_2O} C_6H_5-\overset{CH_3}{\overset{|}{\underset{OH}{\underset{|}{C}}}}-CH_3$$

$$\xrightarrow[\Delta]{\text{Con. }H_2SO_4} C_6H_5-\overset{CH_3}{\overset{|}{C}}=CH_2 \xrightarrow{\text{Ozonolysis}}$$

'B'

$C_6H_5COCH_3 + HCHO$

$$C_6H_5COCH_3 \xrightarrow{3I_2+4NaOH} CHI_3$$

36. (c) By oxidation of tertiary alcohol with stronger oxidising agents ketones may be formed along with carboxylic acid.

$$(CH_3)_3\,COH \xrightarrow{4[O]} CH_3COCH_3 + CO_2 + 2H_2O$$

$$(CH_3)_3\,COH \xrightarrow{8[O]} CH_3COOH + 2CO_2 + 3H_2O$$

37. (c) $CH_3-\overset{O}{\overset{\|}{C}}-H + HCN \longrightarrow CH_3-\underset{CN}{\underset{|}{\overset{OH}{\overset{|}{C}}}}-H \xrightarrow{\text{Hydrolysis}} CH_3-\underset{OH}{\underset{|}{CH}}-COOH$

2-Hydroxy propanol

(As it has a chiral C-atom thus it is optically active).

38. (c) Tollen's reagent is used to detect of aldehydes. Tollen's reagent is an ammonical solution of silver nitrate. When aldehyde is added to Tollen's reagent, silver oxide is reduced to metallic silver which deposits as mirror.

$$RCHO + Ag_2O \longrightarrow RCOOH + 2Ag$$

39. (a) $R-\overset{O}{\overset{\|}{C}}-R' \xrightarrow[KCN]{HCN} R-\underset{R'(A)}{\underset{|}{\overset{OH}{\overset{|}{C}}}}-CN \xrightarrow[LiAlH_4(B)]{\text{Reduction by}} R-\underset{R'}{\underset{|}{\overset{OH}{\overset{|}{C}}}}-CH_2NH_2$

40. (c) Electron withdrawing substituent increases the acidity by increasing the ionic character of –O–H by inductive effect. Electronegativity decreases in the order. $F > Cl > Br$ and hence –I effect also decreases in the same order, therefore the correct option is $FCH_2COOH > ClCH_2COOH > BrCH_2COOH > CH_3COOH$

41. (a) Carbonyl compounds (aldehydes and ketones) are obtained by the oxidation of 1° and 2° alcohols respectively. Among the given options, only (a) is 2° alcohol hence it can be oxidized to ketone.

$$H_3C\overset{OH}{\overset{|}{C}H}CH_3 \xrightarrow{\text{oxidation}} H_3C\overset{O}{\overset{\|}{C}}CH_3$$

2 – hydroxypropane   Acetone

42. (b) Aldehydes containing no α-hydrogen atom on warming with 50% NaOH or KOH undergo disproportionation i.e. self oxidation - reduction known as Cannizzaro's reaction.

$$2HCHO + NaOH \xrightarrow{50\%\,NaOH} HCOONa + CH_3OH$$

43. (a)

$$3\,HCHO \text{ (aq. solution)} \underset{}{\overset{\text{On keeping}}{\rightleftharpoons}} \text{Trioxane (cyclic } (CH_2O)_3\text{)}$$

Trioxane
(meta formaldehyde)

44. (a) Ketones on oxidation give carboxylic acids with lesser number of carbon atoms, i.e.,

$$CH_3COCH_3 \xrightarrow{[O]} CH_3COOH + CO_2 + H_2O$$

45. (b) $2HCOONa \xrightarrow[360^\circ]{\Delta} \underset{COONa}{\overset{COONa}{|}} + H_2$

sod.oxalate

46. (b) Chromic acid and $KMnO_4$ will cleave the molecule at the site of double bond while Cu at 300°C cannot oxidise $COCH_3 \longrightarrow COOH$. The only reagent suitable for this conversion is NaOI or NaOH + I_2 (iodoform test):

$$(CH_3)_2C = CHCOCH_3 \xrightarrow{NaOI} (CH_3)_2C = CHCOOH + CHI_3$$

47. (b) Pinacolone is 3,3-dimethyl-2 butanone.

$$CH_3-\underset{CH_3}{\underset{|}{\overset{CH_3}{\overset{|}{C}}}}-\underset{O}{\underset{\|}{C}}-CH_3$$

48. (b) When benzaldehyde is refluxed with aqueous alcoholic potassium cyanide, two molecules of benzaldehyde condense together to form benzoin

$$C_6H_5-\overset{H}{\overset{|}{\underset{O}{\underset{\|}{C}}}} + \overset{O}{\overset{\|}{\underset{H}{\underset{|}{C}}}}-C_6H_5 \xrightarrow[\Delta]{KCN\ (alc)} C_6H_5-\overset{H}{\overset{|}{\underset{OH}{\underset{|}{C}}}}-\overset{O}{\overset{\|}{C}}-C_6H_5$$

Benzoin

49. (d) Dihydrogen sodium phosphate (NaH_2PO_4) does not have a lone pair of electrons on the P atom. As such it can not act as a nucleophile and hence does not react with aldehydes and ketones.

50. (c) When glycerol is heated with oxalic acid following reaction occurs.

$$\begin{matrix} CH_2OH \\ | \\ CHOH \\ | \\ CH_2OH \end{matrix} + \begin{matrix} HOOC \\ | \\ HOOC \\ \text{oxalic acid} \end{matrix} \xrightarrow[-H_2O]{100-110°C} \begin{matrix} CH_2OC.COOH \\ | \\ CHOH \\ | \\ CH_2OH \end{matrix}$$

$$\xrightarrow{-CO_2} \begin{matrix} CH_2CHO \\ | \\ CHOH \\ | \\ CH_2OH \end{matrix} \xrightarrow{H_2O} \begin{matrix} CH_2OH \\ | \\ CHOH \\ | \\ CH_2OH \end{matrix} + \underset{\text{Formic acid}}{HCOOH}$$

51. (d) **Claisen condensation:** Intermolecular condensation of esters containing α-hydrogen atom in presence of strong base to form β-keto ester

$$\underset{\text{ethylacetate}}{CH_3COOC_2H_5} + H.CH_2.CO.OC_2H_5$$

$$\underset{\longleftarrow}{\xrightarrow{C_2H_5ONa}} \underset{\substack{\text{Ethyl acetoacetate} \\ (\beta\text{-ketoester})}}{CH_3\overset{O}{\overset{\|}{C}}.CH_2COOC_2H_5} + C_2H_5OH$$

52. (b) The ester having α hydrogen atom show Claisen condensation reaction. We know that ethyl benzoate ($C_6H_5COOC_2H_5$) does not contain α-hydrogen. Therefore $C_6H_5COOC_2H_5$ does not undergo Claisen self-condensation.

53. (b) $CH_3-\underset{H}{\underset{|}{C}}=O + H_2N-NH-\overset{O}{\overset{\|}{C}}-NH_2 \xrightarrow{(-H_2O)}$

$$\underset{\textbf{acetaldehyde semicarbazone}}{CH_3-CH=N-NH-\overset{O}{\overset{\|}{C}}-NH_2}$$

54. (a)

$$C_6H_5-C\equiv CH \xrightarrow{H_2SO_4/HgSO_4} \underset{\textbf{enolform}}{C_6H_5-\overset{OH}{\overset{|}{C}}=CH_2} \rightleftharpoons \underset{\textbf{Ketoform}}{C_6H_5-\overset{O}{\overset{\|}{C}}-CH_3}$$

55. (c) In $CH_2=CH-CHO$ due to — M effect of — CHO group polarization of electron takes place as follows

$$CH_2=CH-\underset{H}{\underset{|}{C}}=O \leftrightarrow \overset{+}{C}H_2-CH=\underset{H}{\underset{|}{C}}-\overset{-}{O}$$

Hence partial polarization is represented as

$$\overset{\delta+}{C}H_2=CH-CH=\overset{\delta-}{O}$$

56. (a)

$$\underset{\text{Aldehyde}}{R-\overset{O}{\overset{\|}{C}}-H} + \underset{\text{Hydrazine}}{NH_2-NH_2} \rightarrow \underset{\text{Aldehyde hydrazone}}{R-\underset{H}{\underset{|}{C}}=N-NH_2}$$

57. (d) $CH_3-\overset{H}{\overset{|}{C}}=O+HCN \rightarrow CH_3-\overset{H}{\overset{|}{\underset{CN}{\underset{|}{C}}}}-OH$

$$\xrightarrow{H^+/H_2O} \underset{\text{50\% L-Lactic acid}}{OH-\overset{H}{\overset{|}{\underset{COOH}{\underset{|}{C}}}}-CH_3} + \underset{\text{50\% D-Lactic acid}}{CH_3-\overset{H}{\overset{|}{\underset{COOH}{\underset{|}{C}}}}-OH}$$

58. (d) Out of given compound only acetaldehyde can form optical active hydroxy acid as it contains one asymmetric carbon atom as marked below :

$$CH_3-\overset{O}{\overset{\|}{C}}-H \xrightarrow{HCN} CH_3-\overset{OH}{\overset{|}{\underset{H}{\underset{|}{C}}}}-CN$$

$$\xrightarrow{\text{hydrolysis}} CH_3-\overset{OH}{\overset{|}{\underset{COOH}{\underset{|}{C^*}}}}-H$$

59. (c) Aldehydes are more reactive than ketones due to +I effect of $-CH_3$ group. There are two $-CH_3$ group in acetone which reduces +ve charge density on carbon atom of carbonyl group. More hindered carbonyl group too becomes less reactive. So in the give case CH_3CHO is the right choice.

60. (a) It is an example of Claisen condensation. The product is acetoacetic ester.

$$CH_3\overset{O}{\overset{\|}{C}}-\boxed{OC_2H_5+H}-CH_2-\overset{O}{\overset{\|}{C}}-OC_2H_5 \longrightarrow$$

$$\underset{\text{(ethyl acetoacetate)}}{CH_3-\overset{O}{\overset{\|}{C}}-CH_2-\overset{O}{\overset{\|}{C}}-OC_2H_5}$$

27. AMINES

ANSWER KEY

1	(a)	11	(c)	21	(a)	31	(b)	41	(d)	51	(d)
2	(c)	12	(d)	22	(c)	32	(d)	42	(d)	52	(c)
3	(d)	13	(c)	23	(b)	33	(c)	43	(b)	53	(a)
4	(b)	14	(c)	24	(a)	34	(d)	44	(c)	54	(c)
5	(d)	15	(b)	25	(c)	35	(a)	45	(d)	55	(b)
6	(c)	16	(d)	26	(a)	36	(c)	46	(c)	56	(a)
7	(c)	17	(d)	27	(b)	37	(a)	47	(b)	57	(a)
8	(b)	18	(c)	28	(d)	38	(d)	48	(b)	58	(c)
9	(b)	19	(d)	29	(b)	39	(b)	49	(c)	59	(a)
10	(a)	20	(b)	30	(b)	40	(b)	50	(d)	60	(b)

Hints & Solutions

1. (a) Nitrobenzene is reduced to aniline in the presence of Sn/HCl.

NO_2 (benzene ring) $\xrightarrow{Sn/HCl}$ NH_2 (benzene ring)

Nitrobenzene Aniline

Hence, option (a) is correct.

2. (c)

Phthalimide $\xrightarrow{KOH}$ N-Potassium phthalimide ($\bar{N}K^+$) (a nucleophile)

$\xrightarrow[S_N2\ rxn]{RX/DMF}$ N-alkyl phthalimide (NR) $\xrightarrow{RNH_2}$

Phthalic acid (COOH, COOH) + H_2N-R Primary amine

3. (d) This is Hofmann bromamide reaction in which 1°amine is obtained having one C atom less than the starting amide.

4. (b) In aniline the lone pair on N is involved in delocalization with benzene ring and is not available for protonation. Methyl amine is a stronger base than ammonia because +I effect of methyl group increases electron density on N making it more basic than NH_3.

$\ddot{N}H_2$ (aniline) $\longrightarrow$ $\overset{\oplus}{N}H_2$ (ring with ⊖)

Cannot be protonated. least basic

$CH_3 \rightarrow NH_2$

+I Effect increases basicity.

5. (d)

(a) NH_2 on benzene (aniline)

(b) NH_2 on benzene with NO_2 para

(c) $NHCOCH_3$ on benzene

(d) CH_2NH_2 on benzene

In (a), (b) and (c) nitrogen is directly attached with benzene ring whereas in (d) nitrogen is not directly linked with benzene ring. So lone pair of this compound is relatively more free to be donated. So (d) is most basic. In other rings substituent behaves as electron withdrawing group which decreases the basicity of a niline.

6. (c) **Carbylamine reaction :**

$$R\text{–}NH_2 + 3KOH + CHCl_3 \longrightarrow R\text{–}N{\equiv}C + 3KCl + 3H_2O$$

7. (c) $C_2H_5NH_2 + HONO \longrightarrow \overset{\oplus}{C_2H_5} + N_2$

The carbonium ion $\overset{\oplus}{C_2H_5}$ may react with H_2O to form ethylalcohol or lose a proton to form alkene.

8. (b) A mixture of conc. HNO_3 and conc. H_2SO_4 introduces one $-NO_2$ group at 55°C and two nitro groups at 100°C.

benzene $\xrightarrow[100°C]{\text{conc. } HNO_3 + \text{conc. } H_2SO_4}$ m-Dinitrobenzene

benzene $\xrightarrow[55°C]{\text{conc. } HNO_3 + \text{conc. } H_2SO_4}$ Nitrobenzene

9. (b) $p\text{-}CH_3C_6H_4N_2^+Cl^- \xrightarrow{H_2O} p\text{-}CH_3C_6H_4OH + N_2 + HCl$ (p- cresol)

10. (a) $C_6H_5NH_2 \xrightarrow[\text{HCl (273-278K)}]{NaNO_2}$

$C_6H_5N_2^+Cl^- + H\text{–}C_6H_4\text{–}N(CH_3)_2 \longrightarrow C_6H_5\text{–}N{=}N\text{–}C_6H_4\text{–}N(CH_3)_2$

11. (c) Only 1° amides (i.e. $RCONH_2$) in the present case undergo Hoffmann bromamide reaction.

$RCONH_2 + Br_2 + 4KOH$

$\longrightarrow R-NH_2 + 2KBr + K_2CO_3 + 2H_2O$

(Hoffman's bromamide reaction)

12. (d) Aniline when treated with acetic anhydride forms acetanilide (nucleophilic substitution)

$$C_6H_5NH_2 \xrightarrow{CH_3\text{–}\overset{O}{\overset{\|}{C}}\text{–}O\text{–}\overset{O}{\overset{\|}{C}}\text{–}CH_3} C_6H_5NH\text{–}\overset{O}{\overset{\|}{C}}\text{–}CH_3 + CH_3\text{–}\overset{O}{\overset{\|}{C}}\text{–}OH$$

13. (c) Benzylamine is more basic than aniline. The reason is that in aniline, the lone pair of nitrogen is conjugated with benzene ring so it is not available readily for others. On the other hand in Benzylamine, nitrogen is not directly attached with ring so lone pairs are not conjugated with ring.

14. (c) When ethyl amine reacts with chloroform in the presence of alcoholic potash, then ethylisocynide is formed. This reaction is known as carbyl amine reaction.

$C_2H_5-NH_2 + CHCl_3 + 3KOH \longrightarrow$

Ethyl amine Chloroform

$C_2H_5-NC + 3KCl + 3H_2O$

Ethyl isocyanide (bad smelling compound)

15. (b) The alkyl groups are electron releasing group (+ I), thus increases the electron density around the nitrogen thereby increasing the availability of the lone pair of electrons to proton or lewis acid and making the amine more basic. Hence more the no. of alkyl group more basic is the amine. Therefore the correct order is

$NH_3 < CH_3NH_2 < (CH_3)_2N$

16. (d) **Note :** Aromatic amines are less basic than aliphatic amines. Among aliphatic amines the order of basicity is 2° > 1° > 3°. The electron

density is decreased in 3° amine due to crowding of alkyl group over N atom which makes the approach and bonding by a proton relatively difficult. Therefore the basicity decreases. Further Phenyl group show – I effect, thus decreases the electron density on nitrogen atom and hence the basicity.

∴ Dimethylamine (2° aliphatic amine) is strongest base among given choices.

∴ The correct order of basic strength is

Dimethylamine > Methyl amine > Trimethyl amine > Aniline.

17. (d) $CH_3CN \xrightarrow{\text{Reduction}} CH_3-CH_2-NH_2$

$\downarrow$ HONO

$CH_3-CH_2-OH + N_2 + H_2O$

18. (c) $CH_3CH_2NH_2 + CHCl_3 + 3KOH \longrightarrow$

$CH_3CH_2NC + 3KCl + 3H_2O$

In this reaction, bad smelling compound ethyl isocyanide (CH_3CH_2NC) is produced. This equation is known as **carbyl amine reaction**.

19. (d) p-Toluidine (NH_2, CH_3) $\xrightarrow{(CH_3CO)_2O}$ (A) ($NHCOCH_3$, CH_3) $\xrightarrow[CH_3COOH]{Br_2}$ (B) ($NHCOCH_3$, Br, CH_3) $\xrightarrow{H^+/H_2O}$ (C) (NH_2, Br, CH_3)

20. (b) $HONO_2 + H_2SO_4$

$\longrightarrow NO_2^+ + H_2O + HSO_4^-$

Nitric acid acts as a base by accepting a proton.

21. (a) $CH_3CONH_2 \xrightarrow[Br_2]{NaOH} CH_3NH_2$

(Hofmann bromamide reaction)

22. (c) Aryl amines do not produce phenol on treatment with nitrous acid.

23. (b) When aromatic primary amines are treated with HNO_2 (an ice cold solution of $NaNO_2$ and HCl dilute), diazo salts are obtained.

$C_6H_5NH_2 + HNO_2 \xrightarrow[0-5°C]{NaNO_2/HCl} C_6H_5\overset{+}{N}\equiv N\overset{-}{Cl}$

Aniline (Aromatic primary amine) — Benzene diazonium chloride (Diazo salt)

24. (a) 1° amines (aliphatic and aromatic) react with $COHCl_3$/KOH to yield isocyanide (foul smelling) This is known as carbylamine test which is not given by 2° & 3° amines.

25. (c) The conversion of a primary amide to a primary amine containing one carbon atom less than the original amide on heating with a mixture of Br_2 in presence of NaOH or KOH (i.e. NaOBr or KOBr) is called Hoffmann bromamide reaction or Hoffmann degradation of amides. For example,

$\underset{\text{Acetamide}}{CH_3CONH_2} + Br_2 + 4NaOH \xrightarrow{\Delta}$

$\underset{\text{Methylamine}}{CH_3NH_2} + 2NaBr + Na_2CO_3 + 2H_2O$

26. (a) Chloroform on warming with an alcoholic solution of primary amines and KOH gives carbylamine having offensive smell.

$CHCl_3 + RNH_2 + 3KOH \xrightarrow{\text{warm}}$

$\underset{\text{isocyanide}}{RN \equiv C} + 3KCl + 3H_2O$

27. (b) CH_3CN can be converted into acetic acid by hydrolysis.

$CH_3CN \xrightarrow{H^+/H_2O} CH_3COOH$

28. (d) Carbylamine test is for primary amine for example

$CH_3NH_2 + CHCl_3 + 3KOH$

$\longrightarrow CH_3NC + 3KCl + 3H_2O$

29. (b) $R-CONH_2 \xrightarrow{LiAlH_4} R-NH_2$

30. (b) $\underset{\text{Benzamide (a)}}{C_6H_5CONH_2}$ $\underset{\text{Butamine (b)}}{CH_3CH_2CH_2CH_2NH_2}$

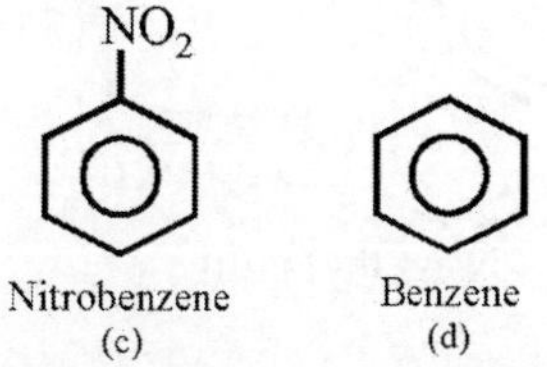

Nitrobenzene (c) Benzene (d)

Electron donar are bases. $-NH_2$ is electron dontaing group. So, butanamine is most basic among the given choices.

31. (b) Acetamide and ethylamine can be distinguished by heating with NaOH solution. Acetamide evolves NH_3 but ethylamine does not.

$$CH_3CONH_2 + NaOH \xrightarrow{\Delta} CH_3COONa + NH_3$$

$$CH_3CH_2NH_2 + NaOH \longrightarrow \text{No reaction.}$$

32. (d) Aniline $+ 3Br_2 \rightarrow$ 2, 4, 6 tribromoaniline $\xrightarrow[\text{\& dil HCl}]{NaNO_2}$

Diazonium salt (N_2Cl) $\xrightarrow[(-NaCl)]{NaBF_4}$ Diazonium tetra fluro borate ($\overset{+}{N}_2\overset{-}{BF_4}$) $\xrightarrow[-N_2,\ -BF_3]{\Delta}$ 2,4,6 tribromofluorobenzene

33. (c) Nitrobenzene (A) $\xrightarrow[\text{reduction}]{Sn+HCl}$ Aniline $\xrightarrow{CHCl_3+KOH}$ (B) $N\equiv C$ (C) $\xrightarrow[Na/C_2H_5OH]{\text{Reduction}}$ N-methylaniline ($NH-CH_3$)

34. (d)

Aniline (NH_2) $\xrightarrow[HCl]{NaNO_2}$ $\overset{\oplus}{N_2}\,\overset{\ominus}{Cl}$ Arene diazonium salt

(follows S_N1 path) $\downarrow$ CuCN

$C_6H_5CN \xrightarrow{H_2/Ni} C_6H_5CH_2NH_2$

$\downarrow HNO_2$

$C_6H_5CH_2-OH$

Benzyl alcohol

35. (a) Reduction of alkyl isocyanides in presence of $LiAlH_4$ yields secondary amines containing methyl as one of the alkyl group.

$$R-N \equiv C + 4[H] \xrightarrow{LiAlH_4} \underset{2°\text{amine}}{R-NH-CH_3}$$

e.g., $CH_3-N \equiv C + 4[H] \xrightarrow{LiAlH_4} \underset{\text{dimethyl amine}}{CH_3-NH-CH_3}$

whereas, alkyl cyanides give 1° amine on reduction.

36. (c) Aniline (NH_2) $\xrightarrow[H_2SO_4]{Na_2Cr_2O_7}$ p-benzoquinone

37. (a) All lower aliphatic amines are stronger bases than ammonia because of +I (inductive) effect. The alkyl groups, which are electron releasing groups, increase the electron density around the nitrogen thereby increasing the availability of the lone pair of electrons to proton or Lewis acids and making the amine more basic (larger K_b).

38. (d) Nitrous acid reacts differently with aliphatic and aromatic amines in cold.

$$\underset{1°\text{ Amine}}{C_2H_5NH_2} + HONO \longrightarrow \underset{\text{Alcohol}}{C_2H_5OH} + N_2$$

$$C_6H_5NH_2 + \underset{\underset{(NaNO_2 + HCl)}{\uparrow}}{HONO} \xrightarrow{\text{Cold}}$$

$$\left[C_6H_5N^{\oplus}\equiv N\right]Cl^{\ominus}$$

Benzene diazonium chloride

39. (b) $CH_3-\overset{\overset{O}{\|}}{C}-\overset{\overset{H}{|}}{N}-H \xrightarrow[-H_2O]{P_2O_5\Delta} CH_3C\equiv N$

$$\xrightarrow[-NH_4Cl]{2H_2O\ +\ HCl} CH_3COOH \text{ (Acetic acid)}$$

$$\xrightarrow[-2H_2O\ +\ Na_2CO_3]{2NaOH\ (CaO)} CH_4 \text{ (Methane)}$$

40. (b) Wurtz reaction is for the preparation of hydrocarbons from alkyl halide.

$$RX + 2Na + XR \longrightarrow R-R + 2NaX$$

41. (d) Ethyl isocyanide on hydrolysis form primary amines.

$$CH_3CH_2N \equiv C + H_2O \xrightarrow{H^+} CH_3CH_2NH_2 + HCOOH$$

Therefore it gives only one mono chloroalkane.

42. (d) All aliphatic amines are stronger bases than NH_3 and among different ethylamines order of basictity is 2° > 3° > 1°. Thus, the correct order is (d) i.e.,

$$(C_2H_5)_2NH > (C_2H_5)_3N > C_2H_5NH_2 > NH_3.$$

This anomolous behaviour of tertiary ethyl amine is due to steric factors i.e., crowding of alkyl groups cover nitrogen atom from all sides and thus makes the approach and bonding by a lewis acid relatively difficult which results the maximum steric strain in tertiary amines. The electrons are there but the path is blocked resulting the reduction in its basicity.

43. (b) Methyl cyanide on treatment with conc. HCl give acetamide.

$$\underset{\text{Acetonitrile}}{CH_3C\equiv N} \xrightarrow{\text{Conc.HCl}} \underset{\text{Acetamide}}{CH_3CONH_2}.$$

44. (c) Benzaldehyde reacts with primary aromatic amine to form schiff's base

$$\underset{\text{Benzaldehyde}}{C_6H_5CH=O} + \underset{\text{Aniline}}{C_6H_5NH_2} \longrightarrow \underset{\text{Benzylidene aniline}}{C_6H_5CH=NC_6H_5} + H_2O$$

This is know as Schiff's base reaction.

45. (d) RDX is prepared by treating urotropine with fuming nitric acid. When the inner bridge system is destroyed by oxidation and the peripheral nitrogen atom are nitrated, it forms cyclonitrite (or RDX).

46. (c) The mecahnism of Hoffmann bromide reaction is

(i) $RCONH_2 + Br_2 \rightarrow RCONHBr + HBr$

(ii) $RCONHBr + HBr \xrightarrow{OH^-} R-CO\overset{\ominus}{\ddot{N}}Br + H_2O$

(iii) $R-\underset{\underset{O}{\|}}{C}-\overset{\ominus}{\ddot{N}}-Br \xrightarrow{K^+} R-\underset{\underset{O}{\|}}{C}-\ddot{N}\!: + KBr$

(iv) $O=\overset{\overset{R}{|}}{C}-N\!: \rightleftharpoons R-\ddot{N}=C=O$ (rearrangement)

(v) $R-\ddot{N}=C=O + 2KOH \rightarrow RNH_2 + K_2CO_3$

47. (b) 3-methoxybenzonitrile ($C\equiv N$, OCH_3) $+ CH_3MgBr \longrightarrow$

$$H_3C-C=NMgBr \text{ (on ring with } OCH_3) \xrightarrow{H_3O^+} H_3C-C=NH \text{ (on ring with } OCH_3)$$

$$\xrightarrow{H_3O^{\oplus}} COCH_3 \text{ (on ring with } OCH_3)$$

48. (b)

$$\text{(CH}_3)_2\text{CH–CO–CH}_3 \xrightarrow{CH_3NH_2} \text{(CH}_3)_2\text{CH–C(=NCH}_3\text{)–CH}_3 \xrightarrow[\text{[H]}]{LiAlH_4/H_2O} \text{(CH}_3)_2\text{CH–CH(NHCH}_3\text{)–CH}_3$$

49. (c) Electrolytic reduction of Nitroalkane in weakly acidic medium give aniline

$$C_6H_5NO_2 \xrightarrow[\text{(weakly acidic medium)}]{\text{Electrolytic reduction}} C_6H_5NH_2$$

Aniline

Whereas in strongly acidic medium it gives p-hydroxyaniline

$$C_6H_5NO_2 \xrightarrow[\text{of conc. } H_2SO_4]{\text{electrolytic reduction in presence}} C_6H_5NHOH$$

(Strongly acidic medium)

$$\xrightarrow{\text{rearrangement}} p\text{-}HOC_6H_4NH_2$$

p-Hydroxyaniline

50. (d) Secondary amine with ($NaNO_2$ + HCl) gives a nitroso product

$$C_6H_5NHCH_3 \xrightarrow[HCl]{NaNO_2} C_6H_5N(CH_3)–N=O$$

N– Nitroso – N– methylaniline

51. (d) (1) $CH_3CH_2-CH_2-CH_2-NH_2$

(ii) with KOH (alcohol) and $CHCl_3$ produces bad smell

(2) $CH_3C\equiv CH$

(ii) gives white ppt with ammonical $AgNO_3$

(3) $CH_3\ CH_2\ COOCH_3$

(i) alkaline hydrolysis

(4) $CH_3\ CHOH-CH_3$

(iv) with Lucas reagent cloudiness appears after 5 minutes

52. (c)

$$m\text{-}BrC_6H_4COOH \xrightarrow{SOCl_2} m\text{-}BrC_6H_4COCl \xrightarrow{NH_3} m\text{-}BrC_6H_4CONH_2 \xrightarrow[+Br_2]{NaOH} m\text{-}BrC_6H_4NH_2$$

53. (a) When nitro compound is reduced with a neutral reducing agent (Zn dust + NH_4Cl) the corresponding hydroxyl amine is formed

$$C_6H_5NO_2 + 4[H] \xrightarrow{\text{Zn dust}+NH_4Cl} C_6H_5NHOH$$

Phenyl hydroxyl amine

54. (c) $C_2H_5I + NH_3 \longrightarrow C_2H_5NH_2 + HI$

Ethyl iodide

$$C_2H_5OH + NH_3 \longrightarrow C_2H_5NH_2 + H_2O$$

55. (b) $C_2H_5NH_2 \xrightarrow{HNO_2} \underset{(A)}{C_2H_5OH}$

$$\xrightarrow{PCl_5} \underset{(B)}{C_2H_5Cl} \xrightarrow{NH_3} C_2H_5NH_2$$

NBS - N - bromo succinamide

56. (a) Primary amides react with nitrous acid to form carboxylic acids with the evolution of nitrogen gas.

$$\underset{\text{Ethanamide}}{CH_3CONH_2} + HONO \longrightarrow \underset{\text{Ethanoic acid}}{CH_3COOH} + N_2 + H_2O$$

In this reaction, nitrogen is eliminated quantitatively and its volume can be easily measured. Therefore, this reaction is used for the ***quantitative estimation of primary amides***.

57. (a) When Aniline reacts with chloroform in the presence of caustic potash (KOH), then a bad smelling compound phenyl isocynide is formed.

$$\underset{\text{Aniline}}{C_6H_5NH_2} + \underset{\text{Chloroform}}{CHCl_3} + 3KOH \longrightarrow C_6H_5N \equiv C + 3KCl + 3H_2O$$

Phenyl isocynide (bad smelling)
(carbylamine Reaction)

58. (c)

$$\underset{\text{aniline}}{C_6H_5NH_2} \longrightarrow \underset{\text{acetanilide}}{C_6H_5NH\text{–}COCH_3}$$

$$C_6H_5NH_2 + (CH_3CO)_2O \xrightarrow{H_2SO_4} C_6H_5NH-\overset{O}{\overset{\|}{C}}-CH_3 + CH_3COOH$$

59. (a) When ethylamine is heated with sodium metal, then hydrogen gas is evolved.

$$2C_2H_5NH_2 + 2Na \longrightarrow 2C_2H_5NHNa + H_2 \uparrow$$

Hence, option (a) is correct.

28. BIOMOLECULES

ANSWER KEY

1	(a)	11	(b)	21	(b)	31	(d)	41	(b)	51	(d)
2	(a)	12	(b)	22	(a)	32	(b)	42	(a)	52	(c)
3	(a)	13	(c)	23	(b)	33	(b)	43	(b)	53	(a)
4	(b)	14	(d)	24	(c)	34	(b)	44	(a)	54	(d)
5	(b)	15	(d)	25	(c)	35	(b)	45	(d)	55	(b)
6	(a)	16	(a)	26	(a)	36	(c)	46	(c)	56	(c)
7	(b)	17	(c)	27	(d)	37	(b)	47	(d)	57	(d)
8	(d)	18	(b)	28	(d)	38	(b)	48	(c)	58	(b)
9	(c)	19	(d)	29	(d)	39	(b)	49	(a)	59	(a)
10	(a)	20	(c)	30	(d)	40	(c)	50	(b)	60	(b)

Hints & Solutions

1. (a) Lactose is a disaccharide.
2. (a) ***Manufacture*** - By hydrolysis of starch with hot dil. mineral acids

$$\underset{\text{Starch}}{(C_6H_{10}O_5)_n} \rightarrow \underset{\text{Dextrin}}{(C_6H_{10}O_5)_n H_2O} \rightarrow$$

$$\underset{\text{Maltose}}{C_{12}H_{22}O_{11}} \rightarrow \underset{\text{Glucose}}{2C_6H_{12}O_6}$$

3. (a) Glucose contains an aldehyde group. It is oxidised into acidic group by bromine water and gluconic acid is formed

$$CH_2OH-(CHOH)_4-CHO \xrightarrow{(O)}$$
$$CH_2OH-(CHOH)_4-COOH$$

$$Br_2 + H_2O \longrightarrow 2HBr + O$$

4. (b) Cellulose is a linear polymer of β–D– glucose in which C_1 of one glucose unit is connected to C_4 of the other through β–D glucosidic linkage. It does not undergo hydrolysis easily. However on heating with dilute H_2SO_4 under pressure. It does undergo hydrolysis to give only D– glucose.

$$(C_6H_{10}O_5)\,n + nH_2O \xrightarrow{H+} \underset{\text{D–Glucose}}{nC_6H_{12}O_6}$$

5. (b) There are 20 amino acids in man out of which 10 amino acids are essential amino acids. These essential amino acids are supplied to our bodies by food which we take because they cannot be synthesised in the body .These are (i) valine (2) leucine (3) Isoleucine (4) Phenyl alanine (5) Threonine (6) Methionine (7) Lysine (8) Tryptho phone (9) Arginine (10) Histidine.
6. (a) Due to resonance C — N bond in protein acquires double bond character and is smaller than usual C — N bond.

$$-\overset{\overset{\ddot{O}:}{||}}{C}-\ddot{N}H- \longleftrightarrow -\overset{\overset{O^-}{|}}{C}=\underset{\oplus}{N}H-$$

7. (b) The α-helix structure is formed when the chain of α-amino acids coils as a right handed screw (called α-helix) because of the formation of hydrogen bonds between amide groups of the same peptide chain, i.e., NH group in one unit is linked to carbonyl oxygen of the third unit by hydrogen bonding. This hydrogen bonding between different units is responsible for holding helix in a position.
8. (d) The function of enzymes in the living system is to catalyse biochemical reactions which occur in living systems. e.g. invertase, pepsin, amylase.

$$\underset{\text{(polymer)}}{\text{Sucrose}} \xrightarrow{\text{Invertase}} \underset{\text{(monomer)}}{\text{glucose + fructose}}$$

$$\underset{\text{(polymer)}}{\text{Starch}} \xrightarrow{\text{amylase}} \underset{\text{(monomer)}}{\text{glucose}}$$

Option (d) is correct.

9. (c) It is found in liver, egg, milk, meat, and fish. Minute amounts are probably present in all animal cells. Peculiarly, unlike other vitamins, B_{12} is not found in significant amounts in green plants.
10. (a) The pyrimidine bases present in DNA are cytosine and thymine.
11. (b) In RNA, the sugar is D–ribose and base is uracil where as in DNA, the sugar is D-2 de oxyribose and the nitrogenous base is thymine.
12. (b) DNA consists of two polynucleotide chains, each chain forms a right handed spiral with ten bases in one turn of the spiral. The two chains coil to double helix and run in opposite direction held together by hydrogen bonding.

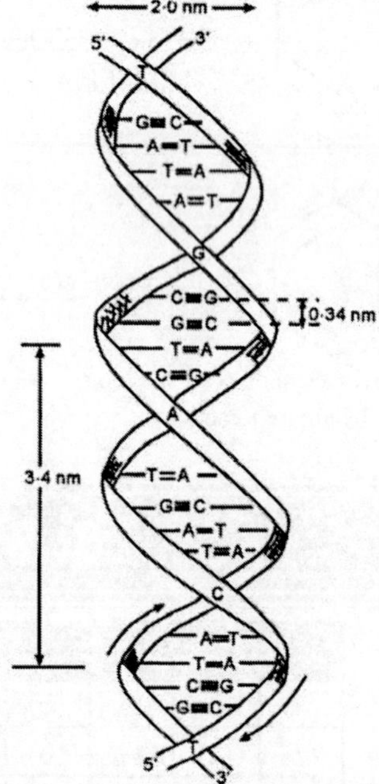

13. (c) RNA contains cytosine and uracil as pyrimidine bases while DNA has cytosine and thymine. Both have the same purine bases i.e., Guanine and adenine.
14. (d) The specific rotation of a freshly prepared solution of glucose changes with time and finally attains a constant value. This phenomenon is called mutarotation.
15. (d) Tollen's reagent is reduced by glucose due to aldehydic group and gives grey colour as silver metal.
    Option (d) is correct.
16. (a) Proteins and peptides are linked by peptide

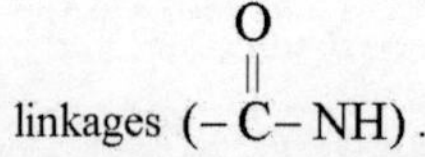

linkages $(-\overset{\overset{O}{||}}{C}-NH)$.

17. (c) Cellulose is a polymer of carbohydrate which is essential constituent of plant cell.

18. (b) ∝ - Amino acid is the building block unit of protein which is formed by polymerisation of amino acid through peptide linkage.

$$\left[-NH-\underset{\underset{O}{\|}}{C}-\right]$$

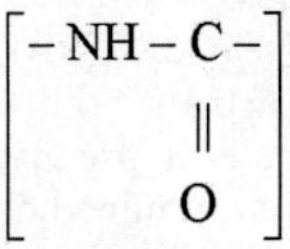

Option (b) is correct.

19. (d) $Hb + CO \longrightarrow HbCO$ (stable compound)
carboxy haemoglobin

$Hb + O_2 \rightleftharpoons HbO_2$
oxyhaemoglobin (unstable).

20. (c) Fructose is the sweetest sugar.

21. (b) Chlorophyll contain Mg.

22. (a) We know that cellulose $(C_6H_{12}O_6)_n$ is the chief constituent of cell walls of plants. It is the most abundant organic substance found in nature. It is a polymer of glucose with 3500 repeat units in a chain.

23. (b) Enzymes are made up of protein with specific structure.

24. (c) The bond formed between two amino acids by the elimination of a water molecule is called a peptide linkage or bond. The peptide bond is simply another name for amide bond.

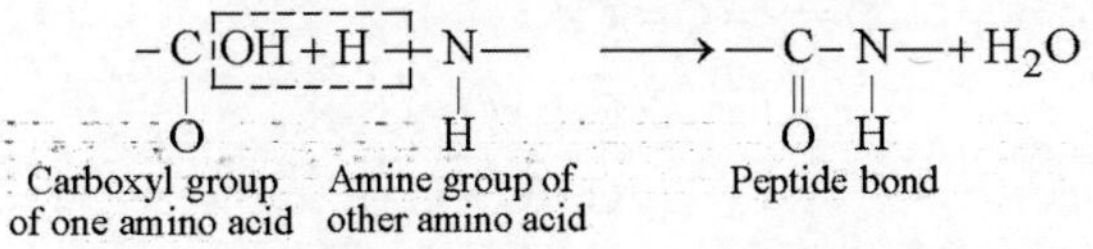

$$-\underset{\underset{O}{\|}}{C}\boxed{OH + H}-\underset{H}{\underset{|}{N}}- \longrightarrow -\underset{\underset{O}{\|}}{C}-\underset{H}{\underset{|}{N}}- + H_2O$$

Carboxyl group of one amino acid; Amine group of other amino acid; Peptide bond

The product formed by linking amino acid molecules through peptide linkages. —CO—NH—, is called a peptide.

25. (c) Cell membranes (Plasma membranes) constitutes bilayer of phospholipid with embedded proteins. In humans, lipids accounts for upto 79% of cell membrance.

26. (a) Vitamins are organic substances which does not provide energy but are essential for healthy growth and proper functioning of body. Vitamins are not synthesized inside human body but they are essential part of our diet.

27. (d)

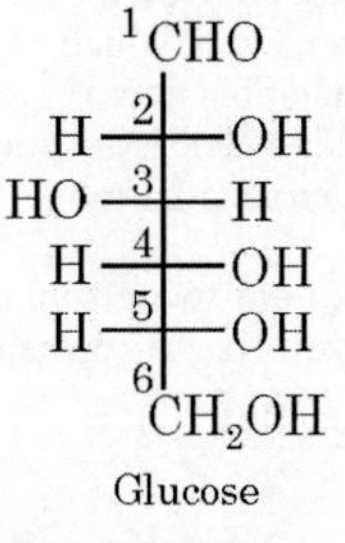

$$\begin{array}{c} {}^{1}CHO \\ H-\overset{2}{C}-OH \\ HO-\overset{3}{C}-H \\ H-\overset{4}{C}-OH \\ H-\overset{5}{C}-OH \\ {}^{6}CH_2OH \end{array}$$

Glucose

28. (d) The most common disaccharide, Lactose has the molecular formula $C_{12}H_{22}O_{11}$.

29. (d) The disaccharides are sugars which on hydrolysis give two moles of the same or different monosaccharides. Sucrose, maltose and lactose $(C_{12}H_{22}O_{11})$ are the common examples.

30. (d) In childhood, lack of vitamin A retards growth and hence like other vitamins, it is also said to be a growth promoting factor. In mild deficiency it leads to night blindness or nyctalopia. Its prolonged deficiency leads to xerophthalmea.

31. (d) Glucose being an aldose responds to Tollen's test while fructose, although a ketose, undergoes rearrangement in presence of basic medium (provided by Tollen's reagent) to form glucose, which then responds to Tollen's test.

32. (b) The two isomeric forms (α – and β –) of D-glucopyronose differ in configuration only at C–1; hence these are called anomers.

33. (b) Insulin is a biochemically active peptide harmone secreted by pancreas.

34. (b) In DNA and RNA heterocyclic base and phosphate ester are at C_1' and C_5' respectively of the sugar molecule.

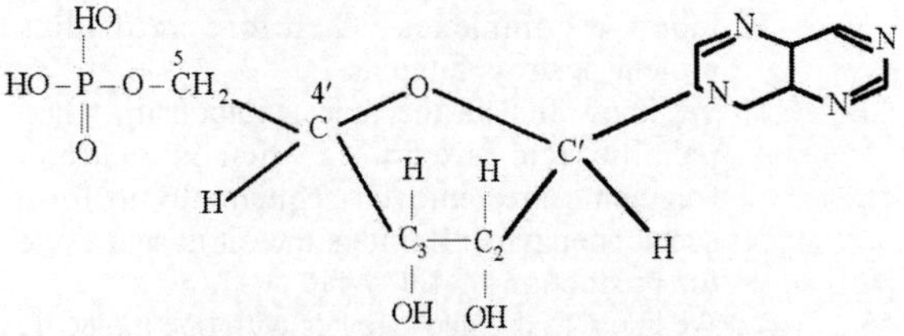

35. (b) The secondary structure of a protein refers to the shape in which a long peptide chain can exist. There are two different conformations of the peptide linkage present in protein, these are α-helix and β-conformation. The α-helix always has a right handed arrangement. In β-conformation all peptide chains are streched out to nearly maximum extension and then laid side by side and held together by intermolecular hydrogen bonds. The structure resembles the pleated folds of drapery and therefore is known as β-pleated sheet.

36. (c) Since $\alpha-D-(+)-$glucose and β–D–(+) glucose differ in configuration at C – 1 atom so they are anomers.
NOTE : Anomers are those diastereomers that differ in configuration at C – 1 atom.
i.e., (c) in the correct answer.

37. (b) Xanthoproteic reaction: On treatment with concentrated nitric acid, certain proteins give yellow colour. This yellow colour is the same that is formed on the skin when the latter comes in contact with the concentrated nitric acid.

38. (b) Glucose and fructose both are reduced by Fehling's solution, Tollen's reagent and Bendict's solution. Therefore, these three reagents can not be used to distinguish between glucose and fructose.

39. (b) Enzymes being biocatalyst can increase the rate of a reaction upto 10 million times. Even very small amount can accelerate a reaction.
40. (c) Basically digestion means hydrolysis.
41. (b) α -D glucose and β–D glucose are the isomers which differ in the orientation (configuration) of H and OH groups around C_1 atom.

```
 H—C¹—OH ┐        HO—C¹—H ┐
   |     |           |     |
 H—C²—OH |         H—C²—OH |
   |     O           |     O
HO—C³—H  |        HO—C³—H  |
   |     |           |     |
 H—C⁴—OH |         H—C⁴—OH |
   |     |           |     |
 H—C⁵————┘         H—C⁵————┘
   |6                |6
  CH₂OH             CH₂OH
α– D - glucose    β– D - glucose
```

42. (a) When antigens enter in to the body cells and destroy them, then antibodies being proteins are synthesised in the body and combine with antigens and destroy these antigens by forming inactive complexes. Therefore antibodies protein destroy antigens.
44. (a) We know that in the lipid metabolism, when palmitic acid is oxidised, then two carbon fragments are removed sequentially to form acetyl coenzyme. It enters the citric acid cycle for production of 130 ATP.
45. (d) We know that glucose reacts with one molecule of phenyl hydrazine to give phenyl hydrazone. When warmed with excess of phenylhydrazine, the secondary alcoholic group adjacent to the aldehyde group is oxidised by another molecule of phenylhydrazine to a ketonic group. With this ketonic group, the third molecule of phenylhydrazine condenses to glucosazone. Therefore the value of X is 3

```
CHO + H₂NNHC₆H₅           CH = NNHC₆H₅
|                          |
CHOH                       CHOH
|              warm        |
(CHOH)₃      ———————→      CH(OH)₃
|                          |
CH₂OH                      CH₂OH
Glucose                    Glucose Phenylhydrozone
                              | H₂NNHC₆H₅
                              ↓
CH = NNHC₆H₅  ←H₂NNHC₆H₅——  CH = NNHC₆H₅
|               – H₂O        |
C = NNHC₆H₅                  C = O
|                            |
CH(OH)₃                      (CHOH)₃
|                            |
CH₂OH                        CH₂OH
Glucosazone                  keto compound of glucose
                             phenyle hydra zone
```

46. (c) $C_6H_{12}O_6 + 6O_2 + 2ATP \longrightarrow$

$$6CO_2 + 6H_2O + 38\,ATP \text{ Molecules}$$

Net total number of ATP molecules evolved = 36 molecules.
47. (d) Progesterone (Gestogens) is a steroid hormone, which controls the development and maintainance of pregnancy. Thryoxine and Adrenaline are Amine hormones.
48. (c) Glucose contain aldehyde group. Hence it give positive fehling solution test.
49. (a) Starch is also know as amylum which occurs in all green plants. A molecule of starch $(C_6H_{10}O_5)_n$ is built of a large number of α-glucose ring joined through oxygen-atom.
50. (b) All proteins are not found in L-form but they may be present in form of D or L
51. (d) It is a common pathway for both the aerobic & anaerobic respiration in which 1 glucose molecule is converted to 2 molecules of pyruvate.
52. (c) Phospholipids are derivatives of glycerol in which two of the hydroxyl groups are esterified with fatty acids while the third is esterified with some derivative of phosphoric acid with some alcohol such as choline, ethanolamine, serine or inositol.

```
      R″
      |
      O
      |
O = P – O – CH₂        O
      |      |         ‖
      OH     CH – O – C –R′
             |
             CH₂O – C –R
                     ‖
                     O
```

53. (a) The sequence of bases in mRNA are read in a serial order in groups of three at a time. Each triplet of nucleotides (having a specific sequence of bases) in known as codon. Each codon specifies one amino acid. Further since, there are four bases. therefore, 4^3 = 64 triplets or codons are possible.
54. (d) Insulin helps in converting glucose to glycogen
55. (b) Triglycerides are lipids, hence these are hydrolysed by *lipases* to glycerol and fatty acids.
56. (c) Haemoglobin acts as an oxygen carrier in the blood since it reacts with oxygen to form unstable oxyhaemoglobin which easily breaks to give back haemoglobin and oxygen.
57. (d) Testorterone and Adrenaline are steroid harmone, Thyroxin is non-steroided harmone glucagon is peptide harmone.
58. (b) Pepsin and Trypsin are two enzymes involved in the process (hydrolysis of proteins)

$$\text{Proteins} \xrightarrow[\text{Stomach}]{\text{Pepsin/ HCl}}$$

Proteases and Peptones

$\xrightarrow[\text{(Pancreatic juice Intestine)}]{\text{Trypsin}\quad\text{Chemotrypsin}}$ Peptides

$\xrightarrow[\text{(Intistine)}]{\text{Peptidases}}$ Amino acids

59. (a) Thyroxine is an amino hormone.
60. (b) The DNA sequence that codes for a specific protein is called a Gene and thus every protein in a cell has a corrosponding gene.

29. POLYMERS

ANSWER KEY											
1	(d)	**11**	(c)	**21**	(d)	**31**	(b)	**41**	(d)	**51**	(c)
2	(c)	**12**	(b)	**22**	(a)	**32**	(c)	**42**	(b)	**52**	(b)
3	(b)	**13**	(c)	**23**	(c)	**33**	(a)	**43**	(c)	**53**	(a)
4	(b)	**14**	(d)	**24**	(a)	**34**	(a)	**44**	(d)	**54**	(d)
5	(c)	**15**	(d)	**25**	(d)	**35**	(b)	**45**	(a)	**55**	(b)
6	(b)	**16**	(c)	**26**	(b)	**36**	(a)	**46**	(a)	**56**	(d)
7	(a)	**17**	(c)	**27**	(d)	**37**	(c)	**47**	(c)	**57**	(d)
8	(b)	**18**	(a)	**28**	(c)	**38**	(b)	**48**	(d)	**58**	(d)
9	(d)	**19**	(a)	**29**	(d)	**39**	(b)	**49**	(c)	**59**	(a)
10	(a)	**20**	(b)	**30**	(a)	**40**	(d)	**50**	(b)	**60**	(d)

Hints & Solutions

2. (c) Polythene is a linear polymer
3. (b) Nylon is a condensation polymer
4. (b) Elastomers are the polymers having very weak intermolecular forces of attraction between the polymer chain. The weak forces permit the polymer to be streched.
5. (c) Bakelite is a thermosetting polymer. On heating it is infusible and cannot be remoulded.
6. (b) Nylon is a polyamide polmer.
7. (a) Polyesters are condensation polymers of a diabasic acid and a diol. e.g., Terylene

$$HO.CH_2CH_2OH + n\ HOOC-C_6H_4-COOH \xrightarrow{\Delta}$$

$$\text{+}O\ CH_2CH_2-O-\overset{O}{\overset{\|}{C}}-C_6H_4-\overset{O}{\overset{\|}{C}}\text{+}_n$$

Ethylene glycol Terephthalic acid

Terylene

8. (b) When phenol is treated with formaldehyde in presence of a basic catalyst, it undergoes condensation polymerization to form either a linear or a cross-linked polymer called Phenol-formaldehyde resin or bakelite.

9. (d) $nCH_2 = CH - \overset{Cl}{\overset{|}{C}} = CH_2 \xrightarrow{K_2S_2O_8}$
Chloroprene

$$\text{+}CH_2 - CH = \overset{Cl}{\overset{|}{C}} - CH_2\text{+}_n$$

Neoprene

10. (a) Natural rubber is prepared by the polymerisation of isoprene.

$$n\left(\begin{matrix} CH_2 = C - CH = CH_2 \\ | \\ CH_3 \end{matrix}\right) \longrightarrow$$

Isoprene

$$\left(\begin{matrix} -CH_2 - C = CH - CH_2 - \\ | \\ CH_3 \end{matrix}\right)_n$$

Poly-isoprene

Option (a) is correct.

11. (c) Nylon-66 is prepared by the reaction between adipic acid and hexamethylene diamine.

12. (b) PVC is a polymer of $CH_2 = CH.Cl$

$$n\,CH_2 = CHCl \longrightarrow \left\{\begin{matrix} -CH_2 - CH - \\ | \\ Cl \end{matrix}\right\}_n$$

Hence, option (b) is correct.

13. (c) We know that polytetrafluoroethylene or teflon is a tough material, resistance to heat and bad conductor of electricity. Therefore it is used for coating the cookware to make them non-sticky.

14. (d) The given compound is a copolymer of hexamethyline diamine and adipic acid. It is actually Nylon-66.

15. (d) Copolymer of adipic acid (6C) and hexamethylene diamine (6C).

$$n\,HOOC(CH_2)_4COOH + nH_2N(CH_2)_6NH_2 \rightarrow$$
$$\underset{\text{Adipic acid}}{}\qquad \underset{\text{Hexamethylene diamine}}{}$$

$$\underset{\text{Nylon 66}}{\left(\overset{O}{\overset{\|}{C}} -(CH_2)_4- \overset{O}{\overset{\|}{C}} -NH-(CH_2)_6-NH \right)_n}$$

It has high tenacity and elasticity. It is resistant to abrasion and not affected by sea water. It is used for reinforcement of rubber tyres, manufacture of parachute, safety belts, carpets and fabrics.

16. (c) Nylon is a general name for all synthetic fibres forming polyamides.

17. (c) Teflon is polymer of $CF_2 = CF_2$.

18. (a) $n\,CF_2 = CF_2 \xrightarrow{\text{Polymerisation}} \underset{\text{Teflon}}{[-CF_2-CF_2-]_n}$

19. (a) Neoprene is a polymer of chloroprene (2 – chloro – 1, 3 – butadiene).

20. (b) Polyethylene (Polyethene). It is addition polymer of ethylene.

21. (d) Polytetrafluoroethylene (Teflon). It is addition polymer of tetrafluoroethylene, $F_2C{=}CF_2$.

22. (a) Buna S, SBR (Styrene-butadiene rubber). It is a copolymer of 75% butadiene ($CH_2 = CH - CH = CH_2$) and 25% styrene ($C_6H_5 - CH = CH_2$).

23. (c) All other three polymers are obtained by addition polymerisation

29. (d) DNA is a biopolymer.

30. (a) Terylene is prepared by condensing dimethylterephthalate and ethylene glycol in presence of a weak base, (calcium acetate). Methanol is eliminated during condensation.

$$\underset{\text{Dimethylterephthalate}}{nH_3COOC-C_6H_4-COOCH_3} + \underset{\text{Ethylene glycol}}{nHOCH_2-CH_2OH}$$

$$\xrightarrow[300^\circ C]{-CH_3OH} \underset{\text{Tervlene}}{\left[-\underset{\|}{\overset{}{C}}-C_6H_4-\underset{\|}{C}-OCH_2-OCH_2- \right]_n}$$
$$\quad O \qquad\qquad O$$

31. (b) Acrilan is a polyacrylonitrile (PAN).

32. (c) Polystyrene is made up by polymerization of $C_6H_5-CH=CH_2$, in which

$C_6H_5-\underset{|}{C}H=CH_2^-$ is the repeating unit.

33. (a) Addition of monomers follows isoprene rule

$$\underset{H}{} \; (CH_3)_2C=CH_2 \; \underset{T}{} \quad \underset{H}{} \; (CH_3)_2C=CH_2 \; \underset{T}{}$$

$$\rightarrow {}^{\oplus}\underset{CH_3}{\overset{CH_3}{\overset{|}{\underset{|}{C}}}}-CH_2-\underset{CH_3}{\overset{CH_3}{\overset{|}{\underset{|}{C}}}}-CH_2 \sim\sim$$

34. (a) Polymerisation starts either by condensation or addition reactions between monomers. Condensation polymers are formed by the combination of monomers with the elimination of simple molecules.Where as the addition polymers are formed by the addition together of the molecules of the monomer or monomers to form a large molecule without elimination of any thing.

35. (b) Buna – N is a copolymer of butadiene ($CH_2 = CH - CH = CH_2$) and acrylonitrile ($CH_2 = CHCN$).

36. (a) Monomer of $\left[-\underset{CH_3}{\overset{CH_3}{\overset{|}{\underset{|}{C}}}}-CH_2- \right]_n$ polymer is 2-methyl propene.

37. (c) Vit B_{12} also called Cyanocobaltamine, is anti-pernicious anaemia vitamin. It is $C_{63}H_{84}O_{14}N_{14}PCo$.

38. (b)

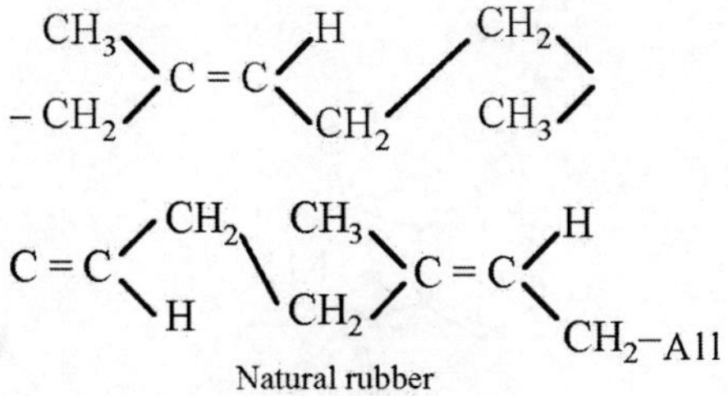

```
CH3         H      CH2
    >C = C<    /       \
-CH2         CH2    CH3/

     CH2   CH3          H
C = C<   \     >C = C<
     H    CH2          CH2–All
```

Natural rubber
(All cis configuration)

statements except (b) are correct

39. (b) PMMA is used in bullet proof glass
40. (d) Orlon is a trade name of polyacrylonitrile
42. (b) Nylon-66 is a polyamide having –CONH– ; oxygen of CO of amide linkage is linked to H of NH of the other –CONH– linkage
43. (c) Ebonite is a hard highly vulcanized rubber, containing 20-30%, rubber
44. (d) The concerned chemical reactions are

(i) $\underset{64\text{kg}}{CaC_2} + 2H_2O \rightarrow Ca(OH)_2 + \underset{\text{Ethyne, 26kg}}{C_2H_2}$

(ii) $C_2H_2 + H_2 \rightarrow \underset{\text{Ethylene, 28kg}}{C_2H_4}$

(iii) $\underset{\substack{n\times 28\text{kg} \\ \text{or } 28\text{kg}}}{nC_2H_2} \rightarrow \underset{\substack{n\times 28\text{kg polythene} \\ \text{or } 28\text{kg}}}{[-CH_2-CH_2-]_n}$

Thus 64 kg of CaC_2 gives 26 kg of acetylene which in turn gives 28 kg of ethylene whose 28 kg gives 28 kg of the polymer, polythene.

45. (a) Molecules of a polymer, being large in size, scatter light
48. (d) Glyptal is an example of a step growth polymer
57. (d) $n(CF_2 = CF_2) \xrightarrow[\Delta]{(NH_4)_2SO_4} \underset{\substack{\text{PTFE} \\ \text{(X)}}}{-\!\!\left(CF_2 - CF_2\right)\!\!-}$

Here X is polytetrafluoroethylene. So, none of these i.e., option (d) is correct choice.

58. (d) Chloro ethylene (vinyl chloride) polymerises in presence of peroxide to form polyvinyl chloride (PVC)

$$\underset{\text{Chlorethylene}}{nCH_2 = \underset{Cl}{\underset{|}{CH}}} \xrightarrow{\text{Peroxides}} \underset{\text{PVC}}{\left(CH_2 - \underset{Cl}{\underset{|}{CH}}\right)_n}$$

59. (a) PVC is an example of thermoplastic polymer.
60. (d) Rayon is a manufactured regenerated cellulosic fiber. Rayon is produced from naturally occurring polymers and therefore it is not a truly synthetic fiber, nor is it a natural fiber. It is known by the names *viscose rayon* and *art silk* in the textile industry. So , option (d) is the correct choice.

30. CHEMISTRY IN EVERYDAY LIFE

ANSWER KEY

1	(d)	**9**	(d)	**17**	(a)	**25**	(c)	**33**	(b)	**41**	(a)	**49**	(b)
2	(d)	**10**	(d)	**18**	(b)	**26**	(d)	**34**	(c)	**42**	(b)	**50**	(b)
3	(b)	**11**	(a)	**19**	(c)	**27**	(c)	**35**	(b)	**43**	(b)	**51**	(b)
4	(d)	**12**	(c)	**20**	(b)	**28**	(d)	**36**	(a)	**44**	(c)	**52**	(b)
5	(c)	**13**	(c)	**21**	(b)	**29**	(c)	**37**	(a)	**45**	(c)	**53**	(a)
6	(c)	**14**	(c)	**22**	(d)	**30**	(d)	**38**	(c)	**46**	(b)	**54**	(d)
7	(b)	**15**	(b)	**23**	(a)	**31**	(c)	**39**	(b)	**47**	(a)	**55**	(c)
8	(d)	**16**	(a)	**24**	(c)	**32**	(c)	**40**	(a)	**48**	(b)		

Hints & Solutions

2. (d) The function of enzymes in the living system is to catalyse biochemical reactions which occure in living systems. e.g. invertase, pepsin, amylase.

$$\underset{\text{(polymer)}}{\text{Sucrose}} \xrightarrow{\text{Invertase}} \underset{\text{(monomer)}}{\text{glucose + fructose}}$$

$$\underset{\text{(polymer)}}{\text{Starch}} \xrightarrow{\text{amylase}} \underset{\text{(monomer)}}{\text{glucose}}$$

Option (d) is correct.

3. (b) Diphenyl hydramine also known as (Banadry) is an antihistamine.
4. (d) Equanil is an important medicine used in depression and hypertension.
5. (c) An analgesic drugs is one which relieves or decrease the pain e.g., analgin aspirin (belongs to non-narcotics and morphine, codein, heroin (belongs to narcotics class)
6. (c) Broad spectrum antibiotics act on different antigens.
7. (b) Alkaline hydrolysis of esters is known as saponification.

$$R-COOR' + NaOH \longrightarrow R'OH + RCOONa$$

8. (d) Commercial detergent are the sodium salts of long chain (linear) alkyl substituted benzene sulphonic acids (LAB) and are most widely used. The most common is *sodium dodecylbenzene sulphonate.*

$$CH_3(CH_2)_{10}CH_2-C_6H_4-SO_2O^-Na^+$$

10. (d) It is used as detergents
11. (a) Antipyretics are those compounds which reduce the body temperature in fever. Examples are aspirin (acetyl salicylic acid), paracetamol, phenacetin, novalgin and analgin.
16. (a) A. Fleming discovered penicillin in 1929.
17. (a) Acetyl salicylic acid is also called acetoxy benzoic acid
18. (b) Aspirin is antipyretic i.e., drug which is responsible for lowering the temperature of Feverish organism to normal, other antipyretic drugs are paracetamol, Phenacetin.
19. (c) Tranquilizers reduce anxiety and tension they are also called psychototropic drugs. These are two type
   (i) Sedative the drugs used for violent and mentaly agitated patietn e.g.., Equanil and diazepam.
   (ii) Antidepressant- The drug are used to patients who are highly depressed and lose self confidence e.g. tofranil vitalin, amphetamine etc.
20. (b) paracetamol is an antipyretic.
26. (d) The mixture of chloroxylenol and terepeneol are dettol which is used as antiseptic.
27. (c) We know that N-acetyl-para-aminophenol (or paracetamol) is an antipyretic which can also be used as an analgesic to relieve pains.
28. (d) Cobalt (60) isotope is used in the treatment of cancer.
29. (c)

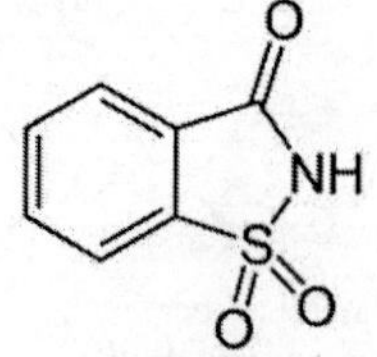

Saccharine

31. (c) Linear alkylbenzenesulphonates (LAB) carrying phenyl chains at secondary positions are biodegradable
36. (a)

OH, H_3C, CH_3, Cl

Chloroxylenol

(4-chloro-3, 5-dimethylphenol)

37. (a) These drugs produce sleep and are habit forming common example of hypnotic drugs are luminal and Saconal.
38. (c) Arsenic drugs are poisonous for syphilis.
39. (b) It is acetyl salicylic acid i.e., aspirin analgesic and antipyretic.
40. (a) Amoxillin is semisynthetic modification of Penicillin
41. (a) Insulin is an antidiabatic drug.
42. (b) Sodium lauryl sulphate is a detergent
45. (c) Any oils which are good for eating or cooking, can be used in making soap. One of the best is said to be Coconut oil. Groundnut, Shea butter, Cocoa butter, Sun flower and many other vegetable oils are also used.
46. (b) Sodium or potassium salts of palmitic acid ($CH_3(CH_2)_{14}COO^-Na^+$) and stearic acid ($CH_3(CH_2)_{16}COO^-Na^+$) are used as soaps.
47. (a) Paracetamol is an antipyretic. It is an effective substitute for aspirin, due to its analgesic and antipyretic properties. However, unlike aspirin, it is not a very effective anti-inflammatory agent. It is well tolerated, lacks many of the side effects of aspirin, and is available over-the-counter, so it is commonly used for the relief of fever, headaches, and other minor aches and pains.
48. (b) The term **broad-spectrum antibiotic** refers to an antibiotic with activity against a wide range of disease-causing bacteria. It also means that it acts against Gram-positive and Gram-negative

bacteria. Penicillins G, bactericidal against susceptible organisms, is a broad spectrum antibiotic.

49. (b) Bithional is a well known antiseptic, added in soaps to reduce odours produced by bacterial decomposition of organic matter of skin.

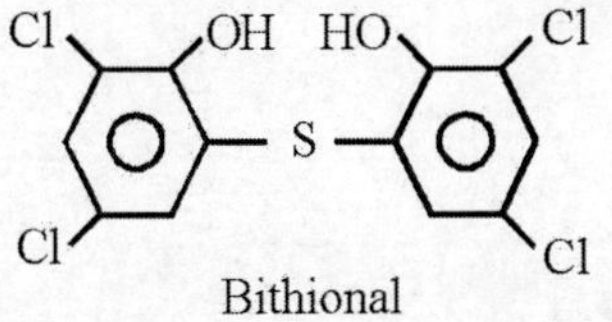

Bithional

50. (b) Chloramin - 'T' is used as antiseptic.

51. (b) Soaps are actually salts of higher fatty acids.

Example; $C_{17}H_{35}COONa$
(sodium stearate)

52. (b) Chloramphenicol is an example of antibiotic.

53. (a) Penicillin is an antibiotic.

54. (d) Urotropine is used as antibiotic for urinary tract infection.

55. (c) **Chloramphenicol** is a broad spectrum drug. [Broad spectrum antibiotics are medicines effective against gram positive as well as gram negative bacteria, *e.g.*, tetracycline, chloramphenicol, etc.]

BIOLOGY

1-278

BIOLOGY

Diversity in the Living World

CONCEPTS

A living organism is a self reproducing system capable of growing and maintaining its integrity by the expenditure of energy.

Common features of living organism are as follows :

- Living organism possesses definite shape and size.
- The real living matter in living being is **protoplasm** which is the physical basis of life.
- All living organisms are made up of one or more structural units called **cells**.
- Living organisms acquire and use energy for metabolism.
- Living organism whether prokaryotes or eukaryotes, respond to the external and internal stimuli and adapt themselves to the changes in the environment.
- Any permanent and irreversible increase in number, mass or overall size of cells of an organism or its parts is called growth. It is the final product of successful metabolism.
- All living organisms possess the capability to produce offspring similar to them.
- Organisms reproduce by sexual as well as asexual means.

DIVERSITY IN THE LIVING WORLD

Biodiversity refers number and types of wide variety of plants and animals present on earth.

Nomenclature

- Nomenclature is defined as system of naming objects (plants and animals).

Identification

- Nomenclature is possible when organisms are described correctly. The description of organism which makes possible its naming is called as identification.

BINOMIAL NOMENCLATURE

- In 1773, Swedish botanist **Carolus Linnaeus** formulated the system of binomial nomenclature in his book '*Species plantarum*'.
- In binomial system, each name is expressed in two parts *i.e.*, **generic** name and **specific** name.
- First letter of the generic name starts with a capital letter and of the species name with a small letter. Both of these words are written in italics and if hand written, it is underlined. For example, scientific name of wheat is *Triticum aestivum*. Here the generic name of wheat plant is *Triticum* and the specific name is *aestivum.*

Classification

- It is the process by which anything is grouped into convenient categories based on some easily observable characters.

Taxon

- Taxon is a group of real organisms contained in any category. A natural taxon refers to group of similar, genetically related individuals having certain characters distinct from those of other group. e.g., tiger (species), rose (genus), eagle (family), reptile (order), fish (class), roundworm (phylum).
- Thus, reptile is taxon but reptilia are category; *Zea mays* (maize) are taxon.

Taxonomy

- Taxonomy is the branch of biology that deals with the framing of laws and principles of classifying the organisms on the basis of their evolutionary relationships.

Systematics

- The word systematic is derived from the latin word '*systema*' which means systematic arrangement of organism.

Hierarchy of Categories

- In the modern classification, the various categories are arranged in a hierarchy. Hierarchy refers to various levels of kinship.
- The hierarchial system of classification was introduced by **Linnaeus**.

  Kingdom → Phylum or Division → Class → Order → Family → Genus → Species

Species

- Species is most important category in taxonomic hierarchy. It is the basic unit of evolution. Species is defined as "the smallest real basic unit of taxonomy which is reproductively isolated from other group of individuals". For example *Mangifera indica, Solanum tuberosum* and *Panthera leo*; *indica, tuberosum* and *leo* are specific epithets.

Genus

- Genus is a group of closely related species that are alike in broad features of their organisation. Potato, brinjal, etc; all come under genus *Solanum.*

Family

- Family is a group of related genera having several common characters. Among animals for example, genus *Panthera*, comprising lion, tiger, leopard is put along with genus, *Felis* (cat) in the family Felidae.

Order

- Generally, order and other higher taxonomic categories are identified based on the aggregates of characters. Thus, the family solanaceae is placed in the order Polymoniales alongwith four other related families (Convolvulaceae, Boraginaceae, Hydrophyllaceae and Polymoniaceae).

Class

- A class is made of one or more related orders. For example, the class dicotyledoneae (dicotyledons) of flowering plants contains all dicots which are grouped into several orders (*e.g.*, rosales, passiflorales, polemoniales, sapindales, ranales etc).

Division or Phylum

- It is a category higher than that of class.
- The term phylum is used for animals while division is commonly employed for plants. The phylum chordata of animals contains not only the class mammalia but also aves (birds), reptilia (reptiles), amphibia (amphibians), cyclostomata, chondirchthyes, etc.

Kingdom

- It is the highest taxonomic category. All plants are included in kingdom plantae. All animals are included in kingdom animalia.

TAXONOMIC AIDS

- Botanical gardens, herbaria, zoological parks, museums are several aids that help in taxonomical studies called as taxonomical aids.
- Monographs (provide information for reversionary work for taxa), manuals (having complete information about keys, description of families, genus and species) and periodicals (provide information about new additions and update information from time to time) also acts as taxonomical aids.

Herbarium

- Herbarium is a collection of pressed and dried plant specimens that are preserved on paper sheets.

Botanical Gardens

- In botanical garden, various plants groups are grown for scientific study, conservation, public education, aesthetics, and recreation. The famous botanical gardens are at Kew (England), Indian Botanical Garden, Howrah (India) and at National Botanical Research Institute, Lucknow (India).

Museum

- Museum is a building used for the preservation, storage and exhibition of inanimate objects. Biological museums have collections of preserved plant and animals specimens.

zoological park

- Zoological park or zoological garden or zoo is a place where wild animals are kept in protected environment under human care. These animals are kept for public exhibition.

Key

- Key refers to a set of alternate characters in such a manner that helped for identification of plants and animals by selecting and eleminating the characters according to their presence or absence in the organism.
- **Bracketed Key:** In this key, pairs of contrasting characters are used for identification of the specimens.
- **Couplet :** The keys are based on contrasting characters generally in a pair called couplet.
- **Lead :** Each statement in a key is called a lead.

Exercise

1. The living organisms can be unexceptionally distinguished from the non-living things on the basis of their ability for
   (a) interaction with the environment and progressive evolution
   (b) reproduction
   (c) growth and movement
   (d) responsiveness to touch
2. Biology is concerned with
   (a) DNA
   (b) biomass
   (c) living organisms
   (d) life as recorded for fossils
3. Study of form and structure of organisms is
   (a) Ecology (b) Taxonomy
   (c) Anatomy (d) Morphology
4. Which of the following are characteristics of living things?
   (a) They reproduce
   (b) They acquire energy
   (c) They respond to stimuli
   (d) All of the above
5. Plants differ from animals in having
   (a) definite lifespan
   (b) localized growth
   (c) indefinite lifespan
   (d) localized growth and indefinite lifespan
6. Energy is utilized by cells for
   (a) reproduction
   (b) growth
   (c) biochemical activities
   (d) All of these
7. The first organisms to appear on earth were more like plants because
   (a) plants are simpler
   (b) plants are more
   (c) plants do photosynthesis
   (d) None of the above
8. The basic unit of classification is ?
   (a) genus (b) Species
   (c) order (d) All of these
9. Group of organisms that closely resemble each other and freely interbreed in nature, constitute a–
   (a) species (b) genus
   (c) family (d) taxon
10. Plant nomenclature means :
    (a) to give names to plants without any rules
    (b) nomenclature of plants under the international rules
    (c) nomenclature of plants in local language
    (d) nomenclature of plants in English language
11. Taxonomy refers to –
    (a) plant classification (b) plant nomenclature
    (c) plant affinity (d) All of these
12. Which of the following is a correct name ?
    (a) *Solanum Tuberosum*
    (b) *Solanum Tuberosum*
    (c) *Solanum Tuberosum Linn*
    (d) All of the above
13. Systematics deals with –
    (a) classification (b) nomenclature
    (c) plant description (d) plant exploration
14. In taxonomy the first step is :
    (a) identification (b) nomenclature
    (c) classification (d) affinities
15. For higher plants, flowers are chiefly used as a basis of classification, because –
    (a) these show a great variety in colour
    (b) they can be preserved easily
    (c) reproductive parts are more conservative than vegetative parts
    (d) None of these
16. Which of the following statements regarding nomenclature is correct?
    (a) Generic name always begins with capital letter whereas specific name with small letter
    (b) Scientific name should be printed in italics
    (c) Scientific name when typed or handwritten should be underlined
    (d) All of the above
17. Which is the most important but generally not used criteria for the identification of the species?
    (a) Interbreeding (b) Morphology
    (c) Genetic material (d) None
18. Herbarium is –
    (a) a garden where medicinal plants are grown
    (b) garden where herbaceous plants are grown
    (c) dry garden
    (d) chemical to kill plants

19. The biological concept of species is mainly based on –
(a) morphological features
(b) morphology and method of reproduction
(c) method of reproduction only
(d) reproductive isolation

20. Which of the following is a species ?
(a) *Tamarindus* (b) *Indicus*
(c) *Indica* (d) *Tamarindus indicus*

21. Practical significance of taxonomy is –
(a) classification
(b) to understand diversity
(c) to understand evolution
(d) identification of organisms

22. Sequence of taxonomic categories is
(a) Class – Phylum – Tribe – Order – Family – Genus – Species
(b) Division – Class – Family – Tribe – Order – Genus – Species
(c) Division – Class – Order – Family – Tribe – Genus – Species
(d) Phylum – Order – Class – Tribe – Family – Genus – Species

23. Binomial nomenclature means
(a) one name given by two scientists
(b) one scientific name consisting of a generic and specific epithet
(c) two names, one latinised, other of a person
(d) two names of same plant

24. The most important feature of all living systems is to
(a) utilize oxygen to generate energy
(b) replicate the genetic information
(c) produce gametes
(d) utilize solar energy for metabolic activities

25. First life on earth was
(a) Cyanobacteria
(b) Chemoheterotrophs
(c) Autotrophs
(d) Photoautotrophs

26. Biosystematics aims at
(a) identification and arrangement of organisms on the basis of their cytological characteristics
(b) the classification of organisms based on broad morphological characters
(c) delimiting various taxa of organisms and establishing their relationships
(d) the classification of organisms based on their evolutionary history and establishing their phylogeny on the totality of various parameters from all fields of studies

27. Angiosperms have dominated the land flora primarily because of their
(a) power of adaptability in diverse habitat
(b) property of producing large number of seeds
(c) nature of self - pollination
(d) domestication by man

28. Which of the following is not true for a species?
(a) Members of a species can interbreed.
(b) Gene flow does not occur between the populations of a species.
(c) Each species is reproductively isolated from every other species.
(d) Variations occur among members of a species.

29. ICBN stands for
(a) International Code of Botanical Nomenclature
(b) International Congress of Biological Names
(c) Indian Code of Botanical Nomenclature
(d) Indian Congress of Biological Names.

30. Biological organisation starts with
(a) cellular level
(b) organismic level
(c) atomic level
(d) submicroscopic molecular level

31. Which one of the following aspects is an exclusive characteristic of living things ?
(a) Isolated metabolic reactions occur *in vitro*
(b) Increase in mass from inside only
(c) Perception of events happening in the environment and their memory
(d) Increase in mass by accumulation of material both on surface as well as internally.

32. A Prediction made by a scientist based on his observations is known as
(a) Law (b) Theory
(c) Principle (d) Hypothesis

33. Organisms which display properties of both living and non - living are
(a) Viruses (b) Diatoms
(c) Lichens (d) Bacteria

34. The highest level of organisation among the following is
(a) ecosystem (b) landscape
(c) population (d) community

35. The basis of life is
(a) lipid (b) protein
(c) nucleic acid (d) nucleoprotein

36. About how long ago was the earth formed?
(a) 4.6 billion years ago
(b) 10 billion years ago
(c) 3.0 billion years ago
(d) 20 billion years ago

37. Which is responsible for origin of life?
(a) Spontaneous generation
(b) Special creation
(c) Catastrophy
(d) Chemosynthesis

38. What is evolution?
(a) Development of DNA from nucleotides
(b) Development of organism through time
(c) Development of a cell from chemicals
(d) None of these

39. Which is basis of evolution?
(a) Cell (b) Individual
(c) Population (d) Species

40. Fossils are
(a) fovea in the retina of vertebrate eye
(b) remains of organisms present in the rocks
(c) the fossa present in the bones
(d) foramen through which nerves pass

41. A scientist kept 72 generations of *Drosophila* in darkness. Even after that the first flies had normal eyes. This disapproves the theory of
(a) natural selection (b) acquired characters
(c) use and disuse (d) synthetic theory

42. Which of the following statement is incorrect?
(a) Life processes are consequences of reactions that in an organism.
(b) Living organisms are made of inorganic and orgnic compounds.
(c) Life comes form pre-existing life.
(d) Genes are not responsible for the stability or change ability of species.

43. Fossils are studied for
(a) tracing evolutionary history of organisms
(b) studying extinct organisms
(c) providing jobs to scientist
(d) Both 'a' and 'b'

44. An important criterion for modern day classification is
(a) resemblances in morphology
(b) anatomical and physiological traits
(c) breeding habits
(d) presence or absence of notochord

45. Linnaeus is credited with
(a) Binomial Nomenclature
(b) theory of biogenesis
(c) discovery of microscope
(d) discovery of blood circulation

46. The high boiling point of water is advantageous to living organisms because
(a) the environment seldom reaches the boiling point of water
(b) organisms can easily boil off enough water to keep themselves cool
(c) it allows organisms to spread heat evenly throughout their bodies
(d) organisms can absorb a great deal of heat before they reach the boiling point from organisms and population

47. If there was no CO_2 in the earth's atmosphere the temperature of earth's surface would be
(a) higher than the present form environmental issues
(b) less than the present
(c) the same
(d) dependent on the amount of oxygen in the atmosphere from environmental issues

48. The practical purpose of classification of living organisms is to
(a) explain the origin of living organisms
(b) trace the evolution of living organisms
(c) name the living organisms
(d) facilitate identification of unknown organisms

49. As we go from species to kingdom in a taxonomic hierarchy, the number of common characteristics
(a) will decrease
(b) will increase
(c) remain same
(d) may increase or decrease

50. Genus represents
(a) an individual plant or animal
(b) a collection of plants or animals
(c) group of closely related species of plants or animals
(d) None of these

51. The taxonomic unit 'Phylum' in the classification of animals is equivalent to which hierachial level in classification of plants
(a) Class (b) Order
(c) Division (d) Family

52. Botanical gardens and zoological parks have
(a) Collection of endemic living species only
(b) Collection of exotic living species only
(c) Collection of endemic and exotic living species
(d) Collection of only local plants and animals

53. Taxonomic key is one of the taxonomic tools in the identificaton and classification of plants and animals. It is used in the prepatation of
(a) monographs (b) flora
(c) Both a & b (d) None of these

54. All living organisms are linked to one another because
(a) they have common genetic material of the same type.
(b) they share common genetic material but to varying degrees
(c) all have common cellular organization
(d) All of the above

55. Match the following and choose the correct option :

A.	Family	i.	*tuberosum*
B.	Kingdom	ii.	Polymoniales
C.	Order	iii.	*Solanum*
D.	Species	iv.	Plantae
E.	Genus	v.	Solanaceae

(a) i-D, ii-C, iii-E, iv-B, v-A
(b) i-E, ii-D, iii-B, iv-A, v-C
(c) i-D, ii-E, iii-B, iv-A, v-C
(d) i-E, ii-C, iii-B, iv-A, v-D

56. Which of the following statement(s) is/are correct?
(a) Only living organisms grow.
(b) Plants grow only up to a certain age.
(c) The growth in living organisms is from inside.
(d) All of these

57. Which one of the following is not a correct statement?
(a) Botanical gardens have collection of living plants for reference.
(b) A museum has collection of photographs of plants and animals.
(c) Key is taxonomic aid for identification of specimens.
(d) Herbarium is a store house that contains dried, pressed and preserved plant specimens.

58. **Statement 1 :** Taxon and category are different things.
Statement 2 : Category shows hierarchical classification.
(a) Statement- 1 is True, Statement-2 is True, Statement-2 is a correct explanation for Statement -1
(b) Statement -1 is True, Statement -2 is True; Statement-2 is NOT a correct explanation for Statement - 1
(c) Statement - 1 is True, Statement- 2 is False
(d) Both the Statements are False.

59. Which one of the following statements is correct about Biodiversity ?
(a) Biodiversity is the occurrence of varied type of organisms on earth.
(b) Each different kind of plant, animal or organism represents a species.
(c) The number of species that are known and described range between 1.7–1.8 million.
(d) All of these

60. Read the following statements
(i) Lower the taxon, more are the characteristics that the members within the taxon share.
(ii) Order is the assemblage of genera which exhibit a few similar characters.
(iii) Cat and dog are included in the same family Felidae.
(iv) Binomial Nomenclature was introduced by Carolus Linnaeus.

Which of the following statements are not correct ?
(a) (i), (ii) and (iii) (b) (ii), (iii) and (iv)
(c) (i) and (iv) (d) (ii) and (iii)

CHAPTER 2

Biological Classification

CONCEPTS

- **Biological classification** was first proposed by **Aristotle** who used simple morphological characters to classify plants and animals.
- **Linnaeus** initially separated plants and animals in **two kingdoms** *i.e.*, Kingdom Plantae and Kingdom Animalia.
- In **three kingdom** system, kingdom Protista was included along with kingdom Plantae and kingdom Animalia created by **Haeckel**.
- **Copeland** classified organisms into **four kingdoms**; Monera (all prokaryotes), Protista (unicellular eukaryotes), Metaphyta (multicellular plants) and Metazoa (multicellular eukaryotes).
- Most accepted system of classification is Five system classification which was given by Whittaker.

BASIC FEATURES OF WHITTAKER'S FIVE KINGDOMS

	Kingdom	Cellular Organisation	Movement	Nutrition	Reproduction
1.	**Monera** (All Prokaryotes)	Unicellular, without nucleus or membranous organelle.	By flagella (tubulin-dynein system)	Absorptive or photosynthetic	Asexual
2.	**Protista** (Protozoans, unicellular algae)	Unicellular, eukaryote with nucleus and membranous organelles.	By flagella, cilia, pseudopodia and mucilage propulsion	Absorptive, photosynthetic & holozoic	Both sexual and asexual
3.	**Fungi** (Multicellular decomposers)	Multicellular eukaryote coenocytic, no plastids, cells wall of cellulose, chitin.	Non-motile	Heterotrophic (saprophytic/ Parasitic)	Asexual and sexual both
4.	**Plantae** (All plants)	Multicellular, higher organisation eukaryotes, cellulosic cell wall, plastids present.	Non-motile	Autotrophic or photosynthetic	Asexual and sexual both
5.	**Animalia** (All animals)	Multicellular, higher organization, eukaryotes without cell wall and chlorophyll.	Highly motile with all type of motile machinery	Heterotrophic (holozoic or saprozoic)	Both sexual and asexual but in higher forms only sexual

KINGDOM MONERA (THE PROKARYOTES)

- This kingdom comprises bacteria (mycoplasma, archaebacteria, eubacteria, actinomycetes, etc.) and cyanobacteria.
- The earliest living organisms may have been thermophiles.
- The proteobacteria is the largest group in terms of the number of species. They are sometimes referred to as the *purple bacteria*. Mitochondria were derived from proteobacteria by endosymbiosis. *E. coli*, *Yersinia pestis*, *Vibrio cholerae*, and *Salmonella typhimurium* are all proteobacteria.

Archaebacteria

- Most archaea live in environments that are extreme in one way or another: temperature, salinity, oxygen concentration, or pH. The archaea lack peptidoglycan in their cell walls.
- All methanogens are obligate anaerobes. Methanogens release approximately 2 billion tons of methane gas into Earth's atmosphere. Approximately a third of this methane comes from methanogens in the guts of grazing herbivores. *Methanopyrus* lives on the ocean bottom near volcanic vents. These can live at 110ºC.

Bacteria

- Bacteria are the simplest, most primitive unicellular organisms.
- Each bacterial cell contains a single chromosomes (nuclear material) that is not enclosed in a nuclear membrane.
- According to shape, bacteria are grouped into four types – Coccus or spherical shaped, bacillus or rod-shaped bacteria, spirillum or spiral/curved shaped bacteria, vibrio or comma-shaped bacteria.
- Autotrophic nutrition consists of manufacture of organic materials from inorganic raw materials with the help of energy obtained from outside sources. It is of two types, chemosynthesis and photosynthesis. The bacteria performing these modes of nutrition are respectively called chemoautotrophs and photo-autotrophs.
- **Photoautotrophic Bacteria:** The bacteria possess photosynthetic pigments of two types, *bacteriochlorophyll* and *bacteriophaeophytin*.
- **Chemoautotrophic Bacteria :** They are able to manufacture their organic food from inorganic raw materials with the help of energy derived from exergonic chemical reactions involving oxidation of an inorganic substance present in the external medium.
- Bacteria reproduce both asexually as well as sexually. **Asexual reproduction** in bacteria is by binary fission which takes place in favourable conditions or by endospore formation.
  Sexual reproduction is not associated with gamete formation but takes place through the transfer of naked DNA. Hence referred to as sexual recombination or genetic recombination which may be- Conjugation, Transformation, Transduction.

Cyanobacteria

- Cyanobacterial cells are larger and more elaborate than bacteria. Cell structure is typically prokaryotic – one envelope organisation with peptidoglycan wall, naked DNA, 70S ribosomes and absence of membrane bound structures like endoplasmic reticulum, mitochondria, golgi bodies, plastids, lysosomes and sap vacuoles.
- Heterocyst is a large-sized pale coloured thick-walled cell which occurs in terminal, intercalary or lateral position in filamentous cyanobacteria, *e.g.*, *Nostoc*.

Mycoplasma

- Mycoplasmas have the least amount of DNA and are the smallest bacteria. They lack cell walls. Some are 0.2 mm in size. They may have the minimum amount of DNA necessary to code for the essential properties of a living cell.

KINGDOM PROTISTA

- Protista include all unicellular eukaryotic microorganisms.

Chrysophytes

- They are a group of diatoms (*Triceratium, Pieurosigma, Navicuia, Cymbella, Amphipleura*) and golden brown photosynthetic microscopic protists.
- Diatoms are very good **pollution indicators.**
- The siliceous frustules of diatoms do not decay easily. They pile up at the bottom of water reservoirs and form big heaps called **diatomite** or **diatomaceous** earth.

Dinoflagellates

- Dinoflagellates are mostly marine and photosynthetic.
- Very often, red dinoflagellates (*e.g.*, *Gonyaulax*) undergo such rapid multiplication that they make the sea appear red (red tides).

Euglenoids

- It includes *Euglena*-like flagellates which have plant like characteristics (chlorophyll) in addition to some animal characteristics. They ingest food particles and carry on photosynthesis.
- An osmoregulatory **contractile vacuole** is found.
- Instead of a cell wall, they have a protein rich layer called pellicle which makes their body flexible.

Slime Moulds

- They are unicellular, saprophytic organism that lack cell walls and have the plasma membrane as their outer covering.
- They move through pseudopodia.
- The body moves along decaying twigs and leaves engulfing organic material.

PROTOZOA

- Classification of protozoa is mainly based on locomotory organelles. Protistan protozoa are divisible into four phyla—zooflagellata, sarcodina, sporozoa and ciliata.

Zooflagellates

- They are protozoan protists which possess one to several flagella for locomotion.
- The body is covered by a firm pellicle. Cyst formation is also present.
  Examples : *Trichonympha, Lophomonas, Giardia, Trypanosoma, Leishmania* and *Trichomonas.*

Sarcodines

- They develop pseudopodia for locomotion and ingestion of food.
- The body is without periplast (with a covering of only plasmalemma) or have a shell.
- *Entamoeba histolytica* causes the disease known as **amoebic dysentery** or **amoebiasis.** *Entamoeba coli* is commensal parasite and does not produce any disease. *Entamoeba gingivalis or mouth amoeba* is found in tartar of teeth and in pus pockets of bleeding gums. The condition is called as **pyorrhoea.**

Sporozoan

- They are develop from paristic mode of life (usually endoparasites).
- The body is covered by elastic pellicle or cuticle and have a single nucleus contractile vacuoles are absent.
- *Plasmodium vivax* causes benign tertian fever. *Plasmodium falciparum* causes malignant tertian fever or cerebral malaria or pernicious malaria or black-water fever. *Plasmodium ovale* causes benign tertian fever.

KINGDOM FUNGI (MULTICELLULAR DECOMPOSERS)

- Fungi are achlorophyllus organisms and hence they live as heterotrophs *i.e.*, as parasites and saprophytes.
- Fungi are non-vascular, non-seeded, non-flowering, multicellular decomposers and mineralisers of organic wastes and help in recycling of matter in the biosphere.
- In true fungi the plant body is **thallus**. It may be non-mycelial or mycelial.
- Hyphae are usually branched tube like structure bounded by a cell-wall of chitin.

Fungi have three types of reproduction.

1. **Vegetative reproduction :** Budding, Fission, Oidia, Chlamydospores, Fragmentation.
2. **Asexual reproduction**
   Asexual reproduction in fungi may be of the following types.
   (a) **Zoospore formation :** They may be uniflagellate (*e.g., Synchytrium*) or biflagellate (*e.g., Saprolgenia*)
   (b) **Aplanospore formation :** They are non-flagellate spores that develop inside sporangia *e.g., Rhizopus, Mucor.*
   (c) **Conidia** : Conidia are non-motile, thin walled exogenous spores produced on a conidiophore *e.g., Aspergillus* and *Penicillium.*
3. **Sexual reproduction**
   Sexual reproduction is reduced in fungi and take place by two fusing gametes. It includes 3 stages:
   (a) **Plasmogamy** or fusion of two protoplasts
   (b) **Karyogamy** or fusion of two nuclei
   (c) **Meiosis** or reduction division

Classification of Fungi

Group	Examples	Morphology	Asexual Reproduction	Sexual Reproduction	Miscellaneous Information
Zygomycota	Zygote fungi, black bread mould	Coenocytic hyphae, chitosan cell walls	Spores formed in sporangia on aerial hyphae	Hyphal extensions fuse, karyogamy, resistant zygosporangia	In soil and decaying matter
Ascomycota	Sac fungi, yeast, *Penicillium*	Septate hyphae, chitin cell walls, some unicellular	Abundant spores in chains or clusters, called conidia	Dikaryotic hyphae in ascocarp, karyogamy in asci, ascospores	Plant parasites, lichens, genetic studies, decomposers
Basidiomycota	Club fungi, mushrooms, puffballs, shelf fungi, rusts	Dikaryotic hyphae with complex pores, mycelium extensive, chitin	Form conidia less commonly than ascomycetes	Basidiocarps, karyogamy in basidia, forms basidiospores	Edible mushrooms, plant parasites, mycorrhizae
Deuteromycota	Fungi imperfecti *Alternaria* Trichoderma	Moulds, rapidly growing hyphae	Form conidia	No sexual stage, may have parasexuality	Non-phylogenetic form division, some produce antibiotics

KINGDOM PLANTAE

- In this kingdom are included multicellular, eukaryotic, photosynthetic organisms (kingdom of multicellular producers or Metaphyta).
- Main characters of this kingdom are :
  - Cellulosic cell wall.
  - Presence of chloroplasts.
  - Photosynthetic mode of nutrition (oxygenic photosynthesis), *e.g.*, different types of algae (green, brown, red algae), bryophytes, pteridophytes, gymnosperms and angiosperms.
  - Reproduction is both asexual and sexual.

KINGDOM ANIMALIA

- The organisms are made up of eukaryotic cells.
- The cells do not contain cell wall, but contain only cell membrane.
- They do not perform photosynthesis but have heterotrophic nutrition.
- They have the power of locomotion.

VIRUSES

- The term 'virus' has been derived from latin, which means poison or venom or viscous fluid.
- They are obligate parasites, *i.e.*, can live inside living host only.
- They have either RNA or DNA.
- They have character of both living and non-living.
- **Bacteriophage :** Bacterial viruses or bacteriophages (viruses that infect the bacteria) are usually double stranded DNA viruses.

VIROIDS

- **T.O.Diener** (1971) discovered some new infections agents, which are still smaller than viruses.
- Viroids contain only very low molecular weight RNA and not protein coat.
- Viroids cause potato spindle tuber disease (PSTV), *Chrysanthemum stunt, Citrus exocortis*, cucumber pale fruit etc.

LICHENS

- Lichens shows symbiotic relationships.
- In lichens, there are 2 components; i.e., algal partner called phycobiont and fungal partner called mycobiont.
- Lichens are economically important as food and fodders, dyes, as a source of medicines etc.
- They are very good pollution indicators – they do not grow in polluted areas.

Exercise

1. Plant decomposers are
   (a) Monera and Fungi
   (b) Fungi and Plants
   (c) Protista and Animalia
   (d) Animalia and Monera
2. What is true for archaebacteria ?
   (a) All halophiles (b) All photosynthetic
   (c) All fossils (d) Oldest living beings
3. Decomposers are organisms that
   (a) elaborate chemical substances, causing death of tissues
   (b) operate in living body and simplifying organic substances of cells step by step
   (c) attack and kill plants as well as animals
   (d) operate in relay terms, simplifying step by step the organic constituents of dead body
4. Mycorrhiza represents
   (a) antagonism (b) endemism
   (c) symbiosis (d) parasitism
5. Photosynthetic bacteria have pigments in
   (a) leucoplasts (b) chloroplasts
   (c) chromoplasts (d) chromatophores
6. Which one belongs to Monera?
   (a) *Amoeba* (b) *Escherichia*
   (c) *Gelidium* (d) *Spirogyra*
7. *Escherichia coli* is used extensively in biological research as it is
   (a) easily cultured
   (b) easily available
   (c) easy to handle
   (d) easily multiplied in host
8. Shell of Diatom is made up of :
   (a) $CaCO_3$
   (b) silica
   (c) magnesium carbonate
   (d) calcium phosphate
9. Bacteria are considered plant because they :
   (a) are green in colour
   (b) have rigid cell wall
   (c) have chlorophyll
   (d) have stomata
10. Whittaker is famous for –
    (a) two kingdom classification
    (b) four kingdom classification
    (c) five kingdom classification
    (d) distinguishing in Bacteria & blue green Algae
11. Fungi are always
    (a) autotrophic (b) heterotrophic
    (c) parasitic (d) saprophytic
12. 'Comma' shaped bacteria are known as
    (a) coccus (b) spiral
    (c) lophotrichus (d) vibrio
13. When fungi feed on dead organic matter, they are known as
    (a) dimorphic (b) parasites
    (c) saprophytes (d) None of these
14. Yeast is divided under the class
    (a) basidiomycetes (b) deuteromycetes
    (c) ascomycetes (d) zygomycetes
15. Mycorrhiza is
    (a) a symbiotic association of plant roots and certain fungi
    (b) an association of algae with fungi
    (c) a fungus parasitising root system of higher plants
    (d) an association of *Rhizobium* with the roots of leguminous plants
16. What is true for cyanobacteria?
    (a) Oxygenic with nitrogenase
    (b) Oxygenic without nitrogenase
    (c) Non-oxygenic with nitrogen
    (d) Non-oxygenic without nitrogenase
17. Enzymes are absent in
    (a) cyanobacteria (b) viruses
    (c) algae (d) fungi
18. Virus envelope is known as:
    (a) capsid (b) virion
    (c) nucleoprotein (d) core
19. There exists a close association between the alga and the fungus within a lichen. The fungus
    (a) provides protection, anchorage and absorption for the alga
    (b) provides food for the alga
    (c) fixes the atmospheric nitrogen for the alga
    (d) releases oxygen for the alga
20. Which bacteria is utilized in gober gas plant?
    (a) Methanogens
    (b) Nitrifying bacteria
    (c) Ammonifying bacteria
    (d) Denitrifying bacteria
21. Lichens indicate SO_2 pollution because they
    (a) show association between algae and fungi
    (b) grow faster than others
    (c) are sensitive to SO_2
    (d) flourish in SO_2 rich environment

22. Absorptive heterotrophic nutrition is exhibited by
(a) algae (b) fungi
(c) bryophytes (d) pteridophytes
23. Name the organisms which do not derive energy directly or indirectly from sun
(a) chemosynthetic bacteria
(b) pathogenic bacteria
(c) symbiotic bacteria
(d) mould
24. In *Amoeba* and *Paramoecium* osmoregulation occurs through
(a) pseudopodia (b) nucleus
(c) contractile vacuole (d) general surface
25. Bacteria lack alternation of generation because there is
(a) neither syngamy nor reduction division
(b) distinct chromosomes are absent
(c) no conjugation
(d) no exchange of genetic material
26. If all ponds and puddles are destroyed, the organism likely to be destroyed is
(a) *leishmania* (b) *trypanosoma*
(c) *ascaris* (d) *plasmodium*
27. Nitrogen fixer soil organisms belong to
(a) mosses (b) bacteria
(c) green algae (d) soil fungi
28. Which of the following organism possesses characteristics of both a plant and an animal?
(a) Bacteria (b) *Euglena*
(c) *Mycoplasma* (d) *Paramoecium*
29. The function of contractile vacuole, in protozoa, is
(a) locomotion (b) food digestion
(c) osmoregulation (d) reproduction
30. Excretion in *Amoeba* occurs through
(a) lobopodia (b) uroid portion
(c) plasma membrane (d) contractile vacuole
31. The organism, used for alcoholic fermentation, is
(a) *Pseudomonas* (b) *Penicillium*
(c) *Aspergillus* (d) *Saccharomyces*
32. Sex factor in bacteria is
(a) chromosomal replicon
(b) f-replicon
(c) RNA
(d) sex-pilus
33. Which one of the following is not true about lichens?
(a) Their body is composed of both algal and fungal cells
(b) Some form food for reindeers in arctic regions
(c) Some species can be used as pollution indicators
(d) These grow very fast at the rate of about 2 cm per year
34. Which one of the following statements about viruses is correct?
(a) Viruses possess their own metabolic system.
(b) Viruses contain either DNA or RNA.
(c) Viruses are facultative parasites.
(d) Viruses are readily killed by antibiotics.
35. The main role of bacteria in the carbon cycle involves
(a) photosynthesis
(b) chemosynthesis
(c) digestion or breakdown of organic compounds
(d) assimilation of nitrogenous compounds
36. A few organisms are known to grow and multiply at temperatures of 100–105°C. They belong to
(a) marine archaebacteria
(b) thermophilic sulphur bacteria
(c) hot-spring blue-green algae (cyanobacteria)
(d) thermophilic, subaerial fungi
37. Transfer of genetic information from one bacterium to another in the transduction process is through
(a) conjugation
(b) bacteriophages released from the donor bacterial strain
(c) another bacterium
(d) physical contact between donor and recipient strain
38. Which one of the following micro-organisms is used for production of citric acid in industries?
(a) *Penicillium citrinum*
(b) *Aspergillus niger*
(c) *Rhizopus nigricans*
(d) *Lactobacillus bulgaris*
39. Due to which of the following organisms, yield of rice has been increased ?
(a) *Anabaena* (b) *Bacillus popilliae*
(c) *Sesbania* (d) *Bacillus polymyxa*
40. In prokaryotes, the genetic material is
(a) linear DNA with histones
(b) circular DNA with histones
(c) linear DNA without histones
(d) circular DNA without histones
41. A virus can be considered a living organism because it
(a) responds to touch stimulus
(b) respires
(c) reproduces (inside the host)
(d) can cause disease

42. The most thoroughly studied fact of the known bacteria-plant interactions is the
   (a) cyanobacterial symbiosis with some aquatic ferns
   (b) gall formation on certain angiosperms by *Agrobacterium*
   (c) nodulation of *Sesbania* stems by nitrogen fixing bacteria
   (d) plant growth stimulation by phosphate-solubilising bacteria
43. Which of the following statements regarding cilia is not correct ?
   (a) Cilia are hair-like cellular appendages.
   (b) Microtubules of cilia are composed of tubulin.
   (c) Cilia contain an outer ring of nine doublet microtubules surrounding two single microtubules.
   (d) The organized beating of cilia is controlled by fluxes of Ca^{2+} across the membrane.
44. The thalloid body of a slime mould (Myxomycetes) is known as
   (a) plasmodium (b) fruiting body
   (c) mycelium (d) protonema
45. Which pair of the following belongs to Basidiomycetes?
   (a) Puffballs and *Claviceps*
   (b) *Peziza* and stink borns
   (c) *Morchella* and mushrooms
   (d) Agaricus and lstilago
46. In the light of recent classification of living organisms into three domains of life (bacteria, archaea and eukarya), which one of the following statements is true about archaea?
   (a) Archaea resemble eukarya in all respects.
   (b) Archaea have some novel features that are absent in other prokaryotes and eukaryotes.
   (c) Archaea completely differ from both prokaryotes and eukaryotes.
   (d) Archaea completely differ from prokaryotes.
47. Which one of the following organisms is not an example of eukaryotic cells ?
   (a) *Paramoecium caudatum*
   (b) *Escherichia coli*
   (c) *Euglena viridis*
   (d) *Amoeba proteus*
48. In eubacteria, a cellular component that resembles eukaryotic cells is :
   (a) plasma membrane (b) nucleus
   (c) ribosomes (d) cell wall
49. Organisms called methanogens are most abundant in a :
   (a) sulphur rock (b) cattle yard
   (c) polluted stream (d) hot spring
50. During rainy seasons, the ground become slippery because of the dense growth of :
   (a) algae (b) cyanobacteria
   (c) bacteria (d) archaebacteria
51. Incorrect statement about "Fungi imperfecti" is :
   (a) no sexual phase
   (b) species prey on nematodes
   (c) include toadstool and puffball
   (d) include *Aspergillus* used for making soya sauce.
52. Which of the following process needs bacteriophage?
   (a) Transduction (b) Translation
   (c) Transformation (d) Conjugation
53. In bacteria respiration occurs in :
   (a) cytoplasmic membrane
   (b) mitochondria
   (c) nuclear membrane
   (d) endoplasmic reticulum
54. Edible part of mushroom is
   (a) basidiocarp (b) secondary mycelium
   (c) primary mycelium (d) tertiary mycelium
55. An organism, which is a connective link between animals and plants, is
   (a) virus (b) bacteria
   (c) *euglena* (d) *amoeba*
56. Slime moulds in the division myxomycota (true slime moulds) have
   (a) pseudoplasmodia.
   (b) spores that develop into free living amoeboid cells.
   (c) spores that develop into flagellated gametes.
   (d) feeding stages consisting of solitary individual cells.
57. Flagella of prokaryotic and eukaryotic cells differ in
   (a) type of movement and placement in cell
   (b) location in cell and mode of functioning
   (c) micro-tubular organization and type of movement
   (d) micro-tubular organization and function
58. HIV has a protein coat and genetic material :
   (a) ss RNA (b) ds RNA
   (c) ss DNA (d) ds DNA
59. In nodules on the roots of leguminous plants, nitrogen is fixed by :
   (a) *rhizobium* (b) *azospirillum*
   (c) *azotobacter* (d) None of these
60. Which of the following statement is true for mycoplasma?
   (a) Presence of cell wall (b) Presence of nucleus
   (c) Absence of cell wall (d) Definite shape

CHAPTER 3

Plant Kingdom

CONCEPTS

PLANT KINGDOM

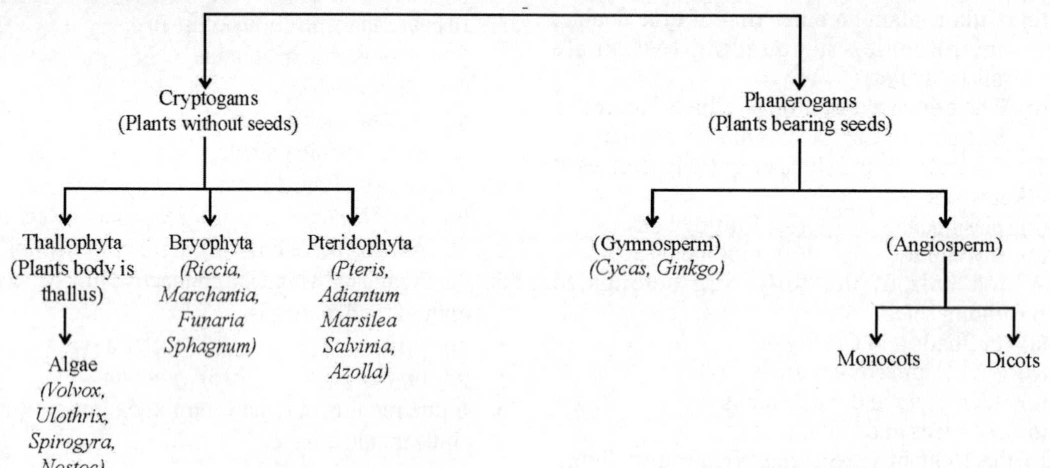

Plant kingdom includes eukaryotic, autotrophic or photosynthetic and non-motile organisms. They have well defined cellulosic cell wall. Plastids are present.

Types of Classification

- **Artificial classification system :** It was used by **Linneaus**. The artificial classification system was based on vegetative characters or on the androecium structure.
- **Natural classification system :** It was based on natural affinities among organisms. Both external and internal features were taken into account. It was used by George **Bentham** and Joseph Dalton Hooker.
- **Phylogenetic classification system :** This system of classification is based upon evolutionary relationship and uses morphological characters, origin and evolution of the different organisms. It was proposed by **Hutchinson**.
- **Numerical taxonomy (Phenetics) or Quantitative taxonomy:** This system uses numerical methods for evaluating the similarities and differences between the species. This uses maximum number of characters, without giving extra emphasis on anyone.
- **Cytotaxonomy** is based on cytological information like chromosome number, structure, behaviour and **chemotaxonomy** uses the chemical constituents of the plant to resolve confusions, are also used by taxonomists these days.

ALGAE

The main characterstics of algae are given below :

- Algae are haploid gametophytic, eukaryotic, autotrophic, chlorophyllous, cryptogamic, non-vascular thallophytic aquatic plants. They are primary producers of food and occupy 3/4 of the surface of earth and account for 90% of total photosynthesis. Marine algae are main producers.

- Algae are plants because they have chlorophyll *a*, cellulosic cell wall, starch as reserve food.
- In multicellular reproductive organs (in some brown algae), all the cells are fertile and take part in reproduction. There is no jacket of sterile cells surrounding the reproductive organs.
- Asexual reproduction occurs by unicellular mitospores. Sex organs are unicellular **non-jacketed** and called **gametangia**.
- Fertilization requires water.
- Zygote sheds and develops into perennating thick walled zygospore.
- True alternation of generations is absent (exception— *Ectocarpus* that shows isomorphic alternation of generations).
- Life cycle mostly haplontic type.

TABLE 1. Divisions of Algae and their Main Characteristics

Classes Name	Common	Major Pigments Food	Stored	Cell Wall	Flagellar Number and Position of Insertions	Habitat
Chlorophyceae	Green algae	Chlorophyll *a*, *b*	Starch	Cellulose	2-8, equal, apical	Fresh water, brackish salt water,
Phaeophyceae	Brown algae	Chlorophyll *a*, *c*, fucoxanthin	Mannitol, laminarin	Cellulose and algin	2, unequal, lateral	Fresh water (rare) brackish water, salt water
Rhodophyceae	Red algae	Chlorophyll *a*, *d*, phycoerythrin	Floridean starch	Cellulose	Absent	Fresh water (some), brackish water, salt water (most)

Economic Importance of Algae

- Certain marine brown and red algae produce large amounts of hydrocolloids (water holding substances), *e.g.*, **algin** (brown algae) and **carrageen** (red algae) which are used commercially.
- **Agar,** one of the commercial products obtained from *Gelidium* and *Gracilaria* are used to grow microbes and in preparations of ice-creams and jellies.
- *Chlorella* and *Spirullina* are unicellular algae, rich in proteins and are used as food supplements even by space travellers.

BRYOPHYTES

- Bryophytes are also called amphibians of the plant kingdom because these plants can live in soil but are dependent on water for sexual reproduction. They usually occur in damp, humid and shaded localities.
- The plant body of bryophytes is more differentiated than that of algae. It is thallus-like and prostrate or erect, and attached to the substratum by unicellular or multicellular rhizoids. They lack true roots, stem or leaves. They may possess root-like, leaf-like or stem-like structures.
- The main plant body of the bryophyte is haploid. It produces gametes, hence is called a **gametophyte.**
- The male sex organ is called **antheridium.** They produce biflagellate **antherozoids**.
- The female sex organ called **archegonium** is flask-shaped and produces a single egg.
- Zygotes do not undergo reduction division immediately. They produce a multicellular body called a sporophyte.
- Some cells of the sporophyte undergo reduction division (meiosis) to produce haploid spores.

Liverworts (Marchantia)

- Grow in moist, shady habits such as banks of streams, marshy ground, damp soil, bark of trees and deep in woods.
- The plant body is haploid (n), gametophytic, small, dorsoventrally flattened, thallose, dichotomously branched fixed by unicellular, unbranched rhizoids.
- On the dorsal side, gemma cups are found. In gemma cups, asexual buds **gemmae** are formed.
- The sporophyte is differentiated into a foot, seta and capsule. After meiosis, spores are produced within the capsule.

Moss (Funaria)

- Found in moist, shady places in patches on damp soils, moist rocks, burnt humid soil, damp walls or on tree trunks.
  The main body of *Funaria* is gametophyte and is of two forms.
- Juvenile form (Creeping protonema): Protonema is a green filamentous, branched structure.
- Adult form (Leafy gametophore): This is the main plant body of moss, consisting of a small erect axis 'stem' surrounded by spirally arranged 'leaves and numerous rhizoids at the base.

Economic Importance of Bryophytes

- Species of *Sphagnum,* a moss, provide peat that have long been used as fuel, and because of their capacity to hold water as packing material for trans-shipment of living material.
- Mosses form dense mats on the soil, they reduce the impact of falling rain and prevent soil erosion.

PTERIDOPHYTA

- Evolutionarily, they are the first terrestrial plants to possess vascular tissues – xylem and phloem.
- The main plant body is a sporophyte which is differentiated into true root, stem and leaves. These organs possess well-differentiated vascular tissues.
- The sporophytes bear sporangia that are subtended by leaf-like appendages called **sporophylls.**
- The spores germinate to give rise to inconspicuous, small but multicellular, free-living, mostly photosynthetic thalloid gametophytes called **prothallus.**
- The gametophytes bear male and female sex organs called antheridia and archegonia, respectively.
- Zygote thereafter produces a multicellular well-differentiated sporophyte which is the dominant phase of the pteridophytes.

Classification of Pteridophytes

Subphylum I – Psilophyta

- These plants represent rootless sporophytes which mainly constitute fossils. The example of living form is *Psilotum*.

Subphylum II – Lycophyta

- These are club mosses, differentiated into root, stems and leaves.

Subphylum III – Arthophyta

- Silica particles are deposited in their stem *e.g.*, *Equisetum* (horsetail).

Subphylum IV – Filicophyta

- Stems usually rhizome
- Homosporus *e.g.*, *Dryopteris*, *Pteris* or heterosporous *e.g.*, *Marsilea.*

GYMNOSPERMS

- Gymnosperms are plants which bear naked seeds *i.e.*, the ovules and the seeds that develop from these ovules after fertilization are not enclosed in fruit wall.
- Most plants are perennial and woody. Plants develop well developed tap root system.
- Roots in some genera have fungal association in the form of **mycorrhiza** (*Pinus*), while in some other (*Cycas*) small specialised roots called coralloid roots are associated with N_2– fixing cyanobacteria.
- The gymnosperms are heterosporous; they produce haploid microspores and megaspores. The two kinds of spores are produced within sporangia that are borne on sporophylls which are arranged spirally along an axis to form lax or compact strobili or **cones**.
- The male or female cones or strobili may be borne on the same tree (*Pinus*) or on different trees (*Cycas*).
- In gymnosperms the male and the female gametophytes do not have an independent free-living existence. They remain within the sporangia retained on the sporophytes.
- Following fertilisation, zygote develops into an embryo and the ovules into seeds. These seeds are not covered.

ANGIOSPERMS (FLOWERING PLANTS)

- Angiosperms are seed bearing, flowering vascular plants in which seeds are enclosed in fruits. The term angiosperm means 'enclosed seeds' as seeds (ovules) are found enclosed in the ovary wall.
- The flower is the most characteristic structure of the angiosperms.

- The male sex organ in a flower is the stamen. Each stamen consists of a slender filament with an anther at the tip. The anthers, following meiosis, produce pollen grains.
- The female sex organ in a flower is the pistil or the carpel. Pistil consists of an ovary enclosing one to many ovules. Within ovules are present highly reduced female gametophytes termed **embryo sacs.**

  Each embryo-sac has a three-celled **egg apparatus** – one **egg cell** and two **synergids,** three **antipodal** cells and two **polar nuclei**. The polar nuclei eventually fuse to produce a diploid secondary nucleus.
- Pollen grain, after dispersal from the anthers, are carried by wind or various other agencies to the stigma of a pistil. This is termed as **pollination**.
- The pollen tubes enter the embryo-sac where two male gametes are discharged. One of the male gametes fuses with the egg cell to form a zygote (syngamy). The other male gamete fuses with the diploid secondary nucleus to produce the triploid primary endosperm nucleus (PEN). Because of the involvement of two fusions, this event is termed as **double fertilisation**, and event unique to angiosperms.

S.No.	Dicotyledons	Monocotyledons
1.	The plants have two cotyledons.	The plants have one cotyledon.
2.	Leaves have reticulate venation.	Leaves possess parallel venation.
3.	Stomata are generally kidney-shaped.	Stomata are commonly dumb-bell shaped.
4.	Tap root system is more common. Adventitious roots are less common.	Adventitious roots are more common.
5.	In stem the vascular bundles are arranged in a single ring.	The vascular bundles are scattered in the stem.
6.	Vascular bundles of the stem are conjoint, collateral and open.	Vascular bundles of the stem are conjoint, collateral and closed.
7.	Tetramerous or pentamerous flowers, i.e. having four or five members in each floral whorls.	Trimerous flowers having three members in each floral whorls.

PLANT LIFE CYCLE AND ALTERNATION OF GENERATION

- In plants, both haploid and diploid cells can divide by mitosis.
- Life cycle of any sexually reproducing plant consists of two morphological phases *i.e.*, haploid **gametophyte** and diploid **sporophyte** that regularly alternate with each other, this phenomenon is termed as **alternation of generation**.
- **Haplontic life cycle:** In haplontic life cycle there is no true alternation of generation as sporophytic generation is represented only by one celled zygote. There is no free living sporophyte. This life cycle is found in algae like *Chlamydomonas, Volvox, Ulothrix, Spirogyra.*
- **Diplontic life cycle:** In diplontic life cycle, diploid sporophyte is dominant vegetative phase. It bears sex organs that produce gametes through meiosis. There is no true alternation of generation. This type of life cycle is found in all seed bearing plants *i.e.*, gymnosperm and angiosperm. Alga *Fucus* shows diplontic life cycle.
- **Haplo-diplontic life cycle:** The haploid gametophyte produces gametes in sex organs. The diploid sporophyte produces spores after meiosis. These spores on germination gives rise to gametophytic phase. It represents true alternation of generation. It is found in bryophytes, pteridophytes and in some algae like *Ectocarpus, Polysiphonia.*

Exercise

1. Both chlorophyll *a* and *b* are present in
(a) rhodophyceae (b) phaeophyceae
(c) chlorophyceae (d) None of these
2. Bryophytes can be separated from algae, because they
(a) are thalloid forms
(b) have no conducting tissue
(c) possess archegonia
(d) contain chloroplast
3. Female reproductive part of bryophytes is
(a) Antheridium (b) Oogonium
(c) Archegonium (d) Sporangium
4. Which of the following is called amphibians of plant kingdom ?
(a) Bryophytes (b) Pteridophytes
(c) Gymnosperms (d) Algae
5. Bryophytes are of
(a) great economic value
(b) no value at all
(c) great ecological importance
(d) a lot of aesthetic value
6. Common characteristic between bryophytes and pteridophytes is
(a) vascularisation
(b) terrestrial habit
(c) water for fertilization
(d) independent sporophyte
7. The plant group that produces spores and embryo but lacks vascular tissues and seeds is
(a) pteridophyta (b) rhodophyta
(c) bryophyta (d) phaeophyta
8. A plant having seeds but lacking flowers and fruits belongs to
(a) pteridophytes (b) mosses
(c) ferns (d) gymnosperms
9. Pteridophytes differ from mosses/bryophytes in possessing
(a) independent gametophyte
(b) well developed vascular system
(c) archegonia
(d) flagellate spermatozoids
10. In Chlorophyceae, sexual reproduction occurs by
(a) isogamy and anisogamy
(b) isogamy, anisogamy and oogamy
(c) oogamy only
(d) anisogamy and oogamy
11. Bryophytes are dependent on water because
(a) water is essential for fertilization for their homosporous nature
(b) water is essential for their vegetative propagation
(c) the sperms can easily reach upto egg in the archegonium
(d) archegonium has to remain filled with water for fertilization
12. Which of the following is without exception in angiosperms?
(a) Presence of vessels
(b) Double fertilisation
(c) Secondary growth
(d) Autotrophic nutrition
13. Agar is commercially obtained from
(a) red algae (b) green algae
(c) brown algae (d) blue-green algae
14. Brown algae is characterised by the presence of
(a) phycocyanin (b) phycoerythrin
(c) fucoxanthin (d) haematochrome
15. Chloroplast of *Chlamydomonas* is
(a) stellate (b) cup-shaped
(c) collar-shaped (d) spiral
16. Sexual reproduction involving fusion of two cells in *Chlamydomonas* is
(a) isogamy (b) homogamy
(c) somatogamy (d) hologamy
17. Prothallus (gametophyte) gives rise to fern plant (sporophyte) without fertilization. It is
(a) apospory (b) apogamy
(c) parthenocarpy (d) parthenogenesis
18. Moss peristome takes part in
(a) spore dispersal (b) photosynthesis
(c) protection (d) absorption

19. Which one of the following is a living fossil?
(a) *Pinus* (b) *Opuntia*
(c) *Ginkgo* (d) *Thuja*

20. Blue-green algae belong to
(a) eukaryotes (b) prokaryotes
(c) rhodophyceae (d) chlorophyceae

21. Dichotomous branching is found in
(a) Fern (b) *Funaria*
(c) Liverworts (d) *Marchantia*

22. Floridean starch is found in
(a) chlorophyceae (b) rhodophyceae
(c) myxophyceae (d) cyanophyceae

23. *Cycas* have two cotyledons but not included in angiosperms because of
(a) naked ovules
(b) seems like monocot
(c) circinate ptyxis
(d) compound leaves

24. Peat moss is used as a packing material for sending flowers and live plants to distant places because
(a) it is hygroscopic
(b) it reduces transpiration
(c) it serves as a disinfectant
(d) it is easily available

25. Conifers differ from grasses in the
(a) lack of xylem tracheids
(b) absence of pollen tubes
(c) formation of endosperm before fertilization
(d) production of seeds from ovules

26. If you are asked to classify the various algae into distinct groups, which of the following characters you should choose?
(a) Nature of stored food materials in the cell
(b) Structural organization of thallus
(c) Chemical composition of the cell wall
(d) Types of pigments present in the cell

27. Algae have cell wall made up of:
(a) cellulose, galactans and mannans
(b) hemicellulose, pectins and proteins
(c) pectins, cellulose and proteins
(d) cellulose, hemicellulose and pectins

28. In angiosperms, functional megaspore develops into
(a) embryo sac (b) ovule
(c) endosperm (d) pollen sac

29. In gymnosperms like *Pinus* and *Cycas*, the endosperm is
(a) triploid (b) haploid
(c) diploid (d) tetraploid

30. Protonema occurs in the life cycle of
(a) *Riccia* (b) *Funaria*
(c) Equisetum (d) *Spirogyra*

31. Which of the following cannot fix nitrogen?
(a) *Nostoc* (b) *Azotobacter*
(c) *Spirogyra* (d) *Anabaena*

32. An alga very rich in protein is
(a) *spirogyra* (b) *ulothrix*
(c) *oscillatoria* (d) *chlorella*

33. Which one of the following plants is monoecious?
(a) *Pinus* (b) *Cycas*
(c) Fucus (d) *Marchantia*

34. Which type of ribosomes are found in *Nostoc* cells ?
(a) 50S (b) 60S
(c) 70S (d) Eukaryotic

35. *Nostoc* is a –
(a) green alga (b) yellow-green alga
(c) blue-green alga (d) red alga

36. The attachment structure/s found in lichens –
(a) Rhizomes (b) Hold fast
(c) Both of these (d) Rhizosphere

37. Bioindicators are –
(a) lichens tracing the presence of pollution
(b) fossil lichens
(c) special type of litmus paper
(d) None of the above

38. Which commonly known as 'Peat moss' or 'Bog moss' ?
(a) *Selaginella* (b) *Equisetum*
(c) *Laminaria* (d) *Sphagnum*

39. Which is commonly known as 'Liverwort' ?
(a) *Funaria* (b) *Marchantia*
(c) *Anthoceros* (d) *Sphagnum*

40. Sporophyte is dependent on gametophyte in
(a) bryophytes (b) gymnosperms
(c) angiosperms (d) pteridophytes

41. The rhizophore is not a true root because it is
(a) endogenous (b) exogenous
(c) mesogenous (d) endarch

42. Seed habit first originated in
(a) certain pteridophytes
(b) certain pines
(c) certain monocots
(d) certain dicots

43. Which of the following plants exhibit independent alternation of generation ?
(a) Angiosperms (b) Gymnosperms
(c) Pteridophytes (d) Bryophytes

44. 'Maiden hair fern' is –
(a) *Dryopteris* (b) *Azolla*
(c) *Adiantum* (d) *Pteris*

45. Sporocarp is a reproductive structure of –
(a) some algae
(b) some aquatic ferns having sori
(c) angiosperms having spores
(d) bryophytes

46. Prothallus means –
(a) Immature gametophyte
(b) Immature sporophyte
(c) Immature archegonium
(d) None of these

47. Fruits are not formed in Gymnosperms because of –
(a) absence of pollination
(b) absence of seed
(c) absence of fertilization
(d) absence of ovary

48. In Gymnosperms, seeds are said to be naked because they–
(a) do not have ovule
(b) do not have ovary
(c) do not contain endosperm
(d) do not contain starch

49. The embryo sac in an angiosperm is a –
(a) megasporangium
(b) megaspore mother cell
(c) megagametophyte
(d) megaspore

50. The role of double fertilization in angiosperms is to produce-
(a) endosperm (b) integuments
(c) cotyledons (d) endocarp

51. Which of the following is a "resurrection" plant?
(a) *Lycopodium* (b) *Selaginella*
(c) *Rhynia* (d) *Marsilia*

52. Algae used in space programme is :
(a) *fucus* (b) *volvox*
(c) *chlamydomonas* (d) *chlorella*

53. Which of the following is not the feature of gymnosperms?
(a) Parallel venation
(b) Perennial plants
(c) Distinct branches (long and short branches)
(d) Xylem with vessels

54. In *Pinus,* male cone bears a large number of
(a) ligules (b) anthers
(c) microsporophylls (d) megasporophylls

55. In gymnosperms, ovules are
(a) covered (b) naked
(c) single coat (d) imbricate

56. *Laminaria* (kelp) and *Fucus* (rock weed) are the examples of
(a) red algae (b) brown algae
(c) green algae (d) golden brown algae

57. Tallest angiosperm is
(a) *Eucalyptus* (b) red wood tree
(c) oak tree (d) *Pinus*

58. Sea weeds are important source of
(a) chlorine (b) fluorine
(c) iodine (d) fromine

59. Which of the following is an fatty oil yielding plant?
(a) Sunflower (b) *Acacia*
(c) *Butea* (d) *Casuarina*

60. The red colour of rhodophyta is due to the preponderance:
(a) phycobilins (b) phycocyanin
(c) phycoerythrin (d) None of these

Animal Kingdom

CONCEPTS

- Animals are the most diverse groups of organisms. Multicellular, heterotrophs characterised by mobility, sensory and nervous systems.

BASIS OF CLASSIFICATION

Arrangement of cell, body symmetry, nature of coelom, pattern of digestive, circulatory and reproductive systems are used as the basis of animal classification.

Levels of Organization (Grades of Organization)

- **Cellular level :** Example— sponges.
- **Tissue level :** Example— coelenterates.
- **Organ level :** Example - member of platyhelminthes.
- **Organ-system level :** Example - annelids, arthropods, molluscs, echinoderms and chordates.

The digestive system is of two types:

(a) **Incomplete digestive system :** There is only single opening to the outside of the body that serve as both mouth and anus *e.g. Hydra*.

(b) **Complete digestive system :** It has two openings mouth and anus *e.g.* Human.

The circulatory system is also of two types.

(a) In **closed** blood vascular system, the blood circulates inside the blood vessels without ever coming in direct contact with the body cells, *e.g.*, annelids, chordates.

(b) In **open** blood vascular system, the blood flows in open spaces like lacunae and sinuses. It baths the cells directly, *e.g.*, arthropods, molluscs.

Symmetry

- **Asymmetrical :** The body cannot be divided in two equal halves on any plane through the centre. *e.g. Amoeba*
- **Radial symmetry :** Any plane passing through a central axis of the body divides the organism into two identical halves. Examples are coelentrates, ctenophores and echinoderms.
- **Bilateral Symmetry:** Body can be divided into two equal halves on only one plane. Examples are annelids, arthropods, human etc.

Diploblastic and Triploblastic Organisation

Diploblastic animals : Cells are arranged in two embryonic layers, an external ectoderm and internal endoderm *e.g.* coelenterates, sponges.

Triploblastic animals : Animals in which developing embryo has the third germinal layers, mesoderm, in between the ectoderm and endoderm are called triploblastic animals. *e.g.* platyhelminthes to chordates.

Coelom

- The cavity between body wall and alimentary canal lined with mesodermal epithelia on both sides .
- **Acoelomates :** The animals which do not have coelom *e.g.*, sponges, coelenterates, ctenophorans and flatworms.
- **Pseudocoelomates :** Body cavity is not lined by mesoderm continuously, but it is present as scattered pouches in between the ectoderm and endoderm. Aschelminthes (round worms) are psuedocoelomates.
- **Eucoelomates (Coelomates)**. The animals which possess true coelom. True coelom is found in annelids, echinoderms and chordates.

Notochord

- It is a mesodermally derived rod-like structure formed on the dorsal side during embryonic development of some animals.

- Animals with notochord are called chordates and those animals which do not form this structure are called non-chordates, *e.g.*, porifera to echinoderms.

PHYLUM-PORIFERA

- Sponges are aquatic, mostly marine, solitary or colonial and sessile.
- Body is porous having two types of pores such as (i) ostia, which is inhalent pores (ii) osculum, which is exhalent pores.
- Most of the sponges are asymmetrical.
- **Choanocytes** or collar cells line the spongocoel and the canals.
- Canal system helps in nutrition, excretion, circulation, transport of gametes and fertilization.
- Sexes are not separate (hermaphrodite).
- Sponges reproduce asexually mainly by fragmentation and gemmule formation, however if reproduces sexually by the formation of gametes.
- Examples of some sponges are : *Sycon* (scypha), *Spongilla* (fresh water sponge) and *Euspongia* (bath sponge)

PHYLUM-COELENTERATA (CNIDARIA)

- All are aquatic and are mostly marine (exception-*Hydra* are found in fresh-water), solitary or colonial, sessile, or free-swimming and radially symmetrical animals.
- Coelenterates are dipoblastic animals.
- Both sexual and asexual reproduction is found.

  Example-*Physalis* (Portuguese man of war), *Adamsia* (Sea anemone), *Pennatula* (Sea-pen), *Gorgonia* (Sea-fan) and *Meandrina* (Brain coral).

PHYLUM-CTENOPHORA

- These are diploblastic, radial symmetrical animals with tissue level of organization.
- Bioluminescence and power of regeneration is well-marked.
- As comb like plates are present, hence, ctenophores are called comb jellies or sea walnuts.
- Reproduction takes place only by sexual means.
- Examples–*Hormiphora* (sea walnut), *Pleurobranchia* (sea gooseberry), *Cestum* (venus girdle), *Beroe.*

PHYLUM-PLATYHELMENTHES

- These are mostly endoparasites, bilateral symmetrical, triploblastic and acoelomate animals with organ level of organisation.
- Excretory organs are flame cells.
- Examples– *Taenia* (Tape worm), *Fasciola* (liver fluke).

PHYLUM-ASCHELMINTHES

- They may be free-living, aquatic and terrestrial or parasitic in plants and animals.
- They have organ-system level of body organisation, bilaterally symmetrical, triploblastic and pseudocoelomate animals.
- Sexes are separate (dioecious). Examples: *Ascaris* (Round worm), *Wuchereria* (filarial worm), *Ancylostoma* (Hook worm), *Enterobius* (Pin worm)

PHYLUM-ANNELIDA

- It is characterised by metameric segmentation forming ring like segments.
- They are bilaterally symmetrical, triploblastic, coelomate animals with organ-level of organisation. Annelids are the first animals to have a true coelom, which coelomic fluid, with serves as a hydrostatic skeleton.
- Blood vascular system is closed type.
- Nephridia helps in osmoregulation and excretion.

  Example: *Neries*, *Pheretima* (Earthworm) and *Hirudinaria* (Blood sucking leech).

PHYLUM-ARTHROPODA

- They are bilateral symmetry, triploblastic animals, which have organ-system level of organisation.
- Respiratory organs are gills or book gills tracheae or book-lungs.
- Excretory organs are either green glands or malpighian tubules.
- Circulatory system is of open type.
- Sensory organs like antennal and eyes are present.

  Example: *Apis* (Honey bee), *Bombyx* (Silkworm), *Laccifer* (Lac insect)

  Peripatus (walking worm), which is also "connecting link" between phylum Annelida and phylum Arthropoda. *Limulus* (king crab), is a living fossil.

PHYLUM-MOLLUSCA

- They are aquatic (marine or fresh water), or terrestrial having an organ-system level of organisation.
- The body of molluscs is unsegmented with a distinct head, muscular foot and visceral hump mostly. It is covered by a calcareous shell.
- Blood vascular system is open type.
- The mouth contains a rasping organ for feeding, called radula.
- They are dioecious generally, but some are hermaphrodite.
- Ex. *Pila* (apple snail, amphibous) *Octopus* (devil fish), *Loligo* (sea squid).

PHYLUM-ECHINODERMATA

- All existing echinoderms are marine, generally live at sea bottom.
- Body is bilateral symmetrical in larval state, but radial in adults, triplobastic with organ-system level of organisation.
- Presence of water vascular system is the characteristic feature of phylum echinodermata.
- Circulatory system is open type.
- Ex. *Asterias* (star fish), *Cucumaria* (commonly called as sea cucumber), *Antedon* (water lily or feather star).

PHYLUM-HEMICHORDATA

- They are bilaterally symmetrical, triploblastic, and entrocoelous animals.
- Circulatory system is open type.
- Respiration is by gills.
- Excretory organ is **proboscis gland**.
- Sexes are separate or unite and reproduction is mostly sexual.
- Ex. *Balanoglossus* (acorn worm or tongue worm), *Saccoglossus.*

PHYLUM-CHORDATA

- The fundamental four characters of phylum chordata are presence of notochord, a dorsal hollow nerve cord, paired pharyngeal gill slits and post anal tail either in the embryonic or adult stage.
- They are bilaterally symmetrical, triploblastic and coelomate with organ system level of organisation.
- They posses a closed circulatory system.
- Ventral myogenic heart is made of cardiac muscle.
- Blood with RBC containing haemoglobin as respiratory pigment.
- Bone and cartilage are the skeletal tissue.
- Phylum Chordata is divided into three sub-phylums:

Sub-phylum 1: Urochordata

- Urochordate adults are generally sedentary.
- Notochord is present in the tail of the larva only, but disappears in the adult.

  Examples: *Herdmania* (sea squirt), *Ascidia, Botryllus* (colonial urochordate), *Pyrosoma* (bioluminescent colonial urochordate).

Sub-phylum 2: Cephalochordata

- Both adult and larva are motile.
- It shows progressive metamorphosis *i.e*, changes of larval body occurs from less developed larva to better developed adult.

  Example: *Branchiostoma* (= Amphioxus).

Sub-phylum 3: Vertebrata or Craniata

- Notochord is only present in the embryonic stage, but replaced by a cartilaginous or bony vertebral column in the adult forms.
- Heart is ventrally situated with two, three or four chambers.
- Circulatory system is closed type consisting of blood and lymphatic systems.
- Sexes are separate (unisexual) except hag fish. Asexual reproduction is absent.

Subphylum vertebrata is divided into two sections:

Section 1 Agnatha (The jawless vertebrates)

Class : Cyclostomata

- Mouth jawless suctorial and round.
- All living members are ectoparasites on some fishes.
- Their body is devoid of scales and paired fins.

  Ex. *Petromyzon* (lamprey), *Lcanpetm* (lamprey ectoparasites on sharks), Myxine (hag fish).

Section 2 Gnathostomata (The jawed vertebrates)

- Embryonic notochord is usually replaced in adult by a vertebral column.
- Paired fins or limbs and nostrils are present.

SUPERCLASS : PISCES (Bear fins)

Class : Chondrichthyes

- They have a cartilagenous skeleton.
- Air bladder is not present in these animals. Hence, they have to swim constantly to avoid sinking.
- Some of them possess electric organs *e.g.* Torpedo.

- They are cold-blooded (poikilothermous) animals.
- Sexes are separate.
- They are oviparous or ovoviviparous.
- Examples: *Scoliodon* (Dog fish), *Trygon* (Sting ray).

Class : Osteichthyes

- They have a bony skeleton.
- Air bladder is present which regulates buoyancy.
- They are cold-blood animals.
- Sexes are separate.
- They are mostly oviparous and development is direct except in *Anguilla* where development is indirect with a larva.
- Examples : Marine – *Exocoetus* (Flying fish), *Hippocampus* (Sea horse), *Lophius* (Angler fish), Fresh water fishes – *Labeo* (Rohu), *Catla* (Katla).

SUPERCLASS : TETRAPODA (Bear limbs)

Class : Amphibia

- Adapted for both water and land life.
- Respiration by gills, skin, lungs and buccal epithelium.
- A tympanum represents the ear.
- Heart is 3 chambered (2 auricles and 1 ventricle).
- Alimentary canal, urinary and reproductive tract opens into common chamber called cloaca which opens to the exterior.
- They are oviparous and development indirect through distinct larval stage called tadpole. Exambles : *Bufo* (Toad), *Rana* (Frog), *Hyla* (Tree frog), *Salamandra* (Salamander), *Ichthyophis* (Limbless amphibial).

Class: Reptilia

- The class name refers to their creeping or crawling mode of locomotion.
- Dry and horny skin with epidermal scales prevent loss of water from the body. Snakes and lizards shed their scales as skin cast.
- Heart is 3 chambered but 4 chambered in crocodiles; separation of ventricle is only partial.
- They do not have external ear openings.
- They are oviparous ; Development direct. Examples: Crocodilus (Crocodile), Bangarus (Krait)

Class: Aves

- Birds are bipedal feathered animals.
- They are warm-blooded (homoiothermous).
- Fore-limbs are modified into wings for flight except some flightless bird such as Ostrich.
- Endoskeleton is fully ossified (bony) and the long bones are hollow with air cavities (pneumatic).
- Respiration is by lungs.
- Heart is four-chambered.
- Sexes are separate. All birds are oviparous & lay eggs.
- Four embryonic membranes (*e.g.*, chorion, amnion, allantois and yolk sac) are formed.
- Brain is better developed than that of reptiles
- Each ear consists of three parts : external, middle and internal.
  Examples : Corvus (crow), Pavo (Peacock)

Class: Mammalia

- These are warm blooded (homiothermous) animals having hair and mammary glands.
- Different types of teeth (heterodont) are present and are embedded in the sockets of jaws (thecodont).
- External ears or pinnae are present. 3 ear ossicles (incus, malleus and stapes) present in the tympanic cavity.
- Respiration is by lungs.
- Heart is four chambered. Red blood corpuscles are without nucleus.
- Sexes are separate and fertilization is internal.
- They are viviparous with few exceptions and development is direct.

Example :

Oviparous – *Tachyglossus* = *Echidna* (spiny Anteater).

Viviparous – *Pteropus* (Flying fox), *Camelus* (Camel), *Macaca* (Monkey), *Rattus* (Rat), *Canis* (Dog), *Felis* (Cat), *Elephas* (Elephant), *Eqqus* (Horse), *Delphinus* (Common dolphin), *Balaenoptera* (Blue whale), *Panthera tigris* (Tiger), *Panthera Leo* (Lion).

Exercise

1. All chordates possess
   (a) exoskeleton
   (b) limbs
   (c) skull
   (d) axial skeletal rod of notochord
2. A common characteristic of all vertebrates is
   (a) presence of skull
   (b) division of body into head, neck, trunk and tail
   (c) presence of two pairs of functional appendages
   (d) body is covered with an exoskeleton
3. Pneumatic bones are expected to be found in
   (a) pigeon (b) house lizard
   (c) frog's tadpole (d) flying fish
4. The long bones are hollow and connected by air passages these are characteristics of
   (a) mammals
   (b) reptiles
   (c) birds
   (d) all land vertebrates
5. Cause of mimicry is
   (a) concealment
   (b) attack (offence)
   (c) protection (defence)
   (d) both (b) and (c)
6. Animals/organisms floating on the surface of water are
   (a) plankton (b) pelagic
   (c) benthos (d) neritic
7. Earthworms are
   (a) useful
   (b) harmful
   (c) more useful than harmful
   (d) more harmful
8. A chordate character is
   (a) gills
   (b) spiracles
   (c) postanal tail
   (d) chitinous exoskeleton
9. Malpighian tubules are
   (a) excretory organs of insects
   (b) excretory organs of annelids
   (c) respiratory organs of insects
   (d) respiratory organs of annelids
10. Skin is a respiratory organ in
    (a) lizards
    (b) birds
    (c) primitive mammals
    (d) frog
11. Silk thread is obtained from silk moth during
    (a) pupal state (b) larval state
    (c) nymph state (d) adult state
12. Flight muscles of birds are attached to
    (a) clavicle (b) keel of sternum
    (c) scapula (d) coracoid
13. Classification of porifera is based on
    (a) branching (b) spicules
    (c) reproduction (d) symmetry
14. Metamorphosis of insects is regulated through hormone
    (a) pheromone (b) thyroxine
    (c) ecdysone (d) All of the above
15. Which one occurs in *Echinodermata* ?
    (a) Bilateral symmetry
    (b) Radical symmetry
    (c) Porous body
    (d) Soft skin
16. A. *Periplaneta americana* is nocturnal, omnivorous, household pest.
    R. It is because it acts as scavenger.
    (a) A is true but R is false
    (b) A is false but R is true
    (c) Both A and R are true and R is correct explanation of A
    (d) Both A and R are true but R is not correct explanation of A
17. Bull Frog of India is
    (a) *Rana tigrina* (b) *R. sylvatica*
    (c) *R. ecutesbeiana* (d) *R. esculenta*
18. The cervical vertebrae in humans is
    (a) same as in whale
    (b) more than that in rabbit
    (c) double than that of horse
    (d) less than that in giraffe
19. What is common in whale, bat and rat ?
    (a) Absence of neck
    (b) Muscular diaphragm between thorax and abdomen

(c) Extra-abdominal testes to avoid high temperature of body
(d) Presence of external ears

20. Gorilla, chimpanzee, monkeys and humans belongs to the same
(a) species (b) genus
(c) family (d) order

21. Tracheae of cockroach and mammal are similar in having
(a) paried nature
(b) noncollapsible walls
(c) ciliated inner lining
(d) origin from head

22. Budding is a normal mode of asexual reproduction in
(a) starfish and *Hydra*
(b) *Hydra* and sponges
(c) tapeworm and *Hydra*
(d) sponge and starfish

23. Radial symmetry is often exhibited by animals having
(a) one opening of alimentary canal
(b) aquatic mode of living
(c) benthos/sedentary living
(d) ciliary mode of feeding

24. Coelom derived from blastocoel is known as
(a) enterocoelom (b) schizocoel
(c) pseudocoelom (d) haemocoelom

25. Special character of Coelenterates is
(a) polymorphism
(b) nematocytes
(c) flame cells
(d) hermaphroditism

26. Point out the non-parasite
(a) Tapeworm (b) Mosquito
(c) Leech (d) Sea anemone

27. What is true about all sponges without exception ?
(a) They are all marine
(b) They have flagellated collar cells
(c) They have a mixed skeleton consisting of spicules and spongin fibres
(d) They reproduce only asexually by budding

28. In desert grasslands which type of animals are relatively more abundant?
(a) Arboreal (b) Aquatic
(c) Fussorial (d) Diurnal

29. In which of the following animal post anal tail is found?
(a) Earthworm
(b) Lower invertebrate
(c) Scorpion
(d) Cobra

30. Which one of the following characters is not typical of the class Mammalia?
(a) Thecodont dentition
(b) Alveolar lungs
(c) Ten pairs of cranial nerves
(d) Seven cervical vertebrae

31. In contrast to Annelids the Platyhelminthes show
(a) absence of body cavity
(b) bilaterial symmetry
(c) radial symmetry
(d) presence of pseudocoel

32. In which one of the following sets of animals do all the four give birth to young ones?
(a) Platypus, penguin, bat, hippopotamus
(b) Shrew, bat, cat, kiwi
(c) Kangaroo, hedgehog, dolphin, , loris
(d) Lion, bat, whale, ostrich

33. Which one of the following is not a characteristic of phylum annelida?
(a) Closed circulatory system
(b) Segmentation
(c) Pseudocoelom
(d) Ventral nerve cord

34. Which one of the following groups of animals is bilaterally symmetrical and triploblastic?
(a) Aschelminthes (round worms)
(b) Ctenophores
(c) Sponges
(d) Coelenterates (Cnidarians)

35. Which one of the following pairs of animals comprises 'jawless fishes'?
(a) Mackerals and rohu
(b) Lampreys and hag fishes
(c) Guppies and hag fishes
(d) Lampreys and eels

36. Which one of the following have the highest number of species in nature?
(a) Fungi (b) Insects
(c) Birds (d) Angiosperms

37. Which one of the following statement is totally wrong about the occurrence of notochord, while the other three are correct?
(a) It is present only in larval tail in Ascidians.
(b) It is replaced by a vertebral column in adult frog.
(c) It is absent throughout life in humans from the very beginning.
(d) It is present throughout life in *Amphioxus.*

38. Frogs differ from humans in possessing
(a) paired cerebral hemispheres
(b) hepatic portal system
(c) nucleated red blood cells
(d) thyroid as well as parathyroid

39. Sponges are characterized by
(a) amoeboid cells (b) choanocytes cells
(c) pigmented cells (d) gland cells

40. Which of the following animal is not a insect ?
(a) Ticks (b) Honey bee
(c) Beetle (d) Wasp

41. Which of the following characteristic is shared by all arthropods ?
(a) Complete metamorphosis
(b) Wings
(c) Jointed appendages
(d) Asexual reproduction

42. Member of Echinodermata has a specific system, which is not found in other phylum, it is
(a) canal system
(b) water vascular system
(c) respiratory system
(d) reproductive system

43. Cleavage in mammals
(a) holoblastic equal (b) holoblastic unequal
(c) superficial (d) discoidal

44. Which of the following is closest relative of man ?
(a) Chimpanzee (b) Gorilla
(c) Orangutan (d) Gibbon

45. Uricotelism is found in
(a) fishes and fresh water protozoans
(b) birds, reptiles and insects
(c) frogs and toads
(d) mammals and birds

46. One of the following is a very unique feature of the mammalian body
(a) presence of diaphragm
(b) four chambered heart
(c) rib cage
(d) homeothermy

47. Presence of gills in the tadpole of frog indicates that
(a) fishes evolved from frog like ancestors
(b) frogs will have gills in future
(c) frogs evolved from gilled ancestors
(d) fishes were amphibious in the past

48. Which of the following is a chordate feature, not shared by the non-chordates ?
(a) Metamerism (b) Axial organization
(c) Bilateral symmetry (d) Pharyngeal gill slits

49. Which of the following is not found in birds ?
(a) Hind limb (b) Fore limb
(c) Pelvic girdle (d) Pectoral girdle

50. Herpetology is the study of
(a) fishes (b) reptiles
(c) insects (d) birds

51. Venom of cobra attacks
(a) digestive system (b) respiratory system
(c) nervous system (d) circulatory system

52. Which of the following is present in man but absent in frogs ?
(a) Salivary gland (b) Adrenal gland
(c) Thyroid gland (d) Pancreas

53. Which of the following is present in molluscs but not in echinodermates ?
(a) Flame cells (b) Malpighian tubules
(c) Kidney (d) None of these

54. Dance pattern in honey bees is for
(a) courtship (b) communication
(c) recreation (d) None of these

55. The methods of dispersal in *Amoeba* is
(a) locomotion (b) sporulation
(c) encystment (d) binary fission

56. The largest invertebrate animal is
(a) whale (b) giant squid
(c) shrew (d) python

57. Spider web is formed by a fluid secreted by its
(a) abdominal gland (b) salivary gland
(c) cephalothorax (d) None of these

58. Which of the following group of animals maintains high and constant body temperature such as mammals ?
(a) Reptiles (b) Amphibians
(c) Birds (d) Fishes

59. Insects have
(a) 2 pairs of legs (b) 3 pairs of legs
(c) 4 pairs of legs (d) 1 pair of legs

60. Which type of respiratory organs are present in spiders and scorpions ?
(a) Book lungs (b) Gills
(c) Gill books (d) Lungs

Morphology of Flowering Plants

CONCEPTS

- Angiosperms are flowering, fruit bearing, spermatophytic plants that are well adapted to life on land. These plants are the most advanced and evolved group that exists today.
- They all have enclosed seeds and bear flowers.
- Angiosperms lack archegonium which is replaced by pistil. They have double fertilization and endosperm is of triploid nature.
- The male gametes are nonmotile and carried by pollen tube to avoid dependence on water for fertilization.

PLANT MORPHOLOGY

- Plant morphology refers to the study of external form and structure of plants.

The Root

- It is the underground system, usually below the soil and originates from the radicle.
- Generally it is positively geotropic, negatively phototropic and positively hydrotropic.
- It lacks nodes, internodes, leaves and buds.
- In few hydrophytes a root cap is absent and instead root pockets are present. A root pocket differs from a cap in its mode of development. In *Pandanus* (screwpine), multiple caps are present.
- The primary functions of root are fixation of plant firmly on soil, absorption of water and conduction of mineral nutrients.
- Roots also prevent soil erosion.

Region of the root

- **Root cap (calyptra):** Root cap is derived from calyptrogen. It covers root tip and protects it against friction from soil particles.
- **Region of active cell division or meristematic region:** The growing point produces new cells for root cap and basal parts.
- **Region of elongation:** The cells elongate rapidly due to vacuolization. This zone is responsible for growing of root in length.
- **Region of root hair:** Outer cells just above the region of elongation give rise to lobular unbranched unicellular root hairs for increasing absorptive area.
- **Region of cell maturaton:** The outermost layer of this region has thick walled or impermeable cells. Its only function is to anchor the plant firmly in the soil.

MODIFICATION OF ROOTS

Modified Tap Roots

- **Fusiform :** It is swollen in the middle and is tapering at both the ends *e.g.* radish.
- **Napiform :** It is much swollen at the upper end and abruptly tapering towards the lower end *e.g.* sugar beet.
- **Conical :** The conical root is broadest at the top (base) and is gradually tapering towards the lower end *e.g.* carrot.
- **Tuberous root :** Tap root is irregularly swollen except at the base *e.g.* tapioca.
- **Nodulated :** In legumes, the roots are symbiotically associated with the nitrogen-fixing bacterium, *Rhizobium leguminosarum.* This results in the formation of several nodules in the branches of the roots *e.g.* rhizobium.

Modified Branched Root

- **Pneumatophores :** Pneumatophores or respiratory roots are short, vertical and negatively geotropic (grow in an upward direction) that occur in certain halophytes, which grow in saline marshes (mangroves). They bear numerous minute pores or special lenticels (pneumatothodes) in their terminal parts that help in taking atmospheric oxygen which is completely absent in the saline water *e.g.* rhizophora.

Modified for Mechanical Functions

- **Prop root :** Plants are profusely branched and branches are thick and heavy. From these branches roots arise and hang downwards in the air and later penetrete the soil and function as prop (for support) to the branch *e.g.* radish*e.g.* banyan.
- **Stilt root :** In some plants roots arise from lower nodes of stem and enter the soil and become stronger *e.g.* screwpine.
- **Climbing root :** arise from nodes and help the plant in climbing *e.g.* betel.
- **Clinging root :** Arise from node of a stem and pierce the stem of host plant and thus help in fixing the plant to the host *e.g.* orchid.
- **Floating root :** Spongy aerial roots filled with air help the plant in floating *e.g.,* Jussiaea.
- **Contractile root :** Roots are meant for keeping storage stems at a proper depth in the soil *e.g.,* onion.
- **Root thorn :** Roots near the soil become hard like thorns. These thorns serve as protective organ *e.g.,* coconut.

Modified Adventitious Root

Modified for Physiological Functions

Storage root

- **Assimilatory root :** Develop chlorophyll and perform photosynthesis *e.g.,* Tinospora.
- **Parasitic root :** Roots arise from the stem and enter the host plant developing contact with xylem and phloem of the host stem *e.g.,* Cuscuta.
- **Reproductive root :** Vegetative buds arise on adventitious roots which after separation from parent plant form new plants *e.g.,* Ipomea.
- **Epiphytic root :** These aerial hanging roots present in epiphytes (plants growing on other plants for space only) *e.g.,* Orchids.
- **Tuberous root :** Arise from nodes of stem and one root out of bunch become tuberous and fleshy for storage of food *e.g.,* sweet potato.
- **Fasciculated root (*Asparagus*) :** Arise in bunches (fascicles) from lower nodes of stem and become thick and fleshy for storage of food *e.g.,* asparagus.
- **Nodulose root :** Terminal portions swell up *e.g.,* Curcuma amada.
- **Moniliform root :** Due to food storage these roots swells at different places *e.g.,* dioscorea.
- **Annulated root :** Swelling at different places takes place in such a way that it looks like stack of rings *e.g.,* Ipecac.

SHOOT SYSTEM

- It is a direct prolongation of the plumule.
- It is negatively geotropic and positively phototropic.
- It is distinguishable into nodes and internodes and lacks a terminal cap.
- The space between two successive nodes is called as internode.
- Stem facilitates conduction of water, mineral and food material. It also produces and supports leaves and reproductive structure.

Modified Stems

Underground

- **Tuber:** Irregularly shaped swollen stem. Stores plenty of food. The 'eyes' on potato tuber indicate axillary buds at the nodes *e.g.* potato.
- **Bulb :** Having a discoid stem. Bears large and fleshy scaly leaves which store food *e.g.* onion.
- **Rhizome :** grows horizontally forward under soil surface. Nodes and internodes are very distinct *e.g.* ginger.
- **Corm :** Highly swollen vertical stem. Adventitious roots are borne by basal nodes only *e.g.* Amorphophallus.

Sub-aerial

- **Runner :** have long and thin internodes and branches creep over the surface of soil *e.g.* Oxalis.
- **Offset :** This stem is like a runner but the internodes are thicker and shorter *e.g.* Pistia.
- **Stolon :** Arise from nodes of underground portions of stem and grow in the horizontal direction. After growing for sometime the apical region of the branches comes out of the soil, grows into air and roots arise from lower side of nodes *e.g.* Mentha.
- **Sucker :** A branch moves obliquely or sometimes grows horizontally under the surface of soil and then comes out of the soil growing obliquely upwards *e.g.* Chrysanthemum.

Aerial or Metamorphosed

- **Thorn :** They are stiff, sharp structures which have lost their growing point and become hard *e.g. Duranta.*

- **Stem-tendril :** Thread-like sensitive structures which can coil around a support and help the plant in climbing *e.g.* grape.
- **Phylloclade :** They are flattened green stems of unlimited growth which have taken over the function of photosynthesis *e.g.* Opuntia.
- **Cladode :** They are green stems of limited growth (usually one internode long) which have taken over the function of photosynthesis from the leaves.

LEAF

The leaf is a specialised organ of photosynthesis, transpiration and gaseous exchange.

Parts of Leaf or Phyllopodium

- **Leaf Base:** It is the point of attachment of the leaf petiole to the stem.
- **Stipules :** Stipules are a pair of lateral outgrowths arising from the leaf base
- **Petiole :** It is the stalk of the leaf.
- **Lamina :** Lamina is usually flattened green photosynthetic part of leaf.

Venation

Venation is the arrangement of the veins and the veinlets in a leaf.

- **Reticulate Venation:** The veins are arranged in a net like manner, *e.g.*, dicots.
- **Parallel Venation:** Here the veins are arranged parallel-to each other, *e.g.*, monocots.

Types of the Leaves

- **Simple leaf :** In simple leaf lamina is not divided completely into distinct leaflets.
- **Compound leaf :** In compound leaf, incision of leaf blade goes down to the rachis so the leaf is broken up into number of segments called **leaflets**. The compound leaves are of two types- pinnate and palmate.
  - **Pinnate :** The segmentation of the leaf-blade is towards the mid-rib, so that the leaflets are borne laterally by midrib or rachis.
  - **Palmate :** In palmate compound leaves the segmentation of the leaf extends from the apical margin to the petiole so that the leaflets are articulated to the tip of the petiole like fingers on a palm.

Phyllotaxy

The arrangement of mature leaves on the stem or its branches is called phyllotaxy.
It is of three types :

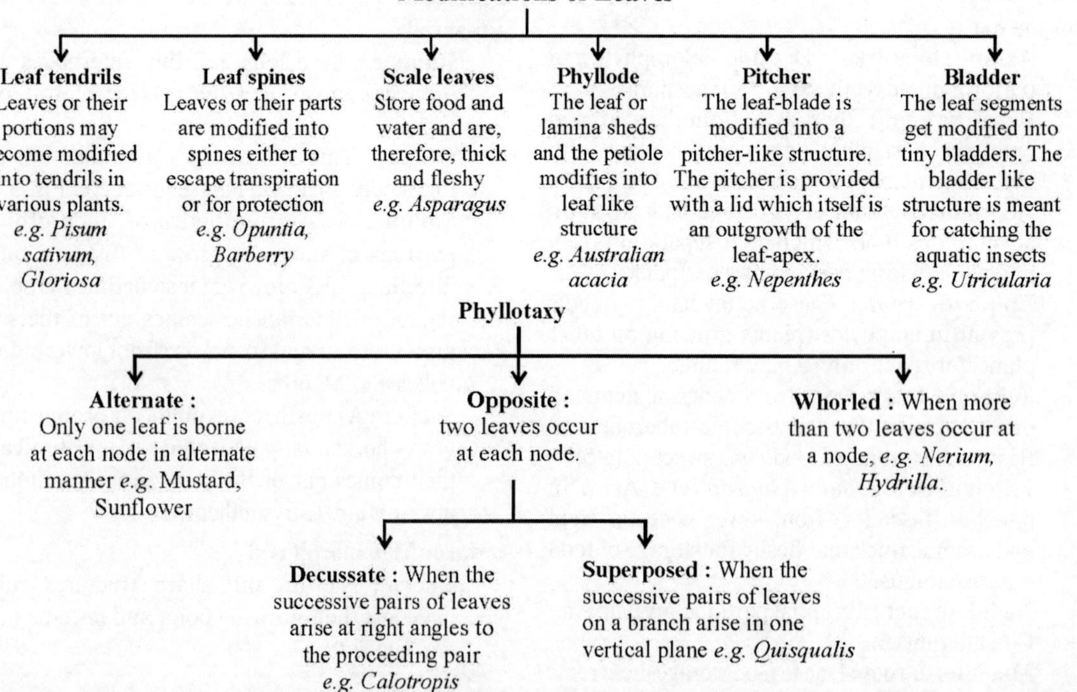

INFLORESCENCE

- An inflorescence is the mode of arrangement of flowers on peduncle or mother axis.

Types of Inflorescence

I. Racemose (Indefinite)
(Main axis grows indefinitely bearing flowers in Acropetal order)

II. Cymose (Definite)
(Main axis terminates into flower & flowers arise in Basipetal order)

RACEMOSE

Main axis elongated

- **Raceme :** Pedicellate (Stalked) flowers occur in an acropetal order on the mother axis. *e.g.*, larkspur.
- **Spike :** Sessile flowers occur in an acropetal order on the elongated mother axis. *e.g.*, bottle brush.
- **Catkin :** Unisexual flowers occur very close to one another on a short axis *e.g.*, Mulberry. The whole inflorescence falls down as a single unit.
- **Spadix :** Small unisexual flowers occur in different groups on a thick axis. All the flowers are enclosed in a large showy bract called spathe e.g., Banana, arum lily.
- **Spikelet :** Unit of compound inflorescence of grasses (members of family Poaceae)

Main axis shortened

- **Corymb :** Pedicels of lower flowers more elongated. All flowers are at the same level *e.g. Iberis*.
- **Umbel :** Axis reduced, pedicels are nearly equal in length *e.g. Coriandrum*.

Main axis reduced or flattened

- **Capitate :** Sessile flowers on suppressed axis forming globose structure *e.g. Mimosa, Acacia*
- **Capitulum or Head :** Sessile flowers on a flattened disc or receptacle *e.g.* Compositae.

CYMOSE

Typical cymes

- **Monochasial or uniparous cyme**
  In this form the main axis ends in a flower and produces only one lateral branch at a time.
  (i) **Helicoid cyme :** *Eg. Heliotropium, Bigonia, Junius.*
  (ii) **Scorpioid cyme :** *Eg. Drosera, Fressia, Hyoscymus.*
- **Dichasial or Biparous cyme :** Eg., *Ixora, Tectona grandis* (teak), *Jasminium, Caryophyllaceae* (Dianthus)
- **Polychasial or multiparous cyme :** Eg., *Calotropis, Hamelia patens.*

Special cymes

- **Cyathium :** Single female flower is surrounded by many male flowers. It has a cup shaped involucre *e.g., Euphorbia, Phyllanthus.*
- **Verticilaster :** Whorls of flowers around the axis at nodes in condensed cymes *e.g., Labiatae* family.
- **Hypanthodium :** Receptacle is fleshy and cup-shaped. Flowers are sessile and unisexual. Female flower are found on the lower side and the male flowers near the aperture *e.g. Ficus* spp.
- **Scapigerous :** In *Allium cepa* (Onion) the leafless flowering axis is known as **scape.** It bears cluster of flowers that form a head which is covered by one or more bracts called **spaths** *e.g.*, Onion, garlic.

FLOWER

A flower is a reproductive shoot which bears sex organs and performs following functions: (i) production of pollen and ovules. (ii) pollination (iii) fertilization (iv) development of fruit, and seeds and their dispersal.

Sex-distribution

1. **Bisexual, perfect or monoclinous:** A flower with both sexual organs, *i.e.*, stamens and pistils.
2. **Unisexual, Imperfect or diclinous :**
   (a) **Monoecious :** Both male and female flowers are present in the same plant.
   (b) **Dioecious :** Male and female flowers are present in different plants.
   (c) **Polygamous :** possess more than two types of flowers *e.g.* Mango, cashewnut, trees possess three types of flowers–bisexual, staminate and neuter.

Arrangement of Floral Parts

- **Acyclic:** Floral organs are arranged in spirals, *e.g., Magnolia.*
- **Cyclic:** Floral organs are arranged in whorls, *e.g., Solanum.*
- **Hemicyclic:** Floral organs are partly in spiral and partly in whorls, *e.g., Anona.*
- **Bracts :** It is a leaf like structure at the base of pedicel in whose axil a flower arises. The flower is described as **bracteate** when the bract is distinct. It is called **ebracteate** when the bract is missing.

Symmetry of Flower

- **Actinomorphic or regular flower :** Flower are radially symmetrical i.e., divisible into two equal halves by any radial plane. *e.g.* Malvaceae.
- **Zygomorphic or irregular flower :** Flower are bilaterally symmetrical i.e., divisible into two equal halves by a single verticle plane. *e.g.* Papilionaceae, Compositeae, Caesalpineae, Poaceae.

Asymmetic flower : *e.g.* Orchid.

Position of the Flower

- **Hypogynous :** When the calyx, corolla and androecium arise from below the ovary, the flower is described as hypogynous. The ovary thus becomes **superior**, *e.g.*, cruciferae, liliaceae.
- **Perigynous :** When the sepals, petals and stamens appear to be arising from middle of the ovary, the flower is described as perigynous. In these flowers, the ovary is **semi-inferior**, *e.g., Saxifraga.*
- **Epigynous :** When the sepals, petals and stamens appear to be arising from the top of the ovary, the flower is described as epigynous. In these flowers, the ovary is **inferior**, *e.g.*, myrtaceae, (*Callistemon*), umbelliferae (*Coriandrum*), compositae, cucurbitaceae, *etc.*
- **Perianth:** Collective terms when petals and sepals are not distinguished.
- **Tepal:** A segment of perianth.
- **Polyphyllous:** In such flowers tepals are free.
- **Gamophyllous:** In such flowers tepals are fused.

Calyx

The outermost non-essential but protective whorl of the flower is the calyx. It is represented by the sepals.

- **Polysepalous :** Sepals are free.
- **Gamosepalous :** Sepals are wholly or partially united.

Corolla – Aestivation

- **Valvate :** Arrangement without overlapping *e.g.*, Custard apple, *Acacia.*
- **Twisted :** All are in or out, *e.g.*, china-rose.
- **Imbricate :** One sepal/petal completely external, one is completely internal and rest overlap one another *e.g., Cassia.*
- **Quincuncial :** A form of imbricate where two petals or sepals are external, two are internal and one partly internal and partly external *e.g.*, Guava.
- **Vexillary** or **Descending Imbricate :** Consists of 5 petals. Odd posterior petal is largest and outermost. It overlaps a pair of smaller lateral petals, which in turn overlap a boat shaped keel *e.g.*, Pea.

Shape of Corolla

- **Cruciform :** Four petals arranged in th form of cross(X). Each petal has a claw and limb *Eg.* Crucifereae.
- **Tubular :** *Eg.* Sunflower
- **Papilionaceous:** *Eg.* Pulses.
- **Bilabiate :** *Eg. Ocimum*

5. **Ligulate :** *Eg.* Ray floret of sunflower.

Androecium

- **Monadelphous :** Stamens fused to form a single bundle (only filaments fused and anthers free) *e.g.*, China-rose.
- **Diadelphous,** When the filaments are united in two bundles and the anthers remain free, *e.g., Pea.*
- **Polydelphous:** When the filaments are united into more than two bundles but anthers are free, *e.g., Citrus.*
- **Syngenesious:** Anthers are united to form one bundle, *e.g., Helianthus.*
- **Synandrous :** Total stamens are united *e.g.* Cucurbitaceae
- **Didynamous:** Four stamens in two pairs, a pair of shorter and a pair of long stamens, *e.g., Oscimum.*
- **Tetradynamous**: Six stamens–four long and two short, *e.g. Brassica.*
- **Epipetalous :** (fusion of stamens with petals) *e.g.* malvaceae, solanaceae, compositeae.

9. **Versatile anther :** *Eg.* Poaceae

Gynoecium

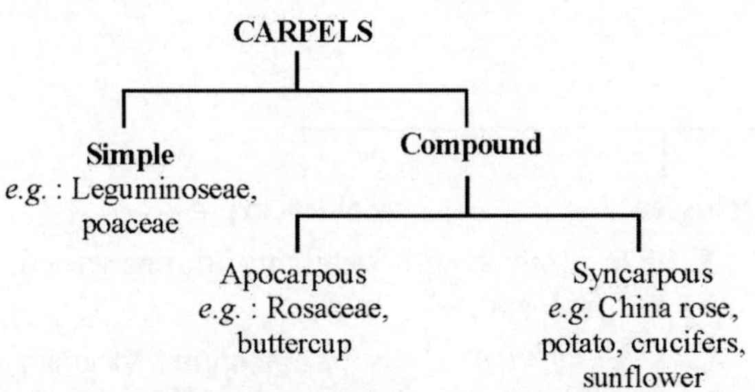

Chambers of Ovary

- **Unilocular :** Eg. Poaceae, Compositeae, Papilionaceae.
- **Bilocular :** Eg. Cruciferеae, Solanaceae.
- **Trilocular :** Eg. Liliaceae, Cucurbitaceae.
- **Pentalocular :** Eg. Malvaceae.

Note : Gynobasic style : Eg. *Salvia, Ocimum* (Labiatae).

Placentation

- **Marginal :** One chambered ovary has placenta forms a ridge on the ventral suture of the ovary and ovules are borne,*e.g.* Pea.
- **Parietal :** One chambered ovary having placentae on the inner wall *e.g.* Papaya, mustard.
- **Axile :** Several chambered ovary having placenta on the central axis ovules are attached to it in a multilocular ovary *e.g.* China-rose, *Petunia.*
- **Free central :** The ovary is one-chambered. The placenta arises from the base of the ovary in the form of a swollen central axis. The ovules are borne all over the surface of the placenta, *e.g,* Primrose
- **Basal :** The ovary is one chambered. The placenta is very small and arises directly on the thalamus. Only one ovule is borne at the base of the ovary, *e.g.* Compositeae, poaceae.

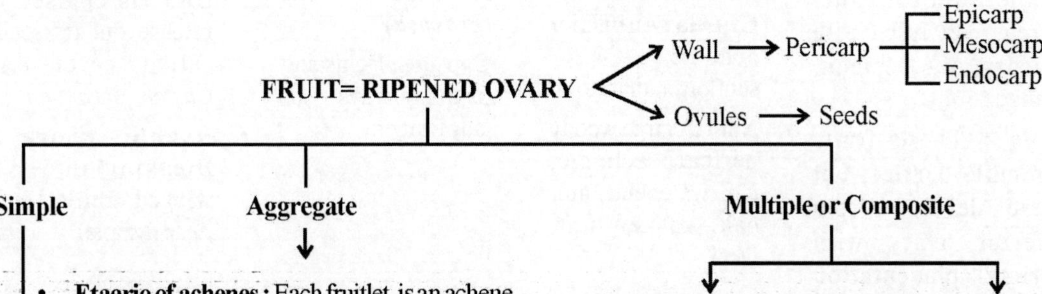

- **Etaerio of achenes :** Each fruitlet is an achene. The thalamus is fleshy and becomes red on maturation and is the edible part in Strawberry, *e.g. Clematis, Ranunculus,* Lotus
- **Etaerio of follicles :** Many follicles (fruitlets) make an etaerio which are arranged on the enlarged thalamus in a bunch, *e.g. Calotropis*
- **Etaerio of drupes :** Many small drupes, developed from different carpels are arranged collectively (in groups) on the fleshy thalamus, *e.g. Rubus*
- **Etaerio of berries :** It is an aggregate fruit consist of small berries e.g., in custard apple, many berries are developed on fleshy thalamus, *e.g.* custard apple, *Artabotrys.*

- **Sorosis :** These fruits develop from spike, spadix or catkin inflorescence. In *Ananas sativus* (Pine apple) too, rachis, bracts and perianth are the edible parts, *e.g.* Mulberry (Morus), Pineapple, Jack fruit (kathal).
- **Syconus :** It develops from hypanthodium infloresence. Receptacle becomes fleshy and many achenes develop from the pistillate flowers. Develops from Hypan-thodium, *e.g.* Banyan

Flesh Simple Fruit
(Pericarp fleshy)

- **Berry :** Develop from mono or multicarpellary, syncarpous, superior ovary. Epicarp in these fruits makes the ring of fruit. Mesocarp is fleshy and endocarp is thin like a membrane, *e.g.* tomato, guava.
- **Hesperidium :** Develop from multicarpellary, syncarpous, superior ovary having axile placentation. Mesocarp is in the form of white, fibrous part fused with epicarp, *e.g.* lemon, orange
- **Pepo :** These fruits resemble berries, but these develop from inferior ovary with parietal placentation Epicarp makes a hard rind, *e.g.* cucumber, watermelon
- **Pome :** It is a false fruit as it, develops from the fleshy thalamus. True fruit remains inside the swollen thalamus, *e.g.* apple, pear.

Dry Simple Fruit
(Pericarp dry)

Indehiscent

- **Achene :** One seeded developed from superior ovary, fruit wall and seed coat free, *e.g.*, Hogweed (*Boerhaavia*), *Mirabilis jalapa*
- **Caryopsis :** A superior acheme, fruit wall and seed coat fused *e.g.*, rice, wheat, maize
- **Cypsela :** An inferior acheme, *e.g.*, sunflower, marigold
- **Nut :** A large woody pericarp acheme, *e.g.*, water chestnut, oak, cashew-nut, litchi

Dehiscent

- **Follicle :** Dehiscing by ventral suture, *e.g.*, *Caloptropis.*
- **Legume or Pod :** Dehiscing by both the sutures. *e.g.*, *pea.*
- **Siliqua :** Dehiscing from base to apex, e.g. mustard to expose replum with attached seeds (*Brassica*)
- **Capsule :** Dehiscing in many ways *e.g.*, lady fiinger.

Schizocarpic

Dehiscing into one seeded parts :

- **Lomentum :** A legnum breaking into one seeded parts *e.g.*, *Acacia, Mimosa.*
- **Cremocarp :** Capsule breaking into two mericarp *e.g.* *Coriander*, Fennel.
- **Regma :** Single seeded fruitlets dehisce to release out its seeds called cocci *e.g.* *Castor, Geranium*
- **Double samra :** Breaking into two winged fruitlets *e.g.* *Acer* (Maple)

THE SEED

Seed is the ripended ovule which contains an integumented embryo or miniature plant with adequate reserve food for future development of the embryo.

Dicotyledonous Seed

The outer or the only seed coat (if one is present) is called **testa**, while the inner one is named as **tegmen**. The surface of the seed possesses a fine pore at one end is called **micropyle**.

Just behind the micropyle is present a dark spot called **hilum**, which is a scar left on the seed coat when it is detached from the fruit wall.

Monocotyledonous Seed

- In maize grain the seed coat is fused with the **pericarp**.
- The endosperm has one to three layered peripheral proteins layer called aleurone layers which separates the embryo with endosperm. The cotyledon is also called **scutellum** in cereals. It is laterally attached to the embryonal axis. The lower end of the axis is called the **radicle** which has a protective sheath termed **coleorhiza**. The upper end of the axis is called the **plumule** which is covered by **coleoptile**.

There are two types of seeds *i.e.*, albuminous and exalbuminous.

- **Albuminous (endospermic) seeds :** In these seeds food is stored in the endosperm, *e.g.*, corn, wheat, caster, onions, *etc.*
- **Exalbuminous (non-endospermic) seeds :** They usually store reserve food material in cotyledons *e.g.*, bean, gram and pea.

DESCRIPTION OF SOME IMPORTANT FAMILIES

FABACEAE

- It is commonly called as legume family. It is 4th largest (after compositae, orchidaceae and gramineae) and 2nd most valuable (after gramineae) family of plant kingdom that provides maximum timber.

Habit – Annual or biennial, herb, shrub or tree.
Root – Tap root system.
Stem – Erect or creeping, solid or weak.
Leaf – Alternate or whorled, stipulate, compound, reticulate venation.

Floral Characters

Inflorescence : Racemose
Flower : Bisexual, zygomorphic
Calyx : Sepals five, gamosepalous; imbricate aestivation.
Corolla : Petals five, polypetalous, papilionaceous, consisting of a posterior standard, two lateral wings, two anterior ones forming a keel (enclosing stamens and pistil), vexillary aestivation.
Androecium : Ten, diadelphous, anther dithecous.
Gynoecium : Ovary superior, mono carpellary, unilocular with many ovules, style single.
Fruit : Legume; seed: one to many, non-endospermic
Floral formula : % ⚥ $K_{(5)}\ C_{1+2+(2)}\ A_{(9)+1}\ \underline{G}_1$

Floral Diagram of Pisum sativum :

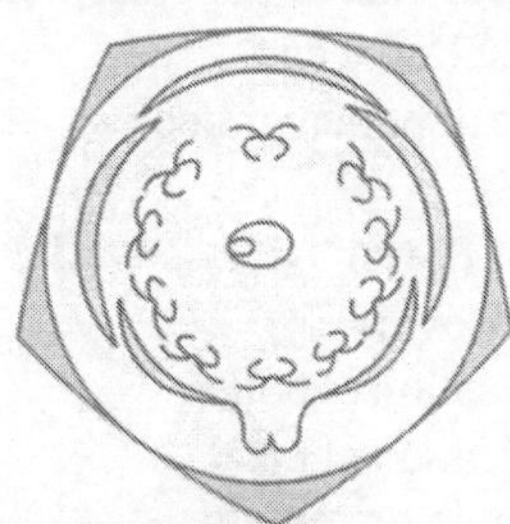

Economic Importance

- Various kinds of pulses are obtained from the plants of this family *e.g.*, garden pea—edible pea *(Pisum sativum)*, chick pea or gram *(Cicer arietinum)*, Lobia *(Vigna inensis)*, Arhar *(Cajanus cajan)* etc.
- Ground nut *(Arachis hypogea)*—Its embryo is used in preparation of vegetable oil. Soyabean *(Glycine max)*—It is rich in protein.
- *Sesbania, Trifolium* is used as fodder.
- Roots of *Glycyrrhiza glabra* (liquorice = mulaithi) are used in sore throat.
- Roots of plants of this family are nodulated that contain nitrogen fixing bacteria *Rhizobium*.

SOLANACEAE

- It is a large family commonly called as 'potato family'.

Vegetative Characters :

- Plants mostly, herbs, shrubs and small trees.
  Stem : Herbaceous rarely woody, aerial; erect, cylindrical, branched, solid or hollow, hairy or glabrous, underground stem in potato (*Solanum tuberosum*).
  Leaves : Alternate, simple, rarely pinnately compound, exstipulate; venation reticulate.

Floral Characters :

- **Inflorescence** – Axillary, solitary or cymose in Datura.
- **Flower** – Ebracteate, actinomorphic, hermaphrodite, pentamerous and hypogynous.
- **Calyx** – Sepals 5, gamosepalous, valvate, persistent.
- **Corolla** – Petals 5, gamopetalous, valvate or twisted.
- **Androecium** – Stamens 5, epipetalous anthers.
- **Gynoecium** – Bicarpellary, syncarpous, ovary superior, bilocular, many ovules in each locule.
- **Fruit** – Generally berry, capsule in Datura.
- **Floral Formula** – $E_{Br.}$ ⊕ ⚥ $K_{(5)}\ \overbrace{C_{(5)}A_{(5)}}\ \underline{G}_{(2)}$
  Many plants belonging to this family are source of food (tomato, brinjal, potato), spice (chilli), medicine (belladonna, *ashwagandha*), fumigatory (tobacco) and of ornamental values such as *Petunia*.

Floral Diagram of *Solanum nigrum* (Makai) :

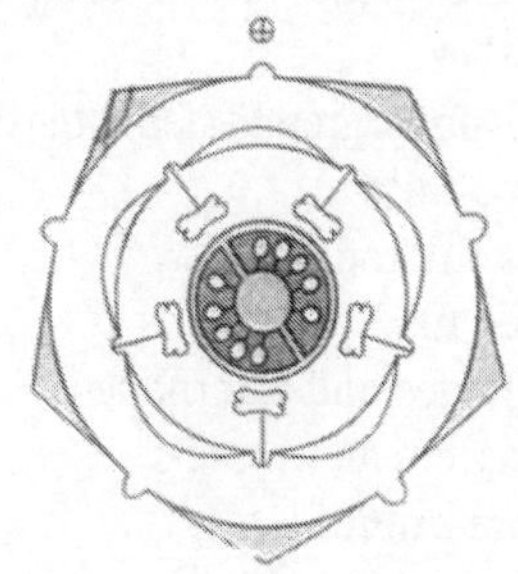

LILIACEAE

It is commonly called as 'Lily family'.

Vegetative Characters:

Habit—Usually perennial herb as *Polygonatum,* sometimes small shrubs as *Ruscus.*
Root—Adventitious, fibrous.
Stem—Erect or climbing as *Smilax,* branched or unbranched, herbaceous.
Leaf— Exstipulate, alternate or whorled, parallel venation.

Floral Characters:

Inflorescence—Raceme, umbel or sometimes solitary.
Flower—Pedicellate, bracteate, hermaphrodite, actinomorphic, trimerous, hypogynous.
Perianth— Tepal six (3 + 3) perianth in two whorls, imbricate aestivation,. **Androecium**— six (3 + 3) stamens in two whorls. **Gynoecium**—Tricarpellary, syncarpous, superior ovary, axile placentation, trilocular, two or more ovules in each locule.
Fruit—Capsule or rarely berry.

Floral formula — Br. ⊕ ⚥ $P_{3+3}\ A_{3+3}\ \underline{G}_{(3)}$

Economic Importance

Allium cepa, A. sativum and Asparagus racemosus are used as vegetables.
Leaves of *Allium cepa* (onion) are used in dental and ear troubles. The leaves of *Allium sativum* (lahasun) are used in cough and fever.
Colchicine, an alkaloid obtained from *Colchicum autumnale* is used to induce polyploidy in plants.

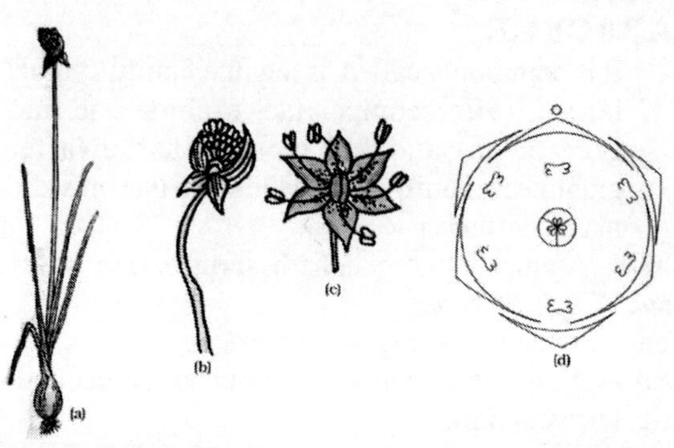

Exercise

1. Monocot plants are characterised by the presence of
   (a) tap roots (b) fibrous roots
   (c) annulated roots (d) stilt roots
2. Prop roots of Banyan tree are meant for
   (a) respiration
   (b) absorption of water in soil
   (c) retention of water in soil
   (d) providing support to big tree
3. Root is distinguishable from stem in
   (a) having a root cap
   (b) having root hairs
   (c) absence of nodes and internodes
   (d) All of these
4. Ginger is an underground stem. It is distinguished from root because
   (a) it lacks chlorophyll
   (b) it stores food
   (c) it has nodes and internodes
   (d) it has xylem and vessels
5. Rachis is present in
   (a) pinnate compound leaf
   (b) palmate compound leaf
   (c) both correct
   (d) both wrong
6. Parallel venation occurs in
   (a) monocots (b) dicots
   (c) all angiosperms (d) ferns
7. The most important function of inflorescence is to help in
   (a) forming large number of fruits
   (b) attracting insects for cross-pollination
   (c) dispersal of seeds
   (d) release of pollen grains
8. Thalamus is
   (a) base of flower
   (b) base of ovary
   (c) modification of pollen
   (d) modification of petal
9. Floral formula represents
   (a) position of flower
   (b) symmetry of a flower
   (c) functions of a flower
   (d) diagrammatic notation of floral characters

10. A plant with both male and female flowers borne over it is
(a) Monoecious (b) Dioecious
(c) Unisexual (d) Bisexual

11. A flower zygomorphic when
(a) Any transverse section divides it into two equal halves
(b) Only one transverse section divides it into two equal halves
(c) Every vertical section passing through its centre divides it into two equal halves
(d) Only one vertical section passing through its centre divides it into two equal halves

12. Which one of the following is an endospermic seed ?
(a) Pea (b) Bean
(c) Gram (d) Castor

13. Fruit of Groundnut is
(a) legume (b) caryopsis
(c) berry (d) nut

14. Floral formula of tomato/tobacco is
(a) $\oplus ⚥ K_{4-5} A_{10} \underline{G}_{(2)}$
(b) $\oplus ⚥ K_{2+2} C_4 A_{2+4} \underline{G}_1$
(c) $\oplus ⚥ P_2 A_3 \underline{G}_1$
(d) $\oplus ⚥ K_{(5)} \widehat{C_{(5)} A_5} \underline{G}_{(2)}$

15. Vivipary is characteristic of
(a) mesophytes (b) xerophytes
(c) hygrophytes (d) halophytes

16. Fruit of *Mangifera indica* is
(a) Berry (b) Drupe
(c) Capsule (d) Siliqua

17. Botanical name of Cauliflower is
(a) *Brassica oleracea var. capitata*
(b) *Brassica campesteris*
(c) *Brassica oleracea var. botrytis*
(d) *Brassica oleracea var. gemmifera*

18. Pulses are obtained from
(a) fabaceae (b) asteraceae
(c) poaceae (d) solanaceae

19. A perennial plant differs from biennial in
(a) having underground perennating structure
(b) having asexual reproductive structures
(c) being tree species
(d) not dying after seasonal production of flowers

20. Plant having column of vascular tissues bearing fruits and having a tap root system is
(a) monocot (b) dicot
(c) gymnosperm (d) succulent

21. Cymose is
(a) thalamus (b) fruit
(c) inflorescence (d) ovary

22. Diadelphous stamens are found in family
(a) Malvaceae (b) Solanaceae
(c) Fabaceae (d) Liliaceae

23. Which one of the following is a true fruit?
(a) Apple (b) Pear
(c) Cashew nut (d) Coconut

24. Which part of the coconut produces coir?
(a) Seed coat (b) Mesocarp
(c) Epicarp (d) Pericarp

25. Most plants are green in colour because
(a) the atmosphere filters out all the colours of the visible light spectrum except green
(b) green light is the most effective wavelength region of the visible spectrum in sunlight for photosynthesis
(c) chlorophyll is least effective in absorbing green light
(d) green light allows maximum photosynthesis

26. Floral features are chiefly used in Angiosperms identification because
(a) flowers are of various colours
(b) flowers can be safely pressed
(c) reproductive parts are more stable and conservative than vegetative parts
(d) flowers are nice to work with

27. What is eye of potato?
(a) Axillary bud (b) Accessory bud
(c) Adventitious bud (d) Apical bud

28. In a cereal grain the single cotyledon of embryo is represented by
(a) scutellum (b) prophyll
(c) coleoptile (d) coleorhiza

29. The technical term used for the androecium in a flower of China rose (*Hibiscus rosa sinensis*) is:
(a) monadelphous (b) diadelphous
(c) polyandrous (d) polyadelphous

30. Sweet potato is homologous to
(a) potato (b) *Colocasia*
(c) ginger (d) turnip

31. Racemose is a type of –
(a) Flower (b) Inflorescence
(c) Stem (d) All

32. Function of stem is –
(a) bear leaves and branches
(b) conduction of water & minerals
(c) conduction and storage of food
(d) All

33. Which is not a stem modification ?
(a) Rhizome of ginger (b) Corm of *Colocasia*
(c) Pitcher of *Nepenthes* (d) Tuber of potato
34. Main function of leaf is –
(a) manufacture of food (b) exchange of gases
(c) both 1 and 2 (d) None of the above
35. Arrangement of leaves on a stem branch is –
(a) venation (b) vernation
(c) infloresecence (d) phyllotaxy
36. A modification of leaf is –
(a) Tendrils (b) Phylloclade
(c) Cladode (d) Corm
37. The fourth whorl of flower is of
(a) petals (b) stamens
(c) carpels (d) sepals
38. Capsule is the fruit of –
(a) Cruciferae (b) Malvaceae
(c) Liliaceae (d) Solanaceae
39. Fruit of brinjal –
(a) Berry (b) Hesperidium
(c) Drupe (d) Pome
40. A true fruit develops from –
(a) ovary (b) thalamus
(c) petals (d) receptacle
41. Pneumatophores are found in –
(a) the vegetation which is found in marshy and saline lake
(b) the vegetation which found in saline soil
(c) xerophytes
(d) epiphytes
42. Stilt roots occur in –
(a) groundnut (b) rice
(c) sugar cane (d) wheat
43. Major source of protein is –
(a) *Oryza sativa*
(b) *Cicer arietinum*
(c) *Beta vulgaris*
(d) *Rhizobium leguminosarum*
44. Which of the following genera is characterised by the production of geocarpic fruits ?
(a) *Cucurbita* (b) *Pisum*
(c) *Glycine* (d) *Arachis*
45. An edible inflorescence is –
(a) *Brassica rapa* (b) Brassica juncea
(c) *Raphanus sativus* (d) *Brassica oleracea*
46. Which is correct pair for edible part ?
(a) Tomato - Thalamus (b) Maize - Cotyledons
(c) Guava - Mesocarp (d) Date palm - Pericarp
47. Which is insectivorous plant?
(a) *Balanophora* (b) *Orobanche*
(c) *Rafflesia* (d) *Nepenthes*
48. Water logged soil is
(a) physically wet (b) physiologically dry
(c) physiologically wet (d) physically dry
49. Buds of garlic have
(a) greatly reduced stems
(b) no stem
(c) extensive root system
(d) no leaves
50. Imparipinnate leaf is the one where
(a) leaflets are borne in pairs
(b) leaflets are small
(c) leaflets are large
(d) rachis is terminated by an odd leaflet
51. The edible part in Gram is :
(a) mesocarp (b) endosperm
(c) cotyledons (d) aril
52. Flower is complete when it has
(a) calyx, corolla, androecium & gynaecium
(b) calyx & corolla
(c) androecium & gynaecium
(d) corolla, androecium & gynaecium
53. Sucking roots are present in
(a) *Tubisoire* (b) *Betel*
(c) *Cuscuta* (d) *Orchids*
54. Which of the following does not have stomata ?
(a) Hydrophytes
(b) Mesophytes
(c) Xerophytes
(d) Submerged hydrophytes
55. Leaves arise from which part of plant?
(a) Rhizome (b) Stem
(c) Internode (d) Node
56. Which of the following is not an insectivorous plant ?
(a) *Drosera* (b) *Nepenthes*
(c) *Opuntia* (d) *Utricularia*
57. Rhizome which grows vertically upwards are
(a) corms (b) stolon
(c) bulbils (d) root stock
58. A plant which lives on another plant but do not take food or anything from plant is called
(a) endophyte (b) epiphyte
(c) parasite (d) host
59. Wheat is which of the following types of fruit?
(a) Berry (b) Nut
(c) Caryopsis (d) Legume (pod)
60. Botanical name of banana is
(a) *Musa paradisica* (b) *Phaseolus vulgaris*
(c) *Ricinus communis* (d) *Ananas sativus*

Anatomy of Flowering Plants

CONCEPTS

A group of structurally similar or dissimilar cells that perform a common function and have a common origin is called a **tissue**.

- **TISSUES**
  - **Meristematic**
    - **on the basis of plane of division**
      - **Mass Meristem (Young, in all planes)**
      - **Plate Meristem (Anticlinal division in two planes)**
      - **Rib Meristem (Perpendicular / anticlinal divisions in 1 plane, occasional periclinal divisions)**
    - **on the basis of histogen**
      - **Dermatogen (Epidermis)**
      - **Periblem (Hypodermis, cortex, endodermis)**
      - **Plerome (Stele-pericycle, phloem, xylem, cambium and pith)**
    - **on the basis of development**
      - **Primary meristem (shoot and root apex)**
      - **Secondary meristem (interfascicular cambium, cork cambium)**
    - **on the basis of position**
      - **Apical meristem (shoot apex, root apex)**
      - **Lateral meristem (vascular cambium, cork cambium)**
      - **Intercalary meristem (as in grasses)**
  - **Permanent**
    - **Simple**
      - **Parenchyma (chlorenchyma, palisade and spongy parenchyma, aerenchyma, idioblast)**
      - **Collenchyma (angular, lamellar, lacunior)**
      - **Sclerenchyma (fibres, sclereids)**
    - **Complex**
      - **Xylem (tracheids, vessels, wood fibres, wood parenchyma)**
      - **Phloem (sieve cells, sieve tubes, companion cells phloem fibres, phloem parenchyma)**
    - **Secretory**
      - **Lacticiferous tissue (lateral cells, latex vessels)**
      - **Glandular tissue**

MERISTEMATIC TISSUE

Cells which are capable of active cell division are called meristematic cells.

Meristematic cells are simple, immature, and isodiametric (same length, width and height, mostly rounded/ oval/ polygonal or rectangular). They have dense cytoplasm with prominent nuclei.

PERMANENT TISSUES

- Permanent tissues are formed as a result of division and differentiation in meristematic tissues.

Simple permanent tissue

- These tissues are a collection of similar cells that perform a common function. These tissues are of 3 types:

Parenchyma

- Parenchyma cells are living, thin-walled, isodiametric (all sides equal) and appear oval or spherical in shape. Sometimes, they become polygonal as a result of pressure of the neighbouring cells.
- The cell wall is thin and made up of cellulose. It helps in storage of food, conduction of substances, provides turgidity to softer parts of plants.
- **Chlorenchyma** (parenchymatous cells that develop chloroplasts) participates in photosynthesis.

  Palisade parenchyma: It is a type of chlorenchyma in which cells are elongated.

  Spongy parenchyma: It is also a type of chlorenchyma with cells of irregular shapes and sizes.

  Aerenchyma : A parenchyma with well developed intercellular spaces which forms a connected system throughout the entire plant.

Collenchyma

- It is living mechanical tissue, found beneath the epidermis (*i.e.*, hypodermis) of herbaceous dicot stem. The intercellular spaces in this tissue are absent because in intercellular spaces at the corner of cells thickenings of cellulose and pectin develop due to which the cell wall become rigid and thick at corners.
- It provides mechanical support to the organs and due to its peripheral position in stems and resists bending and pulling action of wind.

Sclerenchyma

- These are dead, mechanical and act as skeleton in plants.
- It is hardest plant tissue, madeup of highly thick walled cells with no nucleus and no cytoplasm. This uniform thicknening is made up of mainly lignin and cellulose or both.
- Two types of cells are included in sclerenchyma *i.e.*,

Sclerenchymatous fibres

- Fully developed fibres are long and tapering at the ends.
- They provide great strength and flexibility.
- The fibres are present in hypodermis of monocot stem, in pericycle of many dicots, in secondary wood, in vascular bundle sheath in monocot stems, and hypodermis of many leaves.

Stone cells (Sclereids)

- They are lignified, extremely thick walled so that the lumen of the cells is almost obliterated.
- It provides grittiness to fleshy fruits.
- They form part of seed coat in some members of leguminoseae.

COMPLEX TISSUES

- A complex tissue can be defined as a collection of different types of cells performing a common function.

Xylem (Wood)

- Xylem is made of 4 types of cells *i.e.*,

Tracheids

- The cells are long, tubular with tapering, chisel like ends. The cell wall is lignified.
- The tracheids are dead (without cytoplasm and nucleus) at maturity.
- The main function of tracheids is to conduct water but due to their thick and hard walls they also provide mechanical support.

Tracheae (Xylem vessels)

- Vessels is an elongated tube like structure that is composed of row of cells placed one above the other with their transverse walls absent.
- The walls of vessels are lignified and hard and less thicker than tracheids.

Wood (Xylem) fibres

- They are long, narrow and tapering at both ends.
- They provide only mechanical support.

Wood (Xylem) parenchyma

- These are thin walled (cellulose) living cells.
- They store food material and also involved in short distance lateral transport of sap.

Phloem (Bast)

- Phloem is also a conducting tissue that transports food materials, usually from leaves to other part of the plants.
- It is a complex tissue made up of 4 kinds of cells in angiosperms, these are

Sieve element

- The sieve elements in angiosperms are sieve tubes arranged one above the other in distinct linear rows and have sieve plates (oblique or transverse perforated septa) on their end walls. These are associated with companion cells.

Companion cells

- Companion cell is a long, narrow, thin walled cell that is usually attached to the lateral side of sieve tube.
- They are supposed to help in maintaining pressure gradient in sieve tubes. They are absent in gymnosperms and pteridophytes.

Phloem fibres

- These are sclerenchymatous dead cells that provide mechanical strength. They have thick wall with simple pits.
- These are absent in leaves, roots and primary phloem.

Phloem parenchyma

- These cells are living, thin walled parenchyma cells that store food, latex, tannin, crystals, *etc.*
- They are absent in monocots and in some dicots (*e.g., Ranunculus*).

THE TISSUE SYSTEM

Each system usually consists of only one tissue or an association of tissues which perform a common function. There are three main tissue systems :

The Epidermal Tissue System

Epidermis

- It is the outermost layer of all organs of primary plant body.
- It is made up of elongated, compactly arranged single layer of cells.
- The outer wall of epidermal cells contain a fatty substance called cutin and form **cuticle** on the epidermis. Cuticle prevents the loss of water. It is absent in roots.

Stomata

- Stomata are tiny pores that are found on the epidermis of leaves, stem and some fruits.
- Stomata helps in gaseous exchange at the time of respiration and photosynthesis. Stomata are composed of two bean-shaped epidermal cells, called *guard cells* which encloses stomatal pore. In grasses guard cells are dumb-bell shape.
- Unlike other epidermal cells, the guard cells have chloroplasts and the outer walls of guard cells (away from the stomatal pore) are thin and the inner walls (towards the stomatal pore) are highly thickened.
- The guard cells control the opening and closing of stomata, which in turn, regulate the loss of water from the leaves to the atmosphere through transpiration.

Epidermal Appendages (Trichomas and Hairs)

- In shoot system, they are usually multicellular. They may be soft or stiff, branched or unbranched. Sometimes they are secretory.
- They help in retarding the rate of transpiration.
- Root hairs help in absorption of water and minerals from soil.

Ground (Fundamental) Tissue System

- All tissues except epidermis and vascular bundle constitute the ground tissue.
- It consists of simple tissues such as parenchyma, collenchyma and sclerenchyma.

VASCULAR TISSUE SYSTEM

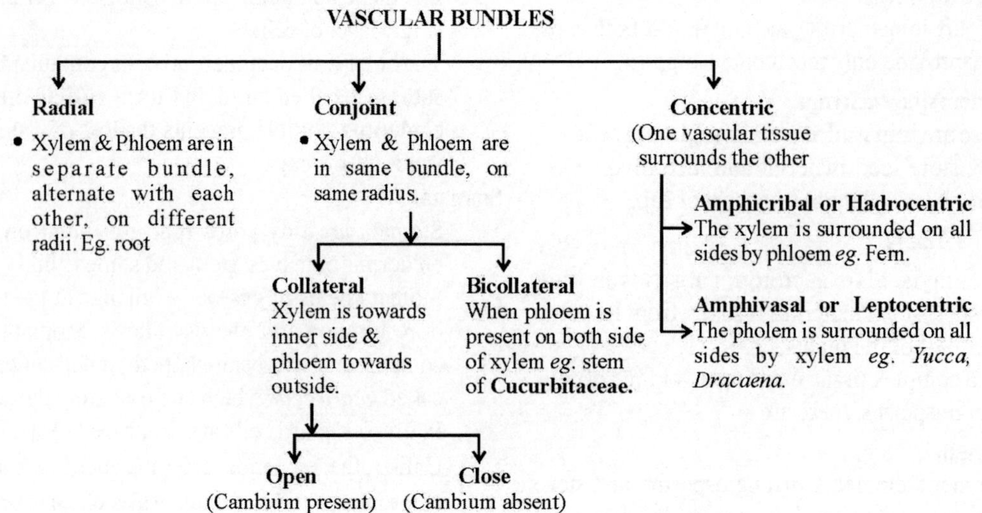

Radial, Concentric V.B. = always close; Bicollateral V.B. = always open; Collateral V.B. = may be open or close

Difference between dicot and monocot roots :

	Characters	Dicot roots	Monocot roots
(1)	Vascular bundles	2-6	More than six
(2)	Pith	Absent or very small	Large
(3)	Pericycle	Forms cambium	No cambium
(4)	Secondary growth	Occurs	Does not occur

Differences between internal structure of dicot and monocot stem :

	Point of Differences	Dicot stem	Monocot stem
(1)	Trichomes	Distributed uniformly	Usually absent
(2)	Hypodermis	Collenchymatous	Sclerenchymatous
(3)	The concentric arrangement of tissues	Present	Absent
(4)	Ground Tissue	Differentiated into – (a) cortex (b) endodermis (c) pericycle (d) pith (e) medullary ray	Undifferentiated, homogenous, parenchymatous
(5)	Stele	Eustele	Atactostele
(6)	Stomata have	kidney shaped guard cell	dumbell shaped guard cell
(7)	Vascular bundles • Arrangement • Type • Bundle cap is present. • Arrangement of primary xylem	 In a ring Conjoint, collateral, open Sclerenchymatous bundle cap replaced by bundle sheath. In chains or rows	 Scattered Conjoint, collateral, close Bundle cap absent, In a Y-shaped manner

• Protoxylem cavities • Shape of V.B. • Size of V.B.	Absent Wedge-shaped Similar size	Present Oval or rounded Different sizes – smaller towards the outside and larger towards the centre
• Phloem parenchyma • Outline of vessel	Present Polygonal	Absent Oval or rounded
(8) Secondary growth	Present	Absent

Differences between dicot and monocot leaf :

	Dicot leaf	Monocot leaf
(1)	Dorsiventral	Isobilateral
(2)	Hypostomatic	Amphistomatic
(3)	Motor cells or bulliform cells are absent.	Bulliform cells are present in upper epidermis.
(4)	Mesophyll is differentiated into palisade and spongy parenchyma.	Mesophyll not differentiated and all cells are like spongy parenchyma.
(5)	Above and below large vascular bundles, patches of parenchymatous or collenchymatous cells are present.	Sclerenchymatous cells patches are present above and below the large vascular bundles.

SECONDARY GROWTH

- The increase in thickness due to the formation of secondary tissue by the **activity of cambium** and **cork cambium**.

Secondary Growth in Dicot Stem In Stelar Region

- **Formation of complete cambium ring :** Fascicular cambium and interfascicular cambium constitute the ring.
- The cells of cambium divide periclinally.
- As a result of division of cambium cell two daughter cells are produced. One of these daughter cells retains meristemetic activity while other differentiate into either secondary xylem or secondary phloem.
- The outwardly cut cambium produces secondary phloem while the inwardly cut produces secondary xylem.
- Cambium ring consists of two type of cells – **fusiform initials** and **ray initials**.
- **Fusiform initials** cut off **vascular elements** while the **ray initials** cut off cells that form **parenchymatous medullary ray**.

Annual Rings

- Wide lumen vessels wood and narrow lumen vessels wood constitute an **annual ring**.
- Wide lumen vessels wood is known as **spring** or **early wood** and produced in spring when growth activities are maximum.
- Narrow lumen vessels wood is known as **autumn wood or winter wood** or **late wood**.

Dendrochronology

- **Determination of age of a tree** by counting annual rings is known as **dendrochronology**.
- **Increment Borer** an instrument is used in age determination.

Ring Porous and Diffuse Porous Wood

- Wood with large xylem vessels, arranged in a ring is known as **ring porous** wood, while wood with uniform distribution of small sized vessels is known as **diffuse porous**.
- **Examples : Ring porous** - *Dalbergia sisso, Cassia fistula.*
- **Diffuse porous** - *Azadirachta indica, Mangifera indica.*

Sapwood (Alburnum) and Heart Wood (Duramen)

Sap wood

- Lies in peripheral region of old stem and creamy in colour.
- Not plugged with resins and tyloses and takes active part in transport.

Heart wood

- Lies in central region of old stem and dark in colour.
- Plugged with resins and tyloses and takes part in transport.

Secondary Growth In Extra Stelar Region

- Due to outward pressure of secondary tissues produced in stelar region, the cortical cells of extrastelar region becomes meristematic and forms a continuous ring of cambium, known as **cork-cambium** or **Phellogen**.
- The cells formed by the cork cambium to its inner face are known as **secondary cortex (Phelloderm)** and the cells formed by the cork cambium to its outerface are known as **cork (Phellem)**.
- Secondary cortex, cork and cork cambium commonly called as **periderm**.
- Cork cells are suberized and dead.

SECONDARY GROWTH IN DICOT ROOTS

There is no cambium in dicot root before secondary growth and secondary growth starts with the formation of **cambium stripes**.

In Stelar Region

- Waxy cambium develops as a result of meristematic activity of parenchyma cells just infront of phloem and behind of the xylem.
- Due to meristematic activity of cambium secondary xylem and secondary phloem form and later cambial ring becomes circular.

In Extra Stelar Region

- Cortex cells act as cork cambium and as a result of its acitivity Periderm forms.

Note :

- Secondary growth in Monocot stem – *e.g.*, *Dracaena, Yucca, Aloe, Agave,* etc.
- Abnormal secondary growth – *e.g.*, *Bignonia, Aristolochia, Amaranthus, Boerhaavia, Mirabilis,* etc.

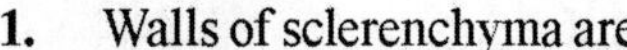

Exercise

1. Walls of sclerenchyma are
   (a) rigid (b) lignified
   (c) pactinised (d) suberised
2. Collenchyma differs from sclerenchyma
   (a) retaining protoplasm at maturity
   (b) having thick walls
   (c) having wide lumen
   (d) being meristematic
3. Sieve tubes are better suited for translocation, because
   (a) possess broader lumen and perforated cross walls
   (b) are broader than long
   (c) possess bordered pits
   (d) possess no end walls
4. Collenchyma differs from parenchyma in having
   (a) living protoplasm
   (b) cellulose walls
   (c) vacuoles
   (d) pectin deposits at corners
5. The cells or tissues of plants which have lost the power of division are called
   (a) permanent tissue
   (b) promeristem tissue
   (c) protoderm tissue
   (d) meristematic tissue
6. Which is absent in plants?
   (a) Cell wall
   (b) Vascular system
   (c) Nervous system
   (d) Epidermal tissue system
7. The xylem fibres are classified into
   (a) protoxylem and metaxylem
   (b) primary and secondary fibres
   (c) fibre tracheids and fibres
   (d) long and short fibres
8. Chloroplast containing parenchyma is called
   (a) chlorenchyma (b) collenchyma
   (c) aerenchyma (d) amylanchyma
9. Cuticle is secreted by
   (a) epidermis (b) endodermis
   (c) both (a) and (b) (d) hypodermis
10. In dicot root
    (a) vascular bundles are scattered and with cambium

(b) vascular bundles are arranged in a ring and have cambium
(c) xylem and phloem radially arranged
(d) xylem is always endarch

11. Well developed pith is found in
(a) monocot stem and dicot root
(b) monocot and dicot stems
(c) dicot stem and dicot root
(d) dicot stem and monocot root

12. Exarch xylem is found in
(a) root (b) stem
(c) leaf (d) rachis

13. Which of the following statements is/are not true?
A. Cork cambium is otherwise called phellogen
B. Cork is otherwise called phellem
C. Secondary cortex is otherwise called periderm
D. Cork cambium, cork and secondary cortex are collectively called phelloderm
(a) C and D (b) A and B
(c) B and C (d) B and D
(e) A and D

14. Intrafasciular cambium is situated in
(a) outside the vascular bundles
(b) in medullary rays
(c) inside the vascular bundles
(d) in between the vascular bundles

15. In roots and stems, secondary growth takes place after the formation of
(a) cambium (b) sclerenchyma
(c) cork (d) bark

16. The vessel elements of angiosperms differ from other elements of xylem in having
(a) simple pits on their radial walls
(b) bordered pits on their lateral walls
(c) simple and bordered pits on their end walls
(d) simple perforation on their end walls

17. Function of storage is performed by
(a) Parenchyma (b) Sclerenchyma
(c) Phloem (d) All of the above

18. Which meristem helps in increasing girth?
(a) Lateral meristem
(b) Intercalary meristem
(c) Primary meristem
(d) Apical meristem

19. Cork is formed from
(a) cork cambium (phellogen)
(b) vascular cambium
(c) phloem
(d) xylem

20. Which is the correct fact about diffuse or ring porous wood?
(a) Ring porous wood, carries more water for short period
(b) Diffuse porous wood carries more water
(c) Ring porous wood carries more water when need is higher
(d) Diffuse porous wood is less specialised but conducts water rapidly throughout

21. Monocot leaves possess
(a) intercalary meristem
(b) lateral meristem
(c) apical meristem
(d) mass meristem

22. Collenchyma occurs in the stem and petioles of
(a) Xerophytes (b) Monocots
(c) Dicot herbs (d) Hydrophytes

23. Pericycle of roots produces
(a) mechanical support
(b) lateral roots
(c) vascular bundles
(d) adventitious buds

24. What is true about a monocot leaf ?
(a) Reticulate venation
(b) Absence of bulliform cells from epidermis
(c) Mesophyll not differentiated into palisade and spongy tissues
(d) Well differentiated mesophyll

25. Vascular cambium produces
(a) primary xylem and primary phloem
(b) secondary xylem and secondary phloem
(c) primary xylem and secondary phloem
(d) secondary xylem and primary phloem

26. Where do the casparian bands occur?
(a) Epidermis (b) Endodermis
(c) Pericycle (d) Phloem

27. An organised and differentiated cellular structure having cytoplasm but no nucleus is
(a) vessels
(b) xylem parenchyma
(c) sieve tubes
(d) tracheids

28. Which is correct about transport or conduction of substances?
(a) Inorganic food moves up through phloem
(b) Organic food moves up through phloem
(c) Inorganic food moves upwardly and downwardly through xylem
(d) Organic food moves upwardly and downwardly through phloem

29. A bicollateral vascular bundle is characterised by
(a) Phloem being sandwitched between xylem
(b) Transverse splitting of vascular bundle
(c) Longitudinal splitting of vascular bundle
(d) Xylem being sandwitched between phloem

30. Bordered pits are found in
(a) sieve cells (b) vessel wall
(c) companion cells (d) sieve tube wall

31. What is not true about sclereids?
(a) These are parenchyma cells with thickened lignified walls
(b) These are elongated and flexible with tapered ends
(c) These are commonly found in the shells of nuts and in the pulp of guava, pear, etc
(d) These are also called the stone cells

32. A leaf primordium grows into the adult leaf lamina by means of
(a) apical meristem
(b) lateral meristem
(c) marginal meristems
(d) at first by apical meristem and later largely by marginal meristems.

33. Which of the following meristems is responsible for extrastellar secondary growth in dicotyledonous stem?
(a) Intrafascicular cambium
(b) Interfascicular cambium
(c) Intercalary meristem
(d) Phellogen

34. Main function of lenticel is
(a) transpiration (b) guttation
(c) gaseous exchange (d) transportation

35. Vessels are found in
(a) all angiosperms and some gymnosperms
(b) most of angiosperms and few gymnosperms
(a) all angiosperms, all gymnosperms and some pteriodophyta
(d) all pteridophyta

36. Which of the following statements is true?
(a) Vessels are multicellular with narrow lumen
(b) Tracheids are multicellular with narrow lumen
(c) Vessels are unicellular with wide lumen
(d) Tracheids are unicellular with wide lumen

37. In a longitudinal section of a root, starting from the tip upward, the four zones occur in the following order:
(a) Root cap, cell division, cell enlargement, cell maturation
(b) Root cap, cell division, cell maturation, cell enlargement
(c) Cell division, cell enlargement, cell maturation, root cap
(d) Cell division, cell maturation, cell enlargement, root cap

38. The most abundant element present in the plants is
(a) Carbon (b) Nitrogen
(c) Manganese (d) Iron

39. For a critical study of secondary growth in plants. Which one of the following pairs is suitable?
(a) Teak and pine
(b) Deodar and fern
(c) Wheat and maiden hair fern
(d) Sugarcane and sunflower.

40. Passage cells are thin walled cells found in
(a) phloem elements that serve as entry points for substance for transport to other plant parts
(b) testa of seeds to enable emergence of growing embryonic axis during seed germination
(c) central region of style through which the pollen tube grows towards the ovary
(d) endodermis of roots facilitating rapid transport of water from cortex to pericycle.

41. In barley stem, vascular bundles are:
(a) closed and scattered
(b) open and in a ring
(c) closed and radial
(d) open and scattered

42. The chief water conducting elements of xylem in gymnosperms are:
(a) vessels (b) fibres
(c) transfusion tissue (d) tracheids

43. Heartwood differs from sapwood in:
(a) presence of rays and fibres
(b) absence of vessels and parenchyma
(c) having dead and non–conducting elements
(d) being susceptible to pests and pathogens

44. The cork cambium, cork and secondary cortex are collectively called:
(a) phelloderm (b) phellogen
(c) periderm (d) phellem

45. Function of companion cells is
(a) providing energy to sieve elements for active transport
(b) providing water to phloem
(c) loading of sucrose into sieve elements by passive transport
(d) loading of sucrose into sieve elements

46. Histogens are components of –
(a) apical meristem (b) intercalary meristem
(c) lateral meristem (d) secondary meristem

47. A tissue is a group of cells which are –
(a) similar in origin, but dissimilar in form and function.
(b) dissimilar in origin, form and function.
(c) dissimilar in origin, but similar in form and function.
(d) similar in origin, form and function.

48. Embryo of a seed is made up of –
(a) meristematic tissue
(b) parenchyma
(c) collenchyma
(d) sclerenchyma

49. The tracheids differ from vessels in having –
(a) thick wall
(b) bordered pit
(c) presence of pitted end wall
(d) spiral thickening

50. In submerged hydrophytes, stomata are found –
(a) On upper surface of leaf
(b) On lower surface of leaf
(c) On both surface of leaf
(d) No where on the plant

51. Apical, intercalary and lateral meristems are differentiated on the basis of-
(a) origin (b) function
(c) position (d) development

52. Aerenchyma is helpful in plants by –
(a) Giving flexibility to plants
(b) Giving mechanical strength to plants
(c) Promoting photosynthesis
(d) Providing buoyancy in hydrophytes

53. Wood is-
(a) primary phloem (b) primary xylem
(c) secondary xylem (d) secondary phloem

54. Largest number of chloroplast is found in -
(a) palisade tissue (b) spongy tissue
(c) transfusion tissue (d) bundle sheath cells

55. From where the commercial jute fibres are obtained-
(a) interxylary fibres (b) xylem fibres
(c) phloem fibres (d) None of these

56. How does autumn wood differ from spring wood ?
(a) Broad vessels & tracheids
(b) Narrow vessels & tracheids
(c) Red colour of xylem
(d) Cambium

57. P-protein belongs to-
(a) xylem parenchyma
(b) trichomes
(c) sieve tube elements
(d) tracheids and vessels

58. Trees at sea do not have annual rings because –
(a) soil is sandy
(b) there is climatic variation
(c) there is no marked climatic variation
(d) there is enough moisture in the atmosphere

59. Best method to determine the age of tree is to
(a) measure its diameter
(b) count number of leaves
(c) count number of annual rings at base stem
(d) number of branches

60. The difference between alburnum and duramen is the presence of :
(a) companion cells (b) living cells
(c) ray cells (d) lignified cells

Structural Organisation in Animals

CONCEPTS

Cells, tissues, organs and organ systems distribute their work in a way that exhibit division of labour *i.e.* they divide their work among each other so as to carry out major activities of the body in an easy and perfect manner.

ANIMAL TISSUES

The structure of the cells vary according to their function. Therefore, the tissues are different and are broadly classified into four types:

Epithelial Tissue

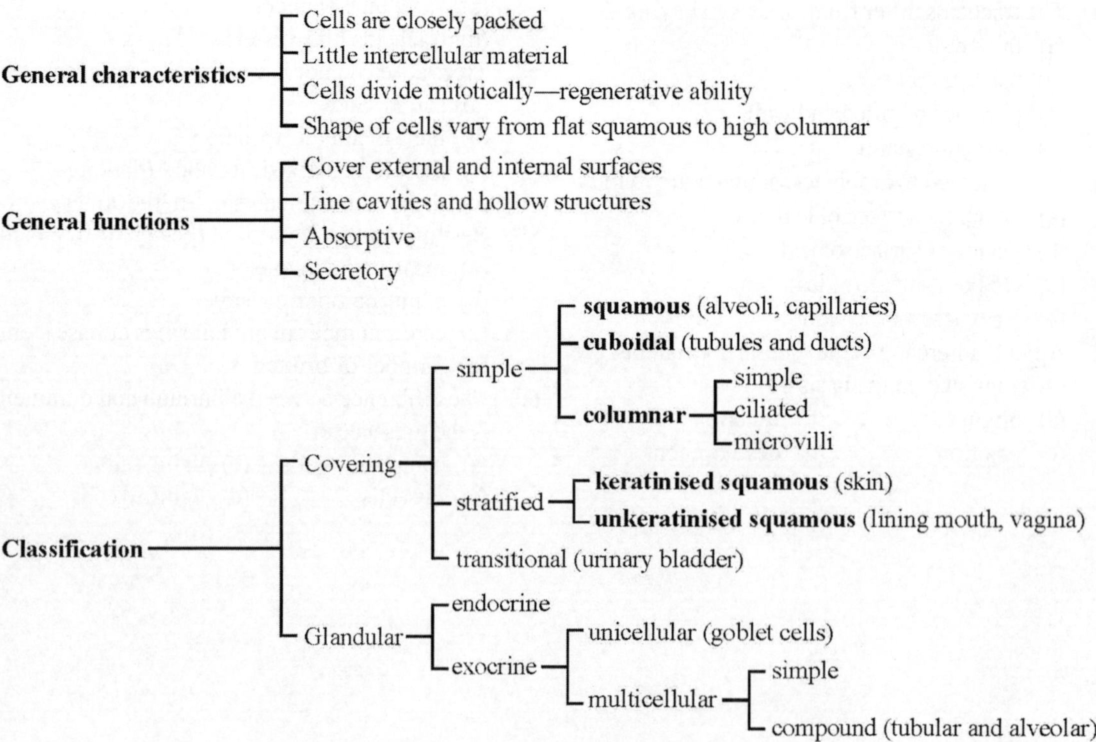

connective Tissue

- General composition
  - Cells : heterogenous; separated by matrix
  - Fibres : collagen, elastic and reticular
  - Matrix : fluid, gel, semi-solid

- **Connective tissue**
  - loose : general connective tissue of body
  - reticular : supports lymphoid tissue
  - dense fibrous : strong, forms tendons, ligaments, etc.
  - elastic : possesses extension and recoil properties, *e.g.* lungs
- **Adipose tissue**
  - composition
    - adipocytes (fat cells)
    - loose connective tissue matrix
  - functions as
    - protective packing material, *e.g.* kidneys
    - heat insulator (skin)
    - energy store
- **Cartilage**
  - composition
    - matrix
    - chondrocytes (cartilage cells)
    - fibres
  - type
    - hyaline : covers articular surfaces, forms framework, *e.g.* larynx elastic, ear.
    - fibrocartilage, *e.g.* intervertebral discs
- **Adipose tissue**
  - composition
    - cells (osteocytes)
    - matrix : solid because it contains bone salts
  - compact : framework for body
- **Haemopoietic tissue**
  - red bone marrow : produces blood cells
  - circulating blood : blood cells suspended in blood plasma

Blood

- Plasma
- Formed Elements of blood
  - Erythrocytes (RBC)
  - Leucocytes (WBC)
    - Granulocytes
      - Eosinophils
      - Basophils
      - Neutrophils
    - Agranulocytes
      - Monocytes
      - Lymphocytes
  - Blood Platelets

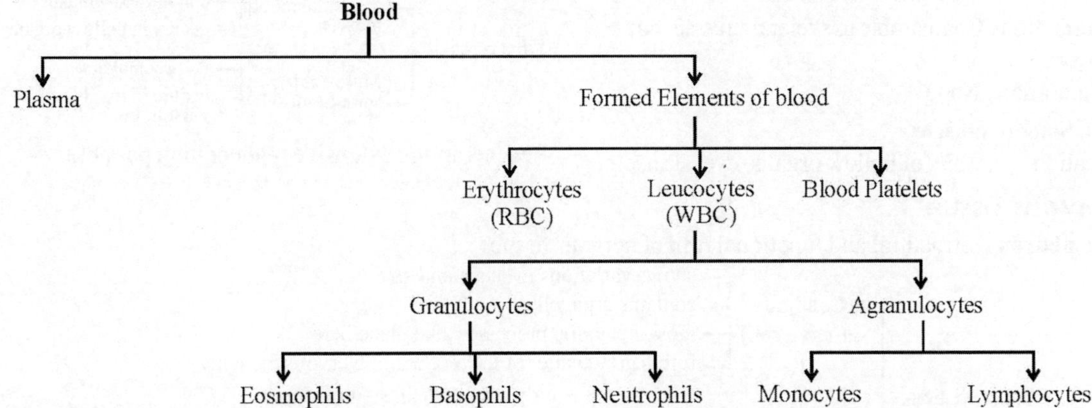

Some important points

- pH of blood - 7.3 to 7.4 (slightly alkaline)
- Proteins constitute about 7% to 8% part of plasma. These mainly include albumins, globulins, prothrombin and fibrinogen.
- RBCs of amphibians are oval, biconvex and nucleated.
- RBCs of mammals are round, biconcave and non-nucleated except camels and Ilamas, in which shape of RBC is oval.
- RBCs of mammals lose their nucleus at the stage of maturation of RBCs.
- Number of RBCs in male -5 to 5.5 millions/mm^3 and in female -4.5 to 5.0 millions/mm^3.

- WBCs are colourless, nucleated and motile cells of blood.
- Number of WBCs about 8000/mm^3 in blood.
- Number of blood platelets = 3 lakhs/mm^3 of blood.
- The mineral necessary for coagulation of blood is calcium.

Muscle Tissue

(i) Skeletal muscle

Location : Skeletal musculature

Functions : Movements of the body

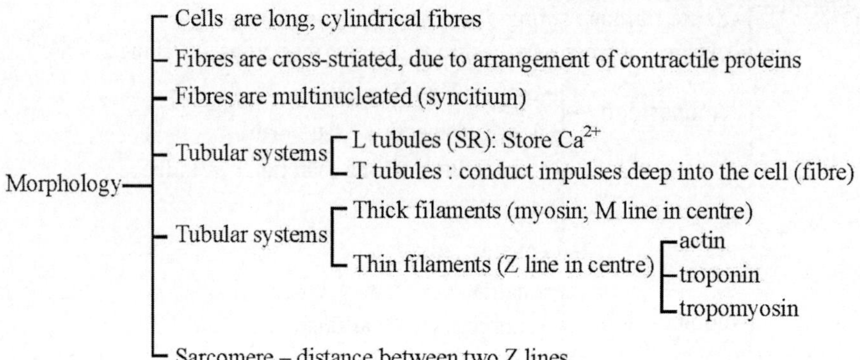

(ii) Cardiac muscle

Location : Cardiac wall

Function : Pumps blood

Control : Involuntary; autonomic nervous system; inherent rhythmicity due to pacemaker cells.

Morphology –
- Cells shorter, branched, join one another
- Intercalated discs
- Single nucleus/cell
- Striated fibres

Contraction : Comparable to skeletal muscle, but slower

Regneration : None

(iii) Smooth muscle

Location : Walls of hollow organs, eye, skin, etc.

Function : Contraction of hollow muscular organs

Control : By pacemaker cells and autonomic nervous sytem – contractions involuntary

Morphology –
- Fibres shorter with single nucleus
- Fibres non-striated
- Contractile proteins = actin and myosin
- No troponin; contains calmodulin
- Ca^{2+} flows into cell from extracellular space

Contraction : Slow

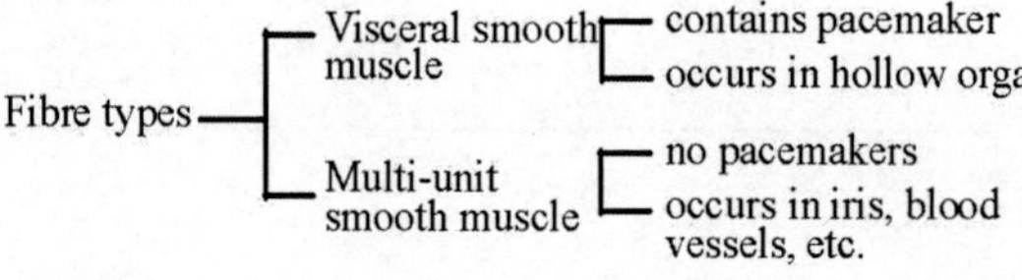

Regeneration : extensive regeneration possible

Nervous Tissue

The neuron – Structural and functional unit of nervous tissue.

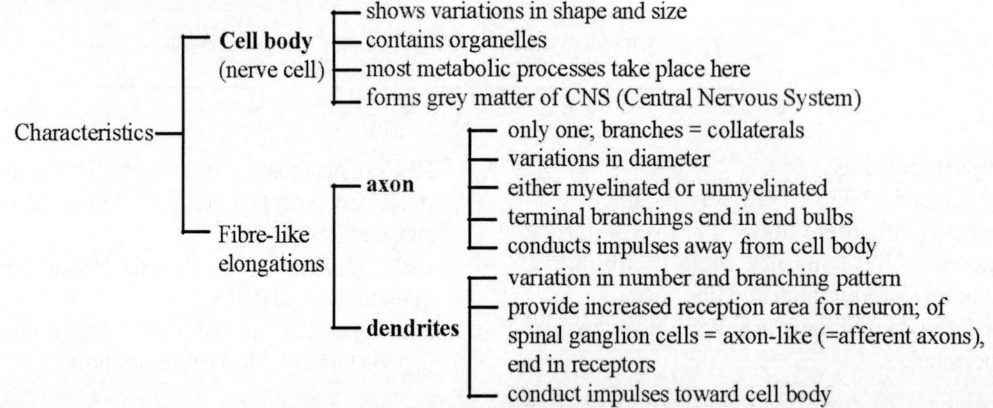

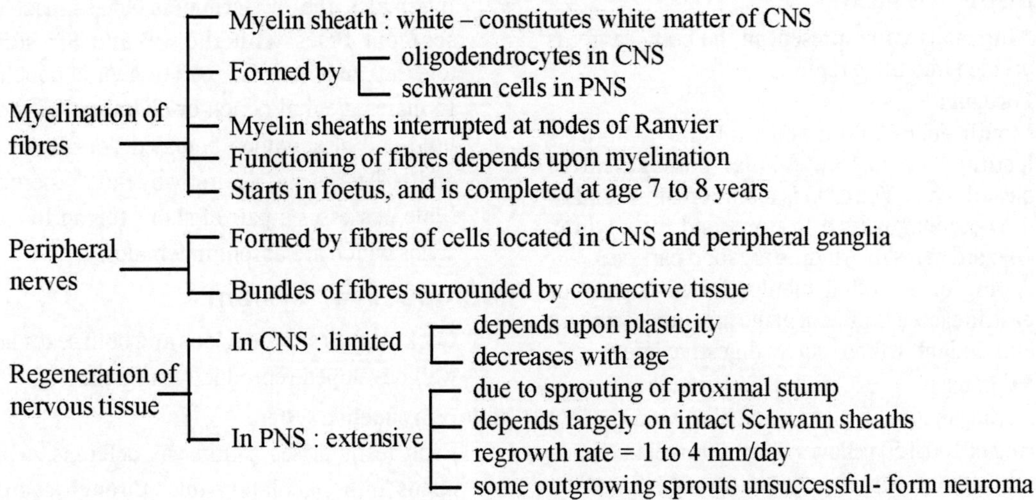

Neuroglia

Act as supporting tissue and are involved in neural nutrition, defence, excitation and inhibition.

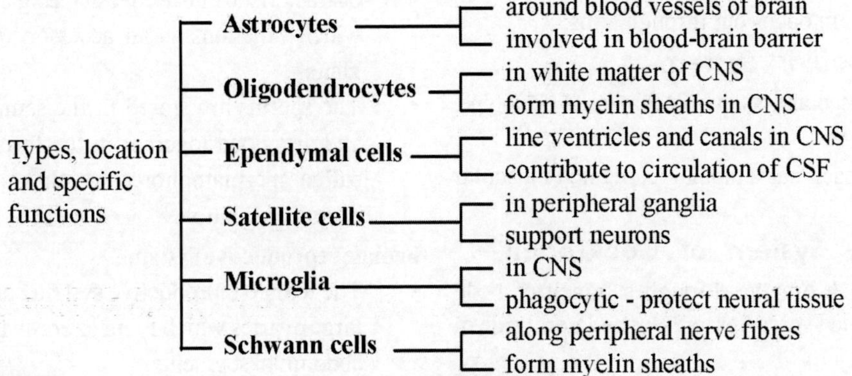

COCKROACH

It belongs to phylum arthropoda; its scientific name is - *Periplaneta americana.*

Morphology

- It is about 40 mm (1.5 inches) long with a reddish-brown body and have a yellowish margin on the body region behind the head.
- The entire body is covered by a hard chitinous exoskeleton (brown in colour).

Head

- It is formed by the fusion of six segments and shows great mobility in all directions due to flexible neck.

Eye

- Each compound eye is made up of 2,000 hexagonal ommatidia with the help of several ommatidia a cockroach can receive several images of an object.

Antennae

- A pair of thread like antennae arises from membranous sockets lying in front of eyes.
- Male cockroaches have special sense receptors that allow them to smell the odours produced by a female.

Mouth

- Anterior end of head bears appendages forming biting and chewing type of mouth parts. Mouth parts consist of a labrum (upper lip), a pair of mandibles, a pair of maxillae and a labium (lower lip).

Thorax

- The head is connected with thorax by short extension of the prothorax known as the neck.
- Forewings (mesothoracic) called tegmina which are opaque dark and leathery which cover the hind wings when at rest.
- The hind wings are transparent, membranous and are used in flight.

Anatomy

- Alimentary canal present in the body cavity is divided into three regions :

  Foregut :

  Mouth opens into a short tubular pharynx - leading to a narrow tubular passage called **oesophagus.** The crop is a section at the end of the oesophagus used to store food for a while. Gizzard helps in grinding the food particles.

  A ring of 6-8 blind tubules called hepatic or gastric caecea is present at the junction of foregut and midgut, which secrete digestive juice.

  Midgut :

  At the junction of midgut and hindgut is present ring of 100-150 yellow coloured thin filamentous **malpighian tubules.** It cleans waste out of the blood.

  Hindgut :

  It is differentiated into ileum, colon and rectum. The rectum opens out through anus.

Blood Vascular System

- Blood vascular system in cockroach is open type.

Heart

- A cockroach has a tubular elongated muscular heart.

Respiratory System of Cockroach

- Cockroach breathe through a bunch of little holes called **spiracles** present on lateral side of its body.
- Exchange of gases takes place at the tracheoles by diffusion.

Excretory System

- Excretion is performed by malpighian tubules.
- They absorb nitrogenous waste products and convert them into uric acid excreted through the hindgut. Hence, it is also called **uricotelic**.

Nervous System of Cockroach

- Its nervous system consists of series of fused, segmentally arranged ganglia which is joined by paired longitudinal connectives on the ventral side.
- In cockroach, the sense organs are antennae, eyes, maxillary palps, labial palps, anal cerci, etc.

Abdomen

- The abdomen in both males and females consists of 10 segments.
- In females, the 7th sternum is boat shaped. The segment fuses with the 8th and 9th sterna together and forms a brood or genital pouch.
- In males, genital pouch or chamber lies at the hind end of abdomen bounded dorsally by 9th and 10th terga and ventrally by the 9th sternum.
- Male possesses a pair of short, thread like ana styles which are absent in females.

Reproductive System

- Cockroaches are **dioecious** and both sexes have well developed reproductive organs.

Male reproductive system

- Each testis arises a thin vas deferens, which opens into ejaculatory duct through seminal vesicles.
- At the 6th - 7th abdominal segments, a characteristic mushroom shaped gland is present which functions as an accessory reproductive gland.
- The sperms are stored in the seminal vesicles and are glued together in the form of bundles called spermatophores which are discharged during copulation.

Female reproductive system

- The female reproductive system consists of two large ovaries which lying laterally in the 2nd-6th abdominal segments.
- Each ovary is formed of a group of eight ovarian tubules or ovarioles which contain a chain of developing ova.

Fertilization

- A pair of spermatheca is present in the 6th segment which opens into the genitals chamber.
- Sperms are transferred through spermatophores and fertilisation occur.
- There fertilised eggs are encased in capsules called oothecae.

Development

- The development of *P. americana* is paurometabolous *i.e.* there is development through nymphal stage.

Exercise

1. Compound tissue is defined as –
   (a) similar types of cells held together by connective tissue
   (b) different types of cells which are different in structures and functions
   (c) different types of cells performing one function
   (d) similar cells at different regions performing many functions
2. Microvilli of epithelial cells –
   (a) increase surface area
   (b) protect the cells
   (c) engulf the foreign matter
   (d) give movements to the cells
3. Tendon is made up of –
   (a) yellow fibrous connective tissue
   (b) adipose tissue
   (c) modified white fibrous tissue
   (d) areolar tissue
4. Which of the following cell contains fat droplets?
   (a) Macrophages (b) Plasma cells
   (c) Adipose cells (d) Leucocytes
5. In Camel, the hump is mainly made of tissue
   (a) Areolar (b) Adipose
   (c) Muscular (d) Skeleton
6. 60% inorganic components are present in
   (a) cartilage (b) bone
   (c) connective tissue (d) None of these
7. Inter-calated discs are present in –
   (a) cardiac muscles (b) skeletal muscles
   (c) unstriated muscles (d) ligament
8. Diagnostic feature of insects is
   (a) three pairs of legs (b) compound eye
   (c) chitinous body (d) two pairs of wings
9. Young cockroach is called –
   (a) maggot (b) ephyra
   (c) nymph (d) pupa
10. Formation of platelets is known as –
    (a) Haemopoiesis (b) Thrombopoiesis
    (c) Haemolysis (d) None of these
11. The major constituent of connective tissue is –
    (a) collagen (b) carbohydrate
    (c) lipid (d) cholesterol
12. The cell junctions called tight, adhering and gap junctions are found in
    (a) muscular tissue
    (b) connective tissue
    (c) epithelial tissue
    (d) neural tissue
13. Cells of squamous epithelium are
    (a) tall with elongated nuclei
    (b) cube like
    (c) flat and tile like
    (d) columnar or cuboidal in shape
14. Epithelial tissue performs the following functions
    (a) Protection, secretion, absorption, respiration
    (b) Protection, secretion, sensation, absorption
    (c) Protection, secretion, absorption, digestion
    (d) None of these
15. Compound squamous epithelium is found in
    (a) stomach (b) intestine
    (c) trachea (d) pharynx
16. The main function of ligament is
    (a) joining of two bones
    (b) joining of muscles
    (c) joining of muscle to bone
    (d) joining of muscle to nerves
17. Below the skin, the fat is in the form of
    (a) lipoproteins (b) adipose tissue
    (c) mucous layer (d) lymphoid tissue
18. The function of nervous tissue is
    (a) irritability (b) sensibility
    (c) responsiveness (d) contraction
19. The summer sleep of frog is termed as
    (a) neoteny (b) aestivation
    (c) paedogenesis (d) hibernation
20. Croaking of frog is
    (a) hunger call (b) danger call
    (c) musical tone (d) sex call for female
21. Inner surface of fallopian tubes, bronchi and bronchioles are lined by
    (a) Squamous epithelium
    (b) Ciliated epithelium
    (c) Columnar epithelium
    (d) Cubical epithelium
22. The most abundant tissue in the body is
    (a) nervous (b) muscular
    (c) vascular (d) epithelial

23. Which of the following is regarded as a unit of nervous tissue?
(a) Axons (b) Dendrites
(c) Neurons (d) Myelin sheath
24. Lymph differs from blood in possessing
(a) only WBC
(b) more RBC and WBC
(c) more RBC and few WBC
(d) more WBC and few RBC
25. Mineral found in red pigment of vertebrate blood is
(a) magnesium (b) iron
(c) calcium (d) copper
26. Characteristics of smooth muscle fibres are
(a) spindle-shaped, unbranched, unstriated, uninucleate and involuntary
(b) spindle shaped, unbranched, unstriped, multinucleate and involuntary
(c) cylindrical, unbranched, unstriped, multinucleate and involuntary
(d) cylindrical, unbranched, striated, multinucleate and voluntary
27. Component of blood responsible for producing antibodies is
(a) thrombocytes (b) monocytes
(c) erythrocytes (d) lymphocytes
28. Formation of cartilage bones involves
(a) deposition of bony matter by osteoblasts and resorption chondroblasts
(b) deposition of bony matter by osteoclasts and resorption by chondroblasts
(c) deposition of bony matter by osteoclasts only
(d) deposition of bony matter by osteoblasts only
29. Blood capillary consists of
(a) endothelium only
(b) endothelium and thin coat of connective tissue
(c) endothelium and thin coat of muscle fibres
(d) endothelium and thin coat of connective tissue plus muscle fibres
30. Protein present in the matrix of cartilage is known as
(a) chondrin (b) casein
(c) cartilagin (d) ossein
31. Muscle fibres are composed of numerous fine fibrils, called
(a) myofibrils (b) sarcomere
(c) Z-band (d) cross bridges
32. A piece of bone such as femur of frog if kept in dilute HCl for about a week will
(a) assume black colour
(b) shrink in size
(c) turn flexible
(d) crack into pieces
33. Which cells do not form layer and remain structurally separate?
(a) Epithelial cells (b) Muscle cells
(c) Nerve cells (d) Gland cells
34. Earthworms have no skeleton but during burrowing, the anterior end becomes turgid and acts as a hydraulic skeleton. It is due to
(a) coelomic fluid (b) blood
(c) gut peristalsis (d) setae
35. Which one of the following is correct pairing of a body part and the kind of muscle tissue that moves it ?
(a) Biceps of upper arm–Smooth muscle fibres
(b) Abdominal wall–Smooth muscle
(c) Iris–Involuntary smooth muscle
(d) Heart wall–Involuntary unstriated muscle
36. If a live earthworm is pricked with a needle on it outer surface without damaging its gut, the fluid that comes out is
(a) coelomic fluid (b) haemolymph
(c) slimy mucus (d) excretory fluid
37. Which one of the following correctly describes the location of some body parts in the earthworm (*Pheretima*)?
(a) Four pairs of spermathecae in 4 – 7 segments.
(b) One pair of ovaries attached at intersegmental septum of 14th and 15th segments.
(c) Two pairs of testes in 10th and 11th segments.
(d) Two pairs of accessory glands in 16 – 18 segments.
38. The kind of tissue that forms the supportive structure in our pinna (external ears) is also found in
(a) nails (b) ear ossicles
(c) tip of the nose (d) vertebrae
39. Which of the following is correctly stated as it happens in the common cockroach ?
(a) Malpighian tubules are excretory organs projecting out from the colon
(b) Oxygen is transported by haemoglobin in blood
(c) Nitrogenous excretory product is urea
(d) The food is ground by mandibles and gizzard
40. The breakdown of detritus into smaller particles by earthworm is a process called
(a) humification (b) fragmentation
(c) mineralisation (d) catabolism

41. Consider the following four statements (A-D) related to the common frog *Rana tigrina*, and select the correct option stating which ones are true (T) and which ones are false (F)

Statements :

(A) On dry land it would die due to lack of O_2 its mouth is forcibly kept closed for a few days

(B) It has four- chambered heart

(C) On dry land, it turns uricotelic from ureotelic

(D) Its life-history is carried out in pond water

Codes :

	(A)	(B)	(C)	(D)
(a)	T	F	F	T
(b)	T	T	F	F
(c)	F	F	T	T
(d)	F	T	T	F

42. Blood platelets are not present in the blood of –
(a) fishes (b) amphibians
(c) reptiles (d) All of the above

43. Percentage of protein is more in –
(a) lymph (b) blood
(c) plasma (d) W.B.C.

44. In man, the number of RBC per cubic mm of blood is –
(a) 1 to 3 millions
(b) 5 to 5.4 millions
(c) Less than one million but more than 50 thousands
(d) Less than 50 thousands

45. "All or None" rule can't be implicated on
(a) non straited muscles
(b) cardiac muscles
(c) straited muscles
(d) All of the above

46. Rhythmical contraction is noticed in
(a) cardiac muscles (b) striped muscles
(c) unstriped muscles (d) in all the muscles

47. Four healthy people in their twenties got involved in injuries resulting in damage and death of a few cells of the following. Which of the cells are least likely to be replaced by new cells –
(a) liver cells
(b) neurons
(c) malpighian layer of the skin
(d) osteocytes

48. Photoreceptor in *Pheretima* are –
(a) on ventral side of skin
(b) on dorsal side of skin
(c) on both
(d) in clitellum

49. Insects excrete nitrogen as
(a) uric acid (b) guanine
(c) urea (d) ammonia

50. Which part of cockroach has both exoskeleton and endoskeleton?
(a) Head (b) Thorax
(c) Both of these (d) None of these

51. Male and female cockroaches can be distinguished externally through
(a) anal styles in male
(b) anal cerci in female
(c) anal style and antennae in female
(d) both (b) and (c)

52. Carotene pigment is found in the cells of :
(a) dermis (b) epidermis
(c) adipose cell (d) both 'b' and 'c'

53. One very special feature in the earthworm *Pheretima* is that
(a) fertilisation for eggs occurs inside the body
(b) the typhlosole greatly increases the effective absorption area of the digested food in the intestine
(c) the S- shaped setae embedded in the integument are the defensive weapons used against the enemies
(d) it has a long dorsal tubular heart

54. The process of bone formation is called –
(a) Ossification (b) Calcification
(c) Calcination (d) None of these

55. Blood is a –
(a) epithelial tissue (b) muscular tissue
(c) connective tissue (d) supportive tissue

56. Uninucleate muscles are –
(a) smooth muscles (b) cardiac muscles
(c) involuntary muscles (d) All of the above

57. Multinucleate muscle cells are –
(a) striped muscles (b) smooth muscles
(c) skeletal muscles (d) All of the above

58. Muscles get fatigued due to accumulation of-
(a) Adenosine triphosphate
(b) CO_2
(c) Lactic acid
(d) Phosphate molecules

59. Calciferous glands are present in Earthworm in –
(a) pharynx & gizzard
(b) stomach & buccal cavity
(c) pharynx & stomach
(d) only in stomach

60. Excretory organs of Earthworm are-
(a) coelom (b) flame cells
(c) nephridia (d) gizzard

Cell : The Unit of Life

CONCEPTS

- All living organisms are constituted of structural and functional units called **cell**.
- **Robert Hooke** in 1665 coined the word 'cell'.
- **Anton von Leeuwenhoek** first saw and described a live cell.
- **Robert Brown** later had discovered the nucleus.
- Cell theory was proposed by **Schleiden** and **Schwann** in 1855 to explain the concept of the cellular nature of living organism.

The three fundamental facts of the cell theory are as follows:

- All living organisms (animals, plants and microbes) are made up of cells and cell products.
- All metabolic reactions in unicellular and multicellular organisms take place in cell.
- Cells originate only from pre-existing cells.

PROKARYOTIC CELLS

- Prokaryotic cells are morphologically most primitive cells.
- A single membrane surrounds the cell. It is without membrane bound nucleus.
- Prokaryotic cells are devoid of membrane bound organelles like plastids, mitochondria and advanced (9+2) flagella.
- They reproduce asexually by transverse binary fission or budding.
- Prokaryotic cells are represented by bacteria, cyanobacteria (blue green algae) mycoplasma and PPLO (pleuro-pneumonia like organisms).
- **Gram-positive bacteria:** These bacteria are stainable by the Gram stain (*e.g.*, *Diplococcus pneumoniae*, etc.). The cell wall of Gram-positive bacteria contains polysaccharides and mucopeptides remain extensively linked to one another that provides considerable strength to the wall.
- **Gram-negative bacteria**: These are not stained by Gram stain (*e.g.*, *Escherichia coli*, *etc.*). The cell walls of Gram-negative bacteria contain a similar network but have in addition an outer layer of lipids, complex with protein and polysaccharides.
- Bacterial cell is covered by an (a) outer slime layer or capsule, (b) a middle cell wall and (c) an inner cell membrane.

Cell wall

- The plasma membrane is covered by a strong, rigid cell wall.

Plasma membrane

- The plasma membrane is semi-permeable in nature and interacts with the outside world.

Ribosomes & Inclusion Bodies

- Bacterial ribosomes are slightly smaller than the ribosomes of eukaryotes (they sediment at 70S as opposed to 80S) that occur freely in the cytoplasm.
- Reserve material in prokaryotic cells are stored in the cytoplasm in the form of inclusion bodies.

Mesosomes

- In some bacteria (*e.g.*, *Bacillus subtilis*) the plasma membrane giving out certain invaginations or in foldings called mesosomes, in the cytoplasm.
- Mesosomes plays role in separation of two daughter nucleoids and in the formation of a new cell plate during cell division.

Genetic Material

- In bacteria, a true nucleus with nuclear envelope, nucleolus and histone protein is absent. It is represented by a single-naked, **double stranded, circular,** ring like highly folded supercoiled DNA with no free ends. Such a nucleus without nuclear envelope is called **nucleoid**.
- Many prokaryotes have small, circular DNA molecules called **plasmids**.

- Locomotion occurs by **flagella** that protrude from an intracellular "basal granule" through the plasma membrane and cell-wall.
- Respiration may be aerobic or anaerobic.
- Nutrition may be autotrophic or heterotrophic.

EUKARYOTIC CELLS

- Eukaryotic cells are characterized by extensive compartmentalisation of cytoplasm through the presence of membrane bound organelles. *e.g.* mitochondria, endoplasmic reticulum, golgi complex, *etc.*, and a true nucleus.
- A eukaryotic cell consists of the following components:

CELL WALL

- The cell wall is a non-living, semi-rigid, external protective covering of the cell.
- Cell wall is entirely lacking in animals.
- It is made up of cellulose secreted by the cell itself.
- The middle lamella functions as cement between adjacent cells. It is chiefly composed of calcium pectate.
- Plasmodesmata are cytoplasmic protoplasmic bridges between two adjacent cells.

Functions

- It acts as a barrier to entry of viral, bacterial and fungal pathogens.
- It helps in transport of materials and metabolites in and out of the cell.

CELL MEMBRANE

- The cell membrane is a living, thin, elastic and semi-permeable membranous covering of plant and animal cells.
- The lipids are arranged within the membrane with the polar head towards the outer sides and the hydrophobic tails towards the inner part.

Fluid mosaic model of plasma membrane

- **S.J.Singer** and **G. Nicolson** in 1972 proposed the most accepted model of membrane structure
- The lipid bilayer is a semifluid in which the proteins are dispersed to give mosaic appearance to the whole membrane.
- Lipids are amphipathic, *i.e.*, they are structurally asymmetric with polar hydrophilic and non-polar hydrophobic group.

Two kinds of proteins are found in the membrane:

- **Peripheral or extrinsic proteins :** They function as carrier for transport of material.
- **Integral or intrinsic proteins :** Intrinsic proteins form channels which provide a rapid passage to water soluble materials across the plasma membrane.

Functions

- One of the most important function of plasma membrane is the transport of the molecules across it.
- Many molecules can move briefly across the membrane without any requirement of energy. This is called as **passive transport**.
- Movement of water by diffusion is called **osmosis**.
- A few ions or molecules are transported across the membrane against their concentration gradient. Such type of transport is called as **active transport** because it is energy dependent process in which ATP is utilised, *e.g.*, Na^+/K^+ pump.
- Endomembrane system includes endoplasmic reticulum, golgi complex, lysosomes and vacuoles.

ENDOPLASMIC RETICULUM (ER)

- Endoplasmic reticulum is vast reticulum or network of inter-connected tubules and vesicles. There are two types of endoplasmic reticulum *i.e.*,
- **Smooth or agranular ER** – They do not have attached ribosomes on their surface. The SER is well developed in those cells which are actively concerned with the synthesis and secretion of steroid hormone. It is also concerned with the synthesis of lipid *etc.*
- **Rough or granular ER** – They bear ribosomes on their surface. The cells which are active in protein synthesis have RER in abundance.

Functions

- Mechanical support :
- Intracellular exchange :
- Lipid secretion :
- Detoxification :

GOLGI APPARATUS

- Golgi apparatus or Golgi complex is a stack of flattened, membrane bounded, parallely arranged organelles that occur in the association of endoplasmic reticulum in the cytoplasmic matrix.
- Each Golgi complex is composed of many lamellae, tubules, vesicles and cisternae of 0.5 μm to 1.0 μm in diameter.
- Golgi body is present in all cells of eukaryotes but absent in mature RBCs of mammals, sieve tubes, bryophytes *etc.*

Function

- The golgi apparatus principally performs the function of packaging materials to be delivered either to the intra-cellular targets or secreted outside the cell.

LYSOSOMES

- Lysosomes are tiny spheroid or irregular-shaped, membrane bounded vesicles.
- Lysosomes are popularly called "suicide bags". They are also called as lytic bags, scavenger of cell or disposal units.

VACUOLES

- The cytoplasm of many plants and some animal cells (*i.e.*, ciliate protozoans) contains numerous small or large-sized, hollow, liquid-filled structures called vacuoles.
- In plant cells, the vacuoles can occupy up to 90 percent of the volume of the cell. The vacuole is bound by a single membrane called **tonoplast**.

Functions

- It is the storage, transmission of the materials and the maintenance of internal pressure of the cell. These vacuoles contain water, phenol, flavonoids, anthocyanins (blue and red pigments), alkaloids and storage products, such as sugars and proteins.

MITOCHONDRIA

- Mitochondria are also called as powerhouse of cells. They are large-sized, rounded or rod-like, double-membranous structures.
- The two membrane have their own specific enzymes associated with the mitochondrial mitochondria are the sites of aerobic respiration functions. Cristae contain different components of respiratory chain phosphorylation.

Functions

- Mitochondria are the sites of aerobic respiration. They produce cellular energy in the form of ATP.

PLASTIDS

- Plastids are found in plants and few protists (*Euglena*). They may be colourless (leucoplastids) or coloured (chromoplastids and chloroplasts).
- Chloroplasts is double membranous structure. Each chloroplast has two parts *i.e.*,

Stroma

- Stroma contains proteins, DNA, ribosomes and fewer amounts of enzymes.
- Due to presence of enzymes, the dark reaction of photosynthesis takes place in stroma.

Grana

- In chloroplast several membranes are arranged in the form of stacks of coins. 20 to 50 membranous sac like structure called thylakoids. These thylakoids are stacked one above the other to form grana.
- Due to presence of chlorophyll in granum lamella, the light reaction of photosynthesis takes place in granum.

Functions

- The chloroplasts help in the biosynthesis of food stuffs by the process of photosynthesis.

RIBOSOMES

- Ribosomes are smallest cell organelle
  There are two types of ribosomes *viz.*,
  - Prokaryotic or 70S ribosomes
  - Eukaryotic or 80S ribosomes

Function

- They are the sites of protein synthesis.

CYTOSKELETON

- An elaborate network of filamentous proteinaceous structures present in the cytoplasm is collectively referred to as the **cytoskeleton.**

Cilia and Flagella

- Cilia and flagella are contractile filamentous or hair-like processes of the cytoplasm, which help in cell movement, create food currents, excretion, secretion and transportation.
- **Centrosome** is an organelle usually containing two cylindrical structures called centrioles.
- The centrioles form the basal body of cilia or

flagella, and spindle fibres that give rise to spindle apparatus during cell division in animal cells.

NUCLEUS

- Nucleus is centrally located, spherical and largest component of the all eukaryotic cell.

Structure of Nucleus

- A typical nucleus consists of four structures: (i) nuclear membrane, (ii) nucleoplasm (iii) chromatin and (iv) the nucleolus.
- The nucleus is bounded by two unit membranes that are separated by a 100-500Å wide perinuclear space.
- The fluid filled in nucleus is known as nucleoplasm or **karyolymph.**

Chromatin

In interphase nucleus the genetic material (DNA) is seen as the thread like structure called **chromatin** or **chromatin fibres**. During cell divisions (mitosis and meiosis) chromatin fibres become condensed to form thick structures called as the **chromosomes.**

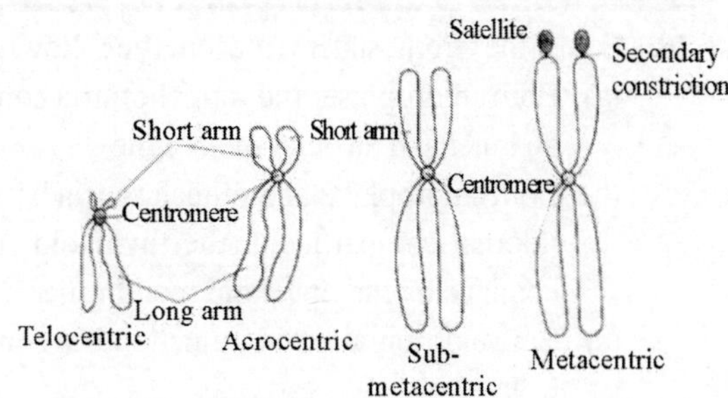

Fig. : Types of chromosomes based on the position of centromere

Nucleolus

- Nucleolus is a conspicuous, darkly stained spherical body that is found in nucleoplasm.
- It is site of active ribosomal RNA synthesis.

MICROBODIES

- Microbodies are single unit membrane bound, minute spherical structures that contain various enzymes.

Exercise

1. Difference between the prokaryotic and eukaryotic cells in having
   (a) cell wall (b) nuclear membrane
   (c) ribosome (d) None of these
2. Who proposed the theory that "cells arise only from the pre-existing cells"?
   (a) Mohl (b) Virchow
   (c) Haeckel (d) Brown
3. The main difference between plant and animal cell is
   (a) animal cells lack cell wall
   (b) plant cell has no cell wall
   (c) animal cell has a rigid cell wall
   (d) plant cells lack cell membrane
4. A mature plant cell has
   (a) cell wall and protoplasm
   (b) protoplasm and vacuole
   (c) vacuole and cell wall
   (d) protoplasm, cell wall and vacuole
5. Ion carriers are located in
   (a) nucleus (b) cell wall
   (c) cellular space (d) plasma membranes
6. The type of cell junction which facilitates cell to cell communication is
   (a) tight junction (b) adhering junction
   (c) gap junction (d) desmosomes
7. The size of mitochondrion is
   (a) 5 – 10 μ (b) 50 – 100 μ
   (c) 0.5 – 1.0 μ (d) 150 – 300 μ
8. Green pigment (Chlorophyll) present in plants is
   (a) chromoplast (b) chloroplast
   (c) ribosome (d) lysosome
9. A flattened disc-like sac in a chloroplast is called a
   (a) Loculus (b) Thylakoid
   (c) Stroma (d) Margin

10. Select the wrong statement from the following
   (a) Both chloroplasts and mitochondria contain an inner and an outer membrane
   (b) Both chloroplasts and mitochondria have an internal compartment, the thylakoid space bounded by the thylakoid membrane
   (c) Both chloroplasts and mitochondria contain DNA
   (d) The chloroplasts are generally much larger than mitochondria
11. Main function of amyloplast is
   (a) water absorption (b) light absorption
   (c) fat accumulation (d) starch accumulation
12. Golgi body is associated with
   (a) packaging and storage of material
   (b) cell plate formation
   (c) secretion of different substance
   (d) All of the above
13. Centrioles and centrosomes are present in cells of
   (a) bacteria (b) cyanobacteria
   (c) green plants (d) animals
14. Function of centriole is
   (a) formation of spindle fibres
   (b) formation of nucleolus
   (c) initiation of cell division
   (d) formation of cell plate
15. The function of centrosome is
   (a) inhibition of cell division
   (b) initiates cell division
   (c) to increase protein synthesis
   (d) None of these
16. Nucleolus is found in
   (a) protoplasm (b) nucleus
   (c) cytoplasm (d) None of these
17. The function of nucleolus is the synthesis of
   (a) DNA (b) m-RNA
   (c) r-RNA (d) t-RNA
18. Chromosomes whose arms are equal are called
   (a) Metacentric (b) Acrocentric
   (c) Concentric (d) Acentric
19. According of fluid mosaic model, plasma membrane is composed of
   (a) phospholipids and oligosaccharides
   (b) phospholipids and hemicellulose
   (c) phospholipids and integral proteins
   (d) phospholipids, extrinsic proteins and intrinsic proteins
20. Organelles can be separated from cell homogenate through
   (a) chromatography
   (b) X-rays diffraction
   (c) differential centrifugation
   (d) auto-radiography
21. Plasma membrane is made of
   (a) proteins and carbohydrates
   (b) proteins and lipids
   (c) proteins, lipids and carbohydrates
   (d) proteins, some nucleic acids and lipids
22. Nucleoproteins are synthesised in
   (a) nucleoplasm (b) nuclear envelope
   (c) nucleolus (d) cytoplasm
23. The latest model for plasma membrane is
   (a) lamellar model
   (b) unit membrane model
   (c) fluid mosaic model
   (d) molecular lipid model
24. Electron microscope has a high resolution power. This is due to
   (a) electromagnetic lenses
   (b) very low wavelength of electron beam
   (c) low wavelength of light source used
   (d) high numerical aperture of glass lenses used
25. Resolution power is the ability to
   (a) distinguish two close points
   (b) distinguish two close objects
   (c) distinguish amongst organelles
   (d) magnify image
26. Fluid mosaic model of cell membrane was put forward by
   (a) Danielli and Davson
   (b) Singer and Nicolson
   (c) Garner and Allard
   (d) Watson and Crick
27. Cell wall shows
   (a) complete permeability
   (b) semipermeability
   (c) differential permeability
   (d) impermeability

28. Ribosomes are the centre for
(a) respiration (b) photosynthesis
(c) protein synthesis (d) fat synthesis

29. All plastids have similar structure because they can
(a) store starch, lipids and proteins
(b) get transformed from one type to another
(c) perform same function
(d) be present together

30. Names of Schleiden and Schwann are associated with
(a) protoplasm as the physical basis of life
(b) cell theory
(c) theory of cell lineage
(d) nucleus functions as control centre of cell

31. Cell recognition and adhesion occur due to biochemicals of cell membranes named
(a) proteins (b) lipids
(c) proteins and lipids
(d) glycoproteins and glycolipids

32. Golgi apparatus is absent in
(a) higher plants (b) yeast
(c) bacteria and blue-green algae
(d) None

33. Membranous bag with hydrolytic enzymes which is used for controlling intracellular digestion of macro -molecules is
(a) endoplasmic reticulum
(b) nucleosome
(c) lysosome (d) phagosome

34. In plant cells, peroxisomes are associated with
(a) photorespiration (b) phototropism
(c) photoperiodism (d) photosynthesis

35. Inner membrane convolutions of a mitochondrion are known as
(a) lamellae (b) thylakoids
(b) grana (d) cristae

36. Mitochondrial cristae are sites of
(a) breakdown of macromolecules
(b) protein synthesis
(c) phosphorylation of flavoproteins
(d) oxidation-reduction reactions

37. The prokaryotic flagella possess
(a) unit membrane enclosed fibre
(b) protein membrane enclosed fibre
(c) '9 + 2' membrane enclosed structure
(d) helically arranged protein molecules

38. The function of rough endoplasmic reticulum is
(a) fat synthesis (b) lipid synthesis
(c) protein synthesis (d) steroid synthesis

39. Lysosomes have a high content of
(a) hydrolytic enzymes(b) lipoproteins
(c) polyribosomes (d) DNA ligases

40. Which of the following organelle has single membrane?
(a) Nucleus (b) Cell wall
(c) Mitochondria (d) Vacuole

41. Microtubules absent in
(a) mitochondria (b) centriole
(c) flagella (d) spindle fibres

42. Element necessary for middle lamella is
(a) Ca (b) Zn
(c) K (d) Cu

43. Mitotic spindle is mainly composed of which protein?
(a) Actin (b) Myosin
(c) Tubulin (d) Myoglobin

44. Flagella of prokaryotic and eukaryotic cells differ in
(a) type of movement and placement in cell
(b) location in cell and mode of functioning
(c) microtubular organization and type of movement
(d) microtubular organization and function

45. Centromere is required for
(a) movement of chromosomes towards poles
(b) cytoplasmic cleavage
(c) crossing over
(d) transcription

46. Vacuole in a plant cell
(a) is membrane-bound and contains storage proteins and lipids
(b) is membrane-bound and contains water and excretory substances
(c) lacks membrane and contains air
(d) lacks membrane and contains water and excretory substances

47. Plasmodesmata are
(a) locomotory structures
(b) membranes connecting the nucleus with plasmalemma
(c) connections between adjacent cells
(d) lignified cemented layers between cells

48. Cytoskeleton is made up of:
(a) callose deposits
(b) cellulosic microfibrils
(c) proteinaceous filaments
(d) calcium carbonate granules

49. Which one of the following has its own DNA?
(a) Mitochondria (b) Dictyosome
(c) Lysosome (d) Peroxisome

50. Which one of the following is not considered as a part of the endomembrane system ?
(a) Golgi complex (b) Peroxisome
(c) Vacuole (d) Lysosome

51. The middle lamella is composed of –
(a) pectates (b) cellulose
(c) lignin (d) proteins

52. Endocytosis includes –
(a) Phagocytosis (b) Pinocytosis
(c) Both (d) None

53. Which one of the following is not a constituent of cell membrane?
(a) Cholesterol (b) Glycolipids
(c) Proline (d) Phospholipids

54. 70S type of ribosomes is found in –
(a) prokaryotic cells
(b) prokaryotic cells, chloroplasts and mitochondria
(c) mitochondria
(d) nucleus and mitochondria

55. Hereditary characters are due to –
(a) chromosomes (b) gene
(c) blood (d) placenta

56. Telomerase is an enzyme which is a –
(a) RNA
(b) ribonucleoprotein
(c) DNA
(d) simple protein

57. The cell wall of algae contains :
(a) galactans (b) pectin
(c) chitin (d) muramic acid

58. Membrane surrounding vacuole is called :
(a) tonoplast (b) nucleoplast
(c) protoplast (d) vacuoplast

59. 'Power house' of the cell is called
(a) nucleus (b) ribosome
(c) mitochondria (d) golgi complex

60. The largest cell organelles in the cell are:
(a) plastids (b) golgi bodies
(c) mitochondria (d) chromosomes

CHAPTER 9

Biomolecules

CONCEPTS

- All the carbon compounds that we get from living tissues can be called **biomolecules**.

PRIMARY & SECONDARY METABOLITES

Metabolites can be defined as those biomolecules which are either utilized in metabolic functions or are synthesized by the cellular machinery. These are of two types:

Primary metabolites

- Primary metabolites are found in all the cell types.
- Examples of primary metabolites are amino acids, carbohydrates, nucleotides, fats, *etc*.

Secondary metabolites

- Secondary metabolites are found only in plant, fungal and microbial cells.

Table : Some Secondary Metabolites

Pigments	Carotenoids, Anthocyanins, etc.
Alkaloids	Morphine, Codeine, etc.
Terpenoides	Monoterpenes, Diterpenes, etc.
Essential oils	Lemon grass oil, etc.
Toxins	Abrin, Ricin
Lectins	Concanavalin A
Drugs	Vinblastin, curcumin, etc.
Polymeric substances	Rubber, gums, cellulose

AMINO ACIDS

Amino acids are building blocks of proteins.

Basic structure

- Amino acids are organic acids containing atleast one amino (NH_2) group and one carboxylic acid (COOH) group, attached to same carbon atom or α-Carbon. Hence they are known α amino acids.
- The other valencies of carbon are satisfied by one H and one R group as side chain which may be polar or nonpolar; aliphatic or aromatic; simple or heterocyclic compound.
- 'R' group may vary from a H-atom to any large aromatic compound; hence the type of amino acids depends upon R-group.

CLASSIFICATION OF AMINO ACIDS

They are usually classified according to the nature of their sidechains or R groups.

1. Acidic amino acid — e.g., glutamic acid.
2. Basic amino acid — e.g., lysine
3. Neutral amino acid — e.g., valine
4. Aromatic amino acid — e.g., tyrosine, phenylalanine, tryptophan

$$\underset{(A)}{H_3^+N-\overset{\overset{R}{|}}{C}H-COOH} \rightleftharpoons \underset{(B)}{H_3^+N-\overset{\overset{R}{|}}{C}H-COO^-}$$

$$\rightleftharpoons \underset{(C)}{H_2N-\overset{\overset{R}{|}}{C}H-COO^-}$$

B is called zwitterionic form.

Essential amino acids

- They can be obtained only from plant source. These are alanine, valine, histidine, isoleucine, lysine, methionine, phenylalanine, threonine and tryptophan.

Non-essential amino acids

- They can be synthesised by organisms in the body and thus these are not required component of diet. E.g., tyrosine, serine, aspartic acid, glutamic acid, proline, alanine, *etc*.

PROTEINS

- They are nitrogenous macromolecules, composed of many amino acids. There are 20 types of amino acids that form proteins that is why they are **heteropolymers** and not homopolymer (formed of only one type of monomer units).

Biological importance of protein

- **Structural proteins :** Proteins that serve as building material of cells and tissues called as structural proteins. *E.g.*, α-keratin (hair, nails, etc.), ossein (bone), chondrin (cartilage), collagen, etc.
- **Functional proteins :** Certain proteins like enzymes, some coenzymes and hormones (insulin, parathormone) act as regulator of body functions and metabolism.
- **Transport proteins :** These proteins serves as physical carriers of other functionally important smaller molecules. Eg., Haemoglobulin, myoglobulin etc.
- **Contractile proteins :** These proteins help in movement and locomotion. Eg., myosin and actin of muscle fibre.
- **Protective proteins :** These proteins acts as antibodies and provide immunity to the body. *E.g.*, antibodies.
- **Storage proteins :** These includes ovalbumen of egg white, glutins of wheat, casein of milk, etc.

STRUCTURE OF PROTEINS

Primary Structure

- A protein is imagined as a line, the left end represented by the first amino acid and the right end represented by the last amino acid.

Secondary Structure

- In proteins, only right handed helices are observed. Other regions of the protein thread are folded into other forms. This folding is known as secondary structure.

Tertiary Structure

- The peptide chain with its secondary structure may be further folded and twisted about itself, gives 3-dimensional structure.

Quaternary Structure

- Some proteins are an assembly of more than one polypeptide or subunits. The manner in which these individual folded polypeptides or subunits are arranged with respect to each other (*e.g.*, linear string of spheres, spheres arranged one upon each other in the form of a cube or plate etc.) is called the quaternary structure of a protein.
- For example, the globin part of haemoglobin is formed of two α and two β chains.

NUCLEOTIDES

- Nucleotides are building blocks of nucleic acids (DNA and RNA). Each deoxynucleotide is made by following three moieties.
- **Pentose sugar :** It is ribose in RNA and 2'-deoxyribose sugar in DNA.
- **Nitrogenous bases :** There are two types of nitrogenous bases *i.e.*, 9-membered, double ringed **purines** (adenine or guanine) and 6-membered, single ringed **pyrimidines** (thymine or cytosine). The nitrogenous base molecules are joined to the sugar molecules by **glycosidic bonds** which develop between 1' carbon of the sugar and the nitrogen at position 1 in case of pyrimidine base and at position 9 in case of purine base.
- **Phosphoric acid :** The phosphoric acid molecule is linked to number 3 and 5 carbon atoms of the deoxyribose molecules by an ester linkage.

Nucleic Acids

These are polymeric compounds of nucleotides so are called polynucleotides. Nucleotides are formed of C, H, O, N and P.

A polynucleotide chain is formed by inter linking of large number of nucleotides forming **5'-3' phospho diester** bonds.

These are two classes of nucleic acids.

- Deoxyribose nucleic acid or DNA
- Ribose nucleic acid or RNA

DNA : DNA is the genetic material and forms molecular basis of heredity. DNA is mainly found in nucleus of all living cells except some viruses *e.g.* (Tobacco Mosaic Virus) TMV.

Watson and Crick Structural Model of DNA or B-Form of DNA

- The backbone of each strand of DNA molecule is made up of deoxyribonucleotide (phosphate + deoxyribose sugar) units that are joined in a chain by phosphodiester bonds which link the

5' carbon of the deoxyribose of one mononucleotide unit with the 3' carbon of the deoxyribose of the next mononucleotide unit.

- The two polynucleotide strands are held together by hydrogen bonds between specific pairs of purines and pyrimidines.
- Both polynucleotides strands remain separated by 20 Å distance.
- The coiling of double helix is **right handed** and a complete turn occurs every 34 Å. As each base pair occupies a distance of 3.4 Å, 10 base pairs are involved per complete turn. (The base pairs are rotated 36° with respect to each adjacent pair).
- The orientation of G ≡ C base pair and A=T base pair in DNA is such that it leaves two external grooves, a deep wide one, called **major groove** and a shallow narrow one, called **minor groove**:

FATTY ACIDS AND GLYCEROL

Fatty Acids

They are organic molecules madeup of long hydrocarbon chain which contains terminal carboxylic group (COOH). The hydrocarbon chain (R group) may be methyl (CH_3) or ethyl (C_2H_5) or higher member of CH_2 groups (1 to 19 carbon).

- **Saturated fatty acids**
  There is no double bond in the carbon chain of these fatty acids. *E.g.*, butyric acid (C_4), palmitic (C_{16}) and stearic (C_{18}) acids.
- **Unsaturated fatty acids**
  They contain double bonds in their carbon-chains because the valencies of some carbon atoms are not saturated with hydrogen. These are also called fats and oils based on melting point.

POLYSACCHARIDES

- Each polysaccharide molecule is composed of many molecules of monosaccharides (*e.g.*, glucose, fructose or galactose).
- Polysaccharides have two subclasses:

Homopolysaccharides (Homoglycans)

Homopolysaccharides are composed of many molecules of the same monosaccharide.

- Starch $(C_6H_{10}O_5)_n$, is a plant polysaccharide stored in grams, legumes, tubers like potato, and roots like carrot.
- Glycogen $(C_6H_{10}O_5)_n$, is stored only in liver and muscle cells.
- Cellulose is a polymer of glucose.
- Insulin is a polymer of fructose.

Heteropolysaccharides (Heteroglycans)

- **Pectins** are structural heteroglycans of the middle lamella of plant cell walls.
- **Cellulose** is main component of cotton fibres and paper formed from paper pulp.
- **Chitin** is structural polysaccharide of cell wall of fungi and exoskeletal plates of arthropods.
- **Peptidoglycan** forms cell wall of bacteria and cynobacteria.
- **Agar** is used in tissue culture, laxative and food.

DYNAMIC STATE OF BODY CONSTITUENTS – CONCEPT OF METABOLISM

A living system is characterised by number of chemical reactions catalyzed by catalytic proteins known as enzymes.

Turn over is the number of biomolecules produced per unit time, while the sum of all the chemical reactions going on inside a living system is called metabolism.

Metabolic Basis for Living

- **Anabolic pathways :** Energy consuming reactions in which simpler structures forms complex structures *e.g.* conversion of acetic acid into cholestrol.
- **Catabolic pathways :** Energy releasing reaction in which complex structure breaks and form simpler structure forms catabolic pathway *e.g.*, respiration.

Role of ATP (Adenosine Triphosphate) in metabolism

ATP is commonly called energy-cheque because when required, it is hydrolysed and chemical energy is released which is used in various endergonic (energy-dependent) metabolic functions.

ENZYMES

- An enzyme may be defined as 'complex biological catalyst present in living organism to regulate the various physiological processes of the body'. Enzymes, through their active site, catalyse reactions at a high rate.

Chemical Reactions

Physical change or physical process

- It refers to a change in shape without breaking of bonds.

Chemical change or chemical process

- It refers to the transformations in which bonds are broken and new bonds are formed for *e.g.*, hydrolysis of starch into glucose.
- Enzymes are the proteins with three dimensional structure including an "active site", convert a substrate (S) into a product.
  It can depicted as $S \rightarrow P$.
- Most of the chemical reactions do not start automatically because the reactant molecules have an **energy barrier** to become reactive.
- Therefore, an external supply of energy is needed for the start of the chemical reaction. It is called **activation energy**.
- Activation energy increases the kinetic energy of the system and brings about forceful collisions between the reactants.
- Enzymes lower the activation energy required for a reaction.

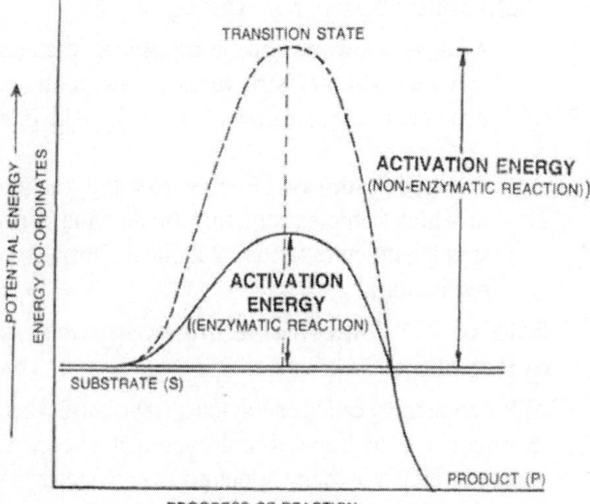

Fig. Lowering of activation energy by enzyme in the energy relations of a chemical reaction.

Nature of Enzyme Action

Lock-and-key hypothesis theory (Fischer's hypothesis): According to this model the active site of the enzyme already exists in the proper conformation even in absence of substrates. Thus, the active site by itself provides a rigid pre-shaped template, fitting with the size, form and groups of the substrate.

Induced fit theory (Koshland's hypothesis): This model proposes flexibility of the three-dimensional form of the active site. It proposes that inspite of the complete primary structure, the active site does not possess the fully active and complementary conformation in absence of the substrate.

This flexibility of the active site may explain complex saturation kinetics, competitive inhibition and allosteric modulation.

Substrate concentration

- Increase in substrate concentration increases the rate of reaction.
- A stage is reached where velocity is maximum.
- It does not increase further by increasing the substrate concentration.
- At this stage the enzyme molecules become fully saturated and no active site is left free to bind additional substrate molecules.

pH

- Each enzyme acts best at a particular pH called the optimum pH – the activity declines both above and below that pH.

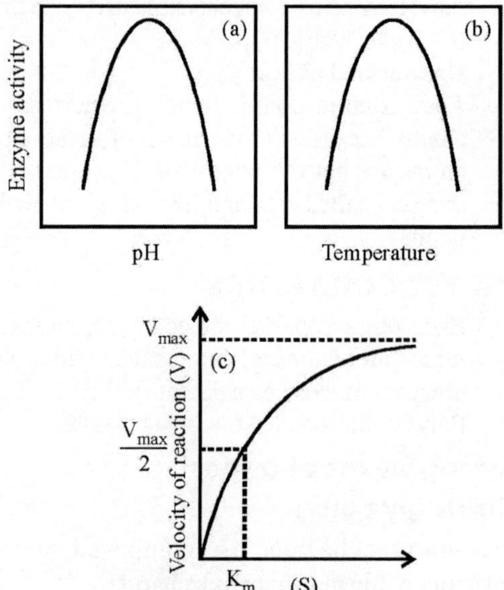

Fig. Effect of change in :
(a) pH (b) Temperature and
(c) Concentration of substrate on enzyme activity.

Temperature

- Each enzymes shows the highest activity at particular temperature called the **optimum temperature**

CLASSIFICATION OF ENZYMES

Oxidoreductases

- Enzyme of this class catalyzes oxidations and reductions of their substrates.
- Dehydrogenases, oxidases (O_2 acts as acceptor), oxygenases and peroxidases are the trivial names of many enzymes found in this class of enzymes.

Transferases

- These enzymes transfer a particular group from one substrate to another. *E.g.*, transaminases, transphosphorylases, *etc.*

Hydrolases

- These enzymes cleave their substrates by the hydrolysis of a covalent bond. *E.g.*, acetylcholinesterase, lipase.

Lyases

- They cleave a covalent bond in the substrate to change it to more than one product, but the cleavage does not involve any hydrolysis. *E.g.*, decarboxylases.

Isomerases

- They change their substrates to their isomers by intramolecular rearrangements and are classified according to the type of isomerization. *E.g.*, triose phosphate isomerase.

Ligases

- These enzymes bind two substrate molecules together by forming bonds. *E.g.*, acetyl CoA synthetases.

Co-Factors

- Some enzymes are composed of only polypeptide chains but there are many enzymes that require additional chemical compound for their activity. These compounds are called as **cofactor**.

Prosthetic Groups

- These are complex organic molecules that are tightly and permanently bound to apoenzyme. Haeme is a prosthetic group.

Co-Enzymes

- These are also organic molecules but there association with apoenzyme is only transient, usually occuring during course of catalysis. Eg., NAD or Nicotinamide adenine dinucleotide (contains vitamin niacin-B_5), FAD or (Flavin adenine dinucleotide (from Riboflavin-B_2).

Metal ions

- Metal ions are always inorganic in nature. Zinc is a cofactor for the proteolytic enzyme carboxypeptidase.

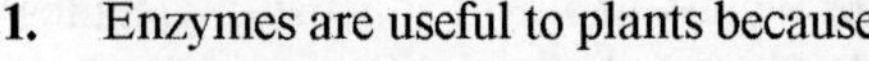

Exercise

1. Enzymes are useful to plants because
   (a) they are building block of chlorophyll
   (b) they are essential for the metabolic processes
   (c) they enhance absorption of water and salts
   (d) they are responsible for paratonic movements
2. NADP is
   (a) a coenzyme (b) a part of tRNA
   (c) an enzyme (d) a part of rRNA
3. The most important property of an enzyme is its
   (a) composition
   (b) thermal denaturation
   (c) specificity
   (d) solubility
4. Starch and cellulose are the compounds made up of many units of
   (a) simple sugar (b) fatty acid
   (c) glycerol (d) amino acid
5. Which of the following is a purine ?
   (a) Prokaryotes (b) Guanine
   (c) Thymine (d) Uracil
6. In DNA guanine pairs with
   (a) cytosine (b) thymine
   (c) uracil (d) adenine
7. Nucleic acids were discovered by
   (a) Watson and Crick (b) Khorana
   (c) Wilkins (d) Miescher

8. The similarity between DNA and RNA is that both are
(a) double stranded
(b) having similar sugars
(c) polymers of nucleotides
(d) having similar pyrimidines

9. The base pairs of DNA are correctly shown as
(a) $A \equiv T$ and $C = G$
(b) $A = T$ and $C = G$
(c) $A = T$ and $C \equiv G$
(d) $A \equiv T$ and $C \equiv G$

10. A nucleotide is formed of
(a) purine, pyrimidine and phosphate
(b) purine, sugar and phosphate
(c) nitrogen base, sugar and phosphate
(d) pyrimidine, sugar and phosphate

11. The basic unit of nucleic acid is
(a) pentose sugar (b) nucleoid
(c) nucleoside (d) nucleotide

12. Adenine is
(a) purine (b) pyrimidine
(c) nucleoside (d) nucleotide

13. Which is wrong about nucleic acids ?
(a) DNA is single stranded in some viruses
(b) RNA is double stranded occasionally
(c) Length of one helix is 45Å in B-DNA
(d) One turn of Z-DNA has 12 bases

14. An enzyme brings about
(a) decrease in reaction time
(b) increase in reaction time
(c) increase in activation energy
(d) reduction in activation energy

15. Two free ribonucleotide units are interlinked with
(a) peptide bond
(b) covalent bond
(c) hydrogen bond
(d) phosphodiester bond

16. In double helix of DNA, the two DNA strands are
(a) coiled around a common axis
(b) coiled around each other
(c) coiled differently
(e) coiled over protein sheath

17. RNA does not possess
(a) uracil (b) thymine
(c) adenine (d) cytosine

18. Which is not consistent with double helical structure of DNA ?
(a) A=T, C=G
(b) Density of DNA decreases on heating
(c) A + T/C + G is not constant
(d) Both A and B

19. Enzymes having slightly different molecular structure but performing identical activity are
(a) holoenzymes (b) isoenzymes
(c) apoenzymes (d) coenzymes

20. A polysaccharide, which is synthesized and stored in liver cells, is
(a) lactose (b) galactose
(c) arabinose (d) glycogen

21. In which one of the following groups, all the three are examples of polysaccharides?
(a) Starch, glycogen, cellulose
(b) Sucrose, maltose, glucose
(c) Glucose, fructose, lactose
(d) Galactose, starch, sucrose

22. Most diverse macromolecules, found in the cell both physically and chemically are
(a) proteins (b) carbohydrates
(c) nucleic acids (d) lipids

23. DNA synthesis can be specifically measured by estimating the incorporation of radiolabelled
(a) uracil
(b) adenine
(c) thymidine
(d) deoxyribose sugar

24. The enormous diversity of protein molecules is due mainly to the diversity of
(a) amino groups on the amino acids
(b) R groups on the amino acids
(c) amino acid sequences within the protein molecule
(d) peptide bonds

25. Cellulose, the most important constitutent of plant cell wall is made of
(a) unbranched chain of glucose molecules linked by a-1, 4 glycosidic bond
(b) branched chain of glucose molecules linked by b-1, 4 glycosidic bond in straight chain and a-1, 6 glycosidic bond at the site of branching
(c) unbranched chain of glucose molecules linked by b-1, 4 glycosidic bond
(d) branched chain of glucose molecules linked by a-1, 6 glycosidic bond at the site of branching

26. Lactose is composed of
(a) Glucose + Glucose
(b) Glucose + Fructose
(c) Fructose + Galactose
(d) Glucose + Galactose

27. The secondary structure of that portion of an integral protein that is buried in the lipid bilayer of a cell membrane is
(a) -strand (b) -bend
(c) random coil (d) -helix
28. Length of one turn of the helix in a -form DNA is approximately
(a) 3.4 nm (b) 2 nm
(c) 0.34 nm (d) 20 nm
29. Feedback inhibition of an enzymatic reaction is caused by
(a) end product (b) substrate
(c) enzyme (d) rise in temperature
30. Which is an essential amino acid?
(a) Serine (b) Aspartic acid
(c) Glycine (d) Phenylalanine
31. Lipids are insoluble in water because lipid molecules are
(a) hydrophilic (b) hydrophobic
(c) neutral (d) zwitter ions
32. Collagen is
(a) fibrous protein (b) globular protein
(c) lipid (d) carbohydrate
33. The catalytic efficiency of two different enzymes can be compared by the
(a) formation of the product
(b) pH optimum value
(c) K_m value
(d) molecular size of the enzyme
34. Antiparallel strands of a DNA molecule means that
(a) the phosphate groups of two DNA strands, at their ends share the same position
(b) the phosphate groups at the start of two DNA strands are in opposite position (pole)
(c) one strand turns clockwise
(d) one strand turns anti-clockwise
35. Carbohydrates, the most abundant bio-molecules on earth, are produced by :
(a) some bacteria, algae and green plant cells
(b) fungi, algae and green plant cells
(c) all bacteria, fungi and algae
(d) viruses, fungi and bacteria
36. About 98 percent of the mass of every living organism is composed of just six elements including carbon, hydrogen, nitrogen, oxygen and
(a) sulphur and magnesium
(b) magnesium and sodium
(c) calcium and phosphorus
(d) phosphorus and sulphur
37. A competitive inhibitor of succinic dehydrogenase is
(a) malonate (b) oxaloacetate
(c) α-ketoglutarate (d) malate
38. Carrier ions like Na^+ facilitate the absorption of substances like:
(a) amino acids and glucose
(b) glucose and fatty acids
(c) fatty acids and glycerol
(d) fructose and some amino acids
39. Which is a disaccharide?
(a) Galactose (b) Fructose
(c) Maltose (d) Dextrin
40. Which element is normally absent in proteins?
(a) C (b) N
(c) S (d) P
41. Protein most abundant in human body is –
(a) collagen (b) myosin
(c) actin (d) albumin
42. Characteristic feature of haemoglobin –
(a) reversible union with oxygen
(b) red Colour
(c) presence of Cu
(d) presence of globulin protein
43. Milk protein is –
(a) lactogen (b) myosin
(c) casein (d) pepsin
44. The amino acids which are not synthesized in the body are called –
(a) non-essential (b) essential
(c) deaminated (d) All of them
45. In India the best source for proteins in herbivorous persons is –
(a) pulses (b) potato
(c) egg (d) meat
46. Which is sweet in taste, but is not sugar?
(a) Maltose (b) Saccharine
(c) Lactose (d) Glucose
47. Ligase enzyme is used for –
(a) denaturation of DNA
(b) splitting DNA into small bits
(c) joining bits of DNA
(d) digestion of lipids
48. Which functional group participates in disulphide bond formation in proteins ?
(a) Thioether (b) Thiol
(c) Thioester (d) Thiolactone

49. Which of the following is the example of acidic amino acid?
(a) Lysine (b) Glutamic acid
(c) Aspartic acid (d) (b) and (c) both

50. Which of the following biomolecule is insoluble in water ?
(a) α-Keratin (b) Haemoglobin
(c) Ribonuclease (d) Adenine

51. Framework elements in plants are
(a) N, Ca, Mg (b) C, H, N, O
(c) Mg, Cu, Fe (d) C, O, Cu

52. In food, starch is detected by its
(a) granular form
(b) white appearance
(c) energy reserve
(d) blue reaction with iodine

53. The smallest RNA is
(a) tRNA
(b) mRNA
(c) rRNA
(d) chromosomal RNA

54. In isolated DNA kept at 82-90°C
(a) two strands uncoil & separate
(b) fragmentation occurs
(c) thymine is replaced by uracil
(d) structure is stabilised

55. Hydrolysis of starch occurs with the help of
(a) sucrose (b) amylase
(c) peptidase (d) lipase

56. Adenosine diphosphate contains
(a) one high energy bond
(b) two high energy bond
(c) three high energy bond
(d) four high energy bond

57. Which enzyme converts glucose into alcohol?
(a) Zymase (b) Diastase
(c) Invertase (d) Lipase

58. Antibodies in our body are complex
(a) steroids (b) prostaglandins
(c) glycoproteins (d) lipoproteins

59. Decreasing order of amount of organic compound in animal body –
(a) carbohydrate, protein, fat, and nucleic acid
(b) protein, fat, nucleic acid and carbohydrate
(c) protein, fat, carbohydrate and nucleic acid
(d) carbohydrate, fat, protein and nucleic acid

60. If base order in one chain of DNA is "ATCGA" then how many number of H-bonds found in DNA duplex ?
(a) 20 (b) 12
(c) 10 (d) 11

CHAPTER 10

Cell Cycle and Cell Division

CONCEPTS

All cells reproduce by dividing into two, with each parental cell giving rise to two daughter cells each time they divide. These newly formed daughter cells can themselves grow and divide, giving rise to a new cell population that is formed by the growth and division of a single parental cell and its progeny.

CELL CYCLE

- Cell cycle consists of two basic phases. There is a long non-dividing growing I-phase (Interphase) and a short-dividing M-phase.

PHASES OF CELL CYCLE

Interphase :

- The interphase is the longest phase of the mitotic cycle, lasts more than 95% of the duration of cell cycle. It has following three sub-phases:
- **G_1 phase:** It (G stands for gap) includes the synthesis of substrate and enzyme necessary for DNA synthesis. During G_1 phase the cell metabolically active and continuously grows but does not replicate its DNA.
- **S phase:** During S or synthesis phase, replication or duplication of chromosomal DNA molecules takes place. During this time the amount of DNA per cell doubles.

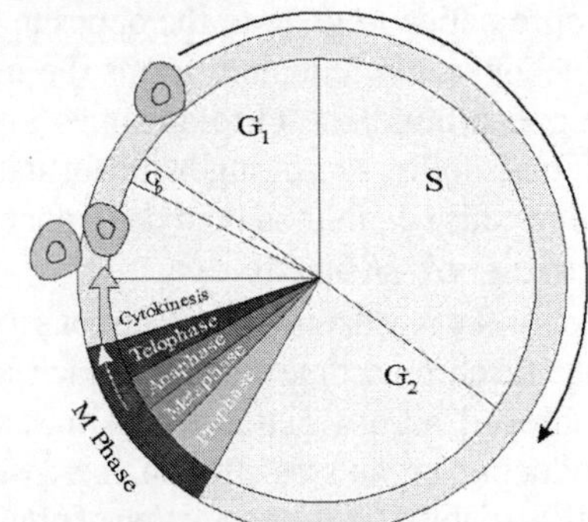

A diagrammatic view of cell cycle indicating formation of two cells from one cell

- **G_2 phase:** During the G_2 phase, proteins are synthesized in preparation for mitosis while cell growth continues.
- **G_0 phase :** Cells that do not divide further exit G_1 phase to enter an inactive stage is called quiescent stage (G_0) of the cell cycle.

M PHASE (MITOSIS PHASE)

It is also called as equational division as the number of chromosomes in the parent and progeny cells is the same. It is usually divided into several stages or phases.

Prophase

- Onset of chromosome coiling and condensation.
- Two pairs of centrioles formed by duplication.
- Initiation of assembly of spindle formation begins.

Metaphase

- At metaphase complete disintegration of the nuclear envelope take place the alignment of chromosome and completion of the spindle apparatus takes place.
- Morphology of chromosome can be observed and studied.

Anaphase

- Mitotic anaphase begins with centromere division. The two daughter chromatids, now free of each other, move towards their respective poles.

Telophase

- The nucleolus condense and reappear. The spindle fibres disperse. The nuclear envelope is assembled around the chromatin mass endoplasmic reticulum reform again.

Cytokinesis

- Karyokinesis (division of nucleus into two) is followed by cytokinesis *i.e.*, division of cytoplasm into two daughter cells, which starts by the end of telophase.

Significance of Mitosis

The mitosis has the following significance for living organisms:

- Helps in the maintenance of equilibrium in the amount of DNA and RNA in the cell.
- Provides the opportunity for the growth and development to organs and the body of the organisms.
- The gonads and the seed cells depend on the mitosis for the increase in their number.

MEIOSIS

Meiosis involves two sequential cycles of nuclear and cell division called **meiosis I** and **meiosis II** but only a single cycle of DNA replication.

Meiosis I

- 7Meiosis starts after an interphase which is not very different from that of an mitotic interphase.

Prophase I

- The first prophase is the longest stage of the meiotic division. It includes following substages:

Leptotene

- The chromosomes become gradually visible under the light microscope.

Zygotene

- The pairing of homologous chromosomes takes place.
- The pairing of the homologous chromosomes is known as **synapsis**.

Pachytene

- This stage is characterised by the appearance of recombina-tion nodules, the sites at which crossing over occurs between non-sister chromatids of the homologous chromosomes.
- Crossing over is the exchange of genetic material between two homologous chromosomes.

Diplotene

- At Diplotene further thickening and shortening of chromosome takes place. The synaptic forces between paired homologues come to an end (synaptonemal complex appears to have dissolved) and the chromosome starts separating.
- The participating chromatids of the paired homologous chromosomes physically joined at one or more discrete points having X-shaped structure called **chiasmata**.
- **Diakinesis :** At this stage the bivalent chromosomes become more condensed and evenly distributed in the nucleus. The nucleolus disappears, the nuclear envelope breaks down. During diakinesis, the terminalisation of chiasmata take place.

Metaphase I

- At metaphase I spindle fibres attach themselves to chromosomes and chromosomes align at the equator.

Anaphase I

- Homologous chromosome with its two chromatids moves towards the opposite poles of the cell and separate from each other.

Telophase I

- In telophase, the nuclear membrane and nucleolus reappear.

Meiosis II

- **Prophase II**
  The nuclear membrane and the nucleolus disappear. The chromosomes with two chromatids become short, thick and compact.
- **Metaphase II**
  The chromosomes get arranged on the equator of the spindle. The microtubules of the spindle are attached with the kinetochores of the chromosomes.
- **Anaphase II**
  The daughter chromosomes move towards the opposite poles due to the splitting of centromere and stretching of internal microtubules of the spindle.
- **Telophase II**
  The chromatids migrate to the opposite poles. The endoplasmic reticulum forms the nuclear envelope around the chromosomes and the cytokinesis follows resulting in the formation of tetrad of cells *i.e.*, four haploid daughter cellls.

Significance of Meiosis

- Meiosis is the mechanism by which conservation of specific chromosome number of each species is achieved across generations in sexually reproducing organisms. It also increases the genetic variability in the population of organisms

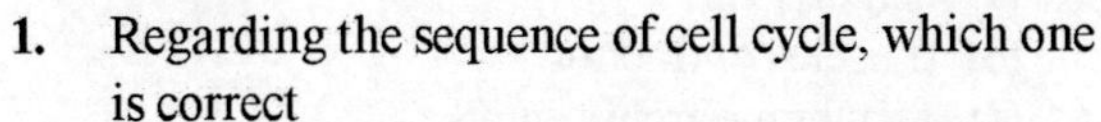

Exercise

1. Regarding the sequence of cell cycle, which one is correct
(a) G_1, G_2, S and M (b) S, G_1, G_2 and M
(c) G_1, S, G_2 and M (d) G_2, S, G_1 and M
2. G_1, G_2 and S phases are seen in which phase of cell cycle
(a) Metaphase (b) Prophase
(c) Anaphase (d) Interphase
3. Condensation of chromosomes occurs in
(a) prophase I (b) prophase II
(c) anaphase (d) metaphase
4. Pick out the correct statements
(A) Mitosis takes place in the somatic cells and meiosis takes place in the germ cells
(B) During mitosis, the DNA replicates once for one cell division and in meiosis the DNA replicates twice for two cell divisions
(C) Mitosis and meiosis occur both in sexually and asexually reproducing organisms
(a) A only (b) B only
(c) C only (d) A and B
5. Which is not the character of mitosis ?
(a) Leptotene (b) Zygotene
(c) Pachytene (d) All of the above
6. Synaptonemal complex is formed during
(a) meiosis (b) amitosis
(c) mitosis (d) cytokinesis
7. Recombinant nodules are found during which of the following.
(a) Anaphase (b) Prophase
(c) Telophase (d) Metaphase
8. Bivalents in meiosis are
(a) tetrad
(b) pairs of non-homologous chromosomes
(c) pairs of several chromatids
(d) pairs of homozygous chromosomes
9. Mitosis is the process by which eukaryotic cells
(a) expose the genes for protein synthesis
(b) become specialized in structure and function
(c) multiply
(d) grow
10. Recombination of genes occur at
(a) prophase in mitosis
(b) prophase i in meiosis
(c) prophase ii in meiosis
(d) metaphase ii in meiosis
11. Which out of the following is not a divisional stage ?
(a) Telophase (b) Interphase
(c) Metaphase (d) Prophase
12. Crossing over is advantageous because it brings about
(a) variation (b) linkage
(c) inbreeding (d) stability
13. Prophase of reduction division is divided into number of stages. The correct chronological sequence is
(a) Leptotene – pachytene – zygotene – diplotene – diakinesis
(b) Leptotene – diplotene – pachytene – zygotene – diakinesis
(c) Leptotene – zygotene – diplotene – pachytene – diakinesis
(d) Leptotene – zygotene – pachytene – diplotene – diakinesis
14. Prophase is longer in
(a) mitosis (b) meiosis
(c) equal in both (d) amitosis
15. Karyokinesis differ from cytokinesis because it involves
(a) division of cytoplasm
(b) division of the nucleus and cytoplasm
(c) division of the nucleus
(d) division of the cell
16. Which is not true for anaphase ?
(a) Golgi body and ER are reformed
(b) Chromosomes move to opposite poles
(c) Spindle poles move farther apart
(d) Centromeres split and chromatide separate
17. Which of the following has centrioles
(a) Chromosomes (b) Spindle fibres
(c) Centrosomes (d) Centromeres

18. In which phase proteins for spindle fibre formation are synthesized ?
(a) G_1 phase (b) G_2 phase
(c) S-phase (d) Anaphase
19. Best stage to observe shape, size and number of chromosomes is
(a) interphase (b) metaphase
(c) prophase (d) telophase
20. In cell cycle, DNA replication takes place in
(a) G_1 phase (b) G_2 phase
(c) mitotic metaphase (d) S phase
21. Meiosis I is reductional division. Meiosis II is equational division due to
(a) pairing of homologous chromosomes
(b) crossing over
(c) separation of chromatids
(d) disjunction of homologous chromosomes
22. Mitotic anaphase differs from metaphase in possessing
(a) same number of chromosomes and same number of chromatids
(b) half number of chromosomes and half number of chromatids
(c) half number of chromosomes and same number of chromatids
(d) same number of chromosomes and half number of chromatids
23. In meiosis, the daughter cells differ from parent cell as well as amongst themselves due to
(a) segregation, independent assortment and crossing over
(b) segregation and crossing over
(c) independent assortment and crossing over
(d) segregation and independent assortment
24. Meiosis is evolutionary significant because it results in
(a) genetically similar daughters
(b) four daughter cells
(c) eggs and sperms
(d) recombinations
25. The exchange of genetic material between chromatids of paired homologous chromosomes during first meiotic division is called
(a) transformation (b) chiasmata
(c) crossing over (d) synapsis
26. Which one of the following structures will not be common to mitotic cells of higher plants?
(a) Cell plate (b) Centriole
(c) Centromere (d) Spindle fibres
27. During cell division in apical meristem the nuclear membrane appears in
(a) metaphase (b) anaphase
(c) telophase (d) cytokinesis
28. Microtubule is involved in the
(a) cell division
(b) muscle contraction
(c) membrane architecture
(d) DNA recognition
29. During cell division, the spindle fibres attach to the chromosome at a region called
(a) chromocentre (b) kinetochore
(c) centriole (d) chromomere
30. Best material for the study of mitosis in laboratory is
(a) anther (b) root tip
(c) leaf tip (d) ovary
31. If you are provided with root-tips of onion in your class and are asked to count the chromosomes, which of the following stages can you most conveniently look into?
(a) Metaphase (b) Telophase
(c) Anaphase (d) Prophase
32. At what stage of the cell cycle are histone proteins synthesized in a eukaryotic cell?
(a) During G-2 stage of prophase
(b) During S-phase
(c) During entire prophase
(d) During telophase
33. Synapsis occurs between
(a) mRNA and ribosomes
(b) spindle fibres and centromere
(c) two homologous chromosomes
(d) a male and a female gamete
34. During mitosis ER and nucleolus begin to disappear at
(a) late prophase (b) early metaphase
(c) late metaphase (d) early prophase
35. Which type of chromosome will appear 'L'-shaped during anaphase ?
(a) Telocentric (b) Acrocentric
(c) Metacentric (d) Submetacentric
36. A contractile mid body forms during cytokinesis in –
(a) animals (b) higher plants
(c) fungi (d) algae
37. Amitosis is characteristic of –
(a) higher plants (b) higher animals
(c) bryophyta (d) lower organisms
38. Which division maintains genetic similarity ?
(a) Mitosis (b) Meiosis
(c) Amitosis (d) Reduction division

39. Cell division in blue green algae is similar to that in
(a) bacteria (b) brown algae
(c) green algae (d) higher plants
40. Gap between meiosis - I and II is called
(a) interphase (b) interkinesis
(c) diakinesis (d) metakinesis
41. Each chromosome composed of one chromatid in
(a) Anaphase-I (b) Anaphase-II
(c) Metaphase-I (d) Metaphase-II
42. Which of the following statement is correct ?
(a) DNA is synthesized through out the cell cycle
(b) Cell division is inhibited by cytokinin
(c) Chromosome are condensed at S-stage
(d) Only extra chromosomal DNA is replicated at any stage of cell cycle
43. Many cells function properly and divide mitotically though they do not have
(a) plasma membrane (b) cytoskeleton
(c) mitochondria (d) plastids
44. Chiasmata are formed as a result of
(a) exchange of parts of paired homologous chromosome
(b) exchange of part of unpaired non-homologous chromosome
(c) duplication of parts of paired homologous chromosome
(d) loss of parts of unpaired non-homolog us chromosome
45. In plants, active mitosis occurs in
(a) cambium (b) leaf tip
(c) root base (d) mid vein
46. Chromosome exhibit high level of coiling at which phase of karyokinesis
(a) Prophase (b) Metaphase
(c) Telophase (d) Interphase
47. Meiosis does not occurs in
(a) ovule (b) anther
(c) microsporangia (d) shoot tip
48. Replication of DNA in meiosis occur during
(a) S - phase
(b) S - phase and leptotene
(c) S - phase and zygotene
(d) All of these
49. Reason of chromosomal movement in Anaphase
(a) Astral rays
(b) Centrioles
(c) Kinetochore
(d) Kinetochore and spindle fibres
50. In which of the following stage, the chromosome is single thin and like long thread ?
(a) Leptotene (b) Zygotene
(c) Pachytene (d) Diakinesis
51. Segregation of mendelian factor (Aa) occurs during
(a) diplotene (b) anaphase I
(c) zygotene/Pachytene(d) anaphase II
52. Which one of the following precedes reformation of the nuclear envelope during M phase of the cell cycle ?
(a) Decondensation from chromosomes, and reassembly of the nuclear lamina
(b) Transcription from chromosomes, and reassembly of the nuclear lamina
(c) Formation of the contractile ring, and formation of the phragmoplast
(d) Formation of the contractile ring, and transcription from chromosomes
53. Cell division or mitosis is normal process in a living cell, but sudden and abnormal mitosis in an organ will frequently result in
(a) zygote (b) cancer
(c) new organ (d) gastrula
54. Centrosome is not present in
(a) cells of higher plants
(b) cells of lower plants
(c) cells of higher animals
(d) cells of lower animals
55. An angiospermic plant has 24 chromosomes in 'microspore mother cell'. The number of chromosomes in its endosperm will be
(a) 12 (b) 24
(c) 36 (d) 48
56. At anaphase - II of meiosis each chromosome contains
(a) 4 - DNA (b) 3 - DNA
(c) 2 - DNA (d) 1 - DNA
57. Genetic information is transferred from zygote to all body cell by
(a) meiosis (b) amitosis
(c) endomitosis (d) mitosis
58. How many chromosome shall be present in a diploid cell at mitotic anaphase if its egg cell has ten chromosome ?
(a) 10 (Ten) (b) 20 (Twenty)
(c) 30 (Thirty) (d) 40 (Forty)
59. Which does not occurs in prophase?
(a) Hydration of chromatin
(b) Dehydration of chromatin
(c) Appearance of chromosome
(d) Disappearance of nuclear membrane and nucleolus
60. Slipping of chiasmata towards the ends of bivalent is called
(a) terminalisation (b) diakinesis
(c) interkinesis. (d) heteropycnosis

CHAPTER 11

Transport in Plants

CONCEPTS

- Plants lack both interstitial fluid as well as a regular circulation system.
- Organic compounds are exported to all other parts including every living cell, growing points, fruits and storage organs.
- A complex traffic of materials is going on in flowering plants, some moving to different directions, some passing out in polarised manner, with most organs, receiving some substances and giving out some others.

MEANS OF TRANSPORT

Diffusion :

- Diffusion is a random movement of individual molecules from a region of higher concentration to a region of lower concentration.
- **Diffusion rates** are affected by concentration gradient, membrane permeability, temperature and pressure.
- The extent to which a membrance permits or restricts the movement of a substance is called membrane permeability.

Facilitated Diffusion :

- The substance that have a hydrophilic moiety, find it difficult to pass through membrane. The movement of such molecules are facilitate, for which proteins provide site at which such molecule cross membrane. This is called as facilitated diffusion.
- The special proteins that help in movement of substances across the membranes do not require expenditure of ATP energy.

Passive Symports and Antiports :

- In symport, both molecules cross the membrane in the same direction.
- In antiport, both molecules cross the membrane in opposite directions.
- In uniport, molecules moves across a membrane independent of other molecule.

Active Transport :

- Active transport uses energy to pump molecules against a concentration gradient.
- Active transport is carried out by membrane bound proteins.

PLANTS-WATER RELATIONS

- Water acts as an excellent solvent and help in the uptake and distribution of mineral nutrients and other solutes.

Water Potential :

The potential energy of water is referred to as *water potential*.

- It is measured in term of pressure.
- The common measurement unit of water potential is ***Pascal***, Pa (1 Megapascal, MPa = 10 bars) and represented by Greek letter, *Psi* (ψ).
- It is a relative quantity expressed as the difference between potential of a solution in a given state and potential of the same solution in a standard state.
- Water potential of any system can be represented as

$$\Psi_w = \Psi_s + \Psi_p$$

Osmosis :

- Osmosis is the diffusion of water through a semipermeable membrane.
- Osmotic pressure and osmotic potential are numerically equal, but osmotic potential has a negative sign.

$\Psi_s = -\pi$ [when Ψ_s = Osmotic potential of solution;

π = osmotic pressure].

- Osmosis depends on two factors
  (i) concentration of dissolved solutes in a solution
  (ii) pressure difference.

Plasmolysis :

- If a turgid cell is placed in a solution that has more solutes, it exerts a higher osmotic pressure and water will move out.
- **Isotonic solution :** When concentration of outer solution (in which cell is placed) is equal to concentration of cell sap.
- **Hypotonic solution :** When concentration of outer solution is lower than concentration of cell sap.
- **Hypertonic solution :** When concentration of outer solution is higher than concentration of cell sap.

Imbibition :

- It is a type of diffusion by which movement of water takes place along a diffusion gradient.
- Two conditions necessary for imbibition are –
  (i) water potential gradient between surface of the adsorbent and the liquid imbibed,
  (ii) affinity between the adsorbent and imbibed liquid.

Factors influencing the rate of imbibition are nature of imbibant, surface area of imbibant, temperature, concentration of solutes, pH of imbibant

LONG DISTANCE TRANSPORT OF WATER

Transport of substances over longer distances through the vascular tissue, *i.e.*, xylem and pholem, is called **translocation**. It occurs through **mass flow**. The direction of translocation is essentially **unidirectional** in case of water. It is **multidirectional**, in case of minerals and organic solutes.

Absorption of Water Follows Two Pathways :

- **Apoplastic** movement of water occurs exclusively through cell wall; without crossing any membrane.
- **Symplastic** movement occurs from cell to cell through plasmodesmata.

WATER MOVEMENT UP A PLANT

Root Pressure :

- Mineral ions from soil are taken up by the roots and get deposited in the xylem vessels.
- Accumulations of ions leads to lowering of osmotic potential.
- The low water potential of xylem sap compared to the soil solution attracts water and water enters the root.

Capillarity :

- Water is first taken in due to the force of adhesion between water and the walls of thin xylem vessels (tracheids).
- As water flows upward along the wall, strong cohesive forces between water molecules come into play to pull the water upward.

Cohesion theory :

- Proposed by ***Henry Dixon*** 1914.
- Evaporation of water from the leaf to atmosphere decreases the water potential of the epidermal cells.
- Water moves from adjacent cells along water potential gradient.
- Thus it form a continuous water column from root to leaves due to cohesive and adhesive properties of water molecules.
- Uptake of water in this way is also referred to an *transpirational pull.*

TRANSPIRATION :

- Loss of water in the form of water vapour from plant through the small pores (stomata) present on leaves is called transpiration.

Significance of transpiration :

- If transpirational loss of water is higher, then plants undergo water stress. This declines the growth in plants.
- Helps in movement of xylem sap and absorption of minerals nutrients.

Factors affecting transpiration :

(i) Humidity – Humidity is high, the rate of transpiration is slow.

(ii) Rate of evaporation doubles with every rise in temperature by 10° C.

(iii) High wind speed increases temperature.

The opening & closing of the stomatal pore is brought about by changes in the volume & shape of the guard cells. When the guard cells absorb water from the surrounding cells, they become turgid & when the water move out of guard cell, they become flaccid. An increase in turgidity causes the outer thinner wall of guard cell to stretch in much more amount than the inner thick wall. This result in the opening of the pore. Thus when the guard cells are turgid, stomata open & when flaccid, the stomata close.

Classification of plant on the basis of distribution of stomata on the sides of leaf :

1. Apple or mulberry type - Stomata only on one side on dorsal (lower) sides.
2. Potato type - Stomata on both sides of leaf, more on dorsal side of leaf.
3. Oat type - No. of stomata equal on both sides.
4. Waterlily type - Stomata only found on one side of leaf, only on upper (ventral) side.
5. *Potamogeton type* - Stomata absent *e.g.* - submerged hydrophytes.

Classification of leaf or plant on the basis of duration of opening & closing of stomata :

1. Alfalfa type - Stomata open whole day & closed whole night.
2. Succulent type - Stomata closed whole day & open whole night.
3. Potato type - Stomata open whole day & night but closed for a short period after sunset.
4. Barley type - Stomata closed whole day & night but it open for a short period at mid-day.
5. *Equisetum type* - Stomata always open.

3 TYPES OF TRANSPIRATION

(a) Stomatal transpiration - 50 – 97% (through stomata)

(b) Cuticular - Through cuticle (3 – 10%).

(c) Lenticular - Through lenticels (0.1%).

Guttation :

- Removal of H_2O in the form of liquid is called exudation of water or guttation. It mainly takes place during warm & moist night.

UPTAKE AND TRANSPORT OF MINERAL NUTRIENTS

Uptake of Mineral Ions

- Plants obtain their supply of carbon and most of their oxygen form CO_2 of atmosphere, hydrogen from water while the rest are minerals which are picked up individually from the soil.
- The movement of ions from soil to interior of root is against concentration gradient and requires an **active transport.** Specific **ion pumps** occur in the membrane of root hairs.
- For active transport, ATPase are present over the plasma membranes of root epidermal cells. They establish an **electrochemical proton gradient** for supplying energy for movement of ions.

Translocation of Mineral Ions

Mineral elements pass up xylem in both inorganic and organic form. They reach the area of their sink, namely young leaves, developing flowers, fruits and seeds, apical and lateral meristems and individual cells for storage. Minerals are unloaded at fine vein endings through diffusion. They are picked up by cells through active uptake.

Phloem transport Flow From Source to Sink

- It is the long distance movement of organic substances from the **source** or supply end (region of manufacture or storage) to the region of utilization or **sink**.
- The source-sink relationship is variable, the direction of movement of organic solutes in phloem can be upwards or downwards *i.e.*, *bidirectional.*

The pressure flow or Mass Flow

Hypothesis

It was put forward by **Munch (1930).** According to this hypothesis, organic substances move from the region of high osmotic pressure to the region of low osmotic pressure in a mass flow due to the development of a gradient of turgor pressure.

Exercise

1. The physical process involved in the release of molecular oxygen from leaves is
(a) diffusion (b) transpiration
(c) osmosis (d) capillarity

2. Osmosis is the diffusion of a solution of a weaker concentration when both are separated by semipermeable membrane. What is the error in the statement ?
(a) The movement of solvent molecule is not specified
(b) There is no mention of DPD
(c) Behaviour of semipermeable membrane is not specified
(d) The exact concentration of solutions are not indicated

3. If a plant cell is immersed in water, the water continues to enter the cell until the
(a) concentration of the salts is the same inside the cell as outside
(b) cell bursts
(c) concentration of water is the same inside the cell as outside
(d) diffusion pressure deficit is the same inside the cell as outside

4. If a cell swells, after being placed in solution, the solution is
(a) neutral (b) hypotonic
(c) hypertonic (d) isotonic

5. Osmosis means
(a) solute from low concentration to higher
(b) solute from higher concentration to low
(c) solvent from low concentration of solution to higher concentration of solution
(d) solvent from higher concentration solution to low concentration solution

6. Which of the following statement is not correct ?
(a) Plants absorb excess quantity of water
(b) Plants take small quantity of mineral salts through soil water
(c) Water and inorganic salts may also simultaneously by root hairs
(d) Plant absorb only one thing at a time water or inorganic salts

7. Which of the following is a rapid type of absorption?
(a) Passive absorption (b) Active absorption
(c) Salt absorption (d) Root absorption

8. The form of water absorbed by plant's root system, from the soil is :
(a) hygroscopic water (b) gravitational water
(c) capillary water (d) All of these

9. The pathway of water from soil upto the secondary xylem
(a) Soil → root hair → cortex → endodermis → pericycle → protoxylem → meta xylem
(b) Metaxylem → protoxylem → pericycle → cortex → endodermis → soil → root hair
(c) Cortex → root hair → endodermis → pericycle → protoxylem → metaxylem
(d) Pericycle → soil → root hair → cortex → endodermis → protoxylem → metaxylem

10. Humus in soil is necessary for plant growth because it
(a) increases aeration and water absorption capacity of soil
(b) makes soil compact
(c) makes soil sterile
(d) decreases rate of percolation

11. Water will be absorbed by root hairs when
(a) concentration of salts in the soil is high
(b) concentration of solutes in the cell sap is high
(c) the plant is rapidly respiring
(d) they are separated from the soil by a semipermeable membrane

12. Water in plants is transported by ascent of sap takes place through
(a) cambium (b) phloem
(c) xylem (d) epidermis

13. The metal ion involved in the stomatal regulation is
(a) iron (b) magnesium
(c) zinc (d) potassium

14. Transpiration from plants would be most rapid when
(a) there is lot of humidity in atmosphere
(b) the air is more humid
(c) there is excess rain fall
(d) environmental conditions are dry

15. Wilting of a plant result from excessive
(a) respiration (b) photosynthesis
(c) absorption (d) transpiration

16. Excessive loss of water causes wilting of leaves, it can be prevented by
(a) keeping the plant in bright light
(b) spraying the plant with alcohol
(c) applying vaseline on the leaf surface
(d) adding high amounts of fertilizers to the soil

17. Increase in CO_2 concentration around leaf results in
(a) rapid opening of stomata
(b) partial closure of stomata
(c) complete closure of stomata
(d) no effect on stomatal opening

18. Which of the following wall of guard cells is thick ?
(a) Outer (b) Inner
(c) Sidewall (d) All the three

19. Which of the atmospheric factor act as anti-transpirant ?
(a) SO_2 (b) CO
(c) CO_2 (d) All pollutant gases

20. Water potential is equal to
(a) Ψ_s + O.P. (b) Ψ_s = T.P.
(c) $\Psi_p + \Psi_w$ (d) $\Psi_s + \Psi_p$

21. Stomata open and close due to
(a) circadian rhythm
(b) genetic clock
(c) pressure of gases inside the leaves
(d) turgor pressure of guard cells

22. The principal pathway of water translocation in angiosperms is
(a) sieve cells
(b) sieve tube elements
(c) xylem vessel system
(d) xylem and phloem

23. Guttation is mainly due to
(a) root pressure (b) osmosis
(c) transpiration (d) imbibition

24. Active and passive transports across cell membrane differ in
(a) passive transport is nonselective
(b) passive transport is along the concentration gradient while active transport is due to metabolic energy
(c) active transport is more rapid
(d) passive transport is confined to anions while active transport is confined to cations

25. Bidirectional translocation of minerals takes place in
(a) xylem (b) phloem
(c) parenchyma (d) cambium

26. The water potential and osmotic potential of pure water are
(a) zero and zero (b) 100 and 100
(c) zero and 100 (d) 100 and zero

27. If a cell A with DPD 4 bars is connected to cell B, C, D whose OP and TP are respectively 4 and 4, 10 and 5 and 7 and 3 bars, the flow of water will be
(a) A and D to B and C
(b) A to B,C and D
(c) B to A, C and D
(d) C to A, B and D

28. Which of the following plant is found to have minimum transpiration ?
(a) Cactus (b) Hydrilla
(c) Mango (d) Guava

29. Stomata of a plant open due to
(a) influx of calcium ions
(b) influx of potassium ions
(c) efflux of potassium ions
(d) influx of hydrogen ions

30. The ability of the Venus fly trap to capture insects is due to
(a) specialised "muscle-like" cells
(b) chemical stimulation by the prey
(c) a passive process requiring no special ability on the part of the plant
(d) rapid turgor pressure changes

31. The translocation of organic solutes in sieve tube members is supported by
(a) P-proteins
(b) mass flow involving a carrier and ATP
(c) cytoplasmic streaming
(d) root pressure and transpiration pull

32. Transpiration differs from evaporation in
(a) rate of water loss
(b) transpiration is a physiological process while evaporation is a physical process
(c) transpiration is a physical process while evaporation is a physiological process
(d) frequency of water loss

33. Excessive supply of chemical fertilizers often causes death of crop plants due to
(a) exosmosis (b) endosmosis
(c) imbibition (d) turgidity

34. In rainy season, door gets swelled due to
(a) imbibition (b) diffusion
(c) transpiration (d) respiration

35. At what time the stomatal transpiration in plants is almost completely stopped?
(a) Morning (b) Night
(c) Evening (d) None of these

36. Which one explains ascent of sap ?
(a) Cohesion tension theory of Dixon & Jolly
(b) Starch-sugar inter-conversion
(c) Photosynthesis
(d) None

37. DPD is abbreviated form of
(a) Daily Photosynthetic Deficit
(b) Daily Phosphorus Deficit
(c) Daily Pressure Deficit
(d) Diffusion Pressure Deficit

38. Good soil is
(a) which holds whole of the water that enters into it
(b) which allows percolating of the water slowly from it
(c) which allows water to pass very quickly from it
(d) which allows limited amount of water to retain into it.

39. Guard cells differ from epidermal cells in having
(a) mitochondria (b) vacuoles
(c) cell wall (d) chloroplasts

40. Turgor pressure become equal to the wall pressure when
(a) water leaves the cell
(b) water enters the cell
(c) no exchange of water takes place
(d) solute goes from cell into water

41. Which one of the following doesn't help in molecule transport?
(a) Diffusion (b) Osmosis
(c) Surface tension (d) Active transport

42. Which of the following is not performed by root hairs ?
(a) Water uptake (b) Oxygen uptake
(c) Mineral uptake (d) CO_2 uptake

43. Movement of ions or molecules against the electrochemical gradient is called
(a) diffusion
(b) pinocytosis
(c) Brownian movement
(d) active transport

44. When stomata open, the pH of guard cells
(a) increases (b) decreases
(c) remains same (d) both 'a' and 'b'

45. Conversion of starch to organic acids is required for
(a) stomatal opening (b) stomatal closing
(c) stomatal formation (d) stomatal activity

46. Plants die from prolonged water-logging because
(a) soil nutrients become very dilute
(b) root respiration stops
(c) cell sap in the plants becomes too dilute
(d) nutrients leach down due to excess water

47. Water potential of actively absorbing cells is
(a) always +ve (b) always – ve
(c) always 0 (d) always > 1

48. The rupture and fractionation do not usually occur in the water column in vessel/tracheids during the ascent of sap because of
(a) lignified thick walls
(b) cohesion and adhesion
(c) weak gravitational pull
(d) transpiration pull

49. Carbohydrates are commonly found as starch in plant storage organs. Which of the following five properties of starch (A–E) make it useful as a storage material?
(A) Easily translocated
(B) Chemically non-reactive
(C) Easily digested by animals
(D) Osmotically inactive
(E) Synthesized during photosynthesis
The useful properties are
(a) (B) and (C) (b) (B) and (D)
(c) (A), (C) and (E) (d) (A) and (E)

50. Leaves which appear wilted in the day time recover at night because
(a) light is essential for photosynthesis
(b) the stomata close down, temperature decrease, transpiration is reduced and the plant is able to absorb more water from the soil
(c) respiration and translocation of organic substance both increase
(d) the plant is sleeping because of dark conditions

51. What is action spectrum of transpiration ?
(a) Green and U.V. (b) Blue and yellow
(c) Blue and far red (d) Blue and red

52. Root pressure is maximum when
(a) transpiration is high and absorption is very low
(b) transpiration is very low and absorption is very high
(c) absorption is very high and transpiration is very high
(d) absorption is low and transpiration is also very low

53. Which of the following plant dies rapidly?
(a) Hollow (b) Girdled plant
(c) Pruned plant (d) Deciduous

54. Path of water movement from soil to xylem is
(a) metaxylem – protoxylem – cortex – soil – root hair
(b) cortex – root hair – endodermis – pericycle – protoxylem – metaxylem
(c) soil – root hair – cortex – endodermis – pericycle – protoxylem – metaxylem
(d) pericycle – soil – root hair – cortex – endodermis – protoxylem – metaxylem

55. The plant ash is an indication of
(a) mineral salts absorbed by the plant
(b) organic matter of the plant
(c) both the mineral salts and organic matter
(d) None of the above

56. A semi-permeable membrane allows the diffusion of
(a) solutes (b) solvent
(c) both 'a' & 'b' (d) None

57. Mango dipped in conc. NaCl solution will
(a) burst (b) contract
(c) swell (d) no effect

58. Which of the following theories gives the latest explanation for closure of stomata?
(a) Starch glucose theory
(b) Munch theory
(c) ABA theory
(d) Active K^+ transport theory

59. Water lost in guttation is :
(a) pure water (b) impure water
(c) in vapour form (d) either 'a' and 'b'

60. The membrane which allows passage of certain substances more readily than others is termed as
(a) permeable
(b) selectively permeable
(c) semipermeable
(d) impermeable

CHAPTER 12

Mineral Nutrition

CONCEPTS

- All living organisms require macromolecules like carbohydrates, proteins, fats and minerals for the development.
- Only the green plants can prepare their own food from simpler substances, rest of the living organisms depend on them directly or indirectly for food.
- **Hydroponics :** The system of growing plants in soilless culture (also called solution culture or *tank farming*) is known as *hydroponics*.

ESSENTIAL MINERAL ELEMENTS

- The criteria for essentiality of an element are -
  - The element must be absolutely necessary for supporting normal growth and development.
  - The requirement of the element must be specific and not replaceable by any element.
  - The element must be directly involved in the metabolism of plants.
- **Macro nutrients -** Carbon, hydrogen, oxygen, nitrogen, phosphorous, sulphur, potassium, helium, magnesium and silicon.
- **Micro nutrients -** Manganese, copper, molybdenum, zinc, boron and chlorine. In some plants cobalt, vanadium and nickels may be essential.

ROLE OF ESSENTIAL ELEMENTS (MACRO & MICRO) AND THEIR DEFICIENCY SYMPTOMS

S. No.	Name of Element	In which form they are absorbed	Functions	Deficiency Symptoms
1.	Nitrogen	NO_2^-, NO_3^- or NH_4^+	Major contituent of	Chlorosis (yellowing of older leaves)proteins, nucleic acids, vitamins and minerals.
2.	Phosphorous	($H_2PO_4^{2-}$, or HPO_4^{2-})	Constituent of cell membrane, nucleic acids, nucleotides and some proteins.	Delay in seed germination purple or red spots on leaves
3.	Potassium	K^+	Involved in protein synthesis, closing & opening of stomata. Maintenance of turgidity of cells.	Chlorosis in intervernial area, loss of apical dominance
4.	Calcium	Ca^{++}	Used in synthesis of cell wall (middle lamalla)	Stunted growth, necrosis of meristematic regions.
5.	Magnesium	Mg^{++}	Activate enzymes in respiration, photo -synthesis, DNA and RNA synthesis.	Chlorosis between leaf veins, necrosis on older leaves.

			Constituents of the porphyrin ring of chlorophyll structure.	
6.	Sulphur	SO_4^{2-}	Constituent of amino acids like cystine and methionine.	Chlorosis of younger leaves, stunted growth anthocyanin accumulation.
7.	Iron	Fe^{3+}	Constiuent of feredoxin and cytochromes.	Chlorosis, initiates in intravenous regions and then in the complete leaf.
8.	Zinc	Zn^{2+}	Neccessary for synthesis of auxin.	Malformed leaves, inter-veinal chlorosis, stunted growth.
9.	Copper	Cu^{2+}	Associated with certain enzymes involved in redox reactions.	Necrosis of tip of young leaves which extends towards leaf base.
10.	Boron	BO_3^{3-} and $B_4O_7^{2-}$	Necessary for uptake and utilization of Ca^{2+}, pollen generation.	Death of root and shoot tips, abscission of flowers.
11.	Chlorine	Cl^-	Determine solute concentration (with Na^+ and K^+) and anion-cation balance in cells, essential for photolysis of water.	Stunted root growth, reduced fruiting.

MECHANISM OF ABSORPTION OF ELEMENTS :

Passive Absorption :

- It is the process of absorption of minerals by physical processes not involving direct expenditure of metabolic energy.
- It occurs through ion channels (transmembrane protein)

Ion Exchange :

- Both cations and anions gets absorbed on the surface of cell wall.

Donnan Equilibrium :

- Donnan equilibrium is said to the attained if the product of anions and cations of internal solution becomes equal to the products of anions and cations of external solution.

Mass Flow Hypothesis :

- According to this hypothesis mass flow of ions occur along with absorption of water as an result of transpirational pull.

Active Absorption :

- It is the process of movement of ions against metabolic energy.

Ion Traffic Into the Root :

- In apoplast pathway diffusion and bulk flow of water occur from cell to cell through spaces between cell wall polysaccharides.
- In symplast pathway, ions enter the xylem vessels through the following pathway : cytoplasm of cortex $\rightarrow$ endodermis of pericycle (through $\rightarrow$ plasmodesmata vessels)

TRANSLOCATION OF SOLUTES :

- By feeding plants with radioisotopes, it has been proved that inorganic substances move up the plant through xylem these substances move along with water by transpiration pull.

Soil as Reservoir of Essential Elements

Soil provides anchorage, air, water and minerals to the plants growing in it. Soil also has a large number of microorganisms that not only decompose organic remains but also release the minerals bound in organic matter. Nitrogen fixing bacteria increase the usable nitrogen content of soil.

METABOLISM OF NITROGEN :

Nitrogen Cycle

- Nitrogen is found in essential compounds like proteins, nucleic acids, growth regulators, and many vitamins.

Fixation of N_2

- The atmospheric N_2 is fixed into organic combinationsles (amino acids, proteins, nucleic acids, vitamins etc.) in living organism *via* NH_4^+.
- Ammonia is rapidly converted first to nitrites (by *Nitrobacter*) by the process *nitrification.*
$$2NH_3 + 3O_2 \rightarrow 2NO_2^- + 2H^+ + 2H_2O$$
$$2NO_2^- + O_2 \rightarrow 2NO_3^-$$
- Nitrate is then either available to the plant, or converted to nitrogen gas in the process of denitrification (by *Pseudomonas*).
- Fixation is done by both free living *e.g. Azotobacters, Clostridium,* Cyanobacteria like *Nostoc, Anabena* and symbiotic bacteria *Rhizobium.*
- Nodules act as the site for N_2 fixation. It contains leghaemoglobin (a pink pigment) and enzyme nitrogenase (Mo-Fe protein).
- During this process, the N_2 atmospheric(dinitrogen) is reduced by the addition of hydrogen atoms to ammonia.
$N_2 + 8e^- + 8H^+ + 16\ ATP \rightarrow 2NH_3 + H_2 + 16ADP + 16Pi$
- Amino acids are formed from ammonia.
- The amino with amides are linked by peptide bonds to form polypeptide chain.
- One or more polypeptide chain form protein.

Heterotrophic Nutrition in Plants

- **Parasites** - Plants that obtained their food from other living tissue by penetrating their haustoria into the vascular bundles of the host plant. *E.g. Cuscuta.*
- **Saprophytes** - Plants grow on dead and decaying matter. *E.g. Monotropa* in some bacteria fungi.
- **Symbionts** - Two organism live in close physical association and are of natural benefit to each other. *E.g. Lichens, Mycorrhiza.*
- **Insectivorous plants** - Plants obtain part of their nourishment from soil and part by catching insects. The amino acids this obtained supplement their nitrogen requirement. *E.g. Drosera, Utricularia, Nepenthes.*

Exercise

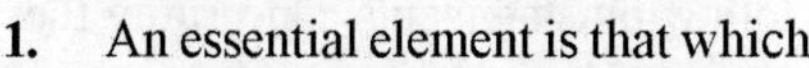

1. An essential element is that which
(a) improve health of the plant
(b) is irreplaceable and indispensable for growth of plants
(c) is found in plant ash
(d) is available in the soil
2. Tracer elements are
(a) micro elements (b) macro elements
(c) radio isotopes (d) vitamins
3. K, N, Ca, Mg deficiency causes
(a) chlorosis (b) leaf curl
(c) exanthema (c) little leaf
4. Essential macro elements are
(a) absorbed from soil
(b) manufactured during photosynthesis
(c) produced by enzymes
(d) produced by growth hormones
5. In nodulous plant leghaemoglobin helps in
(a) Scavanging of oxygen
(b) Nodule formation
(c) Help reduction of nitrite
(d) Dimide formation
6. A plant that manufactures its own food is
(a) autotroph (b) parasite
(c) epiphyte (d) saprophyte
7. Which of the group of elements is not essential for a normal plant ?
(a) K, Ca, Mg (b) Fe, Zn, Mn, B
(c) Pb, I, Na (d) Mg, Fe, Mo

8. Hydroponics is a technique in which plants are grown in
(a) green house
(b) water saturated sand
(c) balanced nutrient solution
(d) purified distilled water

9. For chlorophyll formation a plant needs
(a) Fe, Ca & light (b) Fe, Mg & Light
(c) Ca, K & light (d) Mn & Cu

10. The amino acid having S in its composition is
(a) cystine (b) cysteine
(c) methionine (d) All

11. Which elements are considered as balancing elements?
(a) Ca & K (b) C & H
(c) N & S (d) Mg & Fe

12. The group of mineral nutrients known as frame work elements are
(a) N, S, P (b) C, H, O
(c) Mg, Fe, Zn (d) Zn, Mn, Cu

13. Which element is required in comparatively least quantity for the growth of plant?
(a) Zn (b) N
(c) P (d) Ca

14. Which mineral nutrients are called critical element for crops ?
(a) N, P, K (b) C, H, O
(c) N, S, Mg (d) K, Ca, Fe

15. The mineral nutrient mainly concerning with apical meristematic activity is
(a) K (b) Ca
(c) N (d) S

16. Chlorosis disease is caused by
(a) Zn - deficiency (b) Cu - deficiency
(c) Mo - deficiency (d) S - deficiency

17. Which of the following does NPK denote ?
(a) Nitrogen, Potassium, Kinetin
(b) Nitrogen, Protein, Kinetin
(c) Nitrogen, Protein, Potassium
(d) Nitrogen, Phosphorus, Potassium

18. Plants absorb mineral salts from the soil solution through
(a) a semipermeable membrane into the cytoplasm
(b) perforations at the apex of root hair cells
(c) the cell wall which is semipermeable
(d) None of these

19. Mineral salts which are absorbed by the roots from the soil are in the form of
(a) very dilute solution
(b) dilute solution
(c) concentrated solution
(d) very concentrated solution

20. Which is free ion present in a cell ?
(a) P (b) K
(c) Fe (d) B

21. Carrier protein helped in a
(a) active absorption of ions
(b) passive ions absorption
(c) water absorption
(d) vaporization

22. Active uptake of minerals depends upon
(a) active water absorption
(b) transpiration
(c) photorespiration
(d) dephosphorylation

23. Which element is related with Khaira disease of paddy & auxin synthesis ?
(a) Fe (b) Zn
(c) B (d) Cu

24. Generally plants absorbed N_2 in the form of –
(a) NO_2^- (b) NO_3^-
(c) $N \equiv N$ (d) HNO_2

25. Sulphur is major constituents of
(a) Sugars (b) Vitamins
(c) Hormones (d) Fats

26. Which is essential for N_2 metabolism ?
(a) B (b) Mo
(c) Cu (d) Mg

27. Which one of the following mineral element plays an important role in biological nitrogen fixation?
(a) Copper (b) Manganese
(c) Zinc (d) Molybdenum

28. All of the following statements concerning the Actinomycetous filamentous soil bacterium *Frankia* are correct except the *Frankia* –
(a) can induce root nodules on many plant species
(b) cannot fix nitrogen in the free-living state
(c) like *Rhizobium*, it usually infects, its host plant through root hair deformation and stimulates cell proliferation in the host's cortex
(d) forms specialized vesicles in which the nitrogenase is protected from oxygen by a chemical barrier involving triterpenoids

29. Minerals absorbed by roots move to the leaf through
(a) xylem (b) phloem
(c) sieve tubes (d) None of the above

30. Which one is an essential mineral, not constituent of any enzyme but stimulates the activity of many enzymes?
(a) Zn (b) Mn
(c) K (d) Mg

31. Mycorrhiza is a symbiotic relationship between roots of higher plants and
(a) virus (b) fungi
(c) bacteria (d) blue green algae

32. Which of the following can fix atmospheric nitrogen ?
(a) *Albugo* (b) *Cystopus*
(c) *Saprolegnia* (d) *Anabaena*

33. Which of the following is not caused by deficiency of mineral nutrition?
(a) Necrosis
(b) Chlorosis
(c) Etiolation
(d) Shortening of internodes

34. Which of the following is a free living aerobic nonphotosynthetic nitrogen-fixer?
(a) *Rhizobium* (b) *Azotobacter*
(c) *Azospirillum* (d) *Nostoc*

35. Enzyme involved in nitrogen assimilation is
(a) nitrogenase (b) nitrate reductase
(c) transferase (d) transaminase

36. Boron in green plants assist in
(a) sugar transport
(b) activation of enzymes
(c) acting as enzyme cofactor
(d) photosynthesis

37. A free living nitrogen-fixing cyanobacterium which can also form symbiotic association with the water fern *Azolla* is
(a) *Tolypothrix* (b) *Chlorella*
(c) *Nostoc* (d) *Anabaena*

38. Manganese is required in
(a) plant cell wall formation
(b) photolysis of water during photosynthesis
(c) chlorophyll synthesis
(d) nucleic acid synthesis

39. Which one of the following elements in plants is not remobilised?
(a) Phosphorus (b) Calcium
(c) Potassium (d) Sulphur

40. Nitrifying bacteria
(a) oxidize ammonia to nitrates
(b) convert free nitrogen to nitrogen compounds
(c) convert proteins into ammonia
(d) reduce nitrates to free nitrogen

41. The function of leghaemoglobin in the root nodules of legumes is
(a) inhibition of nitrogenase activity
(b) oxygen removal
(c) nodule differentiation
(d) expression of nif gene

42. Importance of micro-elements was recognized late due to
(a) their toxicity
(b) presence as contaminants in macronutrients
(c) their absence in plant ash
(d) leakage from roots

43. Active transport from outside to inside of molecule across a member requires
(a) cyclic AMP (b) acetylcholine
(c) ATP (d) phospholipid

44. Rotation of crops is useful for
(a) increasing mineral quantity of soil
(b) increasing fertility of soil
(c) high protein content
(d) denitrification of crops

45. Which one of the following essential elements can land plants normally obtain directly from air?
(a) Hydrogen (b) Carbon
(c) Nitrogen (d) Phosphorus

46. Insectivorous plants grow in soils which are deficient in
(a) nitrogen (b) carbohydrate
(c) vitamin C (d) calcium

47. Macro-element found in phosphorylation reactions is
(a) sulphur (b) magnesium
(c) calcium (d) phosphorous

48. The deficiencies of micronutrients not only affects growth of plants but also vital functions such as photosynthetic and mitochondiral electron flow. Among the list given below, which group of three elements shall affect most, both photosynthetic and mitochondrial electron transport ?
(a) Co, Ni, Mo (b) Ca, K, Na
(d) Mn, Co, Ca (d) Cu, Mn Fe

49. Fall of immature leaf is due to the deficiency of
(a) sulphur (b) phosphorus
(c) sodium (d) zinc

50. Which one of these do plants require for the formation of ATP ?
(a) N, Ca (b) N, P
(c) N, Cu (d) K

51. Which element essential stability of chromosome structure ?
(a) Zn (b) Ca
(c) Mo (d) Fe

52. The major portion of the dry weight of plants comprises of
(a) nitrogen, phosphorus and potassium
(b) calcium, magnesium and sulphur
(c) carbon, nitrogen and hydrogen
(d) carbon, hydrogen and oxygen

53. Farmers in a particular region were concerned that pre-mature yellowing of leaves of a pulse crop might cause decrease in the yield. Which treatment could be most beneficial to obtain maximum seed yield ?
(a) Removal of all yellow leaves and spraying the remaining green leaves with 2,4,5-trichlorophenoxy acetic acid.
(b) Application of iron and magnesium to promote synthesis of chlorophyll
(c) Frequent irrigation of the crop
(d) Treatment of the plants with cytokinins alongwith a small dose of nitrogenous fertilizer.

54. Which of the following is a non-leguminous plant with nodules containing nitrogen-fixing micro-organism ?
(a) *Cicer arietinum*
(b) *Alnus*
(c) *Crotalaria juncea*
(d) *Cycas revoluta*

55. Guard cells help in
(a) transpiration
(b) guttation
(c) fighting against infection
(d) protection against grazing

56. An adaptation for better gaseous exchange in plant leaves is
(a) hair on lower surface
(b) multiple epidermis
(c) waxy cuticle
(d) stomata on lower surface away from direct sun rays

57. Transpiration is least in
(a) good soil moisture
(b) high wind velocity
(c) dry environment
(d) high atmospheric humidity

58. What statement can be cited for 10% sodium chloride solution and 10% sugar solution present?
(a) Both have equal OP
(b) The concentration of sodium chloride solution will be less than concentration of sugar solution
(c) The OP of sugar solution will be higher than OP of sodium chloride solution
(d) DPD of sodium chloride solution will be higher than DPD of sugar solution

59. Pieces of beet root do not lose their colour in cold water, but do so in boiling water because :
(a) the cell wall is killed in boiling water
(b) hot water can enter the cells readily
(c) the plasma membrane gets killed in boiling water and becomes permeable
(d) the pigment is not soluble in cold water

60. The following percentage of water absorbed by herbaceous plants is lost in transpiration
(a) 80% (b) 60%
(c) 40% (d) 99%

CHAPTER 13 Photosynthesis

CONCEPTS

- Green plants capture energy from sun and convert it into chemical energy (food) with the help of CO_2 taken from atmosphere and water from soil.
- Photosynthesis is actually **oxidation reduction process** in which water is oxidised and CO_2 is reduced to carbohydrates.
- The reduction of CO_2 to carbohydrates needs **assimilatory powers, *i.e.*, ATP** and **$NADPH_2$**.
- Photosynthesis occurs particularly in specialised cells called mesophyll cell. These cells contain chloroplast, which is the actual sites for photosynthesis.

PIGMENTS OF PHOTOSYNTHESIS

Chlorophyll

- Large molecule composed of four 5 membered rings called **pyrrole rings** and a central core of magnesium. A side chain, called the **phytol chain** extends from one of the pyrrole ring.

Carotenoids

- These are of red, orange and yellow colour.
- They act as antenna complexes and harvest light from different regions of the spectrum other than chlorophyll, which is then funneled into the reaction centres (chl *a*) for conversion into electrical energy.

LIGHT REACTION

- Light reaction occurs in **grana fraction** of chloroplast.
- **Assimilatory powers** (ATP and $NADPH_2$) are mainly produced in this light reaction.

Two Pigment Systems

PS-I is essential for producing the reduced **NADP** (*i.e.*, $NADPH_2$) used to reduce CO_2 whereas PS-II oxidises water and releases O_2.

8 quanta of light are required to evolve one molecule of O_2.

$$\textbf{Quantum yield} = \frac{1}{8} = 0.125$$

The two forms of chlorophyll '*a*' are chl *a* 683 (P_{680}) and chl *a* 703 (P_{700}) with peak absorption at 683 and 703 respectively are anchored in thylakoids membranes. They are the **reaction centres**.

Red Drop

Sudden drop or fall in rate of photosynthesis or in quantum yield above 680 nm (far red light) is called **red drop**.

Emerson Enhancement Effect

Emerson *et al* (1957) discovered that red light (above 680 nm), which is photosynthetically inefficient can be made efficient by supplementing it with a beam of shorter wave length (red beam below 680 nm). Further the quantum yield in combined beam (far red + red) is more than sum total of quantum yields in two separate beams.

ELECTRON TRANSPORT

It was first formulated by **Hill** (1939). Electron transport chain is a series of electron carrier over which electrons pass in a downhill journey releasing energy at every step that is used in generating an electrochemical proton gradient which helps in synthesising ATP.

Cyclic and Non-Cyclic Photophosphorylation

Non-cyclic Photophosphorylation

- When PS II (P_{680}) absorbs light, it is excited & emits electrons.
- The electrons run down an electron transport chain consisting of plastoquinone, cytochrome complex and plastocyanin to P_{700}.

- The electrons, generated from splitting of water, are then passed to the oxidised P_{680}.P_{680} thus restores its electron.
- It then transfers electron to ferredoxin and NADP - reductase to reduce NADP to $NADPH_2$.
- $NADPH_2$ is then utilized in reduction of CO_2 to carbohydrates in the carbon reaction. Since the electron lost by PSII do not come back to it, in the ATP formation, is called non-cyclic photophosphorylation.

Cyclic Photophosphorylation

- If light of wavelengths greater than 680 nm fall photosystem I is activated, but PS II remains inactivated.
- Electron from PS I are not passed to NADP from electron acceptor.
- Instead, electron are transferred back to P700.
  - This downhill movement result in the formation of ATP from ADP and termed as cyclic photophosphorylation.
  - O_2 is not released (no splitting of water) and $NADPH_2$ is not produced.

Chemiosmotic Hypothesis

- Electron transport, produces a proton gradient.
- concentration of protons decreases in matrix or stroma region while their concentration in thylakoid lumen rises resulting in decrease in pH.
- The proton gradient is broken down due to movement of protons through transmembrane channels, CF_0 of ATPase ($CF_0 - CF_1$ particle).
- CF_0 provides facilitated diffusion to H^+ or protons. As protons move to the outer side of APT, they bring about conformational changes in CF_1 particle of ATPase or coupling factor. The transient CF_1 particle of ATPase enzyme form ATP from ADP and inorganic phosphate.
- One molecule of ATP is formed when $2H^+$ pass through ATPase.

Carbon Reaction of 3 Types

	Salient features	C_3-Plants	C_4-Plants	CAM-Plants
1.	Cycle	C_3-Cycle	C_4-Cycle	CAM Cycle
2.	Kranz Anatomy	Absent	Present	Absent
3.	Carboxylating enzyme	RuDP Carboxylase	PEP in Mesophyll and RuDP in Bundle sheath	PEP in dark and RuDP in Light
4.	Initial CO_2 acceptor	RuDP	PEP	PEP
5.	First stable product	3-PGA	Oxaloacetate	Oxaloacetate
6.	Optimum temperature	15 - 25^{0}C	30 - 40^{0}C	35^{0}C
7.	CO_2 Compensation point (PPM CO_2)	50 - 150	0 - 10	0 - 5 in dark
8.	Na^+ requirement as micro nutrient	No	Yes	Yes
9.	Photorespiration	Present	Not easily	Detectable late
10.	O_2 Concentration of air	Above 21% inhibited	No effect	Inhibited
11.	ATP and $NADPH_2$ for reduction of 1 mole CO_2. CO_2 : ATP : $NADPH_2$	1 : 3 : 2	1 : 5 : 2	1 : 65 : 2
12.	Photosynthetic efficiency dry matter production (tones/hectare/years)	22 ± 0.3	39 ± 17	Lower than both C_3 and C_4 plants
13.	Rate of Photosynthesis in full sunlight	15 -35 mg of CO_2 per dm^2 of leaf area/hour	40 to 80 mg of CO_2 per dm^2 of leaf area/hr	–
14.	ATP are required for synthesis	18 A.T.P.	30 A.T.P.	30 A.T.P.

of 1 glucose molecule

Factors Affecting Photosynthesis

- **Temperature** - At 35°C photosynthesis operates fast. The rate of Photosynthesis becomes double from 0°C to 30°C.
- **Light** - The rate of photosynthesis increases with increase in light intensity. At compensation point, the rate of respiration is equal to rate of photosynthesis.
- **CO_2** - The rate of photosynthesis increases with increase in supply of CO_2. High CO_2 concentration reduces the rate of photosynthesis.
- **H_2O** - Plants only utilises 1 - 3% of absorbed water.

Principle of Limiting Factor

- **Blackman (1905)** states that the factor under consideration the other factor remaining constant, the rate of photosynthesis increases from a minimum to an optimum. Any further increase does not affect the rate. If any of the other factors which is kept constant is increased the rate increases.

PHOTORESPIRATION

- Light stimulated oxidation of photosynthetic intermediates to CO_2 is known as photorespiration. The course of photorespiration is related to chloroplasts, peroxisomes and mitochondria.
- First product of photorespiration is P-glycolate and CO_2 is liberated during glycine and serine conversion in mitochondria.
- This is a wasteful process & occurs in C_3 plants.

Exercise

1. Chlorophyll is present
   (a) in the grana of chloroplasts
   (b) on the surface of chloroplasts
   (c) dispersed through out the chloroplasts
   (d) in the stroma of chloroplasts
2. Solar energy is converted into ATP in
   (a) mitochondria (b) chloroplast
   (c) ribosome (d) peroxisome
3. Conditions necessary for photosynthesis are
   (a) light and suitable temperature
   (b) chlorophyll and water
   (c) carbon dioxide
   (d) All of the above
4. The first step in photosynthesis is
   (a) joining of three carbon atoms to form glucose
   (b) formation of ATP
   (c) ionization of water
   (d) excitement of an electron of chlorophyll by a photon of light
5. The ultimate gain of light reaction is
   (a) ATP & $NADPH_2$ (b) $NADPH_2$
   (c) only ATP (d) only O_2
6. Splitting of water in photosynthesis is called
   (a) dark reaction (b) photolysis
   (c) electron transfer (d) phototropism
7. Which one of the following pigment does not occur in the chloroplast ?
   (a) Carotene (b) Xanthophyll
   (c) Chlorophyll 'b' (d) Anthocyanin
8. Chlorophyll contains
   (a) Fe (b) Mg
   (c) K (d) Mn
9. Which pigment is water soluble ?
   (a) Chlorophyll (b) Carotene
   (c) Anthocyanin (d) Xanthophyll
10. Chloroplasts fix
    (a) carbon dioxide (b) oxygen
    (c) nitrogen (d) hydrogen
11. Greatest producers of organic matter are
    (a) crop plants
    (b) forests
    (c) plants of the land area
    (d) phytoplankton of oceans
12. Kranz anatomy is typical of
    (a) C_4 plants (b) C_3 plants
    (c) C_2 plants (d) CAM plants
13. Photosynthetic pigments found in the chloroplasts occur in
    (a) thylakoid membranes
    (b) plastoglobules
    (c) matrix
    (d) chloroplast envelope

14. All types of plastids possess essentially the same structure because they
(a) perform the same function
(b) store food materials like starch, fat and protein
(c) occur in aerial parts
(d) can transform from one form to another

15. Chlorophyll *a* occurs in
(a) all photosynthetic autotrophs
(b) in all higher plants
(c) all oxygen liberating autotrophs
(d) all plants except fungi

16. During monsoon, the rice crop of eastern states of India shows lesser yield due to limiting factor of
(a) CO_2 (b) light
(c) temperature (d) water

17. Dark reactions of photosynthesis occur in
(a) granal thylakoid membranes
(b) stromal lamella membranes
(c) stroma outside photosynthetic lamellae
(d) periplastidial space

18. Translocation of carbohydrate nutrients usually occurs in the form of
(a) glucose (b) maltose
(c) starch (d) sucrose

19. Photorespiration is favoured by
(a) high O_2 and low CO_2
(b) low light and high O_2
(c) low temperature and high O_2
(d) low O_2 and high CO_2

20. Photosynthetically active radiation is represented by the range of wavelength of
(a) 340 – 450 nm (b) 400 – 700 nm
(c) 500 – 600 nm (d) 400 – 950 nm

21. Which enzyme is most abudantly found on earth?
(a) Catalase (b) RuBisCO
(c) Nitrogenase (d) Invertase

22. The process of photophosphorylation take place in
(a) chloroplast (b) ribosomes
(c) mitochondria (d) cell-wall

23. PS-I occurs in
(a) appressed part of granal thylakoids
(b) appressed and non appressed part of grana thylakoids
(c) stroma
(d) stroma thylakoids and non-appressed part of granal thylakoids

24. Photophosphorylation means synthesis of
(a) ATP from ADP (b) NADP
(c) ADP from ATP (d) PGA

25. Which occurs during the light reaction of photosynthesis?
(a) Chlorophyll is produced
(b) Water splits to form $2H^+$ & O_2
(c) CO_2 is given off as a waste
(d) Sugar is formed from CO_2 and water

26. In photosynthesis, hydrogen is transferred from the light reactions to dark reactions by
(a) DPN (b) DNA
(c) ATP (d) NADP

27. Which of the following element is a component of ferredoxin ?
(a) Copper (b) Manganese
(c) Zinc (d) Iron

28. Carbon dioxide acceptor in C_4 plants is
(a) phosphoenol pyruvic acid (PEP)
(b) ribulose-1, 5-diphosphate
(c) NADP
(d) ribulose-5-phosphate

29. What is the first stable intermediate product of photosynthesis?
(a) Glucose
(b) Formaldehyde
(c) Phosphoglyceric acid
(d) Malate

30. Which of the following is C_4 plants ?
(a) Maize (b) *Atriplex*
(c) Sugarcane (d) All of the above

31. C_4 plants are adapted to
(a) hot and dry climate
(b) temperate climate
(c) cold and dry climate
(d) hot and humid climate

32. C_4 plants are found among
(a) only gramineae
(b) only monocot
(c) only dicot
(d) monocots as well as dicots

33. In case of C_4 pathway, in the first step is
(a) CO_2 combines with RuDP
(b) CO_2 combines with PGA
(c) CO_2 combines with PEP
(d) CO_2 combines with RMP

34. The enzyme which catalyzes the photosynthetic C_4 cycle is
(a) RUDP carboxylase (b) PEP carboxylase
(c) Carbonic anhydrase (d) None of these

35. Number of ATP molecules required for regeneration phase of RuBP during synthesis of 1 glucose molecule
(a) 6 (b) 12
(c) 18 (d) 30

36. The rate of photosynthesis does not depend upon
(a) light duration (b) light intensity
(c) light quality (colour) (d) temperature

37. Which one of the following have high CO_2 compensation point ?
(a) C_2 plants (b) C_3 plants
(c) C_4 plants (d) Alpine herbs

38. Solarisation is a process in which ?
(a) sugar are formed with the help of solar energy
(b) chlorophyll is formed
(c) destruction of chlorophyll and ultimate death of protoplasmic components
(d) mobilization of light energy.

39. Accumulation of food in assimilatory cells results in
(a) increase in the rate of photosynthesis
(b) decrease in the rate of photosynthesis
(c) no effect
(d) may increase or decrease

40. Photorespiration in C_3 plants starts from
(a) phosphoglycerate (b) glycerate
(c) glycine (d) phosphoglycolate

41. Etiolated plants are formed due to lack of
(a) light (b) Hg
(c) Fe (d) Mg

42. Generally CO_2 is not limiting for hydrophytes because
(a) mesophytes plants fix H_2S in their photosynthesis
(b) these plants take CO_2 from water in the form of HCO_3^-
(c) glucose is not required for their respiration
(d) All the above

43. What is called "warburg's effect" on photosynthesis?
(a) Low rate of process due to O_2 supply
(b) Low rate of the process due to CO_2 supply
(c) Both (a) & (b)
(d) None of the above

44. Bacterial photosynthesis involves
(a) PS-I only (b) PS-II only
(c) Both PS-I & PS-II (d) None of the above

45. The first acceptor of electrons from an excited chlorophyll molecule of photosystem II is
(a) quinone
(b) cytochrome
(c) iron-sulphur protein
(d) ferredoxin

46. In the leaves of C_4 plants, malic acid formation during CO_2 fixation occurs in the cells of
(a) epidermis (b) mesophyll
(c) bundle Sheath (d) phloem

47. During photosynthesis when PGA is changed into phosphoglyceraldehyde, which of the following reaction occur?
(a) Oxidation (b) Reduction
(c) Electrolysis (d) Hydrolysis

48. $NADP^+$ is reduced to NADPH in
(a) PS-I
(b) PS-II
(c) Calvin cycle
(d) Noncyclic photophosphorylation

49. Which technique has helped in investigation of Calvin cycle?
(a) X-ray crystallography
(b) X-ray technique
(c) Radioactive isotope technique
(d) Intermittent light

50. Ferredoxin is a constituent of
(a) PS I (b) PS II
(c) Hill reaction (d) P_{680}

51. Maximum solar energy is trapped by
(a) planting trees
(b) cultivating crops
(c) growing algae in tanks
(d) growing grasses

52. Which one occurs both during cyclic and noncyclic modes of photophosphorylation?
(a) Involvement of both PS I and PS II
(b) Formation of ATP
(c) Release of O_2
(d) Formation of NADPH

53. Chlorophyll '*a*' molecule at its carbon atom 3 of the pyrrole ring II has one of the following
(a) aldehyde group (b) methyl group
(c) carboxyl group (d) magnesium.

54. Which one of the following statements about cytochrome P-450 is wrong?
(a) It contains iron
(b) It is an enzyme involved in oxidation reaction
(c) It is a coloured cell
(d) It has an important role in metabolism

55. How many turns of Calvin cycle yield one molecule of glucose ?
(a) Eight (b) Two
(c) Six (d) Four

56. Fixation of one CO_2 molecule through Calvin cycle requires
(a) 1 ATP and $2NADPH_2$
(b) 2ATP and $2NADPH_2$
(c) 3ATP and $2NADPH_2$
(d) 2ATP and $1NADPH_2$

57. Which pigment system is inactivated in red drop?
(a) PS -I and PS -II (b) PS-I
(c) PS -II (d) None of these

58. Cytochrome is
(a) metallo - flavo protein
(b) Fe-containing porphyrin pigment
(c) glycoprotein
(d) lipid

59. Stomata of CAM plants
(a) never open
(b) are always open
(c) open during the day and close at night
(d) open during the night and close during the day

60. Which fractions of the visible spectrum of solar radiations are primarily absorbed by carotenoids of the higher plants?
(a) Violet and blue (b) Blue and green
(c) Green and red (d) Red and violet

Respiration in Plants

CONCEPTS

The phenomenon of breaking of the C–C bond of complex organic molecules through oxidation and releasing of energy for cellular use, is called respiration.

$C_6H_{12}O_6 + 6O_2 \rightarrow 6CO_2 + 6H_2O + \text{energy}\ (2870\ \text{kg})$

GLYCOLYSIS

- The scheme of glycolysis was given by Gustav **Embden**, Otto **Meyerhof** and **J. Parnas**, hence it is referred to as the EMP pathway.
- It occurs in the cytoplasm of the cell.

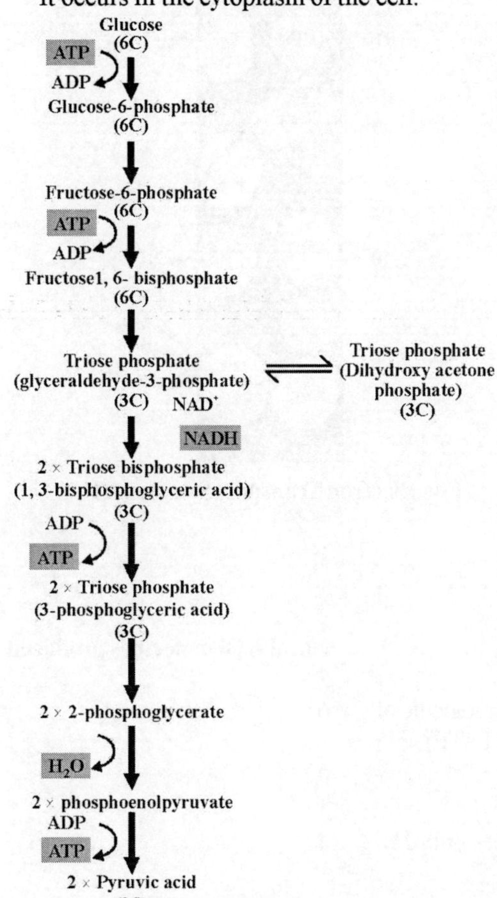

Fig. Steps of Glycolysis

Fermentation

- When O_2 is limiting, NADH and pyruvic acid begin to accumulate.
- Under this condition, plants carry out fermentation (anaerobic respiration), leading to the formation of CO_2 and either ethanol or lactic acid.

Aerobic Respiration

- Before pyruvic acid enters the Kreb's cycle, it is first decarboxylated and then oxidised by enzyme pyruvate dehydrogenase (oxidative decarboxylation).
- The remaining - 2C acetate unit is then readily accepted by a sulphur containing compound, coenzyme - A (CoA) to form acetyl CoA. This is the connecting link between glycolysis and Kreb's cycle.

$$\text{Pyruvic acid} + \text{CoA} \xrightarrow[Mg^{2+}]{NAD^+ \rightarrow NADH} \text{Acetyl CoA} + CO_2$$

CITRIC ACID CYCLE OR TRICARBOXYLIC ACID CYCLE OR KREB'S CYCLE

It occurs in mitochondrial matrix.

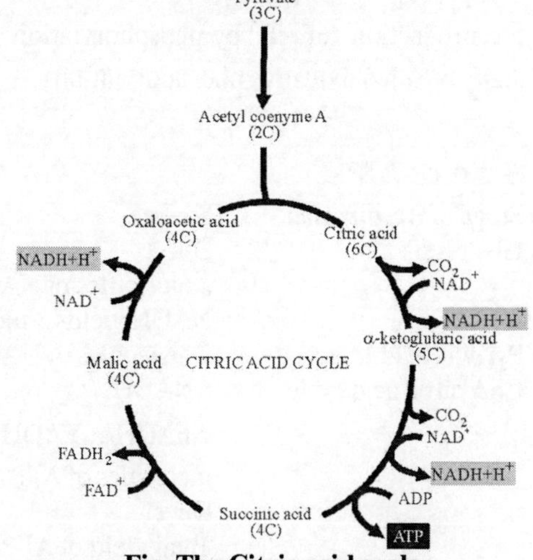

Fig. The Citric acid cycle

ELECTRON TRANSPORT SYSTEM (ETS) AND OXIDATIVE PHOSPHORYLATION

- The metabolic pathway through which the electrons passes from one carrier to another, is called the electron transport system (ETS).
- It operates in inner mitochondrial membrane.
- Electrons from NADH are oxidised by an NADH dehydrogenase (complex - I) and electrons are then transferred to ubiquinone located within the inner membrane.
- Ubiquinone also receive reducing equivalents *via* $FADH_2$ through the activity of the enzyme, succinate dehydrogenase (complex - II), in the citric acid cycle.
- The reduced ubiquinone (ubiquinol) is then oxidised with the transfer of electrons to cytochrome *c via* cytochrome bc_1 complex (complex III).
- Complex IV refers to cytochrome *c* oxidase complex containing cytochromes *a* and a_3 and two copper centres.
- When the electrons pass form one carrier to another *via* complex I to IV in the electron transport chain, they are coupled to ATP synthetase (complex V) for the production of ATP from ADP and inorganic phosphate.
- The process by which oxygen effectively allows the production of ATP by phosphorylation of ADP, is called oxidative phosphorylation.
- Complex V consists of two major components F_1 and F_0.
- F_1 headpiece is a peripheral membrane protein complex and contain the site for synthesis of ATP from ADP and inorganic phosphate.
- F_0 is an integral membrane protein complex that forms the channel through which protons cross inner membrane.

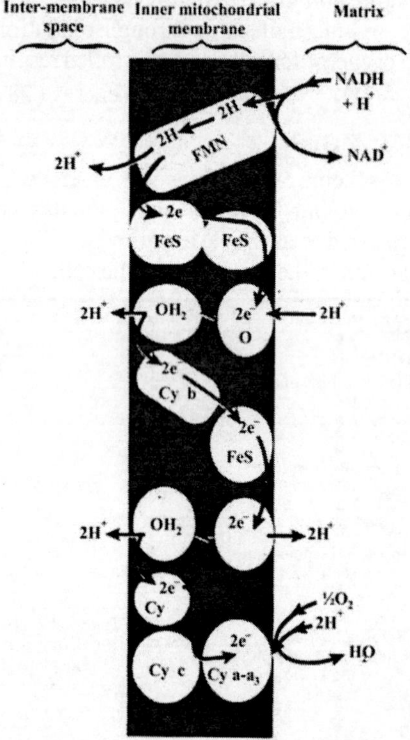

Fig. Electron Transport System (ETS)

Net Gain of ATP

Stage of Respiration	Source	No. of ATP molecules produced
Glycolysis	Direct	2
	2 molecules of NADH (one molecule of NADH yields 3 molecules of ATP)	6
Pyruvic acid to acetyl	2 molecules of NADH	6
CoA citric acid cycle	6 NADH	18
	2 $FADH_2$ ($FADH_2$ produces only 2 molecules of ATP)	4
	Direct	2
	Total yield of ATP molecules	38

Amphibolic pathway

- Respiratory pathway is involved in both anabolism and catabolism it is called as an amphibolic pathway.

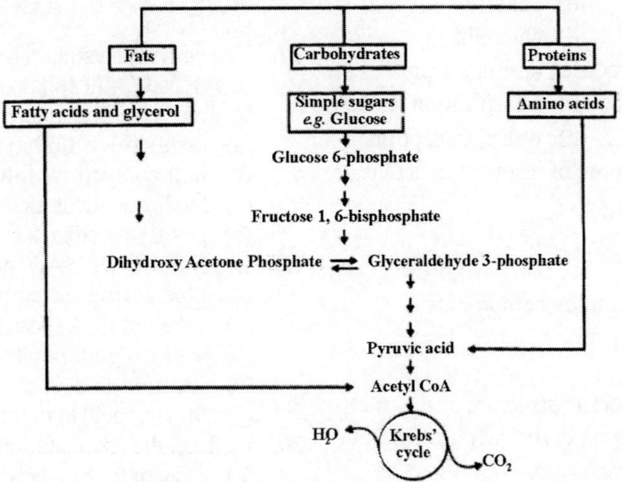

Respiratory Quotient (RQ) H_3 - Ratio between the volume of CO_2 evolved and O_2 absorbed during respiration.

R.Q = 1	Glucose
RQ = >1	Organic acid
RQ = < 1	Fats and Protein
RQ = 0	Succulents
RQ = ∞	Anaerobic respiration.

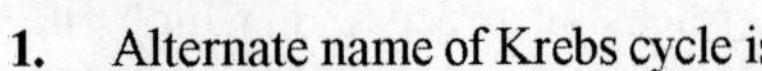

Exercise

1. Alternate name of Krebs cycle is
   (a) glyoxylate cycle (b) glycolate cycle
   (c) citric acid cycle (d) EMP Pathway
2. Respiration in plants
   (a) occur only during day
   (b) result in the formation of vitamins
   (c) is characteristic of all living cells
   (d) often requires CO_2
3. What is the energy coin of a cell ?
   (a) DNA (b) RNA
   (c) ATP (d) Minerals
4. RQ is
   (a) C/N (b) N/C
   (c) CO_2/O_2 (d) O_2/CO_2
5. Krebs cycle occurs in
   (a) mitochondria (b) cytoplasm
   (c) chloroplasts (d) ribosomes
6. End product of glycolysis is
   (a) acetyl CoA
   (b) pyruvic acid
   (c) glucose-1-phosphate
   (d) fructose-1-phosphate
7. Number of carbons in pyruvic acid is
   (a) 6 (b) 3
   (c) 2 (d) 1
8. Glycolysis occurs in
   (a) grana (b) cytosol
   (c) stroma (d) mitochondria
9. End product of fermentation of sugars is
   (a) H_2O & CO_2 (b) CO_2
   (c) alcohol & CO_2 (d) alcohol
10. Synthesis of ATP in mitochondria takes place
    (a) in the matrix
    (b) at the outer membrane
    (c) at the cristae
    (d) in the intracristal space

11. The commonest living, which can respire in the absence of O_2 is
(a) fish (b) yeast
(c) potato (d) chlorella

12. In plants energy is produced during the process of
(a) photosynthesis (b) transpiration
(c) respiration (d) water absorption

13. A very important feature of respiration is that
(a) it liberates energy
(b) it provides O_2
(c) it utilizes CO_2
(d) it synthesizes complex compounds

14. Enzymes are made up of
(a) edible proteins
(b) proteins with specific structure
(c) nitrogen containing carbohydrates
(d) carbohydrates

15. Energy required for start of biochemical reaction is
(a) potential energy (b) entropy
(c) activation energy (d) kinetic energy

16. Respiration is which type of process ?
(a) Catabolic (b) Metabolic
(c) Anabolic (d) None

17. Apparatus to measure rate of respiration and RQ is
(a) auxanometer (b) potometer
(c) respirometer (d) manometer

18. Respiratory quotient (RQ) for fatty acid is
(a) >1 (b) <1
(c) 1 (d) 0

19. Fermentation is anaerobic production of
(a) protein and acetic acid
(b) alcohol, lactic acid or similar compounds
(c) ethers and acetones
(d) alcohol and lipoproteins

20. Site of respiration in bacteria is
(a) episome (b) ribosome
(c) mesosome (d) microsome

21. Net gain of ATP molecules during aerobic respiration is
(a) 36 molecules (b) 38 molecules
(c) 40 molecules (d) 48 molecules

22. Chemiosmotic theory of ATP synthesis in the chloroplasts and mitochondria is based on
(a) membrane potential
(b) accumulation of Na ions
(c) accumulation of K ions
(d) proton gradient

23. Terminal cytochrome of respiratory chain which donates electrons to oxygen is
(a) Cyt. b (b) Cyt. c
(c) Cyt. a_1 (d) Cyt. a_3

24. The enzymes hexokinase which catalyses glucose to glucose-6-phosphate in glycolysis is inhibited by glucose-6-phosphate. This is an example of
(a) competitive inhibition
(b) non-competitive inhibition
(c) feedback allosteric inhibition
(d) positive feedback.

25. In Krebs cycle FAD participates as electron acceptor during the conversion of
(a) succinyl CoA to succinic acid
(b) $\alpha-$ keto glutarate to succinyl CoA
(c) succinic acid to fumaric acid
(d) fumaric acid to malic acid

26. In alcoholic fermentation
(a) oxygen is the electron acceptor
(b) triose phosphate is the electron donor while acetaldehyde is the electron acceptor
(c) triose phosphate is the electron donor while pyruvic acid is the electron acceptor
(d) there is no electron donor

27. All enzymes of TCA cycle are located in the mitochondrial matrix except one which is located in inner mitochondrial membranes in eukaryotes and in cytosol in prokaryotes. This enzyme is
(a) isocitrate dehydrogenase
(b) malate dehydrogenase
(c) succinate dehydrogenase
(d) lactate dehydrogenase.

28. The energy-releasing process in which the substrate is oxidised without an external electron acceptor is called
(a) fermentation (b) photorespiration
(c) aerobic respiration (d) glycolysis

29. In germinating seeds fatty acids are degraded exclusively in the
(a) proplastids (b) glyoxysomes
(c) peroxisomes (d) mitochondria

30. The chemiosmotic coupling hypothesis of oxidative phosphorylation proposes that adenosine triphosphate (ATP) is formed because
(a) high energy bonds are formed in mitochondrial proteins
(b) ADP is pumped out of the matrix into the intermembrane space
(c) a proton gradient forms across the inner membrane
(d) there is a change in the permeability of the inner mitochondrial membrane toward adenosine diphosphate (ADP)

31. The formation of Acetyl Co-A from pyruvic acid is the result of its
(a) reduction
(b) dehydration
(c) phosphorylation
(d) oxidative decarboxylation
32. Enzyme concerned with transfer of electrons are
(a) hydrolasase (b) dehydrogenase
(c) transaminase (d) protease
33. During which stage in the complete oxidation of glucose are the greatest number of ATP molecules formed from ADP
(a) Conversion of pyruric acid to acetyl Co A
(b) Electron transport chain
(c) Glycolysis
(d) Krebs cycle
34. Link between glycolysis & TCA cycle is
(a) pyruvic acid (b) acetyl Co-A
(c) citric acid (d) None
35. Pyruvate dehydrogenase complex is used in converting of
(a) pyruvate to glucose
(b) glucose to pyruvate
(c) pyruvic acid to lactic acid
(d) pyruvate to acetyl Co-A
36. Enzymes, vitamins and hormones can be classified into a single category of biological chemicals, because all of these
(a) enhance oxidative metabolism
(b) are conjugated proteins
(c) are exclusively synthesized in the body of a living organism
(d) help in regulating metabolism
37. The respiration of young maturing seeds is quite high but as water content decreases during maturation, the rate of respiration
(a) remains high (b) stops completely
(c) increases steadily (d) decreases steadily
38. Respiration differs from combustion in
(a) liberating more energy as compared to combustion
(b) liberating all the energy at once in contrast to combustion
(c) liberation of energy gradually in steps in contrast to combustion
(d) carbohydrates take part as the combustion substance
39. At a temperature above 35°C
(a) rate of photosynthesis will decline earlier than that of respiration
(b) rate of respiration will decline earlier than that of photosynthesis
(c) there is no fixed pattern
(d) both decline simultaneously
40. Which of the following enzymes is absent in mitochondria?
(a) Aconitase
(b) Malic dehydrogenase
(c) Fumarase
(d) Hexokinase
41. Which of the following statements is true?
(a) All enzymes are proteins
(b) All proteins are enzymes
(c) All enzymes are not proteins
(d) All enzymes and hormones are proteins
42. Which of the following is not a co-enzyme?
(a) NAD (b) NADP
(c) FAD (d) ATP
43. Which one of the following is first step of glycolysis?
(a) Breakdown of glucose
(b) Phosphorylation of glucose
(c) Conversion of glucose
(d) Dehydrogenation of glucose
44. Last electron receptor in respiration is
(a) CO_2 (b) O_2
(c) H_2 (d) NADH
45. Peroxisomes are rich in
(a) reductive enzymes
(b) lytic enzymes
(c) oxidative enzymes
(d) catalytic enzymes
46. Oxysomes are present on
(a) cell wall
(b) outer membrane of mitochondria
(c) inner membrane of mitochondrial cristae
(d) cytoplasmic membrane
47. Two enzymes common to EMP pathway and C_3 cycle are
(a) aldolase and triose phosphate isomerase
(b) aldolase and enolase
(c) cytochrome oxidase and enolase
(d) phosphoglyceromutase and triose phosphate isomerase

48. During anaerobic digestion of organic waste, such as in producing biogas, which one of the following is left undegraded ?
(a) Cellulose (b) Lipids
(c) Lignin (d) Hemi-cellulose

49. Which one of the following concerns photophosphorylation ?
(a) $AMP + \text{Inorganic } PO_4 \xrightarrow{\text{Light energy}} ATP$
(b) $ADP + AMP \xrightarrow{\text{Light energy}} ATP$
(c) $ADP + \text{Inorganic } PO_4 \xrightarrow{\text{Light energy}} ATP$
(d) $ADP + \text{Inorganic } PO_4 \longrightarrow ATP$

50. In glycolysis, during oxidation electrons are removed by
(a) ATP
(b) glyceraldehyde-3-phosphate
(c) NAD^+
(d) molecular oxygen

51. Aerobic respiratory pathway is appropriately termed
(a) parabolic (b) amphibolic
(c) anabolic (d) catabolic

52. Respiration is
(a) endothermic process
(b) exothermic process
(c) anabolic process
(d) endergonic process

53. Fermentation is performed by
(a) all micro-organisms
(b) all fungi
(c) all bacteria
(d) some fungi & bacteria

54. In succulents, RQ < 1 due to
(a) incomplete oxidation
(b) incomplete reduction
(c) complete reduction
(d) complete oxidation

55. During respiration in yeast
(a) water and CO_2 are end products
(b) CO_2, C_2H_5OH and energy are end products
(c) H_2S, $C_6H_{12}O_6$ and energy are end products
(d) CO_2 is the only end product

56. Energy for ATP synthesis is obtained by
(a) oxygen ion gradient
(b) hydrogen ion gradient
(c) nitrogen ion gradient
(d) all

57. When 2-pyruvic acids forms two lactic acids by anaerobic respiration then
(a) One ATP is lost (b) 3 ATPs lost
(c) 6 ATPs lost (d) None

58. Which of the following statements regarding mitochondrial membrane is not correct ?
(a) The inner membrane is highly convoluted forming a series of infoldings
(b) The outer membrane resembles a sieve
(c) The outer membrane is permeable to all kinds of molecules
(d) The enzymes of the electron transfer chain are embedded in the outer membrane

59. Out of 36 ATPs molecules produced per glucose molecule during respiration
(a) 2 are produced outside glycolysis and 34 during respiratory chain
(b) 2 are produced outside mitochondria and 34 inside mitochondria
(c) 2 during glycolysis and 34 during Krebs cycle
(d) All are formed inside mitochondria

60. End products of aerobic respiration are
(a) sugar and oxygen
(b) water and energy
(c) carbon dioxide, water and energy
(d) carbon dioxide and energy

Plant Growth and Development

CONCEPTS

Growth can be broadly defined as "permanent and irreversible increase in size of living structure which is accompanied by an increase in dry weight and the amount of protoplasm".

PLANT GROWTH GENERALLY IS INDETERMINATE

There are three meristematic regions in plant *i.e.*, apical, intercalary and lateral.

- **Apical meristem :** The two important growing points are present in the root apex and stem apex called apical meristems and growth due to them is known as **primary growth**.
- **Intercalary meristems :** Intercalary meristems are actually a part of apical meristems, which get separated from the apex by intervening permanent tissues. In some plants, increase in length is due to intercalary meristems *e.g.*, bamboo, mint etc.
- **Lateral meristems :** The lateral meristem (cambium) add new tissues for growth in thickness. This type of growth is called as **secondary growth**.

GROWTH IS MEASURABLE

The growth in plants can easily be measured in terms of increase in length or girth, increase in weight, (both fresh and dry), and volume of an organ of plant. The increase in number of cells (especially in algae, yeast and bacteria), also gives an idea about the rate of growth.

Phases of Growth

- **Meristematic phase :** This phase has constantly dividing cells and is restricted to the apical meristems, both at the root and shoot tips. They have large nuclei which are centrally located.
- **Elongation phase :** It lies just behind the formative phase and is aimed at the enlargement of cells. The rich amount of solutes are present in these cells cause absorption of water.
- During cell enlargement a large vacuole appears in the centre and the protoplasm occupies the peripheral region of cell.
- **Maturation phase :** During this phase, the cells start maturing and differentiating to obtain a permanent shape and size such as fibres, tracheids, vessels, sieve tubes, cork cells, collenchyma, sclerids etc.

Growth Rate

If growth rate of plant or plant part is plotted against time, a typical sigmoid or S-shaped growth curve is obtained. The rate of plant growth is slow in the initial stages or in phase of cell division (lag phase), but increases at a very rapid rate to maximum during phase of cell enlargement (logarithmic or log or exponential phase). The growth again slows down due to the limitation of nutrients during the phase of cell maturation (stationary phase).

- Growth rate can show two types of increase *i.e.*, arithmatic increase and geometrical growth.

Arithmatic Increase

$$L_t = L_0 + rt$$

Where

L_t = length at time 't';

L_0 = length at time 'zero';

r = growth rate /elongation per unit time;

t = time of growth.

Geometrical Growth

$$W_1 = W_o e^{rt}$$

Where,

W_1 = Final size (weight, height, number etc.)
W_o = Initial size at the beginning of the period;
e = base of natural logarithms;
r = growth rate;
t = time of growth.

Conditions for Growth

The proper growth of plant is influenced by external and internal factors viz., supply of nutrients, water, oxygen, suitable temperature and light. The force of gravity, and also light, determines the direction of root and shoot growth.

DIFFERENTIATION, DEDIFFERENTIATION AND REDIFFERENTIATION

Development is growth accompanied by differentiation.

Differentiation

- The cells are derived from root apical and shoot apical meristems and cambium differentiate and mature to perform specific functions. This act leading to maturation is called as differentiation.

Dedifferentiation

- The living differentiated cells loose their capacity of division. These cells may regain there capacity to divide under certain conditions. This phenomenon is called as dedifferentiation.

Redifferentiation

- The products of dedifferentiated cells which lose the capability to divide but mature to perform specific functions are called redifferentiated cells.

Plasticity

- The plant may follow different pathways in response to environment or phases of life. It leads to formation of various different structures.

Heterophylly

- Heterophylly occurs due to plasticity. In plants like cotton, coriander and *Delphinium* leaves of juvenile plants differ in shape from the shapes of leaves produced in mature plants. Heterophylly due to plasticity is present in butter cup (*Ranunculus*) due to environment.

PLANT GROWTH REGULATORS

- Plant growth regulators are small, simple molecules of diverse chemical composition. They could be indole compounds (Indole acetic acid), adenine derivatives (kinetin), carotenoid derivatives (abscisic acid), terpenes (gibberelic acid) or gases (ethylene).

Auxins

- Auxins are defined as "an organic compound characterized by its capacity in low concentration (below 10^{-3}M or 0.001M) to induce elongation in shoot cells and inhibition of elongation of root cells.
- The best known universal natural auxin is Indole acetic acid (IAA).
- Synthetic auxins are indole butyric acid (IBA), indole propionic acid (IPA), 2,4- dichlorophenoxy acetic acid-(2,4 D); naphthalene acetic acid (NAA).

Role of auxin in the growth and development of plants

- At the removal of apical bud the lateral buds grow vigorously. It shows that apical bud suppresses the growth of lateral bud (axillary bud) just below it. This is known as **apical dominance**.
- Initiating and promoting cell division in certain tissues such as cambium.
- Promotes elongation of stem and coleoptile.
- Auxin controls the preharvest drop of fruit such as apple, pear and citrus by retardation of abscission.
- Parthenocarpy refers to development of seedless fruit that occurs without pollination and fertilization.
- Auxin generally inhibits flowering but in case of pineapple spray of auxins induces early flowering.

Gibberellins

- Chemically, they are known as gibberellic acid. They contain a gibbane ring system and have specific biological properties.
- GA_3 was one of the first gibberellin to be discovered and remains the most intensively studied form. All GAs are acidic in nature. They occur in various plant organs such as roots, stems, leaves, buds, immature seeds and callus tissues of higher plants.

Physiological Effects of Gibberellins

- The most typical and striking effect of gibberellin is on the elongation of stem. The internodes increase in length.

- Gibberellin break dormancy in potato tubers and in buds of trees.
- Gibberellin causes the plants to bolt and flower.
- Gibberellins have been found be more effective than auxins in causing parthenocarpic development of fruits *e.g.*, tomatos, apples and pears.
- Gibberellin induces *de novo* synthesis of various hydrolytic enzymes viz., α- amylase, ribonuclease etc. in aleurone cells of barely grains.
- Gibberellin causes maleness in plants.

Cytokinins

- **Miller** in 1954 isolated the first crystals of a 'cell division inducing substance' from the autoclaved herring sperm DNA. Since this substance has specific effect on cytokinesis it is called as kinetin (a modified form of adenine).

Physiological Effects of Cytokinins

- Permanent cell division occurs only in presence of cytokinins. It also plays important role in causing expansion of cells.
- Cytokinins play a part in intiation of organ formation in variety of tissue cultures.
- **Richmond** and **Lang** (1957) reported the delay of senescence (chlorophyll disappearance and protein degradation) for several days in the detached leaves of *Xanthium* when they were treated with kinetin.
- Cytokinins not only breaks dormancy but also promotes the germination of seeds.

Abscisic acid

- **Carns** and **Addicott** (1963) isolated two substances Abscisin I and Abscisin II from the cotton balls, that were responsible for accelerated abscission of leaves.
- Abscisic acid (ABA) is a naturally occurring growth regulator that is found in all vascular plants and some mosses. It is absent in bacteria, algae, most of fungi and liverworts.

Physiological Effects of Abscisic Acid

- ABA inhibits seed germination and growth of excised embryos.
- Causes abscission of leaves and flower.
- ABA induces dormancy in various plants.
- ABA stimulates stomatal closure by inhibiting the K^+ uptake by guard cell.
- ABA increases tolerance of plants to various kinds of stresses.

Ethylene

- Ethylene is only **gaseous hormone** ($CH_2 = CH_2$) that is synthesized in large amounts by tissues undergoing senescence and ripening fruits.

Physiological Effects of Ethylene

- Ethylene inhibits elongation of stem, causes swelling of nodes and nullifies geotropism.
- It inhibits the stem elongation; it is associated with radial enlargement of tissues that results in swelling of plant parts.
- It is highly effective in inducing fruit ripening when it is produced in large amount which coincides with respiratory **climactric** *i.e.*, a brief rise to a very high level of respiration. This rise indicates the beginning of senescence and death.

PHOTOPERIODISM

The term photoperiodism is used by **Garner** and **Allard**, 1920 for the ability of plant to detect and respond the relative length of day and night to which the plant is exposed. The site of perception of light/dark duration are leaves.

The plants are classified into three groups according to their photoperiods viz.,

Short Day Plants

- Short day plants require a relatively long period of uninterrupted darkness for flowering.
- The length of night is more important than the length of day. Thus, short day plants are also called as 'long night plants'.
- Most of the plants are short day plants that include *Cosmos*, *Dahlia*, *Chrysanthemum*, rice, etc.

Long Day Plants

- In long day plants, flowering will be initiated only when the day length is longer than critical day length (varies from 4 to over 18 hrs).

- Long day plants can flower under short day conditions, if a flash of light is given to the plants during the long dark period.
- The long day plants are wheat, barley, sugar beat, larkspur, etc.

Intermediate or Day Neutral Plants

- These plants do not require specific photoperiods to flower.
- Cucumber, sunflower, tobacco, tomato etc. are some examples of intermediate day plants.

VERNALISATION

- In some plants early flowering is induced by pretreatment of seeds with a certain low temperature. This phenomenon is known as vernalisation.
- The site for vernalization is meristematic cells in the bud, apex of embryo, developing leaves etc.

Devernalisation

- If a vernalized seed or plant is kept at high temperature, the effect of low temperature treatment is completely removed.
- Devernalized plants can be again vernalized by low temperature treatment.

Significance of Vernalisation

- Vernalisation shortens the vegetative period of the plant.
- It increases the cold resistance of the plants.

SEED DORMANCY

- **Dormancy** may be defined as the inactive state of the seed in which the growth of the embryo is temporarily suspended for a specific length of time.
- The seed dormancy may be due to many causes. Some of which are as follows :
- **Impermeability of seed coats to oxygen**, *e.g.*, *Xanthium*.
- **Impermeability of seed coats to water**, *e.g.*, many plants of legumes.
- **Hard seed coat**, which does not allow proper growth of developing embryo, *e.g.*, mustard.
- **Germination inhibitors :** Some plants produce such chemical compounds that inhibit the germination of their own seeds, *e.g.*, tomato.
- Other chemical inhibitors are abscissic acid, phenolic acid, para-ascorbic acid, etc.

Methods of Breaking Dormancy

- **Mechanical scarification:** Weakening of hard seed coat with anything of sharp edge, *e.g.*, pieces of glass, knives, sand paper or vigorous shaking.
- **Chemical scarification :** Treating the seeds with dilute acids, fat solvents, etc.
- **Chilling treatment :** Low temperature treatment, alternating low and high temperature treatments.
- **To neutralize the effect of inhibitors :** Treating the seeds with KNO_3, thiourea, gibberellin, ethylene, chlorohydrin, etc.

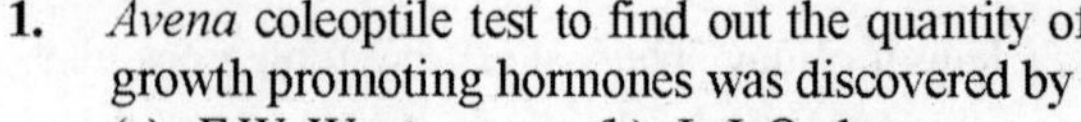

Exercise

1. *Avena* coleoptile test to find out the quantity of growth promoting hormones was discovered by
   (a) F.W. Went (b) L.J. Oudus
   (c) K.V. Thimann (d) F. Skoog
2. In plants, growth is
   (a) restricted to certain regions or structure
   (b) irreversible
   (c) change in size
   (d) All the above
3. Growth is primarily affected by two climatic factors which are
   (a) light and temperature
   (b) temperature and relative humidity
   (c) light and wind
   (d) rainfall and temperature
4. Apical dominance means
   (a) suppression of growth of apical buds by axillary buds
   (b) suppression of growth of axillary buds by the presence of apical bud
   (c) stimulation of growth of axillary buds by removal of apical bud
   (d) inhibition of growth of axillary buds by removal of apical bud

5. "Foolish Seeding" disease of rice led to the discovery of
(a) IAA (b) GA
(c) ABA (d) 2,4-D

6. The maximum growth rate occurs in
(a) exponential phase (b) stationary phase
(c) senescent phase (d) lag phase

7. Phytochrome is involved in
(a) phototropism (b) photorespiration
(c) photoperiodism (d) geotropism

8. Which one increases in the absence of light?
(a) Uptake of minerals
(b) Uptake of water
(c) Elongation of internodes
(d) Ascent of sap

9. Phytohormones are
(a) chemicals regulating flowering
(b) chemicals regulating secondary growth
(c) hormones regulating growth from seed to adulthood
(d) regulators synthesised by plants and influencing physiological processes

10. The response of different organisms to environmental rhythms of light and darkness is called
(a) phototaxis (b) photoperiodism
(c) phototropism (d) vernalization

11. Plant bends towards the light because
(a) they need light for photosynthesis
(b) they need light for respiration
(c) light attract them
(d) cell on the shaded side elongate more

12. Thigmotropism is the response of the plant to
(a) water (b) gravity
(c) contact (d) light

13. A plant cell has potential to develop into a full plant. This is called
(a) totipotency (b) gene cloning
(c) tissue culture (d) regeneration

14. Opening & closing of flowers represent a kind of
(a) nastic movements
(b) tropic movements
(c) nutation movements
(d) autonomic movements

15. Which of the following is not naturally occurring plant hormone?
(a) 2,4-D (b) GA_2
(c) Gibberellin (d) IAA

16. Leaf fall occurs when the content of
(a) auxin increases
(b) auxin decreases
(c) abscisic acid decreases
(d) gibberellic acid decreases

17. Auxanometer is meant for measuring
(a) respiratory activity
(b) photosynthetic activity
(c) growth activity
(d) osmotic pressure

18. The biological activity of IAA is tested by
(a) α-amylase test
(b) avena curvature test
(c) soyabean callus test
(d) xanthium leaf disc test

19. An enzyme that can stimulate germination of barley seeds is
(a) protease (b) invertase
(c) α-amylase (d) lipase

20. Movement of leaves of sensitive plant, *Mimosa pudica* are due to
(a) thermonasty (b) seismonasty
(c) hydrotropism (d) chemonasty

21. Cut or excised leaves remain green for long if induced to root or dipped in
(a) gibberellins (b) cytokinins
(c) auxins (d) ethylene

22. Hormone primarily connected with cell division is
(a) IAA (b) NAA
(c) cytokinin/Zeatin (d) gibberellic acid

23. Which of the following hormones can replace vernalisation?
(a) Auxin (b) Cytokinin
(c) Gibberellins (d) Ethylene

24. Phototropic and geotropic movements are linked to
(a) gibberellins (b) enzymes
(c) auxin (d) cytokinins

25. Abscisic acid controls
(a) cell division
(b) leaf fall and dormancy
(c) shoot elongation
(d) cell elongation and wall formation

26. Highest auxin concentration occurs
(a) in growing tips
(b) in leaves
(c) at base of plant organs
(d) in xylem and phloem

27. Which is employed for artificial ripening of banana fruits?
(a) Auxin (b) Coumarin
(c) Ethylene (d) Cytokinin
28. In short day plants, flowering is induced by
(a) photoperiod less than 12 hours
(b) photoperiod below a critical length and uninterrupted long night
(c) long night
(d) short photoperiod and interrupted long night
29. Bananas can be prevented from over-ripening by
(a) maintaining them at room temperature
(b) refrigeration
(c) dipping in ascorbic acid solution
(d) storing in a freezer
30. Flowering dependent on cold treatment is
(a) cryotherapy (b) cryogenics
(c) cryoscopy (d) vernalisation
31. Movement of auxin is
(a) centripetal (b) basipetal
(c) acropetal (d) both b and c
32. The pigment, that absorbs red and far red light in plants, is
(a) xanthophyll (b) cytochrome
(c) phytochrome (d) carotene
33. Ethylene gas is used for
(a) growth of plants
(b) delaying fruit's abscission
(c) ripening of fruits
(d) stopping the leaf abscission
34. If a tree, flowers thrice in a year (Oct., Jan. and July) in Northern India, it is said to be
(a) photosensitive but thermoinsensitive
(b) thermosensitive but photoinsensitive
(c) photo and thermosensitive
(d) photo and thermoinsensitive
35. A plant hormone used for inducing morphogenesis in plant tissue culture is
(a) gibberellins (b) cytokinins
(c) ethylene (d) abscisic acid
36. The method that renders the seed coat permeable to water so that embryo expansion is not physically retarded is
(a) vernalization (b) stratification
(c) denudation (d) scarification
37. What reason will you assign for coconut milk used in tissue culture?
(a) Gibberellins (b) Cytokinins
(c) Auxins (d) Ethylene
38. What breaks bud dormancy of potato tuber?
(a) Gibberellin (b) IAA
(c) ABA (d) Zeatin
39. Glycolate induces opening of stomata in
(a) presence of oxygen
(b) low CO_2 concentration
(c) high CO_2 concentration
(d) absence of CO_2
40. Differentiation of shoot is controlled by
(a) high gibberellin : cytokinin ratio
(b) high auxin : cytokinin ratio
(c) high cytokinin : auxin ratio
(d) high gibberellin : auxin ratio
41. Plants deficient of element zinc, show its effect on the biosynthesis of plant growth hormone
(a) abscisic acid (b) auxin
(c) cytokinin (d) ethylene
42. Cell elongation in internodal regions of the green plants takes place due to
(a) indole acetic acid (b) cytokinins
(c) gibberellins (d) ethylene
43. Opening of floral buds into flowers, is a type of
(a) autonomic movement of variation
(b) paratonic movement of growth
(c) autonomic movement of growth
(d) autonomic movement of locomotion
44. One of the synthetic auxin is
(a) IAA (b) GA
(c) IBA (d) NAA
45. Which one of the following acids is a derivative of carotenoids ?
(a) Indole-3 -acetic acid
(b) Gibberellic acid
(c) Abscisic acid
(d) Indole butyric acid
46. Which is a weedicide ?
(a) IAA (b) 2,4-D
(c) IBA (d) NAA
47. A potted plant placed near the window bends outwardly due to
(a) greater O_2 availability to the tip
(b) more auxin concentration on the shaded side
(c) greater light availability to tip
(d) availability of necessary warmth to the tip
48. Somaclonal variation can be obtained by

(a) application of colchicine
(b) irradiation with gamma rays
(c) tissue culture
(d) hybridisation

49. Sensation of phototropism is perceived by
(a) root tip (b) shoot tip
(c) axillary bud (d) node

50. Which of the following ion is pulled out in apoplast by auxin during growth ?
(a) K^+ (b) H^+
(c) Na^+ (d) Mg^{2+}

51. Haploid plants can be obtained from
(a) leaf culture (b) bud culture
(c) root culture (d) anther culture

52. The histogen layer present at the apex of root tip is
(a) dermatogen (b) calyptrogen
(c) periblema (d) plerome

53. Rotenone is
(a) an insect hormone
(b) a bioherbicide
(c) a natural insecticide
(d) a natural weedicide

54. Which of the following controls flowering in long day plants and sex expression in certain species?
(a) Auxin (b) Cytokinin
(c) Gibberellin (d) Ethylene

55. Ageing is retarded by
(a) ABA (b) Cytokinin
(c) GA (d) Ethylene

56. Which is essential for root hair growth ?
(a) Zn (b) Ca
(c) Mo (d) S

57. One set of a plant was grown at 12 hours day and 12 hours night period cycles and it flowered. While in the other set night phase was interrupted by flash of light and it did not flower. Under which one of the following categories will you place this plant
(a) Short day (b) Long day
(c) Darkness neutral (d) Day neutral

58. Prunning of plants promotes branching, because the axillary buds get sensitized to
(a) ethylene (b) gibberellin
(c) cytokinin (d) IAA

59. Parthenocarpic tomato fruits can be produced by
(a) raising the plants from vernalized seeds
(b) treating the plants with phenylmercuric acetate
(c) removing androecium of flowers before pollen grains are released
(d) treating the plants with low concentrations of gibberellic acid and auxins

60. Movement due to external stimuli is known as
(a) autonomic movement
(b) paratonic movement
(c) ameboid movement
(d) excretory movement

Digestion and Absorption

CONCEPTS

- The food that we consume contains highly complex substances like proteins, carbohydrates and fats.
- They must be broken down into simpler absorbable forms so that they can be easily absorbed and transported to various parts of our body through blood. This task is accomplished by the digestive system.

The **human digestive** system consists of two parts :

ALIMENTARY CANAL

- In human beings, the alimentary canal is a muscular tube, about 9 metres long that passes through the body's ventral cavity.

The Mouth

Its main function is to receive food and start mechanical digestion by mastication of food.

Teeth :

- In human beings, there are two jaws, a fixed (upper) jaw and a movable (lower) jaw, with the tongue in between. Both jaws bear teeth, which are used to chew or masticate food into smaller pieces.

Dental Formula for Adult human

$$\frac{\text{Upper jaw}}{\text{Lower jaw}} = \frac{\text{I C Pm M}}{\text{I C Pm M}} = \frac{\mathbf{2123}}{\mathbf{2123}}$$

Salivary Secretions

- Saliva is slightly alkaline in nature containing about 99% water and about 1% salt, mucus and an enzyme called salivary amylase (also known as ptyalin).
- Saliva moistens food particles that help in swallowing.

Pharynx

- The next part is pharynx. The pharynx is an area at the back of the throat that connect the nasal and oral cavities with the larynx and oesophagus.
- The larynx or voice box present at the entrance of the wind pipe is pulled upward to regulate the movement of food into the oesophagus.

Oesophagus

- It is a tube like structure about 25 cm in length.
- Food passes through oesophagus by peristalsis.

Stomach

- The stomach is a J-shaped muscular bag like structure.
- It contains a ring of muscles called pyloric sphincter that controls the release of food from the stomach into the small intestine.

Gall bladder

- Stores bile

Small intestine

- Completes digestion
- Absorbs nutrients

Large intestine

- Reabsorbs water, ions,vitamins
- Stores wastes

Rectum

- Expels wastes

Anus

- Opening for waste elimination

DIGESTIVE GLANDS

There are **three** sets of **salivary glands** present in the oral cavity of man.

- **Parotid gland :** It is the largest salivary gland and lies in the region of ear.
- **Submaxillary gland :** It lies in the posterior part of lower jaw.
- **Sub-lingual gland :** It is the smallest and is located below the tongue.

Liver

- Liver is the largest gland in the human body and weighs about 1.5 kg in an adult human beings.
- Liver regulates blood sugar level.
- It detoxifies toxic and metallic poisons and excretes them.

Pancreas

- The pancreas has an exocrine function of producing pancreatic juice that aids in digestion. The pancreas is closely associated with small intestine.

DIGESTION OF FOOD

Digestion of Carbohydrates

Name of the Digestive juice	Name of the enzymes	Substrate	End product
Saliva	Ptyalin (Salivary amylase)	Starch	Maltose
Pancreatic iuice	Amylopsin (pancreatic amylase)	Starch Glycogen	Maltose and Glucose
Intestinal juice	Sucrase (invertase), Maltase, Lactase	Sucrose; Maltose, Lactose	Glucose and fructose, Glucose, Glucose and galactose

Digestion of Proteins

Name of the Digestive juice	Name of the enzymes	Substrate	End product
Gastric iuice	Pepsin, Rennin	Proteins, Casein	Proteoses and peptones, Calciumcaseinate
Pancreatic iuice	Trypsin, Chymotrypsin, Carboxyl peptidases	Proteins Proteins Peptides	Proteoses and peptides Peptides Amino acid.
Intestinal juice	Amino peptidase, Dipeptidase	Peptides	Amino acids, Amino acids

Digestion of Fats

Name of the Digestive juice	Name of the enzymes	Substrate	End product
Gastric iuice	Gastric lipase (functionless in the adult stomach)	Butterfat molecules of milk	Fatty acids and glycerol
Pancreatic iuice	Steapsin (pancreatic lipase)	Emulsified lipids	Fatty acids and glycerol
Bile juice	No lipases. But only bile salts, sodium taurocholate and sodium glycocholate	Lipids	Emulsified lipids

Digestion of Nucleic acids

Name of the Digestive juice	Name of the enzymes	Substrate	End product
Pancreatic iuice	Deoxyribo polynucleotidase Ribo polynucleotidase	DNA RNA	DNA nucleotides, nucleotides, RNA
Intestinal juice	Nucleotidase, Nucleotidase	Nucleotides	Nucleosides and inorganic phosphate, Pentose sugars, purines and pyrimidines

Basal Metabolism

Basal metabolic rate (B.M.R.) is defined as the amount of heat produced in the body in a given time in complete state of physical and mental rest at 20°C room temperature.

BMR is raised with the following :

- In certain pathological conditions, *e.g.*, hyperthyroidism, fever, severe anaemia malignancy (leukaemia). and cardiac disorders causing over-work of the heart.

BMR is decreased in the following cases :

- Hypothyroidism.
- During hypothermia.
- During starvation

Vitamin required by the body

Vitamin	Chemical Name	Function In Body	Deficiency Disease
B_1	Thiamine pyrophosphate	Part of coenzyme for respiration	**Beri-beri:** nerve and heart disorders
B_2	Riboflavin	Part of coenzyme FAD needed for respiration	**Ariboflavinosis:** skin and eye disorders
B_{12}	Cyanocobalamin	Coenzyn me needed for making red blood cells, bone, blood and nerve changes	**Pernicious anaemia**
B_5	Nicotinic acid ('neacin')	Part of coenzymes NAD NADP used in respiration	**Pellagra:** skin, gut and nerve disorders
C	Ascorbic acid	Not precisely known	**Scurvy:** degeneration of skin teeth and blood vessels.
A	Retinol	Not fully known but forms part of visual pigment, rhodopsin	**Xeropthalmia:** 'dry eyes'
D	Cholecalciferol	Stimulates calcium absorption by small intestine, needed for proper bone growth	**Rickets:** bone deformity
E	Tocopherol	Not precisely known	**Infertility**
K	Phylloquinone	Involved in blood clotting	**Possible haemorrage**

Inorganic Elements in the Human Diet

Element	Common ions	Functions in human body
Calcium	Ca^{2+}	Calcium ions are needed for stability of cell membranes, as cofactors for some enzymes and are involved in muscle contraction and blood clotting.
Phosphorus	H_2PO_4	Bones component of many organic molecules like DNA, RNA and ATP.
Potassium Sodium Chlorine	K^+ Na^+ Cl^-	These ions are important in determining the balance of electrical charges in body fluids.
Iron	Fe^{2+}, Fe^{3+}	Component of haemoglobin and cytochrome molecules.
Iodine	I^-	Component of hormone thyroxin.
Copper Manganes Zinc	Cu^{2+} Mn^{2+} Zn^{2+}	Trace elements as enzyme cofactors, for example, Cu^{2+} is co-factor for cytochrome oxidase.

DISORDERS OF DIGESTIVE SYSTEM

Indigestion

- It is the condition in which the food is not properly digested leading to a feeling of fullness.

Diarrhoea

- It refers to the abnormal frequency of bowel movement and increased liquidity of the faecal discharge, absorption of food is impaired.

Vomiting

- It is the ejection of stomach contents through the mouth; this reflex action is controlled by the vomit centre in the medulla.

Jaundice

- It is due to viral infection, where liver is affected and digestion of fats is impaired.
- The eyes and skin turn yellow due to the deposit of bile pigments.

Ulcers

- An ulcer is a crater like lesion or erosion in a membrane. Ulcers that develop in the alimentary canal exposed to acidic gastric juice are called **peptic ulcers.**

Cirrhosis

- Cirrhosis refers to a deserted or scarred liver as a result of chronic inflammation.
- The symptoms of cirrhosis include jaundice, oedema in the legs, uncontrolled bleeding, and increased sensitivity to drugs.

Hepatitis

Hepatitis refers to inflammation of the liver and can be caused by viruses, drugs, and chemicals, including alcohol, clinically several types are recognized.

PEM

- Protein-energy malnutrition (PEM) may affect large sections of the population during drought, famine and political turmoil.
- Marasmus is produced by a simultaneous deficiency of proteins and calories. In Marasmus, protein deficiency impairs growth and replacement of tissue proteins; extreme emaciation of the body and thinning of limbs results, the skin becomes dry, thin and wrinkled. Growth rate and body weight decline considerably.
- Kwashiorkar is produced by protein deficiency unaccompanied by calorie deficiency. Like marasmus, kwashiorkor shows wasting of muscles, thinning of limbs, failure of growth and brain development.

Exercise

1. The food that gives more calories per unit mass of food is
   (a) protein (b) carbohydrates
   (c) fat (d) water
2. Essentially the work digestion means
   (a) breaking food for energy
   (b) building of proteins from amino acids
   (c) changing organic molecules
   (d) breaking complex organic molecules into smaller ones
3. In human beings, carbohydrate is stored as glycogen in
   (a) liver & muscles (b) liver
   (c) muscles (d) spleen
4. Emulsification of fat is carried out by
   (a) bile pigments (b) bile salts
   (c) HCl (d) pancreatic juice
5. Digestion of proteins, fats and carbohydrate is completed in
   (a) stomach (b) duodenum
   (c) ileum (d) none of these
6. Which of the following teeth appear only once in the life of a man ?
   (a) Incisors (b) Canines
   (c) Premolars (d) Molars
7. Which is the element that hardens the tooth enamel
   (a) Calcium (b) Fluorine
   (c) Iodine (d) Sodium
8. Major function of HCl of gastric juice is
   (a) providing acidic medium for pepsin
   (b) killing micro-organisms
   (c) dissolve food
   (d) facilitate absorption of food
9. The incisor teeth are meant for
   (a) biting & cutting
   (b) munching
   (c) munching and chewing
   (d) chewing

10. Which of the following substance can be assimilated unchanged ?
(a) Vitamin (b) Starch
(c) Proteins (d) Lipids
11. The secretions that mix with food in the small intestine are
(a) saliva, gastric juice & bile
(b) gastric juice, bile & pancreatic juice
(c) bile, pancreatic juice & intestinal juice
(d) pancreatic juice, intestinal juice and gastric juice
12. Jaundice is a disease of
(a) kidney (b) liver
(c) pancreas (d) duodenum
13. In human beings cellulose is digested by
(a) enzymes
(b) symbiotic bacteria
(c) symbiotic protozoans
(d) none of the above
14. Thiamine is
(a) enzyme (b) hormone
(c) deficiency disease (d) vitamin
15. Digestion is
(a) absorption of diffusible food
(b) absorption of water
(c) throwing out of non diffusible food substances.
(d) conversion of non-diffusible ingredients into diffusion state.
16. Succus entericus is
(a) swollen area between ileum & rectum
(b) intestinal juice
(c) any swelling in gut
(d) vermiform appendix
17. The structure which prevents the entry of food into respiratory tract is
(a) glottis (b) pharynx
(c) epiglottis (d) larynx
18. Release of pancreatic juice is stimulated by
(a) enterokinase (b) cholecystokinin
(c) trypsinogen (d) secretin
19. Secretin stimulates production of
(a) saliva (b) gastric juice
(c) bile (d) pancreatic juice
20. In man the zymogen or chief cells are mainly found in
(a) cardiac part of stomach
(b) pyloric part of stomach
(c) duodenum
(d) fundic part of stomach
21. Pancreatic juice and hormones of pancreas are produced by
(a) same cells
(b) same cells at different times
(c) statement is wrong
(d) different cells
22. Vitamin K is required for
(a) change of prothrombin to thrombin
(b) synthesis of prothrombin
(c) change of fibrinogen to fibrin
(d) formation of thromboplastin
23. Kupffer's cells occur in
(a) spleen (b) kidney
(c) brain (d) liver
24. Renin acts on
(a) milk changing casein into calcium paracaseinate at 7.2 – 8.2 pH
(b) proteins in stomach
(c) fat in intestine
(d) milk changing casein into calcium paracaseinate at 1 – 3 pH.
25. The enzyme enterokinase helps in the conversion of
(a) pepsinogen into pepsin
(b) trypsinogen into trypsin
(c) caesinogen into caesin
(d) proteins into polypeptides
26. A dental disease characterized by mottling of teeth is due to the presence of a certain chemical element in drinking water. Which of the following is that element ?
(a) mercury (b) chlorine
(c) fluorine (d) boron
27. The contraction of gall bladder is due to
(a) gastrin (b) cholecystokinin
(c) secretin (d) enterogastrone
28. Stool of a person is whitish grey coloured due to malfunction of which of the following organ?
(a) Pancreas (b) Spleen
(c) Kidney (d) Liver
29. During prolonged fasting, in what sequence are the following organic compounds used up by the body ?
(a) First proteins, next lipids and lastly carbohydrates
(b) First carbohydrates, next fats and lastly proteins
(c) First fats, next carbohydrates and lastly proteins
(d) First carbohydrates, next proteins and lastly lipids

30. Epithelial cells of the intestine involved in food absorption have on their surface
(a) pinocytic vesicles
(b) microvilli
(c) zymogen granules
(d) phagocytic vesicles

31. Angiotensinogen is a protein produced and secreted by
(a) endothelial cells (lining the blood vessels)
(b) liver cells
(c) juxtaglomerular (JG) cells
(d) macula densa cells

32. What will happen if the secretion of parietal cells of gastric glands is blocked with an inhibitor?
(a) Gastric juice will be deficient in chymosin
(b) Gastric juice will be deficient in pepsinogen
(c) In the absence of HCI secretion, inactive pepsinogen is not converted into the active enzyme pepsin
(d) Enterokinase will not be released from the duodenal mucosa and so trypsinogen is not converted to trypsin

33. If for some reason our goblet cells are non functional, this will adversely affect
(a) production of somatostatin
(b) secretion of sebum from the sebaceous glands
(c) maturation of sperms
(d) smooth movement of food down the intestine

34. Which one of the following correctly represents the normal adult human dental formula ?
(a) $\frac{3}{3},\frac{1}{1},\frac{3}{1},\frac{1}{1}$ (b) $\frac{1}{1},\frac{1}{1},\frac{3}{1},\frac{3}{3}$
(c) $\frac{1}{1},\frac{1}{1},\frac{1}{1},\frac{3}{3}$ (d) $\frac{3}{3},\frac{1}{1},\frac{3}{3},\frac{3}{3}$

35. The centre of hunger or centre which regulates the amount of food we eat or out appetite is located in
(a) Medula (b) Cerebrum
(c) Hypothalamus (d) Alimentary canal

36. Gastric juice contains which of the following enzymes?
(a) Pepsin and renin
(b) Amylase, pepsin and lipase
(c) Amylase and pepsin
(d) Insulin and glucagon

37. Absorption of a solution against a concentration gradient by the intestinal epithelium is accompanied by?
(a) Osmotic pressure
(b) Active transport
(c) Brownian movement
(d) Donnan equilibrium

38. Hydrolysis of phospholipids yields
(a) glycerol, phosphoric acid and fatty acids
(b) glycerol, phosphoric acid and nitrogen base
(c) glycerol and fatty acids
(d) acetyl Co A

39. When teeth are different in shape, size & function then these are called
(a) acrodont
(b) pleurodont
(c) homodont
(d) heterodont

40. Secretion of hormone cholecystokinin is for
(a) controlling blood pressure
(b) iron
(c) bile functions
(d) release of insulin

41. Trypsinogen is produced by
(a) liver (b) duodenum
(c) stomach (d) pancreas

42. Saliva contains
(a) enterokinase (b) ptyalin
(c) chymotrypsin (d) lipase

43. Weight of human liver is
(a) 6.0 kg (b) 5.0 kg
(c) 3.0 kg (d) 1.5 kg.

44. Bile secreted by liver passes into gall bladder through
(a) hepatopancreatic duct
(b) hepatic duct
(c) cystic duct
(d) hepato-gall duct

45. Chief cells of the gastric glands secreting zymogen have well developed
(a) smooth ER (b) rough ER
(c) mitochondria (d) lysosome

46. Which of the following, food reserves would be first used in a starving person?
(a) Muscle proteins (b) Skin adipose tissue
(c) Liver glycogen (d) Liver proteins

47. After absorption of digested lipids by lymphatics of small intestine, these become milky, then these lipoprotein droplets are called as
(a) chyme (b) chyle
(c) chylomicron (d) micelles

48. Parotid salivary glands are present
(a) below the tongue
(b) below the ears
(c) in the angle between two jaws
(d) below the eye orbits

49. A good source of lipase is
(a) pancreatic juice (b) gastric juice
(c) bile (d) succus entericus

50. Trypsin differs from pepsin in digestion of proteins in
(a) acidic medium (b) alkaline medium
(c) neutral medium (d) different pH ranges

51. Insulin deficiency causes
(a) diabetes insipidus (b) diabetes mellitus
(c) goitre (d) all of these

52. Amino groups of surplus amino acids are turned into urea in
(a) liver (b) spleen
(c) pancreas (d) kidneys

53. Which one is not a high-energy phosphate compound ?
(a) ATP (b) Creatine phosphate
(c) GTP (d) AMP

54. Human being is hungry when
(a) food cannot meet energy requirement
(b) stomach is empty
(c) food can meet energy requirement of the body
(d) food has been digested

55. Water is largely absorbed in
(a) stomach (b) oesophagus
(c) small intestine (d) colon

56. What is common among amylase, rennin & trypsin?
(a) They are all proteins
(b) They are proteolytic enzymes
(c) They are produced in stomach
(d) They are act at pH lower than 7

57. Crypts of Lieberkuhn are present in
(a) intestine (b) stomach
(c) oesophagus (d) all of these

58. This bile product is not used in the solubilization of fats
(a) bile salts (b) cholesterol
(c) lecithin (d) bile pigments

59. Curdling of milk in small intestine takes place due to
(a) trypsin (b) rennin
(c) ptyline (d) chymotrypsin

60. The fastest acting enzyme, in the biological kingdom is
(a) lipase (b) amylase
(c) peptidase (d) carbonic anhydrase

Breathing and Exchange of Gases

CONCEPTS

Respiration is a catabolic process. It may be divided into two steps :

- Breathing or external respiration involves inhalation of air containing oxygen and exhalation of carbon dioxide produced during oxidation of food.
- Internal or cellular respiration involves release of energy by oxidation of food (glucose), and conversion of the energy released into biologically usable form, such as ATP.

HUMAN RESPIRATORY SYSTEM:

- Human respiratory system consists of external nostrils, nasal cavity, nasopharynx, larynx, trachea, bronchiole and lungs.
- **Larynx** is a small box and form connection between pharynx and trachea.
- **Epiglottis**, a large leaf like cartilagenous structure guards the opening of larynx.
- In each lung, the **bronchous** divide and redivide to form *secondary bronchi, tertiary bronchi, bronchioles* and ultimately into many alveoral ducts that lead into *alveoli* or *air sac.*
- A double layered pleural membrane encloses the lungs–the outer layer is attached to the wall of thoracic cavity.
- The space between two membranes contains a fluid secreted by its wall and reduces the friction and makes movement of lungs easy.

MECHANISM OF BREATHING :

- Breathing involves two stages :
- During **inspiration**, the upward movement of rib cage occurs by external intercostal muscles present between the ribs, along with the assistance of few other adjacent muscles.

  The diaphragm also contracts pulling the lower surface of the lungs downward then the volume of lungs increases, pressure inside decreases coming inhalation of air or inspiration.
- During **expiration**, the downward movement of rib cage is facilitated by the internal intercostal muscles. The diaphragm also show relaxation and the lungs are compressed. The pressure inside the lungs increased and air moves out.

Respiratory Volumes and Capacities

- **Tidal Volume :** Volume of air inspired or expired during a normal respiration. A healthy man can inspire or expire approximately 6000 to 8000 mL of air per minute.
- **Inspiratory Reserve Volume :** Additional volume of air, a person can inspire by a forcible inspiration. This averages 2500 mL to 3000 mL.
- **Expiratory Reserve Volume :** Additional volume of air, a person can expire by a forcible expiration. This averages 1000 mL to 1100 mL.
- **Residual Volume :** Volume of air remaining in the lungs even after a forcible expiration. This averages 1100 mL to 1200 mL.

  By adding up a few respiratory volumes described above, one can derive various pulmonary capacities, which can be used in clinical diagnosis.
- **Inspiratory Capacity :** Total volume of air a person can inspire after a normal expiration. This includes tidal volume and inspiratory reserve volume.
- **Expiratory Capacity :** Total volume of air a person can expire after a normal inspiration. This includes tidal volume and expiratory reserve volume.
- **Functional Residual Capacity :** Volume of air that will remain in the lungs after a normal expiration. This includes ERV+RV.
- **Vital Capacity :** The maximum volume of air a person can breathe in after a forced expiration.

This includes ERV, TV and IRV or the maximum volume of air a person can breathe out after a forced inspiration.

- **Total Lung Capacity :** Total volume of air accommodated in the lungs at the end of a forced indpiration. This includes RV, ERV, TV and IRV or vital capacity + residual volume.

EXCHANGE OF GASES

- The process of diffusion is directly proportional to the pressure caused by the gas alone. The pressure exerted by an individual gas is called *partial pressure.* It is represented by pO_2 , pCO_2 , pN_2 for oxygen, carbon dioxide and nitrogen respectively.

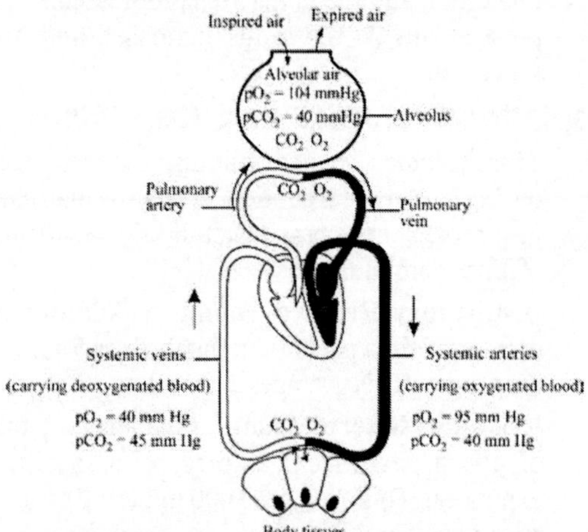

Fig. Diagrammatic representation of exchange of gases at the alveolus and the body tissues with blood and transport of oxygen and carbon dioxide

TRANSPORT OF GASES :

- 97% of oxygen is transported from the lungs to the tissues in combination with haemoglobin ($Hb + O_2 \longrightarrow HbO_2$, oxyhaemoglobin).
- 3% is transported in dissolved condition by the plasma.
- At lower pO_2 oxygen is released from haemoglobin.
- This relationship is expressed by plotting the O_2 saturation of blood against the pO_2 of oxygen. An S–shaped curve, called *oxygen dissociation curve*, is obtained.

There are three ways of transport of CO_2.

- 5%–7% (approximately) of CO_2 is transported, being dissolved in the plasma of blood.
- CO_2 react with the water of RBC to form *carbonic acid* (H_2CO_3) by the enzyme carbonic anhydrase (present in RBC).
- CO_2 reacts with amine radicals (NH_2) of haemoglobin molecule and forms a carbamino–haemoglobin ($HbCO_2$) molecule. Nearly 23% of CO_2 is transported through this mode.

REGULATION OF RESPIRATION :

- The respiratory rhythm is controlled by a number of groups of neuron located bilaterally in the medulla oblongata. These are *respiratory centres.*
- Dorsal respiratory group is present in the dorsal portion of medulla oblongata and its nervous signal is transmitted to the diaphragm to generate basic respirating rhythm.
- The ventral respiratory group is located anterolateral to the dorsal respiratory group. Under enhanced respiratory drive, the respiratory signal of this group fulfill the demand by regulating both inspiration and expiration.
- The pneumotoxic centre is located dorsally in the upper pons. It controls the switch off point of inspiration.

DISORDERS OF RESPIRATORY SYSTEM

Bronchial Asthma :

- It is characterised by the spasm of smooth muscles of the wall of bronchiole.
- It is caused due to hypersensitivity of bronchioles to any foreign substance.

Emphysema :

- It is an inflation of bronchiole, which results into loss of elasticity of these parts. Lung size increase ultimately.
- Cigarette smoking and chronic bronchitis are two reason.

Occupational Lung Disease :

- It is caused because of the exposure of potentially harmful substances persuntion in the environment, where people work.
- Two common occupational diseases are – silicosis and asbestosis.
- It is characterised by fibrosis of upper part of lungs.

Exercise

1. Vocal cords are situated at
(a) Pharynx (b) Larynx
(c) Glottis (d) Bronchial tube
2. Which is a common passage in swallowing food and breathing ?
(a) Larynx (b) Gullet
(c) Glottis (d) Pharynx
3. The structure which prevents the entry of food into respiratory tract is
(a) Pharynx (b) Larynx
(c) Glottis (d) Epiglottis
4. Skin is an accessory organ of respiration in
(a) human (b) frog
(c) rabbit (d) lizard
5. Air is breathed through
(a) Trachea — lungs — larynx — pharynx — alveoli
(b) Nose — larynx — pharynx — bronchus — alveoli — bronchioles
(c) Nostrils — pharynx — larynx — trachea — bronchi — bronchioles — alveoli
(d) Nose — mouth — lungs
6. Lungs have a large number of alveoli for
(a) having spongy texture and proper shape
(b) more surface area for diffusion of gases
(c) more space for increasing volume of inspired air
(d) more nerves supply
7. Oxygen dissociation curve of haemoglobin is
(a) sigmoid (b) hyperbolic
(c) linear (d) hypobolic
8. Rate of respiration is directly affected by
(a) concentration of carbon dioxide
(b) oxygen in trachea
(c) concentration of oxygen
(d) diaphragm expansion
9. One haemoglobin carry how many molecule of O_2 ?
(a) 4 (b) 2
(c) 6 (d) 8
10. One of the following is not respiratory pigment ?
(a) Anthocyanin
(b) Haemoglobin
(c) Haemoerythrin
(d) Haemocfdnin
11. Lungs in addition to the respiratory function also help in
(a) excretion
(b) temperature regulation
(c) pH regulation
(d) maintaining balance of body
12. Cellular respiration depends upon the
(a) availability of carbohydrates in cells
(b) concentration of O_2 in atmosphere
(c) presence of nitrogen with O_2 in air
(d) transport of O_2 to the cells
13. Which have no specific organ for respiration but respire ?
(a) Rabbit (b) Cockroach
(c) Earthworm (d) Frog
14. Intercostal muscles occur in
(a) abdomen (b) thigh
(c) ribs (d) diaphragm
15. Carbon dioxide is transported from tissues to respiratory surface by only
(a) plasma and erythrocytes
(b) plasma
(c) erythrocytes
(d) erythrocytes and leucocytes
16. The exchange of gases in the alveoli of the lungs takes place by
(a) simple diffusion (b) osmosis
(c) active transport (d) passive transport
17. Which one of the following organs in the human body is most affected due to shortage of oxygen?
(a) Intestine (b) Skin
(c) Kidney (d) Brain
18. What is vital capacity of our lungs?
(a) Inspiratory reserve volume plus tidal volume
(b) Total lung capacity minus expiratory reserve volume
(c) Inspiratory reserve volume plus expiratory reserve volume
(d) Total lung capacity minus residual volume
19. The function of nasal cavity & nasopharynx is to
(a) warm the inspired air
(b) moisten the inspired air
(c) filter out the dust particles from the inspired air
(d) All of the above

20. In expiration, diaphragm becomes
(a) flattened (b) relaxed
(c) straightened (d) arched

21. Oxygenated blood from lungs is carried to the heart by
(a) pulmonary artery (b) pulmonary vein
(c) coronary vein (d) pre-cavals

22. A person met with an accident and died instantly without any injury to heart, brain, stomach and kidney.
One of the following is a reason for his death
(a) Intestine got twisted
(b) RBC became coagulated
(c) Stomach stopped digestion
(d) Diaphragm got punctured

23. Which energy is consumed in breathing ?
(a) Mechanical (b) Chemical
(c) Bioelectrical (d) Physical

24. The disease that occurs when the haemoglobin content of the blood goes down is
(a) plurisy (b) emphysema
(c) anaemia (d) pneumonia

25. Volume of air breathed in and out during normal breathing is called
(a) vital capacity (b) IRV
(c) ERV (d) tidal volume

26. Although much CO_2 is carried in blood, yet blood does not become acidic, because
(a) it is absorbed by the leucocytes
(b) blood buffers play an important role in CO_2 transport.
(c) it combines with water to form H_2CO_3 which is neutralized by $NaCO_3$
(d) it is continuously diffused through tissues and is not allowed to accumulate

27. The quantity 1500 ml in the respiratory volumes of a normal human adult refers to
(a) maximum air that can be breathed in and breathed out
(b) residual volume
(c) expiratory reserve volume
(d) total lung capacity

28. Trachea & bronchi possess
(a) incomplete cartilaginous rings
(b) complete cartilaginous rings
(c) thick muscular walls
(d) thick fibrous walls

29. Oxygen carrying capacity of blood is
(a) 20% (b) 30%
(c) 40% (d) 50%

30. Carbonic anhydrase is mostly active in
(a) RBC (b) WBC
(c) blood plasma (d) blood platelets

31. Vital capacity of lungs of an average human is
(a) 3000-4500 ml (b) 1500-1800 ml
(c) 2000-2500 ml (d) 500-1000 ml

32. In the tissues, high concentrations of carbon-dioxide
(a) increases the affinity of haemoglobin to both oxygen and hydrogen
(b) increases the affinity of haemoglobin to oxygen but decreases its affinity to hydrogen
(c) decreases the affinity of haemoglobin to oxygen but increases its affinity to hydrogen
(d) decreases the affinity of haemoglobin to both oxygen and hydrogen

33. The alveolar epithelium in the lungs is
(a) nonciliated columnar
(b) nonciliated squamous
(c) ciliated columnar
(d) ciliated squamous

34. The carbon dioxide is transported via blood to lungs as
(a) dissolved in blood plasma
(b) in the form of carbonic acid only
(c) in combination with haemoglobin only
(d) carbaminohaemoglobin and as carbonic acid

35. In alveoli of the lungs, the air at the site of gas exchange, is separated from the blood by
(a) alveolar epithelium only
(b) alveolar epithelium and capillary endothelium
(c) alveolar epithelium, capillary endothelium and tunica adventitia
(d) alveolar epithelium, capillary endothelium, a thin layer of tunica media and tunica adventitia

36. When CO_2 concentration in blood increases, breathing becomes
(a) shallower and slow
(b) there is no effect on breathing
(c) slow and deep
(d) faster and deeper

37. People living at sea level have around 5 millions RBC per cubic millimeter of their blood whereas those living at an altitude of 5400 metres have around 8 millions. This is because at high altitude
(a) atmospheric O_2 level is less and hence more RBCs are needed to absorb the required amount of O_2 to survive
(b) there is more UV radiation which enhances RBC production
(c) people eat more nutritive food, therefore more RBCs are formed
(d) people get pollution - free air to breathe and more oxygen is available

38. The breathing controlling centre of medulla oblongata is mainly under
(a) chemical control (b) physical control
(c) neural control (d) All of the above

39. In which form CO_2 is carried in blood ?
(a) Sodium bicarbonate
(b) Sodium carbonate
(c) Potassium carbonate
(d) Magnesium carbonate

40. The normal rate of respiration in man per minute is about
(a) 10 to 15 times (b) 16 to 20 times
(c) 19 to 21 times (d) 23 to 25 times

41. The transport of oxygen by the blood takes place by
(a) leucocytes (b) thrombocytes
(c) plasma (d) erythrocytes

42. O_2 is transported in vertebrates as
(a) dissolved in plasma
(b) complexed with haemoglobin
(c) dissolved in cytoplasm of erythrocytes
(d) absorbed over the RBC

43. The covering of the lung is called
(a) pericardium (b) perichondrium
(c) pleural membrane (d) peritoneum

44. Diaphragm present in mammals is
(a) membrane between external & middle ear
(b) membrane around the brain
(c) partition between the thoracic & abdominal cavities
(d) membrane around lungs

45. Percentage of oxygen supplied by haemoglobin is
(a) 97% (b) 100%
(c) 49% (d) 3%

46. On high mountain difficulty in breathing is due to
(a) decrease in partial pressure of O_2
(b) decrease in amount of O_2
(c) increase in CO_2 concentration
(d) all of the above

47. During inspiration, the diaphragm
(a) expands
(b) shows no change
(c) contracts and flattens
(d) relaxes to become dome-shaped

48. 500 ml respiratory volume in a normal adult human is related to
(a) residual volume
(b) total lung capacity
(c) respiratory reserve volume
(d) tidal volume

49. Body tissues obtain oxygen from haemoglobin because of its dissociation in tissues caused by
(a) low oxygen concentration and high carbon dioxide concentration
(b) low oxygen concentration
(c) low carbon dioxide concentration
(d) high carbon dioxide concentration

50. Which is false ?
(a) Blood from Right side of heart is carried to lungs by pulmonary artery
(b) Pleura is double covering of kidney
(c) Pancreas is both exocrine and endocrine gland
(d) Scurvy is due to vitamin C deficiency

51. If the thoracic wall but not lungs is punctured
(a) the lungs get inflated
(b) the man dies as the lungs get collapsed
(c) the breathing rate decreases
(d) the breathing rate increases

52. What is the cause for the movement of oxygen through the alveolar blood capillaries of lungs ?
(a) Difference in the O_2 tension and partial pressure of these chambers
(b) Partial pressure of CO_2
(c) Union of O_2 with haemoglobin
(d) All of the above

53. One of the following is a difference between pulmonary respiration of frog and human
(a) diaphragm and ribs play role in respiration in frog
(b) lungs are respiratory organs
(c) respiration occurs due to pressure gradient in human
(d) None of the above

54. How the transport of O_2 and CO_2 by blood happens ?
(a) with the help of RBCs and blood plasma
(b) with the help of RBCs and WBCs
(c) with the help of WBCs and blood serum
(d) with the help of platelets and corpuscles

55. The breathing rate gets increased by the increase in the content of
(a) nitrogen (b) oxygen
(c) carbon dioxide (d) carbon monoxide

56. Blood analysis of a patient reveals an unusually high quantity of carboxyhaemoglobin content. Which of the following conclusions is most likely to be correct? The patient has been inhaling polluted air containing unusually high content of
(a) carbon disulphide (b) chloroform
(c) carbon dioxide (d) carbon monoxide

57. Which two of the following changes (a–d) usually tend to occur in the plain dwellers when they move to high altitudes (3,500 m or more)?
(A) Increase in red blood cell size
(B) Increase in red blood cell production
(C) Increased breathing rate
(D) Increase in thrombocyte count
Changes occurring are:
(a) (B) and (C) (b) (C) and (D)
(c) (A) and (D) (d) (A) and (B)

58. Which one of the following is a possibility for most of us in regard to breathing, by making a conscious effort?
(a) One can breathe out air totally without oxygen
(b) One can breathe out air through eustachian tubes by closing both the nose and the mouth
(c) One can consiously breathe in and breathe out by moving the diaphragm alone, without moving the ribs at all
(d) The lungs can be made fully empty by forcefully breathing out all air from them

59. Which one of the following mammalian cells is not capable of metabolising glucose to carbon-dioxide aerobically?
(a) Unstraited muscle cells
(b) Liver cells
(c) Red blood cells
(d) White blood cells

60. For proper transport of O_2 & CO_2 blood should be
(a) slightly acidic (b) strongly acidic
(c) strongly alkaline (d) slightly alkaline

Body Fluids and Circulation

CONCEPTS

- The distribution of food and oxygen to all parts of the body as well as the removal of body wastes is performed by a transport system within the body of all living organisms.
- The transport of hormones is also carried by a transport system in our body.

BLOOD

- Blood is a special connective tissue consisting of a fluid matrix, plasma and formed elements.

Plasma

- It is extra cellular, yellow coloured viscous fluid and alkaline in nature.
- It contains 91- 92% water, 77% protein, 0.9 % organic constituents, 0.1% glucose and various other organic and inoganic substances.
- Albumins, globulins and fibrogens are the important types of proteins present in plasma.

Blood Corpuscles

- There are three types of corpuscles-erythrocytes or red blood corpuscles or RBC, leucocytes or white blood corpuscles or WBC and thrombocytes or blood platelets.
- The number of WBC has a range of 7000 - 10,000/mm^3.
- They are grouped into two categories :
- **Granulocytes** are the cells containing granules.
  - **Neutrophils :** Nucleus has two to seven lobes and responsible for protection against infection.
  - **Eosinophils :** Nucleus is bilobed to three lobes and are significant in allergic reaction.
- **Agranulocytes** are the cells without granules.
  - **Lymphocytes :** manufacture serum globulin and play a key role in immunological reactions.
  - **Monocytes :** are phagocytic in nature.
  - **Thrombocytes** or platelets ranges between 2,50,000 – 4,00,000/mm^3 and play an important role in blood coagulation.

Blood Groups

- **ABO grouping :** It is based on the presence or absence of two surface antigens on the RBCs namely A and B.

Table : Blood Groups and Donor Compatibility

Blood Group	Antigens on RBCs	Antibodies in Plasma	Donor's Group
A	A	anti-B	A, O
B	B	anti-A	B, O
AB	A, B	nil	AB, A, B, O
O	nil	anti-A, B	O

Rh Grouping

- Another antigen, the Rh antigen are also observed on the surface of RBCs of majority of humans (Rh$^+$ individuals).
- A special case of Rh incompatibility has been observed between Rh$^-$ blood of a pregnant mother with Rh$^+$ blood of foetus.

LYMPH (TISSUE FLUID)

- It is a colourless fluid containing specialised lymphocytes which are responsible for the immune responses of the body.

Coagulation of Blood

- Blood exhibits coagulation or clotting in response to injury or trauma.
- The clot that is formed at the site of injury is mainly network of threads called fibrin, in which dead and damaged elements of blood are trapped.

- Fibrins are formed by the conversion of inactive fibrinogens in the plasma by enzyme thrombin.
- Ca^{++} plays a very important role in coagulation.

CIRCULATORY PATHWAYS

The circulatory patterns are of two types –

- **Open circulatory system** is present in arthropods and molluscs in which blood pumped by the heart passes through large vessels into open spaces or body cavities called sinuses.
- Annelids and chordates have a **closed circulatory system** in which the blood pumped by the heart is always circulated through a closed network of blood vessels.
- All vertebrates possess a muscular chambered heart. Fishes have a 2-chambered heart with an atrium and a ventricle. Amphibians and the reptiles (except crocodiles) have a 3-chambered heart with two atria and a single ventricle, whereas crocodiles, birds and mammals possess a 4-chambered heart with two artia and two ventricles.

Human Circulatory System

- It is a hollow muscular structure situated behind the sternum and between the lungs in the thoracic cavity.
- It remains enclosed in double walled sac called pericardium.
- Human heart consists of four chambers : two upper thin layered atria and two lower thick layered ventricles.
- The right atrium receives deoxygenated blood from all parts of the body, except the lungs, through superior and inferior vena cava.
- Pulmonary veins bring oxygenated blood to the left atrium from the lungs.
- From right ventricle deoxygenated blood goes to the lungs through the pulmonary artery for purification.
- Left ventricles pumps oxygenated blood to aorta from where it is supplied to the coronary arteries and systematic circulation of the body.
- In between atria and ventricles there are atrioventricular valves.
- Right AV valve or tricuspid valve is present in between right atrium and right ventricle.
- In between left atrium and left ventricle bicuspid valve present.
- The pointed ends of the flaps of AV valves are attached to the ventricular wall by chordae tendinae.

Pumping Action of Human Heart

- Heart muscles continuously generate impulses in a manner that causes rhythmic contraction and relaxation of the heart chamber in a specific sequence.
- The sinoatrial node or SA node situated in the upper lateral wall of the right atrium express the ability of self excitation and control the rate of entire heartbeat.
- The action potential generated in SA node stimulates the AV node through internodal pathway.
- From AV nodes, two specialised muscle fibres (left and right), called bundle branches originate and run into the wall of left and right ventricles.

CARDIAC CYCLE

In the cardiac cycle, the phase of contraction is called **systole**, and phase of relaxation is **diastole**.

- **Atrial systole : -** Blood continuously enters into right atrium through superior and inferior vena cava and coronary sinus. The pulmonary veins bring blood to left atrium from lungs.
- **Ventricular filling : -** Nearly one third of the ventricle is filled when atrioventricular valves open.
- **Ventricular systole : -** When atrial systole approaches its end the ventricles contract and result in the closing of atrioventricular valve.
- **Ventricular diastole : -** During this phase, ventricular pressure falls, semilunar valves close to prevent the back flow of blood, AV valve open to allow the flow of blood from both atria.

Heart Beat and Pulse

- The human heart beats at the rate of about 72-80 per minute in the resting condition.
- A wave of destination that passes along the arteries immediately after ventricular system, is called atrial phase.

Heart Sound

- The rhythmic closer opening of the valves covers the sound of heart beat.
- The first sound *lubb* (dimension 0.16-0.90 sec) is created by closer of atrioventricular valves.
- The second sound *dubb* (duration 0.10 sec) is created by the closure of semilumar valves.

Electrocardiograph

- ECG is the graphic record of electronic current produced by the excitation of cardiac muscles.
- A normal electrocardiogram is composed of a P wave, QRS complex and T wave.
- P wave indicate the depolarisation of the atria.
- QRS complex expresses the ventricular depolarisation.
- T wave indicate an repolarisation of ventrcle.

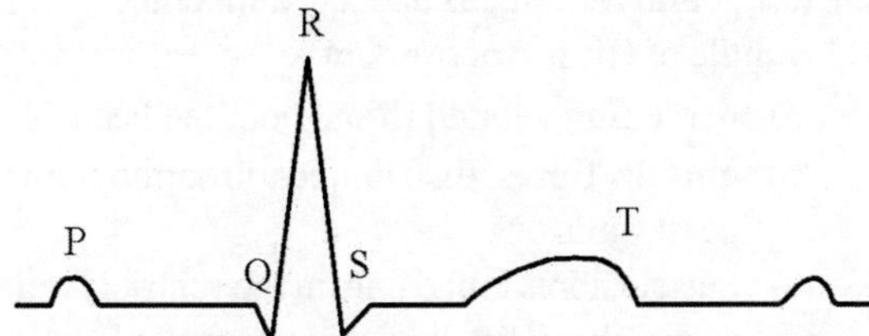

Fig. Diagrammatic presentation of a standard ECG

DOUBLE CIRCULATION :

- Each artery and vein consists of three layers.
  - Tunica intima
  - The tunica media
  - The tunica externa
- The system of blood vessels that ensures the supply of oxygenated blood from left ventricles to all the body organs and return of deoxygenated blood to the right atrium is called **systematic circulation**.
- This circulation starts from the aortic arch arising from left ventricle and terminates as superior and inferior vena cava openning into the right atrium.
- The aortic arch before turning downward to form abdominal aorta , gives of three branches that supply blood to the head region and arms.
- Abdominal aorta caters the need of visceral organs and lower exterminates.
- Regular oxygenation of the blood returning to right atrium is necessary.
- The deoxygenated blood enters the lungs from right ventricles and comes back to the left auricle after oxygenation. This is called **pulmonary circulation**.
- Besides these two, coronary circulation is another pathway in which blood is supplied to the heart muscles and returning to right atrium.
- In some cases, a veins carrying deoxygenated blood from a system of capillaries, divides again into a second capillary system in the tissue before the blood can finally return to the heart, such vein is called portal vein. This system is called portal system *e.g.*, - hepatic portal system.

REGULATION OF CARDIAC ACTIVITY

- Normal activities of the heart are regulated intrinsically, *i.e.*, auto regulated by specialised neural (nodal tissues) hence the heart is called **myogenic**.
- A special neural centre in the medulla oblangata can moderate the cardiac function through autonomic nervous system (ANS).
- On the other hand, parasympathetic neural signals decrease the rate of heart beat, speed of contraction of action potential and thereby the cardiac output.
- Adrenal medullary hormones can also increase the cardiac output.

DISORDERS OF CIRCULATORY SYSTEM

Hypertension

- A continuous or sustained rise in arterial presence is known as hypertension.
- High blood pressure compels heart to work excessive and then can tend to congestive heart disease.

Atherosclerosis

- In this, calcium salts precipitated with cholesterol of the forming or formed opaque making the wall of arteries stiff and rigid and is referred to as the hardening of the arteries.
- It may lead to heart attack and death.

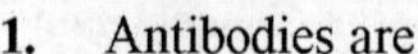

Exercise

1. Antibodies are
   (a) components of blood plasma
   (b) RBC
   (c) substances to inactivate antigens
   (d) WBC which destroy bacteria
2. Lymph differs from blood in
   (a) absence of RBC (b) absence of WBC
   (c) excess of water (d) absence of protein
3. In blood
   (a) WBCs are more than RBCs
   (b) RBCs are more than WBCs
   (c) RBCs are less than platelets
   (d) Platelets are less than WBCs
4. Pulse pressure is
   (a) difference between systolic and diastolic pressure
   (b) systolic pressure
   (c) pressure in the great veins
   (d) diastolic pressure
5. Removal of calcium from freshly collected blood would
   (a) cause delayed clotting
   (b) prevent clotting
   (c) cause immediate clotting
   (d) prevent destruction of haemoglobin
6. Which one engulfs pathogens rapidly?
   (a) Acidophils (b) Monocytes
   (c) Basophils (d) Neutrophils
7. Blood group AB has
   (a) no antigen
   (b) no antibody
   (c) neither antigen nor antibody
   (d) both antigen and antibody
8. An adult human with average health has systolic and diastolic pressures as
   (a) 70 mm Hg and 120 mm Hg
   (b) 120 mm Hg and 80 mm Hg
   (c) 50 mm Hg and 80 mm Hg
   (d) 80 mm Hg and 80 mm Hg
9. What is true about leucocytes ?
   (a) Their sudden fall in number is indication of blood cancer
   (b) These are produced in thymus
   (c) These are enucleated
   (d) These can squeeze out through the capillary walls
10. What is correct for blood group 'O'?
    (a) No antigens but both a and b antibodies are present
    (b) A antigen and b antibody
    (c) Antigen and antibody both absent
    (d) A and B antigens and a, b antibodies
11. Bundle of His is a network of
    (a) nerve fibres found throughout the heart
    (b) muscle fibres distributed throughout the heart walls
    (c) muscle fibres found only in the ventricle wall
    (d) nerve fibres distributed in ventricles
12. Largest number of WBCs are
    (a) eosinophils (b) basophils
    (c) neutrophils (d) monocytes
13. Rh factor is named after
    (a) monkey (b) *Drosophila*
    (c) rat (d) man
14. Sympathetic nervous system
    (a) decreases heart beat
    (b) increases heart beat
    (c) not related to heart beat
    (d) controls heart beat
15. Human/mammalian/elephant heart is
    (a) myogenic (b) neurogenic
    (c) cardiogenic (d) digenic
16. Which one of the following is the main graveyard of RBC ?
    (a) Bone marrow (b) Spleen
    (c) Liver (d) Kidney
17. Blood circulation take following course in heart of man
    (a) Left auricle - left ventricle
    (b) Right auricle -left ventricle
    (c) Left auricle -left ventricle -lungs - right auricle- right ventricle
    (d) None of them
18. If blood cells are removed the liquid left behind is called
    (a) only water (b) serum
    (c) plasma (d) only haemoglobin

19. To reach the left side of heart the blood must pass through
(a) sinus venosus (b) kidneys
(c) liver (d) lungs
20. Interchange of materials in the tissue occurs between
(a) blood capillaries and tissue cells
(b) lymph and tissue fluid
(c) veins and tissue cells
(d) blood and lymph
21. Characteristics of cardiac muscles are that they
(a) contract quickly and get fatigued
(b) contract quickly and do not get fatigued
(c) contract slowly and get fatigued
(d) contract slowly and do not get fatigued
22. Chordae tendinae are found in
(a) ventricles of brain
(b) ventricles of heart
(c) auricles of heart
(d) connection between bone
23. Largest lymphatic organ in the body is –
(a) liver (b) kidney
(c) spleen (d) pancreas
24. A portal system is that in which
(a) a vein begins from an organ and ends in heart
(b) an artery breaks up in an organ & restarts by the union of its capillaries
(c) the blood from gut is brought into kidneys before it is poured into heart
(d) a vein breaks up in an organ into capillaries & restarts by their union as a new vein in the same organ
25. A vein differ from an artery in having
(a) strong muscular walls
(b) narrow lumen
(c) valves to control direction of blood flow opposite to heart
(d) valves to control direction of blood flow towards heart
26. Blood does not clot in the absence of vitamin K because it is essential
(a) components for blood
(b) components of platelets
(c) for synthesis of prothrombin in liver
(d) for synthesis fibrinogen in liver
27. The thickening and loss of elasticity of arterial wall are called
(a) arthritis (b) arteriosclerosis
(c) aneurysm (d) both b and c
28. In which of the following reptiles four chambered heart is present ?
(a) Lizard (b) Snake
(c) Scorpion (d) Crocodile
29. Child death may occur in the marriage between
(a) Rh^+ man and Rh^+ woman
(b) Rh^+ man and Rh^- woman
(c) Rh^- man and Rh^- woman
(d) Rh^- man and Rh^+ woman
30. Tricuspid valve is found in between
(a) sinus venosus and right auricle
(b) right auricle and right ventricle
(c) left ventricle and left auricle
(d) ventricle and aorta
31. A person with blood group A requires blood. The blood group which can be given is
(a) A and B (b) A and AB
(c) A and O (d) A, B, AB and O
32. 'Dup' sound is produced during closure of
(a) semilunar valves (b) bicuspid valve
(c) tricuspid valve (d) both b and c
33. Closed circulatory system occurs in
(a) cockroach (b) tadpole/fish
(c) mosquito (d) housefly
34. Pacemaker of heart is
(a) AV node (b) bundle of His
(c) SA node (d) purkinje fibres
35. A child of blood group O cannot have parents of blood groups
(a) AB and AB/O (b) A and B
(c) B and B (d) O and O
36. Antigens are present
(a) inside the nucleus
(b) on cell surface
(c) inside the cytoplasm
(d) on nuclear membrane
37. The correct route through which pulse-making impulse travels in the heart is
(a) AV node → bundle of His → SA node → Purkinje fibres → heart muscles
(b) AV node → SA node → Purkinje fibres→ bundle of His → heart muscles
(c) SA node → Purkinje fibre → bundle of His → AV node → heart muscles
(d) SA node → AV node → bundle of His → Purkinje fibre → heart muscles

38. Which one of the following statements about blood constituents and transport of respiratory gases is most accurate ?
(a) RBCs transport oxygen whereas WBCs transport CO_2
(b) RBCs transport oxygen whereas plasma transports only CO_2
(c) RBCs as well as WBCs transport both oxygen and CO_2
(d) RBCs as well as plasma transport both oxygen and CO_2

39. Pulmonary artery differ from pulmonary vein in having
(a) no endothelium (b) valves
(c) large lumen (d) thick muscular walls

40. Systemic heart refers to
(a) the two ventricles together in humans
(b) the heart that contracts under stimulation from nervous system
(c) left auricle and left ventricle in higher vertebrates
(d) entire heart in lower vertebrates

41. Which one of the following has an open circulatory system ?
(a) *Periplaneta* (b) *Hirudinaria*
(c) *Octopus* (d) *Pheretima*

42. The most active phagocytic white blood cells are:
(a) neutrophils and eosinophils
(b) lymphocytes and macrophages
(c) eosinophils and lymphocytes
(d) neutrophils and monocytes

43. The most popularly known blood grouping is the ABO grouping. It is named ABO and not ABC, because "O" in it refers to having
(a) overdominance of this type on the genes for A and B types
(b) one antibody only - either anti - A or anti- B on the RBC
(c) no antigens A and B on RBCs
(d) other antigens besides A and B on RBCs

44. There is no DNA in
(a) mature RBCs
(b) a mature spermatozoan
(c) hair root
(d) an enucleated ovum

45. In a standard ECG which one of the following alphabets is the correct representation of the respective activity of the human heart?
(a) S - start of systole
(b) T - end of diastole
(c) P - depolarisation of the atria
(d) R - repolarisation of ventricles

46. Globulins contained in human blood plasma are primarily involved in
(a) osmotic balance of body fluids
(b) oxygen transport in the blood
(c) clotting of blood
(d) defence mechanisms of body

47. Which one of the following plasma proteins is involved in the coagulation of blood ?
(a) An albumin (b) Serum amylase
(c) A globulin (d) Fibrinogen

48. Arteries are best defined as the vessels which :
(a) supply oxygenated blood to the different organs
(b) break up into capillaries which reunite to form one visceral organ
(c) break up into capillaries which reunite to form a vein
(d) carry blood from one visceral organ to another visceral organ

49. Which one of the components of ECG in human is correctly interpreted below?
(a) Complex QRS- one complete pulse
(b) Peak T - initiation of total cardiac contraction
(c) Peak P and peak R together-systolic and diastolic blood pressures
(d) Peak P- initiation of left atrial contraction only

50. Which of the following ion is present in blood ?
(a) HCO_3^- (b) K^+
(c) Cl^- (d) OH^-

51. Maximum surface area of circulating system is seen in
(a) heart (b) capillaries
(c) arterioles (d) veins

52. The pH of blood is
(a) between 7-8 (b) between 2-4
(c) between 12-14 (d) between 2-5

53. Universal blood recipient is
(a) blood group O (b) blood group AB
(c) blood group A (d) blood group B

54. QRS complex is related to

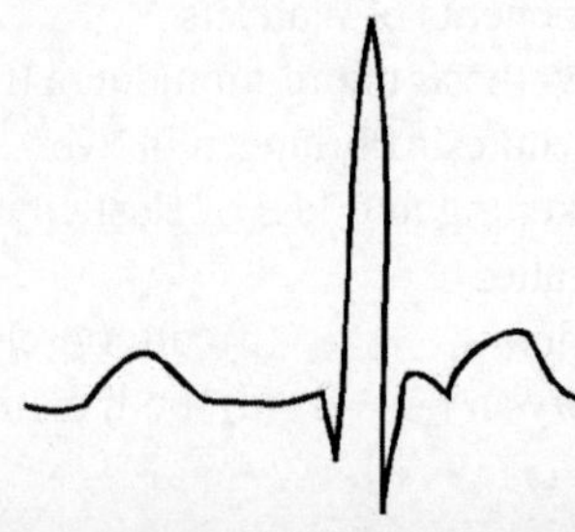

(a) ventricular contraction
(b) auricular contraction
(c) ventricular relaxation
(d) auricular relaxation

55. Heparin is connected with
(a) coagulation
(b) prevention of blood coagulation
(c) dilution of blood
(d) buffering of blood

56. Antibodies required for fighting toxins & pathogens are usually produced by
(a) RBC (b) platelets
(c) WBC (d) All the above

57. The name of disease connected with Rh factor is
(a) haemophilia (b) erythropoiesis
(c) myxoedoma (d) erythroblastosis

58. Number of leucocytes present in one mm^3 of blood is
(a) 5000 – 10,000
(b) 2000 – 3000
(c) 8000 – 13000
(d) 1 million to 15 millions

59. Which of the following is not the main function of lymph glands ?
(a) Forming WBC (b) Forming RBC
(c) Forming antibodies (d) Destroying bacteria

60. Which of the following factors has little effect on blood flow in arteries ?
(a) Heart beat
(b) Blood pressure
(c) Skeletal muscle contraction
(d) Total cross-sectional area of vessels

Excretory Products and Their Elimination

- **Excretion** is the elimination of metabolic waste products from the animal body to regulate the composition of the body fluids and tissues.
- Depending on the availability of water, animals excrete the nitrogeneous wastes in the various forms like ammonia, urea or uric acid.
- The process of excreting ammonia is - **Ammonotelism**. Kidney plays a minor role in the elimination of ammonia *e.g.*, teleost fishes, tadpoles, aquatic soft bodied invertebrates. Organism undergoing ammonotelism are called **ammonotelic**.
- The process of excreting urea is - **Ureotelism**. Examples are mammals, many terrestrial adult amphibians and cartilaginous fishes (shark).
- The process of elimination of uric acid is - **Uricotelism** Examples are land snails, insects, birds and many reptiles.

HUMAN EXCRETORY SYSTEM

Human excretory system consists of -

- **Kidneys** - Paired bean shaped glandular organs located in the abdominal cavity one on either side of the vertebral column.
- Two excretory tubes are **ureters** one from each kidney opens into a **urinary bladder**.
- The urine collected in the urinary bladder passes into a muscular tube called **urethra** from which it is thrown out through an opening at its ends the urinary opening.
- Each kidney has nearly one million complex tubular structures called **nephrons**, which are the functional units of kidney. These filter the blood to produce urine.

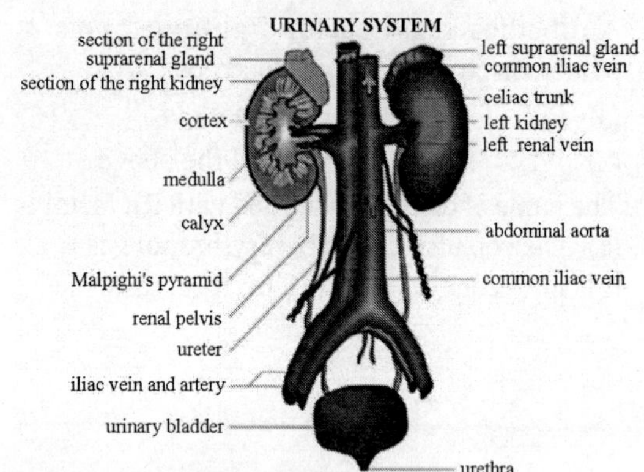

Fig. Urinary System

Urine Formation

The formation of urine involves three processes :

- Glomerular filtration
- Reabsorption
- Secretion

Ultra Filtration

- This forms the **primary urine** which is isosmotic to blood from which it is derived.
- Later the compositon of primary urine is filtered by reabsorption of useful substances (*e.g.*, glucose) into the blood and by the secretion of unwanted substances (*e.g.*, urea) from the blood into the tubular fluid.
- The concentration of primary urine depends upon the length of loop of henle.

Functions of the Tubules

Proximal and Distal Convoluted Tubules

- Main substance reabsorbed in the proximal convulated tubule are glucose, water and salt.
- Some water and salts are also reabsorbed in the distal tubule.
- Dangerous substances in the blood are also secreted into the distal tubule, from the proximal tubule the filtrate, now containing urea, sodium and chloride ions and water, moves into the loops of Henle.

- Loop of Henle - plays a role in the transport of ions & water & the concentrating of urine.
- The loop of Henle is a U-shaped loop that extends from the cortex to the medulla of the kidney and named after it discoverer **F.G. J. Henle.**
- An increasing salt concentration is then created in the ascending limb with the highest concentration deepest in the medulla, because at all the places in loop of Henle salt is actively removed but in the ascending loop it is not allowed to diffuse back in. There is a decreasing salt concentration in the ascending limb.
- The water is reabsorbed by diffusion because of the higher salt concentration on the outside in the medulla of one kidney.
- The filtrate which now consist mainly of urea and other harmful substances as well as varying amount of water depending on the need for it in the body is excreted *via* the urethra.

Mechanism of Concentration of the Filtrate

(Counter Current Mechanism)

- Counter- current multiplier system is an active process which is responsible for the production of concentrated urine in the collecting ducts of the nephrons.
- Sodium and chloride ions are actively pumped from the ascending limb of the loop but water is retained, since the ascending limb is impermeable of water.
- Fluid passing from the loop of Henle to the distal tubule is less concentrated than that entering the loop but because of the high osmotic pressure in the medulla water diffusses out of the collecting ducts producing a concentrated urine.

Regulation of Kidney Function

- The osmolarity of the urine is regulated by nervous and hormonal control of water and salt reabsorption in the kidneys.

Organs Involved are :

- **Hypothalamus** - Urine volume is regulated by the hormone **ADH (antidiuretic hormone),** which is produced by the hypothalamus, and stored & released by the posterior lobe of the pituitary gland in response to an increase in osmotic concentration of the blood caused by dehydration.

Juxtaglomerular Apparatus or JGA -

- JGA operates through multihormonal system -

Renin-Angiotensin-Aldosterone System (RAAS)

- Decrease in blood volume → decrease in pressure → cells of Juxtaglomerular apparatus secretes renin →renin converts angiotensinogen in plasma to angiotensin →enzymes in lungs converts angiotensin to angiotensin II → Blood vessels constricts and aldosterone is secreted by the adrenal gland → aldosterone increase sodium and water reabsorption.
- This leads to an increase in the blood volume and pressure and brings GPR back to normal.
- **Atrial wall of Heart** - An increase in the blood flow (volume) and pressure of the atria of the heart causes the release of Atrial Natriuretic Factor (ANF).
- It works antagonistically to the renin angiotensin system. It is a response to an increase in blood volume and pressure.

Micturition

- Micturition also called urination is the process of disposing urine from the urinary bladder through the urethra to the oustide of the body.

Role of other organs in excretion

- **Lungs** - Responsible for elimination of CO_2 & water vapour.
- **Liver** - Secretes bile, degraded steroid hormones, drugs and certain vitamins.
- **Skin** - Undergo excretion of certain substances through glands present in it *e.g.*, sweat, hydrocarbons waxes.

DISORDERS OF THE EXCRETORY SYSTEM

- Malfunctioning of kidneys can lead to accumulation of urea in blood, a condition called **uremia**, which is highly harmful and may lead to kidney failure.
- In such patients, urea can be removed by a process called **hemodialysis**. Blood drained from a convenient artery is pumped into a dialysing unit called **artificial kidney**.
- **Renal calculi :** Stone or insoluble mass of crystallized salts (oxalates, *etc.*) formed within the kidney.
- **Glomerulonephritis :** Inflammation of glomeruli of kidney.

Exercise

1. Main function of kidney is
(a) passive absorption
(b) ultrafiltration
(c) selective reabsorption
(d) both (b) & (c)

2. Excretory product in mammals is
(a) urea (b) uric acid
(c) ammonia (d) None

3. In man the two kidneys lie
(a) at the level of ovaries
(b) at the same level
(c) left kidney at a higher level than the right one
(d) right kidney at a higher level than the left one

4. The afferent and efferent vessels are
(a) arterial in nature
(b) venous in nature
(c) one is arterial and the other is venous
(d) None of the above

5. A notch present on the medial side of kidney is known as
(a) ureter (b) pelvis
(c) hilus (d) pyramid

6. Functional & structural unit of kidney is
(a) nephron (b) seminiferous tubule
(c) acini (d) None

7. Excretory material are formed in
(a) kidney (b) rectum
(c) liver (d) everybody cell

8. Nephron is also known as
(a) juxta glomerular tubule
(b) seminiferous tubule
(c) uriniferous tubule
(d) all

9. Nitrogenous waste products are eliminated mainly as
(a) urea in tadpole and ammonia in adult frog
(b) ammonia in tadpole and urea in adult frog
(c) urea in both tadpole and adult frog
(d) urea in tadpole and uric acid in adult frog

10. A terrestrial animal must be able to
(a) excrete large amounts of water in urine
(b) conserve water
(c) actively pump salts out through the skin
(d) excrete large amounts of salts in urine

11. Proximal and distal convoluted tubules are parts of
(a) seminiferous tubules
(b) nephron
(c) oviduct
(d) vas deferens

12. If kidney fail to reabsorb water, the effect on tissue would
(a) remain unaffected
(b) shrink and shrivel
(c) absorb water from blood plasma
(d) take more O_2 from blood

13. The enteronephric nephridia of earthworm are concerned with
(a) osmoregulation
(b) excretion of nitrogenous wastes
(c) digestion
(d) respiration

14. A person is undergoing prolonged fasting. His urine will be found to contain abnormal quantities of
(a) fats (b) amino acids
(c) glucose (d) ketones

15. Bowman's glands are located in the
(a) anterior pituitary
(b) female reproductive system of cockroach
(c) olfactory epithelium of our nose
(d) proximal end of uriniferous tubules

16. Excretion is
(a) removal of essential substances not required by body
(b) removal of useless substances present in excess
(c) formation of substances having some role in body
(d) all

17. Malphigian body is found in
(a) kidneys of vertebrates & is involved in formation of urine
(b) vertebrates & is involved in filtration of blood
(c) vertebrates & involved in formation of salts
(d) insects & involved in excretion

18. Glomerulus filtrate possesses
(a) blood plasma without proteins
(b) blood plasma with cell proteins
(c) blood plasma with urea
(d) plasma minus cell

19. Which one has nothing to do with nitrogeneous excretion?
(a) Skin (b) Kidneys
(c) Liver (d) Lungs

20. Diuresis is the condition in which
(a) excretion of volume of urine increases.
(b) excretion of volume of urine decreases
(c) the kidneys fail to excrete urine
(d) the water balance of the body is disturbed

21. The most toxic excretory product is
(a) CO_2 (b) amino acids
(c) urea (d) ammonia

22. Part not belonging to uriniferous tubule is
(a) glomerulus
(b) henle's loop
(c) distal convoluted tubule
(d) collecting duct

23. Urea is formed in liver cells from
(a) ammonia and nitrogen
(b) ammonia and carbon dioxide
(c) ammonia carbon, dioxide and aspartic acid
(d) ammonia and carbon monoxide

24. Vasa rectae are peritubublar capillaries around
(a) posterior part of alimentary canal
(b) PCT
(c) loop of Henle
(d) DCT

25. Micturition is
(a) removal of urea from blood
(b) removal of uric acid
(c) passing out urine
(d) removal of fats

26. Glomerular filtration rate would be decreased by
(a) constriction of the efferent arteriole
(b) an increase in afferent arteriolar pressure
(c) compression of the renal capsule
(d) an increase in renal blood flow

27. Diameter of the renal afferent vessel is
(a) same as that of efferent
(b) smaller than that of efferent
(c) larger than that of efferent
(d) there is no efferent vessel

28. Which secretes renin ?
(a) Stomach
(c) Juxta cortical cell
(b) Juxta glomerular body
(d) Juxta medullary cell

29. Podocytes are present in
(a) afferent arteriole (c) peritubular network
(b) efferent arteriole (d) bowman's cup

30. Which excretory material is least toxic ?
(a) Ammonia
(b) Urea
(c) Uric acid
(d) Trimethylamine oxide

31. Which regulates reabsorption of salts from glomerular filtrate ?
(a) Oxytocin (b) Vasopressin
(c) Gluco corticoides (d) Mineral corticoids

32. Which one of the following body functions is not performed by kidneys ?
(a) Excretion
(b) Osmoregulation
(c) Regulation of blood volume
(d) Destruction of dead blood corpuscles

33. Angiotensinogen is a protein produced and secreted by
(a) juxtaglomerular (JG) cells
(b) macula dense cells
(c) endothelial cells (cells lining the blood vessels)
(d) liver cells

34. A person who is on a long hunger strike and is surviving only on water, will have
(a) less amino acids in his urine
(b) more glucose in his blood
(c) less urea in his urine
(d) more sodium in his urine

35. Water reabsorption in kidney is controlled by
(a) aldosterone (b) ADH
(c) GH (d) oxytocin

36. Ultra filteration is determined by
(a) glomerular hydrostatic pressure
(b) colloid osmotic pressure of blood
(c) capsular hydrostatic pressure
(d) All of the above

37. Urinary bladder is absent in
(a) reptiles (b) aves
(c) amphibians (d) mammals

38. Which one is related to urine concentration in mammals?
(a) Testosterone hormone
(b) Antidiuretic hormone
(c) Oxytocin hormone
(d) All of these

39. The kidneys resemble contractile vacuoles of protozoans in
(a) expelling of excess water
(b) expelling out excess glucose
(c) expelling out urea & uric acid
(d) expelling out salts

40. Loop of Henle is connected with
(a) excretory system
(b) nervous system
(c) reproductive system
(d) muscular system

41. The blood vessel taking blood into Bowman's capsule is
(a) afferent arteriole (b) efferent arteriole
(c) renal vein (d) renal portal vein

42. In which part of the excretory system of mammals you can first use the term urine for fluid it contains?
(a) Bowman's capsule
(b) Loop of Henle
(c) Collecting tubule
(d) Urinary tubule

43. Formation of concentrated (hyperosmotic) urine in vertebrates generally depends on
(a) length of the proximal convoluted tubule
(b) length of Henle's loop
(c) podocytes of Bowman's capsule
(d) capillary network forming glomerulus

44. Cause of glomerular filtration is
(a) dialysis (b) osmosis
(c) high pressure (d) diapedesis

45. Urea is synthesised in
(a) kidneys (b) lungs
(c) intestine (d) liver

46. Occurrence of excess urea in blood due to kidney failure is
(a) renal calculi (b) uremia
(c) uricotelism (d) ureotelism

47. The mechanism of urine formation in nephron involves
(a) ultra-filtration (b) absorption
(c) diffusion (d) osmosis

48. Glomerular filtrate contains glucose in comparison to plasma
(a) more (b) same
(c) less (d) nil

49. In diabetes mellitus, the patient drinks more water as there is urinary loss of
(a) salt (b) insulin
(c) protein (d) glucose

50. The part of the neprhon of kidney where Cl^- is absorbed along with Na^+ is
(a) ascending limb of Henle's loop
(b) descending limb of Henle's loop
(c) DCT
(d) PCT

51. When a fresh-water protozoan possessing a contractile vacuole, is placed in a glass containing marine water, the vacuole will
(a) increase in number
(b) disappear
(c) increase in size
(d) decrease in size

52. Earthworms are
(a) uricotelic when plenty of water is available
(b) uricotelic under conditons of water scarcity
(c) ammonotelic when plenty of water is available
(d) ureotelic when plenty of water is available

53. Consider the following four statements (i-iv) about certain desert animals such as kangaroo, rat
(i) They have dark colour and high rate of reproduction and excrete solid urine
(ii) They do not drink water, breathe at a slow rate to conserve water and have their body covered with thick hairs
(iii) They feed on dry seeds and do not require drinking water
(iv) They excrete very concentrated urine and do not use water to regulate body temperature

Which two of the above statements for such animals are true?
(a) (iii) and (iv) (b) (ii) and (iii)
(c) (iii) and (i) (d) (i) and (ii)

54. What will happen if the stretch receptors of the urinary bladder wall are totally removed?
(a) Micturition will continue
(b) Urine will continue to collect normally in the bladder
(c) There will be no micturition
(d) Urine will not collect in the bladder

55. Which one of the following statements in regard to the excretion by the human kidneys is correct?
(a) Descending limb of Loop of Henle is impermeable to water
(b) Distal convoluted tubule is incapable of reabsorbing HCO_3
(c) Nearly 99 per cent of the glomerular filtrate is reabsorbed by the renal tubules
(d) Ascending limb of Loop of Henle is impermeable to electrolytes

56. Consider the following four statements (i-iv) regarding kidney transplant and select the two correct ones out of these.
(i) Even if a kidney transplant is proper the recipient may need to take immune suppresants for a long time
(ii) The cell mediated immune response is responsible for the graft rejection
(iii) The B-lymphocytes are responsible for rejection of the graft
(iv) The acceptance or rejection of a kidney transplant depends on specific interferons
The two correct statements are:
(a) (ii) and (iii) (b) (iii) and (iv)
(c) (i) and (iii) (d) (i) and (ii)

57. Which one of the following is not a part of a renal pelvis?
(a) Peritubular capillaries
(b) Convoluted tubules
(c) Collecting ducts
(d) Loops of Henle

58. A large proportion of oxygen is left unused in the human blood even after its uptake by the body tissues. This O_2
(a) acts as a reserve during muscular exercise
(b) raise the pCO_2 of blood to 75 mm of Hg
(c) is enough to keep oxyhaemoglobin saturation at 96%
(d) helps in releasing more O_2 to the epithelial tissues

59. Which one of the following correctly explains the function of a specific part of a human nephron ?
(a) Podocytes : create minute spaces (slite pores) for the filtration of blood into the Bowman's capsule
(b) Henle's loop : most reabsorption of the major substances from the glomerular filtrate
(c) Distal convoluted tubule : reabsorption of K^+ ions into the surrounding blood capillaries
(d) Afferent arteriole : carries the blood away from the glomerular towards renal vein.

60. Which one of the following statement is correct with respect to kidney function regulation ?
(a) When someone drinks lot of water, ADH release is suppressed.
(b) Exposure to cold temperature blood flow stimulates formation of Angiotensin II.
(c) An increase in glomerular blood flow stimulates formation of Angiotensin II.
(d) During summer when body loses lot of water by evaporation, the release of ADH is suppressed.

Locomotion and Movement

CONCEPTS

- The movement of an individual from one place to another is called **Locomotion.** It is one of the most important characteristics of living organisms.

TYPES OF MOVEMENT

Movements of External Body Parts

- Movements of limbs, appendages, head and trunk help to maintain equilibrium of the body.
- Movements of eye balls help to see the object, however, movement of the pinna of the ear enable the animal to collect sound vibrations.

Movements of Internal Organs

- Food and urine are moved forward by the movements of the muscles (peristalsis) present in the digestive and urinary tracts respectively.
- Inspiration and expiration involves the movements of the diaphragm, ribs, etc.
- Movements of uterine wall enable the mother to give birth to a child.

MUSCLES

Muscle is a specialized tissue, that arises from the mesoderm.

Types of Muscles :

Based on their location, there are three types of muscles :

Skeletal Muscles :

- Skeletal muscles help in movement and support skeleton.
- They have striped appearance under the microscope and hence are called **striated muscles**.

  They provide power and strength to the body.

Smooth Muscles :

- These muscles are found in the hollow parts of the body like stomach, intestine, blood vessels, reproductive tract etc.
- Smooth muscles are involuntary muscles, *i.e.* we cannot consciously control these muscles. These muscles are under the control of autonomous nervous system.
- They help in transportation of food through the digestive tract and gametes through the genital tract.

Cardiac muscles (A Heart muscle)

- Cardiac muscles are stripped, like skeletal muscles.
- Cardiac muscles contract automatically to squeeze the walls of the heart inward.

Structure of Contractile Protein

ACTIN : All eukaryotic species contain actin.

- Each F actin is a polymer of monomeric 'G' (Globular) actins and arranged as a chain of repeating units, forming two strands of an alpha helix.
- Actin strands are interwoven by rod shaped proteins termed as **Tropomyosin**, which consists of 6-7 actin molecules. Tropomyosin contains a complex proteins, **Troponin** which remains attached to Tropomyosin at regular intervals, which serves as a binding site for Ca^{++} during **excitation - contraction coupling**.

Myosin : There are about 300 molecules of myosin per thick filament, called as meromyosins.

Each meromyosins consists of -

- A globular head with a short arm called as heavy meromyosin (HMM). It is an active ATPase enzyme and has binding sites for ATP and active sites for actin.
- A tail called as light meromyosin (LMM)

Mechanism of Muscle Contraction

- Muscle contraction occurs by a sliding filament mechanism where by actin filaments slide inward among the myosin filaments.

- The contraction of a muscle is dependent on a precisely controlled passing of sodium and calcium ions in and out of a muscle fibre. Contraction is stimulated by a chemical called as **acetylcholine** released from an attached nerve.

SKELETAL SYSTEM

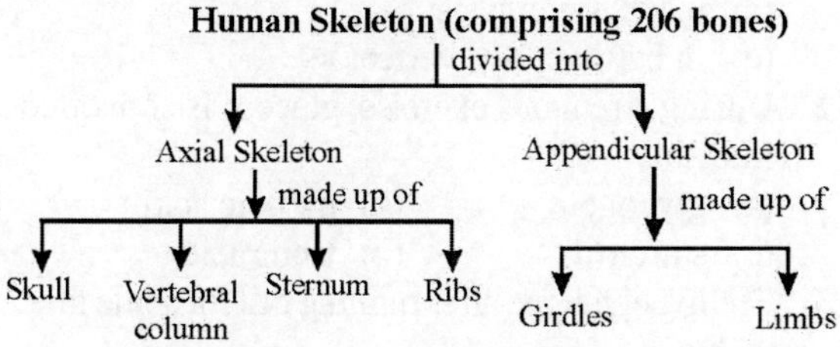

Human Skeleton (comprising 206 bones)
divided into
- Axial Skeleton — made up of: Skull, Vertebral column, Sternum, Ribs
- Appendicular Skeleton — made up of: Girdles, Limbs

Axial Skeleton : Skeleton which occurs in the mid axial or longitudinal part of the body.

(i) **Skull** is made up of 29 bones. It is composed of
- **Cranium (8 bones)** : Frontal - 1; Parietal - 2; Occipital - 1; Temporal - 2; Sphenoid - 1; Ethmoid - 1.
- **Facial bones (14 in number)** : Nasal - 2; Maxillae - 2; Zygomatic - 2; Lacrymals - 2; Mandibles - 1; Inferior turbinals - 2; Vomer - 1; Palatines - 2.
  Hyoid Tongue bone - 1
- **Ear ossicles (6 bones)** : Malleus -2; Incus - 2; Stapes - 2.

(ii) **Vertebral column :** 33 in babies, 26 in adults. Grouped into 5 categories :
Cervical - 7; Thoracic - 12; Lumber - 5; Sacral - 5; Coccygeal - 4 (fused in adults).

(iii) **Sternum** : Composed of 3 parts → Manubrium, body of sternum and xiphoid process .

(iv) **Ribs** : They are twelve pairs. First seven pairs are true ribs. The 8th, 9th and 10th ribs are called false ribs or vertebrochondrial ribs. The last 11th and 12th pairs are called floating ribs.

Appendicular Skeleton : Present laterally or attached to the axial skeleton.

(i) **Girdles :** 2 types - pectoral and pelvic.
Pectoral girdle : made of two parts - clavicle and scapula.
Pelvic girdle : made of three bones - ilium, pubis and ischium.

(ii) **Limb bones** : Hind limbs and fore limbs - both made up of 30 bones each.
Fore limbs : Humerus (1); Radius-Ulna (2); Carpals (8); Metacarpals (5); Phalanges (14); Phalanges formula = 2, 3, 3, 3, 3. **Hind limbs** : Femur (1); Tibia-Fibula (2); Patella (1); Tarsals (7); Metatarsals (5); Phalanges (14).

Joints

- A joint is a location at which two bones make contact and is essential for all types of movements, involving the bony parts of the body.
- **Fibrous joints** - The bones are united by dense connective tissue consisting of collagen fibres, which run between the bones.
- This type of joint is present in flat skull bones and these joints are immovable joints *i.e.* they do not allow movements.
- **Cartilagenous joints (Slightly movable joints)** - Bones are united with each other by cartilage.
- Present in between the adjacent vertebral in the vertebral column.

Synovial Joints - Movable Joints : They are characterised by the presence of a closed space or **cavity** between the bones.

- This kind of joint are classified into six major categories.
  - **Plane** (gliding joint) : Present between carpals. Only sliding motion in all direction is allowed.
  - **Hinge joint** : Present between Knee joint
  - **Pivot joint** : Present between atlas and axis
  - **Ellipsoidal joint**
  - **Saddle joint** : Present between carpal and metacarpal
  - **Ball and Socket joint** : Present between humerus and pectoral girdle.

DISORDERS OF MUSCULAR AND SKELETAL SYSTEM

- **Myasthenia gravis** - Auto immune disorder. It affects neuromuscular transmission.
- **Muscular dystrophy** - Progressive skeletal muscle weakness, defects in muscle proteins, the death of muscle cells and tissue.
- **Rheumatoid Arthritis :** Inflammation of synovial membrane.
- **Osteoarthritis :** Degeneration of articular cartilage.
- **Gout :** Caused by excess formation of uric acid and their deposition in the joints.
- **Osteoporosis :** Low bone mass, increased fragility and proneness to fracture.

Exercise

1. Long bones function in
(a) support
(b) support, erythrocyte and leucocyte synthesis
(c) support and erythrocyte synthesis
(d) erythrocyte formation
2. We move our hand while walking for is
(a) faster movement
(b) balancing
(c) increasing blood circulation
(d) relieving tension
3. Bones are mainly made up of
(a) calcium and phosphorous
(b) calcium and sulphur
(c) calcium and magnesium
(d) calcium and iron
4. Which ion is essential for muscle contraction?
(a) Na^+ (b) K^+
(c) Ca^{++} (d) Cl^-
5. Immediate source of energy for muscle contraction is
(a) glucose (b) GTP
(c) creatine phosphate (d) ATP
6. In mammals the lower jaw is made of
(a) maxilla (b) dentary
(c) mandible (d) ethmoid
7. Which one is a ball and socket joint?
(a) Knee joint
(b) Elbow joint
(c) Humerus and pectoral girdle
(d) Skull and atlas
8. Correct order of ear ossicles in rabbit is
(a) incus, stapes, malleus
(b) malleus, incus, stapes
(c) malleus, stapes , incus
(d) incus, malleus, stapes
9. In old age stiffness of joints is due to the
(a) hardening of bones
(b) inefficiency of muscles
(c) decrease in synovial fluid
(d) enlargement of bones
10. The functional unit of contractile system of a striated muscle is
(a) sarcomere (b) Z-band
(c) cross bridge (d) myofibril
11. Inter-articular disc occurs in
(a) wall of heart
(b) wall of liver
(c) pubic symphysis
(d) in between two vertebrae
12. During strenuous exercise, glucose is converted into
(a) glycogen (b) pyruvic acid
(c) starch (d) (d) lactic acid
13. Feeling of fatigue after running fast for some time is due to
(a) loss of energy
(b) accumulation of lactic acid in muscles
(c) formation of succinic acid
(d) formation of biuret crystals
14. Tendon is made up of
(a) adipose tissue
(b) modified white fibrous tissue
(c) areolar tissue
(d) yellow fibrous connective tissue
15. Ligament is a
(a) modified yellow elastic fibrous tissue
(b) inelastic white fibrous tissue
(c) modified white fibrous tissue
(d) None of these
16. Which of the following is the contractile protein of a muscle?
(a) Myosin (b) Tropomyosin
(c) Actin (d) Tubulin
17. The number of floating ribs, in the human body, is
(a) 6 pairs (b) 5 pairs
(c) 3 pairs (d) 2 pairs
18. Ankle joint is
(a) pivot joint (b) ball and socket joint
(c) hinge joint (d) gliding joint
19. Elbow joint is
(a) ball and socket joint
(b) hinge joint
(c) suture joint
(d) gliding joint
20. Cartilagenous joints –
(a) permit slight movements
(b) are found in symphysis
(c) are found in the bodies of vertebrae
(d) All of the above

21. Hinge joints
(a) are synovial joints
(b) permit movement in one direction
(c) are found in knee
(d) All of the above

22. Joints between carpals are
(a) fibrous joints (b) cartilagenous joints
(c) angular joints (d) gliding joints

23. Myoglobin occurs in
(a) blood (b) liver
(c) spleen (d) muscles

24. Sarcomere is distance between-
(a) two I-bands (b) A and I-bands
(c) two Z-lines (d) Z and A bands

25. Which one is incorporated in muscles fibres ? Which gives a reddish appearance?
(a) Acetylcholine (b) Myoglobin
(c) Histone (d) Cytochrome

26. Joint between atlas and axis is
(a) pivot joint (b) saddle joint
(c) angular joint (d) hinge joint

27. Acetabulum is part of
(a) pelvic girdle (b) pectoral girdle
(c) fore arm (d) upper arm

28. Synovial fluid is present in
(a) spinal cavity
(b) cranial cavity
(c) freely movable joints
(d) fixed joints

29. Bones of pelvic girdle forms a cavity in which head of the femur is fitted are
(a) ilium only
(b) ilium and ischium
(c) ilium, ischium and pubis
(d) ischium and pubis

30. Sutures present between various bones of skull are
(a) cartilagenous joints
(b) synovial joints
(c) hinge joints
(d) fibrous joints

31. Which are involved in muscle contraction?
(a) Ca^{2+} & Mg^{2+} (b) Ca^{2+} & Na^+
(c) Na^+ & K^+ (d) Mg^{2+} & K^+

32. Largest synovial joint is
(a) hip joint (b) knee joint
(c) shoulder joint (d) ankle joint

33. Acromion process is part of
(a) vertebral column (b) pelvic girdle
(c) femur (d) pectoral girdle

34. The number of lumbar vertebral column in the human body, is
(a) 6 (b) 5
(c) 3 (d) 2

35. Which statement is correct for muscle contraction?
(a) Length of H-line decreases
(b) Length of A-band remains constant
(c) Length of I-band increases
(d) Length of two Z-lines increase

36. Three of the following pairs of the human skeletal parts are correctly matched with their respective inclusive skeletal category and one pair is not matched. Identify the non-matching pair.

Pairs of skeletal parts	**Category**
(a) Sternum and ribs	Axial skeleton
(b) Clavicle and glenoid cavity	Pelvic girdle
(c) Humerus and ulna	Appendicular skeleton
(d) Malleus and stapes	Ear ossicles

37. End of long bones are covered with
(a) muscles (b) cartilage
(c) adipose tissue (d) bone marrow

38. In mammals, the number of cervical vertebrae is always
(a) 7 (b) 9
(c) 1 (d) 12

39. Which one is bone of the fore limb?
(a) Humerus (b) Femur
(c) Tibia (d) Fibula

40. Thoracic cage is made up of
(a) ribs, vertebral column & diaphragm
(b) ribs, diaphragm & sternum
(c) vertebral column, diaphragm & sternum
(d) ribs, vertebral column & sternum

41. Which one has the maximum glycogen ?
(a) Liver (b) Muscles
(c) Nerves (d) Kidneys

42. Number of bones of face is
(a) 12 (b) 30
(c) 40 (d) 14

43. How many ribs are present in human beings ?
(a) 6 pairs (b) 9 pairs
(c) 12 pairs (d) 15 pairs

44. Cranium of man is made up of
(a) 8 bones (b) 12 bones
(c) 10 bones (d) 16 bones

45. The last two pairs of ribs are named floating ribs because
(a) Their sternal parts are attached to the sternum directly
(b) Their sternal parts are attached on the 7th pair of ribs
(c) Their sternal parts remain free and do not even reach the sternum
(d) They float in the body cavity

46. Patella is associated with
(a) elbow (b) knee
(c) neck (d) wrist

47. Skull of mammals is
(a) Amphicondylous (b) Condylless
(c) Dicondylous (d) Monocondylous

48. The dark bands (Black bands) (A-bands) of a skeletal muscle are known as
(a) Isotropic bands (b) Anisotropic bands
(c) Intercalated disc (d) Cross bridges

49. Muscles of the heart are
(a) voluntary, striated (b) voluntary, smooth
(c) involuntary, striated (d) involuntary, smooth

50. The dark bands in a myofibril are due to overlapping of
(a) only thick bands
(b) only thin bands
(c) both thick and thin bands
(d) None of the above

51. Number of bones in human body is
(a) 260 (b) 206
(c) 306 (d) 203

52. Malleus is a part of
(a) fore limbs of vertebrates
(b) reproductive organs of cockroach
(c) auditory ossicles of middle ear of rabbit
(d) skull of frog.

53. In main the axial skeleton is made up of
(a) 80 bones (b) 100 bones
(c) 103 bones (d) 106 bones

54. Which of the following is an example of appendicular skeleton ?
(a) Bones of skull
(b) Bones of vertebal column
(c) Ribs
(d) Bones of fore and hind limbs

55. Major protein in the thick filament of skeletal muscle fibre is
(a) tropomyosin (b) myosin
(c) actin (d) troponin

56. In man ribs are attached to
(a) clavicle (b) ileum
(c) sternum (d) scapula

57. The 8th and 9th ribs are known as false ribs because their external portions are attached to
(a) xiphisternum
(b) costa of 7th rib
(c) they have no costa
(d) they are not true ribs

58. Which one is odd pair ?
(a) Femur-humerus (b) Tibia-radius
(c) Axis-atlas (d) Tarsal-carpals

59. Sarcolemma is a membrane found over in
(a) nerve fibre (b) cardiac muscle
(c) skeletal muscle fibre (d) heart

60. Upon stimulation of skeletal muscles calcium is immediately made available for binding to troponin from
(a) Blood
(b) Lymph
(c) Sarcoplasmic reticulum
(d) Bone

Neural Control and Co-ordination

CONCEPTS

Coordination is the process through which two or more organs interact and complement the functions of each other.

NEURAL SYSTEM

- The neural sysem is the control system of the body which consists of highly specialized cells called **neurons**. The neurons detect and receive information from different sense organs (receptors) in the form of stimuli and transmit the stimuli to the central neural system (CNS) through sensory nerve fibres.

HUMAN NEURAL SYSTEM

Human neural system is divisible into two main parts:

- **Central Neural System :** It is a hollow, dorsally placed structure lying along the mid-dorsal axis of the body. It comprises the brain and spinal cord.
- **Peripheral Neural System :** The nerves arising from the central neural system constitute the peripheral neural system. The nerves originate from the brain and spinal cord and are known as cranial nerves and spinal nerves respectively.
- The nerve fibres of PNS are divided into two groups, namely

  The **afferent nerve fibres** transmit sensory impulses from tissues/organs to the CNS and form the sensory or afferent pathway. The **efferent nerve fibres** transmit motor impulses from CNS to the concerned tissues/organs and form the motor or efferent pathways.
- The peripheral nervous system is again divided into two subdivisions :

  Somatic neural system : It controls the movements of the body by acting on the skeletal muscles.
- **Autonomic Neural System :** It controls and coordinates such organs which are under involuntary control. It consists of **sympathetic** and **parasympathetic** nervous system.
- Visceral nervous system is the part of the peripheral nervous system that comprises the whole complex of nerves, fibers, ganglia, and plexuses by which impulses travel from the central nervous system to the viscera and from the viscera to the central nervous system.

NEURONS (NERVE CELLS)

- A neuron is a structural and funtional unit of the neural tissue and hence, the neural system.
- A neuron consists of main cell body and cytoplasmic processes arising from it.

Cell Body (*Cyton or Soma*)

- Like a typical cell, it consists of cytoplasm, nucleus and cell membrane.
- The cytoplasm has mitochondria, golgi apparatus, rough endoplasmic reticulum, ribosomes, lysosomes, fat globules, pigment granules, neurofibrils, neurotubules and Nissl's granules.
- Cyton is concerned with metabolic maintenance and growth.

Dendrites (Dendrons)

- These are usually shorter, tapering and much branched processes. They may be one to several.
- The dendrites contain neurofibrils, neurotubules and Nissl's granules.
- They conduct nerve impulse towards the cell body and are called afferent processes.

Axon

- Axon is a single, usually very long process of uniform thickness.
- The axon contains neurofibrils and neurotubules but does not have Nissl's garanules, Golgi complex, ribosomes, pigment granules, fat globules, etc.

Generation and Conduction of Nerve Impulse

Conduction of Nerve Impulse Along a Non Myelinated Nerve Fibre

- In a resting nerve fibre (*i.e.*, at resting potential), the axoplasm is electronegative and the exterior (ECF) is electropositive.
- This state of the resting membrane is called polarised state.
- When a threshold stimulus is applied, the resting membrane potential undergoes a change to become action potential, where the interior or axoplasm becomes electropositive and the outside is electronegative.
- Consequent to depolarisation, cations diffuse through the cytoplasm from the electropositive depolarised part of the membrane to the electronegative polarised part.
- Simultaneously, cations from the polarised region diffuse through the ECF to the electronegative region on the outer side.
- This flow of ions depolarises the next region of the membrane producing the action potantial there.
- Repetition of this process makes the action potential flow onwards as the depolarisation proceeds along the membrane.
- In a non-myelinated fibre, these ionic changes are repeated all along its length and hence, action potential flows slowly.
- In a myelinated nerve fibre, the lipid-rich myelin acts as an insulator and depolarisation occurs in the Nodes of Ranvier where myelin is absent.
- Since, the action potential jumps from the Node of Ranvier to another, the conduction becomes faster and such a type of conduction is calle saltatory conduction.
- In a non-myelinated fibre, the depolarisation occurs all along its length and hence, conduction becomes slower.
- **Synapse :** The functional/intercommunicating junction between two neurons, the axon of one neuron and the dendron/dendrite/soma of another neuron, through which impulse is conducted, is called a synapse.

CENTRAL NEURAL SYSTEM

- **Central Nervous system :** It comprises of the brain and spinal cord. The brain is protected by the cranium while spinal cord is protected by the vertebral column.
- **Brain :** It is the anterior most part of the central nervous system (CNS). It is covered by three meninges (*membranes – innermost piamater, middle arachnoid mater and the outer duramater*).

The human brain is divisible into three parts:

- **Forebrain :** It comprises the olfactory lobes, cerebrum and diencephalon. Olfactory lobes are not well developed in man. These lobes recieves information about smell.

  Cerebrum is the largest and complex part. It consists of the left and right hemispheres connected by a bundle of myelinated fibres, called corpus callosum. The outer layer of the cerebrum is called the cortex. The surface of the cortex comprises of many folds (usually elevated part) gyri and depression called sulci. In each cerebral hemisphere three types of functional areas are present.

  – **Sensory areas** – receive impulses from the receptors.

  – **Association areas** – interpret the input, store the input and initiate the response and are involved in memory, learning and reasoning.

  – **Motor areas** – transmit impulses to the effectors.

- **Diencephalon :** The main parts of the diencephalon are epithalamus, thalamus and hypothalamus. Anterior part of the epithalamus is folded to form the anterior choroid plexus behind which is located the pineal body. The thalamus is composed mainly of grey matter which develops from the wall of third ventricles. The thalamus sends efferent fibres to the different sensory areas of the cerebral cortex. It receives afferent fibres from the cerebral cortex, eyes, ears, different parts of the brain etc.

- The hypothalamus is the highest centre of autonomic nervous system. It governs emotional reactions and exercise control over sleep mechanism.
- **Midbrain :** It is formed of corpora quadrigemina and cerebral peduncles. Cerebral penduncles are bundles of fibres connecting the cerebral cortex with other parts of brain and spinal cord.
- **Hind brain :** It comprises of :
  - **Cerebellum :** It comprises of cerebellar hemisphere which shows a branching tree like pattern of arrangement of white matter and hence called arborvitae (tree of life). Cerebellum controls the balance and posture of the body.
  - **Pons varolii -** It is situated in front of the cerebellum below the mid brain and above the medulla oblongata. Functionally, the pons is concered with maintenance of normal rhythm of respiration.
  - **Medulla oblongata -** It is extending from pons varolii and is continuous with the spinal cord. Several medullary centres (reflex centres) are present for controlling the functions of important organs, *e.g.*, cardiac centres (heart), respiratory centre, vasomotor centre (for regulating diameter of blood vessels) and reflex centres (for swallowing, vomiting, peristalsis, secretion and activity of alimentary canal, salivation, coughing etc.)

REFLEX ACTION AND REFLEX ARC

The entire process of response to a peripheral nervous stimulation, that occurs involuntarily, i.e., without conscious effort or thought and requires the involvment of a part of the central nervous system is called a reflex action.

SENSE ORGANS

- The sensory organs detect all types of changes in the environment and send appropriate signals to the CNS, where all the inputs are processed and analysed.
- Signals are then sent to different parts/ centres of the brain.

REVIEW OF REFLEX ARC.

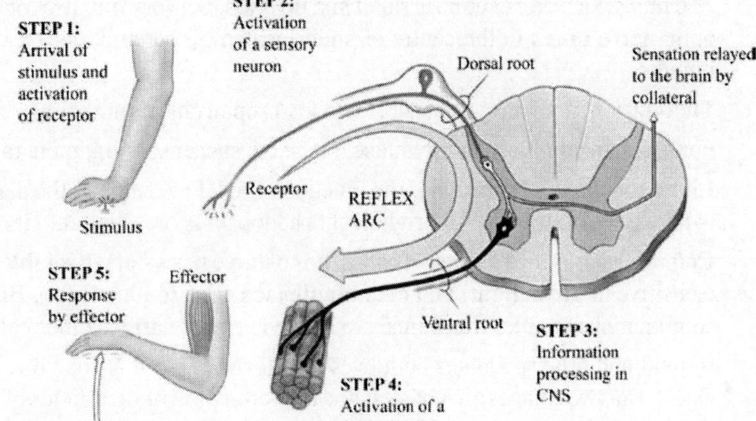

- The nose contains mucus-coated receptors which are specialized for receiving the sense of smell and called olfactory receptors. The tongue detects tastes through taste buds, containing gustatory receptors.

Eye

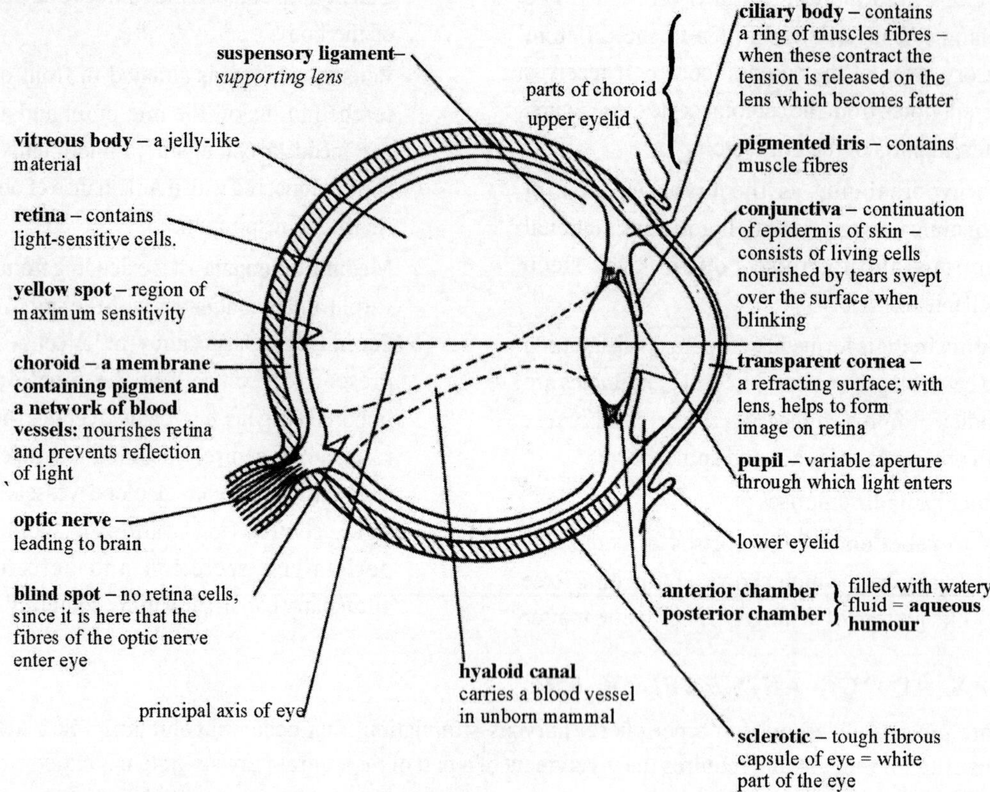

Working of Human Eyes

- Cornea, aqueous humour, lens, vitreous humour are the refractory parts which bend the light rays.
- The image formed on retina is small and inverted. Sensory impulses of the image are carried to the brain by optic nerve fibres to the centre of sight located on cerebral cortex which analyse and co-ordinate the impulse.
- The adjustment of an eye to enable it to focus upon objects at various distances is known as accomodation.
- For seeing near object, ciliary muscle relaxed, suspensory ligament taut, lens becomes less convex.
- **Light and dark adaptation** is the function of rod. For seeing in the dark, high concentration of rhodopsin is necessary in the rods. In bright light rhodopsin is bleached into its constituents (retinol and opsin).
- **Colour vision** is the function of cones. It is of three types - **erythrolable** (sensitive to red light), **chlorolable** (sensitive to green light) and **cyanolable** (sensitive to blue light). Blending of these colours in various combinations enables the animals to have vision for various other colours.
- In man and other primates both eyes face forward, so that they form two separate images of the same object. But two images are not seen due to overlapping of their fields of vision. This type of vision in which the depth or distance is also judged is called binocular or **stereoscopic vision**.

The ear

- Human ears are organs for senses of hearing and balance.
- The ear is a miniature receiver, amplifier and signal processing system.

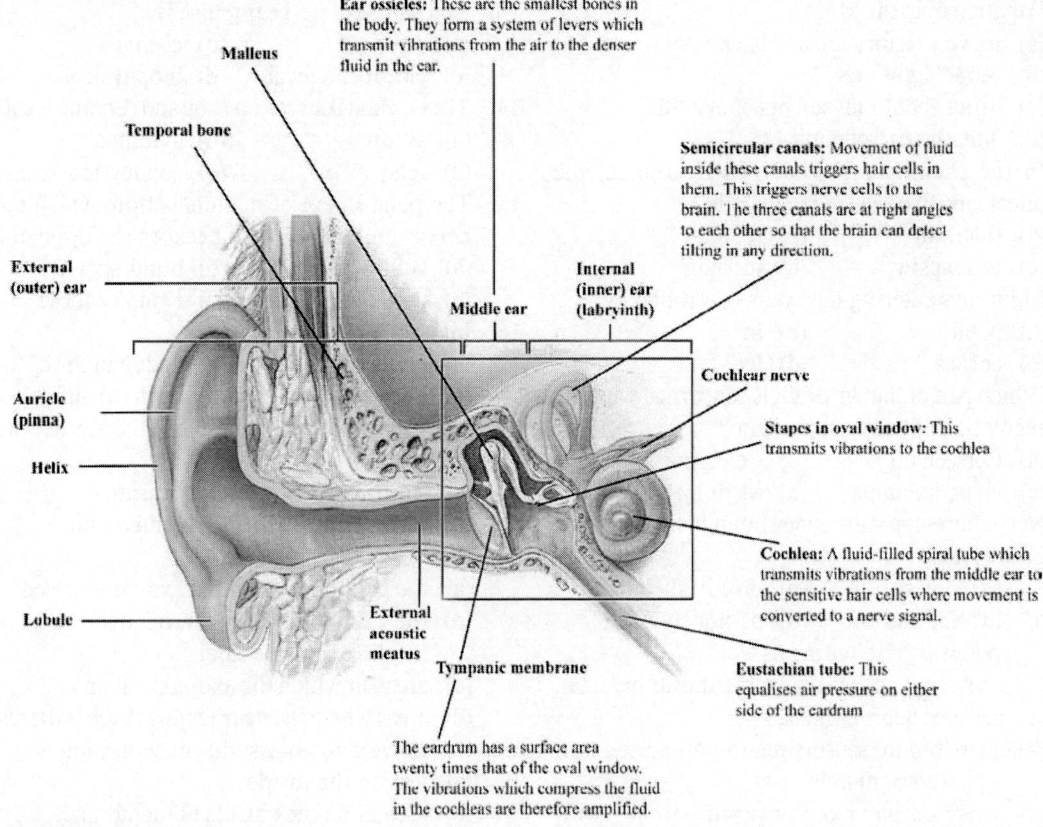

Mechanism of Hearing

- Pinna collects and amplifies sound waves which then pass along the auditory canal to the ear drum.
- Sound waves strike the ear drum and cause vibrations in the thin stretched membrane. The eustachian tube equalizes air pressure on either side of ear drum which allows a free vibration.
- The vibration reaches ear ossicles in the middle ear. Ear ossicles transmit vibrations from the air to the denser fluid in the ear.
- The lever-like action of malleus and incus magnifies the vibration of the stapes.
- The vibrating stapes transmits vibrations to the membrane of the oval window. Round window reduces the movement of fluid in the cochlea.
- Vibrations from oval window transmit to cochlea. This leads to vibration of fluid in cochlear canals.
- Vibrations of fluid in cochlear canals triggers movement of hair cells of organ of corti in cochlea.
- Movement of hair cells is converted to a nerve signal.
- Nerve signal is transmitted to brain *via* auditory nerve and this results in hearing.

Exercise

1. Which one of these processes is found in animals only?
   (a) Nervous system (b) Hormonal control
   (c) Respiration (d) Diffusion

2. Iris is part of
   (a) sclerotic
   (b) choroid/uvea
   (c) choroid and retina
   (d) sclerotic and choroid

3. Function of iris is to
(a) move lens forward and backward
(b) refract light rays
(c) bring about movements of eye lids
(d) alter the size of pupil

4. In the chemistry of vision in mammals, the photosensitive substance is called
(a) sclerotin (b) retinol
(c) rhodopsin (d) melanin

5. Light rays entering the eye is controlled by
(a) pupil (b) iris
(c) cornea (d) lens

6. Which part of human brain is concerned with the regulation of body temperature?
(a) Cerebellum (b) Cerebrum
(c) Hypothalamus (d) Medulla Oblongata

7. Nerve fibres are surrounded by an insulating fatty layer called
(a) adipose sheath (b) myelin sheath
(c) hyaline sheath (d) peritoneum

8. The pathway of reflex arc as :
(a) sense organ, spinal cord, motor neuron, sensory neuron, muscle
(b) sense organ, sensory neuron, motor neuron spinal cord, muscle
(c) sense organ, motor neuron, spinal cord, sensory neuron, muscle
(d) sense organ, motor neuron, spinal cord, sensory neuron, muscle

9. Aqueous humor and vitreous humor are secreted by
(a) iris (b) ciliary body
(c) lens (d) cornea

10. No image formation occurs on blind-spot of retina because
(a) it is not present on the optical axis of the eye
(b) here cones and rods are absent
(c) on this part only cones are present
(d) the nerve fibres of this region do not contribute in the formation of optic chiasma

11. Which one is not a reflex action?
(a) Closing of eye lids against fricking
(b) Release of saliva on seeing sweets
(c) Perspiration due to heat
(d) Obeying the order to run

12. Reflex arc is formed by
(a) brain-spinal cord-muscles
(b) receptor-spinal cord-muscles
(c) muscles-receptor-brain
(d) muscles-spinal cord-receptor

13. The nature of nerve impulse is
(a) physical (b) chemical
(c) electrochemical (d) biophysical

14. The junction between an axon and dendrite is called
(a) cyton (b) synapse
(c) relay (d) conduction zone

15. The point in eye of mammals from which optic nerves and blood vessels leaves the eye ball is
(a) yellow spot (b) blind spot
(c) pars optica (d) None of these

16. Internal ear is filled with
(a) perilymph (b) endolymph
(c) lymph (d) Both (a) and (b)

17. The organ of corti in human is concerned with the sense of
(a) smell (b) hearing
(c) taste (d) equilibrium

18. The nodes of ranvier are
(a) the points in which the axon is exposed
(b) the contact points found over the non-myelinated nerve fiber
(c) areas in which the axons swell up
(d) areas where the neurilemma touches the axon

19. In a nerve cell, potassium concentration is
(a) less on the inside
(b) greater on the outside of membrane
(c) more concentrated inside
(d) equal on both sides of membrane

20. A thin vascular membrane of brain is
(a) arachnoid (b) piameter
(c) durameter (d) epineurium

21. The sense of balance maintained with the help of
(a) hammer (b) stirrup
(c) semicircular canals (d) cochlea

22. A small region on the retina of the eye which contains only cones is called –
(a) area centralis (b) fovea centralis
(c) blind spot (d) ora serrata

23. Afferent nerve fibres carry impulses from
(a) effector organs to CNS
(b) receptors to CNS
(c) CNS to receptors
(d) CNS to muscles

24. The Nissl's granules of nerve cells are made up of
(a) ribosomes (b) proteins
(c) DNA (d) mitochondria

25. In the resting state of the neural membrane, diffusion due to concentration gradients, if allowed, would drive

(a) K^+ into the cell
(b) K^+ and Na^+ out of the cell
(c) Na^+ into the cell
(d) Na^+ out of the cell

26. $Na^+ - K^+$ pump is found in membranes of many cells, like nerve cells. It works against electro chemical gradient and involve an integral protein ATPase. For each molecule of ATP used
(a) 3 ions of Na^+ are pumped out and two K^+ are taken in.
(b) 3 ions of Na^+ are taken in and $2K^+$ are pumped out.
(c) 2 ions of Na^+ are thrown out and $3K^+$ are absorbed.
(d) 3 ions of K^+ are absorbed and $3Na^+$ are pumped out.

27. Which of the following is non-myelinated nerve fibre?
(a) Autonomic nerves (b) Optic nerves
(c) Cranial nerves (d) Spinal nerves

28. Schwann cell is found around :
(a) axon (b) cyton
(c) dendrite (d) dendron

29. The hind brain consists of
(a) pons + cerebellum
(b) hypothalamus + cerebellum
(c) medulla oblongata + cerebellum
(d) medulla oblongata + cerebellum + pons

30. Neural stimulation in visceral organ in human being is done by
(a) sympathetic and parasympathetic nerves and is under involuntary action
(b) sympathetic nerves and is under voluntary action
(c) sympathetic and parasympathetic nerves and is under voluntary action
(d) parasympathetic nerves and is under voluntary action

31. The correct sequence of meninges from outer to the inner sides is
(a) arachnoid – piamater – duramater
(b) arachnoid – duramater – paiamater
(c) piamater – arachnoid – duramater
(d) duramater – arachnoid – piamater

32. The nervous strip connecting both the cerebral hemispheres is
(a) corpus callosum (b) corpus albicans
(c) corpus stratum (d) corpus spongiosum

33. Which of the following is an example of reflex action?
(a) To shoot the bird after aiming
(b) Watering of the mouth on seeing good edibles
(c) To obey the order
(d) To read the story

34. Which is thickened to form organ of corti?
(a) Reissner's membrane
(b) Basilar membrane
(c) Tectorial membrane
(d) All of the above

35. Reflexes for maintaining vital functions like blood pressure are localised in
(a) hind brain (b) mid brain
(c) fore brain (d) cerebrum

36. The membranes enclosing the brain and spinal cord are known as
(a) meninges (b) meningitis
(c) nephron (d) axon

37. The arachnoid membrane covers the
(a) spinal cord (b) otic capsule
(c) piamater (d) None of the above

38. Bipolar nerve cells and ganglion cells are found in the
(a) sclerotea (b) cochlea
(c) retina (d) cristae

39. The dendrite carries impulses
(a) toward the cyton
(b) away from cyton
(c) across the body
(d) from one neuron to another

40. Nerve impulse is
(a) carried by afferent and efferent fibres
(b) is not carried by any afferent or efferent fibres
(c) is carried away by afferent fibres and brought about by efferent fibres
(d) is brought by afferent fibres and brought about by efferent fibres

41. The potential difference between outside and inside of a nerve before excitation is known as
(a) resting potential (b) action potential
(c) spike potential (d) reaction potential

42. Sodium-potassium pump is
(a) a hormone (b) an enzyme
(c) a protein carrier (d) an organelle

43. Axon part of the neuron is highly modified for which of the following function?
(a) Reception of stimuli from neurons
(b) Conduction for neurons
(c) Reception of internal stimuli
(d) Conduction of impulse away from neurons

44. Brain depends on blood for the supply of
(a) oxygen and ATP
(b) oxygen and electrolytes
(c) oxygen and glucose
(d) ATP and glucose

45. Physiological state of body is controlled by
(a) thalamus (b) cerebellum
(c) hypothalamus (d) All of the above

46. Transmission of nerve impulse, across the synapse is accomplished by
(a) release of ions
(b) release of neurotransmitters
(c) movement of water
(d) movement of Na^+ and K^+

47. Some cell in our body can be over a feet long. They are
(a) muscle cells (b) bone cells
(c) nerve cells (d) gland cells

48. Breathing is controlled by
(a) lungs (b) trachea
(c) medulla oblongata (d) hypothalamus

49. Nerve cells do not possess
(a) neurilemma (b) sarcolemma
(c) dendrite (d) axon

50. That part of the brain which is involved in interpreting an input, storing input information and initiating a response in the light of similar past experience is
(a) motor area (b) sensory area
(c) association area (d) pons

51. Choroid plexus is a network of
(a) capillaries (b) muscle fibres
(c) nerves (d) lymph versels

52. White matter is composed of
(a) ependyma (b) nerve cells
(c) nerve fibres (d) None of these

53. If a person has lost his memory in an accident, the following part of the brain have got injured.
(a) Diencephalon (b) Medulla oblongata
(c) Cerebellum (d) Cerebrum

54. Which one of the following is responsible for the control of reflex action?
(a) Sensory nerves
(b) Motor nerves
(c) Sympathetic nervous system
(d) Central nervous system

55. Which one of the following statement is correct?
(a) Neither hormones control neural activity nor the neurons control endocrine activity
(b) Endocrine glands regulate neural activity, but not vice-versa
(c) Neurons regulate endocrine activity, but not vice-versa
(d) Endocrine glands regulate neural activity

56. Which one of the following occurs without the help of brain?
(a) Cranial reflex (b) Spinal reflex
(c) Efferent reflex (d) Afferent reflex

57. During the course of evolution which part of the brain has shown maximum increase in size?
(a) Mid brain (b) Fore brain
(c) Hind brain (d) All of the above

58. Nissl's granules are absent in
(a) axon (b) cyton
(c) dendron (d) Both 'a' and 'b'

59. Depolarization of a stimulated nerve is maintained by
(a) Ca^{++} (b) Cl^-
(c) Mg^{++} (d) K^+

60. Which of the following is not a part of middle ear?
(a) Malleus (b) Incus
(c) Stapes (d) Cornea

CHAPTER 22 Chemical Co-ordination and Regulation

CONCEPTS

- Both endocrine and neural systems are meant for internal communication and regulation of the animal body.
- In the vertebrate body, glands may be classified on the basis of presence or absence of ducts. These are exocrine glands and endocrine glands.
- **Exocrine glands** (glands with ducts) : The secretions of these glands are carried by ducts to a particular organ for some metabolic activity, *e.g.*, salivary glands, liver etc.
- **Endocrine glands** (ductless glands) or glands of internal secretions : These glands have no ducts and their secretions get absorbed into the immediate surrounding blood circulation to reach the specific organs to initiate a particular metabolic change.
- The endocrine glands which secrete only hormones are called the **holocrine glands** (*e.g.*, thyroid, parathyroids, adrenals, pituitary gland).
- The glands which have dual functions (secretion of hormones and some other functions) are termed the **heterocrine glands**. (*e.g.*, pancreas, testes, ovaries etc.,).

CHEMICAL COORDINATION IN ANIMAL (HORMONES)

Endocrine Gland	Hormone	Principal action	Disorders
Thyroid	Thyroxine (T_4) and Triiodothyronine (T_3)	Maintains calcium level normal in the body.	Cretinism, myxoedema goiter
	Calcitonin	Increases rate of metabolism in the body.	
Parathyroid	Parathormone (PTH)	Increases plasma calcium	Parathyroid tetany osteoporosis
Adrenal gland (medulla)	Adrenaline and Noradrenaline	Increases heart beat, blood sugar and also constricts blood vessel	
Adrenal cortex	Mineralocorticoids (aldosterone)	Increases reabsorption of sodium and excertion of potassium	Addison's disease Adrenal virilism
	Glucocorticoids (cortisol)	Increases blood sugar and affects carbohydrate, fat and protein metabolism	Cushing's syndrome

Hypothalamus	ARH	Regulates corticotropin secretion	
	TRH	Thyrotropin secretion	
	SRH	Stimulates secretion of gonadotropins	
	Growth hormone releating factor (GRH)	Regulates secretion of prolactin	
	Prolactin releasing hormone (PRH) and Prolactin inhibitory hormone (PIH)	Control secretion of MSH	
Pituitary gland, anterior lobe	Growth hormone	Stimulates general growth	Pituitary dwarfism, gigantism, Acromegaly
	Prolactin	Stimulates milk production and secretion	
	Follicle stimulating hormone (FSH)	Stimulates ovarian follicle and spermatogenesis	
	Luteinizing hormone (LH)	Stimulates corpus luteum and ovulation in females and interstitial cell in males	
	Thyroid stimulating hormone (TSH)	Stimulates thyroid gland to secrete hormones	
	Adrenocorticotropic hormone (ACTH)	Stimulates adrenal cortex to secrete glucocorticoids	
Intermediate lobe	Melanocyte stimulating hormone (MSH)	Growth and development of melanocyte	
Posterior lobe	Oxytocin	Contraction of uterine muscles and mammary gland cells	
	Vasopressin (ADH)	Promotes reabsorption of water from collecting ducts of kidneys	Diabetes insipidus

Pancreas

- Located posterior to stomach, close to duodenum.
- **Endocrine Pancreas :** Consists of islets of Langerhans. The islet of Langerhans have two main types of cells.

ENDOCRINE PANCREATIC SECRETIONS:

	NAME OF THE CELLS	PRODUCT	FUNCTION
1.	Beta (β) cells	Insulin and Amylin	Lower blood sugar level.
2.	Alpha (α) cells	Glucagon	Raise blood sugar level.

- **Exocrine Pancreas :** Produces enzymes that breaks down digestive food.

EXOCRINE PANCREATIC SECRETIONS:

	NAME OF THE CELLS	CELL PRODUCING	PRODUCING SIGNAL
1.	Bicarbonate ions	Center acinar cells	Secretin
2.	Digestive enzymes	Basophilic	CCK Pancreatic, amylase, trypsin etc.

Diseases of Pancreas :

Diabetes Mellitus : Increased sugars level in blood.

Testes

- Primary sex organ as well as endocrine glands.

Located outside the body as they are suspended by spermatic cord and within the scrotum.

Function : Produces a group of hormones called androgens mainly testosterone.

- Androgen regulates the development, maturation and functions of the male accessory sex organs.

Ovary

- Egg producing reproductive organs found in females.
- Located in the lateral wall of the pelvis in a region called ovarian fussa. It lies behind urethra.

Functions : Ovary produces one ovum during each menstrual cycle. It produces 2 groups of steroid hormones called.

(i) **Estrogen**
- Stimulating growth and activities of female secondary sex organs.
- Also regulates female sexual behaviour.

(ii) **Progesterone**
- Supports Pregnancy.
- Production of milk.

HORMONES OF HEART AND KIDNEY AND GASTRO INTENSTINAL TRACT

	HORMONES	SECRETED BY	FUNCTION
1.	Atrial Natriuretic factor (ANF)	Atrial wall of heart	Control Blood pressure.
2.	Erythropoietin of kidney	Juxtaglomerular cells	Stimulates formation of RBC.
3.	Gastrin secretes cholecystokinin (CCK) and Gastric Inhibitory peptides (GIP)	Gastrointestinal tract	i. **Gastrin -** Stimulates secretion of HCl. ii. **Secretin -** Stimulates secretion of walls and bicarbonate ions. iii. **CCK -** acts on pancreas and gall bladder and secretion of pancreatic enzymes and bile juice respectively. iv. **GIP -** Inhibits gastric secretion.
4.	Growth factors tissues.	Other non-endocrine	Essential for normal growth of tissues and their repairing/regeneration.

MECHANISM OF HORMONE ACTION :

Hormones produce their effects on target tissues by binding to specific proteins called hormone receptors.

Two Types of Hormone Receptors :

	LOCATION OF RECEPTOR	CLASSES OF PEPTIDES	PRINCIPLES OF MECHANISM OF ACTIN
1.	Cell surface receptors (plasma membrane bound receptors)	Proteins and peptides catecholamines and eicosonoids.	Generation of second messengers which alter one activity of other molecules usually enzymes. This happens due to the formation of hormone receptors complex which leads to conformation change in the large cells, as result they produce second messengers.
2.	Intracellular receptors (Present inside the cell cytoplasm/molecules)	Steroids and thyroids.	After transcriptional activity of responsible genes.

Other classes of hormones based on their chemical nature are -

(i) Iodothyronies (thyroid hormones)

(ii) Amino acid derivatives

Example : Epinephrine

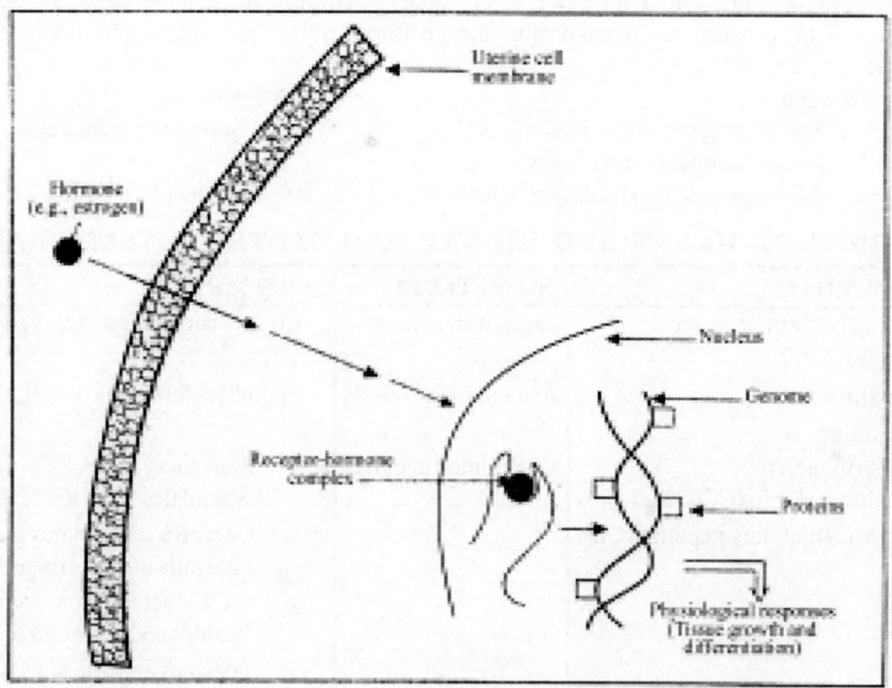

(b)

Exercise

1. Main similarity between hormone and enzyme is
   (a) both act at particular pH
   (b) both are proteins
   (c) both are required in small amounts
   (d) both can be used again and again
2. Largest amount of iodine is found in
   (a) adrenals (b) liver
   (c) thyroid (d) testes
3. Estrogen is secreted by
   (a) liver (b) spleen
   (c) ovaries (d) pituitary
4. Androgens are secreted by
   (a) pituitary (b) testes
   (c) ovaries (d) thyroid
5. Insulin by chemical nature is
   (a) carbohydrate (b) protein
   (c) steroid (d) lipid
6. Melatonin is a hormone produced by
   (a) adrenal gland (b) pituitary gland
   (c) pineal gland (d) thymus gland
7. Chemically hormones are
   (a) biogenic amines only
   (b) proteins, steroids and biogenic amines
   (c) proteins only
   (d) steroids only
8. Which one of the following is not a second messenger in hormone action ?
   (a) Calcium (b) Sodium
   (c) cAMP (d) cGMP
9. According to the accepted concept of hormone action, if receptor molecules are removed from target organs, then the target organ will
   (a) not respond to the hormone
   (b) continue to respond to hormone without any difference
   (c) continue to respond to the hormone but in the opposite way
   (d) continue to respond to the hormone but will require higher concentration
10. Hormone secreted by pituitary gland are
    (a) all steroids
    (b) some steroids, some proteinoids
    (c) simple protein
    (d) proteins or glycoproteins
11. Hormones influence body parts for bringing about
    (a) proper growth
    (b) coordination
    (c) stimulation of body parts
    (d) proper functioning of body parts
12. Hypophysis is the alternate name of
    (a) pituitary (b) thyroid
    (c) thymus (d) pineal gland
13. Receptors for protein hormones are found
    (a) inside nucleus (b) inside cytoplasm
    (c) on surface of ER (d) on cell surface
14. Neurohypophysis secretes
    (a) vasopressin & growth hormone
    (b) oxytocin & estrogen
    (c) vasopressin & oxytocin
    (d) vasopressin & estrogen
15. Fight and flight hormone is
    (a) adrenaline (b) thyroxine
    (c) ADH (d) oxytocin
16. Epinephrine is
    (a) secreted from pancreas and decreases heart beat
    (b) secreted from adrenal medulla and increases heart beat
    (c) secreted from adrenal medulla and decreases heart beat
    (d) secreted from pancreas and increases heart beat
17. Epinephrine and norepinephrine together is known as
    (a) steroid (b) protein
    (c) catecholamine (d) None
18. M S H is secreted by in human
    (a) anterior lobe at pituitary
    (b) middle lobe of pituitary
    (c) posterior lobe of pituitary
    (d) endostyle
19. Thymosin stimulates
    (a) milk secretion (b) erythrocytes
    (c) T-lymphocytes (d) melanocytes
20. Norepinephrine hormone is secreted from
    (a) zona glomerulosa (b) zona fasiculata
    (c) zona reticularis (d) medulla of adrenal

21. Glucagon is secreted by
(a) leydig cells
(b) islets of langerhans
(c) corpus luteum
(d) glisson's capsule

22. A hormone with seat of activity in liver-changing glucose into glycogen is produced by
(a) pituitary (b) thymus
(c) parathyroid (d) pancreas

23. Ovulation in mammals occurs mainly under the influence of
(a) TSH and ACTH (b) FSH and LH
(c) TSH and STH (d) MTH and ACTH

24. Which gland is both exocrine as well as endocrine ?
(a) Pituitary (b) Mammary gland
(c) Thyroid (d) Pancreas

25. Leydig cells are meant for
(a) formation of sperm
(b) to produce progesterone
(c) to produce testosterone
(d) nutrition of sperm

26. Secretion of estrogen is controlled by
(a) HCG (b) progesterone
(c) LH (d) FSH

27. Immediate cause of induction of ovulation in human female is large in plasma surge of
(a) progesterone (b) LH
(c) FSH (d) estradiol

28. Stimulation of uterine contraction during child birth is brought about by
(a) adrenaline (b) progesterone
(c) oxytocin (d) prolactin

29. Role of thymus in *Homo sapiens* is chiefly concerned with
(a) reproduction (b) immunology
(c) calcium balance (d) blood coagulation

30. In diabetes disease the urine contains
(a) salt (b) fat
(c) protein (d) sugar

31. Which hormone secretion is under nervous control ?
(a) Adrenal cortex (b) Anterior pituitary
(c) Posterior pituitary (d) Pineal body

32. Mammals born without a thymus gland fail to manufacture?
(a) B - lymphocytes (b) T - lymphocytes
(c) plasma cells (d) basophils

33. Both adrenaline and cortisol are secreted in response to stress. Which of the following statements is true for both of these hormones ?
(a) They act to increase blood glucose
(b) They are secreted by the adrenal cortex
(c) Their secretion is stimulated by adrenocorticotropin
(d) They are secreted into the blood within seconds of the onset of stress

34. Glucagon is secreted by
(a) β (beta) cells of islets of langerhans
(b) α (alpha) cells of islets of langerhans
(c) β cells of pancreas
(d) adrenal cortex

35. Secretin stimulates the activity of
(a) liver (b) gastric gland
(c) pancreas (d) gall-bladder

36. Which hormone stimulates contraction of gall bladder ?
(l) CCK-PZ (b) ACTH
(c) LTH (d) FSH

37. Which hormone possesses anti-insulin effect?
(a) Cortisol (b) Calcitonin
(c) Oxytocin (d) Aldosterone

38. Occurrence of Leydig's cells and their secretion are
(a) ovary and estrogen
(b) liver and cholesterol
(c) pancreas and glucagon
(d) testes and testosterone

39. ADH or vasopressin is
(a) enzyme that hydrolyses peptides
(b) hormone secreted by pituitary that promotes reabsorption of water from glomerular filtrate
(c) hormone that promotes glycogenolysis
(d) energy rich compound connected with muscle contraction

40. The mammalian corpus luteum produces
(a) estrogen
(b) progesterone
(c) luteotropic hormone
(d) luteinizing hormone

41. Calcitonin is a thyroid hormone which
(a) lowers calcium level in blood
(b) elevates calcium level in blood
(c) has no effect on calcium
(d) elevates potassium level in blood

42. Parathormone deficiency produces muscle ramps or tetany as a result of
(a) lowered blood Ca^{2+}
(b) enhanced blood Na^{+}
(c) enhanced blood glucose
(d) enhanced blood Ca^{2+}

43. Mainly which type of hormones control the menstrual cycle in human beings?
(a) FSH
(b) LH
(c) FSH, LH, estrogen
(d) Progesterone

44. Acromegaly is caused by
(a) excess of GH
(b) excess of thyroxin
(c) deficiency of thyroxin
(d) excess of adrenaline

45. The 24 hour (diurnal) rhythm of our body such as the sleep-wake cycle is regulated by the hormone
(a) calcitonin (b) prolactin
(c) adrenaline (d) melatonin

46. Ductless gland which pour their secretion into the blood are termed
(a) endocrine (b) holocrine
(c) apocrine (d) exocrine

47. In case of stoppage of functioning of islets of Langerhans which hormone will be in short supply & what will be its effect?
(a) Adrenaline-heart beat increases
(b) Insulin-blood glucose level increases
(c) Thyroxine-retardation of growth
(d) Cortisone-tetany

48. Deficiency of ADH or vasopressin will result in
(a) increased volume of urine
(b) decreased volume of urine
(c) excessive secretion of urochrome
(d) change in pH from acidic to alkaline range

49. Growth hormone (GH or STH) works properly
(a) with thyroxine
(b) without thyroxine
(c) without adrenaline
(d) without insulin

50. Hormone connected with osmoregulation is
(a) oxytocin (b) parathormone
(c) androgen (d) vasopressin

51. Insulin is secreted by
(a) α-cells of pancreas
(b) β-cells of pancreas
(c) spleen
(d) mucosa of oesophagus

52. Undersecretion of cortisol (corticoids) produces a disease known as
(a) addison's disease
(b) haemophilia
(c) anaemia
(d) mental retardation

53. Thyroid deficiency leads to in infants
(a) hypothyroidism (b) myxedema
(c) cretinism (d) thyrotoxicosis

54. Mammalian prolactin is secreted by
(a) adenohypophysis (b) neurohypophysis
(c) adrenal cortex (d) adrenal medulla

55. Which is gonadotropic hormone ?
(a) GH (b) MSH
(c) ADH (d) FSH & LH

56. FSH is important for
(a) formation of sperm
(b) fertilisation
(c) oviposition or egg laying
(d) None of these

57. Intercellular communication in multicellular organism occurs through?
(a) Digestive system only
(b) Respiratory system only
(c) Nervous system only
(d) Both nervous and endocrine system

58. ADH acts on the
(a) collecting tubule of kidney
(b) loop of Henle
(c) collecting ducts of testes
(d) None of the above

59. Name the hormone associated with the metabolism of phosphorus and calcium
(a) glucocorticoids
(b) mineralocorticoids
(c) calcitonin
(d) glucagon

60. A trial natriuretic hormone is produced by
(a) kidney (b) heart
(c) duodenum (d) thyorid gland

Reproduction in Organisms

CONCEPTS

The period from birth to death is known as **life span**. The life span of the organisms varies from one organism to other organism *e.g.* the life span of human+s is 65 years on an average while that of Mayfly it is about one day.

REPRODUCTION

Reproduction enables the continuity of the species, generation after generation. Genetic variation is created and inherited during reproduction. Living organisms reproduce by three different modes of reproduction which are:

A. Asexual Reproduction

- In this type of reproduction, the new organisms are produced without the formation and fusion of gametes. In this case, all the genes and genetic contents are received by the offspring from one parent.
- Some common examples are : Bacteria, *Amoeba*, *Hydra*, *Paramoecium*, *Pencillium*, *Bryophyllum* etc.
- Asexual reproduction takes place by following methods :

(i) Binary fission (ii) Multiple fission
(iii) Budding (iv) Spore formation

(i) Binary fission : It is one of the most common modes of asexual reproduction in which the parent cell divides into two small and approximately equal sized individuals, *e.g.*, Bacteria, *Amoeba*.

Modes of binary fission are : Simple binary fission (*e.g.*, *Amoeba*), longitudinal binary fission (*e.g.*, *Euglena*), transverse binary fission (*e.g.*, *Paramecium*) and oblique binary fission (*e.g.*, *Ceratium*).

(ii) Multiple fission : In this division, parent cell divides into many small daughter individuals simultaneously, *e.g.*, *Plasmodium*, *Amoeba*, etc.

(iii) Budding : In this type of asexual reproduction, formation of a daughter organism takes place from a small projection *i.e.*, bud arising from the parent body, *e.g.*, *Hydra*, Yeast etc.

(iv) Spore formation : Fungi and simple plants such as algae reproduce through spores. Spores after certain time give rise to young ones. In different organisms, the spores are of different kinds : zoospores, sporangiospores, chlamydospores, conidia, etc., e.g., *Phytophthora*, *Rhizopus*.

B. Vegetative Propagation

- It is a type of asexual reproduction in plants. In plants, the units of vegetative propogation like runner, rhizome, sucker, tuber, bulb called vegetative propagules all are capable of producing new offsprings.

Natural Methods of Vegetative Propagation

- A portion of the plant gets detached from the body of the mother plant. The detached portion may be a part of stem, leaf, root, etc., which develops into a new independent plant under suitable environmental conditions.

(a) Roots – E.g., Sweet potato, Yam, Dahlia etc.

(b) Underground stem
- Suckers – *Chrysanthemum*
- Rhizomes – Ginger
- Bulbs – Onion
- Tubers – Potato

(c) Creeping stems
- Runners – *Cyanodon*, *Oxalis*
- Stolons – *Fragaria*
- Offset – *Pistia*

(d) Aerial stem – *Opuntia*

(e) Leaves – *Bryophyllum*

(f) Bulbils – *Oxalis*, *Allium*

Artificial Methods of Vegetative Propagation

- Several methods of vegetative propagation are man made and developed by plant growers and horticulturists for commercial production of crops. They are called artificial methods.
- Some of the artificial methods are –
  (a) **Cuttings** – The small piece of any plant (stem, root or leaf) used for propagation is called cutting *e.g.*, *Begonia*, Tamarind.
  (b) **Layering** – The development of adventitious roots is induced before the plant part is detached from the parent plant.
    (i) Mound layering – Jasmine, Strawberry.
    (ii) Air layering – Litchi, Orange.
  (c) **Grafting** – A new variety produced by joining parts of two different plants is called **grafting**. The rooted shoot of one plant, called **stock**, is joined with a piece of shoot of another plant known as **scion**.
    (i) Tongue or whip grafting
    (ii) Wedge grafting
    (iii) Crown grafting
    (iv) Side grafting
    (v) Bud grafting
  Examples – Mango, Roses, Apple, Citrus etc.
  (d) **Propagation by plant tissue culture** – *Orchids, Gladiolus.*

C. Sexual Reproduction

- In this case, the fusion of male and female gametes takes place. Sexual reproduction results in offspring that are not identical to the parents or to themselves.
- **Unisexual Reproduction :** When male and female gametes are produced from male and female parents respectively, *e.g.*, Man, Cow, etc.
- **Bisexual Reproduction :** When both male and female gametes are produced from same individual, *e.g.*, earthworm, tapeworm.
  There are **three phases in an organism's life** – Juvenile phase, Reproductive phase and Old phase.

1. **Juvenile phase** or **Vegetative phase :** All organisms have to reach a certain phase of growth and maturity in their life, before they can reproduce sexually. This growth period is known as juvenile phase or vegetative phase. Its duration varies in different organisms.
2. **Reproductive phase :** When the organism starts reproducing in their life cycle, then this is known as **reproductive phase**.
   - Some plants flower seasonally whereas others flower throughout the year.
   - Among animals, such as birds living in nature lay eggs seasonally whereas birds in captivity lay eggs throughout the year.
   - Placental females show changes in ovaries, accessory reproductive ducts and hormones during reproductive phase.
   - Oestrus cycle is the term for cyclical changes during reproductive phase in non-primates mammals like cows, sheep, deers, dogs, tiger, etc. While such changes in primates is called menstrual cycle like humans, apes, monkeys, etc.
3. **Senescence or Old phase :** In this phase, there is progressive deterioration in the body that ultimately leads to death.

Events in Sexual Reproduction can be grouped into three distinct stages

A. **Pre-fertilization events :** The event before the fusion of gametes is known as pre-fertilization events.

Gametogenesis :

- Homothallic and monoecious terms are used to denote the bisexual or hermaphrodite condition and heterothallic and dioecious terms are used to describe unisexual condition.
- Cucurbits and coconuts are bisexual plants whereas papaya and date palm are unisexual plants.
- Earthworm, sponge, tapeworm, and leech bear both male and female reproductive organs, and are called **hermaphrodite** or bisexual animals while cockroach is a unisexual animal.

Gametic transfer :

- Pollen grains produced in the anthers are transferred to the stigma by a process, known as **pollination**.

Pollination is of two types :

(i) **Self pollination** – Transfer of pollen grains from anther to stigma occurs within the same flower, *e.g.* Pea.

(ii) **Cross-pollination** – Transfer of pollen grains from anthers of male flower to the stigma of female flower of different flowers through the agency of air, water or animals particularly insects.

B. Fertilization : It is the complete and permanent fusion of two gametes from different or same parent to form a diploid zygote. This process is also called **syngamy**.

External fertilization and external development

- In this type of fertilization, parents release sperms and eggs into the surrounding water, where fertilization takes place and zygotes develop into offspring, *e.g.*, *Obelia*, *Nereis*, *Labeo* and Frog.

Internal fertilization and internal development

- In this case, fertilization process and the development of zygote take place internally, *e.g.*, Mammals.
- There are some organisms in which the female gamete undergoes development to form new organisms without fertilization. This process is known as parthenogenesis, *e.g.* rotifers, honeybees and some lizards.

Syngamy versus Fertilization :

- Syngamy is the fusion of two entire gametes to form a zygote including the fusion of nuclei while fertilization includes all the associated events that ultimately lead to syngamy.

C. Post-fertilization events :

- The events after the formation of zygote are known as post-fertilization events.
- Two haploid gametes fuse to form a diploid zygote (2n).
- After the formation of zygote, it develops into an embryo & process is known as embryogenesis.
- On the basis of a development of zygote, the animals are of two types : **Oviparous** and **Viviparous**.

The oviparous animals lay fertilized or unfertilized eggs in a safe place in the environment. These eggs are covered by hard calcareous shell. After passing incubation period, the young ones hatch out from the eggs. For example, reptiles & birds. The viviparous animals give birth to young ones. For example – Majority of mammals, including humans.

Exercise

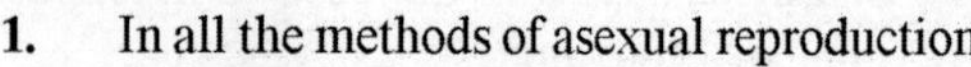

1. In all the methods of asexual reproduction
   (a) offsprings produced are genetically identical to the parents
   (b) offsprings produced are genetically different from the parents
   (c) offsprings produced may or may not be identical to the parents
   (d) None of the above
2. Common method of asexual reproduction is by
   (a) regeneration (b) budding
   (c) archeocytes (d) gemmulation
3. In sexual reproduction, offsprings resemble the parents
   (a) structurally but not functionally
   (b) functionally but not structurally
   (c) both structurally and functionally
   (d) neither structurally nor functionally
4. Asexual reproduction takes place in
   (a) higher animals (b) lower animals
   (c) plants (d) all of these
5. As a result of binary fission number of individuals produced by one fission is
   (a) two (b) three
   (c) four (d) five
6. An alternation of asexual and sexual generations where both the generations are diploid and the haploid stage is represented only by the gametes is called
   (a) binary fission
   (b) metagenesis
   (c) paedogenesis
   (d) parthenogenesis
7. Which mode of reproduction present in *Amoeba* ?
   (a) Binary fission only
   (b) Binary fission and multiple fission
   (c) Binary fission and conjugation

(d) Multiple fission only

8. Binary fission is found in
(a) *Amoeba* (b) *Paramoecium*
(c) *Planaria* (d) all of these

9. What is true in the process of fertilization?
(a) Only one sperm reaches the egg and enters it
(b) The entry of sperm activates the egg for completing meiosis
(c) Two haploid nuclei fuse and immediately divide to produce two nuclei which are again haploid
(d) Only the acrosome of the sperm enters the egg

10. On fertilization of egg nucleus with sperm nucleus
(a) first maturation is completed
(b) second maturation is completed
(c) embryo is formed
(d) first polar body is formed

11. The function of egg cell is/are
(a) It supplies a haploid set of chromosomes to the future embryo
(b) It provides most of the cytoplasm to the embryo
(c) It supplies food reserves to the embryo
(d) All the above

12. Development of an egg without fertilization is called
(a) gametogenesis (b) metagenesis
(c) oogenesis (d) parthenogenesis

13. Cells become variable in morphology and function in different regions of the embryo. The process is
(a) differentiation (b) metamorphosis
(c) organisation (d) rearrangement

14. In an egg, the type of cleavage is determined by
(a) shape and size of the sperm
(b) size and location of the nucleus
(c) amount and distribution of yolk
(d) number of egg membranes

15. The "Eyes" of the potato tuber are
(a) root buds (b) flower buds
(c) shoot buds (d) axillary buds

16. Consider the following statements and choose the correct option
(i) The genetic constitution of a plant is unaffected in vegetative propagation
(ii) Rhizome in ginger serves as an organ of vegetative reproduction
(iii) Totipotency of cells enables us to micropropagate plants
(a) Statements (i) and (ii) are true
(b) Statements (ii) and (iii) are true
(c) Statement (ii) alone is true
(d) All the three statements [(i), (ii) and (iii)] are true

17. The part which is grafting on stalk of another tree is called
(a) graft (b) bulbil
(c) bud (d) scion

18. Plants identical to mother plants can be obtained from
(a) seeds (b) stem cutting
(c) both (a) and (b) (d) none of these

19. Among the following which one is not a method of vegetative propagation?
(a) Budding (b) Layering
(c) Sowing (d) Tissue culture

20. One of the plants using 'Foliar adventitious buds' is a method for vegetative propagation is
(a) banana (b) ginger
(c) *Bryophyllum* (d) *Calocasia*

21. Stem cuttings are commonly used for the propagation of
(a) banana (b) rose
(c) mango (d) cotton

22. A few statements describing certain features of reproduction are given below
(i) Gametic fusion takes place
(ii) Transfer of genetic material takes place
(iii) Reduction division takes place
(iv) Progeny have some resemblance with parents
Select the options that are true for both asexual and sexual reproduction from the options given below
(a) i and ii; (b) ii and iii;
(c) ii and iv; (d) i and iii.

23. The term 'clone' cannot be applied to offspring formed by sexual reproduction because
(a) offsprings do not possess exact copies of parental DNA

(b) DNA of only one parent is copied and passed on the offsprings
(c) Offsprings are formed at different times
(d) DNA of parents and offsprings are completely different.

24. *Amoeba* and Yeast reproduce asexually by fission and budding respectively, because they are
(a) microscopic organisms
(b) heterotrophic organisms
(c) unicellular organisms
(d) uninucleate organisms

25. A few statements with regard to sexual reproduction are given below
(i) sexual reproduction does not always require two individuals
(ii) sexual reproduction generally involves gametic fusion
(iii) meiosis never occurs during sexual reproduction
(iv) external fertilisation is a rule during sexual reproduction
Choose the correct statements from the options below
(a) i and iv (b) i and ii
(c) ii and iii (d) i and iv

26. Given below are few statements related to external fertilization.
(i) The male and female gametes are formed and released simultaneously
(ii) Only few gametes are released into the medium
(iii) Water is the medium in a majority of organisms exhibiting external fertilization
(iv) Offsprings formed as a result of external fertilization have better chance of survival than those formed inside an organism
Choose the correct statements
(a) iii and iv (b) i and iv
(c) ii and iv (d) i and iv

27. The statements given below describe certain features that are observed in the pistil of flowers.
(i) Pistil may have many carpels
(ii) Each carpel may have more than one ovule
(iii) Each carpel has only one ovule
(iv) Pistil have only one carpel
Choose the statements that are true from the options below
(a) i and ii (b) i and iii
(c) ii and iv (d) iii and iv

28. Appearance of vegetative propagules from the nodes of plants such as sugarcane and ginger is mainly because
(a) nodes are shorter than internodes
(b) nodes have meristematic cells
(c) nodes are located near the soil
(d) nodes have non-photosynthetic cells

29. Differentiation in morphology of the two sexes of the same species is called
(a) hermaphrodite (b) heteromorphosis
(c) sexual dimorphism (d) unisexual

30. The croaking sound made by frog is
(a) hunger call
(b) just a musical note
(c) signaling call of danger
(d) sex call for female partner

31. Which type of reproduction is found in *Hydra* ?
(a) Polyembryony (b) Sexual and asexual
(c) Parthenogenesis (d) Encystment

32. Gemmule formation in Sponges is helpful in
(a) parthenogenesis
(b) sexual reproduction
(c) only dissemination
(d) asexual reproduction

33. One of the minute cell which separates from the animal egg during maturation is known as
(a) primary spermatogonia
(b) secondary oogonia
(c) primary oogonia
(d) polar bodies

34. The final event in the process of fertilization is
(a) structural change in male and female gametes
(b) egg activation
(c) amphimixis
(d) organizational change in egg cytoplasm

35. A sperm without acrosome, cannot
(a) get food supply
(b) move
(c) penetrate into the seminal vesicle
(d) penetrate into the ovum

36. The size of the egg chiefly depends upon the

(a) size of the animal
(b) amount of yolk reserves in it
(c) amount of food taken by the mother
(d) all the above

37. When more than one sperm establish contact and penetrate into the egg, then it is called
(a) monospermy (b) polyspermy
(c) both (a) and (b) (d) none of these

38. Which type of stimulus is used to achieve parthenogenic development?
(a) Chemical
(b) Change in pH
(c) Pricking with microneedle
(d) All the above

39. What is common between vegetative reproduction and apomixis?
(a) Both are applicable to only dicot plants
(b) Both bypass the flowering phase
(c) Both occur round the year
(d) Both produce progeny identical to the parent

40. What is true for cleavage?
(a) Size of embryo increases
(b) Size of cells decrease
(c) Size of cells increase
(d) Size of embryo decreases

41. During regeneration, modification of an organ to other organ is known as
(a) Morphogenesis
(b) Epimorphosis
(c) Morphallaxis
(d) Accretionary growth

42. Exponential growth of cells is a characteristic feature of
(a) tissue culture cells
(b) multicellular organisms
(c) unicellular organisms
(d) embryo

43. In *Rhizopus*, sexual fusion takes place between
(a) two gametangia (b) two gametes
(c) two hyphae (d) two sporangia

44. The main difference between animal and plant cell is that
(a) plant cell lacks rigid cell wall
(b) animal cell lacks rigid cell wall
(c) plant cell possesses small vacuoles
(d) animal cell possesses large vacuoles

45. A root is adventitious when it is
(a) swollen
(b) growing in marshy places
(c) formed from plumule
(d) modified for storage

46. A modification of bud that can bear leaves, flowers, fruits & branches is
(a) thorn (b) spine
(c) prickle (d) trichome

47. Identify the wrong statement regarding post-fertilisation development
(a) The ovary wall develops into pericarp
(b) The outer integument of ovule develops into tegmen
(c) The fusion nucleus (triple nucleus) develops into endosperm
(d) The ovary develops into fruit

48. Dicot embryo consists of
(a) radicle and plumule
(b) radicle, plumule, cotyledons and sometimes endosperm
(c) radicle, plumule, cotyledons and tegmen
(d) radicle, plumule, cotyledons, tegmen and testa

49. The parthenogenesis refers to a fruit
(a) with seeds after pollination
(b) without seeds after pollination
(c) with viable seeds after fertilization
(d) seedless without fertilization

50. Which of the following is parthenocarpic fruit ?
(a) Orange (b) Papaya
(c) Pomegranate (d) Apple

51. Endosperm in Angiosperm (flowering plants) is
(a) haploid (b) diploid
(c) triploid (d) polyploid

52. The ovary after fertilization is converted into
(a) embryo (b) endosperm
(c) fruit (d) seed

53. Syngamy means
(a) fusion of similar spores
(b) fusion of dissimilar spores
(c) fusion of cytoplasm
(d) fusion of gametes

54. Fertilization of egg takes place inside
(a) anther (b) stigma
(c) pollen tube (d) embryo sac

55. Milky water of green coconut is
(a) liquid nucellus
(b) liquid of female gametophyte
(c) liquid endosperm
(d) liquid embryo

56. In which one pair both the plants can be vegetative propagated by leaf pieces?
(a) *Bryophyllum* and *Kalanchoe*
(b) *Chrysanthemum* and *Agave*
(c) *Agave* and *Kalanchoe*
(d) *Asparagus* and *Bryophyllum*

57. Read the following statements and select the correct ones.
(i) Conidia are the asexual propagules restricted to kingdom Fungi.
(ii) A piece of potato tuber having at least one eye (or node) is capable of giving rise to a new plant.
(iii) Ginger propagates vegetatively with the help of its underground roots.
(iv) Fleshy buds which take part in vegetative propagation are called bulbils, present in *Dioscorea*, *Agave*, etc.
(a) (ii) and (iii) (b) (i) and (iv)
(c) (i), (ii) and (iv) (d) (i), (ii) and (iii)

58. Match the items in column-I with those in column-II and choose the correct option.

	Column-I		Column-II
A.	Binary fission	I.	Algae
B.	Zoospore	II.	*Amoeba*
C.	Conidium	III.	*Hydra*
D.	Budding	IV.	*Penicillium*
E.	Gemmules	V.	Sponge

(a) A – I; B – IV; C – V; D – III; E – II
(b) A – II; B – I; C – IV; D – III; E – V
(c) A – II; B – IV; C – III; D – V; E – I
(d) A – I; B – IV; C – III; D – II; E – V

59. Which one of the following is correctly matched?
(a) Onion – Bulb
(b) Ginger – Sucker
(c) *Chlamydomonas* – Conidia
(d) Yeast – Zoospores

60. Find out the wrongly matched pair.
(a) Tuber – Potato
(b) Rhizome – Ginger
(c) Bulbil – *Agave*
(d) Leaf buds – Banana

CHAPTER 24 Sexual Reproduction in Flowering Plants

CONCEPTS

FLOWER

- Flowers are highly modified shoots, bearing nodes and modified floral leaves, which are meant essentially for sexual reproduction in plant.
- Flowers, usually have four different kinds of floral members arranged in four whorls (*i.e.*, calyx, corolla, androecium and gynoecium).
- Androecium and gynoecium are the two innermost whorl of flower in which sexual reproduction take place, anther bears spores in sporangia and ovary contains ovules which bear embryo-sac with egg.

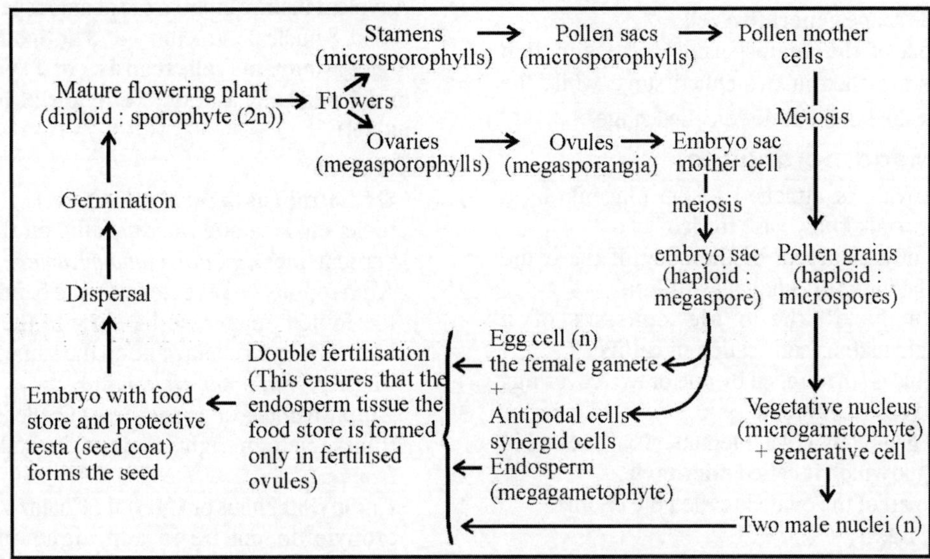

- The study of the development of embyro from zygote after fertilization is called **embryology**.
- Father of Embryology - **Hanstein** (1870)
- Father of Indian Embryology - **Prof. P. Maheshwari**.
- Embryology depends upon the sporogenesis; gametogenesis; pollination, fertilization and post-fertilization.

Microsporangium

- Each microsporangium is surrounded by four layers that are as follows :
  (i) **Epidermis** – Protective in function.
  (ii) **Endothecium** – Cells posses bands of callose and help in dehiscence of anthers at maturity.
  (iii) **Middle layers** – Cells are crushed or where persistent contain starch and other reserve material to be consumed by the growing microspore mother cells.
  (iv) **Tapetum** – Single celled innermost layer surrounding the sporgenous tissue, having dense cytoplasm and are multinucleated. Its main function is to provide nutrition to the developing microspore mother cells and pollen grains.

Microsporogenesis

- Each cell of sporogenous tissue serve as microspore mother cell (MMC).
- These MMCs undergo meiosis and form microspore tetrad and become haploid.
- As anthers mature microspores of the tetrad separate from each other and develops into pollen grains.

Pollen Grains

- These are male reproductive bodies of a flower.
- **Exine** is composed of sporopollenin which is resistant to physical and biological decomposition. It protect the pollens from environmental extremes.
- **Intine** is thin, delicate, cutin and cellulosic layer lying internal to exine.
- Mature pollen grain contains two cells – a large, nucleated **vegetative cell** which is irregular in shape with abundant food reserve and a small spindle shape **generative cell**.
- In most of the angiospermic species, pollen grains are shed at two-celled stage while the others shed them at three-celled stage.

Ovule (Megasporangium)

- Each ovule is attached to the placenta by a slender stalk known as "**funicle**"
- Point of attachment of the body of the ovule with the funicle is known as "**hilum**"
- The body of the ovule consists of a parenchymatous cells called **nucellus**.
- Nucellus is surrounded by one or two coverings termed as **integuments**.
- Small gap is left at the integuments at one end of the ovule which is called **micropyle**.
- Basal part of the ovule is called the **chalaza**.

Megasporogenesis

- The process of formation of megaspore from megaspore mother cell (MMC) is known as megasporogenesis.
- This MMC undergoes meiosis to form '4' haploid cells called **megaspores**.

Female gametophyte (Embryo sac)

- Out of the four haploid megaspores, either few of them degenerate or all take part in embryo sac formation.
- Depending upon the number of megaspores taking part in formation of female gametophyte, it is classified into following 3-types :
  (i) **Monosporic type**
  (ii) **Bisporic type**
  (iii) **Tetrasporic type**
- Female gametophyte development in monosporic type :

Functional Megaspore $\xrightarrow{\text{Mitosis}}$ '2' Nuclei $\xrightarrow{\text{Mitosis}}$ '4' Nuclei $\xrightarrow{\text{Mitosis}}$ '8' Nucleate Embryo sac

- A typical mature embryo sac of angiosperm is 7-celled, 8 nucleate structure *i.e.*, **3 antipodal cells, 3-egg apparatus cells (consists of 2 synergids and 1 egg cell) and one central cell (2 polar nuclei)**.

Forms of Ovules

1. **Orthotropous** or **Straight :** Ovule is erect. Funicle, chalaza and micropyle lie on the same vertical line, *e.g.*, *Polygonum, Rumex*.
2. **Anatropous** or **Inverted :** Ovule bends along the funicle micropyle lie close to the hilum. Micropyle and chalaza lie on the same vertical line, *e.g.*, *Solanum, Helianthus*.
3. **Amphitropous** or **Transverse :** Ovule is placed transversely or right angle to its stalk, *e.g.*, *Lemna*.
4. **Campylotropous** or **Curved :** Chalaza and micropyle do not lie in a straight line, *e.g. Capparis,* Mustard, *Pisum*.
5. **Circinotropous.** In this ovule is inverted at an angle of 360° so that ovule again becomes orthotropous coiled around by funiculus, *e.g.*, *Opuntia*.

Types of Embryo sac –

Embryo sac	M.M.C.	Spores Tetrad	F. Megaspore	Degenerate Spore	Example
1. Monosporic	1 (2n)	4 (n)	1 (n)	3 (n)	*Polygonum*
2. Bisporic	1 (2n)	4 (n)	2 (n)	2 (n)	*Allium*
3. Tetrasporic	1 (2n)	4 (n)	4 (n)	Zero	*Adoxa; Plumbago.*

POLLINATION

- Transfer of pollen grains from anther to stigma of gynoecium.

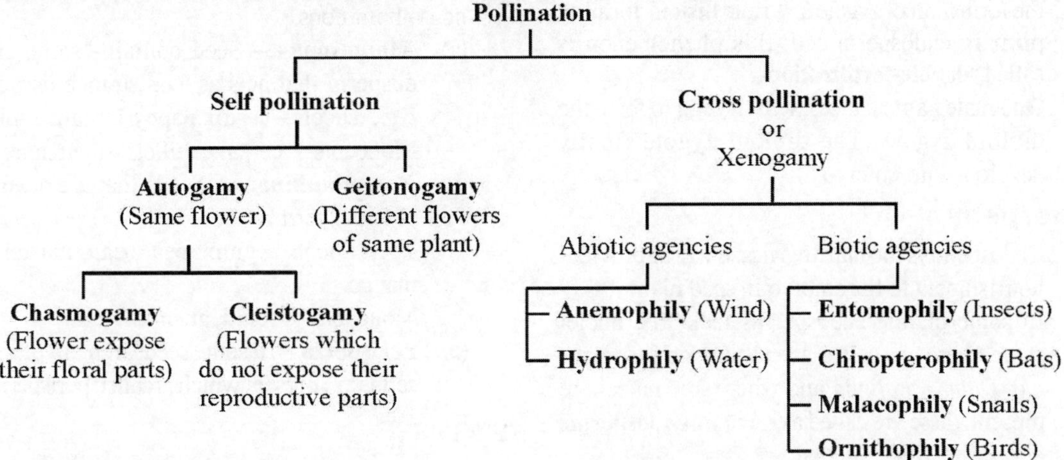

Outbreeding Devices

- Majority of flowers are bisexual, hence self pollination is a common phenomenon.
- Plants are modified variously to ensure cross-pollination :
  (i) **Self sterility (self-incompatibility)** – Plants in which pollen from same flower is incapable of bringing about fertilization, due to the presence of similar self-sterile gene, *e.g.*, tobacco, potato.
  (ii) **Dichogamy** – Maturation of male and female sex organs at different times in a bisexual flowers, *e.g.*, cruciferae family.
  (iii) **Herkogamy** – Male or female sex organs themselves prove as a barrier to prevent self pollination, *e.g.*, *Cyprepedium*, orchids.
  (iv) **Dicling or Unisexuality** – Effectively prevents self-pollination. Presence of unisexual flowers in plants that prevents self pollination, *e.g.*, castor, maize and papaya.
  (v) **Heterostyly** – Presence of styles of different heights in flower of many plants to prevent self-pollination, *e.g.*, *Primula vulgaris.*
  (vi) **Prepotency** – Few pollen grains germinate more rapidly over the stigma of another flower than on the same flower, *e.g.*, apple, grapes.

Pollen-Pistil Interaction

- All the events from landing of pollen-grains on stigma till pollen tube enters ovule are termed as pollen-pistil interaction.
- Pistil and pollen mutual chemical secretions determine the compatibility of pollen grains. If incompatibility is there, then either germination of pollen grain is inhibited or pollen tube abort in style, but if compatible pollen tube reaches the style and causes fertilisation.

Artificial Hybridization

- Dusting of pollen from a chosen plant on to the stigma of the selected plant artificially for obtaining commercially superior varieties is called artificial hybridization.
- **Process involved in artificial hybridisation :**
  (i) Anthers are removed before dehiscence from bisexual flowers by using forceps, this phenomenon is called **emasculation** but if flower is unisexual emasculation is not required.
  (ii) Now, the flower containing only stigma is covered with butter paper, this is called **bagging**, which is done to prevent contamination of stigma from other pollens.

DOUBLE FERTILIZATION

- Fusion of male and female gametes is called fertilization.
- One of these male gametes fuses with egg to form diploid zygote (2n) while the other fuses with two polar nuclei of the central cell to produce triploid primary endosperm nucleus (PEN) (3n). Since, the latter involves fusion of three haploid nucleus, therefore it is called **triple fusion**.

- In some angiosperms, two types of fusion occur in the same embryo sac "syngamy" leading to the formation of zygote & "triple fusion" forming primary endosperm cell, this phenomenon is called "**double-fertilization**".
- One male gamete fuses with the egg to form the diploid zygote. The diploid zygote finally develops into embryo.

Endosperm

- Endosperm is the nutritive tissue which provides nourishment to the embryo in seed plant.
- In some plants, endosperms lack free nuclei stage, these are called as **cellular endosperms** *e.g. Datura, Petunia* and where free nuclei are present, these are called as **nuclear endosperms** *e.g.* maize, wheat, rice etc.

Embryo

- Zygote formed after fertilization of egg cell secretes a cellulose wall around itself.
- The upper cell or embryo cell enlarges and divides repeatedly to form eight cells that are arranged in two tiers – **epibasal** (terminal) and **hypobasal** (near the suspensor). The former form two cotyledons and the plumule in dicots and one cotyledon in monocots. The latter form the hypocotyl except its tip.
- Portion of embryonal axis above cotyledon is **epicotyl** while below it is **hypocotyl**.
- A **mature embryo in monocotyledons** has a single cotyledon called **'scutellum'**. At lower portion of scutellum, portion of embryonal axis has the radicle, which bears apical meristem and root cap at the lower end. The radicle and root cap is enclosed in **coleorhiza**, which is the undifferentiated lower part of the proembryo. It comprises of shoot apex with some leaf primordium enclosed in a hollow follicular structure, the **coleoptile**.

SEED

- Seed is a fertilized ovule.
- During seed formation, endosperm is consumed in varying amount by the developing embryo, Depending on this seed may be albuminous or non-albuminous.

(a) **Albuminous** – Seed contains some endosperm that acts as food storage tissue.
e.g., Dicots – castor, poppy, custard apple.
Monocots – cereals, millets, sugarcane.

(b) **Non-albuminous** – Seeds that are devoid of endosperm.
e.g., Dicots – gram, pea, bean, mustard, mango.
Monocots – orchid, alisma.

(c) **Perisperm** – In some seeds, remain of nucellus are present which is called perisperm.

FRUITS

- After fertilization ovary begins to grow and gradually matures into fruit.
- **True fruit :** Normally it is a ripe ovary that grows into the true fruit, *e.g*, cucumber, tomato, coconut.
- **False fruit :** When floral parts other than ovary particularly thalamus or even calyx form part of the fruit, *e.g.*, apple, cashewnut, strawberry.
- **Parthenocarpic fruit :** In some plants, ovary may grow into fruit without fertilization, *e.g.*, Oranges, watermelon, banana, grapes.

APOMIXIS AND POLYEMBRYONY

- Plants where usual sexual reproduction is replaced by a type of asexual reproduction called **apomixis** which does not involve meiosis and syngamy, *e.g.*, citrus, cactaceae, grasses, *Parthenium*, etc.
- The normal process of cell cycle involving meiosis and fertilisation is called **amphimixis**.
- **Polyembryony** - Given by **Leeuwenhoek** (1917) in Orange
More than one embryo in a seed.
- Polyembryony may be formed by cleavage of zygote or formation of embryo by the cells of the embryo sac other than egg usually synergids or by development of more than one embryo sac within the same ovule.

Exercise

1. Which of the following statement is correct ?
   (a) Flower is a modified root
   (b) Flower is a modified shoot
   (c) Flower is a modified leaf
   (d) Flower is a modified inflorescence
2. Ovule of an angiosperm is technically equivalent to
   (a) a megasporangium
   (b) a megagametangium
   (c) a microspore
   (d) a megaspore
3. Which one is female gametophyte ?
   (a) Embryo (b) Embryo sac
   (c) Endosperm (d) Pistil
4. Filiform apparatus are found in
   (a) antipodal cell
   (b) egg cell
   (c) secondary nucleus
   (d) synergids
5. Perisperm is
   (a) persistent nucellus in seed
   (b) ovule wall
   (c) ovule coat
   (d) fossil of haustoria
6. Self-pollination means
   (a) transfer of pollen from anthers to stigma in the same flowers
   (b) transfer of pollen from one flower to another on the different plants
   (c) occurrence of male and female sex organs in the same flowers
   (d) germination of pollen grains
7. Chasmogamy refers to the condition where
   (a) flowers remains closed
   (b) flowers absent
   (c) flowers open
   (d) flowers gamopetalous
8. When pollen grains of a flower are transferred to stigma of another flower on a different plant, the process is called
   (a) geitonogamy (b) xenogamy
   (c) autogamy (d) homogamy
9. The main embryo developed from the structure formed as a result of
   (a) double fertilization
   (b) triple fusion
   (c) syngamy
   (d) fusion of two polar nuclei of an embryo sac
10. After fertilization the seed is developed from
    (a) ovule
    (b) ovary
    (c) embryo
    (d) endosperm
11. Double fertilization means
    (a) fusion of male gametes and ovum
    (b) fusion of two polar bodies
    (c) a male gamete fused with egg and second male gamete fused with secondary nucleus
    (d) all the above
12. Pollen tube enters into embryosac of angiosperm through the
    (a) egg cell
    (b) any one of the synergids
    (c) any one of the antipodal cells
    (d) none of these
13. The fusion product of polar nuclei and male gamete is
    (a) nucellus
    (b) primary endosperm nucleus
    (c) zygote
    (d) secondary nucleus
14. Haploid, diploid and triploid conditions respectively can be traced in
    (a) egg, nucellus, endosperm
    (b) antipodal, egg, endosperm
    (c) endosperm, nucellus, synergids
    (d) antipodal, synergids & integuments
15. In albuminous seed, the food is stored in
    (a) testa (b) plumule
    (c) cotyledon (d) endosperm
16. Seed is not a morphological equivalent of spore because
    (a) spore is not produced through fertilization
    (b) spore does not possess embryo & endosperm
    (c) spore is a haploid structure
    (d) all the above

17. In non endospermic seed or ex-albuminous seed
(a) the endosperm may be completely used up by developing embryo
(b) no endosperm form
(c) endosperm is starchy
(d) endosperm is oily

18. Male gametophyte of angiosperms/monocots is
(a) microsporangium (b) nucellus
(c) microspore (d) stamen

19. Sperm and egg nuclei fuse due to
(a) base pairing of their DNA and RNA
(b) formation of hydrogen bonds
(c) mutual attraction
(d) attraction of their protoplasts

20. Pollination occurs in
(a) bryophytes and angiosperms
(b) pteridophytes and angiosperms
(c) angiosperms and gymnosperms
(d) angiosperms and fungi

21. Double fertilization is characteristic of
(a) angiosperms (b) pteridophytes
(c) gymnosperms (d) bryophytes

22 The polyembryony commonly occurs in
(a) *Citrus* (b) turmeric
(c) tomato (d) potato

23. In angiosperms, triple fusion is required for the formation of
(a) embryo (b) endosperm
(c) seed coat (d) fruit wall

24. Wind pollinated flowers are
(a) small, brightly coloured, producing large number of pollen grains
(b) small, producing large number of dry pollen grains
(c) large producing abundant nectar and pollen grains
(d) small, producing nectar and dry pollen grains

25. Aleurone layer helps in
(a) storage of food in endosperm
(b) protection of embryo
(c) utilization of stored food
(d) all

26. In the legume seed, food is stored in
(a) cotyledons (b) endosperm
(c) perisperm (d) seed coats

27. Flower is complete when it has
(a) calyx, corolla, androecium & gynoecium
(b) calyx & corolla
(c) androecium & gynoecium
(d) corolla, androecium & gynoecium

28. A bisexual flower which always remains close is
(a) chasmogamous (b) dichogamous
(c) geitonogamous (d) cleistogamous

29. Female gametophyte of angiosperm mostly is
(a) 7 celled (b) 8 celled
(c) 9 celled (d) one celled

30. Emasculation is concerned with
(a) hybridization (b) clonal selection
(c) mass selection (d) pure line selection

31. The egg apparatus usually consists of
(a) egg and antipodals cells
(b) egg and central cell
(c) egg and two synergids
(d) egg and one synergid

32. Microsporogenesis is a synonym for
(a) spermatogenesis
(b) development of pollen grains
(c) development of male gametophyte
(d) development of female gametophyte

33. An example of a naturally occurring parthenocarpic fruit is
(a) guava (b) mango
(c) banana (d) apple

34. In a seed of maize, scutellum is considered as cotyledon because it
(a) protects the embryo
(b) contains food for the embryo
(c) absorbs food materials and supplies them to the embryo
(d) converts itself into a monocot leaf

35. Dicot embryo consists of
(a) radicle and plumule
(b) radicle, plumule, cotyledons and sometimes endosperm
(c) radicle, plumule, cotyledons and tegmen
(d) radicle, plumule, cotyledons, tegmen and testa

36. Identify the wrong statements regarding post-fertilisation development
(a) The ovary wall develops into pericarp
(b) The outer integument of ovule develops into tegmen
(c) The fusion nucleus (triple nucleus) develops into endosperm
(d) The ovule develops into fruit

37. Which of the following floral parts forms pericarp after fertilization?
(a) Nucellus (b) Outer integument
(c) Ovary wall (d) Inner integument

38. Triple fusion is characteristic of
(a) thallophytes (b) bryophytes
(c) pteridophytes (d) angiosperms

39. Triple fusion in angiosperm is the fusion of second sperm with

(a) antipodal cell and one synergid cell
(b) two antipodal cells
(c) two synergid cells
(d) two polar nuclei

40. The cell that divides to form two male nuclei in angiosperms is
(a) vegetative cell (b) generative cell
(c) tube cell (d) antheridial cell

41. Through which route the pollen tube enters the ovule?
(a) Chalaza (b) Micropyle
(c) Funiculus (d) All of these

42. Fusion of two dissimilar gametes is called
(a) fertilization (b) pollination
(c) self pollination (d) self fertilization

43. The ovary after fertilization is converted into
(a) embryo (b) endosperm
(c) fruit (d) seed

44. Part of the gynoecium which receives the pollen is called
(a) ovary (b) ovule
(c) style (d) stigma

45. Pollination by water is seen in
(a) *Nelumbium* (b) *Vallisneria*
(c) *Eichhornia* (d) *Nymphaea*

46. Which one of the following is an example of cleistogamy?
(a) Herkogamy (b) Dichogamy
(c) Self sterility (d) All of these

47. One advantage of cleistogamy is
(a) it leads to greater genetic diversity
(b) seed dispersal is more efficient and widespread
(c) seed set is not dependent on pollinators
(d) each visit of a pollinator results in transfer of hundreds of pollen grains

48. What is the direction of micropyle in anatropous ovule?
(a) Left (b) Right
(c) Upward (d) Inverted

49. The point of attachment of funicle with chalazal end is called
(a) placenta (b) integument
(c) nucellus (d) hilum

50. Egg apparatus of angiosperm consists of
(a) 1 egg cell and 2 synergids
(b) 1 egg cell, 2 synergids 3 antipodal cells
(c) 3 antipodals cells only
(d) secondary nucleus and egg cell

51. The function of innermost layer of pollen sac, tapetum is
(a) dehiscence
(b) mechanical strength
(c) protective
(d) nutritive

52. In a pollen grain, large irregularly shaped nucleus present in
(a) generative cell (b) tube cell
(c) vegetative cell (d) none of these

53. Which of the following is correct statement?
(a) Gametes are diploid
(b) Spores are invariably haploid
(c) Spores and gametes are invariably haploid
(d) Gametes are invariably haploid

54. The odd one is
(a) micropyle (b) embryo sac
(c) nucellus (d) pollen grain

55. In plants meiosis occurs in
(a) anther (b) root tip
(c) cambium (d) pollen grains

56. The anther wall consists of four wall layers where
(a) endothecium lies inner to middle layers
(b) tapetum lies just inner to endothecium
(c) tapetum lies next to epidermis
(d) middle layers lie between endothecium and tapetum

57. What does the filiform apparatus do at the entrance into ovule?
(a) It brings about opening of the pollen tube
(b) It guides pollen tube from a synergid to egg
(c) It helps in the entry of pollen tube into a synergid
(d) It prevents entry of more than one pollen tube into the embryo sac

58. Which one of the following statements is not true?
(a) Pollen grains are released from anthers at 2-celled state
(b) Sporogenous cell directly behaves as the megaspore mother cell
(c) Megaspore divides twice to form an eight nucleate embryosac
(d) Egg and synergids always lie near the micropylar end of ovule

59. Exine of pollen grain is composed of
(a) pectocellulose (b) lignocellulose
(c) sporopollenin (d) lignin

60. The pollen grain is
(a) an immature male gametophyte
(b) a mature male gametophyte
(c) partially developed male gametophyte
(d) last stage of male gametophyte

Human Reproduction

CONCEPTS

- **Reproduction** is the ability of living organisms to produce a new offspring similar to themselves.
- Reproduction are broadly classified as **asexual reproduction** and **sexual reproduction**.
- Human are **unisexual** and **viviparous**.
- The major reproductive events in human beings are
  - (i) **Gametogenesis** – Formation of gametes.
  - (ii) **Insemination** – Transfer of sperms into female genital tract.
  - (iii) **Fertilisation** – Fusion of male and female gametes leading to formation of zygote.
  - (iv) **Implantation** – Formation and development of blastocyst and its attachment to the uterine wall.
  - (v) **Gestation** – Embryonic development; gestation is the time from conception to birth.
  - (vi) **Parturition** – Delivery of baby (the process of birth).

MALE REPRODUCTIVE SYSTEM

Anatomical Components

1. Primary sex organs : A pair of testis
2. Duct system : Involved in the storage, maturation, conveyance and ejection of spermatozoa
3. Accessory sex glands : Seminal vesicles, prostate, cowper's glands
4. External genitalia : penis and scrotum.

Testes

Anatomical aspects →
- Number : two; total mass = 80g
- Origin : urogenital ridge; intra-abdominally
- Differentiation : controlled by H-Y antigen on Y chromosome
- Located in the scrotum, suspended by spermatic cord
- Descend from abdomen to scrotum
- Surrounded by connective tissue capsule (tunica albuginea)
- Subdivided into 250 lobules by connective tissue septa
- Each lobule consists of 2 seminiferous tubules composed of sex and sertoli cells
- Leydig cells are located extratubularly and secrete sex hormones called androgens (testosterone).

Duct system

Segments : Include the tubuli recti → rete testis → vasa efferentia → epididymis → vas deferens → ductus ejaculatorius → urethra

Construction : Lumen lined by epithelial cells; walls contain muscle fibres.

Functions →
- Serve as store and incubator for the maturing spermatozoa
- Convey spermatozoa and eject semen

Accessory glands

Seminal vesicles →
- 5 cm long sac-like glands lined by epithelium
- Walls contain muscle fibres; excretory ducts open in ductus ejaculatorius
- Contribute major portion of the seminal fluid
- Provide volume, fructose, electrolytes and prostaglandins

Prostate glands →
- A chestnut-sized gland surrounding first part of urethra
- Constituted of fibrous capsule, muscle fibres and glandular epithelium
- Secretion ejected during ejaculation
- Adjusts vaginal pH and stimulates sperm motility
- Prostatism common in elderly males

Cowper's glands →
- This gland is comparable to the bertholin's gland of females
- Comprises ~ 5% of semen secretion

External Genitalia

Scrotum →
- A skin-covered pouch containing a muscle layer in its wall : Tunica dartos
- Provides optimal temperature for spermatogenesis
- Consists of three masses of erectile tissue containing muscle fibres and sinuses

Penis →
- Three segments are distinguished : Root, body and glans
- Contains urethra which serves as common canal for urine and semen
- Erection
  - effected by parasympathetic fibres
  - makes copulation and insemination possible

Semen (Ejaculate) :

- It is an organic fluid, that may contain spermatozoa.
- The process that results in the discharge of semen is called **ejaculation**
- Volume : 2 to 5 ml; contains 40 to 120 million sperms/ml at pH 7.4

FEMALE REPRODUCTIVE SYSTEM

Anatomical Components

1. Ovaries (gonads) : Primary sex organs
2. Tubular system : Uterine tubes, uterus and vagina }
3. External genitalia : Hymen, vestibule, labia, clitoris }
4. Vestibular glands

Ovaries

Number : Two; combined mass = 15g

Location : Intrapelvic, either side of uterus on posterior surface of broad ligament

Structural characteristics →
- Consists of connective tissue, sex cells, blood vessels, etc.
- Tissues arranged in two zones : Medulla and cortex
- Medulla or central part consists of connective tissue
- Cortex contains connective tissue and sex cells in different stages of development
- Ovary surrounded by a capsule (tunica albuginea)
- A single layer of germ cells covers ovarian surface - No peritoneal cover
- Sex cells
  - Germ cells
  - Oocytes : germ cell surrounded by single layer of epithelial cells
  - Primary follicles : oocyte surrounded by layers of epithelial cells
  - Secondary follicles : oocyte surrounded by pellucida and granulosa cells
  - Graafian follicle : large oocyte surrounded by pellucida, corona radiata, an antrum, theca interna and theca externa
  - Corpus luteum : formed from granulosal and theca interna cells

Secondary Sex Organs

Oviducts (Fallopian Tubes)

Location →
- One on either side of the uterus
- Lateral end of tube opens into the peritoneal cavity near the ovary
- Medial end opens in upper part of uterus cavity

Anatomy →
- Each tube = 100 mm long
- Lateral end is dilated and possesses fimbriae
- Wall : three layers - inner (ciliated epithelium), middle (muscle), outer (peritoneum)

Functions →
- Convey ovum towards uterus
- Fertilisation occurs in one of the tubes (ectopic tubal pregnancy not uncommon)

Regions →
- **Infundibulum :** bears motile, finger like processes called **fimbriae** helps in collection of the ovum after ovulation
- **Ampulla :** site of fertilization
- **Isthmus :** narrow thick walled, straight part follows the ampulla.
- **Utrine part :** passes through uterine wall and communicates with the uterine cavity

Uterus ("Womb")

Location →
- In the pelvic cavity between the bladder anteriorly and rectum posteriorly
- Almost completely covered by peritoneum - forms broad ligament laterally

Anatomy →
- A hollow, pear-shaped, muscular organ that opens in the vagina below
- Parts : fundus, corpus and cervix (inferior) + tubes
- Orientation
  - position : antiversion/antiflexion; rests on the bladder
  - held in position by broad ligament, round ligament, etc.
- Wall
  - arranged in 3 layers : endometrium, myometrium, perimetrium
  - endometrium : connective tissue, tubular glands, inner lining of ciliated columnar epithelium-becomes stratified squamous epithelium. Blood supply by straight and spiral arteries
  - myometrium : smooth muscle - sensitive to sex hormones, oxytocin, prostaglandins
  - Perimetrium : the peritoneal cover of the uterus - does not cover ovary.

Function →
- Receives embryo and presents optimal conditions for its survival and growth
- Acts as feeder, respirator, protector, "sink", etc., for embryo/foetus
- Expels the foetus at its birth

Vagina

Location : a fibromuscular tube stretching from the uterus to the vestibule

Anatomy →
- Cervix protrudes into its upper end; distal end partly closed by hymen
- Wall
  - arranged in 3 layers : mucosa, smooth muscle layer, elastic tissue
  - mucosa lined by stratified squamous epithelium
  - epithelium lining undergoes cyclic changes under influence of sex hormones

External Genitalia

Labia majora : two sagittally-placed folds of adipose tissue covered by skin
Labia minora : two folds of skin lying deep to labia majora
Clitoris : the female homologue of the male penis
Vulva : the region of the external genital organs of the female
Vestibular glands : secretion lubricates vaginal mucosa

Spermatogenesis

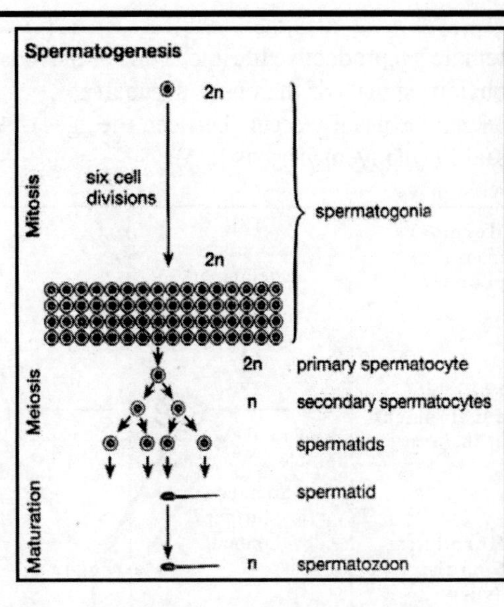

- Production of mature sperm cells.
- It starts at the age of puberty.
- Occurs in seminiferous tubules.
- All 4 products of meiosis I and II develop into mature sperm.
- Each spermatozoan or sperm is a long, flagellated, motile cell.
- Sperm consists of a **head** comprising nucleus and acrosome, a **neck** containing two centrioles, a **middle piece** containing mitochondria and a tail.
- Acrosome contains hydrolysing enzymes for the sperm penetration. The nucleus is the seat of genetic information & contain haploid set of chromosomes.
- Tail helps sperm in swimming (also called motility) which is essential for male fertility.
- The neck contains many mitochondria to provide ATP for movement of the flagellum.

Hormonal control of Spermatogenesis

Hypothalamus $\xrightarrow{\text{GnRH}}$ Anterior pituitary

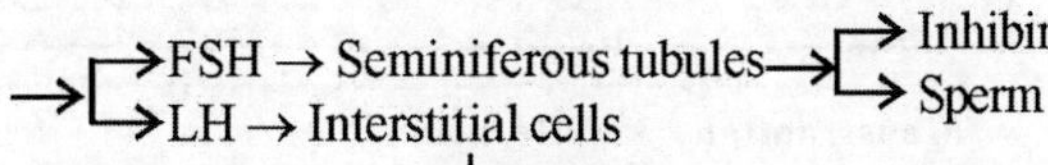

Oogenesis

- During meiotic divisions, cytokinesis is unequal, forming one large cell that becomes the egg, and 3 small polar bodies that soon degenerate.
- At birth and puberty, primary oocytes enlarge, and in the presence of FSH, the first meiotic division produces a secondary oocyte and the first polar body.
- Only after fertilization does the second meiotic division occur to form the mature ovum.

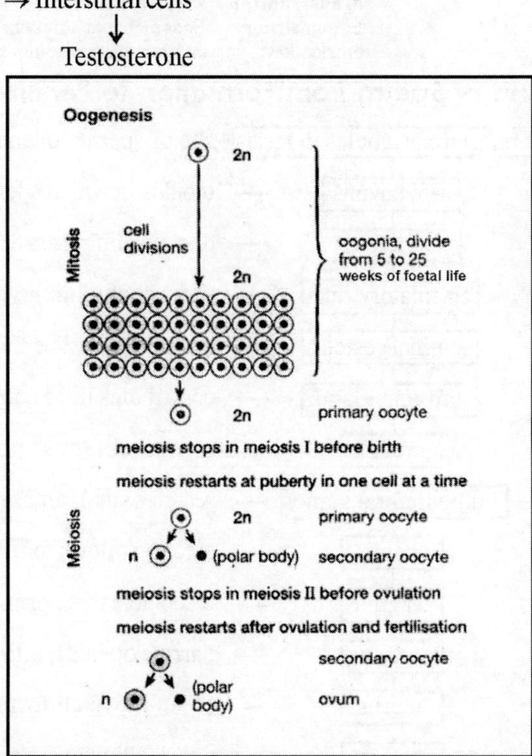

MENSTRUAL CYCLE

Menstruation begins at puberty and menstrual cycles continue until menopause, approximately 36 to 40 years later. The day on which discharge of blood from the vagina begins is taken as the first day of the cycle; the cycle ends on the last day prior to the next menstrual flow. The cycle normally runs for about 1 lunar month, or 28 days, but it can vary from 22 to 35 days.

Menarche

- The epoch of the first menstrual flow, or period, is called menarche; it marks the beginning of a female's reproductive life and is one of the most obvious signals of the onset of puberty.

- Menarche usually occurs between the ages of 9 and 17 (the average age is 12.5).
- Around 45 years of age the maturation of Graafian follicles stops and due to suppression of estrogen secretion menstrual cycle discontinues. This phase is called **menopause**.

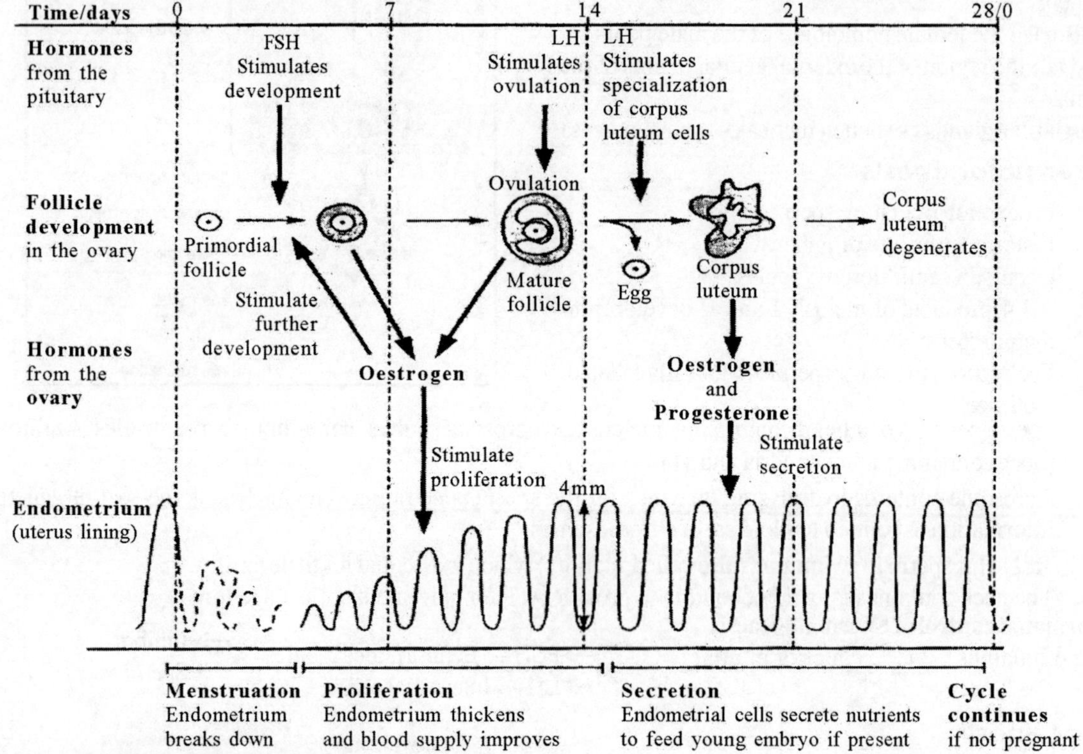

Path of Sperm from Formation to Fertilization

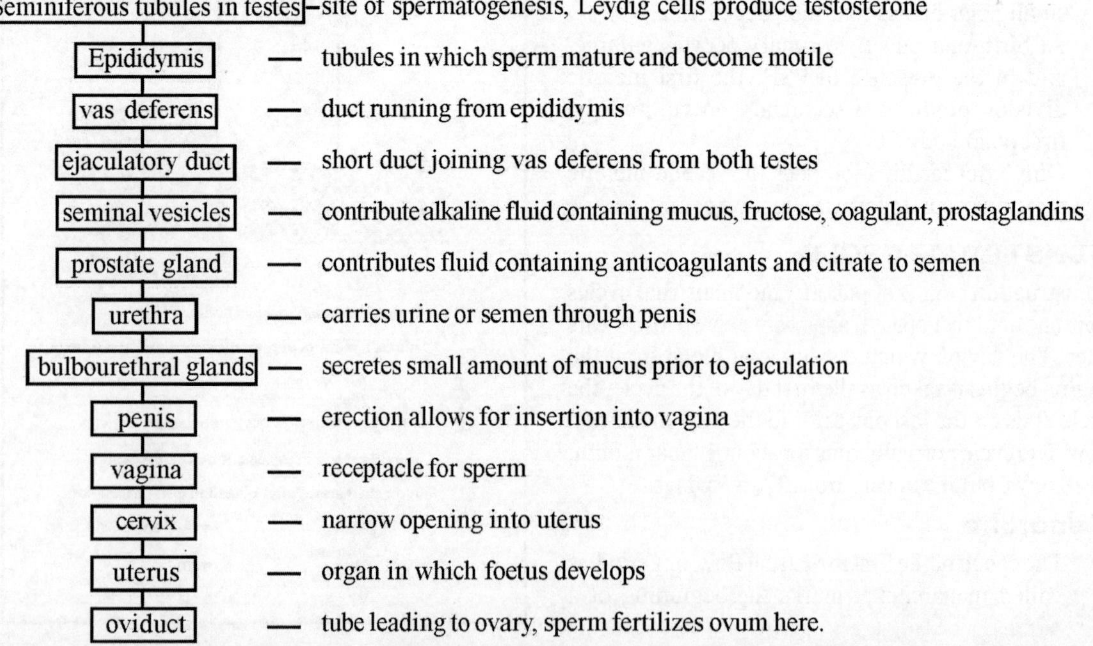

Sequence of Key Events in the Fertilization

Event	Function
Sperm contacts egg; hydrolytic enzymes released from acrosome	Allows acrosomal process to penetrate jelly coat
Protein on acrosomal process attaches to receptor on vitelline layer	Assures that egg is fertilized by sperm of same species
Enzymes digest through vitelline layer	Brings plasma membranes in contact
Sperm and egg plasma membranes fuse; sperm nucleus enters egg cytoplsm	Two haploid nuclei in same egg; membrane fusion causes ion channels to open
Ion channels open, Na^+ flows into egg, and membrane depolarizes	Depolarization creates fast block to polyspermy
Calcium released in egg cytoplasm	Triggered by depolarization, calcium release causes cortical reaction
Exocytosis by cortical granules; fertilization membrane forms	Enzymes loosen vitelline layer, swells by osmotic uptake of water, forms slow block to polyspermy
pH rises, rates of cellular respiration and protein synthesis increase	Activation of egg, synthesis of proteins from mRNA stockpiled in egg
Sperm and egg nucleus fuse; DNA replication begins	Prepares for first cell division, ready for development.

EMBRYONIC DEVELOPMENT

- Cleavage occur into the zygote, converting it into a 16-32 celled **morula**.
- The morula undergoes **compaction** and forms two types of cells–peripheral cells form trophoblast and inner mass cells which develops into **embryo**. Trophoblast cells secrete fluid into the morula to form a blastocoel and the embryo is called **blastocyst**.
- At this stage, implantation into uterus takes place.
- Following this gastrulation (formation of 3 layered embryo), neurulation and organogenesis occur.

Foetal Membranes :

- Amnion, chorion, allantois and yolk sac are four types of foetal membrones.
- Amnion provides fluid medium for the developing embryo and acts as shock absorber.
- Yolk sac gradually shrinks.
- Chorion and allantois form the placenta.

Parturition

- Expulsion of the young one at the end of gestation is known as parturition. Gestation period is 280 days in humans.
- The signals for parturition originate from **fully developed foetus & placenta** and induce mild uterine contractions called **foetal ejection reflex,** which in turn triggers the release of **oxytocin** hormone from post.
- First step in labour is the opening of cervix.
- Due to powerful contractions amnion ruptures releasing amniotic fluid and then baby follows.
- Umbilical cord is cut and then placenta and remains of the cord are expelled.
- Infants lungs expand and it starts breathing.

Lactation

- The mammary glands also undergo certain development during pregnancy under the influence of hormones like prolactin (hPL) and progesterone.
- They start producing milk towards the end of pregnancy by the process called lactation.
- The milk that comes out of the mammary glands of the mother just after child birth, is called **colostrum**; it is rich in nutrients and certain antibodies for the baby.

Exercise

1. Freshly released human egg has
(a) one Y-chromosome
(b) one X-chromosome
(c) two X-chromosomes
(d) one X-chromosome & one Y-chromosome

2. Fertilization is
(a) union of diploid spermatozoan with diploid ovum to form diploid zygote
(b) union of haploid sperm with haploid ovum to form haploid zygote
(c) union of haploid sperm with haploid ovum to form diploid zygote
(d) union of diploid sperm with haploid ovum to form triploid zygote

3. A mature sperm has
(a) a pair of flagella
(b) a nucleus, an acrosome and a centriole
(c) a nucleus, an acrosome, a pair of centrioles
(d) a nucleus, an acrosome, a pair of centrioles and a tail

4. In human beings, the developing embryo resides in
(a) fallopian tubes (b) uterus
(c) ovaries (d) vagina

5. Spermatogenesis is promoted by
(a) estrogen (b) progesterone
(c) testosterone (d) oxytocin

6. Sperm cells are produced in
(a) seminiferous tubules
(b) interstitial cells
(c) epididymis
(d) prostate gland

7. Human sperm moves by
(a) cilia (b) flagella
(c) basal body (d) nucleosome

8. Cauda epididymis leads to
(a) vas efferens (b) vas deferens
(c) ejaculoatory duct (d) rete testis

9. The nutritive medium for the ejaculated sperms is given by
(a) seminal fluid (b) vaginal fluid
(c) uterine lining (d) fallopian tube

10. In female rabbit, the expanded proximal part of the oviduct is known as
(a) uterus (b) vagina
(c) ovary (d) fimbricated funnel

11. Ovulation in mammals is caused by
(a) FSH and TSH (b) FSH and LH
(c) FSH (d) LTH and LH

12. Testis produce
(a) sperms (b) eggs
(c) seeds (d) spores

13. The process of releasing the ripe female gamete (ovum) from the ovary is called
(a) parturition (b) ovulation
(c) fertilization (d) implantation

14. In which phase of cell division is oogonia arrested
(a) Anaphase II
(b) Prophase I
(c) Interphase
(d) Both prophase I and II

15. Oogenesis comprises
(a) multiplication phase
(b) growth phase
(c) maturation phase
(d) all the above

16. Location and secretion of Leydig's cells are
(a) liver – cholesterol
(b) ovary – estrogen
(c) testes – testosterone
(d) pancreas – glucagon

17. Egg is liberated from ovary in
(a) secondary oocyte stage
(b) primary oocyte stage
(c) oogonial stage
(d) mature ovum stage

18. Middle piece of mammalian sperm possesses
(a) mitochondria and centriole
(b) mitochondria only
(c) centriole only
(d) nucleus and mitochondria

19. The growth of corpus luteum is initiated by
(a) human chorionic gonadotropin
(b) follicle stimulating hormone
(c) luteinizing hormone
(d) prolactin

20. At the end of first meiotic division, male sperm differentiates into
(a) secondary spermatocyte
(b) primary spermatocyte
(c) spermatogonium
(d) spermatids

21. Exponential growth of cells is a characteristic feature of
(a) tissue culture cells
(b) multicellular organisms
(c) unicellular organisms
(d) embryo

22. The correct sequence of spermatogenetic stages leading to the formation of sperms in a mature human testes is
(a) spermatogonia - spermatocytes - spermatids - sperms
(b) spermatids - spermatocytes - spermatogonia - sperms
(c) spermatogonia -spermatids-spermatocytes - sperms
(d) spermatocyts - spermatogonia - spermatids - sperms

23. Which one of the followings is the most likely root cause why menstruation is not taking place in regularly cycling human female?
(a) Maintenance of the hypertrophical endometrial lining
(b) Maintenance of high concentration of sex hormones in the blood stream
(c) Retention of well -developed corpus luteum
(d) Fertilisation of the ovum

24. Sertoli cells are found in
(a) ovaries and secrete progesterone
(b) adrenal cortex and secrete adrenaline
(c) seminiferous tubules and provide nutrition to germ cells
(d) pancreas and secrete cholecystokinin

25. Vasa efferentia are the ductules leading from
(a) testicular lobules to rete testis
(b) rete testis to vas deferens
(c) vas deferens to epididymis
(d) epididymis to urethra

26. The signals for parturition originate from
(a) placenta only
(b) placenta as well as fully developed foetus
(c) oxytocin released from maternal pituitary
(d) fully developed foetus only

27. The part of fallopian tube closest to the ovary is
(a) isthmus (b) infundibulum
(c) cervix (d) ampulla

28. Spermatozoa are nourished during their development by
(a) sertoli cells
(b) interstitial cells
(c) connective tissue cells
(d) none of the above

29. Correct sequence in development is –
(a) fertilization → zygote → cleavage → morula → blastula → gastrula
(b) fertilization → zygote → blastula → morula → cleavage → gastrula
(c) fertilization → cleavage → morula → zygote → blastula → gastrula
(d) cleavage → zygote → fertilization → morula → Blastula → Gastrula

30. Polar body is produced during the formation of
(a) sperm (b) secondary oocyte
(c) oogonium (d) spermatocytes

31. The head of mature mammalian sperm is made of
(a) an acrosome
(b) elongated nucleus covered by acrosome
(c) two centrioles and an axial filament
(d) nucleus, acrosome, cytoplasm and mitochondrial sheath

32. Which part of sperm provides energy it to fertilize the egg?
(a) Sperm head (b) Acrosome
(c) Sperm tail (d) Middle piece

33. In which stage of development the embryonic cells form the germinal layers by the movement ?
(a) Morula (b) Blastula
(c) Gastrula (d) Nerula

34. What is true for cleavage ?
(a) Size of embryo increases
(b) Size of cells decrease
(c) Size of cells increase
(d) Size of embryo decreases

35. Seminal vesicle is present at the junction of
(a) prostate & urethra
(b) prostate & vas deferens
(c) prostate & cowper's gland
(d) vas deferens & testes

36. Which layer develops first during embryonic development?
(a) Ectoderm (b) Mesoderm
(c) Endoderm (d) Both (b) & (c)

37. Clitoris in mammals is
(a) homologous to penis
(b) analogous to penis
(c) functional penis in female
(d) non functional

38. Maturation of sperm before penetration is called
(a) spermatogenesis (b) spermiogenesis
(c) capacitation (d) spermatid

39. Which of the following is an accessory reproductive gland in male mammals?
(a) Prostate gland
(b) Gastric gland
(c) Mushroom shaped gland
(d) Inguinal gland

40. There are some special types of cells found in the seminiferous tubules known as sertoli cells. These are
(a) germinal cells (b) reproductive cells
(c) somatic cells (d) protective cells

41. Spermatogenesis is under the regulatory influence of
(a) ADH (b) FSH
(c) LH (d) STH

42. Seminal plasma in humans is rich in
(a) fructose, calcium, certain enzymes
(b) fructose and calcium but has no enzymes
(c) glucose and certain enzymes but has no calcium
(d) fructose and certain enzymes but poor in calcium

43. Supporting cells found in between the germinal epithelium of testes are called
(a) interstitial cells of leydig
(b) sertoli cells
(c) granular cells
(d) phagocytes

44. Inhibition of secretion of which of the following homone is necessary for disintegration of corpus luteum?
(a) LH (b) Progesterone
(c) Testosterone (d) FSH

45. Graafian follicles are characteristically found in the
(a) thyroid of mammal (b) ovary of frog
(c) testis of mammal (d) ovary of mammal

46. Which hormone level reaches peak during luteal phase of menstrual cycle ?
(a) Luteinizing hormone
(b) Progesterone
(c) Follicle stimulating hormone
(d) Estrogen

47. Withdrawl of which of the following hormones is the immediate cause of menstruation ?
(a) FSH-RH (b) Progesterone
(c) Estrogen (d) FSH

48. Which one of the following is the correct matching of the events occuring during menstrual cycle ?
(a) Ovulation : LH and FSH attain peak level and sharp fall in the secretion of progesterone
(b) Proliferative phase : Rapid regeneration of myometerium and maturation of Graafian follicle
(c) Development of corpus luteum : Secretory phase and increased secretion of progesterone
(d) Menstruation : Breakdown of myometrium and ovum not fertilised

49. FSH and LH is
(a) vasopressin (b) vitamin
(c) gonadotropin (d) enzyme

50. Some important events in the human female reproductive cycle are given below. Arrange the events in a proper sequence
A : Secretion of FSH
B : Growth of corpus luteum
C : Growth of the follicle and oogenesis
D : Ovulation
E : Sudden increase in the levels of LH
(a) $C \rightarrow A \rightarrow D \rightarrow B \rightarrow E$
(b) $A \rightarrow C \rightarrow E \rightarrow D \rightarrow B$
(c) $A \rightarrow D \rightarrow C \rightarrow E \rightarrow B$
(d) $B \rightarrow A \rightarrow C \rightarrow D \rightarrow E$

51. Vaginal orifice, urethral orifice and clitoris are protected by
(a) labia majora (b) labia minora
(c) vulva (d) anus

52. Ovulation takes place in/on
(a) ovary (b) about the 14th day
(c) both (a) and (b) (d) none of these

53. The cavity present in the graafian follicle is
(a) amniotic cavity (b) archenteron
(c) antrum (d) ostium

54. Proliferation of endometrium of uterus in controlled by
(a) relaxin (b) oxytocin
(c) progesterone (d) luteinizing

55. In human females ova are produced in
(a) ovarian follicles (b) oviduct
(c) uterus (d) vagina

56. The menstrual cycle in normal adult woman is of
(a) 48 days (b) 38 days
(c) 18 days (d) 28 days

57. Parturition duct in female is called
(a) uterus (b) oviduct
(c) vagina (d) cervix

58. In mammals the female secondary (accessory) sexual characters are developed by the hormone
(a) relaxin (b) estrogens
(c) progesterone (d) gonadotropins

59. What helps in the penetration of egg by the sperm
(a) Fertilizin
(b) Antifertilizin
(c) Sperm lysin
(d) Fertilization membrane

60. On fertilization of egg nucleus with sperm nucleus
(a) First maturation is completed
(b) Second maturation is completed
(c) Embryo is formed
(d) First polar body is formed

CHAPTER 26 Reproductive Health

CONCEPTS

- **Reproductive health** refers to as a total well being in physical, emotional, behavioural and social aspects of reproduction.
- Health is affected by
  - Genetic disorders and deficiencies with which an individual is born.
  - Infections.
  - Life style that includes - food and water, rest, exercise and habits.

REPRODUCTIVE HEALTH-PROBLEMS AND STRATEGIES

- Reproductive and Child Health Care (RCH) programmes, are in operation.
- The main aim of such programmes is to create awareness among people about reproductive organs, accessory organs of reproduction, secondary sexual characters, adolescence and associated changes, safe and hygienic sexual practices, sexually transmitted diseases (STDs).
- Knowledge about available birth control methods, care of pregnant mothers, post-natal care of the mother and the child, importance of breast-feeding, equality of sex etc. among fertile couples and those in marriageable age group and other goals of RCH programmes.

POPULATION EXPLOSION

- The rapid increase in human population size over a relatively short period is called **human population explosion**.
- Increased health facilities and improvement in technology leading to better living conditions, have an explosive impact on the growth of population.
- The probable reasons are
  - decline in death and increase in longevity.
  - decline in maternal mortality rate (MMR).
  - decline in infant mortality rate (IMR).
  - increase in the number of people in the reproductive age.

Consequences of Overpopulation

- Overpopulation has resulted in serious problems like depletion of natural resources, various socio-economic problems, energy crisis and eco-degradation.
- Socio-economic problems are
  - shortage of food supply
  - space
  - unemployment
  - essential goods and poverty
- Energy crisis – Increase in population has created greater demand for energy – fuel, wood, fossil fuel etc.
- Eco-degradation are
  - pollution, and
  - sanitation

Population Control

- Man has started realising his fate and has initiated plans to check this rate of increase by adopting following measures:
  - Planned control of population
  - Education
  - Increasing marriageable age
  - Family planning

BIRTH CONTROL

- A variety of methods are known for birth control. The birth control methods which deliberately prevent fertilization are referred to as **contraception**.
- These methods are of two types – temporary and permanent.

Temporary Methods Include

(i) Natural methods

(ii) Barrier method

- **Coitus Interruptus** involves the withdrawal of the penis by the male before the ejaculation so that semen is not deposited in the vagina & there is no fertilization.

- **Abstinence** is a way to avoid conception by abstain from sexual intercourse.
- Spermicidal creams, jellies and foams if introduced into the vagina before intercourse they adhere to the mucous membrane & immobilise and kill sperms by inhibiting oxygen uptake.
- **Condom** (Nirodh) is a rubber sheath to cover the erect penis.

  It is also a safeguard against AIDS and other sexual diseases.
- **Diaphragms** and **cervical cap** are rubber plastic covers that are fitted on the cervix & check the entry of sperms into the uterus.
- **Intrauterine devices** are plastic or metal devices placed in the uterus. They contain either copper or a common synthetic hormone used in oral contraceptives. These include loop, copper-T, spiral, etc. They prevent fertilization of the egg or implantation of egg.
- Commonly used IUD's are Lippes loop and copper-T.
- **Oral contraceptives (birth control pills)** check ovulation by inhibiting the secretion of FSH and LH that are necessary for ovulation.

Permanent Methods

- **Vasectomy** in male involves a cut in the scrotal sac, cutting or burning of the vas deferens (tubes that carry sperm) and blocking both the cut ends. Vasectomy prevents the passage of sperm into seminal fluid by blocking the vas deferens.
- Female sterilization (called **tubectomy**) prevents fertilization by interrupting the passage through fallopian tube. A small part of the fallopian tube is removed or tied up through a small incision in the abdomen or through vagina.

Medical Termination of Pregnancy

- Medical termination of pregnancy (MTPs) or induced abortion is one of the most widely used methods of fertility control in the world.
- Certain pills acts as abortants. They function by inducing menstruation which checks the implantation of the zygote or detaches the implanted egg.

SEXUALLY TRANSMITTED DISEASES (STDS) :

- Sexually transmitted diseases (STDs) are the diseases or infections which are transmitted through sexual intercourse with infected persons.

	Disease		Causative Agent
(i)	Syphilis	–	*Treponema pallidum*
(ii)	Gonorrhoea	–	*Neisseria gonorrhoeae*
(iii)	Chlamydiases	–	*Chlamydia trachomatis*
(iv)	Trichomoniasis	–	*Trichomonas vaginalis*
(v)	AIDS	–	HumanImmuno-deficiency Virus (HIV)
(vi)	Hepatitis-B	–	Hepatitis virus
(vii)	Genital Herpes	–	Herpes simplex virus, Human papilloma virus

Infertility

- A large number of couples are infertile, *i.e.*, they are unable to produce children inspite of unprotected sexual cohabitation.

Assisted Reproductive Technologies (ART)

- Test-tube baby programme
- Gamete Intra Fallopian Transfer (GIFT)
- Intra-Cytoplasmic Sperm Injection (ICSI)
- Artificial Insemination Technique

- **Test - Tube Baby Programme**

  – In this method, ova from the wife or a donor female and the sperms from the husband or a donor are allowed to fuse under simulated conditions (as that of the body) *in vitro* fertilisation.

  – The zygote or early embryo is transferred into the uterus or fallopian tube; it is called **Embryo Transfer (ET)**.
- **Gamete Intra Fallopian Transfer (GIFT)**

  – This method involves the transfer of an ovum collected from a donor female into another female, who cannot produce ova, but can provide suitable conditions for fertilization and further development of the foetus upto parturition.
- **Intra Cytoplasmic Sperm Injection (ICSI)**

  – In this method, the sperm is directly injected into the ovum to form an embryo in the laboratory and then embryo transfer is carried out.
- **Artificial Insemination**

  – In this method, the semen collected from the husband or a healthy donor, is artifically introduced into the vagina or into the uterus (intra uterine insemination)

Amniocentesis

- It is a prenatal diagnostic technique in which a sample of amniotic fluid from the womb of a pregnant woman, is taken during the early stages of foetal development and the cells are cultured and analysed.
- By this method the chromosomal abnormalities, the sex of the foetus and development disorders could be detected; it is misused for destroying the normal female foetuses.
- Determination of sex by amniocentesis has been banned.

Exercise

1. Main cause of population explosion in the world is
   (a) excellent job facilities
   (b) increase in agricultural production
   (c) excellent health care
   (d) fewer battles and wars
2. The impact of human population is directly related to
   (a) standard of living
   (b) food supply and housing
   (c) health and medical care
   (d) all of the above
3. Marriageable age in India is (in years)
   (a) male 15, female 16 (b) male 25, female 20
   (c) male 18, female 21 (d) male 21, female 18
4. The rate at which new born individuals are joining the population by reproduction is known as
   (a) Natality (b) Fertility
   (c) Contractability (d) None
5. Which of the following means of reducing birth-rate is not legitimate in India?
   (a) Deprivation of the expectant mother from fundamental rights
   (b) Medical termination of pregnancy
   (c) Use of contraceptives
   (d) Use of surgical methods
6. In India, human population is heavily weighed towards the younger age groups as a result of
   (a) long life-span of many individuals and of high birth rate
   (b) short life-span of many individuals and low birth rate
   (c) long life-span of many individuals and of a high birth rate
   (d) short life-span and low birth rate
7. In some tribes, the population does not grow appreciably because of
   (a) low fecundity
   (b) high rate of infant mortality
   (c) illiteracy
   (d) limited food supply
8. Which of the following factors regulate human life with reference to population density?
   (a) Availability of blood, housing and health facilities
   (b) Urbanisation
   (c) Climatic conditions
   (d) All the above
9. For better survival of the human population which of the following steps is most important?
   (a) Reduction in the use of various resources
   (b) Afforestation
   (c) Conservation of wild life
   (d) Ban on mining activity
10. To which population category India belongs?
    (a) High birth rate and high mortality rate
    (b) Low birth rate and low mortality rate
    (c) Low birth rate and high mortality rate
    (d) High birth rate and low mortality rate
11. Which one of the following factors has contributed most in the rapid rise of human population in the present century?
    (a) Increase in birth rate
    (b) Decrease in death rate of old people
    (c) Decrease in infantile mortality
    (d) Polygamy
12. In India, the population growth
    (a) like some animal species, human population is likely to reach zero population growth level
    (b) can be reduced by natural calamities and birth control methods
    (c) like many animal species, it is also going to take sigmoid curve
    (d) can be controlled by adopting national family welfare programmes
13. Which of these is used to control human population?
    (a) Estrogen + progesterone
    (b) IUDs and MTP
    (c) Tubectomy and Vasectomy
    (d) All of these
14. What is the function of copper-T?
    (a) Checks mutation
    (b) Stops fertilization
    (c) Stops zygote formation
    (d) Stops oblituation of blastocoel

15. The method of sterilisation is
(a) Lippes loop (b) IUD
(c) Vasectomy (d) Tubectomy

16. Surgical removal or cutting and ligation of the ends of oviduct is known as
(a) tubectomy (b) oviductomy
(c) vasectomy (d) ovarioctomy

17. Progesterone pill helps in preventing pregnancy by not allowing
(a) ova formation (b) fertilization
(c) implantation (d) none of these

18. A contraceptive pill contains
(a) progesterone and estrogen
(b) oxytocin
(c) relaxin
(d) none of these

19. Surgical removal of uterus is called
(a) vasectomy (b) tubectomy
(c) hysterectomy (d) anatomy

20. Which of the following birth control measure can be considered as the safest?
(a) The rhythm method
(b) The use of physical barriers
(c) Termination of unwanted pregnancy
(d) Sterilization techniques

21. The success of birth control programmes in controlling population growth is dependent on
(a) use of contraceptives
(b) tubectomy
(c) vasectomy
(d) acceptability of the above by the people

22. Trade name of weekly oral contraceptive pill is
(a) Mala (b) Saheli
(c) Mala A (d) Mala D

23. Amniocentesis involves the analysis of
(a) amnion
(b) body fluid of amniotes
(c) amino acids of protein
(d) amniotic fluid

24. Which one of the following is tested by the technique of amniocentesis?
(a) Biochemical abnormalities in the foetus
(b) Errors of metabolism in the foetus
(c) Chromosomal abnormalities in the foetus
(d) All of the above

25. Which of the following is a method for birth control?
(a) IUDs (b) GIFT
(c) HTF (d) IVE-ET

26. In amniocentesis, the fluid is taken from
(a) foetal blood
(b) mother's blood
(c) body fluid of mother
(d) fluid surrounding foetus

27. Cu ions released from copper - releasing Intra Uterine Devices (IUDs)
(a) make uterus unsuitable for implantation
(b) increase phagocytosis of sperms
(c) suppress sperm motility
(d) prevent ovulation

28. Which one of the following is the most widely accepted method of contraception in India, as at present ?
(a) Cervical caps
(b) Tubectomy
(c) Diaphragms
(d) IUDs (Intra uterine devices)

29. Medical termination of pregnancy (MTP) is considered safe up to how many weeks of pregnancy?
(a) Eight weeks (b) Twelve weeks
(c) Eighteen weeks (d) Six weeks

30. The technique called gamete intrafallopian transfer (GIFT) is recommended for those females
(a) who cannot produce an ovum
(b) who cannot retain the foetus inside uterus
(c) whose cervical canal is too narrow to allow passage for the sperms
(d) who cannot provide suitable environment for fertilisation

31. Given below are four methods (A-D) and their modes of action (p-s) in achieving contraception. Select their correct matching from the four options that follow:

	Method		**Mode of action**
A.	The pill	(p)	Prevents sperms reaching cervix
B.	Condom	(q)	Prevents implantation
C.	Vasectomy	(r)	Prevents ovulation
D.	Copper T	(s)	Semen contains no sperms

Matching:
(a) A-(r), B-(p), C-(s), D-(q)
(b) A-(s), B-(p), C-(q), D-(r)
(c) A-(r), B-(s), C-(p), D-(q)
(d) A-(q), B-(r), C-(p), D-(s)

32. Legally acceptable term of abortion is
(a) MTP (b) MMTP
(c) MTTP (d) none

33. Test tube baby means a baby born when
(a) it develops from a non-fertilized egg
(b) it developes in a test tube
(c) it is developes through tissue culture method
(d) the ovum is fertilised externally and thereafter implanted in the uterus

34. The mechanical measure of population control includes
(a) Condom only (b) Diaphragm only
(c) IUD only (d) All of the above

35. The chemical method of contraception includes
(a) jellies only
(b) creams and foams only
(c) oral contraceptives only
(d) all of the above

36. Test tube babies are produced by
(a) fertilising the egg removed from the body of the female with the husband's sperm outside *in vitro* culture. The zygote is transferred back
(b) external fertilisation and development *in vitro* culture till 32 cells stage before putting the embryo back into mother's uterus
(c) complete development of a baby *in vitro*
(d) development upto 32 cells stage and transplanting embryo in the uterus of a surrogate mother

37. Surrogate mother is used for
(a) induction of lactation
(b) artificially inseminated female
(c) future mother with transplanted embryo
(d) artificial insemination and development of embryo

38. Action of contraceptive is
(a) prevent the ovulation only
(b) prevention of ovulation and fertilization only
(c) prevention of ovulation, fertilization and implantation only
(d) prevent the rapid passing of eggs in oviduct

39. Intensely lactating mothers do not generally conceive due to the
(a) suppression of gonadotropins
(b) hyper secretion of gonadotropins
(c) suppression of gametic transport
(d) suppression of fertilisation

40. Sterilisation techniques are generally fool proof methods of contraception with least side effects. Yet, this is the last option for the couples because
(i) it is almost irreversible
(ii) of the misconception that it will reduce sexual urge / drive
(iii) it is a surgical procedure
(iv) of lack of sufficient facilities in many parts of the country
Choose the correct option:
(a) i and iii (b) ii and iii
(c) i and iv (d) i, ii, iii and iv

41. Emergency contraceptives are effective if used within
(a) 72 hours of coitus
(b) 72 hours of ovulation
(c) 72 hours of menstruation
(d) 72 hours of implantation

42. Choose the right one among the statements given below
(a) IUDs are generally inserted by the user herself
(b) IUDs increase phagocytosis reaction in the uterus
(c) IUDs suppress gametogenesis
(d) IUDs once inserted need not be replaced

43. Condoms are one of the most popular contraceptives because of the following reasons
(a) these are effective barriers for insemination
(b) they do not interfere with coital act
(c) these help in reducing the risk of sexually transmitted diseases.
(d) All of the above

44. Choose the correct statement regarding the ZIFT procedure
(a) Ova collected from a female donor are transferred to the fallopian tube to facilitate zygote formation
(b) Zygote is collected from a female donor and transferred to the fallopian tube
(c) Zygote is collected from a female donor and transferred to the uterus
(d) Ova collected from a female donor and transferred to the uterus

45. Contraceptive 'pills' are used by
(a) male (b) female
(c) both (d) none

46. Commonest STD in India is
(a) syphillis (b) gonorrhoea
(c) AIDS (d) herpes

47. Barrier method of contraceptive includes all except
(a) condom (b) dutch cap
(c) CuT (d) diaphragm

48. Which of the following disease is curable ?
(a) HIV infection (b) Hepatitis-B
(c) Genital herpes (d) None

49. Which of the following step is necessary for building up a healthy reproductive society ?
(a) Introduction of sex education in schools
(b) Proper information about reproductive organs, adolescent and safe sexual practices
(c) To create awareness among people about reproduction related aspects
(d) All of these

50. Natality is the characteristic of a population which means
   (a) the total number of individuals present per unit area at a given time
   (b) the increase in number of individuals in a population under given environmental conditions
   (c) loss of individuals due to death in a population under given environmental conditions
   (d) the movement of individuals into and out of population

51. Consider the statements given below regarding contraception
   (a) Medical Termination of Pregnancy (MTP) during first trimester is generally safe
   (b) Generally chances of conception are nill until mother breast-feeds the infant upto two years
   (c) Intra uterine devices like copper-T are effective contraceptives
   (d) Contraception pills may be taken upto one week after coitus to prevents conception

52. What is the full form of AIDS?
   (a) Acquired Immune Deficiency Syndrome
   (b) Acquired Immune Disease Syndrome
   (c) Acquired Immune Deficiency System
   (d) Acquired Immune Defective System

53. Which one is the natural method of avoiding chances of ovum and sperms meeting?
   (a) Periodic abstinence (b) IUDs
   (c) Vasectomy (d) Tubectomy

54. Which one of the followings is not a STDs?
   (a) AIDS (b) Syphillis
   (c) Gonorrhoea (d) Cancer

55. In which year Government of India legalised Medical termination of pregnancy (MTP)?
   (a) 1971 (b) 1972
   (c) 1973 (d) 1974

56. In which year MTP was legalised in India?
   (a) 1971 (b) 1951
   (c) 1981 (d) 1923

57. The Test-tube Baby Programme employs which one of the following techniques
   (a) Intra Cytoplasmic Sperm Injection (ICSI)
   (b) Intra Uterine Insemination (IUI)
   (c) Gamete Intra Fallopian Transfer (GIET)
   (d) Zygote Intra Fallopian Transfer (ZIFT)

58. Read the following 4-statements and mark the option that has both correct statements.
   (i) MTP was legalized in 1971.
   (ii) Inability to conceive or produce children even after 2 years of unprotected sexual cohabitation is called infertility.
   (iii) Surgical method of contraception prevents gamete formation.
   (iv) MTPs are relatively safe up to 14 weeks of pregnancy.
   (a) (i) and (ii) (b) (ii) and (iii)
   (c) (iii) and (iv) (d) (i) and (iii)

59. Consider the statements given below regarding contraception and answer as directed thereafter.
   (i) Medical Termination of Pregnancy (MTP) during first trimester is generally safe.
   (ii) Generally chances of conception are nil until mother breast-feeds the infant upto two years.
   (iii) Intrauterine devices like copper-T are effective contraceptives.
   (iv) Contraception pills may be taken upto one week after coitus to prevent conception.

   Which two of the above statements are correct?
   (a) (i) and (ii) (b) (ii) and (iii)
   (c) (iii) and (iv) (d) (i) and (iii)

60. Choose the correct statements –
   (i) According to the WHO, reproductive health is total well-being in the physical, social, emotional, behavioural aspects of reproduction.
   (ii) According to the WHO, reproductive health is total well being in the physical, social and emotional aspects of reproduction.
   (iii) A reproductively healthy society has people with physically and functionally normal reproductive organs.
   (iv) Reproductively healthy societies have abnormal sex-related emotional and behavioural interactions.
   (a) (i), (ii) and (iii) (b) (ii) and (iv)
   (c) (i) and (iii) (d) (i) only

Heredity and Variation

CONCEPTS

- Study of heredity and variation is called **genetics**.
- Term genetics was given by - **Bateson**.
- Father of genetics - **Gregor Johann Mendel**.
- Father of experimental genetics - **Thomas Hunt Morgan**.
- Father of human genetics - **Archibald Garrod**.
  - **Heredity or Inheritance :** The transfer of characters from one generation to another is called heredity or inheritance.
  - **Variation :** Variation is the differences in characters exhibited between parents and offsprings and among themselves.
- **Variation is of two types :** (a) Environmental variations (b) Genetic variations

(a) Environmental Variations

- Variations due to environment are also called
  Eco phenotype.
- The ability of individual to change the phenotype under the influence of environmental factors is called **phenotypic plasticity**.
- Somatic variations are non heritable.

(b) Genetic Variations

- It involves change in genetic materials.
- Genetic variations are inheritable; genetic variations arise due to - (i) Recombination of genes (ii) Mutations
  Genetic variations are of two kinds :- (i) Continuous variations (ii) Discontinuous variations

(i) Continuous Variations

- Continuous variation refers to the small indistinct differences from the average.
- This variation is characteristic of quantitative characters such as height in man, skin colour in man.

(ii) Discontinuous Variations

- This variation is caused due to mutation.
- Discontinuous variation are meristic *e.g.*, Polydactyly.

SOME IMPORTANT TERMS

1. **Gene :** It is segment of DNA. It is basic unit of heredity.
2. **Allele :** According to Mendel each genetic character is controlled by a pair of unit factors. Alleles of a gene are found on homologous chromosomes.
3. **Locus :** The point on a chromosome where an allele is located.
4. **Homozygous condition :** If both the alleles of a gene are similar. It is pure for a trait (TT), (tt).
5. **Heterozygous condition :** If both the alleles of a gene are dissimilar. It is impure for the trait (Tt).
6. **Hemizygous condition :** If an allele of a gene is missing.
7. **Dominant trait :** Trait which is expressed in heterozygous condition *e.g.* (Tt), (TT).
8. **Recessive trait :** Trait which is not expressed in heterozygous condition (Tt). Eg, tt.
9. **Phenotype :** External appearance of a trait. E.g., Tall, dwarf, fair, black etc.
10. **Genotype :** Genetic make up of a trait. E.g., TT, tt, Tt.
11. **F_1 Generation :** F_1 or first filial generation is the generation of hybrids produced from a cross between the genetically different individuals called parents.
12. **F_2 Generation :** F_2 or second filial generation is the generation of individuals which arises as a result of inbreeding or interbreeding amongst individuals of F_1 generation.
13. **Pure line :** Same as true breeding or pure breeding forms or traits.

14. **Emasculation :** Removal of anther before fertilization.
15. **Hybrid :** Offspring in heterozygous condition. E.g., Tt
16. **Hybridization :** Crossing of male and female in order to produce hybrid offspring. E.g., Tt
17. **Back cross :** It is cross which is performed between hybrid and one of its parents.
18. **Test cross :** Test cross is crossing of offspring with unknown dominant phenotype with the individual homozygous recessive for the trait.
19. **Monohybrid cross :** It is a cross between two organisms of a species which is made to study the inheritance of a single pair of alleles or factors of a character.
20. **Monohybrid ratio :** Monohybrid ratio is usually 3 : 1 (phenotypic ratio) or 1 : 2 : 1 (genotype ratio) in which 25% of the individuals carry the recessive trait, 25% pure dominant and 50% have hybrid dominant trait.
21. **Dihybrid cross :** It is a cross between two organisms of a species which is made to study the inheritance of two pairs of factors or alleles of two genes.
22. **Dihybrid ratio :** Dihybrid ratio is 9 : 3 : 3 : 1 (phenotypic ratio) where 9/16 first recessive and second dominant and 1/16 carry both the recessive traits.
23. **Genome :** It is the complete set of chromosomes where every chromosome is represented as in a gamete. Haploid cell has single genome or diploid cell has two set of genomes while polyploids have multiple sets of genomes.
24. **Heterosis :** The hybrid which has superior quality than parents. The term hybrid vigour or Heterosis was given by **Shull**.
25. **Gene bank :** Genetic preservation of extinct species.

MENDELISM

- Mendel's excellent experiments, valid analysis and formation of laws of inheritance are collectively called Mendelism.
- Mendel conducted cross hybridization experiments on Garden Pea plant (*Pisum sativum*) and studied transmission of characters that had 2 opposing traits. *Eg.*, character is height and traits are tall (T) and dwarf (t).
- **Gregor Mendel** was the first individual to apply a modern scientific approach to the study of heredity. Mendel proposed two basic principles of trasmission genetics. The first was the **Principle of segregation**, which claimed that each trait was specified by paired hereditary determinants (alleles of genes) that separate from each other during gamete formation. This law is also called **Law of purity of gametes or Law of splitting of hybrids.**
- If genes are located on different chromosomes, then the alleles of each gene are transmitted to egg cells and sperm cells independently of each other.
- Mendel's second basic conclusion was the **Principle of independent assortment**, which stated that the segregation of one pair of genes-controlling a given trait - was not influenced by the segregation of other gene pairs. The chormosome theory provided a physical basis for the principle of independent assortment. Genes located on different chromosomes move to gametes independently of each other during meiosis. The principle of independent assortment holds because chromosomes line up independently of each other at the metaphase plate during meiosis I. Thus, gametes receive random combinations of chromosomes.

POST-MENDELIAN ERA - OTHER PATTERNS OF INHERITANCE

Gene Interaction (Cause modification in basic mendelian ratio.)

↓ ↓

Intragenic Interaction

→ Incomplete dominance
→ Multiple alleles
→ Co-dominance
→ Lethal alleles

Intergenic Interaction

→ Complementary genes
→ Supplementary genes
→ Epistatic genes —— → Dominant epistasis; → Recessive epistasis
→ Duplicate genes or Pseudoalleles

Intragenic Interaction

Incomplete Dominance

- Red × White = Pink in F_1
  (RR) (rr) (Rr)
- Not blending of alleles, because in F_2 you get 1 red : 2 pink : 1 white *e.g.* - Four-O-'clock (*Mirabilis jalapa*) and Snapdragon (*Antirrhinum majus*)

Multiple Allele, Co-dominance and ABO Blood Group

- More than two alternative forms of a gene occupying the same locus on a chromosome are known as multiple alleles.
- Multiple alleles reside always at the same locus at the same chromosome.
- No crossing over occurs between alleles of multiple allele series.

Human Blood Groups & Multiple Allele

- There are four phenotypes of Blood **-A, B, AB** and **O** produced by three different alleles $\mathbf{I^A}$, $\mathbf{I^B}$ **& i** of a gene.
- The allele $\mathbf{I^A}$ & $\mathbf{I^B}$ are equally dominant and do not interfere with expression of each other hence the allele $\mathbf{I^A\,I^B}$ are said to be co-dominant because both are expressed in the phenotype **AB.**

NOTE : Co-dominant as well as dominant recessive inheritance is found in case of the gene for the blood group in human beings.

Lethal Alleles (2 : 1)

- Such gene which cause death in an organism are called lethal gene (in homozygous condition). It was first reported in Mice body colour by **Cuenot**.
- Lethal gene of *Drosophila* is curly wings (Cy), plum eyes (Pm) and stubbles (Sub). It may be dominant or recessive.

Intergenic Interaction

- Complementary genes 9 : 7 (in *Lathyrus odoratus* by **Bateson and Punnet)**
- Supplementary gene 9 : 3 : 4
- Dominant epistasis 12 : 3 : 1
- Recessive epistasis 9 : 3 : 4
- Duplicate genes 15 : 1
- Suppressor gene 13 : 3

CHROMOSOMAL THEORY OF INHERITANCE

- Chromosome theory was formulated after rediscovery of Mendel's work by three scientists (**de Vries**, **Correns** and **Tschermak**) in 1900. Infact it was correns who summarised Mendel's conclusions in the familiar term of two principles & coined the term **factor**.
- Three postulates of chromosomal theory of inheritance are
  (i) Both chromosomes and genes were found in pairs in diploid cell.
  (ii) 2 alleles of a gene pair are also located on homologous sites on homologous chromosomes. Homologous chromosomes separate and gene pairs (allele), segregate during meiosis.
  (iii) Paired condition of both chromosome and genes were restored during fertilization.

LINKAGE AND RECOMBINATION

- Experimental verification of the chromosomal theory of inheritance by Thomas Hunt Morgan & his colleagues led to discovering the 'basis of variation that sexual reproduction produced.
- Morgan studied eye colour in *Drosophila.*
- Normal eye colour in flies is red and mutant is white eyed.

Phenotype

Red eyed female : Red eyed male : White eyed male
50 : 25 : 25

- This criss cross pattern of inheritance is of sex linked (X-linked genes)
- **Linkage** is the phenomenon of certain genes staying together during inheritance through generations without any change or separation due to their being present on the same chromosomes.
- Linkage in the genes can be identified by test cross.
- **Linked genes** refers to genes on the same chromosomes and assorted together or show very low recombination.
- **Sex linkage** is the phenotypic expression of an allele that is related to the gender of the individual & is directly tied to the sex chromosomes.
- Genes that are present on the X & Y chromosomes are called **sex linked genes**.
- The rearrangements of linked genes due to crossing over is known as **recombination**. Recombination also occurs due to chance separation of chromosomes during gametogenesis & their random coming together during fertilization.

SEX DETERMINATION

- **Henking** discovered X body in spermatogenesis of few insect and it was given name of X chromosome. Due to involvment of X and Y chromosomes in determination of sex, they were called **sex chromosome.**

- Rest of the chromosomes which determine other metabolic character of the body are called **autosomes**.
- Chromosomal determination of sex varies in different organisms.

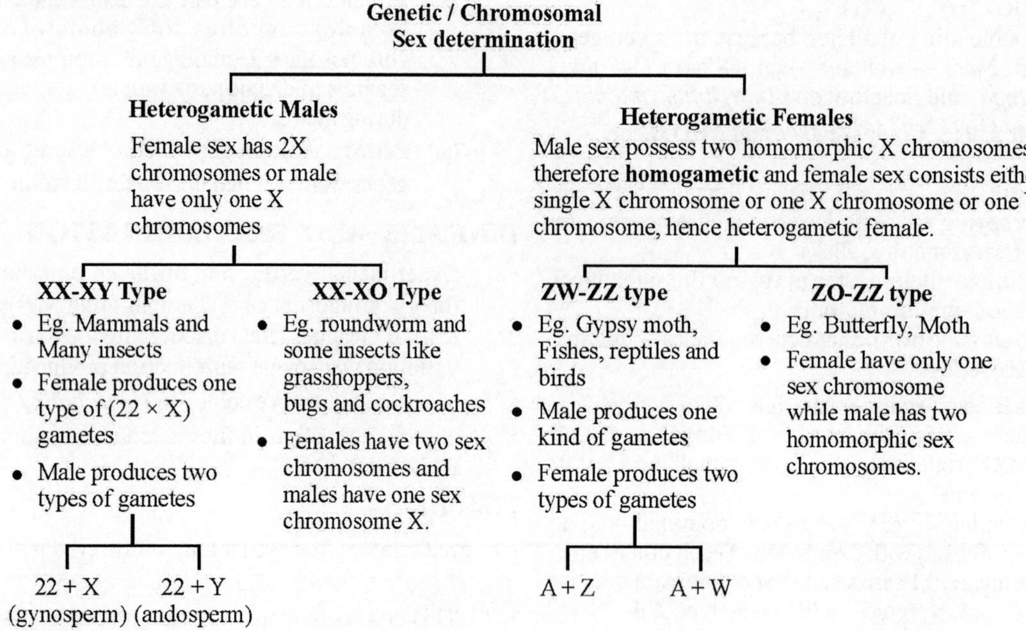

MUTATION

- Phenomenon that results in alteration of DNA sequence and consequently results in change in genotype and phenotype of an organism is called **mutation**.
- **Mutagens** are various chemical and physical factors that induce mutations, *e.g.*, UV radiations, carcinogenic chemicals like nicotine, nitric oxide (NO).

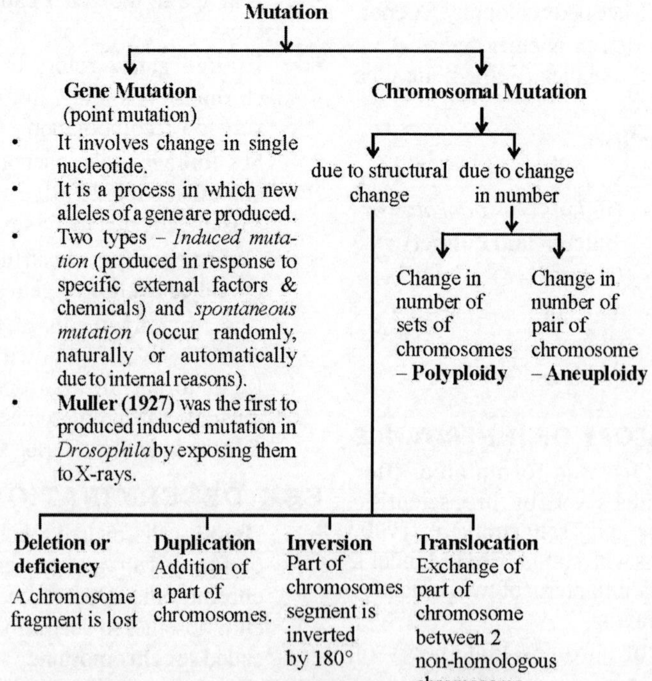

GENETIC DISORDER

- A genetic disorder is a disease that is caused by an abnormality in an individual's DNA.
- Genetic disorder may be grouped into two categories – **Mendelian disorders & chromosomal disorders.**

Mendelian disorders

- Mendelian disorders are chiefly determined by alteration or mutation in the single gene.
- It can be studied by pedigree analysis.

(I) Haemophilia

(i) It is an X linked recessive disease.

(ii) A protein involved in clotting of blood is affected in an affected individual; if person gets a cut, will result in non-stop bleeding.

(iii) Females are heterozygous and carriers of haemophilia.

(II) Sickle cell anaemia

(i) It is an autosomal recessive trait.

(ii) It is transmitted when both the patients are carrier/heterozygous for gene.

(iii) It is due to inheritance of defective allele coding for β-globin. It results in the transformation of Hb^A into Hb^S in which glutamic acid is replaced by valine at 6th position in each of two β-chains of haemoglobin.

(iv) 2 alleles $\underline{Hb^A}$, $\underline{Hb^S}$ $\rightarrow$ Sickle cell anaemia.

(v) It is an excellent example of single mutation.

(III) Phenylketonuria

(i) Phenylketonuria is an inborn error of metabolism.

(ii) It is an autosomal recessive trait.

(iii) Affected individual lacks enzyme phenylalanine hydroxylase that converts amino acid phenylalanine to tyrosine.

(iv) It is characterized by severe mental retardation, hypopigmentation of skin & hair, eczema, etc.

PEDIGREE ANALYSIS

- **Pedigree analysis** is the diagrammatic illustration depicting the passage of a certain traits through successive generations in a family linkage.
- It helps to study the pattern of inheritance of a dominant or recessive trait.
- It helps genetic counsellors to advice couples about the possibility of having genetic defective childrens.

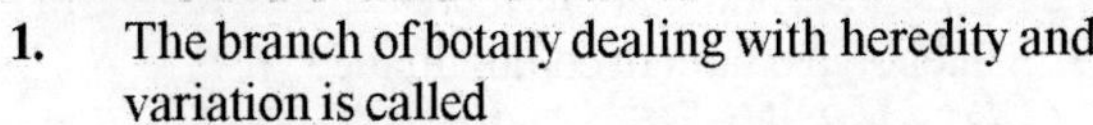

Exercise

1. The branch of botany dealing with heredity and variation is called
   (a) geobotany (b) sericulture
   (c) genetics (d) evolution
2. The resemblance of individuals to their progenitors is called
   (a) heredity (b) genetics
   (c) evolution (d) none of these
3. Mendel proposed which of following term for hereditary units?
   (a) Factor (determiner)
   (b) Genome
   (c) Genetic particle
   (d) None of the above
4. Organism with two different allele is
   (a) heterozygous and homozygous
   (b) heterozygous for the allele
   (c) homozyous for the allele
   (d) none of these
5. Which of the following Mendel has selected for his experiment?
   (a) Garden pea (b) Pigeon pea
   (c) Sweet pea (d) Moong
6. Mendel chose pea plants because
   (a) they were cheap
   (b) they were having seven pairs of contrasting characters
   (c) they were easily available
   (d) of great economic importance
7. Alleles are form of
   (a) genes (b) chromosomes
   (c) DNA (d) none
8. The male human is represented by sex chromosomes
   (a) XX (b) XO
   (c) XY (d) YY
9. Mutation rates are affected by
   (a) temperature
   (b) X-rays
   (c) gamma and beta radiation
   (d) all of the above

10. Mutation caused by a mutagen is
(a) natural (b) chemical
(c) spontaneous (d) induced

11. A dihybrid condition is
(a) tt Rr (b) Tt rr
(c) tt rr (d) Tt Rr

12. Mendel's last law is
(a) segregation
(b) dominance
(c) independent assortment
(d) polygenic inheritance

13. Which one of the following traits of garden pea studied by Mendel was a recessive feature ?
(a) Round seed shape
(b) Axial flower position
(c) Green seed colour
(d) Green pod colour

14. An organism's genetic constitution is called its
(a) genotype
(b) phenotype
(c) holotype
(d) none of these

15. Mendel formulated some laws which are known as
(a) laws of germplasm
(b) laws of origin of species
(c) laws of recapitulation
(d) laws of inheritance

16. An organism with two identical alleles is
(a) dominant (b) hybrid
(c) heterozygous (d) homozygous

17. A gene is said to be dominant if
(a) it expresses its effect only in homozygous state
(b) it expresses its effect only in heterozygous condition
(c) it expresses its effect both in homozygous and heterozygous condition
(d) it never expresses its effect in any conditions

18. Which pair is chromosomal aberration ?
(a) Duplication & Transduction
(b) Duplication & Transversion
(c) Duplication and Translocation
(d) All

19. Mutation is
(a) generally recessive
(b) generally dominant
(c) always dominant
(d) always recessive

20. Haemophilia is
(a) autosomal (b) Y-linked
(c) Z-linked (d) X-linked

21. The main aim of plant breeding is
(a) to produce improved varieties
(b) to make soil fertile
(c) to control pollution
(d) to become more progressive

22. In case of incomplete dominance the monohybrid ratio of phenotypes in F_2 generation is
(a) 1 : 2 : 1 (b) 3 : 1 : 1
(c) 9 : 3 : 3 : 1 (d) 2 : 3 : 1

23. Mendel did not include in his laws
(a) Segregation (b) Dominance
(c) Purity of gametes (d) Linkage

24. Mendel had a difficulty in explaining the linked characters due to
(a) law of dominance
(b) law of segregation
(c) law of independent assortment
(d) all the above

25. Which of the character is dominant in pea plant ?
(a) Wrinkled seeds containing tall plants
(b) Red flower containing dwarf plants
(c) Both (a) and (b)
(d) Neither (a) nor (b)

26. Linked gene shows
(a) always parental combination
(b) sometimes new combinations
(c) always new combination
(d) new combinations more

27. Phenotypic ratio in co-dominance
(a) 1 : 2 : 1 (b) 3 : 1
(c) 2 : 1 (d) 2 : 1 : 3

28. Which of the following is exception to Mendel's laws ?
(a) Linkage
(b) Incomplete dominance
(c) Co-dominance
(d) All of the above

29. AB - blood group shows
(a) co-dominance
(b) incomplete
(c) mixed inheritance
(d) composite inheritance

30. In *Mirabilis* & *Antirrhinum* plant the appearance of the pink hybrid (Rr) between cross of a red (RR) and white (rr) flower parent indicates
(a) incomplete dominance
(b) segregation
(c) dominance
(d) heterosis

31. RR (red) is crossed with ww (white). All Rw offsprings are pink. This indicates that R-gene is
(a) hybrid
(b) incompletely dominant
(c) recessive
(d) mutant

32. What contribute to the success of Mendel ?
(a) Qualitative analysis of data
(b) Observation of distinct inherited traits
(c) His knowledge of biology
(d) Consideration of one character at one time

33. Haemophilia is more common in males because it is a
(a) recessive character carried by Y-chromosome
(b) dominant character carried by Y-chromosome
(c) Dominant trait carried by X-chromosome
(d) Recessive trait carried by X-chromosome

34. In Down's syndrome of a male child, the sex complement is
(a) XO (b) XY
(c) XX (d) XXY

35. An allele is dominant if it is expressed in
(a) both homozygous and heterozygous states
(b) second generation
(c) heterozygous combination
(d) homozygous combination

36. Genes located on Y-chromosome are
(a) mutant genes (b) sex-linked genes
(c) autosomal genes (d) holandric genes

37. Test cross involves
(a) crossing between two F_1 hybrids
(b) crossing the F_1 hybrid with a double recessive genotype
(c) crossing between two genotypes with dominant traits
(d) crossing between two genotypes with recessive traits

38. Phenotype of an organism is the result of
(a) cytoplasmic effects and nutrition
(b) environmental changes and sexual dimorphism
(c) genotype and environmental interactions
(d) mutations and linkages

39. Which kind of disease is "Down's syndrome"?
(a) Sex-linked (b) Viral
(c) Bacterial (d) Autosomal

40. Which one is genetically similar to haemophilia?
(a) Albinism (b) Colour blindness
(c) Night blindness (d) None of these

41. Which is sex linked character ?
(a) Baldness (b) Colour blindness
(c) Diabetes (d) Polio

42. The number of autosomes in human female ovum is
(a) 11 (b) 12
(c) 22 (d) 23

43. In man, which of the following genotypes and phenotypes may be the correct result of aneuploidy in sex chromosomes
(a) 22 pairs + XXY males
(b) 22 pairs + XX females
(c) 22 pairs + XXXY females
(d) 22 pairs + Y females

44. Which of the following statements is not correct for sickle cell anaemia?
(a) Sixth amino acid (from NH_2 terminal) of β chain is glutamic acid
(b) Erythrocytes elongates and form sickle shaped cells
(c) Individual homozygous haemoglobin S allele HB_β^S/Hb_β^S
(d) β chain of haemoglobin S contains valine.

45. Male XX and female XY sometime occur due to
(a) deletion
(b) transfer of segments in X and Y chromosomes
(c) aneuploidy
(d) hormonal imbalance

46. In a dihybrid cross between RRYY and rryy parents, the number of RrYy genotypes in F_2 generation
(a) 4 (b) 3
(c) 2 (d) 1

47. In 1900 A.D. three biologists independently discovered Mendel's principles. They are
(a) De Vries, Correns and Tschermak
(b) Sutton, Morgan and Bridges
(c) Avery, McLeod and McCarthy
(d) Bateson, Punnet and Bridges

48. Which one of the following represents a test cross ?
(a) Ww × WW (b) Ww × Ww
(c) Ww × ww (d) WW × WW

49. How many different kinds of gametes will be produced by a plant having the genotype AABbCC?
(a) Nine (b) Two
(c) Three (d) Four

50. How many pairs of contrasting characters in pea pod were chosen by Mendel?
(a) 2 (b) 3
(c) 4 (d) 7

51. What type of gametes will form by genotype RrYy?
(a) RY, Ry, rY, ry (b) RY, Ry, ry, ry
(c) Ry, ry, Yy, ry (d) Rr, RR, Yy, YY

52. Mendel's law of segregation is applicable to
(a) dihybrid cross only
(b) monohybrid cross only
(c) both dihybrid and monohybrid cross
(d) dihybrid but not monohybrid

53. Mendel did not recognize the linkage phenomenon in his experiments because
(a) there were many chromosomes to handle
(b) characters he studied were located on different chromosomes
(c) he did not have powerful microscope
(d) he studied only pure plants

54. Recombination is involved in the process of
(a) cytokinesis
(b) spindle formation
(c) crossing over
(d) chromosome duplication

55. Exchange of genetic material between chromatids of homologous chromosomes during meiosis is called
(a) synapsis (b) chiasmata
(c) transformation (d) crossing over

56. The four daughter cells derived from a single meiosis differ from each other due to
(a) difference in chromosome number
(b) crossing over only
(c) independent assortment of chromosomes only
(d) crossing over as well as independent assortment of chromosomes

57. Inheritance of ABO blood grouping is an example of
(a) dominance
(b) co-dominance
(c) incomplete dominance
(d) both (a) & (b)

58. A person affected by disease having chromosome complement XXX is called/having
(a) klinefilter's syndrome
(b) down's syndrome
(c) super female
(d) turner's syndrome

59. Point (Gene mutation) mutation involves
(a) insertion
(b) change in single base pair
(c) duplication
(d) deletion

60. Which of the following disorder is not hereditary?
(a) Haemophilia (b) Cataract
(c) Sickle-cell anaemia (d) Colour blindness

CHAPTER 28

Molecular Basis of Inheritance

CONCEPTS

- **Nucleic acids** are the building blocks of genetic material. These are long polymers of nucleotides.
- **Genetic material** is that substance which not only controls the formation and expression of traits in an organism but can replicate & pass from a cell to its daughter cell or from one generation to next.

DNA (DEOXYRIBONUCLEIC ACID)

- DNA is a long chain polymer of deoxyribonucleotides.
- Nucleotide is made up of 3 chemical groups →

  $\underbrace{\textbf{A nitrogen base moiety + pentose sugar}}_{\text{Nucleoside}}$ **+ phosphate group**

- Nitrogenous base are of two types – **purines** (9 membered double rings with nitrogen at 1, 3, 7 & 9th positions) and **pyrimidines** (6 membered rings with nitrogen at 1 and 3rd position).
- Purines are of **two types** – adenine (A) and guanine (G) and pyrimidines are of **three types** – thymine (T), cytosine (C) and uracil (U).
- A characteristics that differentiate DNA from RNA is that DNA contains all of the nitrogen bases except uracil and RNA contains all of the nitrogen bases except thymine.
- Nitrogen base link to phosphate sugar through N-glycoside linkage and form nucleotides.
- The backbone of a DNA strand/chain is build up of alternate deoxyribose & phosphoric acid groups. The phosphate group is connected to carbon 5' of the sugar residue of its own nucleotide & carbon 3' of the sugar residue of the next nucleotide by **phosphodiester bonds**.
- **Wilkins-Franklin** carried out X ray diffraction (X-ray crystallography) on the basis of which **Watson and Crick** suggested 3-D structure of DNA in 1953.

Features of 3-Dimensional double helical structure of DNA

- DNA is composed of 2 strands wound around each other to form a double helix, with base inside and sugar phosphate backbone on outside.
- 2 DNA strands are organized in antiparallel and complementary arrangement [*i.e.*, 2 strands run in opposite orientation (one in 5' – 3' and allies in 3' – 5')].
- Adenine pairs with thymine with 2 hydrogen bonds and guanine pairs with cytosine with three hydrogen bonds.
- The helix is generally right handed *i.e.* its turn run clockwise looking along the helix axis. The pitch of helix is **3.4 nm** (1 nm = 10^{-9}m) and there are **10bp** in each turn.
- Distance between 2 bp in helix = 0.34 nm.
- Diameter of helix is **20 Å**
- According to **Chargaff's rule** the percentage of adenine (A) is equal to percentage of thymine (T) and the percentage of guanine (G) is equal to the percentage of cytosine (C).
- The concept of central dogma in molecular biology was proposed by **Francis Crick** (1958). It proposes unidirectional or one way flow of information from DNA to RNA & then to protein.

  $$\text{DNA} \xrightarrow{\text{Transcription}} \text{mRNA} \xrightarrow{\text{Translation}} \text{Polypeptide (protein)}$$

Prokaryotic Chromosome/DNA Packaging

- DNA of bacteria such as *E.coli* is a circular double stranded molecule which contain 4.6 million base pairs.
- The circular DNA is packaged into **nucleoid** where it is organized into loops or domains bound to a central protein.

Eukaryotic DNA Packaging

- DNA is organized into bead structure called **nucleosome**.
- There is a set of positively charged, basic proteins called **histones** which are rich in basic amino acids – lysine and arginine. They have positively charged side chains.
- Histones organize into unit of 8 molecules called **histone octamer**.
- Negatively charged DNA is wrapped around this positively charged octamer to form nucleosome.
- One nucleosome (DNA + histone octamer) attaches to other nucleosome with the help of linker DNA associated with H_1 protein.
- DNA takes 1.8 left handed turns around octamer.
- A typical nucleosome contains 200 bp of DNA helix.
- Packaging of chromatin to chromosomes occurs with the help of additional set of proteins called **NHC (non-histone chromosomal) proteins.**
- NHC are high molecular weight protein having amino acids - tyrosine & tryptophan.
- On the basis of packaging, chromatin is classified into 2 types –
  (a) **Euchromatin**
    (i) Transcriptionally active.
    (ii) Loosely packed, rich in gene concentration.
    (iii) Lightly stained.
  (b) **Heterochromatin**
    (i) Transcriptionally inactive.
    (ii) Densely packed.
    (iii) Darkly stained.

ESTABLISHMENT OF DNA AS GENETIC MATERIAL

- **Frederick Griffith** (in 1928), a British Medical officer described the phenomenon of bacterial transformation.
- He carried out experiment with ***Streptococcus pneumoniae*** (bacterium causing pneumonia) which is used to infect mice.
- It has **2 types of colonies** on culture plate – **S** (**smooth** due to mucous polysaccharide which makes it infective) and **R** (**rough** strain, mucous polysaccharide absent).
- **Steps of experiment are –**
  – S strain (Virulent) → injected into mice → Mice die
  – R strain (Non-virulent) → injected into mice → Mice survive
  – S strain (Heat killed) → injected into mice → Mice survive
  – S strain (Heat killed) + R strain (Live) → injected into mice → Mice die

Establishment of Biochemical Nature of Transforming Genetic Material

- **Ostwald Avery, Colin Macleod** and **Mc Carty** in 1933-34 revealed the chemical nature of the transforming substance to be DNA.
- They also observed that protein digesting enzyme (Protease) and RNA digesting (RNase) have no effect on transformation whereas digestion of DNA with DNase would not cause transformation.
- This finding **established that transforming genetic material is DNA**.

Hershey-Chase Experiment

- Hershey & Chase (1962) discovered that DNA is the genetic material of bacteriophage.
- Infection of bacteria with virus called **bacteriophage**.

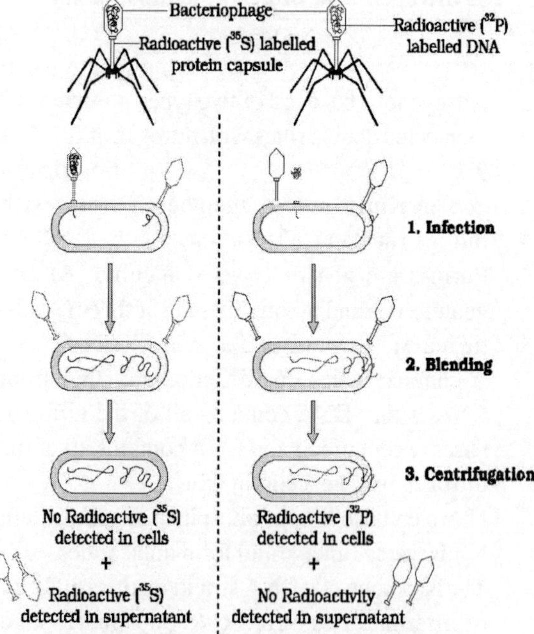

Fig. The Harshey Chase experiment

REPLICATION

- DNA replication or DNA synthesis is the process of copying a double stranded DNA prior to cell division.

- Replication of DNA is found to be semiconservative and its experimental proof has been provided by **Meselson** and **Stahl** by conducting following experiment :
  (a) Culturing *E.coli* for many generations in N^{15} containing culture.
  (b) N^{15} was incorporated as nitrogenous base in newly synthesized DNA.
  (c) Heavy DNA separated from normal by centrifugation in CsCl (cesium chloride) density gradient.
  (d) Culturing of DNA from *E.coli* cells having N^{15} in medium with N^{14}.
  (e) Extraction of DNA from *E.coli* cells in step (d) and to measure density of ds DNA in CsCl centrifugation.
  (f) Cycles repeated for another 40 minutes in normal culture.
- **Results of Experiment**
  - Resultant DNA was double stranded and a hybrid with one strand having N^{15} isotope and other having N^{14} had intermediate density.
  - DNA extracted after recycling of procedure produced 2 types of DNA– hybrid and light.

Mechanism of DNA Replication

- DNA polymerase is responsible for the template directed polymerisation of deoxyribonucleotide triphosphate.
- **Requirements of enzyme are**
  - All four dNTPs (dATP, dGTP, dTTP and dCTP) and Mg^{2+}.
  - DNA template to be copied by DNA polymerase.
  - A primer with a free 3' OH end that enzyme can be extended by enzyme.

Steps of replication

- DNA double helix is unwound by helicase enzyme and single stranded (ss) DNA is stabilized by SSB-single stranded binding protein.
- DNA topoisomerase I allows the helix to unwind without causing extensive rotation of the chromosome.
- DNA topoisomerase II separates the 2 daughter DNA circles following replication.
- DNA replication begins at a specific site called **ori (origin)** which has recognition site for DNA polymerase, which also provides site for attachment of RNA primer.
- dsDNA is antiparallel, one strand runs in 5'-3' direction and its complementary runs in 3'-5'. New DNA strand is made against each template strand. DNA polymerase makes DNA only in 5'-3' direction. So are template strand with 3'-5' orientation, new DNA strand is made in continuous piece in correct 5'-3' direction. This DNA is called **leading strand**.
- On template strand which has 5' → 3' orientation, DNA polymerase synthesizes short pairs of new DNA (about 1000 nucleotide long) in 5' → 3' direction and then joins these piece together. These small fragments are called **okazaki fragments** and new DNA strand made in this **discontinuous manner** is called **lagging strand**. Okazaki fragments are joined by means of DNA ligase.

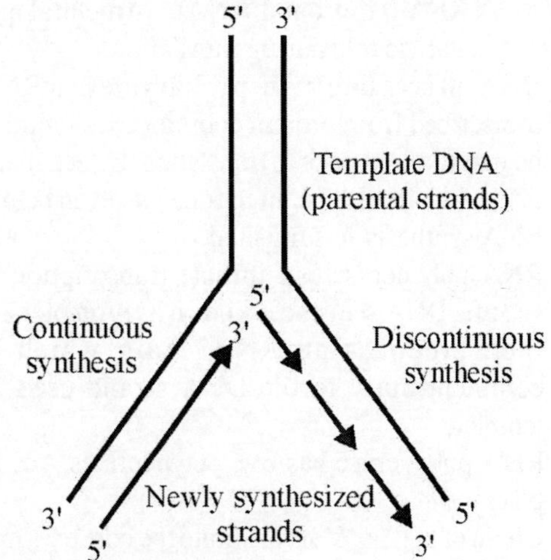

TRANSCRIPTION

- Process of copying genetic information from one strand of DNA to RNA is called **transcription**.
- The strand of DNA with polarity 3' → 5' act as **template** strand and the DNA strand with polarity 5' → 3' act as **coding strand**.
- Transcription is carried out by DNA dependent RNA polymerase.
- Transcription unit has 3 regions in the DNA.
  (i) A promoter – where RNA polymerase binds.
  (ii) A structural gene – which will undergo transcription.

(iii) A terminator – a site where transcription will end.

- Segments of gene represented in mRNA are called **exons** whereas the ones not represented is called **introns**. Introns usually contains palindromic base sequence and constitutes junk/ satellite DNA.

RNA & Process of Transcription

- RNA is a non hereditary nucleic acid except in some viruses (retro viruses).
- RNA is a polymer of ribonucleotide & is made up of pentose ribose sugar, phosphoric acid & nitrogenous base (A, U, C, G).
- The 2' OH group of ribonucleotides is a reactive group, that makes RNA as a catalyst.
- On the basis of molecular size & function, three forms of non genetic RNA are –
  - **mRNA (messenger RNA)** – provides template for translation.
  - **tRNA (transfer RNA)** – brings aminoacids and reads genetic code.
  - **rRNA (ribosomal RNA)** – structural and catalytic role during translation.
- **RNA processing : In prokaryotes**, mRNA transcribed from protein coding genes requires no modification prior to translation. In fact, many mRNA molecules begin to translate even before RNA synthesis has finished.
- RNA polymerase can initiate transcription at specific DNA sequences known as promoters. It then produces an RNA chain which is complementary to the DNA strand used as template.
- RNA polymerase has five polypeptides – σ, β, β', α and ω.
- Chain of β, β', α & ω constitute the core enzyme.
- σ or sigma factor recognises the promoter region while the remaining core enzymes takes part in transcription.
- Rho factor helps in elongation and termination of the RNA synthesis.
- **Transcription Bubble**
  - Region of unwound DNA undergoing transcription.
  - RNA transcript forms a transient DNA-RNA hybrid then pulls away from template as transcription proceeds.
  - DNA is wound ahead of transcription bubble and after the transcription complex has passed, DNA rewinds.
- Transcription continues till a termination sequence in form of GC rich/palindromic region is reached. DNA made from this palindrome is self complementary and basepair internally to form hairpin structure. Termination sequence is located downstream/towards 3' end of coding strand.
- In **prokaryotes :** No splitting event takes place. But in **eukaryotes** – primary transcript has both exons and introns and introns are non functional. Now it is subjected to process called splicing where **introns are removed** and exons are **joined in defined manner**.
- **Capping :** All unusual nucleotide (methyl guanosine triphosphate) is added to 5' end of hnRNA.
- **Tailing :** Adenylate residues [200-300] are added at 3' end of a template in independent manner.
- **RNA splicing :** Intron sequences are removed by process known as splicing and ligates the ends of exon sequences together.
- **RNA editing :** The sequence of an mRNA molecule may be changed after synthesis and processing by RNA editing.

GENETIC CODE

- Transfer of genetic information from a polymer of nucleotides to a polymer of amino acids is called **translation**. This is accomplished with the help of genetic code which is row of **three consecutive nucleotides** – coding for 20 amino acids.
- **Salient features of genetic code are**

  (a) **Triplet nature :** There are 3 stop codon – UAA, UAG UGA.

  (b) **Start codon :** AUG codes for methionine is first codon to be read by ribosome in mRNA.

  (c) Degeneracy of codon

  (d) Wobble position

  (e) Universality of code

  (f) **Open reading frame :** DNA sequence that has run of codon starting with codon ATG and ending with TGA, TAA, TAG These runs of codons are called ORF.

  (g) **Continuity of codons :** Codons are read in mRNA in a continuous fashion.

  (h) **Specificity of codon :** One codon codes for only one amino acid.

tRNA : The Adapter Molecule

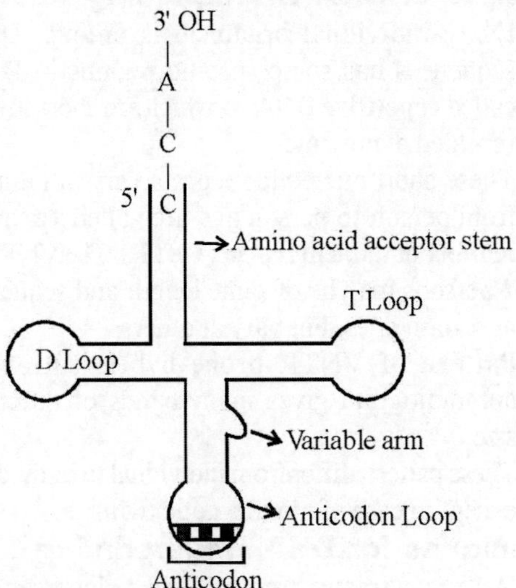

- **Structure of tRNA :**
  (i) **Anticodon arm** contains in its loop 3 nucleotides of anticodon which will form base pair with complementary codon in mRNA during translation.
  (ii) **D/DHU arm with D loop** contains dihydrouracil – an unusual pyrimidine.
  (iii) **T/TψC arm :** Pseudouracil (ψ) in sequence T ψ C.
  (iv) **Variable/arm :** With 3-21 nucleotide.
  (v) **Amino acid acceptor stem :** Site where amino acid attaches at 3'OH group of 3'-CCA sequence.
- tRNA helps to transport amino acids from the surrounding cytoplasm to the site of protein synthesis.

TRANSLATION

- The process of decoding of the message from mRNA to protein with the help of tRNA, ribosome and enzyme is called translation (protein synthesis).
- The main step in protein synthesis are **activation, initiation, elongation & termination of polypeptide chain.**
- During translation, mRNA is read in 5' → 3' direction
- **Aminoacyl tRNA** carry specific amino acid and recognises the corresponding codons in **tRNA with amino acid.**
  It requires ATP activate tRNA and form aminoacyl tRNA.
- **Initiation :**
  – It requires initiation factors.
  – Each ribosome has 3 binding sites for tRNA :
  A site – Incoming amino acyl tRNA binds.
  P site – tRNA linked to growing polypeptide chain is bound.
  E site – Which bind diacylated tRNA to be release from ribosome.
- **Elongation**
  – **Amino acyl tRNA binding:** Acylated tRNA corresponding to 2nd codon binds to A site as ribosome in presence of elongation factor, using GTP.
  – **Peptide bond formation :** Formation of peptide bond between C terminal of amino acyl tRNA at **P** site and amino acyl tRNA at **A site.**
  – **Translocation :** Deacyclated tRNA moves from P → E site.
  – The cycle is repeated till all the codon have been exposed to A-site.
- **Termination**
  – When one of 3 termination/stop codons [UAG, UAA and UGA] are positioned in A site, for which there are no tRNA, termination of translation occurs in presence of release factor (RF) which causes cleaving of bond between polypeptide and tRNA in P site.
  – Polypeptide leaves ribosome followed by mRNA and now tRNA ready to start next cyle of translation.

REGULATION OF GENE EXPRESSION

- Gene expression is the mechanism at the molecular level by which a gene is able to express itself in the phenotype of an organisms.
- One of the best studied example of gene expression is *Lac* **operon.**
- Model of lac operon was suggested by **Jacob** and **Monod** in 1961.
- In *lac* operon, a polycistronic structural gene is regulated by a common **promoter & regulatory genes.**
- It consists of **one regulatory gene (the *i* gene)** & **three structural genes (*Z, Y & A*).**
- The i gene codes for the repressor of the *lac* operon.
- Structural genes produces 3 enzymes for the degradation of lactose to glucose & galactose.

- *Z* produces β-galactosidase for splitting lactose into glucose & galactose.
- Y-produces β-galactoside permease (membrane bound proteins) which is required in entry of the lactose/galactose.
- *A* produces β-galactoside transacetylase enzyme that transfers an acetyl group from acetyl CoA to β-galactosides.

Mechanism of Regulation of *Lac* Operon

- Few molecules of β-galactosidase in cell before induction converts lactose to allolactose which turns on transcription of these three genes in the *lac* operon.
- Allolactose is an inducer.
- The repressor of operon is synthesized from I gene. The repressor protein binds to operator region of the operon and prevents RNA polymerase from transcribing the operon.

HUMAN GENOME PROJECT (HGP)

- HGP established whole genome of human with the help of genetic engineering techniques, cloning techniques and bioinformatics.
  The goal of projects are :
  (i) To develop ways of mapping the human genome at increasing fine level of expression.
  (ii) To store this information in database and develop tools for data analysis.
  (iii) To address the ethical, legal and social issues [ELSI] that may arise from the project.

Implications of Human Genome Project [HGP]

- Prepared sequencing more than 1200 genes that are responsible for common cardiovascular ailments, endocrine disease like diabetes, neurological disorder like Alzheimer, deadly coma and many more.
- HGP holds a key to design drugs, genetically modified diets and finding our genetic identity.
- Many non-human model organisms such as bacteria, yeast, *Caenorhabolitis elegans* (a free living non pathogenie nematode), *Drosophila* (fruit fly), plants (rice and *Arabidopies*) have also been sequenced.

DNA FINGERPRINTING

- The technique of DNA fingerprinting was processed and reflected by British geneticist **Dr. Alec Jeffrey** in 1984.

Principle of DNA Fingerprinting

- Every individual organism is unique. DNA sequences has some specific regions in DNA called **repetitive DNA** → which are short stretch repeated many times.
- These short nucleotide repeats vary in number from person to person and are called **variable number of tandem repeat (VNTR)**. The VNTR of 2 persons may be of same length and sequence at certain sites, but vary at others.
- Pattern of VNTR probe hybridisation in autoradiogram gives many bands of different size.
- These pattern differ from individual to individual except in case of monozygotic twins.

Techniques for DNA Fingerprinting

- DNA is extracted from cells like blood/semen/skin/follicle root of hair in a high speed refrigerated centrifugal.
- Amplification of DNA by PCR.
- Restriction digestion of sample DNA.
- Electrophoresis through gel (agarose polymer).
- Visualization of DNA fragments by staining with dye and observation under UV lamps.
- Double stranded DNA is split into ssDNA using alkaline chemicals.
- Separated DNA sequences are transferred to nylon or nitrocellulose sheet over gel. This is called **southern blotting.**
- Nylon sheet is immersed into bath having probe/marker.
- Probe/marker are radioactive, synthetic DNA fragments of known sequences are added.
- The probe targets a specific nucleotide sequences which are complementary to VNTR sequences and hybridize there.
- X-ray film is exposed to nylon sheet containing radioactive probe. Dark band appear at probe sites which are scanned using scanner.

Application of DNA Fingerprinting

- Identification of criminals.
- Determination of paternity in disputes/legal cases.
- Verification of migrants from one country to another.

Exercise

1. A nucleotide is formed of
   (a) purine, pyrimidine and phosphate
   (b) purine, sugar and phosphate
   (c) nitrogenous bases, sugar and phosphate
   (d) pyrimidine, sugar and phosphate
2. The process of translation is
   (a) ribosome synthesis
   (b) protein synthesis
   (c) DNA synthesis
   (d) RNA synthesis
3. Transformation experiment was first performed on which bacteria?
   (a) *E.coli*
   (b) *Diplococcus pneumoniae*
   (c) *Salmonella*
   (d) *Pasteurella pestis*
4. Which one of the following pairs of nitrogenous bases of nucleic acids, is wrongly matched with the category mentioned against it?
   (a) Thymine, Uracil - Pyrimidines
   (b) Uracil, Cytosine - Pyrimidines
   (c) Guanine, Adenine - Purines
   (d) Adenine, Thymine - Purines
5. Duplication of DNA is called
   (a) replication (b) transduction
   (c) transcription (d) translation
6. Ligase helps in
   (a) removal of few genes
   (b) translation
   (c) inserting few genes in DNA
   (d) bringing transversion in chromosomes
7. DNA is found primarily
   (a) in cell nucleus
   (b) outside the cell nucleus
   (c) in cell cytoplasm
   (d) none of these
8. The one which is capable of self replication is
   (a) DNA (b) RNA
   (c) enzyme (d) protein
9. Replication of DNA is in
   (a) $3' \rightarrow 5'$ direction
   (b) $2' \rightarrow 5'$ direction
   (c) Both $3' \rightarrow 5'$ and $5' \rightarrow 3'$ direction
   (d) none of these
10. Purines of DNA are represented by
    (a) uracil and thymine
    (b) guanine and adenine
    (c) uracil and cytosine
    (d) thymine and cytosine
11. A nucleoside differs from a nucleotide in not having
    (a) phosphate
    (b) sugar
    (c) nitrogen base
    (d) phosphate and sugar
12. What is the type of coiling in DNA ?
    (a) right-handed (b) left-handed
    (c) zig-zag (d) opposite
13. 'Operon model' for gene regulation in bacteria was proposed by
    (a) Jacob and Monad (b) Barry Commoner
    (c) Crick (d) Watson and Crick
14. What is the correct sequence of processes involved in central dogma?
    (a) Replication, transcription, translation
    (b) Replication, translation, transcription
    (c) Translation, replication, transcription
    (d) Transcription, replication, translation
15. Which amino acids are present in histones?
    (a) Lysine and histidine
    (b) Valine and histidine
    (c) Arginine and lysine
    (d) Arginine and histidine
16. DNA replication is
    (a) conservative and discontinuous
    (b) semiconservative and semidiscontinuous
    (c) semiconservative and discontinuous
    (d) conservative
17. The process of transfer of genetic information from DNA to RNA/formation of RNA from DNA is
    (a) transversion (b) transcription
    (c) translation (d) translocation
18. *Escherichia coli* fully labelled with ^{15}N is allowed to grow in ^{14}N medium. The two strands of DNA molecule of the first generation bacteria have
    (a) different density and do not resemble parent DNA
    (b) different density but resemble parent DNA
    (c) same density and resemble parent DNA
    (d) same density but do not resemble parent DNA

19. Nucleotide arrangement in DNA can be seen by
(a) X-ray crystallography
(b) electron microscope
(c) ultracentrifuge
(d) light microscope

20. During DNA replication, the strands separate by
(a) DNA polymerase
(b) topoisomerase
(c) unwindase/helicase
(d) gyrase

21. Because most of the amino acids are represented by more than one codon, the genetic code is
(a) overlapping (b) wobbling
(c) degenerate (d) generate

22. Who proved that DNA is basic genetic material?
(a) Griffith
(b) Watson
(c) Boveri and Sutton
(d) Hershey and Chase

23. Which is not involved in protein synthesis ?
(a) Transcription (b) Initiation
(c) Elongation (d) Termination

24. DNA template sequence of CTGATAGC is transcribed over mRNA as
(a) GUCTUTCG (b) GACUAUCG
(c) GAUTATUG (d) UACTATCU

25. In *Escherichia* coli, *lac* operon is induced by
(a) lactose
(b) promoter gene
(c) β -galactosidase
(d) I-gene

26. Reverse transcriptase is
(a) RNA dependent RNA polymerase
(b) DNA dependent RNA polymerase
(c) DNA dependent DNA polymerase
(d) RNA dependent DNA polymerase

27. In split genes, the coding sequences are called
(a) introns (b) operons
(c) exons (d) cistrons

28. An environmental agent, which triggers transcription from an operon, is a
(a) regulator
(b) inducer
(c) depressor
(d) controlling element

29. The *lac* operon is an example of
(a) arabinose operon
(b) inducible operon
(c) repressible operon
(d) overlapping genes

30. Okazaki fragments are seen during
(a) transcription (b) translation
(c) replication (d) transduction

31. The basis for DNA fingerprinting is
(a) occurrence of restriction fragment length polymorphism (RFLP)
(b) phenotypic differences between individuals
(c) availability of cloned DNA
(d) knowledge of human karyotype

32. Three codons causing chain termination are
(a) TAG, TAA, TGA
(b) GAT, AAT, AGT
(c) AGT, TAG, UGA
(d) UAG, UGA, UAA

33. A mutation at one base of the first codon, of a gene, produces a non-functional protein. Such a mutation is called
(a) nonsense mutation
(b) missense mutation
(c) frameshift mutation
(d) reverse mutation

34. Different mutations referrable to the same locus of a chromosome give rise to
(a) multiple alleles (b) pseudoalleles
(c) polygenes (d) oncogenes

35. The eukaryotic genome differs from the prokaryotic genome because
(a) genes in the former case are organized into operons
(b) the DNA is complexed with histones in prokaryotes
(c) repetitive sequences are present in eukaryotes
(d) the DNA is circular and single stranded in prokaryotes

36. In which direction m-RNA is synthesised on DNA template ?
(a) $5'-3'$ (b) $3'-5'$
(c) Both (a) and (b) (d) Any

37. In a mutational event, when adenine is replaced by guanine, it is a case of
(a) frame shift mutation
(b) transcription
(c) transition
(d) transversion

38. The okazaki fragments in DNA chain growth
(a) polymerize in the 3′ to 5′ direction and forms replication fork
(b) prove semi-conservative nature of DNA replication
(c) polymerize in the 5′ to 3′ direction and explain 3′ to 5′ DNA replication
(d) result in transcription

39. In the DNA molecule
(a) the total amount of purine nucleotides and pyrimidine nucleotides is not always equal
(b) there are two strands which run parallel in the $5' \rightarrow 3'$ direction
(c) the proportion of adenine in relation to thymine varies with the organism
(d) there are two strands which run antiparallel one in $5' \rightarrow 3'$ direction and other in $3' \rightarrow 5'$

40. What is **not** true for genetic code?
(a) It is nearly universal
(b) It is degenerate
(c) It is unambiguous
(d) A codon in mRNA is read in a non contiguous fashion

41. Removal of introns and joining the exons in a defined order in a transcription unit is called
(a) tailing (b) transformation
(c) capping (d) splicing

42. The enzyme used in polymerase chain reaction is
(a) Taq - polymerase (b) RNA polymerase
(c) ribonuclease (d) endouclease

43. DNA probe is used for
(a) DNA finger printing
(b) detection of pathogenic bacteria
(c) medical genetics to find out whether a person carries a particular gene or not
(d) all of the above

44. Tryptophan (Trp) operon is
(a) inducible system
(b) repressible system
(c) controlled by regulator genes
(d) controlled by three structural genes

45. An amino acid binds to tRNA at the
(a) 5' end (b) 3' end
(c) anticodon (d) any of these

46. Genetic information in a DNA molecule is coded in the
(a) number of bases
(b) sequence of nucleotides
(c) length of DNA
(d) number of nucleosides

47. Antiparallel strands of a DNA molecule means that
(a) The phosphate groups at the start of two DNA stands are in opposite position (pole)
(b) One strand turns clockwise
(c) One strands turns anti-clockwise
(d) The phosphate groups of two DNA strands, at their ends, share the same postion

48. One turn of the helix in a B-form DNA is approximately
(a) 3.4 nm (b) 2 nm
(c) 20 nm (d) 0.34 nm

49. True replication of DNA is possible due to
(a) hydroden bonding
(b) phosphate backbone
(c) complementary base pairing rule
(d) none of these

50. The part of DNA molecule that varies among DNA molecule is
(a) phosphate molecule
(b) nitrogen base
(c) sugar molecule
(d) all of these

51. Mode of DNA replication in *E. coli* is
(a) conservative and unidirectional
(b) semi conservative and unidirectional
(c) conservative and bidrectional
(d) semi conservative and bidirectional

52. When DNA replication starts
(a) the leading strand produces Okazaki fragments
(b) the hydrogen bonds between the nucleotides of two strands break
(c) the phosphodiester bonds between the adjacent nucleotides break
(d) the bonds between the nitrogen base and deoxyribose sugar break

53. What is false about *t* RNA?
(a) It binds with an amino acid at its $5'$ end
(b) It has five double stranded regions
(c) It has a codon at one end which recognizes the anticodon on mesenger RNA
(d) It looks like clover leaf in the three dimensional structure

54. Which from of RNA has a structure resembling clover leaf?
(a) *m* RNA (b) *t* RNA
(c) *r* RNA (d) *hn* RNA

55. Which is the initial step in *m*-RNA maturation process?
(a) polyadenylation
(b) $5'$ capping
(c) splicing
(d) endonucleolytic cleavage

56. In Operon concept, regulator gene functions as
(a) repressor (b) regulator
(c) inhibitor (d) all of these

57. Operon is
(a) a set of closely linked genes regulating a metabolic pathways in prokaryotes
(b) the sequence of three nitrogen bases determing a single amino acid
(c) the sequence of nitrogen bases in *m*RNA which codes for a single amino acid
(d) a gene responsible for switching on or off other genes

58. Terminator gene
(a) helps in terminating flowering
(b) helps in terminating seed germination
(c) used in hybridisation
(d) none of these

59. Anticodon is
(a) paired triplet of bases on *m*RNA
(b) unpaired triplet of bases on *r*RNA
(c) paired triplet of bases on *r*RNA
(d) an unpaired triplet of bases in an exposed position of *t*RNA

60. A triplet codon means
(a) a sequence of three nitrogen bases on *m*RNA
(b) a sequence of three nitrogen bases in *t*RNA
(c) a sequence of three nitrogen bases in *r*RNA
(d) the presence of only three nitrogen bases in *m*RNA

CHAPTER 29

Evolution

CONCEPTS

Evolution is the formation of newer types of organisms from the pre-existing ones through modification. Evolution is therefore, often called descent with modification.

ORIGIN OF LIFE

- Origin of life is the process by which living organisms developed from inanimate matter,
- Clouds of dust and gases, chiefly **helium** and **hydrogen**, spread through the space.
- Enormous heat and pressure were produced by condensation due to which thermonuclear reaction initiated and the main condensed mass were converted into sun.
- Earth originated 4600 million years ago.
- Life appeared 3.7 billion years ago. This is indicated by the discovery of microfossils of cyanobacteria like organisms.

Theories for Origin of Life

- **Theory of spontaneous generation (Abiogenesis/autogenesis)**
  - Life originated from non-living things in spontaneous manner.
  - Given by **Aristotle**.
- **Theory of Biogenesis**
  - Life originated from pre-existing life.
  - Given by **Francisco Redi**.
  - Primitive atmosphere was reducing atmosphere because hydrogen atoms combined with all available oxygen atoms to form water, leaving no free oxygen.
  - Water & ammonia were probably the first compound molecules of primitive earth.
  - **Chemical evolution** involves the synthesis of simple organic molecules.
  - With slight lowering of the surface temperature of earth, the lighter elements interacted to form water, methane (the first organic compound), ammonia, CO_2, HCN etc.
  - The early compounds interacted & produced simple organic compounds such as sugar, nitrogenous bases, amino acids, glycerol etc. with the help of external sources (UV rays, cosmic rays, lightning etc.) acting on the mixtures.
  - **Oparin – Haldane** theory was supported experimentally by **Miller-Urey** experiment in 1953 with the help of simulation experiments. Miller & Urey tried to create conditions existing in primitive atmosphere – in **Spark Discharge Apparatus** – a large flask containing mixture of CH_4, NH_3, H_2 and H_2O with electric source and boiling of water for a week.
  - **Findings of experiment:** Miller observed dark condensed liquid which was analysed. Analysis reports concluded, that it was a mixture of amino acids like alanine, glycine, glutamic acid, aspartic acid, valine and leucine and number of other organic compounds like HCN, aldehyde and cyanocompounds.

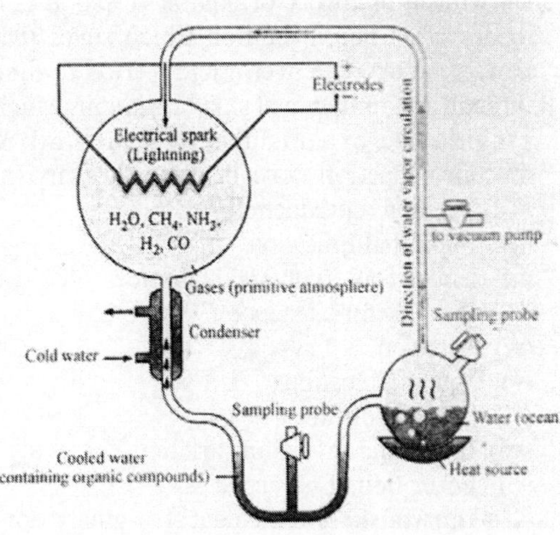

Fig. Diagrammatic representation of Miller's experiment

- Free atoms
  ↓
- Origin of simple inorganic compounds
  ↓
- Origin of early organic compounds
  ↓
- Origin of simple organic compounds
  ↓
- Origin of complex organic compounds
  ↓
- Origin of coacervates
  ↓
- Eobiont (proto cells)
  ↓
- Prokaryotes
  ↓
- Eukaryotes

Flow chart : Main steps in origin of life according to modern theory of origin of life

- Change in organism with time is **organic evolution**. It states that present day complex organism have originated during course of time from earlier simple form of life by process of gradual changes occuring in sequential manner.

THEORIES OF EVOLUTION

• Darwinism

- Darwinism is the term coined for the explanation offered by **Darwin** for the origin of species by **Natural selection** in 1858.
- His findings were based on **Malthus** essays on the 'principles of population'. He said that food increases in arithmetic progression while number of animals increases in geometric progression.
- Darwinism or theory of natural selection is a theory of organic evolution which states that new species evolve over a long period of time through accumulation of small variations which provide the organism with structural & functional superiority over others in their survival & differential reproduction.
  (i) Rapid multiplication
  (ii) Limited environmental resources
  (iii) Struggle for existence
  (iv) Variation
  (v) Natural selection
  (vi) Survival of fittest
  (vii) Inheritance of useful variation
  (viii) Formation of new species
- **Neo-Darwinism** is refinement of original theory of natural selection to remove objections.
- According to neo-Darwinism both mutation & natural selection are responsible for evolution.

Genetic Equilibrium (Hardy Weinberg Law)

- Law states that "both gene frequency and genotype frequency" will remain constant from generation to generation in an infinitely large interbreeding population in which mating is random and no selection, migration or mutation occurs.
- **Hardy Weinberg formula or binomial expression is given as**
  $p^2 + 2pq + q^2 = 1$ for 2 alleles A and a.
  where, p = gene frequency of allele A.
  q = gene frequency of allele a.
  p^2 = genotype frequency of genotype AA.
  q^2 = genotype frequency of genotype aa.
  pq = genotype frequency of genotype Aa.
- **Significance of Hardy Weinberg Law :** Hardy Weinberg principle gives the geneticists a tool to determine when evolution is occurring.
- It shows that any change in genotype due to mutation, genetic drift, gene flow, assortative mating, natural selection lead to change in gene and genotypic frequency which cause evolution leading to speciation.
- **Genetic drift** refers to chance elimination of gene(s) of certain traits when a section of population migrates or dies of natural calamity. It eliminates certain alleles and fixes other alleles.
- Genetic drift in a new colony is called **founder effect**.
- Genetic drift always influences frequencies of alleles & is inversely proportional to the size of population.
- Natural calamity like earthquake greatly reduces the size of population, killing the individuals randomly. The genetic pool of surviving population decreases. This condition of reduced genetic variability is called **bottleneck effect**.

EVIDENCES OF EVOLUTION

- There are many types of evidences which supports the process of organic evolution.

Types of Evolution

(i) Convergent evolution - The characters of different animals become similar due to similarity of habitat *e.g.* - Fish & whale; bat & birds.

(ii) Divergent evolution - The characters of related organisms become different due to difference in habitat *e.g.* - Whale and Bat.

(iii) Parallel evolution - The character difference remains unchanged due to similar environmental pressure.*e.g.* - Horse, zebra and deer.

Vestigial Organs

- The useless and functionless degenerate structures which were large and functional in some other animals.
- They are rudimentary, relatively smaller in size and do not function.
- Examples : Vermiform appendix, coccyx, pinna muscle, wisdom tooth in man, rudiments of hindlimbs in python and boa.

Atavism

- The sudden reappearance of those ancestral characteristics in an organism or in the organisms of a group, which do not occur normally.
- Eg.: Cervical fistula, tail, hairs on the body & face etc.

TABLE : GEOLOGICAL TIME SCALE (TO BE READ FROM BELOW TO UPWARDS)

Era	Period Age in million years	Epoch Age in million years	Some Important Fauna (Animals) & Flora (Plants)
Coenozoic (Era of Modern life)	Quarternary	Recent (Holocene) 0.01	Modern man dominant, modern mammals, bird, fishes, insects. Rise of herbaceous plants, decline of woody plants.
		Pleistocene 1.8	Ice ages, humans appear, Evolution of human society and culture.
	Tertiary	Pliocene (Age of mammals) 5	Ape-like ancestors of humans appear, Adaptive radiation of flowering plants, origin of man
		Miocene 23	Continued radiation of mammals and angiosperms.
		Oligocene 34	Origin of most modern mammalian orders including apes.
		Eocene 57	Angiosperm dominance increases, further increase in mammalian diversity, origin of horse.
		Palaeocene 65	Major radiation of mammals, birds and pollinating insects primitive primates appear.
Mesozoic (Era of Medieval life)	Cretaceous 144		Extinction of dinosaurs and toothed birds. Angiosperms (flowering plants) appear.
	Jurassic 208 (Age of Reptiles)		Dinosaurs dominant, origin of toothed birds (first birds), Gymnosperms were dominant.
	Triassic 245		Origin of dinosaurs and mammals.
Palaeozoic (Era of Ancient life)	Permian 285		Extinction of trilobites. Origin of mammal-like reptile and most modern order of insects.
	Carboniferous 360 (Age of Amphibians)		Origin of reptiles, amphibians dominant, extensive forests of vascular plants, first seed plants .
	Devonian 408 (Age of fishes)		Origin of amphibians, diversification of bony fishes, origin of gymnosperms.
	Silurian 438		Origin of jawed fishes, colonization of land by plants and arthropods, origin of vascular plants.
	Ordovician 505 (Age of invertebrates)		Origin of first vertebrates, e.g. Ostracoderms (jawless fishes), Invertebrates and marine algae abundant.
	Cambrian 544		Origin of most invertebrate phyla, origin of trilobites.
Precambrian (From 570 to 4600 million years ago)	570		Diverse algae.
	700		Origin of first animals
	1500		Oldest eukaryotic fossils
	2500		Oxygen begins accumulating in atmosphere
	3500		Oldest definite fossils known (prokaryotes)
	4600		Approximate origin of earth

* some authors divide precambrian into three eras, namely Proterozoic, Archaeozoic and Azoic.

Embryological Evidence

- In 1828, **Von Baer** put forward Baer Law. **E. Haeckel** later called it the **biogenetic law (1868)**. The view was summarised as ontogeny recapitulates phylogeny. Ontogeny is the life history of an organism while phylogeny is the evolutionary history of the race of that organisms.
- **Embryological evidences are**
  - Mosses and ferns are most complex than algae. But protonema of mosses resemble certain green algae. This suggest their evolutionary relationship.
  - Change of a fish like embryo of human to a seventh month human foetus resembling a baby ape to a human baby.

Phylogeny of horse

- Evolutionary sequence of horse is given below –

Eohippus (Eocene period)
↓
Mesohippus (Oligocene period)
↓
Merychippus (Miocene period)
↓
Pliohippus (Pliocene period)
↓
Equus (Pliocene period)

- Merychippus (ruminant horse) marked the completion of transition from browsing to grazing.
- Modern horse, Equus is the direct descendent of pliohippus.

MUTATION THEORY BY HUGO DE VRIES (1901)

- Experiments on evening prime rose ***(Oenothera lamarckiana)***
- Mutation is a heritable change in genetic material which alters the phenotype of an organism.

Salient Features of Mutation Theory

(i) Mutation appears from time to time among organisms of a naturally breeding species or population. The organisms with mutation are called mutants. These mutants are clearly distinct from their parents.
(ii) Mutations are heritable and form new species.
(iii) Mutation arise suddenly in one step *i.e.* new species arise suddenly in one step & not gradually.
(iv) Mutations occur in all possible directions and may be advantageous or disadvantageous.
(v) Unsuitable mutants are destroyed by natural selection.

ADAPTIVE RADIATION AND BIOLOGICAL EVOLUTION

- Adaptive radiation is a special evolutionary pattern, characterized by a rapid increase in number of kinds of closely related species.
- A **good example of adaptive radiation** is found among the **Finches of Galapagos Islands**. Another example is **Australian Marsupials**.
- Biological evolution is the process of change over time, in the heritable characteristics, or traits of a population organisms.
- Natural selection, one of the processes that drives evolution results from the difference in reproductive success between individual.
- Two significant concepts of Dawinian theory are **branching descent & natural selection**.

ORIGIN AND EVOLUTION OF MAN

- Many experts believe that *Australopithecus garhi* or a similar species gave rise to the genus *Homo*.
- Early hominids–members of the genus *Homo*–lived contemporaneously with *australopithecines* for perhaps a half million years.
- Two major changes accompanied the evolution of *Homo* from *Australopithecus*:
  - An increase in body size
  - A doubling of brain size
- The oldest fossil remains of a member of the genus *Homo* were discovered in Tanzania. It was named *H. habilis*.
- Primitive tools were found with the fossils.
- Members of *H. habilis* likely used these tools to dig into the hard, dry soils to exploit food resources.
- *Homo erectus* is the only other known extinct species of the genus *Homo*.
- *H. erectus* may have exterminated *H. habilis*.
- Members of *H. erectus* used fire for cooking and for hunting large animals.
- They made tools that were probably used for digging, capturing animals, cleaning and cutting meat, scraping hides, and cutting wood.
- *H. erectus* was replaced in tropical regions by *Homo sapiens* about 200,000 years ago.

Brains Steadily Became Larger

- Evolutionary trends that were present in the transition from *Australopithecus* to *H. erectus* continued during the evolution of *H. sapiens*.
- Larger brains evolved as social lives became increasingly complex.

- Several types of *H. sapiens* existed during the mid-Pleistocene epoch, all of which were skilled hunters of large mammals.
- During this time, a distinctly human trait emerged: rituals and a concept of life after death.
- Neanderthals were a group of *H. sapiens* that were widespread in Europe and Asia between 75,000 and 30,000 years ago. They were stocky, powerfully built humans with massive skulls and large brains. They hunted large mammals and manufactured a variety of tools.
- Many scientists believe that they were exterminated by a more modern form of *H. sapiens*, known as the Cro-Magnons.
- Cro-Magnon people made sophisticated tools and created paintings of large mammals. Cro-Magnon people spread across Asia and reached North America perhaps as early as 20,000 years ago.

GENE MIGRATION

Members of a population might enter a new population and interbreed there. They will add new allels to the local gene pool of the host population. This is called gene migration. (It may be in case of inter-specific fertile cross).

Types of Selection

(i) **Stabilizing Selection :** Intermediate characteristics are favoured. (Fig. 1)

(ii) **Directional Selection :-** One of the two possible extremes is favoured. (Fig. 2)

(iii) **Disruptive Selection :-** Natural selection moves characteristics toward both extremes, and intermediate phenotypes become rarest. (Fig. 3)

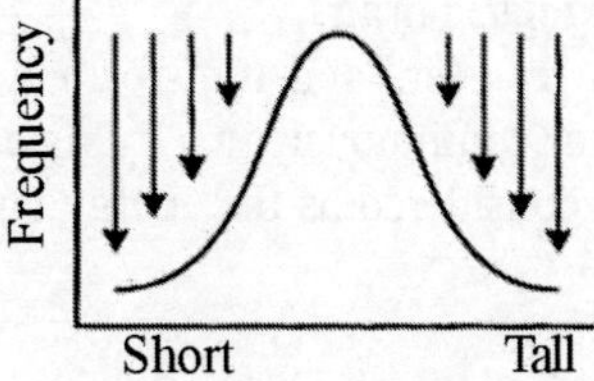

Fig. 1 : Height Variation in Humans

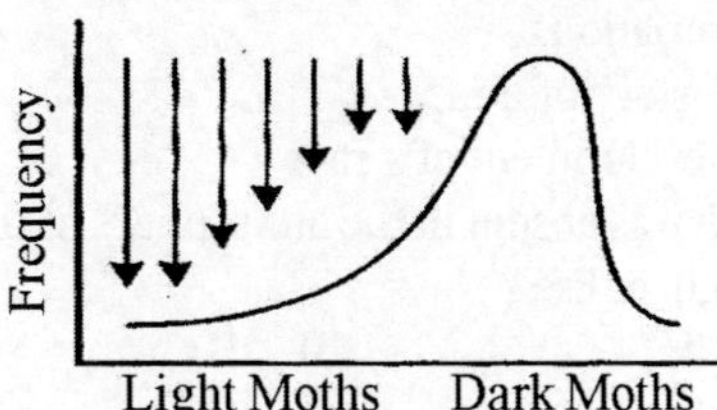

Fig. 2 : Industrial Melanism

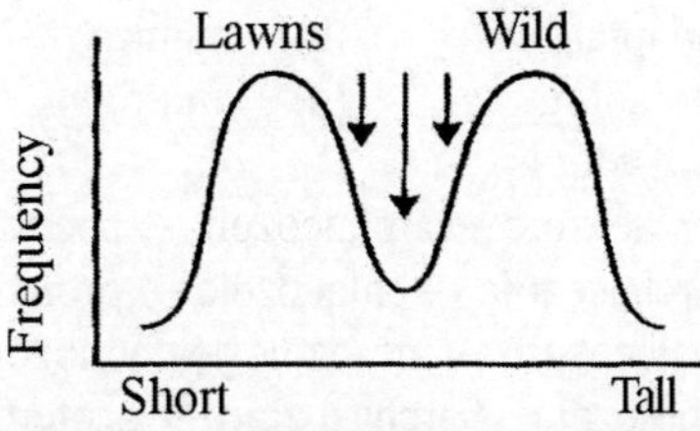

Fig. 3 : Height variation in weeds of lawns and in the wild

Exercise

1. Scientific name of man is
   (a) *Canis familiaris* (b) *Homo habilis*
   (c) *Homo erectus* (d) *Homo sapiens*
2. Which of the following changes for man in the course of evolution is probably useless?
   (a) Development of opposable thumb
   (b) Loss of tail
   (c) Development of being erect
   (d) Development of cranial capacity
3. The primitive atmosphere of earth contained water vapours, hydrogen, ammonia and
   (a) CO_2 (b) O_2
   (c) N_2 (d) CH_4
4. Life originated in
   (a) air (b) soil
   (c) mountains (d) water
5. The book 'Origin of Species' is written by
   (a) Aristotle (b) Darwin
   (c) Watson (d) Lamarck
6. Lamarck's theory of evolution is also called
   (a) theory of special creation
   (b) inheritance of acquired characters
   (c) survival of the fittest
   (d) theory of special creation
7. Basic elements of life are
   (a) C, H, O, N (b) C, P, H, S
   (c) H, O, N, S (d) P, C, O, N

8. Which of the following statements is incorrect?
(a) The living and non-living are made up of the same elements
(b) Life originated in air
(c) Energy transfers and transformations do not occur continuously in a living cell
(d) ATP is considered as the energy currcncy of a cell

9. Evolution is
(a) progressive development of a race
(b) history and development of a race alongwith variations
(c) history of a race
(d) development of a race

10. Which was absent in the atmosphere at the time of origin of life ?
(a) NH_3 (b) H_2
(c) O_2 (d) CH_4

11. Phenomenon of organisms resembling others for escaping from enemies is
(a) adaptation (b) mimicry
(c) homology (d) analogy

12. Correct order is
(a) palaeozoic → archaeozoic → coenozoic
(b) archaeozoic → palaeozoic → proterozoic
(c) palaeozoic → mesozoic → coenozoic
(d) mesozoic → archaeozoic → proterozoic

13. Evolutionary convergence is development of
(a) common set of characters in group of different ancestory
(b) dissimilar characters in closely related groups
(c) common set of characters in closely related groups
(d) random mating

14. Theory of natural selection dwells on
(a) role of environment in evolution
(b) natural selection acting on favourable variations
(c) changes in gene complex resulting in heritable variations
(d) none of these

15. Genetic drift is change of
(a) gene frequency in same generation
(b) appearance of recessive genes
(c) gene frequency from one generation to next
(d) none of the above

16. Frequency of a character increases when it is
(a) recessive (b) dominant
(c) inheritable (d) adaptable

17. Homologous organs are
(a) wings of insects and bat
(b) gills of fish and lungs of rabbit
(c) pectoral fins of fish and fore limbs of horse
(d) wings of grosshopper and crow

18. The homologous organs are those that show similarity in
(a) size (b) origin
(c) function (d) appearance

19. The kind of evolution in which two species of different geneology come to resemble one another closely, is termed as
(a) progressive evolution
(b) convergent evolution
(c) parallel evolution
(d) retrogressive evolution

20. Which one of the following is regarded as the direct ancestor of modern man ?
(a) *Homo erectus* (b) *Ramapithecus*
(c) *Homo habilis* (d) *Australopithecus*

21. Common origin of man and chimpanzee is best shown by
(a) banding pattern in chromosomes number 3 and 6
(b) cranial capacity
(c) binocular vision
(d) dental formula

22. Genetic drift operates only in
(a) smaller populations
(b) larger populations
(c) mendelian populations
(d) island populations

23. Life cannot originate from inorganic materials at present because of
(a) a very high amount of oxygen in the atmosphere
(b) very low atmospheric temperature
(c) absence of raw materials
(d) high degree of environmental pollution

24. Which of the following primate is the closest relative of humans?
(a) Rhesus monkey (b) Orangutan
(c) Gorilla (d) Gibbon

25. Which one of the following features is closely related with the evolution of humans?
(a) Loss of tail
(b) Shortening of jaws
(c) Binocular vision
(d) Flat nails

26. Which of the following is correct order of evolutionary history of man?
(a) Peking man, *Homo sapiens*, Neanderthal, Cro-magnon
(b) Peking, Neanderthal, *Homo sapiens*, Cro-magnon
(c) Peking man, Heidelberg man, Neanderthal, Cro-magnon
(d) Peking man, Neanderthel, *Homo sapiens*, Heidelberg man

27. Industrial melanism is an example of
(a) defensive adaptation of skin against ultraviolet radiations
(b) drug resistance
(c) darkening of skin due to smoke from industries
(d) protective resemblance with the surroundings

28. In a random mating population in equilibrium, which of the following brings about a change in gene frequency in a non-directional manner?
(a) Migration (b) Mutations
(c) Random drift (d) Selection

29. Darwin in his 'Natural Selection Theory' did not believe in any role of which one of the following in organic evolution ?
(a) Discontinuous variations
(b) Parasites and predators as natural enemies
(c) Survival of the fittest
(d) Struggle for existence

30. Presence of gills in the tadpole of frog indicates that
(a) fishes were amphibious in the past
(b) fishes evolved from frog-like ancestors
(c) frogs will have gills in future
(d) frogs evolved from gilled ancestors

31. An important evidence in favour of organic evolution is the occurrence of
(a) analogous and vestigial organs
(b) homologous organs only
(c) homologous and analogous organs
(d) homologous and vestigial organs

32. Among the human ancestors the brain size was more that 1000 cc in
(a) *Homo erectus*
(b) *Ramapithecus*
(c) *Homo habilis*
(d) *Homo neanderthalensis*

33. Which one of the following statement is correct?
(a) There is no evidence of the existence of gills during embryogenesis of mammals
(b) All plant and animal cells are totipotent
(c) Ontogeny repeats phylogeny
(d) Stem cells are specialize cells.

34. The concept of chemical evolution is based on
(a) interaction of water, air and clay under intense heat
(b) effect of solar radiation on chemicals
(c) possible origin of life by combination of chemicals under suitable environmental conditions
(d) crystallization of chemicals

35. One of the important consequences of geographical isolation is
(a) preventing speciation
(b) speciation through reproductive isolation
(c) random creation of new species
(d) no change in the isolated fauna

36. What was the most significant trend in evolution of modern man (*Homo sapiens*) from his ancestors ?
(a) Upright posture
(b) Shortening of jaws
(c) Binocular vision
(d) Increasing brain capacity

37. Darwin's theory of natural selection was based on
(a) inheritance of acquired characters
(b) mutation
(c) enormous rate of reproduction in organisms, struggle for existence and survival of the fittest
(d) changes due to the use and disuse of organ.

38. There would be no evolution if
(a) the inheritance of acquired characters did not take place
(b) somatic variations were not inheritable
(c) genetic variations were not found among members of population
(d) somatic variations would not transform into germinal variations

39. If population of a species is transferred to more suitable environment then it will show
(a) protection against enemies
(b) more individuals would survive
(c) rate of reproduction increases
(d) unlimited food would be available

40. Snakes do not have legs because
(a) legs are lost during their entry in tunnels
(b) legs are lost during evolution
(c) the ancestors of reptiles did not have legs
(d) there are no legs in lizards

41. Ancestor of man who first time showed bipedal movement
(a) Cro-magnon (b) *Australopithecus*
(c) Java apeman (d) Peking man

42. *Homo erectus* is biological name of
(a) Modern man
(b) Neanderthal man
(c) Java man
(d) Peking and java man

43. *Homo erectus* differed from cro-magnon man in having
(a) jaws protruding out
(b) tool making
(c) sloping jaws
(d) arts and paintings

44. The first ancestor of man whose fossils have been discovered was
(a) *Pithecanthropus* (b) *Zinjanthropus*
(c) *Australopithecus* (d) Neanderthal man

45. What was the cranial capacity of java man ?
(a) 400 cc (b) 650 cc
(c) 900 cc (d) 1450 cc

46. Evolution of man was possible because our ape like ancestors
(a) showed bipedal movement on open land
(b) used fire
(c) felt difficulty in nutrition
(d) developed community hunting

47. Cranial capacity of cro-magnon man was
(a) 900 cc (b) 1075 cc
(c) 1450 cc (d) 1600 cc

48. Greatest advantage of bipedal movement is
(a) fore arms becoming free for carrying out order of brain
(b) greater speed
(c) support the body properly
(d) loss of weight

49. Which fossil man had cranial capacity almost equal to modern man ?
(a) *Australopithecus* (b) Java ape man
(c) Neanderthal man (d) Peking man

50. Speciation occurs when
(a) populations are isolated by geographical barrier
(b) populations are reproductively isolated
(c) both of these
(d) none of these

51. Struggle for existence and survival of fittest are related to
(a) Lamarckism (b) Darwinism
(c) Oparin's theory (d) Mendelism

52. Gene frequency in a population remains stable unless and until there is
(a) random drift (b) random mating
(c) selection (d) mutation

53. Which is most important in speciation ?
(a) Geographical isolation
(b) Reproductive isolation
(c) Ecological isolation
(d) Ethological isolation

54. Successful adaptation simply mean
(a) an increase in fitness
(b) producing offsprings
(c) moving to a new place
(d) evolving new characters

55. Theory of Pangenesis was proposed by
(a) Lamarck (b) Darwin
(c) Hugo de Vries (d) Dobzhansky

56. The key point of Lamarck's theory was or Lamarck's theory on evolution is stated as
(a) inheritance of acquired characters
(b) natural selection
(c) survival of fittest
(d) decent with change

57. One major criticism against Darwin's theory of evolution is that
(a) it does not explain presence of vestigial organs found in many mammals
(b) it over estimates the reproductive capacity of organism
(c) it does not takes into account variation with heredity
(d) none

58. Process of evolution is
(a) discontinuous
(b) discontinuous in the past but continuous presently
(c) continuous in the past but discontinuous presently
(d) continuous

59. Which is true about gene mutation?
(a) A mutation that is detrimental in one environment may prove to be beneficial in another environment
(b) Mutation arises when needed
(c) Process of mutation is not affected by low intensity X-rays
(d) Mutation is always disadvantageous

60. The sum of all genes held by the members of population is called
(a) genotype (b) gene pool
(c) gene flow (d) genome

CHAPTER 30

Health and Disease

CONCEPTS

- **Health** is a state of complete physical, social and mental well being and not merely the absence of disease or infirmity.
- **Disease** is any condition which interferes with the normal structure and function of the body that is manifested by a characteristics sets of symptoms and sign.
- Agents that cause disease are called pathogens.
- Pathogens can enter one body by various means, multiply and interfere with normal vital activities resulting in morphological and functional damage.
- Balanced diet, personal hygiene & regular exercise are very important to maintain good health.

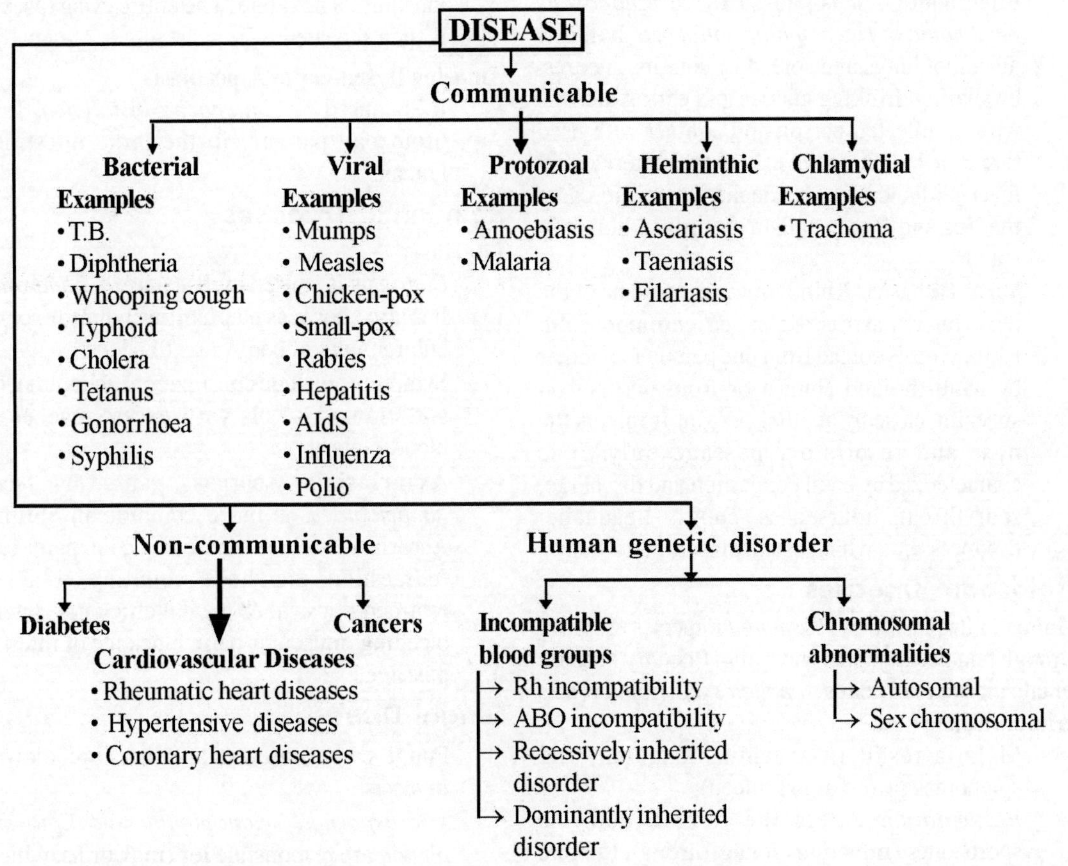

COMMON HUMAN DISEASES

Diseases can be broadly classified into two categories:

(a) **Congenital diseases** (inborn diseases) are present from the birth of the individual and are heritable. These may be due to some metabolic disorder or mutations (gene or chromosomal).

(b) **Acquired diseases** are those which the individuals acquire or develop after birth and are non-inheritable. Such diseases may be communicable (or infectious) or non-communicable.

Bacterial Diseases

- **Typhoid :** Typhoid is caused by *Salmonella typhi*. Typhoid spreads through food, milk and water contaminated with intestinal discharges either directly or through flies and personal hygiene. Symptoms are high fever (39° to 40° C) headache, weakness, loss of apetite, skin rash, constipation, stomach, ache, etc. Typhoid is diagonsed with **widal test**.
- **Pneumonia :** It is caused by *Streptococcus pneumonia* or *Haemophilus influenza*. It infects alveoli of lungs and spread by coughs, sneezes, by sharing drinking glasses and eating utensils with an infected person and contact with used tissue or handkerchiefs. It is characterized by fever chills, cough, headache. In severe cases the lips and finger nails may turn to bluish in colour.
- **Viral Diseases** : Rhinoviruses causes one of the most infectious disease called **common cold.** Rhinovirus is spread from one person to another by hand to hand contact or from one person sneezing close by another person. It infects the nose and respiratory passage only. It is characterized by nasal congestion and discharge, soar throat, hoarseness, cough, headache, tiredness, etc., which usually last for 3 to 7 days.

Protozoan Diseases

Malaria : It is caused by *Plasmodium* species & spread through female *Anopheles* mosquito. Primary host are female mosquito of genus *Anopheles* and humans acts as intermediate host.

- Malaria result in anaemia, toxaemia and splenomegaly (enlarged spleen).
- *Plasmodium* enters the human body as sporozoites (**infectious form**) through the bite of infected female *Anopheles* mosquito. *Plasmodium* sporozoites enters the bloodstream and travel to liver where they divide repeatedly & other attack the red blood cells resulting in their rupture.
- The rupturing of red blood cells cause the symptoms of fever and chills (which is recurring every 3 to 4 day), after releasing a toxic substance called **haemozoin**.
- A mosquito biting an infected host at this stage can ingest **merozoites** (formed inside the red blood cells). If this happens, then merozoites enter the mosquito's stomach and become male & female **gametocytes**. This kicks off a sexual reproductive cycle where the separate male & female gametocytes unite together to form a single celled zygote, which grows to become an **oocyst**. This oocyst eventually divides, releasing a multitude of (asexual free-swimming) sporozoites that migrate to the mosquito's head & salivary glands, where they are ready to pass into the human blood stream during the mosquito's next bite. The entire cycle (asexual) is then repeated.

Amoebic Dysentery or Amoebiasis

- It is caused by *Entamoeba histolytica.* It is a protozoan parasite in the large intestine of human.

Helminthic Diseases

Filariasis

- Filariasis is caused by *Wuchereria bancrofti*.
- It is also known as **elephantiasis** due to excessive enlargement of body parts like legs.
- Symptoms include constipation, abdominal pain and cramps, stools with excess mucous and blood clots.
- **Ascariasis :** Ascariasis, caused by *Ascaris lumbricoides*, is more common in children. Ascariasis is characterized by colic pain, fever, indigestion, diarrhoea, vomiting, weakness, retarded physical & mental efficiency, internal bleeding, muscular pain, blockage of intestinal passage.

Fungal Diseases

- Fungi causes diseases and these are known as **mycosis**.
- *Microsporum*, *Trichophyton* and *Epidermophyton* are responsible for ringworms, which is characterized by appearance of dry scaly lesions of the skin, nails and scalps.

Sexually Transmitted Diseases

- Sexually transmitted diseases (STDs) are a group of communicable diseases that are transmitted mainly by sexual contact. STDs are caused by a wide range of bacterial, viral, protozoal and fungal agents.

STDs	Pathogen
Syphilis	*Treponema pallidum* (bacterium)
Gonorrhoea	*Neisseria gonorrhoea* (bacterium)
Vaginitis	*Candida albicans* (fungus)
AIDS	Human immuno deficiency virus (virus)

- **Diabetes mellitus** is caused by the deficiency of insulin hormone and is characterized by excessive concentration of sugar in the blood and urine. **Diabetes insipidus**, on the other hand, is characterized by excessive urination, urine being sugar-free and is caused by the deficiency of ADH.
- **Arthritis** is a degenerative disease in which inflammation of the joints leading to pain occurs. It may be of three types - **rheumatoid, osteoarthritis or gout**. Rheumatoid arthritis is due to inflammation of the synovial membrane, osteoarthritis is due to degeneration of cartilage at the joints, and gout is due to accumulation of uric acid crystals in the synovial joints.

Cancers

- **Cancers** are not inheritable.
- The non-regulated growth of the cells that accompanies cellular transformation produces **tumours or neoplasms**, each tumour being the product of proliferation of a single abnormal cell.
- Tumours are of two types – **benign** and **malignant**.
- **Benign tumour** cells is a large localized mass of abnormal tissue enclosed in connective tissue which does not spread to distant sites.
- **Malignant tumour** cells are cancer cells that spread to and take up residence in neighbouring tissues - a condition called **metastasis**.
- It is believed that all cells carry cancer-producing genes called **proto-oncogenes** which are activated to oncogenes under certain conditions. These oncogenes multiply rapidly and form a malignant tissue.
- Agents that cause cancers are called **carcinogens**. UV radiations, smoking, mustard gas, soot, viruses, coal tar, aflatoxins and industrial pollutants are known to be carcinogenic.
- **Cancer can be detected by**
  - biopsy and histopathological studies of the tissue.
  - blood and bone marrow tests for increased cell counts as in Leukaemia.
  - use of techniques like radiography, magnetic resonance imaging (MRI) and computed tomography (CT) for cancer of internal organs.
  - use of antibodies against cancer-specific antigens.
  - applying principles of molecular biology to detect genes in individuals with inherited susceptibility to certain cancers.

Treatment of Cancer

- Therapy used in the treatment of cancer are surgery, radiotherapy, chemotherapy, immuno-therapy, hormonal therapy, etc.

AIDS

- **AIDS** is caused by Human Immunodeficiency Virus (HIV) which is retrovirus.

Life Cycle of HIV

- The virus after getting into the body of a person, enters the macrophages.
- The RNA replicates and DNA is formed by reverse transcriptase.
- The viral DNA gets incorporated with the host cell DNA and directs the infected cell to produce virus particles.
- The macrophages continue to produce virus particles.
- The virus then enters the helper T-lymphocytes (T_H), replicates and forms progeny viruses.
- The progeny viruses released in the blood attack other helper T-lymphocytes in the body of the infected persons.
- The person becomes easily infected by bacteria like mycobacterium, viruses and even parasites like Toxoplasma.
- The person is unable to protect himself/herself against any infection.
- AIDS is diagnosed by **ELISA** (Enzyme Linked Immunosorbent Assay).

Prevention of AIDS

- National AIDS Control Organisation (NACO) and non-governmental organisation are trying their best to educate people about AIDS.
- World Health Organisation (WHO) has started a number of programmes to prevent spreading HIV infection; some such steps include :
  - ensuring use of disposable needles and syringes.
  - checking blood for HIV.
  - free distributions of condoms and advocating safe sex.
  - controlling drug abuse.
  - promoting regular check-up for HIV in susceptible populations, etc.
- Treatment with anti-retroviral drugs is only partially effective; they can only prolong the life of the patient and cannot prevent death.
- **Genetic counselling** helps people to understand the nature and risk of genetic disease. Counselling includes diagnosis, estimate of the occurrence or recurrence of a particular trait or any genetic defect, and assists in coping with a particular problem. The chief areas where a genetic counsellor is useful include : (i) inherited disorders, (ii) strategies for overcoming sterility or infertility in either parent, (iii) maternal foetus incompatibilities of blood groups, and (iv) the incidence of forms of mental and physical abnormalities in the family.
- **Amniocentesis** is one of the techniques that is used to detect genetic disorders in the developing foetus.
- **Chorionic villus sampling (CVS)** is another method of prenatal diagnosis. Amniocentesis can be performed after 15 - 16 weeks of pregnancy. CVS can be performed 8 weeks earlier in pregnancy than amniocentesis. CVS is thus preferred to amniocentesis as abortion at an early stage can be carried out much more safely if so required.

IMMUNE SYSTEM

- **Defensive cells and proteins**
  - White blood cells play many defensive roles.
  - Immune system proteins bind pathogens or signal other cells.
- **Innate defenses**
  - Barriers and local agents defend the body against invaders.
  - Innate defenses include chemical and cellular processes.
- **Specific Defenses: The Immune Response**
  - Four features characterize the immune response.
  - There are two interactive immune responses.
  - Clonal selection accounts for the characteristic features of the immune response.
  - Immunological memory and immunity result from clonal selection.
  - Animals distinguish self from nonself and tolerate their own antigens.
- **B Cells: The Humoral Immune Response**
  - Some B cells develop into plasma cells.
  - Antibodies share a common structure, but may be of different classes.
  - Hybridomas produce monoclonal antibodies.
- **T Cells: The Cellular Immune Response**
  - T cell receptors are found on two types of T cells.
  - The major histocompatibility complex encodes proteins that present antigens to the immune system.
  - Helper T cells and MHC II proteins contribute to the humoral immune response.
  - Cytotoxic T cells and MHC I proteins contribute to the cellular immune response.
  - MHC molecules underlie the tolerance of self.
  - MHC molecules are responsible for transplant rejection.
- **The Genetic Basis of Antibody Diversity**
  - Antibody diversity results from DNA rearrangement and other mutations.
  - B cell produces a specific heavy chain.
  - The constant region is involved in class switching.
- **Disorders of the Immune System**
  - An inappropriately active immune system can cause problems.
  - AIDS is an immune deficiency disorder.
  - HIV infection and replication occur in T_H cells.
  - Treatments for HIV infection rely on knowledge of its molecular biology.

Immunity

- The term **immunity** refers to the specific resistance exhibited by the host towards infections by micro-organisms (pathogens) and their products.

- Immunity against infectious disease is of two types

Innate or Natural Immunity

- **Innate immunity** is developed in an individual without having the disease or immunization, *e.g.*, recreation of sweat glands contain certain chemical substances which prevent the entry of micro-organisms.
- It is present from birth.
- It is the general defence of body including the following four mechanism —
  (i) *Phagocytosis* of invaders by leucocytes & tissue macrophages (called *cellular barrier*).
  (ii) Resistance of skin to invading micro-organisms (called *physical barrier*).
  (iii) Destruction of micro-organisms swallowed with food by the HCl of gastric juice & by digestive enzymes and tear from eye (called *physiological barriers*).
  (iv) Virus infected cells secrete proteins (called interferons) which protect non-infected cells from further viral infection (*cytokine barriers*)

Acquired Immunity

- It is a third line defence and developed by an animal in response to a disease caused by infections of microbes.
- The resistance against infectious disease that an individual acquires during life is known as **acquired immunity.**
- **Active acquired immunity** is the resistance developed by an individual as a result of an antigenic stimulus (entry of pathogens into the body). It affords durable and effective protection.
- The resistance that is transmitted to recipient in a ready-made form is known as **passive acquired immunity**. There is no active role of immune system of the recipient.

Antibody Mediated Immunity

- B cells produce specialized proteins called antibodies (immunoglobulin) which are glycoproteins. Each immunoglobulin molecule is made up of 4 polypeptide chains, help together in the form of a Y, the tips of the two upper arms bind to the antigens to form antigen-antibody complex. The antibody molecule may remain bound to the cell membrane or may remain free.
- The free antibodies perform three functions : (a) Agglutination of particulate matter. (b) Opsonisation or coating of bacteria to facilitate phagocytosis. (c) Neutralisation of toxins released by bacteria.

Classification of Antibodies :

(a) **Ig A** – Protects from inhaled or ingested pathogens.
(b) **Ig D** – Present on lymphocyte surface as receptors, activation of B cells.
(c) **Ig E** – Mediator in allergic response.
(d) **Ig G** – Stimulation of phagocytes and complement system, passive immunity to foetus.
(e) **Ig M** – Activation of B cells.

Clonal Selection : The T and B cells have a specific receptor on each of them. When the receptor interacts with the antigenic determinant specific to it, the lymphocyte is activated and it divides to form a clone of cells. These cells are transformed into effector cells, *i.e.*, B cells and T cells, This phenomenon is known as clonal selection.

Primary and Secondary Immune Response : The immune response as a result of the first encounter of the individual, with an antigen is called primary immune response. It takes longer to develop. The secondary response is the immune response, produced by memory cells in a subsequent encounter of an animal with the same antigen.

Immunization

- Inoculation of vaccines to prevent diseases is called **immunization**.
- **Vaccines** are preparations of living or killed micro-organisms or their products. Vaccines are of two types – live vaccines and killed vaccines.
- **Genetic vaccines** involve one or more genes from diseases carrying agent (pathogen) and splicing these gene into plasmids (closed rings of DNA). These rings are then delivered into small groups of cell, often by infection into muscle cells or by propulsion into *via* so called '**gene gun**'.

Allergy & Autoimmunity

- Allergy is an important side effect of immunity.
- Autoimmune disease result when the immune system attack and destroys 'self' cells and molecules. Eg., rheumatoid arthritis.

DRUGS AND ALCOHOL ABUSE

Major groups of psychotrophic drugs, their examples and effects

Types of drug	Examples	Effects
1. Sedatives and transquilizers (depressants)	Benzodiazephines (*e.g.* Valium), Barbiturates	(i) Depress brain activity (ii) Produce feelings of calmness, relaxation, and drowsiness
2. Opiate narcotics	Opium, Morphine, Heroin, Penthidine, Methadone	(i) Suppress brain function (ii) Relieve intense pain (iii) Causes loss of weight, sterility and lack of interest in work.
3. Stimulants	Caffeine (very mild), Cocaine, Amphetamines	(i) Stimulates the nervous system (ii) Make a person more wakeful (iii) Increase alertness and activity (iv) Produce excitement
4. Hallucinogens	LSD, Marijuana, Charas, Bhang, Hashish	(i) Alter thoughts, feelings and perceptions.

- **Common problems of adolescents**
  - (a) Acne
  - (b) Hypochondria - undue concern about health.
  - (c) Neurasthenia - inability to concentrate on or enjoy things. It may lead to irritability, fatigue, insomnia.
  - (d) Phobias - intens dread of things
  - (e) Post traumatic stress disorder due to traumas like rape, etc.
  - (f) Addiction to drugs, tobacco.
  - (g) Physiological aberrations like absence of monthly period in females.

Mental Health

- It refers to the ability to balance feelings, desires, ambition and ideas and to accept the realities of life. Mental illness prevents any such balance. It is reflected in the following ways :

  (i) Depression, (ii) Insomnia, (iii) Excessive sleeping, (iv) Compulsive actions, (v) Feeling of hopelessness, (vi) Thoughts of suicide, (vii) Unreasonable phobias, (viii) Partial or complete loss of memory, (ix) Self destructive behaviour like gambling, drug abuse, over eating, extreme dieting, (x) Delusions and hallucinations, (xi) Vocational and social dysfunctioning.

- **Psychological Disorders**
  - (a) **Psychosis** is a severe type of mental illness in which the patient loses touch with reality.
  - (b) **Neurosis** is a milder type of mental illness in which patients show prolonged emotional reactions.
  - (c) **Epilepsy :** It is a mental disorder characterised by recurrent unconsciousness or fits with generalised convulsions due to abnormal nerve impulses from the over-activated neurons in the brain.
  - (d) **Schizophrenia :** characterised by distorted thoughts; disturbed emotions with rapid shift from one extreme to the other; laughing or crying at inappropriate moments; incoherent bizarre behaviour lasting for a week or more.
  - (e) Borderline Personality Disorder (BPO)
- Drug dependence is the habitual non medical use of a drug.
- **Drug abuse :** Drug abuse is on the rise among the youth due to several factors, like tension and peer group pressure. A sound psychological and reproductive health is essential for the population to become a resource.
- **Adolescence :** This is a period of rapid growth with physical and mental development. It extends from puberty to complete sexual maturity. Changes taking place during adolescence include accelerated (a) Physical growth and development of reproductive organs, (b) Emotional turbulence with mood shifts, (c) Self assertion.

Exercise

1. Vaccines are
   (a) treated bacteria or viruses or one of their proteins
   (b) MHC (major histocompatibility complex) proteins
   (c) curative medicines
   (d) monoclonal antibodies
2. Which one of the following acts as a physiological barrier to the entry of micro-organisms in human body ?
   (a) Epithelium of urogenital tract
   (b) Tears
   (c) Monocytes
   (d) Skin
3. Malaria'a parasite completes its life-cycle in
   (a) one host (b) two host
   (c) three host (d) reservoir host
4. 'Malaria' a common disease world wide is caused by a
   (a) bacterium (b) virion
   (c) protozoa (d) fungi
5. Cancer is related to
   (a) uncontrolled growth of tissues
   (b) non-malignant tumour
   (c) controlled division of tissues
   (d) none of these
6. Broad spectrum antibiotics are one which attack
   (a) pathogens as well as host
   (b) a wide range of pathogens
   (c) only pathogens
   (d) only host
7. Which of the following is a pair of viral diseases?
   (a) Common cold, AIDS
   (b) Dysentery, common cold
   (c) Typhoid, tuberculosis
   (d) Ringworm, AIDS
8. Antibodies in our body are complex
   (a) steroids (b) prostaglandins
   (c) glycoproteins (d) lipoproteins
9. Vaccines are prepared from immune
   (a) vitamins (b) blood
   (c) serum (d) plasma
10. Malaria is spread by
    (a) male *Culex* (b) female *Culex*
    (c) male *Anopheles* (d) female *Anopheles*
11. Which of the following diseases is due to an allergic reaction?
    (a) Goitre (b) Skin cancer
    (c) Hay fever (d) Enteric fever
12. HIV infects
    (a) RBC (b) T - helper cells
    (c) B - cells (d) basophils
13. Cancerous cells can easily be destroyed by radiations due to
    (a) rapid cell divisions (b) lack of nutrition
    (c) fast mutation (d) lack of oxygen
14. Short-lived immunity acquired from mother to foetus across placenta or through mother's milk to the infant is categorised as
    (a) active immunity
    (b) passive immunity
    (c) cellular immunity
    (d) innate non-specific immunity
15. Antigen binding site in an antibody is found between
    (a) two light chains
    (b) two heavy chains
    (c) one heavy and one light chain
    (d) either between two light chains or between one heavy and one light chain depending upon the nature of antigen
16. Cells involved in immune mechanism are
    (a) erythrocytes (b) lymphocytes
    (c) eosinophils (d) thrombocytes
17. A cell coded protein formed in response to infection with most animal viruses is
    (a) antigen (b) antibody
    (c) interferon (d) histone
18. Obstacle to large scale transplantation of organs is
    (a) insufficiency of organ donors
    (b) immunological rejection of foreign bodies
    (c) religious or ethnic considerations
    (d) lack of effective surgical techniques
19. Carcinoma refers to
    (a) malignant turm ours of the connective tissue
    (b) malignant tumours of the skin or mucous membrane
    (c) malignant tumours of the colon
    (d) benign tumours of the connective tissue

20. What is true about T-lymphocytes in mammals ?
(a) There are three main types-cytotoxic T -cells, helper T -cells and suppressor T - cells
(b) These originate in lymphoid tissues
(c) They scavenge damaged cells and cellular debris
(d) These are produced in thyroid

21. Antibodies are the macromolecule of
(a) fats (b) protein
(c) carbohydrate (d) nucleic acid

22. Choose the wrong statement regarding AIDS
(a) AIDS is an immunodeficiency disease
(b) It is caused by a retrovirus, HIV
(c) HIV selectively infects and kill B-lymphocytes
(d) Retroviruses have RNA genomes that replicate via DNA intermediate

23. Only one of the following four ways through which AIDS can spread
(a) Infected needles and syringes
(b) Through mosquito bites
(c) Looking after AIDS patient
(d) Shaking hands, coughing, sneezing, hugging

24. Active immunity development is related to
(a) natural killer cells
(b) memory cells
(c) helper T cells
(d) suppressor T cells

25. Rejection of tissue of organ transplants is brought about mainly by
(a) cytotoxic T cells (b) NK cells
(c) suppressor T cells (d) B cells

26. Cells obtained from cancerous tumours are known as
(a) hybridomas (b) myelomas
(c) lymphocytes (d) monoclonal cells

27. Which immunoglobin (Ig) is produced in primary immune response?
(a) IgA (b) IgE
(c) IgG (d) IgM

28. Expand ELISA
(a) Enzyme linked immuno - sorbent assay
(b) Enzyme linked ion sorbent assay
(c) Enzyme linked inductive assay
(d) None of the above

29. Characters of acquired immunity are
(a) specificity of antigen
(b) difference between self & non-self
(c) retains memory
(d) all of these

30. T-cells are lymphocytes which produce the cellular immunity. These are developed from
(a) thymus
(b) liver
(c) spleen
(d) endothelium of blood vessels

31. The immune system which works against self is
(a) self immune system
(b) autoimmunity
(c) specific immunity
(d) none of the above

32. The immunity obtained after the body has recovered from a disease is
(a) active immunity
(b) passive immunity
(c) both
(d) none of these

33. AIDS virus has
(a) single stranded RNA
(b) double stranded RNA
(c) Single stranded DNA
(d) Double stranded DNA

34. Humoral immunity is due to
(a) B-lymphocytes
(b) T-lymphocytes
(c) L-Lymphocytes
(d) P- Lymphocytes

35. 'ELISA' test is done for the diagnosis of
(a) anthrax (b) HIV
(c) hepatitis (d) malaria

36. Sensitivity to any allergen is related to
(a) deviation from the process of immunity
(b) age of the person
(c) eating habit
(d) rise in environmental temperature

37. A molecule that elicits an immune response is called
(a) antibody (b) antigen
(c) mutagen (d) carcinogen

38. AIDS can be transmitted by
(a) mosquito bite
(b) hand shake
(c) courtship
(d) All of the above

39. The factor responsible for cirrhosis of liver is
(a) sugar (b) vitamins
(c) fats and oils (d) alcoholism

40. Full form of AIDS is
(a) Anti immune deficiency syndrome
(b) Auto immune deficiency syndrome
(c) Acquired immuno deficiency syndrome
(d) Acquired immune disease symptom

41. Which one of the following is an Indian medicinal plant ?
(a) *Saccharum officinarum*
(b) *Rauwolfia serpentina*
(c) *Oryza sativa*
(d) *Solanum melongena*

42. Hypersensitivity to an allergen is associated with
(a) aberrant functioning of the immune mechanism
(b) increase in ambient temperature
(c) age of the individual
(d) food habits

43. The use of *Cannabis* products (bhang, ganja, charas, marijuana or hashish) causes
(a) depression of brain activity and feeling of calmness
(b) alters thoughts, perceptions and feelings
(c) suppresses brain function and relieves pain
(d) stimulates the nervous system and increases alertness and activity

44. Typhoid fever is caused by
(a) *Salmonella* (b) *Shigella*
(c) *Escherichia* (d) *Giardia*

45. During blood typing agglutination indicates that the
(a) RBC carry certain antigens
(b) plasma contains certain antigens
(c) RBC carry certain antibodies
(d) plasma contains certain antibodies

46. Use of anti-histamines and steroids give a quick relief from
(a) nausea (b) cough
(c) headache (d) allergy

47. Select the correct statement from the ones given below?
(a) Barbiturates when given to criminals make them tell the truth
(b) Morphine is often given to persons who have undergone surgery as a pain killer
(c) Chewing tobacco lowers blood pressure and heart rate
(d) Cocaine is given to patients after surgery as it stimulates recovery

48. Widal test is used for the diagnosis of
(a) malaria (b) pneumonia
(c) tuberculosis (d) typhoid

49. Which one of the following is categorised as a parasite in true sense ?
(a) The female *Anopheles* bites and sucks blood from humans
(b) Human foetus developing inside the uterus draws nourishment from the mother
(c) Head louse living on the human scalp as well as laying eggs on human hair
(d) The cuckoo (koel) lays its eggs in crow's nest

50. Select the correct statement with respect to diseases and immunisation
(a) If due to some reason B-and T-lymphocytes are damaged, the body will not produce antibodies against a pathogen
(b) Injection of dead / inactivated pathogens cause passive immunity
(c) Certain protozoans have been used to mass produce hepatitis B vaccine
(d) Injection of snake antivenom against snake bite is an example of active immunisation

51. Fever in malaria is due to
(a) entry of sporozoites into blood capillaries
(b) entry of merozoites into liver cells
(c) release of merozoites from red blood cells
(d) entry of cryptomerozoites into red blood cells

52. The process of spreading the cancerous cells to distant site is know as
(a) hyperstasis (b) metastasis
(c) parastasis (d) parasitesis

53. Benign tumour is the one which
(a) shows metastasis
(b) differentiated and capsulated
(c) undifferentiated and noncapsulated
(d) differentiated and noncapsulated

54. The genes concerned with the production of cancer are called
(a) cancer genes (b) carcino genes
(c) carcinomas (d) oncogenes

55. Antibodies are produced by
(a) leucocytes (b) lymphocytes
(c) erythrocytes (d) none of these

56. Amoebiasis in human is acquired through
    (a) viral infection
    (b) contamination of food with cysts
    (c) accidental consumption of amoebic trophozoites
    (d) none of the above
57. This drug produces hallucinations
    (a) cocaine (b) heroin
    (c) hashish (d) marijuana
58. Common cold is not cured by antibiotics because it is
    (a) caused by a virus
    (b) caused by a gram-positive bacterium
    (c) caused by a gram-negative bacterium
    (d) not an infectious disease
59. At which stage of HIV infection does one usually show symptoms of AIDS?
    (a) When the infecting retrovirus enters host cells
    (b) When viral DNA is produced by reverse trancriptase
    (c) When HIV replicates rapidly in helper T-lymphocytes and damages large number of these cells
    (d) Within 15 days of sexual contact with an infected person
60. A person likely to develop tetanus is immunised by administering
    (a) preformed antibodies
    (b) wide spectrum antibiotics
    (c) weakened germs
    (d) dead germs

Strategies for Enhancement in Food Production

CONCEPTS

- The strategies for enhancement in food production aim at
  - increasing the amount of food obtained from animals.
  - increasing the yield of agricultural crops.

ANIMAL HUSBANDRY

- Animal husbandry deals with the care, breeding & management of domesticated animals that are useful to humans.

Dairy Farming

- In dairy farming, those animals are housed from which we obtain milk and milk products.
- The proper management of dairy farms include
  - processes and systems that increase yield and improve quality of milk.
  - selection of good breeds having high yielding potential, combined with resistance to diseases is very important.
  - providing proper fodder and water to them.

Poultry Farming

- Poultry is a rearing of domesticated fowls (chickens), ducks, geese, turkeys, guinea fowls and pigeons.
- Management procedures that are employed in poultry farm management involves the following steps :
  (i) Selection of disease free and suitable breeds.
  (ii) Proper and safe farm conditions.
- Poultry birds exclusively grown for meat are called **broilers, layers** are for egg production, **cockerel** for young male fowls and **rooster** are mature male fowls.

Fisheries

- Fisheries is an industry devoted to the rearing, breeding, catching and marketing of fishes.
- **Pisciculture** is the rearing, breeding and catching of fishes.
- **Aquaculture** is rearing and management of useful aquatic plants and animals like fishes, oysters, mussels and prawns, etc.
- Fishes are reared and caught for food which is rich in proteins, vitamin A & D. Marine animals are also richer in iodine. Shark liver oil and cod liver oil are very good sources of vitamin A and D.

Bee keeping or Apiculture

- Apiculture is rearing and breeding of honeybees for the production of honey.
- The commonest species of honeybee is **Apis indica**.
- The products obtained from honey bees are honey, bee wax and bee venom.

ANIMAL BREEDING

- Animal breeding is the production of new breeds of domesticated animals with improved traits.

Inbreeding : Mating between the closely related animals of same breed.

- The steps of selective mating during inbreeding–

  Superior male and superior female within the same breed are mated → among the progeny produced → again the superior males and superior females are identified→selected and mated → the same step is repeated in the next progeny and the process in carried out for 4 – 6 generations.
- The improvement in traits occur due to homozygosity.
- The harmful recessive genes which have come together due to homozygosity are eliminated through selection.
- Reduction in fertility after 4 – 6 matings is called inbreeding depression. This is overcome by

breeding the superior animal with the animal of an unrelated breed.

Out-breeding : Mating between the animals which are not closely related.

Out-crossing : Mating between the animals of the same breed which do not have a common ancestor.

Cross-breeding : Mating between the superior animals of different breeds of the same species.

Interspecific Hybridisation

Mating between the animals of two different species.

- The progeny produced is sterile *e.g.*, mule.
- Mule is produced by mating between male ass and female horse (Mare).
- Hinny is produced by mating male horse and female ass.
- Other examples of interspecific hybrids
  - Liger (Male lion and female tiger)
  - Tiglion (Male tiger and female lion)
  - Zonkey (Male zebra and female donkey)

Hybrid Vigour

- Superiority of hybrid in traits over either of its parents.
- It is also known as **heterosis**.

Artificial Insemination

- Semen of a selected male artificially deposited into the vagina of a selected female.
- Semen is preserved at low temperature (–196°C) in liquid nitrogen.
- Helps to overcome several problems of normal mating.
- **MOET (Multiple Ovulation Embryo Transfer)** technique is a programme which improves the chances of successful production of hybrids.
- MOET involves the following steps –
  (i) **Super ovulation :** It involves the stimulation of healthy and high milk yielding breed of female animals by injecting gonadotrophic hormones (FSH & LH) to release more eggs from the ovaries.
  (ii) Super ovulated female is mated with a superior male (yielding high quality lean meat with less lipids) by natural breeding or is artificially inseminated with the semen of a superior male to promote internal fertilization.
  (iii) **Embryo transfer :** Embryo at 8-32 celled stages can be recovered non surgically from the female and then transferred into the surrogate females for implantation and further development.
- MOET has been successfully used in cows, sheep, rabbit, buffaloes, mares etc.

PLANT BREEDING

- **Plant breeding** refers to the modification and improvement of genetic material of plants resulting in the development of crops which are more beneficial to human beings.
- Steps of plant breeding are as follows:

Plants with superior traits are identified

↓

Cross hybridization between the selected plants

↓

Plants with superior traits among the progeny are selected

↓

Again cross hybridization between the superior hybrids is carried out while eliminating the plants with less beneficial traits.

Domestication

- It is the process of bringing a species under human management. Plant breeding is concerned with improvement of crops and animal breeding with improvement in domestication of animals.

Improved varieties

- A variety is a group of plants that have similar genotype and it differs in one or more characters from other varieties of the same crop.

Development of new varieties

- It consists of the following steps : (a) Creation of genetic variation, (b) Selection, (c) Evaluation and release as a variety, (d) Seed multiplication and distribution to farmers.
- **Steps in variety development :-**
  (a) Creation of genetic variation
  (i) Variation - genetic or induced by man.
  (ii) Induced by domestication, germplasm cell, plant intro, hybridization, polyploidy, mutation, genetic engineering.
  (b) Selection
  (i) Self - pollinated crops - isolates homozygous
  (ii) Cross pollinated crops - isolates both homo and hetero zygous.
  (c) Evaluation and release of varieties
  ICAR-carries out evaluation. Performance of new compared with old variety.
  (d) Multiplication of Improved seed

- **Process of collection and conservation of germ plasm**
  (i) Germ plasm decrease due to demand of only good and non-wild variety.
  (ii) Collection maintained in form of seeds at low temperture.
  (iii) Have to be occasionally grown to get fresh seeds to prevent spoiling.
  (iv) Crop taken from one area into new area, where not previously grown → plant introduction.
  (v) Quarantine-examination of all introductions for presence of weeds, insects and disease causing organisms.
- **Mutation Breeding**
  (i) Mutation can be induced by short wave electromagnetic radiations (UV radiation, X-rays, cosmic rays), ionising radiations (Gamma rays obtained from ^{60}Co and ^{137}Caesium) and certain chemicals like nitrous acid, ethyl methane sulfonate (EMS), nitromethyl urea and nitrosoethyl urea.
  (ii) Atomica-2 and Reimei is high yielding variety of rice produced by gamma radiation.
  (iii) Amber coloured Sharbati - Sonora and Pusa - Lerma varieties of wheat have been produced by gamma radiation from red grained variety (Sonora-64 and Rojo-64A)

Table : Crop varieties bred by hybridization & selection for disease resistance to fungi, bacteria & viral disease.

	Crop	Variety	Resistance to diseases
1.	Wheat	Himgiri	Hill bunt & leaf and stripe rust
2.	Cauliflower	Pusa snowball K-1 Pusa shubra	**Blight black rot** Black rot and curl
3.	Brassica	Pusa Swarnim (Karan rai)	White rust
4.	Cowpea	Pusa Komal	Bacterial blight
5.	Chilli	Pusa Sadabahar	Chilly mosaic virus, Tobacco mosaic virus and leaf curl.

- **Polyploid Breeding :**
  (i) Autotetraploidy is found in potato, ground nut, coffee, tomato, plum, strawberry and grapes.
  (ii) Tetraploids are sterile and propagated vegetatively e.g.- Banana, Seedless water melon, Sugar beet etc.
  (iii) Wheat, bread wheat, new world long stapled cotton, tobacco, etc. are alloploids.
- **Disease resistance** is the ability of plants to withstand, oppose, lessen or overcome the attack of pathogen.
- Examples of insect pest resistance crops bred by hybridization are
  (i) **Pusa Gaurav** variety of *Brassica* is resistant to aphids.
  (ii) **Pusa Sawani & Pusa A-4** varieties of Okra (Bhindi) are resistant to shoot and fruit borer.
  (iii) **Pusa sem 2** and **Pusa sem 3** varieties of flat bean are resistance to aphids & fruit borer.
- **Biofortification** is the improvement of nutritional quality of crops to improve public health.
- In plant breeding, biofortification aims at developing new crops with improved
  - protein content and quality
  - lipid content and quality
  - vitamin & mineral content
- Some biofortified vegetables developed by Indian Agricultural Research Institute, New Delhi are –
  - Vitamin A enriched carrot, spinach, pumpkin;
  - Vitamin C enriched mustard, tomato;
  - Iron and calcium enriched spinach and bathua;
  - Protein enriched beans-lablab, french beans.

SINGLE CELL PROTEIN (SCP)

- Single cell protein refers to the dried microbial cells or total protein extracted from pure microbial cell culture (algae, bacteria, yeasts and filamentous fungi) which can be used as food supplement to humans or animals.
- Advantages of SCP are
  - It has high protein and low fat content.
  - It is good source of vitamins particularly B-complex. E.g. Mushrooms and Yeasts.
  - SCP organisms grow faster and produce large quantities of SCP from relatively small area of land and time.

TISSUE CULTURE

- Plant cells and plant organs can be cultured *in vitro* (e.g., in test tube or any other container) on a suitable medium (*e.g.*, agar medium), is called plant tissue culture.
- A plant part excised from its original location and used for initiating a culture is known **as explant**. The explants, media, vessels and instruments used in tissue culture must be sterilised to make them free from any microbes.
- Culture medium should be provided with nutrition that is required for growth and development of explants. Culture of explants on an agar medium provided with 2, 4-D yields callus cultures.
- Plant cells can also be grown in a liquid medium as a suspension of single cells as well as groups.
- A constant agitation of suspension cultures is essential.
- Complete plantlets can be obtained from agar and suspension cultures.
- Explants that contain pre-existing shoot meristems and produce shoots from them are called **meristem culture**.
- Haploids may be produced in many plant species when anthers are cultured on a suitable medium, this is called **anther culture**. Chromosome numbers of haploid plants can be doubled (diploid, 2n) to produce homozygous lines within 2 to 3 years.

Tissue culture depending on *in vitro* growth

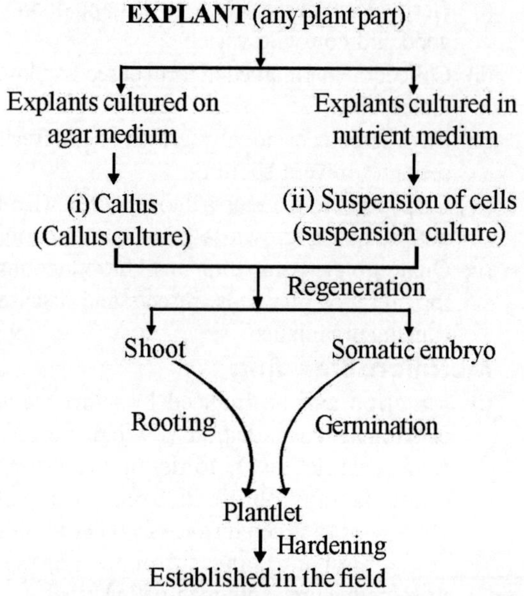

EXPLANT (any plant part)

↓

Explants cultured on agar medium → (i) Callus (Callus culture)

Explants cultured in nutrient medium → (ii) Suspension of cells (suspension culture)

↓ Regeneration

Shoot → Rooting → Plantlet

Somatic embryo → Germination → Plantlet

Plantlet

↓ Hardening

Established in the field

- A hybrid produced by fusion of somatic cells of two species is called somatic hybrid, and process is known as **somatic hybridisation**.
- Cell walls can be digested with a combination of pectinase and cellulase. The plant cells lacking cell walls are called **protoplasts**. The fusion of protoplasts can produce somatic hybrids.

Embryo Culture

- Embryos taken out from seeds are grown on culture medium. The embryos might be obtained from
  - seeds of interspecific hybrid plants (embryo rescue).
  - seeds which lack stored food *e.g.*, orchids.
  - seeds which are dormant.
  - seeds of rare plants.

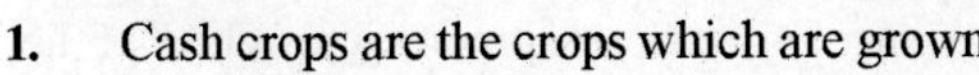

Exercise

1. Cash crops are the crops which are grown
   (a) alternating with other crops to increase the yield
   (b) for their high price to benefit the farmers
   (c) to initiate the germination of parasitic plants before the main crop
   (d) to increase the capacity of parasitic seedlings to infect the main crop
2. The quickest method of plant breeding is
   (a) introduction (b) selection
   (c) hybridization (d) mutation breeding
3. The new varieties of plants are produced by
   (a) introduction & mutation
   (b) selection & introduction
   (c) selection & hybridization
   (d) mutation & selection
4. Silk is produced by
   (a) larva (b) adult moth
   (c) both (a) and (b) (d) cocoon/pupa
5. Domestication of honey bees is called
   (a) sericulture (b) apiculture
   (c) tissue Culture (d) pisciculture
6. What is the best pH of soil for cultivation of plants?
   (a) 3.4 – 5.4 (b) 6.5 – 7.5
   (c) 4.5 – 8.5 (d) 5.6 – 6.5
7. Three crops that contribute maximum to global food grain production are
   (a) wheat, rice and maize
   (b) wheat, rice and barley
   (c) wheat, maize and sorghum
   (d) rice, maize and sorghum
8. The branch of biology which deals with improvement of plant variety
   (a) plant breeding (b) eugenics
   (c) agrology (d) serendipity
9. Production of plant without fertilization is done by
   (a) vegetative propagation
   (b) transplantation
   (c) grafting
   (d) none of these
10. Pure line breed refers to
    (a) heterozygosity only
    (b) homozygosity only
    (c) heterozygosity and linkage
    (d) homozygosity and self-assortment
11. The best way to increase the yield of wheat in India is
    (a) to sow seeds of improved varities
    (b) to use tractors to till the soil
    (c) to reduce the quantity of ration consumers
    (d) to remove weeds from wheat fields
12. Aims of plant breeding are to produce
    (a) disease-free varieties
    (b) high yielding varieties
    (c) early maturing varieties
    (d) all of the above
13. Selection is the method of
    (a) plant physiology (b) plant breeding
    (c) genetics (d) cytology
14. Largest producer of sugarcane is
    (a) China (b) India
    (c) Pakistan (d) USA
15. Mule is a product of
    (a) breeding
    (b) mutation
    (c) hybridization
    (d) interspecific hybridization
16. Silk worm is a
    (a) fly (b) worm
    (c) moth (d) beetle
17. An explant is
    (a) dead plant
    (b) part of the plant
    (c) part of the plant used in tissue culture
    (d) part of the plant that expresses a specific gene

18. Protoplast is
(a) another name for protoplasm
(b) an animal cell
(c) a plant cell without a cell wall
(d) a plant cell

19. The science of rearing, feeding, care, breeding and utilization of animals is called
(a) animism (b) veterinary science
(c) animal husbandry (d) dairy science

20. The milk of which one of these has more fat content
(a) Cow (b) Buffalo
(c) Sheep (d) Goat

21. In crop improvement programme, haploids are important because they
(a) require one half of nutrients
(b) are helpful in study of meiosis
(c) grow better under adverse conditions
(d) form perfect homozygous

22. One of the major difficulties in the biological control of insect pests is the
(a) practical difficulty of introducing the predator to specific areas
(b) method is less effective as compared with the use of insecticides
(c) predator does not always survive when transferred to a new environment
(d) the predator develops a preference to other diets and may itself become a pest

23. Which one of the following is an exotic Indian fish ?
(a) *Catla catla*
(b) *Heteropneustes fossilis*
(c) *Cyprinus carpio*
(d) *Labeo rohita*

24. Of the world's top five crops in terms of annual production
(a) all five belong to the family Poaceae (Gramineae)
(b) four belong to Poaceae (Graminaeae) and one to Leguminosae
(c) four belong to Poaceae (Gramineae) and one to Solanaceae
(d) three belong to Poaceae (Gramineae), one to Leguminosae and one to Solanaceae

25. The reason why vegetatively reproducing crop plants are best suited for maintaining hybrid vigour is that
(a) once a desired hybrid has been produced there are few chances of losing it
(b) they have a longer life span
(c) they are more resistant to diseases
(d) they can be easily propagated

26. Biological control component is central to advanced agricultural production. Which of the following is used as a third generation pesticide?
(a) Insect repellants
(b) Pheromones
(c) Pathogens
(d) Insect hormone analogues

27. Which one of the following statement is correct in relation to honey bees?
(a) *Apis indica* is the largest wild honey bee in India
(b) Honey is predominantly sucrose and arabinose
(c) Beewax is a waste product of honey bees
(d) Communication in honey bees was discovered by Karl Von Frisch

28. In tissue culture medium, the embryoids formed from pollen grains is due to
(a) cellular totipotency
(b) organogenesis
(c) double fertilization
(d) test – tube culture

29. Cellular totipotency is demonstrated by
(a) only bacterial cells
(b) only gymnosperm cells
(c) all plant cells
(d) all eukaryotic cells

30. India's wheat yield revolution in the 1960s was possible primarily due to
(a) hybrid seeds
(b) increased chlorophyll content
(c) mutations resulting in plant height reduction
(d) quantitative trait mutations

31. The technique of obtaining large number of plantlets by tissue culture method is called
(a) plantlet culture
(b) organ culture
(c) micropropagation
(d) macropropagation

32. In order to obtain virus-free plants through tissue culture the best method is
(a) embryo rescue
(b) anther culture
(c) meristem culture
(d) protoplast culture

33. *Triticale,* the first man-made cereal crop, has been obtained by crossing wheat with
(a) pearl millet (b) sugarcane
(c) barley (d) rye

34. Somaclones are obtained by
(a) plant breeding
(b) irradiation
(c) genetic engineering
(d) tissue culture

35. Breeding of crops with high level of minerals, vitamins and proteins is called
(a) somatic hybridisation
(b) biofortification
(c) biomagnification
(d) micropropagation

36. An improved variety of transgenic basmati rice
(a) does not require chemical fertilizers and growth hormones
(b) gives high yield and is rich in vitamin A
(c) is completely resistant to all insect pests and diseases of paddy
(d) gives high yield but has no characteristic aroma

37. In crop improvement programmes, virus-free clones can be obtained through
(a) grafting (b) hybridization
(c) embryo (d) shoot apex culture

38. Desired improved varieties of economically useful crops are raised by
(a) migration (b) biofertilizer
(c) hybridization (d) natural selection

39. Plants can be disease resistant by
(a) breeding with their wild relatives
(b) colchicine treatment
(c) hormone treatment
(d) heat treatment

40. Which of the following is not used for crop improvement ?
(a) Inbreeding (b) Introduction
(c) Hybridization (d) Mutations

41. Which of the following effect is produced by colchicine ?
(a) Duplication of DNA
(b) Duplication of chromosomes
(c) Formation of spindle fibres
(d) Hinderance in the formation of middle wall

42. Which of the following is generally used for induced mutagenesis in crop plants ?
(a) X-rays
(b) UV (260 nm)
(c) Gamma rays (from cobalt 60)
(d) Alpha particles

43. Heterosis means
(a) hybrid vigour
(b) hybrids are weak
(c) hybrids are weak as well as vigorous
(d) hybrids are neither weak nor vigorous

44. The *Triticum aestivum* (wheat) is
(a) haploid (7 chromosomes)
(b) diploid (14 chromosomes)
(c) tetraploid (30 chromosomes)
(d) hexaploid (42 chromosomes)

45. The plant most useful for man kind
(a) coconut (b) cotton
(c) mustard (d) ground nut

46. In tissue culture medium, the embryoid formed from pollen grain is due to
(a) cellular totipotency
(b) organogenesis
(c) double fertilization
(d) test tube culture

47. Honey bee after discovering the new source of nector/honey can convey this information but
(a) cannot convey the direction
(b) can convey the direction by round or tail waving dance
(c) can convey the direction by round dance only
(d) cannot convey the direction by tail waving dance only

48. Parthenocarpic tomato fruits can be produced by
(a) treating the plants with low concentrations of gibberellic acid and auxins
(b) raising the plants from vernalized seeds
(c) treating the plants with phenylmercuric acetate
(d) removing androecium of flowers before pollen grains are released

49. Farmers in a particular region were concerned that pre-mature yellowing of leaves of a pulse crop might cause decrease in the yield. Which treatment could be most beneficial to obtain maximum seed yield ?
(a) Treatment of the plants with cytokinins along with a small dose of nitrogenous fertilizer
(b) Removal of all yellow leaves and spraying the remaining green leaves with 2, 4, 5 - trichlorophenoxy acetic acid
(c) Application of iron and magnesium to promote synthesis of chlorophyll
(d) Frequent irrigation of the crop

50. Crop plants grown in monoculture are
(a) free from intraspecific competition
(b) characterised by poor root system
(c) highly prone to pests
(d) low in yield

51. The most likely reason for the development of resistance against pesticides in insects damaging a crop is
(a) random mutation
(b) genetic recombination
(c) directed mutation
(d) acquired heritable change

52. Haploid plants are preferred over diploids for mutation study because in haploids
(a) recessive mutation expresses immediately
(b) induction of mutation is easier
(c) culturing is easier
(d) dominant mutation expresses immediately

53. The long-term prospects for a truly human civilization depend in a large measure on
(a) the ability of humanity to moderate its fecundity
(b) increasing the food production
(c) colonization of underpopulated areas
(d) control of human diseases

54. Super-ovulation and embryo transplantation are meant for improving
(a) human race
(b) livestock
(c) poultry
(d) plants

55. Artificial breeding of cattle is brought about by
(a) artificial insemination
(b) superovulation and embryo
(c) homozygotic twinning
(d) all the above

56. The transgenic animals are those which have
(a) foreign DNA in some of its cells
(b) foreign DNA in all its cells
(c) foreign DNA and RNA in some of its cells
(d) foreign DNA and RNA in all its cells

57. Inbreeding is carried out in animal husbandry because it
(a) increases vigour
(b) improves the breed
(c) increases heterozygosity
(d) increases homozygosity

58. Sonalika and Kalyan sona are varieties of
(a) wheat (b) rice
(c) millet (d) tobacco

59. Fungicides and antibiotics are chemicals that
(a) enhance yield and disease resistance
(b) kill pathogenic fungi and bacteria, respectively
(c) kill all pathogenic microbes
(d) kill pathogenic bacteria and fungi respectively.

60. Two plants growing in different seasons and different geographical area, can produce hybrid by
(a) pollen culture
(b) tissue culture
(c) somatic embryogenesis
(d) *in vitro* synthesis

Microbes in Human Welfare

CONCEPTS

- Microbes can be grown on culture nutritive media to form colonies which are used for mass production of certain desirable products and for scientific study of metabolism, genetics and structure of micro-organisms.

Microbes in Household and Industrial Products

1. **Dairy Industry**

   Lactic acid bacteria causes coagulation of milk protein casein. They also improve the nutritional quality of curd by increasing vitamin B_{12}.

   Lactobacillus lacti : Milk fermentation
   Lactobacillus acidophilus : Cheese production
   Streptococcus thermophilus : Yoghurt production
   Saccharomyces cerevisiae : Bread production
   S. lactis, L. bulgaricus : Kefir production
   Lactose fermenting yeast : Butter milk production

2. **Fermented food Industry**

Fermented food	***Starting product***	***Micro-organisms***
Sauerkraut	Cabbage	*Enterobacter*
Pickles	Cucumber	*Lactobacillus, Streptococcus*
Green olives	Olives	*L. plantarum, L. brevis*
Sausage	Beef and pork	*Micrococcus, Pediococcus*

3. **Acid and alcoholic industry**

Product	***Micro-organism***	***Uses***
Acetone-butanol	*Clostridium acetobutylicum*	Solvent
Dihydroxy acetone	*Gluconobacter*	Solvent
Lactic acid	*Lactobacillus delbrueckii* *L. bulgaricus*	Textile and laundry delining hides
Dextran (high mol. wt. polysaccharide)	*Leuconostoc mesenteroid*	Molecular sieve

4. **Vitamins**

Sorbose	*Gluconobacter*	Ascorbic acid
Cobalamine B_{12}	*Propionibacterium*	As food supplement

5. **Amino acid production**

Glutamic acid	*Brevibacterium spp*	Food additive
Lysine	*Micrococcus*	Animal feed additive
Microbial protein (Single cell protein)	Methane oxidising bacteria	Food supplement

6. **Enzyme production**

Bacterial protease	*Bacillus subtilis*	Bating hides, desizing fibres, spot remover, tenderizing meat(softening)
Bacterial amylase	*Bacillus subtilis*	Modifying starch desizing fabric

7. **Medicinal use of bacteria**

	Streptokinase – Streptodornase	*Streptococcus*	Dissolution of blood clots
	Insulin, Interferon	Recombinant DNA variety of *E. coli*	Human therapy
8.	**Commercial product of yeast**		
	Product	***Micro-organism***	***Uses***
	• Baker's yeast, beer, wine, bread	*Saccharomyces cerevisiae*	Baking and brewing industry
	• Soya sauce	*Saccharomyces rouxii*	Food condinent
	• Ethanol	*S. cerevisiae*	Fuel, solvent
	• Riboflavin	*Eremothecium*	Vitamin supplement
	• SCP [single cell protein]	*S. lipolytica*	Food supplement
Industrial application of molds		***Product***	***Micro-organism Uses***
	• Citric acid	*Aspergillus niger*	Food product preservative
	• Fumaric acid	*Rhizopus nigricans*	Resin production
	• Gluconic acid	*A. niger*	Pharmaceuticals
	• Itaconic acid	*A. terreus*	Wetting agent
	• Gibberellic acid	*Fusarium moniliformes*	Setting of fruit, seed ripening
	• Lactic acid	*Rhizopus*	Food and pharmaceuticals
	Enzymes		
	• Pectinase	*A. wentii*	Clarifying agent in fruit juice industry
	• Protease	*A. aureus*	
	11-γ-hydroxy progesterone production (therapeutic use)	*Rhizopus nigricans*	Intermediate for hormone
	• Invertase	*Mucor*	Preparation of adhesives, candy making

- Bottled fruit juices are clarified and clearer as compared to home made due to use of **pectinase** and **protease** enzyme.
- **Streptokinase** produced by bacterium *Streptococcus is* used as clot buster for removing blood clot in patients suffering from myocardial infractions.
- **Cyclosporin A**, a bioactive immunosuppressive agent is used in organ transplant patient, produced by *Trichorderma polysporum*.
- **Statin** is blood cholesterol lowering agent produced by yeast *Monascus* and it acts by competitively inhibiting enzyme responsible for cholesterol synthesis.
- **Oil spill** is cleaned by genetically engineered a species of *Pseudomonas poteta* so that it has the capacity to metabolize four major hydrocarbons of petroleum – camphor, octane, xylene and napthalene.
- *Thiobacillus thiooxidans* and *thiobacillus ferroxidans*, when grown in presence of copper ore produce acid and affect oxidation process known as **leaching** and improves recovery of metal from ore and is non polluting to the atmosphere.

PRODUCTION OF ANTIBIOTICS

- Antibiotics are chemical substances, which are produced by some microbes and can kill or retard growth of other disease causing microbes.
- **Alexander Fleming** while working on *Staphylococci* plates observed that in one of unwashed culture plate, it could not grow because of chemical produced by mould *Penicilliun notatum* grown on it which produces chemical **penicillin**.

MICROBES IN SEWAGE TREATMENT

Wastewater treatment is necessary before wastewater is disposed off without producing any undesirable or harmful effect.

Thus, wastewater treatment process are carried out by municipal treatment plant as a community process which is as follows :

1. **Primary (1°) Treatment :** This is a physical process that involves physical removal of particles from the sewage through filtration and sedimentation. So, it remove coarse solids and settleable solids in sedimentation units (tanks, basin, or mechanical devices) by which particulate material is concentrated as **sludge**. The sludge and liquid effluent are processed separately. The effluent from the primary settling tank is taken to secondary.

2. **Secondary (2°) treatment :** Secondary treatment involves the oxidation of organic material in effluent by **microbial** method. These are –
   (i) Activated sludge process.
   (ii) **Filteration** by intermittent sand filters and trickling filter.

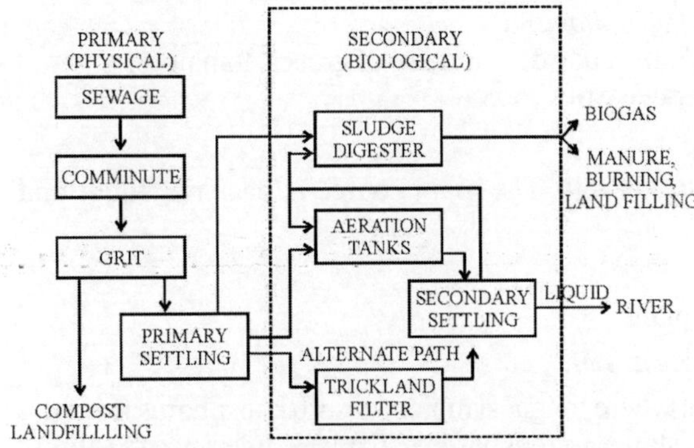

Flow chart of sewage treatment

- **In activated sludge process,** aeration of sewage in large aeration and agitation tank is done to form flocs and floccules. When they are vigorously aerated, flocculation occurs and this contain large number of actively metabolizing bacteria, yeast, mould and protozoa. This growth is effective in oxidation of organic compound. This significantly reduces **BOD (Biochemical oxygen demand).** BOD is measure of organic content in H_2O. Greater is BOD, in H_2O, more is its polluting potential.
- **Anaerobic sludge digester :** Solid which accumulates during sedimentation are pumped into sludge digester. This sludge is digested with the help of anaerobic and facultative bacteria. These microorganisms bring about a decrease in organic solid by degrading them to soluble substances and gaseous products which has 60-70% methane and 20-30% CO_2. This gas mixture is used as **biogas** for heating purpose and for operating power.
- **Ganga Action Plan** and **Yamuna Action Plan** are a part of this strategy of sewage **treatment**.

MICROBES IN PRODUCTION OF BIOGAS

- Produced by **methanogens**, which are anaerobic bacteria growing on cellulosic material and produce large amount of methane alongwith CO_2 and H_2 e.g., *Methanococcus, Methanobacterium.*
- Formed in
  (a) Anaerobic sludge during 2° treatment.
  (b) Humus of cattle
- A **biogas plant** has a large concrete or brick lined airtight cylindrical tank called digester. It has a charge pit for passage of slurry into digester, a floating gas holder of metal with an outlet for gas and a pit for removal of sludge or manure. The biogas thus produced is used for cooking and lighting.

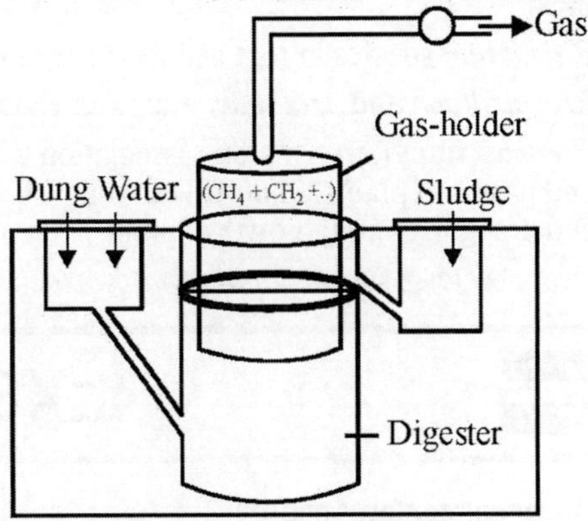

Fig. A typical biogas plant

- Biogas is also called gobar gas because excrete of cattle has cellulose and cellulose digesting, methane releasing methanogens.

Microbes as a Biocontrol Agents

- Use of predators, animal composite, microorganisms such as bacteria/virus to control insect pest, pathogen and weed is called biological control and biological agents are called biocontrol agents.

- **Bt Cotton** is genetically modified cotton having *Bacillus thuringiensis* genes. Bacterium *Bacillus thuringiensis* (Bt) is used as microbial control agent in order to control butterfly caterpillars. It is sprayed on plants like brassica and fruit trees which are attacked by insect larvae. In the gut of larvae, Bt release toxin and insect larvae are killed. It kills only caterpillars, leaving all other insects unharmed.

  Bt cotton has *B. thuringiensis* toxin genes and such plants are resistant to attack by insect pest.

	Micro-organism	*Target organism*
A.	**Insect control**	
	(i) Granulosis viruses [GV] Nucleo-polyhedrosis virus [NPV]	Rice borer bollworm, cotton leaf worm, cabbage hopper larvae
	(ii) **Bacteria** *Bacillus thuringiensis, Bacillus sphaericus*	*Lepidopteran* and *Dipteran* larvae, cockroach, culex and *Anopheles*
	(iii) **Fungi** *Beauveria* spp., *Verticillium metarrhizium*	Beetle, caterpillar, aphids, white flies, hopper, butterfly, mosquito
	(iv) **Protozoa** *Molaemba locusta*	Grasshopper
B.	**Weed control** by bacteria,	*Phytophthora palmivora, Colletotrichum glucospirodes.*
C.	**Disease control** by Fungi	

Fungi	***Disease controlled***
Trichoderma irridal, Ampelomyces spp.	→ Control fungal pathogen such as *Phaseolina* and *Macrophomina*
Agrobacterium radio bacteria	→ Control powdery mildew of grapes, tomatoes, strawbarries, etc.

MICROBE AS BIOFERTILISERS

Biofertilisers are organisms that enrich nutrient quality of soil. The main source is bacteria, fungi and cyanobacterium.

Examples:

- ***Rhizobium*** species in root nodules of leguminous plant.
- *Azospirillium* and *Azotobacter* are free living species in soil.
- *Gleomus* (fungi) in symbiotic association with plant where fungal symbiont absorb phosphorus from soil and passes to plants. Plants in turn show better resistance to root borne pathogens, tolerance to salinity and draught, overall growth of plant.
- Cyanobacteria such as *Anabaena*, *Nostoc* and *Oscillatoria* (autotrophic micorbes) in paddy field.

Exercise

1. Gobar gas contains mainly
   (a) $CH_4 + CO_2$ (b) $CH_4 + O_2$
   (c) $CO_2 + H_2$ (d) $CO_2 + SO_2$
2. Antibiotics are mostly obtained from
   (a) fungi (b) actinomycetes
   (c) cyanobacteria (d) (a) & (b)
3. Bioenergy is the energy obtained from
   (a) biological sources (b) coal
   (c) petroleum (d) natural gases
4. The example of non-renewable source of energy is
   (a) alcohol fuel (b) petroleum
   (c) biogas (d) fuel wood
5. Beer is obtained from
   (a) molasses (b) grapes
   (c) barley (d) rye

6. What are the advantage(s) of gobar gas over conventional utilization ?
   (a) More efficient source of energy
   (b) Used as good fertilizer
   (c) Reduces the chances of spreading of pathogens
   (d) All the above
7. Which is not a character of good fire wood ?
   (a) Highly combustible
   (b) Soft wood
   (c) Non resinous and non smoky
   (d) High caloric value
8. Indicator of water pollution
   (a) *E. coli* (b) *Chlorella*
   (c) *Beggiatoa* (d) *Ulothrix*
9. Non-conventional energy source is
   (a) tidal energy (b) biogas
   (c) geothermal energy (d) all of these
10. Petro-crops are plants
    (a) grown near oil fields
    (b) used in refining crude oil
    (c) whose fossil remains formed crude oil
    (d) from which petrol like fuels can be derived
11. A good fodder
    (a) contains high dry matter
    (b) is free from disease and pest
    (c) has nutrient without toxicity
    (d) all of these
12. Non-renewal energy gas is
    (a) natural gas
    (b) geothermal energy
    (c) gobar gas
    (d) biogas
13. The residue left after methane production from cattle dung is
    (a) burnt
    (b) burried in land fills
    (c) used as manure
    (d) used in civil construction
14. Domestic cooking gas cylinder is filled with
    (a) alcohol
    (b) diesel oil
    (c) liquid petroleum gas
    (d) coal gas
15. Which of the following is included in biopesticide?
    (a) Viruses and bacteria
    (b) Viruses, bacteria and fungi
    (c) Viruses, bacteria, fungi, protozoa and mites
    (d) viruses, bacteria, fungi and protozoa
16. Which one thing is not true about antibiotics?
    (a) The term "antibiotic" was coined by Selman Waksman in 1942
    (b) First antibiotic was discovered by Alexander Flemming
    (c) Each antibiotic is effective only against one particular kind of germ
    (d) Some persons can be allergic to a particular antibiotic
17. Which one of the following is non-symbiotic biofertilizer?
    (a) *Azotobacter* (b) *Anabaena*
    (c) *Rhizobium* (d) VAM
18. Farmers have reported over 50% higher yields of rice by using the biofertilizer
    (a) *Azolla pinnata*
    (b) *Cyanobacteria*
    (c) *Legume-Rhizobium* symbiosis
    (d) Mycorrhiza
19. The aquatic fern, which is an excellent biofertiliser is
    (a) *Azolla* (b) *Salvinia*
    (c) *Marsilia* (d) *Pteridium*
20. Which of the following is pair of bio-fertilizers?
    (a) *Azolla* and blue gree algae
    (b) *Nostoc* and legumes
    (c) *Rhizobium* and grasses
    (d) *Salmonella* and *E. coli*
21. Two microbes found to be very useful in genetic engineering are
    (a) *Vibrio cholerae* and a tailed bacteriophage
    (b) *Diplococcus* sp. and *Pseudomonas* sp.
    (c) Crown gall bacterium and *Caenorhabditis elegans*
    (d) *Escherichia coli* and *Agrobacterium tumefaciens*
22. Human insulin is being commercially produced from a transgenic species of
    (a) *Escherichia* (b) *Mycobacterium*
    (c) *Rhizobium* (d) *Saccharomyces*
23. A common biocontrol agent for the control of plant diseases is
    (a) *Frankia*
    (b) *Bacillus thuringiensis*
    (c) *Glomus*
    (d) *Trichoderma*
24. The common nitrogen fixer in paddy fields is
    (a) *Rhizobium* (b) *Azospirillum*
    (c) *Oscillatoria* (d) *Frankia*

25. Which one of the following is not used in organic farming?
(a) *Glomus* (b) Earthworm
(c) *Oscillatoria* (d) Snail

26. Which one of the following help in absorption of phosphorus from soil by plants?
(a) *Glomus* (b) *Rhizobium*
(c) *Frankia* (d) *Anabaena*

27. Which one of the following is a wrong matching of a microbe and its industrial product, while the remaining three are correct ?
(a) Yeast - statins
(b) *Acetobacter aceti* - acid
(c) *Clostridium butylicum* - lactic acid
(d) *Aspergillus niger* - citric acid

28. Conversion of sugar into alcohol during fermentation is due to the direct action of
(a) temperature
(b) microorganisms
(c) zymase
(d) concentration of sugar solution

29. Rennin used in cheese industry is
(a) antibiotic (b) enzyme
(c) alkaloid (d) inhibitor

30. Which one of the microorganism is used for production of citric acid in industries
(a) *Lactobacillus bulgaricus*
(b) *Penicillium citrinum*
(c) *Aspergillus niger*
(d) *Rhizopus nigricans*

31. Which one of the following is used in the manufacture of alcohol ?
(a) Bacteria (b) Water mould
(c) Yeast (d) Slime mould

32. Which one of the following is used in industrial preparation of ethanol ?
(a) *Lactobacillius* (b) *Azotobacter*
(c) *Penicillium* (d) *Saccharomyces*

33. Biogas consists of
(a) carbon monoxide, methane and hydrogen
(b) carbon dioxide, methane and hydrogen
(c) carbon monoxide, ethane and hydrogen
(d) carbon dioxide, ethane and hydrogen

34. First fermented acid is
(a) gluconic acid (b) lactic acid
(c) fumaric acid (d) all the above

35. Which is not a character of good antibiotic ?
(a) No side effects for host
(b) Destroy microflora of alimentary canal of host
(c) Broad spectrum antibiotics
(d) Immediate reaction

36. What is mode of bacterial resistance against antibiotics ?
(a) Development of thick mucilaginous layer
(b) Alteration of cell membrane
(c) Mutation in bacteria
(d) All the above

37. Cheese and yoghurt are products of
(a) pasteurisation (b) distillation
(c) dehydration (d) fermentation

38. Sewage purification by
(a) microbes (b) fertilizers
(c) antibiotics (d) plants

39. Which one produce gas by decomposing the gobar (dung) in gobar gas ?
(a) Fungus
(b) Virus
(c) Methanogenic bacteria
(d) Algae

40. *Escherichia coli* is used as an indicator organism to determine pollution of water with
(a) heavy metals
(b) faecal matter
(c) industrial effluents
(d) pollen of aquatic plants

41. During anaerobic digestion of organic waste, such as in producing biogas, which one of the following is left undergraded ?
(a) Lipids (b) Lignin
(c) Hemi-cellulose (d) Cellulose

42. With the exception of water, which one of the following is possibly the most important accessory chemical substance in industrial process ?
(a) Petroleum (b) Rubber
(c) Ethanol (d) Liquid nitrogen

43. Biogas can be a good substitute for
(a) fuel wood (b) petroleum and oil
(c) coal (d) charcoal

44. Major autotrophic biomass in oceans is contributed by
(a) forests
(b) algae and phytoplanktons
(c) crops
(d) none of these

45. Energy plantation refers to
(a) setting up new electricity plants
(b) growth of fuel wood trees
(c) manufacture of more generators
(d) erection of more dams

46. What type of fuel are coal, petrol and natural gas ?
(a) Biofuels (b) Electrical fuels
(c) Fossil fuels (d) Liquid fuels
47. The fuel wood crisis can be overcome by
(a) Efficient forest extractions
(b) Afforestation
(c) More efficient heat transfer
(d) A combination of these
48. In developing countries, the heaviest demand on forests is for
(a) fuel wood (b) furniture wood
(c) fruits (d) none of these
49. Fuel wood is
(a) source of petroleum products
(b) renewable resource of energy
(c) source of biogas
(d) source of alcohol
50. The vitamin whose content increases following the conversion of milk into curd by lactic acid bacteria is
(a) vitamin C (b) vitamin D
(c) vitamin B_{12} (d) vitamin E
51. Wastewater treatment generates a large quantity of sludge, which can be treated by
(a) digesters (b) activated sludge
(c) chemicals (d) oxidation pond
52. Match the following list of bacteria and their commercially important products

Bacterium	Product
(A) *Aspergillus niger*	(p) Lactic acid
(B) *Acetobacter aceti*	(q) Butyric acid
(C) *Clostridium butylicum*	(r) Acetic acid
(D) *Lactobacillus*	(s) Citric acid

Choose the correct match
(a) A→q, B→r, C→s, D→p
(b) A→p, B→r, C→s, D→q
(c) A→s, B→r, C→q, D→p
(d) A→q, B→s, C→r, D→p
53. Mycorrhiza does not help the host plant in
(a) enhancing its phosphorus uptake capacity
(b) increasing its tolerance to drought
(c) enhancing its resistance to root pathogens
(d) increasing its resistance to insects
54. Which one is the most important role of micro organism for the well - beings of humans?
(a) sewage treatment
(b) production of methane
(c) biological control of plant diseases
(d) conversion of milk to curd
55. Activated sludge should have the ability to settle quickly so that it can
(a) be rapidly pumped back from sedimentation tank to aeration tank
(b) absorb pathogenic bacteria present in waste water while sinking to the bottom of the settling tank
(c) be discarded and anaerobically digested
(d) absorb colloidal organic matter
56. The term antibiotic was coined by
(a) Alexander Flemming
(b) S. Waksman
(c) Louis Pasteur
(d) Edward Jenner
57. Biogas is produced by anaerobic breakdown of biomass of agricultural waste by methanogenic bacteria. It is a
(a) one step process
(b) two steps process
(c) three steps process
(d) multi steps process
58. Mechanism of action of antibiotics are
(a) disruption of cell wall
(b) inhibition of DNA/RNA synthesis
(c) disruption of plamalemma
(d) All the above
59. *Saccharomyces* is used in the formation of
(a) ethanol (b) methanol
(c) acetic acid (d) antibiotics
60. Maximum utilization of biotechnological techniques has been made in the field of
(a) industries (b) medicines
(c) agriculture (d) biogas production

CHAPTER 33

Biotechnology : Principles and Processes

CONCEPTS

Biotechnology deals with large scale production and marketing of products and processes using living organisms, cells or enzymes.

- Modern biotechnology using genetically modified organisms was made possible only when man learnt to alter the chemistry of DNA and construct recombinant DNA. This key process is called recombinant DNA technology of genetic engineering.
- **Examples of Biotechnology** are GMO [Genetically modified organisms], IVF (*in vitro* fertilization), Test tube babies, DNA vaccines and gene therapy.
- **2 main techniques** in biotechnology are
  (i) Genetic engineering
  (ii) Application of genetic engineering in production of bulk of products like antibiotics, vaccines and enzymes.
- **Genetic engineering** can be defined as the generation of new combination of heritable material by the insertion of desired genes or DNA sequences, produced by whatever means outside the cell, into any carrier system so as to allow their incorporation into a host organism in which they do not normally occur but in which they are able to perform normal behaviour and propagation.
- **Three techniques** of genetic engineering are:
  (i) rDNA (recombinant DNA technology)
  (ii) Gene cloning/gene amplification
  (iii) Gene therapy
- **Tools of (recombinant DNA technology) rDNA** technology are
  (i) Vector – Plasmid, Bacteriophage, Cosmid, BAC, YAC
  (ii) Host – with Ori C [origin of replication]
  (iii) Restriction endonuclease – with well defined recognition site
- **Steps of rDNA technology are**
  (i) Isolation of desired gene by cutting with molecular scissors – "Restriction Endonuclease".
  (ii) Introduction of gene into vector (cloning vehicle).
  (iii) Absorption of vector by host organism in culture media.
  (iv) Production of transformed cells.
  (v) Screening of transformed cells for desired result.
- **Vectors** are cloning vehicles required to transfer DNA of interest from one organism to another.
- **Plasmids** are extra-chromosomal, circular, double stranded autonomously replicating sequence in bacterial cell.
- **Bacteriophage** is virus which infect bacteria. They have special sticky "*cos*" site to accept foreign DNA.
- **Cosmid** can be defined as the hybrid vectors derived from plasmids which contain *cos* site of phage 1.
- **BAC** (Bacterial Artificial Chromosome) : Episome + Foreign DNA.
- **YAC** (Yeast Artificial Chromosome) : Yeast DNA + Marker gene from competent bacteria.
- **Marker sequences** are the ones whose phenotype gives observable change, *e.g.*, Tetracycline resistance gene, alcohol fermentation gene.
- **Palindromic** sequence are the ones which read same in one orientation on both the strand.
  For e.g., 5' — GAATTC — 3'
  3' — CTTAAG — 5'
  is palindrome as it reads same on both strand in $5' \rightarrow 3'$ direction. Same is true for $3' \rightarrow 5'$ direction.

RESTRICTION ENZYMES

- Enzymes that recognizes a specific sequence of double stranded DNA and cut the DNA at that site are called **restriction enzymes**. They are often referred as **molecular scissors**.

- **Restriction endonuclease** was found by **Arber** in 1963 in bacteria. First was *EcoRI* found in colon bacterium ***Escherichia coli***. Recognition sequence of *EcoRI* is 5' —GAATTC— 3'.
- **Nomenclature of Restriction Endonuclease.**
  1st letter – Generic name
  2nd letter – Species name
  3rd letter – Strain name
  4th letter – Order in which enzyme is isolated from that strain of bacteria [Roman Numeral]
  e.g., EcoRY13 → stands for ***Escherichia* coli** strain **RY numbered 13.**
- **Endonuclease** makes cuts at specific positions within the DNA and **exonuclease** makes cut/ removes nucleotides from ends of DNA.
- **Desirable properties of cloning vector are**
  (i) High copy number
  (ii) Presence of origin of replication (*Ori*)
  (iii) Presence of selectable marker
  (iv) Presence of unique recognition site in cloning site
  (v) Ability to sustain in bacterial cell/low incompatibility quotient.
- ***Agrobacterium tumefaciens*** is cloning vector for plants which is **T-DNA** (transforming DNA) to transform normal plant cell into tumour and direct them to form mass of cells called **galls** which start secreting chemical required by bacteria.
- **Bacteria can be transformed** in 3 ways
  (i) by making it competent by treating with divalent cations like Ca^{2+}, which make its wall more permeable to fully in DNA.
  (ii) By giving **heat shock treatment** and **incubating** with foreign DNA at **cool temperature.**
  (iii) By **microinjections** – animals cells.
  (iv) **Gene gun/bolistic guns** – transforming plant cells.

Polymerase Chain Reaction

- *In Vitro* synthesis of DNA of interest can be done by PCR (Polymerase chain reaction). In this reaction, multiple copies of the gene (or DNA) or interest is synthesised *in vitro* using two sets of primers (small chemically synthesised oligonucleotides that are complementary to the regions of DNA) and the enzyme DNA polymerase. The enzyme extends the primers using the nucleotides provided in the reaction and the genomic DNA as template. Polymerase chain reaction has three steps : (i) Denaturation; (ii) Primer annealing; (iii) Extension of primers

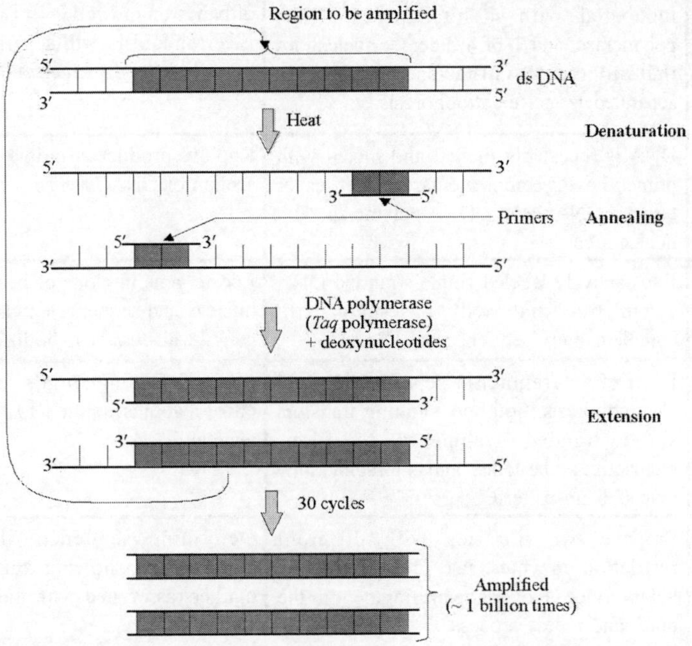

Fig. : Polymerase chain reaction (PCR) : Each cycle has three steps : (i) Denaturation; (ii) Primer annealing; (iii) Extension of primers

BIOREACTORS

- The product from rDNA can be obtained at large scale in **Bioreactors** – where from an inlet, large volume of culture medium is fed and from an outlet, used media is thrown out.
- **2 types of Bioreactor**
  - (i) Simple stirred - tank bioreactor
  - (ii) Sparged stirred - tank bioreactor
- **Downstream processing** is separation and purification of products acheived from bioreactors. The product is converted into formulation for usage, tested for human use, undergoes quality control and then marketed.

DNA Technology

Technique or tool	Brief Description	Use in DNA Technology
Restriction enzymes	Bacterial enzymes that cut DNA at restriction sites, creating "sticky ends" that can base-pair with other fragments.	Cut DNA sequences that can then join to make recombinant DNA (seal with ligase); form restriction fragments used for RFLP analysis and sequencing DNA.
Gel electrophoresis	Mixture of molecules applied to slab of gel in electric field; molecules separate, moving at different rates due to charge and size.	Separate restriction fragments into pattern of distinct bands. Fragments can be sequenced, and put into gene libraries.
cDNA (complementary DNA)	mRNA isolated from cell is treated with reverse transcriptase to produce a complementary DNA strand, which then produces a double-stranded DNA gene.	Creates artificial gene lacking introns which is easier to clone in bacteria, produces cDNA library of the genes that are active in a cell.
DNA sequencing by **Sanger** method	Single-stranded DNA fragments are incubated with 4 nucleotides, DNA polymerase, and 1 of 4 dideoxy nucleotides that interrupt synthesis. Samples are separated by gel electrophoresis.	Sequence of nucleotides is read from the four sets of bands on the gel. Electrophoresis will separate bands that differ by as little as one nucleotide.
Polymerase chain reaction	DNA is repeatedly melted and mixed with primers having complementary sequences for targeted DNA section, DNA polymerase, and nucleotides.	Rapidly produce multiple copies of a gene or section of DNA *in vitro*.
Labeled probe and auto-radiography	Radioactively labeled single-stranded DNA or mRNA that will base-pair with complementary sequences of DNA.	Locate gene in clone of bacteria, identify similar nucleic acid sequences, make cytological map of genes with *in situ* hybridization.
RFLP analysis	Restriction fragments separated by gel electrophoresis, Southern blotting transfers single-stranded fragments to filter, radioactive probe added, and autoradiography reveals band pattern.	Make DNA fingerprints for forensic use, map chromosomes using RFLP markers, diagnose genetic diseases.
Chromosome walking	Prepare two libraries with different restriction enzymes, use probe from one library to locate overlapping fragment in the other and repeat process.	Determine sequence of fragments on a chromosome, showing rough order of genes or markers associated with those fragments.

Exercise

1. Biotechnology is
   (a) a science to culture bacteria
   (b) application of biological organisms to study genetics
   (c) application of biological organisms, systems to manufacture industrial products useful to mankind
   (d) all of the above
2. Genetic engineering aims at
   (a) destroying wild gene
   (b) preserving defective gene
   (c) curing human disease by introducing new gene
   (d) all of the above
3. Which of the enzymes are used to join bits of DNA ?
   (a) Ligase (b) Primase
   (c) DNA polymerase (d) Endonuclease
4 Common bacteria used in genetic engineering is
   (a) *E. coli* (b) *Diplococcus*
   (c) *Rhizobium* (d) *Spirillum*
5. Important objective of biotechnology in agriculture section is
   (a) to produce pest resistant varieties of plant
   (b) to increase the nitrogen content
   (c) to decrease the seed number
   (d) to increase the plant weight
6. Which structure involved in genetic engineering?
   (a) Plastid (b) Plasmid
   (c) Codon (d) None
7. The raising new plants from a plant tissue through tissue culture is termed as
   (a) micropropagation (b) micro grafting
   (c) macro economic (d) macro consumer
8. DNA repairing is done by
   (a) ligase
   (b) DNA polymerase III
   (c) DNA polymerase II
   (d) DNA polymerase I
9. Genetic engineered bacteria is used for production of
   (a) insulin (b) glucagon
   (c) penicillin (d) all of these
10. Transgenic crops are modified through genetic engineering to develop natural resistance to insect pests. Which one of the pair is transgenic plant?
    (a) Tomato and wheat
    (b) Tomato and rice
    (c) Tobacco and tomato
    (d) Maize and sugarcane
11. DNA or RNA segment tagged with a radioactive molecule is called
    (a) vector (b) probe
    (c) clone (d) plasmid
12. Gel electrophoresis is used for
    (a) cutting of DNA into fragments
    (b) separation of DNA fragments according to their size
    (c) construction of recombinant DNA by joining with cloning vectors
    (d) isolation of DNA molecule
13. Restriction endonucleases are enzymes which
    (a) make cuts at specific positions within the DNA molecule
    (b) recognize a specific nucleotide sequence for binding of DNA ligase
    (c) restrict the action of the enzyme DNA polymerase
    (d) remove nucleotides from the ends of the DNA molecule
14. There is a restriction endonuclease called *EcoRI*. What does co. part in it stand for ?
    (a) colon (b) coelom
    (c) coenzyme (d) *coli*
15. A genetic clone is
    (a) a plant produced by asexual means
    (b) hybrid produced by sexual means
    (c) homozygous plant produced by sexual means
    (d) heterozygous plant produced by sexual means

16. Introduction of food plants developed by genetic engineering is not desirable because
(a) economy of developing countries may suffer
(b) these products are less tasty as compared to the already existing products
(c) this method is costly
(d) there is danger of entry of viruses and toxins with introduced crop

17. The linking of antibiotic resistance gene with the plasmid vector became possible with
(a) DNA ligase (b) endonucleases
(c) DNA polymerase (d) exonucleases

18. Polyethylene glycol method is used for
(a) biodiesel production
(b) seedless fruit production
(c) energy production from sewage
(d) gene transfer without a vector

19. Which one of the following is used as vector for cloning genes into higher organisms?
(a) Baculovirus
(b) *Salmonella typhimurium*
(c) *Rhizopus nigricans*
(d) Retrovirus

20. Which one of the following palindromic base sequences in DNA can be easily cut at about the middle by some particular restriction enzyme?
(a) 5'............CGTTCG............3'
3'............ATGGTA............5'
(b) 5'............GATATG............3'
3'............CTACTA............5'
(c) 5'............GAATTC............3'
3'............CTTAAG............5'
(d) 5'............CACGTA............3'
3'............CTCAGT............5'

21. Agarose extracted from sea weeds finds use in
(a) spectrophotometry
(b) tissue culture
(c) PCR
(d) gel electrophoresis

22. *Bacillus thuringiensis* forms protein crystals which contain insecticidal protein
(a) binds with epithelial cells of midgut of the insect pest ultimately killing it
(b) is coded by several genes including the gene *cry*
(c) is activated by acid pH of the foregut of the insect pest
(d) does not kill the carrier bacterium which is itself resistant to this toxin

23. PCR (Polymerase Chain Reaction) method is useful for
(a) amplification of DNA for forming billions of copies of itself
(b) monoclonal antibody production
(c) hybridoma production
(d) all of the above

24. Collection of organisms as bacteria that all contain the same piece of recombinant DNA is called
(a) gene cloning (b) cell cloning
(c) organ cloning (d) all of these

25. *Agrobacterium tumefaciens* is used in genetics engineering for
(a) DNA-mapping
(b) DNA-modification
(c) vector
(d) DNA fingerprinting

26. A genetically engineered bacteria used for clearing oil spills is
(a) *Escherischia coli*
(b) *Bacillus subtilis*
(c) *Agrobacterium tumifaciens*
(d) *Pseudomonas putida*

27. Genetic engineering is
(a) study of extra nuclear genes
(b) manipulation of genes by artificial method
(c) manipulation of RNA
(d) manipulation of enzymes

28. Which of the following enzymes cut the DNA molecule at specific nucleotide sequence ?
(a) Restriction endonuclease
(b) DNA - ligase
(c) RNA - polymerase
(d) Exonuclease

29. Restriction endonucleases are used in genetic engineering because
(a) they can degrade harmful proteins
(b) they can join DNA fragments
(c) they can cut DNA at variable sites
(d) they can cut DNA at specific base sequences

30. When the genotype of an organism is improved by the addition of foreign gene, the process is called
(a) tissue culture
(b) genetic diversity
(c) genetic engineering
(d) plastic surgery

31. Which vector is commonly used in the transfer of gene in a crop plant ?
    (a) Plasmids of *B. subtilis*
    (b) Bacteriophages
    (c) Ti-plasmids of *Agrobacterium*
    (d) *E. coli*
32. Taq - polymerase which is used to amplification of DNA related with
    (a) hybridoma technique
    (b) PCR-technique
    (c) gene cloning
    (d) r-DNA technology
33. First cloned animal
    (a) dolly sheep (b) polly sheep
    (c) molly sheep (d) dog
34. Genetic engineering is possible because
    (a) we can cut DNA at specific sites by endonuclease like DNAase
    (b) restriction endonuclease purified from bacteria is well understood
    (c) the phenomenon of transduction in bacteria is well understood
    (d) we can see DNA by electron microscope
35. Construction of recombinant DNA involves
    (a) cleaving and rejoining DNA segments with ligase alone
    (b) cleaving DNA segments with ligase and rejoining them with endonuclease
    (c) cleaving DNA segment with endonuclease and rejoining them with ligase
    (d) cleaving and rejoining DNA segments with endonuclease alone
36. Which one of the following bacteria has found extensive use in genetic engineering work in plants?
    (a) *Clostridium septicum*
    (b) *Xanthomonas citri*
    (c) *Bacillus coagulens*
    (d) *Agrobacterium tumefaciens*
37. Genetically engineered insulin can be obtained by
    (a) recombinant DNA technique with the help of *E. coli*
    (b) two coded insulin genes separated then incorporated into bacteria
    (c) the extraction of cow's and pig's pancreas
    (d) technique not developed till now
38. The process of joining together different DNA fragments is often referred to as
    (a) transcription (b) cloning
    (c) gene splicing (d) DNA amplification
39. It is now possible to breed plants and animals with desired characters through
    (a) genetic engineering
    (b) chromosome engineering
    (c) gel electrophoresis
    (d) tissue culture
40. A distinct mechanism that usually involves a short segment of DNA with remarkable capacity to move from one location in a chromosome to another, this is called
    (a) DNA replication
    (b) DNA transposition
    (c) DNA hybridization
    (d) DNA recombination
41. Production of a human protein in bacteria by genetic engineering is possible because
    (a) bacterial cell can carry out the RNA splicing reactions
    (b) the human chromosome can replicate bacterial cell
    (c) the mechanism of gene regulation is identical in humans and bacteria
    (d) the genetic code is universal
42. Which one of the following pairs of terms/names mean one and the same thing ?
    (a) Gene pool- genome
    (b) Codon - gene
    (c) Cistron -triplet
    (d) DNA fingerprinting-DNA profiling
43. The transfer of genetic material from one bacterium to another through the mediation of a vector like virus is termed as
    (a) transduction (b) conjugation
    (c) transformation (d) translation
44. In agarose gel electrophoresis, DNA molecules are generally
    (a) positively charged
    (b) negatively charged
    (c) neutral
    (d) both (a) and (b)

45. The most important feature in a plasmid to be used as a vector is
(a) origin of replication (*ori*)
(b) presence of a selectable marker
(c) presence of sites for restriction endonuclease
(d) its size

46. The main technique involved in agricultural biotechnology is called
(a) tissue culture (b) transformation
(c) plant breeding (d) DNA replication

47. Which one of the following pair is incorrect ?
(a) Plasmid - small piece of extrachromosomal DNA in bacteria
(b) Interferon - an enzyme that interferes with DNA replication
(c) Cosmid - A vector for carrying large DNA fragments into host cells
(d) Myeloma - antibody producing tumour cells.

48. Large number of bacteria produced from a single bacterium is called
(a) deme (b) population
(c) clones (d) offspring

49. Restriction enzyme '*Eco RI*' nicks at one of the following sites on DNA
(a) 5' G↓AATTC 3' (b) 5'GG↑CC3'
(c) 5'A↓AGCTT 3' (d) None above

50. Biotechnology is the modern branch of biology which deals with
(a) genetic engineering (b) biochemistry
(c) microbiology (d) all the above

51. Which of the following is not a genetic vector ?
(a) Plasmid (b) Phage
(c) Cosmid (d) Virusoid

52. 'Cloning' is meant for/to
(a) production of gene in *E. coli*
(b) preserve the genotype of organism
(c) replace the original gene
(d) all of the above

53. Which of these best shows that plant cells are totipotent ?
(a) A shoot culture for the purpose of micropropagation
(b) Flower meristem culture for the purpose of somatic embryos
(c) Protoplasmic culture for the purpose of genetic engineering of plants
(d) Leaf, stem and root culture for the purpose of cell suspension cultures

54. 'Axenic culture' is
(a) culture of tissue
(b) growing of shrubs
(c) growing of tall trees
(d) culture of tissue free from contamination

55. The enzyme, which helps to cut one strand of DNA duplex to release tension of coiling of two strands is
(a) DNA ligase
(b) DNA polymerase I
(c) topoisomerase
(d) helicase or unwindase

56. Plasmids are suitable vectors for gene cloning because
(a) these are small circular DNA molecules which can integrate with host chromosomal DNA
(b) these are small circular DNA molecules with their own replication at origin site
(c) these can shuttle between prokaryotic and eukaryotic cells
(d) these often carry antibiotic resistance genes

57. Which of the following is/are pesticide?
(a) DDT (b) BHC
(c) Aldrin (d) All of the above

58. Use of biology in industrial process and for improving quality of life is called
(a) genetic engineering
(b) eugenics
(c) microbiology
(d) biotechnology

59. Product of biotechnology is
(a) transgenic crop (GM crop)
(b) humulin
(c) biofertilizer
(a) all of these

60. Antibiotics belong to the category of
(a) steroids (b) toxins
(c) medicines (d) tonics

Biotechnology and its Applications

CONCEPTS

Biotechnology mainly deals with industrial scale production of bio-pharmaceuticals and biologicals using genetically modified microbes, fungi, plants and animals.

Applications of Biotechnology are

- GM crop production
- Transgenics < Animals / Plants
- Therapeutics
- Gene therapy
- Molecular diagnosis

GM CROP PRODUCTION – GENETICALLY MODIFIED CROP

(i) Purpose of introduction of GM crop was to minimise use of fertilisers and chemicals so that their harmful effect on environment could be reduced.

(ii) GMO are genetically modified organisms. Plants, bacteria, fungi and animals whose genes have been altered are called GMO.

(iii) Examples GM crops are **Bt cotton** and **Pest resistant plants.**

Bt cotton : It is a pest resistant plant which resist attack of *Lepidopterans* insects. The plant Bt toxin gene which cause death of insect larvae by causing cell lysis and swelling of epithelium of midgut.

Bt Toxin :

- Bt toxin is biologically produced by bacterium called *Bacillus thuringiensis* (Bt)
- This toxin is insecticidal protein crystal produced in bacteria (inactive form in bacteria) during a particular phase of growth.
- Inactive protein (protoxin) is converted into active form of toxin due to alkaline pH of gut of insect which solubilise crystals.
- The activated toxin binds to surface of midgut epithelial cell and creates pores that cause cell swelling and lysis and eventually death of the insect.
  - These Bt toxin genes are isolated from bacteria and incorporated into several crop plants like cotton which has these lepidopteran insects as pests.
  - Choice of gene depends on targeted crop and insect species as most Bt toxins are insect group specific.
  - **Bt toxin** is coded by gene called **CRY genes**.
  - *Cry IAc* and *Cry II Ab* – control **cotton bacterium**

    Cry IAb and *Cry II Ac* – control **corn borer worm**

RNA Interference Technology

- Prevention of nematode *Meloidegyne incognitoa* parasitism on roots of tobacco plants by RNA interference technology (RNAi).
- RNAi is a method of cellular defense in eukaryotes.
- This is silencing of mRNA due to complementary double stranded RNA that prevents translation of mRNA [silencing].
- Source of dsDNA is **retrovirus** having RNA genome or **transposons** (mobile genetic material).
- Using ***Agrobacterium*** as a vector, nematode specific DNA is introduced into host plant which produces dsRNA. They initiate RNAi and silence specific mRNA when go inside nematode. Thus, parasite could not survive in transgenic host expressing specific dsRNA acting as **RNAi**.
- Transgenic plants are organisms containing a foreign gene from a different species received by genetic engineering .

- **Examples of Transgenic plants are**

Transgenic plants	Character introduced
Flavr savr tomato	Long shelf life
Soyabean, corn	Herbicide tolerance
Brinjal, tomato	Insect resistance
Tobacco, potato	Viral resistance
Golden rice	Rich in vitamin A

Therapeutics

Insulin production

- Insulin is a hormone used for diabetics patient.
- Insulin was extracted from pancreas of cattle, pigs in limited quantity.
- Gene for insulin is introduced in bacteria and these genetically engineered bacteria produce insulin in bulk as they express gene for insulin along with their genome.
- Insulin is synthesized as prohormones, immature form with 2 chains A and B joined by C-peptide. Removal of C makes it mature.
- Insulin chain A and B are produced separately in plasmid of *E.coli*, exhausted and combined by creating disulfide bond to make it human insulin (active form) *celled humulin*.

OTHER DNA / BIOTECHNOLOGY / THERAPEUTIC PRODUCTS

Therapeutic product	Application
Human insulin	Treatment of diabetes
Human growth hormone	Treatment of dwarfism
Calcitonin	Treatment of rickets
Interferon	Treatment of viral infections
Interleukins	Treatment of immune related disorders

Gene Therapy

- Gene therapy is **alteration of specific genes** to modify the patient's genotype for the treatment of hereditary diseases like haemophilia and phenylketonuria – using ELISA.
- The first clinical gene therapy was given to 4 year old in 1990 with ADA deficiency.
- **Treatment of SCID** [Severely combined immuno deficiency disease] : SCID patient has a defective gene for enzyme adenosine deaminase [ADA]. Patient do not have T lymphocyte and don't mount immune response against pathogens. So in gene therapy–
  - A WBC is extracted from bone marrow of sufferer and a good copy of human gene encoding ADA is introduced by **retrovirus** as vector.
  - This leads to WBC with a good copy of gene, in addition to existing faulty copy.
  - The treated cells are reinjected to patient's bone marrow and cells play normal role in immune system.

Medical Diagnosis of Diseases

- **PCR** is a gene amplification technique that has found wide spread use in medicine & molecular biology.
- **ELISA** is a laboratory test, which can quickly determine antigen or antibody levels in blood or other fluids.
- A single stranded DNA/RNA attached with radioactive molecule is called **molecular probe**. Molecular probe are used for detection of gene by complementary hybridising in cloned cell, followed by **autoradiography**.

Transgenic Animals

These are the animals that had DNA manipulated to possess and express an extra gene.

- They receive foreign genes to produce highly valuable material.
- **Hirudin** is a protein that stops blood clotting. The gene is then transferred into *Brassica napus*, where hirudin accumulates in seeds. The hirudin is purified & used as medicine.
- Transgenic sheep produce clotting factor for hemophilics.
- Transgenic goat produce protein to dissolve clot, cure for coronary thrombosis.
- Transgenic cattle, sheep and fish grow most quickly and produce more meat.

Transgenic animals	Useful application
1. Cow, Sheep, Goat	Therapeutic: human proteins in their milk.
2. Pig	Organ transplantation without risk of rejection.
3. Rabbits	Molecular farming or gene farming.
4. Mouse	Contains a human gene that causes breast cancer. This enables the researchers to study the very early development of cancer.
5. Fish (Common Carp, Cat-fish , Salmon, Gold fish)	They contain human growth hormone (hGH). They attain a size twice to that shown by non transgenic fish.

ETHICAL ISSUES

- Manipulation of living organism by human race is unethical and should not go unregulated.
- GM organisms can have unpredictable result when they are introduced into ecosystems.
- Modification/use of living organism for food and medicine production creates patent problem.

For reduction of these ethical issues Indian government has set up organization GEAC [Genetic engineering approval committees], that make decision regarding validity of GM research and safety of introducing GMO in ecosystem.

Biopiracy

- **Biopiracy** is use of bioresource by multinational companies and other organisation without proper authorization from countries and people concerned without compensatory payment. It is due to lack of awareness and strong patent legalities that Basmati rice has been patented by an american company by crossing Indian variety to new dwarf variety and claiming as novel / inventions.

Practical problems associated with genetic engineering :

1. Biological warfare.
2. Production of drug resistant pathogens.
3. Ethical problem of examining someone's genetic map.
4. Introduction of super-weed due to introduction of faulty gene.
5. Harmless intestinal bacteria can turn into cancer causing bacterial forms.

Exercise

1. Golden rice is a transgenic crop of the future with the following improved trait.
   (a) insect resistance
   (b) high lysine (essential amino acid) content
   (c) high protein content
   (d) high vitamin-A content
2. Which one of the following is commonly used in transfer of foreign DNA into crop plants?
   (a) *Meloidogyne incognita*
   (b) *Agrobacterium tumefaciens*
   (c) *Penicillium expansum*
   (d) *Trichoderma harzianum*
3. Transgenic plants are the ones
   (a) generated by introducing foreign DNA into a cell and regenerating a plant from that cell
   (b) produced after protoplast fusion in artificial medium
   (c) grown in artificial medium after hybridization in the field
   (d) produced by a somatic embryo in artificial medium
4. The genetically-modified (GM) brinjal in India has been developed for
   (a) insect-resistance
   (b) enhancing shelf life
   (c) enhancing mineral content
   (d) drought-resistance

5. Which one of the following techniques made it possible to genetically engineer living organism ?
(a) Recombinant DNA techniques
(b) X-ray diffraction
(c) Heavier isotope labelling
(d) Hybridization

6. Transgenic animal has
(a) foreign DNA in all of its cells
(b) foreign RNA in all of its cells
(c) foreign DNA in some of the cells
(d) both (b) and (c)

7. First transgenic plant
(a) potato (b) tomato
(c) tobacco (d) maize

8. Important objective of biotechnology in agriculture section is
(a) to produce pest resistant varieties of plants
(b) to increase the nitrogen content
(c) to decrease the seed number
(d) to increase the plant weight

9. Which one of the following is not an antibiotic ?
(a) Streptomycin (b) Citric acid
(c) Griseofulvin (d) Cephalosporin

10. Which of these can preserve food stuffs ?
(a) Salt and sugar (b) Sugar and vinegar
(c) Vinegar (d) All of the above

11. Crown gall disease in plants is caused by
(a) Ti-plasmid (b) Pi-plasmid
(c) bacteria (d) virus

12. Lactic acid bacteria (LAB) at suitable temperature converts milk to curd, which improves its nutritional quality by enhancing vitamin
(a) A (b) B
(c) C (d) D

13. Antibiotics belong to the category of
(a) steroids (b) toxins
(c) medicines (d) tonics

14. Product of biotechnology is
(a) transgenic crop (GM crop)
(b) humulin
(c) biofertilizer
(d) all of these

15. Flavr Savr is the transgenic variety of
(a) cotton (b) rice
(c) tomato (d) potato

16. Genetically engineered bacteria have been successfully used in the commercial production of
(a) human insulin (b) testosterone
(c) thyroxine (d) melatonin

17. Maximum application of animal cell culture technology today is in the production of
(a) edible proteins (b) insulin
(c) interferons (d) vaccines

18. ELISA is used to detect viruses where the key reagent is
(a) RNase
(b) alkaline phosphatase
(c) catalase
(d) DNA probe

19. *Bacillus thuringiensis* (Bt) strains have been used for designing novel
(a) biofertilizers
(b) bio-metallurgical techniques
(c) bio-mineralization processes
(d) bioinsecticidal plants

20. A trangenic food crop which may help in solving the problem of night blindness in developing countries is
(a) flavr savr tomatoes (b) starlink maize
(c) Bt soybean (d) golden rice

21. Main objective of production/use of herbicide resistant GM crops is to
(a) eliminate weeds from the field without the use of manual labour
(b) eliminate weeds from the field without the use of herbicides
(c) encourage eco-friendly herbicides
(d) reduce herbicide accumulation in food articles for health safety

22. What is true about Bt toxin?
(a) Bt protein exists as active toxin in the *Bacillus*
(b) The activated toxin enters the ovaries of the pest to sterilise it and thus prevent its multiplication.
(c) The concerned *Bacillus* has antitoxins.
(d) The inactive protoxin gets converted into active form in the insect gut

23. Some of the characteristics of Bt cotton are
(a) long fibre and resistance to aphids
(b) medium yield, long fibre and resistance to beetle pests
(c) high yield and production of toxic protein crystals which kill dipteran pests
(d) high yield and resistance to bollworms

24. Consider the following statements (A-D) about organic farming.
(A) Utilizes genetically modified crops like Bt cotton

(B) Uses only naturally produced inputs like compost
(C) Does not use pesticides and urea
(D) Produces vegetables rich in vitamins and minerals

Which of the above statements are correct?
(a) (B), (C) and (D) (b) (C) and (D) only
(c) (B) and (C) only (d) (A) and (B) only

25. The new strain of bacteria produced by biotechnology in alcohol industry is
(a) *Escherichia coli*
(b) *Saccharomyces cerevisiae*
(c) *Bacillus subtilis*
(d) *Pseudomonas putida*

26. The name of drug used in cancer treatment produced by biotechnology is
(a) interferon
(b) human growth hormone (HGH)
(c) thyroid stimulating hormone (TSH)
(d) insulin

27. The prerequisites for biotechnological production of antibiotics is
(a) to search an antibiotic producing micro-organism
(b) to isolate the antibiotic gene
(c) to join antibiotic gene with *E.coli* plasmid
(d) all of the above

28. Recombinant DNA technology prepared vaccine is called
(a) first generation vaccine
(b) second generation vaccine
(c) third generation vaccine
(d) synthesized vaccine

29. *Cry* protein is obtained from
(a) *Bacillus thuringiensis*
(b) *Bacillus subtilis*
(c) *Clostridium welchii*
(d) *E. coli*

30. Which of following is not related with bioweapons?
(a) *Bacillus anthracis*
(b) *Clostridium botulinum*
(c) *E. coli*
(d) *Yersinia pestis*

31. Use of transgenic plants as biological factories for the production of special chemicals is called
(a) molecular farming (b) molecular genetics
(c) molecular mapping (d) dry farming

32. In tissue culture medium the embryoids formed from pollen grain is due to
(a) cellular totipotency (b) organogenesis
(c) double fertilization (d) test-tube culture

33. Maximum utilization of the techniques of biotechnology has been made in the field of
(a) agriculture (b) medicines
(c) industries (d) biogas production

34. An example of gene therapy is
(a) production of injectable hepatitis B vaccine
(b) production of vaccines in food crops like potatoes which can be eaten
(c) introduction of gene for adenosine deaminase in persons suffering from severe combined immuno deficiency
(d) production of test tube babies by artificial insemination and implantation of fertilized egg

35. Find the incorrect statement
(a) Gene therapy is a genetic engineering technique used to treat diseases at molecular level by replacing defective genes with normal genes
(b) Calcitonin is a medically useful recombinant product in the treatment of infertility
(c) Bt toxin is a biodegradable insecticide obtained from *Bacillus thuringiensis*
(d) *Trichoderma* species is a biocontrol agent for fungal diseases of plants

36. Which of the following is false for Bt transgenic plant ?
(a) Disease resistance
(b) Prepared by *Bacillus thuringiensis*
(c) It is recombinant type
(d) No such plant is known

37. In plant biotechnology, root tumours are induced in plant using the bacterium
(a) *Agrobacterium rhizogenes*
(b) *Agrobacterium basilis*
(c) *Rhizobium*
(d) None of these

38. Curding of milk takes place by
(a) *Streptococcus lactis*
(b) *Streptococcus thermophilus*
(c) *Lactobacillus lactis*
(d) All of the above

39. Highest number of antibiotics are produced by
(a) *Bacillus* (b) *Pencillium*
(c) *Streptomyces* (d) *Cephalosporium*

40. Conversion of sugar into alcohol during fermentation is due to the direct action of
(a) temperature
(b) micro-organisms
(c) concentration of sugar solution
(d) zymase
41. Biopiracy is related to
(a) bioresearches
(b) traditional knowledge
(c) biomolecules and genes discovered
(d) all of these
42. Interferons are
(a) antiviral proteins
(b) complex proteins
(c) anti-bacterial proteins
(d) anti-cancer proteins
43. Which one of the following is not correctly matched ?
(a) *Spirulina* - Single cell protein
(b) *Rhizobium* - Biofertilizer
(c) *Streptomyces* - Antibiotic
(d) *Serratia* - Drug addiction
44. The fruit juices turn bitter in taste if they are kept in open place for sometime, because of
(a) bacteria of the atmosphere react with the juice
(b) fermentation of the juice by yeast
(c) some internal factors
(d) all the above three statements are correct
45. *Bacillus thuringiensis* is used to control
(a) bacterial pathogens
(b) fungal pathogens
(c) nematodes
(d) insect and pests
46. Which one of the following is now being commercially produced by biotechnological procedures ?
(a) Nicotine (b) Morphine
(c) Quinine (d) Insulin
47. Breeding for disease resistance requires
(a) a good source of resistance
(b) planned hybridisation
(c) disease test
(d) All of the above
48. The abbreviation 'Bt' in 'Bt' toxin stands for
(a) biotechnology
(b) biotoxin
(c) toxin released by *Bacillus thuringiensis*
(d) toxins released by bacteria
49. Recombinant DNA technology can be used to produce quantities of biologically active form of which one of the following products in *E. coli* ?
(a) Luteinizing hormone (b) Ecdysome
(c) Rifamycin (d) Interferon
50. In industries microorganisms are used for the manufacture of
(a) gluconic acid (b) malic acid
(c) glycerol (d) castor oil
51. Pasteurised milk is
(a) sterile and will not turn sour
(b) free from pathogenic bacteria
(c) not free from bacteria
(d) both (a) and (b)
52. Yeast is a
(a) prokaryote (b) eukaryote
(c) heterokaryote (d) None of these
53. Enzymes can be immobilised by
(a) cross-linking enzyme molecules
(b) covalently attaching to a solid support
(c) entrapping them in gel
(d) All of the above
54. Streptococcus is employed in the manufacturing of
(a) wine (b) bread
(c) cheese (d) All of these
55. Which of the molecules listed below is a product of fermentation of glucose by yeast ?
(a) $(C_6H_{10}O_5)n$ (b) C_2H_5OH
(c) $C_6H_{12}O_6$ (d) CH_3OH
56. C-peptide of human insulin is
(a) a part of mature insulin molecule.
(b) responsible for formation of disulphide bridges.
(c) removed during maturation of pro-insulin to insulin.
(d) responsible for its biological activity.
57. Which of the following is a correct statement?
(a) "*Bt*" in *Bt*-cotton indicates that it is genetically modified organism produced through biotechnology.
(b) Somatic hybridization involves fusion of two complete plant cells carrying desired genes.
(c) The anticoagulant hirudin is being produced from transgenic *Brassica napus* seeds.
(d) "*Flavr Savr*" variety of tomato has enriched the production of ethylene which improves its taste.

58. Select the correct statement.

(a) Genetic engineering works only on animals and has not yet been successfully used on plants.

(b) There are no risks associated with DNA technology.

(c) The first step in PCR is heat which is used to separate both the strands of target DNA.

(d) DNA from one organism will not bond to DNA from another animal.

59. Read the following statements and choose the correct option.

(i) Gene therapy has been tested on large number of patients with a wider variety of inherited genetic disorders, and in numerous cases it has produced a complete cure.

(ii) Genetic engineering has been used to mass produce insulin for curing the diabetes.

(iii) DNA hybridization is the base pairing of DNA from two different sources.

(iv) Genetic engineering is a technique of plant breeding.

(a) (i) and (ii) (b) (ii) and (iii)
(c) (i), (ii) and (iii) (d) All of these

60. Following are the steps in one type of gene therapy.

(i) Inject engineered cells into patients bone marrow.

(ii) Viral DNA carring the normal allele inserts into chromosome.

(iii) Let retrovirus infect bone marrow cells that have removed from patient and cultured.

(iv) Insert RNA version of normal allele into retrovirus.

The correct sequence is –

(a) (i), (ii), (iii) and (iv) (b) (iv), (iii), (ii) and (i)
(c) (i), (ii), (iv) and (iii) (d)(iv), (iii), (i) and (ii)

Organisms and Environment

CONCEPTS

- The idea about ecology was first started by **Relter**.
- The term ecology was given and defined by **Haeckel** as : **Ecology** is the study of the interactions between the organism and their environment.

LEVELS OF ECOLOGICAL ORGANISATION

- **Individual :** The first level of study in ecology is the individual.
- **Population :** The collection of individuals of a given species is called population.
- **Community :** The interacting groups of populations of various species constitute a community.
- **Ecosystem :** A biological community and the physical environment associated with its constitute ecosystem.
- **Biome :** A major ecological community or complex of communities that extends over a large geographical area.
- **Species** is a group of organisms that resembles each other more than they resemble to any other organism, and that can breed among themselves and produces fertile offspring. Species cannot be subdivided into two or more species.
- The smallest level of organisation is an **individual** of species and the largest level of organisation is **biosphere**.
- The rotation of earth around the sun and the tilt of its axis cause yearly variations in the intensity and duration of temperature which results in different seasons.
- These variations together with other variations results in the formation of **major biomes** such as rainforest, desert, taiga, tundra, alpine regions in various parts of the world.
- **In India, the major types of biomes are**
  (a) **Tropical rainforest :** Includes parts of Assam, Meghalaya, etc.
  (b) **Deciduous forest :** Includes parts of Indo-gangetic plains, etc.
  (c) **Desert :** Includes Rajasthan
  (d) **Sea coast :** Includes coastal parts of Kerela, Tamil Nadu, etc.
- **Habitat :** It is a specific place or locality where an organism lives.
- The key elements that lead to so much variation in the conditions of different habitats can be subdivided into two categories:
  (a) **Abiotic components** – Includes all the physical components, such as soil, temperature, water, light, etc.
  (b) **Biotic components** – Includes all the living organisms.
- **Ecotone :** It is the marginal vegetation present in between two well established habitat. Eg. - Estuary is a water body present near the sea which is connected both to fresh water and sea water.

Types of plant Communities

(a) **Lithophyte :** Plants growing on rocks. Eg - lichens.
(b) **Chasmophyte :** Plants is in rock crevices. Eg- Bryophytes (moss).
(c) **Oxylophytes :** Plants growing in acidic soil or water. Eg - Drosera.
(d) **Halophyte :** Plants growing in saline soil or water. Eg - Marine plants.
(e) **Psammophyte :** Plants growing in sand. Eg - Cactus.
(f) **Cryophyte :** Plants growing in snow. Eg - Deodar.
(g) **Alpine vegetation :** Plants growing on mountain tops. Eg - Deodar.

(h) **Mangrove vegetation :** Plants growing in marshy area.

(i) **Cerute vegetation :** Plants growing on dead logs. Eg - Mushrooms.

(j) **Chersophytes vegetation :** Plants growing on waste land. Eg - Prickly poppy (Argemone).

ENVIRONMENT

- It is the sum total of physical and biotic conditions influencing the behaviour of the organism.
- The life supporting environment of planet earth is the **biosphere**. It is composed of three chief media - air, water & soil, which are the components of 3 major subdivisons of the biosphere - atmosphere, hydrosphere & lithosphere respectively.

Atmosphere

- The multilayered gaseous envelope .
- It is divided into 5 distinct layers troposphere, stratosphere, mesosphere, ionosphere & exosphere.

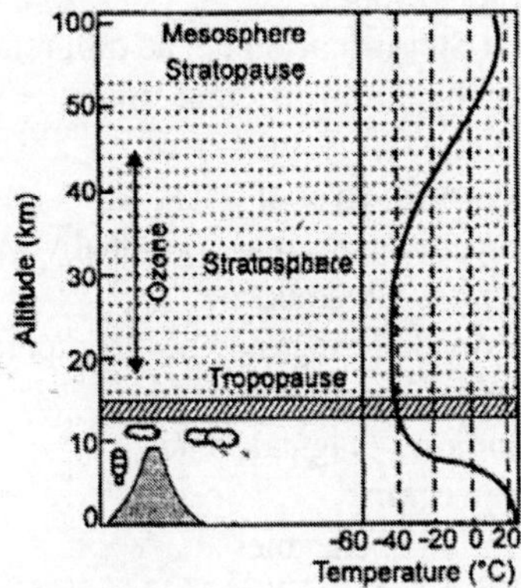

Water Resource

- It is the major component of the **hydrosphere** and covers about 3/4th of the earth's surface.
- The total volume of water present on earth is about 1.4 billion cu km.
- Out of this 97% is sea water which cannot be used directly. Only 3% is fresh water.
- Out of 3% about 77% is stored in ice caps, about 22.5% is ground water. Only about 0.5% is present in river & lakes which is available for direct use.
- Out of the total water evaporated from sea, about 90% falls back in them but about 10% comes on the ground and serves as additional resource.
- Aquatic animals which can tolerate only a narrow range of salinity are called **stenohaline**.
- Aquatic organisms having wide range of salt tolerance are called **euryhaline**.

Lithosphere

The solid component of biosphere.

Land resource is the major component of the lithosphere and covers about 1/4th of the earth's surface. Out of the total landmass, about 36% is occupied by roads, cities industries & human dwelling.

- About 28 to 29% is covered by forest. [The ideal forest cover should be 33 to 35% i.e. about 1/3rd of the landmass].
- About 22% (Natural waste land) is occupied by marshy area, **swamps, desert** etc.
- Sahara deserts is the largest natural wasteland in world.

Soil

- This soil is a mixture of inorganic mineral particles derived from weathering of rock and organic matter consisting of humus.
- The physical properties of soil depend on colour, texture, porosity and water retaining capacity.
- The chemical properties of soil depend on salt content, organic and inorganic nutrients. The quality of soil is depleted mainly by salination, acidification, water logging and soil erosion.
- The removal of top soil from any area is known as **soil erosion**. It takes place mainly by wind & by water. For checking soil erosion, large scale plantation should be done. The plants act as soil binder and wind breakers (wind screen).
- In hilly areas, soil erosion can be checked by **Terrace farming and contour farming.**

Soil Profile

- The soil profile is the vertical section of the crust of soil. By studying soil profile, the nature of different soil layers can be known. The layer are reffered as horizon.

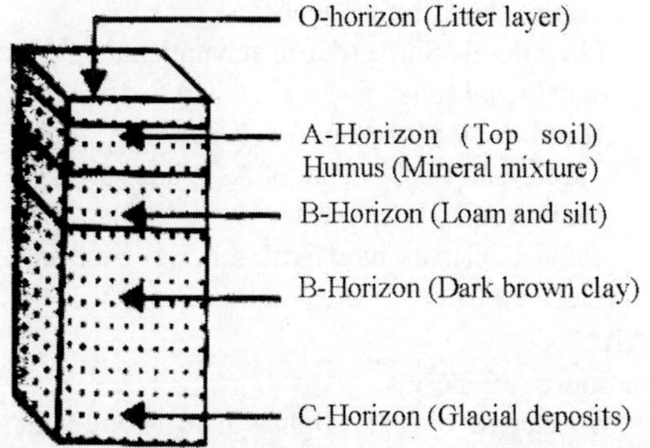

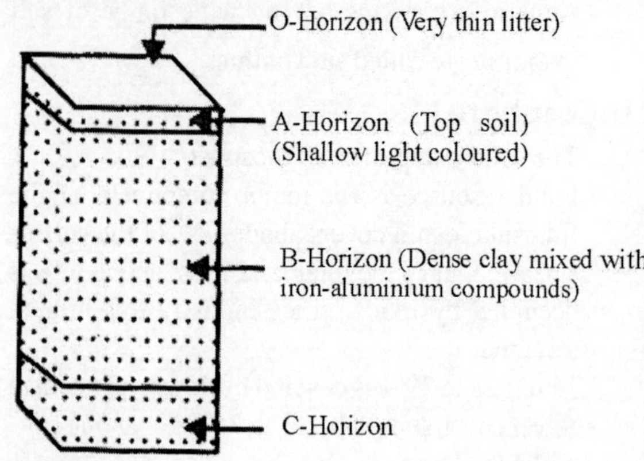

Soil Water

(a) **Run of water**

(b) **Ground water**

(c) **Gravitational water**

(d) **Capillary water :** It is the water which is present in the space between the soil particle. The spaces between the soil particles are known as interstitices. It is the only water available to plants.

- The capillary capacity of soil is also known as **field capacity**. The capillary capacity + gravitational water is referred as storage capacity.

- **On the basis of the size of particles, soil is of following types :**
  (i) Clay - upto .062 mm
  (ii) Silt - .002 to .02 mm
  (iii) Fine sand - .02 to .2 mm
  (iv) Coarse sand - .2 to 2 mm
  (v) Fine gravel - 2 to 5 mm
  (vi) Coarse gravel - 5 and more
- **On the basis of their origin and formation, the soil is of following type :**

(i) **Sedimentary or residual soil** are formed by active degeneration of parental rock material and remain at the same place where they have been formed.

(ii) **Transported or shifted soil** - There are brought to any area by different agencies :

Illuvial soil : Shifted due to gravitational influence or colluvial soil.

Alluvial soil : Is by running water.

Glacial soil : Is by glaciers. As Edian soil.

Aeolian soil : Is by wind.

Alluvial soil in the most fertile soil. Eg - Sand dunes, volcanic ash.

Light

- Source of energy
- Wavelength of visible light = 3900 to 7600 nm

Photoperiodism : It is a phenomena showing the impact of duration of light upon flowering in plants. On this basis, plants have been divided into 2 categories :
(a) Long day plants (b) Short day plants

Phototropism : Growth of plants affected by light.

U.V. radiations are of 3 types
(1) UV–C (100nm – 280nm)
(2) UV–B (280nm – 320nm)
(3) UV–A (320nm – 390nm).

Effect on Animals

- U.V. rays induce mutation.
- Affects skin pigmentation.

Temperature

- Temperature determines the configuration of distribution of organism.
- **Eurythermal organisms** - Which can tolerate very large fluctuation in temp. E.g. - man, lizard.
- **Stenothermal organisms** - Which can tolerate very small fluctuation in temperature. Eg. - Fish, snail.
- Aestivation and Hibernation : Dormancy during extreme temperature.

  Thermal Stratification : The differentiation of aquatic habitat into different strata

(a) **Epilimnion**
- Uppermost layer of water.
- Temperature decreases gradually with depth.

(b) **Metalimnion/Thermocline**
- Intermediate region where temperature fall sharply with depth.

(c) **Hypolimnion :** Stagnant water.

Effect of Temperature

(i) Denaturation of enzymes

(ii) **Rate of metabolism :** Metabolism first increase in range of 17° to 48°C.

(iii) **Colouration :** Gloger's rule : Inhabitants of warm climate have darker pigmentation.

(iv) **Morphology**

- **Bergman's Rule**
  - Warm blooded - Larger in cold climates.
  - Cold blooded - Smaller in cold regions.

  Smaller has less surface area to conserve heat.
- **Allen's Rule :** Extremities (fingers, tail, ears, legs) smaller in cold climates.
- **Rensch's Rule :** Broader wings of birds in warmer climate.
- **Jordan's Rule :** Greater no. of vertebrae in colder climate.
- **Cyclomorphosis :** It is the seasonal change of body form with temperature. E.g., Daphnia

(Crustacean) develops a helmet during summer & disappears during winter.

- **Dormancy :** During this phase the metabolic rate of the organism gets reduced. E.g., Hibernation, aestivation.

(v) **Precipitation :** The moisture falling on an area in liquid or frozen form is termed as precipitation. Average annual rainfall of India - 117cm - Highest in world.

(vi) **Humidity :** The atmospheric moisture in the form of invisible vapour is called humidity. It affects transpiration & absorption of water.

- **Range of Tolerance :** The range between too little and too much is called the range of tolerance. Organisms are abundant in the central optimal range.
- **Acclimatisation** is the gradual physiological adjustment of the organisms to changing new environmental conditions.
- **Ecological Adaptations :** They are special characteristics of plants and animals that enable them to survive in the environmental conditions.

Plant Adaptations

(a) **Adaptation to Light Regime :** Two kinds of plants are found - light adapted plants (Heliophytes) have high optimal temperature for photosynthesis while shade adapted plants (Sciophytes) have low photosynthetic and metabolic activities. For example, ferns.

(b) **Adaptations to Water Scarcity :** Plants are called xerophytes.

(i) Some plants complete their life cycle in the rainy season and survive the dry season in the form of seeds - Ephemerals.

(ii) Some plants have deep tap roots, e.g. Acacia, palms.

(iii) Mostly leaves are small and leathery and with a waxy cuticle having sunken stomata to reduce transpiration.

(iv) In succulent xerophytes, leaves are reduced to spines and stem becomes fleshy to store food and water, e.g., Opuntia, Cacti.

(v) Most grasses show C_4 pathway of photosynthesis to achieve higher rate of photosynthesis and reduce photorespiration.

(vi) Many succulents show Crassulacean Acid Metabolism (CAM pathway) of photosynthesis, i.e., they close their stomata during day and open during night.

(c) **Adaptations to Aquatic Environment :** Plants are hydrophytes,

(i) possess air space or aerenchyma which gives plants buoyancy and also helps in transport of oxygen, e.g., water hyacinth.

(ii) roots may be poorly developed (Hydrilla, Vallisneria) or may be absent (Ceratophyllus).

(iii) may have inflated petiole which helps in floating.

(d) **Adaptations to Saline Environment :** Plants are halophytes,

(i) may store water and dilute the salt concentration.

(ii) mangroves have high level of salt concentration and so the plants excrete salt.

(iii) some plant species accumulate glycerol in their cells for osmoregulation - helps them to tolerate saline conditions.

(iv) other adaptations are vivipary and pneumatophores.

(e) **Adaptations to Oligotrophic Soils** : These are soils low in nutrients. Plants possess mycorrhizae which helps in efficient absorption of nutrients.

- **Adaptation in Animals :** Some adaptations are migration, camouflage, mimicry, hibernation and aestivation.

(a) Adaptation to Water Scarcity : Animals are adapted to reduce water loss, e.g. Kangaroo rat excretes solid urine and can live without drinking water from birth to death.

(b) Adaptation to Arid Climate : Camel uses water economically and can tolerate wide fluctuation in body temperature.

(c) Adaptation to Cold : Some animals show cold hardening, e.g., spiders.

(i) Some animals have ice nucleating proteins, i.e., ice formation can take place in extra nuclear spaces.

(ii) Some animals can decrease the freezing point of their body fluids by accumulating glycerol or antifreeze proteins. e.g. fishes in antarctic regions.

POPULATION

It is the total number of individual of a particular species inhabiting a particular area at a particular time.

Demography : Study of population.

Population Characteristics

(1) **Population density :** It is number of individuals per unit area of environment.

(2) **Natality :** This is addition of individuals in a population due to birth.

(3) **Mortality :** It is the rate of death of individuals in a given population.

(4) **Population Dispersion**

(i) **Emigration** - One way outward movement.

(ii) **Immigration** - One way inward movement.

(iii) **Migration** : It is a cyclical movement with respect to and weather that during life history of an animal at definite intervals, and always includes a return trip from where it began. It is a two way movement, e.g. Birds.

- Fishes (Osteichthyes) -
- Catadromous - Fresh to sea eg. Anguilla
- Anadromous - Sea to Fresh eg. Hilsa.

(5) **Age Structure :** The proportion of individuals of various age group in a population forms the age structure.

Pre reproductive > Reproductive > Post reproductive → Expanding population

Pre reproductive = Reproductive > Post reproductive → Stable population

Pre reproductive < Reproductive < Post reproductive → Decline population.

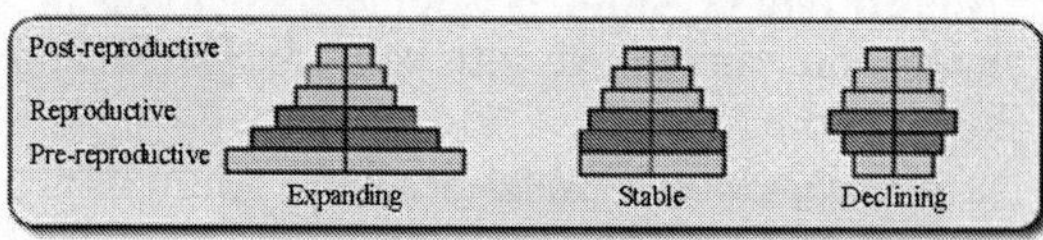

Fig. Representation of age pyramids for human population

(6) **Biotic Potential :** The inherent maximum capacity of an organism to reproduce or increase in number is termed as biotic potential. Nature keeps a check on it.

Carrying Capacity : The maximum population size that can be supported by the environment is called carrying capacity.

Population growth : Current population – Initial population + Birth + Immigration – Death – Emigration.

Population growth forms : Populations have characteristic pattern of increase with time, which is called population growth forms.

There are two main type of population growth forms

(1) **J - Shaped :** The population values when plotted against time gives a J - shaped growth curve, and at the peak of the population, growth ceases abruptly due to environmental resistance (eg. - growth of insects during rainy season) - infection of bacteria (HIV in man).

(2) **Sigmoid or S - Shaped form :** It shows an initial gradual increase in the population size and then it accelerates and finally slows to a nearly constant level.

Asymptote : The peak constant level of population.

Replacement Level : It is the number of children a couple must produce to replace themselves. It is always slightly higher than 2 (2.1 in developed countries & 2.7 in developing countries) since some children will die.

Demographic Transition : If birth rate and death rates were equal, a zero population growth rate would result, which is called demographic transition.

Births, deaths, and movements drive population dynamics

- Ecologists measure rates at which events such as birth and death take place in a population.
- A mathematical representation of population dynamics can be constructed using the following variables:
- N_1 = the number of individuals at time 1
- N_0 = the number of individuals at time 0
- B = the number of individuals born between time 0 and time 1
- D = the number of individuals that died
- I = the number of individuals that immigrated
- E = the number of individuals that emigrated
- Hence, $N_1 = N_0 + B - D + I - E$
- If the rates of the variables of the equation are measured over many time intervals, a population's density variation can be determined.
- Average annual growth rate (in per cent) =

$$\frac{P_2 - P_1}{P_1 \times N} \times 100$$

where,

P_1 is population size in the previous census;

P_2 is population size in the present census,

N is number of years between the two census.

- **Different stages of the demographic transition:** Demographic transition might occur in all countries as they become developed, but it may take many decades.

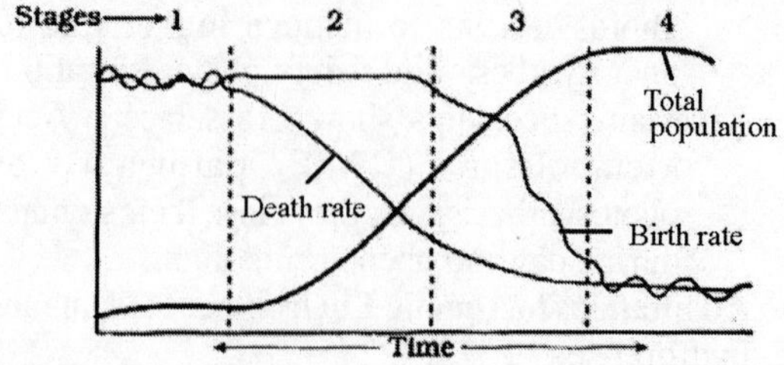

BIOTIC COMMUNITY

All the living entities of an ecosystem form a single biotic component called the community. (Plants + Consumers + decomposer)

Basic Attribute of Community

- Co-occurence of species
- Homeostasis/equilibrium

Characteristics of a Community

(1) **Species diversity :** Various species of plants & animals live in a community & exibit species diversity. Deserts have low species diversity while evergreen forests have high diversity.

(2) **Dominance :** Among several species present in a community, a few exert a major controlling influence by virtue of their size, numbers or activities. These are called dominant species. Communities are named after the name of dominant species (e.g. - grassland Community) or by the name of important environmental factor (e.g. desert Community).

(3) **Physiognomy :** It is the external appearence or the look of the community. It is mainly decided by the external morphological appearence of the dominant species.

(4) **Stratification -** The communities exibit a structure or recognizable pattern in spatial arrangement of members of the communities.

e.g. - A deep pond can be divided vertically into 3 different strata -

(a) **Littoral zone -** The topmost layer - Clear light available.

(b) **Limnetic zone -** Intermediate layer -Diffused light available.

(c) **Profundal zone** - Bottom layer - No light available.

The oceanic region consists of 2 zones

(a) Bathyal zone - The region of continental slope

(b) Abyssal region - Area of deep ocean

In terms of light penetration, the ocean region is vertically divided into 2 regions

(a) Euphotic zone/light compensation zone - light is present

(b) Aphotic zone - permanently dark zone.

The oceanic region also includes two more vertical zone -

(a) Benthic - The bottom zone (Sea bottom) e.g. of benthic community - sponges, crab, sea anemone.

(b) Pelagic - The whole body of water.

Community of pelagic zone includes

(i) Phytoplankton - Algae, diatoms

(ii) Zooplankton - Jelly fish, Snail

(iii) Nekton - Free swimming animals independent of water movement. Eg. - fish, whale.

(iv) Neuston - Minute organisms that float or swim on surface of water (Planktons → Anywhere suspended)

(5) **Keystone species :** The species having much greater influence on community characteristics, relative to their low abundance or biomass, are called key stone species. Extinction of keystone species produces a threat for the existence of community. E.g., - Fig trees provide fruits to different animals in tropical rain forests at the time of food scarcity.

(6) **Link Species :** A link species is one which assists the dominant species of the community to perform some important physiological function. e.g. Mycorrhiza (fungi + plants) assists in absorption of minerals and water.

(7) **Ecotone and Edge Effect :** The place where two major communities meet & blend together is called an ecotone. The variety & density of life is often greatest in this transitional area and this is known as edge effect.

BIOLOGICAL INTERACTIONS

- Biological interactions can be of two types:
  (i) **Intraspecific interaction :** Between members of a species.
  (ii) **Interspecific interaction :** Between members of different species.
- Some species of insects and frogs are cryptically-coloured to avoid being detected easily by the predator. This is called **camouflage**.

Amensalism

It is a relationship in which one species is harmed whereas the other is unaffected.

Competition

- Competition is best defined as a process in which the fitness of one species is significantly lower in the presence of another species.
- **Gause's 'competitive exclusion principle'** states that two closely related species competing for the same resources cannot co-exist indefinitely and the competitively inferior one will be eliminated eventually.

Parasitism

- Parasitism is a relationship between two species

in which one benefits on the host (harm) of the other. It is always an one side relationship for the parasite which is always benefitted from the host.

Commensalism

- The interaction in which one species benefits and the other is neither harmed nor benefited is known as **commensalism**.

Mutualism

- Mutualism interaction is a positive reciprocal relationship between two different species.
- Anabaena (a nitrogen fixing blue green alga) is associated with water fern Azolla.

Predation

- Predation is an interaction between members in which one population adversely affects the other by direct attack (capture, kill and eat) but is nevertheless dependent on other. The former is called **predators** & the latter is called **prey**

Exercise

1. Which one of the following is not abiotic factor?
(a) Temperature (b) Light
(c) Soil (d) Plants
2. A mutually beneficial association necessary for survival of both partners is
(a) mutualism/symbiosis
(b) commensalism
(c) amensalism
(d) both (a) and (b)
3. Soil best suited for plant growth is
(a) clay (b) loam
(c) sandy (d) gravel
4. Sunken stomata is the characteristic feature of
(a) hydrophyte (b) mesophyte
(c) xerophyte (d) halophyte
5. Which part of the world has a high density of organisms?
(a) Grasslands
(b) Savannahs
(c) Deciduous forests
(d) Tropical rain forests
6. In a population unrestricted reproductive capacity is called as
(a) biotic potential (b) fertility
(c) carrying capacity (d) birth rate
7. Choose the correct sequence of stages of growth curve for bacteria
(a) Lag, log, stationary, decline phase
(b) Lag, log, stationary phase
(c) Stationary, lag, log, decline phase
(d) Decline, lag, log phase
8. The maximum growth rate occurs in
(a) stationary phase (b) senescent phase
(c) lag phase (d) exponential phase
9. The formula for exponential population growth is
(a) $dN/rN = dt$ (b) $rN / dN = dt$
(c) $dN / dt = rN$ (d) $dt / dN = rN$
10. Soil is composed of
(a) mineral + water + wir
(b) mineral + organic matter + air + water
(c) mineral + organic matter + air + water
(d) organic matter + water
11. Biotic factors are
(a) chemical factors of soil which effect life
(b) physical factors of soil which effect life
(c) all living organisms which influence other organisms
(d) atmospheric factors which influence life
12. 'Eco' term refers as
(a) biosphere (b) environment
(c) organisms (d) plants
13. In commensalism
(a) population of commensal and host remains unaffected
(b) population of commensal may increase while that of host remains unaffected
(c) population of both commensal and host increases
(d) population of commensal increases while the population of host gradually decreases
14. Species is a
(a) group immediately below a phylum
(b) closely related interbreeding population
(c) taxonomic division of similar genera
(d) closely related non-breeding population

15. Ecology takes into account only
(a) environmental factors
(b) plant adaptations
(c) effect of environment on plants
(d) All of the above

16. Competition for light, nutrients and space is most severe between
(a) closely related organism growing in different niches
(b) closely related organisms growing in the same area/niche
(c) distantly related organisms growing in the same habitat
(d) distantly related organisms growing in different niches

17. Homeostasis is
(a) tendency of biological systems to change with change in environment
(b) tendency of biological systems to resist change
(c) disturbance of self regulatory system and natural controls
(d) biotic materials used in homeopathic medicines

18. Which one is true?
(a) Commensalism when none of the interacting populations affect each other
(b) Symbiosis when the interaction is useful to both the populations
(c) Symbiosis when neither populations affects each other
(d) Commensalism when the interaction is useful to both the populations

19. A fertile agricultural soil appears deep coloured at the surface as compared to soil one metre down. The reason for colour of top soil is
(a) more moisture
(b) rich in organic matter
(c) rich in iron, calcium and magnesium
(d) recent formation

20. Soil particles size determine its
(a) texture
(b) field capacity
(c) water holding capacity
(d) soil flora

21. The sum total of the populations of the same kind of organisms constitute
(a) colony (b) genus
(c) community (d) species

22. Association of animals when one species is harmed and the other one is unaffected
(a) colony (b) mutualism
(c) commensalism (d) amensalism

23. Xeric environment is characterised by
(a) precipitation
(b) low atmospheric humidity
(c) extremes of temperature
(d) high rate of vapourisation

24. Animals that can tolerate a narrow range of salinity are
(a) stenohaline (b) euryhaline
(c) anadromous (d) catadromous

25. In a food chain, the largest population is that of
(a) producers
(b) decomposers
(c) secondary consumers
(d) primary consumers

26. In India, human population is heavily weighed towards the younger age groups as a result of
(a) short life span of many individuals and low birth rate
(b) long life span of many individuals and low birth rate
(c) short life span of many individuals and high birth rate
(d) long life span of many individuals and high birth rate

27. Human population growth in India
(a) tends to follow a sigmoid curve as in case of many other animal species
(b) tends to reach a zero population growth as in case of some animal species
(c) can be reduced by permitting natural calamities and enforcing birth control measures
(d) can be regulated by following the national programme of family planning

28. The nature of climax community ultimately depends on
(a) climate
(b) bed rock
(c) soil organisms
(d) pool of available nutrients

29. In increasing order of organizational complexity, which one of the following is the correct sequence?
(a) Population, species, community, ecosystem
(b) Population, variety, species, ecosystem
(c) Population, ecosystem, species, community
(d) Species, variety, ecosystem, community

30. Keystone species in an ecosystem are those
(a) present in maximum number
(b) that are most frequent
(c) attaining a large biomass
(d) contributing to ecosystem properties

31. What is the most important factor for the success of animal population?
(a) Natality
(b) Unlimited food
(c) Adaptability
(d) Inter-species activity

32. Two opposite forces operate in the growth and development of every population. One of them relates to the ability to reproduce at a given rate. The force opposing to it is called
(a) mortality
(b) fecundity
(c) biotic control
(d) environmental resistance

33. Which of the following communities is more vulnerable to invasion by outside animals and plants ?
(a) Tropical evergreen forests
(b) Oceanic island communities
(c) Mangroves
(d) Temperate forests

34. The semilog of per minute growing bacteria is plotted against time. What will be the shape of graph?
(a) Sigmoid
(b) Hyperbolic
(c) Ascending straight line
(d) Descending straight line

35. What is true about keystone species ?
(a) A species which makes up only a small proportion of the total biomass of a community, yet has a huge impact on the community's organization and survival
(b) A common species that has plenty of biomass, yet has a fairly low impact on the community's organization
(c) A rare species that has minimal impact on the biomass and on other species in the community
(d) A dominant species that constitutes a large proportion of the biomass and which affects many other species

36. In which one of the following habitats does the diurnal temperature of soil surface vary most?
(a) Shrub land (b) Forest
(c) Desert (d) Grassland

37. Certain characteristic demographic features of developing countries are
(a) high fertility, low or rapidly falling mortality rate, rapid population growth and a very young age distribution
(b) high fertility, high density, rapidly rising mortality rate and a very young age distribution
(c) high infant mortality, low fertility, uneven population growth and a very young age distribution
(d) high mortality, high density, uneven population growth and a very old age distribution

38. Geometric representation of age structure is a characteristic of
(a) population (b) landscape
(c) ecosystem (d) biotic community

39. The population of an insect species shows an explosive increase in numbers during rainy season followed by its disappearance at the end of the season. What does this show?
(a) The food plants mature and die at the end of the rainy season
(b) Its population growth curve is of J-type
(c) The population of its predators increases enormously
(d) S-shaped or sigmoid growth of this insect.

40. Two plants can be conclusively said to belong to the same species if they
(a) have more than 90 per cent similar genes
(b) look similar and possess identical secondary metabolites
(c) have same number of chromosomes
(d) can reproduce freely with each other and form seeds

41. A high density of elephant population in an area can result in
(a) intra specific competition
(b) inter specific competition
(c) predation on one another
(d) mutualism

42. What is true about the isolated small tribal populations?
(a) There is a decline in population as boys marry girls only from their own tribe
(b) Hereditary disease like colour blindness do not spread in the isolated population
(c) Wrestlers who develop strong body muscles in their life time pass this character on to their progeny
(d) There is no change in population size as they have a large gene pool

43. If the number of fauna increases than the flora of an ecosystem then it is danger because it increases
(a) percentage of radioactive fallout
(b) percentage of carbondioxide
(c) percentage of oxygen
(d) percentage of diseases

44. Which of the following is correct ?
(a) Number of extinct species exceeds the number of living species
(b) Number of living species exceeds the number of extinct species
(c) Number of taxonomic species exceeds the number of biological species
(d) Number of taxonomic species is equal to the number of biological species

45. The geographic limit within which a population exists is called
(a) niche (b) ecosystem
(c) habitat (d) biome

46. Biotic potential refers to
(a) increase of population under optimum conditions
(b) increase of population under given conditions
(c) increase of population under natural conditions
(d) increase of population under climatic conditions

47. The carrying capacity of a population is determined by its
(a) birth rate
(b) death rate
(c) limiting resource
(d) reproductive status

48. When both partners/components are affected negatively, the nature of interaction is
(a) commensalism (b) predation
(c) competition (d) amensalism

49. Small fish get stuck near the bottom of a shark and derives its nutrition from it. This kind of association is called as
(a) symbiosis (b) commensalism
(c) predation (d) parasitism

50. The possible beneficial affect of grazing animals is the
(a) removal of wild animals
(b) eradication of weeds
(c) removal of wild plants
(d) addition of their excrete to the soil

51. Which biological factor is responsible for poor vegetation in deserts?
(a) Grazing by animals and goats
(b) Low rainfall
(c) Poor fertility of soil
(d) Native mankind

52. Which of the following is a climatic factor?
(a) Oxygen (b) Temperature
(c) Soil structure (d) Altitude

53. The sun loving plants are referred to as
(a) halophytes (b) heliophytes
(c) heterotrophs (d) sciophytes

54. Deforestation has an alarming effect on
(a) increase in grazing area
(b) sunlight
(c) weed control
(d) soil erosion or desertification of habitat

55. Which symbol represents intrinsic rate of increase ?
(a) N_t (b) N_o
(c) r (d) e

56. Find out the wrong statement.
(a) Habitat includes both biotic and abiotic factors.
(b) Abiotic and biotic components interact constantly with each other.
(c) Abiotic components alone can characterize the habitat of an organism.
(d) Major abiotic factors includes temperature, water, light and soil.

57. Read the following statements and choose the correct option.
(i) Light is essential for life to exist on the earth.
(ii) Many species of small plants under the canopy to tall trees in forest show optimal

use of available light due to having large sized antenna and higher number of thylakoids.

(iii) UV rays are not harmful to many organisms.

(iv) Photoperiodic requirement is essential for many plants for flowering.

(v) Red algae can live in deeper water of sea because of having pigment, phycoerythrin.

(a) (i) and (iii) are correct

(b) (i), (iii) and (iv) are correct

(c) only (iii) is correct

(d) (i), (ii), (iv) and (v) are correct

58. (i) Thermoregulation energetically least expensive process for many organisms like shrews and humming birds.

(ii) 99% animals and nearly all plants cannot maintain their constant internal environment.

(iii) During the course of evolution, the costs and heights of maintaining a constant internal environment are discarded.

(iv) In aquatic animals, the osmotic concentration of the body fluids change with that of the ambient water osmotic concentration.

(a) (i) and (ii) are wrong

(b) (iii) and (iv) are wrong

(c) (i) and (iii) are wrong

(d) (ii) and (iii) are wrong

59. Match the following

	Population		Example
A.	Predation	I.	*Cuscuta* and hedge plants
B.	Commensalism	II.	*Balanus* and *Chathamalus*
C.	Parasitism	III.	Cactus and Moth
D.	Competition	IV.	Orchid and Mango

(a) A – III, B – IV, C – I, D – II

(b) A – IV, B – III, C – II, D – I

(c) A – I, B – III, C – II, D – IV

(d) A – III, B – IV, C – II, D – I

60. Match list I with list II and choose the correct option.

	List I		List II
A.	Pacific salmon fish	I	Verhulst - pearl logistic growth
B.	$N_t = N_0 e^{rt}$	II	Breed only once in life time
C.	Oyster	III	Exponential growth
D.	$dN/dt = rN\left[\frac{K-N}{K}\right]$	IV	A large number of small sized offsprings

(a) A – IV; B – III; C – I; D – II

(b) A – III; B – IV; C – I; D – II

(c) A – III; B – I; C – IV; D – II

(d) A – II; B – III; C – IV; D – I

CHAPTER 36

Ecosystem

Any ecological unit that includes all the organisms which interact among themselves and with the physical environment, so that a flow of energy leads to clearly defined trophic structure, biotic diversity and material cycle within the system, is known as ecosystem.

STRUCTURE OF ECOSYSTEM

An ecosystem has 2 major components :

A. **Abiotic Components**

(a) Climatic conditions - Soil, temp, light, water.

(b) Inorganic substances - Nitrogen, sulphur, phosphorus

(c) Organic substances - Carbohydrate, protein, lipid.

Liebig's law of Limiting Factor : Normal growth and reproduction of an organism depends upon the essential material and environmental factors present in the least favourable amount.

B. **Biotic Components**

(a) **Producers :** The organism which produce their own food.

- Terrestrial ecosystem- Herbs, shrubs, tree.
- Aquatic ecosystem - 2 types of producers
- Phytoplanktons - Minute, floating or suspended lower plants (algae) distributed throughout the photic zone.
- Macrophytes - Rooted large sized plants.

(b) **Consumers :** They are the phagocytic heterotrophs which depend for their nutrition on the organic manufactured by producers, the green plants. They are of following 3 types :

(i) Herbivores (Primary Consumers) - These animals feed directly on living plants or plant remains.

(ii) Carnivore order - 1 (Secondary Consumers) - These carnivores feed on herbivores, *e.g.* frog, fish.

(iii) Carnivore order - 2 (Tertiary Consumerrs) - These are carnivores feeding on other carnivores. (Eagle feeding on snake)

(c) **Decomposers :** They bring about the decomposition of dead organic matter of producers as well as consumers. They help in returning the mineral elements to the abiotic phase and help in continuing biogeochemical cycle, *e.g.* - bacteria, fungi.

FUNCTION OF ECOSYSTEM

The basic function of ecosystem to maintain the flow of energy and the cycling of nutrients. In ecosystem, energy flow is always unidirectional. *i.e.* from Sun → Producer → Consumer → Decomposer.

Productivity : The rate of production *i.e.* the amount of organic matter accumulated in any unit time. It is of following types :

(A) **Primary Productivity :** The rate at which radiant energy is stored by producers.

(i) Gross Primary Productivity - It refers to the total rate of photosynthesis including the organic matter used up in respiration.

(ii) Net Primary Productivity - It is the rate of storage of organic matter in plant tissue in excess of the respiratory utilization.

(B) **Secondary Productivity :** It is the rate of energy storage at consumer's level - herbivores, carnivores & decomposers.

(C) **Net Productivity :** It is the rate of storage of organic matter not used by heterotrophs or consumers, *i.e.* equivalent to net primary production minus consumption by the heterotrophs during unit period (year).

Ecological efficiency :

1. **Photosythetic efficiency** $= \dfrac{\text{GPP}}{\text{Incedent total solar reaction}} \times 100$

2. **Net production efficiency** $= \dfrac{\text{NPP}}{\text{Gross primary productivity}} \times 100$

3. **Assimilation efficiency**

$$= \frac{\text{Food energy assimilated}}{\text{Food energy ingested}} \times 100$$

4. **Ecological efficiency**

$$= \frac{\text{Energy / biomass Production at a trophic level}}{\text{Energy / biomass production at previous trophic level}} \times 100$$

ENERGY FLOW

Food Chains : The pattern of eating & being eaten forms a linear chain called food chain, which can always be traced back to the producers. Each step of food chain is known as trophic level. (Plants - 1st trophic level)

Producers → Herbivores → Small fish → Large fish

Plants → Grasshopper → Lizard → Hawk

Ecological Pyramids : Concept of ecological pyramid was given by Elton.

There occurs a regular pattern of change in the properties (like number, energy and biomass) of the organisms across different trophic level. It is the graphical representation of food chain.

Types of Ecological Pyramid

1. Pyramid of Number - upright or inverted
2. Pyramid of Biomass - upright or inverted

- **Biomass may be defined as the total weight of dry matter present in the organism at any particular time.**

3. Pyramid of energy - Always upright

Lindeman's ten percent law : The food chain efficiency is only 10% *i.e.* the organism at a particular trophic level consumes 90% of the total energy which it receives from the preceeding trophic level and passes only the 10% remaining energy to the next level.

ECOLOGICAL SUCCESSION

- Ecological succession includes a series of changes that a biotic community undergoes in maturation towards a stable or climax condition.

Steps in Succession

(1) **Nudation** : This is the development of a bare area without any form of life. This may be due to landslide, erosion, or fire.

(2) **Invasion** :It is the successful establishment of a species in an area (Ecesis). Plants which are the first to colonize in an area are called the pioneers.

(3) **Reaction** :It is the mechanism of the ***modification of the environment*** through the influence of living organisms on it. As a result, the environment gets modified and becomes unsuitable (pH, light condition) for the existing community. So, this community gets replaced by other. The whole sequence of communities that replaces one another in a given area is called a sere.

(4) **Stabilization (Climax)** : Finally the final terminal community mantains itself in equilibrium with the climate of the area. This community is not replaced and is known as ***climax community***.

Types of Succession

(1) **Primary succession** : If an area is colonized by organisms for the first time, the succession is called primary succession. Eg. - Newly exposed rock ; Modified desert.

(2) **Secondary succession** : If succession starts on the area previously colonized, but has been cleared off by some environmental force (fire, flood, lightening), it is called secondary succession. Eg. - forest devastated by fire.

Example of Succession

(1) **Hydrarch** : Succession which begins in water bodies like ponds.

Pioneer stage → Submerged stage → Floating stage → Reed swamp stage → Marsh-meadow stage → Wood land stage → Forest stage (climax)

(2) **Xerarch :** Successions initiated on bare rocks, sand dunes, rocky slopes etc. where there is extreme scarcity of water, are termed as xerarch.

(1) Lichen Stage - The pioneer stage.

(2) Moss Stage - Grow in rock cervices where small amount of soil has accumulated.

(3) Herb Stage - eg. - Poa, Solidago.

(4) Shrub Stage - eg. - Phytocarpus.

(5) Forest Stage - Climax Stage.

BIOGEOCHEMICAL CYCLES

More or less circular pathways through which the chemical elements, including all the essential elements of protoplasm, circulate in the biosphere from environment to organisms and back to the environment are known as biogeochemical cycle.

Water Cycle

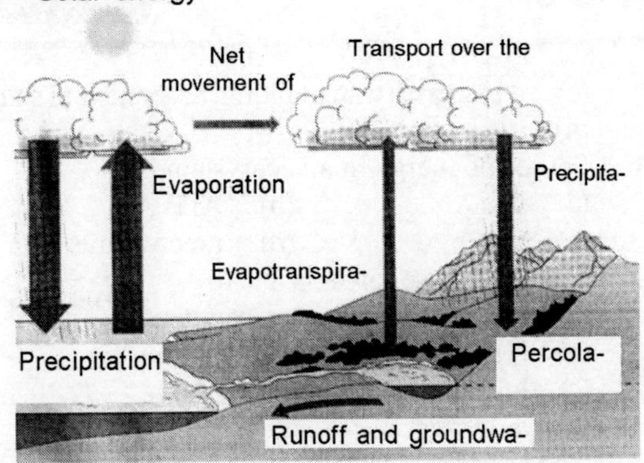

Phosphorus Cycle :

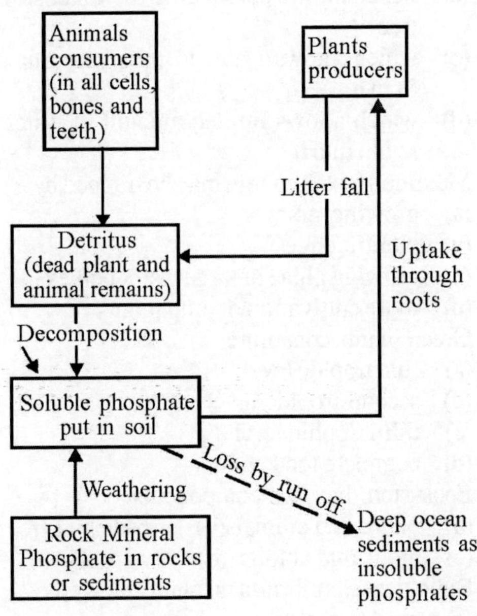

Carbon Cycle :

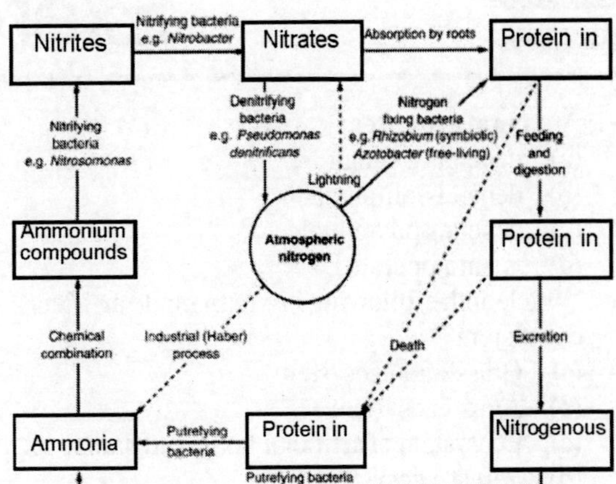

Nitrogen Cycle :

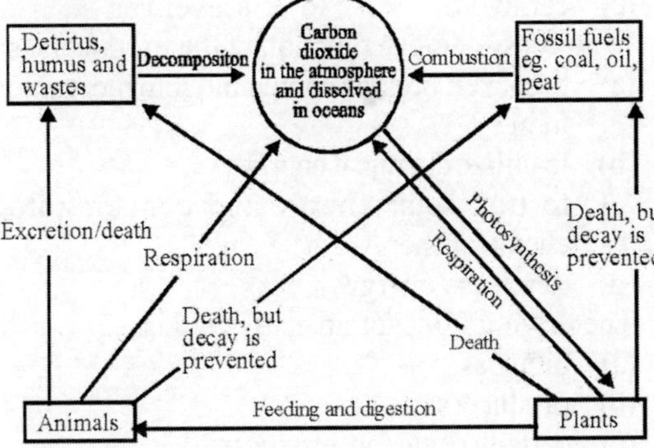

Sulphur Cycle :

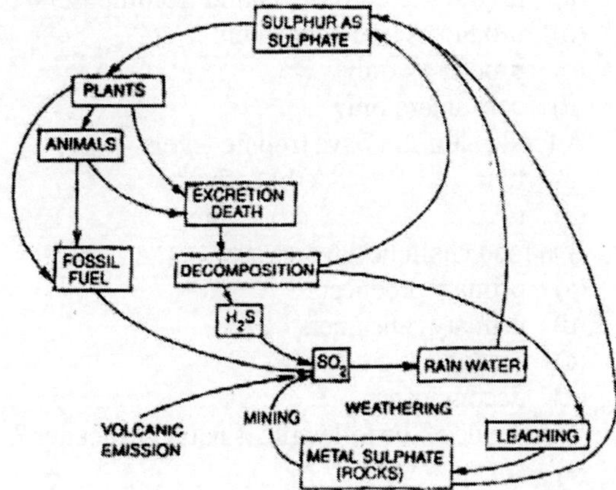

Exercise

1. A natural ecosystem
   (a) depends on man
   (b) depends on plants
   (c) depends on animals
   (d) is auto operated
2. Which of the following is a man made artificial ecosystem?
   (a) Grassland ecosystem
   (b) Agro ecosystem
   (c) Ecosystem of artificial lakes and dams
   (d) Forest ecosystem
3. Grasshopper in grassland is a
   (a) producer (b) herbivore
   (c) carnivore (d) scavenger
4. In an ecosystem the function of the producers is
   (a) to convert organic compound into inorganic form
   (b) to utilize chemical energy
   (c) to trap solar energy and convert into chemical energy
   (d) to release energy
5. The driving force for an ecosystem is
   (a) biomass
   (b) producers
   (c) carbohydrates in producers
   (d) solar energy
6. Biotic components of an ecosystem include
   (a) producers, consumers and decomposers
   (b) producers and consumers
   (c) producers only
   (d) consumers only
7. A food chain can have trophic levels
   (a) four (b) three
   (c) two (d) one
8. In a food chain herbivores are
   (a) primary producers
   (b) primary consumers
   (c) secondary consumers
   (d) decomposers
9. Which one of the following is nature's cleaner ?
   (a) consumers
   (b) producers
   (c) decomposers and scavengers
   (d) symbionts
10. Humus is good for plant growth because
    (a) it improves physical condition of soil
    (b) it makes soil porous
    (c) it increases water holding & aeration of soil
    (d) all
11. Source of energy in an ecosystem is
    (a) sun (b) ATP
    (c) sugar (d) green plants
12. Tropical dense forest is due to
    (a) low rainfall & low temperature
    (b) high rainfall & low temperature
    (c) low rainfall & high temperature
    (d) all
13. Good soil is
    (a) which holds whole of the water that enters into it
    (b) which allows percolating the water slowly from it
    (c) which allows water to pass very quickly from it
    (d) which allows limited amount of water to retain into it
14. Maximum solar energy may be trapped by
    (a) growing grasses
    (b) planting trees
    (c) growing algae in large water bodies
    (d) more cultivation of crop plants
15. Green plants constitute
    (a) first trophic level
    (b) second trophic level
    (c) third trophic level
    (d) complete food chain
16. Ecosystem has two components
    (a) plants and animals(b) weeds and trees
    (c) biotic and abiotic (d) frog and men
17. Biological equilibrium is found among the
    (a) producer plants
    (b) consumers and producers
    (c) decomposers
    (d) producers, consumers and decomposers
18. The correct path of energy flow in an ecosystem is
    (a) Produers $\rightarrow$ Carnivores $\rightarrow$ Herbivores $\rightarrow$ Decomposers
    (b) Producers $\rightarrow$ Herbivores $\rightarrow$ Carnivores $\rightarrow$ Decomposers
    (c) Herbivores $\rightarrow$ Carnivores $\rightarrow$ Producers $\rightarrow$ Decomposers
    (d) Herbivores $\rightarrow$ Producers $\rightarrow$ Carnivores $\rightarrow$ Decomposers

19. The first link in any food chain is always a green plant because
(a) they are widely distributed
(b) they are firmly fixed to the soil
(c) they alone have a capacity to fix atmospheric CO_2 in the presence of sunlight
(d) all of the above

20. Trophic levels are formed by
(a) organisms linked in food chain
(b) only plants
(c) only animals
(d) only carnivores

21. Pyramid of numbers in a grassland/true ecosystem is
(a) always inverted (b) always upright
(c) both (a) and (b) (d) spindle-shaped

22. Food chain in which micro-organisms breakdown the food formed by primary producers is
(a) parasitic food chain
(b) detritus food chain
(c) consumer food chain
(d) predator food chain

23. Pick up the correct food chain
(a) Grass $\rightarrow$ Chameleon $\rightarrow$ Insect $\rightarrow$ Bird
(b) Grass $\rightarrow$ Fox $\rightarrow$ Rabbit $\rightarrow$ Bird
(c) Phytoplankton $\rightarrow$ Zooplankton $\rightarrow$ Fish
(d) Fallen leaves $\rightarrow$ Bacteria $\rightarrow$ Insect larvae

24. Study of inter-relationships between organisms and their environment is
(a) ecology (b) ecosystem
(c) phytogeography (d) ethology

25. Pyramid of numbers deals with number of
(a) species in an area
(b) individuals in a community
(c) individuals in a trophic-level
(d) subspecies in a community

26. Pyramid of numbers in a pond ecosystem is
(a) irregular (b) inverted
(c) upright (d) spindle shaped

27. The pyramid which cannot be inverted in a stable ecosystem is that of
(a) biomass (b) number
(c) energy (d) all of the above

28. Bulk CO_2 fixation occurs in
(a) crop plants
(b) oceans
(c) tropical rain forests
(d) temperate forests

29. If we completely remove the decomposers from an ecosystem, its functioning will be adversely affected, because
(a) energy flow will be blocked
(b) herbivores will not receive solar energy
(c) mineral movement will be blocked
(d) rate of decomposition will be very high

30. The primary succession refers to the development of communities on a
(a) fleshly cleared crop field
(b) forest clearing after devastating fire
(c) pond, freshly filled with water after a dry phase
(d) newly-exposed habitat with no record of earlier vegetation

31. Which of the following is the most stable ecosystem?
(a) Forest (b) Desert
(c) Mountain (d) Ocean

32. In a biotic community, the primary consumers are
(a) carnivores (b) omnivores
(c) detritivores (d) herbivores

33. In a food chain, the largest population is that of
(a) decomposers
(b) producers
(c) primary consumers
(d) tertiary consumers

34. Which of the following ecosystems have highest rate of gross primary production?
(a) Grasslands
(b) Mangroves
(c) Coral reefs
(d) Equatorial rain forest

35. The rate at which light energy is converted to the chemical energy of organic molecules is the ecosystem's
(a) net primary productivity
(b) gross primary productivity
(c) net secondary productivity
(d) gross secondary productivity

36. Upper part of sea/aquatic ecosystem contains
(a) plankton
(b) nekton
(c) plankton and nekton
(d) benthos

37. What is true about ecosystem?
(a) Primary consumers are least dependent upon producers
(b) Primary consumers out-number producers
(c) Producers are more than primary consumers
(d) Secondary consumers are the largest and most powerful

38. The greatest biomass of autotrophs in the oceans is that of
(a) sea grasses and slime moulds
(b) free floating microalgae, cyanobacteria and nanoplankton
(c) benthic brown algae,coastal red algae and diatoms
(d) benthic diatoms and marine viruses

39. An ecosystem which can be easily damaged but can recover after some time if damaging effect stops will be having
(a) low stability and high resilience
(b) high stability and low resilience
(c) low stability and low resilience
(d) high stability and high resilience

40. Which one of the following is not used for construction of ecological pyramids?
(a) Number of individuals
(b) Rate of energy flow
(c) Fresh weight
(d) Dry weight

41. About 70% of total global carbon is found in
(a) grasslands (b) agroecosystems
(c) oceans (d) forests

42. Consider the following statements concerning food chains
(i) removal of 80% tigers from an area resulted in greatly increased growth of vegetation
(ii) removal of most of the carnivores resulted in an increased population of deers
(iii) the length of food chains is generally limited to 3-4 trophic levels due to energy loss
(iv) the length of food chains may vary from 2 to 8 trophic levels
Which two of the above statement are correct ?
(a) ii and iii (b) iii and iv
(b) i and iv (d) i and ii

43. Mass of living matter at a trophic level in an area at any time is called
(a) standing crop (b) deteritus
(c) humus (d) standing state

44. Which one of the following statements for pyramid of energy is incorrect, whereas the remaining three are correct ?
(a) Its base is broad
(b) It shows energy content of different trophic level organisms
(c) It is inverted in shape
(d) It is upright in shape

45. Both, hydrarch and xerarch successions lead to
(a) medium water conditions
(b) xeric conditions
(c) highly dry conditions
(d) excessive wet conditions

46. A pond is a
(a) a biome
(b) a community of plants and animals
(c) a natural ecosystem
(d) none of the above

47. Which one statement is correct regarding man made ecosystem ?
(a) They are highly efficient
(b) They are poor in diversity
(c) They are vulnerable to drought, floods and diseases
(d) All are correct

48. The largest vegetational group is known as
(a) ecosystem (b) association
(c) consociation (d) formation

49. The largest ecosystem of the world is
(a) forest (b) grassland
(c) great lake (d) ocean

50. Dynamic aspect of an ecosystem is
(a) producer and energy flow
(b) producer and mineral cycles
(c) consumer and mineral cycles
(d) energy flow and mineral cycles

51. Most important determinant of marine ecosystem is
(a) rainfall
(b) geographical location
(c) depth of distance from the shore
(d) all the above

52. The resource which regulates the flow of energy in desert ecosystem is the availability of
(a) light (b) water
(c) minerals (d) heat

53. Grassland with scattered trees are called
(a) rain forest
(b) evergreen forest
(c) savanna
(d) deciduous forest

54. Ecosystem is
(a) always open
(b) always closed
(c) both open and closed depending upon community
(d) both open and closed depending upon biomass

55. Zone of transition present between two types of communities for special ecological interest is called
(a) ecotype (b) ecotone
(c) border effect (d) niche

56. Succession in saline soil is
(a) mesosere (b) lithosere
(c) psammosere (d) halosere

57. Pneumatophores are found in vegetation of which ecosystem
(a) desert
(b) tropical rain forest
(c) mangroves of sunderban
(d) taiga

58. Animals which occupy the same trophic level are
(a) tiger and bees
(b) deer and bees
(c) snake and earth worm
(d) crow and cow

59. Which is not correct for ancient plants ?
(a) They have photosynthetic pigment
(b) They are primitive algae like
(c) They use H_2S as hydrogen source
(d) They release O_2 as by product

60. Stability of ecosystem depends upon
(a) primary productivity
(b) interchange between producers and consumers
(c) number of producers
(d) number of consumers

CHAPTER 37 Biodiversities and its Conservation

CONCEPTS

- **Biodiversity** means the sum total of all the species including plants, animals and micro-organisms on the earth.
- The 3 richest region of biodiversity in India are : (1) Western Ghats (2) North East (3) Kashmir.
- More than 60% of total biodiversity are found in tropical rain forest of the world

LEVELS OF BIODIVERSITY

(1) **Genetic Diversity :** It refers to variations of genes within species.

(2) **Species Diversity :** It refers to variety of species per unit area.

(3) **Community and ecosystem diversity :** It has three perspectives

(a) **Alpha diversity :** Diversity of organisms sharing the same community (intra community diversity).

(b) **Beta diversity :** Diversity among the members of different communities. (inter community diversity).

(c) **Gamma diversity :** Diversity over the total landscape or geographical area.

- Biodiversity varies with change in lattitude or altitude. Biodiversity increases, as we move from high to low lattitude as (*i.e.*, from poles to equator).
- **Species–area relationship – Alexander Von Humboldt**, a German Naturalist and Geographer has observed that within a region species richness increased and area for 1 wide varieties of taxa turns out to be rectangular hyperbola.

  The relation between species richness and area having wide variety of taxa is rectangular hyperbola limit. On a logarithmic scale, the relationship is a straight line, described as equation

$$\log S = \log C + Z \log A$$

where, S = Species richness

A = Area

Z = Slope of line (Regression coefficient)

C = Y-intercept

Z for small area = $0.1 \rightarrow 0.2$

Z for large area = $0.6 \rightarrow 1.2$

e.g., for **Z in small taxonomic group area** =

1. Plants in Britain
2. Birds in California
3. Mollusc in New York

Z for large taxonomic area = Frugivorous birds and mammals in tropical forest of different continent.

- Biodiversity is useful for modern agriculture in 3 ways

(i) as a source of new crops.

(ii) as a source material for breeding improved varieties.

(iii) as a source for more biodegradable pesticides.

- Economic values of biodiversity are: **direct value** and **indirect value**
- **Direct value:** Products of biodiversity harvested by people *e.g.*, important drugs from plants like anticancer drug, Taxol from bark of *Taxus brevifolia.*
- **Indirect value:** These are assigned to benefits provided by biodiversity that do not harvest or destroy the resource, *e.g.*,

(i) **Non-consumptive use of value:** Includes ecosystem biodiversity, protection of water and soil resources, waste management, ecotourism, climate monitoring.

(ii) **Option value:** The value of species is its potential to provide an economic benefit to human society in future, *e.g.*, environmental biotechnology.

(iii) **Existence value:** Assigned to protection of wild life. People and government contribution to conservation organisation constitute this.

CAUSES OF LOSS OF BIODIVERSITY

- **Disappearance of species** is a natural process and is called extinction and it is of 3 types
  (i) **Natural extinction:** Some species disappear and are replaced by other species when environmental conditions change. This loss of species which occurred in part at a very slow rate, is called natural or background extinction.
  (ii) **Mass extinction:** Disappearance of large number of species because of some catastrophe is called mass extinction.
  (iii) **Anthropogenic extinction** : Disappearance or extinction of species due to human activity is called anthropogenic extinction.
- **Habitat destruction :** The factor causing habitat destruction are the large industrial and commercial activities, associated with a global economy, such as mining, cattle roaching, fishing, forestry and dam construction. Rain forest, tropical dry forest, wetlands, mangroves and grasslands are threatened habitat and leading to desertification.
- **Habitat fragmentation :** Habitat fragmentation is the process where a large, continuous area of habitat is both, reduced in area and divided into two or more fragments.
- **Habitat degradation and pollution**
- **Introduction of exotic species :** Disease causing microorganism, if introduced to new virgin areas may cause epidemics and native species are eliminated completely.
- **Over exploitation :** Over exploitation threatens 1/3rd of endangered vertebrates in the world, as well as other species. Even if a species is not completely eliminated by over exploitation, the population size may become too small that the species is unable to recover.
- **Co-extinction**

Reasons to conserve biodiversity is

(i) **Narrowly Utilitarian approach**

– **Economic benefits :** By providing natural food, construction material, medicinal drugs. With increased **bioprospecting** (exploration of molecular, genetics and species level diversity for products of economic importance), we can expect enormous benefits from rich biodiversity.

(ii) **Broadly utilitarian approach:** Biodiversity maintains various ecosystem services that nature provides, *e.g.*, Amazon forest is estimated to produce through photosynthesis 20% of total oxygen in Earth's atmosphere.

(iii) **Ethical approach :** Each species has a right to exist. Each species has its own intrinsic survival value. All species are interdependent. The loss of one species may have far reaching consequence for other members of community.

- **Pattern of Biodiversity is Defined in 2 Terms:**

(i) **Lattitudinal gradients:** The diversity of plants and animal is not uniform throughout the world but shows a rather uneven distribution.
- In general, species diversity decreases as we move away from equator to poles.
- Tropics have greater species diversity than temperate or polar areas.
- Tropics Amazonian rain forest in South America has the greatest biodiversity on earth. It has
  - 40,000 species of plants
  - 3,000 of fishes
  - 1300 of birds
  - 427 of mammals
  - 378 of reptiles
  - 1,25,000 invertebrates
  - 2 million insect species yet to be discovered.

(ii) **Species area relationship:** While exploring South America jungles, the naturalist and geographer – Alexander Von Humbolt observed that within a region, species richness increases with increased explored area.

Conservation of Biodiversity

- There are 2 basic methods of biodiversity conservation:
  (i) *In situ* (on site)
  (ii) *Ex situ* (off site)

***In situ* conservation**

- *In situ* method provides protection to total ecosystem. It includes protection of a group of typical ecosystems through a network of protected areas.
- Example of protected areas are National Park, wild life sancturies and biosphere reserve.

***Ex situ* conservation**

- *Ex situ* conservation includes botanical gardens, zoos, pollen seed, seedling, tissue culture and DNA banks.
- In this, gametes of threatened species can be preserved in viable and fertile conditions for long period using cryopreservation and propagation of plants by tissue culture methods.
- Germplasm banks, also called gene banks are established. In Such **gene banks** species of seeds, pollen grains, vegetative propagative parts [culm, bulb, tuber] are collected and stored in gene bank.
- **Cryopreservation** is an *in vitro* conservation technique by which vegetatively propagated crops like potato are preserved in liquid nitrogen at a temperature of –196°C.
- **IUCN** and **WWF** are leading International Organisations concerned with biodiversity conservation.
- IUCN is International Union for the Conservation of Nature and Natural Resources is now called world conservation union.
- **Red List** prepared by IUCN is catalogue of taxa that are facing risk of extinction. The list has 8 categories of species :

(i) Extinct
(ii) Extinct in wild
(iii) Critically endangered
(iv) Endangered
(v) Vulnerable
(vi) Lower risk
(vii) Data deficient
(viii) Not evaluated

Sanctuaries and National Parks : A sanctuary or a National park may be defined as an area declared by state, for the purpose of protecting, propagating or developing wild life therein, or it's natural environment for their scientific, educational & recreational value.

National Park

- No Human activity permitted
- Not for specific animal

Sanctuary

- Human activity allowed on limited scale with permission
- For specific animal.

Biosphere Reserves

- Launched in 1975 as a part of UNESCO's "Man & Biosphere" programme.
- Biosphere reserves are special category of protected area wherein people are an integral component of the system.
- **It consists of 3 zones**

Transitional zone (Area of

Buffer zone (Human activity like

Core zone (Undisturbed by human

Exercise

1. Biodiversity of a geographical region represents
   (a) endangered species found in the region
   (b) the diversity in the organisms found in the region
   (c) genetic diversity present in the dominant species of the region
   (d) species endemic to the region
2. Conservation is
   (a) protection of natural resources
   (b) management of natural resources
   (c) proper use of natural resources
   (d) all the above
3. "Biosphere" is called as
   (a) part of atmosphere where the living organisms cannot live
   (b) part of atmosphere where the living organisms can live
   (c) both (a) and (b)
   (d) none of the above
4. Which is preserved in national park ?
   (a) Flora (b) Fauna
   (c) Both (a) and (b) (d) None of these
5. Which of the following animal has become almost extinct in India ?
   (a) Wolf (b) Rhinoceros
   (c) Hippopotamus (d) Cheetah

6. The species which are very few in number in the world are known as
(a) living fossils (b) endangered
(c) threatened (d) rare
7. Which of the following is the main factor of desertification?
(a) Tourism (b) Irrigated agriculture
(c) Over-grazing (d) All of these
8. Water is a resource
(a) nondegradable and nonmaintainable
(b) degradable and maintainable
(c) renewable
(d) non-renewable
9. Soil conservation is
(a) conversion of sterile soil into fertile one
(b) aeration of soil
(c) erosion of soil
(d) protection against loss
10. Deforestation will decrease
(a) soil erosion (b) land slides
(c) soil fertility (d) rainfall
11. A non-renewable resource is
(a) non-renewable non-conventional energy source
(b) non-renewable conventional energy source
(c) renewable non-conventional energy source
(d) renewable conventional energy source
12. "Chipko movement" is concerned with
(a) animal breeding (b) plant breeding
(c) tiger project (d) plant conservation
13. Endemic plants are those which are
(a) cosmopolitan in distribution
(b) restricted to certain area
(c) found in arctic region
(d) gregarious in habit
14. Wildlife is
(a) all biota excluding man, domestic animals and cultivated crops
(b) all vertebrates of reserve forests
(c) all animals of reserve forests
(d) all animals and plants of reserve forest
15. National bird of india is
(a) hornbill (b) black swan
(c) peacock (d) house sparrow
16. A high density of a protected animal in a national park can result into
(a) mutualism
(b) intraspecific competition
(c) emigration
(d) predation
17. An institution where valuable plant materials likely to become irretrievably lost in the wild or in cultivation is preserved in a viable condition is known as
(a) genome (b) herbarium
(c) gene library (d) gene bank
18. Reason of diversity in living beings is due to
(a) mutation
(b) long term evolutionary change
(c) gradual change
(d) short term evolutionary change
19. Which group of vertebrates comprises the highest number of endangered species ?
(a) Birds (b) Mammals
(c) Fishes (d) Reptiles
20. Diversification in plant life appeared
(a) due to long periods of evolutionary changes
(b) due to abrupt mutations
(c) suddenly on earth
(d) by seed dispersal
21. One of the most important functions of botanical gardens is that
(a) they provide a beautiful area for recreation
(b) one can observe tropical plants there
(c) they allow *ex-situ* conservation of germ plasm
(d) they provide the natural habitats for wildlife
22. Which one of the following is an example of *ex-situ* conservation?
(a) Wildlife sanctuary (b) Seed bank
(c) Sacred groves (d) National park
23. Alpha diversity is present
(a) within community
(b) between community
(c) ranges of communities
(d) all the above
24. Biodiversity is determined by
(a) number of individuals in an area
(b) species richness
(c) evenness
(d) both (b) and (c)
25. Extinction of numerous wild animals is due to
(a) hunting (b) deforestation
(c) overcrowding (d) scarcity of food
26. New approach of conservation is the establishment of
(a) sanctuaries (c) reserve forests
(c) national parks (d) biosphere reserves

27. Genetic diversity in agricultural crops is threatened by
(a) introduction of high yielding varieties
(b) intensive use of fertilizers
(c) etensive intercropping
(d) intensive use of biopesticides

28. Which of the following is not done in a wild life sanctuary?
(a) Fauna is conserved
(b) Flora is conserved
(c) Soil and flora are utilised
(d) Hunting is prohibited

29. Those species whose populations have been seriously depleted and whose ultimate security is not assured are known as
(a) threatened species (b) endangered species
(c) vulnerable species (d) rare species

30. A collection of plants and seeds having diverse alleles of all the genes of a crop is called
(a) herbarium (b) germplasm
(c) gene library (d) genome

31. Which one of the following has maximum genetic diversity in India?
(a) Mango (b) Wheat
(c) Tea (d) Teak

32. Which one of the following is not observed in biodiversity hotspots?
(a) Endemism
(b) Accelerated species loss
(c) Lesser inter-specific competition
(d) Species richness

33. In your opinion, which is the most effective way to conserve the plant diversity of an area?
(a) By tissue culture method
(b) By creating biosphere reserve
(c) By creating botanical garden
(d) By developing seed bank

34. If the forest cover is reduced to half, what is most likely to happen on a long term basis ?
(a) Tribals living in these areas will starve to death
(b) Cattle in these and adjoining areas will die due to lack of fodder
(c) Large areas will become deserts
(d) Crop breeding programmes will suffer due to a reduced availability of variety of germplasm

35. Deforestation does not lead to
(a) quick nutrient cycling
(b) soil erosion
(c) alteration of local weather conditions
(d) destruction of natural habitat of wild animals

36. If the Bengal tiger becomes extinct
(a) hyenas and volves will become scare
(b) the wild area will be safe for man and domestic animals
(c) its gene pool will be lost forever
(d) the population of beautiful animals like deers will be stabilized

37. Red data book provides data on
(a) red flowered plants
(b) red coloured fishes
(c) endangered plants and animals
(d) red eyed birds

38. Habitat loss, fragmentation, over exploitation, alien species invasion and co-extinction are causes for
(a) population explosion
(b) migration
(c) biodiversity loss
(d) pollution

39. Animals species should be preserved mainly because
(a) they are lovely creatures
(b) they are useful to mankind
(c) man cannot recreate a species of animals if it be destroyed
(d) zoologists want to study them

40. Which of the following is not a cause for loss of biodiversity?
(a) Destruction of habitat
(b) Invasion by alien species
(c) Keeping animals in zoological parks
(d) Over-exploitation of natural resources

41. Which one of the following is not a major characteristic feature of biodiversity hot spots?
(a) Large number of species
(b) Abundance of endemic species
(c) Large number of exotic species
(d) Destruction of habitat

42. In India, we find mangoes with different flavours, colours, fibre content, sugar content and even shelf life. The large variation is an account of
(a) species diversity (b) induced mutations
(c) genetic diversity (d) hybridization

43. The most biodiversity rich zone in India
(a) Gangetic planes (b) Trans himalayas
(c) Western ghats (d) Central india

44. Which of the following types of animals does man chiefly protect ?
(a) Harmless animals
(b) Economically useful
(c) Those likely to perish
(d) Feeble animals

45. Which of the following is mainly responsible for the extinction of wild life ?
(a) Hunting for flesh
(b) Destruction of habitats
(c) Pollution of air and water
(d) All of these

46. Which is the first national park established in India?
(a) Bandipur national park
(b) Corbett national park
(c) Kanha national park
(d) Periyar national park

47. A threatened species is
(a) only endangered species
(b) only vulnerable species
(c) endangered and rare species
(d) endangered, vulnerable and rare species

48. The wild life protection act was enacted by the Indian Parliament in
(a) 1962 (b) 1972
(c) 1983 (d) 1989

49. Which gas is used in cryopreservation ?
(a) Nitrogen (b) Oxygen
(c) Hydrogen (d) Carbon dioxide.

50. Which of the following is a sacred grove ?
(a) Aravali hills of Rajasthan
(b) Khasi hills in Meghalaya
(c) Western Ghat of Karnataka
(d) All of these

51. What accounts for low biodiversity at the poles?
(a) Short days (b) Low productivity
(c) Severe climate (d) All of these

52. Which of the following regions of our country are known for their rich biodiversity ?
(a) Western ghats and eastern himalayas
(b) Western ghats and deccan plateau
(c) Eastern himalayas and gangetic plane
(d) Trans himalayas and deccan peninsula

53. Red list of plant species of India is formed by
(a) Botanical Survey of India
(b) Zoological Survey of India
(c) Geological Survey of India
(d) None of these

54. Which of the following is true for national parks
(a) Tourism is allowed in buffer zone
(b) No human activity is allowed
(c) Hunting allowed in core zone
(d) Cattle grazing is allowed in buffer zone

55. Which step is required for better survival of human beings ?
(a) Conservation of wild life
(b) Afforestation
(c) Ban on mining
(d) Reduced utilization of resources

56. Amazon rain forest, called the 'Lungs of the Planet', contribute following percentage of oxygen by photosynthesis, to earth atmosphere
(a) 20% (b) 35%
(c) 42% (d) 50%

57. Which one of the following pairs of geographical areas show maximum biodiversity in our country?
(a) Sunderbans and Rann of Kutch
(b) Eastern Ghats and West Bengal
(c) Eastern Himalaya and Western Ghats
(d) Kerala and Punjab

58. The total number of national parks and sanctuaries in India are
(a) 82 national parks and 410 sanctuaries
(b) 83 national parks and 421 sanctuaries
(c) 85 national parks and 450 sanctuaries
(d) 87 national parks and 460 sanctuaries

59. Mark the correct statement.
(a) Amazonian rain forest has greatest biodiversity on earth.
(b) According to Robert May estimates, the global species diversity is 7 million.
(c) Biodiversity is the greatest in tropics.
(d) All of these.

60. Modern *Ex Situ* conservation includes
(i) cryopreservation techniques
(ii) *in vitro*-fertilisation
(ii) plants can be propagated using tissue culture methods
(a) Only (i) (b) Only (iii)
(c) (ii) and (iii) (d) All of these

CHAPTER 38

Environmental Issues

CONCEPTS

Pollution is an undersirable change in physical, chemical or biological characteristics of environment which adversely affects the biological species including Man.

Waste Products are of 2 Categories

(1) **Biodegradable waste :** These are such waste substances which are acted upon by microorganisms and broken into simpler components. E.g. - Most of the organic waste.

(2) **Non - biodegradable waste :** These are such waste substances which are not acted upon by micro-organisms and remain in the same form for a long period of time. E.g. - Polythene, glass, DDT, etc.

AIR POLLUTION

- Air pollution is the presence of substances that are not present naturally in the atmosphere.

Important Air Pollutants and their Impact

(1) **SO_2, H_2S**
- Lichens are most sensitive to SO_2.
- Interveinal blade damage in plants.
- Eye irritation, destroys bronchial cilia.
- Causes acid rain thus decreasing the pH of soil.

(2) **Carbon Monoxide**
- Reduces O_2 carrying capacity of blood by forming carboxy haemoglobin.
- Increases the rate at which fatty materials are deposited in the arteries.

(3) **Nitrogen oxides**
- Collapse of leaves.
- Nitrogen oxides reacts with hydrocarbons like methane, ethane, toluene, etc. to form peroxyacetyl nitrate or PAN ($C_2H_3O_5N$). PAN bocks the Hill reaction "in photosynthesis".

Photochemical smog : The mixture of PAN with ozone and various organic radicals.

(4) **Ozone :** $NO_2 \rightarrow NO + O$, $O_2 + O \rightarrow O_3$.
- Causes premature senescence in plants.
- Damage pulmonary organ in animals.

(5) **Hydrocarbons**

(a) 3, 4 benzopyrene - Causes lung cancer.

(b) Methyl Isocyanate (MIC)
- Leaked out accidently from a pesticide factory "Union Carbide" on December 2, 1984.
- Killed around 3000 people.

(6) **Cadmium**
- Causes cancer of liver and lungs.
- itai - itai (ouch ouch) disease (painful joints)

Control of Air Pollution : 3 Steps

1. **Separation of pollutants**
   - Certain plants like *Phaseolus vulgaris, Ficus variegata* can fix carbon monoxide and some plants *(Pinus, Quercus, Juniper)* can metabolise nitrogen oxides.
   - Good quality fuel like lead free or low sulphur should be used in motor vehicles.
   - Installation of equipments like **gravity settling tanks** or porous filters and **electrostatic precipitator**. They can minimize air pollution. The precipitator removes smaller particles.

2. **Avoiding pollutants**
   - Use of high temperature incinerator can reduce production of ash.
   - Complete electrification of rail track.
   - Replacement of non - renewable source of energy by renewable ones.

3. **Conservation of pollutant to harmless material**
   - Electronic precipitator has electrode wires that are maintained at several thousand volts, which produce a ray of electrons. They attach to dust particles giving them a net negative charge.
   - The collecting plates attract charged dust particles. (+vely charged)
   - With the help of scrubber, sulphur dioxide are removed from collecting plates.
   - Motor vehicle equipped with catalytic converter should use unleaded petrol because lead in petrol inactivates catalyst.

WATER POLLUTION

- The term water pollution refers to any type of aquatic contamination between the following 2 extremes -
- A highly enriched, over productive biotic community, such as river or lake with nutrients from sewage or fertilizer (cultural eutrophication) or,
- A body of water poisoned by toxic chemicals which eliminate living organisms or even exclude all forms of life.

Sludge = Solid waste + water

Scum = Liquid waste + water

Silt = Organic waste + water

Causes of Water Pollution

(1) **Sewage**

(2) **Industrial waste**

(3) **Agrochemicals**
   - Methaemoglobinuria or Blue baby syndrome : The surface run off water from agricultural fields contain high percentage of nitrates. When it enters the body of foetus, it reacts with the haemoglobin and forms methaemoglobin which has a highly reduced oxygen carrying capacity.
   - Excess of fluoride - causes skeletal fluorosis (teeth and skeletal deformity)
   - Arsenic - Black foot disease
   - Mercury - Minimata disease (numbness of limbs, lips and tongue)

(4) **Thermal pollution or Calefaction :**
   - The source is the heat from thermal & nuclear power plants.
   - Gives thermal shocks which affect the aquatic life.

Biological Magnification (Amplification) : It is the increase in the effect of any non degradable chemical as it passes on in the food chain. E.g. - Polychlorinated biophenyl, DDT.

Biological (Biochemical) Oxygen Demand (BOD) : It is the measure of oxygen required by aerobic decomposers for biochemical degradation of organic materials (biodegradable) in water. This demand of oxygen is directly proportional to increasing input of organic wastes in water.

Chemical Oxygen Demand : It is the amount of oxygen required for oxidation of total organic matter (biodegradable + non biodegradable) present in water.

- Drinking water should have BOD less than 1mg of O_2/litre.

General Remedies to Cure Pollution :

- Biodegradable wastes should be stored in anaerobic tanks and then changed into manures.
- Non-biodegradable substances such as plastic polythene etc. should be disposed off by **land fill method/deep fill method**.
- Use of chemical fertilizers should be replaced by biofertilizers.
- Use of chlorofluorocarbons should be stopped.
- Large scale tree plantation.

NOISE POLLUTION

- Noise is an undesired sound. Sound pollution starts from 80 decibel.
- A constant exposure to noise level of 80 db causes :
  - Mental irritation
  - Hypertension
  - Temporary deafness
- A constant exposure to noise level of 100 db causes : temporary or permanent deafness.
- A constant exposure to noise level of 140 db (highly dangerous) causes :
  - Permanent deafness
  - Physical injury to ear
  - Abortion

Organic Farming

- Integrated approach, a cyclical, zero waste procedure, where waste products from one process are cycled in as nutrients for other processes. This allows maximum utilization of resource and increases efficiency of production.

Radioactive Wastes

- This waste is generated as a result of generation of electricity from nuclear energy in nuclear reactors. Radiation emitted out by waste is lethal at high doses and cause mutation and genetic disorder at high doses which can be transmitted generation after generation.

Remedy/Measure for Radioactive Waste

- Nuclear waste, after sufficient pre-treatment should be stored in suitably shielded container buried within the rocks, about 500 m deep below the earth surface.

GREENHOUSE EFFECT AND GLOBAL WARMING

- Greenhouse gases such as CO_2, CH_4, N_2O and CFC's present in atmosphere radiate part of radiowave radiations emitted by earth back to the earth. This downflux is called **greenhouse flux** which keeps the earth warm and phenomenon is called greenhouse effect.

Effect of Greenhouse Effect

- Temperature of earth has increased by 0.6°C in last 3 decades. This increase in temperature is leading to deleterious change in environment and resulting in climatic changes (**El Nino Effect**), leading to increased melting of polar ice caps as well as of other places like Himalayas snow caps.
- **CO_2 fertilization effect** – Increase in the growth rate of plants in response to elevated concentrations of CO_2 is known as carbondioxide fertilisation effect. **Kyotoprotocol** (1997) was signed to reduce the emission of green house gases. Under the protocol, industrialized countries as a whole will cut their overall CO_2 emission by at least 5.2% below 1990 level.
- Change of species diversity from tropics to temperate, equator to pole.
- Eruption of plant disease and pests, and explosive growth of weeds.

Global Warming

- The mean annual global temperature is 14°C. Any significant rise in this temperature is regarded as global warming. Major green house gases are CO_2, Methane, Chlorofluorocarbons (refrigerators, sprays), Nitrous oxide (NO_2).

Incineration - Burning of waste in presence of O_2.

OZONE DEPLETION

- Dobson unit is the unit for measurement of O_3 level in stratosphere.
  - 1 DV = 0.1 mm of compressed O_3 at NTP.
  - Normal O_3 level should be greater than 400 DV.

Major O_3 Depleting Substances

- Chlorofluorocarbon (C_1 component destroys O_3), Chloroform, CCl_4, Methane, N_2O (Nitrous oxide).
  - O_3 layer protects life from harmful U.V. radiations.
  - 3 forms of U.V. radiations :

(a) U.V. C - 100 nm to 280 nm
  - Completely absorbed by O_3.

(b) U.V. B - 280 nm to 315 nm
  - O_3 layer transforms it into infrared.
  - Thinning of O_3 layer leads to more penetration of U.V. B.

  Harmful impact of U.V. B :

  * Cataract * Skin Cancer (Melanoma)

(c) U.V. A - 315 nm to 400 nm.
  - Reaches the surface of earth
  - harmless.

- Montreal protocol (1987) → Stop use of ozone depleting substances.
- London protocol (1990)
- Chernobyl tragedy - Radioactive pollution - Russia.

NATURAL RESOURCES

Any component of the natural environment that can be utilized by man to promote his welfare is termed as natural resources.

Some examples of natural resources : Land, soil, water, forests, grasslands etc. are some examples of natural resources.

Step involved to improve erosion effected soil :

- Stabilizing of the soil to prevent further erosion.
- Restoring soil fertility.

Advantages of Recycling and Reusing

- Removal of mineral resources.
- Saving unspoiled land from disruption of mining.
- Reducing area of solid waste.
- Reducing energy consumption and pollution.

Solar Energy

- Direct solar energy - Radiant energy
- Indirect solar energy obtained from materials that have incorporated previously the sun's radiant energy.

Water present in the Earth : Oceans contain about 97.5 % of Earth's water in saline condition. The rest 2.5% is fresh water. Most of which is frozen as polar or glacial ice (1.97%). Remaining fresh water occur as ground water (0.5)% and in lakes and river (0.02%), soil (0.01%) and atmosphere (0.001%). Thus only a small fraction of fresh water is available for human consumption.

Water agent approaches to provide a sustainable **supply of high-quality water**

(i) Construction of dams to ensure year round supply of water and generation of electricity.
(ii) Desalination of sea water, making it fit for drinking.
(iii) Diversion of water bodies *eg.* through canal to increase the natural supply to a particular area.
(iv) Regular drinking and distillation of water bodies.

Wetland Conservation Programmes based on :

- Preparation of wetland inventories.
- Identification of wetlands of critical importance for their protection.
- Checking waste disposal in wetland.
- Reduction of excess inflow nutrients by keeping them under plant cover.

Types of Fresh Water Wetlands

- **Marshes** - Where grass like plant dominate.
- **Swamps** - Where trees or shrubs dominate
- **Riverine** - Periodically flooded forests found in low land along stream.

Forms of Forests :

1. **Productive :** Production of wood, fruits and a variety of compounds. Eg. Resin, latex etc.
2. **Protective :** Conservation of soil and water.
3. **Regulative :** Absorption storage and release of gases (CO_2, O_2), water, mineral.

Legal provision to safeguard wildlife : Laws :

1. Forest Act, 1927. 2. Wild life forest act 1972 **Agro forestry :** Agro forestry includes variety of land uses. When woody species are grown in combination with herbaceous crops either at the same time or in the time sequences.

It consists of :

- **Transfer system :** Involving growing crops between rows of planted trees (Sal etc.)
- **Jhum :** Shifting cultivation involving felling and burning of forests followed by cultivation for few years and abanding it to allow regulation.

Exhaustible resources further classified : Exhaustible resources can either be renewable or non-renewable .

Renewable resources : Mostly biotic; can be continuously regenerated. If consumption of these resources continues to exceed their rate of renewable, the quality becomes effected. They may even get totally exhausted.

Example : Forests $\rightarrow$ yield timber, plant products.

Rangelands $\rightarrow$ Sustain grazing animals for milk.

Wild life maintains food chain.

Non-renewable resources : Cannot be regained or reconstructed once used up *e.g.*, biological species. Fossil fuels like coal, petroleum, gas and metal.

Exercise

1. Green house gases include
   (a) CO_2, CFC, CH_4 & NO_2
   (b) CO_2, O_2, N_2, NO_2 & NH_3
   (c) CH_4, N_2, CO_2 & NH_3
   (d) CFC, CO_2, NH_3 & N_2
2. Which gas of the atmosphere holds up ultraviolet rays ?
   (a) O_2 (b) O_3
   (c) N_2 (d) CH_4
3. Lichens are important in the studies on atmospheric pollution because they
   (a) can grow in polluted atmosphere
   (b) can readily multiply in polluted atmosphere
   (c) efficiently purify the atmosphere
   (d) are very sensitive to pollutants like sulphur dioxide
4. Full form of BOD is
   (a) Biological organism death
   (b) Biotic oxygen demand
   (c) Biochemical oxygen demand
   (d) Biological organisation day
5. The expanded form of DDT is
   (a) dichloro diphenyl trichloroethane
   (b) dichloro diethyl trichloroethane
   (c) dichloro dipyrydyl trichloroethane
   (d) dichloro diphenyl tetrachloroacetate
6. Most harmful environmental pollutants are
   (a) biodegradable compounds
   (b) corrosive agents
   (c) non biodegradable chemicals
   (d) all of the above
7. Domestic waste contains
   (a) non-biodegradable pollutants
   (b) biodegradable pollutants
   (c) hydrocarbons
   (d) none of the above
8. Air pollution effects are usually found on
   (a) leaves (b) flowers
   (c) stems (d) roots
9. The most common indicator organism that represents polluted water is
   (a) *E. coli*
   (b) *Pseudomonas. typhi*
   (c) *Chlorella*
   (d) *Entamoeba*
10. A renewable exhaustible natural resource is
    (a) coal (b) petroleum
    (c) minerals (d) forest
11. Which one is not a pollutant normally?
    (a) Hydrocarbons
    (b) Carbon dioxide
    (c) Carbon monoxide
    (d) Sulphur dioxide
12. Major aerosol pollutant in jet plane emission is
    (a) sulphur dioxide
    (b) carbon monoxide
    (c) methane
    (d) fluorocarbon
13. Ultraviolet radiations from sunlight causes a reaction that produces
    (a) fluorides (b) carbon monoxide
    (c) sulphur dioxide (d) ozone
14. The CO_2 content in the atmospheric air is about
    (a) 0.034% (b) 0.34%
    (c) 3.34% (d) 6.5%
15. DDT is
    (a) a non-degradable pollutant
    (b) a bio-degradable pollutant
    (c) not a pollutant
    (d) an antibiotic
16. Acid rains are produced by
    (a) excess emissions of NO_2 and SO_2 from burning fossil fuels
    (b) excess production of NH_3 by industry and coal gas
    (c) excess release of carbon monoxide by incomplete combustion
    (d) excess formation of CO_2 by combustion and animal respiration
17. Green house effect is warming due to
    (a) infra-red rays reaching earth
    (b) moisture layer in atmosphere
    (c) increase in temperature due to increase in carbon dioxide concentration of atmosphere
    (d) ozone layer of atmosphere
18. Gas released during Bhopal tragedy was
    (a) methyl isocyanate
    (b) potassium isothiocyanate
    (c) sodium isothiocyanate
    (d) ethyl isothiocyanate
19. Most hazardous metal pollutant of automobile exhausts is
    (a) mercury (b) cadmium
    (c) lead (d) copper
20. Atmosphere of big/metropolitan cities is polluted most by

(a) automobile exhausts
(b) pesticide residues
(c) household wastes
(d) radioactive fall-out

21. Which gas contributes most to green house effect?
(a) CFC (b) Freon
(c) CO_2 (d) CH_4

22. Drawback of DDT as pesticide is
(a) it becomes ineffective after sometime
(b) it is less effective than others
(c) it is not easily/rapidly degraded in nature
(d) its high cost

23. Fish die in water bodies polluted by sewage due to
(a) pathogens
(b) clogging of gills by silt
(c) reduction in oxygen
(d) foul smell

24. The Taj mahal is threatened due to the effect of
(a) oxygen (b) hydrogen
(c) chlorine (d) sulphur dioxide

25. When huge amount of sewage is dumped into a river, its BOD, will
(a) increase (b) decrease
(c) sharply decrease (d) remain unchanged

26. In coming years, skin related disorders will be more common due to
(a) pollutants in air
(b) use of detergents
(c) water pollution
(d) depletion of ozone layer

27. Formation of ozone hole is maximum over
(a) India (b) Europe
(c) Antarctica (d) Africa

28. How does carbon monoxide, a poisonous gas emitted by automobiles, prevent transport of oxygen into the body tissues?
(a) By destroying the haemoglobin
(b) By forming a stable compound with haemoglobin
(c) By obstructing the reaction of oxygen with haemoglobin
(d) By changing oxygen into carbon dioxide.

29. Which one of the following organisms is used as indicator of water quality?
(a) *Chlorella* (b) *Azospirillum*
(c) *Escherichia* (d) *Beggiatoa*

30. Which of the following is the use of lichens in case of pollution?
(a) They treat the polluted water
(b) They act as bioindicators of pollution
(c) They promote pollution
(d) Lichens are not related with pollution

31. Green house effect refers to
(a) cooling of earth
(b) trapping of UV rays
(c) production of cereals
(d) warming of earth

32. Which of the following is a secondary pollutant?
(a) CO
(b) CO_2
(c) Peroxy acetyl nitrate (PAN)
(d) Aerosol

33. A major component of gobar gas is
(a) ammonia (b) methane
(c) ethane (d) butane

34. Which one of the following pairs is mismatched?
(a) Fossil fuel burning - release of CO_2
(b) Nuclear power - radioactive wastes
(c) Solar energy - green house effect
(d) Biomass burning - release of CO_2

35. Biochemical Oxygen Demand (BOD) in a river water
(a) has no relationship with concentration of oxygen in the water
(b) gives a measure of *Salmonella* in the water
(c) increases when sewage gets mixed with river water
(d) remains unchanged when algal bloom occurs

36. "Good ozone" is found in the
(a) mesosphere (b) troposphere
(c) stratosphere (d) ionosphere

37. In the environment, ozone is known for its
(a) harmful effects (b) useful effects
(c) both (a) and (b) (d) inert nature

38. All organic substance found in the biosphere necessarily contain
(a) silica (b) carbon
(c) oxygen (d) nitrogen

39. Ozone layer is formed in which zone of atmosphere
(a) mesosphere (b) stratosphere
(c) troposphere (d) ionosphere

40. In case CO_2 of earth's atmosphere disappears, the temperature of earth's surface would
(a) increase
(b) decrease
(c) depend on oxygen concentration
(d) remain the same

41. BOD test of biological oxygen demand test is made for measuring
(a) air pollution (b) water pollution
(c) noise pollution (d) soil pollution

42. In BOD test oxygen plays an important role to
(a) destroy inorganic matter
(b) destroy pollution
(c) destroy waste organic matter
(d) none of these

43. Water pollution can be stopped best by
(a) treating effluents to remove injurious chemicals
(b) rearing more fishes
(c) cultivating useful water plants
(d) spraying with ddt

44. Photochemical smog is related to the pollution of
(a) soil (b) water
(c) air (d) noise

45. Which constituent of the atmosphere is likely to change if the forest cover is removed ?
(a) O_2 level is increased
(b) CO_2 level is increased
(c) O_2 level is significantly increases
(d) CO_2 level is significantly decreased

46. Which of the following does not cause pollution ?
(a) Thermal power project
(b) Nuclear energy project
(c) Automobile
(d) Hydro electric project

47. Which of the following is most harmful pollutant ?
(a) NO_2 (b) CO_2
(c) SO_2 (d) CO

48. Major aerosol pollutant in jet emission is
(a) sulphur dioxide
(b) carbon monoxide
(c) methane
(d) fluorocarbon

49. Non-biodegradable pollutants are created by
(a) nature
(b) excessive use of resources
(c) humans
(d) natural disasters

50. Which of the following causes biomagnification?
(a) SO_2 (b) Mercury
(c) DDT (d) Both b & c

51. Among the following which one causes more indoor chemical pollution?
(a) burning coal
(b) burning cooking gas
(c) burning mosquito coil
(d) room spray

52. The major source of noise pollution, world wide is due to
(a) office equipments
(b) transport system
(c) sugar, textile and paper industries
(d) oil refineries and thermal power plants

53. Green house effect means
(a) increase in the temperature of earth due to high concentration of O_2
(b) increase in the temperature of earth due to high concentration of NO_2
(c) increase in the temperature of earth due to high concentration SO_2
(d) increase in temperature of earth due to high concentration of CO_2

54. Which of the following is not an air pollutant ?
(a) water vapour (b) carbon monoxide
(c) methane (d) oxygen

55. Smog is a common pollutant in places having
(a) high temperature
(b) low temperature
(c) excessive SO_2 in the air
(d) excessive ammonia in the air

56. Sewage water can be purified for recycling with the action of
(a) aquatic plants (b) penicillin
(c) micro-organisms (d) fishes

57. CFC are not recommended to be used in refrigerators because they
(a) increase temperature
(b) affect environment
(c) affect aquatic life
(d) affect human body

58. The main cause of water pollution is
(a) soap
(b) industrial effluents
(c) smoke
(d) ammonia

59. Sound becomes hazardous noise pollution at level
(a) above 30 dB (b) above 80 dB
(c) above 100 dB (d) above 120 dB

60. Which important green house gas other than methane is being produced from the agricultural fields?
(a) Sulphur dioxide (b) Ammonia
(c) Nitrous oxide (d) O_3

1. DIVERSITY IN THE LIVING WORLD

ANSWER KEY

1	(b)	11	(d)	21	(d)	31	(c)	41	(b)	51	(c)
2	(c)	12	(c)	22	(c)	32	(d)	42	(d)	52	(c)
3	(d)	13	(c)	23	(b)	33	(a)	43	(d)	53	(c)
4	(d)	14	(a)	24	(b)	34	(b)	44	(b)	54	(d)
5	(b)	15	(c)	25	(b)	35	(c)	45	(a)	55	(a)
6	(d)	16	(d)	26	(d)	36	(a)	46	(b)	56	(c)
7	(c)	17	(a)	27	(a)	37	(d)	47	(b)	57	(b)
8	(b)	18	(c)	28	(b)	38	(b)	48	(d)	58	(a)
9	(a)	19	(d)	29	(a)	39	(d)	49	(a)	59	(d)
10	(b)	20	(d)	30	(d)	40	(b)	50	(c)	60	(d)

Hints & Solutions

1. **(b)** There are several factors and processes which differentiate living beings with non-living beings like reproduction, respiration, growth, etc. But among them reproduction is the only difference which differentiate without any exception living being with non-living beings.

4. **(d)** Life is a unique, complex organisation of molecules, expressing itself through chemical reactions which lead to growth, development, responsiveness, adaptation and reproduction.

8. **(b)** The basic unit of classification is species.

18. **(c)** Herbarium is dry garden.

22. **(c)** With the discovery of more and more organisms, sometimes, it becomes difficult to adjust an organism to the traditional categories. Therefore to make taxonomic position of a species more precise, the various obligate categories in hierarchial classification are explained below

    Division ⟶ Class ⟶ Order ⟶ Family —→ Genus ⟶ Species

23. **(b)** Binomial Nomenclature means the scientific name of any organism consists of a generic epithet and a specific epithet. Generic name starts with capital letter and specific name with small letter. Both the names are written in Italics or underlined in manuscripts.

24. **(b)** Replication of the genetic information causes transfer of genetic information from one generation to the next.

25. **(b)** First living beings were formed in the environment of sea having abundant organic molecules. They absorbed the organic materials for the sake of nutrtion and hence were chemoheterotrophs.

26. **(d)** Biosystematics is the study of diversity of organism and all their comparative and evolutionary relationships.

27. **(a)** Angiosperms have adapted themselves to all kinds of habitat - terrestrial, aquatic, tropical, deciduous and alpine. Self - pollination is seen in very few angiosperms. Production of large number of seeds ensure that at least some will germinate. Not all plants have been domesticated by man.

28. **(b)** Gene flow is the spread of genes through populations as effected by movements of individual and their propagules, e.g., seeds spores etc. Gene flow ensures that all populations of a given species share a common gene pool, i.e., it reduces difference between populations. The interruption of gene flow between populations is a pre-requisite for the formation of new species.

29. **(a)** ICBN (International Code of Botanical Nomenclature) - It is one of the code of nomenclature which is independent of zoological and bacteriological nomenclature. The code applies equally to names of taxonomic groups treated as plants whether or not these groups were originally so treated.

30. **(d)** Biological organisms starts with submicroscopic moleculer level like viruses, bacteria, etc.
    These organisms are unable to be seen by naked eyes without the help of microscope or even electron microscope.

31. **(c)** All living things have an ability to respond to their environment, that is also called stimulation.

34. **(b)** The highest level of organisation is the landscape which is delimited by a particular climate or geographical area (lowest-population).

36. **(a)** Earth came into existence about 4.6 billion years ago from a large spinning cloud of gas and dust.
43. **(d)** The study of fossils is known as palaeonto-logy. Fossils are the remains or impressions of the hard parts of the past individuals in the strata of the earth. Fossils are studied for tracing evolutionary history of organisms and studying extinct organisms.
44. **(b)** Modern day classification is new systematics or biosystematics which includes all the characteristics of organisms gathered from the study of different sections like Physiology, Ecology, Anatomy, Biochemistry, Cytology, etc,.
45. **(a)** Binomial Nomenclature.
46. **(b)** This is one of the reason for organisms being Homeostatic (constant body temperature).
47. **(b)** CO_2 is green house gas which traps the solar radiation i.e., heats up the earth's atmosphere.
48. **(d)** Biological classification is the scientific arrangement of organisms in a hierarchial series of groups and subgroups on the basis of similarities and differences in their traits. It helps in building evolutionary pathways and in identifying new oganisms.
57. **(b)** Biological museums have the collection of preserved animals and plants specimens for study and reference. Biological museums are generally set up in educational institutes, *i.e.* School and colleges. The specimens may be preserved dry or in preservative solution in jars or containers. The insects are preserved in insect boxes after collecting, killing and pinning. The larger animals, like mammals and birds are stuffed and are chemically treated for long term preservation. Museums also have collection of animal skeletons.
58. **(a)** A category is a rank or level in the hierarchial classification of organisms. Taxon is a unit in classification which may represent any level of grouping of organisms based on certain common characteristics. There is some confusion in the use of taxon and category, for example Bryophyta is a taxon while division is a category.
60. **(d)** Order being higher category is the assemblage of families which exhibit a few similar characteristic. Dog (*Canis familaris*) and Cat (*Felis domesticus*) belong to two different families—Cancideae and Felidae respectively.

2. BIOLOGICAL CLASSIFICATION

ANSWER KEY

1	(a)	11	(d)	21	(c)	31	(d)	41	(c)	51	(c)
2	(d)	12	(d)	22	(b)	32	(b)	42	(b)	52	(a)
3	(d)	13	(c)	23	(a)	33	(d)	43	(a)	53	(a)
4	(c)	14	(c)	24	(c)	34	(b)	44	(a)	54	(a)
5	(d)	15	(a)	25	(a)	35	(c)	45	(d)	55	(c)
6	(b)	16	(a)	26	(d)	36	(a)	46	(b)	56	(c)
7	(a)	17	(b)	27	(b)	37	(b)	47	(b)	57	(c)
8	(b)	18	(a)	28	(b)	38	(b)	48	(a)	58	(a)
9	(b)	19	(a)	29	(c)	39	(a)	49	(b)	59	(a)
10	(c)	20	(a)	30	(d)	40	(d)	50	(b)	60	(c)

Hints & Solutions

1. **(a)** Plant decomposers are Bacteria (Kingdom Monera) and Fungi.
2. **(d)** Archaebacteria are the most ancient and halogenous group of bacteria and are called living fossils.
3. **(d)** Decomposers act on dead bodies converting complex organic constituents into simple soluble form. They obtain their energy from dead organisms. Decomposers themselves are organisms (often fungi, earthworms, or bacteria) that breakdown organic materials to gain nutrients and energy. Decomposers accelerate the natural process of decomposition. They supply the required nutrients for other trophic levels. The fungi on the tree are decomposers.

When an organism dies, it leaves behind nutrients that are locked tightly together. A scavenger might eat the carcass, but its faces would still contain a considerable amount of unused energy and nutrients. Decomposers will later induce further breakdown. This last step releases raw nutrients (such as nitrogen, phosphorus and magnesium) in a form usable to plants, which quickly incorporate the chemicals into their own cells. This process greatly increases the nutrient-load of an ecosystem, in turn allowing for greater biodiversity.

4. **(c)** Mycorrhiza is symbiotic association between fungi and roots of higher plants. Mycorrhiza form wooly covering of fungal hyphae on the surface and remain in upper layers. It is of two types ecto and endomycorrhiza. In some endomycorrhiza, the fungal hyphae develop some organs called vesicles within the root cortical cells, such mycorrhizae are called VAM (vesicular arbuscular mycorrhizae) . It is meant for phosphate nutrition.
5. **(d)** Chloroplasts, chromoplasts and leucoplasts are the types of plastids found in higher plants. Chloroplasts contain the green pigment chlorophyll. Chromoplasts contain coloured pigments. Leucoplasts are colourless and store starch granules.
6. **(b)** Whittaker (1969) divided organisms into five kingdoms : Monera, Protista, Fungi, Plantae and Animalia. Kingdom Monera includes all prokaryotes– Mycoplasma, bacteria, actino-mycetes, blue-green algae, archaebac-teria, methanogens. *Escherichia* is bacteria, *Amoeba*, *Gelidium* come under Protista, *Spirogyra* is algae.
7. **(a)** *Escherichia coli*, lives as a symbiont in human intestine.
8. **(b)** Cell wall of diatom is deposited with silica. After the death of diatoms, cell wall is resistant to decomposition, piles up at the floor of water body and forms diatomaceous earth.
9. **(b)** Bacterial cell wall is peculiar in being made up of a compound called Peptidoglycan.
Peptidoglycan $\rightarrow$ Heteropolysaccharide alternative amino-sugar molecules $\rightarrow$ N-acetyl glucosamine and N-acetyl muramic acid.
Chains of peptidoglycan are cross linked by short chains of peptides.

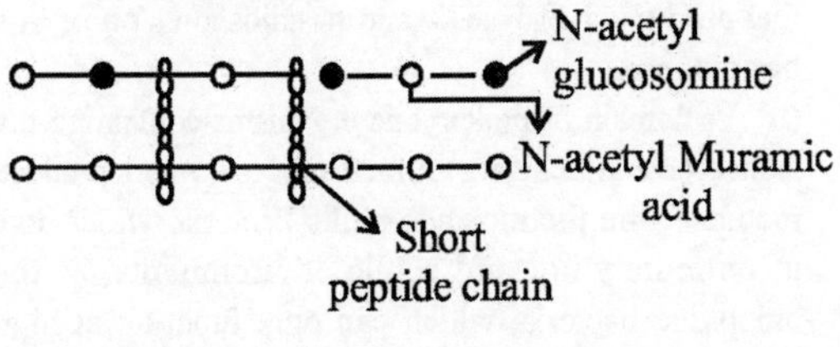

N-acetyl glucosamine
N-acetyl Muramic acid
Short peptide chain

11. **(d)** Plants which can make their own food are autotrophic. Plants which can not make their own food and depend on others are heterotrophic. Parasitic and saprophytic are types of heterotrophic plants. Parasites depend on living organisms and saprophytes depend on dead and decaying matter. Fungi can not make their own food and depend on dead and decaying matters.
12. **(d)** Bacteria posses various forms and shapes and are of 4 different types - coccus (round), bacillus (rod shaped), vibrio (comma shaped) and spirillum (spiral like cork screw). When the shape of a bacteria is like a comma then it is said to be vibrio, *e.g.*, *Vibrio cholerae*.
13. **(c)** When fungi feed on dead organic matter, they are known as saprophytes.
14. **(c)** True yeast bear asci which are not organized into ascocarps. e.g., Schizosaccharomyces, Saccharomyces, Saccharomycodes, Zygosac-charomyces, Nematospora. Therefore, yeast are classified under ascomycetes.
15. **(a)** Association of algae and fungi is referred to as lichen. Symbiotices association of *Rhizobium* with roots of leguminous plants is referred to as symbiosis. Mycorrhiza is symbiotic association between fungi and roots of higher plants. The fungal partner of mycorrhiza obtains food from roots of the higher plants and in return supplies mineral elements to it.
16. **(a)** Cyanobacteria are oxygenic photoautotroph. Many members perform nitrogen fixation.
17. **(b)** Enzymes are absent in viruses because they are unable to transmit their nucleic acid from one host cell to another.
18. **(a)** Virus envelope is known as capsid. The capsid is composed of protein subunits called capsomere.
19. **(a)** Lichens (coined by Theophrastus) are composite or dual organisms which are formed by a fungus partner or mycobiont (mostly ascomycetes) and an algal partner (mostly blue green algae). Fungus forms the body of lichen as well as its attaching and absorbing structures. Algae performs photosynthesis and provides food to the fungus.
20. **(a)** *Methanobacillus* (methanogen) occurs in marshes and also in dung. It produces CH_4 gas under anaerobic condition and is utilized in gobar gas plant.
21. **(c)** Lichens typically grow in harsh environments in nature, most lichens, especially epiphytic fruticose species and those containing cyanobacteria, are sensitive to manufactured pollutants. Hence, they have been widely used as pollution indicator organisms. Lichens are symbiotic associations of a fungus with a photosynthetic partner (called a photobiont or phycobiont) that can produce food for the lichen from sunlight.
22. **(b)** Fungi are nutritionally saprophytes, which grow on dead and decaying matter . They secrete enzyme to the external medium where digestion takes place and digested food absorbed by the body surface. They convert complex organic constituents of dead body into simple soluble forms. That is why Fungi are decomposers.
23. **(a)** Chemosynthetic bacteria does not require sunlight as a source of energy directly or indirectly. The energy

for the synthesis of food is obtained by the oxidation of certain inorganic substances present in the medium. The chemical energy obtained from oxidation reaction is trapped in ATP molecules. The ATP is used in CO_2 assimilation.

24. (c) Contractile vacuole in *Amoeba* and *Paramoecium* maintain the water balance of the cell. This is known as osmoregulation.
25. (a) In bacteria, asexual reproduction through binary fission is the most common method of multiplication. Sexual reproduction which comprises of syngamy and meiosis is entirely absent. Hence no gamete formation takes place. In sexually reproducing organism alternation of generation occurs.
26. (d) *Plasmodium* transmitted through female *Anopheles* mosquito to human and the most favourite dwelling place for *Anopheles* is water or moist places like ponds and puddles.
27. (b) Only some bacteria and blue green algae (cynobacteria) has capacity to fix atmospheric nitrogen.
   • Bacteria : *Azotobacter, Rhizobium, Frankia* etc.
   •Blue green algae : *Nostoc, Oscillatoria, Anabaena,* etc.
28. (b) The *Euglena* is an organism, which possesses both the characteristics of plants and animals, as it can move with a flagella and also contains chlorophyll. Its nutrition is mixotrophic.
29. (c) In protozoa, the function of contractile vacuole is the removal of excretory substances, carbon dioxide etc. It is very essential to regulate water content i.e., osmoregulation.
30. (d) The contractile vacuole is supposed to assist excertion in *Amoeba*, as its watery contents possess traces of carbon dioxide and urea. The CO_2 diffuses directly through plasmalemma.
31. (d) Fermentation refers to the breakdown of organic substances, particularly carbohydrates, under anaerobic conditions. *Saccharomyces cerevisiae* is used for fermenting carbohydrate. The ethyl alcohol and carbon dioxide are formed at the end of the process.
32. (b) Sex-factor or F-factor in bacteria results in high frequency conjugation. It allows bacteria to produce sex pilus necessary for conjugation.
33. (d) Lichens are composite dual organisms formed by the association between a fungus and a photosynthetic symbiont. The bulk of lichen body is formed of fungus.
34. (b) Viruses have either DNA or RNA as the genetic material. Viruses having RNA as the genetic material are known as Retroviruses.
35. (c) Bacteria acts as decomposers in the carbon cycle.
36. (a) These are archaebacteria which can tolerate high temperature.
37. (b) Transfer of genetic information from one bacterium to another by physical contact is called conjugation while if it takes place by some other medium like virus than it is called transduction.
38. (b) *Aspergillus niger* is known as guinea pig of kingdom fungi. The main source of commercial preparation of citric acid is the fermentation of sugar with *Aspergillus niger*.
39. (a) *Azolla* is a freshwater fern harbouring a blue green alga - *Anabaena* in its leaf cavities. The alga fixes atmospheric nitrogen and releases nitrogenous compounds in leaf cavities. This symbiotic system is the main source of algal biofertilizer in rice fields. When rice paddies are flooded in the spring, they can be inoculated with *Azolla*, which then quickly multiplies to cover the water, suppressing weeds. As the plants die, they contribute nitrogen to the rice plants, and as the rice paddy dries out, the *Azolla* all eventually die, making an exceptional green manure.
40. (d) In prokayotes nucleoid consist of double stranded circular DNA without histone protein.
41. (c) Virus is an ultra microscopic nucleoprotein entity which becomes active only inside a living cell. It resembles living beings due to presence of genetic material and reproduction.
42. (b) This phenomenon has been successfully used in genetic engineering to produce disease resistant varieties of plants.
43. (a) Eukaryotic cilia are made up of tubulin protein. Cilia have a core of 2 microtubules and around there are 9 doublet microtubules. These are not at all hair like structures.
44. (a) The thalloid body of slime moulds is made up of multinucleated cell which lacks septa in between and hence it is a multinucleated single celled mass called plasmodium.
45. (d) The class Basidiomycetes includes those members that produce their basidia and basidiospores on or in a basidiocarp.
46. (b) A domain of prokaryotic organisms containing the archaebacteria including the methanogens, which produce methane; the thermoacido-philic bacteria, which live in extremely hot and acidic environments, & the halophilic bacteria, which can only function at high salt concentrations are abundant in the world's oceans.
47. (b) *E. coli* is a prokaryotic celled gram negative bacterium.
48. (a) Eubacteria are prokaryotic but eubacteria are enclosed by plasma membrane like eukaryotic cells.
49. (b) *Methanogens* are archaebacteria abundant in cattle yard and paddy fields.
50. (b) Some common blue green algae are *Spirulina, Oscillatoria, Nostoc, Anabaena etc.*
51. (c) Toad stool Puffball are the members of Basidiomycetes (the club fungi).

52. **(a)** Transduction is the transfer of DNA from one organism to another through bacteriophage.

53. **(a)** In prokaryotes, enzymes for respiration are said to be located on cell membrane.

54. **(a)** The mushrooms belonging to the basidiomycetes are edible. The edible mushrooms are *Agaricus campestris* and *Agaricus rodmani*. The basidiocarp is loosely arranged over the stalk and it is the vegetative phase in the life of *Agaricus*. The basidiocarp which is the button stage in the edible part of the fungi, is loosely attached to the substratum with the help of the hyphae.

55. **(c)** *Euglena* is connecting link between animal and plants because –

(1) It is autotrophic due to the presence of chloroplast – plant character.

(2) It can move from one place to other with the help of flagella – animal like.

56. **(c)** Slime moulds in the division of myxomycota have spores that develop into flagellated gametes.

57. **(c)** Flagella of prokaryotic and eukaryotic cells differ in micro-tubular organization and type of movement. The flagella in eukaryotic cells are composed of several filaments and are far more complex than the flagella in prokaryotic cells.

58. **(a)** HIV (Human Immunodeficiency Virus) is rounded in shape. It has a protein coat and single stranded RNA as its genetic material. There are two single strands of RNA. Enzyme reverse transcriptase and proteins P-24, P-18 and P-15 are also present.

59. **(a)** *Rhizobia* are soil bacteria that fix nitrogen (diazotrophy) after becoming established inside root nodules of legumes (Fabaceae). The rhizobia cannot independently fix nitrogen, and requires a plant host. Morphologically they are generally gram negative, motile, non-sporulating rods.

3. PLANT KINGDOM

ANSWER KEY

1	(c)	11	(c)	21	(d)	31	(c)	41	(b)	51	(b)
2	(c)	12	(b)	22	(b)	32	(d)	42	(a)	52	(d)
3	(c)	13	(a)	23	(a)	33	(a)	43	(c)	53	(d)
4	(a)	14	(c)	24	(a)	34	(c)	44	(c)	54	(c)
5	(c)	15	(b)	25	(c)	35	(c)	45	(b)	55	(b)
6	(c)	16	(a)	26	(d)	36	(c)	46	(d)	56	(b)
7	(c)	17	(b)	27	(a)	37	(a)	47	(d)	57	(a)
8	(d)	18	(a)	28	(a)	38	(c)	48	(b)	58	(c)
9	(b)	19	(c)	29	(b)	39	(b)	49	(c)	59	(a)
10	(b)	20	(b)	30	(b)	40	(a)	50	(a)	60	(c)

Hints & Solutions

2. **(c)** The femal sex organ is archegonia. It is flask shaped structure distinguishable into a long neck and a globular swollen venter.

3. **(c)** Female sex organ is called archegonium which is flask shaped with a tubular neck and a swollen venter.

5. **(c)** They are called pollution indicators. They prevent soil erosion by running water.

6. **(c)** Require water for fertilization as their sperms are motile.

7. **(c)** Bryophytes are the plants which produces spores and embryo but no vascular tissues are present, although primitive type of conducting tissues are present in the form of hadrome and leptome.

•Rhodophytes & phaeophytes are algae produces spores, no embryo & vascular tissues.

•Pteridophytes have spores, embryo and true vascular tissues.

8. **(d)** Gymnosperms are vascular land plants having naked ovules i.e., ovules are not enclosed in a ovary hence, flower is absent. Ovules after fertilization produces naked seeds. Pteridophytes (ferns) and mosses do not produce seeds.

9. **(b)** Well developed vascular system present in the members of Pteridophytes but absent in mosses as the plant body is sporophyte which is distinguished into true root, stem and leaves.

Independent gametophyte, archegonia, flagellate spermatozoids are present in moss and pteridophyte both.

10. **(b)** In chlorophyceae, sexual reproduction takes place by all the three processes

1. Isogamy – Fusion of morphologically and physiologically similar gametes.
2. Anisogamy – Morphologically similar but physiologically dissimilar gametes.
3. Oogamy – Fusing gametes are dissimilar in all respect.

11. **(c)** Bryophytes require a thin film of water on the surface of their substratum during sexual reproduction for the following reasons :
(i) dehiscence of antheridia requires absorption of water from outside
(ii) dehiscence of archegonium
(iii) swimming of sperms to reach the archegonia

12. **(b)** Vessel less angiosperms are *Wintera*, *Trochodendron* etc. Secondary growth is absent in some angiosperms. Angiospermic plants are autotrophic in nutrition. But some angiosperms are heterotrophic in nutrition. The 4 special modes of nutrition and their examples are:
(i) Saprophytic : e.g. *Neottia,* *Monotropa.*
(ii) Symbiotic – e.g. Mycorrhiza-between fungus and roots of higher plants.
(iii) Parasitic – *Cuscuta.*
(iv) Insectivorous plant – *Nepenthes.*
Double fertilization is characteristic of all angiosperms.

13. **(a)** The Agar is obtained from several members of red algae such as *Gracilaria*, *Gelidium*, *Chondrus* etc. Agar gels are extensively used for growing micro-organisms.

14. **(c)** Phycoerythrins are accessory photosynthetic pigments along with phycocyanins. Haematochrome is a pigment in the blood. Brown algae is brown due to fucoxanthin.

15. **(b)** Shape and number of chloroplast in different members of algae is different
Chlamydomonas – cup shaped, 1/cell
Zygnema – Stellate, 2/cell
Spirogyra – Spiral, 1/ cell
Ulothrix – Collar shaped, 1/cell

16. **(a)** Isogamy refers to a form of sexual reproduction involving gametes of the same size. Since both gametes look alike, they cannot be classified as "male" or "female." Instead, organisms undergoing isogamy are said to have different mating types, most commonly noted as "+" and "–" strains. Fertilization occurs when "+" and "–" gametes fuse to form a zygote. There are several types of isogamy. Both gametes may be flagellated and thus motile. This type occurs for example in algae such as *Chlamydomonas.*

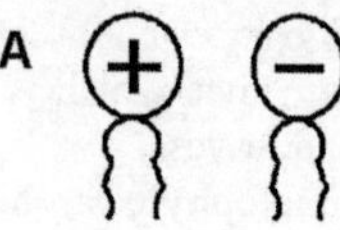

A. Isogamy of motile cells.
B. Isogamy of non-motile cells.

17. **(b)** When prothallus of a fern give rise to sporophyte directly from somatic cell without forming gametes it is Apogamy. Such type of sporophyte is haploid in nature.
Development of gametophyte directly from sporophyte without meiosis and without forming spores, is Apospory.

18. **(a)** Peristomes are teeth present in the capsule of moss. These are in two rows and helps in the dispersal of the spores due to hygroscopic nature.

19. **(c)** Ginkgophyta is an ancient group of which the sole living representative or living fossil is *Ginkgo biloba* or Maiden hair tree which is native of South China.

20. **(b)** Blue green algae or cyanobacteria come under Prokaryotes. Prokaryotes lack a well defined nucleus surrounded by nuclear membrane. Eukaryotes have well organised nucleus. Rhodophyceae & Chlorophyceae are members of algae which come under eukaryotes.

21. **(d)** In mosses branching is lateral but extra-axillary (*Funaria*). *Marchantia* is a dorsiventral dichotomously branched thalloid liverwort. In *Sphagnum* branching is profuse.

22. **(b)** Reserve food in red algae (Rhodophyceae) is Floridean starch. In green algae (Chlorophyceae), reserve food is starch. In brown algae, (Phaeophyceae) reserve food is laminarin and mannitol. In (cyanophyceae), blue green algae reserve food is cyanophycean starch.

23. **(a)** In *Cycas* (gymnosperm) the seed does not occur inside a fruit but lies naked or exposed on the surface of megasporophyll.

24. **(a)** Peat moss is used wherever we require to retain water for a long because Peat Mosses are hygroscopic in nature and they absorb the moisture from the atmosphere and this moisture keep the living materials and flowers fresh for a long time.

25. **(c)** Conifers (Gymnosperms) differ from grasses (angiosperms) because in gymnosperms the female gametophyte is actually endosperm which is made before fertilization while in grasses endosperm is a tissue formed by the fertilization of second male gamete to polar nuclei. Moreover in gymnosperms the endosperm is a haploid tissue while in angiosperms it is triploid.

26. **(d)** It will be types of pigment present in the cell like Rhodophyceae shows presence of phycoerythrin, chlorophyceae shows presence of phycocyanin, etc.

27. **(a)** Algae possess a definite cell wall containing cellulose, galactans and mannans.

28. **(a)** During megagametogenesis functional megaspore (mostly chalazal) gives rise to embryo sac. This is the mature female gametophyte generation.

29. **(b)** In Gymnosperms like *Cycas*/*Pinus*, endo-sperms of female gametophyte is haploid and formed before fertilizaiton.

In angiosperms endosperm is triploid and develops after fertilization.

30. **(b)** In *Funaria*, the spore germinates to form a green multicellular structure called as protonema which later on develops into gametophyte. Presence of protonema is characteristic feature of moss.

31. **(c)** *Spirogyra* is a member of green algae which cannot fix atmospheric nitrogen. Only *Nostoc* and *Anabaena* (Blue green algae) and *Azotobacter* (bacteria) can fix nitrogen.

32. **(d)** *Chlorella* can be grown to provide human food rich in proteins, lipids, carbohydrates, vitamins and minerals.

Chlorella has been researched as a potential source of food and energy, because its efficiency of photosynthesis can reach 8%, this is comparable with other highly efficient crops such as sugarcane. It is high in protein and other essential nutrients. When dried, it is about 45 percent protein, 20 percent fat, 20 percent carbohydrate, 5 percent fiber, and 10 percent minerals and vitamins.

33. **(a)** *Pinus* plant is monoecious i.e. both male and female cone are present in same plant but on different branches. *Pinus*, with over 100 species, is the largest genus of conifers and the most widespread genus of trees in the Northern Hemisphere. *Pinus* are also extensively planted in temperate regions of the Southern Hemisphere. Pines have a relatively rich fossil record.

51. **(b)** In *Selaginella*, some species roll up in dry season to form brown balls which may get uprooted and dispersed. They turn green with the availability of water. Such species are called as Resurrection plants.

52. **(d)** *Chlorella* is an algae which is used by space travellers as food. This algae belonging to chlorophyceae is rich in proteins and is used as a food supplement.

53. **(d)** Gymnosperm have tracheids and do not possess xylem with vessels. The tracheids conduct water in the gymnosperms.

54. **(c)** The male cone bears 60–135 microsporophylls in a spiral manner. It is comparable to the male flower of angiosperm. These micros-porophylls produce microspores with wings or saccus. They are liberated out in large numbers from a cluster of cones. They form a pale yellow cloud phenomenon called "Shower of sulphur" or "Shower of golden dust". Since the pollen is dispersed through air (Anemophily), more pollen has to be produced in order to reach the female cone and hence in *Pinus*, male cone bears large numbers of microsporophylls.

55. **(b)** Gymnosperms are the plant group in which ovules are not enclosed in the ovary. It means ovules are naked. Hence, gymnosperms are regarded as naked seed plants.

56. **(b)** *Laminaria* (kelp) and *Fucus* (rock weed) are marine algae. They are the members of class-phaeophyceae (brown) algae.

57. **(a)** Tallest angiospermic plant is *Eucalyptus* but the tallest plants of all category is red wood tree (*Sequoia*) → a member of gymnosperm.

58. **(c)** Kelps are called as sea weeds, *e.g.*, *Laminaria*, a member of brown algae. They are good sources of iodine.

59. **(a)** Sunflower seeds give fatty oils that have large amount of linoleic acid and do not contain linolenic acid. Its oil is used in cooking and cosmetics.

60. **(c)** Rhodophytes or red algae possess chlorophyll 'a' (like all photosynthetic organisms), carotenoids (carotene and xanthophyll) and phycobilins (phycoerythrin, phycocyanin and allophyco-cyanin). The red colour is imparted by phycoerythrin.

4. ANIMAL KINGDOM

ANSWER KEY

1	(d)	**12**	(b)	**23**	(c)	**34**	(a)	**45**	(b)	**56**	(b)
2	(a)	**13**	(b)	**24**	(c)	**35**	(b)	**46**	(a)	**57**	(a)
3	(a)	**14**	(c)	**25**	(b)	**36**	(b)	**47**	(c)	**58**	(c)
4	(c)	**15**	(b)	**26**	(d)	**37**	(c)	**48**	(d)	**59**	(b)
5	(a)	**16**	(c)	**27**	(b)	**38**	(c)	**49**	(b)	**60**	(a)
6	(a)	**17**	(a)	**28**	(c)	**39**	(b)	**50**	(b)		
7	(a)	**18**	(a)	**29**	(d)	**40**	(a)	**51**	(c)		
8	(c)	**19**	(b)	**30**	(c)	**41**	(c)	**52**	(a)		
9	(a)	**20**	(d)	**31**	(a)	**42**	(b)	**53**	(c)		
10	(d)	**21**	(b)	**32**	(c)	**43**	(a)	**54**	(b)		
11	(a)	**22**	(b)	**33**	(c)	**44**	(a)	**55**	(c)		

Hints & Solutions

1. (d) All chordate possess
— Notochord at some stage of life
— Dorsal tubular nervous system
— Pharyngeal gill slits at same stage of life cycle.
2. (a) Vertebrates are also known as Craniata due to presence of skull in all its members.
3. (a) Birds have pneumatic bones.
4. (c) Hollow bones are characteristic adaptive features of birds. It reduces their body weight and is a major flight adaptation.
5. (a) Mimicry is resemblance of an organism to its natural surroundings, a non living object or another organism for concealing itself from its natural predators or preys.
6. (a) Organisms passively floating on the surface of water are planktons. If floating organisms are animals then it is zooplankton and if plant – phytoplankton.
7. (a) Earthworms are very useful to human as they make the soil porous and increase its fertility through worm casting. They are called as Nature's 'plough man'.
8. (c) Chordates are featured by the presence of the following characters
   - Notochord (cephalic or head region)
   - Pharyngeal gill slits
   - Postanal tail.
9. (a) Malpighian tubules are fine, unbranched, yellow tubules that lie more or less free in the haemocoel. They open into alimentary canal. They absorb nitrogenous waste product, thus acts as excretory organs in insects.
10. (d) Skin infact is a secondary or accessory respiratory organ in frog when frog undergoes the hibernation or aestivation i.e., winter sleep or summer sleep & when it goes into the water for a long period.
11. (a) Caterpillar larva is a voracious eater, it gets matured and by secreting silk thread changes into pupa & it encloses into silk cocoon.
12. (b) In birds flight muscles or breast muscles are attached to the large sternum or breast bone. The wings are powered by these flight muscles.
13. (b) The terms porifera was given by Grant, the phylum includes animals with pores in their body. Its classification based on skeleton or spicules.
14. (c) Metamorphosis of insects is regulated through ecdysone hormone.
15. (b) Larval stage of Echinoderms have bilateral symmetry and the adults have radial symmetry, pentamerous i.e., body parts arranged in five or multiples of five.
16. (c) *Periplaneta americana* (cockroach) cannot act as a scavanger, it does not feed upon dead animals.
17. (a) *Rana tigrina* – Bull frog; Order - Anura Amphibia.
18. (a) The number of cervical vertebrae are same in man & whale that is 7 in number.
19. (b) Diaphragm is present in mammals as a muscular seperator between thorax and abdominal region.
20. (d) Gorilla, chimpanzee, monkeys and man belongs to same order- Primates of class mammalia.
21. (b) Tracheal tube of cockroach and human have non-collapsable wall. Each tracheal tube is lined with cuticle which prevents collapsing.
22. (b) Budding is an asexual mode of reproduction in *Hydra* & sponges. Bud is formed as an outgrowth on the body surface, then detached to form new animal.
23. (c) In radial symmetry – body is cut longitudinally through any radius into two equal halves. Present in sessile animals *i.e.* starfish that live in the sea bottom.
24. (c) Coelom or body cavity, when derived from blastocoel is pseudocoelom as in nematodes, rotifers. In this type of cavity, mesoderm is lined by only one side *i.e.*, towards the body wall. Schizocoel is true coelom derived from mesodermal segments.
25. (b) Presence of nematocytes or stinging cells are the characteristic feature of the phylum Coelenterata. It serves the function of offence and defence by producing toxin to kill the prey & enemy.
26. (d) Sea anemone (*Adamsia*) is marine animal attached to the empty shell of gastropods occupied by hermit crab. They live as symbionts, not as parasites.

   Tapeworm, Leech, Mosquitoes] are parasites

27. (b) Sponges are marine with the exception of family spongilidae, the members of which occur in fresh water *e.g. Spongilla*. Their endoskeleton is made of either spicules or spongin fibres or both. Sponges reproduce asexually by fragmentation or by budding or through gemmules and reduction bodies. Sponges also exhibit sexual reproduction.
28. (c) Fussorial are the animals which live beneath the soil (in low temperature) and are found abundant in desert grasslands.
29. (d) Post anal tail is present in chordates at some stage in the life.
30. (c) Mammals have 12 pairs of cranial nerves.
31. (a) Platyhelminthes includes flat warms.
There is no body cavity. The animals are therefore acoelomates. Annelids also have bilateral symmetry. They are coelomate having a perivisceral cavity divided into compartments.
32. (c) Penguin, kiwi & ostrich all belong to class Aves of chordata (*i.e.* birds) and they do not give birth to their young ones, they are oviparous while kangaroo, hedgehog, dolphin, loris all belong to class mammalia and are viviparous.

33. (c) Pseudocoelomate is any invertebrate animal whose body cavity is a pseudocoel, a cavity between the gut and the outer body wall derived from a persistent blastocoel, rather than a true coelom. Pseudocoelomate animals include the Rotifera and Nematoda.
34. (a) Aschelminthes is bilaterally symmetrical and triploblastic. These are mostly aquatic, free living or parasitic. Their body is three layered which is ectoderm, mesoderm and endoderm.
35. (b) Lampreys and hag fishes are unusual, jawless fish that comprise the order Cyclostomata, so named because of the circular shape of the mouth. The brains of lampreys and hag fishes differ a lot, but they also show a large number of similarities, as do all craniate brain.
36. (b) In nature insects have the highest number of species (30 millions). Fungi have 100,000, angiosperms have 260000 and birds have 10,000 species.
37. (c) Notochord is a flexible rod like structure that forms the main support of the body in the lowest chordates. It is not absent in humans through out their life. Notochord is present in embryonic stage and get changed or replaced by vertebral column in the adult.
38. (c) Human possesses enucleated RBC in mature state. But frog blood has both white and red blood cells which are nucleated. Frog cells do not lack platelets.
48. (d) Chordates show the presence of nerve cord, notochord and pharyngeal gill slits.
49. (b) In birds, forelimbs are modified as wings for flying. Therefore, the forelimb is not found in birds.uwuw
50. (b) Study of fishes – Ichthyology
Study of reptiles – Herpetology
Study of insects – Entomology
Study of birds – Ornithology
51. (c) Venom of cobra attacks nervous system in man. The poison reaches the brain through blood and inactivates the brain cells and simultaneouly body's whole nervous system is collapsed.
52. (a) Salivary gland is present in man because digestion starts in buccal cavity. It is absent in frog.
53. (c) One or two pairs of sac like kidneys are present in Mollusca but entirely absent in the phylum Echinodermata.
54. (b) Earnest Spytzner was first to draw attention to the fact that bees communicate by means of definite movements now called "bee dances".
55. (c) *Amoeba* is dispersed in the form of cysts blown by wind. *Amoeba* protrudes out of the cyst with the return of favourable conditions.
56. (b) *Loligo* (gaint squid), the largest invertebrate, is a member of the phylum Mollusca of class cephalopoda.
57. (a) Abdomen of the spiders contains silkgland which spin silk for many purposes including formation of spider web for capturing the prey.
58. (c) Animals which maintain high and constant body temperature are called warm blooded animals. They are also called homeothermic animals. Birds are the first vertebrate to have warm blood. They are homeothermous.
59. (b) Class insecta has 3 pairs of legs located on the thoracic segments. It is the characteristic feature of class Insecta.
60. (a) The respiratory organ in spiders and scorpions constitute the book lungs. The book lungs are the respiratory organs in terrestrial forms. The exchange of gases also takes place through the general body surface. The book lungs open to the exterior through the spiracles or stigmata.

5. *MORPHOLOGY OF FLOWERING PLANTS*

ANSWER KEY

1	(b)	11	(d)	21	(c)	31	(b)	41	(a)	51	(c)
2	(d)	12	(d)	22	(c)	32	(d)	42	(c)	52	(a)
3	(d)	13	(a)	23	(d)	33	(c)	43	(b)	53	(c)
4	(c)	14	(d)	24	(b)	34	(a)	44	(d)	54	(d)
5	(a)	15	(d)	25	(c)	35	(d)	45	(d)	55	(d)
6	(a)	16	(b)	26	(c)	36	(a)	46	(d)	56	(c)
7	(b)	17	(c)	27	(c)	37	(c)	47	(d)	57	(d)
8	(a)	18	(a)	28	(a)	38	(c)	48	(b)	58	(b)
9	(d)	19	(d)	29	(a)	39	(a)	49	(a)	59	(c)
10	(a)	20	(b)	30	(d)	40	(a)	50	(d)	60	(a)

Hints & Solutions

2. (d) Prop or pillar or columnar roots: They are pillar-like roots which give extra mechanical support to heavy stem branches. *e.g.*, Banyan tree.
5. (a) It is the most familiar and wide spread type of compound leaf in which the rachis is elongated and bears two rows of simple or divided leaflets. The leaflets arranged alternately or in pairs along with the rachis.
6. (a) In parallel venation (striate venation), veins and veinlets run parallel to each other. Parallel venation is the characteristic feature of monocots.
7. (b) The most important function of inflorescence is it makes flower more conspicuous to pollinating agents (insects/birds) so that chances of cross-pollination are high.
8. (a) The broadened base of flower which lies at the tip of pedicel is called torus or thalamus or receptacle.
9. (d) Floral formula is a symbiotic and numerical way to provide information about : number, position, cohesion and adhesion of floral parts, symmetry of the flower and sexuality of the flower.
10. (a) A plant bearing flowers of both sexes, i.e., staminate and pistillate flowers, e.g., ricinus, zea mays (maize) and cucurbits.
12. (d) Castor is an endospermic (Albuminous seed). In albuminous seeds, embryo not consumed all endosperm. So, it persists in the mature seed. In these seeds food stored in endosperm.
13. (a) Legume is a dehiscent dry fruit produced from a monocarpellary superior ovary. Caryopsis is dry and one seeded fruit which developed from a superior monocarpellary ovary. In this the pericarp is closed and fused with seed coat. It is a characteristic of family Poaceae.
14. (d) Tomato and tobacco both belong to the family Solanaceae. Solanaceae has some identifying characteristics-bicarpellary, syncarpous superior ovary, axile plancentation, fruit is berry or capsule.
15. (d) Special type of seed germination is characteristic of halophytes growing in saline conditions is vivipary. The seed germination occurs inside the fruit while it is attached to the parent plant.
16. (b) Drupe is a fleshy one seeded, indehiscent fruit *e.g., Mangifera indica*.
17. (c) Cauliflower has the scientific name *Brassica oleracea* belonging to the variety *botrytis*.
18. (a) Pulses which are economically important belong to the family Fabaceae.
19. (d) Perennials survive for more than two years and produce flowers and fruits during specific seasons.
20. (b) Dicots possess tap root system.
23. (d) The fruit is a mature or ripened ovary. When a fruit develops exclusively from the ovary, it is said to be true fruit. When in addition to the ovary, some other floral parts also participate in the formation of fruits, then it is known as false fruit. Apple, pear, cashewnut, mulberry, etc are all false fruits.
24. (b) Coir is a coarse fibre extracted from the fibrous outer shell of a coconut. Coir fibres are found between the husk and the outer shell of a coconut. The individual fibre cells are narrow and hollow, with thick walls made of cellulose. They are pale when immature but later become hardened and yellowed as a layer of lignin is deposited on their walls.
25. (c) The leaves appear green because of the pigment chlorophyll which does not absorb green light rather reflects it back.
26. (c) Reproductive parts (floral features) are more stable and conservative than vegetative parts, which exhibit changes due to the environmental factors quite readily. So floral features are more reliable.
27. (c) Axillary bud is the bud arising in the axil of branch. Apical bud arises at the apex. Adventitious bud arises from any location other than its usual location.
28. (a) Single cotyledon of embryo in cereal grain is represented by scutellum. Coleoptile represent the covering of stem. Coleorhiza represents the covering of root.
29. (a) *Hibiscus* belongs to Malvaceae family. It possess numerous stamens that are called monadelphous.
30. (d) Sweet potato and turnip both are roots and become modified for special functions, the most common being the formation of tuberous roots for food storage.
40. (a) A true fruit develops from ovary.
48. (b) Water logged soil is physiologically dry.
When soil is waterlogged, plants literally drown. Water fills all the air spaces between the soil particles and this prevents oxygen from reaching the roots.
49. (a) Bud of garlic have greatly reduced stems. A corm is an underground, vertical, compressed stem with reduced scale leaves. Garlic bulbs consist mostly of individual segments called cloves.
50. (d) Imparipinnate leaf is one where rachis is terminated by an odd leaflet. Pinnate with a single terminal leaflet.
52. (a) Flower is complete when it has calyx, corolla, androecium and gynaecium.
53. (c) Sucking roots are present in *Cuscuta reflexa*. It is non-chlorophyllous parasitic plant. These roots have specialized sucking organ called as haustorium that absorbs required food material from the host plant.
54. (d) Submerged hydrophytes do not have stomata (a protective mechanism in aquatic plants against water logging of internal cells and tissues).
55. (d) The leaf is green, lateral appendage, borne on the node of the stem and bears bud in its axil, exogenous in origin and develops from the leaf primordia of shoot apex. The main function of the leaf is to synthesize food in the process of photosynthesis.
57. (d) Rhizome is a prostrate, branched horizontally growing stems having nodes and internodes. Vertical rhizome is also called root stock.

58. **(b)** Epiphytes are those plants which live on another plant for shelter but do not take food or any thing from that plant.

59. **(c)** Wheat (*Triticum aestivum*) is caryopsis type of fruit. Caryopsis is simple fruit which develops from monocarpellary, unilocular ovary. The fruit is small and one seeded. The pericarp is fused with the seed coat. Such fruits are found in family *Gramineae* or *Poaceae* eg. wheat, rice, maize etc. The fruit is indehiscent.

60. **(a)** Botanical name of banana is *Musa paradi-sica.*

6. ANATOMY OF FLOWERING PLANTS

ANSWER KEY

1	(b)	12	(a)	23	(b)	34	(a)	45	(d)	56	(b)
2	(a)	13	(a)	24	(c)	35	(b)	46	(a)	57	(c)
3	(a)	14	(c)	25	(b)	36	(d)	47	(d)	58	(c)
4	(d)	15	(a)	26	(b)	37	(a)	48	(a)	59	(c)
5	(a)	16	(d)	27	(c)	38	(a)	49	(c)	60	(b)
6	(c)	17	(a)	28	(d)	39	(a)	50	(d)		
7	(c)	18	(a)	29	(d)	40	(d)	51	(c)		
8	(a)	19	(a)	30	(b)	41	(a)	52	(d)		
9	(a)	20	(c)	31	(a)	42	(d)	53	(c)		
10	(c)	21	(a)	32	(d)	43	(c)	54	(a)		
11	(d)	22	(c)	33	(d)	44	(c)	55	(c)		

Hints & Solutions

1. **(b)** They posses hard and extremely thick secondary walls due to uniform deposition of lignin.

2. **(a)** The cells of collenchyma contain protoplasm and are living without intercellular spaces. When sclerenchyma consist of thick-walled dead cells. In the beginning the cells are living and have protoplasm but due to deposition of impermeable secondary walls (lignin) they become dead, thick and hard.

4. **(d)** In collenchyma cell walls show localized thickenings due to presence of approximately 45% pectin, 35% hemicellulose and 20% cellulose.

5. **(a)** Permanent tissues are made up of mature cells which have lost the capacity to divide and have attained a permanent shape, size and function, due to division and differentiation in meristematic tissue.

6. **(c)** Nervous system is present in animals.

7. **(c)** Xylem fibres are two types : libriform fibres (thickwalled with simple pits) and fibre tracheids (thin walled with reduced bordered pits).

8. **(a)** Chloroplast containing parenchyma (chlorenchyma) are mostly present in leaf.

9. **(a)** Waxy outermost lining, which is secreted by epidermis.

10. **(c)** The xylem and phloem strands alternate with each other separated by parenchymatous cells. Such kinds of vascular bundles are called radial and found mainly in roots.

11. **(d)** Pith or medulla occupies the central part in dicot stem and monocot root.

16. **(d)** The wall area bearing perforation is called perforation plate. Usually the perforations in vessels are present in the end walls but they may be present in the lateral wall too.

17. **(a)** Parenchyma storage of food materials e.g., carrot, beet root, etc.

18. **(a)** Lateral meristem occurs on the sides of stem and help in increasing girth of stem and root. It divides only periclinally or radially and is responsible for increase in girth or diameter.

19. **(a)** Phellogen present in outer cortical cells produces cork or phellem on the outer side which consists of dead and compactly arranged rectangular cells that possess suberised cell walls which causes them to become buoyant. Phellogen also cuts off cells on innerside called as phelloderm or secondary cortex.

20. **(c)** Ring porous wood is more advanced than diffuse porous wood as it provides for better translocation when requirement of plant is high.

21. **(a)** Intercalary meristems are derived from apical meristems and separated from the same by permanent cells. They are responsible for localised growth.

22. **(c)** Collenchyma provides mechanical strength to young dicot stems, petioles and leaves.

23. **(b)** Pericycle in roots is active in the formation of root branches or lateral roots.

24. **(c)** In monocot leaves, the mesophyll cells are undifferentiated.

25. **(b)** Vascular cambium produces secondary xylem on the inner side and secondary phloem to the outer side.

26. **(b)** Endodermis or innermost layer of cortex has casparian strips in roots. It is called starch sheath in dicot stems. It separates cortex from stele. The cell walls are thickened at the corners in angular collenchyma.

27. **(c)** Internally, sieve tubes possess peripheral layer of cytoplasm and are devoid of nucleus. They are living cells. Cell walls of sieve tubes are thickened than surrounding parenchyma cells.

28. **(d)** Sieve tubes are long distance channels for transport of organic nutrients. The movement of the nutrients are bidirectional.

29. **(d)** Bicollateral vascular bundles have phloem in both outer and inner side of xylem. These type of bundles occur in cucurbitaceae.

30. **(b)** The walls of xylem vessels are lignified and bordered pits are common in the walls.

31. **(a)** Sclereids are small bundles of sclerenchyma tissue in plants that form durable layers, such as the cores of apples and the gritty texture of pears. Sclereids are variable in shape. The cells can be isodiametric, prosenchymatic, forked or fantastically branched. The cell walls fill nearly all the cell's volume. The shell of many seeds like those of nuts as well as the stones of drupes like cherries or plums are made up from sclereids.

32. **(d)** A leaf primordium grows into the adult leaf lamina by means of first by apical meristem and later by marginal meristems.

33. **(d)** Intrafascicular cambium occurs inside the vascular bundles in between xylem and phloem. Interfascicular cambium develops in the form of strips at the level of intrafascicular cambium of vascular bundles. Intercalary meristem lies in between areas of permanent tissues.

34. **(a)** Lenticels are pores present in woody stem through which transpiration or loss of water vapour takes place. Lenticel formation begins during the development of the first periderm. In the stem, they usually appear below a stoma or group of stomata. It should also be noted that lenticels can be present on fruits such as apples and pears.

35. **(b)** Generally gymnosperms do not have vessels but a few gymnosperms having vessels are *Ephedra*, *Smilax*, etc. Most of angiosperms have vessels except of few e.g., *Trochodendron*.

36. **(d)** Vessels are elongated, multicellular water conducting channels with wide lumen formed by end to end fusion of a large number of vessel elements. Tracheids are elongated dead cells with tapering ends having lignified walls and large or wide lumen. Their main function is conduction of water and minerals from root to leaf.

37. **(a)** Root shows the following regions
Root cap or calyptra - caplike protective covering over tip of the root
Meristematic region is the subapical position
Zone of elongation-receives news cells from the growing point
Root hair zone - is the zone of differentiation
Zone of mature cells-having thick walled impermeable cells.

38. **(a)** Manganese and iron are micronutrients required in trace amounts. Concentration of nitrogen in dry matter- 15000 mg/gm. Concentration of carbon in dry matter is 45,000 mg / gm.

39. **(a)** Teak and pine is most suitable for the study of critical secondary growth because in secondary growth, secondary tissues are formed from lateral meristem which is well developed in these two cases and secondary growth occurs in gymnosperms and dicots.

40. **(d)** The innermost layer of the cortex is called endodermis. It comprises a single layer of barrel shaped cells without any intercellular spaces. In roots thick walled endodermal cells are interrupted by thin walled passage cells or transfusion tissue.

41. **(a)** In barley stem, vascular bundles are closed and scattered. They are open only for a hour in the day time and never open at night. e.g Cereals.

42. **(d)** Tracheids are chief water conducting elements of xylem in gymnosperms. They are devoid of protoplasm and hence dead. The wall constituting the tracheids is hard, thick and lignified. These are elongated cells with tapering ends.

43. **(c)** Heartwood differs from sapwood in having dead and non-conducting elements. In old trees, the inner region that comprises dead elements with highly lignified walls is called heartwood. Heartwood does not conduct water but gives mechanical support to the stem. On the other hand, the peripheral region, which is lighter in colour are called sapwood. It is involved in the conduction of water and minerals from root to leaf.

44. **(c)** Phellem, phellogen and phelloderm are collectively called periderm.

45. **(d)** Function of companion cell is to load sugar and amino acids into sieve elements. These cells use transmembrane proteins to take up by active transport.

47. **(d)** A tissue is a group of cells which are similar in origin, form and function.

59. **(c)** By counting the number of annual rings at base stem of a tree the age of the tree can be determined.

60. **(b)** The wood, which is in fact the secondary xylem, is divided into 2 types in old trees outer, light loloured alburnum or sap wood and inner dark coloured duramen or heart wood. Sap wood is functional while heart wood is non–functional.

7. STRUCTURAL ORGANISATION IN ANIMALS

ANSWER KEY											
1	(c)	11	(a)	21	(b)	31	(a)	41	(a)	51	(a)
2	(a)	12	(c)	22	(b)	32	(c)	42	(d)	52	(a)
3	(c)	13	(c)	23	(c)	33	(c)	43	(b)	53	(b)
4	(c)	14	(b)	24	(a)	34	(a)	44	(b)	54	(a)
5	(b)	15	(d)	25	(b)	35	(b)	45	(d)	55	(c)
6	(b)	16	(a)	26	(a)	36	(a)	46	(a)	56	(d)
7	(a)	17	(b)	27	(d)	37	(c)	47	(b)	57	(c)
8	(a)	18	(a)	28	(a)	38	(c)	48	(b)	58	(c)
9	(c)	19	(b)	29	(a)	39	(d)	49	(a)	59	(d)
10	(b)	20	(d)	30	(a)	40	(b)	50	(c)	60	(c)

Hints & Solutions

2. **(a)** Microvilli of epithelial cells increase surface area.

3. **(c)** Tendon is made up of modified white fibrous tissue.

13. **(c)** Cells of squamous epithelium are flattened scale like and attached to each other like a mosaic.

14. **(b)** Epithelial tissues have a widespread distribution throughout the body and serve several important functions such as protection, secretion, sensation and absorption.

16. **(a)** **Ligament** is a modified yellow elastic fibrous connective tissue and connects bone to bone.
Tendon is non-elastic but flexible tissue consists of parallel bundles of collagenous fibres. It joins the muscles to bones.

22. **(b)** Because there are about 639 muscles which constitutes the bulk of body, 40% of total weight and 50% of soft parts.

23. **(c)** Neurons are structural and functional units of nerve tissue. Their number is quite high, more than 1×10^{12} in brain alone.

24. **(a)** Lymph differs from blood in that it contains less protein and no red blood cells. Lymph contains white blood cells.

25. **(b)** Mineral found in red pigment of vertebrate blood is iron.

26. **(a)** Characteristics of smooth muscle fibres.
- cells range from 5 to 10 μm in diameter and 30 to 200 μm in length
- spindle-shaped
- single, centrally located nucleus
- smooth muscle tissue occurs within almost every organ, forming sheets, bundles or sheaths around other tissues.

Smooth muscle differs from both skeletal and cardiac muscle tissues in structure and function. Sarcomeres or myofibrils are not present - therefore not striated, i.e., smooth.

27. **(d)** Lymphocyte is a type of white blood cells in the vertebrate immune system.
Lymphocytes are responsible for making antibodies.

28. **(a)** Deposition of bony matter by osteoblasts and resorption chondroblasts.

29. **(a)** The wall of blood vessels (arteries & veins) are made up of mainly 3 layers – the outer most tunica externa has a lose connective tissue, collagen fibres, lymph vessels and the middle tunica media has circular smooth muscle fibres, some elastic fibres and the inner most tunica interna is a single layered endothelium of polyhedral squamous cells. If the arteries reach organs they form arterioles which are thinner and mainly consist smooth muscles and when they subdivide in the tissues they make capillaries. The wall of capillaries are very thin (usually less than one micron) and have numerous minute pores and made up of only endothelium. Exchange of material takes place between blood and tissue fluid across the endothelial membrane of capillaries through active diffusion.

30. **(a)** Chondrin is a gelatin-like substance and can be obtained by boiling cartilage in water. The cartilage is a connective tissue that contains cells embedded in a matrix of chondrin.

32. **(c)** When a bone is kept in dilute acids like HCl for a few days, the calcium salts get dissolved leaving only the soft and flexible organic matrix. This process is called Decalcification. If a bone is kept in KOH it remains unaffected and only the surrounding muscles and connective tissues get dissolved so that the bone now appears clean.

33. **(c)** Only nerve cells do not form layers. These also remain structurally separate from each other. Nerve cells or neurons are the cells specialized to conduct an electrochemical current. Nerve tissue is made up of these cells and supporting cells. Epithelial tissue is made of one or a few layers of compactly arranged cells overlying a basement membrane. Glands are epithelial cells specialized to elaborate and release chemicals for use in the body. Muscle tissue is made of elongated and contractile cells called myocytes.

34. **(a)** Earthworms have no skeleton but during burrowing, anterior end becomes turgid and acts as a hydraulic skeleton. It is due to coelomic fluid. The coelom is filled with an alkaline milky coelomic fluid. This coelomic fluid contains water, salts, proteins and 4 types of cells namely amoebocytes, mucocytes, circular cells and chloragogen cells.

35. **(b)** The structure of the abdominal wall is similar in principle to the thoracic wall. There are three layers, an external, internal and inner-most layer. The vessels and nerves lie between the internal and innermost layers. The abdomen can be divided into quadrants or nine abdominal regions. Pain felt in these regions may be considered to be direct or referred abdominal wall made up of smooth muscles.

36. **(a)** If a live earthworm is pricked with a needle on its outer surface without damaging its gut, the fluid that comes out is coelomic fluid because coelomic fluid lies between body wall and alimentary canal. Earthworms are helpful to man as bait for fishing and loosen the soil for roots to grow. A earthworm's wastes help to fertilize the soil.

37. **(c)** In *Pheretima* two pairs of testes sacs are situated in the 10th and 11th segments. Each testes sac of the 10th segment encloses a testes and a seminal funnel. Each testes sac of the 11th segment encloses a testes, a seminal vesicle and a seminal funnel.

38. **(c)** The kind of tissue that forms the supportive structure in our pinna (external ears) is also found in tip of the nose. External pinna is made up of elastic cartilage. It is passage towards middle ear.

39. **(d)** Cockroach takes the food after grounding by its mandibles and gizzard.

40. **(b)** The break down of detritus into smaller particles by earthworm is known as fragmen-tation.

41. **(a)** (A) Dry skin cause ceased cutaneous respiration
(B) Three chambered heart.
(C) Frog never becomes uricotelic
(D) External fertilization occurs in water

52. **(a)** Some organism have the power to change body colour with the help of pigment carotene in chromatophores present in dermis.

53. **(b)** The intestine of an earthworm has a peculiar fold of tissue that forms a pocket in the intestine that is visible when viewing a cross section. This space, which runs almost the whole length of the body dorsally is known as the typhlosole and it serves to increase the surface area of the intestine for absorption.

57. **(c)** Multinucleate muscle cells are skeletal muscles.

8. CELL : THE UNIT OF LIFE

ANSWER KEY															
1	(b)	9	(b)	17	(c)	25	(b)	33	(c)	41	(a)	49	(a)	57	(a)
2	(b)	10	(b)	18	(a)	26	(b)	34	(a)	42	(a)	50	(b)	58	(a)
3	(a)	11	(d)	19	(d)	27	(a)	35	(d)	43	(c)	51	(a)	59	(c)
4	(d)	12	(d)	20	(c)	28	(c)	36	(d)	44	(c)	52	(c)	60	(a)
5	(d)	13	(d)	21	(c)	29	(b)	37	(d)	45	(a)	53	(c)		
6	(c)	14	(a)	22	(d)	30	(b)	38	(c)	46	(b)	54	(b)		
7	(c)	15	(d)	23	(c)	31	(d)	39	(a)	47	(c)	55	(b)		
8	(b)	16	(b)	24	(b)	32	(c)	40	(d)	48	(c)	56	(b)		

Hints & Solutions

1. **(b)** The prokaryotic cells do not have nuclear membrane while eukaryotic cell have well organised nuclear membrane.

2. **(b)** Virchow gave 'Omnis cellula a cellula' theory.

5. **(d)** Surface of plasma membrane has the ion carriers which participates in exchange of ions.

8 **(b)** The green colour of plants is due to chlorophyll, which is found in chloroplast.

14. **(a)** During cell division spindle fibres attached on centriole .

18. **(a)** Metacentric chromosomes are 'V' shaped in which centromere is situated in the mid portion of chromatids.

19. **(d)** According to fluid mosaic model proposed by Singer and Nicolson (1972, 74), plasma membrane is

composed of phospholipid bilayer and proteins. Protein molecules occur at places both inside and outside of lipid bilayer. The internal proteins are called intrinsic proteins and the external proteins are called extrinsic proteins.

20. **(c)** Centrifuge is an instrument used to create a centrifugal force. The homogenate of the cell organelles is allowed to undergo centrifugation by which different cell organelles get separated. The process is known as differential centrifugation.

21. **(c)** Plasma membrane is made up of proteins, lipids and carbohydrates. Carbohydrates are covalently linked to both lipid and protein.

22. **(d)** Nucleoproteins are conjugated proteins. They are deoxyribonucleoproteins and ribonuc-leoproteins synthesised in cytoplasm of the cell.

23. **(c)** Fluid mosaic model is the most recent model of a biomembrane proposed by Singer and Nicolson (1972, 74).

24. **(b)** The resolution is a power of a microscope to distinguish between two particles that are closely situated. Electron microscope has high resolution power due to very low wavelength of electron beam.

25. **(b)** The ability to distinguish two close objects is called resolution power.

26. **(b)** Fluid mosaic model of cell membrane was put forward by Singer and Nicolson.

27. **(a)** Cell wall shows complete permeability because it helps in the transport of substances in and out of the cell.

28. **(c)** Ribosomes are the workhouses of protein biosynthesis, the process of translating messenger RNA (mRNA) into protein. The mRNA comprises a series of codons that dictate to the ribosome the sequence of the amino acids needed to make the protein. Using the mRNA as a template, the ribosome translates each codon of the mRNA, pairing it with the appropriate amino acid. This is done using molecules of transfer RNA (tRNA) containing a complementary anticodon on one end and the appropriate amino acid on the other.

29. **(b)** All plastids are similar in structure. So they can easily get transformed from one type to another. Depending upon the colours there are three types of plastids- leucoplasts, chromoplasts and chloroplasts.

30. **(b)** Cell theory was proposed by Schleiden and Schwann (1839) according to which "All living organisms are composed of cells." Cell theory is also known as cell principle, cell doctrine etc.

31. **(d)** Cell recognition and adhesion occur due to biochemicals of cell membranes named glycoproteins and glycolipids. These are formed due to the small carbohydrate molecules present on lipids and extrinsic proteins.

32. **(c)** Golgi complex is absent in prokaryotic cell i.e bacteria and blue green algae. It is present in all eukaryotic cells except sieve tubes of plants, sperms of bryophytes and pteridophytes and red blood corpuscles.

33. **(c)** Lysosomes are membranous bag with hydrolytic enzymes which is used for controlling intracellular digestion of macro-molecules.

34. **(a)** In plants peroxisomes are associated with photorespiration. Peroxisomes are ubiquitous organelles in eukaryotes that function to rid the cell of toxic substances. They have a single membrane that separate their contents from the cytosol (the internal fluid of the cell) and that contain membrane proteins critical for various functions, such as importing proteins into the organelles and aiding in proliferation. Unlike lysosomes, which are formed in the secretory pathway, peroxisomes usually self-replicate by enlarging and then dividing, although there is some indication that new ones may be formed directly.

35. **(d)** The inner membrane of mitochondria is infolded to form involutions called cristae. They are meant for increasing the active area of the inner membrane.

36. **(d)** Mitochondrial cristae possess small particles called elementary particles which contain ATP- ase. Therefore, they are the centes of ATP synthesis during oxidative-phosphorylation.

37. **(d)** The prokaryotic flagellum is single stranded made up of several parallel protein fibrils and helical in shape while in eukaryotes it is made up of axonema and sheath.

38. **(c)** Rough endoplasmic reticulum bears ribosomes on its outer surface. These ribosomes take part in protein synthesis.

39. **(a)** Lysosomes are single membrane bound vesicular organelles rich in acid hydrolases. They are involved in phagocytosis.

41. **(a)** Microtubules are formed of dimers of tubulin. Microtubules are present in centriole, flagella, spindle fibres but not in mitochondria.

42. **(a)** Zinc is an activator of enzymes like carbonic anhydrase. Copper takes part in electron transport as plastocyanin. Potassium maintains membrane permeability.

43. **(c)** Mitotic spindles are composed of micro-tubules which are made of tubulin dimers. Actin and myosin are the contractile proteins of muscle fibres. Myoglobin is oxygen carrying pigment found in muscle.

44. **(c)** Prokaryotes have simple flagella without microtubules whereas eukaryotes have complex cilia and flagella which consists of microtubules arranged in 9 + 2 fashion (an outer ring of nine pairs surrounding 1 center pair). Further in prokaryotes the arrangement is 9 + 0.

45. **(a)** During anaphase APC (anaphase promoting complex) develops. It degenerates proteins binding the two chromatids in the region of centromere.
As a result the centromere of each chrom-osome divides. This converts the two chromatids into daughter chromosomes.

46. **(b)** Vacuole in a plant cell is membrane bound and contains water and excretory substance. Vacuole is a

space within the cytoplasm of living cell that is filled with air, water or other liquid, cell sap or food particles. In plant cells, there is usually one large vacuole bounded by a single layered membrane (tonoplast or vacuole membrane); animal cells usually have several small vacuoles.

47. (c) Plasmodesmata are connections between adjacent cells. Plasmodesmata are narrow channels that act as intercellular cytoplasmic bridges to facilitate communication and transport of materials between plant cells. Plasmodesmata are formed during cell division, when traces of the endoplasmic reticulum become caught in the new wall that divides the parent cell.

48. (c) The cytoskeleton is made up of three kinds of protein filaments actin filaments (also called microfilaments), intermediate filaments and microtubules whose major constituents are actin and tubulin respectively.

49. (a) Mitochondria has its own DNA. It is as structure within cytoplasm of eukaryotic cells that carries out aerobic respiration. It is the site of Kreb's cycle and ETS.
Therefore, it is also called as cell's energy production site.

50. (b) Except peroxisome the remaining three and ER are the parts of endomembrane system.

58. (a) Tonoplast also shows fluid mosaic model and is similar to plasma membrane.

59. (c) Power house of the cell is mitochondria as it is the site for oxidation of food material (cellular respiration) to release energy.

60. (a) Plastids (chloroplasts) are the largest cell organelles. They may have a diameter of 4-6 micron. In a eukaryotic cell only nucleus is larger than the chloroplasts.

9. BIOMOLECULES

ANSWER KEY

1	(b)	11	(d)	21	(a)	31	(b)	41	(a)	51	(b)
2	(a)	12	(a)	22	(a)	32	(a)	42	(a)	52	(d)
3	(c)	13	(c)	23	(c)	33	(c)	43	(c)	53	(a)
4	(a)	14	(d)	24	(c)	34	(a)	44	(b)	54	(a)
5	(b)	15	(d)	25	(c)	35	(a)	45	(a)	55	(b)
6	(a)	16	(a)	26	(d)	36	(c)	46	(b)	56	(a)
7	(d)	17	(b)	27	(d)	37	(a)	47	(c)	57	(a)
8	(c)	18	(c)	28	(a)	38	(a)	48	(b)	58	(c)
9	(c)	19	(b)	29	(a)	39	(c)	49	(d)	59	(c)
10	(c)	20	(d)	30	(d)	40	(d)	50	(a)	60	(b)

Hints & Solutions

1. (b) Because enzymes work as biocatalysts in metabolic process.

3. (c) Because a particular enzyme can catalyse only a particular type of reaction.

4. (a) Simple sugar *i.e.,* monosaccharide. These are the simplest carbohydrates and are building units of complex carbohydrates. i.e., Starch and cellulose.

10. (c) Nucleic acids (DNA and RNA) are repeating units of nucleotides i.e. polynucleotides. Each nucleotide comprises of nitrogenous heterocyclic bases viz. purines or pyrimidines, pentose sugar and phosphoric acid.

11. (d) Nucleic acids (DNA and RNA) comprise of polynucleotides. Each nucleotide consists of nitrogenous bases, pentose sugar and phosphoric acid.

12. (a) The nitrogenous bases are constituents of DNA. The nitrogenous bases may be purines or pyrimidines. Purines are 9 membered, double ringed structures e.g. adenine or guanine.

13. (c) Length of one helix is 34Å in B-DNA.

14. (d) Enzymes accelerate chemical reaction by lowering the activation energy.

15. (d) In a phosphodiester bond, linkage takes place between the phosphate group of one nucleotide and the hydroxyl group of sugar (ribose or deoxyribose) of the adjacent nucleotide.

16. (a) The two chains comprising the DNA are sprially coiled around a common axis to form a regular, right handed double helix.

17. (b) In RNA, pyrimidine thymine is replaced by uracil.

18. (c) The ratios for the (A+ G)/ (T+C) are constant at approximately 1.0 for the different organisms. Each of these organisms contains a double stranded genome. The percentage of guanine and cytosine are almost

equal to each other, similarly the percentages of adenine and thymine are almost equal to each other as well. In other words, the percentage of purines should be equal to the percentage of pyrimidines for double stranded DNA. This means that (A + G) = (C + T)

The (A+T) / (C + G) ratios are not constant. The relative numbers of (A + T) base pairs and (G + C) base pairs are unique to each organism and can vary between the different species.

19. **(b)** There are certain enzymes which have slightly different molecular structure but have similar catalytic function such enzymes are known as isoenzymes. Holoenzymes is the active compound formed by combination of a coenzyme and an apoenzyme. Apoenzyme is the protein component of an enzyme, to which the coenzyme attaches to form an active enzyme. Coenzyme are organic non-protein molecules that bind with the protein molecule (apoenzyme) to form the active enzyme (holoenzyme).
20. **(d)** Glycogen is a polysaccharide, which is synthesized and stored in the liver. It is released into the blood by breakdown of simple glucose, and energy is released.
21. **(a)** Polysaccharides are complex carbohydrates formed by polymerization of a large number of units called monosaccharides, starch is a glucosan polysaccharide and is the major reserve food of animals. Cellulose is the structural polysaccharide of plant cell wall. Sucrose is a disaccharide formed of one unit of D-glucose and one unit of D-fructose. Maltose is a disaccharide formed of two molecules of glucose held together, lactose is disaccharide formed through condensation of D-glucose and D-glucose. Glucose and fructose are monosaccharides.
22. **(a)** Proteins are polymers of amino acids. There are only 20 amino acids which can be arranged in different orders in a polypeptide chain to form a wide array of proteins.
23. **(c)** Thymidine is used to assess the rate of DNA synthesis, since thymine is unique to DNA.
24. **(c)** There are only 20 amino acids . These can be arranged in different order in the polypeptide chain to form a diverse array of proteins.
25. **(c)** Cellulose is the structural polysaccharide of plant cell walls. They are long chain polymers of -glucose units. The adjacent -glucose units are joined by linkages.
26. **(d)** Lactose is a disaccharide that consists of -D-galactose and -D-glucose molecules bonded through a 1-4 glycosidic linkage. Lactose makes up around 2-8% of the solids in milk. The name comes from the Latin word for milk, plus the -ose ending used to name sugars.
27. **(d)** Secondary structure is development of new stearic relationships amongst the amino acids through formation of intra-and inter polypeptide hydrogen bonds.
28. **(a)** Length of one turn of DNA helix is 34Å.

$1\text{Å} = 10^{-1}$ nm

$$\therefore 34\text{Å} = 10^{-1} \times 34 = \frac{34 \times 1}{10} = 3.4\,\text{nm}$$

30. **(d)** Essential amino acids are those which cannot be synthesized in the body from precursors. These amino acids must be present in our diet. They are valine, leucine, isoleucine, lysine, methionine, phenylalanine, tryptophan and threonine.
31. **(b)** Water attracting molecules are called hydrophilic. Water repelling molecules are called hydrophobic. Amino acids carry simultaneously positive and negative charges. Such substances are called zwitter ions. Lipids are compounds of C, H, O but the ratio of H and O is more than 2 : 1 that is the ratio of oxygen is lesser than carbohydrates. Lipids are insoluble in water but soluble in non-polar solvents such as benzene, chloroform, etc.

    Commonest lipid found in a cell is phospholipid. It contains a hydrophilic (polar) head and a hydrophobic (non- polar tail).
32. **(a)** Collagen is a major fibrous protein of connective tissue occuring as white fibres produced by fibroblast.
33. **(c)** K_m (Michealis Menten constant). It is defined as that substrate concentration at which under optimum conditions the rate of an enzyme catalysed reaction reaches half the maximum rate. K_m is inversely proportional to affinity of enzyme for its substrate.
34. **(a)** Antiparallel strands of DNA have two strands having phosphate groups at same positions in space.
35. **(a)** Some bacteria (such as Rhodopseudomonas), algae and green plant cells produce carbohyd-rates.
36. **(c)** It is calcium and phosphorus.
37. **(a)** Malonate is a powerful inhibitor of cellular respiration, because it binds to the active site of the succinate dehydrogenase in the citric acid cycle but does not react, thereby competes with succinate, since it does not have the -CH_2-CH_2- group as in succinate which is required for dehydrogenation. For the oxidative phosphory-lation reaction, malonate is an inhibitor for complex II which again contains succinate dehydrogenase.
38. **(a)** Massive ions like Na^+ facilitate the absorption of substances like amino acids and glucose through Co-transport.
43. **(c)** Milk protein is Casein.
47. **(c)** Ligase enzyme is used for joining bits of DNA.
51. **(b)** C, H, N, O are the frame work elements in plants.
52. **(d)** With Iodine starch turns blue black. Starch is detected by Iodine test.
53. **(a)** tRNA constitute about 15% of total RNA. It is the smallest RNA with 70-85 nucleotides and sedimentation coefficient of 4S.
54. **(a)** At high temperature two strands uncoil & separate.
55. **(b)** Amylase is a group of closely related enzymes that degrade starch.
56. **(a)** ATP has 2 phosphate bonds, ADP has 1 high

energy phosphate bond while AMP has no phosphate bond.

57. (a) Zymase is an enzyme complex ("mixture") that catalyzes glycolysis, the fermentation of sugar into ethanol and carbon dioxide. As the conversion takes place, the reaction will gradually slow down. They occur naturally in yeasts. Zymase was first isolated from the yeast cell in 1897 by a German chemist named Eduard Buchner who fermented sugar in the laboratory without living cells, leading to 1907 Nobel Prize in Chemistry.

58. (c) In case of any infection our body's immune system produces antibodies to kill or react against the germs. The antibodies are made up of carbo-hydrates and proteins i.e. they are glycoproteins. Antibodies are also called immunoglobulins.

10. CELL CYCLE AND CELL DIVISION

ANSWER KEY

1	(c)	11	(b)	21	(c)	31	(a)	41	(b)	51	(b)
2	(d)	12	(a)	22	(d)	32	(b)	42	(d)	52	(b)
3	(a)	13	(d)	23	(a)	33	(c)	43	(d)	53	(b)
4	(a)	14	(b)	24	(d)	34	(d)	44	(a)	54	(a)
5	(d)	15	(c)	25	(c)	35	(d)	45	(a)	55	(a)
6	(a)	16	(a)	26	(b)	36	(a)	46	(b)	56	(d)
7	(b)	17	(c)	27	(c)	37	(d)	47	(d)	57	(d)
8	(a)	18	(b)	28	(a)	38	(a)	48	(c)	58	(d)
9	(c)	19	(b)	29	(b)	39	(a)	49	(d)	59	(a)
10	(b)	20	(d)	30	(a)	40	(b)	50	(a)	60	(a)

Hints & Solutions

3. (a) Condensation of chromosomes occurs in the prophase-I. After that chromosome moves toward the poles.

5. (d) Leptotene, zygotene, pachytene, diplotene and diakinesis all are successive stages of meiosis.

6. (a) Synaptonemal complex is formed during meiotic prophase first.

7. (b) Knot like structure 'Chromomeres' found in the stage leptotene of prophase first of meiosis.

8. (a) Bivalent formation occurs in pachytene of meiosis. In this stage, two chromatids of homologous chromosome (Bivalent) later on forms a cross and now showing a tetravalent or tetrad stage.

10. (b) The crossing over of homologous chromosome occurs in pachytene of prophase first of meiosis. It is known as recombination.

19. (b) During metaphase, the centomeres of the chromosome lie on the equatorial plate according to their size and spatial arrangement. So it is the best time to count the number and study the morphology of chromosomes.

20. (d) G_1 phase, also called Gap I phase is characterized by increase in cell size. In the S phase or synthetic phase DNA molecules replicate. G_2 is the second growth phase, also called. Gap II where there is intensive formation of RNAs and proteins. In the mitotic metaphase, the chromosomes are arranged at the equatorial plate.

21. (c) Meiosis II is an equational division which is meant for maintaining the haploid number, separating the two chromatids of a chromosome.

22. (d) In mitotic anaphase, the sister chromatids separate from each other and begin to move towards the opposite poles.

23. (a) Due to segregation, independent assortment and crossing over at the time of meiosis, daughter cells exhibit variation.

24. (d) Meiosis provides a chance for new combinations of chromosomes mainly by the phenomenon of crossing over and random distribution of homologous chromosomes between daughter cells.

25. (c) The points of attachment between homologous chromosomes after their separation in diplotene are called chiasmata. The process of pairing of homologous chromosomes is called synapsis, the phenomenon by which DNA isolated from one type of cell, when introduced into another type, is able to bestow some of the properties of the former to the latter is known as transformation.

26. (b) Plants cells do not have centrioles. Cell plate appears during cytokinesis in dividing cells. Centromere holds the sister chromatids in a

chromosome. Spindle fibres appear during metaphase.

27. **(c)** Metaphase is characterized by the appearance of spindle fibres. Anaphase involves movement of chromatids towards the poles. Cytokinesis marks the division of the cytoplasm of the parent cell into daughter cells.
28. **(a)** The spindle fibres involved in cell division is made of microtubules.
29. **(b)** Kinetochore is the proteinaceous covering of centriole, to which spindle fibers attach.
30. **(a)** Anther cells are used to study mitosis.
31. **(a)** Chromosomes are most distinct in the metaphase stage. In the telophase stage they regain their coiled composition. In anaphase the chromatids separate and move towards opposite poles. In prophase stage the chromosomes appear thread like and individual chromatid cannot be seen.
32. **(b)** Histone proteins are synthesized during S-phase of cell cycle.
33. **(c)** Synapsis (also called syndesis) is the pairing of two homologous chromosomes that occurs during prophase I (zygotene stage) of meiosis. The two chromosome move together and pairing of corresponding points along their lengths as they lie side to side. The resulting structure is called a bivalent.
34. **(d)** During mitosis, endoplasmic reticulum and nucleus begins to disappear at early prophase. Also, nucleus and cell become spherical and DNA molecules condense to form shortened chromosome.
37. **(d)** Amitosis is characteristic of lower organisms.
38. **(a)** Mitosis division maintains genetic similarity.
50. **(a)** During leptotene sub-stage of prophase I of meiosis, chromosomes appear long, thin, beaded thread like appearance.
51. **(b)** The paired homologous chromosomes separate in Anaphase I of meiosis I so that each gamete receives one chromosome of each homologous pair.
52. **(b)** At the beginning of M phase or mitotic phase the nuclear membrane and nucleolus disappear. The decondensation of chromosomes occur. But when nuclear envelop restart forming then nuclear lamina assemble and chromosomes start condensing.
53. **(b)** Sudden, abnormal and uncontrolled mitosis in cells of an organ results in cancer.
54. **(a)** Centrosome are present in animal cells and cells of lower plants where flagellated structures are present in the life cycle. They are absent in higher plants. Centrosome help in cell division by forming spindle poles.
55. **(a)** Microspore mother cell is 2n = 24

n = 12

Endosperm is 3n. Therefore 3n = 36.

11. TRANSPORT IN PLANTS

ANSWER KEY

1	(a)	11	(b)	21	(d)	31	(a)	41	(c)	51	(d)
2	(a)	12	(c)	22	(c)	32	(b)	42	(d)	52	(b)
3	(d)	13	(d)	23	(a)	33	(a)	43	(d)	53	(b)
4	(b)	14	(d)	24	(b)	34	(a)	44	(a)	54	(c)
5	(c)	15	(d).	25	(b)	35	(b)	45	(a)	55	(a)
6	(d)	16	(c)	26	(a)	36	(a)	46	(b)	56	(b)
7	(a)	17	(b)	27	(c)	37	(d)	47	(b)	57	(b)
8	(c)	18	(b)	28	(a)	38	(b)	48	(b)	58	(d)
9	(a)	19	(c)	29	(b)	39	(d)	49	(b)	59	(b)
10	(a)	20	(d)	30	(d)	40	(c)	50	(b)	60	(b)

Hints & Solutions

4. **(b)** If a cell swells, after being placed in solution, the solution is called Hypotonic.
10. **(a)** Humus in soil is necessary for plant growth because it increases aeration and water absorption capacity of soil.
12. **(c)** Water in plants is transported by ascent of sap takes place through xylem.
15. **(d)** Wilting of a plant result from excessive transpiration.
20. **(d)** Water potential is defined as the difference between chemical potential of water at any point in a system and that of pure water under standard conditions. The unit of measurement of water potential is pascal, Pa (1 mega pascal, Mpa = 10 bars). It is represented by greek letter Psi (Ψ). Water potential has two components solute potential which is always negative and pressure potential which is usually positive. Therefore water potential Ψw is expressed as:

$$\Psi w = \Psi s + \Psi p$$

21. **(d)** Turgor pressure is the pressure that develops in a cell due to osmotic diffusion of water inside it and is responsible for pushing the membrane against cell wall. Stomata open under conditions of increased turgor

pressure of guard cell and stomata get closed under conditions of decreased turgor pressure of guard cells. When turgid, they swell and bend outward. As a result, the stomatal aperture opens. When they are flaccid, the tension from the wall is released and stomatal aperture closes.

22. **(c)** Water movement occurs through the trachaery elements of xylem.
23. **(a)** Loss of water from margins of leaves is called guttation. Guttation occurs when trans-piration rate is very low as compared to rate of water absorption. Due to which root pressure is developed and water is pushed out through stomata like pores called hydathodes. It occurs in more saturated atomosphere.
24. **(b)** Active transport always involves the expenditure of energy for the movements of substances against concentration gradient.
25. **(b)** Minerals can move upward and downward (bidirectional movement) as well as laterally in phloem. Phloem plays an important role in translocation of minerals.
26. **(a)** Osmotic potential is the potential of a solution to cause water movement into it across a semi- permeable membrane. Water potential is the tendency of water to leave a system.
27. **(c)** D.P.D. or suction pressure (S.P.) = O.P. – T. P.
    Hence D.P.D. for A = 4 bars, B = 0 bars, C = 5 bars, D = 4 bars
    We know that direction of movement of water is from lower D.P.D. (S.P.) to higher D.P.D. (S.P.).
    flow will be from B to A, C, and D.
28. **(a)** Cactus is a xerophytic plant and have sunken stomata to reduce the rate of transpiration, confined to lower epidermis.
29. **(b)** During day time, due to photosynthesis, malic acid forms which breaks to H^+ and malate. H^+ move out of guard cells and K^+ enter forming potassium malate which makes guard cells turgid and stomata opens.
30. **(d)** Bending of tentacles in Venus fly trap or *Drosera* after coming in contact with an insect is thigmonastic or chemonastic movement of variation or turgor movements.
    The mechanism by which the trap snaps shut involves a complex interaction between elasticity, turgor and growth. In the open, untripped state, the lobes are convex , but in the closed state, the lobes are concave. It is the rapid flipping of this bistable state that closes the trap, but the mechanism by which this occur is still poorly understood. When the trigger hairs are stimulated, an action potential involving calcium ions is generated, which propagates across the lobes and stimulates cells in the lobes and in the midrib between them.
31. **(a)** P-protein is the supporting factor for the translocation of organic solutes in sieve tubes.
32. **(b)** Transpiration differs from evaporation because it is a physiological process while evaporation is a physical process.
33. **(a)** Excessive supply of chemical fertilizers often causes death of crop plants due to exosmosis.
34. **(a)** In rainy season, door gets swelled due to the phenomenon of imbibition. It is the process of absorption of water without forming a solution.
35. **(b)** Stomata of most of the plants close during night and hence stomatal transpiration is stopped at night.
36. **(a)** This theory was proposed by Dixon and Jolly (1894). According to this theory leaves loose water into atmosphere by transpiration. This in turn decreases water potential of mesophyll cells. These cells draw water from adjacent cells having high water potential. This potential gradient extends upto sap of xylem. As a result, a tension is created in xylem elements of leaf which is transmitted down to the xylem elements of root. This puts the water in xylem to a great pull or tension. This is known as transpiration pull or transpiration tension. The water molecules within a column are joined to one another by a strong mutual force of attraction called cohesion force.
37. **(d)** DPD is abbreviated form of diffusion pressure deficit. The term was given by Meyer.
38. **(b)** Good soil allows percolating of the water slowly from it
39. **(d)** Guard cells contain chloroplast while epidermal cells do not. The guard cells are bean-shaped in surface view, while the epidermal cells are irregular in shape. The guard cells contain chloroplasts, so they can manufacture food by photosynthesis (The epidermal cells do not contain chloroplasts).
40. **(c)** Cell placed in a hypotonic solution becomes turgid due to endosmosis. Water exerts a pressure on the walls of the cell called turgor pressure. Cell wall being a rigid and hard structure also exerts a pressure on cytoplasm in response to turgor pressure when TP = WP, DPD = 0 (No net exchange of water).
41. **(c)** Diffusion – Movement of solid, liquid, gases particles from high conc. to low conc.
    Osmosis – Special case of diffusion which is applicable to liquid only.
42. **(d)** Root hairs absorb water, minerals and oxygen but they do not absorb CO_2. CO_2 intake takes place in leaves.
43. **(d)** The energy mediated transport of biomolecules across cell membranes is termed as active transport. This process requires the expenditure of the cellular energy in the form of ATP since it involves movement against an electrical or concentration gradient. The random movement of particles in a fluid is termed as Brownian movement. Pinocytosis is a form of endocytosis operating in living cells wherein extracellular fluids are engulfed into larger vesicles. Diffusion is the movement of molecules from region of their higher concentration to region of lower concentration.
44. **(a)** When Stomata open, the pH of guard cells increases. Stomata contains a pore which is surrounded by kidney shaped guard cells. During day time CO_2 is used in the photo synthesis which increases the pH of guard cells on the contrary, at night there is increase in CO_2 and decrease in pH of guard cells.
45. **(a)** Starch is converted to malic acid (organic acid) through the action of an enzyme phosphoe-nolpyruvate carboxylase (PEPCase) under condition of high pH when most of the CO_2 is utilized due to high photosynthetic rate during daytime.

46. **(b)** Under water logged conditions roots die due to lack of air for respiration.

47. **(b)** Water potential is regarded as the tendency of water to leave a system. Pure water has the highest possible water potential which is zero. All solutions have a water potential lower than that of water.

48. **(b)** The rupture and fractionation do not usually occur in the water column in vessel/ tracheids during the ascent of sap because of cohesion and tension. The water molecules have a great mutual attraction with each other or in other words we can say that they have tremendous cohesive power which is sometimes as much as 350 atmospheres. Thus, the transpiration pull develops a negative pressure in the uppermost xylem cells. It is transmitted from there into the xylem of stems, and from there into the xylem of stems, and from there to the roots.

49. **(b)** Carbohydrates are commonly found as starch in plant storage organs. It is chemically non-reactive and osmotically inactive polysaccharides of much greater molecular weight. Carbohydrates performs a vital role in living organisms. Starch and other polysaccharides serve as energy storage in plants, particularly in seeds, tubers, *etc.* which provide a major energy source for animals, including humans.

52. **(b)** When root pressure is maximum then transpiration is very low and absorption of water and minerals are very high.

53. **(b)** For life, food is essential for every cell. When girdling takes place there is no movement of food (phloem sap) towards the root & hence the plant will be die.

54. **(c)** Path of water movement from soil to xylem is:
soil → root hair → cortex → endodermis → pericycle → protoxylem → metaxylem

55. **(a)** Ash is the residue of burned plant parts like; bark, wood, sawdust, leaves, woody debris, pulp, husk, hulls, fronds, and other plant debris. Ash has been used for soil liming (Risse, 2002) and for traditional pest control to some crawling pests (HDRA, 2002; Stoll, 2000). The minerals required in the greatest amounts by plants (also called macroelements) are those containing the elements potassium, phosphorous and in particular nitrogen.

56. **(b)** A semipermeable membrane, also termed a selectively-permeable membrane, a partially-permeable membrane or a differentially-permeable membrane, is a membrane that will allow certain molecules or ions to pass through it by diffusion and occasionally specialized "facilitated diffusion." The rate of passage depends on the pressure, concentration, and temperature of the molecules or solutes on either side, as well as the permeability of the membrane to each solute.

57. **(b)** Hypertonic solutions contain a high concen-tration of solute relative to another solution (e.g. the cell's cytoplasm). When a mango is placed in a hypertonic solution, the water diffuses out of the cell, causing the mango to shrivel.

58. **(d)** Active K^+ transport theory gives the the latest explanation for the closure of stomata. Uptake of K^+ into the cells and subsequentally the loss of K+. The loss of these solutes causes a reduction in osmotic pressure, thus making the cell flaccid and so closing the stomatal pores.

59. **(b)** Water lost in guttaion is impure water. Guttation takes place through special structures called hydathodes. These are usually found on the margins and tips of leaves. The water which is lost now contains organic and inorganic salts and is impure water.

60. **(b)** The membrane which allows passage of certain substances more readily than others is termed as selectively permeable. It is a membrane in cells that only allows certain things in . This allows certain molecules or ions to pass through it by diffusion & occassionally specialized facilitated diffusion. The rate of passage depends on the pressure, concentration, and temperature of the molecules of solutes on either side, as well as the permeability of the membrane to each solute.

12. MINERAL NUTRITION

ANSWER KEY

1	(b)	11	(a)	21	(a)	31	(b)	41	(b)	51	(b)
2	(c)	12	(b)	22	(d)	32	(d)	42	(b)	52	(d)
3	(a)	13	(a)	23	(b)	33	(c)	43	(c)	53	(b)
4	(a)	14	(b)	24	(b)	34	(b)	44	(b)	54	(b)
5	(a)	15	(b)	25	(b)	35	(a)	45	(b)	55	(a)
6	(a)	16	(a)	26	(b)	36	(a)	46	(a)	56	(d)
7	(c)	17	(d)	27	(d)	37	(d)	47	(d)	57	(d)
8	(c)	18	(a)	28	(b)	38	(b)	48	(d)	58	(d)
9	(b)	19	(a)	29	(a)	39	(b)	49	(d)	59	(c)
10	(d)	20	(b)	30	(c)	40	(a)	50	(b)	60	(d)

Hints & Solutions

1. (b) Essential elements are indispensable for growth of plants. Plants are not able to grow normally and hence unable to complete their life cycle in the absence of essential element. It cannot be replaced by other element and it plays direct role in metabolism of plant.

2. (c) Tracer elements are radioactive isotopes of elements, which are used to detect various metabolic pathways in plants e.g., C^{14}, N^{15}, P^{32}, S^{35} etc.

6. (a) Autotrophs are green plants capable of synthesizing their own food from raw material (CO_2, water and sunlight). Thus, they are independent and self nourishing.

9. (b) For chlorophyll formation a plant needs Fe, Mg & light.

22. (d) Active uptake of minerals depends upon dephosphorylation.

29. (a) Along with water, minerals move upwards through the trachaery elements of xylem.

30. (c) Potassium is not a constituent of any enzyme but is an activator of enzymes. Potassium is an essential mineral and is found in more amounts in actively dividing cells such as of buds, young leaves, root tips, etc. It has some other roles like involved in regulating stomatal movement and protein synthesis.

31. (b) Symbiotic relationship between roots of higher plants and fungi is called mycorrhiza, e.g. *Pinus* roots associated with fungal hyphae.

32. (d) *Anabaena* is a nitrogen fixing blue-green alga composed of barrel-shaped cells held in a gelatinous matrix. So, it can fix atmospheric nitrogen.

33. (c) Etiolation is a plant response to stimuli. When plant seeds are grown in insufficient light, either in partial or complete absence of light, the seedlings will have long, weak stems, smaller and fewer leaves because of longer internodes, and a pale yellow colour (chlorosis). This is due to a lack of chlorophyll. The plant's increase in height helps to reach a possible source of light faster.

34. (b) *Rhizobium* is a symbiotic nitrogen fixer. *Azospirillum* is a nitrogen fixing fungus and *Nostoc* is anaerobic.

35. (a) A molecule of N_2 is changed to ammonia by nitrogenase. Transaminase transfers amino group from one substrate to another.

36. (a) Manganese is an activator of enzymes. Copper is essential for photosynthesis. Molybdenum is a cofactor of enzymes. Boron assists in sugar transport.

37. (d) The freshwater fern *Azolla* harbours *Anabaena* (a blue green alga) in its leaf cavities. *Chlorella* is simply a green alga. *Nostoc* is a free living cyanobacteria involved in N_2 fixation.

38. (b) Manganese is essential for many plant functions. Some of them are
   - The assimilation of carbon dioxide in photosynthesis.
   - It is involved in the Hill Reaction where water is split during photosynthesis.
   - It aids in the synthesis of chlorophyll and in nitrate assimilation.
   - Manganese activates fat forming enzymes.
   - It functions in the formation of riboflavin, ascorbic acid, and carotene.
   - It functions in electron transport during photosynthesis.

39. (b) Calcium is not remobilized from the leaves to the fruits, like potassium, phosphorus and sulphur.

40. (a) The nitrifying bacteria bring about transformations of ammonia to nitrite and of nitrite to nitrate.

41. (b) Leghaemoglobin is the haemoglobin like red pigments found in root nodules of legumes and reported to function as an oxygen-carrying pigment in symbiotic nitrogen fixation.

42. (b) Importance of micro-elements was recognized late due to their presence as contaminants in macronutrients.

43. (c) In active transport expenditure of energy required, so it requires ATP.

44. (b) Crop rotation or Crop sequencing is the practice of growing a series of dissimilar types of crops in the same area in sequential seasons for various benefits such as to avoid the build up of pathogens and pests that often occurs when one species is continuously cropped. Crop rotation also seeks to balance the fertility demands of various crops to avoid excessive depletion of soil nutrients. A traditional component of crop rotation is the replenishment of nitrogen through the use of green manure in sequence with cereals and other crops. It is one component of polyculture. Crop rotation can also improve soil structure and fertility by alternating deep-rooted and shallow-rooted plants.

45. (b) Land plants normally obtain carbon directly from air.

46. (a) An insectivorous plant, also called a carnivorous plant, captures prey items, such as insects, spiders, crustaceans, mites, and protozoans, as a nitrogen source. Many insectivorous species live in freshwater bogs, where nitrogen is not present in available form, because the pH of the water is extremely acid. The forms of entrapment by these types of plants are modified leaves.

48. (d) Copper (Cu), manganese (Mn) and iron (Fe) are essential micronutrients which are required for photosynthetic and mitochondrial electron transport. So, their deficiency will affect growth of the plants by affecting photosynthetic and mitochondrial electron transport.

49. (d) Deficiency of zinc causes fall of immature young leaves due to formation of abscission layer, which is prevented by auxin and zinc plays an important role in auxin formation.

50. (b) ATP (Adenosine triphosphate) is known as energy currency. It stores energy. It is a nucleotide, made of nitrogenous base (adenine), ribose sugar and phosphate (PO_4). Hence, it requires N and P.

55. **(a)** Guard cells help in transpiration. Transpiration is the evaporation of water from the aerial parts of plants, especially leaves but also stems, flowers and roots. Stomatal opening allowing the diffusion of carbon dioxide and oxygen during photosynthesis.
56. **(d)** Stomata present on lower surface of plant leaves effectively exchange gases without allowing much loss of water from the leaves.
57. **(d)** Transpiration is a process of loss of water in the form of vapours from the aerial parts of the plants. Transpiration is inversely proportional to the atmospheric humidity. The rate of transpiration would be higher when humidity is low.

13. PHOTOSYNTHESIS

ANSWER KEY

1	(a)	12	(a)	23	(d)	34	(b)	45	(a)	56	(c)
2	(b)	13	(a)	24	(a)	35	(a)	46	(b)	57	(c)
3	(d)	14	(d)	25	(b)	36	(a)	47	(b)	58	(b)
4	(d)	15	(c)	26	(d)	37	(b)	48	(d)	59	(d)
5	(a)	16	(b)	27	(d)	38	(c)	49	(c)	60	(a)
6	(b)	17	(c)	28	(a)	39	(b)	50	(a)		
7	(d)	18	(d)	29	(c)	40	(d)	51	(c)		
8	(b)	19	(a)	30	(d)	41	(a)	52	(b)		
9	(c)	20	(b)	31	(a)	42	(a)	53	(b)		
10	(a)	21	(b)	32	(d)	43	(a)	54	(a)		
11	(d)	22	(a)	33	(c)	44	(a)	55	(c)		

Hints & Solutions

1. **(a)** Chlorophyll is present in the grana of chloroplasts.
5. **(a)** The ultimate gain of light reaction is ATP & $NADPH_2$.
6. **(b)** Splitting of water in photosynthesis is called photolysis.
10. **(a)** Chloroplasts fix carbon dioxide.
11. **(d)** Phytoplanktons (algae) of oceans are the highest producers of organic matter.
12. **(a)** C_4 plants have a characteristic leaf anatomy called kranz anatomy with two distinct choloroplasts containing cells mesophyll and bundle sheath cells.
13. **(a)** Photosynthetic pigments occur on photosynthetic thylakoids of chloroplasts. A stack of thylakoids is called granum.
14. **(d)** Depending upon their colour, plastids are of three main types – leucoplasts, chromoplasts and chloroplasts.
15. **(c)** Chlorophyll *a* is found in all photosynthetic plants except bacteria. Hence, it is termed as universal photosynthetic pigment.
16. **(b)** According to law of limiting factors given by Blackman, when a process depends upon a number of factors which are separate, the rate of the process is limited by the pace of the slowest factor. Rate of yield is dependent on light as photosynthesis is dependent on light. During monsoon, the light is dim and so this reduces rate of photosynthesis and hence yield.
17. **(c)** Dark reaction occurs in the stroma or matrix of chloroplasts. It is independent of light but depends on the products of light reaction (ATP and NADPH–assimilatory process).
18. **(d)** Translocation of organic solutes occur through sieve tubes in the form of sucrose. Transportable form of sugar is sucrose.
19. **(a)** Photorespiratory loss of CO_2 occurs when RuBisCO starts functioning as an oxygenase instead of carboxylase under conditions of high O_2 and low CO_2. It involves three organelles chloroplast, mitochondria and peroxisomes. Half of the photosynthetically fixed carbon (in the form of RuBP) may be lost into the atmosphere through this process and no ATP formation occurs.
20. **(b)** Photosynthesis takes place only in the visible part (400 – 700 nm wavelength) of electromagnetic radiations. Hence this component comprises the photosynthe-tically active radiation.
21. **(b)** RuBisCO is the enzyme involved in Calvin cycle. Nitrogenase catalyses nitrogenation. Invertase catalyses breaking of sucrose to glucose and fructose.
24. **(a)** Photophosphorylation means synthesis of ATP

from ADP.

34. (b) PEP carboxylase catalyzes the photosyn-thetic C_4 cycle.

38. (c) Solarisation is a process in which destruction of chlorophyll and ultimate death of protoplasmic components.

44. (a) Bacterial photosynthesis involves PS-I only.

46. (b) In the leaves of C_4 plants, malic acid formation during CO_2 fixation occurs in the cells of Mesophyll.

48. (d) Non cyclic photophosphorylation is not only connected with ATP synthesis but also production of NADPH. It involves two photosystems—PSI and PSII. PSII (680) receives electrons from photolysis of water and transfer it to PSI (700) to form NADPH from $NADP^+$. The hydrogen required for reduction comes from breaking of water molecules.

49. (c) In 1954-55 Calvin traced the pathway of carbon fixation by using radioactive isotope method by using $^{14}CO_2$ and gave the Calvin cycle.

50. (a) In PSI, when the electron passes from ferredoxin and plastoquinone, the electron loses sufficient energy to form ATP from ADP and inorganic phosphate.

51. (c) Algae are the most efficeint photosyn-thesizing organisms.

52. (b) Cyclic photophosphorylation involves only PSI and there occurs only ATP formation.

53. (b) Chlorophyll A has a methyl group attached to the third carbon, which is replaced by an aldehyde group in chlorophyll B.

54. (a) Cytochrome P-450 contains copper.

55. (c) Each turn of Calvin cycle generates one carbon atom hence six turns of the cycle is required to generate one molecule of hexose sugar glucose.

56. (c) 2 ATPs are required during conversion of PGA to 1, 3-diphosphoglyceric acid and 1 ATP during conversion of glyceraldehyde phosphate to ribulose bisphosphate. $2NADH_2$ molecules are utilised for converting 1, 3-diphosphoglyceric acid to glyceraldehyde phosphate.

57. (c) The reduction in the rate of photosynthesis in monochromatic light of more than 680 nm is called red drop.

58. (b) Cytochrome is Fe containing porphyrin pigment. These are electron transferring proteins often regarded as enzymes.

59. (d) To conserve water, the stomata of CAM plants open during night to take up CO_2 and store it in the form of malic acid which is utilised during the time of photosynthesis.

60. (a) Carotenoids have three absorption peaks in the blue-violet range of the spectrum.
In photosynthetic organisms, carotenoids play a vital role in the photosynthetic reaction centre. They either participate in the energy-transfer process, or protect the reaction center from auto-oxidation. In non-photosynthe-sizing organisms, carotenoids have been linked to oxidation-preventing mechanisms.

14. RESPIRATION IN PLANTS

ANSWER KEY

1	(c)	11	(b)	21	(a)	31	(d)	41	(a)	51	(b)
2	(c)	12	(c)	22	(d)	32	(b)	42	(d)	52	(b)
3	(c)	13	(a)	23	(d)	33	(b)	43	(b)	53	(d)
4	(c)	14	(b)	24	(c)	34	(b)	44	(b)	54	(a)
5	(a)	15	(c)	25	(c)	35	(d)	45	(c)	55	(b)
6	(b)	16	(a)	26	(b)	36	(d)	46	(c)	56	(b)
7	(b)	17	(c)	27	(c)	37	(d)	47	(a)	57	(c)
8	(b)	18	(b)	28	(d)	38	(c)	48	(c)	58	(d)
9	(c)	19	(b)	29	(b)	39	(a)	49	(c)	59	(b)
10	(c)	20	(c)	30	(c)	40	(d)	50	(c)	60	(c)

Hints & Solutions

4. (c) Respiratory Quotient

$$RQ = \frac{\text{Volume of } CO_2 \text{ evolved}}{\text{Volume of } O_2 \text{ consumed}}$$

5. (a) The enzyme involved in Krebs cycle are localized in the mitochondrial matrix.

6. (b) Pyruvic acid is the end product of glycolysis.

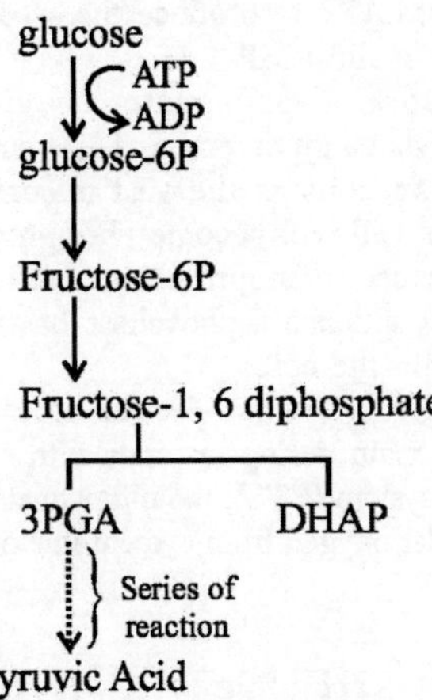

7. (b) Number of carbons in pyruvic acid is 3. It is also called veto pyruvic acid. During glycolysis pyruvic acid is formed by the oxidation of glucose.

$$\underset{\text{Pyruvic acid}}{CH_3\,CO\,.\,COOH}$$

8. (b) Glycolysis is the breakdown of glucose to 2-molecules of pyruvate. It occurs in cell cytoplasm (cytosol) and is the common step in both anaerobic and aerobic respiration. The total gain is of 8 ATPs.

9. (c) The end product of fermentation of sugars is alcohol and CO_2. Sugars are the most common substrate of fermentation, and typical examples of fermentation products are ethanol, lactic acid, and hydrogen. However, more exotic compounds can be produced by fermentation, such as butyric acid and acetone. Yeast carries out fermentation in the production of ethanol in beers, wines and other alcoholic drinks, along with the production of large quantities of carbon dioxide. Fermentation occurs in mammalian muscle during periods of intense exercise where oxygen supply becomes limited, resulting in the creation of lactic acid.

10. (c) ATP synthesis in mitochondria (oxidative phosphorylation) occurs in F_1 particles or oxysomes present on cristae.

17. (c) Respirometer is an instrument for measuring respiratory quotient as well as rate of respiration.

18. (b) Respiratory quotient (RQ) is defined as the ratio of the molecules of carbon dioxide given out to the number of oxygen molecules taken in during respiration. Its value for fatty acids is always less than unity.

19. (b) Fermentation is referred to as the decomposition of organic compounds, anaerobically by enzymes yielding alcohol, organic acids, gases etc.

20. (c) Mesosomes are the invaginations of the plasma membrane that can form into vesicles. They are found to be present in both gram-positive and gram-negative bacteria. Mesosomes may play a role in cell wall formation during cell division and/or chromosome replication and distribution and/or electron transfer systems of respiration.

21. (a) 38 molecules of ATP produced during aerobic respiration. Out of these 2 ATP are used in link reaction. Hence net gain is of 36 ATP.

22. (d) The enzyme responsible for oxidative phosphorylation is ATP synthase. ATP synthase is located in the F_1 component of $F_0 - F_1$ or elementary particles. ATP synthase becomes active in ATP formation only where there is a proton gradient having higher concentration of H^+ or protons on the F_0 side as compared to F_1 side. Increased proton concentration is produced in the outer chamber of outer surface of inner mitochondrial memberane by the pushing of protons with the help of energy liberated by passage of electrons from one carrier to another.

23. (d) Terminal cytochorome is cyt a_3. cyt a_3 posseses two copper centers. It helps in transfer of electrons to oxygen.

24. (c) Competitive inhibition is the reversible inhibition of enzyme activity due to presence of substrate analogues. Non competitive inhibition is the reduction of enzyme activity by a factor that has no real structural similarity with the substrate. Allosteric inhibition is reversible non-competitive inhibition occurring in case of allosteric enzymes. Here the inhibitors are the products or intermediates of reactions catalyzed by the enzymes. Hence it is also called end product inhibition or feedback inhibition.

25. (c) In the conversion of succinyl Co A to succinic acid, energy liberated during hydrolysis of succinyl CoA is used in synthesis of GTP. For the conversion of α-ketoglutarate to succinyl CoA, NAD^+ and CoA are required. Conversion of fumaric acid to malic acid involves simple hydration. FAD participates in conversion of succinic acid to fumaric acid.

26. **(b)** In alcohol fermentation triose phosphate is the electron donor while acetaldehyde is the electron acceptor.
27. **(c)** Succinate:ubiquinone oxidoreductase, also known in mitochondria as Complex II, provides a link between the citric acid cycle and the membrane-bound electron-transport system. The membrane extrinsic, water-soluble domain, known as succinate dehydrogenase (SDH), contains the fumarate/succinate active site with a covalently bound FAD group and three iron-sulfur clusters: $[2Fe\text{-}2S]^2$ +/1+, $[4Fe\text{-}4S]^2$+/1+, and $[3Fe\text{-}4S]^1$ +/0 . The enzyme catalyzes the interconversion of fumarate and succinate, and is closely related to fumarate reductase.
28. **(d)** Glycolysis takes place in all body cells and is of two types
   (a) Anaerobic glycolysis-from glycogen or glucose to lactic acid in muscles.
   (b) Aerobic glycolysis-from glycogen or glucose to pyruvic acid (all cells of body)
29. **(b)** In germinating seeds, fatty acids are degraded exclusively in the glyoxysomes.
   Glyoxylate cycle occurs in tissues rich in fats, such as those of germinating seeds; the enzymes involved in the cycle, which have not been found in mammals, are contained in organelles called glyoxysomes.
30. **(c)** Chemiosmotic theory postulated by the British biochemist Peter Mitchell (1920-22) to explain the formation of ATP in the mitochondrial electron transport chain. As electrons are transferred along the electron carrier system in the inner mitochondrial membrane, hydrogen ions (protons) are actively transported into the space between the inner and outer mitochondrial membranes, which thus contains a higher concentration of protons than the matrix. This creates an electrochemical gradient across the inner membrane, through which protons move back into the matrix.
31. **(d)** The formation of acetyl Co-A from pyruvic acid is the result of its oxidative decarboxylation.
37. **(d)** The respiration of young maturing seeds is quite high but as water content decreases during maturation, the rate of respiration decreases steadily.
38. **(c)** It is the process of cellular oxidation, in which breakdown of food in the presence of oxygen to release in carbon dioxide and energy which is used in cellular activities.
39. **(a)** The optimum temperature for photosyn-thesis is 10°-25°C for C_3 plants and 30°-45°C for C_4 plants.
40. **(d)** Hexokinase is absent in the mitochondria. Since it is enzyme of glycolysis which occurs in cytoplasm.
41. **(a)** We know that all enzymes are proteins, but all proteins are not enzymes. The enzymes are proteins made up of one or more than one polypeptide chains.
42. **(d)** ATP is not a co-enzyme. Adenosine-5'-triphosphate (ATP) is a multifunctional nucleotide, the most significant function of which in cell biology is as a "molecular currency" of intracellular energy transfer. In this role, ATP transports chemical energy within cells for metabolism. It is produced as an energy source during the processes of photosynthesis and cellular respiration and consumed by many enzymes and a multitude of cellular processes including biosynthetic reactions, motility and cell division. In signal transduction pathways, ATP is used as a substrate by kinases that phosphorylate proteins and lipids, as well as by adenylate cyclase, which uses ATP to produce the second messenger molecule cyclic AMP.
43. **(b)** Glucose 6-phosphate is glucose sugar phosphorylated on carbon 6. This compound is very common in cells as the vast majority of glucose entering a cell will become phosp-horylated in this way. Because of its prominent position in cellular chemistry, glucose 6-phosphate has many possible fates within the cell.
44. **(b)** Oxygen is the last electron acceptor in electron transport chain during respiration. In electron transport system (ETS), the ultimate electron acceptor is molecular oxygen from cytochrome oxidase, to form H_2O.

$$\frac{1}{2}O_2 + 2H^+ + 2e^- \longrightarrow H_2O$$

45. **(c)** (i) Peroxisomes are the micro bodies found in plants as well as animal cells.
   (ii) contain oxidative enzymes like urate oxidase, D-aminoacid oxidase. The reactions catalysed by these enzymes produce H_2O_2 (hydrogen peroxide) which is immediately metabolized by another enzyme called catalase.
46. **(c)** Oxysomes are f_0-f_1 particles.
47. **(a)** EMP pathway or glycolysis is the first and anaerobic phase of cellular respiration. C_3 cycle or Calvin – Benson cycle is the dark reaction or Black reaction of photosynthesis. Enzymes aldolase and triose phosphate isomerase take part in both.
48. **(c)** Lignin is a complex polymer of phenylpropane units, which are cross-linked to each other with a variety of different chemical bonds.
49. **(c)** Photophosphorylation refers to addition of phosphate in the presence of light.
50. **(c)** ATP is the energy currency in all cells. Glyceraldehyde 3-phosphate is reduced during glycolysis. Molecular oxygen is the terminal electron acceptor in ETS.
51. **(b)** All energy-releasing pathways whether aerobic (requiring oxygen) or anaerobic (not requiring oxygen) begin with a pathway called glycolysis, which occurs in the cytoplasm (cytosol). Aerobic respiratory pathway is appropriately termed amphibolic. Aerobic respiration is the main energy-releasing pathway leading to ATP formation. It occurs in the mitochondria. Aerobic respiration yields thirty-six ATPs.

52. **(b)** The process in which energy is released is called exothermic process. Since energy is released during respiration so its an exothermic process. About 673 kcal energy is released during aerobic respiration and 28 kcal during anaerobic respiration.

$$C_6H_{12}O_6 + 6O_2 \longrightarrow 6CO_2 + 6H_2O + \underset{\text{(energy)}}{673\text{ kcal}} \uparrow$$

53. **(d)** Fermentation is performed by some fungi and bacteria. Fermentation is the process of deriving energy from the oxidation of organic compounds, such as carbohydrates, using an endogenous electron acceptor, which is usually an organic compound. This is in contrast to cellular respiration, where electrons are donated to an exogenous electron acceptor, such as oxygen, via an electron transport chain. Fermentation does not necessarily have to be carried out in an anaerobic environment, however. For example, even in the presence of abundant oxygen, yeast cells greatly prefer fermentation to oxidative phosphorylation, as long as sugars are readily available for consumption.

54. **(a)** In succulents, RQ <1 due to incomplete oxidation. The respiratory quotient (or RQ or respiratory coefficient), is a unitless number used in calculations of basal metabolic rate (BMR) when estimated from carbon dioxide production. Such measurements, like measurements of oxygen uptake, are forms of indirect calorimetry. It is measured by using Ganong's respirometer.

55. **(b)** In yeast, during anaerobic respiration CO_2, C_2H_5OH and energy are formed.

$$C_6H_{12}O_6 \xrightarrow{\text{Enzymes}} 2CO_2 + 2C_2H_5OH + 50\text{kcal}$$

56. **(b)** Based on chemiosmotic theory.

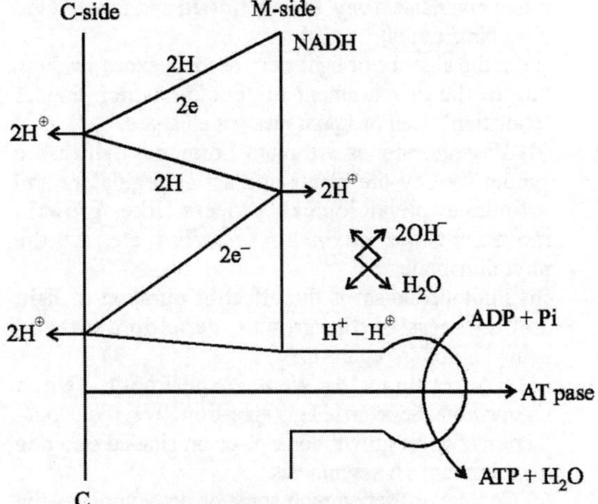

Chemiosmotic Theory

59. **(b)** 2 ATPs are produced during glycolysis in cytoplasm. 34 ATPs are produced by Krebs cycle, oxidative phosphorylation, ETC.

60. **(c)** Aerobic respiration is an enzymatically controlled release of energy in a stepwise catabolic process of complete oxidation of organic substrate into CO_2 and H_2O with O_2 acting as terminal oxidant.

15. PLANT GROWTH AND DEVELOPMENT

ANSWER KEY

1	(a)	11	(d)	21	(b)	31	(d)	41	(b)	51	(d)
2	(d)	12	(c)	22	(c)	32	(c)	42	(c)	52	(b)
3	(a)	13	(a)	23	(c)	33	(c)	43	(c)	53	(a)
4	(b)	14	(a)	24	(c)	34	(d)	44	(d)	54	(c)
5	(b)	15	(a)	25	(b)	35	(b)	45	(c)	55	(b)
6	(a)	16	(b)	26	(a)	36	(d)	46	(b)	56	(b)
7	(c)	17	(c)	27	(c)	37	(b)	47	(b)	57	(a)
8	(c)	18	(b)	28	(b)	38	(a)	48	(c)	58	(c)
9	(d)	19	(c)	29	(c)	39	(b)	49	(b)	59	(d)
10	(b)	20	(b)	30	(d)	40	(c)	50	(b)	60	(b)

Hints & Solutions

6. (a) The maximum growth rate occurs in exponential phase.
7. (c) Borthwick & Hendricks (1952) gave the concept of phytochrome. Phytochrome is a coloured protein pigment located in plasma membrane to perceive the photoperiodic stimulus in leaves. It may exist in two interconvertable forms Pr and Pfr (red and far red light absorbing forms).
8. (c) In the absence of light stem become excessive long due to the development of 'etiolin' which causes 'etiolation'. Cell of internodes get enlarged.
9. (d) Phytohormones are plant hormones which are synthesized by the plants and act as regulators and influence physiological process like growth, movements *etc. auxins*, *gibberellins etc.* are the phytohormones.
10. (b) Photoperiodism is the effect of duration of light and darkness on the growth, development and behaviour of an organism.
11. (d) According to Went experiment, when monochromatic beam of light spread on *Avena coleoptile* then maximum growth took place on shaded side due to maximum IAA-synthesis.
12. (c) Contact or friction with some object stimulates the movement of a plant part in a specific direction *eg.* tendril in the family – *Cucurbitaceae*.
13. (a) Any cell which has an ability to develop into a complete organism is totipotent.
14. (a) Nastic movements are non-directional responses to stimuli (e.g. temperature, humidity, light irradiance). The movement can be due to changes in turgor or changes in growth. Nastic movements differ from tropic movements in that the direction of tropic responses depends on the direction of the stimulus, whereas the direction of nastic movements is independent of the stimulus. The rate or frequency of these responses increases as intensity of the stimulus increases. An example of such a response is the opening and closing of flowers (photonastic response). Nastic responses are usually associated with plants.
16. (b) Leaf fall occurs when the content of auxin decreases.
20. (b) *Mimosa pudica* (touch me not) leaves show seismonastic movement in response to physical contact. The movement is due to turgor changes due to ejection of water by cells of pulvinus (swollen leaf base) on touching it. It causes folding and drooping of leaves.
21. (b) Because cytokinins help cells in retaining chlorophyll for longer time. Cytokinin delay the senescence (ageing) of leaves and other organs.
22. (c) The most important function of cytokinins (zeatin is one of the cytokinins present in *Zea mays* - the corn) is the promotion of cell division. These are now established as true cell division factors.
23. (c) Vernalization is the low temperature treatment for flowering in plants. Plants are treated (seeds and seedlings) with very low temperature to promote germination.
24. (c) Phototropic movements are the plant responses towards light. Generally plant shoots and flowers show positive phototropism while geotropism is the plant response to the gravity of earth. Generally roots are positively geotropic. All these movements are linked to auxin.
25. (b) Abscisic acid as the name suggests cause abscission due to which leaves fall. It also acts as a growth inhibitor & induce bud dormancy in various plants.
26. (a) Highest concentration of auxins occurs at the growing tips was proved by Went in 1928 by his experiment on *Avena* coleoptile. It was demonstrated that auxins are synthesized at the tips and then translocated downwards where they cause growth.
27. (c) Ethylene is a plant hormone which has the most remarkable effect of stimulation of fruit ripening.
28. (b) A short day plant is one that flowers on photoperiods shorter than the critical day length. It generally flowers when treated by a short day and uninterrupted long night (more than 12 hrs.)
29. (c) Ascorbic acid is an organic acid with antioxidant properties. Its appearance is white to light yellow crystals or powder. It is water soluble. The L-enantiomer of ascorbic acid is commonly known as Vitamin C. Ascorbic acid is easily oxidized and so is used as a reductant in photographic developer solutions and as a preservative.
30. (d) Vernalization is the method of inducing early flowering in plants by pretreatment of their seeds at low temperature.
31. (d) Auxins are synthesized at the tip of shoot & then transported towards the base i.e. basipetal. In all segments of root at all temperatures, the movement of Auxin is polarised acroketalles.
32. (c) The pigment phytochrome is involved in flowering and many other light-mediated processes. It absorbs red and far red light in plants for photosynthesis.
33. (c) Ethylene is a gaseous plant hormone which plays an important role in fruit ripening. In the presence of ethylene there is an increase in the rate of respiration.
34. (d) In northern India, during these months there is observed maximum variation in light intensity and duration and temperature. Hence only plants that are photo-and thermoinsensitive can flower during these months.
35. (b) Cytokinins are used in tissue culture to induce shooting of the cultural callus.
36. (d) Scarification is the breaking of the seed coat barriers which can be done by mechanical or chemical methods.
37. (b) The medium in tissue culture provides all the essential nutrients required for growth and division of callus. Coconut milk contains cytokinins which is useful in tissue culture.
38. (a) IAA promotes apical dominance. ABA induces dormancy in buds.
39. (b) During day time, due to photosynthesis CO_2 concentration is low.

40. **(c)** Differentiation of root is controlled by high auxin concentration while in tissue culture auxin concentration is made high to promote rooting.
41. **(b)** Zinc is one of the component of Auxin.
42. **(c)** Gibberellins induces elongation of internodes.
43. **(c)** Opening of floral buds into flowers, is a type of autonomic movement of growth which is due to epinasty.
44. **(d)** NAA is a synthetic auxins. Some of the most widely used weed killers are synthetic auxins. Synthetic auxin analogs include 1 naphthale-neacetic acid (NAA), 2, 4-dichlorophenoxy acetic acid (2,4-D) and 2,4,5-trichlorophenoxy acetic acid (2,4,5- T).
45. **(c)** Abscisic acid (ABA), also known as abscisin II and dormin, is a plant hormone. It functions in many plant developmental processes, includsing bud dormancy. Abscisic acid is a derivative of carotenoids. It was called "abscisin II" originally because it was thought to play a major role in abscission of fruits. At about the same time another group was calling it "dormin" because they thought it had a major role in bud dormancy. The name abscisic acid (ABA) was coined by a compromise between the two groups.
46. **(b)** 2, 4-D is used for killing broad leaved weeds (generally dicot).
47. **(b)** A potted plant placed near the window bends outwardly due to maximum concentration of auxin on the back side.
48. **(c)** Somaclonal variations are differences in plants that are raised from the callus by tissue culture. If these variations are of economic value, *e.g.* induce tolerance of pests, diseases, etc. such plants are selected and multipled. This technique has been used in wheat, rice, potato and tomato.
49. **(b)** Sensation of phototropism is perceived by shoot tip. The term meristem, shoot tip, meristem tip are often interchanged. The term shoot tip to refer to an apical tip ranging from 1-3 cm.
50. **(b)** Hydrogen ion is pulled out in apoplast by auxin during growth.
51. **(d)** Anthers produce haploid male gametes in angiosperms called pollengrain (n) which can be cultured to raise haploid plants.
52. **(b)** Differential divisions in various parts of root apical meristem gives rise to 3-4 regions – Procambium, ground meristem and calyptrogen. Calyptrogen gives rise to root cap.
53. **(a)**

Some **natural insecticides sources**

Rotenones	Roots of *Derris Elliptica*
Pyrethroids	*Chrysanthemum cinerarifolium*
Azadirachtin	*Azadirachta indica*

54. **(c)** Gibberellins promote the formation of male flowers on genetically female plants of *Cannabis*. They can also replace female flowers with male flowers on monoecious plants of cucurbits.
55. **(b)** Functions of cytokinin
    (i) Delays senescence of leaves and other organs.
    (ii) In presence of auxin, cytokinins bring about division even in permanent cells.
    (iii) Cell division in callus, and differentiation of cells.
56. **(b)** Some of the deficiency symptoms of calcium are
    (i) decomposition of root epidermis
    (ii) terminates growth of meristematic region
    (iii) younger leaves become wilted.
58. **(c)** Prunning of plants promotes branching, because the axillary buds get sensitized to cytokinin.
59. **(d)** Parthenocarpic tomato fruits can be produced by treating the plants with low concentrations of gibberellic acid and auxins.

16. DIGESTION AND ABSORPTION

ANSWER KEY

1	(c)	11	(c)	21	(d)	31	(b)	41	(d)	51	(b)
2	(d)	12	(b)	22	(b)	32	(c)	42	(b)	52	(a)
3	(a)	13	(d)	23	(d)	33	(d)	43	(d)	53	(d)
4	(b)	14	(d)	24	(d)	34	(c)	44	(c)	54	(b)
5	(c)	15	(d)	25	(b)	35	(c)	45	(b)	55	(d)
6	(c)	16	(b)	26	(c)	36	(a)	46	(c)	56	(a)
7	(b)	17	(c)	27	(b)	37	(b)	47	(c)	57	(a)
8	(a)	18	(b)	28	(d)	38	(a)	48	(b)	58	(b)
9	(a)	19	(d)	29	(b)	39	(d)	49	(a)	59	(b)
10	(a)	20	(d)	30	(b)	40	(c)	50	(b)	60	(d)

Hints & Solutions

3. (a) Glycogen is stored in liver & muscles in human beings. Carbohydrates are used primarily as source of chemical energy to be metabolized immediately into glucose or stored as glycogen. The synthesis of glycogen is called glycogenesis.
4. (b) Bile salts viz sodium glycolate & sodium taurocholate (both are organic salts present in bile) reduce the surface tension of large fat globules and break them to fine minute fat droplets. This process is known as emulsification of fat.
5. (c) Digestion of proteins, fats and carbohydrate is completed in Ilieum.
8. (a) Oxyntic cells of stomach secrete HCl, which provide acidic medium and kill micro-organisms.
14. (d) Thiamine is vitamin B_1. Deficiency of Thiamine is responsible for Beri-Beri disease.
15. (d) Hydrolysis of food substance is called Digestion. Digestion is the process of metabolism whereby a biological entity processes a substance, in order to chemically and mechanically convert the substance into nutrients.
16. (b) Intestinal juice (succus entericus) refers to the clear to pale yellow watery secretions from the glands lining the small intestinal walls. Secretion is stimulated by the mechanical pressure of partly digested food in the intestine. Its function is to complete the process begun by pancreatic juice; the enzyme trypsin exists in pancreatic juice in the inactive form trypsinogen, it is activated by the intestinal enterokinase in intestinal juice. Trypsin can then activate other protease enzymes and catalyze the reaction pro-colipase ® colipase. Colipase is necessary, along with bile Salts, to enable lipase function. Intestinal juice also contains hormones, digestive enzymes, mucus, substances to neutralize hydrochloric acid coming from the stomach and trypsin which further digests polypeptides into amino acids, completing protein digestion.
17. (c) Epiglottis prevents the entry of food into respiratory tract.
18. (b) Cholecystokinin (in fact it is cholecystokinin - pancreozymin CCK-PZ) is a hormone which is secreted by the duodenal wall. It causes the release of bile by the contraction of the gall bladder & increases the enzyme release process of pancreatic cells. The pancreatic juice enterokinase is an 'activator enzyme' which is present in intestinal juice (succus entericus) and converts the inactive proenzyme trypsinogen to active trypsin enzyme. Secretin is also a hormone and causes the release of sodium bicarbonate in the pancreatic juice (it is said that the production of pancreatic juice is stimulated by secretin & its release is stimulated by CCK-PZ).
19. (d) Pancreatic juice secretion is regulated by the hormone secretin, which is produced by the walls of the duodenum upon detection of acid food, proteins, and fats.
20. (d) Zymogen or chief cells are mainly found in the fundic part of the stomach. They are the most abundant among the gastric glands & secrete proenzymes, propepsin and prorenin. (They are also present in intestine but their main site is stomach).
21. (d) Pancreas consists mainly of large lobules / acini / alveoli and secretes pancreatic juice. In between the acini, here and there i.e. randomly spread endocrine cells are present known as Islets of Langerhans which consist of α, β, and F cells & secrete insulin, glucagon, pancreatic polypeptide and somatostatin hormones.
22. (b) Vitamin K (Phylloquinone = Nepthaquinone) also known as 'antihaemorrhagic factor'. This vitamin is essential for the synthesis of prothrombin in liver cells which is essential for normal blood clotting. It is also synthesized by the colon bacteria.
23. (d) Kupffer's cells or stellate cells are active phagocytic cells present in the liver. They engulf the warn out and dead RBCs & break out haemoglobin. They are also infact hepatic macrophages.
24. (d) Renin acts on milk protein casein & changes it into paracasein spontaneously precipitated into insoluble calcium paracaseinate, forming solid curd or coagulated milk. Renin is present in gastric juice and hence works in highly acidic pH (due to HCl) in stomach (about 1-3 pH) and absent in adult human being.
25. (b) The enzyme enterokinase (enterope-ptidase) helps in the conversion of trypsinogen (inactive form of another enzyme) into trypsin (active form of the enzyme) for protein digestion in the intestine.
26. (c) The excess of fluorine in water causes a condition called fluorosis, which is identified by mottling of teeth (yellowish streaks) and abnormal bones liable to fracture etc.
27. (b) Secretin released by duodenum inhibits movements in gastric intestinal tract. Enterog-asterone inhibits digestive activity of stomach. Gastrin stimulates secretion of HCl.
28. (d) Stool colour is due to the bile pigments biliverdine and bilirubin secreted by liver. If stool colour is whitish grey that means secretion of biliveridin and bilirubin is not proper from the liver.
29. (b) Carbohydrates are basic source of energy, proteins provide material for growth and development.
30. (b) Presence of microvilli on the epithelial cells of intestine increases the surface area for absorption of food.
31. (b) Angiotensinogen is a protein secreted by liver cells.
32. (c) If the secretion of parietal cells of gastric glands is blocked with an inhibitor, in the absence of HCl

secretion, inactive pepsinogen is not converted into the active enzyme pepsin.

33. **(d)** Goblet cells, found in the intestinal mucosal epithelium, secrete mucus. The mucus lubricates the food for an easy passage. So, if for some reasons, goblet cells become non-functional, it will adversely affect the smooth movement of food down the intestine.

34. **(c)** The adult dental formula of human is

$$\text{incisor } \frac{1}{1}, \text{ canine } \frac{1}{1}, \text{ premolar } \frac{1}{1},$$

$$\text{molar } \frac{3}{3}.$$

39. **(d)** When teeth are different in shape, size & function then these are called heterodont.

44. **(c)** Bile secreted by liver passes into gall bladder through cystic duct.

50. **(b)** Trypsin functions in alkaline medium while pepsin functions in the acidic medium.

$$\text{Peptones} \xrightarrow{\text{Trypsin}} \text{Peptide}$$
$$\text{Protein} \xrightarrow{\text{Pepsin}} \text{Peptones}$$

51. **(b)** Insulin deficiency causes diabetes mellitus. It results from hypo-secretion of insulin. The symptoms are excessive urine production, excessive thirst and excessive eating.

52. **(a)** Surplus amino acids undergo deamination producing ammonia, which is later turned into urea through a series of cyclic reactions called ornithine cycle in liver.

53. **(d)** ATP and GTP both are biphosphate nucleotide and contain three phosphate groups, linearly linked together by covalent bond. The hydrolysis of triphosphate nucleotide first gives diphosphate nucleotide and then, monophosphate nucleotide yielding a large amount of energy (30.6 kJ mol^{-1}) in each step. AMP is one of the such mono-phosphate nucleotide so it is not a high energy phosphate compound.

55. **(d)** Maximum H_2O absorption takes place in colon.

56. **(a)** All are enzymes which are made up of proteins.

57. **(a)** Crypts of Lieberkuhn are the intestinal glands which secrete digestive enzymes and mucus.

58. **(b)** Cholesterol is a soft, waxy substance found among the lipids (fats) in the bloodstream and in all our body's cells. It's an important part of a healthy body because it's used to form cell membranes, some hormones and is needed for other functions. But a high level of cholesterol in the blood —hypercholesterolemia — is a major risk factor for coronary heart disease, which leads to heart attack.

59. **(b)** Rennin is an enzyme present in gastric juice. It converts the soluble milk protein (Casein) into insoluble calcium paracaseinate. This is called curdling of milk. Renin is a hormone secreted by the kidney.

60. **(d)** The fastest acting enzyme in the biological kingdom is carbonic anhydrase. Carbon dioxide produced by the tissues diffuses into the blood stream and passes into red blood cells where it combines with water to form carbonic acid. This prosses is catalysed by the enzyme carbonic anhydrase found in red blood cells. This is the fastest acting enzyme.

17. BREATHING AND EXCHANGE OF GASES

ANSWER KEY

1	(b)	11	(c)	21	(b)	31	(a)	41	(d)	51	(b)
2	(d)	12	(d)	22	(d)	32	(c)	42	(b)	52	(a)
3	(d)	13	(c)	23	(b)	33	(b)	43	(c)	53	(c)
4	(b)	14	(c)	24	(c)	34	(d)	44	(c)	54	(a)
5	(c)	15	(a)	25	(d)	35	(b)	45	(a)	55	(c)
6	(b)	16	(a)	26	(b)	36	(d)	46	(a)	56	(d)
7	(a)	17	(d)	27	(b)	37	(a)	47	(c)	57	(a)
8	(a)	18	(d)	28	(a)	38	(a)	48	(d)	58	(b)
9	(a)	19	(d)	29	(a)	39	(a)	49	(d)	59	(c)
10	(a)	20	(d)	30	(a)	40	(a)	50	(b)	60	(d)

Hints & Solutions

1. **(b)** Inside the larynx there are vocal cords. These are two pairs of fold of mucous membrane that extend into the lumen of the larynx from the sides.
2. **(d)** The pharynx provides passage to both air and food.
3. **(d)** It serves to close the glottis during deglutition thereby preventing entry of food into respiratory tract.
4. **(b)** Frog has lungs as its main respiratory organs but during hibernation & aestivation and during its habitat in water it respires through skin.
5. **(c)** The pathway of inhaled air is - Nostrils - pharynx (common passage for food & air) - larynx (voice box) - trachea (the wind pipe) - bronchi (2 for each side lungs) - bronchioles (give arise to alveolar ducts) - alveoli (the exchange site for gases in the form of small sacs or pouches).
7. **(a)** It is the relationship between the percentage saturation of haemoglobin (by volume) in the blood and the oxygen tension (in partial tension) pO_2 of the blood.
   It is usually a sigmoid plot. Haemoglobin molecules can bind up to four oxygen molecules in a reversible way. The shape of the curve results from the interaction of bound oxygen molecules with incoming molecules. The binding of the first molecule is difficult. However, this facilitates the binding of the second and third molecules, and it is only when the fourth molecule is to be bound that the difficulty increases, partly as a result of crowding of the haemoglobin molecule, partly as a natural tendency of oxygen to dissociate.
13. **(c)** In earthworm, there is no specific organ for respiration. Haemoglobin is dissolved in the plasma.
14. **(c)** Intercostal muscles (external intercostal & internal intercostal) are attached with the ribs which help in the movement of rib cage during breathing.
15. **(a)** CO_2 from the respiratory tissues to the lungs is transported by the blood in 3 ways
   (i) In dissolved state or as a physical solution : Very small amount physically dissolved in plasma (7% i.e. $\cong$ 0.3 ml of CO_2 by each 100 ml of blood)
   (ii) Bicarbonate ions : $\cong$70% (i.e. $\cong$2.5 ml per 100 ml of blood) CO_2 diffuses in plasma & then into RBCs where it (in the presence of carbonic anhydrase) combines with H_2O to form carbonic acid which is almost spontaneously dissociated into hydrogen ions & bicarbonate ions.
   (iii) Carbaminohaemoglobin : $\cong$23% (i.e. $\cong$1 ml of CO_2 per 100 ml of blood) combines with haemoglobin forming an unstable compound.
16. **(a)** Gaseous exchange between blood (dissolved phase) and alveolar air (gaseous phase) across respiratory membrane occurs by simple diffusion. Obviously, it depends upon the concentration gradient (= partial pressure) of the concerned gases in blood and alveolar air.
17. **(d)** Brain is the most vital organ. It stops functioning in the absence of O_2.
18. **(d)** Vital capacity of lungs is largest possible expiration after largest possible inspiration that is greatest. Volume of air can be exchanged in single respiration or amount of air breathe in and out with greatest possible efforts.

$$VC = TRV + TV + ERV$$
$$= 3000 + 500 + 1100 = 4600 \text{ ml.}$$

24. **(c)** The disease that occurs when the haemoglobin content of the blood goes down is anaemia.
26. **(b)** Due to the buffering action of blood much amount of CO_2 is carried out by blood yet blood does not become acidic.
27. **(b)** The total volume of air that can be expelled from the lungs after maximum inspiration and then expiring to the maximum is known as the vital capacity (3-5*l*). The volume of air that remains inside lungs at the end of maximum forceful expiration is the residual volume (1.5 *l*). Expiratory reserve volume is the maximum extra volume of air that can be expired by forceful expiration after a normal tidal expiration (1 *l*). Total lung capacity is the maximum volume of air that can be contained in the lungs after a maximal inspiration (5 *l*).
28. **(a)** Incomplete cartilaginous rings are present around trachea & bronchi.
29. **(a)** At 15 gm / 100 ml of haemoglobin, the oxygen carrying capacity would be 19.4 ml. Blood leaving tissue carries about 14.4 ml of O_2/100 ml.
30. **(a)** Carbonic anhydrase catalyses the reaction between CO_2 and H_2O and thus facilitates transfer of CO_2 from tissues to blood and from blood to lung alveoli. It is present in RBC and kidney cells.

$$CO_2 + H_2O \xrightarrow[\text{anhydrase}]{\text{Carbonic}} H_2CO_3 \rightleftharpoons H^{\oplus} + HCO_3^-$$

31. **(a)** Vital capacity is the amount of air one can inhale and exhale with maximum effort. It varies from 3000-4500 ml.

$$\boxed{TLC = VC + RV}$$

32. **(c)** The high concentration of CO_2, called hyperkapnia leads to decrease in the affinity of haemoglobin to oxygen. This eventually leads to an obligatory decrease in the partial pressure of oxygen known as hypoxaemia. It also increase the affinity of haemoglobin to hydrogen resulting in respiratory acidosis.
33. **(b)** The very thin alveolar wall (about 0.0001 mm) is composed of moist, nonciliated, squamous epithelial cells.
34. **(d)** Carbon dioxide is transported via blood to lungs mostly as carbaminohaemoglobin and carbonic acid. It is released in lungs in exchange with oxygen.
35. **(b)** Alveoli are the site of the respiratory exchange of gases. Oxygen from the alveolar air diffuses through the alveolar epithelium and the capillary endothelium

into the capillary blood and carbon dioxide diffuses in the opposite direction.

36. **(d)** The breathing becomes faster and deeper in order to oxygenate the blood at a fast rate. Shallow and slow breathing occurs during rest.

37. **(a)** At the altitude of 5400 meters the low atmospheric pressure of O_2 will be too low so the solubility of oxygen in the blood will be very less hence the oxygen carried by each RBC will be too less. But to fulfill the oxygen requirement of the body blood has to carry more oxygen to the body tissue and this is done by the increased number of RBCs.

38. **(a)** The breathing controlling centre of medulla oblongata is mainly under chemical control.

42. **(b)** Oxygen combines with haemoglobin to form oxyhaemoglobin that is transported in the vertebrates through the blood.

44. **(c)** Diaphragm, an important feature of the mammals, is a partition between thoracic and abdominal cavities.

45. **(a)** About 97% of oxygen is carried in combination with haemoglobin of the erythrocytes to form oxyhaemoglobin.

46. **(a)** At high altitude there is decrease in partial pressure of oxygen due to which the person feels difficulty in breathing.

47. **(c)** During inspiration diaphragm contracts and flattens which increases the lung volume and lowers the pressure inside it.

48. **(d)** In a normal adult human, 500 ml respiratory volume is related to tidal volume. It is the air that normally goes in and out of lungs during breathing.

56. **(d)** Carboxyhaemoglobin is the stable product formed by the association of CO and Hb in the blood. The association of carbon dioxide and haemoglobin forms carbamino-haemoglobin.

57. **(a)** At high altitude, the body undergoes numerous changes in order to increase oxygen delivery to cells and improve the efficiency of oxygen usage. The early changes include increased breathing rate and increased red blood cell production.

58. **(b)** Eustachian tube connect middle ear cavity (tympanic cavity) with nasopharynx, approxi-mately it is 35 mm long in adults, which helps in breathing out by closing both the nose and mouth.

59. **(c)** Since RBCs do not have mitochondria so they can respire only anaerobically.

60. **(d)** For proper transport of O_2 & CO_2 blood should be slightly alkaline. Haemoglobin can carry more O_2 in slightly alkaline blood.

18. BODY FLUIDS AND CIRCULATION

ANSWER KEY

1	(c)	12	(c)	23	(c)	34	(c)	45	(c)	56	(c)
2	(a)	13	(a)	24	(b)	35	(a)	46	(d)	57	(d)
3	(b)	14	(b)	25	(c)	36	(b)	47	(d)	58	(a)
4	(a)	15	(a)	26	(c)	37	(d)	48	(b)	59	(b)
5	(b)	16	(b)	27	(b)	38	(d)	49	(c)	60	(c)
6	(d)	17	(a)	28	(d)	39	(d)	50	(a)		
7	(b)	18	(c)	29	(b)	40	(c)	51	(b)		
8	(b)	19	(d)	30	(b)	41	(a)	52	(a)		
9	(d)	20	(a)	31	(c)	42	(d)	53	(b)		
10	(a)	21	(b)	32	(a)	43	(c)	54	(a)		
11	(c)	22	(b)	33	(b)	44	(a)	55	(b)		

Hints & Solutions

5. **(b)** Because the process of blood clotting starts when prothrombinase (an enzyme formed by thromboplastin) catalyzes the conversion of prothrombin (an inactive protein in the blood plasma) into thrombin (active protein) and this reaction takes place in the presence of Ca^{2+}. Now this thrombin converts fibrinogen into fibrin which makes fibrin threads on polymerization & with blood corpuscles it makes blood clot so if Ca^{2+} are removed from blood it'll prevent clotting. It may also be done by adding some anticoagulants e.g. sodium & potassium oxalates in blood as they precipitate Ca^{2+} present in the blood.

6. **(d)** Neutrophils are a type of leucocyte (WBCs) that can take all types of stain (acid-basic-Neutral). It is most abundant (60–70% of total WBCs) and most active type of WBCs i.e., they are the most actively phagocytic in nature.

7. (b) Blood group A – antigen-A & antibody 'b' or Anti A or a,
Blood group B – antigen-B & antibody a or Anti B or b
Blood group AB – antigen-A & B and no antibody.
Blood group O – no antigen & antibodies a & b both.

8. (b) An adult human with average health has systolic and diastolic pressures as 120 mm Hg and 80 mm Hg respectively.

9. (d) Leucocytes are colourless, nucleated amoeboid cells found in blood which are devoid of haemoglobin and are capable of coming out of blood capillaries through the process of diapedesis. Fall of WBC count is called leucopenia, and occurs due to folic acid deficiency and AIDS, etc.

10. (a) Blood group 'O' is known as the universal donor since, it does not contain antigens.

11. (c) The atrioventricular bundle (AV bundle), on entering the ventricles splits into right and left branches. These branches then enter the heart muscles. They branch extensively and are called Purkinje fibres. They conduct the cardiac impulse to the ventricular wall.

12. (c) Neutrophils are largest in number and constitute about 62-75% of leucocytes. They eat harmful germs and are therefore, phagocytic in nature.

Neutrophil

13. (a) Rh factor is named after the Rhesus monkey. Experiments by Karl Landsteiner and Alexander S. Wiener, showed that rabbits, when immunised with rhesus monkey red cells, produce an antibody that also agglutinates the red blood cells of many humans.

14. (b) Sympathetic nervous system increases rate of heart beat. In sympathetic nervous system preganglionic nerve small while post ganglionic nerve is long. Sympathetic nervous system generally show sympathy.

15. (a) The heart beat is initiated by a path of modified heart muscle in man and elephant. The impulse of contraction originates itself in the heart that is why myogenic in origin.

27. (b) The thickening and loss of elasticity of arterial wall are called arteriosclerosis.

29. (b) Rh factor was discovered by Karl Landsteiner. A child of Rh^+ man will be Rh^+ whether the mother is Rh^+ or Rh^-. If the mother is Rh^+ then there will be no problem but if mother is Rh^- so when the blood of Rh^+ child (in womb) mixes with the blood of Rh^- mother then some antibodies in mother's blood are formed against Rh^+ factor which coagulate the womb blood causing death. If birth takes place then there is a possibility of child death in early years. This in known as *erythroblastosis foetalis*. In most cases the Ist pregnancy may succeed but after that it fails.

30. (b) Tricuspid valve is present between right auricle and right ventricle. It has 3 membranous flaps to check the flow of blood in reverse direction. Sinus venosus and right auricle is guarded by sinu-atrial value. The opening of left auricle and left ventricle is guarded by bicuspid valve. The opening of ventricle and aorta is guarded by semilunar valves.

31. (c) A has Antigen A & antibody b so it can have the same blood group or O has no any antigen & antibodies a & b so, it is called universal donor.

32. (a) Heart produces two types of sounds. 1st sound : 'Lubb' produced by closure of bicuspid and tricuspid valves. 2nd sound : 'Dup' produced by closure of semilunar valves present at the base of aorta to avoid back flow of blood into the ventricle.

33. (b) Closed circulatory system means blood flows in some definite pipelines in some specific direction in more organized manner than that of open system in which blood does not flow in vessels.

34. (c) Sino-Auricular node (SA node) present in the walls of right auricle has a myogenic initiation of heartbeat in a regular fashion and controls the pace of heartbeat called pacemaker.

35. (a) The inheritance of blood group is determined by some allelic genes which are I^A, I^B & I^o. In one individual two allelic genes are present. I^a & I^b are dominant while I^o is recessive in heterozygous condition. So, if a child has a blood group O he/she must have I^oI^o & he/she cannot have the parents of blood group AB & AB/O.

36. (b) Antigens are foreign proteins, which stimulate specific immune response (antibody) against itself when introduced into the body. They are present on the surface of cell wall.

37. (d) The pulse making impulse travels in the heart in the order of SA node → AV node → bundle of His → Purkinje fibres → heart muscles.

38. (d) RBCs as well as plasma both carry oxygen and CO_2 in dissolved and chemically combined form.

39. (d) Arteries have thick wall and narrow lumen while veins have thin walls and large lumen. Arteries do not have valves.

40. (c) Higher vertebrates exhibit double circulation of blood namely systemic and pulmonary circulation. Systemic circulation involves the distribution of oxygenated blood via the aorta from the heart to all parts of the body, and the collection of deoxygenated blood from all parts of the body through the great veins into the heart. Pulmonary circulation denotes the route impure blood takes from the heart to the lungs for purification and then back to the heart, the left auricle receives oxygenated blood and passes it to the left ventricle. The left ventricle pumps the pure blood to all parts of the body through aorta. Hence these two chambers involved in systemic circulation are referred to as systemic heart.

41. (a) Periplaneta (*P. americana,* the Indian Cockroach) is an insect & insects do not have closed blood circulation.

Their blood (called haemolymph) even does not serve for respiration. They do not possess blood vessels in this circulatory system and hence it is an open system.

42. **(d)** The most active phagocytic white blood cells are neutrophils and monocytes.
Neutrophil is a type of WBC (leucocyte) that has a lobed nucleus and granular cytoplasm. Neutrophils engulf bacteria and release various substances such as lysozyme and oxidizing agents.
Monocyte is the largest form of WBC in vertebrates. It has a kidney shaped nucleus and is actively phagocytic, ingesting bacteria and cells debris.

43. **(c)** The ABO blood group system is the most important blood type system (or blood group system) in human blood transfusion. It is named ABO and not ABC, because blood group O (or blood group zero in some countries) individuals do not have either A or B antigens on the surface of their RBCs, but their blood serum contains IgM anti-A antibodies and anti-B antibodies against the A and B blood group antigens. Therefore, a group O individual can receive blood only from a group O individual, but can donate blood to individuals of any ABO blood group (i.e., A, B, O or AB).

44. **(a)** Red blood cells are the most common type of blood cell and delivering oxygen to the body tissues via the blood. There is no DNA in mature RBC. The reticulocyte is the immediate precursor of the mature RBC and, within 24 hours of release into the peripheral circulation, evolves into the mature RBC.

45. **(c)** In a standard ECG, the P-wave is a small upward wave that indicates the depolarisation of the atria. This is caused by the activation of SA node.

46. **(d)** Globulins contained in human blood plasma are primarily involved, in defence mechanism of body. Globulin is one of the two types of serum proteins, the other being albumin. Globulins can be divided into three fractions based on their electrophoretic mobility. Most of the alpha and beta globulins are synthesized by the liver, whereas gamma globulins are produced by lymphocytes and plasma cells in lymphoid tissue.

47. **(d)** Fibrinogen, the protein of blood plasma is converted to insoluble protein fibrin during the clotting process. The fibrinogen free fluid obtained after removal of the clot, called blood serum is plasma minus fibrinogen.

48. **(b)** Arteries are the vessels which break up into capillaries which reunite to form one visceral organ.

49. **(c)** Peak P-causes diastolic phase in ventricle while R-Peak causes systole in ventricle means diastolic and systolic phases represented by P & R.

50. **(a)** HCO_3^- is the ion present in blood. 23% of CO_2 is transported in the blood in the form of carbaminohaemoglobin. 70% of CO_2 is transported in the form of bicarbonate ions HCO_3^-.

$$CO_2 + H_2O \longrightarrow H_2CO_3$$

Carbonic acid is unstable which gives H^+ and HCO_3^- ions. H^+ combines with haemoglobin, HCO_3^- remains in the blood and carries CO_2.

51. **(b)** Maximum surface area of circulating system is seen in capillaries. The blood vessels arteries and veins divide to form arterioles and venules which further divide to form thin walled vessels called capillaries. Capillaries are composed of endothelium only. The nutrients, hormones, gases can diffuse through the wall of capillaries.

52. **(a)** Blood is an opaque, mobile connective tissue fluid which has salty taste. pH of blood is between 7-8. Hence, the blood possesses slightly alkaline pH.

53. **(b)** Blood group AB individuals have both A and B antigens on the surface of their RBCs, and their blood plasma does not contain any antibodies against either A or B antigen. Therefore, an individual with type AB blood can receive blood from any group with AB being preferable. Hence, blood group AB is known as universal recipient.

54. **(a)** QRS complex is related with ventricular contraction. It forms a curve complex and begins after a fraction of second of the P wave. It begins as a small downward deflection (Q) and continues as large upright (R) and a triangular wave ending as downward wave (S) at its base.

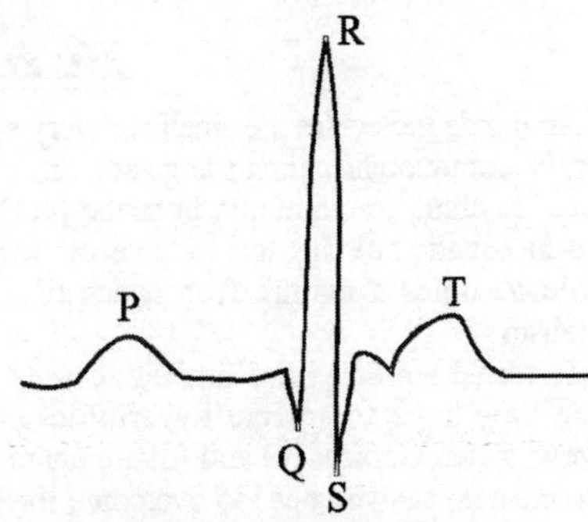

ECG of a single heart beat

55. **(b)** Heparin is an anticoagulant secreted by basophilic type of granulocyte (WBC). It is formed in liver and mast cells common.

56. **(c)** Lymphocytes are the types of WBCs that help in producing antibodies to fight the antigens or toxins produced by germs.

57. **(d)** Erythroblastosis is the disease related with Rh-factor. It occurs in the foetus due to Rh incompatibility between the mother and the foetus.

58. **(a)** The number of leucocytes per microlitre of blood is called TLC or total leucocyte count. This varies from 5000 to 8000 per cubic millimeter of blood in humans.

59. **(b)** Lymph is described as blood minus RBC. It plays an important role in defence against microbes, forms WBC but it does not form erythrocytes or RBCs.

60. **(c)** Skeletal muscle contraction has nothing to do with the blood flow in arteries whereas heart beat, blood pressure and total cross sectional area of vessels are directly or indirectly associated with it.

19. EXCRETORY PRODUCTS AND THEIR ELIMINATION

ANSWER KEY											
1	(d)	12	(b)	23	(b)	34	(c)	45	(d)	56	(d)
2	(a)	13	(b)	24	(c)	35	(a)	46	(b)	57	(c)
3	(c)	14	(d)	25	(c)	36	(d)	47	(a)	58	(a)
4	(a)	15	(c)	26	(c)	37	(b)	48	(b)	59	(a)
5	(c)	16	(b)	27	(c)	38	(b)	49	(d)	60	(a)
6	(a)	17	(a)	28	(d)	39	(a)	50	(a)		
7	(d)	18	(a)	29	(d)	40	(a)	51	(d)		
8	(c)	19	(d)	30	(c)	41	(a)	52	(c)		
9	(b)	20	(a)	31	(d)	42	(c)	53	(a)		
10	(b)	21	(d)	32	(d)	43	(b)	54	(c)		
11	(b)	22	(d)	33	(d)	44	(c)	55	(c)		

Hints & Solutions

9. **(b)** Ammonia molecules are small and very soluble in water. In ammonotelic animals large amount of H_2O is required to eliminate ammonia from the body, so it is found in aquatic animals like tadpole of frog. When tadpole becomes a mature frog it acquires mainly ureotelism.
10. **(b)** Terrestial animals have limited access to water. Hence they have to secrete hypertonic urine and conserve water. Choices (c) and (d) are characteristics of marine animals who need to overcome the problem of preventing entry of salts and loss of water. Choice (a) is characteristic of fresh water animals.
11. **(b)** Nephron is the excretory unit of human excretory system. Each nephron has a bowman's Capsule, a proximal Convoluted tubule (PCT), loop of henle (descending & ascending limbs) and distal convoluted tubule (DCT) which then enter into collecting duct.
12. **(b)** If kidneys fail to reabsorb water the urine will be diluted causing polyuria (frequent urination). Then the body tissues get dehydrated & the cell will loose water content & shrink.
13. **(b)** The nephridia are the excretory organs. The nephridia in earthworm are – septal nephridia, pharyngeal nephridia and integumentary nephridia. The septal nephridia do not discharge excretory fluid to the exterior rather it pour them into the intestine. Hence these are also called enteronephric nephridia.
14. **(d)** Under fasting conditions and in uncontrolled diabetes, the liver produces large amount of ketone bodies like acetoacetate, dihydroxy butyrate. The presence of excess ketone bodies in blood or urine is termed ketonemic and ketonuria respectively.
15. **(c)** Bowman's glands are located in the olfactory epithelium of our nose.
16. **(b)** Excretion is the removal of metabolic waste products from the animal body to regulate the composition of the body fluids and tissues.
17. **(a)** Malphigian body is present in Bowman's capsule. This visceral layer surrounds the glomerulus and is composed of special type of cells called podocytes.
18. **(a)** Glomerulus filtrate contains a large amount of water and dissolved substances. Its filterate mainly contains blood plasma without proteins and fats.
19. **(d)** The lungs are concerned with respiratory process. They help in taking in oxygen *i.e.,* inspiration and giving out of carbon dioxide *i.e.,* expiration.
20. **(a)** The production of watery urine in large quantities is called diuresis. It causes the osmotic pressure of the blood to rise end is counteracted by antidiuretic hormone.
21. **(d)** Ammonia is highly soluble in water and the most toxic excretory product. Thus, its concentration must be kept very low in the blood.
23. **(b)** Urea is formed in liver cells from ammonia and carbon dioxide.
31. **(d)** Mineralocorticoids regulates reabsorption of salts from glomerular filtrate.
33. **(d)** Angiotensinogen is a protein produced and secreted by liver cells.
35. **(a)** Water reabsorption in the kidney is under the control of ADH and aldosterone. ADH controls the urine concentration by changing the permeability of the collecting tubule whereas aldosterone does the same by retaining Na^+ ions in the interstitial tissues.
36. **(d)** Ultrafiltration in uriniferous tubules is determined by glomerular hydrostatic pressure (GHP = 60 mmHg),

blood colloidal osmotic pressure (BCOP = 32 mm Hg) and capsular hydrostatic pressure (CHP = 18 mm Hg).

37. (b) Aves (or birds) do not have urinary bladder. (exceptions - ostrich, rhea). Their urine is semi-solid uric acid. Uric acid goes down to cloaca and is excreted out alongwith faeces.

38. (b) Vasopressin hormone is also known as anti diuretichormone (ADH). It is secreted by the posterior lobe of pituitary gland. It helps in the reabsorption of water from glomerular filtrate thereby making the urine concentrate. The deficiency of this hormone results in the excretion of dilute and excessive urine. This condition is known as diabetes insipidus.

39. (a) The contractile vacuole mainly expels surplus water in protozoans. The fresh water protozoans gain water by osmosis from their hypotonic environment.

40. (a) Loop of Henle is a U-shaped structure found in nephron which absorbs water. It is connected to the excretory system.

41. (a) An afferent arteriole enters each Bowman's capsule to form the glomerulus. An efferent arteriole arises from the glomerulus.

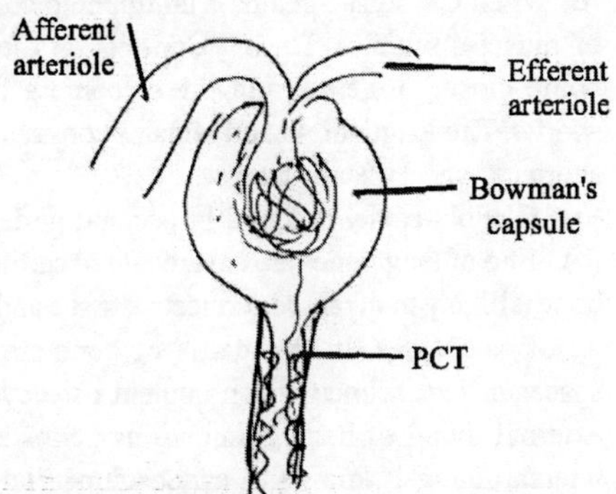

42. (c) When the filtrate reaches the collecting tubules then it can be said as urine. The filtrate enters the collecting duct where further reabsorption of water takes place. The filtrate is now hypertonic to blood plasma. Sodium is reabsorbed under the influence of aldosterone. The filtrate is called urine.

44. (c) The diameter of afferent blood vessels of glomerulus is wider than the efferent blood vessels. This difference creates pressure gradient in glomerulus which is responsible for glomerular filtration.

45. (d) Liver converts toxic substances into less toxic ones. Toxic ammonia (NH_3) is converted into much less toxic urea which is excreted in urine. Urea is synthesized in liver by urea cycle or ornithine cycle and transported to kidneys for excretion in urine.

46. (b) Uremia is a disease in which excess of urea gets collected in the blood. There is high concentration of non-protein nitrogens that include urea, uric acid, creatinine and some other compounds.

47. (a) The mechanism of urine formation takes place in three steps : (i) Ultrafiltration; (ii) Selective reabsorption and (iii) Tubular secretion. Ultrafiltration takes place through basement membrane. Selective reabsorption takes place in (PCT + Loop of Henle + DCT). Tubular secretion is supporting mechanism for filtration and reabsoprtion.

48. (b) The amount of water, urea, glucose salts and amino acids in glomerular filterate is same as in plasma i.e., it is isotonic with plasma.

49. (d) Diabetes mellitus is a diseased condition where more of glucose is excreted in urine due to insufficiency of insulin.

50. (a) In ascending limb of Henle's loop, small amounts of Cl^- is reabsorbed. Some Na^+ is also reabsorbed by diffusion because of the electrostatic attraction of the reabsorbed Cl^- ions.

51. (d) A fresh water protozoan when placed in marine water medium, faces a condition of preventing entry of salts and loss of water. Hence the contractile vacuole will decrease in size.

52. (c) Some animals have dual behaviour of excretion eg. earthworm. It excretes ammonia (hence called ammonotelic) when enough water is available but excretes urea (ureotelic) in dry conditions.

53. (a) Characteristics of certain desert animals such as kangaroo, rat are
- they feed on dry seeds and do not require drinking water.
- they excrete very concentrated urine and do not use water to regulate body temperature.

54. (c) Micturition is same as urination. Urination or micturition is the act of passing urine which is a reflex phenomenon. As urine accumulates in bladder the stretch receptors are activated that pass the stimulus to the spinal cord. In the absence of stretch receptors the urine would get collected and probably overflow.

55. (c) Urine formation involves three main process called, glomerular filtration, reabsorption and secretion. A comparison of the volume of the filtrate formed per day (which is 180 litres per day) with that of urine released (about 1.5 litres) suggests that nearly 99 percent of the glomerular filtrate is reabsorbed by the renal tubules. The descending limb of loop of Henle is permeable to water but impermeable to electrolytes. The ascending limb is impermeable to water but allows transport of electrolytes.
Reabsorption of sodium ions and water takes place in distal convoluted tubule.
Also, it is capable of reabsorption of HCO_3.

56. (d) Tissue and blood group matching are essential before undertaking kidney transplant. Even if kidney transplant is proper, the recipient may need to take immuno suppresant all his/her life. The ability of body to differentiate self and nonself and the cell-mediated immune response is responsible for graft rejection.

57. (c) Renal pelvis is the innermost portion of kidney. So, collecting ducts are not a part of renal pelvis.

58. (a) A large portion of oxygen is left unused in the human blood even after its uptake by the body tissues. This O_2 acts as a reserve during muscular exercise.

59. (a) Glome podocytes are highly specialized cells with a complex cytoarchitecture plays a major role in establishing the selective permeability of glomerular filtration barrier.

60. (a) When some one drinks lots of water, kidney release of ADH is suppressed.

20. LOCOMOTION AND MOVEMENT

ANSWER KEY													
1	(b)	10	(a)	19	(b)	28	(c)	37	(b)	46	(b)	55	(b)
2	(b)	11	(d)	20	(d)	29	(c)	38	(a)	47	(c)	56	(c)
3	(a)	12	(d)	21	(d)	30	(d)	39	(a)	48	(b)	57	(b)
4	(c)	13	(b)	22	(d)	31	(a)	40	(d)	49	(c)	58	(c)
5	(d)	14	(b)	23	(d)	32	(b)	41	(b)	50	(c)	59	(c)
6	(c)	15	(a)	24	(c)	33	(b)	42	(d)	51	(b)	60	(c)
7	(c)	16	(a)	25	(b)	34	(b)	43	(c)	52	(c)		
8	(b)	17	(d)	26	(a)	35	(b)	44	(a)	53	(a)		
9	(c)	18	(c)	27	(a)	36	(b)	45	(c)	54	(d)		

Hints & Solutions

3. **(a)** Bone is the solid rigid strong connective tissue. Its matrix consists of ossein protein & mineralization of matrix occurs by calcium- phosphate salts. This gives rigidity & strength to bones.
5. **(d)** Immediate source of energy for muscle contraction is ATP.
7. **(c)** Humerus and pectoral girdle is a ball and socket joint.
8. **(b)** A chain of three ear ossicles or three little ear bones is present in the cavity of middle ear of rabbit. These are present in a sequence as given below :

   Malleus ⟶ Incus ⟶ Stapes
   (hammer) (anvil) (stirrup)
14. **(b)** Tendons are made of inelastic white fibrous connective tissue. A tendon (or sinew) is a tough band of fibrous connective tissue that connects muscle to bone or muscle to muscle and is designed to withstand tension. Tendons are similar to ligaments except that ligaments join one bone to another. Tendons and muscles work together and can only exert a pulling force.
15. **(a)** Ligament contains both yellow elastin fibres and collagen fibres. The latter are arranged in a manner to allow for stretching. Ligaments help in binding bones to bones.
16. **(a)** Actin and tropomyosin are part of thin filaments of skeletal muscle. Tubuline is presents in microtubules. Myosin is muscle protein.
17. **(d)** The last two pairs *i.e.,* 11th and 12th pairs ribs remain free anteriorly, hence, they are called as floating ribs.
26. **(a)** Joint between atlas and axis is pivot joint.
29. **(c)** Bones of pelvic girdle forms a cavity in which head of the femur is fitted are ilium, ischium and pubis.
31. **(a)** Ca^{2+} & Mg^{2+} are involved in muscle contraction.
35. **(b)** When Ca^{+} ions combine with troponin contraction of muscles initiates. During Contraction the Z lines come closer together and the sarcomere becomes shorter. The length of A band remains constant. I bands shortens and H-band narrows.
36. **(b)** Glenoid cavity is found in pectoral girdle.
37. **(b)** End of long bones are covered with cartilage. The bone is highly mineralised connective tissue and is made up of osteocytes. In the matrix of bone are present spaces called lacunae which contain osteocytes. The external living of bone is known as periosteum and internal lining is known as endoostium. End of long bones are lined by hyaline cartilage or articular cartilage.
38. **(a)** All mammals possess 7-cervical vertebrae except : Seacow & Sloth. First cervical vertebra is known as atlas and the second cervical vertebra is called as axis.
39. **(a)** The bones of fore limb are humerus and radio-ulna while that of hind limb are femur and tibio-fibula. The humerus bone is elongated and has a proximal expanded round head for articulation with the glenoid cavity of the scapula. At the sides of head there are two tuberosities, a long on the outerside and a small on the innerside.
40. **(d)** Rib cage consists of vertebral column (dorsal), sternum (ventral) and ribs (lateral).
42. **(d)**

Nasal	=	2
Maxillae	=	2
Zygomatic	=	2
Mandible	=	1
Lacrimal	=	2
Palatines	=	2
Inferior chonchae	=	2
Vomer	=	1
		14

43. (c) 12 pairs ribs are present in human beings, 1st to 7th pairs are called true ribs, 8th, 9th, 10th pairs are called false ribs, and last two pairs of ribs are called floating ribs.
44. (a) The cranium is formed by 8 bones. (1 frontal bone, 2 parietal, 2 temporal, 1 occipital, 1 sphenoid, 1 ethmoid).
46. (b) Patella is flat bone. Patella is also called knee cap.
48. (b) Each myofibril shows dark and light bands. The dark bands of the myofibril are termed as A-bands or anisotropic bands.
51. (b) Human skeleton is mainly formed of bones and cartilages. It is formed of 206 bones in adult man.
53. (a) It forms the main axis. It consists of skull - 22 bones, vertebral column - 33 vertebrae, sternum - 1 and ribs - 24.
54. (d) Appendicular skeleton consists of pectoral and pelvic girdles and bones of arms and legs.
55. (b) Myosin filaments are thicker and confined to the A-bands only. They are composed of myosin protein
56. (c) The first seven pair are attached directly to the sternum 8th, 9th and 10th pairs are attached indirectly to the sternum, and the last two pairs of ribs are called floating ribs because their anterior ends are not attached to either the sternum or the cartilage of another ribs.
57. (b) The eighth, ninth and tenth pairs of ribs are called false ribs. They articulate by cartilage with costal cartilage of the seventh rib.
59. (c) Sarcolemma is present on the outside of skeletal muscle fibre. In the region of attachment, the sarcolemma is folded to form finger like projections.

21. NEURAL CONTROL AND CO-ORDINATION

ANSWER KEY

1	(a)	11	(c)	21	(c)	31	(d)	41	(a)	51	(a)
2	(c)	12	(b)	22	(b)	32	(a)	42	(c)	52	(c)
3	(d)	13	(c)	23	(b)	33	(b)	43	(d)	53	(d)
4	(c)	14	(b)	24	(a)	34	(b)	44	(c)	54	(d)
5	(a)	15	(b)	25	(c)	35	(a)	45	(c)	55	(d)
6	(c)	16	(d)	26	(a)	36	(a)	46	(b)	56	(b)
7	(b)	17	(b)	27	(a)	37	(c)	47	(c)	57	(b)
8	(c)	18	(a)	28	(a)	38	(c)	48	(c)	58	(a)
9	(b)	19	(c)	29	(d)	39	(a)	49	(b)	59	(d)
10	(b)	20	(b)	30	(a)	40	(d)	50	(c)	60	(d)

Hints & Solutions

2. (c) Iris consists of two layers, outer thickened layer is continuation with choroid and inner simple cuboidal epithelium layer continued from retina.
3. (d) Iris controls the size of pupil by contracting or relaxing of its circular muscles. It controls the amount of light entering in eye.
4. (c) Melanin is the pigment which gives colour to the skin. Retinol is the other name for vitamin A. Sclerotin is the component of the carapace in crustaceans.
5. (a) Pupil is the opening which controls the amount of light entering in eye. When light intensity is high, it decreases in size and when light intensity is low it dilates to allow more light in the eye to make eye enable to see the object.
6. (c) Hypothalamus part of human brain is concerned with the regulation of body temperature. The hypothalamus is a portion of the brain that contains a number of small nuclei with a variety of functions. The hypothalamus is small cone-shaped structure, projects downward, ending in the pituitary.
11. (c) Reflex action is the immediate involuntary response to stimulus. It includes sudden action of body parts due to heat stimulus, burn stimulus, pricking, sneezing, coughing yawning, etc. Perspiration is not a reflex action. It is also known as sweating. Sweating is loss of water from sweat glands of the skin. Sweat includes some salts and urea. Its for temperature regulation.
12. (b) The sensory nerve fibres bring sensory impulses from the receptor organ to the central nervous system. The motor nerve fibres relay the motor impulses from the central nervous system to the effector organ. This

path of reflex action is called the reflex arc.

13. (c) Nerve impulse is electrochemical in nature. The membrane or ionic theory of nerve impulse was proposed by Hodgkin and Huxley. The theory states that electrical events in the nerve fibre are governed by the differential permeability of membranes to sodium & potassium. These are regulated by the electric field across the membrane.

14. (b) The junction between an axon and dendrite is known as synapses through which nerve impulse is transmitted from one neuron to other. A synapse consists of swelling at the end of nerve fibre called synaptic knob and small synaptic vesicle containing neuro-transmitter–acetylcholine for the transmission of nerve impulse.

15. (b) Blind spot is the place found close to the yellow spot in the human eye from where optic nerve exits and goes to the brain.

16. (d) The internal ear is called the labyrinth which consists of two parts
 (i) bony labyrinth filled with fluid called perilymph and
 (ii) a membranous labyrinth, contains a fluid called endolymph.

17. (b) The organ of corti in human is concerned with the sense of hearing. It rests on the basilar membrane, which are stimulated by sound waves.

18. (a) The nodes of ranvier are the points in which the axons is exposed.

23. (b) Afferent nerve fibres – sensory nerve fibres which carry senses from receptors to brain or spinal cord (CNS).
 Efferent nerve fibres – motor nerve fibres carry orders or responses from CNS to effectors (muscles and glands).

24. (a) Prominent granules (Nissl's granules) observed in the cytoplasm of the cyton are believed to be made of ribosomes.
 A Nissl body (or Nissl's granule or tigroid body) is a large granular body found in nerve cells. It was named after Franz Nissl, German neurologist (1860-1919). Nissl bodies can be demonstrated by selective staining, which was developed by Nissl and the stain used was an aniline stain to label extranuclear RNA granules. These granules are rough endoplasmic reticulum (with ribosomes) and are the site of protein synthesis. Nissl bodies show changes under various physiological conditions and in pathological conditions they may dissolve and disappear (karyolysis).

25. (c) In a resting state of the neural membrane, Na^+ concentration is higher on the outer side and K^+ concentration is more within the cell. This concentration gradient is maintained by voltage gated channels. Hence, if diffusion is allowed Na^+ would enter the cell and K^+ would leave.

26. (a) The process of movements of sodium and potassium ions takes place against an electro-chemical gradient. This process involves energy by the breaking of ATP bonds. It also involves an integral protein ATPase. For one molecule of ATP used, 3 ions of Na^+ are pumped out and two K^+ ion are taken in.

27. (a) Non-medullated fibres are present in autonomic nerves.

28. (a) Schwann cells are found around axon of myelinated fibers. They secrete myelin sheath around axon.

29. (d) Hind brain has two distinct parts
 (i) metencephalon (medulla oblongata)
 (ii) myelencephalon (cerebellum), transverse band of white matter called pons varoli.

31. (d) Central nervous system comprises brain and spinal cord both are surrounded by the three protective membranes these are duramater, arachnoid piamater.

32. (a) It coordinates the activity of both the cerebral hemispheres.

35. (a) Medulla oblongata of hind brain contains centres that regulate heart beat, blood pressure, breathing, swallowing, salivation, sneezing, vomiting and coughing and some other involuntary movements.

36. (a) Brain and spinal cord are surrounded by three meninges-duramater, piamater and arachnoid in mammals.

37. (c) Arachnoidmater lies between the duramater and piamater.

39. (a) The dendrites contain neurofibrils, neurotubules and Nissl's granules. They conduct nerve impulse towards the cell body (cyton).

43. (d) Axons are modified for the conduction of impulses away from CNS.

45. (c) Hypothalamus controls the physiological state of body such as eating, drinking, body temperature, etc.

46. (b) Transmission of nerve impulse, across a synapse is accomplished by neurotransmitters because of synapse. It separates one neuron from other.

47. (c) Certain neurons or nerve cells may almost equal the length of body itself.

48. (c) Medulla oblongata contains centre for the autonomic reflex control of respiration, heart beat and these centres are called vital centres.

51. (a) The thin roof of III ventricle (diocoel) is called epithalamus. The anterior part of epithalamus contains a highly coiled bunch of blood capillaries called anterior Choroid plexus. The dorsal surface of metacoel also contains highly coiled bunch of blood capillaries called post choroid plexus.

52. (c) A dense network of white medullated nerve fibres constitute the white matter.

53. (d) A mammal, from whose brain cerebrum is removed loses all responses that depend upon consciousness and memory.

54. (d) Reflex action is an immediate involuntary response to a stimulus without the intervention of brain, controlled by spinal cord of central nervous system.

56. **(b)** It is under the sole control of spinal cord.

58. **(a)** Axon is a single long process of uniform thickness. It contains neurofibrils and neurotubules but lack nissle's granules, golgi complex, ribosomes, etc.

59. **(d)** Efflux of K^+ causes repolarization. Therefore it is the movement of K^+ that maintains the depolarized state.

22. CHEMICAL CO-ORDINATION AND REGULATION

ANSWER KEY											
1	(c)	**12**	(a)	**23**	(b)	**34**	(b)	**45**	(d)	**56**	(a)
2	(c)	**13**	(d)	**24**	(d)	**35**	(c)	**46**	(a)	**57**	(d)
3	(c)	**14**	(c)	**25**	(c)	**36**	(a)	**47**	(b)	**58**	(a)
4	(b)	**15**	(a)	**26**	(d)	**37**	(a)	**48**	(a)	**59**	(c)
5	(b)	**16**	(b)	**27**	(b)	**38**	(d)	**49**	(a)	**60**	(b)
6	(c)	**17**	(c)	**28**	(c)	**39**	(b)	**50**	(d)		
7	(b)	**18**	(a)	**29**	(b)	**40**	(b)	**51**	(b)		
8	(b)	**19**	(c)	**30**	(d)	**41**	(a)	**52**	(a)		
9	(a)	**20**	(d)	**31**	(c)	**42**	(a)	**53**	(c)		
10	(d)	**21**	(b)	**32**	(b)	**43**	(c)	**54**	(a)		
11	(b)	**22**	(d)	**33**	(a)	**44**	(a)	**55**	(d)		

Hints & Solutions

4. **(b)** Androgens are secreted by testes.

6. **(c)** Melatonin is a hormone produced by pineal gland.

7. **(b)** Hormones are chemical messengers formed by endocrine cells. Chemically hormones are of the following types: Amines–composed of amino group e.g., Melatonin.

Amino acids –e.g. thyroxine
Peptides – e.g. insulin
Steroids – e.g. aldosterone

8. **(b)** In heart cells AMP acts as secondary messenger which is made in the response of adrenaline and it stimulates Ca^{2+} ions to come out from the sarcoplasmic reticulum of muscle fibres which causes muscle contraction.
cGMP (Cyclic Guanosine Monophosphate) which acts as secondary messenger works on the action of acetylcholine, increase in flow of Ca^{2+} into muscle fibres & hence causes muscle relaxation. There is no role of sodium in hormonal action.

9. **(a)** The hormones in blood act on specific sites or receptors of target organ. So if we remove the receptor molecule from the target organs, there will be no response.

10. **(d)** 7 protein hormones secreted by Adenohy-pophysis are:

→ STH
→ Prolactin
→ FSH
→ LH
→ ACTH
→ TSH
→ MSH

and some non-peptide hormones secreted by Neurohypophysis are :
(a) Vasopressin (ADH);
(b) Oxytocin (Pitocin)

11. **(b)** Hormones influence body parts for bringing about co-ordination.

12. **(a)** Pituitary is known as hypophysis cerebri. Its name pituitary was given by Vesalins. Muller's gland of amphioxus and subneural gland of hardmania is homologous to pituitary of vertebrates.

13. **(d)** Receptor for protein hormone are found on cell surface and steroid on the nuclear surface.

14. **(c)** Oxytocin is called birth hormone and milk ejecting hormone while vasopressin also called pitressin has two functions antidiuretic effect and pressor effect. Both of them are secreted by posterior lobe of pituitary.

15. **(a)** Adrenaline hormone secreted by adrenal medulla is also called fight, flight and fright hormone because it is secreted when somebody gets excited during a fight or feel fear of something. It is also called emergency hormone because it is secreted at the time of emergency.

37. **(a)** Cortisol is secreted by the middle region of adrenal cortex. It increases the blood glucose level (which is anti-insulin effect) by converting proteins & fats into glucose.

38. **(d)** Leydig's cells are present in testes and they secrete male hormone testosterone.
Small groups of polygonal, large cells present in the connective tissue in between the seminiferous tubules are known as Leydig's cells are interstitial cells. They secret male sex hormone testosterone.

39. **(b)** ADH : Antidiuretic hormone–vasopressin is secreted by neurohypophysis (posterior lobe) of pituitary gland. This hormone increases the reabsorption of water from the urine and hence urine gets concentrated and ECF gets diluted. Diabetes insipidus caused by the hyposecretion of ADH.

40. **(b)** Mammalian corpus luteum produces progesterone. It is essential for the continuation of pregnancy.

41. **(a)** Calcitonin is a polypeptide hormone which lowers calcium and phosphate level of plasma by inhibiting bone degradation and stimulating their uptake by bone parathyroid hormone elevates calcium level in blood.

42. **(a)** Parathormone released by the parathyroid gland elevates the level of Ca^{2+} in blood. The deficiency of this hormone lowers blood Ca^{2+}. As a result, the excitability of muscles and nerves increases producing tetany sustained contraction.

43. **(c)** Estrogens are steroid hormones secreted by growing ovarian follicles. During menstrual cycle a negative feedback prevents the over secretion of estrogen. FSH stimulates maturation of Graafian follicles. LH stimulates ovulation and development of corpus luteum.

44. **(a)** Deficiency of thyroxin causes cretinism in infants and myxoedema in adults. Excess of adrenaline does not cause any disorder. Acromegaly is caused by over secretion of STH or somatotrophic hormone (Growth hormone).

45. **(d)** Melatonin is a hormone that regulates the 24 hour rhythm of our body such as the sleep-wake cycle *etc.* Melatonin is produced by pineal gland in the brain, which is responsible for operation and regulation of the biological clock in mammals.

46. **(a)** Endocrine glands are ductless glands which pour their secretion directly into blood. Their secretion is called hormone. The endocrine glands which secrete only hormones are called holocrine glands e.g. thyroid, parathyroid, etc. The glands which secrete hormone and perform other functions also are called heterocrine glands e.g. pancreas, testes, ovaries, etc, exocrine glands are glands with ducts e.g. salivary glands.

47. **(b)** If islets of Langerhans stops functioning insulin & glucagon will be in short supply. So, the blood glucose level would increase.

48. **(a)** ADH will enhance retention of water in Henle's loop.

49. **(a)** Thyroxine also enhances growth by increasing metabolism. So, GH works properly with thyroxine. Thyroxine is secreted by thyroid gland, GH by pituitary.

51. **(b)** There are four kinds of cells in Islets of Langerhans :
(i) α-cells produce glucagon
(ii) β-cells produce insulin
(iii) Delta cells produce somatostatin
(iv) F-cell/PP cells or pancreatic polypeptide cells produce pancreatic polypeptide.

52. **(a)** Corticoids are hormones secreted by adrenal cortex. It includes mineralocorticoid, gluco-corticoid and gonadocorticoid. Addison's disease is caused by the deficiency of minerals corticoid and glucocorticoid. Symptom of this disease are low blood sugar, low plasma Na^+, high plasma K^+, increased urinary Na^+, nausea, vomitting diarrhoea and a bronze – like pigmentation of skin.

53. **(c)** Due to hyposecretion of thyroxine cretinism is caused in infants.

54. **(a)** Prolactin hormone, growth hormone, follicle stimulating hormone & LH, TSH and adenocorticotropic hormones are secreted by adenohypo-physis. Neurohypophysis secretes oxytocin & vasopressin.

55. **(d)** FSH is the gametokinetic factor for growth of ovarian follicle in female and spermatogenesis in male. FSH activates Leyding's cells of testis for secretion of androgen in male. In females, LH causes rupture of ovarian follicle.

56. **(a)** Follicle stimulating hormone (FSH) is a steroid hormone or sex hormone. It is concerned with gamete formation or gametogenesis.

57. **(d)** Intercellular communication in multicellular organism is through nervous and endocrine system.

58. **(a)** ADH or Antidiuretic hormone or Vasopressin is secreted by the posterior lobe of pituitary. ADH increases water reabsorption in the collecting tubule of kidney. ADH also stimulates contraction of involuntary muscles in the wall of intestine, gall bladder, urinary bladder and blood vessels.

59. **(c)** Calcitonin hormone is secreted by the parafollicular cells of thyroid gland. It is hypocalcaemic and hypophosphatemic i.e., tends to decrease the plasma levels of calcium and phosphate ions. It does so by preventing the mobilisation from bones. It is antagonist to parathormone and its deficiency leads to osteoporosis (light and porous bones) due to unopposed mobilisation from bones by parathormone.

60. **(b)** The Atrial natriuretic hormone is produced by heart.
The Atrial natriuretic horomone is produced by heart which helps in regulating the sodium and water balance of the body.

23. REPRODUCTION IN ORGANISMS

ANSWER KEY															
1	(a)	9	(b)	17	(d)	25	(b)	33	(d)	41	(a)	49	(d)	57	(c)
2	(b)	10	(c)	18	(b)	26	(b)	34	(c)	42	(d)	50	(b)	58	(b)
3	(b)	11	(b)	19	(c)	27	(a)	35	(d)	43	(a)	51	(c)	59	(a)
4	(b)	12	(d)	20	(c)	28	(b)	36	(b)	44	(b)	52	(c)	60	(d)
5	(a)	13	(a)	21	(b)	29	(c)	37	(b)	45	(c)	53	(d)		
6	(b)	14	(c)	22	(c)	30	(d)	38	(d)	46	(a)	54	(d)		
7	(b)	15	(d)	23	(a)	31	(b)	39	(d)	47	(b)	55	(c)		
8	(d)	16	(d)	24	(c)	32	(d)	40	(b)	48	(b)	56	(a)		

Hints & Solutions

2. **(b)** Most common forms of asexual reproduction in animals are fission and budding.
12. **(d)** Development of an egg (ovum) into complete individual without fertilization by a sperm is known as parthenogenesis.
13. **(a)** Cells become variable in shape, size & getting their specialization for the formation of particular tissue or organ in future foetus. They place themselves at some specific regions in embryo for further organogeny.
14. **(c)** The amount of yolk and its distribution determines the type and depth of cleavage. Cleavage is effected due to yolk resulting in different types of segmentation or cleavage.
15. **(d)** The axillary buds of the potato tuber are called "eyes" in common language. They are found at the nodes of the stem tuber.
19. **(c)** Sowing is related with sexual reproduction.
20. **(c)** Foliar (on the leaf) adventitious buds are formed at place other than nodes. e.g., *Bryophyllum*.
21. **(b)** Stem cutting of atleast one year old stem are employed in case of rose for propagation.
31. **(b)** Both sexual and asexual type of reproduction are found in *Hydra*.
35. **(d)** Acrosome contains an enzyme hyaluronidase which breaks the covering of the ovum.
39. **(d)** Vegetative reproduction and apomixis both are asexual methods of reproduction, which give the progeny genetically similar to parents.
40. **(b)** Cleavage involves repeated cell division without increase in size. But cells size decrease.
41. **(a)** Morphallaxis is a mechanism of regeneration involving reorganization of body cells. In epimorphosis new cells proliferate from the surface of the wound to form the missing structure. In accretionary growth some specialized cells retain the ability to divide and produce new cells to replace the worn-out.
42. **(d)** Embryo growth graph is 'J' shaped.
43. **(a)** In *Rhizopus*, sexual reproduction takes place between two gametangia. Sexual reproduction takes place by conjugation between two multinucleate but single celled gametangia. The gametes are isogamous and non-motile.
44. **(b)** In animal cell only the cell membrane is its outer covering. Cell wall is absent.
45. **(c)** A root is adventitious when it is formed from plumule. Root growing in an unusual location e.g. from a stem. Root is usually underground organ that lacks buds or leaves or nodes; absorbs water and mineral salts; usually it anchors the plant to the ground.
46. **(a)** A modification of bud that can bear leaves, flowers , fruits and branches is thorn. Thorn, a sharp structure in plants.
53. **(d)** Syngamy is fusion of one of the two male gametes with egg to produce diploid zygote (oospore).
54. **(d)** Because egg is the part of embryo sac.
55. **(c)** In *Cocos nucifera* (coconut) milky endosperm is found in which many nuclei, vitamins and growth hormone e.g., cytokinins, auxin and induced cytokinin is found.
56. **(a)** Leaves of a number of plants develop or possess adventitious buds for vegetative propagation *e.g.*, *Bryophyllum*, *Kalanchoe*, *Adiantum caudatum*, etc.
58. **(b)** Asexual reproduction takes place in many ways:
    (i) binary fission, *e.g., Amoeba*.
    (ii) spore formation - by motile zoospores, *e.g., Ulothrix, Chlamydomonas* or by non motile conidia, *e.g., Penicillium.*
    (iii) budding - by external budding, *e.g., Hydra* or by gemmules (internal buds), *e.g.*, Sponges.
59. **(a)** Onion - Bulb - Undeground stem , Ginger - Rhizome, *Chlamydomonas* - Zoospore, Yeast - ascospores
60. **(d)** Banana is vegetatively propagated by modified stem (Rhizomes).

24. SEXUAL REPRODUCTION IN FLOWERING PLANTS

ANSWER KEY

1	(b)	11	(c)	21	(a)	31	(c)	41	(d)	51	(d)
2	(a)	12	(b)	22	(a)	32	(b)	42	(a)	52	(c)
3	(b)	13	(b)	23	(b)	33	(c)	43	(c)	53	(c)
4	(d)	14	(a)	24	(b)	34	(c)	44	(d)	54	(d)
5	(a)	15	(d)	25	(c)	35	(b)	45	(b)	55	(a)
6	(a)	16	(d)	26	(a)	36	(b)	46	(c)	56	(d)
7	(c)	17	(d)	27	(a)	37	(c)	47	(c)	57	(c)
8	(b)	18	(c)	28	(d)	38	(d)	48	(d)	58	(c)
9	(c)	19	(d)	29	(a)	39	(d)	49	(d)	59	(c)
10	(a)	20	(c)	30	(a)	40	(b)	50	(a)	60	(c)

Hints & Solutions

5. **(a)** Perisperm is persistent nucellus in seed.

9. **(c)** The main embryo developed from the structure formed as a result of syngamy.

11. **(c)** Double fertilization means, a male gamete fused with egg and second male gamete fused with secondary nucleus.

15. **(d)** In albuminous seed, the food is stored in endosperm.

18. **(c)** Microspore is haploid, uninucleate, minute spore produced in large numbers as a result of meiosis in microspore mother cells inside the microsporangia. These are the first cell of gametophytic generation in angiosperms.

19. **(d)** Sperm or male gamete fuses with egg or female gamete because of the attraction of their protoplast. Some chemicals are secreted by ovule or egg wall to attract the male gamete or sperm.

20. **(c)** Pollen grains are found only in gymnosperms and angiosperms.

21. **(a)** Double fertilization is a characteristic feature of angiosperms in which one male gamete fuses with egg (called syngamy) and other male gamete fuses with 2 polar nuclei (called triple fusion) to form endosperm.

22. **(a)** Polyembryony refers to the formation of more than one embryo within a seed of a flowering plant. e.g. *Citrus*.

23. **(b)** In angiosperms, one male gamete fuses with the egg resulting in the formation of a diploid zygote. The other male gamete fuses with the two polar nuclei forming the primary endosperm nucleus. This type of fertilization is called double fertilization.

24. **(b)** Pollination by wind is called anemophily. Wind pollinated flowers are small in size, producing large number of dry pollen grains. Pollens are small, dry and light in weight. Grasses are anemophilous plants.

25. **(c)** Aleurone layer helps in utilization of food. Aleurone is a protein found in protein granules of maturing seeds and tubers. The term is also used for the outermost cell layer of the seed coat, the aleurone layer.

26. **(a)** In the legume seed, food is stored in cotyledons. Legume vegetables have large, relatively fragile seed. Food reserves are stored in the pair of cotyledons, sometimes called the seed leaves.

27. **(a)** Flower is complete when it has calyx, corolla, androecium and gynoecium. A flower is regarded as a modified stem with shortened internodes and bearing, at its nodes, structures that may be highly modified leaves. In essence, a flower structure forms on a modified shoot or axis with an apical meristem that does not grow continuously (growth is determinate).

28. **(d)** Cleistogamous flower is found in pea, where the flower remains closed unless pollination is accomplished. Thus, such flowers favour self pollination but artificial cross pollination for the experiments can be easily carried out in such flowers. This was one of the reason why Mendel selected pea plant for the studies of Genetics.

29. **(a)**

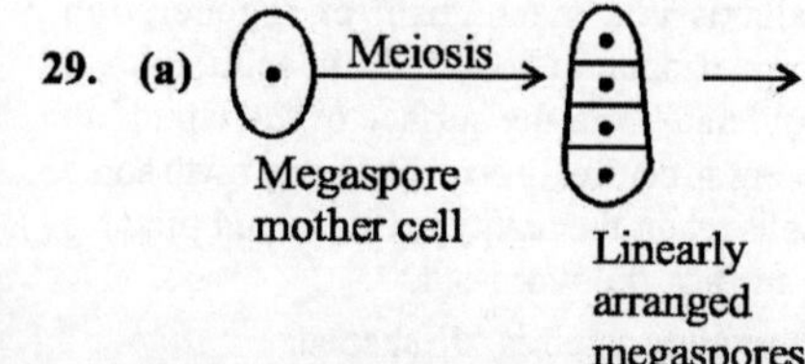

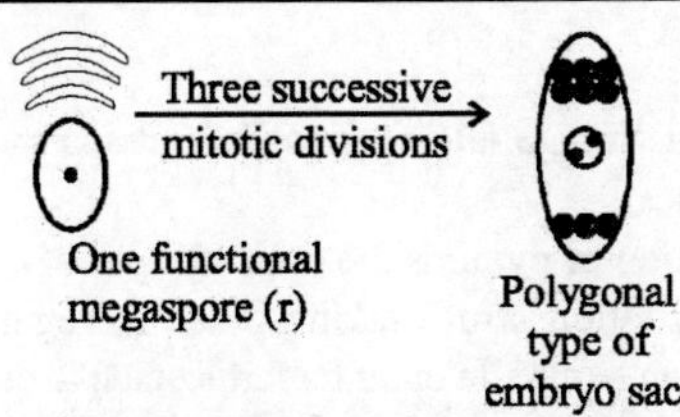

Polygonum is most common type of embryo eac containing 8 nuclei but 7 cells – 3 micropylar, 3 chalazal and one central.

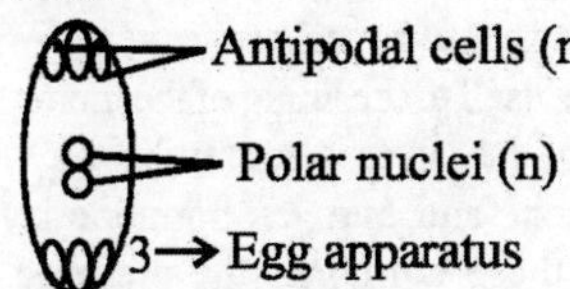

Mature embryo sac

30. **(a)** Emasculation is defined as removal of anthers from the flower before maturity to avoid self pollination and to favour cross pollination. Emasculation is an essential step of the process of hybridization.

31. **(c)** Typical angiospermic embryo sac is seven celled 8 nucleate structure, as shown in the figure.

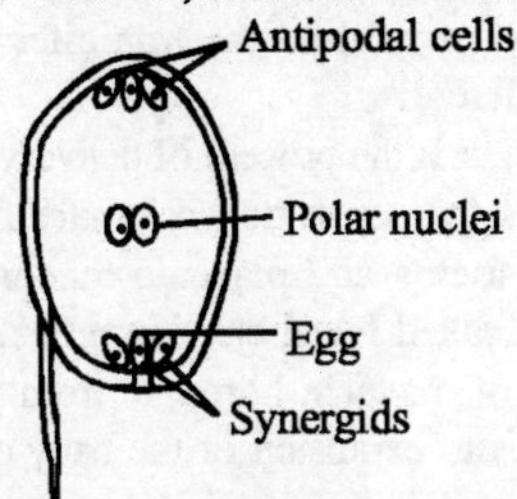

Egg is always with two synergids.

32. **(b)** Microsporogenesis is the development of pollen grains. In this, primary sporogenous cells found in the anther divide to give microspore mother cells. The meiotic division of microspore mother cell gives a tetrad of pollen grains (4-pollen grains).

40. **(b)** Generative cell divides into two male gametes, if it has not divided already.

41. **(d)** All three types can be route of the pollen tube enters in the ovule as chalazogamy mesogamy and porogamy.

45. **(b)** *Vallisneria* is submerged, dioecious, fresh water plant hence pollinated in water.

46. **(c)** In cleistogamy, bisexual flowers never open therefore the pollen grains may only pollinate the stigma of the same flower e.g., *Commelina bengalenis* (day flower).

48. **(d)** When the funicle lies parallel to the body of the ovule and micropyle, the body of the ovule has rotated by 180°, the ovule is called anatropous.

50. **(a)** Egg apparatus of angiosperm is 3 celled. It has one egg cell and two synergids on either side of egg cell. 3 antipodals are present on chalazal side. The secondary nucleus is binucleate, is present in the centre of embryo sac.

54. **(d)** Pollen grains are not part of ovule.

55. **(a)** Meiosis occurs in pollen mother cells of anther.

59. **(c)** Exine is made up of sporopollenin (derived from carotenoids).

60. **(c)** Pollen grain is partially developed male gametophyte because the rest of the development is completed on stigma when pollen grains start to germinate and produces pollen tube having two male nuclei.

25. HUMAN REPRODUCTION

ANSWER KEY

1	(b)	11	(b)	21	(d)	31	(b)	41	(b)	51	(c)
2	(c)	12	(a)	22	(a)	32	(d)	42	(a)	52	(c)
3	(d)	13	(b)	23	(d)	33	(c)	43	(b)	53	(c)
4	(b)	14	(b)	24	(c)	34	(b)	44	(b)	54	(d)
5	(c)	15	(d)	25	(b)	35	(b)	45	(d)	55	(a)
6	(a)	16	(c)	26	(b)	36	(c)	46	(b)	56	(d)
7	(b)	17	(a)	27	(b)	37	(a)	47	(b)	57	(c)
8	(b)	18	(a)	28	(a)	38	(c)	48	(c)	58	(b)
9	(a)	19	(c)	29	(a)	39	(a)	49	(c)	59	(c)
10	(a)	20	(a)	30	(b)	40	(d)	50	(b)	60	(c)

Hints & Solutions

1. (b) Human female has XX chromosomes. When eggs formed each egg carries 22 autosomes & one X chromosome. In human male each sperm carries 22 autosomes & either X or Y-chromosomes.
2. (c) Fertilization is union of haploid sperm with haploid ovum to form diploid zygote.
3. (d) A mature sperm has a nucleus, an acrosome, a pair of centrioles and a tail.
4. (b) In humans fertilization takes place in fallopian duct and the embryo develops in uterus.
8. (b) Vas deferens is larger duct arises from cauda epididymis and reach upto seminal vesicles.
9. (a) Seminal fluid has a pH about 7.4 and contains fructose, citrate, ascorbic acid, prostaglandins and various enzymes. The fructose is a source of energy for the spermatazoa.
11. (b) Graafian follicle develops under influence of FSH of anterior pituitary. Its follicular cells secrete estrogen. Rising level of estrogen decreases production of FSH and stimulates secretion of LH. The two cause the mature Graafian follicle to rise to the ovarian surface and burst open releasing ovum.
15. (d) Oogenesis is process of formation develo-pment and maturation of haploid ova from diploid germinal cells.
16. (c) Leydig's cells are present in testes and secrete male hormone testosterone.
17. (a) Egg is liberated from the ovary (ovulation) at secondary oocyte stage when the meiosis II is arrested in secondary oocyte. It will be induced by the sperm.
18. (a) Sperm has (i) head in which acrosome & nucleus are present
    (ii) middle piece in which it contains many mitochondria
    (iii) A long & highly motile tail.
    (A short neck is also present between middle piece & head which has centrioles and may sometimes regarded as a part of middle piece).
19. (c) Luteinizing hormone causes ovulation. The remaining cells of ovarian follicle are stimulated by LH to develop corpus luteum.
20. (a) After first meiotic division primary spermatocyte (diploid) gives rise to two secondary spermatocytes (haploid).
21. (d) Embryo growth graph is 'J' shaped.
22. (a) The formation of sperms in mature human testes completed in three important phases in gametogenesis-multiplication, growth and maturation phase. During these phases primarily germ cells change into spermatogonia then spermatocytes during growth phase which change into spermatids and spermatids into sperms.
23. (d) Fertilization of ovum is the most likely root cause why menstruation is not taking place in regularly cycling human female. In order for fertilization to occur, one of these sperm must attach itself to the ovum and penetrate its outer surface. The fertilized ovum will then continue travelling down the fallopian tube, taking several days until it reaches the uterus. When it arrives, it attaches itself to the lining of the uterus and continues to grow. If the ovum is not fertilized it will die and the progesterone and estrogen hormone levels will drop. Without these hormones the prepared uterus lining breaks up and is released through the vagina. This is known as a 'period'.
24. (c) Sertoli cells are found in the walls of seminiferous tubules of the testes. They anchor and provide nutrition to the developing germ cells especially the spermatids.
25. (b) Vasa efferentia are ductules leading from rete testis to vas deferens. The rete testis is an anastomosing network of tubules located in the hilum of the testicles that carries sperm from the seminiferous tubules to the vasa efferentia.
26. (b) Parturition is the process of delivery of the foetus. The signals for parturition originate from the fully developed foetus and placenta that induce uterine contraction called foetal ejection reflex. This triggers the release of oxytocin hormone from pituitary. This in turn leads to expulsion of the baby out of uterus.
27. (b) The part of fallopian tube closest to the ovary is infundibulum. Infundibulum possess finger-like projections called fimbriae that help in collection of ovum after ovulation. It leads to wider part of oviduct called ampulla. The last part of oviduct is isthmus that has a narrow lumen and joins the uterus.
35. (b) Seminal vesicle secretes semen which is alkaline and contain fructose. It is present at the junction of prostate & vas deferens.
36. (c) After fertilization, embryonic development starts. Zygote undergoes rapid mitotic divisions to form blastula. Blastula forms gastrula and the process is called gastrulation. During gastrulation three germ layers are formed ectoderm, mesoderm and endoderm. But endoderm is formed first. Endoderm differentiates into primitive gut.
37. (a) Clitoris in female is homologous to penis of male.
40. (d) Each seminiferous tubules is lined by a germinal epithelium formed of two types of cells germ or spermatogenic cells and sertoli or nurse cells. Sertoli cells are a few pyramidal shaped cells wich provide nutrition to developing sperms.
43. (b) The germinal epithelium lining of the seminiferous tubules is made of two kinds of cells. A few larger cell

columnar supporting cells are called sertoli cells or sustentacular cells or nurse cells.

44. **(b)** Progesterone is a hormone produced primarily by the corpus luteum of the ovary but also by the placenta. The action of the hormone is to cause endometrial development preparatory for the reception and nutrition of the embryo. It suppresses ovulation and the secretion of LH by pituitary. During pregnancy, progesterone production by corpus luteum continues through, till near term. If pregnancy has not occurred, the output of both estrogen and progesterone suddenly fall on or about the twenty eight days of the menstrual cycle, menstrual flow starts and the uterine endometrium starts sloughing.

45. **(d)** Cortex contains many small and large spherical or oval ovarian follicles. A fully matured follicle is called graafian follicle.

53. **(c)** The cavity of graafian follicle is antrum or follicular cavity

57. **(c)** Vagina serves as birth canal during parturition.

58. **(b)** Estrogens are produced by the theca interna cells of Graafian follicles. Estrogens regulate growth and development of female accessory reproductive organs.

26. REPRODUCTIVE HEALTH

ANSWER KEY

1	(c)	8	(d)	15	(d)	22	(b)	29	(b)	36	(b)	43	(d)	50	(b)	57	(d)
2	(d)	9	(c)	16	(a)	23	(d)	30	(a)	37	(c)	44	(b)	51	(a)	58	(a)
3	(d)	10	(d)	17	(a)	24	(c)	31	(a)	38	(b)	45	(c)	52	(a)	59	(d)
4	(a)	11	(c)	18	(a)	25	(a)	32	(a)	39	(a)	46	(b)	53	(a)	60	(c)
5	(a)	12	(b)	19	(c)	26	(d)	33	(d)	40	(d)	47	(c)	54	(d)		
6	(b)	13	(d)	20	(d)	27	(c)	34	(d)	41	(a)	48	(d)	55	(a)		
7	(b)	14	(b)	21	(d)	28	(d)	35	(d)	42	(b)	49	(d)	56	(a)		

Hints & Solutions

1. **(c)** The main cause of population explosion in the world is excellent health care as a better medical facilities control of insect vectors of fatal diseases and epidemics and proper care of new born children and their mothers.

4. **(a)** Number of offsprings produced per unit of population by reproduction is called Natality or birth rate.

6. **(b)** India has high birth rate, but average life span is lower (62 year) as compared to 75 – 80 in advanced countries.

9. **(c)** Conservation of wild life conserves the ecosystem in its natural form. Future of man depends on ecological balance.

14. **(b)** The copper - T is also known as intra-uterine device. It is a loop which is placed in the uterus and prevents fertilization and implantation.

15. **(d)** The method of sterilization is tubectomy (female) or vasectomy (male). While loop, IUD and diaphragm are temporary way to keep time gap between two children.

16. **(a)** Sterilization provides a permanent and sure birth control. It is called vasectomy in man and tubectomy in woman.

18. **(a)** A contraceptive pill contains progesterone and estrogen hormone which prevents ovulation and helps in birth control.

19. **(c)** Surgical removal of uterus is called hysterectomy. Vasectomy is surgery of seminal tube or tubectomy is surgery of fallopian tube. Anatomy is the internal structure.

23. **(d)** Amniocentesis involves the analysis of amniotic fluid is withdrawn. It contains foetal skin cells and a number of proteins especially enzymes. The cells can be cultured *in vitro* for further examination.

24. **(c)** Amniocentesis is a technique used for determining the sex of the foetus and a process to determine any hereditary disease in the embryo.

25. **(a)** Intra utrine devices (Copper MIT) inhibit fertilization and help in birth control.

26. **(d)** In amniocentesis the fluid is taken from amniotic fluid, which is immediate surrounding of foetus.

27. **(c)** Cu ions released by copper releasing intra Uterine Devices suppress sperm motility.
Intra-uterine devices are inserted by doctors in the uterus through vagina. They are available as the non-medicated IUDs, copper releasing IUDs and hormone releasing IUDs.

28. **(d)** Intra uterine device (IUD) is a method of contraception in India.

29. **(b)** Medical termination of pregnancy is considered safe up to twelve weeks of pregnancy.

30. **(a)** Gamete intrafallopian transfer (GIFT) is recommended for those females who cannot produce an ovum. In this process, the eggs of the donor woman

are removed and in a form of mixture with sperm transferred into fallopian tube of another woman who cannot produce ovum, but can provide suitable environment for fertilization. Thus in GIFT, site of fertilization is fallopian tube, not laboratory.

31. **(a)** A. The pill — Prevents ovulation
B. Condom — Prevents sperm reaching cervix
C. Vasectomy — Semen contains no sperms
D. Copper-T — Prevent implantation.

34. **(d)** The mechanical measure of population control includes condom (Nirodh), Diaphragm and cervical cap and Intra uterine devices (IUDs).

37. **(c)** Surrogate mother is woman who bears a baby for another couple.

56. **(a)** Medical Termination of Pregnancy (MTP) was legalised in 1971 Government of India with some strict conditions to avoid its misuse.

57. **(d)** Test tube baby programme employs zygote intrafallopion transfer (ZIFT) technique. In this technique fusion of ovum and sperm is done outside the body of woman to form zygote which is allowed to divide forming 8 blastomeres, then it is transfered the fallopion tube of the woman.

27. HEREDITY AND VARIATION

ANSWER KEY

1	(c)	12	(c)	23	(d)	34	(b)	45	(d)	56	(d)
2	(a)	13	(c)	24	(c)	35	(a)	46	(a)	57	(d)
3	(a)	14	(a)	25	(d)	36	(d)	47	(a)	58	(c)
4	(b)	15	(d)	26	(b)	37	(b)	48	(c)	59	(b)
5	(a)	16	(d)	27	(a)	38	(c)	49	(b)	60	(b)
6	(b)	17	(c)	28	(d)	39	(d)	50	(d)		
7	(a)	18	(c)	29	(a)	40	(b)	51	(a)		
8	(c)	19	(a)	30	(a)	41	(b)	52	(c)		
9	(d)	20	(d)	31	(b)	42	(c)	53	(b)		
10	(d)	21	(a)	32	(d)	43	(a)	54	(c)		
11	(d)	22	(a)	33	(d)	44	(a)	55	(d)		

Hints & Solutions

1. **(c)** Genetics is the study of principles and mechanism of heredity and variations.

2. **(a)** Heredity is the transmission of genetic characters from parents to the offspring.

3. **(a)** Mendel factor now known as gene (Johannsen, 1905) is a unit of inheritance.

7. **(a)** Allele is an alternative form of a gene for example, T and t alleles represent tall and short stem height of pea plant.

10. **(d)** Mutation induced by a mutagen are called as induced mutations.

11. **(d)** Dihybrid condition involves simultaneous inheritance of two pairs of mendelian factors or genes.

12. **(c)** According to the law of independent assortment, alleles of different characters located in different pairs of homologous chromosomes are independent of one another in their separation or segregation during gamete formation.

13. **(c)** Yellow seed colour is dominant over green seed colour.

16. **(d)** Homozygous individuals have identical mendelian factors or genes for a character (TT or tt). They are always pure for a particular trait (character).

17. **(c)** A dominant gene would lead to the expression of its phenotype irrespective of the fact whether its allelic gene is dominant or recessive.

18. **(c)** Duplication and Translocation is chromosomal abberrations. Duplication and translocation both are mutation.

19. **(a)** Mutation is generally recessive. Mutation is a process which produces a heritable alteration in the genotype. Mutations are classified into two main types : (a) Chromosomal mutation; (b) Point mutation.

20. **(d)** Haemophilia is a X-linked disease. It is also known as Royal's disease. Haemophilia is due to 8th factor of blood clotting (AHG) Anti Haemophilic Globin factor which is also known as AHF (Anti Haemophilic Factor).

32. **(d)** Consideration of one character with two contrasting conditions at one time contributed to Mendel's success in proposing the laws of inheritance of traits. His choice of experimental plant as pea and use of statistical methods also contributed to his success.

33. (d) Haemophilia is a sex linked recessive trait carried by X chromosome (also known as bleeder's disease).

34. (b) Down's syndrome is due to trisomy of 21st chromosome and is an autosomal abnormality.

35. (a) Dominant allele expresses itself in the homozygous as well as heterozygous condition.

36. (d) Holandric genes are present on the differential region of the Y-chromosome. The differential regions carry completely sex linked genes and they do not undergo crossing over.

37. (b) A cross of F_1 hybrid with its recessive homozygous parent is called the test cross. It is done to determine the genotype of a given plant. If the given plant has homozygous dominant traits then on test cross it gives all dominant trait plants but if it is heterozygous dominant than it gives dominant and recessive phenotypes in 1 : 1 ratio.

38. (c) Phenotypex is the appearance shows by organism while genotype is the gene complement that it has from its ancestors. These genes only show their effect in phenotype but environment also plays an important role in this hence phenotype is a result of genotype and environmental interaction.

39. (d) Down's syndrome is an autosomal trait. It is due to trisomy of 21st pair of autosomal chromosome. Trisomy is due to non disjunction of chromosome pair. Down's syndrome is also known as mangolism characteristics of disease are flat face, slanting eye, short neck, permanent opened mouth, protuding tongue etc.

40. (b) Colour blindness is genetically similar to haemophilia. Colour blindness or haemophilia is always inherited by female and these are due to recessive genes of female. Colour blindness also known as daltonism and haemophilia is known as royal's disease.

41. (b) Colourblindness is a sex linked character. Colourblindness is caused by a single recessive gene in man. Sex linked character usually found in man. The character which are linked together and associated with sex. Chromosome in inheritance are known as sex linked inheritance or sex linked character.

42. (c) In human female ovum, number of autosome are 22 pairs.

43. (a) An addition or loss of chromosomes is more or less than an entire set of chromosomes, is known as aneuploidy.

44. (a) In sickle cell anaemia, the β peptide chain of haemoglobin is having valine at position 6 (instead of glutamic acid). The affected RBCs, under conditions of oxygen stress, become sickle shaped. This results in impaired transport of oxygen by blood.

45. (d) Symptoms of hormone imbalance are caused primarily by the incorrect relationship between progesterone and estrogen levels in the body. The two female hormones, estrogen and progesterone, exist in a delicate balance. Variations in that balance can have a dramatic effect on your health, resulting in symptoms of hormone imbalance. The amounts of these hormones that the woman's body produces from month to month can vary, depending on factors such as stress, nutrition, exercise and most importantly — ovulation or the lack of ovulation.

47. (a) Mendel died before his work could be appreciated by the rest of the scientific community. In 1900 three botanists, Correns of Germany, De Vries of the Netherlands and Tschermak of Austria rediscovered his work after reaching similar conclusions independently.

50. (d) Mendel in his experiment considered total 7 characters (3 characters of seed i.e., seed shape, seed colour, cotyledon colour, 2 characters of pod i.e., plant height and position of pods on the stem).

51. (a) RrYy is a dihybrid, so four type of gametes are formed as RY, Ry, rY, ry.

55. (d) Crossing over is a process by which chromatids of homologous exchange portions of genetic material, usually during tetrad formation of meiosis. Crossing over occurs at pachytene substage of prophase I of meiosis but visible at diplotene substage of prophase I.

28. MOLECULAR BASIS OF NHERITANCE

ANSWER KEY

1	(c)	12	(a)	23	(d)	34	(a)	45	(b)	56	(a)
2	(b)	13	(a)	24	(b)	35	(c)	46	(b)	57	(a)
3	(b)	14	(a)	25	(a)	36	(a)	47	(a)	58	(b)
4	(d)	15	(c)	26	(d)	37	(c)	48	(a)	59	(d)
5	(a)	16	(b)	27	(c)	38	(c)	49	(c)	60	(a)
6	(c)	17	(b)	28	(b)	39	(d)	50	(b)		
7	(a)	18	(a)	29	(b)	40	(d)	51	(d)		
8	(a)	19	(a)	30	(c)	41	(d)	52	(b)		
9	(c)	20	(c)	31	(a)	42	(a)	53	(a)		
10	(b)	21	(c)	32	(d)	43	(a)	54	(b)		
11	(a)	22	(d)	33	(b)	44	(b)	55	(b)		

1. (c) Nucleotide is a unit of DNA, which is formed of nitrogenous bases (purines & pyrimidines), sugar (pentose) & phosphate.
2. (b) Protein synthesis occurs over ribosomes which are also referred to as protein factories.
3. (b) Transformation is change in genetic material of an organism by obtaining genes from outside.
4. (d) Purine is an organic nitrogenous base sparingly soluble in water, that gives rise to a group of biologically important derivatives, notably adenine and guanine, which occur in nucleotides and nucleic acids (DNA and RNA).
6. (c) Ligase helps in inserting few genes in DNA.
7. (a) DNA occur mainly in nucleus, forming major chemical proportion of chromosomes. Some amount of DNA is also present in cytoplasm (mitochondria and plastids).
11. (a) In a nucleotide, purine or pyrimidine nitrogenous base is joined by deoxyribose pentose sugar (D), which is further linked with phosphate (P) group to form nucleotides.
13. (a) Operon model was given by Jacob and Monad (1961) for regulation of protein synthesis in prokaryotes.
15. (c) Histones are low molecular weight proteins with high proportion of positively charged basic amino acids arginine and lysine.
16. (b) In semiconservative replication of DNA in each replica one half is the old strand and the other half is a new strand synthesized over it.
17. (b) Central dogma of protein synthesis explains a unidirectional or one way flow of information from DNA to RNA (transcription) and from RNA to polypeptide (translation).
18. (a) After one generation in ^{14}N medium DNA of bacteria settled at a level intermediate between heavy and light bands due to semi conservative DNA replication.
19. (a) In 1953 Wilkins obtained very fine X-ray crystallographic pictures of DNA from which Watson and Crick developed the double helix model of DNA.
20. (c) Unwindase or helicase takes part in separation of two DNA strands. In prokaryotes, helicase or unwindase is assisted by gyrase in this function.
21. (c) All amino acids are specified by more than one codon (except tryptophan and methionine). Hence they are degenerate.
22. (d) Hershey and Chase (1952) worked on *Escherichia coli* and conclusively proved that DNA is the genetic material.
23. (d) Protein synthesis stops when a nonsense codon (UAA, UGA or UAG) reaches the amino acyl or A site.
24. (b) During transcription, from the DNA template complementary mRNA is formed and thymine is replaced by uracil.
25. (a) *Lac* operon is an inducible operon system which regulates genetic material. The genetic material remains switched off normally but becomes operational in the presence of inducer lactose.
26. (d) The phenomenon of making DNA over RNA genome through enzyme reverse transcriptase is called reverse transcription.
27. (c) The region of a gene, which becomes part of m-RNA and code for different regions of the proteins are called exons.
28. (b) The inducer is an environmental agent, which triggers transcription from an operon. The inducer or effector molecule are small sized which can bind to a regular protein.
29. (b) The *lac* operon consists of a promotor gene, an operator gene and structural genes. There are three structural genes designated as z, y and a, which code for the enzymes β-galactosidase, *lac* permease and transacetylase respectively. It is an example of inducible operon.
30. (c) Replication of DNA is discontinuous over the lagging strand over which only small stretches of DNA are built due to opposite running of DNA template. The small stretches of DNA are called okazaki fragments.
31. (a) DNA fingerprinting is a technique to identify a person on the basis of his DNA which is specific for every individual. This technique for first time was developed by Alec Jeffreys and his colleagues at Leicester University in U.K. DNA of an individual carries specific sequences of nucleotides which are repeated many times throughout the length of DNA. This reveals polymorphism in DNA. These are inherited. Each individual inherits these repeats from his/her parents which are used as genetic markers in personal identity test.
32. (d) Termination codons are the stop signals which when encountered cause termination of polypeptide synthesis. They are UAA, UAG and UGA. Codons are determined by the sequence of bases on the DNA. Thymine is not present on RNA.
33. (b) A mutation bringing about early stoppage of polypeptide formation is called nonsense mutation. Frameshift mutation is a type of gene mutation where the reading of codons is changed due to insertion or deletion of nucleotides.
34. (a) Multiple alleles are multiple alternatives or alleles of the same gene which occur in the population of same species . Polygene is a gene, single dominant allele of which express only a unit of trait . Polygenes are the genes controlling quantitative inheritances. Oncogenes are the cancer causing genes.
35. (c) DNA is complexed with histones in eukaryotes.
36. (a) The enzyme polymerase can synthesize the bases only in $5' \rightarrow 3'$ direction.

37. **(c)** In transition substitution, a purine is replaced by another purine base (A with G or vice versa). In transversion substitution, a purine is replaced by a pyrimidine base or vice versa. Frameshift mutation is a type of mutation where the reading of codons is changed due to insertion or deletion of nucleotides. Transition is the formation of RNA over the template of DNA.

38. **(c)** Okazaki fragments in DNA are linked up by the enzyme DNA ligase. Replication always occur in 5′ - 3′ direction. Okazaki fragments synthesized on 3′ - 5′ DNA template, join to form lagging strand which grows in 3′ - 5′ direction.

39. **(d)** In the DNA molecule, there are two strands which run antiparallel one is 5′ - 3′ direction and other in 3′ -5′ direction, the two chains are held together by hydrogen bonds between their bases. Adenine (A), a purine of one chain his exactly opposite thymine (T), a pyrimidine of the other chain. Similarly, cytosine (C), a pyrimidine lies opposite guanine (G), a purine. This allows a sort of lock & key arrangment between large sized purine & small sized pyrimidine. It is strengthened by the appearance of hydrogen bonds between the two.

40. **(d)** The genetic code consists of 64 triplets of nucleotides. These triplets are called codons. With three exceptions, each codon encodes for one of the 20 amino acids used in the synthesis of proteins that produces some redundancy in the code. Most of the amino acids being encoded by more than one codon. The genetic code can be expressed as either RNA codons or DNA codons.

41. **(d)** Splicing is the removal of introns and joining the exons in a defined order in a transcription unit. In molecular biology, splicing is a modification of RNA after transcription, in which introns are removed and exons are joined.

43. **(a)** DNA probe is used for DNA finger printing.

44. **(b)** Tryptophan (Trp) operon is repressible system.

53. **(a)** Amino acid binds with 3′ end of *m*-RNA

54. **(b)** A tRNA molecule as proposed by R.W. Holley in 1965, has the resemblance of a clover leaf that results from self-folding and base pairing, creating paired stems and unpaired loops.

56. **(a)** Regulator gene produces a repressor that binds to operator gene and stops the working of the latter.

29. EVOLUTION

ANSWER KEY

1	(d)	12	(c)	23	(a)	34	(c)	45	(c)	56	(a)
2	(b)	13	(a)	24	(c)	35	(b)	46	(a)	57	(c)
3	(d)	14	(c)	25	(a)	36	(d)	47	(d)	58	(d)
4	(d)	15	(c)	26	(c)	37	(c)	48	(a)	59	(a)
5	(b)	16	(d)	27	(d)	38	(c)	49	(c)	60	(b)
6	(b)	17	(c)	28	(c)	39	(b)	50	(b)		
7	(a)	18	(b)	29	(a)	40	(b)	51	(b)		
8	(b)	19	(b)	30	(d)	41	(b)	52	(d)		
9	(b)	20	(a)	31	(d)	42	(d)	53	(b)		
10	(c)	21	(a)	32	(d)	43	(a)	54	(d)		
11	(b)	22	(a)	33	(c)	44	(c)	55	(b)		

Hints & Solutions

2. **(b)** Any change in the development of tail during the course of evolution in man is useless.

3. **(d)** The primitive atmosphere of the earth contained : H_2O, H_2, NH_3 & CH_4. The ratio was (2 : 2 : 1) of : NH_3, CH_4 and hydrogen.

4. **(d)** According to Oparin & Haldane life originated in H_2O. The primitive atmosphere being a reducing atmosphere had no free O_2. H_2 and O_2 combined to form H_2O. Torrential rains resulted in rushing dam of water with salts. Thus, oceans were formed. The ocean water with lots of salts was called hot dilute soup or prebiotic soup. The complex molecules helped in the formation of heterotrophs or coacervates which due to mutation led to the formation of photosynthetic plants. Thus, life originated in water.

5. **(b)** 'Origin of Species' book was written by Charles Robert Darwin. During his world journey (1831–1936) as a naturalist in H.M.S. Beagle ship, he visitied many parts

of the world. On the basis of the observations of this journey, he wrote the book thereby describing his theory of natural selection.

6. (b) Lamarck's theory of evolution is also called theory of inheritance of acquired characters.
He explained his theory in four propositions :
(i) Every organism or its organs have tendency to increase in size
(ii) Due to environmental change, the need of new organs or modification in organs arises
(iii) The organs which are in constant use are better developed
(iv) All the characters which are acquired in the life are passed on to the next generation

7. (a) Various elements are present in the body of plants and animals. Some of them constitute the living matter while some are required for the proper metabolic functioning. Of these elements C, H, O and N are the basic elements of life which make the major portion of living matter or protoplasm.

8. (b) Life originated in water instead of air. Living and non-living things both are made up of the same elements. The difference lies in the various combinations between these elements leading to the formation of different molecules. Energy transfer and transformation take place simultaneously in a cell with ATP being the cellular energy currency.

9 (b) The term evolution was given by Herbert Spencer that is "Descent with modifi-cations". Evolution helps us to understand the history of life. It is a process in which something changes into a different and usually more complete or better form over time and in response to environment. This results in descendents becoming different from ancestors.

10. (c) Oxygen gas was not present due to reducing atmosphere , it only came to existence after the evolution of photosynthesis process.

11. (b) Adaptation is the evolutionary process in which population becomes better suited to its habitat. Mimicry is an example of long term adaptation, in which there is a close resemblance of an organism to the surrounding in order to avoid detection. e.g. Leaf insect *Phyllium.* The term mimicry was introduced by Bates (1862).
Homology is phenomenon in which origin of organs are same but functions are different. e.g. hand of human and fore limb of horse.
Analogy - origin is different but functions are same of two different organs.
e.g. wings of birds, wings of insects.

12. (c) The correct geological time scale of earth is
Precambrian → Palaeozoic →
Mesozoic → Coenozoic
A geological time scale is a diagram that details the history of earths geology, noting major events like the formation of the earth, the first life forms and mass extinctions. The first geological time scale was proposed in 1913 by the british geologist Arthur Holmes (1890-1965).
The history of the earth has been subdivided into eras, eras into periods and periods into major divisions.

13. (a) Evolutionary convergence is a type of evolutionary pattern in which unrelated organisms/different species develop common set of characters / adaptive functional structures (analogous structures). They do not share common ancestors, related to one another in only the most distant sense.

14. (c) In 1859 Charles Darwin published his theory of natural selection in the book 'Origin of Species'. The theory is based on five important observations
1. All species have great reproductive potential
2. Most populations are stable in size
3. Natural resources are limited
4. Individuals of a population have variations.
5. Most of the variations are heritable.
Species with variations selected by nature.

15. (c) Genetic drift is also known as Sewall Wright effect. It refers to the random changes in gene frequencies in a population by chance, either due to intensive inbreeding or death of a small section of population by natural calamity. Its effect is more marked in small isolated population.

17. (c) Organs that are similar in fundamental structure but different in functions are "homologous organs", Richard Owen, introduced the term homologous. Pectoral fins of fish and fore limbs of horse similar in structure but different in functions are homologous organs.
Rest of the organs compared in the question are analogous organs.

18. (b) Homologous organs have same basic structure and origin but they differ in their external appearance and functions.

19. (b) Progressive evolution is development of organisms with more elaborate and specialized structures from those having less elaborate features e.g. amphibians from reptiles. Retrogressive or degenerative evolution is development of simpler forms from more complex ones. Such evolution has occurred in case of vestigial organs, parasitic forms, and in reduction of overspecialized structures such as wings in flightless birds. Parallel evolution is formation of similar traits in related groups of organisms independently due to similar requirement e.g. running of two toed deer and one toed horse. Evolution of wings in insects and birds serve as example of convergent evolution.

22. (a) Genetic drift can operate only on smaller populations where its fluctuation can be observed in the proportion of allele distribution in the presence of external disturbances.
Genetic drift is the random change in allele number and frequency in a gene pool.

23. (a) Synthesis of organic molecules from simple molecules is no longer possible because of the oxidizing atmosphere.

24. (c) Chimpanzees and gorillas are our closest relatives among the living primates.

25. (a) Loss of a prehensile tail is associated with the gradual development of errect posture and bipedal gait.

26. (c) Peking man had a cranial capacity of about 1000cc. Heidelberg man is regarded as ancestor of Neanderthal man. Cro-Magnon is the form of modern man living in Europe.

27. (d) The replacement of grey coloured moth by dark coloured melanic species due to industrial smoke is called industrial melanism. Initially on the lichen covered tree trunks, the mutant moths was more conspicuous due to its black colouration and was therefore more susceptible to predation by birds. The large scale burning of coal during industrial revolution resulted in the deposition of sooty particulate matter on tree trunks. Hence the grey coloured moths became more conspicuous to predatory

birds. Thus the population of black moths increased considerably.

28. (c) Migration refers to the movement of individuals into and out of population. Mutation refers to random and sudden heritable variations or changes arising in the genetic constitution.

29. (a) In any population, there is always competition for space and food resources.

30. (d) Amphibians have moved onto terrestrial habitat but they still need to go back to water for reproduction. The presence of gills in tadpole indicate that amphibians have evolved from fishes. Fishes are completely aquatic so the other three options are incorrect.

31. (d) Homologous organs represent the divergent evolution. These are the organs belonging to different organisms of different species may be for dissimilar functions but they all have the same basic structure. On the other hand vestigial organs show that how a species evolved from its ancestors through the ages according to the use and disuse of organs. These are the organs which were sometime functional in humans (or may be any other organisms) but during the course of evolution their utility gradually decline and now they are present as non-functional and vestigial organs.

33. (c) Ontogeny repeats phylogeny.

34. (c) The concept of chemical evolution is based on possible origin of life by combination of chemicals under suitable environmental conditions. The distribution of elements in the cosmos is the result of many processes in the history of the universe.

35. (b) Speciation takes place via reproductive isolation which is the most important consequence of geographical isolation.

36. (d) The most significant trend in evolution of modern man (*Homo sapiens*) from his ancestors is development of brain capacity.

37. (c) Darwin's theory of natural selection was based on enormous rate of reproduction in organisms, struggle for existence and survival of the fittest.

42. (d) *Homo erectus* is biological name of Peking and Java man.

44. (c) The first ancestor of man whose fossils have been discovered was *Australopithecus*.

47. (d) Cranial capacity of Cro-magnon man was 1600 cc.

56. (a) Evolution is a continuous process taking place due to changes in the environment. This also leads to new needs and then acquired characters creating a new species.

57. (c) Darwin had not taken into account variation and heredity. Mutation in DNA leads to variation in species, which is followed by inheritance of the character.

59. (a) The process of evolution will not take place without a mutation & consequently speciation will not take place. Hence, mutation is very important and necessary for the process of evolution.

60. (b) Gene pool is the total gene content of all the individuals in a population. These members share the same gene pool and interbreed.

30. HEALTH AND DISEASE

ANSWER KEY

1	(a)	10	(d)	19	(b)	28	(a)	37	(b)	46	(c)	55	(b)
2	(d)	11	(c)	20	(a)	29	(d)	38	(c)	47	(b)	56	(b)
3	(b)	12	(b)	21	(b)	30	(a)	39	(d)	48	(d)	57	(c)
4	(c)	13	(a)	22	(c)	31	(b)	40	(c)	49	(c)	58	(a)
5	(a)	14	(b)	23	(a)	32	(a)	41	(b)	50	(a)	59	(c)
6	(b)	15	(c)	24	(b)	33	(a)	42	(a)	51	(c)	60	(c)
7	(a)	16	(b)	25	(a)	34	(a)	43	(b)	52	(b)		
8	(c)	17	(c)	26	(b)	35	(b)	44	(a)	53	(b)		
9	(c)	18	(b)	27	(d)	36	(a)	45	(a)	54	(d)		

Hints & Solutions

1. (a) Vaccine contains dead, attenuated form or antigen of a pathogen which can be injected to provide immunity towards that pathogen. Monoclonal antibodies are homogenous immunological reagents of defined specificity, so that these can be utilized for diagnosis and screening with certainty.

2. (d) Skin has a metabolic function to prevent non-resident bacteria from developing.

6. (b) There are two types of antibiotics – narrow spectrum and broad spectrum. Broad spectrum are those antibiotics which attack wide range of pathogens.

7. (a) Common Cold, AIDS is a pair of viral diseases. Viral diseases are very common type of infectious disease. Viruses are the smallest life-form existing, since they are not even a single cell. It is almost like they are not alive at all. They are small strands of DNA-like cell material. A virus consists mostly of RNA and cannot survive without host cells.

8. (c) Whenver our body gets attacked by some foreign invadors, our body's immune system produces some

chemicals to kill or to react against the invader. These chemicals are actually made up of carbohydrates & proteins *i.e.* glycoproteins called antibody.

11. **(c)** Hay fever is due to some fungal spores sensitivity, which is an allergic disease with symptoms of bronchial asthma and skin rashes and also with increase in eosinophil (white cells) of blood.
13. **(a)** Cancerous cells can easily be destroyed by radiations due to rapid cell division.
16. **(b)** Lymphocytes are the part of specific defence system called immune system.
17. **(c)** A special defence system works specially against viral infection. It has no effect on micro-organism. Cells invaded by a virus produce an antiviral protein called interferon (IFN). The latter is released from the infected cell and on reaching the nearby non-infected cells it makes them resistant to the virus infection.
18. **(b)** The major obstacle in transplantation of organs is that the recipient body does not accept the donor's organ. The body defence mechanism reject & treat the transplanted organ as a foreign particle and reacts actively.
20. **(a)** There are three main types-cytotoxic T - cells, helper T -cells and suppressor T - cells
21. **(b)** Antibodies are the macromolecules of protein.
29. **(d)** Acquired immunity is obtained during the life of an individual to a particular micro-organism due to previous infection, vaccination or inoculation of antiserum. It has four unique characteristics :
    (i) Specificity : It is specific for each & every type of pathogen.
    (ii) Diversity : It operates against the whole diversity of pathogenic organisms, their toxins & pollutants.
    (iii) Difference between self & non-self : It can differentiate molecules & cells of foreign origin from those of self.
    (iv) Memory : The immunity against a pathogen developed during an initial infection is retained by the individual so that second encounter with the pathogen invites a heightened immune response.
41. **(b)** *Rauwolfia serpentina* belong to family Apocynaceae, its roots yield a chemical useful for high blood pressure.
42. **(a)** Allergy means all kinds of reactions manifested in the body due to hypersensitivity of substances called allergens. Allergy reactions occur due to liberation of histamine by tissue cells.
43. **(b)** Tranquillizers bring about depression of brain activity and feeling of calmness. Stimulants stimulate the nervous system, and increase alertness and activity. Hallucinogens (*Cannabis* products) alters thoughts, perceptions and feelings.
44. **(a)** Typhoid fever is caused by *Salmonella typhi* (*S. typhosa*). Bacterial genus *Shigella* causes shigellosis or bacillary dysentry. *Escherichia coli* is a facultative anaerobe found in the intestine of human beings.
45. **(a)** Antibodies bind to specific antigens to produce large insoluble complexes which render them harmless and facilitate their destruction by other cells of the immune system this process is called agglutination.
46. **(c)** The use of anti-histamines and steroids give a quick relief from headache. Antihistamines are drugs that block the action of histamine (a compound released in allergic inflammatory reactions) at the H_1 receptor sites, responsible for immediate hypersensitivity reactions such as sneezing and itching.
47. **(b)** Morphine is potent opioid analgesic that is often given to persons (who have undergone surgery) as a pain killer. It is mainly used to relieve severe and persistent pain. It is administrated by mouth, injection or suppositories.
48. **(d)** Widal test is used for the diagnosis of typhoid. It is an agglutination test for the presence of antibodies against the *Salmonella* organisms, which cause typhoid fever.
49. **(c)** Head louse is an obligate ectoparasite of human scalp and as well as laying eggs on human hair.
50. **(a)** B and T-lymphocytes produce antibodies against pathogen in the body, if due to some reasons B and T - lymphocytes are damaged the body will not produce antibodies against a pathogen. Each B cell and T cell is specific for a particular antigen.
55. **(b)** Antibodies are produced by lymphocytes. Lymphocytes are agranulocytes found in blood. They are non-phagocytic with spherical nucleus. They participate in body's immunity by producing antibodies. Such kind of immunity is called antibodies mediated immunity.
56. **(b)** The trophozoites of *Entamoeba histolytica* live inside the cells of host (man). These trophozoites later on develop a cyst around them and through faeces they come out of the host's body. Contamination of food with these cysts and their ingestion leads to the infection of *amoebiasis*.
57. **(c)** Hashish is a preparation of *Cannabis* composed of the compressed trichomes collected from the *Cannabis* plant. It contains the same active ingredients but in higher concentrations than other parts of the plant such as the buds or the leaves. Psychoactive effects vary between types of hashish but are usually the same as those of other *Cannabis* preparations such as marijuana.
58. **(a)** Common cold is due to rhinovirus. Recent studies found that antibiotics do not kill this virus.
59. **(c)** When HIV replicates rapidly in helper T-lymphocytes and damages large number of these cells, at this stage infected persons start showing symptoms of AIDS.
60. **(c)** Tetanus toxoid is a vaccine consisting of growth products of *Clostridium tetani* treated with

formaldehyde serving as an active immunising agent. Hence it is weakened germs. The first tetanus toxoid (inactivated toxin) was produced in 1924 and was used successfully to prevent tetanus in the armed services during world war II. In the mid-1940s, tetanus vaccine was combined with diphtheria toxoid and inactivated pertussis vaccine to make the combination DTP vaccine for routine childhood immunization.

31. STRATEGIES FOR ENHANCEMENT IN FOOD PRODUCTION

ANSWER KEY											
1	(b)	11	(a)	21	(d)	31	(a)	41	(b)	51	(a)
2	(d)	12	(d)	22	(d)	32	(c)	42	(c)	52	(a)
3	(c)	13	(b)	23	(c)	33	(d)	43	(a)	53	(d)
4	(a)	14	(b)	24	(c)	34	(d)	44	(d)	54	(b)
5	(b)	15	(d)	25	(a)	35	(b)	45	(a)	55	(d)
6	(b)	16	(c)	26	(b	36	(b)	46	(a)	56	(b)
7	(a)	17	(c)	27	(d)	37	(d)	47	(b)	57	(d)
8	(a)	18	(c)	28	(a)	38	(c)	48	(a)	58	(a)
9	(a)	19	(c)	29	(c)	39	(a)	49	(c)	59	(b)
10	(b)	20	(b)	30	(c)	40	(a)	50	(c)	60	(b)

Hints & Solutions

6. **(b)** The best pH of soil for cultivation of plants is 6.5 – 7.5.

7. **(a)** Three crops that contribute maximum to global food grain production are wheat, rice and maize, which belong to the family *Poaceae* (*Graminae*).

12. **(d)** The aims of plant breeding are increase in yield of seeds, resistance to diseases, insects, pests and earliness or lateness in maturity period or change in maturity behavior.

15. **(d)** Mule is an inter specific hybrid between female horse and male donkey. Mule is sterile hybrid.

16. **(c)** Silk is produced by silk worm insect called as silk moth.

21. **(d)** In crop improvement programme, haploids are very important because they form perfect homozygous which are helpful to get a pure line.

22. **(d)** Biological control refers the use of living organisms like natural predators and parasites to reduce pest population and control diseases. However they may become pest as they can develop a preference to other diets.

23. **(c)** *Catla catla* and *Labeo rohita* are the two Indian major carps whereas *Heteropneustes* is a catfish. *Cyprinus* is the exotic breed.

24. **(c)** The top five crops are potato (which belongs to the family *Solanaceae*) and wheat, rice, corn, barley (which belongs to the family *Poaceae*).

25. **(a)** In vegetative reproduction the same genotype can be propagated from one generation to the next.

26. **(b)** Inorganic substances, oils, plant extracts used as insecticides are called first generation pesticides. Synthetic organic compounds are known as second generation pesticides. Insect hormones are third generation pesticides. Pheromones are the chemical substances which when released into an animal's surroundings, influence the behaviour or development of other individuals of the same species.
Insect hormone analogues help in trapping the insects by confusing them as they mimic the odour of pheromones.

27. **(d)** Karl Von Frisch, carried out many experiments and determined that when a foraging bee returns to the hive, it performs a waggle dance. Honey is predominantly glucose and fructose. *Dorsata* is a biger bee than *Apis indica* (a medium sized bee). Bees wax is secreted by special wax glands to make compartments.

28. **(a)** Cellular totipotency is the ability, shown by many living cells to form all types of tissue that constitute the mature organisms.

29. **(c)** Cellular totipotency is exhibited only by plant cells wherein the plant cells retain the ability to deactivate and form the new dividing cells.

30. (c) Green revolution was possible due to hybridization. Increased chlorophyll content has no effect on yield of crops. Hybrid plants were dwarf because of selection. Quantitative trait mutations are not involved in hybridization.
32. (c) Meristem culture is done for the development of virus free plants. Meristematic tissue cell can be taken either from shoot or root tip.
33. (d) *Triticale* is the first man made cereal which is made by the crossing of wheat (*Triticum aestivum*) and rye (*Secale cereale*).
34. (d) Genetic variation present among plant cells during tissue culture is called somaclonal variation. The term somaclonal variation is also used for the genetic variation present in plants regenerated from a single culture. This variation has been used to develop several useful varieties against resistance to diseases and pests, better yield, better quality etc.
35. (b) Breeding of crops with high levels of minerals, vitamins and minerals is called biofortification. This is most practical aspect to improve the health of people.
36. (b) An improved variety of transgenic basmati rice gives high yield and is rich in vitamin A.
38. (c) In hybridization, two or more plants of unlike genotypes are crossed together to get offsprings with new desirable combinations of characters as a result of genetic recombinations.
40. (a) Introduction, hybridization and mutations are methods of plant breeding.
42. (c) Gamma rays (from cobalt 60) is generally used for induced mutation in crops plants.
48. (a) Parthenocarpy is the development of fruits without prior fertilization which results in the formation of seedless fruits. In some plant species, parthenocarpic fruits are produced naturally or they may be induced by treatment of the unpollinated flowers with auxin.

    Removal of androecium before pollen release is called emasculation which is helpful in preventing unwanted pollination. Vernalized seeds are the chilled treated seeds for breaking dormancy. Phenyl mercuric acetate is an antitranspirant.

    Gibberellins and auxins are known to induce parthenocarpy in plants. If a tomato plant is treated with a low concentration of auxin and gibberellic acid it'll produce fruits without fertilization i.e. parthenocarpic fruits.
49. (c) Because iron promotes the formation of chlorophyll and magnesium is an integral part of chlorophyll molecule so in the absence of these nutrients plants show chlorosis & leaves get yellowing. Application of iron and magnesium to promote the synthesis of chlorophyll can cure the disease.
50. (c) Crop plants grown in monoculture are highly prone to pests.
51. (a) According to Darwin resistance to pesticides has a genetic basis due to presence of certain alleles. In the presence of pesticides, the insects devoid of the allele for resistance die while the ones having them survive. They are selected by nature.
52. (a) Mutations are generally recessive so they are not expressed in their heterozygous condition. If plant is haploid then mutation can express themselves.
53. (d) The long-term prospects for a truly human civilization depend in a large measure on control of human diseases.

32. MICROBES IN HUMAN WELFARE

ANSWER KEY

1	(a)	11	(d)	21	(d)	31	(c)	41	(b)	51	(d)
2	(d)	12	(a)	22	(a)	32	(d)	42	(a)	52	(c)
3	(a)	13	(c)	23	(d)	33	(b)	43	(a)	53	(d)
4	(b)	14	(c)	24	(b)	34	(b)	44	(b)	54	(c)
5	(c)	15	(c)	25	(d)	35	(b)	45	(b)	55	(a)
6	(d)	16	(c)	26	(a)	36	(d)	46	(c)	56	(b)
7	(b)	17	(a)	27	(c)	37	(d)	47	(a)	57	(c)
8	(a)	18	(a)	28	(c)	38	(a)	48	(a)	58	(a)
9	(d)	19	(a)	29	(b)	39	(c)	49	(a)	59	(a)
10	(d)	20	(a)	30	(c)	40	(b)	50	(c)	60	(b)

Hints & Solutions

1. **(a)** Gobar gas contains mainly $CH_4 + CO_2$.
16. **(c)** 'Each antibiotic is effective only against one particular kind of germ' is not correct.
17. **(a)** *Azotobacter* is a free living nitrogen fixing bacterium, *Rhizobium* is symbiotic and found in the root nodules of leguminous plants, *Anabaena* occurs in assoction with *Azolla*. VAM is vesicular arbuscular mycorrhiza.
18. **(a)** Farmers have reported over 50% higher yields of rice by using the biofertilizer *Azolla*.
19. **(a)** *Azolla* is a freshwater fern harbouring a bluegreen alga - *Anabaena* in its leaf cavities. The alga fixes atmospheric nitrogen and releases nitrogenous compounds in leaf cavities. This symbiotic system is the main source of algal biofertilizer in rice fields.
20. **(a)** *Azolla* and blue green algae- *Anabaena* form biofertilizer in rice fields.
21. **(d)** *Escherichia coli* is a bacterium found in human colon. On this bacterium, scientists have made extensive genetic experiments to make some vital chemicals like insulin. Another bacterium is *Agrobacterium tumefaciens* which causes crown gall in plants is extensively used for genetic experiments.
22. **(a)** Human insulin is being commercially produced from a transgenic species of *Escherichia coli*. *E. coli* is a bacterium that is commonly found in the lower intestine of warm blooded animals. The bacteria can also be grown easily and its genetics are comparatively simple and easily manipulated, making it one of the best studied prokaryotic model organisms, and an important species in biotechnology.
23. **(d)** A common biocontrol agent for control of plant diseases is *Trichoderma*. *Trichoderma* is a tree living fungus that exerts biocontrol over several plant pathogens for the control of plant diseases. It is the natural method of pest and pathogen control.
24. **(b)** The common nitrogen fixer in paddy field is *Azospirillum*. It is an anaerobic bacteria that forms loose association with roots of paddy crops.
25. **(d)** Organic farming involves use of organic wastes and other biological materials along with beneficial microbes to release nutrients to crop to increase the soil fertility in an ecofriendly, and pollution free environment. *Glomus*, earthworm and *Oscillatoria* can be used in organic farming while snail cannot.
26. **(a)** *Glomus aggregatum* is a mycorrhizal fungus used as a soil inoculant in agriculture and horticulture. Its purpose is to increase the surface area of roots for nutrient absorption like phosphorus.
27. **(c)** *Clostridium butylicum* industrially produces butyric acid.
33. **(b)** Biogas consists of carbon dioxide, methane and hydrogen.
40. **(b)** *Escherichia coli* is used as an indicator organism to determine pollution of water with faecal matter.
45. **(b)** Energy plantation mean grow more trees for fuel wood, solar energy stored, safe ecologically, renewable etc.
48. **(a)** Wood which is used as source of energy is called fuel wood and about 2 billions people in world are dependent upon wood as a source of fuel.
60. **(b)** Modern biotechnology has the potential to make the 21st century an era of the most profound and wide ranging progress in the history of medicine. Compared the advent of antibiotics, which created a revolution in the treatment of infections caused by bacteria to advances in genetic science have the potential to create a revolution in the diagnosis and treatment of many diseases and inherited disorders.

33. BIOTECHNOLOGY : PRINCIPLES AND PROCESSES

ANSWER KEY

1	(c)	11	(b)	21	(d)	31	(c)	41	(d)	51	(d)
2	(c)	12	(b)	22	(a)	32	(b)	42	(d)	52	(b)
3	(a)	13	(a)	23	(a)	33	(a)	43	(a)	53	(a)
4	(a)	14	(d)	24	(a)	34	(a)	44	(b)	54	(d)
5	(a)	15	(a)	25	(c)	35	(c)	45	(a)	55	(c)
6	(b)	16	(d)	26	(d)	36	(d)	46	(a)	56	(b)
7	(a)	17	(a)	27	(b)	37	(a)	47	(b)	57	(d)
8	(d)	18	(d)	28	(a)	38	(c)	48	(c)	58	(d)
9	(a)	19	(d)	29	(d)	39	(a)	49	(a)	59	(d)
10	(c)	20	(c)	30	(c)	40	(b)	50	(d)	60	(c)

Hints & Solutions

7. **(a)** Living cells or tissues or organs are cultured in aseptic conditions in nutrient medium to produce plantlet which is disease resistant. This is known as tissue culture or micropropagation.
8. **(d)** DNA repairing is done by the involvement of three enzymes – DNA repair nucleases, DNA polymerase I and DNA ligase. Both DNA polymerase I and DNA ligase have important general roles in DNA metabolism. Both functions in DNA replication as well as DNA repair.
9. **(a)** Genetic engineered bacteria is used for production of insulin. On 5th July 1983, the american firm Eli Lilly launched humulin (1st genetically engineered human insulin).
10. **(c)** Tobacco and tomato, the transgenic plants possess a gene or genes that have been transferred from a different species. Although, DNA of another species can be integrated in a plant genome by natural processes, the term "transgenic plants" refers to plants created in a laboratory using recombinant DNA technology. The aim is to design plants with specific characteristics by artificial insertion of genes from other species or sometimes entirely different kingdoms.
11. **(b)** DNA or RNA segment tagged with a radioactive molecule is called probe. They are used to detect the presence of complementary sequences in nucleic acid samples. Probes are used for identification and isolation of DNA and RNA.
12. **(b)** Gel electrophoresis is a technique to separation of DNA fragments according to their size. DNA is negatively charged so in gel tank when electric passed, DNA move towards positive electrode.
13. **(a)** Restriction endonucleases are enzymes that make cuts at specific positions within the DNA molecule. They acts as molecular scissors. They recognise specific base sequence at palindrome sites in DNA duplex and cut its strands.
14. **(d)** *EcoRI* is an endonuclease enzyme isolated from strains of *E.coli* and a part of restriction modified system. So co part stands for *coli*.
15. **(a)** A genetic clone is a plant produced by asexual means.
16. **(d)** Plants developed by genetic engineering are called transgenic plants or genetically modified crops from which genetically modified food is produced. For their production micro-organisms (bacteria, virus) are used. So, by consuming them there is a danger of entry of viruses and toxins causing different types of allergies and other health hazards to human beings.
17. **(a)** The linking of antibiotic resistance gene with the plasmid vector became possible with DNA ligase. DNA ligase is an enzyme that is able to join together two portions of DNA and therefore plays an important role in DNA repair. DNA ligase is also used in recombinant DNA technology as it ensures that the foreign DNA is bound to the plasmid into which it is incorporated.
18. **(d)** Direct gene transfer is the transfer of naked. DNA into plant cells but the presence of rigid plant cell wall acts as a barrier to uptake. Therefore protoplasts are the favoured target for direct gene transfer. Polyethylene glycol mediated DNA uptake is a direct gene transfer method that utilizes the interaction between polyethylene glycol, naked DNA, salts and the protoplast membrane to effect transport of the DNA into the cytoplasm.
19. **(d)** Retrovirus has the ability to transform normal cells into cancerous cells. Hence, it can used as a vector for cloning desirable genes into animal cells.
20. **(c)** Palindromic sequences in DNA molecule are group of bases that forms the same sequence when read in both forward and backward direction. In the given question, only option (c) represent a palindromic sequence.
21. **(d)** In gel electrophoresis agarose extracted from sea weed used as gel agarose, made of 0.7% gel show good resolution of large DNA and 2% gel will show good resolution of small fragments.
22. **(a)** *Bacillus thuringiensis* produces a large amount of crystalline protein during sporulation. In the cell, toxins are formed along with the spore and are referred to as parasporal body. The bacteria are capable of entering the insect's blood and using the host insect to reproduce. The proteins from ingested spores are activated by gut, high pH and the polypeptide toxins destroy gut epithelial cells and kill the pest.
46. **(a)** The tissue culture is a latest method of crop improvement. Besides also used in manufacture of antibiotics, alkaloids and dyes.
51. **(d)** Virusoid is not a genetic vector. Virusoids are circular single stranded RNA's dependent on plant viruses for replication & encapsidation.
52. **(b)** Cloning is the production of an organism with exactly similar genetic make up as in the mother individual. Cloning is done to preserve genotype of an individual. This is achieved by cell culture, tissue culture or genetic engineering.
53. **(a)** Totipotency of plant cells are best explained by tissue culture. When a cell or group of cells are grown in a culture medium it gives rise to new plantlet. Every cell of a plant has capacity to gives rise a new plant.
54. **(d)** Axenic is a method of culture of isolated plant cells, tissues or organs in an artificial, nutritive medium.
55. **(c)** During the process of replication the enzyme that helps to cut one strand of DNA duplex to release the tension of coiling of two strands is topoisomerase.
56. **(b)** A plasmid is a DNA molecule separate from the chromosomal DNA and capable of autonomous replication. It is typically circular and double-stranded. It usually occurs in bacteria, and is sometimes found in eukaryotic organisms. Plasmids used in genetic

engineering are called vectors. They are used to transfer genes from one organism to another and typically contain a genetic marker conferring a phenotype that can be selected for or against, along with their independent replication site.

57. **(d)** DDT, BHC (Benzene Hexachloride) – chlorinated hydrocarbons
Aldrin – Carbamate (reversibly inhibits acetylcholenesterase).

58. **(d)** Use of biology in industrial process and for improving quality of life is called as biotechno-logy. It is the application of discoveries in biology for the production of useful organisms and their products.

34. BIOTECHNOLOGY AND ITS APPLICATIONS

ANSWER KEY															
1	(d)	9	(b)	17	(d)	25	(d)	33	(b)	41	(d)	49	(d)	57	(c)
2	(b)	10	(d)	18	(b)	26	(a)	34	(c)	42	(a)	50	(a)	58	(c)
3	(a)	11	(a)	19	(d)	27	(d)	35	(b)	43	(d)	51	(d)	59	(b)
4	(a)	12	(b)	20	(d)	28	(b)	36	(d)	44	(b)	52	(b)	60	(b)
5	(a)	13	(c)	21	(d)	29	(a)	37	(a)	45	(d)	53	(d)		
6	(a)	14	(d)	22	(d)	30	(c)	38	(d)	46	(d)	54	(c)		
7	(c)	15	(c)	23	(d)	31	(a)	39	(a)	47	(d)	55	(b)		
8	(a)	16	(a)	24	(c)	32	(a)	40	(d)	48	(c)	56	(c)		

Hints & Solutions

1. **(d)** Golden rice is a transgenic crop of the future with high Vit. A content. Millions of people suffer from Vit. A deficiency which leads to vision impairment. Transgenic rice has been developed which is capable of synthesizing beta-carotene, the precursor of Vitamin A. The rice variety is now being crossed into adapted varieties with field tests possible in an year or two.

2. **(b)** *Agrobacterium tumefaciens* is the causal agent of crown gall disease (the formation of tumours) in over 140 species of dicots. This disease caused by a DNA plasmid (T_i plasmid) carried by bacterium and transferred to the plant cells. T_i plasmid has widely used in plant engineering as a vector in order to inject gene in host plant to form transgenic plant.

3. **(a)** Transgenic plants are the ones generated by introducing foreign DNA into a cell and regenerating a plant from that cell.

4. **(a)** The genetically modified brinjal in India has been developed for insect resistance. Bt brinjal is a transgenic brinjal that is developed by inserting a crystal gene from the *Bacillus thuringiensis* into the brinjal's genome. This process of insertion is accomplished using *Agrobacterium* mediated recombination.

5. **(a)** Recombinant DNA technology is the process joining together two DNA molecules from two different species that are inserted into a host organism to produce new genetic combination.

11. **(a)** Crown gall disease in plants is caused by Ti-plasmid.

16. **(a)** Human insulin has 53 amino acids in two polypeptides (A and B) connected by two S -S (disulphide) linkages. In 1983, American company Eli Lilly prepared two DNA sequences corresponding to A and B insulin chains. When introduced in plasmids of *E.coli*, insulin chains were formed. They were extracted and fused to produce humulin (human insulin).

18. **(b)** ELISA test is a technique which can detect any amount of an antibody or antigen with the help of an enzyme. The commonly used enzymes are alkaline phosphatase and peroxidase.

19. **(d)** Bt cotton, a transgenic crop variety has been introduced in India. The Bt cotton variety contains a foreign gene obtained from *Bacillus thuringiensis*. This bacterial gene protects cotton from the ball worm, a major pest of cotton.

20. **(d)** Golden rice is a transgenic food crop which may help in solving the problem of night blindness in developing countries. Golden rice or miracle rice is rich in vitamin A or β-carotene and iron and decaffeinated coffee are also valuable achievements of gene transfer technology.

21. **(d)** Main objective of production/use of herbicide resistant GM crops is to reduce herbicide accumulation in food articles for health safety. GM plants have been useful in many ways. Genetic modifications has made crops more tolerant to abiotic stresses, reduced reliance on chemical pesticides, enhanced nutritional value of food.

22. **(d)** About Bt toxin, it is true, that the inactive protoxin gets converted into active form in the insect gut. There are several advantages in expressing Bt toxins in transgenic Bt crops. The level of toxin expression can be very high, thus delivering sufficient dosage to the pest. The toxin expression is contained within the plant system and hence only those insects that feed on the crop perish. The toxin expression can be modulated by using tissue-specific promoters and replaces the use of synthetic pesticides in the environment.

23. **(d)** Bt toxin genes were isolated from *Bacillus thuringiensis* and incorporated into cotton plant to form a genetically modified crop called Bt cotton. Bt cotton has high yield and resistance to bollworms.

24. **(c)** Organic farming is the form of agriculture that relies on the techniques like crop rotation, green manure, compost and biological pest control.

28. **(b)** Recombinant DNA technology prepared vaccine is called second generation vaccine.

33. **(b)** Maximum utilization of the techniques of biotechnology has been made in the field of medicines.

42. **(a)** Interferons are active proteins (antiviral proteins) that are effective against most viruses.

50. **(a)** In industries microorganisms are used for the manufacture of gluconic acid.

53. **(d)** By cross-linking enzyme molecules or by covalently attaching them to a solid support or by entrapping them in gel, metabolic activities of enzymes are stopped.

55. **(b)**

$$C_{12}H_{22}O_{11} + H_2O \xrightarrow[\text{Invertase}]{\text{Yeast}} \underset{\text{(Glucose)}}{C_6H_{12}O_6} + \underset{\text{(Fructose)}}{C_6H_{12}O_6}$$

$$C_6H_{12}O_6 \xrightarrow[\text{Zymase}]{\text{Yeast}} \underset{\text{(Ethyl alcohol)}}{2C_2H_5OH} + 2CO_2$$

56. **(c)** Insulin is made up of 51 amino acids arranged in two polypeptide chains, A having 21 amino acids and B with 30 amino acids that are linked together by disulphide bridges. In mammals, including humans, insulin is synthesised as a pro-hormone (like a pro-enzyme, the pro-hormone also needs to be processed before it becomes a fully mature and functional hormone) which contains an extra stretch called the C peptide. This C peptide is not present in the mature insulin and is removed during maturation into insulin. The main challenge for production of insulin using rDNA techniques was getting insulin assembled into a mature form.

35. ORGANISMS AND ENVIRONMENT

ANSWER KEY

1	(d)	11	(c)	21	(d)	31	(c)	41	(a)	51	(b)
2	(a)	12	(b)	22	(d)	32	(d)	42	(a)	52	(b)
3	(b)	13	(b)	23	(b)	33	(b)	43	(b)	53	(b)
4	(c)	14	(b)	24	(a)	34	(c)	44	(a)	54	(d)
5	(d)	15	(d)	25	(a)	35	(a)	45	(c)	55	(c)
6	(a)	16	(b)	26	(c)	36	(c)	46	(a)	56	(c)
7	(a)	17	(b)	27	(d)	37	(a)	47	(c)	57	(d)
8	(d)	18	(b)	28	(a)	38	(a)	48	(c)	58	(c)
9	(c)	19	(b)	29	(a)	39	(b)	49	(b)	59	(a)
10	(c)	20	(a)	30	(d)	40	(d)	50	(d)	60	(d)

Hints & Solutions

2. **(a)** Symbiosis or mutualism is a mutually beneficial association necessary for the survival of both the partners.

3. **(b)** Loam soil is best suited for plant growth because it possesses good aeration, nutritive salts and good water retaining capacity.

4. **(c)** Sunken stomata is the characteristic feature of xerophytes which help in reducing loss of water from leaf surface. These type of stomata are found in oleander or pine.

5. **(d)** Tropical rain forests occur in equatorial and subequatorial regions. The forests receive all the external inputs for optimum plant growth. Due to abundant plant growth, a large number of animals live in tropical rain forests. In grasslands fire occur periodically which prevent tree growth. In savannahs periods of drought are common. In deciduous forests are found predominantly broad leaved hard wood deciduous trees.
6. **(a)** Carrying capacity refers to the maximum number of individuals that can be sustained by the environment. Birth rate refers to number of births per unit population.
7. **(a)** The growth curve for bacteria is sigmoid. First phase is the phase of slow growth- lag phase. Second phase is the period of accelerated growth - log phase. Third phase is the phase when growth rate becomes stable-stationary phase. Finally growth rate declines.
8. **(d)** In the sigmoid growth curve growth is slow in the lag phase. Growth rate remains stable during the stationary phase and it declines during the senescent phase. Growth increases logarithmically during the exponential phase.
9. **(c)** The formula of exponential growth is where is the rate of change in population size, r is the biotic potential and N is the population size.
14. **(b)** Species is a group of organisms which interbreed freely and produce fertile offsprings in nature.
15. **(d)** Ecology is the branch of biology that deals with the inter relationships amongst organisms and interactions between organisms and their environment.
16. **(b)** Competition is most severe between the members of a population belonging to same niche.
17. **(b)** Homeostasis is the property of an open system, especially living organisms, to regulate its internal environment to maintain a stable, constant condition, by means of multiple dynamic equilibrium adjustments, controlled by interrelated regulation mechanisms.
18. **(b)** When the interaction is useful for both the populations.
19. **(b)** High organic matter darkens the colour of soil.
20. **(a)** Soil particles size defines its texture. The mineral particles present in soil, vary enormously in size.
21. **(d)** A species is a group of individuals of same kind of phenotypic characters and can interbreed easily.
23. **(b)** Xeric environment is found in deserts where temperature remains high in the day and low during the night and air humidity is very low.
24. **(a)** Stenohaline: 'Steno' meaning narrow and 'haline' meaning salt. A fish cannot handle a wide fluctuation in salt content in water. Many fresh water fishes tend to be stenohaline and die in environments of high salinity such as the ocean. Fish living in coastal estuaries and tide pools are often euryhaline (tolerant to changes in salinity) as are many species which have life cycle requiring tolerance to both fresh water and sea water environments such as *Salmon*.
25. **(a)** Number of individuals decreases at successive trophic levels.
26. **(c)** In India, there is short life span of individuals and high birth rate as compared to other countries. That is why, the human population is heavily weighed towards the younger age groups.
27. **(d)** Human population growth in India can be regulated by following the national programme of family planning.
28. **(a)** In ecological succession, climax community is a stable, self perpetuating biotic community developing towards the end of biotic succession in an area and is in harmony with the climate of the region.
29. **(a)** Population is an aggregation or grouping of individuals of the same species at the same time in a particular area. Species is a grouping of individuals of one or more populations resembling each other in important morphological, anatomical and biochemical characters and can potentially interbreed. Biotic community is an assemblage of interdependent, and ecosystem is a self sufficient and self regulating segment of nature comprising of a biotic community and its physical environment, both interacting and exchanging materials. The basic level of ecological organisation starts with the individual, the next more complex levels are populations, species, community and then ecosystem.
30. **(d)** The keystone species in an ecosystem are those who are the main contributors to the ecosystem.
31. **(c)** Natality refers to the rate at which new individuals are born. Adaptability refers to the capacity of the organisms to adapt themselves to their environment.
32. **(d)** Mortality is the average number of natural deaths per unit population per unit time. Fecundity refers to the biotic potential of the organism. Environmental resistance is the sum total of all limiting factors that prevents a population to realize its full potential.
33. **(b)** This is so because oceanic islands are normally cut off from other habitats.
34. **(c)** Semilog of per minute growing bacterium when plotted against time, would yield ascending straight line.
35. **(a)** A keystone species is the one which makes up only a small proportion of the total biomass of a community, yet has a huge impact on the community's organization and survival.
36. **(c)** Desert shows maximum difference between day and night temperature.
37. **(a)** Demography is the study of population in all aspects. Fertility refers to the number of children per couple. Mortality rate is the average number of natural deaths per unit population per unit time. Age distribution refers to the proportionate occurrence of

individuals of the three age groups. Developing countries usually have high rate of population growth, because of increasing fertility and declining mortality.

38. **(a)** Geometric representation of age structure is a characteristic of population.

39. **(b)** If a population (*e.g* reindeer population) is allowed to grow in a predator free environment, the population grows beyond carrying capacity and there occurs population crash due to sudden shortage of food. Such growth curves also occur in insect populations during rainy season, and in algal blooms.

40. **(d)** When the members of a species are inter - fertile and produce fertile offsprings.

41. **(a)** It is competition between individuals of same species. The intraspecific competition may be very severe because all the members of a species have similar requirements of food, habitat mate, *etc* and they also have similar adaptations to get their needs.

42. **(a)** There is a decline in population as boys marry girls only from their own tribe in the isolated small tribal populations.

46. **(a)** Biotic potential is the maximum capacity of a population to reproduce under ideal conditions (environmental).

47. **(c)** The carrying capacity of a population is determined by its limiting resources. Carrying capacity is the upper limit of an ecosystem upto which it can provide the basic needs to the population under given circumstances.

49. **(b)** Commensalism is an association between two organisms in which one is benefitted without the other being harmed e.g. - sucker fish has a hold fast and attached itself underside of shark and thus remains protected and also get food left over when shark is feeding on its prey.

57. **(d)** An overexposure to UV-B radiation can cause sunburn and some form of skin cancer. In humans, prolonged exposure to solar UV-radiation may result in acute and chronic health effects on the skin, eye and immune system. Moreover, UV-C can cause adverse effects that can variously be mutagenic or carcinogenic.

58. **(c)** Thermoregulation is energetically expensive for many organisms particularly true for small animals like shrews and humming birds. During the course of evolution, the costs benefits of maintaining a constant internal environment are taken into consideration.

36. ECOSYSTEM

ANSWER KEY

1	(d)	11	(a)	21	(b)	31	(d)	41	(c)	51	(c)
2	(b)	12	(b)	22	(b)	32	(d)	42	(a)	52	(a)
3	(b)	13	(b)	23	(c)	33	(b)	43	(a)	53	(c)
4	(c)	14	(c)	24	(a)	34	(d)	44	(c)	54	(c)
5	(c)	15	(a)	25	(c)	35	(b)	45	(a)	55	(b)
6	(a)	16	(c)	26	(c)	36	(a)	46	(c)	56	(d)
7	(a)	17	(d)	27	(c)	37	(c)	47	(d)	57	(c)
8	(b)	18	(b)	28	(b)	38	(b)	48	(d)	58	(b)
9	(c)	19	(c)	29	(c)	39	(a)	49	(d)	59	(d)
10	(d)	20	(a)	30	(d)	40	(c)	50	(d)	60	(b)

Hints & Solutions

4. **(c)** In an ecosystem the function of the producers is to trap solar energy and convert into chemical energy.

10. **(d)** Humus is rich source of organic matter and it makes soil porous and increases its water holding capacity.

11. **(a)** Source of energy in ecosystem is sun. Energy from the sun in the form of sunlight, supports almost all life on earth via photosynthesis, and drives the earth's climate and weather.

12. **(b)** Tropical rainforests are generally found near the equator. They are common in Asia,Australia, Africa, South America, Central America, and on many of the Pacific Islands. Within the world wildlife fund's biome classification scheme, tropical rainforests are considered a type of tropical wet forest (or tropical moist broadleaf forest) and may also be referred to as lowland equatorial evergreen rainforests. Minimum normal annual rainfall between 1,750 millimetres (69 in) and 2,000 millimetres (79 in) occur in this climate region. Mean monthly temperatures exceed 18 °C during all months of the year.

13. **(b)** Good soil is essential to a healthy garden. One of the most important parts of indoor and outdoor gardening is the soil you provide for your plants and flowers. One of the most important parts of indoor and outdoor gardening is the soil you provide for your plants and flowers.
14. **(c)** Because in large water bodies the number of (producers) for food.
15. **(a)** Because plants have the capacity of synthesizing food (producers).
16. **(c)** The ecosystem comprises abiotic and biotic components. The entire living community comprising plants and animals constitute the biotic component whereas the entire physical environment forms the abiotic component.
17. **(d)** An ecosystem should always maintains a balance, if primary consumers in an ecosystem are absent, then over crowding occur. It results in competition and consequently number of producers will decrease to near normal.
20. **(a)** The trophic structure in any ecosystem is a kind of producer consumer arrangement. Here each food level is called trophic level.
21. **(b)** Pyramid of numbers in a grassland ecosystem is always upright. Pyramids of numbers are upright because numbering of organisms decrease at successively higher trophic levels. Trophic level refers to the organisms, position in the food chain.
22. **(b)** Detritus food chain being with dead organic matter with dead plant parts, animals and their excretory products which is being acted upon by decomposers such as saprophytes to obtain energy needed for their survival.
23. **(c)** Phytoplankton → Zooplankton → Fish. This is a type of food chain in aquatic ecosystem.
24. **(a)** German biologist Ernst Haeckel coined the term 'ecology', the study of inter-relationship between organisms and their environment. Ecosystem, consist of different communities of organism and the physical environment with which they interact. Phytogeography is the study of geographic distribution of plants species on earth.
25. **(c)** Pyramid of number represents the number of individuals on each trophic level.
26. **(c)** Pyramid of numbers of pond ecosystem is upright.
27. **(c)** Energy pyramid can not be inverted because energy always flows in one direction only. According to 10% law of energy transfer given by Lindemann (1972), during transfer of energy from one trophic level to other there is always loss in respiration and other activities only 10% of energy is transferred to next trophic level. So the amount of energy flow decreases with successive trophic levels.
28. **(b)** It due to majority of phytoplanktons.
29. **(c)** The decomposers act on the dead organic matter and break them down into simpler compounds and minerals.
30. **(d)** The primary succession means the development of communities only on a newly exposed habitat with no record of earlier vegetation.
31. **(d)** The oceans cover about 2/3 of the whole surface of our earth. Thus it is the most stable ecosystem, because of buffering action by water.
32. **(d)** In a biotic community the herbivores (goat, deer) are those animals, which consume the primary producers (green plants). Hence they are primary consumers.
33. **(b)** Producers are the direct and indirect source of food for all the trophic levels. They are always more in number.
34. **(d)** Coral reefs are often called "rainforest of the sea". It forms some of the most diverse ecosystem on earth. Coral reef ecosystem have the highest grass primary productivity in the sea.
35. **(b)** The rate at which organic molecules are formed in a green plants is called gross productivity.
36. **(a)** Plankton → Organisms passively floating in upper water.
    Nekton → Actively swimming
    Benthos → Lead sedentary life upon the sea bottom
37. **(c)** In an ecosystem producers (green plants) are always more than consumers (herbivores, carnivores etc.)
38. **(b)** The greatest biomass of autotrophs in the oceans is that of free floating microalgae, cyanobacteria and nanoplankton.
39. **(a)** An ecosystem having low stability can be easily damaged. An ecosystem having high resilience will take less time to recover.
40. **(c)** Fresh weight is not used for the construction of ecological pyramids because the total fresh weight does not change into energy. Hence we can say that fresh weight is not continuous in the trophic levels.
41. **(c)** About 70% of total global carbon is found in oceans. This oceanic reservoir regulates the amount of carbon dioxide in the atmosphere. Atmosphere contains only about one percent of total global carbon.
42. **(a)** Food chain is the transfer of energy from green plants (primary producers), through a sequence of organisms occupies in a food chain is known as its trophic level. Therefore, statements b and c are correct.
43. **(a)** A standing crop is the quantity or total weight or energy content of the organism, which are in a particular location at a particular time.
44. **(c)** An energy pyramid is the graphical representation of the trophic level (nutritionals) by which the incoming solar energy is transferred into an ecosystem. It can never be inverted in shape.
45. **(a)** Hydrarch succession takes place in wetter areas and xerarch succession takes place in dry areas. So, both hydrarch and xerarch successions lead to medium water conditions.
48. **(d)** The largest vegetational group is known as formation.

53. (c) Grassland with scattered trees are called savanna.
55. (b) Zone of transition present between two types of communities for special ecological interest is called ecotone.
58. (b) Deer and bees occupy same trophic level. Both are primary consumers or consumers of first order. They get their food directly from green plants *i.e.* they are herbivores. Alton 1939 have given the name key industry animals to them because many other organisms of ecosystem depend on them for their food e.g. deer, cow, insects etc.
59. (d) Ancient plants were photosynthetic but they were not using water as source of electrons. Hence, evolution of oxygen was absent.
60. (b) Stability of the ecosystem mainly depends upon interaction between two components of the ecosystems, the producers and the consumers. Interactions between the two lead to development of different trophic levels and flow of energy in the ecosystems. The balance in the ecosystem is maintained by the relationship between the producers and the consumers. They maintain a constant ratio in the evironment since primary consumers eat producers. The latter are eaten by secondary consumers these are then by tertiary consumers and finally all by decomposers. So, their number do not increase immensely.

37. BIODIVERSITIES AND ITS CONSERVATION

ANSWER KEY															
1	(b)	9	(d)	17	(d)	25	(b)	33	(d)	41	(d)	49	(a)	57	(c)
2	(d)	10	(d)	18	(b)	26	(d)	34	(c)	42	(c)	50	(d)	58	(b)
3	(b)	11	(b)	19	(b)	27	(a)	35	(a)	43	(c)	51	(d)	59	(d)
4	(c)	12	(d)	20	(a)	28	(c)	36	(b)	44	(c)	52	(a)	60	(d)
5	(d)	13	(b)	21	(c)	29	(b)	37	(c)	45	(d)	53	(a)		
6	(d)	14	(a)	22	(b)	30	(c)	38	(c)	46	(b)	54	(b)		
7	(c)	15	(c)	23	(a)	31	(b)	39	(c)	47	(d)	55	(d)		
8	(c)	16	(b)	24	(d)	32	(b)	40	(c)	48	(b)	56	(a)		

Hints & Solutions

1. (b) Biodiversity is the number of variety of organism found within a specified geographic region.
6. (d) Rare species are species with naturally small population, either localised or thinly scattered.
7. (c) Desertification is the process leading to the formation of desert. The main factor of desertification of land is over-grazing by herbivorous animals.
8. (c) Water is a renewable resource but can become non renewable if they are not used properly.
9. (d) Soil conservation is to conserve fertile soil from the losses like, heavy rainfall, drainage, high wind, flood, drought etc.
10. (d) Transpiration will be decreased, air temperature will increase and water content will decrease hence rainfall decreases.
11. (b) Non-renewable sources of energy are generally conventional (traditional) sources.
12. (d) "Chipko movements" is concerned with plant conservation.
13. (b) Species restricted to small area are called endemic.
14. (a) Wild life comprises all the living organisms in their natural habitats which are neither cultivated nor tamed.
16. (b) Competition between organisms of the same species is called intraspecific competition. Mutualism is obligatory and mutually beneficial relationship occurring between two interacting organisms. Predation is the act of capturing, killing and eating. It is a negative interaction between two organisms. Migration is a permanent outward movement of individuals from a population for settling into a new area.
17. (d) An institution where valuable plant materials likely to become irretrievably lost in the wild or in cultivation is preserved in a viable condition is known as gene bank.
18. (b) Diversity is due to the long term evolutionary changes. Adaptability to continuous changes in environmental conditions is important for natural selection of variants and variations generation after generation leading to emergence of diverse descendants.
20. (a) According to Darwin gradual accumulation of small variations give rise to new species. Abrupt mutations do not give rise to new species . Plants arose by gradual adaptation to diverse habitats and not

suddenly on earth. Seeded plants are just one group in the plant kingdom.

21. **(c)** Ex-situ conservation means off site conservation. It is the process of protecting an endangered species of plant or animal by removing it from an unsafe or threatened habitat and placing it or part of it under the care of man. Botanical garden is one of the method of ex-situ conservation.

22. **(b)** *Ex-situ* conservation is the conservation of selected organism in places outside their natural homes. They include off site collection and gene banks. *In situ* conservation, on the other hand, is the conservation of endangered species in their natural habitat. Biosphere reserves, national parks, wildlife sanctuaries and sacred groves all are examples of *in situ* conservation.

23. **(a)** Alpha diversity is present within community.

29. **(b)** Those species whose populations have been seriously depleted and whose ultimate security is not assured are known as endangered species.

30. **(c)** Gene library contains DNA fragments representing the entire genome of an organism. So collection of alleles of the genes of a crop is called gene library.

31. **(b)** Wheat genetics is more complicated than that of most other domesticated species. Some wheat species are diploid, with two sets of chromosomes, but many are stable polyploids, with four sets of chromosomes (tetraploid) or six (hexaploid).

32. **(b)** Lesser inter-specific competition is not observe in biodiversity hotspots, levels of species richness and high degree of endemism. Initially 25 biodiversity hotspots were identified but subsequently nine more have of biodiversity hotspots in the world to 34. These hotspots are also regions of accelerated habitat loss.

33. **(d)** Tissue culture method is used to propagate plant. Creating biosphere reserve conserves both flora and fauna. Botanical gardens can conserve only a few specific varieties of plants.

34. **(c)** If the forest cover is reduced to half, large areas will become deserts.

35. **(a)** Deforestation causes the depletion of wood and nutrients from an ecosystem and hence nutrient cycling gets slowed down.

37. **(c)** The list of endangered species of plants and animals is given in red data book.

43. **(c)** Largest region is deccan, peninsula and most biodiversity rich region is western ghats (4%) with a very large number of endemic amphibian species.

48. **(b)** The wild life protection act was passed by the Indian parliament in 1972. This act has the provisions for conservation of wild life, nationalisation of bioreserves and establishment of national parks and sanctuaries.

53. **(a)** Botanical Survey of India (BSI) has its head quarters at Hawrah (Kolkata). BSI is connected with plant exploration and writing up of regional flora and also preparation of flora of India and red list of plant species.

58. **(b)** There are 83 national parks and 421 sanctuaries found in India. Their total area is about 1,41,295 sq. km. It is about 4% of India's total geographic area.

38. ENVIRONMENTAL ISSUES

ANSWER KEY

1	(a)	11	(b)	21	(c)	31	(d)	41	(b)	51	(c)
2	(b)	12	(d)	22	(c)	32	(c)	42	(c)	52	(b)
3	(d)	13	(d)	23	(c)	33	(b)	43	(a)	53	(d)
4	(c)	14	(a)	24	(d)	34	(c)	44	(c)	54	(d)
5	(a)	15	(a)	25	(a)	35	(c)	45	(b)	55	(b)
6	(c)	16	(a)	26	(d)	36	(c)	46	(b)	56	(c)
7	(b)	17	(c)	27	(c)	37	(c)	47	(c)	57	(b)
8	(a)	18	(a)	28	(b)	38	(b)	48	(d)	58	(b)
9	(a)	19	(c)	29	(c)	39	(b)	49	(c)	59	(b)
10	(d)	20	(a)	30	(b)	40	(b)	50	(d)	60	(c)

Hints & Solutions

3. **(d)** Lichens are important in the studies on atmospheric pollution because they are very sensitive to pollutants like sulphur dioxide.

9. **(a)** *E. coli* lives in the human intestine. If they are present in water it indicates that the water is polluted. *E.coli* coliform count test is done.

10. **(d)** Exhaustible resources are those natural resources that are likely to be exhausted due to their continuous use. Forest is a renewable exhaustible resource that can be again and again produced. Coal, petroleum and minerals are non-renewable exhaustible natural resources that cannot be again produced.
11. **(b)** CO_2 in its normal fraction in air (about 0.03%) is actually not a pollutant but above this it is considered as a pollutant. At the elevated level it becomes a cause for global waring.
12. **(d)** Fluorides of carbon is the major pollutant from jet plane emission.
13. **(d)** Ultraviolet radiations cause photochemical splitting of oxygen molecules in the stratosphere. The nascent oxygen combines with molecular oxygen to form ozone. In troposphere nascent oxygen comes from nitrogen dioxide.
14. **(a)** The content of CO_2 in atmospheric air is 0.034%. The main contributors to air are N_2 and O_2.
15. **(a)** DDT cannot be broken down or converted to harmless state by natural processes of oxidation or biological decomposition. Hence it is non biodegradable.
16. **(a)** NO_2 and SO_2 after reacting with water in atmosphere produce HNO_3 & H_2SO_4 which fall down as acid rain.
17. **(c)** Increase in the concentration of CO_2 do not allow the earth's radiations to go out of earth's atmosphere and contributes in increasing temperature of earth.
18. **(a)** MIC $\rightarrow$ (Methyl isocyanate) was the deadly gas leaked from the fertilizer plant of union carbide company in bhopal.
19. **(c)** Lead (Pb) is the most hazardous metal pollutant present in automobile exhaust.
20. **(a)** Automobile exhaust is the main cause of air pollution in metro cities.
21. **(c)** CO_2 is the major green house gas causing green house effect which is released in the atmosphere by burning fossils (combustion) or by activities like animal respiration. Deforestation adds to the increase in CO_2 in atmosphere because trees are cut down which are the big absorber of CO_2.
22. **(c)** In 1962, American biologist Rachel Carson published the book *Silent Spring*, which alleged that DDT caused cancer and harmed bird reproduction by thinning egg shells. This followed the principles of biological magnification, killing higher level organisms like the birds. DDT is a persistent organic pollutant with a reported half life of between 2-15 years, and is immobile in most soils.
23. **(c)** Sewage kills plant life and also use oxygen for degradation, so oxygen content of water is lowered and hence animal life in water body starts decreasing.
24. **(d)** The Taj mahal is threatened by environmental pollution, especially by acid rain due to sulphur dioxide emitted from mathura refinery.
25. **(a)** By dumping of huge amount of sewage, the oxygen levels are depleted, which are reflected in terms of BOD values of water. The number of microbes also increases tremendously and these also consume most of the oxygen. Thus BOD of the river water will increase.
26. **(d)** Increase in air pollution has resulted in gradual depletion of the ozone layer. This has exposed the earth's surface to the harmful ultraviolet radiations of the sun. Exposure to UV rays increases the risk of skin cancer and other skin related disorders.
27. **(c)** Ozone hole is an area with extreme reduction in the ozone concentration. It was discovered over Halley Bay in Antarctica by Farman *et. al* (1985), members of British Antarctic Survey Team.
28. **(b)** Carbon monoxide, when inhaled, combines with blood haemoglobin to form carboxy-haemoglobin at a rate 210 times faster than the rate at which oxygen forms oxyhaemoglobin. Thus, respiration is impaired.
30. **(b)** Lichens are sensitive to sulphur dioxide pollution, hence they serve as bioindicators of pollution.
31. **(d)** Green house effect is the rise in atmospheric temperature due to increase in concentration of green house gases.
The greenhouse effect, discovered by Joseph Fourier in 1824 and first investigated quantitatively by Svante Arrhenius in 1896, is the process in which the emission of infrared radiation by an atmosphere warms a planet's surface. In the case of the Earth, without these greenhouse gases its surface would be up to 30°C cooler. The name comes from an incorrect analogy with the way in which greenhouses are heated by the sun in order to facilitate plant growth. In addition to the Earth, Mars and especially Venus have greenhouse effects. The maximum green house gases are emitted by USA.
32. **(c)** Primary pollutant is that which persists in the form in which it is released in the environment e.g. CO. A secondary pollutant is one which is formed from another pollutant due to change or reaction e.g. PAN (Peroxy acetyl nitrate).
33. **(b)** Methane forms a major part of gobar-gas. Ammonia is toxic. Ethane and butane are not major components of gobar gas.
34. **(c)** Solar energy coming to the earth is not responsible for green house effect. It is the increase in green house gases in atmosphere like CO_2 which is released by complete combustion of fossil fuels or biomass in industries or transportation vehicles that prevent the reradiation of infrared radiation from the earth and result in increase in temperature of the earth.
35. **(c)** Biochemical oxygen demand (bod) in a river water increases when sewage gets mixed with river water. Biochemical oxygen demand or biological oxygen demand (BOD) is a chemical procedure for determining how fast biological organisms use up oxygen in a body of water. It is used in water quality management and assessment, ecology and environmental science. BOD

considered as an indication of the quality of a water source.

36. **(c)** Good ozone occurs naturally in the earth's upper atmosphere 10 to 30 miles above earth's surface (stratosphere) where it forms a protective layer that shield us from the sun's harmful ultraviolet rays.

37. **(c)** In the environmental, ozone is known for its both harmful and useful effects.

40. **(b)** In case CO_2 of earth's atmosphere disappears, the temperature of earth's surface would decrease.

43. **(a)** Water pollution can be stopped best by treating effluents to remove injurious chemicals.

48. **(d)** Major aerosol pollutant in jet emission is fluorocarbon.

56. **(c)** Because in the presence of oxygen micro-organisms (bacteria) oxidise the complex organic components to inorganic forms.

57. **(b)** CFCs reacts with ozone and cause its depletion. That is why CFCs are not recommended to be used in refrigerators.

59. **(b)** Above 80 dB sound becomes hazardous.

60. **(c)** Some of the important green house gases are CO_2, CFCs, Nitrous oxide, Ozone, Methane etc. Nitrous oxide is produced by denitrifying bacteria acting on artificial fertilisers applied to poorly aerated soil.